AQA
A Level Further Maths

Year 1 +
Year 2

Series Editor
David Baker

Authors
Brian Jefferson, David Bowles, Eddie Mullan, Garry Wiseman,
John Rayneau, Katie Wood, Mike Heylings, Rob Wagner

OXFORD
UNIVERSITY PRESS

OXFORD
UNIVERSITY PRESS

Great Clarendon Street, Oxford, OX2 6DP, United Kingdom

Oxford University Press is a department of the University of Oxford.

It furthers the University's objective of excellence in research, scholarship, and education by publishing worldwide. Oxford is a registered trade mark of Oxford University Press in the UK and in certain other countries.

© Oxford University Press 2018

The moral rights of the authors have been asserted.

First published in 2018

British Library Cataloguing in Publication Data
Data available

978 0 19 841291 5

10 9 8 7 6 5

Paper used in the production of this book is a natural, recyclable product made from wood grown in sustainable forests.
The manufacturing process conforms to the environmental regulations of the country of origin.

Printed and bound by CPI Group (UK) Ltd, Croydon, CR0 4YY

Acknowledgements

Authors
Brian Jefferson, David Bowles, Eddie Mullan, Garry Wiseman, John Rayneau, Katie Wood, Mike Heylings, Rob Wagner

Editorial team
Dom Holdsworth, Ian Knowles, Matteo Orsini Jones, Felicity Ounsted

With thanks also to Geoff Wake, Matt Woodford, Brian Brooks, Deb Dobson, Katherine Bird, Linnet Bruce, Keith Gallick, Jim Newall, Laurie Luscombe, Susan Lyons and Amy Ekins-Coward for their contribution.

Index compiled by Milla Hills

Although we have made every effort to trace and contact all copyright holders before publication, this has not been possible in all cases. If notified, the publisher will rectify any errors or omissions at the earliest opportunity.

Cover: Aleksandr Kurganov/Shutterstock; mexrix/Shutterstock

p3, p30, p71, p176, p202, p483, p507, p524, p527, p558, p660(b), p665, p686, p689, p724, p750, p753, p774(m), p774(b), p810(b), p833(t), p852, p855, p881, p900, p905, p934, p939, p958 Shutterstock; **p35** josemoraes/iStockphoto; **p75** Atosan/Shutterstock; **p122** Manzotte Photography/Shutterstock; **p122** Soleil Nordic/Shutterstock; **p125** BEST-BACKGROUNDS/Shutterstock; **p143** MarkLG/Shutterstock; **p179** science photo/Shutterstock; **p205** Dolomites-image/iStockphoto; **p230** Andyd/iStockphoto; **p230** Steve Mann/Dreamstime; **p235** FatCamera/iStockphoto; **p256** selensergen/iStockphoto; **p256** Tatiana Shepeleva/Shutterstock; **p261** choja/iStockphoto; **p272** Roman023_photography/Shutterstock; **p272** Annette Shaff/Shutterstock; **p275** Wafuefotodesign/Dreamstime.com; **p315** Alphotographic/iStockphoto; **p341** El Nariz/Shutterstock; **p341** El Nariz/Shutterstock; **p345** kynny/iStockphoto; **p409** Pavel L Photo and Video/Shutterstock; **p426** turtix/Shutterstock; **p431** xieyuliang/Shutterstock; **p467** Monkey Business Images/Shutterstock; **p479** Rvector/Shutterstock; **p479** Zvonimir Atletic/Shutterstock.com; **p503** NEW YORK PUBLIC LIBRARY/SCIENCE PHOTO LIBRARY; **p561, p603** iStockphoto; **p609** Mikephotos/Dreamstime; **p660(t), p727, p774(t), p779, p810(t), p810(m)** iStockphoto; **p833(b), p837, p855** iStockphoto.

MECHANICS

This book has been specifically created for those studying the AQA 2017 Further Mathematics AS and A Level. It has been written by a team of experienced authors and teachers, and it's packed with questions, explanation and extra features to help you get the most out of your course.

Every section starts by covering the basic **Fluency and skills**.

Support for when and how to use **calculators** is available throughout this book.

Worked examples provide a model answer and commentary to practice questions.

There is a Fluency and skills exercise for each section, to practise the skills before moving on to the Reasoning and problem-solving section.

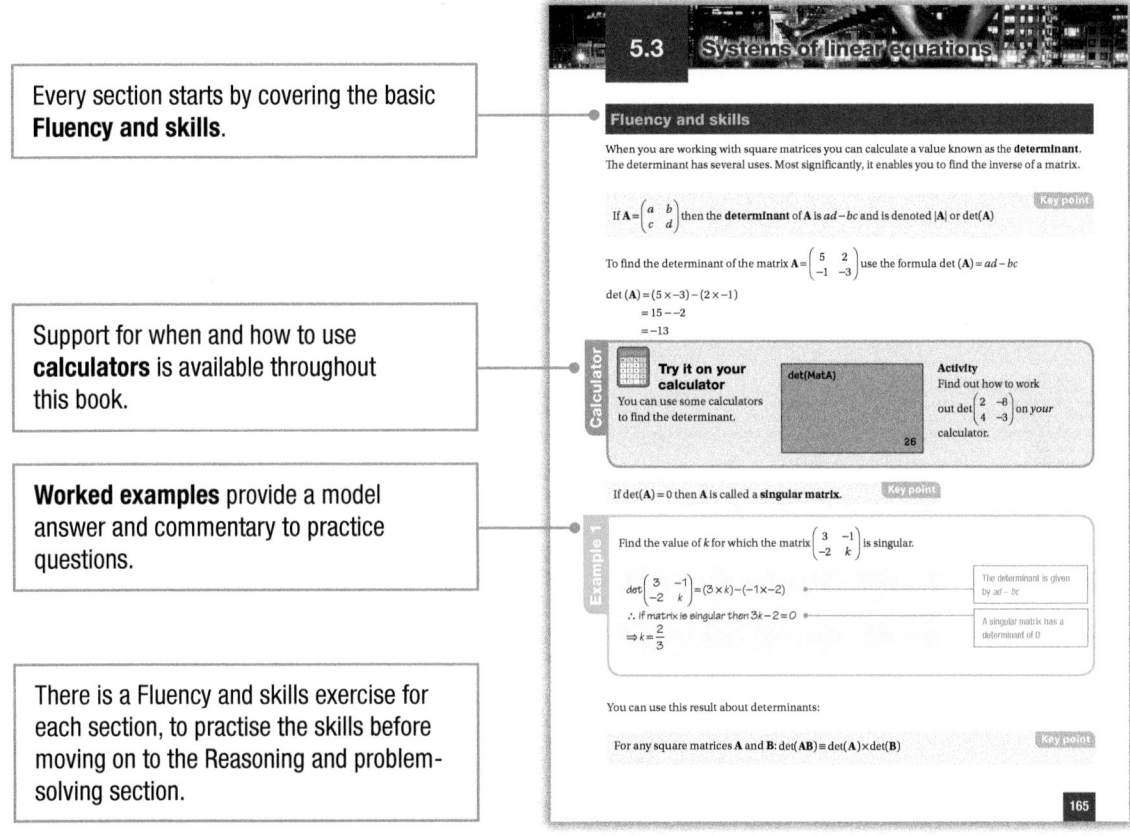

On the chapter **Introduction page**, the **Orientation box** explains what you should already know, what you will learn, and what this leads to.

At the end of every chapter, an **Exploration page** gives you an opportunity to explore the subject beyond the specification.

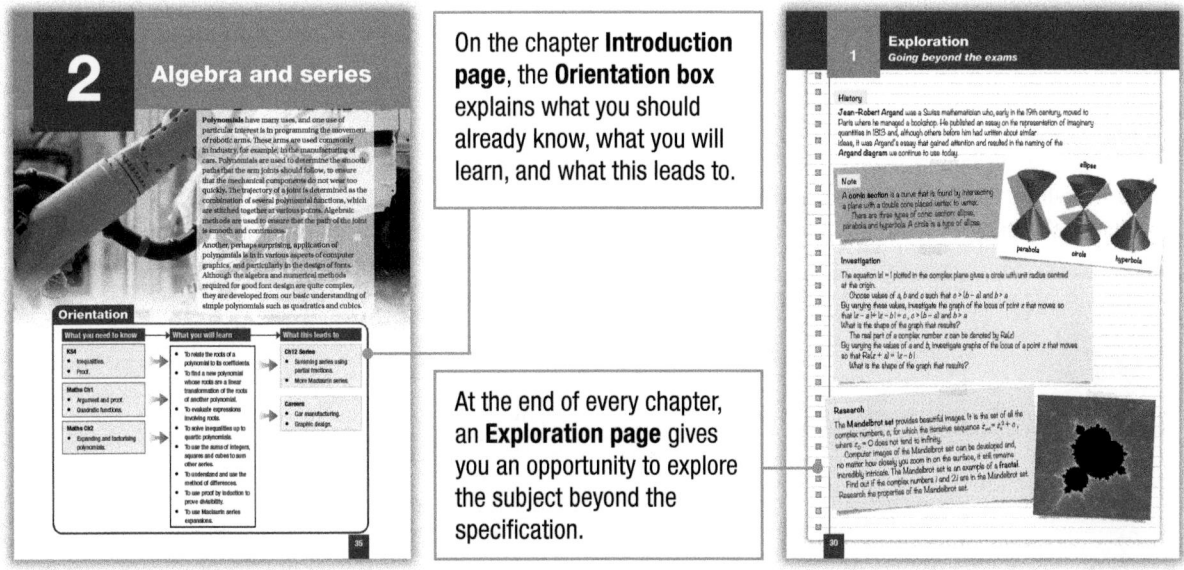

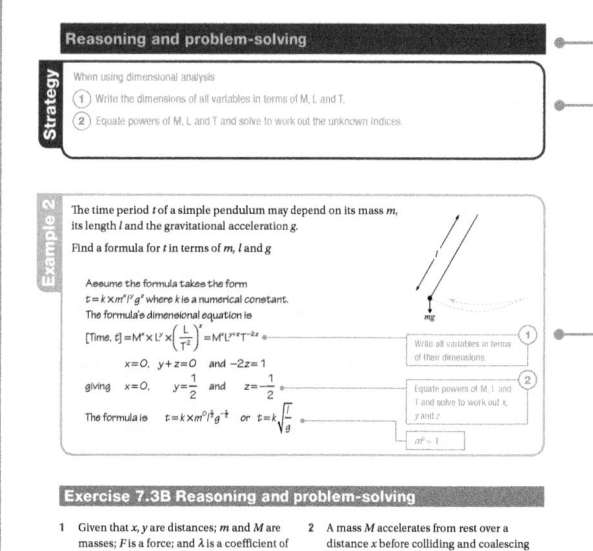

Reasoning and problem-solving

Strategy

When using dimensional analysis

① Write the dimensions of all variables in terms of M, L and T.

② Equate powers of M, L and T and solve to work out the unknown indices.

Example 2

The time period t of a simple pendulum may depend on its mass m, its length l and the gravitational acceleration g.

Find a formula for t in terms of m, l and g.

Assume the formula takes the form
$t = k \times m^x l^y g^z$ where k is a numerical constant.
The formula's dimensional equation is

$[\text{Time}, t] = M^x \times L^y \times \left(\dfrac{L}{T^2}\right)^z = M^x L^{y+z} T^{-2z}$ ①

Write all variables in terms of their dimensions.

$x = 0$, $y+z = 0$ and $-2z = 1$

$\text{giving} \quad x = 0, \quad y = \dfrac{1}{2} \quad \text{and} \quad z = -\dfrac{1}{2}$ ②

Equate powers of M, L and T and solve to work out x, y and z

The formula is $\quad t = k \times m^0 l^{\frac{1}{2}} g^{-\frac{1}{2}} \quad$ or $\quad t = k \sqrt{\dfrac{l}{g}}$

$M^0 = 1$

Exercise 7.3B Reasoning and problem-solving

1 Given that x, y are distances; m and M are masses; F is a force; and λ is a coefficient of elasticity, find the dimensions of P and Q in the following equations, and suggest what each of P and Q might represent.

$P = \sqrt{\dfrac{4Fmx}{3M(m+M)}} \qquad Q = \sqrt{\dfrac{\lambda(x-y)}{2mxy}}$

2 A mass M accelerates from rest over a distance x before colliding and coalescing with a mass m. After the collision, they are brought to rest over a distance y by a constant resisting force R. It is suggested that $R = \pi\left[(m+M)g + \dfrac{M^2y}{(m+M)x}g\right]$

Determine whether this formula is dimensionally consistent.

226 Forces and energy Dimensional analysis

Fluency techniques are built on in the **Reasoning and problem-solving** section.

Strategy boxes help build problem-solving techniques.

Circled numbers show how each step of an example is linked to the strategy box.

The questions in each exercise increase in difficulty. For extra challenge, watch out for the last question or two in each problem-solving (B) exercise.

Exercise 5.3B Reasoning and problem-solving

1 Use matrices to solve these pairs of simultaneous equations.

a $\quad 4x - y = 11$
$\quad 2x + 3y = -5$

b $\quad x - 5y = 0$
$\quad 2x - 8y = -2$

c $\quad 3x + 6y = 12$
$\quad x - 2y = 2$

d $\quad 4x + 6y = 1$
$\quad -8x + 3y = 3$

2 Use matrices to find a solution to these systems of linear equations.

a $\quad x + y - 2z = 3$
$\quad 2x - 3y + 5z = 4$
$\quad 5x + 2y + z = -3$

b $\quad 4x + 6y - z = -3$
$\quad 2x - 3y = 2$
$\quad 8y + 4z = 0$

3 Given that $\begin{pmatrix} a & 3 \\ -2a & -1 \end{pmatrix}\begin{pmatrix} x \\ y \end{pmatrix} = \begin{pmatrix} 5 \\ 10a \end{pmatrix}$, find expressions in terms of a for x and y Simplify your answers.

4 You are given the equations $(k+3)x - 2y = k-1$ and $kx + y = k$

Use matrices to find expressions in terms of

10 If $\mathbf{A}^2 = 2\mathbf{I}$, write the 2×2 matrix $\mathbf{A}$

11 Simplify each of these expressions involving the non-singular matrices $\mathbf{A}$ and $\mathbf{B}$

a $(\mathbf{AB})^{-1}\mathbf{A}$ b $\mathbf{B}(\mathbf{A}^{-1}\mathbf{B})^{-1}$

12 If $\mathbf{A} = \mathbf{C}^{-1}\mathbf{BC}$, prove that $\mathbf{B} = \mathbf{CAC}^{-1}$

13 Prove that if $\mathbf{ABA}^{-1} = \mathbf{I}$ then $\mathbf{B} = \mathbf{I}$

14 Prove that $(\mathbf{ABC})^{-1} = \mathbf{C}^{-1}\mathbf{B}^{-1}\mathbf{A}^{-1}$ for non-singular matrices $\mathbf{A}$, $\mathbf{B}$ and $\mathbf{C}$

15 Prove that if $\mathbf{P}$ is self-inverse then $\mathbf{P}^2 = \mathbf{I}^{-1}$

16 Prove that if $\mathbf{PQP} = \mathbf{I}$ for non-singular matrices $\mathbf{P}$ and $\mathbf{Q}$, then $\mathbf{Q} = (\mathbf{P}^{-1})^2$

17 A point P is transformed by the matrix $\mathbf{T} = \begin{pmatrix} 2 & 1 \\ -3 & -4 \end{pmatrix}$ and the coordinates of the image of P are $(9, -11)$

Find the coordinates of P

PURE

Assessment sections at the end of each chapter test everything covered within that chapter.

3   Assessment

1 a Sketch the graph of $y = \dfrac{x+5}{x-7}$, clearly labelling any intercepts with the coordinate axes. **[4 marks]**

b Write down the equations of the asymptotes to the curve. **[2]**

c Find the coordinates of the points where the curve $y = \dfrac{x+5}{x-7}$ meets the line $y = -x$ **[3]**

2 a Sketch the curve $y = \dfrac{5x^2}{x^2+1}$ **[3]**

b Find the points of intersection between the curve $y = \dfrac{5x^2}{x^2+1}$ and the line $y = 1$ **[3]**

3 a Express the Cartesian coordinates $(\sqrt{3}, -3)$ as polar coordinates, giving the exact value of r and writing θ in terms of π where $-\pi < \theta \leq \pi$ **[4]**

11 a Sketch the curve $y = \dfrac{x^2 - 5x + 4}{x^2+1}$ **[3]** b Solve the inequality $\dfrac{x^2 - 5x + 4}{x^2+1} > 1 - x$ **[1]**

12 A rational function is defined by $f(x) = \dfrac{3x^2 + 5x - 8}{x^2 + 2x - 6}$
A sketch of the curve $y = f(x)$ is shown.

a For what values of x is the function undefined? Leave your answer as a surd. **[2]**

b Where does the curve of the function cut the coordinate axes? **[3]**

c For what values of x is $f(x) > 1$? **[5]**

13 a Show that the equation $xy - 5y + x - 9 = 0$ can be written as $(x-5)(y+1) = k$ where k is a constant to be found. **[2]**

1 Complex numbers 1

Complex numbers are a crucial tool in the design of modern electrical components, such as motherboards. Electrical engineers use them to simplify many of their calculations. This helps them to analyse varying currents and voltages in electrical circuits. When the current flows in one direction constantly, the resistance can often be calculated using a simple formula. However, when the direction of the current is alternating, this formula doesn't work. Therefore, engineers express the quantities as complex numbers, which makes it easier to understand the processes involved and perform the necessary calculations.

Complex numbers are used in many other fields such as chemistry, economics, and statistics. Their unique properties provide powerful ways of solving and interpreting complicated equations that can be applied in a wide variety of contexts. From some of the deepest mysteries in number theory to the signal processing used in playing digital music, these so-called imaginary numbers have led to a surprising array of advances in the real world.

Orientation

What you need to know

KS4
- Use the quadratic formula.

Maths Ch1
- Solve simultaneous equations.

Maths Ch3
- Work with sine and cosine functions.

What you will learn

- To calculate with complex numbers in the form $a + bi$
- To understand and use the complex conjugate.
- To solve quadratic, cubic and quartic equations with real coefficients.
- To convert between modulus-argument form and the form $a + bi$ and calculate with numbers in modulus-argument form.
- To sketch and interpret Argand diagrams.

What this leads to

Ch16 Complex numbers 2
- Exponential form.
- The Euler formula $e^{i\theta} = \cos \theta + i\sin \theta$
- De Moivre's formula.
- Roots of unity.

Careers
- Electrical engineering.

1.1 Properties and arithmetic

Fluency and skills

Some equations, including those deriving from real-world situations, have no real solutions. For example, up to this point, you have been unable to solve equations such as $x^2 = -1$. You know that the solutions are $x = \pm\sqrt{-1}$, but this is not a real number and is difficult to manipulate. In order to solve this problem, mathematicians denoted the square root of negative one by i. This means that the solutions of the equation $x^2 = -1$ can be written as $\pm$i. Although this number is known as 'imaginary', it means that all polynomial equations do indeed have solutions.

A good analogy is negative numbers: these can be hard to grasp in isolation. For example, the concept of 'minus one apple' is a difficult one, but it's useful in sums such as 3 apples − 1 apple = 2 apples. In exactly the same way, imaginary numbers are essential in many calculations and have many real-world applications.

> **Key point**
>
> The **imaginary number** i is defined as $i = \sqrt{-1}$

You can solve the equation $x^2 = -9$ by square-rooting both sides to give $x = \pm\sqrt{-9}$

Then, because $\sqrt{-9}$ can be written as $\sqrt{9}\sqrt{-1}$, you can use the fact that $i = \sqrt{-1}$ to give $x = \pm 3$i

Notice that once you have defined the square root of minus one, the square root of all other negative numbers can be written in terms of i

For example, to solve the equation $(x-3)^2 = -5$, you first square-root both sides to give $x - 3 = \pm\sqrt{-5}$, and since $\sqrt{-5} = \sqrt{5}i$, this gives the solutions $x = 3 \pm \sqrt{5}i$

Numbers with both real and imaginary parts are called complex numbers.

> **Key point**
>
> **Complex numbers** can be written in the form $a + bi$ where $a, b \in \mathbb{R}$. The set of complex numbers is denoted $\mathbb{C}$

Complex numbers can be added, subtracted and multiplied by a constant in the same way as algebraic expressions.

Example 1

Simplify the expression $3(4-7i) - 2(3-2i)$

$$3(4-7i) - 2(3-2i) = (12-21i) - (6-4i)$$
$$= 6 - 17i$$

Multiply real and imaginary parts by the constant.

Simplify real parts: $12 - 6 = 6$

Simplify imaginary parts: $-21i + 4i = -17i$

When multiplying complex numbers together, you will use the fact:

$i^2 = -1$ (since $i^2 = (\sqrt{-1})^2$)

Example 2

Solve the equation $(x+7)^2 = -16$

$x + 7 = \pm\sqrt{-16}$ — Square-root both sides of equation.

$x = -7 \pm \sqrt{-16}$

$= -7 \pm \sqrt{16}\sqrt{-1}$

$= -7 \pm 4i$ — Since $\sqrt{-1} = i$

Calculator

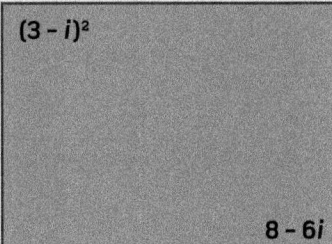

Try it on your calculator

Some calculators enable you to manipulate complex numbers.

$(3 - i)^2$

$8 - 6i$

Activity

Find out how to work out $(3 - i)^2$ on *your* calculator. Use your calculator to find these expressions.

a $(2+i)(5-7i)$ **b** $4i(2-8i)$
c $(30+5i) \div (2-i)$

To rationalise the denominator in surd form such as $\dfrac{1}{a+\sqrt{b}}$, you multiply the numerator and the denominator by $a - \sqrt{b}$

You can use a similar method to simplify fractions with complex denominators.
Doing this will always change the denominator into a positive real number.

To simplify $\dfrac{1}{a+bi}$, multiply the numerator and denominator by $a - bi$
(this is called the **complex conjugate**).

Example 3

Simplify $\dfrac{1+3i}{1-2i}$

$\dfrac{1+3i}{1-2i} = \dfrac{(1+3i)(1+2i)}{(1-2i)(1+2i)}$ — Multiply the numerator and denominator by the complex conjugate.

$= \dfrac{1+5i+6i^2}{1-4i^2}$ — The imaginary parts of the denominator cancel each other out.

$= \dfrac{-5+5i}{5} = -1+i$ — Use $i^2 = -1$

Write in the form $a + bi$

1 Solve these equations.

  a $x^2 = -25$ **b** $x^2 = -121$

  c $x^2 = -20$ **d** $x^2 + 8 = 0$

  e $z^2 = -9$ **f** $z^2 + 12 = 0$

2 Simplify these expressions, giving your answers in the form $a + bi$ where $a, b \in \mathbb{R}$

  a $(2 + 3i) + (5 - 9i)$

  b $(5 - 7i) - (12 + 3i)$

  c $3(6 - 9i)$ **d** $3(2 + 10i) + 5(4 - i)$

  e $4 - 9(7i + 5)$ **f** $2(6 - 2i) - 3(2i - 5)$

3 Write each of these expressions in the form $a + bi$ where $a, b \in \mathbb{R}$

  a $(2 + 3i)(i + 5)$ **b** $(7 - i)(6 - 3i)$

  c $i(8 - 3i)$ **d** $(9 - 4i)^2$

4 Fully simplify each of these expressions.

  a i^3 **b** i^4 **c** i^5

  d $(2i)^3$ **e** $(3i)^4$ **f** $2i^2(5i - 9)^2$

5 Simplify these fractions, giving your answers in the form $a + bi$ where $a, b \in \mathbb{R}$

  a $\dfrac{3}{2 + i}$ **b** $\dfrac{2i}{1 - 5i}$ **c** $\dfrac{1 + 7i}{3 - i}$

  d $\dfrac{i + 3}{2i - 1}$ **e** $\dfrac{6 + 3i}{i - \sqrt{2}}$ **f** $\dfrac{\sqrt{2}i - \sqrt{6}}{\sqrt{3} - i}$

6 You are given that $z_1 = 3i - 2$, $z_2 = 4 + i$

Calculate these expressions, fully simplifying your answers.

  a $z_1 + z_2$ **b** $z_1 z_2$

  c $\dfrac{z_1}{z_2}$ **d** $\dfrac{z_2}{z_1}$

Reasoning and problem-solving

Strategy

To solve equations involving imaginary numbers

  (1) Write all the numbers and expressions in the form $a + bi$

  (2) Equate real parts and imaginary parts on both sides of the equation or identity.

  (3) Solve the equations simultaneously.

Example 4

Find real numbers a and b such that $(a + 5i)(2 - i) = 9 + bi$

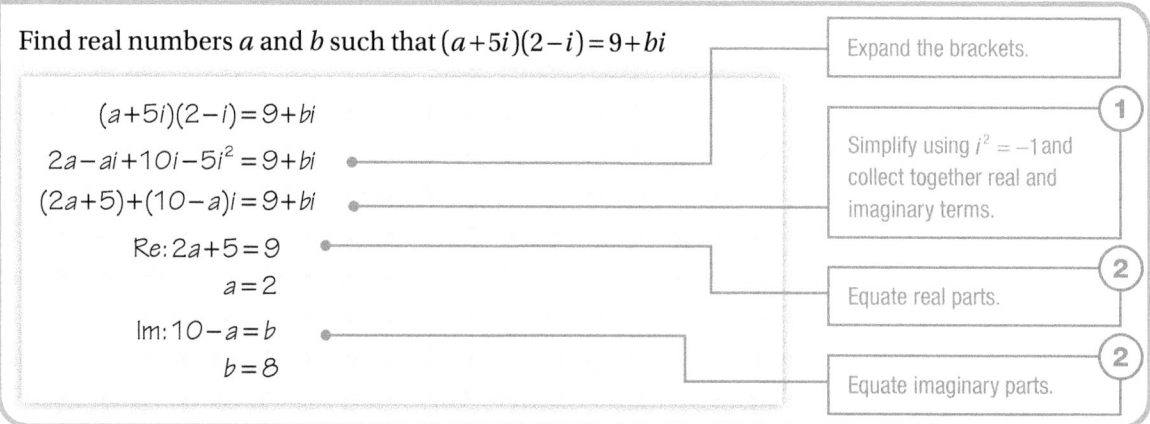

$$(a + 5i)(2 - i) = 9 + bi$$
$$2a - ai + 10i - 5i^2 = 9 + bi$$
$$(2a + 5) + (10 - a)i = 9 + bi$$
$$\text{Re}: 2a + 5 = 9$$
$$a = 2$$
$$\text{Im}: 10 - a = b$$
$$b = 8$$

Expand the brackets. **(1)**

Simplify using $i^2 = -1$ and collect together real and imaginary terms.

Equate real parts. **(2)**

Equate imaginary parts. **(2)**

Example 5 | PURE

Find the complex numbers z such that $z^2 = 5+12i$

Let $z = a+bi$ where $a, b \in \mathbb{R}$, then $(a+bi)^2 = 5+12i$

You need to find the complex number z such that $z^2 = 5+12i$

$a^2 + 2abi + b^2i^2 = 5+12i$

Simplify, using $i^2 = -1$ ①

$(a^2 - b^2) + 2abi = 5+12i$

Real: $a^2 - b^2 = 5$

Imaginary: $2ab = 12 \Rightarrow a = \dfrac{6}{b}$

Equate real parts and equate imaginary parts. ②

$\left(\dfrac{6}{b}\right)^2 - b^2 = 5$

Substitute for a and solve simultaneously. ③

$36 - b^4 = 5b^2$

$b^4 + 5b^2 - 36 = 0$

Solve the quadratic in b^2

$b^2 = 4 \text{ or } -9 \Rightarrow b = \pm 2$

b is a real number so $b^2 = 4$

$a = \dfrac{6}{b} \Rightarrow a = \pm 3 \Rightarrow z = 3+2i \text{ or } z = -3-2i$

Give both possible answers.

Exercise 1.1B Reasoning and problem-solving

1 Find the real numbers a and b such that $(2a+i)^2 = 35 - bi$

2 Find the values $a, b \in \mathbb{R}$ that satisfy the equation $(2-bi)(3+4i) = a-13i$

3 Solve the equation $z(2-5i) = 13-18i$ for $z \in \mathbb{C}$

4 Solve the equation $(x+3i)(10-7i) = 34+6i$ for a complex number x

5 Find the complex numbers z_1 and z_2 such that $z_1 + z_2 = 11-3i$ and $z_2 - z_1 = 5+7i$

6 Find the complex numbers z and w that satisfy the equations $z+2w = 6i$ and $3z-4w = 20+23i$

7 Find $z_1, z_2 \in \mathbb{C}$ such that $2z_1 - 3z_2 = 10+8i$ and $5z_1 - \dfrac{1}{2}z_2 = 4+6i$

8 Find the complex numbers w in each of these cases.
   a $w^2 = 30i - 16$ b $w^2 = -3-4i$
   c $w^2 - 1 = 20(1-i)$

9 Calculate the square roots of $2 - 4\sqrt{2}i$

10 Solve these simultaneous equations, where w and z are complex numbers.
   $w^2 + z^2 = 0$
   $z - 3w = 10$

11 Solve each of these equations to find $z \in \mathbb{C}$
   a $z^2 - 2z = -50$ b $z^2(1+i) = 7-17i$

12 Simplify each of these expressions, giving your answer in the form $a+bi$ in each case.
   a $(1+2i)^4$ b $(2-5i)^5$
   c $(3i-1)^3(1+i)$

13 Find $a, b \in \mathbb{R}$ such that $(a+i)^4 = 28+bi$

Full A Level

Fluency and skills

Using complex numbers, the quadratic formula $x = \dfrac{-b \pm \sqrt{b^2 - 4ac}}{2a}$ can be used to solve all quadratic equations of the form $ax^2 + bx + c = 0$, with $a, b, c \in \mathbb{R}$

When the discriminant $(b^2 - 4ac)$ is negative the solutions will contain imaginary numbers.

Because the $\pm$ in the formula gives two solutions, you can see that:

> **Key point**
>
> If $z = a + bi$ is a solution of a quadratic equation with **real coefficients** then the complex conjugate, $z^* = a - bi$, will also be a solution.

Example 1

Given that $z = 5 - 3i$, find

a z^* **b** zz^* **c** $z + z^*$

a $z^* = 5 + 3i$ — Write the complex conjugate of z

b $zz^* = (5 - 3i)(5 + 3i)$ — Since $i^2 = -1$

$\qquad = 25 + 15i - 15i - 9i^2$

$\qquad = 25 + 9$ — zz^* will always be a real number.

$\qquad = 34$

c $z + z^* = 5 - 3i + 5 + 3i$

$\qquad = 10$ — $z + z^*$ will always be a real number.

If you know the roots of a quadratic equation, then you can find the original equation by multiplying the factors together.

Example 2

Find a quadratic equation with real coefficients that has one root of $2 + 7i$

Roots are $2 + 7i$ and $2 - 7i$ — The complex conjugate will also be a root.

So the factors are $x - (2 + 7i)$ and $x - (2 - 7i)$ — Ensure you know the difference between a *root* and a *factor*.

Equation is $\quad (x - (2 + 7i))(x - (2 - 7i)) = 0$

$x^2 - x(2 - 7i) - x(2 + 7i) + (2 + 7i)(2 - 7i) = 0$

$\qquad\qquad x^2 - 4x + 53 = 0$ — Since $-49i^2 = 49$

Complex roots for any real polynomial equation will always occur in **complex conjugate pairs**. Therefore, a polynomial equation will always have an even number of complex solutions. This implies that a cubic equation will always have at least one real root (as there cannot be three complex roots). You can use these facts to help you solve cubic and quartic equations.

Example 3

The quartic equation $3x^4 + ax^3 + bx^2 + cx + d = 0$ has solutions $-3, \dfrac{2}{3}, 5+i$ and $5-i$. Find the values of the real constants a, b, c and d

$x+3$ and $3x-2$ are factors of $3x^4 + ax^3 + bx^2 + 242x + c$ → Since -3 and $\dfrac{2}{3}$ are solutions of the equation.

Therefore $(x+3)(3x-2) = 3x^2 + 7x - 6$ is a factor

$x-(5+i)$ and $x-(5-i)$ are factors of $3x^4 + ax^3 + bx^2 + 242x + c$ → Since $5 + i$ and $5 - i$ are solutions of the equation.

Therefore $(x-(5+i))(x-(5-i)) = x^2 - 10x + 26$ is a factor
So the quartic equation can be written as
$(3x^2 + 7x - 6)(x^2 - 10x + 26)$

$$(3x^2 + 7x - 6)(x^2 - 10x + 26) = 3x^4 - 30x^3 + 78x^2 + 7x^3 - 70x^2$$
$$+ 182x - 6x^2 + 60x - 156$$
$$= 3x^4 - 23x^3 + 2x^2 + 242x - 156$$ → Expand the brackets and simplify.

Therefore, $a = -23, b = 2, c = 242, d = -156$

Exercise 1.2A Fluency and skills

1 Write the complex conjugate of z in each case.

 a $z = 5 - 2i$ b $z = 8 + i$

 c $z = 5i - 6$ d $z = \sqrt{2} - i\sqrt{3}$

 e $z = \dfrac{1}{3} + 4i$ f $z = \dfrac{2}{3}i - 5$

2 Given that $z = 9 - 2i$, calculate

 a zz^* b $z + z^*$

 c $z - z^*$ d $\dfrac{z}{z^*}$

 e $(z^*)^*$ f $\dfrac{z^*}{z}$

3 Given that $w = -\sqrt{6} + \sqrt{2}i$, calculate

 a ww^* b $w + w^*$

 c $w - w^*$ d $\dfrac{w}{w^*}$

 e $w^2 + (w^*)^2$ f $(w + w^*)^2$

4 Solve each of these quadratic equations.

 a $x^2 + 5x + 7 = 0$ b $x^2 - 3x + 5 = 0$

 c $2x^2 + 7x + 7 = 0$ d $3x^2 - 10x + 9 = 0$

5 Find a quadratic equation with solutions

 a $x = 7$ and $x = -4$

 b $x = 3 + 5i$ and $x = 3 - 5i$

 c $x = -1 - 9i$ and $x = -1 + 9i$

 d $x = -5 + 4i$ and $x = -5 - 4i$

6 A quadratic equation has a solution of $z = \sqrt{3} + i$

 a Write down the other solution of the equation.

 b Find a possible equation.

7 Find a quadratic equation where one solution is given as

 a $2 + i$ b $4 - 3i$ c $7i - 1$

 d $-5 - 2i$ e $a + 3i$ f $5 - bi$

8 Find a cubic equation with solutions

   a $x=-5$, $x=2+3i$ and $x=2-3i$

   b $x=2$, $x=i-1$ and $x=-i-1$

   c $x=0$, $x=\sqrt{3}+2i$ and $x=\sqrt{3}-2i$

   d $x=3$, $x=-\sqrt{2}-i$ and $x=-\sqrt{2}+i$

9 A cubic equation has solutions of $z=-\dfrac{1}{2}$ and $z=6-2i$

   a Write down the other solution of the equation.

   b Find a possible equation that has these solutions.

10 You are given that $f(x)=x^3+9x^2+25x+25$ and $f(-5)=0$

   a Show that the other solutions of the equation $f(x)=0$ satisfy $x^2+4x+5=0$

   b Solve the equation $f(x)=0$

11 Given that $1+8i$ is a root of $x^4+4x^3+66x^2+364x+845=0$

   a Show that two of the other solutions to the quartic equation satisfy $x^2+6x+13=0$

   b Find all the solutions of the equation $x^4+4x^3+66x^2+364x+845=0$

12 Use the solutions of these cubic equations to find the values of the real constants a, b and c

   a $x^3+ax^2+bx+c=0$ with solutions

     $x=-3$, $x=1+3i$

   b $ax^3+bx^2+27x+c=0$ with solutions

     $x=-6i-1$, $x=5$

13 Use the solutions of these quartic equations to find the values of the real constants a, b and c

   a $x^4+ax^3+bx^2+8x+c=0$ with solutions

     $x=1+2i$, $x=2$

   b $x^4-8x^3+ax^2+bx+c=0$ with solutions

     $x=1+i$, $x=3-2i$

14 Use the fact that $7i$ is a root to show that the quartic equation $x^4-x^3+43x^2-49x-294=0$ can be written in the form $(x^2+A)(x+B)(x+C)=0$ where A, B and C are constants to be found. You must show your working.

Reasoning and problem-solving

You can find a polynomial equation when given its roots.

To derive a polynomial equation when given its roots

(1) Use the fact that if z is a root of $f(z) = 0$ then z^* will also be a root.

(2) Multiply the factors together.

(3) Simplify and write the equation in descending powers of x or z

A quadratic equation has roots α and β

Show that the equation is $x^2-(\alpha+\beta)x+\alpha\beta=0$

You can apply this result to other problems that involve quadratics.

Roots are α and β so equation is $(x-\alpha)(x-\beta)=0$

     (1) If α and β are roots then $x-\alpha$ and $x-\beta$ must be factors, so multiply these together.

which becomes $x^2-\alpha x-\beta x+\alpha\beta=0$

     (3) Expand brackets then simplify.

$\therefore x^2-(\alpha+\beta)x+\alpha\beta=0$

     You can also use this equation and the factor theorem to prove its factors are α and β

Example 5

Two of the roots of the equation $ax^4 + bx^3 + cx^2 + dx + e = 0$ are $3 - 2i$ and $4i - 1$

Find the values of a, b, c, d and e

| ① | If $3 - 2i$ and $4i - 1$ are roots, then so are their complex conjugates. |

$3 + 2i$ and $-4i - 1$ are also roots.

Roots $3 - 2i$ and $3 + 2i$ give the quadratic factor $x^2 - 6x + 13$

Similarly, roots $4i - 1$ and $-4i - 1$ give the quadratic factor $x^2 + 2x + 17$

| ② | Multiply the factors together or use the result from Example 4 |

$(x^2 - 6x + 13)(x^2 + 2x + 17) = x^4 - 4x^3 + 18x^2 - 76x + 221$

So $a = 1$, $b = -4$, $c = 18$, $d = -76$, $e = 221$

| ③ | Now multiply the two quadratic factors together and simplify the equation. |

You can solve a cubic equation by using the factor theorem to find the first solution, then dividing by the factor found and solving the remaining quadratic equation.

Strategy 2

To find solutions of polynomial equations with real coefficients

(1) Use the fact that if z is a root of f$(z) = 0$ then z^* will also be a root.

(2) Factorise f(z) using the factor you know.

(3) Equate coefficients to find the value of unknown constants.

Example 6

$f(z) = z^4 + az^3 - 15z^2 + bz - 18 = 0$ where a and b are integers.

Use the fact that $f(1 - i) = 0$ to solve the equation $f(z) = 0$ and find the values of a and b

| ① | Since $i - 1$ is a solution, $i + 1$ will also be a solution. |

$f(1 - i) = 0$ implies that $z = 1 - i$ is a solution of $f(z) = 0$

$z = 1 - i$ and $z = 1 + i$ are both solutions

So $z - (1 - i)$ and $z - (1 + i)$ are factors of $f(z)$

Therefore, $z^2 - 2z + 2$ is a factor of $z^4 + az^3 - 2z^2 + bz - 18 = 0$

| | Multiply the factors or use the result that $z^2 - (\alpha + \beta)z + \alpha\beta = 0$ has solutions α and β |

$z^4 + az^3 - 15z^2 + bz - 18 = (z^2 - 2z + 2)(z^2 + kz - 9)$ for some real k

$\quad = z^4 + kz^3 - 9z^2 - 2z^3 - 2kz^2 + 18z + 2z^2 + 2kz - 18$

$\quad = z^4 + (k - 2)z^3 + (-7 - 2k)z^2 + (18 + 2k)z - 18$

| ② | Factorise $f(z)$, use the fact that the constant term is -18 and $2 \times (-9) = -18$ |

$z^2 : -15 = -7 - 2k \Rightarrow 2k = 8$

$\qquad\qquad\quad \Rightarrow k = 4$

| ③ | Equate coefficients of z^2 to find the value of k |

So $(z^2 - 2z + 2)(z^2 + 4z - 9) = 0$

The solutions of $z^2 + 4z - 9 = 0$ are $z = -2 \pm \sqrt{13}$

$z : b = 18 + 2k = 18 + 2(4) = 26$

| ③ | Equate coefficients of z to find the value of b |

$z^3 : a = k - 2 = 4 - 2 = 2$

So the quartic equation is $z^4 + 2z^3 - 15z^2 + 26z - 18 = 0$

It has two real solutions, $z = -2 \pm \sqrt{13}$ and two complex solutions, $z = 1 \pm i$

| ③ | Equate coefficients of z^3 to find the value of a |

You can prove results involving complex numbers by manipulating them in the form $a + bi$

To prove results involving complex numbers

(1) Write complex numbers in the form $a + bi$

(2) Use the fact that if $z = a + bi$ then $z^* = a - bi$

(3) Manipulate in the usual way to prove the result.

Example 7

Prove that $(z + w)^* \equiv z^* + w^*$ for all $z, w \in \mathbb{C}$

Let $z = a + bi$ and $w = c + di$ | Write in the form $a + bi$ (1)
with $a, b, c, d \in \mathbb{R}$

$(z + w)^* = ((a + c) + (b + d)i)^*$

$\qquad = (a + c) - (b + d)i$ | Find complex conjugate. (2)

$\qquad = (a - bi) + (c - di)$ | Rearrange equation. (3)

$\qquad = z^* + w^*$ as required.

Exercise 1.2B Reasoning and problem-solving

1 Prove these equations for all $z, w \in \mathbb{C}$

  a $(zw)^* = z^* w^*$

  b $(z^*)^* = z$

  c $\left(\dfrac{z}{w}\right)^* = \dfrac{z^*}{w^*}$

2 Prove that, for all $z \in \mathbb{C}$

  a $z + z^*$ is real,

  b $z - z^*$ is imaginary,

  c zz^* is real.

3 Find the possible complex numbers z such that $z + z^* = 6$ and $zz^* = 58$

4 Find the possible complex numbers w such that $w - w^* = 18i$ and $ww^* = 85$

5 Find the complex number z such that

  $z + z^* = 2\sqrt{3}$ and $\dfrac{z}{z^*} = \dfrac{1}{2} - \dfrac{1}{2}\sqrt{3}i$

6 Find the complex number w such that

  $w - w^* = 4i$ and $\dfrac{w}{w^*} = \dfrac{3}{5} + \dfrac{4}{5}i$

7 Given that $x = 7$ is a solution of the equation $x^3 - 9x^2 + 31x + k = 0$

  a Find the value of k

  b Solve the equation fully.

8 Given that $x - 1$ is a factor of the equation $x^3 + 2x^2 + 5x + k = 0$, find the value of k and solve the equation.

9 Find all the solutions of the equation $x^4 - 4x^3 + 4x^2 - 4x + 3 = 0$ given that $x^2 - 4x + 3$ is a factor of $x^4 - 4x^3 + 4x^2 - 4x + 3$
You must show your working.

10 The curve of $y = x^4 - 6x^3 + ax^2 + bx + c$ is shown where a, b, and c are real constants.

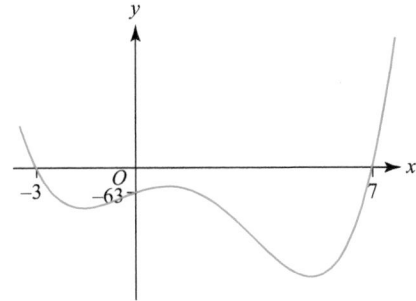

Find all the solutions of the equation
$x^4 - 6x^3 + ax^2 + bx + c = 0$

11 The equation $x^4 + kx^3 + kx^2 - 110x - 111 = 0$ has a root of $i - 6$

   a Find the value of k

   b Solve the equation.

12 Given that $5 - i$ is a root of the equation $x^4 + Ax^3 + 28x^2 - 20x + 52 = 0$, find the value of A and solve the equation.

13 Find a quartic equation with real coefficients and repeated root

   a $1 + 3i$

   b $2i - 3$

14 $f(x) = x^3 - 19x^2 + 89x + 109$

   a Write $f(x)$ as the product of a linear and a quadratic factor.

   b Sketch the graph of $y = f(x)$

15 **a** Write the quartic expression
$$4x^4 + 12x^3 - 35x^2 - 300x + 625$$
as the product of two quadratic expressions with real coefficients.

   b Sketch the graph of
$$y = 4x^4 + 12x^3 - 35x^2 - 300x + 625$$

16 The curve of a quartic equation $y = p(x)$ is shown.

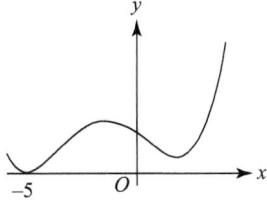

   a What does the graph tell you about the nature of the roots of $p(x) = 0$?

Given that $p(x) = x^4 + 6x^3 + ax^2 + bx + 125$ where a and b are integers

   b Solve the equation $p(x) = 0$

17 If the roots of a cubic equation are a, b and c, and the coefficient of x^3 is 1, write the equation in terms of a, b and c

18 The solutions of the equation $x^3 + 5x^2 + Ax + 12 = 0$ are α, β and γ

Write the value of

   a $\alpha + \beta + \gamma$ **b** $\alpha\beta\gamma$

19 A quartic equation has roots α, β, γ and δ

Given that the coefficient of x^4 is 1, write an expression in terms of α, β, γ and δ for

   a The coefficient of x^3

   b The constant term.

20 $f(z) = az^5 + bz^4 + cz^3 + dz^2 + ez - 841$ where a, b, c, d, and e are real numbers.

The complex number $5 - 2i$ is a repeated root of the equation $f(z) = 0$

Given that $f\left(\dfrac{1}{2}\right) = 0$, find the values of a, b, c, d and e

21 Three solutions of a polynomial equation with real coefficients are -2, $4 - i$ and $3i$

   a State the lowest possible order of the equation.

   b Find an equation of this order with these solutions.

Argand diagrams

Fluency and skills

You represent real numbers visually as points on a number line, but imaginary numbers will not fit onto the real number line. Therefore, we need a new number line of imaginary numbers. Combining these two number lines together gives a plane. So complex numbers can be represented visually as points on a plane.

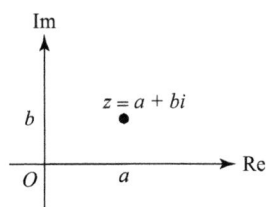

This plane is called an **Argand diagram**. Argand diagrams are used to represent complex numbers graphically.

Key point

The complex number $z = a + bi$ can be represented as the point (a, b) on an Argand diagram.

In an Argand diagram, the horizontal axis is the **real axis** and the vertical axis is the **imaginary axis**.

Example 1

Given that $z = 1 + 3i$, show z and z^* on an Argand diagram.

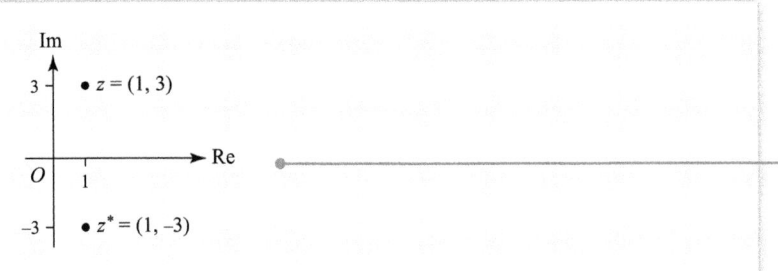

The real axis is labelled 'Re' and the imaginary axis is labelled 'Im'.

z^* is the point z reflected in the real axis.

Example 2

Given that $z = 4 + i$ and $w = -2 + i$, show z, w and $z + w$ on an Argand diagram.

Complex numbers can also be shown as position vectors.

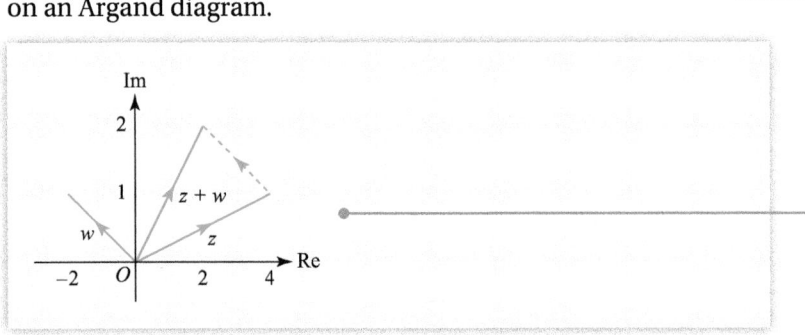

The sum can be shown as a vector addition on the Argand diagram.

Ensure that you label each vector clearly.

Example 3

Given that $z = 1 + 2i$, show z, $3z$ and $-2z$ on an Argand diagram.

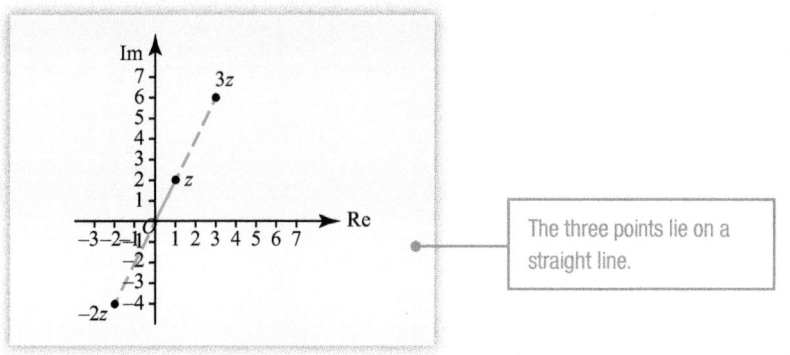

The three points lie on a straight line.

Exercise 1.3A Fluency and skills

1 Write the complex numbers represented by the vectors $\overrightarrow{OA}, \overrightarrow{OB}, \overrightarrow{OC}, \overrightarrow{OD}, \overrightarrow{OE}$ and $\overrightarrow{OF}$, as shown in the Argand diagram, in the form $a + bi$

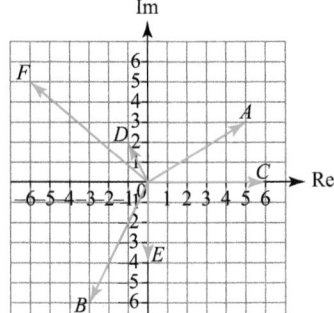

2 The Argand diagram shows the complex numbers u, v, w and z

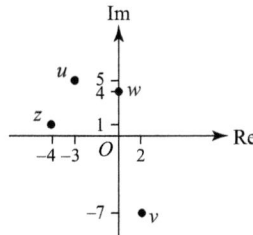

Write the complex numbers u, v, w and z in Cartesian form.

3 $z_1 = 5 - 8i$, $z_2 = 2 + 4i$
Show z_1, z_2 and $z_1 + z_2$ on an Argand diagram.

4 $z = 2 - 7i$, $w = 6i - 4$
Show z, w and $z - w$ on an Argand diagram.

5 Show the vector addition $z = (3 + 5i) + (2 - 7i)$ on an Argand diagram.

6 Show the vector subtraction $z = (6 - 2i) - (2 - 5i)$ on an Argand diagram.

7 Given that $w = \sqrt{3} + 2i$, draw w and w^* on an Argand diagram and describe the geometric relationship between them.

8 Given that $z^* = -7 + i$, draw z and z^* on the same Argand diagram and describe the geometric relationship between them.

9 Solve each of these equations and plot the solutions on an Argand diagram. Describe the geometric relationship between the two solutions in each case.

  a $x^2 = -16$ b $x^2 = -80$

10 a Solve the equation $(z + 1)^2 = -3$
   b Plot the solutions on an Argand diagram.
   c Describe the geometric relationship between the two solutions.

11 a Solve the equation $(1 - z)^2 = -25$
   b Plot the solutions on an Argand diagram.
   c Describe the geometric relationship between the two solutions.

12 For each value of z, show z and iz on an Argand diagram. What transformation maps z to iz in each case?

  a $z = -3 - 4i$ b $z = 2 - 6i$

13 For the complex numbers given, draw w and $i^2 w$ on the same Argand diagram and describe the geometrical relationship between them.

  a $w = 5 - 7i$ b $w = -2 + 3i$

Strategy

To solve geometric problems involving complex numbers

1. Draw an Argand diagram.

2. Use rules for calculating area, lengths and angles.

3. Use the fact that the product of gradients of perpendicular lines is -1

4. Fully define transformations.

Example 4

$f(x) = 2x^4 + x^3 - 15x^2 + 100x - 250$

a Find all the roots of the equation $f(x) = 0$ given that $x + 5$ is a factor of $f(x)$

b Find the area of the quadrilateral formed by the points representing the four roots.

c Prove that the quadrilateral contains two right angles.

a $2x^4 + x^3 - 15x^2 + 100x - 250 = (x+5)(2x^3 - 9x^2 + 30x - 50) = 0$

$\therefore 2x^3 - 9x^2 + 30x - 50 = 0$

Solve to give roots $x = \dfrac{5}{2}, 1 \pm 3i$

Use long division or an inspection method to factorise.

Some calculators have an equation solver. This is the easiest method to use to find the three remaining roots.

Show the roots on an Argand diagram. ①

b Area $= 2\left(\dfrac{1}{2} \times 7.5 \times 3\right) = 22.5$ square units

The quadrilateral formed is a kite which we can split into two triangles. ②

c Gradient of $AB = \dfrac{3}{6} = \dfrac{1}{2}$

Gradient of $BC = \dfrac{-3}{1.5} = -2$

$\dfrac{1}{2} \times -2 = -1$ so $\angle ABC$ is a right angle.

Similarly for $\angle ADC$

The product of the gradients is -1 ③

Example 5

You are given $z = 1 - 3i$

The points A, B and C are represented by the complex numbers z, z^2 and $z^2 - 4z$ respectively.

a Find the complex number $z^2 - 4z$ in the form $a + bi$

b Describe the transformation that maps line segment OA to CB

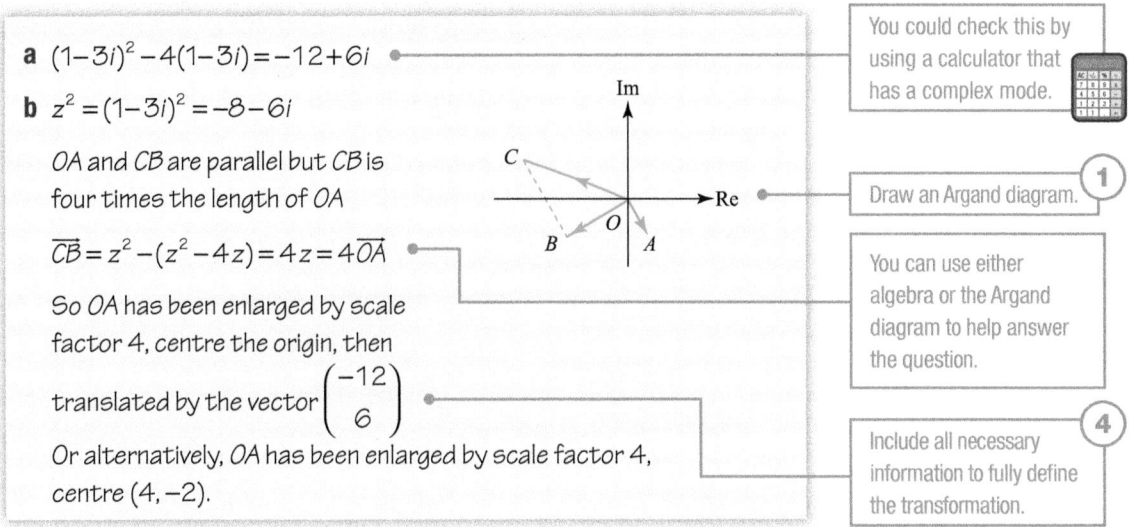

a $(1 - 3i)^2 - 4(1 - 3i) = -12 + 6i$ •————

You could check this by using a calculator that has a complex mode.

b $z^2 = (1 - 3i)^2 = -8 - 6i$

OA and CB are parallel but CB is four times the length of OA

Draw an Argand diagram. ①

$\overrightarrow{CB} = z^2 - (z^2 - 4z) = 4z = 4\overrightarrow{OA}$ •————

You can use either algebra or the Argand diagram to help answer the question.

So OA has been enlarged by scale factor 4, centre the origin, then

translated by the vector $\begin{pmatrix} -12 \\ 6 \end{pmatrix}$ •————

Include all necessary information to fully define the transformation. ④

Or alternatively, OA has been enlarged by scale factor 4, centre $(4, -2)$.

Exercise 1.3B Reasoning and problem-solving

1 $z_1 = 5 - 2i$, $z_2 = \dfrac{20 + 21i}{z_1}$

    a Find z_2 in the form $a + bi$

    b Show the points A and B representing z_1 and z_2 respectively on an Argand diagram.

    c Show that AOB is a right angle.

2 **a** Show the three roots of the equation $x^3 + x^2 + 6x - 8 = 0$ on an Argand diagram.

    b What type of triangle is formed by the points representing the three roots?

    c Find the exact area of the triangle.

3 The points A, B and C on an Argand diagram represent the solutions of the cubic equation
$x^3 - 9x^2 + 16x + 26 = 0$
Calculate the area of triangle ABC

4 **a** Given that the expression
$x^4 + x^3 + 3x + 9$ can be written in the form
$(x^2 + 3x + 3)(x^2 + Ax + B)$, find the values of A and B

    b Solve the equation $x^4 + x^3 + 3x + 9 = 0$

    c Show the roots of $x^4 + x^3 + 3x + 9 = 0$ on an Argand diagram.

    d What type of quadrilateral is formed by the points representing the roots?

    e Find the area of the quadrilateral.

5 The quartic equation
$x^4 + 8x^3 + 40x^2 + 96x + 80 = 0$
has a repeated real root.

    a Show that the repeated root is $x = -2$

    b Calculate the other solutions of the equation.

    c Show all the solutions on an Argand diagram.

6 The points A, B, C and D represent the solutions of the quartic equation
$x^4 - 3x^3 + 10x^2 - 6x - 20 = 0$

    a Use the factor theorem to find two real solutions.

    b Calculate the area of quadrilateral $ABCD$

7 The solutions of the cubic equation $x^3 - x^2 + 9x - 9 = 0$ are represented by the points A, B and C on an Argand diagram.

  a Prove that triangle ABC is isosceles.

  b Calculate the area of triangle ABC

8 You are given that $w = 5 - 2i$

  The points A, B and C represent the complex numbers w, $w + iw$ and iw respectively.

  a Prove that $OABC$ is a square.

  b Find the area of the square.

9 The points P, Q and R represent the complex numbers z, z^* and $2z + z^*$, where $z = 2 - 4i$

  a Show the points P, Q and R on an Argand diagram.

  b Describe the transformation that maps OP to QR

10 The points A, B and C are represented by the complex numbers z, z^3 and $z^3 - 2z$ where $z = 2 + i$

  a Find the complex number $z^3 - 2z$ in the form $a + bi$

  b Describe the transformation that maps line segment OA to CB

11 The solutions of a quartic equation are $z = a \pm bi$ and $w = c \pm di$, where $a, b, c, d \in \mathbb{C}$, $b \neq 0$, $d \neq 0$

  Find the area of the quadrilateral formed by the roots on an Argand diagram.

12 Given that $z = 8 + 12i$, show the complex numbers z, zi, zi^2 and zi^3 on an Argand diagram.

  Describe the transformations that map z to each of the other points.

13 The points A and B represent the complex numbers z and iz respectively. Prove that OA is perpendicular to OB

14 The points P and Q represent the complex numbers $4 + 6i$ and $-3 + 4i$ respectively. Find the area of the triangle OPQ

15 Find the area of the shape formed by the solutions of the equation $x^5 - x^4 + 18x^3 - 18x^2 + 81x - 81 = 0$

16 $f(z) = z^4 - 4z^3 + 56z^2 - 104z + 676$

  The equation $f(z) = 0$ has complex roots z_1 and z_2 which are represented on an Argand diagram by the points A and B. Given that $f(z) = 0$ has no real roots, calculate the length of AB

17 The complex numbers w and z are such that $w = 3 + 2i$ and $z = 12 - 5i$

  a Show that $\dfrac{z}{w} = 2 - 3i$

  b Mark on an Argand diagram the points A and B representing the numbers w and $\dfrac{z}{w}$, respectively.

  c Show that triangle OAB is right-angled.

  d Hence calculate the area of triangle OAB

18 $f(z) = z^4 - az^3 + 47z^2 - bz + 290$ where $a, b \in \mathbb{R}$ are constants.

  Given that $z = 2 - 5i$ is a root of the equation $f(z) = 0$, show all the roots of $f(z) = 0$ on an Argand diagram.

19 $f(z) = az^4 + bz^3 + cz^2 + dz + e$ where $a, b, c, d, e \in \mathbb{R}$ are constants.

  The solutions of $f(z) = 0$ are plotted on an Argand diagram.

  a Describe the shape formed by the points representing the solutions of $f(z) = 0$ when the equation has

  i Precisely two (distinct) real roots,

  ii Precisely one real root,

  iii No real roots.

  Given that the points representing the solutions of $f(z) = 0$ form a square,

  b Calculate the values of the constants a, b, c, d and e when

  i $z = 2i$ is a root of $f(z) = 0$

  ii $z = 1 + 3i$ is a root of $f(z) = 0$

Fluency and skills

You have seen how the complex number $z = a + bi$ can be represented on an Argand diagram.

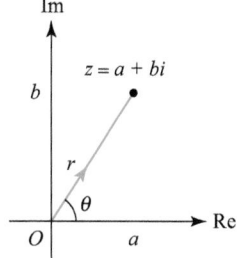

The length of the vector representing z is known as the **modulus** of z and written $|z|$

The angle between the positive real axis and the vector representing z is known as the **(principal) argument** of z and is written $\arg(z)$

Key point

The modulus of the complex number $z = a + bi$ is given by $|z| = \sqrt{a^2 + b^2}$

Key point

Write $\arg(z) = \theta$ where $-\pi < \theta \le \pi$

Example 1

Find the argument and modulus of the complex number $w = -3 + 3i$

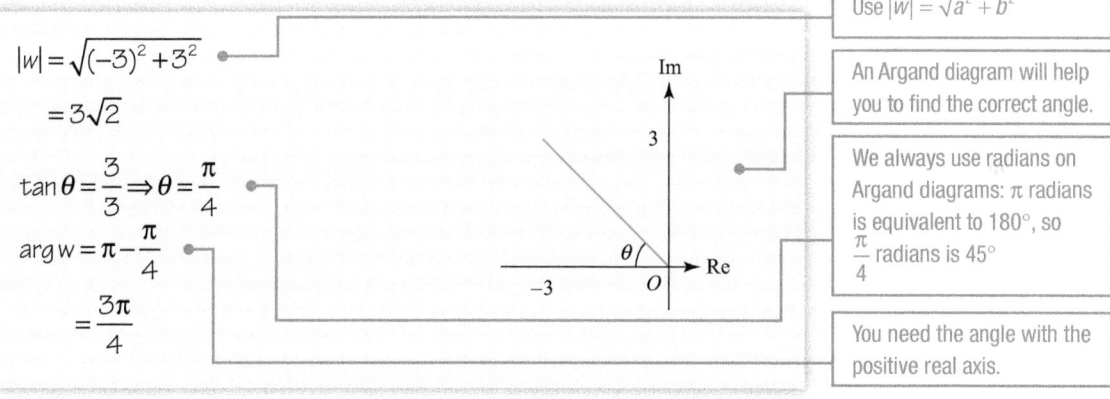

$|w| = \sqrt{(-3)^2 + 3^2}$

$\quad = 3\sqrt{2}$

$\tan\theta = \dfrac{3}{3} \Rightarrow \theta = \dfrac{\pi}{4}$

$\arg w = \pi - \dfrac{\pi}{4}$

$\quad = \dfrac{3\pi}{4}$

Use $|w| = \sqrt{a^2 + b^2}$

An Argand diagram will help you to find the correct angle.

We always use radians on Argand diagrams: π radians is equivalent to $180°$, so $\dfrac{\pi}{4}$ radians is $45°$

You need the angle with the positive real axis.

Using $w = -3 + 3i$ from the previous example, if you wished to find the modulus of $-2w$ then it is clear from the diagram that $|-2w| = |-2| |w|$

The argument of $-2w$ can also be seen to be given by $\arg(-2w) = \arg(w) + \pi$ which is $\arg w + \arg(-2)$

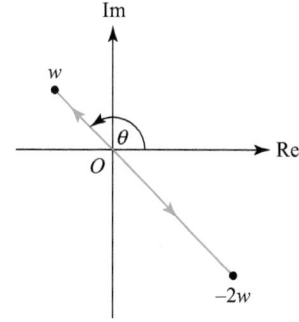

In fact, these results can be generalised for any two complex numbers.

Key point

$$|z_1 z_2| = |z_1||z_2| \text{ and } \left|\frac{z_1}{z_2}\right| = \frac{|z_1|}{|z_2|} \text{ for all } z_1, z_2 \in \mathbb{C}$$

Key point

$$\arg(z_1 z_2) = \arg(z_1) + \arg(z_2) \text{ and}$$
$$\arg\left(\frac{z_1}{z_2}\right) = \arg(z_1) - \arg(z_2) \text{ for all } z_1, z_2 \in \mathbb{C}$$

You can quote these results and will learn how to prove them in the next section.

Calculator

Try it on your calculator

Some calculators can be used to find the argument of a complex number. Work out how to find the argument of $5 - 6i$ on *your* calculator.

$$\arg(5 - 6i)$$

$$-0.8760580506$$

Example 2

Given that $z_1 = \sqrt{3} + i$ and $z_1 z_2 = -4 - 4i$, find the argument and modulus of z_2

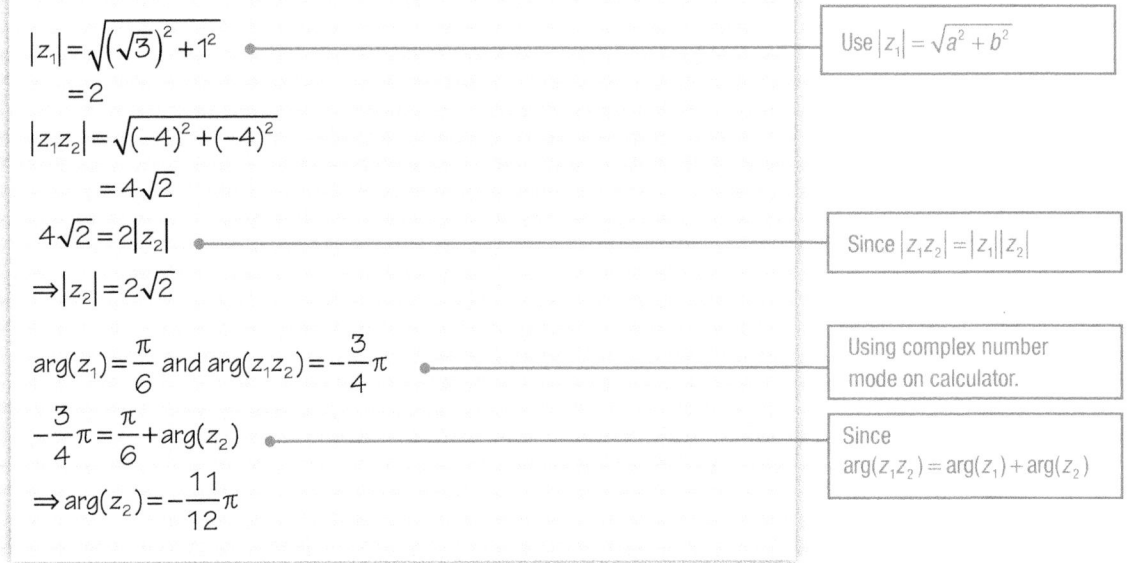

$$|z_1| = \sqrt{\left(\sqrt{3}\right)^2 + 1^2}$$
$$= 2$$

Use $|z_1| = \sqrt{a^2 + b^2}$

$$|z_1 z_2| = \sqrt{(-4)^2 + (-4)^2}$$
$$= 4\sqrt{2}$$

$$4\sqrt{2} = 2|z_2|$$
$$\Rightarrow |z_2| = 2\sqrt{2}$$

Since $|z_1 z_2| = |z_1||z_2|$

$$\arg(z_1) = \frac{\pi}{6} \text{ and } \arg(z_1 z_2) = -\frac{3}{4}\pi$$

Using complex number mode on calculator.

$$-\frac{3}{4}\pi = \frac{\pi}{6} + \arg(z_2)$$

Since $\arg(z_1 z_2) = \arg(z_1) + \arg(z_2)$

$$\Rightarrow \arg(z_2) = -\frac{11}{12}\pi$$

Instead of the form $z = a + bi$, sometimes known as Cartesian form, you can write complex numbers in **modulus** and **argument** form.

You can see in the diagram that the real component of r is $r\cos\theta$ and the imaginary component is $r\sin\theta$

Key point

The **modulus-argument form** of the complex number $z = a + bi$ is given by $z = r(\cos\theta + i\sin\theta)$ where r is the modulus of z and θ is the argument.

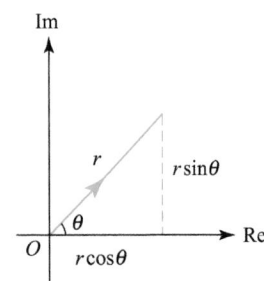

Example 3

Write the number $z = 7 - i$ in modulus-argument form.

The argument is measured in radians, where π radians is equal to $180°$

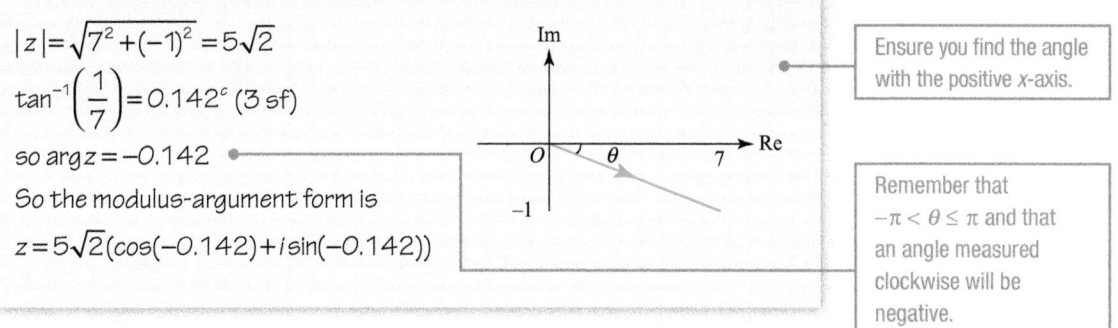

$$|z| = \sqrt{7^2 + (-1)^2} = 5\sqrt{2}$$

$$\tan^{-1}\left(\frac{1}{7}\right) = 0.142^c \text{ (3 sf)}$$

so $\arg z = -0.142$

So the modulus-argument form is

$$z = 5\sqrt{2}(\cos(-0.142) + i\sin(-0.142))$$

Ensure you find the angle with the positive x-axis.

Remember that $-\pi < \theta \leq \pi$ and that an angle measured clockwise will be negative.

Calculator

Try it on your calculator

Some calculators can be used to convert to and from modulus-argument form. Find out how to convert $\sqrt{3} + i$ to modulus-argument form on *your* calculator. Also find out how to enter a number in modulus-argument form and convert back to the form $a + bi$

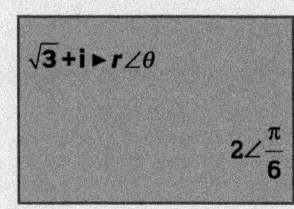

$\sqrt{3} + i \blacktriangleright r\angle\theta$

$2\angle\dfrac{\pi}{6}$

You can draw **loci** in an Argand diagram. These are the set of points that obey a given rule.

Example 4

Sketch the locus of points that satisfy $|z| = 4$

This will be a circle, centre the origin and radius 4

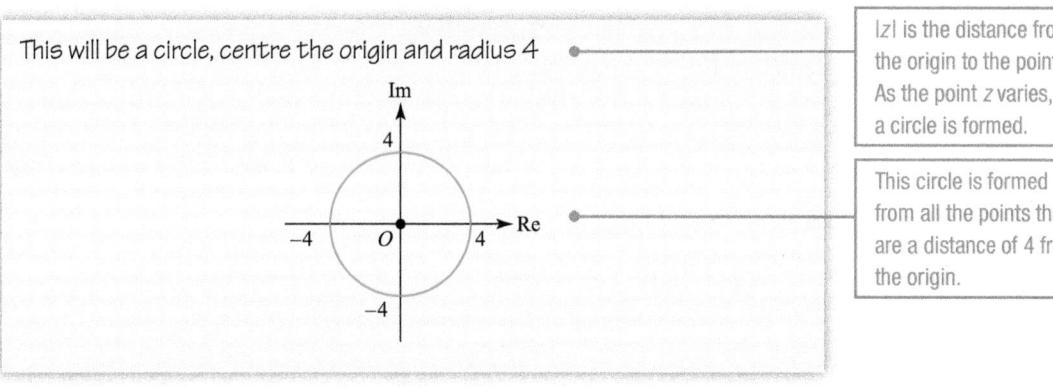

$|z|$ is the distance from the origin to the point z. As the point z varies, a circle is formed.

This circle is formed from all the points that are a distance of 4 from the origin.

If we have any fixed point z_1, then

Key point

The locus of points satisfying $|z - z_1| = r$ will be a circle centre z_1 and radius r

Example 5

Sketch the locus of points that satisfy $|z-2+3i|=2$

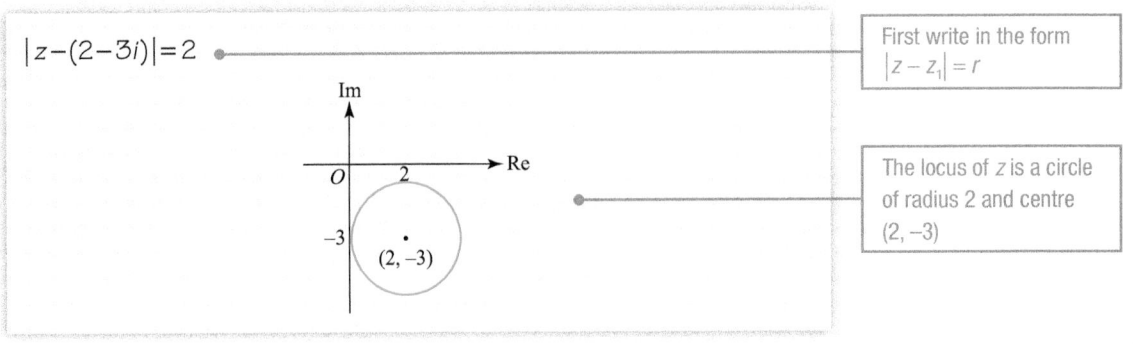

$|z-(2-3i)|=2$ — First write in the form $|z-z_1|=r$

The locus of z is a circle of radius 2 and centre $(2, -3)$

Example 6

Sketch the locus of points satisfying $\arg(z)=\dfrac{2\pi}{3}$

This will be a line that makes an angle of $\dfrac{2\pi}{3}$ with the positive real axis. — $\arg(z)$ is the angle between the positive real axis and the line representing z

Notice that the line ends at the origin.

> **Key point**
>
> The locus of points satisfying $\arg(z-z_1)=\theta$ is a **half-line** from the point z_1 at an angle of θ to the positive real axis.

Example 7

Sketch the locus of z where $\arg(z+2+3i)=\dfrac{\pi}{6}$

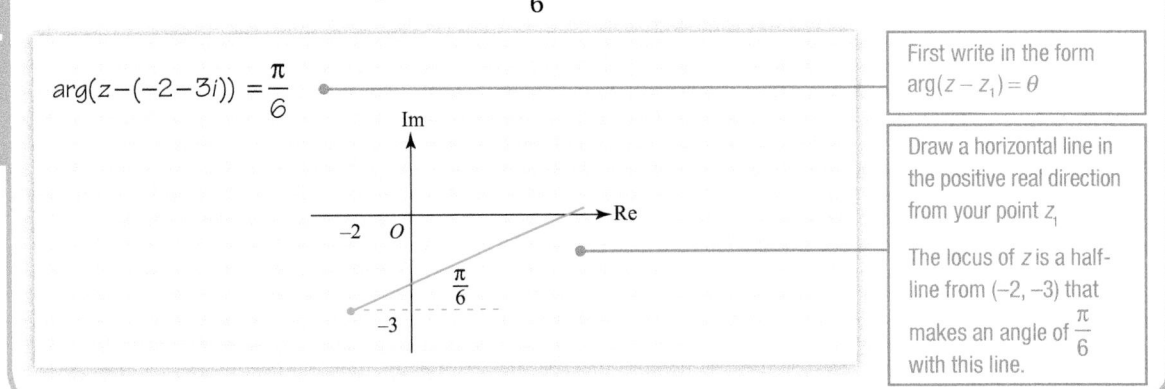

$\arg(z-(-2-3i))=\dfrac{\pi}{6}$ — First write in the form $\arg(z-z_1)=\theta$

Draw a horizontal line in the positive real direction from your point z_1

The locus of z is a half-line from $(-2, -3)$ that makes an angle of $\dfrac{\pi}{6}$ with this line.

The locus of points satisfying $|z-z_1|=|z-z_2|$ is the perpendicular bisector of the line joining z_1 and z_2

Key point

This is because the locus includes all points that are equidistant from the fixed points z_1 and z_2

Example 8

Sketch the locus of points satisfying $|z-2|=|z+3i|$

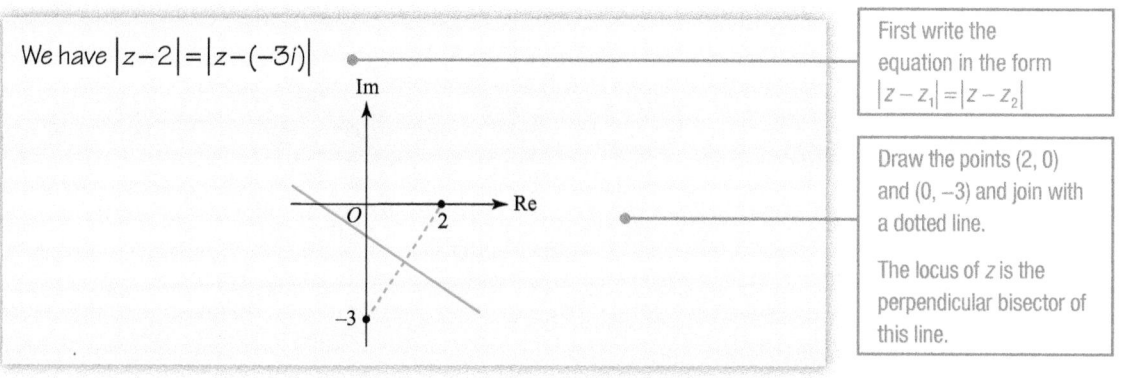

We have $|z-2|=|z-(-3i)|$

First write the equation in the form $|z-z_1|=|z-z_2|$

Draw the points (2, 0) and (0, −3) and join with a dotted line.

The locus of z is the perpendicular bisector of this line.

Exercise 1.4A Fluency and skills

1 Find the modulus and argument of each complex number.

a $12+5i$ b $4-3i$

c $-3i$ d $-6-8i$

e $-1+7i$ f $-2-i$

g $-\sqrt{2}-\sqrt{2}i$ h $\sqrt{3}-\sqrt{6}i$

2 Verify in each case that $|zw|=|z||w|$ and $\left|\dfrac{z}{w}\right|=\dfrac{|z|}{|w|}$

a $z=1+3i, w=-5+2i$

b $z=-2-i, w=\sqrt{5}i$

c $z=-\sqrt{3}+6i, w=1-\sqrt{3}i$

3 In each case, verify that $\arg(zw)=\arg z+\arg w$ and $\arg\left(\dfrac{z}{w}\right)=\arg z-\arg w$

a $z=1+i, w=3+\sqrt{3}i$

b $z=i, w=2-2i$

c $z=\sqrt{3}-3i, w=-2i$

4 Given that $z=1-\sqrt{3}i$ and $zw=\sqrt{3}+3i$, find the modulus and the argument of w

5 Given that $z_1=-1-i$ and $\dfrac{z_1}{z_2}=3\sqrt{2}+\sqrt{6}i$, find the modulus and the argument of z_2

6 Given that $z=-2\sqrt{3}+6i$ and $\dfrac{w}{z}=\sqrt{3}-i$, find the modulus and the argument of w

7 Write each of these complex numbers in Cartesian form.

a $3\left(\cos\left(\dfrac{\pi}{2}\right)+i\sin\left(\dfrac{\pi}{2}\right)\right)$

b $5(\cos(-\pi)+i\sin(-\pi))$

c $10\left(\cos\left(\dfrac{5\pi}{6}\right)+i\sin\left(\dfrac{5\pi}{6}\right)\right)$

d $\sqrt{3}\left(\cos\left(-\dfrac{2\pi}{3}\right)+i\sin\left(-\dfrac{2\pi}{3}\right)\right)$

8 Write each of these numbers in modulus-argument form.

a $z=3+3i$ b $z=1-\sqrt{3}i$

c $z=-2\sqrt{3}-2i$ d $z=-4+9i$

9 Sketch the locus of z in each case.

a $|z|=7$ b $|z-2|=5$

c $|z-i|=3$ d $|z-(1+2i)|=2$

e $|z-3+5i|=5$ f $|z+4-2i|=4$

10 Sketch the locus of z in each case.

a $\arg z = -\dfrac{\pi}{4}$ **b** $\arg(z-3) = \dfrac{\pi}{2}$

c $\arg(z+i) = \dfrac{3\pi}{4}$ **d** $\arg(z-2i) = -\dfrac{\pi}{6}$

e $\arg(z-2+i) = \dfrac{5\pi}{6}$ **f** $\arg(z-4-i) = -\dfrac{2\pi}{3}$

g $\arg(z+5-7i) = -\dfrac{\pi}{3}$

11 Sketch the locus of z in each case.

a $|z| = |z+4|$ **b** $|z-2i| = |z|$

c $|z-2i| = |z+2|$ **d** $|z+6+2i| = |z+6|$

e $|z+4-i| = |z-5+2i|$

Reasoning and problem-solving

You need to be able to prove the results: $\left|\dfrac{z}{w}\right| = \dfrac{|z|}{|w|}$, $|zw| = |z||w|$, $\arg\left(\dfrac{z}{w}\right) = \arg z - \arg w$ and $\arg(zw) = \arg z + \arg w$

Strategy 1

To prove results about modulus and argument

(**1**) Write complex numbers in modulus-argument form.

(**2**) Simplify powers of i

(**3**) Split into real and imaginary parts.

(**4**) Use the addition formulae for sine and for cosine.

Example 9

Prove that $\left|\dfrac{z}{w}\right| = \dfrac{|z|}{|w|}$ and $\arg\left(\dfrac{z}{w}\right) = \arg z - \arg w$ for all $z, w \in \mathbb{C}$

Let $z = |z|(\cos A + i \sin A)$ and $w = |w|(\cos B + i \sin B)$ ←   Write both numbers in modulus-argument form. (**1**)

Then $\dfrac{z}{w} = \dfrac{|z|(\cos A + i \sin A)}{|w|(\cos B + i \sin B)}$

$= \dfrac{|z|(\cos A + i \sin A)(\cos B - i \sin B)}{|w|(\cos B + i \sin B)(\cos B - i \sin B)}$ ←   Expand the brackets.

$= \dfrac{|z|(\cos A \cos B - i \cos A \sin B + i \cos B \sin A - i^2 \sin A \sin B)}{|w|(\cos^2 B - i^2 \sin^2 B)}$ ←   Use the fact that $i^2 = -1$ (**2**)

$= \dfrac{|z|(\cos A \cos B + \sin A \sin B + i(\cos B \sin A - \cos A \sin B))}{|w|(\cos^2 B + \sin^2 B)}$ ←   Separate real and imaginary parts. (**3**)

$= \dfrac{|z|(\cos(A-B) + i \sin(A-B))}{|w|(1)}$

$= \dfrac{|z|}{|w|}(\cos(A-B) + i \sin(A-B))$ ←   Use $\cos(A-B)$ $= \cos A \cos B + \sin A \sin B$ and $\sin(A-B)$ $= \cos A \sin B - \sin A \cos B$ (**4**)

Therefore the number $\dfrac{z}{w}$ has modulus $\dfrac{|z|}{|w|}$ and argument $A - B$

So $\left|\dfrac{z}{w}\right| = \dfrac{|z|}{|w|}$ and $\arg\left(\dfrac{z}{w}\right) = A - B = \arg z - \arg w$

You know how to find the Cartesian equation of certain loci by drawing the graph and using known equations of circles or lines. However, it is possible to find the Cartesian equation of any locus by setting $z = x + iy$ and finding a relationship between x and y

To find the Cartesian equation of a locus

① Write z as $x + iy$

② Calculate the modulus.

③ Use tan to form an equation from the argument.

④ Rearrange to the required form.

See Maths Ch1.5

For a reminder of the equations of lines and circles.

Example 10

Find the Cartesian equations of these loci.

a $|z - 3 + 4i| = \sqrt{5}$

b $\arg(z + i) = -\dfrac{\pi}{6}$

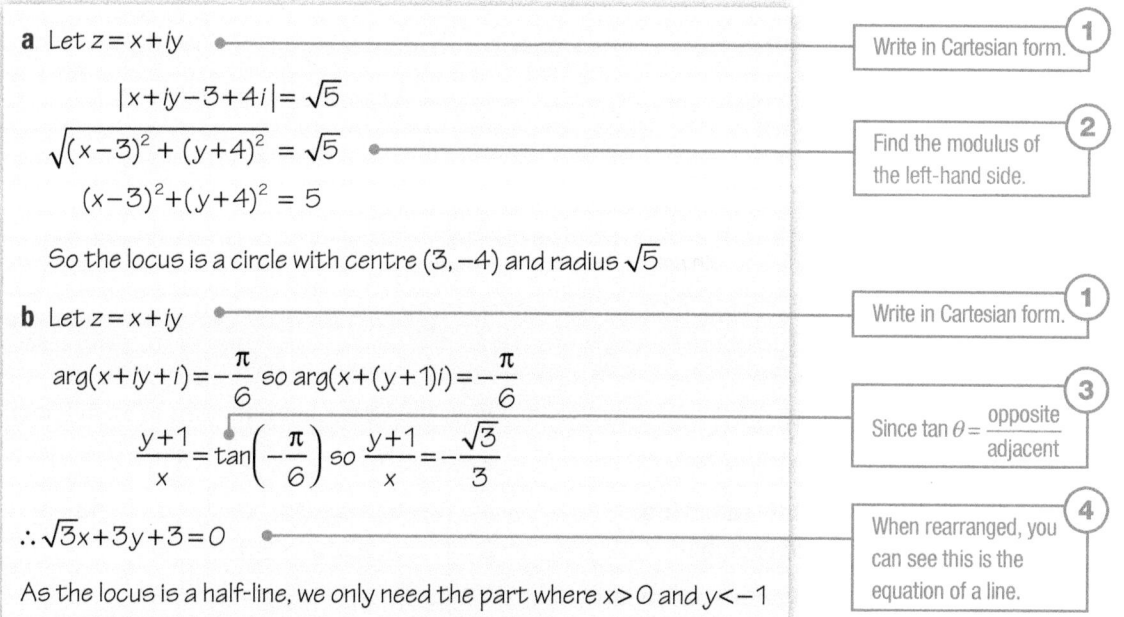

a Let $z = x + iy$

$$|x + iy - 3 + 4i| = \sqrt{5}$$

$$\sqrt{(x-3)^2 + (y+4)^2} = \sqrt{5}$$

$$(x-3)^2 + (y+4)^2 = 5$$

So the locus is a circle with centre $(3, -4)$ and radius $\sqrt{5}$

Write in Cartesian form. ①

Find the modulus of the left-hand side. ②

b Let $z = x + iy$

$$\arg(x + iy + i) = -\frac{\pi}{6} \text{ so } \arg(x + (y+1)i) = -\frac{\pi}{6}$$

$$\frac{y+1}{x} = \tan\left(-\frac{\pi}{6}\right) \text{ so } \frac{y+1}{x} = -\frac{\sqrt{3}}{3}$$

$$\therefore \sqrt{3}x + 3y + 3 = 0$$

As the locus is a half-line, we only need the part where $x > 0$ and $y < -1$

Write in Cartesian form. ①

Since $\tan \theta = \dfrac{\text{opposite}}{\text{adjacent}}$ ③

When rearranged, you can see this is the equation of a line. ④

To find a region bounded by a locus

① Sketch the locus of the boundary of the region.

② Test a point to see if it is inside the region or not.

③ Shade the correct area.

Example 11

Shade these sets of points.

a $\{z \in \mathbb{C} : |z-5| \geq 3\}$

b $\left\{z \in \mathbb{C} : 0 < \arg(z-4i) < \dfrac{3\pi}{4}\right\}$

$\{z \in \mathbb{C} : |z-5| \geq 3\}$ is the set of complex numbers z such that $|z-5| \geq 3$

See Maths Ch 1.7
For a reminder of set notation.

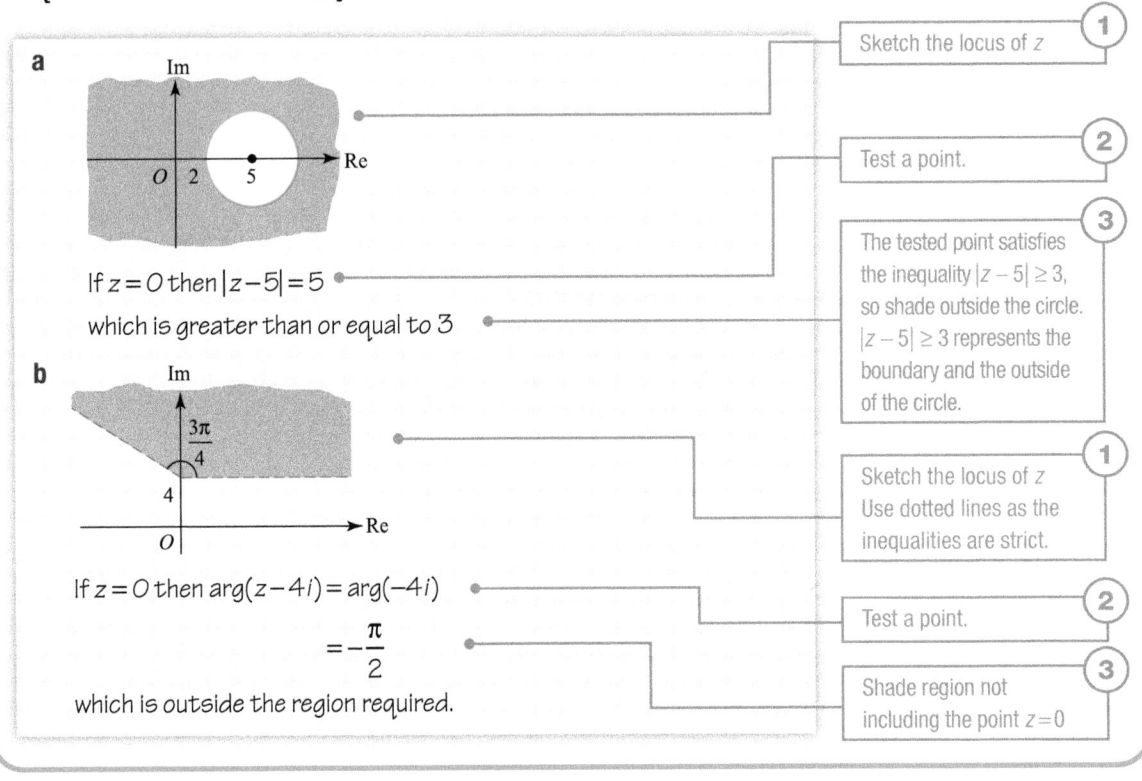

a

If $z = 0$ then $|z-5| = 5$

which is greater than or equal to 3

Sketch the locus of z ①

Test a point. ②

③ The tested point satisfies the inequality $|z-5| \geq 3$, so shade outside the circle. $|z-5| \geq 3$ represents the boundary and the outside of the circle.

b

If $z = 0$ then $\arg(z-4i) = \arg(-4i)$

$= -\dfrac{\pi}{2}$

which is outside the region required.

① Sketch the locus of z Use dotted lines as the inequalities are strict.

② Test a point.

③ Shade region not including the point $z = 0$

Exercise 1.4B Reasoning and problem-solving

1 Prove that $|zw| = |z||w|$ and $\arg(zw) = \arg z + \arg w$ for all $z, w \in \mathbb{C}$

2 Show that the locus of points satisfying $|z+3-2i| = 4$ is a circle and sketch it.

3 a Show that the locus of points satisfying $|z-2-i| = 2$ is a circle.

   The circle touches the imaginary axis at the point A and crosses the real axis at B and C

   b Calculate the exact area of triangle ABC

4 Find the Cartesian equation of each locus.

   a $|z-5| = |z|$ b $|z+2| = |z-2i|$

   c $|z-4i| = |z+2|$ d $|z+1-i| = |z-3|$

   e $|z-5+i| = |z+i-2|$

   f $|z-7+4i| = |z+6-3i|$

5 In each case, find the Cartesian equation of the line on which the half-line lies.

   a $\arg(z-3i) = \dfrac{\pi}{4}$ b $\arg(z+5) = \dfrac{\pi}{2}$

   c $\arg(z+2-i) = \dfrac{\pi}{3}$ d $\arg(z-4+i) = \dfrac{2\pi}{3}$

6 Find the Cartesian equation of the locus of points satisfying $|z-2|=|z+3-i|$

7 Find the Cartesian equation of the locus of $|z-\sqrt{2}+i|=|z-1-\sqrt{2}|$

8 Find the Cartesian equation of the locus of $|z-2|=|4i-z|$

9 Shade the region that satisfies

   a $|z-4i|\le 3$ b $|z+2-i|\ge 1$

   c $4\le|z|\le 10$ d $2<|z-5+2i|<5$

10 Find the area of the region that satisfies $\sqrt{7}\le|z+3-7i|\le 7$

11 Shade the region represented by each inequality.

   a $0\le\arg z\le\dfrac{\pi}{3}$ b $-\dfrac{5\pi}{6}<\arg(z-3i)<0$

   c $\dfrac{\pi}{2}\le\arg(z-2-i)<\dfrac{3\pi}{4}$

12 Find the area of the region that satisfies $|z-5+2i|<8$ and $0\le\arg(z-5+2i)\le\dfrac{\pi}{2}$

13 Shade the region that satisfies

   a $|z-3|\ge|z+5|$ b $|z-i|<|z-3i|$

   c $|z-2-4i|\le|z+8-4i|$

   d $|z-3-i|>|z-5+3i|$

   e $|z-1+2i|\le|z-3-2i|$ and $|z|>|z-2|$

14 Sketch and shade the region that satisfies both $|z-3+3i|\le 3$ and $-\dfrac{\pi}{4}\le\arg z\le 0$

15 Sketch and shade the region that satisfies both $|z|\le|z-8i|$ and $|z-2i|\ge 8$

16 Sketch and shade the region that satisfies both $\dfrac{\pi}{3}<\arg(z-2)<\dfrac{2\pi}{3}$ and $|z-2i|<|z-4i|$

17 Shade on an Argand diagram the set of points $\{z\in\mathbb{C}:|z+i-2|\le 1\}$

18 Shade on an Argand diagram the set of points $\{z\in\mathbb{C}:|z+2i|>|z-2|\}$

19 Shade on an Argand diagram the set of points $\left\{z\in\mathbb{C}:|z|\ge 4\right\}\cap\left\{z\in\mathbb{C}:-\dfrac{\pi}{3}<\arg(z)<\dfrac{\pi}{3}\right\}$

20 Shade on an Argand diagram the set of points $\left\{z\in\mathbb{C}:|z-3i|>|z+5i|\right\}\cap\left\{z\in\mathbb{C}:-\pi\le\arg(z)\le-\dfrac{\pi}{2}\right\}$

Full A Level

21 a Use algebra to show that the locus of points satisfying $|z+3|=2|z-6i|$ is a circle, then sketch it.

   b Shade the region that satisfies $|z+3|\le 2|z-6i|$ and $|z-1-3i|\le 20$

22 The point P represents a complex number z on an Argand diagram such that $|z-3+i|=3$

   a State the Cartesian equation of the locus of P

   The point Q represents a complex number z on an Argand diagram such that $|z+2-i|=|z-1+2i|$

   b Find the Cartesian equation of the locus of Q

   c Find the complex number that satisfies both $|z-3+i|=3$ and $|z+2-i|=|z-1+2i|$, giving your answer in surd form.

23 Find the complex number that satisfies both $|z-3i|=4$ and $\arg(z-3i)=-\dfrac{\pi}{4}$

1 Summary and review

Chapter summary

- The imaginary number i is defined as $i = \sqrt{-1}$
- Complex numbers written in the form $a + bi$ can be added, subtracted and multiplied in the same way as algebraic expressions.
- Powers of i should be simplified: $i^2 = -1$, $i^3 = -i$ and so on.
- The complex conjugate of the number $z = a + bi$ is $z^* = a - bi$
- Fractions with a complex number in the denominator can be simplified by multiplying the numerator and the denominator by the complex conjugate of the denominator.
- Complex roots of polynomial equations with real coefficients occur in conjugate pairs.
- The complex number $z = a + bi$ can be represented by the point (a, b) on an Argand diagram.
- The modulus of a complex number $z = a + bi$ is given by $|z| = \sqrt{a^2 + b^2}$
- The (principal) argument of a complex number is the angle between the vector representing it and the positive real axis. Write $\arg z = \theta$ where $-\pi < \theta \le \pi$
- $|z_1 z_2| = |z_1||z_2|$ and $\left|\dfrac{z_1}{z_2}\right| = \dfrac{|z_1|}{|z_2|}$ for all $z_1, z_2 \in \mathbb{C}$
- $\arg(z_1 z_2) = \arg(z_1) + \arg(z_2)$ for all $z_1, z_2 \in \mathbb{C}$
- The modulus-argument form of the complex number $z = a + bi$ is given by $z = r(\cos\theta + i\sin\theta)$ where r is the modulus of z and θ is the argument.
- The locus of points satisfying $|z - z_1| = r$ will be a circle, centre z_1 and radius r
- The locus of points satisfying $\arg(z - z_1) = \theta$ is a half-line from the point z_1 at an angle of θ to the positive real axis.
- The locus of points satisfying $|z - z_1| = |z - z_2|$ is the perpendicular bisector of the line joining z_1 and z_2

Check and review

You should now be able to...	Try Questions
✔ Add, subtract, multiply and divide complex numbers in the form $a + bi$	1
✔ Understand and use the complex conjugate.	1–3
✔ Solve quadratic, cubic and quartic equations with real coefficients.	2–4
✔ Calculate the modulus and the argument of a complex number.	5
✔ Convert between modulus-argument form and the form $a + bi$	6, 7
✔ Multiply and divide numbers in modulus-argument form.	8
✔ Sketch and interpret Argand diagrams.	9
✔ Construct and interpret loci in the Argand diagram.	10–14

1 Given that $z = 5 - 4i$ and $w = -2 - 3i$, find each of these in the form $a + bi$

  a $z + w$ b $3w$

  c $2z - w$ d zw

  e $z^2 + w^2$ f $(z + w)^2$

  g z^* h w^*

  i $\dfrac{z}{w}$ j $\dfrac{3w}{2z}$

  k $2 \div z$ l $w^* \div 3i$

2 $f(x) = ax^2 + bx + c = 0$ where a, b and c are real numbers.

  Given that $f(x)$ has a root $x = -7 + 2i$

  find possible values of a, b and c

3 $f(z) = z^3 + az^2 + 93z - 130$ where a is a real number.

  Given that $f(2) = 0$,

  a Find the value of a

  b Solve the equation $f(z) = 0$

4 Solve the quartic equation
$$x^4 + ax^3 - ax^2 + bx + 169 = 0$$
given that a and b are real constants and that $3 + 2i$ is one solution of the equation.

5 Calculate the modulus and the argument of these complex numbers.

  a $2 + 9i$ b $3 - 3i$

  c $7i$ d $-2i$

  e $-1 + 4i$ f $-3 - 4i$

6 Write these complex numbers in modulus-argument form.

  a $8 + 6i$ b $-12 + 5i$

  c $-2 - 2i$ d $\sqrt{3} - i$

  e $\sqrt{5}\left(\cos\left(\dfrac{\pi}{3}\right) - i\sin\left(\dfrac{\pi}{3}\right)\right)$

7 Write these in the form $a + bi$

  a $2\left(\cos\left(\dfrac{\pi}{6}\right) + i\sin\left(\dfrac{\pi}{6}\right)\right)$

  b $\sqrt{3}\left(\cos\left(-\dfrac{\pi}{4}\right) + i\sin\left(-\dfrac{\pi}{4}\right)\right)$

8 Given that
$$z = \sqrt{6}\left(\cos\left(-\dfrac{\pi}{3}\right) + i\sin\left(-\dfrac{\pi}{3}\right)\right)$$
and
$$w = \sqrt{3}\left(\cos\left(\dfrac{\pi}{6}\right) + i\sin\left(\dfrac{\pi}{6}\right)\right)$$
find these in modulus-argument form.

  a zw b $\dfrac{z}{w}$

  c $\dfrac{w}{z}$

9 Given that $z_1 = 5 - 2i$ and $z_2 = -2 + 3i$, draw these complex numbers on the same Argand diagram.

  a z_1 b z_2

  c $z_1 + z_2$ d $z_1 - z_2$

10 Sketch these loci on separate Argand diagrams and give the Cartesian equation of each.

  a $|z| = 7$ b $|z - 8| = 5$

  c $|z + 3 - i| = 3$ d $|z - 2 - 3i| = 2$

11 Sketch these loci on separate Argand diagrams.

  a $\arg z = \dfrac{\pi}{6}$ b $\arg(z - 3) = -\dfrac{\pi}{3}$

  c $\arg(z + 2i) = \dfrac{2\pi}{3}$ d $\arg(z + 1 - i) = -\dfrac{\pi}{4}$

12 Sketch these loci on separate Argand diagrams and give the Cartesian equation of each.

  a $|z| = |z - 6|$ b $|z + 2i| = |z - 8i|$

  c $|z - 2i| = |z + 4|$ d $|z - 1 - i| = |z + 1 + i|$

13 Sketch and shade the region satisfying each inequality.

  a $|z - 5 + 2i| \le 2$ b $\dfrac{\pi}{2} \le \arg(z - i) < \dfrac{5\pi}{6}$

  c $3 \le |z + 3 - 5i| \le 5$ d $|z - 6i| < |z + 6|$

14 Sketch and shade the region satisfying both of these inequalities.

  $|z - 7| \ge 7$ and $-\dfrac{\pi}{2} \le \arg(z - 7) \le 0$

History

Jean-Robert Argand was a Swiss mathematician who, early in the 19th century, moved to Paris where he managed a bookshop. He published an essay on the representation of imaginary quantities in 1813 and, although others before him had written about similar ideas, it was Argand's essay that gained attention and resulted in the naming of the **Argand diagram** we continue to use today.

Note

A **conic section** is a curve that is found by intersecting a plane with a double cone placed vertex to vertex.
There are three types of conic section: ellipse, parabola and hyperbola. A circle is a type of ellipse.

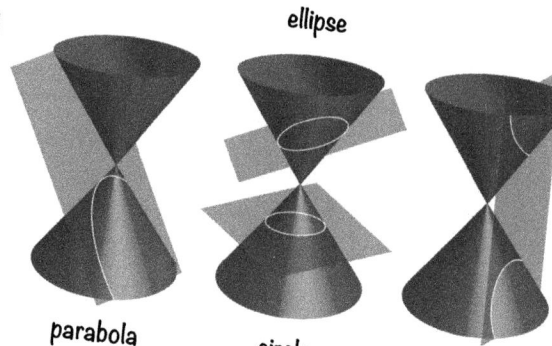

ellipse

parabola   circle   hyperbola

Investigation

The equation $|z| = 1$ plotted in the complex plane gives a circle with unit radius centred at the origin.
Choose values of a, b and c such that $c > (b - a)$ and $b > a$
By varying these values, investigate the graph of the locus of point z that moves so that $|z - a| + |z - b| = c$, $c > (b - a)$ and $b > a$
What is the shape of the graph that results?
The real part of a complex number z can be denoted by $\text{Re}(z)$
By varying the values of a and b, investigate graphs of the locus of a point z that moves so that $\text{Re}(z + a) = |z - b|$
What is the shape of the graph that results?

Research

The **Mandelbrot set** provides beautiful images. It is the set of all the complex numbers, c, for which the iterative sequence $z_{n+1} = z_n^2 + c$, where $z_0 = 0$ does not tend to infinity.
Computer images of the Mandelbrot set can be developed and, no matter how closely you zoom in on the surface, it still remains incredibly intricate. The Mandelbrot set is an example of a **fractal**.
Find out if the complex numbers i and $2i$ are in the Mandelbrot set.
Research the properties of the Mandelbrot set.

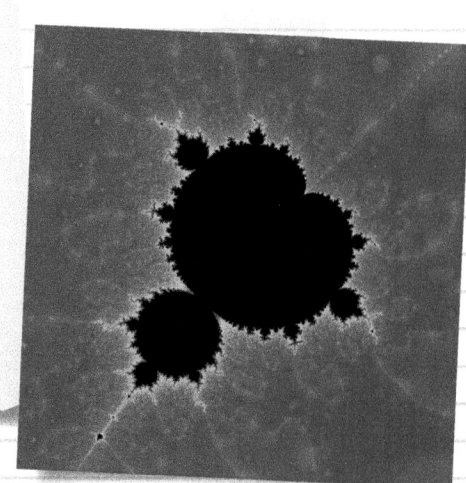

1 The complex numbers z and w are given as $z = 9 - 11i$ and $w = 7 + 3i$

    a Calculate these expressions, giving the answers in the form $a + bi$

        i $w - z$ **ii** wz **iii** $\dfrac{w}{z}$ **[5 marks]**

    b Calculate the modulus and the argument of z, giving your answers to 3 significant figures.   **[3]**

2 The quadratic equation $x^2 + 2x + 5 = 0$ has complex roots α and β

    a Find α and β **[2]**

    b Show α and β on an Argand diagram.   **[2]**

    c Describe the transformation that maps α to β **[1]**

3 $z = \dfrac{5 + 2i}{3 + i}$

    a Show that $z = a + bi$ where a and b are constants to be found.   **[3]**

    b Calculate the value of

        i $z + z^*$ **ii** $z - z^*$ **iii** zz^* **[4]**

    c **i** Draw z and iz on the same Argand diagram.

        ii Describe the geometric relationship between z and iz **[3]**

4 Find the values of a and b such that $(a + bi)^2 = -15 + 8i$ **[6]**

5 Find the values of a and b such that $(a + bi)^2 = 1 + 2\sqrt{2}i$ **[6]**

6 $f(x) = x^3 - 2x^2 + kx + 26$ for some real constant k

    a Given that $x = 2 - 3i$ is a solution of $f(x) = 0$, find all the roots of $f(x)$ **[3]**

    b Find the area of the triangle formed by the three roots of $f(x)$ **[2]**

7 Given that $5 - i$ is a root of the equation $x^3 + ax^2 + bx - 182 = 0$

    a Find the other two roots,   **[4]**

    b Calculate the values of a and b **[3]**

8 Solve the equation $z - 2z^* = -6 + 27i$ **[3]**

9 Solve the simultaneous equations

$$z - 2w = 8 + 5i$$
$$2z - 3w = 13 + 3i$$

    [4]

10 Given that $z = 3 + 3\sqrt{3}i$ and $w = -\sqrt{2} + \sqrt{2}i$

    a Calculate the modulus and argument of z **[3]**

    b Calculate the modulus and argument of w **[3]**

c Hence, or otherwise, find the value of

   i $|zw|$ **ii** $\left|\dfrac{z}{w}\right|$

   iii $\arg(zw)$ **iv** $\arg\left(\dfrac{z}{w}\right)$ **[6]**

11 Given that $z = 4+\sqrt{2}i$ and $zw = 5\sqrt{2}-2i$

  a Find w in modulus-argument form.   **[6]**

  The points A and B represent w and zw on an Argand diagram.

  b Calculate the distance AB, giving your answer as a surd.   **[2]**

12 a Write the complex number $z = \sqrt{3}-i$ in modulus-argument form.   **[3]**

  The complex number w has argument $-\dfrac{\pi}{12}$

  b Find the argument of

    i zw **ii** $\dfrac{z}{w}$ **[4]**

  c Find $|w|$ given that $|zw| = 10$ **[2]**

13 a Sketch the locus of points that satisfy

    i $|z-4i| = 2$ **ii** $|z+3| = |z-3i|$ **[4]**

  b Find the Cartesian equation of the locus of points drawn in part **a**.   **[3]**

14 a Show that the locus of points satisfying $|z+3-4i| = 4$ is a circle.   **[3]**

  b Sketch the locus of points satisfying $|z+3-4i| = 4$ **[2]**

  c Shade in the region that satisfies $|z+3-4i| \le 4$ **[2]**

15 For the locus of points satisfying $|z+5| = |z-i|$

  a Sketch the locus,   **[2]**

  b Find the Cartesian equation.   **[3]**

16 Find the square roots of the complex number $39-80i$ **[6]**

17 Solve the equation $z^2 = 4-2\sqrt{5}i$ **[6]**

18 The solutions of the quartic equation $x^4 - 14x^3 + ax^2 - bx + 58 = 0$ are represented on an Argand diagram by the points P, Q, R and S

  Given that a and b are real constants and that one solution of the equation is $w = 7+3i$

  calculate the area of the quadrilateral $PQRS$ **[6]**

19 Two solutions of a cubic equation with real coefficients are $x = -3$ and $x = i-4$

  a State the third solution of the equation.   **[1]**

  b Find a possible equation with these solutions.   **[3]**

20 Two solutions of a quartic equation with real coefficients are $z = -2+i$ and $z = 3-5i$

  a State the two other solutions of the equation.   **[2]**

  b Find a possible equation with these solutions.   **[3]**

21 A quartic equation with real coefficients $ax^4 + bx^3 + cx^2 + dx + e = 0$ has exactly two distinct solutions, one of which is $1 + 5i$

Find the values of a, b, c, d and e **[4]**

22 Solve the simultaneous equations

$z^2 - w^2 = -6$

$z + 2w = 3$ **[6]**

23 Solve the simultaneous equations.

$z^2 + w^2 = 30$

$z + 3w = 20$ **[4]**

24 The equation $2x^5 + x^4 + 36x^3 + 18x^2 + 162x + 81 = 0$ has a repeated quadratic root.

  a Show that $x = 3i$ is a solution of the equation. **[2]**

  b Fully factorise the equation. **[3]**

25 The curve of $y = x^4 - 4x^3 - x^2 + 6x + 18$ is shown.

Find all the solutions of the equation
$x^4 - 4x^3 - x^2 + 6x + 18 = 0$ **[4]**

26 A quadratic equation has roots α and β, where α has argument $\dfrac{5\pi}{6}$ and modulus 3

  a Find β in modulus-argument form. **[2]**

  b Calculate

    i $|\alpha\beta|$ **ii** $\left|\dfrac{\alpha}{\beta}\right|$ **iii** $\arg(\alpha\beta)$ **[6]**

  c Find the quadratic equation with roots α and β **[3]**

27 The complex numbers z_1 and z_2 are given by $z_1 = 3 + ai$ and $z_2 = 2 - i$ where a is an integer.

  a Find $\dfrac{z_1}{z_2}$ in terms of a **[3]**

Given that $\left|\dfrac{z_1}{z_2}\right| = \sqrt{18}$

  b Find the possible values of a **[5]**

28 Given that $w = \dfrac{1}{2}\left(\cos\dfrac{\pi}{3} - i\sin\dfrac{\pi}{3}\right)$

  a Write w in modulus-argument form, **[1]**

  b Find the complex number z such that $\arg(wz) = -\dfrac{\pi}{6}$ and $\left|\dfrac{w^2}{z}\right| = 3$

Write your answer in exact Cartesian form. **[3]**

29 Describe the locus shown in terms of z **[2]**

30 a Sketch the locus of points that satisfy $\arg(z-2i)=\dfrac{\pi}{4}$ **[3]**

b Find the Cartesian equation of the locus drawn in part **a**.   **[3]**

31 Sketch and shade the region satisfying

$0 < \arg(z+1+3i) \leq \dfrac{2\pi}{3}$ **[5]**

32 Sketch and shade the region satisfying

$|z-i| \geq 1$ and $|z+i| \leq 2$ **[4]**

33 The complex numbers $4-3i$ and $-2-i$ represent the points A and B respectively.

Find the area of the triangle OAB **[5]**

34 The locus of the complex number z is a half-line as shown.

Describe the locus in terms of z **[3]**

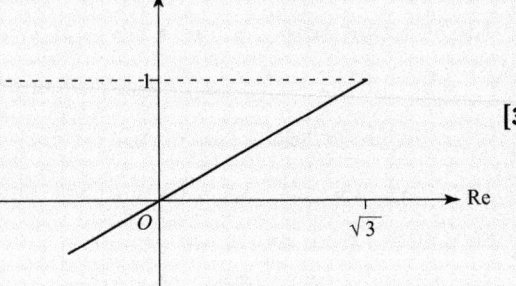

35 The point P represents a complex number z on an Argand diagram such that

$|z+2| = 3|z-2i|$

Show that, as z varies, the locus of P is a circle and state the radius of the circle and the coordinates of the centre.   **[6]**

36 The point P represents a complex number z on an Argand diagram such that

$|z-1| = 1$

The point Q represents a complex number z on an Argand diagram such that

$\arg(z-i) = -\dfrac{\pi}{4}$

a Sketch the loci of P and Q on the same axes.   **[4]**

b Find the complex number that satisfies both

$|z-1| = 1$ and $\arg(z-i) = -\dfrac{\pi}{4}$ **[6]**

37 a Find the possible complex numbers, z, that satisfy both $|z+3i| = |z-i|$ and $|z-2+i| = 4$ **[4]**

b Sketch the region that satisfies both $|z+3i| > |z-i|$ and $|z-2+i| \geq 4$ **[5]**

38 a Shade the region that satisfies $1 \leq |z-3i| \leq 3$ **[4]**

b Find the exact area of the shaded region.   **[2]**

39 a Shade the region that satisfies both

$|z+i-3| \leq 2$ and $-\dfrac{\pi}{2} \leq \arg(z+i) \leq 0$ **[4]**

b Find the exact area of the shaded region.   **[2]**

2 Algebra and series

Polynomials have many uses, and one use of particular interest is in programming the movement of robotic arms. These arms are used commonly in industry, for example, in the manufacturing of cars. Polynomials are used to determine the smooth paths that the arm joints should follow, to ensure that the mechanical components do not wear too quickly. The trajectory of a joint is determined as the combination of several polynomial functions, which are stitched together at various points. Algebraic methods are used to ensure that the path of the joint is smooth and continuous.

Another, perhaps surprising, application of polynomials is in in various aspects of computer graphics, and particularly in the design of fonts. Although the algebra and numerical methods required for good font design are quite complex, they are developed from our basic understanding of simple polynomials such as quadratics and cubics.

Orientation

What you need to know	What you will learn	What this leads to
KS4 • Inequalities. • Proof. **Maths Ch1** • Argument and proof. • Quadratic functions. **Maths Ch2** • Expanding and factorising polynomials.	• To relate the roots of a polynomial to its coefficients. • To find a new polynomial whose roots are a linear transformation of the roots of another polynomial. • To evaluate expressions involving roots. • To solve inequalities up to quartic polynomials. • To use the sums of integers, squares and cubes to sum other series. • To understand and use the method of differences. • To use proof by induction to prove divisibility. • To use Maclaurin series expansions.	**Ch12 Series** • Summing series using partial fractions. • More Maclaurin series. **Careers** • Car manufacturing. • Graphic design.

Fluency and skills

See Maths Ch 2.3 For a reminder of the factor theorem.

The roots of polynomials can be found using the factor theorem and long division. You are now going to learn how to transform one polynomial into another polynomial with roots that are related in some way.

Quadratic equations

Let the quadratic equation $ax^2 + bx + c = 0$ have roots $x = \alpha$ and $x = \beta$

Dividing through by a gives

$$x^2 + \frac{b}{a}x + \frac{c}{a} = 0$$

Since $x = \alpha$ and $x = \beta$ are the roots of this quadratic, you can write the equation in the form

$$(x - \alpha)(x - \beta) = 0$$

Expanding the brackets gives

$$x^2 - (\alpha + \beta)x + \alpha\beta = 0$$

Comparing the two versions of the quadratic equation gives

$$x^2 + \frac{b}{a}x + \frac{c}{a} \equiv x^2 - (\alpha + \beta)x + \alpha\beta = 0$$

So, comparing the coefficients for x and the constant gives $(\alpha + \beta) = -\frac{b}{a}$ and $\alpha\beta = \frac{c}{a}$

For a quadratic equation:

Key point

The sum of the roots $= \alpha + \beta = -\frac{b}{a}$ and the product of the roots $= \alpha\beta = \frac{c}{a}$

This also shows that all quadratics can be written in the form
$x^2 - (\text{sum of the roots})x + (\text{product of the roots}) = 0$

Example 1

The roots of $x^2 - 7x + 12 = 0$ are $x = \alpha$ and $x = \beta$. Without finding the values of α and β separately,

a   Write down the values of $\alpha + \beta$ and $\alpha\beta$

b   Hence find the quadratic equations whose roots are   i α^2 and β^2 ii $\frac{1}{\alpha}$ and $\frac{1}{\beta}$

a $a = 1$, $b = -7$ and $c = 12$, so

$\alpha + \beta = -(-7) = 7$ and $\alpha\beta = 12$

$(\alpha + \beta) = -\frac{b}{a}$ and $\alpha\beta = \frac{c}{a}$

(Continued on the next page)

b **i** The new equation must be in the form

$$x^2 - (\alpha^2 + \beta^2)x + \alpha^2\beta^2 = 0$$

Now $\alpha^2 + \beta^2 = (\alpha + \beta)^2 - 2\alpha\beta$

So $\alpha^2 + \beta^2 = 7^2 - 2 \times 12 = 25$

and $\alpha^2\beta^2 = (\alpha\beta)^2 = 12^2 = 144$

Hence the new equation is $x^2 - 25x + 144 = 0$

Box: $x^2 - $ (sum of roots)$x + $ (product of roots) $= 0$

Box: Write $\alpha^2 + \beta^2$ and $\alpha^2\beta^2$ in terms of $\alpha + \beta$ and $\alpha\beta$

Box: Substitute using $\alpha + \beta = 7$ and $\alpha\beta = 12$

ii The new equation must be

$$x^2 - \left(\frac{1}{\alpha} + \frac{1}{\beta}\right)x + \frac{1}{\alpha\beta} = 0$$

$$\left(\frac{1}{\alpha} + \frac{1}{\beta}\right) = \frac{\alpha + \beta}{\alpha\beta} = \frac{7}{12} \text{ and } \frac{1}{\alpha\beta} = \frac{1}{12}$$

Hence the new equation is $x^2 - \dfrac{7}{12}x + \dfrac{1}{12} = 0$

or $12x^2 - 7x + 1 = 0$

Box: Write $\left(\dfrac{1}{\alpha} + \dfrac{1}{\beta}\right)$ and $\dfrac{1}{\alpha\beta}$ in terms of $\alpha + \beta$ and $\alpha\beta$

Cubic equations

There are similar relationships between the coefficients of x and the roots of the equation for higher-order equations (cubics, quartics, and so on).

Let the cubic equation

$ax^3 + bx^2 + cx + d = 0$ have roots $x = \alpha$, $x = \beta$ and $x = \gamma$

Let $ax^3 + bx^2 + cx + d = 0 \equiv (x - \alpha)(x - \beta)(x - \gamma) = 0$

$$\equiv (x^2 - \alpha x - \beta x + \alpha\beta)(x - \gamma) = 0$$

$$\equiv x^3 - \alpha x^2 - \beta x^2 + \alpha\beta x - x^2\gamma + \alpha\gamma x + \beta\gamma x - \alpha\beta\gamma = 0$$

$$\equiv x^3 - (\alpha + \beta + \gamma)x^2 + (\alpha\beta + \beta\gamma + \gamma\alpha)x - \alpha\beta\gamma = 0$$

Dividing $ax^3 + bx^2 + cx + d = 0$ by a gives $x^3 + \dfrac{b}{a}x^2 + \dfrac{c}{a}x + \dfrac{d}{a} = 0$

Hence

$$(\alpha + \beta + \gamma) = -\frac{b}{a} \qquad (\alpha\beta + \beta\gamma + \gamma\alpha) = \frac{c}{a} \qquad \text{and } \alpha\beta\gamma = -\frac{d}{a}$$

Example 2

The roots of the equation $x^3 - 7x^2 + 3x + 2 = 0$ are α, β and γ. Find the values of

a $\alpha^2 + \beta^2 + \gamma^2$ **b** $\alpha^3 + \beta^3 + \gamma^3$ **c** $(\alpha+2)(\beta+2)(\delta+2)$

Box: Consider the expression $(\alpha + \beta + \gamma)^2$ to find a way to write $\alpha^2 + \beta^2 + \gamma^2$

a $(\alpha + \beta + \gamma)^2 = \alpha^2 + 2\alpha\beta + 2\alpha\gamma + \beta^2 + 2\beta\gamma + \gamma^2$

$= (\alpha^2 + \beta^2 + \gamma^2) + 2(\alpha\beta + \beta\gamma + \alpha\gamma)$

$7^2 = (\alpha^2 + \beta^2 + \gamma^2) + 2(3)$

$\alpha^2 + \beta^2 + \gamma^2 = 49 - 6 = 43$

Box: Substitute $\alpha + \beta + \gamma = 7$ and $\alpha\beta + \beta\gamma + \alpha\gamma = 3$ and rearrange.

(*Continued on the next page*)

b α is a root of the equation, which means

$\alpha^3 - 7\alpha^2 + 3\alpha + 2 = 0$, so by rearranging $\alpha^3 = 7\alpha^2 - 3\alpha - 2$

The same can be done with the other roots.

$\beta^3 = 7\beta^2 - 3\beta - 2$ $\gamma^3 = 7\gamma^2 - 3\gamma - 2$

Now find the sum of these expressions.

$\alpha^3 + \beta^3 + \gamma^3 = 7\alpha^2 - 3\alpha - 2 + 7\beta^2 - 3\beta - 2 + 7\gamma^2 - 3\gamma - 2$

$= 7(\alpha^2 + \beta^2 + \gamma^2) - 3(\alpha + \beta + \gamma) - 6$ ⟵ Substitute $\alpha^2 + \beta^2 + \gamma^2 = 43$ and $\alpha + \beta + \gamma = 7$

$= 7(43) - 3(7) - 6 = 274$

c $(\alpha + 2)(\beta + 2)(\delta + 2) = \alpha\beta\delta + 2(\alpha\beta + \alpha\delta + \beta\delta) + 4(\alpha + \beta + \delta) + 8$ ⟵ Expand the brackets.

$= -2 + 2(3) + 4(7) + 8$ ⟵ Use $\alpha\beta\delta = -2$, $\alpha\beta + \alpha\delta + \beta\delta = 3$ and $\alpha + \beta + \delta = 7$

$= 40$

Quartic equations

The same method can also be used to find the relationships between the roots, α, β, γ and δ, and the coefficients of the quartic equation $ax^4 + bx^3 + cx^2 + dx + e = 0$

$(\alpha + \beta + \gamma + \delta) = -\dfrac{b}{a}$ i.e. $\Sigma\alpha = -\dfrac{b}{a}$

$(\alpha\beta + \alpha\gamma + \alpha\delta + \beta\gamma + \beta\delta + \gamma\delta) = \dfrac{c}{a}$ i.e. $\Sigma\alpha\beta = \dfrac{c}{a}$

$(\alpha\beta\gamma + \beta\gamma\delta + \gamma\delta\alpha + \delta\alpha\beta) = -\dfrac{d}{a}$ i.e. $\Sigma\alpha\beta\delta = -\dfrac{d}{a}$

$\alpha\beta\gamma\delta = \dfrac{e}{a}$

Key point

$\alpha + \beta + \gamma + \delta$ is often abbreviated to $\Sigma\alpha$

$\alpha\beta + \beta\gamma + \gamma\delta + \delta\alpha$ is often abbreviated to $\Sigma\alpha\beta$

$\alpha\beta\gamma + \beta\gamma\delta + \gamma\delta\alpha + \delta\alpha\beta$ is often abbreviated to $\Sigma\alpha\beta\gamma$

Exercise 2.1A Fluency and skills

1 Find the sum and product of the roots of these equations.

a $x^2 - 5x + 9 = 0$ **b** $x^2 + 6x + 7 = 0$

c $x^2 - 8x - 12 = 0$ **d** $x^2 + 10x - 5 = 0$

e $3x^2 - 12x + 8 = 0$ **f** $4x^2 + x + 6 = 0$

2 Write the sum, the sum of the products in pairs and the product of the roots of these equations.

a $x^3 + 4x^2 - 9x - 14 = 0$

b $x^3 - 7x^2 - 11x + 12 = 0$

c $x^3 - 13x^2 + 22x - 26 = 0$

d $2x^3 + 5x^2 + 17x - 21 = 0$

e $4x^3 - x^2 + 3x + 8 = 0$

f $\dfrac{1}{2}x^3 + \dfrac{3}{8}x^2 - \dfrac{3}{4}x + \dfrac{5}{16} = 0$

3 A quadratic equation $3x^2 + kx - 4 = 0$ has roots α and β

Find an expression in terms of k for

a $\alpha + \beta$ **b** $(\alpha + 1)(\beta + 1)$

c $\alpha^2 + \beta^2$ **d** $(\alpha - 4)(\beta - 4)$

e $\alpha^3 + \beta^3$ **f** $\dfrac{1}{\alpha} + \dfrac{1}{\beta}$

4 A cubic equation $p(x)=0$ has roots α, β and γ

Use the facts that $\alpha+\beta+\gamma=5$, $\alpha\beta\gamma=-2$ and $\alpha\beta+\alpha\gamma+\beta\gamma=4$ to write down a possible expression for $p(x)$

5 These equations have roots α, β, γ and δ

a $x^4-13x^2+36=0$

b $x^4-4x^3-7x^2+22x+24=0$

Find the values of

i $\alpha^2+\beta^2+\gamma^2+\delta^2$ **ii** $\dfrac{1}{\alpha}+\dfrac{1}{\beta}+\dfrac{1}{\gamma}+\dfrac{1}{\delta}$

6 The cubic equation $2x^3+ax^2+bx+c=0$ has roots α, β and γ

Find an expression in terms of a, b and c for

a $\alpha+\beta+\gamma$ **b** $\alpha\beta+\alpha\gamma+\beta\gamma$

c $\alpha\beta\gamma$ **d** $\alpha^2+\beta^2+\gamma^2$

e $\alpha^3+\beta^3+\gamma^3$ **f** $(1+\alpha)(1+\beta)(1+\gamma)$

g $\dfrac{1}{\alpha}+\dfrac{1}{\beta}+\dfrac{1}{\gamma}$ **h** $(\alpha\beta)^2+(\alpha\gamma)^2+(\beta\gamma)^2$

7 The quartic equation $ax^4+bx^3+cx^2+dx+e=0$ has roots α, β, γ and δ. Show that

a $\alpha+\beta+\gamma+\delta=-\dfrac{b}{a}$

b $\alpha\beta+\alpha\gamma+\alpha\delta+\beta\gamma+\beta\delta+\gamma\delta=\dfrac{c}{a}$

c $\alpha\beta\gamma+\alpha\beta\delta+\alpha\gamma\delta+\beta\gamma\delta=-\dfrac{d}{a}$

d $\alpha\beta\gamma\delta=\dfrac{e}{a}$

8 The quartic equation $3x^4+kx^3+7x^2-2x+k=0$ has roots α, β, γ and δ

Find an expression in terms of k for

a $\alpha^2+\beta^2+\gamma^2+\delta^2$ **b** $\dfrac{1}{\alpha}+\dfrac{1}{\beta}+\dfrac{1}{\gamma}+\dfrac{1}{\delta}$

Reasoning and problem-solving

Sometimes it is useful to transform an equation into another whose roots are related in a simple way to the roots of the original equation.

> **Key point**
>
> If the roots are transformed in a linear way, so that $y=mx+c$, then you transform the equation by substituting $x=\dfrac{y-c}{m}$
>
> If the new roots are reciprocals, so that $y=\dfrac{1}{x}$, then you transform the equation by substituting $x=\dfrac{1}{y}$

Strategy 1

To solve a question involving the transformation of one polynomial into another

1. Rewrite the transformation $y=mx+c$ as $x=\dfrac{y-c}{m}$

2. Substitute $\dfrac{y-c}{m}$ for x in the original polynomial and simplify to produce the transformed equation.

Example 3

The roots of the equation $x^3 - 2x^2 - x + 2 = 0$ are α, β and γ

a Write the value of $\alpha\beta\gamma$

b **i** Write down a cubic equation with roots $\alpha - 2$, $\beta - 2$ and $\gamma - 2$

ii By considering the product of the roots of this new equation, find the value of
$\alpha\beta\gamma(1 + \beta\gamma)(1 + \alpha\gamma)(1 + \alpha\beta)$

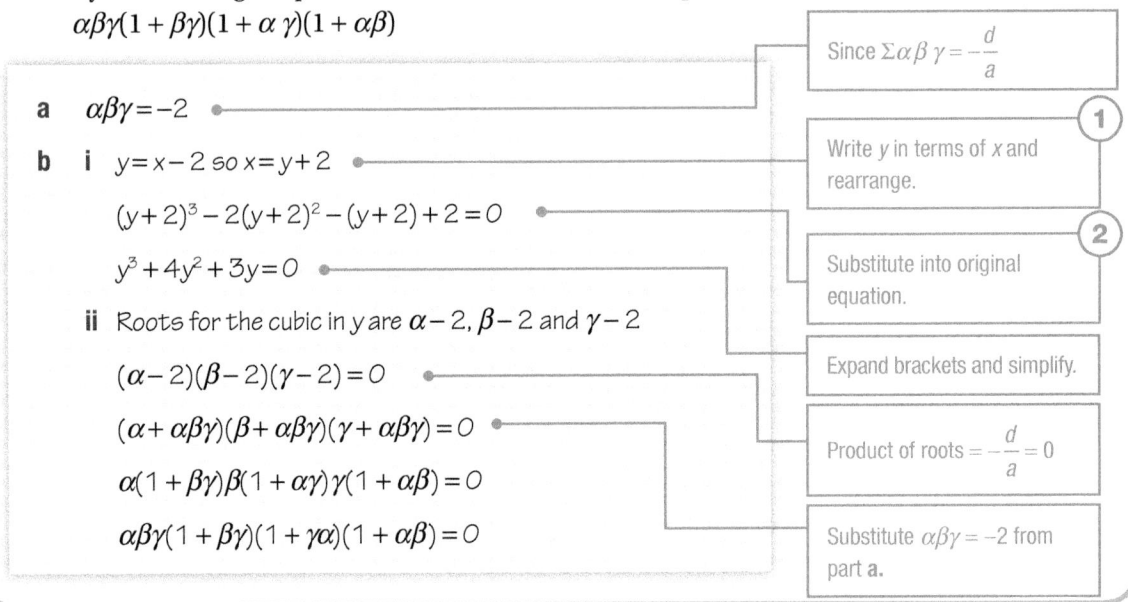

a $\alpha\beta\gamma = -2$

Since $\Sigma\alpha\beta\gamma = -\dfrac{d}{a}$ ①

b **i** $y = x - 2$ so $x = y + 2$

Write y in terms of x and rearrange. ②

$(y+2)^3 - 2(y+2)^2 - (y+2) + 2 = 0$

Substitute into original equation.

$y^3 + 4y^2 + 3y = 0$

ii Roots for the cubic in y are $\alpha - 2$, $\beta - 2$ and $\gamma - 2$

Expand brackets and simplify.

$(\alpha - 2)(\beta - 2)(\gamma - 2) = 0$

$(\alpha + \alpha\beta\gamma)(\beta + \alpha\beta\gamma)(\gamma + \alpha\beta\gamma) = 0$

Product of roots $= -\dfrac{d}{a} = 0$

$\alpha(1 + \beta\gamma)\beta(1 + \alpha\gamma)\gamma(1 + \alpha\beta) = 0$

$\alpha\beta\gamma(1 + \beta\gamma)(1 + \gamma\alpha)(1 + \alpha\beta) = 0$

Substitute $\alpha\beta\gamma = -2$ from part **a**.

To solve problems with unknown roots and coefficients

① Use a coefficient you know the value of to form an equation.

② Solve this equation to find the roots.

③ Use this root to find the value of the others.

④ Use the roots with the rules learnt to find the other coefficients in the equation.

Example 4

The quartic equation $2x^3 + px^2 + qx - 50 = 0$ has roots α, $\dfrac{50}{\alpha}$, $\alpha + \dfrac{50}{\alpha} + \dfrac{5}{2}$

Solve the equation and find the values of p and q

$\dfrac{50}{2} = \alpha\left(\dfrac{50}{\alpha}\right)\left(\alpha + \dfrac{50}{\alpha} + \dfrac{5}{2}\right)$

Use fact that product of roots equals $-\dfrac{d}{2}$ since you know. ①

$\Rightarrow 25 = 50\left(\alpha + \dfrac{50}{\alpha} + \dfrac{5}{2}\right)$

The value of d cancels the factor of α

$\Rightarrow \dfrac{1}{2} = \alpha + \dfrac{50}{\alpha} + \dfrac{5}{2}$

Simplify and multiply through by α to give a quadratic equation.

$\Rightarrow \alpha + \dfrac{50}{\alpha} + 2 = 0$

$\Rightarrow \alpha^2 + 2\alpha + 50 = 0$

$\Rightarrow \alpha = -1 \pm 7i$

Solve the quadratic to find α ②

(*Continued on the next page*)

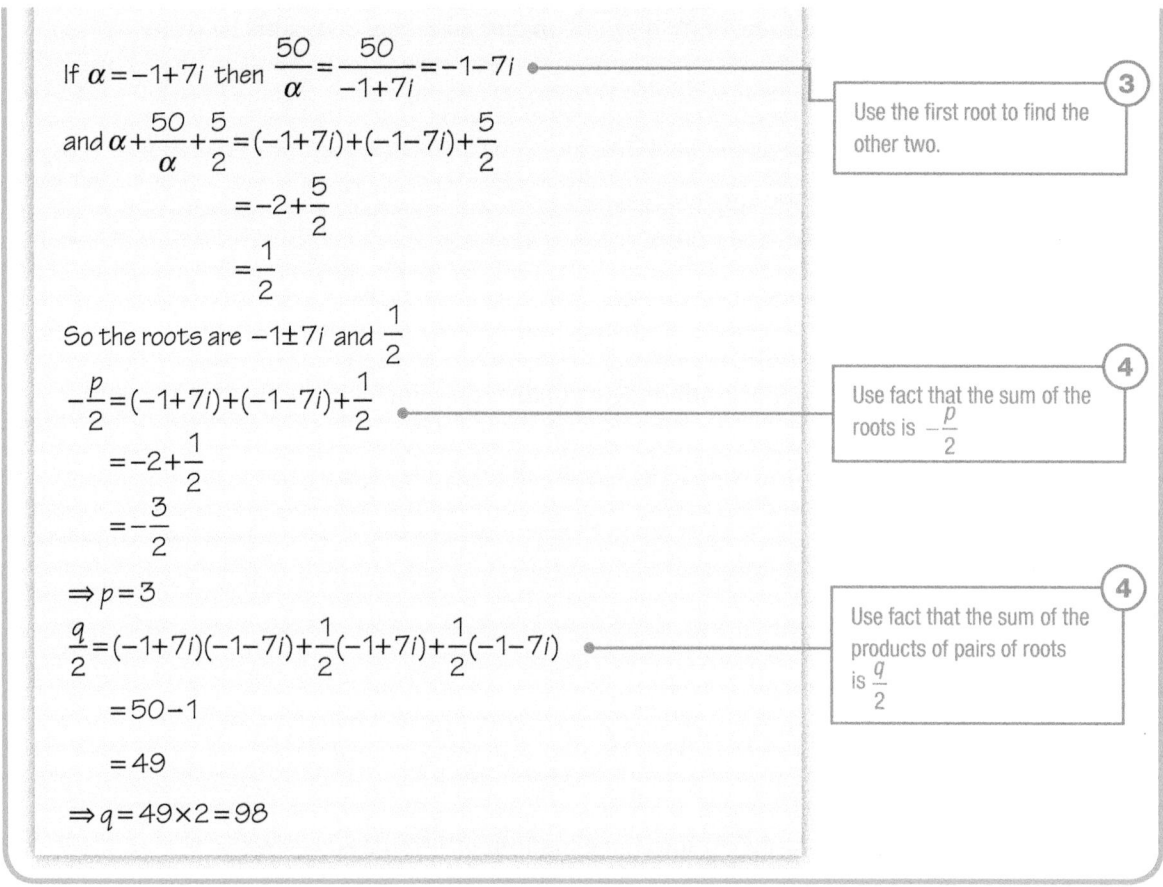

If $\alpha = -1 + 7i$ then $\dfrac{50}{\alpha} = \dfrac{50}{-1+7i} = -1-7i$ •————————

③ Use the first root to find the other two.

and $\alpha + \dfrac{50}{\alpha} + \dfrac{5}{2} = (-1+7i) + (-1-7i) + \dfrac{5}{2}$

$\qquad = -2 + \dfrac{5}{2}$

$\qquad = \dfrac{1}{2}$

So the roots are $-1 \pm 7i$ and $\dfrac{1}{2}$

$-\dfrac{p}{2} = (-1+7i) + (-1-7i) + \dfrac{1}{2}$ •————————

④ Use fact that the sum of the roots is $-\dfrac{p}{2}$

$\qquad = -2 + \dfrac{1}{2}$

$\qquad = -\dfrac{3}{2}$

$\Rightarrow p = 3$

$\dfrac{q}{2} = (-1+7i)(-1-7i) + \dfrac{1}{2}(-1+7i) + \dfrac{1}{2}(-1-7i)$ •————————

④ Use fact that the sum of the products of pairs of roots is $\dfrac{q}{2}$

$\qquad = 50 - 1$

$\qquad = 49$

$\Rightarrow q = 49 \times 2 = 98$

Exercise 2.1B Reasoning and problem-solving

1 Find the polynomial whose roots are

    a Reciprocals of, **b** Triple

    the roots of $x^3 + 3x^2 + 5x + 1 = 0$

2 The equation $x^3 - 2x^2 - x + 2 = 0$ has roots α, β and γ

    a Find a cubic equation with roots of $\alpha - 2$, $\beta - 2$ and $\gamma - 2$

    b Solve this cubic equation.

    c Use the substitution $y = x - 2$ to find the roots of the original equation in x

3 The equation $x^3 - 7x^2 + 9x + 11 = 0$ has roots α, β and γ

    Without calculating the roots, find an equation with roots of $\dfrac{\alpha}{2} - 1, \dfrac{\beta}{2} - 1$ and $\dfrac{\gamma}{2} - 1$

4 The equation $x^4 + 6x^3 + 7x + 8 = 0$ has roots α, β, γ and δ

    Without calculating the roots, find an equation with roots of $2\alpha - 1$, $2\beta - 1$, $2\gamma - 1$ and $2\delta - 1$

5 The equation $x^3 - 9x^2 + 6x + k = 0$ has roots α, β, γ and δ

    a Find an equation with roots of $\alpha - 4$, $\beta - 4$, $\gamma - 4$ and $\delta - 4$

    b If $k = 56$, demonstrate how you can use your solution to part **a** to find the solutions to $x^3 - 9x^2 + 6x + k = 0$

6   The cubic equation $x^3 + mx^2 + nx - 50 = 0$ where m and n are real constants has roots

$\alpha, \dfrac{10}{\alpha}$ and $\alpha + \dfrac{10}{\alpha} - 1$

   a   Solve the equation $x^3 + mx^2 + nx - 50 = 0$

   b   Calculate the values of the constants m and n

7   The equation $4x^3 + kx^2 - 11x + 119 = 0$ has roots α, β and γ

   a   Find the value of k when $\alpha + \beta + \gamma = -6$

For this same value of k

   b   Write down a cubic equation with integer coefficients which has roots $\dfrac{1}{\alpha}, \dfrac{1}{\beta}$ and $\dfrac{1}{\gamma}$

8   The quartic equation $4x^4 - 44x^3 + px^2 + qx + 4165 = 0$ has roots

α, β, γ and δ where $\alpha < 0$

Given that $\alpha = \beta$ and $\gamma = \dfrac{85}{\delta}$

   a   Solve the equation,

   b   Find the values of p and q

9   Find the value of b if the equations $x^2 + 6x + b = 0$ and $x^2 + 4x - b = 0$ have a common root that is not equal to 0

10 α, β and γ are the roots of $x^3 = 4x + 3$
    By substituting $x = \alpha$, $x = \beta$ and $x = \gamma$ into the equation, prove that $\alpha^3 + \beta^3 + \gamma^3 = 9$

11 α, β and γ are the roots of $2x^3 - 2x^2 - 6x - 3 = 0$
    By considering the sum of the roots, find a suitable substitution to transform this equation into a polynomial with roots $\alpha + \beta$, $\beta + \gamma$ and $\gamma + \alpha$

12 The equation $x^3 - 4x^2 - x + 7 = 0$ has roots α, β and γ

   a   Write down the value of $\alpha + \beta + \gamma$

   b   Use the substitution $x = 4 - y$ to find a cubic equation in y

   c   By considering the product of the roots for this equation, find the value of $(\beta + \gamma)(\alpha + \gamma)(\alpha + \beta)$

The rules for solving inequalities are very similar to those for solving equations, but there are two differences.

Key point

- Answers to inequalities are a range of values rather than individual values.
- When you multiply or divide by a negative number you reverse the inequality sign.

You need to be particularly careful to avoid incorrect assumptions when working with inequalities that involve negative numbers.

For example, to solve the inequality $\dfrac{2x-1}{-5} < 3+x$ you can multiply

both sides of the equation by -5 but you must remember to reverse the direction of the inequality sign:

so $\dfrac{2x-1}{-5} < 3+x$ becomes $2x-1 > -15-5x$

rearrange to give $\qquad\qquad 7x > -14$

so the solution is $\qquad\qquad x > -2$ (a range of values)

Example 1

If $a < b$, is $a^2 < b^2$? Explain your answer.

Not always.

If $a = 2$ and $b = 3$ then $2^2 < 3^2$ —————— Since $4 < 9$

but if $a = -3$ and $b = 2$ then $(-3)^2 > 2^2$ even though $-3 < 2$ —————— Since $9 > 4$

If an expression can be positive or negative, you can consider each case separately.

For example, to solve the inequality $\dfrac{5}{x+2} > 1$ you could consider the

cases when $x + 2$ is positive and $x + 2$ is negative separately.

If $x + 2$ is positive then $x > -2$ and $5 > x+2 \Rightarrow x < 3$ so $-2 < x < 3$

If $x + 2$ is negative then $5 < x+2 \Rightarrow x > 3$, but this cannot be true since we know that $x + 2$ is negative so $x < -2$

Hence the solution is $-2 < x < 3$

An alternative method is to multiply both sides of the equation by $(x+2)^2$ since you know this will be a positive number.

Example 2

Solve the inequality $\dfrac{18x-135}{x-3} \le 21$

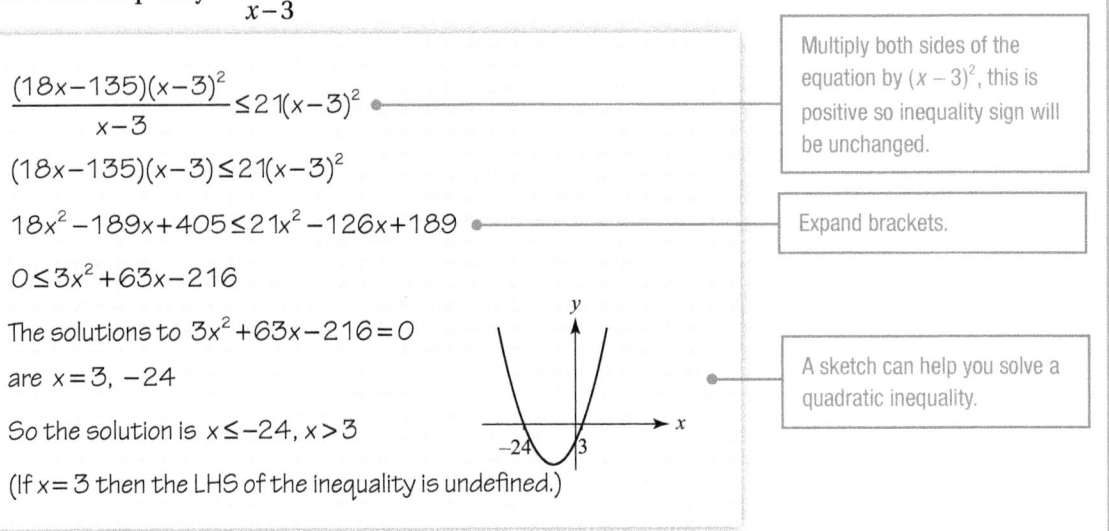

$\dfrac{(18x-135)(x-3)^2}{x-3} \le 21(x-3)^2$

Multiply both sides of the equation by $(x-3)^2$, this is positive so inequality sign will be unchanged.

$(18x-135)(x-3) \le 21(x-3)^2$

$18x^2 - 189x + 405 \le 21x^2 - 126x + 189$

Expand brackets.

$0 \le 3x^2 + 63x - 216$

The solutions to $3x^2 + 63x - 216 = 0$

are $x = 3, -24$

A sketch can help you solve a quadratic inequality.

So the solution is $x \le -24, x > 3$

(If $x = 3$ then the LHS of the inequality is undefined.)

If the inequality in Example 2 had been $\dfrac{18x-135}{x-3} < 0$, you can multiply both sides by $(x-3)$, and use the same argument as in Example 2

Case 1: $x > 3$

Solve $18x - 135 < 0$ to get $x < 7.5$

Case 2: $x < 3$

Solve $18x - 135 > 0$ to get $x > 7.5$, which isn't possible.

So, $\dfrac{18x-135}{x-3} < 0$ when $3 < x < 7.5$

Alternatively, to work out when $\dfrac{18x-135}{x-3}$ is negative, you could consider the signs of the numerator and denominator for different values of x

	$x < 3$	$3 < x < 7.5$	$x > 7.5$
$18x - 135$	−ve	−ve	+ve
$x - 3$	−ve	+ve	+ve
$\dfrac{18x-135}{x-3}$	+ve	−ve	+ve

If you need to solve an inequality of the form $f(x) \ge 0$ or $f(x) \le 0$, remember to include values where $f(x) = 0$ (where the graph intersects the x-axis).

So, $\dfrac{18x-135}{x-3} < 0$ when $3 < x < 7.5$

You could also sketch a graph of the function $y = \dfrac{18x - 135}{x - 3}$ and find

the values of x where $y < 0$ and the graph lies below the x-axis.

Example 3

a Sketch the graph of $f(x) = (x + 3)(x + 1)(x - 4)$

b Use your sketch to solve   **i** $f(x) = 0$ **ii** $f(x) > 0$ **iii** $f(x) < 0$

> You can use a graphical calculator to help you sketch this graph.
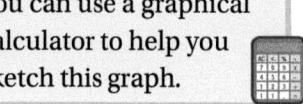

a

> Use where the graph intersects, is above and is below the x-axis to solve the inequalities.

b i $(x + 3)(x + 1)(x - 4) = 0$ when $x = -3$ or $x = -1$ or $x = 4$

ii $(x + 3)(x + 1)(x - 4) > 0$ when $-3 < x < -1$ or $x > 4$

iii $(x + 3)(x + 1)(x - 4) < 0$ when $x < -3$ or $-1 < x < 4$

Example 4

a For the function $f(x) = \dfrac{5x + 15}{x - 1}$, use an algebraic method to find

the values of x where

   i $f(x) = 0$ **ii** $f(x) > 0$ **iii** $f(x) < 0$

b Using graph plotting software, sketch a graph of $y = f(x)$ to confirm your answers.

> You can use a graphical calculator to help you sketch this graph.

a i $\dfrac{5x + 15}{x - 1} = 0$ when $5x + 15 = 0$ so when $x = -3$

> $f(x) = 0$ when numerator $= 0$

ii $\dfrac{5x + 15}{x - 1} > 0$ when numerator and denominator have the same sign.

If $x > -3$, numerator is positive.

For denominator to also be positive, $x > 1$

So $x > 1$

> x must satisfy $x > -3$ and $x > 1$

If $x < -3$, numerator is negative.

For denominator to also be negative $x < 1$

So $x < -3$

> x must satisfy $x < -3$ and $x < 1$

$\dfrac{5x + 15}{x - 1} > 0$ when $x > 1$ or $x < -3$

> Combine the results.

(*Continued on the next page*)

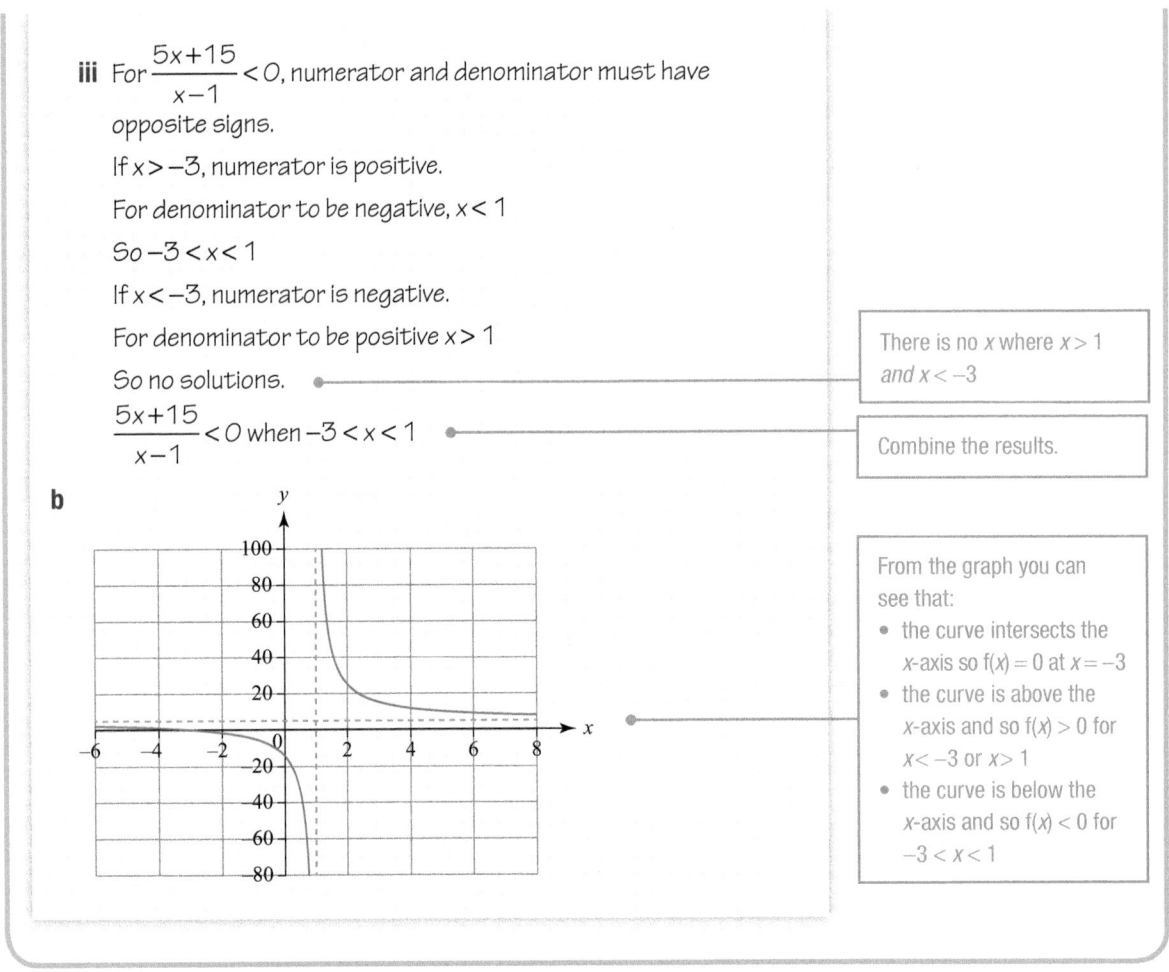

iii For $\dfrac{5x+15}{x-1} < 0$, numerator and denominator must have opposite signs.

If $x > -3$, numerator is positive.

For denominator to be negative, $x < 1$

So $-3 < x < 1$

If $x < -3$, numerator is negative.

For denominator to be positive $x > 1$

So no solutions. •————— There is no x where $x > 1$ and $x < -3$

$\dfrac{5x+15}{x-1} < 0$ when $-3 < x < 1$ •————— Combine the results.

b

From the graph you can see that:
- the curve intersects the x-axis so $f(x) = 0$ at $x = -3$
- the curve is above the x-axis and so $f(x) > 0$ for $x < -3$ or $x > 1$
- the curve is below the x-axis and so $f(x) < 0$ for $-3 < x < 1$

In Example 4, for $x < -3$ the numerator is negative and for $x > -3$ it is positive so $x = -3$ is known as a **critical value**. Similarly, $x = 1$ is the critical value for the denominator.

Exercise 2.2A Fluency and skills

1 By trying values for a, b, x and y, show that each of these statements are sometimes, but not always, true.

a If $a < b$ then $\dfrac{1}{a} < \dfrac{1}{b}$

b If $a < x$ and $b < y$ then $a - b < x - y$

c If $a < x$ and $b < y$ then $ab < xy$

2 Solve these inequalities algebraically.

a $\dfrac{x-1}{x} < 1$

b $\dfrac{2x-3}{3x} \le 1$

c $\dfrac{x-1}{x+2} \ge -1$

d $\dfrac{2x+5}{x-2} > 3$

3 Solve these inequalities by sketching a graph in each case.

a $x(x+4)(x-5) > 0$

b $(x+5)(x+4)(x-5) < 0$

c $(2x+3)(3x-4)(5x-9) \le 0$

d $(x+4)(x-2)^2 \ge 0$

e $(3x-5)(2x+3)^2 > 0$

f $(x^2-25)(3x+7) < 0$

g $(x-3)(x^2-x-12) \le 0$

4 Solve these inequalities.

a $x(x+1)(x-2)(x-9) > 0$

b $(x+2)(x+1)(x-3)(x-7) < 0$

c $(2x+1)(4x-5)(3x-4)(x-11) \leq 0$

d $(x-1)^2(x-3)^2 \geq 0$

e $(2x-7)^2(3x+2)^2 > 0$

f $(x^2-49)(x^2-5) < 0$

g $(x^2-3x-10)(x^2-x-12) \leq 0$

5 For what values of x are the following functions positive?

a $\dfrac{x-2}{x}$

b $\dfrac{x-2}{x+3}$

c $\dfrac{x-2}{x-5}$

d $\dfrac{(x+1)(x+2)}{x}$

e $\dfrac{(x+3)(x-4)}{(x+5)}$

f $\dfrac{(2x-3)(3x+4)}{(x-7)}$

g $\dfrac{(4-x)(5+2x)}{(2x+3)}$

h $\dfrac{(3x-2)(2x+4)}{2x-3}$

6 For what values of x are the following functions negative?

a $\dfrac{x+1}{x}$

b $\dfrac{x+3}{x+4}$

c $\dfrac{x-3}{x+2}$

d $\dfrac{(x+3)(x-5)}{x}$

e $\dfrac{(x+7)(x+4)}{(x+6)}$

f $\dfrac{(4x-3)(2x+5)}{(x+6)}$

g $\dfrac{(9-2x)(6+5x)}{(2x-1)}$

h $\dfrac{(2x+5)(x-2)}{x+1}$

Reasoning and problem-solving

Strategy

To solve an inequality algebraically

1. Factorise and simplify using the same rules as for equations, but remember to change the direction of the inequality sign if you multiply or divide by a negative value.

2. Use graphs or critical values to investigate where functions are positive or negative.

3. Solve the inequality by considering all possibilities and cases.

Example 5

a Solve the inequality $x^4 - 37x^2 + 36 \geq 0$ algebraically.

b Using a graphical calculator or graph sketching software on your computer, sketch a graph to confirm your answer.

a $f(x) \equiv x^4 - 37x^2 + 36$ is a quadratic in x^2

$f(x) \equiv (x^2-1)(x^2-36)$ ◄——————— Factorise the polynomial. ①

$\equiv (x-1)(x+1)(x-6)(x+6)$

(*Continued on the next page*)

Solve $(x-1)(x+1)(x-6)(x+6) \geq 0$

$(x+6)(x+1)(x-1)(x-6) \geq 0$ •————

The critical values are $x = -6, -1, 1$ and 6 •————

When $x < -6$, f(x) is positive. •————

When $-6 < x < -1$, f(x) is negative. •————

When $-1 < x < 1$, f(x) is positive. •————

When $1 < x < 6$, f(x) is negative. •————

When $x > 6$, f(x) is positive. •————

Hence $x^4 - 37x^2 + 36 \geq 0$ when

$x \leq -6$ or $-1 \leq x \leq 1$ or $x \geq 6$ •————

b

It might be helpful to re-order the brackets into order of size of the root.

②

At these values one factor $= 0$
This means f(x) $= 0$ and the graph crosses the x-axis.

All four factors are negative.

Three negative and one positive factor.

Two negative and two positive factors.

One negative and three positive factors.

All four factors are positive.

③

Include the critical points as the inequality is ≥ 0 and so includes 0

②

$x^4 - 37x^2 + 36 \geq 0$ when the curve intersects or is above the x-axis.
You can see that this is when $x \leq -6$ or $-1 \leq x \leq 1$ or $x \geq 6$

It is sometimes necessary to find inequalities for functions with real roots.

Example 6

The function $f(x) = \dfrac{x^2 - 4x + 4}{x}$ intersects the straight line $y = k$

a Form a quadratic equation in x and k

b Hence find the values of k for which f(x) has real roots.

c Use a graphical calculator or graph sketching software on your computer to confirm your answer.

a $k = \dfrac{x^2 - 4x + 4}{x}$

$kx = x^2 - 4x + 4$ •————

$x^2 - (k+4)x + 4 = 0$

①

Multiply both sides by x and simplify to form a quadratic equation in x

(Continued on the next page)

b The function has real roots if $b^2 - 4ac \geq 0$ ●————

$(k+4)^2 \geq 4 \times 1 \times 4$ ●————

$k^2 + 8k + 16 - 16 \geq 0$

$k^2 + 8k \geq 0$

$k(k+8) \geq 0$

$k = \dfrac{x^2 - 4x + 4}{x}$ has real roots if $k \geq 0$ or $k \leq -8$ ●————

The discriminant must be ≥ 0

Substitute values and solve the inequality.

②③ k and $(k + 8)$ must either both be positive or both be negative.

c

② The graph of $y = \dfrac{x^2 - 4x + 4}{x}$ shows that x does not exist if $-8 \leq y \leq 0$

So x can only have real roots if $y \geq 0$ or $y \leq -8$

You can solve double inequalities using the same methods.

Example 7

Find the values of x for which $0 \leq \dfrac{4+x}{3-x} < 1$

Consider $0 \leq \dfrac{4+x}{3-x}$

Critical values are $x = -4$ and $x = 3$

For $x < -4$, $\dfrac{4+x}{3-x}$ is negative. ●————

For $-4 < x < 3$, $\dfrac{4+x}{3-x}$ is positive.

For $x > 3$, $\dfrac{4+x}{3-x}$ is negative.

So $\dfrac{4+x}{3-x} \geq 0$ when $-4 \leq x < 3$ ●————

Now consider $\dfrac{4+x}{3-x} < 1$

When $x < 3$

$4 + x < 3 - x$ ●————

$x < -\dfrac{1}{2}$

When $x > 3$

$4 + x > 3 - x$ ●————

$x > -\dfrac{1}{2}$, but we have the more restrictive condition $x > 3$

So $\dfrac{4+x}{3-x} < 1$ when $x < -\dfrac{1}{2}$ or $x > 3$

Full solution: $0 \leq \dfrac{4+x}{3-x} < 1$ when $-4 \leq x < -\dfrac{1}{2}$ ●————

② Consider the signs of the numerator and denominator for each set of values.

Do not include 3 as $\dfrac{4+x}{3-x}$ is not defined at $x = 3$

$(3 - x)$ is positive.

① $(3 - x)$ is negative so reverse the inequality.

③ Find the values which satisfy both inequalities.

1 The diagram shows the graph of a cubic function, $f(x) = Ax^3 + Bx^2 + Cx + D$

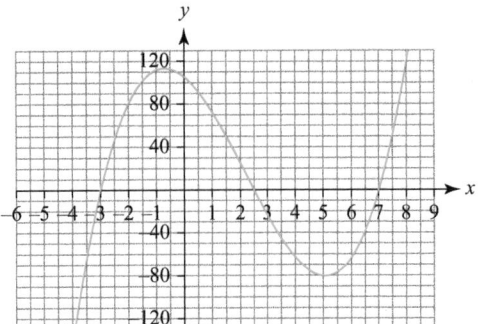

  a Write down the coordinates of all the intercepts with the coordinate axes.

  b Hence write down and expand the equation of the function.

  c Write down the values of A, B, C and D.

  d Estimate and write down the coordinates of the turning points.

  e Write down the ranges of values of x where $f(x) < 0$

2 The diagram shows the graph of a quartic function, $f(x) = Ax^4 + Bx^3 + Cx^2 + Dx + E$

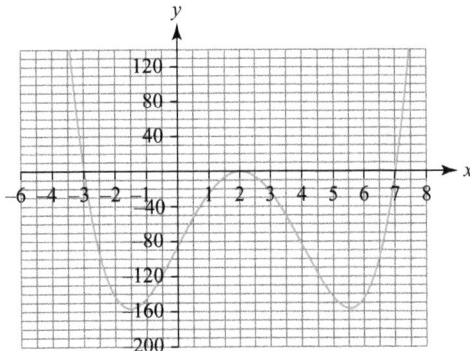

  a Write down the coordinates of all the intercepts with the coordinate axes. Explain why there are only three x-intercepts even though this quartic has four real roots.

  b Hence write down and expand the equation of the function.

  c Write down the values of A, B, C, D and E

  d Estimate and write down the coordinates of the turning points.

  e Write down the ranges of values of x where $f(x) > 0$

3 Solve the inequality $x^3 - 8x^2 + 12x \geq 0$

  Use a graphical calculator to confirm your answer with a sketch.

4 Solve the inequality $x^4 - 20x^2 + 64 \leq 0$

  Use a graphical calculator to confirm your answer with a sketch.

5 Solve these inequalities.

  a $\dfrac{x+3}{x-2} < x+3$ **b** $\dfrac{x-1}{x+4} > x - \dfrac{1}{4}$

  c $\dfrac{x-5}{x-4} \geq 5 - x$ **d** $\dfrac{2x-3}{x+1} \leq x+9$

6 In an experiment, three students recorded the temperature of their liquid as $(x+3)\,°C$, $(x-2)\,°C$ and $(x-7)\,°C$.

  For what values of x is the product of these temperatures positive?

7 $f(x) \equiv x^4 - 34x^2 + 225$. Find the range of values of x for which $f(x) \leq 0$

8 Find the range of values for which

$$10 - 2x < \frac{8}{x}$$

9 **a** Find the range of values of y for which the function $y = \dfrac{x^2 - 3x + 2}{x}$ has real roots.

  b Use a graphical calculator or graph sketching software on your computer to confirm your results.

10 **a** Find the range of values of x for which the function $y = \dfrac{2x^2 - 5x - 12}{x+2}$ is positive.

  b Use a graphical calculator or graph sketching software to confirm your answer.

11 **a** For all values of θ, $-1 \leq \sin\theta \leq 1$ Hence find the values of x such that $\sin\theta = \dfrac{x-3}{x+3}$ has a solution in θ

  b Confirm your answer using a sketch.

12 If x is any positive number, prove that the sum of the number and its reciprocal cannot be less than 2

Full A Level

Fluency and skills

If you want to write the sum of a series of n terms $u_1, u_2, \ldots,$ $u_r \ldots u_{n-1}, u_n$, you can use sigma notation, Σ

$\sum_{r=1}^{r=n} u_r$ or $\sum_1^n u_r$ means 'find the sum of all the terms from u_1 to u_n.

u_r is the general term and, if all terms are defined algebraically, u_r is a function of the variable.

> **Key point**
>
> If $S_n = u_1 + u_2 + \ldots + u_r + \ldots + u_{n-1} + u_n$, then $S_n = \sum_1^n u_r$

> Σ means 'find the sum of all these terms'.

> S_n means 'the sum of these n terms'.

If $u_r = r^2$ then $S_n = 1^2 + 2^2 + 3^2 + \ldots + r^2 \ldots + n^2 = \sum_1^n r^2$

You need to know the formulae for the sums of integers, squares and cubes.

Finding the sum of integers from 1 to n

$$\sum_1^n r = 1 \quad + \quad 2 \quad + \quad 3 \quad + \ldots + \quad (n-2) \quad + \quad (n-1) \quad + \quad n$$

Write the same sequence in reverse:

$$\sum_1^n r = n \quad + (n-1) + (n-2) + \ldots + \quad 3 \quad + \quad 2 \quad + \quad 1$$

Add the two together:

$$2\sum_1^n r = (n+1) + (n+1) + (n+1) + \ldots + (n+1) \quad + \quad (n+1) \quad + (n+1)$$

Every term is equal to $n+1$. Try it with some numbers.

There are n lots of $(n+1)$ so $2\sum_1^n r = n(n+1)$ and so $\sum_1^n r = \dfrac{n(n+1)}{2}$

> **Key point**
>
> $$\sum_1^n r = 1 + 2 + \ldots + n = \frac{n(n+1)}{2}$$

> You will be expected to remember and be able to quote without proof, the formulae for Σr

Example 1

Find the value of $2 + 4 + 6 + 8 + \ldots + 40$

$2 + 4 + 6 + 8 + \ldots + 40 = 2(1 + 2 + 3 + 4 + \ldots + 20)$
— Rewrite as a sum of integers.

$$= 2\sum_1^{20} r$$

$$= 2 \times \frac{20(20+1)}{2} = 2 \times \frac{20 \times 21}{2}$$
— Substitute $n = 20$ into the formula.

$$= 420$$

Sum of squares

You can derive the formula for the sum of squares using the method of differences. You will see how to do this later in this section.

Key point

$$\sum_{1}^{n} r^2 \equiv \frac{n(n+1)(2n+1)}{6}$$

Sum of cubes

Compare sums of r and r^3 for different values of n

n	Σr	Σr^3
1	1	1
2	$1+2=3$	$1+8=9$
3	$1+2+3=6$	$1+8+27=36$
4	$1+2+3+4=10$	$1+8+27+64=100$
5	$1+2+3+4+5=15$	$1+8+27+64+125=225$

For each value of n, $\Sigma r^3 = (\Sigma r)^2$

Hence $\Sigma n^3 = (\Sigma n)^2 = \left(\frac{n(n+1)}{2} \right)^2 = \frac{n^2(n+1)^2}{4}$

Key point

$$\sum_{1}^{n} r^3 = \frac{n^2(n+1)^2}{4}$$

Example 2

Evaluate $\displaystyle\sum_{5}^{n} r(r+4)$

$\displaystyle\sum_{5}^{n} r(r+4) = 5(9) + 6(10) + \ldots + n(n+4)$

Write out some of the sequence to help you understand it. This is the same as $\displaystyle\sum_{1}^{n} r(r+4)$ without the first four terms.

$= \displaystyle\sum_{1}^{n} r(r+4) - \sum_{1}^{4} r(r+4)$

$= \displaystyle\sum_{1}^{n} r^2 + 4\sum_{1}^{n} r - (1(5) + 2(6) + 3(7) + 4(8))$

$= \left[\dfrac{n}{6}(n+1)(2n+1) \right] + 4\dfrac{n(n+1)}{2} - 70$

Use the standard formulae.

$= n(n+1)\left[\dfrac{1}{6}(2n+1) + \dfrac{4}{2} \right] - 70$

$= n(n+1)\left[\dfrac{2n+1+12}{6} \right] - 70$

$= \dfrac{n(n+1)(2n+13) - 420}{6}$

$= \dfrac{(n-4)(2n^2 + 23n + 105)}{6}$

Simplify.

Example 3

a Find a formula for the sum to n terms of the series

$1 \times 2 \times 4 + 2 \times 3 \times 5 + 3 \times 4 \times 6 + ...$

b Find the sum of the first 10 terms and check your formula.

a The general term is $r(r+1)(r+3)$

$$\sum_1^n r(r+1)(r+3) \equiv \sum_1^n (r^3 + 4r^2 + 3r)$$

Rewrite in terms of sums you know.

$$\equiv \sum_1^n r^3 + 4\sum_1^n r^2 + 3\sum_1^n r$$

$\Sigma 4n^2 \equiv 4\Sigma n^2$ and $\Sigma 3n \equiv 3\Sigma n$

$$\equiv \frac{n^2(n+1)^2}{4} + 4\left[\frac{n}{6}(n+1)(2n+1)\right] + 3\frac{n(n+1)}{2}$$

Substitute formulae for Σn^3, Σn^2 and Σn, and simplify.

$$\equiv \frac{n(n+1)}{12}\left[3n(n+1) + 8(2n+1) + 18\right]$$

$$\equiv \frac{n(n+1)}{12}\left[3n^2 + 19n + 26\right]$$

$$\equiv \frac{n(n+1)(n+2)(3n+13)}{12}$$

b $$\sum_1^{10} r(r+1)(r+3) \equiv \frac{10 \times 11}{12}(3 \times 10^2 + 19 \times 10 + 26)$$

Find the sum using the formula.

$$= 4730$$

$$(1 \times 2 \times 4) + (2 \times 3 \times 5) + ... + (10 \times 11 \times 13)$$

Calculate the first 10 terms and add to check your formula.

$$= 8 + 30 + ... + 1430 = 4730$$

If a series has alternate positive and negative terms, $\Sigma f(x)$ can be written with a $(-1)^n$ multiple in it.

For example, $-1 + 2 - 3 + 4 - ...$ for n terms, is expressed as $\sum_1^n (-1)^n r$

The method of differences

There is no general method for summing series.

Sometimes it is not possible to express Σn as an algebraic expression.

For example $1 + \dfrac{1}{2} + \dfrac{1}{3} + \dfrac{1}{4} + \dfrac{1}{5} + ... + \dfrac{1}{n}$

However, if the general term of a function can be expressed as $f(r+1) - f(r)$, you can find the sum of the series using the **method of differences**.

Example 4

Find the sum to n terms of the series $\dfrac{1}{r} - \dfrac{1}{r+1}$

$$\sum_{1}^{n} \frac{1}{r(r+1)} = \frac{1}{1} - \frac{1}{2}$$
$$+ \frac{1}{2} - \frac{1}{3}$$
$$+ \frac{1}{3} - \frac{1}{4}$$
$$+ \frac{1}{n} - \frac{1}{(n+1)}$$

Write the 'differences' for successive terms vertically and eliminate terms wherever possible.

Thus $\displaystyle\sum_{1}^{n} \frac{1}{r(r+1)} = 1 - \frac{1}{(n+1)} = \frac{n}{(n+1)}$

Collect up the remaining terms and simplify.

Example 5

The expression $\dfrac{18}{(3r-2)(3r+1)(3r+4)}$ can be written as $\dfrac{1}{(3r-2)} - \dfrac{2}{(3r+1)} + \dfrac{1}{(3r+4)}$

Find the sum to n terms of the series $\dfrac{18}{(3r-2)(3r+1)(3r+4)}$

$$\sum_{1}^{n} \frac{18}{(3r-2)(3r+1)(3r+4)} \equiv \sum_{1}^{n}\left[\frac{1}{(3r-2)} - \frac{2}{(3r+1)} + \frac{1}{(3r+4)} \right]$$

$$\equiv \quad \frac{1}{1} - \frac{2}{4} + \frac{1}{7}$$
$$= \quad \frac{1}{4} - \frac{2}{7} + \frac{1}{10}$$
$$= \quad \frac{1}{7} - \frac{2}{10} + \frac{1}{13}$$
$$= \quad \frac{1}{10} - \frac{2}{13} + \frac{1}{16}$$

$$= \quad \frac{1}{3n-8} - \frac{2}{3n-5} + \frac{1}{3n-2}$$
$$= \quad \frac{1}{3n-5} - \frac{2}{3n-2} + \frac{1}{3n+1}$$
$$= \quad \frac{1}{3n-2} - \frac{2}{3n+1} + \frac{1}{3n+4}$$

Write the 'differences' for successive terms vertically and eliminate terms wherever possible.

Hence $\displaystyle\sum_{1}^{n} \frac{18}{(3r-2)(3r+1)(3r+4)} \equiv \frac{1}{1} - \frac{2}{4} + \frac{1}{4} + \frac{1}{3n+1} - \frac{2}{3n+1}$
$$+ \frac{1}{3n+4}$$

Collect up the remaining terms and simplify.

$$\equiv \frac{3}{4} - \frac{1}{3n+1} + \frac{1}{3n+4}$$

Exercise 2.3A Fluency and skills

1 Write out these series.

a $\displaystyle\sum_{1}^{5} r^3$ b $\displaystyle\sum_{4}^{n} r^2$

c $\displaystyle\sum_{1}^{n}(r^2-2r)$ d $\displaystyle\sum_{1}^{n}\frac{1}{m+2}$

e $\displaystyle\sum_{1}^{6}(-1)^r r^3$ f $\displaystyle\sum_{n-3}^{n} r(r+1)$

2 Use Σ notation to write these sums.

a $1+3+5+\dots+31$

b $1^5+2^5+3^5+\dots+n^5$

c $1+\dfrac{1}{2}+\dfrac{1}{3}+\dfrac{1}{4}+\dots+\dfrac{1}{n+1}$

d $3-6+9-12+\dots-42$

e $1\times3+2\times4+3\times5+\dots$ for n terms

f $\dfrac{1\times4}{3}-\dfrac{3\times5}{4}+\dfrac{5\times6}{5}-\dfrac{7\times7}{6}+\dots$ for

   n terms

3 Use standard results to prove each of these results

a $\displaystyle\sum_{r=1}^{n}(r^2+6r-3)=\frac{1}{6}n(2n^2+21n+1)$

b $\displaystyle\sum_{r=1}^{n}(2r^2-7r)=\frac{1}{6}n(n+1)(4n-19)$

c $\displaystyle\sum_{r=1}^{n}(8-5r^2)=-\frac{1}{6}n(10n^2+15n-43)$

d $\displaystyle\sum_{r=1}^{n}(r+1)^2=\frac{1}{6}n(2n^2+9n+13)$

e $\displaystyle\sum_{r=1}^{n}(r^3+2r+3)=\frac{1}{4}n(n^3+2n^2+5n+16)$

f $\displaystyle\sum_{r=1}^{n}(2r^3+3r^2)=\frac{1}{2}n(n+1)(n^2+3n+1)$

g $\displaystyle\sum_{r=1}^{n}(r^3-2r^2+3r)=\frac{1}{12}n(n+1)(3n^2-5n+14)$

h $\displaystyle\sum_{r=1}^{n}(r+2)^3=\frac{1}{4}n(n^3+10n^2+37n+60)$

4 a How many terms are there in the series

   $\displaystyle\sum_{1}^{3n}(r^2-r+3)$?

   b Write and simplify the $(2n+1)$th term.

5 Find the sums of these series.

a $1+2+3+\dots+3n$

b $1^2+2^2+3^2+\dots+(2n-1)^2$

c $1^3+2^3+3^3+\dots+(2n-1)^3$

d $1+3+5+\dots+(2n-1)$

e $1\times2+2\times3+3\times4+\dots+(n)(n+1)$

f $0+2+6+12+\dots+(n^2-n)$

6 a Find a formula for the sum to n terms of the series

   $0\times1+1\times2+2\times3+\dots+(n-1)n$

   b Find the sum of the first 10 terms and check your formula.

7 Find the rth term and the sum to n terms of these series.

a $1\times2\times3+2\times3\times4+3\times4\times5+\dots$

b $1\times4+3\times5+5\times6+\dots$

c $1\times3\times5+2\times4\times6+3\times5\times7+\dots$

d $2+6+10+16+\dots$

e $2+5+10+17+\dots$

8 Evaluate

a $\displaystyle\sum_{5}^{n}(2r+3)$ b $\displaystyle\sum_{3}^{n-2}(r^2-r)$

c $\displaystyle\sum_{n}^{2n+1}(r^3+3r)$

9 Find the sum of each series, using the method of differences where appropriate.

a $\displaystyle\sum_{1}^{n}(2-3r)$

b $\displaystyle\sum_{1}^{n}\left(\frac{1}{r}-\frac{1}{r+1}\right)$

c $\displaystyle\sum_{1}^{n}\left(\frac{1}{r}-\frac{1}{r+2}\right)$

d $\displaystyle\sum_{1}^{n}\left(\frac{1}{r}-\frac{1}{r+3}\right)$

e $\displaystyle\sum_{1}^{n}\left(\frac{1}{2r-1}-\frac{1}{2r+1}\right)$

f $\displaystyle\sum_{1}^{n}\left(\frac{1}{r+2}-\frac{2}{r+3}+\frac{1}{r+4}\right)$

Strategy

To solve problems involving sums of series

① Rearrange the expression into multiples of integers, squares and cubes, or into the form $f(x+1) - f(x)$

② Substitute standard formulae or terms as required.

③ Simplify the resulting expression for the sum.

Example 6

a Simplify the expression $(2r+1)^3 - (2r-1)^3$

b Hence, use the method of differences to show that $\sum_1^n r^2 \equiv \dfrac{n(n+1)(2n+1)}{6}$

a $(2r+1)^3 - (2r-1)^3 \equiv (8r^3 + 12r^2 + 6r + 1) - (8r^3 - 12r^2 + 6r - 1)$

$$\equiv 24r^2 + 2$$

b $24r^2 + 2 \equiv (2r+1)^3 - (2r-1)^3$

$$r^2 \equiv \frac{1}{24}[(2r+1)^3 - (2r-1)^3 - 2]$$

$$\sum_1^n r^2 \equiv \frac{1}{24}(3^3 - 1^3 - 2)$$

$$+ \frac{1}{24}(5^3 - 3^3 - 2)$$

$$+ \frac{1}{24}(7^3 - 5^3 - 2)$$

$$+ \frac{1}{24}(9^3 - 7^3 - 2)$$

$$+ \frac{1}{24}[(2n-1)^3 - (2n-3)^3 - 2]$$

$$= \frac{1}{24}[(2n+1)^3 - (2n-1)^3 - 2]$$

② Write the 'differences' for successive series of terms vertically and eliminate terms wherever possible.

Hence $\sum_1^n r^2 \equiv \frac{1}{24}[(2n+1)^3 - 1^3 + n \times (-2)]$

③ Collect up the remaining terms and simplify.

$$\equiv \frac{1}{24}[8n^3 + 12n^2 + 6n + 1 - 1 - 2n]$$

$$\equiv \frac{1}{24}[8n^3 + 12n^2 + 4n]$$

$$\equiv \frac{n}{6}[2n^2 + 3n + 1]$$

$$\equiv \frac{n}{6}(n+1)(2n+1)$$

Example 7

PURE

a Show that $\dfrac{2}{r(r+1)(r+2)}$ can be written as $\dfrac{1}{r} - \dfrac{2}{r+1} + \dfrac{1}{r+2}$

b Hence or otherwise find the sum to n terms of the series

$$\frac{1}{3} + \frac{1}{12} + \frac{1}{30} + \frac{1}{60} + \dots + \frac{1}{n \times (n+1) \times (n+2)}$$

c Use your result to deduce $\displaystyle\sum_{1}^{\infty} \frac{2}{n(n+1)(n+2)}$

a $\dfrac{1}{r} - \dfrac{2}{r+1} + \dfrac{1}{r+2} \equiv \dfrac{(r+1)(r+2) - 2r(r+2) + r(r+1)}{r(r+1)(r+2)}$

> Form a single fraction and simplify.

$$\equiv \frac{r^2 + 3r + 2 - 2r^2 - 4r + r^2 + r}{r(r+1)(r+2)}$$

$$= \frac{2}{r(r+1)(r+2)}$$

b $S_n = \dfrac{1}{1} - \dfrac{2}{2} + \dfrac{1}{3}$

$+ \dfrac{1}{2} - \dfrac{2}{3} + \dfrac{1}{4}$

$+ \dfrac{1}{3} - \dfrac{2}{4} + \dfrac{1}{5}$

$+ \dfrac{1}{4} - \dfrac{2}{5} + \dfrac{1}{6}$

$+ \dfrac{1}{5} - \dfrac{2}{6} + \dfrac{1}{7}$

$\downarrow \quad \downarrow \quad \downarrow$

$+ \dfrac{1}{n-2} - \dfrac{2}{n-1} + \dfrac{1}{n}$

$+ \dfrac{1}{n-1} - \dfrac{2}{n} + \dfrac{1}{n+1}$

$+ \dfrac{1}{n} - \dfrac{2}{n+1} + \dfrac{1}{n+2}$

> **2** Write the 'differences' for successive series of terms vertically and eliminate terms wherever possible.

Hence $S_n = \dfrac{1}{1} - \dfrac{2}{2} + \dfrac{1}{2} + \dfrac{1}{n+1} - \dfrac{2}{n+1} + \dfrac{1}{n+2}$

$$= \frac{1}{2} - \frac{1}{n+1} + \frac{1}{n+2}$$

> **3** Collect up the remaining terms and simplify.

c $\displaystyle\sum_{1}^{n} \frac{2}{r(r+1)(r+2)}$ can be written as $\dfrac{1}{2} - \dfrac{1}{n+1} + \dfrac{1}{n+2}$

Therefore, when $n \to \infty$

$$\frac{1}{2} - \frac{1}{n+1} + \frac{1}{n+2} \to \frac{1}{2} - 0 - 0 = \frac{1}{2}$$

Hence $S_\infty = \dfrac{1}{2}$

> You could also find S_∞ by saying: As n gets very large,
> $$\frac{n(n+1)}{2(n+1)(n+2)} \to \frac{n^2}{2n^2} = \frac{1}{2}$$
> Hence, $\displaystyle\lim_{n \to \infty} \frac{n(n+1)}{2(n+1)(n+2)} = \frac{1}{2}$

1 The sum to n terms of a series is $2n^2 + 4n$

  a By considering S_1, S_2 and S_3, find the first three terms of the series.

  b By considering S_{n-1} and S_n, find the nth term.

2

Admiral Nelson's ship has a triangular pile of cannonballs stacked on deck. The top cannon ball is supported by three cannonballs in the layer underneath it. These three are supported by six cannonballs in the layer underneath them, and so on. There are n layers of cannon balls.

  a How many cannon balls are there in the bottom layer?

  b How many cannonballs are there altogether?

  c Calculate the weight of the pile of 'six pounder' balls contained in a stack of 10 layers. A 'six pounder' cannon ball weighs approximately 2.722 kg.

3

The Spanish Armada's cannon balls are in square pyramids. The base is a square with $2n$ cannonballs in each side, the layer above has $(2n-1)$ cannonballs in each side, and so on. There are n layers in total. How many cannon balls are there in this pyramid?

4 **a** Write down the first four terms and the nth term of the series whose general term is $\left(\dfrac{1}{r^2} - \dfrac{1}{(r+1)^2} \right)$

b Find $\displaystyle\sum_1^n \left(\dfrac{1}{r^2} - \dfrac{1}{(r+1)^2} \right)$ and hence find the sum to infinity.

5 Write down the general term in the series
$$S = 1(m) + 2(m-1) + 3(m-2) + \dots + (m)(1)$$
and find $\displaystyle\sum_1^m S$

6 **a** Show that
$$\dfrac{1}{(2r-1)^2} - \dfrac{1}{(2r+1)^2} \equiv \dfrac{8r}{((2r)^2 - 1)^2}$$

  b Hence find $\displaystyle\sum_1^n \dfrac{8r}{((2r)^2 - 1)^2}$ and deduce the value of $\displaystyle\sum_1^n \dfrac{r}{((2r)^2 - 1)^2}$

7 **a** Show that $r(r+1) - r(r-1) \equiv 2r$

  b Use this information to show that $\displaystyle\sum_1^n r \equiv \dfrac{n(n+1)}{2}$

8 **a** Simplify the expression $(2r+1)^3 - (2r-1)^3$

  b Hence show that $\displaystyle\sum_1^n r^2 \equiv \dfrac{n(n+1)(2n+1)}{6}$

9 **a** Use standard results to find a formula for the sum of the series
$$1^2 + 3^2 + 5^2 + \dots + (2r-1)^2$$

  b Confirm your result by finding the formula in a different way.

10 Show that $\dfrac{2(x-4)}{x(x+2)(x+4)}$ can be written

as $-\dfrac{1}{x} + \dfrac{3}{x+2} - \dfrac{2}{x+4}$

11 Show that
$$2r^2 + 3r + \dfrac{1}{r} - \dfrac{1}{r+1} \equiv \dfrac{2r^4 + 5r^3 + 3r^2 + 1}{r(r+1)}$$
and hence find $\displaystyle\sum_1^n \dfrac{2r^4 + 5r^3 + 3r + 1}{r(r+1)}$

Proof by induction

Fluency and skills

There are a number of different methods of directly proving a mathematical statement. **Proof by induction** is a method that is generally used to prove a mathematical statement for all natural numbers (positive integers).

> Other methods of direct proof are **proof by deduction** and **proof by exhaustion**.

It is a powerful method that can be used in many different contexts. The principle behind proof by induction is that if you prove a statement is true when $n = 1$, and if you can prove it is true for $n = k + 1$ by assuming it is true for any $n = k$, then you can deduce that the statement must be true for all $n \in \mathbb{N}$

Key point

The three key steps to a proof by induction are:

1 Prove the statement is true for $n = 1$
2 Assume the statement is true for $n = k$ and use this to prove the statement is true for $n = k + 1$
3 Write a conclusion.

> $\mathbb{N}$ means the set of natural numbers: these are the positive integers 1, 2, 3, ...

You must always explain the steps fully and end your proof with a conclusion.

Example 1

Prove by induction that $\displaystyle\sum_{r=1}^{n} 4^{r-1} = \frac{1}{3}(4^n - 1)$ for all $n \in \mathbb{N}$

When $n = 1$, $\displaystyle\sum_{r=1}^{n} 4^{r-1} = 4^0 = 1$

> Calculate the LHS of the statement when $n = 1$

and $\dfrac{1}{3}(4^n - 1) = \dfrac{1}{3}(4 - 1)$

> Calculate the RHS of the statement when $n = 1$

$= \dfrac{1}{3}(3) = 1$

> Since LHS = RHS, the statement is true for $n = 1$

So the statement is true when $n = 1$

Assume the statement is true for $n = k$ and consider $n = k + 1$

$$\sum_{r=1}^{k+1} 4^{r-1} = \sum_{r=1}^{k} 4^{r-1} + 4^{k+1-1}$$

> Write the sum to the $(k+1)$th term as a sum to the kth term plus the $(k+1)$th term.

$$= \frac{1}{3}(4^k - 1) + 4^k$$

> Since you are assuming the statement is true for $n = k$, you can replace $\displaystyle\sum_{r=1}^{k} 4^{r-1}$ with $\dfrac{1}{3}(4^k - 1)$

(Continued on the next page)

$$= \frac{1}{3}(4^k - 1 + 3(4^k)) = \frac{1}{3}(4(4^k) - 1)$$

Collect like terms and then use index laws.

$$= \frac{1}{3}(4^{k+1} - 1)$$

This is $\frac{1}{3}(4^n - 1)$ with n replaced by $k + 1$

So the statement is true when $n = k + 1$

The statement is true for $n = 1$ and by assuming it is true for $n = k$ it is shown to be true for $n = k + 1$

Therefore, by mathematical induction, it is true for all $n \in \mathbb{N}$

You must always write a conclusion.

Example 2

Prove by induction that $\sum_{r=1}^{n} r^2 = \frac{1}{6}n(n+1)(2n+1)$ for all $n \in \mathbb{N}$

Calculate the LHS of the statement when $n = 1$

When $n = 1$, $\sum_{r=1}^{n} r^2 = 1^2 = 1$

Calculate the RHS of the statement when $n = 1$

and $\frac{1}{6}n(n+1)(2n+1) = \frac{1}{6} \times 1 \times (1+1)(2 \times 1 + 1) = \frac{1}{6}(2)(3)$

$$= 1$$

Since LHS = RHS, the statement is true for $n = 1$

So the statement is true when $n = 1$

Assume statement is true for $n = k$ and substitute $n = k + 1$ into the formula:

Write the sum to the $(k+1)$th term as sum to the kth term plus the $(k + 1)$th term.

$$\sum_{r=1}^{k+1} r^2 = \sum_{r=1}^{k} r^2 + (k+1)^2$$

$$= \frac{1}{6}k(k+1)(2k+1) + (k+1)^2$$

Since you are assuming the statement is true for $n = k$, you can replace $\sum_{r=1}^{k} r^2$ with $\frac{1}{6}k(k+1)(2k+1)$

$$= \frac{1}{6}(k+1)[k(2k+1) + 6(k+1)]$$

$$= \frac{1}{6}(k+1)(2k^2 + k + 6k + 6)$$

$$= \frac{1}{6}(k+1)(2k^2 + 7k + 6)$$

$$= \frac{1}{6}(k+1)(k+2)(2k+3)$$

Look for common factors – avoid multiplying out brackets unless necessary.

$$= \frac{1}{6}(k+1)((k+1)+1)(2(k+1)+1)$$

This is $\frac{1}{6}n(n+1)(2n+1)$ with n replaced by $k + 1$

So statement is true when $n = k + 1$

The statement is true for $n = 1$ and by assuming it is true for $n = k$ it is shown to be true for $n = k + 1$

Therefore, by mathematical induction, it is true for all $n \in \mathbb{N}$

You must always write a conclusion.

Exercise 2.4A Fluency and skills

1 Use proof by induction to prove these statements for all $n \in \mathbb{N}$

a $\displaystyle\sum_{r=1}^{n} 1 = n$ 　　　　 **b** $\displaystyle\sum_{r=1}^{n} r = \frac{1}{2}n(n+1)$

c $\displaystyle\sum_{r=1}^{n}(2r+3) = n(n+4)$

d $\displaystyle\sum_{r=1}^{n} r(r+1) = \frac{1}{3}n(n+1)(n+2)$

e $\displaystyle\sum_{r=1}^{n}(r-1)^2 = \frac{1}{6}n(n-1)(2n-1)$

f $\displaystyle\sum_{r=1}^{n}(r+1)(r-1) = \frac{1}{6}n(2n+5)(n-1)$

2 Use induction to prove these sums for all natural numbers, n

a $4 + 9 + 16 + 25 + \ldots + (n+1)^2$
$= \dfrac{1}{6} n(2n^2 + 9n + 13)$

b $1 + 5 + 25 + 125 + \ldots + 5^{(n-1)} = \dfrac{1}{4}(5^n - 1)$

3 Prove by induction that $\displaystyle\sum_{r=1}^{n} r^3 = \frac{1}{4}n^2(n+1)^2$ for all $n \in \mathbb{N}$

4 Prove by induction that $\displaystyle\sum_{r=1}^{2n} r = n(2n+1)$ for all $n \in \mathbb{N}$

5 Prove by induction that
$\displaystyle\sum_{r=1}^{2n} r^2 = \frac{1}{3}n(2n+1)(4n+1)$ for all $n \in \mathbb{N}$

6 Use induction to prove these statements for all $n \in \mathbb{N}$

a $\displaystyle\sum_{r=1}^{n} 2^r = 2(2^n - 1)$ 　 **b** $\displaystyle\sum_{r=1}^{n} 3^r = \frac{3}{2}(3^n - 1)$

c $\displaystyle\sum_{r=1}^{n} 4^r = \frac{4}{3}(4^n - 1)$ 　 **d** $\displaystyle\sum_{r=1}^{n} 2^{r-1} = 2^n - 1$

e $\displaystyle\sum_{r=1}^{n} 3^{r-1} = \frac{1}{2}(3^n - 1)$

f $\displaystyle\sum_{r=1}^{n} \left(\frac{1}{2}\right)^r = 1 - \left(\frac{1}{2}\right)^n$

7 Prove by induction that $\displaystyle\sum_{r=1}^{n} \frac{1}{r(r+1)} = \frac{n}{n+1}$ for all $n \in \mathbb{N}$

8 Prove by induction that $\displaystyle\sum_{r=2}^{n} \frac{1}{r(r-1)} = \frac{n-1}{n}$ for all $n \in \mathbb{N}$

9 Prove by induction that
$\displaystyle\sum_{r=1}^{n} \frac{1}{r^2 + 2r} = \frac{n(3n+5)}{4(n+1)(n+2)}$ for all $n \in \mathbb{N}$

Reasoning and problem-solving

As well as proving sums of series, you can use proof by induction in other contexts such as to prove an algebraic expression is divisible by a given integer.

Strategy

To prove an expression is divisible by a particular integer

(1) Substitute $n = 1$ into the expression and show the result is divisible by the integer you are using.

(2) Assume the expression is divisible by the integer for $n = k$ and substitute in $n = k+1$

(3) Separate the part you know to be divisible by the integer given.

(4) Write the conclusion.

Example 3

Prove that $n^3 + 2n$ is divisible by 3 for all $n \in \mathbb{N}$

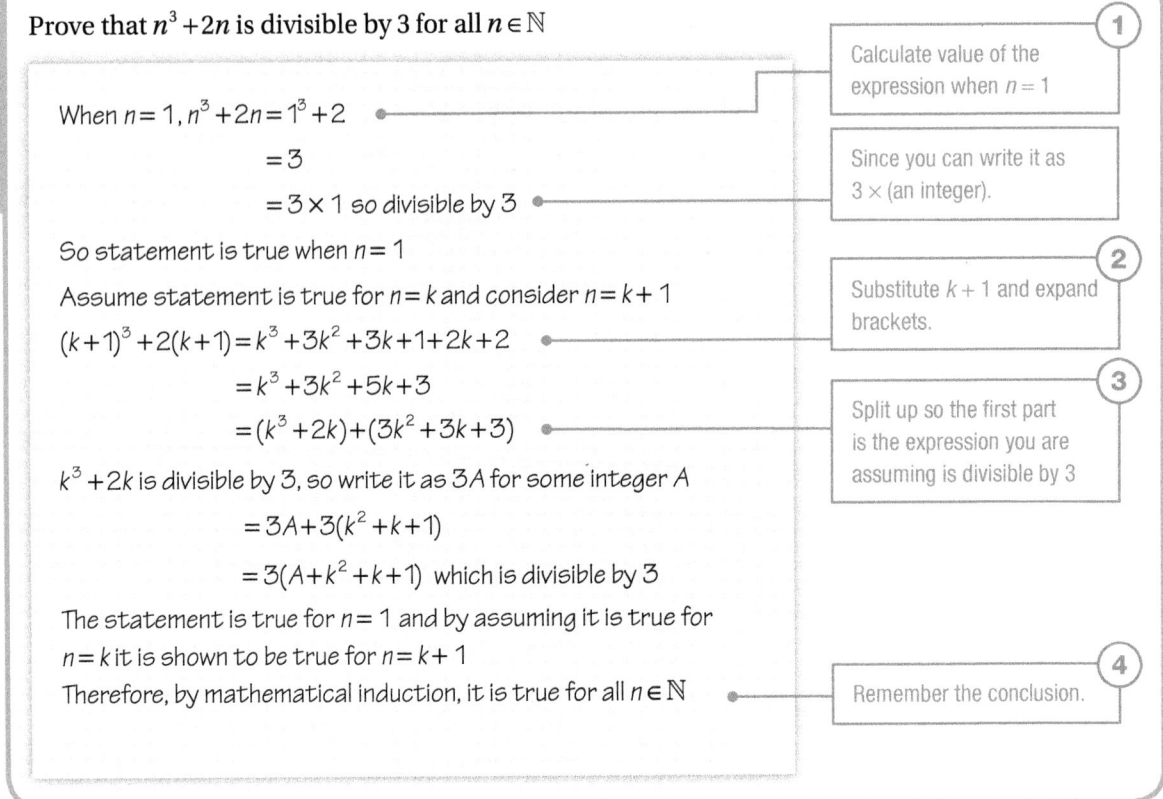

When $n = 1$, $n^3 + 2n = 1^3 + 2$

$\qquad = 3$

$\qquad = 3 \times 1$ so divisible by 3

So statement is true when $n = 1$

Assume statement is true for $n = k$ and consider $n = k + 1$

$(k+1)^3 + 2(k+1) = k^3 + 3k^2 + 3k + 1 + 2k + 2$

$\qquad = k^3 + 3k^2 + 5k + 3$

$\qquad = (k^3 + 2k) + (3k^2 + 3k + 3)$

$k^3 + 2k$ is divisible by 3, so write it as $3A$ for some integer A

$\qquad = 3A + 3(k^2 + k + 1)$

$\qquad = 3(A + k^2 + k + 1)$ which is divisible by 3

The statement is true for $n = 1$ and by assuming it is true for $n = k$ it is shown to be true for $n = k + 1$

Therefore, by mathematical induction, it is true for all $n \in \mathbb{N}$

① Calculate value of the expression when $n = 1$

Since you can write it as $3 \times$ (an integer).

② Substitute $k + 1$ and expand brackets.

③ Split up so the first part is the expression you are assuming is divisible by 3

④ Remember the conclusion.

Example 4

Given that $\mathrm{f}(n) = 2^{n+1} + 3^{2n-1}$

prove by induction that $\mathrm{f}(n)$ is a multiple of 7 for all positive integers n

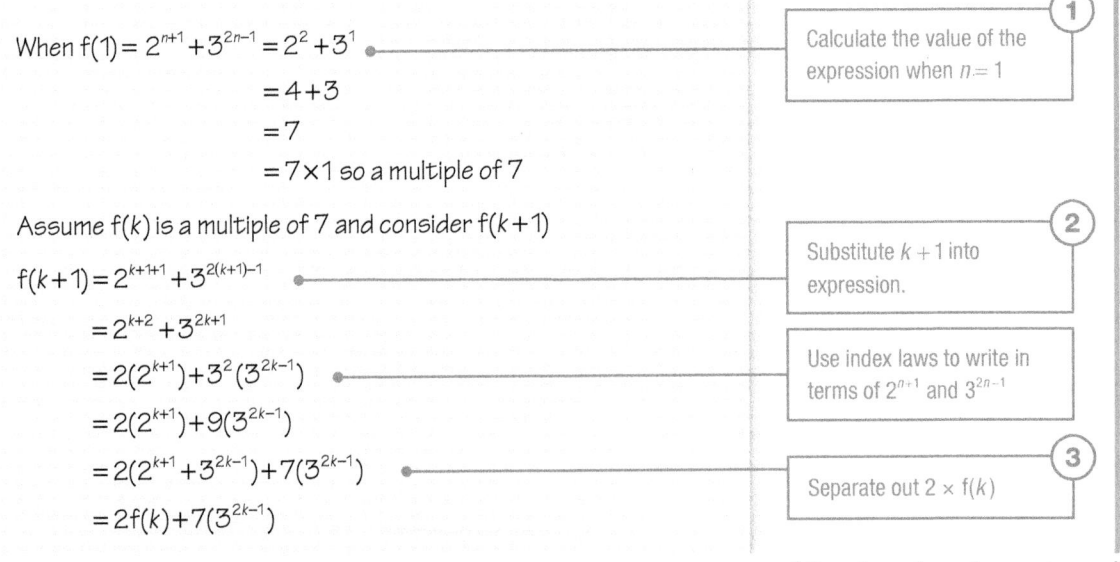

When $\mathrm{f}(1) = 2^{n+1} + 3^{2n-1} = 2^2 + 3^1$

$\qquad = 4 + 3$

$\qquad = 7$

$\qquad = 7 \times 1$ so a multiple of 7

Assume $\mathrm{f}(k)$ is a multiple of 7 and consider $\mathrm{f}(k+1)$

$\mathrm{f}(k+1) = 2^{k+1+1} + 3^{2(k+1)-1}$

$\qquad = 2^{k+2} + 3^{2k+1}$

$\qquad = 2(2^{k+1}) + 3^2(3^{2k-1})$

$\qquad = 2(2^{k+1}) + 9(3^{2k-1})$

$\qquad = 2(2^{k+1} + 3^{2k-1}) + 7(3^{2k-1})$

$\qquad = 2\mathrm{f}(k) + 7(3^{2k-1})$

① Calculate the value of the expression when $n = 1$

② Substitute $k + 1$ into expression.

Use index laws to write in terms of 2^{n+1} and 3^{2n-1}

③ Separate out $2 \times \mathrm{f}(k)$

(*Continued on the next page*)

which is a multiple of 7 since $f(k)$ and $7(3^{2k-1})$ are both multiples of 7

$f(1)$ is a multiple of 7 and by assuming that $f(k)$ is a multiple of 7 it is shown that $f(k+1)$ is a multiple of 7

Therefore, by mathematical induction, $f(n)$ is a multiple of 7 for all positive integers n ●————————————— Write the conclusion.   ④

Exercise 2.4B Reasoning and problem-solving

1 Use proof by induction to prove these statements for all $n \in \mathbb{N}$

   a $n^2 + 3n$ is divisible by 2

   b $5n^2 - n$ is divisible by 2

   c $8n^3 + 4n$ is divisible by 12

   d $11n^3 + 4n$ is divisible by 3

2 Prove by induction that $7n^2 + 25n - 4$ is divisible by 2 for all $n \in \mathbb{N}$

3 Prove by induction that $n^3 - n$ is divisible by 3 for all $n \geq 2$, $n \in \mathbb{N}$

4 Prove by induction that $10n^3 + 3n^2 + 5n - 6$ is divisible by 6 for all $n \in \mathbb{N}$

5 Use proof by induction to prove these statements for all $n \in \mathbb{N}$

   a $6^n + 9$ is divisible by 5

   b $3^{2n} - 1$ is divisible by 8

   c $2^{3n+1} - 2$ is divisible by 7

6 Prove by induction that $5^n - 4n + 3$ is a multiple of 4 for all $n \in \mathbb{N}$

7 Prove by induction that $3^n + 2n + 7$ is a multiple of 4 for all $n \in \mathbb{N}$

8 Prove by induction that $7^n - 3n + 5$ is a multiple of 3 for all $n \in \mathbb{N}$

9 Use proof by induction to prove these statements for all $n \in \mathbb{N}$

   a $\displaystyle\sum_{r=n+1}^{2n} r^2 = \frac{1}{6}n(2n+1)(7n+1)$

   b $\displaystyle\sum_{r=n}^{2n} r^3 = \frac{3}{4}n^2(5n+1)(n+1)$

   c $8^n - 5^n$ is divisible by 3 for all $n \in \mathbb{N}$

Fluency and skills

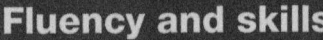

It is sometimes useful to be able to express a function as a series of terms. The **Maclaurin series** for a function $f(x)$ is based on three assumptions:

- $f(x)$ can be expanded as a **convergent** infinite series of terms
- each of the terms in $f(x)$ can be differentiated
- each of the differentiated terms has a finite value when $x = 0$

> A convergent series is one where an infinite number of terms has a finite sum.

With these assumptions, you can write

$f(x) \equiv a_0 + a_1 x + a_2 x^2 + a_3 x^3 + a_4 x^4 + a_5 x^5 + a_6 x^6 + \dots + a_r x^r + \dots$ where $a_0, a_1, a_2, \dots$ are constants.

> The proof of these assumptions is outside the scope of this chapter.

Differentiating repeatedly and substituting $x = 0$, gives

$$f(x) \equiv a_0 + a_1 x + a_2 x^2 + a_3 x^3 + a_4 x^4 + a_5 x^5 + \dots + a_r x^r + \dots$$

So $f(0) \equiv a_0$

$$f'(x) \equiv a_1 + 2a_2 x + 3a_3 x^2 + 4a_4 x^3 + 5a_5 x^4 + \dots + ra_r x^{r-1} + \dots$$

So $f'(0) \equiv a_1$

$$f''(x) \equiv 2 \times 1 a_2 + 3 \times 2a_3 x + 4 \times 3a_4 x^2 + 5 \times 4a_5 x^3 + \dots + r(r-1)a_r x^{r-2} + \dots$$

So $f''(0) \equiv 2! a_2$

$$f'''(x) \equiv 3 \times 2 \times 1 a_3 + 4 \times 3 \times 2a_4 x + 5 \times 4 \times 3a_5 x^2 + \dots + r(r-1)(r-2)a_r x^{r-3} + \dots$$

So $f'''(0) \equiv 3! a_3$

Continuing this process gives the definition of the Maclaurin series.

Key point

$$f(x) \equiv f(0) + x f'(0) + \frac{x^2}{2!} f''(0) + \frac{x^3}{3!} f'''(0) + \dots + \frac{x^r}{r!} f^r(0) + \dots$$

This is the definition of a Maclaurin series.

A Maclaurin series can be found for any function that meets the conditions given above. You should be able to recognise and use the Maclaurin series for the following functions and know the values of x for which they are valid.

> You do not need to be able to prove the conditions for validity of these Maclaurin series.

Key point

$$(1 + x)^n = 1 + nx + \frac{n(n-1)}{2!} x^2 + \dots + \frac{n(n-1)\dots(n-1+r)}{r!} x^r + \dots \text{ for } -1 < x < 1, \, n \in \mathbb{R}$$

$$e^x = 1 + x + \frac{x^2}{2!} + \dots + \frac{x^r}{r!} + \dots \text{ for all } x$$

$$\ln(1 + x) = x - \frac{x^2}{2} + \frac{x^3}{3} - \dots + (-1)^{r+1} \frac{x^r}{r} + \dots \text{ for } -1 < x \le 1$$

$$\sin x = x - \frac{x^3}{3!} + \frac{x^5}{5!} - \dots + (-1)^r \frac{x^{2r+1}}{(2r+1)!} + \dots \text{ for all } x$$

$$\cos x = 1 - \frac{x^2}{2!} + \frac{x^4}{4!} - \dots + (-1)^r \frac{x^{2r}}{(2r)!} + \dots \text{ for all } x$$

Example 1

Use the Maclaurin expansion of $(1+x)^n$ to find the first five terms of the series for $f(x) = \dfrac{1}{(1-x)^2}$

$(1+x)^n = 1 + nx + \dfrac{n(n-1)}{2!}x^2 + \dots + \dfrac{n(n-1)\dots(n-1+r)}{r!}x^r + \dots$

This expansion will be given in your formulae booklet.

$\dfrac{1}{(1-x)^2} \equiv 1 + 2x + \dfrac{(-2)(-3)}{2}x^2 + \dfrac{(-2)(-3)(-4)}{6}(-x)^3$

$+ \dfrac{(-2)(-3)(-4)(-5)}{24}x^4 + \dots$

$\equiv 1 + 2x + 3x^2 + 4x^3 + 5x^4 + \dots$

For the expansion of $\dfrac{1}{(1-x)^2}$, $n = -2$ and x is replaced by $-x$

If n is not a positive integer, the expansion of $(1+x)^n$ is valid for $-1 < x < 1$

This means that in Example 1, the expansion for $(1-x)^{-2}$ is valid when $-1 < -x < 1$

This can be rearranged to $-1 < x < 1$

Example 2

Write down the range of values of x that make the Maclaurin series for these functions valid.

a $\ln(1-3x)$ b $\ln(1+x^2)$ c $\dfrac{1}{3+x}$

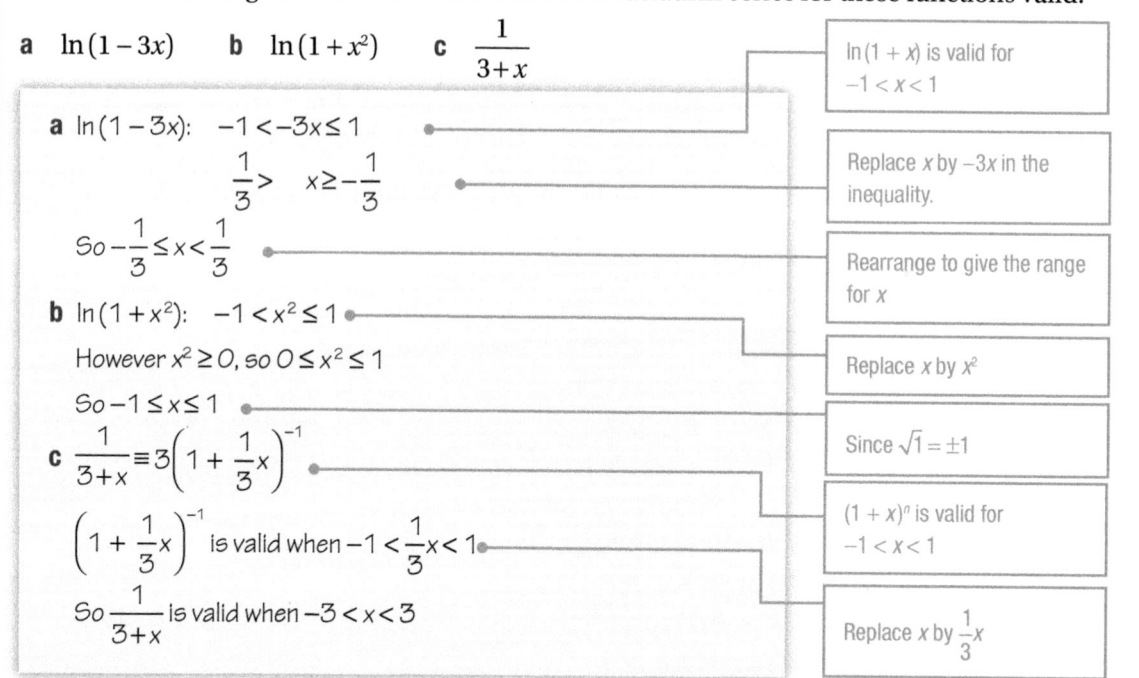

a $\ln(1-3x)$: $-1 < -3x \le 1$

$\dfrac{1}{3} > x \ge -\dfrac{1}{3}$

So $-\dfrac{1}{3} \le x < \dfrac{1}{3}$

b $\ln(1+x^2)$: $-1 < x^2 \le 1$

However $x^2 \ge 0$, so $0 \le x^2 \le 1$

So $-1 \le x \le 1$

c $\dfrac{1}{3+x} \equiv 3\left(1 + \dfrac{1}{3}x\right)^{-1}$

$\left(1 + \dfrac{1}{3}x\right)^{-1}$ is valid when $-1 < \dfrac{1}{3}x < 1$

So $\dfrac{1}{3+x}$ is valid when $-3 < x < 3$

$\ln(1+x)$ is valid for $-1 < x < 1$

Replace x by $-3x$ in the inequality.

Rearrange to give the range for x

Replace x by x^2

Since $\sqrt{1} = \pm 1$

$(1+x)^n$ is valid for $-1 < x < 1$

Replace x by $\dfrac{1}{3}x$

Exercise 2.5A Fluency and skills

1 Use the Maclaurin expansion to find the first five terms of these functions.

a e^{2x} b e^{-x} c e^{-3x}

d $e^{\frac{x}{2}}$ e $e^{\frac{-x}{3}}$

2 Use the Maclaurin expansion to

i   Find the first five terms of these functions,

ii   Find the range of values of x for which each expansion is valid.

a $\ln(1-x)$ b $\ln(1+2x)$ c $\ln(1-3x)$

d $\ln\left(1+\dfrac{x}{2}\right)$ e $\ln\left(1-\dfrac{x}{3}\right)$

3 Use the Maclaurin expansion to find the first four terms of these functions.

a $\sin 2x$ b $\sin\dfrac{x}{2}$ c $\sin(-x)$

d $\sin(-3x)$ e $\sin\dfrac{3x}{2}$

4 Use the Maclaurin expansion to find the first four terms of these functions.

a $\cos 4x$ **b** $\cos \dfrac{x}{3}$ **c** $\cos(-x)$

d $\cos(-2x)$ **e** $\cos \dfrac{-x}{2}$

a $(1+x)^{\frac{1}{2}}$ **b** $\sqrt{(1-x)}$

c $\sqrt[3]{(1+x)}$ **d** $(1+2x)^{\frac{3}{2}}$

e $(1-3x)^{\frac{5}{2}}$ **f** $\dfrac{1}{1-x}$ **g** $\dfrac{3}{1+x}$

5 For each function

i Find the first four terms of the Maclaurin expansion,

ii Find the range of values of x for which the expansion is valid.

6 Use the Maclaurin series to write down the first four terms in these expansions.

a $\sin(x^2)$ **b** $\sin(\sqrt{x})$

c $\cos(x^2)$ **d** $\cos(\sqrt{x})$

Reasoning and problem-solving

Strategy

To find series expansions for compound functions

(1) Use standard series to expand each part of the function.

(2) Combine the series to give the expansion of the compound function.

(3) Consider the validity of the values of x

Example 3

Expand $\dfrac{\sin x}{\sqrt{(1+x)}}$ as far as the term in x^5 and give the range of x for which the series is valid.

$\dfrac{1}{\sqrt{(1+x)}} \equiv (1+x)^{\frac{-1}{2}}$

$(1+x)^n = 1 + nx + \dfrac{n(n-1)}{2!}x^2 + \dfrac{n(n-1)(n-2)}{3!}x^3 + \dots$

$(1+x)^{-\frac{1}{2}} = 1 + \left(-\dfrac{1}{2}\right)x + \left(\dfrac{3}{4}\right)\left(\dfrac{x^2}{2!}\right) + \left(-\dfrac{15}{8}\right)\left(\dfrac{x^3}{3!}\right)$

$\qquad + \left(\dfrac{105}{16}\right)\left(\dfrac{x^4}{4!}\right) + \left(\dfrac{945}{32}\right)\left(\dfrac{x^5}{5!}\right) + \dots$

$\qquad = 1 - \dfrac{1}{2}x + \dfrac{3x^2}{8} - \dfrac{5x^3}{16} + \dfrac{35x^4}{128} - \dfrac{63x^5}{256} + \dots$

> **(1)** Replace n by $-\dfrac{1}{2}$ in the expansion for $(1+x)^n$

$\sin x = x - \dfrac{x^3}{3!} + \dfrac{x^5}{5!} - \dots$

> **(1)** Use the standard expansion for $\sin x$

$\dfrac{\sin x}{\sqrt{(1+x)}} \equiv (\sin x)(1+x)^{-\frac{1}{2}}$

$\qquad = \left(x - \dfrac{x^3}{6} + \dfrac{x^5}{120}\right)\left(1 - \dfrac{1}{2}x + \dfrac{3x^2}{8} - \dfrac{5x^3}{16} + \dfrac{35x^4}{128} - \dfrac{63x^5}{256} + \right)$

$\qquad = x - \dfrac{1}{2}x^2 + \dfrac{3x^3}{8} - \dfrac{5x^4}{16} + \dfrac{35x^5}{128} + \dots - \dfrac{x^3}{6} + \dfrac{x^4}{12} - \dfrac{x^5}{16}$

$\qquad + \dots + \dfrac{x^5}{120} + \dots$

> **(2)** Multiply the expansions to find the expansion for the compound function. Only multiply out the terms that give you powers up to x^5

(Continued on the next page)

Hence $\dfrac{\sin x}{\sqrt{(1+x)}} \equiv x - \dfrac{1}{2}x^2 + \dfrac{5x^3}{24} - \dfrac{11x^4}{48} + \dfrac{421x^5}{1920} + \ldots$

The expansion for $(1+x)^{-\frac{1}{2}}$ is valid for $-1 < x < 1$

The expansion for $\sin x$ is valid for all values of x

So, the expansion for $\dfrac{\sin x}{\sqrt{(1+x)}}$ is valid for $-1 < x < 1$

③ Choose the range that is valid for both series.

Exercise 2.5B Reasoning and problem-solving

1 Use the Maclaurin series to find the first three non-zero terms of these functions.

    a $e^x \sin x$ **b** $e^{\sin x}$

    c $\sin x + \cos x$ **d** $\cos^2 x$

2 Find the first three non-zero terms in the Maclaurin expansion of $x \cos x$. Give the values of x for which the expansion is valid.

3 Find the first three non-zero terms in these Maclaurin expansions. Find the range of values of x for which each expansion is valid.

    a $(1+3x)^{\frac{1}{2}}$ **b** $\ln\left(\dfrac{1-x}{1+x}\right)$

    c $\dfrac{x}{1-x}$ **d** $\dfrac{\cos 2x}{1+x}$

4 a Find the first four non-zero terms in the expansion of

    i $\sqrt{1+\dfrac{1}{x}}$ **ii** $\ln\left(1+\dfrac{1}{x}\right)$

   b Find the range of values of x for which the expansion of $\ln\left(1+\dfrac{1}{x}\right)$ is valid.

5 a Make use of known series expansions to obtain these expansions.

    i $\dfrac{1}{2}(e^x - e^{-x})$ (up to the term in x^5)

    ii $\cos x$ (up to the term in x^4)

    iii $\ln(1+\cos x)$ (use your answer to part **ii** and expand up to the term in x^{12})

   b Find the range of values of x for which the expansion of $\ln(1+\cos x)$ is valid.

6 a Use the Maclaurin series to approximate $e^{-0.6}$ (to 4 sf).

   b **i** By writing 3.5 as $3\left(1+\dfrac{1}{6}\right)$, using $\ln 3 \approx 1.09861$ and the expansion of $\ln(1+x)$, find the value of $\ln(3.5)$ correct to 5 sf.

    ii Explain why just substituting $x = 2.5$ does not give a valid answer.

7 At any point x, the gradient of exponential function e^{kx} is ke^{kx}

This means that the derivative of e^{kx} is ke^{kx}

   a Use this to derive the Maclaurin series for e^{2x} from first principles using the definition of the Maclaurin series.

   b Check your answer by substituting into the standard formula for the expansion of e^x

   c Substitute $x = \dfrac{1}{4}$ in your first five terms and hence obtain an approximation for $\sqrt{e}$

8 a The differential of $\sin x$ is $\cos x$ and the differential of $\cos x$ is $-\sin x$
Use these facts and the definition of the Maclaurin series to find the first four terms of the function $f(x) = \sin x$

   b Substitute $x = \dfrac{\pi}{2}$ into your expansion and find the percentage error between that and the exact value of $\sin \dfrac{\pi}{2}$

Chapter summary

- For the quadratic equation $ax^2 + bx + c = 0$, the sum of the roots $= -\dfrac{b}{a}$ and the product of the roots $= \dfrac{c}{a}$

- If the roots of the cubic equation $ax^3 + bx^2 + cx + d = 0$ are $\alpha,\ \beta$ and γ, then

$$(\alpha + \beta + \gamma) = -\frac{b}{a} \qquad (\alpha\beta + \beta\gamma + \gamma\alpha) = \frac{c}{a} \qquad \text{and} \qquad \alpha\beta\gamma = -\frac{d}{a}$$

- If the roots of the quartic equation $ax^4 + bx^3 + cx^2 + dx + e = 0$ are $\alpha,\ \beta,\ \gamma$ and δ, then

$$(\alpha + \beta + \gamma + \delta) = -\frac{b}{a} \quad (\alpha\beta + \beta\gamma + \gamma\delta + \delta\alpha) = \frac{c}{a} \quad (\alpha\beta\gamma + \beta\gamma\delta + \gamma\delta\alpha + \delta\alpha\beta) = -\frac{d}{a} \quad \text{and} \quad \alpha\beta\gamma\delta = \frac{e}{a}$$

- If roots are transformed by a linear amount so $y = mx + c$, then substitute $x = \dfrac{y - c}{m}$

- To solve inequalities, follow the same rules as in solving equations, but remember that
  - answers to inequalities come as a range of values
  - when you multiply or divide by a negative number you reverse the inequality sign.

- $\displaystyle\sum_{1}^{n} u_r$ means 'find the sum of all the terms $u_1,\ u_2,\ ...,\ u_r,\ ...\ u_{n-1},\ u_n$'

- $\displaystyle\sum_{1}^{n} r \equiv 1 + 2 + 3 + ... + n = \frac{n(n+1)}{2}$

- $\displaystyle\sum_{1}^{n} r^2 \equiv \frac{n}{6}(n+1)(2n+1)$

- $\displaystyle\sum_{1}^{n} r^3 \equiv \frac{n^2(n+1)^2}{4}$

- If a series has alternate positive and negative terms, include $(-1)^{r+1}$ in $\Sigma f(r)$
  For example, $1 - 2 + 3 - 4 + ... +$ for n terms, can be written as $\displaystyle\sum_{1}^{n} (-1)^{r+1}$

- If you can find a function $f(x)$ such that the general term, u_r can be expressed as $f(r+1) - f(r)$, you can find the sum of the series using the method of differences.

- To prove a statement by induction you must first prove the result for $n = 1$ then assume the statement is true for $n = k$ and use this fact to prove the statement is true for $n = k + 1$. Always remember to write a conclusion.

- You should recognise the following Maclaurin series and their range of validity:
  - $(1 + x)^n = 1 + nx + \dfrac{n(n-1)}{2!} x^2 + ... + \dfrac{n(n-1)...(n-1+r)}{r!} x^r + ...$ for all x if n is a positive integer,
    otherwise $-1 < x < 1,\ n \in \mathbb{R}$
  - $e^x = 1 + x + \dfrac{x^2}{2!} + ... + \dfrac{x^r}{r!} + ...$ for all x
  - $\ln(1 + x) = x - \dfrac{x^2}{2} + \dfrac{x^3}{3} - ... + (-1)^{r+1} \dfrac{x^r}{r} + ...$ for $-1 < x \leq 1$
  - $\sin x = x - \dfrac{x^3}{3!} + \dfrac{x^5}{5!} - ... + (-1)^r \dfrac{x^{2r+1}}{(2r+1)!} + ...$ for all x
  - $\cos x = 1 - \dfrac{x^2}{2!} + \dfrac{x^4}{4!} - ... + (-1)^r \dfrac{x^{2r}}{(2r)!} + ...$ for all x

Check and review

1 Write down the sum, the sum of the products in pairs and the product of the roots of these equations.

a $x^3 + 5x^2 + 2x + 1 = 0$

b $x^3 + 9x^2 - 11x - 8 = 0$

2 Find the polynomial whose roots are double the roots of $x^3 - 4x^2 + 6x - 2 = 0$

3 The cubic equation $2x^3 - 35x^2 + kx + 510 = 0$ has solutions α, β and γ

  a Use the fact that $\alpha\beta + \alpha\gamma + \beta\gamma = 51$ to find the value of k

  b Find a cubic equation with roots $\dfrac{1}{\alpha} + 1$, $\dfrac{1}{\beta} + 1$ and $\dfrac{1}{\gamma} + 1$

4 The cubic equation $x^3 + ax^2 + bx + 2 = 0$ has roots $\alpha, \dfrac{1}{\alpha}$ and $\alpha + \dfrac{1}{\alpha} - 2$

  a Solve the equation,

  b Find the values of a and b

5 The roots of the equation $x^3 - x^2 - 3x + 1 = 0$ are α, β and γ

  a Write down the value of $\alpha\beta\gamma$

  b **i** Find a cubic equation with roots $\alpha - 1$, $\beta - 1$, $\gamma - 1$

    ii By considering the product of the roots of this new equation, find the value of $\alpha\beta\gamma(1 + \beta\gamma)(1 + \gamma\alpha)(1 + \alpha\beta)$

6 The equation $2x^3 - 6x^2 + 3x - 1 = 0$ has roots α, β and γ

Find the values of

  a $\dfrac{1}{\alpha} + \dfrac{1}{\beta} + \dfrac{1}{\gamma}$

  b $(2 - \alpha)(2 - \beta)(2 - \gamma)$

  c $\alpha^2 + \beta^2 + \gamma^2$

7 Solve these inequalities algebraically.

  a $\dfrac{3x - 15}{8x} \geq 0$ **b** $\dfrac{x}{x - 4} < 0$

  c $\dfrac{x - 1}{x + 2} > 4$ **d** $\dfrac{4x}{2x - 5} \geq -4$

  e $\dfrac{x - 3}{x - 2} < 3 - x$ **f** $\dfrac{2x - 3}{x + 4} \geq 3$

8 Solve these inequalities.

  a $x(2x + 3)(3x - 2) < 0$

  b $(x + 6)(x + 1)(x - 8) > 0$

  c $(3x + 8)(x - 1)(2x - 5)(x - 8) \leq 0$

  d $(x^2 + 6x + 5)(x^2 - 10x + 21) \leq 0$

9 Solve these inequalities.

a $1 \le \dfrac{2x-5}{x-2} \le 3$

b $-8 \le \dfrac{3x+1}{x+4} \le 2$

10 Write out these sequences.

a $\displaystyle\sum_{3}^{10} r$

b $\displaystyle\sum_{n}^{n+4} r^2$

c $\displaystyle\sum_{1}^{n} (r^2 + 5r)$

d $\displaystyle\sum_{1}^{n} \dfrac{1}{m-3}$

e $\displaystyle\sum_{2}^{10} (-1)^r r^2$

f $\displaystyle\sum_{n-1}^{n+3} 2r(3r-2)$

11 Find the sums of these series to n terms.

a $3 - 2r$

b $2r^2 - 4r$

c $5r + r^3$

12 a Find $\displaystyle\sum_{n}^{2n+1} (r^3 - 3r + 2)$

b Verify your solution using $n = 6$

13 Find the sums of these series to n terms.

a $\dfrac{1}{r+1} - \dfrac{1}{r+2}$

b $\dfrac{1}{r+2} - \dfrac{1}{r+3}$

14 Use the method of differences to find the sum to n terms of the series

$$\frac{1}{2} + \frac{1}{6} + \ldots + \left(\frac{1}{n} - \frac{1}{n+1}\right)$$

Hence write down the sum to infinity.

15 Show that $t^2(t+1)^2 - t^2(t-1)^2 \equiv 4t^3$

Hence show that $\displaystyle\sum_{1}^{n} t^3 = \dfrac{n^2(n+1)^2}{4}$

16 The expression $\dfrac{2n+1}{n^2(n+1)^2}$ can be written as

$\dfrac{1}{n^2} - \dfrac{1}{(n+1)^2}$. Use the method of differences

on $\left(\dfrac{1}{n^2} - \dfrac{1}{(n+1)^2}\right)$ to show that

$$\sum_{1}^{k}\left(\frac{1}{n^2} - \frac{1}{(n+1)^2}\right) = \frac{k(k+2)}{(k+1)^2}$$

17 Prove by induction that $3^{2n+1} + 1$ is divisible by 4 for all $n \in \mathbb{N}$

18 Prove by induction that $2^{2n} - 3n + 2$ is divisible by 3 for all $n \in \mathbb{N}$

19 Prove by induction that $6^n - 1$ is divisible by 5 for all $n \in \mathbb{N}$

20 Find the validity of each of these functions.

a $(1+x^2)^{-\frac{1}{2}}$

b $\ln\left(1 - \dfrac{x}{2}\right)$

21 Find the first three non-zero terms in these Maclaurin expansions.

a $\ln(1 + x^2)$

b $(1+x)^{-\frac{1}{3}} \sin 2x$

22 Use the Maclaurin expansion of e^x to find the first four non-zero terms of $e^{\frac{x}{3}}$ and hence find an approximate value of $\sqrt[3]{e}$ to 4 dp.

Did you know?

The **Riemann zeta function** is an infinite summation formula.
It is given by the formula

$$\zeta(s) = \sum_{n=1}^{\infty} \frac{1}{n^s}$$

where s is any complex number with $\text{Re}(s) > 1$

The Riemann zeta function is a famous function in mathematics because it has a link to prime numbers. The **Riemann hypothesis**, proposed by Bernhard Riemann, is a conjecture that makes statements about the zeros of the zeta function. The hypothesis implies results about how the prime numbers may be distributed.

The Riemann hypothesis is yet to be proved, but there is $1 million for the first person who manages it.

Note

ζ is the Greek letter "zeta".

Research

The **Basel problem** was first posed by **Pietro Mengoli** in 1644.
The Basel problem simply asks for the precise summation of the reciprocals of the squares of the natural numbers.
In other words, $\zeta(2)$

Many famous mathematicians failed to solve the problem until **Leonhard Euler**, at the age of 28, solved it in 1734.
Euler found the solution to be

$$\sum_{n=1}^{\infty} \frac{1}{n^2} = \frac{\pi^2}{6}$$

Find out how Euler solved the problem.

After Euler, **Augustin-Louis Cauchy** came up with an alternative proof for the result.
Find out Cauchy's approach to the problem.

Leonhard Euler

Information

The sum of the natural numbers is a divergent sequence. This can be shown by considering increasingly longer finite sums of the natural numbers:

$$\sum_{n=1}^{100} \frac{1}{n} = 5050 \qquad \sum_{n=1}^{1000} \frac{1}{n} = 500\,500 \qquad \sum_{n=1}^{10000} \frac{1}{n} = 50\,005\,000$$

As the length of the sequence increases, so does the sum. Therefore as the length tends to infinity, so does the sum.

Research

In 1913, the Indian mathematician **Ramanujan** claimed to be able to show that

$$\sum_{n=1}^{\infty} \frac{1}{n} = -\frac{1}{12}$$

Find out about the proof he offered, and how it related to the Riemann zeta function.

71

2 Assessment

1 The quadratic equation $3x^2 + 4x + 2 = 0$ has roots α and β. Work out the value of $\alpha + \beta$
Select the correct answer.

  a $\dfrac{2}{3}$ b $-\dfrac{4}{3}$ c $\dfrac{4}{3}$ d $\dfrac{2}{3}$ **[1 mark]**

2 The quartic equation $x^4 - 4x^2 + x - 2 = 0$ has roots α, β, γ and δ. Work out the value of $\sum \alpha\beta\delta$
Select the correct answer.

  a –3   b 3   c –1   d 1   **[1]**

3 Solve the inequality $(x-1)(x+3) > 0$
Select the correct solution.

  a $0 < x < -3$ or $x < 1$ b $x < -3$ or $0 < x < 1$ c $-3 < x < 1$ or $x > 0$ d $-3 < x < 0$ or $x > 1$ **[1]**

4 Given that $(1+x)^n \approx 1 + nx + \dfrac{n(n-1)}{2!}x^2 + \ldots + \dfrac{n(n-1)\ldots(n-1+r)}{r!}x^r + \ldots$

  a Find the expansion of $\dfrac{1}{(1+5x)^3}$ in ascending order up to the term in x^3 **[4]**

  b State the range of values of x for which this expansion is valid.   **[1]**

5   a Use the Maclaurin expansion of e^x to find the first four terms in the expansion of $e^{-\frac{x}{2}}$ **[3]**

  b Use your expansion to estimate the value of $e^{0.05}$ **[3]**

6 Express the sum $\sum_{r=1}^{n} (6r^2 + 2r)$ as a polynomial in n **[3]**

7 A cubic equation is given by $4x^3 + 3x = 1 + 6x^2$

  a Use the substitution $u = 2x - 1$ to form a cubic in u **[3]**

  b Solve this cubic equation in u **[2]**

  c Hence, or otherwise, solve the original equation in x **[2]**

8 Solve the inequality $1 + \dfrac{x+2}{-5} > x - \dfrac{x-2}{3}$ **[3]**

9   a Express $\sum_{r=1}^{n} (3r^2 - 2r - 1)$ as a factorised polynomial in n **[4]**

  b Hence evaluate the sum to $n = 16$ **[1]**

10 The nth term of a sequence is defined by $u_n = (2n-1)(n+3)$

  a Express the sum to n terms of the sequence as a function of n **[4]**

  b Evaluate the sum of the first 15 terms.   **[2]**

  c Find the sum of the 16th to the 20th term.   **[3]**

11 a Form a polynomial function, f(x), whose roots are 2, 3 and −4 and whose curve passes through the point (1, 5) **[3]**

b The function $y = $ f(x) undergoes the transformation described by $y = \dfrac{1}{2}x + 1$
Find the equation of the transformed polynomial, g(x) **[3]**

12 The roots of the cubic $2x^3 - x^2 - 3x + 2 = 0$ are α, β and γ

a Write down the values of $\alpha\beta\gamma$, $\sum\alpha\beta$ and $\sum\alpha$ **[3]**

b Find the cubic equation with roots $2\alpha - 1$, $2\beta - 1$ and $2\gamma - 1$ **[4]**

13 a Use the Maclaurin expansion of $\ln(1+x)$ to find the expansion of $\ln(1-x^2)$ up to the term in x^4 **[3]**

b State the range of values of x for which this series is valid. **[2]**

14 Find the first four non-zero terms in the series expansion of $\dfrac{\sin\left(\dfrac{x}{3}\right)}{e^x}$ **[5]**

15 A cubic function is given by f(x) = $2x^3 + 3x^2 - 8x + 3$

a Evaluate f(1) and write down a linear factor of f(x) **[1]**

b Factorise f(x) fully. **[3]**

c Hence solve the inequality f(x) < 0 **[2]**

16 Prove by induction that $\displaystyle\sum_{r=1}^{n} 4^{r-1} = \frac{1}{3}(4^n - 1)$ for all $n \in \mathbb{N}$ **[6]**

17 Prove by induction that $4n^3 + 8n$ is divisible by 12 for all $n \in \mathbb{N}$ **[6]**

18 Prove by induction that $4^{2n+1} - 1$ is a multiple of 3 for all $n \in \mathbb{N}$ **[6]**

19 A function is given by f(x) = $\dfrac{(x-3)(x+4)}{x-5}$

a Sketch the curve $y = $ f(x) and find the set of values for x for which the function is positive. **[3]**

b Find the set of values for which f(x) ≤ −8 **[4]**

20 a Show that $\dfrac{1}{x} - \dfrac{1}{x+1} = \dfrac{1}{x(x+1)}$ **[1]**

b Find an expression in n for $\displaystyle\sum_{r=1}^{n} \frac{1}{r(r+1)}$ **[3]**

c Hence evaluate the sum $\dfrac{1}{1\times2} + \dfrac{1}{2\times3} + \dfrac{1}{3\times4} + \dfrac{1}{4\times5} + ... + \dfrac{1}{99\times100}$ **[2]**

d Evaluate $\displaystyle\sum_{r=50}^{99} \frac{1}{r(r+1)}$ **[2]**

21 a Prove that $\dfrac{1}{x^2-1} = \dfrac{1}{2}\left(\dfrac{1}{x-1} - \dfrac{1}{x+1}\right)$; $x > 1$ **[2]**

b Find an expression in n for $\displaystyle\sum_{r=2}^{n} \frac{1}{r^2-1}$ **[4]**

c Evaluate $\dfrac{1}{3} + \dfrac{1}{8} + \dfrac{1}{15} + \dfrac{1}{24} + ... + \dfrac{1}{399}$ **[3]**

d Find $\displaystyle\sum_{r=2}^{\infty} \frac{1}{r^2-1}$ **[2]**

22 Given that $\displaystyle\sum_{r=1}^{n}1=n$, that $\displaystyle\sum_{r=1}^{n}r=\frac{n}{2}(n+1)$ and that $(r+1)^3-r^3=3r^2+3r+1$

find an expression for $\displaystyle\sum_{r=1}^{n}r^2$ in terms of n **[6]**

23 The quartic equation $x^4+x+1=0$ has four roots α, β, γ and δ

  a Find the quartic equation with integer coefficients which has roots 3α, 3β, 3γ and 3δ **[3]**

  b By considering the substitution $u=x^2$, or otherwise, find the quartic equation with roots α^2, β^2, γ^2 and δ^2 **[3]**

24 Solve the inequality $\dfrac{x+1}{2x-2}\geq 2x+2$ **[7]**

25 The function $\mathrm{f}(x)=\dfrac{x^2+x+4}{x}$ intersects the straight line $y=k$

  a Form a quadratic equation in x and k and hence find the set of values of k for which $\mathrm{f}(x)$ has real roots. **[4]**

  b Hence show that the equation $\sin\theta=\dfrac{x^2+x+4}{x}$ has no solutions. **[3]**

26 **a** Expand $\ln\left(\dfrac{1+4x}{\sqrt{1-2x}}\right)$ up to the term in x^3 **[5]**

  b State the range of values of x for which this expansion is valid. **[2]**

27 **a** Prove that $\dfrac{1}{x(x+1)(x+2)}=\dfrac{1}{2}\left(\dfrac{1}{x}-\dfrac{2}{x+1}+\dfrac{1}{x+2}\right)$ **[2]**

  b Hence find an expression in n for $\displaystyle\sum_{r=1}^{n}\frac{1}{r(r+1)(r+2)}$ **[4]**

  c Evaluate $\dfrac{1}{11\times12\times13}+\dfrac{1}{12\times13\times14}+\dfrac{1}{13\times14\times15}+...+\dfrac{1}{20\times21\times22}$ **[3]**

28 **a** Show that $2x+\dfrac{3}{x}-\dfrac{3}{x+2}=\dfrac{2x^3+4x^2+6}{x(x+2)}$ **[2]**

  b Hence find, as a single fraction, an expression for $\displaystyle\sum_{r=1}^{n}\frac{2r^3+4r^2+6}{r(r+2)}$ **[6]**

29 **a** Prove by induction that $\displaystyle\sum_{r=2}^{n}\frac{1}{r^2-1}=\frac{3n^2-n-2}{4n(n+1)}$ for all $n\geq 2\in\mathbb{N}$ **[8]**

  b Explain why $\displaystyle\sum_{r=1}^{n}\frac{1}{r^2-1}$ does not exist. **[1]**

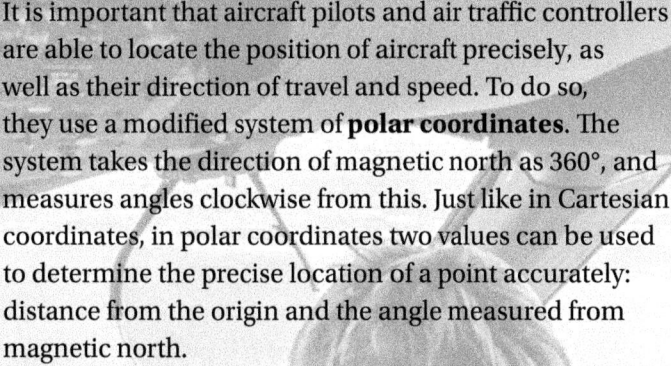

3 Curve sketching 1

It is important that aircraft pilots and air traffic controllers are able to locate the position of aircraft precisely, as well as their direction of travel and speed. To do so, they use a modified system of **polar coordinates**. The system takes the direction of magnetic north as 360°, and measures angles clockwise from this. Just like in Cartesian coordinates, in polar coordinates two values can be used to determine the precise location of a point accurately: distance from the origin and the angle measured from magnetic north.

Polar coordinates are useful in many situations when working with phenomena on or near the Earth's surface. For example, meteorologists use them to model existing weather systems and predict future weather, pilots of ocean-going tankers use them for navigation, and cartographers use them to map the Earth.

Orientation

What you need to know	What you will learn	What this leads to
KS4 • Quadratic functions. • Cartesian coordinates.	• To sketch graphs of linear and quadratic rational functions, finding asymptotes, turning points, limits, etc. • To solve inequalities using graph sketches. • To use polar coordinates and to convert between Cartesian and polar coordinates. • To use the equations of parabolas, ellipses and hyperbolae. • To use the definitions and graphs of hyperbolic functions and their inverses.	**Ch18 Curve sketching 2** • Relationships between zeros and asymptotes. • Combined transformations. • Hyperbolic functons.
Maths Ch2 • Curve sketching.		**Ch19 Integration 2** • Integrating hyperbolic functions.
		Careers • Aircraft pilot. • Cartography.

Fluency and skills

When solving problems involving rational functions, you may find it helpful to sketch a graph.

When sketching a graph of a function of the form $\dfrac{ax+b}{cx+d}$

- find the intercepts where the curve crosses the axes,
- find any asymptotes,
- consider what happens when x and y become very large and approach infinity.

When sketching a graph you might not need all of these techniques. Just use the relevant ones for each question.

> For $y = \mathrm{f}(x)$, vertical asymptotes occur at values where the denominator is zero.
>
> To find horizontal asymptotes, rearrange to $x = \mathrm{f}(y)$ and find the values of y which make the denominator zero.

Example 1

Find the key features (intercepts and asymptotes) and sketch the graph of $y = \dfrac{(x+1)}{(x-4)}$

$y = \dfrac{(x+1)}{(x-4)}$

When $x = 0$, $y = -\dfrac{1}{4}$ and when $y = 0$, $x = -1$

So the curve crosses the axes at $\left(0, -\dfrac{1}{4}\right)$ and $(-1, 0)$

> Find the intercepts by substituting $x = 0$ and $y = 0$

When $x - 4 = 0$, $x = 4$

So $x = 4$ is the vertical asymptote.

> The vertical asymptote is where the denominator is zero.

As x becomes very large $y \to \dfrac{x}{x} \to \dfrac{1}{1} \to 1$

So $y = 1$ is the horizontal asymptote.

> Consider large values for x

Putting $x = 1000$ gives a value just greater than 1, so the curve is above the asymptote, and tends to 1 as x tends to ∞. As x tends to 4 from the right, y tends to ∞. As x tends to ∞, y tends to 1 from above.

> Looking at specific large and small values tells you on which side of the asymptote the curve lies.

Putting $x = -1000$ gives a value just less than 1, so the curve is below the asymptote, and tends to 1 as x tends to $-\infty$. Since the curve passes through $\left(0, -\dfrac{1}{4}\right)$, y must therefore tend to $-\infty$ as x tends to 4 from the left.

> Consider large values for y

As y becomes very large $x \to \dfrac{4}{1} = 4$

(This is another indication that $x = 4$ is an asymptote.)

> Sketch the graph using the key features to help you.
>
> Asymptotes are usually marked with dashed or dotted lines.

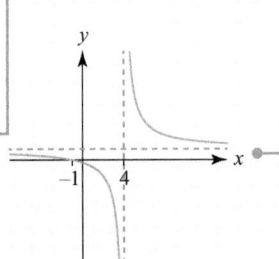

> Check your sketch by plotting the curve using a graphical calculator or graph plotting software.

You can of course draw the graph using a graphical calculator and use this to read off the key features directly. However, you should be able to work them out algebraically because you may not be able to plot all curves using a calculator.

Example 2

Find the key features and sketch the graph of $y = \dfrac{(x+a)}{(x-2)}$ where $a > 0$

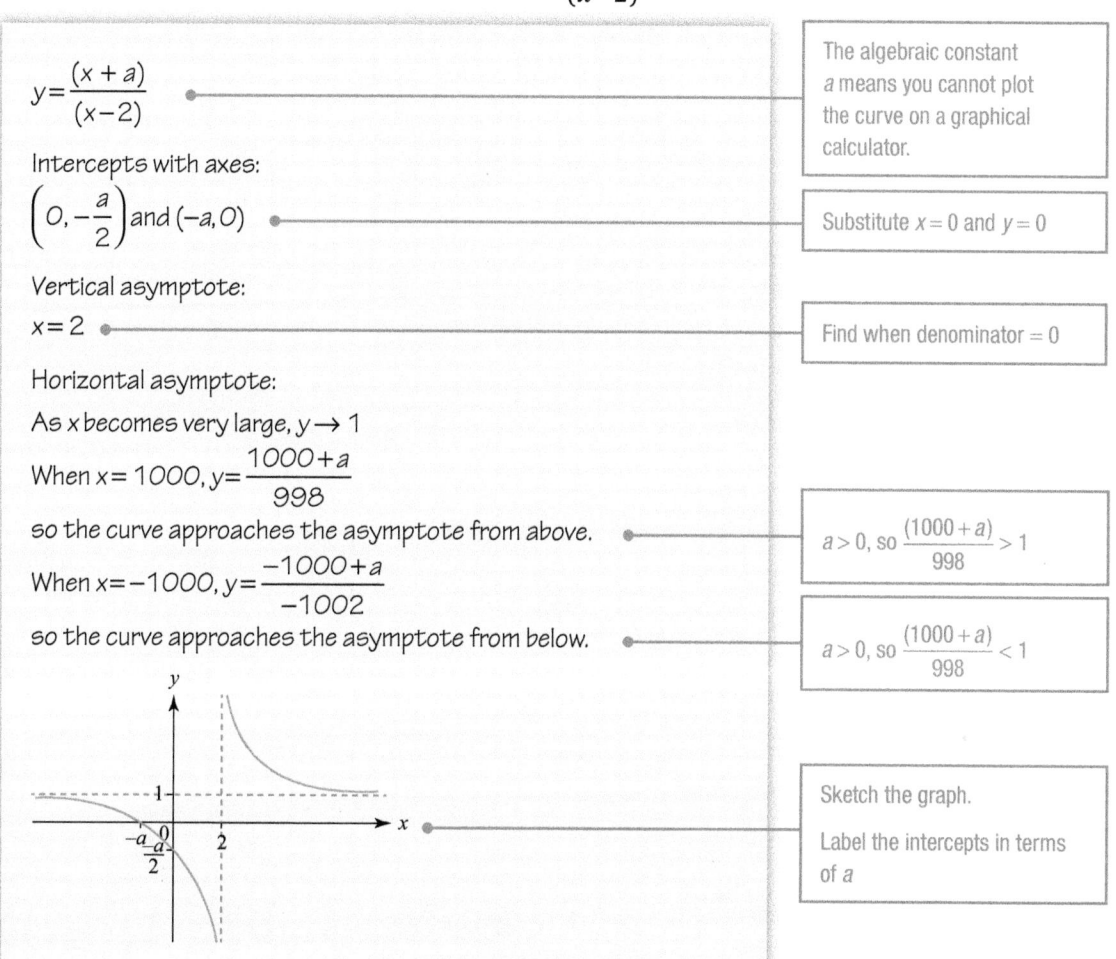

$y = \dfrac{(x+a)}{(x-2)}$

| The algebraic constant a means you cannot plot the curve on a graphical calculator. |

Intercepts with axes:

$\left(0, -\dfrac{a}{2}\right)$ and $(-a, 0)$

| Substitute $x = 0$ and $y = 0$ |

Vertical asymptote:

$x = 2$

| Find when denominator $= 0$ |

Horizontal asymptote:

As x becomes very large, $y \to 1$

When $x = 1000$, $y = \dfrac{1000 + a}{998}$

so the curve approaches the asymptote from above.

| $a > 0$, so $\dfrac{(1000 + a)}{998} > 1$ |

When $x = -1000$, $y = \dfrac{-1000 + a}{-1002}$

so the curve approaches the asymptote from below.

| $a > 0$, so $\dfrac{(1000 + a)}{998} < 1$ |

| Sketch the graph. |
| Label the intercepts in terms of a |

Example 3

a Identify the key features and sketch the graphs of $y = \dfrac{x}{(x-5)}$ and $y = 5x - 24$

b Find the points of intersection.

a For $y = \dfrac{x}{(x-5)}$:

when $x = 0$, $y = 0$ and when $y = 0$, $x = 0$

| Substitute $x = 0$ and $y = 0$ |

When $x - 5 = 0$, $x = 5$, so $x = 5$ is a vertical asymptote

| Find the vertical asymptote. |

(*Continued on the next page*)

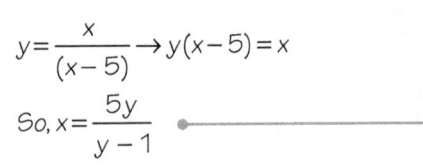

$$y = \frac{x}{(x-5)} \rightarrow y(x-5) = x$$

So, $x = \dfrac{5y}{y-1}$

Rearrange to the form $x = \dots$ to find the horizontal asymptote.

When $y - 1 = 0$, $y = 1$, so $y = 1$ is a horizontal asymptote.

As x becomes very large $y \rightarrow \dfrac{1}{1} = 1$

You can substitute large values of x and y to see from which side the curve approaches the asymptote.

$y = 5x - 24$ is a straight line with gradient 5 and y-intercept -24

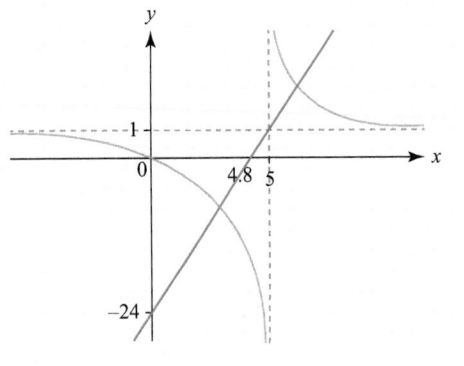

Draw a sketch and label the key features.

b The curve and the straight line intersect when

$$\frac{x}{(x-5)} = 5x - 24$$

Rearrange the equation to form a quadratic.

$$x^2 - 10x + 24 = 0$$

$$(x-4)(x-6) = 0$$

Factorise.

So the curve and line intersect when $x = 6$ and $x = 4$

Points of intersection are $(4, -4)$ and $(6, 6)$

You can use a graphical calculator to draw the graph and check your answer.

Exercise 3.1A Fluency and skills

Wherever possible, you should use a graphical calculator or graph sketching software on your computer to check your answers to the questions in this exercise.

1 For each of these graphs

   i Determine any intercepts with the axes,

   ii Find any vertical and horizontal asymptotes,

   iii Draw a sketch.

  a $y = \dfrac{x+1}{x-2}$ **b** $y = \dfrac{x-1}{x+2}$

  c $y = \dfrac{x+5}{x-3}$ **d** $y = \dfrac{6x-5}{8x+3}$

  e $y = \dfrac{x+12}{5x-12}$ **f** $y = \dfrac{7x-10}{3x-20}$

2 Find any asymptotes and intercepts with the axes for each of following graphs.

  a $y = \dfrac{x+a}{2x-1}$ **b** $y = \dfrac{3x+4}{x+b}$

  c $y = \dfrac{2x-a}{3x-5}$ **d** $y = \dfrac{x-a}{2x+b}$

3 Identify the key features, sketch each pair of equations and find the points of intersection. Show all your working.

  a $y = \dfrac{x+3}{x-4}$ and $y = 2$

  b $y = \dfrac{x+6}{x-4}$ and $y = 6$

c $y = \dfrac{x+12}{5x-12}$ and $y = 0$

d $y = \dfrac{4-3x}{2x-3}$ and $y = -8$

e $y = \dfrac{x+5}{x+1}$ and $y = 2x+1$

f $y = \dfrac{2x-6}{x+5}$ and $y = 3-x$

g $y = \dfrac{10-6x}{x+3}$ and $y = 7x+15$

h $y = \dfrac{8x+48}{3x-8}$ and $y = 24-x$

4 By sketching a curve and a line, solve these inequalities graphically. Show all your working and label all the key features.

a $\dfrac{x-2}{x+5} < 8$

b $\dfrac{x-15}{x+6} \leq 4$

c $\dfrac{4x-1}{3x+4} < 2$

d $\dfrac{3x+5}{2x-3} \geq 11$

Reasoning and problem-solving

Sometimes, solving an inequality algebraically can be quite complicated. You can use a graphical method to solve the inequality or to check your algebraic result.

When solving inequalities graphically

1. Investigate the key features to find the shape of the graph of the functions.
2. Sketch the graphs and find the points of intersection.
3. Use the graphs to write down the ranges of the variable which solve the inequality.

Example 4

Solve the inequality $\dfrac{18x-135}{x-3} > 21 - x$ graphically.

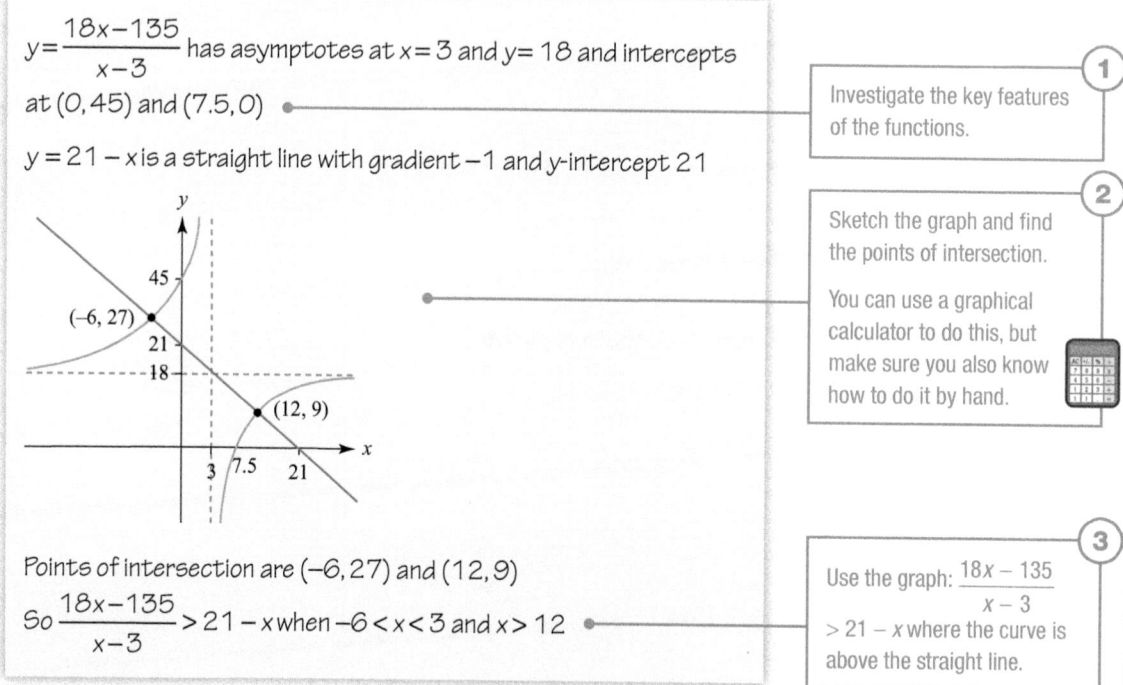

$y = \dfrac{18x-135}{x-3}$ has asymptotes at $x = 3$ and $y = 18$ and intercepts at $(0, 45)$ and $(7.5, 0)$

$y = 21 - x$ is a straight line with gradient -1 and y-intercept 21

① Investigate the key features of the functions.

② Sketch the graph and find the points of intersection.

You can use a graphical calculator to do this, but make sure you also know how to do it by hand.

Points of intersection are $(-6, 27)$ and $(12, 9)$

So $\dfrac{18x-135}{x-3} > 21 - x$ when $-6 < x < 3$ and $x > 12$

③ Use the graph: $\dfrac{18x-135}{x-3} > 21 - x$ where the curve is above the straight line.

Example 5

Following an experiment, student A's results fit the graph of $y(x-3)=2$, whilst student B's results fit with $y(x+2)=3$

The students plotted their graphs of y against x on the same grid.

a Find the point of intersection of their graphs.

b Using a graphical calculator or graph sketching software on your computer, find the range of values of x where the graph of student A gives higher results than that of student B.

c Confirm your results algebraically.

a $y = \dfrac{2}{x-3} = \dfrac{3}{x+2} \to 2(x+2) = 3(x-3)$

$2x+4 = 3x-9$

$x = 13$

The point of intersection is $(13, 0.2)$

b $y = \dfrac{2}{x-3}$ has asymptotes at $x=3$ and $y=0$ ● — Investigate the key features. ①

$y = \dfrac{3}{x+2}$ has asymptotes at $x=-2$ and $y=0$

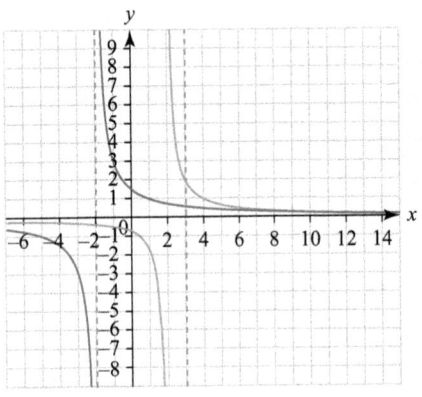

● — Sketch the graphs. ②

The blue graph is $y = \dfrac{2}{x-3}$

The red graph is $y = \dfrac{3}{x+2}$

The results of student A are higher than student B when the blue graph is above the red graph.

$\dfrac{2}{x-3} > \dfrac{3}{x+2}$ when $x < -2$ or when $3 < x < 13$

See Ch 2.2

For a reminder on solving inequalities algebraically.

c Case 1: both denominators positive or both negative

$2(x+2) > 3(x-3)$ if $(x+2)(x-3) > 0$

    i.e. if $x > 3$ or $x < -2$

$2x+4 > 3x-9$

$13 > x$

So, $x < -2$ or $3 < x < 13$

(Continued on the next page)

Case 2: one denominator positive and one negative

$2(x+2) < 3(x-3)$ if $(x+2)(x-3) < 0$

i.e. if $-2 < x < 3$

$2x+4 < 3x-9$

$13 < x$

So no solution since $-2 < x < 3$

> Remember to reverse the inequality sign when you multiply by a negative number.

Exercise 3.1B Reasoning and problem-solving

Use a graphical calculator or graph plotting software to answer the questions in this exercise.

1 Solve these inequalities graphically, labelling all the intercepts, asymptotes and points of intersection.

a $\dfrac{x+5}{x+1} > 2x+1$　　**b** $\dfrac{2x-6}{x+5} < 3-x$

c $\dfrac{x+8}{x-4} \geq 2x-9$　　**d** $\dfrac{2x+6}{2x-3} \geq 14-2x$

e $\dfrac{10-6x}{x+3} \leq 7x+15$　　**f** $\dfrac{8x+48}{3x-8} > 24-x$

2 Solve these inequalities graphically, labelling all the intercepts, asymptotes and points of intersection.

a $-2 < \dfrac{x+2}{x-4} < 3$　　**b** $-9 \geq \dfrac{2x+5}{2x-5} \geq 0$

c $-1 \leq \dfrac{3x-5}{2x-5} \leq 1$　　**d** $2-x < \dfrac{x-12}{x-6} \leq 6-\dfrac{x}{2}$

e $11-3x \leq \dfrac{2x-18}{x-3} \leq \dfrac{9-x}{3}$

f $\dfrac{2}{x+3} > \dfrac{3}{x-1}$　　**g** $\dfrac{5}{x+4} \leq \dfrac{3}{x-2}$

3 John says that the ratio of $2x$ to $x+2$ is always less than 1. Use graphs to explain why John is wrong.

4 Sketch two appropriate graphs to solve the inequality $\dfrac{3}{x-1} - x - 1 \geq 0$

5 For all real values of θ, $-1 \leq \sin\theta \leq 1$
Find the values of x for which $\sin\theta = \dfrac{x-5}{x+5}$ is satisfied by real values of θ

6 For all real values of θ, $-1 \leq \cos\theta \leq 1$
Find the values of x for which $\cos\theta = \dfrac{x+1}{1-x}$ is satisfied by real values of θ

7 a Use a graphical calculator or graph sketching software to sketch the graph of $y = \dfrac{2x}{x+1} - \dfrac{x-3}{x-5}$

b Show that the graph crosses the y-axis at $y=-0.6$

c Show that the asymptotes of the graph are $x=-1$, $x=5$ and $y=1$

8 When is $\dfrac{x}{x+1} \geq \dfrac{x+1}{x}$?

9 You can solve $x^2 - 2x - 15 = 0$ by sketching, on the same grid, the graph of $y=x-2$ and an appropriate curve.

a Give the equation of an appropriate curve and draw a sketch to solve the equation.

b Verify your solution by solving $x^2 - 2x - 15 = 0$ algebraically.

10 Solve the equation $x^3 - x^2 - 6x = 0$ by using a graphical calculator or graph sketching software to sketch, on the same grid, the graph of $y=x^2$ and an appropriate curve of the form $\dfrac{Ax}{x+B}$
State the values of A and B

11 Three fractions are $\dfrac{1}{x}$, $\dfrac{x}{2}$ and $\dfrac{3}{x+2}$
Sketch suitable graphs on the same grid and calculate the exact ranges of values of x when each fraction is the largest of the three.

Quadratic rational functions

Fluency and skills

You can sketch graphs of quadratic rational functions using similar techniques to those you learned in Section 3.1.

When sketching a graph of a function of the form $\dfrac{ax^2+bx+c}{dx^2+ex+f}$

See Ch18.4
To learn about oblique asymptotes

- find any asymptotes,
- find the intercepts where the curve crosses the axes,
- consider what happens when x and y approach infinity,
- find any limitations on the possible ranges of y by solving the function for y and/or x and use this to find any maximum or minimum points on the curve.

> You need to be able to find the key features of a graph using algebraic techniques, but make sure you also know how to plot graphs on a graphical calculator to check your answers.

Example 1

For the graph of $y=\dfrac{x^2-16}{x^2+4}$

a Find the intercepts where the curve crosses the axes,

b Show that there is no vertical asymptote, but that there is a horizontal asymptote at $y=1$

a When $x=0$, $y=\dfrac{-16}{4}=-4$

When $y=0$, $x^2-16=0$ (provided $x^2+4\neq0$)

So the curve crosses the axes at $(0,-4)$ $(4,0)$ and $(-4,0)$

> Find the intercepts by substituting $x=0$ and $y=0$

b $x^2+4\neq0$ for any real value of x

so there is no vertical asymptote.

Rearranging: $y(x^2+4)=x^2-16$

$$4y+16=x^2(1-y)$$

$$x=\sqrt{\dfrac{4y+16}{1-y}}$$

There is a horizontal asymptote at $y=1$

> A vertical asymptote occurs when the denominator is zero.

> Rearrange to $x=\ldots$ and find when the denominator is zero to find any horizontal asymptote.

Alternatively,

As x gets very large $y\to\dfrac{x^2}{x^2}=1$

So $y=1$ is a horizontal asymptote.

> You can also find horizontal asymptotes by considering large values of x

Let $y=k$ so $\dfrac{x^2-16}{x^2+4}=k\Rightarrow x^2-16=kx^2+4k$

$\Rightarrow(1-k)x^2=4k+16\Rightarrow x^2=\dfrac{4k+16}{1-k}$

Since $x\geq0$, $k\leq1$. However, as $x\to\infty$, $\dfrac{x^2}{x^2}\to1$

so $y=1$ is a horizontal asymptote.

Hence $y\leq1$

> Sketch the graph using a graphical calculator to confirm your values for the intercepts and asymptote.

Example 2

Find the range of x and y values for which the graph of $y = \dfrac{x^2 + 6x + 9}{x^2 + 3x + 3}$ exists.

Use this information to find any turning points. Check your answers using a graphical calculator.

x values

The graph exists for all values of x except for any vertical asymptotes.

Vertical asymptotes occur when the denominator $= 0$

Since $x^2 + 3x + 3$ has no real solutions, there is no vertical asymptote.

This means the graph exists for all real values of x

y values

Consider where the graph crosses the line $y = k$ and deduce allowable values for k

Let $y = k$

Rearrange the equation •————————

> Solve for x to find the limits on the range of y

$$k(x^2 + 3x + 3) = x^2 + 6x + 9$$

$$(k-1)x^2 + 3(k-2)x + 3(k-3) = 0$$

x can only take real values when $b^2 - 4ac \geq 0$

$$9(k-2)^2 - 4(k-1)(3)(k-3) \geq 0$$

$$-3k^2 + 12k \geq 0$$

$$3k(k-4) \leq 0$$ •————————

> Change the direction of the inequality since you have multiplied by -1

So the curve only exists for $0 \leq k \leq 4$

since $y = k$, $0 \leq y \leq 4$

Turning points

As x gets very large $y \to \dfrac{x^2}{x^2} = 1$, so $y = 1$ is a horizontal asymptote.

$y = 1$ is the only horizontal asymptote and $0 \leq y \leq 4$

So $y = 0$ and $y = 4$ must be either minimum or maximum points.

When $y = 4$ •————————

> Substitute $y = 4$ into the equation and solve.

$$3x^2 + 6x + 3 = 0 \text{ so } 3(x+1)^2 = 0$$

Hence the maximum point is $(-1, 4)$

When $y = 0$

$$x^2 + 6x + 9 = 0 \text{ (provided } x^2 + 3x + 3 \neq 0)$$

So, $(x+3)^2 = 0$ and the minimum point is $(-3, 0)$

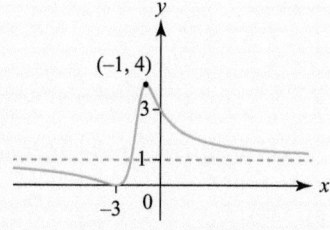

> Check your answers with a graphical calculator.
>
> Notice that, in some circumstances, a curve can cross an asymptote.

Example 3

Identify the key features and sketch the graph of $y = \dfrac{x^2 - x - 6}{x^2 - x - 12}$

Intercepts

When $x = 0$, $y = \dfrac{-6}{-12} = \dfrac{1}{2}$

When $y = 0$, $x^2 - x - 6 = 0$ (provided $x^2 - x - 12 \neq 0$)

$\quad\quad (x - 3)(x + 2) = 0$

$\quad\quad\quad x = 3 \text{ or } x = -2$

(Check: $(3)^2 - (3) - 12 \neq 0$ and $(-2)^2 - (-2) - 12 \neq 0$)

So, the curve crosses the axes at $\left(0, \dfrac{1}{2}\right)$, $(-2, 0)$ and $(3, 0)$

> Substitute $x = 0$ and $y = 0$

Asymptotes

$\quad\quad x^2 - x - 12 = 0$

$\quad\quad (x + 3)(x - 4) = 0$

So, $x = -3$ and $x = 4$ are vertical asymptotes.

As $x \to \infty$, $\dfrac{x^2}{x^2} \to 1$, so $y = 1$ is a horizontal asymptote.

Substituting $x = \pm 1000$, say, shows the graph approaches the asymptote from above on both sides of the graph.

> The denominator is zero at a vertical asymptote.

> Consider large values of x

Range

Consider where the graph crosses the line $y = k$ and deduce allowable values for k

Let $y = k$

Rearranging the equation:

$k(x^2 - x - 12) = x^2 - x - 6$

$\Rightarrow (k - 1)x^2 - (k - 1)x + (6 - 12k) = 0$

x can only take real values when

$\quad\quad (k-1)^2 - 4(k-1)(6 - 12k) \geq 0$

$\quad\quad\quad\quad 49k^2 - 74k + 25 \geq 0$

$\quad\quad\quad\quad (49k - 25)(k - 1) \geq 0$

$49k - 25 \geq 0$ or $k - 1 \geq 0$

So

$k \geq \dfrac{25}{49}$ or $k \geq 1$

So the curve exists for $k \leq \dfrac{25}{49}$ or $k \geq 0$

Since $k = y$

$y \leq \dfrac{25}{49}$ or $y \geq 1$

> Solve for x to find the limits on the range of y

> However, because of the asymptote, y does not exist when $\dfrac{25}{49} < y \leq 1$

(Continued on the next page)

Turning points

When $y = 1$

$y = 1$ is an asymptote as x tends to $\pm\infty$

When $y = \dfrac{25}{49}$

$$\left(\frac{25}{49} - 1\right)x^2 + \left(1 - \frac{25}{49}\right)x + \left(6 - 12 \times \frac{25}{49}\right) = 0$$

$$\frac{-24}{49}x^2 + \frac{24}{49}x - \frac{6}{49} = 0$$

$$4x^2 - 4x + 1 = 0$$

$$(2x - 1)^2 = 0$$

So there is a turning point at $x = \dfrac{1}{2}$

This shows that $\left(\dfrac{1}{2}, \dfrac{25}{49}\right)$ is a maximum point.

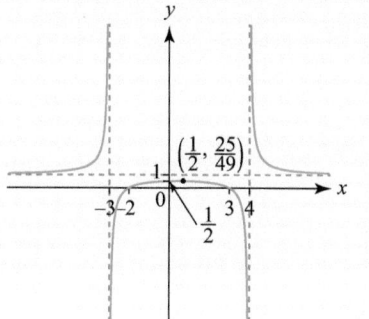

> Use these features to sketch the graph.
>
> You should use a graphical calculator to check your sketch.

Exercise 3.2A Fluency and skills

Use a graphical calculator or graph sketching software to check your answers in this exercise.

1 Find any asymptotes and points of intersection with the axes for each graph.

  a $y = \dfrac{x^2 - 9}{x^2 + 9}$ **b** $y = \dfrac{x^2 + 9}{x^2 - 9}$

  c $y = \dfrac{x^2}{x^2 + 1}$ **d** $y = \dfrac{x^2 + 5x}{x^2 + 5}$

  e $y = \dfrac{x^2 - 3x + 4}{x^2 + 4x}$ **f** $y = \dfrac{x^2 - 3x - 4}{x^2 + 4x}$

2 For each of these functions, determine the ranges of x and y values for which the graph exists.

  a $y = \dfrac{x^2}{x^2 + 4x - 3}$ **b** $y = \dfrac{x^2 - 3x}{x^2 - 3x + 2}$

  c $y = \dfrac{x^2 - 3x - 4}{x^2 - 3x + 1}$ **d** $y = \dfrac{2 - x}{2 + x^2}$

3 For each of these curves, find any local maximum and minimum points.

  a $y = \dfrac{x^2}{x + 1}$ **b** $y = \dfrac{12x}{x^2 + 4}$

  c $y = \dfrac{6}{x^2 + 4}$ **d** $y = \dfrac{5}{x^2 + 4x}$

4 Sketch these curves.

  a $y = \dfrac{x^2 - 2x}{x^2 - 4x + 8}$ **b** $y = \dfrac{x^2 - 8x}{x^2 - 5x - 6}$

  c $y = \dfrac{x^2 - 4x}{x^2 - 5x + 6}$ **d** $y = \dfrac{x^2 - 4x + 4}{x^2 - 5x - 6}$

Reasoning and problem-solving

Strategy

To sketch a curve of the form $y = \dfrac{ax^2 + bx + c}{dx^2 + ex + f}$

(1) Find any points of intersection with the axes and any asymptotes.

(2) Investigate the areas where y and x have real values and determine any maxima/minima.

(3) Sketch the curve and use a graphical calculator to check your answer.

Example 4

Sketch the curve $y = \dfrac{x^2 + x - 6}{x^2 - x - 6}$

Consider the function $y = \dfrac{x^2 + x - 6}{x^2 - x - 6}$

This can be written as $y = \dfrac{(x+3)(x-2)}{(x-3)(x+2)}$

Therefore there are critical values at $x = -3, -2, 2$ and 3

Considering the sign of y between and on either side of these values we obtain:

$x < -3$	$-3 < x < -2$	$-2 < x < 2$	$2 < x < 3$	$x > 3$
y negative	positive	negative	positive	negative

Hence, every time a critical value of x is passed, y changes from one side of the x-axis to the other.

Also, when $x = 0$, $y = 1$ so the curve passes through $(0, 1)$

There are vertical asymptotes at $x = -2$ and 3 (you can see this by considering the denominator of the curve).

There is a horizontal asymptote because $y \to 1$ as $x \to \infty$

And the graph crosses this horizontal asymptote once as can be seen by solving

$\dfrac{x^2 + x - 6}{x^2 - x - 6} = 1$

$\Rightarrow x^2 + x - 6 = x^2 - x - 6$

$\Rightarrow x = 0$

(Continued on the next page)

Although we cannot say that y is continuous, the curve actually passes through every value of y

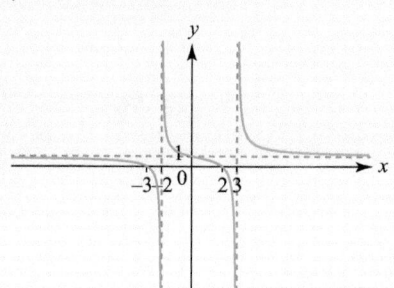

Exercise 3.2B Reasoning and problem-solving

Use a graphical calculator or graph sketching software to answer the questions in this exercise where appropriate.

1 Prove that $-\dfrac{1}{2} \le \dfrac{x}{x^2+1} \le \dfrac{1}{2}$ and hence sketch the graph of the function.

2 Prove that $0 \le \dfrac{x^2}{x^2+1} \le 1$ and use this to sketch the graph of the function.

3 Prove that $y = \dfrac{x^2}{x^2-1}$ cannot lie between 0 and 1 and use this to sketch the graph.

4 Prove that $y = \dfrac{x^2+5}{x-2}$ has a minimum value of 10 and a maximum value of -2, and explain this apparent contradiction using a sketch graph.

5 Sketch the graph of the function $\dfrac{2x^2+3x-2}{x^2-2x+2}$

6 a Find the greatest value of c for which the value of the expression $\dfrac{x^2-4x+4}{x^2+c}$ is 5

   b Sketch the graph of the expression for this value of c

7 Calculate the range of values of C for which the equation $\dfrac{1-3x}{x^2+x} = C$ has real roots. Hence find the coordinates of the local maximum and minimum points on the curve.

8 Show that $y = \dfrac{x^2+A}{x^2-B^2}$ has real solutions except for the range $-\dfrac{A}{B^2} < y < 1$. Hence sketch the graph of $y = \dfrac{x^2+4}{x^2-3^2}$ and find the coordinates of the maximum turning point.

9 a Sketch on the same grid the graphs of $y = \dfrac{4}{x^2-x}$ and $y^2 = \dfrac{4}{x^2-x}$

   b Find the exact values of x and y at the points of intersection of the graphs and hence solve the inequality $\dfrac{4}{x^2-x} > \sqrt{\dfrac{4}{x^2-x}}$

   c Explain why $y^2 = \dfrac{4}{x^2-x}$ has no value between $x = 0$ and $x = 1$

10 a Prove that $y = \dfrac{x^2-A}{x^2-4x-A}$ has real roots if $A \ge 0$

    b i Sketch the graph when $A = 4$ and show there is no turning point.

       ii Solve the equation $\dfrac{x^2-4}{x^2-4x-4} = 0$

11 If $y = \dfrac{kx}{x^2-x+1}$, find the minimum and maximum values of y in terms of k Hence sketch the curve when $k = 6$

3.3 Polar coordinates

Fluency and skills

A point can be defined by its (x, y) coordinates, but you could also define it by a distance from the origin and an angle as shown in the diagram.

The distance of P from the origin is 2, and the angle it makes with the positive x axis is $\dfrac{\pi}{3}$

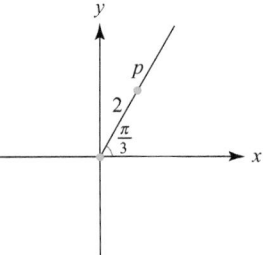

These two numbers describe the point perfectly, and they are called the polar coordinates of the point P, written $\left(2, \dfrac{\pi}{3}\right)$.

However, the angle is not unique: there is an infinite number of ways to represent a point in polar coordinates. For example,

$$\left(2, \frac{\pi}{3}\right) = \left(2, -\frac{5\pi}{3}\right) = \left(2, \frac{7\pi}{3}\right) = \left(2, -\frac{11\pi}{3}\right)$$

The letters r and θ are usual for the distance and the angle.

It is possible to use a negative value of r, by going backwards from the origin:

$$\left(2, \frac{\pi}{3}\right) = \left(-2, \frac{4\pi}{3}\right) = \left(-2, -\frac{2\pi}{3}\right)$$

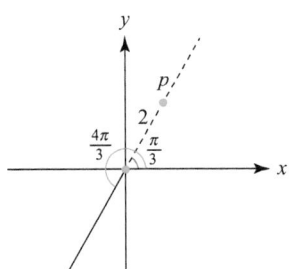

Every point, however, can only be written in one way if we always choose $r \geq 0$ and $-\pi < \theta \leq \pi$ or $0 \leq \theta < 2\pi$

The point $(0, 0)$ is called the pole, and the positive x axis, which is where the angle is measured from, is called the initial line.

> **Key point**
>
> r is the distance from the pole to the point.
>
> θ is the angle between the initial line and the line segment from the pole to the point, measured counterclockwise.

You can convert polar (r, θ) coordinates into Cartesian (x, y) coordinates. In the example above, the x coordinate is $2\cos\dfrac{\pi}{3}$ and the y coordinate is $2\sin\dfrac{\pi}{3}$

The same method will always work, so

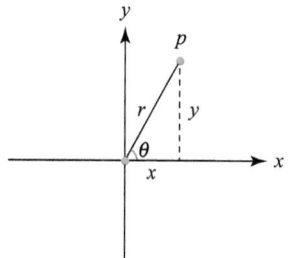

Key point

To convert from polar to Cartesian, use

$$x = r\cos\theta$$

$$y = r\sin\theta$$

To convert from Cartesian to polar, use

$$r^2 = x^2 + y^2$$

$$\tan\theta = \frac{y}{x}$$

but you should draw a diagram to ensure that you choose the right value of θ

Example 1

Convert the polar coordinates $\left(5, -\dfrac{3\pi}{4}\right)$ into Cartesian coordinates.

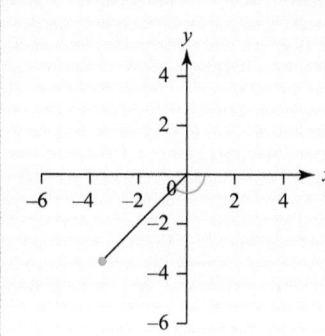

$$x = r\cos\theta = 5\cos\left(-\frac{3\pi}{4}\right) = -\frac{5}{\sqrt{2}}$$

$$y = r\sin\theta = 5\sin\left(-\frac{3\pi}{4}\right) = -\frac{5}{\sqrt{2}}$$

so the Cartesian coordinates are

$$\left(-\frac{5}{\sqrt{2}}, -\frac{5}{\sqrt{2}}\right)$$

Example 2

Convert the Cartesian coordinates $(-2\sqrt{3}, 2)$ into polar coordinates.

First, draw a sketch.

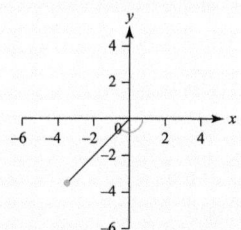

$$r^2 = x^2 + y^2 = 12 + 4 = 16$$

$$\Rightarrow r = 4$$

$$\tan\theta = \frac{y}{x} = -\frac{1}{\sqrt{3}}$$

$$\Rightarrow \theta = -\frac{\pi}{6}, \frac{5\pi}{6}$$

but the sketch shows which value to use, so polar coordinates are

$$\left(4, \frac{5\pi}{6}\right)$$

PURE

An equation involving x and y can be represented as a graph. So can an equation involving r and θ

Example 3

Draw the graph

$r = 3(1 - \sin\theta),\ 0 \le \theta < 2\pi$

This means that for any value of θ in the range 0 to 2π, you can calculate the value of r and plot that point. It is a good idea to make a table of values using easy values of θ Then you can join the points up to make a smooth curve.

θ	0	$\dfrac{\pi}{6}$	$\dfrac{\pi}{3}$	$\dfrac{\pi}{2}$	$\dfrac{2\pi}{3}$	$\dfrac{5\pi}{6}$	π	$\dfrac{7\pi}{6}$	$\dfrac{4\pi}{3}$	$\dfrac{3\pi}{2}$	$\dfrac{5\pi}{3}$	$\dfrac{11\pi}{6}$	2π
r	3	1.5	0.40	0	0.40	1.5	3	4.5	5.6	6	5.6	4.5	3

Polar graph paper can be useful: the lines represent points with the same θ coordinate, and the circles represent points with the same r coordinate.

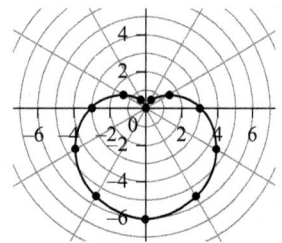

Example 4

a Draw the graph
$r = 5\sin 3\theta,\ 0 \le \theta < 2\pi$

b Draw the graph
$r = 5\sin 3\theta,\ 0 \le \theta < 2\pi,\ r \ge 0$

a You could draw up a table of values as before, but you do not usually do this for (x, y) graphs, and you don't always need to for (r, θ) graphs.

First, notice that $-1 \le \sin 3\theta \le 1$, so $-5 \le r \le 5$

It is useful to find the values of θ that give r these maximal and minimal values:

$r = 5 \Rightarrow \sin 3\theta = 1 \Rightarrow 3\theta = \dfrac{\pi}{2}, \dfrac{5\pi}{2}, \dfrac{9\pi}{2}, \ldots$

$\Rightarrow \theta = \dfrac{\pi}{6}, \dfrac{5\pi}{6}, \dfrac{3\pi}{2} \cdot \ldots$

Also,

$r = -5 \Rightarrow \sin 3\theta = -1 \Rightarrow 3\theta = \dfrac{3\pi}{2}, \dfrac{7\pi}{2}, \dfrac{11\pi}{2}, \ldots$

$\Rightarrow \theta = \dfrac{\pi}{2}, \dfrac{7\pi}{6}, \dfrac{11\pi}{6}, \ldots$

(Continued on the next page)

And it would also be a good idea to find where r is zero:

$r = 0 \Rightarrow \sin 3\theta = 0 \Rightarrow 3\theta = 0, \pi, 2\pi, 3\pi, 4\pi, 5\pi, 6\pi, \ldots$

$\Rightarrow \theta = 0, \dfrac{\pi}{3}, \dfrac{2\pi}{3}, \pi, \dfrac{4\pi}{3}, \dfrac{5\pi}{3}, 2\pi, \ldots$

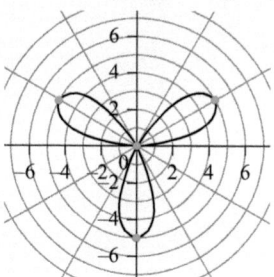

As θ increases from 0, r increases from 0 until it reaches 5, when $\theta = \dfrac{\pi}{6}$

Then r decreases until it reaches 0 when $\theta = \dfrac{\pi}{3}$

This makes the top right loop on the curve.

When θ is between $\dfrac{\pi}{3}$ and $\dfrac{2\pi}{3}$, r is negative, going from 0 to −5 and back to 0 again. This makes the bottom loop.

Then when θ is between $\dfrac{2\pi}{3}$ and π, r goes from 0 to 5 and back to 0 again, making the top left loop.

When θ goes above π, the curve traces over the same path that it has already completed.

b Here, you are only interested in the positive values of r (or zero), so you only need values of θ for which $5\sin 3\theta$ is positive or zero. This gives you the top right loop when

$0 \leq \theta \leq \dfrac{\pi}{3}$, the top left loop when $\dfrac{2\pi}{3} \leq \theta \leq \pi$, and the bottom loop when $\dfrac{4\pi}{3} \leq \theta \leq \dfrac{5\pi}{3}$.

However, if the equation had been

$$r = 5\sin 3\theta, \ 0 \leq \theta < \pi, \ r \geq 0$$

with θ only going as far as π, then the bottom loop would be missing.

Key point

Pay careful attention to the range of values of θ and r specified in the question.

Example 5

a Draw the graph

$r = 1 - 2\sin\theta,\ 0 \leq \theta < 2\pi$

b Draw the graph

$r = 1 - 2\sin\theta,\ 0 \leq \theta < 2\pi,\ r \geq 0$

a First notice that

$-1 \leq 1 - 2\sin\theta \leq 3$

$\Rightarrow -1 \leq r \leq 3$

When $r = 0$, $1 - 2\sin\theta = 0 \Rightarrow \theta = \dfrac{\pi}{6}, \dfrac{5\pi}{6}$

When $r = -1$, $1 - 2\sin\theta = -1 \Rightarrow \theta = \dfrac{\pi}{2}$

When $r = 3$, $1 - 2\sin\theta = 3 \Rightarrow \theta = \dfrac{3\pi}{2}$

When $r = 1$, $1 - 2\sin\theta = 1 \Rightarrow \theta = 0, \pi$

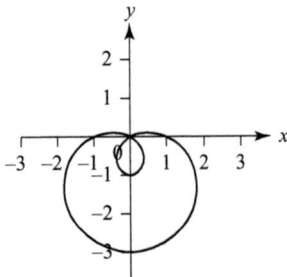

When $0 < \theta < \dfrac{\pi}{6}$, r decreases from 1 to 0

When $\dfrac{\pi}{6} < \theta < \dfrac{5\pi}{6}$, r decreases from 0 to -1 and increases back to 0

When $\dfrac{5\pi}{6} < \theta < 2\pi$, r increases from 0 to 3 and decreases back to 1

b As in example 4b, you are not interested in the values of θ for which r is negative, so the smaller inner loop is missing.

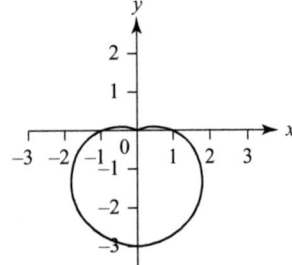

Example 6

Draw these graphs:

a $r = 4$ **b** $\theta = \dfrac{5\pi}{6}$ **c** $\theta = \dfrac{5\pi}{6}, r \geq 0$ **d** $r = \theta, r \geq 0$

a This graph includes all points that are 4 units away from the pole, so it is a circle radius 4 with centre at the pole.

b This graph includes all points on the line that forms an angle $\dfrac{5\pi}{6}$ with the positive x axis, so it is the line $y = -\dfrac{\sqrt{3}}{3}x$

c In this case, r cannot take negative values, so this graph is only the half line.

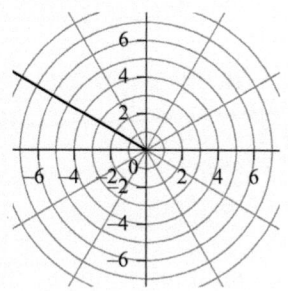

d As θ gets bigger, so does r

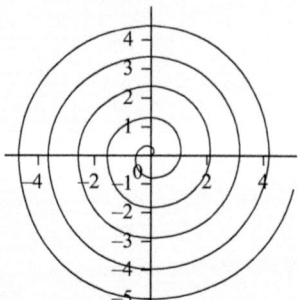

Exercise 3.3A Fluency and skills

1 Draw a polar chart similar to the one shown and plot the given points.

Convert each set of polar coordinates to their Cartesian equivalent.

a $\left(2, \dfrac{\pi}{3}\right)$ **b** $\left(4, \dfrac{4\pi}{3}\right)$ **c** $\left(2, \dfrac{5\pi}{3}\right)$

d $\left(3, \dfrac{2\pi}{3}\right)$ **e** $\left(3, -\dfrac{5\pi}{6}\right)$ **f** $\left(3, -\dfrac{2\pi}{3}\right)$

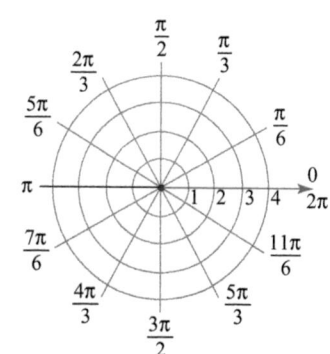

2 Convert these Cartesian coordinates to polar form (r, θ) where $0 < \theta \leq 2\pi$. Give your answers to 3 sf where appropriate.

   a $(2\sqrt{3}, 2)$ **b** $(-3, 3\sqrt{3})$ **c** $(4, -4\sqrt{3})$ **d** $(-5, -12)$

3 Convert these Cartesian coordinates to polar form (r, θ) where $-\pi < \theta \leq \pi$

   Give r as a surd and angles to 3 sf where appropriate.

   a $(-2\sqrt{3}, -2)$ **b** $(5, -5\sqrt{3})$ **c** $(-3, 4)$ **d** $(1, 6)$

4 State the polar equation of the curve $x^2 + y^2 = 5$ and describe the curve.

5 Describe the curve $r = 4$ and state its Cartesian equation.

6 Sketch the graphs for each of these polar equations and find the Cartesian equation of the line they lie on.

   a $\theta = \dfrac{\pi}{4}, r \geq 0$ **b** $\theta = \dfrac{\pi}{2}, r \geq 0$ **c** $\theta = \dfrac{3\pi}{4}, r \geq 0$ **d** $\theta = -\dfrac{\pi}{6}, r \geq 0$

7 **a** Plot the curve $r = \theta$ in the domain $0 \leq \theta \leq 6\pi$ using step-sizes of $\dfrac{\pi}{4}$

    b Plot $r = \dfrac{10}{2 + \cos\theta}$ in the domain $0 \leq \theta \leq 2\pi$

8 Sketch these polar curves for $0 \leq \theta \leq 2\pi$ and state the maximum and minimum value of r in each case.

   a **i** $r = \cos 2\theta, r \geq 0$ **ii** $r = \cos 2\theta, r \in \mathbb{R}$

   b $r = \sin 4\theta, r \geq 0$

   c **i** $r = 2\cos 3\theta, r \geq 0$ **ii** $r = 2\cos 3\theta, r \in \mathbb{R}$

   d **i** $r = 4\sin 2\theta, r \geq 0$ **ii** $r = 4\sin 2\theta, r \in \mathbb{R}$

   e $r = 1 + \cos\theta$

   f $r = 4 + \sin\theta$

   g $r = 3 - 2\cos\theta$

   h **i** $r = 3 - 5\cos\theta, r \geq 0$ **ii** $r = 3 - 5\cos\theta, r \in \mathbb{R}$

   i $r = 2\theta$

   j **i** $r^2 = 4\sin 2\theta, r \geq 0$ **ii** $r^2 = 4\sin 2\theta, r \in \mathbb{R}$

Reasoning and problem-solving

An equation that is simple in (r, θ) coordinates can be complicated in (x, y) coordinates, and vice versa. You can always convert one into the other, though.

Key point

To convert from polar to Cartesian, use

$$r^2 = x^2 + y^2$$

$$\tan\theta = \frac{y}{x}$$

$$\cos\theta = \frac{x}{r} = \pm\frac{x}{\sqrt{x^2+y^2}}$$

$$\sin\theta = \frac{y}{r} = \pm\frac{y}{\sqrt{x^2+y^2}}$$

[Choose + or − depending on whether r is positive or negative]

To convert from Cartesian to polar, use

$$x = r\cos\theta$$

$$y = r\sin\theta$$

The difficulty sometimes lies in simplifying the answer.

Example 7

Convert the equation $\dfrac{x^2}{3} + \dfrac{y^2}{4} = 1$ to a polar equation.

$$\frac{(r\cos\theta)^2}{3} + \frac{(r\sin\theta)^2}{4} = 1$$

$$\Rightarrow 4r^2\cos^2\theta + 3r^2\sin^2\theta = 12$$

$$\Rightarrow r^2 = \frac{12}{4\cos^2\theta + 3\sin^2\theta}$$

This is a perfectly good answer, although we could also write:

$$r^2 = \frac{12}{3(\cos^2\theta + \sin^2\theta) + \cos^2\theta}$$

$$\Rightarrow r^2 = \frac{12}{3 + \cos^2\theta}$$

Notice that this implies that r could be positive or negative. So long as the range of values of θ is big enough, however, it makes no difference whether we include the negative values of r or not.

Example 8

Convert the equation $r = \dfrac{\tan\theta}{\cos\theta}$ to a Cartesian equation.

When $r > 0$

$$\sqrt{x^2 + y^2} = \frac{y}{x} \div \frac{x}{\sqrt{x^2 + y^2}}$$

$$= \frac{y\sqrt{x^2 + y^2}}{x^2}$$

$$\Rightarrow 1 = \frac{y}{x^2}$$

$$\Rightarrow y = x^2$$

When $r < 0$

$$-\sqrt{x^2 + y^2} = \frac{y}{x} \div \frac{x}{-\sqrt{x^2 + y^2}}$$

$$\Rightarrow y = x^2$$

So this is the Cartesian equation of the graph.

We can also find the points of intersection of two curves given their polar equations without having to convert into Cartesian equations.

Example 9

a Find the points of intersection of the curves

$r = \sin 2\theta \ \ r \geq 0$ and $r = \cos 2\theta \ \ r \geq 0$

b Find the points of intersection of the curves $r = \sin 2\theta$ and $r = \cos 2\theta$, where r can be any real number.

In this question, only the values of **θ** for which r is positive are considered for each curve.

At the points of intersection, $\sin 2\theta = \cos 2\theta$

$$\Rightarrow \tan 2\theta = 1$$

$$\Rightarrow 2\theta = \frac{\pi}{4}, \frac{5\pi}{4}, \frac{9\pi}{4}, \frac{13\pi}{4} \dots$$

$$\Rightarrow \theta = \frac{\pi}{8}, \frac{5\pi}{8}, \frac{9\pi}{8}, \frac{13\pi}{8} \dots$$

However, r is not positive on both curves for all these values of **θ**

When $\theta = \dfrac{\pi}{8}$ or $\theta = \dfrac{9\pi}{8}$, $r = \sin 2\theta = \cos 2\theta = \dfrac{\sqrt{2}}{2}$

When $\theta = \dfrac{5\pi}{8}$ or $\theta = \dfrac{13\pi}{8}$, $r = \sin 2\theta = \cos 2\theta = -\dfrac{\sqrt{2}}{2} < 0$

So the only solutions are

$$\left(\frac{\sqrt{2}}{2}, \frac{\pi}{8} \right), \left(\frac{\sqrt{2}}{2}, \frac{9\pi}{8} \right)$$

(Continued on the next page)

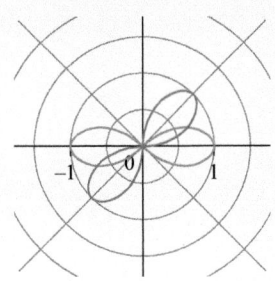

Drawing the curves will help you to visualise what is going on, but you need to choose values of θ that make r positive on each curve:

$$0 \le \theta \le \frac{\pi}{2} \text{ or } \pi \le \theta \le \frac{3\pi}{2} \Rightarrow r = \sin 2\theta \ge 0$$

$$-\frac{\pi}{4} \le \theta \le \frac{\pi}{4} \text{ or } \frac{3\pi}{4} \le \theta \le \frac{5\pi}{4} \Rightarrow r = \cos 2\theta \ge 0$$

b At the points of intersection, $\sin 2\theta = \cos 2\theta$

$$\Rightarrow \tan 2\theta = 1$$

$$\Rightarrow 2\theta = \frac{\pi}{4}, \frac{5\pi}{4}, \frac{9\pi}{4}, \frac{13\pi}{4} \dots$$

$$\Rightarrow \theta = \frac{\pi}{8}, \frac{5\pi}{8}, \frac{9\pi}{8}, \frac{13\pi}{8} \dots$$

This gives 4 points of intersection. However, drawing the curves shows 8 points of intersection. Where do the other 4 come from? The problem is that polar coordinates are not unique. In this case, notice that the point $\left(\frac{\sqrt{2}}{2}, \frac{3\pi}{8} \right)$ is on the curve $r = \sin 2\theta$

But this point is the same as $\left(-\frac{\sqrt{2}}{2}, \frac{11\pi}{8} \right)$, which is on the curve $r = \cos 2\theta$, and at this point, $\tan 2\theta = -1$

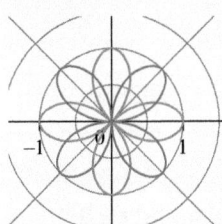

This means you also need to find the solutions to
$\tan 2\theta = -1$

$$\Rightarrow 2\theta = \frac{3\pi}{4}, \frac{7\pi}{4}, \frac{11\pi}{4}, \frac{15\pi}{4} \dots$$

$$\Rightarrow \theta = \frac{3\pi}{8}, \frac{7\pi}{8}, \frac{11\pi}{8}, \frac{15\pi}{8} \dots$$

giving 8 points of intersection:

$$\left(\frac{\sqrt{2}}{2}, \frac{\pi}{8} \right), \left(\frac{\sqrt{2}}{2}, \frac{3\pi}{8} \right), \left(\frac{\sqrt{2}}{2}, \frac{5\pi}{8} \right), \left(\frac{\sqrt{2}}{2}, \frac{7\pi}{8} \right), \left(\frac{\sqrt{2}}{2}, \frac{9\pi}{8} \right), \left(\frac{\sqrt{2}}{2}, \frac{11\pi}{8} \right), \left(\frac{\sqrt{2}}{2}, \frac{13\pi}{8} \right), \left(\frac{\sqrt{2}}{2}, \frac{15\pi}{8} \right)$$

Perhaps the easiest way to deal with this is to sketch the two curves, and use the idea of symmetry to see where all the solutions lie.

1   Find the polar equation for

   a   The line passing through the origin with gradient 1

   b   The line passing through the origin with gradient 2

   c   The positive y-axis

   d   The positive x-axis

   e $y = 5$

   f $x = 3$

   g $y = x + 1$

   h $y = mx + c$

   i   The circle centre the origin, radius 5

   j   The circle with centre $(5, 12)$ and radius 13

   k   The circle with centre $(3, 0)$ passing through the origin

   l   The circle with centre $(0, 4)$ passing through the origin

   m   The circle with centre $(5, 5)$ passing through the origin.

   n $y = x^2$

   o $x^2 + y^2 + 2xy = 5$

2   Convert these polar equations into Cartesian form and describe the graphs.

   a $r = \dfrac{1}{\sin\theta + \cos\theta}$ b $r = \dfrac{5}{\sin\theta - 2\cos\theta}$

   c $r = \dfrac{2}{\cos\theta}$ d $r = \dfrac{-2}{3\sin\theta}$

   e $r = 4\sin\theta$ f $r = 8\cos\theta + 6\sin\theta$

3   Find the points of intersection between these pairs of curves for $-\pi < \theta \le \pi$ and $r \ge 0$
   Give angles in terms of π

   a $r = 2$, $r = 3 - 2\sin\theta$

   b $r = \cos 4\theta$, $r = \dfrac{1}{2}$

   c $r = \sin 2\theta$, $r = \cos 2\theta$

4   Show algebraically that any spiral of the form $r = a\theta$ will intersect any circle centred on the origin precisely once.

5 For each pair of equations, the domain is $0 \leq \theta \leq 2\pi$, $r \geq 0$ unless otherwise stated.

   i   Sketch the curves on the same graph,

   ii Find their points of intersection.

   a $r = \sqrt{3}\sin\theta$, $r = \cos\theta$

   b $r = \theta$, $r = \theta - \sin\theta$

   c $r = 2 + \cos\theta$, $r = \cos^2\theta$

   d $r = \sin^2\theta$, $r = \cos\theta$

   e $r = -\sin 3\theta$, $r = \sqrt{3}\cos 3\theta$, $r \geq 0$

   f $r = -\sin 3\theta$, $r = \sqrt{3}\cos 3\theta$, $r \in \mathbb{R}$

6 The family of curves $r = a + 4\cos\theta$, $-\pi < 0 \leq \pi$, $r \in \mathbb{R}$ produces different shapes depending on the choice of value for a

   a Show that if a is greater than 4 then the curve never passes through the pole.

   b Prove that the curve is symmetrical about the initial line for any value of a

   c Hence, sketch the curve when

      i $a = 0$ ii $0 < a \leq 4$ iii $4 < a < 8$ iv $a \geq 8$

   d State, in terms of a, the maximum and minimum values of r for the curves in part **c**.

7 Show that the graphs of $r = \sec\theta$ and $r^2 = 2\tan\theta$ intersect only once whether or not negative values of r are included, and find their point of intersection.

8 Find the points of intersection between these pairs of curves for $-\pi < \theta \leq \pi$ and $r \geq 0$
   Give your coordinates to 3 sf.

   a $r = 2\cos 2\theta$, $r = 1 + \sin\theta$

   b $r^2 = 4\cos\theta$, $r = 2 - \cos\theta$

   What is the difference if r can take any value, including negatives?

9 Show that the points of intersection between the curves $r = 2\sin^2\theta$ and $r = \cos 2\theta$ form a rectangle and find its area.

3.4 Parabolas, ellipses and hyperbolas

Fluency and skills

Conic sections are a special category of curve. They are formed by the intersection of a plane and a cone. You can think of this as slicing through the cone. The angle at which the plane intersects the cone determines the type of curve which is formed. For example, a plane parallel to the base of the cone will form a circle and a plane parallel to the curved face will cut through the base and the curved face and form a **parabola**.

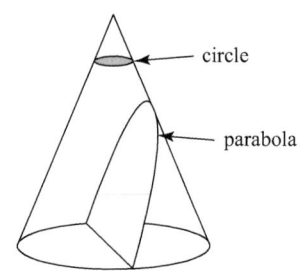

You are already familiar with the parabola $y = Ax^2$ and with transformations of this curve.

Parabolas can also be of the form $x = Ay^2$ which can be rearranged to make y^2 the subject.

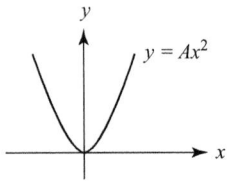 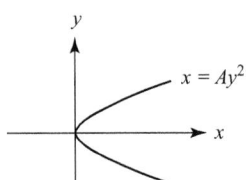

> **Key point**
>
> An equation of the form $y^2 = kx$ describes a parabola with its vertex at the origin.

> You may also see the standard equation of a parabola written as
> $y^2 = 4ax$

Example 1

A curve has equation $y^2 = 24x$

a Sketch the curve, giving the coordinates of any intercepts with the coordinate axes.

b Find the points of intersection between the curve and the line with equation $y = 9 - 2x$

a

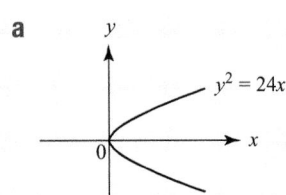

b $(9-2x)^2 = 24x$ — Solve the equations simultaneously.

$81 - 36x + 4x^2 = 24x$

$4x^2 - 60x + 81 = 0$

(Continued on the next page)

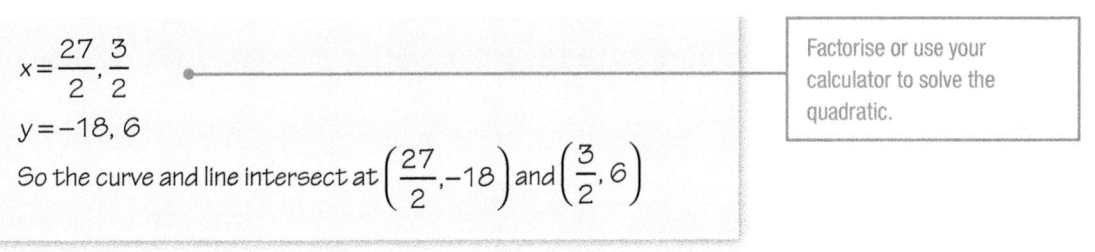

$$x = \frac{27}{2}, \frac{3}{2}$$

$$y = -18, 6$$

So the curve and line intersect at $\left(\frac{27}{2}, -18\right)$ and $\left(\frac{3}{2}, 6\right)$

> Factorise or use your calculator to solve the quadratic.

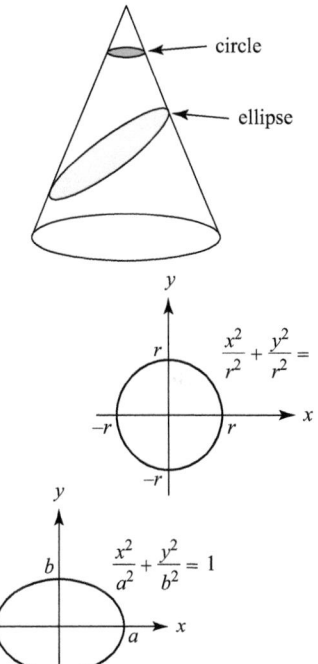

Another type of conic section is an **ellipse**. The ellipse is formed by the intersection of a cone with a plane that cuts through only the curved surface of the cone. If this plane is parallel to the base of the cone then a circle if formed, but in general an ellipse will be formed.

You know that the equation of a circle with centre the origin is $x^2 + y^2 = r^2$

This can be written $\frac{x^2}{r^2} + \frac{y^2}{r^2} = 1$

In contrast, an ellipse does not have a constant radius.

Its equation can be written $\frac{x^2}{a^2} + \frac{y^2}{b^2} = 1$ where $a, b > 0$

Key point

An equation of the form $\frac{x^2}{a^2} + \frac{y^2}{b^2} = 1$ describes an ellipse centred on the origin. The ellipse will pass through the points $(\pm a, 0)$ and $(0, \pm b)$

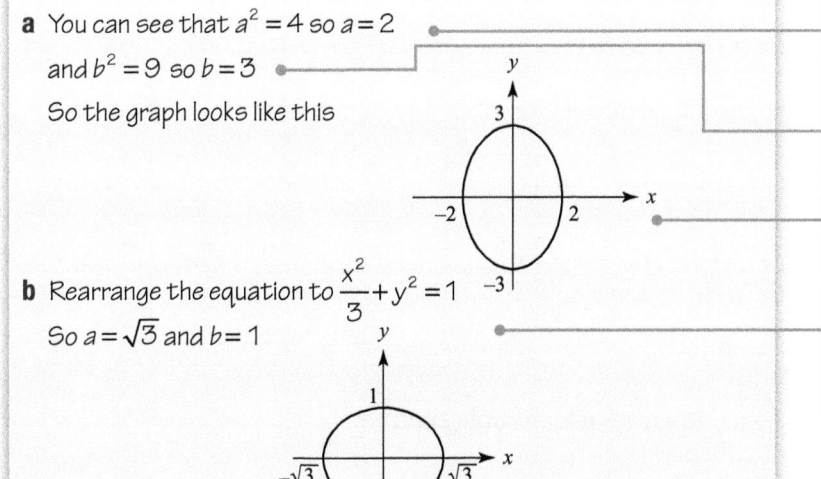

Example 2

Sketch each of these ellipses, clearly labelling their points of intersection with the coordinate axes.

a $\frac{x^2}{4} + \frac{y^2}{9} = 1$ **b** $x^2 + 3y^2 = 3$

a You can see that $a^2 = 4$ so $a = 2$
and $b^2 = 9$ so $b = 3$
So the graph looks like this

> Check by letting $y = 0$, this gives $x^2 = 4 \Rightarrow x = \pm 2$

> Check by letting $x = 0$, this gives $y^2 = 9 \Rightarrow y = \pm 9$

> Try to make your graph approximately to scale.

b Rearrange the equation to $\frac{x^2}{3} + y^2 = 1$
So $a = \sqrt{3}$ and $b = 1$

> Writing the equation in the usual form makes it easier to identify the values of a and b

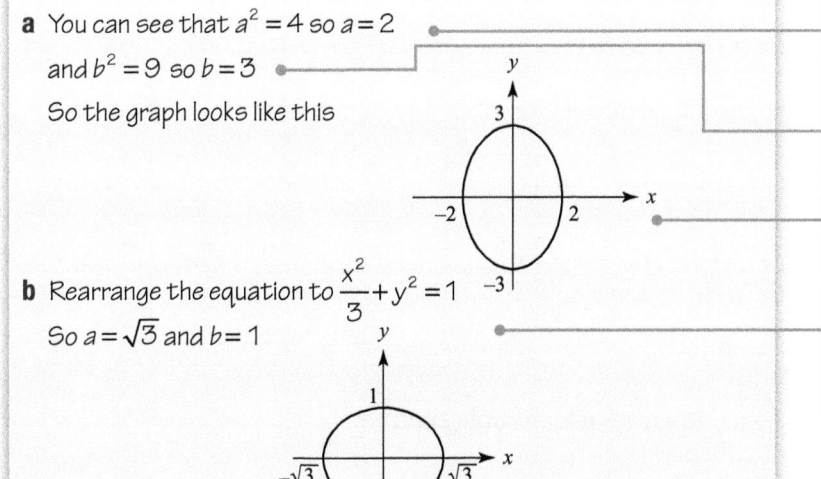

Example 3

A circle with equation $x^2 + y^2 = 9$ is stretched by scale factor 2 in the y-direction to form an ellipse.

Sketch the ellipse and state its equation.

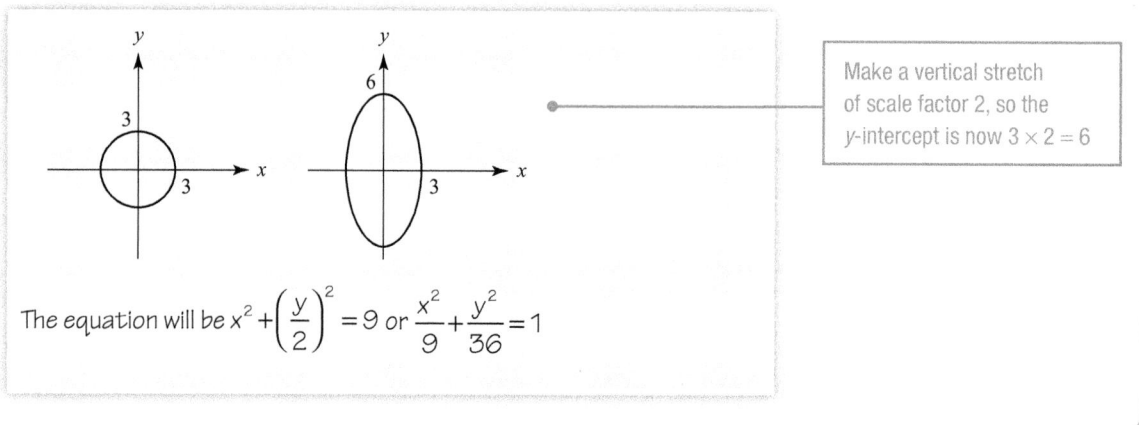

Make a vertical stretch of scale factor 2, so the y-intercept is now $3 \times 2 = 6$

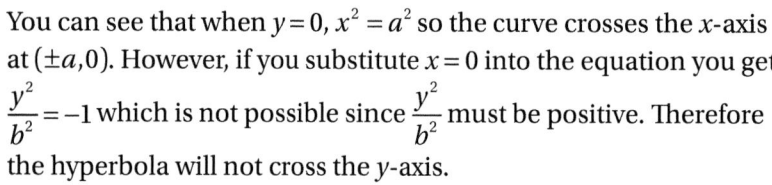

The equation will be $x^2 + \left(\dfrac{y}{2}\right)^2 = 9$ or $\dfrac{x^2}{9} + \dfrac{y^2}{36} = 1$

The final conic section is a **hyperbola**, this can be formed by the intersection of a cone with a plane that is steeper than the curved face of the cone. A double-cone is shown as the plane will intersect both cones.

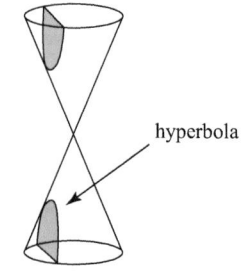

hyperbola

A hyperbola has equation $\dfrac{x^2}{a^2} - \dfrac{y^2}{b^2} = 1$ for $a, b > 0$

Notice that this is very similar to the equation of the ellipse. Only the minus sign is different.

You can see that when $y = 0$, $x^2 = a^2$ so the curve crosses the x-axis at $(\pm a, 0)$. However, if you substitute $x = 0$ into the equation you get $\dfrac{y^2}{b^2} = -1$ which is not possible since $\dfrac{y^2}{b^2}$ must be positive. Therefore the hyperbola will not cross the y-axis.

In fact, since the equation can be rearranged to give $\dfrac{x^2}{a^2} = 1 + \dfrac{y^2}{b^2}$, you can see that $\dfrac{x^2}{a^2} \geq 1$, therefore x is only defined when $|x| \geq a$

As x and y become large, $\dfrac{x^2}{a^2} = 1 + \dfrac{y^2}{b^2}$ becomes $\dfrac{x^2}{a^2} \approx \dfrac{y^2}{b^2}$

So the graph has asymptotes at $y^2 = \dfrac{b^2}{a^2} x^2$, i.e. $y = \pm \dfrac{b}{a} x$

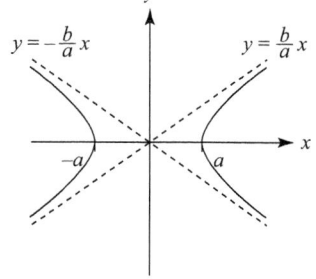

Key point

An equation of the form $\dfrac{x^2}{a^2} - \dfrac{y^2}{b^2} = 1$ describes a hyperbola centred on the origin with asymptotes $y = \pm \dfrac{b}{a} x$

Example 4

Sketch the hyperbola $\dfrac{x^2}{4} - \dfrac{y^2}{12} = 1$ and state the equations of the asymptotes.

In this equation, $a = \sqrt{4} = 2$ and $b = \sqrt{12} = 2\sqrt{3}$

$\dfrac{b}{a} = \dfrac{2\sqrt{3}}{2}$, so the equations of the asymptotes are $y = \pm\sqrt{3}x$

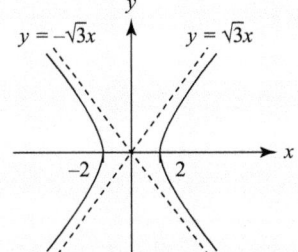

The x-intercepts are at $(\pm 2, 0)$

A hyperbola whose asymptotes are the coordinate axes is called a **rectangular hyperbola**. The equation of a rectangular hyperbola is $xy = c^2$

Key point

An equation of the form $xy = c^2$ describes a rectangular hyperbola centred on the origin with asymptotes $y = 0$ and $x = 0$

Example 5

A curve, C, has equation $xy = 25$

a Sketch C, giving the equations of the asymptotes.

The line with equation $y = 2x + 5$ intersects C at the points A and B

b Calculate the length of the line segment AB

a

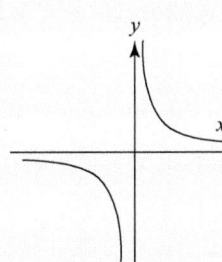

$xy = 25$ represents a rectangular hyperbola.

Asymptotes are $x = 0$ and $y = 0$

b $xy = 25 \Rightarrow y = \dfrac{25}{x}$

$\dfrac{25}{x} = 2x + 5$

Solve simultaneously.

$2x^2 + 5x - 25 = 0$

$x = 2.5, -5 \qquad y = 10, -5$

So A is $(-5, -5)$ and B is $(2.5, -10)$

Length of $AB = \sqrt{7.5^2 + 15^2} = 16.8$

Use Pythagoras' theorem to find the length of AB

1   Sketch each of these curves, clearly labelling their points of intersection with the coordinate axes.

   a $y = 2x^2$ **b** $y^2 = 32x$

   c $\dfrac{x^2}{9} + \dfrac{y^2}{4} = 1$ **d** $\dfrac{x^2}{16} + \dfrac{y^2}{25} = 1$

   e $x^2 + \dfrac{y^2}{7} = 1$ **f** $x^2 + 9y^2 = 36$

   g $9x^2 + 4y^2 = 36$ **h** $4x^2 + y^2 = 8$

2   Sketch the curves with these equations, label the points of intersection with the x-axis and state the equation of the asymptotes in each case.

   a $\dfrac{x^2}{4} - \dfrac{y^2}{9} = 1$ **b** $\dfrac{x^2}{5} - \dfrac{y^2}{3} = 1$

   c $xy = 14$ **d** $x = \dfrac{3}{y}$

   e $x^2 - \dfrac{y^2}{2} = 1$ **f** $3x^2 - y^2 = 6$

   g $x^2 - 5y^2 = 5$ **h** $2x^2 - 5y^2 = 20$

3   A parabola has equation $y^2 = 20x$

   a Sketch the curve.

   b Find the points of intersection between the parabola and the line with equation $y = 2x$

4   An ellipse has equation $\dfrac{x^2}{4} + \dfrac{y^2}{5} = 1$

   a Sketch the curve.

   b Find the points of intersection between the ellipse and the line $y = x + 2$

5   An ellipse has equation $x^2 + 3y^2 = 7$

   a Sketch the curve.

   b Show that the line with equation $2x + 3y = 7$ is a tangent to the ellipse $x^2 + 3y^2 = 7$

6   For each graph, state what type of conic section they are and give a possible equation.

   a

   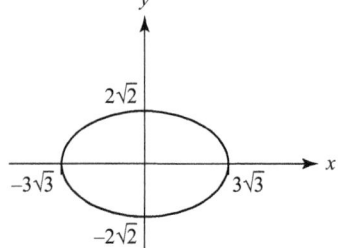

   b

   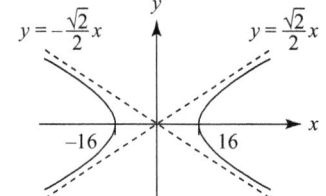

7   The asymptotes of the hyperbola $\dfrac{x^2}{a^2} - \dfrac{y^2}{b^2} = 1$ are $y = \pm 3x$

   Find the equation of the hyperbola given that it passes through the point $(1, 1)$

8   A curve has equation $xy = 16$

   a What name is given to this type of curve?

   b Sketch the curve, giving the equations of the asymptotes.

   The curve intersects the line $y = 3x + 2$ at the points A and B

   c Find the length of the line segment AB

9   A hyperbola has equation $x^2 - y^2 = k^2$

   a Sketch the curve.

   b Show that the asymptotes of the hyperbola are at right angles to each other.

Reasoning and problem-solving

Replacing $f(x)$ by $f(x-c)$ in an equation translates the curve c units in the x-direction. Similarly, replacing $g(y)$ by $g(y-c)$ translates the curve c units in the y-direction.

To sketch translations of conic sections, identify the type of curve involved and find the vertex (for a parabola) or centre (for a hyperbola or an ellipse) of the transformed curve.

Strategy 1

To sketch a translation of a conic section

1. Identify the curve by converting the equation to a standard form.

2. Find the centre or vertex of the transformed curve.

3. Sketch the transformed curve, stating the values of any intercepts and the equations of any asymptotes.

Example 6

Sketch the curve of $y^2 = 2x + 10$

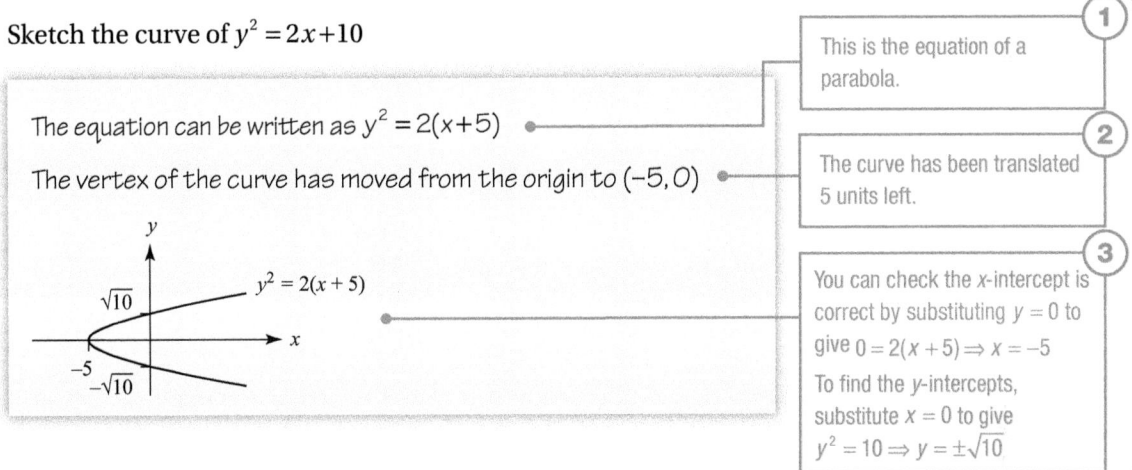

The equation can be written as $y^2 = 2(x+5)$

The vertex of the curve has moved from the origin to $(-5, 0)$

1 This is the equation of a parabola.

2 The curve has been translated 5 units left.

3 You can check the x-intercept is correct by substituting $y = 0$ to give $0 = 2(x+5) \Rightarrow x = -5$

To find the y-intercepts, substitute $x = 0$ to give $y^2 = 10 \Rightarrow y = \pm\sqrt{10}$

Key point

A parabola with equation $(y - y_1)^2 = 4a(x - x_1)$ will have its vertex on the point (x_1, y_1)

Example 7

Sketch the curve of $\dfrac{x^2}{4} + (y-5)^2 = 25$

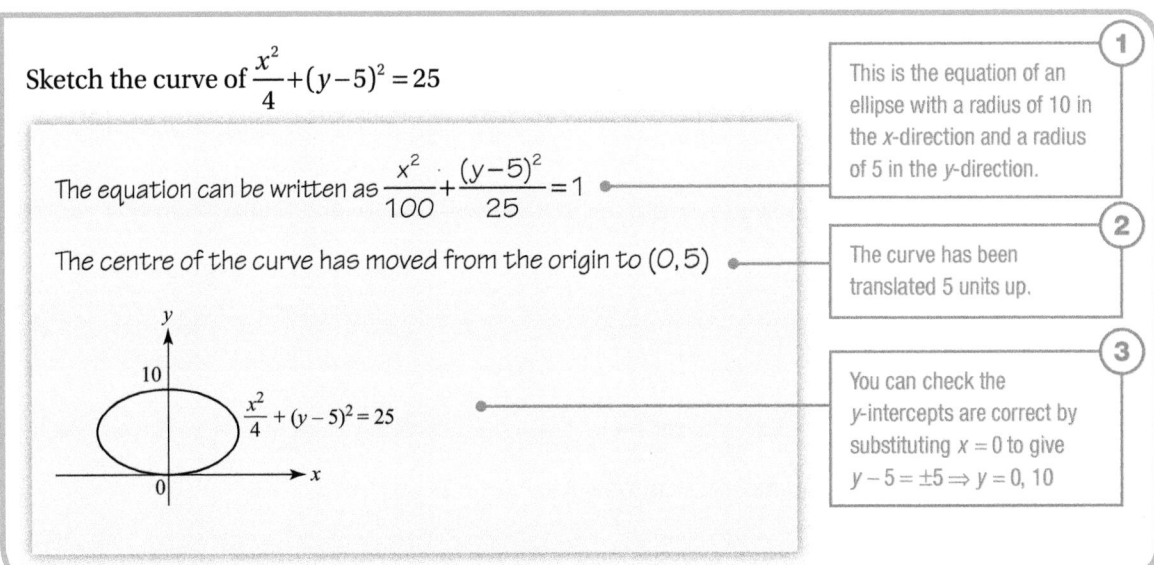

The equation can be written as $\dfrac{x^2}{100} + \dfrac{(y-5)^2}{25} = 1$

The centre of the curve has moved from the origin to $(0, 5)$

1 This is the equation of an ellipse with a radius of 10 in the x-direction and a radius of 5 in the y-direction.

2 The curve has been translated 5 units up.

3 You can check the y-intercepts are correct by substituting $x = 0$ to give $y - 5 = \pm 5 \Rightarrow y = 0, 10$

An ellipse with equation $\dfrac{(x-x_1)^2}{a^2}+\dfrac{(y-y_1)^2}{b^2}=1$ will be centred on **Key point** the point (x_1,y_1) and have radius of a in the x-direction and b in the y-direction.

Example 8

Sketch the curve of $x+xy=4$

The equation can be written as $x(y+1)=4$

The centre of the curve has moved from the origin to $(0,-1)$

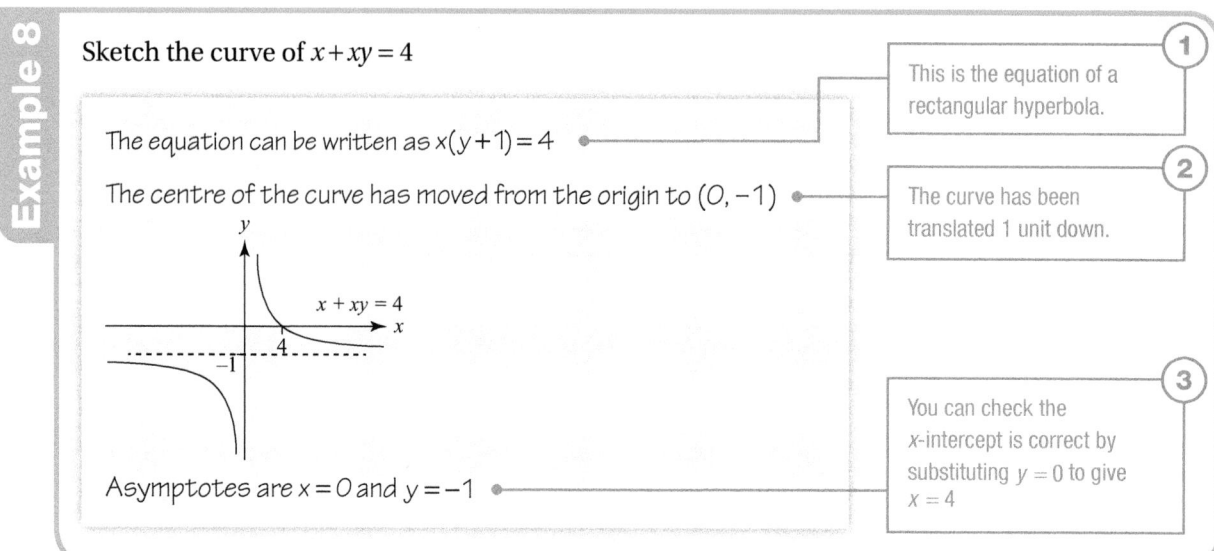

Asymptotes are $x=0$ and $y=-1$

① This is the equation of a rectangular hyperbola.

② The curve has been translated 1 unit down.

③ You can check the x-intercept is correct by substituting $y=0$ to give $x=4$

A rectangular hyperbola with equation $(x-x_1)(y-y_1)=c^2$ **Key point** will be centred on the point (x_1,y_1) and have asymptotes $x=x_1$, $y=y_1$

Example 9

Sketch the curve of $12(x-3)^2-3y^2=48$

The equation can be written as $\dfrac{(x-3)^2}{4}-\dfrac{y^2}{16}=1$

The centre of the curve has moved from the origin to $(3,0)$

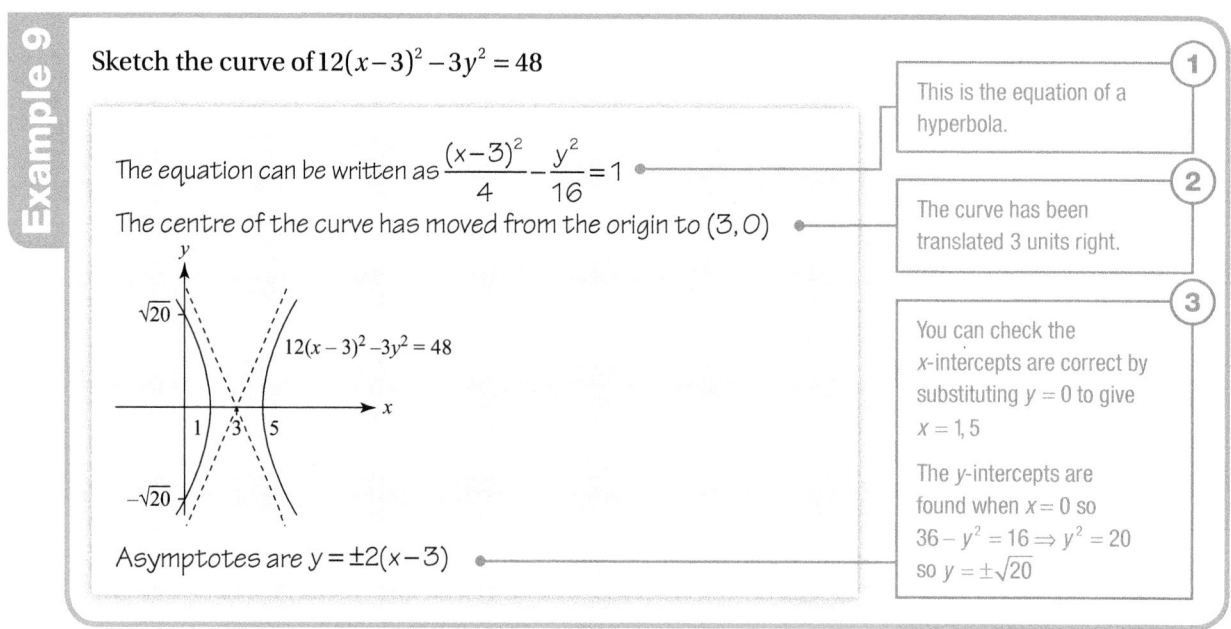

Asymptotes are $y=\pm2(x-3)$

① This is the equation of a hyperbola.

② The curve has been translated 3 units right.

③ You can check the x-intercepts are correct by substituting $y=0$ to give $x=1,5$

The y-intercepts are found when $x=0$ so $36-y^2=16 \Rightarrow y^2=20$ so $y=\pm\sqrt{20}$

A hyperbola with equation $\dfrac{(x-x_1)^2}{a^2}-\dfrac{(y-y_1)^2}{b^2}=1$ will be **Key point** centred on the point (x_1,y_1) and have asymptotes $y-y_1=\pm\dfrac{b}{a}(x-x_1)$

Replacing $f(x)$ by $f\left(\dfrac{x}{k}\right)$ in an equation will stretch the curve by scale factor k in the x-direction. Similarly, replacing $g(y)$ by $g\left(\dfrac{y}{k}\right)$ will stretch the curve by scale factor k in the y-direction.

Strategy 2

To sketch a stretch or a reflection in a coordinate axis of a conic section

(1) Identify the curve by converting the equation to a standard form.

(2) Consider the direction to stretch in or the axis to reflect in.

(3) Sketch the transformed curve, stating the values of any intercepts and the equations of any asymptotes.

Example 10

Sketch the curve of $y^2 = -2x$

> Notice that a stretch or a reflection in a coordinate axis will not affect the vertex.

The equation can be written as $y^2 = 2(-x)$

The curve has been reflected in the line $x = 0$

$y^2 = -2x$

(1) This is the equation of a parabola.

(2) Note that in this case a reflection in the line $y = 0$ would have no effect since $(-y)^2 = y^2$

(3) The x-intercept remains at the origin.

Example 11

The curve with equation $x^2 + 5y^2 = 5$ is stretched by scale factor 3 in the y-direction. Write the equation of the transformed curve and state its points of intersection with the coordinate axes.

Equation of original curve is $\dfrac{x^2}{5} + y^2 = 1$

A stretch of scale factor 3 in the y-direction means equation becomes $\dfrac{x^2}{5} + \left(\dfrac{y}{3}\right)^2 = 1$

Which simplifies to $\dfrac{x^2}{5} + \dfrac{y^2}{9} = 1$

This is an ellipse which intersects the coordinate axes at $(\pm\sqrt{5}, 0), (0, \pm 3)$

(1) This is the equation of an ellipse.

Example 12

The curve with equation $x^2 - 2y^2 = 8$ is stretched by scale factor $\dfrac{1}{2}$ in the x-direction.

Sketch the transformed curve and state the equations of its asymptotes.

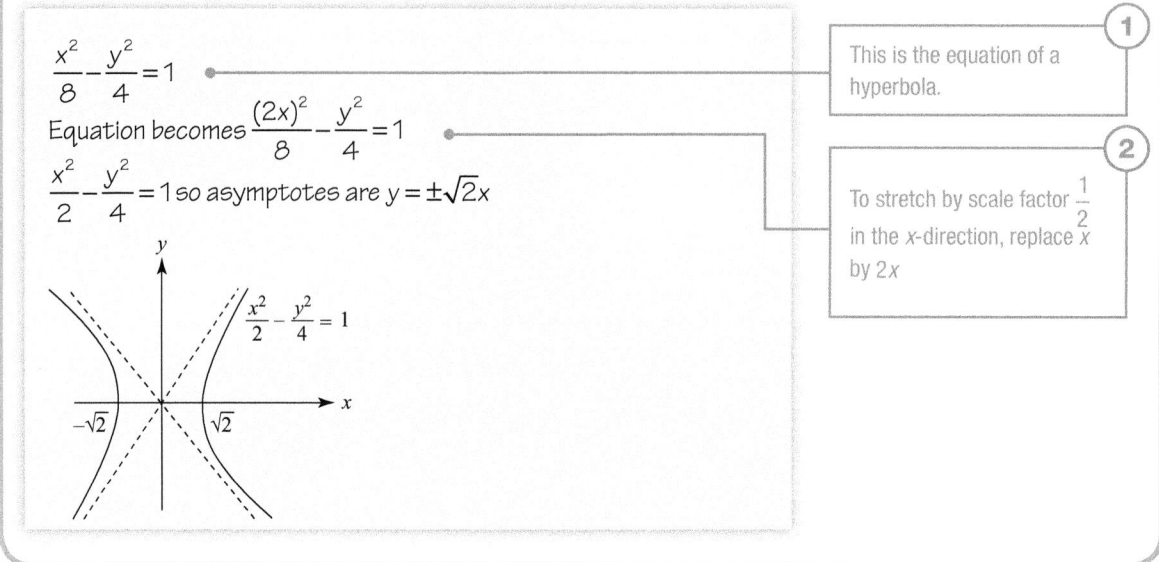

$\dfrac{x^2}{8} - \dfrac{y^2}{4} = 1$ •————————————————

Equation becomes $\dfrac{(2x)^2}{8} - \dfrac{y^2}{4} = 1$ •————————

$\dfrac{x^2}{2} - \dfrac{y^2}{4} = 1$ so asymptotes are $y = \pm\sqrt{2}x$

① This is the equation of a hyperbola.

② To stretch by scale factor $\dfrac{1}{2}$ in the x-direction, replace x by $2x$

You can reflect in the line $y = x$ by reversing the roles of x and y in the equation.

- A parabola of the form $y^2 = kx$ will become $x^2 = ky$
- An ellipse of the form $\dfrac{x^2}{a^2} + \dfrac{y^2}{b^2} = 1$ will become $\dfrac{y^2}{a^2} + \dfrac{x^2}{b^2} = 1$
- A hyperbola of the form $\dfrac{x^2}{a^2} - \dfrac{y^2}{b^2} = 1$ will become $\dfrac{y^2}{a^2} - \dfrac{x^2}{b^2} = 1$ with asymptotes at $x = \pm\dfrac{b}{a}y$
- A rectangular hyperbola of the form $xy = c^2$ will become $yx = c^2$ so no change!

To reflect in the line $y = -x$, you need to replace x with $-y$ and y with $-x$ in the equation.

Example 13

The curve with equation $(x-1)^2 + 36y^2 = 9$ is reflected in the line $y = -x$

Sketch the transformed curve and state its equation.

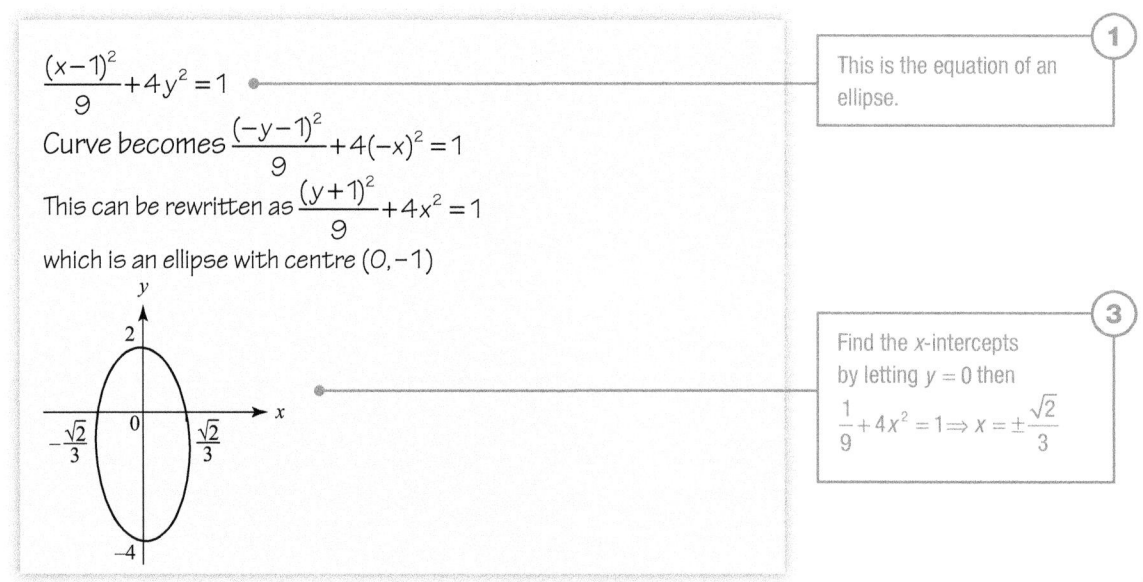

$\dfrac{(x-1)^2}{9} + 4y^2 = 1$ •————————————————

Curve becomes $\dfrac{(-y-1)^2}{9} + 4(-x)^2 = 1$

This can be rewritten as $\dfrac{(y+1)^2}{9} + 4x^2 = 1$

which is an ellipse with centre $(0, -1)$

① This is the equation of an ellipse.

③ Find the x-intercepts by letting $y = 0$ then

$\dfrac{1}{9} + 4x^2 = 1 \Rightarrow x = \pm\dfrac{\sqrt{2}}{3}$

Exercise 3.4B Reasoning and problem-solving

1 Sketch each of these curves, clearly labelling the points of intersection with the coordinate axes in each case.

a $y^2 = 3(x-2)$ **b** $(y-3)^2 = 3x$

c $(y+2)^2 = 8(x-1)$ **d** $y^2 = 4x+8$

e $y^2 + 2y = x$ **f** $\dfrac{x^2}{5} + \dfrac{(y+3)^2}{4} = 1$

g $\dfrac{(3+x)^2}{9} + \dfrac{y^2}{4} = 1$ **h** $x^2 + 9(y-2)^2 = 36$

i $(x-4)^2 + 4(y-2)^2 = 16$

j $4x^2 - 8x + y^2 - 12y + 4 = 0$

2 Sketch each of these curves and state the equations of the asymptotes in each case.

a $x(y+3) = 9$ **b** $(x-4)y = 25$

c $(x+3)(y-3) = 10$ **d** $xy - 2x = 4$

e $xy - 5y = 2$ **f** $\dfrac{x^2}{2} - \dfrac{(y+5)^2}{9} = 1$

g $\dfrac{(x+2)^2}{16} - \dfrac{y^2}{25} = 1$ **h** $x^2 - 4(y-7)^2 = 36$

i $(x-5)^2 - 2(y+2)^2 = 16$

j $x^2 + 6x - 3y^2 + 30y = 147$

3 Sketch each of these curves, clearly labelling the points of intersection with the coordinate axes in each case.

a $y^2 = -5x$ **b** $-xy = 7$

4 Write down the equation of the image of each of these transformations of the curve $y^2 = 8x$ in the form $y^2 = f(x)$

a Stretch by scale factor 3 in the x-direction

b Stretch by scale factor $\dfrac{1}{4}$ in the y-direction

c Reflection in the x-axis

d Reflection in the y-axis.

5 Write down the equation of the image of each transformation of the curve $y^2 = -4x$

a Translation by the vector $\begin{pmatrix} 2 \\ -3 \end{pmatrix}$

b Reflection in the line $y = x$

6 Write the equation of the image of each transformation of the curve $\dfrac{x^2}{12} + \dfrac{y^2}{4} = 1$

Write the equations in the form $\dfrac{x^2}{a^2} + \dfrac{y^2}{b^2} = 1$

a Stretch vertically by scale factor 2

b Stretch horizontally by scale factor –5

c Reflection in the line $y = 0$

d Reflection in the line $y = -x$

7 Write the equation of the curve formed by translating the ellipse $x^2 + 2y^2 = 8$ by the vector $\begin{pmatrix} 3 \\ -1 \end{pmatrix}$

8 For each of these transformations of the curve $x(y-1) = 4$

i Write down the equation of the image,

ii State the equations of the asymptotes.

   a Reflection in the line $x = 0$

   b Translation by the vector $\begin{pmatrix} 5 \\ 2 \end{pmatrix}$

   c Reflection in the line $y = x$

   d Reflection in the line $y = -x$

9 For each of these transformations of the curve $9x^2 - y^2 = 81$

i Write the equation of the image,

ii State the equations of the asymptotes.

   a Stretch vertically by scale factor –3

   b Reflection in the x-axis

   c Translation by the vector $\begin{pmatrix} -1 \\ 4 \end{pmatrix}$

   d Reflection in the line $y = x$

10 A curve has equation $\dfrac{(x-k)^2}{a^2} - (y+k)^2 = a^2$ where a and k are constants, $a, k > 0$

a State the equations of the asymptotes to the curve.

b Sketch the reflection of the curve in the line $y = -x$ and state the equations of the asymptotes of the reflected curve.

Hyperbolic functions 1

Fluency and skills

If you plot all the points of the form $(\sin \theta, \cos \theta)$ where $0 \le \theta \le 2\pi$, you produce a circle (shown by the red curve).

There exists another pair of functions, denoted by $\sinh \theta$ and $\cosh \theta$. If you plot the points $(\sinh \theta, \cosh \theta)$, you produce the curve shown in blue.

This curve is called a hyperbola and so the functions are referred to as **hyperbolic** functions and defined using exponentials.

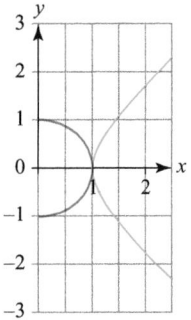

Key point

$$\sinh x = \frac{e^x - e^{-x}}{2} \qquad \cosh x = \frac{e^x + e^{-x}}{2}$$

$$\tanh x = \frac{\sinh x}{\cosh x} = \frac{e^x - e^{-x}}{e^x + e^{-x}}$$

These functions are commonly read as 'shine', 'cosh' and 'tanch' or 'than'.

Find the exact value of

a $\tanh 0$ **b** $\sinh(\ln 3)$

> Use the definition of $\tanh x$

a $\tanh 0 = \dfrac{e^0 - e^{-0}}{e^0 + e^{-0}}$

$= \dfrac{1-1}{1+1}$

$= 0$

b $\sinh(\ln 3) = \dfrac{e^{\ln 3} - e^{-\ln 3}}{2}$

$= \dfrac{e^{\ln 3} - e^{\ln 3^{-1}}}{2}$

> Use the fact that $a \ln b = \ln b^a$

$= \dfrac{3 - \dfrac{1}{3}}{2}$

$= \dfrac{4}{3}$

 Try it on your calculator

You can use a calculator to evaluate hyperbolic functions and inverse hyperbolic functions.

$\sinh(\ln 5)$

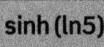

$\dfrac{12}{5}$

Activity

Find out how to evaluate $\sinh(\ln 5)$ on your calculator.

The graph $y = \cosh x$

The curve has a minimum point at $(0, 1)$

So $y \geq 1$

The curve is symmetrical about the y-axis.

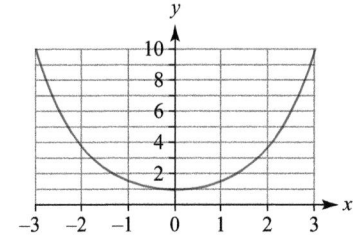

The graph $y = \sinh x$

The curve has rotation symmetry about the origin.

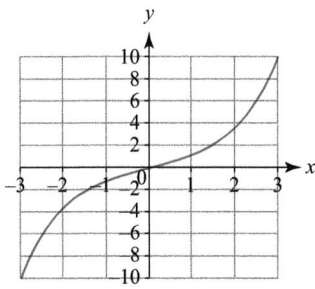

The graph $y = \tanh x$

As $x \to +\infty \Rightarrow e^{-x} \to 0 \Rightarrow \tanh x \to \pm 1$

The curve has asymptotes at $y = 1$ and $y = -1$

So $-1 < y < 1$

The curve has rotational symmetry about the origin.

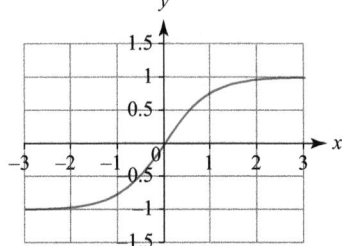

> **Key point**
>
> Graphs of hyperbolic functions can be transformed in the same way as other functions so for $y = f(x)$:
>
> - $y = af(x)$ is $y = f(x)$ stretched vertically by scale factor a
> - $y = f(ax)$ is $y = f(x)$ stretch horizonally by scale factor $\dfrac{1}{a}$
> - $y = a + f(x)$ is $y = f(x)$ translated vertically by a units
> - $y = f(x+a)$ is $y = f(x)$ translated horizontally by $-a$ units.

Example 2

Sketch the graph of $y = 2 + \cosh x$

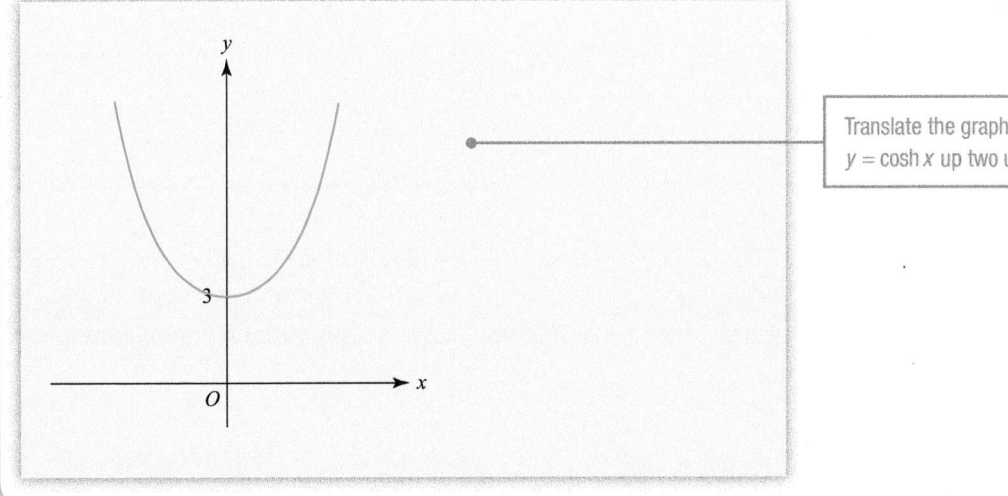

Translate the graph of $y = \cosh x$ up two units.

Example 3

Sketch the graphs of $y = \sinh x$ and $y = \sinh \dfrac{x}{2}$ on the same axes.

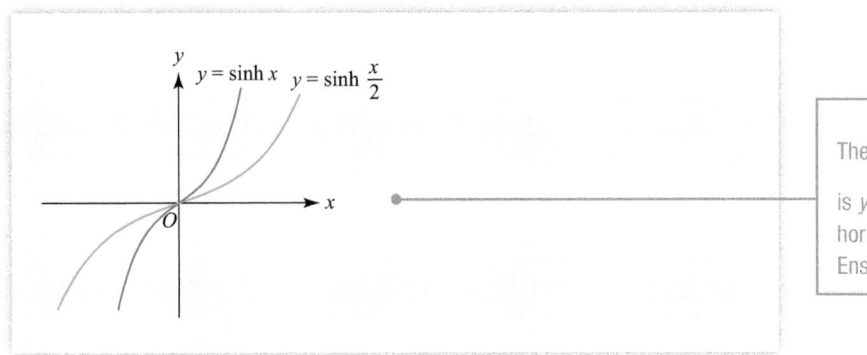

The graph of $y = \sinh \dfrac{x}{2}$ is $y = \sinh x$ stretched horizontally by scale factor 2. Ensure you label each curve.

If you are given the value of $\sinh x$, $\cosh x$ or $\tanh x$ you can use their definitions to find the value of x

Example 4

Solve the equation $\tanh x = \dfrac{1}{2}$

$\dfrac{e^x - e^{-x}}{e^x + e^{-x}} = \dfrac{1}{2}$

Use the definition of $\tanh x$

$2(e^x - e^{-x}) = e^x + e^{-x} \Rightarrow e^x = 3e^{-x}$

$e^{2x} = 3$

Multiply both sides by e^x

$2x = \ln 3 \Rightarrow x = \dfrac{1}{2}\ln 3$

Take logarithms on both sides.

This example could also be done on a calculator using $\tanh^{-1} \dfrac{1}{2}$, however your calculator will give a decimal, not an exact answer.

Example 5

Solve the equation $\sinh x + 2\cosh x = 2$

$\dfrac{e^x - e^{-x}}{2} + \dfrac{2(e^x + e^{-x})}{2} = 2$

Use the definitions of $\sinh x$ and $\cosh x$

$e^x - e^{-x} + 2e^x + 2e^{-x} = 4$

$3e^x - 4 + e^{-x} = 0$

$3e^{2x} - 4e^x + 1 = 0$

Multiply through by e^x to give a quadratic equation in e^x

$(3e^x - 1)(e^x - 1) = 0$

$e^x = 1 \text{ or } e^x = \dfrac{1}{3}$

$x = \ln 1 = 0 \text{ or } x = \ln\left(\dfrac{1}{3}\right)$

Exercise 3.5A Fluency and skills

1 Work out the value of

  a $\sinh 0$ **b** $\cosh(-1)$ **c** $\sinh(\ln 2)$

  d $\cosh(-\ln 3)$ **e** $\cosh 0$ **f** $\tanh(\ln 2)$

2 Sketch these graphs.

  a $y = \sinh(x-2)$ **b** $y = 10 + \sinh x$ **c** $y = \tanh(x-1)$

  d $y = \cosh(x+2)$ **e** $y = 2 + \tanh(x)$ **f** $y = 1 - \cosh(x)$

3 The function f is defined as $f(x) = \sinh(x)$

  a Sketch the graphs of $y = f(x)$ and $y = f(2x)$ on the same set of axes.

  b Solve the equation $f(2x) = 2$, giving your answer to 3 significant figures.

4 Sketch the graphs of $y = \cosh\left(\dfrac{x}{3}\right)$ and $y = \cosh(x)$ on the same set of axes.

5 The function f is defined as $f(x) = \tanh(x)$

  a Sketch the graph $y = -2f(x)$

  b Solve the equation $-2f(x) = 1$, giving your answer to 3 significant figures.

6 Given that $f(x) = \tanh(x+a)$, $a > 0$

  a Sketch the graph of $y = f(x)$

  b Write down the equations of the asymptotes.

7 Given that $g(x) = \tanh x$, $a > 0$

  a Sketch the graph of $y = a + g(x)$

  b Write down the equations of the asymptotes.

8 Given that $f(x) = \cosh x$, $x \in \mathbb{R}$, sketch the graphs of

  a $y = f(x-a)$, for $a > 0$ **b** $y = f(x) + a$, for $a > 0$ **c** $y = af(x)$, for $a < 0$

9 Solve each of these equations, giving your answers to 3 sf.

  a $\sinh x = 5$ **b** $\cosh x = 2$ **c** $\tanh x = -\dfrac{1}{2}$

  d $\cosh(x+1) = 3$ **e** $\sinh(3x) = 4$ **f** $2\tanh(x) + 1 = 2$

10 Solve these equations, leaving your answer as a logarithm in its simplest form where appropriate. Show your working.

  a $\sinh x = 0$ **b** $\cosh x = 1$ **c** $\sinh x = \dfrac{3}{4}$

  d $\cosh x = \dfrac{17}{8}$ **e** $\tanh x = \dfrac{3}{5}$ **f** $\tanh x = \dfrac{40}{41}$

11 Solve these equations, leaving your answer as a logarithm in its simplest form where appropriate.

  a $\cosh x - \sinh x = 2$ **b** $2\sinh x + 3\cosh x = 3$

  c $3 - \sinh x = 2\cosh x$ **d** $\dfrac{1}{\sinh x} - \dfrac{1}{\cosh x} = 4e^x$

Hyperbolic functions have identities which are very similar to the trigonometric ones you already know. You can prove these using the definitions.

> Notice how the sign is different to the trigonometric identity $\cos^2 x + \sin^2 x \equiv 1$

Key point

Learn this important identity: $\cosh^2 x - \sinh^2 x \equiv 1$

Strategy 1

To solve problems using hyperbolic functions

(1) Use the definitions of $\sinh x$, $\cosh x$ and $\tanh x$

(2) Use laws of indices and logarithms as necessary to simplify expressions.

Example 6

Prove that $\cosh(A+B) \equiv \cosh(A)\cosh(B) + \sinh(A)\sinh(B)$

$\cosh(A)\cosh(B) + \sinh(A)\sinh(B)$

$$\equiv \left(\frac{e^A + e^{-A}}{2}\right)\left(\frac{e^B + e^{-B}}{2}\right) + \left(\frac{e^A - e^{-A}}{2}\right)\left(\frac{e^B - e^{-B}}{2}\right)$$

(1) Use the definitions of $\cosh x$ and $\sinh x$

$$\equiv \frac{e^{A+B} + e^{(A-B)} + e^{-(A-B)} + e^{-(A+B)}}{4} + \frac{e^{A+B} - e^{(A-B)} - e^{-(A-B)} + e^{-(A+B)}}{4}$$

(2) Expand the brackets, using laws of indices.

$$\equiv \frac{2e^{A+B} + 2e^{-(A+B)}}{4}$$

$$\equiv \frac{e^{A+B} + e^{-(A+B)}}{2}$$

$$\equiv \cosh(A+B) \text{ as required}$$

Example 7

Prove that $\cosh^2 x - \sinh^2 x \equiv 1$

$$\cosh^2 x - \sinh^2 x \equiv \left(\frac{e^x + e^{-x}}{2}\right)^2 - \left(\frac{e^x - e^{-x}}{2}\right)^2$$

(1) Use the exponential definitions.

$$\equiv \frac{e^{2x} + 2 + e^{-2x}}{4} - \frac{e^{2x} - 2 + e^{-2x}}{4}$$

(2) Simplify the numerator.

$$\equiv \frac{4}{4}$$

$$\equiv 1 \text{ as required}$$

Example 8

Derive the logarithmic form of the inverse function $\cosh^{-1} x$

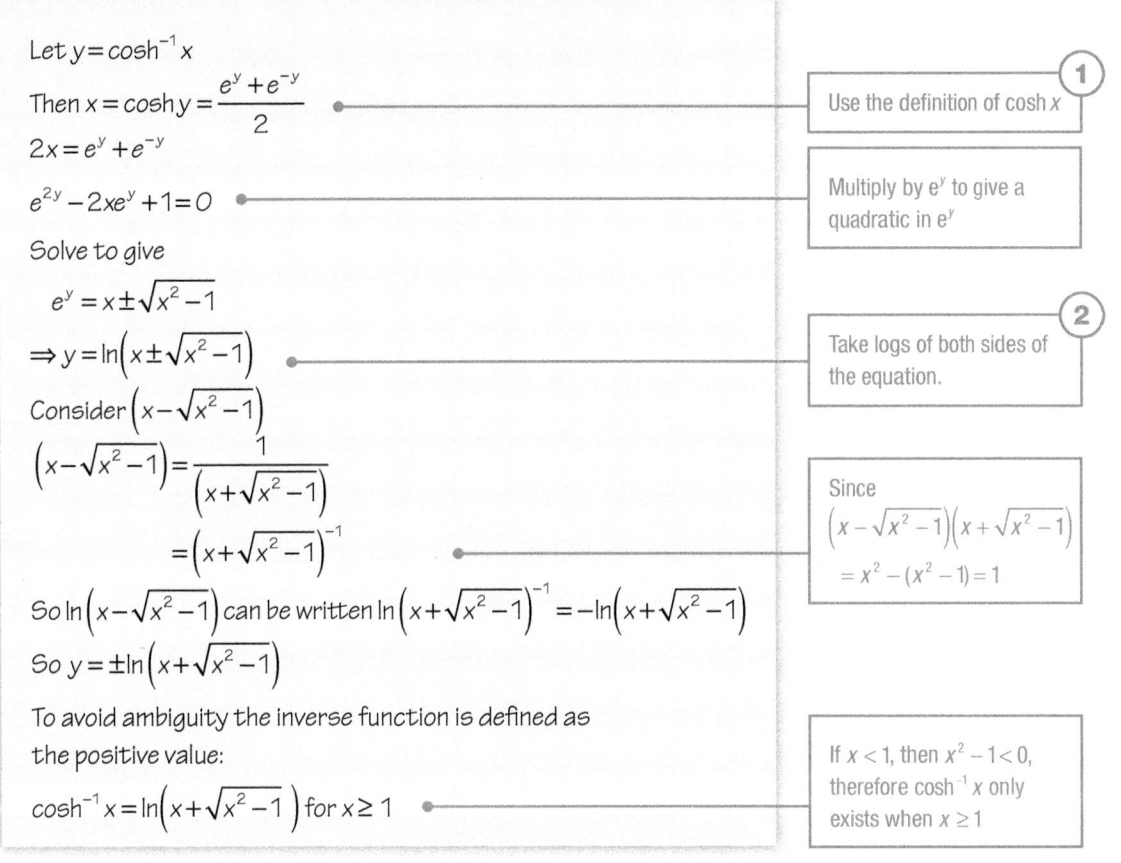

Let $y = \cosh^{-1} x$

Then $x = \cosh y = \dfrac{e^y + e^{-y}}{2}$ — ① Use the definition of $\cosh x$

$2x = e^y + e^{-y}$

$e^{2y} - 2xe^y + 1 = 0$ — Multiply by e^y to give a quadratic in e^y

Solve to give

$e^y = x \pm \sqrt{x^2 - 1}$

$\Rightarrow y = \ln\left(x \pm \sqrt{x^2 - 1}\right)$ — ② Take logs of both sides of the equation.

Consider $\left(x - \sqrt{x^2 - 1}\right)$

$\left(x - \sqrt{x^2 - 1}\right) = \dfrac{1}{\left(x + \sqrt{x^2 - 1}\right)}$

$\quad = \left(x + \sqrt{x^2 - 1}\right)^{-1}$ — Since $\left(x - \sqrt{x^2-1}\right)\left(x + \sqrt{x^2-1}\right) = x^2 - (x^2-1) = 1$

So $\ln\left(x - \sqrt{x^2 - 1}\right)$ can be written $\ln\left(x + \sqrt{x^2 - 1}\right)^{-1} = -\ln\left(x + \sqrt{x^2 - 1}\right)$

So $y = \pm\ln\left(x + \sqrt{x^2 - 1}\right)$

To avoid ambiguity the inverse function is defined as the positive value:

$\cosh^{-1} x = \ln\left(x + \sqrt{x^2 - 1}\right)$ for $x \geq 1$ — If $x < 1$, then $x^2 - 1 < 0$, therefore $\cosh^{-1} x$ only exists when $x \geq 1$

The other inverse hyperbolic functions can be derived in a similar way.

Key point

$\sinh^{-1} x = \ln\left(x + \sqrt{x^2 + 1}\right)$

$\cosh^{-1} x = \ln\left(x + \sqrt{x^2 - 1}\right);\ x \geq 1$

$\tanh^{-1} x = \dfrac{1}{2}\ln\left(\dfrac{1 + x}{1 - x}\right);\ -1 < x < 1$

You can also refer to the inverse functions as $\operatorname{arsinh} x$ or $\operatorname{arcsinh} x$, $\operatorname{arcosh} x$ or $\operatorname{arccosh} x$ and $\operatorname{artanh} x$ or $\operatorname{arctanh} x$

Strategy 2

To solve quadratic equations involving hyperbolic functions

① Use identities to write the equation in terms of a single hyperbolic function.

② Solve the quadratic equation to find the value of $\sinh x$, $\cosh x$ or $\tanh x$

③ Use the definitions of the inverse hyperbolic functions to find the exact values of x

Example 9

Solve the equation $\sinh x + 2\cosh^2 x = 3$

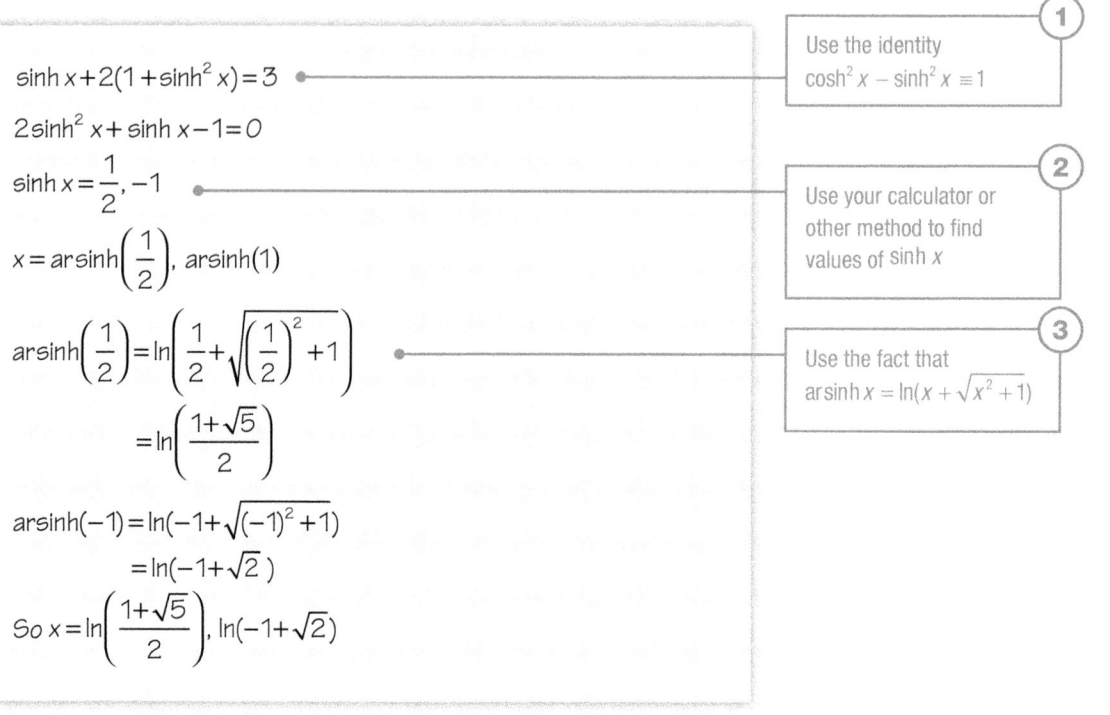

$\sinh x + 2(1 + \sinh^2 x) = 3$ ← Use the identity $\cosh^2 x - \sinh^2 x \equiv 1$ ①

$2\sinh^2 x + \sinh x - 1 = 0$

$\sinh x = \dfrac{1}{2}, -1$ ← Use your calculator or other method to find values of $\sinh x$ ②

$x = \operatorname{arsinh}\left(\dfrac{1}{2}\right), \operatorname{arsinh}(1)$

$\operatorname{arsinh}\left(\dfrac{1}{2}\right) = \ln\left(\dfrac{1}{2} + \sqrt{\left(\dfrac{1}{2}\right)^2 + 1}\right)$ ← Use the fact that $\operatorname{arsinh} x = \ln(x + \sqrt{x^2 + 1})$ ③

$\qquad = \ln\left(\dfrac{1 + \sqrt{5}}{2}\right)$

$\operatorname{arsinh}(-1) = \ln(-1 + \sqrt{(-1)^2 + 1})$

$\qquad = \ln(-1 + \sqrt{2})$

So $x = \ln\left(\dfrac{1 + \sqrt{5}}{2}\right), \ln(-1 + \sqrt{2})$

Exercise 3.5B Reasoning and problem-solving

1 Solve the equations, giving your answers as exact logarithms where appropriate.

  a $\sinh^2 x + \sinh x = 6$

  b $2\cosh^2 x - 5\cosh x + 3 = 0$

  c $3\tanh^2 x - 2 = 0$

  d $\cosh^2(2x) - 3\cosh(2x) = 0$

2 Solve the equations. Give your answers as exact logarithms where appropriate.

  a $\sinh^2 x + \cosh x = 1$

  b $\cosh^2 x + 3\sinh x = 5$

  c $2\sinh x + 3\cosh^2 x = 3$

  d $2\tanh^2 x + 3\tanh x = 2$

  e $\sinh 2x = 3$

  f $\cosh 3x = 4$

  g $2\tanh 3x = 1$

3 Use the definitions of $\sinh x$ and $\cosh x$ to prove that $\cosh^2 x - \sinh^2 x \equiv 1$

4 Use the definitions of $\cosh x$ and $\sinh x$ to prove these identities.

  a $\sinh(A + B) = \sinh(A)\cosh(B) + \sinh(B)\cosh(A)$

  b $\sinh(A - B) = \sinh(A)\cosh(B) - \sinh(B)\cosh(A)$

  c $\sinh(2x) = 2\sinh(x)\cosh(x)$

5 a Use the definitions of $\sinh x$ and $\cosh x$ to prove that $\tanh x = \dfrac{e^{2x}-1}{e^{2x}+1}$

b Hence show that $\tanh(2x) = \dfrac{2\tanh x}{1+\tanh^2 x}$

6 a Use the definition of $\cosh x$ to prove that $\cosh(2x) \equiv 2\cosh^2 x - 1$

b Hence find the exact solutions to the equation $\cosh(2x) + \cosh x = 5$

7 Use a similar method as in Example 8 to prove that $\operatorname{arsinh} x = \ln\left(x + \sqrt{x^2+1}\right)$

Explain clearly why $\operatorname{arsinh} x \neq \ln\left(x - \sqrt{x^2+1}\right)$

8 a Prove that $\operatorname{artanh} x = \dfrac{1}{2}\ln\left(\dfrac{1+x}{1-x}\right)$

b Explain why the formula in part **a** is only valid for $-1 < x < 1$

9 Solve the simultaneous equations

$3\cosh x + 2\sinh x = 1$ and $\cosh^2 x + \sinh x = 3$

10 Solve the simultaneous equations

$\cosh x - 2\sinh^2 x = 3$ and $\cosh x + \sinh x = 4$

11 Solve the equation $\cosh 2x + 7 = 7\cosh x$

12 a Show that $3\sinh\left(\dfrac{x}{2}\right) - \sinh(x) \equiv \sinh\left(\dfrac{x}{2}\right)\left(3 - 2\cosh\left(\dfrac{x}{2}\right)\right)$

b Hence, find the exact solutions to the equation $3\sinh\left(\dfrac{x}{2}\right) - \sinh(x) = 0$

13 a Show that $\dfrac{1}{\cosh(x)+1} + \dfrac{1}{\cosh(x)-1} \equiv -\dfrac{2}{\sinh^2 x}$

b Hence solve the equation $\dfrac{1}{\cosh(x)+1} + \dfrac{1}{1-\cosh(x)} + 8 = 0$

14 a Given that $\sinh(2x) = \cosh^2 x$, show that $\tanh x = \dfrac{1}{2}$

b Hence solve the equation $\sinh(2x) - \cosh^2 x = 0$

15 Solve the equation $\operatorname{arcosh} x = 2\ln x$

16 Given that $\sinh x = 2$, $x > 0$, calculate the exact value of

   a $\cosh x$ **b** $\tanh x$

17 Given that $\cosh x = 3$, $x > 0$, calculate the exact value of

   a $\sinh x$ **b** $\tanh x$

18 Given that $\tanh x = \dfrac{1}{2}$, $x > 0$, calculate the exact value of

   a $\sinh x$ **b** $\cosh x$

19 a Use the definitions to find the derivatives of $\sinh x$ and $\cosh x$

b Find the gradient of both curves when $x = 0$

c Show that for $x > 0$ the curve $y = \sinh x$ is always steeper than $y = \cosh x$

d Find the gradient of the tangent to the curve $y = \tanh x$ at $x = 0$

3 Summary and review

Chapter summary

- When sketching the graph of a function of the form $\dfrac{ax+b}{cx+d}$
  - find the intercepts and any asymptotes,
  - consider what happens when x and y become very large and approach infinity.
- When sketching the graph of a function of the form $\dfrac{ax^2+bx+c}{dx^2+ex+f}$
  - use the above techniques and also find any limitations on the possible ranges of y
  - use the results of these techniques to find any turning points.
- You can use sketches of graphs to solve inequalities.
- r is the distance from the pole to the point.
- θ is the angle between the initial line and the line segment from the pole to the point, measured counterclockwise.
- To convert from polar to Cartesian, use
  $$x = r\cos\theta$$
  $$y = r\sin\theta$$
- To convert from Cartesian to polar, use
  $$r^2 = x^2 + y^2$$
  $$\tan\theta = \frac{y}{x}$$
  but you should draw a diagram to ensure that you choose the right value of θ
- To convert from polar to Cartesian, use
  $$r^2 = x^2 + y^2$$
  $$\tan\theta = \frac{y}{x}$$
  $$\cos\theta = \frac{x}{r} = \pm\frac{x}{\sqrt{x^2+y^2}}$$
  $$\sin\theta = \frac{y}{r} = \pm\frac{y}{\sqrt{x^2+y^2}}$$

  [Choose $+$ or $-$ depending on whether r is positive or negative]
- To convert from Cartesian to polar, use
  $$x = r\cos\theta$$
  $$y = r\sin\theta$$
- The equation of a parabola centred at the origin is $y^2 = 4ax$
- The equation of an ellipse centred at the origin is $\dfrac{x^2}{a^2}+\dfrac{y^2}{b^2}=1$
- The equation of a hyperbola centred at the origin with asymptotes at $y=\pm\dfrac{b}{a}x$ is $\dfrac{x^2}{a^2}-\dfrac{y^2}{b^2}=1$
- A rectangular hyperbola centred at the origin with asymptotes $y=0$ and $x=0$ has equation $xy=c^2$

- The hyperbolic functions are defined using exponentials and have graphs as shown.

  ○ $\sinh x = \dfrac{e^x - e^{-x}}{2}$

  ○ $\cosh x = \dfrac{e^x + e^{-x}}{2}$

  ○ $\tanh = \dfrac{e^x - e^{-x}}{e^x + e^{-x}}$

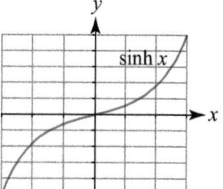

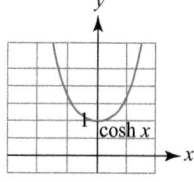

- $\cosh^2 x - \sinh^2 x \equiv 1$
- The inverse hyperbolic functions are

  ○ $\sinh^{-1} x = \ln\left(x + \sqrt{x^2 + 1}\right)$

  ○ $\cosh^{-1} x = \ln\left(x + \sqrt{x^2 - 1}\right),\ x \geq 1$

  ○ $\tanh^{-1} x = \dfrac{1}{2}\ln\left(\dfrac{1+x}{1-x}\right),\ -1 < x < 1$

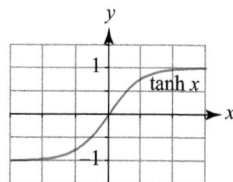

Check and review

You should now be able to...	Try Questions
✔ Graph functions of the form $\dfrac{ax+b}{cx+d}$ and $\dfrac{ax^2+bx+c}{dx^2+ex+f}$, and solve associated inequalities.	1–3
✔ Use quadratic theory to find the range of possible values of a function and find stationary points.	4, 5
✔ Convert between polar and Cartesian coordinates and equations.	6–9
✔ Sketch polar curves and find points of intersection.	10–13
✔ Sketch a parabola with vertex at the origin and an ellipse centred at the origin, giving the coordinates of the points of intersection with the coordinate axes.	14, 15
✔ Sketch a hyperbola centred at the origin, giving the equations of the asymptotes.	16, 17
✔ Sketch single transformations of conic sections.	18–20
✔ Find the equations of conic sections reflected in the lines $y = x$ or $y = -x$	21
✔ Know the definition of $\sinh x$, $\cosh x$ and $\tanh x$ in terms of exponentials; calculate exact values and solve equations.	22, 25, 26
✔ Sketch hyperbolic graphs and transformations.	23
✔ Recall and use the identity $\cosh^2 x - \sinh^2 x \equiv 1$	24
✔ Derive the logarithmic form of the inverse hyperbolic functions and calculate exact values.	26–27

1 Sketch the graph of $y = \dfrac{(x-1)}{(x+4)}$

2 Solve these inequalities graphically, naming any intercepts and vertical and/or horizontal asymptotes.

   a $-3 < \dfrac{x+12}{x-4} < 9$ **b** $7 \geq \dfrac{x-6}{x-12} \geq -5$

3 Sketch the graph of

   a $y = \dfrac{x^2 + 12x + 24}{x^2 + 2x + 2}$ **b** $y = \dfrac{x^2 + 12x + 24}{x^2 + 2x - 3}$

4 Find the values of y for which the graph of the function $y = \dfrac{x^2 - 3x - 4}{x^2 - 3x + 1}$ exists.

5 Without using calculus, find the coordinates of the turning points of the curve with equation $y = \dfrac{x^2 - 5x}{x^2 - 5x + 5}$ and find the maximum value.

6 Convert the sets of polar coordinates to Cartesian coordinates.

   a $\left(5, \dfrac{\pi}{3}\right)$ **b** $\left(5, -\dfrac{\pi}{3}\right)$

7 Convert the sets of Cartesian coordinates to polar coordinates.

   a $\left(2\sqrt{3}, 2\right)$ **b** $\left(-2\sqrt{3}, 2\right)$

   c $\left(-2\sqrt{3}, -2\right)$ **d** $\left(2\sqrt{3}, -2\right)$

8 Express these Cartesian equations in polar form.

   a $y = 3x$ **b** $y = 2x + 1$

   c $x^2 + y^2 = 16$ **d** $(x-1)^2 + (y-1)^2 = 2$

9 Express these polar equations in Cartesian form.

   a   The circle $r = 4 \sin \theta$, $r \geq 0$, stating its centre and radius,

   b   The graph $r = 2 - 4 \sin \theta$, $r \geq 0$

10 For each polar equation

   i   Sketch the half-line it represents,

   ii   Give the Cartesian equation of the line it lies on.

   a $\theta = -\dfrac{3\pi}{4}, r \geq 0$ **b** $\theta = \dfrac{\pi}{3}, r \geq 0$

11 For each of the polar equations

   i   Sketch the graph for $r \geq 0$ and $0 \leq \theta < 2\pi$

   ii   State the maximum and minimum values of r

   a $r = 3$ **b** $r = 7 \cos \theta$

   c $r = 4 \cos 2\theta$ **d** $r = 2 + \sin \theta$

   e $r = 4 - 3 \cos \theta$ **f** $r = 3\theta$

12 Find the points of intersection between the polar curves $r = 2$ and $r = 3 - 2 \cos \theta$

13 Find the points of intersection between the polar curves $r = \sin 2\theta$ and $r = \sqrt{3}\cos 2\theta$ where $r \geq 0$. What difference does it make if r can take any value, including negatives?

14 **a** Sketch the parabola $y^2 = 4x$

   b Find the points of intersection between this curve and the line $y = x - 3$

15 Sketch each of these ellipses, clearly labelling where they cross the coordinate axes.

   a $\dfrac{x^2}{36} + \dfrac{y^2}{49} = 1$ **b** $\dfrac{x^2}{12} + y^2 = 2$

16 Sketch the hyperbola $\dfrac{x^2}{8} - \dfrac{y^2}{4} = 1$, giving the equations of the asymptotes.

17 A rectangular hyperbola has equation $xy = 5$

   a Sketch the curve and state the equations of the asymptotes.

   b Show that the hyperbola does not intersect the line with equation $x + y = 1$

18 Sketch these parabolas, clearly labelling where they cross the coordinate axes.

   a $y^2 = -2x$ **b** $(y-1)^2 = 3x$

19 Sketch these ellipses, clearly labelling where they cross the coordinate axes.

   a $\dfrac{x^2}{3} + \dfrac{(y-2)^2}{4} = 1$ **b** $\dfrac{(x-5)^2}{25} + (y+1)^2 = 1$

20 Sketch these hyperbolas and state the equations of their asymptotes.

   a $\dfrac{x^2}{3} - \dfrac{(y+5)^2}{4} = 1$ **b** $\dfrac{(x+3)^2}{9} - \dfrac{(y-2)^2}{4} = 1$

   c $-xy = 7$ **d** $(x+3)(y-4) = 2$

21 **a** The parabola $\dfrac{x^2}{8} + y^2 = 2$ is reflected in the line $y = x$. Sketch the curve and write down its equation.

   b The hyperbola $6x^2 - 2y^2 = 6$ is reflected in the line $y = -x$. Sketch the curve and write down the equations of its asymptotes.

22 Calculate the exact value of

   a $\cosh(\ln 5)$ **b** $\tanh(-\ln 2)$

23 Sketch these graphs for $a \geq 1$

   a $y = a\cosh x$ **b** $y = \sinh(x-a)$

   c $y = a\tanh x$ **d** $y = \cosh x - a$

24 Solve the equation $\cosh^2 x + \sinh x = 3$

25 Solve these equations, giving your answers as logarithms.

   a $\sinh x + \cosh x = 3$ **b** $\cosh x + 1 = e^x$

26 Use the definition of $\cosh x$ to prove that
   $\operatorname{arcosh}(2x) = \ln\left(2x + \sqrt{4x^2 - 1}\right)$

27 Find the exact solution to the equations

   a $\sinh x = 2$ **b** $\tanh x = -\dfrac{1}{2}$

   Assume $r \geq 0$ in each case.

History

Two mathematicians, the Flemish Gregoire de Saint-Vincent and the Italian Bonaventura Cavalieri, introduced the concepts of the polar coordinate system independently in the mid-seventeenth century. Cavalieri used them to solve a problem involving the area within an Archimedian spiral. The French mathematician Blaise Pascal used polar coordinates to calculate the lengths of parabolic arcs like the ones on this bridge.

Did you know?

Although many bridges are known as suspension bridges, they are actually suspended deck bridges. The cables follow a parabolic curve.

Using a photograph of a famous bridge, such as the Forth Road Bridge, together with graph plotting technology, explore how well you can model the suspension cables using quadratic and hyperbolic functions.

Research

Coordinate systems are used to locate the position of a point in space. Explore coordinate systems that can be used effectively in three-dimensions, for example, the spherical coordinate system commonly used by mathematicians. You should also research cylindrical coordinate systems. These are often used by engineers. Explore when, and why, these systems are used.

Different coordinate systems are used in other subject areas. For example, geographers use a geographical coordinate system, and space scientists and astronomers use a celestial coordinate system. Explore the advantages and disadvantages of these systems.

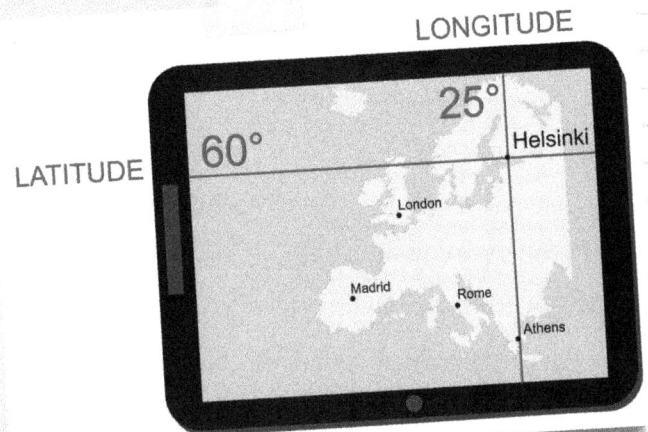

3 Assessment

1 a Sketch the graph of $y = \dfrac{x+5}{x-7}$, clearly labelling any intercepts with the coordinate axes. **[4 marks]**

   b Write down the equations of the asymptotes to the curve. **[2]**

   c Find the coordinates of the points where the curve $y = \dfrac{x+5}{x-7}$ meets the line $y = -x$ **[3]**

2 a Sketch the curve $y = \dfrac{5x^2}{x^2+1}$ **[3]**

   b Find the points of intersection between the curve $y = \dfrac{5x^2}{x^2+1}$ and the line $y = 1$ **[3]**

3 a Express the Cartesian coordinates $\left(\sqrt{3}, -3\right)$ as polar coordinates, giving the exact value of r and writing θ in terms of π where $-\pi < \theta \leq \pi$ **[4]**

   b Sketch the polar graph with equation

   i $r = 9$ **[2]**   ii $\theta = \dfrac{\pi}{6}, r \geq 0$ **[1]**

   c Find the Cartesian equations of the curves in part **b**. **[5]**

4 a Express the polar coordinates $\left(4, -\dfrac{5\pi}{6}\right)$ as exact Cartesian coordinates. **[3]**

   b Find the Cartesian equation of the polar curve

   $r = 2\sin\theta$ **[3]**

   c Sketch the graph of the curve in part **b**. **[3]**

5 a Sketch the graph of $y^2 = 3x$ and state the name given to this type of curve. **[3]**

   The line with equation $y = x + c$ is a tangent to $y^2 = 3x$

   b Calculate the value of c **[4]**

6 Given that $k > 3$, sketch the curve with these equations. Label the points of intersection with the coordinate axes and give the equations of any asymptotes.

   a $\dfrac{x^2}{9} + \dfrac{y^2}{k^2} = 1$ **[3]**   b $\dfrac{x^2}{9} - \dfrac{y^2}{k^2} = 1$ **[4]**

7 Sketch these polar curves.

   a $r = 4\theta$ for $0 \leq \theta \leq 3\pi$ **[2]**   b $r = \cos 3\theta$ **[2]**

   c $r = 1 + \sin\theta$ **[2]**   d $r = 8 + 3\cos\theta$ **[2]**

8 a Use the exponential definition of $\sinh x$ to show that $\sinh(\ln 2) = \dfrac{3}{4}$ **[3]**

   b Solve the equation $\sinh x = 3$, giving your solution to 1 decimal place. **[3]**

9 a Sketch the graph of $y = \tanh(x+1)$ **[2]**

   b Write down the equations of the asymptotes to the curve in part **a**. **[2]**

   c Use the exponential definitions of $\sinh x$ and $\cosh x$ to show that $\tanh x = \dfrac{e^{2x}-1}{e^{2x}+1}$ **[4]**

   d Hence solve the equation $\tanh x = \dfrac{1}{2}$. Give your answer as a logarithm. **[3]**

10 a Sketch the graph of $y = \dfrac{2+x}{2x-1}$ **[3]** b Solve the inequality $\dfrac{2+x}{2x-1} \geq \dfrac{x}{x+4}$ **[5]**

11 a Sketch the curve $y = \dfrac{x^2 - 5x + 4}{x^2 + 1}$ **[3]** **b** Solve the inequality $\dfrac{x^2 - 5x + 4}{x^2 + 1} > 1 - x$ **[1]**

12 A rational function is defined by $f(x) = \dfrac{3x^2 + 5x - 8}{x^2 + 2x - 6}$

A sketch of the curve $y = f(x)$ is shown.

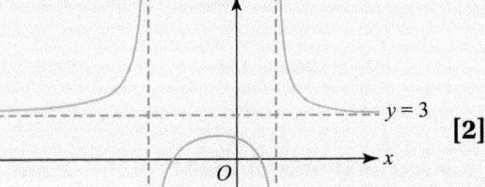

a For what values of x is the function undefined?
Leave your answer as a surd. **[2]**

b Where does the curve of the function cut the
coordinate axes? **[3]**

c For what values of x is $f(x) > 1$? **[5]**

13 a Show that the equation $xy - 5y + x - 9 = 0$ can be written as $(x - 5)(y + 1) = k$ where k is a
constant to be found. **[2]**

b Hence sketch the curve of $xy - 5y + x - 9 = 0$, giving the equations of the asymptotes. **[3]**

14 Sketch the curve with equation $\dfrac{(x+2)^2}{4} + (y - 2)^2 = 1$, labelling any points of intersection
with the coordinate axes. **[3]**

15 Show that the curve $y = \dfrac{1}{x}$ has polar equation $r^2 = 2 \operatorname{cosec} 2\theta$ **[4]**

16 a Give the Cartesian equation of the curve with polar equation $r = 3 \sin 2\theta$ for
$r \geq 0$ and $0 \leq \theta \leq 2\pi$ **[4]**

b Sketch the polar curve found in part **a**. **[2]**

c State the maximum value of $r = 3 \sin 2\theta$ and the values of θ at which it occurs. **[3]**

17 a State the maximum and minimum values of $r = 4 + 3 \cos \theta$ where $0 \leq \theta < 2\pi$ **[2]**

b Sketch the curve $r = 4 + 3 \cos \theta$ **[2]**

18 a Sketch the graphs of $y = \sinh x$ and $y = \sinh 2x$ on the same diagram. **[3]**

b If $y = \sinh x$, use the exponential definition of $\sinh x$ to show that $x = \ln\left(y + \sqrt{y^2 + 1}\right)$ **[5]**

19 The hyperbola with equation $\dfrac{x^2}{10} - \dfrac{y^2}{5} = 1$ is reflected in the line $y = x$

a Write down the equation of the image. **[1]**

b Sketch the image and state the equations of the asymptotes. **[3]**

20 Sketch each transformation of the parabola with equation $y^2 = 12x$ and write down its equation.

a A stretch of scale factor 2 in the y-direction **[2]**

b A translation by the vector $\begin{pmatrix} 3 \\ -8 \end{pmatrix}$ **[2]**

c A reflection in the line $y = -x$ **[2]**

21 a Sketch these curves on the same diagram, where $r \geq 0$

    i $r = \sin 4\theta$ **[2]**   **ii** $r = 2 \sin 3\theta$ **[2]**

b What difference does it make to the diagram if r can take any value? **[3]**

22 a Use the definitions of $\cosh x$ and $\sinh x$ to prove that $\sinh(2x) \equiv 2\sinh x \cosh x$ **[3]**

b Hence find the exact solutions to these equations.

    i $\sinh(2x) - 3 \sinh^2 x = 0$ **ii** $\sinh x \cosh x - \cosh^2(2x) + 4 = 0$ **[10]**

4 Integration 1

In economics many factors, such as currency rates, vary with time. Changes are often not sudden. Economists may model them using functions that, although complex, are developed using fundamental functions with which we are familiar, such as trigonometric functions. An average value of the factor might then be found over a given period by using integration to find the area under the function. This is then equated to the area of a rectangle with the period as base and mean value as height.

This method of finding a mean value using integral calculus has applications in other very different areas. For example, it is used extensively in electrical and electronic engineering in the design of essential circuits that are used everywhere, from delivering electricity to our homes, through to ensuring that all electrical appliances, transport, computers, and so on, work as intended.

Orientation

What you need to know	What you will learn	What this leads to
Maths Ch1 • Finding solutions of simultaneous equations.	• To calculate the mean value of a function. • To find the area enclosed by a curve and lines. • To calculate the volume of revolution when rotated around the x-axis.	**Ch19 Integration 2** • Surface area and volumes of revolution.
Maths Ch4 • Finding definite integrals.	• To calculate the volume of revolution when rotated around the y-axis. • To calculate more complicated volumes of revolution by adding or subtracting volumes.	**Careers** • Architecture.

Mean values

Fluency and skills

You already know that you can use integration to find the area, A, enclosed by the curve $y = f(x)$, the x-axis and the lines $x = a$ and $x = b$: $A = \int_a^b f(x)\,dx$

The Mean Value Theorem is based on the idea that for any continuous function there is always a rectangle with exactly the same area, A. The base of this rectangle will be between a and b. The height, h, of the rectangle is known as the mean value of the function $f(x)$

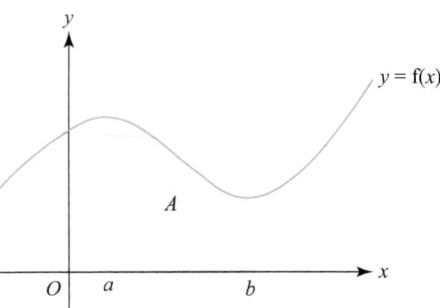

You can calculate the height of the rectangle by dividing its area, found by integration, by its width. So $h = \dfrac{1}{b-a}\int_a^b f(x)\,dx$

Key point

The mean value of a function $f(x)$ in the range $a \le x \le b$ is given by $\dfrac{1}{b-a}\int_a^b f(x)\,dx$

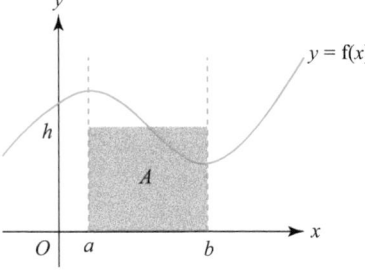

Example 1

Calculate the mean value of the function $f(x) = 2x^3 + 6x - 1$ in the interval $[2, 5]$

Mean value $= \dfrac{1}{5-2}\int_2^5 (2x^3 + 6x - 1)\,dx$

Use the definition of the mean value as $\dfrac{1}{b-a}\int_a^b f(x)\,dx$

$= \dfrac{1}{3}\left[\dfrac{x^4}{2} + 3x^2 - x\right]_2^5$

$= \dfrac{1}{3}\left(\dfrac{625}{2} + 75 - 5\right) - \dfrac{1}{3}(8 + 12 - 2)$

Substitute in limits.

$= \dfrac{243}{2}$

So the mean value of $2x^3 + 6x - 1$ in the interval $[2, 5]$ is $\dfrac{243}{2}$ (or 121.5)

Example 2

PURE

Calculate the mean value with respect to t of the function $t^2(3t+1)(t-2)$ for $1 \leq t \leq 3$

$$t^2(3t+1)(t-2) = t^2(3t^2 - 5t - 2)$$

$$= 3t^4 - 5t^3 - 2t^2$$

> Expand the brackets and use index rules to simplify.

$$\text{Mean value} = \frac{1}{3-1}\int_1^3 (3t^4 - 5t^3 - 2t^2)\,dt$$

> Use the definition of the mean value as $\frac{1}{b-a}\int_a^b f(t)\,dt$ since our function is in terms of the variable t

$$= \frac{1}{2}\left[\frac{3}{5}t^5 - \frac{5t^4}{4} - \frac{2}{3}t^3\right]_1^3$$

$$= \frac{1}{2}\left(\frac{729}{5} - \frac{405}{4} - 18\right) - \frac{1}{2}\left(\frac{3}{5} - \frac{5}{4} - \frac{2}{3}\right)$$

> Substitute in limits.

$$= \frac{209}{15}$$

So the mean value of $t^2(3t+1)(t-2)$ for $1 \leq t \leq 3$ is $\dfrac{209}{15}$ (or 13.9)

 **Try it on your calculator**

Definite integrals can be worked out on your calculator.

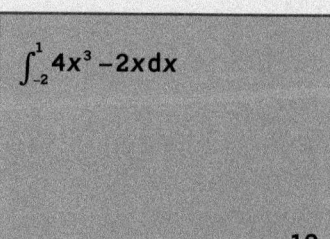
$$\int_{-2}^1 4x^3 - 2x\,dx$$

-12

Activity

Find out how to work out $\int_{-2}^1 4x^3 - 2x\,dx$ on *your* calculator.

Exercise 4.1A Fluency and skills

1 The graph of $y = x^3 - 3x^2 + 6$ is shown.

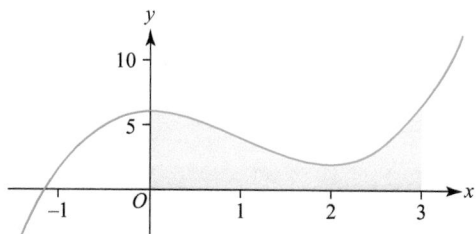

a Calculate the area bounded by the curve, the coordinate axes and the line $x = 3$

b Work out the mean value of $x^3 - 3x^2 + 6$ for $0 \leq x \leq 3$

2 $f(x) = x^2 - 3x + 2$

a Sketch the graph of $y = f(x)$, labelling where the curve crosses the coordinate axes.

b Calculate the area bounded by the curve $y = f(x)$ and the x-axis.

c Work out the mean value of $f(x)$ for each of these intervals.

   i $1 \leq x \leq 2$ **ii** $0 \leq x \leq 1$

3 $g(x) = x^3 + 5x^2$

a Sketch the graph of $y = g(x)$, labelling where the curve crosses the coordinate axes.

b Calculate the area bounded by the curve $y = g(x)$ and the x-axis.

c Work out the mean value of $g(x)$ for each of these intervals.

    i $-5 \leq x \leq 0$ **ii** $0 \leq x \leq 2$

4 Calculate the mean value of the function $f(x) = 5x^4$ in each of these intervals.

    a $[0, 4]$ **b** $[1, 3]$

    c $[-1, 2]$ **d** $\left[-\dfrac{1}{2}, \dfrac{1}{2}\right]$

5 Calculate the mean value of $2x^2 + 3$ for each of these ranges of values of x

    a $0 \leq x \leq 3$ **b** $2 \leq x \leq 6$

    c $-1 \leq x \leq 2$ **d** $-2 \leq x \leq -1$

6 Calculate the mean value of $3x - 8x^3$ for x between

    a   1 and 4   **b**   0 and 3

    c -1 and 1   **d** -3 and -2

7 Show that the mean value of $\dfrac{1}{x^2}$ for $2 \leq x \leq 5$ is $\dfrac{1}{10}$

8 Show that the mean value of $\sqrt{x}$ for $1 \leq x \leq 4$ is $\dfrac{14}{9}$

9 Show that the mean value of $\dfrac{1}{\sqrt{x}}$ for $4 \leq x \leq 9$ is 0.4

10 Given that $f(x) = \dfrac{2\sqrt{x} + 3x}{2x}$

    a   Write $f(x)$ in the form $Ax^c + B$, for constants A, B and c and state their values.

    b   Show that the mean value of $f(x)$ in the interval $[1, 9]$ is 2

11 Given that $g(x) = \dfrac{3x - x^2}{5\sqrt{x}}$

    a   Write $g(x)$ in the form $Ax^c + Bx^d$ for constants A, B, c and d and state their values.

    b   Calculate the exact mean value of $g(x)$ for $2 \leq x \leq 4$

12 Find the mean value of each of these functions of t for the range given.

    a $2t\left(5t^3 - 8t + 1\right)$ for $-1 \leq t \leq 0$

    b $3t\sqrt{t} - 2t^2$ for $\dfrac{1}{4} \leq t \leq 4$

    c $\dfrac{1}{2t\sqrt{t}}$ for $1 \leq t \leq 9$

    d $\dfrac{\left(t^3\right)^{\frac{1}{2}}}{t}$ for $4 \leq t \leq 64$

13 Find an expression for the mean value of $(1 - 2t)^2$ across the interval

    a $0 \leq t \leq T$ **b** $T \leq t \leq T + 2$

    Give your answers as polynomials in T

14 Calculate the mean value of the function $\dfrac{4 + x}{\sqrt{x}}$ for $2 \leq x \leq 8$

    Give your answer in the form $A\sqrt{2}$

15 Derive an expression in terms of a for the mean value of the function $3x^3 - \dfrac{1}{x^3}$ for x in the interval $[a, 2a]$. Give your answer in its simplest form.

16 Find an expression in terms of X for the mean value of the function $\dfrac{1}{2x^4}$, for x in the interval $\left[0, \dfrac{1}{X}\right]$. Simplify your answer.

Reasoning and problem-solving

Example 3

PURE

The velocity of a particle after t seconds is given by $v = \frac{1}{8}t\left(9 - \frac{4}{\sqrt{t}}\right) \mathrm{ms}^{-1}$

a Show that the mean velocity for $1 \le t \le 4$ is $2.03\,\mathrm{ms}^{-1}$ to 3 significant figures.

b Calculate the mean acceleration of the particle over the same time period.

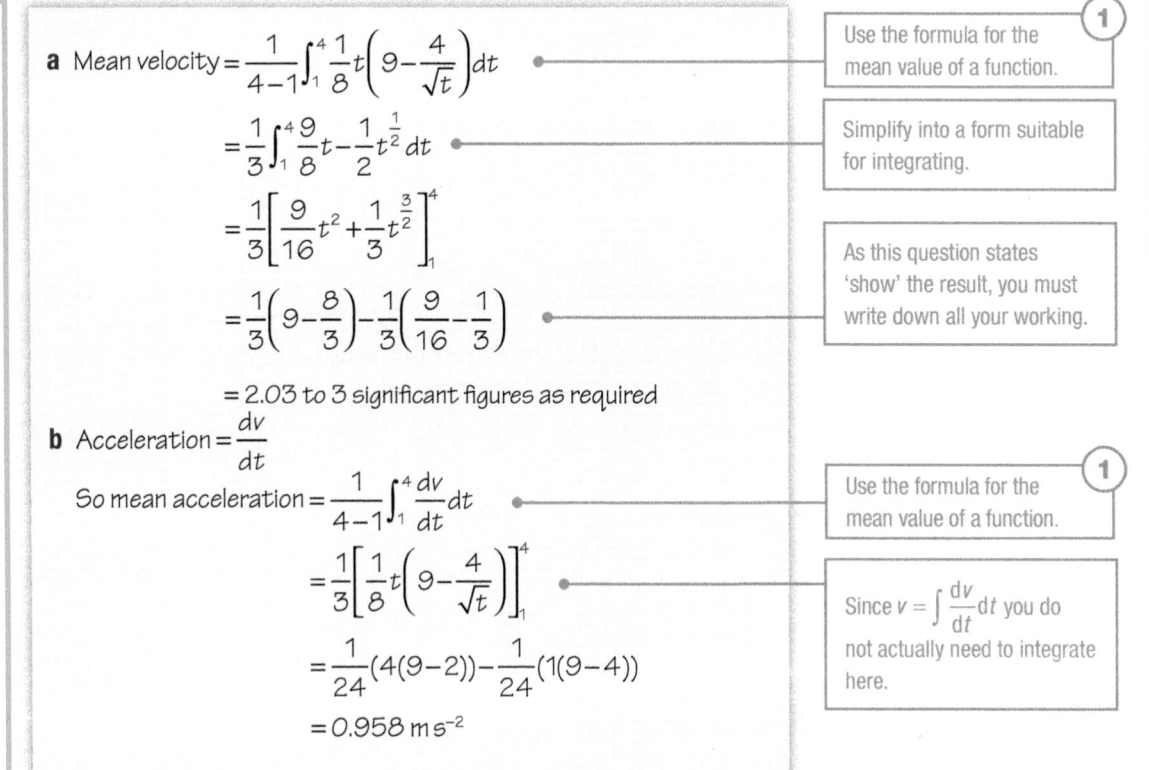

a Mean velocity $= \frac{1}{4-1}\int_1^4 \frac{1}{8}t\left(9 - \frac{4}{\sqrt{t}}\right)dt$

              $= \frac{1}{3}\int_1^4 \frac{9}{8}t - \frac{1}{2}t^{\frac{1}{2}}\,dt$

              $= \frac{1}{3}\left[\frac{9}{16}t^2 + \frac{1}{3}t^{\frac{3}{2}}\right]_1^4$

              $= \frac{1}{3}\left(9 - \frac{8}{3}\right) - \frac{1}{3}\left(\frac{9}{16} - \frac{1}{3}\right)$

              $= 2.03$ to 3 significant figures as required

b Acceleration $= \frac{dv}{dt}$

  So mean acceleration $= \frac{1}{4-1}\int_1^4 \frac{dv}{dt}\,dt$

                     $= \frac{1}{3}\left[\frac{1}{8}t\left(9 - \frac{4}{\sqrt{t}}\right)\right]_1^4$

                     $= \frac{1}{24}(4(9-2)) - \frac{1}{24}(1(9-4))$

                     $= 0.958\,\mathrm{ms}^{-2}$

> **(1)** Use the formula for the mean value of a function.

> Simplify into a form suitable for integrating.

> As this question states 'show' the result, you must write down all your working.

> **(1)** Use the formula for the mean value of a function.

> Since $v = \int \frac{dv}{dt}\,dt$ you do not actually need to integrate here.

> **See Maths Ch7**
> For a reminder on the link between differentiation and kinematics

---

Example 4

The mean value of the function $4x + 7$ in the interval $[a, b]$ is 2 and in the interval $[a, 2b]$ is 3 Calculate the values of a and b

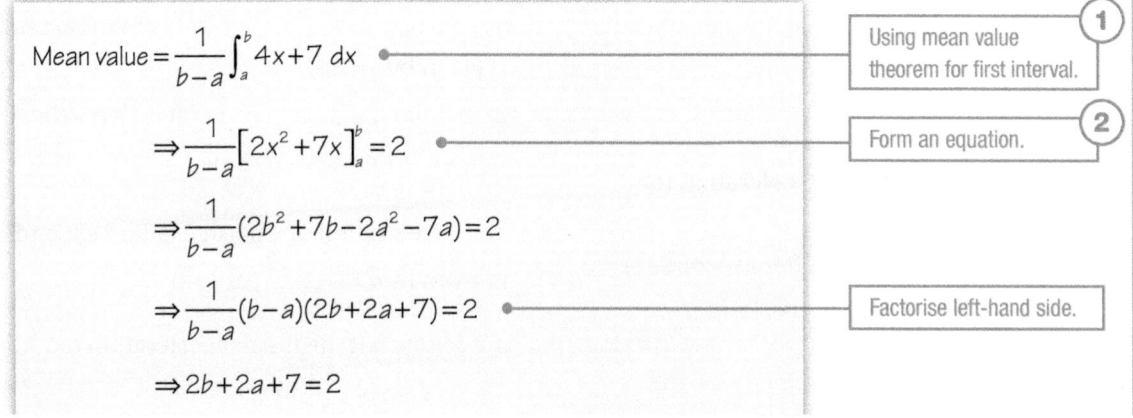

Mean value $= \frac{1}{b-a}\int_a^b 4x + 7\,dx$

    $\Rightarrow \frac{1}{b-a}\left[2x^2 + 7x\right]_a^b = 2$

    $\Rightarrow \frac{1}{b-a}(2b^2 + 7b - 2a^2 - 7a) = 2$

    $\Rightarrow \frac{1}{b-a}(b-a)(2b + 2a + 7) = 2$

    $\Rightarrow 2b + 2a + 7 = 2$

> **(1)** Using mean value theorem for first interval.

> **(2)** Form an equation.

> Factorise left-hand side.

(*Continued on the next page*)

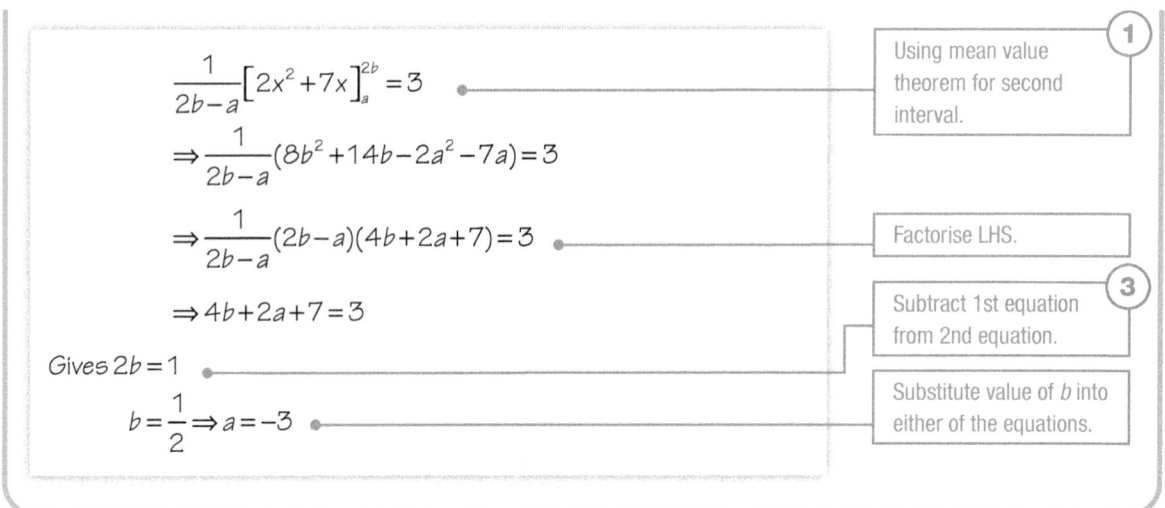

$$\frac{1}{2b-a}\left[2x^2+7x\right]_a^{2b}=3$$

Using mean value theorem for second interval. ①

$$\Rightarrow \frac{1}{2b-a}(8b^2+14b-2a^2-7a)=3$$

$$\Rightarrow \frac{1}{2b-a}(2b-a)(4b+2a+7)=3$$

Factorise LHS.

$$\Rightarrow 4b+2a+7=3$$

Subtract 1st equation from 2nd equation. ③

Gives $2b=1$

$$b=\frac{1}{2} \Rightarrow a=-3$$

Substitute value of b into either of the equations.

Example 5

Given that $f(x)=\dfrac{2x+1}{\sqrt{x}}$, show that the mean value of $f(x)$ for x in the interval $2 \le x \le 8$ is $\dfrac{31}{9}\sqrt{2}$

$$f(x)=\frac{2x+1}{x^{\frac{1}{2}}}$$

Write in simplified index form.

$$=2x^{\frac{1}{2}}+x^{-\frac{1}{2}}$$

Mean value $=\dfrac{1}{8-2}\displaystyle\int_2^8\left(2x^{\frac{1}{2}}+x^{-\frac{1}{2}}\right)dx$

Use the definition of the mean value.

$$=\frac{1}{6}\left[\frac{4}{3}x^{\frac{3}{2}}+2x^{\frac{1}{2}}\right]_2^8$$

$$=\frac{1}{6}\left(\frac{64}{3}\sqrt{2}+4\sqrt{2}\right)-\frac{1}{6}\left(\frac{8}{3}\sqrt{2}+2\sqrt{2}\right)$$

$$=\frac{31}{9}\sqrt{2}\text{ as required}$$

Exercise 4.1B Reasoning and problem-solving

1   The velocity of a particle after t seconds is given by $v=\dfrac{2}{5}t^3-\dfrac{1}{5}t$ ms^{-1}

 a   Calculate the mean velocity for $1 \le t \le 3$

 b   Show that the mean acceleration for $1 \le t \le 3$ is 5 m s^{-2}

2   The velocity of a particle after t seconds is given by $v=\dfrac{3t^2-5}{5}$ ms^{-1}

 a   Show that the mean velocity over the first 5 seconds is 4 ms^{-1}

 b   Calculate the mean acceleration over the first 5 seconds.

3   The displacement of a particle after t seconds is given by $s=2t\sqrt{t}$

 Show that the mean acceleration over the range $\dfrac{1}{2} \le t \le 1$ is $6-3\sqrt{2}$ ms^{-2}

4   The acceleration of a particle after t seconds is given by $a=\dfrac{t}{10}-\dfrac{10}{t^3}$ for $t>0$

 a   Show that the mean acceleration for $1 \le t \le 3$ is $-\dfrac{91}{45}$ m s^{-2}

 b   Given that after 1 second the particle is travelling at 5 m s^{-1}, calculate the mean velocity for $1 \le t \le 3$

5 Show that the mean value of a straight line $y = mx + c$ in the interval $[a, b]$ is given by $\dfrac{m(a+b)}{2} + c$

6 Show that the mean value of the curve $y = x^2$ for $0 \le x \le a$ is given by $\dfrac{a^2}{3}$

7 $g(x) = x^3 + 4x^2 - 5x$

  a Sketch the graph of $y = g(x)$, labelling where the curve crosses the coordinate axes.

  b Calculate the area bounded by the curve $y = g(x)$ and the x-axis.

  c Find the mean value of $g(x)$ for values of x in each of these intervals

  i $[-5, 0]$ **ii** $[0, 1]$

8 $g(x) = \sqrt{x}$

  The mean value of $g(x)$ in the interval $[1, 4]$ is double the mean value of $g(x)$ in the interval $[0, k]$ for $k > 0$

  Find the exact value of k

9 $f(x) = 2x^2 + a$

  The mean value of $f(x)$ in the interval $[-3, 3]$ is -2. Calculate the value of a

10 The mean value of the function $f(x) = x - 5$ for $0 \le x \le a$ is -1.5

  Calculate the value of a

11 The mean value of the function $g(x) = x^2 + 2x - 1$ for $0 \le x \le X$ is 17

  Calculate the possible values of X

12 Given $h(x) = 2x + 1$

  The mean value of $h(x)$ for $a \le x \le b$ is 4 and the mean value of $h(x)$ for $a \le x \le 2b$ is 8

  Calculate the values of a and b

13 Is the mean value of the function $x^4 - 2x^3 + 3x - 5$ greater in the range $[1, 3]$ or in the range $[-2, -1]$? Show your working and state how much bigger the greater mean value is.

14 Particle A has a speed at time t seconds given by $t^2 + t$ and particle B has a speed at time t seconds given by $\sqrt{t}$

  Which particle has the fastest average speed over the range $0 \le t \le \dfrac{1}{2}$? Show all your working and write down precisely how much quicker this particle's average speed is.

15 Show that the mean value of the function $\ln x$ in the interval $\dfrac{1}{2} \le x \le 3$ is $a \ln b + c$ where a, b and c are constants to be found.

16 $f(x) = \dfrac{x}{4x^2 - 3}$

  Find an expression for the mean value of $f(x)$ in the interval $[k, k+1]$. Give your answer as a single logarithm.

17 The graph of $y = A + \dfrac{1}{x - B}$ is shown.

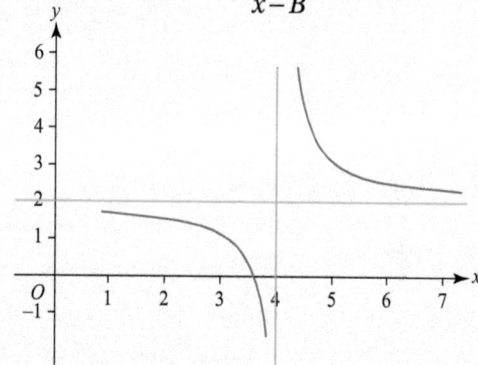

  a Calculate the mean value of the function for $5 \le x \le 7$

  b Is it possible to calculate the mean value of the function for $0 \le x \le 7$?

  Explain your answer.

18 Calculate the mean value of each of these functions for the interval given.

  a $\sin x$ for $\left[0, \dfrac{\pi}{3}\right]$ **b** $\cos 2x$ for $\left[\dfrac{\pi}{6}, \dfrac{\pi}{3}\right]$

  c e^{3x} for $[0, 1]$ **d** $2xe^x$ for $[0, 3]$

Fluency and skills

You can use calculus to find the volumes of 3D shapes. You can also generate the formulae for shapes like cones and spheres.

Look at this graph of $y = f(x)$. If the section of this curve between $x = a$ and $x = b$ is rotated 360° around the x-axis it will form a solid.

Since the curve has rotated through 360°, the cross-section of this solid will always be a circle with radius y.

Now consider thin strips of thickness δx. Each of these strips is approximately a cylinder so it will have a volume of $\delta V = \pi y^2 \delta x$

The volume of the whole solid is the sum of these thin strips. Using integration to sum these strips gives the formula

$$V = \int_a^b \pi y^2 \, dx$$

> **Key point**
>
> The volume of the solid formed by rotating the curve $y = f(x)$ between $x = a$ and $x = b$ a full turn around the x-axis is given by $V = \int_a^b \pi y^2 \, dx$

Because this is an integral with respect to x, you need the expression in terms of x only, so replace the y^2 with the function in terms of x

Example 1

Find the volume formed when the area enclosed by the curve with equation $y = x^3 + 1$, the coordinate axes and the line $x = 2$ is rotated

a 360° around the x-axis,   **b** 180° around the x-axis.

a $V = \pi \int_0^2 \left(x^3 + 1\right)^2 dx$ •   Using $V = \int_a^b \pi y^2 \, dx$ where $y = x^3 + 1$

$\quad = \pi \int_0^2 x^6 + 2x^3 + 1 \, dx$ •   Expand and simplify before attempting to integrate.

$\quad = \pi \left[\dfrac{x^7}{7} + \dfrac{x^4}{2} + x \right]_0^2$

$\quad = \pi \left(\dfrac{128}{7} + 8 + 2 - 0 \right)$

$\quad = \dfrac{198}{7} \pi$ cubic units

b Rotating through 180° produces exactly $\dfrac{1}{2}$ of the shape in part

a so the volume in this case is $\dfrac{1}{2}\left(\dfrac{198}{7} \pi \right) = \dfrac{99}{7} \pi$ cubic units.

You could also be asked to calculate the volume when a curve is rotated around the y-axis. In this case, the roles of x and y are swapped.

Therefore you need the function to be $x = f(y)$ and the limits to be $y = a$ and $y = b$. Then the volume is $V = \int_a^b \pi x^2 \, dy$

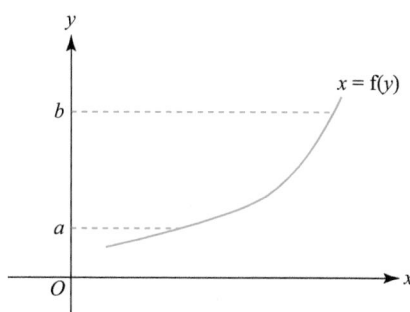

Key point

The volume of the solid formed by rotating the curve $x = f(y)$ between $y = a$ and $y = b$ a full turn around the y-axis is given by $V = \int_a^b \pi x^2 \, dy$

Example 2

The region A is enclosed by the curve with equation $y = x^5$, the y-axis and the lines $y = 0.5$ and $y = 1$

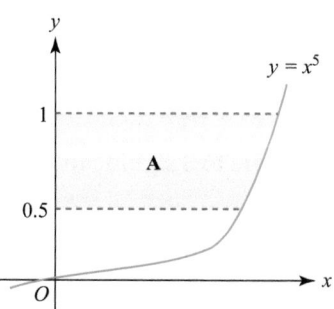

Calculate the volume when region A is rotated 2π radians around the y-axis.

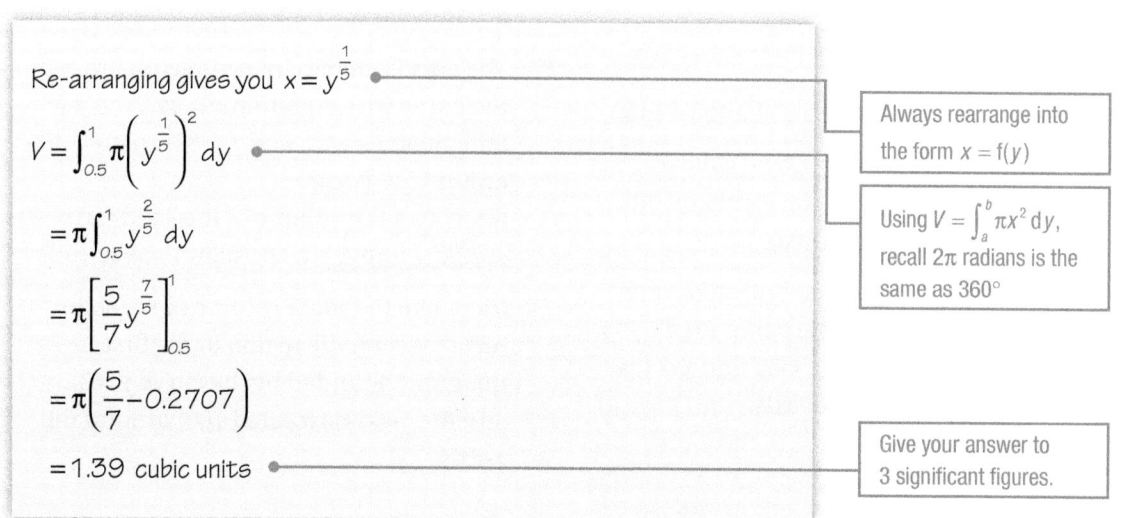

Re-arranging gives you $x = y^{\frac{1}{5}}$ — Always rearrange into the form $x = f(y)$

$V = \int_{0.5}^{1} \pi \left(y^{\frac{1}{5}} \right)^2 dy$

$= \pi \int_{0.5}^{1} y^{\frac{2}{5}} \, dy$ — Using $V = \int_a^b \pi x^2 \, dy$, recall 2π radians is the same as 360°

$= \pi \left[\frac{5}{7} y^{\frac{7}{5}} \right]_{0.5}^{1}$

$= \pi \left(\frac{5}{7} - 0.2707 \right)$

$= 1.39$ cubic units — Give your answer to 3 significant figures.

1 Calculate the volume formed when the region bounded by the curve $y = 3x^2$, the x-axis and the lines $x = 1$ and $x = 4$ is rotated $360°$ around the x-axis.

2 The region R is enclosed by the curve $y = 8 - x^3$ and the coordinate axes.

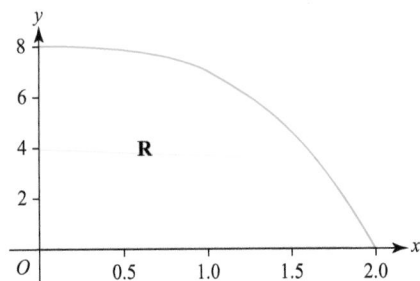

a Show that the area of R is 12 square units.

The region R is rotated $360°$ around the x-axis.

b Calculate the volume of the solid formed.

3 The shaded region is enclosed by the curve $y = x^4 - x^3$ and the x-axis.

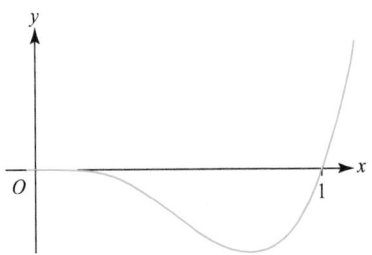

a Calculate the area of the shaded region.

b Calculate the volume of the solid formed when the shaded region is rotated 2π radians around the x-axis.

4 Find the volume of the solid formed when the region bounded by the curve $y = 4x^2 - 5$, the x-axis and the lines $x = 0.5$ and $x = 1$ is rotated $180°$ around the x-axis.

5 Calculate the volume of the solid formed when the region bounded by the curve $x = 2y^3 - y$ and the lines $y = 1$ and $y = 2$ is rotated 2π radians around the y-axis.

6 The region R is enclosed by the curve $y = \dfrac{1}{x^2}$, the x-axis and the lines $x = \dfrac{1}{2}$ and $x = 1$

Show that the volume of the solid formed when R is rotated 2π radians around the x-axis is $\dfrac{7}{3}\pi$

7 The region A is enclosed by the curve $y = \dfrac{2 + \sqrt{x}}{5x^2}$, the x-axis and the lines $x = 1$ and $x = 3$

a Show that the area of A is $\dfrac{10 - 2\sqrt{3}}{15}$

b Calculate the volume of the solid formed when the region A is rotated

i $360°$ around the x-axis,

ii $90°$ around the x-axis.

Give your answers to 3 significant figures.

8 Calculate the volume of the solid formed when the area bounded by the curve with equation $y = 4 - x^2$, the positive x-axis and the positive y-axis is rotated $360°$ around

a The x-axis,   b The y-axis.

9 The area bounded in the first quadrant by the curve with equation $y = 4x^4$, the y-axis and the lines $y = 2$ and $y = 8$ is rotated $360°$ around the y-axis.
Show that the volume of the solid formed is $A\sqrt{2}\pi$ where A is a constant to be found.

10 Calculate the exact volume of the solid formed when the region in the first quadrant bounded by the curve $y = 3x - x^3$ and the x-axis is rotated $180°$ around the x-axis.

Strategy

To solve problems involving volumes of revolution

(1) Sketch a graph. Note carefully which axis you are rotating the graph about.

(2) Choose the correct version of the formula for a volume of revolution, either

$V = \int_a^b \pi x^2 \, dy$ or $V = \int_a^b \pi y^2 \, dx$

(3) Add or subtract another volume where necessary.

Example 3

The region R is enclosed by the curve $y = \sqrt{x}$ and the lines $y = x - 6$, $x = 9$ and $y = 0$

Calculate the volume of the solid formed when R is rotated 360° around the x-axis.

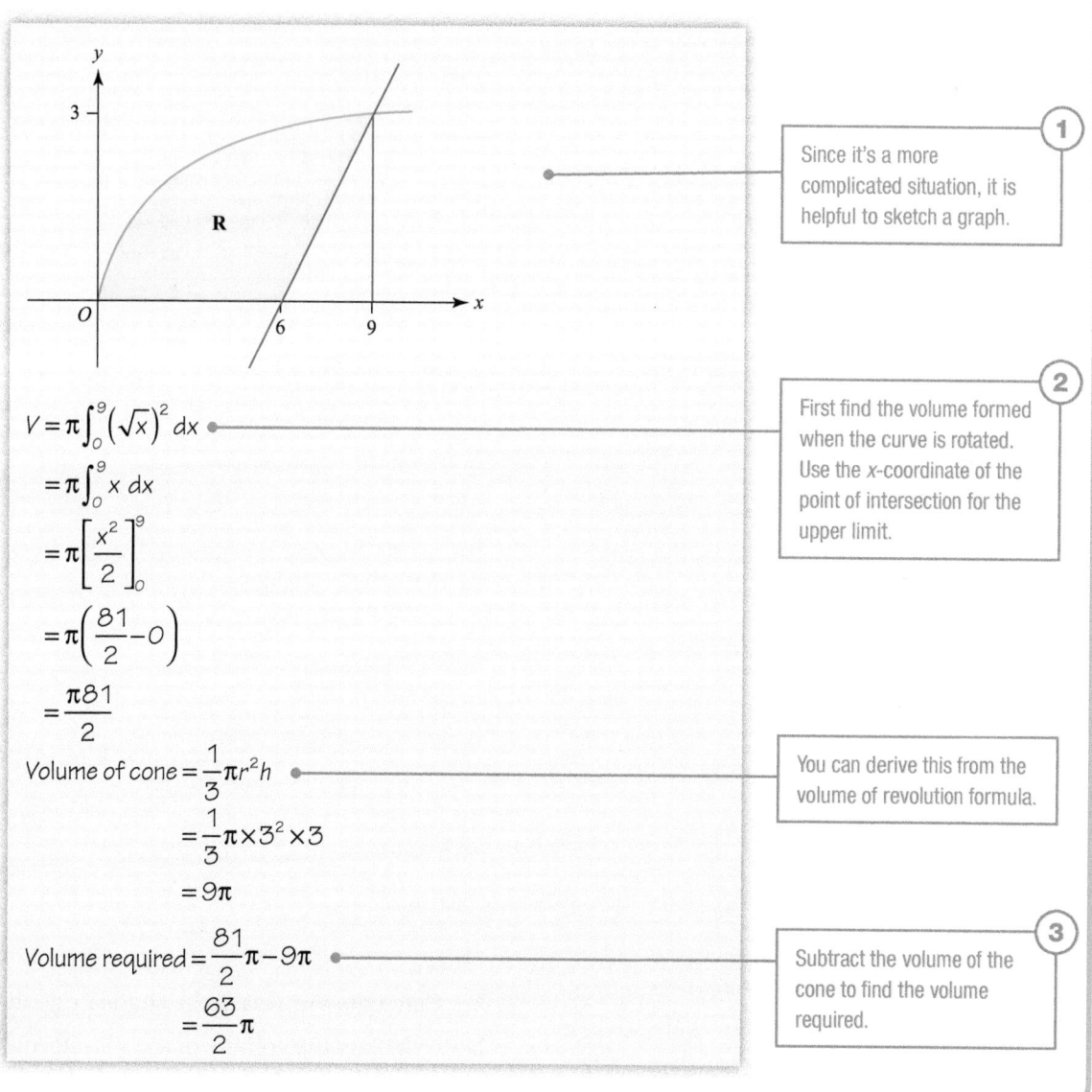

1 Since it's a more complicated situation, it is helpful to sketch a graph.

$V = \pi \int_0^9 \left(\sqrt{x}\right)^2 dx$

$= \pi \int_0^9 x \, dx$

$= \pi \left[\dfrac{x^2}{2}\right]_0^9$

$= \pi \left(\dfrac{81}{2} - 0\right)$

$= \dfrac{\pi 81}{2}$

2 First find the volume formed when the curve is rotated. Use the x-coordinate of the point of intersection for the upper limit.

Volume of cone $= \dfrac{1}{3}\pi r^2 h$

$= \dfrac{1}{3}\pi \times 3^2 \times 3$

$= 9\pi$

You can derive this from the volume of revolution formula.

Volume required $= \dfrac{81}{2}\pi - 9\pi$

$= \dfrac{63}{2}\pi$

3 Subtract the volume of the cone to find the volume required.

You can use volumes of revolution to prove some standard volume formulae.

Example 4

By rotating the curve $y = \sqrt{r^2 - x^2}$ between $-r$ and r around the x-axis, show that the volume of a sphere is $\frac{4}{3}\pi r^3$

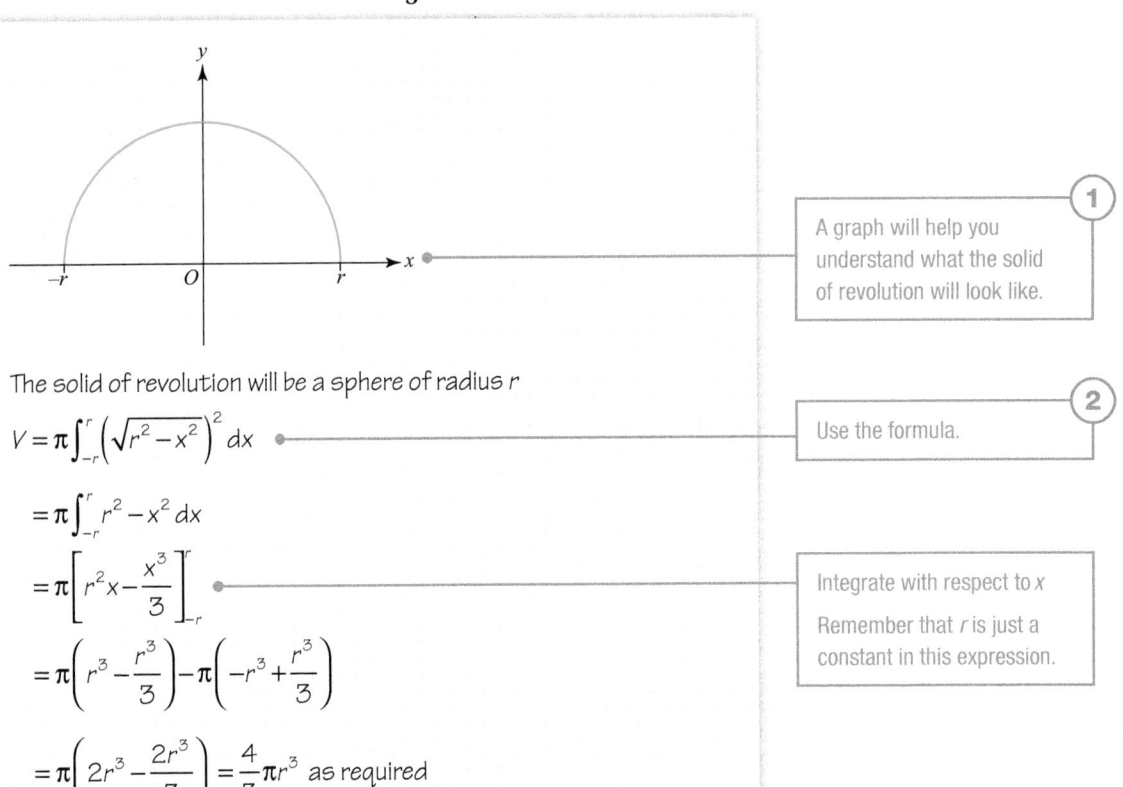

The solid of revolution will be a sphere of radius r

$$V = \pi \int_{-r}^{r} \left(\sqrt{r^2 - x^2} \right)^2 dx$$

$$= \pi \int_{-r}^{r} r^2 - x^2 \, dx$$

$$= \pi \left[r^2 x - \frac{x^3}{3} \right]_{-r}^{r}$$

$$= \pi \left(r^3 - \frac{r^3}{3} \right) - \pi \left(-r^3 + \frac{r^3}{3} \right)$$

$$= \pi \left(2r^3 - \frac{2r^3}{3} \right) = \frac{4}{3}\pi r^3 \text{ as required}$$

① A graph will help you understand what the solid of revolution will look like.

② Use the formula.

Integrate with respect to x
Remember that r is just a constant in this expression.

Exercise 4.2B Reasoning and problem-solving

1 The shaded region is bounded by the curve $y = 9x - 8x^2$, the x-axis and the line $y = 2x - 1$

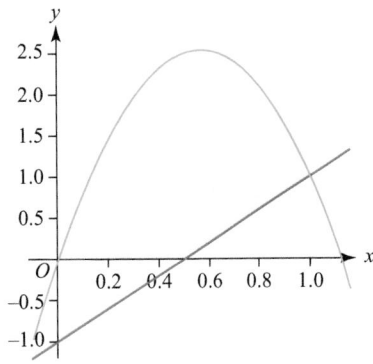

a Calculate the area of the shaded region.

b Calculate the volume of the solid formed when the shaded region is rotated $360°$ around the x-axis. Give your answer in terms of π

2 The shaded region is bounded by the curve $y = 10 - x^3$, the y-axis and the line $y = 2$

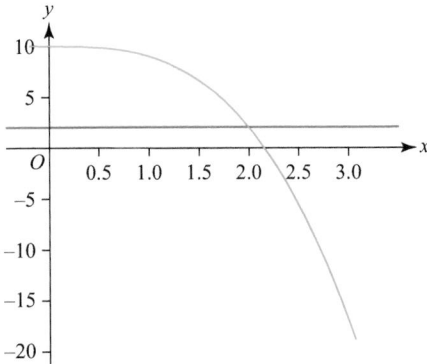

a Calculate the area of the shaded region.

b Calculate the volume of the solid formed when the shaded region is rotated $360°$ around the x-axis. Give your answer in terms of π

3 The region R is bounded in the first quadrant by the curve with equation $y = 16 - x^4$, the line $y = 7$ and the y-axis.

    a Show that the area of R is $\dfrac{36}{5}\sqrt{3}$ square units.

    b Calculate the volume of the solid formed when R is rotated 360° around the x-axis.

4 The region A, to the right of the y-axis, is bounded by the curve with equation $y = 8 + 2x - x^2$, the line $y = x - 4$ and the y-axis.

    a Show that the area of A is $\dfrac{104}{3}$ square units.

    b Calculate the volume of the solid formed when A is rotated 180° around the x-axis.

5 The region R is bounded by the curve $y = \sqrt{x}$, the line $y = x - 2$ and the x-axis.

    a Calculate the area of R.

    R is rotated 180° around the x-axis.

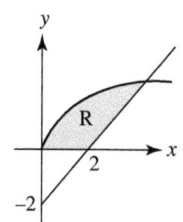

    b Calculate the volume of the solid formed.

6 By rotating the line $y = ax$ between $x = 0$ and $x = b$ 360° around the x-axis, show that the volume of a cone is $\dfrac{1}{3}\pi r^2 h$, where r is the radius of the base and h is the height.

7 Use a volume of revolution to show that the volume of a cylinder is $\pi r^2 h$, where r is the radius and h is the height.

8 Use a volume of revolution to derive a formula for the volume of a hemisphere of radius r

9 The section above the x-axis between $x = 1$ and $x = 3$ of the curve with equation $y = \dfrac{Ax}{\sqrt{x}}$ is rotated 2π radians around the x-axis and the volume of the solid formed is $\dfrac{8}{3}\pi$

    a Calculate the value of A

    b Calculate the volume when the same section of the curve is rotated 2π radians around the y-axis.

10 The parabola $y^2 = 28x$ intersects the line $x = k$ at the points A and B
The area bounded by the parabola and the line segment AB is $130\dfrac{2}{3}$ square units.

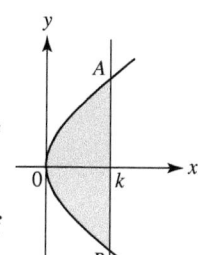

    a Find the coordinates of A and B

The shaded region is rotated 180° around the x-axis.

    b Calculate the volume of the solid formed.

11 Calculate the volume of the solid formed when each of these curves is rotated 360° around the x-axis between the limits shown. Give each of your answers in its simplest form.

    a $y = \dfrac{1}{2\sqrt{x}}$, $x = 2$ and $x = 5$

    b $y = \sec x$, $x = \dfrac{\pi}{6}$ and $x = \dfrac{\pi}{3}$

    c $y = (2x - 5)^3$, $x = 3$ and $x = 3.5$

    d $y = 3e^{2x}$, $x = -\dfrac{1}{4}$ and $x = 0$

12 Find the volume of the solid formed when the region in question **2** is rotated 180° around the y-axis. Give your answer in terms of π

Chapter summary

- The mean value of a function $f(x)$ in the range $a \le x \le b$ is given by $\dfrac{1}{b-a}\int_a^b f(x)dx$
- The area enclosed between the curve with equation $y = f(x)$, the x-axis and the lines $x = a$ and $x = b$ is given by $\int_a^b y\,dx$
- The volume of the solid formed by rotating the curve $y = f(x)$ between $x = a$ and $x = b$ a full turn around the x-axis is given by $V = \int_a^b \pi y^2\,dx$
- The volume of the solid formed by rotating the curve $x = f(y)$ between $y = a$ and $y = b$ a full turn around the y-axis is given by $V = \int_a^b \pi x^2\,dy$

Check and review

You should now be able to...	Try Questions
✔ Calculate the mean value of a function.	1–5
✔ Find the area enclosed by a curve and lines.	7, 10, 12
✔ Calculate the volume of revolution when rotated around the x-axis.	6–8, 11, 12
✔ Calculate the volume of revolution when rotated around the y-axis.	9–11
✔ Calculate more complicated volumes of revolution by adding or subtracting volumes.	11–13

1 Find the mean value of the function $f(x) = 8x^3 + x^2 - 4$ for $1 \le x \le 3$

2 Show that the mean value of the function $g(x) = \dfrac{1+x}{\sqrt{x}}$ in the interval $[2, 8]$ is $A\sqrt{2}$ where A is a constant to be found.

3 Work out the mean value of $4x^3 - \dfrac{16}{x^3}$ in the interval $\dfrac{1}{2} \le x \le 2$

4 Show that the mean value of the function $\dfrac{2-x}{2x^3}$ for x in the interval $[a, 1]$ is $\dfrac{1}{2a^2}$ and $a > 0$

5 The velocity of a particle after t seconds is given by $v = \dfrac{1}{20}(5t^4 - 3t^2)\,\mathrm{m\,s}^{-1}$

 a Calculate the mean velocity over the first 4 seconds.

 b Calculate the mean acceleration over the first 4 seconds.

6 Calculate the volume of the solid formed when the line with equation $y = 1 - 3x$ between $x = 2$ and $x = 5$ is rotated 2π radians around the x-axis.

7 The region R is enclosed by the curve with equation $y = 3 - \dfrac{1}{x^2}$, the x-axis and the line $x = 2$

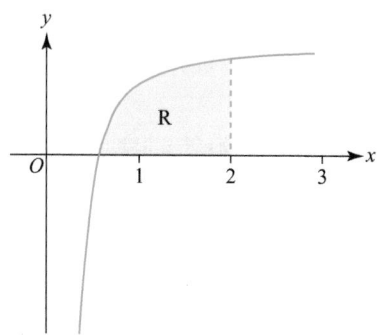

a Show that the area of R is $\dfrac{13 - 4\sqrt{3}}{2}$ square units.

The region R is rotated 360° around the x-axis.

b Calculate the volume of the solid formed.

8 Calculate the volume of the solid formed when the curve with equation $y = 5\sqrt{x}$ between $x = 1$ and $x = 3$ is rotated 180° around the x-axis.

9 Show that the volume of revolution when $x = 4y^2$ between $y = 0$ and $y = \dfrac{1}{2}$ is rotated 2π radians around the y-axis is $\dfrac{\pi}{10}$ cubic units.

10 The shaded region is bounded by the curve with equation $x = y^4 + 1$, $y \geq 0$, the coordinate axes and the line with equation $y = 2$

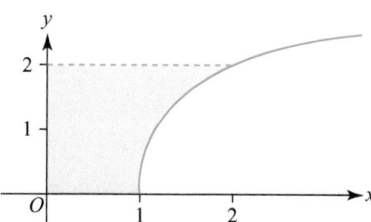

a Show that the area of the shaded region is $\dfrac{42}{5}$ square units.

The region is rotated 360° around the y-axis.

b Calculate the volume of the solid formed.

11 The region A, bounded by the curve with equation $y = 2 - \sqrt{x}$, the line $y = 2 - \dfrac{x}{2}$ and the y-axis, is rotated 360° around the x-axis.

a Sketch a graph showing the region A.

b Calculate the volume of revolution when region A is rotated 2π radians around the x-axis.

c Calculate the volume of revolution when region A is rotated 2π radians around the y-axis.

12 The shaded region is bounded by the curve with equation $y = 5x - 4x^2$, the line $y = 1$ and the x-axis.

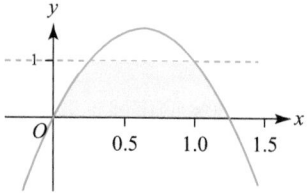

a Calculate the area of the shaded region.

b Calculate the volume of the solid formed when the shaded region is rotated 180° around the x-axis.

13 The region A is bounded by the curve with equation $y = 2\sqrt{x}$ and the line $y = x$

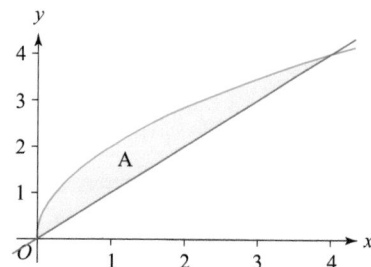

Calculate the volume of revolution when the shaded region is rotated 2π radians around

a The x-axis, **b** The y-axis.

Investigation

A different method of finding the volume of a solid by considering the revolution of a function, relies on considering the solid to be composed of thin cylindrical shells.

Each cylindrical shell is formed by rotating a vertical strip about the y-axis. The inner surface area of this cylinder is a rectangle with width $2\pi x$ and height y, so it has area $2\pi xy$

This means that the volume, δv, of a cylinder of width δx is given by $\delta v = 2\pi xy\delta x$

Summing these cylinders between $x = x_1$ and $x = x_2$, and letting δx

tend to zero, results in the formula $v = 2\pi \int_{x_2}^{x_2} xy\,dx$

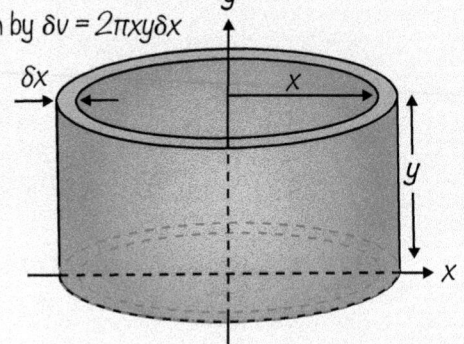

Convince yourself that this formula is correct by using it to find
a The volume of a cylinder of radius a and height h
b The volume of a cone of radius a and height h

Information

Gabriel's horn is a geometric solid that has a finite volume but an infinite surface area. This is sometimes known as the painter's paradox as although the horn would contain a finite volume of paint you would never have enough paint to paint its outside.

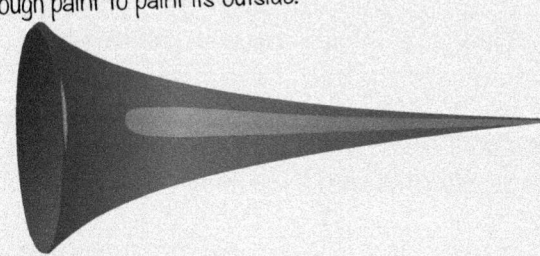

The solid is formed by rotating
the function $f(x) = \frac{1}{x}$ about the

x-axis for $x > 1$

Have a go

When carrying out an integral with no limit on the upper bound, it is usual practice to take an arbitrary value, say $x = a$, and then consider what happens as a tends to infinity.

Try the method described above to calculate the finite volume of Gabriel's horn.

Challenge

The formula for a **surface area of revolution** that is found by rotating $y = f(x)$ from $x = a$ to $x = b$ about the x-axis is given by

$$A = 2\pi \int_a^b y\sqrt{1+\left(\frac{dy}{dx}\right)^2}\,dx$$

Use this formula to show that Gabriel's horn has an infinite surface area.

1   Calculate the mean value of these functions for the limits shown.

    a x^2+3x+1 for $0 \le x \le 2$ **b** $x^3\left(2-x^3\right)$ for $-1 \le x \le 1$ **[8 marks]**

2   You are given that $f(x)=\dfrac{x^3-3\sqrt{x}}{2x^2}$

    a   Write $f(x)$ in the form $Ax^m + Bx^n$ where A, B, m and n are constants to be found.   **[2]**

    b   Calculate the mean value of $f(x)$ in the interval $[1, 4]$.   **[4]**

3   Find the mean value of the function $g(x)=x\sqrt{x}+\dfrac{3}{x^2}$ for $1 \le x \le 3$

    Give your answer in the form $a+b\sqrt{3}$ where a and b are constants to be found.   **[4]**

4   The velocity of a particle after t seconds is given by $v=5+3t^2\,\text{m s}^{-1}$

    a   Calculate the mean velocity for $1 \le t \le 5$ **[4]**

    b   Calculate the mean acceleration for $1 \le t \le 5$ **[4]**

5   The mean value of the function $1+2x^3$ in the interval $[0, a]$ is 109

    Evaluate a **[5]**

6   The shaded region is bounded by the curve $y = x^2 - 2x$, the x-axis and the line $x = 4$

    a   Find the value of a **[2]**

    b   Calculate the area of the shaded region.   **[3]**

    The shaded region is rotated one full turn around the x-axis.

    c   Calculate the volume of the solid formed.   **[4]**

7   Calculate the volume of revolution when the region bounded by the curve $y = 2\sqrt{x}$, the x-axis and the line $x = 3$ is rotated $360°$ about the x-axis. Give your answer in terms of π **[4]**

8   The region R is bounded by the curve $y = \dfrac{2}{x} + x^3$, the x-axis and the lines $x = 1$ and $x = 2$

    Calculate the volume of revolution when R is rotated $360°$ around the x-axis.   **[4]**

9   Calculate the volume of revolution when the region bounded by the curve $y = x^2(5-x)$ and the x-axis is rotated $180°$ around the x-axis. Give your answer to 3 significant figures.   **[5]**

10   The acceleration of a particle is given by $a=\dfrac{3}{2}t^2 - \dfrac{2}{t^3}$

    Given that after 1 second the particle is travelling at $3\,\text{m s}^{-1}$, calculate the mean velocity between 1 and 4 seconds after starting.   **[5]**

11   $f(x)=ax^2 + bx^5$

    The mean value of the function $f(x)$ in the interval $[1, 2]$ is 35 and the mean value of $f(x)$ in the interval $[0, 3]$ is 99

    Calculate the values of a and b **[5]**

12   The curve with equation $y = x - k\sqrt{x}$ between $x = 0$ and $x = 1$ is rotated 2π radians around the x-axis and the volume of revolution is $\dfrac{11}{15}\pi$. Calculate the possible values of k **[5]**

13 The mean value of $f(x) = Ax(2x^2 - 1)^3$ in the interval $\left[\frac{1}{2}, 1\right]$ is 15

Calculate the value of A [4]

14 The shaded region is part of an ellipse with

equation $\dfrac{x^2}{9} + \dfrac{y^2}{4} = 1$

   a Write the values of a and b in the diagram. [2]

   b Calculate the volume when the shaded area is
rotated π radians about the x-axis. [5]

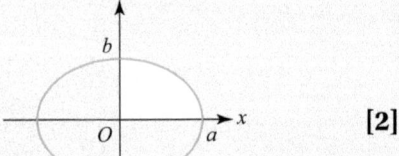

Full A Level

15 a Sketch the graphs of $y = 5 - |x|$ and $y = |x^2 - 25|$ on the same axes. [4]

   b The enclosed region bounded by $y = 5 - |x|$ and $y = |x^2 - 25|$ is rotated $360°$ around
the x-axis. Calculate the volume of the solid formed. [5]

16 $g(x) = e^{2x} + x^3 + 2$

Show that the mean value of the function $g(x)$ in the interval $[0, 2]$ is $\frac{1}{4}e^4 + k$, where k is a
constant to be found. [4]

17 Calculate the mean value of the function $f(x) = \dfrac{1}{2x - 3}$ for x in the range $a \le x \le 3$

Give your answer in its simplest form in terms of a [4]

18 Calculate the mean value of each of these functions for the limits shown

   a $\sin 2x$ for $0 \le x \le \dfrac{\pi}{6}$ b $\sin^2 x$ for $\dfrac{\pi}{3} \le x \le \dfrac{\pi}{2}$ [9]

19 Calculate the mean value of the function $f(t) = \dfrac{t}{2}\ln t$ for t in the region $2 \le t \le 4$

Give your answer in the form $A + B\ln 2$ where A and B are constants to be found. [6]

20 The region R is bounded by the curve with equation $= xe^{\frac{x}{2}}$, the x-axis and the line $x = 2$

   a Calculate the area of R [4]

   b Calculate the volume of revolution when R is rotated $360°$ around the x-axis. [6]

21 The curve C is defined by the parametric equations

   $y = 3t^2 + 1, x = 5t$

The region bounded by C, the x-axis and the lines $x = 5$ and $x = 10$ is rotated π radians
around the x-axis. Calculate the volume of revolution of the solid formed. [6]

22 a Sketch the graph of $y = \cosh 2x$ [2]

The region R is bounded by the curve $y = \cosh 2x$ and the lines $x = \pm 1$

   b Calculate the volume of the solid formed when R is rotated $90°$ about the x-axis. [6]

23 The region R is bounded by the curve with equation $y = \tan x$, the x-axis and the line $x = \dfrac{\pi}{4}$

R is rotated $k\pi$ radians around the x-axis and the volume of the solid formed is $\dfrac{\pi}{3} - \dfrac{\pi^2}{12}$

Calculate the value of k [6]

24 The curve C is defined by the parametric equations

   $x = 2\sin\theta, y = \sin 2\theta, 0 \le \theta \le \dfrac{\pi}{2}$

Calculate, in terms of π, the volume of the solid formed when the region bounded by C and
the y-axis is rotated $360°$ around the y-axis. [8]

5 Matrices 1

The mathematics of matrices is used in a diverse range of applications. For example, civil and structural engine engineers find them useful because they can be used to find solutions to **systems of linear simultaneous** equations with multiple unknowns. When analysing the forces in, for example, the design of a new bridge, the mathematics can involve systems of equations where many forces need to be found. The use of matrices, together with computer software, provides a powerful method that ensures that the equations can be solved and the forces involved calculated. This allows more complex structures to be built than was previously the case.

Matrix algebra methods can be used in other branches of engineering. For example, to find the current flowing in various parts of electrical circuits, to determine traffic flow in models of road networks, the production of different chemicals in an industrial process, and so on.

Orientation

What you need to know

KS4
- Transforming objects by reflection, rotation and enlargement.

Maths Ch1
- Solving simultaneous equations.

Maths Ch6
- Using vectors.

What you will learn

- To identify the order of a matrix.
- To add, subtract and multiply matrices by a scalar or a conformable matrix.
- To apply linear transformations given as matrices, describe transformations given as matrices and write linear transformations as a matrix.
- To find invariant points and lines of linear transformations.
- To calculate determinants and inverses of matrices.
- To use matrices to solve systems of linear equations.

What this leads to

Ch22 Further Matrices
- Eigenvalues and eigenvectors.
- 3×3 systems.
- Matrix diagonalisation.

Ch12 Graphs and networks
- Matrix formulation of Prim's algorithm.

Ch14 Linear programming and game theory
- Zero sum games.

5.1   Properties and arithmetic

Fluency and skills

See Maths Ch6

For a reminder of vectors.

Matrices are a way of representing information in a form that can be manipulated mathematically. You will have already used vectors, which are a particular sort of matrix.

> **Key point**
>
> A **matrix** with n rows and m columns has **order** $n \times m$

For example, the matrix $\begin{pmatrix} 5 & 3 & -1 \\ -4 & 0 & 5 \end{pmatrix}$ is said to have order 2×3

The matrix $\begin{pmatrix} 2 \\ 7 \\ -5 \end{pmatrix}$ is said to have order 3×1

Computers use matrices to carry out an operation, such as addition or subtraction, and to carry it out on multiple numbers simultaneously. This is particularly used in the processing of computer graphics.

> **Key point**
>
> If two matrices have the same order then they can be added or subtracted by adding or subtracting their corresponding **elements**.

You can think of putting the two matrices on top of each other and adding or subtracting the elements that match.

> **Key point**
>
> To multiply a matrix by a constant, you should multiply each of its elements by that constant.

Example 1

You are given that $\mathbf{A} = \begin{pmatrix} 2 & 5 \\ 0 & 4 \\ -1 & -3 \end{pmatrix}$, $\mathbf{B} = \begin{pmatrix} 2 & -3 \\ 0 & 5 \end{pmatrix}$ and $\mathbf{C} = \begin{pmatrix} 6 & 15 \\ 0 & 12 \\ -3 & -9 \end{pmatrix}$. Find, if possible,

   a $\mathbf{A} + \mathbf{B}$ **b** $\mathbf{A} + \mathbf{C}$ **c** $2\mathbf{B}$ **d** $\mathbf{C} - 3\mathbf{A}$

   a   Not possible as **A** and **B** are not of the same order.

   b $\mathbf{A} + \mathbf{C} = \begin{pmatrix} 2 & 5 \\ 0 & 4 \\ -1 & -3 \end{pmatrix} + \begin{pmatrix} 6 & 15 \\ 0 & 12 \\ -3 & -9 \end{pmatrix}$

       $= \begin{pmatrix} 2+6 & 5+15 \\ 0+0 & 4+12 \\ -1+-3 & -3+-9 \end{pmatrix}$ Add corresponding elements of **A** and **C**

       $= \begin{pmatrix} 8 & 20 \\ 0 & 16 \\ -4 & -12 \end{pmatrix}$

(Continued on the next page)

c $\quad 2B = 2\begin{pmatrix} 2 & -3 \\ 0 & 5 \end{pmatrix}$

$\quad = \begin{pmatrix} 2 \times 2 & 2 \times -3 \\ 2 \times 0 & 2 \times 5 \end{pmatrix}$ — Multiply every element of **B** by 2

$\quad = \begin{pmatrix} 4 & -6 \\ 0 & 10 \end{pmatrix}$

d $\quad C - 3A = \begin{pmatrix} 6 & 15 \\ 0 & 12 \\ -3 & -9 \end{pmatrix} - 3\begin{pmatrix} 2 & 5 \\ 0 & 4 \\ -1 & -3 \end{pmatrix}$

$\quad = \begin{pmatrix} 6 - 3 \times 2 & 15 - 3 \times 5 \\ 0 - 3 \times 0 & 12 - 3 \times 4 \\ -3 - 3 \times -1 & -9 - 3 \times -3 \end{pmatrix}$ — Multiply every element of **A** by 3 then subtract from the corresponding element in **C**

$\quad = \begin{pmatrix} 0 & 0 \\ 0 & 0 \\ 0 & 0 \end{pmatrix}$ — This is called a zero matrix as all its elements are 0

The **zero matrix**, **0**, is a matrix, of any order, with all elements equal to zero. **Key point**

Matrices can only be multiplied together if the number of columns in the first matrix is the same as the number of rows in the second matrix.

If two matrices can be multiplied, then it is said that they are **conformable for multiplication**. **Key point**

When multiplying matrices, to find the element in the nth row and mth column, you must multiply the first term of the nth row in the first matrix by the first term of the mth column in the second matrix, then the second term of the nth row by the second term of the mth column and so on, then add these terms together.

The product of an $n \times m$ matrix and an $m \times p$ matrix has **order** $n \times p$ **Key point**

To find the first term in a matrix multiplication, look at the first row of the first matrix and the first column of the second matrix.

$$\begin{pmatrix} a & d \\ b & e \\ c & f \end{pmatrix}\begin{pmatrix} w & y \\ x & z \end{pmatrix} = \begin{pmatrix} aw+dx & \ldots \\ \ldots & \ldots \\ \ldots & \ldots \end{pmatrix}$$

Then consider the first row and second column.

$$\begin{pmatrix} a & d \\ b & e \\ c & f \end{pmatrix}\begin{pmatrix} w & y \\ x & z \end{pmatrix} = \begin{pmatrix} aw+dx & ay+dz \\ \ldots & \ldots \\ \ldots & \ldots \end{pmatrix}$$

Then move on to using the second row of the first matrix. Continue in this way until the last term which is found by considering the final row of the first matrix and the final column of the second.

$$\begin{pmatrix} a & d \\ b & e \\ c & f \end{pmatrix}\begin{pmatrix} w & y \\ x & z \end{pmatrix} = \begin{pmatrix} aw+dx & ay+dz \\ bw+bx & by+ez \\ cw+fx & cy+fz \end{pmatrix}$$

Example 2

If $\mathbf{A} = \begin{pmatrix} 3 & 0 \\ -1 & 2 \\ 7 & -4 \end{pmatrix}$, $\mathbf{B} = \begin{pmatrix} 5 & 1 \\ -3 & 0 \end{pmatrix}$ find, if possible, the products

a **AB** **b** **BA** **c** **$\mathbf{B}^2$**

a $\mathbf{AB} = \begin{pmatrix} 3 & 0 \\ -1 & 2 \\ 7 & -4 \end{pmatrix} \begin{pmatrix} 5 & 1 \\ -3 & 0 \end{pmatrix}$

$= \begin{pmatrix} (3 \times 5)+(0 \times -3) & (3 \times 1)+(0 \times 0) \\ (-1 \times 5)+(2 \times -3) & (-1 \times 1)+(2 \times 0) \\ (7 \times 5)+(-4 \times -3) & (7 \times 1)+(-4 \times 0) \end{pmatrix}$

To find the term in the 3rd row and the 2nd column: multiply the first term from the 3rd row of matrix **A** by the first term from the 2nd column of matrix **B**. Then multiply the second term from the 3rd row of matrix **A** by the second term in the 2nd column of matrix **B**. Finally, add these values together.

$= \begin{pmatrix} 15 & 3 \\ -11 & -1 \\ 47 & 7 \end{pmatrix}$

Notice how the product of a 3×2 matrix and a 2×2 matrix is a 3×2 matrix.

b Not possible as **B** has 2 columns but **A** has 3 rows.

c $\mathbf{B}^2 = \begin{pmatrix} 5 & 1 \\ -3 & 0 \end{pmatrix} \begin{pmatrix} 5 & 1 \\ -3 & 0 \end{pmatrix}$

$\mathbf{B}^2$ means $\mathbf{B} \times \mathbf{B}$
It does not mean square each element of the matrix.

$= \begin{pmatrix} (5 \times 5)+(1 \times -3) & (5 \times 1)+(1 \times 0) \\ (-3 \times 5)+(0 \times -3) & (-3 \times 1)+(0 \times 0) \end{pmatrix} = \begin{pmatrix} 22 & 5 \\ -15 & -3 \end{pmatrix}$

Key point

It is only possible to find $\mathbf{A}^2$ if **A** is a **square matrix**, that is, it has the same number of rows as columns.

Try it on your calculator

Some calculators can be used to add, subtract and multiply matrices.

MatA × MatB

$\begin{bmatrix} 15 & 3 \\ -11 & -1 \\ 47 & 7 \end{bmatrix}$

Activity
Find out how to work out $\begin{pmatrix} 3 & 0 \\ -1 & 2 \\ 7 & -4 \end{pmatrix} \begin{pmatrix} 5 & 1 \\ -3 & 0 \end{pmatrix}$ on *your* calculator.

Key point

The **transpose** of a matrix is formed by swapping the rows and columns.

For example, the transpose of the matrix $\begin{pmatrix} 1 & -2 & 4 \\ 0 & 5 & -6 \end{pmatrix}$ is $\begin{pmatrix} 1 & 0 \\ -2 & 5 \\ 4 & -6 \end{pmatrix}$: the first row of the matrix became the first column of the transpose and the second row of the matrix became the second row of the transpose.

You can use this result about transposes:

Key point

Give two $n \times n$ matrices **A** and **B**: $(\mathbf{AB})^{\mathrm{T}} \equiv \mathbf{B}^{\mathrm{T}}\mathbf{A}^{\mathrm{T}}$

Example 3

The matrix $\mathbf{A} = \begin{pmatrix} 1 & 0 & -3 \\ k & 2 & 4 \\ 5 & -6 & 0 \end{pmatrix}$ and $\mathbf{B} = \begin{pmatrix} 2 & k & 0 \\ -3 & 1 & -5 \\ 0 & 1 & -1 \end{pmatrix}$

Without finding $\mathbf{AB}$, find $(\mathbf{AB})^{\mathrm{T}}$ in terms of k

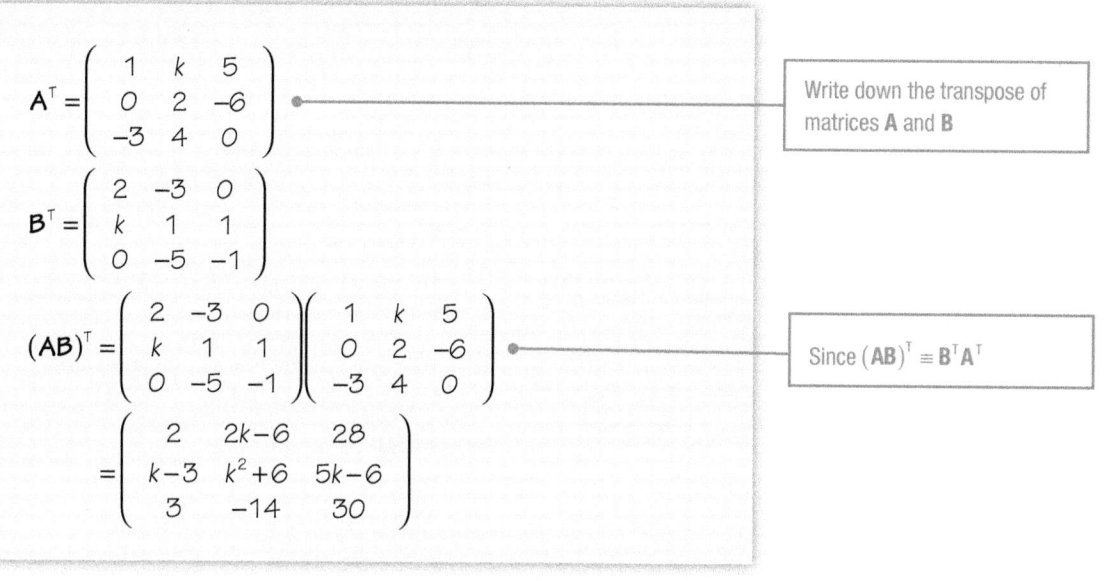

$\mathbf{A}^{\mathrm{T}} = \begin{pmatrix} 1 & k & 5 \\ 0 & 2 & -6 \\ -3 & 4 & 0 \end{pmatrix}$ Write down the transpose of matrices $\mathbf{A}$ and $\mathbf{B}$

$\mathbf{B}^{\mathrm{T}} = \begin{pmatrix} 2 & -3 & 0 \\ k & 1 & 1 \\ 0 & -5 & -1 \end{pmatrix}$

$(\mathbf{AB})^{\mathrm{T}} = \begin{pmatrix} 2 & -3 & 0 \\ k & 1 & 1 \\ 0 & -5 & -1 \end{pmatrix}\begin{pmatrix} 1 & k & 5 \\ 0 & 2 & -6 \\ -3 & 4 & 0 \end{pmatrix}$ Since $(\mathbf{AB})^{\mathrm{T}} \equiv \mathbf{B}^{\mathrm{T}}\mathbf{A}^{\mathrm{T}}$

$= \begin{pmatrix} 2 & 2k-6 & 28 \\ k-3 & k^2+6 & 5k-6 \\ 3 & -14 & 30 \end{pmatrix}$

Exercise 5.1A Fluency and skills

1 a Write down the order of these matrices.

     i $\begin{pmatrix} 5 & -3 \\ 1 & 0 \\ 9 & 11 \end{pmatrix}$ **ii** $\begin{pmatrix} 9 \\ 3 \\ -2 \end{pmatrix}$

     iii $\begin{pmatrix} 3 & -7 \\ -2 & 0 \end{pmatrix}$ **iv** $\begin{pmatrix} -3 & 0 & 5 & 8 \\ 2 & -1 & 4 & 0 \end{pmatrix}$

   b Which of the matrices in part **a** is a square matrix?

2 Calculate

   a $\begin{pmatrix} 5 & -1 & 2 \\ 0 & 6 & 4 \end{pmatrix} + \begin{pmatrix} -1 & 4 & 0 \\ 5 & -7 & 2 \end{pmatrix}$

   b $\begin{pmatrix} 4 & -2 \\ 8 & -5 \end{pmatrix} - \begin{pmatrix} 8 & 4 \\ 7 & -2 \end{pmatrix}$

   c $4\begin{pmatrix} -1 & 0 & 4 \\ 2 & 5 & -3 \\ 8 & -6 & 0 \end{pmatrix}$

   d $\begin{pmatrix} -6 & 14 \\ 3 & 9 \\ 2 & -4 \end{pmatrix} + \dfrac{1}{2}\begin{pmatrix} 8 & -4 \\ 6 & 3 \\ -2 & 0 \end{pmatrix}$

3 If $\mathbf{A} = \begin{pmatrix} 9 & -4 \\ 0 & -2 \end{pmatrix}$, $\mathbf{B} = \begin{pmatrix} 8 & 4 \\ 2 & -6 \\ -2 & 0 \end{pmatrix}$,

     $\mathbf{C} = \begin{pmatrix} -5 & -2 \\ 7 & 0 \\ 3 & 1 \end{pmatrix}$, $\mathbf{D} = \begin{pmatrix} 0 & 1 \\ -5 & 4 \end{pmatrix}$

   a Calculate if possible or, if not, explain why.

     i $\mathbf{A} + \mathbf{D}$ **ii** $\mathbf{A} + \mathbf{B}$

     iii $\mathbf{B} - \mathbf{C}$ **iv** $3\mathbf{B}$

   b Show that

     i $5\mathbf{A} - \mathbf{D} = \begin{pmatrix} 45 & -21 \\ 5 & -14 \end{pmatrix}$

     ii $2\mathbf{C} + 7\mathbf{B} = 2\begin{pmatrix} 23 & 12 \\ 14 & -21 \\ -4 & 1 \end{pmatrix}$

4 Calculate these matrix products.

   a $\begin{pmatrix} -3 & 0 \\ 1 & 4 \end{pmatrix}\begin{pmatrix} 2 & 5 & -1 & 6 \\ -3 & 1 & 0 & -4 \end{pmatrix}$

b $\begin{pmatrix} -2 & 1 \\ 0 & 9 \\ -5 & 0 \end{pmatrix}\begin{pmatrix} -6 & 3 \\ 4 & 8 \end{pmatrix}$

b $\begin{pmatrix} a \\ 2 \\ -a \end{pmatrix}(3-a)$

c $\begin{pmatrix} 5 & -3 & 0 \\ 2 & -4 & 1 \end{pmatrix}\begin{pmatrix} -3 \\ -5 \\ -1 \end{pmatrix}$

c $(-1 \quad a \quad 0)\begin{pmatrix} 3a & a \\ 4 & 2 \\ -1 & 5 \end{pmatrix}$

d $(9 \quad -2)\begin{pmatrix} 0 & 1 & -2 \\ 4 & 7 & -4 \end{pmatrix}$

d $\begin{pmatrix} a & 3 \\ -2 & a \end{pmatrix}^2$

e $\begin{pmatrix} 4 \\ 3 \end{pmatrix}(-1 \quad 2 \quad 5)$

8 Given $\mathbf{A} = \begin{pmatrix} 2 & 5 \\ -1 & 2 \end{pmatrix}$, show that

f $\begin{pmatrix} 1 & 3 & -1 \\ -2 & 0 & 2 \\ 0 & -3 & 1 \end{pmatrix}^2$

$\mathbf{A}^3 = \begin{pmatrix} -22 & k \\ -7 & -22 \end{pmatrix}$, stating the value of k

5 Show that $\begin{pmatrix} 3 & 2 & 0 \\ -1 & 0 & -2 \end{pmatrix}\begin{pmatrix} 4 & 1 \\ 0 & 3 \\ -3 & 0 \end{pmatrix} = \begin{pmatrix} 12 & 9 \\ 2 & -1 \end{pmatrix}$

9 The matrix $\mathbf{A} = \begin{pmatrix} 3 & k \\ 2 & 0 \end{pmatrix}$ and $\mathbf{B} = \begin{pmatrix} k & 4 \\ 0 & -2 \end{pmatrix}$

Find $(\mathbf{AB})^{\mathrm{T}}$ in terms of k

6 If $\mathbf{A} = \begin{pmatrix} 3 \\ 7 \end{pmatrix}$, $\mathbf{B} = \begin{pmatrix} 0 & 2 \\ -1 & 3 \end{pmatrix}$,

10 The matrix $\mathbf{A} = \begin{pmatrix} 4 & 0 & k \\ -1 & 3 & 0 \\ 0 & 2 & 1 \end{pmatrix}$ and

$\mathbf{C} = \begin{pmatrix} 2 & 9 & 4 \\ -3 & 0 & -5 \end{pmatrix}$, $\mathbf{D} = \begin{pmatrix} -1 \\ 5 \\ 2 \end{pmatrix}$,

$\mathbf{B} = \begin{pmatrix} 5 & 1 & 3 \\ 3 & 0 & -1 \\ 2 & k & 2 \end{pmatrix}$

find, if possible, or if the calculation is not possible, explain why.

Find $(\mathbf{AB})^{\mathrm{T}}$ in terms of k

a $\mathbf{AB}$ **b** $\mathbf{CD}$ **c** $\mathbf{BCD}$

d $\mathbf{CB}$ **e** $\mathbf{B}^2\mathbf{A}$ **f** $\mathbf{C}^2$

7 Calculate these matrix products, simplifying each element where possible.

a $\begin{pmatrix} 2a & 3 \\ 1 & -1 \end{pmatrix}\begin{pmatrix} 2 & 4 & 3 \\ a & 2 & -a \end{pmatrix}$

Reasoning and problem-solving

To solve problems involving matrix arithmetic

1. If two matrices are of equal order, then equate corresponding elements.

2. Solve the equations simultaneously to find the values of unknowns.

3. Use subscript notation for the elements of general matrices, for example, a_1, a_2 etc.

You can solve problems involving unknown elements using basic algebra.

Example 4

Given that $\begin{pmatrix} x & 5 \\ 3 & 3y \end{pmatrix}\begin{pmatrix} 1 \\ 5 \end{pmatrix} = \begin{pmatrix} y \\ x \end{pmatrix}$, find the values of x and y

Multiply the matrices on the left-hand side of the equation.

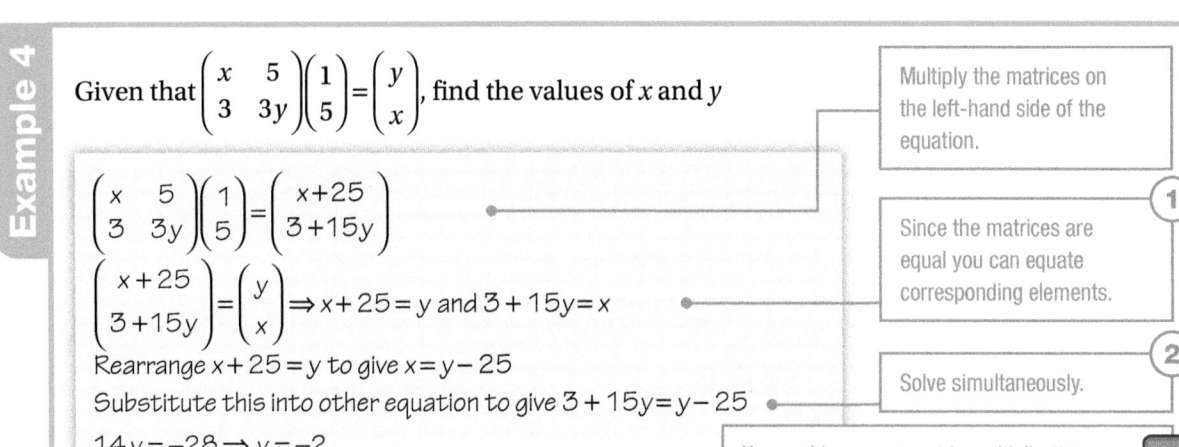

$$\begin{pmatrix} x & 5 \\ 3 & 3y \end{pmatrix}\begin{pmatrix} 1 \\ 5 \end{pmatrix} = \begin{pmatrix} x+25 \\ 3+15y \end{pmatrix}$$

(1)

Since the matrices are equal you can equate corresponding elements.

$$\begin{pmatrix} x+25 \\ 3+15y \end{pmatrix} = \begin{pmatrix} y \\ x \end{pmatrix} \Rightarrow x+25 = y \text{ and } 3+15y = x$$

Rearrange $x+25 = y$ to give $x = y-25$

(2)

Solve simultaneously.

Substitute this into other equation to give $3+15y = y-25$

$14y = -28 \Rightarrow y = -2$

You could now use matrix multiplication on your calculator to check the answer.

$\therefore x = -2-25 = -27$

Alternatively, you may be given a system of linear equations and have to write it in matrix form.

You will be shown later how to use matrices to solve 2×2 systems of equations.

Example 5

Write a matrix equation for each of these systems of simultaneous equations.

a $x+7y = -5$
   $3x-y = 29$

b $3x-2y-5z = -3$
   $x+7y+4z = 30$
   $5x-9z = 13$

(1)

Check by equating the individual elements.

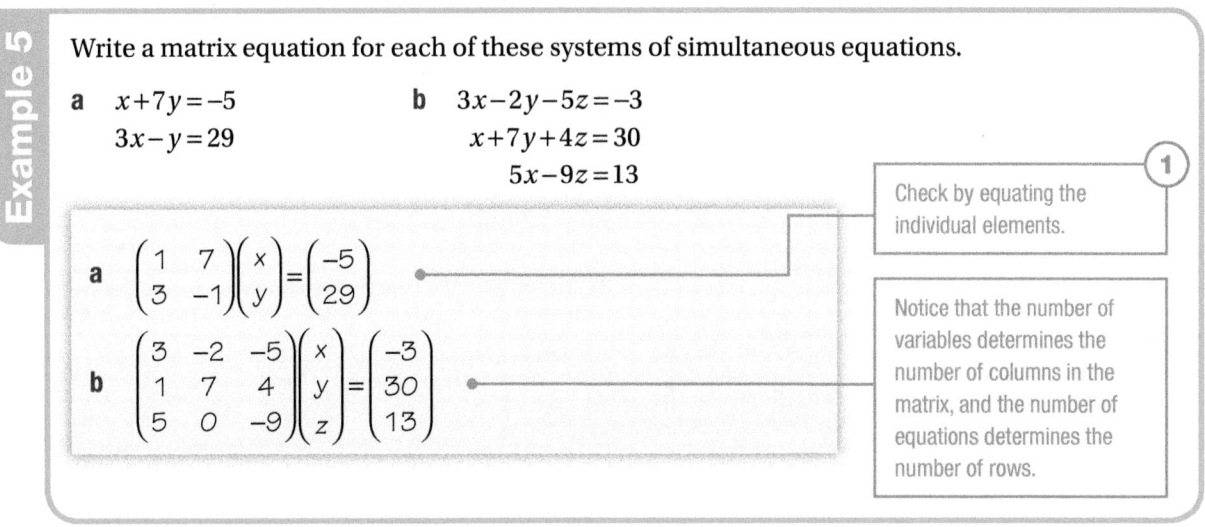

a $\begin{pmatrix} 1 & 7 \\ 3 & -1 \end{pmatrix}\begin{pmatrix} x \\ y \end{pmatrix} = \begin{pmatrix} -5 \\ 29 \end{pmatrix}$

b $\begin{pmatrix} 3 & -2 & -5 \\ 1 & 7 & 4 \\ 5 & 0 & -9 \end{pmatrix}\begin{pmatrix} x \\ y \\ z \end{pmatrix} = \begin{pmatrix} -3 \\ 30 \\ 13 \end{pmatrix}$

Notice that the number of variables determines the number of columns in the matrix, and the number of equations determines the number of rows.

Example 6

Use the result $(\mathbf{AB})^{\top} \equiv \mathbf{B}^{\top}\mathbf{A}^{\top}$ to prove that $(\mathbf{MPR})^{\top} \equiv \mathbf{M}^{\top}\mathbf{P}^{\top}\mathbf{R}^{\top}$ for all matrices $\mathbf{M}$, $\mathbf{P}$ and $\mathbf{R}$ that are conformable for multiplication.

$$(\mathbf{MPR})^{\top} \equiv ((\mathbf{MP})\mathbf{R})^{\top}$$

> Since matrix multiplication is associative.

$$\equiv \mathbf{R}^{\top}(\mathbf{MP})^{\top}$$

> Use result $(\mathbf{AB})^{\top} \equiv \mathbf{B}^{\top}\mathbf{A}^{\top}$ with $\mathbf{A} = \mathbf{MP}$ and $\mathbf{B} = \mathbf{R}$

$$\equiv \mathbf{R}^{\top}\mathbf{P}^{\top}\mathbf{M}^{\top} \text{ as required}$$

> Since $(\mathbf{MP})^{\top} \equiv \mathbf{P}^{\top}\mathbf{M}^{\top}$

Matrix addition is both **associative**, so $\mathbf{A} + (\mathbf{B} + \mathbf{C}) = (\mathbf{A} + \mathbf{B}) + \mathbf{C}$, and **commutative**, so $\mathbf{A} + \mathbf{B} = \mathbf{B} + \mathbf{A}$

Matrix multiplication is associative, that is, $(\mathbf{AB})\mathbf{C} = \mathbf{A}(\mathbf{BC})$, but it is not commutative since, in general, $\mathbf{AB} \neq \mathbf{BA}$

Matrix multiplication is also **distributive**, that is, $\mathbf{A}(\mathbf{B} + \mathbf{C}) = \mathbf{AB} + \mathbf{AC}$

Example 7

Prove the associative property for matrix multiplication of 2×2 matrices.

Let $A = \begin{pmatrix} a_1 & a_2 \\ a_3 & a_4 \end{pmatrix}$, $B = \begin{pmatrix} b_1 & b_2 \\ b_3 & b_4 \end{pmatrix}$ and $C = \begin{pmatrix} c_1 & c_2 \\ c_3 & c_4 \end{pmatrix}$

> ③ Use subscript notation for the elements of general 2×2 matrices.

> **See Ch2** For a reminder of proof by induction

Then $(AB)C = \left[\begin{pmatrix} a_1 & a_2 \\ a_3 & a_4 \end{pmatrix} \begin{pmatrix} b_1 & b_2 \\ b_3 & b_4 \end{pmatrix} \right] \begin{pmatrix} c_1 & c_2 \\ c_3 & c_4 \end{pmatrix}$

$$= \begin{pmatrix} a_1 b_1 + a_2 b_3 & a_1 b_2 + a_2 b_4 \\ a_3 b_1 + a_4 b_3 & a_3 b_2 + a_4 b_4 \end{pmatrix} \begin{pmatrix} c_1 & c_2 \\ c_3 & c_4 \end{pmatrix}$$

> First find $\mathbf{AB}$

$$= \begin{pmatrix} (a_1 b_1 + a_2 b_3)c_1 + (a_1 b_2 + a_2 b_4)c_3 & (a_1 b_1 + a_2 b_3)c_2 + (a_1 b_2 + a_2 b_4)c_4 \\ (a_3 b_1 + a_4 b_3)c_1 + (a_3 b_2 + a_4 b_4)c_3 & (a_3 b_1 + a_4 b_3)c_2 + (a_3 b_2 + a_4 b_4)c_4 \end{pmatrix}$$

> Now multiply by $\mathbf{C}$

$$= \begin{pmatrix} a_1 b_1 c_1 + a_2 b_3 c_1 + a_1 b_2 c_3 + a_2 b_4 c_3 & a_1 b_1 c_2 + a_2 b_3 c_2 + a_1 b_2 c_4 + a_2 b_4 c_4 \\ a_3 b_1 c_1 + a_4 b_3 c_1 + a_3 b_2 c_3 + a_4 b_4 c_3 & a_3 b_1 c_2 + a_4 b_3 c_2 + a_3 b_2 c_4 + a_4 b_4 c_4 \end{pmatrix}$$

> Expand the brackets.

$$= \begin{pmatrix} a_1(b_1 c_1 + b_2 c_3) + a_2(b_3 c_1 + b_4 c_3) & a_1(b_1 c_2 + b_2 c_4) + a_2(b_3 c_2 + b_4 c_4) \\ a_3(b_1 c_1 + b_2 c_3) + a_4(b_3 c_1 + b_4 c_3) & a_3(b_1 c_2 + b_2 c_4) + a_4(b_3 c_2 + b_4 c_4) \end{pmatrix}$$

> Factorise out terms from matrix $\mathbf{A}$

$$= \begin{pmatrix} a_1 & a_2 \\ a_3 & a_4 \end{pmatrix} \begin{pmatrix} b_1 c_1 + b_2 c_3 & b_1 c_2 + b_2 c_4 \\ b_3 c_1 + b_4 c_3 & b_3 c_2 + b_4 c_4 \end{pmatrix}$$

$$= \begin{pmatrix} a_1 & a_2 \\ a_3 & a_4 \end{pmatrix} \left[\begin{pmatrix} b_1 & b_2 \\ b_3 & b_4 \end{pmatrix} \begin{pmatrix} c_1 & c_2 \\ c_3 & c_4 \end{pmatrix} \right]$$

$$= A(BC) \text{ as required}$$

You can apply the method of proof by induction to matrices.

Strategy 2

To solve induction problems involving matrices

1. Check the base case.

2. Assume the statement is true for $n = k$

3. Use $A^{k+1} = A \times A^k$ to show the statement is true for $n = k + 1$

4. Write a conclusion.

Example 8

Prove by induction that $\begin{pmatrix} 1 & 0 \\ 2 & 1 \end{pmatrix}^n = \begin{pmatrix} 1 & 0 \\ 2n & 1 \end{pmatrix}$ for all positive integers n

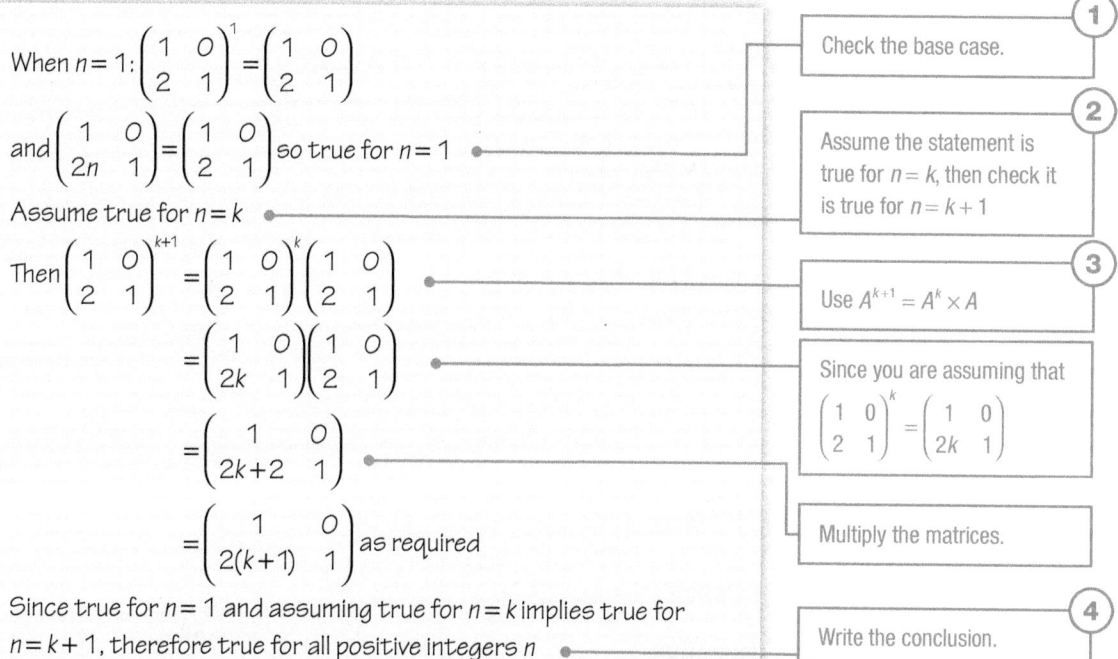

When $n = 1$: $\begin{pmatrix} 1 & 0 \\ 2 & 1 \end{pmatrix}^1 = \begin{pmatrix} 1 & 0 \\ 2 & 1 \end{pmatrix}$

and $\begin{pmatrix} 1 & 0 \\ 2n & 1 \end{pmatrix} = \begin{pmatrix} 1 & 0 \\ 2 & 1 \end{pmatrix}$ so true for $n = 1$ — Check the base case. ①

Assume true for $n = k$ — Assume the statement is true for $n = k$, then check it is true for $n = k + 1$ ②

Then $\begin{pmatrix} 1 & 0 \\ 2 & 1 \end{pmatrix}^{k+1} = \begin{pmatrix} 1 & 0 \\ 2 & 1 \end{pmatrix}^k \begin{pmatrix} 1 & 0 \\ 2 & 1 \end{pmatrix}$ — Use $A^{k+1} = A^k \times A$ ③

$= \begin{pmatrix} 1 & 0 \\ 2k & 1 \end{pmatrix}\begin{pmatrix} 1 & 0 \\ 2 & 1 \end{pmatrix}$ — Since you are assuming that $\begin{pmatrix} 1 & 0 \\ 2 & 1 \end{pmatrix}^k = \begin{pmatrix} 1 & 0 \\ 2k & 1 \end{pmatrix}$

$= \begin{pmatrix} 1 & 0 \\ 2k+2 & 1 \end{pmatrix}$ — Multiply the matrices.

$= \begin{pmatrix} 1 & 0 \\ 2(k+1) & 1 \end{pmatrix}$ as required

Since true for $n = 1$ and assuming true for $n = k$ implies true for $n = k + 1$, therefore true for all positive integers n — Write the conclusion. ④

Strategy 3

To solve problems involving tables by using matrices

1. Convert from tabular form into a matrix of suitable order.

2. Identify any relevant vectors.

3. Multiply the matrix by the vector to perform the necessary data analysis.

4. Write a conclusion, if required.

Example 9

The table shows the probabilities of a spring day being rainy in four UK cities.

	March	April	May
London	0.32	0.3	0.29
Edinburgh	0.38	0.33	0.37
Cardiff	0.42	0.37	0.36
Belfast	0.45	0.38	0.38

(Continued on the next page)

a Write a matrix, **A**, to represent the table of information and a vector, **x**, to represent the number of days in each month.

b Show how you can use your matrices to calculate the total number of rainy days expected in each city over the three months.

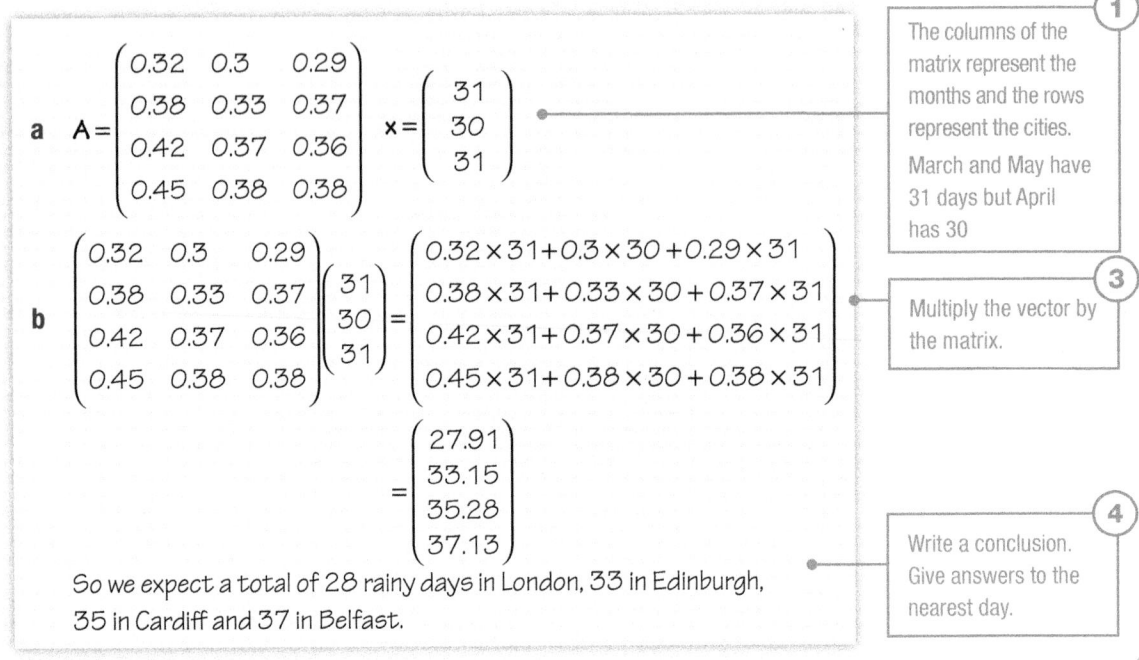

a $A = \begin{pmatrix} 0.32 & 0.3 & 0.29 \\ 0.38 & 0.33 & 0.37 \\ 0.42 & 0.37 & 0.36 \\ 0.45 & 0.38 & 0.38 \end{pmatrix}$ $x = \begin{pmatrix} 31 \\ 30 \\ 31 \end{pmatrix}$

The columns of the matrix represent the months and the rows represent the cities. March and May have 31 days but April has 30

b $\begin{pmatrix} 0.32 & 0.3 & 0.29 \\ 0.38 & 0.33 & 0.37 \\ 0.42 & 0.37 & 0.36 \\ 0.45 & 0.38 & 0.38 \end{pmatrix} \begin{pmatrix} 31 \\ 30 \\ 31 \end{pmatrix} = \begin{pmatrix} 0.32 \times 31 + 0.3 \times 30 + 0.29 \times 31 \\ 0.38 \times 31 + 0.33 \times 30 + 0.37 \times 31 \\ 0.42 \times 31 + 0.37 \times 30 + 0.36 \times 31 \\ 0.45 \times 31 + 0.38 \times 30 + 0.38 \times 31 \end{pmatrix}$

Multiply the vector by the matrix.

$$= \begin{pmatrix} 27.91 \\ 33.15 \\ 35.28 \\ 37.13 \end{pmatrix}$$

Write a conclusion. Give answers to the nearest day.

So we expect a total of 28 rainy days in London, 33 in Edinburgh, 35 in Cardiff and 37 in Belfast.

Exercise 5.1B Reasoning and problem-solving

1 Find the values of a, b, c and d such that
$$\begin{pmatrix} 1 & a \\ 3 & 7 \end{pmatrix} + \begin{pmatrix} -2 & 4 \\ b & -5 \end{pmatrix} = \begin{pmatrix} c & 1 \\ -2 & d \end{pmatrix}$$

2 Find the values of a, b and c such that
$$\begin{pmatrix} 3 & a \\ 1 & -2 \end{pmatrix} \begin{pmatrix} 5 & b \\ 4 & 1 \end{pmatrix} = \begin{pmatrix} 3 & 9 \\ c & 2 \end{pmatrix}$$

3 Find the values of x and y in each case.

a $\begin{pmatrix} x & 3 \\ 2 & 0 \end{pmatrix} \begin{pmatrix} 2 & 5 \\ y & -y \end{pmatrix} = \begin{pmatrix} 10 & -17 \\ 4 & 10 \end{pmatrix}$

b $\begin{pmatrix} x & -1 \\ y & 2 \end{pmatrix}^2 = \mathbf{0}$

4 Solve each of these equations to find the possible values of x

a $\begin{pmatrix} 3 & x & -2 \\ 4 & 0 & 0 \end{pmatrix} \begin{pmatrix} 1 & 1 \\ 5x & 0 \\ x & 5 \end{pmatrix} = \begin{pmatrix} 6 & -7 \\ 4 & 4 \end{pmatrix}$

b $(7 \quad x \quad 4) \begin{pmatrix} x & 3 & 0 \\ x & 0 & 2 \\ 5 & -1 & x \end{pmatrix} = (10 \quad 17 \quad 6x)$

c $\begin{pmatrix} 5 & x \\ -x & -5 \end{pmatrix}^2 = 16 \begin{pmatrix} 1 & 0 \\ 0 & 1 \end{pmatrix}$

d $\begin{pmatrix} 0 & x \\ x & 1 \end{pmatrix}^3 = \begin{pmatrix} 4 & -10 \\ -10 & 9 \end{pmatrix}$

5 Calculate the values of a, b and c in this matrix equation.
$$\begin{pmatrix} -1 & 2 & 5 \\ 4 & 1 & 0 \\ 0 & 3 & 0 \end{pmatrix} \begin{pmatrix} a \\ b \\ c \end{pmatrix} = \begin{pmatrix} -3 \\ 25 \\ -9 \end{pmatrix}$$

6 Calculate the values of x, y and z in this matrix equation.
$$\begin{pmatrix} 2 & 0 & 0 \\ 0 & 4 & 6 \\ -1 & 3 & 1 \end{pmatrix} \begin{pmatrix} x \\ y \\ z \end{pmatrix} = \begin{pmatrix} 6 \\ 10 \\ 8 \end{pmatrix}$$

7 Calculate the values of a, b and c in this matrix equation.
$$\begin{pmatrix} 1 & 3 & 2 \\ 5 & 0 & 0 \\ 7 & -2 & 1 \end{pmatrix} \begin{pmatrix} a \\ b \\ c \end{pmatrix} = \begin{pmatrix} 5 \\ -5 \\ 3 \end{pmatrix}$$

8 Write a matrix equation for each of these systems of simultaneous equations.

a $2x + 5y = 6$
   $6x - y = 3$

b $x + y - z = -4$
   $2x + y + z = 4$
   $3x + 2y + 2z = 10$

9 Prove by induction that $\begin{pmatrix} 1 & 5 \\ 0 & 1 \end{pmatrix}^n = \begin{pmatrix} 1 & 5n \\ 0 & 1 \end{pmatrix}$ for all positive integers n

10 Prove by induction that
$$\begin{pmatrix} 5 & 4 \\ 0 & 1 \end{pmatrix}^n = \begin{pmatrix} 5^n & 5^n - 1 \\ 0 & 1 \end{pmatrix}$$
for all positive integers n

11 Prove by induction that
$$\begin{pmatrix} -2 & -1 \\ 9 & 4 \end{pmatrix}^n = \begin{pmatrix} 1 - 3n & -n \\ 9n & 3n + 1 \end{pmatrix}$$
for all positive integers n

12 Prove by induction that
$$\begin{pmatrix} 1 & 4 \\ 0 & 2 \end{pmatrix}^n = \begin{pmatrix} 1 & 4(2^n - 1) \\ 0 & 2^n \end{pmatrix}$$
for all positive integers n

13 Prove that $\mathbf{A} + (\mathbf{B} + \mathbf{C}) = (\mathbf{A} + \mathbf{B}) + \mathbf{C}$ for any 3×2 matrices $\mathbf{A}$, $\mathbf{B}$ and $\mathbf{C}$

14 Prove that matrix addition is commutative for any 2×2 matrices.

15 Prove by counter-example that matrix multiplication is not commutative.

16 Prove that $\mathbf{A}(\mathbf{B} + \mathbf{C}) = \mathbf{AB} + \mathbf{AC}$ for any 2×2 matrices $\mathbf{A}$, $\mathbf{B}$ and $\mathbf{C}$

17 Prove that $\mathbf{A}\begin{pmatrix} 1 & 0 \\ 0 & 1 \end{pmatrix} = \begin{pmatrix} 1 & 0 \\ 0 & 1 \end{pmatrix}\mathbf{A}$ for any 2×2 matrix $\mathbf{A}$

18 Prove that $\mathbf{B}\begin{pmatrix} 1 & 0 & 0 \\ 0 & 1 & 0 \\ 0 & 0 & 1 \end{pmatrix} = \begin{pmatrix} 1 & 0 & 0 \\ 0 & 1 & 0 \\ 0 & 0 & 1 \end{pmatrix}\mathbf{B}$ for any 3×3 matrix $\mathbf{B}$

19 The profit (in £1000) made by three employees for a company over the four quarters of a year is shown in the table.

	Q1	Q2	Q3	Q4
Employee A	8	14	15	17
Employee B	6	11	7	9
Employee C	9	18	19	12

The company plans to pay bonuses of 5% of profit from Q1, 2% from Q2, 3% from Q3 and 1% from Q4

a Write a matrix to represent the table of information and a vector to represent the percentage paid in bonuses.

b Show how you can use your matrices to calculate the amount of bonus paid to each of the employees.

c Write down the total amount of bonuses paid to all three employees.

20 The stationery requirements of a group of Maths teachers is given in the table.

	Red pens	Pencils	Rulers	Board pens	Paper clips
Teacher 1	5	20	7	6	50
Teacher 2	4	15	15	8	100
Teacher 3	12	4	2	30	50

Red pens cost 12p each, pencils cost 8p, rulers 18p, board pens 30p and paper clips 1p

a Show how you can use matrices to calculate the expenditure on each type of stationery by each teacher.

b Write down the total cost of all the stationery required.

21 Given that
$$\begin{pmatrix} a & b \\ c & d \end{pmatrix}\begin{pmatrix} e & f \\ g & h \end{pmatrix} = \begin{pmatrix} 1 & 0 \\ 0 & 1 \end{pmatrix},$$

a Verify that $e = \dfrac{d}{ad - bc}$ and $g = -\dfrac{c}{ad - bc}$

b Find similar expressions for f and h in terms of a, b, c and d

22 Prove by induction that
$$\begin{pmatrix} 0 & 1 \\ 2 & 0 \end{pmatrix}^{2n} = 2^n \begin{pmatrix} 1 & 0 \\ 0 & 1 \end{pmatrix}$$ for all $n \in \mathbb{N}$

Fluency and skills

Matrices can be used to represent certain transformations such as some rotations, reflections and enlargements. Transformations that can be described in this way are known as **linear**.

In order to find the image of a vector $\begin{pmatrix} x \\ y \end{pmatrix}$ under a transformation $\mathbf{T} = \begin{pmatrix} a & b \\ c & d \end{pmatrix}$, we **pre-multiply** the vector by the matrix, so $\begin{pmatrix} x' \\ y' \end{pmatrix} = \begin{pmatrix} a & b \\ c & d \end{pmatrix}\begin{pmatrix} x \\ y \end{pmatrix}$

A point (x, y) can also be represented by the vector $\begin{pmatrix} x \\ y \end{pmatrix}$

To find the image of a point $A(1, -3)$ under the transformation

$\mathbf{T} = \begin{pmatrix} 1 & 0 \\ 0 & -1 \end{pmatrix}$,

multiply the vector $\begin{pmatrix} 1 \\ -3 \end{pmatrix}$ by $\mathbf{T}$

$\begin{pmatrix} x' \\ y' \end{pmatrix} = \begin{pmatrix} 1 & 0 \\ 0 & -1 \end{pmatrix}\begin{pmatrix} 1 \\ -3 \end{pmatrix} = \begin{pmatrix} 1 \\ 3 \end{pmatrix}$

This is a reflection in the x-axis.

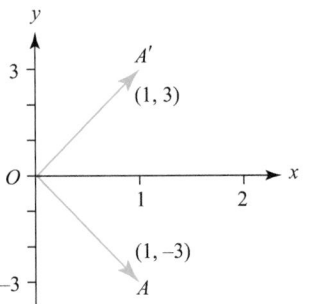

The matrix $\begin{pmatrix} 1 & 0 \\ 0 & -1 \end{pmatrix}$ represents a reflection in the x-axis.

The matrix $\begin{pmatrix} -1 & 0 \\ 0 & 1 \end{pmatrix}$ represents a reflection in the y-axis.

The matrix $\begin{pmatrix} 0 & 1 \\ 1 & 0 \end{pmatrix}$ represents a reflection in the line $y = x$

The matrix $\begin{pmatrix} 0 & -1 \\ -1 & 0 \end{pmatrix}$ represents a reflection in the line $y = -x$

You can find matrices that represent particular transformations by considering the effect the transformation will have on a pair of vectors.

Example 1

Find the 2×2 matrix that represents a rotation of 90° anticlockwise about the origin.

Let the matrix be given by $\begin{pmatrix} a & b \\ c & d \end{pmatrix}$

Choose two points and their images after the rotation.

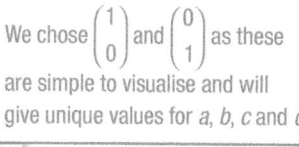

We chose $\begin{pmatrix} 1 \\ 0 \end{pmatrix}$ and $\begin{pmatrix} 0 \\ 1 \end{pmatrix}$ as these are simple to visualise and will give unique values for a, b, c and d

$(1, 0)$ moves to $(0, 1)$ and $(0, 1)$ moves to $(-1, 0)$

So $\begin{pmatrix} a & b \\ c & d \end{pmatrix}\begin{pmatrix} 1 \\ 0 \end{pmatrix} = \begin{pmatrix} 0 \\ 1 \end{pmatrix}$ and $\begin{pmatrix} a & b \\ c & d \end{pmatrix}\begin{pmatrix} 0 \\ 1 \end{pmatrix} = \begin{pmatrix} -1 \\ 0 \end{pmatrix}$

So from the first equation: $a = 0$, $c = 1$

From the second equation: $b = -1$ and $d = 0$

Therefore the transformation matrix is $\begin{pmatrix} 0 & -1 \\ 1 & 0 \end{pmatrix}$

Notice that the first column of the matrix is the image of $\begin{pmatrix} 1 \\ 0 \end{pmatrix}$ and the second column is the image of $\begin{pmatrix} 0 \\ 1 \end{pmatrix}$

Key point

The matrix $\begin{pmatrix} a & b \\ c & d \end{pmatrix}$ represents the transformation

that maps $\begin{pmatrix} 1 \\ 0 \end{pmatrix}$ to $\begin{pmatrix} a \\ c \end{pmatrix}$ and $\begin{pmatrix} 0 \\ 1 \end{pmatrix}$ to $\begin{pmatrix} b \\ d \end{pmatrix}$

It follows from this fact that the matrix $\begin{pmatrix} 1 & 0 \\ 0 & 1 \end{pmatrix}$ maps all points to themselves.

In number multiplication the identity is 1 since $1 \times a = a \times 1 = a$

Key point

The matrix $\mathbf{I} = \begin{pmatrix} 1 & 0 \\ 0 & 1 \end{pmatrix}$ is known as the **identity matrix**

and has the property that $\mathbf{AI} = \mathbf{IA} = \mathbf{A}$ for all 2×2 matrices

This can be generalised for any square matrix, in particular $\mathbf{I}_3 = \begin{pmatrix} 1 & 0 & 0 \\ 0 & 1 & 0 \\ 0 & 0 & 1 \end{pmatrix}$

To find the 2×2 matrix that represents a stretch of scale factor 3 parallel to the x-axis, look at what

happens to the vectors $\begin{pmatrix} 1 \\ 0 \end{pmatrix}$ and $\begin{pmatrix} 0 \\ 1 \end{pmatrix}$

A stretch 'parallel to the x-axis' can be thought of a stretch 'along the x-axis' so only the x-coordinates are affected.

The image of $\begin{pmatrix} 1 \\ 0 \end{pmatrix}$ under this transformation is $\begin{pmatrix} 3 \\ 0 \end{pmatrix}$

The image of $\begin{pmatrix} 0 \\ 1 \end{pmatrix}$ under this transformation is $\begin{pmatrix} 0 \\ 1 \end{pmatrix}$

Therefore the transformation matrix is $\begin{pmatrix} 3 & 0 \\ 0 & 1 \end{pmatrix}$

The matrix $\begin{pmatrix} k & 0 \\ 0 & 1 \end{pmatrix}$ represents a stretch of scale factor k parallel to the x-axis.

The matrix $\begin{pmatrix} 1 & 0 \\ 0 & k \end{pmatrix}$ represents a stretch of scale factor k parallel to the y-axis.

Transformations can be applied to several points at once by forming a matrix from their position vectors.

Example 2

Find the image of the points $(2, -1)$, $(-4, 5)$ and $(-3, 0)$ under the transformation described by the matrix $\begin{pmatrix} 2 & 0 \\ 0 & 2 \end{pmatrix}$. Describe the transformation geometrically.

$$\begin{pmatrix} 2 & 0 \\ 0 & 2 \end{pmatrix}\begin{pmatrix} 2 & -4 & -3 \\ -1 & 5 & 0 \end{pmatrix} = \begin{pmatrix} 4 & -8 & -6 \\ -2 & 10 & 0 \end{pmatrix}$$

Ensure you always pre-multiply by the transformation matrix.

This is an enlargement of scale factor 2, centre the origin.

The matrix $\begin{pmatrix} k & 0 \\ 0 & k \end{pmatrix}$ represents an enlargement of scale factor k, centre the origin.

See Maths Ch3.1
For a reminder of exact values of trig functions.

To find the matrix that represents a 60° rotation anticlockwise about the origin, look at what happens to the vectors $\begin{pmatrix} 1 \\ 0 \end{pmatrix}$ and $\begin{pmatrix} 0 \\ 1 \end{pmatrix}$

The image of $\begin{pmatrix} 1 \\ 0 \end{pmatrix}$ under this transformation is $\begin{pmatrix} \cos 60° \\ \sin 60° \end{pmatrix}$

and the image of $\begin{pmatrix} 0 \\ 1 \end{pmatrix}$ is $\begin{pmatrix} -\sin 60° \\ \cos 60° \end{pmatrix}$

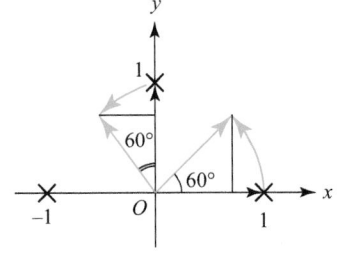

Therefore the matrix is $\begin{pmatrix} \cos 60° & -\sin 60° \\ \sin 60° & \cos 60° \end{pmatrix} = \begin{pmatrix} \dfrac{1}{2} & -\dfrac{\sqrt{3}}{2} \\ \dfrac{\sqrt{3}}{2} & \dfrac{1}{2} \end{pmatrix}$

This result can be generalised for any angle.

The matrix $\begin{pmatrix} \cos\theta & -\sin\theta \\ \sin\theta & \cos\theta \end{pmatrix}$ represents an anticlockwise rotation by angle θ about the origin. Positive angles always give anticlockwise rotations.

Transformations can also be combined and you can use the product of the matrices to represent the combined transformation.

If the transformation represented by matrix **A** is followed by the **Key point**
transformation represented by matrix **B**, then **BA** is the combined transformation.

Notice that if you wish to apply transformation **A** to a vector first, you should multiply by **A** then by
B hence **BA**

Example 3

Find the single matrix that represents a rotation of 270° anticlockwise followed by a reflection
in the y-axis.

Rotation of 270°: $\begin{pmatrix} 1 \\ 0 \end{pmatrix}$ is transformed to $\begin{pmatrix} 0 \\ -1 \end{pmatrix}$ and $\begin{pmatrix} 0 \\ 1 \end{pmatrix}$ is transformed to $\begin{pmatrix} 1 \\ 0 \end{pmatrix}$

Therefore $\mathbf{A} = \begin{pmatrix} 0 & 1 \\ -1 & 0 \end{pmatrix}$

Reflection in the y-axis: $\begin{pmatrix} 1 \\ 0 \end{pmatrix}$ is transformed to $\begin{pmatrix} -1 \\ 0 \end{pmatrix}$ and $\begin{pmatrix} 0 \\ 1 \end{pmatrix}$ is transformed to $\begin{pmatrix} 0 \\ 1 \end{pmatrix}$

Therefore $\mathbf{B} = \begin{pmatrix} -1 & 0 \\ 0 & 1 \end{pmatrix}$

So combined transformation $= \mathbf{BA}$ | The order of the matrices is important.

$$= \begin{pmatrix} -1 & 0 \\ 0 & 1 \end{pmatrix}\begin{pmatrix} 0 & 1 \\ -1 & 0 \end{pmatrix}$$

$$= \begin{pmatrix} 0 & -1 \\ -1 & 0 \end{pmatrix}$$ | This is a reflection in the line $y = -x$

You may recall learning about 3D coordinates in GCSE Maths. Points in 3D space have 3D position
vectors, and you will learn more about these in your Further Maths course.

See Ch 6.1
For vectors
in Further
Maths.

You can use 3×3 matrices to represent transformations in 3D.

To reflect in the plane $x = 0$, the 'mirror' is a plane through the
y-axis and extended in the positive and negative z-directions.

The image of the vector $\begin{pmatrix} 1 \\ 0 \\ 0 \end{pmatrix}$ will be $\begin{pmatrix} -1 \\ 0 \\ 0 \end{pmatrix}$

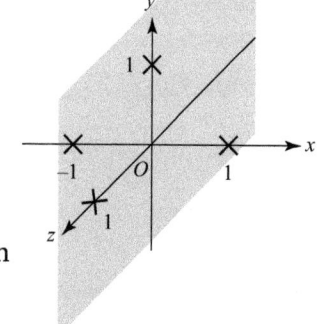

The vectors $\begin{pmatrix} 0 \\ 1 \\ 0 \end{pmatrix}$ and $\begin{pmatrix} 0 \\ 0 \\ 1 \end{pmatrix}$ are in the 'mirror' plane so are unchanged when
reflected in x

Therefore the transformation matrix is $\begin{pmatrix} -1 & 0 & 0 \\ 0 & 1 & 0 \\ 0 & 0 & 1 \end{pmatrix}$

The matrices for reflection in the planes $y = 0$ and $z = 0$ can be found using the same method.

The 3D reflection matrices you need to use are as follows.

Reflection in $x = 0$: $\begin{pmatrix} -1 & 0 & 0 \\ 0 & 1 & 0 \\ 0 & 0 & 1 \end{pmatrix}$

Reflection in $y = 0$: $\begin{pmatrix} 1 & 0 & 0 \\ 0 & -1 & 0 \\ 0 & 0 & 1 \end{pmatrix}$

Reflection in $z = 0$: $\begin{pmatrix} 1 & 0 & 0 \\ 0 & 1 & 0 \\ 0 & 0 & -1 \end{pmatrix}$

You also need to know about 3D rotation around one of the coordinate axes.

This diagram illustrates rotating the point $(1, 0, 0)$ by angle θ anticlockwise around the z-axis. Notice how the z-coordinate of the point is unchanged so this is actually the same as rotating around the origin in the 2D case.

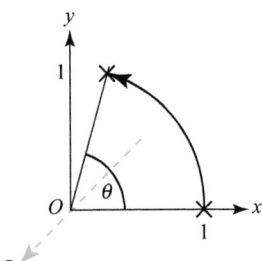

Therefore you can represent anticlockwise rotation around the z-axis by the matrix

Notice how this is the 2×2 matrix for anticlockwise rotation around the origin.

$\begin{pmatrix} \cos\theta & -\sin\theta & 0 \\ \sin\theta & \cos\theta & 0 \\ 0 & 0 & 1 \end{pmatrix}$

Consider an anticlockwise rotation around the x-axis or the y-axis in a similar way.

The 3D rotation matrices you need to use are as follows.

Rotation around the x-axis: $\begin{pmatrix} 1 & 0 & 0 \\ 0 & \cos\theta & -\sin\theta \\ 0 & \sin\theta & \cos\theta \end{pmatrix}$

Rotation around the y-axis: $\begin{pmatrix} \cos\theta & 0 & \sin\theta \\ 0 & 1 & 0 \\ -\sin\theta & 0 & \cos\theta \end{pmatrix}$

Rotation around the z-axis: $\begin{pmatrix} \cos\theta & -\sin\theta & 0 \\ \sin\theta & \cos\theta & 0 \\ 0 & 0 & 1 \end{pmatrix}$

Example 4

Find the 3×3 matrix, **T**, that represents a rotation of $45°$ anticlockwise around the x-axis and find the image of the point $P(1, 4, -2)$ under **T**

$$T = \begin{pmatrix} 1 & 0 & 0 \\ 0 & \cos 45° & -\sin 45° \\ 0 & \sin 45° & \cos 45° \end{pmatrix}$$

$$= \begin{pmatrix} 1 & 0 & 0 \\ 0 & \dfrac{\sqrt{2}}{2} & -\dfrac{\sqrt{2}}{2} \\ 0 & \dfrac{\sqrt{2}}{2} & \dfrac{\sqrt{2}}{2} \end{pmatrix}$$

$$P' = \begin{pmatrix} 1 & 0 & 0 \\ 0 & \dfrac{\sqrt{2}}{2} & -\dfrac{\sqrt{2}}{2} \\ 0 & \dfrac{\sqrt{2}}{2} & \dfrac{\sqrt{2}}{2} \end{pmatrix} \begin{pmatrix} 1 \\ 4 \\ -2 \end{pmatrix}$$

Pre-multiply the vector representing point P by the matrix **T**

$$= \begin{pmatrix} 1 \\ 2\sqrt{2} + \sqrt{2} \\ 2\sqrt{2} - \sqrt{2} \end{pmatrix}$$

$$= \begin{pmatrix} 1 \\ 3\sqrt{2} \\ \sqrt{2} \end{pmatrix}$$

So the actual point is $(1, 3\sqrt{2}, \sqrt{2})$

Since you are transforming a point you should give the final answer as coordinates.

Exercise 5.2A Fluency and skills

1   The matrix **B** represents a stretch of scale factor 5 parallel to the x-axis and a stretch of scale factor -2 parallel to the y-axis.

  a   Use a diagram to show the transformation of the vectors $\begin{pmatrix} 0 \\ 1 \end{pmatrix}$ and $\begin{pmatrix} 1 \\ 0 \end{pmatrix}$ under matrix **B**

  b   Write down matrix **B**

2   The matrix **A** represents a rotation of $180°$

  a   Use a diagram to show the transformation of the vector $\begin{pmatrix} 1 \\ 0 \end{pmatrix}$ under this transformation.

  b   Write down the image of the vector $\begin{pmatrix} 0 \\ 1 \end{pmatrix}$ under this transformation.

  c   Write down the matrix **A**

3 Find the 2×2 matrix which represents each of these transformations.

   a Reflection in the line $y = x$

   b Rotation of $30°$ anticlockwise about the origin

   c Enlargement of scale factor 5, centre the origin

   d Stretch parallel to the y-axis of scale factor 4

4 A triangle has vertices $(0, -2)$, $(2, 3)$ and $(-4, 3)$

   a Find the image of these points under the transformation represented by the matrix

$$T = \begin{pmatrix} 2 & 0 \\ 0 & 1 \end{pmatrix}$$

   b Describe geometrically the transformation represented by matrix T

5 The matrix $\dfrac{\sqrt{2}}{2}\begin{pmatrix} 1 & 1 \\ -1 & 1 \end{pmatrix}$ represents an anticlockwise rotation of θ degrees, centre the origin. Find the value of θ

6 A transformation is given by $A = \begin{pmatrix} 1 & 0 \\ 0 & 0.5 \end{pmatrix}$

   a Find the image of the vectors $\begin{pmatrix} 6 \\ -2 \end{pmatrix}$ and $\begin{pmatrix} -1 \\ 4 \end{pmatrix}$ under the transformation represented by matrix A

   b Describe geometrically the transformation represented by matrix A

7 Write the transformation represented by each of these matrices.

   a $\begin{pmatrix} 3 & 0 \\ 0 & 3 \end{pmatrix}$ b $\begin{pmatrix} 1 & 0 \\ 0 & 2 \end{pmatrix}$

   c $\dfrac{\sqrt{2}}{2}\begin{pmatrix} -1 & -1 \\ 1 & -1 \end{pmatrix}$ d $\begin{pmatrix} 0 & 1 \\ -1 & 0 \end{pmatrix}$

   e $\begin{pmatrix} -1 & 0 \\ 0 & 1 \end{pmatrix}$ f $\begin{pmatrix} \cos 20° & \sin 20° \\ -\sin 20° & \cos 20° \end{pmatrix}$

8 Given the transformation matrices

$$A = \begin{pmatrix} 1 & 1 \\ 0 & 1 \end{pmatrix} \text{ and } B = \begin{pmatrix} 0 & -1 \\ -1 & 1 \end{pmatrix},$$

   a Find the matrix representing

      i Transformation A followed by transformation B

     ii Transformation B followed by transformation A

   b Comment on your answers to part a.

9 Give the transformation matrices

$$A = \begin{pmatrix} 1 & 0 \\ 0 & -1 \end{pmatrix}, B = \begin{pmatrix} 0 & 1 \\ -1 & 0 \end{pmatrix}$$

   a Describe the transformations A and B geometrically.

   b Find the matrix representing

      i Transformation A followed by B

     ii Transformation B followed by A

   c Describe both transformations in part b geometrically as a single transformation.

10 Find the single 2×2 matrix that represents each of these combinations of transformations.

   a A rotation of $45°$ anticlockwise about the origin followed by a reflection in the y-axis

   b A reflection in the line $y = -x$ followed by an enlargement of scale factor 2 about the origin

   c A stretch parallel to the x-axis of scale factor -2 followed by a clockwise rotation of $90°$ about the origin.

11 A square has vertices at $(2, 5)$, $(2, 1)$, $(-3, 1)$ and $(-3, 5)$

The square is rotated $225°$ anticlockwise about the origin and then stretched by a scale factor of -2 in the y-direction.

Find the vertices of the image of the square following these transformations.

12 a Write a matrix A that represents a rotation of $135°$ anticlockwise about the origin.

   b Show that $A^2 = \begin{pmatrix} 0 & 1 \\ -1 & 0 \end{pmatrix}$

   c Show that A^2 represents a clockwise rotation of $90°$ about the origin.

   d Describe the transformation represented by A^3

13 The rotation represented by the 2×2 matrix **A** is such that $\mathbf{A}^2 = \mathbf{I}$ and $\mathbf{A} \neq \mathbf{I}$

a Write down a possible matrix **A** and describe the transformation fully.

The rotation represented by the 2×2 matrix **B** is such that $\mathbf{B}^3 = \mathbf{I}$ and $\mathbf{B} \neq \mathbf{I}$

b Write down a possible matrix **B** and describe the transformation fully.

14 Matrix **P** represents a reflection in the x-axis and matrix **Q** represents a stretch of scale factor 3 parallel to the y-axis.

a Find the single matrix that represents transformation **P** followed by transformation **Q**

b Describe geometrically the single transformation that could replace this combination of transformations.

15 The 2×2 matrix **M** represents an enlargement around the origin of scale factor k, $k > 0$, followed by an anticlockwise rotation of θ, $0 \leq \theta \leq 180$ around the origin.

a Write **M** in terms of k and θ

The image of the point $(1, -2)$ under **M** is $(2, 6)$.

b Find the values of k and θ

16 A transformation is represented by the matrix $\mathbf{T} = \begin{pmatrix} 1 & 0 & 0 \\ 0 & -1 & 0 \\ 0 & 0 & 1 \end{pmatrix}$

a Find the image of the points $(4, -1, 0)$ and $(2, 5, 3)$ under **T**

b Describe this transformation geometrically.

17 Find the 3×3 matrix that represents each of these transformations.

a A rotation of $30°$ anticlockwise around the x-axis

b A reflection in $z = 0$

c A rotation of $45°$ anticlockwise around the y-axis.

18 The matrix $\begin{pmatrix} 1 & 0 & 0 \\ 0 & \dfrac{1}{2} & \dfrac{\sqrt{3}}{2} \\ 0 & -\dfrac{\sqrt{3}}{2} & \dfrac{1}{2} \end{pmatrix}$ represents an anticlockwise rotation of θ degrees around one of the coordinate axes.

a Find the value of θ and state which axis the rotation is around.

b Find the image of the vector $3\mathbf{i} + 4\mathbf{j} - \mathbf{k}$ under this transformation.

19 The matrix $\dfrac{1}{2}\begin{pmatrix} 1 & \sqrt{3} & 0 \\ -\sqrt{3} & 1 & 0 \\ 0 & 0 & 2 \end{pmatrix}$ represents an anticlockwise rotation of θ around one of the coordinate axes. Find the value of θ and state which axis the rotation is around.

20 The point $A(3, 7, -2)$ is transformed by a transformation matrix **M** to the point $A'(3, 7, 2)$

a Describe this transformation geometrically.

b Write down the transformation matrix **M**

21 Describe the transformation represented by each of these matrices.

a $\begin{pmatrix} 1 & 0 & 0 \\ 0 & 0 & -1 \\ 0 & 1 & 0 \end{pmatrix}$ **b** $\begin{pmatrix} -1 & 0 & 0 \\ 0 & 1 & 0 \\ 0 & 0 & 1 \end{pmatrix}$

c $\begin{pmatrix} -1 & 0 & 0 \\ 0 & -1 & 0 \\ 0 & 0 & 1 \end{pmatrix}$ **d** $\begin{pmatrix} -\dfrac{1}{2} & 0 & \dfrac{\sqrt{3}}{2} \\ 0 & 1 & 0 \\ -\dfrac{\sqrt{3}}{2} & 0 & -\dfrac{1}{2} \end{pmatrix}$

22 The 3×3 matrix **P** represents a clockwise rotation of $135°$ around the y-axis.

a Find **P** in its simplest form.

b Find the image of the point $(\sqrt{2}, 0, 1)$ under the transformation represented by **P**

23 The 3×3 matrix **T** represents an anticlockwise rotation of $330°$ around the z-axis.

a Find **T** in its simplest form.

b Find the image of the point $(\sqrt{3}, 0, 1)$ under the transformation **T**

Reasoning and problem-solving

A point which is unaffected by a transformation is known as an invariant point.

Given a transformation matrix **T** and a position vector **x**, if **Tx** = **x** then **x** represents an **invariant point**.

You can be asked to find invariant points for a specific transformation.

Strategy

To find invariant points and lines

1. Write vectors in a general form.
2. Pre-multiply a vector by a transformation matrix to form a system of simultaneous equations.
3. Solve simultaneous equations to find either an invariant point or a line of invariant points.
4. Equate coefficients to find equations of invariant lines.

Example 5

Find any invariant points under the transformation given by $\begin{pmatrix} 3 & 1 \\ -1 & 2 \end{pmatrix}$

Let $\begin{pmatrix} x \\ y \end{pmatrix}$ be the invariant point.

> Write a general vector to represent the invariant point. ①

Then $\begin{pmatrix} 3 & 1 \\ -1 & 2 \end{pmatrix}\begin{pmatrix} x \\ y \end{pmatrix} = \begin{pmatrix} x \\ y \end{pmatrix}$

$\therefore 3x + y = x \Rightarrow 2x + y = 0$

and $-x + 2y = y \Rightarrow y = x$

> Form equations. ②

Therefore $3x = 0 \Rightarrow x = 0$ and $y = 0$

So the invariant point is $(0, 0)$

> In this case only the point (0, 0) is invariant. ③

For any linear transformation, (0, 0) is an invariant point.

Some transformations have a **line of invariant points** where every point satisfying a certain property is invariant.

For example, consider a reflection in the line $y = x$

All points on the line will be unaffected by the transformation.

Therefore $y = x$ is a line of invariant points.

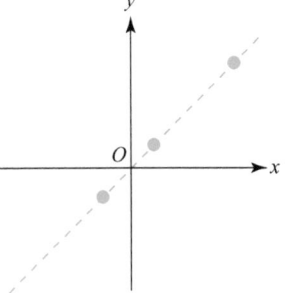

If every point on a line is mapped to itself under a transformation then it is known as a **line of invariant points**.

Example 6

Find the invariant points under the transformation given by $\begin{pmatrix} 2 & 3 \\ -1 & -2 \end{pmatrix}$

Let $\begin{pmatrix} x \\ y \end{pmatrix}$ be an invariant point. $\bullet$ ———

Then $\begin{pmatrix} 2 & 3 \\ -1 & -2 \end{pmatrix}\begin{pmatrix} x \\ y \end{pmatrix} = \begin{pmatrix} x \\ y \end{pmatrix}$

$\therefore 2x + 3y = x \Rightarrow x = -3y$ $\bullet$ ———

and $-x - 2y = y \Rightarrow x = -3y$

So there is a line of invariant points given by $x = -3y$ $\bullet$ ———

> **1** Write a general vector to represent the invariant point.

> **2** Form equations.

> **3** If the two equations simplify to the same thing then there is a line of invariant points.

In some cases, every point on a line will map to another point on the same line.

For example, consider a rotation of 180° around the origin. The point $(0, 0)$ is invariant but if you take any line through the origin then every other point on that line will map to a different point on the same line. So any line through the origin will be an invariant line.

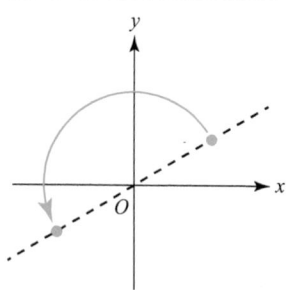

> **Key point**
>
> If every point on a line is mapped to another point on the same line then it is known as an **invariant line**. Lines of invariant points are a subset of invariant lines.

To find the equations of invariant lines use $y = mx + c$ to rewrite your general vector and $\begin{pmatrix} x' \\ y' \end{pmatrix}$ for the image of that vector. You should then use $y' = mx' + c$ to find the possible values of m and c

Example 7

Find the equations of the invariant lines of the transformation given by $\begin{pmatrix} 1 & 1 \\ 0 & 3 \end{pmatrix}$

$\begin{pmatrix} 1 & 1 \\ 0 & 3 \end{pmatrix}\begin{pmatrix} x \\ mx+c \end{pmatrix} = \begin{pmatrix} x' \\ y' \end{pmatrix}$ $\bullet$ ———

$x + mx + c = x'$

$3(mx + c) = y'$ $\bullet$ ———

So $3(mx + c) = m(x + mx + c) + c$ $\bullet$ ———

$(3m - m - m^2)x + 3c - mc - c = 0$ $\bullet$ ———

$(2m - m^2)x + 2c - mc = 0$

$2m - m^2 = 0 \Rightarrow m = 0 \text{ or } 2$ $\bullet$ ———

$2c - mc = 0$ $\bullet$ ———

Therefore when $m = 0$, $c = 0$

and when $m = 2$ you have $2c - 2c = 0$ so c can be any value.

Therefore the invariant lines are $y = 0$ and $y = 2x + c$ (for any c). $\bullet$ ———

> **1** Write the vector in a general form in terms of x

> **2** Form simultaneous equations.

> Using $y' = mx' + c$

> Collect terms in x

> **4** Equate coefficients of x

> **4** Equate constant terms.

> State the invariant lines.

1 Give the equation of the line of invariant points under the transformation given by each of these matrices.

a $\begin{pmatrix} 1 & -2 \\ 0 & 3 \end{pmatrix}$ b $\begin{pmatrix} 5 & 2 \\ 4 & 3 \end{pmatrix}$

2 Show that the origin is the only invariant point under the transformation given by each of these matrices.

a $\begin{pmatrix} 3 & -2 \\ 2 & 3 \end{pmatrix}$ b $\begin{pmatrix} 2 & 0 & 1 \\ 0 & 3 & -2 \\ 1 & 0 & -4 \end{pmatrix}$

3 In each of these cases in 2D, decide whether or not the order the transformations are applied affects the final image. Justify your answers.

a A reflection in the y-axis and a stretch parallel to the y-axis

b A rotation about the origin and an enlargement with centre the origin

c A reflection in the line $y = x$ and a stretch along the x-axis

4 Describe the combination of transformations that will be represented by each of these transformation matrices.

a $\begin{pmatrix} 3 & 0 \\ 0 & -2 \end{pmatrix}$ b $\begin{pmatrix} 0 & 4 \\ 4 & 0 \end{pmatrix}$

5 Find the equations of the invariant lines under each of these transformations.

a $\begin{pmatrix} 2 & 0 \\ 1 & -3 \end{pmatrix}$ b $\begin{pmatrix} 2 & -1 \\ -3 & 0 \end{pmatrix}$

6 a Suggest two different transformations that have an invariant line given by the equation $x = 0$

b Which of your transformations in part **a** has a line of invariant points given by $x = 0$?

7 The invariant lines under transformation **A** are given by the equations $x + y = c$ where c can be any value. Describe a possible transformation **A** geometrically.

8 The invariant lines under transformation **B** are given by the equations $y = kx$ where k can be any value. Describe transformation **B** geometrically.

9 Find the equations of the invariant lines under the transformation given by each of these matrices.

a $\begin{pmatrix} 1 & 2 \\ 2 & -1 \end{pmatrix}$ b $\begin{pmatrix} 3 & 0 \\ 0 & 3 \end{pmatrix}$ c $\begin{pmatrix} -\dfrac{3}{5} & \dfrac{4}{5} \\ \dfrac{4}{5} & \dfrac{3}{5} \end{pmatrix}$

10 a For each of these transformations, find either the invariant point or the equations of the invariant lines as appropriate.

i Reflection in the x-axis

ii Rotation of 90° around the origin

iii Stretch of scale factor 2 parallel to the x-axis

iv Reflection in the line $y = -x$

b For each of the invariant lines in part **a**, explain whether or not it is a line of invariant points.

11 A transformation is represented by the matrix $\mathbf{T} = \begin{pmatrix} -2 & 3 \\ -3 & 4 \end{pmatrix}$

a Find the line of invariant points under transformation **T**

b Show that all lines of the form $y = x + c$ are invariant lines of the transformation **T**

12 The matrix representing transformation **A**, a reflection in the line $x = y$, followed by transformation **B** is given by $\begin{pmatrix} -1 & 0 \\ 0 & 1 \end{pmatrix}$. Find the matrix that represents the transformation **B** followed by the transformation **A**

13 Give the invariant points, lines and planes for each of these transformations.

a $\begin{pmatrix} -1 & 0 & 0 \\ 0 & 1 & 0 \\ 0 & 0 & 1 \end{pmatrix}$ b $\begin{pmatrix} 1 & 0 & 0 \\ 0 & 0 & 1 \\ 0 & 1 & 0 \end{pmatrix}$

c $\begin{pmatrix} 0 & 1 & 0 \\ 0 & -1 & 0 \\ 0 & 0 & 1 \end{pmatrix}$ d $\begin{pmatrix} 1 & 0 & 0 \\ 1 & 0 & 0 \\ 0 & 0 & 1 \end{pmatrix}$

Full A Level

Fluency and skills

When you are working with square matrices you can calculate a value known as the **determinant**. The determinant has several uses. Most significantly, it enables you to find the inverse of a matrix.

> **Key point**
>
> If $\mathbf{A} = \begin{pmatrix} a & b \\ c & d \end{pmatrix}$ then the **determinant** of A is $ad - bc$ and is denoted $|\mathbf{A}|$ or $\det(\mathbf{A})$

To find the determinant of the matrix $\mathbf{A} = \begin{pmatrix} 5 & 2 \\ -1 & -3 \end{pmatrix}$ use the formula $\det(\mathbf{A}) = ad - bc$

$$\det(\mathbf{A}) = (5 \times -3) - (2 \times -1)$$
$$= 15 - -2$$
$$= -13$$

Try it on your calculator

You can use some calculators to find the determinant.

det(MatA)

26

Activity

Find out how to work out $\det\begin{pmatrix} 2 & -8 \\ 4 & -3 \end{pmatrix}$ on *your* calculator.

> **Key point**
>
> If $\det(\mathbf{A}) = 0$ then A is called a **singular matrix**.

Example 1

Find the value of k for which the matrix $\begin{pmatrix} 3 & -1 \\ -2 & k \end{pmatrix}$ is singular.

$\det\begin{pmatrix} 3 & -1 \\ -2 & k \end{pmatrix} = (3 \times k) - (-1 \times -2)$ 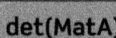   The determinant is given by $ad - bc$

$\therefore$ if matrix is singular then $3k - 2 = 0$ A singular matrix has a determinant of 0

$\Rightarrow k = \dfrac{2}{3}$

You can use this result about determinants:

> **Key point**
>
> For any square matrices A and B: $\det(\mathbf{AB}) \equiv \det(\mathbf{A}) \times \det(\mathbf{B})$

Example 2

The 2×2 matrices **A**, **B** and **C** are such that $\mathbf{C} = \mathbf{AB}$

Given that $\mathbf{C} = \begin{pmatrix} 2 & -4 \\ 3 & -5 \end{pmatrix}$ and $\det(\mathbf{A}) = \dfrac{1}{4}$, calculate the determinant of **B**

$\det(\mathbf{C}) = 2 \times (-5) - (-4) \times 3$ ● —————— Use $ad - bc$ to calculate the determinant of **C**

$\qquad = -10 + 12$

$\qquad = 2$

So $\dfrac{1}{4} \times \det(\mathbf{B}) = 2$ ● —————— Use the result $\det(\mathbf{AB}) \equiv \det(\mathbf{A}) \times \det(\mathbf{B})$

$\qquad \Rightarrow \det(\mathbf{B}) = 2 \times 4$

$\qquad\qquad = 8$

Recall that $\mathbf{I} = \begin{pmatrix} 1 & 0 \\ 0 & 1 \end{pmatrix}$ is the identity matrix.

The inverse of a matrix, **A**, is $\mathbf{A}^{-1}$ where $\mathbf{AA}^{-1} = \mathbf{A}^{-1}\mathbf{A} = \mathbf{I}$

Key point

If $\mathbf{A} = \begin{pmatrix} a & b \\ c & d \end{pmatrix}$ then the inverse matrix is given by $\mathbf{A}^{-1} = \dfrac{1}{\det(\mathbf{A})} \begin{pmatrix} d & -b \\ -c & a \end{pmatrix}$

Key point

Notice how this definition will not work for a matrix which is singular. Singular matrices do not have an inverse. Therefore, only non-singular matrices will have an inverse.

Example 3

Find the inverse of the matrix $\mathbf{B} = \begin{pmatrix} 2 & 5 \\ -1 & 4 \end{pmatrix}$

$\det(\mathbf{B}) = (2 \times 4) - (5 \times -1)$ ● —————— First calculate the determinant using $ad - bc$

$\qquad = 8 - -5$

$\qquad = 13$

$\mathbf{B}^{-1} = \dfrac{1}{13}\begin{pmatrix} 4 & -5 \\ 1 & 2 \end{pmatrix}$ ● —————— Use $\mathbf{B}^{-1} = \dfrac{1}{\det(\mathbf{B})}\begin{pmatrix} d & -b \\ -c & a \end{pmatrix}$

Checking answer:

$\mathbf{BB}^{-1} = \dfrac{1}{13}\begin{pmatrix} 2 & 5 \\ -1 & 4 \end{pmatrix}\begin{pmatrix} 4 & -5 \\ 1 & 2 \end{pmatrix}$ ● —————— You can check your answer by verifying that $\mathbf{BB}^{-1} = \mathbf{I}$

$\qquad = \dfrac{1}{13}\begin{pmatrix} 13 & 0 \\ 0 & 13 \end{pmatrix}$

$\qquad = \begin{pmatrix} 1 & 0 \\ 0 & 1 \end{pmatrix}$

$\qquad = \mathbf{I}$

Try it on your calculator

Some calculators can be used to find the inverse of a matrix.

MatA⁻¹

$$\begin{bmatrix} -4 & 3 \\ -3 & 2 \end{bmatrix}$$

Activity

Find out how to work out the inverse of $\begin{pmatrix} 2 & -3 \\ 3 & -4 \end{pmatrix}$ on *your* calculator.

Using the result $\det(\mathbf{AB}) \equiv \det(\mathbf{A}) \times \det(\mathbf{B})$ with $\mathbf{B} = \mathbf{A}^{-1}$ you can see that

$$\det(\mathbf{AA}^{-1}) \equiv \det(\mathbf{A}) \times \det(\mathbf{A}^{-1})$$

Then, since $\mathbf{AA}^{-1} = \mathbf{I}$ (the identity matrix), you have

$$\det(\mathbf{I}) \equiv \det(\mathbf{A}) \times \det(\mathbf{A}^{-1})$$

Think about the 2×2 identity matrix, $\mathbf{I}_2 = \begin{pmatrix} 1 & 0 \\ 0 & 1 \end{pmatrix}$, the determinant of this is $\det(\mathbf{I}_2) = 1 - 0 = 1$

The same will be true for an identity matrix of any size (you do not need to prove this fact).

Therefore, $1 \equiv \det(\mathbf{A}) \times \det(\mathbf{A}^{-1})$

$$\Rightarrow \det(\mathbf{A}^{-1}) \equiv \frac{1}{\det(\mathbf{A})}$$

You can use this result about the determinant of an inverse:

Key point

For any square, non-singular matrix $\mathbf{A}$: $\det(\mathbf{A}^{-1}) \equiv \dfrac{1}{\det(\mathbf{A})}$

Example 4

Given that the matrix $\mathbf{C}$ is self-inverse, deduce the possible values of $\det(\mathbf{C})$

C is self-inverse so $C^{-1} = C$

Therefore $\det(C^{-1}) = \det(C)$

$\Rightarrow \det(C) = \dfrac{1}{\det(C)}$ •──────── Use the result $\det(\mathbf{C}^{-1}) \equiv \dfrac{1}{\det(\mathbf{C})}$

$\Rightarrow [\det(C)]^2 = 1$

$\Rightarrow \det(C) = \pm 1$

1 Show that $\det\begin{pmatrix} -7 & 3 \\ 5 & -4 \end{pmatrix} = 13$

2 Calculate the determinants of each of these matrices.

a $\begin{pmatrix} 2 & 4 \\ 5 & 1 \end{pmatrix}$ **b** $\begin{pmatrix} -3 & 2 \\ 0 & 1 \end{pmatrix}$

c $\begin{pmatrix} 4a & 3 \\ -a & -2 \end{pmatrix}$ **d** $\begin{pmatrix} 3\sqrt{2} & -\sqrt{2} \\ -2\sqrt{2} & \sqrt{2} \end{pmatrix}$

3 Show that $\det\begin{pmatrix} a+b & 2a \\ 2b & a+b \end{pmatrix} = (a-b)^2$

4 Find the possible values of k given that
$\det\begin{pmatrix} 7 & k \\ -k & 2-k \end{pmatrix} = 2$

5 Decide whether each of these matrices is singular or non-singular. You must show your working.

a $\begin{pmatrix} 3 & 1 \\ -6 & 2 \end{pmatrix}$ **b** $\begin{pmatrix} 2 & 3 \\ 4 & 6 \end{pmatrix}$

c $\begin{pmatrix} -a & -2 \\ -2a & 4 \end{pmatrix}$ **d** $\begin{pmatrix} \sqrt{3} & 3 \\ -2\sqrt{3} & -6 \end{pmatrix}$

6 Find the values of x for which each of these matrices is singular.

a $\begin{pmatrix} x & 1 \\ 5 & 2 \end{pmatrix}$ **b** $\begin{pmatrix} x & -3 \\ -1 & x \end{pmatrix}$

c $\begin{pmatrix} 2x & x \\ -3 & x \end{pmatrix}$ **d** $\begin{pmatrix} 4 & -x \\ -x & x \end{pmatrix}$

7 Given that $\mathbf{A} = \begin{pmatrix} y+3 & y+5 \\ -1 & y-1 \end{pmatrix}$, calculate the possible values of y for which the matrix $\mathbf{A}$ has no inverse.

8 Given that $\mathbf{A} = \begin{pmatrix} -1 & 2 \\ 0 & 5 \end{pmatrix}$ and $\mathbf{B} = \begin{pmatrix} 0 & -3 \\ 2 & 4 \end{pmatrix}$

a Calculate the values of

   i $\det(\mathbf{A}) \times \det(\mathbf{B})$ **ii** $\det(\mathbf{A}) + \det(\mathbf{B})$

b Show that

   i $\det(\mathbf{AB}) = -30$ **ii** $\det(\mathbf{A}+\mathbf{B}) = -7$

9 The square matrices $\mathbf{A}$ and $\mathbf{B}$ have determinants $\dfrac{1}{3}$ and -6 respectively. Calculate the determinant of

a $\mathbf{AB}$ **b** $\mathbf{A}^2$

10 The 2×2 matrices $\mathbf{A}$, $\mathbf{B}$ and $\mathbf{C}$ are such that $\mathbf{C} = \mathbf{AB}$

Given that $\mathbf{C} = \begin{pmatrix} 3 & -2 \\ 4 & 8 \end{pmatrix}$, and $\det(\mathbf{A}) = -4$, calculate

a $\det(\mathbf{B})$ **b** $\det(\mathbf{A}^2)$

11 The non-singular 2×2 matrix, $\mathbf{M}$, has determinant 7

Calculate $\det(\mathbf{M}^{-1})$

12 The matrix $\mathbf{M} = \begin{pmatrix} 2 & a \\ 3 & 5 \end{pmatrix}$ has determinant k

Given that $\det(\mathbf{M}^{-1}) = \dfrac{1}{5}$

a State the value of k

b Calculate the value of a

13 Find the inverse of each of these matrices.

a $\begin{pmatrix} 7 & 2 \\ 4 & 5 \end{pmatrix}$ **b** $\begin{pmatrix} -4 & 3 \\ 8 & -1 \end{pmatrix}$

c $\begin{pmatrix} 2x & 5 \\ -x & -3 \end{pmatrix}$ **d** $\begin{pmatrix} 4\sqrt{5} & -\sqrt{5} \\ 2\sqrt{5} & \sqrt{20} \end{pmatrix}$

14 Show that the inverse of $\begin{pmatrix} 6 & 1 \\ -9 & -1 \end{pmatrix}$ is $\begin{pmatrix} -\dfrac{1}{3} & -\dfrac{1}{3} \\ 3 & 2 \end{pmatrix}$

15 $\mathbf{A} = \begin{pmatrix} b & 2a \\ b & 3a \end{pmatrix}$

    a Find the inverse of **A**, writing each element in its simplest form.

    b Show that $\det(\mathbf{A}^{-1}) = \dfrac{1}{\det(\mathbf{A})}$ for the matrix **A** given.

16 Show that the matrix $\begin{pmatrix} 3 & -2 \\ 4 & -3 \end{pmatrix}$ is self-inverse.

17 Find the values of a such that the matrix $\begin{pmatrix} a & -3 \\ 2 & -a \end{pmatrix}$ is self-inverse.

18 Find the value of a such that the matrix $\begin{pmatrix} 1 & 3-a \\ a-5 & 2-a \end{pmatrix}$ is self-inverse.

19 $\mathbf{B} = \begin{pmatrix} x+1 & -x \\ x-1 & 2x \end{pmatrix}$

    a Given that the determinant of **B** is 10, find the possible values of x

    b Write the inverse of **B** in terms of x

20 $\mathbf{T} = \begin{pmatrix} 2 & 0 \\ 0 & 2 \end{pmatrix}$

    a Show that $\mathbf{T}^{-1} = \dfrac{1}{2}\mathbf{I}$

    b Describe the transformations represented by **T** and $\mathbf{T}^{-1}$ geometrically.

21 **a** Write the matrix representing rotation of $135°$ anticlockwise about the origin.

    b Show that the inverse of this matrix is $\dfrac{1}{\sqrt{2}}\begin{pmatrix} -1 & 1 \\ -1 & -1 \end{pmatrix}$

22 For each of these transformations, explain whether it is always, sometimes or never self-inverse. For those which are sometimes self-inverse, state when this occurs.

    a Reflection in a coordinate axis

    b Enlargement centre the origin

    c Rotation about the origin

    d Reflection in the line $y = \pm x$

Reasoning and problem-solving

Matrices can be used to solve systems of linear equations.

Take, for example, the system of equations $ax + by = x'$

$$cx + dy = y'$$

If $\mathbf{M} = \begin{pmatrix} a & b \\ c & d \end{pmatrix}$ then $\mathbf{M}\begin{pmatrix} x \\ y \end{pmatrix} = \begin{pmatrix} a & b \\ c & d \end{pmatrix}\begin{pmatrix} x \\ y \end{pmatrix} = \begin{pmatrix} ax+by \\ cx+dy \end{pmatrix}$ so the equations can be written using matrices

as $\mathbf{M}\begin{pmatrix} x \\ y \end{pmatrix} = \begin{pmatrix} x' \\ y' \end{pmatrix}$

Then you can pre-multiply both sides by $\mathbf{M}^{-1}$ to give $\mathbf{M}^{-1}\mathbf{M}\begin{pmatrix} x \\ y \end{pmatrix} = \mathbf{M}^{-1}\begin{pmatrix} x' \\ y' \end{pmatrix}$

As $\mathbf{M}^{-1}\mathbf{M} = \mathbf{I}$, this can be rewritten as $\begin{pmatrix} x \\ y \end{pmatrix} = \mathbf{M}^{-1}\begin{pmatrix} x' \\ y' \end{pmatrix}$ and you can calculate the values of x and y

Strategy

To solve problems involving systems of linear equations

① Rewrite a system of linear equations using matrices.

② Pre-multiply or post-multiply a matrix by its inverse.

③ Use the fact that $\mathbf{A}\mathbf{A}^{-1} = \mathbf{A}^{-1}\mathbf{A} = \mathbf{I}$

Example 5

Use matrices to solve the simultaneous equations $2x+3y=11$
$$4x-y=-6$$

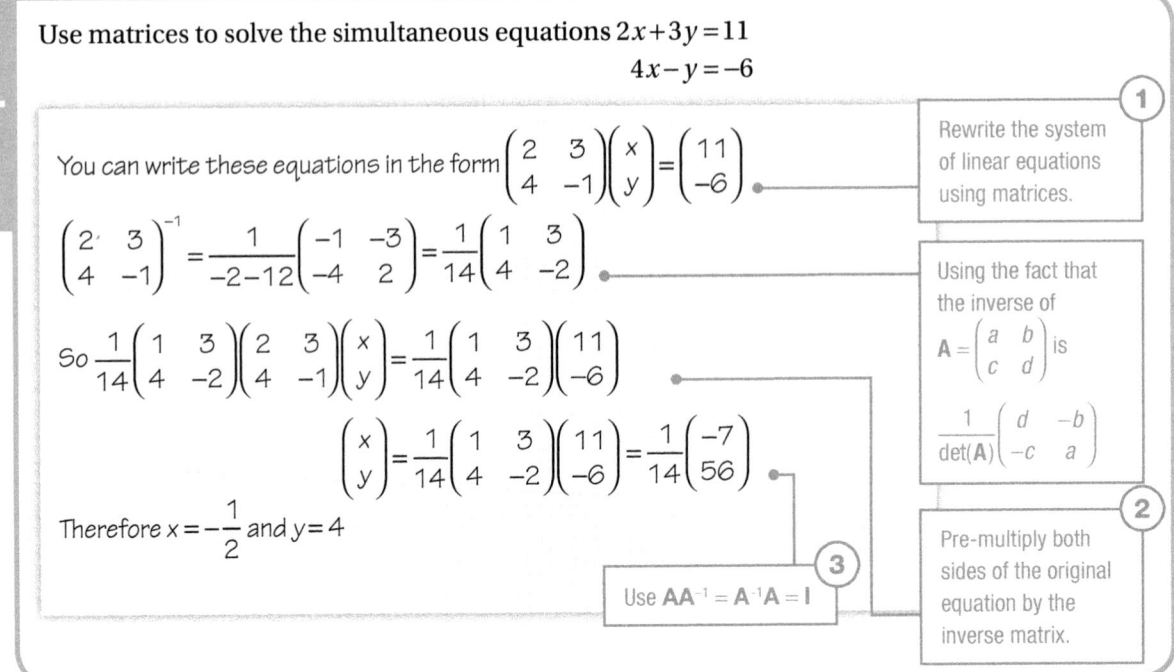

You can write these equations in the form $\begin{pmatrix} 2 & 3 \\ 4 & -1 \end{pmatrix}\begin{pmatrix} x \\ y \end{pmatrix}=\begin{pmatrix} 11 \\ -6 \end{pmatrix}$

$\begin{pmatrix} 2 & 3 \\ 4 & -1 \end{pmatrix}^{-1}=\dfrac{1}{-2-12}\begin{pmatrix} -1 & -3 \\ -4 & 2 \end{pmatrix}=\dfrac{1}{14}\begin{pmatrix} 1 & 3 \\ 4 & -2 \end{pmatrix}$

So $\dfrac{1}{14}\begin{pmatrix} 1 & 3 \\ 4 & -2 \end{pmatrix}\begin{pmatrix} 2 & 3 \\ 4 & -1 \end{pmatrix}\begin{pmatrix} x \\ y \end{pmatrix}=\dfrac{1}{14}\begin{pmatrix} 1 & 3 \\ 4 & -2 \end{pmatrix}\begin{pmatrix} 11 \\ -6 \end{pmatrix}$

$\begin{pmatrix} x \\ y \end{pmatrix}=\dfrac{1}{14}\begin{pmatrix} 1 & 3 \\ 4 & -2 \end{pmatrix}\begin{pmatrix} 11 \\ -6 \end{pmatrix}=\dfrac{1}{14}\begin{pmatrix} -7 \\ 56 \end{pmatrix}$

Therefore $x=-\dfrac{1}{2}$ and $y=4$

(1) Rewrite the system of linear equations using matrices.

Using the fact that the inverse of $\mathbf{A}=\begin{pmatrix} a & b \\ c & d \end{pmatrix}$ is $\dfrac{1}{\det(\mathbf{A})}\begin{pmatrix} d & -b \\ -c & a \end{pmatrix}$

(2) Pre-multiply both sides of the original equation by the inverse matrix.

(3) Use $\mathbf{AA}^{-1}=\mathbf{A}^{-1}\mathbf{A}=\mathbf{I}$

You can construct matrix proofs by pre-multiplying or post-multiplying both sides of an equation.

Example 6

Prove that $(\mathbf{AB})^{-1}=\mathbf{B}^{-1}\mathbf{A}^{-1}$

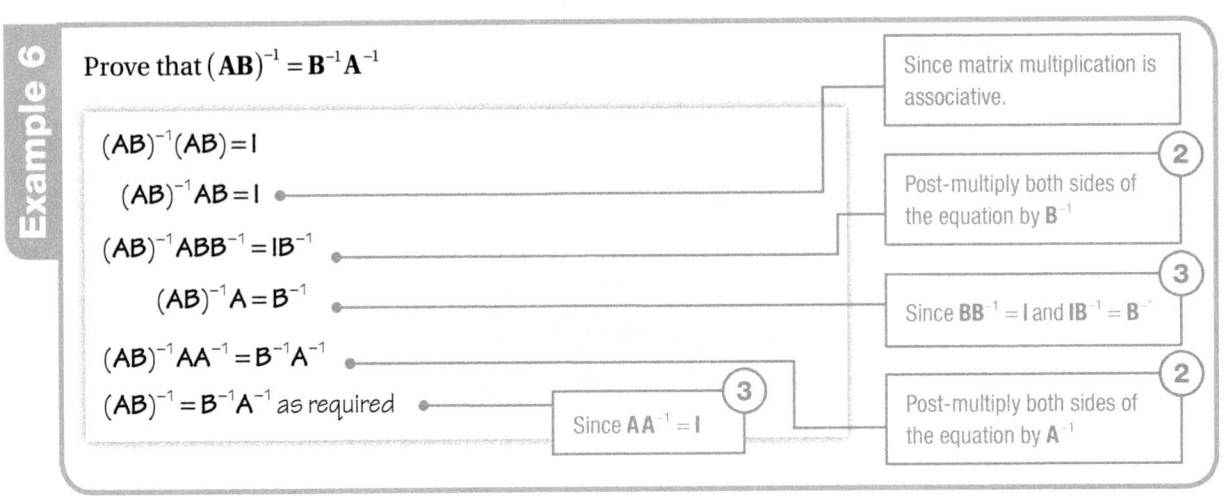

$(\mathbf{AB})^{-1}(\mathbf{AB})=\mathbf{I}$

$(\mathbf{AB})^{-1}\mathbf{AB}=\mathbf{I}$

$(\mathbf{AB})^{-1}\mathbf{ABB}^{-1}=\mathbf{IB}^{-1}$

$(\mathbf{AB})^{-1}\mathbf{A}=\mathbf{B}^{-1}$

$(\mathbf{AB})^{-1}\mathbf{AA}^{-1}=\mathbf{B}^{-1}\mathbf{A}^{-1}$

$(\mathbf{AB})^{-1}=\mathbf{B}^{-1}\mathbf{A}^{-1}$ as required

Since matrix multiplication is associative.

(2) Post-multiply both sides of the equation by $\mathbf{B}^{-1}$

(3) Since $\mathbf{BB}^{-1}=\mathbf{I}$ and $\mathbf{IB}^{-1}=\mathbf{B}^{-1}$

(2) Post-multiply both sides of the equation by $\mathbf{A}^{-1}$

(3) Since $\mathbf{AA}^{-1}=\mathbf{I}$

Key point

If $\mathbf{A}$ and $\mathbf{B}$ are non-singular matrices then $(\mathbf{AB})^{-1}=\mathbf{B}^{-1}\mathbf{A}^{-1}$

1   Use matrices to solve these pairs of simultaneous equations.

   a $4x - y = 11$
          $2x + 3y = -5$

   b $x - 5y = 0$
          $2x - 8y = -2$

   c $3x + 6y = 12$
          $x - 2y = 2$

   d $4x + 6y = 1$
          $-8x + 3y = 3$

2   Use matrices to find a solution to these systems of linear equations.

   a $x + y - 2z = 3$
          $2x - 3y + 5z = 4$
          $5x + 2y + z = -3$

   b $4x + 6y - z = -3$
          $2x - 3y = 2$
          $8y + 4z = 0$

3   Given that $\begin{pmatrix} a & 3 \\ -2a & -1 \end{pmatrix}\begin{pmatrix} x \\ y \end{pmatrix} = \begin{pmatrix} 5 \\ 10a \end{pmatrix}$,

   find expressions in terms of a for x and y
   Simplify your answers.

4   You are given the equations $(k+3)x - 2y = k - 1$ and $kx + y = k$

   Use matrices to find expressions in terms of k for x and y

5   Given that $\mathbf{A}\begin{pmatrix} -7 & -2 \\ 0 & 1 \end{pmatrix} = \begin{pmatrix} 14 & 8 \\ 7 & 0 \end{pmatrix}$, find the matrix $\mathbf{A}$

6   Given that $\begin{pmatrix} 3 & -5 \\ -1 & 2 \end{pmatrix}\mathbf{B} = \begin{pmatrix} 12 & -1 \\ -4 & 0 \end{pmatrix}$, find the matrix $\mathbf{B}$

7   Given that $\mathbf{P}\begin{pmatrix} 2a & b \\ -a & -b \end{pmatrix} = \begin{pmatrix} a & 2a \\ a & -b \end{pmatrix}$, find an expression for $\det(\mathbf{P})$ in terms of a and b
   Fully simplify your answer.

8   If $\mathbf{A} = \begin{pmatrix} 3 & 5 \\ 2 & 4 \end{pmatrix}$ and $\mathbf{AB} = \begin{pmatrix} 6 & 8 \\ -2 & 0 \end{pmatrix}$, find $\mathbf{B}$

9   If $\mathbf{B} = \begin{pmatrix} 2 & 3 \\ -3 & -1 \end{pmatrix}$ and $\mathbf{AB} = \begin{pmatrix} 10 & 15 \\ 11 & 13 \end{pmatrix}$, find the matrix $\mathbf{A}$

10 If $\mathbf{A}^2 = 2\mathbf{I}$, write a possible 2×2 matrix $\mathbf{A}$

11 Simplify each of these expressions involving the non-singular matrices $\mathbf{A}$ and $\mathbf{B}$

   a $(\mathbf{AB})^{-1}\mathbf{A}$ **b** $\mathbf{B}(\mathbf{A}^{-1}\mathbf{B})^{-1}$

12 If $\mathbf{A} = \mathbf{C}^{-1}\mathbf{BC}$, prove that $\mathbf{B} = \mathbf{CAC}^{-1}$

13 Prove that if $\mathbf{ABA}^{-1} = \mathbf{I}$ then $\mathbf{B} = \mathbf{I}$

14 Prove that $(\mathbf{ABC})^{-1} = \mathbf{C}^{-1}\mathbf{B}^{-1}\mathbf{A}^{-1}$ for non-singular matrices $\mathbf{A}$, $\mathbf{B}$ and $\mathbf{C}$

15 Prove that if $\mathbf{P}$ is self-inverse then $\mathbf{P}^2 = \mathbf{I}^{-1}$

16 Prove that if $\mathbf{PQP} = \mathbf{I}$ for non-singular matrices $\mathbf{P}$ and $\mathbf{Q}$, then $\mathbf{Q} = (\mathbf{P}^{-1})^2$

17 A point P is transformed by the matrix $\mathbf{T} = \begin{pmatrix} 2 & 1 \\ -3 & -4 \end{pmatrix}$ and the coordinates of the image of P are $(9, -11)$

   Find the coordinates of P

18 Prove that $\det(\mathbf{AB}) = \det(\mathbf{A})\det(\mathbf{B})$ for any non-singular 2×2 matrices $\mathbf{A}$ and $\mathbf{B}$

19 Prove that $\det(\mathbf{A}^{-1}) = [\det(\mathbf{A})]^{-1}$ for any non-singular 2×2 matrix $\mathbf{A}$

20 **a** Write the matrix that represents

      i   a rotation of angle θ anticlockwise around the origin,

      ii a rotation of angle θ clockwise around the origin.

   b Hence, deduce the identity $\sin^2 \theta + \cos^2 \theta \equiv 1$

21 By considering a square with one vertex at the origin, show that, under any stretch, $\mathbf{T}$, in the x- or y-direction (or both),

   a The image of the square will be a rectangle,

   b The area of the image will be the area of the original square multiplied by the determinant of $\mathbf{T}$

Chapter summary

- A matrix with m rows and n columns has order $m \times n$
- Matrices of the same order can be added by adding corresponding elements.
- A matrix can be multiplied by a constant by multiplying each of its elements by that constant.
- Two matrices are conformable for multiplication if the number of columns in the first matrix is the same as the number of rows in the second matrix.
- The product of an $n \times m$ matrix and an $m \times p$ matrix has order $n \times p$
- Under the transformation given by the matrix $\begin{pmatrix} a & b \\ c & d \end{pmatrix}$ the vector $\begin{pmatrix} 1 \\ 0 \end{pmatrix}$ is transformed to $\begin{pmatrix} a \\ c \end{pmatrix}$ and the vector $\begin{pmatrix} 0 \\ 1 \end{pmatrix}$ is transformed to $\begin{pmatrix} b \\ d \end{pmatrix}$
- The matrix representing transformation $\mathbf{A}$ followed by transformation $\mathbf{B}$ is given by the product $\mathbf{BA}$
- The identity matrix is a matrix with ones along the leading diagonal and zeros elsewhere, so in 2D, $\mathbf{I} = \begin{pmatrix} 1 & 0 \\ 0 & 1 \end{pmatrix}$
- The zero matrix is a matrix with 0 as every element, so in 2D, $\mathbf{0} = \begin{pmatrix} 0 & 0 \\ 0 & 0 \end{pmatrix}$
- The invariant points of the matrix $\mathbf{T}$ can be found by solving $\mathbf{T}\begin{pmatrix} x \\ y \end{pmatrix} = \begin{pmatrix} x \\ y \end{pmatrix}$
- The invariant lines of the matrix $\mathbf{T}$ can be found by solving $\mathbf{T}\begin{pmatrix} x \\ mx+c \end{pmatrix} = \begin{pmatrix} x' \\ mx'+c \end{pmatrix}$ to find the values of m and c
- The determinant of a 2×2 matrix is $\det\begin{pmatrix} a & b \\ c & d \end{pmatrix} = ad - bc$
- A matrix is singular if its determinant is zero.
- Under a transformation $\mathbf{T}$, area of image = area of original $\times |\det(\mathbf{T})|$
- If $\det(\mathbf{T}) > 0$ then the transformation represented by $\mathbf{T}$ preserves orientation.
- If $\mathbf{A} = \begin{pmatrix} a & b \\ c & d \end{pmatrix}$ then its inverse is $\mathbf{A}^{-1} = \dfrac{1}{\det(\mathbf{A})}\begin{pmatrix} d & -b \\ -c & a \end{pmatrix}$
- For any non-singular matrix $\mathbf{A}$, $\mathbf{AA}^{-1} = \mathbf{A}^{-1}\mathbf{A} = \mathbf{I}$

You should now be able to...	Try Questions
✔ Identify the order of a matrix.	1
✔ Add and subtract matrices of the same order and multiply matrices by a constant.	1, 2
✔ Multiply conformable matrices.	1, 2
✔ Apply a linear transformation given as a matrix to a point or vector.	3, 4
✔ Describe transformations given as matrices geometrically.	3, 4
✔ Write a linear transformation given geometrically as a matrix.	5, 6
✔ Find the matrix that represents a combination of two transformations.	7
✔ Find invariant points and lines.	8, 9
✔ Calculate the determinant of a 2×2 matrix.	10
✔ Understand what is meant by a singular matrix.	11
✔ Find the inverse of a 2×2 matrix.	12–14
✔ Use matrices to solve systems of linear equations.	15
✔ Describe the geometrical significance of solutions.	3, 4

PURE

1   $A = \begin{pmatrix} 6 & 2 \\ -4 & 12 \end{pmatrix}$, $B = \begin{pmatrix} 0 & 2 & -1 \\ 5 & 4 & 0 \end{pmatrix}$, $C = \begin{pmatrix} -1 & -5 \\ 4 & 0 \\ 0 & 2 \end{pmatrix}$ and $D = \begin{pmatrix} 8 & -2 & 3 \\ 0 & -1 & 0 \end{pmatrix}$

   a   State the order of each of the matrices.

   b   Find each of these matrices.

     i $B + D$

     ii $3C$

     iii $D - 2B$

     iv $\dfrac{1}{2}A$

   c   Calculate these matrix products where possible. If it is not possible, explain why not.

     i AB

     ii BA

     iii BC

     iv CA

     v AC

     vi CD

     vii A^2

     viii C^2

2 Given that $\mathbf{A} = \begin{pmatrix} a & 3 \\ a & 2 \end{pmatrix}$, $\mathbf{B} = \begin{pmatrix} -1 & a & 3 \\ 0 & 1 & a \\ 2a & -3 & 0 \end{pmatrix}$ and $\mathbf{C} = \begin{pmatrix} a & -1 \\ 0 & 3 \\ -2 & 2a \end{pmatrix}$,

find each of these matrices in terms of a

a $\mathbf{CA}$

b $\mathbf{BC}$

c $\mathbf{A}^2$

d $\mathbf{B}^2$

3 a Apply each of these transformations to the point $(-1, 2)$ and describe the effect of the transformation geometrically.

i $\begin{pmatrix} 0 & 1 \\ 1 & 0 \end{pmatrix}$ **ii** $\begin{pmatrix} 7 & 0 \\ 0 & 1 \end{pmatrix}$ **iii** $\begin{pmatrix} 0 & -1 \\ 1 & 0 \end{pmatrix}$ **iv** $\begin{pmatrix} -\dfrac{\sqrt{3}}{2} & \dfrac{1}{2} \\ -\dfrac{1}{2} & -\dfrac{\sqrt{3}}{2} \end{pmatrix}$

b A triangle has area 5 square units. Give the area of the image of the triangle under each of the transformations given in part **a**.

c Explain which of the transformations in part **a** preserves orientation.

4 Apply each of these transformations to the point $(3, -2, 4)$ and describe the effect of the transformation geometrically.

a $\begin{pmatrix} 1 & 0 & 0 \\ 0 & \dfrac{\sqrt{2}}{2} & -\dfrac{\sqrt{2}}{2} \\ 0 & \dfrac{\sqrt{2}}{2} & \dfrac{\sqrt{2}}{2} \end{pmatrix}$

b $\begin{pmatrix} 1 & 0 & 0 \\ 0 & 1 & 0 \\ 0 & 0 & -1 \end{pmatrix}$

5 Write down a 2×2 matrix to represent each of these transformations.

a Reflection in the line $x = 0$

b Rotation of $315°$ anticlockwise around the origin

c Enlargement of scale factor -3, centre the origin.

6 Write down a 3×3 matrix to represent each of these 3D transformations.

a Rotation of $45°$ anticlockwise around the y-axis

b Reflection in the plane $y = 0$

7 Write down the single 2×2 matrix that represents each of these transformations.

a Rotation of $270°$ anticlockwise around the origin followed by a stretch of scale factor 2 parallel to the y-axis

b Reflection in the line $y = -x$ followed by reflection in the x-axis

c Reflection in the x-axis followed by rotation of $150°$ anticlockwise about the origin.

8 Find the invariant points under each of these transformations.

a $\begin{pmatrix} -1 & 0 \\ 4 & -1 \end{pmatrix}$

b $\begin{pmatrix} 2 & -3 \\ -1 & 4 \end{pmatrix}$

9 Find all of the invariant lines under each of these transformations.

a $\begin{pmatrix} 0 & 2 \\ 3 & 1 \end{pmatrix}$

b $\dfrac{1}{5}\begin{pmatrix} -4 & 3 \\ 3 & 4 \end{pmatrix}$

10 Work out the determinant of each of these matrices.

a $\begin{pmatrix} 2 & -1 \\ 4 & -3 \end{pmatrix}$

b $\begin{pmatrix} 3a & b \\ -2a & b \end{pmatrix}$

11 Calculate the values of x for which these matrices are singular.

a $\begin{pmatrix} 3x & 4 \\ 5 & 5 \end{pmatrix}$

b $\begin{pmatrix} x & -1 \\ 4 & -2x \end{pmatrix}$

12 Find the inverse, if it exists, of each of these matrices. If it does not exist, explain why not.

a $\begin{pmatrix} 6 & 3 \\ -1 & -2 \end{pmatrix}$

b $\begin{pmatrix} 0 & 1 \\ 3 & -2 \end{pmatrix}$

c $\begin{pmatrix} 3 & -6 \\ -4 & 8 \end{pmatrix}$

d $\begin{pmatrix} 5a & 2b \\ 7a & 3b \end{pmatrix}$

13 Given that $\mathbf{A} = \begin{pmatrix} 5 & 7 \\ 1 & -4 \end{pmatrix}$ and $\mathbf{AB} = \begin{pmatrix} 10 & 22 \\ 2 & -1 \end{pmatrix}$, work out matrix $\mathbf{B}$

14 Given that $\mathbf{C} = \begin{pmatrix} 9 & -2 \\ 5 & 1 \end{pmatrix}$ and $\mathbf{BC} = \begin{pmatrix} -17 & -11 \\ -60 & 7 \end{pmatrix}$, work out matrix $\mathbf{B}$

15 Use matrices to solve each pair of simultaneous equations.

a $x+6y=7$
$3x-y=-17$

b $5x-7y=8$
$10x+3y=-1$

History

The Russian mathematician **Andrey Andreyevich Markov** (1856 – 1922) used matrices in his work on **stochastic processes**. Stochastic processes are collections of random variables. They can be used to show how a system might change over time.

There are now many applications of **Markov processes** including random walks, the Gambler's ruin problem, modelling queues arriving at an airport, exchange rates and the PageRank algorithm.

ICT

Here is an example of a Markov process that helps to predict long-term weather probabilities.

In a simplified model the weather on any one day can either be dry or wet.

Assume that the weather today is dry so the probabilities of dry and wet are given by $W_0 = [1 \quad 0]$

Let $P = \begin{bmatrix} 0.75 & 0.25 \\ 0.6 & 0.4 \end{bmatrix}$ where $P = \begin{bmatrix} P(\text{dry tomorrow} \mid \text{dry today}) & P(\text{wet tomorrow} \mid \text{dry today}) \\ P(\text{dry tomorrow} \mid \text{wet today}) & P(\text{wet tomorrow} \mid \text{wet today}) \end{bmatrix}$

It follows that the weather tomorrow is given by

$W_1 = W_0 P = [1 \quad 0] \begin{bmatrix} 0.75 & 0.25 \\ 0.6 & 0.4 \end{bmatrix} = [0.75 \quad 0.25]$

The weather the next day would be given by

$W_2 = W_1 P = [0.75 \quad 0.25] \begin{bmatrix} 0.75 & 0.25 \\ 0.6 & 0.4 \end{bmatrix} = [0.7125 \quad 0.2875]$

Using a spreadsheet, find (to 4 decimal places) the 1×2 matrix that the weather probabilities tend towards.

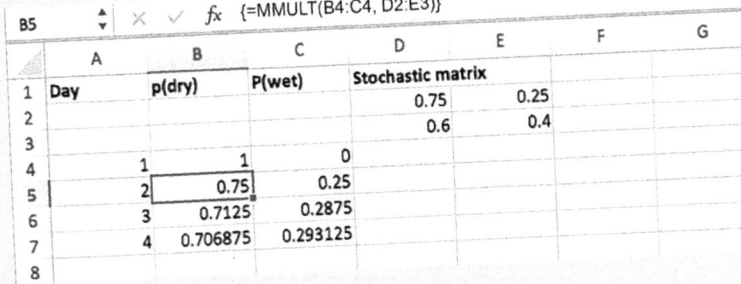

B5	⇕	× ✓	f_x {=MMULT(B4:C4, D2:E3)}				
	A	B	C	D	E	F	G
1	Day	p(dry)	P(wet)	Stochastic matrix			
2				0.75	0.25		
3				0.6	0.4		
4	1	1	0				
5	2	0.75	0.25				
6	3	0.7125	0.2875				
7	4	0.706875	0.293125				
8							

Research

Matrices can be used to solve systems of linear equations using a method now known as **Gaussian elimination**. The method was first invented in China before being reinvented in Europe in the 1700s.

It is named after **Carl Friedrich Gauss** after the adoption, by professional computers, of a specialised notation that Gauss devised.

Find out about Gaussian elimination.

Karl Friedrich Gauß.

1. Given that $\mathbf{M} = \begin{pmatrix} 5 & 2 \\ -3 & 1 \end{pmatrix}$ and $\mathbf{N} = \begin{pmatrix} -4 & 3 \\ 0 & 2 \end{pmatrix}$, calculate

    a $3\mathbf{M}$ **b** $-2\mathbf{N}$ **c** $\mathbf{M} - \mathbf{N}$ **d** $\mathbf{MN}$ **[4 marks]**

2. You are told that $\mathbf{M} = \begin{pmatrix} 1 & 2 \\ 4 & -3 \end{pmatrix}$ and $\mathbf{N} = \begin{pmatrix} 2 & 7 \\ 5 & 1 \end{pmatrix}$

    a   Calculate $\mathbf{MN}$ **b**   Calculate $\mathbf{NM}$ **c**   What does this say about $\mathbf{M}$ and $\mathbf{N}$?   **[3]**

3. Given $\mathbf{M} = \begin{pmatrix} 2 & 7 \\ 9 & 11 \end{pmatrix}$, calculate

    a $\mathbf{MI}$ **b** $\mathbf{0} + 2\mathbf{M}$ **c** $\mathbf{M}^2$ **[3]**

4. You are told that $\begin{pmatrix} a & -1 \\ 2 & 3 \end{pmatrix}\begin{pmatrix} b & 5 \\ 6 & 2 \end{pmatrix} \equiv \begin{pmatrix} -2 & 18 \\ 20 & c \end{pmatrix}$ is an identity. Find a, b and c **[2]**

5. Each of these matrices represents a transformation. Fill in the blanks in each sentence.

    a   The matrix $\begin{pmatrix} 6 & 0 \\ 0 & 6 \end{pmatrix}$ describes an _____ with scale factor _____.

    b   The matrix $\begin{pmatrix} 0 & -1 \\ 1 & 0 \end{pmatrix}$ describes a rotation about _____ by _____ in the anticlockwise direction.

    c   The matrix $\begin{pmatrix} 1 & 0 \\ 0 & -1 \end{pmatrix}$ describes a _____ in _____.   **[6]**

6. $\mathbf{M}$ is the matrix which describes a rotation by 135° anticlockwise about the origin. $\mathbf{N}$ is the matrix which describes a rotation by 45° clockwise about the origin.

    a   Write matrices $\mathbf{M}$ and $\mathbf{N}$

    b   Calculate $\mathbf{MN}$

    c   What effect does $\mathbf{MN}$ have and how does it relate to $\mathbf{M}$ and $\mathbf{N}$?   **[5]**

7. Calculate any invariant points for these matrices.

    a $\begin{pmatrix} 2 & -1 \\ 2 & 3 \end{pmatrix}$

    b $\begin{pmatrix} 5 & 2 \\ -4 & -1 \end{pmatrix}$

    c $\begin{pmatrix} 1 & 0 \\ 2 & 7 \end{pmatrix}$ **[6]**

8. Find the invariant lines under these transformations.

    a   Reflection in the y-axis in 2D space

    b   Rotation by 180° clockwise about the z-axis in 3D space.   **[2]**

9 The matrix $\mathbf{M} = \begin{pmatrix} 4 & 1 \\ 3 & 7 \end{pmatrix}$ has the inverse $\mathbf{M}^{-1} = \dfrac{1}{25}\begin{pmatrix} 7 & -1 \\ -3 & 4 \end{pmatrix}$

Find the solution to the equation $\begin{pmatrix} 4 & 1 \\ 3 & 7 \end{pmatrix}\begin{pmatrix} x \\ y \end{pmatrix} = \begin{pmatrix} 10 \\ 13 \end{pmatrix}$ **[2]**

10 The matrix $\mathbf{M} = \begin{pmatrix} 3 & 10 \\ 7 & 5 \end{pmatrix}$ has the inverse $\mathbf{M}^{-1} = \dfrac{1}{55}\begin{pmatrix} -5 & 10 \\ 7 & -3 \end{pmatrix}$

Find the solution to the equation $\begin{pmatrix} 3 & 10 \\ 7 & 5 \end{pmatrix}\begin{pmatrix} x \\ y \end{pmatrix} = \begin{pmatrix} 42 \\ 43 \end{pmatrix}$ **[2]**

11 The determinant of the matrix $\begin{pmatrix} 3 & 1 \\ 6 & a \end{pmatrix}$ is $3a - 6$

    a For what value of a does the matrix $\begin{pmatrix} 3 & 1 \\ 6 & a \end{pmatrix}$ have no inverse?

    b For that value of a find any invariant points under $\begin{pmatrix} 3 & 1 \\ 6 & a \end{pmatrix}$ **[4]**

12 Calculate the determinants of these matrices.

    a $\begin{pmatrix} 6 & 0 \\ 3 & 2 \end{pmatrix}$

    b $\begin{pmatrix} 1 & 5 \\ 2 & 10 \end{pmatrix}$ **[4]**

13 Find the inverses of these matrices.

    a $\begin{pmatrix} 9 & 16 \\ 5 & 9 \end{pmatrix}$

    b $\begin{pmatrix} 4 & -2 \\ 5 & 3 \end{pmatrix}$ **[8]**

14 a Find the inverses of $\mathbf{A} = \begin{pmatrix} 3 & 1 \\ 5 & 2 \end{pmatrix}$ and $\mathbf{B} = \begin{pmatrix} -2 & 4 \\ 1 & -1 \end{pmatrix}$

    b Calculate $\mathbf{B}^{-1}\mathbf{A}^{-1}$

    c Find the inverse of $\mathbf{B}^{-1}\mathbf{A}^{-1}$

    d How does this relate to $\mathbf{A}$ and $\mathbf{B}$? **[6]**

6 Vectors 1

Vectors are used extensively in engineering. One area of engineering where they are particularly important is in the design of electrical circuits. The size of circuit varies from the design of heavy engineering plants through to the design of miniature electronic circuits. Vectors can be used to represent current and voltage in an alternating current (A. C.) circuit. These vectors are considered to be rotating to provide different sinusoidal waves and, when two or more are brought together, they can be added using vector addition that takes into account not only the magnitude of the voltage, but also the phase angle(s) between them.

The applications of vector methods are also used in all branches of mechanical, civil and structural engineering. A particularly important use is to model forces. They are applied to calculate forces throughout buildings at the design stage to ensure that buildings are structurally sound and can withstand extreme weather, earthquakes, etc.

Orientation

What you need to know

KS4
- The equation of a line.

Maths Ch1
- Parallel and perpendicular lines.
- Points of intersection.
- Solving simultaneous equations.

Maths Ch6
- Position vector and displacements.
- Magnitude and direction.
- i, j form.

What you will learn

- To write and use the equation of a line in Cartesian and vector form.
- To decide whether lines are intersecting, parallel or skew and determine any points of intersection.
- To calculate the scalar product and use it to find angles and show lines are perpendicular.
- To find points of intersection, angles and calculate distances between points and lines.

What this leads to

Ch23 Further vectors
- The vector product.
- The scalar triple product.
- Lines and planes.

Ch8 Momentum
- Momentum and impulse in two dimensions.

6.1 The vector equation of a line

Fluency and skills

See Maths Ch6.1
For a reminder of magnitude and direction of vectors.

Vectors are used to describe magnitude and direction in 2D. In this section, you will learn how to express the equation of a 2D line and a 3D line in terms of vectors, as well as how to convert these vector equations to Cartesian form by using parametric equations.

In 3D, the vector $\begin{pmatrix} x \\ y \\ z \end{pmatrix}$ can also be written $x\mathbf{i} + y\mathbf{j} + z\mathbf{k}$ where $\mathbf{i}$, $\mathbf{j}$ and $\mathbf{k}$ are perpendicular unit vectors.

Vectors can also be used to describe points, lines and planes in 2D or 3D. Suppose you have a line that passes through the point with position vector $\mathbf{a}$ and is parallel to vector $\mathbf{b}$

Then the general point on this line will have position vector $\mathbf{r}$ where $\mathbf{r} = \mathbf{a} + \lambda\mathbf{b}$. You can see this from the triangle in the diagram. λ is a scalar quantity. It can take any value and determines how far along the line you move.

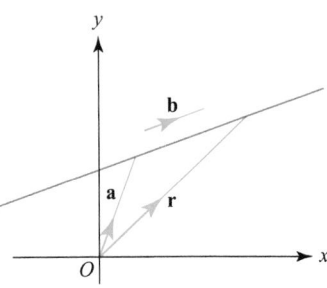

See Maths Ch6.1
For a reminder of finding parallel vectors.

Therefore, in either 2D or 3D:

Key point

$\mathbf{r} = \mathbf{a} + \lambda\mathbf{b}$ is the vector equation of a line which is parallel to the vector $\mathbf{b}$ and which passes through the point with position vector $\mathbf{a}$

Example 1

Work out the vector equation of the line passing through the points with position vectors $\begin{pmatrix} 1 \\ -3 \\ 2 \end{pmatrix}$ and $\begin{pmatrix} 3 \\ 0 \\ -1 \end{pmatrix}$

First, find a vector in the direction of the line.

$\begin{pmatrix} 3 \\ 0 \\ -1 \end{pmatrix} - \begin{pmatrix} 1 \\ -3 \\ 2 \end{pmatrix} = \begin{pmatrix} 2 \\ 3 \\ -3 \end{pmatrix}$ is parallel to the line

To find a vector parallel to the line, subtract the position vector of a point on the line from the position vector of another point on the line.

Therefore the equation of the line is $\mathbf{r} = \begin{pmatrix} 1 \\ -3 \\ 2 \end{pmatrix} + \lambda\begin{pmatrix} 2 \\ 3 \\ -3 \end{pmatrix}$

You can use either of the position vectors given.

Or, alternatively, $\mathbf{r} = \begin{pmatrix} 3 \\ 0 \\ -1 \end{pmatrix} + \lambda\begin{pmatrix} 2 \\ 3 \\ -3 \end{pmatrix}$

Suppose you have a line which is parallel to vector $\mathbf{b} = (b_1, b_2, b_3)$ and passes through the point with position vector $\mathbf{a} = (a_1, a_2, a_3)$

Let the general point on this line be $\mathbf{r} = (x, y, z)$. Then the equation of the line is

$(x, y, z) = (a_1, a_2, a_3) + \lambda(b_1, b_2, b_3)$

Therefore the line can be written using **parametric equations**: $x = a_1 + \lambda b_1, y = a_2 + \lambda b_2, z = a_3 + \lambda b_3$

These can be rearranged to give $\dfrac{x - a_1}{b_1} = \dfrac{y - a_2}{b_2} = \dfrac{z - a_3}{b_3} (= \lambda)$, which are the **Cartesian equations of the line**.

> **Key point**
>
> The **Cartesian equations of a line** parallel to the vector $\mathbf{b} = (b_1, b_2, b_3)$ and passing through the point with position vector $\mathbf{a} = (a_1, a_2, a_3)$ are $\dfrac{x - a_1}{b_1} = \dfrac{y - a_2}{b_2} = \dfrac{z - a_3}{b_3}$

Example 2

The Cartesian equations of a line are $\dfrac{x-3}{2} = \dfrac{y+1}{4} = \dfrac{z-2}{-5}$

Give the vector equation of the line.

The direction vector is $\mathbf{b} = \begin{pmatrix} 2 \\ 4 \\ -5 \end{pmatrix}$

The position vector is $\mathbf{a} = \begin{pmatrix} 3 \\ -1 \\ 2 \end{pmatrix}$

So the vector equation is $\mathbf{r} = \begin{pmatrix} 3 \\ -1 \\ 2 \end{pmatrix} + t \begin{pmatrix} 2 \\ 4 \\ -5 \end{pmatrix}$

> Since the Cartesian equations are of the form
> $$\dfrac{x - a_1}{b_1} = \dfrac{y - a_2}{b_2} = \dfrac{z - a_3}{b_3}$$

> Use $\mathbf{r} = \mathbf{a} + t\mathbf{b}$

Example 3

Give the Cartesian equations of the line with vector equation $\mathbf{r} = \mathbf{i} + 2\mathbf{j} - \mathbf{k} + s(3\mathbf{i} - \mathbf{k})$

The direction vector is $\begin{pmatrix} 3 \\ 0 \\ -1 \end{pmatrix}$

The position vector is $\begin{pmatrix} 1 \\ 2 \\ -1 \end{pmatrix}$

The line will lie on the plane $y = 2$

Therefore the Cartesian equations of the line are

$\dfrac{x-1}{3} = \dfrac{z+1}{-1}, y = 2$

> Use $\mathbf{r} = \mathbf{a} + t\mathbf{b}$

> Since the j-component in the direction vector is zero, don't write $\dfrac{y-2}{0}$

You may have a line that lies on one of the planes $x = a_1$, $y = a_2$ or $z = a_3$.

For example, $\dfrac{x - a_1}{b_1} = \dfrac{y - a_2}{b_2}, z = a_3$ lies on the plane $z = a_3$, so the $\mathbf{k}$ component of the direction vector will be zero so the vector equation is $\mathbf{r} = (a_1, a_2, a_3) + s(b_1, b_2, b_3)$.

To determine whether or not a point lies on a line, you can make an equation by using $\mathbf{r} = \mathbf{a} + \lambda \mathbf{b}$ and then substitute the position vector of the given point for $\mathbf{r}$. Then, by considering the components of $\mathbf{i}$, $\mathbf{j}$ and $\mathbf{k}$ separately, you can determine whether or not there is a unique value of λ

Example 4

Establish whether or not the point $(-5, 5, 6)$ lies on the line $\mathbf{r} = \begin{pmatrix} 3 \\ 1 \\ 0 \end{pmatrix} + \lambda \begin{pmatrix} 4 \\ -2 \\ -3 \end{pmatrix}$

$\begin{pmatrix} -5 \\ 5 \\ 6 \end{pmatrix} = \begin{pmatrix} 3 \\ 1 \\ 0 \end{pmatrix} + \lambda \begin{pmatrix} 4 \\ -2 \\ -3 \end{pmatrix}$

Form an equation and substitute in the position vector of the point in place of $\mathbf{r}$

$\mathbf{i}$ components: $-5 = 3 + 4\lambda \Rightarrow \lambda = -2$

$\mathbf{j}$ components: $5 = 1 - 2\lambda \Rightarrow \lambda = -2$

$\mathbf{k}$ components: $6 = -3\lambda \Rightarrow \lambda = -2$

Find the value of λ arising from each of the components.

Since you get the same value of λ for each component, $(-5, 5, 6)$ does lie on the line.

If λ was not the same for all three equations, then you would conclude that the point is not on the line.

Exercise 6.1A Fluency and skills

1 Write a vector equation of the line that is parallel to the vector $\begin{pmatrix} 1 \\ -2 \\ 0 \end{pmatrix}$ and passes through the point with position vector $\begin{pmatrix} 0 \\ 4 \\ 3 \end{pmatrix}$

2 Write a vector equation of the line that is parallel to the vector $5\mathbf{i} - 3\mathbf{j} + \mathbf{k}$ and passes through the point with position vector $7\mathbf{i} - \mathbf{j} + 2\mathbf{k}$

3 Write a vector equation of the line that passes through the points with position vectors $\begin{pmatrix} -7 \\ 2 \\ -5 \end{pmatrix}$ and $\begin{pmatrix} 4 \\ -9 \\ -2 \end{pmatrix}$

4 Write a vector equation of the line that passes through the points with position vectors $2\mathbf{i} + 3\mathbf{k}$ and $2\mathbf{i} + \mathbf{j} - \mathbf{k}$

5 Write Cartesian equations of the line that is parallel to the vector $\begin{pmatrix} 0 \\ 4 \\ -1 \end{pmatrix}$ and passes through the point with position vector $\begin{pmatrix} 5 \\ 2 \\ -7 \end{pmatrix}$

6 Write the Cartesian equations of the line that is parallel to the vector $8\mathbf{i} - 2\mathbf{j} + 3\mathbf{k}$ and passes through the point with position vector $4\mathbf{i} - 2\mathbf{j}$

7 Write the Cartesian equations of the line that passes through the points with position vectors $\begin{pmatrix} 9 \\ 4 \\ -3 \end{pmatrix}$ and $\begin{pmatrix} 1 \\ 0 \\ 5 \end{pmatrix}$

8 Write the Cartesian equations of the line that passes through the points with position vectors $5\mathbf{j} + \mathbf{k}$ and $4\mathbf{i} - \mathbf{j} - \mathbf{k}$

9 Convert each of these vector equations to Cartesian form.

a $\mathbf{r} = \begin{pmatrix} 0 \\ 1 \\ -2 \end{pmatrix} + \lambda \begin{pmatrix} -4 \\ 0 \\ 8 \end{pmatrix}$ b $\mathbf{r} = \begin{pmatrix} -3 \\ 0 \\ 7 \end{pmatrix} + \lambda \begin{pmatrix} 5 \\ -1 \\ 3 \end{pmatrix}$

10 Convert each of these Cartesian equations into the form $\mathbf{r} = \mathbf{a} + \lambda \mathbf{b}$

a $\frac{x-5}{2} = \frac{y-1}{-3} = \frac{z+3}{1}$ b $\frac{x+2}{4} = \frac{y}{-2} = \frac{5-z}{3}$

11 Establish whether the point $(3, 2, 1)$ lies on each of these lines.

a $\mathbf{r} = \begin{pmatrix} 1 \\ -4 \\ 3 \end{pmatrix} + \lambda \begin{pmatrix} 1 \\ 3 \\ -1 \end{pmatrix}$ b $\mathbf{r} = \begin{pmatrix} 5 \\ 2 \\ 0 \end{pmatrix} + \lambda \begin{pmatrix} 2 \\ 0 \\ 1 \end{pmatrix}$

12 Establish whether the point $(-2, 0, 4)$ lies on each of these lines.

a $\frac{x-1}{3} = \frac{y+3}{-3} = \frac{z-0}{4}$ b $\frac{x+8}{3} = \frac{y+2}{1} = \frac{z-6}{-1}$

Reasoning and problem-solving

A vector equation of a line is not unique. There are lots of points you can use to set the vector from the origin and you can use any vector which is parallel to the line to give its direction, so there are many correct vector equations for any one line. You need to be able to decide whether or not two equations represent the same line.

Strategy 1

To check whether two equations represent the same line

1. Check whether their direction vectors are equivalent.

2. Check whether they have a point in common.

3. If they have the same direction and a point in common then they must represent the same line.

Example 5

Decide which of these equations represent the same line.

$$L_1 : \mathbf{r} = \begin{pmatrix} 9 \\ 1 \\ -2 \end{pmatrix} + \lambda \begin{pmatrix} -2 \\ -1 \\ 1 \end{pmatrix} \qquad L_2 : \mathbf{r} = \begin{pmatrix} 3 \\ 0 \\ 4 \end{pmatrix} + \lambda \begin{pmatrix} 4 \\ 2 \\ -2 \end{pmatrix} \qquad L_3 : \frac{x-1}{2} = \frac{y+3}{1} = \frac{z-2}{-1}$$

All three lines have the same direction since $\begin{pmatrix} -2 \\ -1 \\ 1 \end{pmatrix} = -\begin{pmatrix} 2 \\ 1 \\ -1 \end{pmatrix}$ and $\begin{pmatrix} 4 \\ 2 \\ -2 \end{pmatrix} = 2\begin{pmatrix} 2 \\ 1 \\ -1 \end{pmatrix}$

If a vector is a multiple of another vector then they are parallel. (1)

Take the point $(1, -3, 2)$ which you know lies on line L_3

Alternatively, you could pick a point you know lies on line L_1 or on line L_2 (2)

Check if this point lies on L_1: $\begin{pmatrix} 1 \\ -3 \\ 2 \end{pmatrix} = \begin{pmatrix} 9 \\ 1 \\ -2 \end{pmatrix} + \lambda \begin{pmatrix} -2 \\ -1 \\ 1 \end{pmatrix}$

$9 - 2\lambda = 1 \Rightarrow \lambda = 4$
$1 - \lambda = -3 \Rightarrow \lambda = 4$
$-2 + \lambda = 2 \Rightarrow \lambda = 4$

Since the same value of λ is found for each of the components. (2)

Therefore $(1, -3, 2)$ lies on line L_1

So L_1 and L_3 are equations of the same line.

Since they have the same direction and have a point in common. (3)

Check if this point lies on L_2: $\begin{pmatrix} 1 \\ -3 \\ 2 \end{pmatrix} = \begin{pmatrix} 3 \\ 0 \\ 4 \end{pmatrix} + \lambda \begin{pmatrix} 4 \\ 2 \\ -2 \end{pmatrix}$

$3 + 4\lambda = 1 \Rightarrow \lambda = -\frac{1}{2}$
$2\lambda = -3 \Rightarrow \lambda = -\frac{3}{2}$
$4 - 2\lambda = 2 \Rightarrow \lambda = 1$

So $(1, -3, 2)$ does not lie on L_2

Therefore L_2 does not represent the same line as L_1 and L_3

L_2 is parallel to the other lines but does not have a point in common. (3)

If you have two lines in 2D then they must either intersect or be parallel. However, with two lines in 3D they could also be **skew**, that is, they pass one another on different planes so do not intersect but are also not parallel.

For example, in the cuboid shown:

- *AB* and *DC* are parallel
- *AB* and *AD* intersect
- *AB* and *GH* are skew (they will never intersect however far they are extended, but they are not parallel)

You already know that lines are parallel if their direction vectors are multiples of each other. To decide whether or not two lines intersect, you must attempt to solve their equations simultaneously and see if they are consistent.

Strategy 2

To establish if lines are parallel, intersecting or skew

(1) Check whether their direction vectors are multiples of each other, in which case they are parallel. You can then use strategy 1 above to determine whether or not they are the same line.

(2) If the lines are not parallel, attempt to solve their equations simultaneously.

(3) If the solution is consistent then the lines intersect and you can find the point of intersection.

(4) If the solutions leads to an inconsistency then the lines are skew.

Example 6

For each pair of lines, decide whether they are parallel, intersecting or skew.

If they intersect, find their point of intersection.

a $L_1 : \mathbf{r} = 2\mathbf{i} - \mathbf{j} + \lambda(3\mathbf{i} + \mathbf{j} - 2\mathbf{k})$, $\quad L_2 : 3\mathbf{i} + 2\mathbf{j} - \mathbf{k} + \mu(2\mathbf{i} - \mathbf{k})$

b $L_1 : \mathbf{r} = \begin{pmatrix} 1 \\ 0 \\ -8 \end{pmatrix} + \lambda \begin{pmatrix} 2 \\ -2 \\ -3 \end{pmatrix}$, $\quad L_2 : \mathbf{r} = \begin{pmatrix} -4 \\ 7 \\ -1 \end{pmatrix} + \mu \begin{pmatrix} 3 \\ -1 \\ -5 \end{pmatrix}$

a They are not parallel. ● — (1) Since their direction vectors are not multiples of each other.

$$\begin{pmatrix} 2 \\ -1 \\ 0 \end{pmatrix} + \lambda \begin{pmatrix} 3 \\ 1 \\ -2 \end{pmatrix} = \begin{pmatrix} 3 \\ 2 \\ -1 \end{pmatrix} + \mu \begin{pmatrix} 2 \\ 0 \\ -1 \end{pmatrix}$$

● — You may find it easier to work with column vectors.

$\mathbf{i} : 2 + 3\lambda = 3 + 2\mu \Rightarrow \mu = \dfrac{3\lambda - 1}{2}$ ● — (2) Equating the components of **i**

$\mathbf{j} : -1 + \lambda = 2 \Rightarrow \lambda = 3$ ● — (2) Equating the components of **j**

$\mathbf{k} : -2\lambda = -1 - \mu \Rightarrow \mu = -1 + 2\lambda$ ● — (2) Equating the components of **k**

Therefore, from the **i** components, $\mu = \dfrac{3 \times 3 - 1}{2} = 4$ ● — (2) Solve simultaneously.

But from the **k** components, $\mu = -1 + 2 \times 3 = 5$ ●

The equations are inconsistent so there is no point of intersection. Therefore the lines are skew. — (4) An inconsistency as μ cannot be both 4 and −5

(Continued on the next page)

b They are not parallel.

$$\begin{pmatrix} 1 \\ 0 \\ -8 \end{pmatrix} + \lambda \begin{pmatrix} 2 \\ -2 \\ -3 \end{pmatrix} = \begin{pmatrix} -4 \\ 7 \\ -1 \end{pmatrix} + \mu \begin{pmatrix} 3 \\ -1 \\ -5 \end{pmatrix}$$

> (1) Since their direction vectors are not multiples of each other.

i: $1 + 2\lambda = -4 + 3\mu$

j: $-2\lambda = 7 - \mu$

k: $-8 - 3\lambda = -1 - 5\mu$

Adding equations from i and j components gives $1 = 3 + 2\mu \Rightarrow \mu = -1$

> (2) Solve simultaneously.

Therefore, from the i components, $\lambda = \dfrac{-5 + 3 \times -1}{2} = -4$

and from the k components, $\lambda = \dfrac{7 - 5 \times -1}{-3} = -4$

Therefore the lines intersect.

> (4) Since the equations are consistent.

To find point of intersection, substitute $\mu = -1$ into L_2:

$$\begin{pmatrix} -4 \\ 7 \\ -1 \end{pmatrix} - 1 \begin{pmatrix} 3 \\ -1 \\ -5 \end{pmatrix} = \begin{pmatrix} -7 \\ 8 \\ 4 \end{pmatrix}$$

> Alternatively substitute $\lambda = -4$ into L_1

So point of intersection is $(-7, 8, 4)$

Strategy 3

To apply a matrix transformation to a line

(1) Form a vector for a general point on the line.

(2) Pre-multiply this vector by the transformation matrix.

(3) Write the equation of the image.

> **See Ch5.2**
> For a reminder of using matrices for linear transformations.

Example 7

A line can be transformed by a linear transformation to another line or to a point.

The line with equation $\mathbf{r} = \mathbf{i} - \mathbf{j} + \lambda(\mathbf{i} + 3\mathbf{j})$ is transformed by the matrix $\mathbf{T} = \begin{pmatrix} 1 & -2 \\ 2 & 0 \end{pmatrix}$

Work out the equation of the image of the line.

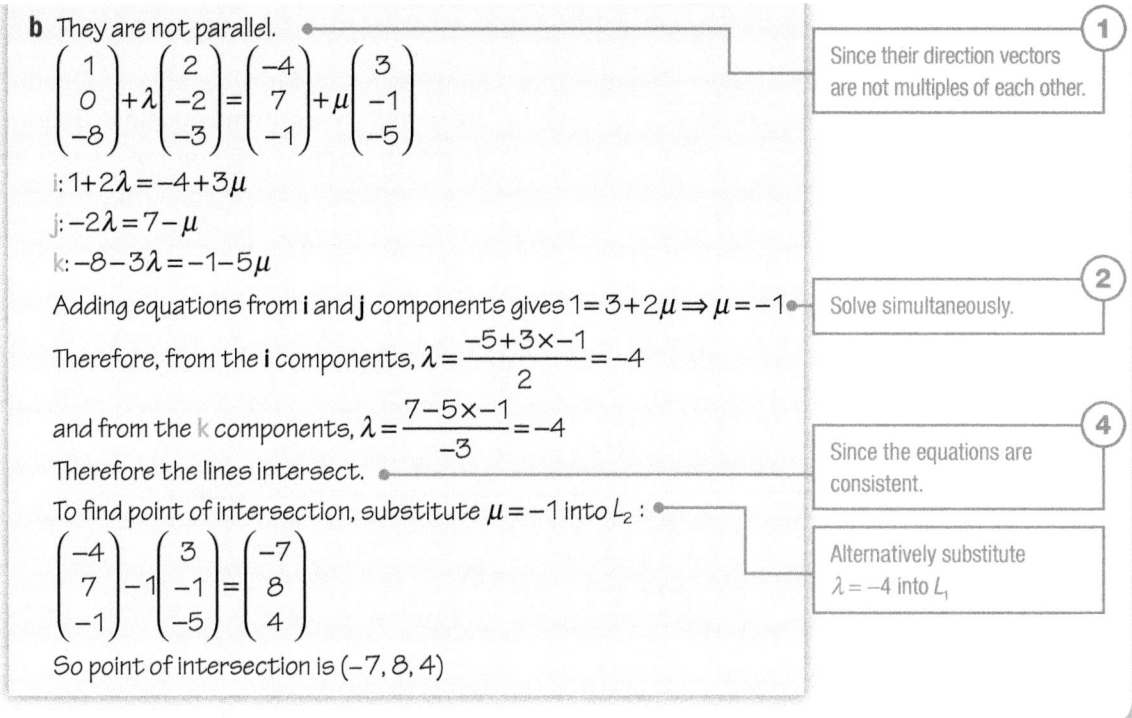

A general point on the line is $\begin{pmatrix} 1 + \lambda \\ -1 + 3\lambda \end{pmatrix}$

> (1) Combine to a single column vector. From the line equation you know that the x component is $1 + \lambda$ and the y component is $-1 + 3\lambda$

$$\begin{pmatrix} 1 & -2 \\ 2 & 0 \end{pmatrix} \begin{pmatrix} 1 + \lambda \\ -1 + 3\lambda \end{pmatrix} = \begin{pmatrix} 1 + \lambda - 2(-1 + 3\lambda) \\ 2(1 + \lambda) + 0 \end{pmatrix}$$

$$= \begin{pmatrix} 3 - 5\lambda \\ 2 + 2\lambda \end{pmatrix}$$

> (2) Pre-multiply by **T**

So the equation of the image is $\mathbf{r} = \begin{pmatrix} 3 \\ 2 \end{pmatrix} + \lambda \begin{pmatrix} -5 \\ 2 \end{pmatrix}$ or alternatively write $\mathbf{r} = 3\mathbf{i} + 2\mathbf{j} + \lambda(-5\mathbf{i} + 2\mathbf{j})$

> (3) Write the equation of the image in the form required.

1 For each pair of equations, decide whether or not they represent the same line. Show your working clearly.

a $\quad \mathbf{r} = \begin{pmatrix} 1 \\ 5 \end{pmatrix} + \lambda \begin{pmatrix} 4 \\ -3 \end{pmatrix}$ and $\mathbf{r} = \begin{pmatrix} 2 \\ 3 \end{pmatrix} + \mu \begin{pmatrix} -2 \\ 6 \end{pmatrix}$

b $\quad \mathbf{r} = \begin{pmatrix} 4 \\ -1 \end{pmatrix} + \lambda \begin{pmatrix} 1 \\ -1 \end{pmatrix}$ and $\mathbf{r} = \begin{pmatrix} 1 \\ 2 \end{pmatrix} + \mu \begin{pmatrix} -1 \\ 1 \end{pmatrix}$

c $\quad \mathbf{r} = 2\mathbf{i} + 7\mathbf{j} + \lambda(\mathbf{i} - \mathbf{j})$ and $y = 9 - x$

2 For each pair of equations, decide whether or not they represent the same line. Show your working clearly.

a $\quad \mathbf{r} = \begin{pmatrix} -3 \\ 3 \\ 3 \end{pmatrix} + \lambda \begin{pmatrix} 2 \\ 0 \\ -1 \end{pmatrix}$ and $\mathbf{r} = \begin{pmatrix} 1 \\ 3 \\ -1 \end{pmatrix} + \mu \begin{pmatrix} -4 \\ 0 \\ 2 \end{pmatrix}$

b $\quad \mathbf{r} = \begin{pmatrix} 2 \\ 3 \\ 0 \end{pmatrix} + \lambda \begin{pmatrix} 1 \\ 1 \\ -1 \end{pmatrix}$ and $\mathbf{r} = \begin{pmatrix} 2 \\ 5 \\ -2 \end{pmatrix} + \mu \begin{pmatrix} 1 \\ -1 \\ 1 \end{pmatrix}$

See Ch5.2
For a reminder of writing transform-ations as matrices.

c $\quad \mathbf{r} = 2\mathbf{j} - \mathbf{k} + \lambda(\mathbf{i} + 2\mathbf{j} - 3\mathbf{k})$ and $\mathbf{r} = 5\mathbf{i} + 6\mathbf{j} - 10\mathbf{k} + \mu(2\mathbf{i} + 4\mathbf{j} - 6\mathbf{k})$

d $\quad \dfrac{x-2}{1} = \dfrac{y-3}{3} = \dfrac{z+1}{-2}$ and $\dfrac{x+3}{-1} = \dfrac{y+12}{-3} = \dfrac{z-9}{2}$

e $\quad \dfrac{x+5}{-1} = \dfrac{y-0}{2} = \dfrac{z-1}{2}$ and $\mathbf{r} = \begin{pmatrix} 1 \\ -12 \\ 11 \end{pmatrix} + \lambda \begin{pmatrix} 2 \\ -4 \\ -4 \end{pmatrix}$

3 Find the point of intersection between these pairs of lines.

a $\quad \mathbf{r} = \begin{pmatrix} -1 \\ 1 \end{pmatrix} + \lambda \begin{pmatrix} -2 \\ 3 \end{pmatrix}$ and $\mathbf{r} = \begin{pmatrix} 1 \\ 7 \end{pmatrix} + \mu \begin{pmatrix} 4 \\ -3 \end{pmatrix}$

b $\quad \mathbf{r} = -4\mathbf{i} + 14\mathbf{j} + \lambda(-\mathbf{i} + 2\mathbf{j})$ and $\mathbf{r} = \begin{pmatrix} 4 \\ 1 \end{pmatrix} + \mu \begin{pmatrix} -2 \\ 3 \end{pmatrix}$

c $\quad \mathbf{r} = \begin{pmatrix} 0 \\ 2 \end{pmatrix} + \lambda \begin{pmatrix} -1 \\ 3 \end{pmatrix}$ and $y = 2x - 3$

4 For each pair of lines, decide whether they are parallel, skew or intersecting. If they are intersecting, find their point of intersection.

a $\quad \mathbf{r} = \begin{pmatrix} 7 \\ -6 \\ -2 \end{pmatrix} + \lambda \begin{pmatrix} 3 \\ 0 \\ -1 \end{pmatrix}$ and $\mathbf{r} = \begin{pmatrix} 3 \\ 4 \\ 0 \end{pmatrix} + \mu \begin{pmatrix} -2 \\ 5 \\ 1 \end{pmatrix}$

b $\quad \mathbf{r} = \mathbf{i} + 2\mathbf{k} + \lambda(3\mathbf{i} + 4\mathbf{j} - 2\mathbf{k})$ and $\mathbf{r} = 3\mathbf{i} + \mathbf{j} - \mathbf{k} + \mu(-6\mathbf{i} - 8\mathbf{j} + 4\mathbf{k})$

c $\quad \dfrac{x+5}{-1} = \dfrac{y}{2} = \dfrac{z-1}{2}$ and $\dfrac{x-2}{1} = \dfrac{y+9}{3} = \dfrac{z-5}{-2}$

d $\quad \dfrac{-4-x}{3} = \dfrac{y-2}{-1} = \dfrac{z}{5}$ and $\dfrac{x-6}{2} = \dfrac{y+8}{-2} = \dfrac{5-z}{-1}$

5 Two lines, L_1 and L_2, have equations $\dfrac{x+10}{5} = \dfrac{y-5}{-1} = \dfrac{z-5}{2}$ and $\mathbf{r} = \begin{pmatrix} 3 \\ 3 \\ 5 \end{pmatrix} + \lambda \begin{pmatrix} -3 \\ 0 \\ 4 \end{pmatrix}$ respectively.

Given that L_1 and L_2 intersect at the point A, calculate the length of $\overrightarrow{OA}$

6 Find the image of the line $\mathbf{r} = \begin{pmatrix} 1 \\ 3 \end{pmatrix} + \lambda \begin{pmatrix} -2 \\ 1 \end{pmatrix}$ under the transformation given by the matrix $\begin{pmatrix} 2 & 1 \\ 0 & -2 \end{pmatrix}$

7 Find the image of the line $\mathbf{r} = \begin{pmatrix} 2 \\ -1 \end{pmatrix} + \lambda \begin{pmatrix} 2 \\ 1 \end{pmatrix}$ under a rotation of 90° anticlockwise about the origin.

See Ch5
For a reminder of writing transform-ations as matrices.

8 Find the image of the line $\mathbf{r} = \begin{pmatrix} 3 \\ 0 \\ -1 \end{pmatrix} + \lambda \begin{pmatrix} 1 \\ -1 \\ -2 \end{pmatrix}$ under each of these transformations.

a Rotation 90° anticlockwise around the x-axis.

b Rotation 180° around the y-axis.

c Reflection in the plane $z = 0$

9 Find the image of the line $\dfrac{x-1}{2} = \dfrac{y+3}{-1} = \dfrac{5-z}{-4}$ after a reflection in the plane $y = 0$

Fluency and skills

A very important concept in the application of vectors is the **scalar product**. As its name suggests, the result of finding the scalar product of two vectors will be a scalar quantity.

> **Key point**
>
> The scalar product $\mathbf{a} \cdot \mathbf{b}$ of vectors $\mathbf{a}$ and $\mathbf{b}$ is defined as
> $\mathbf{a} \cdot \mathbf{b} = |\mathbf{a}||\mathbf{b}|\cos\theta$ where θ is the angle between the vectors $\mathbf{a}$ and $\mathbf{b}$
> The vectors $\mathbf{a}$ and $\mathbf{b}$ are both directed away from the angle (or both directed towards the angle) and $0 \le \theta \le 180°$

See Maths Ch6.1
For a reminder of magnitude notation.

Example 1

Find the value of

a $\mathbf{i} \cdot \mathbf{i}$ **b** $\mathbf{k} \cdot \mathbf{j}$ **c** $\mathbf{i} \cdot (\mathbf{j} + \mathbf{k})$

a $\mathbf{i} \cdot \mathbf{i} = 1 \times 1 \times \cos 0°$
     $= 1$

b $\mathbf{k} \cdot \mathbf{j} = 1 \times 1 \times \cos 90°$
     $= 0$

c $\mathbf{i} \cdot (\mathbf{i} + \mathbf{k}) = \mathbf{i} \cdot \mathbf{i} + \mathbf{i} \cdot \mathbf{k}$
       $= 1 + 1 \times \cos 90°$
       $= 1$

$\mathbf{i}$ is the unit vector in the positive x-direction so has magnitude 1

$\mathbf{k}$ and $\mathbf{j}$ are the unit vectors in the positive z- and y-directions respectively so the angle between them is 90°

Since the scalar product is distributive.

The scalar product is sometimes called the **dot product**. The dot product is a distributive operation. If vectors $\mathbf{a}$ and $\mathbf{b}$ are perpendicular then $\mathbf{a} \cdot \mathbf{b} = 0$. This is because $\cos 90° = 0$

> **Key point**
>
> If you have two vectors $\mathbf{a} = a_1\mathbf{i} + a_2\mathbf{j} + a_3\mathbf{k}$ and $\mathbf{b} = b_1\mathbf{i} + b_2\mathbf{j} + b_3\mathbf{k}$, then the scalar product is $\mathbf{a} \cdot \mathbf{b} = a_1b_1 + a_2b_2 + a_3b_3$

Example 2

Given $\mathbf{a} = \begin{pmatrix} 1 \\ -3 \\ 2 \end{pmatrix}$ and $\mathbf{b} = \begin{pmatrix} -4 \\ -1 \\ 5 \end{pmatrix}$, calculate

a The scalar product $\mathbf{a} \cdot \mathbf{b}$ **b** The angle between vectors $\mathbf{a}$ and $\mathbf{b}$

a $\mathbf{a} \cdot \mathbf{b} = \begin{pmatrix} 1 \\ -3 \\ 2 \end{pmatrix} \cdot \begin{pmatrix} -4 \\ -1 \\ 5 \end{pmatrix}$

       $= (1 \times -4) + (-3 \times -1) + (2 \times 5)$

       $= 9$

Using $\mathbf{a} \cdot \mathbf{b} = a_1b_1 + a_2b_2 + a_3b_3$

(Continued on the next page)

b You need $|\mathbf{a}| = \sqrt{1^2 + 3^2 + 2^2} = \sqrt{14}$

and $|\mathbf{b}| = \sqrt{4^1 + 1^2 + 5^2} = \sqrt{42}$

> To find the magnitude of a 3D vector find the sum of the squares of each component and take the square root.

$\cos\theta = \dfrac{\mathbf{a}\cdot\mathbf{b}}{|\mathbf{a}||\mathbf{b}|}$

> From rearranging
> $\mathbf{a}\cdot\mathbf{b} = |\mathbf{a}||\mathbf{b}|\cos\theta$

$= \dfrac{9}{\sqrt{14}\sqrt{42}}$

$= \dfrac{3\sqrt{3}}{14}$

$\theta = 68.2°$ (1 decimal place)

> Using $\cos^{-1}\left(\dfrac{3\sqrt{3}}{14}\right)$

Calculator

Try it on your calculator

You can use a calculator to find the scalar product.

VetA · VetB

-6

Activity

Find out how to work out

$\begin{pmatrix} 2 \\ 3 \\ -1 \end{pmatrix} \cdot \begin{pmatrix} -1 \\ 0 \\ 4 \end{pmatrix}$ on *your* calculator.

Exercise 6.2A Fluency and skills

1 Find the value of

  a $\mathbf{j}\cdot\mathbf{j}$ **b** $\mathbf{i}\cdot\mathbf{k}$ **c** $2\mathbf{i}\cdot\mathbf{j}$ **d** $\mathbf{j}\cdot(\mathbf{i}+\mathbf{j})$

  e $\mathbf{i}\cdot(2\mathbf{i})$ **f** $(\mathbf{i}+\mathbf{j})\cdot(\mathbf{i}-\mathbf{j})$

  g $(\mathbf{j}+\mathbf{k})\cdot(\mathbf{i}+\mathbf{j})$ **h** $(\mathbf{i}+\mathbf{j}+\mathbf{k})\cdot(\mathbf{i}+\mathbf{j}+\mathbf{k})$

2 Calculate these scalar products.

  a $\begin{pmatrix} 2 \\ 4 \\ -1 \end{pmatrix} \cdot \begin{pmatrix} 5 \\ -3 \\ -6 \end{pmatrix}$ **b** $(2\mathbf{i}-\mathbf{j}+3\mathbf{k})\cdot(\mathbf{i}+4\mathbf{j}-\mathbf{k})$

  c $\begin{pmatrix} \sqrt{2} \\ 1 \\ 3 \end{pmatrix} \cdot \begin{pmatrix} 2 \\ 0 \\ -\sqrt{2} \end{pmatrix}$ **d** $(5\mathbf{i}+\mathbf{k})\cdot(2\mathbf{j}-3\mathbf{k})$

3 Find an expression in terms of a for the dot product of $(2a\mathbf{i}-\mathbf{j}+a\mathbf{k})$ and $(3a\mathbf{i}-a\mathbf{j}-5\mathbf{k})$

4 Find an expression in terms of k of the dot product of $\begin{pmatrix} 2 \\ k \\ -3 \end{pmatrix}$ and $\begin{pmatrix} k \\ -1 \\ 2 \end{pmatrix}$

5 Calculate the angle between the vectors

  a $\begin{pmatrix} 5 \\ 1 \\ 3 \end{pmatrix}$ and $\begin{pmatrix} 2 \\ -2 \\ 4 \end{pmatrix}$

  b $(5\mathbf{i}+\mathbf{j}-2\mathbf{k})$ and $(4\mathbf{i}+3\mathbf{j}-\mathbf{k})$

  c $\begin{pmatrix} 2\sqrt{3} \\ 4 \\ \sqrt{3} \end{pmatrix}$ and $\begin{pmatrix} -1 \\ \sqrt{3} \\ 2 \end{pmatrix}$ **d** $(3\mathbf{j}+2\mathbf{k})$ and $(4\mathbf{i}-\mathbf{j})$

6 Calculate the cosine of the acute angle between each pair of vectors.

  Give your answers in terms of a where $a > 0$

  a $\begin{pmatrix} a \\ 1 \\ 0 \end{pmatrix}$ and $\begin{pmatrix} 1 \\ 0 \\ a \end{pmatrix}$

  b $3a\mathbf{i}-4a\mathbf{j}+5\mathbf{k}$ and $5a\mathbf{j}+5\mathbf{k}$

7 Show that the vectors $\begin{pmatrix} 6 \\ -4 \end{pmatrix}$ and $\begin{pmatrix} -8 \\ -12 \end{pmatrix}$ are perpendicular.

8 Show that the vectors $\begin{pmatrix} 4 \\ -1 \\ 5 \end{pmatrix}$ and $\begin{pmatrix} -2 \\ -3 \\ 1 \end{pmatrix}$ are perpendicular.

9 Show that the vectors $3\mathbf{i}-5\mathbf{j}-2\mathbf{k}$ and $2\mathbf{i}+4\mathbf{j}-7\mathbf{k}$ are perpendicular.

10 Show that the vectors $2\mathbf{i}+\mathbf{k}$ and $4\mathbf{i}-\mathbf{j}-3\mathbf{k}$ are not perpendicular.

11 Find the value of a for which the vectors
$$\begin{pmatrix} 3a \\ 2 \\ -2 \end{pmatrix} \text{ and } \begin{pmatrix} 2 \\ a \\ 4 \end{pmatrix} \text{ are perpendicular.}$$

12 Find the value of b for which the vectors $(2b\mathbf{i}+\mathbf{j}-\mathbf{k})$ and $(3\mathbf{i}-b\mathbf{j}+10\mathbf{k})$ are perpendicular.

13 Find the values of c for which the vectors
$$\begin{pmatrix} c \\ 3 \\ -1 \end{pmatrix} \text{ and } \begin{pmatrix} -c \\ c \\ 2 \end{pmatrix} \text{ are perpendicular.}$$

Reasoning and problem-solving

Strategy 1

To find the obtuse angle between two intersecting lines

(**1**) Identify the direction vector of each line, **a** and **b**

(**2**) Use $\cos\theta = \left| \dfrac{\mathbf{a}\cdot\mathbf{b}}{\|\mathbf{a}\|\|\mathbf{b}\|} \right|$ to find the acute angle between the lines.

(**3**) Subtract the acute angle from 180° to find the obtuse angle between the lines.

Example 3

Given that L_1 and L_2 have equations $L_1 : \mathbf{r} = \begin{pmatrix} 1 \\ -1 \\ 2 \end{pmatrix} + \lambda \begin{pmatrix} 1 \\ -2 \\ 1 \end{pmatrix}$ and $L_2 : \dfrac{x+4}{2} = \dfrac{y+3}{-1} = \dfrac{z-1}{1}$ and that

they intersect at the point A, find the obtuse angle between L_1 and L_2

The direction vectors are $\begin{pmatrix} 1 \\ -2 \\ 1 \end{pmatrix}$ and $\begin{pmatrix} 2 \\ -1 \\ 1 \end{pmatrix}$ (**1**) Identify direction vectors.

$\begin{pmatrix} 1 \\ -2 \\ 1 \end{pmatrix} \cdot \begin{pmatrix} 2 \\ -1 \\ 1 \end{pmatrix} = 2+2+1 = 5$

$\left\| \begin{pmatrix} 1 \\ -2 \\ 1 \end{pmatrix} \right\| = \sqrt{1^2+2^2+1^2} = \sqrt{6}$ (**2**) Find the magnitude of the two direction vectors.

$\left\| \begin{pmatrix} 2 \\ -1 \\ 1 \end{pmatrix} \right\| = \sqrt{2^2+1^2+1^2} = \sqrt{6}$

$\cos\theta = \dfrac{5}{\sqrt{6}\sqrt{6}}$ (**2**) First find the acute angle.

$\theta = 33.6°$

So obtuse angle is $180-33.6 = 146.4°$ (1 decimal place)   (**3**) Then, subtract from 180° to find the obtuse angle.

To prove properties of the scalar product

1. Use general vectors, for example $\mathbf{a} = a_1\mathbf{i} + a_2\mathbf{j} + a_3\mathbf{k}$

2. Use the fact that if $\mathbf{a}$, $\mathbf{b}$ are perpendicular vectors then $\mathbf{a} \cdot \mathbf{b} = 0$

3. Use the fact that $\mathbf{i} \cdot \mathbf{i} = \mathbf{j} \cdot \mathbf{j} = \mathbf{k} \cdot \mathbf{k} = 1$

Example 4

Show that $\mathbf{a} \cdot \mathbf{b} = \mathbf{b} \cdot \mathbf{a}$ for any 3D vectors $\mathbf{a}$ and $\mathbf{b}$

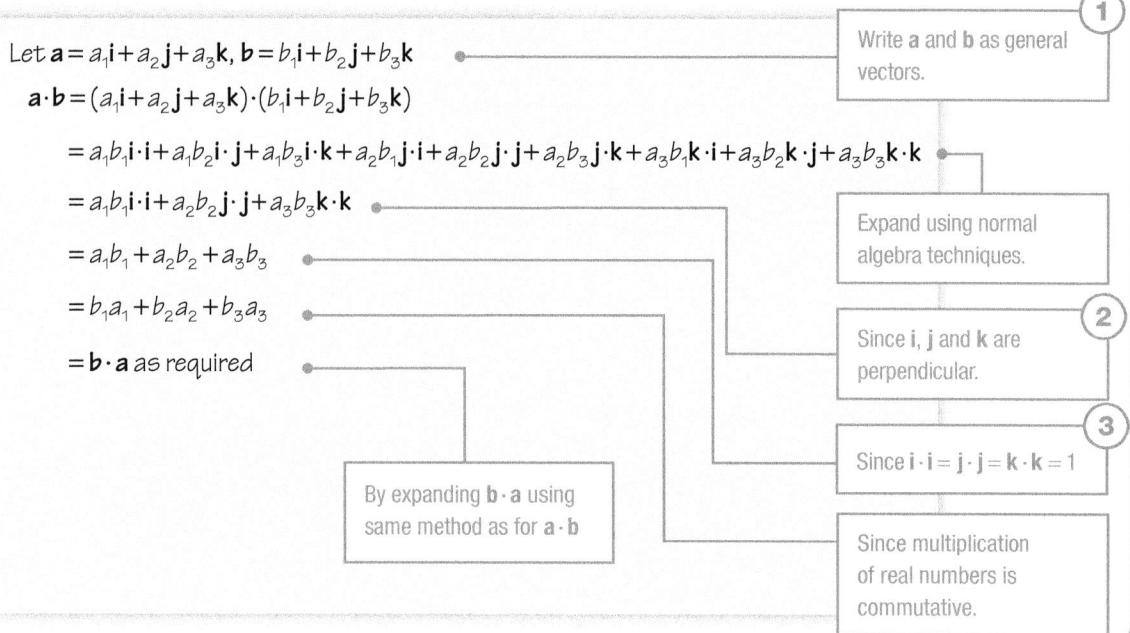

Let $\mathbf{a} = a_1\mathbf{i} + a_2\mathbf{j} + a_3\mathbf{k}$, $\mathbf{b} = b_1\mathbf{i} + b_2\mathbf{j} + b_3\mathbf{k}$ — Write $\mathbf{a}$ and $\mathbf{b}$ as general vectors. **①**

$\mathbf{a} \cdot \mathbf{b} = (a_1\mathbf{i} + a_2\mathbf{j} + a_3\mathbf{k}) \cdot (b_1\mathbf{i} + b_2\mathbf{j} + b_3\mathbf{k})$

$= a_1b_1\mathbf{i} \cdot \mathbf{i} + a_1b_2\mathbf{i} \cdot \mathbf{j} + a_1b_3\mathbf{i} \cdot \mathbf{k} + a_2b_1\mathbf{j} \cdot \mathbf{i} + a_2b_2\mathbf{j} \cdot \mathbf{j} + a_2b_3\mathbf{j} \cdot \mathbf{k} + a_3b_1\mathbf{k} \cdot \mathbf{i} + a_3b_2\mathbf{k} \cdot \mathbf{j} + a_3b_3\mathbf{k} \cdot \mathbf{k}$

$= a_1b_1\mathbf{i} \cdot \mathbf{i} + a_2b_2\mathbf{j} \cdot \mathbf{j} + a_3b_3\mathbf{k} \cdot \mathbf{k}$ — Expand using normal algebra techniques.

$= a_1b_1 + a_2b_2 + a_3b_3$ — Since $\mathbf{i}$, $\mathbf{j}$ and $\mathbf{k}$ are perpendicular. **②**

$= b_1a_1 + b_2a_2 + b_3a_3$ — Since $\mathbf{i} \cdot \mathbf{i} = \mathbf{j} \cdot \mathbf{j} = \mathbf{k} \cdot \mathbf{k} = 1$ **③**

$= \mathbf{b} \cdot \mathbf{a}$ as required — By expanding $\mathbf{b} \cdot \mathbf{a}$ using same method as for $\mathbf{a} \cdot \mathbf{b}$ — Since multiplication of real numbers is commutative.

Exercise 6.2B Reasoning and problem-solving

1. Show that the lines $= \begin{pmatrix} 3 \\ -2 \end{pmatrix} + \lambda \begin{pmatrix} -1 \\ 3 \end{pmatrix}$ and $\mathbf{r} = 2\mathbf{i} - \mathbf{j} + \mu(6\mathbf{i} + 2\mathbf{j})$ are perpendicular.

2. Show that the lines $\mathbf{r} = \begin{pmatrix} 5 \\ -2 \\ 4 \end{pmatrix} + \lambda \begin{pmatrix} 6 \\ 7 \\ -5 \end{pmatrix}$ and $\mathbf{r} = \begin{pmatrix} 1 \\ 0 \\ -3 \end{pmatrix} + \lambda \begin{pmatrix} 2 \\ -1 \\ 1 \end{pmatrix}$ are perpendicular.

3. Use the scalar product to show that the lines $\mathbf{r} = \begin{pmatrix} 1 \\ 1 \\ 2 \end{pmatrix} + \lambda \begin{pmatrix} 3 \\ -2 \\ 6 \end{pmatrix}$ and $\dfrac{x-5}{2} = \dfrac{y+7}{6} = \dfrac{z}{1}$ are perpendicular.

4. Show that the lines with equations $\mathbf{r} = \begin{pmatrix} 6 \\ 6 \end{pmatrix} + \lambda \begin{pmatrix} 2 \\ 1 \end{pmatrix}$ and $\mathbf{r} = \begin{pmatrix} 4 \\ 0 \end{pmatrix} + \mu \begin{pmatrix} 3 \\ -1 \end{pmatrix}$ intersect and find the acute angle between them.

5. Show that the lines with equations $\mathbf{r} = \begin{pmatrix} 2 \\ 1 \\ -3 \end{pmatrix} + s \begin{pmatrix} 0 \\ 2 \\ 1 \end{pmatrix}$ and $\mathbf{r} = \begin{pmatrix} 5 \\ 11 \\ 3 \end{pmatrix} + t \begin{pmatrix} 3 \\ -4 \\ -1 \end{pmatrix}$ intersect and find the acute angle between them to the nearest degree.

6 Given that they intersect, find the obtuse angle between the lines with equations

$$\mathbf{r}=\begin{pmatrix}4\\0\\12\end{pmatrix}+\lambda\begin{pmatrix}0\\2\\5\end{pmatrix} \text{ and } \frac{x-1}{3}=\frac{y+2}{-6}=\frac{z+10}{2}$$

7 Show that the lines with equations
$\mathbf{r}=(2\mathbf{i}-\mathbf{j}+3\mathbf{k})+\lambda(\mathbf{i}+\mathbf{j}-\mathbf{k})$ and
$\mathbf{r}=(2\mathbf{i}-2\mathbf{j}+\mathbf{k})+\mu(-\mathbf{i}+3\mathbf{k})$ intersect and find the acute angle between them to 3 significant figures.

8 Find the values of a for which the lines

$$\mathbf{r}=\begin{pmatrix}3\\-1\\2\end{pmatrix}+\lambda\begin{pmatrix}2\\a\\0\end{pmatrix} \text{ and } \mathbf{r}=\mu\begin{pmatrix}3\\0\\\sqrt{3}\end{pmatrix} \text{ meet at a } 60°$$

angle.

9 Find the values of a for which the lines

$$\mathbf{r}=\begin{pmatrix}1\\4\\4\end{pmatrix}+\lambda\begin{pmatrix}-2\\a\sqrt{2}\\2\end{pmatrix} \text{ and } \frac{x-3}{1}=\frac{z-1}{-1}, y=-2$$

meet at a 135° angle.

10 The lines with equations $\dfrac{x}{4}=\dfrac{y+1}{8}=\dfrac{z-3}{1}$ and $\dfrac{x-1}{2}=\dfrac{y}{6}=\dfrac{z+2}{11}$ intersect at the point A

   a Find the coordinates of the point A

   b Calculate the exact cosine of the acute angle between the two lines at the point A

11 The acute angle between the vectors
$\mathbf{a}=\mathbf{i}-k\mathbf{j}$ and $\mathbf{b}=\mathbf{i}+\mathbf{j}$ is 60°

   Calculate the possible values of k

12 The acute angle between the vectors $\begin{pmatrix}1\\k\\2\end{pmatrix}$

   and $\begin{pmatrix}0\\-1\\1\end{pmatrix}$ is 45°

   Calculate the possible values of k

13 You are given that $\mathbf{a}=3\mathbf{i}-\mathbf{k}$ and $\mathbf{b}=\mathbf{i}-2\mathbf{j}+2\mathbf{k}$

   Calculate the exact area of triangle ABC

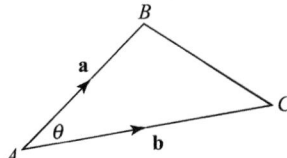

14 The points A, B and C have position vectors

$$\begin{pmatrix}4\\0\\1\end{pmatrix}, \begin{pmatrix}5\\-2\\0\end{pmatrix} \text{ and } \begin{pmatrix}3\\-1\\1\end{pmatrix} \text{ respectively.}$$

   Calculate the exact area of triangle ABC

15 Given that $\mathbf{a}=a_1\mathbf{i}+a_2\mathbf{j}+a_3\mathbf{k}$
   and $\mathbf{b}=b_1\mathbf{i}+b_2\mathbf{j}+b_3\mathbf{k}$, show that
   $\mathbf{a}\cdot\mathbf{b}=a_1b_1+a_2b_2+a_3b_3$

16 Show that the dot product is distributive, that is, $\mathbf{a}\cdot(\mathbf{b}+\mathbf{c})=\mathbf{a}\cdot\mathbf{b}+\mathbf{a}\cdot\mathbf{c}$ for any 3D vectors $\mathbf{a}$, $\mathbf{b}$ and $\mathbf{c}$

17 a Prove that $\mathbf{a}\cdot\mathbf{a}=|\mathbf{a}|^2$ for any vector $\mathbf{a}$

   b Hence prove the cosine rule:
   $|\mathbf{a}|^2=|\mathbf{b}|^2+|\mathbf{c}|^2-2|\mathbf{b}||\mathbf{c}|\cos\theta$ for the triangle shown.

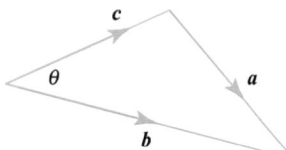

18 a Show that the line $\dfrac{x+2}{-4}=\dfrac{y-5}{5}=\dfrac{z-3}{3}$
   cuts the x-axis and find the intercept's coordinates.

   b Calculate the acute angle between the x-axis and the line.

19 Three lines have equations as follows:
$L_1: \mathbf{r}=6\mathbf{i}-3\mathbf{j}+\lambda(\mathbf{i}+\mathbf{k})$, $L_2: \mathbf{r}=s(3\mathbf{i}-\mathbf{j}+\mathbf{k})$
and $L_3: \mathbf{r}=t(\mathbf{j}+2\mathbf{k})$

   The lines L_1 and L_2 intersect at the point A,
   L_1 and L_2 intersect at the point B,
   and L_2 and L_3 intersect at the point C as shown.

   Calculate the exact area of triangle ABC

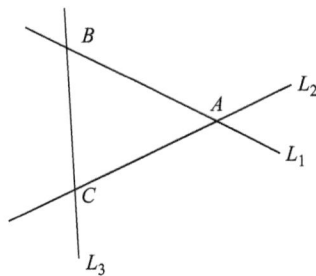

6.3   Finding distances 1

Suppose you wish to find the shortest distance from a line $\mathbf{r} = \mathbf{a} + \lambda\mathbf{b}$ to a given point, C, with position vector $\mathbf{c}$

The shortest possible distance will be CD, which is perpendicular to the original line.

Therefore, the scalar product $(\mathbf{c} - (\mathbf{a} + \lambda\mathbf{b})) \cdot \mathbf{b} = 0$
since $DC = \mathbf{c} - (\mathbf{a} + \lambda\mathbf{b})$.

This allows you to find the point of intersection between DC and the line $\mathbf{r} = \mathbf{a} + \lambda\mathbf{b}$. You will then be able to find the length of DC and hence the shortest distance from the point to the line.

Example 1

Calculate the shortest distance from the point $(3, 3, 1)$ to the line with equation $\mathbf{r} = \begin{pmatrix} 2 \\ 0 \\ -6 \end{pmatrix} + \lambda \begin{pmatrix} -3 \\ 0 \\ -1 \end{pmatrix}$

$$\begin{pmatrix} 3 \\ 3 \\ 1 \end{pmatrix} - \left[\begin{pmatrix} 2 \\ 0 \\ -6 \end{pmatrix} + \lambda \begin{pmatrix} -3 \\ 0 \\ -1 \end{pmatrix} \right] = \begin{pmatrix} 1 + 3\lambda \\ 3 \\ 7 + \lambda \end{pmatrix}$$

Find a vector that passes through the point and the line.

$$\begin{pmatrix} 1 + 3\lambda \\ 3 \\ 7 + \lambda \end{pmatrix} \cdot \begin{pmatrix} -3 \\ 0 \\ -1 \end{pmatrix} = 0$$

Since the vectors are perpendicular, their scalar product is zero.

$$\Rightarrow -3 - 9\lambda - 7 - \lambda = 0 \quad \Rightarrow \lambda = -1$$

$$\begin{pmatrix} 2 \\ 0 \\ -6 \end{pmatrix} - 1 \begin{pmatrix} -3 \\ 0 \\ -1 \end{pmatrix} = \begin{pmatrix} 5 \\ 0 \\ -5 \end{pmatrix}$$

This is the position vector of the point of intersection of the line with the perpendicular through $(3, 3, 1)$

$$\text{Distance} = \sqrt{(5 - 3)^2 + (0 - 3)^2 + (-5 - 1)^2}$$

Find the distance from $(3, 3, 1)$ to the point of intersection.

$$= 7$$

You already know how to find the point of intersection between two lines, if it exists. If two lines do not intersect, you can instead find the shortest possible distance between them. In the case of parallel lines, this is straightforward as the distance between them is constant.

If two parallel lines have equations $\mathbf{r} = \mathbf{a} + \lambda\mathbf{b}$ and $\mathbf{r} = \mathbf{c} + \mu\mathbf{b}$ then, since their separation is always the same, you can chose any point on either one of the lines (for example, the point with position vector $\mathbf{a}$ on the line $\mathbf{r} = \mathbf{a} + \lambda\mathbf{b}$) and use the method above to find the shortest distance from a point to a line.

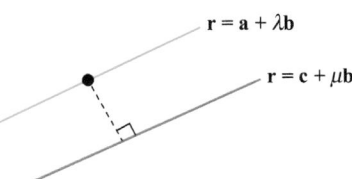

Example 2

PURE

Lines l_1 and l_2 have equations $l_1 : \mathbf{r} = 2\mathbf{i} - \mathbf{j} + 8\mathbf{k} + s(4\mathbf{i} + 2\mathbf{j} - \mathbf{k})$ and
$$l_2 : \mathbf{r} = 3\mathbf{j} + \mathbf{k} + t(-12\mathbf{i} - 6\mathbf{j} + 3\mathbf{k})$$

Calculate the shortest distance between l_1 and l_2

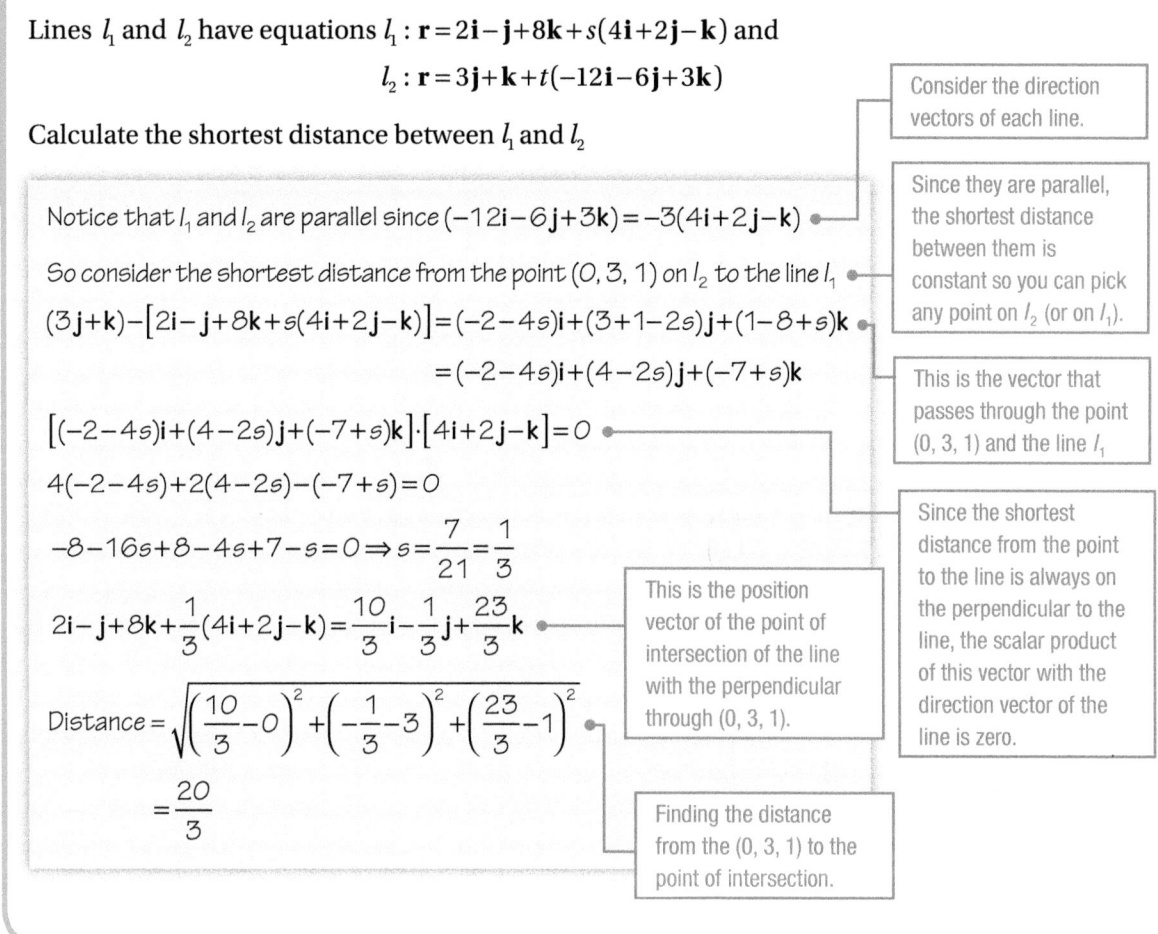

Consider the direction vectors of each line.

Notice that l_1 and l_2 are parallel since $(-12\mathbf{i} - 6\mathbf{j} + 3\mathbf{k}) = -3(4\mathbf{i} + 2\mathbf{j} - \mathbf{k})$

Since they are parallel, the shortest distance between them is constant so you can pick any point on l_2 (or on l_1).

So consider the shortest distance from the point $(0, 3, 1)$ on l_2 to the line l_1

$(3\mathbf{j} + \mathbf{k}) - [2\mathbf{i} - \mathbf{j} + 8\mathbf{k} + s(4\mathbf{i} + 2\mathbf{j} - \mathbf{k})] = (-2 - 4s)\mathbf{i} + (3 + 1 - 2s)\mathbf{j} + (1 - 8 + s)\mathbf{k}$

$= (-2 - 4s)\mathbf{i} + (4 - 2s)\mathbf{j} + (-7 + s)\mathbf{k}$

This is the vector that passes through the point $(0, 3, 1)$ and the line l_1

$[(-2 - 4s)\mathbf{i} + (4 - 2s)\mathbf{j} + (-7 + s)\mathbf{k}] \cdot [4\mathbf{i} + 2\mathbf{j} - \mathbf{k}] = 0$

$4(-2 - 4s) + 2(4 - 2s) - (-7 + s) = 0$

$-8 - 16s + 8 - 4s + 7 - s = 0 \Rightarrow s = \dfrac{7}{21} = \dfrac{1}{3}$

Since the shortest distance from the point to the line is always on the perpendicular to the line, the scalar product of this vector with the direction vector of the line is zero.

$2\mathbf{i} - \mathbf{j} + 8\mathbf{k} + \dfrac{1}{3}(4\mathbf{i} + 2\mathbf{j} - \mathbf{k}) = \dfrac{10}{3}\mathbf{i} - \dfrac{1}{3}\mathbf{j} + \dfrac{23}{3}\mathbf{k}$

This is the position vector of the point of intersection of the line with the perpendicular through $(0, 3, 1)$.

$\text{Distance} = \sqrt{\left(\dfrac{10}{3} - 0\right)^2 + \left(-\dfrac{1}{3} - 3\right)^2 + \left(\dfrac{23}{3} - 1\right)^2}$

$= \dfrac{20}{3}$

Finding the distance from the $(0, 3, 1)$ to the point of intersection.

You can also find the perpendicular distance between skew lines by solving simultaneous equations.

Example 3

Find the minimum distance between the skew lines with equations

$\mathbf{r} = 2\mathbf{i} + \mathbf{j} + \lambda(-\mathbf{j} + 2\mathbf{k})$ and $\mathbf{r} = \mathbf{j} - 2\mathbf{k} + \mu(\mathbf{i} + 2\mathbf{j})$

A general point, P, on the first line has position vector, $\overrightarrow{OP} = \begin{pmatrix} 2 \\ 1 - \lambda \\ 2\lambda \end{pmatrix}$.

A general point, Q, on the second line has position vector

$\overrightarrow{OQ} = \begin{pmatrix} \mu \\ 1 + 2\mu \\ -2 \end{pmatrix}$,

Therefore, the vector joining these two points is

$\overrightarrow{PQ} = \begin{pmatrix} \mu \\ 1 + 2\mu \\ -2 \end{pmatrix} - \begin{pmatrix} 2 \\ 1 - \lambda \\ 2\lambda \end{pmatrix} = \begin{pmatrix} \mu - 2 \\ 2\mu + \lambda \\ -2 - 2\lambda \end{pmatrix}$

Use $\overrightarrow{PQ} = \overrightarrow{OQ} - \overrightarrow{OP}$

(Continued on the next page)

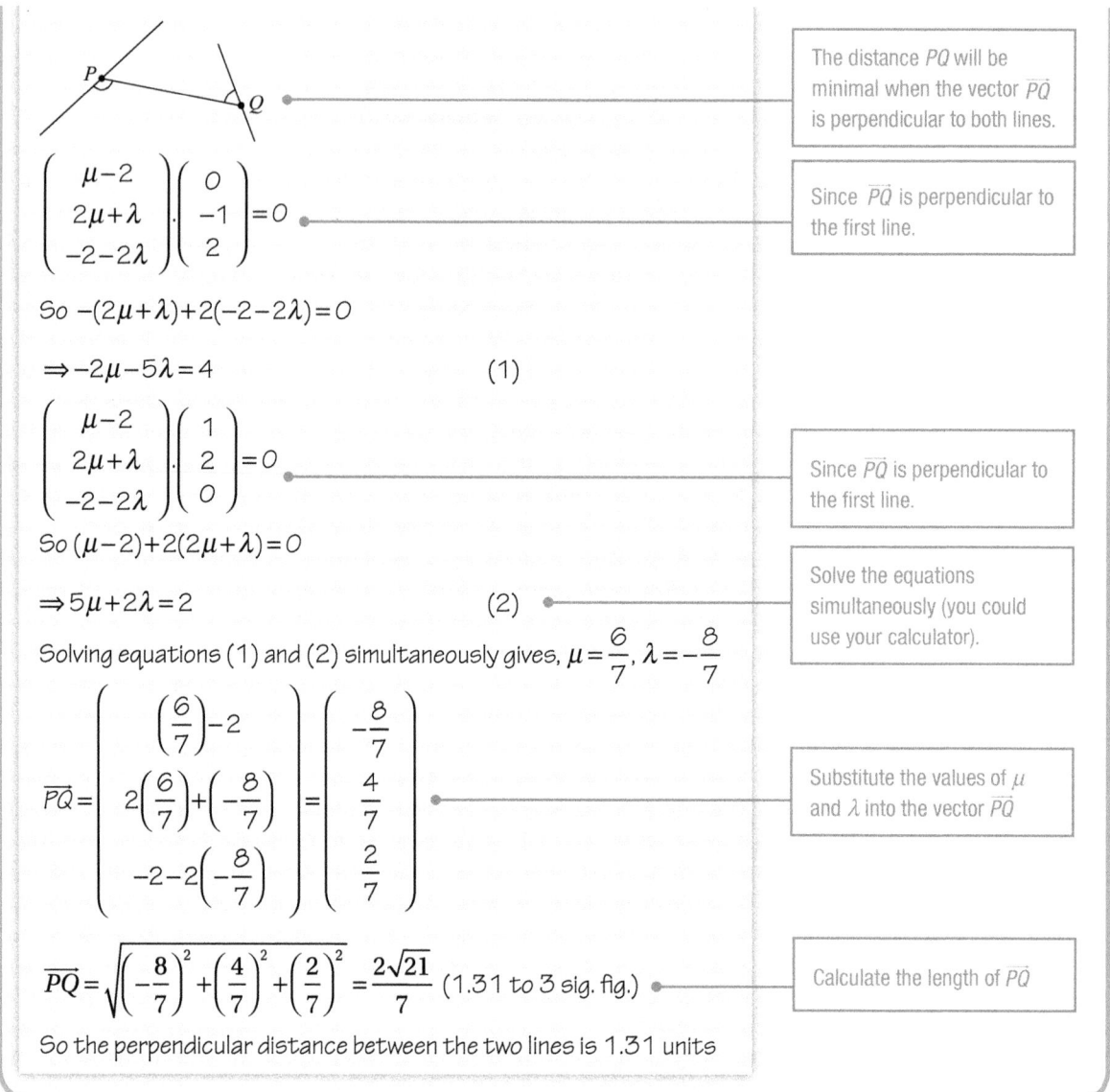

$$\begin{pmatrix} \mu-2 \\ 2\mu+\lambda \\ -2-2\lambda \end{pmatrix} \cdot \begin{pmatrix} 0 \\ -1 \\ 2 \end{pmatrix} = 0$$

> The distance PQ will be minimal when the vector $\vec{PQ}$ is perpendicular to both lines.

> Since $\vec{PQ}$ is perpendicular to the first line.

So $-(2\mu+\lambda)+2(-2-2\lambda)=0$

$\Rightarrow -2\mu-5\lambda=4$ (1)

$$\begin{pmatrix} \mu-2 \\ 2\mu+\lambda \\ -2-2\lambda \end{pmatrix} \cdot \begin{pmatrix} 1 \\ 2 \\ 0 \end{pmatrix} = 0$$

> Since $\vec{PQ}$ is perpendicular to the first line.

So $(\mu-2)+2(2\mu+\lambda)=0$

$\Rightarrow 5\mu+2\lambda=2$ (2)

> Solve the equations simultaneously (you could use your calculator).

Solving equations (1) and (2) simultaneously gives, $\mu=\dfrac{6}{7}$, $\lambda=-\dfrac{8}{7}$

$$\vec{PQ}=\begin{pmatrix} \left(\dfrac{6}{7}\right)-2 \\ 2\left(\dfrac{6}{7}\right)+\left(-\dfrac{8}{7}\right) \\ -2-2\left(-\dfrac{8}{7}\right) \end{pmatrix} = \begin{pmatrix} -\dfrac{8}{7} \\ \dfrac{4}{7} \\ \dfrac{2}{7} \end{pmatrix}$$

> Substitute the values of μ and λ into the vector $\vec{PQ}$

$$\overline{PQ}=\sqrt{\left(-\dfrac{8}{7}\right)^2+\left(\dfrac{4}{7}\right)^2+\left(\dfrac{2}{7}\right)^2}=\dfrac{2\sqrt{21}}{7} \text{ (1.31 to 3 sig. fig.)}$$

> Calculate the length of $\vec{PQ}$

So the perpendicular distance between the two lines is 1.31 units

Exercise 6.3A Fluency and skills

1   Calculate the shortest distance between the line with equation $\mathbf{r}=6\mathbf{i}-4\mathbf{j}+\mathbf{k}+s(2\mathbf{i}-\mathbf{k})$ and the point $(-4, 2, 6)$

2   Find the shortest distance between the line with equation $\mathbf{r}=\begin{pmatrix} 1 \\ 3 \\ -2 \end{pmatrix}+\lambda\begin{pmatrix} 1 \\ -2 \\ 0 \end{pmatrix}$ and the point $(0, 5, 1)$

3   Find the shortest distance between the line with equations $\dfrac{x-3}{2}$, $y=1$, $z=-1$ and the point $(0, 3, 1)$

4   Calculate the shortest distance between the line with equation $\mathbf{r}=\mu(\mathbf{i}+\mathbf{j}-\mathbf{k})$ and the point with position vector $2\mathbf{i}-3\mathbf{j}+\mathbf{k}$

5 Calculate the shortest distance between the point $(2, 1, 0)$ and the line $x = y = 2z$

6 Calculate the shortest distance between these pairs of parallel lines.

a $\mathbf{r} = 5\mathbf{i} + \mathbf{k} + s(\mathbf{i} - \mathbf{j})$ and $\mathbf{r} = 2\mathbf{i} + \mathbf{j} - 7\mathbf{k} + t(\mathbf{i} - \mathbf{j})$

b $\mathbf{r} = \begin{pmatrix} 2 \\ 3 \\ 5 \end{pmatrix} + \lambda \begin{pmatrix} 3 \\ 1 \\ -2 \end{pmatrix}$ and $\mathbf{r} = \begin{pmatrix} -6 \\ 1 \\ -1 \end{pmatrix} + \mu \begin{pmatrix} 9 \\ 3 \\ -6 \end{pmatrix}$

c $x = y + 3 = \dfrac{z - 6}{-2}$ and $x + 1 = y - 4 = \dfrac{z - 8}{-2}$

d $\mathbf{r} = \begin{pmatrix} 0 \\ 4 \\ -1 \end{pmatrix} + \lambda \begin{pmatrix} -3 \\ 5 \\ 1 \end{pmatrix}$ and $\dfrac{x}{3} = \dfrac{y + 3}{-5} = -z - 1$

e $\mathbf{r} = (2\mathbf{i} - 3\mathbf{j} + 11\mathbf{k}) + s(\mathbf{i} - 5\mathbf{j} - \mathbf{k})$ and $2 - x = \dfrac{y - 1}{5} = z$

7 Calculate the length of the perpendicular between these pairs of skew lines

a $\mathbf{r} = \begin{pmatrix} 3 \\ 0 \\ -1 \end{pmatrix} + \lambda \begin{pmatrix} 1 \\ 1 \\ -1 \end{pmatrix}$ and $\mathbf{r} = \begin{pmatrix} 0 \\ 2 \\ 0 \end{pmatrix} + \mu \begin{pmatrix} 0 \\ 0 \\ 1 \end{pmatrix}$

b $\mathbf{r} = \mathbf{i} - \mathbf{j} + s(\mathbf{i} - \mathbf{k})$ and $\mathbf{r} = \mathbf{j} + 2\mathbf{k} + t(2\mathbf{i} - \mathbf{j})$

c $\mathbf{r} = \mathbf{i} - 4\mathbf{j} + 2\mathbf{k} + s\mathbf{k}$ and $\dfrac{x - 3}{2} = \dfrac{y + 1}{1} = \dfrac{z - 2}{-1}$

Reasoning and problem-solving

An alternative approach to finding the minimum distance between a line and a point is to use calculus. Suppose you have the line $\mathbf{r} = \mathbf{a} + t\mathbf{b}$ and a point with position vector $\mathbf{c}$. Then you need to find the minimum value of $x = |\mathbf{a} + t\mathbf{b} - \mathbf{c}|$ which can be found by using $\dfrac{\mathrm{d}x}{\mathrm{d}t} = 0$

Strategy 1

To find the minimum distance between a line and a point using calculus

(**1**) Find the vector joining the point given to a general point on the line.

(**2**) Find an expression for x, the length of this vector.

(**3**) Use $\dfrac{\mathrm{d}x}{\mathrm{d}t} = 0$ to calculate the value of t which minimises x

(**4**) Substitute this value of t into your expression for the length.

Example 4

Find the minimum distance between the line with equation $\mathbf{r} = \mathbf{i} + \mathbf{j} + t(\mathbf{j} - \mathbf{k})$ and the point $(2, 1, -2)$

You are interested in the vector $\mathbf{i} + \mathbf{j} + t(\mathbf{j} - \mathbf{k}) - (2\mathbf{i} + \mathbf{j} - 2\mathbf{k})$

$$= -\mathbf{i} + t\mathbf{j} + (2 - t)\mathbf{k}$$

① Find the vector joining the point to a general point on the line.

The length of this vector is

$$x = \sqrt{1^2 + t^2 + (2 - t)^2} = \sqrt{2t^2 - 4t + 5}$$

② Find the magnitude of the vector.

Therefore $x^2 = 2t^2 - 4t + 5$

It is simpler to consider the expression for x^2

$$\frac{d(x^2)}{dt} = 4t - 4$$

Minimising x^2 will also minimise x

$$\frac{dx}{dt} = 0 \Rightarrow \frac{d(x^2)}{dt} = 0$$

$$\Rightarrow t = 1$$

③ Use the derivative to calculate t

$$x = \sqrt{1^2 + 1^2 + (2 - 1)^2} = \sqrt{3}$$

④ Substitute t into the expression.

A similar process can be used to find the perpendicular distance between a pair of parallel lines.

Example 5

Lines l_1 and l_2 have equations $l_1 : \dfrac{x-1}{4} = \dfrac{y+1}{2} = \dfrac{z+2}{-1}$ and $l_2 : \mathbf{r} = \begin{pmatrix} 1 \\ 3 \\ 0 \end{pmatrix} + \lambda \begin{pmatrix} -2 \\ -1 \\ 0.5 \end{pmatrix}$

Use calculus to calculate the shortest distance between l_1 and l_2
Give your answer to 3 significant figures.

Since a line with the Cartesian equations
$$\frac{x - a_1}{b_1} = \frac{y - a_2}{b_2} = \frac{z - a_3}{b_3}$$
has vector equation
$$\mathbf{r} = \begin{pmatrix} a_1 \\ a_2 \\ a_3 \end{pmatrix} + \lambda \begin{pmatrix} b_1 \\ b_2 \\ b_3 \end{pmatrix}.$$

The vector equation of l_1 is $\mathbf{r} = \begin{pmatrix} 1 \\ -1 \\ -2 \end{pmatrix} + \lambda \begin{pmatrix} 4 \\ 2 \\ -1 \end{pmatrix}$

So l_1 and l_2 are parallel since $\begin{pmatrix} 4 \\ 2 \\ -1 \end{pmatrix} = -2 \begin{pmatrix} -2 \\ -1 \\ 0.5 \end{pmatrix}$

This is important since the method only works if they are parallel.

So consider the shortest distance from the point $(1, 3, 0)$ on l_2 to the line l_1

Since they are parallel, the shortest distance between them is constant so you can pick any point on l_2 (or on l_1).

(Continued on the next page)

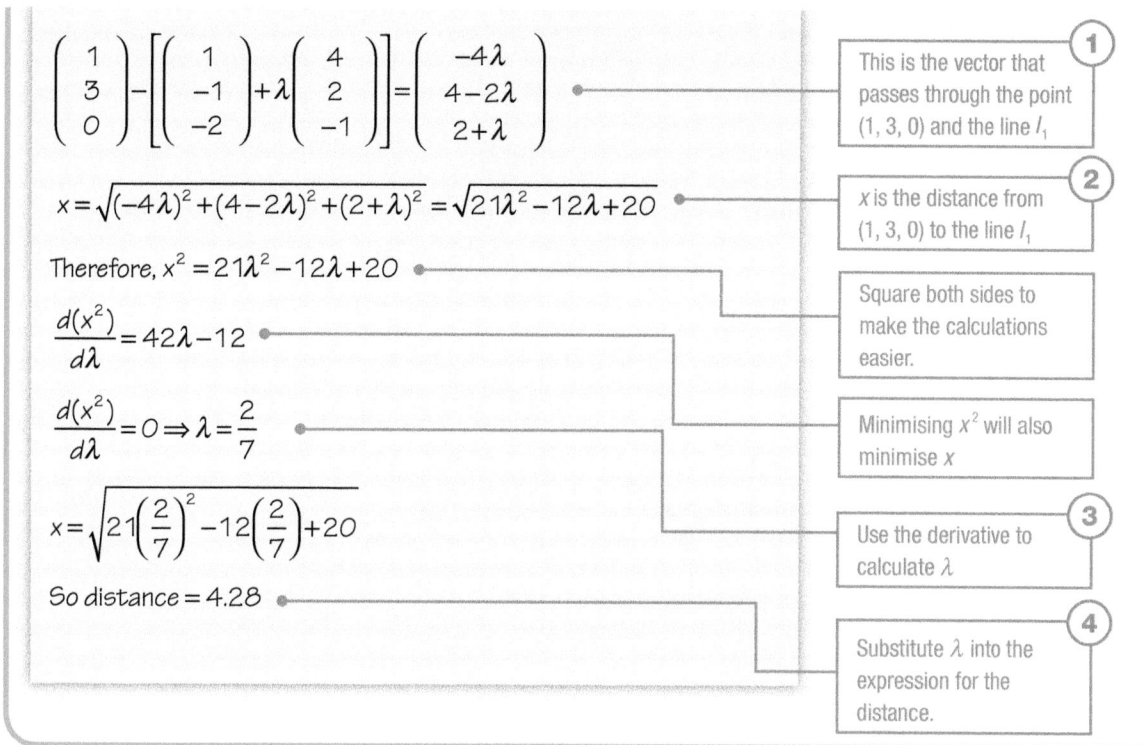

$$\left[\begin{pmatrix} 1 \\ 3 \\ 0 \end{pmatrix} - \left[\begin{pmatrix} 1 \\ -1 \\ -2 \end{pmatrix} + \lambda\begin{pmatrix} 4 \\ 2 \\ -1 \end{pmatrix}\right]\right] = \begin{pmatrix} -4\lambda \\ 4-2\lambda \\ 2+\lambda \end{pmatrix}$$

(1) This is the vector that passes through the point (1, 3, 0) and the line l_1

$$x = \sqrt{(-4\lambda)^2 + (4-2\lambda)^2 + (2+\lambda)^2} = \sqrt{21\lambda^2 - 12\lambda + 20}$$

(2) x is the distance from (1, 3, 0) to the line l_1

Therefore, $x^2 = 21\lambda^2 - 12\lambda + 20$

Square both sides to make the calculations easier.

$$\frac{d(x^2)}{d\lambda} = 42\lambda - 12$$

$$\frac{d(x^2)}{d\lambda} = 0 \Rightarrow \lambda = \frac{2}{7}$$

Minimising x^2 will also minimise x

$$x = \sqrt{21\left(\frac{2}{7}\right)^2 - 12\left(\frac{2}{7}\right) + 20}$$

(3) Use the derivative to calculate λ

So distance = 4.28

(4) Substitute λ into the expression for the distance.

You can use the perpendicular vector to find the reflection of a point in a line. When a point A is reflected in a line l to give the image A', then l will be the perpendicular bisector of the line segment AA'. You can use this fact to find the coordinates of A'.

Strategy 2

To find the image of a point reflected in a line

(1) Write the vector $\overrightarrow{AM}$

(2) Use the fact that $\overrightarrow{AM}$ is perpendicular to the mirror line so the scalar product of $\overrightarrow{AM}$ with the direction vector of the line is zero

(3) Use the fact that $\overrightarrow{AA'} = 2\overrightarrow{AM}$

Example 6

Find the coordinates of the image of the point $A(-1, 5, 2)$ after it is reflected in the line with

equation $\mathbf{r} = \begin{pmatrix} 0 \\ 3 \\ -2 \end{pmatrix} + t\begin{pmatrix} 1 \\ 1 \\ -1 \end{pmatrix}$

A general point on the line has position vector $\overrightarrow{OM} = \begin{pmatrix} t \\ t-3 \\ 2-t \end{pmatrix}$

So the vector $\overrightarrow{AM} = \begin{pmatrix} t \\ t+3 \\ -2-t \end{pmatrix} - \begin{pmatrix} -1 \\ 5 \\ 2 \end{pmatrix} = \begin{pmatrix} t+1 \\ t-2 \\ -4-t \end{pmatrix}$

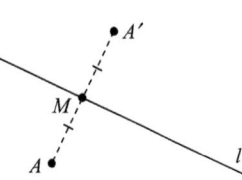

(1) Using $\overrightarrow{AM} = \overrightarrow{OM} - \overrightarrow{OA}$

(Continued on the next page)

$$\begin{pmatrix} t+1 \\ t-2 \\ -4-t \end{pmatrix} \cdot \begin{pmatrix} 1 \\ 1 \\ -1 \end{pmatrix} = 0$$

② Since $\overrightarrow{AM}$ is perpendicular to the line.

$$\Rightarrow (t+1)+(t-2)-(-4-t)=0$$
$$\Rightarrow 3t=-3$$
$$\Rightarrow t=-1$$

So $\overrightarrow{AM} = \begin{pmatrix} -1+1 \\ -1-2 \\ -4-(-1) \end{pmatrix} = \begin{pmatrix} 0 \\ -3 \\ -3 \end{pmatrix}$

Therefore, $\overrightarrow{AA'} = 2\begin{pmatrix} 0 \\ -3 \\ -3 \end{pmatrix} = \begin{pmatrix} 0 \\ -6 \\ -6 \end{pmatrix}$

③ Since $\overrightarrow{AA'} = 2\overrightarrow{AM}$

So $\overrightarrow{OA'} = \begin{pmatrix} -1 \\ 5 \\ 2 \end{pmatrix} + \begin{pmatrix} 0 \\ -6 \\ -6 \end{pmatrix} = \begin{pmatrix} -1 \\ -1 \\ -4 \end{pmatrix}$

Using $\overrightarrow{OA'} = \overrightarrow{OA} + \overrightarrow{AA'}$

So the coordinates of the image A' are $(-1, -1, -4)$

Exercise 6.3B Reasoning and problem-solving

1   Use calculus to find the shortest distance between the point $(1, 5, -7)$ and the line with equation

$$\mathbf{r} = \begin{pmatrix} -1 \\ 9 \\ -5 \end{pmatrix} + \lambda \begin{pmatrix} 0 \\ 3 \\ 1 \end{pmatrix}$$

2   Use calculus to find the shortest distance between the point $(-4, 0, 2)$ and the line with equation
$$\frac{x+2}{-1} = \frac{y+2}{2} = \frac{z-1}{1}$$

3   Use calculus to find the shortest distance between these lines.
$$\frac{y+1}{6} = \frac{z}{2}, \ x = 3 \text{ and } \mathbf{r} = \begin{pmatrix} -1 \\ 1 \\ 5 \end{pmatrix} + \lambda \begin{pmatrix} 0 \\ 3 \\ 1 \end{pmatrix}$$

4   Determine whether each of these pairs of lines intersect.

If they intersect, find their point of intersection. If they do not intersect, find the minimum distance between them.

a $\mathbf{r} = 2\mathbf{i} - \mathbf{k} + s(\mathbf{i} + \mathbf{j} - 2\mathbf{k})$ and $\dfrac{(x-1)}{5} = \dfrac{(y+5)}{5} = \dfrac{z}{-10}$

b $\mathbf{r} = \mathbf{i} + 5\mathbf{j} + 4\mathbf{k} + s(-2\mathbf{i} + \mathbf{j} + 4\mathbf{k})$ and $\mathbf{r} = \mathbf{i} + 2\mathbf{k} + t(3\mathbf{i} + \mathbf{j} - 5\mathbf{k})$

c $\mathbf{r} = 2\mathbf{i} + \mathbf{k} + \lambda(\mathbf{i} + 2\mathbf{j} - \mathbf{k})$ and $\mathbf{r} = \mathbf{i} - 8\mathbf{j} + \mu(\mathbf{i} + 3\mathbf{k})$

5   Find the reflection of the point $A(-4, 3, -2)$ in the line $\mathbf{r} = -2\mathbf{i} + 5\mathbf{j} - \mathbf{k} + t(\mathbf{i} + \mathbf{j} - \mathbf{k})$

6   Find the reflection of the point $A(3, 0, -4)$ in the line $\mathbf{r} = \mathbf{i} + 3\mathbf{j} - 2\mathbf{k} + t(\mathbf{i} - 2\mathbf{j} - 2\mathbf{k})$

7   The points A, B and C are $(2, 6, -4)$, $(0, -2, 3)$ and $(-5, -17, 13)$ respectively.

The line L_1 passes through the origin and point A

Line L_2 passes through the points B and C

Calculate the shortest distance between L_1 and L_2

8   Which of these lines comes closest to the origin?

$$L_1: \mathbf{r} = \begin{pmatrix} 5 \\ -2 \\ -4 \end{pmatrix} + \lambda \begin{pmatrix} 1 \\ 2 \\ 1 \end{pmatrix} \quad \text{or} \quad L_2: \mathbf{r} = \begin{pmatrix} -1 \\ 3 \\ 2 \end{pmatrix} + \lambda \begin{pmatrix} 1 \\ 1 \\ 0 \end{pmatrix}.$$

Fully explain your reasoning.

9   Which of these two points lie closest to the line $x - 4 = \dfrac{y+3}{3} = \dfrac{7-z}{2}$?

$A: (4, 1, 6)$ or $B: (6, -2, 6)$

Fully explain your reasoning.

10 L_1 and L_2 have equations $\mathbf{r} = \mathbf{i} + t\mathbf{j}$ and $\mathbf{r} = \mathbf{j} - \mathbf{k} + s(\mathbf{i} + \mathbf{k})$ respectively and intersect at the point A

The line L_3 is parallel to the vector $\mathbf{i} + \mathbf{j}$ and passes through the point $(2, 1, 0)$.

Calculate the shortest distance from A to the line L_3

6 Summary and review

Chapter summary

- The vector equation of a line parallel to the vector **b** and passing through the point with position vector **a** is $\mathbf{r} = \mathbf{a} + \lambda\mathbf{b}$
- The Cartesian equations of a line parallel to the vector $\mathbf{b} = b_1\mathbf{i} + b_2\mathbf{j} + b_3\mathbf{k}$ and passing through the point with position vector $\mathbf{a} = a_1\mathbf{i} + a_2\mathbf{j} + a_3\mathbf{k}$ are $\dfrac{x-a_1}{b_1} = \dfrac{y-a_2}{b_2} = \dfrac{z-a_3}{b_3}$
- Two lines in 3D can be parallel, intersecting or skew
- The scalar product is defined as $\mathbf{a} \cdot \mathbf{b} = |\mathbf{a}||\mathbf{b}|\cos\theta$ where θ is the angle between the vectors **a** and **b**
- If you have two vectors $\mathbf{a} = a_1\mathbf{i} + a_2\mathbf{j} + a_3\mathbf{k}$ and $\mathbf{b} = b_1\mathbf{i} + b_2\mathbf{j} + b_3\mathbf{k}$, then the scalar product is defined as $\mathbf{a} \cdot \mathbf{b} = a_1b_1 + a_2b_2 + a_3b_3$
- If vectors **a** and **b** are perpendicular then the scalar product, $\mathbf{a} \cdot \mathbf{b} = 0$
- The acute angle, θ, between two lines $\mathbf{r} = \mathbf{a} + \lambda\mathbf{b}$ and $\mathbf{r} = \mathbf{c} + \lambda\mathbf{d}$ can be found using $\cos\theta = \left|\dfrac{\mathbf{b} \cdot \mathbf{d}}{|\mathbf{b}||\mathbf{b}|}\right|$
- The shortest distance between a point and a line can be found by using the fact that it will be perpendicular to the line.
- The shortest distance between two lines can be found by using the fact that it will be perpendicular to both lines.

Check and review

You should now be able to...	Try Questions
✔ Write and use the equation of a line in Cartesian and vector form.	1, 2
✔ Decide whether two lines are intersecting, parallel of skew.	3
✔ Find the coordinates of the point of intersection of two lines.	3
✔ Calculate the scalar product of two vectors.	4, 5
✔ Use the scalar product to find the angle between two vectors.	5, 6
✔ Use the scalar product to show that two lines are perpendicular.	7
✔ Calculate the shortest distance from a point to a line.	8
✔ Calculate the shortest distance between two parallel lines.	9, 10
✔ Calculate the shortest distance between two skew lines.	11

1 A line is parallel to the vector $2\mathbf{i}-3\mathbf{j}+\mathbf{k}$ and passes through the point $(1, -1, 4)$

   Write the equation of the line in

   a   Cartesian form,

   b   Vector form.

2 A line passes through the points $(3, 0, 5)$ and $(-2, 4, 7)$

   Write the equation of the line in

   a   Cartesian form,

   b   Vector form.

3 For each pair of lines, decide whether they are parallel, skew or intersecting. If they are intersecting, find their point of intersection.

   a $\mathbf{r} = \begin{pmatrix} 1 \\ 3 \\ -2 \end{pmatrix} + s\begin{pmatrix} -2 \\ 4 \\ -1 \end{pmatrix}$ and

       $\mathbf{r} = \begin{pmatrix} -4 \\ 0 \\ 2 \end{pmatrix} + t\begin{pmatrix} 5 \\ -3 \\ 3 \end{pmatrix}$

   b $\mathbf{r} = 4\mathbf{i} - \mathbf{j} + \lambda(2\mathbf{i} + 3\mathbf{j} - \mathbf{k})$ and
       $$\frac{1-x}{10} = \frac{1-y}{15} = \frac{z+3}{5}$$

   c $\dfrac{x-2}{3} = \dfrac{y+1}{2} = \dfrac{z-3}{-1}$ and
       $$\frac{x-1}{-4} = \frac{y+3}{-3} = \frac{z-6}{2}$$

   d $\mathbf{r} = -5\mathbf{i} + 2\mathbf{j} + 4\mathbf{k} + \lambda(5\mathbf{i} - 3\mathbf{j} + \mathbf{k})$ and
       $\mathbf{r} = 12\mathbf{i} - 3\mathbf{j} + 6\mathbf{k} + \mu(2\mathbf{i} + 4\mathbf{j} - \mathbf{k})$

4 Calculate the value of $\begin{pmatrix} -3 \\ 7 \\ 4 \end{pmatrix} \cdot \begin{pmatrix} -2 \\ 5 \\ -1 \end{pmatrix}$

5 For the vectors $\mathbf{a} = 2\mathbf{i} - \mathbf{j} + 3\mathbf{k}$ and $\mathbf{b} = 3\mathbf{i} + 4\mathbf{j} - 5\mathbf{k}$, calculate

   a $\mathbf{a} \cdot \mathbf{b}$

   b   The acute angle between $\mathbf{a}$ and $\mathbf{b}$

6 Find the obtuse angle between the line $\mathbf{r} = 5\mathbf{i} + 3\mathbf{j} + 2\mathbf{k} + s(\mathbf{i} + \mathbf{j} - 2\mathbf{k})$ and the line
   $$\frac{x-1}{2} = \frac{y+3}{3} = z$$

7 Show that the lines $\dfrac{x-7}{-2} = \dfrac{y+2}{8} = \dfrac{z-3}{-4}$ and
   $\mathbf{r} = \begin{pmatrix} 5 \\ 0 \\ -6 \end{pmatrix} + \lambda\begin{pmatrix} -4 \\ 2 \\ 6 \end{pmatrix}$ are perpendicular.

8 Calculate the shortest distance from the point $(5, -2, 4)$ to the line with equation
   $\mathbf{r} = \begin{pmatrix} 5 \\ 0 \\ -1 \end{pmatrix} + s\begin{pmatrix} 1 \\ 3 \\ -2 \end{pmatrix}$

9 The lines l_1 and l_2 have equations
   $l_1 : \mathbf{r} = (-5\mathbf{i} - \mathbf{j} + 2\mathbf{k}) + s(3\mathbf{i} - 6\mathbf{k})$ and
   $l_2 : \mathbf{r} = (-4\mathbf{i} + 3\mathbf{k}) + t(-2\mathbf{i} + 4\mathbf{k})$.

   a   Show that l_1 and l_2 are parallel.

   b   Find the distance between l_1 and l_2

       Give your answer to 3 significant figures.

10 Use calculus to find the distance between the lines l_1 and l_2 where l_1 has equation
   $\mathbf{r} = \begin{pmatrix} 0 \\ 2 \\ -4 \end{pmatrix} + t\begin{pmatrix} 1 \\ 5 \\ -2 \end{pmatrix}$

   and l_2 is the line through the origin which is parallel to l_1
   Give your answer to 3 significant figures.

11 The lines l_1 and l_2 have equations
   $\mathbf{r} = \begin{pmatrix} 3 \\ -2 \\ 4 \end{pmatrix} + s\begin{pmatrix} 2 \\ 1 \\ -1 \end{pmatrix}$ and

   $\mathbf{r} = \begin{pmatrix} 4 \\ -1 \\ 5 \end{pmatrix} + t\begin{pmatrix} 1 \\ 1 \\ -2 \end{pmatrix}$ respectively.

   a   Show that l_1 and l_2 are skew lines.

   a   Calculate the minimum distance between l_1 and l_2

Investigation

A **median** of a triangle is the line joining one of the vertices to the midpoint of the opposite side.
The scalar product of vectors can be used to calculate the length of a median.
Consider a triangle defined by the vectors **a**, **b** and **c**,
and the median that cuts side **c** is vector **p**

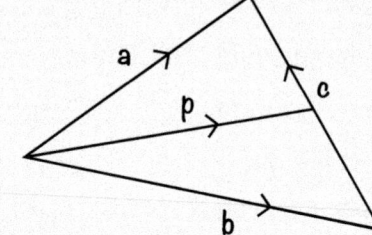

Vector **a** can be expressed as $\mathbf{p} + \dfrac{\mathbf{c}}{2}$ and
b can be expressed as $\mathbf{p} - \dfrac{\mathbf{c}}{2}$

Find **a·a** + **b·b** in terms of **p** and **c**

Hence, find an expression for the median in terms of the lengths of the sides of a triangle.
Use it to find the lengths of the medians of a triangle with side lengths 3, 4 and 5

Did you know?

The word vector is derived from the Latin word vector, meaning
"one who carries or conveys".

Have a go

When **R** is a vector representing a constant resultant force acting on a body which is moved
thorough a displacement represented by the vector **r**, the work done is given by **R·r**
    Using this vector expression, explain why it is a waste of energy to push a heavy object in a
direction near perpendicular to the direction in which you wish the body to travel.
    To ensure that more than half of a person's efforts are of use, what would you advise about
the direction in which they should push?

History

The historical development of vectors relies on many underpinning
ideas in both pure and applied mathematics.

    One important development was that of complex numbers and
their geometrical representation. This led mathematicians to
wonder how to devise a system that allowed analysis of
three-dimensional space.

Isaac Newton

    Alongside such developments, Newton and others were busy considering
how to understand and analyse forces and motion. In his famous Principia
Mathematica, Newton wrote, "A body, acted on by two forces simultaneously, will describe the
diagonal of a parallelogram in the same time as it would describe the sides by those forces separately".

    As you will now see, such thinking relating to a parallelogram of forces was very close to what
we might now represent by vectors.

6 Assessment

1 **a** Write down a vector equation of the line that passes through the point with position vector $2\mathbf{i} + 4\mathbf{j} - \mathbf{k}$ and is parallel to the vector $\mathbf{i} + \mathbf{j} - 2\mathbf{k}$ **[1 mark]**

 b Show that the point $(-1, 1, 5)$ lies on the line. **[3]**

2 **a** Find the vector equation of the line that passes through the points with position vectors $2\mathbf{i} + \mathbf{j}$ and $3\mathbf{i} + \mathbf{k}$ **[2]**

 b Show that the point $(8, -5, 7)$ does not lie on the line. **[3]**

3 Find the vector equation of the line that passes through the point with position vector $\begin{pmatrix} 2 \\ 1 \\ 0 \end{pmatrix}$

and is parallel to the line with equation $\mathbf{r} = \begin{pmatrix} 2 \\ 4 \\ -3 \end{pmatrix} + \lambda \begin{pmatrix} 8 \\ -5 \\ 3 \end{pmatrix}$ **[2]**

4 The position vectors of points A and B are $\begin{pmatrix} 7 \\ 1 \\ -3 \end{pmatrix}$ and $\begin{pmatrix} -2 \\ 8 \\ -1 \end{pmatrix}$ respectively.

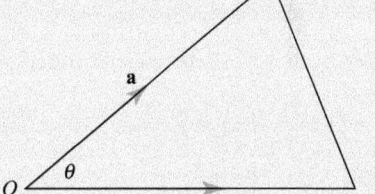

 a Calculate $\overrightarrow{OA} \cdot \overrightarrow{OB}$ **[2]**

 b Find the size, in degrees, of the acute angle AOB **[3]**

5 Calculate the cosine of the acute angle between the

vectors $\begin{pmatrix} 4 \\ 3 \\ 5 \end{pmatrix}$ and $\begin{pmatrix} -1 \\ -2 \\ 1 \end{pmatrix}$

Give your answer in the form $a\sqrt{3}$ where a is a constant to be found. **[4]**

6 Show that the vectors $-4\mathbf{i} + \mathbf{k}$ and $2\mathbf{i} - \mathbf{j} + 8\mathbf{k}$ are perpendicular. **[3]**

7 Given that $\mathbf{a} = \begin{pmatrix} 4 \\ -3 \\ -7 \end{pmatrix}$ and $\mathbf{b} = \begin{pmatrix} -1 \\ 2 \\ -2 \end{pmatrix}$, calculate $\mathbf{a} \cdot \mathbf{b}$ **[2]**

8 Find a vector which is perpendicular to both $\begin{pmatrix} 6 \\ 0 \\ 4 \end{pmatrix}$ and $\begin{pmatrix} -2 \\ 3 \\ 1 \end{pmatrix}$ **[2]**

9 Find the Cartesian equation of this line

 $\mathbf{r} = 2\mathbf{i} - 3\mathbf{k} + s(-\mathbf{i} + 4\mathbf{j} + 3\mathbf{k})$ **[2]**

10 The point $(2, 7, -3)$ lies on the line with equation $\dfrac{x-a}{2} = \dfrac{y-b}{-3} = \dfrac{z-c}{1}$

 a Write down the values of a, b and c **[1]**

 b Find the equation of the line in vector form. **[2]**

11 Calculate the perpendicular distance between the lines with equations
 $\mathbf{r} = 3\mathbf{i} - \mathbf{j} + \mu(-2\mathbf{i} + 3\mathbf{j} - \mathbf{k})$ and $\mathbf{r} = 2\mathbf{i} + \mathbf{j} - \mathbf{k} + \lambda(-2\mathbf{i} + 3\mathbf{j} - \mathbf{k})$ **[5]**

12 Is the point $(-2, 4, 7)$ closer to the line with equation $\mathbf{r} = 5\mathbf{i} + 2\mathbf{k} + \lambda(\mathbf{i} - 2\mathbf{j} + 4\mathbf{k})$ or
 the line parallel to the vector $2\mathbf{i} + \mathbf{j}$ which passes through the origin? **[10]**

13 For these pairs of lines, state and justify whether they are skew, parallel or intersecting.

If they are intersecting find their point of intersection.

a $\mathbf{r} = \begin{pmatrix} 3 \\ -1 \\ 5 \end{pmatrix} + \lambda \begin{pmatrix} 2 \\ 1 \\ -1 \end{pmatrix}$ and $\mathbf{r} = \begin{pmatrix} 13 \\ 1 \\ 6 \end{pmatrix} + \mu \begin{pmatrix} 4 \\ 1 \\ 0 \end{pmatrix}$ **[5]**

b $\mathbf{r} = 3\mathbf{i} + \mathbf{j} - \mathbf{k} + s(\mathbf{i} - 2\mathbf{j})$ and $\mathbf{r} = 5\mathbf{i} + 2\mathbf{j} - 3\mathbf{k} + t(-2\mathbf{i} + 4\mathbf{j})$ **[2]**

c $\mathbf{r} = \mathbf{i} + 2\mathbf{j} + 3\mathbf{k} + \lambda(2\mathbf{i} + \mathbf{j} - \mathbf{k})$ and $\mathbf{r} = 6\mathbf{i} - \mathbf{k} + \mu(3\mathbf{i} + \mathbf{k})$ **[4]**

14 a Find the Cartesian equation of the line with vector equation
$\mathbf{r} = 2\mathbf{i} - 3\mathbf{k} + s(-\mathbf{i} + 4\mathbf{j} + 3\mathbf{k})$ **[2]**

b Does this represent the same line as $\dfrac{x-1}{-1} = \dfrac{y-4}{4} = \dfrac{z+6}{-3}$? Explain how you know. **[3]**

15 Which of these equations, if any, represent the same line?

A: $\dfrac{x-4}{2} = \dfrac{y-7}{-1} = \dfrac{z-2}{5}$ **B:** $\mathbf{r} = \begin{pmatrix} 8 \\ 5 \\ 8 \end{pmatrix} - \lambda \begin{pmatrix} 2 \\ -1 \\ 5 \end{pmatrix}$ **C:** $\mathbf{r} = \begin{pmatrix} 10 \\ 4 \\ 13 \end{pmatrix} + \lambda \begin{pmatrix} 2 \\ -1 \\ 5 \end{pmatrix}$ **[3]**

16 Calculate the shortest distance from the point $(5, -1, 6)$ to the line with equation

$\mathbf{r} = \begin{pmatrix} 7 \\ -3 \\ 9 \end{pmatrix} + t \begin{pmatrix} 2 \\ 1 \\ -2 \end{pmatrix}$ **[4]**

17 Calculate the perpendicular distance between the lines with equations
$\mathbf{r} = 3\mathbf{i} - \mathbf{j} + \mu(-2\mathbf{i} + 3\mathbf{j} - \mathbf{k})$ and $\mathbf{r} = 2\mathbf{i} + \mathbf{j} - \mathbf{k} + \lambda(-2\mathbf{i} + 3\mathbf{j} - \mathbf{k})$ **[5]**

18 The lines l_1 and l_2 have Cartesian equations $\dfrac{x+1}{2} = \dfrac{y-4}{7} = \dfrac{1+z}{-1}$ and $\mathbf{r} = \begin{pmatrix} -5 \\ 4 \\ 1 \end{pmatrix} + t \begin{pmatrix} -4 \\ -2 \\ 2 \end{pmatrix}$

a Calculate the acute angle of l_1 and l_2 **[4]**

b Find the point of intersection of l_1 and l_2 **[4]**

19 The line l_1 is given by $\mathbf{r} = 6\mathbf{i} - 3\mathbf{j} + \mathbf{k} + s(-\mathbf{i} + \mathbf{j} - 2\mathbf{k})$ and the line l_2 is given by
$\dfrac{x-4}{-5} = \dfrac{y-2}{2} = \dfrac{z+5}{-8}$

The point A is the intersection of l_1 and l_2 and the point B has coordinates $(1, -2, 7)$

Calculate the area of triangle OAB **[8]**

20 The line l_1 has equation $\mathbf{r} = \lambda(2\mathbf{i} + \mathbf{j} - \mathbf{k})$ and the line l_2 has equation $\mathbf{r} = \mathbf{j} - 3\mathbf{k} + \lambda(\mathbf{i} - \mathbf{k})$

a Show that the lines l_1 and l_2 are skew, **[3]**

b Calculate the perpendicular distance between l_1 and l_2 **[7]**

The image of the point $A(2, -3, 4)$ when reflected in l_1 is A'.

c Find the coordinates of A' **[7]**

d Calculate the area of the triangle OAA' **[4]**

7 Forces and energy

If you've ever ridden a bike down a hill, you'll have noticed that you gain speed without needing to use any energy. Cycling uphill, on the other hand, is hard work and needs a lot of energy to gain any amount of speed and height. The exchange of energy, the work that needs to be done, and the power necessary to do this are all important considerations for vehicle designers who strive for fuel efficiency.

This exchange of energy is a key principle in **regenerative braking,** where kinetic energy is converted to electrical energy, by essentially reversing the drive motor during braking so that it becomes a generator. With electronic technology on the rise, the uses of regenerative braking are becoming ever-more widespread: in all kinds of electric trains, buses and cars.

Orientation

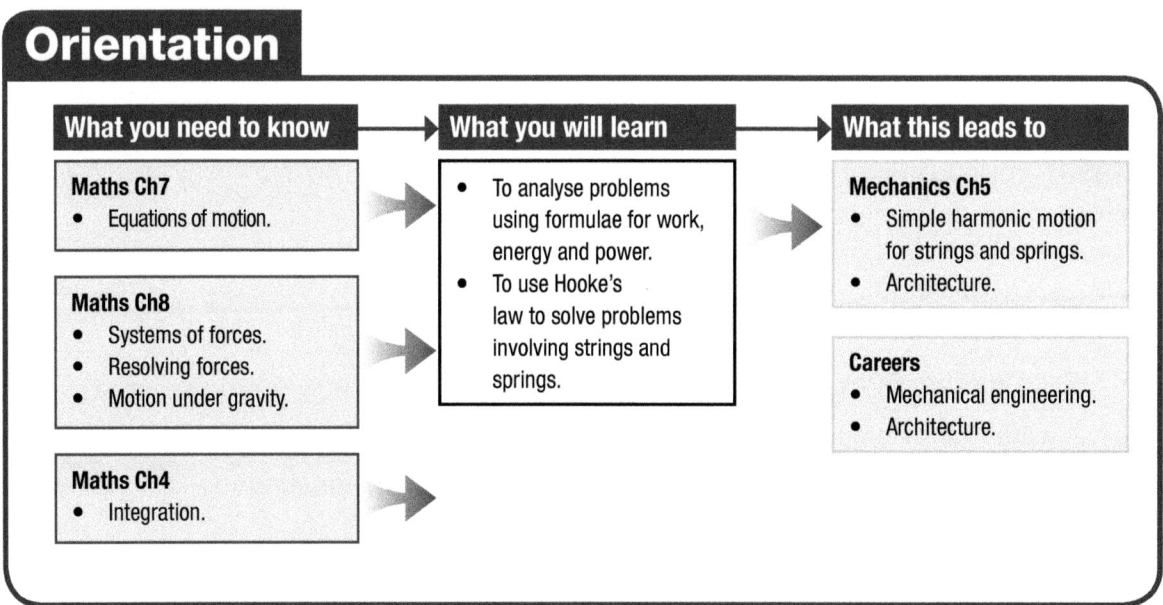

What you need to know	What you will learn	What this leads to
Maths Ch7 • Equations of motion.	• To analyse problems using formulae for work, energy and power. • To use Hooke's law to solve problems involving strings and springs.	**Mechanics Ch5** • Simple harmonic motion for strings and springs. • Architecture.
Maths Ch8 • Systems of forces. • Resolving forces. • Motion under gravity.		**Careers** • Mechanical engineering. • Architecture.
Maths Ch4 • Integration.		

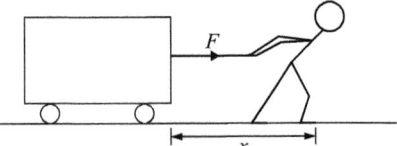

Fluency and skills

If you want to move an object, you have to push or pull it and so do some work. How much work you do depends on the force F you exert and the distance x you move the object *in the direction of the force.*

> **Key point**
>
> **Work done** by a constant force = force × distance
> $$= F \times x$$
> when F and x are in the same direction.

The units of work are **joules** (J) when force is in N and distance in m.

> Joules (J) are the same as N m.

When the force F is not constant, but varies over the distance x moved in the direction of F, the work done by F over a small distance δx is given by $F \times \delta x$

> **Key point**
>
> Summing over the total distance x gives:
>
> the total work done by a variable force $= \int_0^x F \, dx$

Full A Level

A third scenario is when a constant force F acts at an angle θ to the horizontal forward motion. The object does not move vertically, so the vertical component of the force, $F \sin \theta$, does no work. Its horizontal component $F \cos \theta$ moves a distance x in the direction of motion and so does work equal to $F \cos \theta \times x$

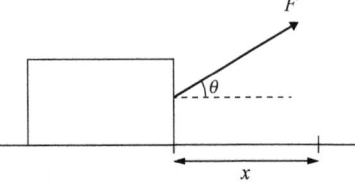

> **Key point**
>
> **Work done** by a constant force F is $Fx \cos \theta$ when the angle between F and x is θ.

Full A Level

Example 1

A force F pulls a trolley a horizontal distance x. Find the work done by F if

a F acts horizontally over 4 metres with a constant magnitude of 10 N

b F acts horizontally and varies so that $F = (10 - 2x)$ N for $0 \le x \le 4$ metres

c F acts at 60° to the horizontal over 4 metres with a constant magnitude of 10 N   **Full A Level**

(Continued on next page)

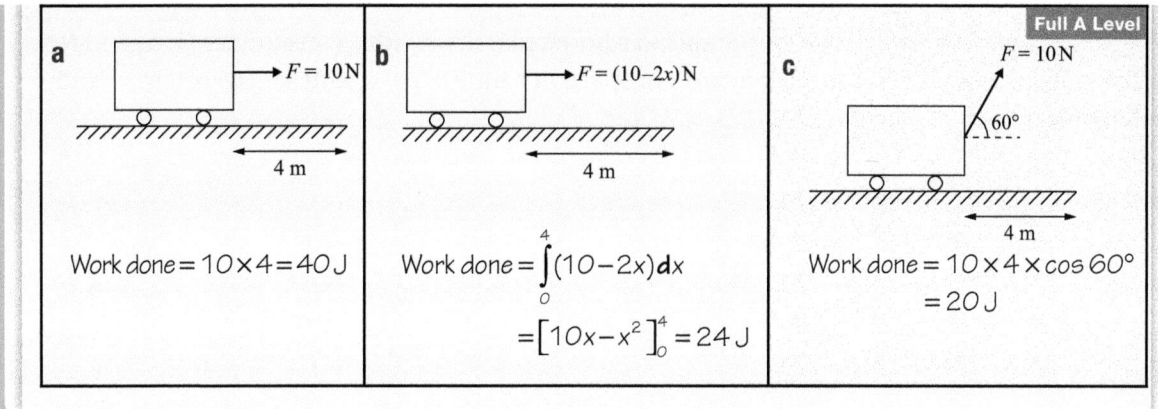

a

Work done $= 10 \times 4 = 40\,\text{J}$

b

Work done $= \int_{0}^{4}(10-2x)\,\mathbf{d}x$

$= \left[10x - x^2\right]_{0}^{4} = 24\,\text{J}$

c

Work done $= 10 \times 4 \times \cos 60°$

$= 20\,\text{J}$

If you drag an object along a rough floor for a distance x, four forces are acting on the object.

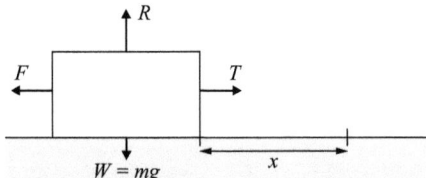

The weight W and the reaction R between object and floor do no work because they act at right-angles to the motion. The object is dragged by a force T (for example, the tension in a rope or the tractive force of an engine) against a frictional force F. These forces act in the same straight line as the motion of the object, so the work done *by* the dragging force over distance x is $T \times x$, and the work done *against* the frictional force is $F \times x$ (if T and F are constant). Use the word *by* when the force does work and the word *against* when work is done against a resisting force such as friction.

Power is the rate at which work is done.

If an engine is set to work for a period of time, then the average power of the engine is equal to $\dfrac{\text{work done}}{\text{time taken}}$

The units of power are **watts** (W) when work is in J and time in s.

If a constant force FN pulls an object in its direction at a constant speed $v\,\text{m s}^{-1}$, the object moves a distance of v metres every second. For example, consider a train where an engine is pulling coaches at a steady speed. The work done every second by the engine is $F \times v$ joules, so the power of the engine is $F \times v$ watts.

> **Key point**
>
> The power of a constant force F moving at a steady speed v is equal to $F \times v$, where power is in W, force in N and speed in m s^{-1}
>
> If F or v are variable, $F \times v$ gives the power at that particular instant.

The letter F is often used for any general force and also for a frictional force.

The letter R is often used for a reaction and also for a resisting force.

The pulling force of an engine is usually called the **tractive force** or **driving force**. It is often represented by the letter T or P

Example 2

A person pulls a crate with a rope parallel to a horizontal floor with a constant tension of 15 N. The crate passes points A and B that are 4 m apart, in a time of 1.6 s, with speeds of 2 m s^{-1} and 3 m s^{-1} as shown.

Find

a The average power exerted over the 4 metres,

b The power exerted when at point B

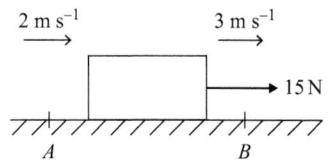

a Average power $= \dfrac{\text{work done}}{\text{time taken}} = \dfrac{15 \times 4}{1.6} = 37.5\,\text{W}$

b Power exerted at $B = F \times v = 15 \times 3 = 45\,\text{W}$.

Example 3

A metal ingot of weight 80 N is pulled at a constant speed of 0.2 m s^{-1} for 3 metres across a rough horizontal floor by a rope parallel to the floor. There is a constant frictional force F of 30 N. Find

a The work done by each force acting on the ingot

b The power exerted by the tension T in the rope.

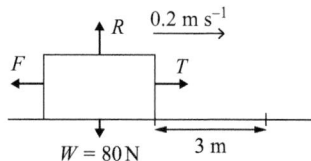

There is no vertical acceleration because there is no vertical motion, and there is no horizontal acceleration because the horizontal motion is at constant speed. So there is no resultant force, and hence,

resolving vertically,   reaction R = weight W = 80 N

resolving horizontally,   tension T = friction F = 30 N

a There is no vertical distance moved, so the work done by R and W is zero.

The horizontal distance moved = 3 m

The work done by tension T = 30 × 3 = 90 N

The work done against friction F = 30 × 3 = 90 N

b Power generated by tension = $T \times v$ = 30 × 0.2 = 6 W

Exercise 7.1A Fluency and skills

1 A constant force F pulls an object a distance x in its own direction. Find the work done by the force when

   a $F = 20$ N, $x = 4$ m

   b $F = 50$ N, $x = 12$ m

   c $F = 0.3$ kN, $x = 0.5$ km

2 Each of the three contestants A, B and C, in a strong-man competition drags a weight a distance x along level ground by exerting a constant tension T in a rope which is parallel to the ground. Use this data to find which man does the most work.

	A	B	C
Tension, T	500 N	300 N	800 N
Distance, x	20 m	35 m	15 m

3 For the strong men in question 2, *A* completed his task in a time of 20 s, *B* in 25 s and *C* in 30s. Using the relationship

$$\text{power} = \frac{\text{work done}}{\text{time taken}}, \text{find which man exerted}$$

the most power.

4 A car's engine exerts a constant driving force *F* of 800 N as it increases its speed from 20 m s^{-1} to 30 m s^{-1}. Using the relationship *power* = *F* × *v*, find the power exerted by the engine at the start and end of its motion.

5 A cyclist pedals his bike to create a constant driving force of 50 N as he climbs a hill, starting with a speed of 14 m s^{-1} and finishing with a speed of 3 m s^{-1}

What power is he exerting at the start and end of his climb?

6 Two cranes, *P* and *Q*, each lift loads of 2000 N to a height of 25 m. *P* takes 10 s and *Q* takes 15 s. How much work does each crane do and what is the rate of working (that is, the power) of each crane?

7 A car engine works at a rate of 20 kW. Find its driving force if it travelling at

  a 20 m s^{-1} **b** 120 km h^{-1}

8 Two equally powerful athletes start a race at the same time. They finish the race with speeds of 10 m s^{-1} and 15 m s^{-1}. If their final power output is 300 W, what tractive force is each then producing?

9 Find the work done and power required when

  a A crate is moved steadily for 6 m horizontally in 4 s by a horizontal rope under a tension of 120 N

  b A spring balance which reads 20 N steadily lifts a suitcase 3 m upwards in 1.5 s.

10 A winch uses a rope to lift three boxes *A*, *B* and *C* vertically, one after the other, at constant speeds for distances and times as given in the table. For each box find, in terms of *g*,

  a The constant tension *T* in the rope

  b The work done by the winch

  c The power exerted by the winch.

Box	*A*	*B*	*C*
Mass, kg	12	4	0.5
Distance, m	3	0.5	0.2
Time, s	4	4	2

11 A variable horizontal force *F* N pulls an object along a horizontal table through a distance *x* of 4 m. Find the work done by the force if

  a $F = 8 - 2x$ **b** $F = 16 - x^2$

  c $F = x^2 - x + 2$ **d** $F = \frac{1}{2}x + 1$

12 A horizontal rope pulls a trolley 6 m along a horizontal floor. A graph is drawn to show how the tension *T* in the rope varies with distance. Calculate the work done by the tension in each case

a 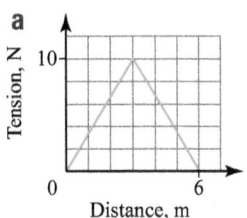   **b**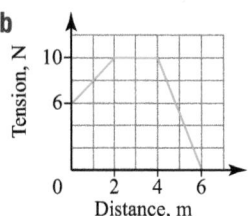

13 A boy pulls a block on a level track for a distance *x* by a string inclined at an angle θ to the horizontal. If the tension *T* in the string is constant, find the work done by the boy in each of these three cases:

	Tension, *T*	Distance, *x*	Angle, θ
a	10 N	8 m	60°
b	5 N	4 m	45°
c	0.2 N	0.6 m	30°

14 The three blocks in question 13 were initially at rest when the boy began to pull them. Find the power exerted by the boy in each case at the instant when the speed is

  a 2 m s^{-1} **b** 5 m s^{-1} **c** 4 m s^{-1}

15 A mass at the point (*x*, 0) moves along the *x*-axis of a horizontal *x*-*y* plane under a force $F = (2x + 1)$ N which acts in the plane at 30° to the *x*-axis. Find the power supplied by force *F* if the speed of the mass is 5 m s^{-1} when *x* = 4 m.

Full A Level

To solve problems involving work, energy and power

(1) Draw a clearly labelled diagram to show the information you are given.

(2) When writing an energy equation, ensure you have the gains and losses balanced correctly.

The **energy** of a body is its ability to do work. There are different kinds of energy such as electrical, nuclear and heat. In this chapter you will study two kinds: **kinetic energy** (due to a body's motion) and **potential energy** (due to a body's position).

When you apply a force F to a body of mass m which is initially at rest, it moves with acceleration a. After travelling a distance s, it has velocity v

Newton's second law gives $\qquad\qquad\qquad F = ma$

The kinematic equation $v^2 = u^2 + 2as$ gives $\quad v^2 = 2as$ because $u = 0$

So the work done by the force F is $\qquad\quad F \times s = ma \times \dfrac{v^2}{2a} = \dfrac{1}{2}\, mv^2$

The energy of the body due to the work done to give it this velocity v is called its kinetic energy.

Key point

The **kinetic energy** (KE) of a body with mass m and velocity v is equal to $\dfrac{1}{2}\, mv^2$

When you release a mass m from height h, it falls downwards freely as a result of the gravitational force F which is equal to its weight mg

Newton's second law gives $F = ma$, so $mg = ma$, and a is the acceleration due to gravity, g

So the work done by F is $F \times h = mgh$

Before being released, the mass had energy with the potential to do this work. The energy it possessed due to its height is called its gravitational potential energy.

Key point

The **gravitational potential energy** (GPE) of a body with mass m at height h is equal to mgh

The **principle of conservation of mechanical energy** says that the total mechanical energy (the sum of KE and GPE) of a system remains constant, provided no work or energy is lost to friction or impacts.

For example, when an object falls freely under gravity, the total energy is constant because the loss in GPE when falling equals its gain in KE, provided no energy is lost to air resistance. However, if energy is lost, you would have, for this example, the energy equation:

Loss of GPE = Gain in KE + Work done against air resistance

Example 4

An 4 kg object, which is initially at rest, falls 12 m freely under gravity to hit the ground. Take $g = 9.81$ m s^{-2}

a Find its final speed v if
   i Air resistance is neglected
   ii Air resistance is modelled by a constant force, F of 5 N.
b Describe a limitation to this model and explain how it might be made more realistic.

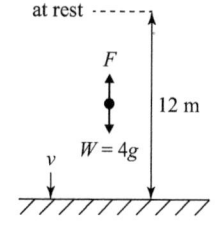

a i The energy equation is   GPE lost = KE gained

$$4 \times g \times 12 = \frac{1}{2} \times 4 \times v^2$$

Final velocity, $v = \sqrt{24g} = 15.3$ m s^{-1}

ii The energy equation is   GPE lost = KE gained + Work done against resistance

$$4 \times g \times 12 = \left(\frac{1}{2} \times 4 \times v^2\right) + (5 \times 12)$$

$$2v^2 = 48g - 60 = 410.9$$

Final velocity, $v = \sqrt{205.4} = 14.3$ m s^{-1}

b It is common experience that air resistance increases with speed. The model is limited by having a constant resistance, F
A more realistic model would have F dependant on speed.

Note that part **a i** of Example 4 could also be easily solved using the kinematic equation $v^2 = u^2 + 2as$ with $a = g$. In part **b**, the acceleration, a could be found first using Newton's 2nd law. (To better understand how to use Newton's 2nd law in this way, see Example 6.) In problems where acceleration is not constant, the energy equation is likely to be the easier method.

Example 5

A car of mass 800 kg starts from rest at the foot of a slope with a constant tractive force, T N. It climbs the slope, travelling 500 m in 50 s to reach a speed of 20 m s^{-1} as it rises a vertical height of 4 m.

Stating your assumptions and taking $g = 10$ m s^{-2}, calculate

a The gain in its GPE and KE

b The tractive force, T

c The average power of the engine over the whole journey

d The power of the engine at the end of the journey.

(*Continued on next page*)

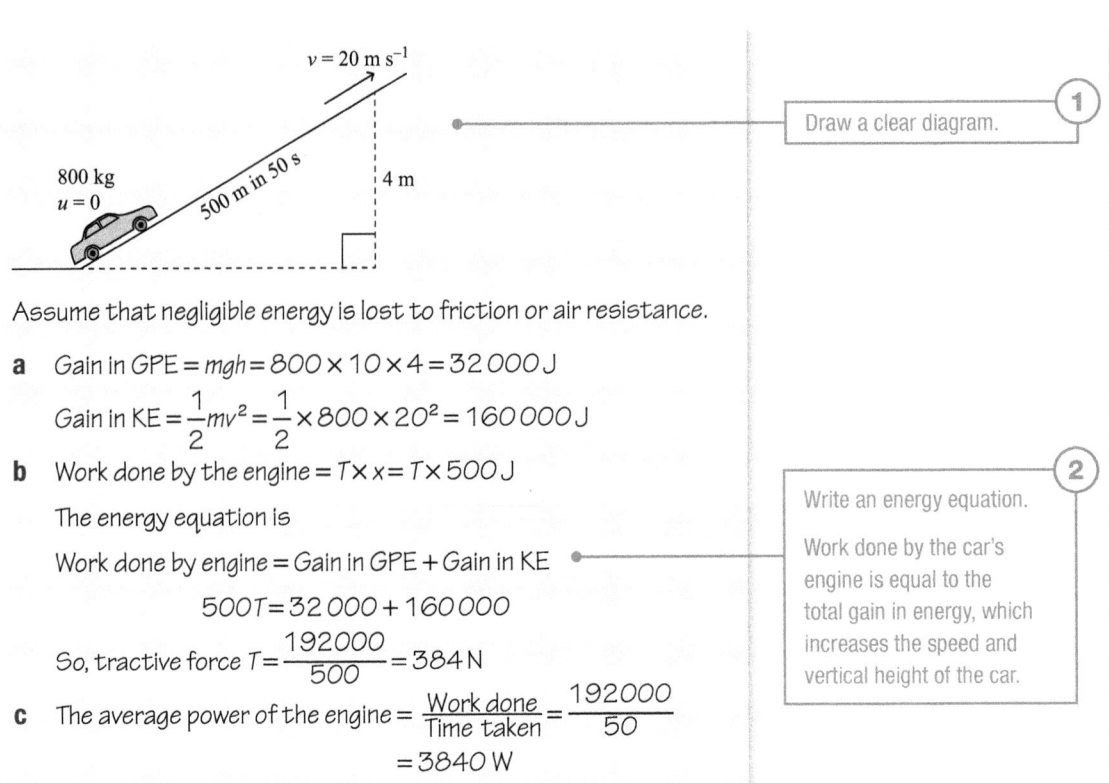

①
Draw a clear diagram.

Assume that negligible energy is lost to friction or air resistance.

a Gain in GPE $= mgh = 800 \times 10 \times 4 = 32\,000\,\text{J}$

  Gain in KE $= \dfrac{1}{2}mv^2 = \dfrac{1}{2} \times 800 \times 20^2 = 160\,000\,\text{J}$

b Work done by the engine $= T \times x = T \times 500\,\text{J}$

  The energy equation is

  Work done by engine = Gain in GPE + Gain in KE

  $$500T = 32\,000 + 160\,000$$

  So, tractive force $T = \dfrac{192\,000}{500} = 384\,\text{N}$

②
Write an energy equation.

Work done by the car's engine is equal to the total gain in energy, which increases the speed and vertical height of the car.

c The average power of the engine $= \dfrac{\text{Work done}}{\text{Time taken}} = \dfrac{192\,000}{50}$

  $= 3840\,\text{W}$

d Instantaneous power at the end $= T \times v = 384 \times 20 = 7680\,\text{W}$

Example 6

A car of mass 1 tonne travels from rest up a slope at 30° to the horizontal with a constant acceleration against a constant resistance R of 400 N. On reaching the top of the slope, it has a speed of 10 m s⁻¹ and its engine is working at a rate of 58 kW. Calculate the length of the slope. Take $g = 9.8\,\text{m s}^{-1}$

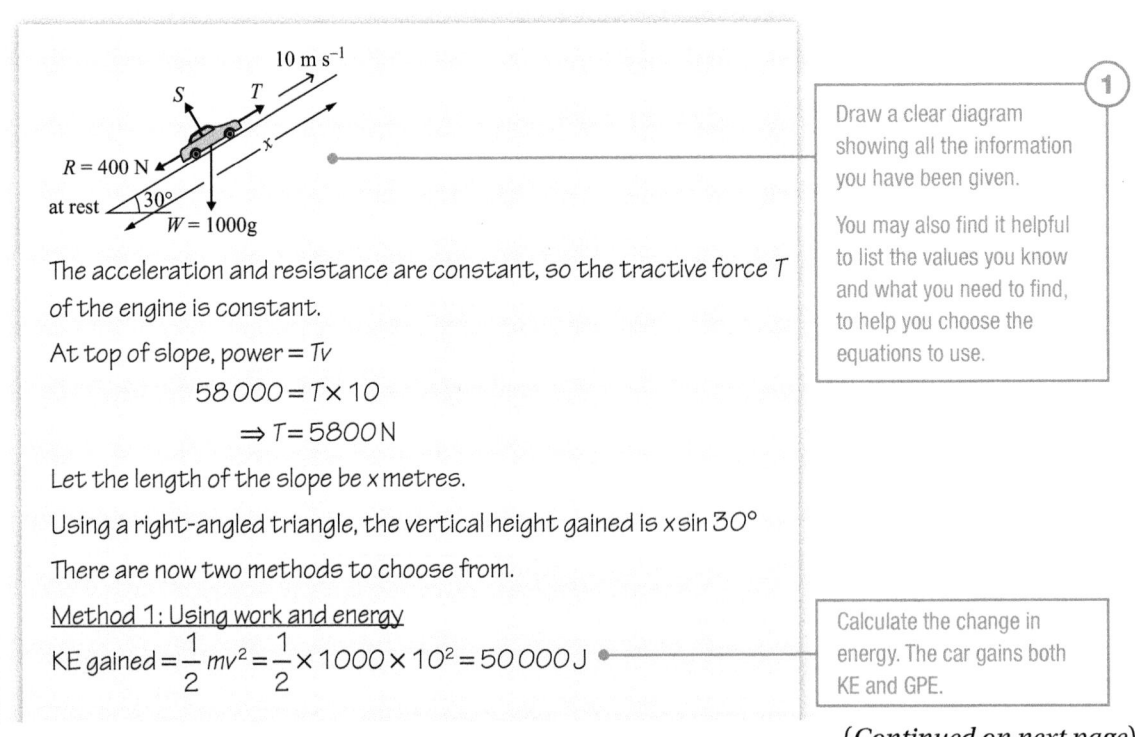

①
Draw a clear diagram showing all the information you have been given.

You may also find it helpful to list the values you know and what you need to find, to help you choose the equations to use.

The acceleration and resistance are constant, so the tractive force T of the engine is constant.

At top of slope, power $= Tv$

$$58\,000 = T \times 10$$

$$\Rightarrow T = 5800\,\text{N}$$

Let the length of the slope be x metres.

Using a right-angled triangle, the vertical height gained is $x \sin 30°$

There are now two methods to choose from.

Method 1: Using work and energy

KE gained $= \dfrac{1}{2}mv^2 = \dfrac{1}{2} \times 1000 \times 10^2 = 50\,000\,\text{J}$

Calculate the change in energy. The car gains both KE and GPE.

(Continued on next page)

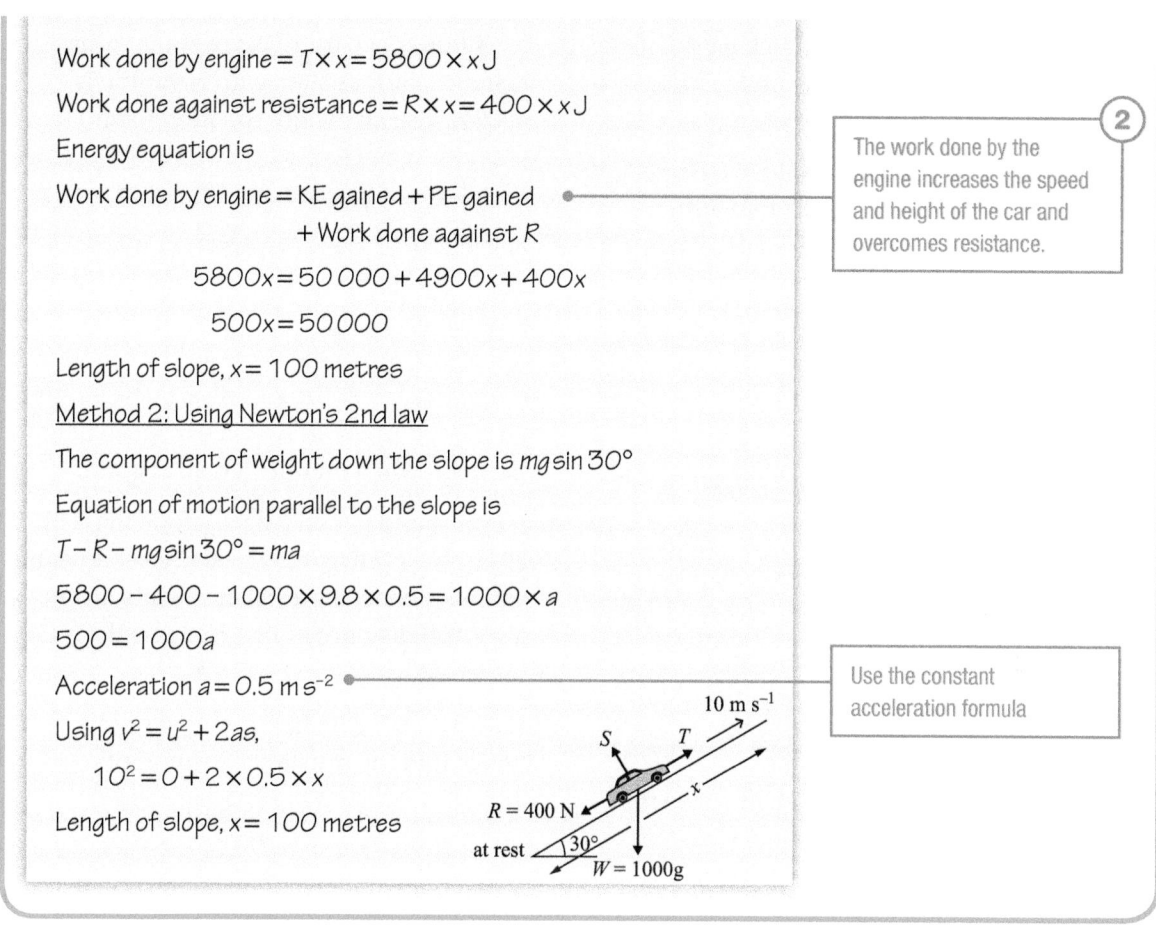

Work done by engine $= T \times x = 5800 \times x$ J

Work done against resistance $= R \times x = 400 \times x$ J

Energy equation is

Work done by engine $=$ KE gained + PE gained + Work done against R

The work done by the engine increases the speed and height of the car and overcomes resistance.

$5800x = 50\,000 + 4900x + 400x$

$500x = 50\,000$

Length of slope, $x = 100$ metres

Method 2: Using Newton's 2nd law

The component of weight down the slope is $mg \sin 30°$

Equation of motion parallel to the slope is

$T - R - mg \sin 30° = ma$

$5800 - 400 - 1000 \times 9.8 \times 0.5 = 1000 \times a$

$500 = 1000a$

Acceleration $a = 0.5$ m s^{-2}

Use the constant acceleration formula

Using $v^2 = u^2 + 2as$,

$10^2 = 0 + 2 \times 0.5 \times x$

Length of slope, $x = 100$ metres

Exercise 7.1B Reasoning and problem-solving

1 A ball of mass 0.5 kg drops 12 m vertically from rest to hit the ground. Use an energy equation to find its speed on impact if

  a Air resistance is negligible,

  b Air resistance is a constant force of magnitude 4 N.

2 A rocket of mass 5 kg is fired vertically upwards with an initial speed of 24 m s^{-1}

  Use an energy equation to find its maximum height if

  a Air resistance is negligible,

  b Air resistance is a constant 4 N.

3 A 5 kg box with an initial speed of 10 m s^{-1} moves 4 m across a rough horizontal floor under a constant frictional force F and comes to rest. Find the value of F

4 A person pulls a crate for 6 m on a horizontal floor using a horizontal rope with a constant tension of 12 N. The motion is resisted by a constant frictional force F of 4 N. The crate's final velocity is 2 m s^{-1}

  Find

  a The work done by the tension,

  b The work done against friction,

  c The final power exerted by the person at the end of the motion.

5 **a** A lorry of mass 2 tonnes accelerates from rest to a speed of 10 m s^{-1} over 400 m on a horizontal road with negligible resistance. Find

    i The gain in its KE,

    ii The work done by the engine,

    iii The engine's tractive force.

b If the lorry had experienced a constant resistance of 50 N, how much work would the engine now do and what would be the tractive force?

6 A train of mass 50 tonnes increases its speed from $10\,\text{m s}^{-1}$ to $20\,\text{m s}^{-1}$ over a horizontal distance of 500 metres against a resistance of 2000 N due to wind and friction.

    a Calculate the increase in the train's KE.

    b Calculate the work done by the engine and the engine's tractive force.

    c What is the initial and final power of the engine?

7 **a** A brick of mass 1.2 kg falls from rest from the top of a building 30 metres tall.

      i What is the GPE lost and the KE gained just before hitting the ground?

      ii Calculate the brick's speed on impact.

    b If the brick is thrown down from the top of the building with an initial speed of $14\,\text{m s}^{-1}$, what is its total KE and its speed just before impact?

    c If the brick is thrown upwards from the top of the building with a speed of $14\,\text{m s}^{-1}$, what is its speed on impact with the ground?

    d If the brick experiences a constant resistance of 3 N in falling from rest, what it its velocity when striking the ground?

8 A train of mass 100 tonnes accelerates steadily from rest over 450 metres horizontally to reach a speed of $15\,\text{m s}^{-1}$

Resistance is negligible.

    a Use an energy equation to find the maximum power exerted by the train's engine.

    b Find the acceleration of the train and use Newton's 2nd law to check your answer.

9 The same train as in Question 8 reaches the same speed from rest over the same distance, but resistance to its motion is 4000 N. Use an energy equation to find the maximum power now exerted by its engine. Check your answer using Newton's 2nd law.

10 Particles P of mass 2 kg and Q of mass 5 kg are connected by a light inextensible string passing over a light smooth pulley. They are initially at rest at the same level with the string taut when the system is released. Use an energy equation to find the speed of the particles when they are 3 metres apart.

11 A cyclist and her bike have a total mass of 80 kg. The cyclist free-wheels 200 m downhill against a constant resistance while dropping a vertical distance of 8 m. Her speed increases from $3\,\text{m s}^{-1}$ to $10\,\text{m s}^{-1}$

Calculate the gain in kinetic energy and hence find the resistance to her motion.

12 The maximum power of the engine of a 5 tonne coach is 40 kW. When the resistance to motion is 1500 N, find the maximum speed it can reach on a horizontal road.

13 A fountain uses a pump to raise 3000 litres of water through 4 metres every minute and expels it at a speed of $8\,\text{m s}^{-1}$

Find the power of the pump.

14 Water is discharged at $6\,\text{m s}^{-1}$ through a circular nozzle of radius 4 cm, having been raised by 2 m. Find the power of the pump to the nearest watt.

15 A train of mass 1200 kg with velocity $v\,\text{ms}^{-1}$ is resisted by a force of $600 + v^2$ N. The power of its engine is $1.5v$ kW. When on level ground, calculate

    a The train's maximum speed,

    b Its acceleration when moving at $10\,\text{m s}^{-1}$

16 The total mass of an engine and its train is 450 000 kg. The resistance to their motion is $\dfrac{v^2}{4}$ N per 1000 kg at a speed of $v\,\text{m s}^{-1}$

If the maximum power of the engine is 1125 kW, find its greatest speed on horizontal ground.

17 A cyclist on a level road has a power output of 75 W when travelling at a constant speed $v\,\text{m s}^{-1}$ against a resistive force $R = kv^2\,\text{N}$. Calculate the value of the constant k if the cyclist is traveling at 24 km h^{-1}

18 The resistance R to the motion of a car is directly proportional to its speed v, such that $R = kv$

   a A car of mass 1500 kg has a maximum speed of 45 m s^{-1} on level ground when its engine has a power output of 8 kW. Find the value of k

   b Calculate the car's acceleration when it moves at 20 m s^{-1} on a level road if the power of its engine is then 6 kW.

19 An object of mass 8 kg is dragged along a rough horizontal floor from rest for a distance s m by a variable force T where $T = 36 - s^2\,\text{N}$. Find its speed and the power exerted by the force T after the object has moved 4 m, given that the magnitude of the frictional force is a quarter that of the normal reaction to the floor.

20 A car of mass 1.2 tonnes travels up a slope at 30° to the horizontal at a constant speed of 25 m s^{-1} for 300 metres.

   a If the slope is smooth, calculate

   i The increase in the car's energy,

   ii The tractive force of its engine,

   iii Its power output.

   b If the slope is rough and the car experiences a constant frictional force down the slope of 3000 N, calculate the tractive force and power of its engine.

21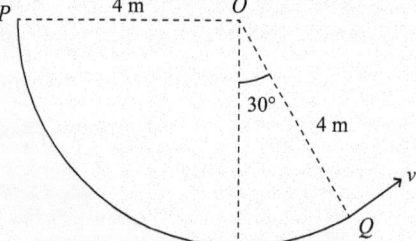

A skateboarder of mass 60 kg begins at rest at point P. He travels down a ramp shaped as an arc of a circle, centre O, with radius 4 m. Calculate his speed v when he leaves the ramp at Q when there is

   a No resistance to motion,

   b A constant resistance to motion of 48 N.

22 For the coach described in question 12, what is its maximum speed up a slope which makes an angle of $\sin^{-1}\dfrac{1}{98}$ to the horizontal?

23 A car of mass 900 kg pulls a trailer of mass 200 kg and the resistances to their motion are constant at 200 N and 80 N respectively. When travelling on a horizontal road at a maximum speed of $28\,\mathrm{m\,s^{-1}}$, the car's engine is working at maximum power. The car and trailer travel with speed $8\,\mathrm{m\,s^{-1}}$ at full power up a hill which makes an angle of $\arcsin\dfrac{1}{40}$ to the horizontal.

Calculate

a   The acceleration,

b   The tension in the tow-bar.

24 A smooth ring of mass m is threaded on to a fixed vertical circular hoop with centre O and radius r

The ring is projected with velocity u from its lowest point.

a   Show that if the ring just reaches the highest point Q, then $u^2 = 4gr$

b   If the ring passes a point P (where the acute angle $QOP = \theta$) with velocity $\dfrac{1}{2}u$, prove that $\cos\theta = \dfrac{3u^2}{8gr} - 1$ and that $\dfrac{8gr}{3} < u^2 < \dfrac{16gr}{3}$

25 For the cyclist in question 17 above, calculate the steepest gradient she can climb at $12\,\mathrm{km\,h^{-1}}$ when working at the same rate, given that her total mass is 110 kg.

26 A racing car of mass 1500 kg, travelling at speed $v\,\mathrm{m\,s^{-1}}$, experiences a resistance $R = 36v\,\mathrm{N}$. The car travels down a slope inclined at 2° to the horizontal. If the maximum power of its engine is 120 kW, find the maximum speed of the car down the slope.

27 The object in question 19 is now dragged along a floor that is tilted upwards at 5° to the horizontal. All other given data stays the same. Calculate the speed of the object up the slope and the power exerted by the dragging force after the object has moved 4 metres.

28 A train of mass 240 000 kg is pulled from rest up a slope of $\arcsin\dfrac{1}{250}$. The resistance to motion is $2000g\,\mathrm{N}$. The tractive force T of its engine varies with the distance s travelled, as in this table. Take $g = 9.8\,\mathrm{m\,s^{-2}}$

$s\,(\mathrm{m})$	0	25	50	75	100	125	150
$T\,(\times 10^4\,\mathrm{N})$	10	13	12	11	9	7	6

a   Use a numerical method to calculate an approximate value for the work done in pulling the train 150 m up the slope.

b   Calculate the speed of the train after it has travelled 100 m and the power of its engine at this point to the nearest watt.

7.2 Hooke's law

Fluency and skills

In many situations, it is not sensible to model **strings** as inextensible. You are now going to adopt a new model which assumes that strings are **light** and **elastic** and that they recover their **natural length** after being stretched.

A **spring** is also modelled as light and elastic. As well as recovering its natural length after being stretched, it also recovers it after being compressed. A string cannot be compressed; it simply goes slack.

For a string or stretched spring, the difference between the stretched length l' and the natural length l is the **extension** x. For a compressed spring, the difference in length is called a **compression**, x

Hooke's law states that the **tension** T in an elastic string or spring is proportional to its extension x

$T \propto x$ where $x = l' - l$

So, $T = k \times x$ where the constant k is the **stiffness** of the string or spring.

Stiffness depends on the material from which the string or string is made and also on its natural length. Stiffness k is equal to $\dfrac{\lambda}{l}$ where the constant λ is called the **modulus of elasticity**.

> In reality, if the stretch goes beyond the **elastic limit**, the string or spring is permanently deformed and Hooke's law does not apply.

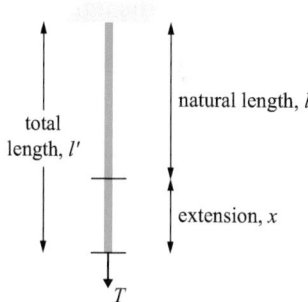

> The units of λ are newtons.

Key point

Hooke's law gives $T = \dfrac{\lambda}{l}x$ where $x = l' - l$

To calculate the work required to create an extension X from its natural length, imagine a small extension δx when the tension is T

The small amount of work required is force $\times$ distance $= T \times \delta x$
To calculate the total work needed, integrate for the whole extension.

There are two methods.

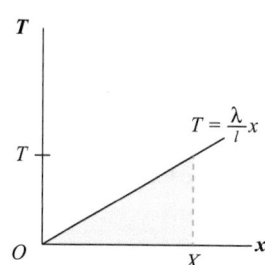

Using calculus

Work required $= \displaystyle\int_0^X T \, dx$

$= \dfrac{\lambda}{l} \displaystyle\int_0^X x \, dx = \dfrac{\lambda}{l} \left[\dfrac{x^2}{2} \right]_0^X$

$= \dfrac{1}{2} \times \dfrac{\lambda}{l} \times X^2 = \dfrac{\lambda}{2l} X^2$

Using a graph

Work required $= \displaystyle\int_0^X T \, dx$

$=$ Area of shaded triangle

$= \dfrac{1}{2} \times X \times T = \dfrac{1}{2} \times X \times \dfrac{\lambda}{l} \times X$

$= \dfrac{\lambda}{2l} X^2$

217

The work done to stretch a string or spring is stored as **elastic potential energy** (EPE) in the string or spring.

The elastic potential energy, EPE, stored in a string or spring when its natural length l is extended by a distance x is

$$\text{EPE} = \frac{\lambda}{2l}x^2$$

Key point

This formula has the same shape as $\frac{1}{2}mv^2$ for kinetic energy.

To increase the extension from x_1 to x_2, the energy required is

$$\frac{\lambda}{2l}x_2^2 - \frac{\lambda}{2l}x_1^2 = \frac{\lambda}{2l}(x_2^2 - x_1^2) = \frac{\lambda}{l}(x_2 + x_1)(x_2 - x_1)$$

Example 1

A string with natural length 5 m and modulus of elasticity 10 N is extended 0.4 m.

a What is the stiffness of the string?

b Calculate the tension in the string and the EPE stored in it.

c If the string is extended a further 0.3 m, how much extra EPE is stored in it?

a Stiffness $= \dfrac{\lambda}{L} = \dfrac{10}{5} = 2\,\text{Nm}^{-1}$

b From Hooke's Law, tension $T = \dfrac{\lambda}{l}x = \dfrac{10}{5} \times 0.4 = 0.8\,\text{N}$

   EPE stored $= \dfrac{\lambda}{2l}x^2 = \dfrac{10}{2\times5} \times 0.4^2 = 0.16\,\text{J}$

c Increase in EPE $= \dfrac{\lambda}{2l}(x_2^2 - x_1^2) = \dfrac{10}{2\times5} \times (0.7^2 - 0.4^2) = 0.33\,\text{J}$

Example 2

Two strings OA and OB are tied to an object O which is held on a smooth table between two fixed points A and B which are 1.0 m apart.

OA has natural length 0.4 m and modulus of elasticity 12 N. OB has natural length 0.3 m and modulus of elasticity 18 N.

Find the extensions of the strings and their total EPE.

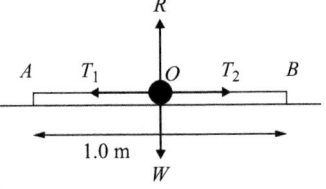

For vertical equilibrium,   reaction $R =$ weight W

For horizontal equilibrium,   tension $T_1 =$ tension T_2

Hooke's law gives:

for OA, $T_1 = \dfrac{\lambda_1}{l_1}x_1 = \dfrac{12}{0.4}x_1 = 30x_1$

for OB, $T_2 = \dfrac{\lambda_2}{l_2}x_2 = \dfrac{18}{0.3}x_2 = 60x_2$

The object is held in equilibrium, so the two horizontal forces balance.

Use Hooke's law for OA and OB

(Continued on next page)

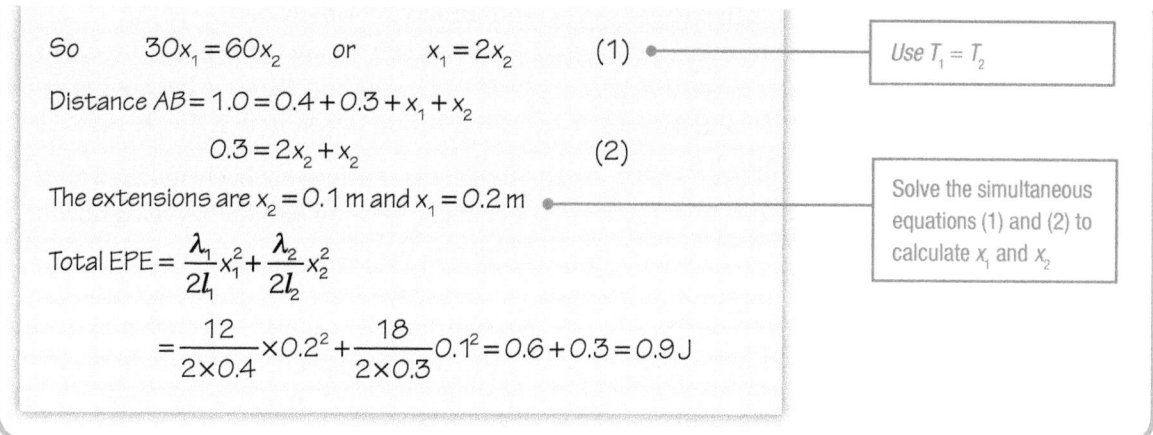

So $30x_1 = 60x_2$ or $x_1 = 2x_2$ (1) •

Distance $AB = 1.0 = 0.4 + 0.3 + x_1 + x_2$

$0.3 = 2x_2 + x_2$ (2)

The extensions are $x_2 = 0.1\,m$ and $x_1 = 0.2\,m$ •

Total EPE $= \dfrac{\lambda_1}{2l_1}x_1^2 + \dfrac{\lambda_2}{2l_2}x_2^2$

$= \dfrac{12}{2 \times 0.4} \times 0.2^2 + \dfrac{18}{2 \times 0.3}0.1^2 = 0.6 + 0.3 = 0.9\,J$

> Use $T_1 = T_2$

> Solve the simultaneous equations (1) and (2) to calculate x_1 and x_2

MECH

Exercise 7.2A Fluency and skills

1 Five different strings A to E of natural length l m and modulus of elasticity λ N are extended by x m

   a Copy and complete this table, where T is the tension in each string and EPE is the elastic potential energy stored in the string.

	l, m	x, m	λ, N	T, N	EPE, J
A	2	0.5	8		
B	3	1	6		
C	1.5		15	4	
D	1.2	0.6		4.5	
E	2.5		10		1.28

   b Find the stiffness of each string A to E.

2 A force of 15 N extends an elastic string by 0.2 metres. What will be the extension when the force is 30 N?

3 A force of 20 N compresses a spring by 5 cm. How far will the spring be compressed under a force of 10 N?

4 An light elastic string of natural length 2 m with a modulus of elasticity of 50 N is extended by a force of 20 N.

   Calculate

   a Its extension,

   b The work done by the force,

   c The elastic potential energy stored in the extended string.

5 A spring of natural length 1.2 m with $\lambda = 25$ N, where λ is the modulus of elasticity, is compressed by a force of 15 N.

   Calculate

   a The length of the compressed string,

   b The spring's elastic potential energy when compressed.

6 An elastic spring of natural length 1.5 m hangs from a ceiling with an object of mass 8 kg attached at its lower end. If the modulus of elasticity, λ, is 120 N and the object hangs in equilibrium, calculate

   a The tension in the spring,

   b The extension of the spring.

7 Two fixed points P and Q are 3 m apart on a smooth horizontal table. Two horizontal strings PO and QO are attached to an object O which sits between P and Q

   The natural length and value of λ are 0.9 m and 24 N respectively for PO, and are 0.6 m and 32 N respectively for QO

   Calculate the extensions of the two strings and the tensions in them when O is in equilibrium.

8   Two springs OX and OY are attached to an object O which is in equilibrium on a smooth horizontal table between two fixed points X and Y which are on the table and lie 1 m apart. OX has natural length 0.8 m and $\lambda = 15$ N. OY has natural length 1.2 m and $\lambda = 18$ N. Calculate the compressed lengths of the two springs and their total EPE.

9   Two light vertical springs AB and BC are joined at B with end A fixed to a horizontal table, and with C directly above B

    A mass of 5 kg, which is attached to C, compresses both springs, and rests in equilibrium. The natural length and modulus of elasticity of AB are 0.4 m and 90 N, and those of BC are 0.2 m and 80 N. Find the distance between A and B and the total EPE stored in the springs.

10 Two vertical strings are each attached to a ceiling by one end so that they hang down. Together they hold a mass m in equilibrium at their two lower ends. They have the same natural length l and their moduli of elasticity are λ_1 and λ_2 respectively. Show that the extension of each string is $\dfrac{mgl}{\lambda_1 + \lambda_2}$ and calculate the tension in each string.

Reasoning and problem-solving

To solve problems involving extensions in strings and springs

(1) Sketch a clear diagram to show all information.

(2) Use Hooke's law to calculate the energy stored in the string or spring.

(3) Write an energy equation which balances the changes in energy and any work done.

The principle of conservation of mechanical energy says that the total mechanical energy of a system remains constant, provided there is no external work done by resisting forces such as friction.

This implies that any increase in one form of energy is balanced by a decrease in another form. When you write an energy equation, you may find it easier to think in terms of increases and decreases of KE, GPE and EPE rather than the total energy.

Example 3

A vertical elastic string, of natural length $AB = 1$ m with $\lambda = 49$ N, has A fixed to a ceiling and B fixed to a mass of 2 kg. Take $g = 9.8$ m s^{-2}

a The mass is lowered gently until in equilibrium at point O. Calculate the extension when at O

It is then pulled down a further 20 cm to point P and released.

b Calculate its velocity v when passing O and the greatest height that it reaches. Describe its subsequent motion.

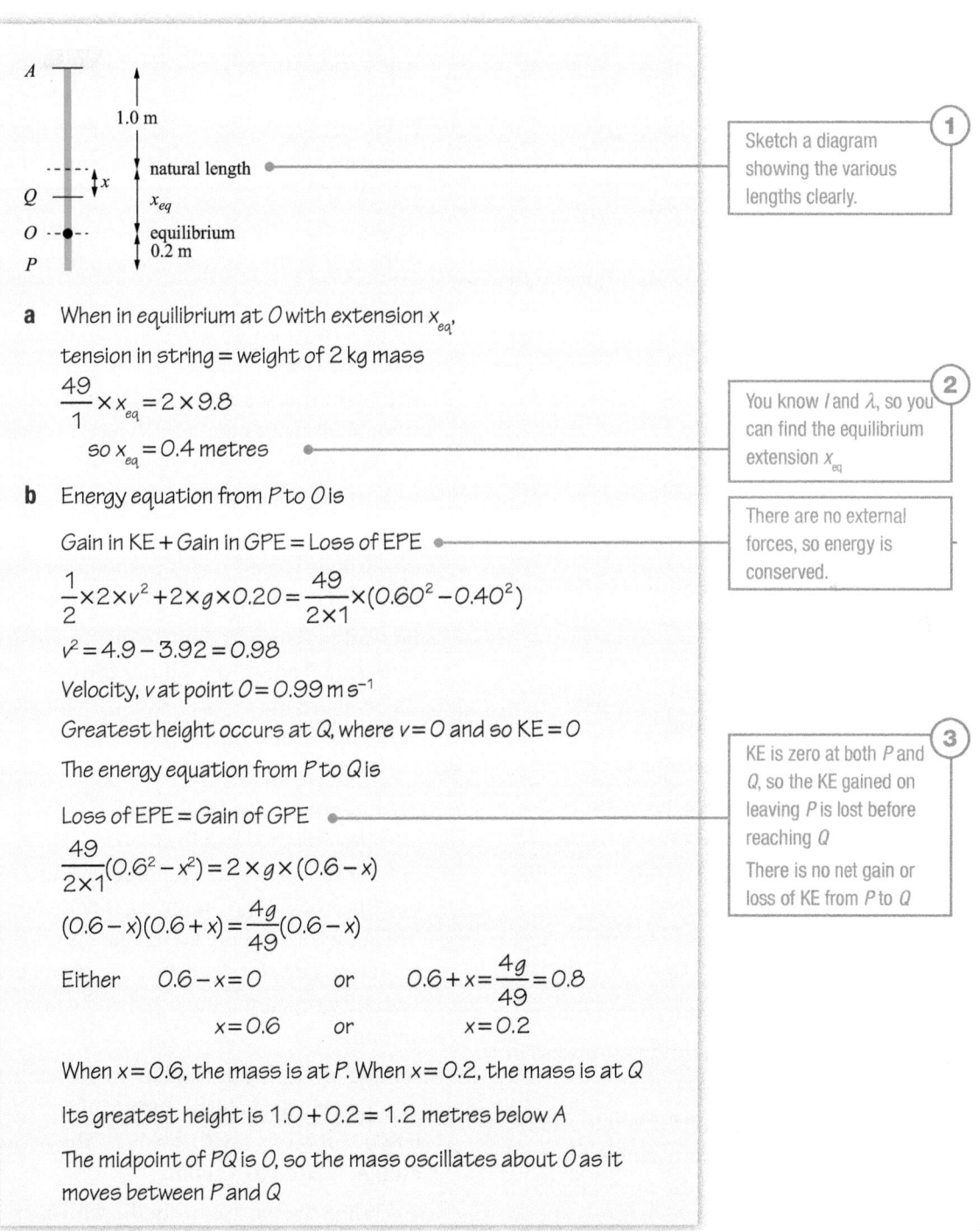

A
1.0 m
natural length
Q x
x_{eq}
O
equilibrium
0.2 m
P

> **1** Sketch a diagram showing the various lengths clearly.

a When in equilibrium at O with extension x_{eq},

tension in string = weight of 2 kg mass

$$\frac{49}{1} \times x_{eq} = 2 \times 9.8$$

so $x_{eq} = 0.4$ metres

> **2** You know l and λ, so you can find the equilibrium extension x_{eq}

b Energy equation from P to O is

Gain in KE + Gain in GPE = Loss of EPE

> There are no external forces, so energy is conserved.

$$\frac{1}{2} \times 2 \times v^2 + 2 \times g \times 0.20 = \frac{49}{2 \times 1} \times (0.60^2 - 0.40^2)$$

$v^2 = 4.9 - 3.92 = 0.98$

Velocity, v at point $O = 0.99$ m s^{-1}

Greatest height occurs at Q, where $v = 0$ and so KE = 0

The energy equation from P to Q is

Loss of EPE = Gain of GPE

> **3** KE is zero at both P and Q, so the KE gained on leaving P is lost before reaching Q
>
> There is no net gain or loss of KE from P to Q

$$\frac{49}{2 \times 1}(0.6^2 - x^2) = 2 \times g \times (0.6 - x)$$

$$(0.6 - x)(0.6 + x) = \frac{4g}{49}(0.6 - x)$$

Either $0.6 - x = 0$ or $0.6 + x = \dfrac{4g}{49} = 0.8$

$x = 0.6$ or $x = 0.2$

When $x = 0.6$, the mass is at P. When $x = 0.2$, the mass is at Q

Its greatest height is $1.0 + 0.2 = 1.2$ metres below A

The midpoint of PQ is O, so the mass oscillates about O as it moves between P and Q

In all questions, take the value of g to be $9.8 \, m \, s^{-2}$ unless told otherwise. λ is the modulus of elasticity.

1 How much work is done to stretch a light spring from a length of $2 \, m$ to a length of $3 \, m$, if its natural length is $1.5 \, m$ and its modulus of elasticity is $25 \, N$?

2 An elastic spring of natural length $0.5 \, m$ with $\lambda = 80 \, N$ hangs from a ceiling. One end of an elastic string of natural length $0.6 \, m$ with $\lambda = 120 \, N$ is attached to the spring's lower end. An object of mass $4 \, kg$ hangs freely in equilibrium from the lower end of the string.

Calculate

a The tension in the spring and the string,

b The total extension of the spring and the string,

c The total EPE stored in the spring and the string.

3 Two thin light vertical springs, A and B, stand side by side with their lower ends fixed to a horizontal table. A mass of $5 \, kg$ is placed on their upper ends, compressing both springs so that the mass is in equilibrium. If the natural length and modulus of elasticity of A are $0.8 \, m$ and $80 \, N$, and those of B are $1.5 \, m$ and $60 \, N$, find

a The compressed length of the springs,

b The tension in each spring,

c The total EPE in the springs.

4 a A model railway has a horizontal track which ends at a buffer made from a light, horizontal spring. In testing the spring, a mass of $2 \, kg$ compresses it by $2 \, cm$. Subsequently, a truck of mass $0.5 \, kg$ runs into the buffer with a speed of $0.60 \, m \, s^{-1}$
Calculate the stiffness of the spring, $\dfrac{\lambda}{l}$
How much does the buffer compress in bringing the truck to rest?

b A light, vertical spring is fixed to a floor. It is compressed by a distance a when a mass m rests on it. If the mass falls from rest onto the spring from a height of $\dfrac{3a}{2}$

above the top of the spring, calculate the greatest compression of the spring in the subsequent motion.

5 A mass P of $2 \, kg$ on a rough horizontal table is tied to a light horizontal elastic string, of natural length $2 \, m$ and stiffness $25 \, N \, m^{-1}$, whose other end is attached to a fixed point Q on the table. The mass is held such that $PQ = 3 \, m$. What is the initial acceleration of the mass and what is the distance PQ when the mass comes to rest, if the magnitude of the frictional force is three-fifths that of the normal reaction on mass P?

6 An elastic string of natural length $2 \, m$ with $\lambda = 80 \, N$ has one end tied to a $3 \, kg$ mass and the other end to a $5 \, kg$ mass. The string is stretched so that the masses are at rest $4 \, m$ apart on a smooth horizontal surface. They are released and move towards each other. Find their speeds at the moment when the string goes slack.

7 A light string of natural length l and modulus of elasticity λ has an initial extension of x_1
It is stretched further until its extension is x_2
Prove that the work done to increase the extension from x_1 to x_2 equals the product of the increase in the extension with the mean of the initial and final tensions.

8 A light horizontal spring, of natural length $2 \, m$ and modulus of elasticity $50 \, N$, has one end attached to a fixed point A on a horizontal table. The other end is attached to a mass of $0.2 \, kg$. The spring is compressed by 0.4 metres and the $0.2 \, kg$ mass is released from rest. How far does the mass travel before coming to instantaneous rest, if the table is smooth?

9 A mass of $1 \, kg$ is attached to the lower end of a light vertical elastic spring, of natural length $AB = 1.5 \, m$ with $\lambda = 49 \, N$. The other end, A, is fixed to a ceiling.

a When the mass is in equilibrium at point P, what is the extension in the string?

b The mass is now pulled down to point Q which is 30 cm below P, and released from rest. Find its velocity v when it reaches P and the greatest height to which it rises above P. Describe its subsequent motion.

c The mass is pulled down further to point R so that it is 50 cm below P. When released from rest, how close does it get to point A?

d If the string was replaced by a light spring with the same natural length AB and modulus of elasticity λ, how would your answer to part **b** be different? How close does the mass now get to A in this case?

10 A light elastic rope of natural length 25 m with $\lambda = 1200$ N has one end tied to a bungee jumper of mass 60 kg and the other end tied to a bridge. The jumper falls from rest off the point of attachment to the bridge and reaches a speed of v m s^{-1} after dropping a distance x m.

a Show that $30v^2 = 1788x - 24x^2 - 15000$ provided $x > k$, for constant k

State the value of k and calculate the maximum value of v

b Calculate the maximum value of x

11 A light elastic string of natural length a and modulus of elasticity λ has two masses m_1 and m_2 attached to its ends. They rest on a smooth horizontal table a distance a apart. The mass m_1 is struck in the direction of the string so that it has an initial velocity u away from m_2

Show that the maximum extension of the string is $\sqrt{\dfrac{m_1 m_2 u^2 a}{(m_1 + m_2)\lambda}}$

12 An object of 4 kg lies on a plane which slopes at 30° to the horizontal. A spring, of natural length 1.5 m with $\lambda = 40$ N, is attached at one end to the top of the plane with the other end attached to the object.

a If the plane is smooth, calculate the length of the spring when the object rests in equilibrium.

b If the plane is so rough that the frictional force is a fifth of the normal reaction on the object, calculate the distance along the plane in which the object can be in equilibrium.

13 A light elastic string of natural length 3 m hangs vertically from a fixed point and is stretched to a length of 4.5 m by a mass of 15 kg. The same mass is now attached to the midpoint of the same string which is stretched horizontally between two points X and Y which are 4 metres apart.

a On release, the mass falls vertically from rest. Calculate its velocity after it falls 1.5 m

b Verify that, if the mass is gently lowered until it hangs in equilibrium, its distance below the horizontal through X and Y is 1.1 m

14 A light rod AB of length 2 m is smoothly hinged to a vertical wall at A with B higher than A. A light elastic string, of natural length 1 m with $\lambda = 100$ N, is stretched from B to point C on the wall 2 m above A

A mass of m kg is fixed to B

a If angle $CAB = 60°$, calculate the value of m when the mass is in equilibrium.

b If B is placed to coincide with C and the mass m is released from rest in this position with a negligibly small horizontal velocity so that AB rotates about A, what is the maximum value of angle CAB in the subsequent motion?

7.3   Dimensional analysis

Fluency and skills

Using dimensional analysis to examine formulae can serve two purposes.

Firstly, you can use it to check whether a formula is valid or not. This process looks at the dimensional consistency of the terms in the formula. For example, it is not possible to add together a term with units in metres to a term with units in kilograms, because the dimensions of the terms are not consistent within the formula.

Secondly, dimensional analysis can be used to help build a possible formula between variables which are suspected to be related. For example, the density of seawater and the speed of waves might affect the height of the waves and a formula might exist to connect them.

Most quantities can be expressed in terms of three basic dimensions: mass M, length L and time T. Dimensions are not the same as units, but units can sometimes help you work out dimensions. For example, speed = distance ÷ time, the units of speed are $m\,s^{-1}$, and the dimensions of speed are $\dfrac{L}{T}$ or LT^{-1}

You write 'dimensions of' using square brackets, so $[\text{speed}] = LT^{-1}$

> Some quantities, such as pure numbers and angles, have no dimensions and do not appear in dimensional equations.

Example 1

a   Given that kinetic energy $KE = \dfrac{1}{2}mv^2$, work out the dimensions of energy.

b   Show that the equation $GPE = mgh$ is dimensionally consistent.

a $[KE] = [m] \times [v]^2 = M \times \left(\dfrac{L}{T}\right)^2 = M\,L^2\,T^{-2}$

b   On the left-hand side, $[GPE] = [KE] = M\,L^2\,T^{-2}$

> All forms of energy have the same dimensions.

On the right-hand side, $[m] \times [g] \times [h] = M \times LT^{-2} \times L = M\,L^2\,T^{-2}$

Both sides have the same dimensions, so the equation is dimensionally consistent.

1 Which of these quantities are dimensionless?

 a Speed of a car

 b The ratio of height to length for a bus

 c $\sin 60°$

 d Speed of rotation in radians per second

 e Angle of rotation in radians

 f The ratio of circumference to diameter for a circle

 g Temperature

2 a Write the dimensions of velocity and acceleration.

 b Use Newton's 2nd Law to write the dimensions of force and so write the dimensions of work.

 c Write the dimensions of momentum.

3 Find the dimensions of

 a volume   b density   c power

 d energy   e impulse   f pressure

4 Show that the dimensions of the term $2as$ in the kinematic equation $v^2 = u^2 + 2as$ are the same as the dimensions v^2 and u^2

5 Show that both of the terms ut and $\frac{1}{2}at^2$

 in the equation $s = ut + \frac{1}{2}at^2$ have the dimension L

6 Boyle's law gives the equation $pv = k$, where p and v are the pressure and volume of a gas. What are the dimensions of k?

7 Einstein said energy E and mass m are equivalent. If $E = k \times m$, work out the dimensions of k and show that they are the same as those of (velocity)²

8 Given that A is an area, a is acceleration, t is time, x is distance, u, v are velocities, F is force and m is mass

 i Make k the subject of each formula

 ii Find the dimensions of k

 a $A = kt$ b $F = kv^2$

 c $F = kxt$ d $a = kmu$

 e $v = \dfrac{kxF}{ma}$ f $\dfrac{u}{t} = k \times at$

9 The tension T in a string over a pulley which connects two masses m_1 and m_2 is given by $T = \dfrac{2mM}{m+M}g$

 Show that the dimensions of $\dfrac{2mM}{m+M}g$ are the same as the dimensions of force.

10 The resisting force F on a ball of radius r falling vertically with velocity v through a fluid of viscosity η is given by $F = 6\pi r v \eta$ Make η the subject of the formula and so find the dimensions of viscosity η

11 Newton's law of gravitation is $F = \dfrac{Gm_1m_2}{r^2}$

 Work out the dimensions of the constant G

12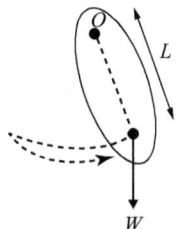

 A compound pendulum of weight W and pivot O has a length L and a time period t,

 where $t = 2\pi\sqrt{\dfrac{I}{mgL}}$

 Find the dimensions of the variable I and state its SI units.

13 Which of these formulae are dimensionally inconsistent?

 s, h are lengths; u, v are velocities; a, g are accelerations; F is force; and t is time.

 a $s = ut + \dfrac{1}{2}at^3$ b $Ft = m\sqrt{a}$

 c $v^2 = \dfrac{2as}{t}$ d $\dfrac{1}{2}mv^2 = mgh$

 e $F = \dfrac{m_1m_2g}{m_1+m_2}$ f $s = \dfrac{1}{2}t(u+v)$

MECH

Reasoning and problem-solving

Strategy

When using dimensional analysis

(1) Write the dimensions of all variables in terms of M, L and T.

(2) Equate powers of M, L and T and solve to work out the unknown indices.

Example 2

The time period t of a simple pendulum may depend on its mass m, its length l and the gravitational acceleration g.

Find a formula for t in terms of m, l and g

Assume the formula takes the form
$t = k \times m^x l^y g^z$ where k is a numerical constant.
The formula's dimensional equation is

$$[\text{Time, } t] = M^x \times L^y \times \left(\frac{L}{T^2}\right)^z = M^x L^{y+z} T^{-2z}$$ •——

$$x = 0, \quad y + z = 0 \quad \text{and} \quad -2z = 1$$

giving $\quad x = 0, \qquad y = \frac{1}{2} \quad \text{and} \quad z = -\frac{1}{2}$ •——

The formula is $\quad t = k \times m^0 l^{\frac{1}{2}} g^{-\frac{1}{2}} \quad$ or $\quad t = k\sqrt{\dfrac{l}{g}}$ •——

① Write all variables in terms of their dimensions.

② Equate powers of M, L and T and solve to work out x, y and z

$m^0 = 1$

Exercise 7.3B Reasoning and problem-solving

1   Given that x, y are distances; m and M are masses; F is a force; and λ is a coefficient of elasticity, find the dimensions of P and Q in the following equations, and suggest what each of P and Q might represent.

$$P = \sqrt{\frac{4Fmx}{3M(m+M)}} \qquad Q = \sqrt{\frac{\lambda(x-y)}{2mxy}}$$

2   A mass M accelerates from rest over a distance x before colliding and coalescing with a mass m. After the collision, they are brought to rest over a distance y by a constant resisting force R. It is suggested

that $R = \pi\left[(m+M)g + \dfrac{M^2 y}{(m+M)x}g\right]$

Determine whether this formula is dimensionally consistent.

3 A small mass m is attached to a fixed point O on a table by a string of length r. It travels on a circular path about O with a speed v. Find a formula for tension T in the string of the form

$$T = k \times m^a \times v^b \times r^c$$

where k is a numerical constant, clearly stating the values of a, b and c

4 The velocity v of sound in a gas with density ρ and pressure P has the form $v = k \times \rho^x \times P^y$ where k is a numerical constant. Work out the values of x and y and rewrite the formula using your answers.

5 The lifting force F on an aeroplane's wing depends on its speed v, its surface area A and the density of the air r. If $F = k \times v^x A^y r^z$ where k is a numerical constant, calculate the values of x, y and z. Write the formula using your answers.

6 "There's a hole in my bucket, dear Liza", said Henry, as he sees liquid leaking through the hole at the bottom with a speed v. Henry says that v will depend on the depth of the liquid, d and the gravitational acceleration, g. Liza agrees with him but says that it might also depend on the density of the liquid, ρ. So, their two models are

$$v \propto d^a g^b \rho^c \quad \text{and} \quad v \propto d^x g^y$$

where a, b, c, x and y are constants.

Which model would you accept? Give your reasons, stating the values of the constants to give dimensional consistency.

7 This formula is used in the design of jet engines: $F = mgI - pA$ where F, m, p and A are force, mass, pressure and area, and g is acceleration due to gravity. What are the dimensions of the symbol I?

8 The velocity v of the wave on a violin string may depend on the string's tension T, mass m and length L. Construct a formula for v which uses these variables.

9 The speed v at which waves travel across the ocean depends on the density ρ of the water, the distance λ between the wave crests and the gravitational constant g. Construct a formula for v in terms of ρ, λ and g

There is a dimensionless quantity in the equation.

10 The flow F, in m³ per second, of liquid down a tube depends on the tube's radius rm, its length Lm, its viscosity η kg m⁻¹ s⁻¹ and the drop in pressure PN per m² along the tube.

a Construct a formula for the flow in the form $F = k \times r^{4x} L^{-x} \eta^y P^z$

b For a particular tube with $r = 0.05$, $L = 6.0$, $\eta = 0.01$, $P = 3600$ and $F = 0.15$, show that $k = 0.4$

Chapter summary

- Work done by a constant force F moving a distance x in its direction is $F \times x$ and, for a variable force, work done $= \int F \, dx$
- Energy is gained by an object when work is done on it. Mechanical energy is either kinetic energy or potential energy. The units of both energy and work are joules (J).
- Kinetic energy, $KE = \frac{1}{2} mv^2$, relates to a body's motion.
- Gravitational potential energy, $GPE = mgh$, relates to a body's vertical height in Earth's gravity.
- Elastic potential energy, $EPE = \frac{\lambda}{2l} x^2$ relates to the elasticity of a string or spring.
- Power is the rate of doing work. For a constant force F moving with a speed v, power $= F \times v$ The unit of power is watts.
- When there is no external input of work or energy, the total energy of a system is conserved. So, gains in one form of energy are balanced by losses in another form of energy.
- Hooke's law states that the tension T in a string or spring is proportional to its extension (or, for a spring, its compression), x, where $T = \frac{\lambda}{l} x$
- The stiffness of a string or spring is equal to $\frac{\lambda}{L}$
- Dimensional analysis can help you to construct a formula involving various quantities or to check the consistency of a formula.

Check and review

You should now be able to...	Try Questions
✔ Analyse problems using formulae for work, energy and power.	1–9, 12–13
✔ Use Hooke's law to solve problems involving strings and springs.	8, 9, 13
✔ Construct equations and check them using dimensional analysis.	10, 11, 14

In all questions, take the value of g to be $9.8 \, \text{m s}^{-2}$ unless told otherwise.

1 A car has a mass of 1200 kg. It starts from rest and reaches a speed of $24 \, \text{m s}^{-1}$ after travelling 320 metres. Assuming that resistances to motion are negligible, use an energy equation to calculate the tractive force of its engine.

2 Reindeer pull a sleigh of mass 300 kg horizontally over 100 m on ice from rest with a constant acceleration of $0.5 \, \text{m s}^{-2}$

Assuming negligible resistance to motion, calculate

a The horizontal force pulling the sleigh,

b The work done by the reindeer,

c The gain in KE of the sleigh,

d The final power exerted by the reindeer.

3 A box of mass 200 kg is winched, with constant acceleration, 12 metres vertically for 10 seconds from rest by a rope. Calculate

a Its acceleration,   **b** Its final velocity,

c The gain in its

  i KE,   **ii** GPE,

d The work done by the winch,

e The tension in the rope,

f The final power of the winch.

4 A fountain pumps 1500 litres of water 2 metres vertically every minute and expels it at a speed of $5\,\text{m}\,\text{s}^{-1}$

Assuming no loss of energy, find the power of the pump.

5 A cyclist of total mass 90 kg has a speed of $10\,\text{m}\,\text{s}^{-1}$ at point P at the top of a hill. She free-wheels downhill to point Q which is a vertical distance of 50 m below P, reaching Q with a speed of $24\,\text{m}\,\text{s}^{-1}$

The distance PQ by road is 360 m. There is a constant resistance to motion. Calculate its value.

6 A 2 kg mass is dragged along a rough horizontal floor from rest for a distance s m by a variable force F where $F = 24 + 2s - s^2$ N. If the magnitude of the frictional force is two-sevenths that of the normal reaction, what is the speed of the mass after moving 3 m? What is the power of the force at this point?

7 The resistance to a car's motion when moving with speed v is kv, where k is a constant and the maximum power of its engine is P

Its maximum speed down a slope is V

Its maximum speed up the same slope is $\frac{1}{2}V$

Prove that $V = \sqrt{\dfrac{2P}{k}}$

8 A mass of 12 kg is attached to one end of a light elastic string of natural length 1.5 m whose upper end is fixed to a ceiling. If the modulus of elasticity, λ, is 120 N and the mass is in equilibrium, calculate

a The tension in the string,

b Its extension,

c The EPE stored in it.

9 Two fixed points X and Y are 2 m apart on a smooth table. Two horizontal strings OX and OY are attached to an object O which lies between X and Y. The natural length and modulus of elasticity for OX are 0.6 m and 16 N respectively, and for OY are 0.4 m and 20 N respectively. Find the extensions of the two strings and the tensions in them when O is in equilibrium.

10 Work out whether these formulae are dimensionally consistent given that m is mass; s and l are lengths; u and v are velocities; a and g are accelerations; F is force and t is time.

a $s = vt - \dfrac{1}{2}at^2$ **b** $Ft = m(v^2 - u^2)$

c $t^2 = \pi \dfrac{l}{g}$ **d** $Fs = m \times a \times t$

11 Check the dimensional consistency of these equations.

a $Fs = \dfrac{1}{2}mv^2$ **b** Power $= mgv$

12 The forward thrust T on the blade of a turbine depends on the area A swept by the blade, the speed v of the air and density ρ of the air. Given that $T = kA^x v^y \rho^z$ where k is a dimensionless constant and that $T = 40$ when $A = 7$, $v = 13$ and $\rho = 1.2$, work out the values of x, y, z and k and write the formula for T

13 A lorry of mass 3 tonnes travels up a slope where there is no resistance to motion at an angle of $\arcsin \dfrac{1}{140}$ to the horizontal. Its speed increases steadily from $15\,\text{m}\,\text{s}^{-1}$ to $25\,\text{m}\,\text{s}^{-1}$ over a distance of 500 m. Find the total increase in its energy, the tractive force of its engine and its final power output.

14 An elastic string has a natural length 4 m and a modulus of elasticity of $20g$ N. The string is stretched horizontally between two points A and B that are 6 m apart, and a mass of 15 kg is attached to its midpoint. The mass drops vertically after being released from rest. Calculate its velocity after it falls 3 m.

History

Robert Hooke was a remarkable scientist who lived in the 17th century. He lends his name to Hooke's law of elasticity, devised in 1660, but he carried out research in a wide range of fields. Perhaps most notably, he illustrated his observations through a microscope in the book *Micrographia: or Some Physiological Descriptions of Minute Bodies Made by Magnifying Glasses*. His amazing drawings were the first major publication of the Royal Society in 1665. Far from the world of microscopes, he also explored the stars and planets with telescopes, drawing some very detailed sketches of Mars.

Investigation

A section of a rollercoaster can be modelled as a ball, with mass 100 kg and negligible radius, rolling around the inside of a vertical circle with radius 10 m. The ball needs to be travelling at 10 m s⁻¹ at the top of the circle for it to make a complete loop-the-loop without leaving the track. Investigate what speed the ball has to be travelling as it enters the circle. How would the situation change if the ball were fixed to the circle, so that it couldn't fall off?

Note

The SI unit for energy is the same as the unit of work – the joule (J). This is named in honour of James Prescott Joule. 1 joule is the work done when lifting an object with a weight of 1 newton by 1 metre.

Research

Look up different landing vehicles that have visited other planets or moons in the Solar System. How much do they weigh, how much energy do they need to leave the surface, and in what way is the energy affected by the size of the moon or planet?

Investigation

In a bungee jump, people are tied to an elastic rope of a carefully calculated length, so that when they fall their head just dips into the water below. Investigate how the length of the rope is varied, based on the weight of the jumper as a fraction of the typical male adult.

10 c
AIR MAIL

UNITED STATES

FIRST MAN ON THE MOON

In all questions, take the value of g to be $9.8\,\text{m s}^{-2}$ unless told otherwise. λ is the modulus of elasticity.

1   A man raises a load of 20 kg through a height of 6 m using a rope and simple pulley. His maximum power output is 180 W. Assuming the rope and pulley are light and smooth

    a   How much work does he do in completing the task?   **[2 marks]**

    b   What is the shortest time in which he could complete it?   **[2]**

2   A child of mass 25 kg moves from rest down a slide. The total drop in height is 4 m.

    a   Assuming that there is no resistance to motion, calculate the speed of the child at the bottom of the slide.   **[3]**

    b   In fact, the child reaches the bottom travelling at $6\,\text{m s}^{-1}$. The length of the slide is 6 m. Calculate

       i   The work done against friction,   **[2]**

       ii   The frictional force (assuming it to be constant).   **[2]**

3   A horizontal force is applied to a 6 kg body so that it accelerates uniformly from rest and moves across a horizontal plane in a straight line against a total resistance to motion of 30 N. After it has travelled 16 m, it has a speed of $4\,\text{m s}^{-1}$

Calculate

    a   The applied force,   **[2]**

    b   The total work done by the force.   **[2]**

4   Two elastic strings, AB and BC, are joined at B, and the other ends are fixed to points D and E on a smooth horizontal table. AB has natural length 85 cm, and modulus of elasticity 45 N. BC has natural length 45 cm and modulus of elasticity 65 N. The distance DE is 2.4 m. Calculate the stretched lengths of the strings.   **[4]**

5   Particles A and B, of mass 0.5 kg and 1.5 kg respectively, are connected by a light inextensible string of length 2.5 m. Particle A rests on a smooth horizontal table at a distance of 1.5 m from its edge. The string passes directly over the smooth edge of the table and particle B hangs suspended. The system is held at rest with the string taut and then released. Calculate the speed of A when it reaches the edge of the table.   **[3]**

6   A pump, working at 3 kW, raises water from a tank at $1.2\,\text{m}^3\,\text{min}^{-1}$ and emits it through a nozzle at $15\,\text{m s}^{-1}$

Calculate the height through which the water is raised.
(Density of water = $1000\,\text{kg m}^{-3}$)   **[4]**

7   A particle is moved along a straight line from O to A on a horizontal plane by a force F N. The length $OA = 5$ m. Find the work done by the force if

    a $F = 12$ N   **[2]**

    b $F = (12 - 2x)$ N where x m is the distance of the particle from O **[3]**

8   The resistance to motion of a car is proportional to its speed. A car of mass 1000 kg has a maximum speed of 45 m s^{-1} on the level when its power output is 8 kW. Calculate its acceleration when it is travelling on the level at 20 m s^{-1} and its engine is working at 6 kW.   [6]

9   The time, T, taken for a satellite to complete a circular orbit of radius r around the Earth (radius R) is given by

$$T^2 = \frac{4\pi^2 r^3}{gR^2}$$

Show that this formula is dimensionally consistent.   [3]

10 A cyclist has a maximum power P

The resistance to motion is kv and in addition there is a wind blowing which exerts a constant force W

The cyclist's maximum speed against the wind is V and their maximum speed with the wind is $2V$

Show that $V = \sqrt{\dfrac{P}{2k}}$ [5]

11 A ball of mass 500 grams is fastened to one end of a light, elastic rope, whose natural length is 3 m and whose modulus of elasticity is 90 N. The other end of the rope is fastened to a bridge. The ball is held level with the fixed end, and is released from rest.

a   Calculate the speed of the ball when the rope just becomes taut.   [2]

b   How far below the bridge is the lowest point reached by the ball?   [4]

12 A particle of mass 2 kg is suspended from a point A on the end of a light spring of natural length 1 m and modulus of elasticity 196 N.

a   Calculate the length of the spring when the particle hangs in equilibrium.   [3]

The particle is now pulled down a further distance of 0.5 m and released from rest.

b   Calculate the distance below A at which the particle next comes instantaneously to rest, assuming that the spring can compress to that point without the coils touching.   [5]

c   Calculate the highest position reached by the particle if, instead of the spring, an elastic string of the same natural length and modulus of elasticity had been used.   [2]

13 An elastic string OA has natural length 2 m and modulus of elasticity 40 N. O is a fixed point on a rough horizontal plane, and a particle of mass 2 kg is attached to the string at A. The particle is projected horizontally from O with speed v ms^{-1}. It is subject to a constant resistance force of 10 N. It reaches a maximum distance x m from O, then rebounds and comes to rest exactly at O. Calculate

a   The value of x [7]

b   The value of v [2]

14 The frequency, f vibrations per second, with which a guitar string vibrates is thought to depend on the length, l, of the string, the density, ρ, of the steel and the tension, T, applied to it. Assuming this to be the case, find a possible formula for f in terms of l, ρ, T and a dimensionless constant k [7]

15 Two particles of equal mass are connected by an elastic string of length l and modulus of elasticity λ. They lie at rest on a smooth horizontal surface. One particle is set in motion with speed u

The maximum extension, x, reached by the string is given by

$$x = ku\sqrt{\frac{l}{\lambda}}$$

Use dimensional analysis to find the dimensions of k [4]

16 A catapult is made by attaching a light elastic string of natural length 10 cm to points A and B, which are a horizontal distance of 6 cm apart. The modulus of elasticity of the string is 5 N. A stone of mass 10 grams is placed at the centre of the string, which is then pulled back until the stone is 25 cm below the centre of AB. Calculate the greatest speed reached by the stone when it is released. State any assumptions you make in your solution. [6]

17 An object of mass 4 kg is released from rest at a point O and falls under gravity. It is subject to air resistance of magnitude $2.5x$ N, where x m is the distance it has fallen from O. It hits to ground at A, where $OA = 5$ m. Calculate

   a   The work done against resistance in travelling from O to A [3]

   b   The speed with which it hits the ground. [3]

18 A particle is placed at a distance r from the centre of a rough horizontal disc. The disc rotates at n revolutions per second. The maximum value of n for the particle to remain in place is given by $n = \dfrac{\mu}{2\pi}\sqrt{\dfrac{g}{r}}$, where μ is a dimensionless constant.

   Show that this formula is dimensionally consistent. [3]

19 A car working at a rate P watts has a maximum speed $V\,\text{m s}^{-1}$ when travelling on a horizontal road and experiences a resistance proportional to the square of its speed. At what rate, in terms of P, would the car have to work to double its maximum speed? [4]

20 A car of mass 1 tonne is towing a trailer of mass 400 kg on a level road. The resistance to motion of the car is 400 N and of the trailer is 300 N. At a certain instant, they are travelling at $10\,\text{m s}^{-1}$ and the power output of the engine is 10.5 kW. Calculate the tension in the coupling between the car and the trailer. [4]

21 An object of mass m kg is accelerated from rest at a constant rate up a vertical tube of height h m. When it emerges, it rises a further $3h$ m before coming instantaneously to rest. Show that the average rate of working while the projectile is in the tube is $2m\sqrt{6g^3h}$ watts. [8]

22 A particle is projected vertically into the air. Assuming that the greatest height, h, reached by the particle depends only on its initial velocity, v, and the acceleration due to gravity, g, use dimensional analysis to predict the formula for h in terms of v, g and a dimensionless constant k [5]

23 Liquid flows through a pipe because there is a pressure difference between the ends of the pipe. The rate of flow, $V\,\text{m}^3\,\text{s}^{-1}$, depends upon this pressure difference, $p\,\text{N m}^{-2}$, the viscosity of the liquid, $\eta\,\text{kg m}^{-1}\,\text{s}^{-1}$, the length of the pipe, l m, and the radius of the pipe, r m.

   a   Explain why it would be impossible to use dimensional analysis to determine the four unknowns in the formula $V = K\eta^{\alpha}r^{\beta}p^{\gamma}l^{\delta}$ [2]

   b   It is decided to use the pressure gradient $\dfrac{p}{l}$ instead of p and l separately. Use dimensional analysis to obtain a possible formula relating the rate of flow, V, to the viscosity of the liquid, the radius of the pipe and the pressure gradient. [7]

24 A car of mass 900 kg moves on a level road against a resistance proportional to its speed. Its power output is 6 kW and its maximum speed is 40 m s^{-1}. Calculate its acceleration when its speed is 20 ms^{-1}. [5]

25 A train of mass 40 tonnes accelerates uniformly. At point A it has speed 18 kmh^{-1} and has doubled its speed by the time it reaches B, where $AB = 600$ m. It is subject to a constant resistance of 3000 N. Calculate

    a The work done by the engine in travelling from A to B [3]

    b The power output of the engine when it is at the midpoint of AB [5]

26 A block of mass 4 kg is placed on a rough slope inclined at 30° to the horizontal. It is projected up the slope with initial speed 15 ms^{-1} and comes to rest 12 m further up the slope. Calculate the magnitude of the friction force, assuming it to be constant. [4]

27 A ship can travel at a constant speed of 30 kmh^{-1} when its engines are working at 15 MW. The resistance to its motion varies as the square of its speed. What power would the engines need to exert if it had to travel at a constant speed of 36 kmh^{-1}? [6]

28 A light elastic rope OA has natural length 10 m and modulus of elasticity 80 N. The end O is fixed to the ground and an object of mass 2 kg is attached to A

The object is projected vertically upwards from O with initial speed 40 ms^{-1}.

    a Assuming that there is no air resistance, what is the maximum height reached by the object? [4]

    b If, in fact, there is a constant air resistance of 15 N, find the maximum height reached by the object. [4]

    c In each of the cases **a** and **b** find the speed of the object when it returns to O [3]

29 An elastic string of natural length 2 m and modulus of elasticity kmg is stretched between two points A and B, where A is 4 m vertically above B. A particle of mass m is attached to the midpoint of the string and is gently lowered a distance d until it is in equilibrium.

    a Show that $d = \dfrac{1}{2k}$ provided $k \geq 0.5$ [4]

    b Explain what happens if $k < 0.5$ and find the value of d in this case. [3]

30 In resisted sprint training athletes attach themselves to a wall by means of a bungee cord and then run as far as possible away from the wall until the cord pulls them back. Ashley has mass 80 kg and he is attached to the wall at O by a light elastic rope of natural length 10 m and modulus of elasticity 1600 N. He can exert a constant forward force of 400 N.

    a What is the furthest that Ashley can stand from the wall? [3]

Ashley starts from rest at O and uses his maximum force to accelerate away from the wall.

    b What is the greatest speed he achieves? [4]

    c How far does he get from the wall before the rope pulls him back? [4]

8 Momentum

Forensic scientists use the principles of conservation of momentum, and momentum change during impacts, to construct models of collisions between two vehicles. Data from the accident scene, such as measurements of tyre skid marks and vehicle compression, are used to perform calculations that can reconstruct collisions and make an expert judgement of what really happened. These calculations are used in court cases to determine who was really at fault during a collision, and so it's vital that the calculations involved are accurate and give a true reflection of the incident.

Vehicle collisions are just one area in which principles relating to momentum are vitally important. Other areas include space engineering and rocket science, the design of theme park rides, and railway engineering.

Orientation

What you need to know	What you will learn	What this leads to
Maths Ch7 • Equations of motion. **Maths Ch8** • Resolving forces. • Newton's laws of motion. **Ch7 Work energy and power** • Kinetic energy.	• To use the momentum equation and Newton's experimental law properly. • To find the size of an impulse for constant and variable forces. • To apply these concepts to problems in two dimensions using vectors.	**Careers** • Forensic science. • Automotive testing. • Mechanical engineering.

Fluency and skills

When a particle of mass m moves with velocity $\mathbf{v}$, its **momentum** is equal to the mass multiplied by the velocity. If it is acted on by a constant force $\mathbf{F}$ for a time t, then the **impulse** acting on the particle is equal to the force multiplied by the time during which the force acts.

> **Key point**
>
> Momentum $= m \times \mathbf{v}$ Impulse $= \mathbf{F} \times t$

The units of momentum and impulse are the same. They are newton seconds ($N\,s$).

$\mathbf{F}$ and $\mathbf{v}$ are vectors, so momentum and impulse are also vectors.

If the force $\mathbf{F}$ is constant and the particle's velocity increases from $\mathbf{u}$ to $\mathbf{v}$ then its acceleration is given by $\mathbf{a} = \dfrac{\mathbf{v} - \mathbf{u}}{t}$

Newton's 2nd law gives $\mathbf{F} = m\mathbf{a} = \dfrac{m(\mathbf{v} - \mathbf{u})}{t} = \dfrac{m\mathbf{v} - m\mathbf{u}}{t}$

> **Key point**
>
> $\mathbf{F}t = m\mathbf{v} - m\mathbf{u}$
>
> Impulse = Change in momentum

Before impact

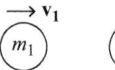
During impact

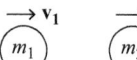
After impact

Bodies A and B with masses m_1 and m_2 and initial velocities $\mathbf{u}_1$ and $\mathbf{u}_2$ collide directly. During the collision, forces act on the bodies for a short time t

If the force acting on B is $\mathbf{F}$, then Newton's 3rd law states that an equal and opposite force acts on A

After the collision, A and B separate with velocities $\mathbf{v}_1$ and $\mathbf{v}_2$

For A, $-\mathbf{F}t = m_1\mathbf{v}_1 - m_1\mathbf{u}_1$

For B, $\mathbf{F}t = m_2\mathbf{v}_2 - m_2\mathbf{u}_2$

So $-(m_1\mathbf{v}_1 - m_1\mathbf{u}_1) = m_2\mathbf{v}_2 - m_2\mathbf{u}_2$

Rearrange

$$m_1\mathbf{u}_1 + m_2\mathbf{u}_2 = m_1\mathbf{v}_1 + m_2\mathbf{v}_2$$

Total initial momentum = Total final momentum

A **direct** collision means that bodies collide along their common line of travel. If they are moving towards each other, the direct collision is head-on. Most collisions in this section are direct.

> **Key point**
>
> The **principle of conservation of linear momentum** states that, when no *external* forces are present, the total momentum of a system of particles is unchanged by collisions between them.

'**Linear**' indicates that motion is considered as acting in a straight line.

Example 1

A body P of mass 1 kg moving with velocity 5 m s^{-1} collides directly with another body Q of mass 2 kg moving towards P with velocity 4 m s^{-1}

After the collision, Q is at rest. Find the final velocity of P

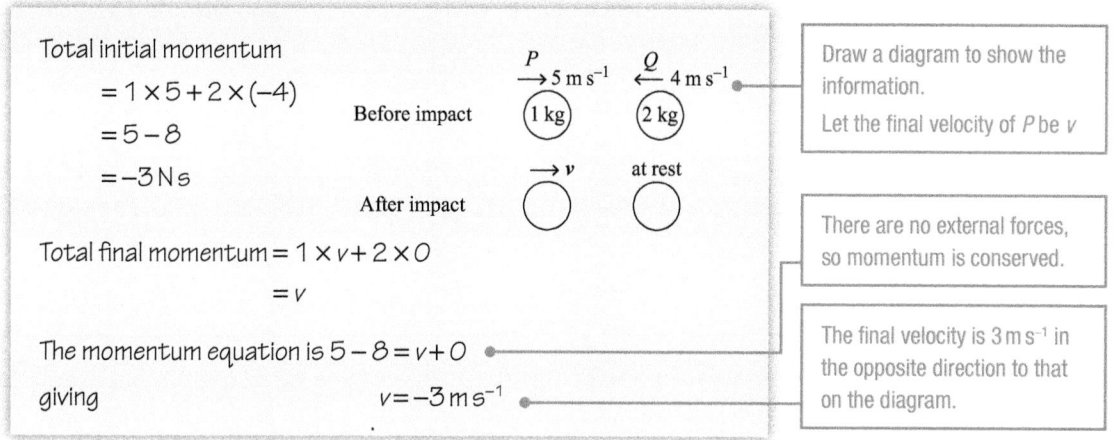

Total initial momentum

$$= 1 \times 5 + 2 \times (-4)$$
$$= 5 - 8$$
$$= -3 \, Ns$$

Before impact

Total final momentum $= 1 \times v + 2 \times 0$
$$= v$$

The momentum equation is $5 - 8 = v + 0$

giving $\qquad v = -3 \, m \, s^{-1}$

Draw a diagram to show the information.
Let the final velocity of P be v

There are no external forces, so momentum is conserved.

The final velocity is 3 m s^{-1} in the opposite direction to that on the diagram.

Example 2

A particle P of mass 1 kg with a velocity $2\mathbf{i} - 3\mathbf{j}$ m s^{-1} collides and coalesces with a particle Q of mass 4 kg and velocity $\mathbf{i} + 2\mathbf{j}$ m s^{-1} They move off together with a common velocity $\mathbf{v}$

When two particles **coalesce**, they combine into one particle.

Calculate the speed $|\mathbf{v}|$ and the angle which $\mathbf{v}$ makes with the $\mathbf{i}$-direction.

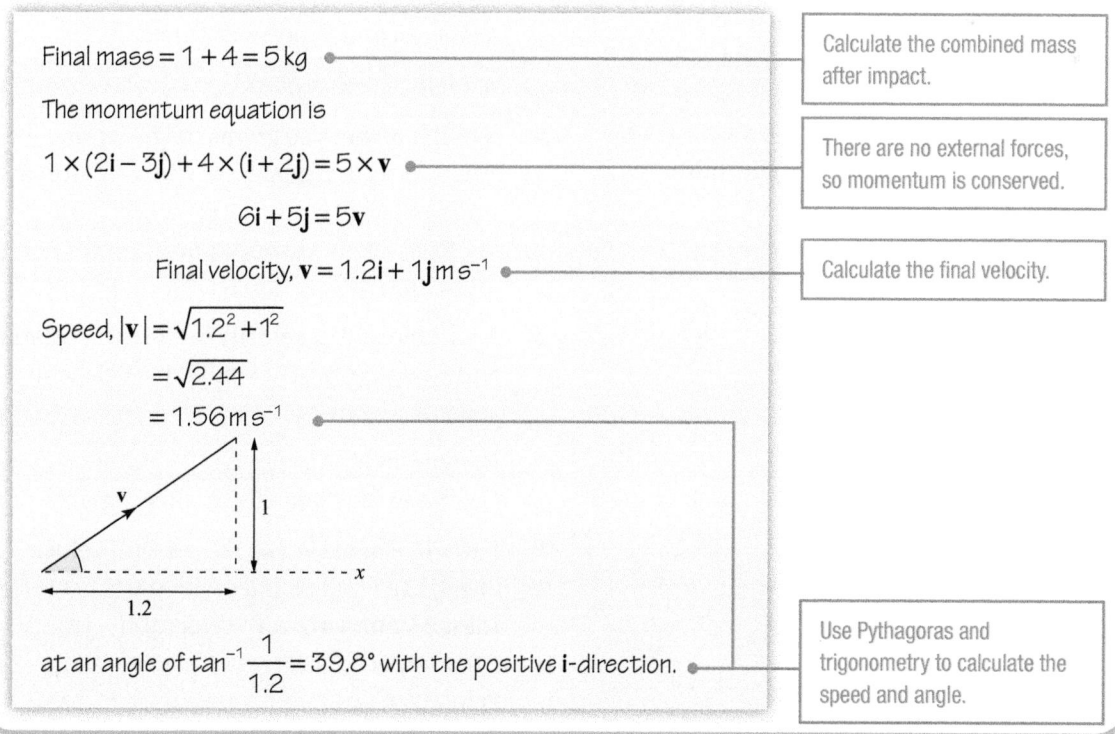

Final mass $= 1 + 4 = 5 \, kg$

The momentum equation is

$$1 \times (2\mathbf{i} - 3\mathbf{j}) + 4 \times (\mathbf{i} + 2\mathbf{j}) = 5 \times \mathbf{v}$$

$$6\mathbf{i} + 5\mathbf{j} = 5\mathbf{v}$$

Final velocity, $\mathbf{v} = 1.2\mathbf{i} + 1\mathbf{j} \, m \, s^{-1}$

Speed, $|\mathbf{v}| = \sqrt{1.2^2 + 1^2}$
$$= \sqrt{2.44}$$
$$= 1.56 \, m \, s^{-1}$$

at an angle of $\tan^{-1} \dfrac{1}{1.2} = 39.8°$ with the positive $\mathbf{i}$-direction.

Calculate the combined mass after impact.

There are no external forces, so momentum is conserved.

Calculate the final velocity.

Use Pythagoras and trigonometry to calculate the speed and angle.

Example 3

A child of mass 30 kg travels in a horizontal straight line on a skateboard of mass 5 kg with a velocity of 1.4 ms⁻¹

The child jumps off the skateboard with an initial horizontal backwards velocity of 0.2 ms⁻¹ while the skateboard continues forwards. All motion is in the same straight line. Calculate the final velocity, v, of the skateboard.

Total initial momentum $= (30 + 5) \times 1.4 = 49\,\text{Ns}$ ●——— Total mass $= 30 + 5 = 35\,\text{kg}$

Total final momentum $= (30 \times -0.2) + (5 \times v) = 5v - 6\,\text{Ns}$

The momentum equation is $49 = 5v - 6$ ●——— There are no external forces, so momentum is conserved.

Final velocity of skateboard, $v = \dfrac{55}{5} = 11\,\text{ms}^{-1}$

Exercise 8.1A Fluency and skills

1

P 2 kg $\qquad$ Q 4 kg

Before impact $\bigcirc \rightarrow u_1 \qquad \bigcirc \rightarrow u_2$

After impact $\bigcirc \rightarrow v_1 \qquad \bigcirc \rightarrow v_2$

Two particles P and Q, with masses 2 kg and 4 kg respectively, move along a straight line. They collide directly, travelling with velocities u_1 and u_2 before impact and with velocities v_1 and v_2 after impact. Calculate the values, x, of the unknown velocities in this table. All velocities are given in m s⁻¹

	u_1	u_2	v_1	v_2
a	3	2	2	x
b	6	1	x	3.5
c	5	4	4	x
d	10	−2	x	4
e	1	−4	−4	x

2

Y m_1 $\qquad$ Z m_2

Before impact $\bigcirc \rightarrow u_1 \qquad \bigcirc \rightarrow u_2$

$Y + Z$

After impact $\bigcirc \rightarrow v$

Two particles Y and Z, with masses m_1 kg and m_2 kg respectively, move along a straight line with velocities u_1 and u_2 m s⁻¹

They impact directly, coalesce, and then travel with velocity v m s⁻¹

Calculate v in each case.

	m_1	m_2	u_1	u_2
a	4	6	3	4
b	7	5	4	−2
c	2	8	5	−3

3 Particle A of mass 4 kg moves at 5 m s⁻¹ towards particle B which is at rest. After a direct impact, A is at rest and B moves at 2 m s⁻¹

What is the mass of B?

4 A bullet of mass 40 grams travelling at 80 m s⁻¹ hits and coalesces with a stationary wooden target of mass 2 kg which is free to move. Find their common speed v immediately after impact.

5 A particle with a velocity $4\mathbf{i} - 6\mathbf{j}$ m s⁻¹ collides and coalesces with a similar particle with velocity $\mathbf{i} + 2\mathbf{j}$ m s⁻¹

Both particles have mass 500 grams. Calculate their velocity, $\mathbf{v}$, after impact.

6 Particle P of mass 2 kg moves with velocity $3\mathbf{i} + 3\mathbf{j}$ m s⁻¹ and collides with particle Q of mass 3 kg moving with velocity $4\mathbf{i} - \mathbf{j}$ m s⁻¹ After impact, P is at rest. Calculate the velocity of Q

7 Two particles A and B have masses 3 kg and 4 kg and velocities $\begin{pmatrix} 4 \\ -2 \end{pmatrix}$ m s⁻¹ and $\begin{pmatrix} 3 \\ 4 \end{pmatrix}$ m s⁻¹

respectively. They collide and, after the collision, the velocity of A is $\begin{pmatrix} 0 \\ -4 \end{pmatrix}$ m s^{-1}

Calculate the velocity of B after the collision.

8 Particle P has mass 2 kg and velocity $5\mathbf{i} - \mathbf{j}$ m s^{-1}

Particle Q has mass 4 kg and velocity $-\mathbf{i} - 4\mathbf{j}$ m s^{-1}

They collide and coalesce. Calculate their velocity after impact.

9 Particle S of mass 4 kg travels with velocity $5\mathbf{i} + 4\mathbf{j}$ m s^{-1} and collides with particle T of mass 2 kg travelling with velocity $-2\mathbf{i} + \mathbf{j}$ m s^{-1}

After impact, the velocity of T is double the velocity of S. Calculate their final velocities.

10 An unexploded mine that is at rest suddenly explodes into two pieces. The pieces travel in opposite directions at speeds of 10 ms^{-1} and 16 ms^{-1}

If the first piece has a mass of 2 kg, find the mass of the other piece.

11 An object of mass 3 kg is moving horizontally at 120 ms^{-1} when it explodes into two pieces which travel in the same straight horizontal line. One piece continues forwards at 200 ms^{-1} and the other travels backwards at 80 ms^{-1} Find the masses of the two pieces.

Reasoning and problem-solving

Strategy

To solve collision problems involving the conservation of momentum

(1) Draw and label a diagram showing the situation before and after the collision.

(2) Check that there are no external forces involved, so that momentum is conserved.

(3) Write momentum equations, ensuring correct signs for velocities, and solve them.

Example 4

Three particles A, B and C, with masses 1 kg, 4 kg and 12 kg respectively, are positioned in a straight line.

Particles B and C are at rest and particle A is moving towards B with a speed of 10 m s^{-1}

After A and B collide, particle A rebounds backwards and B moves towards C with twice the speed of A

After B and C collide, they move in opposite directions with the same speed.

Show that there are no more collisions between the particles, assuming no other forces act on them.

> The * in the diagram indicates a collision.

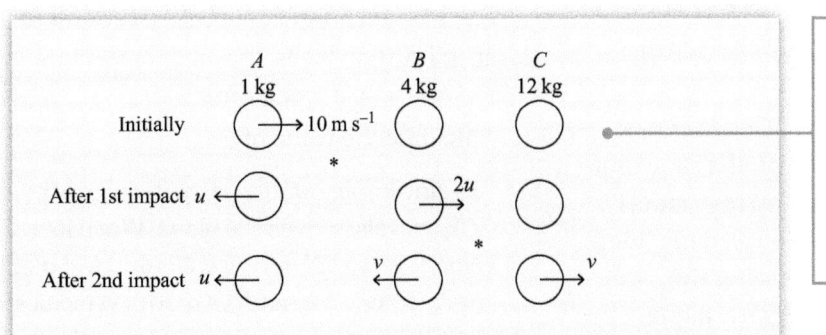

> (1)(2) Draw a clearly-labelled diagram showing the velocities of the particles at different stages.
>
> There are no external forces so momentum is always conserved.

(Continued on the next page)

Momentum equation when A strikes B is

$1 \times 10 + 0 = 1 \times (-u) + 4 \times 2u$

$\Rightarrow \qquad 7u = 10$

$\Rightarrow \qquad u = 1\frac{3}{7}\,\mathrm{m\,s^{-1}}$

Momentum equation when B strikes C is

$4 \times 2u + 0 = 4 \times (-v) + 12 \times v$

$\Rightarrow \qquad 8v = 8u$

$\Rightarrow \qquad v = u = 1\frac{3}{7}\,\mathrm{m\,s^{-1}}$

After these two collisions, A and B are moving in the same direction with the same speed and in the opposite direction to C Hence, there are no further collisions.

③ Write a momentum equation for each collision. Note the negative signs.

Exercise 8.1B Reasoning and problem-solving

1   A hammer of mass 10 kg travelling vertically downwards with speed 24 m s^{-1} strikes the top of a vertical post of mass 2 kg and does not rebound. Find the speed of the hammer and post immediately after impact.

2   An arrow, of mass 250 grams, travelling at 25 m s^{-1}, hits directly and coalesces with a stationary target of mass 9.0 kg which can move freely. Calculate their velocity immediately after impact.

3   Two cars of mass 900 kg and 1200 kg are travelling in opposite directions on a straight road with speeds of 135 km h^{-1} and 90 km h^{-1} when they collide head-on and lock together. What is their speed in km h^{-1} immediately after impact?

4   A bullet of mass 40 grams is fired horizontally with a speed of 600 m s^{-1} into a target of mass 2 kg suspended from a vertical string. The bullet becomes embedded in the target. Find their common speed immediately after the impact.

5   Two railway trucks with masses 2 tonnes and 3 tonnes are travelling on the same line at speeds of 5 m s^{-1} and 2 m s^{-1} respectively. They collide and become coupled together. Find their common velocity after the collision if they were initially travelling

   a   In the same direction,

   b   In opposite directions.

6   Masses of 3 kg and 5 kg are joined by a stretched elastic string and held apart on a smooth horizontal table. At a certain time after they are released from rest, the lighter mass has speed 12 m s^{-1}

   What is the speed of the other mass at this time?

7   When a gun is fired, the bullet shoots forwards and the gun is free to recoil backwards. The bullet has mass 50 grams and the gun has mass 2 kg. If the speed of the bullet is 250 m s^{-1} find the speed at which the gun recoils.

   Given that the explosion creates an impulse, explain why the principle of conservation of momentum can still be used.

8   A soldier fires a shell of mass 10 kg. When travelling horizontally at 120 m s^{-1}, it explodes into two pieces which also initially travel horizontally. Immediately after the explosion, one piece of mass 4 kg travels backwards with speed 40 m s^{-1} whilst the other piece continues forward with a speed v m s^{-1}

   Calculate the value of v

9   At a November bonfire party, a rocket of mass 0.5 kg travels in a vertical plane. When its velocity is $6\mathbf{i} + 20\mathbf{j}$ m s^{-1}, it splits into two pieces with masses 0.3 kg and 0.2 kg with velocities $-2\mathbf{i} + 18\mathbf{j}$ m s^{-1} and $a\mathbf{i} + b\mathbf{j}$ m s^{-1} respectively. Calculate a and b

10 Two particles, P of mass 3 kg and Q of mass 2 kg, lie on a smooth horizontal table connected by a light inextensible string which is slack. P is at rest and Q is moving away from P with a speed of $8\,\text{m}\,\text{s}^{-1}$

Calculate their common velocity immediately after the string becomes taut.

11 Three spheres A, B and C, with masses 5 kg, 4 kg and 3 kg respectively, lie in a straight line on a smooth horizontal surface with B and C at rest. Sphere A moves towards B with a speed of $18\,\text{m}\,\text{s}^{-1}$

On impact, A and B coalesce and move towards C. On impact, they coalesce with C and move with velocity $v\,\text{m}\,\text{s}^{-1}$

Calculate v

12 Three particles X, Y and Z, with masses 4 kg, 2 kg and 6 kg respectively, lie in a straight line on a smooth horizontal surface, with X and Z at rest. Y moves towards Z at $10\,\text{m}\,\text{s}^{-1}$ and collides with Z. After impact, Y and Z both move with speed $u\,\text{m}\,\text{s}^{-1}$ in opposite directions. Y now collides and coalesces with X and they continue with speed $v\,\text{m}\,\text{s}^{-1}$

Calculate u and v and explain why there is or is not another collision.

13 Three particles R, S and T, of mass 6 kg, 3 kg and 5 kg respectively, lie in a straight line. S is between R and T and they are on a smooth horizontal surface. R and S are joined by a light, inextensible, slack string. R and T are at rest. S moves away from R with a speed of $10\,\text{m}\,\text{s}^{-1}$ and the string becomes taut before S reaches T

  a Calculate the speed of R after the string becomes taut.

  b S now collides with T and comes to rest. Calculate the speed of T after this collision. Explain why there is a further collision between R and S

14 Three particles A, B and C, of mass 2 kg, 3 kg and 1 kg respectively, are in line with B and C at rest, B between A and C, and A moving on a smooth horizontal surface towards B with velocity u

The collision between A and B brings A to rest. B now collides with C

After this collision, B and C move in the same direction, with C moving twice as fast as B

Find the final velocities of B and C in terms of u

15 Particle A of mass 10 kg has speed $5\,\text{m}\,\text{s}^{-1}$ It collides directly on a smooth horizontal surface with particle B of mass m kg moving in the opposite direction at $2\,\text{m}\,\text{s}^{-1}$ After the collision, A continues in the same direction with a speed of $3\,\text{m}\,\text{s}^{-1}$ Show that, if there are no further collisions, $m \le 4$

16 A particle of mass 4 kg has a velocity of $\mathbf{i} - \mathbf{j}\,\text{m}\,\text{s}^{-1}$ It collides with a second particle which has a mass of m kg and a velocity of $2\mathbf{i} - 3\mathbf{j}\,\text{m}\,\text{s}^{-1}$ After the impact, the first particle is at rest and the second particle has a velocity of $4\mathbf{i} + a\mathbf{j}\,\text{m}\,\text{s}^{-1}$ Find the values of m and a

17 Particles A and B have masses 2 kg and m kg and velocities $4\mathbf{i} + 3\mathbf{j}$ and $p\mathbf{i} + q\mathbf{j}\,\text{m}\,\text{s}^{-1}$ respectively. They collide and coalese. If, after impact,

  a They move parallel to the x-axis, calculate the value of m given that $q = -1$

  b They move at 45° to the x-axis, calculate the final speed given that $q = 3$

    Use the value of m you found in part **a**.

18 A particle of mass 3 kg has a velocity of $4\,\text{m}\,\text{s}^{-1}$ in the x-direction. A second particle has mass m kg and velocity $6\,\text{m}\,\text{s}^{-1}$ in the y-direction. They collide, coalesce and begin to move together at 45° to their initial directions. Calculate the magnitude of their final velocity.

19 A smooth sphere A of mass 4 kg and centre C_1 has a velocity of $6\,\text{m}\,\text{s}^{-1}$ at 60° to the x-axis. A smooth sphere B of mass 2 kg is at rest with centre C_2 on the x-axis. A strikes B so that the line of centres C_1C_2 lies on the the x-axis. After impact, A moves at right angles to the x-axis. Explain why B moves along the x-axis and calculate the final velocities of A and B

Fluency and skills

When two bodies with known masses and velocities collide and rebound, their velocities after impact are unknown. You need two equations to find them. One is the momentum equation. The other is given by **Newton's experimental law**. It involves the speed at which the gap between two bodies changes.

> **Key point**
> $$\frac{\text{Speed of separation}}{\text{Speed of approach}} = e$$

The constant e is the **coefficient of restitution**. It depends on the **elasticity** of the bodies and has a value in the range $0 \leq e \leq 1$

$e = 0$ if the impact is **inelastic** (that is, with no rebound)

$e = 1$ if the impact is **perfectly elastic**.

> **Key point**
> If you say the velocities of the objects before impact are u_1 and u_2 and the velocities after impact are v_1 and v_2 then you can use the following equation.
> $$e = \frac{v_2 - v_1}{u_1 - u_2}$$

The **speed of approach** is the component of the speed along the line of impact at which the gap decreases before impact.

The **speed of separation** is the component of the speed along the line of impact at which the gap increases after impact.

A golf ball has $e \approx 0.8$ for an impact with a hard object, like a golf club.

Example 1

A particle P with mass 4 kg and velocity 20 m s^{-1} collides directly with particle Q with mass 12 kg travelling with velocity 4 m s^{-1} in the opposite direction.

Given $e = \frac{1}{2}$, calculate the velocities v_1 and v_2 after impact.

P 4 kg 20 m s^{-1} Q 12 kg

Before impact

After impact $\to v_1$ 4 m s^{-1} $\to v_2$

The momentum equation is

$4 \times 20 + 12 \times (-4) = 4v_1 + 12v_2$

$\qquad v_1 + 3v_2 = 8 \qquad\qquad$ **(1)**

The speed of approach $= 20 + 4 = 24$ m s^{-1}

The speed of separation $= v_2 - v_1$

Newton's equation is

$\dfrac{v_2 - v_1}{24} = \dfrac{1}{2} \qquad \Rightarrow \qquad v_2 - v_1 = 12 \qquad$ **(2)**

(1) + **(2)** gives $4v_2 = 20$

$\qquad v_2 = 5$

Substituting in **(1)** gives $v_1 = 8 - 15 = -7$

So, P moves at 7 m s^{-1} and Q moves at 5 m s^{-1} and both change direction after impact.

> Work out the speed of approach and the speed of separation. Take care with the signs.

> Solve **(1)** and **(2)** simultaneously to find v_1 and v_2

The particle in this diagram strikes a smooth, fixed plane with a velocity **u** and rebounds with velocity **v**

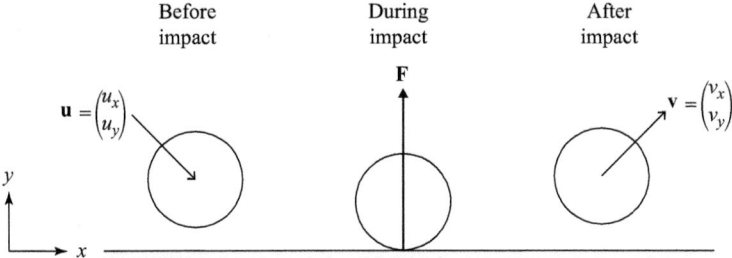

As the plane is smooth, there is no force parallel to the plane and therefore there is no change in the particle's momentum parallel to the plane. So, $u_x = v_x$

Upon impact, the force **F** acts entirely perpendicular to the plane. Momentum is not conserved in this direction (because an external force acts), but Newton's experimental law still applies perpendicular to the plane, so $e = \dfrac{v_y}{u_y}$

MECH

Example 2

A ball strikes a smooth fixed plane with a velocity of $\mathbf{u} = 20\sqrt{3}\mathbf{i} - 20\mathbf{j}$

The coefficient of restitution, e, is 0.8

Find the final velocity $\mathbf{v} = v_x\mathbf{i} + v_x\mathbf{j}$ after impact and the angle θ it makes with the plane.

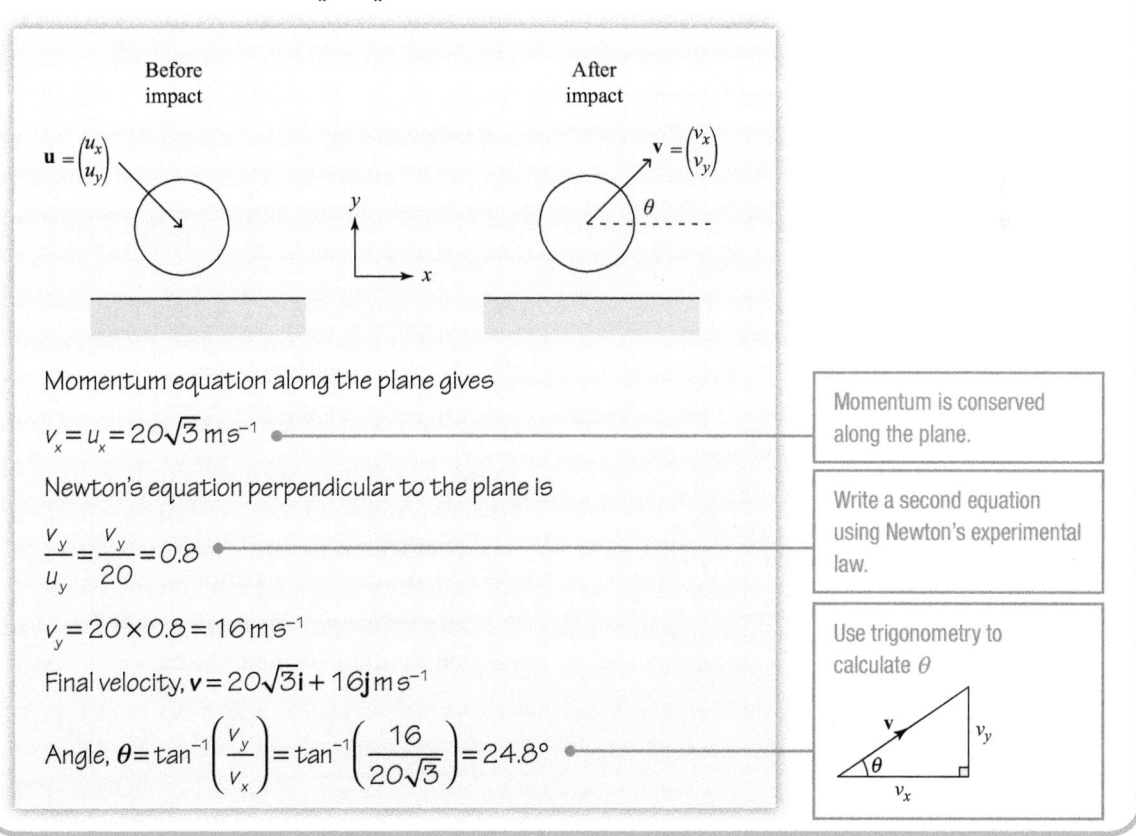

Momentum equation along the plane gives

$v_x = u_x = 20\sqrt{3}\,\mathrm{m\,s^{-1}}$ ●—————————————

Newton's equation perpendicular to the plane is

$\dfrac{v_y}{u_y} = \dfrac{v_y}{20} = 0.8$ ●—————————————

$v_y = 20 \times 0.8 = 16\,\mathrm{m\,s^{-1}}$

Final velocity, $\mathbf{v} = 20\sqrt{3}\mathbf{i} + 16\mathbf{j}\,\mathrm{m\,s^{-1}}$

Angle, $\theta = \tan^{-1}\left(\dfrac{v_y}{v_x}\right) = \tan^{-1}\left(\dfrac{16}{20\sqrt{3}}\right) = 24.8°$ ●———————

Momentum is conserved along the plane.

Write a second equation using Newton's experimental law.

Use trigonometry to calculate θ

1   Two particles P and Q, with masses 2 kg and 4 kg respectively, move in a straight line and collide directly. They travel with velocities u_1 and u_2 before impact and with velocities v_1 and v_2 after impact. In each case, calculate their velocities after impact.

	u_1	u_2	e
a	5	3	$\frac{1}{2}$
b	6	-2	$\frac{3}{4}$
c	-1	-10	$\frac{2}{3}$
d	5	-4	$\frac{1}{3}$
e	8	-2	$\frac{1}{4}$

2   Two particles R and S, with masses 3 kg and 6 kg respectively, move in a straight line and collide directly. They travel with velocities u_1 and u_2 before impact and with velocities v_1 and v_2 after impact. In each case, calculate the value of v_2 and the coefficient of restitution e

	u_1	u_2	v_1
a	16	4	2
b	5	2	2.5
c	-6	4	2
d	4	-1	-2

3   Two dodgem cars at a fairground collide head-on from opposite directions and rebound. Their masses, including passengers, are 125 kg and 150 kg. Their initial speeds are 0.6 m s^{-1} and 0.5 m s^{-1} respectively. The coefficient of restitution is 0.3

Calculate their speeds just after impact.

4   An ice-hockey puck of mass 800 grams skims across the ice at 15 m s^{-1}

It hits an identical puck, which is stationary, and both move on in the same straight line. If $e = 0.6$, find their velocities immediately after the collision.

5   Two toy trains travel on the same track in the same direction and collide. The front and rear trains have masses of 0.15 kg and 0.24 kg and initial speeds of 5.0 cm s^{-1} and 8.0 cm s^{-1} respectively. After impact, the rear train has a speed of 6.0 cm s^{-1} in the same direction. Find the value of e

6   A sphere strikes a smooth horizontal plane with a velocity $\mathbf{u} = u_x\mathbf{i} - u_y\mathbf{j}$

It rebounds with a velocity $\mathbf{v} = v_x\mathbf{i} + v_y\mathbf{j}$

The coefficient of restitution is e

In each case, find the final velocity $\mathbf{v} = v_x\mathbf{i} + v_y\mathbf{j}$ and the angle which it makes with the plane.

	u_x	u_y	e
a	4	6	$\frac{1}{2}$
b	3	4	0.25
c	2	6	$\frac{1}{3}$
d	1.2	9.0	0.8

7   A particle moving with velocity $3\mathbf{i} - 4\mathbf{j}$ m s^{-1} strikes a smooth horizontal surface. If the coefficient of restitution is 0.5, find its velocity $\mathbf{v} = v_x\mathbf{i} + v_y\mathbf{j}$ after impact. Also calculate the angle through which the direction of the particle changes.

Reasoning and problem-solving

To solve a collision problem involving elasticity

(1) Draw and label a diagram showing the situation before and after the collision.

(2) Check whether there are any external forces and whether momentum is conserved.

(3) Write the momentum equation and Newton's equation, taking care with signs, and solve.

Example 3

Two spheres P of mass 1 kg and Q of mass 3 kg lie on a smooth horizontal plane with the line PQ at right-angles to a vertical wall. P moves at $10\,\text{m s}^{-1}$ and collides directly with Q, which is initially at rest.

Q then hits the wall and rebounds. If the coefficient of restitution between the spheres is 0.4 and between Q and the wall is e, show that P and Q collide again if $e > \dfrac{1}{7}$

First impact

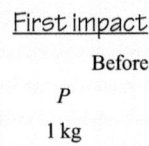
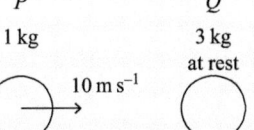
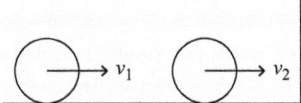

Before 1st impact — P 1 kg, $10\,\text{m s}^{-1}$; Q 3 kg at rest.

After 1st impact — $P \to v_1$; $Q \to v_2$.

> Draw a clearly labelled diagram showing the velocities.
>
> There is no external force at first impact so momentum is conserved.

Momentum equation is:

$$10 + 0 = v_1 + 3v_2 \quad \textbf{(1)}$$

Newton's equation is:

$$\frac{v_2 - v_1}{10} = 0.4 \quad \Rightarrow \quad v_2 - v_1 = 4 \quad \textbf{(2)}$$

> Write the two equations and solve simultaneously.

(1) and **(2)** give $v_1 = -0.5$ and $v_2 = 3.5$

P is now moving away from the wall and Q towards the wall.

Second impact

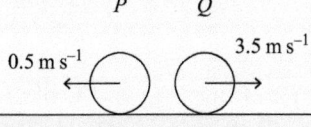
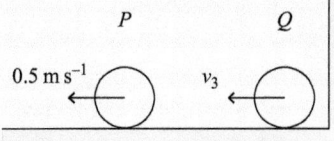

Before 2nd impact — P $0.5\,\text{m s}^{-1}$; Q $3.5\,\text{m s}^{-1}$.

After 2nd impact — P $0.5\,\text{m s}^{-1}$; Q v_3.

> Draw another diagram showing the velocities.
>
> The wall produces an external force on Q so momentum is not conserved.

Newton's equation is:

$$\frac{v_3}{3.5} = e \quad \text{so } v_3 = 3.5e$$

> Write only one equation since momentum is not conserved for Q

P and Q are now both moving away from the wall.

Q collides again with P if $3.5e > 0.5$ or $e > \dfrac{1}{7}$

Example 4

Particles P, Q and R have masses 3 kg, 2 kg, 1 kg and velocities 6 m s^{-1}, 4 m s^{-1}, 2 m s^{-1} respectively. They move in a straight line in this order and P strikes Q with perfect elasticity. A second collision between Q and R has a coefficient of restitution, e

Find the possible values of e if there are no more collisions.

> The * in the diagram indicates a collision.

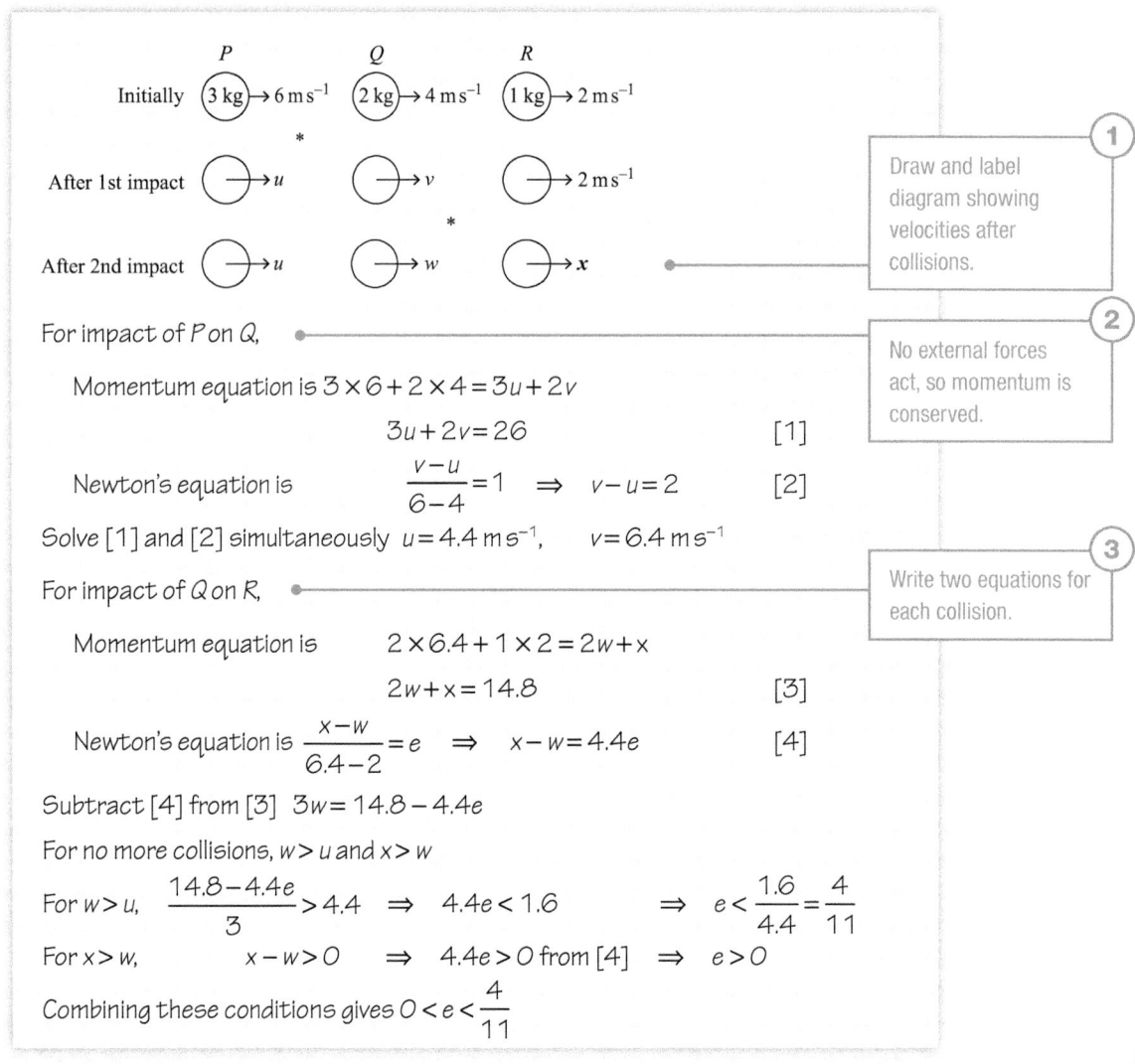

1 Draw and label diagram showing velocities after collisions.

2 No external forces act, so momentum is conserved.

For impact of P on Q,

Momentum equation is $3 \times 6 + 2 \times 4 = 3u + 2v$

$$3u + 2v = 26 \qquad [1]$$

Newton's equation is $\dfrac{v - u}{6 - 4} = 1 \;\Rightarrow\; v - u = 2 \qquad [2]$

Solve [1] and [2] simultaneously $u = 4.4$ m s^{-1}, $v = 6.4$ m s^{-1}

3 Write two equations for each collision.

For impact of Q on R,

Momentum equation is $\qquad 2 \times 6.4 + 1 \times 2 = 2w + x$

$$2w + x = 14.8 \qquad [3]$$

Newton's equation is $\dfrac{x - w}{6.4 - 2} = e \;\Rightarrow\; x - w = 4.4e \qquad [4]$

Subtract [4] from [3] $\quad 3w = 14.8 - 4.4e$

For no more collisions, $w > u$ and $x > w$

For $w > u$, $\quad \dfrac{14.8 - 4.4e}{3} > 4.4 \;\Rightarrow\; 4.4e < 1.6 \qquad \Rightarrow\; e < \dfrac{1.6}{4.4} = \dfrac{4}{11}$

For $x > w$, $\qquad x - w > 0 \;\Rightarrow\; 4.4e > 0$ from [4] $\;\Rightarrow\; e > 0$

Combining these conditions gives $0 < e < \dfrac{4}{11}$

Exercise 8.2B Reasoning and problem-solving

1 A ball of mass m kg is dropped vertically onto a stone floor from a height of 4 metres and rebounds to a height of 3 metres. Using $g = 9.8$ m s^{-1}, calculate

   a The speed at which it hits the floor and the speed at which it leaves the floor,

   b The coefficient of restitution, e, between the ball and the floor.

2 Two spheres P of mass 2 kg and Q of mass 4 kg lie on a smooth horizontal plane. P is at rest. Q moves away from P at 10 m s^{-1} towards a vertical wall where $e = \dfrac{1}{2}$

   Q hits the wall and rebounds to hit and coalesce with P

   Calculate the final speed of P and Q

3 Two balls Y and Z, of mass $4\,\text{kg}$ and $m\,\text{kg}$ respectively, lie on a smooth horizontal surface between two parallel walls on a line at right angles to them. Y moves at $2\,\text{m s}^{-1}$ towards one wall and Z moves at $4\,\text{m s}^{-1}$ towards the other wall. They bounce off the walls, where $e = \dfrac{1}{3}$ and collide with each other on their return. After this collision, both balls are at rest. Calculate m

4 A particle with velocity $6\mathbf{i} - 4\mathbf{j}\,\text{m s}^{-1}$ collides with a smooth horizontal plane which contains the x-axis. Its velocity after impact is $a\mathbf{i} + 2\mathbf{j}\,\text{m s}^{-1}$

Write the value of a and find the coefficient of restitution, e

5 A ball moves with velocity $5\mathbf{i} - 3\mathbf{j}$ towards a smooth horizontal plane which contains the x-axis. The coefficient of restitution e is 0.8 Calculate the speed of the ball immediately after impact and the angle it makes with the plane.

6 Two spheres R and S, of mass $2\,\text{kg}$ and $m\,\text{kg}$ respectively, lie on a smooth horizontal plane. S moves with speed u away from R (which is at rest) towards a wall fixed at right-angles to its path, rebounds from the wall and returns towards R. The coefficient of restitution for all impacts is 0.75. Calculate the value of m such that, after S impacts on R, it is at rest.

7 Two spheres, A of mass $1\,\text{kg}$ and B of mass $3\,\text{kg}$, are joined by a light slack string on a smooth horizontal plane. A is at rest. B moves away from A with speed u to strike and rebound from a wall at right angles to its path, with the string still slack. Given that $e = 0.5$ for all collisions, find in terms of u

  a The velocities of A and B after they collide,

  b The common speed of A and B after the string becomes taut.

8 Sphere B of mass $2\,\text{kg}$ lies on a smooth horizontal plane between sphere A of mass $30\,\text{kg}$ and a fixed vertical wall. A is at rest and B moves towards A with speed u

The value of e is 0.6 for all collisions. Show that B is at rest after its second collision with A

9 Three particles P, Q and R, with masses $6\,\text{kg}$, $4\,\text{kg}$ and $8\,\text{kg}$ respectively, lie in a straight line on a smooth horizontal surface with Q and R at rest. P moves towards Q with velocity $15\,\text{m s}^{-1}$

If $e = \dfrac{3}{5}$ for all collisions, find their velocities after the third collision. Explain why there are no more collisions.

10 A rubber ball of mass m is dropped 5 metres from rest onto a horizontal floor. Taking $g = 10\,\text{m s}^{-2}$, show that the ball's speed just before the first impact is $10\,\text{m s}^{-1}$

Given $e = 0.5$, calculate the time between the first and second impacts. Use a geometric progression to find the total time for the ball to come to rest after being released.

11 Three particles X, Y and Z, of masses $2\,\text{kg}$, $4\,\text{kg}$ and $4\,\text{kg}$ respectively, lie in a straight line on a smooth horizontal plane. X and Y are joined by a light slack string. With X and Z at rest, Y moves towards Z at $6\,\text{m s}^{-1}$ and the string becomes taut before Y reaches Z

If $e = \dfrac{1}{2}$ for all impacts, calculate the common speed of X and Y

  a Before Y strikes Z,

  b After the string becomes taut for a second time.
    Explain why there are no more collisions.

12 A ball drops vertically and strikes a smooth slope with a speed of $10\,\text{m s}^{-1}$

After impact, the ball moves horizontally. If the slope makes an angle of $30°$ with the horizontal, calculate the coefficient of restitution.

13 A ball falls freely under gravity for 10 metres until it strikes a smooth plane that is inclined at $45°$ to the horizontal. If $e = 0.2$, calculate the speed and direction of the ball immediately after impact. Use $g = 10\,\text{m s}^{-1}$

MECH

Full A Level

247

Fluency and skills

You already know that the impulse **I** of a constant force **F** acting on a body for a time t is equal to the change in momentum of the body.

> **Key point**
>
> Impulse, $\mathbf{I} = \mathbf{F} \times t$
>
> $\qquad = m\mathbf{v} - m\mathbf{u}$

I, **F**, **v** and **u** are all vectors.

F can be very large and t very small, as in a collision.

F can be constant over time, as when a jet of water hits a fixed surface.

F can also vary with time, in which case you need to use calculus to find the impulse.

> **Key point**
>
> $$\mathbf{I} = \int_0^t \mathbf{F}dt = \int_0^t m\frac{d\mathbf{v}}{dt}\, dt = \left[m\mathbf{v}\right]_u^v = m\mathbf{v} - m\mathbf{u}$$

$\mathbf{F} = m\dfrac{d\mathbf{v}}{dt}$ comes from Newton's 2nd law of motion.

In each case, impulse = change in momentum.

Example 1

A ball of mass 0.2 kg moves with velocity $-8\,\text{m s}^{-1}$ horizontally towards a bat. It is struck by the bat, reverses its direction and moves away at $12\,\text{m s}^{-1}$

Find the impulse of the bat on the ball. What is the impulse of the ball on the bat?

Impulse equation for the ball alone is

$I = mv - mu$

$I = 0.2 \times 12 - 0.2 \times (-8)$

$\quad = 2.4 + 1.6 = 4.0\,\text{N s}$

The impulse of the bat on the ball is $4\,\text{N s}$

From Newton's 3rd law of motion, the impulse of the ball on the bat is $4\,\text{N s}$ in the opposite direction.

Apply the impulse equation for ball alone. The positive direction is to the right in the diagram.

Example 2

Water flows horizontally from a pipe of cross-section 0.02 m² at a speed of 15 m s⁻¹ It strikes a fixed vertical wall.

Find the force *F* on the wall. (Density of water = 1000 kg m⁻³)

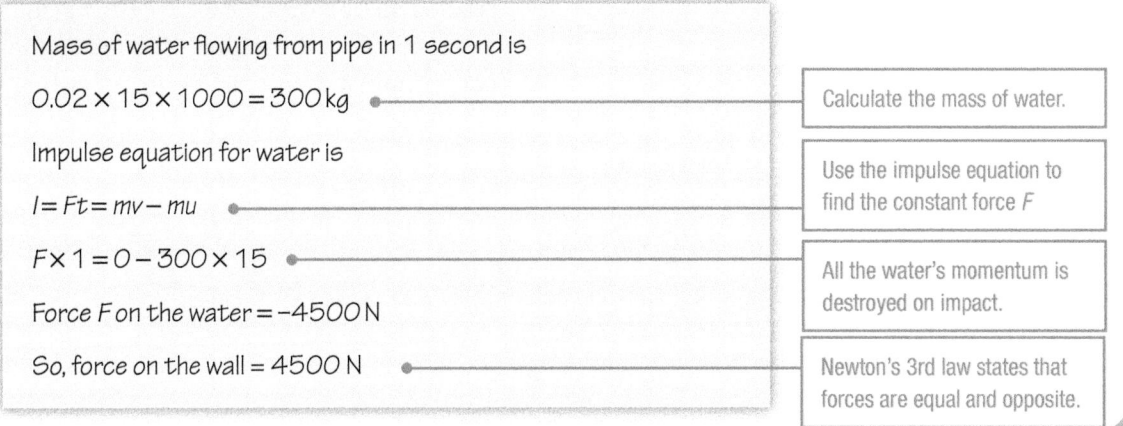

Mass of water flowing from pipe in 1 second is

$0.02 \times 15 \times 1000 = 300\,kg$ • ─────────── Calculate the mass of water.

Impulse equation for water is

$I = Ft = mv - mu$ • ─────────── Use the impulse equation to find the constant force *F*

$F \times 1 = 0 - 300 \times 15$ • ───────────

Force *F* on the water = $-4500\,N$ ─────────── All the water's momentum is destroyed on impact.

So, force on the wall = $4500\,N$ • ─────────── Newton's 3rd law states that forces are equal and opposite.

Example 3

A particle of mass 2 kg has an initial velocity *u* of 3 m s⁻¹

It is subject to a force $F = 8t - 3t^2$ newtons for 3 seconds where $0 \le t \le 3$

Find its speed *v* after 3 seconds.

Impulse $I = \int_0^3 F\,dt = \int_0^3 8t - 3t^2\,dt$ • ─────────── Use calculus.

$= \left[4t^2 - t^3 \right]_0^3 = 36 - 27 - 0 = 9\,Ns$

Impulse equation is

$I = mv - mu$ • ─────────── Solve to find *v*

$9 = 2v - 2 \times 3$

Final velocity, $v = 7.5\,m\,s^{-1}$

Exercise 8.3A Fluency and skills

1 A ball of mass *m* kg falls vertically and hits a horizontal floor with a velocity *u* m s⁻¹ It rebounds with velocity *v* m s⁻¹

   Find the impulse of the floor on the ball when

   a $m = 2, u = 12, v = -8$

   b $m = 6, u = 25, v = -10$

2 Two identical particles *A* and *B* of mass *m* kg collide head-on. Particle *A* has an initial velocity *u* m s⁻¹ and a final velocity *v* m s⁻¹

   Find the impulse that *A* exerts on *B* when

   a $m = 3, u = 8, v = 2$

   b $m = 5, u = 6, v = -3$

3 **a** A golf club hits a ball at rest with a horizontal impulse of 48 N s. If the ball has a mass of 50 grams, find its initial velocity.

   b A cricket bat strikes a ball of mass 0.15 kg horizontally with an impulse of 36 N s

   If the ball is moving horizontally towards the bat with a speed 8 m s⁻¹, what is the ball's final velocity?

MECH

249

4   Water flows from a circular pipe of radius r with a speed $v\,\text{m s}^{-1}$ and strikes a fixed wall at right angles without any rebound. Given the density of water is $1000\,\text{kg m}^{-3}$, find the force that the water exerts on the wall, when

    a $r = 2\,\text{cm}$, $v = 12\,\text{m s}^{-1}$

    b $r = 15\,\text{mm}$, $v = 20\,\text{m s}^{-1}$

5   Find the force on each square metre of ground when $1.2\,\text{cm}$ of rain falls in 2 hours, striking the ground vertically with a speed of $72\,\text{m s}^{-1}$

    ($1\,\text{m}^3$ of water has a mass of $1000\,\text{kg}$.)

6   A $5\,\text{kg}$ particle with an initial velocity of $2\,\text{m s}^{-1}$ is acted on by a variable force F (in N s) for 4 seconds. Find the impulse I acting on the particle and its final velocity, $v\,\text{m s}^{-1}$, when

    a $F = 20t - 6t^2$

    b $F = 3t^2 - 8t + 1$

Reasoning and problem-solving

To solve problems involving impulses

(1) Draw and label a diagram to show the information.

(2) Consider the situation and briefly explain your strategy and method.

(3) Write and solve equations to calculate the required values.

A $2\,\text{kg}$ ball with velocity $\mathbf{u} = 3\mathbf{i} - 4\mathbf{j}$ impacts a smooth plane, which is parallel to the x-axis. Given $e = 0.5$, calculate the final velocity $\mathbf{v} = v_x\mathbf{i} + v_y\mathbf{j}$ of the ball and the impulse $\mathbf{I}$ of the plane on the ball.

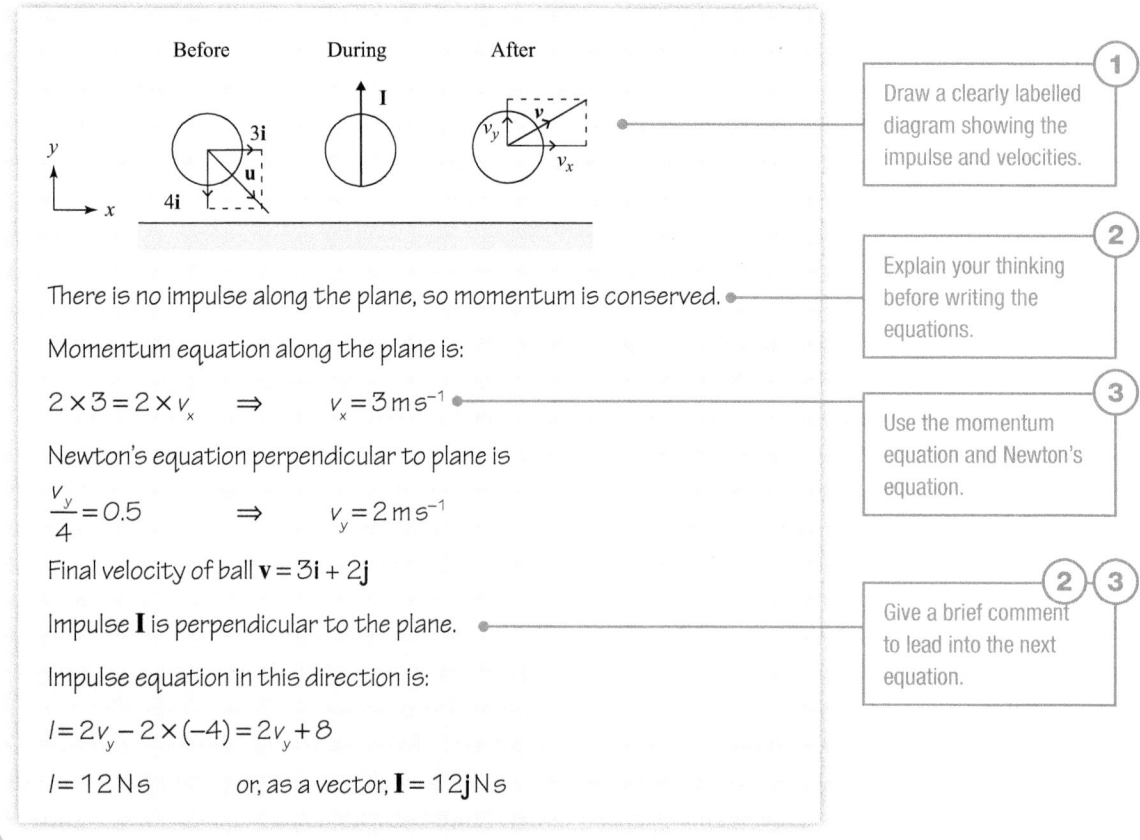

Draw a clearly labelled diagram showing the impulse and velocities. **(1)**

There is no impulse along the plane, so momentum is conserved.

Explain your thinking before writing the equations. **(2)**

Momentum equation along the plane is:

$2 \times 3 = 2 \times v_x \quad \Rightarrow \quad v_x = 3\,\text{m s}^{-1}$

Use the momentum equation and Newton's equation. **(3)**

Newton's equation perpendicular to plane is

$\dfrac{v_y}{4} = 0.5 \quad \Rightarrow \quad v_y = 2\,\text{m s}^{-1}$

Final velocity of ball $\mathbf{v} = 3\mathbf{i} + 2\mathbf{j}$

Impulse $\mathbf{I}$ is perpendicular to the plane.

Give a brief comment to lead into the next equation. **(2)(3)**

Impulse equation in this direction is:

$I = 2v_y - 2 \times (-4) = 2v_y + 8$

$I = 12\,\text{N s}$ or, as a vector, $\mathbf{I} = 12\mathbf{j}\,\text{N s}$

Example 5

Two spheres A and B of masses $4\,\text{kg}$ and $2\,\text{kg}$ respectively are connected by an inelastic string on a smooth horizontal plane. B is struck by an external impulse $\mathbf{J} = 24\mathbf{i} + 30\mathbf{j}\,\text{N s}$

a Show that B begins to move at $75°$ to the line AB

b Calculate the impulsive tension in the string, $\mathbf{I}$

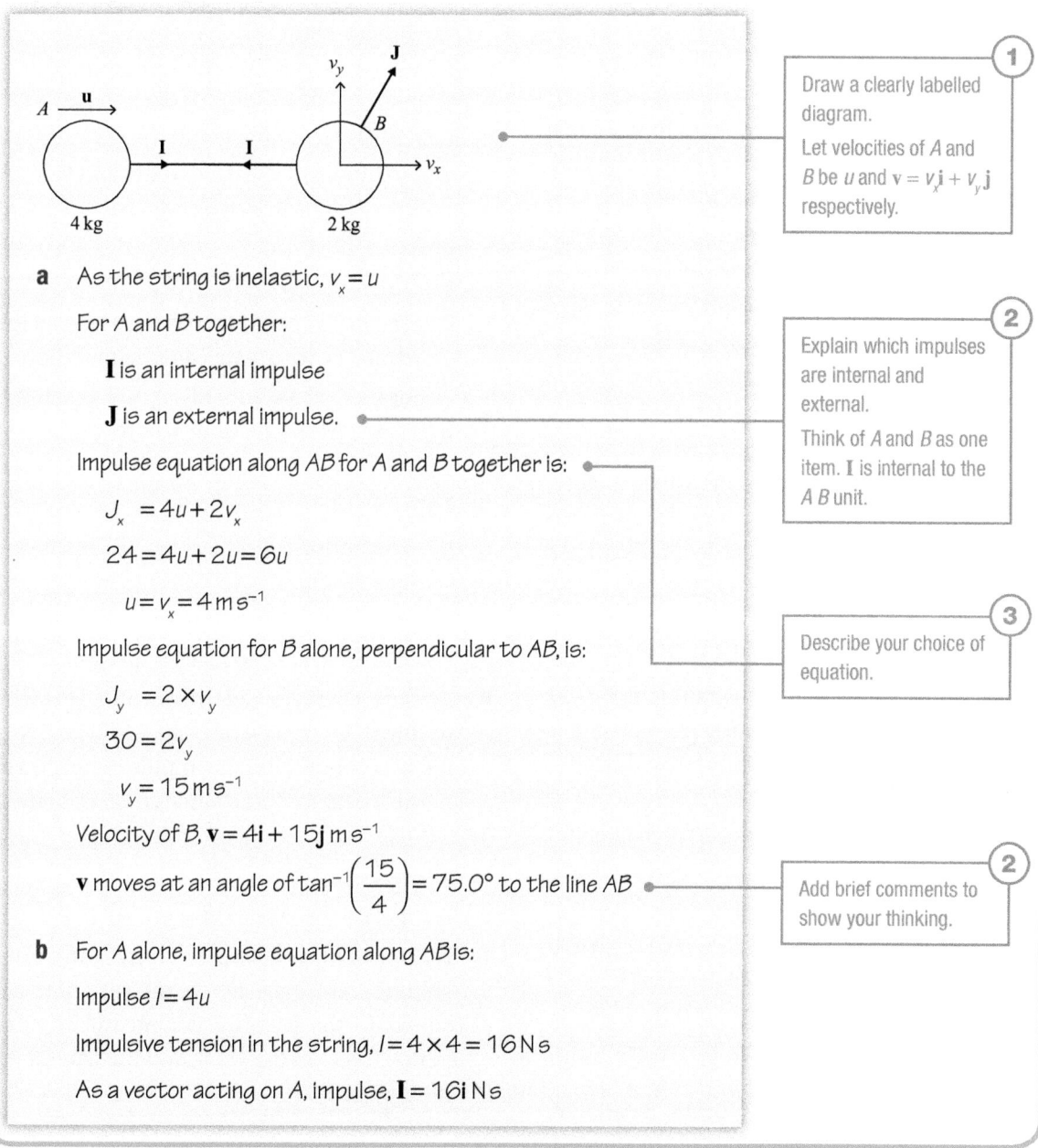

① Draw a clearly labelled diagram.

Let velocities of A and B be u and $\mathbf{v} = v_x\mathbf{i} + v_y\mathbf{j}$ respectively.

a As the string is inelastic, $v_x = u$

For A and B together:

 $\mathbf{I}$ is an internal impulse

 $\mathbf{J}$ is an external impulse.

Impulse equation along AB for A and B together is:

 $J_x = 4u + 2v_x$

 $24 = 4u + 2u = 6u$

 $u = v_x = 4\,\text{m s}^{-1}$

Impulse equation for B alone, perpendicular to AB, is:

 $J_y = 2 \times v_y$

 $30 = 2v_y$

 $v_y = 15\,\text{m s}^{-1}$

Velocity of B, $\mathbf{v} = 4\mathbf{i} + 15\mathbf{j}\,\text{m s}^{-1}$

$\mathbf{v}$ moves at an angle of $\tan^{-1}\left(\dfrac{15}{4}\right) = 75.0°$ to the line AB

② Explain which impulses are internal and external.

Think of A and B as one item. $\mathbf{I}$ is internal to the $A\,B$ unit.

③ Describe your choice of equation.

② Add brief comments to show your thinking.

b For A alone, impulse equation along AB is:

Impulse $I = 4u$

Impulsive tension in the string, $I = 4 \times 4 = 16\,\text{N s}$

As a vector acting on A, impulse, $\mathbf{I} = 16\mathbf{i}\,\text{N s}$

You could check the answer to Example 5 part **b** by considering the impulse equation for B alone in the direction AB

$J_x - I = 2 \times v_x$ giving $I = 24 - 8 = 16\,\text{N s}$ in the direction towards A as in the diagram.

As a vector acting on B, impulse, $\mathbf{I} = -16\mathbf{i}$

1 Particle A of mass 2 kg is moving at $12\,\text{m s}^{-1}$ and collides directly with particle B of mass 3 kg which is moving at $6\,\text{m s}^{-1}$ in the same direction. After impact, A and B coalesce. Calculate the impulse of A on B

2 A 5 kg ball with velocity $\mathbf{u} = 2\mathbf{i} - 3\mathbf{j}$ strikes a smooth plane containing the x-axis. If the coefficient of restitution is $e = 0.2$, calculate the final velocity $\mathbf{v} = v_x\mathbf{i} + v_y\mathbf{j}$ of the ball and the impulse $\mathbf{I}$ of the plane on the ball.

3 A sphere with velocity $\mathbf{u} = -4\mathbf{i} + 3\mathbf{j}$ impacts a smooth plane containing the y-axis. The coefficient of restitution is $e = 0.6$
If the sphere's mass is 6 kg, calculate its velocity $\mathbf{v} = v_x\mathbf{i} + v_y\mathbf{j}$ after impact and the impulse $\mathbf{I}$ of the plane on the sphere.

4 Two smooth spheres collide with their line of centres parallel to the x-axis. Before impact, sphere A of mass 4 kg has a velocity $\mathbf{u} = 2\mathbf{i} + 3\mathbf{j}\,\text{m s}^{-1}$ and sphere B of mass 2 kg is at rest.

  a Explain why B moves off in the direction of the x-axis after impact.

  b If the coefficient of restitution is $e = 0.5$, find their velocities after impact and the magnitude of the impulse during impact.

5 When two smooth spheres, P of mass 2 kg and Q of mass 4 kg, collide, their line of centres is parallel to the y-axis. Before impact, the velocity of P is $\mathbf{u} = 4\mathbf{i} + 6\mathbf{j}$ and Q is at rest. Given the coefficient of restitution is $e = 0.4$, find their velocities immediately after impact and the magnitude of the impulse during impact.

6 A 3 kg mass travels for 5 s under a variable force F where, after t seconds, $F = 3t^2 + 2\,\text{N}$
Calculate

  a The impulse on the mass during the 5 s,

  b The final velocity of the mass if its initial velocity is
    i zero   ii $4\,\text{m s}^{-1}$

7 A variable force $F\,\text{N}$ acts for 10 s on a 5 kg mass with an initial speed of $2\,\text{m s}^{-1}$
If $F = 2t + 3$ at a time t s, find

  a The impulse on the mass during the 10 seconds,

  b The final velocity of the mass,

  c The time for the mass to reach half its final velocity.

8 A ball of mass 0.2 kg is hit horizontally at $10\,\text{ms}^{-1}$ to strike a vertical wall at right angles. It is in contact with the wall for 0.25 s and the force of the wall on the ball is $2t^2(1 - 4t)\,\text{kN}$ where $0 \le t \le 0.25$
Find

  a The impulse of the wall on the ball,

  b The ball's speed at the instant when it loses contact with the wall after rebounding.

9 A mass of 0.75 kg, initially at rest, is acted on by a force F for 4 seconds where
$F = t + t^3\,\text{N}$ at a time t s
Calculate

  a The impulse on the mass during the 4 seconds,

  b Its speed at the end of the 4 seconds,

  c The time taken to reach a speed of $33\,\text{m s}^{-1}$

10 An object of mass 4 kg with an initial velocity of $10\,\text{ms}^{-1}$ is acted on by a variable force $F\,\text{N}$ for 12 seconds. Find the impulse acting on the particle over this time and its final velocity, if

$$F = \begin{cases} \dfrac{1}{2}t^2 & \text{for } 0 \le t \le 4 \\ 12 - t & \text{for } 4 \le t \le 12 \end{cases}$$

11 A truck of mass 400 kg moving at $5\,\text{m s}^{-1}$ on a smooth track experiences the variable force shown by this graph. Calculate the impulse of this force on the truck and the truck's final velocity.

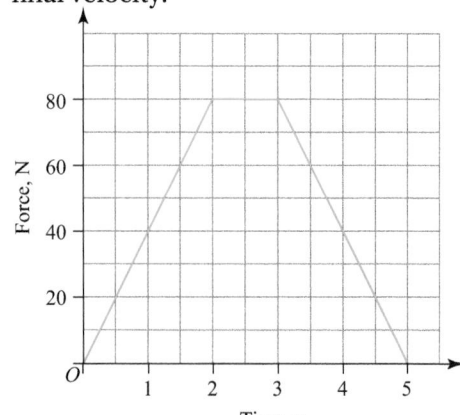

12 A box of mass 20 kg is pushed from rest over a smooth horizontal surface by the variable force shown in the diagram.

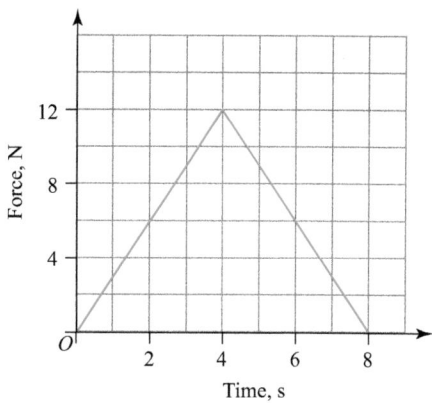

Calculate

a The impulse on the box,

b The box's final velocity,

c The time when its velocity is $0.6\,\text{m s}^{-1}$

13 A bullet of mass m and speed u strikes and embeds in a wooden disc of mass M moving in the same direction with speed v
Find the impulse of the bullet on the disc.

14 Two spheres P and Q, of mass 3 kg and 4 kg, on a smooth table are joined by a slack inelastic string. Q moves at $21\,\text{m s}^{-1}$ directly away from P which is at rest. At the moment the string becomes taut, find the velocity of both spheres and the magnitude of the impulsive tension in the string.

15 A shell of mass 1.25 kg is fired horizontally with a speed of $380\,\text{m s}^{-1}$ from a gun of mass 50 kg. The gun is brought to rest by a constant horizontal force acting over a distance of 1 metre. Calculate

a The impulse of the explosion and the initial speed of recoil of the gun,

b The magnitude of the force.

16 A smooth ball P of mass 6 kg moving at $10\,\text{m s}^{-1}$ collides with a smooth ball Q of mass 12 kg that is at rest. The direction of motion of P makes an angle of 30° with their line of centres. Given the coefficient of restitution is $e = 0.75$, find the velocities of the spheres after impact and the magnitude of the impulse during their contact.

17 A smooth ball of mass 2 kg with speed $4\sqrt{3}\mathbf{i} + 4\mathbf{j}\,\text{m s}^{-1}$ strikes a second smooth ball of mass 4 kg with speed $\sqrt{3}\mathbf{i} + \mathbf{j}\,\text{m s}^{-1}$
Just before impact, they are moving in parallel directions and the direction of motion makes an angle of 30° with the line of their centres. Given the coefficient of restitution is $e = \dfrac{1}{3}$, calculate their velocities after impact and the magnitude of the impulse between them during impact.

18 Two identical smooth spheres of mass m move with the same speed u
They collide when their centres are on the positive x-axis and their directions of motion are at 30° and 60°, respectively, to the x-axis. Find their speeds after impact and the magnitude of the impulse between them during impact if the spheres are perfectly elastic.

19 Two particles P and Q of masses m and M are at rest and connected by a light rod that lies in the direction $\mathbf{i}$
P is struck a blow of impulse $\mathbf{I} = x(\cos\theta\mathbf{i} + x\sin\theta\mathbf{j})$ which creates an impulsive thrust in the rod. Immediately after the impulse, the velocity of P is $v(\cos\phi\mathbf{i} + v\sin\phi\mathbf{j})$.

a Show that $\tan\phi = \dfrac{m+M}{M}\tan\theta$

b Comment on what happens to the motion of P when
   i M is much larger than m,
   ii M is zero.

20 Two equal mass particles, A and B, are joined by an inextensible string. Particle A is at rest. Particle B moves with velocity $u\left(\dfrac{\sqrt{3}}{2}\mathbf{i} + \dfrac{1}{2}\mathbf{j}\right)$. At the instant the string becomes taut it lies in the direction $\mathbf{j}$
Find the magnitude of the impulse in the string and the magnitude of the velocity of B at the instant the string becomes taut if

a A is held fixed, **b** A is free to move.

Chapter summary

- The momentum of a body is its mass multiplied by its velocity; momentum $= m \times \mathbf{v}$; the units of momentum are N s
- The principle of conservation of momentum applies when no external forces are present. It says that the total momentum is constant.
  So, total initial momentum = total final momentum
- When two bodies collide, Newton's experimental law gives the equation

$$\frac{\text{Speed of separation}}{\text{Speed of approach}} = e$$

$$e = \frac{v_2 - v_1}{u_1 - u_2}$$

- The constant e is known as the coefficient of restitution.
  If $e = 0$, the impact is inelastic.
  If $e = 1$, the bodies are perfectly elastic.
  In general, $0 < e < 1$
- When a force $\mathbf{F}$ acts on a body for a time t, the impulse $\mathbf{I} = \mathbf{F} \times t$ when $\mathbf{F}$ is constant or $\mathbf{I} = \int \mathbf{F} dt$ when $\mathbf{F}$ is variable.
- An impulse creates a change in momentum. The impulse equation is
  Impulse of force = Change in momentum produced
- Velocity, momentum, force and impulse are all vectors, so equations involving them must take account of their directions.

Check and review

You should now be able to...	Try Questions
✔ Use the conservation of momentum equation and Newton's equation appropriately.	1–14
✔ Find the size of an impulse for constant and variable forces.	6–9, 12–15
✔ Apply these concepts to problems in two-dimensions using vectors.	2–3, 8, 12

1 Particle A of mass 2 kg moves with velocity $10 \, \text{m s}^{-1}$

It collides directly with another particle B of mass 4 kg moving towards A with velocity $8 \, \text{m s}^{-1}$

After the collision, B is at rest. Find the final velocity of A

2 A sphere of mass 2 kg has a velocity $5\mathbf{i} + 3\mathbf{j} \, \text{m s}^{-1}$

It collides and coalesces with a sphere of mass 4 kg with a velocity $2\mathbf{i} - 3\mathbf{j} \, \text{m s}^{-1}$
Their common velocity after the collision is $\mathbf{v}$
Calculate the speed $|\mathbf{v}|$ and the angle which $\mathbf{v}$ makes with the x-axis.

3 A particle of mass 5 kg has a velocity of $4\,\text{m s}^{-1}$ in the x-direction. A second particle has a mass m kg and a velocity of $2\,\text{m s}^{-1}$ in the y-direction. They collide, coalesce and begin to move together at 45° to their initial directions. Find their final velocity and the value of m

4 A particle with mass 2 kg and speed $20\,\text{m s}^{-1}$ collides head-on with a particle with mass 6 kg and speed $6\,\text{m s}^{-1}$

If the coefficient of restitution is $e = \dfrac{1}{2}$, calculate their speeds after impact.

5 A ball, which is initially at rest, drops from a height of 9 m onto a horizontal plane and rebounds to a height of 1 m. The ball then continues to bounce. Calculate the value of e and the total distance it has travelled when it hits the ground for the third time. (Use $g = 10\,\text{m s}^{-2}$)

6 A railway truck of mass 10 tonnes and speed $3\,\text{m s}^{-1}$ strikes and couples with a lighter truck of mass 5 tonnes which is at rest on the same line. Find their common speed after impact and the impulse in the coupling between them during impact.

7 Sphere A of mass 3 kg is moving at $12\,\text{m s}^{-1}$ and collides directly with sphere B of mass m kg, which is moving at $2\,\text{m s}^{-1}$ in the same direction. After the impact, A moves at $2\,\text{m s}^{-1}$ in the opposite direction and B continues at $6\,\text{m s}^{-1}$

Calculate the values of m and e and the size of the impulse during the collision.

8 A smooth wall lies along the x-axis. A 4 kg ball with velocity $\mathbf{u} = 5\mathbf{i} - 2\mathbf{j}\,\text{m s}^{-1}$ strikes the wall where the coefficient of restitution, $e = 0.5$. Calculate the final velocity $\mathbf{v} = v_1\mathbf{i} + v_2\mathbf{j}\,\text{m s}^{-1}$ of the ball and the impulse $\mathbf{I}$ of the wall on the ball.

9 Water flows horizontally from a pipe of cross-section $0.03\,\text{m}^2$ at a speed of $25\,\text{m s}^{-1}$ It strikes a fixed vertical wall and does not rebound. Take the density of water as $1000\,\text{kg m}^{-3}$ and find the force F on the wall.

10 A particle of mass 2 kg has an initial velocity of $9\,\text{m s}^{-1}$

A force $F = 3t^2 - 6t + 2\,\text{N}$ acts on the particle for 3 s in the same direction as the velocity. Find the impulse on the particle and its speed after 3 seconds.

11 Three particles A, B and C with masses of 1 kg, 2 kg and 4 kg respectively are in the same line on a smooth horizontal surface in this order with A at rest. B and C collide head-on with velocities of $10\,\text{m s}^{-1}$ and $8\,\text{m s}^{-1}$ respectively. Given that $e = 0.5$ for this and any subsequent impacts, show that B then collides with A

Find all three velocities after this second collision and explain why there are no more collisions.

12 Two spheres, P of mass 2 kg and Q of mass 4 kg, lie at rest on the x-axis with P nearer the origin. They are connected by a straight string. Q is struck by an impulse $\mathbf{I} = 20\mathbf{i} + 24\mathbf{j}\,\text{N s}$. Find the impulsive tension in the string and the spheres' velocities immediately after the blow.

13 A ball of mass 1 kg strikes a smooth fixed plane with a velocity of $20\,\text{m s}^{-1}$ at 30° to the plane where the coefficient of restitution is $e = 0.4$. Find the angle that the final velocity makes with the wall and also the impulse on the ball during the collision.

14 Four identical balls of mass m lie at rest on a smooth horizontal surface at the points $(0, 0)$, $(0, a)$, $(a, 0)$ and (a, a). The balls are joined together by four inelastic strings of length a

The ball at (a, a) is struck by an impulse

$$I\left(\frac{1}{\sqrt{2}}\mathbf{i} + \frac{1}{\sqrt{2}}\mathbf{j}\right)$$

Find the initial velocities of the four balls in terms of m and I

Clearly state any assumptions that you have made.

15 A mass of 0.5 kg, initially at rest, is acted on by a variable force F for 4 s where $F = 2t + 4\,\text{N}$ and t is the time in seconds. Find

a The impulse acting on the mass,

b The final speed of the mass,

c The time for the mass to reach half its final speed.

Information

The term momentum refers to a 'quantity of motion of a moving body'. It is borrowed from Latin, where it means 'movement' or 'moving power'. The concept of momentum was first introduced by the French mathematician Descartes, who is perhaps most famous for his development of Cartesian graphs.

Investigation

A Newton's cradle is a device that is named after Sir Isaac Newton, and can be used to demonstrate conservation of momentum and energy. The device consists of a number of spheres (usually 5 or 6) suspended so that, when at rest, their centres are all at the same height. When a sphere at the end is lifted and released, it strikes the stationary spheres – a force is transmitted through the stationary spheres and pushes the last sphere out and upward.

Investigate, using ideas of conservation of momentum and energy, what will happen if first one ball is lifted at the end of the cradle and released. Then think about the case for two balls, three balls, and four balls.

History

Sir Isaac Newton first presented his three laws of motion in the "Principia Mathematica Philosophiae Naturalis" in 1686. His second law defines a **force** to be equal to the rate of change of momentum with respect to time, with momentum being defined to be the product of the mass, m, and its velocity, v. An alternative form is $F = ma$

Research

In physics, **angular momentum** is the rotational equivalent of linear momentum. It is an important quantity in mechanics because it is a conserved quantity, and so the total angular momentum of a system remains constant unless acted on by an external **torque**. A torque is the rotational equivalent of a force and provides a twist to an object.

Ideas of angular momentum and torque are important in understanding the motion of a gyroscope. Research gyroscopes and write a report on what they are, how they work, and how they can be used.

8 Assessment

1 An object of mass 7 kg, travelling at a speed of $4\,\text{m s}^{-1}$, is acted on by a constant force in its direction of travel which increases its speed to $10\,\text{m s}^{-1}$ in the same direction. Calculate

   a The impulse exerted on the object, **[2 marks]**

   b The force involved if the process took 0.35 seconds. **[2]**

2 A particle of mass 2 kg is travelling at $8\,\text{m s}^{-1}$ in a straight line. A variable braking force, acting along the same straight line, is applied so that, after t s, the magnitude of the force is $2t\,\text{N}$. Calculate the time taken for the particle to come to rest. **[3]**

3 A particle of mass 5 kg is travelling with a velocity of $(4\mathbf{i} + \mathbf{j})\,\text{m s}^{-1}$ when it is subjected to an impulse of $(2\mathbf{i} - 7\mathbf{j})\,\text{N s}$. Calculate the new velocity of the particle. **[3]**

4 A particle of mass 3 kg has velocity $(2\mathbf{i} - 3\mathbf{j})\,\text{m s}^{-1}$

   It is acted on by a constant force of $(-\mathbf{i} + 2\mathbf{j})\,\text{N}$, which changes its velocity to $0.5\mathbf{i}\,\text{m s}^{-1}$

   For how long does the force act? **[4]**

5 A bullet of mass 0.1 kg is fired horizontally at a block of wood of mass 2 kg, which is stationary and free to move. The bullet enters the block travelling at $100\,\text{m s}^{-1}$
   Calculate the subsequent speed of the block if

   a The bullet passes through the block and emerges travelling at $40\,\text{m s}^{-1}$, **[2]**

   b The bullet becomes embedded in the block. **[2]**

6 A railway truck of mass $3m$, travelling at a speed of $2v$, collides with another truck of mass $4m$, which is travelling with a speed of v

   The trucks become coupled together. Find, in terms of v, the common speed of the trucks if, before impact, they were travelling

   a In the same direction, **[2]**

   b In opposite directions. **[2]**

7 A particle A, of mass 10 kg, is moving at $5\,\text{m s}^{-1}$ when it collides with a particle B, of mass m kg, travelling in the opposite direction at $2\,\text{m s}^{-1}$

   After the collision, A travels in the same direction as before but with its speed reduced to $3\,\text{m s}^{-1}$

   a If $m = 3$, find the velocity of B after the collision. **[3]**

   b Show that the value of m cannot be greater than 4 **[4]**

8 Smooth uniform spheres A and B, of equal size, have masses of 3 kg and 2 kg respectively and move on a smooth horizontal surface. A has velocity $(4\mathbf{i} + 3\mathbf{j})\,\text{m s}^{-1}$ and B has velocity $(-\mathbf{i} - 5\mathbf{j})\,\text{m s}^{-1}$

   They collide when the line joining their centres is parallel with $\mathbf{j}$

   The coefficient of restitution between the spheres is 0.5

a   Calculate the velocities of A and B after the collision.   **[5]**

b   Calculate the impulse received by B **[2]**

c   Calculate the angle through which the motion of A has been deflected by the collision.   **[2]**

9   A sledgehammer of mass 6 kg, travelling at $20\,\text{m s}^{-1}$, strikes the top of a post of mass 2 kg and stays in contact with the post.

a   Calculate the common speed of the hammer and post immediately after impact.   **[2]**

b   The post and hammer are brought to rest in 0.02 s by the action of a resistive force R from the ground. By modelling R as constant, find its magnitude.   **[2]**

c   If, in fact, the force is given by $R = k(1 + 2t)\,\text{N}$, where t s is the time from the moment of impact, find the value of the constant k given that post is again brought to rest in 0.02 s   **[3]**

10   An object of mass 3 kg has velocity $(3\mathbf{i} + 2\mathbf{j})\,\text{m s}^{-1}$

It collides with another object, which has a mass of 2 kg and a velocity of $(\mathbf{i} - \mathbf{j})\,\text{m s}^{-1}$

After the impact, the first object has a velocity of $(2\mathbf{i} + \mathbf{j})\,\text{m s}^{-1}$

Calculate the velocity of the second object after the collision.   **[2]**

11   Particles A and B have masses of 3 kg and 2 kg respectively. They are connected by a light inextensible string. The particles lie at rest on a smooth horizontal surface. The coefficient of restitution between the particles is 0.5. A is projected towards B with velocity $10\,\text{m s}^{-1}$

a   Calculate

     i   The velocities of the particles after the collision,   **[4]**

     ii   The common velocity of the particles after the string becomes taut.   **[2]**

b   **i**   Explain why the answer to **a ii** is independent of e, provided $e > 0$ **[1]**

     ii   Describe what would happen if $e = 0$ **[1]**

12   A particle A, with mass 2 kg and velocity $10\,\text{m s}^{-1}$, is moving on a smooth horizontal surface. It catches up and collides with a second particle B, with mass 1 kg which is traveling with velocity $5\,\text{m s}^{-1}$ along the same straight line. After the impact, B then goes on to collide head-on with a vertical wall. The coefficient of restitution between A and B is 0.5, and between B and the wall is 0.75

a   Calculate the velocities of A and B after they collide for the first time.   **[4]**

b   Calculate the velocities of the particles after they collide for a second time.   **[4]**

c   Describe what happens after the second impact.   **[4]**

13   Three particles A, B and C have masses 3 kg, 2 kg and 1 kg respectively. They are moving, in that order, along a straight line with velocities of $3\,\text{m s}^{-1}$, $2\,\text{m s}^{-1}$ and $1\,\text{m s}^{-1}$ respectively. The collision between A and B, which happens first, is perfectly elastic. The second collision, between B and C, has coefficient of restitution e

a   Find the velocities of A and B after the first collision.   **[4]**

b   Find, in terms of e, the velocity of B after the second collision.   **[3]**

c   Hence show that there will be no more collisions if $e \leq \dfrac{4}{11}$ **[3]**

14 The diagram shows a snooker table *ABCD* from above. The surface is horizontal. The sides *AB* and *AD* are parallel to the **i**- and **j**-directions respectively, as shown.

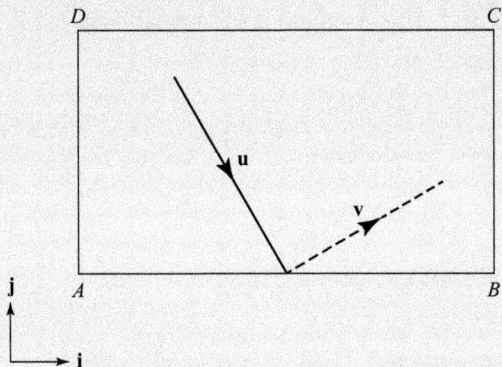

A ball of mass 0.16 kg, travelling with velocity **u** = $(5\mathbf{i} - 7\mathbf{j})$ m s^{-1}, strikes the cushion *AB* (which is assumed to be smooth) and rebounds with velocity **v** m s^{-1}

The coefficient of restitution between the ball and the cushion is 0.7

a i Explain why the *x*-component of **v** is 5 m s^{-1} **[1]**

   ii Calculate the *y*-component of **v**   **[2]**

  iii Find the angle through which the direction of the ball has been changed.   **[4]**

b The ball is in contact with the cushion for 0.1 s

   The force exerted by the cushion has magnitude $kt^2(3 - 8t)$ N, where *t* s is the time since first contact. Find the value of the constant *k*   **[4]**

15 The diagram shows particles *A*, *B* and *C* of masses 1 kg, 2 kg and 2 kg respectively, which are initially at rest in a straight line.

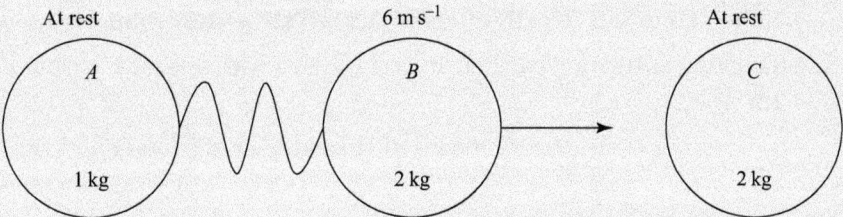

A and *B* are connected by a light inextensible string which is slack. *B* is then propelled towards *C* at 6 m s^{-1}

The coefficient of restitution between *B* and *C* is 0.5. The string becomes taut before *B* reaches *C*

a Find the common speed of *A* and *B* at the instant after the string becomes taut.   **[2]**

b Find the impulse on *C* when *B* collides with it.   **[4]**

c Show that, after the string becomes taut for a second time, there are no more collisions.   **[4]**

16 Particles of mass *m* and 2*m*, both travelling with speed *u*, collide head on. After the collision, one particle is travelling at twice the speed of the other. Find their speeds after impact in terms of *u*   **[3]**

17 Particle A of mass m, moving at $7\,\mathrm{m\,s^{-1}}$, catches up and collides with particle B, of mass km, moving at $1\,\mathrm{m\,s^{-1}}$

After the collision, the speed of B is twice the speed of A. The coefficient of restitution is 0.75
Find the two possible values of k **[5]**

18 A particle A, of mass 4 kg, traveling at $6\,\mathrm{m\,s^{-1}}$ catches up and collides with a particle B, of mass m kg, travelling at $1\,\mathrm{m\,s^{-1}}$

The particles coalesce and move with speed $3\,\mathrm{m\,s^{-1}}$

  a   Calculate the value of m **[2]**

  b   What would their common velocity after the collision have been if they had been travelling in opposite directions before the collision?   **[2]**

19 A tennis player strikes a ball so that its path is exactly reversed. The ball approaches the racket at $35\,\mathrm{m\,s^{-1}}$ and leaves at $45\,\mathrm{m\,s^{-1}}$

The mass of the ball is 90 g. Find the magnitude of the impulse exerted on the ball.   **[2]**

20 A gun of mass 500 kg, which is free to move, fires a shell of mass 5 kg horizontally at a speed of $200\,\mathrm{m\,s^{-1}}$

Find the initial speed of recoil of the gun.   **[3]**

21 Two particles A and B, of mass 4 kg and 2 kg respectively, are moving with respective speeds of $1\,\mathrm{m\,s^{-1}}$ and $10\,\mathrm{m\,s^{-1}}$ directly towards a fixed vertical wall. B hits the wall, rebounds and then collides with A. The coefficient of restitution between B and the wall is 0.4, and between the particles is 0.2

  a   Show that B is brought to rest when it collides with A **[3]**

  b   Calculate the speed at which A is travelling after its collision with B **[2]**

22 An object of mass 4 kg, travelling with velocity $(5\mathbf{i} + 2\mathbf{j})\,\mathrm{m\,s^{-1}}$, is struck by a second object of mass 6 kg and velocity $\mathbf{v}$

The two objects coalesce. If their common velocity after impact is $(2\mathbf{i} - 4\mathbf{j})\,\mathrm{m\,s^{-1}}$, find $\mathbf{v}$ **[3]**

23 A particle A, of mass 3 kg travelling at $10\,\mathrm{m\,s^{-1}}$, catches up and collides with a second particle, B, of mass 5 kg, travelling at $2\,\mathrm{m\,s^{-1}}$

After the collision, A is moving in the same direction with its speed reduced to $4\,\mathrm{m\,s^{-1}}$

Calculate

  a   The new speed of B **[2]**

  b   The coefficient of restitution.   **[2]**

24 A particle of mass 6 kg, travelling at $8\,\mathrm{m\,s^{-1}}$, is brought to rest in a head-on collision with a second particle of mass 4 kg

If $e = 0.3$, calculate the initial and final velocities of the second particle.   **[4]**

25 A body of mass 5 kg is travelling with velocity $(2\mathbf{i} + 3\mathbf{j})\,\mathrm{m\,s^{-1}}$ when it receives an impulse $\mathbf{I}\,\mathrm{N\,s}$. This changes its velocity to $(4\mathbf{i} + 7\mathbf{j})\,\mathrm{m\,s^{-1}}$

Find $\mathbf{I}$ and show that it has a magnitude of $10\sqrt{5}\,\mathrm{N\,s}$ **[4]**

26 A particle of mass 8 kg is travelling with velocity $2\,\mathrm{m\,s^{-1}}$ when it is acted upon for a period of 4 s by a force $F = (3t^2 + 1)\,\mathrm{N}$ in the same direction as the velocity. Find

  a   The impulse received by the particle,   **[2]**

  b   Its final velocity.   **[2]**

9 Circular motion 1

The analysis of circular motion has been particularly significant in making advances in medical sciences because of the design of clinical centrifuges for use in medical laboratories. These allow technicians to separate serum, urea, blood samples and so on into their constituent parts. Measurements can then be made that support quick and accurate diagnoses of patients' medical problems. This ensures that they can be treated as soon as possible, which increases the likelihood of a quick return to good health.

Understanding of circular motion has applications in other areas of life. For example, a good understanding of circular motion is significant in the design of theme park rides, and in improving athletic performance.

Orientation

What you need to know	What you will learn	What this leads to
Maths Ch7 • Equations of motion.	• To use equations describing circular motion. • To analyse horizontal circular motion.	**Chapter 24 Circular motion 2**

Fluency and skills

When a point mass P moves in a circle with centre O and radius r, the distance s it travels is the arc length and θ is the angle through which the radius OP rotates.

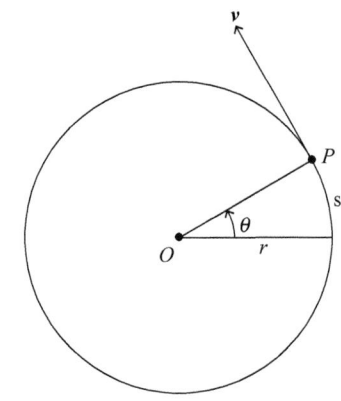

The arc length, $s = \dfrac{\theta}{2\pi} \times 2\pi r = r\theta$, with θ in radians.

The **linear velocity**, v of P along the tangent is the rate of change of s and the **angular velocity**, ω about O is the rate of change of θ

So: $\quad \omega = \dfrac{d\theta}{dt} \quad$ and $\quad v = \dfrac{ds}{dt} = r\dfrac{d\theta}{dt} = r\omega$

In this chapter, P moves at constant speed, so the magnitude of v is constant but its direction changes. Angular velocity ω is also constant and is measured in **radians per second** (rad s^{-1}), although revolutions per minute (rpm) is sometimes used.

The time period T to make one complete revolution is

$$\frac{\text{Angular displacement, radians}}{\text{Angular speed, rad/sec}} = \frac{2\pi}{\omega} \text{ seconds}$$

Key point

For circular motion at constant speed with θ in radians

$$s = r\theta \qquad \omega = \frac{d\theta}{dt} \qquad v = r\omega \qquad T = \frac{2\pi}{\omega}$$

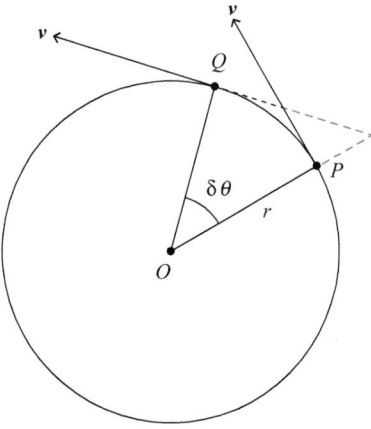

As the mass travels around the circle, it moves from P to Q through a small angle $\delta\theta$ in a time δt

At P and Q, the velocity along the tangent is v

The magnitude of this tangential velocity is constant.

However, the change in velocity along $PO = v\sin\delta\theta - 0 \approx v\delta\theta$

So the acceleration towards $O = \dfrac{\text{change in velocity}}{\text{time taken}} = \dfrac{v\delta\theta}{\delta t}$

As $\delta\theta \to 0$, the acceleration along PO tends to $v\dfrac{d\theta}{dt}$

For a small angle $\delta\theta$ in radians, $\sin\delta\theta = \delta\theta$

where angular velocity $\dfrac{d\theta}{dt} = \omega = \dfrac{v}{r}$ Hence $a = v\dfrac{d\theta}{dt} = \dfrac{v^2}{r} = r\omega^2$

In summary, there is no linear acceleration along the tangent, but there is a linear acceleration of $r\omega^2 = \dfrac{v^2}{r}$ towards the centre of the circle. You can think of this result as an acceleration that changes the direction of the particle but not its speed.

A body moving on a circular path of radius r with constant angular velocity ω about the centre has

Key point

- a constant speed $v = r\omega$ along the tangent

- an acceleration $a = r\omega^2 = \dfrac{v^2}{r}$ towards the centre.

Example 1

A cyclist is mending his bike. He places it upside down with the wheel spinning 200 times every minute. A piece of grit is stuck in the tyre.

If the diameter of the tyre is 60 cm, calculate

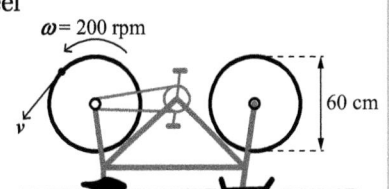

a   The angular speed of the wheel in radians per second,

b   The linear tangential speed of the grit in metres per second,

c   The acceleration of the grit towards the centre of the wheel.

a   Angular speed, $\omega = 200$ revs per minute

$$= 200 \times \frac{2\pi}{60} \text{ rad s}^{-1} = 20.9 \text{ rad s}^{-1}$$

Each revolution is 2π radians.

b   Linear speed of the grit, $v = r\omega$

$$= 0.30 \times 20.9 = 6.3 \text{ m s}^{-1}$$

Radius is needed in metres.

c   Acceleration of the grit towards the centre, $a = r\omega^2$

$$= 0.3 \times 20.9^2 = 130 \text{ m s}^{-2} \text{ to 2 sf}$$

Exercise 9.1A Fluency and skills

1   This table shows various speeds measured in different units. Copy and complete the table.

		rpm	rad/min	rad s^{-1}
a	Vinyl record	33		
b	Roundabout	9		
c	Wind turbine			10
d	Car engine			260

2   The London Eye, with a diameter of 120 m, makes a full revolution in 27 minutes. Calculate the magnitude of its angular velocity (in rad s^{-1}) and its linear velocity (in m s^{-1}) of a point on its circumference.

3   The hands of the Big Ben clock in London are 2.7 m and 4.2 m long. For each hand, calculate the angular velocity (in rad s^{-1}) and the speed of the tip of the hand (in m s^{-1}).

4   Given that the Earth's radius is 6371 km, calculate its angular speed in rad s^{-1} and work out the speed, in km h^{-1}, relative to the centre of the Earth, of a point which lies on the equator.

5   A mass is attached to a fixed point, P, on a smooth horizontal plane by a string 1.5 m long. The mass moves in a circular path about P, with the string taut, at a speed of 6 m s^{-1}

Calculate the angular velocity of the mass and its acceleration towards P

6   A fairground carousel rotates 10 times every minute. A child is 3 metres from the centre. Calculate the linear speed of the child and her acceleration towards the centre of the carousel.

7 The crankshaft of a car engine has a radius of 3 cm and rotates at a steady speed of 2000 rpm. Calculate the linear speed (in m s^{-1}) of a point on its circumference and the acceleration (in m s^{-2}) of that same point towards the centre of the shaft.

8 A spin dryer has a maximum angular speed of 1200 rpm and a drum with a diameter of 48 cm. What is the linear speed of a sock on the circumference of the drum when the drum is spinning at maximum speed? What is the sock's acceleration towards the centre of the drum?

Reasoning and problem-solving

Since a body moving in a circle has an acceleration of $r\omega^2$ (or $\dfrac{v^2}{r}$) towards the centre, Newton's 2nd law requires that there is a force equal to $m \times r\omega^2$ (or $m\dfrac{v^2}{r}$) towards the centre.

The force can be provided in different ways. For example, when a car rounds a horizontal bend, friction between tyres and road provides the force; when clothing is spinning in a spin dryer, the reaction between drum and clothes provides the force. This force F to the centre is called the **centripetal force**.

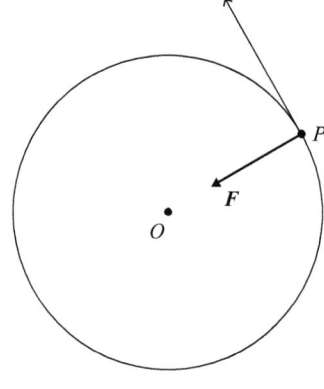

Key point

The equation of motion towards the centre is:

Centripetal force $= mr\omega^2 = m\dfrac{v^2}{r}$

Strategy

To solve problems involving motion in a circle

1. Draw a clear diagram showing all the forces acting on the moving object.

2. Write and solve an equation for motion towards the centre of the circle using Newton's 2nd law

Example 2

The ends of a 2 m length of string are attached to a fixed point O and a 3 kg mass. The mass rotates about O on a smooth table.

If the string breaks when the tension is greater than 530 N, find the maximum angular velocity in rpm.

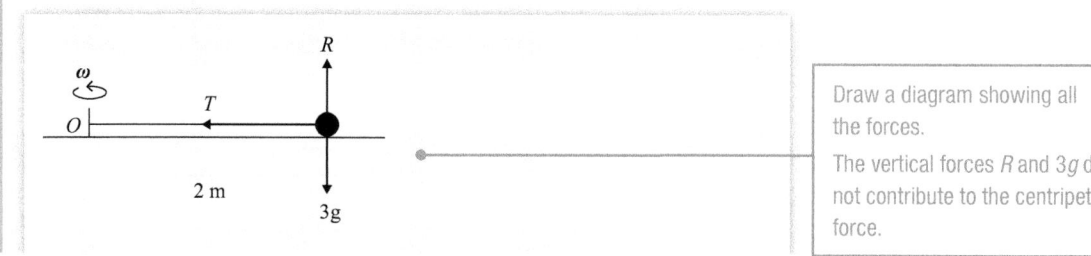

Draw a diagram showing all the forces.

The vertical forces R and $3g$ do not contribute to the centripetal force.

(*Continued on the next page*)

Let ω be the maximum angular velocity.

On the point of breaking, the equation of motion is

$$530 = m \times r\omega^2 = 3 \times 2 \times \omega^2$$

Maximum angular velocity, $\omega = \sqrt{\dfrac{530}{3\times2}} = 9.4\,\text{rad s}^{-1}$

$$= \dfrac{9.4 \times 60}{2\pi}$$

$$= 90\,\text{rpm to 2 significant figures}$$

> ② Use Newton's second law for motion towards the centre of the circle O

Exercise 9.1B Reasoning and problem-solving

1. A 2 kg mass is attached to a light, inextensible string of length 1.5 m, the other end of which is fixed to a point on a smooth horizontal table. The mass moves on the table in a circle at a speed of $3\,\text{m s}^{-1}$ with the string taut. Calculate the tension in the string.

2. An inextensible string 1.5 m long connects a particle P of mass 0.5 kg to a fixed point O on a smooth horizontal table. If the particle moves on the table in a circle about O with the string taut and completes 90 revolutions every minute, calculate the tension in the string.

3. The engine of a toy train has a mass of 0.8 kg

   It runs on a circular horizontal track of radius 1.2 m

   a What produces the force that enables the engine to travel in a circle?

   b If the engine has an angular speed of 10 rpm, calculate the force towards the centre of the circle, in newtons.

4. A spring balance is used to measure force. One end of a spring balance is attached to a fixed point O on a horizontal smooth surface. The other end is attached to a 5 kg mass which rotates at a steady speed in a circle of radius 20 cm about O fifty times each minute. What is the reading on the spring balance?

5. A car of mass 900 kg is driven round a circular bend on an icy road of radius 20 m The frictional force between the tyres and road cannot exceed 250 N. What is the maximum angular speed of the car if no skidding occurs? What is the reading on the car's speedometer (which shows kilometres per hour) in this case?

6. A particle of mass 100 grams rests on a rough horizontal turntable at a distance of 20 cm from the centre. The turntable rotates at a constant angular velocity. The particle is on the point of moving when the frictional force is 0.5 N. What is its maximum angular velocity, in rpm, for which the particle will not slip?

7. A railway engine of mass 50 tonnes travels on a horizontal track at $18\,\text{km h}^{-1}$ round a circular arc of radius 0.3 km. Calculate the total sideways force exerted on its wheels by the track.

8. A mass m on a smooth horizontal table is attached by an inextensible string through a small smooth hole in the centre of the table to an equal mass which hangs freely under the table. The mass on the table moves on a circular path, of radius r, around the hole. Its speed is sufficient to hold the hanging particle at rest. Show that the required linear speed is $\sqrt{rg}$

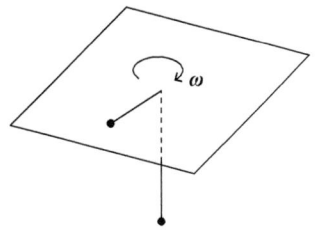

Fluency and skills

When circular motion takes place in a horizontal plane, the centripetal force is horizontal.

The force of gravity has no effect on horizontal motion and so does not contribute to the centripetal force.

Key point

A body moving on a circular path of radius r with constant angular velocity ω about the centre has

- a constant speed $v = r\omega$ along the tangent

- an acceleration $a = r\omega^2 = \dfrac{v^2}{r}$ towards the centre

- a centripetal force $F = mr\omega^2 = m\dfrac{v^2}{r}$

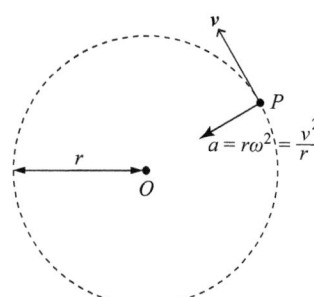

Example 1

Calculate the maximum speed (in km h^{-1}) for a car of 900 kg to round a bend of radius 60 m on a level road if

a The maximum frictional force between the tyres and road is 1440 N

b The coefficient of friction between the car's tyres and the road is 0.5   **Full A Level**

Take $g = 9.8 \text{ m s}^{-2}$

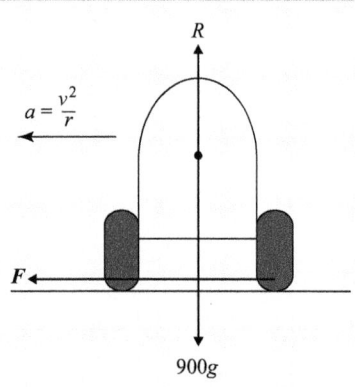

Draw a diagram showing all the forces.

The total reaction of the ground on the car through its tyres has a vertical component, R, and a frictional component, F

To prevent side-slipping, F acts towards the centre of the bend.

a   Let v be the maximum speed of the car.

At maximum speed, the horizontal equation of motion is:

$$F = m \times \frac{v^2}{r}$$

$$1440 = 900 \times \frac{v^2}{60}$$

R does not contribute to the centripetal force

$$v^2 = \frac{1440 \times 60}{900} = 96$$

Maximum speed $v = \sqrt{96} \text{ m s}^{-1} = \dfrac{\sqrt{96} \times 60 \times 60}{1000} = 35.3 \text{ km h}^{-1}$

(*Continued on the next page*)

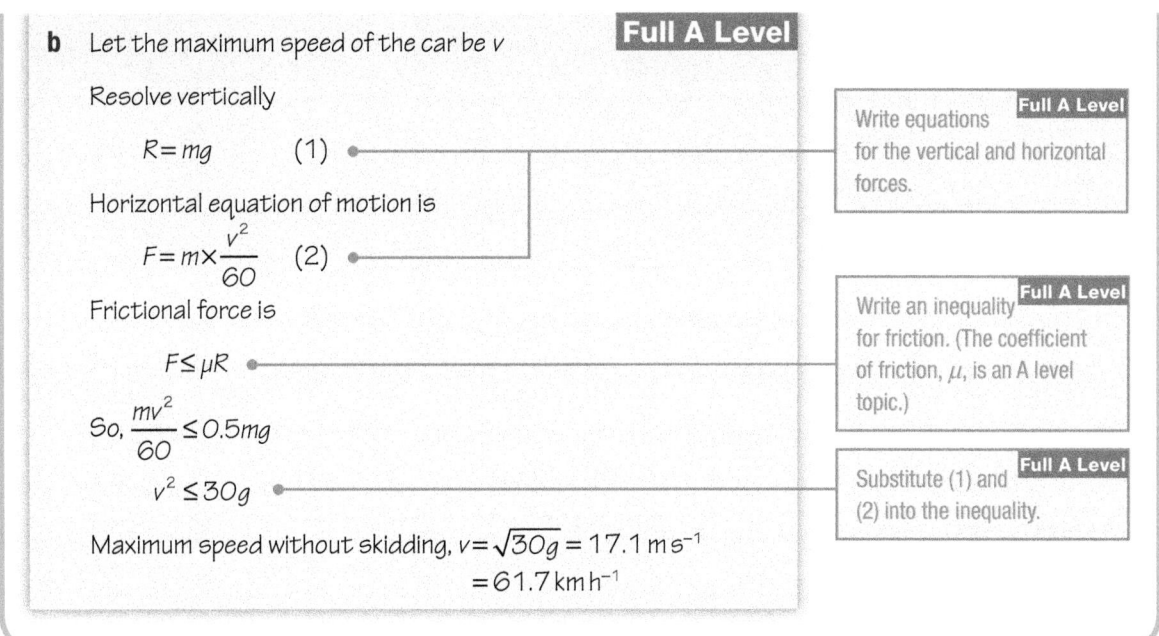

b Let the maximum speed of the car be v **Full A Level**

Resolve vertically

$$R = mg \qquad (1)$$

> **Full A Level**
> Write equations for the vertical and horizontal forces.

Horizontal equation of motion is

$$F = m \times \frac{v^2}{60} \qquad (2)$$

Frictional force is

$$F \le \mu R$$

> **Full A Level**
> Write an inequality for friction. (The coefficient of friction, μ, is an A level topic.)

So, $\dfrac{mv^2}{60} \le 0.5mg$

$$v^2 \le 30g$$

> **Full A Level**
> Substitute (1) and (2) into the inequality.

Maximum speed without skidding, $v = \sqrt{30g} = 17.1\,\mathrm{m\,s^{-1}}$
$$= 61.7\,\mathrm{km\,h^{-1}}$$

MECH

Exercise 9.2A Fluency and skills

1. A particle of mass 4 kg is attached to a fixed point O on a smooth horizontal table by a light inextensible string 0.6 m long. Calculate the tension in the string when the particle travels in a circle once every second about point O

2. A particle of mass 200 grams on a smooth horizontal table is attached by a string of length 80 cm to a fixed point on the table. What is the linear speed at which it rotates if the tension in the string is 0.5 N?

3. A string 0.6 m long can just support a mass of 10 kg without breaking. With one end fixed to a point on a smooth horizontal table, a mass of 2 kg rotates about the fixed point. Calculate the greatest number of revolutions the 2 kg mass can complete each minute without breaking the string. Use $g = 9.8\,\mathrm{m\,s^{-1}}$

4. A smooth circular table of radius 80 cm has a raised vertical rim. A small ball travels round the table in contact with the rim at a speed of $3\,\mathrm{m\,s^{-1}}$

   If the mass of the ball is 40 grams, calculate the force exerted by the rim on the ball.

5. A light rod OMA of length $2a$ has two equal masses m attached to it, one at its midpoint M and the other at its end A. The rod's other end O is fixed to a point on a smooth horizontal table. The rod and masses are in contact with the table and rotate about O with angular velocity ω

   Calculate the tensions in the two halves of the rod in terms of m, a and ω

6. Three equal masses m are attached to points P, Q and R on a light rod OR, revolving round O in a horizontal plane, such that OP, PQ and QR are equal. Prove that the stresses in the three parts of the rod are in the ratio $3:5:6$

7. A particle of mass m on a rough turntable is on the point of moving when it is a distance r from the centre and the turntable is rotating at $\omega\,\mathrm{rad\,s^{-1}}$

   If the particle were at a distance of $2r$ from the centre, what angular speed would cause it to be on the point of moving?

To solve problems involving motion in a horizontal circle

(1) Draw a clear diagram showing all the forces acting on the moving object.

(2) Decide which forces contribute to the centripetal force and write an equation for motion towards the centre of the circle.

(3) Write other equations as necessary and solve them simultaneously.

Example 2

A particle of mass m is tied to a fixed point O on a smooth horizontal table by an elastic string of natural length 0.3 m and modulus of elasticity $\lambda = mg$

The particle moves at a constant angular speed of 20 rpm about point O

Calculate the extension of the string if $g = 9.8 \text{ m s}^{-2}$

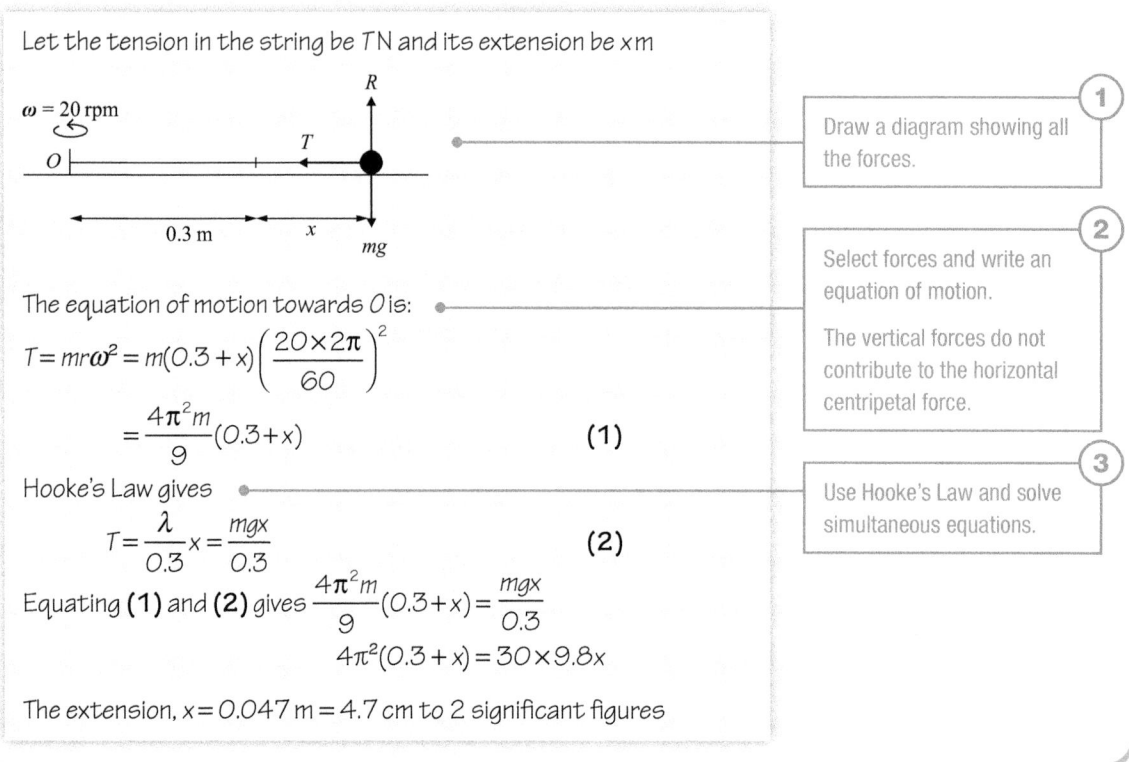

Let the tension in the string be T N and its extension be x m

$\omega = 20$ rpm

0.3 m x

Draw a diagram showing all the forces. (1)

The equation of motion towards O is:

$$T = mr\omega^2 = m(0.3 + x)\left(\frac{20 \times 2\pi}{60}\right)^2$$

$$= \frac{4\pi^2 m}{9}(0.3 + x) \qquad (1)$$

Select forces and write an equation of motion. (2)

The vertical forces do not contribute to the horizontal centripetal force.

Hooke's Law gives

$$T = \frac{\lambda}{0.3}x = \frac{mgx}{0.3} \qquad (2)$$

Use Hooke's Law and solve simultaneous equations. (3)

Equating (1) and (2) gives $\dfrac{4\pi^2 m}{9}(0.3 + x) = \dfrac{mgx}{0.3}$

$$4\pi^2(0.3 + x) = 30 \times 9.8x$$

The extension, $x = 0.047 \text{ m} = 4.7 \text{ cm}$ to 2 significant figures

Exercise 9.2B Reasoning and problem-solving

Throughout this exercise use $g = 9.8 \text{ m s}^{-2}$ unless told otherwise.

1   An elastic string with a natural length of 0.6 m and modulus of elasticity λ of 120 N has one end attached to a fixed point O on a smooth horizontal table and the other end attached to a 2 kg mass. Calculate the extension in the string when the mass rotates at 40 rpm about point O

2   A spring balance with a natural length of 50 cm has one end attached to a fixed point O and the other end to a mass of 5 kg

The mass rotates in circles about O at a speed of 8 rpm on a smooth horizontal surface when the spring balance reads 4 N. Calculate the modulus of elasticity of the spring balance.

3 Two masses m_1 and m_2 are connected by a taut string of length x

The two masses lie along a radius of a rough horizontal disc at distances x and $2x$ respectively from the centre O and rotate with uniform angular velocity ω about O. If the coefficient of friction between the disc and the particles is μ, show that they are both on the point of slipping if

$$\omega = \sqrt{\frac{\mu g(m_1 + m_2)}{x(m_1 + 2m_2)}}$$

4 Ten light spokes, each 40 cm long, connect the axle of a wheel to its light rim. A mass of 800 grams is attached to the end of each spoke. The wheel rotates at 100 rpm in a horizontal plane. Find the tension in each spoke and the total kinetic energy of the arrangement. Why does the tension in the rim not affect the tension in the spokes?

5 An elastic string of natural length 1.0 m and modulus of elasticity of 80 N has one end fixed to point O on a smooth horizontal table. The other end is attached to a mass of 3 kg which rotates about O on a circular path of radius 1.2 m. Calculate the linear speed of the mass and the total energy stored as kinetic and elastic energy.

6 A smooth ring is fixed to a point O on a smooth horizontal surface so it can rotate and act as a pivot. A light rod of length 80 cm is free to slide through the ring. The rod has masses of 10 kg and 25 kg fixed to its two ends. The rod rotates about O at a constant speed of 15 revs per second. How far is the rod's centre from O?

7 A car of mass m rounds a corner of radius 60 m with a coefficient of friction of 0.4 between its tyres and the horizontal road. Find the maximum speed of the car if there is no slipping. Does this speed depend on the mass of the car?

8 A car of mass 800 kg is travelling at 36 km h^{-1} and rounds a bend of radius 250 m on a horizontal road without skidding. Calculate the smallest value of the coefficient of friction between the tyres and the road.

9 A level turntable of radius 12 cm rotates at 33 rpm

Calculate the least coefficient of friction if a small mass stays on the turntable wherever it is placed.

10 A rough horizontal circular disc rotates about a vertical axis through its centre O

It makes two full revolutions every second. Prove that the greatest distance that a particle can be placed from O so that it is stationary relative to the disc is 6.2μ cm where μ is the coefficient of friction between the disc and particle.

11 Before its launch, a satellite on the Earth's surface has a weight of mg. When orbiting the Earth at a height h, it experiences a gravitational attractive force F which is inversely proportional to the square of its distance from the centre of the Earth.

a If the Earth's radius is R, show that the constant of proportionality equals mgR^2

b If the satellite takes 90 minutes to complete each circuit of the Earth's circumference, take R as 6370 km and calculate its angular velocity and height above the Earth's surface.

c A 'geostationary' satellite appears to be stationary in the sky above a fixed point on the equator. Calculate the height of a geostationary satellite above the Earth's surface.

Chapter summary

- When an object moves along the arc of a circle of radius r, its linear displacement, s, and the angle through which it moves, θ, respectively, are related by $s = r\theta$
- An object's linear velocity v along the tangent and its angular velocity ω about the centre of the circle are related by $v = r\omega$
- When ω is constant, the time period T for one complete revolution is $\dfrac{2\pi}{\omega}$
- The SI units for angular velocity ω are radians per second (rad s^{-1}), but revolutions per minute (rpm) are also common units.
- When v and ω are constant, there is no linear acceleration along the tangent and no angular acceleration. There is, however, an acceleration a towards the centre of the circle where
$a = r\omega^2 = \dfrac{v^2}{r}$
- The acceleration towards the centre is the result of a force in this direction called the centripetal force. You can use Newton's 2nd law to write an equation of motion towards the centre involving the centripetal force.

Check and review

You should now be able to...	Try Questions
✔ Change units and convert linear to angular motion and vice versa.	1–5
✔ Write equations of motion involving a centripetal force.	6–16
✔ Solve problems involving horizontal circular motion.	8–15

1 A washing machine has two speed settings: 800 rpm and 1200 rpm. What are these speeds in rad s^{-1}?

2 A satellite travels round the Earth twenty times each day. Calculate its angular velocity in rad s^{-1}

3 A model train runs on a circular track of radius 0.8 metres. If the train's linear speed is 0.6 m s^{-1}, what is its angular speed in rad s^{-1} and its acceleration towards the centre of the track in m s^{-2}?

4 A car rounds a bend at a steady speed of 36 km h^{-1}

If the bend has a radius of 300 m and turns the car through an angle of 60°, calculate the car's angular velocity and the time it takes to round the bend.

5 A fairground ride has ten 'pods' equally spaced around the edges of a light horizontal wheel of radius 5 m

Each pod has a mass of 600 kg

What is the total kinetic energy stored in the pods when they are empty and rotating at 12 rpm?

6 A spin-dryer rotates at 1000 rpm and has a radius of 15 cm. When an item of clothing with a mass of 0.5 kg is spinning horizontally on the circumference of the drum, what is the reactive force of the drum on the item?

7 The Earth is 1.5×10^8 km from the Sun. Calculate the angular speed of the Earth around the Sun in rad s^{-1}

Given that the mass of the Earth is 6.0×10^{24} kg, calculate the centripetal force on the Earth towards the Sun.

8 One end of a light, inextensible string of length 2 m is attached to a fixed point on a smooth horizontal table. The other end is tied to a mass of 5 kg which rests on the table. The mass moves in a horizontal circle at a speed of 3 m s^{-1} with the string taut. Calculate the tension in the string.

9 A railway engine of mass 50 tonnes travels on a horizontal track round a circular arc of radius 400 m. If the maximum lateral force towards the centre of the arc that the rails can withstand is 10^5 N, calculate the maximum speed that the engine can travel round the bend.

10 A car of mass 1500 kg travels around a bend, which can be modelled as a quarter of the circumference of a circle with radius 15 m. The car will skid if the tyres have to provide a frictional force greater than 200 N. What is the shortest time in which the car can go around the bend without skidding?

11 An elastic spring, fixed at one end O, is attached to a mass of 0.5 kg at its other end. The mass rotates about O on a smooth horizontal surface. The spring has a natural length l of 40 cm and a modulus of elasticity λ of 80 N. Calculate the linear speed at which the mass travels if the spring is extended by 5 cm.

12 An elastic string has a natural length of 0.8 m and modulus of elasticity $\lambda = 100$ N. Its ends are attached to a fixed point O and to a 2 kg mass. Calculate the extension in the string when the mass rotates at 4 rad s^{-1} about O on a smooth horizontal table. What is the total of the kinetic and elastic energy in this arrangement?

13 A smooth curved wire forms a fixed horizontal hoop of radius r on which a bead of mass m is threaded. If the bead describes circles at a constant angular speed ω, calculate the magnitude and direction of the total reaction between the bead and the hoop.

14 The force of gravity F between an object of mass m and the Earth of mass M is given by $F = \dfrac{GmM}{d^2}$ where
$G = 6.67 \times 10^{-11}$ N m^{2}kg^{-2}, $M = 5.98 \times 10^{24}$ kg and d is the distance of the object from the centre of the Earth. Taking the radius of the Earth as 6370 km, calculate the time for a satellite to orbit the Earth if its height above the Earth's surface is 250 km.

15 A horizontal turntable of radius 20 cm rotates at 45 rpm. When a particle is placed on the turntable 15 cm from the centre, it is on the point of slipping. What is the coefficient of friction between the particle and the turntable?

16 A car of mass 900 kg travelling at 54 km h^{-1} rounds a bend of radius 40 m on a horizontal road. Calculate the smallest value of the coefficient of friction if there is no sideways skidding.

Did you know?

Gradians are a measure of angle. The gradian or grad is a unit of measurement equivalent to 1/400 of a complete turn. Gradians are commonly used as the measure of angle in surveying, mining and geology.

Investigation

On a fairground ride, the rider sits in a seat which is rotated at a speed so that, eventually, it swings outwards. As the speed increases, the angle that the chains make with the vertical also increases. The situation can be modelled with the passenger and the seat modelled as a particle, and the chains holding the seat modelled as an inextensible string in tension, so that the passenger is rotating about a vertical axis.

Investigate the situation, answering questions such as:

- Is the passenger's mass significant? (What happens to empty seats?)
- Is the length of the supporting chains significant?
- How far from the vertical do passengers swing out?

Investigation

Imagine that you place several coins of the same type on a horizontal turntable at different distances from the centre of rotation and that you can vary the speed of the turntable. Which coin will slip first? The one nearest or furthest away from the centre? What changes if the coins are of different mass?
Explain your answers!

Take $g = 9.8 \, \text{ms}^{-2}$ unless otherwise stated.

1   A particle of mass 2 kg is attached by a light, inextensible string of length 1.2 m to a fixed point on the surface of a horizontal, smooth table. The particle travels in a circle on the table at a speed of $2.5 \, \text{m s}^{-1}$

    Calculate the tension in the string.   **[2 marks]**

2   A string of length 80 cm can just support a suspended mass of 40 kg without breaking. A 2 kg mass is attached to the string and the other end of the string is fastened to a point on the surface of a smooth, horizontal table. The mass is made to move in a circle on the table. Calculate, in revolutions per minute, the maximum angular speed at which the mass can revolve without breaking the string.   **[4]**

3   A particle of mass 2 kg travels in a circle about a point O on a smooth horizontal plane. It is attached to O by means of a light elastic string of natural length 50 cm and modulus of elasticity 200 N. The speed of the particle is $2\sqrt{3} \, \text{m s}^{-1}$

    Calculate the radius of the circle.   **[5]**

4   Particles A and B, of mass 0.5 kg and 0.2 kg respectively, are connected together by means of a light inextensible string. The string is threaded through a small smooth hole, O, in a smooth horizontal table. Particle B hangs at rest 20 cm below the table surface, while particle A describes a circle about O at an angular speed of $2.8 \, \text{rad s}^{-1}$

    Calculate the length of the string.   **[5]**

5   A car of mass m travels over a hump-backed bridge which has a cross-section in the form of an arc of a circle of radius 25 m. The speed of the car at the highest point of the bridge is $V \, \text{m s}^{-1}$

    Calculate the greatest value that V can take if the car is not to leave the surface of the road at this point.   **[4]**

6   A smooth wire has a bead of mass 0.005 kg threaded onto it. It is then bent round to form a circular hoop of radius 0.2 m. The hoop is fastened in a horizontal position, whilst the bead travels round it at a constant speed of $1 \, \text{m s}^{-1}$

    Work out the magnitude and direction of the reaction between the hoop and the bead.   **[7]**

7   The gravitational attraction between objects of mass M and m that lie a distance r apart has magnitude $\dfrac{GMm}{r^2}$, where the universal gravitational constant $G = 6.67 \times 10^{-11} \, \text{N m}^2\text{kg}^{-2}$

    Make the modelling assumptions that the Earth is fixed and the Moon travels in a circular orbit around it. The distance of Moon from the Earth is 384 000 km and it orbits once every 27.3 days. Use this information to estimate the mass of the Earth.   **[4]**

8   A particle of mass 2 kg travels in a circle on a smooth horizontal plane about a point O

    The particle is attached to O by means of a light elastic string of natural length 1 m and modulus of elasticity 300 N. The total energy of the system is 42 J. Calculate

    a   The radius of the circle,   **b**   The speed of the particle.   **[8]**

9   Particles A and B each have mass m

    Particle A lies on a smooth horizontal table and is connected to B by a light inextensible string of length $4a$ which passes through a small hole O in the table. Particle A travels in a circle of radius r about O, and B hangs at rest. Calculate the difference in total energy (potential and kinetic) between the situations $r = a$ and $r = 3a$ [8]

10 A car of mass m is travelling on a horizontal road round a bend of radius $50\,\text{m}$
    The coefficient of friction between the car and the road is 0.9

    a   If the speed of the car is $45\,\text{km}\,\text{h}^{-1}$ find, in terms of m, the frictional force acting on the car.   [2]

    b   Calculate the maximum speed at which the car could travel without skidding.   [4]

11 A particle P, of mass $3\,\text{kg}$, is placed on a rough, horizontal turntable at a distance of $0.8\,\text{m}$ from its centre O

    The coefficient of friction between the particle and the turntable is 0.4

    a P is connected to O by a light, inextensible string of length $0.8\,\text{m}$. The turntable is set in motion and its speed gradually increased until the tension in the string is $50\,\text{N}$. Calculate the angular speed of the turntable at this time.   [2]

    b   The process is then repeated with P connected to O by a light, elastic string of natural length $0.8\,\text{m}$ and modulus of elasticity $200\,\text{N}$. The speed of the turntable is gradually increased from zero to $4\,\text{rad}\,\text{s}^{-1}$.

        Calculate the length of the string at this time.   [3]

12 A fairground ride is made from a large hollow cylindrical chamber placed with its axis vertical. The riders stand against the inside wall of the cylinder, and the chamber rotates about its axis. When it reaches top speed, the floor is lowered, leaving the riders supported only by the friction between themselves and the wall. The radius of the cylinder is $3\,\text{m}$ and the coefficient of friction between a rider and the wall is $\dfrac{1}{3}$

    What is the minimum angular speed at which the floor can be lowered so that riders do not slide down the wall?   [4]

13 A particle is placed on a horizontal turntable at a distance r from its centre. The turntable is set in motion and the particle is on the point of slipping when the turntable is rotating with angular speed ω

    The coefficient of friction between the particle and the turntable is μ
    Find an expression for ω in terms of μ, r and g [3]

14 A rough, horizontal disc rotates at 2 revolutions per second. A particle is to be placed on the disc. The coefficient of friction between the particle and the disc is μ

    Find, in terms of μ, the maximum distance from the centre at which the particle can be placed without it slipping.   [3]

15 Two particles, each of mass m, are attached to the ends of a light, inextensible string. The string passes through a hole in the centre of a rough, horizontal turntable. One particle is placed on the turntable at a distance a from its centre, and the other hangs freely below the turntable. The coefficient of friction between the particle and the turntable is μ

    The contact between the string and the hole is smooth. The turntable is rotating with angular speed ω and the suspended particle is stationary. Show that the ratio between the maximum and minimum values of ω is $\sqrt{\dfrac{1+\mu}{1-\mu}}$ [5]

10 Discrete and continuous random variables

Drivers are required by law to be insured against accidents because the outcome can be expensive for themselves and for other people involved. To calculate the insurance payments to be made by an individual driver, many factors that relate to their circumstances are taken into account. For example, their age, how many years they have been driving, where they live and the age and type of car they will be driving are considered. These factors all have an estimated probability of having an accident associated with them and these probabilities are used to calculate the cost of premiums to drivers.

This is one area where actuaries use expected values to inform the likelihood of events occurring and to calculate insurance payments for travel, homes against theft, property against fire and health of both people and animals. The associated mathematics is used extensively throughout industry and business.

Orientation

What you need to know

Maths Ch9
- Calculate the mean.
- Calculate the variance.

Maths Ch10
- To model a probability distribution using the binomial distribution.

What you will learn

- To calculate expectation values for given random variables.
- To use Poisson probability models.
- To derive probability generating functions for given random variables and use them to calculate summary statistics.

What this leads to

Ch 3 Continuous random variables
- Calculating the mean of a continuous random variable.
- Calculating the variance of a continuous random variable.

Fluency and skills

See Maths
Ch 10.1

For a
reminder
about discrete
random
variables.

A probability distribution for a random experiment shows how the total probability of 1 is distributed between all the possible outcomes. If a fair coin is thrown three times, the probability distribution for the number of heads is shown in this table.

Number of heads	0	1	2	3
Probability	$\dfrac{1}{8}$	$\dfrac{3}{8}$	$\dfrac{3}{8}$	$\dfrac{1}{8}$

The number of heads can take four values only so the variable is discrete. Since its value is determined by the outcome of a random experiment, it is an example of a **discrete random variable**. Sometimes the distribution is defined using a probability distribution function, for example,

$$P(X=x) = \frac{x^2}{30}; x = 1, 2, 3, 4$$

As with all distributions, you can define measures of location and spread.

The median and mode of a discrete random variable can be calculated from tables of probabilities or distribution functions.

The median value of a random variable divides the probability distribution into two equal parts.

> **Key point**
>
> The median of a discrete random variable is given by M where
>
> $$P(X \le M) \ge \frac{1}{2} \text{ and } P(X \ge M) \ge \frac{1}{2}$$

Solving each of these inequalities for M may give two distinct values. If this is the case, to find the median you should calculate the arithmetic mean of these values.

The median value of a random variable divides the probability distribution into two equal parts. For a discrete random variable, to find the median value you find the arithmetic mean of two values either side of the median.

> **Key point**
>
> The median of a discrete random variable is given by
>
> $$M = \frac{x_1 + x_2}{2}, \text{ where } x_1 \text{ and } x_2 \text{ are given by } P\{X \ge x_1\} \ge \frac{1}{2} \text{ and } P\{X \le x_2\} \ge \frac{1}{2}$$

For sample data, the mode has been defined as the most commonly occurring value. An equivalent definition is used for random variables.

> **Key point**
>
> For a discrete random variable, the mode is defined as the value with the greatest probability.

Some random variables may have more than one modal value; others may have none.

Example 1

A random variable, X, has the probability distribution shown in the table. Write down the mode and median of X

x	−1	2	3	5
$P\{X=x\}$	0.05	0.55	0.3	0.1

Mode = 2

Median = 2

$P\{X \le 2\} \ge \dfrac{1}{2}$ and $P\{X \ge 2\} \ge \dfrac{1}{2} \rightarrow M = 2$

The mode is the x-value with the greatest probability.

Example 2

The number of raisins in a cake is modelled by a random variable, X, with distribution function

$$P\{X = x\} = \begin{cases} \dfrac{e^{-2} 2^x}{x!}; & x = 1, 2, 3, 4 \\ a; & x = 4 \\ 0; & \text{otherwise} \end{cases}$$

Find the value of a and write down the median value of X

If $p_i = P(X = i)$, then $p_0 = 0.1353$, $p_1 = 0.2707$

$p_2 = 0.2707$, $p_3 = 0.1804$

$a = 1 - (0.1353 + \ldots + 0.1804) = 0.143 \ (3dp)$

Median = 2

Using $P(X = x) = \dfrac{e^{-2} 2^x}{x!}$

When you throw a fair dice, the probability of getting a six is $\dfrac{1}{6}$

If you throw the dice 600 times this probability would predict that you would get a six one hundred times. The expected value of the number of sixes obtained if you throw a fair dice 600 times is 100

The expected value of a random variable and the arithmetic mean are closely related. If you throw a fair dice N times and the random variable, X, denotes the score on a randomly chosen throw, then the distribution of X can be written as

The **expected value** of a random variable is the number of occurrences of the particular event that you would expect to get if the relative frequency obtained in an experiment equals the number of occurrences predicted by probability.

x	1	2	3	4	5	6
$P\{X=x\}$	$\dfrac{1}{6}$	$\dfrac{1}{6}$	$\dfrac{1}{6}$	$\dfrac{1}{6}$	$\dfrac{1}{6}$	$\dfrac{1}{6}$

The expected number of each outcome is $\dfrac{N}{6}$ and therefore the

expected total score is $1 \times \dfrac{N}{6} + 2 \times \dfrac{N}{6} + 3 \times \dfrac{N}{6} + 4 \times \dfrac{N}{6} + 5 \times \dfrac{N}{6} + 6 \times \dfrac{N}{6}$

The expected (or mean) value of X, the score *per throw*, is therefore

The $\forall$ symbol means 'for all', so $\forall x$ means 'for all x'

$$\dfrac{1 \times \dfrac{N}{6} + 2 \times \dfrac{N}{6} + 3 \times \dfrac{N}{6} + 4 \times \dfrac{N}{6} + 5 \times \dfrac{N}{6} + 6 \times \dfrac{N}{6}}{N} = 1 \times \dfrac{1}{6} + 2 \times \dfrac{1}{6} + 3 \times \dfrac{1}{6} + 4 \times \dfrac{1}{6} + 5 \times \dfrac{1}{6} + 6 \times \dfrac{1}{6}$$

$$= 3\dfrac{1}{2}$$

This gives a general result for discrete random variables.

The **expected value** or **mean value** of a discrete random variable is given by

$$E[X] = \sum_{\forall x} x P(X = x)$$

Key point

Since this is a mean, it is sometimes given the symbol μ, a Greek m

Example 3

A random variable, S, has the probability distribution shown in the table. Find the expected value of S

s	-1	0	1	2
$P\{S = s\}$	$\dfrac{1}{2}$	$\dfrac{1}{4}$	a	a

$$\frac{1}{2} + \frac{1}{4} + 2a = 1 \Rightarrow a = \frac{1}{8}$$

Total probability on sample space equals 1

$$E[S] = \sum_{\forall s} sP(S = s) = -1 \times \frac{1}{2} + 0 \times \frac{1}{4} + 1 \times \frac{1}{8} + 2 \times \frac{1}{8}$$

$$= -\frac{1}{8}$$

You will sometimes need to find the expected (mean) value of a function, $g(X)$, of a random variable. This can be done by replacing the x by $g(x)$ in the expression above.

The **expected value** of a function, $g(X)$, of a discrete random variable is given by

$$E[g(X)] = \sum_{\forall x} g(x) P(X = x)$$

Key point

For example, for a random variable X, you could work out the expected value of X^2 by multiplying the values of X^2 by the corresponding probabilities and summing the results.

Example 4

A random variable, X, has the probability distribution shown in the table.

Find the expected value of

a X **b** X^2

x	-1	2	3	5
$P\{X = x\}$	0.2	0.3	0.4	0.1

a $E[X] = \sum_{\forall x} xP(X = x)$

$= -1 \times 0.2 + 2 \times 0.3 + 3 \times 0.4 + 5 \times 0.1$

$= 2.1$

b

x	-1	2	3	5
x^2	1	4	9	25
$P\{X = x\}$	0.2	0.3	0.4	0.1

Work out the values of X^2

$E[X^2] = \sum_{\forall x} x^2 P(X = x)$

$= 1 \times 0.2 + 4 \times 0.3 + 9 \times 0.4 + 25 \times 0.1$

$= 7.5$

To measure the spread of a distribution you calculate the **variance**, which is the 'expected value of the squared deviations from the mean'.

You can write the variance of X as $\mathrm{Var}(X)$ or σ^2

Key point

The variance of a random variable X is given by

$$\mathrm{Var}(X) = E[(X - \mu)^2]$$
$$= \sum_{\forall x} (x - \mu)^2 P(X = x)$$

To make calculations simpler, you usually use an alternative form for variance.

$$\mathrm{Var}(X) = E[X^2] - \mu^2$$
$$= \sum_{\forall x} x^2 P(X = x) - \mu^2$$

Key point

The square root of the variance, σ, is the **standard deviation** of the distribution.

STATISTICS

Example 5

A random variable, X, has the probability distribution shown in the table.

x	1	2	3	4
$P\{X = x\}$	0.3	0.2	0.4	0.1

Find the expected value and variance of X

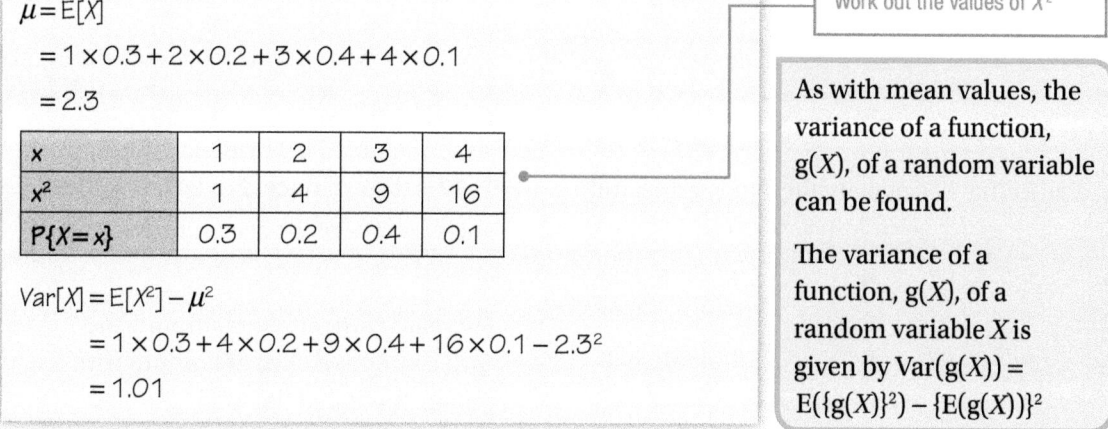

Work out the values of X^2

$$\mu = E[X]$$
$$= 1 \times 0.3 + 2 \times 0.2 + 3 \times 0.4 + 4 \times 0.1$$
$$= 2.3$$

x	1	2	3	4
x^2	1	4	9	16
$P\{X = x\}$	0.3	0.2	0.4	0.1

$$\mathrm{Var}[X] = E[X^2] - \mu^2$$
$$= 1 \times 0.3 + 4 \times 0.2 + 9 \times 0.4 + 16 \times 0.1 - 2.3^2$$
$$= 1.01$$

As with mean values, the variance of a function, $g(X)$, of a random variable can be found.

The variance of a function, $g(X)$, of a random variable X is given by $\mathrm{Var}(g(X)) = E(\{g(X)\}^2) - \{E(g(X))\}^2$

Consider the classic random experiment where a fair dice is rolled once. Since the dice is fair, each of the six outcomes, integers 1–6, have an equal probability of occurring. This probability is $\frac{1}{6}$

Therefore, if X is a random variable for the score obtained when the dice is rolled once, then its probability distribution is given by

$$P\{X = x\} = \frac{1}{6}; x = 1, 2, \ldots, 6$$

This is a special case of a discrete uniform distribution. In general, the x-values don't have to be consecutive integers.

If a random variable, U, has a uniform distribution taking values 1, 2,..., n, it follows that $P\{U = u\} = k$; $u = 1, 2, \ldots, n$, where k is some constant. But $P\{S\} = 1$, where S is the set of all outcomes.

Therefore $nk = 1$ and so $k = \frac{1}{n}$

A discrete uniform random variable, U, taking values $u = 1, 2, \ldots, n$, has a probability distribution given by $P\{U = u\} = \dfrac{1}{n}$; $u = 1, 2, \ldots, n$

You can use this rule to calculate the mean and variance of a discrete uniform distribution.

A discrete uniform random variable, U, with distribution function $P\{U = u\} = \dfrac{1}{n}$; $u = 1, 2, \ldots, n$ has a mean and variance given by $\mu = \dfrac{n+1}{2}$ and $\sigma^2 = \dfrac{n^2 - 1}{12}$

Example 6

An ordinary fair dice is rolled once. Calculate the mean, μ_X, and variance, $\sigma_X{}^2$, of X, the number scored.

$$P\{X = x\} = \frac{1}{6}; \; x = 1, 2, \ldots, 6$$

$$\mu_X = \frac{n+1}{2} = \frac{6+1}{2} = \frac{7}{2}$$

$$\sigma_X{}^2 = \frac{n^2 - 1}{12} = \frac{6^2 - 1}{12} = \frac{35}{12}$$

Exercise 10.1A Fluency and skills

1   For each of the following probability distributions, calculate the expectation and variance of the random variables.

x	1	4	9	16	25	36
$P\{X=x\}$	$\dfrac{1}{6}$	$\dfrac{1}{6}$	$\dfrac{1}{6}$	$\dfrac{1}{6}$	$\dfrac{1}{6}$	$\dfrac{1}{6}$

y	-1	0	1	2
$P\{Y=y\}$	$2a$	$2a$	a	a

2   For each of the following probability distributions, find the mean and the variance.

a $P\{X = a\} = \dfrac{1}{4}$; $a = 1, 2, 3, 4$

b $P\{X = b\} = \dfrac{b}{3}$; $b = 1, 2$

c $P\{X = c\} = \begin{cases} \dfrac{1}{c}; & c = 2, 3 \\[2mm] \dfrac{1}{6}; & c = 4 \end{cases}$

3   A fair ten-sided dice with faces numbered 1–10 is thrown once. Prove that the mean and variance of the score obtained are $\dfrac{11}{2}$ and $\dfrac{33}{4}$

4   Five cards are numbered from 1 to 5 Two cards are drawn at random with replacement and the total, T, noted. Find the expected value of T

5   A four-sided spinner with sides numbered 1–4 is biased so that odd numbered outcomes are each twice as likely as even numbered outcomes. Find the expected value and variance of the score on a randomly chosen spin.

6   A spinner can land on numbers 1–9 with equal probability. Show that the expected score for one throw is 5 and find the variance of this score.

7   A fair dice is thrown once. Find the mean value of the **square** of the score.

8 A six-sided dice with faces numbered 1–6 is biased so that the probability of any score is proportional to the square of the score. Find the expected value of S, the score on a randomly chosen throw.

9 A fair dice is thrown once. If R is the reciprocal of the score, find the mean value of R

10 X is a random variable for the score on a biased tetrahedral dice. Its probability distribution is

x	1	2	3	4
$P\{X=x\}$	0.4	0.2	0.2	0.2

  a Find the mean and variance.

  b Find the mean and variance of the random variable Y, where $Y = 2X^2$

11 A number is chosen at random from the integers 1 to 30

  a Find the expected number of prime numbers.

  b Explain what this means for a set of repeated trials of this experiment.

12 A fair coin is tossed until a head or four tails appear.

  a Give the probability distribution of X, the number of throws required.

  b What is the expected number of throws?

13 The number of made-to-measure suits ordered per month from a tailor has the following discrete probability distribution.

x	0	1	2	3
$P\{X=x\}$	0.18	0.39	0.31	0.12

Find the mode and the median of X

14 A discrete random variable X has probability distribution function given by

$$P\{X=x\} = \begin{cases} \dfrac{x}{12}; & x = 1, 2, 3 \\ k; & x = 4, 5, 6 \\ 0; & \text{otherwise} \end{cases}$$

  a Find the value of k

  b Write down the modal value of X

15 Prove that a discrete uniform random variable, U, with distribution function $P\{U=u\} = \dfrac{1}{n}$; $u = 1, 2, \dots, n$ has a mean and variance given by $\mu = \dfrac{n+1}{2}$ and $\sigma^2 = \dfrac{n^2-1}{12}$

Reasoning and problem-solving

Strategy

To solve a problem involving expectation

① Find the probability distribution of the random variable, either in the form of a function or as a table.

② Use standard formulae to find the mean and variance of the random variable.

③ Where necessary, find the expectation value of a function of X, $g(X)$.

④ Give your conclusion.

You will sometimes need to calculate the expectation of linear functions of random variables.

Example 7

An ordinary six-sided dice is thrown once. X is a random variable for the score obtained. The random variable Y is defined by the equation $Y = 2X - 3$. By finding the possible values of Y, show that

a $E[Y] = 2E[X] - 3$ **b** $\text{Var}[Y] = 4\text{Var}[X]$

(*Continued on the next page*)

a Since X takes values 1–6, Y takes values $-1, 1, 3, 5, 7, 9$

x	1	2	3	4	5	6
y	-1	1	3	5	7	9
P	$\dfrac{1}{6}$	$\dfrac{1}{6}$	$\dfrac{1}{6}$	$\dfrac{1}{6}$	$\dfrac{1}{6}$	$\dfrac{1}{6}$

1 Find the probability distribution of the random variable in the form of a table.

$$E[X] = 1 \times \frac{1}{6} + 2 \times \frac{1}{6} + 3 \times \frac{1}{6} + 4 \times \frac{1}{6} + 5 \times \frac{1}{6} + 6 \times \frac{1}{6}$$

$$= 3.5$$

2 Use standard formulae to find the mean of the random variable.

$$2E[X] - 3 = 2 \times 3.5 - 3$$

$$= 4$$

$$E[Y] = -1 \times \frac{1}{6} + 1 \times \frac{1}{6} + 3 \times \frac{1}{6} + 5 \times \frac{1}{6} + 7 \times \frac{1}{6} + 9 \times \frac{1}{6}$$

$$= 4$$

Therefore $E[Y] = 2E[X] - 3$, as required.

4 Give your conclusion.

b $Var(Y) = E[Y^2] - \{E(Y)^2\}$

$$= (-1)^2 \times \frac{1}{6} + 1^2 \times \frac{1}{6} + 3^2 \times \frac{1}{6} + 5^2 \times \frac{1}{6} + 7^2 \times \frac{1}{6} + 9^2 \times \frac{1}{6} - 4^2$$

$$= \frac{35}{3}$$

2 Use standard formulae to find the variance of the random variable.

$Var(X) = E[X^2] - \{E(Y)^2\}$

$$= 1^2 \times \frac{1}{6} + 2^2 \times \frac{1}{6} + 3^2 \times \frac{1}{6} + 4^2 \times \frac{1}{6} + 5^2 \times \frac{1}{6} + 6^2 \times \frac{1}{6} - 3.5^2$$

$$= \frac{35}{12}$$

Therefore $Var[Y] = 4Var[X]$, as required.

4 Give your conclusion.

This example illustrates a general result.

Key point

If X and Y are two discrete random variables and $Y = aX + b$, where a and b are constants, then

$E(Y) = aE(X) + b$ and $Var(Y) = a^2Var(X)$

You can also write this as
$E(aX + b) = aE(X) + b$
and $Var(aX + b) = a^2Var(X)$

Example 8

A random variable, S, has probability distribution

s	1	2	3
$P\{S = s\}$	$\dfrac{1}{2}$	a	$2a$

where a is a positive constant.

a Find the value of a

The variable T is related to S by the equation $T = -3S + 6$

b Find the mean and variance of S and hence the mean and variance of T

(Continued on the next page)

a $3a + \dfrac{1}{2} = 1 \Rightarrow a = \dfrac{1}{6}$

① Find the probability distribution.

b $E(S) = 1 \times \dfrac{1}{2} + 2 \times \dfrac{1}{6} + 3 \times \dfrac{1}{3}$

$\qquad = \dfrac{11}{6}$

② Use standard formulae to find the mean of the random variable.

$Var(S) = 1^2 \times \dfrac{1}{2} + 2^2 \times \dfrac{1}{6} + 3^2 \times \dfrac{1}{3} - \left(\dfrac{11}{6}\right)^2$

$\qquad = \dfrac{29}{36}$

② Use standard formulae to find the variance of the random variable.

$T = -3S + 6$

$E(T) = -3E(S) + 6$

$\qquad = -3 \times \dfrac{11}{6} + 6$

$\qquad = \dfrac{1}{2}$

$Var(T) = 9Var(S)$

$\qquad = 9 \times \dfrac{29}{36}$

$\qquad = \dfrac{29}{4}$

Example 9

Two independent random variables, R and S, have distributions as follows.

$$f(r) = \begin{cases} k(r+1); & 0 < r < 1 \\ 0; & \text{otherwise} \end{cases}$$

$$f(r) = \begin{cases} ms^2; & 1 < s < 2 \\ 0; & \text{otherwise} \end{cases}$$

where k and m are constants.

a Find the values of k and m

b Find the expected value and variance of the function $2R + 3S$

① Use the 'area equals 1' property to find any unknown constants.

a $k\displaystyle\int_0^1 (r+1)\,dr = 1 \rightarrow k = \dfrac{2}{3}; \quad m\displaystyle\int_1^2 s^2\,ds = 1 \rightarrow m = \dfrac{3}{7}$

b $E(2R + 3S) = E(2R) + E(3S) = 2E(R) + 3E(S)$

② $2R$ and $3S$ are independent random variables.

$\qquad = 2 \times \dfrac{2}{3} \times \displaystyle\int_0^1 r(r+1)\,dr + 3 \times \dfrac{3}{7} \times \int_1^2 s^3\,ds = \dfrac{10}{9} + \dfrac{135}{28}$

$\qquad = 5.93\,(2dp)$

$Var(2R + 3S) = Var(2R) + Var(3S) = 4Var(R) + 9Var(S)$

$\qquad = 4 \times \left(\dfrac{2}{3} \times \displaystyle\int_0^1 r^2(r+1)\,dr - \left(\dfrac{5}{9}\right)^2\right) + 9 \times \left(\dfrac{3}{7} \times \int_1^2 s^4\,ds - \left(\dfrac{45}{28}\right)^2\right)$

④ Find mean and variance of the functions of random variables.

$\qquad = \dfrac{26}{81} + \dfrac{2619}{3920} = 0.99\,(2dp)$

It is possible to find the mean and variance of sums of independent random variables.

For two discrete random variables, X and Y, the expected value of the sum, $X + Y$, is given by $E(X + Y) = E(X) + E(Y)$

This simple result is surprisingly difficult to prove but the following example makes it plausible.

Example 10

The following table gives the bivariate probability distribution for the variables X and Y

		Y			
		1	**2**	**3**	**P{X=x}**
X	**1**	0.15	0.1	0.05	0.3
	2	0.2	0.1	0.4	0.7
	P{Y=y}	0.35	0.2	0.45	1

For example, $P\{X = 1 \text{ and } Y = 1\} = 0.15$

a Write down the probability distribution of the random variable T given by $T = X + Y$

b By finding the expected values of the three variables, show that $E[T] = E[X] + E[Y]$

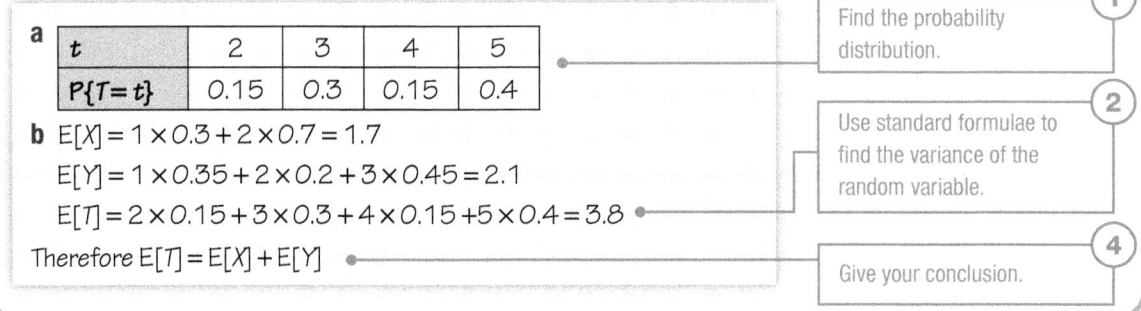

a

t	2	3	4	5
P{T=t}	0.15	0.3	0.15	0.4

b $E[X] = 1 \times 0.3 + 2 \times 0.7 = 1.7$

$E[Y] = 1 \times 0.35 + 2 \times 0.2 + 3 \times 0.45 = 2.1$

$E[T] = 2 \times 0.15 + 3 \times 0.3 + 4 \times 0.15 + 5 \times 0.4 = 3.8$

Therefore $E[T] = E[X] + E[Y]$

① Find the probability distribution.

② Use standard formulae to find the variance of the random variable.

④ Give your conclusion.

A similar result applies to the variance of sums of independent random variables.

For two independent random variables, X and Y, the variance of the sum, $X + Y$, is given by $\text{Var}(X + Y) = \text{Var}(X) + \text{Var}(Y)$

The proof of this result is found in question **10** of Exercise 10.1B

Example 11

West End FC Under 11s football team played a set of home and away games. During home games, the number of goals scored had a mean of 3.1 and a standard deviation of 1.2

During away games, these statistics were 2.7 and 1.4 respectively. Find the mean and standard deviation of the total number of goals scored. State any necessary assumptions made, indicating to which part of the solution the assumption relates.

Let X and Y be random variables for the numbers of goals scored during home and away games respectively.

$E(X + Y) = E(X) + E(Y) = 3.1 + 2.7 = 5.8$

$\text{Var}(X + Y) = \text{Var}(X) + \text{Var}(Y) = 1.2^2 + 1.4^2 = 3.4$

Standard deviation of $X + Y = 1.84$ (2dp).

For the variance result, assume that X and Y are independent.

1 A random variable X has this probability distribution.

x	1	2	3	4	5	6
$P\{X=x\}$	$\dfrac{1}{6}$	$\dfrac{1}{6}$	$\dfrac{1}{6}$	$\dfrac{1}{6}$	a	b

Given that the mean of X is $\dfrac{52}{15}$, find the values of a and b

2 R is a random variable with this distribution.

r	−3	−2	−1	0	1	2	3
$P\{R=r\}$	$\dfrac{1}{7}$	$\dfrac{1}{7}$	$\dfrac{1}{7}$	$\dfrac{1}{7}$	$\dfrac{1}{7}$	$\dfrac{1}{7}$	$\dfrac{1}{7}$

A single value of R is chosen at random.

a Find the mean and standard deviation of R

b What is the probability that it will be one standard deviation or closer to the mean?

3 The proportion of boys born in the UK is 0.51 of all live births. Four UK births are selected at random.

a Write down all possible permutations of gender and hence calculate the probability distribution for X, a random variable for the number of girls born in the four births.

b Find the mean and variance of X

4 Jenny and John play a dice game. John says that he will get at least one six in four throws of a fair dice and will pay Jenny £1 if he fails.

a What is the probability of John winning any given game?

b How much should Jenny pay to enter the game so that in the long run neither player can expect to win anything? Give your answer to the nearest 1p.

5 A bag contains 1 red and 2 black balls. A ball is drawn from the bag and then, without the first ball being replaced, a second ball is drawn. By first finding the probability distribution, find the expected value of B, the number of black balls drawn.

6 **a** Find the probability of getting a double six in 24 throws of a pair of fair dice.

b In a dice game, you throw the pair of dice 3 times and receive 10 counters for every double six you throw. How many counters should you pay to enter the game if you want to break even in the long run? Give your answer to the nearest integer.

7 A random variable X has the probability distribution function

$$P\{X=x\}=\begin{cases} \dfrac{1}{5n}; & x=1,2,3\ldots,5n \\ 0; & \text{otherwise} \end{cases}$$

where n is a positive integer.

a Prove that $E[X]=\dfrac{(1+5n)}{2}$

b Find in terms of n an expression for the variance of X

You may assume that

$$\sum_{i=1}^{n} i = \frac{n(n+1)}{2}$$

and that

$$\sum_{i=1}^{n} i^2 = \frac{n}{6}(n+1)(2n+1)$$

8 A circle of radius r is inscribed in a square of side $2r$

Three darts are randomly dropped onto the square.

D is a random variable for the number of darts that land within the circle.

a Find $E[D]$

(Assume that the probability of a dart landing in the circle is equal to $\dfrac{\text{area of circle}}{\text{area of square}}$)

b One dart is dropped randomly onto the square 100 times and on 71 occasions the dart lands within the circle. Estimate the value of π

9 X and Y are two independent random variables. Given that $E[X]=3.2$, $Var[X]=12.1$, $E[Y]=22$ and $Var[Y]=9$, find the mean and standard deviation of $X+Y$

10 Prove that, for two independent random variables, $Var(X+Y)=Var(X)+Var(Y)$. You may assume that for independent random variables, $E[XY]=E[X]E[Y]$

11 a A random variable X has mean μ and variance σ^2

Write down the mean and variance of

i aX

ii $bX-c$

where a, b and c are constants.

b In a dice game, you throw an ordinary, fair dice once and receive in tokens five times the score obtained. You pay 20 tokens to enter the game. Find the mean and variance of your net winnings (after allowing for the entry cost).

12 A random variable, S, has probability distribution

s	1	2	3
$P\{S=s\}$	$\dfrac{3}{8}$	$2k$	k

where k is a positive constant.

a Find the value of k

The variable R is related to S by the equation $R = 4S - 3$

b Find the mean and variance of S and hence the mean and variance of R

13 A random variable X has mean 4, variance 8 and an independent variable Y has mean 5, variance 12

Write down the mean and variance of the variables S and T, where $S = X + Y$ and $T = X - Y$

14 A fair dice with faces numbered 2, 4, 6, 8, 10 and 12 is thrown once and a fair eight-sided spinner with sides numbered 1–8 is spun once. If X and Y are random variables for the scores on the dice and the spinner respectively, find the expected value and variance of $X + Y$

15 An ordinary pack of cards has the jacks, queens and kings removed. Two cards are chosen at random, with replacement, from the remaining cards. Find the mean and variance of the sum of the scores.

10.2 The Poisson distribution

Fluency and skills

The Poisson distribution can be used to model the occurrence of certain types of random events in time or space. Here, the word 'random' is used in a very particular way. In this context, events are random if the occurrence of an event at a particular point in time or space is independent of occurrences elsewhere. Further, the probability of an event happening in a small interval of a given size is the same at all times or positions in space and events can't occur simultaneously.

Key point

If X is the number of random events that occur in a given interval of time or space and λ is the mean number of occurrences in that interval, then its probability distribution function is given by

$$P(X=x) = \frac{e^{-\lambda}\lambda^x}{x!}; x = 0, 1, 2,...$$

X is said to follow a Poisson distribution, parameter λ. For a Poisson distribution, the mean = variance.

Some examples of random variables that may follow a Poisson distribution are the number of radioactive disintegrations in five-second intervals, the number of stars of a given type in a given volume of space, the number of telephone calls you receive in a one-hour interval of time. It is worth noting that although the Poisson distribution function gives non-zero probabilities for outcomes 0, 1, 2..., (that is, there is no upper limit on the value of a Poisson variable), in practice, there will almost always be an upper limit on the variable.

> Random events in continuous time or space constitute a **Poisson process**.

Example 1

A Geiger counter detects radioactive disintegrations at a mean rate of 23 per minute. Use the Poisson distribution function with $\lambda = 23$ to find the probability that, in a randomly chosen minute, there will be

a 20 disintegrations,

b Between 21 and 23 inclusive disintegrations.

Check your answers using the Poisson function of your calculator.

> Let X be a random variable for the number of disintegrations in a randomly chosen minute.

a $P(X=20) = \dfrac{e^{-23}\lambda^{20}}{20!}$

$= 0.072 \ (3 \ dp)$

> Substitute $\lambda = 23$ and $x = 20$ into $P(X = x) = \dfrac{e^{-\lambda}\lambda^x}{x!}$

b $P(21 \le X \le 23) = \displaystyle\sum_{x=21}^{23} \dfrac{e^{-23}23^x}{x!}$

$= \dfrac{e^{-23}23^{21}}{21!} + \dfrac{e^{-23}23^{22}}{22!} + \dfrac{e^{-23}23^{23}}{23!}$

$= 0.079 + 0.083 + 0.083$

$= 0.245 \ (3 \ dp)$

> Substitute $x = 21$, $x = 22$ and $x = 23$, into $P(X = x) = \dfrac{e^{-23}23^x}{x!}$

> Use your calculator to check your answers.

Example 2

A random variable X has a Poisson distribution with parameter $\lambda = 2$

a Copy the following table and use the probability distribution function of X to complete it.

x	0	1	2	3	4	5	6	>6
$P\{X=x\}$	0.135	0.271					0.012	0.005

b Use the values $x = 0, 1, 2, ..., 6$ to estimate the mean and variance of X

a

x	0	1	2	3	4	5	6	>6
$P\{X=x\}$	0.135	0.271	0.271	0.180	0.090	0.036	0.012	0.005

b $E(X) \displaystyle\sum_{x=0}^{6} x P\{X=x\}$

$= 0 \times 0.135 + 1 \times 0.271 + 2 \times 0.271 + 3 \times 0.18 + 4 \times 0.09 + 5 \times 0.036$
$+ 6 \times 0.012$

$= 1.965$

$Var(X) \displaystyle\sum_{x=0}^{6} x^2 P\{X=x\} - \mu^2$

$= 0^2 \times 0.135 + 1^2 \times 0.271 + 2^2 \times 0.271 + 3^2 \times 0.18 + 4^2 \times 0.09$
$+ 5^2 \times 0.036 + 6^2 \times 0.012 - 1.965^2$

$= 1.886$

You can use the Poisson distribution function to calculate exact values of the mean and variance of the distribution.

Key point

If X has a probability distribution function given by

$$P(X=x) = \frac{e^{-\lambda} \lambda^x}{x!}; \quad x = 0, 1, 2, ...$$

then the mean and variance of X are, respectively, given by

$$E\{X\} = \sum_{x=0}^{\infty} x P\{X = x\} = \lambda$$
$$Var\{X\} = \sum_{x=0}^{\infty} x^2 P\{X = x\} - \lambda^2 = \lambda$$

Example 3

a Write down the mean and variance of a random variable X with a Poisson distribution, $\lambda = 2$

b Find the probability that a randomly chosen value of X will be within 1 standard deviation of the mean.

a $X \sim Po(\lambda = 2)$

mean $= 2$, variance $= 2$

b Standard deviation $\sqrt{2} = 1.414$ (3 dp).

Standard deviation is the square root of the variance.

$P(2 - 1.414 < X < 2 + 1.414) = P(0.586 < X < 3.414)$

$= P(X = 1 \text{ or } 2 \text{ or } 3)$

$= 0.271 + 0.271 + 0.180$

$= 0.722$

Compare your answers to part **a** with the estimates found in Example 2 above.

The key point above shows that for a Poisson random variable, mean equals variance. This result can be used to investigate whether or not data is likely to come from a Poisson distributed population.

Example 4

The following famous data gives information about the random variable X, the number of deaths from horse kicks for 10 units of the Prussian Army over 20 years (that is, 200 observations) in the last decades of the nineteenth century.

Number of deaths, X	0	1	2	3	4
Frequency	109	65	22	3	1

By finding the mean and variance of X, decide whether this data is likely to follow a Poisson distribution.

Mean $= 0.610$ Variance $= 0.608$ (3dp)

Use a calculator to find these values.

Mean approximately equal to variance $\rightarrow$ data likely to be Poisson distributed.

Exercise 10.2A Fluency and skills

1   A random variable X has a Poisson distribution, mean 4. Find the probability that X takes values 0, 1, 2, 3, > 3

2   Random events occur at a rate of three per ten seconds. Find the probability that there will be at least two events in any randomly chosen ten-second interval.

3   A random variable R is Poisson distributed, parameter 4

Find the probability that R takes values 0 or greater than 4

4   A random variable X has a Poisson distribution with parameter $\lambda = 1.1$

   a   Copy this table and use the probability distribution function of X to complete it.

x	0	1	2	3	4	>4
$P\{X=x\}$	0.333	0.366				

   b   Sketch the probability distribution function of X for $x = 0, 1, 2, 3, 4$

5   Random events occur at an average rate of 10 per 12 minutes. Find the probability that there will be less than three events in any randomly chosen 12-minute interval.

6   Radioactive decays occur at a rate of 34 per 12-second interval. Find the probability that, in two randomly chosen 12-second intervals, there will be at least 30 disintegrations in exactly one of the intervals.

7   A household receives on average four telephone calls per day. What is the probability that, in a randomly chosen week (of seven days), there is at least one call every day? State any assumptions you make about the appropriate probability distribution.

8   Emails arrive at your account randomly at a rate of four per two hours. Find the probability that, in a randomly chosen two-hour period, you will receive at least three emails.

9   There are on average four hazelnuts in a 200 g chocolate bar. What is the probability that, in five randomly chosen bars, there is at least one hazelnut in each bar? State any distributional or other assumptions you make.

10 A fabric-weaving machine produces cloth with flaws, which occur at a rate of 3.4 per square metre. A customer orders a $2\,m^2$ piece of cloth and insists that it must be flawless. Find the probability that the first piece of cloth produced for the customer meets her requirements.

11 Cars pass a traffic checkpoint at a rate of two per hour. Find the probability that, in two consecutive hours, there is a total of four cars passing. You should treat the two one-hour intervals separately.

12 The number of cars passing a rural traffic survey station in 200 four-minute intervals is shown in the table below.

Number of cars, X	0	1	2	3	4	5	6	7	8
Frequency	24	49	52	34	19	11	7	3	1

By finding the mean and variance of X, decide whether the cars are likely to be passing the station randomly.

Reasoning and problem-solving

Strategy

To solve a problem involving the Poisson distribution

(1) Ensure that the conditions for a Poisson distribution apply to the problem.

(2) Use the probability distribution function to define the required probabilities.

(3) Use a calculator, statistical tables or the Poisson distribution function to find probabilities.

(4) Where required, use standard results to find the mean and variance of the distribution.

(5) Give your conclusion.

Example 5

Events occur randomly at a mean rate of μ per hour. Find, in terms of μ, the probability that three events occur in a randomly chosen **two** hours.

$X_T = X_1 + X_2$

$P\{X_T = 3\} = P\{X_1 = 3, X_2 = 0 \text{ or } X_1 = 2, X_2 = 1 \text{ or } X_1 = 1, X_2 = 2 \text{ or } X_1 = 0, X_2 = 3\}$

$P\{X_T = 3\} = \dfrac{e^{-\mu}\mu^3}{3!} \times \dfrac{e^{-\mu}\mu^0}{0!} + \dfrac{e^{-\mu}\mu^2}{2!} \times \dfrac{e^{-\mu}\mu^1}{1!} + \dfrac{e^{-\mu}\mu^1}{1!} \times \dfrac{e^{-\mu}\mu^2}{2!} + \dfrac{e^{-\mu}\mu^0}{0!} \times \dfrac{e^{-\mu}\mu^3}{3!}$

$= \dfrac{e^{-2\mu}\mu^3}{3!} + \dfrac{e^{-2\mu}\mu^3}{2!} + \dfrac{e^{-2\mu}\mu^3}{2!} + \dfrac{e^{-2\mu}\mu^3}{3!}$

$= e^{-2\mu}\mu^3 \left(\dfrac{1}{3!} + \dfrac{1}{2!} + \dfrac{1}{2!} + \dfrac{1}{3!} \right)$

$= e^{-2\mu}\mu^3 \times \dfrac{4}{3}$

$= \dfrac{e^{-2\mu}(2\mu)^3}{3!}$

Conditions for a Poisson distribution apply to the problem. (1)

X_1 is the number of events in the first hour, X_2 is the number of events in the second hour and X_T is the total number of events in the 2 hours.

A total of 3 events in 2 hours means 3 in the first and 0 in the second or 2 in the first and 1 in the second...

Use the probability distribution function to define the required probabilities. (2)

Note that this is the probability distribution function of a Poisson distribution, mean 2μ

This suggests the following result:

X_i; $i = 1, 2,..., n$ is a set of n independent and identically distributed random variables each with a Poisson distribution, mean λ. If $X_T = \sum_{i=1}^{n} X_i$, then X_T has a Poisson distribution with mean $n\lambda$

In n equal intervals of time or space, the total number of random events occurring equals the sum of the number of events in each of the separate intervals. Hence, you can apply the above result.

If events occur randomly at a rate of λ per given interval of time or space, then the number of events in n such intervals is a Poisson random variable with mean $n\lambda$

Example 6

Cars pass a traffic checkpoint at a rate of two per hour.
a Explain why it is necessary to assume that the checkpoint is, for example, on a quiet rural road or a motorway with free overtaking, in order to use a Poisson distribution.
b Assuming that the conditions in part **a** are met, find the probability, that in a two-hour period, a total of more than two cars pass.

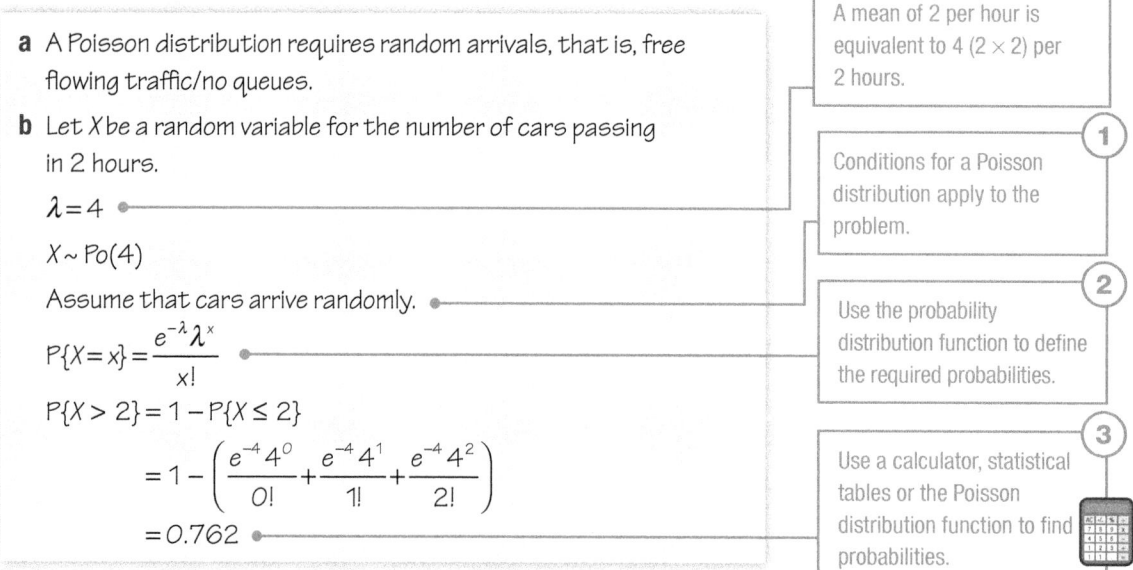

a A Poisson distribution requires random arrivals, that is, free flowing traffic/no queues.

b Let X be a random variable for the number of cars passing in 2 hours.

$\lambda = 4$

$X \sim Po(4)$

Assume that cars arrive randomly.

$P\{X = x\} = \dfrac{e^{-\lambda} \lambda^x}{x!}$

$P\{X > 2\} = 1 - P\{X \le 2\}$

$= 1 - \left(\dfrac{e^{-4} 4^0}{0!} + \dfrac{e^{-4} 4^1}{1!} + \dfrac{e^{-4} 4^2}{2!} \right)$

$= 0.762$

A mean of 2 per hour is equivalent to 4 (2×2) per 2 hours.

① Conditions for a Poisson distribution apply to the problem.

② Use the probability distribution function to define the required probabilities.

③ Use a calculator, statistical tables or the Poisson distribution function to find probabilities.

Suppose that a number of identical red objects are scattered randomly over a large area A with an average density of r objects per unit area. Let X be a random variable for the number of these objects in a randomly chosen unit area. Suppose, further, that green objects are scattered in a similar way over the same area with an average density of s objects per unit area and with Y objects in a randomly chosen unit area. From the theory in the previous section, X and Y will each have Poisson distributions with parameters r and s respectively.

Now, considering all objects as a single collection, these will be randomly scattered over the area with a mean number per unit area of $r + s$

That is, the total number of objects in a randomly chosen unit area, $X + Y$, will follow a Poisson distribution, parameter $r + s$

> **Key point**
>
> If X and Y are two independent random variables that can be modelled using the Poisson distribution with distributions given by $X \sim \text{Po}(\lambda_x)$ and $Y \sim \text{Po}(\lambda_y)$, then $X + Y \sim \text{Po}(\lambda_x + \lambda_x)$

Example 7

The number of cars passing a point on a quiet rural road between 7am and 8am is given by the random variable X

The random variable Y gives the same statistic for the interval between 8am and 9am.

a Explain why the random variables could each be modelled by the Poisson distribution. The means of X and Y are 21 and 34 respectively.

b Assuming that the distribution in part **a** provides a good model, find the probability that the total number of cars passing the point between 7am and 9am is between 50 and 60 inclusive.

a On a quiet road, cars are likely to arrive randomly.

b $X \sim \text{Po}(21) \ Y \sim \text{Po}(32)$

T is the total number of cars passing the point between 7 am and 9am

$T = X + Y$

$T \sim \text{Po}(53)$

$P\{50 \leq T \leq 60\} = P\{T \leq 60\} - P\{T < 50\} = 0.848 - 0.322 = 0.53 \ (2dp)$

Exercise 10.2B Reasoning and problem-solving

1   Random events occur at a rate of three per ten seconds. Find the probability that there will be at most 33 events in any randomly chosen two-minute interval.

2   Cars pass a traffic checkpoint at a rate of two per minute. Find the probability that, in two consecutive minutes, a total of four cars pass.

3   Shoppers arrive at a supermarket at an average rate of seven every minute. Find the probability that, in a randomly chosen one-hour interval, at least 400 shoppers will arrive.

4   At a certain set of traffic lights there is on average one road traffic accident (RTA) every nine days. Find the probability that, in a randomly chosen month (of 31 days), there is at most one accident.

5 In the radioactive decay of Americium-241, alpha particles are emitted at a rate of 20 per ten-second interval. Find the probability that, in a randomly chosen one-minute interval, there will be at least 100 particles emitted.

6 Strontium-90 emits on average 100 beta particles in 10 seconds. Find the probability that, in a randomly chosen 90-second interval, there will be no fewer than 910 particles emitted.

7 I receive on average 14 emails per day. Assuming that their arrival constitutes a random process, what is the probability that in a randomly chosen seven-day period

    a I receive at least twelve emails every day,

    b I receive a total of at least 84 emails?

8 At a weather station during light rain, raindrops land on a one square metre detector plate at a rate of 25 per minute. Assuming that their arrival constitutes a random process, what is the probability that, in a randomly chosen ten-minute period, there will be no rain detected on a $10\,\text{cm} \times 20\,\text{cm}$ plate. State any assumptions you make.

9 Silver chain is made by linking circles of silver. Quality management rules require that the first three 5 m lengths of chain produced in a day should have at most one flaw each, otherwise the machine must be adjusted. Find the probability that no adjustment is necessary if imperfections in the chain occur at a rate of 3.2 per ten metres.

10 Suppose that there exist two types of star, A and B. The number of each type in a given volume of space follows a Poisson distribution with means 1.8 and 2.1 respectively. Assuming that these random variables are independent, find the probability that the volume contains a total of three stars.

11 Cars at a given point on the M62 motorway early in the morning travel either towards or away from Leeds with mean rates of 123 and 111 per 10 minutes respectively. Assuming that traffic is flowing freely, find the probability that the total number of cars passing the point in ten minutes is over 250

Fluency and skills

You know that, for discrete random variables, you give a probability distribution function by stating the probabilities of each value the variable can take. This is not possible for continuous variables because you cannot list all the individual values. In the same way as you use histograms instead of bar charts for continuous variables, probability density functions replace probability distribution functions for continuous variables.

You use a **probability density function** (pdf) of a random variable, X, to find the probability that a randomly chosen x-value will be between two values. To do this, you find the area under the function between these values. You can do this by integrating.

Key point

If X is a random variable with probability density function f(x), then $P\{a < X < b\} = \int_{x=a}^{x=b} f(x)\mathrm{d}x$

This is similar to histograms where area gives frequency.

Since the area under the probability density function gives a probability, it follows that the probability of X taking any exact value is zero; that is, $P\{X = x\} = 0$

Probability density functions must obey two properties.

- $\int_{-\infty}^{+\infty} f(x)\mathrm{d}x = 1$ (total probability is represented by total area).
- $f(x) \geq 0$ (all probabilities are greater than or equal to zero).

Example 1

A random variable, X, has a probability density function given by

$f(x) = \dfrac{3}{8}x^2$ for $0 < x < 2$

a Sketch the curve $y = f(x)$ and verify that f(x) is a valid pdf. On your diagram, shade the area representing $P\{X > 1\}$

b Find the value of $P\{X > 1\}$

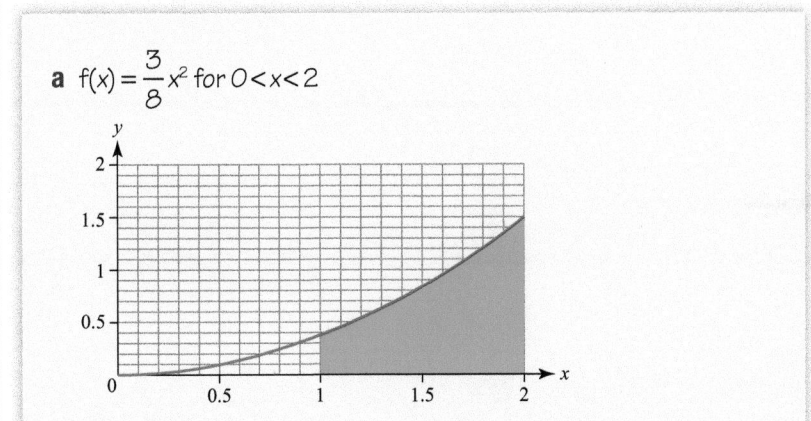

a $f(x) = \dfrac{3}{8}x^2$ for $0 < x < 2$

(*Continued on the next page*)

$f(x) > 0$ for $0 < x < 2$

All probabilities are greater than zero.

$$\int_0^2 \frac{3}{8}x^2\,dx = \left[\frac{x^3}{8}\right]_0^2$$

$$= 1 - 0$$

$$= 1$$

Total probability is represented by total area.

So $f(x)$ is a valid pdf.

b $P\{X > 1\} = \int_1^2 \frac{3}{8}x^2\,dx$

$$= \left[\frac{x^3}{8}\right]_1^2$$

$$= 1 - \frac{1}{8}$$

$$= \frac{7}{8}$$

Example 2

A random variable, X, has probability density function

$$f(x) = \begin{cases} -kx(x-4); & 0 < x < 4 \\ 0; & \text{otherwise} \end{cases}$$

where k is some positive constant.

Find the value of k

$$\int_0^4 -kx(x-4)\,dx = 1$$

For $f(x)$ to be a pdf, the total probability rule must apply.

$$\int_0^4 (-kx^2 + 4kx)\,dx = 1$$

Integrate.

$$\left[-k\frac{x^3}{3} + 2kx^2\right]_0^4 = 1$$

$$-\frac{64}{3}k + 32k = 1$$

$$\frac{32}{3}k = 1$$

$$k = \frac{3}{32}$$

For a continuous random variable, the median is the value that divides the total probability, and therefore the area under the corresponding pdf, into two halves.

Key point

If X is a continuous random variable with probability density function f(x), then the median value of X, denoted by M, say, is given by the equation

$$\int_{-\infty}^{M} f(x)dx = \frac{1}{2}$$

$\int_{-\infty}^{M} f(x)dx$ gives the probability up to the median value and is therefore, by definition, equal to $\frac{1}{2}$

In practice, the lower limit of the integral will be the smallest value of x

Example 3

A random variable, X, has a probability density function given by

$$f(x) = \begin{cases} \dfrac{1}{2}x; \ 1 < x < \sqrt{5} \\ \\ 0; \ \text{otherwise} \end{cases}$$

Find the median of X

$$\int_{-\infty}^{M} f(x)dx = \frac{1}{2}$$

$$\int_{1}^{M} \frac{x}{2}dx = \frac{1}{2}$$

$$\left[\frac{x^2}{4}\right]_{1}^{M} = \frac{1}{2}$$

$$\frac{M^2}{4} - \frac{1}{4} = \frac{1}{2}$$

$$M = \sqrt{3}$$

You can also calculate the lower and upper quartiles from the probability density function. They are defined by the equations

$$P(X < Q_1) = \frac{1}{4} \text{ and } P(X < Q_3) = \frac{3}{4}$$

Key point

The lower quartile (Q_1) and upper quartile (Q_3) are given by

$$\int_{-\infty}^{Q_1} f(x)dx = \frac{1}{4}$$

$$\int_{-\infty}^{Q_3} f(x)dx = \frac{3}{4}$$

where f(x) is the pdf.

STATISTICS

The mode of a continuous random variable is the value of the variable at a local maximum. This can be found by considering a sketch of the probability density function or by differentiation.

For a continuous random variable with probability density function f(x), the mode can be found from the equation

$$\frac{\mathrm{df}(x)}{\mathrm{d}x} = 0$$

A random variable, X, has probability density function

$$f(x) = a(x^2 - x - 2); \quad -1 < x < 1$$

where a is a constant.

a Find the value of a

b By differentiating f(x), or otherwise, find the mode of X

$f(x) = a(x^2 - x - 2); \quad -1 < x < 1$

a $\displaystyle\int_{-1}^{1} a(x^2 - x - 2)\mathrm{d}x = 1$

$$\left\{ a\left(\frac{x^3}{3} - \frac{x^2}{2} - 2x \right) \right\}_{-1}^{1} = 1 \;\rightarrow\; -\frac{10a}{3} = 1$$

$$a = -\frac{3}{10}$$

b $\dfrac{\mathrm{df}(x)}{\mathrm{d}x} = 0$ at modal value

$$\left(-\frac{3}{10} \right)(2x - 1) = 0$$

$$x = \frac{1}{2}$$

> The modal value can also be found from the graph of f(x); the graph cuts the x-axis at -1 and 2 and the modal value is midway, $x = \dfrac{1}{2}$

You can often find results for continuous random variables by considering the discrete equivalents. So the formulae for the expected (mean) value and variance of continuous variables are found by first considering their discrete versions. Changing Σ to $\int$ and P{$X = x$} to f(x)dx give the following results.

If X is a random variable with mean μ and variance σ^2, **Key point**
then

$$\mu = \mathrm{E}(X) = \int_{-\infty}^{+\infty} x\mathrm{f}(x)\mathrm{d}x$$

$$\sigma^2 = \mathrm{Var}[X] = \mathrm{E}[(X-\mu)^2] = \int_{-\infty}^{\infty} (x-\mu)^2 \mathrm{f}(x)\mathrm{d}x$$

For ease of calculation, the variance formula can be written

$$\mathrm{Var}[X] = \mathrm{E}[X^2] - \mu^2 = \int_{-\infty}^{\infty} x^2\mathrm{f}(x)\mathrm{d}x - \mu^2$$

The continuous versions of E[X] and Var[X] are found from their discrete equivalents by changing the Σ-operators into integrals because integration is the continuous equivalent of the discrete summation, Σ

Example 5

A random variable, X, has a probability density function given by

$$\mathrm{f}(x) = \begin{cases} \dfrac{1}{2}x; & 1 < x < \sqrt{5} \\ \\ 0; & \text{otherwise} \end{cases}$$

Find the mean, μ, and variance of X

$$\mu = \int_{-\infty}^{+\infty} x\mathrm{f}(x)\,dx$$

$$= \int_{x=1}^{\sqrt{5}} \frac{x^2}{2}\,dx$$

Substitute in f(x) and the limits.

$$= \left[\frac{x^3}{6}\right]_1^{\sqrt{5}}$$

Integrate.

$$= \frac{\left(\sqrt{5}\right)^3}{6} - \frac{1^3}{6}$$

$$= 1.70$$

$$\sigma^2 = \int_{-\infty}^{\infty} x^2\mathrm{f}(x)\,dx - \mu^2$$

$$= \int_1^{\sqrt{5}} \frac{x^3}{2}\,dx - 1.70^2$$

Substitute in f(x), μ and the limits.

$$= \left[\frac{x^4}{8}\right]_1^{\sqrt{5}} - 1.70^2$$

Integrate.

$$= \frac{\left(\sqrt{5}\right)^4}{8} - \frac{1^4}{8} - 1.70^2$$

$$= 0.12$$

The mean value and variance of a function, g(X), of X can be found by replacing x by g(x)

If X is a random variable with probability density function f(x), then the mean value and variance of g(X) are given by

$$E(g(X)) = \int_{-\infty}^{\infty} g(x)f(x)dx$$

$$Var(g(X)) = E(\{g(X)\}^2) - \{E(g(X))\}^2$$
$$= \int_{-\infty}^{\infty} \{g(x)^2\}f(x)dx - \{E(g(X))\}^2$$

The above result can be used with the following for the expectation of sums of functions of random variables.

For a random variable X, the mean value of the function g(X) + h(X) is given by E(g(X) + h(X)) = E(g(X)) + E(h(X))

Example 6

A continuous random variable X has probability density function $f(x) = \dfrac{x+1}{6}$ for x between 1 and 3 Find the expected value of

a $2X^2$

b $X + \dfrac{15}{X^3}$

a $E(2X^2) = \int_1^3 2x^2 \left(\dfrac{x+1}{6}\right)dx$

$= \dfrac{1}{3}\int_1^3 (x^3 + x^2)dx$

$= \dfrac{1}{3}\left[\dfrac{x^4}{4} + \dfrac{x^3}{3}\right]_1^3$

$= 9.56 \ (2 \ dp)$

b $E\left(X + \dfrac{15}{X^3}\right) = \int_1^3 \left(x + \dfrac{15}{x^3}\right)\left(\dfrac{x+1}{6}\right)dx$

$= \dfrac{1}{6}\int_1^3 (x^2 + x + 15x^{-2} + 15x^{-3})dx$

$= \dfrac{1}{6}\left[\dfrac{x^3}{3} + \dfrac{x^2}{2} - 15x^{-1} - \dfrac{15}{2}x^{-2}\right]_1^3$

$= \dfrac{44}{9}$

Or use

$E\left(X + \dfrac{15}{X^3}\right) = E(X) + E\left(\dfrac{15}{X^3}\right)$

1 A random variable, X, has a probability density function given by

$$f(x) = \begin{cases} \dfrac{1}{2}x; & 1 < x < a \\ 0; & \text{otherwise} \end{cases}$$

   a Sketch a graph of the probability density function.

   b Prove that $a = \sqrt{5}$ and hence calculate $P(X > 1.5)$

2 A random variable, X, has a pdf given by

$$f(x) = \begin{cases} kx^2; & 0 < x < 2 \\ 0; & \text{otherwise} \end{cases}$$

   Find the value of k and $P\{X < 1\}$

3 A random variable, X, has probability density function

$$f(x) = \begin{cases} -ax(x-6); & 0 < x < 6 \\ 0; & \text{otherwise} \end{cases}$$

   where a is some positive constant.

   Find the value of a

4 A random variable, X, has probability density function

$$f(x) = \begin{cases} ax(2x-1); & 0 < x < \dfrac{1}{2} \\ 0; & \text{otherwise} \end{cases}$$

   where a is a constant.

   a Find the value of a

   b Calculate the values of

      i The median,

      ii The lower and upper quartiles,

      iii The interquartile range of X

5 **a** Sketch the probability density function of X, given by

$$f(x) = \begin{cases} a - |x|; & -a < x < a \\ 0; & \text{otherwise.} \end{cases}$$

   b Show that $a = 1$

   c Find the exact values of the lower and upper quartiles and the interquartile range.

6 The random variable Y has a pdf given by
$$f(y) = \dfrac{2y+1}{k} \text{ for } 0 < y < 4$$

   a Find the value of k

   b Show that the first quartile is 1.79 (3 sf) and find the second quartile.

   c Find $E(Y)$, $E(Y^2)$ and $\text{Var}(Y)$

7 A continuous random variable X has pdf given by

$$f(x) = \begin{cases} \dfrac{a(2x+1)}{4}; & 0 < x < 5 \\ 0; & \text{otherwise.} \end{cases}$$

   a Show that the value of a is $\dfrac{2}{15}$ and find the mean and variance of X

   b Show that in three randomly chosen x-values, the probability of only one being above 4 is $\dfrac{4}{9}$

8 A continuous random variable X has probability density function $f(x) = \dfrac{2x+1}{18}$ for x between 1 and 4 Find the expected value of

   a X^2

   b $X + \dfrac{1}{X^3}$

9 A random variable, X, has probability density function

$$f(x) = \begin{cases} ax(x-5); & 0 < x < 5 \\ 0; & \text{otherwise} \end{cases}$$

   where a is some positive constant.

   a Find the value of a and $P\{1 < X < 3\}$

   b Write down the value of $P\{X = 2\}$ and explain your answer.

   c By differentiation, show that the mode of X is 2.5

10 A random variable, X, has probability density function

$$f(x) = \begin{cases} x(x-4); & 4 < x < a \\ 0; & \text{otherwise} \end{cases}$$

where a is a constant.

a Find the value of a

b Calculate the median value of X

c Find the interquartile range of X

11 A random variable, X, has a probability density function given by

$$f(x) = \begin{cases} \dfrac{1}{4}x; & 0 < x < a \\ 0; & \text{otherwise} \end{cases}$$

a Sketch a graph of the probability density function.

b Prove that $a = 2\sqrt{2}$ and hence show that the median of X is 2

12 For every parcel handled by a delivery company, an estimated time of delivery is given. Among parcels which are delivered late, the number of hours late can be modelled by a random variable T with probability density function

$$f(t) = \begin{cases} \dfrac{t}{5}; & 0 < t < 1 \\ -\dfrac{1}{45}t + \dfrac{2}{9}; & 1 < t < 10 \\ 0; & \text{otherwise} \end{cases}$$

a Sketch the graph of $f(t)$

b Verify that $f(t)$ is a valid probability density function.

c Write down the mode of T

d Find the median of T

Reasoning and problem-solving

To solve a problem involving continuous random variables

(1) Use the 'area equals 1' property to find any unknown constants.

(2) Use integration to find probabilities and the cumulative distribution function.

(3) Calculate summary statistics for location and spread.

(4) Find mean and variance of functions of random variables.

You can find the expectation value and variance of a linear function of a random variable X from the probability density function. Suppose Y is a linear function of X, that is, $Y = aX + b$, where a and b are constants.

$$E(Y) = E(aX + b)$$

$$= \int_{-\infty}^{\infty} (ax + b)f(x)\,dx$$

$$= \int_{-\infty}^{\infty} (axf(x) + bf(x))\,dx$$

$$= a\int_{-\infty}^{\infty} xf(x)\,dx + b\int_{-\infty}^{\infty} f(x)\,dx$$

$$= aE(X) + b$$

since $E(X) = \int_{-\infty}^{\infty} xf(x)\,dx$ and $\int_{-\infty}^{\infty} f(x)\,dx = 1$

Discrete and continuous random variables Continuous distributions 1

$$\text{Var}[Y] = \text{Var}(aX + b)$$

$$= \text{E}\left((aX+b)^2\right) - \left(\text{E}(aX+b)\right)^2 \text{ (Definition of variance)}$$

$$= \text{E}\left((aX+b)^2\right) - \left(a\text{E}(X)+b\right)^2 \text{ (Using the result above)}$$

$$= \text{E}\left(a^2X^2 + 2abX + b^2\right) - \left((a\text{E}(X))^2 + 2ab\text{E}(X) + b^2\right)$$

$$= a^2\text{E}\left(X^2\right) + 2ab\text{E}(X) + b^2 - \left(a^2\left(\text{E}(X)\right)^2 + 2ab\text{E}(X) + b^2\right)$$

(Using the result for sums of functions)

$$= a^2\left(\text{E}\left(X^2\right) - \left(\text{E}(X)\right)^2\right)$$

$$= a^2\text{Var}(X)$$

Key point

If X and Y are two continuous random variables and $Y = aX + b$, where a and b are constants, then

$$\text{E}(Y) = a\text{E}(X) + b$$

$$\text{Var}(Y) = a^2\text{Var}(X)$$

These results can also be written as
$\text{E}(aX + b) = a\text{E}(X) + b$ and
$\text{Var}(aX + b) = a^2\text{Var}(X)$.

Example 7

A random variable, X, has a pdf given by $f(x) = \begin{cases} a(x-4); & 4 < x < 6 \\ 0; & \text{otherwise} \end{cases}$

a Find the value of a and the mean and variance of X

The random variable Y is related to X by the equation $Y = -2X + 1$

b Find the mean and variance of Y

a $f(x) = a(x - 4)$

$$\int_4^6 a(x-4)\,dx = 1$$

$$\left[a\left(\frac{x^2}{2} - 4x\right)\right]_4^6 = 1$$

$$a[(18 - 24) - (8 - 16)] = 1$$

$$2a = 1$$

$$a = \frac{1}{2}$$

①

Use the 'area equals 1' property to find any unknown constants.

(Continued on the next page)

Therefore $f(x) = \dfrac{1}{2}(x-4)$

$$E(X) = \int_4^6 \frac{1}{2}x(x-4)dx$$

$$= \left[\frac{1}{2}\left(\frac{x^3}{3} - 2x^2\right)\right]_4^6$$

$$= \left[\frac{x^3}{6} - x^2\right]_4^6$$

$$= (36-36) - \left(\frac{64}{6} - 16\right)$$

$$= \frac{16}{3}$$

$$Var(X) = \int_4^6 \frac{1}{2}x^2(x-4)dx - \left(\frac{16}{3}\right)^2$$

$$= \left[\frac{1}{2}\left(\frac{x^4}{4} - \frac{4x^3}{3}\right)\right]_4^6 - \left(\frac{16}{3}\right)^2$$

$$= \left[\frac{x^4}{8} - \frac{2x^3}{3}\right]_4^6 - \left(\frac{16}{3}\right)^2$$

$$= (162-144) - \left(32 - \frac{128}{3}\right) - \frac{256}{9}$$

$$= 18 + \frac{32}{3} - \frac{256}{9}$$

$$= \frac{2}{9}$$

3 Calculate summary statistics for spread.

$Y = -2X + 1$

$$E(Y) = -2E(X) + 1$$

$$= -2 \times \frac{16}{3} + 1$$

$$= -\frac{29}{3}$$

$$Var(Y) = 4Var(X)$$

$$= 4 \times \frac{2}{9}$$

$$= \frac{8}{9}$$

4 Find mean and variance of linear functions of random variables.

As with discrete variables, the expected (mean) value and variance of the sum of two independent continuous random variables can be found.

Key point

For two independent continuous random variables, X and Y, the expected value and variance of the sum, $X + Y$, is given by

$$E(X + Y) = E(X) + E(Y)$$

$$Var(X + Y) = Var(X) + Var(Y)$$

Example 8

Two independent random variables, R and S, have distributions as follows.

$$f(r) = \begin{cases} k(r+1); & 0 < r < 1 \\ 0; & \text{otherwise} \end{cases}$$

$$f(s) = \begin{cases} ms^2; & 1 < s < 2 \\ 0; & \text{otherwise} \end{cases}$$

where k and m are constants.

a Find the values of k and m

b Find the expected value and variance of the function $2R + 3S$

(Continued on the next page)

a $k\int_{0}^{1}(r+1)dr = 1 \rightarrow k = \frac{2}{3}; m\int_{1}^{2}s^2 ds = 1 \rightarrow m = \frac{3}{7}$

> Use the 'area equals 1' property to find any unknown constants.

b $E(2R+3S) = E(2R) + E(3S) = 2E(R) + 3E(S)$

> 2R and 3S are independent random variables.

$$= 2 \times \frac{2}{3} \times \int_{0}^{1} r(r+1)dr + 3 \times \frac{3}{7} \times \int_{1}^{2} s^3 ds$$

$$= \frac{10}{9} + \frac{135}{28} = 5.93 \,(2\,dp)$$

$Var(2R+3S) = Var(2R) + Var(3S) = 4Var(R) + 9Var(S) =$

$$= 4 \times \left(\frac{2}{3} \times \int_{0}^{1} r(r+1)dr - \left(\frac{5}{9}\right)^2 \right) + 9 \times \left(\frac{3}{7} \times \int_{1}^{2} s^4 ds - \left(\frac{45}{28}\right)^2 \right)$$

$$= \frac{26}{81} + \frac{2619}{3920} = 0.99 \,(2\,dp)$$

> Find mean and variance of the functions of random variables.

Exercise 10.3B Reasoning and problem-solving

1 A random variable, X, has a pdf given by

$$f(x) = \begin{cases} kx^2; & -1 < x < 3 \\ 0; & \text{otherwise} \end{cases}$$

a Find the value of k and the mean and variance of X

The random variable Y is related to X by the equation $Y = -2X + 1$

b Find the mean and variance of Y

2 In a maths exam, marks over several years were found to be modelled by a probability density function given by $f(x) = kx(60 - x)$ for $0 < x < 60$, where k is a positive constant.

a Find the value of k and $P\{X > 50\}$

One year the exam board decided to scale the marks according to the equation $y = 0.9x + 3$, where y is the scaled mark.

b What values do marks of 50 and 10 become after scaling?

c Calculate the change in the mean mark as a result of the scaling.

3 A random variable, X, has probability density function

$$f(x) = \begin{cases} ax(x - b); & 0 < x < 3 \\ 0; & \text{otherwise} \end{cases}$$

where a and b are constants.

a Given that the mean of X is $\frac{7}{4}$, find the constants a and b

b Show that the variance of X equals $\frac{43}{80}$

c If μ and σ are the mean and standard deviation of X, respectively, find the probability that a randomly chosen value of X will lie between $\mu + \sigma$ and $\mu - \sigma$. Give your answer to 2dp.

4 Demand for coal, in tonnes per week, from a wholesaler can be modelled by a probability density function

$$f(x) = \frac{3x(100 - x)}{500000} \text{ for } 0 < x < 100$$

Show that the median demand is 50 tonnes per week.

5 The median of a random variable with probability density function $f(x) = ax(2-x)$; $0 < x < 2$, $a > 0$ is M

  a Find the value of a

  b Show that M satisfies the equation $M^3 - 3M^2 + 2 = 0$

  c Hence show that $M = 1$

6 Two independent random variables, X and Y, have distributions as follows.

$$f(x) = \begin{cases} 2x; & 0 < x < 1 \\ 0; & \text{otherwise} \end{cases}$$

$$f(y) = \begin{cases} my^2; & 1 < y < 3 \\ 0; & \text{otherwise} \end{cases}$$

where m is a constant.

  a Find the values of m

  b Find the expected value and variance of the function $3X + 2Y$

7 A continuous random variable X has probability density function

$$f(x) = \begin{cases} \dfrac{1}{6}x^2 + 1; & 0 < x < a \\ 0; & \text{otherwise} \end{cases}$$

  a Find the value of a

  b Find the values of
  i $E(3X + 4)$, **ii** $Var(2X - 1)$

8 A random variable, X, has a probability density function given by

$$f(x) = \begin{cases} 2x^3; & 0 < x < a \\ 0; & \text{otherwise} \end{cases}$$

  a Show that $a = 2^{\frac{1}{4}}$

  b Find the mean and variance of X

  c Find the mean and variance of the function $2X - 3$

Give your answers to 2 decimal places.

9 Two continuous random variables, X and Y, have means and standard deviations given by μ_X and σ_X, μ_Y and σ_Y. Write down expressions for the mean and variance of $X + Y$, stating any assumptions you make.

10 Anouk travels to work by bicycle and train. The probability density function of X, the time spent on the train in minutes, is given by

$$f(x) = \begin{cases} a(x^2 - 12x); & 0 < x < 12 \\ 0; & \text{otherwise} \end{cases}$$

  a Find the value of a

  b Find the mean and variance of X

  The mean and standard deviation of the cycle ride are 7 minutes and 2 minutes respectively. The total journey time, T, is the sum of her train and cycle journey times plus 2 minutes spent walking.

  c Find the mean and variance of T

Chapter summary

- The expected value or mean value and variance of a discrete random variable are given by

$$\mu = \mathrm{E}[X] = \sum_{\text{all } x} x \mathrm{P}\{X = x\}$$

$$\sigma^2 = \mathrm{Var}(X) = \mathrm{E}[(X - \mu)^2] = \sum_{\text{all } x} (x - \mu)^2 \mathrm{P}\{X = x\} = \sum_{\text{all } x} x^2 \mathrm{P}\{X = x\} - \{\mathrm{E}(X)\}^2$$

- The expected value and variance of a function, $g(X)$, of a discrete random variable are given by

$$\mathrm{E}[g(X)] = \sum_{\text{all } x} g(x) \mathrm{P}\{X = x\}$$

$$\mathrm{Var}(g(X)) = \mathrm{E}(\{g(X)\}^2) - \{\mathrm{E}(g(X))\}^2 = \sum_{\text{all } x} \{g(x)^2\} \mathrm{P}\{X = x\} - \{\mathrm{E}(g(X))\}^2$$

- If X and Y are two random variables and $Y = aX + b$, where a and b are constants, then $\mathrm{E}(Y) = a\mathrm{E}(X) + b$ and $\mathrm{Var}(Y) = a^2 \mathrm{Var}(X)$

- If X and Y are two independent random variables, the expected value and variance of the sum, $X + Y$, are given by $\mathrm{E}(X + Y) = \mathrm{E}(X) + \mathrm{E}(Y)$ and $\mathrm{Var}(X + Y) = \mathrm{Var}(X) + \mathrm{Var}(Y)$ respectively.

- A discrete uniform random variable, U, taking values $u = 1, 2, \ldots, a$, has probability distribution given by $P\{U = u\} = \dfrac{1}{n}$; $u = 1, 2, \ldots, n$

- A discrete uniform random variable, U, with distribution function $P\{U = u\} = \dfrac{1}{n}$; $u = 1, 2, \ldots, n$ has a mean and variance given by $\mu = \dfrac{n+1}{2}$ and $\sigma^2 = \dfrac{n^2 - 1}{12}$

- If X is the number of random events that occur in a given interval of time or space and λ is the mean number of occurrences in that interval, then its probability distribution function is given by

$$P\{X = x\} = \frac{e^{-\lambda} \lambda^x}{x!}; x = 0, 1, 2\ldots$$

and X is said to follow a Poisson distribution, parameter λ.

- If X has a probability distribution function given by $P\{X = x\} = \dfrac{e^{-\lambda} \lambda^x}{x!}$; $x = 0, 1, 2\ldots,$

then the mean and variance of X are, respectively, given by $\mathrm{E}(X) = \lambda$ and $\mathrm{Var}(X) = \lambda$

- If X_i; $i = 1, 2, \ldots, n$ is a set of n independent and identically distributed random variables each with a Poisson distribution, mean λ and if $X_T = \sum_{i=1}^{n} X_i$, then X_T has a Poisson distribution with mean $n\lambda$

- If events occur randomly at a rate of λ per given interval of time or space, then the number of events in n such intervals is a Poisson random variable with mean $n\lambda$

- If X is a random variable with density function $f(x)$, then $P(a < X < b) = \displaystyle\int_{x=a}^{x=b} f(x) \mathrm{d}x$

- Probability density functions must obey two properties.

$$\int_{-\infty}^{+\infty} f(x)dx = 1 \text{ and } f(x) \geq 0 \text{ for all } x$$

- If X is a continuous random variable with probability density function $f(x)$, then the median value of X is given by the equation $\int_{-\infty}^{M} f(x)dx = \frac{1}{2}$

- The lower and upper quartiles, Q_1 and Q_3, are given by the formulae $\int_{-\infty}^{Q_1} f(x)dx = \frac{1}{4}$ and $\int_{-\infty}^{Q_3} f(x)dx = \frac{3}{4}$, where $f(x)$ is the probability density function.

- If X is a random variable with mean μ and variance σ^2, then

$$\mu = E(X) = \int_{-\infty}^{+\infty} xf(x)dx$$

$$\sigma^2 = \text{Var}[X] = \int_{-\infty}^{\infty} (x-\mu)^2 f(x)dx = \int_{-\infty}^{\infty} x^2 f(x)dx - \mu^2$$

- If X is a random variable with density function $f(x)$, then the mean value and variance of $g(X)$ are given by

$$E(g(X)) = \int_{-\infty}^{\infty} g(x)f(x)dx$$

$$\text{Var}(g(X) = E(\{g(X)\}^2) - \{E(g(X))\}^2 = \int_{-\infty}^{\infty} \{g(x)\}^2 f(x)dx - \{E(g(X))\}^2$$

Check and review

You should now be able to...	Review Questions
✔ Find the expected value and variance of random variables and their functions.	1, 2, 3
✔ Apply the Poisson distribution to suitable random experiments.	4
✔ Find probabilities and the mean and variance of continuous random variables and their functions.	5
✔ Find the mean and variance of sums of independent random variables and their functions.	6

1 Calculate the expected value and variance for this probability distribution.

x	1	3	5	7	9	11
$P(X=x)$	$\frac{1}{6}$	$\frac{1}{6}$	$\frac{1}{6}$	$\frac{1}{6}$	$\frac{1}{6}$	$\frac{1}{6}$

2 A six-sided dice with faces numbered 1–6 is biased so that the probability of any score is proportional to the cube of the score. Find the expected value of C, the score on a randomly chosen throw.

3 Random events occur at an average rate of five per four minutes. Find the probability that there will be fewer than two events in any randomly chosen four-minute interval.

4 A firm receives, on average, 18 letters per day. Assuming that their arrival constitutes a random process, what is the probability that in five randomly chosen days

a There are at least twelve letters every day,

b There is a total of at least 90 letters?

5 A continuous random variable X has probability density function given by

$$f(x) = \begin{cases} \dfrac{a(x+3)}{5}; & 0 < x < 2 \\ 0; & \text{otherwise} \end{cases}$$

a Show that the value of a is $\dfrac{5}{8}$ and find the mean and variance of X

b A sample of four x-values is taken at random. Find the probability that exactly one value will be greater than one.

6 Two independent random variables, S and T, have probability density functions as follows:

$$f(s) = \begin{cases} as; & 1 < s < 2 \\ 0; & \text{otherwise} \end{cases}$$

$$f(t) = \begin{cases} 6t(1-t); & 0 < t < 1 \\ 0; & \text{otherwise} \end{cases}$$

where a is a constant.

a Find the values of a

b Find the expected value and variance of the function $S + 2T$

7 X is a random variable for the score on a randomly chosen throw of a fair, 10-sided dice. Find the expected values and variances of

a $2X$

b $X - 1$

Did you know?

The Poisson distribution is named after the French Mathematician, Geometer and Physicist, Simeon Poisson. Poisson is also well known for his partial differential equation that is frequently written as $\nabla^2 \phi = f$, and widely used in mechanical engineering and theoretical physics.

> Life is good for only two things: doing mathematics and teaching it.
> – Simeon Poisson

ICT

Many discrete distributions can be simulated in spreadsheet packages. For instance, the distribution of 100 rolls of a fair dice can be simulated in Excel.

Use the function =RANDBETWEEN(1,6) to generate a random number between 1 and 6. Then copy this formula into 100 (say) cells in the A column. Use the COUNTIF function to count how many of each number appears in the list. (An example formula can be seen in cell D3.) Then draw a graph of the results.

Try to simulate the following situations using a similar process:
- 100 rolls on a twelve-sided dice;
- 100 rolls on two six-sided dice;
- 100 rolls on a biased six-sided dice (for example, even numbers are twice as likely as odd numbers).

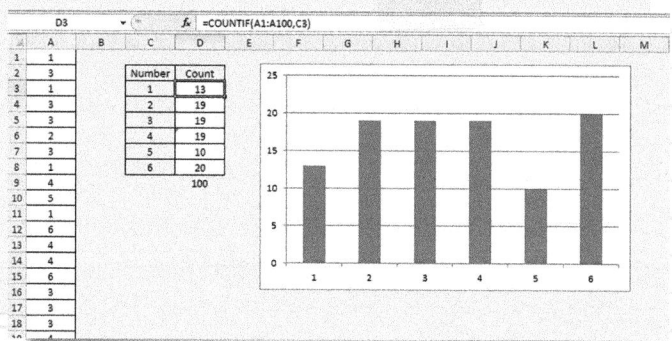

ICT

The Excel function =POISSON(x, mean, cumulative) can be used to calculate the probability of x events, with the mean average number of events per interval.

For example, by 2017 the mean average number of goals Ipswich Town scored against Norwich City per match was 1.455. Find the probability of
(a) Ipswich Town scoring 5 goals in one game against Norwich City,
(b) Ipswich Town scoring 4 or fewer goals in one game against Norwich City.
Using Excel,
(a) = POISSON(5, 1.455, false) = 0.0127 (to 3 significant figures).
(b) = POISSON(4, 1.455, true) = 0.983 (to 3 significant figures).

1   The probability distribution of X and Y are as follows.

$$P(X = x) = \begin{cases} \dfrac{x}{k}; & x = 1, 2, 3, 4 \\ 0; & \text{otherwise} \end{cases}$$

y	1	2	3	4
$P\{Y = y\}$	0.3	a	a	b

Given that $P\{X = 4\} = P\{Y = 4\}$

   a   Find the value of k, a and b **[5 marks]**

   b   Which discrete random variable has the highest value on average? Explain how you know.   **[4]**

2   This table shows the probability distribution of N, the number of people living in a household as estimated from a sample of 200 households.

n	1	2	3	4	5
$P\{N = n\}$	$5k$	$6k$	$7k$	k	k

   a   Find the value of k **[2]**

   b   Calculate the median and mode of N **[2]**

   c   Calculate the expected number of people per household.   **[2]**

   d   Calculate the variance.   **[3]**

3   The continuous random variable X is such that $X \sim \text{Po}(3)$

   a   Calculate each of these probabilities.

     i $P\{X \leq 5\}$ **[1]**

     ii $P\{X = 3\}$ **[2]**

     iii $P\{4 < X \leq 7\}$ **[2]**

     iv $P\{X < \lambda\}$ **[2]**

   b   State the variance of X **[1]**

4   A shop sells apples at the rate of 2.5 per minute on average.

   a   Explain what assumptions you would have to make in order to model the sale of apples as a Poisson distribution.   **[2]**

   b   Use the Poisson distribution to find the probability that in a one-minute period

     i   No apples are sold,   **[2]**

     ii   Fewer than three apples are sold,   **[2]**

     iii   More than four apples are sold.   **[2]**

   c   Calculate the probability that, over a three-minute period, more than ten apples are sold. **[3]**

5 A random variable X has mean 4 and standard deviation 3

  a Calculate $E(X^2)$ **[2]**

The random variable M is defined as $M = X + 2Y$, where X and Y are independent random variables.

  b Use the information that $E(M) = 11$ and $Var(M) = 19$ to find the mean and standard deviation of the random variable Y **[4]**

6 The continuous random variable X has probability density function given by

$$f(x) = \begin{cases} kx; & 0 \le x \le 6 \\ 0; & \text{otherwise.} \end{cases}$$

  a Calculate the value of k **[4]**

  b Find each of these probabilities.

    i $P\{X < 3\}$ **[2]**   **ii** $P\{X > 1\}$ **[2]**   **iii** $P\{1.5 < X < 5\}$ **[2]**

  c Calculate the expectation of X **[3]**

  d Calculate the variance of X **[4]**

7 The continuous random variable X has pdf given by

$$f(x) = \begin{cases} \dfrac{1}{9}(4x - x^2); & 0 \le x \le 3 \\ 0; & \text{otherwise} \end{cases}$$

  a Sketch the probability density function of X **[2]**

  b Write down the mode of X **[1]**

  c Calculate the median of X **[4]**

8 The table gives the bivariate probability distribution for the variables X and Y

		Y		
		10	**20**	**P{X=x}**
	1	0.2	0.1	0.3
X	**2**	0.15	0.05	0.2
	3	0.3	0.2	0.5
P{Y=y}		0.65	0.35	1

  a Write down the probability distribution of the random variable T, where $T = X + Y$ **[3]**

  b Find $E(T)$ **[1]**

  c Without doing any further calculations, explain why $E(T) = E(X) + E(Y)$ but $Var(T) \ne Var(X) + Var(Y)$ **[3]**

9 Given that $X \sim Po(7)$ and $Y \sim Po(3)$, where X and Y are independent random variables,

  a Write down the value of

    i $E(X + Y)$ **ii** $Var(X + Y)$ **[2]**

  b Calculate $P\{X + Y > 8\}$ **[3]**

  c Find the probability that $X = 5$ and $Y = 3$ **[3]**

10 The discrete uniform variable U is defined over the interval $[2, 7]$

 a Calculate $E(U)$ and $Var(U)$ [3]

 b Find $P\{X > 3.5\}$ [2]

11 The discrete random variable V has probability density function $f(v) = \begin{cases} \dfrac{v^2}{30}; & v = 1, 2, 3, 4 \\ 0; & \text{otherwise} \end{cases}$

 a Calculate the median of V [2]

 b Calculate $E(5V + 3)$ and $Var(5V + 3)$ [7]

12 Given that the random variable X is such that X $Po(5.5)$

 a Calculate the range of values of a such that $P\{X < a\} > 0.7$ [2]

 b Calculate the range of values of b such that $P\{X > b\} < 0.1$ [2]

 c Calculate the probability that the value of X is within one standard deviation of the mean. [5]

A second, independent, random variable Y is such that Y $Po(2)$

 d Calculate each of these probabilities.

   i $P\{X + Y < 9\}$ [2] ii $P\{X - Y \geq 2\}$ [3] iii $P\{(2X - Y)\} > E(2X - Y)$ [4]

13 The probability density function of Y is given by $f(y) = \begin{cases} ky^3; & 0 \leq y \leq 2 \\ 0; & \text{otherwise} \end{cases}$

 a Calculate the variance of Y [8]

 b Calculate the interquartile range of Y [7]

The continuous random variable W is defined by $W = 4Y - 3$

 c Find the expectation and variance of W [4]

14 The discrete random variable T has probability density function $f(t) = \begin{cases} \dfrac{1}{A}; & t = 1, 2, 3 \\ \dfrac{t}{A}; & t = 4, 5 \\ 0; & \text{otherwise} \end{cases}$

 a Calculate $P\{T > E(T)\}$ [5]

 b Find $Var(7 - 3T)$. [4]

15 The probability density function of X is given by $f(x) = \begin{cases} \dfrac{1}{n}; & x = 1, 2, \ldots n \\ 0; & \text{otherwise} \end{cases}$

 a Write down an expression for $E(X)$ [1]

 b Prove that the variance of X is given by the formula $Var(X) = \dfrac{1}{12}(n^2 - 1)$ [7]

16 The number of emails a person receives during the hours of 9 am to 5 pm is modelled by a Poisson distribution. Emails are received at a rate of three every ten minutes.

 a Calculate the probability of receiving more than seven emails in a ten-minute period. [3]

 b Calculate the probability of receiving between five and eight emails (inclusive) over half an hour. [3]

 c Calculate the probability of receiving at least three emails every ten minutes for an hour. [5]

17 A continuous random variable has probability density function $f(x) = \begin{cases} ax+b; & 0 \le x \le 2 \\ 0; & \text{otherwise} \end{cases}$

Given that $E(X) = \dfrac{13}{12}$

   a   Calculate the values of a and b **[6]**

   b   Calculate the median of X **[4]**

The continuous random variable Y is defined by $Y = AX + 2$

   c   Write down an expression for the mean of Y **[1]**

Given that the variance of Y is 47

   d   Calculate the values of the constant A **[4]**

18 X and Y are two independent random variables. $E(X) = 12$ and $E(X+2Y) = 15$

   a   Find the expectation of Y **[2]**

The random variables X and Y have standard deviations of 1.7 and 0.5 respectively.

   b   Calculate $\text{Var}(X + 2Y)$ **[4]**

19 A continuous random variable $f(x)$ is defined by $f(x) = \begin{cases} k(9 - x^2); & 0 \le x \le 3 \\ 0; & \text{otherwise} \end{cases}$

   a   Show that $k = \dfrac{1}{18}$ **[4]**

   b   Calculate the expected value of $X^2 + X$ **[4]**

20 A continuous random variable $f(x)$ is defined by $f(x) = \dfrac{1}{9}(2 + 3x^2)$ for x between 1 and 2

Find $E\left(X + \dfrac{2}{X^3}\right)$ **[5]**

21 The continuous random variable R has probability density function

$$f(r) = \begin{cases} k\left(\dfrac{1}{2} + r\sin r\right); & 0 \le r \le \dfrac{\pi}{3} \\ 0; & \text{otherwise} \end{cases}$$

   a   Calculate the value of k **[5]**

   b   Find $E\left(\sqrt{3}R + 2\right)$, giving your answer in terms of π **[7]**

22 Two independent random variables, R and S, have distributions

$$P(R = r) = \begin{cases} Ar^2; & r = 1, 2, 3, 4 \\ 0; & \text{otherwise} \end{cases}$$

and $f(s) = \begin{cases} \dfrac{1}{8}\left(s^2 - \dfrac{1}{s^2}\right); & 1 < s < k \\ 0; & \text{otherwise} \end{cases}$

   a   Show that $k = 3$ and find the value of A **[3]**

   b   Calculate $E(4R + S^3)$ **[7]**

11 Hypothesis testing and contingency tables

The principles that underpin hypothesis testing were adopted centuries ago in response to concerns that the contractors responsible for minting coins might be tempted to syphon some of the precious metals used in the coins for their own use. The process used to prevent this became known as the *Trial of the Pyx,* which was held at regular intervals. This involved careful inspection of a sample of coins from every batch of a certain weight from the mint. The sample was compared with control coins that were stored in Pyx Chapel in Westminster Abbey in a wooden box called the 'pyx'.

This comparison of data sets underpins modern day hypothesis testing in a wide range of different areas. Such methods are used throughout manufacturing industries to control quality; they are used in medical trials of new drugs or surgical techniques and in social psychology to better understand human behaviour.

Orientation

What you need to know	What you will learn	What this leads to
Maths Ch11 • Conduct and interpret a hypothesis test.	• To test the parameter of a Poisson distribution. • To find the probability **A Level** of making type I and type II errors. • To perform a χ^2 test to evaluate a distribution model. • To use a contingency table to calculate expected frequencies and χ^2 contributions. • To calculate the number of degrees of freedom and the critical value or the p-value of the test statistic. • To perform a test for association.	• Hypothesis testing for correlation.

11.1 Hypothesis testing and errors

Fluency and skills

Look back at Section 10.2 for a reminder on the Poisson distribution

You can use the Poisson distribution to model the probability of a number of independent, random events that occur over a fixed time if you know the value of λ (lambda), the parameter that gives the average number of events.

> **Key point**
>
> In the absence of any other information, the best estimate of λ is the average rate taken from a random sample.

If you have an estimate and then take a new sample, a hypothesis test will tell you if the sample could reasonably fit your estimate for λ

The **null hypothesis** is that the parameter takes a fixed value $H_0 : \lambda = \lambda_0$

The **alternative hypothesis** is either simply that λ_0 is wrong, $H_1 : \lambda \neq \lambda_0$ (a 2-tailed test), or that λ is higher or lower than λ_0, $H_1 : \lambda < \lambda_0$ or $H_1 : \lambda > \lambda_0$ (a 1-tailed test).

> A single observation is a sample of size 1

Example 1

A Poisson distribution is believed to have a parameter of 13.2

A sample is taken to test whether or not this is the case. State the null and alternative hypotheses for this test.

> The null hypothesis is $H_0 : \lambda = 13.2$ and the alternative hypothesis is $H_1 : \lambda \neq 13.2$

As with any hypothesis test you need a **test statistic**.

> **Key point**
>
> The **test statistic** when testing a Poisson distribution is the total number of times the event occurs in a sample of size n

If this test statistic is sufficiently different from $n\lambda$ then you have sufficient evidence to reject the null hypothesis. You can compare the test result to a critical value, or you can compare the p-value of the test statistic to the significance level.

> The total number of events in the sample is compared with the number you would expect on average in that sample. The Poisson distribution only allows integer numbers of outcomes, which is why you can't compare the average number of events to λ

Example 2

A Poisson distribution is believed to have parameter 8.6, $X \sim \mathrm{Po}(8.6)$. The hypotheses $H_0 : \lambda = 8.6$ and $H_1 : \lambda \neq 8.6$ are tested at the 10% level.

a Calculate the critical values for a sample of size 5

A sample of size 5 is taken which has an average rate of 10.4

b State, with a reason, whether the null hypothesis is accepted or rejected.

> Find out how to use your calculator to work out cumulative probabilities.

> Look back at Section 10.3 for a reminder on how to calculate individual probabilities using $P(X = n) = \frac{\lambda^n e^{-\lambda}}{n!}$

a $Y \sim \mathrm{Po}(43)$

> As $43 = 8.6 \times 5$ (sample size × average rate)

$P(Y \leq 32) = 4.97\%$ and $P(Y \leq 33) = 6.93\%$

$P(Y \geq 54) = 5.86\%$ and $P(Y \geq 55) = 4.39\%$

So the critical values are 32 and 55

b $5 \times 10.4 = 52.5$ lies between the critical values so the null hypothesis is accepted.

> Recall that the critical region is bounded by the critical values.

Example 3

A Poisson distribution is believed to have parameter 6.1, $X \sim \mathrm{Po}(6.1)$

The hypotheses $H_0 : \lambda = 6.1$ and $H_1 : \lambda > 6.1$ are tested at the 5% level.

A sample is taken with results 3, 4, 4, 6, 7, 8, 9, 10, 13, and 14

a Calculate the total number of occurrences, n

b Consider $Y \sim \mathrm{Po}(61)$, the Poisson distribution for the number of successes over 10 samples. Calculate the probability of obtaining a result at least this extreme if the parameter is truly 6.1 (i.e., find $P(Y \geq n)$, the p-value for the test statistic).

c State, with a reason whether the null hypothesis is accepted or rejected.

a $3 + 4 + 4 + 6 + 7 + 8 + 9 + 10 + 13 + 14 = 78$

b $P(Y \geq 78) = 0.02032$

> Find the p-value.

c Reject the null hypothesis since the p-value is less than the significance level.

There are two types of possible error you should consider when hypothesis testing.

> **Key point**

A **type I error** is when a null hypothesis which is true is rejected.

A **type II error** is when a null hypothesis which is false is accepted.

> Calculating the probability of a type II error is covered at A Level.

The probability of making a type I error is equal to the significance of the hypothesis test if the probability distribution is continuous. For discrete distributions, the probability of making a type I error may be less than the significance of the hypothesis test.

Example 4

A Poisson distribution is believed to have parameter 1.3, $X \sim Po(1.3)$

The hypotheses $H_0: \lambda = 1.3$ and $H_1: \lambda > 1.3$ are tested at the 5% level.

A sample of size 8 is taken.

a Find the critical value.

b Find the probability of making a type I error.

> A type I error occurs when a result at least as extreme as the critical value is found but the true value of λ is 1.3

a $Y \sim Po(10.4)$ so $P(Y \geq 16) = 0.06405$ and $P(Y \geq 17) = 0.03681$

Therefore, the critical value is 17

> Since $0.06405 > 0.05$ and $0.05 > 0.03681$

b $P(\text{type I error}) = 0.03681$

> $P(\text{type I error}) = P(Y \geq 17)$

A type I error is often called a **false positive**, where a result that should be ignored attracts attention. A type II error is often called a **false negative**, where a result is overlooked which could have been important. Both types of error can have serious consequences, depending on the context.

Example 5

A kitchen in a nut-free school looks at the ingredients it is purchasing to ensure that they comply with regulations. The null hypothesis for each ingredient is that it contains no trace of nuts.

a Explain what a type I error represents and why the kitchen might not worry about making one.

b Explain what a type II error represents and why the kitchen should worry a lot about making one.

a A false positive here is that the kitchen suspects an ingredient contains nuts when it actually doesn't. This might cost a bit of money, but at least the students with allergies are safe.

b A false negative would be the kitchen believing an ingredient doesn't contain nuts when it actually does. This has potentially fatal consequences and should be avoided at all reasonable costs.

Exercise 11.1A Fluency and skills

1 A Poisson distribution with assumed parameter 5.31 is being tested to see if this is an overestimate.

a State the null and alternative hypotheses.

A sample of size 12 is taken to test at the 10% level.

b Find any critical values.

2 A Poisson distribution is believed to have parameter 3.4, $X \sim Po(3.4)$. The hypotheses $H_0: \lambda = 3.4$ and $H_1: \lambda \neq 3.4$ are tested at the 10% level.

a Calculate the critical values for a sample of size 9

A sample of size 9 is taken which has an average rate of 3

b State, with a reason whether the null hypothesis is accepted or rejected.

3   A Poisson distribution models a situation where the parameter is usually found to be 72

A new sample is taken to see if this has changed.

a   Explain why the hypotheses are $H_0: \lambda = 72$ and $H_1: \lambda \neq 72$

The sample is of size 6 and a total of 482 outcomes occur. The critical values at the 5% significance level are 391 and 474

b   Determine the conclusion of the test.

c   Actually a type I error has been made. What does this mean?

4   A Poisson distribution models a situation where the parameter is usually found to be 12.5

A new sample is taken to see if this has been reduced.

a   Explain why the hypotheses are $H_0: \lambda = 12.5$ and $H_1: \lambda < 12.5$

The sample is of size 14 and a total of 155 outcomes occur. The p-value of this statistic is 6.81%

b   Determine the conclusion of the test at the 5% level.

c   Actually a type II error has been made. What does this mean?

5   A Poisson distribution with assumed parameter 8.3 is being tested to see if this is a good estimate or not.

a   State the null and alternative hypotheses.

A sample of size 9 is taken to test at the 10% level. A total of 87 events occur in the whole sample.

b   Calculate the p-value of this result.

c   State, with a reason whether the null hypothesis is accepted or rejected.

6   A Poisson distribution is believed to have parameter 14.2, $X \sim \text{Po}(14.2)$. The hypotheses $H_0: \lambda = 14.2$ and $H_1: \lambda > 14.2$ are tested at the 5% level. A sample is taken with results 7, 8, 10, 11, 14, 14, 15, 23, 27, 34

a   Calculate the total number of occurrences, n

b   Consider $Y \sim \text{Po}(142)$, the Poisson distribution for the number of successes over 10 samples. Calculate the probability of obtaining a result at least this extreme if the parameter is truly 14.2 (i.e., find $P\{Y \geq n\}$, the p-value for the test statistic).

c   State, with a reason whether the null hypothesis is accepted or rejected.

7   A Poisson distribution with assumed parameter 3.9 is being considered. The hypotheses $H_0: \lambda = 3.9$ and $H_1: \lambda > 3.9$ are tested at the 5% level using a sample of size 15

a   Calculate any critical values.

b   State the probability of making a type I error.

8   A Poisson distribution is being considered. The hypotheses $H_0: \lambda = 6.3$ and $H_1: \lambda < 6.3$ are tested at the 5% level using a sample of size 20

A total of 108 events occur in the whole sample.

a   Calculate the p-value of this result.

b   State, with a reason whether the null hypothesis is accepted or rejected.

c   Calculate the probability of making a type I error.

9   A medical test to detect the early presence of a disease is being designed.

a   Explain why a type I error would be tolerated.

b   Explain why a type II error would preferably be avoided.

10 A dice manufacturer is designing new 12-sided dice and wants to see if they are fair. Each redesign can be a costly process.

a   Explain why they would want to reduce the probability of a type I error.

b   Why might they not be too bothered about a type II error?

11 Describe your own situations in which

a   A type I error is preferably avoided,

b   A type II error is preferably avoided.

Strategy

To solve problems involving hypothesis testing

(1) Conduct a test.

(2) Find any critical values.

(3) Accept or reject the null hypothesis.

(4) Interpret the result in context.

Example 6

The Poisson distribution with a parameter of 2.8 is thought to be a good model for the number of goals scored per football match. Due to recent developments in defensive tactics it is believed that the average number of goals conceded has dropped.

The hypotheses $H_0 : \lambda = 2.8$ and $H_1 : \lambda < 2.8$ are tested at the 5% level.

A random sample of 24 matches is taken.

a Find any critical values.

The sample contains 49 goals in total.

b State, with a reason, whether the null hypothesis is accepted or rejected. Determine the conclusion to the hypothesis test in context.

a The critical value is 53 as $P(X \leq 53) = 4.34\%$ and $P(X \leq 54) = 5.69\%$

b Since 49 is less than the critical value, there is sufficient evidence to reject the null hypothesis.

There are, on average fewer goals conceded per match now.

(1)(2) Conduct a test and find the critical values.

(3) Reject the null hypothesis.

(4) Interpret the result in context.

Type I and II errors are not exclusive to Poisson distributions. The definitions apply to any hypothesis test.

This is a binomial distribution.

Example 7

A dice is rolled 27 times to test the hypotheses $H_0 : p = \dfrac{1}{6}$ and $H_1 : p < \dfrac{1}{6}$ at the 5% level, where p is the probability of rolling a six. The critical value is found to be one six. Calculate the probability of making a type I error.

$X \sim B\left(27, \dfrac{1}{6}\right)$ and $P(X \leq 1) = 0.04659$

The probability of making a type I error is the probability of the critical region.

(2) Find the critical value.

(4) Interpret the result in context.

1 The number of phone calls a call centre receives in a 10-minute period can be modelled by a Poisson distribution with parameter $\lambda = 6.8$

The hypotheses $H_0 : \lambda = 6.8$ and $H_1 : \lambda \neq 6.8$ are being tested at the 5% level and a sample of seven 10-minute periods are taken.

    a Find any critical values.

The seven periods contain a total of 59 calls.

    b State, with a reason, whether the null hypothesis is accepted or rejected. Determine the conclusion to the hypothesis test in context.

    c Define a type I error and when one occurs in a hypothesis test.

    d Determine the probability of making a type I error.

2 The number of points scored per game by a basketball team is believed to be well modelled by a Poisson distribution with parameter 102.3

Due to recent training, it is believed that the average number of points scored has increased. The hypotheses $H_0 : \lambda = 102.3$ and $H_1 : \lambda > 102.3$ are being tested at the 5% level. A random sample of 20 matches is taken.

    a Find any critical values.

The sample contains a total of 2098 goals.

    b State, with a reason whether the null hypothesis is accepted or rejected. Determine the conclusion of the hypothesis test in context.

3 The number of accidents occurring in a nursery is measured over one month by the owner. After 41 accidents are measured, they take steps to reduce the number of accidents. During the next month only 38 occur.

    a Assuming the occurrence of accidents follows a Poisson distribution, state the null and alternative hypotheses to test whether the number of accidents has been reduced.

    b Complete the hypothesis test at the 10% level and state the conclusion.

    c Describe in words what it would mean in context if a type II error had been made.

4 Over a single season, 1064 goals were scored by all the teams collectively in a football league during the 380 matches played. On the final day of the season, 37 goals were scored in the 10 matches played that day.

    a The number of goals per match is modelled by a Poisson distribution. Calculate the expected number of goals scored in 10 matches that season.

    b State the null and alternative hypotheses if it is believed that the number of goals on the last day was exceptionally large.

    c Determine the conclusion to the test at the 10% significance level.

    d Describe in words what it would mean in context if a type I error had been made.

5 The number of accidents per month at a sports centre can be modelled by a Poisson distribution with parameter $\lambda = 0.8$ but new safety procedures have been put in place. The hypotheses $H_0 : \lambda = 0.8$ and $H_1 : \lambda < 0.8$ are being tested at the 5% level. A sample of one year is used and there were five accidents.

    a Determine the p-value of the result.

    b State, with a reason whether the null hypothesis is accepted or rejected. Determine the conclusion of the hypothesis test in context.

STATISTICS

6 The number of customers coming into a town book shop in a half-hour period can be modelled by a Poisson distribution with parameter $\lambda = 15$

The hypotheses $H_0 : \lambda = 15$ and $H_1 : \lambda > 15$ are being tested at the 5% level and a sample of twelve 30-minute periods are taken. Over those periods a total of 204 customers came in.

a Find the p-value of this outcome.

b State, with a reason, whether the null hypothesis is accepted or rejected. Determine the conclusion of the hypothesis test in context.

7 The number of light bulbs in a shopping centre which burn out each month can be well modelled by a Poisson distribution with parameter $\lambda = 28$

The hypotheses $H_0 : \lambda = 28$ and $H_1 : \lambda \neq 28$ are being tested at the 10% level. A sample of four random months in a period of several years is taken and a total of 99 light bulbs were replaced in that time.

a Find the p-value of this outcome.

b State, with a reason, whether the null hypothesis is accepted or rejected. Determine the conclusion of the hypothesis test in context.

8 A coin is suspected of being biased. The hypotheses are $H_0 : p = \dfrac{1}{2}$ and $H_1 : p \neq \dfrac{1}{2}$ at the 5% level, where p is the probability of flipping a tail. The test is performed at the 5% level and the coin is flipped 16 times.

a Find any critical values.

b Determine the probability of making a type I error.

9 The probability of success for a binomial distribution is tested. It is hypothesised that the true value is 0.45 and a sample of size 12 is taken.

a Determine the probability of making a type I error if the significance level is

   i 10%   **ii** 5%   **iii** 1%

b Find also the probability of making a type II error in each case if the true value is actually 0.6

Correlation describes linear relationships between variables, but correlation cannot describe relationships between variables that do not have a scale. There can be many different relationships between variables. A relationship is known as an **association**.

Below is data about regional preferences for types of meat. A sample of 266 people from the north, east and south of a city were asked which meat they liked most out of beef, chicken, or pork. The **observed frequencies** are shown in the **contingency table** below. Contingency tables are used in lots of research when looking for association between two variables.

Observed frequencies	North	East	South	Totals
Beef	32	9	44	84
Chicken	41	53	20	114
Pork	8	32	28	68
Totals	81	94	88	266

> A cell tells you how many of the total have both of the qualities that its row and column indicate. For example, there are 41 people from the north of the city who like chicken most.

It looks as though there might be regional preferences, but how can you be sure? The first stage is to work out what the table would look like if the variables were independent.

You would expect $266 \times \dfrac{81}{266} \times \dfrac{84}{266}$ or 25.6 (to 3 sf) people to be from the north of the city with beef as their favourite.

Using this method, the table below shows all the expected frequencies (values given to 3sf).

Expected frequencies	North	East	South
Beef	25.6	29.7	28.7
Chicken	34.7	40.3	39.0
Pork	20.7	24.0	23.3

> If the variables are independent then the probability of liking beef and coming from the north of the city is the product of two probabilities. This gives the proportion of people who like beef and are from the north.

The next stage is to set up a hypothesis to test. The **null hypothesis** being considered is H_0: there is no association between the variables. The **alternative hypothesis** is H_1: there is some association. This example is to be tested at the 5% significance level.

> Note that you don't test for positive or negative association, just the presence of any association.

The X^2 (**chi squared**) **statistic** gives a way of measuring how different the observed values are from the expected values.

The X^2 statistic is a calculation involving the observed frequencies $\{O_i\}$ and the expected frequencies $\{E_i\}$.

Key point

$$X^2 = \sum_i \frac{(O_i - E_i)}{E_i}$$

The individual X^2 contributions to this calculation are shown in the table below.

X^2 contributions	North	East	South
Beef	1.61	14.41	7.08
Chicken	1.14	4.01	9.26
Pork	7.80	2.64	0.965

This sum of the individual contributions gives a value of $X^2 = 48.9$ (all values to 3sf).

The final stage is to decide whether you have sufficient evidence to reject the null hypothesis. To do this, you compare $X^2 = 48.9$ to the χ^2 distribution with four degrees of freedom.

Key point

The χ^2 distribution is a continous distribution but the observed values are discrete, so it is an approximation.
For this test to work accurately, it is required that every expected frequency is greater than 5

Note that observed frequencies may be 5 or lower.

The χ^2 distribution (also chi squared, but this one is lower case) with k degrees of freedom is the distribution of a sum of the squares of k independent standard normal random variables.

If the X^2 statistic is larger than the critical value found using the χ^2 distribution, or if the p-value is less than the significance level, then you have sufficient evidence to reject the null hypothesis.

You find the critical value by calculating the value of x for which $P\{X \geq x\}$ is equal to the significance level.

For $X \sim \chi^2_4$, $P\{X \geq 0.95\}$ the critical value is 9.49

Since the calculated value of $X^2 = 48.9$ is larger than the critical value, you have sufficient evidence to reject the null hypothesis.

For an $m \times n$ contingency table you should use the χ^2 distribution with $(m-1) \times (n-1)$ degrees of freedom.

Example 1

A test is performed to see if there is any association between age and movie genre preferences. The contingency table shows the recorded data.

a Draw a table of expected frequencies.

b Draw a table of X^2 contributions and calculate the X^2 test statistic.

c Calculate the number of degrees of freedom and hence the p-value of the test statistic.

d Determine the conclusion of the test.

O_i	Action	Comedy	Romance	Totals
< 25 years	41	34	27	102
25–50 years	35	38	32	105
> 50 years	7	12	19	38
Totals	83	84	78	245

(Continued on the next page)

a

E_i	Action	Comedy	Romance
< 25 years	34.6	35.0	32.5
25–50 years	35.6	36.0	33.4
> 50 years	12.9	13.0	12.1

> The expected frequencies are calculated and given in this table to 3 sf.
>
> For example, the cell that corresponds to Action and < 25 years is $245 \times \dfrac{83}{245} \times \dfrac{102}{245} = 34.6$ (3 sf)

b

	Action	Comedy	Romance
< 25 years	1.20	0.03	0.92
25–50 years	0.01	0.11	0.06
> 50 years	2.68	0.08	3.938

The X^2 contributions are provided in this table to 3 sf.

The total X^2 statistic is 9.03 (3 sf).

> For example, $\dfrac{(41-34.6)^2}{34.6} = 1.2$ is the X^2 for Action and < 25 years.

c The 3 × 3 table has $(3-1) \times (3-1) = 4$ degrees of freedom. The p-value is 6.04%.

> The p-value is $P(X \geq 9.03)$ where $X \sim \chi^2_4$. 9.03 is used as it is the test statistic.

d The p-value is larger than the significance level, so there is insufficient evidence to reject the null hypothesis.

When expected frequencies are below five but some of the options are alike, the similar options can be grouped and their frequencies pooled. After doing so, the expected frequencies should be recalculated to ensure that there are now at least five and the number of degrees of freedom should be calculated to match the grouped table. Even if only a single expected frequency is below 5, there must be complete rows and columns after grouping.

Example 2

Voters in three constituencies are polled about their voting intentions in an upcoming election. Their options are the Red and Green parties, who have similar policies to one another but are both very different to those of the Blue party. The intentions are produced as follows to see if there is any association at the 5% level between location and intention.

O_i	Red	Green	Blue	Totals
Uppersville	9	8	85	102
Middleton	51	7	42	100
Workby	40	6	13	59
Totals	100	21	140	261

a Produce an observed frequency table using a relevant grouping, and explain why you have made that grouping.

b Calculate the number of degrees of freedom.

After grouping, the χ^2 statistic is 65.3

The critical value is 5.99

c Determine the conclusion of the test.

(Continued on the next page)

a

O_i	Red and Green	Blue	Totals
Uppersville	47.3	54.7	102
Middleton	46.4	53.6	100
Workby	27.4	31.6	59
Totals	121	140	261

The Red and Green columns have been grouped because, with the initial data, the Green share in Workby had an expected frequency below 5. The Red and Green columns were combined because they were said to represent similar options.

b Now there are $(3 - 1) \times (2 - 1) = 2$ degrees of freedom.

c There is sufficient evidence to suggest that there is association between the location the voters live in and their voting intention.

Note that all values are individually correct to 1 dp. The exact values, rather than these rounded ones, should be used in further calculations.

Having grouped the data, any conclusion we draw must treat Red and Green intentions together.

When you have a 2×2 contingency table, the method explained above changes slightly and you must calculate the X^2 statistic using **Yates' correction**.

Key point

$$X^2_{\text{Yates}} = \sum \frac{(|O_i - E_i| - 0.5)^2}{E_i}$$

Example 3

The lunch choices of two year groups at a school are tested for association. A sample of 63 students in year 12 and year 13 are asked whether they prefer fish 'n' chips or salad.

a Draw a table of expected frequencies (correct to 3 sf).

b Draw a table of X^2 contributions and calculate the X^2 test statistic.

c At the 10% level the critical value is 2.71. Determine the conclusion of the test.

O_i	Fish 'n' chips	Salad	Totals
Year 12	19	12	31
Year 13	14	18	32
Totals	33	30	63

a

E_i	Fish 'n' chips	Salad
Year 12	16.2	14.8
Year 13	16.8	15.2

b

E_i	Fish 'n' chips	Salad
Year 12	0.32	0.35
Year 13	0.31	0.34

The test statistic is 1.30 (3 sf)

c The test statistic is less than the critical value. There is insufficient evidence to reject the null hypothesis.

The X^2 contributions (3 sf) have been calculated using Yates' correction.

The sum of the individual contributions.

Exercise 11.2A Fluency and skills

For each of the contingency tables in questions 1 to 4

  a Draw a table of expected frequencies.

  b Draw a table of X^2 contributions and calculate the X^2 test statistic.

  c Calculate the number of degrees of freedom and hence the critical value at the 5% level.

  d Calculate the p-value of the X^2 statistic.

  e Determine the conclusion of the test.

1

	A	B	C	Totals
X	57	92	5	154
Y	7	32	13	52
Z	75	100	23	198
Totals	139	224	41	404

2

	A	B	C	D	Totals
W	16	13	16	10	55
X	8	13	8	17	46
Y	17	7	17	17	58
Z	8	15	18	18	59
Totals	49	48	59	62	218

3

	A	B	C	Totals
X	79	57	85	221
Y	61	72	62	195
Totals	140	129	147	416

4 (Remember to use Yates' correction.)

	A	B	Totals
X	70	124	194
Y	50	45	95
Totals	120	169	289

5 A test is performed to see if there is any association between the colour of a hen and how many eggs it lays per week. The contingency table shows the recorded data.

O_i	0–1	2–3	4+	Totals
Brown	6	13	16	35
White	17	20	19	56
Totals	23	33	35	91

  a Draw a table of expected frequencies.

  b Draw a table of X^2 contributions and calculate the X^2 test statistic at the 10% significance level.

  c Calculate the number of degrees of freedom and hence the p-value of the test statistic.

  d Determine the conclusion of the test.

6 A snack shop tests meal deal choices for association. A sample of sales is checked for sandwich and drink choices.

	Chicken	BLT	Ham and cheese	Egg mayo	Totals
Orange	31	28	19	6	84
Apple	19	29	17	8	73
Water	24	21	20	9	74
Totals	74	78	56	23	231

  a Draw a table of expected frequencies.

  b Draw a table of X^2 contributions and calculate the X^2 test statistic.

  c At the 10% level the critical value is 10.6 Determine the conclusion of the test.

7 Three classes of students are polled about their preferences for an upcoming year group lunch trip. A teacher wants to determine whether there is any association between class and food preference.

O_i	Hamburger	Cheeseburger	Pizza	Hot Dog	Totals
Class A	4	5	8	6	23
Class B	2	7	6	9	24
Class C	8	2	9	4	23
Totals	14	14	23	19	70

  a Calculate a table of expected frequencies.

  b State which groupings should be made, if any, and calculate the number of degrees of freedom.

c With grouping, the test statistic is 2.63 Determine whether the result is significant at the 5% level and state the conclusion in context.

8 Two identical twins, Frieda and Helga, and their friend Hans keep records of the snacks they eat over a month to see if there's any association between person and preference at the 5% level.

O_i	Chocolate Bars	Apples	Packets of Crisps	Totals
Frieda	15	2	11	28
Helga	12	3	14	29
Hans	10	9	21	40
Totals	37	14	46	97

a Draw a table of expected frequencies.

b Explain why Frieda and Helga's values should be grouped together.

c The X^2 statistic is 6.52

Find the critical value and determine the conclusion of the test.

Reasoning and problem-solving

Example 4

An airline takes a survey of passengers of various heights to gauge views on the amount of legroom available.

	>1.80 m	<1.50 m	Between 1.50 m and 1.80 m	Totals
Too little	67	32	304	403
Too much	19	48	415	482
About right	48	51	379	478
Totals	134	131	1098	1363

a Perform a test for association at the 5% level.

b Comment on the association.

a

	>1.80m	<1.50m	Between 1.50m and 1.80m
Too little	39.6	38.7	324.6
Too much	47.4	46.3	388.3
About right	47.0	45.9	385.1

This is the table of expected frequencies.

	>1.80m	<1.50m	Between 1.50m and 1.80m
Too little	18.92	1.17	1.31
Too much	17.00	0.06	1.84
About right	0.02	0.56	0.096

This is the table of X^2 contributions.

The X^2 statistic is 41.0 (3 sf).

The critical value at the 5% level is $\chi^2_4 = 9.49$

There is sufficient evidence to reject the null hypothesis.

(1) Perform a test for association.

(Continued on the next page)

b There is association between height and opinions on how much legroom is available.

> Identify positive association. **2**

The large contributions from people over 1.80 m show that their opinions differ most from what is expected. Looking at the observed and expected frequencies there is evidence for a strong positive association between being tall and believing there is too little legroom. Similarly, there is a strong negative association between being tall and believing there is too much legroom.

> Identify positive or negative association by comparing observed and expected values for large X^2 contributions. **2**

The number of short people who believe there is too much legroom is very close to what is expected.

> Identify where the observed and expected frequencies closely match for small X^2 contributions. **3**

The contingency tables will not always be provided explicitly. You must be able to construct them from the question. Observed values must be absolute frequencies, but information can also be given as proportions or percentages.

A climbing centre tracks its customers' usage to better tailor its facilities. Of the 98 who climb using ropes, $\frac{1}{7}$ are under 18, $\frac{4}{7}$ are over 65 and the rest are in between (middle). The under 18s form 15% of the boulderers (non-rope climbing), 25% of the boulderers are over 65 and the other 60% are in between.

a Create a contingency table containing this information.

b Perform a hypothesis test to determine whether there is any association between age and climbing style preference at the 5% level.

c Comment on any association.

a

O_i	Under-18s	Middle	Over-65s	Totals
Ropes	14	28	56	98
Bouldering	9	36	15	60
Totals	23	64	71	158

b The expected frequencies are

Expected	Under-18s	Middle	Over-65s
Ropes	14	39.7	44
Bouldering	8.7	24.3	27

The X^2 contributions are

X^2	Under-18s	Middle	Over-65s
Ropes	0.005	3.446	3.249
Bouldering	0.01	5.629	5.31

which give a total X^2 statistic of 17.6

There are 2 degrees of freedom so the critical value is 5.99

The result is significant so there is sufficient evidence to suggest that there is association between age and preference of climbing style.

c The under-18s are in line with what you'd expect. The over 65s have a relative preference for roped climbing and the middle ages have a relative preference for bouldering.

For each of the following situations

  a   Perform a test for association at the 5% level,

  b   Comment on the association.

1   A group of people is surveyed to see if there is any association between their eye colour and their favourite colour.

Observed	Blue eyes	Green eyes	Brown eyes	Totals
Blue	99	37	38	174
Red	34	29	33	96
Yellow	32	34	27	93
Totals	165	100	98	363

2   Live-action role players are asked which ancient tribes they most enjoy playing and which weapon they most like to carry.

Observed	Axe	Sword	Bow	Spear	Totals
Celts	10	29	20	34	93
Vikings	12	24	10	31	77
Huns	0	14	31	24	69
Totals	22	67	61	89	239

3   In a coastal town, the colours of rocks are compared with whether they were found on the beach or by the river.

Observed	Grey	Brown	Totals
Beach	76	64	140
River	107	59	166
Totals	183	123	306

4   A restaurant is looking for association between choices of starter and main course from its menu.

Observed	Pizza	Pasta	Totals
Soup	87	121	208
Salad	104	96	200
Totals	191	217	408

  a   Perform a test for association at the 10% level.

  b   Comment on the association.

5   A toy company tests its new range of coloured monster toys for association between child gender and their preference of monster colour.

Observed	Red	Blue	Yellow	Totals
Boys	75	71	57	203
Girls	58	79	60	197
Totals	133	150	117	400

  a   Perform a test for association at the 1% level.

  b   Comment on the association.

6   A crowd is observed at a concert. People are categorised by the colour of their hair and the colour of their clothes. The person in charge of measurements tells you the following:

'318 people attended in total. There were 96 people with black hair, of whom one third wore colourful clothes, one quarter wore light-coloured clothes and the rest wore dark-coloured clothes. Of the 120 people with brown hair, 55% wore dark-coloured clothes, 35% wore colourful clothes and the remaining 10% wore light-coloured clothes. The other people had blonde hair and were equally-divided among the different colours of clothing.'

  a   Construct a contingency table using the given information.

  b   Perform a hypothesis test at the 5% level to determine if there is any association between hair colour and choice of clothing colour.

  c   Comment on the association.

Full A Level

7 The number of yellow cards acquired per season by football teams from different regions of the country is collected over three seasons to identify whether there is any association between the two. You are told the following:

- The most yellow cards a southern team earned in a season was 68 with the second highest being 60. No team from the South equalled that amount.
- One third of the entries for the Midlands are in the 60–69 range.
- The North-West teams earned one third of the entries in the 0–59 range.
- A total of 55 entries come from London teams.

a Copy and complete the following table using the information provided.

O_i	0–59	60–69	70–79	80+	Totals
London		14		1	
Midlands			13	2	48
North-East	5	12	7		
North-West		21	10		61
South					10
Totals	90	65	40	5	200

b Explain why the table should be redrawn with columns showing the number of teams who earned 70 + cards.

c State which region should be excluded from further analysis and explain why.

d State the number of degrees of freedom. The test statistic is 12.78 and the critical value at the 5% level is 12.59, making the result significant.

e The largest contribution (3.45) is from the North-East in the 0–59 range and the smallest (0.00) is from the Midlands in the 60–69 range. Interpret this in context.

8 For a 2×2 contingency table Yates' correction can be written as

$$\chi^2_{\text{Yates}} = \frac{N\left(|ad-bc|-\dfrac{N}{2}\right)^2}{T_X T_Y T_A T_B}$$

for tables in the form

Observed	A	B	Totals
X	a	b	T_X
Y	c	d	T_Y
Totals	T_A	T_B	N

For every 2×2 contingency table in this section, check that this formula agrees with your value.

STATISTICS

Full A Level

11.3 Confidence intervals

Fluency and skills

Suppose you want to know the mean and variance of the heights of all the adults in a country. The most accurate values possible would be obtained by measuring every adult, but that is completely impractical. Instead, you would find a sample and calculate the mean and variance of this sample.

For a set of n data values $\{x_i\}$ the **sample mean is**

> **Key point**
>
> $$\bar{x} = \frac{\sum x_i}{n}$$

and the **sample variance is**

> **Key point**
>
> $$s^2 = \frac{1}{n-1}\sum (x_i - \bar{x})^2$$

These are known to be unbiased estimators of the population mean μ and population variance σ^2. While the expected values of these random variables are equal to the population parameters, that doesn't mean that they are necessarily likely to have values close to them.

It can be shown that, for a large sample from a population, the distribution of the sample mean is distributed as

$$\bar{X} \approx N\left(\mu, \frac{\sigma^2}{n}\right)$$

So, if you find a number of sample means, they will themselves follow the normal distribution and their mean is expected to be equal to the population mean. So the sample mean is increasingly likely to be close to the population mean as the size of the sample increases.

> Notice the $n-1$ in the denominator of the sample variance. Using n gives a value which is expected to underestimate the population variance, so the estimator is biased.

> An unbiased estimator for a population parameter is a random variable whose expected value is equal to the parameter.

> This is due to the central limit theorem, knowledge of which is not required for this course.

> When the sample size equals the population size the values will obviously match exactly.

Example 1

A population follows a normal distribution with mean and variance 6.2
A sample of size 50 is taken. Give a model of the distribution of the sample mean.

The sample mean is distributed as

$$N\left(6.2, \frac{6.2}{50}\right) = N(6.2, 0.124)$$

Quite how close the sample mean is to the population mean depends on the population variance, which is likely to be unknown.

Although the sample variance is an unbiased estimator of the population variance, it is unlikely to be very accurate for small samples. A good rule of thumb is that a sample of at least size 30 is 'large enough' to obtain a reasonable estimate of the population variance. Then the sample variance can be used in place of σ^2

The sample data can be used to generate a **confidence interval**. This is a range of values in which it is likely that the true population mean appears.

> **Key point**
>
> A $p\%$-**confidence interval** is generated from a sample. It is expected before generation that the population mean μ will fall into this interval with probability $p\%$.

The $p\%$-confidence interval generated by a sample of size n is

$$\bar{x} - z \times \frac{s}{\sqrt{n}} < \mu < \bar{x} + z \times \frac{s}{\sqrt{n}}$$

where z is calculated from p, $\bar{x}$ is the sample mean and s^2 is the sample variance.

> **Key point**
>
> The standard deviation of the sample mean, $\frac{s}{\sqrt{n}}$, is called the **standard error**.

You can never be certain that your interval contains the population mean and the more confident you want to be, the larger the interval becomes (z increases as p increases).

You can calculate these yourself. $z = \Phi^{-1}\left(\frac{1+p}{2}\right)$, where Φ^{-1} is the inverse of the standard normal distribution.

p	z
99%	2.576
98%	2.326
95%	1.96
90%	1.645

It is not possible to give the probability that the true mean lies in an interval once it has been generated. The $p\%$ probability only applies before a sample is taken.

Example 2

A sample of size 36 is taken from a population whose standard deviation is 20.4 in order to generate a 95% confidence interval for the mean of the population.

a What is the probability that the confidence interval will contain the population mean?

The sample mean is 13.6

b Find the confidence interval.

---

a 95%

b The number of standard errors is $\left|\phi^{-1}(0.975)\right| = 1.96$

The interval is given by

$$13.6 - 1.96 \times \frac{20.4}{\sqrt{36}} < \mu < 13.6 + 1.96 \times \frac{20.4}{\sqrt{36}}$$

which simplifies to $6.94 < \mu < 20.3$ (3 sf)

Find out how to use your calculator to calculate the interval.

When the population variance is known it can be used without needing to be estimated from the sample. In this case, the p%-confidence interval is

$$\bar{x} - z \times \frac{\sigma}{\sqrt{n}} < \mu < \bar{x} + z \times \frac{\sigma}{\sqrt{n}}$$

where z is calculated from p as before.

A 90% confidence is to be created for a normal distribution whose variance is known to be 19.5

A sample of size 72 is taken and the sample mean is 19.3

a Calculate the standard error.

b Determine the value of z

Use tables or your calculator.

c Find the confidence interval.

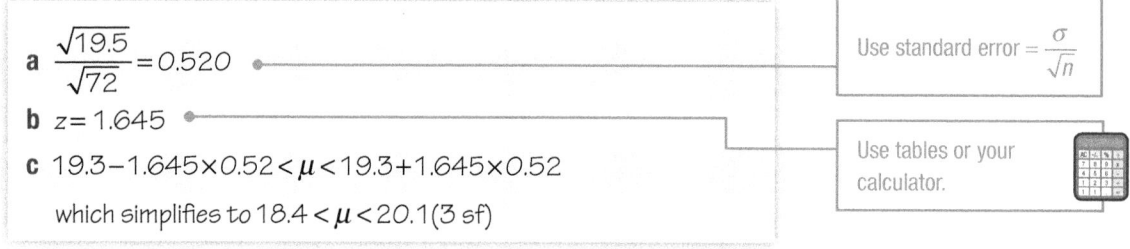

a $\dfrac{\sqrt{19.5}}{\sqrt{72}} = 0.520$ •   Use standard error $= \dfrac{\sigma}{\sqrt{n}}$

b $z = 1.645$ •

c $19.3 - 1.645 \times 0.52 < \mu < 19.3 + 1.645 \times 0.52$ Use tables or your calculator.

which simplifies to $18.4 < \mu < 20.1\,(3\,\text{sf})$

Exercise 11.3A Fluency and skills

1 Find 95% confidence intervals for samples with these summary statistics.

a $n = 58$, $\bar{x} = 91.4$, $s^2 = 14.2$

b $n = 41$, $\bar{x} = 0.122$, $s^2 = 0.00563$

c $n = 103$, $\bar{x} = 12.7$, $s^2 = 126.1$

2 Find 99% confidence intervals for samples with these summary statistics.

a $n = 14$, $\bar{x} = 41.5$, $\sigma^2 = 192.5$

b $n = 31$, $\bar{x} = -5.76$, $\sigma^2 = 91.5$

3 A sample of size 19 is drawn from a population whose variance is $\sigma^2 = 217.4$ The sample mean is 62.5 and the sample variance is $s^2 = 205.1$

a Explain why the population variance is used in finding a confidence interval.

b Find a 95% confidence interval and a 99% confidence interval using the sample data.

4 The heights of ten pole vaulters are measured to determine if they are taller than the general population. The following heights are recorded.

| 182.0 | 191.5 | 190.4 | 195.3 | 171.7 |
| 183.9 | 185.8 | 181.5 | 200.1 | 190.3 |

a Determine the mean height of the sample.

b If the population standard deviation is 6.8 cm, find a 95% confidence interval for the heights of pole vaulters.

5 The reaction times, in milliseconds, of 30 people to a stimulus are measured. Construct a 95% confidence interval for the mean reaction time.

209	207	216	207	196	205
211	206	203	207	228	204
195	217	206	220	216	217
223	228	225	217	210	218
207	211	233	226	205	192

6 The number of bees in a swarm is well approximated by a normal distribution with variance 6 250 000

    a Find the probability that a sample of size 22 has a mean that lies within 2000 of the true mean

    b Find the confidence interval for a sample whose mean value is 3700

7 For $X \sim N(317, 53^2)$ find

    a $P\{X \leq 213.12\}$

    b $P\{X \geq 420.88\}$

Hence give a 95% confidence interval for a sample of size 36 with mean 317 and standard deviation 318

8 A statistician is planning a survey to find the true mean value of something believed to be well modelled by a normal distribution with standard deviation 98

Find the minimum size of the sample required for the following intervals to have a range of at most size 50

    a 99% interval

    b 95% interval

    c 90% interval

Reasoning and problem-solving

Strategy

When estimating population parameters

  (1) Make sure your sample is representative to avoid bias.

  (2) Use the sample mean and sample variance as unbiased estimators.

  (3) Generate a confidence interval for the population mean.

  (4) Answer the question in context.

Confidence intervals are a useful tool in hypothesis testing. When the true mean is unknown, a sample mean and a measure of population variance provide a test statistic for deciding whether a proposed true mean is a reasonable estimate.

Example 4

It is thought that the true mean of a population is 12.9
This is tested at the 10% significance level and 20 samples are taken to generate 20 corresponding confidence intervals. All 20 of them contain 12.9
What does this suggest?

It is expected that 18 of the intervals contain the true mean. All 20 of them contain the number 12.9, which does not automatically suggest that 12.9 is the true mean but it does not rule it out. It would be reasonable to say that 12.9 is likely to be close to the true mean.

    (4) Answer the question in context.

It is important to remember that a $p\%$-confidence interval can only be said to be $p\%$ likely to contain the true mean before it is generated. If you take repeated samples and form confidence intervals, then you would expect $p\%$ of them to contain μ

Example 5

A 90% confidence interval is generated for the average height of a human adult. The sample gives the interval 187cm to 204cm. Explain why it is not reasonable to say that this interval is 90% likely to contain the true average human adult height.

Experience suggests that 187cm is a very tall height.

It does not seem likely that the average human adult height is even larger than this.

> **1**
> Make sure your sample is representative to avoid bias.

> **4**
> Answer the question in context. Perhaps the sample was taken from a basketball team or a town with very tall people.

Increasing the sample size decreases the width of the confidence interval because the size of the standard error is decreased. A larger sample gives a more accurate estimate of the mean.

As the standard error is based on $\sqrt{n}$ though, a sample four times as large will halve the width of the interval. This can be costly in real life.

Exercise 11.3B Reasoning and problem-solving

1 A statistician wishes to know the mean of a normal distribution whose variance is unknown. They take two samples. The first, of size 51, has a mean of 61.3 and a variance of 91.6

The second, of size 39, has a mean of 60.5 and a variance of 105.1

a Generate 95% confidence intervals for the two separate samples.

The combined samples have a variance of 97.6

b Calculate the size and sample mean of the combined samples and hence generate a 95% confidence interval.

c Explain why the statistician is more likely to use the interval just generated over the previous two.

2 A population is normally distributed and it is suspected, but not known, that its variance is 432

A sample of size 32 is taken whose mean is 1071 and variance is 1905

a Generate confidence intervals from the sample using the sample variance and suspected population variance.

b Explain which interval should be used in estimating the true mean.

3 The lengths of wood cut by a machine are thought to be well modelled by a normal distribution. The differences from the intended mean length of 61 pieces of wood are given in this table.

Value	<−0.3	−0.3	−0.2	−0.1	−0.05	0	0.05	0.1	0.15	>0.15
f	0	4	8	12	12	9	5	7	4	0

a Estimate the mean length difference of the sample.

b Estimate the standard deviation of the sample.

c Construct a 95% confidence interval for the mean difference.

4   The weights (in kg) of the 73 members of a bridge club are taken to estimate the mean weight of a bridge player. The values are grouped and given in this table.

Weight (kg)	50 to 65	65 to 75	75 to 85	85 to 95	95 to 105	105 to 110
f	5	13	20	26	8	1

a   Estimate the mean and standard deviation of the weights of the bridge players.

b   Calculate the expected frequencies and comment on the assumption that the weights are normally distributed with your estimated mean and standard deviation.

c   Construct a 95% confidence interval using the sample.

5   The masses of 10-year old boys are thought to be modelled by a normal distribution. Forty boys are weighed by their doctors and the results are given here.

32.2	29.8	36.4	39.6	26.9	28.2	41.9	40.5
32.0	32.6	32.8	34.0	17.4	33.8	29.6	24.1
28.9	34.9	31.5	27.7	27.3	38.1	33.7	36.4
16.1	32.4	33.8	29.9	25.0	46.7	30.0	33.3
28.9	29.8	37.8	22.6	28.6	33.0	32.5	28.8

a   Estimate the population variance using these values.

b   The 95%-confidence interval generated by this sample is $29.4 \le \mu \le 33.1$

Someone claims that there is a 95% chance that the true mean of a 10-year old boy lies in this range. Explain why they are incorrect.

STATISTICS

Chapter summary

- A Poisson distribution has a parameter, λ, which is the average rate that the event occurs.
- When you don't know the value of λ you can estimate it as $\lambda = \lambda_0$ and then perform a hypothesis test:

  The null hypothesis is $H_0: \lambda = \lambda_0$

  The alternative hypothesis is either:
    - $H_1: \lambda \neq \lambda_0$, a 2-tailed test
    - $H_1: \lambda < \lambda_0$, a 1-tailed test, or
    - $H_1: \lambda > \lambda_0$, a 1-tailed test.

- If the null hypothesis is true, then the distribution of the number of events in a sample of size n is a Poisson distribution with parameter λn. If the number of events in the sample is sufficiently unlikely, then you reject the null hypothesis.
- Sometimes a null hypothesis which is true is rejected due to an unlikely result. This is called a type-1 error.

  The test gives evidence for or against λ_0 being a reasonable estimate of the parameter of the distribution.

- A contingency table shows the observed frequencies, $\{O_i\}$, from a sample which splits data according to two discrete variables.
- In a test for association the null hypothesis is that there is no association between the variables and the alternative hypothesis is that there is some association.
- You should calculate a table of expected frequencies, $\{E_i\}$, to show what the sample would look like if the two variables were independent.
- The X^2 test statistic can be calculated from the observed and expected frequencies.

$$X^2 = \sum_i \frac{(O_i - E_i)^2}{E_i}$$

- For a 2×2 contingency table, or when an observed frequency is less than 5, **Full A Level** the test statistic is calculated using Yates' correction.

$$X^2{}_{\text{Yates}} = \sum \frac{(|O_i - E_i| - 0.5)^2}{E_i}$$

- The χ^2 distribution with k degrees of freedom is the distribution of a sum of the squares of k independent standard normal random variables.
- The X^2 test statistic can be compared with the appropriate χ^2 distribution to test for significance. For an $m \times n$ contingency table you should use the χ^2 distribution with $(m-1) \times (n-1)$ degrees of freedom.
- If the statistic is larger than the critical value, or if the p-value is less than the significance level, then the null hypothesis H_0 is rejected.
- Large X^2 contributions highlight association and you should compare observed and expected frequencies so that you can comment on the association.

- A $p\%$-confidence interval is generated from a sample. It is expected that $p\%$ of intervals generated in that way will contain the population mean. For a population with unknown mean μ and known variance σ, that has a sample with mean $\bar{x}$, the confidence interval is

$$\bar{x} - \left| \phi^{-1}\left(\frac{1+p}{2}\right) \right| \times \frac{\sigma}{\sqrt{n}} < \mu < \bar{x} + \left| \phi^{-1}\left(\frac{1+p}{2}\right) \right| \times \frac{\sigma}{\sqrt{n}}$$

- The value $\dfrac{\sigma}{\sqrt{n}}$ is known as the **standard error**.

Check and review

You should now be able to...	Try Questions
✔ Test the parameter of a Poisson distribution.	1, 2
✔ Calculate the probability of making a type I error.	2
✔ Interpret the result of the test in context.	3
✔ Use a contingency table to calculate expected frequencies and X^2 contributions.	4, 7, 10
✔ Calculate the number of degrees of freedom, and the critical value or the p-value of the test statistic.	5, 6
✔ Perform a test for association.	6, 7
✔ Interpret the result of a test for association in context and describe the association.	7
✔ Generate confidence intervals.	8, 9
✔ Use Yates' correction appropriately. Full A Level	10

1   A Poisson distribution is believed to have parameter 1.9 and a test is performed to see if it is higher than this.

   a   State the null and alternative hypotheses.

   A sample of size 19 is taken to test at the 5% level. A total of 48 events occur in the whole sample.

   b   Calculate the p-value of this result.

   c   State, with a reason whether the null hypothesis is accepted or rejected.

2   It is thought that $X \sim \text{Po}(0.75)$. The hypotheses $H_0 : \lambda = 0.75$ and $H_1 : \lambda < 0.75$ are tested at the 5% level. A sample of size 29 is taken.

   a   Find the critical value.

   b   Find the probability of making a type I error.

3   The number of road accidents on a country road per month is well modelled by a Poisson distribution with parameter 1.3

   It is hoped that new traffic calming measures will reduce this. The hypotheses $H_0 : \lambda = 1.3$ and $H_1 : \lambda < 1.3$ are being tested at the 10% level. The number of accidents in next 8 months form a sample.

   a   Find any critical values.

   There are 7 accidents in those 8 months.

   b   State, with a reason whether the null hypothesis is accepted or rejected. Determine the conclusion of the hypothesis test in context.

4   For this contingency table

   a   Draw a table of expected frequencies,

   b   Draw a table of X^2 contributions and calculate the X^2 test statistic.

O_i	A	B	C	Totals
X	103	71	35	209
Y	96	84	41	221
Z	111	58	49	218
Totals	310	213	125	648

5   For this contingency table, calculate

   a   The number of degrees of freedom,

   b   The critical value at the 5% significance level.

O_i	A	B	C	D	E	Totals
X	29	37	25	87	41	219
Y	14	39	93	20	53	219
Z	10	5	2	7	28	52
Totals	53	81	120	114	122	490

6   The heights and birthplaces of UK adults are tested for association at the 5% level. A large sample is taken with 24 degrees of freedom. The X^2 statistic is found to be 34.91

   Calculate the p-value of this statistic and determine the conclusion of the test in context.

7   A car designer is testing public opinion on four new models (A, B, C, D) for association with preference for colour. Using the contingency table

O_i	A	B	C	D	Totals
Red	100	84	60	80	324
Blue	55	64	57	78	254
Silver	30	36	27	23	116
Totals	185	184	144	181	694

   a   Draw a table of expected frequencies,

   b   Draw a table of X^2 contributions and calculate the X^2 test statistic.

There are 6 degrees of freedom so the critical value at the 5% level is 11.07

   c   Determine the conclusion of the test and comment on the association.

8   A sample of size 36 is taken from a normal distribution with unknown variance. The sample mean is 14.7 and the sample variance is 29.5

   a   Generate a 95% confidence interval for the true mean of the distribution.

   The hypotheses $H_0 : \mu = 16.5$ and $H_1 : \mu \neq 16.5$ are tested at the 5% level.

   b   Determine the conclusion of the test given the sample of size 36

   c   Determine the probability of making a type I error.

9   Find 95% confidence intervals for samples with the following summary statistics.

   a $n = 25, \Sigma x = 100, \Sigma x^2 = 1600$

   b $n = 61, \Sigma x = 301, \Sigma x^2 = 1486$

10 Given this contingency table and its table of expected values, calculate the X^2 test statistic using Yates' correction.

O_i	A	B	Totals
X	24	36	60
Y	60	24	84
Totals	84	60	144

E_i	A	B
X	35	25
Y	49	35

Full A Level

Did you know?

The theory behind confidence intervals is based around dealing with uncertainty in results that are derived from data that is taken from only a randomly selected subset of a population. Confidence intervals treat their bounds as random variables and the parameter as a fixed value.

Bayesian inference provides an alternative explanation in the form of the theory of **credible intervals**. In contrast to confidence intervals, Bayesian intervals treat their bounds as fixed and the estimated parameter as a random variable.

> Life is complicated, but not uninteresting.
> - Jerzy Neyman

Research

A confusion matrix is a contingency table that reports on the number of false positives, false negatives, true positives and true negatives. The F1 score is used to measure the accuracy of a test.

For example, imagine a piece of software that was being used to identify whether 16 photographs were of males or females.

It could achieve these results:

		Actual class	
		Male	Female
Predicted class	Male	5	6
	Female	2	3

The software achieved 5 true-positives, 6 false-positives, 2 false-negatives and 3 true-negatives. Research how this information can be used to calculate the F1 score.

1   A Geiger counter indicates radioactive decay by a series of clicks, one click per decay. For one sample of radioactive material, the number of decays over a randomly chosen 10-second interval is observed to be 6

The sample is thought to be of a material with a mean number of decays of 2.7 per 10-second period. A scientist wishes to test the hypothesis that the source is actually a material with a greater mean.

   a   State the null and alternative hypotheses for this test.   **[1 mark]**

   b   Perform the test for a significance level of 5%.   **[4]**

2   Using traditional thread, a machine produces parachute material with an average of 3.2 flaws per $5\,m^2$. It is claimed that a new type of thread will reduce the number of flaws. In a random sample of $10\,m^2$ of the new material, two flaws are found.

   a   State the null and alternative hypotheses to test this claim.   **[1]**

   b   Perform the test for a significance level of 5%.   **[4]**

   c   Explain what is meant by a type 1 error and give the probability that a type 1 error is not made in the test in part **b**.   **[2]**

3   A business receives on average 12 telephone calls per hour throughout the day. The day after a newspaper advertising campaign, it receives 90 calls between 10am and 4pm.

   a   Using a significance level of 5%, test the hypothesis that the campaign has increased the rate of telephone calls.   **[4]**

   b   State the probability of a type 1 error for your test and explain what this means.   **[2]**

   (You may assume that the number of calls in a given period follows a Poisson distribution.)

4   At an accident blackspot, the rate of road traffic accidents is known to be 3.2 per month. During the month after traffic calming measures are put in place, only 2 accidents are recorded.

   a   Explain why a Poisson distribution may be appropriate to test whether the accident rate has been reduced.   **[1]**

   b   Carry out this test using a significance level of 5%. You should state clearly your null and alternative hypotheses.   **[4]**

   c   What is the largest number of accidents in a month that would result in a conclusion that the rate of accidents had been reduced?   **[2]**

5 A school is recording how students spend their free time. A group of 100 students were asked to name their main pastime outside school. The results by gender, are as follows.

		Pastime				
		Reading	Sport	Online	Television	Other
Gender	Male	6	12	20	7	14
	Female	7	17	10	6	1

a Find the expected frequencies based upon the hypothesis that there is no association between pastime and gender. **[2]**

b Test at a significance level of 10% whether females and males differ in their choice of pastime. **[5]**

6 A teacher wishes to assess her students. She decides to use two assessment methods, a traditional problem-solving exam (A) and a multiple choice exam (B).

		Result		
		Fail	Pass	Distinction
Exam	A	7	18	8
	B	7	14	6

The teacher believes that there is no significant difference in exam performance between the two exams.

a Complete the following table of expected frequencies on the assumption that she is correct. **[2]**

		Result			
		Fail	Pass	Distinction	Total
Exam	A				
	B				
	Total				

b Test the teacher's belief at a significance level of 5%. **[5]**

7 A researcher is asked to assess whether patient safety is affected by weekend staffing levels. She obtains data for nursing staff drug errors and the day of the week on which they occurred. The drug errors are classified as type A: errors with no significant consequences to the patient, type B: errors with minor consequences and type C: errors with serious consequences.

		Weekday	Weekend	Total
Error	Type A	45	18	63
	Type B	24	8	32
	Type C	14	5	19
	Total	83	31	114

Test at a significance level of 5% whether the timing (weekday or weekend) of a drug error affects its seriousness. **[7]**

8 A batch of 37 brands of olive oil are assessed for quality by a panel of experts and rated either outstanding, good or unsatisfactory. This judgement is recorded with the price per litre of the oil.

		Quality			
		Outstanding	Good	Unsatisfactory	Total
Price per litre	< £7	5	5	7	17
	≥ £7	10	7	3	20
	Total	15	12	10	37

Test at a significance level of 5% whether there is an association between the price of olive oil and its quality. **[7]**

9 A random variable X has a binomial distribution, parameters $n = 12$ and p, a constant. The value of p was known to be 0.3 but is now thought to have increased.

a Write down the null and alternative hypotheses in a test of this belief. **[2]**

b Complete the following table given that the null hypothesis in part **a** is correct. Give your answers to 4 dp. **[2]**

x	6	7	8	9
$P\{X \geq x\}$				

c What values of X would suggest that the belief is incorrect? Use a significance level of 5%. **[1]**

d Describe the meaning of the term 'type I error' and work out its probability for the test above. **[2]**

10 Fasting blood sugar levels of 15 healthy adult females had a mean of 4.4 mmol/L.

a Assuming these to be a random sample from a normal population with standard deviation 0.6 mmol/L, find a 95% confidence interval for the population mean. **[2]**

It is proposed that the population from which this sample was taken had a mean of 4.6 mmol/L.

b Use your answer to part **a** to determine whether or not to reject this hypothesis. You should give a reason for your answer. **[2]**

11 A sample of size 60 is taken from a normal population with unknown mean and variance.

a Explain why it is acceptable in this situation to use the **sample** variance when using normal distribution theory to find a confidence interval for the population mean. **[2]**

b Given that the sample mean and variance are 14 and 36, respectively, find an approximate 90% confidence interval for the population mean. **[2]**

12 Graphs and networks 1

To ensure fast internet connectivity in towns and cities, and throughout buildings like schools and airports, cable networks are used. Often during installation some existing conduits for services such as lighting can be used, whereas in other areas entirely new conduits will have to be put in place, and costs along certain routes may be greater than others. It is important that the cabling for such networks is done efficiently and at minimum cost. To ensure that costs are kept to a minimum, the maths of minimum spanning trees is used.

Similar calculations are used in other areas of design of networks, for example, in planning transportation networks, electricity distribution and water supply.

Orientation

What you need to know	What you will learn	What this leads to
• The construction and language of graphs.	• To use Kruskal's algorithm to find the minimum spanning tree for a network given as a diagram. • To use Prim's algorithm to find the minimum spanning tree for a network given as a diagram or as a table. • To find the shortest path between vertices in a network using Dijkstra's algorithm. • To solve the route inspection problem for a network with odd order vertices. • To find upper and lower bounds for the travelling salesman problem.	**Ch13 Critical path analysis** • Activity networks. **Ch28 Graphs and networks 2** • Optimising network flows.

Full A Level

Fluency and skills

You use the term **graph** for a diagram involving a set of points and interconnecting lines. Each point is called a **vertex** (plural **vertices**) and a line joining two points is called an **edge**.

> **Key point**
>
> A graph consists of a number of points (vertices) connected by a number of lines (edges). Note that it is possible to have a vertex that does not have an edge connected to it.

What matters is which vertex is connected to which. The shape or layout of the diagram is irrelevant. These three diagrams show the same set of vertices and connections – they are effectively the same graph (check that you can see this).

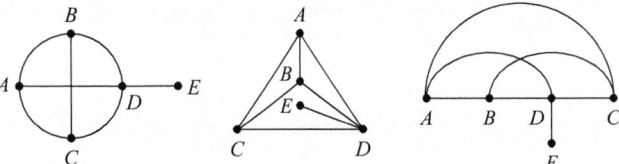

Graphs like these are called **isomorphic** (of the same form).

A graph is **connected** if you can travel from any vertex to any other vertex (perhaps passing through others on the way).

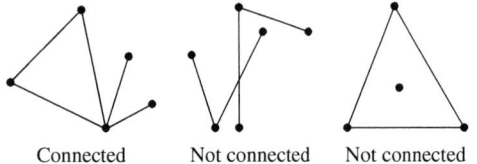

Connected   Not connected   Not connected

Two or more edges may connect the same pair of vertices. These are **multiple edges**. There may be a **loop** connecting a vertex to itself.

Loop   Multiple edges

A graph with no loops or multiple edges is a **simple** graph.

The graph formed by using only some of the vertices and edges of a graph is a **subgraph** of the original graph.

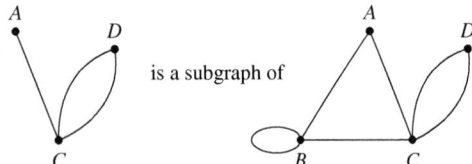

is a subgraph of

If you add extra vertices along the edges of a graph the result is called a **sub-division** of the graph.

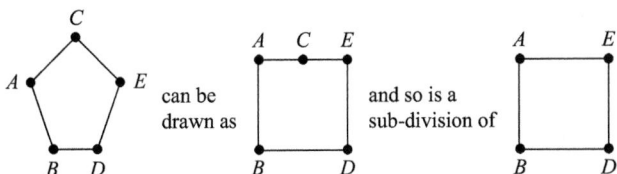

If a simple graph has an edge connecting all possible pairs of vertices, it is a **complete** graph. Complete graphs have their own notation.

The complete graph with n vertices is called K_n

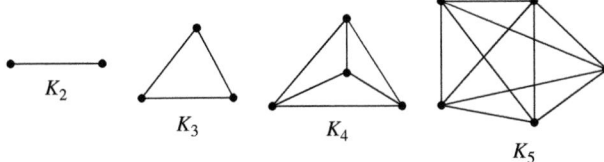

In some graphs the vertices belong to two distinct sets, and each edge joins a vertex in one set to a vertex in the other. Such a graph is called a **bipartite graph**. The graph on the right graph is a bipartite graph. The two sets of vertices are $\{A, B, C\}$ and $\{D, E\}$.

If every possible edge in a bipartite graph is present, it is a **complete bipartite graph**. Again there is a special notation.

The complete bipartite graph connecting m vertices to n vertices is called $K_{m,n}$

The number of edges meeting at a vertex is called the **degree** or **order** of the vertex. The degrees of the vertices in this graph are shown in brackets.

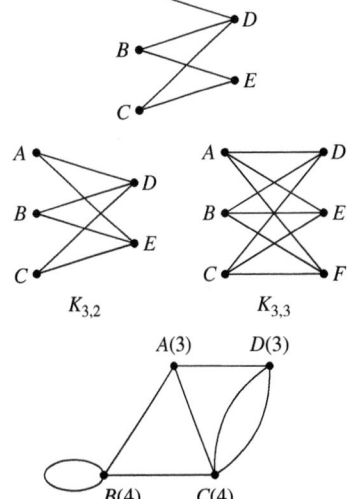

DISCRETE

Example 1

Draw a graph with four vertices – one with degree 4, one with degree 2 and two with degree 1

The solution is not unique. Here are two possible graphs which fit the requirements.

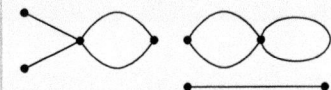

There is at least one other connected solution and one other disconnected solution. You might like to try to find these.

a Draw a graph with vertices labelled with the integers 2, 3, ..., 9

Draw an edge between two vertices only if the two numbers have no common factor.

b Write down the number of edges and the total of the degrees of the vertices. State the relationship between these quantities.

a

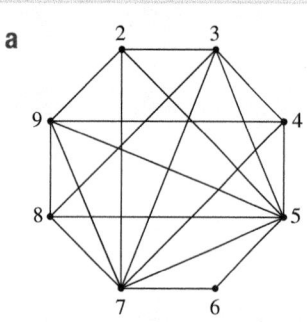

b Number of edges = 19

Degrees are 4, 5, 4, 7, 2, 7, 4 and 5, so the total of degrees = 38

Total of degrees = 2 × number of edges

The relationship in the above example is true for all graphs, because each edge contributes 1 to the degree of the vertex at each of its two ends.

> **Key point**
>
> Total number of degrees = 2 × number of edges

A lemma is a fact which can be taken for granted in subsequent work (you can quote it without proof).

This is called the **handshaking lemma** because two hands (vertices) are involved in one handshake (edge).

Here are two drawings of the complete graph K_4

The second version is preferable because none of the edges cross. If you can draw a graph in this way is said to be **planar**, and drawing it with no crossing edges is called drawing it 'in the plane'.

> **Key point**
>
> A graph is planar if it can be drawn in a plane in such a way that no two edges meet each other except at a vertex.

When a graph is drawn in the plane, the areas created by its edges are called **faces**. The area outside the graph is also a face, sometimes called the **infinite face**.

If a connected graph drawn in the plane has V vertices, E edges and F faces, these values are related by **Euler's formula**:

> Euler's formula **Key point**
>
> $V + F - E = 2$

Example 3

Show that Euler's formula holds for this graph.

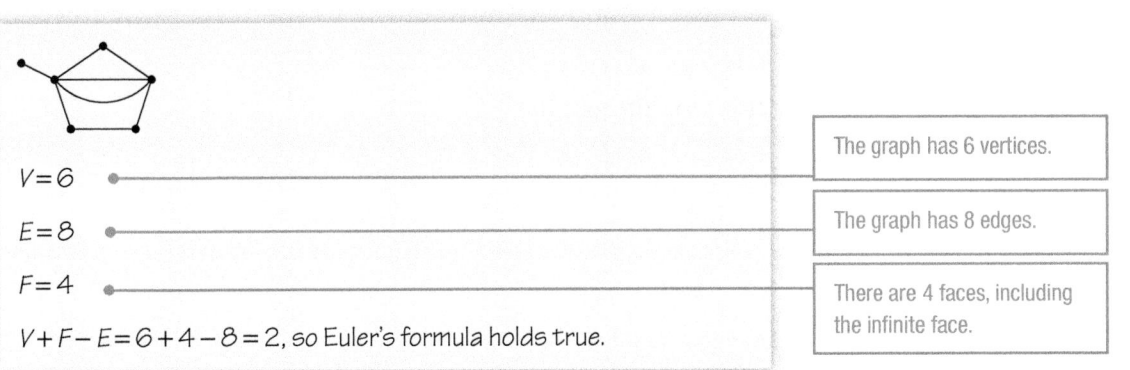

$V = 6$ — The graph has 6 vertices.

$E = 8$ — The graph has 8 edges.

$F = 4$ — There are 4 faces, including the infinite face.

$V + F - E = 6 + 4 - 8 = 2$, so Euler's formula holds true.

To prove Euler's formula consider any connected plane graph such as the one shown.

You remove edges one by one without making it disconnected.

If you remove an edge such as AB, then $E \rightarrow (E-1)$ and $V \rightarrow (V-1)$

If you remove an edge such as BE, one of the two edges DE or the loop at C, then $E \rightarrow (E-1)$ and $F \rightarrow (F-1)$

(In the diagram AB and DE have been removed.)

In both cases, the value of $V + F - E$ is unchanged.

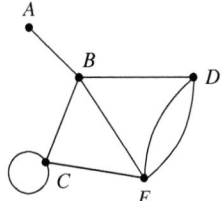

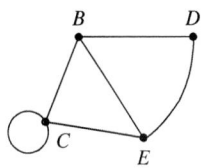

Eventually, you reduce the graph to a single vertex, so $V = 1$, $F = 1$ and $E = 0$

This gives $V + F - E = 2$

This value did not change as the graph reduced, so for the original graph $V + F - E = 2$

In addition to drawing a diagram, there are two ways in which you can record a graph. One way is to list its vertices and edges (the *vertex set* and *edge set*), but, more usually, you construct an **adjacency matrix** showing the vertices and the number of connections between them.

Example 4

a List the vertex set and edge set for this graph.

b Draw an adjacency matrix for this graph.

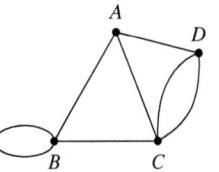

a $\{A, B, C, D\}$, $\{AB, AC, AD, BB, BC, CD\}$

b

	A	B	C	D
A	0	1	1	1
B	1	2	1	0
C	1	1	0	2
D	1	0	2	0

Count the number of direct routes between each pair of vertices.

There are 2 routes from B to itself because you can go round the loop in either direction.

DISCRETE

349

The adjacency matrix for a bipartite graph will contain two square blocks of zeros provided the two sets of vertices are kept separate in the table. You can then replace it by a more convenient reduced form.

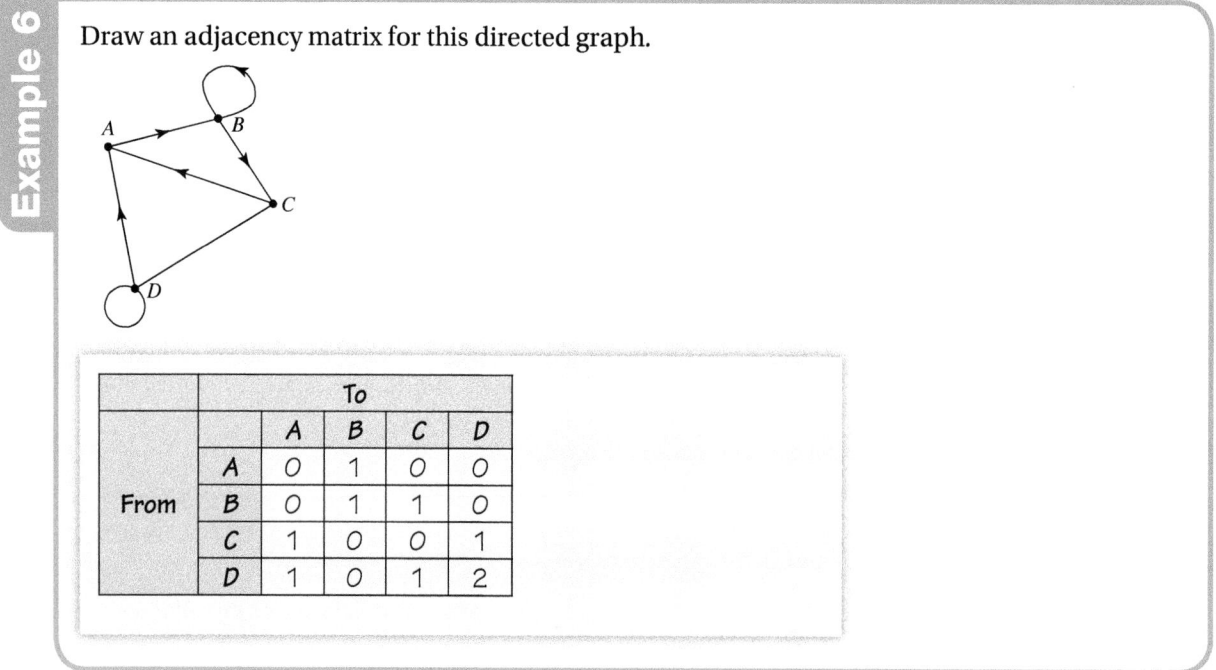

Example 5

Draw an adjacency matrix for this bipartite graph.

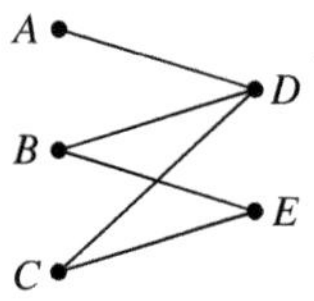

	A	B	C	D	E
A	O	O	O	1	O
B	O	O	O	1	1
C	O	O	O	1	1
D	1	1	1	O	O
E	O	1	1	O	O

This is the full adjacency matrix.

	D	E
A	1	O
B	1	1
C	1	1

This is the reduced form.

Sometimes one or more edges of a graph are 'one-way streets'. These edges are called **directed edges** and the graph is a **directed graph** (or **digraph**). You can still use an adjacency matrix, but this will no longer be symmetrical and you need to include 'from' and 'to' to show the direction.

Example 6

Draw an adjacency matrix for this directed graph.

		To			
		A	B	C	D
	A	O	1	O	O
From	B	O	1	1	O
	C	1	O	O	1
	D	1	O	1	2

For every simple graph G there is a corresponding graph G', called the **complement of G**. This is formed by drawing those edges which do not exist in G and removing the edges which do. If you combine G and G' you get a complete graph.

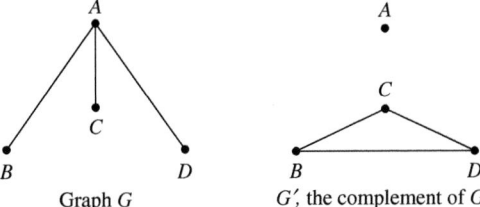

Graph G G', the complement of G

You find the adjacency matrix for G' by changing zeros to ones and ones to zeros in the adjacency matrix of G (ignoring the leading diagonal because no loops are added).

Example 7

Draw the adjacency matrix for this graph, G, and hence find the adjacency matrix for the complement, G'. Draw G'.

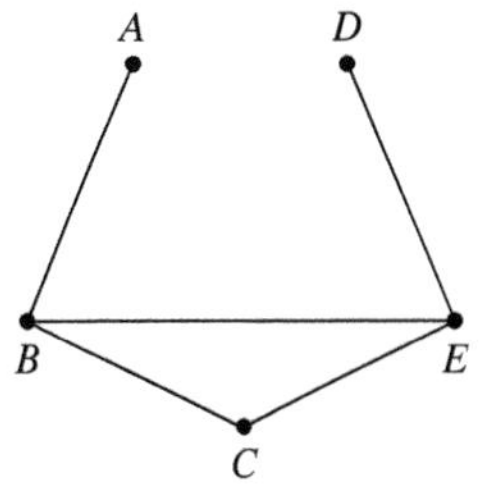

The adjacency matrix for G is:

	A	B	C	D	E
A	0	1	0	0	0
B	1	0	1	0	1
C	0	1	0	0	1
D	0	0	0	0	1
E	0	1	1	1	0

The adjacency matrix for G' is:

	A	B	C	D	E
A	0	0	1	1	1
B	0	0	0	1	0
C	1	0	0	1	0
D	1	1	1	0	0
E	1	0	0	0	0

Change 1 to 0 and 0 to 1 (except the leading diagonal).

The complement, G', looks like this:

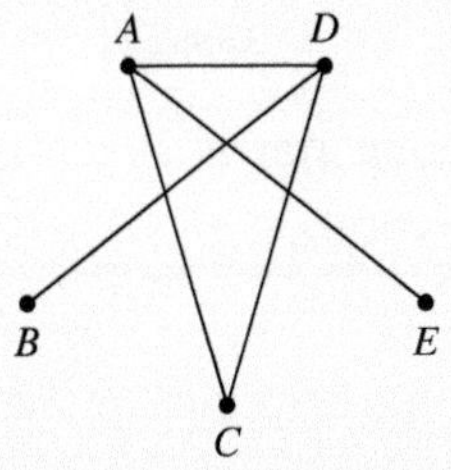

A network, or **weighted graph**, is a graph with a number associated with each arc. This number is called the **weight** of the arc. Weights may correspond to distances, times, costs or many other things. A network may be **directed** (some or all the arcs are 'one-way streets'). You can represent a network using a **distance matrix**, with entries showing the weights of the arcs.

When talking about networks, use the terms 'node' and 'arc'.

DISCRETE

351

Example 8

This network shows the travel times (in minutes) between five Somerset towns.

Draw a distance matrix to represent this network.

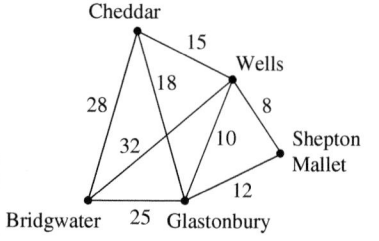

Abbreviate the towns as B, C, G, S and W.

	B	C	G	S	W
B	–	28	25	–	32
C	28	–	18	–	15
G	25	18	–	12	10
S	–	–	12	–	8
W	32	15	10	8	–

Exercise 12.1A Fluency and skills

1 Draw a connected graph which has

  a One vertex of order 3 and three of order 1

  b One vertex of order 1 and three of order 3

  c Two vertices of order 3 and two of order 1

  d Four vertices, each with a different order

2 The table shows the number of vertices (*V*), edges (*E*) and faces (*F*) in a planar graph. In each case, fill in the missing value and draw a possible layout for the graph.

	V	E	F
a	5	7	
b	8		6
c		9	7

3 Two of the three graphs *A*, *B* and *C* are sub-graphs of the following graph.

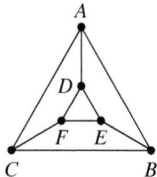

  a Identify the odd one out.

  b For each of the others, list a possible vertex set and edge set.

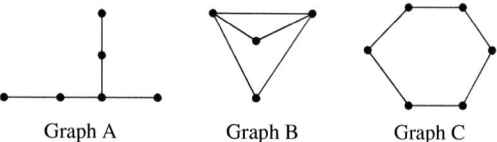

Graph A   Graph B   Graph C

4 One of these graphs is a sub-division of the complete graph K_4

Identify which one, and list the 'extra' vertices.

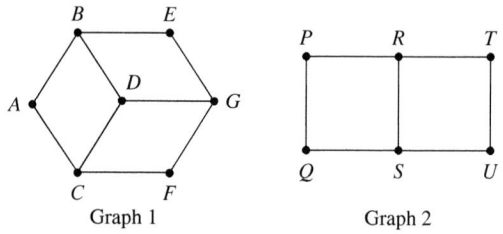

Graph 1   Graph 2

5 Write down the adjacency matrix corresponding to this graph.

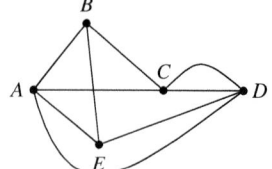

6 Draw graphs corresponding to the following adjacency matrices.

a

	A	B	C	D
A	0	1	1	1
B	1	0	0	1
C	1	0	0	0
D	1	1	0	0

b

	A	B	C	D
A	0	1	2	0
B	1	0	0	1
C	2	0	0	0
D	0	1	0	2

7 **a** Write down the adjacency matrix corresponding to this graph, *G*.

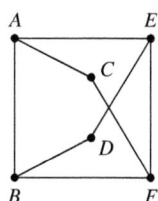

b Construct the adjacency matrix for *G'*, the complement of *G*. Hence draw *G'*.

8 Draw the bipartite graphs with these adjacency matrices.

a

	D	E	F
A	0	1	1
B	1	1	0
C	1	0	1

b

	S	T	U	V
P	1	1	0	1
Q	1	0	1	0
R	1	0	1	1

9 Draw the directed graph corresponding to this adjacency matrix.

		To			
		A	B	C	D
From	A	0	1	2	0
	B	0	1	1	1
	C	1	0	0	1
	D	1	0	1	0

10 Draw the directed network corresponding to this distance matrix.

		To				
		A	B	C	D	E
From	A	–	8	5	–	9
	B	8	–	–	–	–
	C	–	3	–	6	–
	D	2	4	11	–	–
	E	–	–	5	–	–

DISCRETE

Reasoning and problem-solving

Strategy

To solve a problem involving graphs

(1) If necessary model the situation by means of a graph.

(2) Use a diagram or a table, as appropriate.

(3) Ensure that you use the correct graph terms.

(4) Answer the question in context.

One reason for studying graph theory is that graphs and networks give a simplified representation, a mathematical model of a wide variety of situations. A familiar example of a graph used as a model is the map of the London Underground. The geography of the system and the distances between stations are not accurately represented, because the user only needs to know which line to travel on and which station follows which.

Example 9

Ice cream is available in four flavours – strawberry, vanilla, chocolate and mint. George likes strawberry and chocolate, Harriet likes vanilla, chocolate and mint, Isla likes strawberry and mint while Juan only likes chocolate.

a Use a graph to model this situation.

b If only one ice cream of each flavour is available, use your graph to find how they should be distributed so that everyone is happy.

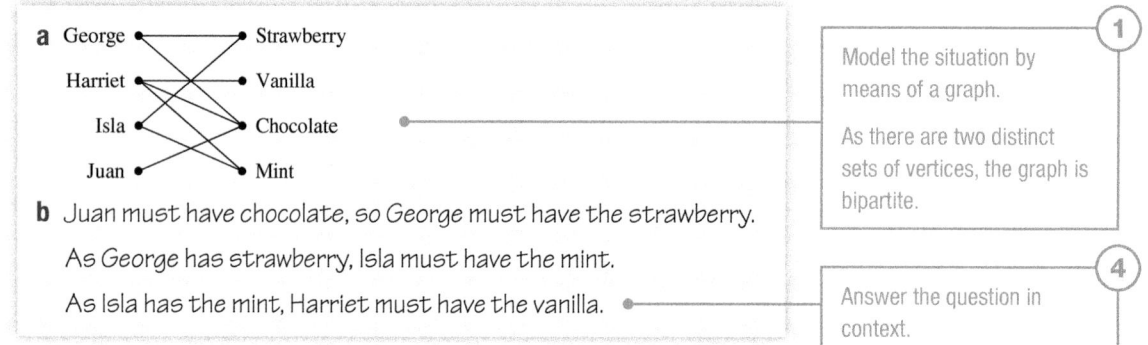

a George — Strawberry
Harriet — Vanilla
Isla — Chocolate
Juan — Mint

b Juan must have chocolate, so George must have the strawberry.

As George has strawberry, Isla must have the mint.

As Isla has the mint, Harriet must have the vanilla.

(1) Model the situation by means of a graph.

As there are two distinct sets of vertices, the graph is bipartite.

(4) Answer the question in context.

If the weights in a network represent direct, straight line distances, then they must obey the **triangle inequality**.

> **Key point**
>
> The triangle inequality states that the sum of the lengths of any two sides of a triangle cannot be less than the length of the third side.
>
> $AB + BC \geq AC$

Of course, if the distances are for 'wiggly' routes or the weights represent other things, such as travel times or costs, the network may not satisfy the triangle inequality.

Example 10

This network shows the travel times, in minutes, between four towns.

a Show that this network does not satisfy the triangle inequality.

b Construct a table of shortest travel times.

c Draw a K_4 graph for the network corresponding to your answer to **b**.

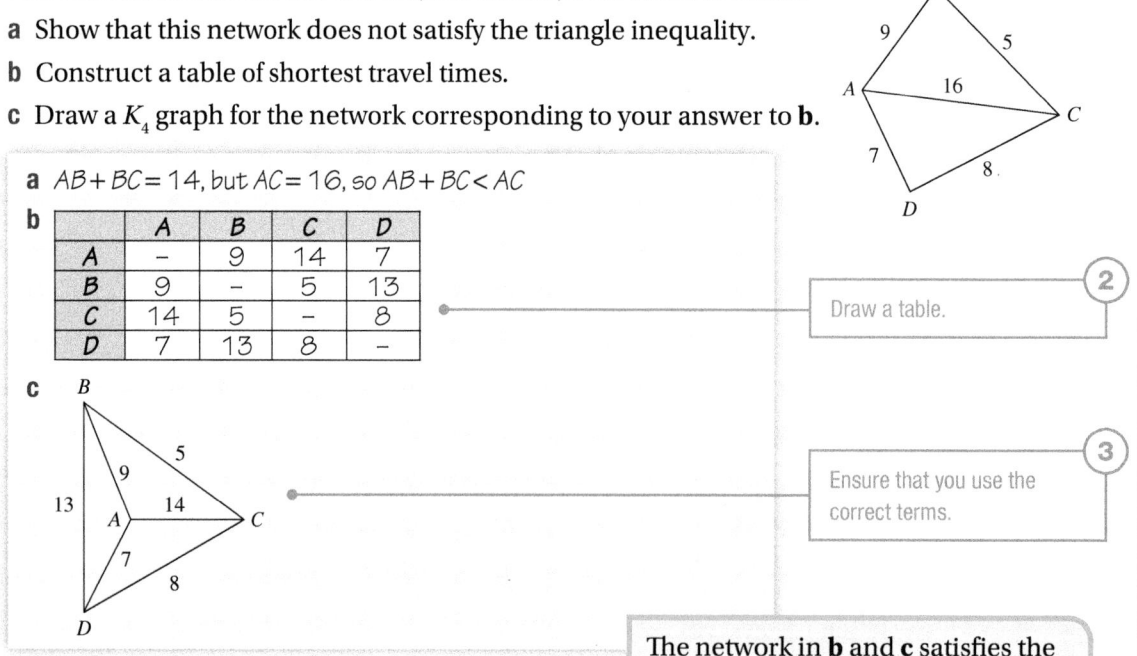

a $AB + BC = 14$, but $AC = 16$, so $AB + BC < AC$

b

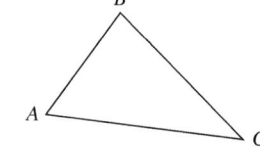

	A	B	C	D
A	–	9	14	7
B	9	–	5	13
C	14	5	–	8
D	7	13	8	–

(2) Draw a table.

c

(3) Ensure that you use the correct terms.

The network in **b** and **c** satisfies the triangle inequality.

1   A team of four runners is to run a $4 \times 100\,\text{m}$ relay race. Dwayne likes to run either the first or second leg. Anton will run any leg other than the first. Kris prefers the third or fourth leg. Jason likes either the first or the last leg.

   a   Draw a bipartite graph to show this information.

   b   List the three possible ways in which the legs could be allocated so that everyone is happy.

2   The diagram shows the direct bus routes offered by an operator, together with the cost, in pounds, of a ticket on each route.

   a   Draw a distance matrix corresponding to this network.

   b   Show that the network does not satisfy the triangle inequality.

   c   Construct a table of cheapest costs between these five towns.

   d   Draw the K_5 graph corresponding to the network in **c**.

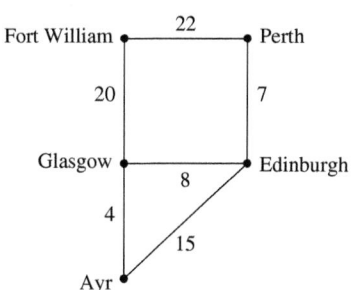

3   a   State in terms of n the degree of each vertex of K_n

   b   Use the handshaking lemma to find the number of edges in K_n in terms of n

4   Alan has four friends – Ben, Candy, Dee and Eli. Ben is friends with Candy and Eli but doesn't know Dee. Candy is friends with Dee and Eli. Dee does not know Eli.

	Alan	Ben	Candy	Dee	Eli
Alan	0	1	1	1	1
Ben	1	0	1	0	1
Candy	1				
Dee	1				
Eli	1				

   a   Copy and complete the adjacency matrix showing these friendships.

   b   Draw the corresponding graph.

   c   Construct the adjacency matrix for the complement of this graph and draw the corresponding graph. What information does this graph provide?

   d   Is the graph K_4 a sub-graph of the graph you drew in **b**? If it is, list the vertices corresponding to this sub-graph and state what it means about the people involved.

5   The diagram shows a road junction at which traffic lights are to be installed.

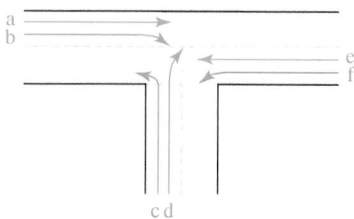

   a   Draw a graph with a vertex for each traffic stream to model which streams can safely flow at the same time.

   b   What is the maximum number of streams which can coexist?

   c   If the lights are to have three phases, which streams should be allowed to go in each phase?

DISCRETE

Fluency and skills

Most problems modelled by graphs and networks involve moving around the graph. Any continuous journey around a graph with no restrictions on repeating edges or vertices is sometimes called a walk, but different types of walk have special names.

- A walk with no repeated edges is called a **trail**. A trail can visit vertices more than once. If a trail returns to its starting vertex, it is a **closed** trail.
- A walk with no repeated edges or vertices (except if you return to the start) is called a **path**. A closed path is called a **cycle**.

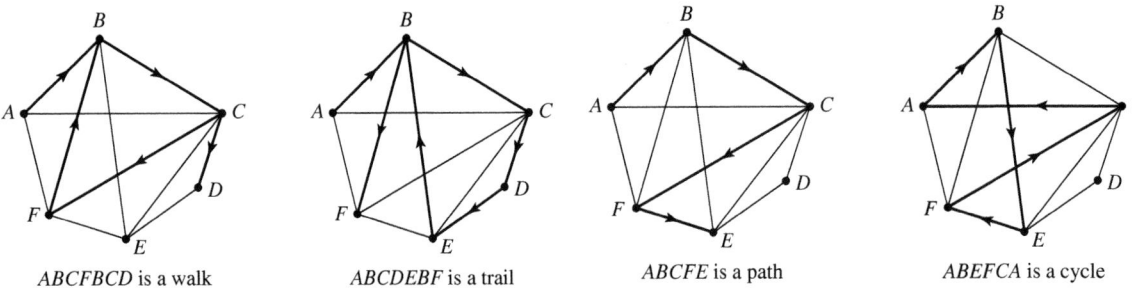

ABCFBCD is a walk   *ABCDEBF* is a trail   *ABCFE* is a path   *ABEFCA* is a cycle

- A connected graph with no cycles is a **tree**.
- A cycle which visits every vertex once and once only is a **Hamiltonian cycle** (sometimes called a **Hamiltonian tour**). A graph which has such a cycle is a **Hamiltonian graph**.

You may have seen puzzles asking you to draw a graph without lifting your pencil or going over a line twice. For example:

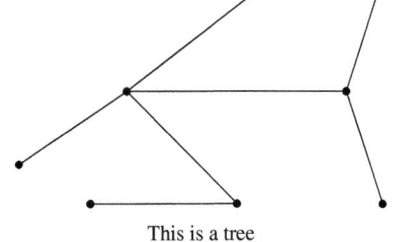

This is a tree

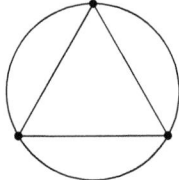 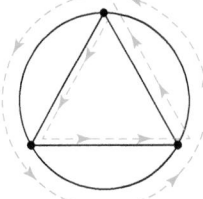

The trail starts and finishes in the same place (it is closed). A graph for which this is possible is called **traversable** or **Eulerian** (pronounced *oi-leerie-ann*).

If a graph can only be drawn by starting and finishing at different points, it is **semi-traversable** or **semi-Eulerian**.

> Named after Leonhard Euler, the 18th century mathematician who studied such graphs.

 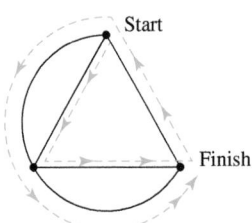

Some graphs are non-traversable (non-Eulerian). For example, try to draw this graph.

You can tell if a graph is traversable by examining whether the degrees of the vertices are odd or even.

> **Key point**
> A graph is traversable (Eulerian) if all vertices are of **even** degree.

A route which traverses the graph is called an **Eulerian trail**. For this graph, *ABCABCA* is an Eulerian trail.

> **Key point**
> A graph is semi-traversable (semi-Eulerian) if it has two odd degree vertices. The two odd vertices have to be the start and finish vertices.

> **Key point**
> A graph with more than two odd degree vertices is non-traversable (non-Eulerian).

The handshaking lemma states that the total of degrees is twice the number of edges. The total of the degrees is therefore an even number, which means the list of degrees must have an even number of odd values.

> **Key point**
> The number of odd vertices in a graph is even.

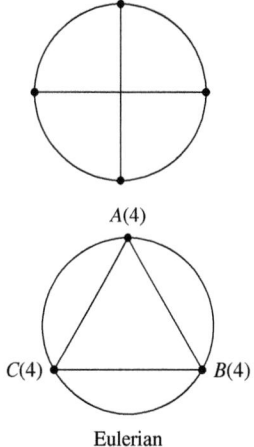

A(4)

C(4)   B(4)

Eulerian

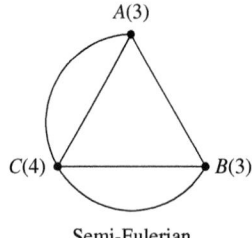

A(3)

C(4)   B(3)

Semi-Eulerian

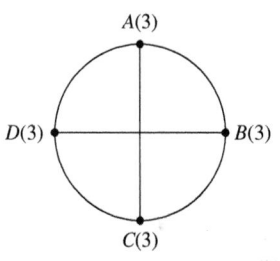

A(3)

D(3)   B(3)

C(3)

Non-Eulerian

DISCRETE

Example 1

a Show that this graph is Hamiltonian.

b Show that the graph is semi-Eulerian. Where would a trail start and finish if it is to traverse the graph?

c Modify the graph to make it Eulerian.

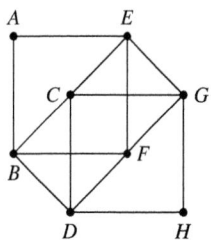

a *AECGHDBA* is a Hamiltonian cycle, so the graph is Hamiltonian.

b The degrees of the vertices are A(2), B(3), C(4), D(3), E(4), F(4), G(4), H(2)

There are two vertices with odd degree, so the graph is semi-Eulerian.

The trail would have to start at *B* and end at *D*, or vice versa.

c The graph will be Eulerian if the edge *BD* is added, making the degrees of *B* and *D* both 4

1 For the graph shown

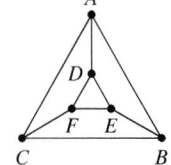

    a Write a path from *A* to *C* which uses

        i 3 edges   **ii** 5 edges.

    b Write a cycle, starting and ending at *B*, which involves (including *B*)

        i 3 vertices   **ii** 5 vertices.

    c Write a Hamiltonian cycle.

    d Explain why *EFCABC* is not a path.

2 Which of these graphs are trees?

    a 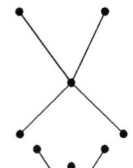   **b**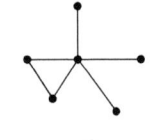

    c    **d**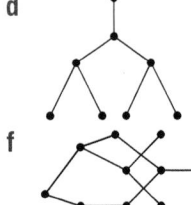

    e   **f**

3 State whether these graphs are Eulerian, semi-Eulerian or neither.

    a 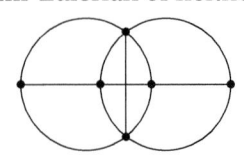   **b**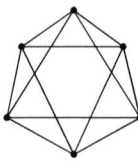

    c 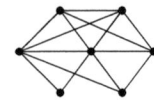   **d**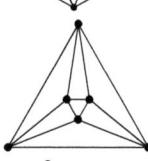

    e    **f**

4 Draw a simple, connected graph with five vertices which is

    a Eulerian but not Hamiltonian

    b Hamiltonian but not Eulerian

    c Neither Hamiltonian nor Eulerian

    d Both Hamiltonian and Eulerian.

5 A simple connected graph has 5 vertices and 5 edges.

    a State the sum of the degrees of its vertices.

    b Draw the graph if exactly one of the vertices is of degree 4. State whether your graph is Eulerian, semi-Eulerian or neither.

    c Draw the graph given that it is Hamiltonian.

Reasoning and problem-solving

Strategy

To solve a problem involving graphs

1 If necessary, model the situation by means of a graph.

2 Use a diagram or a table as appropriate.

3 Ensure that you use the correct graph terms.

4 Answer the question in context.

Example 2

The plan shows a building with five rooms and two doors to the outside. A prize is offered to anyone who can enter the building, pass through every door once only and return to the outside.

By modelling the situation as a graph, investigate whether this is possible.

(Continued on the next page)

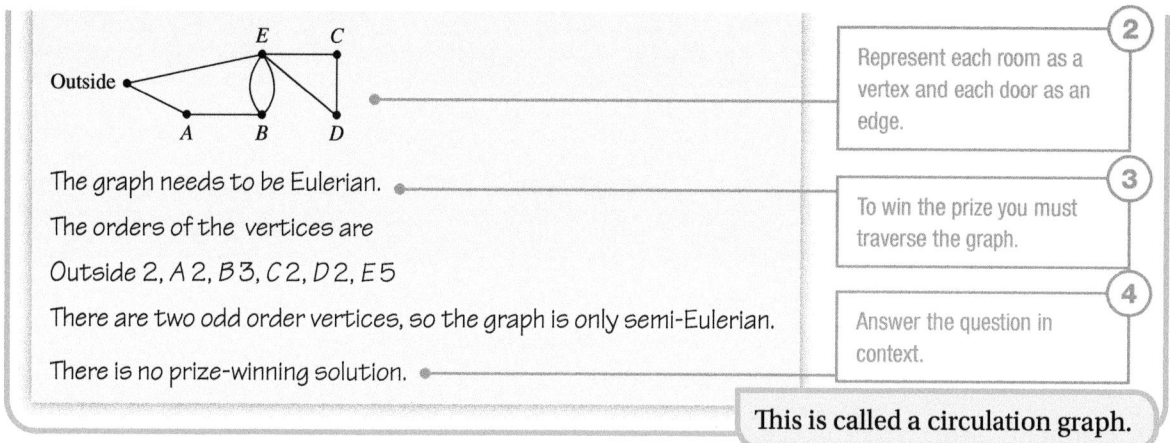

The graph needs to be Eulerian.

The orders of the vertices are

Outside 2, A 2, B 3, C 2, D 2, E 5

There are two odd order vertices, so the graph is only semi-Eulerian.

There is no prize-winning solution.

Represent each room as a vertex and each door as an edge. ②

To win the prize you must traverse the graph. ③

Answer the question in context. ④

This is called a circulation graph.

Exercise 12.2B Reasoning and problem-solving

1 The diagram shows the layout of a building with seven rooms. There are three doors to the outside.

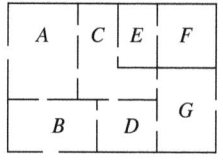

a Draw a circulation graph, with vertices for the rooms and the outside, and edges showing possible movement between them.

b Is your graph Eulerian, semi-Eulerian or neither? Give your reason.

2 The structure of a saturated hydrocarbon molecule is a tree with vertices of order 4 (the carbon atoms) and vertices of order 1 (the hydrogen atoms). Draw two distinct molecules with four carbon atoms. How many hydrogen atoms are required in each case?

3 a Explain why the graph shown is semi-Eulerian.

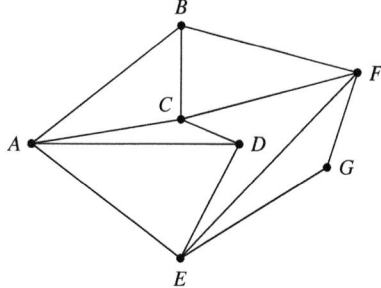

b Which edge would you need to add to the graph to make it Eulerian?

c Give an example of an Eulerian trail for your modified graph.

4 The diagram shows the town of Königsberg.

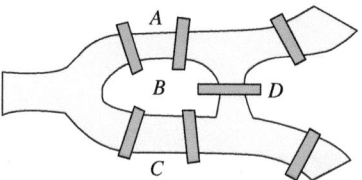

There is an island just downstream from where two rivers meet, and in the eighteenth century there were seven bridges, as shown. The citizens of Königsberg used to spend their Sunday afternoons trying to walk across each bridge once only, ending up at their starting point. By using a graph to model the situation, decide whether the citizens of Königsberg were wasting their time trying to solve this problem. (The Königsberg bridge problem was part of Euler's original study of traversability, published in 1736.)

5 There are only two distinct trees with 4 vertices.

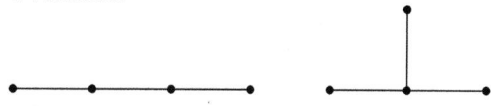

All other such trees are isomorphic to one of these two.

How many distinct trees can you draw with

a 5 vertices **b** 6 vertices?

6 Investigate the conditions under which the complete graph K_n is traversable.

7 Investigate the conditions under which the complete bipartite graph $K_{m,n}$ is Hamiltonian.

DISCRETE

Fluency and skills

If you list the sub-graphs of a given graph, many of them do not contain a cycle; they are trees. Some of these, called **spanning trees** or **connectors**, contain every vertex of the graph.

> **Key point**
> A spanning tree or connector of a graph is a sub-graph that is a tree containing every vertex of the graph.

Here are two possible spanning trees of a graph.

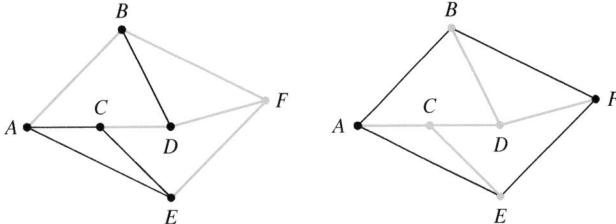

In this example, the graph has six vertices, and each spanning tree has five edges.

A graph with two vertices is spanned by one edge, and each extra vertex requires one extra edge, so, in general, the number of edges in a spanning tree is one fewer than the number of vertices.

> **Key point**
> A spanning tree for a graph with n vertices has $(n-1)$ edges.

For a network (weighted graph), you can find the spanning tree with the lowest total weight.

> **Key point**
> For a network, the spanning tree with the lowest total weight is the minimum spanning tree (MST) or minimum connector.

This is the minimum spanning tree for this network.

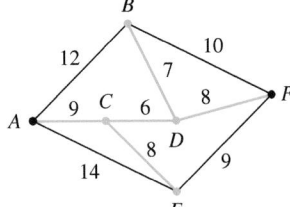

For a small network, the minimum spanning tree is obvious by inspection, but the number of possible spanning trees increases rapidly as the size of the network increases (for K_n there are n^{n-2} possible spanning trees), so, in general, you need an algorithm to find the minimum.

One such algorithm is **Kruskal's algorithm**.

Kruskal's algorithm

Step 1 Choose the arc with minimum weight.

Step 2 Choose the arc with least weight from the remaining arcs, avoiding those that form a cycle with those already chosen.

Step 3 If any nodes remain unconnected, go to **Step 2**

If at any stage there is a choice of arcs, choose at random.

Kruskal's algorithm is an example of a greedy algorithm. At each stage, you make the most advantageous choice without thinking ahead.

DISCRETE

Example 1

Use Kruskal's algorithm to find a minimum spanning tree for the network shown. List the order in which arcs are chosen.

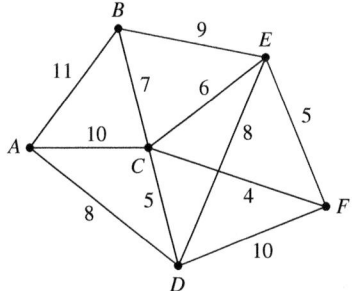

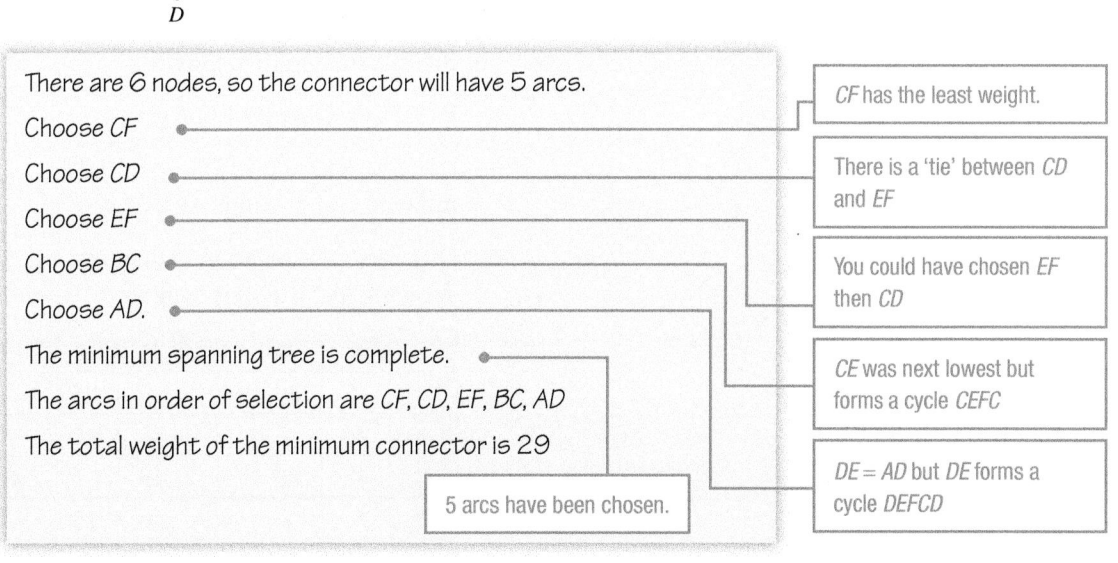

There are 6 nodes, so the connector will have 5 arcs.

Choose CF •————————————————— CF has the least weight.

Choose CD •————————————————— There is a 'tie' between CD and EF

Choose EF •—————————————————

Choose BC •————————————————— You could have chosen EF then CD

Choose AD. •—————————————————

The minimum spanning tree is complete. •————————— CE was next lowest but forms a cycle CEFC

The arcs in order of selection are CF, CD, EF, BC, AD

The total weight of the minimum connector is 29

5 arcs have been chosen.

DE = AD but DE forms a cycle DEFCD

1 Use Kruskal's algorithm to find the minimum spanning tree for each of the networks shown. List the order in which you choose the arcs, and find the total weight of the connector.

a

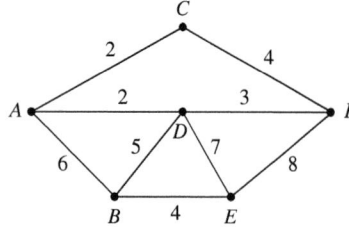

b

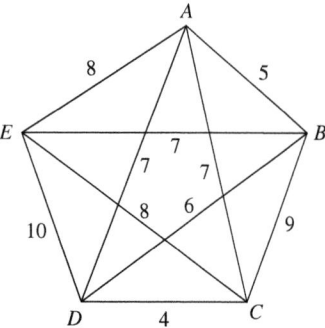

c

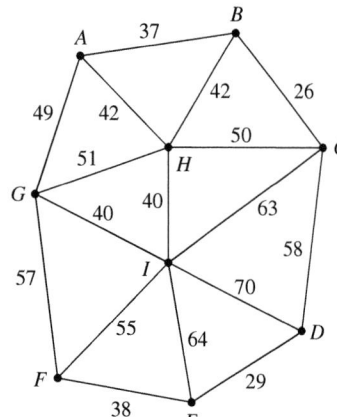

d

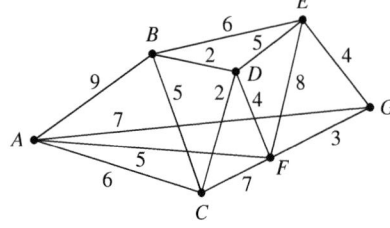

e

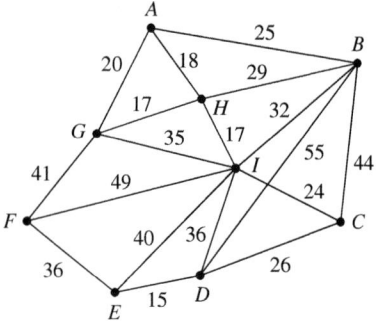

2 Use Kruskal's algorithm to find the two possible minimum spanning trees for the network shown. List the order in which you choose the arcs, and find the total weight of the connector.

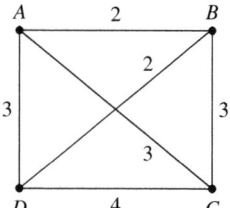

3 Find all possible minimum spanning trees for this network.

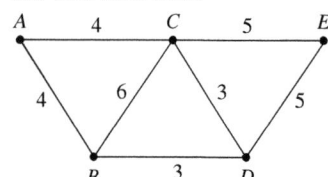

4 For the distance matrix shown,

a Draw the corresponding graph,

b Use Kruskal's algorithm to find the minimum spanning tree,

c Draw a separate labelled diagram showing just the spanning tree.

	A	B	C	D	E	F
A	–	20	42	–	35	–
B	20	–	30	–	40	–
C	42	30	–	45	–	38
D	–	–	45	–	25	55
E	35	40	–	25	–	–
F	–	–	38	55	–	–

Reasoning and problem-solving

To solve a problem using Kruskal's algorithm

(1) If necessary, draw a clear graph.

(2) Apply Kruskal's algorithm, showing clearly the order in which you choose the arcs.

(3) Answer the question in context.

The network represents a country park, with seven picnic sites and a car park joined by rough paths. Distances are in metres.

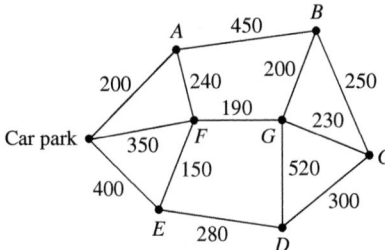

The council plans to upgrade some paths, so that all the picnic sites are accessible by wheelchair. The cost of the upgrade is £55 per metre of path. Use Kruskal's algorithm to decide which paths should be upgraded, and state how much money the project will cost.

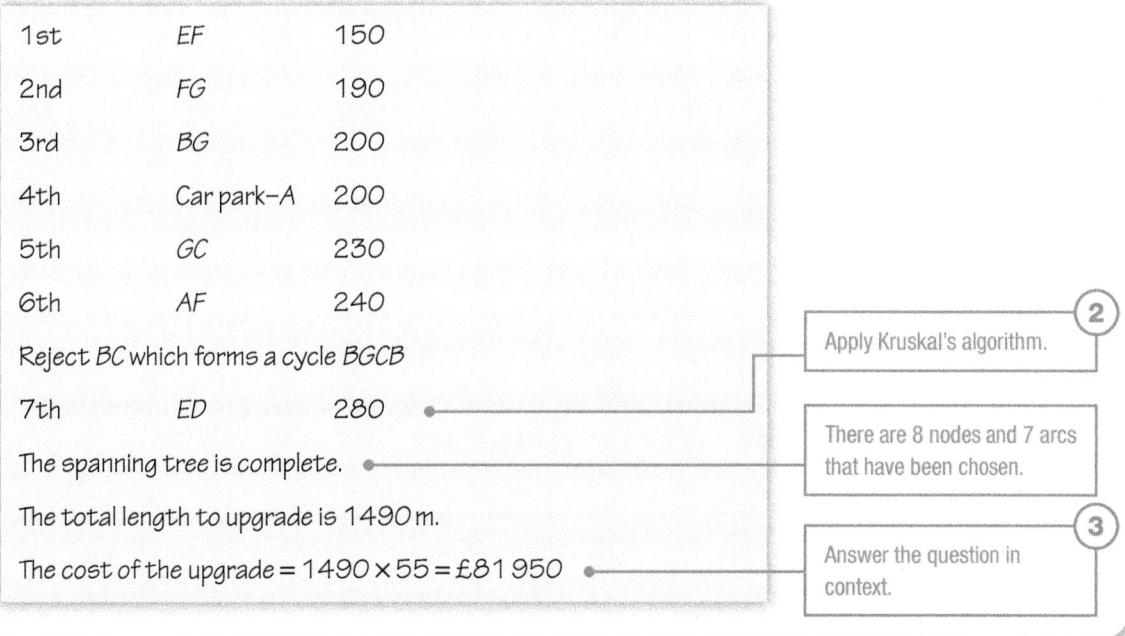

1st	EF	150
2nd	FG	190
3rd	BG	200
4th	Car park–A	200
5th	GC	230
6th	AF	240

Reject BC which forms a cycle BGCB

| 7th | ED | 280 |

The spanning tree is complete.

The total length to upgrade is 1490 m.

The cost of the upgrade = 1490 × 55 = £81 950

Apply Kruskal's algorithm.

There are 8 nodes and 7 arcs that have been chosen.

Answer the question in context.

DISCRETE

363

1   A farmer has five animal shelters on his land and wishes to connect them all to the water supply at the farmhouse. The table shows the distances (in metres) between the farmhouse, F, and the shelters A, B, C, D, and E (some direct connections are not possible).

	A	B	C	D	E	F
A	–	70	100	120	80	50
B	70	–	–	–	70	–
C	100	–	–	60	–	–
D	120	–	60	–	–	80
E	80	70	–	–	–	–
F	50	–	–	80	–	–

Use Kruskal's algorithm to find which connections the farmer should make to achieve the water supply as efficiently as possible. Find the total length of water piping he would need.

2   The diagram shows the roads connecting seven towns. The lengths of the roads are given in miles.

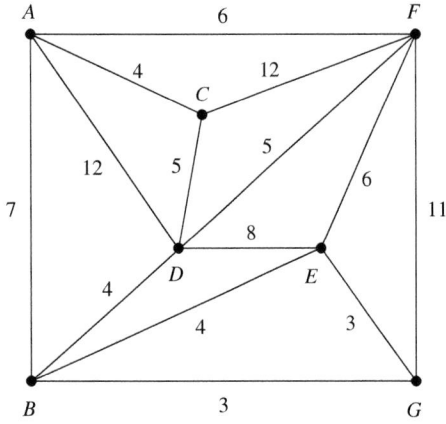

During the winter, snowploughs are used to keep enough roads open so that travel is possible between any two towns.

a   Use Kruskal's algorithm to decide which roads should be kept open to minimise the length of road needing attention. Find the total length of road involved.

b   The ambulance station is situated along the road from A to B, so this must be kept open. Which road would it replace in your answer to part **a**?

3 The table shows the cost, in pounds per thousand words, of translating between languages.

	English	French	German	Italian	Portuguese	Spanish
English	–	25	30	27	38	22
French	25	–	35	22	36	28
German	30	35	–	35	40	32
Italian	27	22	35	–	26	20
Portuguese	38	36	40	26	–	23
Spanish	22	28	32	20	23	–

a Use Kruskal's algorithm to find the minimum spanning tree for this network.

b From your solution to part **a**, state the cheapest sequence of translations to make a document of 10 000 words, originally in German, available in all the languages, and calculate the cost of the operation.

c Calculate the cost of translating the document directly from German into each of the other languages. Suggest reasons why, despite the extra cost, this might be a preferable course of action.

4 The diagram shows an inlet with five islands.

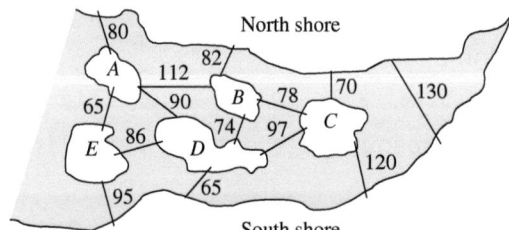

The islands and the mainland are connected by ferries or toll bridges, and the figures shown give the prices, in pounds, of a three month pass for each crossing.

A delivery firm needs to deliver to all areas.

a Draw a network to model the situation.

b Use Kruskal's algorithm to decide which passes the firm should purchase to give them the desired access as cheaply as possible.

Fluency and skills

Kruskal's algorithm cannot easily be adapted to a network stored as a matrix because it is difficult to check whether an arc will form a cycle with those already chosen. This means that it is not well suited to computerisation. The alternative is **Prim's algorithm**.

With Prim's algorithm, the minimum spanning tree is built up from a chosen starting node. Nodes and arcs are added one at a time to a connected sub-graph until the spanning tree is complete.

Prim's algorithm | Key point

Step 1 Choose any node to be the first in the connected set.

Step 2 Choose the arc of minimum weight joining a connected node to an unconnected node. Add this arc to the spanning tree and the node to the connected set.

Step 3 If any unconnected nodes remain, go to **Step 2**

If at any stage there is a choice of arcs, choose at random.

> This is also a **greedy algorithm**. It avoids the problem of forming cycles because it only considers arcs to unconnected nodes, which cannot possibly form a cycle.

Example 1

Starting from *A*, use Prim's algorithm to find a minimum spanning tree for this network. List the order in which arcs are chosen.

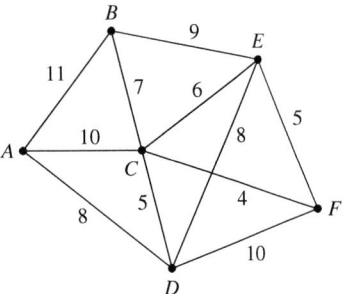

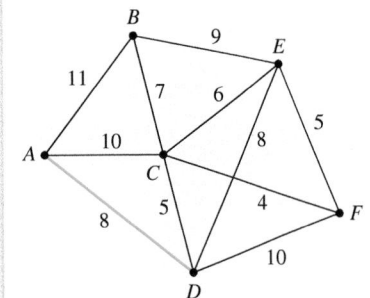

Take *A* as the first node in the connected set.

The arcs to unconnected nodes are *AB*, *AC* and *AD*

The least weight is *AD*, so you add this arc to the connector.

(Continued on the next page)

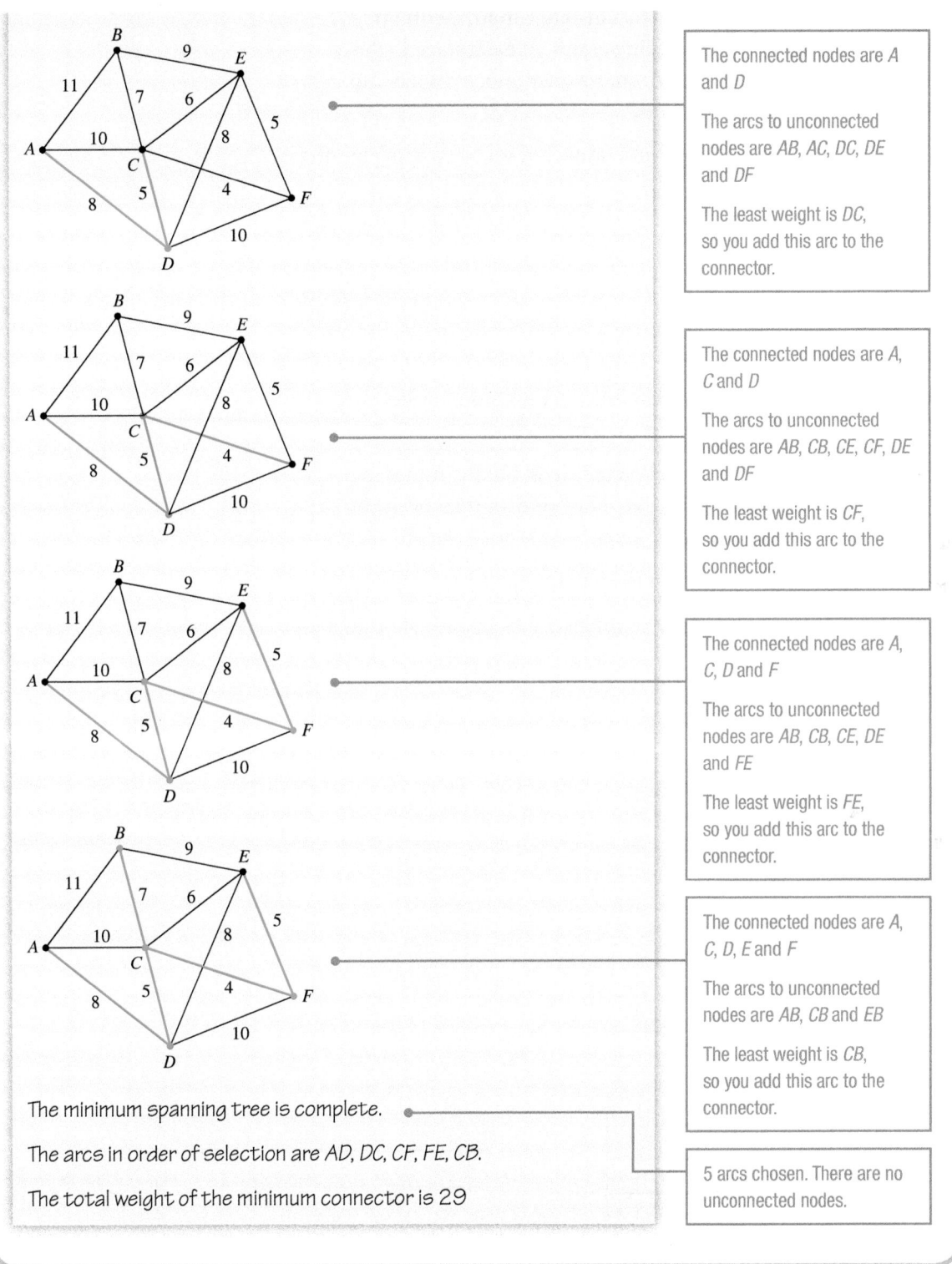

The connected nodes are A and D

The arcs to unconnected nodes are AB, AC, DC, DE and DF

The least weight is DC, so you add this arc to the connector.

The connected nodes are A, C and D

The arcs to unconnected nodes are AB, CB, CE, CF, DE and DF

The least weight is CF, so you add this arc to the connector.

The connected nodes are A, C, D and F

The arcs to unconnected nodes are AB, CB, CE, DE and FE

The least weight is FE, so you add this arc to the connector.

The connected nodes are A, C, D, E and F

The arcs to unconnected nodes are AB, CB and EB

The least weight is CB, so you add this arc to the connector.

5 arcs chosen. There are no unconnected nodes.

The minimum spanning tree is complete.

The arcs in order of selection are AD, DC, CF, FE, CB.

The total weight of the minimum connector is 29

Prim's algorithm can be applied to a distance matrix without the need to draw the corresponding graph. At each stage in the algorithm, you transfer a node from the unconnected set to the connected set. On the matrix you:

- Remove the node from the unconnected set by crossing out the row for that node.
- Add it to the connected set by labelling the column to show the order in which you choose the nodes.

Prim's algorithm on a matrix

Step 1 Select the first node.

Step 2 Cross out the row and number the column for the chosen node.

Step 3 Find the minimum undeleted weight in the numbered columns. Circle this value. The node for this row is the next chosen node.

Step 4 Repeat steps 2 and 3 until all nodes have been chosen.

If there is a choice between two equal weights, choose at random.

Example 2

Use Prim's algorithm to find the minimum spanning tree for the network given by this distance matrix.

	1					
	A	B	C	D	E	F
A	–	12	9	–	14	–
B	12	–	–	7	–	10
C	9	–	–	6	8	–
D	–	7	6	–	–	8
E	14	–	8	–	–	9
F	–	10	–	8	9	–

	1					
	A	B	C	D	E	F
~~A~~	–	~~12~~	~~9~~	–	~~14~~	–
B	12	–	–	7	–	10
C	9	–	–	6	8	–
D	–	7	6	–	–	8
E	14	–	8	–	–	9
F	–	10	–	8	9	–

Cross through row A (removes A from the unconnected set).

Label column A with 1. (A is the first connected node.)

(Continued on the next page)

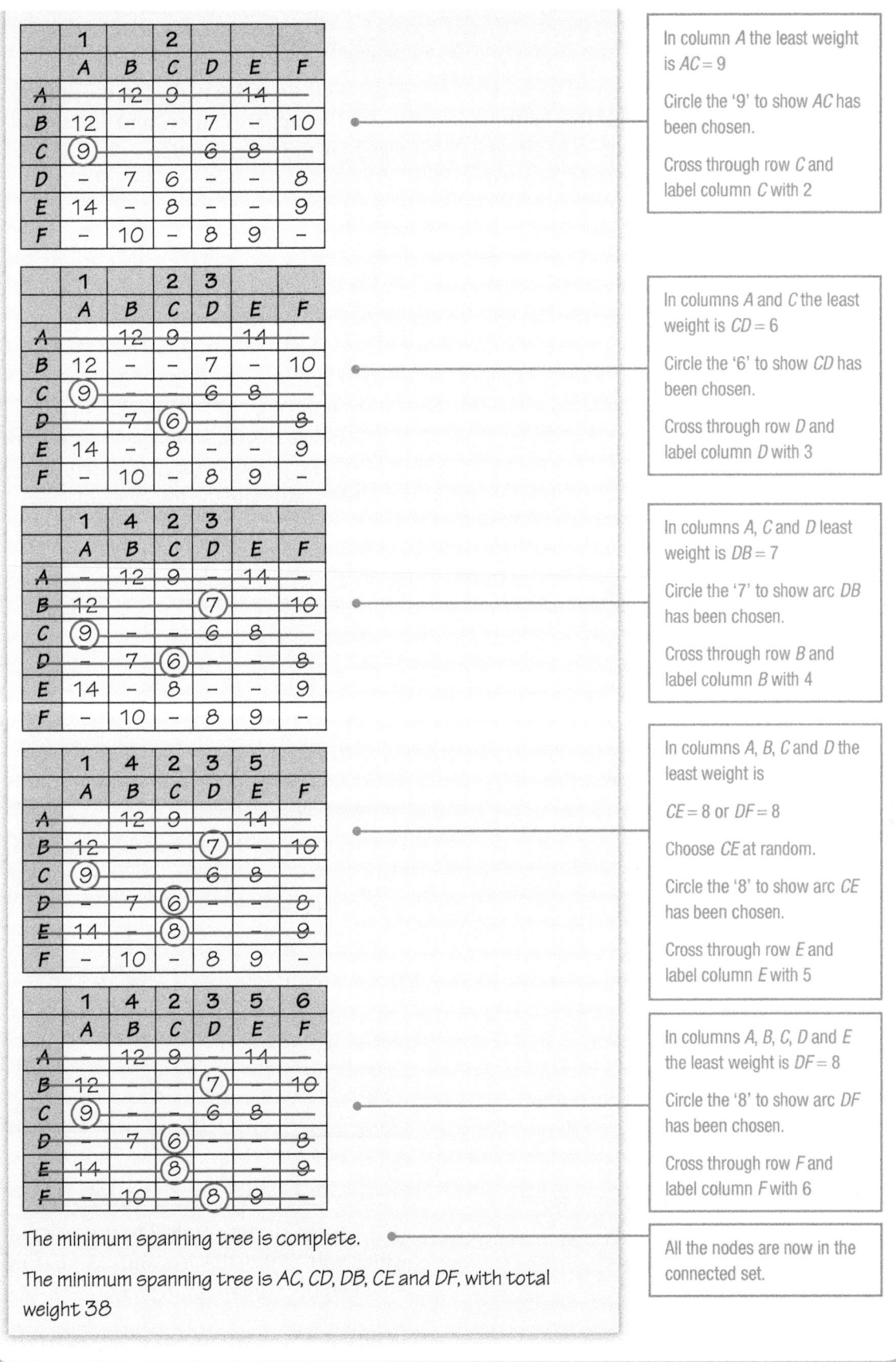

Table 1

	1		2			
	A	B	C	D	E	F
A	–	12	9		14	
B	12	–	–	7	–	10
C	(9)			6	8	
D	–	7	6	–	–	8
E	14	–	8	–	–	9
F	–	10	–	8	9	–

> In column A the least weight is AC = 9
>
> Circle the '9' to show AC has been chosen.
>
> Cross through row C and label column C with 2

Table 2

	1		2	3		
	A	B	C	D	E	F
A	–	12	9		14	
B	12	–	–	7	–	10
C	(9)			6	8	
D	–	7	(6)	–	–	8
E	14	–	8	–	–	9
F	–	10	–	8	9	–

> In columns A and C the least weight is CD = 6
>
> Circle the '6' to show CD has been chosen.
>
> Cross through row D and label column D with 3

Table 3

	1	4	2	3		
	A	B	C	D	E	F
A	–	12	9	–	14	–
B	12		(7)			10
C	(9)	–		6	8	
D	–	7	(6)	–	–	8
E	14	–	8	–	–	9
F	–	10	–	8	9	–

> In columns A, C and D least weight is DB = 7
>
> Circle the '7' to show arc DB has been chosen.
>
> Cross through row B and label column B with 4

Table 4

	1	4	2	3	5	
	A	B	C	D	E	F
A	–	12	9		14	
B	12	–		(7)	–	10
C	(9)			6	8	
D	–	7	(6)	–	–	8
E	14	–	(8)			9
F	–	10	–	8	9	–

> In columns A, B, C and D the least weight is
>
> CE = 8 or DF = 8
>
> Choose CE at random.
>
> Circle the '8' to show arc CE has been chosen.
>
> Cross through row E and label column E with 5

Table 5

	1	4	2	3	5	6
	A	B	C	D	E	F
A	–	12	9		14	
B	12	–		(7)		10
C	(9)	–	–	6	8	
D	–	7	(6)	–	–	8
E	14	–	(8)		–	9
F	–	10	–	(8)	9	–

> In columns A, B, C, D and E the least weight is DF = 8
>
> Circle the '8' to show arc DF has been chosen.
>
> Cross through row F and label column F with 6

The minimum spanning tree is complete.

> All the nodes are now in the connected set.

The minimum spanning tree is AC, CD, DB, CE and DF, with total weight 38

1 Use Prim's algorithm starting from node A to find the minimum spanning tree for the networks shown. List the order in which you choose the arcs, and find the total weight of the connector.

a

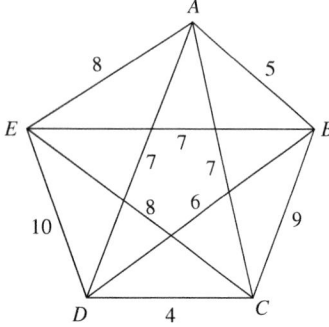

b

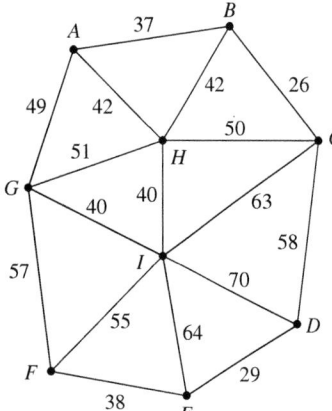

c

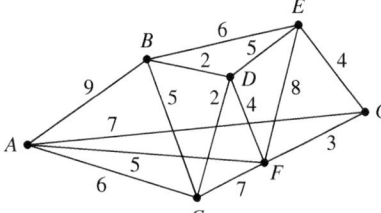

d

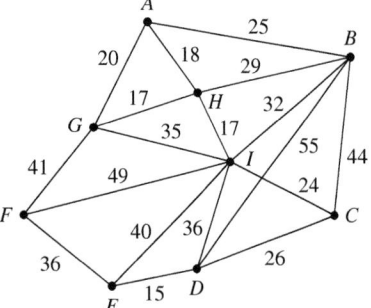

2 Use Prim's algorithm to find the minimum spanning tree for the network shown in this distance matrix.

Use A as the starting node. List the order in which the arcs are chosen, and find the total weight of the spanning tree.

	A	B	C	D	E
A	–	5	9	6	3
B	5	–	11	4	7
C	9	11	–	5	8
D	6	4	5	–	12
E	3	7	8	12	–

3 **a** By applying Prim's algorithm to this distance matrix, show that the network has two possible minimum spanning trees, and state the total weight.

	P	Q	R	S	T	U
P	–	2	5	–	–	9
Q	2	–	1	–	4	7
R	5	1	–	4	–	–
S	–	–	4	–	3	6
T	–	4	–	3	–	4
U	9	7	–	6	4	–

b Draw network diagrams to illustrate the two possible trees.

4 Use Prim's algorithm to find the minimum spanning tree for the network shown in this distance matrix.

	A	B	C	D	E	F	G
A	–	9	17	8	22	16	12
B	9	–	6	15	–	20	14
C	17	6	–	–	12	8	19
D	8	15	–	–	10	–	6
E	22	–	12	10	–	11	–
F	16	20	8	–	11	–	13
G	12	14	19	6	–	13	–

Use A as the starting node. List the order in which you choose the arcs and calculate the total length of the spanning tree.

Reasoning and problem-solving

To solve a problem using Prim's algorithm

1 When working with a graph, list the order in which you choose the arcs.

2 When working with a table, label the columns to show clearly the order in which you choose the arcs.

3 Answer the question in context.

Example 3

The network represents a country park, with seven picnic sites and a car park joined by rough paths. Distances are in metres. The council plans to upgrade some paths, so that all the picnic sites are accessible by wheelchair. The cost of the upgrade is £55 per metre of path.

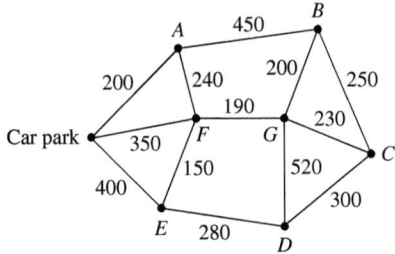

Use Prim's algorithm, starting from the car park, to decide which paths should be upgraded, and state how much money the council should earmark for the project.

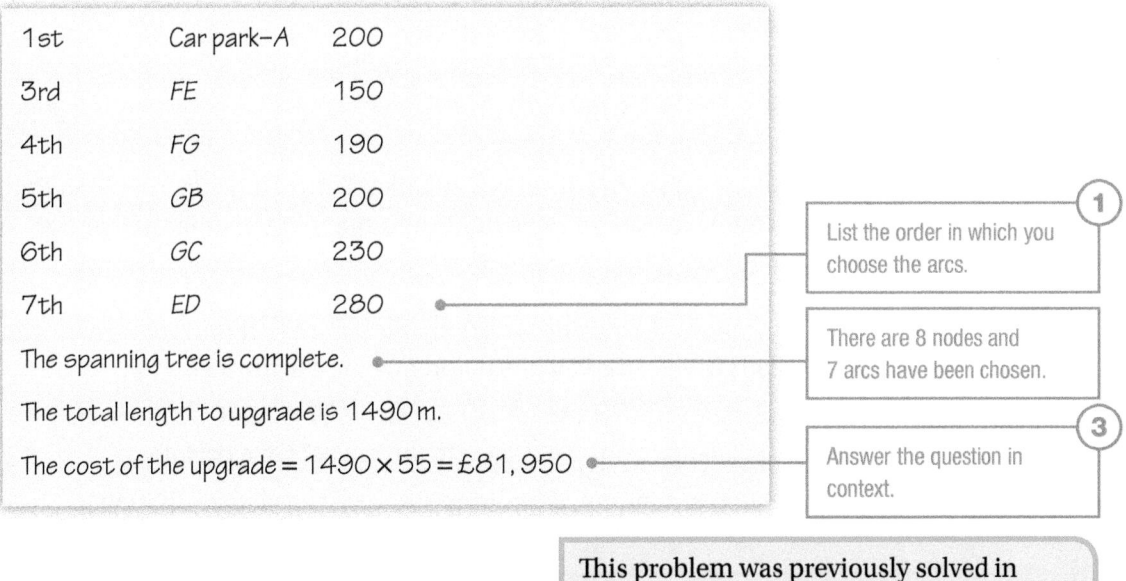

1st	Car park–A	200
3rd	FE	150
4th	FG	190
5th	GB	200
6th	GC	230
7th	ED	280

List the order in which you choose the arcs. **1**

The spanning tree is complete.

There are 8 nodes and 7 arcs have been chosen.

The total length to upgrade is 1490 m.

The cost of the upgrade = 1490 × 55 = £81,950

Answer the question in context. **3**

This problem was previously solved in Section 12.3. Look back and compare Prim's algorithm and Kruskal's algorithm to see how they differ.

1 The table shows the distance by direct rail link between eight towns.

	A	B	C	D	E	F	G	H
A	–	56	20	–	–	–	–	70
B	56	–	–	15	65	–	75	88
C	20	–	–	87	95	–	120	30
D	–	15	87	–	60	–	25	112
E	–	65	95	60	–	30	40	70
F	–	–	–	–	30	–	45	–
G	–	75	120	25	40	45	–	115
H	70	88	30	112	70	–	115	–

It is decided to close some of the links, leaving just enough connections so that it is possible to travel from any town to any other by rail. Use Prim's algorithm to decide which links must be kept so that the amount of track is a minimum. Use town A as the starting node and record the order in which you make the links.

2 a Explain why Prim's algorithm is preferred to Kruskal's algorithm when the network is given in the form of a distance matrix.

	A	B	C	D	E	F
A	–	4	6	12	15	20
B	4	–	5	8	14	25
C	6	5	–	8	12	24
D	12	8	8	–	19	20
E	15	14	12	19	–	11
F	20	25	24	20	11	–

b i The distance matrix shown gives the distances, in miles, between six towns. Using Prim's algorithm, show that there are two possible minimum spanning trees for the network.

ii Draw both minimum spanning trees.

c A bus company only runs services along the roads in the spanning tree.

i How far would a person go when travelling by bus from town A to town F?

ii Where would you suggest that the company should site the bus station if they wish to minimise the greatest distance from the station to the towns? Give reasons.

3 The table shows the walking times, in minutes, between six tourist attractions in a city.

	A	B	C	D	E	F
A	–	4	6	4	3	2
B	4	–	7	6	5	5
C	6	7	–	6	6	5
D	4	6	6	–	3	5
E	3	5	6	3	–	2
F	2	5	5	5	2	–

The information bureau wants to install signposts along a minimum number of routes so that every attraction is accessible and the total walking time is a minimum.

a Use Prim's algorithm to find which routes they should signpost.

b Draw a tree to represent the minimum connector found in **a**.

c What is the greatest time for which a tourist would walk between any two attractions?

4 The diagram shows the roads connecting five villages, with distances in kilometres.

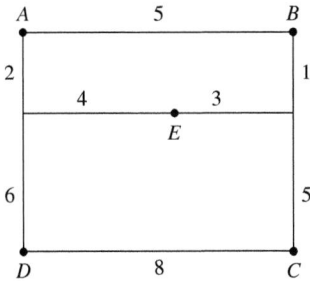

a Construct a table showing the minimum distances by road between the five villages (for example, the distance from A to E is 6 km).

b Draw a network using the distance matrix from part **a**.

c Use Prim's algorithm, starting from node A, to find the minimum connector for the network in part **b**.

As an environmental measure, the local council plans to make some stretches of road 'pedestrians, horses and cycles only', leaving just the minimum unrestricted road so that cars can travel between the five villages.

d Explain why the result you obtained in part **c** does not give the number of kilometres of road which must stay open to cars.

e Using the original diagram, find, by inspection, the roads which should be kept open to cars.

5

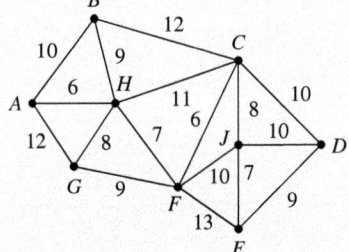

The diagram shows the weight limits (in tonnes) for lorry traffic on roads between nine towns. The council plans to ban lorries on as many roads as possible without stopping lorries from reaching all the towns. By a suitable modification to Prim's algorithm, choose the roads which should remain open to lorries, and state the heaviest lorry which would then have access to all the towns.

Fluency and skills

The route inspection problem is sometimes called the Chinese postman problem (it is the problem rather than the postman that is Chinese. It was originally discussed by the Chinese mathematician Mei-ko Kwan).

The problem is:

For a given network, find the shortest route that travels at least once along every arc and returns to the starting point.

Ideally, you want to travel just once along each arc. You can do this if the network is Eulerian (traversable).

> **Key point**
> If all nodes in the network are even, the network is traversable. The distance travelled is the sum of the weights.

If there are nodes of odd degree, you must repeat some arcs to complete the route. The route inspection problem then becomes:

Which arcs must be repeated to complete the route as efficiently as possible?

Repeating an arc is the same as adding an extra arc to the network. You must choose which arcs to add so that all nodes are made even and the network becomes traversable.

Suppose you must solve the route inspection problem for the network shown on the right.

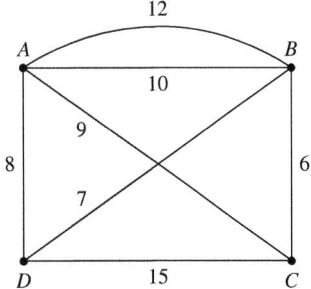

The sum of the weights is 67. The network has two nodes of odd degree, C and D, so the route must travel twice between these nodes. The shortest route from C to D is $CBD = 13$, so you double up on the arcs CB and BD, as shown.

All the nodes now have even degree, so the network is Eulerian. A possible route starting from A is $ABCDABDBCA$

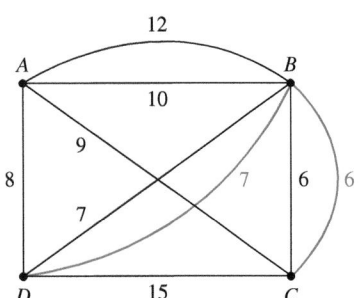

The total distance is $67 + (6 + 7) = 80$

There will always be an even number of odd nodes. If there are more than two, you can pair them together in various ways. You look for the best way to pair them so that the extra arcs have the lowest possible total weight. This gives the **route inspection algorithm:**

Route inspection algorithm

Step 1 Identify the odd nodes.

Step 2 List all possible pairings of the odd nodes.

Step 3 For each pairing find the shortest routes between paired nodes and the total weight of those routes.

Step 4 Choose the pairing with the smallest total.

Step 5 Repeat the shortest route arcs (equivalent to adding extra arcs to the network). The network is now traversable.

The total length of the route = (sum of original weights) + (sum of extra arc weights).

Example 1

Solve the route inspection problem for the network shown. State a possible route.

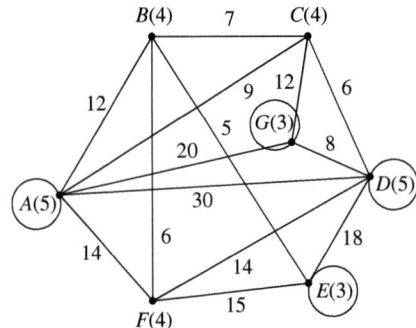

The odd nodes (circled) are A, D, E and G

You can pair these in three ways.

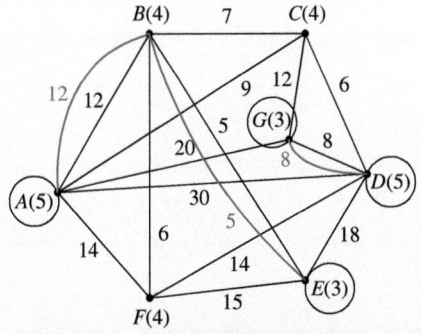

Pairing	Shortest route	Length	Total
AD	AC + CD	15	39
EG	EB + BC + CG	24	
AE	AB + BE	17	25
DG	DG	8	
AG	AG	20	38
DE	DE	18	

The best pairing is AE and DG

The extra arcs are AB, BE and DG

The total distance = 176 + 25 = 201

A possible route, starting from A, is ABCDEFABEBFDGCAGDA

(sum of weights) + (weight of extra arcs).

The main difficulty with the route inspection algorithm is that the number of possible pairings increases very rapidly as the number of odd nodes increases.

2 odd nodes   1 pairing
4 odd nodes   3 pairings
6 odd nodes   15 pairings
8 odd nodes   105 pairings

For a network with n odd nodes, there are
$(n-1) \times (n-3) \times (n-5) \times ... \times 3 \times 1$ possible pairings.

> You will not be expected to deal with more than 4 odd nodes unless there is extra information given to reduce the number of pairs you need to examine.

Exercise 12.5A Fluency and skills

1 This network has a total weight of 41

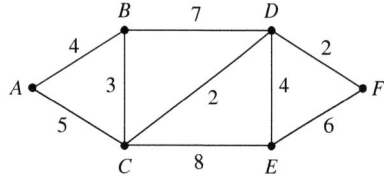

a Explain why a route inspection route starting and finishing at A will need to travel along some arcs more than once.

b Find the arcs that need to be repeated in the route.

c Find the total length of the route.

2 For the network shown,

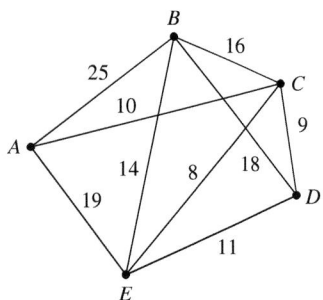

a Find which arcs which should be repeated to solve the route inspection problem,

b State the length of the route,

c State a possible route starting from node A

3 Solve the route inspection problem for each of the networks shown. In each case, show your working clearly, state which arcs will be repeated and give the total weight of the route.

a
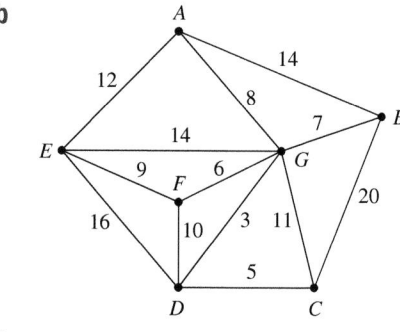

b
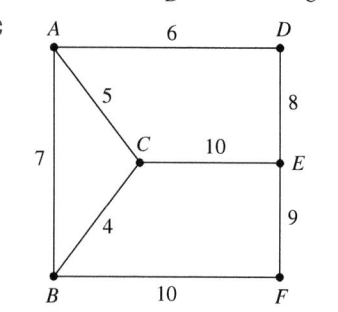

c

d
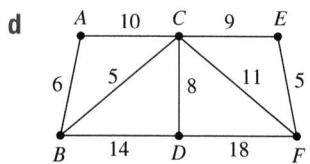

4 This network has a total weight of 82

    a Show that there are two solutions to the route inspection problem. List the arcs that must be repeated in each case.

    b Find the total length of the route.

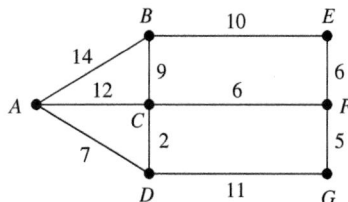

Reasoning and problem-solving

Strategy

To solve a route inspection problem

① Identify nodes with odd degree.

② Find the shortest routes between all pairs of odd nodes.

③ Find which arcs must be added to solve the problem.

④ Answer the question in context.

The inspection route may pass through a particular node several times. Each time, it uses one arc into and one arc out of the node, so, for example, it will pass 3 times through a node of degree 6

Key point

The inspection route passes $\frac{1}{2}n$ times through a node of degree n

Some problems may require different start and finish points. In this case, the network must be made semi-Eulerian.

Example 2

The diagram shows the times, in minutes, that a cleaning vehicle takes to sweep the paths in a park. The total time is 80 minutes. The vehicle is kept at A and must sweep every path and return to A

a Find the minimum time needed to complete the task.

b How many times will the vehicle pass through D?

A change of policy means that the vehicle can start and finish at different points.

c Find the minimum time now needed to complete the task and state the start and finish points.

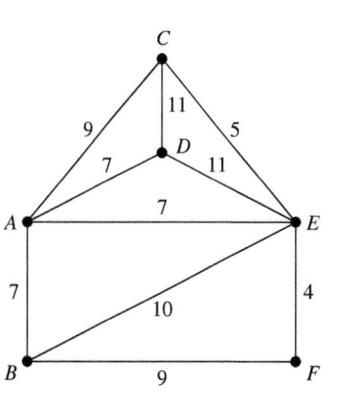

(*Continued on the next page*)

DISCRETE

a The odd nodes are B, C, D and E

Pairing	Shortest route	Length	Total
BC	BE + EC	15	26
DE	DE	11	
BD	BA + AD	14	19
CE	CE	5	
BE	BE	10	21
CD	CD	11	

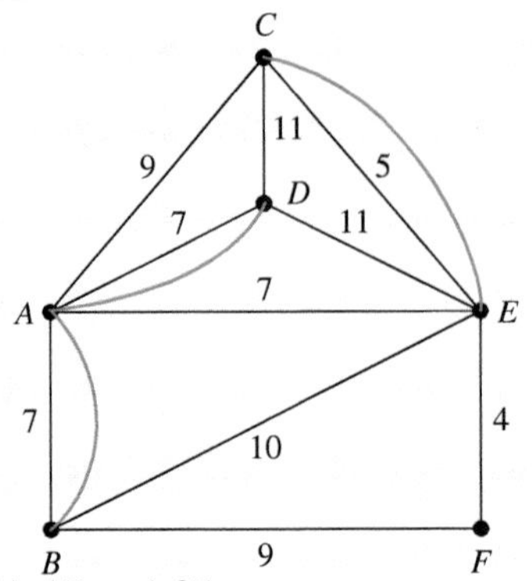

Best pairing is BD and CE, repeating BA, AD and CE

Time needed $= 80 + 19 = 99$ minutes.

b On the modified graph, D had degree 4

Route passes through D twice.

c Route will need to repeat one pair.

The shortest pair is $CE = 5$, so repeat CE

Time needed $= 80 + 5 = 85$ minutes

B and D are still odd nodes. The network is semi-Eulerian.

The route must start at B and finish at D, or vice versa.

① Identify nodes with odd degree.

② Find the shortest routes between all pairs of odd nodes.

③ Find which arcs must be added to solve the problem.

④ Answer the question in context.

(½ × degree)

④ Answer the question in context.

Exercise 12.5B Reasoning and problem-solving

1 The diagram shows the map of the roads on an estate, with distances in metres. All junctions are right angles and all roads are straight apart from the crescent.

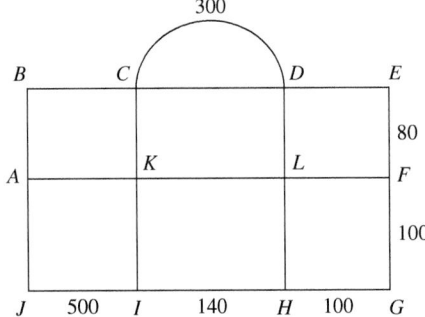

A team of workers is to paint a white line along the centre of each road.

a Assuming that the team enter and leave the estate at A and wish to travel the least possible distance on the estate, which sections of road will they will need to travel along twice? Find the distance they will travel.

b Instead, the team enters the estate at H and leaves at I to go on to another job. How will this affect your results from part **a**?

c What would be the best start and end points for the team if they wish to travel the least possible distance? How far would they go in this case?

2 The table shows the direct road links between six towns, with distances in kilometres.

	A	B	C	D	E	F
A	–	15	–	8	20	–
B	15	–	10	–	6	5
C	–	10	–	9	–	14
D	8	–	9	–	9	4
E	20	6	–	9	–	–
F	–	5	14	4	–	–

Abigail is to do a sponsored cycle ride, travelling at least once along each of these roads and starting and finishing at her home in town A

a Draw the network diagram corresponding to this table.

b Find which roads she should repeat to minimise her journey. State how long her journey will take.

3 The diagram shows the roads connecting five towns, labelled with the distances in kilometres. The total of the distances is 86 km.

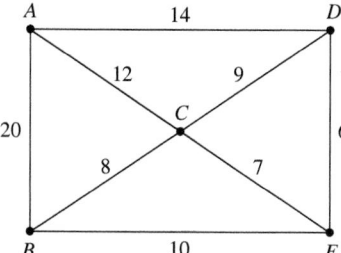

A vehicle starts from A to inspect all the roads for potholes. The vehicle must return to A when the inspection is complete.

a If the vehicle must travel both ways along each road to inspect both carriageways, explain why it can do the job in a minimum distance of 172 km.

b If the vehicle only needs to travel once along a road to inspect it, find the minimum distance it needs to travel, and state which roads must be used twice.

4 The diagram shows the paths in a nature reserve, with lengths in metres. The total length of the paths is 3350 m.

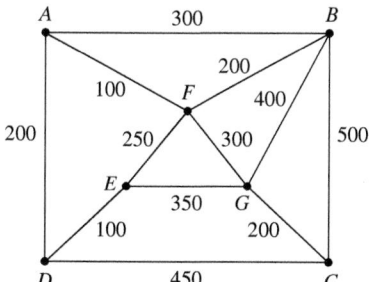

Barry lost a ring during a visit. He went back the next day to look for it.

a What is the minimum distance he needs to walk to search every path, starting and ending at A? Which paths will he travel along twice?

b Barry is sure he didn't go along BG Explain why, nevertheless, he might choose to walk along it.

c Barry's friend agrees to drop him at A and pick him up at another point. Where will he be picked up and how far might he need to walk if he wants to check every path and keep the walking distance to a minimum?

5 The diagram shows the layout of paths in a small pedestrian shopping precinct. Distances are in metres.

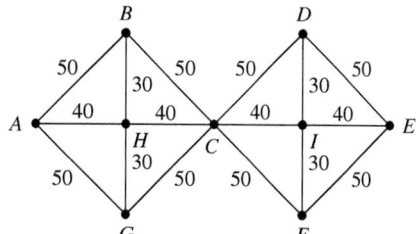

A cleaning cart must drive around the precinct, starting and ending at A

a Find the sections which should be travelled twice to minimise the distance travelled. (Hint: Make use of symmetry.)

b Find the total distance travelled.

c How many times will the cart pass through C?

6 The diagram shows a logo displayed outside a shop. The lengths shown are in cm and the logo is symmetrical, with all angles right angles.

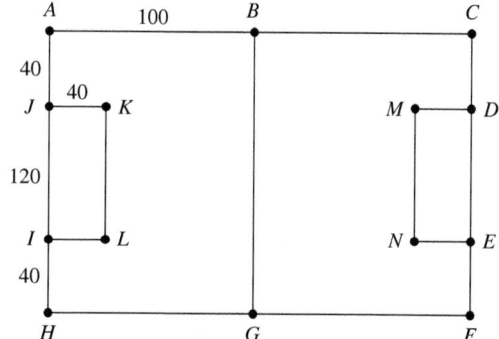

a The logo is outlined with a single rope of LEDs, starting and ending at A

Calculate the minimum length of rope needed for this, and state which sections of the logo will have a double run of rope.

b If the rope could start and end at different points, how much rope would be needed? State the start and end points.

Fluency and skills

A travelling sales person needs to make a number of visits and return to base as efficiently a possible. The travelling salesman problem is named after this situation and can be stated as:

The travelling salesman problem (TSP)   Key point

Find a route that visits every node of a network and returns to the start in the shortest possible distance.

A Hamiltonian cycle (tour) is a closed path that visits every node of a network. The **classical TSP** is to visit every node of a network *once only* and return to the start in the least possible distance: that is, to find a minimum weight Hamiltonian cycle.

However, in a practical situation, the problem may not fit this classical pattern. Firstly, there may not be a Hamiltonian cycle. For example, any complete tour of this graph would need to visit B twice. The graph is not Hamiltonian.

Secondly, the network may not satisfy the triangle inequality, especially if the weights are, for example, travel times.

This network has a Hamiltonian cycle $ABCA$ with total weight 36, but the best practical solution to the problem is $ACBCA$ with total weight 32

To solve a practical TSP with n nodes, you convert it to an equivalent classical TSP.

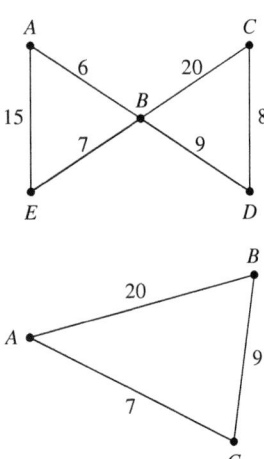

Solving a practical TSP with n nodes   Key point

- Create a complete network, K_n, of shortest distances between pairs of nodes.
- Find a minimum weight Hamiltonian cycle for the complete network (this is the classical problem).
- Interpret the solution in the practical situation.

Example 1

Convert this network to a complete network of shortest distances. Find, by inspection, the solution to the classical TSP for the complete network and hence find a solution to the practical problem.

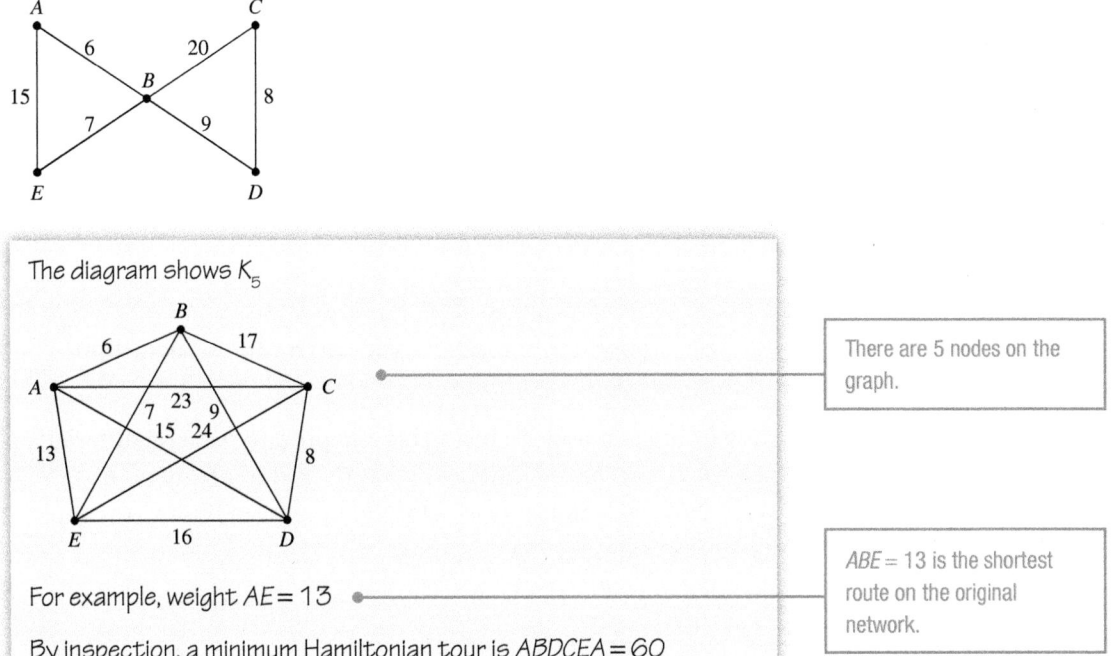

The diagram shows K_5

For example, weight $AE = 13$

By inspection, a minimum Hamiltonian tour is $ABDCEA = 60$

> There are 5 nodes on the graph.

> $ABE = 13$ is the shortest route on the original network.

In the original network:

shortest distance $CE = CDBE = 24$

shortest distance $EA = EBA = 13$

The solution to the practical problem is $ABDCDBEBA = 60$

The only known sure-fire way to find the best tour is to check every Hamiltonian cycle.

From a chosen starting node in K_n, there are $(n-1)$ possible arcs to travel, then $(n-2)$ second arcs and so on. Each tour gets listed twice (forwards and backward) so

Key point

A network with n nodes has $\frac{1}{2}(n-1)!$ possible tours.

This method has exponential order, so even if the network is quite small you have to check a huge number of cycles. Because of this, you need a way to:

- Find a good (not necessarily optimal) solution in a reasonable length of time.
- Decide whether the solution you have found is good enough.

An algorithm which finds a good, but perhaps not perfect solution is called a **heuristic algorithm**. For the TSP a simple heuristic method is the **nearest neighbour algorithm**, which is another example of a greedy algorithm.

The nearest neighbour algorithm

Key point

Step 1 Choose a starting node V

Step 2 From your current position, choose the arc with minimum weight leading to an unvisited node. Travel to that node.

Step 3 If there are unvisited nodes, go to **Step 2**

Step 4 Travel back to V

If at Step 2 there are equal arcs, then you choose one at random. You can rerun the algorithm making the other choice(s) to see if they give better solutions.

You can repeat the algorithm starting from each node in turn. This can produce several different tours from which you select the best. Once you have found a tour, you can start at any node. The 'starting node' referred to in the algorithm is the starting node for the process of finding the tour. It does not have to be the actual start of the salesperson's journey.

Example 2

This network shows the distances, in km, between five towns. Use the nearest neighbour algorithm, starting from each node in turn, to find a route for a travelling salesman, who is based at A and needs to visit every town.

Starting from A, 'nearest' node is C, weight $AC = 8$

From C, 'nearest' unvisited node is D, $CD = 8$

From D, 'nearest' unvisited node is B, $DB = 10$

From B, 'nearest' unvisited node is E, $BE = 15$

No more unvisited nodes, so travel $EA = 14$

The total weight for ACDBEA is $8 + 8 + 10 + 15 + 14 = 55$

Repeat starting from the other nodes:

Start node	Tour(s)	Total weight
A	ACDBEA	55
B	BACDEB	59
C	CABDEC	58
C	CDBAEC	53
D	DCABED	59
E	ECABDE	58
E	ECDBAE	53

There are, in fact, four different tours here.

For a salesman based at A, the best tour found is $ABDCEA = 53$ km.

> For example, *BACDEB* and *DCABED* are the same tour.

For a practical situation, you convert the problem to the classical TSP and then use the nearest neighbour algorithm.

Example 3

The diagram shows the travel times, in minutes, between six locations.

a Draw up a table of shortest travel times.

b Use the nearest neighbour algorithm to find a solution to the travelling salesman problem. List the route the salesperson will take.

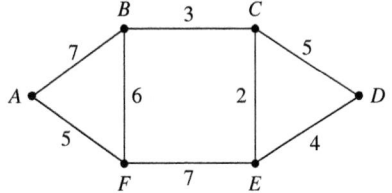

c Show, by inspection, that the solution found in **b** is not optimal.

a These are the minimum travel times, for example A to D is
ABCD = 15

	A	B	C	D	E	F
A	–	7	10	15	12	5
B	7	–	3	8	5	6
C	10	3	–	5	2	9
D	15	8	5	–	4	11
E	12	5	2	4	–	7
F	5	6	9	11	7	–

b Starting from A, 'nearest' node is F, AF = 5

From F, 'nearest' unvisited node is B, FB = 6

From B, 'nearest' unvisited node is C, BC = 3

From C, 'nearest' unvisited node is E, CE = 2

From E, 'nearest' unvisited node is D, ED = 4

All nodes visited, so return to A, DA = 15

The route is AFBCEDA, with total weight = 35

Repeat starting from the other nodes.

Start node	Tour(s)	Total weight
A	AFBCEDA	35
B	BCEDFAB	32
C	CEDBFAC	35
D	DECBFAD	35
E	ECBFADE	35
F	FABCEDF	32

The two 32 minute tours are the same tour. With A as the start node, these give ABCEDFA

The salesperson travels ABCEDEFA, because the shortest route from D to F goes through E

c On the original network, the tour ABCDEFA = 31, so the solution found in **b** is not optimal.

Because there is no efficient algorithm for finding the length, T, of the optimal solution, you need a way of knowing what a 'good' solution is like. To decide if a solution you have found is good enough, you find two values, an **upper bound** and a **lower bound**, such that

lower bound $\leq T \leq$ upper bound.   **Key point**

The closer together the two bounds are, the more accurately you know the length of the optimal tour. You therefore try to find the smallest upper bound and the largest lower bound that you can.

If you have found a tour, it is either optimal or longer. It follows that:

The length of any known tour is an upper bound.   **Key point**

Using the nearest neighbour algorithm to find a tour automatically provides an upper bound, but not necessarily a good one. As an alternative method, you use a minimum spanning tree for the network. You could visit every node by travelling once in each direction along every arc of the minimum spanning tree, so

$(2 \times$ length of minimum spanning tree$)$ is an upper bound.   **Key point**

This is not usually a good upper bound, but you can improve it by using short cuts.

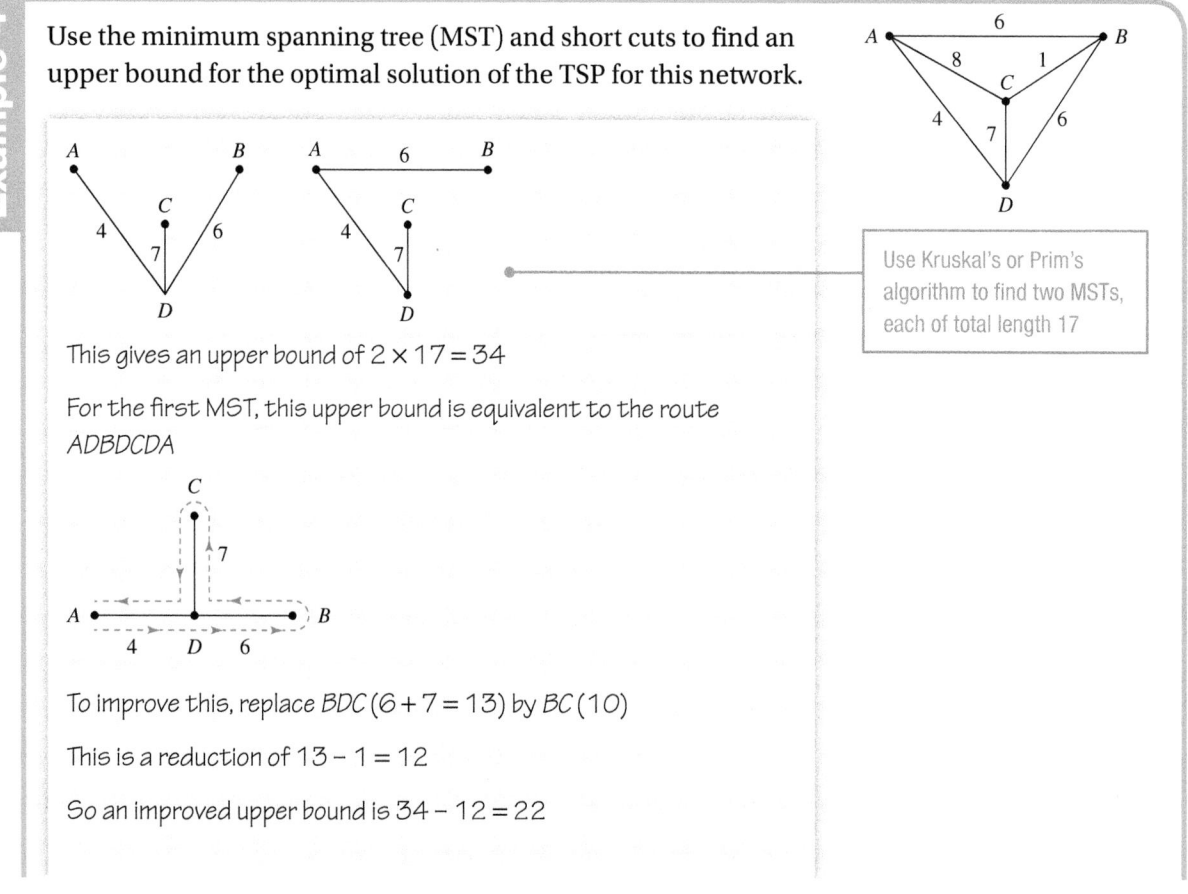

Example 4

Use the minimum spanning tree (MST) and short cuts to find an upper bound for the optimal solution of the TSP for this network.

Use Kruskal's or Prim's algorithm to find two MSTs, each of total length 17

This gives an upper bound of $2 \times 17 = 34$

For the first MST, this upper bound is equivalent to the route ADBDCDA

To improve this, replace BDC $(6 + 7 = 13)$ by BC (10)

This is a reduction of $13 - 1 = 12$

So an improved upper bound is $34 - 12 = 22$

(Continued on the next page)

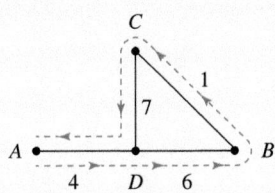

To improve this, replace CDA $(7 + 4 = 11)$ by CA (8)

This is a reduction of $11 - 8 = 3$

So an improved upper bound is $22 - 3 = 19$

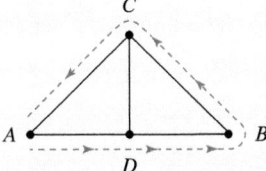

This is a tour, $ADBCA$, so there are no more short cuts.

The best upper bound using this MST is 19

For the second MST, the initial upper bound of 34 is equivalent to the route $BADCDAB$

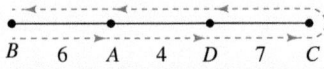

To improve this, return directly from C to B, replacing $CDAB$ $(7 + 4 + 6 = 17)$ by CB (10)

This is a reduction of $17 - 1 = 16$

The improved upper bound is $34 - 16 = 18$

This is a tour $BADCB$, so there are no more short cuts.

The best upper bound using this MST is 18

Of the two values found, the best upper bound $= 18$

> You always want the upper bound to be as small as possible.

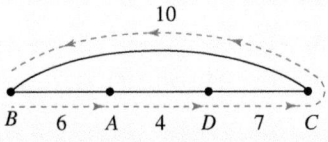

In addition to an upper bound, you need to find a lower bound. Suppose you have the network shown.

The optimal tour enters and leaves A along two of AB, AC, AD and AE

You separate these arcs from the rest of the network. The two arcs involved in the tour must total at least $6 + 7 = 13$

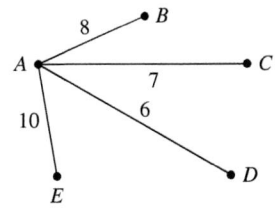

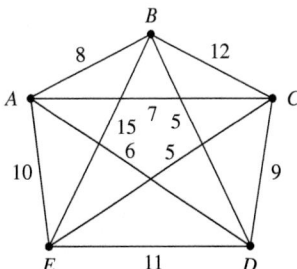

The remaining three arcs of the tour join B, C, D and E

They form a spanning tree for the subgraph shown.

The minimum spanning tree for this subgraph is $5 + 5 + 9 = 19$, so the three arcs in the tour must have a total weight of at least 19

So the complete tour has a total weight of at least $13 + 19 = 32$

This is a lower bound for the tour.

To get the best lower bound available, you could repeat this process, removing each node in turn. Removing B gives a lower bound of 31, removing C gives 33, removing D gives 31 and removing E gives 33 (you should check that you can find these values). You want the lower bound to be as large as possible, so the best lower bound is 33

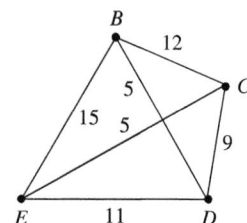

To find a lower bound

Step 1 Choose a node V

Step 2 Identify the two lowest weights, p and q, of the arcs connected to V

Step 3 Remove V and its connecting arcs from the network. Find the total weight, m, of the minimum spanning tree of the remaining subgraph.

Step 4 Calculate lower bound $= p + q + m$

Step 5 If possible, choose another node V and go to Step 2

Choose the largest result obtained as the best lower bound.

Example 5

For the network shown

	A	B	C	D	E
A	–	12	6	7	2
B	12	–	9	7	10
C	6	9	–	4	8
D	7	7	4	–	5
E	2	10	8	5	–

a Find a lower bound for the optimal tour by deleting node A

b Find a second lower bound by deleting node B

c State which of your values is the better lower bound,

d Given that there is an upper bound of 32, write an inequality for T, the optimal tour.

a The shortest arcs from A are $AC = 6$ and $AE = 2$

	1	3	2	4
	B	C	D	E
B	–	9	7	10
C	9	–	④	8
D	⑦	4	–	5
E	10	8	⑤	–

The table shows the graph with A removed.

(Continued on the next page)

The minimum spanning tree is $BD = 7$, $CD = 4$ and $DE = 5$

Use Prim's algorithm.

A lower bound is $(6 + 2) + (7 + 4 + 5) = 24$

b The shortest arcs from B are $BC = 9$ and $BD = 7$

	1	4	3	2
	A	C	D	E
A	–	6	7	2
C	6	–	④	8
D	7	4	–	⑤
E	②	8	5	

The table shows the graph with B removed.

The minimum spanning tree is $AE = 2$, $CD = 4$ and $DE = 5$

A lower bound is $(9 + 7) + (2 + 4 + 5) = 27$

c $27 > 24$, so the better lower bound is 27

d T lies between the lower and upper bounds,

so $27 \leq T \leq 32$

The arcs of the better lower bound, BC, BD, AE, CD and DE, do not form a tour. If they did, you would know that $T = 27$

Key point

If you find a lower bound which forms a tour, then that tour is the optimal solution. If none of the possible lower bounds form a tour, the optimal tour must be greater than the best lower bound.

Exercise 12.6A Fluency and skills

Answer sheet available

1 a Draw up a distance matrix for the complete network of shortest distances corresponding to the network shown.

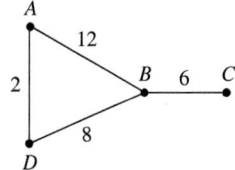

b Show that the nearest neighbour algorithm with the four possible starting nodes leads to two distinct Hamiltonian tours, each with the same total weight.

c Starting from A, list the order in which the nodes would be visited on the original network.

2 a Use the nearest neighbour algorithm with A as the starting node to obtain a Hamiltonian tour of the network shown.

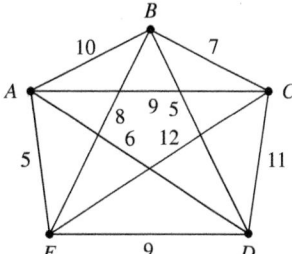

State the length of the tour.

b Show that by using B as the starting node you obtain a shorter tour.

3 The network shown represents the travelling times, in minutes, between four towns.

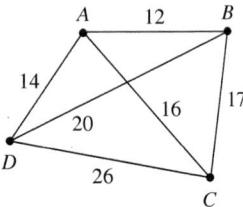

a Find the minimum spanning tree for the network and hence state an upper bound for the optimal solution to the travelling salesman problem for the network.

b By introducing two short cuts, obtain an improved upper bound.

4 The network shows the distances, in km, between six towns. A van based at A needs to deliver to the other five towns and return to A

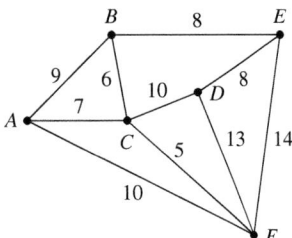

a By finding the minimum spanning tree for the network, obtain an upper bound for the total distance that the van will need to travel.

b Use short cuts to show that the optimal tour is at most 50 km.

5 a Find the two minimum spanning trees for the network shown.

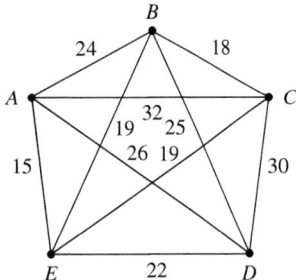

b Using the total weight of the minimum spanning tree, state an upper bound for the optimal solution to the TSP.

c Using short cuts, obtain the best upper bound you can using your minimum spanning trees.

6 a Use Prim's algorithm to find the minimum spanning tree for this network.

	A	B	C	D	E	F
A	–	9	4	7	11	8
B	9	–	12	6	8	14
C	4	12	–	6	10	7
D	7	6	6	–	5	8
E	11	8	10	5	–	13
F	8	14	7	8	13	–

b State an upper bound for the solution to the TSP for this network.

c Using short cuts, obtain an upper bound less than 45

7 a By deleting node A from the network shown, obtain a lower bound for the optimal tour. Explain, by reference to the network, why a tour of this length is not possible.

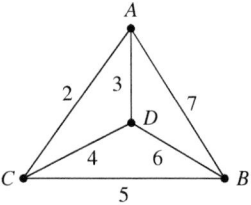

b By deleting node B, obtain another lower bound for the optimal tour and explain why this is, in fact, the length of the optimal tour.

8 By deleting each node in turn from the network shown, obtain the best lower bound available for the solution to the travelling salesman problem.

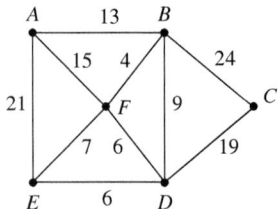

9 The table shows the distances, in km, between five locations.

	A	B	C	D	E
A	–	14	22	18	12
B	14	–	15	32	25
C	22	15	–	20	13
D	18	32	20	–	28
E	12	25	13	28	–

By deleting each node in turn, find the best lower bound for the solution to the travelling salesman problem for this network.

Reasoning and problem-solving

Strategy

To solve travelling salesman problems

1. Convert a practical problem to the classical problem by constructing a complete network of shortest distances.

2. Find possible tours using the nearest neighbour algorithm. Any tour is an upper bound for the optimal tour.

3. Find upper bounds using short cuts on the minimum spanning tree. The smallest upper bound is the best.

4. Find lower bounds by deleting a node and using the MST of the remainder. The largest lower bound is the best.

5. Answer the question in context.

Example 6

The network shows the cost of bus travel, in £, between four towns.

A tourist staying in town C wants to visit all the other towns and return to C

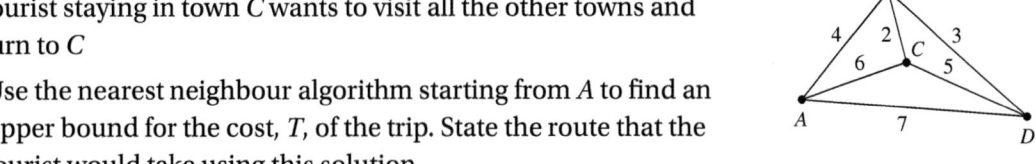

a Use the nearest neighbour algorithm starting from A to find an upper bound for the cost, T, of the trip. State the route that the tourist would take using this solution.

b By deleting first A and then B, find two possible lower bounds for the trip.

c State the best inequality you have for T

a The cheapest routes are shown in this table:

	A	B	C	D
A	–	4	6	7
B	4	–	2	3
C	6	2	–	5
D	7	3	5	–

> For example the cheapest from A to D is $ABD = 7$

From A, the nearest neighbour route is $ABCDA = 4 + 2 + 5 + 7$
$$= 18$$

An upper bound for T is £18. The route from C is $CBDBABC$

(Continued on the next page)

DISCRETE

b The two lowest weight arcs from A are $AB(4)$ and $AC(6)$

The minimum spanning tree of BCD is $BC(2)$ and $BD(3)$

This gives, lower bound $= 4 + 6 + 2 + 3 = 15$

The two lowest weight arcs from B are $BC(2)$ and $BD(3)$

The minimum spanning tree of ACD is $AC(6)$ and $CD(5)$

This gives, lower bound $= 2 + 3 + 6 + 5 = 16$

c The better of the two lower bounds is £16

So the best available inequality for T is $16 \le T \le 18$

> Always look for the largest lower bound.

> In fact, 16 is the largest of the four possible lower bounds and the four arcs involved do not form a tour, so you can say that $16 < T \le 18$

Exercise 12.6B Reasoning and problem-solving

1 A guide is taking a party of tourists from a hotel (H) to four attractions in a city – the museum (M), the art gallery (A), the cathedral (C) and the Guild Hall (G). She estimates the walking times (in minutes) between the various locations as shown in this network.

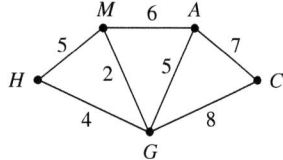

a Draw a complete network of shortest journey times.

b Use the nearest neighbour algorithm starting from A to plan a route for the party. State the total walking time and the order in which their route from the hotel would pass through the nodes of the original network.

c Show, by inspection, that the route found in **b** is not optimal.

2 The table shows the direct distances (in miles) between four locations. There are no other direct links between them.

	A	B	C	D
A	–	5		6
B	5	–		7
C			–	3
D	6	7	3	–

a Fill in the blank cells in the table with the shortest indirect routes available.

b Use the nearest neighbour algorithm with A as the starting node to find a Hamiltonian tour of your completed network. State the total length of the tour and list the order in which it would actually pass through the locations.

3 An orchestra and chorus is available as a whole unit (option A) or as four subgroups (B, C, D and E) offering smaller scale performances. A six-day music festival wants to start and finish with option A, but to have the other four options on the remaining four days. The changeover costs (in £) between the options vary, as shown in this table.

	A	B	C	D	E
A	–	250	300	150	400
B	250	–	120	360	220
C	300	120	–	260	170
D	150	360	260	–	290
E	400	220	170	290	–

a By deleting A, find a lower bound for the overall cost.

b Use the minimum spanning tree and a single short cut to find an upper bound for the cost.

c If the optimal cost is C, use your results to write inequalities satisfied by C

4 The diagram shows the locations of five nesting boxes in a bird reserve. The values shown are the lengths, in metres, of the various sections of path.

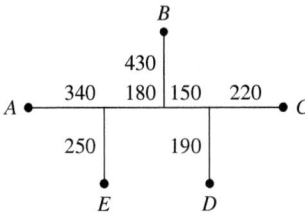

A volunteer wants to visit every nest box, starting and finishing at A

a Draw a complete network of shortest distances.

b The volunteer visits the nest boxes in alphabetical order. By deleting node A, find a lower bound for the TSP for this network and hence show that she could save no more than 120 m by finding the optimal route.

5 The table shows the distances, in km, between seven locations. A distributor based at A needs to deliver bundles of newspapers to shops in each of the other locations.

	A	B	C	D	E	F	G
A	–	5	6	7	4	9	8
B	5	–	4	5	4	8	6
C	6	4	–	9	7	8	9
D	7	5	9	–	9	7	9
E	4	4	7	9	–	6	8
F	9	8	8	7	6	–	9
G	8	6	9	9	8	9	–

a Use Prim's algorithm to find the minimum spanning tree for this network. Hence state an upper bound for the distance the distributor will need to travel.

b Using short cuts, reduce your upper bound to below 50 km.

c By deleting node A, obtain a lower bound for the distributor's journey.

d Using this and your result from part **b**, write inequalities satisfied by his optimal route.

6 The network shows the distances, in miles, between five locations.

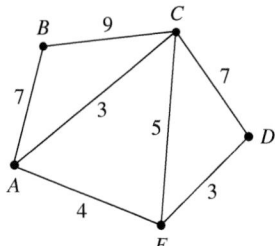

a Draw a complete network showing the shortest distances between the five locations.

b Use the nearest neighbour algorithm starting from A to obtain a possible solution to the travelling salesman problem. State the length of the route and the order in which the nodes would be visited on the original network.

7 A knight's move in chess consists of moving two squares parallel to one side of the board and then one square at right angles to this.

Belinda and Manuel play a game in which one of them places counters on six squares of the board and the other must make a 'knight's tour', starting and ending on one of the marked squares and capturing' each of the other counters en-route, making as few moves as possible. The diagram shows the positions of the counters that Belinda has set up.

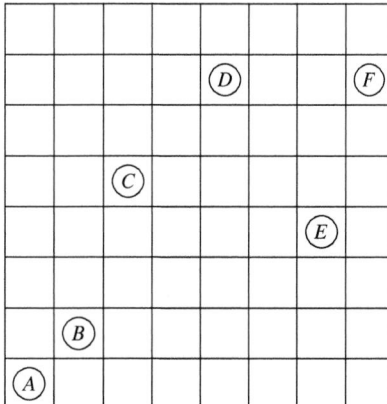

a Draw up a table to show the numbers of moves required to travel between each of these positions.

b Find upper and lower bounds for the number of moves that Manuel will have to make.

Fluency and skills

A directed network can represent the routes along which a commodity flows. The commodity could be electricity, oil, freight, data or many other things. The weight of an arc is its **capacity**, which is the maximum possible flow along that arc. The network is sometimes called a **capacitated network**. Here is a simple example.

The flow enters the network at S 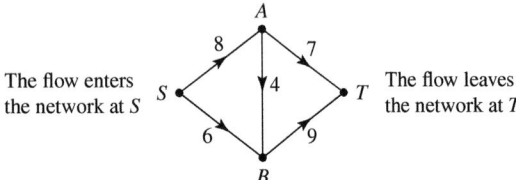 The flow leaves the network at T

At S, all the arcs are directed away from the node. A node like this is called a **source**. It is usual to label a source S

At T, all the arcs are directed towards the node. A node like this is called a **sink**. It is usual to label a sink T

When the commodity flows through the network, there is a non-negative number – the **flow in the arc** – assigned to each arc. Together these form the **flow in the network**. The flow must satisfy certain conditions.

Firstly, the capacity of an arc is the maximum possible flow in that arc.

> **Key point**
> The **feasibility condition** states that the flow in an arc cannot be greater than its capacity.

Secondly, the commodity cannot be 'stored' at a node, so you have the following condition.

> **Key point**
> The **conservation condition** states that, at every node apart from S and T, total inflow = total outflow.

It follows from this second condition that:

> **Key point**
> Total outflow from S = total inflow to T

This total outflow/inflow is called the **value of the flow**.

A feasible flow for the network above is:

At A inflow = outflow = 7

At B inflow = outflow = 9

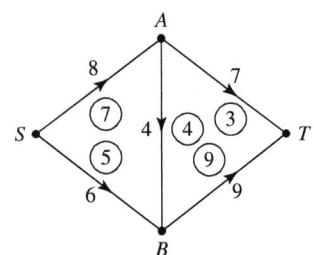

Outflow at S = inflow at T = 12

The value of the flow = 12

You circle the actual flows in the arcs. In the diagram at the bottom of the previous page, in SA, for example, the flow is 7 and the capacity (maximum possible flow) is 8. Notice that in AB and BT you have flow = capacity. These arcs are **saturated**. The other arcs are **unsaturated**.

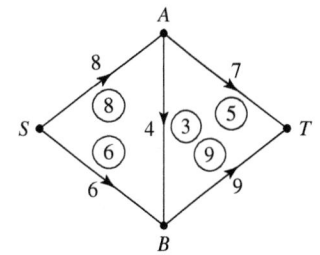

You usually need to find the maximum flow through the network. The network shown has a maximum flow of 14

The diagram on the right is one way in which this flow could take place.

The diagram shows a capacitated network. The circled values represent a feasible flow through the network.

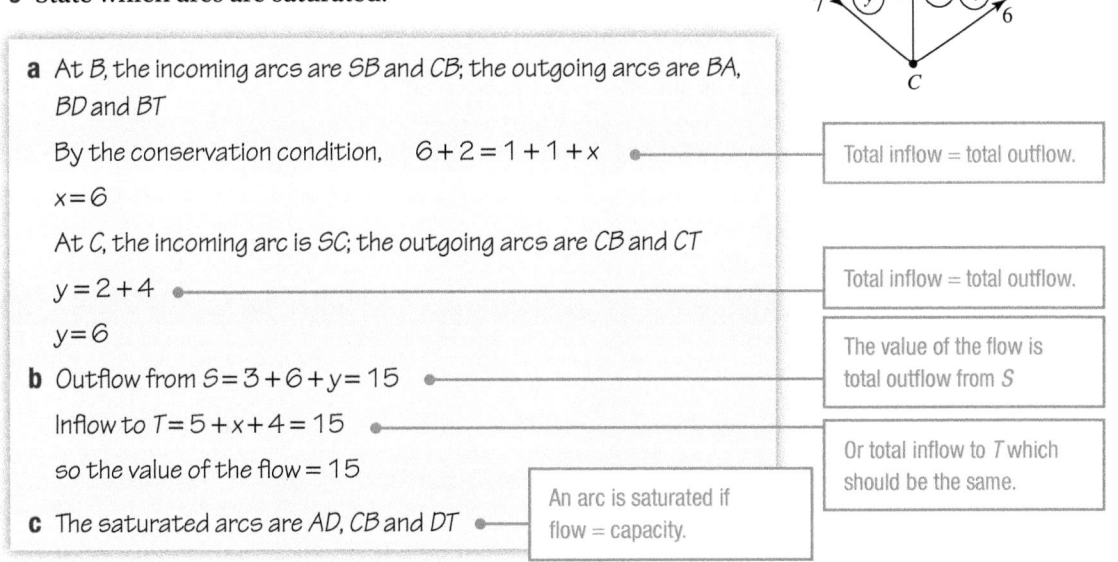

a Find the values of x and y

b State the value of the flow.

c State which arcs are saturated.

a At B, the incoming arcs are SB and CB; the outgoing arcs are BA, BD and BT

By the conservation condition, $6 + 2 = 1 + 1 + x$ ←   Total inflow = total outflow.

$x = 6$

At C, the incoming arc is SC; the outgoing arcs are CB and CT

$y = 2 + 4$ ←   Total inflow = total outflow.

$y = 6$

b Outflow from $S = 3 + 6 + y = 15$ ←   The value of the flow is total outflow from S

Inflow to $T = 5 + x + 4 = 15$ ←   Or total inflow to T which should be the same.

so the value of the flow = 15

c The saturated arcs are AD, CB and DT ←   An arc is saturated if flow = capacity.

You have met the idea of a 'bottleneck' in a road system, where traffic flow is restricted and affects the overall flow through the system.

In the network shown below, a flow of 9 is possible from S to C (3 along SAC, 6 along SBC) and 11 from D to T (4 along DET, 7 along DFT). However, the maximum value of the flow in the network is only 6, which is the most that can pass through the bottleneck CD

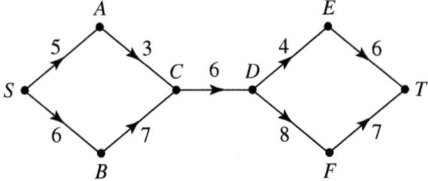

A bottleneck may involve more than one arc. Here the bottleneck is formed by the arcs DE and DF. The maximum value of the flow in the network is 4

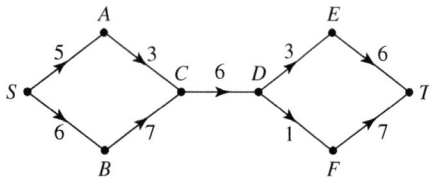

A bottleneck separates the network into two parts, one containing S and the other containing T

The value of the network flow cannot be more than the flow through a bottleneck. The maximum value of the network flow is therefore given by the flow through the 'worst' bottleneck.

This idea is formalised by the concept of a **cut**.

A **cut** is a set of arcs whose removal disconnects the network into two parts, X and Y, with X containing S and Y containing T

The **capacity of a cut** is the sum of the capacities of those arcs of the cut which are directed from X to Y

You describe a cut either by listing the set of arcs in the cut (the **cut set**), or by listing the nodes in the **source set** X and in the **sink set** Y

Example 2

For each cut shown in this diagram

a List the cut set, source set and sink set,

b Find the capacity of the cut.

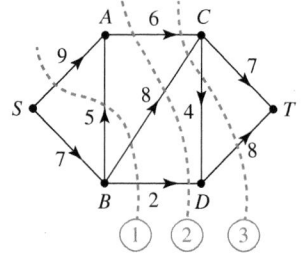

Cut 1 **a** Cut set = {SA, BA, BC, BD}
Source set $X = \{S, B\}$, sink set $Y = \{A, C, D, T\}$

b The capacity of the cut = $9 + 5 + 8 + 3 = 25$

Cut 2 **a** Cut set = {AC, BC, BD}
Source set $X = \{S, A, B\}$, sink set $Y = \{C, D, T\}$

b The capacity of the cut = $6 + 8 + 2 = 16$

Cut 3 **a** Cut set = {AC, BC, CD, DT}
Source set $X = \{S, A, B, D\}$, sink set $Y = \{C, T\}$

b The capacity of the cut = $6 + 8 + 8 = 22$

The arc CD does not contribute to the capacity of cut 3 because it is directed from Y to X

Any flow must cross from set X to set Y. It follows that:

> **Key point**
> The value of any flow is less than or equal to the capacity of any cut.

The maximum flow corresponds to the 'worst bottleneck'. This is the **maximum flow–minimum cut theorem**.

> **Key point**
> **The maximum flow–minimum cut theorem** states that the value of the maximal flow is equal to the capacity of a minimum cut.

DISCRETE

Example 3

List all possible cuts for the network shown, and find their capacities. Hence state the maximal flow for the network.

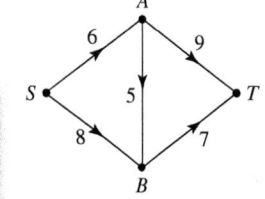

The diagram shows the four possible cuts.

Cut 1 {SA, SB} Capacity $= 6 + 8 = 14$

Cut 2 {SA, AB, BT} Capacity $= 6 + 7 = 13$

Cut 3 {SB, AB, BT} Capacity $= 8 + 5 + 9 = 22$

Cut 4 {AT, BT} Capacity $= 9 + 7 = 16$

The maximal flow $= 13$ •————

The arc AB does not contribute to the capacity of cut 2 because it is directed from Y to X

A useful consequence of the maximum flow–minimum cut theorem is:

> **Key point**
> If you have a flow and a cut such that the value of flow is equal to capacity of cut, then the flow is a maximum and the cut is a minimum.

Example 4

Find, by inspection, a minimum cut for the network shown. Confirm that it is a minimum cut by finding a flow with that value.

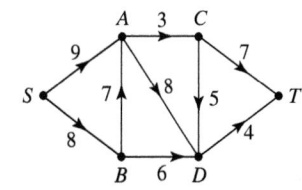

The cut {AC, CD, DT}, shown, appears to be a minimum. •————

The capacity of this cut $= 3 + 4 = 7$

There is a flow of value 7, consisting of 3 along SACT and 4 along SBDT

So this flow is maximal, and the cut is a minimum.

Look for arcs with the lowest weights.

1 For each of the cuts on each of these networks, state

  i The set of arcs (the cut set),

  ii The source set X and the sink set Y,

  iii The capacity.

  a

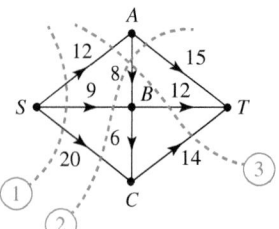

  b

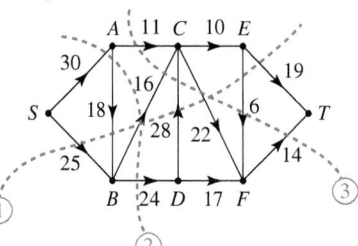

  c

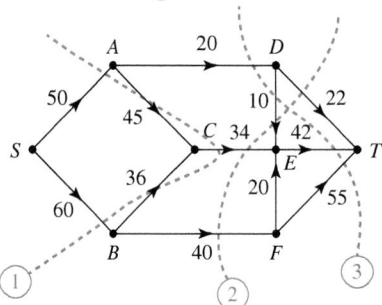

  d

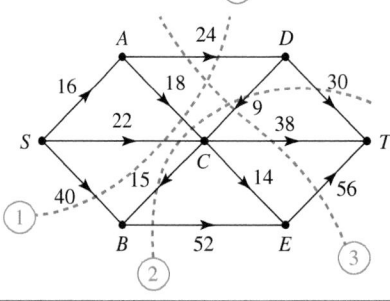

2 For the network in question **1a**

  a Find a cut with a capacity of 35

  b Find a flow with a value of 35

  c What can you deduce from your results in **a** and **b**?

3 Find a minimum cut for the network in question **1b**, and confirm that it is a minimum by finding a flow with that value.

4 For the network shown, find, by inspection, the maximum flow and a minimum cut.

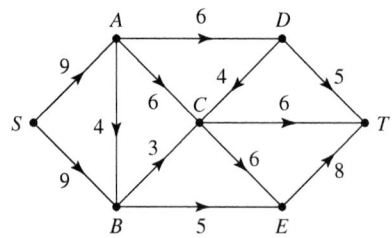

5 Find, by inspection, a minimum cut for the network shown and confirm that it is a minimum by finding a flow with that value.

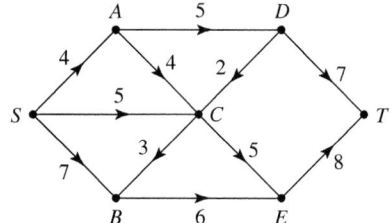

Reasoning and problem-solving

Strategy

To solve problems involving network flow

1 Draw clear diagrams, as necessary.

2 Use the standard notation. For example, circle the actual flow in an arc.

3 Use the maximum flow–minimum cut theorem.

4 Answer the question in context.

In some networks, you will have more than one source and/or sink. In this network, there are no incoming arcs at A or B. A and B are both sources. Similarly, there are no outgoing arcs at G or H. G and H are both sinks.

You deal with multiple sources by connecting them to a dummy **supersource**, S

Each actual source can receive from S as much flow as it needs to supply the network.

Similarly, you deal with multiple sinks by connecting them to a dummy **supersink**, T

Each sink can send to T all the flow it receives from the network.

The possible outflows from A and B are 14 and 15, so the capacities of the dummy arcs SA and SB must be at least 14 and 15

Similarly, the possible inflows to G and H are 14 and 13, so the capacities of the dummy arcs GT and HT must be at least 14 and 13

You now find the maximal flow for the modified network by the usual methods. Once this is found, you can remove the dummy arcs and nodes to leave the solution for the original network.

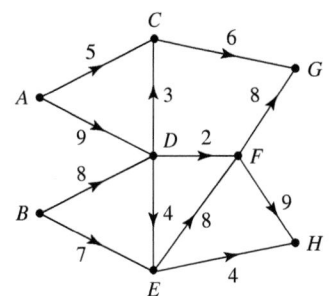

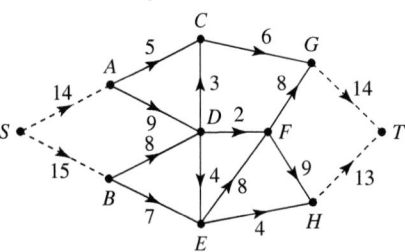

Example 5

a The diagram shows the capacities of the arcs in a network. Show, on the diagram, a flow of 2 along AD and 3 along ABE

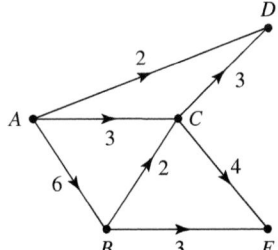

b Identify two additional flows to bring the total flow to 10

c Show that this is the maximal flow.

a D and E are sinks, so introduce a supersink T. Show the given flows on the diagram.

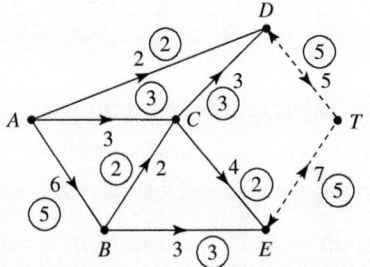

b For example, a flow of 3 along $ACDT$ and 2 along $ABCET$

c The cut $\{AD, AC, BC, BE\}$ has capacity 10

Hence there is a flow and a cut of capacity 10, so this is maximal by the maximum flow–minimum cut theorem.

1 The network shown has a sink, *G*, and two sources.

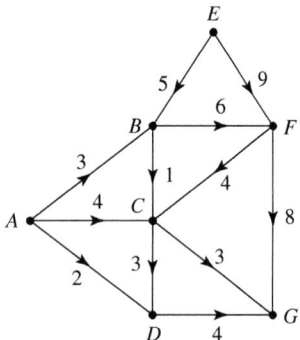

a Identify the sources.

b Introduce a supersource, *S*

c In a certain flow pattern, the arcs *AC*, *AD* and *BC* are saturated. Find the maximum value this flow can have.

d By finding a suitable cut, show that the flow found in **c** is the maximum possible for the network.

2 The network shown represents a system of one-way streets. Traffic enters the system at *A* and *B* and leaves at *I* and *H*. The weights are the maximum traffic flows, in hundreds of cars per hour, which can safely pass along the streets.

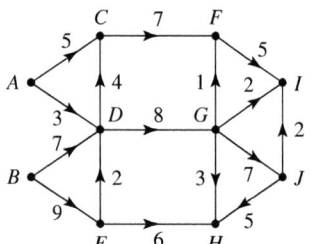

a Draw a diagram with a supersource, *S*, and a supersink, *T*

b State the maximum possible flows along *ACFI*, *BDGJH* and *BEH*

c Find, by inspection, a network flow of value 19

d By finding a cut of capacity 19, show that the flow you found in **c** is maximal.

3 For the network shown,

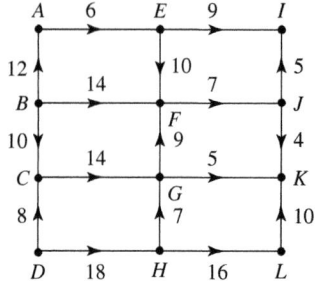

a Identify the source(s) and sink(s),

b Draw the diagram with a supersource and supersink, as necessary,

c Show that the maximum flow cannot be more than 28

d Find, by inspection, a flow with this value.

4 The table shows the flows in the arcs of a network.

		To							
		A	**B**	**C**	**D**	**E**	**F**	**G**	**H**
From	**A**	–	12	8	–	–	–	–	–
	B	–	–	–	10	–	14	–	–
	C	–	11	–	–	–	–	12	–
	D	–	–	–	–	–	–	–	–
	E	–	–	9	–	–	–	15	–
	F	–	–	8	15	–	–	–	7
	G	–	–	–	–	–	15	–	18
	H	–	–	–	–	–	–	–	–

a Identify the sources and sinks.

b Draw the network and add in a supersource, *S*, and a supersink, *T*

c Explain why the maximum flow cannot be more than 44

d Find, by inspection, a flow with this value.

Chapter summary

- A graph consists of points (vertices or nodes) connected by lines (edges or arcs).
- Graphs are isomorphic if they have the same vertices connected in the same way but drawn with a different layout.
- A graph is connected if there is a continuous route between every pair of vertices.
- A simple graph has no loops or multiple edges.
- A graph having some of the vertices and edges of a given graph, G, is a sub-graph of G
- A graph formed by adding vertices of degree 2 along edges of a given graph, G, is a sub-division of G
- A complete graph is a simple graph with all possible pairs of vertices connected. The complete graph with n vertices is called K_n
- In a bipartite graph, the vertices belong to two distinct sets, and each edge joins a vertex in one set to a vertex in the other. If every possible edge is present, it is a complete bipartite graph. The complete bipartite graph connecting m vertices to n vertices is called $K_{m,n}$
- The degree (or order) of a vertex is the number of edges connecting to it.
- Total of orders $= 2 \times$ number of edges, so the number of odd vertices is even.
- An adjacency matrix lists vertices and the number of edges connecting them.
- A network is a graph with a number, called a weight, associated with each arc (you should use node and arc when talking about networks). You can represent it using a distance matrix, with entries showing the weights of the arcs.
- If a graph or network has 'one-way streets', then it is a directed graph (digraph) or directed network.
- Any continuous journey around a graph is a walk. A walk with no repeated edges is a trail. If a trail returns to its starting vertex, it is a closed trail.
- A walk with no repeated edges or vertices (except maybe the start) is a path. A closed path is a cycle.
- A connected graph with no cycles is a tree.
- A Hamiltonian cycle (or tour) visits every vertex of the graph once only. A graph with such a cycle is a Hamiltonian graph.
- A graph is traversable, or Eulerian, if there is a closed trail including every edge once only. The trail is an Eulerian trail. If the trail is not closed, then the graph is semi-traversable or semi-Eulerian.
- A graph is Eulerian if all vertices have even degree. It is semi-Eulerian if it has two odd degree vertices. The two odd vertices are the start and finish vertices. A graph with more than two odd degree vertices is non-traversable.
- A graph is planar if it can be drawn in a plane so that no two edges meet each other except at a vertex.
- When a graph is drawn in the plane, the areas created by its edges are called faces. The area outside the graph is the infinite face.
- A graph drawn in the plane has V vertices, E edges and F faces related by Euler's formula:
$$V + F - E = 2$$

- A spanning tree or connector for a graph with n vertices is a tree with $(n-1)$ edges containing every vertex of the graph.
- For a network, the minimum spanning tree (MST), or minimum connector, is the spanning tree with the lowest total weight.
- Kruskal's algorithm:
  Step 1   Choose the arc with minimum weight.
  Step 2   Choose the arc with least weight from the remaining arcs, avoiding those that form a cycle with those already chosen.
  Step 3   If any nodes remain unconnected, go to Step 2
- Prim's algorithm:
  Step 1   Choose any node to be the first in the connected set.
  Step 2   Choose the arc of minimum weight joining a connected node to an unconnected node. Add this arc to the spanning tree and the node to the connected set.
  Step 3   If any unconnected nodes remain, go to Step 2
- Prim's algorithm on a matrix:
  Step 1   Select the first node.
  Step 2   Cross out the row and number the column for the chosen node.
  Step 3   Find the minimum undeleted weight in the numbered columns. Circle this value. The node for this row is the next chosen node.
  Step 4   Repeat steps 2 and 3 until all nodes have been chosen.
- The route inspection (Chinese postman) problem:
  Find the shortest route that travels at least once along every arc of a network and returns to the start. Which arcs must be repeated?
- Route inspection algorithm:
  Step 1   Identify the odd nodes.
  Step 2   List all possible pairings of the odd nodes.
  Step 3   For each pairing, find the shortest routes between paired nodes and the total weight of those routes.
  Step 4   Choose the pairing with the smallest total.
  Step 5   Repeat the shortest route arcs (equivalent to adding extra arcs to the network). The network is now traversable.
  The total length of the route = (sum of original weights) + (sum of extra arc weights).
- The travelling salesman problem (TSP):
  Find a route which visits every node of a network and returns to the start in the shortest possible distance.
- To solve a practical TSP with n nodes:
  Create a complete network, K_n, of shortest distances between pairs of nodes.
  Find a minimum weight Hamiltonian cycle for the complete network (this is the classical problem).
  Interpret the solution in the practical situation.
- The nearest neighbour algorithm:
  Step 1   Choose a starting node V
  Step 2   From your current position, choose the arc with minimum weight leading to an unvisited node. Travel to that node.
  Step 3   If there are unvisited nodes, go to Step 2
  Step 4   Travel back to V

- For the optimal tour, T, lower bound $\leq T \leq$ upper bound.
- The length of any known tour is an upper bound.
- ($2 \times$ length of minimum spanning tree) is an upper bound which may be improved by using short cuts.
- To find a lower bound:
  Step 1   Choose a node V
  Step 2   Identify the two lowest weights, p and q, of the arcs connected to V
  Step 3   Remove V and its connecting arcs from the network. Find the total weight, m, of the minimum spanning tree of the remaining subgraph.
  Step 4   Calculate lower bound $= p + q + m$
  Step 5   If possible, choose another node V and go to Step 2
- Choose the largest result obtained as the best lower bound.
- The weights in a directed network can represent flows. The weight of an arc is its capacity.
- A node is a source if all arcs are directed away from it.
- A node is a sink if all arcs are directed towards it.
- In a given situation, there is a flow in each arc. These combine to form the flow in the network.
- If the flow in an arc equals its capacity, the arc is saturated.
- Flows satisfy these conditions.
  - The feasibility condition: The flow in an arc cannot be greater than its capacity.
  - The conservation condition: At every node apart from S and T, total inflow = total outflow
- Total outflow at S = total inflow at T
  This total is the value of the flow.
- A cut is a set of arcs whose removal disconnects the network into two parts, X and Y, with X containing S and Y containing T
- The capacity of a cut is the sum of the capacities of those arcs of the cut that are directed from X to Y
- You describe a cut either by listing the set of arcs in the cut (the cut set) or by listing the nodes in the source set X and in the sink set Y
- The maximum flow–minimum cut theorem:
  The value of the maximal flow = the capacity of a minimum cut
- It follows that if a flow and a cut are such that (value of flow) = (capacity of cut), then the flow is maximum and the cut is minimum.
- If a network has multiple sources or sinks, you connect them to a dummy source, S, or a dummy sink, T

You should now be able to...	Review Questions
✔ Draw a graph/network from an adjacency/distance matrix and vice versa.	1
✔ Understand and use the terms related to graphs and networks.	2
✔ Use Euler's formula for planar graphs.	3
✔ Use Kruskal's algorithm to find the minimum spanning tree for a network given as a diagram.	4
✔ Use Prim's algorithm to find the minimum spanning tree for a network given as a diagram or as a table.	5, 6
✔ Solve the route inspection problem for a network with nodes of odd degree.	7
✔ Find an upper bound for the travelling salesman problem using the nearest neighbour algorithm or short cuts applied to the minimum spanning tree.	8
✔ Find a lower bound for the travelling salesman problem by deleting a node and using the minimum spanning tree.	8
✔ Analyse the flow through a network and use the maximum flow–minimum cut theorem to decide if it is maximal.	9

Answer sheet available

1 Draw the graph corresponding to this adjacency matrix:

	A	*B*	*C*	*D*
A	0	1	2	0
B	1	2	1	1
C	2	1	0	1
D	0	1	1	0

2 A simple connected graph is semi-Eulerian and has 5 vertices. Draw the graph if

  a It has exactly one vertex of order 3

  b It is a tree,

  c It has 7 edges.

3 A graph drawn in the plane has 5 vertices and 5 faces.

  a How many edges does the graph have?

  b Draw a graph which satisfies this data.

4 Use Kruskal's algorithm to find the minimum spanning tree for the network shown. List the order in which you choose the arcs and state the total length of the tree.

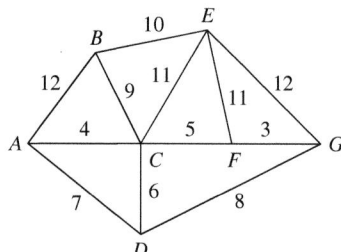

5 Use Prim's algorithm, starting from *B*, to find the minimum spanning for the network in question 4. List the order in which you choose the arcs.

6 Use Prim's algorithm with the table shown, starting from *A*, to find the minimum spanning tree. Show all your working. List the order in which you choose the arcs.

	A	*B*	*C*	*D*	*E*	*F*	*G*
A	–	9	12	4	6	8	14
B	9	–	10	13	8	11	3
C	12	10	–	2	9	15	7
D	4	13	2	–	13	14	4
E	6	8	9	13	–	15	8
F	8	11	15	14	15	–	13
G	14	3	7	4	8	13	–

7 The sum of the weight of the network shown is 104

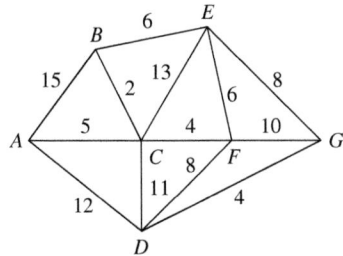

a Find the length of a Chinese postman route through this network.

b State the number of times that the route would pass through *D*

8 The diagram shows the distances between five locations. A delivery van starting from *A* has to visit all five locations and return to base.

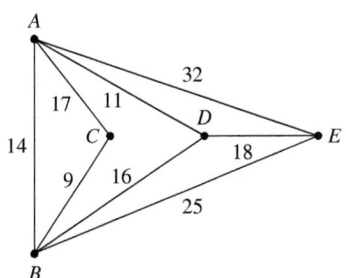

a Draw up a table of shortest distances (found by inspection),

b Use the nearest neighbour algorithm, starting from *A*, to find a possible route. State the length of the route and list the actual route of the van.

c The nearest neighbour algorithm starting from *B* gives a route of length 80 State which of the two values you have is the better upper bound.

d By deleting node *A* from your table, find a lower bound for the optimal tour. Hence write inequalities satisfied by *T*, the length of the optimal tour.

9 **a** Find by inspection a flow of 18 in the network shown.

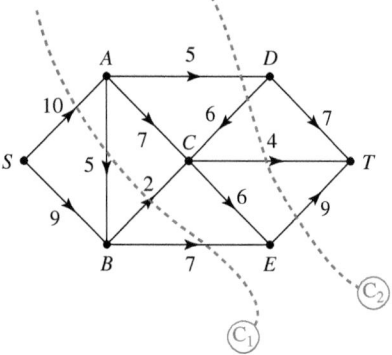

b Find the capacity of the cuts C_1 and C_2

c State, with reasons, the maximum flow in the network.

Did you know?

The Chinese mathematician Meigu Guan is credited for formulating the route inspection problem. His paper was translated into English in 1962 and became known as the 'Chinese postman problem' in his honour.

Investigation

The windy postman problem is a variant of the route inspection problem. In this case, each edge may have one cost for traversing it in one direction and a different cost for traversing it in the other direction.

The cost of traveling each way is given in the brackets as (c_{ij}, c_{ji}) where $i < j$. In this example, the cost for travelling from node 3 to node 4 is 2, and the cost for travelling from node 4 to node 3 is 3.

Solve the problem starting and ending at node 3.

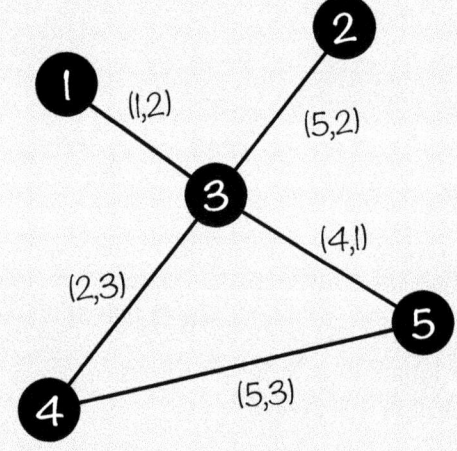

Research

Fleury's algorithm and Hierholzer's algorithm both enable you to find an Eulerian trail around a graph.

- Fleury's algorithm chooses the next edge in the path to be one whose deletion would not disconnect the graph.
- Hierholzer's algorithm requires you to find closed paths and substitute them into other closed paths.

Research how these two algorithms work.

1 a Explain what is meant by a 'tree' in the context of graphs. **[2 marks]**

b Draw the network corresponding to this distance matrix. **[3]**

	A	B	C	D	E
A	–	2	7	4	–
B	2	3	–	9	–
C	7	–	–	6	–
D	4	9	6	–	5
E	–	–	–	5	2

c Give an example of a path on the network with at least four nodes. **[1]**

d Explain whether or not this network is a Hamiltonian graph. **[2]**

e How do you that know this graph is not Eulerian? **[1]**

f Which arc/s would need to be added to change this to an Eulerian graph? **[1]**

2

Graph 1 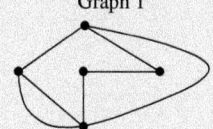 Graph 2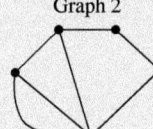

a Are graphs 1 and 2 isomorphic? Explain how you know. **[2]**

b Is graph 1 traversable, semi-traversable or non-traversable? Explain how you know. **[2]**

c Draw a subgraph of graph 2 which is also a tree. **[1]**

d Draw the graph K_4 and explain whether or not it is planar. **[3]**

3 a Draw the adjacency matrix that corresponds to this network. **[4]**

b Write down a semi-Eulerian trail for this graph and state
its weight. **[2]**

c State whether or not the graph is planar and explain how you know. **[2]**

4 a Find a minimum spanning tree for the graph in question 3 using

   i Kruskal's algorithm,   **ii** Prim's algorithm starting at B **[6]**

In each case, list the order in which the arcs are added or rejected.

b Use your minimum spanning tree to state an upper bound for the optimal solution
to the travelling salesman problem. **[3]**

5 For the network in question 3 find the length of the shortest route that travels along each arc at
least once and then returns to the start point. **[6]**

6 The graph gives the distances between five points.

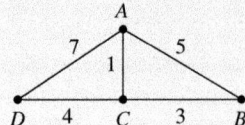

a Write down a Hamiltonian cycle for the graph shown and state its weight. **[2]**

b Draw a complete network of shortest distances. **[2]**

c Use the nearest neighbour algorithm with *A* as the starting node to find a Hamiltonian cycle on the network. **[2]**

d State the order in which the nodes would be visited in the original network. **[2]**

7 This diagram shows a capacitated network with a feasible flow given in circles on each edge.

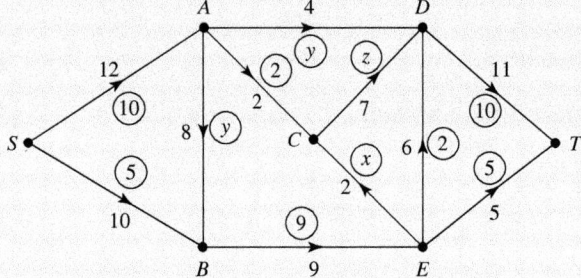

a Find the values of *x*, *y* and *z* **[3]**

b Which of the arcs are saturated? **[2]**

A cut is defined by source set {*S, A, B, E*} and sink set {*C, D, T*}.

c List the arcs in the cut and state its capacity. **[2]**

8 a Draw the graph that corresponds to this adjacency matrix. **[2]**

	A	*B*	*C*
D	1	0	1
E	1	1	1
F	0	0	1

b State the name given to this type of graph. **[1]**

c How many vertices and how many edges must be added to this graph to create the complete graph $K_{4,3}$? **[3]**

d Is the graph $K_{4,3}$ Eulerian, semi-Eulerian or non-Eulerian? Explain how you know. **[2]**

9 **a** Use Prim's algorithm for a matrix, starting at A, to find a minimum spanning tree for the network corresponding to this distance matrix.

	A	B	C	D	E	F
A	–	12	9	–	7	–
B	12	–	4	5	–	1
C	9	4	–	3	8	–
D	–	5	3	–	6	–
E	7	–	8	6	–	2
F	–	1	–	–	2	–

List the order in which you select the arcs and give the weight of the tree. **[4]**

b Show that using Kruskal's algorithm leads to the same minimum spanning tree and sketch the tree. **[4]**

c Solve the route inspection problem, starting at A, for this matrix. Give a possible route and state its length. **[5]**

10 The graph shows the travel times (in minutes) between six towns. A salesperson wishes to visit each town at least once and then return to their starting point.

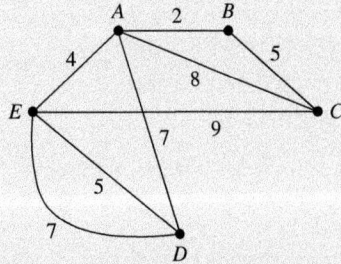

a Use a minimum spanning tree to find an upper bound for the time of the optimal tour. **[4]**

b By finding a shortcut, calculate an improved upper bound. **[2]**

c By removing each node from the network in turn, find the best lower bound for the time of the optimal tour. **[5]**

d Draw a matrix corresponding to the complete network of shortest times. **[2]**

e Use the nearest neighbour algorithm, starting and ending at C, to find a tour that visits each node at least once. **[3]**

f Show that the solution found in **d** is not optimal. **[2]**

11 Here is a capacitated network.

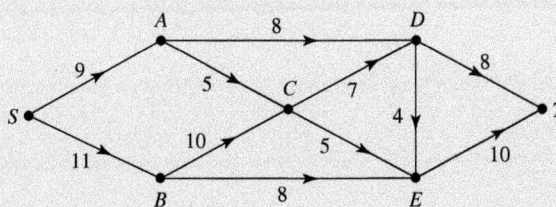

a What is the maximum possible flow along the route $SBCDET$? **[1]**

b Find a cut with a capacity of 28 and write down

   i The cut set,   **ii** The source set,   **iii** The sink set.   **[3]**

c Find, by inspection, a minimum cut and hence state the maximum flow for the network. **[3]**

12 Cables are to be laid to connect to each of the points shown in the network.
The lengths (in metres) between the points are shown.

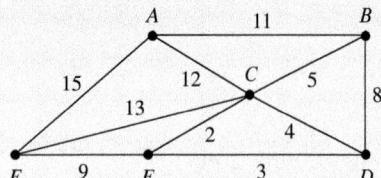

a Use a suitable algorithm to find the minimum length of cable required and state which edges are to be included. State the name of the algorithm used. **[5]**

It is discovered that there is already cable along *AC* and *BD* so this can be used and will not need to be laid.

b Explain which algorithm you chose and how you would adapt it in order to find the minimum length of cable now required. Find the additional length of cable required. **[4]**

c Find a solution to the travelling salesperson problem for this network starting and ending at *A* using only edges in your solution for part **a**. Then use two short cuts to show that 47 is an upper bound for the travelling salesperson problem. **[4]**

d Obtain two different lower bounds for the optimal solution to the travelling salesman problem for this network. **[4]**

e Use your solutions to part **c** and **d** to write the best inequality for the length of the optimal tour. **[2]**

f Show that the nearest neighbour algorithm does not give an optimal tour. **[5]**

13 The network represents a system of roads with the direction of travel and maximum traffic flows given on each edge.

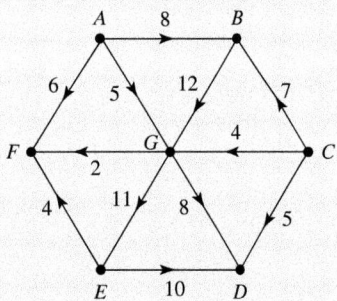

a Copy the network and add on a supersource, *S*, and a supersink, *T*. Write the minimum necessary capacity on each of the dummy edges. **[6]**

b What is the maximum flow along *SABGEFT*? **[1]**

c Given that there exists a possible flow of size 32, show that this flow is maximal. You should state the name of any theorem used. **[4]**

13 Critical path analysis 1

In large construction projects, it is essential to ensure that different contractors, equipment, building materials, and so on, are all in place at the right time. This is a massively complex undertaking. Even ensuring that facilities, such as toilets, are in place for workers at the right part of the construction site at the right time must be considered. Some projects can take many years and ensuring that access is in place to various parts of the site has to be considered. Critical path analysis is used in such situations to determine exactly what needs to be done and when.

Even projects on a much smaller scale, for example building and extension to a house, require some work of this type. Construction projects are just one example of an area in which critical path analysis is used. In almost all projects across areas as diverse as making movies to turning aircraft round at airports, such analysis is necessary to ensure smooth running and efficient use of resources.

Orientation

What you need to know	What you will learn	What this leads to
Chapter 12 Graphs and networks.	• To produce an activity network from a precedence table. • To solve problems involving activity networks. • To identify critical paths and activities.	Chapter 29 Critical path analysis 2.

Fluency and skills

To produce a timetable for the completion of a project, you need:

- A list of the activities involved
- Details of which activities depend on others
- How long each activity will take (its **duration**).

You record this information in a **precedence table** or **dependence table**.

For example, suppose you are planning to paint your bedroom. It might involve the following activities:

A   Remove furniture

B   Remove curtains

C   Remove carpets

D   Wash ceiling

E   Wash walls

F   Paint ceiling

G   Paint walls

H   Replace carpets

I   Replace curtains

J   Replace furniture

Some activities are dependent on others. For example, you can't remove the carpets until you have removed the furniture.

Here is a possible precedence table, with likely durations.

Activity	Duration (minutes)	Depends on
A	15	–
B	10	–
C	15	A
D	20	A, B, C
E	30	A, B, C
F	80	D
G	140	E
H	20	F, G
I	15	F, G
J	15	H

> The timings assume that you have friends to help you, so that, for example, the ceiling can be painted at the same time as the walls.

You can now draw an **activity network**. Each activity is represented by a node, and the arcs show the order of precedence.

Activities *A* and *B* can both start straight away and *A* must finish before *C* can start. Once *A*, *B* and *C* are complete, *D* and *E* can start.

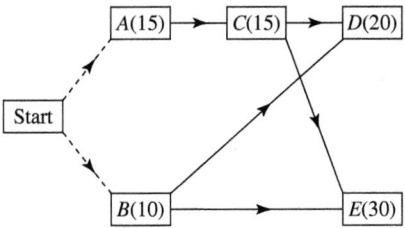

F and *G* must follow *D* and *E*, respectively. These must finish before *H* and *I* can start. *J* can start once *H* has finished.

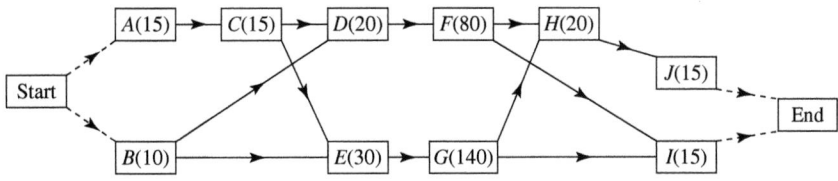

> The 'Start' and 'End' boxes are not strictly needed (they are dummy activities with zero duration) but you may find them helpful.

You can now analyse the timing of the activities. For each activity you need to know:

- What is the **earliest start time**, assuming all preceding activities are completed as soon as possible?
- What is the **latest finish time**, that is, the latest you could finish the activity without increasing the overall length of the project?

It is usual to record these in a box at each node.

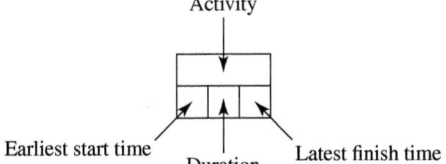

To find the earliest start times you make a **forward pass** through the network.

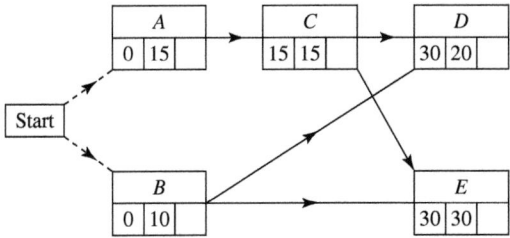

A and *B* can start straight away, so the earliest start time is 0
If *A* starts immediately, it will finish after 15 minutes. *C* cannot start until *A* finishes, so *C* has an earliest start time of 15 minutes.

The earliest that *B* can finish is 10 minutes. The earliest that *C* can finish is 15 + 15 = 30 minutes. *D* and *E* cannot start until both *B* and *C* have finished, so *D* and *E* have earliest start time of 30 minutes.

> Make sure that you understand how these earliest start times were found.

Continuing in this way, the complete forward pass looks like this.

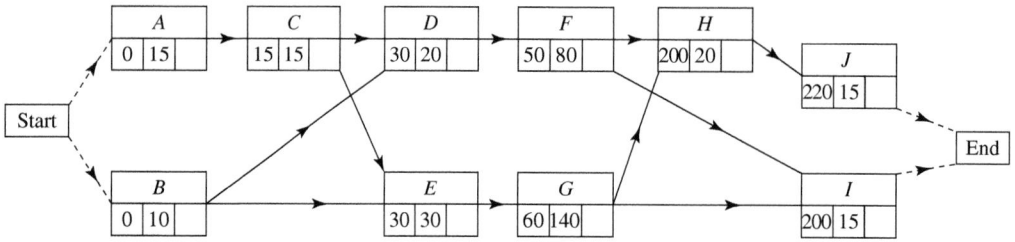

The project finishes when both *I* and *J* end. *I* can finish by
215 minutes and *J* by 235 minutes. This means that the overall
project duration is 235 minutes.

You now make a **backward pass** to find the latest finish time for
each activity consistent with the overall duration of 235 minutes.

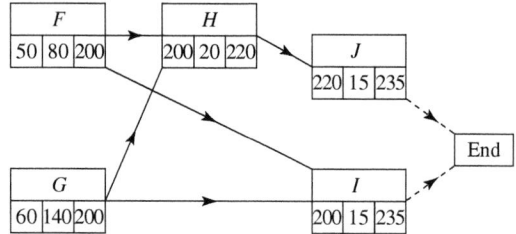

I has a duration of 15 minutes, so to finish by 235 minutes it must
start by 220 minutes at the latest. Similarly, *J* must start by
220 minutes.

J follows on from *H*, so the latest *H* can finish is 220 minutes. *H* has a
duration of 20 minutes, so must start by 200 minutes at the latest.

F and *G* must both finish before *H* and *I* can start. This means that
the latest *F* and *G* can finish is 200 minutes.

> Again, make sure that
> you understand how
> these latest finish times
> were found.

Continuing in this way, the complete backward pass looks like this.

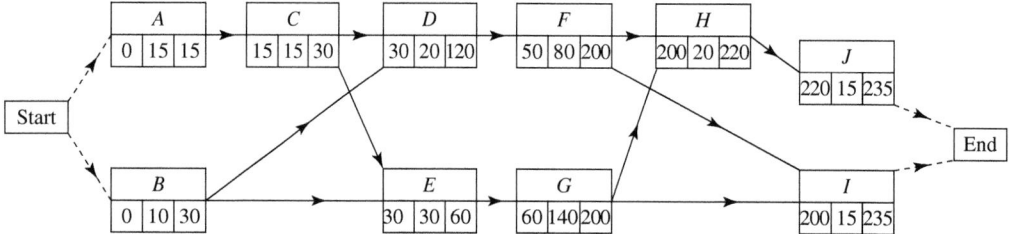

To summarise, an activity may depend on several preceding
activities and be followed by several activities which depend on it.

When doing a forward pass:

Key point

The earliest start time for an activity is the maximum
of the earliest times by which all preceding activities can be
completed.

Completing the forward pass tells you how long the whole project will take.

> **Key point**
> The project duration is the earliest time by which all activities can be completed.

When doing a backward pass:

> **Key point**
> The latest finish time for an activity is the minimum of the latest times by which all following activities can start without affecting the project duration.

Example 1

a Draw an activity network to show the project described by this precedence table.

Activity	Duration (hours)	Depends on
A	3	–
B	6	A
C	8	A
D	10	A
E	7	B
F	7	C
G	9	E, F
H	4	D,F

b Find the minimum project duration and the earliest start and latest finish times for each activity.

c State the latest time at which activity D could start.

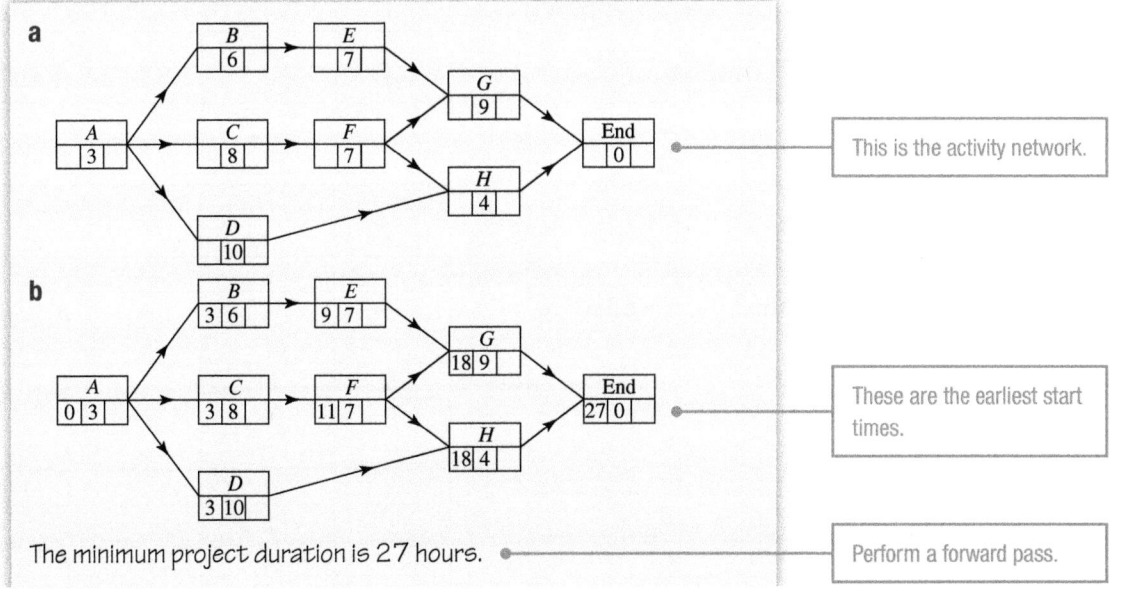

a

This is the activity network.

b

These are the earliest start times.

End 27|0

The minimum project duration is 27 hours.

Perform a forward pass.

(Continued on the next page)

DISCRETE

413

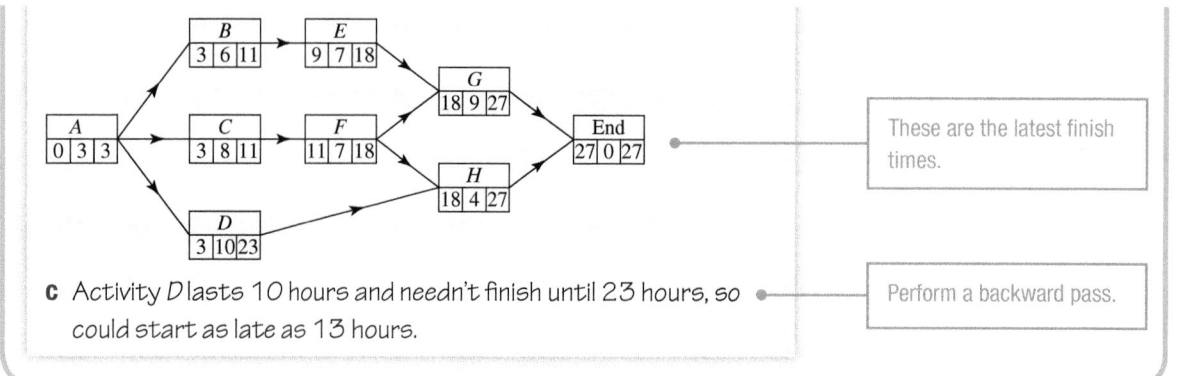

These are the latest finish times.

Perform a backward pass.

c Activity D lasts 10 hours and needn't finish until 23 hours, so could start as late as 13 hours.

Exercise 13.1A Fluency and skills

Answer sheet available

1 This diagram shows part of an activity network. Find the missing values.

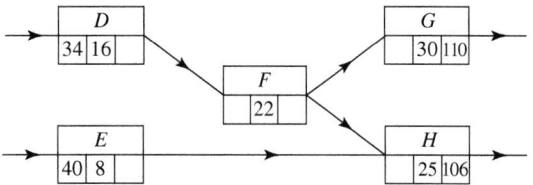

2 Draw an activity network for each of the following precedence tables. Complete a forward and a backward pass to find the project duration and the earliest start times and latest finish times.

a

Activity	Duration	Must be preceded by
A	4	–
B	2	–
C	1	A
D	5	A
E	3	B
F	2	C, E

b

Activity	Duration	Must be preceded by
A	2	–
B	5	–
C	1	A
D	3	B
E	2	B
F	2	C
G	4	D, E

c

Activity	Duration	Must be preceded by
A	3	–
B	2	A
C	3	A
D	4	A
E	5	B, C
F	3	C, D
G	1	D

d

Activity	Duration	Must be preceded by
A	6	–
B	1	–
C	12	–
D	7	A, B
E	3	D
F	4	D
G	1	E, F
H	7	F
I	2	C, G
J	3	I

e

Activity	Duration	Must be preceded by
A	3	–
B	6	A
C	4	A
D	7	A
E	11	B
F	6	B
G	3	C, D, F
H	7	D
I	8	E
J	2	E
K	3	G, I
L	2	G, H, I
M	9	K, L

Reasoning and problem-solving

Strategy

To solve problems involving activity networks

1. Before drawing the activity network, make a rough sketch to help you to produce a clear layout.

2. Draw a clear diagram (a ruler is recommended).

3. If required, do a forward and backward pass to find the project duration and the activity timings.

4. Interpret the relationship between activity durations and timings.

5. Answer the question.

Example 2

a Use an activity network to find the earliest start times and latest finish times for the activities in this project.

Activity	Duration (days)	Depends on
A	2	-
B	3	A
C	5	A
D	8	B
E	7	B
F	3	C
G	4	D
H	1	E, F
I	1	G, H

b State the minimum project duration.

c One of the activities over-runs by 5 days. What effect will this have on the project duration if the delayed activity is

   i D   **ii** E   **iii** F   (assume that the activity starts as soon as possible).

a This is the activity network:

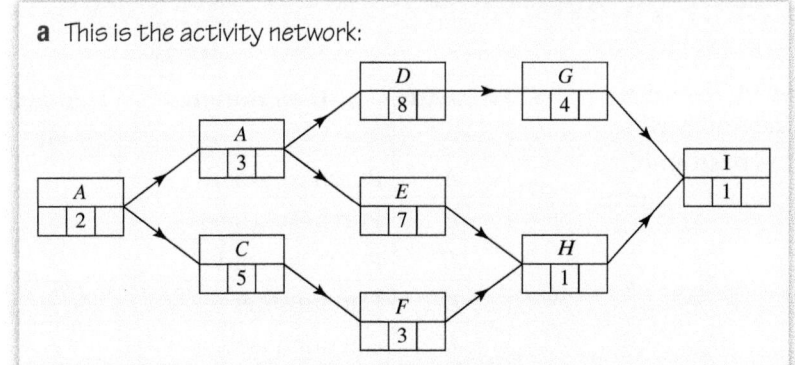

(*Continued on the next page*)

DISCRETE

415

These are the earliest start times:

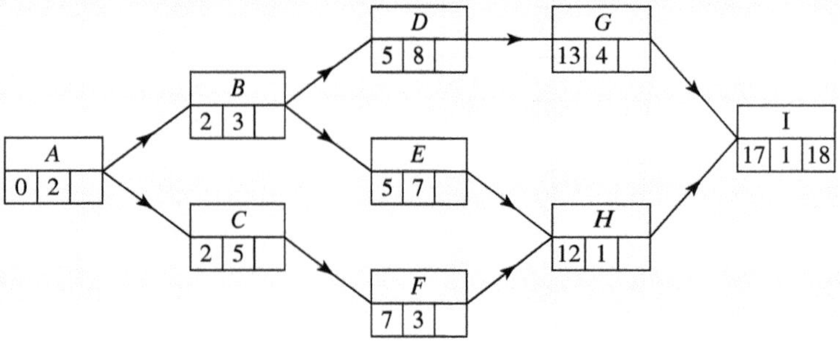

D
5	8

G
13	4

B
2	3

A
0	2

E
5	7

I
17	1	18

Perform a forward pass.

C
2	5

H
12	1

F
7	3

These are the latest finish times:

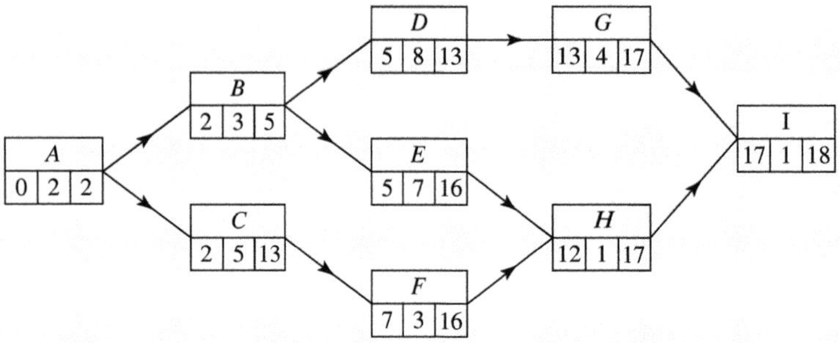

D
5	8	13

G
13	4	17

B
2	3	5

A
0	2	2

E
5	7	16

I
17	1	18

Perform a backward pass.

C
2	5	13

H
12	1	17

F
7	3	16

b The minimum project duration is 18 days.

c i Any increase in D will take it past its latest finish time. If it over-runs by 5 days the whole project takes 23 days.

ii E could take up to 4 extra days without any knock-on effect. If it over-runs by 5 days the whole project takes 19 days.

iii F could take up to 6 extra days without any knock-on effect, so project still takes 18 days.

Exercise 13.1B Reasoning and problem-solving

Answer sheet available

1 The following is a list of tasks for making a wooden trolley for a child. The trolley will have wheels, a square wooden base and four wooden sides with a cord attached to one of them for towing the trolley.

Activity	Description	Depends on
A	Purchase wood	
B	Purchase wheels	
C	Purchase cord	
D	Cut out sides	
E	Cut out base	
F	Attach sides to base	
G	Attach wheels to base	
H	Attach cord	

a Complete the table to show how the activities depend on each other.

b Draw the corresponding activity network.

2 Draw an activity network corresponding to this precedence table.

Activity	Preceding activity
A	–
B	–
C	–
D	A
E	A
F	B
G	B
H	C, D, E, F
I	F
J	F
K	I
L	I
M	G, H, K

3

Activity	Must be preceded by	Duration (days)
A	–	15
B	–	12
C	A	10
D	B	10
E	C, D	8
F	C, D	6
G	E	10
H	E, F	7

a Draw an activity network for this precedence table.

b Perform a forward and a backward pass on your network. State the project duration.

c Find the effect on the project duration if

  i Activity C took 2 days longer than expected,

  ii Activity F took 3 days longer than expected.

4

Activity	Depends on	Duration (hours)
A	–	2
B	A	5
C	A	9
D	B, C	12
E	D	3
F	C	2
G	D, F	5
H	D, F	9
I	E, G, H	5

a Draw an activity network for this precedence table.

b Perform a forward and a backward pass on your network. State the project duration.

c Activity F takes longer than expected. What is the maximum time it could take without affecting the overall project duration?

Fluency and skills

The time available for an activity is the gap between its earliest possible start time and its latest possible finish time. For some activities, the time available is greater than the duration of the activity. The activity has **float**.

Float = (latest finish time – earliest start time) – duration **Key point**

An activity with zero float is a **critical activity**. An increase in its duration, or delay in starting it, will increase the overall project duration.

A critical activity has a float of zero. **Key point**

A sequence of critical activities is a **critical path**.

Example 1

The diagram shows the result of a forward and backward pass on an activity network.

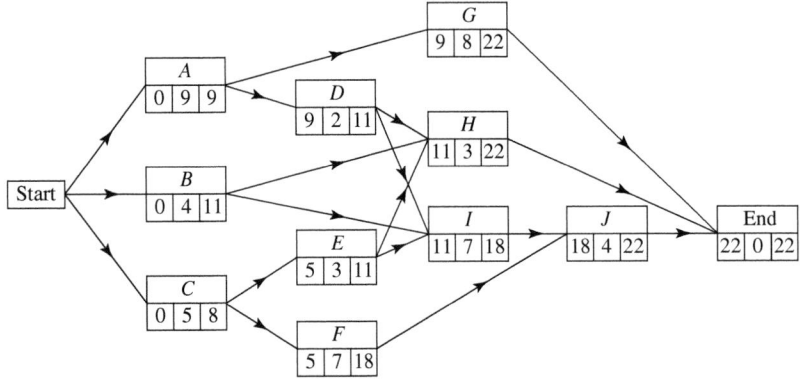

Find the float on each activity and hence identify the critical activities.

This table shows the results.

Activity	Duration	Earliest start time	Latest finish time	Total float
A	9	0	9	0
B	4	0	11	7
C	5	0	8	3
D	2	9	11	0
E	3	5	11	3
F	7	5	18	6
G	8	9	22	5
H	3	11	22	8
I	7	11	18	0
J	4	18	22	0

For example the float for $E = (11 - 3) - 5$ $= 3$

The critical activities are A, D, I and J, shown in red. ●

The sequence of activities $A - D - I - J$ form the critical path.

Example 2

Sometimes there are two or more critical paths.

The diagram shows the result of a forward and backward pass on an activity network.

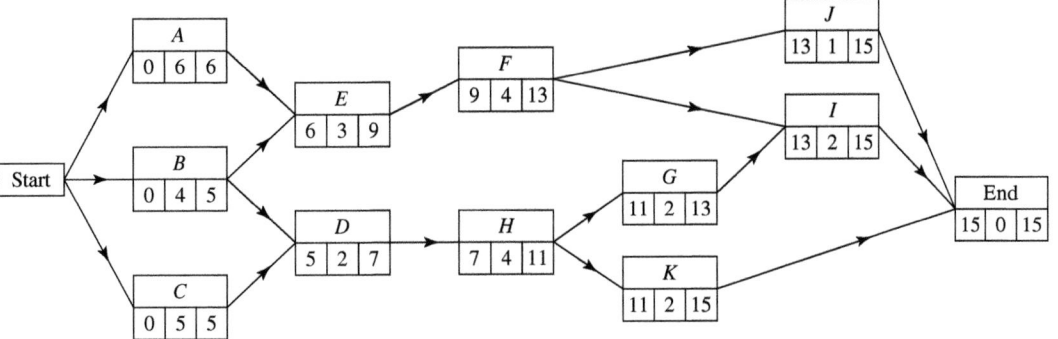

Show that there are two critical paths on this network. Illustrate them on the diagram.

Activity	Duration	Earliest start time	Latest finish time	Total float
A	6	0	6	0
B	4	0	5	1
C	5	0	5	0
D	2	5	7	0
E	3	6	9	0
F	4	9	13	0
G	2	11	13	0
H	4	7	11	0
I	2	13	15	0
J	1	13	15	1
K	2	11	15	2

The critical activities are A, C, D, E, F, G, H and I

These form two paths A-E-F-I and C-D-H-G-I, as shown in the diagram below.

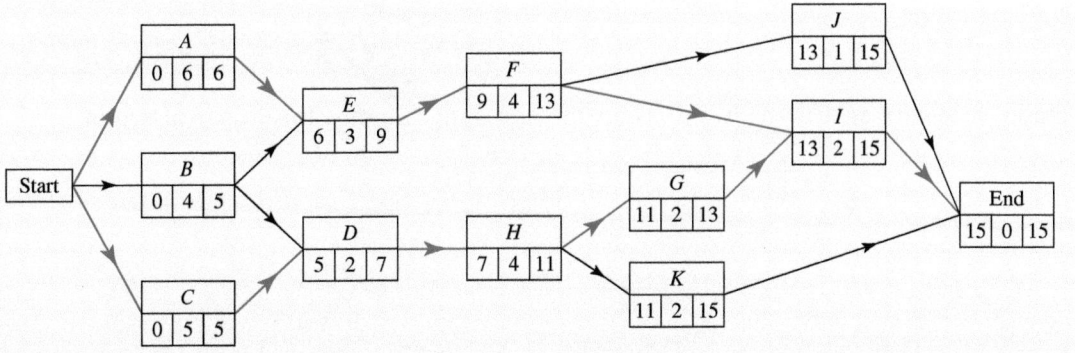

Example 3

This precedence table shows a project with the duration of each activity given in days.

Activity	Preceded by	Duration
A	–	9
B	–	5
C	–	8
D	–	3
E	A	3
F	D	4
G	B, E	10
H	C, F	4
I	B, H	10
J	G, I	2

a Draw an activity network and find the duration of the project.

b Identify the critical path(s).

c You could reduce the duration of any of activities A–F by 1 day at the cost of £100. How much would it cost to reduce the project duration by 1 day, and how would you achieve this?

a This is the network with forward and backward passes completed.

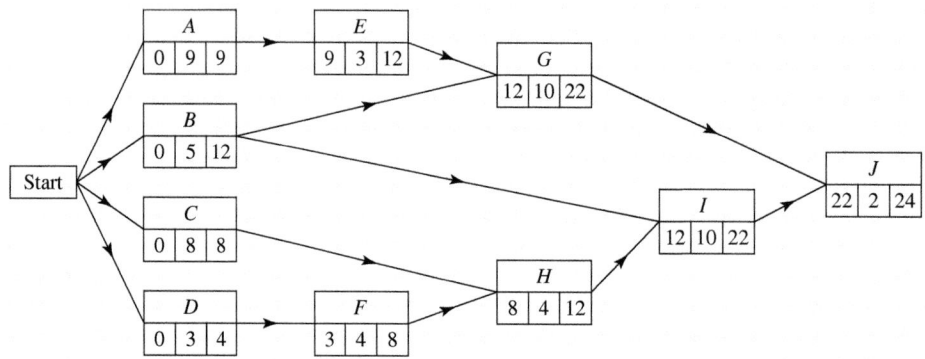

The project duration is 24 days.

b The floats of the activities are A 0, B 7, C 0, D 1, E 0, F 1, G 0, H 0, I 0, J 0. The critical activities (zero float) form two critical paths A–E–G–J and C–H–I–J.

c Both critical paths would need to be shortened by 1 day to have the desired effect on the overall duration. C must be shortened by 1 day, and either A or E by 1 day, with a total cost of £200

Exercise 13.2A Fluency and skills

Answer sheet available

1 Find the float for each activity in this network and hence identify the critical activities.

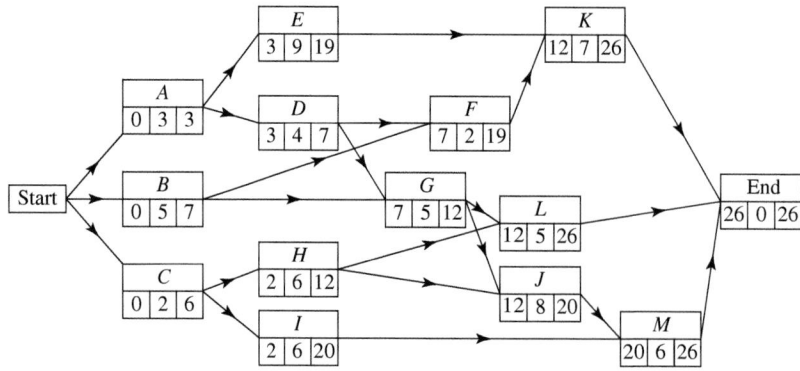

2 For each of the following networks, perform a forward and a backward pass. State the project duration. Find the float for each activity and hence identify the critical activities.

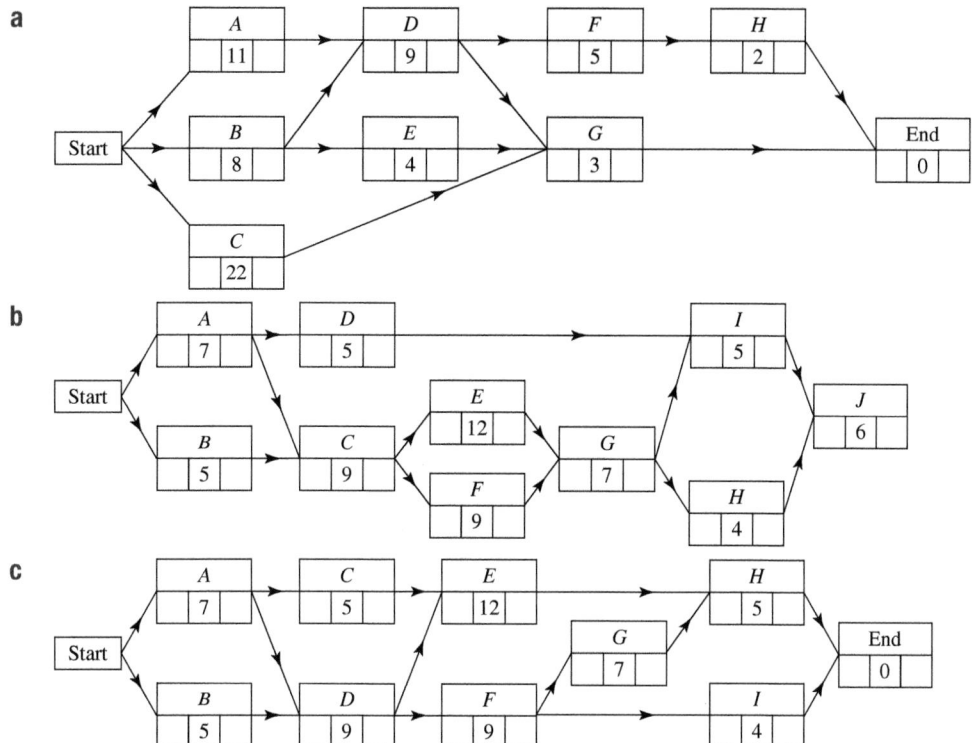

a

A	D	F	H
11	9	5	2

Start

| B | E | G | End |
| 8 | 4 | 3 | 0 |

| C |
| 22 |

b

| A | D | I |
| 7 | 5 | 5 |

Start

| J |
| 6 |

| B | C | E | G | H |
| 5 | 9 | 12 | 7 | 4 |

| F |
| 9 |

c

| A | C | E | H | End |
| 7 | 5 | 12 | 5 | 0 |

Start

| G |
| 7 |

| B | D | F | I |
| 5 | 9 | 9 | 4 |

3 For this precedence table, draw an activity network and use it to find the project duration and the critical activities.

Activity	Duration	Must be preceded by
A	5	–
B	2	–
C	2	A
D	6	A, B
E	4	B
F	2	C, E

4 For this precedence table,

a Draw an activity network and find the project duration,

b Show that there are two critical paths.

Activity	Duration (hours)	Must be preceded by
A	8	–
B	6	A
C	7	A
D	11	A
E	12	B, C
F	8	C, D
G	5	B, D

Reasoning and problem-solving

To solve problems involving critical activities

1. If necessary, draw a neat activity network (sketch it first to get a clear layout).

2. Perform forward and backward passes to find the earliest start times and latest finish times.

3. Calculate the float for each activity and hence identify the critical activities.

4. Answer the question.

Example 4

The network shown represents a building project, with durations in days.

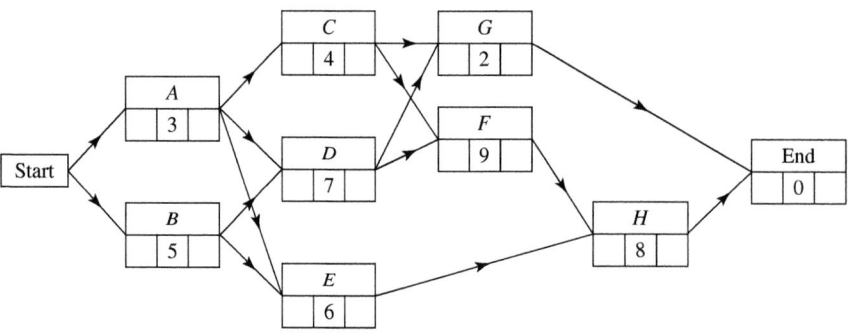

a Find the duration of the project if all goes to plan.

b Identify the critical path.

c In fact, activity C is delayed by 7 days. What effect does this have on the duration of the project?

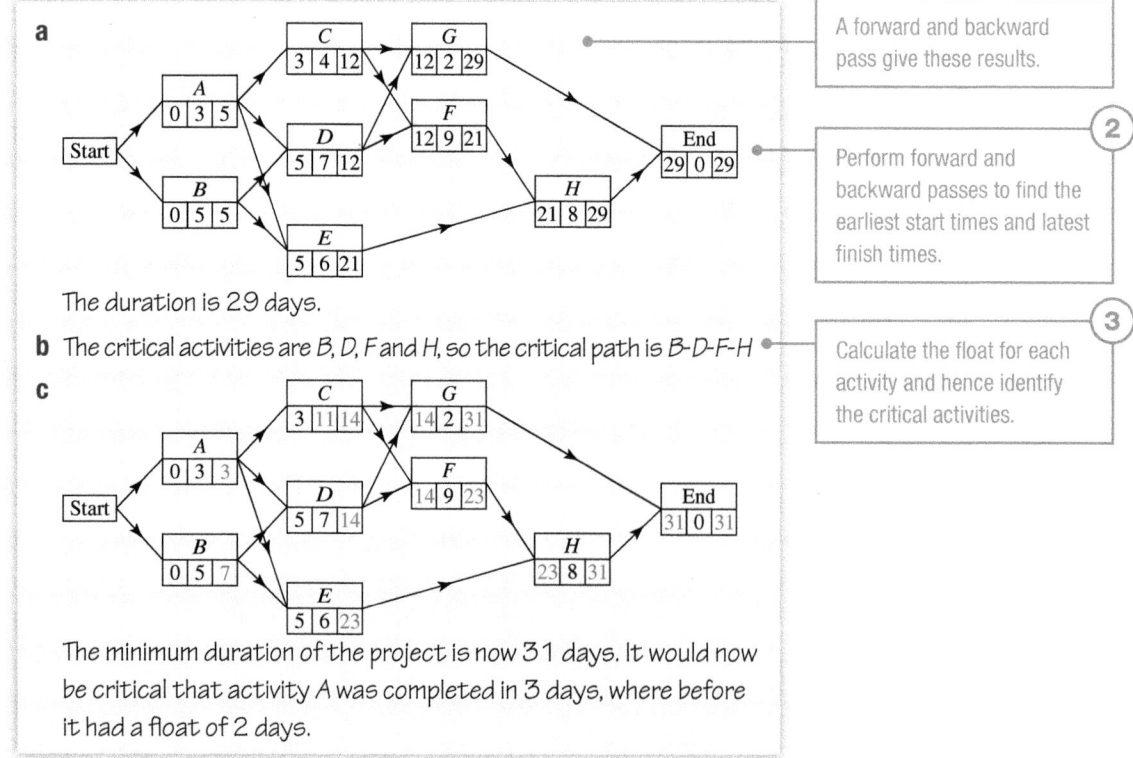

a

The duration is 29 days.

b The critical activities are B, D, F and H, so the critical path is B-D-F-H

A forward and backward pass give these results.

2 Perform forward and backward passes to find the earliest start times and latest finish times.

3 Calculate the float for each activity and hence identify the critical activities.

c

The minimum duration of the project is now 31 days. It would now be critical that activity A was completed in 3 days, where before it had a float of 2 days.

1

Activity	Preceded by	Duration (hours)
A	-	9
B	A	6
C	A	11
D	B	11
E	B	7
F	C	1
G	C	12
H	D	4
I	E. F	10

a Find the duration of this project.

b Find the critical path(s).

c It is possible to reduce the duration of activities C, D and E at a cost of £50 per hour per activity. How much will it cost to reduce the duration of the project by 3 hours and how would this be achieved?

2 Here is a precedence table for a project.

Activity	Preceded by	Duration (days)
A	–	6
B	A	4
C	A	7
D	B	9
E	B	6
F	B, C	3
G	C, E	8
H	D, E	3
I	E. F	5

a Find the duration of the project.

b State the critical activities.

c Find the effect, if any, on the duration of the project if

  i D was delayed by 4 days,

  ii F and I were each delayed by 2 days.

3 Here is a precedence table for a project.

Activity	Must be preceded by	Duration (days)
A	–	12
B	–	9
C	A	6
D	B	8
E	C, D	5
F	C, D	6
G	E	7
H	E, F	4

a Find the duration of the project.

b Identify the critical activities and state the float of the non-critical activities.

c What is the maximum period by which activity B could be delayed without increasing the overall duration of the project?

4 Here is a precedence table for a project.

Activity	Depends on	Duration (hours)
A	–	10
B	A	10
C	–	6
D	A, C	5
E	D	9
F	A	10
G	B, D	6
H	E, F, G	7

a Find the duration of the project and identify the critical activities.

b Activity B can be split into two 5 hour tasks, one of which can be delayed until activity F is complete but is not required before G can start. Investigate whether it is possible to shorten the overall project, justifying your conclusions.

Chapter summary

- A **precedence table** or **dependence table** records activities, their duration and the order in which they can occur.
- In an **activity network**, each activity is represented by a node, and the arcs show the order of precedence.
- You analyse the project to find the **earliest start time** and **latest finish time** for each activity, and these are recorded in a box like this at each node.

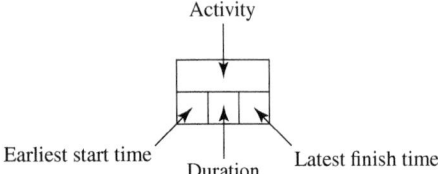

Earliest start time — Duration — Latest finish time

- To find the earliest start times, you make a **forward pass** through the network. The earliest start time for an activity is the maximum of the earliest times by which all preceding activities can be completed.
- The project duration is the earliest time by which all activities can be completed.
- To find the latest finish times, you make a **backward pass** through the network. The latest finish time for an activity is the minimum of the latest times by which all following activities can start without affecting the project duration.
- An activity has **float** if the time available for it (the gap between its earliest start time and its latest finish time) is greater than its duration.
- Float = (latest finish time − earliest start time) − duration
- An activity with zero float is a **critical activity**. An increase in its duration, or delay in starting it, will increase the overall project duration.
- The sequence of critical activities is the **critical path**.

Check and review

You should now be able to...	Review Questions
✔ Draw an activity network corresponding to a given precedence table.	1
✔ Perform a forward pass to find the earliest start time for each activity.	2
✔ Find the project duration.	2
✔ Perform a backward pass to find the latest finish time for each activity.	3
✔ Calculate the float for each activity.	4
✔ Identify the critical activities.	4

1

Activity	Preceded by
A	–
B	–
C	–
D	A, B
E	B, C
F	D
G	D
H	F, G
I	G
J	E, G

Draw an activity network to show this project.

2

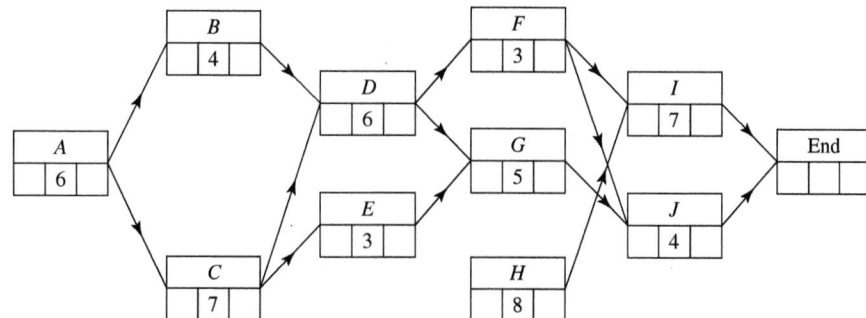

Perform a forward pass on this network to find the earliest start time for each activity (the durations are in hours). State the project duration.

3

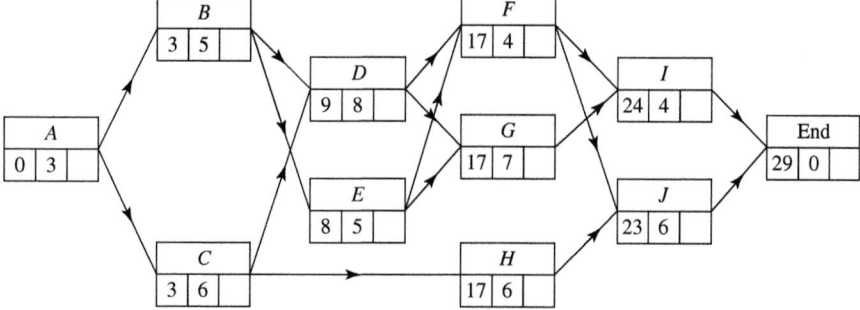

Perform a backward pass on this network to find the latest finish time for each activity (durations are in days).

4

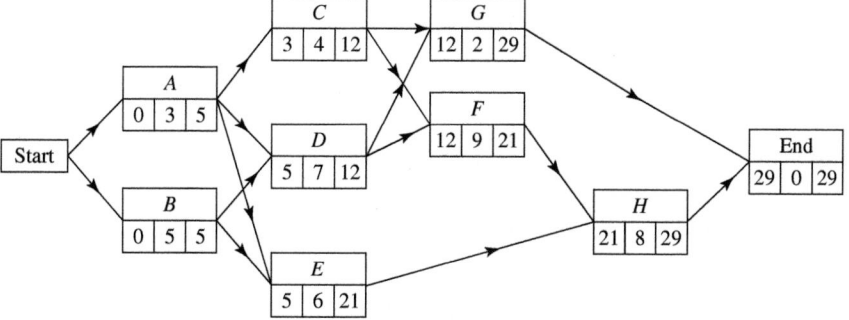

a Calculate the float for each activity in this network (durations are in days).

b State which activities are critical.

History

Henry Laurence Gantt (1861–1919) was an American mechanical engineer who created and popularised the Gantt chart in the 1910s.

Gantt Charts play a key role in the management of many projects. They were used in the construction of the Hoover Dam.

> The Gantt chart, because of its presentation of facts in their relation to time, is the most notable contribution to the art of management made in this generation.
> – Wallace Clark

Research

A resource histogram can be used to help deal with the issue of an activity requiring multiple workers.

Look at this cascade diagram and the resources required for the activities.

Activities A and B each require 4 people and C, D, E and F each require 2 people.

The corresponding resource histogram would be as shown here.

Using the float in the cascade diagram, see if you can level out the resource histogram to minimise the number of people employed and maximise employment time of workers.

1 Draw an activity network to represent this precedence table.

Activity	Immediate predecessor(s)
A	–
B	–
C	A, B
D	C
E	C
F	D, E

[2 marks]

2 Draw an activity network to represent this precedence table.

Activity	Immediate predecessor(s)
A	–
B	–
C	–
D	–
E	A
F	B
G	C
H	D
I	E, F
J	G
K	I, J
L	H, K

[3]

3 Draw an activity network to represent this precedence table.

Activity	Immediate predecessor(s)
A	–
B	A
C	A
D	A
E	B
F	B, C
G	C
H	D
I	E, F
J	F, G
K	H
L	I
M	J, K

[4]

4 Find the earliest start time and latest finish time for each activity in this network.

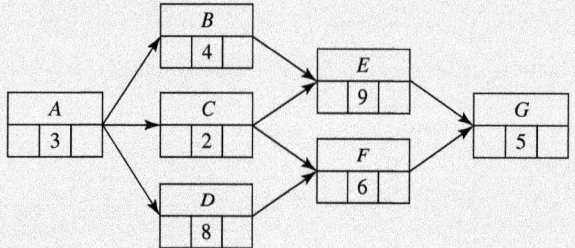

[3]

5 Find the earliest start time and latest finish time for each activity in this network.

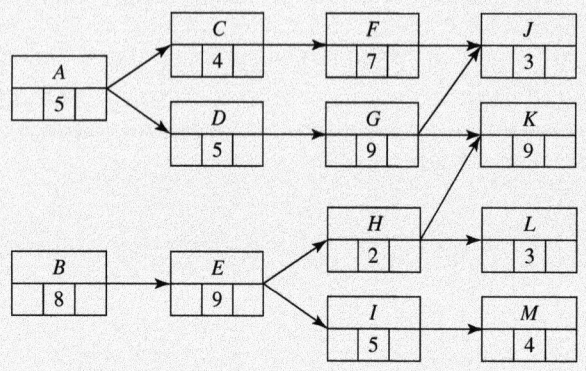

[4]

6 a Draw an activity network to represent this precedence table.

Activity	Duration	Immediate predecessor(s)
A	1	–
B	6	A
C	3	A
D	6	A
E	4	B
F	5	B,C
G	5	D

b Find the earliest start time and latest finish time for each activity in the network. [6]

7 a Draw an activity network to represent this precedence table.

Activity	Duration	Immediate predecessor(s)
A	8	–
B	8	–
C	4	–
D	9	A, B
E	4	B, C
F	8	C
G	9	D, E
H	7	E, F
I	7	G, H

b Find the earliest start time and latest finish time for each activity in the network.

c Find the critical activities. [8]

8 **a** Draw an activity network to represent this precedence table.

Activity	Duration	Immediate predecessor(s)
A	2	–
B	5	–
C	5	–
D	3	A, B
E	9	B
F	3	B, C
G	10	D
H	4	D, E, F
I	7	F
J	5	G, H
K	2	H, I

b Find the earliest start time and latest finish time for each activity in the network.

c Find the critical paths. **[9]**

9 **a** Draw an activity network to represent this precedence table.

Activity	Duration	Immediate predecessor(s)
A	10	–
B	6	–
C	2	–
D	1	–
E	9	A, B
F	15	C, D
G	8	A, E
H	3	B, C
I	6	D, F
J	4	G, H
K	3	H, I

b Find the earliest start time and latest finish time for each activity in the network.

c Find the critical path.

d The duration of task I is changed to x days ($x \geq 6$) without changing the minimum completion time of all the activities. Find a further inequality relating to the maximum value of x **[8]**

10 a Draw an activity network to represent this precedence table.

Activity	Duration	Immediate predecessor(s)
A	7	–
B	8	–
C	11	A, B
D	2	B
E	6	–
F	6	C, D
G	4	E, F
H	1	F
I	6	G, H
J	3	H
K	5	I, J

b Find the earliest start time and latest finish time for each activity in the network.

c Find the critical path.

d Given that A now takes 10 days to complete, list the critical activities.   [8]

11 a Draw an activity network to represent this table of activities.

Activity	Duration	Immediate predecessor(s)
A	9	–
B	3	–
C	2	A
D	5	A, B
E	5	B
F	8	C, D, E
G	4	F
H	1	F
I	7	F
J	6	G, H
K	7	H, I

b Find the earliest start time and latest finish time for each activity in the network.

c Find the critical path.

d Given that G and J each now take 9 days to complete, find the new minimum completion time for the full set of activities.   [9]

14 Linear programming and game theory 1

Linear programming is used to inform managerial decisions in manufacturing and business. This helps ensure that profits are maximised, costs minimised, and the best use is made of resources. Functions are found to represent the boundaries of operational constraints. These are used to identify the region in which operations can take place, and then what is optimal is considered. This is part of what has come to be known as operational research.

The development of computers has assisted with the increasingly complex calculations that are required. This builds on mathematics that was first developed during World War II to ensure that military supplies and personnel were distributed most efficiently. Nowadays, linear programming techniques have a wide range of applications. These include consideration of deploying staff, production, managing stock, marketing and financial management.

Orientation

What you need to know	What you will learn	What this leads to
KS 4 • Inequalities.	• To formulate and solve linear programming problems. • To analyse zero-sum games. • To analyse mixed-strategy games.	• Chapter 30 Applying techniques of linear programming to game theory. • The simplex method.

Fluency and skills

Linear programming is a set of mathematical techniques which help with decision making in a variety of industrial and economic situations. The aim is to choose the best combination of a number of quantities (variables) to achieve the best (optimal) outcome. The objective is typically to maximise profit or minimise cost.

To translate the problem into mathematical terms, you first produce a **linear programming (LP) formulation**. To do this you:

- Identify the quantities you can vary. These are the **decision variables** (or **control variables**).
- Identify the limitations on the values of the decision variables. These are the **constraints**.
- Identify the quantity to be optimised. This is the **objective function**.

> There are two more (trivial) constraints $x \geq 0$ and $y \geq 0$

Example 1

A manufacturer makes two types of muesli, Standard and De Luxe.

1 kg of Standard contains 800 grams of oat mix and 200 grams of fruit mix. 1 kg of De Luxe has 600 grams of oat mix and 400 grams of fruit mix. 1 kg of Standard makes 60 p profit. 1 kg of De Luxe makes 80 p.

They have 3000 kg of oat mix and 1000 kg of fruit mix in stock. They know that they can sell at most 3500 kg of Standard and 2000 kg of De Luxe.

You must decide how much of each type the manufacturer must make in order to maximise the profit. Write this problem as a linear programming formulation.

	Oat mix (kg)	Fruit mix (kg)	Max sales (kg)	Profit (£ per kg)
1 kg Standard	0.8	0.2	3500	0.60
1 kg De Luxe	0.6	0.4	2000	0.80
Availability	3000	1000		

> This table summarises the information.

Suppose they make x kg of Standard and y kg of De Luxe.

> These are the decision variables.

The upper limits on sales give $x \leq 3500$ and $y \leq 2000$

> The constraints are the limitations on sales and the amount of raw materials available.

x kg of Standard uses $0.8x$ kg of oat mix.

y kg of De Luxe uses $0.6y$ kg of oat mix.

There is 3000 kg of oat mix, so $0.8x + 0.6y \leq 3000$

which simplifies to $4x + 3y \leq 15000$

(Continued on the next page)

x kg of Standard uses $0.2x$ kg of fruit mix and y kg of De Luxe uses $0.4y$ kg of fruit mix.

There is 1000 kg of oat mix, so $\quad 0.2x + 0.4y \leq 1000$

which simplifies to $\qquad\qquad x + 2y \leq 5000$

The profit on x kg of Standard and y kg of De Luxe is

$P = 0.6x + 0.8y$

> The function you want to maximise is called the objective function and is often denoted by the letter P

Maximise $\qquad P = 0.6x + 0.8y$

Subject to $\qquad 4x + 3y \ 15\,000$

$\qquad\qquad x + 2y \leq 5000$

$\qquad\qquad x \leq 3500$

$\qquad\qquad y \leq 2000$

> This is the complete linear programming formulation.

$\qquad\qquad x \geq 0, y \geq 0$

It is important that you write out the problem in this form.

Example 2

A paper recycler has two processing plants.

Plant A can process 4 tonnes of waste paper and 1 tonne of cardboard per hour.

Plant B can process 5 tonnes of waste paper and 2 tonnes of cardboard per hour.

It costs £400 per hour to run plant A and £600 per hour to run plant B.

There is a union agreement that each plant must get at least one third of the run time of every consignment.

A consignment of 100 tonnes of waste paper and 35 tonnes of cardboard must be shared between the plants so that costs are minimised.

Write this as a linear programming formulation.

This table summarises the information:

	Waste paper (tonnes)	Cardboard (tonnes)	Cost (£)
Plant A (1 hour)	4	1	400
Plant B (1 hour)	5	2	600
Availability	100	35	

> The decision variables are the running times.

Let x be hours for plant A and y be hours for plant B.

$x \geq \dfrac{1}{3}(x+y)$ and $y \geq \dfrac{1}{3}(x+y)$

> This is the union agreement written algebraically.

which simplify to

$y \leq 2x$ and $x \leq 2y$

$4x + 5y \geq 100$

> Since the plants must process at least 100 tonnes of paper in the time available.

(Continued on the next page)

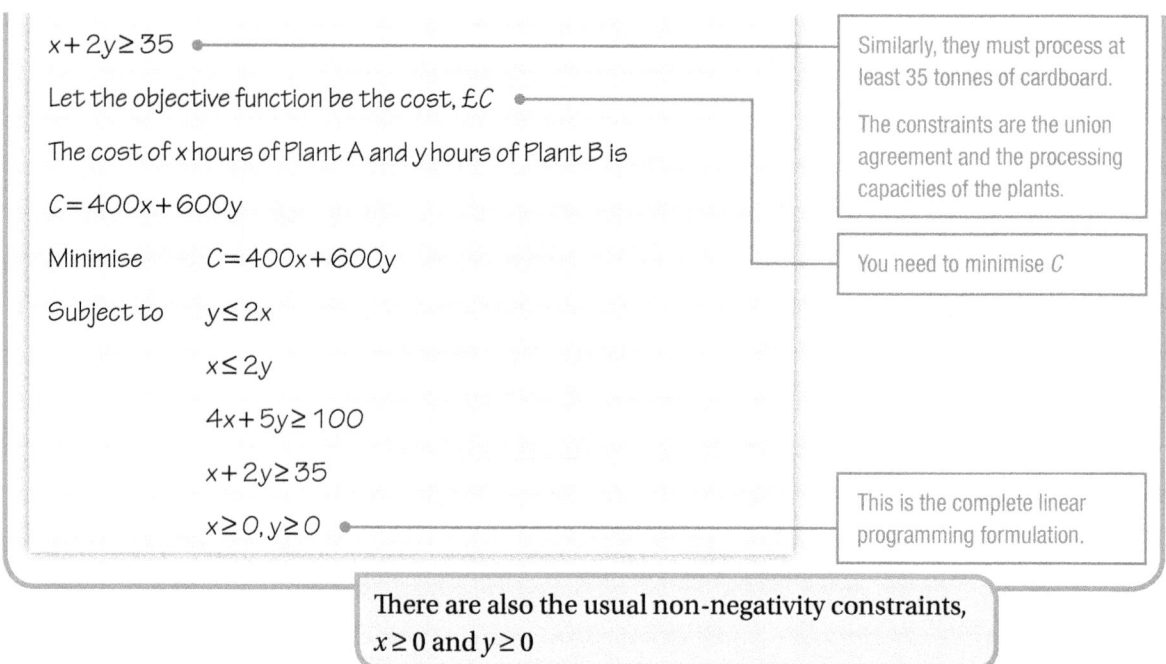

$x + 2y \geq 35$ — Similarly, they must process at least 35 tonnes of cardboard.

Let the objective function be the cost, $£C$ — The constraints are the union agreement and the processing capacities of the plants.

The cost of x hours of Plant A and y hours of Plant B is

$C = 400x + 600y$

Minimise $\quad C = 400x + 600y$ — You need to minimise C

Subject to $\quad y \leq 2x$

$\quad\quad\quad\quad x \leq 2y$

$\quad\quad\quad\quad 4x + 5y \geq 100$

$\quad\quad\quad\quad x + 2y \geq 35$

$\quad\quad\quad\quad x \geq 0, y \geq 0$ — This is the complete linear programming formulation.

There are also the usual non-negativity constraints, $x \geq 0$ and $y \geq 0$

Once you have stated a problem as a linear programming formulation, the next stage is to solve it. If there are just two decision variables, you can use graphical methods.

Consider the muesli problem from Example 1, where x kg of Standard and y kg of De Luxe muesli were produced.

Maximise $\quad P = 0.6\,x + 0.8\,y$

Subject to $\quad x \leq 3500$

$\quad\quad\quad\quad y \leq 2000$

$\quad\quad\quad\quad 4x + 3y \leq 15\,000$

$\quad\quad\quad\quad x + 2y \leq 5000$

$\quad\quad\quad\quad x \geq 0, y \geq 0$

You draw a graph to illustrate the constraints.

First, consider $x \leq 3500$ and $y \leq 2000$ by drawing the lines $x = 3500$ and $y = 2000$

Shade the regions which are **not** needed. This is called shading out.

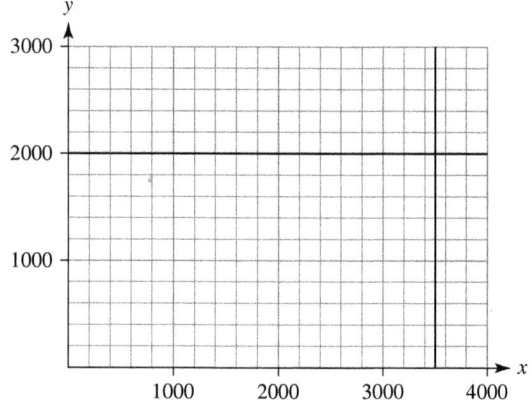

Next, consider $4x + 3y \le 15\,000$ by drawing the line $4x + 3y = 15\,000$

As before, shade the region which does not satisfy the inequality.

Finally, consider $x + 2y \le 5000$, $x \ge 0$ and $y \ge 0$

All allowable combinations of x and y lie in the unshaded region of the graph (including its boundary lines). This is called the **feasible region**.

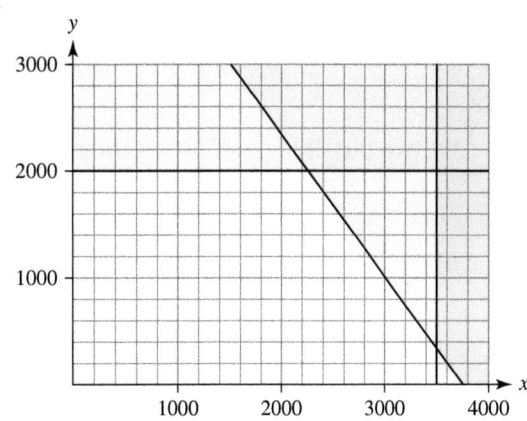

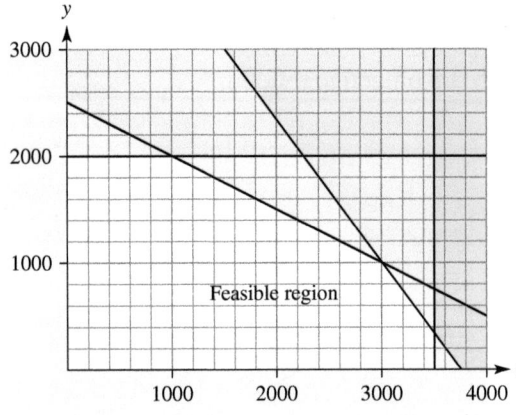

If the inequalities had been $<$ rather than $\le$, the boundaries would not be included. The usual convention is to draw the boundary line dotted for $<$ or $>$, and continuous for $\le$ and $\ge$

The **feasible region** is the set of (x, y) values that satisfy all the constraints. It is the unshaded region on the graph.

You now illustrate the objective function $P = 0.6x + 0.8y$. To do this, choose any values of x and y inside the feasible region. Suppose you chose $x = 1000$, $y = 1000$

This would give $P = 0.6 \times 1000 + 0.8 \times 1000 = 1400$

This plan gives £1400 profit.

There are other production plans giving $P = 1400$, for example $x = 0$, $y = 1750$ or $x = 2000$, $y = 250$
These plans all lie on the line $0.6\,x + 0.8\,y = 1400$

You draw this line on your graph.

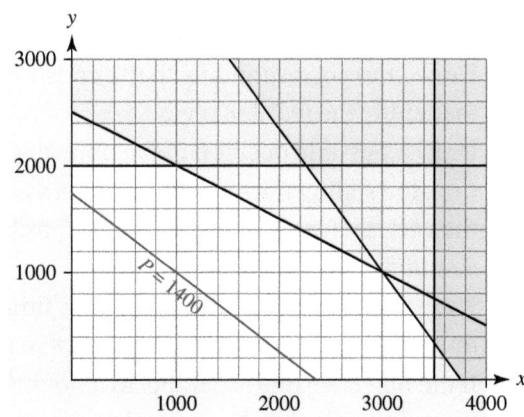

435

Similarly, you could draw the line $0.6x + 0.8y = 1800$ to show all the production plans giving a profit of £1800

These are two possible positions of the **objective line.**

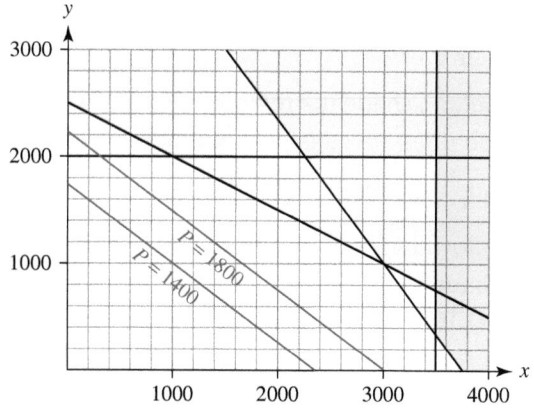

The **objective line** is a line joining all points (x, y) for which the objective function takes a specified value.

Profit P increases as the objective line moves to the right (always keeping the same gradient). As long it crosses the feasible region, there is a production plan giving that profit.

The maximum profit occurs at the extreme position of the line which still includes a point of the feasible region. As you move the line to the right, its last contact with the feasible region will be at a vertex.

In this example, it leaves the feasible region where the lines $x + 2y = 5000$ and $4x + 3y = 15\,000$ intersect.

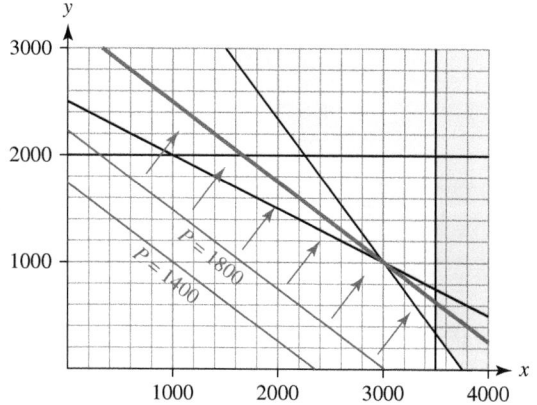

Solving these two equations simultaneously you get the solution $x = 3000$, $y = 1000$ (the point $(3000, 1000)$)
This gives $P = 0.6 \times 3000 + 0.8 \times 1000 = 2600$

So the best production plan is 3000 kg of Standard muesli and 1000 kg of De Luxe muesli, giving a profit of £2600

In summary, to solve a linear programming problem graphically:

- Draw lines corresponding to the constraints and shade out areas to identify the feasible region.
- Draw a possible position of the objective line.
- Imagine sliding the objective line across the graph keeping it parallel to the original. Identify the vertex where it would leave the feasible region. This vertex corresponds to the optimal solution.
- If the choice of vertex is not obvious, find the coordinates of the likely vertices and test which one gives the best value of the objective function.
- If the objective line is parallel to a boundary of the feasible region, then all points on that boundary will give the optimal value for the objective function.

Example 3

This continues the waste paper problem in Example 2

Minimise $\quad C = 400x + 600y$

Subject to $\quad x \le 2y$

$\qquad\qquad y \le 2x$

$\qquad\qquad 4x + 5y \ge 100$

$\qquad\qquad x + 2y \ge 35$

$\qquad\qquad x \ge 0, y \ge 0$

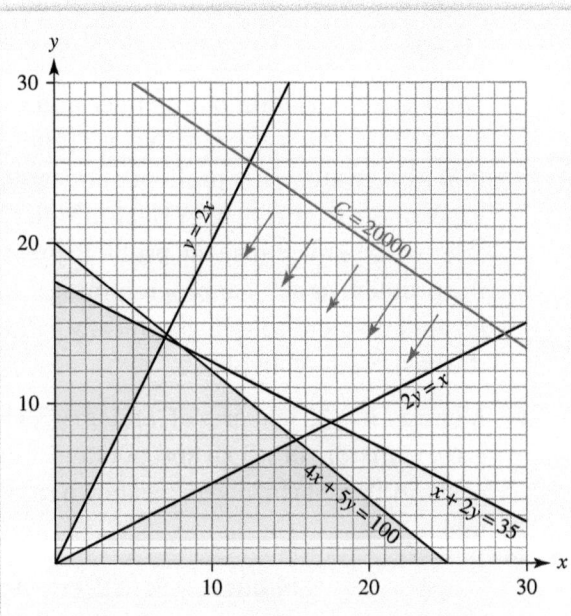

The line shown is the objective line $C = 20\,000$

Solving $y = 2x$ and $4x + 5y = 100$ gives $x = 7\frac{1}{7}, y = 14\frac{2}{7}$

This gives $C = 400 \times 7\frac{1}{7} + 600 \times 14\frac{2}{7} = 11\,428\frac{4}{7}$

Solving $4x + 5y = 100$ and $x + 2y = 35$ gives $x = 8\frac{1}{3}, y = 13\frac{1}{3}$

This gives $C = 400 \times 8\frac{1}{3} + 600 \times 13\frac{1}{3} = 11\,333\frac{1}{3}$

Solving $x + 2y = 35$ and $x = 2y$ gives $x = 17\frac{1}{2}, y = 8\frac{3}{4}$

This gives $C = 400 \times 17\frac{1}{2} + 600 \times 8\frac{3}{4} = 12\,250$

The minimum value of C is $11\,333\frac{1}{3}$, when $x = 8\frac{1}{3}, y = 13\frac{1}{3}$

The merchant should run plant A for $8\frac{1}{3}$ hours and plant B for $13\frac{1}{3}$ hours.

Moving the objective line to the left parallel to itself reduces the value of C Remember that you are trying to minimise C

It is hard to see from the graph which of the three vertices gives the minimum value of C, so find C at each vertex.

DISCRETE

1 A company makes two brands of dog food, A and B, which it sells in 500 g tins. Both brands are a mixture of meat and vegetables. The table shows the amounts of these in each brand and the amounts in stock.

	Meat (kg)	Vegetables (kg)
Brand A	0.2	0.3
Brand B	0.4	0.1
Amount in stock (kg)	8000	6000

A tin of brand A makes 10 p profit, while a tin of brand B makes 15 p. The company wants a production plan to maximise their profit. Taking the number of tins of brands A and B produced as x and y, respectively, express this problem as a linear programming formulation. You do not need to solve the problem.

2 For each of the following linear programming problems, draw a graph showing the feasible region and one possible position of the objective line. Hence find the optimal value of the objective function and the corresponding values of x and y

a Maximise $P = 3x + 2y$
Subject to $3x + 4y \leq 120$
$3x + y \leq 75$
$x \geq 10, y \geq 5$

b Maximise $P = 4x + y$
Subject to $x + y \leq 10$
$2x + y \leq 16$
$x \geq 0, y \geq 0$

c Minimise $C = 5x + 4y$
Subject to $15x + 8y \geq 90$
$y \geq x$
$y \leq 2x$

d Maximise $R = 2x + y$
Subject to $x + y \leq 14$
$x + 2y \leq 20$
$2x + 3y \leq 32$
$x \geq 0, y \geq 0$

3 Roger Teeth Ltd make fruit drinks of two types, Econofruit and Healthifruit, consisting of fruit juice, sugar syrup and water. The proportions of these in the two drinks are shown in the table.

	Fruit juice	Sugar syrup
Econofruit	20%	50%
Healthifruit	40%	30%

There are 20 000 litres of fruit juice and 30 000 litres of sugar syrup in stock (and unlimited water). The profit per litre is 30 p for Econofruit and 40 p for Healthifruit. They wish to maximise the profit. Express the problem as a linear programming formulation and solve it graphically to find the best production plan.

4 A farmer has 75 hectares of land on which to grow a mixture of wheat and potatoes. The costs and profits involved are shown in the table.

	Labour (man-hours per ha)	Fertiliser (kg per ha)	Profit (£ per ha)
Wheat	30	700	80
Potatoes	50	400	100

There are 2800 man-hours of labour and 40 tonnes of fertiliser available. The aim is to maximise the profit. Express the problem as a linear programming formulation and solve it graphically to find the best planting scheme.

To solve a problem involving linear programming

(1) Identify the decision variables and label them $x, y, z\ldots$

(2) Express the constraints as inequalities.

(3) Identify the objective function (to be maximised or minimised).

(4) If the problem has two variables, solve graphically.

(5) Answer the question.

A manufacturer makes three types of dining chair. All chairs pass through three workshops for cutting, assembly and finishing. The cutting workshop runs for 50 hours per week, the assembly workshop for 35 hours per week and the finishing workshop for 40 hours per week. The times in the workshops for each type of chair and the profit gained from each are shown in the table.

> This example has three decision variables, but you can reduce the number of variables to two in the light of extra information. There is also a new type of constraint involved.

Chair type	Cutting (hours)	Assembly (hours)	Finishing (hours)	Profit (£)
A	0.5	0.6	0.75	20
B	2	1	0.75	60
C	0.5	0.4	0.5	20

The aim is to maximise the weekly profit.

a Write this as a linear programming formulation.

b Given that half of the chairs they sell are type C, rewrite the problem with two variables and solve to find the best production plan.

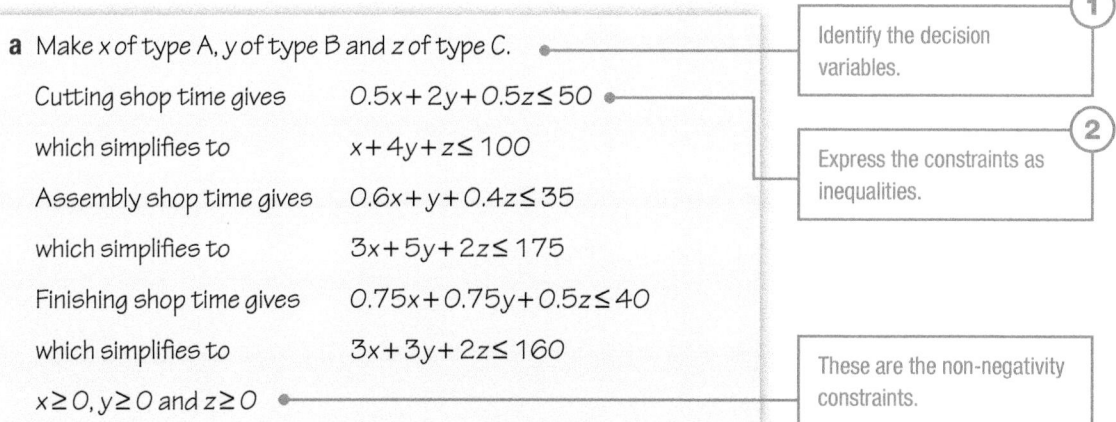

a Make x of type A, y of type B and z of type C. — *Identify the decision variables.* **(1)**

Cutting shop time gives $\quad 0.5x + 2y + 0.5z \le 50$

which simplifies to $\quad x + 4y + z \le 100$

Assembly shop time gives $\quad 0.6x + y + 0.4z \le 35$ — *Express the constraints as inequalities.* **(2)**

which simplifies to $\quad 3x + 5y + 2z \le 175$

Finishing shop time gives $\quad 0.75x + 0.75y + 0.5z \le 40$

which simplifies to $\quad 3x + 3y + 2z \le 160$

$x \ge 0,\, y \ge 0$ and $z \ge 0$ — *These are the non-negativity constraints.*

(*Continued on the next page*)

DISCRETE

There is also the constraint that x, y and z are integers

You can't sell fractions of a chair.

The profit is $\qquad$ $P = 20x + 60y + 20z$

Maximise $\qquad$ $P = 20x + 60y + 20z$

③

This is the objective function.

You need to maximise the profit, $£P$

Subject to $\qquad$ $x + 4y + z \leq 100$

$3x + 5y + 2z \leq 175$

$3x + 3y + 2z \leq 160$

$x \geq 0,\ y \geq 0,\ z \geq 0$

x, y and z are integers.

This is the complete linear programming formulation.

b Given $z = x + y$

Half of the sales are type C

Profit $P = 20x + 60y + 20(x + y) = 40x + 80y$

The constraints are
$x + 4y + (x + y) \leq 100$, so $\qquad$ $2x + 5y \leq 100$

$3x + 5y + 2(x + y) \leq 175$, so $\qquad$ $5x + 7y \leq 175$

$3x + 3y + 2(x + y) \leq 160$, so $\qquad$ $x + y \leq 32$

Substitute $z = x + y$ into the formulation.

$x \geq 0,\ y \geq 0,$ x and y are integers.

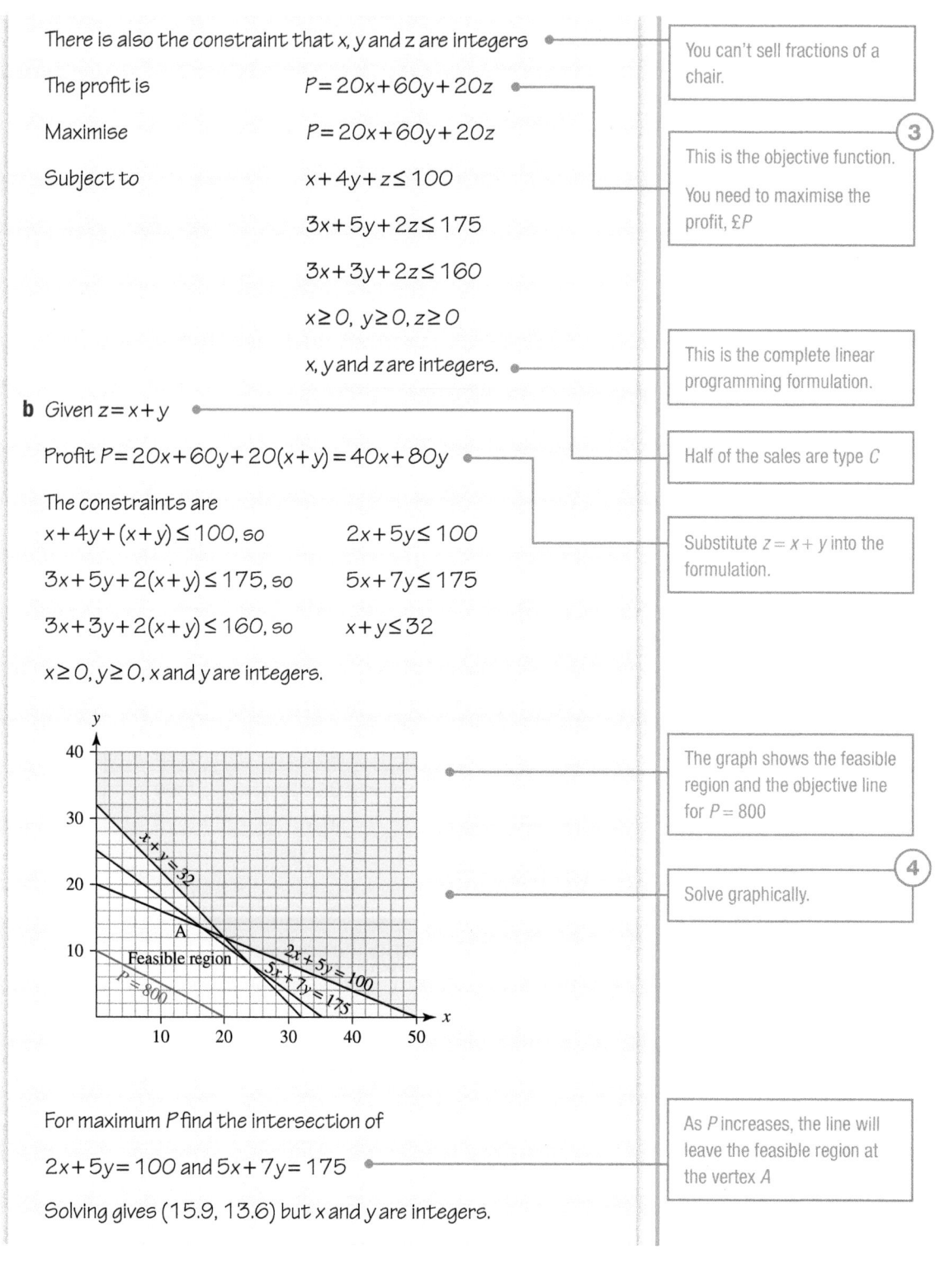

The graph shows the feasible region and the objective line for $P = 800$

④

Solve graphically.

For maximum P find the intersection of

$2x + 5y = 100$ and $5x + 7y = 175$

As P increases, the line will leave the feasible region at the vertex A

Solving gives (15.9, 13.6) but x and y are integers.

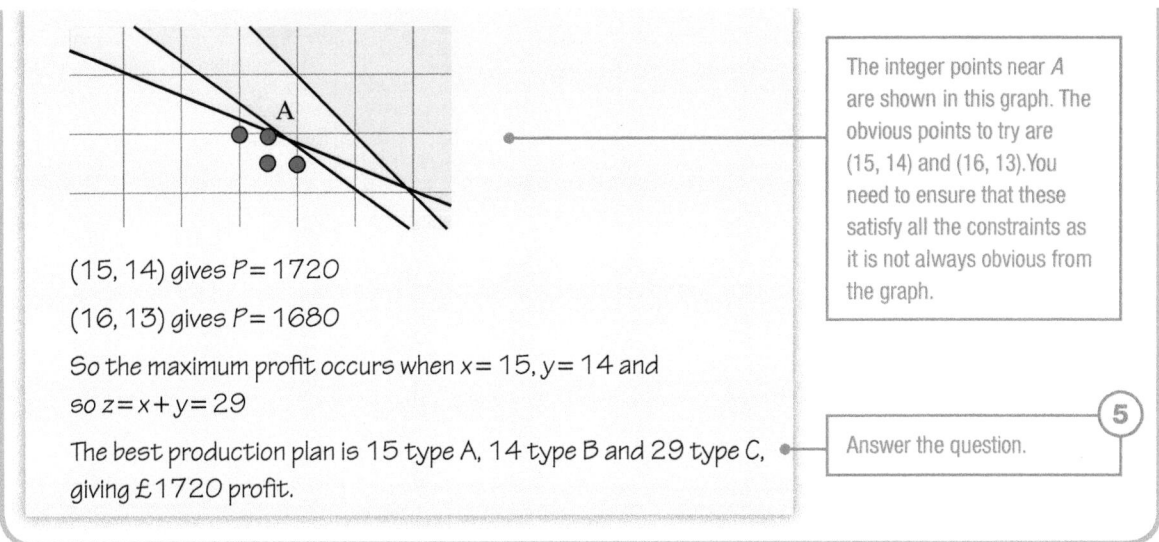

(15, 14) gives $P = 1720$

(16, 13) gives $P = 1680$

So the maximum profit occurs when $x = 15$, $y = 14$ and

so $z = x + y = 29$

The best production plan is 15 type A, 14 type B and 29 type C, giving £1720 profit.

> The integer points near A are shown in this graph. The obvious points to try are (15, 14) and (16, 13). You need to ensure that these satisfy all the constraints as it is not always obvious from the graph.

> (5) Answer the question.

Linear programming can also be used to decide on the most cost-effective proportions to use when blending materials.

Example 5

A company supplying vegetable oil buys from two sources, A and B. The oils are already a blend of olive oil, sunflower oil and other vegetable oils. The table shows the proportions, price and minimum weekly order of these.

	Olive oil	Sunflower oil	Other	Cost (p per litre)	Minimum order (litres)
A	50%	10%	40%	25	35 000
B	20%	60%	20%	20	50 000

The company wants to make a blend with at least 30% olive oil and at least 30% sunflower oil. They want to produce at least 90 000 litres per week, and to minimise the cost.

Write this as a linear programming formulation and solve to find the best blend.

Use x litres of A and y litres of B.

> (1) Identify the decision variables. These are the amounts of the oils to use.

The amount of olive oil is $(0.5x + 0.2y)$ in a total production of $(x + y)$.

At least 30% olive oil gives $\dfrac{0.5x + 0.2y}{x + y} \geq 0.3$

which simplifies to $y \leq 2x$

> (2) Express the constraints as inequalities.

The amount of sunflower oil is $(0.1x + 0.6y)$ in a total production of $(x + y)$

At least 30% sunflower oil gives $\dfrac{0.1x + 0.6y}{x + y} \geq 0.3$

which simplifies to $3y \geq 2x$

The minimum order requirements give $x \geq 35\,000$ and $y \geq 50\,000$

The total production constraint is $x + y \geq 90\,000$

(Continued on the next page)

DISCRETE

The total production constraint is $x + y \geq 90\,000$

$C = 0.25x + 0.2y$

③

The objective function is the cost, £C

You need to minimise C

Minimise $\quad C = 0.25x + 0.2y$

Subject to
$\quad y \leq 2x$
$\quad 3y \geq 2x$
$\quad x \geq 35\,000$
$\quad y \geq 50\,000$
$\quad x + y \geq 90\,000$

This is the complete linear programming formulation.

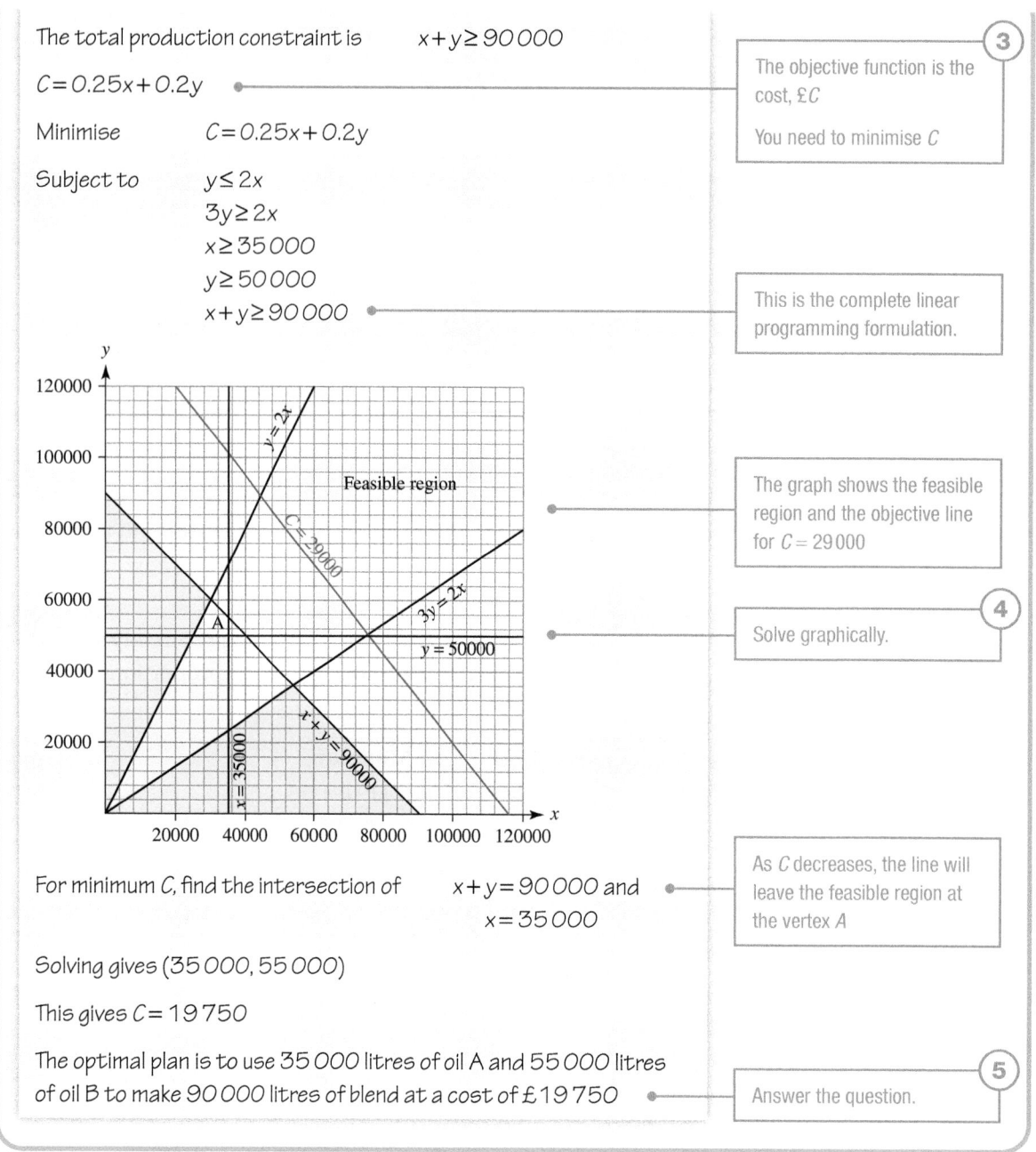

The graph shows the feasible region and the objective line for $C = 29\,000$

④

Solve graphically.

For minimum C, find the intersection of $\quad x + y = 90\,000$ and $\quad x = 35\,000$

As C decreases, the line will leave the feasible region at the vertex A

Solving gives $(35\,000, 55\,000)$

This gives $C = 19\,750$

The optimal plan is to use 35 000 litres of oil A and 55 000 litres of oil B to make 90 000 litres of blend at a cost of £19 750

⑤

Answer the question.

Exercise 14.1B Reasoning and problem-solving

1 A club which has 80 members, is organising a trip. They intend to hire vehicles they can drive themselves and travel in convoy. Only eight of the members are prepared to drive. A car, which can carry five people including the driver, costs £20 per day to hire. A minibus, which can carry 12 people including the driver, costs £60 per day to hire. They wish to minimise the hire costs. Express this problem as a linear programming formulation and hence find the best combination of vehicles to hire.

2 Whisky mac is a mixture of whisky and ginger wine. Whisky is 40% alcohol and costs £12 per litre. Ginger wine is 12% alcohol and costs £5 per litre. A bar-tender wishes to minimise the cost of making a whisky mac,

which must be at least 100 ml of liquid, at least 20% alcohol and contain at most 30 ml of alcohol. Use linear programming to find the amount of each drink in the optimal mixture.

3   A trader buys goods from a warehouse and takes them by van to his shop. He wants shampoo and washing powder. These are packed in cases of the same size. A case of shampoo weighs 6 kg and he can make a profit of £20 per case. A case of washing powder weighs 4 kg and his profit will be £14 per case. His van has room for 60 cases and can carry a maximum load of 280 kg. Find how much of each product he should buy to maximise his profit.

4   An investor has up to £20 000 to invest. She can buy 'safe' bonds yielding 5% interest or 'risky' shares yielding 10% interest. She wants to make at least 8% interest overall, and as she is a cautious investor she wants her investment in bonds to be as great as possible. How much should she invest in each?

5   A building firm has a plot of land with area 6000 m². The intention is to build a mixture of houses and bungalows. A house occupies 210 m² and a bungalow occupies 270 m². Planning regulations limit the total number of dwellings to 25, and insist that there must be no more than 15 of either type. A house makes £20 000 profit, and a bungalow £25 000. Formulate this situation as a linear programming problem and find graphically the best combination of dwellings.

6   There are 50 places on a trip. The group comprises x senior staff, y trainees and z children. There must be at least one adult to every two children, and at least one senior staff member to every trainee. There must be at least five trainees and at least ten senior staff. The cost of the trip is £20 for each senior staff member, £15 for each trainee and £12 for each child. It is required to minimise the total cost.

a   Express this as a linear programming formulation in x, y and z

b   Show that your formulation can be rewritten to minimise the objective function as

$$C = 8x + 3y + 600$$

and write the constraints in terms of the two variables, x and y

c   Solve to find the optimum combination of people.

7   A dog food manufacturer makes three types of chew, each 10 g in weight, from two basic ingredients. The table shows the proportions of these together with the amounts of the two ingredients in stock. Ingredient 1 costs the manufacturer £2 per kg and ingredient 2 costs £1 per kg.

	Ingredient 1	Ingredient 2
Chew A	8 g	2 g
Chew B	6 g	4 g
Chew C	5 g	5 g
In stock	800 kg	400 kg

The manufacturer wants to make 1600 packets of mixed chews. Each must contain 60 chews, and there must be no more than 30 of each type of chew in a packet. Find the best combination of chew types in each packet to minimise the cost.

Fluency and skills

Two players A and B decide to play a game with these rules.

Each person plays a card–king (K), queen (Q) or jack (J). Depending on the cards played, either player A pays player B or player B pays player A an agreed amount.

For example, if they both play a king, player B pays player A 5p. Record this as (5, −5), meaning that player A gains 5p and player B loses 5p.

Similarly, if player A plays a king and player B plays a queen, player A pays player B 4p, Record this as (−4, 4)

The complete set of payments is shown in this table.

		B		
		K	**Q**	**J**
	K	(5, −5)	(−4, 4)	(2, −2)
A	**Q**	(3, −3)	(1, −1)	(4, −4)
	J	(2, −2)	(3, −3)	(−1, 1)

This is a two-person game – there are just two competitors (it could be two teams rather than two people).

On each play, one competitor's gain equals the other's loss. This is called a **zero-sum game**.

> **Key point**
>
> In a **zero-sum** game, the sum of the gains made by the players on each play is zero.

To record a zero-sum game, you only need to put one number in each cell, as the other is just its negative. The table becomes:

		B		
		K	**Q**	**J**
	K	5	−4	2
A	**Q**	3	1	4
	J	2	3	−1

This is player A's **payoff matrix**. It is conventional to record the gains of the player on the left of the table, the **row player**.

Player B's payoff matrix would look like this.

		A		
		K	Q	J
B	K	−5	−3	−2
	Q	4	−1	−3
	J	−2	−4	1

You analyse the game from player A's point of view.

- If she plays a king, the worst outcome is that she loses 4 p.
- If she plays a queen, the worst is that she wins 1 p.
- If she plays a jack, the worst is that she loses 1 p.

		B			Row minimum	
		K	Q	J		
A	K	5	−4	2	−4	
	Q	3	1	4	1	← max = 1
	J	2	3	−1	−1	

These are the minimum values for the rows of the table.

Her **play-safe strategy** is to play a queen, which gives the best guaranteed outcome (win 1 p). This is called a **maximin strategy**, because it maximises her minimum gain.

> **Key point**
>
> A play-safe strategy gives the best guaranteed outcome regardless of what the other player does.

Using the same table, you can analyse the game from player B's point of view by looking at the columns instead of the rows.

- If he plays a king, the worst outcome is that he loses 5 p.
- If he plays a queen, the worst is that he loses 3 p.
- If he plays a jack, the worst is that he loses 4 p.

These are the maximum values for the columns of the table.

		B			Row minimum	
		K	Q	J		
A	K	5	−4	2	-4	
	Q	3	1	4	1	← max = 1
	J	2	3	−1	-1	
Column maximum		5	3	4		

↑
min = 3

His play-safe strategy is to play a queen, which gives the best guaranteed outcome (lose 3 p). This is called a **minimax strategy**, because it minimises his maximum loss.

If both players play safe every time, it is called a **pure-strategy game**. Player A wins 1 p each time. This is the **value of the game** to player A. (The value of the game to player B is −1 p).

If player A knows that player B intends to play a queen, she should play a jack instead of a queen, winning 3 p instead of 1 p. This means the solution is **unstable**.

> **Key point**
>
> A game has a **stable solution** if neither player can gain by changing from their play-safe strategy.

Suppose the payoff matrix is changed, like this.

		B			Row minimum	
		K	**Q**	**J**	**Row minimum**	
A	**K**	4	2	−4	−4	
	Q	3	4	2	2	← max = 2
	J	2	−2	1	−2	
Column maximum		4	4	2		

$\uparrow$
min = 2

Player A's play-safe strategy is still to play a queen while player B's is now to play a jack.

Neither player can gain by changing strategy. If player A plays a queen, then player B's best option is to play a jack, and if player B plays a jack, then player A's best option is to play a queen. The game has a **stable solution**.

The **value of the game** to player A is 2 p. (The stated value of a game is always the value to the row player.)

> A stable solution can also be called a saddle point because the 2 in the table is the lowest value in its row and the highest in its column, just as the centre of a horse's saddle is its lowest point in the nose-to-tail direction and its highest point in the side-to-side direction.

> **Key point**
>
> The value of a game is the payoff to the row player if both players use their best strategy.

Notice that, in this stable solution, the maximum of the row minima is the same as the minimum of the column maxima (both = 2)

These two quantities are always equal in a stable solution.

> **Key point**
>
> A game has a stable solution if
> maximum of row minima = minimum of column maxima

Example 1

This payoff matrix describes a game between players A and B. Player A has three strategies A_1, A_2 and A_3, and player B has four strategies B_1, B_2, B_3 and B_4. Show that the game has a stable solution, and find the value of the game.

		B			
		B_1	B_2	B_3	B_4
A	A_1	8	6	9	6
	A_2	4	3	−2	2
	A_3	10	−1	1	5

(*Continued on the next page*)

		B_1	B_2	B_3	B_4	Row min	
	A_1	8	6	9	6	6	← max = 6
A	A_2	4	3	−2	2	−2	
	A_3	10	−1	1	5	−1	
Column max		10	6	9	6		
			↑		↑		
			min = 6		min = 6		

A's play safe strategy is A_1

B's play safe strategy is either B_2 or B_4

> Find the row minima and choose the maximum of these.

Max of row minima = min of row maxima, so the game has a stable solution.

> Find the column maxima and choose the minimum of these.

The value of the game is 6

In some games, one or more strategies should never be used. You can remove those rows or columns from the payoff matrix.

For example, in this game, A would never use strategy A_2, because A_1 has a better payoff in all three columns. Row 1 **dominates** row 2

		B_1	B_2	B_3
	A_1	7	3	9
A	A_2	4	1	6
	A_3	4	4	−1

Because strategy A_2 will never be used, you can eliminate row 2

		B_1	B_2	B_3
A	A_1	7	3	9
	A_3	4	4	−1

In the revised matrix, B would never use strategy B_1, because its outcomes are the same or worse than those of B_2

Column 2 dominates column 1, so you can eliminate column 1

		B_2	B_3
A	A_1	3	9
	A_3	4	−1

There is now no dominance, so the matrix cannot be simplified any further.

In a payoff matrix, row i dominates row j if, for every column, the value in row $i \geq$ value in row j

Similarly, column i dominates column j if, for every row, the value in column $i \leq$ value in column j

Example 2

Use dominance to simplify this payoff matrix as far as possible. What can be deduced from the result?

		B		
		B_1	B_2	B_3
A	A_1	5	3	3
	A_2	−2	7	−1

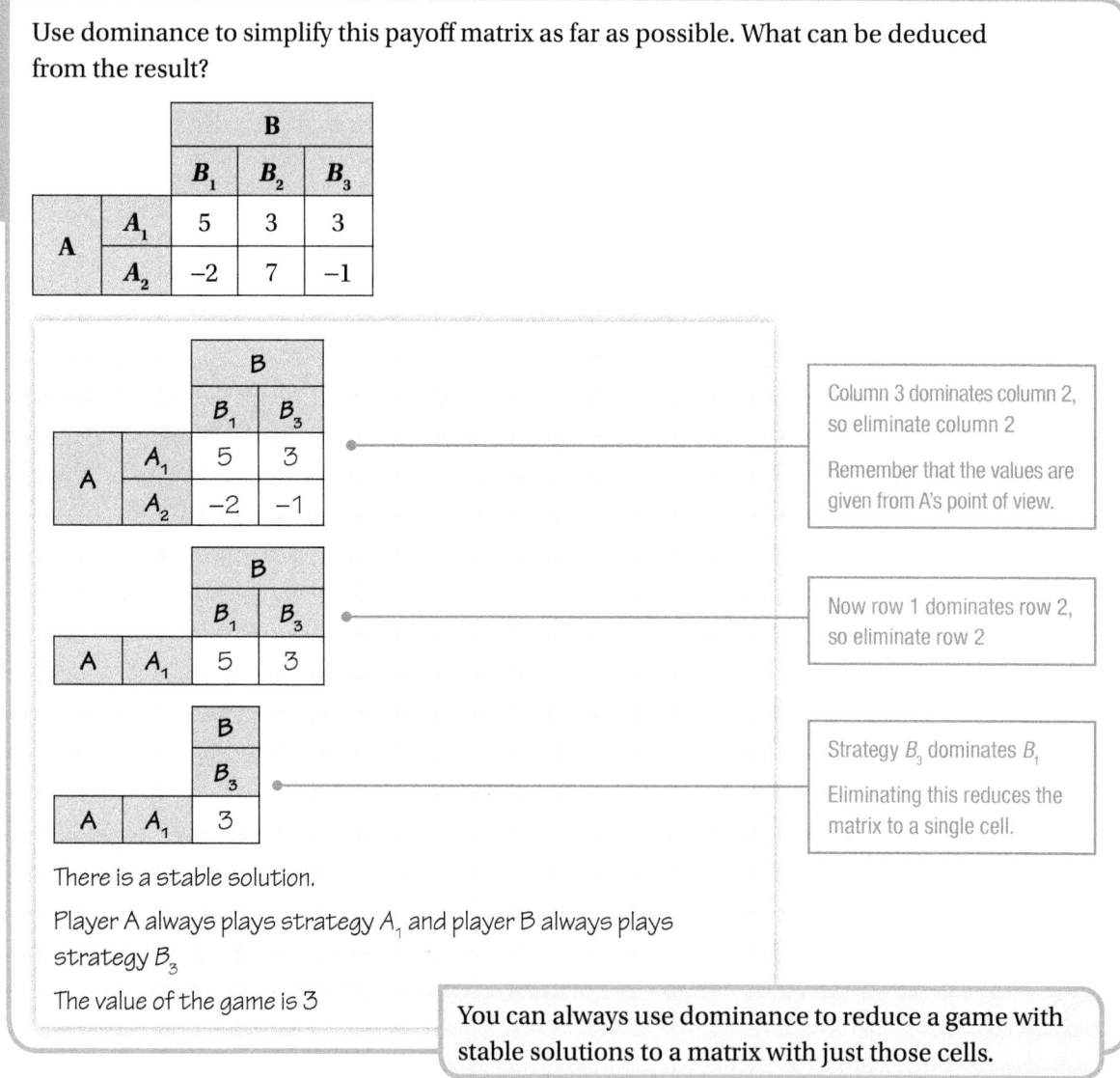

		B	
		B_1	B_3
A	A_1	5	3
	A_2	−2	−1

Column 3 dominates column 2, so eliminate column 2

Remember that the values are given from A's point of view.

		B	
		B_1	B_3
A	A_1	5	3

Now row 1 dominates row 2, so eliminate row 2

		B
		B_3
A	A_1	3

Strategy B_3 dominates B_1

Eliminating this reduces the matrix to a single cell.

There is a stable solution.

Player A always plays strategy A_1 and player B always plays strategy B_3

The value of the game is 3

You can always use dominance to reduce a game with stable solutions to a matrix with just those cells.

1 For each of these payoff matrices, find the play-safe strategy for each player. Determine whether the game has a stable solution and, if so, state the value of the game.

2 For the game shown in question **1d**, write down the payoff matrix for player B.

3 For each of these payoff matrices, use dominance to simplify the problem as far as possible. If there is a stable solution, state the strategies the players should adopt and the value of the game.

a

		B		
		B_1	B_2	B_3
A	A_1	5	3	9
	A_2	4	7	6
	A_3	2	4	5

b

		B		
		B_1	B_2	B_3
A	A_1	9	−2	3
	A_2	−1	−4	2
	A_3	4	6	3

c

		B		
		B_1	B_2	B_3
A	A_1	6	−1	4
	A_2	−1	−3	5

d

		B	
		B_1	B_2
A	A_1	−1	3
	A_2	2	7
	A_3	6	−2

e

		B			
		B_1	B_2	B_3	B_4
A	A_1	4	−2	5	9
	A_2	2	1	3	5
	A_3	3	−1	−8	4

f

		B			
		B_1	B_2	B_3	B_4
A	A_1	3	2	7	10
	A_2	5	5	7	8
	A_3	−3	2	−5	6

a

		B	
		B_1	B_2
A	A_1	−1	2
	A_2	2	3
	A_3	4	−2

b

		B		
		B_1	B_2	B_3
A	A_1	4	−1	4
	A_2	−1	−3	−4

c

		B		
		B_1	B_2	B_3
A	A_1	3	1	6
	A_2	2	5	3
	A_3	0	2	3

d

		B		
		B_1	B_2	B_3
A	A_1	−6	−2	1
	A_2	3	−1	−2
	A_3	−4	3	1

e

		B			
		B_1	B_2	B_3	B_4
A	A_1	−2	2	−3	−1
	A_2	−1	2	1	−1
	A_3	−2	−3	−4	−3

f

		B		
		B_1	B_2	B_3
A	A_1	3	4	2
	A_2	−1	2	1
	A_3	1	2	4
	A_4	2	1	−3

DISCRETE

449

Reasoning and problem-solving

Strategy

To solve problems with zero-sum payoff matrices

1. If necessary, construct the matrix of payoffs for the row player.
2. Check whether the matrix can be reduced by using dominance.
3. Find the maximum of the row minima and the minimum of the row maxima to determine each player's play-safe strategy.
4. Decide whether the game has a stable solution (saddle point) and, if so, find its value.
5. Answer the question in context.

Example 3

Analyse the game shown in the table.

		B			
		B_1	B_2	B_3	B_4
A	A_1	0	−3	5	−9
	A_2	5	−8	−2	10
	A_3	3	10	6	9
	A_4	4	11	−3	2

		B			
		B_1	B_2	B_3	B_4
A	A_2	5	−8	−2	10
	A_3	3	10	6	9
	A_4	4	11	−3	2

Row 3 dominates row 1, so delete row 1

2 Reduce the rows using dominance.

		B		
		B_1	B_2	B_3
A	A_2	5	−8	−2
	A_3	3	10	6
	A_4	4	11	−3

Column 3 dominates column 4, so delete column 4

2 Reduce the columns using dominance.

3 Find row minima and column maxima.

		B			
		B_1	B_2	B_3	
A	A_2	5	−8	−2	−8
	A_3	3	10	6	3
	A_4	4	11	−3	−3
		5	11	6	

Max of row minima = 3 (shown).

Min of column maxima = 5 (shown).

4 Decide whether the solution is stable.

A's play-safe strategy is A_3

B's play-safe strategy is B_1

The solution is not stable.

Max of row minima ≠ min of column maxima.

1 Analyse each of these payoff matrices and state what you can about the game and the players' strategies.

a

		B		
		B_1	B_2	B_3
A	A_1	−4	6	8
	A_2	−2	−8	10
	A_3	−5	0	3

b

		B		
		B_1	B_2	B_3
A	A_1	5	−2	6
	A_2	3	−4	−1
	A_3	−2	−2	3
	A_4	1	−5	−2

c

		B			
		B_1	B_2	B_3	B_4
A	A_1	4	0	1	−2
	A_2	3	1	−3	−1
	A_3	−1	4	3	0
	A_4	−2	2	1	−1

d

		B			
		B_1	B_2	B_3	B_4
A	A_1	3	−1	−1	0
	A_2	2	2	1	1
	A_3	1	1	0	3
	A_4	−1	−2	1	2

e

		B			
		B_1	B_2	B_3	B_4
A	A_1	−2	−5	3	−11
	A_2	13	−10	−4	8
	A_3	5	8	−2	7
	A_4	4	9	−5	0

f

		B				
		B_1	B_2	B_3	B_4	B_5
A	A_1	2	2	0	1	0
	A_2	1	0	−1	1	−2
	A_3	1	3	−1	4	−1
	A_4	4	2	0	1	0

2 X and Y play a game in which X chooses a number from the set {3, 5, 9} and Y a number from the set {2, 6, 7}. The difference between the chosen numbers, ignoring minus signs, is d. If $d > 2$, Y pays X £d; otherwise X pays Y £$2d$

a Construct X's payoff matrix for this game.

b Analyse the game to find the play-safe strategies for X and Y. Explain how you know that the game has no stable solution.

c Modify the game by changing one of the numbers in Y's set, so that the revised game has a stable solution.

3 A game in which there is a fixed number of points awarded on each play can be analysed as a zero-sum game by subtracting half the total score. For example, if there are 10 points available in a game between A and B, possible scores might be (10, 0) or (4, 6). These have the same effect as (5, −5) and (−1, 1), respectively, as, in the first case, A moves 10 points ahead of B and, in the second case, B moves 2 points ahead of A.

The table shows a game in which a total of 6 points is awarded for each play.

		B		
		B_1	B_2	B_3
A	A_1	(4, 2)	(3, 3)	(2, 4)
	A_2	(2, 4)	(0, 6)	(3, 3)
	A_3	(6, 0)	(4, 2)	(5, 1)

a Construct the conventional payoff matrix that is equivalent to this game.

b Find the play-safe strategy for each player and show that the game has a saddle point.

c If the game were played ten times and each player followed a pure strategy, what would be the final score?

Full A Level

DISCRETE

Fluency and skills

Look at the game shown in this table.

		B		
		B_1	B_2	**Row minimum**
A	A_1	−3	5	−3
	A_2	3	−1	−1 ← max = −1
Column maximum		3	5	
		↑ min = 3		

The play-safe strategies are A_2 for A and B_1 for B. If they play a pure-strategy game (always playing safe), A will win 3 every time. If the solution is stable (has a saddle point), there is no advantage to be gained in moving from the play-safe strategy, but in this case the solution is not stable. If B knows that A will play A_2, she will play B_2 instead of the play-safe strategy B_1. However, if A knows that B will do this, he will play A_1 instead of the play-safe strategy A_2

The best approach for each player is to play each strategy some of the time. This is a **mixed-strategy game**.

To analyse a mixed-strategy game, you need to use the idea of **expectation** or **expected payoff**.

Suppose you play a game where the probability of winning is $\frac{1}{4}$

If you win, you get £10

If you lose you pay £5

	Win	Lose
Payoff	£10	−£5
Probability	$\frac{1}{4}$	$\frac{3}{4}$

If you play the game four times, you expect to win once and lose three times.

Your expected total payoff is $£10 + 3 \times (-£5) = -£5$

Dividing by 4, your expected average payoff per game is $\frac{1}{4} \times £10 + \frac{3}{4} \times (-£5) = -£5 \div 4 = -£1.25$

This is the sum of each payoff multiplied by its probability.

If payoffs $x_1, x_2, ..., x_n$ occur with probabilities p_1, $p_2, ..., p_n$, the **expectation** or **expected (mean) payoff** $E(x)$ is given by

$$E(x) = x_1 p_1 + x_2 p_2 + ... + x_n p_n = \sum_{i=1}^{n} x_i p_i$$

You can apply this to the mixed-strategy game in the initial example.

		B	
		B_1	B_2
A	A_1	−3	5
	A_2	3	−1

Suppose player A plays A_1 randomly for a proportion p of plays and A_2 for the remaining $(1-p)$. She needs to decide the best value of p

Let the value of the game be v

If B plays B_1, the possible payoffs for A are:

	A_1	A_2
Payoff	−3	3
Probability	p	$(1-p)$

The expected payoff is $(-3) \times p + 3 \times (1-p) = 3 - 6p$

The value of the game cannot be more than this, so $v \le 3 - 6p$

If B plays B_2, the possible payoffs for A are:

	A_1	A_2
Payoff	5	−1
Probability	p	$(1-p)$

The expected payoff is $5 \times p + (-1) \times (1-p) = 6p - 1$

The value of the game cannot be more than this, so $v \le 6p - 1$

Player A needs to maximise v subject to $v \le 3 - 6p$ and $v \le 6p - 1$

Plot lines $v = 3 - 6p$ and $v = 6p - 1$

For a given value of p, the value of v occurs on the lower of the two lines.

You find the maximum value of v where the two lines intersect, so

$$3 - 6p = 6p - 1$$
$$p = \frac{1}{3}$$

Player A's best strategy is to play A_1 and A_2 randomly with probabilities $\frac{1}{3}$ and $\frac{2}{3}$

This is a linear programming problem, which, in this case, can be solved graphically.

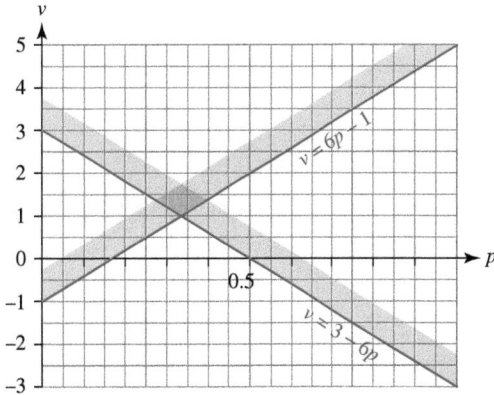

DISCRETE

The value of the game is $v = 3 - 6 \times \dfrac{1}{3} = 1$

You find B's best strategy in the same way.

Suppose B plays B_1 and B_2 with probabilities q and $(1-q)$

If A plays A_1, B's expected payoff is

$$3 \times q + (-5) \times (1-q)) = 8q - 5$$

If A plays A_2, B's expected payoff is

$$(-3) \times q + 1 \times (1-q)) = 1 - 4q$$

The optimal strategy occurs when these are equal:

$$8q - 5 = 1 - 4q \implies q = \frac{1}{2}$$

B's best strategy is to play B_1 and B_2 half of the time each, at random.

The value of the game to B is $8 \times \dfrac{1}{2} - 5 = -1$

> Remember the table shows A's payoffs, and B's payoffs are the negative of these.

> This is $-v$, as you would expect.

Example 1

a Show that this game does not have a stable solution.

		B	
		B_1	B_2
A	A_1	2	−1
	A_2	−2	3

b Find the optimum mixed strategy for player A.

c Find the value of the game.

d Find the optimum mixed strategy for player B.

a

		B		Row minima	Max of row minima
		B_1	B_2		
A	A_1	2	−1	−1	−1
	A_2	−2	3	−2	
Column maxima		2	3		
Min of column maxima		2			

The max of the row minima ≠ min of column maxima, so there is no stable solution.

b Let the value of the game be v and let A play A_1 and A_2 with probabilities p and $(1-p)$

(*Continued on the next page*)

If B plays B_1, then A's expected payoff is $2p + -2(1 - p) = 4p - 2$, so $v \leq 4p - 2$

If B plays B_2, then A's expected payoff is $-p + 3(1 - p) = 3 - 4p$, so $v \leq 3 - 4p$

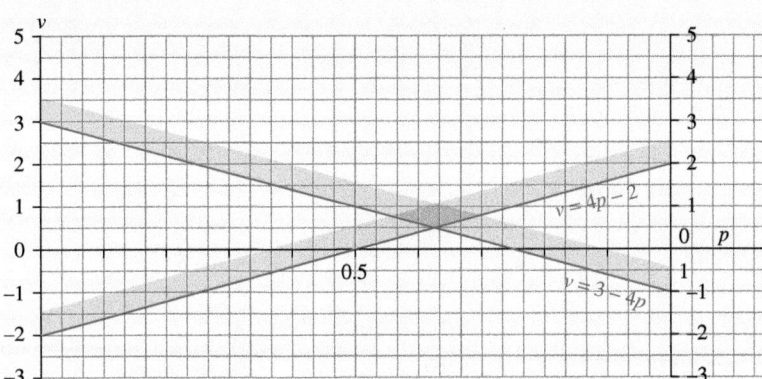

Maximum v occurs when $4p - 2 = 3 - 4p$, so $p = \dfrac{5}{8}$ ●————— Equate to find p

A plays A_1 and A_2 with probabilities $\dfrac{5}{8}$ and $\dfrac{3}{8}$

c The value of the game is $v = 4 \times \dfrac{5}{8} - 2 = \dfrac{1}{2}$

d Let B play B_1 and B_2 with probabilities q and $(1 - q)$

If A plays A_1, B's expected payoff is $-2q + (1 - q) = 1 - 3q$

If A plays A_2, B's expected payoff is $2q - 3(1 - q) = 5q - 3$

The optimal strategy occurs when these are equal:

$1 - 3q = 5q - 3$, so $q = \dfrac{1}{2}$ ●————— Equate to find q

So B plays B_1 and B_2 with equal probability.

As a check, the value to B is $1 - 3 \times \dfrac{1}{2} = -\dfrac{1}{2} = -v$, as expected.

Example 2

The game shown does not have a stable solution.

		B	
		B_1	B_2
A	A_1	−1	5
	A_2	4	−3

The value of the game is v. Analyse the game from the column player's point of view to find their optimal mixed strategy and the value of v

The value of the game to the column player, B, is $(-v)$

Let B play B_1 with probability q

If A plays A_1, then B's expected payoff is $q - 5(1-q) = 6q - 5$
So $(-v) \le 6q - 5$ and so $v \ge 5 - 6q$

If A plays A_2, then B's expected payoff is $-4q + 3(1-q) = 3 - 7q$
So $(-v) \le 3 - 7q$ and so $v \ge 7q - 3$

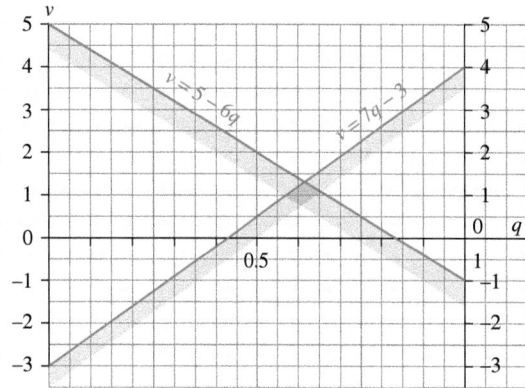

Max $(-v)$ corresponds to minimum v on the graph, that is

where $7q - 3 = 5 - 6q$, so $q = \dfrac{8}{13}$

Hence B plays B_1, B_2 with probabilities $\dfrac{8}{13}$, $\dfrac{5}{13}$ and $v = 1\dfrac{4}{13}$

Alternatively, you could have used value $= -v$ to find q, by putting

$1 - 3q = -\dfrac{1}{2}$ or $5q - 3 = -\dfrac{1}{2}$

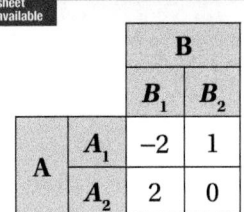

1 For the following games, decide if a mixed strategy is needed. Find the optimal strategy (pure or mixed) for both players and the value of the game.

a

		B	
		B_1	B_2
A	A_1	−2	4
	A_2	3	2

b

		B	
		B_1	B_2
A	A_1	4	2
	A_2	−1	3

c

		B	
		B_1	B_2
A	A_1	5	−3
	A_2	3	2

d

		B	
		B_1	B_2
A	A_1	1	4
	A_2	4	3

2 For each of these games, show that a mixed strategy is necessary. Find the optimal mixed strategy for each player and the value of the game.

a

		B	
		B_1	B_2
A	A_1	2	−2
	A_2	−2	3

b

		B	
		B_1	B_2
A	A_1	7	6
	A_2	5	8

c

		B	
		B_1	B_2
A	A_1	5	2
	A_2	−1	3

d

		B	
		B_1	B_2
A	A_1	−2	1
	A_2	2	0

3 Analyse each of these games from the column player's point of view to find their optimal mixed strategy and the value, v, of the game.

a

		B	
		B_1	B_2
A	A_1	−6	2
	A_2	1	−2

b

		B	
		B_1	B_2
A	A_1	9	7
	A_2	3	10

c

		B	
		B_1	B_2
A	A_1	−5	2
	A_2	1	−4

4 The table shows a game in which player A has three strategies A_1, A_2 and A_3, while player B has four strategies B_1, B_2, B_3 and B_4

		B			
		B_1	B_2	B_3	B_4
A	A_1	−1	3	7	−1
	A_2	1	5	4	2
	A_3	3	2	3	3

a Show that this game has no stable solution.

b Use dominance arguments to reduce the matrix as far as possible.

c Find the optimal mixed strategy for player A and the value of the game.

d Find the optimal mixed strategy for player B.

DISCRETE

Strategy

To solve problems involving two-player zero-sum games

1. If necessary, construct a payoff matrix.

2. Use dominance to reduce the matrix as much as possible.

3. Find the play-safe strategies and decide whether the game has a stable solution.

4. For mixed-strategy games, find the constraints on v, the value of the game, in terms of the probability or probabilities of playing the various strategies.

5. Solve the resulting linear programming problem to find the optimal strategies and the value of the game.

6. Answer the question in context.

You can find the optimum mixed strategy for larger payoff matrices if one of the players has just two strategies, because only two of the other player's options will appear in their best strategy.

Example 3

Find the optimal strategies for the players in this 2×3 game, and find the value of the game.

		B		
		B_1	B_2	B_3
A	A_1	4	2	−1
	A_2	−3	−2	5

The game has no stable solution, so A plays A_1 and A_2 with probabilities p and $(1 − p)$

> **3** Decide whether the game has a stable solution.

If B plays B_1, A's expectation $= 4p − 3(1 − p) = 7p − 3$

If B plays B_2, A's expectation $= 2p − 2(1 − p) = 4p − 2$

If B plays B_3, A's expectation $= −p + 5(1 − p) = 5 − 6p$

> **4** Find the constraints on v

If the value of the game is v, you need to maximise v subject to

$v \leq 7p − 3,\ v \leq 4p − 2,\ v \leq 5 − 6p$

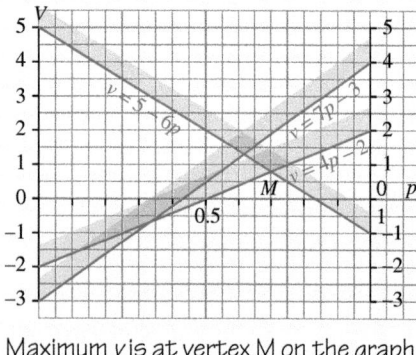

Maximum v is at vertex M on the graph.

(Continued on the next page)

This is the intersection of $v = 4p - 2$ and $v = 5 - 6p$

$4p - 2 = 5 - 6p \implies p = 0.7$

So A plays A_1 and A_2 randomly with probabilities 0.7 and 0.3

The value of the game is $v = 4 \times 0.7 - 2 = 0.8$

The line $v = 7p - 3$ passes above M, so B should not play B_1 because A's payoff would be greater than 0.8

B plays B_2 and B_3 with probabilities q and $(1 - q)$

If A plays A_1, B's expectation is $-2q + (1 - q) = 1 - 3q$

If A plays A_2, B's expectation is $2q - 5(1 - q) = 7q - 5$

The optimal strategy occurs where these lines intersect.

$1 - 3q = 7q - 5 \implies q = 0.6$

So B plays B_2 and B_3 with probabilities 0.6 and 0.4

The value of the game to B is $1 - 3 \times 0.6 = -0.8$, as expected.

In general, in a $2 \times n$ game the column player will only use two of the n available strategies, as the others will give the row player a greater payoff.

In an $n \times 2$ game, you analyse the situation from the point of view of the column player.

DISCRETE

Example 4

Find the optimal strategies for the players in this 3×2 game, and find the value of the game.

		B	
		B_1	B_2
A	A_1	-4	1
	A_2	0	-2
	A_3	1	-4

The game has no stable solution, so B plays B_1 and B_2 with probabilities q and $(1 - q)$

If A plays A_1, B's expectation $= 4q - (1 - q) = 5q - 1$

If A plays A_2, B's expectation $= 0 \times q + 2(1 - q) = 2 - 2q$

If A plays A_3, B's expectation $= -q + 3(1 - q) = 4 - 5q$

If the value of the game (to A) is v, then the value of the game to B is $V = -v$. You need to maximise V subject to $V \leq 5q - 1$, $V \leq 2 - 2q$, $V \leq 4 - 5q$

(*Continued on the next page*)

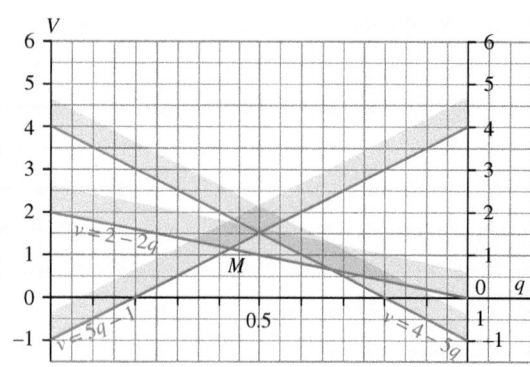

Maximum V is at vertex M on the graph.

This is the intersection of $V = 2 - 2q$ and $V = 5q - 1$

$$2 - 2q = 5q - 1 \quad \Rightarrow \quad q = \frac{3}{7}$$

So B plays B_1 and B_2 with probabilities $\frac{3}{7}$ and $\frac{4}{7}$

The value of the game to B is $V = 2 - 2 \times \frac{3}{7} = 1\frac{1}{7}$

The value of the game is $v = -V = -1\frac{1}{7}$

Line $V = 4 - 5q$ passes above M, so A should not play A_3 because B's payoff would be greater than $1\frac{1}{7}$ ●────────────

A plays A_1 and A_2 with probabilities p and $(1 - p)$

If B plays B_1, A's expectation is $-4p + 0 \times (1 - p) = -4p$

If B plays B_2, A's expectation is $p - 2(1 - p) = 3p - 2$

The optimal strategy occurs where these lines intersect.

$$-4p = 3p - 2 \quad \Rightarrow \quad p = \frac{2}{7}$$

So A plays A_1 $\frac{2}{7}$ and A_2 $\frac{5}{7}$ of the time.

> In general, in an $n \times 2$ game the row player will only use two of the n available strategies, as the others will give the column player a greater payoff.

Exercise 14.3B Reasoning and problem-solving

Answer sheet available

1   Analyse the games shown in these tables to find the optimal strategy for each player and the value of the game.

a

		B		
		B_1	B_2	B_3
A	A_1	−8	−2	1
	A_2	−3	3	4
	A_3	0	−4	−2

b

		B		
		B_1	B_2	B_3
A	A_1	2	1	5
	A_2	1	−4	4
	A_3	5	2	1

c

	B_1	B_2	B_3
A_1	5	2	3
A_2	1	4	2

d

		B		
		B_1	B_2	B_3
A	A_1	3	−1	1
	A_2	−2	4	2

2 The table shows a game between players X and Y, each with four possible strategies.

		Y			
		I	II	III	IV
X	I	1	4	3	2
	II	−1	3	0	3
	III	4	2	5	1
	IV	−2	6	−3	1

a Show that there is no stable solution.

b Find the optimal mixed strategy for each player and the value of the game.

3 The table shows a game with value V.

		B		
		B_1	B_2	B_3
A	A_1	−1	0	−2
	A_2	1	−2	−1
	A_3	0	−1	1

Player A has an optimal mixed strategy in which she plays A_1, A_2 with probabilities p_1 and p_2. Find and simplify three inequalities connecting p_1, p_2 and V. (You do not need to solve the problem.)

4 The table shows a game between players A and B. A has two strategies, A_1 and A_2. B has three strategies, B_1, B_2 and B_3

		B		
		B_1	B_2	B_3
A	A_1	2	5	6
	A_2	5	2	1

a Analyse the game graphically and show that player A's optimal strategy is to play her two strategies at random with equal probability.

b Find the value of the game.

c Explain, with reference to your graph, why, in this game, player B could reasonably make use of all three available strategies.

Let B play his strategies with probabilities q_1, q_2 and q_3, respectively.

d Making use of the known value of the game, write down three equations connecting q_1, q_2 and q_3

e Putting $q_1 = q$, show that $q_2 = 2\frac{1}{2} - 4q$ and find q_3 in terms of q

f Show that $\frac{1}{2} \le q \le \frac{5}{8}$ and find the range of possible values of q_2 and q_3

Chapter summary

- In a linear programming (LP) formulation:
  - The quantities you can vary are the decision variables (control variables).
  - The limitations on the values of the decision variables are the constraints.
  - The quantity to be optimised is the objective function.
- On the graph of the constraints, the feasible region is the set of points satisfying all the constraints.
- The objective line joins all points for which the objective function takes a specified value.
- The optimal solution corresponds to a vertex of the feasible region.
- In a zero-sum game between two players, A and B, the sum of their gains and losses on each play is zero. The gains for the row player, A, are recorded in a pay-off matrix. B's gains are the negative of these entries.
- A play-safe strategy gives the best guaranteed outcome regardless of what the other player does. For A, you find the minimum value in each row. The play-safe strategy corresponds to the maximum of these row minima. For B, you find the maximum value in each column. The play-safe strategy corresponds to the minimum of these column maxima. In a pure-strategy game, both players always play safe.
- The value of a game is the payoff to the row player (A) if both players use their best strategy.
- The game has a stable solution (saddle point) if neither player gains by changing from their play-safe strategy. In this case, max of row minima = min of column maxima.
- In a payoff matrix:

      row i dominates row j if, for every column, value in row $i \geq$ value in row j

      column i dominates column j if, for every row, value in column $i \leq$ value in column j
- A row or column which is dominated can be deleted.
- If a matrix can be reduced by domination to one containing a single value, the game has a stable solution. For an unstable situation, each player should play a mixed-strategy game, that is, play each strategy some of the time.
- In a mixed-strategy game, you aim to maximise the expected payoff, where for payoffs $x_1, x_2, ..., x_n$ that occur with probabilities $p_1, p_2, ..., p_n$, the expected (mean) payoff $E(x)$ is given by

$$E(x) = x_1 p_1 + x_2 p_2 + ... + x_n p_n = \sum_{i=1}^{n} x_i p_i$$

- For a 2×2 game, you assume that the row player plays their first strategy with probability p You write an inequality connecting v, their expected value, with p for each of the column player's strategies. You can solve the resulting linear programming problem graphically.
- For a $2 \times n$ or an $n \times 2$ game, you derive the linear programming formulation working from the point of view of the player with two options. From the resulting graph, you can reject all but two of the lines.

Check and review

You should now be able to...	Try Questions
✔ Use graphical methods to solve a two-variable linear programming problem.	1, 2
✔ Express a worded problem as a linear programming formulation.	2

1 A linear programming problem is:

Minimise $\quad C = 2x + 5y$

Subject to $\quad x + y \leq 20$,

$\qquad\qquad y \geq 7$

$\qquad\qquad y \leq 2x$

$\qquad\qquad y + 3x \geq 25$

a Draw a graph of these constraints. Label the feasible region.

b Draw an objective line.

c Find the values of x and y which give the optimal solution.

2 A manufacturer makes two brands of blackcurrant jam – 'Value' and 'Luxury'. It takes 0.4 kg of blackcurrants and 0.8 kg of sugar to make 1 kg of Value jam, while 1 kg of Luxury jam requires 0.6 kg of blackcurrants and 0.6 kg of sugar. The profit is 10p per kg on Value jam and 12p per kg on Luxury jam. There are 480 kg of blackcurrants and 810 kg of sugar available. The aim is to maximise the profit.

a Express this situation as a linear programming formulation.

b By drawing a suitable graph, find the maximum profit and the corresponding quantities of the two brands of jam.

3 The table shows the payoff matrix for player A in a two-person zero-sum game.

		B	
		B_1	B_2
A	A_1	−1	3
	A_2	4	2

a Find the play-safe strategy for A and the value of the game to her.

b Find the play-safe strategy for B.

c Write down the payoff matrix for B.

4 A two-person zero-sum game is represented by this payoff matrix for player A.

		B		
		I	II	III
A	I	1	0	−2
	II	−1	4	0
	III	2	3	5

a By finding the play-safe strategy for each player, show that the game has a stable solution. Explain the implications of this.

b State the value of the game.

5 The table shows the payoff matrix for player A in a two-person zero-sum game.

		B		
		I	II	III
A	I	5	3	9
	II	4	7	6
	III	2	4	5

a Use dominance arguments to reduce the matrix as far as possible.

b State, with reasons, whether the game has a stable solution.

6 The table shows the payoffs in a game between A and B.

		B		
		B_1	B_2	B_3
A	A_1	3	−1	1
	A_2	−2	4	4

a Show that the game does not have a stable solution.

b Show that the table can be reduced to a 2×2 game.

c A plays strategy A_1 with probability p. Find two inequalities connecting v and p, where v is the value of the game to player A.

d Use a graph to find the optimal value v and the corresponding value of p

e Find B's optimal mixed strategy.

History

The American mathematical scientist, George Dantzig, is credited for developing the simplex algorithm during World War II. It was kept a secret until 1947. When it published, many other industries began to use it. Dantzig is also known for solving two other statistical problems that were unsolved at the time. He had mistaken them for homework after arriving late to a lecture!

ICT

Excel has an add-in called the Solver which can be used to solve linear programming problems. (You may need to select it from the Add-Ins.)

First formulate the following as a linear programming problem.

> A company plans on building a maximum of 12 new shops in a large city. They will build these shops in one of three sizes for each location – a convenience store, a standard store, or a mega store. The convenience store requires £2.5 million to build and 20 employees to operate. The standard store requires £7.125 million to build and 16 employees to operate. The mega store requires £8.275 million to build and 40 employees to operate. The corporation can dedicate £80 million in construction capital, and 280 employees to staff the shops. On average, the convenience store nets £1.6 million annually, the standard store nets £2 million annually, and the mega store nets £2.6 million annually. How many of each shop should they build to maximise revenue?

Once formulated, enter the problem in to Excel. A suggested format is shown on the left. You enter the objective function and constraints in the white cells. (The formula for the objective function in cell D9 will give you a hint of what to do.)

Click on the Solver button (on the Data tab) and complete the pop up box as shown.

Note that the solving method selected must be 'Simplex LP'.

Click on the Solve button and remember to interpret your answers in cells D2, D3 and D4 appropriately.

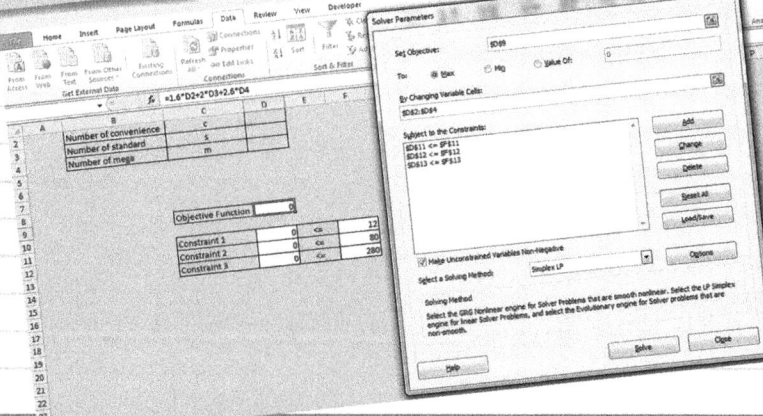

14 Assessment

1 This payoff matrix represents a zero-sum game between two players, P and Q.

		Q		
		Q_1	Q_2	Q_3
P	P_1	6	4	9
	P_2	5	8	7
	P_3	3	5	6

   a Find the play-safe strategy for each player. [4 marks]

   b Determine whether the game has a stable solution [1]

   c State the value of the game. [1]

2 This payoff matrix represents a zero-sum game between two players, M and N.

		N		
		N_1	N_2	N_3
M	M_1	6	6	8
	M_2	4	1	6
	M_3	4	2	−1

   a Use dominance to simplify the problem as far as possible. [4]

   b State what you can deduce from your answer to part **a** and give the value of the game. [3]

3 A linear programming problem is represented by the following constraints:

   Minimise $C = 2x + 3y$
   Subject to $x + y \leq 30$ $\quad y \leq 2x$ $\quad 2y \geq 9$ $\quad y + 2x \geq 28$

   a Draw a graph of these constraints and label the feasible region. [7]

   b Draw an example of the objective function on your graph. [2]

4 This payoff matrix represents a zero-sum game between two players A and B.

		B			
		B_1	B_2	B_3	B_4
A	A_1	14	2	2	7
	A_2	6	2	2	9
	A_3	15	−1	1	0

   a Decide whether this game has a stable solution. [6]

   b Find the value of the game. [1]

5 Connor and Isla are playing a zero-sum game. They are both trying to gain the maximum number of points possible. The table below shows the number of points that Connor scores for each possible strategy that each player can use.

	I_1	I_2	I_3
C_1	2	−5	3
C_2	−1	−3	4
C_3	3	−5	2
C_4	3	−2	−1

   a How many points does Connor score if he uses strategy C_3 and Isla uses strategy I_2? [1]

   b Explain why Isla would never use strategy I_3 [2]

   c Find the play-safe strategy for each player and explain whether the game is stable or not. You must explain your answer. [4]

6 A linear programming problem is represented by the following constraints.

Minimise $P = 3x + 6y + 4z$

Subject to $4x + 2y + 5z \leq 75$

$x + y + z = 24$

$x \geq 8$

$y \geq 9$

$z \geq 2$

a Rewrite these constraints, eliminating the variable z **[6]**

b Represent these constraints graphically, clearly identifying the feasible region.   **[5]**

7 Sanjay and Danny are playing a card game. They each have three coloured cards and choose a card to show simultaneously so that they do not know what the other player will choose. They each start with 20 points and they gain or lose points according to the table.

		Danny		
		Red	Blue	Purple
Sanjay	Yellow	2	−4	−1
	Orange	−4	1	3
	Green	1	−1	1

a Find the play-safe strategy for each player. You must show your working.   **[6]**

b Sanjay thinks that Danny will play safe. Which colour should he choose?   **[1]**

8 A farmer has 20 hectares of land in which he grows cauliflowers and sprouts. The cost per hectare for cauliflowers is £300, whereas the cost for sprouts is £200. The farmer allocates £4800 for these products. The care of the crops requires £10 per hectare for cauliflower and £20 per hectare for sprouts. The farmer has £360 available for crop care. The profit on cauliflowers is £100 per hectare and the profit on sprouts is £125 per hectare.

a Formulate this as a linear programming problem.   **[5]**

b By plotting a suitable graph, calculate the number of hectares of each crop the farmer should plant in order to maximise his profits. State the profit.   **[7]**

9 A company produces two different types of vitamin tablet. The contents of the tablets are shown below.

	Vitamin C	Niacin	Vitamin E
Supervit	200 mg	30 mg	20 mg
Extravit	135 mg	20 mg	40 mg

Supervit tablets cost 15p each, whereas Extravit tablets cost 20p each. In a 30 day month, Jenny wants to spend no more than £6 on tablets. She requires no more than 5400 mg of Vitamin C, at least 700 mg of Niacin and at least 800 mg of Vitamin E.

a Formulate this as a linear programming problem.   **[5]**

b By drawing a suitable graph, find the combination of tablets that Jenny should take to ensure that she has the required amount of vitamins at the minimum cost.   **[9]**

15 Abstract algebra

What might, at first, seem quite abstract algebraic methods that have been developed for their own sake by mathematicians can often come to have important applications. For example, modular arithmetic is used extensively around the world wherever barcodes or universal products codes are used. The bar codes you see being scanned when you are shopping involve 12-digit numbers that are used by the computer system to allocate the current price to charge the customer at the checkout. The computer will also inform other parts of the system about stock levels and automatically re-order stock if necessary to ensure the retailer is always able to cope with customers' needs.

These checking methods that draw on modular arithmetic are used by other everyday systems. For example, they provide one part of online security when credit cards are used, as well as providing a unique code (ISBN) for all books published throughout the world.

Orientation

What you need to know	What you will learn	What this leads to
KS 4 • Sets.	• To use binary operations. • To use Cayley tables. • To use modular arithmetic.	**Careers** • Computer programming.

15.1 Binary operations

Fluency and skills

A fundamental concept in mathematics is the **set**.

A set is a collection of individual objects called **elements**. A set can be finite, for example the set $\{1, 2, 3, 4, 5\}$ has five elements, or infinite, for example the set of natural numbers, $\mathbb{N}$, is the infinite set $\{1, 2, 3, 4, ...\}$

A **binary operation** is a function with two inputs. For example, $+$ on the set $\mathbb{N}$ is a binary operation. The function could be more complicated; for example, the binary function $*$ on the set $\mathbb{N}$ could be defined as $a*b = a^b + 1$

In both these examples, the binary function always produces a member of the set $\mathbb{N}$

> **Key point**
>
> A function is a binary operation if it can be applied to any two elements of a set so that the result is also a member of the set.

$\mathbb{N}$ represents the set of natural numbers; these are the 'counting numbers': $\{1, 2, 3, 4, ...\}$

$\mathbb{Z}$ represents the set of integers (positive or negative) including 0: $\{0, \pm1, \pm2, \pm3, ...\}$

$\mathbb{Z}^+$ is the set of positive integers: $\{1, 2, 3, 4 ...\}$

$\mathbb{Z}^-$ is the set of negative integers: $\{-1, -2, -3, -4, ...\}$

Example 1

Which of these functions are binary operations on the set $\mathbb{Z}$?

a $a*b = a - b$ **b** $a \blacksquare b = \dfrac{a}{b}$

a $a - b$ will always produce a member of $\mathbb{Z}$ so $*$ is a binary operation on $\mathbb{Z}$

> Any two integers can be subtracted to give another integer.

b $\dfrac{a}{b}$ is not, in general, a member of $\mathbb{Z}$ therefore $\blacksquare$ is not a binary operation on $\mathbb{Z}$

For example, if $a = 2$ and $b = 3$ then $a \blacksquare b = \dfrac{2}{3}$ which is not an integer.

> Also, $\dfrac{a}{b}$ is not defined when $b = 0$

> **Key point**
>
> A binary operation, $*$, is **commutative** if $a*b = b*a$ for all a and b

---

Example 2

Which of these binary operations on the set $\mathbb{N}$ are commutative?

a $a*b = a^b$ **b** $a \circ b = a + b$

a In general, $a^b \neq b^a$

For example, $2*5 = 2^5 = 32$ but $5*2 = 5^2 = 25$

> Give a counter-example to prove $*$ is not commutative.

Therefore $*$ is not commutative.

b $a \circ b = b \circ a$ since addition of natural numbers is commutative.

Therefore $a \circ b$ is commutative.

A binary operation, $*$, is **associative** if $(a*b)*c = a*(b*c)$

Example 3

Which of these binary operations on $\mathbb{R}$ is associative?

> $\mathbb{R}$ is the set of real numbers.

a $a \triangleright b = 2a + b$ **b** $a*b = a \times b$

a $(a \triangleright b) \triangleright c = (2a + b) \triangleright c$

> Apply the function to a and b first.

$\qquad = 2(2a + b) + c$

$\qquad = 4a + 2b + c$

$a \triangleright (b \triangleright c) = a \triangleright (2b + c)$

> Now apply to b and c first.

$\qquad = 2a + (2b + c)$

$\qquad = 2a + 2b + c$

> This is not the same as the value of $(a \triangleright b) \triangleright c$

So, in general, $(a \triangleright b) \triangleright c \neq a \triangleright (b \triangleright c)$, therefore $\triangleright$ is not associative.

b $(a*b)*c = (a \times b) \times c$

$\qquad = abc$

$a*(b*c) = a \times (b \times c)$

> Since multiplication of real numbers is associative.

$\qquad = abc$

Therefore $*$ is associative.

Sets sometimes have an **identity** element under a particular binary operation. For example, the identity element of $\mathbb{R}$ under multiplication is 1 since any member of $\mathbb{R}$ multiplied by 1 will be unchanged.

The identity element, e, of a set under an operation $*$ is such that $a*e = e*a = a$ for all values of a in the set.

The identity element for a binary operation is always unique. To prove this, assume that the binary operation $*$ has two identity elements, e and f, in the set S

Then, since f is an identity element, $e*f = e$

Also, since e is an identity element, $e*f = f$

Which implies that $e = f$

Therefore the identity element is unique.

Elements in a set sometimes have **inverses** in the set. For example in the set $\mathbb{R}$ under the operation of multiplication, the inverse of 4 is $\dfrac{1}{4}$ since $4 \times \dfrac{1}{4} = 1$

The inverse of an element combines with it to give the identity. In this case every element of the set except 0 has an inverse.

The inverse, a^{-1}, of an element a under an operation $*$ is such that $a*a^{-1} = a^{-1}*a = e$

An element, a, is a **self-inverse** if $a^{-1} = a$

469

Example 4

The binary operation $*$ is defined by $a*b = a+b$

$\mathbb{Q}$ is the set of rational numbers: those that can be written as $\dfrac{p}{q}$ for some integers p and q

Irrational numbers cannot be written as $\dfrac{p}{q}$

e.g. $\sqrt{2}$

a Find the identity element of the set $\mathbb{Q}$ under $*$

b Which element of $\mathbb{Q}$ is self-inverse under $*$?

a The identity element is $e = 0$ since $a + 0 = 0 + a = a$

Need to show that
$a*e = e*a = a$

b Need to find a such that $a*a = 0$

Since the identity element is
$e = 0$

$a + a = 0 \Rightarrow a = 0$ so 0 is the only self-inverse element of $\mathbb{Q}$ under $*$

Exercise 15.1A Fluency and skills

1 Which of these are binary operations on the set $\mathbb{N}$? Explain your answers.

   a $a \circ b = a - b$　　**b** $a*b = a \times b$

2 Which of these are binary operations on the set $\mathbb{Z}$? Explain your answers.

   a $a*b = a - b$　　**b** $a \diamondsuit b = \dfrac{a}{b^2 + 1}$

3 Which of these are binary operations on the set $\mathbb{Z}^+$? Explain your answers.

   a $a \diamondsuit b = \max(a,b)$

      [where $\max(a, b)$ means the maximum of a and b]

   b $a*b = 2a + b$

4 Which of these are binary operations on the set $\mathbb{Q}$? Explain your answers.

   a $a \circ b = a^b$　　**b** $a \star b = \dfrac{a}{b}$

5 Which of these are binary operations on the set of even integers?

   a $a \star b = \dfrac{a}{2} + b$　　**b** $a*b = 3a - 7b$

6 Explain whether or not each of these binary operations are associative and commutative.

   a $a*b = a - b$ on the set $\mathbb{Q}$

   b $a*b = a + b$ on the set $\mathbb{R}$

   c $a*b = \max(a, b)$ on the set $\mathbb{N}$

   d $a*b = a^2 + b$ on the set $\mathbb{Z}$

   e $a*b = a \times b$ on the set $\mathbb{Z}^+$

   f $a*b = \min(a,b)$ on the set $\mathbb{Z}^-$

   g $a*b = a \times b$ on the set $\mathbb{R}$

7 For each of the sets and binary operations in question **6**

   i Identify the identity element (if it exists),

   ii If the identity exists, explain which of the elements have inverses.

8 For the set of 2×2 matrices with real coefficients

   a Prove that the binary operation defined as matrix multiplication is not commutative.

   b Prove that the binary operation defined as matrix addition is associative and commutative.

   c For each of the binary operations in parts **a** and **b**

     i State the identity element.

     ii Write down the general form of the inverse element.

9 For each of these sets and binary operations, identify the identity element (if it exists) and explain which of the elements have inverses.

   a $a*b = a \times b$ on the set $\{1, 0, -1\}$

   b $a*b = \max(a,b)$ on the set $\{2, 3, 4, 5\}$

   c $a*b = a + b$ on the set of all even numbers

   d $a*b = a \times b$ on the set of all odd numbers.

For a finite set, a table can be drawn showing the result of certain types of binary operation on all possible pairs of elements. This is called a **Cayley table**.

Cayley tables are named after the nineteenth-century mathematician Arthur Cayley.

For example, this is the Cayley table for the binary operation $*$ on the set $\{a, b, c\}$. Notice how the convention is to take the first element from those on the left of the table and the second from the elements on the top of the table.

$*$	a	b	c
a	$a*a$	$a*b$	$a*c$
b	$b*a$	$b*b$	$b*c$
c	$c*a$	$c*b$	$c*c$

Here is the Cayley table for the set $\{1, -1\}$ under multiplication.

$\times$	1	-1
1	1	-1
-1	-1	1

In this Cayley table, you can see that the identity element, e, is 1 This is because all the elements in the column and in the row corresponding to the element 1, are unchanged.

$\times$	1	-1
1	1	-1
-1	-1	1

You can also see that both 1 and -1 are self-inverses since both 1×1 and -1×-1 give the identity element, 1

$\times$	1	-1
1	1	-1
-1	-1	1

Strategy

To find the identity and inverse elements from a Cayley table

1. Look along the columns to find one the same as the initial column of elements; the element at the top of the column is the identity element, e. Also check along the relevant row to make sure it equals the initial row of elements.

2. Find all the instances of e in the table.

3. Work down the rows. For each element, a, at the start of a row the element at the top of the column containing e is the inverse of a

DISCRETE

Example 5

For the Cayley table given,

a Find the identity element,

b Work out which of the elements are self-inverses.

> Notice how the table is symmetrical along its leading diagonal. This indicates that the operation is commutative.

$*$	a	b	c	d
a	b	c	d	a
b	c	d	a	b
c	d	a	b	c
d	a	b	c	d

a $a*d=a, b*d=b, c*d=d$ and $d*d=d$
so d is the identity element.

> The fourth column is a, b, c, d so all of the elements are unchanged, you can also see this in the fourth row of the table. ①

b

$*$	a	b	c	d
a	b	c	d	a
b	c	d	a	b
c	d	a	b	c
d	a	b	c	d

> Find all the times the identity element, d, appears. ②

b and d are self-inverses since $b*b=d$ and $d*d=d$

> You can also see that a is the inverse of c and vice versa. ③

Exercise 15.1B Reasoning and problem-solving

Answer sheet available

1 For each of these Cayley tables
   a Work out the identity element,
   b Find the inverse of each of the elements.

i

	a	b	c
a	b	c	a
b	c	a	b
c	a	b	c

ii

	a	b	c	d
a	a	b	c	d
b	b	c	d	a
c	c	d	a	b
d	d	a	b	c

iii

	A	B	C	D
A	B	A	D	C
B	A	B	C	D
C	D	C	B	A
D	C	D	A	B

iv

	A	B	C	D	E
A	A	B	C	D	E
B	B	C	E	A	D
C	C	E	D	B	A
D	D	A	B	E	C
E	E	D	A	C	B

2 For parts **i** and **ii** of question **1**, verify that the operation is associative.

3 **a** Copy and complete the Cayley table for the binary operation $a \times b$ on the set $\{1, -1, i, -i\}$, where $i = \sqrt{-1}$

$\times$	1	-1	i	$-i$
1				
-1				
i				
$-i$				

b What is the identity element?

c Find the inverse of each element.

d Is the operation associative? Explain your answer.

4 The set S is the set of matrices of the form

$$\begin{pmatrix} \cos\left(\dfrac{n\pi}{2}\right) & -\sin\left(\dfrac{n\pi}{2}\right) \\ \sin\left(\dfrac{n\pi}{2}\right) & \cos\left(\dfrac{n\pi}{2}\right) \end{pmatrix} \text{ where } n = 0, 1, 2, 3$$

a Write down the members of S.

b Explain whether matrix addition is a binary operation on S.

c Copy and complete the Cayley table for matrix multiplication where the elements represent the matrices generated by those values of n

	0	1	2	3
0				
1		2		
2			0	
3	3			

d Write down

   i The identity element,

   ii The inverse of each of the elements.

e Explain the geometrical significance of this set of matrices.

5 The set S is the set of matrices of the form

$$\begin{pmatrix} \cos\left(\dfrac{2n\pi}{5}\right) & \sin\left(\dfrac{2n\pi}{5}\right) \\ -\sin\left(\dfrac{2n\pi}{5}\right) & \cos\left(\dfrac{2n\pi}{5}\right) \end{pmatrix}$$

where $n = 1, 2, \ldots k$.

a What is the smallest value of k for which matrix multiplication is a binary operation on S?

b Construct a Cayley table for matrix multiplication on S using this value of k

c Write down the inverse of each of the matrices.

6 The set S consists of the following transformations in 2D:

   A = reflection in x-axis

   B = reflection in y-axis

   C = reflection in line $y = x$

   D = reflection in line $y = -x$

   E = rotation 90° anticlockwise about origin

   F = rotation 180° about origin

   G = rotation 90° clockwise about origin

   H = no transformation

a Write down the inverse of each element.

b Write down the 2×2 matrix that represents each transformation.

c Construct a Cayley table for the binary operation A ∗ B = transformation A followed by transformation B

d Prove by counter-example that ∗ is not commutative.

7 The set S is

$$\left\{ \begin{pmatrix} -1 & 0 & 0 \\ 0 & 1 & 0 \\ 0 & 0 & 1 \end{pmatrix}, \begin{pmatrix} 1 & 0 & 0 \\ 0 & -1 & 0 \\ 0 & 0 & 1 \end{pmatrix}, \begin{pmatrix} 1 & 0 & 0 \\ 0 & 1 & 0 \\ 0 & 0 & -1 \end{pmatrix}, \right.$$

$$\begin{pmatrix} -1 & 0 & 0 \\ 0 & -1 & 0 \\ 0 & 0 & 1 \end{pmatrix}, \begin{pmatrix} -1 & 0 & 0 \\ 0 & 1 & 0 \\ 0 & 0 & -1 \end{pmatrix}, \begin{pmatrix} 1 & 0 & 0 \\ 0 & -1 & 0 \\ 0 & 0 & -1 \end{pmatrix},$$

$$\left. \begin{pmatrix} -1 & 0 & 0 \\ 0 & -1 & 0 \\ 0 & 0 & -1 \end{pmatrix}, \begin{pmatrix} 1 & 0 & 0 \\ 0 & 1 & 0 \\ 0 & 0 & 1 \end{pmatrix} \right\}$$

a Explain how you know that matrix multiplication is a binary operation on S.

b State the identity element.

c Explain which of the elements of S are self-inverse.

d Show that matrix multiplication is commutative on S.

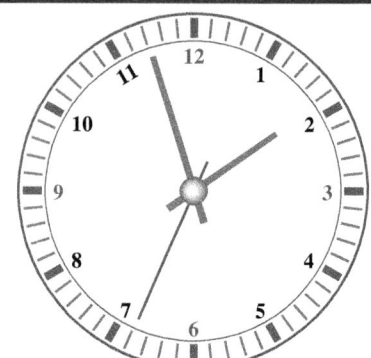

Fluency and skills

When you are using the 12-hour clock, five hours after 11:00 is 4:00 and three hours before 1:00 is 10:00. This is an example of **modular arithmetic**. The modulo in this case is 12

So you could write $11 + 5 = 4 \ (\text{mod } 12)$ and $1 - 3 = 10 \ (\text{mod } 12)$

Every time you reach 12 you start again from 0; 12 is said to be **congruent** to 0 in mod 12

Any modulo can be used for modular arithmetic. For example, using modulo 5, whenever you reach a multiple of 5 you start again from 0

Therefore, 5 is congruent to 0, 6 is congruent to 1 and so on.

0	1	2	3	4
5	6	7	8	9
10	11	12	13	14
15	16	17	18	19

So, for example, $3 + 4 \equiv 2 \ (\text{mod } 5)$ and $4 \times 4 \equiv 1 \ (\text{mod } 5)$

Another way of thinking about this is that $a \ (\text{mod } n)$ is the remainder when a is divided by n

You can use rules to add, subtract, multiply and divide in modular arithmetic.

If $a \equiv b \ (\text{mod } n)$, then a can be written $a = b + kn$ where k is an integer. Similarly, if $c \equiv d \ (\text{mod } n)$ then $c = d + mn$ where m is an integer.

So $a + c = (b + kn) + (d + mn)$

$\qquad = (b + d) + (k + m)n$

Therefore the solution can be written as $b + d \ (\text{mod } n)$ since $k + m$ is an integer.

> You have previously used $\equiv$ to mean 'is identical to' but it can also be used to mean 'is congruent to', as in this example.

Key point

If $a \equiv b \ (\text{mod } n)$ and $c \equiv d \ (\text{mod } n)$ then:

- $a + c = b + d \ (\text{mod } n)$
- $a - c = b - d \ (\text{mod } n)$
- $ac = bd \ (\text{mod } n)$ as long as a, b, c, d, n are integers.

These rules can be used to simplify calculations.

Example 1

Use the rules of modular arithmetic to calculate

a $53+14 \pmod{10}$　　**b** $53 \times 14 \pmod{10}$　　**c** $152^6 \pmod{10}$

> Since $53 \div 10$ has remainder 3 and $14 \div 10$ has remainder 4

a $53 \equiv 3 \pmod{10}$ and $14 \equiv 4 \pmod{10}$

So $53+14 \equiv 3+4 \pmod{10}$

$\equiv 7 \pmod{10}$

> 53, 14, 3, 4, 10 are all integers so can use the rule.

b $53 \times 14 \equiv 3 \times 4 \pmod{10}$

$\equiv 12 \pmod{10}$

$\equiv 2 \pmod{10}$

> Write the answer in its simplest form.

c $152 \equiv 2 \pmod{10}$ therefore $152^6 \equiv 2^6 \pmod{10}$

Therefore $152^6 \equiv 4 \pmod{10}$

> Since $2^6 = 64$ and $64 \equiv 4 \pmod{10}$

> Since $a \equiv b \pmod{n} \Rightarrow$ $a^k \equiv b^k \pmod{n}$ for $a, b,$ $n \in \mathbb{Z}$ and $k \in \mathbb{Z}^+$

Exercise 15.2A Fluency and skills

1 Write each number in the form $a \pmod{n}$ where $0 \le a < n$, using the modulo given.

a $7 \pmod 3$　　　**b** $12 \pmod 7$

c $19 \pmod 4$　　**d** $27 \pmod 9$

e $25 \pmod 6$　　**f** $-5 \pmod 8$

g $-55 \pmod{11}$　**h** $48 \pmod{13}$

2 Use modular arithmetic to work out these calculations. Show your method clearly. Write each answer in the form $a \pmod{n}$ where $0 \le a < n$, using the modulo given.

a $21 + 19 \pmod 3$　　**b** $4 + 25 \pmod 7$

c $68 - 34 \pmod 4$　　**d** $79 - 94 \pmod 9$

e $5 \times 14 \pmod 6$　　**f** $25 \times 37 \pmod 8$

g $82 \times -16 \pmod{11}$　**h** $29 \times 75 \pmod{13}$

3 Use the exponential rule of modular arithmetic to simplify these powers. Give your answers in the form $a \pmod{n}$ where $0 \le a < n$, using the modulo given.

a $37951^{37} \pmod{10}$

b $1736^{26} \pmod 5$

c $982^9 \pmod 5$

d $3219^{48} \pmod 3$

e $3268^{84} \pmod 3$

f $12653^{11} \pmod 6$

Reasoning and problem-solving

You can define binary operations involving modular arithmetic and draw Cayley tables of finite sets under binary operations mod n

To construct a Cayley table for the binary operation * on the set of integers modulo n

（1） Draw the table for the set $\{0, 1, 2, \ldots, (n-1)\}$

（2） Calculate $a * b$ for each pair of integers and write the answers (mod n) in the tables.

（3） Use the fact that $a * e = e * a = a$ for the identity element e

The notation $+_n$ represents addition modulo n, where **Key point**
n is a positive integer

The notation $\times_n$ represents multiplication modulo n, where n is a positive integer

a Draw the Cayley table of the binary operation $+_6$ on the set of integers modulo 6

b State the identity element.

c Work out the inverse of each element.

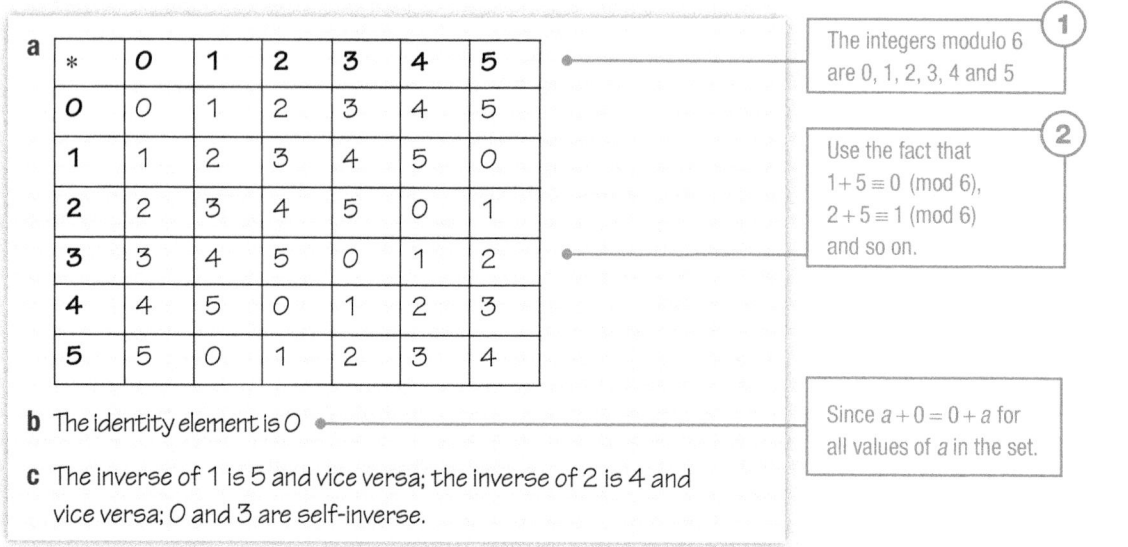

a

*	0	1	2	3	4	5
0	0	1	2	3	4	5
1	1	2	3	4	5	0
2	2	3	4	5	0	1
3	3	4	5	0	1	2
4	4	5	0	1	2	3
5	5	0	1	2	3	4

(1) The integers modulo 6 are 0, 1, 2, 3, 4 and 5

(2) Use the fact that
$1 + 5 \equiv 0 \pmod 6$,
$2 + 5 \equiv 1 \pmod 6$
and so on.

b The identity element is 0

Since $a + 0 = 0 + a$ for all values of a in the set.

c The inverse of 1 is 5 and vice versa; the inverse of 2 is 4 and vice versa; 0 and 3 are self-inverse.

Exercise 15.2B Reasoning and problem-solving

1 Draw the Cayley table for the set of integers modulo 2 under the binary operation $*$ where

$a * b = a - b \pmod 2$

2 Draw the Cayley table for the binary operation

a $+_3$ on the set $\{0, 1, 2\}$

b $+_4$ on the set $\{0, 1, 2, 3\}$

c $\times_5$ on the set $\{1, 2, 3, 4\}$

d $\times_7$ on the set $\{1, 2, 3, 4, 5, 6\}$

3 For each part of questions **1** and **2**

a State the identity element,

b Explain which elements have an inverse.

4 The operation $*$ is defined as $a * b = a + b + a \times b \pmod 5$ on the set $\{0, 1, 2, 3\}$

a Draw the Cayley table for $*$

b **i** State the identity element.

ii Write down the inverse of each of the elements.

5 The operation $\blacklozenge$ is defined as $a \blacklozenge b = a + b + 4 \pmod 6$ on the set $\{0, 1, 2, 3, 4, 5\}$

a Draw the Cayley table for $\blacklozenge$

b **i** State the identity element.

ii Write down the inverse of each of the elements.

Chapter summary

- A binary operation is a function that can be applied to any two elements of a set so that the result is also a member of the set.
- A binary operation, $*$, is commutative if $a*b = b*a$
- A binary operation, $*$, is associative if $(a*b)*c = a*(b*c)$
- The identity element, e, of a set under an operation $*$ is such that $a*e = e*a = a$ for all values of a in the set.
- The inverse, a^{-1}, of an element a under an operation $*$ is such that $a*a^{-1} = a^{-1}*a = e$
- An element, a, is self-inverse if $a^{-1} = a$
- A Cayley table shows the results of applying a binary operation to all possible pairs of elements from a finite set.
- The notation $+_n$ represents addition modulo n, where n is a positive integer
- The notation $\times_n$ represents multiplication modulo n, where n is a positive integer

Check and review

You should now be able to...	Review Questions
✔ Decide if a function on a set is a binary operation.	1
✔ Work out if a binary operation is associative.	2
✔ Work out if a binary operation is commutative.	3, 5
✔ Work out the identity element of a binary operation.	4, 5
✔ Work out the inverses of the elements of a set under a binary operation.	4, 5
✔ Add, subtract and multiply numbers using modular arithmetic.	6
✔ Draw Cayley tables for sets under binary operations.	7, 8

1 Decide whether or not each of these functions is a binary operation. Explain your answers.

a $a+3b$ on the set $\mathbb{R}$

b $(a+1)(b-1)$ on the set $\mathbb{Z}^+$

c $\dfrac{b}{a^2+1}$ on the set $\mathbb{R}$

d $\sqrt{ab}$ on the set $\mathbb{R}$

2 Explain whether or not each of these binary operations is associative.

a $\dfrac{a}{b}$ on the set $\mathbb{R}$

b a^2+b^2 on the set $\mathbb{Z}$

c $\dfrac{ab}{a+b}$ on the set $\mathbb{R}^+$

3 For each of the binary operations in question **2**, explain whether or not it is commutative.

DISCRETE

4 For each of the binary operations in question **2**

    a Find the identity element (if it exists),

    b Write down the inverse of each element where possible.

5 The Cayley table shows the binary operation $*$ over the set $\{A, B, C, D, E\}$

$*$	**A**	**B**	**C**	**D**	**E**
A	B	C	D	E	A
B	C	A	E	A	B
C	D	E	A	B	C
D	E	A	B	C	D
E	A	B	C	D	E

    a State the identity element.

    b Give the inverse of each element.

    c How can you tell from the table that $*$ is commutative?

6 Use modular arithmetic to work out

    a $379 + 612 \pmod 5$

    b $1079 - 351 \pmod 3$

    c $326 \times 249 \pmod 8$

    d $-532 \times 249 \pmod{10}$

7 **a** Copy and complete the Cayley table for the set $\{0, 2, 4, 6, 8\}$ under $+_{10}$

$*$	**0**	**2**	**4**	**6**	**8**
0					
2			8		
4		8			
6					
8					

    b What is the identity element?

    c State the inverse of each of the elements.

8 The set S consists of the matrices

$$A = \begin{pmatrix} -\dfrac{1}{2} & \dfrac{\sqrt{3}}{2} \\ -\dfrac{\sqrt{3}}{2} & -\dfrac{1}{2} \end{pmatrix}$$

$$B = \begin{pmatrix} -\dfrac{1}{2} & -\dfrac{\sqrt{3}}{2} \\ \dfrac{\sqrt{3}}{2} & -\dfrac{1}{2} \end{pmatrix}$$

$$C = \begin{pmatrix} 1 & 0 \\ 0 & 1 \end{pmatrix}$$

    a Copy and complete the Cayley table for matrix multiplication on the set S.

	A	**B**	**C**
A			
B			
C			

    b Write down

      i The identity element,

      ii The inverse of each of the elements.

    c Explain the geometrical significance of this set of matrices.

History

Examples of modular arithmetic have been around for a long time. For example, in the third century Chinese book, Master Sun's Mathematical Manual, the following problem appears:

> We have a number of things, but we do not know exactly how many. If we count them by threes we have two left over. If we count by fives we have three left over. If we count by sevens there are two left over. How many things are there?

Find out how this problem relates to the Chinese Remainder Theorem.

Did you know?

The check digit in an ISBN number (found on all books) is calculated using modulo 10 division.

Research

The Cayley table for the Klein four-group gives the group of permutations in twelve-note music composition. This is a method of composing music devised by Arnold Schoenberg and ensures that all twelve notes in a chromatic scale are given equal importance and hence played the same number of times.

Many pieces of music have been composed using this technique as a basis. Find out more by searching for Schoenberg Opus 23 movement 5.

The initial prime (P) ordering of the twelve notes of a chromatic scale can be subjected to interval-preserving transformations of retrograde (R), inversion (I) and retrograde-inversion(RI) form.

The Cayley table representing these transformations would be:

	P	R	I	RI
P	P	R	I	RI
R	R	P	RI	I
I	I	RI	P	R
RI	RI	I	R	P

1 The binary operation ■ is defined by $a ■ b = a + b^2$ on the set of natural numbers, $\mathbb{N}$

   a Calculate $3 ■ 7$ **[2 marks]**

   b Prove by counter-example that ■ is neither associative nor commutative.   **[4]**

2 Use modular arithmetic to write each of these in the form $a \pmod{10}$ where $0 \leq a < n$

   a $3 + 7$ b $125 - 621$

   c 358×6914 d 2563^4 **[6]**

3 The operation $*$ is defined as $a * b = a^b$

   a Explain whether or not $*$ is a binary operation on each of these sets.

      i $\mathbb{Q}$ ii $\mathbb{R}$ iii $\mathbb{N}$ **[6]**

   b Explain why the operation $*$ on the set $\mathbb{N}$ does not have an identity element. **[2]**

   c Is $*$ associative on $\mathbb{N}$? Explain your answer.   **[2]**

4 The binary operation $*$ is defined as $a * b = \dfrac{ab}{2}$ on the set $\{-2, 2\}$

   a Copy and complete the Cayley table for $*$ **[2]**

$*$	-2	2
-2		
2		

   b Write down the identity element for $*$ **[1]**

   c What is the inverse of each element?   **[1]**

5 The table defines the binary operation $\circ$ on the set $\{S, T, U, V, W, X\}$

$\circ$	S	T	U	V	W	X
S	T	U	S	X	V	W
T	U	S	T	W	X	V
U	S	T	U	V	W	X
V	X	W	V	T	S	U
W	V	X	W	S	U	T
X	W	V	X	U	T	S

   a Use the table to find the identity element.   **[1]**

   b Write down the inverse of V.   **[1]**

   c Which of the elements are self-inverses?   **[1]**

6　**a**　The binary operation $*$ is defined on the set $\{0, 1, 2, 3, 4\}$ as $a * b = a +_5 b$

　　i　Copy and complete the Cayley table for $*$　　　　　　　　　　**[2]**

$*$	0	1	2	3	4
0					
1					
2					
3					
4					

　　ii　What is the identity element?　　　　　　　　　　　　　　**[1]**

　　iii　Write down the inverse of each of the elements.　　　　　　**[2]**

　b　Use modular arithmetic to calculate the remainder when 18^{12} is divided by 5　　**[4]**

7　The binary operation $\blacklozenge$ is defined on the set $\{A, B, C, D, E\}$ by the table.

$\blacklozenge$	A	B	C	D	E
A	D	B	E	A	C
B	E	D	C	B	A
C	B	A	D	C	E
D	C	E	A	D	B
E	A	C	B	E	D

　a　What is A $\blacklozenge$ E?　　　　　　　　　　　　　　　　　　**[1]**

　b　Explain why there is no identity element.　　　　　　　　　　**[3]**

　c　Prove that $\blacklozenge$ is not associative.　　　　　　　　　　　**[3]**

8　The group S consists of the matrices

$$A = \begin{pmatrix} 0 & -1 \\ 1 & 0 \end{pmatrix}, B = \begin{pmatrix} -1 & 0 \\ 0 & -1 \end{pmatrix}, C = \begin{pmatrix} 0 & 1 \\ -1 & 0 \end{pmatrix}, D = \begin{pmatrix} 1 & 0 \\ 0 & 1 \end{pmatrix}$$

　a　Construct a Cayley table for S under matrix multiplication.　　**[3]**

　b　State the inverse of each of the elements.　　　　　　　　　　**[2]**

9　The operation $*$ is defined by $a * b = 2a - b$

　Prove that $*$ is a binary operation on the set of odd integers $\{\pm 1, \pm 3, \pm 5, \pm 7, \pm 9, \ldots\}$　　**[4]**

10 The set S consists of matrices of the form $m_n = \begin{pmatrix} 1 & 0 & 0 \\ 0 & \cos\left(\dfrac{n\pi}{3}\right) & -\sin\left(\dfrac{n\pi}{3}\right) \\ 0 & \sin\left(\dfrac{n\pi}{3}\right) & \cos\left(\dfrac{n\pi}{3}\right) \end{pmatrix}$ for $n = 1, 2, 3, 4, 5, 6$

a Copy and complete the Cayley table for the binary operation of matrix multiplication on S. **[6]**

$\times$	M_1	M_2	M_3	M_4	M_5	M_6
M_1						
M_2						
M_3						
M_4						
M_5						
M_6						

b State the identity element. **[1]**

c What is the inverse of M_2? **[1]**

d Which element is self-inverse? **[1]**

16 Complex numbers 2

Liquids such as water, and gases such as air, are known as fluids. In many ways, the flow of water can be treated the same as the flow of air. The study of such flow is known as fluid dynamics, where complex functions are used to model flow. Aerodynamics is the application of fluid dynamics to the flow of air. Hydrodynamics is the application of fluid dynamics to the flow of liquids. The scientific principles and the underpinning mathematics of fluid dynamics are important in many areas. For example, in the design of vehicles that move in gases and liquids. However, understanding gas and liquid flow is also important when planning how water and gas will reach homes and businesses.

An aircraft that is full of passengers, their luggage, and other cargo can lift off from a runway primarily as a result of the motion of the wing through the air. The mathematics of fluid dynamics allows this motion of the aircraft relative to the air, a fluid, to be analysed in detail. The mathematics relies on complex numbers to provide insight into the flow of the air. It is used by aeronautical engineers when they are designing the wings of an aircraft.

Orientation

What you need to know	What you will learn	What this leads to
Ch1 Complex numbers 1	• How to use exponential form. • How to use de Moivre's theorem. • How to use roots of unity.	**Careers** • Electrical engineering. • Aeronautical engineering. • Mechanical engineering.

Fluency and skills

A complex number $z = a + bi$ can be expressed in modulus–argument form as $z = r(\cos\theta + i\sin\theta)$ where $r = |z|$ and $\theta = \arg z$

See Ch2.5

For a reminder of Maclaurin expansions.

The first few terms of the series expansions of $\cos\theta$ and of $\sin\theta$ are

$$\cos\theta = 1 - \frac{\theta^2}{2!} + \frac{\theta^4}{4!} - \frac{\theta^6}{6!} + \cdots$$

$$\sin\theta = \theta - \frac{\theta^3}{3!} + \frac{\theta^5}{5!} - \frac{\theta^7}{7!} + \cdots$$

Therefore $z = r\left[\left(1 - \frac{\theta^2}{2!} + \frac{\theta^4}{4!} - \frac{\theta^6}{6!} + \cdots\right) + i\left(\theta - \frac{\theta^3}{3!} + \frac{\theta^5}{5!} - \frac{\theta^7}{7!} + \cdots\right)\right]$

$$= r\left(1 + i\theta - \frac{\theta^2}{2!} - \frac{\theta^3 i}{3!} + \frac{\theta^4}{4!} + \frac{\theta^5 i}{5!} - \frac{\theta^6}{6!} - \frac{\theta^7 i}{7!} + \cdots\right)$$

$$= r\left(1 + i\theta + \frac{(i\theta)^2}{2!} + \frac{(i\theta)^3}{3!} + \frac{(i\theta)^4}{4!} + \frac{(i\theta)^5}{5!} + \frac{(i\theta)^6}{6!} + \frac{(i\theta)^7}{7!} + \cdots\right)$$

since $(i\theta)^2 = -\theta$, $(i\theta)^3 = -i\theta$, $(i\theta)^4 = \theta^4$ and so on.

This is the expansion of $e^{i\theta}$

Key point

The formula $re^{i\theta} = r(\cos\theta + i\sin\theta)$ is known as **Euler's formula**.

Key point

So, using Euler's formula, you can write the complex number z in **exponential form** as $z = re^{i\theta}$ where $r = |z|$ and $\theta = \arg z$, $-\pi < \theta \le \pi$

Example 1

Write $2e^{\frac{3\pi i}{4}}$ in the form $a + bi$

To get from exponential form to $a + bi$ form, you need to first convert to modulus–argument form.

$$2e^{\frac{3\pi i}{4}} = 2\left(\cos\left(\frac{3\pi}{4}\right) + i\sin\left(\frac{3\pi}{4}\right)\right) \qquad \text{Use Euler's formula.}$$

$$= 2\left(-\frac{\sqrt{2}}{2} + \frac{\sqrt{2}}{2}i\right)$$

$$= -\sqrt{2} + \sqrt{2}i$$

Example 2

Write $z = 3 - i$ in exponential form.

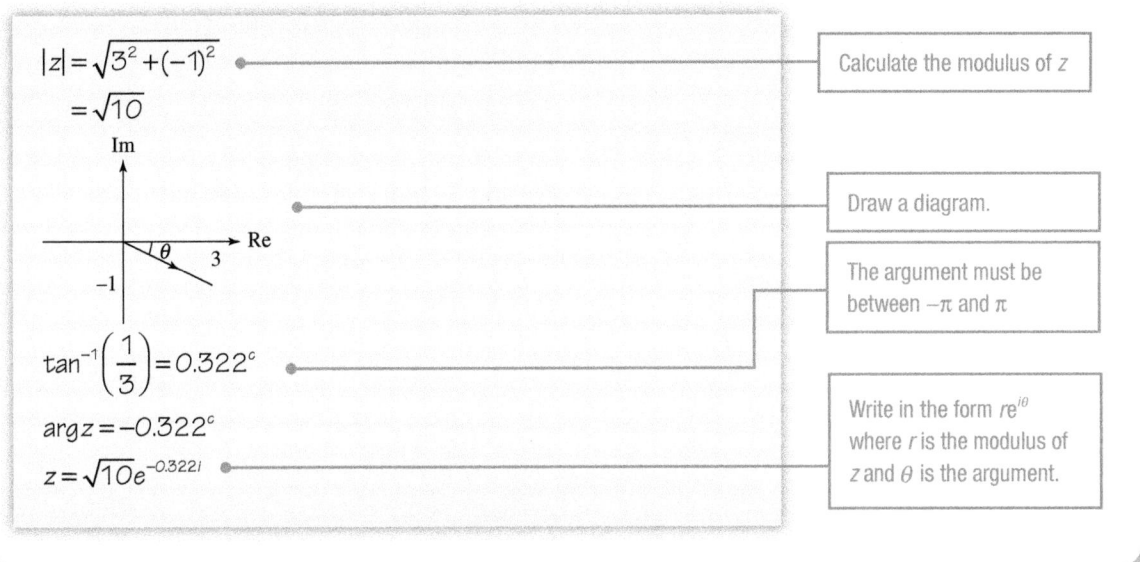

$|z| = \sqrt{3^2 + (-1)^2}$ — Calculate the modulus of z

$= \sqrt{10}$

— Draw a diagram.

— The argument must be between $-\pi$ and π

$\tan^{-1}\left(\dfrac{1}{3}\right) = 0.322^c$

$\arg z = -0.322^c$

$z = \sqrt{10}e^{-0.322i}$ — Write in the form $re^{i\theta}$ where r is the modulus of z and θ is the argument.

Calculator

Try it on your calculator

Calculators can be used to convert to and from modulus–argument form.

Find out how to convert $\sqrt{2}e^{-\frac{\pi}{4}i}$ to Cartesian form on your calculator.

$\sqrt{2}\angle-\dfrac{\pi}{4} \blacktriangleright a + bi$

$1 - i$

Exercise 16.1A Fluency and skills

1 Write each of these numbers in exponential form.

   a $3 + 4i$ b $2 - i$

   c 10 d -5

   e $2i$ f $-6i$

   g $-5 + 12i$ h $-4 - 8i$

   i $\sqrt{3} + i$ j $5 - 5i$

2 Write each of these complex numbers in exponential form.

   a $2\left(\cos\left(\dfrac{\pi}{12}\right) + i\sin\left(\dfrac{\pi}{12}\right)\right)$

   b $4\left(\cos\left(-\dfrac{2\pi}{3}\right) + i\sin\left(-\dfrac{2\pi}{3}\right)\right)$

   c $3\left(\cos\left(\dfrac{5\pi}{6}\right) - i\sin\left(\dfrac{5\pi}{6}\right)\right)$

   d $6\left(\cos\left(\dfrac{\pi}{7}\right) - i\sin\left(\dfrac{\pi}{7}\right)\right)$

   e $\cos\left(\dfrac{7\pi}{5}\right) + i\sin\left(\dfrac{7\pi}{5}\right)$

   f $\sqrt{2}\left(\cos\left(-\dfrac{15\pi}{8}\right) + i\sin\left(-\dfrac{15\pi}{8}\right)\right)$

   g $\sqrt{3}\left(\cos\left(-\dfrac{5\pi}{6}\right) - i\sin\left(-\dfrac{5\pi}{6}\right)\right)$

   h $8\left(\cos\left(-\dfrac{17\pi}{12}\right) - i\sin\left(-\dfrac{17\pi}{12}\right)\right)$

3 Write each of these complex numbers in the form $a+bi$

a $2e^{\frac{\pi}{2}i}$

b $7e^{-\frac{\pi}{3}i}$

c $\sqrt{2}e^{\frac{\pi}{4}i}$

d $e^{-\frac{\pi}{6}i}$

e $\sqrt{8}e^{-\pi i}$

f $\sqrt{3}e^{\frac{5\pi}{6}i}$

4 Given that $z=2e^{\frac{\pi}{3}i}$ and $w=3e^{-\frac{\pi}{3}i}$, calculate the value of

a $|zw|$

b $\left|\dfrac{z}{w}\right|$

c $\arg(zw)$

d $\arg\left(\dfrac{z}{w}\right)$

5 Given that $z=5e^{\frac{2\pi}{7}i}$ and $w=\dfrac{1}{5}e^{-\frac{\pi}{7}i}$, calculate the value of

a $|zw|$

b $\left|\dfrac{z}{w}\right|$

c $\arg(zw)$

d $\arg\left(\dfrac{z}{w}\right)$

6 Given that $z_1=\sqrt{6}e^{-\frac{\pi}{4}i}$ and $z_2=\sqrt{3}e^{-\frac{5\pi}{6}i}$, calculate the value of

a $|z_1 z_2|$

b $\left|\dfrac{z_1}{z_2}\right|$

c $\arg(z_1 z_2)$

d $\arg\left(\dfrac{z_1}{z_2}\right)$

Reasoning and problem-solving

Using the expansions at the beginning of this section, you can define trigonometric functions in terms of sums of exponentials, in particular:

Key point

$$\cos\theta = \frac{e^{i\theta}+e^{-i\theta}}{2} \text{ and } \sin\theta = \frac{e^{i\theta}-e^{-i\theta}}{2i}$$

Strategy

These results can then be used to prove trigonometric identities

(1) Use Euler's formula: $e^{i\theta} \equiv \cos\theta + i\sin\theta$

(2) Use the facts that $\cos(-\theta) \equiv \cos\theta$ and $\sin(-\theta) \equiv -\sin\theta$

(3) Use $\cos\theta \equiv \dfrac{e^{i\theta}+e^{-i\theta}}{2}$ or $\sin\theta \equiv \dfrac{e^{i\theta}-e^{-i\theta}}{2i}$

(4) Use index laws.

Example 3

Prove that $\cos\theta = \dfrac{e^{i\theta}+e^{-i\theta}}{2}$

$e^{i\theta} = \cos\theta + i\sin\theta$ ① Start with Euler's formula.

$e^{-i\theta} = \cos(-\theta) + i\sin(-\theta)$

$\quad\quad = \cos\theta - i\sin\theta$ ② Since $\cos(-\theta)=\cos(\theta)$ and $\sin(-\theta)=-\sin(\theta)$

$e^{i\theta} + e^{-i\theta} = \cos\theta + i\sin\theta + \cos\theta - i\sin\theta$

$\quad\quad\quad = 2\cos\theta$

Therefore $\cos\theta = \dfrac{e^{i\theta}+e^{-i\theta}}{2}$, as required.

Example 4

Prove that $\sin 2\theta \equiv 2\sin\theta\cos\theta$

$$2\sin\theta\cos\theta \equiv 2\left(\frac{e^{i\theta}-e^{-i\theta}}{2i}\right)\left(\frac{e^{i\theta}+e^{-i\theta}}{2}\right)$$

Write $\sin(\theta)$ and $\cos(\theta)$ in terms of exponentials. **3**

$$\equiv \frac{(e^{i\theta}-e^{-i\theta})(e^{i\theta}+e^{-i\theta})}{2i}$$

$$\equiv \frac{e^{2i\theta}+1-1-e^{-2i\theta}}{2i}$$

Expand brackets. **4**

$$\equiv \frac{e^{2i\theta}-e^{-2i\theta}}{2i}$$

$$\equiv \sin 2\theta, \text{ as required}$$

Since $\sin 2\theta = \dfrac{e^{i(2\theta)}-e^{-i(2\theta)}}{2i}$ **3**

Exercise 16.1B Reasoning and problem-solving

1 Use Euler's formula to show that $\sin\theta = \dfrac{e^{i\theta}-e^{-i\theta}}{2i}$

2 A complex number z has modulus 1 and argument θ

   a Show that $z^n + \dfrac{1}{z^n} = 2\cos(n\theta)$ **b** Show that $z^n - \dfrac{1}{z^n} = 2i\sin(n\theta)$

3 Given that $z_1 = r_1 e^{\theta_1 i}$ and $z_2 = r_2 e^{\theta_2 i}$, show that

   a $|z_1 z_2| = |z_1||z_2|$ and $\arg(z_1 z_2) = \arg z_1 + \arg z_2$ **b** $\left|\dfrac{z_1}{z_2}\right| = \dfrac{|z_1|}{|z_2|}$ and $\arg\left(\dfrac{z_1}{z_2}\right) = \arg z_1 - \arg z_2$

4 Use $\cos\theta = \dfrac{e^{i\theta}+e^{-i\theta}}{2}$ and $\sin\theta = \dfrac{e^{i\theta}-e^{-i\theta}}{2i}$ to show that

   a $\sin(A+B) \equiv \sin A\cos B + \sin B\cos A$ **b** $\cos(A+B) \equiv \cos A\cos B - \sin A\sin B$

5 Use exponentials to show that

   a $\cos 2x \equiv \cos^2 x - \sin^2 x$ **b** $\cos^2 x + \sin^2 x \equiv 1$

6 Use exponentials to show that

   a $(\cos\theta + i\sin\theta)^2 \equiv \cos 2\theta + i\sin 2\theta$ **b** $(\cos\theta + i\sin\theta)^n \equiv \cos(n\theta) + i\sin(n\theta)$

7 Given that $z = 4\left(\cos\left(\dfrac{\pi}{9}\right)+i\sin\left(\dfrac{\pi}{9}\right)\right)$ and $w = 3\left(\cos\left(\dfrac{2\pi}{9}\right)+i\sin\left(\dfrac{2\pi}{9}\right)\right)$, show that $zw = 6+6\sqrt{3}i$

8 Given that $z = 8\left(\cos\left(\dfrac{5\pi}{12}\right)+i\sin\left(\dfrac{5\pi}{12}\right)\right)$, show that $z^2 = -32\sqrt{3}+32i$

9 The complex number z is such that $|z| = k$ and $\arg(z) = \theta$ for $k > 0$ and $-\pi < \theta \le \pi$

Another complex number is defined as $w = 1-i$

Find expressions in terms of k and θ for the modulus and the argument of

   a zw **b** $\dfrac{z}{w}$

Fluency and skills

If you write a complex number in the form $z = r(\cos\theta + i\sin\theta)$, where r is a rational number, then you can see that $z^n = [r(\cos\theta + i\sin\theta)]^n$. You can write this as $r^n(\cos\theta + i\sin\theta)^n$

Therefore $z^n = r^n(e^{i\theta})^n$ since $e^{i\theta} = \cos\theta + i\sin\theta$ (using Euler's formula).

You can then use index laws to write $r^n(e^{i\theta})^n = r^n e^{in\theta}$

Using Euler's formula again, this becomes $r^n(\cos(n\theta) + i\sin(n\theta))$

Putting these two results together gives

> **Key point**
>
> $[r(\cos\theta + i\sin\theta)]^n = r^n(\cos(n\theta) + i\sin(n\theta))$, for all integers n, which is known as **de Moivre's theorem**.

> You can prove this result using proof by induction.

De Moivre's theorem can be used to simplify powers of complex numbers.

For example, $(\cos\theta + i\sin\theta)^3 = \cos 3\theta + i\sin 3\theta$

$$\frac{1}{\cos\theta + i\sin\theta} = (\cos\theta + i\sin\theta)^{-1} = \cos(-\theta) + i\sin(-\theta)$$

Example 1

Write each of these numbers in the form $a + bi$

a $\left(\cos\dfrac{\pi}{3} + i\sin\dfrac{\pi}{3}\right)^4$

b $\left(\cos\dfrac{\pi}{4} - i\sin\dfrac{\pi}{4}\right)^6$

a $\left(\cos\dfrac{\pi}{3} + i\sin\dfrac{\pi}{3}\right)^4 = \cos\dfrac{4\pi}{3} + i\sin\dfrac{4\pi}{3}$ ——— Use de Moivre's theorem.

$\qquad = -\dfrac{1}{2} - \dfrac{\sqrt{3}}{2}i$ ——— In the form $a + bi$

b $\left(\cos\dfrac{\pi}{4} - i\sin\dfrac{\pi}{4}\right)^6 = \left(\cos\left(-\dfrac{\pi}{4}\right) + i\sin\left(-\dfrac{\pi}{4}\right)\right)^6$ ——— Using $\cos(-\theta) = \cos\theta$ and $\sin(-\theta) = -\sin\theta$

$\qquad = \left(\cos\left(-\dfrac{3\pi}{2}\right) + i\sin\left(-\dfrac{3\pi}{2}\right)\right)^6$ ——— Needs to be in modulus–argument form and then you can apply de Moivre's theorem.

$\qquad = i$

Example 2

Given the complex number $z = -\sqrt{3} + i$, use de Moivre's theorem to find z^{-2} in the form $a + bi$

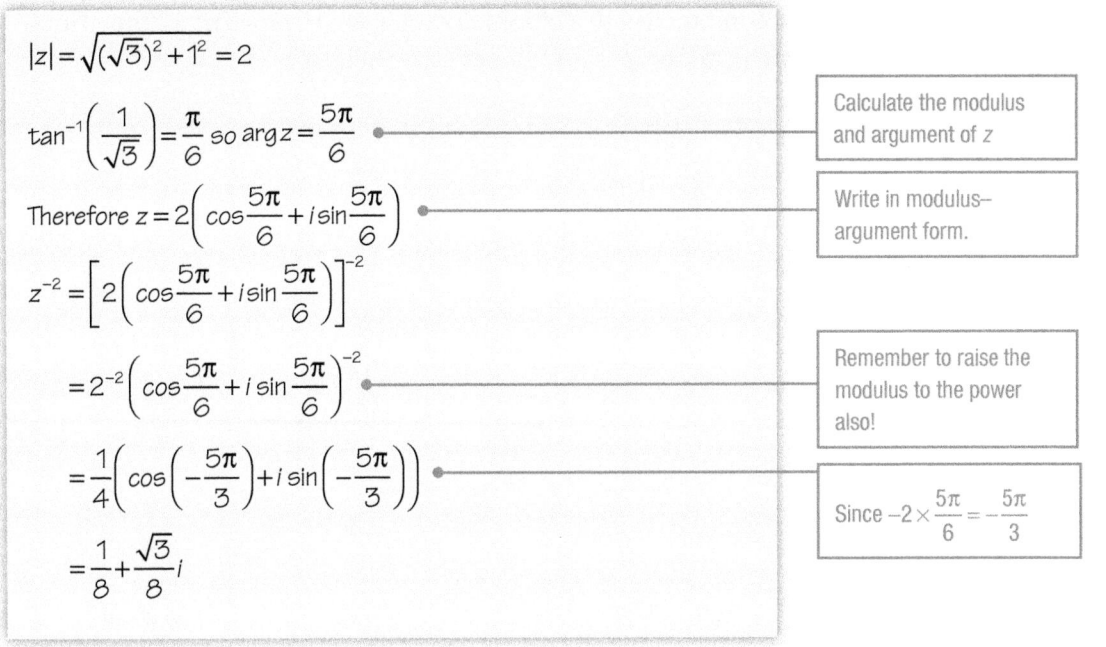

$|z| = \sqrt{(\sqrt{3})^2 + 1^2} = 2$

$\tan^{-1}\left(\dfrac{1}{\sqrt{3}}\right) = \dfrac{\pi}{6}$ so $\arg z = \dfrac{5\pi}{6}$ — Calculate the modulus and argument of z

Therefore $z = 2\left(\cos\dfrac{5\pi}{6} + i\sin\dfrac{5\pi}{6}\right)$ — Write in modulus–argument form.

$z^{-2} = \left[2\left(\cos\dfrac{5\pi}{6} + i\sin\dfrac{5\pi}{6}\right)\right]^{-2}$

$= 2^{-2}\left(\cos\dfrac{5\pi}{6} + i\sin\dfrac{5\pi}{6}\right)^{-2}$ — Remember to raise the modulus to the power also!

$= \dfrac{1}{4}\left(\cos\left(-\dfrac{5\pi}{3}\right) + i\sin\left(-\dfrac{5\pi}{3}\right)\right)$ — Since $-2 \times \dfrac{5\pi}{6} = -\dfrac{5\pi}{3}$

$= \dfrac{1}{8} + \dfrac{\sqrt{3}}{8}i$

Exercise 16.2A Fluency and skills

1 Express each of these numbers in the form $a + bi$

   a $\left(\cos\dfrac{\pi}{3} + i\sin\dfrac{\pi}{3}\right)^6$

   b $\left(\cos\dfrac{\pi}{6} + i\sin\dfrac{\pi}{6}\right)^5$

   c $\left(\cos\left(-\dfrac{\pi}{12}\right) + i\sin\left(-\dfrac{\pi}{12}\right)\right)^4$

   d $\left(\cos\dfrac{\pi}{14} - i\sin\dfrac{\pi}{14}\right)^7$

   e $\left(\cos\dfrac{\pi}{4} + i\sin\dfrac{\pi}{4}\right)^{-3}$

2 Given that $z = 3\left(\cos\left(\dfrac{\pi}{24}\right) + i\sin\left(\dfrac{\pi}{24}\right)\right)$, express in exact Cartesian form

   a z^4 **b** z^{-6}

3 Given that $z = 2\left(\cos\left(\dfrac{\pi}{12}\right) + i\sin\left(\dfrac{\pi}{12}\right)\right)$, express in exact Cartesian form

   a z^3 **b** z^{-2}

   c z^6 **d** z^{-4}

4 Given that $z = 4\left(\cos\left(\dfrac{3\pi}{2}\right) + i\sin\left(\dfrac{3\pi}{2}\right)\right)$, express in exact Cartesian form

   a z^2 **b** z^3

   c $\dfrac{1}{z}$ **d** $16z^{-4}$

5 Given that $z = 1 - i$, use de Moivre's theorem to write the following powers of z in the form $a + bi$

   a z^3 **b** z^7

   c z^{-5} **d** z^{-6}

6 Given that $z = 3i$, use de Moivre's theorem to write the following powers of z in Cartesian form.

   a z^2 **b** z^{-1}

   c z^{-3} **d** $\dfrac{3}{z^3}$

7 Given that $z = -\sqrt{3} + i$, use de Moivre's theorem to write the following in Cartesian form.

   a z^4 **b** z^{-3}

   c z^{-2} **d** $\dfrac{8}{z^6}$

Reasoning and problem-solving

You can use de Moivre's theorem combined with a binomial expansion to prove trigonometric identities involving powers of $\sin\theta$ or $\cos\theta$

To write a power of $\cos\theta$ or $\sin\theta$ as a series involving $\cos(n\theta)$ or $\sin(n\theta)$

(1) Use $\cos\theta \equiv \dfrac{e^{i\theta}+e^{-i\theta}}{2}$ or $\sin\theta \equiv \dfrac{e^{i\theta}-e^{-i\theta}}{2i}$

(2) Write out the binomial expansion.

(3) Use rules of indices to simplify.

(4) Group terms together to write the expression in terms of trigonometric functions.

Prove that $8\cos^4\theta \equiv \cos4\theta + 4\cos2\theta + 3$

$$16\cos^4\theta \equiv (2\cos\theta)^4$$

You want $2\cos\theta$ or $2i\sin\theta$ to start with.

$$\equiv (e^{i\theta}+e^{-i\theta})^4$$

(1) Since $2\cos\theta \equiv e^{i\theta}+e^{-i\theta}$

$$\equiv (e^{i\theta})^4 + 4(e^{i\theta})^3(e^{-i\theta}) + 6(e^{i\theta})^2(e^{-i\theta})^2$$
$$+ 4(e^{i\theta})(e^{-i\theta})^3 + (e^{-i\theta})^4$$

(2) Write the binomial expansion of $(e^{i\theta}+e^{-i\theta})^4$

$$\equiv e^{4i\theta} + 4e^{2i\theta} + 6 + 4e^{-2i\theta} + e^{-4i\theta}$$

$$\equiv (e^{4i\theta}+e^{-4i\theta}) + 4(e^{2i\theta}+e^{-2i\theta}) + 6$$

(3) Simplify using index rules.

$$\equiv 2\cos4\theta + 8\cos2\theta + 6$$

$$8\cos^4\theta \equiv \dfrac{2\cos4\theta + 8\cos2\theta + 6}{2}$$

Remember, this is $16\cos^4\theta$

(4) $e^{4i\theta}+e^{-4i\theta} = 2\cos4\theta$ and $e^{2i\theta}+e^{-2i\theta} = 2\cos2\theta$

$$\equiv \cos4\theta + 4\cos2\theta + 3$$

To write $\cos(n\theta)$ or $\sin(n\theta)$ in terms of powers of $\cos\theta$ or $\sin\theta$

(1) Use de Moivre's theorem.

(2) Write out the binomial expansion.

(3) Simplify powers of i

(4) Equate coefficients of real or imaginary parts.

(5) Use $\cos^2\theta + \sin^2\theta \equiv 1$ to write the expression as powers of either $\cos\theta$ or $\sin\theta$

Example 4

Express $\sin 5x$ in the form $A\sin x + B\sin^3 x + C\sin^5 x$ where A, B and C are constants to be found.

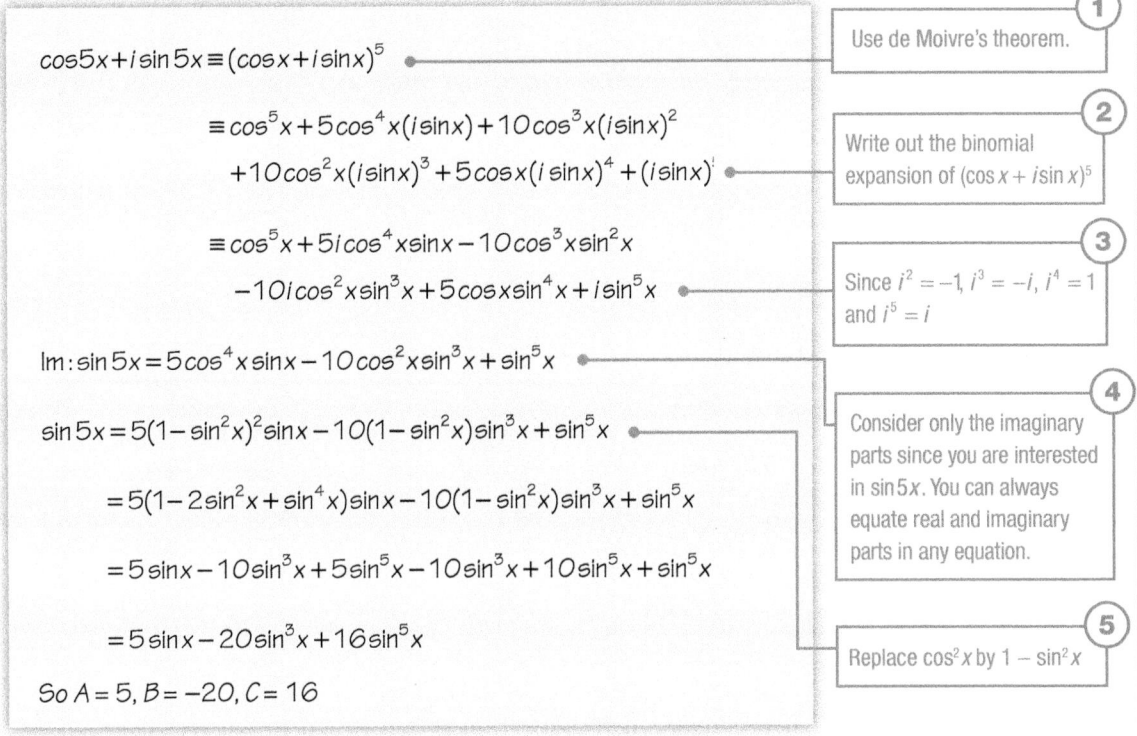

$$\cos 5x + i\sin 5x \equiv (\cos x + i\sin x)^5$$

① Use de Moivre's theorem.

$$\equiv \cos^5 x + 5\cos^4 x(i\sin x) + 10\cos^3 x(i\sin x)^2$$
$$+ 10\cos^2 x(i\sin x)^3 + 5\cos x(i\sin x)^4 + (i\sin x)^5$$

② Write out the binomial expansion of $(\cos x + i\sin x)^5$

$$\equiv \cos^5 x + 5i\cos^4 x\sin x - 10\cos^3 x\sin^2 x$$
$$- 10i\cos^2 x\sin^3 x + 5\cos x\sin^4 x + i\sin^5 x$$

③ Since $i^2 = -1$, $i^3 = -i$, $i^4 = 1$ and $i^5 = i$

$$\text{Im}: \sin 5x = 5\cos^4 x\sin x - 10\cos^2 x\sin^3 x + \sin^5 x$$

$$\sin 5x = 5(1 - \sin^2 x)^2\sin x - 10(1 - \sin^2 x)\sin^3 x + \sin^5 x$$

④ Consider only the imaginary parts since you are interested in $\sin 5x$. You can always equate real and imaginary parts in any equation.

$$= 5(1 - 2\sin^2 x + \sin^4 x)\sin x - 10(1 - \sin^2 x)\sin^3 x + \sin^5 x$$

$$= 5\sin x - 10\sin^3 x + 5\sin^5 x - 10\sin^3 x + 10\sin^5 x + \sin^5 x$$

$$= 5\sin x - 20\sin^3 x + 16\sin^5 x$$

⑤ Replace $\cos^2 x$ by $1 - \sin^2 x$

So $A = 5$, $B = -20$, $C = 16$

Strategy 3

To find the sum of a series involving $\sin(r\theta)$ or $\cos(r\theta)$

① Consider a sum involving $\cos(r\theta) + i\sin(r\theta)$

② Use de Moivre's theorem.

③ Use the formula for the sum of a geometric series.

④ Use formulae for $\sin(A \pm B)$ or $\cos(A \pm B)$

⑤ Select only the real or the imaginary parts as required.

Example 5

Show that $\displaystyle\sum_{r=1}^{n}\sin(r\theta) = \dfrac{\sin\left(\dfrac{(n+1)\theta}{2}\right)\sin\left(\dfrac{n\theta}{2}\right)}{\sin\left(\dfrac{\theta}{2}\right)}$

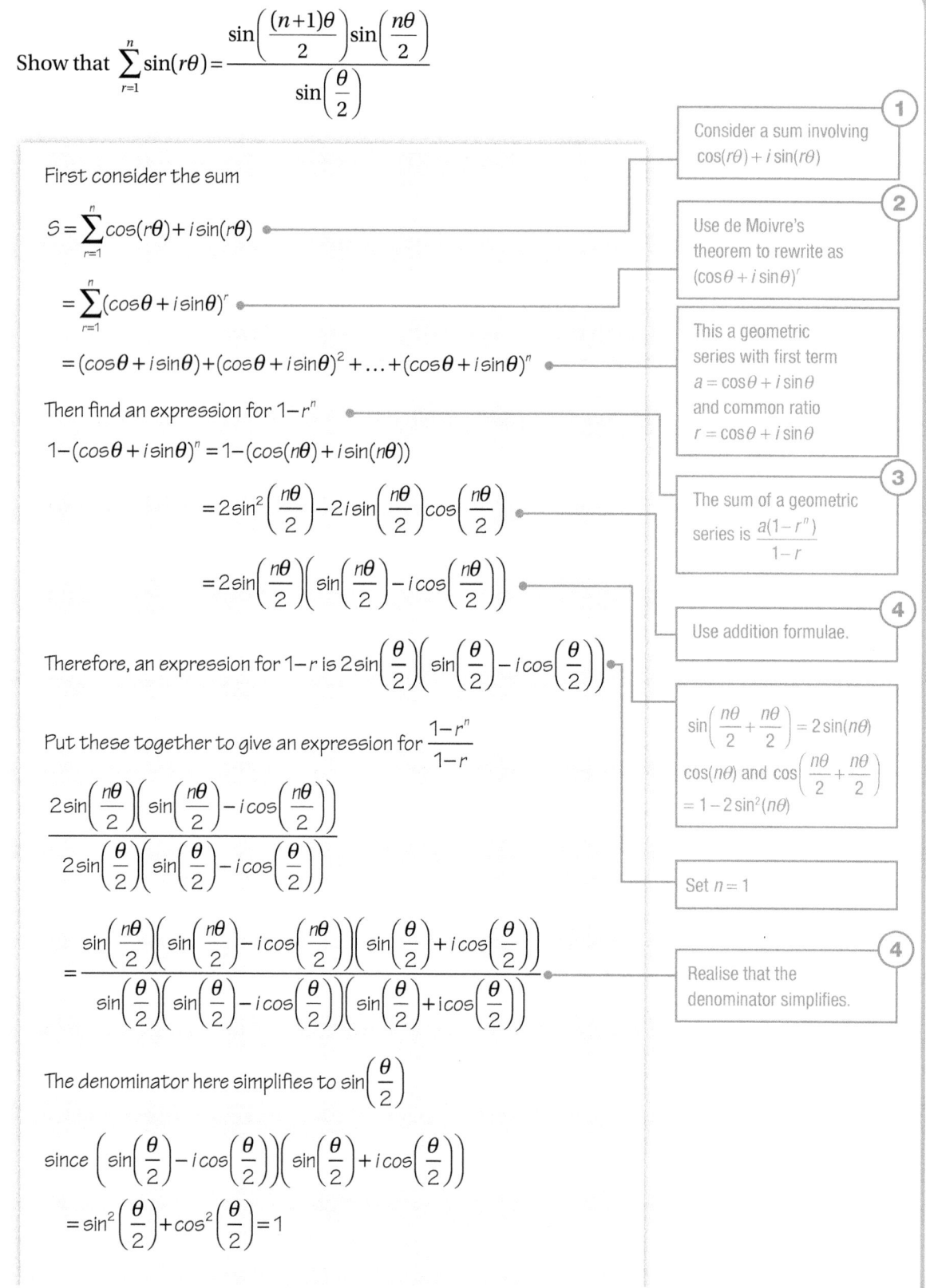

First consider the sum

$$S = \sum_{r=1}^{n}\cos(r\theta)+i\sin(r\theta)$$

(1) Consider a sum involving $\cos(r\theta)+i\sin(r\theta)$

$$= \sum_{r=1}^{n}(\cos\theta+i\sin\theta)^{r}$$

(2) Use de Moivre's theorem to rewrite as $(\cos\theta+i\sin\theta)^{r}$

$$= (\cos\theta+i\sin\theta)+(\cos\theta+i\sin\theta)^{2}+\dots+(\cos\theta+i\sin\theta)^{n}$$

This a geometric series with first term $a = \cos\theta+i\sin\theta$ and common ratio $r = \cos\theta+i\sin\theta$

Then find an expression for $1-r^{n}$

$$1-(\cos\theta+i\sin\theta)^{n} = 1-(\cos(n\theta)+i\sin(n\theta))$$

$$= 2\sin^{2}\left(\frac{n\theta}{2}\right)-2i\sin\left(\frac{n\theta}{2}\right)\cos\left(\frac{n\theta}{2}\right)$$

(3) The sum of a geometric series is $\dfrac{a(1-r^{n})}{1-r}$

$$= 2\sin\left(\frac{n\theta}{2}\right)\left(\sin\left(\frac{n\theta}{2}\right)-i\cos\left(\frac{n\theta}{2}\right)\right)$$

(4) Use addition formulae.

Therefore, an expression for $1-r$ is $2\sin\left(\dfrac{\theta}{2}\right)\left(\sin\left(\dfrac{\theta}{2}\right)-i\cos\left(\dfrac{\theta}{2}\right)\right)$

$$\sin\left(\frac{n\theta}{2}+\frac{n\theta}{2}\right) = 2\sin(n\theta)$$
$$\cos(n\theta) \text{ and } \cos\left(\frac{n\theta}{2}+\frac{n\theta}{2}\right)$$
$$= 1-2\sin^{2}(n\theta)$$

Put these together to give an expression for $\dfrac{1-r^{n}}{1-r}$

$$\dfrac{2\sin\left(\dfrac{n\theta}{2}\right)\left(\sin\left(\dfrac{n\theta}{2}\right)-i\cos\left(\dfrac{n\theta}{2}\right)\right)}{2\sin\left(\dfrac{\theta}{2}\right)\left(\sin\left(\dfrac{\theta}{2}\right)-i\cos\left(\dfrac{\theta}{2}\right)\right)}$$

Set $n=1$

$$= \dfrac{\sin\left(\dfrac{n\theta}{2}\right)\left(\sin\left(\dfrac{n\theta}{2}\right)-i\cos\left(\dfrac{n\theta}{2}\right)\right)\left(\sin\left(\dfrac{\theta}{2}\right)+i\cos\left(\dfrac{\theta}{2}\right)\right)}{\sin\left(\dfrac{\theta}{2}\right)\left(\sin\left(\dfrac{\theta}{2}\right)-i\cos\left(\dfrac{\theta}{2}\right)\right)\left(\sin\left(\dfrac{\theta}{2}\right)+i\cos\left(\dfrac{\theta}{2}\right)\right)}$$

(4) Realise that the denominator simplifies.

The denominator here simplifies to $\sin\left(\dfrac{\theta}{2}\right)$

since $\left(\sin\left(\dfrac{\theta}{2}\right)-i\cos\left(\dfrac{\theta}{2}\right)\right)\left(\sin\left(\dfrac{\theta}{2}\right)+i\cos\left(\dfrac{\theta}{2}\right)\right)$

$$= \sin^{2}\left(\dfrac{\theta}{2}\right)+\cos^{2}\left(\dfrac{\theta}{2}\right) = 1$$

(Continued on the next page)

The numerator expands to give

$$\sin\left(\frac{n\theta}{2}\right)\left[\left(\cos\left(\frac{n\theta}{2}\right)\cos\left(\frac{\theta}{2}\right)+\sin\left(\frac{n\theta}{2}\right)\sin\left(\frac{\theta}{2}\right)\right)\right.$$

$$\left.+i\left(\cos\left(\frac{\theta}{2}\right)\sin\left(\frac{n\theta}{2}\right)-\cos\left(\frac{n\theta}{2}\right)\sin\left(\frac{\theta}{2}\right)\right)\right]$$

$$=\sin\left(\frac{n\theta}{2}\right)\left(\cos\left(\frac{(n-1)\theta}{2}\right)+i\sin\left(\frac{(n-1)\theta}{2}\right)\right)$$

> **4** Use formulae for $\cos\left(\frac{n\theta}{2}-\frac{\theta}{2}\right)$ and $\sin\left(\frac{n\theta}{2}-\frac{\theta}{2}\right)$

Therefore,

$$S=\frac{(\cos\theta+i\sin\theta)\sin\left(\frac{n\theta}{2}\right)\left(\cos\left(\frac{(n-1)\theta}{2}\right)+i\sin\left(\frac{(n-1)\theta}{2}\right)\right)}{\sin\left(\frac{\theta}{2}\right)}$$

> **3** Use $S=\dfrac{a(1-r^n)}{1-r}$

$$\sum_{r=1}^{n}\sin(r\theta)=\frac{\sin\left(\frac{n\theta}{2}\right)\left(\sin\theta\cos\left(\frac{(n-1)\theta}{2}\right)+\sin\left(\frac{(n-1)\theta}{2}\right)\cos\theta\right)}{\sin\left(\frac{\theta}{2}\right)}$$

> **5** Select the imaginary parts of the sum.

$$=\frac{\sin\left(\frac{(n+1)\theta}{2}\right)\sin\left(\frac{n\theta}{2}\right)}{\sin\left(\frac{\theta}{2}\right)}$$

> **4** Use formula for $\sin\left(\frac{(n-1)\theta}{2}+\theta\right)$

For $\sum_{r=1}^{n}\cos(r\theta)$ you would need to consider the real part of the sum.

1 Prove each of these identities.

    a $2\cos^2\theta \equiv \cos 2\theta + 1$ **b** $8\sin^3\theta \equiv 6\sin\theta - 2\sin 3\theta$

    c $4\sin^4\theta \equiv \dfrac{1}{2}\cos 4\theta - 2\cos 2\theta + \dfrac{3}{2}$

2   **a**   Show that $\cos^5\theta \equiv A(10\cos\theta + 5\cos 3\theta + \cos 5\theta)$, where A is a constant to be found.

    b   Hence find $\int \cos^5\theta \; d\theta$

3   **a**   Show that $\sin^6\theta \equiv B(15\cos 2\theta - 6\cos 4\theta + \cos 6\theta - 10)$, where B is a constant to be found.

    b   Hence find $\int \sin^6\theta \; d\theta$

4   **a**   Show that $2\sin^3\theta \equiv \dfrac{3}{2}\sin\theta - \dfrac{1}{2}\sin 3\theta$

    b   Hence solve the equation $3\sin\theta - \sin 3\theta = \dfrac{1}{2}$ for $-\pi \le \theta \le \pi$

5   **a**   Show that $5\cos^4\theta \equiv A\cos 4\theta + B\cos 2\theta + C$, where A, B and C are constants to be found.

    b   Hence solve the equation $\cos 4\theta + 4\cos 2\theta + 3 = 2$ for $-\pi \le \theta \le \pi$

6 Use de Moivre's theorem to prove the following identities.

    a $\sin 2\theta \equiv 2\cos\theta\sin\theta$ **b** $\sin 3\theta \equiv 3\sin\theta - 4\sin^3\theta$

    c $\cos 3\theta \equiv 4\cos^3\theta - 3\cos\theta$ **d** $\sin 4\theta \equiv 4\cos\theta\sin\theta - 8\cos\theta\sin^3\theta$

7 Prove the identities

    a $\cos 6\theta \equiv 32\cos^6\theta - 48\cos^4\theta + 18\cos^2\theta - 1$ **b** $\sin 6\theta \equiv 2\sin\theta\cos\theta(16\sin^4\theta - 16\sin^2\theta + 3)$

8   **a**   Use de Moivre's theorem to show that $\cos 5\theta \equiv 16\cos^5\theta - 20\cos^3\theta + 5\cos\theta$

    b   Hence find 3 solutions to the equation $16x^5 - 20x^3 + 5x = 1$

9   **a**   Use de Moivre's theorem to show that $\cos 4\theta \equiv 8\cos^4\theta - 8\cos^2\theta + 1$

    b   Hence find 4 solutions to the equation $x^4 - x^2 = -\dfrac{1}{16}$

10 Use de Moivre's theorem to show that $\tan 2\theta \equiv \dfrac{2\tan\theta}{1-\tan^2\theta}$

11 Use proof by induction to prove that $\left[r(\cos\theta + i\sin\theta)\right]^n = r^n(\cos n\theta + i\sin n\theta)$ for all positive integers n

12   **a**   Given that $z = \cos\theta + i\sin\theta$, use de Moivre's theorem to show that $2\cos(n\theta) = z^n + \dfrac{1}{z^n}$

    b   Hence show that $4\cos\theta\sin^2\theta \equiv \cos\theta - \cos(3\theta)$

13   **a**   Given that $z = \cos\theta + i\sin\theta$, use de Moivre's theorem to show that $2i\sin(n\theta) = z^n - \dfrac{1}{z^n}$

    b   Hence show that $16\sin^3\theta\cos^2\theta \equiv 2\sin\theta + \sin(3\theta) - \sin(5\theta)$

14 Show that $\displaystyle\sum_{r=1}^{n}\cos(r\theta) = \dfrac{\cos\left(\dfrac{(n+1)\theta}{2}\right)\sin\left(\dfrac{n\theta}{2}\right)}{\sin\left(\dfrac{\theta}{2}\right)}$

Fluency and skills

A fundamental rule in maths is that an equation of order n must have n solutions. So, the equation $z^3 = 1$ must have three roots so there must be three cube roots of 1. One of them is real (1) and two are complex. These are called **the cube roots of unity**.

This is easily seen in the case of $z^4 = 1$. Square root both sides to give $z^2 = \pm 1$

If $z^2 = 1$ then $z = 1$ or $z = -1$

If $z^2 = -1$ then $z = i$ or $z = -i$

So there are four solutions because the equation has order 4

You can use the exponential form of a complex number to solve equations of the form $z^n = 1$ to find the **nth roots of unity**.

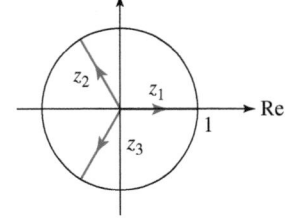

Example 1

a Find all the solutions to the equation $z^6 = 1$, giving your answers in Cartesian form.

b Illustrate the 6th roots of unity on an Argand diagram.

a $z^6 = e^{2k\pi i}$

$z = \left(e^{2k\pi i}\right)^{\frac{1}{6}}$

$= e^{\frac{2k\pi}{6}i}$

> Since the modulus is 1 and the argument of any positive real number is 0 or $\pm 2\pi$ or $\pm 4\pi$ and so on, in general, this could be written as $2k\pi$, where k is an integer.

Consider each of the possible values of k from 0 to 5. Going beyond 5 would just repeat the same values again.

$k = 0: z = e^0 = 1$

$k = 1: z = e^{\frac{2\pi}{6}i} = \frac{1}{2} + \frac{\sqrt{3}}{2}i$

$k = 2: z = e^{\frac{4\pi}{6}i} = -\frac{1}{2} + \frac{\sqrt{3}}{2}i$

$k = 3: z = e^{\frac{6\pi}{6}i} = -1$

$k = 4: z = e^{\frac{8\pi}{6}i} = -\frac{1}{2} - \frac{\sqrt{3}}{2}i$

$k = 5: z = e^{\frac{10\pi}{6}i} = \frac{1}{2} - \frac{\sqrt{3}}{2}i$

> Use the index law.

> There should be 6 solutions to the equation.

(Continued on the next page)

b

Notice how the 6 solutions split a circle into 6 equal sectors each with angle $\dfrac{\pi}{3}$

Key point

The equation $z^n = 1$ has n solutions of the form

$z = e^{\frac{2k\pi i}{n}}$, where $k = 0, 1, 2, ..., n$

This method can be extended to find the nth roots of any complex number.

Example 2

Solve the equation $z^4 = 2\sqrt{3} - 2i$, giving your answers in the form $re^{i\theta}$ where $r > 0$ and $-\pi < \theta \leq \pi$

$\left| 2\sqrt{3} - 2i \right| = \sqrt{(2\sqrt{3})^2 + 2^2} = 4$

Calculate the modulus and argument.

$\arg(2\sqrt{3} - 2i) = -\dfrac{\pi}{6}$

So in, general, the argument is $-\dfrac{\pi}{6} + 2k\pi = \dfrac{(12k-1)\pi}{6}$

$z^4 = 4e^{\frac{(12k-1)\pi}{6}i}$

Write a general term in exponential form.

$z = \left(4e^{\frac{(12k-1)\pi}{6}i} \right)^{\frac{1}{4}}$

$\quad = \sqrt{2}e^{\frac{(12k-1)\pi}{24}i}$

Consider each of the possible values of k from 0 to 3

$k = 0: z = \sqrt{2}e^{-\frac{\pi}{24}i}$

$k = 1: z = \sqrt{2}e^{\frac{11\pi}{24}i}$

$k = 2: z = \sqrt{2}e^{\frac{23\pi}{24}i}$

$k = 3: z = \sqrt{2}e^{\frac{35\pi}{24}i}$

The final solution is not in the interval $-\pi < \theta \leq \pi$

So instead write $z = \sqrt{2}e^{-\frac{13\pi}{24}i}$

Since $-\dfrac{13\pi}{24} = \dfrac{35\pi}{24}$

1 a Solve the equation $z^3 = 1$, giving your answers in Cartesian form.

b Illustrate the cube roots of unity on an Argand diagram.

2 a Solve the equation $z^4 = 1$, giving your answers in Cartesian form.

b Illustrate the 4th roots of unity on an Argand diagram.

3 a Solve the equation $z^5 = 1$, giving your answers in exponential form

b Illustrate the 5th roots of unity on an Argand diagram.

4 Solve each of these equations, giving your solutions in Cartesian form.

a $z^3 = 8$

b $z^3 = i$

c $z^4 = -49$

d $z^3 = -27i$

e $z^6 = -8$

5 Solve each of these equations, giving your solutions in exponential form.

a $z^4 = -16i$ **b** $z^5 = 32i$

6 Solve each of these equations, giving your solutions in exponential form

a $z^3 = 4\sqrt{2} + 4\sqrt{2}i$

b $z^3 = -4\sqrt{2} + 4\sqrt{2}i$

c $z^3 = -4\sqrt{2} - 4\sqrt{2}i$

d $z^3 = 4\sqrt{2} - 4\sqrt{2}i$

7 Solve each of these equations, giving your solutions in modulus–argument form with θ given to 2 decimal places.

a $z^4 = 3\sqrt{5} + 6i$

b $z^4 = -3\sqrt{5} + 6i$

c $z^4 = -6 - 3\sqrt{5}i$

d $z^4 = 6 - 3\sqrt{5}i$

8 Solve each of these equations, giving your solutions in exponential form.

a $z^5 + 243 = 0$

b $32z^5 + 1 = 0$

c $z^6 + 64i = 0$

d $4z^4 + 3i = 4$

9 Solve each of these equations, giving your solutions in Cartesian form.

a $z^4 = 8 + 8\sqrt{3}i$

b $z^4 = -8 + 8\sqrt{3}i$

10 Solve the equation $z^{\frac{3}{2}} = (2 - 2i)^2$, giving your answers in the form $a + bi$

11 a Solve the equation $z^4 = -32 + 32\sqrt{3}i$, giving your solution in Cartesian form.

b Represent the solutions on an Argand diagram.

12 Solve each of these equations, giving your solutions in the form $re^{i\theta}$ where $r > 0$ and $-\pi < \theta \leq \pi$

a $z^3 = 4\sqrt{3} + 4i$

b $z^4 = 3\sqrt{2} - 3\sqrt{2}i$

13 Solve each of these equations, giving your solutions in the form $r(\cos\theta + i\sin\theta)$, where $r > 0$ and $-\pi < \theta \leq \pi$

a $z^3 = \sqrt{2} - \sqrt{6}i$

b $z^6 = -4\sqrt{3} + 4i$

The equation $z^n = 1$ has n solutions of the form $z = e^{\frac{2k\pi i}{n}}$, where $k = 0, 1, 2, ..., n-1$ Using index laws,

you could write this as $z = \left(e^{\frac{2\pi i}{n}} \right)^k$. Therefore, if ω is a complex solution of $z^n = 1$, then

$1, \omega, \omega^2, \omega^3, ..., \omega^{n-1}$ are all distinct solutions.

$$1 + \omega + \omega^2 + \omega^3 + ... + \omega^{n-1} = \left(e^{\frac{2\pi i}{n}} \right)^0 + \left(e^{\frac{2\pi i}{n}} \right)^1 + \left(e^{\frac{2\pi i}{n}} \right)^2 + \left(e^{\frac{2\pi i}{n}} \right)^3 + ... + \left(e^{\frac{2\pi i}{n}} \right)^{n-1}$$

This is a geometric series with first term 1 and common ratio $\omega = e^{\frac{2\pi i}{n}}$

So, using the formula $S_n = \dfrac{a(1-r^n)}{1-r}$, you can see that the sum of the series is $\dfrac{1\left(1 - \left(e^{\frac{2\pi i}{n}} \right)^n \right)}{1 - e^{\frac{2\pi i}{n}}}$

This simplifies to $\dfrac{1(1 - e^{2\pi i})}{1 - e^{\frac{2\pi i}{n}}} = 0$ since $e^{2\pi i} = 1$

Key point

Therefore, $1 + \omega + \omega^2 + ... + \omega^{n-1} = 0$ where ω is an nth root of unity.

You can quote this result.

Strategy 1

To find complex solutions of an equation

(1) Expand brackets or simplify fractions.

(2) Use fact that $\omega^n = 1$ if ω is an nth root of unity.

(3) Use $1 + \omega + \omega^2 + ... + \omega^{n-1} = 0$

Example 3

You are given that ω is a complex 4th root of unity.

a Show that $1 + \omega + \omega^2 + \omega^3 = 0$

b Evaluate $(1+\omega)(1+\omega^2) + \omega^2$

a This is a geometric series with first term 1 and common ratio ω

So $1 + \omega + \omega^2 + \omega^3 = \dfrac{1(1-\omega^4)}{1-\omega}$

Since $S_n = \dfrac{a(1-r^n)}{1-r}$ for a geometric series.

$= \dfrac{1-\omega^4}{1-\omega}$

(2)

$= \dfrac{1-1}{1-\omega}$

Since $\omega^4 = 1$, as ω is a complex 4th root of unity.

$= 0$ as required

b $(1+\omega)(1+\omega^2) + \omega^4 = 1 + \omega + \omega^2 + \omega^3 + \omega^4$

(1)

(2)

$= 1 + \omega + \omega^2 + \omega^3 + 1$

Since $\omega^4 = 1$

$= 0 + 1$

$= 1$

(3)

To solve geometric problems

1. Write z^n in exponential form.

2. Find the roots of an equation.

3. Sketch an Argand diagram.

4. Use Pythagoras' theorem or trigonometry to find lengths and areas.

Example 4

The points A, B and C represent the solutions to the equation $z^3 = 8i$

Calculate the exact area and perimeter of triangle ABC

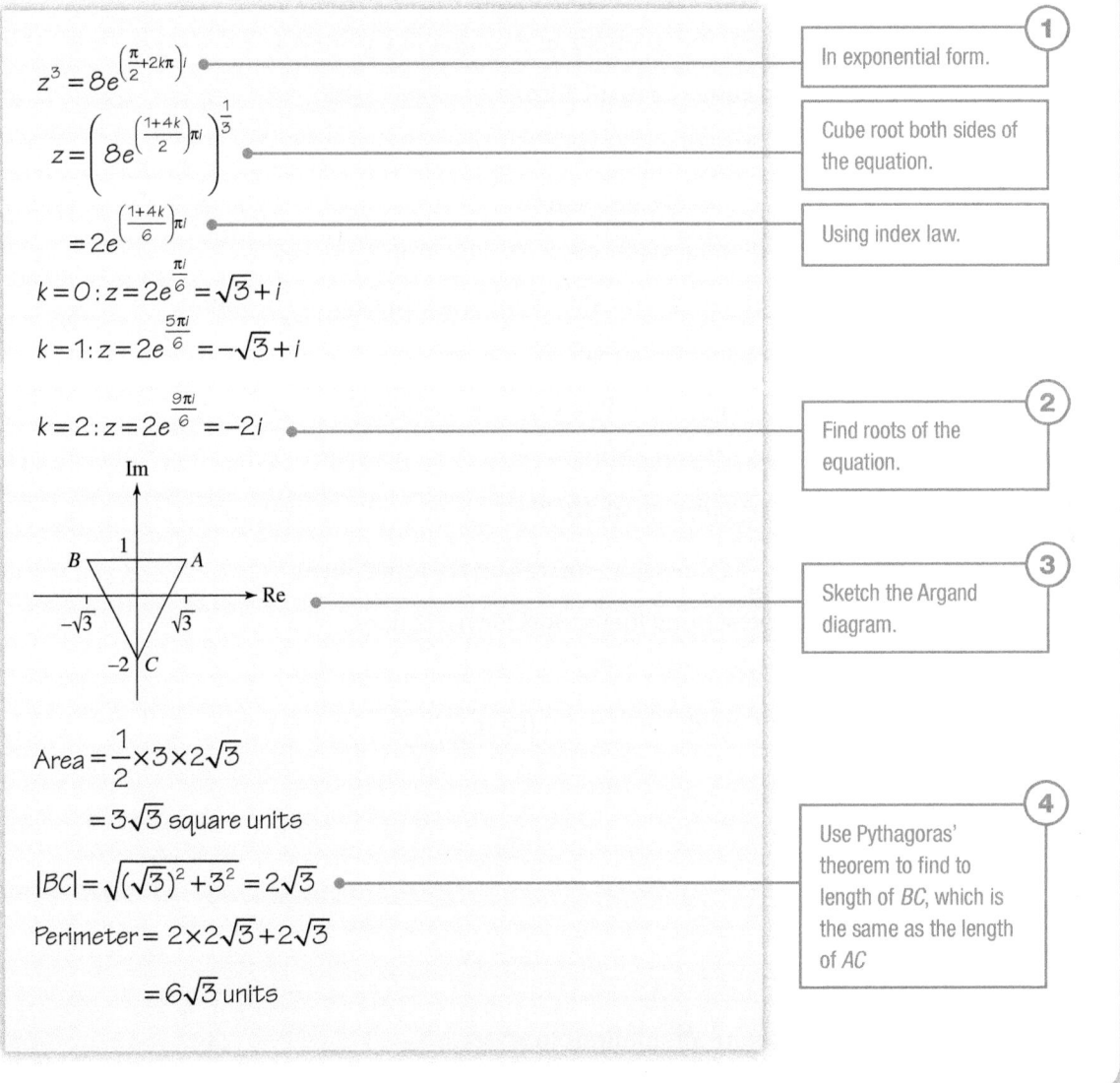

$z^3 = 8e^{\left(\frac{\pi}{2}+2k\pi\right)i}$ — ① In exponential form.

$z = \left(8e^{\left(\frac{1+4k}{2}\right)\pi i}\right)^{\frac{1}{3}}$ — Cube root both sides of the equation.

$= 2e^{\left(\frac{1+4k}{6}\right)\pi i}$ — Using index law.

$k = 0: z = 2e^{\frac{\pi i}{6}} = \sqrt{3} + i$

$k = 1: z = 2e^{\frac{5\pi i}{6}} = -\sqrt{3} + i$

$k = 2: z = 2e^{\frac{9\pi i}{6}} = -2i$ — ② Find roots of the equation.

③ Sketch the Argand diagram.

$\text{Area} = \frac{1}{2} \times 3 \times 2\sqrt{3}$

$= 3\sqrt{3}$ square units

$|BC| = \sqrt{(\sqrt{3})^2 + 3^2} = 2\sqrt{3}$ — ④ Use Pythagoras' theorem to find to length of BC, which is the same as the length of AC

$\text{Perimeter} = 2 \times 2\sqrt{3} + 2\sqrt{3}$

$= 6\sqrt{3}$ units

1 Given that ω is a complex cube root of unity

   a Show that $1+\omega+\omega^2=0$

   b Evaluate the following expressions.

      i $(1+\omega)^2-\omega$ **ii** $(1+\omega)(1+\omega^2)$ **iii** $\omega(\omega+1)$ **iv** $\dfrac{2\omega+1}{\omega-1}+\omega$

2 Given that ω is a complex 5th root of unity

   a Show that $1+\omega+\omega^2+\omega^3+\omega^4=0$

   b Evaluate the following expressions.

      i $\omega(1+\omega)(1+\omega^2)$ **ii** $\dfrac{\omega^2}{\omega+1}+\omega^3+1$ **iii** $\omega(1+\omega+\omega^2+\omega^3)$

3 The points A, B and C represent the solutions to the equation $z^3=-27i$

   a Find the solutions to the equation in the form $a+bi$

   b Calculate the exact

      i Area,   **ii** Perimeter of triangle ABC

4 The points A, B and C represent the solutions to the equation $z^3=-125$

   a Show that the area of triangle ABC is $k\sqrt{3}$, where k is a constant to be found,

   b Calculate the exact perimeter of triangle ABC

5 The points A, B, C and D represent the solutions to the equation $z^4=-4i$

   a State the name of the quadrilateral $ABCD$

   b Calculate the area of the quadrilateral $ABCD$

6 **a** Find the fourth roots of $\dfrac{1}{9}$, giving your answers in Cartesian form.

   b Show that the points representing these roots form a square.

   c Find the area of the square.

7 The points A, B, C, D, E and F are the vertices of the regular hexagon shown.

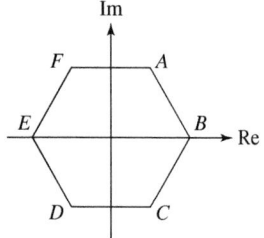

Given that A, B, C, D, E and F are all solutions to $z^n=1$

   a State the value of n

   b Find the coordinates of F

   c Calculate the area of the hexagon.

Chapter summary

- A complex number $z = a + bi$ can be written
  - in modulus–argument form as $z = r(\cos\theta + i\sin\theta)$
  - in exponential form as $z = re^{i\theta}$

  where $r = |z|$ and $\theta = \arg z$, $-\pi < \theta \leq \pi$

- For complex numbers z and w
  - $|zw| = |z||w|$
  - $\left|\dfrac{z}{w}\right| = \dfrac{|z|}{|w|}$
  - $\arg(zw) = \arg z + \arg w$
  - $\arg\left(\dfrac{z}{w}\right) = \arg z - \arg w$

- Euler's formula is $re^{i\theta} = r(\cos\theta + i\sin\theta)$

- $\cos\theta = \dfrac{e^{i\theta} + e^{-i\theta}}{2}$ and $\sin\theta = \dfrac{e^{i\theta} - e^{-i\theta}}{2i}$

- De Moivre's theorem states that $\left[r(\cos\theta + i\sin\theta)\right]^n = r^n(\cos(n\theta) + i\sin(n\theta))$ for all integers, n

- The equation $z^n = 1$ has n solutions of the form $z = e^{\frac{2k\pi i}{n}}$, where $k = 0, 1, 2, \ldots, n$

- Equations of the form $z^n = a + bi$ have solutions of the form $z = r^{\frac{1}{n}} e^{\frac{(\theta + 2k\pi)i}{n}}$ where $|z| = r$, $\arg(z) = \theta$ and $k = 0, 1, 2, \ldots, n$

- $1 + \omega + \omega^2 + \ldots + \omega^{n-1} = 0$ where ω is an nth root of unity.

Check and review

You should now be able to...	Review
✔ Write complex numbers in exponential form.	1, 2
✔ Convert complex numbers from exponential to Cartesian form.	3
✔ Use rules to find the argument and modulus of products of complex numbers.	4
✔ Use rules to find the argument and modulus of quotients of complex numbers.	5
✔ Use de Moivre's theorem to simplify powers of complex numbers.	6, 7
✔ Use de Moivre's theorem to prove trigonometric identities.	8
✔ Simplify powers of trigonometric functions using a binomial expansion.	9
✔ Calculate roots of unity.	10
✔ Calculate the nth root of a complex number.	11–12

1 Write these complex numbers in exponential form.

   a $-3i$ b $1+i$

   c 5 d $-\sqrt{3}-i$

   e $\sqrt{2}-i$ f $-1+\sqrt{3}i$

2 Write each of these complex numbers in exponential form

   a $3\left(\cos\dfrac{\pi}{7}+i\sin\dfrac{\pi}{7}\right)$

   b $\sqrt{2}\left(\cos\dfrac{\pi}{9}+i\sin\dfrac{\pi}{9}\right)$

   c $\sqrt{3}\left(\cos\dfrac{\pi}{8}-i\sin\dfrac{\pi}{8}\right)$

   d $5\left(\cos\left(-\dfrac{\pi}{5}\right)-i\sin\left(-\dfrac{\pi}{5}\right)\right)$

3 Write these complex numbers in the form $a+bi$

   a $7e^{\left(\frac{\pi}{2}\right)i}$ b $6e^{\left(\frac{\pi}{3}\right)i}$

   c $\sqrt{3}e^{\left(-\frac{\pi}{6}\right)i}$ d $\sqrt{2}e^{\left(\frac{3\pi}{4}\right)i}$

   e $\sqrt{6}e^{\left(-\frac{2\pi}{3}\right)i}$ f $2\sqrt{3}e^{\left(\frac{5\pi}{6}\right)i}$

4 Find the argument and modulus of zw in each case.

   a $z=3\left(\cos\dfrac{\pi}{5}+i\sin\dfrac{\pi}{5}\right)$ and

      $w=5\left(\cos\dfrac{\pi}{7}+i\sin\dfrac{\pi}{7}\right)$

   b $z=\sqrt{2}\left(\cos\left(-\dfrac{\pi}{8}\right)+i\sin\left(-\dfrac{\pi}{8}\right)\right)$ and

      $w=\sqrt{6}\left(\cos\dfrac{\pi}{3}+i\sin\dfrac{\pi}{3}\right)$

   c $z=1+\sqrt{3}i$ and $w=3-3i$

   d $z=\dfrac{1}{2}e^{\frac{2\pi}{9}i}$ and $w=4e^{\frac{\pi}{3}i}$

5 Find the argument and modulus of $\dfrac{z}{w}$ in each case.

   a $z=5\left(\cos\dfrac{\pi}{2}+i\sin\dfrac{\pi}{2}\right)$ and

      $w=10\left(\cos\dfrac{\pi}{4}+i\sin\dfrac{\pi}{4}\right)$

   b $z=\sqrt{15}\left(\cos\left(-\dfrac{3\pi}{4}\right)+i\sin\left(-\dfrac{3\pi}{4}\right)\right)$ and

      $w=\sqrt{5}\left(\cos\left(-\dfrac{\pi}{8}\right)+i\sin\left(-\dfrac{\pi}{8}\right)\right)$

   c $z=-5+5i$ and $w=\sqrt{6}-3\sqrt{2}i$

   d $z=16e^{-\frac{2\pi}{11}i}$ and $w=\sqrt{2}e^{\frac{5\pi}{11}i}$

6 Given that $z=8\left(\cos\left(\dfrac{\pi}{2}\right)+i\sin\left(\dfrac{\pi}{2}\right)\right)$, write these powers of z in Cartesian form.

   a z^2 b z^3

7 Given that $w=-2\sqrt{3}-2i$, express w^2 in the form $a+bi$

8 a Use de Moivre's theorem to show that
      $\sin 5\theta \equiv 5\sin\theta - 20\sin^3\theta + 16\sin^5\theta$

   b Hence find 3 solutions to the equation
      $5x-20x^3+16x^5 = 0$
      Give your answers to 3 significant figures.

9 Prove that $\cos^3\theta \equiv A(\cos 3\theta + 3\cos\theta)$, where A is a constant to be found.

10 a Calculate the 8th roots of unity. Give your answers in exact Cartesian form.

    b Draw the roots on an Argand diagram.

11 Solve these equations, giving your answers in Cartesian form.

   a $z^8=16$ b $z^3=i$

   c $z^2=-9i$ d $z^6=-125$

12 Solve the equation $z^4 = -2\sqrt{2}-2\sqrt{2}i$, giving your solutions in the form $re^{i\theta}$ where $r>0$ and $-\pi<\theta\leq\pi$

History

Abraham de Moivre was a French mathematician (1667–1754). He was a contemporary and friend of Isaac Newton and Edmund Halley (the astronomer after whom Halley's comet is named). As well as his work on de Moivre's formula, he wrote a major work on probability *The Doctrine of Chances*.

Halley suggested that de Moivre looked at the astronomical world, and so he also worked on the mathematical ideas associated with centripetal force.

Note

Complex numbers are widely used across many areas of engineering, particularly in electrical/electronic engineering. Engineers use 'j' rather than 'i' to represent the imaginary part of a complex number, so a complex number is written, for example, as $1 + j$

Investigation

Investigate the nth roots of unity for $n = 1, 2, 3, 4, 5, \ldots$ Use a graph plotting package such as GeoGebra to plot these.
- What do you know about the results?
- How does it relate to what you know about geometry?
- What happens to the distance between roots as the series develops?
- What are the connections with series?
- What are the connections with calculus?

Research

The hyperbolic functions $f(x) = \sinh(x)$, $g(x) = \cosh(x)$ are complex analogues of the circular functions $f(x) = \sin(x)$ and $g(x) = \cos(x)$.

Research how the hyperbolic functions relate to the circular functions, in terms of their relationships with circles and hyperbolas.

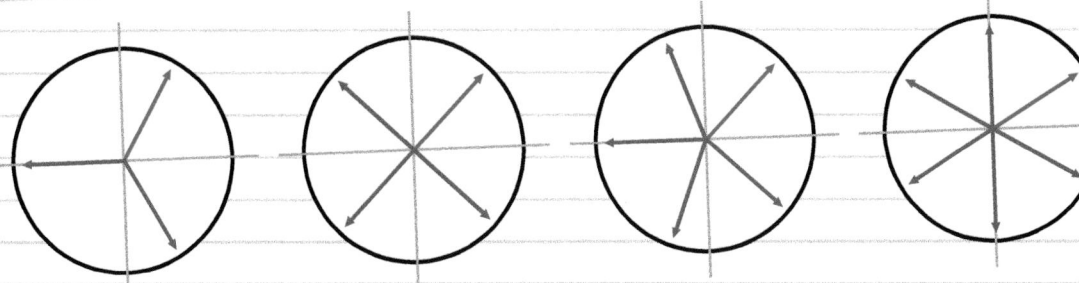

16 Assessment

1 The complex number z is defined by $z = 3 - \sqrt{3}i$

    a Write z in the form $re^{i\theta}$ where r is given as a surd in its simplest form and θ is given as a multiple of π **[4 marks]**

    b Hence work out z^4, giving your answer in modulus–argument form. You must show your working.   **[2]**

2 The complex number z is defined as $z = -k + k\sqrt{3}i$

    a Find the modulus and argument of z in terms of k where appropriate,   **[4]**

    b Given that $k = 2$, use de Moivre's theorem to express z^4 in the form $a + bi$, where a and b are exact constants to be found.   **[4]**

3   **a** Solve the equation $z^3 = -i$, giving your answers in the form $e^{i\theta}$, where $-\pi < \theta \le \pi$ **[4]**

    b Sketch the solutions on an Argand diagram.   **[2]**

4 Given that $z = -2 + 2i$

    a Write z in modulus–argument form,   **[2]**

    b Hence use de Moivre's theorem to find z^{-3} in the form $a + bi$, where a and b are constants to be found.   **[3]**

5 Express $(1 + \sqrt{3}i)^4$ in exact Cartesian form. You must show your working.   **[4]**

6 Use de Moivre's theorem to prove that $(\cos\theta - i\sin\theta)^n = \cos(n\theta) - i\sin(n\theta)$ **[4]**

7   **a** Show that $\sin 5\theta \equiv 5\sin\theta - 20\sin^3\theta + 16\sin^5\theta$ **[5]**

    b Hence solve the equation $5\sin\theta - 20\sin^3\theta + 16\sin^5\theta = 0.5$ for θ in the range $0 < \theta \le 180°$ **[3]**

8   **a** Prove that $32\cos^5 x = 2\cos(5x) + 10\cos(3x) + 20\cos(x)$ **[5]**

    b Hence find $\int \cos^5 x\,\mathrm{d}x$ **[2]**

9   **a** Show that $\sin 3\theta = 3\cos^2\theta\sin\theta - \sin^3\theta$ **[4]**

    b Hence show that $\tan 3\theta \equiv \dfrac{3\tan\theta - \tan^3\theta}{1 - 3\tan^2\theta}$ **[4]**

    c Find the exact value of $\tan 3\theta$, given that $\tan\theta = -2\sqrt{3}$ **[2]**

10   **a** Find the value of n such that $(\sqrt{2})^n = 8$ **[2]**

    b Solve the equation $z^6 + 8 = 0$, giving your answers in the form $r(\cos\theta + i\sin\theta)$, where $r > 0$ is an exact surd and $-\pi < \theta \le \pi$ **[5]**

11 Use exponentials to prove that $\cos(2x) = 1 - 2\sin^2 x$ **[4]**

12 A complex number is given as $z = 2e^{\frac{\pi}{2}i}$

A second complex number is $w = a + bi$, where $\arg(w) = \frac{\pi}{4}$

Calculate the exact value of

   a $\arg(zw)$ **b** $\arg\left(\dfrac{z}{w}\right)$

   c $\arg(w^n)$ **d** $\arg\left(\dfrac{w}{z^2}\right)$ **[7]**

13 a Use de Moivre's theorem to show that

    $\cos 7\theta = \cos\theta\,(64\cos^6\theta - 112\cos^4\theta + 56\cos^2\theta - 7)$ **[5]**

   b Hence solve the equation $64x^6 - 112x^4 + 56x^2 = 7$

    Give your answers to 3 significant figures.   **[4]**

14 Solve the equation $z^8 + 4\sqrt{12} + 8i = 0$, giving your answers the form $re^{i\theta}$

where $r > 0$ and $-\pi < \theta \le \pi$ **[5]**

15 a Prove that $\sin^4\theta \equiv \dfrac{1}{8}(\cos(4\theta) - 4\cos(2\theta) + 3)$ **[4]**

   b Hence find $\int 8\sin^4\theta\,\mathrm{d}\theta$ **[3]**

16 a Solve the equation $z^4 = -6 - 6\sqrt{3}i$, giving your answers in modulus–argument form,

    where r is given to 3 significant figures and θ is given as an exact multiple of π **[5]**

   b Hence solve the equation $(z-1)^4 = -6 - 6\sqrt{3}i$, giving your answers in Cartesian

    form to 2 significant figures.   **[3]**

17 a Use de Moivre's theorem to show that if $z = \cos\theta + i\sin\theta$, then

    $z^n - \dfrac{1}{z^n} \equiv 2i\sin(n\theta)$ **[4]**

   b Hence show that $4\sin^3\theta \equiv 3\sin\theta - \sin(3\theta)$ **[4]**

   c Solve the equation $2\sin(3\theta) - 6\sin\theta = 1$ for θ in the range $0 < \theta \le 2\pi$ **[4]**

18 a Illustrate the solutions to $z^5 = 1$ on an Argand diagram,   **[2]**

   b Given that ω is a complex 5th root of unity, evaluate

    i $\omega(1+\omega)(1+\omega^2)$ **ii** $(1+\omega^5)^2$ **[5]**

19 a Use de Moivre's theorem to show that $\cos 5\theta \equiv 16\cos^5\theta - 20\cos^3\theta + 5\cos\theta$ **[5]**

   b Hence find the general form of all the solutions to the equation

    $\cos 5\theta = 5\cos\theta$ **[4]**

20 a Show that $\omega = \cos\left(\dfrac{2\pi}{7}\right) + i\sin\left(\dfrac{2\pi}{7}\right)$ is one of the seventh roots of unity,   **[3]**

   b Write down the other non-real roots of the equation $z^7 = 1$ in terms of ω **[1]**

   c State the value of $\displaystyle\sum_{r=1}^{6} \omega^r$ **[1]**

21 This question refers to the equation $(z-1)^3 = 1$

   a Verify that $z_1 = \dfrac{1}{2} + \dfrac{\sqrt{3}}{2}i$ is a root of the equation,   **[3]**

   b Find the other two roots of the equation,   **[2]**

   c Produce an Argand diagram showing the three roots of the equation,   **[2]**

   d State the centre and radius of the circle on which the three roots lie.   **[2]**

22 a Solve the equation $z^4 + 16 = 0$

Give your answers in the form $a + bi$ where a and b are real numbers. **[5]**

b The points A, B, C and D in the complex plane represent these four roots.

Find the area of the square $ABCD$ **[2]**

23 a Prove that $\sin 5\theta = 16\sin^5\theta - 20\sin^3\theta + 5\sin\theta$ **[5]**

b Use the result in part **a** to solve the equation $16x^5 - 20x^3 + 5x = 1$ **[4]**

Give your solutions in the form $x = \sin k\pi$, where $0 \le k < 2$

24 a Solve the equation $(z-1)^3 = 8$

Give your solutions in the form $a + bi$, where a and b are real numbers. **[4]**

The points A, B, C in the complex plane represent these three roots.

b Draw the points A, B and C on an Argand diagram. **[2]**

c Calculate the area of the triangle ABC **[3]**

d The triangle ABC is rotated $\dfrac{\pi}{3}$ radians anticlockwise around the origin.

Find the complex numbers that represent the vertices of the triangle after this rotation. **[4]**

25 a Use de Moivre's theorem to show that $z = \dfrac{1}{2}(1+i)$ is a solution to the equation
$\left(\dfrac{z-1}{z}\right)^6 = -1$ **[5]**

b Find the other 5 solutions to the equation.

Give your answers in the form $a + bi$ where a and b are real numbers. **[6]**

The root $z = \dfrac{1}{2}(1+i)$ is represented by the point A in the complex plane.

c Find the coordinates of the image of A following a clockwise rotation of $\dfrac{\pi}{6}$ radians around the origin. **[3]**

26 a Solve the equation $z^3 = 4 - 4\sqrt{3}i$

Give your answers in the form $re^{i\theta}$, where $r > 0$ and $-\pi < \theta \le \pi$ **[6]**

b The roots of the equation $z^3 = 4 - 4\sqrt{3}i$ are represented by the points A, B and C on an Argand diagram.

Calculate the exact area of the triangle ABC **[3]**

c Show that $\sin\left(\dfrac{\pi}{9}\right) - \sin\left(\dfrac{5\pi}{9}\right) + \sin\left(\dfrac{7\pi}{9}\right) = 0$ **[3]**

27 Use proof by induction to show that de Moivre's theorem $(\cos\theta + i\sin\theta)^n = \cos(n\theta) + i\sin(n\theta)$

holds for all integers n (Hint: consider separately the cases when n is positive or negative.) **[11]**

17 Series

Around the world, nations are increasingly moving away from fossil fuels and towards sources of renewable energy, such as wind power. This helps to minimise pollution and depletion of the world's natural resources. The move has led to rapid growth in the number of modern wind turbines that we now see on our skyline, both on land and out at sea. In some places several turbines are clustered together to form wind farms.

The technology that underpins this 'harvesting of the wind' is developed by engineers from across disciplines and, in particular, brings together the expertise of both mechanical and electrical engineers. The design of this technology relies on the mathematics of series.

Orientation

What you need to know

Ch2 Algebra and series
- The method of differences.

What you will learn

- To use partial fractions to sum series.
- To expand functions using the Maclaurin series.
- To evaluate limits using L'Hopital's rule.

What this leads to

Careers
- Mechanical engineering.
- Electrical engineering.

Fluency and skills

If the general term of a function can be expressed as $f(r+1) - f(r)$, you can find the sum of the series using the **method of differences**.

Example 1

> You can prove $\sum_{1}^{n} r \equiv \dfrac{n(n+1)}{2}$ using the formula for an arithmetic progression.

By considering the identity $2r \equiv r(r+1) - r(r-1)$, use the method of differences to prove that $\sum_{1}^{n} r \equiv \dfrac{n(n+1)}{2}$

$$2r \equiv r(r+1) - r(r-1)$$

$$\text{Hence } 2\sum_{1}^{n} r \equiv \quad 2 \qquad\qquad - \qquad 0$$
$$+ \quad 6 \qquad - \qquad 2$$
$$+ \quad 12 \qquad - \qquad 6$$
$$+ \quad (n-1)(n) \quad - \quad (n-1)(n-2)$$
$$+ \quad n(n+1) \quad - \quad n(n-1)$$

$$2\sum_{1}^{n} r \equiv n(n+1) - 0 \equiv n(n+1)$$

$$\text{So } \sum_{1}^{n} r \equiv \frac{n(n+1)}{2}$$

> Write the 'differences' for successive terms vertically and cancel terms wherever possible.

> Collect up the remaining terms and simplify.

In order to express a function as $f(r+1) - f(r)$ you may need to use partial fractions.

You can decompose some functions into their partial fractions.

Start by checking that the degree of the numerator is smaller than the degree of the denominator.

Then split into partial fractions.

Key point

$$\frac{px+q}{(x-a)(x-b)} \equiv \frac{A}{(x-a)} + \frac{B}{(x-b)}$$

You can work out the constants A, B, etc. using substitution or by comparing coefficients (or a combination of these methods).

Example 2

a Find the partial fractions of $\dfrac{1}{r(r+2)}$

b Use your answer to find $\displaystyle\sum_{1}^{n}\dfrac{1}{r(r+2)}$

a Let $\dfrac{1}{r(r+2)} \equiv \dfrac{A}{r} + \dfrac{B}{(r+2)}$

So $\qquad 1 \equiv A(r+2) + Br$

When $r=0$: $\quad 1 = 2A$ so $A = \dfrac{1}{2}$

When $r=-2$: $\quad 1 = -2B$ so $B = -\dfrac{1}{2}$

Hence $\dfrac{1}{r(r+2)} \equiv \dfrac{1}{2r} - \dfrac{1}{2(r+2)} = \dfrac{1}{2}\left(\dfrac{1}{r} - \dfrac{1}{(r+2)}\right)$

| | Multiply through by the denominator. |

| | Substitute values for r to work out A and B |

b $\displaystyle\sum_{1}^{n}\dfrac{1}{r(r+2)} \equiv \dfrac{1}{2}\sum_{1}^{n}\left(\dfrac{1}{r} - \dfrac{1}{(r+2)}\right)$

$$\equiv \dfrac{1}{2}\left[\begin{array}{ccc} 1 & - & \dfrac{1}{3} \\[2mm] + \dfrac{1}{2} & & \dfrac{1}{4} \\[2mm] + \dfrac{1}{3} & & \dfrac{1}{5} \\[2mm] \downarrow & & \downarrow \\[2mm] + \dfrac{1}{(n-2)} & & \dfrac{1}{n} \\[2mm] + \dfrac{1}{(n-1)} & & \dfrac{1}{(n+1)} \\[2mm] + \dfrac{1}{n} & - & \dfrac{1}{(n+2)} \end{array}\right]$$

Write the 'differences' for successive terms vertically and eliminate where possible.

This time the terms that cancel are two rows apart.

$$\equiv \dfrac{1}{2}\left[1 + \dfrac{1}{2} - \dfrac{1}{(n+1)} - \dfrac{1}{(n+2)}\right]$$

$$\equiv \dfrac{3}{4} - \dfrac{(n+2) + (n+1)}{2(n+1)(n+2)}$$

$$\equiv \dfrac{n(3n+5)}{4(n+1)(n+2)}$$

The expression can be factorised and written as a single fraction.

1   Express these functions as the sum of their partial fractions.

 a $\dfrac{6x+10}{(x-5)(x+5)}$

 b $\dfrac{-4x}{(x-3)(x-7)}$

 c $\dfrac{7x+56}{(1-x)(x+6)}$

 d $\dfrac{27x+46}{(x-2)(x+2)(x+3)}$

2   $f(x) \equiv \dfrac{1}{x(x-1)}$

 a   Show that $f(x) - f(x+1) \equiv \dfrac{2}{x(x-1)(x+1)}$

 b   Use the method of differences to find

   $\displaystyle\sum_{2}^{n} \dfrac{1}{x(x-1)(x+1)}$

3   Find the sum of each series.

 a $\displaystyle\sum_{1}^{n} \dfrac{1}{r+1} - \dfrac{1}{r+3}$

 b $\displaystyle\sum_{1}^{n} \dfrac{1}{2r-1} - \dfrac{1}{2r+1}$

 c $\displaystyle\sum_{0}^{n} \dfrac{1}{2r+1} - \dfrac{2}{2r+3} + \dfrac{1}{2r+5}$

4   Use partial fractions to find these sums.

 a $\displaystyle\sum_{2}^{n} \dfrac{1}{r(r-1)}$

 b $\displaystyle\sum_{1}^{n} \dfrac{1}{(r+1)(r+2)}$

 c $\displaystyle\sum_{1}^{n} \dfrac{1}{(r+1)(r+3)}$

 d $\displaystyle\sum_{2}^{n} \dfrac{1}{(r-1)(r+1)}$

Reasoning and problem-solving

Strategy

To sum a series using partial fractions and the method of differences

(1) Express the function using partial fractions.

(2) Use the method of differences to sum the series.

(3) If necessary, consider what happens for large values of n to work out the sum of an infinite number of terms.

Example 3

PURE

Express $\dfrac{16}{r(r+2)(r+4)}$ in partial fractions.

Hence show that $\displaystyle\sum_{r=1}^{\infty}\dfrac{16}{r(r+2)(r+4)}=\dfrac{11}{6}$

$$\frac{16}{r(r+2)(r+4)}\equiv\frac{A}{r}+\frac{B}{r+2}+\frac{C}{r+4}$$

$$16\equiv A(r+2)(r+4)+B(r)(r+4)+C(r)(r+2)$$

Multiply through by the denominator.

$r=0$: $16=8A$ so $A=2$

$r=-2$: $16=-4B$ so $B=-4$

Substitute values of r to eliminate factors.

$r=-4$: $16=8C$ so $C=2$

$$\text{So,}\quad \frac{16}{r(r+2)(r+4)}\equiv\frac{2}{r}-\frac{4}{r+2}+\frac{2}{r+4}$$

$$\sum_{r=1}^{n}\frac{16}{r(r+2)(r+4)}\equiv\quad \frac{2}{1}\quad-\quad\frac{4}{3}\quad+\quad\frac{2}{5}$$

$$+\quad\frac{2}{2}\quad-\quad\frac{4}{4}\quad+\quad\frac{2}{6}$$

$$+\quad\frac{2}{3}\quad-\quad\frac{4}{5}\quad+\quad\frac{2}{7}$$

$$+\quad\frac{2}{4}\quad-\quad\frac{4}{6}\quad+\quad\frac{2}{8}$$

$$+\quad\frac{2}{5}\quad-\quad\frac{4}{7}\quad+\quad\frac{2}{9}$$

$$+\quad\frac{2}{6}\quad-\quad\frac{4}{8}\quad+\quad\frac{2}{10}$$

$$\downarrow\qquad\qquad\downarrow\qquad\qquad\downarrow$$

$$+\quad\frac{2}{n-4}\quad-\quad\frac{4}{n-2}\quad+\quad\frac{2}{n}$$

$$+\quad\frac{2}{n-3}\quad-\quad\frac{4}{n-1}\quad+\quad\frac{2}{n+1}$$

$$+\quad\frac{2}{n-2}\quad-\quad\frac{4}{n}\quad+\quad\frac{2}{n+2}$$

$$+\quad\frac{2}{n-1}\quad-\quad\frac{4}{n+1}\quad+\quad\frac{2}{n+3}$$

$$+\quad\frac{2}{n}\quad-\quad\frac{4}{n+2}\quad+\quad\frac{2}{n+4}$$

Write the 'differences' for successive terms vertically and eliminate where possible.

This time the terms that cancel are several rows apart.

$$\sum_{r=1}^{n}\frac{16}{r(r+2)(r+4)}\equiv\frac{2}{1}-\frac{4}{3}+\frac{2}{2}-\frac{4}{4}+\frac{2}{3}+\frac{2}{4}+\frac{2}{n+1}+\frac{2}{n+2}-\frac{4}{n+1}$$

$$+\frac{2}{n+3}-\frac{4}{n+2}+\frac{2}{n+4}$$

$$\equiv\frac{11}{6}-\frac{2}{n+1}-\frac{2}{n+2}+\frac{2}{n+3}+\frac{2}{n+4}$$

As $n\to\infty$, each of $\dfrac{2}{n+1},\dfrac{2}{n+2},\dfrac{2}{n+3}$ and $\dfrac{2}{n+4}\to 0$

Consider what happens for large n

Hence $\displaystyle\sum_{1}^{\infty}\dfrac{16}{r(r+2)(r+4)}=\dfrac{11}{6}$

1 **a** Show that
$$r(r+1)(r+2)-(r-1)r(r+1) \equiv 3r(r+1)$$

    b Hence find $\displaystyle\sum_{1}^{n} r(r+1)$

2 **a** Find the partial fractions of $\dfrac{1}{4r^2-1}$

    b Hence find $\displaystyle\sum_{1}^{n} \dfrac{1}{4r^2-1}$

    c Find the sum to infinity of this series.

3 $f(r) = \dfrac{1}{(r+1)(r+2)}$

    a Show that $f(r)-f(r+1) \equiv \dfrac{2}{(r+1)(r+2)(r+3)}$

    b Hence find $\displaystyle\sum_{1}^{n} \dfrac{1}{(r+1)(r+2)(r+3)}$

    c Find the sum to infinity of this series.

4 **a** Simplify the expression $r^2(r+1)^2 - r^2(r-1)^2$

    b Use your answer to find a formula for $\displaystyle\sum_{1}^{n} r^3$

5 **a** Use partial fractions to find $\displaystyle\sum_{2}^{n} \dfrac{4}{x^2-1}$

    b Use your result to find the maximum value of $\displaystyle\sum_{2}^{n} \dfrac{1}{x^2-1}$ for $n \geq 2$

6 The rth term of a series, S, is $\dfrac{2r-1}{r(r+1)(r+2)}$

    Show that $\displaystyle\sum_{1}^{\infty} S = \dfrac{3}{4}$

7 **a** Show that $\dfrac{x-1}{x} - \dfrac{x-2}{x-1} \equiv \dfrac{1}{x(x-1)}$

    b Use your answer to find a formula for $\displaystyle\sum_{3}^{n} \dfrac{1}{x(x-1)}$

    c Hence evaluate $\displaystyle\sum_{3}^{\infty} \dfrac{1}{x(x-1)}$

8 Find a formula for $\displaystyle\sum_{2}^{n} \ln\left(\dfrac{r}{r-1}\right)$

9 **a** Express $\dfrac{x-3}{(x-1)(x)(x+1)}$ in its partial fractions.

    b Hence find $\displaystyle\sum_{2}^{n} \dfrac{x-3}{(x-1)(x)(x+1)}$

    c Hence show that the sum to infinity is 0

10 Use partial fractions to show that
$$\sum_{1}^{\infty} \left[\dfrac{1}{r(r+1)(r+2)(r+3)} \right] = \dfrac{1}{18}$$

11 **a** Express $\dfrac{1}{(1-x)(1+2x)}$ in its partial fractions.

    b Hence find the first four terms in the binomial expansion of $\dfrac{\sqrt{1+x}}{(1-x)(1+2x)}$

Fluency and skills

A function $f(x)$ can be expanded using the Maclaurin series given that

- $f(x)$ can be expanded as a **convergent** infinite series of terms
- each of the terms in $f(x)$ can be differentiated
- each of the differentiated terms has a finite value when $x = 0$

A convergent series is one where an infinite number of terms has a finite sum.

Key point

The Maclaurin series, or expansion, for $f(x)$ is

$$f(x) \equiv f(0) + xf'(0) + \frac{x^2}{2!}f''(0) + \frac{x^3}{3!}f'''(0) + \frac{x^4}{4!}f''''(0) + \dots + \frac{x^r}{r!}f^{(r)}(0) + \dots$$

Here are the range of values of x for which the Maclaurin series is valid for different functions.

Function	Range of x where series is valid
e^x	all values of x
$\sin x$	all values of x
$\cos x$	all values of x
$(1+x)^n$	$-1 < x < 1$ for $n \in \mathbb{R}$
$\ln(1+x)$	$-1 < x \leq 1$

You came across the series for these functions in Chapter 2, but now you should be able to derive them yourself.

Example 1

a Explain why a Maclaurin series of $f(x) = \ln(x)$ is not possible.

b Derive the Maclaurin series of $f(x) = \ln(1+x)$

c The first three terms of the series for $\tan x$ are $x + \frac{x^3}{3} + \frac{2x^5}{15}$. Use this series and your answer to part **b** to find the expansion of $\tan(2x)\ln(1+x)$ as far as the term in x^5

a The first constant $f(0) = \ln(0) = -\infty$, which is not finite.

Hence a Maclaurin series of $y = \ln(x)$ is not possible.

b $f(x) = \ln(1+x)$ so $f(0) = \ln(1) = 0$

$f'(x) = (1+x)^{-1} = \dfrac{1}{1+x}$, so $f'(0) = 1$ ← Use the chain rule to differentiate $f(x)$

$f''(x) = -(1+x)^{-2}$ so $f''(0) = -1$

$f'''(x) = 2(1+x)^{-3}$ so $f'''(0) = 2$

$f''''(x) = -6(1+x)^{-4}$ so $f''''(0) = -6 = -3!$

$f'''''(x) = 24(1+x)^{-5}$ so $f'''''(0) = 4!$ ← The pattern is now clear and can be proved, for example, by induction.

(Continued on the next page)

$$f(x) \equiv f(0) + xf'(0) + \frac{x^2}{2!}f''(0) + \frac{x^3}{3!}f'''(0) + \frac{x^4}{4!}f''''(0) + \dots$$

Use the Maclaurin expansion.

$$f(x) = \ln(1+x) \equiv x - \frac{x^2}{2!} + \frac{2x^3}{3!} - \frac{6x^4}{4!} + \frac{24x^5}{5!} - \dots$$

$$\equiv x - \frac{x^2}{2} + \frac{x^3}{3} - \frac{x^4}{4} + \frac{x^5}{5} - \dots$$

c $\tan(2x)\ln(1+x) \equiv \left(2x + \frac{(2x)^3}{3} + \frac{2(2x)^5}{15} + \dots\right) \times \left(x - \frac{x^2}{2} + \frac{x^3}{3} - \frac{x^4}{4} + \frac{x^5}{5} - \dots\right)$

$$\equiv \left(2x + \frac{8x^3}{3} + \frac{64x^5}{15} + \dots\right)\left(x - \frac{x^2}{2} + \frac{x^3}{3} - \frac{x^4}{4} + \frac{x^5}{5} - \dots\right)$$

$$\equiv 2x^2 + x^3\left[2x - \frac{1}{2}\right] + x^4\left[\frac{2}{3} + \frac{8}{3}\right] + x^5\left[-\frac{1}{2} - \frac{8}{6}\right] + \dots$$

You only need to multiply the terms as far as x^5

$$\equiv 2x^2 - x^3 + \frac{10}{3}x^4 - \frac{11}{6}x^5 + \dots$$

Example 2

Derive the Maclaurin series of $f(x) = \ln(2+x)$ by adapting the series for $\ln(1+x)$ you found in Example 1

$$\ln(1+x) \equiv x - \frac{x^2}{2} + \frac{x^3}{3} - \frac{x^4}{4} + \frac{x^5}{5} + \dots$$

$$\ln(2+x) \equiv \ln\left[2\left(1 + \frac{x}{2}\right)\right]$$

Take out a factor of 2

$$\equiv \ln 2 + \ln\left(1 + \frac{x}{2}\right)$$

Use laws of logarithms.

$$\equiv \ln 2 + \left(\frac{x}{2}\right) - \frac{\left(\frac{x}{2}\right)^2}{2} + \frac{\left(\frac{x}{2}\right)^3}{3} - \frac{\left(\frac{x}{2}\right)^4}{4} + \frac{\left(\frac{x}{2}\right)^5}{5} + \dots$$

Expand $\ln\left(1 + \frac{x}{2}\right)$ by replacing x with $\frac{x}{2}$ in the expansion of $\ln(1+x)$

$$\equiv \ln 2 + \frac{x}{2} - \frac{x^2}{8} + \frac{x^3}{24} - \frac{x^4}{64} + \frac{x^5}{160} - \dots$$

See Ch2.5

For the general terms of some well-known Maclaurin series.

General term

Sequences and series can be represented by a formula for the general term, sometimes called the rth term or nth term. For example, the formula $3r - 2$ represents the sequence of terms 1, 4, 7, 10, ...

Sometimes you will only know the first few terms of a sequence, and need to find the rth term. You do this by looking for patterns which you can write as a formula.

Example 3

Use your knowledge of these Maclaurin series to write down their general terms.

a e^{2x} **b** $\ln(1+3x)$

a $e^x = 1 + x + \dfrac{x^2}{2!} + \ldots + \dfrac{x^r}{r!} + \ldots$

So the general term of the sequence for e^{2x} is $\dfrac{(2x)^r}{r!}$ or $\dfrac{2^r x^r}{r!}$

b $\ln(1+x) = x - \dfrac{x^2}{2} + \dfrac{x^3}{3} - \ldots + (-1)^{r+1}\dfrac{x^r}{r} + \ldots$

So the rth term for $\ln(1+3x)$ is $(-1)^{r+1}\dfrac{(3x)^r}{r}$ or $(-1)^{r+1}\dfrac{3^r x^r}{r}$

Example 4

a $f(x) \equiv e^x \cos x$

  i Show that $f''(x) = -2e^x \sin x$

  ii Find $f'''(x)$ and $f''''(x)$

b Use the Maclaurin series to find the first four non-zero terms in the expansion of $f(x)$

a i $f(x) \equiv e^x \cos x$

$f'(x) = e^x(\cos x - \sin x)$ ●————————— Use the product rule to differentiate.

$f''(x) = e^x(\cos x - \sin x - \sin x - \cos x) = -2e^x \sin x$

ii $f'''(x) = -2e^x(\sin x + \cos x)$

$f''''(x) = -2e^x(\sin x + \cos x + \cos x - \sin x) = -4e^x \cos x$

b $f(x) \equiv f(0) + xf'(0) + \dfrac{x^2}{2!}f''(0) + \dfrac{x^3}{3!}f'''(0) + \dfrac{x^4}{4!}f''''(0) + \ldots$

$\equiv e^0 \cos 0 + xe^0(\cos 0 - \sin 0) + \dfrac{x^2}{2!}(-2\,e^0 \sin 0)$ ●————— Substitute values for f(0), f'(0), etc.

$+ \dfrac{x^3}{3!}[-2e^0(\sin 0 + \cos 0)] + \dfrac{x^4}{4!}(-4e^0 \cos 0) + \ldots$

$f(x) \equiv 1 + x(1) + \dfrac{x^2}{2!}(0) + \dfrac{x^3}{3!}(-2) + \dfrac{x^4}{4!}(-4) + \ldots$

$\equiv 1 + x - \dfrac{x^3}{3} - \dfrac{x^4}{6} + \ldots$

Limits

A limiting value, or limit, is a specific value that a function approaches or tends towards as the variable approaches a particular value. This idea is used when differentiating from first principles.

The derivative of a function $f(x)$ is defined as $f'(x) = \lim\limits_{h \to 0} \dfrac{f(x+h)-f(x)}{h}$

It gives the gradient of the curve at any value of x

In this case $f'(x)$ is the limit as h tends to zero, but you can also find a limit approaching other numbers, for example, 0, 1 or $\dfrac{\pi}{2}$

You can sometimes easily work out the limit of a function

$$\lim_{n \to 0}(2+e^n) = 2 + e^0 = 2 + 1 = 3 \qquad \lim_{n \to \frac{\pi}{2}}(1-\cos n) = 1 - \cos\frac{\pi}{2} = 1 - 0 = 1$$

In other cases, you may need to manipulate the function to find the limit. For example, when finding $\lim\limits_{n \to \infty} \dfrac{n+10}{n}$ you must manipulate the expression before you find the limit

$$\lim_{n \to \infty} \frac{n+10}{n} = \lim_{n \to \infty}\frac{n}{n}+\frac{10}{n} = \lim_{n \to \infty}\frac{n}{n}+\frac{10}{n} = 1 + 0 = 1$$

Find these limits.

a $\lim\limits_{n \to \infty} \dfrac{2n+5}{n}$ **b** $\lim\limits_{n \to \infty} \dfrac{2n^2+5}{n}$ **c** $\lim\limits_{n \to \infty} \dfrac{2n+5}{n^2}$ **d** $\lim\limits_{n \to \infty}\left(\dfrac{2n^2+n-35}{5n^2-3n+7}\right)$

a $\lim\limits_{n \to \infty} \dfrac{2n+5}{n} = \lim\limits_{n \to \infty} 2+\dfrac{5}{n}$

> Divide through by n before finding the limit.

As $n \to \infty$, $\dfrac{5}{n} \to 0$

So $\lim\limits_{n \to \infty} \dfrac{2n+5}{n} = 2$

b $\lim\limits_{n \to \infty} \dfrac{2n^2+5}{n} = \lim\limits_{n \to \infty} 2n+\dfrac{5}{n} = \infty$

c $\lim\limits_{n \to \infty} \dfrac{2n+5}{n^2} = \lim\limits_{n \to \infty} \dfrac{2}{n}+\dfrac{5}{n^2} = 0$

d $\lim\limits_{n \to \infty}\left(\dfrac{2n^2+n-35}{5n^2-3n+7}\right) = \lim\limits_{n \to \infty}\left(\dfrac{2+\dfrac{1}{n}-\dfrac{35}{n^2}}{5-\dfrac{3}{n}+\dfrac{7}{n^2}}\right) = \dfrac{2}{5}$

> You should never say $\dfrac{5}{n} = 0$ if $n = \infty$, or $\dfrac{1}{\infty} = 0$, but say '$\dfrac{5}{n}$ tends to 0 as n tends to ∞'. This is because ∞ is not a number. 'n tends to ∞' is just a mathematical way of saying that there is no limit to the size that n can take.

> Divide each term of the numerator and denominator by n^2 before finding the limit.

L'Hopital's rule

Sometimes when you investigate a limit you are not able to evaluate it in the same way as you did in Example 5

For example $\lim\limits_{x \to 0}\left(\dfrac{\sin x}{x}\right)$ gives you $\dfrac{0}{0}$

The expression $\dfrac{\sin x}{x}$ is known as an **indeterminate form** when $x = 0$ and hence the limit is indeterminate.

Indeterminate forms occur when, for an expression of the form $\dfrac{f(x)}{g(x)}$, in the limit as $x \to a$,

$$\lim_{x \to a} f(x) = \lim_{x \to a} g(x) = 0$$

or

$$\lim_{x \to a} f(x) = \pm\infty \quad \text{and} \quad \lim_{x \to a} g(x) = \pm\infty$$

The French mathematician, **l'Hopital**, published a rule which allows you to evaluate limits like this.

His rule says that when the limit of $\dfrac{f(x)}{g(x)}$ is indeterminate, it can usually be found by evaluating the limit of $\dfrac{f'(x)}{g'(x)}$

So $\quad \lim_{x \to 0}\left(\dfrac{\sin x}{x}\right) = \dfrac{0}{0} = \lim_{x \to 0}\left(\dfrac{\cos x}{1}\right) = \dfrac{1}{1} = 1$

> **Key point**
>
> L'Hopital's rule states that
>
> If $\quad \lim\limits_{x \to a} \dfrac{f(x)}{g(x)}$ is indeterminate, where a is any real number or $\pm\infty$,
>
> then $\quad \lim\limits_{x \to a} \dfrac{f(x)}{g(x)} = \lim\limits_{x \to a} \dfrac{f'(x)}{g'(x)}$
>
> This method can be repeated as many times as necessary if $\lim\limits_{x \to a} \dfrac{f'(x)}{g'(x)}$, etc. is still indeterminate.

Example 6

Use l'Hopital's rule to evaluate these limits.

a $\lim\limits_{x \to 1} \dfrac{x^3 - 1}{x^2 - 1}$ **b** $\lim\limits_{x \to 0} \dfrac{(2 - 2\cos x)}{x^2}$

a $\lim\limits_{x \to 1} \dfrac{x^3 - 1}{x^2 - 1} = \dfrac{0}{0}$

$\lim\limits_{x \to 1} \dfrac{f'(x)}{g'(x)} = \lim\limits_{x \to 1} \dfrac{3x^2}{2x} = \lim\limits_{x \to 1} \dfrac{3x}{2} = \dfrac{3 \times 1}{2} = \dfrac{3}{2}$ Apply l'Hopital's rule.

Hence $\lim\limits_{x \to 1} \dfrac{x^3 - 1}{x^2 - 1} = \dfrac{3}{2}$

b $\lim\limits_{x \to 0} \dfrac{(2 - 2\cos x)}{x^2} = \dfrac{0}{0}$

$\lim\limits_{x \to 0} \dfrac{f'(x)}{g'(x)} = \lim\limits_{x \to 0} \dfrac{2\sin x}{2x} = \lim\limits_{x \to 0} \dfrac{\sin x}{x} = \lim\limits_{x \to 0} \dfrac{\sin 0}{0} = \dfrac{0}{0}$

$\lim\limits_{x \to 0} \dfrac{f''(x)}{g''(x)} = \lim\limits_{x \to 0} \dfrac{\cos x}{1} = \lim\limits_{x \to 0} \dfrac{\cos 0}{1} = 1$ Sometimes you have to apply l'Hopital's rule more than once.

Hence $\lim\limits_{x \to 0} \dfrac{(2 - 2\cos x)}{x^2} = 1$

1   Use your knowledge of these Maclaurin series to write down their general terms.

   a e^{-x}

   b $\ln(1-2x)$

   c $\sin\left(\dfrac{x}{5}\right)$

   d $(1+3x)^n$

2   Use the Maclaurin expansion to derive a series for $\sin x$. Include the general term.

3   Use the Maclaurin expansion to derive a series for $\cos x$. Include the general term.

4   Find the first three non-zero terms in the Maclaurin expansion of $f(x) = \ln\left(\dfrac{1-2x}{1+2x}\right)$

5   Derive the first three non-zero terms in the Maclaurin expansion of $f(x) = \cos^2 x$

6   Derive the first three non-zero terms in the Maclaurin expansion of $f(x) = \sqrt{1+3x^2}$

7   Find the first three non-zero terms in the Maclaurin expansion of $f(x) = e^{2x}\sin x$

8   Find the first three non-zero terms in the Maclaurin expansion of $f(x) = e^x\ln(1+x)$

9   Evaluate these limits.

   a $\lim\limits_{x\to\infty}(1+e^{-x})$

   b $\lim\limits_{x\to1}\dfrac{3x+5}{x^2-9}$

   c $\lim\limits_{x\to0}\dfrac{3\cos x}{\sin x+\cos x}$

   d $\lim\limits_{x\to\infty}\dfrac{x^2+2x+3}{2x^2-5}$

   e $\lim\limits_{x\to0}\dfrac{x^2+1}{3x^2+2x-4}$

   f $\lim\limits_{x\to\infty}\dfrac{x^2+1}{3x^2+2x-4}$

   g $\lim\limits_{x\to\infty}\dfrac{3x^2+8}{2-3x^2}$

   h $\lim\limits_{x\to-1}\dfrac{3x^2+8}{2-3x^2}$

10 Evaluate these limits.

   a $\lim\limits_{x\to2}\sqrt{\dfrac{x^3-2x^2+135}{9x^3-12}}$

   b $\lim\limits_{x\to\infty}\sqrt{\dfrac{x^3-2x^2+1}{9x^3+4}}$

   c $\lim\limits_{x\to1}\dfrac{3x+2}{x^2-9}$

   d $\lim\limits_{x\to0}\dfrac{x+2}{\sqrt{2x^2-3x+1}}$

   e $\lim\limits_{x\to\infty}\dfrac{x+2}{\sqrt{2x^2-3x+1}}$

11 Use l'Hopital's method to evaluate these limits.

   a $\lim\limits_{x\to0}\dfrac{\sin^{-1}x}{x}$

   b $\lim\limits_{x\to0}\dfrac{\tan^{-1}x}{x}$

   c $\lim\limits_{x\to\infty}\dfrac{\ln(1+2x)}{x}$

   d $\lim\limits_{x\to\infty}\dfrac{e^{2x}+1}{e^{2x}-1}$

   e $\lim\limits_{x\to0}\dfrac{x}{x+1-\sqrt{x+1}}$

   f $\lim\limits_{x\to0}\dfrac{1-\cos3x+2x\sin5x}{6x^2}$

   g $\lim\limits_{x\to0}\dfrac{e^x\sin5x}{\ln(2x+1)}$

Reasoning and problem-solving

When you investigate a limit and end up with an indeterminate value such as $\dfrac{0}{0}$ or $\dfrac{\infty}{\infty}$ you can use Maclaurin expansions to evaluate the limit as an alternative to using l'Hopital's rule.

To solve problems involving Maclaurin series and limits

(1) Derive the Maclaurin series from first principles or adapt a standard Maclaurin series.

(2) Substitute the series into the expression.

(3) Use the Maclaurin series or l'Hopital's rule to work out the limit of the function.

Example 7

$$f(x) \equiv \frac{4x^4}{3\ln(1+x^2)-3x^2}$$

Adapt the known Maclaurin series for $\ln(1+x)$ in $f(x)$ to calculate $\lim_{x\to 0} f(x)$

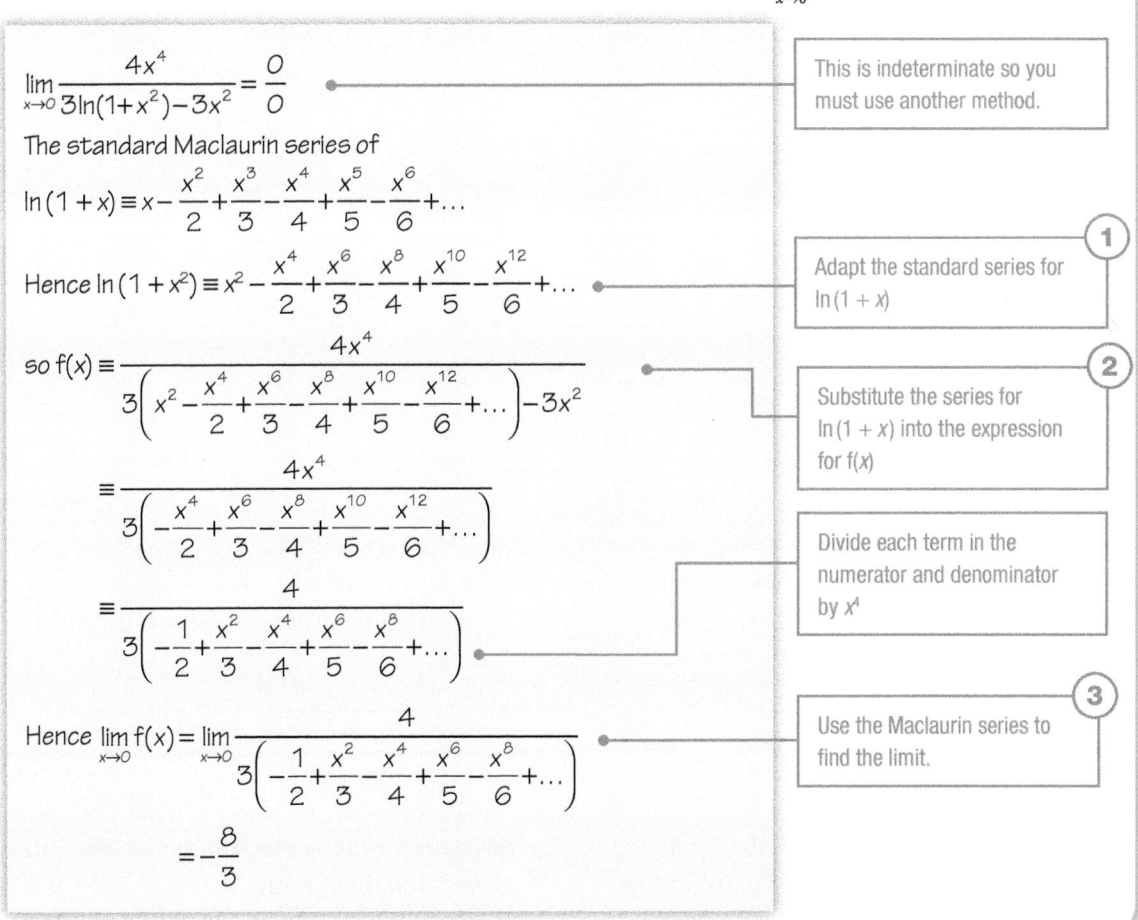

$$\lim_{x\to 0}\frac{4x^4}{3\ln(1+x^2)-3x^2}=\frac{0}{0}$$

This is indeterminate so you must use another method.

The standard Maclaurin series of

$$\ln(1+x)\equiv x-\frac{x^2}{2}+\frac{x^3}{3}-\frac{x^4}{4}+\frac{x^5}{5}-\frac{x^6}{6}+\ldots$$

Hence $\ln(1+x^2)\equiv x^2-\dfrac{x^4}{2}+\dfrac{x^6}{3}-\dfrac{x^8}{4}+\dfrac{x^{10}}{5}-\dfrac{x^{12}}{6}+\ldots$

(1) Adapt the standard series for $\ln(1+x)$

so $f(x)\equiv\dfrac{4x^4}{3\left(x^2-\dfrac{x^4}{2}+\dfrac{x^6}{3}-\dfrac{x^8}{4}+\dfrac{x^{10}}{5}-\dfrac{x^{12}}{6}+\ldots\right)-3x^2}$

(2) Substitute the series for $\ln(1+x)$ into the expression for $f(x)$

$$\equiv\dfrac{4x^4}{3\left(-\dfrac{x^4}{2}+\dfrac{x^6}{3}-\dfrac{x^8}{4}+\dfrac{x^{10}}{5}-\dfrac{x^{12}}{6}+\ldots\right)}$$

Divide each term in the numerator and denominator by x^4

$$\equiv\dfrac{4}{3\left(-\dfrac{1}{2}+\dfrac{x^2}{3}-\dfrac{x^4}{4}+\dfrac{x^6}{5}-\dfrac{x^8}{6}+\ldots\right)}$$

Hence $\lim_{x\to 0}f(x)=\lim_{x\to 0}\dfrac{4}{3\left(-\dfrac{1}{2}+\dfrac{x^2}{3}-\dfrac{x^4}{4}+\dfrac{x^6}{5}-\dfrac{x^8}{6}+\ldots\right)}$

(3) Use the Maclaurin series to find the limit.

$$=-\dfrac{8}{3}$$

Example 8

$$f(x) = \frac{e^{2x} - e^x}{x}$$

a Find the first four terms and the general term of the expansion of $f(x)$

b Use your expansion to find $\lim_{x \to 0} f(x)$

c Use l'Hopital's rule to confirm your answer to part **b**.

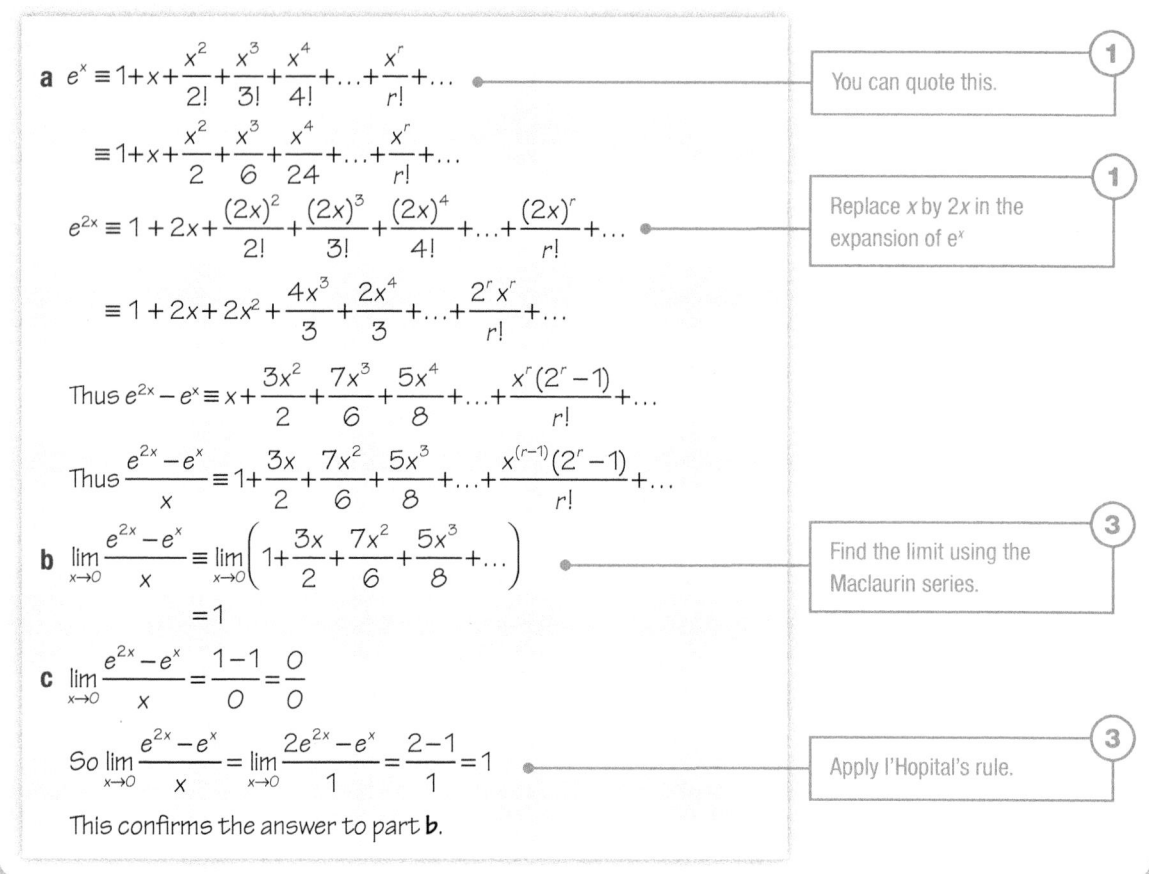

a $e^x \equiv 1 + x + \dfrac{x^2}{2!} + \dfrac{x^3}{3!} + \dfrac{x^4}{4!} + \ldots + \dfrac{x^r}{r!} + \ldots$ ● You can quote this. ①

$\equiv 1 + x + \dfrac{x^2}{2} + \dfrac{x^3}{6} + \dfrac{x^4}{24} + \ldots + \dfrac{x^r}{r!} + \ldots$

$e^{2x} \equiv 1 + 2x + \dfrac{(2x)^2}{2!} + \dfrac{(2x)^3}{3!} + \dfrac{(2x)^4}{4!} + \ldots + \dfrac{(2x)^r}{r!} + \ldots$ ● Replace x by $2x$ in the expansion of e^x ①

$\equiv 1 + 2x + 2x^2 + \dfrac{4x^3}{3} + \dfrac{2x^4}{3} + \ldots + \dfrac{2^r x^r}{r!} + \ldots$

Thus $e^{2x} - e^x \equiv x + \dfrac{3x^2}{2} + \dfrac{7x^3}{6} + \dfrac{5x^4}{8} + \ldots + \dfrac{x^r(2^r - 1)}{r!} + \ldots$

Thus $\dfrac{e^{2x} - e^x}{x} \equiv 1 + \dfrac{3x}{2} + \dfrac{7x^2}{6} + \dfrac{5x^3}{8} + \ldots + \dfrac{x^{(r-1)}(2^r - 1)}{r!} + \ldots$

b $\lim_{x \to 0} \dfrac{e^{2x} - e^x}{x} \equiv \lim_{x \to 0}\left(1 + \dfrac{3x}{2} + \dfrac{7x^2}{6} + \dfrac{5x^3}{8} + \ldots\right)$ ● Find the limit using the Maclaurin series. ③

$= 1$

c $\lim_{x \to 0} \dfrac{e^{2x} - e^x}{x} = \dfrac{1-1}{0} = \dfrac{0}{0}$

So $\lim_{x \to 0} \dfrac{e^{2x} - e^x}{x} = \lim_{x \to 0} \dfrac{2e^{2x} - e^x}{1} = \dfrac{2-1}{1} = 1$ ● Apply l'Hopital's rule. ③

This confirms the answer to part **b**.

Exercise 17.2B Reasoning and problem-solving

1 a Use the binomial theorem to expand
$$\frac{4}{1+4x}$$

b $\displaystyle\int_0^x \frac{4}{1+4y}\, dy = \ln(1+4x)$

Use this and your answer to part **a** to obtain the first four terms in the series of $\ln(1+4x)$

c Check your solution by using the Maclaurin series for $\ln(1+4x)$

2 Use series expansions to determine these limits.

a $\displaystyle\lim_{x \to 0} \frac{(1+2x)^{-3} - 1}{x}$

b $\displaystyle\lim_{x \to 0} \frac{x}{\sin 2x}$

c $\displaystyle\lim_{x \to 0} \frac{1 - \cos 4x}{x^2}$

d $\displaystyle\lim_{x \to 0} \frac{x \ln(1-x)}{e^{x^2} - 1}$

3 Use series expansions where necessary to determine these limits.

a $\displaystyle\lim_{x \to \infty}\left(x - \sqrt{x^2 - 4x}\right)$

b $\displaystyle\lim_{x \to \infty}\left(\sqrt[3]{(x^3 - 2)} - x\right)$

c $2\lim\limits_{x\to\infty}\ln\left(\dfrac{x^4+3}{x^4+2}\right)$

d $\lim\limits_{x\to 0}\dfrac{e^{2x}-3}{4e^{2x}+6}$

e $\lim\limits_{x\to\infty}\dfrac{e^{2x}-3}{4e^{2x}+6}$

4 a Find the first four non-zero terms in the expansion of $\dfrac{\ln(1+x)}{1-x}$

b i Find the first four non-zero terms in the expansion of $\dfrac{\sqrt{(2-x)}}{1+2x}$

ii Write down $\lim\limits_{x\to 0}\dfrac{\sqrt{(2-x)}}{1+2x}$ and confirm the result using your answer from part **i**.

5 a Find the partial fractions of $\dfrac{3(1-2x)}{(x+2)(1+x)}$

b Hence expand $\dfrac{3(1-2x)}{(x+2)(1+x)}$ as a series as far as the term in x^3.

c Write down $\lim\limits_{x\to 0}\left(\dfrac{3(1-2x)}{(x+2)(1+x)}\right)$ and confirm the result using your answer from part **b**.

6 a Use standard Maclaurin expansions to find $\lim\limits_{x\to 0}\dfrac{e^x-e^{2x}}{x}$

b Confirm your answer by evaluating $\lim\limits_{x\to 0}\dfrac{e^x-e^{2x}}{x}$ using l'Hopital's rule.

7 a Use standard Maclaurin expansions to find the first three non-zero terms in the expansion of $(2-x)e^{(2-x)}$ and hence find $\lim\limits_{x\to 0}(2-x)e^{(2-x)}$

b Without using your answer to part **a** write down $\lim\limits_{x\to 0}(2-x)e^{(2-x)}$

8 a Find the first three non-zero terms in the expansion of $\dfrac{x}{e^x-1}$. Hence find $\lim\limits_{x\to 0}\dfrac{x}{e^x-1}$

b Use l'Hopital's rule to confirm your answer.

9 a Write down the first three non-zero terms of the expansion, in ascending powers of x, of $1+e^{-x}$

b Find the first two non-zero terms in the expansion, in ascending powers of x, of $\ln\left(\dfrac{1+e^{-x}}{2-3x}\right)$

c Find $\lim\limits_{x\to 0}\left(\dfrac{\ln\left(\dfrac{1+e^{-x}}{2-3x}\right)}{4x}\right)$

10 a Write down the first three non-zero terms in the expansions of e^{x^2} and $\sin 2x$

b Find the expansion of $\ln\left(\dfrac{\sin 2x}{2x}\right)$ as far as the term in x^4

c Evaluate $\lim\limits_{x\to 0}\dfrac{\ln\left(\dfrac{\sin 2x}{2x}\right)}{(e^{x^2}-1)}$

11 Make use of known series expansions to obtain the expansion of $\dfrac{1}{2}(e^x-e^{-x})$ up to the term in x^5

Hence evaluate $\lim\limits_{x\to 0}\dfrac{6x}{(e^x-e^{-x})}$

12 a Use Maclaurin series to expand $\ln\left(\dfrac{2+x}{1-x}\right)$ up to the term in x^3

b Write down $\lim\limits_{x\to 0}\left(\ln\left(\dfrac{2+x}{1-x}\right)\right)$ and confirm the result using your answer from part **a**.

13 a Show that
$$\ln(\cos x)\equiv -\dfrac{x^2}{2}-\dfrac{x^4}{12}-\dfrac{x^6}{45}+\ldots$$

b Hence find $\lim\limits_{x\to 0}\dfrac{\ln(\cos x)+x^2}{x^2}$

14 Evaluate $\lim\limits_{x\to\infty}(x-\sqrt{x^2-3})$

[Hint: $A^2-B^2\equiv (A-B)(A+B)$]

Chapter summary

- If the general term of a function can be expressed as $f(r+1) - f(r)$, you can find the sum of the series using the method of differences.
- To express a function as $f(r+1) - f(r)$ you may need to use partial fractions.
- To express a function in partial fractions, check that the degree of the numerator is at least one lower than the denominator and then split it up using $\dfrac{px+q}{(x-a)(x-b)} \equiv \dfrac{A}{(x-a)} + \dfrac{B}{(x-b)}$
- You can use substitution, comparing coefficients, or a mixture of the two, to work out the constants A, B, etc.
- The Maclaurin series, or expansion, for a function $f(x)$ is
$$f(x) \equiv f(0) + xf'(0) + \frac{x^2}{2!}f''(0) + \frac{x^3}{3!}f'''(0) + \frac{x^4}{4!}f''''(0) + \ldots + \frac{x^r}{r!}f^{(r)}(0) + \ldots$$
- Maclaurin series are valid for specific ranges of x

Function	Range of x where series is valid
e^x	all values of x
$\sin x$	all values of x
$\cos x$	all values of x
$(1+x)^n$	$-1 < x < 1$ for $n \in \mathbb{R}$
$\ln(1+x)$	$-1 < x \le 1$

- A limiting value, or limit, is a specific value that a function approaches or tends towards as the variable approaches a particular value.
- L'Hopital's rule states that if $\lim\limits_{x \to a} \dfrac{f(x)}{g(x)}$ is indeterminate, then $\lim\limits_{x \to a} \dfrac{f(x)}{g(x)} = \lim\limits_{x \to a} \dfrac{f'(x)}{g'(x)}$

  (where a is any real number, including 0 or $\pm\infty$).
- L'Hopital's rule can be applied to a function repeatedly.
- You can also use Maclaurin expansions to evaluate limits.

Check and review

You should now be able to...	Review Questions
✔ Express a function in terms of its partial fractions.	1, 2
✔ Find the sum of a series using the method of differences.	2, 3
✔ Derive the Maclaurin series for a function and find its range of validity.	4–8
✔ Find the limit of a function.	9
✔ Use l'Hopital's rule.	9

1 Express $\dfrac{7x-41}{(2x+9)(3x-1)}$ in its partial fractions.

2 a Express $\dfrac{4}{x(x+4)}$ in its partial fractions.

   b Hence find $\displaystyle\sum_{1}^{n}\dfrac{4}{x(x+4)}$ and find the sum to infinity.

3 a Show that

$$\dfrac{1}{r^2}-\dfrac{2}{(r+1)^2}+\dfrac{1}{(r+2)^2}\equiv\dfrac{2(3r^2+6r+2)}{r^2(r+1)^2(r+2)^2}$$

   b Use the method of differences to find $\displaystyle\sum_{1}^{n}\dfrac{2(3r^2+6r+2)}{r^2(r+1)^2(r+2)^2}$

   c Hence find the sum to infinity.

4 Use known series expansions to find the first three non-zero terms of the Maclaurin series for $e^x\ln(1+2x)$

5 Use your knowledge of standard Maclaurin series to write down the general terms in the expansion of these series.

   a $e^{\frac{x}{3}}$ **b** $\ln(1-x^2)$ **c** $\cos\left(\dfrac{x}{3}\right)$

   d $\sin(4x+5)$ **e** $\left(1-\dfrac{x}{6}\right)^n$ **f** xe^x

6 Write down the range of values of x for which these series are valid.

   a $(2-x^2)^{-1}$ **b** $\ln\left(1+\left(\dfrac{x^2}{4}\right)\right)$

7 Write down the range of values of x for which these series are valid.

   a $\left(2+\dfrac{x}{3}\right)^{-4}$ **b** $\ln(1-3x)$ **c** $\left(2+\dfrac{x}{3}\right)^{-4}(\ln(1-3x))$

8 Use differentiation and the Maclaurin expansion to find the first three non-zero terms in the expansions of these functions.

   a $e^{2x}\sin x$ **b** $\dfrac{4}{(1+2x)}$ **c** $\dfrac{4}{\ln(2+x)}$

   d $\cos 2x-\sin 2x$ **e** 3^x **f** $\dfrac{2\cos x}{5+\sin x}$

9 Use l'Hopital's rule to evaluate these limits.

   a $\displaystyle\lim_{n\to\infty}\dfrac{4n}{3n+1}$ **b** $\displaystyle\lim_{n\to\infty}\dfrac{n^2+3}{n}$ **c** $\displaystyle\lim_{n\to\infty}\dfrac{(2n+1)^2}{(n-2)^2}$

   d $\displaystyle\lim_{x\to0}\dfrac{e^x}{4x}$ **e** $\displaystyle\lim_{x\to0}\dfrac{x^5+8}{2\sin x}$ **f** $\displaystyle\lim_{n\to0}\dfrac{(8n-4)^2}{3n^2}$

   g $\displaystyle\lim_{x\to1}\dfrac{x^2-1}{x-1}$

Investigate

The Maclaurin series of some even functions.
The Maclaurin series of some odd functions.
Can you explain what you find?

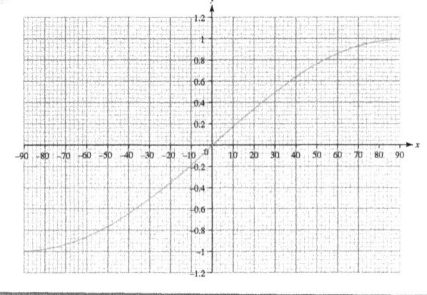

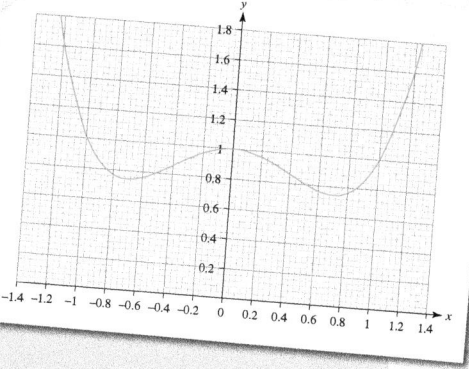

Investigation

The formula for the amount, £P, in a bank account with compound interest for an initial deposit £P_0 is

$$p = p_0 \left(1 + \frac{r}{n} \right)^{nt}$$

where n is the number of times per year that compound interest is added, r is the rate of interest in that period and t is the number of years.

For continuous compounding, $n \to \infty$, so

$$p = \lim_{n \to \infty} p_0 \left(1 + \frac{r}{n} \right)^{nt}$$

To evaluate this it is best to proceed by taking natural logarithms of both sides. L'Hopital's rule will be helpful when finding the limit of the resulting expression. Try this to see if you can arrive at the formula used for continuous compounding:

$$p = p_0 e^{rt}$$

Use this to explore the effect of continuous compounding as opposed to adding compound interest over discrete periods during a year.

Research

Use Maclaurin series to explore the validity of trig identities such as $2 \sin x \cos x = \sin 2x$
Research further identities such as

$$\sin 3x = 3 \sin x - 4 \sin^3 x$$
$$\cos 3x = 4\cos^3 x - 3\cos x$$

17 Assessment

1 a Find an expression in terms of n for $\sum_{r=1}^{n} \dfrac{1}{(r+1)(r+2)}$ [5]

  b Hence, find the value of k for which $\sum_{r=11}^{k} \dfrac{1}{(r+1)(r+2)} = \dfrac{5}{78}$ [3]

2 Use the method of differences to prove that $\sum_{r=1}^{n} r^2 = \dfrac{n}{6}(n+1)(2n+1)$ [6]

3 a Give the first four terms of the expansion of $\sin(2x^2)$. Simplify each term to the form Ax^n [4]

  b State the range of values of x for which the expansion is valid. [1]

  c Use your expansion to find the limit of $\dfrac{x^2}{\sin(2x^2)}$ as $x \to 0$ [3]

4 a Give the first four terms of the expansion of $\ln(1-5x)$. Simplify each term to the form Ax^n [4]

  b State the range of values of x for which the expansion is valid. [1]

  c Use your expansion from part **a** to find the limit of $\dfrac{x}{\ln(1-5x)}$ as $x \to 0$ [3]

5 a Find the expansion of $e^x \sin x$ up to the term in x^5 [4]

  b Use your expansion to estimate the value of $e^{0.3} \sin(0.3)$ [2]

6 a Find an expression in terms of n for $\sum_{r=1}^{n} \dfrac{3}{r(r+1)}$ [5]

  b Hence, show that $\sum_{r=n}^{2n} \dfrac{3}{r(r+1)} = \dfrac{3(n+1)}{n(2n+1)}$ [3]

  c Evaluate $\dfrac{1}{2} + \dfrac{1}{6} + \dfrac{1}{12} + \dfrac{1}{20} + \ldots + \dfrac{1}{930}$ [3]

7 a Use the method of differences to prove that $\sum_{r=2}^{n} \dfrac{1}{r^2-1} = \dfrac{(3n+2)(n-1)}{4n(n+1)}$ [6]

  b Show that $\sum_{r=2}^{\infty} \dfrac{1}{r^2-1} = \dfrac{3}{4}$ [3]

8 a Use Maclaurin's theorem to show that $\cos(3x) = 1 - \dfrac{9}{2}x^2 + \dfrac{27}{8}x^4 - \ldots$ [4]

  b Give the rule for the general term in the expansion of $\cos(3x)$ [2]

9 a Find the expansion of $\sqrt{1-3x}$ up to the term in x^3 [3]

  b Hence find the limit of $\dfrac{2x}{1-\sqrt{1-3x}}$ as $x \to 0$ [3]

10 Find the Maclaurin expansion of $\sec x$ up to the term in x^3 [5]

11 Use Maclaurin's theorem to find the expansion of $e^{\sin 2x}$ up to the term in x^2 [5]

12 **a** Find $\lim\limits_{x\to 0} \dfrac{2x}{e^{3x}-1}$ [3]

   b Explain why l'Hopital's rule can be used in this case. [2]

13 Use l'Hopital's rule to evaluate

   a $\lim\limits_{x\to 0} \dfrac{\sin(x^2)}{3x^2}$ [3]

   b $\lim\limits_{x\to 1} \dfrac{\ln(x^2)}{x-1}$ [3]

14 **a** Show that $\dfrac{1}{r!}-\dfrac{1}{(r+1)!}=\dfrac{r}{(r+1)!}$ [3]

   b Hence use the method of differences to show that $\sum\limits_{r=1}^{n}\dfrac{r}{(r+1)!}=1-\dfrac{1}{(n+1)!}$ [3]

   c Find an expression for $\dfrac{n}{(n+1)!}+\dfrac{n+1}{(n+2)!}+...+\dfrac{2n}{(2n+1)!}$, expressing your answer as a single fraction. [3]

15 **a** Express $\dfrac{2}{r(r+1)(r+2)}$ in partial fractions. [4]

   b Hence use the method of differences to show that $\sum\limits_{r=1}^{n}\dfrac{1}{r(r+1)(r+2)}=\dfrac{n(n+3)}{4(n+1)(n+2)}$ [6]

   c Find $\sum\limits_{r=1}^{\infty}\dfrac{1}{r(r+1)(r+2)}$ [3]

16 Use l'Hopital's rule to evaluate the $\lim\limits_{x\to 0}\dfrac{3-3\cos x}{2x^2}$

   You must explain why l'Hopital's rule is applicable. [5]

17 Use l'Hopital's rule to evaluate $\lim\limits_{x\to 0} x^2\ln(x^2)$

   You must explain why l'Hopital's rule is applicable. [4]

18 Use l'Hopital's rule to evaluate $\lim\limits_{x\to 0}\dfrac{x-x^3-\sin x}{x^3}$ [4]

19 Use Maclaurin's series to find $\lim\limits_{x\to 0}\left(\dfrac{x-3\sin x}{2e^x-3x^2-2}\right)$ [5]

20 Use partial fractions and the method of differences to show that $\sum\limits_{r=1}^{\infty}\dfrac{1}{r(r+1)(r+2)}=\dfrac{1}{4}$ [7]

21 Use partial fractions and the method of differences to find the sum to infinity of the sequence given by $u_n=\dfrac{1}{n^2+9n+20}$ [7]

18 Curve sketching 2

Transforming points or functions on a Cartesian plane is a very useful mathematical idea which is applied in a wide range of occupations, including engineering, operational research and computer animation. This allows users to build on a 'base model' to develop more and varied models without always having to always start from scratch.

Operational researchers use such techniques in their attempts to model complex situations and scenarios to optimise, for example, profits for a company importing and selling high street goods. In their work, as they explore the impact of varying a number of key parameters, ideas of transformations of functions have an important role to play. Although such techniques have been understood for decades, the use of computer technology has allowed such explorations to be carried out at high speed on an hour-by-hour basis.

Orientation

What you need to know	What you will learn	What this leads to
Ch3 Curve sketching 1	• To solve problems involving reciprocal and modulus functions. • To transform graphs of conic sections. • To sketch graphs of inverse hyperbolic functions. • To sketch graphs of reciprocal hyperbolic functions. • To work with graphs with oblique asymptotes.	**Careers** • Operational research. • Engineering. • Computer animation.

Fluency and skills

The graph of $y = \dfrac{1}{f(x)}$ is related to the graph of $f(x)$ in the following ways.

- $\dfrac{1}{f(x)}$ is undefined when $f(x) = 0$, so any roots (or zeros) of

  $y = f(x)$ become vertical asymptotes in $y = \dfrac{1}{f(x)}$

> The values of x where $f(x)$ crosses the x-axis are called the roots or zeros of the function.

- As $f(x) \to 0$ from above $\dfrac{1}{f(x)} \to +\infty$

  as $f(x) \to 0$ from below $\dfrac{1}{f(x)} \to -\infty$

- The sign of $\dfrac{1}{f(x)}$ is the same as the sign of $f(x)$

- As $f(x)$ increases, $\dfrac{1}{f(x)}$ decreases and vice versa

- If $f(x) = \pm 1$ exists, then $\dfrac{1}{f(x)}$ also equals ± 1 and the graphs intersect

- Minimum points become maximum points and vice-versa (except for where these points are zero where they become asymptotes)

- If $f(x) > 1$, then $0 < \dfrac{1}{f(x)} < 1$

  and if $f(x) < -1$, then $-1 < \dfrac{1}{f(x)} < 0$

- If $0 < f(x) < 1$, then $\dfrac{1}{f(x)} > 1$

  and if $-1 < f(x) < 0$, then $\dfrac{1}{f(x)} < -1$

Example 1

$f(x) = 2x^2 + x - 1$

a Sketch the graph of $y = f(x)$

b On the same axes, sketch $y = \dfrac{1}{f(x)}$

a $y = 2x^2 + x - 1$ is a quadratic curve with intercepts $(0, -1)$,

$(-1, 0)$ and $\left(\dfrac{1}{2}, 0\right)$

$y = 2\left(x + \dfrac{1}{4}\right)^2 - \dfrac{9}{8}$

> Completing the square shows there is a minimum point at $\left(-\dfrac{1}{4}, -\dfrac{9}{8}\right)$

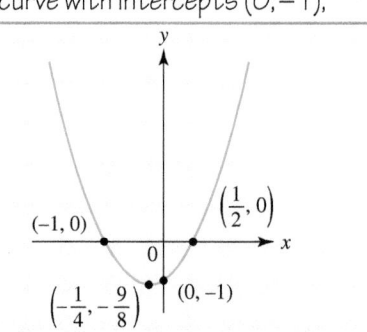

(Continued on the next page)

b f(x) has roots at $x=-1$ and $x=0.5$ so $\dfrac{1}{f(x)}$ is undefined at these values.

Hence $x=-1$ and $x=0.5$ are asymptotes for $\dfrac{1}{f(x)}$

To find the points of intersection

$f(x) = 1$

$y = 2x^2 + x - 1 = 1$

$2x^2 + x - 2 = 0$

$x = \dfrac{-1 \pm \sqrt{17}}{4} \approx 0.78$ and -1.28

$f(x) = -1$

$y = 2x^2 + x - 1 = -1$

$2x^2 + x = 0$

$x = 0$ or $x = -0.5$

So $y = f(x)$ and $y = \dfrac{1}{f(x)}$ intersect at $(0, -1)$, $(-0.5, -1)$,

$(0.78, 1)$ and $(-1.28, 1)$

To find the signs and gradient

f(x) has a minimum point at $x = -0.25$, so $x = -0.25$ becomes a maximum point of $\dfrac{1}{f(x)}$

$y = f(x)$ is increasing for $x > -0.25$ and decreasing for $x < -0.25$

So $y = \dfrac{1}{f(x)}$ decreases for $x > -0.25$ and increases for $x < -0.25$

f(x) and $\dfrac{1}{f(x)}$ are both negative for $-1 < x < 0.5$ and positive elsewhere.

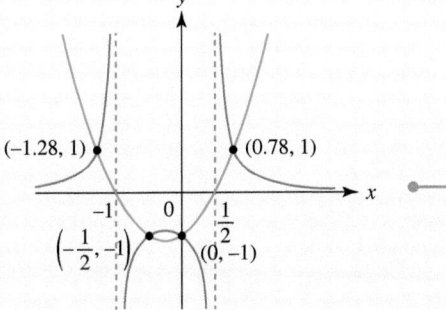

> **Consider the different features** of f(x) and $\dfrac{1}{f(x)}$ in turn.

> $f(x) = \pm 1$ does exist, so solve for x to find where f(x) and $\dfrac{1}{f(x)}$ intersect.

> **Use the information to draw** the graph of $\dfrac{1}{f(x)}$
>
> You can see that for
>
> f(x) > 1, then $0 < \dfrac{1}{f(x)} < 1$
>
> f(x) < -1, then $-1 < \dfrac{1}{f(x)} < 0$
>
> 0 < f(x) < 1, then $\dfrac{1}{f(x)} > 1$
>
> -1 < f(x) < 0, then $\dfrac{1}{f(x)} < -1$

> **Make sure you know how to check your answer using a graphical calculator.**

Sometimes you may not be given the function and instead have to work out the shape of the reciprocal graph by just looking at f(x) and using what you know about the relationship between f(x) and $\dfrac{1}{f(x)}$

Example 2

The graph of $y = f(x)$ is shown.

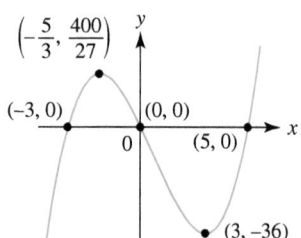

Use this to sketch the graph of $y = \dfrac{1}{f(x)}$

Explain all your working.

f(x) has three roots at $x = -3$, $x = 0$ and $x = 5$ so there are three vertical asymptotes where $f(x) = 0$

The maximum point of f(x) becomes a minimum point of $\dfrac{1}{f(x)}$

f(x) at the maximum point is positive so the minimum point is also positive. $\dfrac{1}{\left(\dfrac{400}{27}\right)} = \dfrac{27}{400}$ so the minimum point is at $\left(-\dfrac{5}{3}, \dfrac{27}{400}\right)$

The minimum point of f(x) becomes a maximum point of $\dfrac{1}{f(x)}$

f(x) at the minimum point is negative so the maximum point is also negative. The value of $\dfrac{1}{f(x)}$ is $-\dfrac{1}{36}$ so the maximum point is at $\left(3, -\dfrac{1}{36}\right)$.

f(x) is decreasing between the maximum and minimum points so $\dfrac{1}{f(x)}$ will increase for these x values.

Elsewhere, f(x) is increasing so $\dfrac{1}{f(x)}$ will decrease.

When f(x) approaches zero $\dfrac{1}{f(x)}$ will approach infinity, so $y = \dfrac{1}{f(x)}$ will have three vertical asymptotes.

When f(x) approaches infinity in either the positive or negative direction, $\dfrac{1}{f(x)}$ will approach zero.

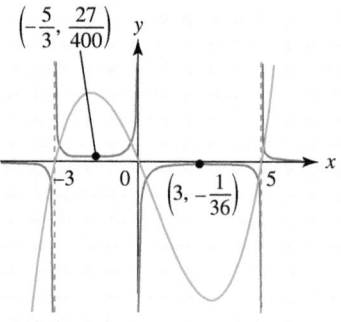

Graphs of the modulus function

$|x|$ is called the **modulus** of x

It is also known as the absolute value of x

The modulus of a real number is always positive. You can think of it as its distance from the origin.

For example, if $x = -3$, then $|x| = 3$

To sketch the graph of $y = |f(x)|$

- Start with a sketch of the graph of $y = f(x)$
- Reflect any negative part of $f(x)$ in the x-axis.

The diagram shows the graphs of $y = x$ (in blue) and $y = |x|$ (in red) drawn on the same axes.

When you carry out the reflection, any minimum turning point below the x-axis will be reflected into a maximum turning point above the x-axis.

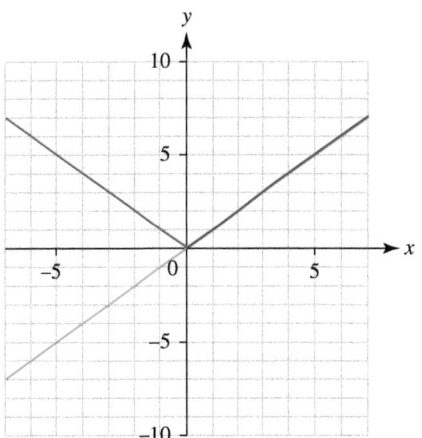

Example 3

Sketch $y = f(x)$ and $y = |f(x)|$ for these functions.

a $f(x) = 2x - 4$ **b** $f(x) = 2x^2 + 3x - 9$

a $y = 2x - 4$ is a straight line with gradient 2

It crosses the x-axis at $(2, 0)$ and crosses the y-axis at $(0, -4)$

so, $y = |2x - 4|$ crosses the axes at $(2, 0)$ and $(0, 4)$

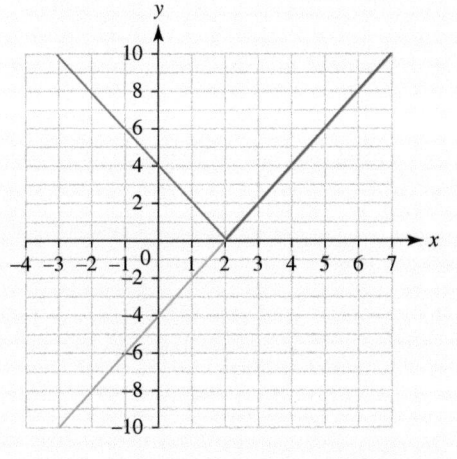

$y = 2x - 4$ is the blue graph and $y = |2x - 4|$ is the red graph.

b $y = 2x^2 + 3x - 9$ is a positive quadratic curve.

It crosses the x-axis at $(-3, 0)$ and $\left(\dfrac{3}{2}, 0 \right)$ and the y-axis at $(0, -9)$

$y = |2x^2 + 3x - 9|$ crosses the x-axis at $(-3, 0)$ and $\left(\dfrac{3}{2}, 0 \right)$ and the y-axis at $(0, 9)$

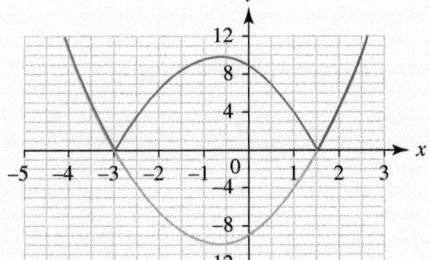

$y = 2x^2 + 3x - 9$ is the blue graph and $y = |2x^2 + 3x - 9|$ is the red graph.

The graph of $y = f(x)$ is shown.

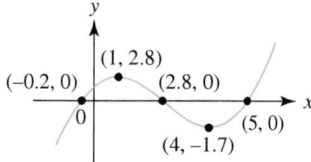

Sketch the graph of $y = |f(x)|$

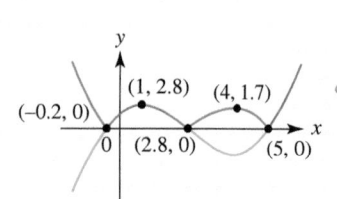

> Reflect the negative parts of the graph in the y-axis.
>
> The minimum point becomes a maximum point. Coordinate point $(4, -1.7)$ becomes $(4, 1.7)$.

Exercise 18.1A Fluency and skills

1 For each graph $y = f(x)$

   i Copy the graph and sketch on the same axes the graph of $y = \dfrac{1}{f(x)}$

   ii On a second copy of the graph, sketch $y = |f(x)|$

a

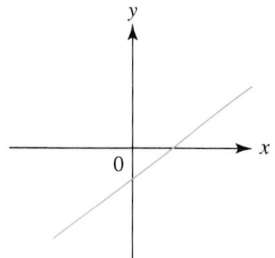

b

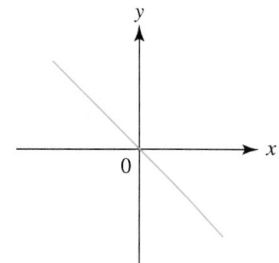

c

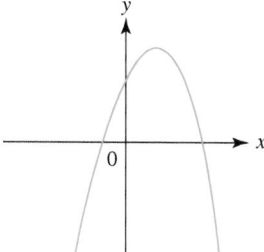

d

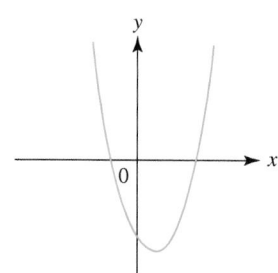

2 For each function

   i Sketch $f(x)$

   ii Sketch $\dfrac{1}{f(x)}$ labelling any asymptotes and points of intersection.

a $y = x$	**b** $y = 2x + 1$
c $y = 3x - 4$	**d** $y = \dfrac{1}{2}x + 3$
e $y = x^2 - x - 6$	**f** $y = x^2 - 8x + 15$
g $y = x^2 + 6x + 8$	**h** $y = -x^2 - x + 12$

3 Sketch |f(x)| for each of the functions in question **2**

4 For each graph $y = f(x)$

  i Copy the graph and sketch on the same axes the graph of $y = \dfrac{1}{f(x)}$

  ii On a second copy of the graph, sketch $y = |f(x)|$

  a

  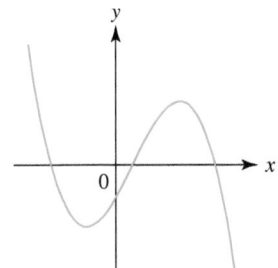

  b

  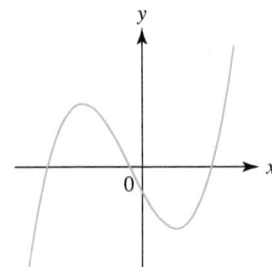

5 For each function

  i Sketch f(x)

  ii Sketch $\dfrac{1}{f(x)}$ labelling any asymptotes and points of intersection,

  iii Sketch |f(x)|

   a $y = x^3 - 4x$

   b $y = x^3 - 2x^2 - 24x$

   c $y = x^3 - 3x^2 - 13x + 15$

6 The diagram shows the graph of $y = f(x)$ and the points P, Q and R

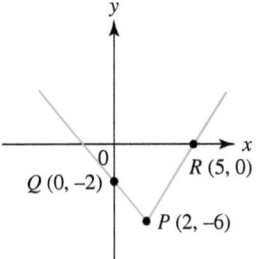

Sketch the graph of $y = |f(x)|$, labelling the transformed points P', Q' and R'

7 The diagram shows the graph of $y = f(x)$ and the points P, Q and R

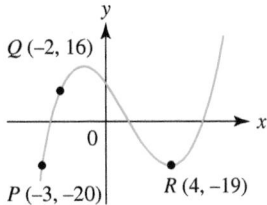

Sketch the graph of $y = |f(x)|$, labelling the transformed points P', Q' and R'

8 The diagram shows the graph of $y = f(x)$ and the points P, Q and R

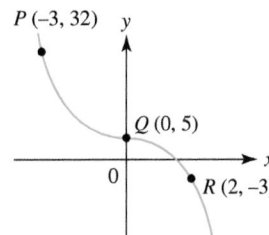

Sketch the graph of $y = |f(x)|$, labelling the transformed points P', Q' and R'

Reasoning and problem-solving

Strategy

To sketch a graph of $y = \dfrac{1}{f(x)}$ or $y = |f(x)|$

1 If it has not been provided, sketch the graph of $y = f(x)$

2 Consider the key features and transform them appropriately, clearly labelling the final graph.

Example 5

The sketch shows part of the graph of f(x)

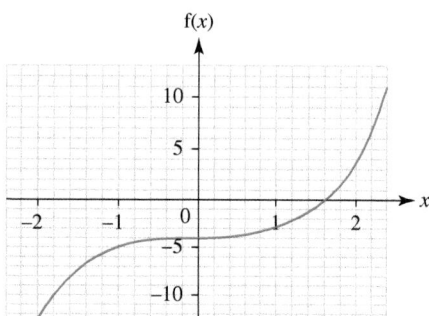

a Find the coordinates of the images of the points $(-2, -12)$, $(0, -4)$ and $(2, 4)$ when f(x) is transformed into $|f(x)|$

b Sketch the graph of $|f(x)|$ and use it to solve $|f(x)| \leq 4$

c Sketch the graph of $\dfrac{1}{f(x)}$ and use it to estimate the solution to $\dfrac{1}{f(x)} < 0$

---

a Under the transformation $f(x) \rightarrow |f(x)|$

$(-2, -12) \rightarrow (-2, 12), \quad (0, -4) \rightarrow (0, 4), \quad (2, 4) \rightarrow (2, 4)$

> ② Negative y-coordinates become positive.
>
> All negative parts of the graph are reflected in the x-axis.

b

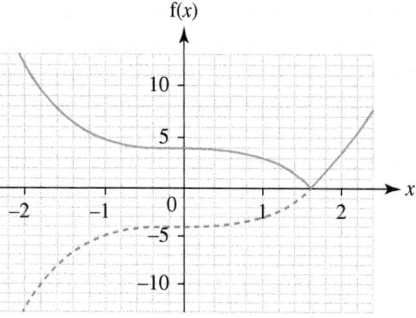

$|f(x)| \leq 4$ when $0 \leq x \leq 2$

> The line $y = 4$ intersects with $|f(x)|$ at $(0, 4)$ and $(2, 4)$

c

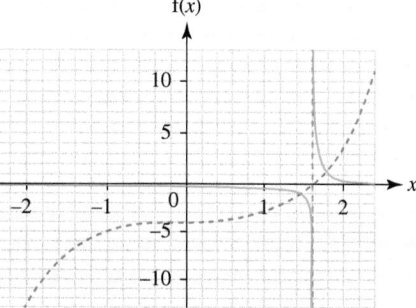

$\dfrac{1}{f(x)} < 0$ when $x < 1.6$ (approximately)

> ② The asymptote is approximately $x = 1.6$
>
> The point where $f(x) = 0$ becomes a vertical asymptote.
>
> As f(x) becomes very large, either positive or negative, $\dfrac{1}{f(x)}$ tends to zero.

1 a Use the relationship between the graphs of $y = \dfrac{1}{f(x)}$ and $y = f(x)$ to sketch, on the same axes

   i $\sin x$ and $\dfrac{1}{\sin x}$ for $-180° \le x \le 180°$

   ii $\cos x$ and $\dfrac{1}{\cos x}$ for $-180° \le x \le 180°$

   iii $\tan x$ and $\dfrac{1}{\tan x}$ for $-180° \le x \le 180°$

b Use your sketches to write down the interval(s) where

   i Both $\sin x$ and $\dfrac{1}{\sin x} \ge 0$

   ii Both $\cos x$ and $\dfrac{1}{\cos x} \ge 0$

   iii Both $\tan x$ and $\dfrac{1}{\tan x} \ge 0$

2 The diagram shows the graph of $y = f(x)$ and points P, Q and R

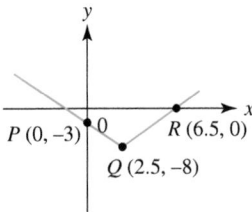

a Sketch the graph of $y = |f(x)|$, labelling the transformed points P', Q' and R'

b $f(x)$ can be written as $a|x + b| + c$

   Write down the values of a, b and c

3 The diagram shows $y = f(x)$ and points A and B

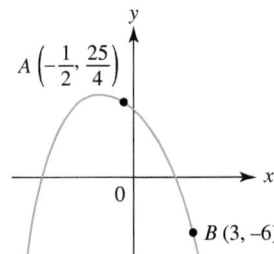

a Write down the coordinates of the points A' and B' when $f(x)$ is transformed into $\dfrac{1}{f(x)}$

b Sketch the graph of $y = \dfrac{1}{f(x)}$ and use it to solve the inequality $\dfrac{1}{f(x)} > 0$

4 $f(x) = \dfrac{2 - x}{x - 3}$

Sketch the graph of $\dfrac{1}{f(x)}$ and use it to solve the inequality $\dfrac{1}{f(x)} < 0$

5 A curve has equation $y = \dfrac{2 - x}{1 - x}$

a **i** Write down the equations of the asymptotes to the curve and the points of intersection with the coordinate axes.

   ii Sketch the curve, indicating clearly the coordinates of the points of intersection of the curve with the coordinate axes.

b Sketch also the curve $y = \left| \dfrac{2 - x}{1 - x} \right|$

c Add the straight line $y = 2 - x$ to each sketch and hence solve the inequalities

$$\dfrac{2 - x}{1 - x} < 2 - x \quad \text{and} \quad \left| \dfrac{2 - x}{1 - x} \right| < 2 - x$$

6 A curve has equation $f(x) = \dfrac{x}{3 + x}$

By drawing suitable curves and lines, solve

a $\dfrac{x}{3 + x} < -x$

b $\left| \dfrac{x}{3 + x} \right| < -x$

Fluency and skills

> **Key point**
>
> An enlargement is equivalent to a stretch by the same scale factor in both the x-direction and the y-direction.
>
> For an enlargement scale factor k replace x with $\dfrac{x}{k}$ and y with $\dfrac{y}{k}$

Example 1

An ellipse with equation $\dfrac{x^2}{4} + \dfrac{y^2}{9} = 1$ is enlarged by scale factor 3

a Find the equation of the transformed curve and state its points of intersection with the axes.

b Sketch the transformed curve.

a $\dfrac{\left(\dfrac{x}{3}\right)^2}{4} + \dfrac{\left(\dfrac{y}{3}\right)^2}{9} = 1$ ● —————————— Substitute for x and y

$\dfrac{x^2}{36} + \dfrac{y^2}{81} = 1$ ● —————————— Simplify to find the equation of the transformed curve.

When $x = 0$, $y^2 = 81 \Rightarrow y = \pm 9 \Rightarrow$ coordinates $(0, 9)$
and $(0, -9)$

When $y = 0$, $x^2 = 36 \Rightarrow x = \pm 6 \Rightarrow$ coordinates $(6, 0)$
and $(-6, 0)$ ● —————————— Substitute $x = 0$ and $y = 0$ to find the coordinates of the points of intersection.

b

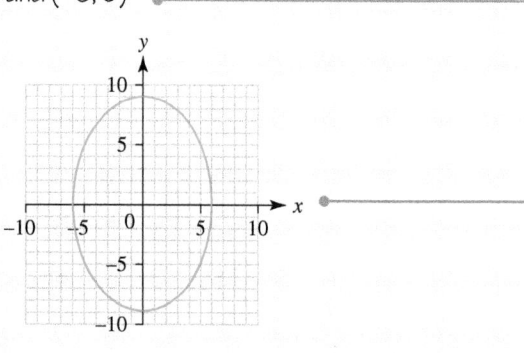

Use the coordinates from part **a** to sketch the curve of the transformed ellipse.

A conic can be rotated through θ radians about the origin. It is convention to describe rotations in an anticlockwise direction. In this course, the rotations are limited to multiples of $\dfrac{\pi}{2}$

Key point

To rotate a conic by $\dfrac{\pi}{2}$ radians, replace x by y and y by $-x$

To rotate a conic by π radians, replace x by $-x$ and y by $-y$

To rotate a conic by $\dfrac{3\pi}{2}$ radians, replace x by $-y$ and y by x

These results come from substituting
$\theta = \dfrac{\pi}{2}$, π and $\dfrac{3\pi}{2}$ into the general transformation formulae:
$$X = x\cos\theta + y\sin\theta$$
$$Y = -x\sin\theta + y\cos\theta$$
where (X, Y) are the coordinates of the transformed point (x, y).

Example 2

The hyperbola $\dfrac{x^2}{9} - \dfrac{y^2}{16} = 1$ is rotated anticlockwise through $\dfrac{3\pi}{2}$ radians about $(0, 0)$.

a Write down the equation of the transformed curve.

b Sketch the transformed curve and find the equations of the asymptotes.

a $\dfrac{(-y)^2}{9} - \dfrac{(x)^2}{16} = 1 \Rightarrow \dfrac{y^2}{9} - \dfrac{x^2}{16} = 1$

Substitute for x and y and simplify.

b
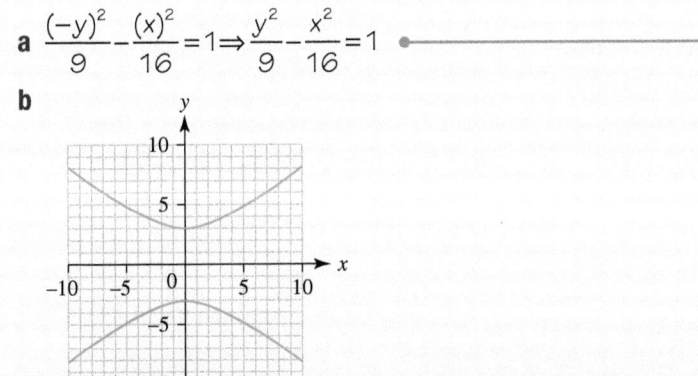

The asymptotes of the original hyperbola are at $y = \pm\dfrac{4}{3}x$

Use the result that the asymptotes to a hyperbola are at $y = \pm\dfrac{b}{a}x$

The asymptotes of the transformed curve are at

$x = \pm\dfrac{4}{3}(-y) \Rightarrow y = \pm\dfrac{3}{4}x$

Apply the same substitution as the one used to transform the conic and rearrange to make y the subject.

Exercise 18.2A Fluency and skills

1 Find the equation of the new curve and sketch its graph when the parabola $y^2 = 12x$ is transformed by

   a An enlargement scale factor 2

   b An enlargement scale factor $\dfrac{1}{3}$

   c A rotation of $\dfrac{\pi}{2}$ radians anticlockwise about $(0, 0)$

   d A rotation of π radians anticlockwise about $(0, 0)$

2 Find the equation of the new curve when the ellipse $\dfrac{x^2}{16} + \dfrac{y^2}{25} = 1$ is transformed in the following ways. In each case, sketch the new graph and write down the coordinates of the points where the curve crosses the axes.

   a An enlargement scale factor $\dfrac{1}{2}$

   b An enlargement scale factor 3

c A rotation of $\dfrac{3\pi}{2}$ radians anticlockwise about $(0, 0)$

d A rotation of π radians anticlockwise about $(0, 0)$

3 Find the equation of the new curve when the hyperbola $\dfrac{x^2}{9} - \dfrac{y^2}{4} = 1$ is transformed in the following ways. In each case, sketch the new graph and write down the equations of the asymptotes.

a An enlargement scale factor 4

b An enlargement scale factor $\dfrac{3}{2}$

c A rotation of $\dfrac{\pi}{2}$ radians anticlockwise about $(0, 0)$

d A rotation of $\dfrac{3\pi}{2}$ radians anticlockwise about $(0, 0)$

4 Find the equation of the new curve when the rectangular hyperbola $xy = 16$ is transformed in the following ways. In each case, sketch the curve and write down the equations of the asymptotes.

a An enlargement scale factor $\dfrac{2}{3}$

b An enlargement scale factor $\dfrac{4}{3}$

c A rotation of π radians anticlockwise about $(0, 0)$

d A rotation of $\dfrac{\pi}{2}$ radians anticlockwise about $(0, 0)$

5 An ellipse with equation $\dfrac{x^2}{4} + \dfrac{y^2}{25} = 1$ is enlarged by scale factor k

The equation of the transformed curve is
$$\dfrac{x^2}{16} + \dfrac{y^2}{100} = \dfrac{1}{9}$$
Find the value of k

6 A hyperbola with equation $\dfrac{x^2}{25} - \dfrac{y^2}{4} = 1$ is enlarged by scale factor k

The equation of the transformed curve is
$$\dfrac{x^2}{36} - \dfrac{25y^2}{144} = 1$$
Find the value of k

7 A parabola with equation $y^2 = 40x$ is rotated anticlockwise through θ radians. The equation of the transformed curve is
$$x^2 = -40y$$
Find the value of θ

Reasoning and problem-solving

You need to be able to find the equation and sketch the graph of a given conic after two or more successive transformations and also identify a sequence of transformations that has been applied to transform one curve into another.

Example 3

The ellipse with equation $x^2 + \dfrac{y^2}{4} = 1$ is enlarged by scale factor 2, then rotated by $\dfrac{\pi}{2}$ radians anticlockwise about (0, 0) before being translated by vector $\begin{pmatrix} 3 \\ 2 \end{pmatrix}$

Find the equation of the new conic in the form $ax^2 + by^2 + cx + dy + e = 0$ where a, b, c, d and e are constants to be found and sketch the graph.

Enlargement; Rotation; Translation.

> ① Identify the order of the transformations.

$$\dfrac{\left(\dfrac{x}{2}\right)^2}{1} + \dfrac{\left(\dfrac{y}{2}\right)^2}{4} = 1 \Rightarrow \dfrac{x^2}{4} + \dfrac{y^2}{16} = 1$$

> ② Use the rule for enlargements to transform the original conic.

$$\dfrac{(y)^2}{4} + \dfrac{(-x)^2}{16} = 1 \Rightarrow \dfrac{x^2}{16} + \dfrac{y^2}{4} = 1$$

$$\dfrac{(x-3)^2}{16} + \dfrac{(y-2)^2}{4} = 1$$

> ③ Apply the second transformation to the new equation.

$$(x-3)^2 + 4(y-2)^2 = 16$$

$$x^2 - 6x + 9 + 4y^2 - 16y + 16 = 16$$

$$x^2 + 4y^2 - 6x - 16y + 9 = 0$$

> ④ Apply the third transformation.

> ⑤ Expand and simplify to give the final answer in the correct form.

> ⑥ A sketch of the final curve is required. You can do this by sketching the curve after each successive transformation or directly from the final answer.

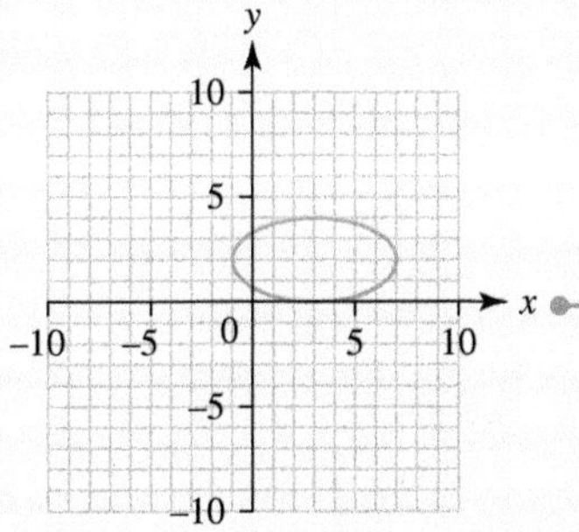

Example 4

A parabola C has equation $y^2 = x$

Describe a sequence of two transformations which maps C onto the curve with equation $y^2 - 6y + x + 11 = 0$

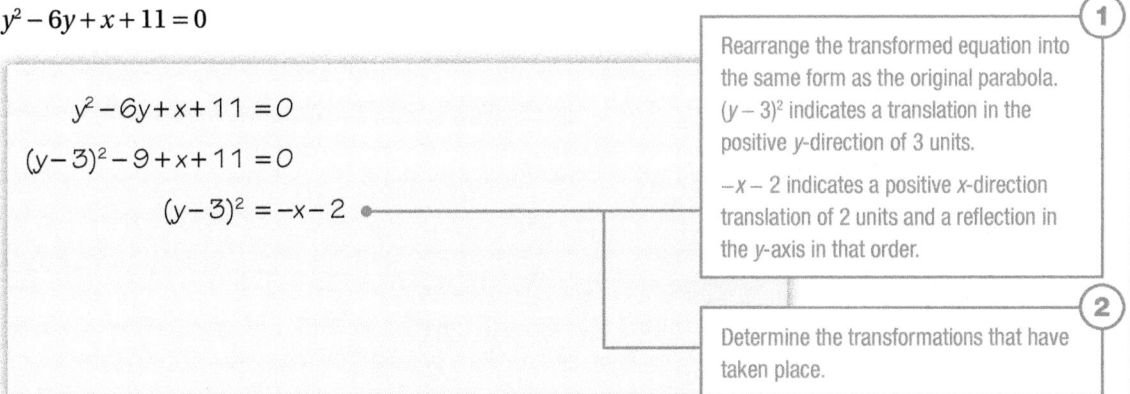

$$y^2 - 6y + x + 11 = 0$$

$$(y-3)^2 - 9 + x + 11 = 0$$

$$(y-3)^2 = -x - 2$$

> ① Rearrange the transformed equation into the same form as the original parabola. $(y-3)^2$ indicates a translation in the positive y-direction of 3 units.
>
> $-x - 2$ indicates a positive x-direction translation of 2 units and a reflection in the y-axis in that order.

> ② Determine the transformations that have taken place.

(Continued on the next page)

A possible sequence with two transformations is a translation using vector $\begin{pmatrix} 2 \\ 3 \end{pmatrix}$, followed by a reflection in the y-axis.

3 Determine the order in which the transformations have taken place.

An alternative way of writing the transformed equation is $(y-3)^2 = -(x+2)$. In this case, the order of transformations is different. The y-direction translation is independent and still 'plus 3 units', but the sequence of transformations on the right-hand side is now a reflection in the y-axis followed by a translation of two units in the *negative* x-direction. Hence the correct sequence of two transformations is now a reflection in the y-axis followed by a translation using vector $\begin{pmatrix} -2 \\ 3 \end{pmatrix}$.

Strategy 2

To find the sequence of transformations that have been used to map one given conic onto another

1 Rearrange the transformed equation into the standard form for the given conic.

2 Determine the transformations that have taken place.

3 Determine, where necessary, the order in which they have taken place.

Example 5

A hyperbola H has equation $x^2 - \dfrac{y^2}{3} = 1$

Describe a sequence of transformations which maps H onto the curve with equation
$12x^2 = y^2 + 48x - 36$

$12x^2 = y^2 + 48x - 36$

$12x^2 - 48x - y^2 = -36$

$x^2 - 4x - \dfrac{y^2}{12} = -3$

$(x-2)^2 - 4 - \dfrac{y^2}{12} = -3$

$(x-2)^2 - \dfrac{\left(\dfrac{y}{2}\right)^2}{3} = 1$

$(x-2)^2$ indicates a translation in the positive x-direction of 2 units.

$\left(\dfrac{y}{2}\right)^2$ indicates a stretch parallel to the y-axis scale factor 2

A possible sequence is a stretch parallel to the y-axis scale factor 2, followed by a translation in the positive x-direction of 2 units.

1 Rearrange the transformed equation into the same form as the original hyperbola.

2 Determine the transformations that have taken place.

3 Determine the order in which the transformations have taken place. Note that in some cases, like this one, the transformations can be applied in *either order*.

1 The hyperbola $\dfrac{x^2}{4} - \dfrac{y^2}{9} = 1$ is rotated by $\dfrac{3\pi}{2}$ radians anticlockwise about $(0, 0)$ and then enlarged by scale factor 3

   a Find the equation of the new conic in the form $ax^2 + by^2 = c$ where a, b and c are constants to be found.

   b State the equations of the asymptotes.

2 The parabola $y^2 = 6x$ is enlarged by scale factor 2, translated by vector $\begin{pmatrix} -1 \\ 2 \end{pmatrix}$ and then rotated through π radians about $(0, 0)$.

   a Find the equation of the new conic in the form $ay^2 + by + cx = d$ where a, b, c and d are constants to be found.

   b Find the coordinates of the turning point.

   c Sketch the graph of the new conic.

3 The ellipse $\dfrac{x^2}{2} + \dfrac{y^2}{3} = 1$ is translated by vector $\begin{pmatrix} 3 \\ -1 \end{pmatrix}$ before being rotated anticlockwise by $\dfrac{\pi}{2}$ radians about the origin and stretched in the x-direction by scale factor 2

   a Show that the equation of the new conic can be written as $x^2 + 6y^2 - 4x - 36y + c = 0$ where c is a constant to be found.

   b Given that the line $y = k$ intersects the new conic at two points, find the possible values of k

4 An ellipse C has equation $\dfrac{x^2}{4} + \dfrac{y^2}{5} = 1$
Describe a sequence of transformations which maps C onto the curve with equation $5x^2 + 16y^2 - 20x + 32y + 16 = 0$

5 A rectangular hyperbola H has equation $xy = 25$

Describe a sequence of transformations which maps H onto the curve with equation $xy - 3y - 9x = 198$

6 An ellipse, E, has equation $\dfrac{x^2}{9} + \dfrac{y^2}{16} = 1$

   a Sketch the ellipse E and state the coordinates of the points where the curve intercepts the coordinate axes.

   b Given that the line $y = c - x$ intersects the ellipse at two distinct points, show that $-5 < c < 5$

   c The ellipse E is translated by vector $\begin{pmatrix} a \\ b \end{pmatrix}$ to form another ellipse with equation $16x^2 + 9y^2 - 128x + 18y + d = 0$

     Find the values of the constants a, b and d

   d Hence find the equation for each of the tangents to the ellipse

     $16x^2 + 9y^2 - 128x + 18y + d = 0$

     which are parallel to the line $y = -x$

7 A hyperbola H has equation $\dfrac{x^2}{4} - \dfrac{y^2}{5} = 1$

   a Show that the line with equation $2x + 3y = 4$ intersects H at two points and find the exact coordinates of these points.

   The hyperbola H is reflected in the line $y = x$

   b Show that the line $2x + 3y = 4$ is now a tangent to the reflected curve.

Fluency and skills

Here are the definitions of the hyperbolic functions, and their inverses.

See Ch3.5

For a reminder of hyperbolic functions.

Key point

$$\sinh x \equiv \frac{1}{2}(e^x - e^{-x}) \qquad \operatorname{arsinh} x \equiv \ln(x + \sqrt{x^2 + 1})$$

$$\cosh x \equiv \frac{1}{2}(e^x + e^{-x}) \qquad \operatorname{arcosh} x \equiv \ln(x + \sqrt{x^2 - 1}), \ x \geq 1$$

$$\tanh x \equiv \frac{e^x - e^{-x}}{e^x + e^{-x}} \qquad \operatorname{artanh} x \equiv \frac{1}{2}\ln\left(\frac{1+x}{1-x}\right), \ -1 < x < 1$$

The domain of the function $f(x) = \sinh x$ is $x \in \mathbb{R}$ and the range is $f(x) \in \mathbb{R}$, so the domain of the function $f^{-1}(x) = \operatorname{arsinh} x$ is $x \in \mathbb{R}$ and the range is $f^{-1}(x) \in \mathbb{R}$

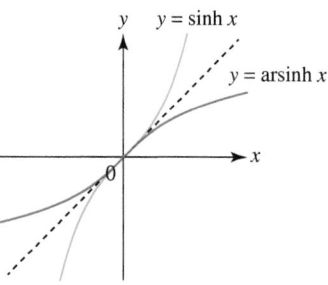

In order to be able to define an inverse you must restrict the domain of $f(x) = \cosh x$ to $x \in \mathbb{R}$, $x \geq 0$. The range of $f(x) = \cosh x$ is $f(x) \in \mathbb{R}$, $f(x) \geq 1$, so the domain of $f^{-1}(x) = \operatorname{arcosh} x$ is $x \in \mathbb{R}$, $x \geq 1$ and the range is $f^{-1}(x) \in \mathbb{R}$, $f^{-1}(x) \geq 0$

The domain of the function $f(x) = \tanh x$ is $x \in \mathbb{R}$ and the range is $f(x) \in \mathbb{R}$, $-1 < f(x) < 1$ since the graph of $y = \tanh x$ has horizontal asymptotes at $y = \pm 1$

Therefore, the graph of $y = \operatorname{artanh} x$ will have vertical asymptotes at $x = \pm 1$. So the domain of $f^{-1}(x) = \operatorname{artanh} x$ is $x \in \mathbb{R}$, $-1 < x < 1$ and the range is $f^{-1}(x) \in \mathbb{R}$

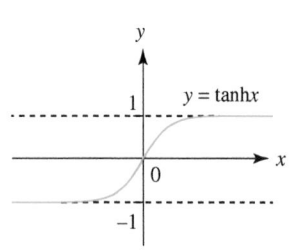

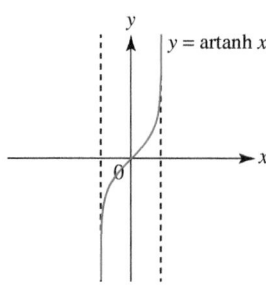

Example 1

Sketch the graph of $y = \operatorname{arcosh}\left(\dfrac{x}{2}\right)$, and state its domain and range.

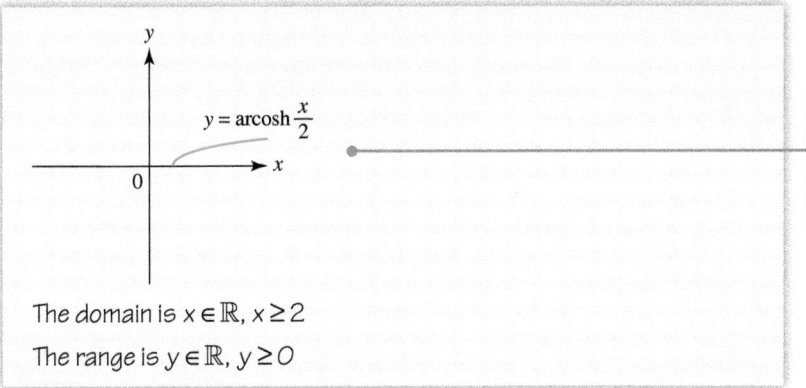

The graph of $y = \cosh^{-1}(x)$ has been stretched by scale factor 2 in the x-direction.

The domain is $x \in \mathbb{R}, x \geq 2$

The range is $y \in \mathbb{R}, y \geq 0$

There are also reciprocal hyperbolic functions, defined in the same way as with trigonometric functions.

Key point

$$\operatorname{cosech} x \equiv \frac{1}{\sinh x} \equiv \frac{2}{e^x - e^{-x}} \quad \text{for } x \in \mathbb{R}, x \neq 0$$

$$\operatorname{sech} x \equiv \frac{1}{\cosh x} \equiv \frac{2}{e^x + e^{-x}} \quad \text{for } x \in \mathbb{R}$$

$$\coth x \equiv \frac{1}{\tanh x} \equiv \frac{e^{2x} + 1}{e^{2x} - 1} \quad \text{for } x \in \mathbb{R}, x \neq 0$$

These functions are commonly read as, 'cosetch', 'setch' and 'coth'.

You can sketch the graphs of these reciprocal functions using the techniques covered earlier in this chapter.

The domain of $f(x) = \operatorname{cosech} x$ is $x \in \mathbb{R}, x \neq 0$

and the range is $f(x) \in \mathbb{R}, f(x) \neq 0$

See Ch18.1

For a reminder of sketching reciprocal graphs.

The domain of $f(x) = \operatorname{sech} x$ is $x \in \mathbb{R}$,

and the range is $f(x) \in \mathbb{R}, 0 < f(x) \leq 1$

The domain of $f(x) = \coth x$ is $x \in \mathbb{R}, x \neq 0$

and the range is $f(x) \in \mathbb{R}, f(x) < -1$ or $f(x) > 1$

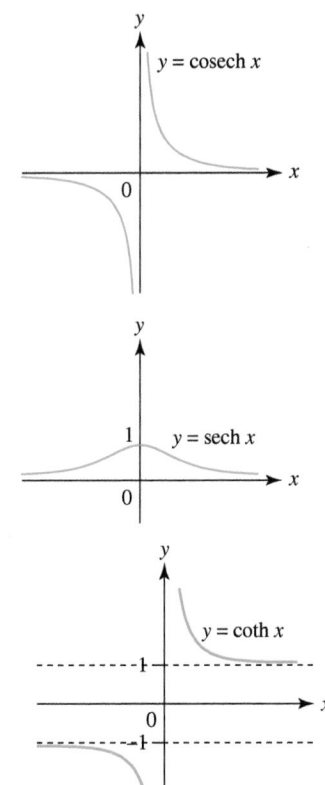

Example 2

Find the exact value of the expression $\operatorname{sech}(\ln\sqrt{3})$. Show your working.

$$\operatorname{sech}(\ln\sqrt{3}) = \frac{2}{e^{\ln\sqrt{3}} + e^{-\ln\sqrt{3}}}$$

$$= \frac{2}{e^{\ln\sqrt{3}} + e^{\ln\left(\frac{1}{\sqrt{3}}\right)}}$$

$$= \frac{2}{\sqrt{3} + \frac{1}{\sqrt{3}}}$$

$$= \frac{\sqrt{3}}{2}$$

Use definition

$\operatorname{sech} x = \dfrac{2}{e^x + e^{-x}}$ since

$-\ln\sqrt{3} = \ln\sqrt{3}^{\,-1} = \ln\left(\dfrac{1}{\sqrt{3}}\right)$

Example 3

Find the exact solution to the equation $1 + \coth x = 4$

$1 + \coth x = 4 \Rightarrow \coth x = 3$

Therefore $\tanh x = \dfrac{1}{3}$

Which gives $x = \dfrac{1}{2}\ln\left(\dfrac{1 + \dfrac{1}{3}}{1 - \dfrac{1}{3}}\right)$

$$= \frac{1}{2}\ln 2$$

$$= \ln\sqrt{2}$$

Rearrange.

Since $\tanh x = \dfrac{1}{\coth x}$

Use the definition of the inverse,

$\tanh^{-1} x = \dfrac{1}{2}\ln\left(\dfrac{1 + x}{1 - x}\right)$

Exercise 18.3A Fluency and skills

1 Find the exact value of each of these expressions and give your answers in their simplest form. Show all your working and do not use a calculator.

   a $\operatorname{sech}(\ln 2)$ b $\operatorname{cosech}(\ln 5)$

   c $\coth(\ln\sqrt{2})$ d $\operatorname{sech}(2\ln 4)$

   e $\operatorname{cosech}\left(\dfrac{1}{2}\ln 5\right)$ f $\coth(\ln 3)$

2 Sketch the graph of $y = f(x)$ for each of these functions and state the domain and range.

   a $y = \operatorname{arcosh}(x+2)$ b $y = \operatorname{arsinh}\left(\dfrac{x}{3}\right)$

   c $y = 1 + \operatorname{artanh}(x)$ d $y = \operatorname{arcosh}(4x)$

   e $y = \operatorname{artanh}(x-2)$ f $y = \operatorname{artanh}(3x)$

3 Sketch the graph of $y = f(x)$ for each of these functions and state the domain and range.

   a $y = \operatorname{sech}(2x)$ b $y = \operatorname{cosech}(x+1)$

   c $y = 1 + \coth(x)$ d $y = \operatorname{sech}(x) - 2$

   e $y = -\operatorname{cosech}(x)$ f $y = 1 - \operatorname{sech}(x)$

4 Solve each of these equations. Give your answers in the form $\ln k$ where k is a constant to be found.

   a $\operatorname{cosech} x = 2$ b $\coth x = 3$

   c $\operatorname{sech} x = \dfrac{1}{\sqrt{2}}$ d $2\coth x = 5$

   e $1 + \operatorname{cosech} x = -3$ f $5 - 6\operatorname{sech} x = 2$

5 Find the exact solutions to these equations.

   a $\operatorname{arsech} x = \ln 7$ b $\operatorname{arcosech} x = 3\ln 2$

   c $\operatorname{arcoth} 3x = \ln\sqrt{3}$ d $2\operatorname{arsech} 5x = \ln 16$

6 Find the exact solutions to the equations

   a $\operatorname{sech}^2 x = \dfrac{3}{4}$ b $3\operatorname{cosech}^2 x = 1$

   c $\coth^4 x - 9 = 0$

Reasoning and problem-solving

The identity $\cosh^2 x - \sinh^2 x \equiv 1$ can be used to prove other identities. For example, dividing both sides of $\cosh^2 x - \sinh^2 x \equiv 1$ by $\cosh^2 x$ leads to the identity $1 - \tanh^2 x \equiv \operatorname{sech}^2 x$

Similarly, dividing by $\sinh^2 x$ leads to the identity $\coth^2 x - 1 \equiv \operatorname{cosech}^2 x$

Key point

$\cosh^2 x - \sinh^2 x \equiv 1$

$1 - \tanh^2 x \equiv \operatorname{sech}^2 x$

$\coth^2 x - 1 \equiv \operatorname{cosech}^2 x \quad$ for $x \neq 0$

> The first identity is given in the formula book. You should memorise the other two and be able to derive them from the first.

> **See Ch3.5** For a reminder of hyperbolic identities.

You can also use the definitions for the hyperbolic functions in terms of exponentials to prove the double angle formulae:

Key point

$\sinh 2x \equiv 2\sinh x \cosh x$

$\cosh 2x \equiv \cosh^2 x + \sinh^2 x$

$\tanh 2x \equiv \dfrac{2\tanh x}{1 + \tanh^2 x}$

> The first two identities are given in the formula book. You should be able to derive the third identity using the fact that
> $$\tanh(2x) \equiv \frac{\sinh(2x)}{\cosh(2x)}$$

For example, to prove the double angle formula for $\cosh(2x)$, use the definitions $\sinh x = \dfrac{e^x - e^{-x}}{2}$ and $\cosh x = \dfrac{e^x + e^{-x}}{2}$ to write

$\cosh^2 x + \sinh^2 x$ as $\left(\dfrac{e^x + e^{-x}}{2}\right)^2 + \left(\dfrac{e^x - e^{-x}}{2}\right)^2$

Expand and use index rules to give $\dfrac{e^{2x} + 2 + e^{-2x}}{4} + \dfrac{e^{2x} - 2 + e^{-2x}}{4}$

which simplifies to $\dfrac{2e^{2x} + 2e^{-2x}}{4} = \dfrac{e^{2x} + e^{-2x}}{2}$

which is the definition of $\cosh 2x$

You can also use the identity $\cosh^2 x - \sinh^2 x \equiv 1$ to rewrite the double angle formula for $\cosh 2x$

Key point

Double angle formulae

$\cosh 2x \equiv 1 + 2\sinh^2 x$ or $\cosh 2x \equiv 2\cosh^2 x - 1$

> These versions are particularly useful when integrating as you will see in the next chapter.

You will need to use these identities to solve equations involving hyperbolic functions.

Strategy 1

To solve equations involving hyperbolic functions

1. Use $\operatorname{sech} x = \dfrac{1}{\cosh x}$, $\operatorname{cosech} x = \dfrac{1}{\sinh x}$ or $\coth x = \dfrac{1}{\tanh x}$

2. Use the definitions of $\cosh x$, $\sinh x$ or $\tanh x$ in terms of exponentials.

3. Form a quadratic in e^x then solve.

4. Find the values of x

Example 4

Solve the equation $4\operatorname{sech} x + \tanh x = 4$

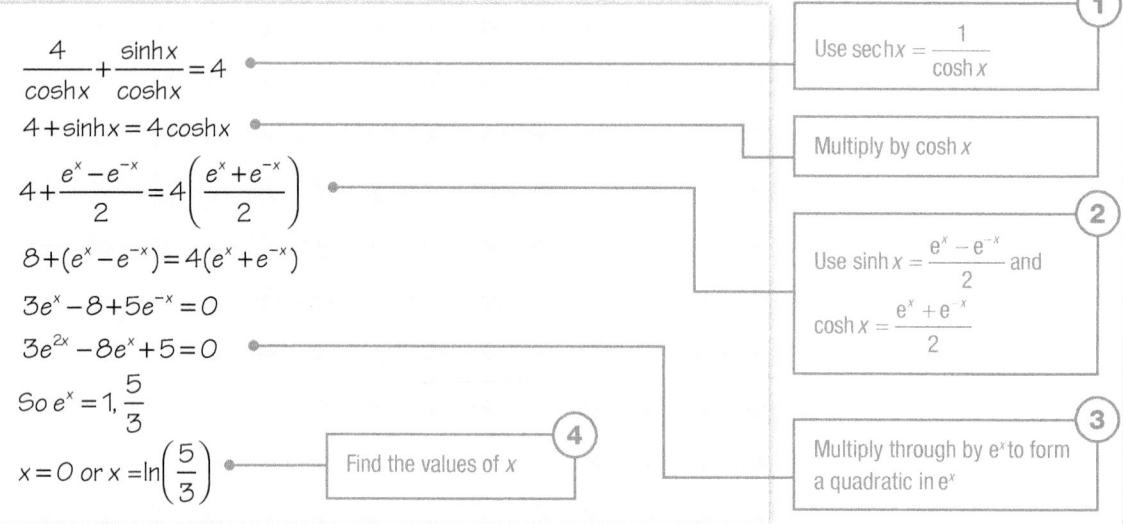

$$\frac{4}{\cosh x} + \frac{\sinh x}{\cosh x} = 4$$

$$4 + \sinh x = 4\cosh x$$

$$4 + \frac{e^x - e^{-x}}{2} = 4\left(\frac{e^x + e^{-x}}{2}\right)$$

$$8 + (e^x - e^{-x}) = 4(e^x + e^{-x})$$

$$3e^x - 8 + 5e^{-x} = 0$$

$$3e^{2x} - 8e^x + 5 = 0$$

$$\text{So } e^x = 1, \frac{5}{3}$$

$$x = 0 \text{ or } x = \ln\left(\frac{5}{3}\right)$$

1 Use $\operatorname{sech} x = \dfrac{1}{\cosh x}$

Multiply by $\cosh x$

2 Use $\sinh x = \dfrac{e^x - e^{-x}}{2}$ and $\cosh x = \dfrac{e^x + e^{-x}}{2}$

3 Multiply through by e^x to form a quadratic in e^x

4 Find the values of x

Strategy 2

To solve quadratic equations involving reciprocal hyperbolic functions

(**1**) Use identities to write the equation in terms of a single hyperbolic function.

(**2**) Solve the quadratic to find the possible values of the reciprocal hyperbolic function.

(**3**) Use the definitions of the reciprocal hyperbolic functions to find the values of $\sinh x$, $\cosh x$ or $\tanh x$

(**4**) Use the definitions of the inverse hyperbolic functions to find the exact values of x

---

Example 5

Find the exact solutions to the equation $\coth^2 x - 2\operatorname{cosech} x = 4$

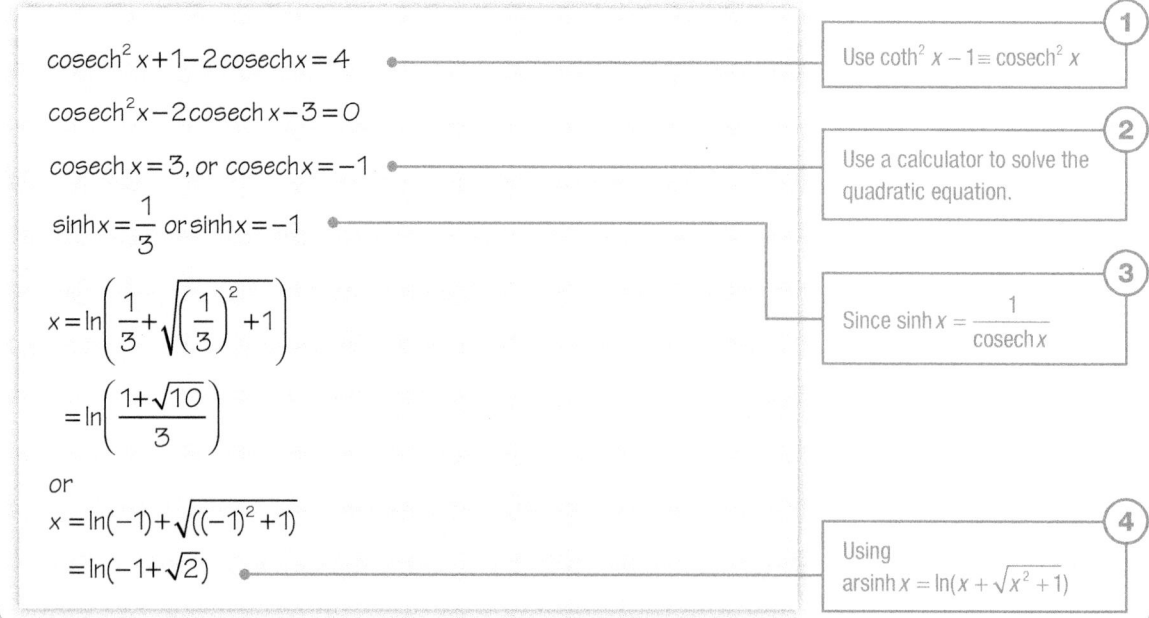

$$\operatorname{cosech}^2 x + 1 - 2\operatorname{cosech} x = 4$$

$$\operatorname{cosech}^2 x - 2\operatorname{cosech} x - 3 = 0$$

$$\operatorname{cosech} x = 3, \text{ or } \operatorname{cosech} x = -1$$

$$\sinh x = \frac{1}{3} \text{ or } \sinh x = -1$$

$$x = \ln\left(\frac{1}{3} + \sqrt{\left(\frac{1}{3}\right)^2 + 1}\right)$$

$$= \ln\left(\frac{1 + \sqrt{10}}{3}\right)$$

or
$$x = \ln(-1) + \sqrt{((-1)^2 + 1)}$$

$$= \ln(-1 + \sqrt{2})$$

1 Use $\coth^2 x - 1 \equiv \operatorname{cosech}^2 x$

2 Use a calculator to solve the quadratic equation.

3 Since $\sinh x = \dfrac{1}{\operatorname{cosech} x}$

4 Using $\operatorname{arsinh} x = \ln(x + \sqrt{x^2 + 1})$

1 Use definitions in terms of exponentials to prove these identities

a $1+\operatorname{cosech}^2 x \equiv \coth^2 x$

b $\operatorname{cosech}2x \equiv \dfrac{1}{2}\operatorname{cosech} x\operatorname{sech} x$

c $\coth x \equiv \dfrac{\coth^2\left(\dfrac{x}{2}\right)+1}{2\coth\left(\dfrac{x}{2}\right)}$

d $\tanh^2\left(\dfrac{x}{2}\right) \equiv \dfrac{1-\operatorname{sech}x}{1+\operatorname{sech}x}$

2 a Show that $\cosh x+2\operatorname{sech} x=3$ can be written as $\cosh^2 x-3\cosh x+2=0$

b Hence, solve the equation $\cosh x+2\operatorname{sech} x=3$

3 a Use the exponential definitions of $\cosh x$ and $\sinh x$ to prove that $\sinh(2x)\equiv 2\sinh x\cosh x$

b Hence, show that $\coth x-\tanh x \equiv 2\operatorname{cosech}(2x)$

c Hence, solve the equation $\coth x-\tanh x=-1$

4 Solve the equation $\operatorname{cosech}x-\operatorname{sech}x=\dfrac{1}{2}e^x$

Give your answer in the form $\ln A$ where A is a constant to be found.

5 a Use the definitions of $\sinh x$ and $\cosh x$ in terms of exponentials to prove that $\sinh^2 x=\dfrac{1}{2}(\cosh(2x)-1)$

b Hence solve the equation $\sinh(x)+\cosh(2x)=1$

6 a Prove that $\operatorname{arcosech} x \equiv \ln\left(\dfrac{1}{x}+\sqrt{\dfrac{1}{x^2}+1}\right)$

b Hence, prove that $\operatorname{arcosech} x+\operatorname{arcosech}(-x)=0$ for all $x\in\mathbb{R}$

7 Prove that $\operatorname{arsech} x \equiv \pm\ln\left(\dfrac{1}{x}+\sqrt{\dfrac{1}{x^2}-1}\right)$

8 Given that $\operatorname{cosech} x=\sqrt{2}$, find the exact values of

a $\cosh x$ **b** $\tanh x$

9 Given that $\operatorname{sech}x=\dfrac{1}{\sqrt{3}}$, find the exact values of

a $\sinh x$ **b** $\coth x$

10 a Use the double angle formulae for sinh and cosh to prove that $\tanh(2x)\equiv\dfrac{2\tanh x}{1+\tanh^2 x}$

b Hence show that $\coth(2x)\equiv\dfrac{1}{2}(\coth x+\tanh x)$

c Solve the equation $\coth x+\tanh x=4$

11 a Prove that $\dfrac{1}{1+\coth x}+\dfrac{1}{1-\coth x}\equiv -2\sinh^2 x$

b Hence solve the equation $\dfrac{1}{1+\coth x}+\dfrac{1}{1-\coth x}=-\dfrac{9}{2}$

Give your answers as simplified logarithms.

12 a On the same diagram sketch the graph with equations $y=\sinh x$ and $y=2\operatorname{sech} x$

b Hence state the number of solutions to the equation $\sinh x=2\operatorname{sech} x$

c Solve the equation $\sinh x=2\operatorname{sech} x$

13 a On the same diagram sketch the graph with equations $y=2\coth x$ and $y=\sinh x$. State the equations of any asymptotes.

b Solve the equation $2\coth x=\sinh x$

Give your solutions to 3 significant figures.

14 Solve the equation $\operatorname{cosech}^2 x+2\operatorname{cosech} x=3$ Give your answers as exact logarithms.

15 Find the exact solutions to the equation $\operatorname{sech}^2 x-\tanh^2 x=1-3\operatorname{sech} x$

16 Solve the equation $6\operatorname{sech}^2 x-\tanh x=4$

Give your answers as logarithms in their simplest form.

17 Find the solution to the equation $3\operatorname{cosech}^2 x+\coth^2 x=4\coth x$

Give your answer in the form $\ln k$, where k is a positive constant to be found.

18 Solve the equation $3\tanh 2x=\coth x$

Give your answers as simplified logarithms.

Fluency and skills

See Ch3.1
For a reminder on finding asymptotes.

You can sketch the graphs of rational functions with horizontal and/or vertical asymptotes.

In this section, you will look at extending this to include rational functions that have **oblique** (or slant) asymptotes.

Consider the graph of the function $y = \dfrac{x}{x^2 - 4}$

There are vertical asymptotes at $x = 2$ and $x = -2$ since the function is undefined for these values of x

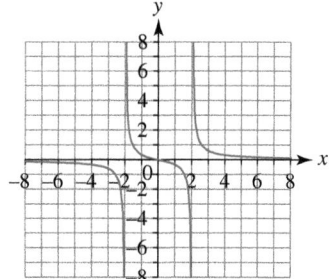

There is also a horizontal asymptote at $y = 0$ since y tends to zero as $x \to \pm\infty$

This happens since as x gets large (either negative or positive) the denominator of the function starts to dominate and 'pulls' the value of the function towards zero.

If the rational function is **improper** (or 'top heavy') this logic does not hold. For large x, the **numerator** of the function starts to dominate.

Consider the graph of the function $y = \dfrac{x^2}{x - 4}$

There is a vertical asymptote when $x = 4$ since the function is undefined for this value of x

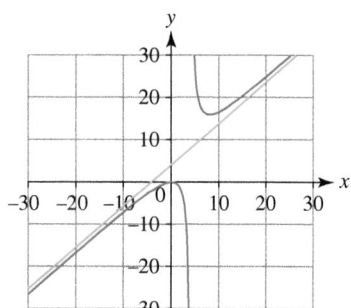

There is also an oblique asymptote as shown.

Using algebraic long division, you can re-write the function as

$$y = x + 4 + \frac{16}{x - 4}$$

Now, as $x \to \pm\infty$ you can see that the fraction part of the functions tends to zero as in the case of proper rational functions, but that leaves the $x + 4$ ('whole') bit. Hence, as $x \to \pm\infty$, y tends to $x + 4$ and there is an oblique asymptote with equation $y = x + 4$

> **Key point**
>
> Oblique asymptotes occur when a rational function is improper.

You find the equation of an oblique asymptote by dividing the function out to find the 'whole' part and setting y equal to this.

Example 1

Find the equations of the asymptotes of the curve $y = \dfrac{x^2 - x}{x+1}$

There is a vertical asymptote at $x = -1$ ●────────────

$$
\begin{array}{r}
x-2 \\
x+1\overline{)x^2 - x} \\
\underline{x^2 + x} \\
-2x \\
\underline{-2x-2} \\
2
\end{array}
$$

Hence $y = x - 2 + \dfrac{2}{x+1}$

There is an oblique asymptote with equation $y = x - 2$ ●────────

> The function is undefined when $x = -1$ since the denominator would be zero.

> Use algebraic long division to divide out the function.

> Use the divided out form to identify the equation of the oblique asymptote.

Exercise 18.4A Fluency and skills

1 Find the equations of the asymptotes for each of these curves.

  a $y = \dfrac{x^2 + 2x - 12}{x-5}$ **b** $y = \dfrac{2x^2}{1-x}$

  c $y = \dfrac{2 - 3x^2}{x-1}$ **d** $y = \dfrac{x^2 + x - 1}{x-2}$

2 Sketch the graphs of each of these functions. Show clearly any asymptotes.

  a $y = \dfrac{x^2 - 1}{x+3}$ **b** $y = \dfrac{2x^2 - 1}{2x+3}$

  c $y = \dfrac{x^2 - 2x + 1}{4 - x}$ **d** $y = \dfrac{x^2 - 4x + 3}{3 - 2x}$

3 Find the equations of the asymptotes for each of these curves. Hence sketch the curves.

  a $y = \dfrac{x^3 - 1}{2x^2 - 1}$ **b** $y = \dfrac{x^3 - x}{x^2 - 4}$

Reasoning and problem-solving

You need to be able to find the value of unknowns in a rational function when you are given the equation of the oblique asymptote.

Typically, this will be the values of a and/or b when the numerator is in the form $ax^2 + bx + c$

To find the value of unknowns in the rational function when you are given the equation of the oblique asymptote

  (1) Multiply the asymptote by the denominator of the rational function to form a quadratic expression in the form $ax^2 + bx + c$

  (2) Read off the values of a and/or b from this quadratic and these will correspond to the values of a and/or b in the rational function.

Example 2

A curve with equation $y = \dfrac{ax^2 + bx + 1}{x - 2}$ has an asymptote $y = 2x + 1$

a Find the values of a and b

b Write down the equation of the other asymptote.

c Without using calculus, find the coordinates of the turning points.

d Sketch the curve.

a $(2x + 1)(x - 2) = 2x^2 - 3x - 2$ •————

Hence $a = 2$ and $b = -3$ •————

The rational function is $y = \dfrac{2x^2 - 3x + 1}{x - 2}$

> **1** Multiply the asymptote by the denominator of the rational function.

b $x = 2$ •————

> **2** Read off the values of a and b

c The range of allowable values of y can be found by setting $y = k$ and solving for k

$$k(x - 2) = 2x^2 - 3x - 1$$

$$0 = 2x^2 - 3x - kx - 1 + 2k$$

$$0 = 2x^2 - (3 + k)x + (2k - 1)$$

Solving for k using $b^2 - 4ac \geq 0$ gives

$$k \geq 5 + 2\sqrt{2} \text{ or } k \leq 5 - 2\sqrt{2}$$

Hence the curve exists for $y \geq 5 + 2\sqrt{2}$ and $y \leq 5 - 2\sqrt{2}$

There will be a minimum point at $y = 5 + 2\sqrt{2}$ and a maximum point at $y = 5 - 2\sqrt{2}$ •————

> This is the value of x for which the function is undefined.

> For a full review of this technique, see Section 3.2, Example 3

To find the x-values for these turning points, substitute the y-values into the original function

$$(5 \pm 2\sqrt{2})(x - 2) = 2x^2 - 3x - 1$$

$$\Rightarrow 2x^2 - (8 \pm 2\sqrt{2})x + (9 \pm 2\sqrt{2}) = 0$$

Solving for x in this equation gives the x-values as $\dfrac{4 \pm \sqrt{2}}{2}$

Hence the turning points are:

Minimum at $\left(\dfrac{4 + \sqrt{2}}{2}, 5 + 2\sqrt{2} \right)$

Maximum at $\left(\dfrac{4 - \sqrt{2}}{2}, 5 - 2\sqrt{2} \right)$

d

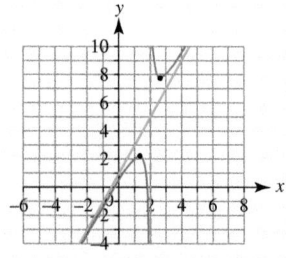

> Draw in both asymptotes and use your answer to part **c** to sketch the sections of the graph.

1 A curve with equation $y = \dfrac{ax^2 + bx - 1}{x + 1}$ has an asymptote $y = 4x - 2$

   a Find the values of a and b

   b Write down the equation of the other asymptote.

   c Without using calculus, find the coordinates of the turning points.

   d Sketch the curve.

2 A curve with equation $y = \dfrac{ax^2 + bx + 1}{x - 2}$ has an asymptote $y = -x - 5$

   a Find the values of a and b

   b Write down the equation of the other asymptote.

   c Without using calculus, find the coordinates of the turning points.

   d Sketch the curve.

3 $f(x) = \dfrac{x^3 - 1}{2(x^2 - 1)}$

   a Show that $f(x)$ can be written in the form $g(x) + \dfrac{bx + c}{2(x + 1)}$ where b and c are constants to be found.

   b Hence write down the equations of the asymptotes to the curve $y = f(x)$

   c Find the coordinates of the turning points on the curve $y = f(x)$

   d Hence sketch the curve $y = f(x)$

4 $f(x) = \dfrac{x^3}{x - 1}$

   a Write $f(x)$ in the form $g(x) + \dfrac{c}{x - 1}$ where c is a constant.

   b Hence write down the equations of the asymptotes to the curve $y = f(x)$

   c Sketch the curve $y = f(x)$

Chapter summary

- Given $y = f(x)$ you can sketch the graph of $y = \dfrac{1}{f(x)}$ by following some simple rules

  - The sign of $\dfrac{1}{f(x)}$ is the same as the sign of $f(x)$
  - Any roots (or zeros) of $y = f(x)$ become vertical asymptotes in $y = \dfrac{1}{f(x)}$
  - As $f(x) \to 0$ then $\dfrac{1}{f(x)} \to \pm\infty$ and vice versa
  - As $f(x)$ increases, $\dfrac{1}{f(x)}$ decreases and vice versa
  - If $f(x) = \pm 1$ exists then $\dfrac{1}{f(x)}$ also equals ± 1 and the graphs intersect
  - Minimum points become maximum points or asymptotes and vice versa

- To sketch the graph of $y = |f(x)|$
  - Sketch $y = f(x)$ and reflect any negative parts of the graph in the x-axis
  - Any minimum turning point below the x axis will be reflected to a maximum turning point above the x axis.

- To transform graphs of conic sections
  - Enlarge by scale factor k by replacing x with $\dfrac{x}{k}$ and y with $\dfrac{y}{k}$
  - Rotate by $\dfrac{\pi}{2}$ by replacing x by y and y by $-x$
  - Rotate by π by replacing x by $-x$ and y by $-y$
  - Rotate by $\dfrac{3\pi}{2}$ by replacing x by $-y$ and y by x

- $\sinh x = \dfrac{e^x - e^{-x}}{2} \Rightarrow \operatorname{arsinh} x = \ln(x + \sqrt{x^2 + 1})$

- $\cosh x = \dfrac{e^x + e^{-x}}{2} \Rightarrow \operatorname{arcosh} x = \ln(x + \sqrt{x^2 - 1}),\ x \geq 1$

- $\tanh x = \dfrac{e^x - e^{-x}}{e^x + e^{-x}} \Rightarrow \operatorname{artanh} x = \dfrac{1}{2}\ln\left(\dfrac{1+x}{1-x}\right), -1 < x < 1$

- Graphs of inverse hyperbolic functions

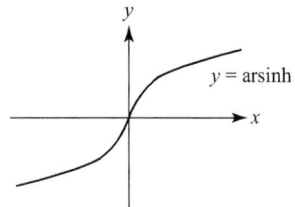

domain is $x \in \mathbb{R}$,
range is $y \in \mathbb{R}$

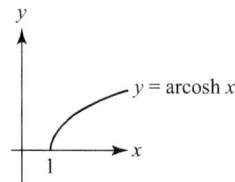

domain is $x \in \mathbb{R}, x \geq 1$
range is $y \in \mathbb{R}, y \geq 0$

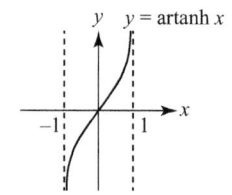

domain is $x \in \mathbb{R}, -1 < x < 1$
range is $y \in \mathbb{R}$

f(x)	Domain of f(x)	Range of f(x)	f⁻¹(x)	Domain of f⁻¹(x)	Range of f⁻¹(x)
$\sinh x$	$x \in \mathbb{R}$	$f(x) \in \mathbb{R}$	$\operatorname{arsinh} x$	$x \in \mathbb{R}$	$f^{-1}(x) \in \mathbb{R}$
$\cosh x$	$x \in \mathbb{R},$ $x \geq 0$	$f(x) \in \mathbb{R}, f(x) \geq 1$	$\operatorname{arcosh} x$	$x \in \mathbb{R},$ $x \geq 1$	$f^{-1}(x) \in \mathbb{R}, f^{-1}(x) \geq 0$
$\tanh x$	$x \in \mathbb{R}$	$f(x) \in \mathbb{R},$ $-1 < f(x) < 1$	$\operatorname{artanh} x$	$x \in \mathbb{R},$ $-1 < x < 1$	$f^{-1}(x) \in \mathbb{R}$

- $\operatorname{cosech} x \equiv \dfrac{1}{\sinh x} \equiv \dfrac{2}{e^x - e^{-x}}$

- $\operatorname{sech} x \equiv \dfrac{1}{\cosh x} \equiv \dfrac{2}{e^x + e^{-x}}$

- $\coth x \equiv \dfrac{1}{\tanh x} \equiv \dfrac{e^{2x} + 1}{e^{2x} - 1}$

- Graphs of reciprocal hyperbolic functions

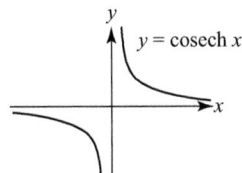

domain is $x \in \mathbb{R}, x \neq 0$,
range is $y \in \mathbb{R}, y \neq 0$

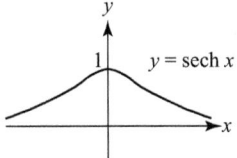

domain is $x \in \mathbb{R}$,
range is $y \in \mathbb{R}, 0 < y \leq 1$

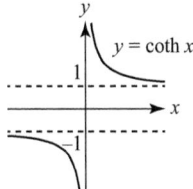

domain is $x \in \mathbb{R}, x \neq 0$
range is $y \in \mathbb{R}, y < -1, y > 1$

f(x)	Domain	Range
$\operatorname{cosech} x$	$x \in \mathbb{R}, x \neq 0$	$f(x) \in \mathbb{R}, f(x) \neq 0$
$\operatorname{sech} x$	$x \in \mathbb{R}$	$f(x) \in \mathbb{R}, 0 < f(x) \leq 1$
$\coth x$	$x \in \mathbb{R}, x \neq 0$	$f(x) \in \mathbb{R}, f(x) < -1 \text{ or } f(x) > 1$

- $\cosh^2 x - \sinh^2 x \equiv 1 \ \Rightarrow \ 1 - \tanh^2 x \equiv \operatorname{sech}^2 x$ and $\coth^2 x - 1 \equiv \operatorname{cosech}^2 x$
- Oblique asymptotes occur when a rational function is improper.
- You find the equation of an oblique asymptote by dividing the function out to find the 'whole' part and setting y equal to this.
- To find the value of unknowns in the rational function when you are given the equation of the oblique asymptote
  - Multiply the asymptote by the denominator of the rational function to form a quadratic expression in the form $ax^2 + bx + c$
  - Read off the values of a and/or b from this quadratic and these will correspond to the values of a and/or b in the rational function.

You should now be able to...	Review Questions
✔ Use graphical and algebraic methods to find and sketch reciprocal and modulus functions.	1, 3, 4–9
✔ Use graphical and algebraic methods to solve inequalities involving reciprocal and modulus functions.	2, 9
✔ Enlarge graphs of conic sections.	10, 11
✔ Rotate graphs of conic sections.	12, 13
✔ Combine transformations of conic sections.	12–15
✔ Sketch graphs of inverse hyperbolic functions and state each domain and range.	16
✔ Sketch graphs of reciprocal hyperbolic functions and state each domain and range.	17
✔ Know, prove and use identities involving hyperbolic functions.	19–20
✔ Solve equations involving hyperbolic functions.	18–21
✔ Understand and use the concept of an oblique asymptote, including finding the equation of an oblique asymptote.	22–23

1 Sketch the graph of $y = \dfrac{(x-1)}{(x+4)}$

2 Solve these inequalities graphically, naming any intercepts and vertical and/or horizontal asymptotes.

    a $-3 < \dfrac{x+12}{x-4} < 9$ **b** $7 \geq \dfrac{x-6}{x-12} \geq -5$

3 Sketch the graphs of

    a $y = \dfrac{x^2+12x+24}{x^2+2x+2}$ **b** $y = \dfrac{x^2+12x+24}{x^2+2x-3}$

4 Find the localised minimum turning point of the function $f(x) = \dfrac{x^2-5x}{x^2-5x+5}$ and find a maximum value.

5 The diagram shows the graph of $y = f(x)$ and the points P, Q and R

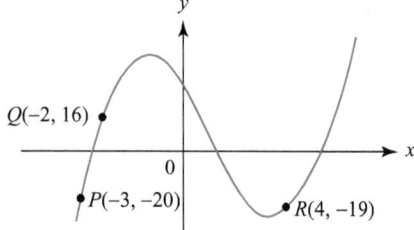

Sketch the graph of $y = |f(x)|$, labelling the transformed points P', Q' and R'

6 Sketch the graphs of $y = |f(x)|$ for these functions.

   a $f(x) = x^2 - 2x - 5$

   b $f(x) = 2x^2 - 3x - 6$

7 Copy each graph $y = f(x)$ and sketch on the same axes the graph of $y = \dfrac{1}{f(x)}$

   a

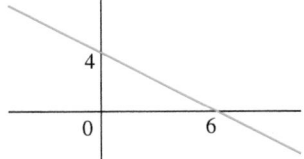

   b

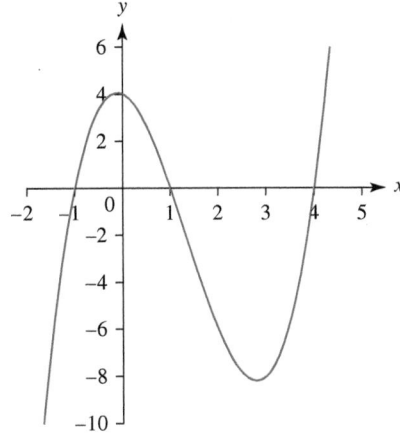

8 For each function

   i   Sketch $y = f(x)$

   ii   Sketch $\dfrac{1}{f(x)}$, labelling any asymptotes and points of intersection.

     a $y = -3x - 2$

     b $y = x^2 + 6x + 5$

     c $y = x^3 + 4$

9 A curve has equation $y = \dfrac{4+x}{2-x}$

   a **i**   Write down the equations of the asymptotes to the curve and the points of intersection with the coordinate axes.

     ii   Sketch the curve, indicating clearly the coordinates of the points of intersection of the curve with the coordinate axes.

   b   Sketch also the curve $y = \left|\dfrac{4+x}{2-x}\right|$

   c   Add the straight line $y = 2 - \dfrac{3}{2}x$ to each sketch and hence solve the inequalities $\dfrac{4+x}{2-x} > 2 - \dfrac{3}{2}x$ and $\left|\dfrac{4+x}{2-x}\right| > 2 - \dfrac{3}{2}x$

10 The ellipse with equation $\dfrac{x^2}{5} + \dfrac{y^2}{9} = 1$ is enlarged by scale factor 3

    a Find the equation of the transformed curve.

    b Sketch the transformed curve, stating the coordinates of the points of intersection with the coordinate axes.

11 The hyperbola with equation $\dfrac{x^2}{64} - \dfrac{y^2}{36} = 1$ is enlarged by scale factor $\dfrac{1}{2}$

    a Find the equation of the transformed curve.

    b Sketch the transformed curve and find the equations of the asymptotes.

12 The parabola with equation $y^2 = 18x$ is rotated $\dfrac{\pi}{2}$ radians anticlockwise about the origin.

    a Find the equation of the transformed curve.

    The curve is then translated by the vector $\begin{pmatrix} -1 \\ 2 \end{pmatrix}$

    b Find the equation of the transformed curve.

    c Sketch the transformed curve, and state the coordinates of the minimum point.

13 The ellipse with equation $\dfrac{x^2}{9} + \dfrac{y^2}{4} = 1$ is rotated $\dfrac{3\pi}{2}$ radians anticlockwise about the origin.

    a Find the equation of the transformed curve.

    The curve is then translated by vector $\begin{pmatrix} 2 \\ 3 \end{pmatrix}$

    b Find the equation of the transformed curve.

    c Sketch the transformed curve.

14 The ellipse with equation $\dfrac{x^2}{4} + \dfrac{y^2}{3} = 1$ is first enlarged by scale factor 5 then translated by vector $\begin{pmatrix} 1 \\ -4 \end{pmatrix}$

    a Find the equation of the transformed curve.

    b Sketch the transformed curve.

15 The parabola with equation $y^2 = 7x$ is reflected in the line $y = -x$ then rotated $\dfrac{\pi}{2}$ radians anticlockwise about the origin.

    a Find the equation of the transformed curve.

    b Sketch the transformed curve.

16 Sketch the graph of $y = f(x)$ for each of these functions and state each domain and range.

    a $y = 1 + \operatorname{arsinh}(x)$

    b $y = \operatorname{arcosh}(2x)$

    c $y = \operatorname{artanh}(x+1)$

17 Sketch the graph of $y = f(x)$ for this function and state the domain and range.

$y = 3\operatorname{sech}(x)$

18 Solve these equations and give your answer in the form $a \ln b$ where a and b are constants to be found.

   a $\coth(x) = 3$

   b $e^x \operatorname{cosech}(x) = -1$

   c $\operatorname{cosech} x - \operatorname{sech} x = 2e^{-x}$

19 **a**   Use exponentials to prove that $\operatorname{cosech}(2x) \equiv \dfrac{1}{2}\operatorname{cosech}(x)\operatorname{sech}(x)$

   b   Hence solve the equation $\operatorname{cosech}(x)\operatorname{sech}(x) = -2$

20 Find the exact solutions to each of these equations.

   a $\operatorname{cosech}^2 x + 3\operatorname{cosech} x = 4$

   b $5\coth x = \operatorname{cosech}^2 x + 1$

21 Show that the equation $\tanh^2 x + \operatorname{sech} x + 5 = 0$ has no real solutions.

22 $f(x) = \dfrac{x^2 - 2x - 1}{x - 5}$

   a   Find the equations of the asymptotes of the curve $y = f(x)$

   b   Hence sketch the curve $y = f(x)$

23 A curve with equation $y = \dfrac{ax^2 + bx - 1}{3x - 1}$ has an asymptote $y = x - \dfrac{1}{3}$

   a   Find the values of a and b

   b   Write down the equation of the other asymptote.

   c   Sketch the curve.

History

The Cartesian coordinate system is named after the French mathematician René Descartes (1596–1650). It is suggested that he developed the system whilst musing about how to best describe the position of a fly on his ceiling: this led to his development of a system of coordinates which referred to an origin at the corner of the ceiling. This development led to the possibility of describing shapes such as circles and parabolas using equations that describe the relationships between their x and y coordinates, thus linking geometry and algebra.

Note

Sine and cosine functions are known as circular functions as they can be defined using the position of a point on a circle. Let a point, P, lie on a unit circle with centre O such that the angle at O measured anticlockwise from the positive x-axis to the line segment OP is θ. As the point moves around the circle, the x- and y-coordinates of P define cosine and sine respectively.

The hyperbolic functions are defined in a similar way using the hyperbola $x^2 - y^2 = 1 \, (x > 1)$ What is similar and what is different?

In particular, notice that the hyperbolic angle is defined in terms of the area enclosed between a line passing through a point on the hyperbola, the hyperbola and the x-axis. (Areas beneath the x-axis are considered negative.)

Explore the graphs of the hyperbolic functions by considering how they can be traced out as the point, P, moves on the hyperbola.

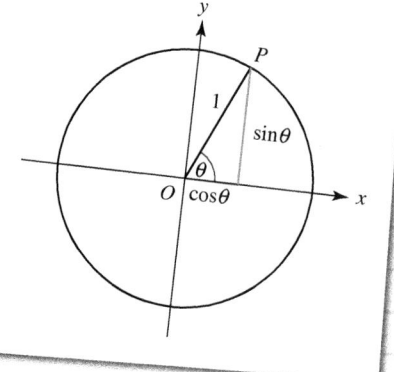

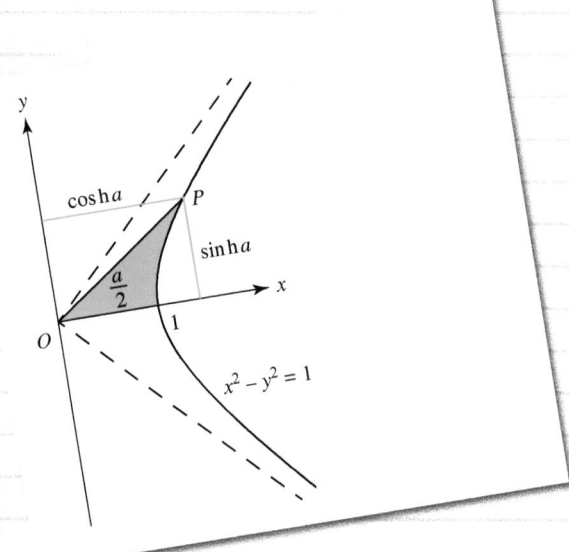

1 $f(x) = 2x^2 - 3x - 2$

   a Sketch the graph of $y = f(x)$ showing clearly the points where the curve crosses the coordinate axes. **[2 marks]**

   b On the same set of axes, sketch the graph of $y = \dfrac{1}{f(x)}$, labelling clearly any asymptotes. **[3]**

   c Calculate the set of values of x for which $f(x) < \dfrac{1}{f(x)}$ **[6]**

2. The graph of $y = f(x)$ is shown in the diagram.

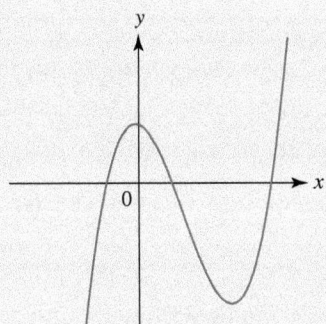

   a Sketch the graph of $y = \dfrac{1}{f(x)}$, showing clearly any asymptotes. **[4]**

   b Sketch the graph of $y = |f(x)|$ **[3]**

3 $f(x) = \dfrac{3-x}{1+x}$

   a Write down the equations of any asymptotes on the graph of $y = f(x)$ and state the coordinates of any points of intersection with the coordinate axes. **[3]**

   b Sketch the graph of $y = |f(x)|$ **[3]**

   c Using your sketch or otherwise, find the set of values of x for which $|f(x)| < 3$ **[2]**

4 The ellipse E has equation $\dfrac{x^2}{4} + \dfrac{y^2}{2} = 1$

   The ellipse is enlarged by scale factor 2 and then translated by vector $\begin{pmatrix} 2 \\ -3 \end{pmatrix}$

   a Show that the new conic has equation $x^2 + 2y^2 - 4x + 12y + k = 0$ where k is a constant to be found. **[5]**

   b Find the equations of the tangents to the new conic which are parallel to the x-axis. **[4]**

5 A parabola P has equation $y^2 = 16x$

   Describe a sequence of transformations that map P onto the curve $x^2 - 2x + 16y + 33 = 0$ **[5]**

6 A hyperbola H has equation $\dfrac{x^2}{9} - \dfrac{y^2}{4} = 1$

    a Show that the line $3y - x = 3$ intersects the hyperbola at two points and find the exact coordinates of these points. **[5]**

    The hyperbola H is stretched by scale factor 2 in the x-direction and then rotated anticlockwise through $\dfrac{\pi}{2}$ radians about the origin to produce the hyperbola H'

    b Sketch the graph of H' **[3]**

    c Write down the equations of the asymptotes to H' **[1]**

7 From the definitions of $\sinh x$ and $\cosh x$ in terms of exponentials,

    a Prove that $\cosh^2 x - \sinh^2 x = 1$ **[3]**

    b Hence, or otherwise, solve the equation $\operatorname{cosech} x = 2(1 + \coth x)$, giving your answer in the form $a \ln b$ where a and b are constants to be found. **[5]**

8 **a** Use exponentials to show that $2\cosh^2 x = \cosh 2x + 1$ **[3]**

    b Hence, or otherwise, solve the equation $5\cosh x = \cosh 2x - 2$, giving your answers in exact logarithmic form. **[4]**

9 **a** Show that $\sinh x + 3\operatorname{cosech} x = 4$ can be written as $\sinh^2 x - 4\sinh x + 3 = 0$ **[2]**

    b Hence find the exact solutions to the equation $\sinh x + 3\operatorname{cosech} x = 4$ **[3]**

10 $f(x) = \operatorname{cosech}(2x + 1)$

    a Sketch the graph of $y = f(x)$, showing clearly the equations of any asymptotes. **[3]**

    b State the domain and range of $f(x)$ **[2]**

    c Show that the coordinates of the point where the graph of $y = f(x)$ intersects the y-axis is $\left(0, \dfrac{2e}{e^2 - 1}\right)$ **[3]**

11 $f(x) = \dfrac{2x^2 - 3}{x + 1}$

    a Find the equations of the asymptotes of the curve $y = f(x)$ **[3]**

    $g(x) = \dfrac{ax^2 + bx - 1}{2x - 1}$

    The curve with equation $y = g(x)$ has the same oblique asymptote as the curve with equation $y = f(x)$

    b Find the values of a and b **[2]**

    c Hence sketch the curve $y = g(x)$ **[3]**

19 Integration 2

Economics is an area of study and work that has become increasingly important. You can see this by the time and space that is devoted to business and the economy in various parts of the media. Many aspects of the work of economists has become more mathematical, as they develop, and use, a range of mathematical models. These models allow them to understand not only what has happened in the past, but to also predict what may happen in the future.

Aspects of calculus are important in the mathematics used by economists. Methods and techniques associated with integration and differentiation allow economists to consider how measurable quantities change with time. For example, supply and demand, in relation to (number of) sales and pricing, are quantities that can be modelled using mathematical functions. Once the model is set up, calculus can be used to work out optimal conditions.

Orientation

What you need to know	What you will learn	What this leads to
Chapter 4 Integration 1	• How to work with improper integrals. • How to differentiate inverse trigonometric functions. • How to differentiate hyperbolic functions. • How to use partial fractions to help with integration. • How to use a reduction formula. • How to use integration to find areas enclosed by a polar curve. • How to use integration to find lengths and areas.	**Careers** • Economics.

19.1 Improper integrals

Fluency and skills

You can find the value of a definite integral by integrating and then substituting in the limits. This value represents a finite area between the curve and an axis.

> **Key point**
>
> An **improper integral** is a definite integral where either:
> - one or both of the limits is $\pm\infty$
> - the integrand (expression to be integrated) is undefined at one of the limits of the integral
> - the integrand is undefined at some point between the limits of the integral.

It is sometimes possible to calculate the value of an improper integral by replacing a limit of $\pm\infty$ with a variable and then considering what happens as that variable tends to $\pm\infty$

> **Key point**
>
> To evaluate an improper integral with a limit of $\pm\infty$ use
> $$\int_a^\infty f(x)dx = \lim_{t\to\infty}\int_a^t f(x)dx \quad \text{or} \quad \int_{-\infty}^a f(x)dx = \lim_{t\to-\infty}\int_t^a f(x)dx$$
> If the limit exists, then the improper integral is called **convergent**.
>
> If the limit does not exist, then the improper integral is called **divergent**.

For example, the integral, $\int_1^\infty \dfrac{1}{x^2}\,dx$ is an improper integral because one of the limits is infinite. To evaluate it, use

$$\int_1^\infty \frac{1}{x^2}dx = \lim_{t\to\infty}\int_1^t \frac{1}{x^2}dx$$

$$= \lim_{t\to\infty}\left(1-\frac{1}{t}\right) \quad \text{since} \int_1^t \frac{1}{x^2}dx = \left[-\frac{1}{x}\right]_1^t = 1-\frac{1}{t}$$

$$= 1 \text{ since } \frac{1}{t}\to 0 \text{ as } t\to\infty$$

Therefore, the improper integral $\int_1^\infty \dfrac{1}{x^2}dx$ is convergent and represents a finite area of 1

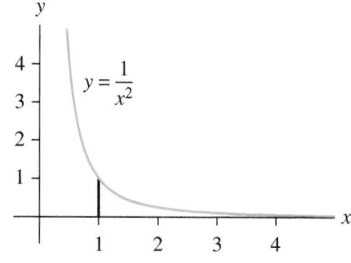

However, if you were to try to evaluate the integral $\int_0^\infty x^2 dx$ then

$$\lim_{t\to\infty}\int_0^t x^2 dx = \lim_{t\to\infty}\left(\frac{t^3}{3}\right) \quad \text{since} \int_0^t x^2 dx = \left[\frac{x^3}{3}\right]_0^t = \frac{t^3}{3}$$

$\dfrac{t^3}{3}\to\infty$ as $t\to\infty$, therefore the improper integral $\int_0^\infty x^2 dx$ is divergent and represents an infinite area.

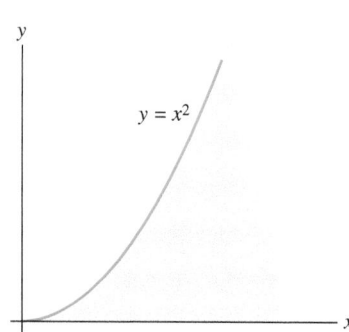

Example 1

Find the value of $\int_{2}^{\infty} \dfrac{2}{x^3}\,dx$

$$\int_{2}^{\infty} \frac{2}{x^3}\,dx = \lim_{t\to\infty}\int_{2}^{t} 2x^{-3}\,dx$$

— Replace ∞ with t

$$= \lim_{t\to\infty}\left[-x^{-2}\right]_{2}^{t}$$

$$= \lim_{t\to\infty}((-t^{-2})-(-2^{-2})) = \lim_{t\to\infty}\left(-\frac{1}{t^2}+\frac{1}{4}\right) = \frac{1}{4}\ \text{since}\ \frac{1}{t^2}\to 0\ \text{as}\ t\to\infty$$

If both the limits are $\pm\infty$, then the integral needs to be split into two integrals, each with one finite limit. Any point can be chosen for this limit, so just choose a convenient value.

> When splitting the integral, you need to use different variables for ∞ and $-\infty$
> This is because both parts of the integral must be convergent for the original integral to exist.

Example 2

a Find the value of $\displaystyle\int_{-\infty}^{\infty} \dfrac{x}{e^{x^2}}\,dx$

b Show that the improper integral $\displaystyle\int_{-\infty}^{\infty} e^{x}$ is divergent.

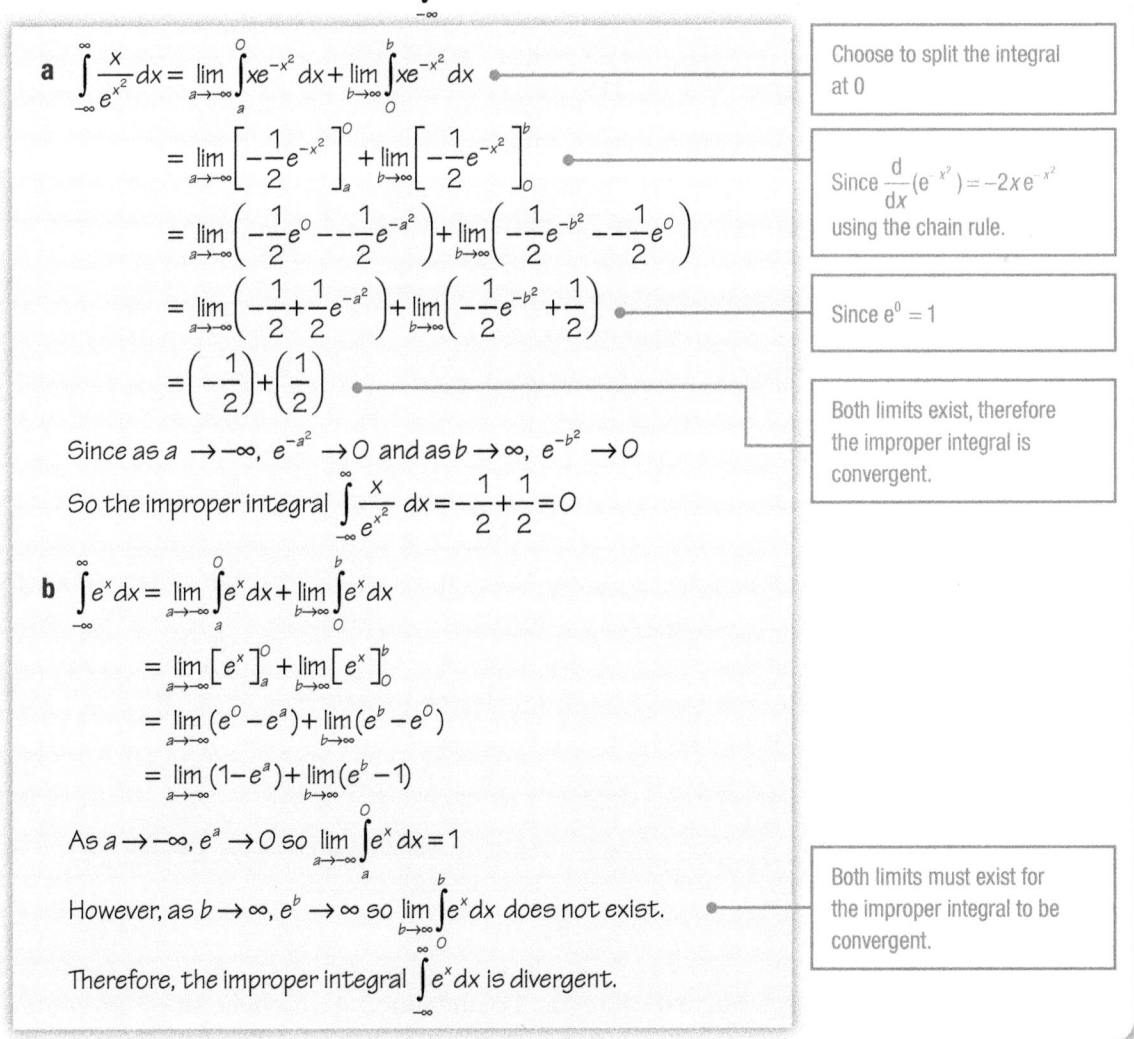

a $\displaystyle\int_{-\infty}^{\infty} \frac{x}{e^{x^2}}\,dx = \lim_{a\to-\infty}\int_{a}^{0} xe^{-x^2}\,dx + \lim_{b\to\infty}\int_{0}^{b} xe^{-x^2}\,dx$

— Choose to split the integral at 0

$$= \lim_{a\to-\infty}\left[-\frac{1}{2}e^{-x^2}\right]_{a}^{0} + \lim_{b\to\infty}\left[-\frac{1}{2}e^{-x^2}\right]_{0}^{b}$$

— Since $\dfrac{d}{dx}(e^{-x^2}) = -2xe^{-x^2}$ using the chain rule.

$$= \lim_{a\to-\infty}\left(-\frac{1}{2}e^{0}--\frac{1}{2}e^{-a^2}\right) + \lim_{b\to\infty}\left(-\frac{1}{2}e^{-b^2}--\frac{1}{2}e^{0}\right)$$

$$= \lim_{a\to-\infty}\left(-\frac{1}{2}+\frac{1}{2}e^{-a^2}\right) + \lim_{b\to\infty}\left(-\frac{1}{2}e^{-b^2}+\frac{1}{2}\right)$$

— Since $e^0 = 1$

$$= \left(-\frac{1}{2}\right)+\left(\frac{1}{2}\right)$$

— Both limits exist, therefore the improper integral is convergent.

Since as $a\to-\infty$, $e^{-a^2}\to 0$ and as $b\to\infty$, $e^{-b^2}\to 0$

So the improper integral $\displaystyle\int_{-\infty}^{\infty} \frac{x}{e^{x^2}}\,dx = -\frac{1}{2}+\frac{1}{2}=0$

b $\displaystyle\int_{-\infty}^{\infty} e^{x}\,dx = \lim_{a\to-\infty}\int_{a}^{0} e^{x}\,dx + \lim_{b\to\infty}\int_{0}^{b} e^{x}\,dx$

$$= \lim_{a\to-\infty}\left[e^{x}\right]_{a}^{0} + \lim_{b\to\infty}\left[e^{x}\right]_{0}^{b}$$

$$= \lim_{a\to-\infty}(e^{0}-e^{a}) + \lim_{b\to\infty}(e^{b}-e^{0})$$

$$= \lim_{a\to-\infty}(1-e^{a}) + \lim_{b\to\infty}(e^{b}-1)$$

As $a\to-\infty$, $e^{a}\to 0$ so $\displaystyle\lim_{a\to-\infty}\int_{a}^{0} e^{x}\,dx = 1$

However, as $b\to\infty$, $e^{b}\to\infty$ so $\displaystyle\lim_{b\to\infty}\int_{0}^{b} e^{x}\,dx$ does not exist.

— Both limits must exist for the improper integral to be convergent.

Therefore, the improper integral $\displaystyle\int_{-\infty}^{\infty} e^{x}\,dx$ is divergent.

Another type of improper integral is where the integrand is undefined at one of the limits of the integral. To evaluate these integrals you replace that limit of integration with a variable as before. You then consider what happens as the variable tends to the original value of the limit of integration.

Find the value of $\displaystyle\int_0^4 \frac{2}{\sqrt{x}}\,dx$

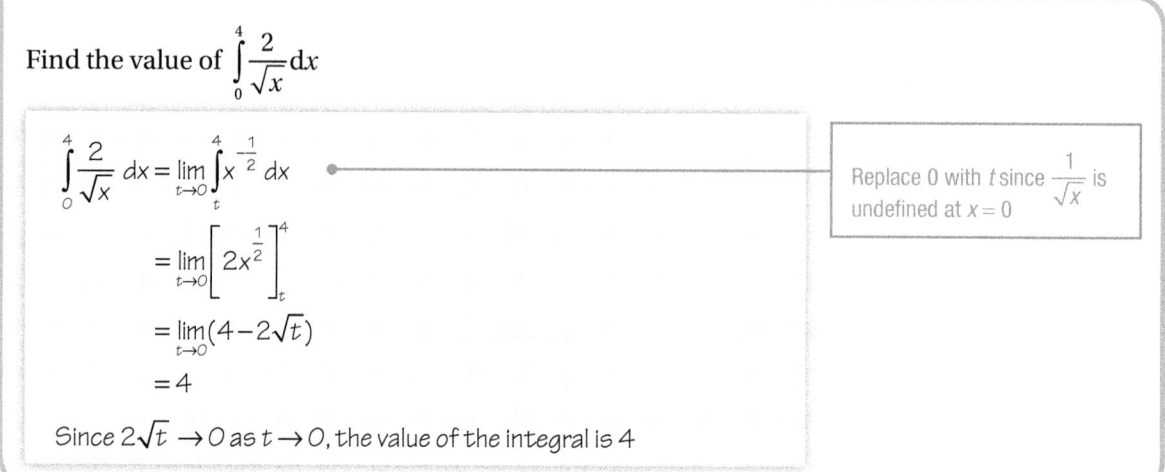

$$\int_0^4 \frac{2}{\sqrt{x}}\,dx = \lim_{t\to 0}\int_t^4 x^{-\frac{1}{2}}\,dx$$

Replace 0 with t since $\dfrac{1}{\sqrt{x}}$ is undefined at $x = 0$

$$= \lim_{t\to 0}\left[2x^{\frac{1}{2}}\right]_t^4$$

$$= \lim_{t\to 0}(4 - 2\sqrt{t})$$

$$= 4$$

Since $2\sqrt{t} \to 0$ as $t \to 0$, the value of the integral is 4

When evaluating integrals, you sometimes need to find the limit of $x^k e^{-x}$ as $x \to \infty$ for some value of k. Since $x^k \to \infty$ but $e^{-x} \to 0$ it is not obvious what will happen to the value of $x^k e^{-x}$. You need to know this result:

Key point

For any real number k, $x^k e^{-x} \to 0$ when $x \to \infty$

This can be proved using the series expansion of e^x but you can simply quote it.

Another common limit is that of $x^k \ln x$ as $x \to 0$. Again, as $x^k \to 0$ and $\ln x \to -\infty$ it is not obvious what the result will be, but the previous result can be used to prove the following:

Key point

For any real number k, $x^k \ln x \to 0$ when $x \to 0+$
(this means that x approaches zero from above as x must be positive for $\ln x$ to be defined).

You can also quote this result.

Find the value of $\displaystyle\int_0^1 2 - \ln x\,dx$

$$\int_0^1 2 - \ln x\,dx = \lim_{t\to 0}\int_t^1 2 - \ln x\,dx$$

Replace 0 with t since $2 - \ln x$ is undefined at $x = 0$

$$= \lim_{t\to 0}\left[2x - (x\ln x - x)\right]_t^1$$

Use integration by parts with $u = \ln x$ and $\dfrac{dv}{dx} = 1$
$$\int \ln x\,dx = x\ln x - \int 1\,dx$$

$$= \lim_{t\to 0}((2 - 1\ln 1 + 1) - (2t - t\ln t + t))$$

$$= \lim_{t\to 0}(3 - 3t + t\ln t)$$

$$= 3$$

Since $3t \to 0$ and $t\ln t \to 0$ as $t \to 0$

So the value of the integral is 3

Using the fact that $t^k \ln t \to 0$ when $t \to 0$, in this case $k = 1$

To evaluate an improper integral where the integrand is undefined at a point between the limits of the integral, you need to split the integral into two parts about the point of discontinuity.

Example 5

PURE

You need to identify the point of discontinuity and split the integral at this point.

Evaluate the improper integral $\int_{-1}^{e} x \ln|x|\, dx$

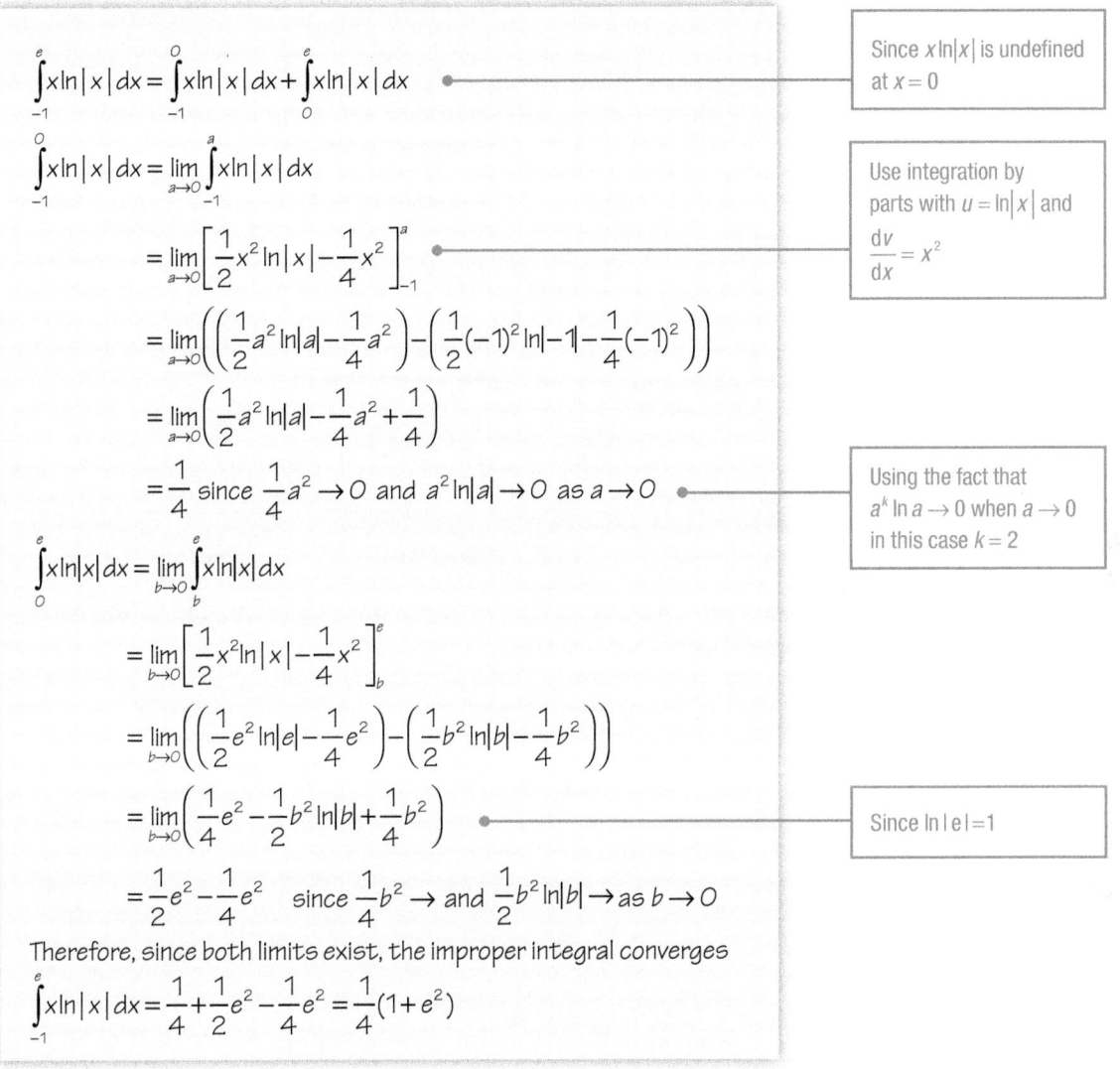

$\int_{-1}^{e} x\ln|x|\,dx = \int_{-1}^{0} x\ln|x|\,dx + \int_{0}^{e} x\ln|x|\,dx$

Since $x\ln|x|$ is undefined at $x=0$

$\int_{-1}^{0} x\ln|x|\,dx = \lim_{a\to 0}\int_{-1}^{a} x\ln|x|\,dx$

Use integration by parts with $u=\ln|x|$ and $\dfrac{dv}{dx}=x^2$

$= \lim_{a\to 0}\left[\dfrac{1}{2}x^2\ln|x|-\dfrac{1}{4}x^2\right]_{-1}^{a}$

$= \lim_{a\to 0}\left(\left(\dfrac{1}{2}a^2\ln|a|-\dfrac{1}{4}a^2\right)-\left(\dfrac{1}{2}(-1)^2\ln|-1|-\dfrac{1}{4}(-1)^2\right)\right)$

$= \lim_{a\to 0}\left(\dfrac{1}{2}a^2\ln|a|-\dfrac{1}{4}a^2+\dfrac{1}{4}\right)$

$= \dfrac{1}{4}$ since $\dfrac{1}{4}a^2 \to 0$ and $a^2\ln|a| \to 0$ as $a \to 0$

Using the fact that $a^k\ln a \to 0$ when $a \to 0$ in this case $k=2$

$\int_{0}^{e} x\ln|x|\,dx = \lim_{b\to 0}\int_{b}^{e} x\ln|x|\,dx$

$= \lim_{b\to 0}\left[\dfrac{1}{2}x^2\ln|x|-\dfrac{1}{4}x^2\right]_{b}^{e}$

$= \lim_{b\to 0}\left(\left(\dfrac{1}{2}e^2\ln|e|-\dfrac{1}{4}e^2\right)-\left(\dfrac{1}{2}b^2\ln|b|-\dfrac{1}{4}b^2\right)\right)$

$= \lim_{b\to 0}\left(\dfrac{1}{4}e^2-\dfrac{1}{2}b^2\ln|b|+\dfrac{1}{4}b^2\right)$

Since $\ln|e|=1$

$= \dfrac{1}{2}e^2-\dfrac{1}{4}e^2$ since $\dfrac{1}{4}b^2 \to$ and $\dfrac{1}{2}b^2\ln|b| \to$ as $b \to 0$

Therefore, since both limits exist, the improper integral converges

$\int_{-1}^{e} x\ln|x|\,dx = \dfrac{1}{4}+\dfrac{1}{2}e^2-\dfrac{1}{4}e^2 = \dfrac{1}{4}(1+e^2)$

Exercise 19.1A Fluency and skills

1 Which of these are improper integrals? Explain your answers.

a $\int_{0}^{5} e^{-x}\,dx$

b $\int_{0}^{2} \ln x\,dx$

c $\int_{1}^{\infty} \dfrac{1}{x^2}\,dx$

d $\int_{-\infty}^{0} \sin x\,dx$

e $\int_{\frac{\pi}{4}}^{\frac{\pi}{2}} \dfrac{1}{\sin x}\,dx$

f $\int_{0}^{\pi} \tan x\,dx$

2 Evaluate each of these improper integrals.

a $\int_{1}^{\infty} \dfrac{1}{x^2}\,dx$

b $\int_{2}^{\infty} \dfrac{3}{x^4}\,dx$

c $\int_{-\infty}^{0} \dfrac{1}{(1-x)^2}\,dx$

d $\int_{-\infty}^{0} \dfrac{1}{(2-3x)^2}\,dx$

e $\int_{0}^{\infty} \dfrac{1}{(x+2)^3}\,dx$

f $\int_{-\infty}^{1} \dfrac{1}{(x-2)^4}\,dx$

g $\int_{0}^{\infty} xe^{-2x}\,dx$

h $\int_{1}^{\infty} \dfrac{\ln x}{x^2}\,dx$

3 Find the exact value of this improper integral.

$$\int_{3}^{\infty}(x-3)e^{-x}dx$$

4 Evaluate this improper integral by first splitting the integral into two parts.

$$\int_{-\infty}^{\infty}\frac{1}{e^{|x|}}dx$$

5 Evaluate each of these improper integrals.

a $\displaystyle\int_{0}^{9}\frac{1}{\sqrt{x}}dx$ **b** $\displaystyle\int_{0}^{27}\frac{1}{x^{\frac{1}{3}}}dx$

c $\displaystyle\int_{2}^{4}\frac{1}{\sqrt{x-2}}dx$ **d** $\displaystyle\int_{0}^{3}\frac{x}{\sqrt{9-x^2}}dx$

e $\displaystyle\int_{0}^{1}\frac{\ln x}{\sqrt{x}}dx$ **f** $\displaystyle\int_{0}^{\ln 2}\frac{e^x}{\sqrt{e^x-1}}dx$

g $\displaystyle\int_{0}^{\frac{\pi}{2}}\frac{\sin x}{\sqrt{\cos x}}dx$ **h** $\displaystyle\int_{0}^{\frac{\pi}{12}}\frac{\cos 2x}{\sqrt{\sin 2x}}dx$

6 Find the exact value of each of these improper integrals.

a $\displaystyle\int_{0}^{e}x\ln x\,dx$ **b** $\displaystyle\int_{0}^{e}x^2\ln x\,dx$

7 Show that $\displaystyle\int_{1-e}^{1}\ln(1-x)dx=0$

8 By splitting the integral into two parts, find the exact value of

$$\int_{-e}^{e}x^2\ln|x|\,dx$$

Reasoning and problem-solving

Many improper integrals are divergent because the areas they represent are not finite. You can use algebra to show that an improper integral is divergent.

Strategy

To decide whether an improper integral converges or diverges

(1) Replace the limit where the integrand is undefined by a variable.

(2) Integrate and substitute in the limits.

(3) Consider the behaviour of the integral as the variable tends towards the original limit.

Example 6

Show that the improper integral $\displaystyle\int_{0}^{5}\frac{1}{x^2}dx$ is divergent.

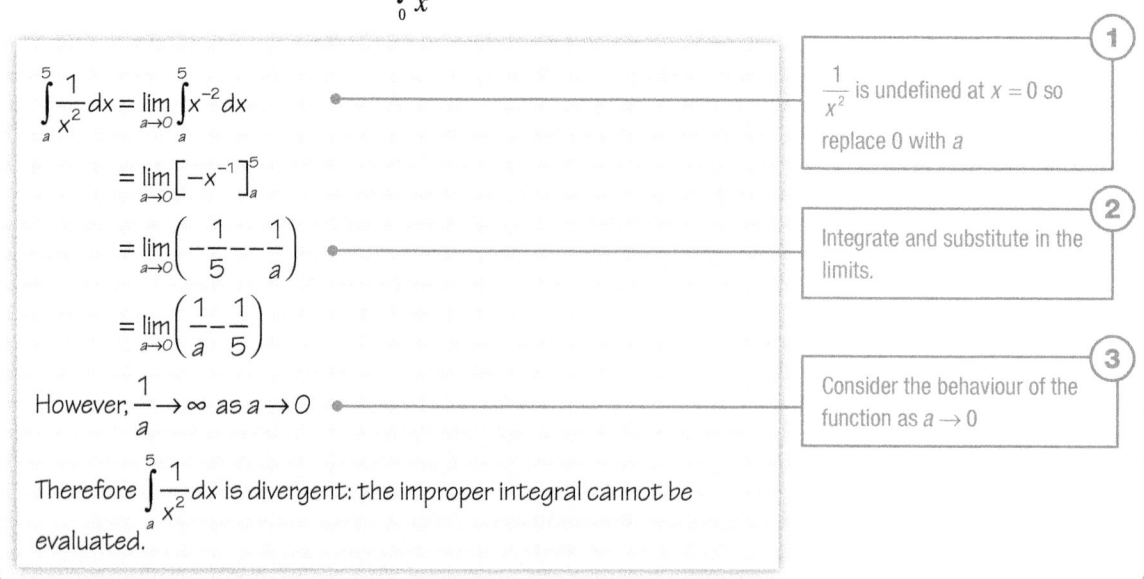

$$\int_{a}^{5}\frac{1}{x^2}dx=\lim_{a\to 0}\int_{a}^{5}x^{-2}dx$$

$$=\lim_{a\to 0}\left[-x^{-1}\right]_{a}^{5}$$

$$=\lim_{a\to 0}\left(-\frac{1}{5}--\frac{1}{a}\right)$$

$$=\lim_{a\to 0}\left(\frac{1}{a}-\frac{1}{5}\right)$$

However, $\dfrac{1}{a}\to\infty$ as $a\to 0$

Therefore $\displaystyle\int_{a}^{5}\frac{1}{x^2}dx$ is divergent: the improper integral cannot be evaluated.

(1) $\dfrac{1}{x^2}$ is undefined at $x=0$ so replace 0 with a

(2) Integrate and substitute in the limits.

(3) Consider the behaviour of the function as $a\to 0$

You may need to consider the behaviour of more complex expressions, such as those involving **rational expressions**.

You do this by dividing the numerator and the denominator by the highest power of the variable.

For example, the expression $\dfrac{x^2+2x}{3x^2-5}$ can be written as $\dfrac{1+\dfrac{2}{x}}{3-\dfrac{5}{x^2}}$ by

dividing the numerator and the denominator by x^2

As $x \to \infty, \dfrac{2}{x} \to 0$ and $\dfrac{5}{x^2} \to 0$, therefore you can see that $\dfrac{1+\dfrac{2}{x}}{3-\dfrac{5}{x^2}} \to \dfrac{1}{3}$

Example 7

Is this improper integral convergent or divergent? $\displaystyle\int_1^\infty \dfrac{2x}{x^2+1} - \dfrac{4x-1}{2x^2-x}\,dx$

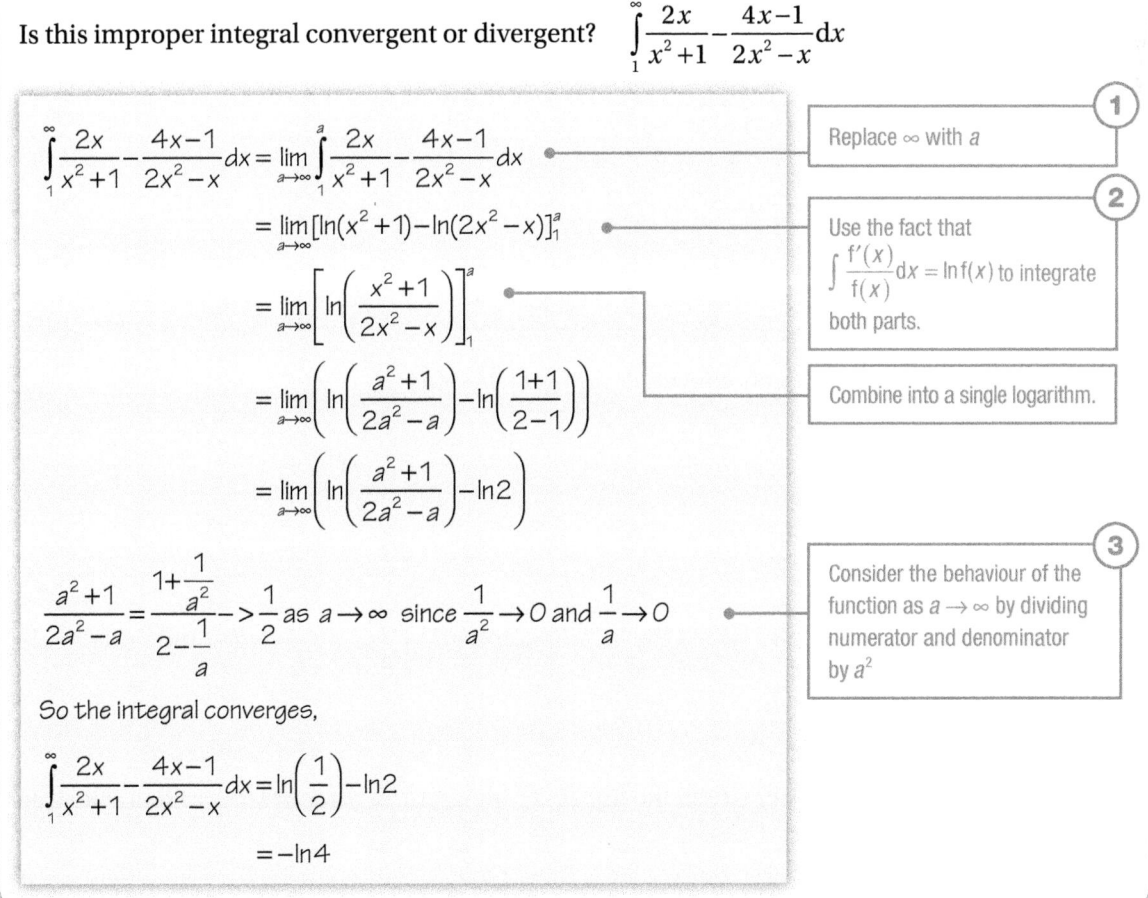

$$\int_1^\infty \dfrac{2x}{x^2+1} - \dfrac{4x-1}{2x^2-x}\,dx = \lim_{a\to\infty}\int_1^a \dfrac{2x}{x^2+1} - \dfrac{4x-1}{2x^2-x}\,dx$$

1 Replace ∞ with a

$$= \lim_{a\to\infty}[\ln(x^2+1)-\ln(2x^2-x)]_1^a$$

2 Use the fact that $\int \dfrac{f'(x)}{f(x)}\,dx = \ln f(x)$ to integrate both parts.

$$= \lim_{a\to\infty}\left[\ln\left(\dfrac{x^2+1}{2x^2-x}\right)\right]_1^a$$

Combine into a single logarithm.

$$= \lim_{a\to\infty}\left(\ln\left(\dfrac{a^2+1}{2a^2-a}\right)-\ln\left(\dfrac{1+1}{2-1}\right)\right)$$

$$= \lim_{a\to\infty}\left(\ln\left(\dfrac{a^2+1}{2a^2-a}\right)-\ln 2\right)$$

$$\dfrac{a^2+1}{2a^2-a} = \dfrac{1+\dfrac{1}{a^2}}{2-\dfrac{1}{a}} \to \dfrac{1}{2} \text{ as } a\to\infty \text{ since } \dfrac{1}{a^2}\to 0 \text{ and } \dfrac{1}{a}\to 0$$

3 Consider the behaviour of the function as $a \to \infty$ by dividing numerator and denominator by a^2

So the integral converges,

$$\int_1^\infty \dfrac{2x}{x^2+1} - \dfrac{4x-1}{2x^2-x}\,dx = \ln\left(\dfrac{1}{2}\right)-\ln 2$$

$$= -\ln 4$$

1 Decide whether or not each of these integrals converges.

If it does converge, find its value. If it diverges, explain why.

a $\displaystyle\int_0^1 \frac{1}{x^4}\,dx$ **b** $\displaystyle\int_1^\infty \frac{1}{x^4}\,dx$

c $\displaystyle\int_0^{\frac{\pi}{4}} \tan x\,dx$ **d** $\displaystyle\int_0^{\frac{\pi}{2}} \tan x\,dx$

e $\displaystyle\int_0^\infty \cos x\,dx$ **f** $\displaystyle\int_1^\infty \frac{1}{x}\,dx$

g $\displaystyle\int_{-\infty}^0 \frac{1}{3-x}\,dx$ **h** $\displaystyle\int_0^7 \frac{1}{\sqrt{7-x}}\,dx$

i $\displaystyle\int_0^7 \frac{1}{(7-x)^2}\,dx$ **j** $\displaystyle\int_2^\infty \frac{1}{x-1}-\frac{2}{2x-1}\,dx$

k $\displaystyle\int_1^\infty \frac{1}{x}-\frac{2x}{x^2+1}\,dx$ **l** $\displaystyle\int_1^\infty \frac{x}{x^2+1}-\frac{2x}{2x^2+1}\,dx$

2 a Show that $\displaystyle\int \frac{x}{x^2+3}-\frac{2}{2x+3}\,dx = \frac{1}{2}\ln\frac{(x^2+3)}{4x^2+12x+9}$

     b Hence show that $\displaystyle\int_0^\infty \frac{x}{x^2+3}-\frac{2}{2x+3}\,dx = \ln k$, where k is a constant to be found.

3 Show that each of these integrals converges and give its value.

a $\displaystyle\int_{-\infty}^0 \frac{6}{3x-2}-\frac{2x}{x^2+4}\,dx$

b $\displaystyle\int_{-\infty}^0 \frac{3-x}{x^2-6x+1}-\frac{8}{5-8x}\,dx$

4 Find the range of values of p for which the improper integral $\displaystyle\int_0^1 \frac{1}{x^p}\,dx$ converges and find its value in terms of p

5 Find the range of values of p for which the improper integral $\displaystyle\int_1^\infty \frac{1}{x^p}\,dx$ converges and find its value in terms of p

19.2 Inverse trigonometric functions

Fluency and skills

See Maths Ch15.5
For a reminder of the chain rule.

To differentiate inverse trigonometric functions you need to use the relationship $\dfrac{dy}{dx} = \dfrac{1}{\frac{dx}{dy}}$

For example, to differentiate $y = \arcsin x$, first rearrange to give $x = \sin y$

Then you know that $\dfrac{dx}{dy} = \cos y$ so $\dfrac{dy}{dx} = \dfrac{1}{\cos y}$

Now rewrite this in terms of x, using the fact that $\cos y = \sqrt{1 - \sin^2 y} = \sqrt{1 - x^2}$

Substituting this gives the result $\dfrac{dy}{dx} = \dfrac{1}{\sqrt{1 - x^2}}$

The derivatives of $\arccos x$ and $\arctan x$ can be derived in a similar way and different functions of x can be used.

Key point

$$\frac{d(\arcsin x)}{dx} = \frac{1}{\sqrt{1 - x^2}} \qquad \frac{d(\arccos x)}{dx} = -\frac{1}{\sqrt{1 - x^2}} \qquad \frac{d(\arctan x)}{dx} = \frac{1}{1 + x^2}$$

Example 1

Differentiate $y = \arctan 3x$ with respect to x

$3x = \tan y \Rightarrow x = \dfrac{1}{3}\tan y$

$\dfrac{dx}{dy} = \dfrac{1}{3}\sec^2 y$

$\dfrac{dy}{dx} = \dfrac{1}{\frac{1}{3}\sec^2 y}$ — Use $\dfrac{dy}{dx} = \dfrac{1}{\frac{dx}{dy}}$

$\phantom{\dfrac{dy}{dx}} = \dfrac{3}{1 + \tan^2 y}$ — Use $1 + \tan^2 y = \sec^2 y$

$\phantom{\dfrac{dy}{dx}} = \dfrac{3}{1 + (3x)^2}$ — Write in terms of x, using the fact that $3x = \tan y$

$\phantom{\dfrac{dy}{dx}} = \dfrac{3}{1 + 9x^2}$

Using the fundamental theorem of calculus, you can use these derivatives to obtain the following integration results:

$$\int \frac{1}{\sqrt{1-x^2}}\,dx = \arcsin x + c \qquad \int \frac{1}{\sqrt{1-x^2}}\,dx = -\arccos x + c \qquad \int \frac{1}{1+x^2}\,dx = \arctan x + c$$

These results can also be derived using a suitable substitution.

> Here you are given the substitution, but you could work out what to use from your knowledge of the derivative of $\arcsin x$

Example 2

Use the substitution $x = \sin u$ to prove that $\int \frac{1}{\sqrt{1-x^2}}\,dx = \arcsin x + c$

$$\frac{dx}{du} = \cos u$$

$$\int \frac{1}{\sqrt{1-x^2}}\,dx = \int \frac{1}{\sqrt{1-\sin^2 u}}\cos u\,du$$ •——— Substitute for x and use $dx = \cos u\,du$

$$= \int \frac{\cos u}{\sqrt{\cos^2 u}}\,du$$ •——— Use the fact that $1 - \sin^2 u = \cos^2 u$

$$= \int 1\,du$$

$$= u + c$$

$$= \arcsin x + c, \text{ as required}$$ Write in terms of x using the fact that $x = \sin u$

Exercise 19.2A Fluency and skills

1 Use the technique in the example above to differentiate these expressions with respect to x

  a $\arccos x$ **b** $\arctan x$

  c $\arcsin 2x$ **d** $\arccos 5x$

  e $\arctan(x-1)$ **f** $2\arcsin x$

  g $3\arccos\left(\dfrac{x}{3}\right)$ **h** $3\arcsin(2-x)$

  i $\arccos x^2$ **j** $x\arcsin x$

2 Prove that $\dfrac{d(\operatorname{arcsec} x)}{dx} = \dfrac{1}{x\sqrt{x^2-1}}$

3 Prove that $\dfrac{d(\operatorname{arccosec} x)}{dx} = -\dfrac{1}{x\sqrt{x^2-1}}$

4 Prove that $\dfrac{d(\operatorname{arccot} x)}{dx} = -\dfrac{1}{x^2+1}$

5 Use the derivatives of $\arccos x$, $\arcsin x$ and $\arctan x$ to find $\dfrac{dy}{dx}$ in each case.

  a $y = e^x \arctan x$ **b** $y = \arccos(3x^2-1)$

  c $y = \sin x \arccos 2x$ **d** $y = (\arcsin x)^2$

  e $y = \arcsin(e^x)$ **f** $e^{\arctan 2x}$

6 Use the substitution $x = \cos u$ to show that

$$\int \frac{1}{\sqrt{1-x^2}}\,dx = -\arccos x + c$$

7 Use the substitution $x = \tan u$ to show that

$$\int \frac{1}{1+x^2}\,dx = \arctan x + c$$

8 Use the substitution $x = \dfrac{1}{3}\sin u$ to integrate

$$\int \frac{1}{\sqrt{1-9x^2}}\,dx$$

9 Use the substitution $x = 5\cos u$ to integrate

$$\int \frac{5}{\sqrt{25-x^2}}\,dx$$

10 Use the substitution $x = 3\tan u$ to integrate

$$\int \frac{1}{9+x^2}\,dx$$

Reasoning and problem-solving

You will not always be told what substitution to use so you will need to choose a suitable one. From the questions in Exercise 19.2A, you can see that these substitutions are often suitable.

Key point

- For an integral involving $\sqrt{a^2 - x^2}$, try the substitution $x = a\sin u$
- For an integral involving $a^2 + x^2$, try the substitution $x = a\tan u$

You may need to do some rearranging of the integral.

For example, $\int \dfrac{1}{1 + \dfrac{x^2}{4}}\,dx$ can be written as $\int \dfrac{4}{4 + x^2}\,dx$. You can then see that a suitable substitution is

$x = 2\tan u$ since the integral involves $a^2 + x^2$, where $a = 2$

Sometimes it is useful to complete the square for the denominator first.

For example, $\int \dfrac{1}{\sqrt{8 + 2x - x^2}}\,dx$ can be written as $\int \dfrac{1}{\sqrt{9 - (x-1)^2}}\,dx$, so a suitable substitution is

$x - 1 = 3\sin u$ since the integral involves $\sqrt{a^2 - x^2}$, where 'x' $= x - 1$ and $a = 3$

Strategy

To find an integral using a trigonometric substitution

(**1**) Rewrite the integrand so it involves either $\sqrt{a^2 - x^2}$ or $a^2 + x^2$

(**2**) Choose the correct substitution, either $x = a\sin u$ or $x = a\tan u$

(**3**) Use integration by substitution to work out the integral and then write the answer in terms of x

Example 3

Work out each of these integrals.

a $\displaystyle\int \dfrac{3}{\sqrt{1 - 9x^2}}\,dx$ **b** $\displaystyle\int \dfrac{9}{x^2 + 6x + 25}\,dx$

a $\displaystyle\int \dfrac{3}{\sqrt{1 - 9x^2}}\,dx = \int \dfrac{1}{\sqrt{\dfrac{1}{9} - x^2}}\,dx$ (**1**) Divide numerator and denominator by 3

Let $x = \dfrac{1}{3}\sin u$

Then $\dfrac{dx}{du} = \dfrac{1}{3}\cos u$ (**2**) $a^2 = \dfrac{1}{9}$ so $a = \dfrac{1}{3}$

$\displaystyle\int \dfrac{1}{\sqrt{\dfrac{1}{9} - \left(\dfrac{1}{3}\sin u\right)^2}}\,\dfrac{1}{3}\cos u\,du = \dfrac{1}{3}\int \dfrac{\cos u}{\sqrt{\dfrac{1}{9} - \dfrac{1}{9}\sin^2 u}}\,du$

$= \dfrac{1}{3}\displaystyle\int \dfrac{\cos u}{\sqrt{\dfrac{1}{9}\cos^2 u}}\,du$ Use $1 - \sin^2 u = \cos^2 u$

(Continued on the next page)

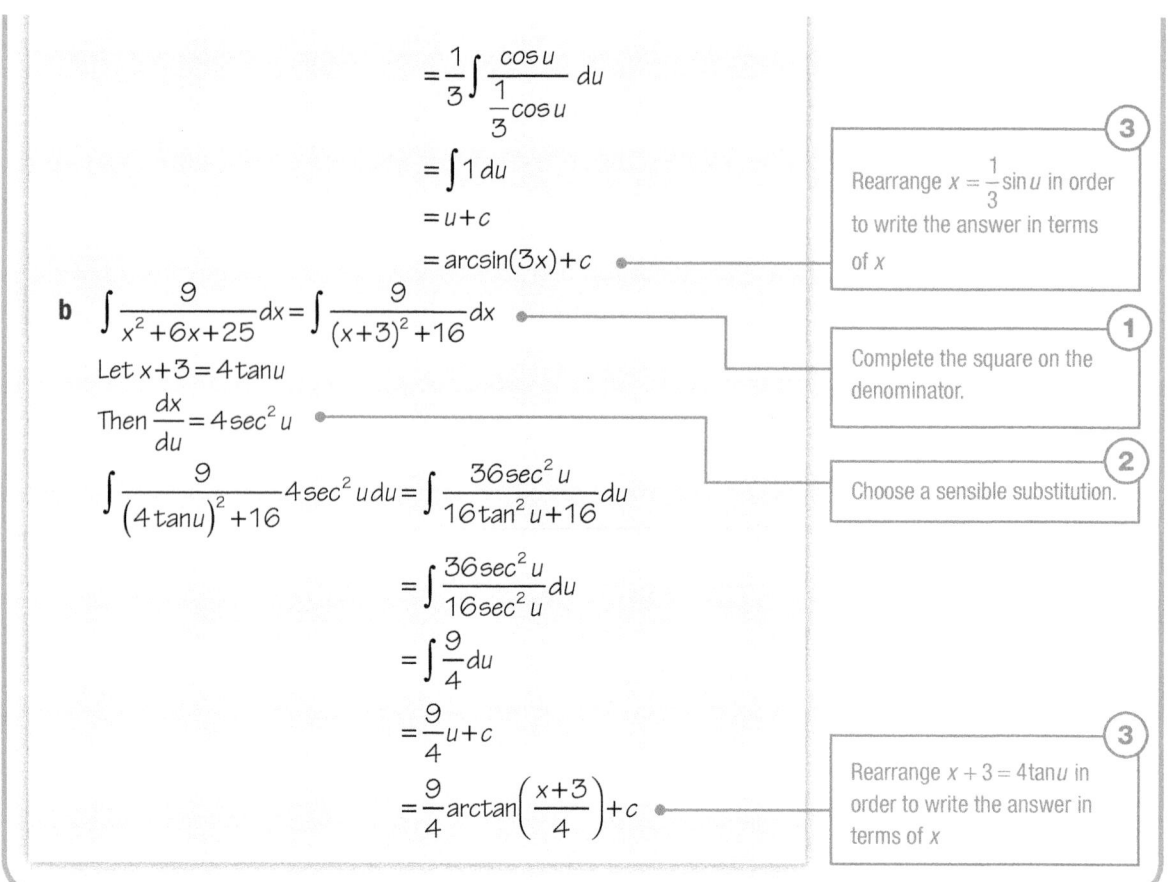

$$= \frac{1}{3}\int \frac{\cos u}{\frac{1}{3}\cos u}\,du$$

$$= \int 1\,du$$

$$= u + c$$

$$= \arcsin(3x) + c$$

3 Rearrange $x = \frac{1}{3}\sin u$ in order to write the answer in terms of x

b $\displaystyle\int \frac{9}{x^2 + 6x + 25}\,dx = \int \frac{9}{(x+3)^2 + 16}\,dx$

1 Complete the square on the denominator.

Let $x + 3 = 4\tan u$

Then $\dfrac{dx}{du} = 4\sec^2 u$

2 Choose a sensible substitution.

$$\int \frac{9}{(4\tan u)^2 + 16}\,4\sec^2 u\,du = \int \frac{36\sec^2 u}{16\tan^2 u + 16}\,du$$

$$= \int \frac{36\sec^2 u}{16\sec^2 u}\,du$$

$$= \int \frac{9}{4}\,du$$

$$= \frac{9}{4}u + c$$

$$= \frac{9}{4}\arctan\left(\frac{x+3}{4}\right) + c$$

3 Rearrange $x + 3 = 4\tan u$ in order to write the answer in terms of x

Exercise 19.2B Reasoning and problem-solving

1 Work out each of these integrals by choosing a suitable substitution.

a $\displaystyle\int \frac{1}{\sqrt{9 - x^2}}\,dx$

b $\displaystyle\int \frac{1}{\sqrt{100 - x^2}}\,dx$

c $\displaystyle -\int \frac{2}{\sqrt{36 - x^2}}\,dx$

d $\displaystyle\int \frac{-4}{\sqrt{8 - x^2}}\,dx$

e $\displaystyle\int \frac{1}{25 + x^2}\,dx$

f $\displaystyle\int \frac{1}{49 + x^2}\,dx$

g $\displaystyle\int \frac{-3}{x^2 + 2}\,dx$

h $\displaystyle\int \frac{6}{x^2 + 6}\,dx$

i $\displaystyle\int \sqrt{1 - x^2}\,dx$

j $\displaystyle\int \sqrt{16 - x^2}\,dx$

2 Work out each of these integrals by first rearranging the integrand then choosing a suitable substitution.

a $\displaystyle\int \frac{1}{1 + \dfrac{x^2}{64}}\,dx$

b $\displaystyle\int \frac{25}{1 + 25x^2}\,dx$

c $\displaystyle\int \frac{2}{\dfrac{x^2}{2} + 4}\,dx$

d $\displaystyle\int \frac{4}{2x^2 + 6}\,dx$

e $\displaystyle\int \frac{1}{\sqrt{1 - \dfrac{x^2}{9}}}\,dx$

f $\displaystyle\int \frac{1}{\sqrt{16 - 4x^2}}\,dx$

g $\displaystyle -\int \frac{1}{\sqrt{1 - \dfrac{x^2}{36}}}\,dx$

h $\displaystyle\int \frac{3}{\sqrt{18 - 9x^2}}\,dx$

i $\displaystyle\int \sqrt{1 - \frac{x^2}{81}}\,dx$

j $\displaystyle\int \sqrt{36 - 9x^2}\,dx$

3 Work out each of these integrals.

a $\displaystyle\int \frac{1}{x^2 + 2x + 5}\,dx$

b $\displaystyle\int \frac{1}{x^2 - 6x + 10}\,dx$

c $\displaystyle\int \frac{1}{x^2 - 14x + 53}\,dx$

d $\displaystyle\int \frac{2}{x^2 + 12x + 38}\,dx$

e $\displaystyle\int \frac{1}{\sqrt{20 - 8x - x^2}}\,dx$

f $\displaystyle\int \frac{1}{\sqrt{1 + 2x - x^2}}\,dx$

g $\displaystyle -\int \frac{2}{\sqrt{8x - x^2}}\,dx$

h $\displaystyle\int \frac{1}{\sqrt{4 - 4x - x^2}}\,dx$

i $\displaystyle\int \sqrt{12 + 4x - x^2}\,dx$

j $\displaystyle\int \sqrt{16 - 6x - x^2}$

4 Calculate the exact value of the integral

$$\int_{-\frac{1}{4}}^{\frac{1}{2}} \frac{1}{\sqrt{8 + 4x - 4x^2}}\,dx$$

Fluency and skills

You can use the exponential definition of hyperbolic functions to work out their derivatives.

For example, $\sinh x = \frac{1}{2}(e^x - e^{-x})$ so $\frac{d(\sinh x)}{dx} = \frac{1}{2}(e^x + e^{-x})$ which is the definition of $\cosh x$

> **See Ch3.5**
> For a reminder about hyperbolic functions.

Key point

$$\frac{d(\sinh x)}{dx} = \cosh x \qquad \frac{d(\cosh x)}{dx} = \sinh x \qquad \frac{d(\tanh x)}{dx} = \text{sech}^2 x$$

These results can be used to find other derivatives.

Example 1

$$\frac{1}{\cosh x} = \text{sech}\, x \qquad \frac{1}{\sinh x} = \text{cosech}\, x \qquad \frac{1}{\tanh x} = \coth x$$

Differentiate $\text{sech}\, x$ with respect to x

$$\text{sech}\, x = (\cosh x)^{-1}$$
$$\frac{d(\text{sech}\, x)}{dx} = -(\cosh x)^{-2}(\sinh x) \qquad \text{Use the chain rule.}$$
$$= -\frac{\sinh x}{\cosh^2 x}$$
$$= -\tanh x \, \text{sech}\, x \qquad \text{Notice the difference in sign compared with the derivative of } \sec x \text{ which is } \tan x \sec x$$

You can also use these derivatives:

Key point

$$\frac{d(\text{sech}\, x)}{dx} = -\text{sech}\, x \tanh x \qquad \frac{d(\text{cosech}\, x)}{dx} = -\text{cosech}\, x \coth x \qquad \frac{d(\coth x)}{dx} = -\text{cosech}^2 x$$

You can integrate $\sinh x$ and $\cosh x$ using the fundamental theorem of calculus:

Key point

$$\int \sinh x = \cosh x + c, \quad \int \cosh x = \sinh x + c$$

To find the integral of $\tanh x$, use the fact that $\tanh x = \frac{\sinh x}{\cosh x}$. Then

$$\int \tanh x \, dx = \int \frac{\sinh x}{\cosh x} dx$$
$$= \ln \cosh x + c \text{ since } \int \frac{f'(x)}{f(x)} dx = \ln f(x) + c$$

Example 2

Work out the exact value of $\int_{0}^{\ln 3} \sinh^2 x \, dx$

Use the identity $\cosh(2x) \equiv 1 + 2\sinh^2 x$

$$\int_{0}^{\ln 3} \sinh^2 x \, dx = \int_{0}^{\ln 3} \frac{1}{2}(\cosh(2x) - 1) dx$$

Use the fact that
$$\sinh^2 x = \frac{1}{2}(\cosh(2x) - 1)$$

$$= \left[\frac{1}{4}\sinh(2x) - \frac{x}{2} \right]_{0}^{\ln 3}$$

Integrate and substitute in the limits.

$$= \frac{1}{4}\left(\frac{e^{2\ln 3} - e^{-2\ln 3}}{2} \right) - \frac{1}{2}\ln 3 - 0$$

$$= \frac{e^{\ln 3^2} - e^{\ln 3^{-2}}}{8} - \frac{1}{2}\ln 3$$

Use rules of logarithms to write $e^{2\ln 3}$ as $e^{\ln 3^2} = 9$ and $e^{-2\ln 3}$ as $e^{\ln 3^{-2}} = \frac{1}{9}$

$$= \frac{9 - \frac{1}{9}}{8} - \frac{1}{2}\ln 3$$

$$= \frac{10}{9} - \frac{1}{2}\ln 3$$

You can check this on your calculator, but it will not give an exact answer.

You can differentiate the inverse hyperbolic functions using a similar method as for inverse trigonometric functions in Section 19.2

Example 3

Differentiate $\text{arcosh}\, x$ with respect to x

Use the identity $\cosh^2 x - \sinh^2 x \equiv 1$

Let $y = \text{arcosh}\, x$

Then $x = \cosh y$

$$\frac{dx}{dy} = \sinh y$$

So $\dfrac{dy}{dx} = \dfrac{1}{\sinh y}$

Use the chain rule: $\dfrac{dy}{dx} = \dfrac{1}{\dfrac{dx}{dy}}$

$$= \frac{1}{\sqrt{\cosh^2 y - 1}}$$

Use the identity $\cosh^2 y - \sinh^2 y \equiv 1$

$$= \frac{1}{\sqrt{x^2 - 1}}$$

The derivative of $\operatorname{arsinh} x$ can be found in a similar way.

$$\frac{d(\operatorname{arcosh} x)}{dx} = \frac{1}{\sqrt{x^2-1}} \qquad \frac{d(\operatorname{arsinh} x)}{dx} = \frac{1}{\sqrt{x^2+1}}$$

Exercise 19.3A Fluency and skills

1 Use the definition of $\cosh x$ and $\sinh x$ to show that $\dfrac{d(\cosh x)}{dx} = \sinh x$

2 a Use exponentials to show that $\dfrac{d(\tanh x)}{dx} = \operatorname{sech}^2 x$

  b Hence show that $\dfrac{d(\coth x)}{dx} = -\operatorname{cosech}^2 x$

3 Use exponentials to show that $\dfrac{d(\operatorname{cosech} x)}{dx} = -\operatorname{cosech} x \coth x$

4 Differentiate these expressions with respect to x

  a $3\operatorname{cosech} x$ **b** $\coth(x+1)$

  c $\sinh 2x$ **d** $\cosh\left(\dfrac{x}{3}\right)$

  e $\tanh(x-2)$ **f** $\sinh x^2$

  g $\cosh(2x-3)$ **h** $\sinh^2 x$

  i $x\tanh x$ **j** $x\sqrt{\operatorname{sech} x}$

5 Calculate these integrals.

  a $\int \operatorname{sech}^2 x\, dx$ **b** $\int \coth x \operatorname{cosech} x\, dx$

  c $\int \cosh 2x\, dx$ **d** $\int \sinh\left(\dfrac{x}{3}\right) dx$

  e $\int \cosh x \sinh^2 x\, dx$ **f** $\int \cosh^2 x\, dx$

  g $\int \coth 4x\, dx$ **h** $\int \operatorname{cosech}^2 5x\, dx$

6 Evaluate these integrals. Give the exact answer in each case.

  a $\displaystyle\int_0^{\ln 2} \sinh(2x)dx$ **b** $\displaystyle\int_0^{\ln 3} \cosh(x-\ln 3)$

  c $\displaystyle\int_{\ln 2}^{\ln 5} \tanh x \operatorname{sech} x\, dx$ **d** $\displaystyle\int_{\ln 2}^{\ln 3} \operatorname{cosech}^2(2x)dx$

7 Show that $\dfrac{d(\operatorname{arsinh} x)}{dx} = \dfrac{1}{\sqrt{x^2+1}}$

8 Show that $\dfrac{d(\operatorname{artanh} 2x)}{dx} = \dfrac{2}{1-4x^2}$

9 Show that $\dfrac{d(\operatorname{arcoth} x)}{dx} = \dfrac{1}{1-x^2}$

10 Show that $\dfrac{d(\operatorname{arcosech} x)}{dx} = \dfrac{1}{-x\sqrt{x^2+1}}$

11 Differentiate the following expressions with respect to x

  You can quote the derivatives of $\operatorname{arsinh} x$ and $\operatorname{arcosh} x$

  a $\operatorname{arcosh} 4x$ **b** $\operatorname{arsinh}\left(\dfrac{x}{5}\right)$

  c $\operatorname{arcosh}(x^2)$ **d** $\operatorname{arsinh} e^x$

  e $\operatorname{arcosh}(\sinh x)$ **f** $e^x \operatorname{arsinh}(x^2-1)$

You can use the derivatives of arsinh x and arcosh x to choose a suitable substitution to use when integrating.

- For an integral involving $\sqrt{x^2+a^2}$, try the substitution $x = a\sinh u$
- For an integral involving $\sqrt{x^2-a^2}$, try the substitution $x = a\cosh u$

You may need to rearrange the integral first, using the same techniques that you saw in Section 19.2

Strategy

To find an integral using a hyperbolic substitution

1. Rewrite the integrand so that it involves either $\sqrt{x^2+a^2}$ or $\sqrt{x^2-a^2}$

2. Choose the correct substitution, either $x = a\sinh u$ or $x = a\cosh u$

3. Use integration by substitution to work out the integral and then write the answer in terms of x

Example 4

Work out each of these integrals

a $\displaystyle\int \frac{6}{\sqrt{36+4x^2}}\,dx$ **b** $\displaystyle\int \frac{1}{\sqrt{x^2-8x-20}}\,dx$

a $\displaystyle\int \frac{6}{\sqrt{36+4x^2}}\,dx = \int \frac{3}{\sqrt{9+x^2}}\,dx$

Divide numerator and denominator by 2 ①

Let $x = 3\sinh u$

$a^2 = 9$ so $a = 3$ ②

Then $\dfrac{dx}{du} = 3\cosh u$

$\displaystyle\int \frac{3}{\sqrt{9+(3\sinh u)^2}}\,3\cosh u\,du = \int \frac{9\cosh u}{\sqrt{9+9\sinh^2 u}}\,du$

$\displaystyle = \int \frac{9\cosh u}{\sqrt{9\cosh^2 u}}\,du$

Use $1+\sinh^2 u = \cosh^2 u$

$\displaystyle = \int \frac{9\cos u}{3\cos u}\,du$

$\displaystyle = \int 3\,du$

$= 3u + c$

$\displaystyle = 3\,\text{arsinh}\left(\frac{x}{3}\right) + c$

Rearrange $x = 3\sinh u$ in order to write the answer in terms of x ③

(Continued on the next page)

b $\displaystyle\int \frac{1}{\sqrt{x^2-8x-20}}dx = \int \frac{1}{\sqrt{(x-4)^2-36}}dx$

Let $x-4 = 6\cosh u$

Then $\dfrac{dx}{du} = 6\sinh u$

$\displaystyle\int \frac{1}{\sqrt{(6\cosh u)^2-36}}6\sinh u\,du = \int \frac{6\sinh u}{\sqrt{36\cosh^2 u-36}}du$

$\qquad\qquad = \displaystyle\int \frac{6\sinh u}{\sqrt{36\sinh^2 u}}du$

$\qquad\qquad = \displaystyle\int \frac{6\sinh u}{6\sinh u}du$

$\qquad\qquad = \displaystyle\int 1\,du$

$\qquad\qquad = u+c$

$\qquad\qquad = \operatorname{arcosh}\left(\dfrac{x-4}{6}\right)+c$

(1) Complete the square on the denominator.

(2) Choose a sensible substitution.

(3) Rearrange $x-4 = 6\cosh u$ in order to write the answer in terms of x

Exercise 19.3B Reasoning and problem-solving

1 Work out each of these integrals.

a $\displaystyle\int \frac{1}{\sqrt{x^2+49}}dx$ **b** $\displaystyle\int \frac{1}{\sqrt{x^2-81}}dx$

c $\displaystyle\int \frac{2}{\sqrt{64+4x^2}}dx$ **d** $\displaystyle\int \frac{-1}{\sqrt{\dfrac{x^2}{9}-1}}dx$

e $\displaystyle\int \frac{2}{\sqrt{x^2+6x+25}}dx$

f $\displaystyle\int \frac{1}{\sqrt{x^2-10x+26}}dx$

g $\displaystyle\int \frac{1}{\sqrt{x^2+14x+24}}dx$

h $\displaystyle\int \frac{3}{\sqrt{x^2-24x+44}}dx$

2 Evaluate these integrals, giving your answers to 3 significant figures.

a $\displaystyle\int_0^1 \frac{1}{\sqrt{6+x^2}}dx$ **b** $\displaystyle\int_6^{12} \frac{\sqrt{27}}{\sqrt{\dfrac{x^2}{12}-3}}dx$

3 Calculate the exact values of these integrals.

a $\displaystyle\int_0^3 \sqrt{\frac{3}{27+3x^2}}dx$ **b** $\displaystyle\int_4^5 \frac{1}{\sqrt{3x^2-48}}dx$

4 Integrate each of these expressions with respect to x

a $\dfrac{x+1}{\sqrt{16+9x^2}}$ **b** $\dfrac{3-x}{\sqrt{\dfrac{x^2}{4}-3}}$

c $\dfrac{x+4}{\sqrt{16-2x^2}}$ **d** $\dfrac{2-5x}{\dfrac{x^2}{7}+7}$

5 Use integration by parts to show that
$$\int \operatorname{arsinh}x\,dx = x\operatorname{arsinh}x - \sqrt{x^2+1}+c$$

6 Work out each of these integrals.

a $\displaystyle\int \operatorname{arcosh}x\,dx$ **b** $\displaystyle\int \operatorname{arcoth}x\,dx$

7 One of these improper integrals converges for $a>0$ and the other does not.

A: $\displaystyle\int_a^{2a} \frac{1}{\sqrt{x^2-a^2}}dx$ **B:** $\displaystyle\int_{2a}^{\infty} \frac{1}{\sqrt{x^2-a^2}}dx$

Explain why one doesn't converge and find the exact value of the integral that does converge.

19.4 Partial fractions

Fluency and skills

See Maths Ch12.5

For a reminder on partial fractions.

Rational functions which have linear factors in the denominator can be split into partial fractions to help you to integrate the function.

If the degree of the numerator is the same as or greater than the degree of the denominator, then this is an **improper** fraction.

Key point

A fraction $\dfrac{f(x)}{g(x)}$ where degree of $f(x) \geq$ degree of $g(x)$ is called an **improper** fraction and can be written in the form $P(x) + \dfrac{Q(x)}{g(x)}$

It may then be possible to write $\dfrac{Q(x)}{g(x)}$ in partial fractions. You can use long division to find $P(x)$ and $Q(x)$

However, if you prefer, you can consider the degree of the numerator and the denominator to decide on a general form for the quotient.

For example, if you have $\dfrac{f(x)}{g(x)}$ where $f(x)$ is a quartic (polynomial of degree 4) and $g(x)$ is a quadratic, then you know that the quotient is of the form $Ax^2 + Bx + C$

Example 1

a Write $\dfrac{30x^3 - 13x^2 + 6x + 6}{15x^2 + x - 6}$ in partial fractions. **b** Hence work out $\displaystyle\int \dfrac{30x^3 - 13x^2 + 6x + 6}{15x^2 + x - 6}\,dx$

a Dividing a cubic by a quadratic will give a linear quotient.

$$\frac{30x^3 - 13x^2 + 6x + 6}{15x^2 + x - 6} = Ax + B + \frac{C}{3x + 2} + \frac{D}{5x - 3}$$

Alternatively, you could use long division.

$$30x^3 - 13x^2 + 6x + 6 = (Ax + B)(15x^2 + x - 6) + C(5x - 3) + D(3x + 2)$$

Multiply both sides by $(3x + 2)(5x - 3)$

Equating coefficients

$x^3 : 30 = 15A \Rightarrow A = 2$ $x^2 : -13 = A + 15B \Rightarrow B = -1$

$x : 6 = -6A + B + 5C + 3D \Rightarrow 5C + 3D = 19$

$1 : 6 = -6B - 3C + 2D \Rightarrow 3C - 2D = 0$

Solve simultaneously to give $C = 2$, $D = 3$

$$\frac{30x^3 - 13x^2 + 6x + 6}{15x^2 + x - 6} = 2x - 1 + \frac{2}{3x + 2} + \frac{3}{5x - 3}$$

b $\displaystyle\int \frac{30x^3 - 13x^2 + 6x + 6}{15x^2 + x - 6}\,dx = \int 2x - 1 + \frac{2}{3x + 2} + \frac{3}{5x - 3}\,dx$

Use the answer from part **a**.

$$= x^2 - x + \frac{2}{3}\ln(3x + 2) + \frac{3}{5}\ln(5x - 3) + c$$

You can also find partial fractions when the denominator includes a quadratic factor which cannot be factorised, for example x^2+5. In these cases, you should use a linear expression $Ax+B$ as the numerator.

Key point

$\dfrac{f(x)}{(\alpha x^2+\beta)(\gamma x+\delta)}$ can be split into partial fractions of the form $\dfrac{Ax+B}{\alpha x^2+\beta}+\dfrac{C}{\gamma x+\delta}$

In some questions you may have to factorise the denominator yourself.

Example 2

Work out $\int \dfrac{2x+12}{(x+1)(x^2+9)}\,dx$

$\dfrac{2x+12}{(x+1)(x^2+9)}=\dfrac{A}{x+1}+\dfrac{Bx+C}{x^2+9}$

$2x+12=A(x^2+9)+(Bx+C)(x+1)$

$\quad\quad\quad = Ax^2+9A+Bx^2+Bx+Cx+C$

Multiply both sides by $(x+1)(x^2+9)$

Equating coefficients

$x^2: 0=A+B$ so $A=-B$

$x: 2=B+C$ (equation 1)

$1: 12=9A+C$ (equation 2)

Or use an alternative method to find the values of A, B and C

Subtract equation 1 from equation 2 to give: $9A-B=10$

$A=-B \Rightarrow -9B-B=10 \Rightarrow B=-1$

Solve the three equations simultaneously.

So $A=1$ and $C=3$

$\int \dfrac{2x+12}{(x+1)(x^2+9)}\,dx = \int \dfrac{1}{x+1}+\dfrac{3-x}{x^2+9}\,dx$

$\quad\quad = \int \dfrac{1}{x+1}+\dfrac{3}{x^2+9}-\dfrac{x}{x^2+9}\,dx$

Split the numerator of the second fraction.

$\quad\quad = \ln(x+1)+\arctan\left(\dfrac{x}{3}\right)-\dfrac{1}{2}\ln\left(x^2+9\right)+c$

Use the substitution $x=3\tan u$ to integrate the second fraction.

which can be written as $\ln\left(\dfrac{x+1}{\sqrt{x^2+9}}\right)+3\arctan\left(\dfrac{x}{3}\right)+c$

Exercise 19.4A Fluency and skills

1 Work out each of these integrals by first expressing the integrand in partial fractions.

a $\int \dfrac{5x^2+14x-42}{(x-2)(x+4)}\,dx$

b $\int \dfrac{2x^2+31x+115}{x^2+14x+49}\,dx$

c $\int \dfrac{5x^3+x^2-46x-24}{x^2-9}\,dx$

d $\int \dfrac{8x^3+92x^2+243x+72}{x^3+11x^2+24x}\,dx$

e $\int \dfrac{x^4-7x^3-20x^2+14x+46}{(x-9)(x+1)}\,dx$

f $\int \dfrac{9x^4}{(x+1)(x-2)^2}\,dx$

g $\int \dfrac{12x^4+7x^3-22x-9}{12-7x+12x^2}\,dx$

PURE

2 Work out each of these integrals by first expressing the integrand in partial fractions.

a $\displaystyle\int \frac{9-7x}{(x+3)(x^2+1)}dx$

b $\displaystyle\int \frac{10+8x}{(x-3)(x^2+25)}dx$

c $\displaystyle\int \frac{x+66}{(x^2+36)(6-x)}dx$

d $\displaystyle\int \frac{7x-1}{(1-x)(x^2+2)}dx$

e $\displaystyle\int \frac{36x+6}{(x+1)(x-2)(x^2+9)}dx$

f $\displaystyle\int \frac{79-28x}{(x-1)^2(x^2+16)}dx$

3 Work out each of these integrals by first expressing the integrand in partial fractions.

a $\displaystyle\int \frac{3x+49}{x^3+49x}dx$ b $\displaystyle\int \frac{x-192}{2x^3+128x}dx$

c $\displaystyle\int \frac{10x+9}{3x^3-x^2+12x-4}dx$

4 a Given that x^2+5 is a factor of $x^4-6x^3+14x^2-30x+45$, write the fraction $\dfrac{20x-46}{x^4-6x^3+14x^2-30x+45}$ in partial fractions.

b Hence work out $\displaystyle\int \frac{20x-46}{x^4-6x^3+14x^2-30x+45}\,dx$

Reasoning and problem-solving

You can also use partial fractions when evaluating improper integrals. The integral will only converge if all parts of it converge.

Example 3

a Show algebraically that the improper integral $\displaystyle\int_{\frac{1}{2}}^{1} \frac{x-9}{2x^3-x^2+8x-4}\,dx$ does not converge.

b Explain whether the improper integral $\displaystyle\int_{1}^{\infty} \frac{x-9}{2x^3-x^2+8x-4}\,dx$ converges.

a $\dfrac{x-9}{2x^3-x^2+8x-4} = \dfrac{x-9}{(2x-1)(x^2+4)}$

This can be written in partial fractions: $\dfrac{x-9}{(2x-1)(x^2+4)} = \dfrac{A}{2x-1}+\dfrac{Bx+C}{x^2+4}$

$x-9 = A(x^2+4)+(Bx+C)(2x-1)$

Consider coefficients.

$x^2 : 0 = A+2B$

$x : 1 = -B+2C$

$1 : -9 = 4A-C$

Solve simultaneously to give $A=-2$, $B=1$, $C=1$

So $\dfrac{x-9}{(2x-1)(x^2+4)} = \dfrac{x+1}{x^2+4}-\dfrac{2}{2x-1}$

> x^2+4 cannot be factorised further so will be a quadratic factor.

(1)

(Continued on the next page)

$\int_{\frac{1}{2}}^{1}\dfrac{x+1}{x^2+4}dx$ is a proper integral so can be calculated directly.

However, $\dfrac{2}{2x-1}$ has a point of discontinuity at $x=\dfrac{1}{2}$ so $\int_{\frac{1}{2}}^{1}\dfrac{2}{2x-1}dx$ is an improper integral.

2

Consider any points of discontinuity for either fraction within the limits given.

$$\int_{\frac{1}{2}}^{1}\dfrac{2}{2x-1}\,dx=\lim_{a\to\frac{1}{2}}\int_{a}^{1}\dfrac{2}{2x-1}\,dx$$

$$=\lim_{a\to\frac{1}{2}}\left[\ln(2x-1)\right]_{a}^{1}$$

$$=\lim_{a\to\frac{1}{2}}(\ln 1-\ln(2a-1))$$

$$=\lim_{a\to\frac{1}{2}}(-\ln(2a-1))$$

3

Consider behaviour as $a\to\dfrac{1}{2}$

$2a-1\to 0$ as $a\to\dfrac{1}{2}$ so $\ln(2a-1)\to\infty$

Therefore, the integral does not converge.

b $\int_{1}^{\infty}\dfrac{x-9}{2x^3-x^2+8x-4}dx=\lim_{a\to\infty}\int_{1}^{a}\dfrac{x}{x^2+4}+\dfrac{1}{x^2+4}-\dfrac{2}{2x-1}dx$

$$=\lim_{a\to\infty}\left[\dfrac{1}{2}\ln(x^2+4)+\dfrac{1}{2}\arctan\left(\dfrac{x}{2}\right)-\ln(2x-1)\right]_{1}^{a}$$

$$=\lim_{a\to\infty}\left(\left(\dfrac{1}{2}\ln(a^2+4)+\dfrac{1}{2}\arctan\left(\dfrac{a}{2}\right)-\ln(2a-1)\right)-\left(\dfrac{1}{2}\ln 5+\dfrac{1}{2}\arctan\dfrac{1}{2}-\ln 1\right)\right)$$

$$=\lim_{a\to\infty}\left(\dfrac{1}{2}\ln\left(\dfrac{a^2+4}{(2a-1)^2}\right)+\dfrac{1}{2}\arctan\left(\dfrac{a}{2}\right)-\dfrac{1}{2}\ln 5-\dfrac{1}{2}\arctan\dfrac{1}{2}\right)$$

$$\dfrac{a^2+4}{(2a-1)^2}=\dfrac{a^2+4}{4a^2-4a+1}$$

$$=\dfrac{1+\dfrac{4}{a^2}}{4-\dfrac{4}{a}+\dfrac{1}{a^2}}\to\dfrac{1}{4}\text{ as }a\to\infty\text{ since }\dfrac{4}{a},\dfrac{1}{a^2}\to 0$$

3

So this part of the integral does converge.

So $\dfrac{1}{2}\ln\left(\dfrac{a^2+4}{(2a-1)^2}\right)\to\dfrac{1}{2}\ln\dfrac{1}{4}$

$\arctan\left(\dfrac{a}{2}\right)\to\dfrac{\pi}{2}$ as $a\to\infty$

3

Consider the graph of $y=\arctan x$, it has a horizontal asymptote at $y=\dfrac{\pi}{4}$

Therefore, the improper integral does converge.

$$\int_{1}^{\infty}\dfrac{x-9}{2x^3-x^2+8x-4}dx=\dfrac{1}{2}\ln\dfrac{1}{4}+\dfrac{1}{2}\left(\dfrac{\pi}{2}\right)-\dfrac{1}{2}\arctan\dfrac{1}{2}-\dfrac{1}{2}\ln 5$$

$$=\dfrac{\pi}{4}-\dfrac{1}{2}\arctan\dfrac{1}{2}-\dfrac{1}{2}\ln 20$$

1   Find the exact value of these improper integrals.

   a $\displaystyle\int_{0}^{\infty}\frac{1}{x^2+7x+12}\,dx$

   b $\displaystyle\int_{1}^{\infty}\frac{10}{3-8x-3x^2}\,dx$

   c $\displaystyle\int_{0}^{\infty}\frac{7x+26}{x^3+9x^2+24x+20}\,dx$

   d $\displaystyle\int_{-\infty}^{0}\frac{9}{8x^2-22x+5}\,dx$

2   One of these improper integrals converges and the other does not. Explain which does not converge and find the value of the integral that does converge.

   A: $\displaystyle\int_{1}^{\infty}\frac{-2x-12}{2x^3-x^2+6x-3}\,dx$

   B: $\displaystyle\int_{0}^{1}\frac{-2x-12}{2x^3-x^2+6x-3}\,dx$

3   **a**   Express $\dfrac{2x+a}{x^3+ax}$ in partial fractions.

   b   Hence show that the improper integral $\displaystyle\int_{0}^{a}\frac{2x+a}{x^3+ax}$ does not converge for any $a>0$

4   For what range of values of a (if any) does each improper integral converge? Explain your answers.

   a $\displaystyle\int_{-\infty}^{0}\frac{3-a}{(x+a)(x+3)}\,dx$

   b $\displaystyle\int_{0}^{\infty}\frac{3-a}{(x+a)(x+3)}\,dx$

19.5 Reduction formulae

Integration by parts can be used to find integrals of the form $\int u \dfrac{dv}{dx} dx$ where you know how to find $\dfrac{du}{dx}$ and v. For example,

$$\int x e^{-x} dx = -x e^{-x} - \int -e^{-x} dx \qquad\qquad u = x, \dfrac{dv}{dx} = e^{-x}$$

$$\qquad\qquad = -x e^{-x} - e^{-x} + c$$

Key point

The formula for integration by parts is

$$\int u \dfrac{dv}{dx} dx = uv - \int v \dfrac{du}{dx} dx$$

You can apply this formula twice, for example,

$$\int x^2 e^{-x} dx = -x^2 e^{-x} - \int -2x e^{-x} dx \qquad\qquad u = x^2, \dfrac{dv}{dx} = e^{-x}$$

$$\qquad\qquad = -x^2 e^{-x} + 2 \int x e^{-x} dx$$

Then apply again to the new integral:

$$\qquad = -x^2 e^{-x} + 2 \left(-x e^{-x} - \int -e^{-x} dx \right) \qquad u = x, \dfrac{dv}{dx} = e^{-x}$$

$$\qquad = -x^2 e^{-x} - 2x e^{-x} - 2e^{-x} + c$$

You could use a similar process for higher powers of x but it would be very time-consuming. Instead, you can find a **reduction formula** which relates $I_n = \int x^n e^{-x} dx$ to $I_{n-1} = \int x^{n-2} e^{-x} dx$

$$I_n = \int x^n e^{-x} dx = -x^n e^{-x} - \int -n x^{n-1} e^{-x} dx \qquad u = x^n, \dfrac{dv}{dx} = e^{-x}$$

$$\qquad\qquad = -x^n e^{-x} + n \int x^{n-1} e^{-x} dx$$

$$\qquad\qquad = -x^n e^{-x} + n I_{n-1}$$

You can now use this formula to find, for example, $I_3 = \int x^3 e^{-x} dx$

$I_3 = -x^3 e^{-x} + 3I_2$, now use the formula again for I_2

$$\quad = -x^3 e^{-x} + 3(-x^2 e^{-x} + 2I_1)$$

You can continue to use the formula until you have an integrand which is simpler to integrate.

$$I_3 = -x^3 e^{-x} - 3x^2 e^{-x} + 6I_1$$

$$\quad = -x^3 e^{-x} - 3x^2 e^{-x} + 6(-x e^{-x} + I_0)$$

$$\quad = -x^3 e^{-x} - 3x^2 e^{-x} - 6x e^{-x} + 6I_0$$

$I_0 = \int e^{-x} dx = -e^{-x}$ so substitute this to find the solution for I_3

$$I_3 = -x^3 e^{-x} - 3x^2 e^{-x} - 6x e^{-x} - 6e^{-x} + c$$

Key point

A **reduction formula** for I_n is an equation that relates I_n to I_{n-1} and/or I_{n-2}
It can be repeatedly applied to reduce the integral to one not requiring integration by parts.

You can use a reduction formula for a definite or an indefinite integral.

Given that $I_n = \int_0^1 (\ln x)^n \, dx$, prove that $I_n = -nI_{n-1}$ and use this formula to evaluate I_4

$I_n = \int_0^1 1.(\ln x)^n \, dx = \left[x(\ln x)^n \right]_0^1 - \int_0^1 nx \left(\frac{1}{x} \right) (\ln x)^{n-1} \, dx$ ●——

Integration by parts with $\frac{dv}{dx} = 1$ and $u = (\ln x)^n$ so that $v = x$ and $\frac{du}{dx} = \frac{1}{x}(\ln x)^{n-1}$

$= 0 - \int_0^1 n(\ln x)^{n-1} \, dx$ ●——

Substituting the limits into $x(\ln x)^n$ gives 0

$= -nI_{n-1}$ as required

$I_4 = -4I_3$

$= -4(-3I_2)$

$= 12(-2I_1)$

$= -24(1I_0)$ ●——

Use the reduction formula repeatedly.

$I_0 = \int_0^1 (\ln x)^0 \, dx = \int_0^1 1 \, dx$

$= [x]_0^1$

$= 1$

Therefore $I_4 = -24I_0 = -24$

After you have used the integration by parts formula, you may need to rearrange the integrand so that it is clearly in the form of I_{n-1} and/or I_{n-2}

Given that $I_n = \int_{-1}^0 x^n \sqrt{x+1} \, dx$, show that $I_n = -nI_{n-1}$ and use this formula to evaluate $\int_{-1}^0 x^3 \sqrt{x+1} \, dx$

$I_n = \int_{-1}^0 x^n (x+1)^{\frac{1}{2}} \, dx = \left[\frac{2}{3} x^n (x+1)^{\frac{3}{2}} \right]_{-1}^0 - \int_0^1 \frac{2}{3} nx^{n-1}(x+1)^{\frac{3}{2}} \, dx$ ●——

Integration by parts with $\frac{dv}{dx} = (x+1)^{\frac{1}{2}}$ and $u = x^n$ so that $v = \frac{2}{3}(x+1)^{\frac{3}{2}}$ and $\frac{du}{dx} = nx^{n-1}$

$= 0 - \frac{2}{3} n \int_{-1}^0 x^{n-1}(x+1)^{\frac{3}{2}} \, dx$ ●——

$= -\frac{2}{3} n \int_{-1}^0 x^{n-1} \left(x(x+1)^{\frac{1}{2}} + (x+1)^{\frac{1}{2}} \right) dx$ ●——

Substituting the limits into $\frac{2}{3} x^n (x+1)^{\frac{3}{2}}$ gives 0

$= -\frac{2}{3} n \int_{-1}^0 x^n \sqrt{x+1} + x^{n-1} \sqrt{x+1} \, dx$

so $\qquad I_n = -\frac{2}{3} nI_n - \frac{2}{3} nI_{n-1}$

Factorising this expression gives $I_n = -\frac{\frac{2}{3} nI_{n-1}}{1 + \frac{2}{3} n}$

which simplifies to $I_n = -\frac{2n}{3+2n} I_{n-1}$

Write $(x+1)^{\frac{3}{2}}$ as $(x+1)(x+1)^{\frac{1}{2}}$ then expand to give $x(x+1)^{\frac{1}{2}} + (x+1)^{\frac{1}{2}}$

(*Continued on the next page*)

So $I_3 = -\dfrac{6}{9}I_2$

$= -\dfrac{6}{9}\left(-\dfrac{4}{7}\right)I_1$

$= -\dfrac{6}{9}\left(-\dfrac{4}{7}\right)\left(-\dfrac{2}{5}\right)I_0$

$= -\dfrac{16}{105}I_0$

Use the reduction formula repeatedly.

$I_0 = \displaystyle\int_{-1}^{0}(x+1)^{\frac{1}{2}}\,dx = \left[\dfrac{2}{3}(x+1)^{\frac{3}{2}}\right]_{-1}^{0}$

$= \dfrac{2}{3}$

Therefore $\displaystyle\int_{-1}^{0} x^3\sqrt{x+1}\,dx = -\dfrac{16}{105}\left(\dfrac{2}{3}\right) = -\dfrac{32}{315}$

Exercise 19.5A Fluency and skills

1 $I_n = \displaystyle\int x^n e^x\,dx$

   a Show that $I_n = x^n e^x - nI_{n-1}$

   b Use this formula to find I_4

2 $I_n = \displaystyle\int x^n e^{-\frac{x}{2}}\,dx$

   a Show that $I_n = 2nI_{n-1} - 2x^n e^{-\frac{x}{2}}$

   b Use this formula to find $\displaystyle\int x^3 e^{-\frac{x}{2}}\,dx$

3 $I_n = \displaystyle\int_0^1 x^n e^{3x}\,dx$

   a Show that $I_n = \dfrac{1}{3}e^3 - \dfrac{n}{3}I_{n-1}$

   b Use this formula to evaluate $\displaystyle\int_0^1 x^4 e^{3x}\,dx$

4 $I_n = \displaystyle\int x(\ln x)^n\,dx$

   a Show that $I_n = \dfrac{x^2}{2}(\ln x)^n - \dfrac{n}{2}I_{n-1}$

   b Use this formula to find $\displaystyle\int x(\ln x)^2\,dx$

5 $I_n = \displaystyle\int_0^1 \dfrac{x^n}{\sqrt{1-x}}\,dx$

   a Show that $I_n = \dfrac{2n}{1+2n}I_{n-1}$

   b Use this formula to find $\displaystyle\int_0^1 \dfrac{x^6}{\sqrt{1-x}}\,dx$

6 $I_n = \displaystyle\int x^n \sin x\,dx$

   a Show that

$$I_n = -x^n \cos x + nx^{n-1}\sin x$$
$$-n(n-1)I_{n-2} \text{ for } n > 1$$

   b Use this formula to find

     i $\displaystyle\int x^4 \sin x\,dx$

     ii $\displaystyle\int x^3 \sin x\,dx$

7 $I_n = \displaystyle\int_0^{\frac{\pi}{4}} x^n \cos 2x\,dx$

   a Show that

$$I_n = \dfrac{1}{2}\left(\dfrac{\pi}{4}\right)^n - \dfrac{n(n-1)}{4}I_{n-2} \text{ for } n > 1$$

   b Use this formula to evaluate

     i $\displaystyle\int_0^{\frac{\pi}{4}} x^6 \cos 2x\,dx$

     ii $\displaystyle\int_0^{\frac{\pi}{4}} x^5 \cos 2x\,dx$

Sometimes the integrand is not in the form of the product of two functions. In these cases, you can split up the integral yourself. For example, you can write $\sin^n x$ as $\sin x \sin^{n-1} x$

Strategy

To find and use a reduction formula

(1) Split up the function if necessary and integrate.

(2) Find an expression for I_n in terms of I_{n-1} and/or I_{n-2}

(3) Use the formula repeatedly until you reach I_0 or I_1

(4) Calculate I_0 or I_1 and use with your formula to find I_n for the value of n required.

Example 3

Find a reduction formula for $I_n = \int_0^\pi \sin^n x \, dx$ and use it to find

a $\int_0^\pi \sin^8 x \, dx$ **b** $\int_0^\pi \sin^7 x \, dx$

$I_n = \int_0^\pi \sin^n x \, dx = \int_0^\pi \sin x \sin^{n-1} x \, dx$

⟵ (1) Split the function.

$\qquad = [-\cos x \sin^{n-1} x]_0^\pi - \int_0^\pi -(n-1)\cos^2 x \sin^{n-2} x \, dx$

⟵ (1) Integration by parts with $\dfrac{dv}{dx} = \sin x$ and $u = \sin^{n-1} x$ so that $v = -\cos x$ and $\dfrac{du}{dx} = (n-1)\sin^{n-2} x \cos x$

$\qquad = (n-1)\int_0^\pi \cos^2 x \sin^{n-2} x \, dx$

$\qquad = (n-1)\int_0^\pi \sin^{n-2} x - \sin^n x \, dx$

⟵ Replace $\cos^2 x$ with $1 - \sin^2 x$

$\qquad = (n-1)(I_{n-2} - I_n)$

$\qquad = (n-1)I_{n-2} - (n-1)I_n$

Rearrange to give $I_n = \dfrac{n-1}{n} I_{n-2}$

⟵ (2) This is your reduction formula.

a Use to find I_8

$I_8 = \dfrac{7}{8} I_6$

$\quad = \dfrac{7}{8}\dfrac{5}{6} I_4$

$\quad = \dfrac{7}{8}\dfrac{5}{6}\dfrac{3}{4} I_2$

$\quad = \dfrac{7}{8}\dfrac{5}{6}\dfrac{3}{4}\dfrac{1}{2} I_0$

⟵ (3) Use the formula repeatedly until I_0 reached.

$\quad = \dfrac{35}{128} I_0$

$I_0 = \int_0^\pi 1 \, dx$

$\quad = [x]_0^\pi$

$\quad = \pi$

⟵ (4) Substitute value of I_0 into formula for I_8

So $I_8 = \dfrac{35}{128}\pi$

(Continued on the next page)

b Use to find I_7

$$I_7 = \frac{6}{7}I_5$$

$$= \frac{6}{7}\frac{4}{5}I_3$$

$$= \frac{6}{7}\frac{4}{5}\frac{2}{3}I_1$$

$$= \frac{16}{35}I_1$$

> 3 — This time, I_1 is reached.

$$I_1 = \int_0^\pi \sin x\,dx = -\left[\cos x\right]_0^\pi$$

$$= -(\cos\pi - \cos 0)$$

$$= -(-1-1)$$

$$= 2$$

$$\text{So } I_7 = \frac{16}{35}(2) = \frac{32}{35}$$

> 4 — Substitute value of I_1 into formula for I_7

Exercise 19.5B Reasoning and problem-solving

1 $I_n = \displaystyle\int_0^{\frac{\pi}{2}} \cos^n x\,dx$

   a Show that $I_n = \dfrac{n-1}{n}I_{n-2}$ for $n > 1$

   b Use this formula to evaluate

      i $\displaystyle\int_0^{\frac{\pi}{2}} \cos^5 x\,dx$ **ii** $\displaystyle\int_0^{\frac{\pi}{2}} \cos^6 x\,dx$

2 $I_n = \displaystyle\int_{\frac{\pi}{2}}^{\pi} \sin^n x\,dx$

   a Show that $I_n = \dfrac{n-1}{n}I_{n-2}$

   b Use this formula to evaluate

      i $\displaystyle\int_{\frac{\pi}{2}}^{\pi} \sin^7 x\,dx$ **ii** $\displaystyle\int_{\frac{\pi}{2}}^{\pi} \sin^8 x\,dx$

3 $I_n = \displaystyle\int \tan^n x\,dx$

   a By writing $\tan^n x$ as $\tan^{n-2} x\tan^2 x$, find a reduction formula for I_n

   b Use this formula to find

      i $\displaystyle\int \tan^6 x\,dx$ **ii** $\displaystyle\int_0^{\frac{\pi}{4}} \tan^7 x\,dx$

4 $I_n = \displaystyle\int_0^1 x^n\sqrt{1+x^2}\,dx$

   a Show that $I_n = \dfrac{2\sqrt{2}-(n-1)I_{n-2}}{2+n}$

   b Use this formula to find

      i $\displaystyle\int_0^1 x^5\sqrt{1+x^2}\,dx$ **ii** $\displaystyle\int_0^1 x^4\sqrt{1+x^2}\,dx$

5 $I_n = \displaystyle\int_0^\pi \dfrac{\cos nx}{\cos x}\,dx$

   a Show that $I_n = -I_{n-2}$

   Hint: write $nx = (n-1)x + x$ and use the addition formula for cos.

   b Use the reduction formula to find

      $\displaystyle\int_0^\pi \dfrac{\cos 7x}{\cos x}\,dx$

   c Hence write down the value of

      i $\displaystyle\int_0^\pi \dfrac{\cos 27x}{\cos x}\,dx$ **ii** $\displaystyle\int_0^\pi \dfrac{\cos 29x}{\cos x}\,dx$

   d Show that $\displaystyle\int_0^\pi \dfrac{\cos nx}{\cos x}\,dx$ does not converge when n is an even number.

19.6 Polar graphs and areas

Fluency and skills

See Ch3.3
For a reminder of polar coordinates.

Curves can be defined in polar form.

For example, $r = 2$ is the circle $x^2 + y^2 = 4$ and $\theta = \dfrac{\pi}{4}$ is a half-line that lies on $y = x$. To sketch polar curves, you work out the value of r for certain values of θ such as $\theta = 0, \dfrac{\pi}{2}, \pi, \dfrac{3\pi}{2}, 2\pi$

Example 1

Sketch the curve $r = 3(1 + \cos\theta)$

When $\theta = 0$, $r = 6$

When $\theta = \dfrac{\pi}{2}$, $r = 3$

When $\theta = \pi$, $r = 0$

When $\theta = \dfrac{3\pi}{2}$, $r = 3$

When $\theta = 2\pi$, $r = 6$

$\theta = 0$

Since $\cos(0) = \cos(2\pi) = 1$,
$\cos\left(\dfrac{\pi}{2}\right) = \cos\left(\dfrac{3\pi}{2}\right) = 0$,
and $\cos(\pi) = -1$

This is a **cardioid.**

The minimum value of r is 0, which occurs when $\theta = \pi$

The maximum value of r is 6, which occurs when $\theta = \dfrac{\pi}{2}$ or $\theta = \dfrac{3\pi}{2}$

Polar curves of the form $r = a\cos(n\theta)$ or $r = b\sin(n\theta)$ will have n 'loops'. Since r is the radius, it must be positive. So you can work out the position of these loops by considering where r is positive.

Example 2

Sketch the curve $r = 2\cos(3\theta)$

$\cos(3\theta) = 0 \Rightarrow 3\theta = \dfrac{\pi}{2}, \dfrac{3\pi}{2}, \dfrac{5\pi}{2}, \dfrac{7\pi}{2}, \dfrac{9\pi}{2}, \dfrac{11\pi}{2}$

$\Rightarrow \theta = \dfrac{\pi}{6}, \dfrac{\pi}{2}, \dfrac{5\pi}{6}, \dfrac{7\pi}{6}, \dfrac{3\pi}{2}, \dfrac{11\pi}{6} \left(-\dfrac{\pi}{6}\right)$

So r is positive for $-\dfrac{\pi}{6} < \theta < \dfrac{\pi}{6}, \dfrac{\pi}{2} < \theta < \dfrac{5\pi}{6}$ and $\dfrac{7\pi}{6} < \theta < \dfrac{3\pi}{2}$

The maximum value of r is 2 and occurs when $\cos(3\theta) = 1$ so when $\theta = 0, \dfrac{2\pi}{3}, \dfrac{4\pi}{3}$

$\theta = \dfrac{\pi}{2}$

$\theta = \dfrac{\pi}{6}$

$\theta = \dfrac{5\pi}{2}$

$\theta = 0$

$\theta = \dfrac{7\pi}{6}$

$\theta = \dfrac{11\pi}{6}$

$\theta = \dfrac{3\pi}{2}$

Consider the graph of $y = \cos(3\theta)$

You don't need to draw all the half-lines shown but you need to be able to calculate where they are.

The area of a sector of a circle is given by the formula area $= \dfrac{1}{2}r^2\theta$

You can use integration to calculate the area enclosed by a polar curve between two half-lines.

Split the area into n slices, each with an angle of $\delta\theta$

The area of each slice can be approximated as the area of a sector of a circle. The formula for the area of a sector of angle θ from a circle of radius r is $A = \dfrac{1}{2}r^2\theta$

Therefore, the area of the slice shown is $\delta A \approx \dfrac{1}{2}\big[f(\theta_i)\big]^2\,\delta\theta$

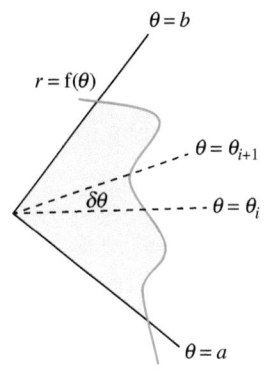

Taking the limit as $n \to \infty$ of the sum of these areas gives the integral $\displaystyle\int_a^b \dfrac{1}{2}\big[f(\theta)\big]^2\,\mathrm{d}\theta$, which can be written $\dfrac{1}{2}\displaystyle\int_a^b r^2\,\mathrm{d}\theta$

Key point

The area of a sector of a polar curve $r = f(\theta)$ between the half lines $\theta = a$ and $\theta = b$ is given by $A = \dfrac{1}{2}\displaystyle\int_a^b r^2\,\mathrm{d}\theta$

Remember that θ must be measured in radians.

Example 3

Calculate the area enclosed by the curve $r = \sin(2\theta)$, where $r > 0$

$\sin(2\theta) = 0 \Rightarrow 2\theta = 0,\ \pi,\ 2\pi,\ 3\pi,\ 4\pi$

$\Rightarrow \theta = 0,\ \dfrac{\pi}{2},\ \pi,\ \dfrac{3\pi}{2},\ 2\pi$

So r is positive for $0 < \theta < \dfrac{\pi}{2},\ \pi < \theta < \dfrac{3\pi}{2}$

Consider the graph of $y = \sin(2\theta)$

A sketch of $r = \sin(2\theta)$ will help you see the required areas.

Calculate the area of one 'loop'.

$A = \dfrac{1}{2}\displaystyle\int_0^{\frac{\pi}{2}} (\sin(2\theta))^2\,\mathrm{d}\theta$

$= \dfrac{1}{2}\displaystyle\int_0^{\frac{\pi}{2}} \sin^2(2\theta)\,\mathrm{d}\theta$

$= \dfrac{1}{4}\displaystyle\int_0^{\frac{\pi}{2}} 1 - \cos(4\theta)\,\mathrm{d}\theta$

Use the identity $\cos 2x = 1 - 2\sin^2 x$

$= \dfrac{1}{4}\left[\theta - \dfrac{1}{4}\sin(4\theta)\right]_0^{\frac{\pi}{2}}$

$= \dfrac{1}{4}\left(\dfrac{\pi}{2} - 0\right) - \dfrac{1}{4}(0 - 0)$

$= \dfrac{\pi}{8}$

Total area $= 2 \times \dfrac{\pi}{8} = \dfrac{\pi}{4}$

Since there are two identical loops.

1 Sketch each of these curves or lines that are given in polar form.

   a $r = 8$ **b** $\theta = \dfrac{\pi}{3}$

   c $\theta = \dfrac{7\pi}{6}$ **d** $r = 2\theta$ for $0 \le \theta \le 2\pi$

2 For each of these polar equations, where $r > 0$

   i State the maximum and the minimum values of r

   ii Sketch the curve.

      a $r = 1 - \cos\theta$ **b** $r = 2 + \sin\theta$

      c $r = 2(1 - \sin\theta)$ **d** $r = 6 + 3\cos\theta$

      e $r = 5 + \sin\theta$ **f** $r = \cos(2\theta)$

      g $r = a\sin(3\theta)$ **h** $r = b\cos(4\theta)$

      i $r^2 = c^2\sin\theta$

3 Calculate the area bounded by the curve with equation $r = 4\cos\theta$ and the half lines $\theta = 0$ and $\theta = \dfrac{\pi}{3}$

4 Calculate the area bounded by the curve with equation $r = \theta$ and the half lines $\theta = 0$ and $\theta = \dfrac{\pi}{2}$

5 Calculate the area bounded by the curve with equation $r = 2\sin\theta$ and the half lines $\theta = \dfrac{\pi}{12}$ and $\theta = \dfrac{5\pi}{12}$

6 Calculate the area bounded by the curve with equation $r^2 = 4\sin\theta$ and the half lines $\theta = \dfrac{\pi}{4}$ and $\theta = \dfrac{3\pi}{4}$

7 Calculate the area enclosed by each of these cardioids.

   a $r = 3 + \cos\theta$ **b** $r = 5 + 2\sin\theta$

   c $r = 1 + \sin\theta$ **d** $r = 3 - 2\cos\theta$

8 Calculate the total area enclosed by these curves where $r > 0$

   a $r = \cos 2\theta$ **b** $r = \sin 4\theta$

   c $r = 4\cos 3\theta$ **d** $r = 5\sin 3\theta$

   e $r^2 = 2\sin 2\theta$ **f** $r^2 = (5 + 2\sin\theta)$

Reasoning and problem-solving

To find the area between two polar curves, you first need to work out where they intersect. As with Cartesian equations, you do this by solving the equations simultaneously.

You can then calculate the area between the two curves by adding the areas of the different parts.

Strategy

To calculate the area between two polar curves

1 Sketch the curves and solve simultaneously to find their point of intersection.

2 Calculate the area of a sector for each curve using $\dfrac{1}{2}\displaystyle\int_a^b r^2\,d\theta$

3 Add the areas together to find the required area.

Example 4

PURE

Find the area bounded by the curves $r = 3\cos\theta$ and $r = 1 + \cos\theta$

To find point of intersection, solve $3\cos\theta = 1 + \cos\theta$

$\Rightarrow \cos\theta = \dfrac{1}{2}$

$\Rightarrow \theta = \pm\dfrac{\pi}{3}$

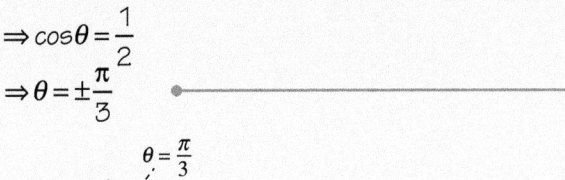

The points of intersection are $\left(\dfrac{3}{2}, \dfrac{\pi}{3}\right)$ and $\left(\dfrac{3}{2}, -\dfrac{\pi}{3}\right)$. Draw the half lines through these points.

The total required area is shaded.

Consider the area labelled A_1 which is enclosed by $1 + \cos\theta$ between the half lines $\theta = 0$ and $\theta = \dfrac{\pi}{3}$

$A_1 = \dfrac{1}{2}\displaystyle\int_0^{\frac{\pi}{3}}(1 + \cos\theta)^2\,d\theta$

Use the formula $A = \dfrac{1}{2}\displaystyle\int r^2\,d\theta$

$= \dfrac{1}{2}\displaystyle\int_0^{\frac{\pi}{3}}1 + 2\cos\theta + \cos^2\theta\,d\theta$

$= \dfrac{1}{2}\displaystyle\int_0^{\frac{\pi}{3}}1 + 2\cos\theta + \dfrac{1}{2}(1 + \cos 2\theta)\,d\theta$

Use the identity $\cos 2x = 2\cos^2 x - 1$

$= \dfrac{1}{2}\left[\theta + 2\sin\theta + \dfrac{1}{2}\theta + \dfrac{1}{4}\sin 2\theta\right]_0^{\frac{\pi}{3}}$

$= \dfrac{1}{2}\left(\dfrac{3}{2}\left(\dfrac{\pi}{3}\right) + \sqrt{3} + \dfrac{\sqrt{3}}{8}\right) - 0$

$= \dfrac{1}{4}\pi + \dfrac{9}{16}\sqrt{3}$

Consider the area labelled A_2 which is enclosed by $3\cos\theta$ between the half lines $\theta = \dfrac{\pi}{3}$ and $\theta = \dfrac{\pi}{2}$

$A_2 = \dfrac{1}{2}\displaystyle\int_{\frac{\pi}{3}}^{\frac{\pi}{2}}(3\cos\theta)^2\,d\theta$

$= \dfrac{1}{2}\displaystyle\int_{\frac{\pi}{3}}^{\frac{\pi}{2}}9\cos^2\theta\,d\theta$

(Continued on the next page)

$$= \frac{9}{4} \int_{\frac{\pi}{3}}^{\frac{\pi}{2}} 1 + \cos 2\theta \, d\theta$$

$$= \frac{9}{4} \left[\theta + \frac{1}{2} \sin 2\theta \right]_{\frac{\pi}{3}}^{\frac{\pi}{2}}$$

$$= \frac{9}{4} \left(\frac{\pi}{2} + 0 \right) - \frac{9}{4} \left(\frac{\pi}{3} + \frac{1}{4} \sqrt{3} \right)$$

$$= \frac{3\pi}{8} - \frac{9}{16} \sqrt{3}$$

$$A_1 + A_2 = \frac{1}{4} \pi + \frac{9}{16} \sqrt{3} + \frac{3\pi}{8} - \frac{9}{16} \sqrt{3}$$

$$= \frac{5}{8} \pi$$

So total area $= 2 \times \left(\frac{5}{8} \pi \right) = \frac{5}{4} \pi$

③ This is the top half of the required area.

Exercise 19.6B Reasoning and problem-solving

1 The diagram shows the graphs of
$r = \cos\theta + \sin\theta$ and $r = 2 + \sin\theta$ for $r > 0$

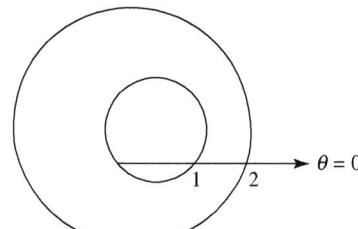

a Prove that the curves do not intersect.

b Calculate the shaded area.

2 The diagram shows the graphs of
$r = \sin\theta\sqrt{2\cos\theta}$ and $r = \sin\theta$, for $r > 0$

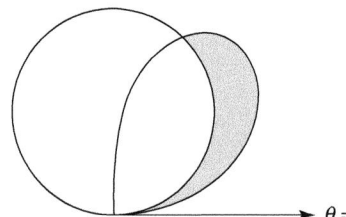

a Find the polar coordinates of the points of intersection of the two curves.

b Calculate the shaded area.

3 The curve shown has polar equation

$r = 2 + \cos(2\theta)$, $0 \leq \theta \leq \frac{\pi}{2}$

At the point A on the curve, $r = 1.5$

Calculate the shaded area.

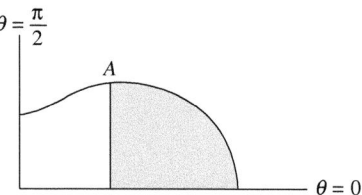

4 Calculate the area bounded by these pairs of curves.

a $r = \cos\theta$, $r = \sqrt{3}\sin\theta$

b $r = 2 - \cos\theta$, $r = 3\cos\theta$

c $r = \sqrt{2} + \sin\theta$, $r = 3\sin\theta$

d $r^2 = 1 - \sin\theta$, $r = \sqrt{2}\sin\theta$

e $r = 5 + 2\cos\theta$, $r = 4$

5 The circle with equation $x^2 + (y-2)^2 = 4$ intersects the curve with polar equation $r = 1 + \sin\theta$ at the points A and B

a Calculate the length of AB

b Calculate the area enclosed by the circle $x^2 + (y-2)^2 = 4$ but not the curve $r = 1 + \sin\theta$

19.7 Lengths and surface areas

Fluency and skills

You can use integration to calculate the length of a curve between two points. Consider two points, A and B, that are close to each other on a curve. If you approximate the length of the curve, s, between these two points by a straight line, then you can use Pythagoras' theorem to see that the length of this line is

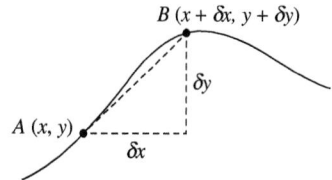

$$(\delta s)^2 = (\delta x)^2 + (\delta y)^2$$

Dividing by $(\delta x)^2$ gives $\dfrac{(\delta s)^2}{(\delta x)^2} = 1 + \dfrac{(\delta y)^2}{(\delta x)^2}$, which can be written as

$$\frac{\delta s}{\delta x} = \sqrt{1 + \left(\frac{\delta y}{\delta x}\right)^2}$$

As $\delta x \to 0$, $\dfrac{\delta s}{\delta x} \to \dfrac{ds}{dx}$ and $\dfrac{\delta y}{\delta x} \to \dfrac{dy}{dx}$; therefore $\dfrac{ds}{dx} = \sqrt{1 + \left(\dfrac{dy}{dx}\right)^2}$

You can integrate this to find the length of the curve.

$$s = \int \sqrt{1 + \left(\frac{dy}{dx}\right)^2}\, dx$$

Key point

The length of the arc of the curve $y = f(x)$ from x_1 to x_2 is given by $s = \displaystyle\int_{x_1}^{x_2} \sqrt{1 + \left(\frac{dy}{dx}\right)^2}\, dx$

Example 1

Calculate the length of the curve $y = 2x\sqrt{x}$ between the points $(0, 0)$ and $(4, 16)$

Write $y = 2x^{\frac{3}{2}}$, then $\dfrac{dy}{dx} = 3\sqrt{x}$

$s = \displaystyle\int_0^3 \sqrt{1 + (3\sqrt{x})^2}\, dx$ Use $s = \displaystyle\int_{x_1}^{x_2} \sqrt{1 + \left(\frac{dy}{dx}\right)^2}\, dx$

$= \displaystyle\int_0^3 \sqrt{1 + 9x}\, dx$

$= \left[\dfrac{2}{27}(1 + 9x)^{\frac{3}{2}}\right]_0^3$ You can check this result using differentiation.

$= \dfrac{2}{27}(1 + 27)^{\frac{3}{2}} - \dfrac{2}{27}(1 + 0)^{\frac{3}{2}}$

$= \dfrac{2}{27}(56\sqrt{7} - 1)$

Example 1

If you are given a curve defined by parametric equations $x = f(t)$, $y = g(t)$, then you need to divide by $(\delta t)^2$ instead.

So $(\delta s)^2 = (\delta x)^2 + (\delta y)^2$ and divide by $(\delta t)^2$ to give

$$\left(\frac{\delta s}{\delta t}\right)^2 = \left(\frac{\delta x}{\delta t}\right)^2 + \left(\frac{\delta y}{\delta t}\right)^2 \text{ which can be written as}$$

$$\frac{\delta s}{\delta t} = \sqrt{\left(\frac{\delta x}{\delta t}\right)^2 + \left(\frac{\delta y}{\delta t}\right)^2}$$

As $\delta t \to 0$, $\dfrac{\delta s}{\delta t} \to \dfrac{ds}{dt}$, $\dfrac{\delta x}{\delta t} \to \dfrac{dx}{dt}$ and $\dfrac{\delta y}{\delta t} \to \dfrac{dy}{dt}$; therefore

$$\frac{ds}{dt} = \sqrt{\left(\frac{dx}{dt}\right)^2 + \left(\frac{dy}{dt}\right)^2}$$

You can then integrate this to find the length of the curve.

$$s = \int \sqrt{\left(\frac{dx}{dt}\right)^2 + \left(\frac{dy}{dt}\right)^2}\, dt$$

Key point

The length of the section of the curve $x = f(t)$, $y = g(t)$

from t_1 to t_2 is given by $s = \displaystyle\int_{t_1}^{t_2} \sqrt{\left(\dfrac{dx}{dt}\right)^2 + \left(\dfrac{dy}{dt}\right)^2}\, dt$

Example 2

Calculate the length of the arc of the curve $x = t^2 - 1$, $y = \dfrac{2}{3}t^3 + 1$ from $t = 0$ to $t = 2$

$$s = \int_0^2 \sqrt{(2t)^2 + (2t^2)^2}\, dt$$

Use the formula $= \displaystyle\int_{t_1}^{t_2} \sqrt{\left(\dfrac{dx}{dt}\right)^2 + \left(\dfrac{dy}{dt}\right)^2}\, dt$

$$= \int_0^2 \sqrt{4t^2 + 4t^4}\, dt$$

with $\dfrac{dx}{dt} = 2t$ and $\dfrac{dy}{dt} = 2t^2$

$$= \int_0^2 2t\sqrt{1 + t^2}\, dt$$

Take out a factor of $\sqrt{4t^2} = 2t$

$$= \left[\frac{2}{3}(1 + t^2)^{\frac{3}{2}}\right]_0^2$$

$$= \frac{2}{3}(1 + 2^2)^{\frac{3}{2}} - \frac{2}{3}(1)^{\frac{3}{2}}$$

Substitute in the limits.

$$= \frac{2}{3}(5\sqrt{5} - 1)$$

See Ch4.2 For a reminder of volumes of revolution. You can use the formula $V = \pi \displaystyle\int_{x_1}^{x_2} y^2\, dx$ to calculate the volume of revolution when $f(x)$ between x_1 and x_2 is rotated through 2π radians around the x-axis.

It is also possible to calculate the surface area of revolution. Consider a small section of the curve rotated through 2π radians around the x-axis. The shape can be approximated by a truncated cone (also known as a frustum).

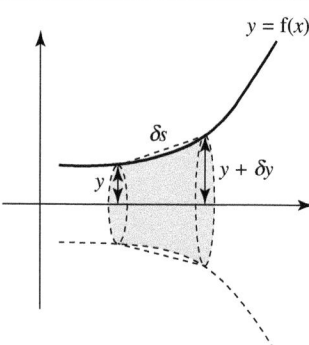

The curved surface area of a truncated cone is given by $S = \pi(r_1 + r_2)l$ where r_1 is the radius of the base, r_2 is the radius of the top and l is the length of the slanted side of the frustum.

Therefore $\delta S = \pi(y + \delta y + y)\delta s = \pi(2y + \delta y)\delta s$

Divide by δx to give $\dfrac{\delta S}{\delta x} = \pi(2y + \delta y)\dfrac{\delta s}{\delta x}$

As $\delta x \to 0$, $\delta y \to 0$, $\dfrac{\delta s}{\delta x} \to \dfrac{ds}{dx}$ and $\dfrac{\delta S}{\delta x} \to \dfrac{dS}{dx}$; therefore

$$\frac{dS}{dx} = 2y\pi\frac{ds}{dx}$$

You can then integrate both sides with respect to x to find the surface area of revolution.

$$S = \int 2y\pi \, ds$$

As $\dfrac{ds}{dx} = \sqrt{1 + \left(\dfrac{dy}{dx}\right)^2}$, you can write this as

$$S = \int 2y\pi\sqrt{1 + \left(\frac{dy}{dx}\right)^2}\,dx$$

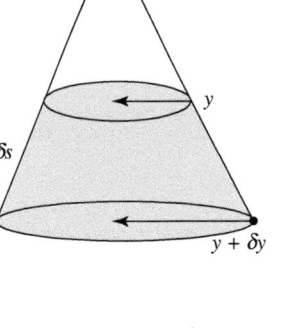

> **Key point**
>
> The surface area of revolution when the curve $y = f(x)$ from x_1 to x_2 is rotated through 2π radians around the x-axis is given by $S = 2\pi\displaystyle\int_{x_1}^{x_2} y\sqrt{1 + \left(\dfrac{dy}{dx}\right)^2}\,dx$

Example 3

The section of the curve with equation $y = x^3$ between $x = 0$ and $x = \dfrac{1}{2}$ is rotated through 2π radians around the x-axis. For the solid formed, calculate the exact value of the curved surface area.

$y = x^3$ so $\dfrac{dy}{dx} = 3x^2$

$s = 2\pi\displaystyle\int_0^{\frac{1}{2}} x^3\sqrt{1 + (3x^2)^2}\,dx$ Use $2\pi\displaystyle\int_{x_1}^{x_2} y\sqrt{1 + \left(\dfrac{dy}{dx}\right)^2}\,dx$

$= 2\pi\displaystyle\int_0^{\frac{1}{2}} x^3(1 + 9x^4)^{\frac{1}{2}}\,dx$

$= 2\pi\left[\dfrac{1}{54}(1 + 9x^4)^{\frac{3}{2}}\right]_0^{\frac{1}{2}}$ Substitute in the limits of 0 and $\dfrac{1}{2}$

$= 2\pi\left(\dfrac{1}{54}\left(\dfrac{125}{64} - 1\right)\right)$

$= \dfrac{61}{1728}\pi$

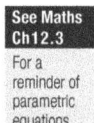
See Maths
Ch12.3

For a reminder of parametric equations.

You can also find a formula for the surface area of revolution when the curve is defined by parametric equations.

Using $\dfrac{ds}{dt} = \sqrt{\left(\dfrac{dx}{dt}\right)^2 + \left(\dfrac{dy}{dt}\right)^2}$ in the formula for surface area of

revolution, $S = \int 2y\pi\,ds$ gives $S = \int 2y\pi \sqrt{\left(\dfrac{dx}{dt}\right)^2 + \left(\dfrac{dy}{dt}\right)^2}\,dt$

Key point

The surface area of revolution when the curve $x = f(t)$, $y = g(t)$ from t_1 to t_2 is rotated through 2π radians around the x-axis is given by

$$S = 2\pi \int_{t_1}^{t_2} y \sqrt{\left(\dfrac{dx}{dt}\right)^2 + \left(\dfrac{dy}{dt}\right)^2}\,dt, \text{ where } y \text{ is replaced by } g(t)$$

Exercise 19.7A Fluency and skills

1 Show that the length of the arc of the curve $y = \dfrac{2}{3}x^{\frac{3}{2}}$ between the points $x = 0$ and $x = 1$ is $A\sqrt{2} + B$, where A and B are constants to be found.

2 Calculate the arc length of the curve $y = x\sqrt{x}$ between $(4, 8)$ and $(9, 27)$, giving your answer to 3 significant figures.

3 Calculate the length of the curve with equation $y = \cosh x$, $\ln 2 \le x \le \ln 4$

4 Calculate the length of the arc of the curve $y = \dfrac{1}{2}\cosh 2x$ between the points $x = 0$ and $x = \ln 2$

5 Calculate the length of the arc of the curve $y = (1 + x)^{\frac{3}{2}}$ between $(-1, 0)$ and $\left(\dfrac{1}{3}, \dfrac{8}{9}\sqrt{3}\right)$

6 a Prove that $\dfrac{d(\ln(\sec x + \tan x))}{dx} = \sec x$

   b Hence calculate the length of the arc of the curve $y = \ln\left(\dfrac{1}{2}\cos x\right)$ between $x = 0$ and $x = \dfrac{\pi}{6}$

7 The curve C is defined by the parametric equations $x = t^2$, $y = 2t$. The section of C between $t = 0$ and $t = 2$ is rotated through $360°$ about the x-axis.

   Calculate the surface area of revolution.

8 A curve is given by the parametric equations $x = \dfrac{t^3}{6} + 1$, $y = t^2$, $0 \le t \le 3$

   Calculate the arc length.

9 A curve C has equation $y = 4x^3$, $0 \le x \le \dfrac{1}{2}$

   The curve C is rotated through 2π radians around the x-axis. Calculate

   a The volume of the solid formed,

   b The curved surface area of the solid formed.

10 The arc of the curve with equation $y = \cosh x$ between $x = 0$ and $x = \ln 3$ is rotated through 2π radians around the x-axis. Calculate

   a The volume of the solid formed,

   b The curved surface area of the solid formed.

Reasoning and problem-solving

You may need to use substitution and other integration techniques to calculate arc lengths and surface areas of revolution.

Strategy

Strategy

To calculate the surface area of revolution

(1) Choose the correct formula for arc length or for surface area, depending on whether the equation of the curve is in parametric or Cartesian form.

(2) Use a suitable substitution, being sure to apply the correct limits.

(3) Use trigonometric or hyperbolic identities.

The hyperbolic identities $\cosh^2 x = \frac{1}{2}(\cosh 2x + 1)$ and $\sinh^2 x = \frac{1}{2}(\cosh 2x - 1)$ are particularly useful for integration.

Example 4

The curve C has equation $y = 2x^2$. Calculate the length of the curve between $(0, 0)$ and $\left(\frac{1}{4}, \frac{1}{8}\right)$

$\dfrac{dy}{dx} = 4x$

So $s = \displaystyle\int_0^1 \sqrt{1 + (4x)^2}\, dt$

Use the formula
$$s = \int_{x_1}^{x_2} \sqrt{1 + \left(\frac{dy}{dx}\right)^2}\, dx$$

$= \displaystyle\int_0^{\frac{1}{4}} \sqrt{1 + 16x^2}\, dt$

Let $x = \dfrac{1}{4}\sinh u$, then $\dfrac{dx}{du} = \dfrac{1}{4}\cosh u$

When $x = 0, u = 0$

Choose a suitable substitution to use and change the limits.

When $x = \dfrac{1}{4}, u = \ln(1 + \sqrt{2})$

$s = \displaystyle\int_0^{\ln(1+\sqrt{2})} \sqrt{1 + 16\left(\frac{1}{4}\sinh u\right)^2}\left(\frac{1}{4}\cosh u\right) du$

$= \dfrac{1}{4}\displaystyle\int_0^{\ln(1+\sqrt{2})} \sqrt{1 + \sinh^2 u}\,(\cosh u)\, du$

Use the hyperbolic identity $\cosh^2 x - \sinh^2 = 1$

$= \dfrac{1}{4}\displaystyle\int_0^{\ln(1+\sqrt{2})} \sqrt{\cosh^2 u}\,(\cosh u)\, du$

$= \dfrac{1}{4}\displaystyle\int_0^{\ln(1+\sqrt{2})} \cosh^2 u\, du$

(Continued on the next page)

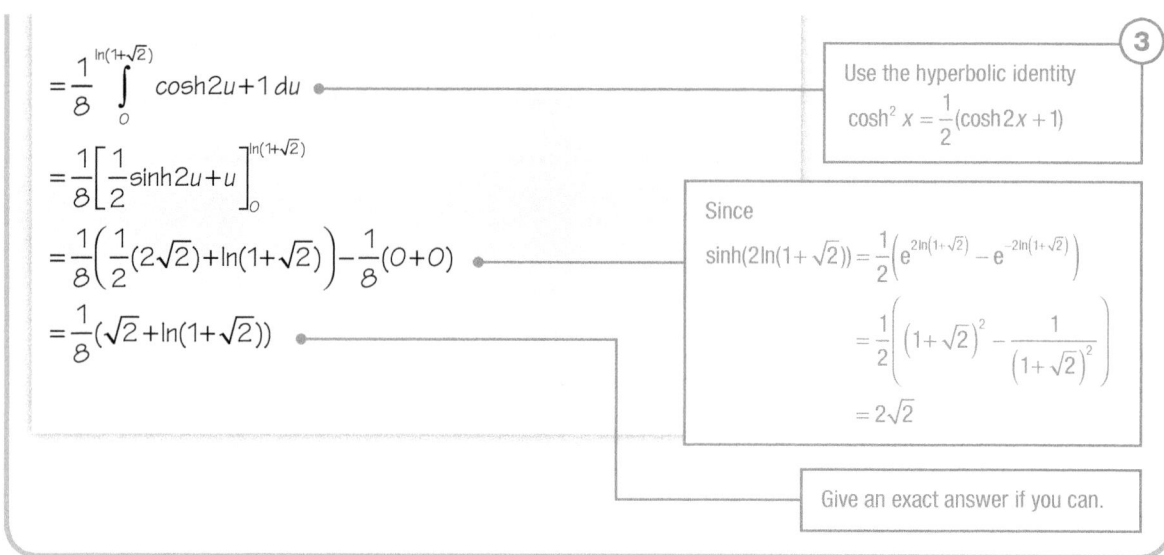

$$= \frac{1}{8} \int_0^{\ln(1+\sqrt{2})} \cosh 2u + 1 \, du$$

Use the hyperbolic identity

$$\cosh^2 x = \frac{1}{2}(\cosh 2x + 1)$$

$$= \frac{1}{8} \left[\frac{1}{2}\sinh 2u + u \right]_0^{\ln(1+\sqrt{2})}$$

$$= \frac{1}{8} \left(\frac{1}{2}(2\sqrt{2}) + \ln(1+\sqrt{2}) \right) - \frac{1}{8}(0+0)$$

Since

$$\sinh(2\ln(1+\sqrt{2})) = \frac{1}{2} \left(e^{2\ln(1+\sqrt{2})} - e^{-2\ln(1+\sqrt{2})} \right)$$

$$= \frac{1}{2} \left((1+\sqrt{2})^2 - \frac{1}{(1+\sqrt{2})^2} \right)$$

$$= \frac{1}{8}(\sqrt{2} + \ln(1+\sqrt{2}))$$

$$= 2\sqrt{2}$$

Give an exact answer if you can.

Exercise 19.7B Reasoning and problem-solving

1 a Use a suitable substitution to show that

$$\int \sqrt{1+16x^2} \, dx$$

$$= \frac{1}{8}\operatorname{arsinh}(4x) + \frac{1}{2}x\cosh(\operatorname{arsinh} 4x) + c$$

b Hence calculate the length of the arc of the curve $y = 2x^2$ between the points $x = 0$ and $x = \frac{1}{4}$

2 A curve is defined by the parametric equations $x = t^2$, $y = 2t$

Calculate the exact arc length from $t = 0$ to $t = 1$

3 A curve has equation $y = 2\sqrt{x}$

a Show that the arc length of the curve between $x = 0$ and $x = 4$ is given by

$$s = \int_0^4 \sqrt{\frac{x+1}{x}} \, dx$$

b Use the substitution $x = \sinh^2 u$ to calculate the exact value of s

4 A curve has parametric equations $x = \cos^3 t$, $y = \sin^3 t$

Calculate the arc length of the curve between $(1, 0)$ and $(0, 1)$

5 The section of the curve $y = 2e^{-x}$ between $(0, 2)$ and $(\ln 2, 1)$ is rotated 2π radians around the x-axis.

a Calculate the volume of revolution.

b Show that the surface area of revolution is given by $4\pi \int_0^{\ln 2} e^{-x} \sqrt{1+4e^{-2x}} \, dx$

c Use the substitution $e^{-x} = \frac{1}{2}\sinh u$ to calculate the surface area of revolution.

6 A circle is defined by the parametric equations $x = r\cos\theta$, $y = r\sin\theta$

Use integration to prove that

a The circumference of the circle is $2\pi r$

b The volume of a sphere is $\frac{4}{3}\pi r^3$

c The surface area of a sphere is $4\pi r^2$

7 A curve has parametric equation $x = \cosh 2\theta$, $y = 2\sinh\theta$, $0 < \theta < \ln 2$

Calculate the surface area of revolution when the curve is rotated $360°$ around the x-axis.

8 a Show that, if $\sin\frac{\theta}{2}$ is positive,

$$\sqrt{1-\cos\theta} = A\sin\frac{\theta}{2},$$ where A is a constant to be found.

b A curve is defined by the parametric equations $x = \theta - \sin\theta$, $y = 1 - \cos\theta$

Calculate the arc length between $\theta = 0$ and $\theta = \pi$

c Calculate the surface area of revolution when the section of the curve between $\theta = 0$ and $\theta = \pi$ is rotated 2π radians around the x-axis.

Chapter summary

- An improper integral is a definite integral where either:
  - one or both of the limits is $\pm\infty$
  - the integrand (expression to be integrated) is undefined at one of the limits of the integral
  - the integrand is undefined at some point between the limits of the integral.
- To evaluate an improper integral, replace the limit where the integrand is undefined with a variable and then consider what happens to the integral as the variable tends to the original value of the limit.
- If the integral is undefined at more than one point, then split it into two integrals.
- Learn and use these important limits:
  - For any real number k, $x^k e^{-x} \to 0$ when $x \to \infty$
  - For any real number k, $x^k \ln x \to 0$ when $x \to 0+$
- Inverse trigonometric functions can be differentiated using the chain rule:
  - $$\frac{d(\arcsin x)}{dx} = \frac{1}{\sqrt{1-x^2}} \qquad \frac{d(\arccos x)}{dx} = -\frac{1}{\sqrt{1-x^2}} \qquad \frac{d(\arctan x)}{dx} = \frac{1}{1+x^2}$$
- Trigonometric functions can be used as substitutions in integration:
  - for an integral involving $\sqrt{a^2 - x^2}$, try the substitution $x = a\sin u$
  - for an integral involving $a^2 + x^2$, try the substitution $x = a\tan u$
- The hyperbolic functions are defined using exponentials:
  - $$\sinh x = \frac{1}{2}(e^x - e^{-x}) \qquad \cosh x = \frac{1}{2}(e^x + e^{-x}) \qquad \tanh x = \frac{e^x - e^{-x}}{e^x + e^{-x}}$$
- Particularly useful hyperbolic identities are:
  - $\cosh^2 x - \sinh^2 x \equiv 1$
  - $\sinh 2x = 2\sinh x \cosh x$ and $\cosh 2x = \cosh^2 x + \sinh^2 x$
- The reciprocal hyperbolic functions are:
  - $$\operatorname{cosech} x = \frac{1}{\sinh x} \qquad \operatorname{sech} x = \frac{1}{\cosh x} \qquad \coth x = \frac{1}{\tanh x}$$
- Hyperbolic functions can be differentiated and integrated:
  - $$\frac{d(\sinh x)}{dx} = \cosh x \qquad \frac{d(\cosh x)}{dx} = \sinh x \qquad \frac{d(\tanh x)}{dx} = \operatorname{sech}^2 x$$
  - $$\frac{d(\operatorname{sech} x)}{dx} = -\operatorname{sech} x \tanh x \qquad \frac{d(\operatorname{cosech} x)}{dx} = -\operatorname{cosech} x \coth x \qquad \frac{d(\coth x)}{dx} = -\operatorname{cosech}^2 x$$
  - $\int \sinh x = \cosh x + c, \int \cosh x = \sinh x + c$
- Inverse hyperbolic functions can be differentiated using the chain rule:
  - $$\frac{d(\operatorname{arcosh} x)}{dx} = \frac{1}{\sqrt{x^2 - 1}} \qquad \frac{d(\operatorname{arsinh} x)}{dx} = \frac{1}{\sqrt{x^2 + 1}}$$
- Hyperbolic functions can be used as substitutions in integration:
  - for an integral involving $\sqrt{x^2 + a^2}$, try the substitution $x = a\sinh u$
  - for an integral involving $\sqrt{x^2 - a^2}$, try the substitution $x = a\cosh u$

- A fraction $\dfrac{f(x)}{g(x)}$ where degree of $f(x) \geq$ degree of $g(x)$ is called an **improper** fraction and can be written in the form $P(x) + \dfrac{Q(x)}{g(x)}$

- $\dfrac{f(x)}{(\alpha x^2 + \beta)(\gamma x + \delta)}$ can be split into partial fractions of the form $\dfrac{Ax + B}{\alpha x^2 + \beta} + \dfrac{C}{\gamma x + \delta}$

- The formula for integration by parts is $\displaystyle\int u\dfrac{dv}{dx}\,dx = uv - \int v\dfrac{du}{dx}\,dx$

- A reduction formula for I_n is an equation that relates I_n to I_{n-1} and/or I_{n-2}. It can be repeatedly applied to reduce the integral to one not requiring integration by parts.

- If a point P has polar coordinates (r, θ), then:
  - r is the length of OP
  - θ is the angle between OP and the initial line.

- A half-line is a straight line that extends infinitely from a point.

- The area of a sector of a polar curve $r = f(\theta)$ between the half lines $\theta = a$ and $\theta = b$ is given by
$$A = \dfrac{1}{2}\int_a^b r^2\,d\theta$$

- The length of the curve $y = f(x)$ from x_1 to x_2 is given by $s = \displaystyle\int_{x_1}^{x_2}\sqrt{1 + \left(\dfrac{dy}{dx}\right)^2}\,dx$

- The length of the curve $x = f(t)$, $y = g(t)$ from t_1 to t_2 is given by $s = \displaystyle\int_{t_1}^{t_2}\sqrt{\left(\dfrac{dx}{dt}\right)^2 + \left(\dfrac{dy}{dt}\right)^2}\,dt$

- The volume of revolution when the curve $y = f(x)$ from x_1 to x_2 is rotated 2π radians around the x-axis is given by $V = \pi\displaystyle\int_{x_1}^{x_2}y^2\,dx$

- The surface area of revolution when the curve $y = f(x)$ from x_1 to x_2 is rotated 2π radians around the x-axis is given by $S = 2\pi\displaystyle\int_{x_1}^{x_2}y\sqrt{1 + \left(\dfrac{dy}{dx}\right)^2}\,dx$

- The surface area of revolution when the curve $x = f(t)$, $y = g(t)$ from t_1 to t is rotated 2π radians around the x-axis is given by $S = 2\pi\displaystyle\int_{t_1}^{t_2}y\sqrt{\left(\dfrac{dx}{dt}\right)^2 + \left(\dfrac{dy}{dt}\right)^2}\,dt$

Check and review

You should now be able to...	Review Questions
✔ Recognise and evaluate an improper integral.	1, 2
✔ Decide whether an improper integral converges.	3, 4
✔ Derive and use the derivatives of inverse trigonometric functions.	5, 6
✔ Use trigonometric functions as substitutions in integration.	7
✔ Derive and use hyperbolic identities.	8, 10

✔ Differentiate hyperbolic functions.	9
✔ Differentiate inverse hyperbolic functions.	10, 11
✔ Use hyperbolic functions as substitutions in integration.	12
✔ Write an improper fraction in the form $P(x) + \dfrac{Q(x)}{g(x)}$	13
✔ Integrate a rational function by first splitting into partial fractions.	13, 14
✔ Derive a reduction formula for indefinite and definite integrals.	15, 16
✔ Use a reduction formula for indefinite and definite integrals.	15, 16
✔ Sketch polar curves.	17
✔ Use integration to calculate the area of a sector of a polar curve.	18
✔ Find the point of intersection between two polar curves.	19
✔ Calculate the area between two polar curves.	19
✔ Calculate the length of a curve given by a Cartesian equation.	20
✔ Calculate the length of a curve given by parametric equations.	21
✔ Calculate the surface area of revolution of a curve given by a Cartesian equation.	22
✔ Calculate the surface area of revolution of a curve given by parametric equations.	23

1 Which of these are improper integrals? Explain your answers.

A: $\displaystyle\int_0^3 \frac{1}{1-x}\,dx$ B: $\displaystyle\int_0^4 \frac{1}{x+2}$ C: $\displaystyle\int_1^\infty \frac{1}{x}\,dx$

2 Evaluate these improper integrals.

a $\displaystyle\int_2^\infty \frac{2}{x^2}\,dx$ b $\displaystyle\int_{-6}^3 \frac{2}{\sqrt{3-x}}\,dx$

3 Show that the integral $\displaystyle\int_0^{2e} x^3 \ln x\,dx$ converges and find its value.

4 Show that the integral $\displaystyle\int_1^\infty \frac{2}{x}\,dx$ does not converge.

5 Prove that $\dfrac{d(\arccos x)}{dx} = -\dfrac{1}{\sqrt{1-x^2}}$

6 Differentiate these expressions with respect to x

a $x^2 \arctan x$ b $\arcsin\left(\dfrac{x^2}{2}\right)$

7 Use a substitution to evaluate each of these integrations.

a $\displaystyle\int \frac{1}{\sqrt{4-x^2}}\,dx$ b $\displaystyle\int \frac{1}{x^2+16}\,dx$

8 Use the definitions of $\sinh x$ and $\cosh x$ to prove these identities.

a $\sinh(A-B) \equiv \sinh A \cosh B - \sinh B \cosh A$

b $\cosh^2 x \equiv \dfrac{1}{2}(1+\cosh 2x)$

9 Differentiate these expressions with respect to x

a $\cosh 2x$ b $x^2 \sinh x$

10 Show that $\dfrac{d(\operatorname{artanh} x)}{dx} = \dfrac{1}{1-x^2}$

11 Differentiate these expressions with respect to x

a $\operatorname{arsinh}(2x^2)$ b $x\operatorname{arcosh}(x-1)$

12 Evaluate these integrals.

a $\displaystyle\int_0^5 \frac{1}{\sqrt{25+x^2}}\,dx$ **b** $\displaystyle\int_3^9 \frac{1}{\sqrt{2x^2-18}}\,dx$

13 a Write $\dfrac{2x^3-x^2-12x+32}{(x-2)(x+3)}$ in the form

$P(x)+\dfrac{Q(x)}{g(x)}$

b Hence evaluate $\displaystyle\int_3^4 \frac{2x^3-x^2-12x+32}{(x-2)(x+3)}\,dx$

14 Calculate the exact value of
$\displaystyle\int_0^3 \frac{x^2+5x+15}{(4-x)(x^2+1)}\,dx$

15 $I_n = \displaystyle\int x^n e^{-2x}\,dx$

a Show that $I_n = -\dfrac{1}{2}x^n e^{-2x} + \dfrac{n}{2}I_{n-1}$

b Use this formula to find I_3

16 $I_n = \displaystyle\int_0^\pi x^n \cos x\,dx$

a Show that
$I_n = -n\pi^{n-1} - n(n-1)I_{n-2}$ for $n>1$

b Use this formula to evaluate

i $\displaystyle\int_0^\pi x^6 \cos x\,dx$ **ii** $\displaystyle\int_0^\pi x^5 \cos x\,dx$

17 For each polar equation

i Sketch the curves for $r\geq 0$ and $0\leq\theta<2\pi$

ii State the maximum and the minimum value of r for for $r\geq 0$ and $0\leq\theta<2\pi$

a $r=8\cos 2\theta$

b $r=-3\sin\theta$

c $r=\dfrac{\theta}{4}$

d $r=7+5\sin\theta$

18 Calculate the total area enclosed within the polar curves for $r\geq 0$ and $0\leq\theta<2\pi$

a $r=3+\cos\theta$ **b** $r=2\sin 4\theta$

19 The graphs of $r=\sqrt{2}+\sin\theta$ and $r=3\sin\theta$ are shown.

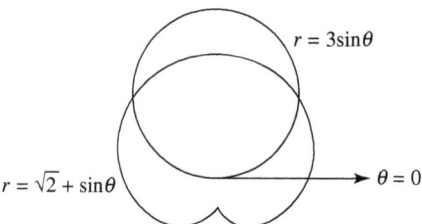

Calculate the shaded area.

20 Calculate the length of the curve $y=(2x+1)^{\frac{3}{2}}$ between the points where $x=3$ and $x=5$

21 A curve is given by the parametric equations

$x=2t,\ y=t^2,\ 0\leq t\leq 2$

Calculate the length of the curve.

22 Calculate the surface area of revolution when the arc of the curve $y=\dfrac{1}{3}x^3$ between $(0,0)$ and $\left(1,\dfrac{1}{3}\right)$ is rotated 2π radians around the x-axis.

23 A curve is given by the parametric equations
$x=\cos\theta,\ y=\sin\theta,\ \dfrac{\pi}{2}\leq\theta\leq\dfrac{3\pi}{4}$

Show that the surface area of revolution when the curve is rotated 360° around the x-axis is $A\pi$, where A is a constant to be found.

Investigation

A geometric solid that has finite volume but infinite surface area is called Gabriel's horn. The solid is formed by rotating the function $f(x) = \frac{1}{x}$ about the x-axis between $x = 1$ and infinity.

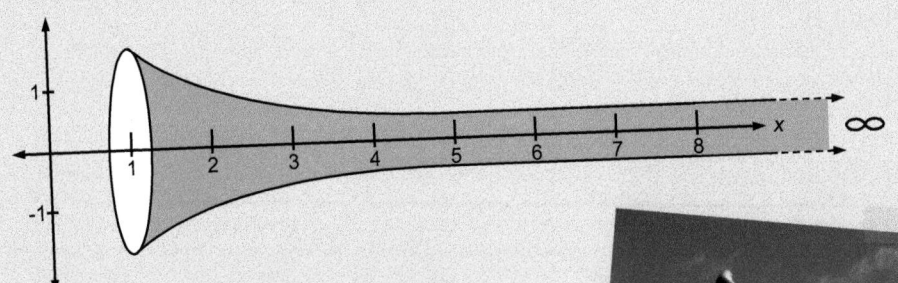

Try this to calculate the finite volume of the solid of revolution.

Research:

- How the formula for the surface area of revolution is derived,
- How the formula for the surface area of revolution leads to the answer of Gabriel's horn having infinite surface area.

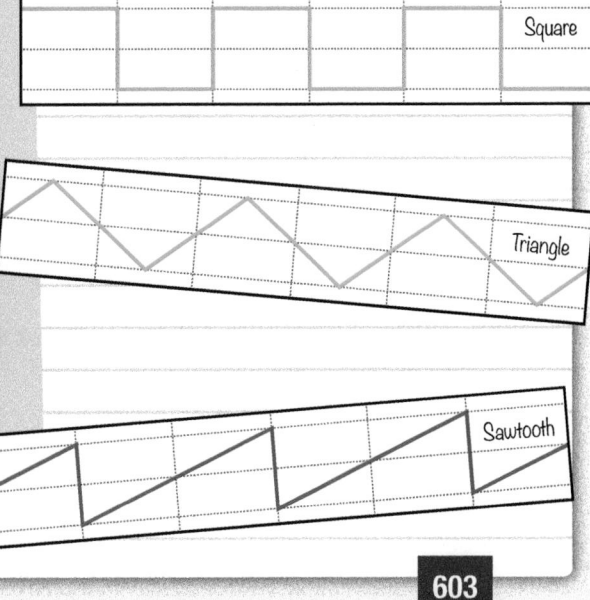

Consider the lower bound of your integrals to be a where $0 \leq a \leq 1$

The results imply that, although the horn could contain a finite amount of paint, this would not be sufficient to paint its surface. This is sometimes known as the painter's paradox.

Investigation

The mean value of a trigonometric wave will be zero over one complete cycle because exactly half of the area enclosed by the function and the horizontal axis is positive and the other half is negative. One way to overcome this problem is to find the root mean square value, that is, the square root of the mean value of the function squared.

Investigate the root mean square of various trigonometric functions such as: $f(x) = A \sin x$ in the interval $0 \leq x \leq 2\pi$

That is find the value of $\sqrt{\frac{1}{2\pi} \int_0^{2\pi} A^2 \sin^2(\omega t)\, dt}$

Root mean square values have many applications in electrical/electronic engineering where engineers work with the flow of alternating and direct current. Other waveforms such as those illustrated may also be used in these fields of engineering.

Investigate their root mean square values.

Square

Triangle

Sawtooth

19 Assessment

1   Find $\int \dfrac{3}{5\sinh x - 4\cosh x}\,dx$ **[10 marks]**

2   **a**   Given $I_n = \int (\ln x)^n\,dx$, show that, for $n \geq 1$, $I_n = x(\ln x)^n - nI_{n-1}$

  b   Hence evaluate $\displaystyle\int_1^4 (\ln x)^3\,dx$ **[10]**

3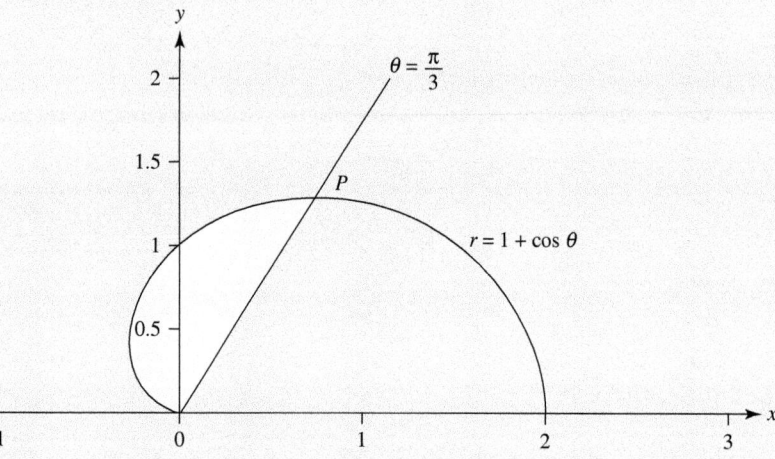

The diagram shows the polar curve with equation $r = 1 + \cos\theta$ for $0 \leq \theta \leq \pi$, along with the line $\theta = \dfrac{\pi}{3}$. The curve and the line intersect at the origin, O, and at the point P

  a   Find the coordinates of the point P

  b   Find the area of the shaded region that is bounded by the line and the curve.   **[8]**

4   **a**   Find $\displaystyle\int_0^1 \dfrac{1}{\sqrt{1-x^2}}\,dx$ **b**   Find $\displaystyle\int_1^2 \dfrac{1}{\sqrt{x^2-1}}\,dx$ **[12]**

5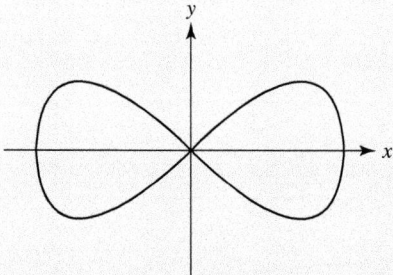

The curve with equation $a^2 y^2 = x^2(a^2 - x^2)$ has two loops, as shown in the diagram above.

  a   Find an expression, in terms of a, for the area of one of the loops.

The curve is rotated through π radians about the x-axis to form two equal solids.

  b   Find the volume of the solids.   **[8]**

6 **a** Given $I_n = \int \tanh^n x \, dx$, show that, for $n \geq 2$, $I_n = I_{n-2} - \dfrac{\tanh^{n-1} x}{n-1}$

   b Hence find an expression for $\int \tanh^8 x \, dx$ **[8]**

7 **a** Sketch the curve with polar equation $r = 4\sin 3\theta$, $0 \leq \theta \leq \pi$

   b Find the area enclosed by one loop of this curve. **[9]**

8 **a** Given $y = \sinh^{-1} x$, prove that $\dfrac{dy}{dx} = \dfrac{1}{\sqrt{1+x^2}}$

   b Show that $\int_0^2 \sinh^{-1} x \, dx = 2\ln(2+\sqrt{5}) + 1 - \sqrt{5}$ **[9]**

9 An arc of a curve is given parametrically by the equations $x = a\cos^3 t$, $y = a\sin^3 t$ for $0 \leq t \leq \dfrac{\pi}{2}$ and $a > 0$

   The points A and B on the curve correspond to the values $t = 0$ and $t = \dfrac{\pi}{2}$ respectively.

   a Find the length of the arc, AB, of the curve.

   b Find the area of the curved surface generated when this arc is rotated through $360°$ about the x-axis. **[11]**

10 Show that $\int_1^4 \dfrac{2x+1}{\sqrt{x^2-2x+10}} \, dx = 6\sqrt{(2)} + 3\,\mathrm{arsinh}(1) - 6$ **[11]**

11 The region bounded by the curve $y = \sin x^2$, the y-axis and the line $y = 1$ is rotated through $360°$ about the y-axis. Find the volume of the solid of revolution. **[8]**

12 **a** Find $\int \dfrac{1}{x^2-4x+13} \, dx$ **b** Evaluate $\int_0^\infty x^2 e^{-x} \, dx$ **[15]**

13 The diagram shows the polar curves with equations $r = \sin 2\theta$ and $r = \cos\theta$ for $0 \leq \theta \leq \dfrac{\pi}{2}$
   The curves intersect at the origin, O, and at the point P

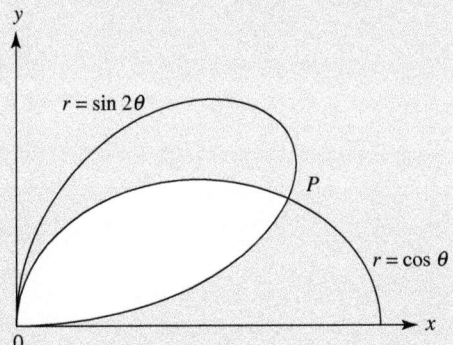

   a Find the polar coordinates of the point P

   b Show that the area of the shaded region that is bounded by the two curves is given
   by $\dfrac{\pi}{8} - \dfrac{3\sqrt{3}}{32}$ **[13]**

14 **a** Find $\int_{\frac{\pi}{3}}^{\frac{\pi}{2}} \dfrac{1}{1+\cos\theta} \, d\theta$ **b** Show that $\int_1^3 \dfrac{3x-x^2}{(x+1)(x^2+3)} \, dx = \dfrac{\pi}{2\sqrt{3}} - \ln 2$ **[16]**

15 An arc, L, of a parabola is given parametrically by the equations $x = at^2$, $y = 2at$, for $0 \leq t \leq 2$

L is rotated through 360° about the x-axis.

a Show that the area of the surface of revolution is given by $8\pi a^2 \displaystyle\int_0^2 t\sqrt{1+t^2}\, dt$

b Find this surface area.

c Show also that the length of L is given by $2a \displaystyle\int_0^2 \sqrt{1+t^2}\, dt$

d Find this length. **[16]**

16 a Given $I_n = \displaystyle\int_0^{\frac{\pi}{2}} \sin^n x\, dx$, show that, for $n \geq 2$, $nI_n = (n-1)I_{n-2}$

The region R is bounded by the curve $y = \sin^4 x$, the line $x = \dfrac{\pi}{2}$ and the x-axis, between $x = 0$ and $x = \dfrac{\pi}{2}$

b Find the area of R.

c Find the volume generated when R is rotated through 360° about the x-axis. **[15]**

17 The curve C has polar equation $r = 1 + \cos\theta$, $0 \leq \theta \leq 2\pi$. The curve D has polar equation $r = 2 - \cos\theta$, $0 \leq \theta \leq 2\pi$. The two curves intersect at the points P and Q

a Find the polar coordinates of the points P and Q

b Sketch, on one diagram, the graphs of C and D.

c Find the area of the region that is both outside D and inside C. **[13]**

18 a Find $\displaystyle\int \dfrac{9x^3 - 4x + 6}{9x^2 - 4}\, dx$

b Use the substitution $x = \sin\theta$ to show that $\displaystyle\int_0^{\frac{1}{2}} \dfrac{24x^2}{\sqrt{1-x^2}}\, dx = 2\pi - 3\sqrt{3}$ **[16]**

19 a Given that $I_n = \displaystyle\int_0^{\frac{\pi}{4}} \dfrac{\sin^{2n+1} x}{\cos x}\, dx$, show that $I_n = I_{n-1} - \dfrac{1}{2^{n+1}\, n}$

b Hence find the exact value of $\displaystyle\int_0^{\frac{\pi}{4}} \dfrac{\sin^7 x}{\cos x}\, dx$ **[11]**

20

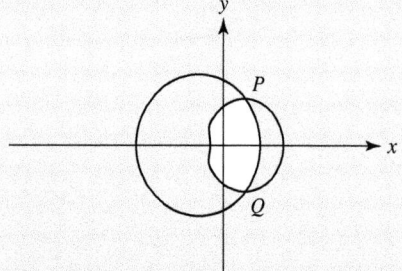

The diagram shows the curves with polar equations $r = a(3 + 2\cos\theta)$ and $r = a(5 - 2\cos\theta)$ for $0 \leq \theta \leq 2\pi$. The curves intersect at the points P and Q

a Find the coordinates of P and Q

b Show that the area of the shaded region that is bounded by the two curves is given by

$$\dfrac{a^2 \left(49\pi - 48\sqrt{3}\right)}{3}$$ **[16]**

21

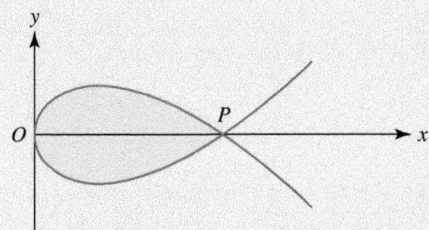

The diagram shows the curve with equation $3y^2 = x(1-x)^2$. The curve meets the x-axis at the origin and at the point P

a   Find the coordinates of P

b   Show that the perimeter of the closed shaded region between O and P is given by $\displaystyle\int_0^1 \frac{1+3x}{\sqrt{3x}}\,dx$

c   Evaluate this perimeter.

The shaded region is rotated through $180°$ about the x-axis.

d   Show that the surface area of the curved surface is $\dfrac{\pi}{3}$ **[14]**

22 a   Find **i** $\displaystyle\int \frac{1}{\sqrt{12-4x-x^2}}\,dx$ **ii** $\displaystyle\int \operatorname{artanh}2x\,dx$

b   Use the substitution $x=4\operatorname{cosech}u$ to show that $\displaystyle\int_1^\infty \frac{1}{x\sqrt{x^2+16}}\,dx = \frac{1}{4}\ln(4+\sqrt{17})$ **[15]**

23 a   Given that $I_n = \displaystyle\int_0^1 x^n\sqrt{1-x^2}\,dx$, show that $I_{2n+1} = \dfrac{2n}{2n+3}I_{2n-1}$ for $n\geq 1$

b   Hence show that $\displaystyle\int_0^1 x^{2n+1}\sqrt{1-x^2}\,dx = \dfrac{2^{2n+1}\,n!\,(n+1)!}{(2n+3)!}$ **[14]**

24

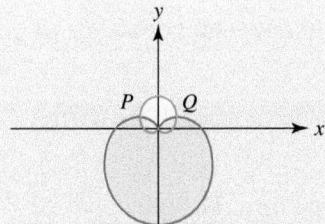

The diagram shows the curves with polar equations $r=3(1-\sin\theta)$ and $r=1+\sin\theta$ for $0\leq\theta\leq 2\pi$
The curves intersect at the pole, and at the points P and Q

a   Find the coordinates of P and Q

b   Prove that the area that is bounded by the blue shape but not the red shape is $9\sqrt{3}-4\pi$ **[15]**

25 a   Find $\displaystyle\int \frac{1}{\sqrt{x^2-2x+10}}\,dx$

b   Show that $\displaystyle\int_{-\infty}^{\infty} \frac{1}{1+x^2}\,dx = \pi$

c   What is wrong with the following argument?

Since $\dfrac{1}{(1+x)^2} > 0$ for all x, clearly $\displaystyle\int_{-a}^{a} \frac{1}{(1+x)^2}\,dx > 0$ for any positive value of a

But $\displaystyle\int_{-a}^{a} \frac{1}{(1+x)^2}\,dx = \left[-\frac{1}{1+x}\right]_{-a}^{a} = -\frac{1}{1+a}+\frac{1}{1-a} = \frac{2a}{1-a^2}$

As $a\to\infty$ $\dfrac{2a}{1-a^2}\to 0$ so $\displaystyle\int_{-\infty}^{\infty}\frac{1}{(1+x)^2}\,dx = 0$, so we have a contradiction.   **[12]**

26

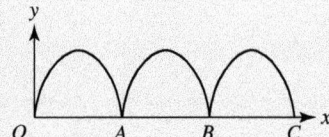

The diagram shows a cycloid with parametric equations $x = \theta - \sin\theta$, $y = 1 - \cos\theta$ for $0 \le \theta \le 6\pi$
The cycloid passes through the origin, and meets the x-axis again at the points A, B and C, as in the diagram.

a   Find the coordinates of the points A, B and C

b   Show that the length of the arc of the cycloid between points O and C is given by

$$3\sqrt{2} \int_0^{2\pi} \sqrt{1 - \cos\theta}\ \mathrm{d}\theta$$

c   Find the value of this arc length.

This arc, between O and C, is rotated through 2π radians about the x-axis.

d   Show that the surface area of the curved surface generated is 64π **[14]**

27 The diagram shows part of the graph of $y = \dfrac{1}{\sqrt{4 - x^2}}$

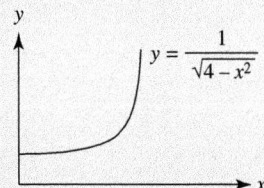

a   Find the area of the region bounded by the curve, the x-axis, the y-axis and the line $x = \sqrt{2}$

b   Find the volume of the solid generated when this region is rotated through $360°$ about the x-axis.

c   Find also the volume generated when this same region is rotated through $360°$ about the y-axis.   **[19]**

Differential equations

Mathematics, and particularly calculus, plays a large part in the science and engineering that underpins space exploration. For centuries, the idea of space travel was only a remote possibility, and the mathematics needed to support the engineering required was only just being developed. The physical principles that these mathematicians worked with are relatively straight forward. For example, rocket launches rely on the application of Newton's third law to the jet engines that are used.

One of the differential equations that space engineers work with is the Tsiolkovsky rocket equation. This is named after the man who developed it, Konstantin Tsiolkovsky, a Russian physicist who was born in 1857 and died in 1935. The equation allows space engineers to calculate how much propellant is necessary to lift a rocket off the ground.

Orientation

What you need to know	What you will learn	What this leads to
Maths Ch16 • Integration and differential equations.	• How to solve first order differential equations. • How to solve second order differential equations. • How differential equations are used to model simple harmonic motion. • How differential equations are used to model damped harmonic motion. • How to work with differential equations that involve more than two variables.	**Careers** • Engineering. • Economics. • Ecology.

Fluency and skills

An equation that involves only a first order derivative, such as $\dfrac{dy}{dx}$, is called a **first order differential equation**. There are some first order differential equations where the terms involving x can be factorised out from the terms involving y. This method is called **separating the variables**. There are a number of methods for solving first order differential equations. One method involves using an **integrating factor**.

Solve the differential equation $x^2 \dfrac{dy}{dx} + 2xy = 4x^3$, giving your answer in the form $y = f(x)$

$$\dfrac{d}{dx}\left[x^2 y\right] = 4x^3$$

The LHS can be rewritten as the derivative of a product.

$$x^2 y = x^4 + c$$

Integrate each side.

$$y = \dfrac{x^4 + c}{x^2}$$

Rearrange to get an expression for y in terms of x

Key point

An expression that can be integrated by spotting that it is a perfect differential of a product is called an **exact equation**.

Suppose instead you had been asked to solve the differential equation $\dfrac{dy}{dx} + \dfrac{2}{x} y = 4x$. This is not as easy because you can no longer spot the product rule on the LHS. However, by multiplying through by x^2, you can transform the equation into an exact equation. The multiplier, x^2, is known as the **integrating factor**. You need to know how to find the integrating factor.

First order differential equations that can be solved using an integrating factor can often be written in the form

$$\dfrac{dy}{dx} + P(x)y = f(x)$$

where $P(x)$ and $f(x)$ are functions involving only the variable x

To find the integrating factor, first notice that if you differentiate a product in the form $e^{\int P(x)\,dx}y$

then

$$\frac{d}{dx}\left[e^{\int P(x)\,dx}y\right] = e^{\int P(x)\,dx}\frac{dy}{dx} + P(x)e^{\int P(x)\,dx}y$$

so if you take the equation $\frac{dy}{dx} + P(x)y = f(x)$ and

multiply throughout by $e^{\int P(x)\,dx}$, the equation becomes

$$e^{\int P(x)\,dx}\frac{dy}{dx} + P(x)e^{\int P(x)\,dx}y = e^{\int P(x)\,dx}f(x)$$

The LHS is now an exact equation, and you have

$$\frac{d}{dx}\left[e^{\int P(x)\,dx}y\right] = e^{\int P(x)\,dx}f(x)$$

which you can solve by integrating each side.

Key point

The expression $e^{\int P(x)\,dx}$ is called the integrating factor.

To solve the equation $\frac{dy}{dx} + \frac{2}{x}y = 4x$, $P(x) = \frac{2}{x}$, and the integrating

factor is $e^{\int \frac{2}{x}\,dx} = e^{2\ln x} = e^{\ln x^2} = x^2$

Multiplying the equation by x^2 gives you $x^2\frac{dy}{dx} + x^2\frac{2}{x}y = x^2 4x$,

i.e. $x^2\frac{dy}{dx} + 2xy = 4x^3$, and you are back to Example 1

Example 2

Solve the differential equation $\frac{dy}{dx} - y\tan x = 3\sin^2 x$, for $0 \le x < \frac{\pi}{2}$

This equation is in the form $\frac{dy}{dx} + P(x)y = f(x)$, where $P(x) = -\tan x$ and $f(x) = 3\sin^2 x$

$e^{\int P(x)\,dx} = e^{\int -\tan x\,dx} = e^{\ln(\cos x)} = \cos x$

Find the integrating factor.

$\cos x\frac{dy}{dx} - y\cos x\tan x = 3\cos x\sin^2 x$

Multiply through by $\cos x$

Since $\cos x\tan x = \cos x\dfrac{\sin x}{\cos x} = \sin x$

$\cos x\frac{dy}{dx} - y\sin x = 3\cos x\sin^2 x$

Use the fact it is now an exact equation.

$\frac{d}{dx}[y\cos x] = 3\cos x\sin^2 x$

Integrate each side. You can leave the solution in implicit form.

$y\cos x = \sin^3 x + c$

A solution that includes a constant of integration is called a **general solution**. Key point

If you can find a value for the constant you obtain a **particular solution**.

Example 3

a Find the general solution to the equation $x\dfrac{dy}{dx} - y = 2x^3$

b Given that $y = 4$ when $x = 1$, find the particular solution.

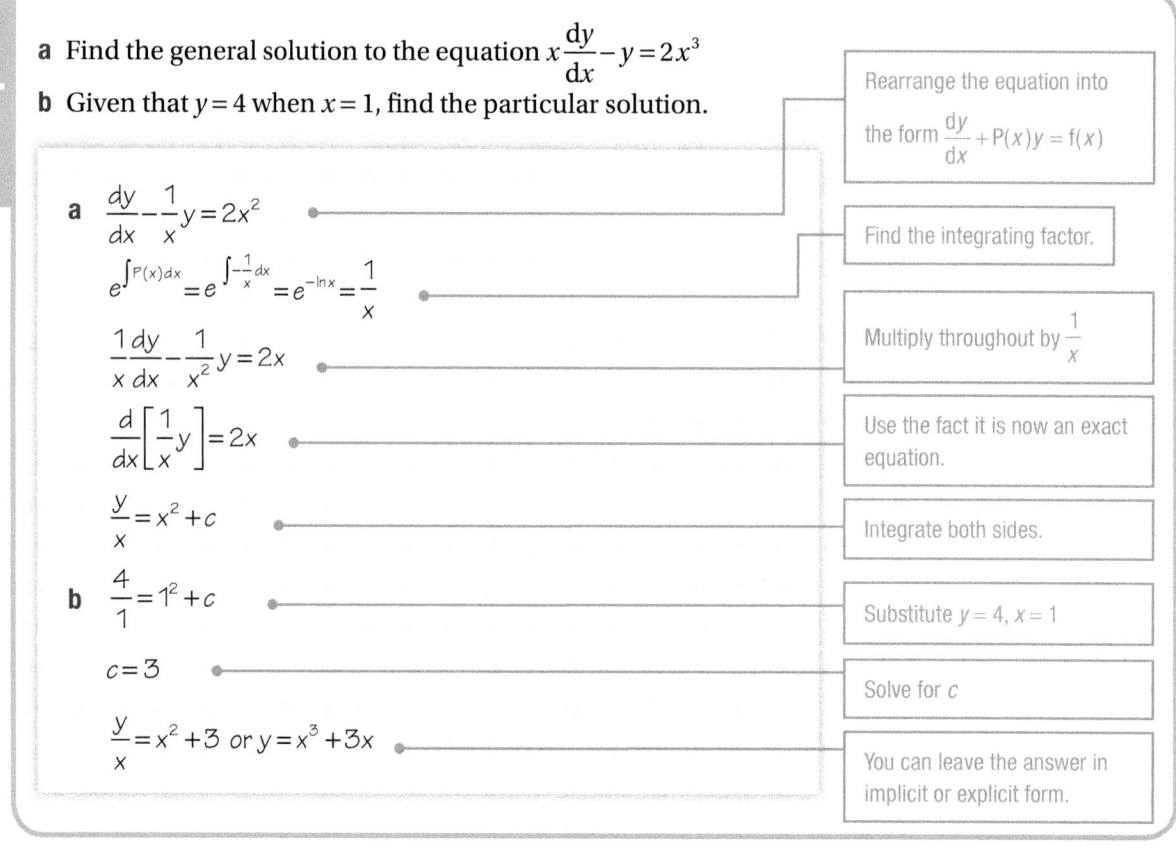

a $\dfrac{dy}{dx} - \dfrac{1}{x}y = 2x^2$ ●————— Rearrange the equation into the form $\dfrac{dy}{dx} + P(x)y = f(x)$

$e^{\int P(x)dx} = e^{\int -\frac{1}{x}dx} = e^{-\ln x} = \dfrac{1}{x}$ ●————— Find the integrating factor.

$\dfrac{1}{x}\dfrac{dy}{dx} - \dfrac{1}{x^2}y = 2x$ ●————— Multiply throughout by $\dfrac{1}{x}$

$\dfrac{d}{dx}\left[\dfrac{1}{x}y\right] = 2x$ ●————— Use the fact it is now an exact equation.

$\dfrac{y}{x} = x^2 + c$ ●————— Integrate both sides.

b $\dfrac{4}{1} = 1^2 + c$ ●————— Substitute $y = 4$, $x = 1$

$c = 3$ ●————— Solve for c

$\dfrac{y}{x} = x^2 + 3$ or $y = x^3 + 3x$ ●————— You can leave the answer in implicit or explicit form.

Sometimes you need to use a given substitution to transform the differential equation into one that can be solved using an integrating factor.

Example 4

a Use the substitution $y = u^2$ to transform the differential equation $\dfrac{dy}{dx} + x^2 y = x\sqrt{y}$ into the differential equation $\dfrac{du}{dx} + \dfrac{1}{2}x^2 u = \dfrac{1}{2}x^2$

b Hence use an integrating factor to find the general solution to the differential equation $\dfrac{dy}{dx} + x^2 y = x\sqrt{y}$ in the form $y = f(x)$

a $\dfrac{dy}{dx} = \dfrac{dy}{du} \times \dfrac{du}{dx} = 2u\dfrac{du}{dx}$ ●————— Find an expression for $\dfrac{dy}{dx}$

Now $\dfrac{dy}{dx} + x^2 y = x\sqrt{y}$ becomes $2u\dfrac{du}{dx} + x^2 u^2 = x^2 u$ ●————— Substitute for $\dfrac{dy}{dx}$, y and $\sqrt{y}$

Hence $\dfrac{du}{dx} + \dfrac{1}{2}x^2 u = \dfrac{1}{2}x^2$ ●————— Divide through by $2u$

(*Continued on the next page*)

b The integrating factor is $e^{\int \frac{1}{2}x^2\,dx} = e^{\frac{1}{6}x^3}$

Find the integrating factor.

$$e^{\frac{1}{6}x^3}\frac{du}{dx} + \frac{1}{2}x^2 e^{\frac{1}{6}x^3}u = \frac{1}{2}x^2 e^{\frac{1}{6}x^3}$$

Multiply through by the integrating factor.

Hence $e^{\frac{1}{6}x^3}u = \int \frac{1}{2}x^2 e^{\frac{1}{6}x^3}\,dx$

$$= e^{\frac{1}{6}x^3} + c$$

Solve the differential equation in u and x

$$e^{\frac{1}{6}x^3}\sqrt{y} = e^{\frac{1}{6}x^3} + c$$

$$y = \left(1 + \frac{c}{e^{\frac{1}{6}x^3}}\right)^2$$

Substitute for y and make y the subject.

Exercise 20.1A Fluency and skills

1 Find the general solution to each of the following differential equations.

a $\dfrac{dy}{dx} + 3y = e^{-x}$

b $\dfrac{dy}{dx} + 2xy = 8x$

c $\dfrac{dy}{dx} + y\tan x = \sin^2 x \cos^2 x$

d $\dfrac{dy}{dx} - \dfrac{y}{x} = \dfrac{x}{x-5}$

e $3x\dfrac{dy}{dx} + y = \sqrt{x}$

f $x\dfrac{dy}{dx} - y = x^2 \ln x$

g $\cos x\dfrac{dy}{dx} - y\sin x = 4\sec^2 x$

h $\dfrac{1}{x}\dfrac{dy}{dx} + 2\dfrac{(x+1)}{x}y = 2e^{(x-1)^2}$

2 Find the particular solution to each of the following differential equations, giving your answers in the form $y = f(x)$

a $\dfrac{dy}{dx} - 2y = 2xe^{3x}$, given $y = -1$ when $x = 0$

b $\dfrac{dy}{dx} + \dfrac{3}{x}y = \dfrac{4}{x^2}$, given $y = 5$ when $x = 1$

c $\dfrac{dy}{dx} + y\cot x = 4\cos^3 x$, given $y = \dfrac{1}{4}$ when $x = \dfrac{\pi}{6}$

d $\dfrac{dy}{dx} - 2y\cos 2x = 2e^{\sin 2x}$, given $y = 1$ when $x = 0$

e $(x+1)\dfrac{dy}{dx} + 2y = \dfrac{14}{(x+1)}$, given $y = 3$ when $x = -2$

f $\sin x \dfrac{dy}{dx} + y\cos x = -3\cos x \sin 2x$, given $y = 2$ when $x = \dfrac{\pi}{4}$

g $x\dfrac{dy}{dx} - y = x^3 \ln x$, given $y = 2$ when $x = 1$

h $\coth x \dfrac{dy}{dx} + y = 5e^{5x}\operatorname{cosech} x$, given $y = 4$ when $x = 0$

3 **a** Find the general solution to the equation $\cos x \dfrac{dy}{dx} + 2y\sin x = \sin^2 x \cos x$

b Given that $y = \dfrac{1}{2}$ when $x = \dfrac{\pi}{4}$, find the particular solution.

4 Find the solution to the differential equation $\dfrac{dy}{dx} - y\tan x = 3x^2 \sec x$, given that $y = 2$ when $x = 0$

5 Find the solution to the differential equation $x\dfrac{dy}{dx} - 2y = x^3 \ln x$, given that $y = 5$ when $x = 1$

6 **a** Find the general solution to the equation $\dfrac{dy}{dx} + 2y = 5\sin x$

b Given that $y = 1$ when $x = 0$, find the particular solution.

7 **a** Given that $u = \dfrac{1}{y^2}$ and $\dfrac{dy}{dx} = y + 2xy^3$, show that $\dfrac{du}{dx} + 2u = -4x$

b Hence find the solution to the differential equation $\dfrac{dy}{dx} = y + 2xy^3$ given that $y = \dfrac{1}{2}$ when $x = 0$

8 **a** Find the general solution of the differential equation $\cos x \dfrac{dy}{dx} + 2y\sin x = e^x \cos^3 x$

b Find the particular solution for which $y = 5$ at $x = 0$, giving your solution in the form $y = f(x)$

9 a Find the general solution of the differential equation $\tan x\dfrac{dy}{dx}+y=12\sin 2x\tan x$

b Hence show that the particular solution to this equation for which $y=2$ at $x=\dfrac{\pi}{4}$, is given by $y\sin x=8\sin^3 x-\sqrt{2}$

10 a Find $\displaystyle\int x^3 e^x\,dx$

b Hence show that the general solution to the differential equation $x\dfrac{dy}{dx}+(x+2)y=2x^2$ is given by $x^2 y=2\,(x^3-3x^2+6x-6\,)+ce^{-x}$

11 a Show that $\displaystyle\int\sec^3 x\,dx=\dfrac{1}{2}\big[\sec x\tan x+\ln(\sec x+\tan x)\big]+c$

b Hence show that the general solution to the differential equation

$x\dfrac{dy}{dx}+(1-x\tan x)y=2\sec^4 x$ is $y=\dfrac{\sec x\tan x+\ln(\sec x+\tan x)+c}{x\cos x}$

12 Find the general solution of the differential equation $x+(t\ln t)\dfrac{dx}{dt}=2te^{2t},\,t>0$

13 a Find the general solution of the differential equation $(1+t)\dfrac{dx}{dt}+3x=\ln(1+t),\,t>-1$

b Show that the particular solution for which $x=\dfrac{8}{9}$ at $t=0$, is given by

$x=\dfrac{3\ln(1+t)-1}{9}+\dfrac{1}{(1+t)^3}$

14 a Given that $z=y^{\frac{1}{2}}$ and $\dfrac{dy}{dx}-6y\tan x=3y^{\frac{1}{2}}$, show that $\dfrac{dz}{dx}-3z\tan x=\dfrac{3}{2}$

b Hence find the general solution to the differential equation $\dfrac{dy}{dx}-6y\tan x=3y^{\frac{1}{2}}$ in the form $y=f(x)$

15 a Use the substitution $v=\dfrac{dx}{dt}$ to transform the differential equation $t\dfrac{d^2x}{dt^2}-\dfrac{dx}{dt}=5t^2$ into $\dfrac{dv}{dt}-\dfrac{1}{t}v=5t$

b Hence find the general solution of the differential equation $t\dfrac{d^2x}{dt^2}-\dfrac{dx}{dt}=5t^2$, giving your answer in the form $y=f(x)$

To answer a question that involves first order differential equations

(**1**) Decide whether to separate the variables or use an integrating factor.

(**2**) Find the general solution of the differential equation.

(**3**) Substitute the given values to find the constant, and hence obtain the particular solution.

(**4**) Answer the question in context.

Example 5

A population of bacteria has an initial size of 100. After t hours, the size of the population is P

The connection between P and t can be modelled by the equation $\dfrac{dP}{dt} = 2(50t - P)$

a Solve this equation to show that $P = 25(2t + 5e^{-2t} - 1)$

b Find the size of the population after 24 hours.

c Prove that the number of bacteria never falls below 40

a $\dfrac{dP}{dt} + 2P = 100t$

Rearrange the equation into the form $\dfrac{dP}{dt} + F(t)P = f(t)$

$e^{\int F(t)dt} = e^{\int 2dt} = e^{2t}$

Find the integrating factor.

$e^{2t}\dfrac{dP}{dt} + 2e^{2t}P = 100te^{2t}$

Multiply throughout by e^{2t}

$\dfrac{d}{dt}\left[e^{2t}P\right] = 100te^{2t}$

Use the fact it is now an exact equation.

$e^{2t}P = \int 100t\,e^{2t}\,dt$

Integrate each side.

$e^{2t}P = 50te^{2t} - \int 50\,e^{2t}dt = 50te^{2t} - 25\,e^{2t} + c$

Use integration by parts.

$100 = -25 + c \Rightarrow c = 125$

$e^{2t}P = 50te^{2t} - 25\,e^{2t} + 125$

Substitute $t = 0$, $P = 100$ to find c

$P = 25(2t + 5e^{-2t} - 1)$

b At $t = 24$, $P = 25[2(24) + 5e^{-2(24)} - 1]$

Substitute $t = 24$

$= 1175$

c $\dfrac{dP}{dt} = 25(2 - 10e^{-2t}) = 0$

Set $\dfrac{dP}{dt} = 0$

$2 - 10e^{-2t} = 0$ gives $t = \dfrac{1}{2}\ln 5$

Solve $\dfrac{dP}{dt} = 0$

$P = 25(\ln 5 + 1 - 1) = 25\ln 5$

By inspection this is a minimum, so the value of P never falls below $25\ln 5 \approx 40$

Substitute $t = \dfrac{1}{2}\ln 5$

Example 6

A raindrop falls vertically through a cloud. Initially the raindrop is at rest. At time t seconds, the velocity, $v\,\text{m s}^{-1}$ of the raindrop, satisfies the equation $(2+t)\dfrac{dv}{dt}+v=10(2+t)$

a Solve this equation to find an expression for v in terms of t

b Find the time at which the velocity of the raindrop is $21\,\text{m s}^{-1}$

c Criticise the model.

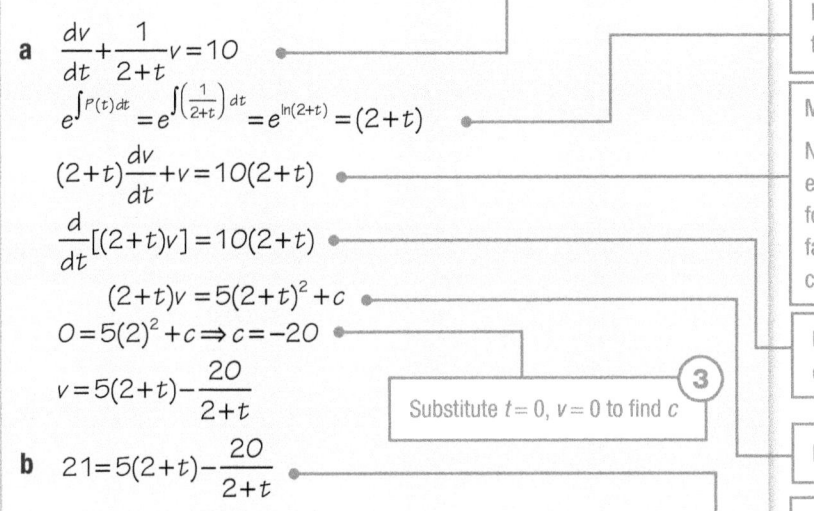

Rearrange the equation into the form $\dfrac{dv}{dt}+P(t)v=f(t)$

a $\dfrac{dv}{dt}+\dfrac{1}{2+t}v=10$

$e^{\int P(t)\,dt}=e^{\int\left(\frac{1}{2+t}\right)dt}=e^{\ln(2+t)}=(2+t)$

$(2+t)\dfrac{dv}{dt}+v=10(2+t)$

$\dfrac{d}{dt}[(2+t)v]=10(2+t)$

$(2+t)v=5(2+t)^2+c$

$0=5(2)^2+c \Rightarrow c=-20$

$v=5(2+t)-\dfrac{20}{2+t}$

b $21=5(2+t)-\dfrac{20}{2+t}$

$5t-11-\dfrac{20}{2+t}=0$

$5t^2-t-42=0$

$(5t+14)(t-3)=0$

So $t=3$

c Under this model, as t increases, v is unlimited, but, in practice, the raindrop would approach a terminal velocity.

1 Decide to use an integrating factor.

Multiply throughout by $(2+t)$

Note that the differential equation was already in the form $[u'v+uv']$. The integrating factor manipulations are here in case you didn't spot that.

Use the fact it is now an exact equation.

3 Substitute $t=0$, $v=0$ to find c

2 Find the general solution.

Substitute $v=21$ to get an equation in t

You only need the positive value of t

4 Answer the question in context.

Example 7

A fuel-filled rocket of mass 25 kg burns fuel at a rate of $2\,\text{kg s}^{-1}$. The rocket is initially at rest. After t seconds, the velocity $v\,\text{m s}^{-1}$ of the rocket, satisfies the equation $(25-2t)\dfrac{dv}{dt}+v=100$ for $t\leq12$

a Show that $v=100-20\sqrt{25-2t}$

b Find the speed of the rocket when $t=12$

c Sketch a graph of v against t

d What happens for $t>12.5$?

(Continued on the next page)

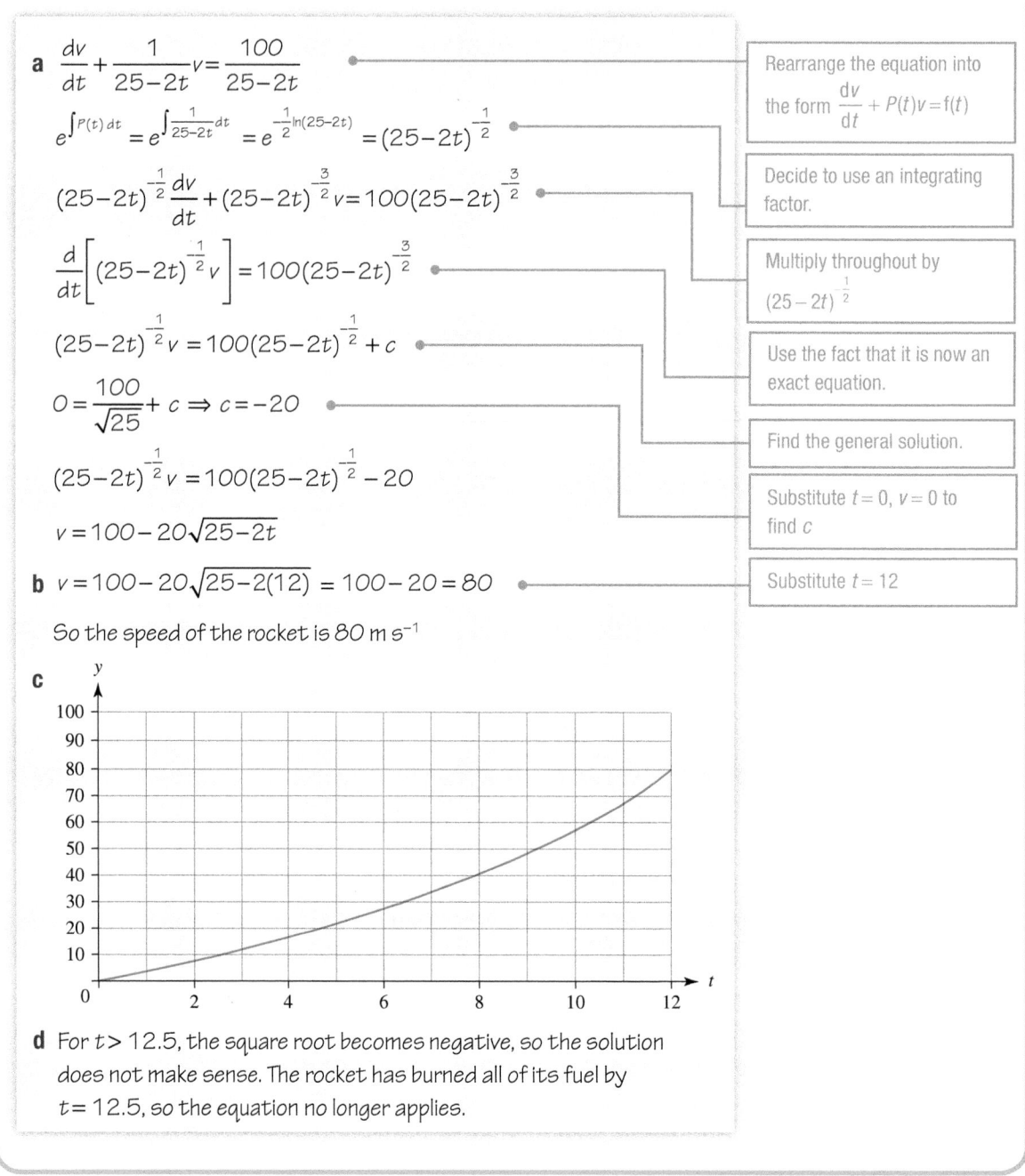

a $\dfrac{dv}{dt}+\dfrac{1}{25-2t}v=\dfrac{100}{25-2t}$

$e^{\int P(t)\,dt}=e^{\int\frac{1}{25-2t}dt}=e^{-\frac{1}{2}\ln(25-2t)}=(25-2t)^{-\frac{1}{2}}$

$(25-2t)^{-\frac{1}{2}}\dfrac{dv}{dt}+(25-2t)^{-\frac{3}{2}}v=100(25-2t)^{-\frac{3}{2}}$

$\dfrac{d}{dt}\left[(25-2t)^{-\frac{1}{2}}v\right]=100(25-2t)^{-\frac{3}{2}}$

$(25-2t)^{-\frac{1}{2}}v=100(25-2t)^{-\frac{1}{2}}+c$

$0=\dfrac{100}{\sqrt{25}}+c\Rightarrow c=-20$

$(25-2t)^{-\frac{1}{2}}v=100(25-2t)^{-\frac{1}{2}}-20$

$v=100-20\sqrt{25-2t}$

b $v=100-20\sqrt{25-2(12)}=100-20=80$

So the speed of the rocket is 80 m s⁻¹

c

d For $t>12.5$, the square root becomes negative, so the solution does not make sense. The rocket has burned all of its fuel by $t=12.5$, so the equation no longer applies.

Rearrange the equation into the form $\dfrac{dv}{dt}+P(t)v=f(t)$

Decide to use an integrating factor.

Multiply throughout by $(25-2t)^{-\frac{1}{2}}$

Use the fact that it is now an exact equation.

Find the general solution.

Substitute $t=0$, $v=0$ to find c

Substitute $t=12$

Exercise 20.1B Reasoning and problem-solving

1 A population has an initial size of 40
After t days the size of the population is P
The connection between P and t can be modelled by the equation

$\dfrac{dP}{dt}=20t-P$

a Solve this equation to show that
$P=20(t+3e^{-t}-1)$

b Find the size of the population after 2 weeks.

c Prove that the size of the population never falls below 21

2 In an electric circuit the current, I amps at time t seconds is modelled by the equation
$$\frac{dI}{dt}+2I=6$$
Initially $I=8$

  a Using an integrating factor, solve this equation to show that $I=3+5e^{-2t}$

  b Show that the current never drops below 3 amps.

  c Sketch a graph of I against t

3 A population of 8 million bacteria is injected into a body. After t days the size of the population in the body is x million where x and t satisfy the differential equation $\dfrac{dx}{dt}=t-x+4$

  a Show that the size of the population initially starts to decline.

  b Find an expression for x in terms of t

  c Find the minimum size of the population.

  d What happens to the population in the long term?

4 A hailstone falls vertically through a cloud. Initially the hailstone is at rest. At time t seconds, the velocity, $v\,\text{m s}^{-1}$, of the hailstone satisfies the equation $\dfrac{dv}{dt}=10-\dfrac{2v}{1+2t}$

  a Solve this equation to find an expression for v in terms of t

  b Find, correct to one decimal place, the time at which the velocity of the hailstone is $24\,\text{m s}^{-1}$.

  c Criticise the model.

5 As part of an industrial process a chemical is dissolved in a solution. t minutes after the start of the process, the mass, $C\,\text{kg}$, of the chemical that has dissolved can be modelled by the equation $\dfrac{dC}{dt}+\dfrac{C}{20+t}=4$
Initially $C=10\,\text{kg}$.

  a Solve this equation to show that
$$C=2(20+t)-\frac{600}{20+t}$$

  b Find the time at which $C=40$

6 Water is draining from a tank. The depth of water in the tank is initially 2 metres, and after t minutes, the depth is x metres. The depth can be modelled by the equation
$$\frac{dx}{dt}=-\frac{1}{2}(x+e^{\frac{1}{2}t}\cos t)$$

  a Solve this equation to find an expression for x in terms of t

  b Find the depth of the water in the tank after five minutes.

7 A particle is projected vertically upwards with a velocity of $13\,\text{m s}^{-1}$. During the motion it absorbs moisture so that when the particle is a distance x metres above the point of projection, its velocity, $v\,\text{m s}^{-1}$ is given by the equation $2v\dfrac{dv}{dx}+\dfrac{2}{1+2x}v^2=-4$

  a Solve this equation to show that
$$v^2(1+2x)=170-(1+2x)^2$$

  b Calculate the greatest height reached by the particle.

8 A population of bacteria grows from an initial size of 10 000. After t years, the size of the population is P. The connection between P and t can be modelled by the equation
$$\frac{dP}{dt}=\frac{P}{1+t}+2$$

  a Solve this equation to show that
$$P=2(1+t)[5000+\ln(1+t)]$$

  b Find the size of the population of bacteria after six years.

9 In fluid dynamics, Bernoulli's equation is used to measure flow. The equation states that
$$\frac{dy}{dx}+py=qy^n$$
where p and q are both functions of x

  a Show that the differential equation
$$x\frac{dy}{dx}+4y=x^4y^2 \qquad \text{(I)}$$
is a Bernoulli equation.

  b Use the substitution $y=\dfrac{1}{v}$ to solve equation (I) given that
$$y=-2 \text{ when } x=2$$

Fluency and skills

An equation that involves a second order derivative, such as $\dfrac{d^2 y}{dx^2}$, but nothing of higher order, is called a **second order differential equation**. Some second order differential equations can be written in the form

$a\dfrac{d^2 y}{dx^2}+b\dfrac{dy}{dx}+cy=f(x)$, where a, b and c are constants.

You can solve these types of equation in two stages. The first stage is to solve the related **homogeneous equation**, that is, the equation

$a\dfrac{d^2 y}{dx^2}+b\dfrac{dy}{dx}+cy=0$

> Homogeneous means '= 0'

> The full formal method to solve this is as follows. You will see later that there is also a much quicker method.

Example 1

Solve the differential equation $\dfrac{d^2 y}{dx^2}+\dfrac{dy}{dx}-6y=0$

Rewrite the equation as $\dfrac{d^2 y}{dx^2}+3\dfrac{dy}{dx}-2\dfrac{dy}{dx}-6y=0$

> This just uses the fact that
> $\dfrac{dy}{dx}=3\dfrac{dy}{dx}-2\dfrac{dy}{dx}$
> This is very much like solving a quadratic equation.

So $\dfrac{d}{dx}\left[\dfrac{dy}{dx}+3y\right]-2\left[\dfrac{dy}{dx}+3y\right]=0$

Substitute $u=\dfrac{dy}{dx}+3y$ to get $\dfrac{du}{dx}-2u=0$

> Since $\dfrac{d^2 y}{dx^2}+3\dfrac{dy}{dx}=\dfrac{d}{dx}\left[\dfrac{dy}{dx}+3y\right]$

So $e^{-2x}\dfrac{du}{dx}-2e^{-2x}u=0$

> Multiplying throughout by the integrating factor, e^{-2x}

$ue^{-2x}=C$

> Integrate each side.

So $u=Ce^{2x}\Rightarrow\dfrac{dy}{dx}+3y=Ce^{2x}$

> Substitute $u=\dfrac{dy}{dx}+3y$

$e^{3x}\dfrac{dy}{dx}+3e^{3x}y=Ce^{5x}$

> Multiply throughout by the integrating factor, e^{3x}

$e^{3x}y=\dfrac{1}{5}Ce^{5x}+A$

> Integrate both sides.

$y=\dfrac{1}{5}Ce^{2x}+Ae^{-3x}$

> Rearrange.

The solution is $y=Ae^{-3x}+Be^{2x}$

> Relabel the constant $\dfrac{1}{5}C$ as B

You don't need to go through all this working every time. By assuming that solutions exist of the form $y = Ce^{mx}$, where C and m are constants, you can cut out several steps of the working.

If $y = Ce^{mx}$, then $\dfrac{dy}{dx} = Cme^{mx}$ and $\dfrac{d^2 y}{dx^2} = Cm^2 e^{mx}$

If you substitute each of these into the differential equation, then $\dfrac{d^2 y}{dx^2} + \dfrac{dy}{dx} - 6y = 0$ becomes $Cm^2 e^{mx} + Cme^{mx} - 6Ce^{mx} = 0$ which simplifies to $m^2 + m - 6 = 0$

This is called the **auxiliary equation**, and when solving these problems, you can use the solution to this equation to complete the first stage of the solution to the differential equation.

So $(m+3)(m-2) = 0 \Rightarrow m = -3$ or $m = 2$, giving solutions $y = Ce^{-3x}$ and $y = Ce^{2x}$

Since either of these solutions satisfies the homogeneous equation, any linear combination of these solutions will also satisfy the homogeneous equation, and the general solution is, therefore, $y = Ae^{-3x} + Be^{2x}$

> As there are two solutions, you need two different constants.

This method only works if the auxiliary equation has real distinct roots. There are two other cases for the roots of the auxiliary equation. The equation could have one repeated root, or the roots could be a complex conjugate pair. These three cases need three different general solutions.

Table 1	
Roots of auxiliary equation $ax^2 + bx + c = 0$	**General solution**
Real distinct roots, m_1 and m_2	$y = Ae^{m_1 x} + Be^{m_2 x}$
Repeated root, m	$y = (Ax + B)e^{mx}$
Complex roots $m \pm in$	$y = e^{mx}(A\cos nx + B\sin nx)$

You will be able to prove why these work in the next exercise.

Example 2

Solve the differential equation $\dfrac{d^2 y}{dx^2} + 7\dfrac{dy}{dx} + 12y = 0$

The auxiliary equation is $m^2 + 7m + 12 = 0$

$(m+3)(m+4) = 0$, so $m = -3$ or $m = -4$

The general solution is $y = Ae^{-3x} + Be^{-4x}$

> Write down and solve the auxiliary equation.

> This is 'real distinct roots' from Table 1

Example 3

Solve the differential equation $\dfrac{d^2 y}{dx^2} - 8\dfrac{dy}{dx} + 16y = 0$

The auxiliary equation is $m^2 - 8m + 16 = 0$

$(m-4)^2 = 0$, so $m = 4$

The general solution is $y = (Ax + B)e^{4x}$

> This is 'repeated root' from Table 1

Example 4

Solve the differential equation $\dfrac{d^2 y}{dx^2} - 4\dfrac{dy}{dx} + 13y = 0$

The auxiliary equation is $m^2 - 4m + 13 = 0$

$(m-2)^2 = -9$, so $m = 2 \pm 3i$

The general solution $y = e^{2x}(A\cos 3x + B\sin 3x)$

> This is 'complex roots' from Table 1

The second stage is used to solve equations that are not homogeneous.

To solve an equation like $\dfrac{d^2 y}{dx^2} + \dfrac{dy}{dx} - 6y = e^{4x}$, the first step is to solve the related homogeneous equation $\dfrac{d^2 y}{dx^2} + \dfrac{dy}{dx} - 6y = 0$

This solution, $y = Ae^{-3x} + Be^{2x}$, is now called the **complementary function**.

> You have already solved this equation in Example 1

You know that this gives the answer 'zero' when substituted into

$$\dfrac{d^2 y}{dx^2} + \dfrac{dy}{dx} - 6y$$

For the second stage you need to look for a function that will give e^{4x} when substituted into $\dfrac{d^2 y}{dx^2} + \dfrac{dy}{dx} - 6y$. This function is called the **particular integral**. Adding together the **complementary function** and the **particular integral** gives you the **general solution**.

> **Key point**
>
> The general solution is obtained by adding together the complementary function and the particular integral.

In this example, the particular integral is easy to find. It is likely to be a multiple of e^{4x}, so try $y = ae^{4x}$

$$y = ae^{4x} \Rightarrow \dfrac{dy}{dx} = 4ae^{4x} \text{ and } \dfrac{d^2 y}{dx^2} = 16ae^{4x}$$

Substituting gives

$$\frac{d^2y}{dx^2} + \frac{dy}{dx} - 6y = 16ae^{4x} + 4ae^{4x} - 6ae^{4x}$$

$$= 14ae^{4x}$$

This gives

$$14ae^{4x} = e^{4x} \text{ and } a = \frac{1}{14}$$

The **complementary function** is $y = Ae^{-3x} + Be^{2x}$ and the

particular integral is $y = \frac{1}{14}e^{4x}$

Adding the two together, gives the **general solution**

$$y = Ae^{-3x} + Be^{2x} + \frac{1}{14}e^{4x}$$

Not all particular integrals are that easy to find. The following table offers suggestions as to what to try.

Table 2	
Form of f(x)	**Form of particular integral**
C – constant	c – constant
$Mx + C$	$mx + c$
$P_n(x)$ – a polynomial of degree n	$p_n(x)$ – a polynomial of degree n
Ce^{mx}	ce^{mx}
$A\cos kx + B\sin kx$	$a\cos kx + b\sin kx$
$A\cosh kx + B\sinh kx$	$a\cosh kx + b\sinh kx$

Example 5

Solve the differential equation $\dfrac{d^2y}{dx^2} - 4\dfrac{dy}{dx} + 4y = 12x - 8$

The auxiliary equation is $m^2 - 4m + 4 = 0$

$(m-2)^2 = 0$, so $m = 2$

So the complementary function is $y = (Ax + B)e^{2x}$ ●————— This is the 'repeated root' type equation from Table 1

For the particular integral try $y = ax + b$ ●————— This is from Table 2

$\dfrac{dy}{dx} = a$, and $\dfrac{d^2y}{dx^2} = 0$

So $\dfrac{d^2y}{dx^2} - 4\dfrac{dy}{dx} + 4y = -4a + 4(ax+b) = 4ax + (-4a + 4b) = 12x - 8$

$4a = 12$, and $-4a + 4b = -8$ ●————— Equate coefficients.

$a = 3$ and $b = 1$

So the particular integral is $y = 3x + 1$ ●

and the general solution is $y = (Ax + B)e^{2x} + 3x + 1$ ●————— Add together the complementary function and the particular integral to get the general solution.

1 Find the general solution to each of the following differential equations.

a $\dfrac{d^2y}{dx^2} - 6\dfrac{dy}{dx} + 8y = 0$

b $\dfrac{d^2y}{dx^2} + 8\dfrac{dy}{dx} + 16y = 0$

c $\dfrac{d^2y}{dx^2} - 4\dfrac{dy}{dx} + 5y = 0$

d $\dfrac{d^2y}{dx^2} + 3\dfrac{dy}{dx} = 0$

e $\dfrac{d^2y}{dx^2} + \dfrac{dy}{dx} - 12y = 0$

f $\dfrac{d^2y}{dx^2} + y = 0$

g $2\dfrac{d^2y}{dx^2} + 5\dfrac{dy}{dx} - 3y = 0$

h $4\dfrac{d^2y}{dx^2} + 4\dfrac{dy}{dx} + 5y = 0$

2 Find the general solution to each of the following differential equations.

a $\dfrac{d^2y}{dx^2} + 2\dfrac{dy}{dx} - 15y = 19 - 30x$

b $\dfrac{d^2y}{dx^2} - 6\dfrac{dy}{dx} + 9y = 3e^{2x}$

c $\dfrac{d^2y}{dx^2} + 2\dfrac{dy}{dx} + 17y = 8\cos 3x - 6\sin 3x$

d $\dfrac{d^2y}{dx^2} - 16y = 32x - 48$

e $\dfrac{d^2y}{dx^2} + 3\dfrac{dy}{dx} - 4y = 4e^{-3x}$

f $\dfrac{d^2y}{dx^2} + 2\dfrac{dy}{dx} + y = x^2 + 4x + 7$

g $3\dfrac{d^2y}{dx^2} - 8\dfrac{dy}{dx} - 3y = 5e^{2x}$

h $4\dfrac{d^2y}{dx^2} + 4\dfrac{dy}{dx} + y = 5x + 18$

3 a Show that the differential equation

$\dfrac{d^2y}{dx^2} - 2a\dfrac{dy}{dx} + a^2 y = 0$ can be written as

$\dfrac{du}{dx} - a \times u = 0$, where a is a constant, and

$u = \dfrac{dy}{dx} - a \times y$

b Solve the differential equation

$\dfrac{du}{dx} - a \times u = 0$

c Hence show that $y = (Ax + B)e^{ax}$, where A and B are constants.

4 If the auxiliary equation has complex conjugate roots $m \pm in$, use Euler's formula to deduce that the general solution $y = Ae^{(m+in)x} + Be^{(m-in)x}$ can be expressed as $y = e^{mx}(\alpha \cos nx + \beta \sin nx)$ for constants α and β

Reasoning and problem-solving

Strategy

To answer a question that involves a second order differential equation

① Write down the auxiliary equation.

② Find the complementary function.

③ Decide on the form of the particular integral.

④ Find the particular integral.

⑤ Write down the general solution.

⑥ Use the information given in the question to find the values of the constants.

⑦ Answer the question(s) in context.

Example 6

Solve the differential equation $\dfrac{d^2y}{dx^2}+2\dfrac{dy}{dx}+10y=13\cos x+16\sin x$, given that $y(0)=1$ and $y'(0)=8$

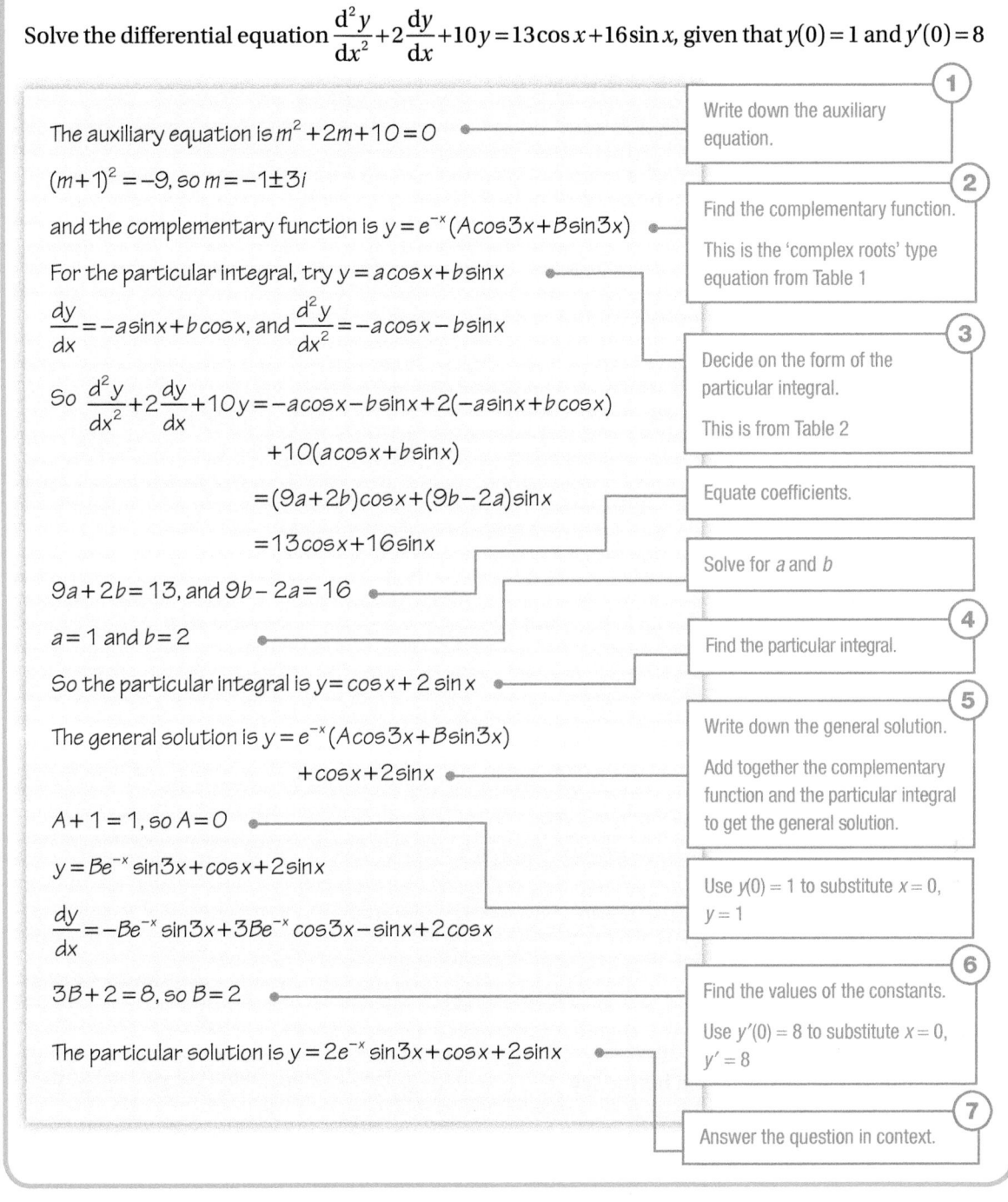

The auxiliary equation is $m^2+2m+10=0$

① Write down the auxiliary equation.

$(m+1)^2=-9$, so $m=-1\pm3i$

and the complementary function is $y=e^{-x}(A\cos3x+B\sin3x)$

② Find the complementary function.

This is the 'complex roots' type equation from Table 1

For the particular integral, try $y=a\cos x+b\sin x$

$\dfrac{dy}{dx}=-a\sin x+b\cos x$, and $\dfrac{d^2y}{dx^2}=-a\cos x-b\sin x$

③ Decide on the form of the particular integral.

This is from Table 2

So $\dfrac{d^2y}{dx^2}+2\dfrac{dy}{dx}+10y=-a\cos x-b\sin x+2(-a\sin x+b\cos x)$

$+10(a\cos x+b\sin x)$

$=(9a+2b)\cos x+(9b-2a)\sin x$

Equate coefficients.

$=13\cos x+16\sin x$

Solve for a and b

$9a+2b=13$, and $9b-2a=16$

$a=1$ and $b=2$

④ Find the particular integral.

So the particular integral is $y=\cos x+2\sin x$

The general solution is $y=e^{-x}(A\cos3x+B\sin3x)$

$+\cos x+2\sin x$

⑤ Write down the general solution.

Add together the complementary function and the particular integral to get the general solution.

$A+1=1$, so $A=0$

$y=Be^{-x}\sin3x+\cos x+2\sin x$

Use $y(0)=1$ to substitute $x=0$, $y=1$

$\dfrac{dy}{dx}=-Be^{-x}\sin3x+3Be^{-x}\cos3x-\sin x+2\cos x$

$3B+2=8$, so $B=2$

⑥ Find the values of the constants.

Use $y'(0)=8$ to substitute $x=0$, $y'=8$

The particular solution is $y=2e^{-x}\sin3x+\cos x+2\sin x$

⑦ Answer the question in context.

Key point

When you have found the values of the constants in the general solution, the final answer is called the **particular solution**.

If the form of the particular integral you try matches the complementary function, it will not work. You simply end up with '0 = 0'. As long as this is not the 'repeated root' case in Table 1 you should simply multiply the function you try for the particular integral by x. In the case of the 'repeated root', multiplying x will also give '0 = 0'. In this case, instead of multiplying by x, you multiply by x^2

These cases are illustrated in Example 7

Example 7

Solve the differential equation $\dfrac{d^2y}{dx^2} - 5\dfrac{dy}{dx} + 6y = 6e^{2x}$, given that $y(0) = 3$ and $y'(0) = -2$

The auxiliary equation is $m^2 - 5m + 6 = 0$

$(m-2)(m-3) = 0$, so $m = 2$ or $m = 3$

So the complementary function is $y = Ae^{2x} + Be^{3x}$

> This is the 'real distinct roots' type equation from Table 1

For the particular integral try $y = axe^{2x}$

$\dfrac{dy}{dx} = ae^{2x} + 2axe^{2x}$, and $\dfrac{d^2y}{dx^2} = 4ae^{2x} + 4axe^{2x}$

> Since e^{2x} is already part of the complementary function, you need to multiply ae^{2x}, from Table 2, by x

So $\dfrac{d^2y}{dx^2} - 5\dfrac{dy}{dx} + 6y = 4ae^{2x} + 4axe^{2x} - 5(ae^{2x} + 2axe^{2x}) + 6axe^{2x}$

$= -ae^{2x}$

> It is no coincidence that the terms in xe^{2x} have cancelled. They always will.

$= 6e^{2x}$

$a = -6$

So the particular integral is $y = -6xe^{2x}$

and the general solution is $y = Ae^{2x} + Be^{3x} - 6xe^{2x}$

$\Rightarrow \dfrac{dy}{dx} = 2Ae^{2x} + 3Be^{3x} - 6e^{2x} - 12xe^{2x}$

> Use $y(0) = 3$ and $y'(0) = -2$ to get equations in A and B

$A + B = 3$, and $2A + 3B - 6 = -2$

so $A = 5$ and $B = -2$

> Solve the equations simultaneously to find A and B

The particular solution is $y = 5e^{2x} - 2e^{3x} - 6xe^{2x}$

Exercise 20.2B Reasoning and problem-solving

1 Solve each of the following differential equations subject to the given boundary conditions.

a $\dfrac{d^2y}{dx^2} - 7\dfrac{dy}{dx} + 6y = 0$, given that $y(0) = 3$ and $y'(0) = 8$

b $\dfrac{d^2y}{dx^2} + 4\dfrac{dy}{dx} + 4y = 0$, given that $y(0) = 1$ and $y'(0) = 2$

c $\dfrac{d^2y}{dx^2} + 25y = 0$, given that $y(0) = 3$ and $y\left(\dfrac{\pi}{10}\right) = 6$

d $\dfrac{d^2y}{dx^2} + \dfrac{dy}{dx} - 6y = 0$, given that $y(0) = 1$ and $y'(0) = 12$

e $\dfrac{d^2y}{dx^2}-6\dfrac{dy}{dx}+9y=0$, given that $y(0)=5$ and $y'(0)=3$

f $\dfrac{d^2y}{dx^2}-\dfrac{dy}{dx}-12y=0$, given that $y(0)=12$ and $y'(0)=6$

g $9\dfrac{d^2y}{dx^2}-12\dfrac{dy}{dx}+4y=0$, given that $y(0)=12$ and $y'(0)=8$

h $4\dfrac{d^2y}{dx^2}-4\dfrac{dy}{dx}+5y=0$, given that $y(0)=6$ and $y'(0)=-1$

2 Solve each of the following differential equations subject to the given boundary conditions.

a $\dfrac{d^2y}{dx^2}-7\dfrac{dy}{dx}+10y=12e^x$, given that $y(0)=13$ and $y'(0)=8$

b $\dfrac{d^2y}{dx^2}-12\dfrac{dy}{dx}+36y=72$, given that $y(0)=3$ and $y(0)=-2$

c $\dfrac{d^2y}{dx^2}+4\dfrac{dy}{dx}+5y=16\cos x$, given that $y(0)=5$ and $y'(0)=-6$

d $\dfrac{d^2y}{dx^2}+2\dfrac{dy}{dx}-8y=30-24x$, given that $y(0)=1$ and $y'(0)=0$

e $9\dfrac{d^2y}{dx^2}-4y=4x^2+2$, given that $y(0)=-8$ and $y'(0)=10$

f $\dfrac{d^2y}{dx^2}-4\dfrac{dy}{dx}+5y=\cos 2x+8\sin 2x$, given that $y(0)=9$ and $y'\left(\dfrac{\pi}{2}\right)=4e^\pi-1$

g $4\dfrac{d^2y}{dx^2}-9y=26+27x-9x^2$, given that $y(0)=8$ and $y'(0)=-9$

h $16\dfrac{d^2y}{dx^2}-8\dfrac{dy}{dx}+y=12e^{-\frac{1}{4}x}$, given that $y(0)=10$ and $y'(0)=2$

3 Given that the differential equation $\dfrac{d^2y}{dx^2}-2\dfrac{dy}{dx}-8y=12e^{-2x}$ has a particular integral of the form $y=axe^{-2x}$, determine the value of the constant a, and find the general solution of the differential equation.

4 Given that the differential equation $\dfrac{d^2y}{dx^2}-2\dfrac{dy}{dx}+y=6e^x$ has a particular integral of the form $y=ax^2e^x$, determine the value of the constant a, and find the general solution of the differential equation.

5 Find the solution of the differential equation $\dfrac{d^2y}{dx^2}+5\dfrac{dy}{dx}-6y=21e^x+12$ for which $y=-2$ and $\dfrac{dy}{dx}=17$ at $x=0$

6 Find the solution of the differential equation $\dfrac{d^2y}{dx^2}-6\dfrac{dy}{dx}+9y=34e^{3x}$ for which $y=3$ and $\dfrac{dy}{dx}=11$ at $x=0$

7 Solve the differential equation $\dfrac{d^3y}{dx^3}-6\dfrac{d^2y}{dx^2}+11\dfrac{dy}{dx}-6y=12e^{4x}$, given that $y(0)=4$, $y'(0)=11$ and $y''(0)=35$

8 a Given $x=e^u$ and $x^2\dfrac{d^2y}{dx^2}-4x\dfrac{dy}{dx}+6y=12$, show that $\dfrac{d^2y}{du^2}-5\dfrac{dy}{du}+6y=12$

b Hence solve the equation $x^2\dfrac{d^2y}{dx^2}-4x\dfrac{dy}{dx}+6y=12$, given that $y(1)=7$ and $y(2)=14$

Fluency and skills

Differential equations can be used to model many different situations. In particular, they can be used to describe things that oscillate. This includes the motion of a weight hanging on the end of a spring, moving up and down, or a pendulum swinging from side to side, or the depth of the water in a harbour as the tide moves in and out. Oscillatory motions like these follow a regular, cyclic pattern. If you plot a displacement–time graph of the motion of an oscillatory system, you will often find that it is a sine graph.

Example 1

a Solve the differential equation $\dfrac{d^2x}{dt^2} = -4x$, given that $x(0) = 4$ and $x\left(\dfrac{\pi}{4}\right) = -4$

b Sketch the solution, and state its amplitude.

a First rewrite the equation as $\dfrac{d^2x}{dt^2} + 4x = 0$

The auxiliary equation is $m^2 + 4 = 0$

$m^2 = -4$, so $m = \pm 2i$

So the general solution is $x = A\cos 2t + B\sin 2t$

$A = 4$ •——————————————————— Use $x(0) = 4$ to substitute $t = 0$, $x = 4$

$B = -4$ •———————————————————

The particular solution is $x = 4\cos 2t - 4\sin 2t$

Use $x\left(\dfrac{\pi}{4}\right) = -4$ to substitute $t = \dfrac{\pi}{4}$, $x = -4$

$$= 4\sqrt{2}\left(\frac{1}{\sqrt{2}}\cos 2t - \frac{1}{\sqrt{2}}\sin 2t\right)$$

$$= 4\sqrt{2}\cos\left(2t + \frac{\pi}{4}\right) \quad •—— \text{Use harmonic form.}$$

b

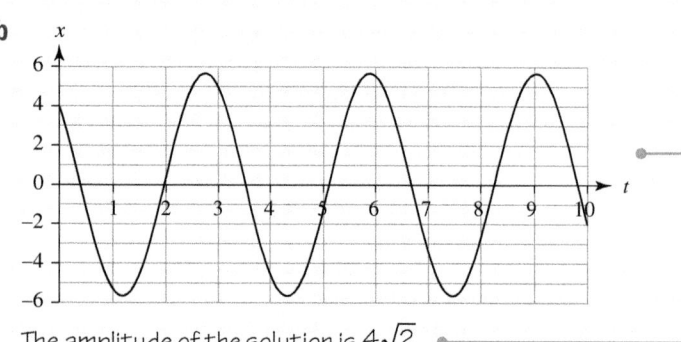

Try plotting this on your graphical calculator.

The amplitude of the solution is $4\sqrt{2}$ •——— The amplitude is the maximum displacement from the centre of the motion.

Motion that satisfies a differential equation of the form $\dfrac{d^2x}{dt^2}=-\omega^2 x$ is called simple harmonic motion (SHM).

There are certain standard results that apply to SHM. These are derived below.

Using $\dfrac{d^2x}{dt^2}=v\dfrac{dv}{dx}$ gives $v\dfrac{dv}{dx}=-\omega^2 x$

$$\dfrac{d^2x}{dt^2}=\dfrac{dv}{dt}=\dfrac{dv}{dx}\times\dfrac{dx}{dt}=\dfrac{dv}{dx}\times v$$

So $\displaystyle\int v\,dv=-\int\omega^2 x\,dx$

$\dfrac{v^2}{2}=-\dfrac{\omega^2 x^2}{2}+c$ (1)

Separate the variables.

Let the amplitude of the motion be a, then

$\dfrac{0^2}{2}=-\dfrac{\omega^2 a^2}{2}+c\Rightarrow c=\dfrac{\omega^2 a^2}{2}$

Use the fact that when $x=a$ the velocity is zero.

So $\dfrac{v^2}{2}=-\dfrac{\omega^2 x^2}{2}+\dfrac{\omega^2 a^2}{2}$

Substitute for c into (1)

i.e. $v^2=\omega^2\left(a^2-x^2\right)$ or $v=\omega\sqrt{a^2-x^2}$

$\dfrac{dx}{dt}=\omega\sqrt{a^2-x^2}$

Use $v=\dfrac{dx}{dt}$

$\displaystyle\int\dfrac{1}{\sqrt{a^2-x^2}}dx=\int\omega\,dt$

Separate the variables, and integrate.

$\sin^{-1}\left(\dfrac{x}{a}\right)=\omega t+\varepsilon$

$x=a\sin(\omega t+\varepsilon)$

Use the standard result $\displaystyle\int\dfrac{1}{\sqrt{a^2-x^2}}dx=\sin^{-1}\left(\dfrac{x}{a}\right)$, where ε is a constant of integration

The value of ε depends on when the clock starts ticking, i.e. where we measure $t=0$ from.

- If $t=0$ when $x=0$, then $\varepsilon=0$, and $x=a\sin\omega t$
- If $t=0$ when $x=a$, then $\varepsilon=\dfrac{\pi}{2}$, and $x=a\sin\left(\omega t+\dfrac{\pi}{2}\right)=a\cos\omega t$

The period of a simple $\sin t$ or $\cos t$ function is 2π. Hence the period of $x=a\sin(\omega t+\varepsilon)$ is $\dfrac{2\pi}{\omega}$

The following results apply to any particle that is moving with simple harmonic motion, i.e. any motion for which $\dfrac{d^2x}{dt^2}=-\omega^2 x$

Standard results

$$\frac{d^2x}{dt^2} = -\omega^2 x \Rightarrow v = \omega\sqrt{a^2 - x^2}$$

$$x = a\cos\omega t \quad \text{or}$$

$$x = a\sin\omega t$$

$$T = \frac{2\pi}{\omega}$$

where a is the amplitude and T is the period.

It is the starting conditions that determine which form of x to use.

Example 2

A particle moves with SHM defined by the equation $x = -16x$

$\left(\text{This is another way of writing } \dfrac{d^2x}{dt^2} = -16x\right)$

The amplitude of the motion is 3 metres.

Find

a The period of the motion,

b The maximum speed of the particle.

a $T = \dfrac{2\pi}{4} = \dfrac{\pi}{2}$ s

Use $T = \dfrac{2\pi}{\omega}$ with

$\omega = \sqrt{16} = 4$

b $v = 4\sqrt{3^2 - 0^2} = 12\,\text{ms}^{-1}$

Use $v = \omega\sqrt{a^2 - x^2}$

with $\omega = 4$ and $x = 0$

The maximum speed occurs at the centre of the motion, i.e. when $x = 0$

Key point

Example 3

A particle is projected from a point O at time $t = 0$, and performs SHM with O as the centre of oscillation.

The motion is of amplitude 5 metres, and the period is $\dfrac{\pi}{3}$ s

Find

a The speed of projection,

b The time it takes for the particle to first reach a point that is 4 metres from O

Use $T = \dfrac{2\pi}{\omega}$ to find ω

a $\dfrac{2\pi}{\omega} = \dfrac{\pi}{3} \Rightarrow \omega = 6$

$v = 6\sqrt{5^2 - 0^2} = 30\,\text{ms}^{-1}$

Use $v = \omega\sqrt{a^2 - x^2}$ with $\omega = 6$, $x = 0$, and $a = 5$

b The equation of motion is $x = 5\sin 6t$

Use $x = a\sin\omega t$, since the motion starts at $x = 0$

$5\sin 6t = 4 \Rightarrow \sin 6t = \dfrac{4}{5}$

$t = 0.155$ s

Solve for t
Remember to use radians.

Exercise 20.3A Fluency and skills

1 Solve each of the following differential equations of SHM, subject to the given initial and boundary conditions.

a $\dfrac{d^2x}{dt^2} = -25x$, given that $x = 7$ when $t = 0$,

and $x = 5$ when $t = \dfrac{\pi}{10}$

b $\dfrac{d^2x}{dt^2} = -4x$, given that $x = 2\sqrt{2}$ when

$t = \dfrac{\pi}{8}$, and $x = 8$ when $t = \dfrac{\pi}{4}$

c $\dfrac{d^2x}{dt^2} = -100x$, given that $x = -5$ and

$\dfrac{dx}{dt} = 30$ when $t = 0$

d $\dfrac{d^2x}{dt^2} = -x$, given that $x = -1$ and $\dfrac{dx}{dt} = 9$

when $t = 0$

e $4\dfrac{d^2x}{dt^2} = -x$, given that $x = \sqrt{2}$ when $t = \dfrac{\pi}{2}$,

and $x = -4\sqrt{2}$ when $t = \dfrac{3\pi}{2}$

f $9\dfrac{d^2x}{dt^2} = -x$, given that $x = 7$ when $t = \dfrac{\pi}{2}$,

and $x = \sqrt{3}$ when $t = \pi$

2 Solve each of the following differential equations of SHM, subject to the given initial and boundary conditions.

a $\dfrac{d^2x}{dt^2} = -16x + 48$, given that $x = 0$ when

$t = 0$, and $x = 5$ when $t = \dfrac{\pi}{8}$

b $\dfrac{d^2x}{dt^2} = -25x + 100$, given that $x = 5$ when

$t = \dfrac{\pi}{10}$, and $x = -1$ when $t = \dfrac{\pi}{5}$

c $\dfrac{d^2x}{dt^2} = -9x + 36$, given that $x = -1$ and

$\dfrac{dx}{dt} = 12$ when $t = 0$

d $\dfrac{d^2x}{dt^2} = -4x + 12$, given that $x = -3$ and

$\dfrac{dx}{dt} = 8$ when $t = 0$

e $16\dfrac{d^2x}{dt^2} + x - 1 = 0$, given that $x = 8$ when

$t = \pi$, and $x = 4$ when $t = 3\pi$

f $4\dfrac{d^2x}{dt^2} + 25x + 50 = 0$, given that $x = 0$ when

$t = \dfrac{2\pi}{5}$, and $x = -2\sqrt{2} - 2$ when $t = \dfrac{\pi}{2}$

3 Solve the differential equation $\dfrac{d^2x}{dt^2} = -25x$,

given that $x(0) = 4$ and $x\left(\dfrac{\pi}{10}\right) = 3$

Hence sketch the graph of x against t

4 Solve the differential equation

$\dfrac{d^2x}{dt^2} = -9x + 18$, given that

$x(0) = 7$ and $x\left(\dfrac{\pi}{2}\right) = 14$

Hence sketch the graph of x against t

5 A particle moves with SHM defined by the equation $\ddot{x} = -9x$. The amplitude of the motion is 2 metres.

Find

a The period of the motion,

b The maximum speed of the particle.

6 A particle moves with SHM defined by the equation $\ddot{x} = -4x$. The maximum speed of the particle is $12 \, \text{m s}^{-1}$

Find

a The period of the motion,

b The amplitude of the motion.

7 A particle moves with SHM defined by the equation $\ddot{x} = -\dfrac{x}{25}$. When $x = 0.4$ metres, the speed of the particle is $0.06 \, \text{m s}^{-1}$.

Find

a The period of the motion,

b The amplitude of the motion,

c The speed of the particle when $x = 0.3$ metres.

8 A particle is projected from a point O and performs SHM with O as the centre of oscillation.

The motion has amplitude 0.9 metres, and the period is π s.

Find

a The speed of projection,

b The time it takes for the particle to first reach a point P, where P is 0.7 metres from O

9 A particle is projected from a point O at time $t = 0$ and performs SHM with O as the centre of oscillation.

The motion is of amplitude 0.6 metres, and the period is $\dfrac{\pi}{2}$ s.

Find

a The speed of projection,

b The time it takes for the particle to first reach a point P, where P is 0.3 metres from O

10 A particle moves with SHM about a centre O The amplitude of the motion is 1.3 metres, and the period is $\dfrac{\pi}{4}$ s.

Find how far the particle is from O when its speed is 4 ms^{-1}.

11 A man on the end of a bungee performs SHM of period 5π seconds and amplitude 10 metres about a centre O. After passing through O the man passes through a point X which is 3 metres below O, and then he passes through a point Y which is 9 metres below O

Find the time taken for the man to travel from X to Y

12 The head of a piston moves with SHM about a centre O. When the piston-head is 1 metre from O its speed is 7.2 ms^{-1}, and when it is 2.4 metres from O its speed is 3 ms^{-1}.

Find

a The amplitude of the motion,

b The period of the motion.

13 A toy on the end of a spring moves with SHM about a centre O. When the toy is 30 cm from O its speed is 0.8 cm s^{-1}, and when it is 40 cm from O its speed is 0.6 cm s^{-1}.

Find

a The amplitude of the motion,

b The period of the motion.

Reasoning and problem-solving

Strategy

To answer a question involving simple harmonic motion

① Draw a force diagram.

② Use Newton's 2nd law to form a second order differential equation.

③ Find the general solution to that differential equation.

④ Use the information given in the question to find the particular solution.

⑤ Answer the question in context.

Example 4

A particle, *P*, of mass 0.06 kg moves in a horizontal straight line under the action of a force directed towards a fixed point *O*. The force varies with the distance of the particle from *O*, and is equal to $(1.5x)$ N, where *x*, measured in metres, is the displacement of *P* from *O*. *P* is initially at rest at a displacement of 10 metres from *O*

a Prove that the motion of *P* is SHM.

b Find the velocity of *P* when *P* is 6 metres from *O*

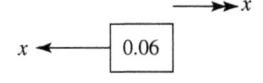

a $-1.5x = 0.06\ddot{x}$

$\ddot{x} = -25x$, which is SHM with $\omega = 5$

The minus sign is because the force acts towards *O*

b $v^2 = 5^2(10^2 - 6^2)$

$v = 40\,\text{ms}^{-1}$

Use $v^2 = \omega^2(a^2 - x^2)$ with $\omega = 5$, $a = 10$, $x = 6$

Example 5

One end of a light elastic spring of unstretched length 2 metres is fixed to a point *O* on a smooth horizontal table. A particle of mass 0.1 kg is attached to the other end of the spring and rests in equilibrium at a point *A* on the table. The particle is then pulled away from *A*, in the direction *OA*, causing the spring to stretch by 20 cm. The force in the spring is directed towards *A*, and has magnitude $10x$ N, where *x*, measured in metres, is the displacement of the particle from *A*. The body is released from rest at time $t = 0$ seconds.

a Show that $\dfrac{d^2x}{dt^2} = -100x$

b Solve this differential equation to show that $x = 0.2\cos(10t)$

c Calculate the distance of the particle from *O* when $t = \dfrac{\pi}{30}$

d Calculate the speed of the particle when $t = \dfrac{\pi}{30}$

a Apply $F = ma$ to get $-10x = 0.1\dfrac{d^2x}{dt^2}$

$\Rightarrow \dfrac{d^2x}{dt^2} = -100x$

b The auxiliary equation is

$m^2 + 100 = 0$

$m^2 = -100$, so $m = \pm 10i$

So the general solution is $x = A\cos 10t + B\sin 10t$

Use $x = 0.2$, $\dot{x} = 0$ at $t = 0$ to get $A = 0.2$, $B = 0$

$\Rightarrow x = 0.2\cos 10t$

c $x = 0.2\cos\left(\dfrac{10\pi}{30}\right) = 0.2\cos\left(\dfrac{\pi}{3}\right) = 0.1$ metres

Substitute $t = \dfrac{\pi}{30}$

d $\dfrac{dx}{dt} = -2\sin 10t = -2\sin\left(\dfrac{10\pi}{30}\right) = -\sqrt{3}$

The speed of the particle is $\sqrt{3}\,\text{ms}^{-1}$

Differentiate *x* to get an expression for the velocity and substitute $t = \dfrac{\pi}{30}$

Example 6

A small box of mass m kg is at rest at a point O on a smooth horizontal table. The box is attached to two identical springs. The other ends of the springs are attached to the points A and B, respectively, which are 3 metres apart on a straight line through O.

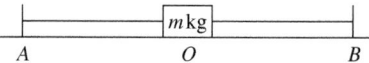

The box is struck so that it moves towards B with an initial velocity of $1.2\,\mathrm{ms^{-1}}$. At a time t seconds after the box is struck, the displacement of the box from O, in the direction OB, is x metres. The force that each of the springs applies to the box, is $(2mx)$ N in the direction BA

a Prove that the motion of the box is SHM.

b Write down the period of the motion.

c Find the minimum distance of the box from B in the subsequent motion.

d State three assumptions that you have made in forming your model.

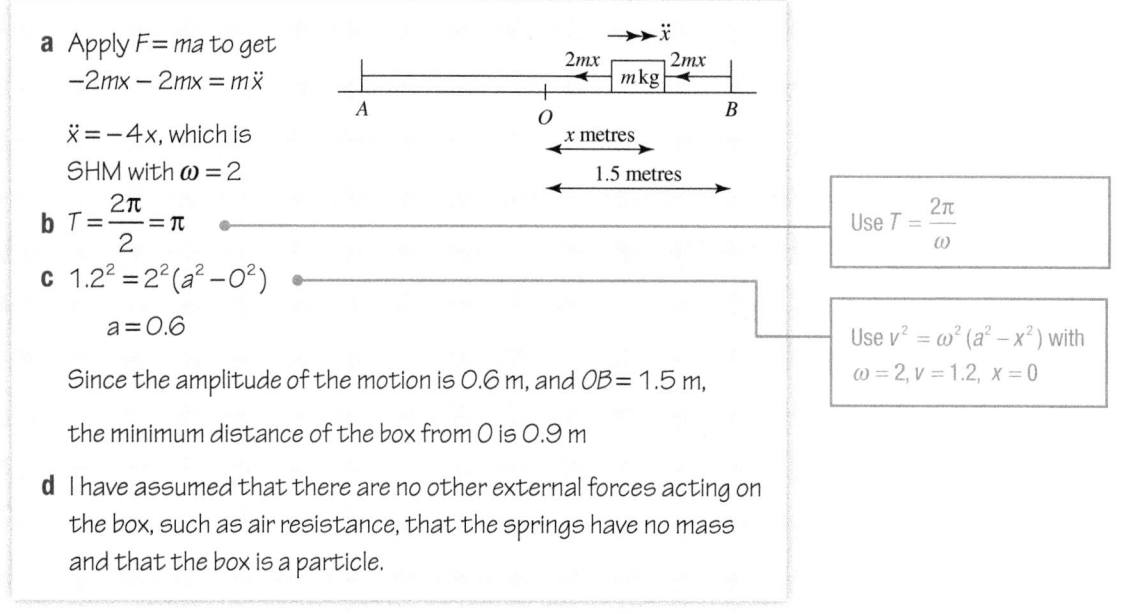

a Apply $F = ma$ to get
$$-2mx - 2mx = m\ddot{x}$$
$\ddot{x} = -4x$, which is
SHM with $\omega = 2$

b $T = \dfrac{2\pi}{2} = \pi$ Use $T = \dfrac{2\pi}{\omega}$

c $1.2^2 = 2^2(a^2 - 0^2)$ Use $v^2 = \omega^2(a^2 - x^2)$ with $\omega = 2, v = 1.2, x = 0$

$a = 0.6$

Since the amplitude of the motion is 0.6 m, and $OB = 1.5$ m,

the minimum distance of the box from O is 0.9 m

d I have assumed that there are no other external forces acting on the box, such as air resistance, that the springs have no mass and that the box is a particle.

Example 7

A particle of mass 0.5 kg rests at a fixed point P, on a smooth horizontal surface, attached to one end on a light elastic spring of natural length 3 metres. The other end of the spring is attached to O, where O is a fixed point on the surface. The particle is pulled a distance 2 metres from P, in the direction OP, and released from rest at time $t = 0$. At time t seconds, the displacement of the particle from O is x metres. The force in the spring is directed towards P and has magnitude $32(x-3)$ newtons.

a Write down the equation of motion of the particle.

b Find an expression for x in terms of t

c Sketch the graph of x against t

d Write down the value of t when the particle is closest to O for the first time.

(*Continued on the next page*)

a $-32(x-3)=0.5\ddot{x}$

$\ddot{x}+64x=192$

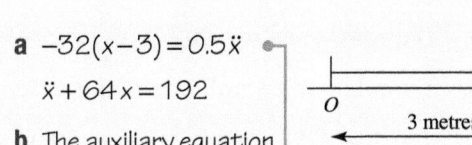

Apply Newton's 2nd law in the direction OP

b The auxiliary equation is $m^2+64=0$

$m^2=-64$, so $m=\pm 8i$

So the complementary function is $x=A\cos 8t + B\sin 8t$

By inspection the particular integral is $x=3$

So the general solution is $x=3+A\cos 8t + B\sin 8t$

Use $x=5$, $\dot{x}=0$ at $t=0$ to get $A=2$, $B=0$

$x=3+2\cos 8t$

c

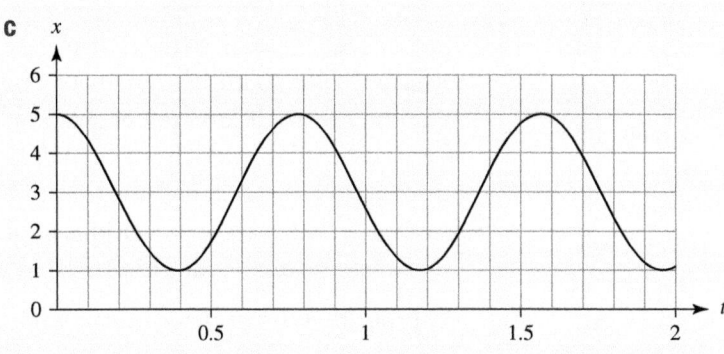

d $\cos 8t = -1$ ———— At the minimum point.

$3+2\cos 8t = 1$

$8t=\pi$

So $t=\dfrac{\pi}{8}$

Key point

This motion is still SHM. It is centred on $x=3$ rather than $x=0$

Exercise 20.3B Reasoning and problem-solving

1 A box B, of mass 5 kg, moves in a horizontal straight line under the action of a force directed towards a fixed point O. The force varies with the displacement of the particle from O, and is equal to $20x$ N, where x, measured in metres, is the distance of B from O. The box is initially at rest at distance of 5 metres from O

a Prove that the motion of B is SHM.

b Write down the period of the motion.

c Calculate the velocity of the box when it is 3 metres from O

2 One end of a light elastic spring of unstretched length 2 metres is fixed to a point O on a smooth horizontal table. A body of mass 200 g is attached to the other end of the spring and rests in equilibrium at a point A on the table. The body is then pulled away from A, a distance 30 cm in the direction OA, and released from rest. The force in the spring is directed towards A, and

635

has magnitude $5x$ N, where x, measured in metres, is the displacement of the body from A at time t seconds after the particle is released.

a   Show that $\dfrac{d^2x}{dt^2} = -25x$

b   Solve this differential equation to show that $x = 0.3\cos(5t)$

c   Calculate the distance of the body from O when $t = \dfrac{\pi}{15}$

d   Calculate also the speed of the body when $t = \dfrac{\pi}{15}$

3   Two points A and B are 1 metre apart on a smooth horizontal table. The point C lies on AB and is 80 cm from A

A light spring of unstretched length 80 cm has one end fastened to the table at A, and the other end is fastened to a body of mass 3 kg which is held at rest at B. When extended, the force in the spring is directed towards C, and is equal to $12x$ N, where x, measured in metres, is the displacement of the body from C. The body is released from rest at B

a   Prove that the motion is SHM.

b   Write down the period of the motion.

c   Calculate the speed of the body when it is 4 cm from B

d   State three assumptions that you have made in forming your model.

4   One end of a light elastic spring of unstretched length 60 cm is fixed to a point O in a smooth horizontal table. A small parcel of mass 100 g is attached to the other end of the spring. The parcel is at rest at the point P on the table, where $OP = 60$ cm. At time $t = 0$ the parcel is struck so that it moves in the direction OP with an initial velocity of $1.2\,\mathrm{m\,s^{-1}}$. The force in the spring is directed towards P and has magnitude $10x$ N, where x, measured in metres, is the displacement of the parcel from P at time t seconds.

a   Show that $\dfrac{d^2x}{dt^2} = -100x$

b   Calculate the amplitude of the motion.

c   Calculate the speed of the parcel when it is 6 cm from P

5   A small package of mass 1.5 kg is at rest at a point O on a smooth horizontal table. The package is attached to two identical springs. The other ends of the springs are attached to the points A and B, respectively, which are 2.4 metres apart on a straight line through O

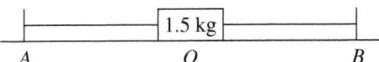

The package is struck so that it moves towards B with an initial velocity of $4\,\mathrm{m\,s^{-1}}$ At a time t seconds after the package is struck, its displacement from O, in the direction OB, is x metres. The force, directed towards A, that each of the springs applies to the package is of magnitude $48x$ N.

a   Prove that the package moves with SHM.

b   Write down the period of the motion.

c   Calculate the time at which the package is first 0.7 metres from B

d   Calculate the speed of the package when it is 0.9 metres from B

6   A light elastic string of unstretched length 4 metres is stretched between two points, A and B, 6 metres apart on a smooth horizontal surface. O is the midpoint of AB A body of mass 0.5 kg is attached to the string at O. The body is pulled 90 cm towards B and released from rest. At a time t seconds after the body is released, its displacement from O is x metres, in the direction OB. The total force, directed towards A, that the string exerts on the body, is $0.32x$ N

a   Show that $\dfrac{d^2x}{dt^2} = -0.64x$

b   Solve this differential equation to show that $x = 0.9\cos(0.8t)$

c Calculate also the speed of the body when $t = \dfrac{5\pi}{12}$

d Explain how the model would need to be changed if, instead of pulling the body 90 cm towards B, it had been pulled 120 cm.

7 A small box of mass m kg is at rest in equilibrium at a point O on a smooth horizontal table. The box is attached to two springs. The other ends of the springs are attached to the points A and B which are 4 metres apart, with $AO = 1$ metre and $OB = 3$ metres.

$$\boxed{\;\; | \quad \fbox{m kg} \quad | \quad\quad\quad\quad | \;\;}$$
$$A \quad\quad O \quad\quad\quad\quad\quad\quad B$$

The box is pulled 20 cm in the direction OB, and released from rest. At a time t seconds after the box is released, its displacement from O, in the direction OB, is x metres. The spring connected to B applies a force of mx N to the box in the direction BO, and the spring connected to A applies a force of $3mx$ N to the box in the direction OA

a Prove that the box moves with SHM.

b Write down the period of the motion.

c Find the speed of the box when it is 10 cm from O

d State three assumptions that you have made in forming your model.

8 A small box of mass 0.1 kg rests at a fixed point P on a large smooth horizontal table, attached to one end of a light elastic spring of natural length 2 metres. The other end of the spring is attached to O, a fixed point on the table. The box is pulled a distance 0.4 metres from P, in the direction OP and released from rest at time $t = 0$. At time t seconds, the displacement of the box from O is x metres. The force in the spring is directed towards P, and has magnitude $\dfrac{2(x-2)}{5}$ newtons.

a Write down the equation of motion of the box.

b Show that $x = 2 + \dfrac{2}{5}\cos(2t)$

c Calculate the maximum speed of the box.

d Calculate the value of t when the box is closest to O for the first time.

9 A small body of mass 0.3 kg rests at a fixed point A, on a smooth horizontal surface, attached to one end on a light elastic spring of natural length 1.5 metres. The other end of the spring is attached to O, a fixed point on the table. At time $t = 0$ the body is struck so that it moves in the direction OA with an initial velocity of $20\,\mathrm{m\,s^{-1}}$. The force in the spring is directed towards A, and has magnitude $15(2x - 3)$ N, where x, measured in metres, is the displacement of the body from P at time t seconds.

a Show that $\ddot{x} + 100x = 150$

b Find an expression for x in terms of t

c Find $\dfrac{\mathrm{d}x}{\mathrm{d}t}$

d Write down the maximum speed of the body.

10 A simple pendulum consists of a bob of mass m kg suspended from a light inextensible string of length ℓ metres. The other end of the string is attached to a fixed point O. The bob oscillates to and fro. At time t seconds, the angle between the string and the downward vertical through O is θ radians.

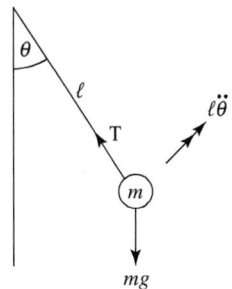

a Show that $-mg\sin\theta = m\ell\ddot{\theta}$

b Using the fact that, for small θ, $\sin\theta \approx \theta$, deduce that this motion is approximately SHM with period $2\pi\sqrt{\dfrac{\ell}{g}}$

Fluency and skills

When a particle is attached to an elastic string or spring, and set in motion, with no extra forces acting on it, the motion is simple harmonic. However, in practice, the amplitude of the oscillations decreases with time. In fact, in some cases, there are no oscillations at all. This is due to other external forces such as friction and air resistance.

Example 1

A particle, P, moves along a horizontal line under the action of a force directed towards a fixed point O. The displacement, x metres, of P from its initial position, at a time t seconds from the start, satisfies the differential equation $\dfrac{d^2 x}{dt^2} + 2\dfrac{dx}{dt} + 17x = 34$

At time $t = 0$ the particle is moving through the point $x = 2$ with velocity $20\,\text{m s}^{-1}$.

a Solve the differential equation to obtain an expression for x in terms of t

b Sketch the graph of x against t

a The auxiliary equation is $m^2 + 2m + 17 = 0$

$(m+1)^2 = -16$, so $m = -1 \pm 4i$

So the complementary function is $x = e^{-t}(A\cos 4t + B\sin 4t)$ •————— This is 'complex roots'.

For the particular integral try $x = c$

$17c = 34$, so $c = 2$

So the particular integral is $x = 2$

The general solution is $x = e^{-t}(A\cos 4t + B\sin 4t) + 2$ •————— Add together the complementary function and the particular integral to get the general solution.

$A + 2 = 2$, so $A = 0$ •—

The solution is now $x = Be^{-t}(\sin 4t) + 2$

$\dfrac{dx}{dt} = -Be^{-t}\sin 4t + 4Be^{-t}\cos 4t$ •————— Use $x(0) = 2$ to get an equation in A and B

$4B = 20$, so $B = 5$ •—

The particular solution is $x = 2 + 5e^{-t}\sin 4t$ •————— Use $x'(0) = 20$ to substitute $t = 0$, $\dot{x} = 20$ to get an equation in A and B

b

Try plotting this on your graphical calculator.

This is called damped harmonic motion. The graph oscillates about $x = 2$. The damping effect is caused by the term $2\dfrac{dx}{dt}$

In practice, one way of thinking of SHM is a particle moving freely on the end of a spring. Damped harmonic motion is a bit like putting the particle and spring in a tub of treacle. If the treacle is thin, the particle will oscillate much as before, with the amplitude slowly decreasing. If the treacle is thick, there might be no oscillations at all, and the particle will just move slowly towards an equilibrium position.

In general, simple harmonic motion is modelled by the differential equation $\dfrac{d^2 x}{dt^2} = -\omega^2 x$ or $\dfrac{d^2 x}{dt^2} + \omega^2 x = 0$

For damped harmonic motion the motion of the particle is subject to a resistive force, which is proportional to the velocity of the particle. This can be modelled by a differential equation of the form $\dfrac{d^2 x}{dt^2} + k\dfrac{dx}{dt} + \omega^2 x = 0$, where $k > 0$. The larger the value of k, the stronger is the resistance to motion (i.e. the thicker the treacle). The larger the value of ω^2, the stiffer the spring. So it is the ratio of $k : \omega^2$ that determines whether the particle will oscillate towards the equilibrium position as in Example 1, or approach the equilibrium position without oscillating.

This can be summarised with reference to the auxiliary equation, $m^2 + km + \omega^2 = 0$

- If $k^2 - 4\omega^2 > 0$, $x = Ae^{m_1 t} + Be^{m_2 t}$ This is called heavy damping
- If $k^2 - 4\omega^2 = 0$, $x = (At + B)e^{mt}$ This is called critical damping
- If $k^2 - 4\omega^2 < 0$, $x = e^{-\frac{kt}{2}}\left(A\sin\dfrac{\alpha t}{2} + B\cos\dfrac{\alpha t}{2}\right)$ This is called light damping where $\alpha^2 = 4\omega^2 - k^2$

Example 2

A particle moves along a horizontal line under the action of a force directed towards a fixed point O. The displacement, x metres of the particle from its initial position, at a time t seconds from the start, satisfies the differential equation $\dfrac{d^2 x}{dt^2} + 5\dfrac{dx}{dt} + 6x = 3$

At time $t = 0$ the particle is moving through the point $x = 1$ with velocity $3\,\text{m s}^{-1}$

a Solve the differential equation to obtain an expression for x in terms of t

b Find the time at which the particle is at rest.

c Sketch the graph of x against t

d Explain whether the type of damping is light, critical or heavy.

a The auxiliary equation is $m^2 + 5m + 6 = 0$

$(m+2)(m+3) = 0$, so $m = -2$ or -3

So the complementary function is $x = Ae^{-2t} + Be^{-3t}$

This is 'real distinct roots'.

(Continued on the next page)

639

For the particular integral, try $x = c$ ●───────────

$6c = 3$, so $c = \dfrac{1}{2}$

So the particular integral is $x = \dfrac{1}{2}$

This is from Table 2

The general solution is $x = Ae^{-2t} + Be^{-3t} + \dfrac{1}{2}$ ●──

$x = -2Ae^{-2t} - 3Be^{-3t}$

$t = 0, x = 1$ gives $A + B + \dfrac{1}{2} = 1$

$t = 0, x = 3$ gives $-2A - 3B = 3$

So $A = 4.5$ and $B = -4$

The particular solution is $x = 4.5e^{-2t} - 4e^{-3t} + 0.5$

Add together the complementary function and the particular integral to get the general solution.

b $\dfrac{dx}{dt} = -9e^{-2t} + 12e^{-3t} = 0$

$9e^{t} = 12$

$t = \ln\left(\dfrac{4}{3}\right) \approx 0.288$

c
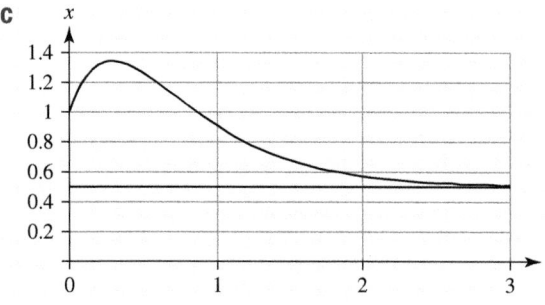

d The auxiliary equation is $m^2 + 5m + 6 = 0$

The discriminant, $\Delta = 5^2 - 4(1)(6) = +1 > 0$

Hence the damping is heavy.

Not all harmonic motion is simple or damped. In the two examples above, one end of the spring has been attached to a fixed point. Suppose, instead, you allow that end to move. It is still possible for the motion of the particle to follow a regular pattern.

Example 3

A box, P, is attached to one end of a light elastic string. The box is initially at rest on a smooth horizontal table when the other end of the spring, A, starts to move away from P with constant velocity $1.2 \, \text{m s}^{-1}$. The displacement, y metres, of P from its initial position at a time t seconds from the start, satisfies the differential equation $\dfrac{d^2x}{dt^2} + 25y = 10t$

a Find an expression for y in terms of t

b Calculate the time at which P is next at rest.

c Sketch the graph of y against t

(*Continued on the next page*)

a The auxiliary equation is $m^2 + 25 = 0$

$m^2 = -25$, so $m = \pm 5i$

So the complementary function is $y = A\cos 5t + B\sin 5t$

For the particular integral, try $y = at + b$

Then $\dot{y} = a$ and $\ddot{y} = 0$

$25(at + b) = 10t$

giving $a = \dfrac{2}{5}$ and $b = 0$

So the particular integral is $y = \dfrac{2}{5}t$

The general solution is $y = A\cos 5t + B\sin 5t + \dfrac{2}{5}t$

$A = 0$

The solution is now $y = B\sin 5t + \dfrac{2}{5}t$

$\dfrac{dy}{dt} = 5B\cos 5t + \dfrac{2}{5}$

$5B + \dfrac{2}{5} = 0$, so $B = -\dfrac{2}{25}$

The particular solution is $y = -\dfrac{2}{25}\sin 5t + \dfrac{2}{5}t$

b $\dfrac{dy}{dt} = -\dfrac{2}{5}\cos 5t + \dfrac{2}{5}$

$-\dfrac{2}{5}\cos 5t + \dfrac{2}{5} = 0$

$\cos 5t = 1$

$t = \dfrac{2\pi}{5}$ s

c

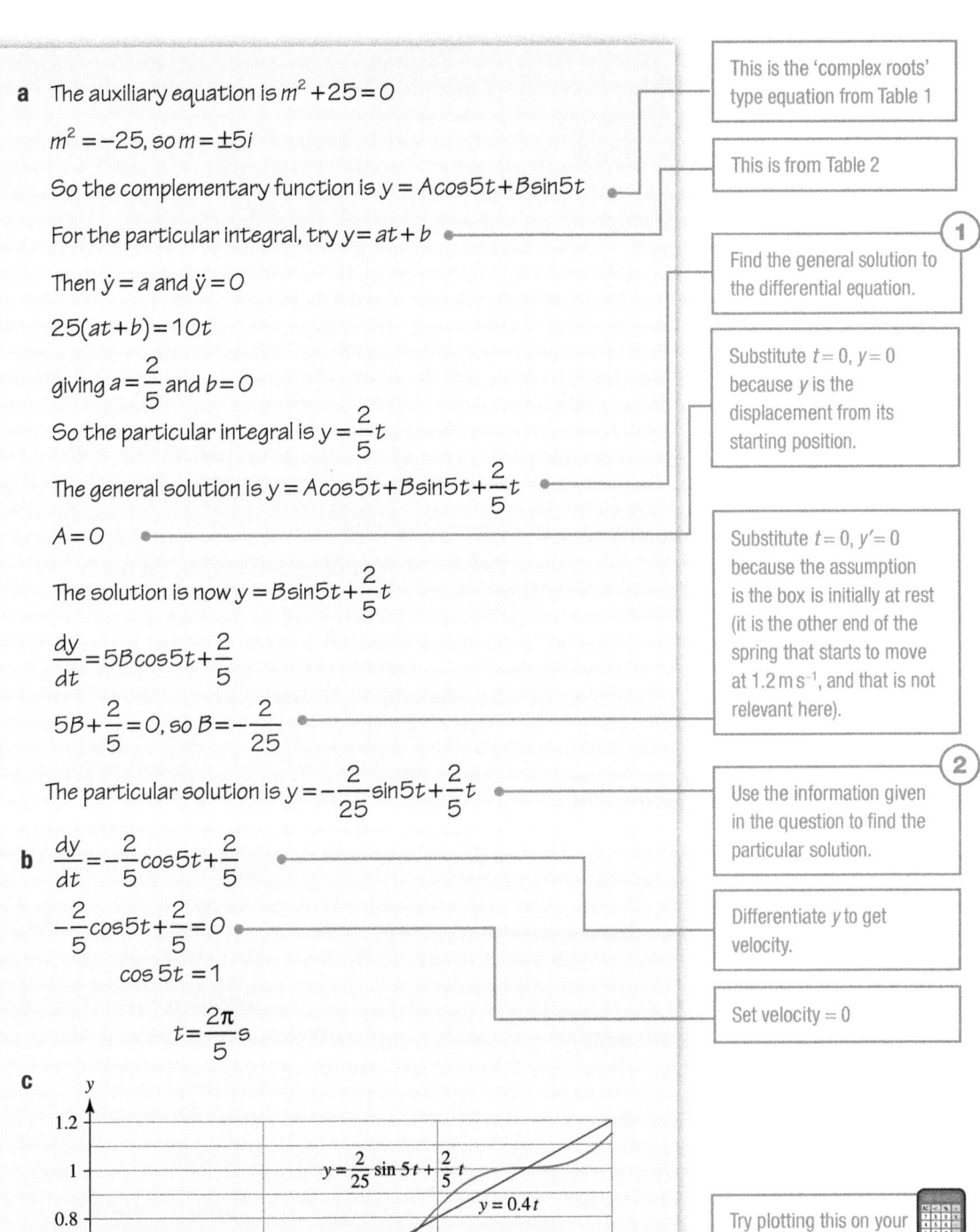

$$y = \frac{2}{25}\sin 5t + \frac{2}{5}t$$

$$y = 0.4t$$

> This is the 'complex roots' type equation from Table 1

> This is from Table 2

> **(1)** Find the general solution to the differential equation.

> Substitute $t = 0$, $y = 0$ because y is the displacement from its starting position.

> Substitute $t = 0$, $y' = 0$ because the assumption is the box is initially at rest (it is the other end of the spring that starts to move at $1.2\,\mathrm{m\,s^{-1}}$, and that is not relevant here).

> **(2)** Use the information given in the question to find the particular solution.

> Differentiate y to get velocity.

> Set velocity $= 0$

> Try plotting this on your graphical calculator.

As you can see from the diagram, the graph oscillates about the line $y = 0.4t$

1 Solve each of the following differential equations subject to the given initial conditions, and classify each type of damping as heavy, critical or light.

a $\dfrac{d^2x}{dt^2}+6\dfrac{dx}{dt}+13x=0$, given that $x=-5$

and $\dfrac{dx}{dt}=3$ when $t=0$

b $\dfrac{d^2x}{dt^2}+6\dfrac{dx}{dt}+8x=0$, given that $x=2$

and $\dfrac{dx}{dt}=-2$ when $t=0$

c $\dfrac{d^2x}{dt^2}+4\dfrac{dx}{dt}+4x=0$, given that $x=-1$

and $\dfrac{dx}{dt}=5$ when $t=0$

d $\dfrac{d^2x}{dt^2}+10\dfrac{dx}{dt}+41x=0$, given that $x=2$

and $\dfrac{dx}{dt}=-6$ when $t=0$

e $2\dfrac{d^2x}{dt^2}+10\dfrac{dx}{dt}+13x=0$, given that $x=6$

and $\dfrac{dx}{dt}=-8$ when $t=0$

f $9\dfrac{d^2x}{dt^2}+6\dfrac{dx}{dt}+x=0$, given that $x=0$

and $\dfrac{dx}{dt}=6$ when $t=0$

2 Solve each of the following differential equations subject to the given initial conditions.

a $\dfrac{d^2x}{dt^2}+10\dfrac{dx}{dt}+9x=-18$, given that $x=0$

and $\dfrac{dx}{dt}=6$ when $t=0$

b $\dfrac{d^2x}{dt^2}+8\dfrac{dx}{dt}+16x=48$, given that $x=5$

and $\dfrac{dx}{dt}=-3$ when $t=0$

c $\dfrac{d^2x}{dt^2}+10\dfrac{dx}{dt}+29x=29$, given that $x=6$

and $\dfrac{dx}{dt}=-21$ when $t=0$

d $\dfrac{d^2x}{dt^2}+4\dfrac{dx}{dt}+5x=20$, given that $x=3$

and $\dfrac{dx}{dt}=2$ when $t=0$

e $\dfrac{d^2x}{dt^2}+4\dfrac{dx}{dt}+4x=12$, given that $x=4$

and $\dfrac{dx}{dt}=9$ when $t=0$

f $\dfrac{d^2x}{dt^2}+2\dfrac{dx}{dt}+2x=10$, given that $x=6$

and $\dfrac{dx}{dt}=0$ when $t=0$

3 Solve the differential equation
$\dfrac{d^2x}{dt^2}+4\dfrac{dx}{dt}+3x=6$, given that $x(0)=5$ and $\dot{x}(0)=-1$, and hence sketch the graph of x against t

4 Solve the differential equation
$\dfrac{d^2x}{dt^2}+2\dfrac{dx}{dt}+5x=20$, given $x(0)=4$ and $\dot{x}(0)=2$, and hence sketch the graph of x against t

5 A particle moves along a horizontal line under the action of a force directed towards a fixed point O. The displacement, x metres of P from O, at a time t seconds from the start, satisfies the differential equation $\dfrac{d^2x}{dt^2}+4\dfrac{dx}{dt}+8x=0$. At time $t=0$ the particle is moving through O with velocity $0.8\,\text{ms}^{-1}$.

a Solve the differential equation to obtain an expression for x in terms of t

b Sketch the graph of x against t

c State whether the type of damping is light, critical or heavy.

6 A particle P is moving in a straight line. At time t seconds, the displacement of P from a fixed point on the line is x metres. The motion of the particle can be modelled by the differential equation $\dfrac{d^2x}{dt^2}+7\dfrac{dx}{dt}+12x=6$ When $t=0$, the particle is moving through the point where $x=1$ with velocity $2\,\text{ms}^{-1}$

a Solve the differential equation to obtain an expression for x in terms of t

b Calculate the time at which the particle is at rest.

c Sketch the graph of x against t

d State whether the type of damping is light, critical or heavy.

7 A particle, P, is attached to one end of a light elastic spring. The other end of the spring is attached to a fixed point A, and P hangs at rest, vertically below A. At time $t = 0$, P is projected vertically downwards with speed u ms^{-1}. The displacement of P downwards from its equilibrium position at time t seconds is x metres. The motion of P can be modelled by the differential equation

$$\frac{d^2 x}{dt^2} + 8\omega \frac{dx}{dt} + 16\omega^2 = 0,$$

where ω is a constant.

a Find an expression for x in terms of u, t and ω

b Find an expression in terms of ω for the time at which P first comes to rest.

c State whether the type of damping is light, critical or heavy.

8 When an alternating electromotive force of $2\cos 2t$ is applied across an electrical circuit, the current, I amps at a time t seconds satisfies the differential equation

$$\frac{d^2 I}{dt^2} + 16I = 36\sin 2t$$

At $t = 0$, $I = 0$ and $\frac{dI}{dt} = 14$

a Find an expression for I in term of t

b Find the time at which the current first returns to zero.

9 A swing door is fitted with a damping device. The angular displacement, θ radians, of the door from its equilibrium position at time t seconds, is modelled by the differential equation $\frac{d^2 \theta}{dt^2} + 6\frac{d\theta}{dt} + 13\theta = 0$

The door starts from rest at an angle of $\frac{\pi}{4}$ to the equilibrium position.

a Find an expression for θ in terms of t

b Sketch the graph of θ against t

c What does the model predict as t becomes large?

10 The differential equation $\frac{d^2 x}{dt^2} + \frac{dx}{dt} + \frac{x}{2} = \frac{3}{4}e^{-\frac{1}{2}t}$ describes the motion of a particle along the x-axis, where x, measured in metres, is the displacement of the particle from the origin at time t seconds.

At time $t = 0$, $x = 5$ and $\frac{dx}{dt} = -\frac{3}{2}$

a Solve this differential equation to get an expression for x in terms of t

b Prove that the particle never reaches the origin.

c Calculate the value of t when the particle first comes to rest.

11 In an experiment, John attaches a small weight, W, to one end of an elastic spring. John holds the other end of the spring, and W hangs at rest at a fixed point A. John then starts to move his end of the spring up and down, causing the weight to oscillate. The vertical displacement, x cm, of W above A, at a time t seconds after John has started to move his end, is given by the differential equation $2\frac{d^2 x}{dt^2} + 5\frac{dx}{dt} + 2x = 10\cos t, t \geq 0$

a Solve this differential equation to find an expression for x in terms of t

b Find the distance of W from A at $t = 2$, making clear whether W is above A or below A at that time.

12

A box, P, is attached to one end of a light elastic string. The box in initially at rest on a rough horizontal table, when the other end of the spring, A, starts to move away from P with constant velocity 0.5 m s^{-1}.

The displacement, x metres, of P from its initial position, at a time t seconds from the start, satisfies the differential equation

$$\frac{d^2 x}{dt^2} + 4x = 2t + 1$$

a Find an expression for x in terms of t

b Calculate the time at which P is next at rest.

c Sketch the graph of x against t

13 A particle, P, moves along a horizontal line under the action of a force directed towards a fixed point O

The displacement, x metres, of P from its initial position, at a time t seconds from the start, satisfies the differential equation

$$\frac{d^2x}{dt^2} + 2\frac{dx}{dt} + 2x = 0$$

At $t = 0$, $x = 1$ and $\dfrac{dx}{dt} = 0$

a Find an expression for x in term of t

b Find the times at which P is at rest.

c Show that the total distance moved by P will not exceed $\coth\left(\dfrac{\pi}{2}\right)$

Strategy

To answer a question involving damped or forced simple harmonic motion

(1) Draw a force diagram.

(2) Use Newton's 2nd law to form a second order differential equation.

(3) Find the general solution to that differential equation.

(4) Use the information given in the question to find the particular solution.

(5) Answer the question in context.

Example 4

A particle, P, of mass 2 kg moves along the positive x-axis under the action of a force directed towards the origin. At time t seconds, the displacement of P from O is x metres, and P is moving away from O with a speed of v m s^{-1}. The force has magnitude $40x$ N. The particle, P, is also subject to a resistive force of magnitude $16v$ N.

a Show that the equation of motion of P is $\dfrac{d^2x}{dt^2} + 8\dfrac{dx}{dt} + 20x = 0$

Given also that $x(0) = 0$ and $x\left(\dfrac{\pi}{4}\right) = 3e^{-\pi}$

b Solve the differential equation, and find an expression for x in terms of t

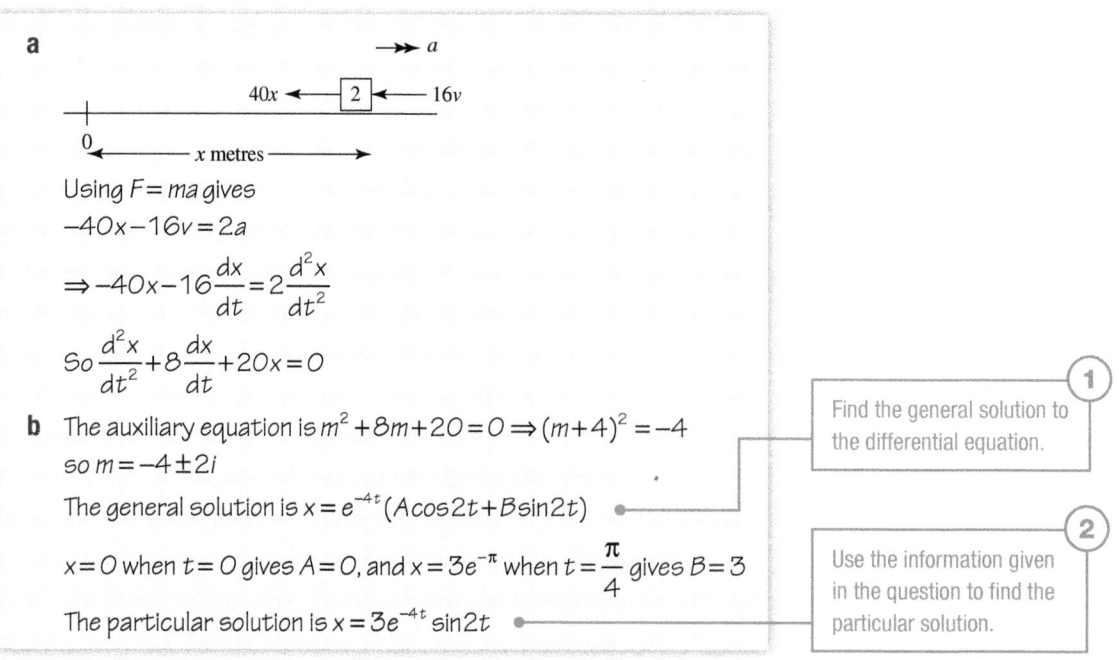

a

$$\longrightarrow a$$

$$40x \longleftarrow \boxed{2} \longleftarrow 16v$$

$$0 \longleftarrow x \text{ metres} \longrightarrow$$

Using $F = ma$ gives

$$-40x - 16v = 2a$$

$$\Rightarrow -40x - 16\frac{dx}{dt} = 2\frac{d^2x}{dt^2}$$

So $\dfrac{d^2x}{dt^2} + 8\dfrac{dx}{dt} + 20x = 0$

b The auxiliary equation is $m^2 + 8m + 20 = 0 \Rightarrow (m+4)^2 = -4$

so $m = -4 \pm 2i$

(1) Find the general solution to the differential equation.

The general solution is $x = e^{-4t}(A\cos 2t + B\sin 2t)$

$x = 0$ when $t = 0$ gives $A = 0$, and $x = 3e^{-\pi}$ when $t = \dfrac{\pi}{4}$ gives $B = 3$

(2) Use the information given in the question to find the particular solution.

The particular solution is $x = 3e^{-4t}\sin 2t$

Example 5

A train of mass m kg runs into the buffers. The buffers can be modelled as an elastic spring that is fixed at one end. At time t seconds after the train first comes into contact with the buffers, the compression of the buffers is x metres, and the magnitude of the force in the spring is given by $m\omega^2 x$ newtons, where ω is a positive constant. The motion of the train is also subject to a resistive force of magnitude mkv newtons, where v m s^{-1} is the speed of the train and k is a positive constant.

a Show that $\dfrac{d^2x}{dt^2} + k\dfrac{dx}{dt} + \omega^2 x = 0$

b At time $t = 0$ the train is travelling with speed U. Given that $k = \dfrac{10\omega}{3}$, find an expression for x in terms of U, ω and t

c Show that the train comes to instantaneous rest when $t = \dfrac{3}{4\omega}\ln 3$

d State three assumptions you have made in forming your model.

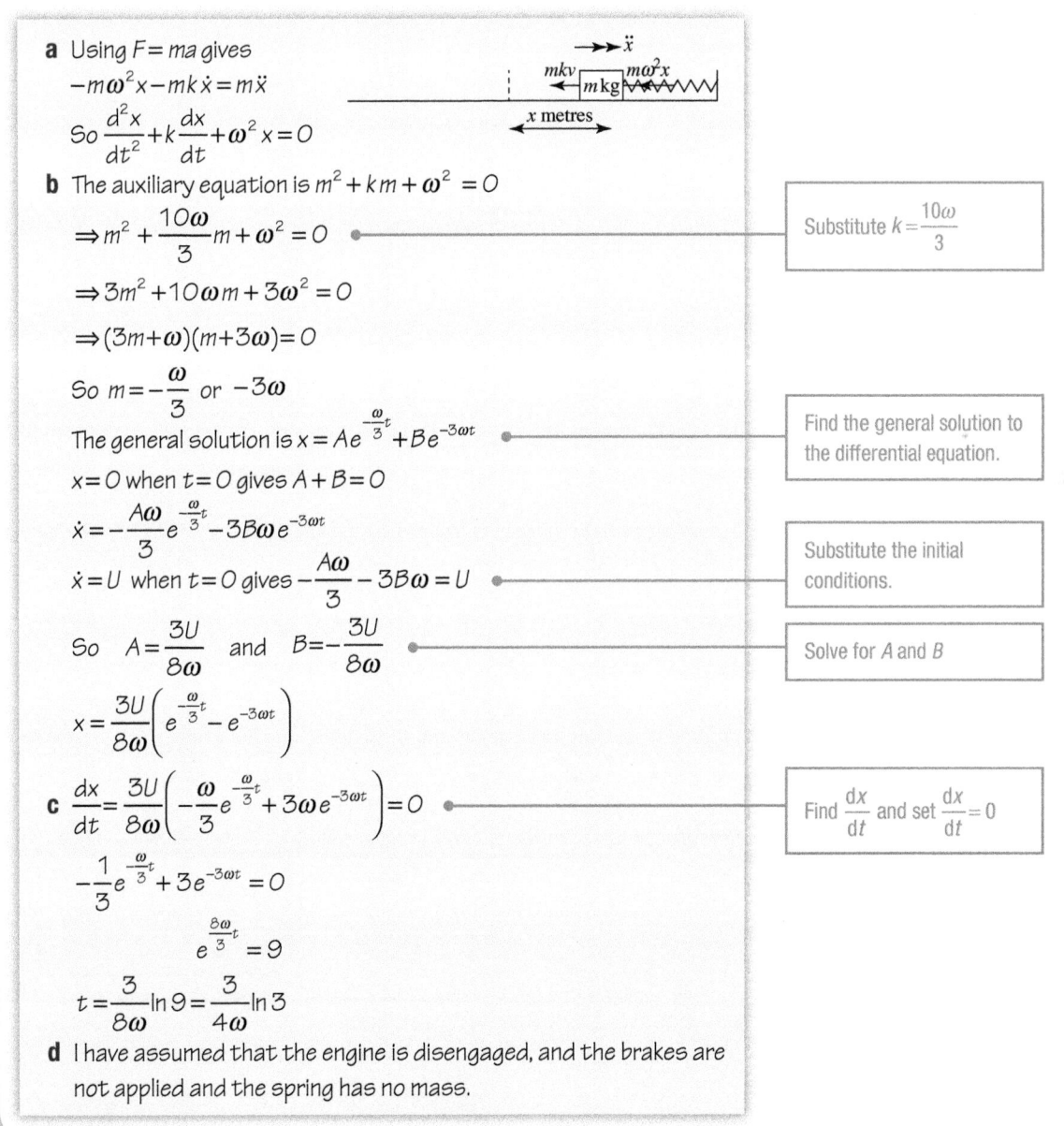

a Using $F = ma$ gives

$-m\omega^2 x - mk\dot{x} = m\ddot{x}$

So $\dfrac{d^2x}{dt^2} + k\dfrac{dx}{dt} + \omega^2 x = 0$

b The auxiliary equation is $m^2 + km + \omega^2 = 0$

$\Rightarrow m^2 + \dfrac{10\omega}{3}m + \omega^2 = 0$ *Substitute $k = \dfrac{10\omega}{3}$*

$\Rightarrow 3m^2 + 10\omega m + 3\omega^2 = 0$

$\Rightarrow (3m + \omega)(m + 3\omega) = 0$

So $m = -\dfrac{\omega}{3}$ or -3ω

The general solution is $x = Ae^{-\frac{\omega}{3}t} + Be^{-3\omega t}$ *Find the general solution to the differential equation.*

$x = 0$ when $t = 0$ gives $A + B = 0$

$\dot{x} = -\dfrac{A\omega}{3}e^{-\frac{\omega}{3}t} - 3B\omega e^{-3\omega t}$

$\dot{x} = U$ when $t = 0$ gives $-\dfrac{A\omega}{3} - 3B\omega = U$ *Substitute the initial conditions.*

So $A = \dfrac{3U}{8\omega}$ and $B = -\dfrac{3U}{8\omega}$ *Solve for A and B*

$x = \dfrac{3U}{8\omega}\left(e^{-\frac{\omega}{3}t} - e^{-3\omega t}\right)$

c $\dfrac{dx}{dt} = \dfrac{3U}{8\omega}\left(-\dfrac{\omega}{3}e^{-\frac{\omega}{3}t} + 3\omega e^{-3\omega t}\right) = 0$ *Find $\dfrac{dx}{dt}$ and set $\dfrac{dx}{dt} = 0$*

$-\dfrac{1}{3}e^{-\frac{\omega}{3}t} + 3e^{-3\omega t} = 0$

$e^{\frac{8\omega}{3}t} = 9$

$t = \dfrac{3}{8\omega}\ln 9 = \dfrac{3}{4\omega}\ln 3$

d I have assumed that the engine is disengaged, and the brakes are not applied and the spring has no mass.

Example 6

A box of mass 0.2 kg is attached to the end of a light elastic string. The other end of the spring is attached to a fixed point. The mass falls freely vertically under gravity, and is travelling with a speed of $7\,\text{m}\,\text{s}^{-1}$ when the string becomes taut for the first time. At time t seconds after the string first becomes taut, the extension of the string is x metres. While the string is taut, the three forces that act on the box are an upwards force in the string of magnitude $2x$ newtons, a resistance force of magnitude $0.4\dot{x}$ newtons, and the weight of the box.
Taking $g=10\,\text{m}\,\text{s}^{-2}$,

a Show that $\ddot{x}+2\dot{x}+10x=10$

b Solve this equation to show that $x=\text{e}^{-t}\left(2\sin 3t-\cos 3t\right)+1$

c Find the time at which particle box first comes to rest.

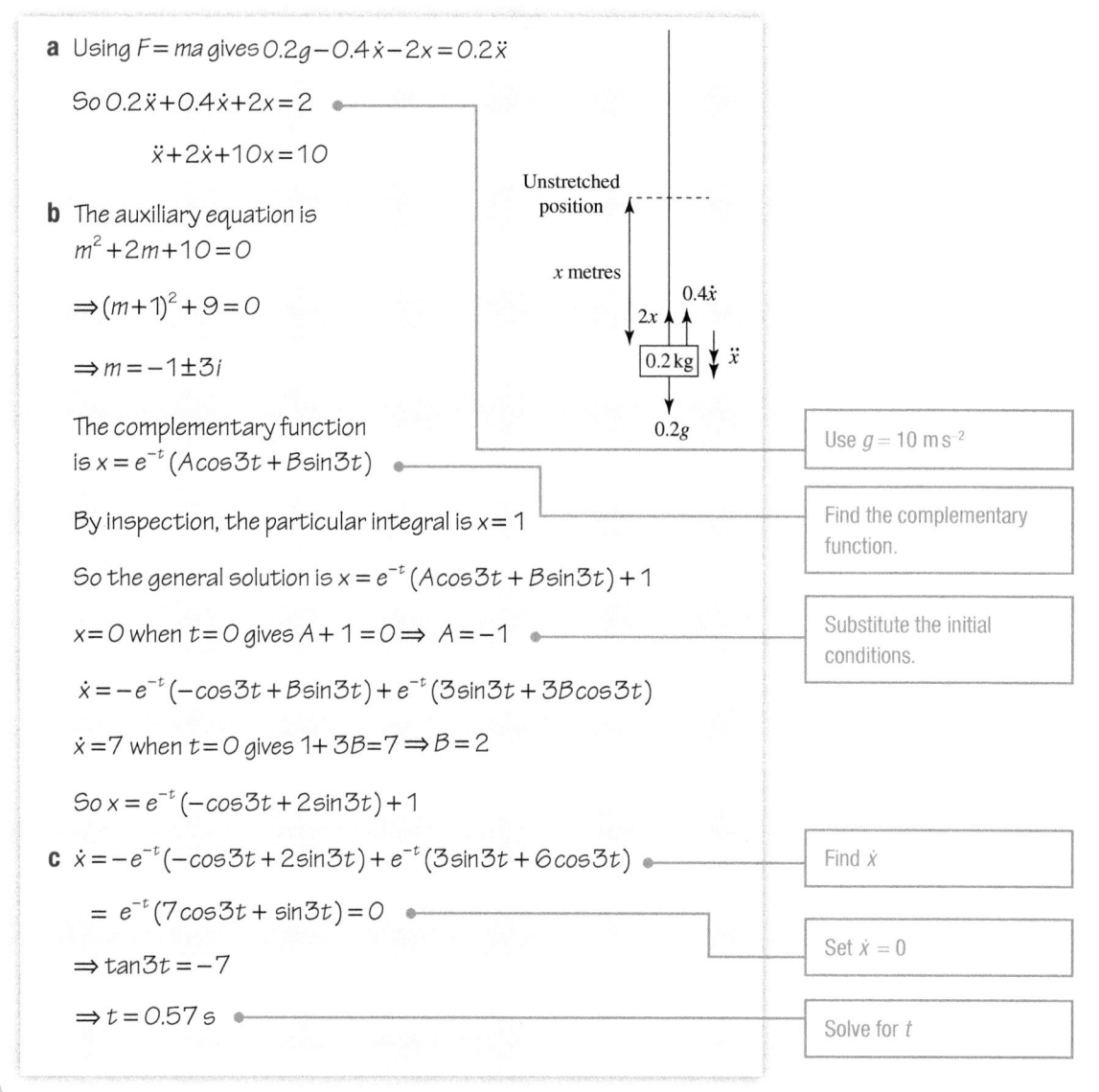

a Using $F=ma$ gives $0.2g-0.4\dot{x}-2x=0.2\ddot{x}$

So $0.2\ddot{x}+0.4\dot{x}+2x=2$

$\ddot{x}+2\dot{x}+10x=10$

b The auxiliary equation is
$m^2+2m+10=0$

$\Rightarrow (m+1)^2+9=0$

$\Rightarrow m=-1\pm 3i$

The complementary function
is $x=\text{e}^{-t}\left(A\cos 3t+B\sin 3t\right)$ — Use $g=10\ \text{m}\,\text{s}^{-2}$

— Find the complementary function.

By inspection, the particular integral is $x=1$

So the general solution is $x=\text{e}^{-t}\left(A\cos 3t+B\sin 3t\right)+1$

$x=0$ when $t=0$ gives $A+1=0\Rightarrow A=-1$ — Substitute the initial conditions.

$\dot{x}=-\text{e}^{-t}\left(-\cos 3t+B\sin 3t\right)+\text{e}^{-t}\left(3\sin 3t+3B\cos 3t\right)$

$\dot{x}=7$ when $t=0$ gives $1+3B=7\Rightarrow B=2$

So $x=\text{e}^{-t}\left(-\cos 3t+2\sin 3t\right)+1$

c $\dot{x}=-\text{e}^{-t}\left(-\cos 3t+2\sin 3t\right)+\text{e}^{-t}\left(3\sin 3t+6\cos 3t\right)$ — Find $\dot{x}$

$=\text{e}^{-t}\left(7\cos 3t+\sin 3t\right)=0$ — Set $\dot{x}=0$

$\Rightarrow \tan 3t=-7$

$\Rightarrow t=0.57\,\text{s}$ — Solve for t

1 A body B, of mass 9 kg moves in a horizontal straight line. At time t seconds, the displacement of B from a fixed point O on the line, is x metres, and B is moving with a velocity of $\dot{x}$ ms^{-1}. Throughout the motion two forces act on B, a restorative force of magnitude $36x$ newtons directed towards O, and a resistance force of magnitude $45\dot{x}$ newtons. At time $t = 0$, $x = -2$ and $\dot{x} = 11$

  a Show that $\ddot{x} + 5\dot{x} + 4 = 0$

  b Find an expression for x in terms of t

  c Hence state the type of damping that occurs.

  d Find the value of t when B passes through O for the first time.

2 A particle P, of mass m kg moves in a horizontal straight line. At time t seconds, the displacement of P from a fixed point O on the line, is x metres, and P is moving with a velocity of $\dot{x}$ ms^{-1}. Throughout the motion two forces act on P, a restorative force of magnitude $25mn^2x$ newtons directed towards O, and a resistance force of magnitude $km\dot{x}$ newtons, where n and k are positive constants.

  a Show that $\ddot{x} + k\dot{x} + 25n^2x = 0$

  Given the damping is critical

  b Find an expression for k in terms of n

  c Find a general expression for x in terms of n and t

3 Two points, A and B, are 2 metres apart on a horizontal table. The point C lies on AB and is 160 cm from A. A light spring of unstretched length 160 cm has one end

fastened to the table at A, and the other end fastened to a body of mass m kg which is held at rest at B. When extended, the force in the spring is directed towards C, and is equal to $\dfrac{4mx}{9}$N, where x, measured in metres, is the displacement of the body from C

The body is released from rest at B. When in motion, two additional resistive forces act on the body, one a constant force of magnitude $\dfrac{4m}{45}$ N, and the other a variable force of magnitude $\dfrac{4m\dot{x}}{3}$ N.

  a Suggest two physical properties that might cause the two resistive forces.

  b Write down an equation of motion for the body as it moves towards A

  c Show that $x = \dfrac{1}{15}\left[(2t+3)e^{-\frac{2}{3}t} + 3\right]$

  d With reference to the type of damping, explain what happens as the value of t increases.

4 A particle P of mass 3 kg is moving on the x-axis. At time t seconds the displacement of P from the origin O, is x metres, and the velocity of P is v ms^{-1}. Three forces act on P, namely, a restoring force towards O, of magnitude $15x$ newtons, a resistance to motion of magnitude $6v$ newtons, and a force of $24e^{-t}$ newtons acting in the direction OP

  a Show that $\dfrac{d^2x}{dt^2} + 2\dfrac{dx}{dt} + 5x = 8e^{-t}$

  Given also that at time $t = 0$, $x = 5$ and $v = 3$

  b Find an expression for x in terms of t

  c Find the value of t for which $x = 0$ for the first time.

5   A particle P of mass m kg is attached to one end of a light elastic string of natural length ℓ metres. P is at rest in equilibrium on a smooth horizontal table. The other end of the string is attached to a fixed point, O, on the table. A time-dependent force is then applied to P in the direction OP. At time t seconds after the force is applied, the extension of the string is x metres. The magnitude of the force in the string is $6m\omega^2 x$, where ω is a constant, and the magnitude of the time-dependent force is $m\omega^2 \ell e^{-\omega t}$. The motion of the particle is opposed by a force of magnitude $5m\omega\dot{x}$.

   a   Write down a differential equation connecting, x, t, ω and ℓ

   b   Solve this equation to show that
   $$x = \frac{\ell}{2}e^{-\omega t}(1-e^{-\omega t})^2$$

   c   Show that the greatest extension of the string is $\dfrac{2\ell}{27}$

6   On a water-ride at a theme park, passengers travel down a course in an open tub. At the end of the course, the tub is brought to rest by running into a retarding device. The retarding device is a large elastic spring that is fixed at the far end. At time t seconds after a tub first comes into contact with the spring, the compression of the spring is x metres, and the magnitude of the force in the spring is $m\omega^2 x$ newtons, where m is the combined mass of the passengers and tub, and ω is a positive constant. The motion of the tub is also subject to a resisting force of magnitude $2mk\dot{x}$ newtons, where $\dot{x}$ ms^{-1} is the velocity of the tub and k is a positive constant.

   a   Show that $\ddot{x} + 2k\dot{x} + \omega^2 x = 0$

   At time $t = 0$ the tub is travelling with speed U as it hits the spring.

   b   Given that $k=\omega$, find an expression for x in terms of U, ω and t

   c   Hence state the type of damping that occurs.

   Health and safety rules state the the tub must come to rest within two seconds of the tub hitting the spring.

   d   Show that $\omega > \dfrac{1}{2}$

   e   State two assumptions that you have made in forming your model.

7   A box of mass 5 kg is attached to the end of a light elastic string. The other end of the string is attached to a fixed point. The box falls vertically under gravity and is travelling with a speed of 3ms^{-1} when the string becomes taut for the first time. At time t seconds after the string first becomes taut, the extension of the string is x metres. While the string is taut, the three forces that act on the box are an upwards force in the string of magnitude $25x$ newtons, a resistive force of magnitude $10\dot{x}$ newtons, and the weight of the box. Taking $g=10$ms^{-2},

   a   Show that $\ddot{x} + 2\dot{x} + 5x = 10$

   b   Solve this equation to get an expression for x in terms of t

   c   Find the time at which the box comes to rest for the first time.

8   A stuntman of mass 80 kg attaches himself to one end of an elastic rope. The other end of the rope is attached to a fixed point on a high bridge. The man throws himself off the bridge and falls vertically under gravity. He is travelling with a speed of $15\,\mathrm{ms}^{-1}$ when the rope becomes taut for the first time. At time t seconds after the rope first becomes taut, the extension of the rope is x metres. While the rope is taut, the three forces that act on the man are an upwards force in the string of magnitude $10x$ newtons, a resistive force of magnitude $40\dot{x}$ newtons, and the weight of the man. Taking $g=10\,\mathrm{ms}^{-2}$,

a   Show that $8\dfrac{\mathrm{d}^2 x}{\mathrm{d}t^2}+4\dfrac{\mathrm{d}x}{\mathrm{d}t}+x=80$

b   Solve this equation to get an expression for x in terms of t

c   Find the time at which the man first comes to rest.

d   State two modelling assumption you have made in answering the question.

9

A box, B, of mass 5 kg is attached to one end of a light elastic spring of natural length 2 metres and modulus 90 N. The box is initially at rest on a smooth horizontal table, with the spring in equilibrium when the other end of the spring, A, starts to move away from B with constant velocity $3\,\mathrm{m\,s^{-1}}$. The air resistance acting on B has magnitude $30v$ N, where $v\,\mathrm{ms^{-1}}$ is the speed of B. At time t seconds, the extension of the string is x metres, and the displacement of B from its initial position is y metres.

a   Show that

   i $x+y=3t$

   ii $\dfrac{\mathrm{d}^2 x}{\mathrm{d}t^2}+6\dfrac{\mathrm{d}x}{\mathrm{d}t}+9x=18$

   iii $x=2-(2+3t)\mathrm{e}^{-3t}$

b   Describe the motion of the box for large values of t

Fluency and skills

So far all of the differential equations you have looked at have involved just two variables. Many differential equations involve more than two variables. Consider, for example, an island populated by foxes and rabbits. You can denote the number of foxes by F, and the number of rabbits by R. Foxes eat rabbits and are known as the **predators**. Rabbits eat grass, and are eaten by the foxes, and are known as the **prey**.

- $\dfrac{dR}{dt}$ depends on R and F. The more rabbits in the population, the more breeding and deaths there are, but the more foxes in the population, the more rabbits are eaten.

- $\dfrac{dF}{dt}$ depends on R and F. The more rabbits in the population, the more food there is for the foxes, but the more foxes in the population, the more breeding.

This gives rise to equations of the type

$$\frac{dR}{dt} = \alpha R - \beta F \quad (1) \qquad \text{and} \qquad \frac{dF}{dt} = \lambda R + \mu F \quad (2)$$

> **Key point**
>
> Problems that involve equations of this type are called **predator–prey** problems.

To solve equations of this type, first differentiate one of the equations with respect to t

Differentiating equation (1) gives

$$\frac{d^2 R}{dt^2} = \alpha \frac{dR}{dt} - \beta \frac{dF}{dt}$$

Now substitute for $\dfrac{dF}{dt}$ from equation (2) to get

$$\frac{d^2 R}{dt^2} = \alpha \frac{dR}{dt} - \beta(\lambda R + \mu F)$$

$$\frac{d^2 R}{dt^2} = \alpha \frac{dR}{dt} - \beta \lambda R - \mu \beta F$$

Now substitute $\beta F = \alpha R - \dfrac{dR}{dt}$ from equation (1) to get

$$\frac{d^2 R}{dt^2} = \alpha \frac{dR}{dt} - \beta\lambda R - \mu\left(\alpha R - \frac{dR}{dt}\right)$$

which rearranges to

$$\frac{d^2 R}{dt^2} - (\alpha+\mu)\frac{dR}{dt} + (\beta\lambda + \alpha\mu)R = 0$$

This is a second order differential equation in R and t that you can solve in the usual way.

Example 1

A system of differential equations is given by

$$\frac{dx}{dt} = x + 4y \qquad (1)$$

$$\frac{dy}{dt} = 2x + 3y - 10 \qquad (2)$$

where $(x, y) = (3, 2)$ when $t = 0$

Find expressions for x and y in terms of t

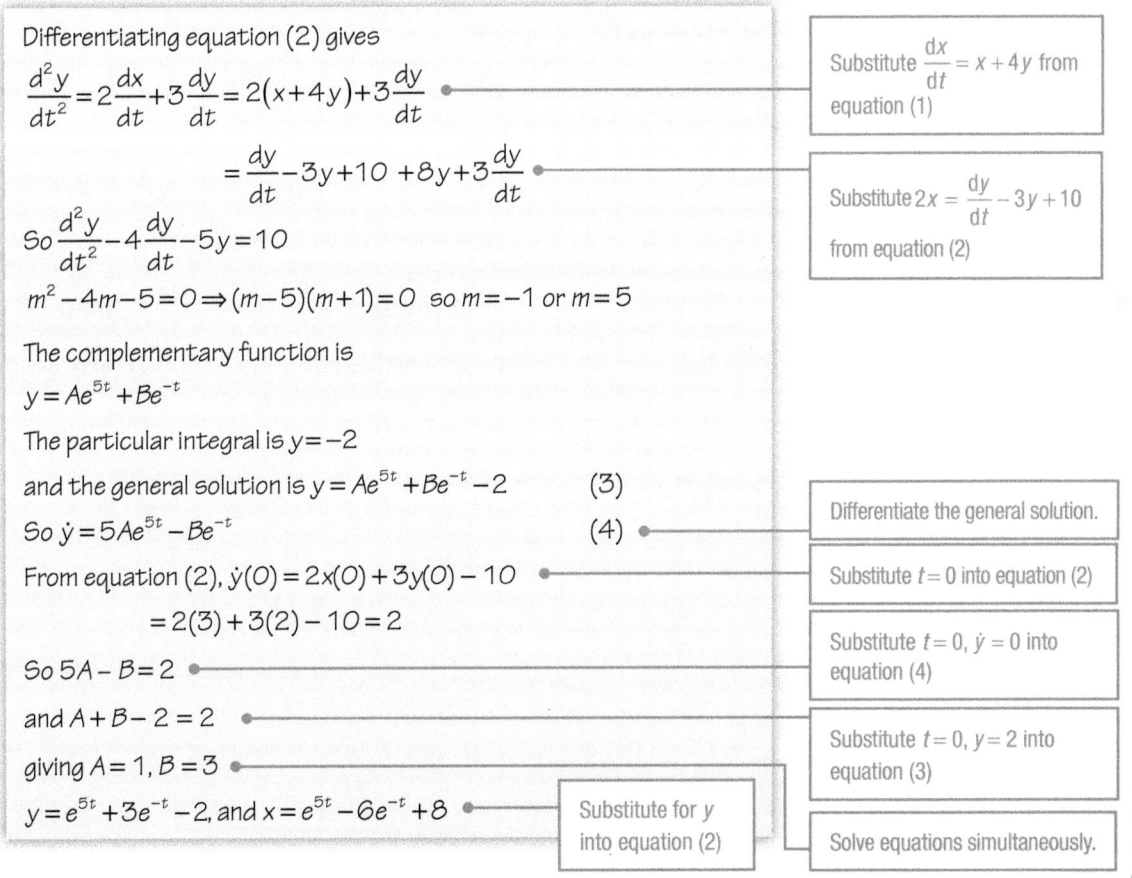

Differentiating equation (2) gives

$$\frac{d^2 y}{dt^2} = 2\frac{dx}{dt} + 3\frac{dy}{dt} = 2(x+4y) + 3\frac{dy}{dt}$$

Substitute $\dfrac{dx}{dt} = x + 4y$ from equation (1)

$$= \frac{dy}{dt} - 3y + 10 + 8y + 3\frac{dy}{dt}$$

Substitute $2x = \dfrac{dy}{dt} - 3y + 10$ from equation (2)

So $\dfrac{d^2 y}{dt^2} - 4\dfrac{dy}{dt} - 5y = 10$

$m^2 - 4m - 5 = 0 \Rightarrow (m-5)(m+1) = 0$ so $m = -1$ or $m = 5$

The complementary function is

$y = Ae^{5t} + Be^{-t}$

The particular integral is $y = -2$

and the general solution is $y = Ae^{5t} + Be^{-t} - 2 \qquad (3)$

So $\dot{y} = 5Ae^{5t} - Be^{-t} \qquad (4)$

Differentiate the general solution.

From equation (2), $\dot{y}(0) = 2x(0) + 3y(0) - 10$

Substitute $t = 0$ into equation (2)

$= 2(3) + 3(2) - 10 = 2$

So $5A - B = 2$

Substitute $t = 0$, $\dot{y} = 0$ into equation (4)

and $A + B - 2 = 2$

giving $A = 1, B = 3$

Substitute $t = 0$, $y = 2$ into equation (3)

$y = e^{5t} + 3e^{-t} - 2$, and $x = e^{5t} - 6e^{-t} + 8$

Substitute for y into equation (2)

Solve equations simultaneously.

PURE

651

1 Solve each of the following systems of differential equations to find expressions for y in terms of t

   a $\dfrac{\mathrm{d}x}{\mathrm{d}t} = x - 2y; \dfrac{\mathrm{d}y}{\mathrm{d}t} = x + 4y$

   b $\dfrac{\mathrm{d}x}{\mathrm{d}t} = -3y; \dfrac{\mathrm{d}y}{\mathrm{d}t} = 3x$

   c $\dfrac{\mathrm{d}x}{\mathrm{d}t} = 5x + 4y; 3\dfrac{\mathrm{d}y}{\mathrm{d}t} = x + 4y$

   d $\dfrac{\mathrm{d}x}{\mathrm{d}t} = 4x - y; \dfrac{\mathrm{d}y}{\mathrm{d}t} = 6x - 3y + 2$

   e $\dfrac{\mathrm{d}x}{\mathrm{d}t} = 2x - 10y; \dfrac{\mathrm{d}y}{\mathrm{d}t} = 5x - 12y$

   f $\dfrac{\mathrm{d}x}{\mathrm{d}t} = -3x + 4y + \cos t; \dfrac{\mathrm{d}y}{\mathrm{d}t} = -2x + y + \sin t$

2 Solve each of the following systems of differential equations, subject to the given boundary conditions, to find expressions for x and y in terms of t

   a $\dfrac{\mathrm{d}x}{\mathrm{d}t} = 2x + 4y; \dfrac{\mathrm{d}y}{\mathrm{d}t} = x - y$; given that $y(0) = 3$ and $\dot{y}(0) = 4$

   b $\dfrac{\mathrm{d}x}{\mathrm{d}t} = 2x - y + 3; \dfrac{\mathrm{d}y}{\mathrm{d}t} = 5x - 4y$; given that $y(0) = -1$ and $\dot{y}(0) = -8$

   c $\dfrac{\mathrm{d}x}{\mathrm{d}t} = x - 5y; \dfrac{\mathrm{d}y}{\mathrm{d}t} = x - 3y$; given that $y(0) = 3$ and $\dot{y}(0) = -5$

   d $\dfrac{\mathrm{d}x}{\mathrm{d}t} = x + 2y + t + 2; \dfrac{\mathrm{d}y}{\mathrm{d}t} = -2x - 3y + 3t$; given that $y(0) = 9$ and $\dot{y}(0) = -2$

   e $\dfrac{\mathrm{d}x}{\mathrm{d}t} = -3x - y - 3; \dfrac{\mathrm{d}y}{\mathrm{d}t} = 2x - y + 2$; given that $y(0) = 5$ and $\dot{y}(0) = 2$

   f $\dfrac{\mathrm{d}x}{\mathrm{d}t} = 7x - 9y + 3\mathrm{e}^{-2t}; \dfrac{\mathrm{d}y}{\mathrm{d}t} = 4x - 5y + \mathrm{e}^{-2t}$; given that $y(0) = 2$ and $\dot{y}(0) = 3$

3 A system of differential equations is given by

   $$\dfrac{\mathrm{d}x}{\mathrm{d}t} = 2x + y \qquad (1)$$

   $$\dfrac{\mathrm{d}y}{\mathrm{d}t} = x + 2y \qquad (2)$$

   where $(x, y) = (3, 1)$ when $t = 0$

   Find expressions for x and y in terms of t

4 A system of differential equations is given by

   $$\dfrac{\mathrm{d}x}{\mathrm{d}t} = -3x - 2y + t \qquad (1)$$

   $$\dfrac{\mathrm{d}y}{\mathrm{d}t} = 2x + y + 3t - 1 \qquad (2)$$

   where $(x, y) = (8, -11)$ when $t = 0$

   Find expressions for x and y in terms of t

5 A system of differential equations is given by

   $$\dfrac{\mathrm{d}x}{\mathrm{d}t} = x + 2y \qquad (1)$$

   $$\dfrac{\mathrm{d}y}{\mathrm{d}t} = y - z \qquad (2)$$

   $$\dfrac{\mathrm{d}z}{\mathrm{d}t} = -x \qquad (3)$$

   At $t = 0$, $x = 0$, $\dfrac{\mathrm{d}x}{\mathrm{d}t} = 4$ and $\dfrac{\mathrm{d}^2 x}{\mathrm{d}t^2} = 5$

   a Show that $\dfrac{\mathrm{d}^3 x}{\mathrm{d}t^3} - 2\dfrac{\mathrm{d}^2 x}{\mathrm{d}t^2} + \dfrac{\mathrm{d}x}{\mathrm{d}t} - 2x = 0$

   b Solve this equation to find an expression for x in terms of t

Reasoning and problem-solving

To solve problems involving a coupled system

① Define any variables that you need.

② Use the information in the question to set up the coupled differential equations.

③ Differentiate one of the equations and use the original equations to eliminate the other variable to obtain a second order differential equation.

④ Solve the second order differential equation and hence solve for both variables.

⑤ Interpret your solution in the context of the question.

Example 2

In a chemical reaction, substance X decays into substance Y, which itself decays.

The rate of decay of X, in grams per hour, is given by twice the amount of substance Y, in grams.

The rate of change of Y, in grams per hour, is given by the amount of substance X, in grams, minus three times the amount of substance Y, in grams.

a Set up two differential equations for x and y, the amounts in grams, of substances X and Y respectively.

b Given that initially $x = 20$ and $y = 0$, solve for x and y at time t hours.

c Prove that there can never be equal amounts of X and Y

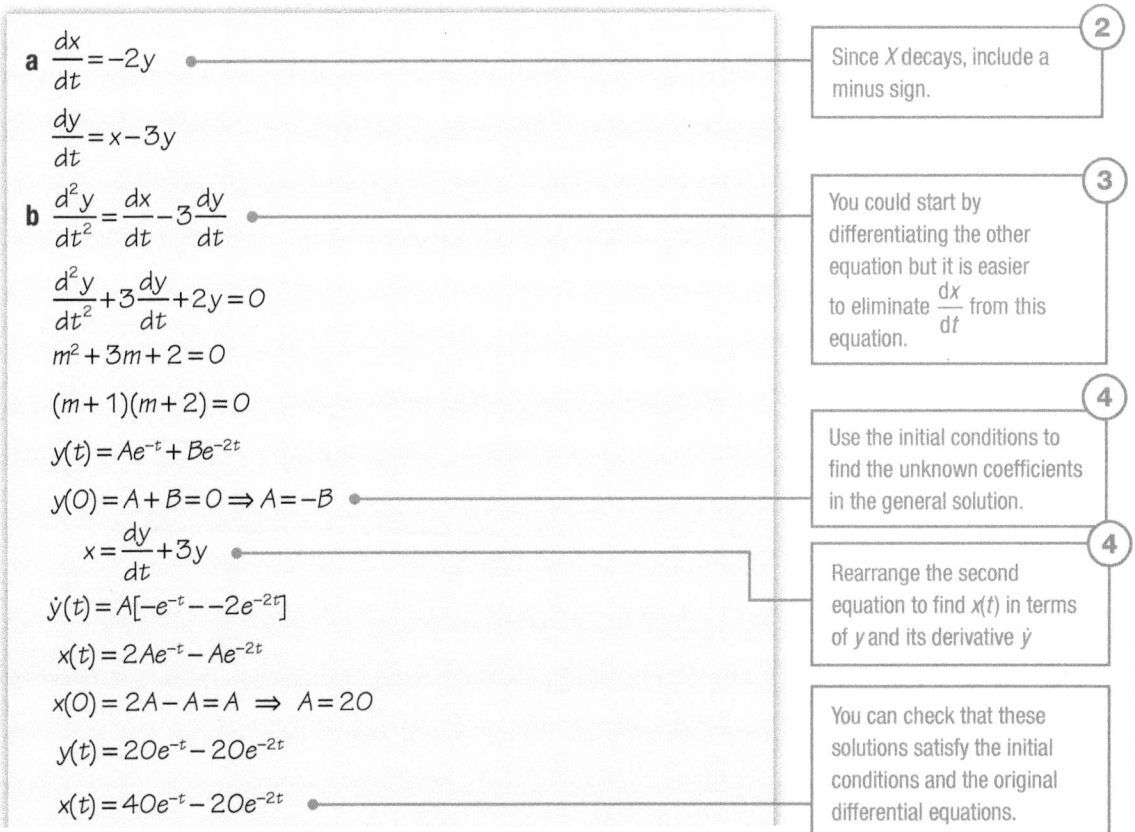

a $\dfrac{dx}{dt} = -2y$

> ② Since X decays, include a minus sign.

$\dfrac{dy}{dt} = x - 3y$

b $\dfrac{d^2y}{dt^2} = \dfrac{dx}{dt} - 3\dfrac{dy}{dt}$

> ③ You could start by differentiating the other equation but it is easier to eliminate $\dfrac{dx}{dt}$ from this equation.

$\dfrac{d^2y}{dt^2} + 3\dfrac{dy}{dt} + 2y = 0$

$m^2 + 3m + 2 = 0$

$(m+1)(m+2) = 0$

$y(t) = Ae^{-t} + Be^{-2t}$

$y(0) = A + B = 0 \Rightarrow A = -B$

> ④ Use the initial conditions to find the unknown coefficients in the general solution.

$x = \dfrac{dy}{dt} + 3y$

> ④ Rearrange the second equation to find $x(t)$ in terms of y and its derivative $\dot{y}$

$\dot{y}(t) = A[-e^{-t} - -2e^{-2t}]$

$x(t) = 2Ae^{-t} - Ae^{-2t}$

$x(0) = 2A - A = A \Rightarrow A = 20$

> You can check that these solutions satisfy the initial conditions and the original differential equations.

$y(t) = 20e^{-t} - 20e^{-2t}$

$x(t) = 40e^{-t} - 20e^{-2t}$

(Continued on the next page)

c Suppose that at time t there are equal amounts of X and Y

$$20e^{-t} - 20e^{-2t} = 40e^{-t} - 20e^{-2t}$$

$$20e^{-t} = 0$$

But, $e^{-t} > 0$ for all t

∴ there can never be equal amounts of X and Y

⑤ Answer the question in context.

Example 3

An isolated island supports populations of sparrowhawks and finches.

- The number of sparrowhawks increases at a rate proportional to the number of finches. When there are 64 finches present, the rate of increase of sparrowhawks is 16 per year.
- If there are no sparrowhawks present, then the finch population would increase by 120% per year.
- If there are sparrowhawks present, then, on average, each sparrowhawk kills 1.44 finches per year.

a Set up two differential equations to model the population of sparrowhawks and finches.

Initially there are 75 sparrowhawks and 120 finches.

b Find the number of sparrowhawks and finches t years later.

c What happens to the populations of finches and sparrowhawks in the distant future?

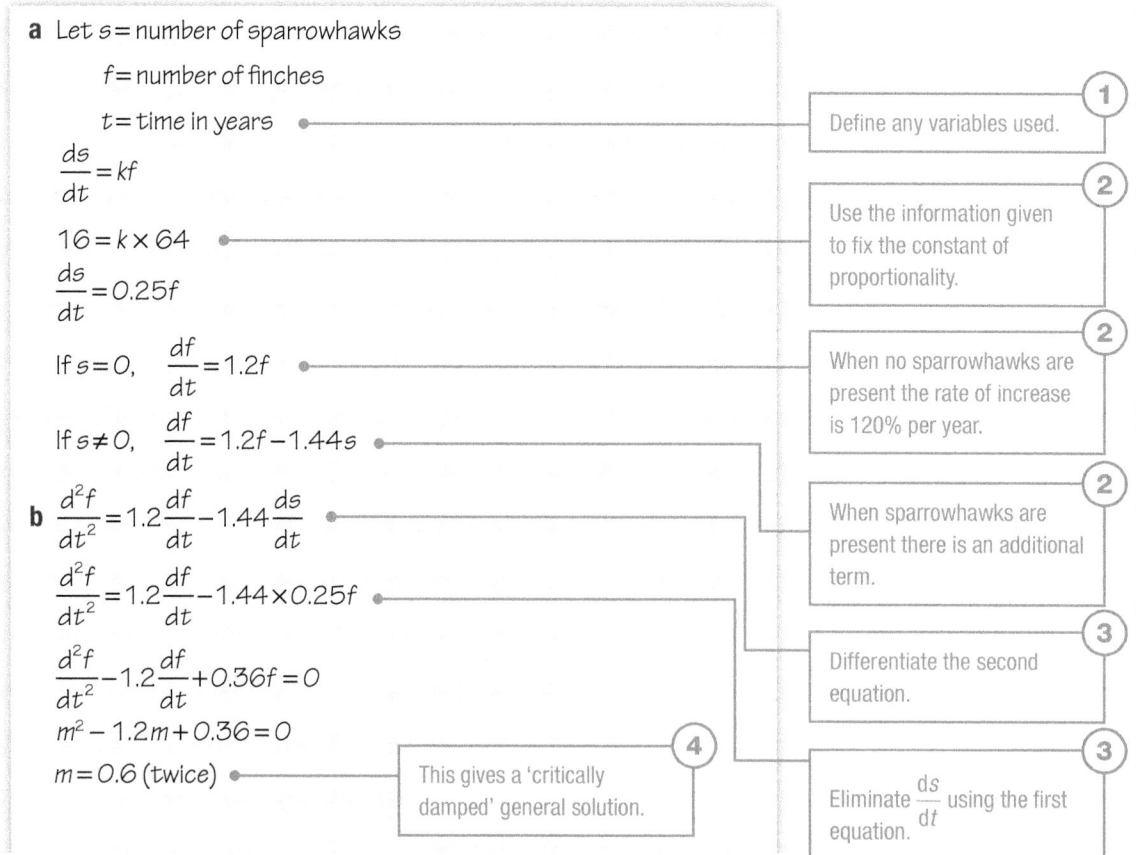

a Let s = number of sparrowhawks

f = number of finches

t = time in years

① Define any variables used.

$$\frac{ds}{dt} = kf$$

$$16 = k \times 64$$

$$\frac{ds}{dt} = 0.25f$$

② Use the information given to fix the constant of proportionality.

If $s = 0$, $\dfrac{df}{dt} = 1.2f$

② When no sparrowhawks are present the rate of increase is 120% per year.

If $s \neq 0$, $\dfrac{df}{dt} = 1.2f - 1.44s$

b $$\frac{d^2f}{dt^2} = 1.2\frac{df}{dt} - 1.44\frac{ds}{dt}$$

② When sparrowhawks are present there is an additional term.

$$\frac{d^2f}{dt^2} = 1.2\frac{df}{dt} - 1.44 \times 0.25f$$

③ Differentiate the second equation.

$$\frac{d^2f}{dt^2} - 1.2\frac{df}{dt} + 0.36f = 0$$

$$m^2 - 1.2m + 0.36 = 0$$

$m = 0.6$ (twice)

④ This gives a 'critically damped' general solution.

③ Eliminate $\dfrac{ds}{dt}$ using the first equation.

(*Continued on the next page*)

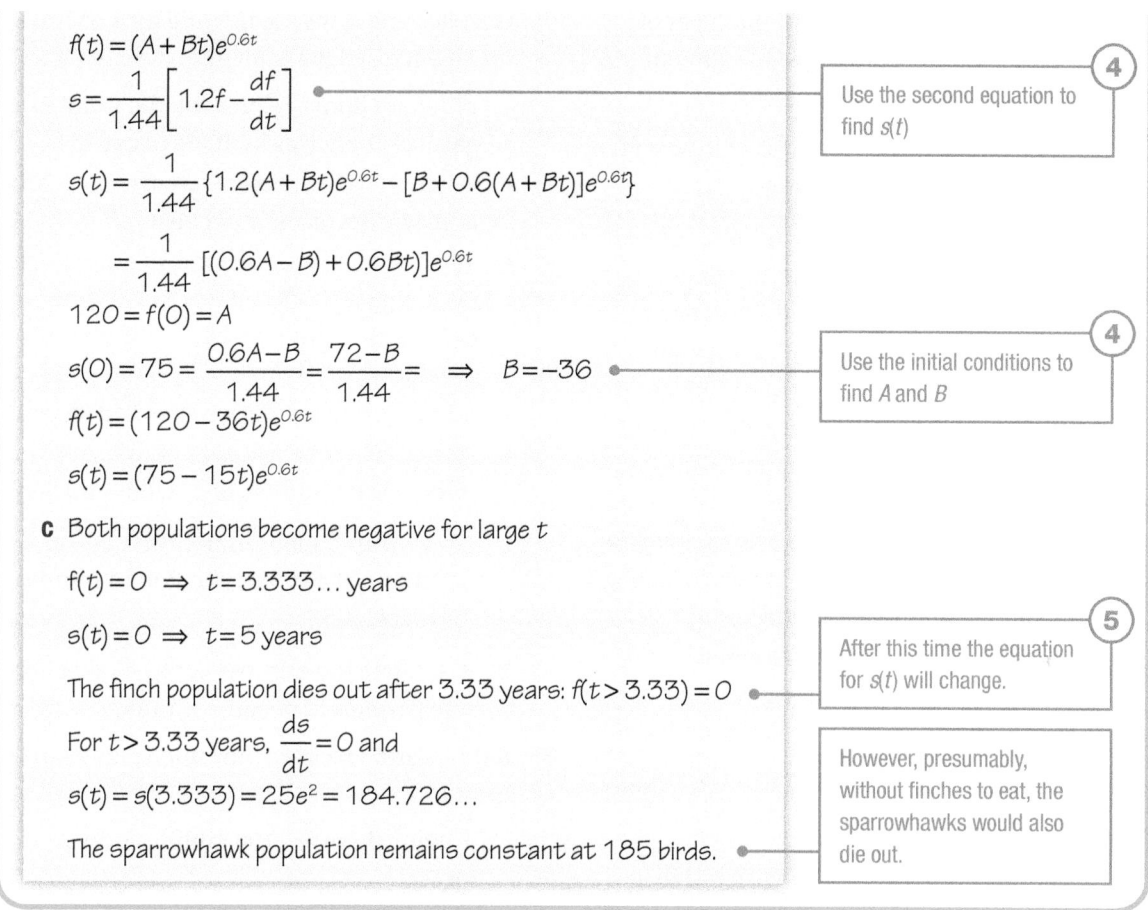

$$f(t) = (A + Bt)e^{0.6t}$$

$$s = \frac{1}{1.44}\left[1.2f - \frac{df}{dt}\right]$$

Use the second equation to find $s(t)$

4

$$s(t) = \frac{1}{1.44}\{1.2(A + Bt)e^{0.6t} - [B + 0.6(A + Bt)]e^{0.6t}\}$$

$$= \frac{1}{1.44}[(0.6A - B) + 0.6Bt]e^{0.6t}$$

$$120 = f(0) = A$$

$$s(0) = 75 = \frac{0.6A - B}{1.44} = \frac{72 - B}{1.44} = \quad \Rightarrow \quad B = -36$$

Use the initial conditions to find A and B

4

$$f(t) = (120 - 36t)e^{0.6t}$$

$$s(t) = (75 - 15t)e^{0.6t}$$

c Both populations become negative for large t

$$f(t) = 0 \Rightarrow t = 3.333\ldots \text{ years}$$

$$s(t) = 0 \Rightarrow t = 5 \text{ years}$$

After this time the equation for $s(t)$ will change.

5

The finch population dies out after 3.33 years: $f(t > 3.33) = 0$

For $t > 3.33$ years, $\dfrac{ds}{dt} = 0$ and

$$s(t) = s(3.333) = 25e^2 = 184.726\ldots$$

The sparrowhawk population remains constant at 185 birds.

However, presumably, without finches to eat, the sparrowhawks would also die out.

Exercise 20.5B Reasoning and problem-solving

1 In a chemical reaction, substance X changes into substance Y which evaporates away.

- The rate of disappearance of X, in grams per hour, is given by three times the amount of substance Y, in grams.
- The rate of change of Y, in grams per hour, is given by the amount of substance X, in grams, minus four times the amount of substance Y, in grams.

a Set up two differential equations for x and y, the amounts, in grams, of substances X and Y respectively.

b Given that initially $x = 40$ and $y = 0$, solve for x and y at time t hours.

c Find the time at which there is precisely four times as much of substance X as there is of substance Y

2 Two declining populations of mollusc species, X and Y, are competing for supremacy.

- The rate of change of the numbers of species X, in thousands per year, is given by three times the amount the numbers of species Y, in thousands, minus twice the number of species X, in thousands.
- The rate of change of the numbers of species Y, in thousands per year, is given by the number of species X, in thousands, minus four times the number of species Y, in thousands.

a Set up two differential equations to model the numbers of molluscs of species X and Y

b Given that initially there are 7000 molluscs of species X and 5000 molluscs of species Y, find the number of molluscs of species X and Y a number of years later.

PURE

c Show that the number of molluscs of species X will always be greater than the number of molluscs of species Y

3 Two species of insect, X and Y, compete for survival in the same environment.

- The rate of change of the number of species X, in millions per month, is given by three times the difference in the numbers of species X and species Y, measured in millions.
- The rate of change of the number of species Y, in millions per month, is given by five times the number of species Y, in millions, minus the numbers of species X, in millions.
- Initially there are 5×10^7 insects of species X and 3×10^7 insects of species Y

a Set up and solve two differential equations for the numbers of insects of species X and Y a number of months later.

b Which species becomes extinct? When does this happen?

4 A chemical reaction occurs between two liquids, A and B

- The rate of change of liquid A, in litres per minute, is given by the amount of liquid B, measured in litres, minus four times the amount of liquid A, measured in litres.
- In addition, liquid A is added at the rate of 7 litres per minute.
- The rate of change of liquid B, in litres per minute, is given by six times the amount of liquid A, measured in litres, minus five times the amount of liquid B, measured in litres.

a Set up two differential equations to model the amounts of liquids A and B

b Given that initially there are 3 litres of liquid A and 14 litres of liquid B, find the amounts of A and B a number of minutes later.

c Show that as the reaction progresses, the ratio of the amount of liquid A to the amount of liquid B tends to a constant ratio. Find this ratio.

5 Gazelles and lions are fighting for survival on an enclosed grass plain.

- When no lions are present, the number of gazelles increases by 100% per year.
- When lions are present, on average, each lion eats one gazelle per year.
- The rate of change of lions, in animals per year, is given by five times the number of lions plus three times the number of gazelles.
- Initially there are 120 gazelles and 80 lions.

a Set up two differential equations to model the numbers of gazelles and lions.

b Solve your equations to find the numbers of gazelles and lions a number of years later.

c Calculate the number of years taken for the gazelles to become extinct.

6 Rabbits and foxes are introduced on to an island.

- Initially there are 100 rabbits and 160 foxes.
- When no foxes are present, the number of rabbits increases by 600% per year.
- When foxes are present, on average, each fox eats one rabbit per year.
- The rate of change of foxes, in animals per year, is given by twice the number of foxes plus three times the number of rabbits.

a Set up and solve two differential equations for the numbers of rabbits and foxes a number of years later.

b Show that, over time, there will be approximately equal numbers of rabbits and foxes.

7 A forest contains bamboo plants and pandas. The pandas like to eat bamboo.

- The rate of change of bamboo plants is proportional to the number of pandas present. Three pandas eat one bamboo plant per year.
- The panda population increases by 20% per year plus 3% of the number of bamboo plants present.

- Initially there are 100 pandas and 2000 bamboo plants.

How many years pass before there are no bamboo plants left in the forest?

8 A small colony of bears feed on fish in a lake.

- When no bears are present, the number of fish would increase at a rate of 20% per year.
- When bears are present, on average, each bear eats a fifth of a fish per year.
- The rate of increase of bears per year is equal to 40% of the number of bears plus 10% of the number of fish.
- Initially there are 4 bears in the colony and 1000 fish in the lake.

Let x represent the number of bears, y represent the number of fish and t represent the time that has passed in years.

a Show that
$$\frac{d^2y}{dt^2} - 0.6\frac{dy}{dt} - 0.1y = 0$$

b Find expressions for x and y in terms of t

c Using the model, how many years pass before there are no fish in the lake?

d Give one criticism of this population model.

9 An Alaskan forest is populated by Canadian lynx and snowshoe hares.

- The rate of increase of lynx per year equals 5% of the number of hares present.
- When no lynx are present, the number of hares would increases at a rate of 120% per year.
- When lynx are present, on average, each lynx eats 1.15 hares per year.
- Initially there are 60 lynx and 500 hares.

a If L represents the number of lynx, H represents the number of hares and t represents the time that has passed in years, explain why
$$\frac{dH}{dt} = 1.2H - 1.15L$$

b Show that $\frac{d^2H}{dt^2} - 1.2\frac{dH}{dt} + 0.0575H = 0$

c Find expressions for the number of lynx and the number of hares at time t years.

d Give one criticism of this model.

10 An African nature reserve is populated by lions and zebra.

- The rate of increase of lions per year equals 20% of the number of zebra present.
- When no lions are present, the number of zebra would increase at a rate of 130% per year.
- When lions are present, on average, each lion kills 1.1 zebra per year.
- Initially there are 100 lions and 1000 zebra.

a If L represents the number of lions, Z represents the number of zebra and t represents the time that has passed in years, show that
$$\frac{d^2Z}{dt^2} - 1.3\frac{dZ}{dt} + 0.22Z = 0$$

b Find expressions for the number of lions and the number of zebras at time t years.

c Show that, for large values of t, the ratio number of lions : number of zebra $= 2 : 11$

11 A radioactive element X decays into a second radioactive element Y which in turn decays into the stable element Z

- The rate of decay of X equals 0.2 times the amount of X present.
- The rate of decay of Y equals 0.1 times the amount of Y present.
- Initially, there are 100 milligrams of X present and no Y or Z

This process can be modelled by the equations
$$\frac{dx}{dt} = -0.2x \qquad \frac{dy}{dt} = 0.2x - 0.1y \qquad \frac{dz}{dt} = 0.1y$$
where x, y and z are the masses of elements X, Y and Z respectively, measured in milligrams, and t is the time, measured in seconds.

a Explain why the equations take these forms.

b Find expressions for x, y and z at time t

c Prove that $x + y + z = 100$ for all t

Chapter summary

- An equation that involves only a first-order derivative, such as $\dfrac{dy}{dx}$, is called a **first order differential equation**.
- An equation of the form $\dfrac{dy}{dx} = F(x)G(y)$ is solved by **separating the variables** to get
$$\int F(x)\,dx = \int \frac{1}{G(y)}\,dy$$
- An equation of the form $\dfrac{dy}{dx} + P(x)y = f(x)$ is solved by multiplying throughout by an **integrating factor** $e^{\int P(x)\,dx}$
- A solution with constant(s) is called a **general solution**. Substitute given value(s) to get a **particular solution**.
- An equation that involves a second order derivative, such as $\dfrac{d^2 y}{dx^2}$, but nothing of higher order, is called a **second order differential equation**.
- To solve an equation of the form $a\dfrac{d^2 y}{dx^2} + b\dfrac{dy}{dx} + cy = f(x)$
  - First solve the **auxiliary equation** $am^2 + bm + c = 0$. There are three cases for the **complementary function** that come from the solution of the auxiliary equation.
  - Next find the **particular integral**.
  - To get the **general solution**, add together the complementary function and the particular integral.
- Differential equations that include two or more variables linked to a common additional variable are called **coupled equations**.
- To solve a system of coupled equations, use differentiation to change the differential equations into a single equation of higher order connecting just two variables.

Check and review

You should now be able to...	Review Questions
✔ Solve a first order differential equation using an integrating factor.	1
✔ Solve a first order differential equation and interpret the solution in context.	2
✔ Set up a first order differential equation from a model and solve it.	3, 4
✔ Solve a second order differential equation of the form $a\dfrac{d^2 y}{dx^2} + b\dfrac{dy}{dx} + cy = f(x)$	5
✔ Solve a second order differential equation and interpret the solution in context.	6
✔ Set up a second order differential equation from a model and solve it.	7
✔ Set up and solve a system of coupled equations.	8

1 Find the solution to the differential equation $\dfrac{dy}{dx}+y\cot x=\cos^3 x$ for which $y=1$ at $x=\dfrac{\pi}{2}$, giving your answer in the form $y=f(x)$

2 As part of an industrial process, salt is dissolved in a liquid. The mass, S kg, of the salt dissolved, t minutes after the process begins, is modelled by the equation
$$\frac{dS}{dt}+\frac{2S}{300-t}=\frac{1}{2}, 0\le t<300$$
Given that $S=0$ when $t=0$,

a Find S in terms of t

b Calculate the maximum mass of salt that the model predicts will be dissolved in the liquid.

3 At time t minutes, the rate of change of temperature of a cooling liquid is proportional to the temperature, $T°C$, of that liquid at that time. Initially $T=100$

a Show that $T=100e^{-kt}$, where k is a positive constant.

Given also that $T=25$ when $t=6$,

b Show that $k=\dfrac{1}{3}\ln 2$

c Calculate the time at which the temperature of this liquid will reach $5°C$.

4 A woman and her bicycle have a combined mass of 120 kg. The woman starts from rest, exerting a constant force of 600 N. When her velocity is v m s^{-1} her motion is subject to a resistive force of $6v^2$ N.

a Show that $100-v^2=20\dfrac{dv}{dt}$

b Calculate the time taken by the woman to reach a speed of 9 m s^{-1} from rest.

c Calculate the distance travelled by the woman in reaching a speed of 9 m s^{-1} from rest.

5 For the differential equation
$$\frac{d^2y}{dx^2}+4\frac{dy}{dx}+13y=90e^{4x}$$

a Find the general solution,

b Find the particular solution for which $y(0)=6$ and $y'(0)=-3$

6

A box, B, is attached to one end of a light elastic string. The box is initially at rest on a smooth horizontal table, when the other end of the string, A, starts to move away from B with constant velocity 2 m s^{-1}. The displacement, y metres, of B from its initial position, at a time t seconds from the start, satisfies the differential equation
$$\frac{d^2y}{dt^2}+16y=8t$$

a Find an expression for y in terms of t

b Calculate the time when B is next at rest.

c Sketch the graph of y against t

7 A particle, P, of mass 4 kg moves along the horizontal x-axis under the action of a force directed towards the origin. At time t seconds, when the displacement of P from the origin is x metres, P is moving with a speed of v m s^{-1} and the force has magnitude $9x$ N. The particle is also subject to a resistive force of $12v$ N.

a Show that $4\dfrac{d^2x}{dt^2}+12\dfrac{dx}{dt}+9x=0$

Initially $x=6$ and $\dfrac{dx}{dt}=-11$

b Show that $x=e^{-\frac{3t}{2}}(6-2t)$

c Write down the time when P passes through the origin.

d Find the speed of P at that time.

8 A particle moves in the plane such that at time $t\ge 0$, it has coordinates (x,y)

- The rate of change of x equals the difference in the x- and y-coordinates minus three times the elapsed time.
- The rate of change of y equals the height of the particle above the line $y=4x+3$
- The particle is initially at $(2,4)$

a Show that
$$\frac{d^2x}{dt^2}-2\frac{dx}{dt}-3x=3t$$

b Find the position of the particle at time t

659

PURE

Research

Bungee jumping was first attempted in 1979, by students jumping off the Clifton suspension bridge in Bristol. Their inspiration came from vine jumping, as practised by the islanders of Vanuatu in the Pacific. The bungee cords used in Northern and Southern hemispheres tend to have different construction, with the Northern hemisphere cords giving a harder, sharper bounce than those in the southern hemisphere.

Try modelling the motion of a bungee jumper using differential equations where appropriate. Research the sorts of values you might use in this motion, including different values for the elasticity of the ropes in the different hemispheres.

Investigation

The diagram shows a potential situation that can be developed as a water feature.

Water flows into the top tank and, once it reaches the holes in the tank, it flows into the middle tank. As the water in the middle tank fills and reaches the holes in this tank, it then flows into the lower tank, which starts to fill. Water continues to fill the lower tank until it overflows.

Investigate this situation using differential equations. Develop your own mathematical model. Assume that that the rate of flow of water out of a tank depends on the volume of water in the tank. In other words, $\frac{dV}{dt} = g(V)$.

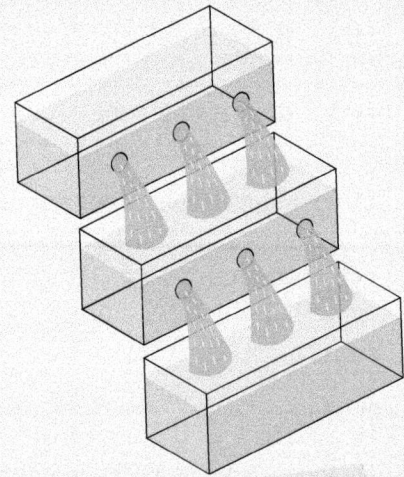

Start your investigation with a function such as
$g(V) = -kV$

Try different functions to develop a design that would be attractive.

Research

Engineers use differential equations when designing a rocket to launch a satellite into space. These equations are derived from Newton's Laws of motion. Important factors that had to be considered when the differential equations were developed included the changing mass of the rocket and the change in the pull of gravity and air resistance as the rocket gets further away from the Earth. Research the use of differential equations to model the motion of a rocket.

20 Assessment

1 Find the general solution to the differential equation $\dfrac{dy}{dx} = e^{x+y}\cos x$ **[5 marks]**

2 Solve the differential equation $\dfrac{1}{\ln x}\dfrac{dy}{dx} + e^{y} = 0$ given that when $x=1, y=0$ **[6]**

3 Given that $y\dfrac{dy}{dx} - x\ln x = 0$

  a Find the general solution to the equation, **[4]**

  b Find the particular solution if when $x=1, y=4$ **[2]**

4 The number of sparrows nesting in a wooded area is thought to be declining. The rate at which the population is changing is known to depend on the population, P, at any given time and on the space available to build nests $(200 - P)$

An initial model predicts that $\dfrac{dP}{dt} = kP + (200 - P)$, where k is a constant of proportionality and t is the number of years.

Initially, there were 550 birds in the area.

  a Taking the value of k to be 5, find the general solution to the differential equation, expressing P as an explicit function of t **[6]**

  b Use the initial conditions to find the particular solution. **[2]**

  c What does the model predict will happen to the number of sparrows in the wood after a number of years? **[2]**

  d Comment on the suitability of the model for predicting the future population of sparrows. **[1]**

5 Find the general solution to the differential equation $x\dfrac{dy}{dx} + 2y = x^2$ **[4]**

6 Solve the differential equation $\dfrac{dy}{dx} - y\tan x = 4x^2$ given that when $x=0, y=1$ **[6]**

7 Given that $2x\dfrac{dy}{dx} + y = \ln x$

  a Find the general solution to the equation. **[5]**

  b Find the particular solution if when $x=e^2, y=e^{-1}$ **[2]**

8 A family of differential equations takes the form

$$2\frac{d^2y}{dx^2} + 8\frac{dy}{dx} + ky = 0 \text{ where } k \text{ is a constant.}$$

Find the general solution to the equation when

a $k = 6$ **[4]**   **b** $k = 8$ **[4]**   **c** $k = 10$ **[4]**

9 Find the particular solution of $2\frac{d^2y}{dx^2} + 3\frac{dy}{dx} - 2y = 0$ given that when $x = 0$, $y = 5$ and $\frac{dy}{dx} = -5$ **[7]**

10 Find the particular solution of $9\frac{d^2y}{dx^2} - 6\frac{dy}{dx} + y = 0$ given that when $x = 0$, $y = 6$ and $\frac{dy}{dx} = 4$ **[7]**

11 Find the particular solution of $\frac{d^2y}{dx^2} + 2\frac{dy}{dx} + 5y = 0$ given that when $x = 0$, $y = 10$ and $\frac{dy}{dx} = 2$ **[7]**

12 Find the general solution of $\frac{d^2y}{dx^2} + \frac{dy}{dx} - 12y = 12x + 1$ **[7]**

13 Find the particular solution of $\frac{d^2y}{dx^2} - 4\frac{dy}{dx} + 13y = 40\sin x$ given that when $x = 0$, $y = 5$ and $\frac{dy}{dx} = -1$ **[7]**

14 The force acting on a compressed spring is given by the equation $m\frac{d^2x}{dt^2} = -\frac{7}{2}\frac{dx}{dt} - (5x - 6e^{-t})$, where m kg is the mass of the spring and x m is the displacement from its initial position at time t seconds.

a If $m = 0.5$ kg, show that the equation can be written as $\frac{d^2x}{dt^2} + 7\frac{dx}{dt} + 10x = 12e^{-t}$ **[2]**

b Find the particular solution of this equation, given that when $t = 0$, $x = 0$ and $\frac{dx}{dt} = 2$ **[11]**

c Comment on the nature of the damping of the spring. Give a reason for your answer.   **[2]**

d Sketch a graph of x against t **[2]**

15 The motion of an air particle when a musical note is played by a wind instrument can be modelled by the equation $\dfrac{d^2x}{dt^2} + 9x - 8\sin t = 0$, where x m is the displacement of the particle from its rest position.

   a Given that when $t = \dfrac{\pi}{2}$, $x = 0$ and $\dfrac{dx}{dt} = 0$, find the particular solution to the differential equation. **[9]**

   b Use the formula $\sin A + \sin B = 2\sin\left(\dfrac{A+B}{2}\right)\cos\left(\dfrac{A-B}{2}\right)$ to rewrite your answer to part **a** and find the times at which the displacement of the particle from its rest position is zero. **[4]**

16 Money placed in a savings account will grow in direct proportion to the amount of money in the bank.

   Initially £1000 is placed in the account. At the end of year 1, there is £1005 in the account. Let £A represent the amount after a time t years.

   a Form a differential equation to model the situation. **[1]**

   b Find the particular solution to the equation. **[6]**

   c How much will be in the bank after 5 years? **[1]**

17 The voltage in a circuit containing a resistor and an inductor connected in series with a power supply is given by the equation

$$V = RI + L\frac{dI}{dt}$$

where I amps is the current, V volts is the voltage of the power supply, C farads is the capacitance of the capacitor and R ohms is the resistance.

   a Show that the equation can be written as $\dfrac{dI}{dt} + \dfrac{RI}{L} = \dfrac{V}{L}$ **[1]**

   b Given that $I = 0$ when $t = 0$, solve the differential equation to find I as a function of t **[9]**

   If $R = 50$ ohms, $V = 100$ V and $L = 10$ H

   c Find the value of I when $t = 0.1$ seconds. **[2]**

   d What happens to the current as t becomes large? **[2]**

   e Sketch a graph of I against t **[2]**

18 A particle moves so that at time t seconds, it is x units from the origin.

Its motion is modelled by $100\dfrac{d^2x}{dt^2} = -25x$

Initially $x = 0$ when $t = 0$. When $t = \pi$, $x = 4$

    a   Find the particular solution of the equation.   **[6]**

    b   Find the value of x when $t = \dfrac{\pi}{3}$ **[1]**

19 A damped oscillating system is modelled by the differential equation

$\dfrac{d^2x}{dt^2} + 0.3\dfrac{dx}{dt} + 0.15^2 x = 0$

    a   Explain why the damping will be critical.   **[3]**

    b   Given that $x = 0.5$ when $t = 0$ and $\dfrac{dx}{dt} = 0.425$ when $t = 0$, solve the differential equation.   **[8]**

    c   Sketch a graph of x against t **[2]**

20 A particle of mass m attached to a spring is subject to a damping force proportional to its velocity given by $-5m\dfrac{dx}{dt}$ and a tension in the spring given by $-6mx$. A disturbing

force $3m\sin 2t$ is applied to the particle.

    a   Show that $\dfrac{d^2x}{dt^2} + 5\dfrac{dx}{dt} + 6x = 3\sin 2t$ **[2]**

    b   Find the complementary function of this differential equation.   **[3]**

    c   Find the particular integral.   **[8]**

    d   Find the particular solution given that $x = 0$ when $t = 0$ and $\dfrac{dx}{dt} = 1$ when $t = 0$ **[6]**

21 A reaction vessel contains two chemicals, X and Y

    •   The rate of increase of X, in grams per minute, is given by the amount of chemical X plus eight times the amount of chemical Y, both in grams.

    •   The rate of decrease of Y, in grams per minute, is given by the amount of chemical X plus five times the amount of chemical Y, both in grams.

    •   Chemical X is added to the reaction vessel at a rate of one gram per minute.

    •   Chemical Y is removed from the reaction vessel at a rate of four grams per minute.

    a   Set up two differential equations to describe the amounts of chemicals X and Y in the reaction vessel.   **[5]**

    b   Initially the vessel contains one gram of Y and no X
        Find the amounts of X and Y at later times.   **[17]**

21

Numerical methods

Those working in computer animation need to understand differential equations. Differential equations are used to describe the motion of objects that are being animated. These equations become complex very quickly and so numerical methods are used to 'solve' them. High-powered computers process the equations rapidly and efficiently so that the animation is smooth and effective.

The mathematical and computing methods that underpin the development of computer animation have applications in many fields throughout science and technology. For example, they can be used to model, and therefore understand, the flow of liquids and gases in aero-engines and nuclear reactors. They can also help designers of cars ensure that their designs are aerodynamic.

Orientation

What you need to know	What you will learn	What this leads to
Maths Ch16 • Differential equations. **Maths Ch17** • Numerical integration.	• How to use Simpson's rule. • How to use Euler's method.	**Careers** • Computer animation. • Aerodynamic engineering.

Fluency and skills

You find the area under a curve by integrating the equation of the curve between appropriate limits.

For example, in the top diagram the area in blue, under the function f(x), between the values $x = 2$ and $x = 7$, is found by calculating $\int_{2}^{7} f(x)\,dx$

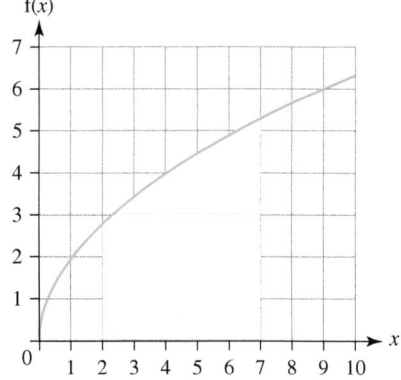

However, there are many functions which you cannot integrate using straightforward methods. Simple examples are $y = \cos\frac{1}{2}x$, $y = e^{x^2}$ and $y = e^{\frac{1}{x}}$

The same problem arises if the graph is from gathered data and so you do not know the equation of the graph.

Even if the functions cannot be integrated, they can still be drawn, and so the area under the curve will exist. For example, the middle diagram shows the curve of $f(x) = e^{\frac{1}{x}}$ with the area beneath the curve between $x = 2$ and $x = 7$ shaded.

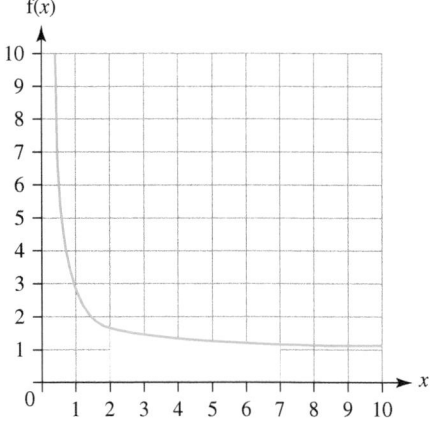

One way of finding an area like this is to use the **mid-ordinate rule**. Using this rule you divide the area required into vertical rectangles where the height of each rectangle is the y value (ordinate) in the middle of the rectangle. For example, in the function shown in the bottom diagram, the area from $x = 0$ to $x = 10$ is divided into 5 rectangles.

The sum of the areas of all the rectangles is an approximation to the area under the curve between $x = 0$ and $x = 10$, i.e. $\int_{0}^{10} f(x)\,dx$

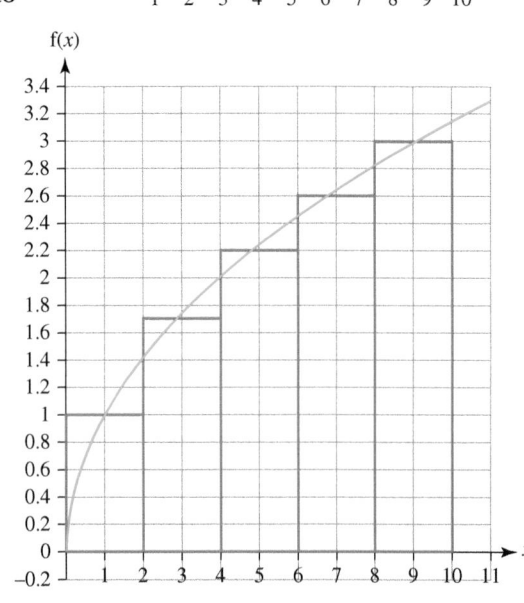

At the top of each rectangle, a small area on the left, above the curve, is added into the area, but this is compensated for by the small area on the right, which is left out. This helps to minimise any errors. The sum of the area of the rectangles is often very close to the actual area under the curve. Estimating the values of the ordinates from the graph at $x = 1, 3, 5, 7$ and 9, we can sum the rectangle areas as
$(2 \times 1) + (2 \times 1.7) + (2 \times 2.2) + (2 \times 2.6) + (2 \times 3) = 21$

The **mid-ordinate rule** states that if you divide the area to be estimated under the function, from a to b, into n rectangles, then the width of each rectangle is $\dfrac{b-a}{n}$

If you then let the ordinates at the centre of each of these rectangles be $y_1, y_2, \ldots, y_n$, then the estimated area is given by the formula $A \quad \left(\dfrac{b-a}{n}\right)(y_1 + y_2 + y_3 + \ \ + y_n)$

In the example shown above, y_1, y_2, y_3, y_4 and y_n are the ordinates when $x = 1, 3, 5, 7$ and 9

Key point

The more rectangles you use the more accurate your estimate will be.

The function shown is $y = \sqrt{x}$ and, using calculus, $\displaystyle\int_0^{10} \sqrt{x}\ dx = 21.1$ (3 s.f.).

Example 1

a Draw the graph of $y = 10x - x^2$

b Divide the area between the curve and the x-axis into 10 rectangles and draw them on your graph.

c Use your diagram to estimate the area between the curve and the x-axis.

d Use calculus to find the exact area between the curve and the x-axis.

e Calculate the relative error between your estimate and the exact area.

a
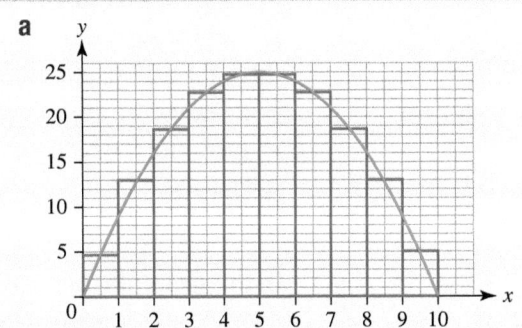

b The width of each rectangle = 1

c The sum of the required ordinates is

$(10 \times 0.5 - 0.5^2) + (10 \times 1.5 - 1.5^2) + (10 \times 2.5 - 2.5^2)$
$+ (10 \times 3.5 - 3.5^2) + (10 \times 4.5 - 4.5^2) + (10 \times 5.5 - 5.5^2)$
$+ (10 \times 6.5 - 6.5^2) + (10 \times 7.5 - 7.5^2) + (10 \times 8.5 - 8.5^2)$
$+ (10 \times 9.5 - 9.5^2)$

Use the mid-ordinate rule.

$= 12.75 + 18.75 + 22.75 + 24.75 + 22.75 + 18.7 + 12.75 + 4.75$

Hence, $A = 162.75$

d $\displaystyle\int_0^{10} (10x - x^2)\, dx = \left[5x^2 - \frac{x^3}{3} \right]_0^{10}$

Integrate.

$= \left[5(10)^2 - \frac{(10)^3}{3} \right] - \left[5(0)^2 - \frac{(0)^3}{3} \right]$

$= 166\frac{2}{3}$

Apply the limits and evaluate.

e Relative error $= \dfrac{\text{actual error}}{\text{true value}}$

$= \dfrac{162.75 - 166.67}{166.67}$

$= -0.0235$ (3 s.f.)

The mid-ordinate rule is a simple formula for estimating the area under a curve. Thomas Simpson, in 1743, published a rule which set out to give a more accurate answer. Simpson's rule minimises the discrepancies at the top of each rectangle by replacing the mid-ordinates with a quadratic curve.

Take three points on this curve, which has equation $y = ax^2 + bx + c$

The area under the parabola is given by

$$A = \int_{-d}^{d} (ax^2 + bx + c)\, dx$$

$$= \left[\frac{ax^3}{3} + \frac{bx^2}{2} + cx \right]_{-d}^{d}$$

$$= \frac{2}{3} ad^3 + 2cd$$

> Just as you can always draw a straight line between two points, you can always draw a quadratic curve between three points.

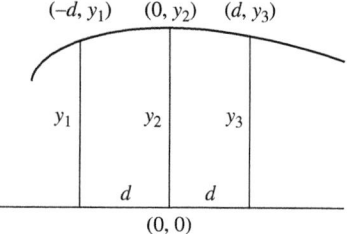

Substituting for the three points, you get

$y_1 = ad^2 - bd + c$ (1)

$y_2 = c$ (2)

$y_3 = ad^2 + bd + c$ (3)

From these equations, if you work out $(1) + 4 \times (2) + (3)$, you get

$y_1 + 4y_2 + y_3 = 2ad^2 + 6c$

Multiplying through by $\dfrac{d}{3}$ you get

$$\frac{d}{3}(y_1 + 4y_2 + y_3) = \frac{d}{3}(2ad^2 + 6c)$$

$$= \frac{2}{3} ad^3 + 2cd$$

Therefore the area under this parabola is $\dfrac{d}{3}(y_1 + 4y_2 + y_3)$

> You don't have to find the actual equation of the parabola which passes through these three points because that is governed by the values y_1, y_2, y_3 and d which you have used.

So an approximation for the area between a curve and the x-axis passing through the points $(-d, y_1)$, $(0, y_2)$ and (d, y_3) is given by the formula $\dfrac{d}{3}(y_1 + 4y_2 + y_3)$

This is known as **Simpson's rule**.

The diagram shows the area under a curve divided into eight 'strips' or, more importantly, four **pairs** of strips.

> In this example there are 3 ordinates covering two equally spaced intervals. However, normally you will need the area under a curve for a larger range of ordinates.

The area under the first two strips is approximated by $\dfrac{d}{3}(y_0 + 4y_1 + y_2)$

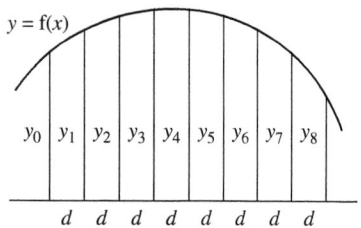

The area under the second two strips is approximated by $\dfrac{d}{3}(y_2 + 4y_3 + y_4)$

The area under the third two strips is approximated by $\dfrac{d}{3}(y_4 + 4y_5 + y_6)$

The area under the last two strips is approximated by $\dfrac{d}{3}(y_6 + 4y_7 + y_8)$

Therefore the area under the whole section shown is approximated by

$$A = \frac{d}{3}(y_0 + 4y_1 + y_2) + \frac{d}{3}(y_2 + 4y_3 + y_4) + \frac{d}{3}(y_4 + 4y_5 + y_6) + \frac{d}{3}(y_6 + 4y_7 + y_8)$$

$$= \frac{d}{3}(y_0 + 4y_1 + 2y_2 + 4y_3 + 2y_4 + 4y_5 + 2y_6 + 4y_7 + y_8)$$

$$= \frac{d}{3}[(y_0 + y_8) + 4(y_1 + y_3 + y_5 + y_7) + 2(y_2 + y_4 + y_6)]$$

This is typically much more accurate than trying to divide the area into just two strips.

> You could remember this as $\frac{d}{3} \times$ [(sum of first and last ordinates) $+ 4 \times$ (sum of the odd ordinates) $+ 2 \times$ (sum of the even ordinates)]

Therefore a generalised version of Simpson's rule is

Key point

If the area required is divided into n strips of equal thickness, **where n is even,**

then $\int_{x_0}^{x_n} f(x)\, dx \simeq \frac{d}{3}[(y_0 + y_n) + 4(y_1 + y_3 + y_5 + \cdots + y_{n-1}) + 2(y_2 + y_4 + y_5 + \cdots + y_{n-2})]$

Example 2

a Using Simpson's rule with ten strips, find the approximate area for the function and range given in Example 1

b Comment on your result.

a $d = 1$ •————————————————————— Define d

x	$10x - x^2$	y_x
0	$10(0) - (0)^2$	0
1	$10(1) - (1)^2$	9
2	$10(2) - (2)^2$	16
3	$10(3) - (3)^2$	21
4	$10(4) - (4)^2$	24
5	$10(5) - (5)^2$	25
6	$10(6) - (6)^2$	24
7	$10(7) - (7)^2$	21
8	$10(8) - (8)^2$	16
9	$10(9) - (9)^2$	9
10	$10(10) - (10)^2$	0

$$A = \frac{d}{3}[(y_0 + y_{10}) + 4(y_1 + y_3 + y_5 + y_7 + y_9) + 2(y_2 + y_4 + y_6 + y_8)]$$

$$= \frac{1}{3}[(0+0) + 4(9 + 21 + 25 + 21 + 9) + 2(16 + 24 + 24 + 16)]$$ •——— Substitute y values.

$$= \frac{1}{3}[(0) + (340) + (160)]$$

$$= 166\frac{2}{3}$$ •————————————————————— Evaluate the area.

Hence $\int_{0}^{10} (10x - x^2)\, dx \simeq 166\frac{2}{3}$

b The function given in Example 1 is, in fact, a quadratic, so the answer should be exact.

1 Use the mid-ordinate rule with the number of rectangles stated to estimate each of these areas to 3 significant figures.

a $\int_{1}^{11} e^{\frac{1}{x}}\, dx$ with 10 rectangles

b $\int_{0}^{2\pi} \cos^2 x\, dx$ with 8 rectangles

c $\int_{\frac{\pi}{4}}^{2\pi} |\ln(x)\sin(x)|\, dx$ with 7 rectangles

d $\int_{-2}^{2} e^{x^2}\, dx$ with 8 rectangles

e $\int_{1}^{11} \left(\dfrac{x}{1+e^x}\right) dx$ with 10 rectangles

f $\int_{-\frac{\pi}{2}}^{\frac{\pi}{2}} \sqrt{\cos x}\, dx$ with 10 rectangles

g $\int_{-\pi}^{\pi} \left(\dfrac{\sin x}{x}\right) dx$ with 10 rectangles

h $\int_{1}^{5} \left(\dfrac{\ln x}{e^x}\right) dx$ with 8 rectangles

2 Use Simpson's rule with an appropriate number of strips to estimate these areas to 4 s.f.

a $\int_{1}^{11} e^{\frac{1}{x}}\, dx$

b $\int_{2}^{4} |\ln(x)\tan(x)|\, dx$

c $\int_{-2}^{2} e^{-x^2}\, dx$

d $\int_{1}^{4} \left(\dfrac{-x}{1-e^x}\right) dx$

e $\int_{0}^{\frac{\pi}{3}} \sqrt{\tan x}\, dx$

f $\int_{\frac{3\pi}{2}}^{\frac{5\pi}{2}} \left(\dfrac{\cos x}{x}\right) dx$

g $\int_{1}^{5} \left(\dfrac{\ln x}{e^{-x}}\right) dx$

3 Use the mid-ordinate rule with an appropriate number of rectangles to estimate these areas to 4 s.f. In each case, also find the exact answer by integration and hence calculate the relative error.

a $\int_{8}^{12} (x^2 - 6x)\, dx$

b $\int_{0}^{\pi} (\sin x)\, dx$

c $\int_{1}^{6} (-x^3 + 5x^2 + 8x - 12)\, dx$

d $\int_{1}^{11} \ln(5x)\, dx$

4 Use Simpson's rule with an appropriate number of strips to estimate these areas to 4 s.f. In each case, also find the exact answer by integration and hence calculate the relative error.

a $\int_{2}^{4} \sec^2 x\, dx$

b $\int_{2}^{12} \left(\dfrac{5}{x}\right) dx$

c $\int_{1}^{7} \left(\dfrac{5}{x^{\frac{3}{2}}}\right) dx$

d $\int_{0}^{10} \left(\dfrac{16x}{x^2 + 1}\right) dx$

Reasoning and problem-solving

Strategy

To find an approximate area under a curve using Simpson's rule

(1) Divide the required area into an appropriate even number of strips of equal width, d

(2) Calculate the values of the ordinates for each x-value.

(3) Complete the area approximation by substituting into the formula

$\dfrac{d}{3} \times [(\text{sum of first and last ordinates}) + 4 \times (\text{sum of odd ordinates}) + 2 \times (\text{sum of even ordinates})]$

Example 3

The cross-section of a vase is circular. The radius r cm, at any point h cm above the base, is shown in this table. The vase is 40 cm high.

h	0	5	10	15	20	25	30	35	40
r	6	5	11	13	15	13	11	5	6

Estimate the volume of the vase in litres to the nearest ml.

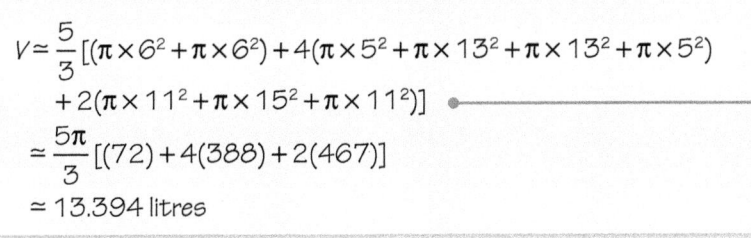

$V \approx \dfrac{5}{3}[(\pi \times 6^2 + \pi \times 6^2) + 4(\pi \times 5^2 + \pi \times 13^2 + \pi \times 13^2 + \pi \times 5^2)$

$\qquad + 2(\pi \times 11^2 + \pi \times 15^2 + \pi \times 11^2)]$

$\approx \dfrac{5\pi}{3}[(72) + 4(388) + 2(467)]$

≈ 13.394 litres

③ Substitute into the formula.

The volume of the vase is found by finding the area under a graph of circular cross-sectional area plotted against height.

Example 4

The equation of a circle, radius 12, is given by the formula $x^2 + y^2 = 144$

Using Simpson's rule in the first quadrant with 12 equal intervals, find an estimate for π correct to 2 d.p.

Find also the relative error, to 1 s.f., of your estimate from the 2 d.p. value of π

$x^2 + y^2 = 144 \Rightarrow y^2 = 144 - x^2$, so $y = \sqrt{144 - x^2}$ and $A = \displaystyle\int_{0}^{12} (\sqrt{144 - x^2})\,dx$

x	0	1	2	3	4	5	6	7	8	9	10	11	12
y	12	11.958	11.832	11.619	11.314	10.909	10.392	9.747	8.944	7.937	6.633	4.796	0

$A \approx \dfrac{1}{3}[(12 + 0) + 4(11.958 + 11.619 + 10.909 + 9.747 + 7.937 + 4.796)$

$\qquad + 2(11.832 + 11.314 + 10.392 + 8.944 + 6.633)]$

$\approx \dfrac{1}{3}[(12) + (227.861) + (98.23)]$

$\approx \dfrac{1}{3}(338.091)$

≈ 112.697

$\dfrac{1}{4} \times \pi \times 12^2 \approx 112.697$

$\qquad 36\pi \approx 112.697$

$\qquad \pi \approx \dfrac{112.697}{36}$

$\qquad \approx 3.13$

Relative error $= \dfrac{3.13 - 3.14}{3.14} = -0.003$

This area is an approximation to a quarter of the area of a circle.

1 The table shows the velocity of a particle over 9 seconds.

Time (s)	1	2	3	4	5	6	7	8	9
Velocity (m s⁻¹)	4.8	19	44	78	122	176	240	313	397

Use Simpson's rule to calculate how far the particle has travelled over this time.

2 a Find the first three non-zero terms in the expansion of $\cosh x = \left(\dfrac{e^x + e^{-x}}{2} \right)$

  b Use Simpson's rule with this expansion to estimate $\displaystyle\int_0^{0.6} \cosh x \, \mathrm{d}x$ to 5 d.p.

3 a Write down the first four terms in the expansion of $\sin x$

  b Use Simpson's rule with this expansion to estimate $\displaystyle\int_0^1 \sin x \, \mathrm{d}x$ to 4 d.p.

4 a Prove by integration that the area between the curve $y = \tan x$ and the coordinate axes between $x = 0$ and $x = \dfrac{\pi}{4}$ is $\ln \sqrt{2}$

  b Use Simpson's rule with 8 strips to calculate an approximation to this value and find the relative error.

5 Use the binomial theorem to expand $(1 + x^2)^{\frac{1}{3}}$ as far as the term in x^6

Hence find $\displaystyle\int_0^{0.6} (1 + x^2)^{\frac{1}{3}} \, \mathrm{d}x$ to 4 d.p.

  a By integration,

  b By using Simpson's rule with six intervals.

6 Use the mid-ordinate rule to estimate $\displaystyle\int_0^{10} \left(\sqrt[3]{1 + \dfrac{x^3}{8}} \right) \mathrm{d}x$ to 5 d.p.

7 a Use the mid-ordinate rule and Simpson's rule to estimate $\displaystyle\int_0^1 \left(\dfrac{24x}{4 + x^2} \right) \mathrm{d}x$

  b Find $\displaystyle\int_0^1 \left(\dfrac{24x}{4 + x^2} \right) \mathrm{d}x$ by integration.

  c Find the relative error for the calculation in part **a**.

8 $\int \dfrac{1}{x}\, \mathrm{d}x = \ln(x)$. Use Simpson's rule with six strips between $x = 1$ and $x = 3$ to find an estimate for $\ln(3)$

9 The coordinates of the centre of mass of a lamina are found by using the formulae

$$\bar{x} = \dfrac{\int xy\, \mathrm{d}x}{\int y\, \mathrm{d}x}, \quad \bar{y} = \dfrac{\int \dfrac{y^2}{2}\, \mathrm{d}x}{\int y\, \mathrm{d}x}$$

A lamina in the shape of the quadrant of the ellipse $\dfrac{x^2}{16} + \dfrac{y^2}{9} = 1$ is formed enclosing the curve and the positive x- and y-axes. Use direct integration to calculate, between appropriate limits, $\int xy\, \mathrm{d}x$ and $\int \dfrac{y^2}{2}\, \mathrm{d}x$ and Simpson's rule with an appropriate number of strips to calculate $\int y\, \mathrm{d}x$ and hence find the coordinates of the centre of mass of the lamina.

Fluency and skills

Up to this point, all the differential equations that you have met could be solved. However, almost all first order differential equations cannot be solved using calculus and algebra. When this is the case, you have to use numerical methods such as Euler's method to find approximate solutions to the differential equation.

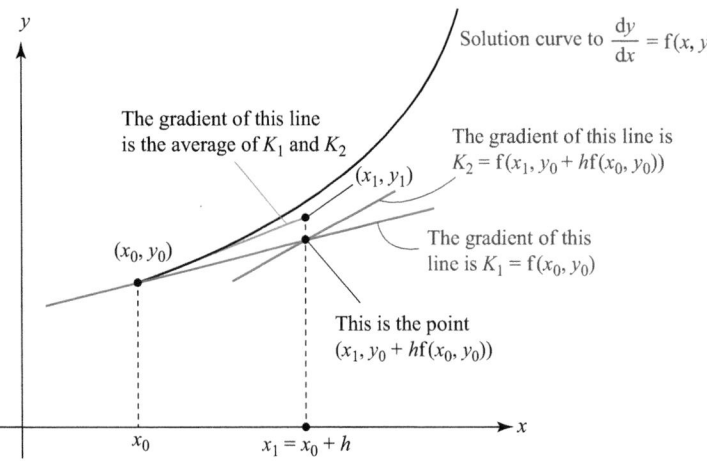

Assume you wish to solve the differential equation $\dfrac{dy}{dx} = f(x, y)$, where you know one point which lies on the curve, say $y = y_0$ when $x = x_0$

Euler's method enables you to estimate successive points (x_1, y_1), (x_2, y_2), (x_3, y_3), … which lie approximately on the solution curve of the differential equation.

Since $\dfrac{dy}{dx} = f(x_0, y_0)$ at the point (x_0, y_0), then the value of $f(x_0, y_0)$ is the gradient of the tangent to the solution curve at that point (shown in red on the diagram). Therefore, if x increases by a small amount, h, then the y value of that straight line will increase by an amount $hf(x_0, y_0)$

This is because on a straight line the change in y is equal to the change in x times the gradient.

If h is small, the solution curve should stay close to the tangent line, and so the point $(x_1, y_1) = (x_0 + h, y_0 + hf(x_0, y_0))$ should be close to the correct solution.

The accuracy is improved the smaller the step length, h and when the start and end points are close.

Clearly, this process can be repeated, so successive approximations to points on the solution curve are given by $x_2 = x_1 + h$, $y_2 = y_1 + hf(x_1, y_1)$; $x_3 = x_2 + h$, $y_3 = y_2 + hf(x_2, y_2)$ and so on.

> **Key point**
>
> If $x_{r+1} = x_r + h$ then $y_{r+1} = y_r + hf(x_r, y_r)$

This is called Euler's method, named after the Swiss mathematician, Leonhard Euler (1707–1783).

Euler approximations are above the curve when it is concave and below when it is convex.

Example 1

PURE

a $\dfrac{dy}{dx} = 4x^3$

Taking (x_0, y_0) as $(1, 1)$ and step length, h, as 0.5, use Euler's method to calculate the approximate value of the curve when $x = 2$

b Find the relative error of this value.

c Change the step length to $h = 0.1$ and find the new approximate value of the curve when $x = 2$ and the resulting relative error.

a

r	x_r	y_r
0	1	1
1	1.5	$1 + 0.5 \times 4 \times 1^3 = 3$
2	2	$3 + 0.5 \times 4 \times 1.5^3 = 9.75$

The relative value required is 9.75

In each calculation,
$y_{r+1} = y_r + hf(x_r, y_r)$.

b $\dfrac{dy}{dx} = 4x^3$ so $F(x) = x^4$

The constant of integration equals 0 due to the point (1, 1) lying on the solution curve.

Hence, when $x = 2$, the accurate value is $2^4 = 16$

Relative error $= \dfrac{9.75 - 16}{16} = -0.390\,625$

c

r	x_r	y_r
0	1	1
1	1.1	$1 + 0.1 \times 4 \times 1^3 = 1.4$
2	1.2	$1.4 + 0.1 \times 4 \times 1.1^3 = 1.9324$
3	1.3	$1.9324 + 0.1 \times 4 \times 1.2^3 = 2.6236$
4	1.4	$2.6236 + 0.1 \times 4 \times 1.3^3 = 3.5024$
5	1.5	$3.5024 + 0.1 \times 4 \times 1.4^3 = 4.6$
6	1.6	$4.6 + 0.1 \times 4 \times 1.5^3 = 5.95$
7	1.7	$5.95 + 0.1 \times 4 \times 1.6^3 = 7.5884$
8	1.8	$7.5884 + 0.1 \times 4 \times 1.7^3 = 9.5536$
9	1.9	$9.5536 + 0.1 \times 4 \times 1.8^3 = 11.8864$
10	2.0	$11.8864 + 0.1 \times 4 \times 1.9^3 = 14.63$

The value required is 14.63

Relative error $= \dfrac{14.63 - 16}{16} = -0.085\,625$

Find out how to use a spreadsheet, your graphical calculator or computer software to do the calculations in part **c** quickly and simply.

Example 2

$\dfrac{dy}{dx} = x + 2y - 3xy$ and $y = 2$ when $x = 1$

Use Euler's method to estimate the value of y when $x = 1.5$ using a step length, h, of 0.1
Give your answer to 4 d.p.

r	x_r	y_r
0	1	2
1	1.1	$2 + 0.1 \times (1 + 2 \times 2 - 3 \times 1 \times 2) = 1.9$
2	1.2	$1.9 + 0.1 \times (1.1 + 2 \times 1.9 - 3 \times 1.1 \times 1.9) = 1.763$
3	1.3	$1.763 + 0.1 \times (1.2 + 2 \times 1.763 - 3 \times 1.2 \times 1.763)$ $= 1.60092$
4	1.4	$1.60092 + 0.1 \times (1.3 + 2 \times 1.60092 - 3 \times 1.3 \times 1.60092)$ $= 1.4267452$
5	1.5	$1.4267452 + 0.1 \times (1.4 + 2 \times 1.4267452 - 3 \times 1.4 \times 1.4267452)$ $= 1.252861256$

$y_5 = 1.2529$ (4 d.p.)

The improved Euler method uses the average of the gradients at the initial point (A) and at the end point of the step (B). As before, the curve has function $F(x)$ and the gradient function at any point is given by $f(x)$

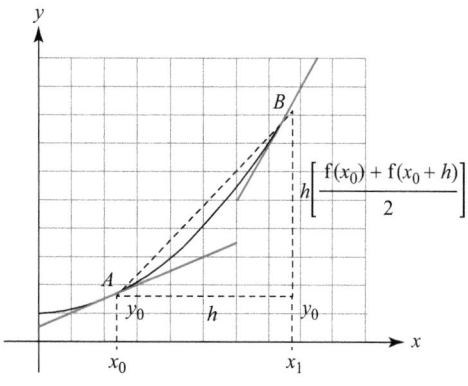

Let A be (x_0, y_0) and B be (x_1, y_1). Therefore, using Euler's method,

$x_1 = x_0 + h$ and $y_1 = y_0 + hf(x_0, y_0)$, so, the improved method gives

$y_1 = y_0 + \dfrac{hf(x_0, y_0) + hf((x_0 + h), (y_0 + hf(x_0, y_0)))}{2}$

If you write k_1 as $hf(x_0, y_0)$ and k_2 as $hf(x_0 + h, y_0 + k_1)$, you get the

improved Euler formula $y_1 = y_0 + \dfrac{1}{2}(k_1 + k_2)$. So, in general,

It is helpful to note that, in effect, working out k_2 is just using the ordinary Euler method on the right-hand boundary.

Key point

$y_{r+1} = y_r + \dfrac{1}{2}(k_1 + k_2)$

$k_1 = hf(x_0, y_0)$ and $k_2 = hf(x_0 + h, y_0 + k_1)$

Example 3

PURE

a Estimate the value of y when $x = 0.2$ for the differential equation $\dfrac{dy}{dx} = 3 + xy$ given that $y = 2$ when $x = 0$

b Use a spreadsheet or your graphical calculator or computer software, to do these calculations quickly and simply.

a $\dfrac{dy}{dx} = 3 + xy$ so $(x_0, y_0) = (0, 2)$ and $h = 0.1$

Step 1

$k_1 = 0.1(3 + (0)(2)) = 0.3$

$k_2 = 0.1f(0.1, 2 + k_1) = 0.1(3 + (0.1)(2 + 0.3))$

$k_2 = 0.1(3.23) = 0.323$

Hence $y_1 = 2 + \dfrac{1}{2}(0.3 + 0.323)$

$\qquad y_1 = 2.3115$

Step 2

$(x_1, y_1) = (0.1, 2.3115)$

$k_1 = 0.1(3 + (0.1)(2.3115)) = 0.323\,115$

$k_2 = 0.1f(0.2, 2.3115 + k_1) = 0.1f(0.2, 2.634\,615)$

$\quad = 0.1(3 + (0.2)(2.634\,615))$

$k_2 = 0.1(3.526\,923) = 0.352\,6923$

Hence $y_2 = 2.3115 + \dfrac{1}{2}(0.323\,115 + 0.352\,6820)$

$\qquad y_2 = 2.649\,403\,65$

$\qquad = 2.649$ (3 d.p.).

b

y_{r+1} $=y_r + \frac{1}{2}(k_1 + k_2)$		$k_1 = hf(x_r, y_r)$		$k_2 = hf(x_r + h, y_r + k_1)$		$f(x) = 3 + xy$		$h = 0.1$
r	x	y (formula)	y	k (formula)	k	k (formula)	k	
0	0		2					
1	0.1	(D3)+0.5*((F4)+(H4))	2.3115	0.1*(3+(B3)*(D3)	0.3	0.1*(3+(B4)*((D3)+(F3))	0.323	
2	0.2	(D4)+0.5*((F5)+(H5))	2.649404	0.1*(3+(B4)*(D4)	0.323 115	0.1*(3+(B5)*((D4)+(F4))	0.352 6923	
3	0.3	(D4)+0.5*((F5)+(H6))	3.020934	0.1*(3+(B5)*(D5)	0.352 988 073	0.1*(3+(B6)*((D5)+(F5))	0.390 071 752	
4	0.4	(D5)+0.5*((F6)+(H7))	3.434479	0.1*(3+(6)*(D6)	0.390 628 007	0.1*(3+(B7)*((D6)+(F6))	0.436 462 463	
5	0.5	(D6)+0.5*((F7)+(H8))	3.899965	0.1*(3+(B7)*(D7)	0.437 379 152	0.1*(3+(B8)*((D7)+(F7))	0.493 592 897	
6	0.6	(D7)+0.5*((F8)+(H9))	4.429313	0.1*(3+(B8)*(D8)	0.494 998 241	0.1*(3+(B9)*((D8)+(F8))	0.563 697 784	
7	0.7	(D8)+0.5*((F9)+(H10))	5.03702	0.1*(3+(B9)*(D9)	0.565 758 77	0.1*(3+(B10)*((D9)+(F9))	0.649 655 012	
8	0.8	(D9)+0.5*((F10)+(H11))	5.7409	0.1*(3+(B10)*(D10)	0.652 591 381	0.1*(3+(B11)*((D10)+(F10))	0.755 168 888	
9	0.9	(D10)+0.5*((F11)+(H12))	6.563044	0.1*(3+(B11)*(D11)	0.759 271 989	0.1*(3+(B12)*((D11)+(F11))	0.885 015 466	

This is a spreadsheet method for calculating these estimates. Since, after the initial 'programming', the spreadsheet takes the drudgery out of the calculations, the table goes up to the value of y at $x = 0.9$

An alternative improved Euler method is sometimes referred to as the midpoint formula since it compares the gradient of the chord PQ on a curve $F(x, y)$ with the gradient of the curve $f(x, y)$ at the midpoint of the chord at A, (x_0, y_0).

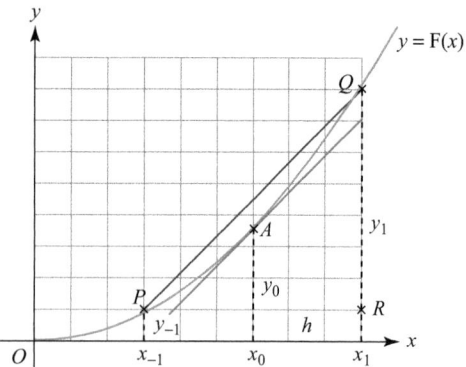

Most questions give you a starting point (x_0, y_0) and a gradient function $f(x, y)$. This means you do not know (x_{-1}, y_{-1}). You will have to use the general Euler formula $y_1 = y_0 + hf(x_0, y_0)$ to create (x_1, y_1).
You can then use
$$y_{r+1} = y_{r-1} + 2hf(x_r, y_r)$$
to create
$y_2 = y_0 + 2hf(x_1, y_1)$ and so on.

Let P be (x_{-1}, y_{-1}) and let Q be (x_1, y_1) and let $x_0 - x_{-1} = x_1 - x_0 = h$

The gradient of the chord $PQ = \dfrac{QR}{PR} = \dfrac{y_1 - y_{-1}}{x_1 - x_{-1}} = \dfrac{y_1 - y_{-1}}{2h}$

This gradient is taken as an approximation to the gradient of the curve at A,

i.e. $\left(\dfrac{dy}{dx}\right)_0 = f(x_0, y_0)$

Hence $f(x_0, y_0) \approx \dfrac{y_1 - y_{-1}}{2h}$

Rearranging this equation gives $y_1 = y_{-1} + 2hf(x_0, y_0)$, or in general

$$y_{r+1} = y_{r-1} + 2hf(x_r, y_r)$$

Key point

Example 4

Use the midpoint method to estimate the value of y when $x = 0.3$ for the differential equation
$$\frac{dy}{dx} = 3 + xy \text{ given that } y = 2 \text{ when } x = 0$$
Use a step length of 0.1

$(x_0, y_0) = (0, 2)$ and $h = 0.1$

$y_1 = 2 + 0.1 \times 3 = 2.3$ •————————

Use $y_1 = y_0 + hf(x_0, y_0)$ to find y_1

$(x_1, y_1) = (0.1, 2.3)$

$y_2 = 2 + 0.2(3 + 0.1 \times 2.3) = 2.646$ •————

Use $y_2 = y_0 + 2hf(x_1, y_1)$ to find y_2

$(x_2, y_2) = (0.2, 2.646)$

$y_3 = 2.3 + 0.2(3 + 0.2 \times 2.646) = 3.00584$

Thus an estimate for y when $x = 0.3$ is 3.01 (3 s.f.)

Exercise 21.2A Fluency and skills

In each of the questions use Euler's method when

1 $\dfrac{dy}{dx} = 2x + 3y$, $x_0 = 0$, $y_0 = 1$, $h = 0.1$ and the endpoint is at $y(0.1)$

2 $\dfrac{dy}{dx} = 3 + xy$, $x_0 = 1$, $y_0 = 1$, $h = 0.1$ and the endpoint is at $y(1.2)$

3 $\dfrac{dy}{dx} = x^2 y$, $x_0 = 0$, $y_0 = 1$, $h = 0.1$ and the endpoint is at $y(0.3)$

4 $\dfrac{dy}{dx} = \cos x$, $x_0 = 0$, $y_0 = 0$, $h = 0.1$ and the endpoint is at $y(0.5)$

5 $\dfrac{dy}{dx} = \sqrt{x}$, $x_0 = 0$, $y_0 = 5$, $h = 1$ and the endpoint is at $y(10)$

Use the improved Euler method when

6 $\dfrac{dy}{dx} = \ln(\sqrt{x} + \sqrt{y})$, $x_0 = 1$, $y_0 = 2$, $h = 1$ and the endpoint is at $y(5)$

7 $\dfrac{dy}{dx} = \sin(x^2)$, $x_0 = 1$, $y_0 = 1$, $h = 0.1$ and the endpoint is at $y(1.4)$

8 $\dfrac{dy}{dx} = \sin^2 x$, $x_0 = 1$, $y_0 = 1$, $h = 0.1$ and the endpoint is at $y(1.4)$

9 $\dfrac{dy}{dx} = (x - y)e^{-x}$, $x_0 = 0$, $y_0 = 5$, $h = 1$ and the endpoint is at $y(5)$

10 $\dfrac{dy}{dx} = ye^{\tan x}$, $x_0 = 0$, $y_0 = 5$, $h = 0.25$ and the endpoint is at $y(1)$

Use the midpoint method on these functions:

11 $\dfrac{dy}{dx} = 2x - 3y$, $x_0 = 0$, $y_0 = 1$, $h = 0.1$ and the endpoint is at $y(0.3)$

12 $\dfrac{dy}{dx} = 3x + xy - 4$, $x_0 = 1$, $y_0 = 1$, $h = 0.1$ and the endpoint is at $y(1.3)$

13 $\dfrac{dy}{dx} = x^2 y^2$, $x_0 = 0$, $y_0 = 1$, $h = 0.1$ and the endpoint is at $y(0.3)$

14 $\dfrac{dy}{dx} = \tan x$, $x_0 = 0$, $y_0 = 0$, $h = 0.1$ and the endpoint is at $y(0.5)$

15 $\dfrac{dy}{dx} = x^{-\frac{1}{2}}$, $x_0 = 1$, $y_0 = 5$, $h = 1$ and the endpoint is at $y(7)$

Reasoning and problem-solving

Strategy

To find an approximate solution to a differential equation

① Identify the differentiated function f(x, y), a starting point (x_0, y_0) and a suitable step value, h

② Calculate $y_1 = y_0 + hf(x_0, y_0)$

③ Continue step 2 using $y_{r+1} = y_r + hf(x_r, y_r)$ until the desired endpoint is reached.

④ Answer the question in context.

Example 5

a Solve the equation $\dfrac{dy}{dx} = y$ where $y = 1$ when $x = 0$

b Use Euler's method with a step length of 0.1 to find, correct to 5 d.p., an approximate value of e.

c Find the relative error.

a $\dfrac{dy}{dx} = y \Rightarrow \int \dfrac{1}{y}\, dy = \int 1 dx \Rightarrow \ln y = x + c$

Since $y = 1$ when $x = 0$, $c = 0$ therefore $y = e^x$

b

r	x_r	y_r
0	0	1
1	0.1	$1 + 0.1(1) = 1.1$
2	0.2	$1.1 + 0.1(1.1) = 1.21$
3	0.3	$1.21 + 0.1(1.21) = 1.331$
4	0.4	$1.331 + 0.1(1.331) = 1.4641$
5	0.5	$1.4641 + 0.1(1.4641) = 1.61051$
6	0.6	$1.61051 + 0.1(1.61051) = 1.771561$
7	0.7	$1.771561 + 0.1(1.771561) = 1.948717$
8	0.8	$1.948717 + 0.1(1.948717) = 2.143589$
9	0.9	$2.143589 + 0.1(2.143589) = 2.357948$
10	1.0	$2.357948 + 0.1(2.357948) = 2.593742$

2 3 Calculate $y_1 = y_0 + hf(x_0, y_0)$ and use $y_{r+1} = y_r + hf(x_r, y_r)$ until the desired endpoint is reached.

$e \approx 2.59374$ (5 d.p.)

4 Answer the question in context.

c Relative error $= \dfrac{2.593742 - 2.71828}{2.71828} = -0.04582$ (5 d.p.)

Example 6

The rate of temperature decrease along a straight copper wire is modelled as $\dfrac{dT}{dx} = -\dfrac{1}{5000} - \dfrac{x}{100}$, where x (cm) is the distance from one end. $T = 100°$ when $x = 0$

a Use Euler's method to estimate the temperature when $x = 10$

Take a step length of 1

b Solve the original equation analytically and hence find the relative error in your estimate.

a

r	x_r	T_r
0	0	100
1	1	99.9998
2	2	99.9896
3	3	99.9694
4	4	99.9392

2 3 Calculate $y_1 = y_0 + hf(x_0, y_0)$ and use $y_{r+1} = y_r + hf(x_r, y_r)$ until the desired endpoint is reached.

(*Continued on the next page*)

5	5	99.899
6	6	99.8488
7	7	99.7886
8	8	99.7184
9	9	99.6382
10	0	99.548

The estimated temperature 10 cm from the end is 99.548°

Answer the question in context.

④

b $\dfrac{dT}{dx} = -\dfrac{1}{5000} - \dfrac{x}{100} \Rightarrow T = \int \left(-\dfrac{1}{5000} - \dfrac{x}{100} \right) dx$

$$= -\dfrac{x}{5000} - \dfrac{x^2}{200} + C$$

When $T = 100$, $x = 0$, hence $C = 100$

$$T = 100 - \dfrac{x}{5000} - \dfrac{x^2}{200}$$

When $x = 10$, $T = 99.498°$

Relative error $= \dfrac{99.548 - 99.498}{99.498} = 0.0005025$ (4 s.f.)

Example 7

The number of cells in a bacterial culture is governed by the formula $n = 1000e^{0.2t}$, where t is in hours.

a How many cells are there initially in the culture?

b After how many hours, to the nearest hour, have the cells in the culture quadrupled?

c Use an improved Euler method to find an estimate for the number of cells after this time.

a Initially $t = 0$, so $n = 1000e^0 = 1000$

b $n = 1000e^{0.2t}$ so $\dfrac{n}{1000} = e^{0.2t}$ and hence

$0.2t = \ln\left(\dfrac{n}{1000}\right)$ or $t = 5\ln\left(\dfrac{n}{1000}\right)$

Thus, when $n = 4000$, $t = 5\ln 4 = 6.93$ hours.

Therefore the number of cells has quadrupled in 7 hours, to the nearest hour.

(Continued on the next page)

c If $n = 1000e^{0.2t}$, then $\dfrac{dn}{dt} = 200e^{0.2t}$

Hence using a starting point of $(0, 1000)$ with a step length of 1, you get

1	1	(D3)+0.5*((F4)+(H4))	1222.140276	200*EXP(0.2*(B3))	200	200*(EXP(0.2*(B4)))	244.280 551 6
2	2	(D4)+0.5*((F5)+(H5))	1493.463 021	200*EXP(0.2*(B4))	244.280 551 6	200*(EXP(0.2*(B5)))	298.364 939 5
3	3	(D4)+0.5*((F6)+(H6))	1824.857 371	200*EXP(0.2*(B5))	298.364 939 5	200*(EXP(0.2*(B6)))	364.423 760 1
4	4	(D5)+0.5*((F7)+(H7))	2229.623 344	200*EXP(0.2*(B6))	364.423 760 1	200*(EXP(0.2*(B7)))	445.108 185 7
5	5	(D6)+0.5*((F8)+(H8))	2724.005 62	200*EXP(0.2*(B7))	445.108 185 7	200*(EXP(0.2*(B8)))	543.656 365 7
6	6	(D6)+0.5*((F9)+(H9))	3327.845 495	200*EXP(0.2*(B8))	543.656 365 7	200*(EXP(0.2*(B9)))	664.023 384 5
7	7	(D6)+0.5*((F10)+(H10))	4065.377 184	200*EXP(0.2*(B9))	664.023 384 5	200*(EXP(0.2*(B10)))	811.039 993 4
8	8	(D6)+0.5*((F11)+(H11))	4966.200 423	200*EXP(0.2*(B10))	811.039 993 4	200*(EXP(0.2*(B11)))	990.606 484 9
9	9	(D6)+0.5*((F12)+(H12))	6066.468 412	200*EXP(0.2*(B11))	990.606 484 9	200*(EXP(0.2*(B12)))	1209.929 493
10	10	(D6)+0.5*((F13)+(H13))	7410.338 768	200*EXP(0.2*(B12))	1209.929 493	200*(EXP(0.2*(B13)))	1477.811 22
11	11		9051.745 728		1477.811 22		1805.002 7
12	12		11056.564 72		1805.002 7		2204.635 276

The number of cells after 7 hours is approximately equal to 4065

> Using formulae in a spreadsheet, you can see clearly that the number of cells has quadrupled in 7 hours.

Exercise 21.2B Reasoning and problem-solving

1 The vertical velocity of a hot-air balloon t s after some ballast has been ejected, is given by the approximate formula $\dfrac{ds}{dt} = (2t - 1)\,\text{m s}^{-1}$.

a The balloon was 150 m above the ground when the ballast was ejected. Use a step size of 0.5 to estimate its height above the ground after 4 s.

b Use integration to find the accurate height.

2 A body, falling freely from rest under gravity, encounters air resistance. The acceleration of the body is given approximately by the formula $\dfrac{dv}{dt} = 10 - 0.5v$. Using a step length of 0.1 estimate the velocity of the body after one second.

3 Water is steadily poured into a hemispherical bowl of radius 10 cm. When the height is h the rate of change of volume against height is given by the formula $\dfrac{dv}{dh} = \pi(20h - h^2)$. Using a step length of 0.5 estimate the volume when $h = 2$ cm.

4 An iron marble runs along a channel under the influence of a magnet. Its motion at time t s is given by the formula $\dfrac{ds}{dt} = 3t^2 - 10t + 18$ where $v = 0$ when $t = 0$

a Use a step length of 1 to find its distance from the start when $t = 6$

b Use integration to find the accurate distance.

5 The acceleration of a body is governed by the equation $\dfrac{dv}{dt} = \dfrac{4}{\sqrt{t}}$ where $v = 4$ when $t = 1$

   a Use a step length of 2 to estimate the velocity of the body when $t = 21$

   b Use integration to find the accurate velocity and find the relative and percentage errors.

6 For a minute after leaving a signal from rest, the speed of a train is given approximately by the formula $\dfrac{dv}{dt} = \dfrac{t(60 - 3t)}{1000}$ where t is the time in seconds and v is the speed in m s^{-1}.

   a Use a step size of 1 to estimate the velocity of the train after 10 s.

   b Use integration to find the accurate velocity.

7 A particle is following simple harmonic motion in a straight line. At any time its rate of change of velocity with respect to its distance from the centre of motion is given by the formula $v\dfrac{dv}{dx} = -x$ and $v = 10$ cm s^{-1} when $x = 0$ cm.

   a Using a step length of 1 estimate the velocity of the particle when $x = 5$

   b Use integration to find the accurate velocity.

8 The ellipse $\dfrac{x^2}{16} + \dfrac{y^2}{9} = 1$ has the differential equation $\dfrac{dy}{dx} = -\dfrac{9x}{16y}$

   a Use an improved Euler method with a step length of 0.5 to estimate a value of y when $x = 3$

   b Find the relative and percentage errors.

9 The acceleration of a body is governed by the equation $\dfrac{dv}{dt} = \dfrac{2}{\sqrt[3]{t}}$ and $v = 4$ when $t = 1$

Use an improved Euler method with a step length of 2 to estimate the velocity of the body when $t = 21$

10 The gradient of a curve at any point (x, y) is given by $\dfrac{dy}{dx} = \dfrac{y}{x^2 + 1}$ and it passes through the point $(1, 2)$

   a Use a step length of 0.2 to estimate the value of y when $x = 2$

   b Use integration to find the accurate value of y

11 A particle runs along a groove. Its motion at time t s is given by the formula $\dfrac{ds}{dt} = t^2 - 5t + 8$ and $v = 0$ when $t = 0$

   a Use the midpoint method with a step length of 1 to find its distance from the start when $t = 5$

   b Use integration to find the accurate distance.

Chapter summary

- To find an approximate area under a curve using the mid-ordinate rule:
  - Divide the required area into an appropriate number of vertical rectangles.
  - Calculate and sum the values of the ordinates at the midpoint of each rectangle.
  - Complete the area approximation by multiplying the sum of the ordinates by the width of each rectangle.
- To find an approximate area under a curve using Simpson's rule:
  - Divide the required area into an appropriate even number of strips of equal width, d
  - Calculate the values of the ordinates for each x-value
  - Complete the area approximation by substituting into the formula:
  - $\frac{d}{3} \times [(\text{sum of first and last ordinates}) + 4 \times (\text{sum of odd ordinates}) + 2 \times (\text{sum of even ordinates})]$
- Euler's method states that $x_{r+1} = x_r + h$, $y_{r+1} = y_r + hf(x_r, y_r)$
- To find an approximate value in an Euler equation:
  - Identify the differentiated function f(x, y), a starting point (x_0, y_0) and a suitable step value, h
  - Calculate $y_1 = y_0 + hf(x_0, y_0)$
  - Continue using $y_{r+1} = y_r + hf(x_r, y_r)$ until the desired endpoint is reached.
- The improved Euler equation is $y_{r+1} = y_r + \frac{1}{2}(k_1 + k_2)$ where $k_1 = hf(x_r, y_r)$ and $k_2 = hf(x_r + h, y_r + k_1)$
- The midpoint formula is $y_{r+1} = y_{r-1} + 2hf(x_r, y_r)$

Check and review

You should now be able to...	Review Questions
✔ Find an approximation for the area under a graph using the mid-ordinate rule.	1
✔ Find an approximation for the area under a graph using Simpson's rule.	2
✔ Find the relative error between an approximation and the accurate answer.	3
✔ Use Euler's step-by-step first order method for equations of the form $y' = f(x, y)$	4
✔ Use improved Euler methods for equations of the form $y' = f(x, y)$	5, 6, 7, 8, 9

1 Use the mid-ordinate rule with an appropriate number of rectangles to estimate these areas to 3 s.f.

a $\displaystyle\int_{1}^{10} e^{\frac{-2}{x}}\,dx$ **b** $\displaystyle\int_{0}^{6} \ln(1+x^2)\,dx$

c $\displaystyle\int_{\frac{\pi}{4}}^{2\pi} e^{\sin x}\,dx$

2 Use Simpson's rule with an appropriate number of strips to estimate these areas to 3 s.f.

a $\displaystyle\int_{1}^{7} e^{\sin(x^3)}\,dx$ **b** $\displaystyle\int_{0}^{\frac{\pi}{3}} \tan^2 x\,dx$

c $\displaystyle\int_{1}^{5} \frac{\sqrt{1+x}}{x^2}\,dx$

3 Questions using Simpson's rule led to these results. Calculate the relative and percentage errors in each case.

  a Approximate result 212.6, accurate answer 220

  b Approximate result 5.97, accurate answer 5.9

  c Approximate result 0.0023, accurate answer 0.01

  d Approximate result 4×10^7, accurate answer 39 450 000

4 Use Euler's method when

  a $\dfrac{dy}{dx} = 10 - 3x$, $x_0 = 0$, $y_0 = 1$, $h = 0.1$, endpoint is at $y(0.4)$

  b $\dfrac{dy}{dx} = 10 - 3y$, $x_0 = 1$, $y_0 = 1$, $h = 0.1$, endpoint is at $y(1.5)$

  c $\dfrac{dy}{dx} = x^2\sqrt{y}$, $x_0 = 3$, $y_0 = 2$, $h = 0.01$, endpoint is at $y(3.05)$

5 Use the improved Euler method when

  a $\dfrac{dy}{dx} = 2x - y + xy$, $x_0 = 3$, $y_0 = 4$, $h = 0.1$, endpoint is at $y(3.6)$

  b $\dfrac{dy}{dx} = \sqrt{xy}$, $x_0 = 0$, $y_0 = 36$, $h = 0.01$, endpoint is at $y(0.06)$

  c $\dfrac{dy}{dx} = \dfrac{e^{x^2}}{y}$, $x_0 = 1$, $y_0 = 4$, $h = 1$, endpoint is at $y(5)$

6 The mass of a block of 10 g of francium-223 is governed by the formula $m \simeq 10e^{-0.0318t}$, where t is in minutes.

  a After t minutes, the mass has halved in size. This is called its 'half life'. Calculate the half life of francium-223.

  b Use an improved Euler method to find out an estimate for the half life.

  c What is the mass after 12 minutes?

7 The gradient of a curve at any point (x, y) is given by $\dfrac{dy}{dx} = \dfrac{y}{1-x^2}$ and it passes through the point $(0, 1)$

  a Use a step length of 0.1 in an improved Euler method to estimate the value of y when $x = 0.6$

  b Use integration to find the accurate value of y

8 Use the midpoint method on these functions.

  a $\dfrac{dy}{dx} = 2xy$, $x_0 = 1$, $y_0 = 1$, $h = 0.1$, endpoint is at $y(1.4)$

  b $\dfrac{dy}{dx} = \tan^2 x$, $x_0 = 0$, $y_0 = 0$, $h = 0.2$, endpoint is at $y(1)$

  c $\dfrac{dy}{dx} = 2x + \ln(1+x)$, $x_0 = 1$, $y_0 = 2$, $h = 0.01$, endpoint is at $y(1.05)$

9 The village of Far Far Away has a population, $n = 1000$

The government needs to implement a massive house building project in order to cope with a predicted population increase of rate $\dfrac{dn}{dt} = 700 \times 1.5^t$ where t is measured in years.

  a Use integration to find the accurate population prediction for the population of Far Far Away after 5 years.

  b Use a midpoint method to find out an estimate for the population of Far Far Away at this time.

Investigation

In the first part of their fall from an aircraft, sky divers free-fall under gravity. If we assume the acceleration due to gravity has a value of 10 ms^{-2}, then the motion of a sky diver can be modelled by the differential equation $\frac{dv}{dt} = 10$ where v ms^{-1} is the velocity of the sky diver (in this case measured vertically downwards) after time t seconds. This assumes that there is no force due to air resistance, whereas in reality this is not the case.

Experience suggests that the force of air resistance increases as speed increases. Experiments in wind tunnels suggest that forces due to air resistance can often be modelled as being proportional to velocity or the square of velocity.

The motion of sky divers, taking air resistance into account, can be modelled by one of these differential equations:

$$\frac{dv}{dt} = 10 - kv \qquad \frac{dv}{dt} = 10 - cv^2$$

Use the Euler step method to explore what would typically happen for a sky diver if you apply each of these models for air resistance in turn. Assume that a sky diver has a speed downwards of zero at $t = 0$

(You may wish to get started by using a value of $k = 0.1$ and a value of $c = 0.01$)

Try different values of k and c

Explore the resulting terminal velocities and how long it takes to reach these. Research typical values and consider which models might be most appropriate.

Investigation

Medicinal drugs typically decay exponentially with time in the body.

The differential equation $dA/dt = -kA$ where A is the amount of drug in the body at time t and k is a constant that primarily depends on the body's physiology, can be used to model the rate of change of concentration of the drug in a body.

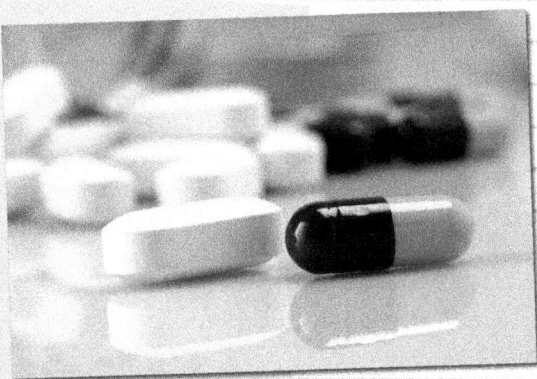

Doctors prescribe drugs to be taken as an initial dose, followed by a repeat dose after a certain time period and for a certain number of days.

Use the Euler step method to investigate how a drug decays in a body over time, and how this is affected by taking repeat doses of the drug every so often. Start by considering the drug to have a half-life of one hour (that is, the drug decays so that its concentration in the body naturally halves every hour).

1

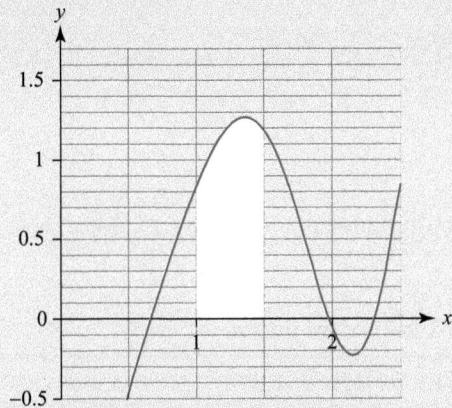

The diagram shows the graph of $y = f(x)$ where $f(x) = \ln x + \sin(x^2)$

The shaded area is bounded by the curve, the x-axis and the lines $x = 1$ and $x = 1.5$

a Use the mid-ordinate rule with five strips to find an estimate for the shaded area. **[5 marks]**

b Given that your estimate in part **a** gives a percentage error of 0.26%, find, correct to four significant figures, the actual area of the shaded region. **[2]**

2

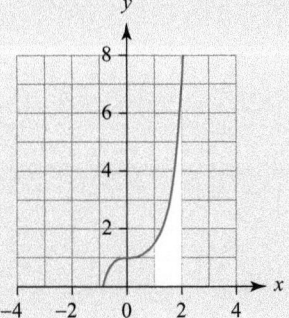

The diagram shows the graph of $y = f(x)$ where $f(x) = \cos(x^2) + x^3$

The shaded area is bounded by the curve, the x-axis and the lines $x = 1$ and $x = 2$

a Use Simpson's rule with six strips to estimate the shaded area. **[5]**

b Suggest how your estimate in part **a** could be improved. **[1]**

3 The diagram shows the design for a new turbine fin. The fin is modelled as the area enclosed by the curve $y = \tan x + x^2$, the x-axis and the lines $x = 0$ and $x = \dfrac{\pi}{4}$ where each unit on the axes represents 10 cm.

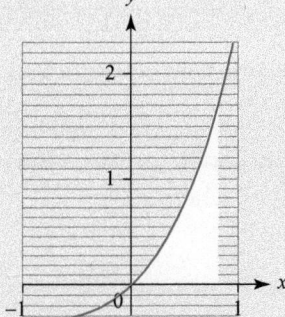

a Use Simpson's rule with five ordinates to estimate the area of the fin, correct to three significant figures. **[4]**

b Find the exact area of the fin, giving your answer in the form $p\pi^3 + q\ln(\sqrt{2})$ where p and q are rational constants to be found. **[4]**

c In light of your answer to part **b**, comment on your answer to part **a**. **[1]**

4 $\dfrac{dy}{dx} = 3x - 1$

Given that $y = 2$ when $x = 1$

a Use Euler's method with a step length of 0.5 to estimate the value of y when $x = 2$ **[4]**

b Use integration to find the exact value of y when $x = 2$ **[4]**

c Hence find the percentage error in using your approximation in part **a**. **[2]**

5 The value, x hundred pounds, of a particular tradeable stock t hours after it is purchased is modelled by the differential equation $\dfrac{dx}{dt} = \dfrac{x^2 - 2t}{2xt - t^2}$

If the stock is worth £500 three hours after it is purchased, use two iterations of Euler's method to estimate, correct to the nearest ten pence, the value of the stock four hours after it is purchased. **[6]**

6 $\dfrac{dy}{dx} = 2 - x^2 y$

Joanna uses an improved Euler method to estimate the particular solution of the differential equation when $x = 1.5$ given that $y = 1$ when $x = 1$

She uses a strip width of 0.25

a Show that after one iteration, Joanna gets a y-value of $\dfrac{579}{512}$ **[4]**

b Complete Joanna's method to find her estimated solution, correct to four decimal places. **[4]**

7 A particle moves such that the rate of change of displacement with respect to time has differential equation $\dfrac{ds}{dt} = t^3 - 4t + 2$

Given that $s = 0$ when $t = 0$,

a Use the midpoint formula with step length 1 to estimate the displacement when $t = 3$ **[5]**

b By solving the differential equation, find the exact value of s when $t = 3$ **[4]**

c Find the relative error in using your approximation in part **a**. **[2]**

8 The velocity, $v\,\text{m s}^{-1}$, of a particle attached to a vertically hanging spring, at the point where the spring becomes taut, is modelled using the differential equation $\dfrac{dv}{dx} = \dfrac{3x - 23}{2v} - 0.01v$ where x is the vertical displacement, in cm, from the point of release. Given that $v = 6$ when $x = 4$, use the midpoint formula with a step length of 0.25 to find the value of v, to 3 decimal places, when $x = 5$ **[6]**

Matrices 2

The mathematical techniques that use matrices, eigenvalues and eigenvectors have uses in computer models in many disciplines. For example, engineering, geology, statistics and financial analysis.

One area where such computer models are used extensively is in weather forecasting.

Weather forecasts help people with everyday decisions. They are also useful for those who take part in outdoor activities such as sailing and hill walking. For people who work outdoors, such as farmers and fishermen, it is important that forecasts are accurate for as far into the future as possible. This is only possible if the use of the mathematics of matrices continues to develop.

Orientation

What you need to know	What you will learn	What this leads to
Ch5 Matrices 1	• How to calculate the determinate of a matrix, and the inverse of a matrix. • How to solve a system of linear equations. • How to work out and use eigenvalues and eigenvectors. • How to use matrix diagonalization.	**Careers** • Computer programming. • Meteorology.

Fluency and skills

See Ch5.3
For a reminder of matrices.

You can find the determinant of the 2×2 matrix $\mathbf{A} = \begin{pmatrix} a & b \\ c & d \end{pmatrix}$ either on your calculator or using the fact that $\det(\mathbf{A}) = ad - bc$

In order to find the determinant of a 3×3 matrix, you need to find the **minors** of the top row of elements in your matrix. Given a 3×3 matrix, the minor of an element is the determinant of the 2×2 matrix remaining when the row and column of the element are crossed out.

For example, to find the minor of a in the matrix $\begin{pmatrix} a & b & c \\ d & e & f \\ g & h & i \end{pmatrix}$, you cross out the row and the column involving a then find the determinant of the matrix $\begin{pmatrix} e & f \\ h & i \end{pmatrix}$ that remains.

Key point

Then $\det \begin{pmatrix} a & b & c \\ d & e & f \\ g & h & i \end{pmatrix} = a \begin{vmatrix} e & f \\ h & i \end{vmatrix} - b \begin{vmatrix} d & f \\ g & i \end{vmatrix} + c \begin{vmatrix} d & e \\ g & h \end{vmatrix}$

Example 1

Find the determinant of the matrix $\begin{pmatrix} 3 & 2 & -1 \\ 0 & 4 & -2 \\ -3 & 1 & 5 \end{pmatrix}$

$\det \begin{pmatrix} 3 & 2 & -1 \\ 0 & 4 & -2 \\ -3 & 1 & 5 \end{pmatrix} = 3 \begin{vmatrix} 4 & -2 \\ 1 & 5 \end{vmatrix} - 2 \begin{vmatrix} 0 & -2 \\ -3 & 5 \end{vmatrix} - 1 \begin{vmatrix} 0 & 4 \\ -3 & 1 \end{vmatrix}$

Use the formula
$a \begin{vmatrix} e & f \\ h & i \end{vmatrix} - b \begin{vmatrix} d & f \\ g & i \end{vmatrix} + c \begin{vmatrix} d & e \\ g & h \end{vmatrix}$

$= 3(20--2)-2(0-6)-1(0--12)$

Take care with negative signs.

$= 66+12-12$

$= 66$

Matrices can be used to perform transformations. Consider a square with vertices at $(0, 0)$, $(1, 0)$, $(0, 1)$ and $(1, 1)$

Under the transformation $\mathbf{T} = \begin{pmatrix} a & 0 \\ 0 & a \end{pmatrix}$ the square is enlarged by scale factor a centre the origin. So the vertices of the image are $(0, 0)$, $(a, 0)$, $(0, a)$ and (a, a). The area of the original square is 1 and the area of the image is a^2, which is also the determinant of $\mathbf{T}$. This result extends to all linear transformations.

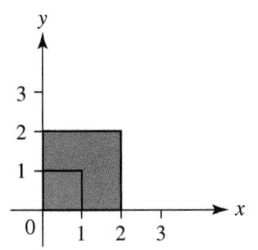

Key point

Under a transformation represented by the 2×2 matrix $\mathbf{T}$

area of image = area of original $\times |\det(\mathbf{T})|$

PURE

Example 2

The triangle with vertices $(3, -1)$, $(4, 2)$ and $(0, 2)$ is stretched by scale factor 2 parallel to the x-axis and stretched by scale factor -3 parallel to the y-axis. Find the area of the image under this transformation.

The transformation is given by $\mathbf{T} = \begin{pmatrix} 2 & 0 \\ 0 & -3 \end{pmatrix}$

$\det(\mathbf{T}) = 2 \times -3 = -6$

The area of the original triangle is $\frac{1}{2} \times 4 \times 3 = 6$

So the area of image is $6 \times |-6| = 36$

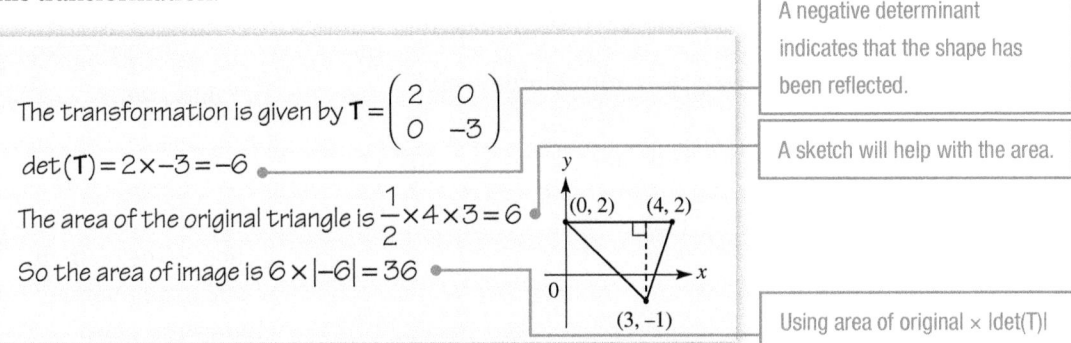

A negative determinant indicates that the shape has been reflected.

A sketch will help with the area.

Using area of original $\times |\det(\mathbf{T})|$

Key point

If $\det(\mathbf{T}) < 0$ then the transformation represented by $\mathbf{T}$ involves a reflection.

If $\det(\mathbf{T}) > 0$ then the orientation of the original shape is preserved.

The same principles can be extended to the volumes of shapes.

Key point

Under a transformation represented by the 3×3 matrix $\mathbf{M}$
volume of image = volume of original $\times |\det(\mathbf{M})|$

A parallelepiped is a prism where each face is a parallelogram.
A cuboid is a special case of a parallelepiped.

Example 3

A cube is transformed by the matrix $\mathbf{M} = \begin{pmatrix} 1 & -2 & -1 \\ 0 & 3 & 2 \\ 2 & 1 & 4 \end{pmatrix}$ to create a parallelepiped.

The volume of the parallelepiped is $36\,\text{cm}^3$.

a Calculate the volume of the original cube.

b Explain whether the transformation $\mathbf{M}$ involves a reflection.

a $\det(\mathbf{M}) = 1(12 - 2) - -2(0 - 4) - 1(0 - 6) = 8$

$36 \div 8 = 4.5$

So the volume of the original cube is $4.5\,\text{cm}^3$.

Volume of original $\div |\det(\mathbf{M})|$

b $\det(\mathbf{M}) > 0$, therefore the transformation $\mathbf{M}$ does not involve a reflection.

You can find the inverse of the 2×2 matrix $\mathbf{A} = \begin{pmatrix} a & b \\ c & d \end{pmatrix}$, either on

your calculator or using the rule $\mathbf{A}^{-1} = \dfrac{1}{\det(\mathbf{A})} \begin{pmatrix} d & -b \\ -c & a \end{pmatrix}$

To find the inverse of a 3×3 matrix, $\mathbf{A}$, you first need to find the **matrix of minors**, $\mathbf{M}$. You do this by replacing each element in $\mathbf{A}$ by its minor.

So if $\mathbf{P} = \begin{pmatrix} a & b & c \\ d & e & f \\ g & h & i \end{pmatrix}$ then the matrix of minors is

$\mathbf{M} = \begin{pmatrix} A & B & C \\ D & E & F \\ G & H & I \end{pmatrix}$ where A is the minor of a, B is the minor

of b and so on.

You then transpose this matrix (swap rows and columns) and

change the sign of alternating elements to give $\begin{pmatrix} A & -D & G \\ -B & E & -H \\ C & -F & I \end{pmatrix}$

> **Key point**
>
> The inverse of $\mathbf{P}$ is $\mathbf{P}^{-1} = \dfrac{1}{\det(\mathbf{P})} \begin{pmatrix} A & -D & G \\ -B & E & -H \\ C & -F & I \end{pmatrix}$

Remember, you can only find the inverse of a non-singular matrix. Singular matrices (which have a determinant of zero) do not have an inverse.

The minor of element a is

$A = \det \begin{pmatrix} e & f \\ h & i \end{pmatrix} = ei - fh$

The sign matrix

$\begin{pmatrix} + & - & + \\ - & + & - \\ + & - & + \end{pmatrix}$

indicates the elements of which you need to change the sign.

Example 4

Find the inverse of $\mathbf{A} = \begin{pmatrix} 2 & 1 & -3 \\ 3 & 1 & -2 \\ 0 & 2 & -1 \end{pmatrix}$, given that it is non-singular.

$\mathbf{M} = \begin{pmatrix} (1\times-1)-(-2\times2) & (3\times-1)-(-2\times0) & (3\times2)-(1\times0) \\ (1\times-1)-(-3\times2) & (2\times-1)-(-3\times0) & (2\times2)-(1\times0) \\ (1\times-2)-(-3\times1) & (2\times-2)-(-3\times3) & (2\times1)-(1\times3) \end{pmatrix}$

Find the minor of every element.

$= \begin{pmatrix} 3 & -3 & 6 \\ 5 & -2 & 4 \\ 1 & 5 & -1 \end{pmatrix}$

$\det(\mathbf{A}) = (2\times3) - (1\times-3) + (-3\times6)$

Use the minors from the top row of the table as

$\det(\mathbf{A}) = 2\begin{vmatrix} 1 & -2 \\ 2 & -1 \end{vmatrix} - 1\begin{vmatrix} 3 & -2 \\ 0 & -1 \end{vmatrix} - 3\begin{vmatrix} 3 & 1 \\ 0 & 2 \end{vmatrix} = -9$

Therefore $\mathbf{A}^{-1} = -\dfrac{1}{9} \begin{pmatrix} 3 & -5 & 1 \\ 3 & -2 & -5 \\ 6 & -4 & -1 \end{pmatrix}$

Use $\mathbf{A}^{-1} = \dfrac{1}{\det(\mathbf{A})} \begin{pmatrix} A & -D & G \\ -B & E & -H \\ C & -F & I \end{pmatrix}$

This could be worked out on a calculator but you do need to know the method.

Exercise 22.1A Fluency and skills

1   Find the determinant of each of these matrices.

a $\begin{pmatrix} 3 & -7 \\ 2 & -4 \end{pmatrix}$ b $\begin{pmatrix} a & 1 \\ -b & 5 \end{pmatrix}$

c $\begin{pmatrix} 6 & 2 & -1 \\ -2 & 4 & 0 \\ 3 & 5 & 1 \end{pmatrix}$ d $\begin{pmatrix} 2a & 3 & a \\ -b & 0 & b \\ 3 & 5 & 1 \end{pmatrix}$

2   Find the values of x for which each of these matrices is singular.

a $\begin{pmatrix} x & -3 \\ 4 & 2 \end{pmatrix}$ b $\begin{pmatrix} -x & x+2 \\ 3x & 4-x \end{pmatrix}$

c $\begin{pmatrix} 1 & 2 & -x \\ x & 4 & -3x \\ 1 & 0 & 2 \end{pmatrix}$ d $\begin{pmatrix} 4 & x & 2 \\ -x & 3 & 3x \\ -2 & x & 1 \end{pmatrix}$

3   A rectangle has area 7 square units.

a   Find the area of the image of the rectangle under each of these transformations.

i $\begin{pmatrix} -2 & 3 \\ 1 & -3 \end{pmatrix}$ ii $\begin{pmatrix} 2 & 1 \\ 5 & 2 \end{pmatrix}$

b   Explain whether or not each of the transformations in part **a** involves a reflection.

4   A triangle is transformed by the matrix $\begin{pmatrix} 1 & -3 \\ 4 & -2 \end{pmatrix}$. The area of the image is 15 square units.

a   Calculate the area of the original triangle.

b   Explain whether or not the triangle has been reflected.

5   A cube with volume 5 cube units is transformed by the matrix $\begin{pmatrix} 1 & 3 & 2 \\ -2 & 1 & 2 \\ -1 & 2 & 0 \end{pmatrix}$

a   Calculate the volume of the image.

b   Explain whether or not the transformation represented by M involves a reflection.

6   Find the inverse of each of these matrices.

a $\begin{pmatrix} 2 & 3 \\ -1 & 0 \end{pmatrix}$ b $\begin{pmatrix} a & 2 \\ a & 2-a \end{pmatrix}$ c $\begin{pmatrix} -2 & -1 & 0 \\ 4 & 0 & 2 \\ 5 & 1 & 3 \end{pmatrix}$

d $\begin{pmatrix} 5 & 1 & 4 \\ 6 & 2 & -3 \\ -2 & -1 & 4 \end{pmatrix}$ e $\begin{pmatrix} -4a & 2 & 1 \\ a & 0 & 2 \\ 3a & -1 & 1 \end{pmatrix}$ f $\begin{pmatrix} a & -2 & 2a \\ 0 & 2 & 3-a \\ -3a & 3 & 1 \end{pmatrix}$

7   Find the inverse of the matrix
    $\mathbf{B} = \begin{pmatrix} 1 & -1 & k \\ k & 0 & 1 \\ 0 & k & 1 \end{pmatrix}$ in terms of k

8 A transformation is represented by the matrix $\mathbf{T} = \begin{pmatrix} 1 & 0 & 0 \\ 0 & \dfrac{\sqrt{2}}{2} & \dfrac{\sqrt{2}}{2} \\ 0 & -\dfrac{\sqrt{2}}{2} & \dfrac{\sqrt{2}}{2} \end{pmatrix}$

The image of the point A under $\mathbf{T}$ is $A'\,(1, -\sqrt{2}, \sqrt{2})$. Use the inverse of $\mathbf{T}$ to find the coordinates of point A

Reasoning and problem-solving

See Ch23.1
For more on equations of planes.

You can use matrices to solve a system of two linear equations in two variables. The solution gives you the point of intersection of two lines.

This method can be extended to three dimensions, where the solution represents the intersection of three planes.

An equation of the form $ax + by + cz = d$ can represent a plane. If you have three different planes then one of these will apply:

- There are no points that lie on all three planes (their equations are said to be inconsistent). This could be because two or three of the planes are parallel or because the three planes form a triangular prism.

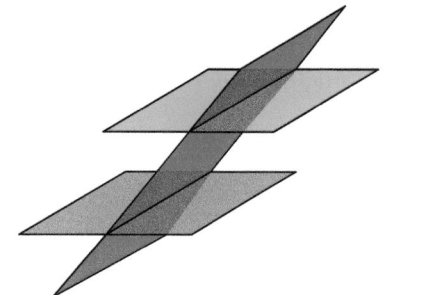 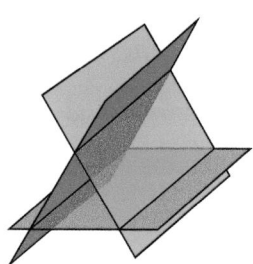

- There are infinitely many solutions as the three planes meet along a line (called a sheaf).

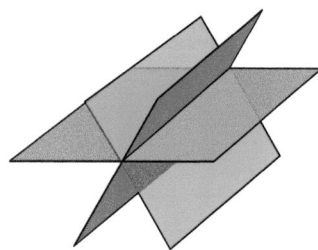

- They intersect at a single point, so there is exactly one solution.

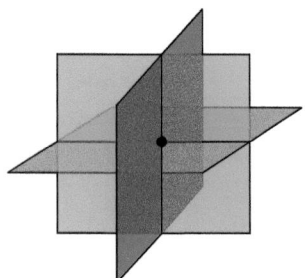

Strategy 1

To solve problems involving systems of linear equations

(1) Rewrite a system of linear equations using matrices.

(2) Pre-multiply or post-multiply a matrix by its inverse.

(3) Use the fact that $\mathbf{AA}^{-1} = \mathbf{A}^{-1}\mathbf{A} = \mathbf{I}$

Example 5

Given that there is a unique solution, use matrices to solve this system of equations:

$$x+y+z=2 \qquad 2x-3y-z=-1 \qquad 3x-2y+2z=11$$

Write the equations in the form $\begin{pmatrix} 1 & 1 & 1 \\ 2 & -3 & -1 \\ 3 & -2 & 2 \end{pmatrix}\begin{pmatrix} x \\ y \\ z \end{pmatrix} = \begin{pmatrix} 2 \\ -1 \\ 11 \end{pmatrix}$

> Rewrite the system of linear equations using matrices.

$$\det\begin{pmatrix} 1 & 1 & 1 \\ 2 & -3 & -1 \\ 3 & -2 & 2 \end{pmatrix} = 1(-6-2)-1(4--3)+1(-4--9) = -10$$

> Calculate the determinant.

$$\begin{pmatrix} 1 & 1 & 1 \\ 2 & -3 & -1 \\ 3 & -2 & 2 \end{pmatrix}^{-1} = -\frac{1}{10}\begin{pmatrix} -8 & -4 & 2 \\ -7 & -1 & 3 \\ 5 & 5 & -5 \end{pmatrix}$$

> Find the inverse.

$$-\frac{1}{10}\begin{pmatrix} -8 & -4 & 2 \\ -7 & -1 & 3 \\ 5 & 5 & -5 \end{pmatrix}\begin{pmatrix} 1 & 1 & 1 \\ 2 & -3 & -1 \\ 3 & -2 & 2 \end{pmatrix}\begin{pmatrix} x \\ y \\ z \end{pmatrix} = -\frac{1}{10}\begin{pmatrix} -8 & -4 & 2 \\ -7 & -1 & 3 \\ 5 & 5 & -5 \end{pmatrix}\begin{pmatrix} 2 \\ -1 \\ 11 \end{pmatrix}$$

> Pre-multiply both sides of the original equation by the inverse matrix.

$$\begin{pmatrix} x \\ y \\ z \end{pmatrix} = -\frac{1}{10}\begin{pmatrix} -8 & -4 & 2 \\ -7 & -1 & 3 \\ 5 & 5 & -5 \end{pmatrix}\begin{pmatrix} 2 \\ -1 \\ 11 \end{pmatrix}$$

> Use $\mathbf{AA}^{-1} = \mathbf{A}^{-1}\mathbf{A} = \mathbf{I}$

$$= -\frac{1}{10}\begin{pmatrix} 10 \\ 20 \\ -50 \end{pmatrix}$$

So $x=-1$, $y=-2$, $z=5$

Strategy 2

To decide if three planes intersect

(1) Check if any of the equations represent parallel planes – if so, they will not intersect.

(2) Write two of the variables in terms of the third.

(3) If the equations are inconsistent then the planes form a triangular prism.

Example 6

Explain why these systems of equations have no solutions.

a $2x-3y-z=2$
$4x-6y-2z=1$
$12x-18y-6z=3$

b $x+y-z=5$
$2x+2y=10$
$x+y+z=7$

> **1**
> Also, notice that
> $12x-18y-6z=3$ is a multiple of $4x-6y-2z=1$ so these represent the same plane.

a $4x-6y-2z=1$ is a multiple of $2x-3y-z=2$ except the constant term, therefore these represent parallel planes. Since parallel planes do not intersect, there are no solutions.

> **2**
> None of the planes are the same or parallel so you need to start trying to simplify the problem by eliminating a variable.

b Adding the first and third equations together gives $2x+2y=12$ however, the second equation is $2x+2y=10$ so the equations are inconsistent and there are no solutions.

> **3**
> The planes form a triangular prism.

The system of equations in Example 6b can be written in matrix form as

$$\begin{pmatrix} 1 & 1 & -1 \\ 2 & 2 & 0 \\ 1 & 1 & 1 \end{pmatrix}\begin{pmatrix} x \\ y \\ z \end{pmatrix} = \begin{pmatrix} 5 \\ 10 \\ 7 \end{pmatrix}$$

You can attempt to solve these by pre-multiplying both sides of the equation by the inverse of

$$\begin{pmatrix} 1 & 1 & -1 \\ 2 & 2 & 0 \\ 1 & 1 & 1 \end{pmatrix}$$

However, for this particular system of equations

$$\det\begin{pmatrix} 1 & 1 & -1 \\ 2 & 2 & 0 \\ 1 & 1 & 1 \end{pmatrix} = 1(2-0)-1(2-0)--1(2-2)=0$$

This implies that there is either no solution (as in this case) or an infinite number of solutions.

> **Key point**
> A system of linear equations represented by the matrix equation $\mathbf{M}x=x'$ has a unique solution if and only if $\det(\mathbf{M}) \neq 0$

To decide on the nature of a system of simultaneous equations

1 Rewrite the system of linear equations using matrices.

2 Calculate a determinant – a non-zero determinant implies a unique solution.

3 Write two of the variables in terms of the third.

4 Check for inconsistencies.

Example 7

You are given a system of simultaneous equations:

$x+3y-7z=8, 2x-2z=-2, x-y+z=-4$

a Decide whether there is a unique solution, an infinite number of solutions or no solutions.

b Describe the geometric significance.

a $\begin{pmatrix} 1 & 3 & -7 \\ 2 & 0 & -2 \\ 1 & -1 & 1 \end{pmatrix} \begin{pmatrix} x \\ y \\ z \end{pmatrix} = \begin{pmatrix} 8 \\ -2 \\ -4 \end{pmatrix}$

> ① Rewrite the system of linear equations using matrices.

$\det \begin{pmatrix} 1 & 3 & -7 \\ 2 & 0 & -2 \\ 1 & -1 & 1 \end{pmatrix} = 1(0-2)-3(2--2)-7(-2-0)$

$= 0$

> ② Since the determinant is zero, there must either be no solution or an infinite number of solutions.

This implies there is not a unique solution.

Subtracting the third equation from the first equation gives

$4y-8z=12 \Rightarrow y=3+2z$

The second equation gives $x=z-1$

> ③ Here y and x are expressed in terms of z, but you could have chosen any of the three variables to write the other two in terms of.

Verify there are no inconsistencies:

$(z-1)+3(3+2z)-7z=8$

$2(z-1)-2z=-2$

$(z-1)-(3+2z)+z=-4$

> ④ You can verify there are no inconsistencies by substituting back into the original equations.

So there is an infinite number of solutions. These lie on the line with equations $y=3+2z, x=z-1$

> You will learn more about writing the equation of lines in 3D in Chapter 23

b The planes meet in a line: they form a sheaf.

Exercise 22.1B Reasoning and problem-solving

1 Use matrices to find the point of intersection between the three planes in each case.

a $x+y-2z=3$

$2x-3y+5z=4$

$5x+2y+z=-3$

b $4x+6y-z=-3$

$2x-3y=2$

$8y+4z=0$

2 Given that $\mathbf{A}\begin{pmatrix} 2 & 0 & -3 \\ 0 & 1 & 4 \\ -5 & 2 & -1 \end{pmatrix} = \begin{pmatrix} 14 & -9 & 9 \\ 25 & -10 & 5 \\ 9 & 1 & 7 \end{pmatrix}$, find the matrix $\mathbf{A}$.

3 Given that $\begin{pmatrix} 5 & -1 & 0 \\ 0 & 2 & -2 \\ 3 & 1 & 4 \end{pmatrix}\mathbf{B} = \begin{pmatrix} -32 & 16 & 15 \\ -4 & 8 & 6 \\ 0 & -12 & -3 \end{pmatrix}$, find the matrix $\mathbf{B}$.

4 If $\mathbf{A} = \begin{pmatrix} 8 & 0 & 2 \\ 4 & 3 & -1 \\ 5 & 0 & 2 \end{pmatrix}$ and $\mathbf{AB} = \begin{pmatrix} -2 & 28 & 14 \\ 13 & 10 & -5 \\ -2 & 19 & 11 \end{pmatrix}$, find the matrix $\mathbf{BA}$.

5 Show that each of these systems of linear equations are inconsistent and explain the geometric significance.

    a $3x - 2y + z = 7$

       $6x - 4y + 2z = 5$

       $x + 3y + 2z = 3$

    b $-2x + 3y - z = 4$

       $6x - 9y + 3z = 7$

       $5x + y + z = 3$

6 A cube is transformed by the matrix $\mathbf{T} = \begin{pmatrix} 1 & 0 & k \\ 0 & 2 & -3 \\ 2-k & -1 & k \end{pmatrix}$ to give a parallelepiped.

    a State the range of values of k for which the transformation preserves the orientation of the cube.

    The point $C(-3, 2, 1)$ is transformed by $\mathbf{T}$ to $C'(-4, 1, -12)$

    b Find the value of k

7 Three planes have equations

    $-7x - 3y + 5z = 4$

    $x + y + z = 0$

    $x + 2y + 4z = 1$

    a Show that the planes do not have a unique point of intersection.

    b Show that a general point on the line is given by $(2\lambda + 1, -3\lambda - 2, \lambda + 1)$

8 For each system of equations, show that the planes that the equations represent form a sheaf.

a
$$x+2z=1$$
$$-2x+y+4z=0$$
$$9x-2y+2z=5$$

b
$$5x-y+z=0$$
$$2x+y-2z=7$$
$$36x+11y-24z=91$$

9 Find the values of a for which this system of simultaneous equations does not have a unique solution.

$$3y+az=4$$
$$-x+ay+2z=1$$
$$x+y=3$$

10 A family keeps rabbits, hamsters and fish as pets.

They initially have 17 pets in total and one more rabbit than hamsters.

Two of the fish die. They then have 8 more fish than hamsters.

a Write a matrix equation to represent this situation.

b Solve the matrix equation and state how many of each pet the family had initially.

11 A property developer owns 12 homes which have a mix of two, three and four bedrooms.

The total number of bedrooms in all her houses is 36 and she has the same number of two-bedroom homes as four-bedroom homes.

a Write a matrix equation to represent this situation.

b Show that there is no unique solution to this problem.

c How many possible solutions are there? Explain your answer.

12 The triangle ABC is stretched by scale factor k parallel to the x-axis then reflected in the line $y=x$. The images of vertices A, B and C are given by $A'(4, -6)$, $B'(0, 0)$ and $C'(4, 2)$

a Work out the area of the original triangle in terms of k

b State the range of values of k for which the transformation preserves the orientation of the triangle.

c Find the coordinates of A, B and C in terms of k

d If the coordinates of C are $(-1, 4)$, find the value of k

13 Three planes have equations:

$\Pi_1 : kx - y - z = 3,$

$\Pi_2 : 2x + ky - z = 4,$
$\Pi_3 : 3x + 2y - 2z = 3$

   a Show that the planes intersect at a single, unique point for all possible values of k

   b Find the coordinates of the point of intersection when $k = 2$

14 a Find the value of k for which the equations

     $x + 2y + kz = 3$

     $x - y - 3z = 0$

     $2x + y + z = -3$

     do not have a unique solution.

   b Assuming k is not equal to the value found in part **a**, find the solution to the equations in terms of k

15 The system of equations

$x - y + z = b$

$y + z = 0$

$x - ay = 4$

does not have a unique solution

   a Find the value of a

   b Given that the planes with these equations form a sheaf, find the value of b

Fluency and skills

The determinant of a 3×3 matrix, $\mathbf{A} = \begin{pmatrix} a & b & c \\ d & e & f \\ g & h & i \end{pmatrix}$ was defined as

$\det(\mathbf{A}) = a \begin{vmatrix} e & f \\ h & i \end{vmatrix} - b \begin{vmatrix} d & f \\ g & i \end{vmatrix} + c \begin{vmatrix} d & e \\ g & h \end{vmatrix}$, which uses the top row to multiply out the determinant.

In fact, any row or column can be used. The sign of the terms is given by the sign matrix

$\begin{pmatrix} + & - & + \\ - & + & - \\ + & - & + \end{pmatrix}$

For example, using, the second row gives

$\det(\mathbf{A}) = -d \begin{vmatrix} b & c \\ h & i \end{vmatrix} + e \begin{vmatrix} a & c \\ g & i \end{vmatrix} - f \begin{vmatrix} a & b \\ g & h \end{vmatrix}$

$= -bdi + cdh + aei - ceg - afh + bfg$

which is the same as using the original definition

$\det(\mathbf{A}) = a \begin{vmatrix} e & f \\ h & i \end{vmatrix} - b \begin{vmatrix} d & f \\ g & i \end{vmatrix} + c \begin{vmatrix} d & e \\ g & h \end{vmatrix}$

$= aei - afh - bdi + bfg + cdh - ceg$

Using, for example, the third column will also give this solution

$\det(\mathbf{A}) = c \begin{vmatrix} d & e \\ g & h \end{vmatrix} - f \begin{vmatrix} a & b \\ g & h \end{vmatrix} + i \begin{vmatrix} a & b \\ d & e \end{vmatrix}$

$= cdh - ceg - afh + bfg + aei - bdi$

Example 1

a Multiply out $\begin{vmatrix} 5 & 0 & 2 \\ 7 & 1 & 3 \\ 6 & 2 & 4 \end{vmatrix}$ using

  i The second column,   **ii** The third row,

b Hence write down the determinant of $\begin{pmatrix} 5 & 7 & 6 \\ 0 & 1 & 2 \\ 2 & 3 & 4 \end{pmatrix}$

a $\begin{vmatrix} 5 & 0 & 2 \\ 7 & 1 & 3 \\ 6 & 2 & 4 \end{vmatrix} = -0 + 1 \begin{vmatrix} 5 & 2 \\ 6 & 4 \end{vmatrix} - 2 \begin{vmatrix} 5 & 2 \\ 7 & 3 \end{vmatrix}$ | Ensure that you use the correct pattern of signs.

$= 1(20 - 12) - 2(15 - 14)$

$= 8 - 2$

$= 6$

(Continued on the next page)

$$\begin{vmatrix} 5 & 0 & 2 \\ 7 & 1 & 3 \\ 6 & 2 & 4 \end{vmatrix} = 6\begin{vmatrix} 0 & 2 \\ 1 & 3 \end{vmatrix} - 2\begin{vmatrix} 5 & 2 \\ 7 & 3 \end{vmatrix} + 4\begin{vmatrix} 5 & 0 \\ 7 & 1 \end{vmatrix}$$

$$= 6(0-2) - 2(15-14) + 4(5-0)$$

$$= -12 - 2 + 20$$

$$= 6$$

> You should get the same answer using any row or column.

b $\begin{pmatrix} 5 & 7 & 6 \\ 0 & 1 & 2 \\ 2 & 3 & 4 \end{pmatrix} = \begin{pmatrix} 5 & 0 & 2 \\ 7 & 1 & 3 \\ 6 & 2 & 4 \end{pmatrix}^T$ so multiplying out $\begin{vmatrix} 5 & 7 & 6 \\ 0 & 1 & 2 \\ 2 & 3 & 4 \end{vmatrix}$

using the second row is the same as multiplying out $\begin{vmatrix} 5 & 0 & 2 \\ 7 & 1 & 3 \\ 6 & 2 & 4 \end{vmatrix}$

using the second column.

Therefore $\det \begin{pmatrix} 5 & 7 & 6 \\ 0 & 1 & 2 \\ 2 & 3 & 4 \end{pmatrix} = 6$

Key point

$|\mathbf{M}| = |\mathbf{M}^T|$ where $\mathbf{M}^T$ is the transpose of matrix $\mathbf{M}$.

From Example 1, you can see that there is some advantage to using a row or column with the most zeros in order to simplify the calculation of the determinant. Therefore, you may wish to manipulate a matrix in order to simplify the calculation of the determinant. You can do this using **row and column operations**.

Key point

These row and column operations can be carried out without affecting the value of the determinant
- Adding or subtracting any multiple of a row to another row.
- Adding or subtracting any multiple of a column to another column.

Key point

However, these row and column operations change the sign of the determinant
- Swapping two rows.
- Swapping two columns.

Finally

Key point

- Multiplying a row or a column of a matrix by a scalar will multiply the determinant by that scalar.
- Dividing a row or a column of a matrix by a scalar will divide the determinant by that scalar.

Example 2

PURE

a Use row and column operations to show that $\begin{vmatrix} 24 & 20 & -1 \\ 9 & 4 & -2 \\ 3 & 6 & 2 \end{vmatrix} = 6 \begin{vmatrix} 0 & 0 & -1 \\ 2 & 5 & 0 \\ 1 & 6 & 2 \end{vmatrix}$

b Evaluate $\begin{vmatrix} 24 & 20 & -1 \\ 9 & 4 & -2 \\ 3 & 6 & 2 \end{vmatrix}$

a $\begin{vmatrix} 24 & 20 & -1 \\ 9 & 4 & -2 \\ 3 & 6 & 2 \end{vmatrix} = \begin{vmatrix} 24 & 20 & -1 \\ 12 & 10 & 0 \\ 3 & 6 & 2 \end{vmatrix}$ —————— Add row 3 to row 2

$= 2 \begin{vmatrix} 24 & 20 & -1 \\ 6 & 5 & 0 \\ 3 & 6 & 2 \end{vmatrix}$ —————— Divide row 2 by 2

$= 2 \begin{vmatrix} 0 & 0 & -1 \\ 6 & 5 & 0 \\ 3 & 6 & 2 \end{vmatrix}$ —————— Subtract $4 \times$ row 2 from row 1

$= 6 \begin{vmatrix} 0 & 0 & -1 \\ 2 & 5 & 0 \\ 1 & 6 & 2 \end{vmatrix}$ —————— Divide column 1 by 3

b $6 \begin{vmatrix} 0 & 0 & -1 \\ 2 & 5 & 0 \\ 1 & 6 & 2 \end{vmatrix} = 6(-1(12-5))$ —————— Use the answer to part **a** and multiply out by the top row.

$= -42$

Exercise 22.2A Fluency and skills

1 Multiply out each of these determinants, using the row or column specified; show your working.

a $\begin{vmatrix} 2 & 5 & 1 \\ 4 & 3 & -1 \\ -3 & 2 & 0 \end{vmatrix}$ using the third column

b $\begin{vmatrix} 0 & 14 & 3 \\ 0 & -1 & 2 \\ 8 & 3 & 1 \end{vmatrix}$ using the first column

c $\begin{vmatrix} 8 & 13 & -2 \\ 0 & 1 & 5 \\ 2 & 5 & 7 \end{vmatrix}$ using the second row

d $\begin{vmatrix} 12 & 4 & 6 \\ 8 & -2 & 5 \\ 0 & 9 & 0 \end{vmatrix}$ using the third row

e $\begin{vmatrix} 0 & b & 2a \\ a & a & 1 \\ 0 & b & 0 \end{vmatrix}$ using the first column

f $\begin{vmatrix} b & a & -b \\ 0 & -3 & 0 \\ a & 2 & a \end{vmatrix}$ using the second row.

2 a Use one or more column operations to write $\begin{vmatrix} 2 & 1 & -3 \\ 4 & 2 & 1 \\ 1 & 1 & 2 \end{vmatrix}$ as a determinant with at least two zero elements,

b Hence find $\det \begin{pmatrix} 2 & 1 & -3 \\ 4 & 2 & 1 \\ 1 & 1 & 2 \end{pmatrix}$

3 **a** Use at least 3 column operations to

simplify $\begin{vmatrix} 6 & 3 & -3 \\ 9 & 2 & -1 \\ -5 & -7 & -2 \end{vmatrix}$ and calculate its

value,

b State the value of det$\begin{pmatrix} 6 & 9 & -5 \\ 3 & 2 & -7 \\ -3 & -1 & -2 \end{pmatrix}$

4 Use row and column operations to write

$\begin{vmatrix} 5 & -2 & 12 \\ -2 & -8 & 24 \\ 4 & 1 & 16 \end{vmatrix}$ as a determinant with at least

two zero elements.

5 Given that the determinant of $\begin{pmatrix} a & 2 & -3 \\ 1 & a & 4 \\ 3 & 2 & 5 \end{pmatrix}$

is 128, calculate each of these determinants.
Justify your answer in each case.

a $\begin{vmatrix} 1 & a & 4 \\ a & 2 & -3 \\ 3 & 2 & 5 \end{vmatrix}$ **b** $\begin{vmatrix} a & 1 & 3 \\ 2 & a & 2 \\ -3 & 4 & 5 \end{vmatrix}$

c $\begin{vmatrix} a+1 & 2+a & 1 \\ 1 & a & 4 \\ 3 & 2 & 5 \end{vmatrix}$ **d** $\begin{vmatrix} a & 2 & -3-4a \\ 1 & a & 0 \\ 3 & 2 & -7 \end{vmatrix}$

e $\begin{vmatrix} a & 6 & -3 \\ 1 & 3a & 4 \\ 3 & 6 & 5 \end{vmatrix}$ **f** $\begin{vmatrix} a & 1 & 3 \\ 1 & \dfrac{a}{2} & -4 \\ 3 & 1 & -5 \end{vmatrix}$

g $\begin{vmatrix} -3 & 2a & 2 \\ 4 & 2 & a \\ 5 & 6 & 2 \end{vmatrix}$ **h** $\begin{vmatrix} a & 8 & -3 \\ 1 & 4a & 4 \\ \dfrac{3}{2} & 4 & \dfrac{5}{2} \end{vmatrix}$

Reasoning and problem-solving

Row and column operations can be used to factorise a determinant.

Key point

A determinant can be factorised by using row or
column operations to obtain rows or columns with a common
factor.

Strategy

To factorise a determinant

1. Take out common factors from rows or columns.

2. Perform row and column operations to create rows or columns where the elements have a common
factor.

3. Multiply out the determinant.

Example 3

a Factorise $\begin{vmatrix} a & b & 1 \\ a^2 & b^2 & 1 \\ a^3 & b^3 & 1 \end{vmatrix}$

b Under what conditions on a and b does this system of linear equations have a unique solution?

$$ax + by + z = 4 \qquad a^2 x + b^2 y + z = -2 \qquad a^3 x + b^3 y + z = 0$$

a
$$\begin{vmatrix} a & b & 1 \\ a^2 & b^2 & 1 \\ a^3 & b^3 & 1 \end{vmatrix} = a \begin{vmatrix} 1 & b & 1 \\ a & b^2 & 1 \\ a^2 & b^3 & 1 \end{vmatrix}$$

> **1** Factor out a from column 1

$$= ab \begin{vmatrix} 1 & 1 & 1 \\ a & b & 1 \\ a^2 & b^2 & 1 \end{vmatrix}$$

> **1** Factor out b from column 2

$$= ab \begin{vmatrix} 0 & 1 & 1 \\ a-1 & b & 1 \\ a^2-1 & b^2 & 1 \end{vmatrix}$$

> **2** Subtract column 3 from column 1

$$= ab \begin{vmatrix} 0 & 0 & 1 \\ a-1 & b-1 & 1 \\ a^2-1 & b^2-1 & 1 \end{vmatrix}$$

> **2** Subtract column 3 from column 2

$$= ab(a-1) \begin{vmatrix} 0 & 0 & 1 \\ 1 & b-1 & 1 \\ a+1 & b^2-1 & 1 \end{vmatrix}$$

> **1** Factor out $a-1$ from column 1. This works since $a^2 - 1 = (a+1)(a-1)$

$$= ab(a-1)(b-1) \begin{vmatrix} 0 & 0 & 1 \\ 1 & 1 & 1 \\ a+1 & b+1 & 1 \end{vmatrix}$$

> **1** Factor out $b-1$ from column 2

$$= ab(a-1)(b-1)[(b+1)-(a+1)]$$
$$= ab(a-1)(b-1)(b-a)$$

> **3** Multiply out the determinant by the top row.

b
$$\begin{pmatrix} a & b & 1 \\ a^2 & b^2 & 1 \\ a^3 & b^3 & 1 \end{pmatrix} \begin{pmatrix} x \\ y \\ z \end{pmatrix} = \begin{pmatrix} 4 \\ -2 \\ 0 \end{pmatrix}$$

> The system of equations can be written as a matrix equation.

$$\det \begin{pmatrix} a & b & 1 \\ a^2 & b^2 & 1 \\ a^3 & b^3 & 1 \end{pmatrix} = ab(a-1)(b-1)(b-a)$$

> From part **a**

$$ab(a-1)(b-1)(b-a) = 0$$

> The system will not have a unique solution if the determinant is zero.

$$\Rightarrow a = 0 \text{ or } b = 0 \text{ or } a = 1 \text{ or } b = 1 \text{ or } a = b$$

Therefore, the system of equations has a unique solution if these conditions are met:

$$a \neq 0, b \neq 0, a \neq 1, b \neq 1, a \neq b$$

1 Use row or column operations to show that

a $\begin{vmatrix} a & b & c \\ b & c & a \\ c & a & b \end{vmatrix} = (a+b+c)(ab+ac+bc-a^2-b^2-c^2)$

b $\begin{vmatrix} a^2 & b^2 & c^2 \\ bc & ca & ab \\ 1 & 1 & 1 \end{vmatrix} = (b-a)(c-a)(c-b)(a+b+c)$

c $\begin{vmatrix} a+b & a+c & b+c \\ c & b & a \\ c^2 & b^2 & a^2 \end{vmatrix} = (b-a)(a-c)(c-b)(a+b+c)$

d $\begin{vmatrix} a & -b & c \\ c-b & a+c & a-b \\ -bc & ac & -ab \end{vmatrix} = (a+b)(c-a)(b+c)(a-b+c)$

2 Use row or column operations to fully factorise the determinant $\begin{vmatrix} 1 & 1 & 1 \\ a & a^2 & a^3 \\ b & b^2 & b^3 \end{vmatrix}$. Show your working clearly.

3 a Use row or column operations to show that $\begin{vmatrix} a & b & c \\ a^2 & b^2 & c^2 \\ a^3 & b^3 & c^3 \end{vmatrix} = abc \begin{vmatrix} 1 & 1 & 1 \\ 0 & b-a & c-a \\ 0 & 0 & (c-b)(c-a) \end{vmatrix}$

b Hence state the values of

i $\begin{vmatrix} 2 & 3 & 4 \\ 4 & 9 & 16 \\ 8 & 27 & 64 \end{vmatrix}$ **ii** $\begin{vmatrix} 5 & 25 & 125 \\ 1 & 1 & 1 \\ 7 & 49 & 343 \end{vmatrix}$

4 a Show that $a^3 - b^3 = (a-b)(a^2+ab+b^2)$

b Fully factorise $\begin{vmatrix} a & 1 & a^3 \\ b & 1 & b^3 \\ c & 1 & c^3 \end{vmatrix}$

c Hence find the conditions on a, b and c under which the following system of linear equations has a unique solution:

$ax+y+a^3z=1$

$bx+y+b^3z=2$

$cx+y+c^3z=3$

5 a Fully factorise the determinant $\begin{vmatrix} a^3 & b^3 & c^3 \\ a^2 & b^2 & c^2 \\ 1 & 1 & 1 \end{vmatrix}$

b Given that $ab+ac+bc>0$, find the conditions on a, b and c under which the following system of linear equations has a unique solution:

$a^3x+b^3y+c^3z=0$

$a^2x+b^2y+c^2z=-1$

$x+y+z=2$

6 a Express $\begin{vmatrix} a & b & a+b-1 \\ a+1 & b+1 & 2 \\ b & a & 1 \end{vmatrix}$ as a product of three linear factors,

b Therefore, find the conditions on a and b under which the system of equations

$ax+(a+1)y+bz=0$

$bx+(b+1)y+az=0$

$(a+b-1)x+2y+z=0$

has a unique solution other than $x=y=z=0$

7 $\mathbf{S}=\begin{pmatrix} a^2 & b^2 & a \\ 1 & c & b \\ a & ac & ab \end{pmatrix}$

Prove that matrix $\mathbf{S}$ is singular for all values of a, b and c

8 $\mathbf{T}=\begin{pmatrix} b & x^2 & x \\ b & a^2 & a \\ x & bx & x \end{pmatrix}$, where $a\neq b$

Given that matrix $\mathbf{T}$ is singular find the possible values of x in terms of a and b

9 $\mathbf{M}=\begin{pmatrix} a+b & x^2 & x \\ a+x & b^2 & b \\ b+x & a^2 & a \end{pmatrix}$, where $a\neq b$

Given that matrix $\mathbf{M}$ is singular find the possible values of x in terms of a and b

10 $\mathbf{A}=\begin{pmatrix} a & a^2-a & a^2+b^2+c^2 \\ b & b^2-b & a^2+b^2+c^2 \\ c & c^2-c & a^2+b^2+c^2 \end{pmatrix}$ where $a\neq b\neq c$

Show that matrix $\mathbf{A}$ is non–singular for all real values of a, b and c

Fluency and skills

A vector whose direction is maintained under a transformation is known as an **eigenvector**. If the matrix used for the transformation is **A** then when the vector **x** is transformed using **A**, the result is a multiple of **x**.

> **Key point**
>
> **Eigenvectors** of the square matrix **A** are non-zero vectors that satisfy the equation $\mathbf{Ax} = \lambda\mathbf{x}$.
> The scalar λ is known as the **eigenvalue**.

You can rearrange the equation $\mathbf{Ax} = \lambda\mathbf{x}$ to give $\mathbf{Ax} - \lambda\mathbf{Ix} = 0$
since $\mathbf{x} = \mathbf{Ix}$

then factorise to give $(\mathbf{A} - \lambda\mathbf{I})\mathbf{x} = 0$

Since **x** is a non-zero vector, it must be the case that the matrix $\mathbf{A} - \lambda\mathbf{I}$ is singular.
Therefore $\det(\mathbf{A} - \lambda\mathbf{I}) = 0$

> You need to write $\lambda\mathbf{x}$ as $\lambda\mathbf{Ix}$ where **I** is an identity matrix with the same dimensions as **A** in order to be able to carry out the subtraction.

> **Key point**
>
> The equation $\det(\mathbf{A} - \lambda\mathbf{I}) = 0$ is the **characteristic equation** of **A** and is used to find the eigenvalues.

Example 1

Find the eigenvalues and corresponding eigenvectors of the matrix $\mathbf{A} = \begin{pmatrix} 3 & -1 \\ 4 & -2 \end{pmatrix}$

$A - \lambda I = \begin{pmatrix} 3 & -1 \\ 4 & -2 \end{pmatrix} - \lambda \begin{pmatrix} 1 & 0 \\ 0 & 1 \end{pmatrix}$

$\quad = \begin{pmatrix} 3-\lambda & -1 \\ 4 & -2-\lambda \end{pmatrix}$

$det(A - \lambda I) = (3-\lambda)(-2-\lambda) - -4$

$\lambda^2 - \lambda - 2 = 0$

$(\lambda - 2)(\lambda + 1) = 0$

$\lambda = 2, -1$

$\begin{pmatrix} 3 & -1 \\ 4 & -2 \end{pmatrix}\begin{pmatrix} a \\ b \end{pmatrix} = 2\begin{pmatrix} a \\ b \end{pmatrix}$

$3a - b = 2a \Rightarrow a = b$

$4a - 2b = 2b \Rightarrow 4a = 4b$

> Solve the equation $\det(\mathbf{A} - \lambda\mathbf{I}) = 0$ to find the possible values of λ

> These are the eigenvalues.

> Use $\mathbf{Ax} = \lambda\mathbf{x}$ with $\mathbf{x} = \begin{pmatrix} a \\ b \end{pmatrix}$ and $\lambda = 2$ to find the possible vectors.

> Both equations give $a = b$

(continued on the next page)

So a possible eigenvector corresponding to the eigenvalue 2 is $\begin{pmatrix} 1 \\ 1 \end{pmatrix}$

$$\begin{pmatrix} 3 & -1 \\ 4 & -2 \end{pmatrix}\begin{pmatrix} a \\ b \end{pmatrix} = -1\begin{pmatrix} a \\ b \end{pmatrix}$$

$3a - b = -a \Rightarrow 4a = b$

$4a - 2b = -b \Rightarrow 4a = b$

So a possible eigenvector corresponding to the

eigenvalue -1 is $\begin{pmatrix} 1 \\ 4 \end{pmatrix}$

> Repeat process with $\lambda = -1$ to find the possible vectors **x**

> Both equations give $4a = b$

> Any non-zero multiples of these will also be eigenvectors.

The same process can be used for any square matrix **A**.

> You will need to be able to find eigenvalues and eigenvectors of 2×2 and 3×3 matrices.

Example 2

Find the eigenvalues and eigenvectors of the matrix $\mathbf{T} = \begin{pmatrix} 2 & 2 & 1 \\ 1 & 1 & 2 \\ 0 & 0 & 2 \end{pmatrix}$

$$\begin{vmatrix} 2-\lambda & 2 & 1 \\ 1 & 1-\lambda & 2 \\ 0 & 0 & 2-\lambda \end{vmatrix} = (2-\lambda)[(2-\lambda)(1-\lambda)-2]$$

$(2-\lambda)(\lambda^2 - 3\lambda) = 0$

$\lambda(2-\lambda)(\lambda - 3) = 0$

$\lambda = 0, 2, 3$

$$\begin{pmatrix} 2 & 2 & 1 \\ 1 & 1 & 2 \\ 0 & 0 & 2 \end{pmatrix}\begin{pmatrix} a \\ b \\ c \end{pmatrix} = 0\begin{pmatrix} a \\ b \\ c \end{pmatrix}$$

$2a + 2b + c = 0$

$a + b + 2c = 0$

$2c = 0 \Rightarrow c = 0$

$\Rightarrow a + b = 0$ so $a = -b$

So a possible eigenvector corresponding to the eigenvalue

0 is $\begin{pmatrix} 1 \\ -1 \\ 0 \end{pmatrix}$

$$\begin{pmatrix} 2 & 2 & 1 \\ 1 & 1 & 2 \\ 0 & 0 & 2 \end{pmatrix}\begin{pmatrix} a \\ b \\ c \end{pmatrix} = 2\begin{pmatrix} a \\ b \\ c \end{pmatrix}$$

> Use the simplest method to calculate the determinant; in this case using the third row to multiply out.

> Solve $\det(\mathbf{A} - \lambda\mathbf{I})$ to find the eigenvalues.

> Use $\mathbf{Ax} = \lambda\mathbf{x}$ with $\mathbf{x} = \begin{pmatrix} a \\ b \\ c \end{pmatrix}$ and $\lambda = 0$ to find the possible vectors **x**

> Repeat process with $\lambda = 2$ to find the possible vectors **x**

(*continued on the next page*)

$2a+2b+c=2a \Rightarrow c=-2b$

$a+b+2c=2b \Rightarrow a+b+2(-2b)=2b \Rightarrow a=5b$

$2c=2c$

So a possible eigenvector corresponding to the eigenvalue

2 is $\begin{pmatrix} 5 \\ 1 \\ -2 \end{pmatrix}$

$\begin{pmatrix} 2 & 2 & 1 \\ 1 & 1 & 2 \\ 0 & 0 & 2 \end{pmatrix} \begin{pmatrix} a \\ b \\ c \end{pmatrix} = 3 \begin{pmatrix} a \\ b \\ c \end{pmatrix}$

> Repeat process with $\lambda = 3$ to find the possible vectors **x**

$2c=3c \Rightarrow c=0$

$2a+2b+c=3a \Rightarrow a=2b$

$a+b+2c=3b \Rightarrow a=2b$

So a possible eigenvector corresponding to the eigenvalue

3 is $\begin{pmatrix} 2 \\ 1 \\ 0 \end{pmatrix}$

The eigenvectors are $\begin{pmatrix} 1 \\ -1 \\ 0 \end{pmatrix}$, $\begin{pmatrix} 5 \\ 1 \\ -2 \end{pmatrix}$ and $\begin{pmatrix} 2 \\ 1 \\ 0 \end{pmatrix}$

Key point

A matrix will always satisfy its own **characteristic equation**.

> This property is known as the Cayley-Hamilton Theorem.

To prove this for any 2×2 matrix, consider $\mathbf{A} = \begin{pmatrix} a & b \\ c & d \end{pmatrix}$

The characteristic equation is $\det(\mathbf{A} - \lambda\mathbf{I}) = 0$ which leads to $(a-\lambda)(d-\lambda) - bc = 0$

Expanding the brackets gives $\lambda^2 - (a+d)\lambda + (ad-bc) = 0$

- Now replace λ by matrix $\mathbf{A}$

$$\begin{pmatrix} a & b \\ c & d \end{pmatrix}^2 - (a+d)\begin{pmatrix} a & b \\ c & d \end{pmatrix} + (ad-bc)\begin{pmatrix} 1 & 0 \\ 0 & 1 \end{pmatrix}$$

> Notice that the $(ad - bc)$ term is multiplied by the identity matrix, **I**
>
> Otherwise you would be trying to add a constant to a matrix which isn't possible.

$$= \begin{pmatrix} a^2+bc & ab+bd \\ ac+cd & bc+d^2 \end{pmatrix} - \begin{pmatrix} a^2+ad & ab+bd \\ ac+cd & ad+d^2 \end{pmatrix} + \begin{pmatrix} ad-bc & 0 \\ 0 & ad-bc \end{pmatrix}$$

$$= \begin{pmatrix} a^2+bc-a^2-ad+ad-bc & ab+bd-ab-bd \\ ac+cd-ac-cd & bc+d^2-ad-d^2+ad-bc \end{pmatrix}$$

$$= \begin{pmatrix} 0 & 0 \\ 0 & 0 \end{pmatrix} \text{ as required.}$$

Example 3

Verify the that the matrix $\mathbf{A} = \begin{pmatrix} 1 & 0 \\ 2 & 3 \end{pmatrix}$ satisfies its characteristic equation.

$\det \begin{pmatrix} 1-\lambda & 0 \\ 2 & 3-\lambda \end{pmatrix} = 0 \Rightarrow (1-\lambda)(3-\lambda) - 0 = 0$

This becomes $\lambda^2 - 4\lambda + 3 = 0$

Replace λ with the matrix $\mathbf{A}$

$\begin{pmatrix} 1 & 0 \\ 2 & 3 \end{pmatrix}^2 - 4 \begin{pmatrix} 1 & 0 \\ 2 & 3 \end{pmatrix} + 3 \begin{pmatrix} 1 & 0 \\ 0 & 1 \end{pmatrix}$

> You need to replace 3 by 3**I** where **I** is the 2 × 2 identity matrix.

$= \begin{pmatrix} 1 & 0 \\ 8 & 9 \end{pmatrix} - \begin{pmatrix} 4 & 0 \\ 8 & 12 \end{pmatrix} + \begin{pmatrix} 3 & 0 \\ 0 & 3 \end{pmatrix}$

$= \begin{pmatrix} 1-4+3 & 0-0+0 \\ 8-8+0 & 9-12+3 \end{pmatrix}$

$= \begin{pmatrix} 0 & 0 \\ 0 & 0 \end{pmatrix}$ as required

> **Key point**
>
> A square matrix is **diagonal** if all its elements except those on the leading diagonal are zero.

You can convert a square matrix into diagonal form. This can be useful when dealing with powers of matrices.

> **Key point**
>
> A matrix, **M**, can be **diagonalised** by finding **P** and **D** such that $\mathbf{M} = \mathbf{PDP}^{-1}$. It can be shown that:
>
> - **D** is a diagonal matrix with the eigenvalues of M along the leading diagonal
> - **P** is a matrix where the columns are the eigenvalues of M
>
> The eigenvectors in the columns of **P** must occur in the same order as their corresponding eigenvalues in **D**

Example 4

Diagonalise the matrix $\mathbf{M} = \begin{pmatrix} 2 & 0 & 0 \\ 0 & 1 & -2 \\ 0 & -2 & 4 \end{pmatrix}$ by finding matrices **P** and **D** such that $\mathbf{M} = \mathbf{PDP}^{-1}$

$\det \begin{pmatrix} 2-\lambda & 0 & 0 \\ 0 & 1-\lambda & -2 \\ 0 & -2 & 4-\lambda \end{pmatrix} = (2-\lambda)[(1-\lambda)(4-\lambda) - 4] = 0$

(continued on the next page)

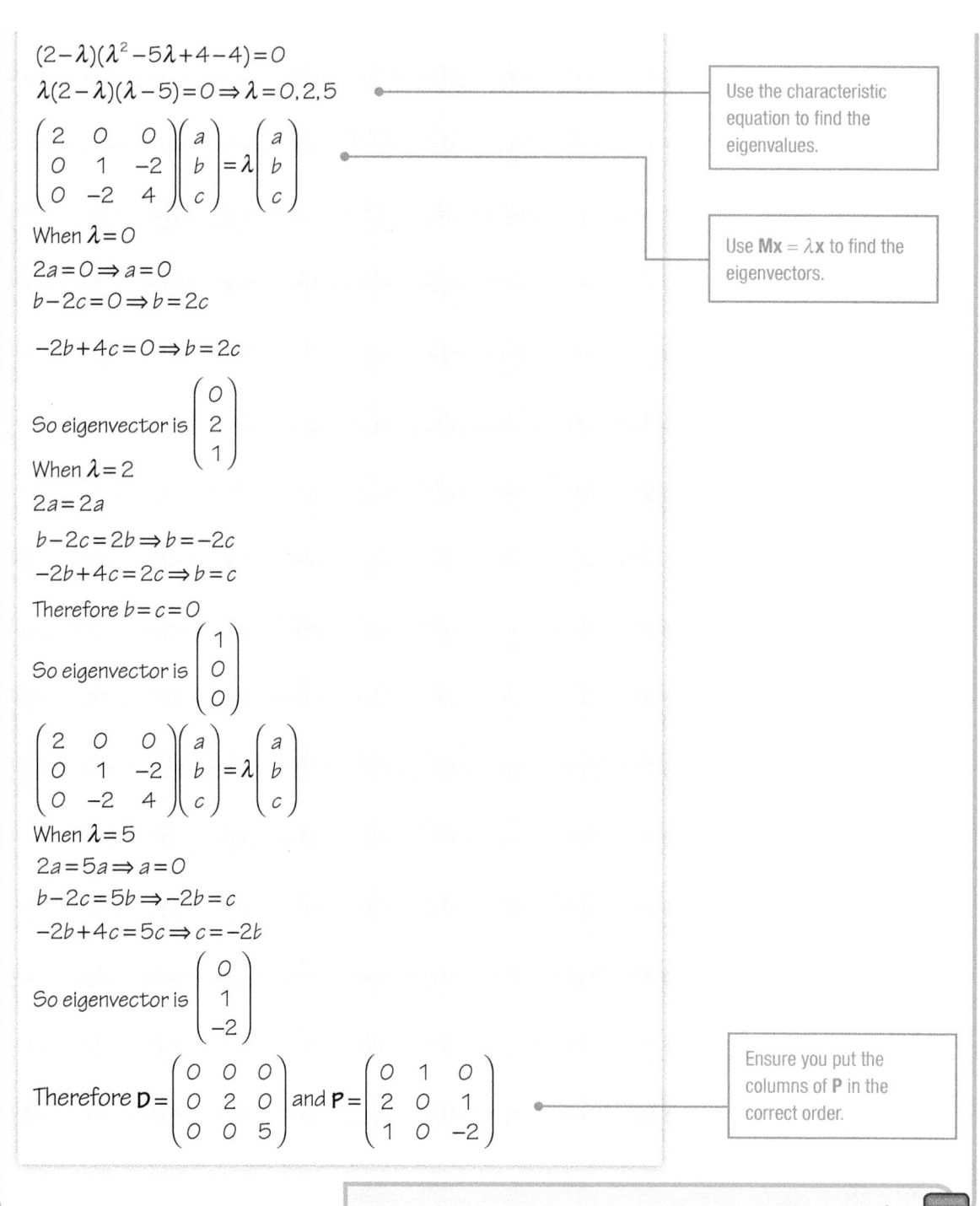

$(2-\lambda)(\lambda^2-5\lambda+4-4)=0$

$\lambda(2-\lambda)(\lambda-5)=0 \Rightarrow \lambda=0,2,5$

$\begin{pmatrix} 2 & 0 & 0 \\ 0 & 1 & -2 \\ 0 & -2 & 4 \end{pmatrix}\begin{pmatrix} a \\ b \\ c \end{pmatrix}=\lambda\begin{pmatrix} a \\ b \\ c \end{pmatrix}$

When $\lambda=0$

$2a=0 \Rightarrow a=0$

$b-2c=0 \Rightarrow b=2c$

$-2b+4c=0 \Rightarrow b=2c$

So eigenvector is $\begin{pmatrix} 0 \\ 2 \\ 1 \end{pmatrix}$

When $\lambda=2$

$2a=2a$

$b-2c=2b \Rightarrow b=-2c$

$-2b+4c=2c \Rightarrow b=c$

Therefore $b=c=0$

So eigenvector is $\begin{pmatrix} 1 \\ 0 \\ 0 \end{pmatrix}$

$\begin{pmatrix} 2 & 0 & 0 \\ 0 & 1 & -2 \\ 0 & -2 & 4 \end{pmatrix}\begin{pmatrix} a \\ b \\ c \end{pmatrix}=\lambda\begin{pmatrix} a \\ b \\ c \end{pmatrix}$

When $\lambda=5$

$2a=5a \Rightarrow a=0$

$b-2c=5b \Rightarrow -2b=c$

$-2b+4c=5c \Rightarrow c=-2b$

So eigenvector is $\begin{pmatrix} 0 \\ 1 \\ -2 \end{pmatrix}$

Therefore $\mathbf{D}=\begin{pmatrix} 0 & 0 & 0 \\ 0 & 2 & 0 \\ 0 & 0 & 5 \end{pmatrix}$ and $\mathbf{P}=\begin{pmatrix} 0 & 1 & 0 \\ 2 & 0 & 1 \\ 1 & 0 & -2 \end{pmatrix}$

Use the characteristic equation to find the eigenvalues.

Use $\mathbf{Mx}=\lambda\mathbf{x}$ to find the eigenvectors.

Ensure you put the columns of **P** in the correct order.

You could use your calculator to check that $\mathbf{M}=\mathbf{PDP}^{-1}$

Exercise 22.3A Fluency and skills

1 a Find the eigenvalues and eigenvectors for each of these matrices.

i $\begin{pmatrix} 2 & 3 \\ 0 & -4 \end{pmatrix}$ ii $\begin{pmatrix} 3 & -1 \\ 4 & -2 \end{pmatrix}$

iii $\begin{pmatrix} 3 & 0 \\ 3 & 7 \end{pmatrix}$ iv $\begin{pmatrix} 4 & -5 \\ -1 & 0 \end{pmatrix}$

b Show that each of the matrices in part **a** satisfies its characteristic equation.

2 Find the eigenvalues and eigenvectors for each of these matrices.

a $\begin{pmatrix} 1 & 0 & -5 \\ 4 & 5 & 1 \\ 0 & 0 & -3 \end{pmatrix}$ **b** $\begin{pmatrix} 1 & 3 & 0 \\ 2 & 0 & 1 \\ 0 & 0 & 2 \end{pmatrix}$

c $\begin{pmatrix} 5 & -3 & 0 \\ 0 & 2 & 0 \\ 0 & -5 & -4 \end{pmatrix}$ **d** $\begin{pmatrix} -7 & 0 & 0 \\ 5 & 4 & 1 \\ 4 & 0 & 9 \end{pmatrix}$

3 Given that $\mathbf{A} = \begin{pmatrix} 1 & 3 & -1 \\ -2 & 3 & 2 \\ 1 & 0 & 2 \end{pmatrix}$

  a Show that 3 is the only real eigenvalue of **A** and find the eigenvector corresponding to it.

  b Show that **A** satisfies its characteristic equation.

4 Given that $\mathbf{A} = \begin{pmatrix} -2 & 6 & -3 \\ 3 & 2 & 6 \\ 2 & 6 & -3 \end{pmatrix}$

  a Show that -7 is an eigenvalue of **A** and find the other two eigenvalues.

  b Find the eigenvector corresponding to the eigenvalue -7

  c Show that **A** satisfies its characteristic equation.

5 Given $\mathbf{A} = \begin{pmatrix} 2 & 0 \\ 0 & 7 \end{pmatrix}$, find matrices **P** and **D** such that $\mathbf{A} = \mathbf{PDP}^{-1}$

6 Given $\mathbf{B} = \begin{pmatrix} 2 & 6 \\ 1 & -3 \end{pmatrix}$, write **B** in the form $\mathbf{B} = \mathbf{PDP}^{-1}$

7 Given $\mathbf{M} = \begin{pmatrix} -4 & 0 & 0 \\ 5 & 2 & 6 \\ 1 & 6 & -3 \end{pmatrix}$, find matrices **P** and **D** such that $\mathbf{M} = \mathbf{PDP}^{-1}$

8 Given $\mathbf{T} = \begin{pmatrix} 7 & -1 & 0 \\ -1 & 7 & 0 \\ 0 & 0 & -7 \end{pmatrix}$, write **T** in the form $\mathbf{T} = \mathbf{PDP}^{-1}$

Reasoning and problem-solving

You can use the diagonalised form of a matrix to calculate powers of a matrix more easily.

If $\mathbf{A} = \mathbf{PDP}^{-1}$ then $\mathbf{A}^n = (\mathbf{PDP}^{-1})^n$

$$= (\mathbf{PDP}^{-1})(\mathbf{PDP}^{-1})(\mathbf{PDP}^{-1})....(\mathbf{PDP}^{-1})$$

$$= (\mathbf{PD})(\mathbf{P}^{-1}\mathbf{P})\mathbf{D}(\mathbf{P}^{-1}\mathbf{P})\mathbf{D}...(\mathbf{P}^{-1}\mathbf{P})(\mathbf{DP}^{-1})$$

$$= (\mathbf{PD})\mathbf{DD}...\mathbf{D}(\mathbf{DP}^{-1})$$

$$= \mathbf{PD}^n\mathbf{P}^{-1}$$

> Since matrix multiplication is associative.

> Since $\mathbf{P}^{-1}\mathbf{P} = \mathbf{I}$

Key point

If $\mathbf{A} = \mathbf{PDP}^{-1}$ then $\mathbf{A}^n = \mathbf{PD}^n\mathbf{P}^{-1}$

You can use this to solve problems involving $\mathbf{A}^n$, since

$$\mathbf{D}^n = \begin{pmatrix} a & 0 & 0 \\ 0 & b & 0 \\ 0 & 0 & c \end{pmatrix}^n = \begin{pmatrix} a^n & 0 & 0 \\ 0 & b^n & 0 \\ 0 & 0 & c^n \end{pmatrix}$$

> You can use proof by induction to prove this result.

Strategy

To calculate powers of a square matrix, **A**

1 Diagonalise the matrix using $\mathbf{A} = \mathbf{PDP}^{-1}$

2 Use $\mathbf{A}^n = \mathbf{PD}^n\mathbf{P}^{-1}$ to find powers of **A**

3 Use the fact that $\begin{pmatrix} a & 0 & 0 \\ 0 & b & 0 \\ 0 & 0 & c \end{pmatrix}^n = \begin{pmatrix} a^n & 0 & 0 \\ 0 & b^n & 0 \\ 0 & 0 & c^n \end{pmatrix}$

Example 5

Given that $\mathbf{A} = \begin{pmatrix} 3 & -1 \\ 0 & 5 \end{pmatrix}$, show that $\mathbf{A}^n = \begin{pmatrix} 3^n & \dfrac{3^n}{2} - \dfrac{5^n}{2} \\ 0 & 5^n \end{pmatrix}$

$det \begin{pmatrix} 3-\lambda & -1 \\ 0 & 5-\lambda \end{pmatrix} = 0 \Rightarrow (3-\lambda)(5-\lambda) = 0$

Find the eigenvalues using the characteristic equation.

$\Rightarrow$ eigenvalues are $\lambda = 3$ and 5

$\begin{pmatrix} 3 & -1 \\ 0 & 5 \end{pmatrix}\begin{pmatrix} a \\ b \end{pmatrix} = 3\begin{pmatrix} a \\ b \end{pmatrix}$

$5b = 3b \Rightarrow b = 0$

$3a - b = 3a$

So eigenvector is $\begin{pmatrix} 1 \\ 0 \end{pmatrix}$

$\begin{pmatrix} 3 & -1 \\ 0 & 5 \end{pmatrix}\begin{pmatrix} a \\ b \end{pmatrix} = 5\begin{pmatrix} a \\ b \end{pmatrix}$

$5b = 5b$

$3a - b = 5a \Rightarrow b = -2a$

So eigenvector is $\begin{pmatrix} 1 \\ -2 \end{pmatrix}$

So $\mathbf{A} = \begin{pmatrix} 1 & 1 \\ 0 & -2 \end{pmatrix}\begin{pmatrix} 3 & 0 \\ 0 & 5 \end{pmatrix}\begin{pmatrix} 1 & \dfrac{1}{2} \\ 0 & -\dfrac{1}{2} \end{pmatrix}$

1 Find $\mathbf{P}^{-1}$ then write in the form $\mathbf{A} = \mathbf{PDP}^{-1}$

$\mathbf{A}^n = \begin{pmatrix} 1 & 1 \\ 0 & -2 \end{pmatrix}\begin{pmatrix} 3 & 0 \\ 0 & 5 \end{pmatrix}^n\begin{pmatrix} 1 & \dfrac{1}{2} \\ 0 & -\dfrac{1}{2} \end{pmatrix}$

2 Use $\mathbf{A}^n = \mathbf{PD}^n\mathbf{P}^{-1}$

$= \begin{pmatrix} 1 & 1 \\ 0 & -2 \end{pmatrix}\begin{pmatrix} 3^n & 0 \\ 0 & 5^n \end{pmatrix}\begin{pmatrix} 1 & \dfrac{1}{2} \\ 0 & -\dfrac{1}{2} \end{pmatrix}$

3 Simplify D^n then multiply the matrices together.

$= \begin{pmatrix} 3^n & 5^n \\ 0 & -2(5^n) \end{pmatrix}\begin{pmatrix} 1 & \dfrac{1}{2} \\ 0 & -\dfrac{1}{2} \end{pmatrix}$

$= \begin{pmatrix} 3^n & \dfrac{3^n}{2} - \dfrac{5^n}{2} \\ 0 & 5^n \end{pmatrix}$ as required

Example 6

PURE

The matrix $\mathbf{B} = \begin{pmatrix} a & a \\ b & b \end{pmatrix}$ where $a \neq -b$ and $b \neq 0$

a Find the eigenvalues and corresponding eigenvectors of $\mathbf{B}$

b Write $\mathbf{B}$ in the form $\mathbf{ADA}^{-1}$ for a diagonal matrix $\mathbf{D}$

c Hence, show that $\mathbf{B}^n = (a+b)^{n-1}\mathbf{B}$

a $\det\begin{pmatrix} a-\lambda & a \\ b & b-\lambda \end{pmatrix} = 0$

$\Rightarrow (a-\lambda)(b-\lambda) - ab = 0$

$\Rightarrow ab - (a+b)\lambda + \lambda^2 - ab = 0$

$\Rightarrow \lambda^2 - (a+b)\lambda = 0$

$\Rightarrow \lambda(\lambda - (a+b)) = 0$

$\Rightarrow \lambda = 0, a+b$

$\begin{pmatrix} a & a \\ b & b \end{pmatrix}\begin{pmatrix} x \\ y \end{pmatrix} = \begin{pmatrix} 0 \\ 0 \end{pmatrix}$

$ax + ay = 0 \Rightarrow x = -y$

So an eigenvector corresponding to the eigenvalue

of 0 is $\begin{pmatrix} 1 \\ -1 \end{pmatrix}$

$\begin{pmatrix} a & a \\ b & b \end{pmatrix}\begin{pmatrix} x \\ y \end{pmatrix} = (a+b)\begin{pmatrix} x \\ y \end{pmatrix}$

$ax + ay = ax + bx \Rightarrow ay = bx$

So an eigenvector corresponding to the eigenvalue of $a+b$ is $\begin{pmatrix} a \\ b \end{pmatrix}$

b $\mathbf{A} = \begin{pmatrix} 1 & a \\ -1 & b \end{pmatrix}$

so $\mathbf{A}^{-1} = \dfrac{1}{a+b}\begin{pmatrix} b & -a \\ 1 & 1 \end{pmatrix}$

and $\mathbf{D} = \begin{pmatrix} 0 & 0 \\ 0 & a+b \end{pmatrix}$

Therefore, $\mathbf{B} = \dfrac{1}{a+b}\begin{pmatrix} 1 & a \\ -1 & b \end{pmatrix}\begin{pmatrix} 0 & 0 \\ 0 & a+b \end{pmatrix}\begin{pmatrix} b & -a \\ 1 & 1 \end{pmatrix}$

c $\mathbf{B}^n = \dfrac{1}{a+b}\begin{pmatrix} 1 & a \\ -1 & b \end{pmatrix}\begin{pmatrix} 0 & 0 \\ 0 & a+b \end{pmatrix}^n\begin{pmatrix} b & -a \\ 1 & 1 \end{pmatrix}$

$= \dfrac{1}{a+b}\begin{pmatrix} 1 & a \\ -1 & b \end{pmatrix}\begin{pmatrix} 0 & 0 \\ 0 & (a+b)^n \end{pmatrix}\begin{pmatrix} b & -a \\ 1 & 1 \end{pmatrix}$

$= \dfrac{1}{a+b}\begin{pmatrix} 0 & a(a+b)^n \\ 0 & b(a+b)^n \end{pmatrix}\begin{pmatrix} b & -a \\ 1 & 1 \end{pmatrix}$

Solve $\det(\mathbf{B} - \lambda\mathbf{I}) = 0$ to find the eigenvalues.

Use $\mathbf{Bv} = 0$ to find the eigenvectors corresponding to the eigenvalue of 0

Use $\mathbf{Bv} = (a+b)\mathbf{v}$ to find the eigenvectors corresponding to the eigenvalue of $(a+b)$

$\mathbf{A}$ is composed of the eigenvectors of $\mathbf{B}$

$\mathbf{D}$ is a diagonal matrix with the eigenvalues on the leading diagonal.

(Continued on the next page)

$$= \frac{1}{a+b} \begin{pmatrix} a(a+b)^n & a(a+b)^n \\ b(a+b)^n & b(a+b)^n \end{pmatrix}$$

$$= \frac{(a+b)^n}{a+b} \begin{pmatrix} a & a \\ b & b \end{pmatrix}$$

$$= (a+b)^{n-1}\mathbf{B}$$

See Ch5.2
For a reminder of invariant line.

Another application of eigenvectors is in finding the direction of an invariant line through the origin.

Key point

The eigenvectors of a matrix $\mathbf{A}$ determine the direction of the invariant lines through the origin of the transformation represented by $\mathbf{A}$

If an eigenvalue of a transformation represented by matrix $\mathbf{A}$ is $\lambda = 1$, then $\mathbf{Av} = \mathbf{v}$ for all corresponding eigenvectors $\mathbf{v}$

Therefore,

Key point

If a transformation given by matrix $\mathbf{A}$ has an eigenvalue of 1 then the corresponding eigenvectors determine the direction of a line of invariant points though the origin.

If you have a repeated eigenvalue then you will sometimes be able to find two non-parallel corresponding eigenvectors. In these cases, the eigenvalue has a whole plane of associated eigenvectors, found by any linear combination of the two non-parallel eigenvectors. So the transformation represented by $\mathbf{A}$ has an invariant plane, namely the plane of eigenvectors. In other cases, there will only be one eigenvector associated with the repeated eigenvalue. In these cases the transformation has an invariant line corresponding to this eigenvector.

Key point

If a transformation given by matrix $\mathbf{A}$ has a repeated eigenvalue and this eigenvalue has two non-parallel corresponding eigenvectors, then $\mathbf{A}$ has an invariant plane defined by a linear combination of two corresponding non-parallel eigenvectors.

Example 7

The matrix $\mathbf{A} = \begin{pmatrix} 4 & -3 & 3 \\ 6 & -5 & 3 \\ 0 & 0 & -2 \end{pmatrix}$ represents a transformation.

Given that the eigenvalues of $\mathbf{A}$ are 1, −2 and −2,

a For each eigenvalue, find a full set of eigenvectors,

b Describe the geometric significance of the eigenvectors of $\mathbf{A}$ in relation to the transformation that $\mathbf{A}$ represents.

(*continued on the next page*)

a For $\lambda = 1$

$$\begin{pmatrix} 4 & -3 & 3 \\ 6 & -5 & 3 \\ 0 & 0 & -2 \end{pmatrix} \begin{pmatrix} x \\ y \\ z \end{pmatrix} = \begin{pmatrix} x \\ y \\ z \end{pmatrix}$$

Use **Av** = **v** to find the eigenvectors corresponding to the eigenvalue of 1

$-2z = z \Rightarrow z = 0$

$4x - 3y + 3z = x \Rightarrow 3x - 3y = 0$

Since $z = 0$

$\Rightarrow x = y$

Therefore, the set of eigenvectors corresponding

to the eigenvalue 1 is the set of all vectors of the form $\alpha \begin{pmatrix} 1 \\ 1 \\ 0 \end{pmatrix}$

Remember, there are infinitely many eigenvectors, all of which are scalar multiples of each other.

For $\lambda = -2$

$$\begin{pmatrix} 4 & -3 & 3 \\ 6 & -5 & 3 \\ 0 & 0 & -2 \end{pmatrix} \begin{pmatrix} x \\ y \\ z \end{pmatrix} = -2 \begin{pmatrix} x \\ y \\ z \end{pmatrix}$$

Use **Av** = -2**v** to find the eigenvectors corresponding to the eigenvalue of -2

$4x - 3y + 3z = -2x \Rightarrow 6x - 3y + 3z = 0$

$\Rightarrow 2x - y + z = 0$

$6x - 5y + 3z = -2y \Rightarrow 6x - 3y + 3z = 0$

This is the same as the first equation.

$-2z = -2z$

This just tells us that z can be any value.

So, we have the equation of a plane $2x - y + z = 0$

$\begin{pmatrix} 0 \\ 1 \\ 1 \end{pmatrix}$ and $\begin{pmatrix} 1 \\ 2 \\ 0 \end{pmatrix}$

Choose any two, non-parallel vectors on the plane.

Therefore, the eigenvectors corresponding

to the eigenvalue -2 are linear combinations of $\begin{pmatrix} 0 \\ 1 \\ 1 \end{pmatrix}$ and $\begin{pmatrix} 1 \\ 2 \\ 0 \end{pmatrix}$,

i.e. $\beta \begin{pmatrix} 0 \\ 1 \\ 1 \end{pmatrix} + \gamma \begin{pmatrix} 1 \\ 2 \\ 0 \end{pmatrix}$

b The eigenvector $\begin{pmatrix} 1 \\ 1 \\ 0 \end{pmatrix}$ gives the direction of a line of invariant

points through the origin.

The set of eigenvectors $\beta \begin{pmatrix} 0 \\ 1 \\ 1 \end{pmatrix} + \gamma \begin{pmatrix} 1 \\ 2 \\ 0 \end{pmatrix}$ represents an invariant plane.

1 Given that $\mathbf{A} = \begin{pmatrix} 1 & 0 \\ 3 & 2 \end{pmatrix}$

    a Work out $\mathbf{A}^6$

    b Find the matrix $\mathbf{A}^n$ in its simplest form.

2 Given that $\mathbf{T} = \begin{pmatrix} 1 & -2 \\ -2 & 4 \end{pmatrix}$

    a Work out $\mathbf{T}^5$

    b Show that $\mathbf{T}^n = 5^{n-1} \begin{pmatrix} 1 & -2 \\ -2 & 4 \end{pmatrix}$

3 Given that $\mathbf{M} = \begin{pmatrix} 5 & 2 \\ 2 & 5 \end{pmatrix}$

    a State matrices $\mathbf{U}$ and $\mathbf{D}$ such that $\mathbf{M} = \mathbf{UDU}^{-1}$

    b Work out the eigenvalues of $\mathbf{M}^3$

    c Write down the eigenvectors of $\mathbf{M}^3$

4 The matrix $\mathbf{B}$ can be written as

$$\mathbf{B} = \begin{pmatrix} 1 & 2 \\ -3 & 1 \end{pmatrix} \begin{pmatrix} 7 & 0 \\ 0 & -7 \end{pmatrix} \begin{pmatrix} \dfrac{1}{7} & -\dfrac{2}{7} \\ \dfrac{3}{7} & \dfrac{1}{7} \end{pmatrix}$$

    a Write down the vector equations of two invariant lines under the transformation represented by matrix $\mathbf{B}$

    b State the eigenvalues and eigenvectors of $\mathbf{B}^2$

5 The matrix $\mathbf{R}$ can be written as

$$\mathbf{R} = \begin{pmatrix} 1 & 2 & 1 \\ 0 & -3 & 1 \\ 1 & 0 & 2 \end{pmatrix} \begin{pmatrix} 3 & 0 & 0 \\ 0 & 2 & 0 \\ 0 & 0 & -4 \end{pmatrix} \begin{pmatrix} 6 & 4 & -5 \\ -1 & -1 & 1 \\ -3 & -2 & 3 \end{pmatrix}$$

    a State the eigenvalues and the eigenvectors of $\mathbf{R}^4$

    b Write down the vector equations of the invariant lines through the origin of the transformation represented by matrix $\mathbf{R}$

6 Given that $\mathbf{A} = \begin{pmatrix} 5 & -1 & 0 \\ 2 & 2 & 0 \\ 0 & 0 & 1 \end{pmatrix}$

    a Work out $\mathbf{A}^3$

    b Find the matrix $\mathbf{A}^n$ in its simplest form.

7 **a** Find the equations of two invariant lines under the transformation $\mathbf{S} = \begin{pmatrix} -3 & 4 \\ 2 & -1 \end{pmatrix}$

    b Which of these lines is also a line of invariant points? Explain your answer.

8 The matrix $\mathbf{M}$ is given by $\mathbf{M} = \begin{pmatrix} a & b \\ b & a \end{pmatrix}$

where a and b are constants and $b \neq 0$

Find the eigenvalues and eigenvectors of $\mathbf{M}$

9 The matrix $\mathbf{M} = \begin{pmatrix} -15 & 24 \\ -8 & 13 \end{pmatrix}$ represents a transformation.

    a Find the eigenvalues and corresponding full sets of eigenvectors of $\mathbf{M}$

    b Describe the geometric significance of the eigenvectors of $\mathbf{M}$ in relation to the transformation $\mathbf{M}$ represents.

10 The matrix $\mathbf{A} = \begin{pmatrix} 3 & -10 & -2 \\ 0 & -1 & 0 \\ 0 & 2 & 1 \end{pmatrix}$ represents a transformation.

    a Find the eigenvalues of $\mathbf{A}$

    b For each eigenvalue, find a full set of eigenvectors.

    c Describe the geometric significance of the eigenvectors of $\mathbf{A}$ in relation to the transformation $\mathbf{A}$ represents.

11 The matrix $T = \begin{pmatrix} 3 & -2 & 1 \\ 1 & 0 & 1 \\ -2 & 4 & 0 \end{pmatrix}$ represents a transformation.

a Use row and column operations to find the eigenvalues of T

b For each eigenvalue, find a full set of eigenvectors.

c Describe the geometric significance of the eigenvectors of T in relation to the transformation T represents.

12 The matrix M is given by $M = \begin{pmatrix} 1 & 1 & 2 \\ 0 & 1 & -1 \\ 0 & 3 & 0 \end{pmatrix}$

a Show that there is only one real eigenvalue of M

b Hence, show that the x–axis is a line of invariant points under the transformation represented by M

13 a Given that the matrix M can be written UDU^{-1} where D is a diagonal matrix, show that $M^n = UD^nU^{-1}$

The 2×2 matrix A has eigenvalues of $\dfrac{3}{5}$ and 1 with corresponding eigenvectors $\begin{pmatrix} 1 \\ 0 \end{pmatrix}$ and $\begin{pmatrix} -1 \\ 2 \end{pmatrix}$

b Given that $A^n \to L$ as $n \to \infty$, find the matrix L

14 Given the matrix $M = \begin{pmatrix} 3 & k+3 \\ k-3 & -3 \end{pmatrix}$, where k is a constant and $k \neq \pm 3$

a Find the eigenvalues and the corresponding eigenvectors of M

b Hence, show that $M^{2n+1} = k^{2n}M$ for all integers n

Chapter summary

- The determinant of a 3×3 matrix is

$$\det\begin{pmatrix} a & b & c \\ d & e & f \\ g & h & i \end{pmatrix} = a\begin{vmatrix} e & f \\ h & i \end{vmatrix} - b\begin{vmatrix} d & f \\ g & i \end{vmatrix} + c\begin{vmatrix} d & e \\ g & h \end{vmatrix}$$

- Under a transformation represented by a 2×2 matrix $\mathbf{T}$,
  area of image = area of original $\times |\det(\mathbf{T})|$
- Under a transformation represented by a 3×3 matrix $\mathbf{T}$,
  volume of image = volume of original $\times |\det(\mathbf{T})|$
- If $\det(\mathbf{T}) < 0$ then the transformation represented by $\mathbf{T}$ involves a reflection.

- If $\mathbf{P} = \begin{pmatrix} a & b & c \\ d & e & f \\ g & h & i \end{pmatrix}$ then the matrix of minors is $\mathbf{M} = \begin{pmatrix} A & B & C \\ D & E & F \\ G & H & I \end{pmatrix}$ where A is the minor of a, that

  is $\det\begin{pmatrix} e & f \\ h & i \end{pmatrix} = (ei - fh)$, B is the minor of b and so on,

  then the inverse matrix is $\mathbf{P}^{-1} = \dfrac{1}{\det(\mathbf{P})}\begin{pmatrix} A & -D & G \\ -B & E & -H \\ C & -F & I \end{pmatrix}$

- A system of linear equations represented by the matrix equation $\mathbf{Mx} = \mathbf{x}'$ has a unique solution if and only if $\det(\mathbf{M}) \neq 0$
- Effect of row and column operations:
  - adding or subtracting any multiple of a row to another row or any multiple of a column to another column will not affect the determinant
  - Swapping two rows or swapping two columns will change the sign of the determinant
  - Multiplying or dividing a row or a column of a matrix by a scalar will multiply or divide the determinant by that scalar.
- Eigenvectors of the square matrix $\mathbf{A}$ are non-zero vectors that satisfy the equation $\mathbf{Ax} = \lambda\mathbf{x}$.
- λ is a scalar known as the eigenvalue.
- The characteristic equation $\det(\mathbf{A} - \lambda\mathbf{I}) = 0$ can be used to find the eigenvalues.
- A matrix will always satisfy its own characteristic equation.
- The eigenvectors of a matrix $\mathbf{T}$ determine the direction of the invariant lines through the origin of the transformation represented by $\mathbf{T}$
- If a transformation given by matrix $\mathbf{T}$ has an eigenvalue of 1 then the corresponding eigenvectors determine the direction of a line of invariant points though the origin.
- If a transformation given by matrix $\mathbf{T}$ has a repeated eigenvalue then $\mathbf{T}$ has an invariant plane defined by a linear combination of two corresponding non-parallel eigenvectors.
- A square matrix is diagonal if all its elements except those on the leading diagonal are zero.

- A 3×3 matrix, **M**, can be diagonalised by finding **P** and **D** such that $\mathbf{M} = \mathbf{PDP}^{-1}$

  - $$\mathbf{D} = \begin{pmatrix} \lambda_1 & 0 & 0 \\ 0 & \lambda_2 & 0 \\ 0 & 0 & \lambda_3 \end{pmatrix}$$ where λ_1, λ_2, λ_3 are the eigenvalues of **M**

  - $$\mathbf{P} = \begin{pmatrix} a_1 & a_2 & a_3 \\ b_1 & b_2 & b_3 \\ c_1 & c_2 & c_3 \end{pmatrix}$$ where $\mathbf{v}_n = \begin{pmatrix} a_n \\ b_n \\ c_n \end{pmatrix}$ is an eigenvector corresponding to the eigenvalue λ_n

- If $\mathbf{M} = \mathbf{PDP}^{-1}$ then $\mathbf{M}^n = \mathbf{PD}^n\mathbf{P}^{-1}$

Check and review

You should now be able to...	Review Questions
✔ Use determinants as scale factors.	1, 4
✔ Understand the implications of a negative determinant.	1, 4
✔ Calculate the determinant of a 3×3 matrix.	2, 3
✔ Find the inverse of a 3×3 matrix.	5
✔ Use matrices to solve systems of linear equations.	6
✔ Identify when a system of 3 simultaneous equations has a unique solution, infinitely many solutions and no solutions, and interpret geometrically.	7
✔ Calculate the determinant of a matrix by multiplying out any row or column.	8
✔ Understand the effect of row and column operations on the determinant of a matrix.	9
✔ Factorise a determinant using row and column operations.	10, 11
✔ Find the eigenvalues and eigenvectors of a 2×2 matrix and a 3×3 matrix.	12, 13, 15
✔ Demonstrate that a matrix satisfies its characteristic equation.	14
✔ Diagonalise a square matrix.	16
✔ Use diagonal form to find the power of a matrix.	17
✔ Understand the geometrical significance of eigenvalues and eigenvectors.	18, 19

1 A triangle has area 8 square units.

a Give the area of each image of the triangle under each of these transformations

   i $\begin{pmatrix} 5 & 0 \\ 0 & 5 \end{pmatrix}$ **ii** $\begin{pmatrix} 2 & 4 \\ 3 & 1 \end{pmatrix}$

   iii $\begin{pmatrix} \dfrac{1}{\sqrt{2}} & -\dfrac{1}{\sqrt{2}} \\ \dfrac{1}{\sqrt{2}} & \dfrac{1}{\sqrt{2}} \end{pmatrix}$ **iv** $\begin{pmatrix} 2a & 3a \\ 4a & -2a \end{pmatrix}$

b Explain which of the transformations involve(s) a reflection.

2 Work out the determinant of each of these matrices.

a $\begin{pmatrix} 3 & -2 & -1 \\ 4 & 0 & 5 \\ 2 & 1 & -3 \end{pmatrix}$ **b** $\begin{pmatrix} a & -3 & 2a \\ 2b & 4 & b \\ 5 & 0 & -1 \end{pmatrix}$

3 Calculate the values of x for which each of these matrices are singular.

a $\begin{pmatrix} 2 & -1 & 1 \\ x & 3 & 0 \\ -x & 4 & 2 \end{pmatrix}$ **b** $\begin{pmatrix} x & 3 & 0 \\ 1 & 2 & 2x \\ 1 & -1 & 2 \end{pmatrix}$

4 A cube is transformed by matrix **T** and the volume of the image is 56 cubic units.

a Give the volume of the original cube when

i $\mathbf{T} = \begin{pmatrix} 3 & 4 & 2 \\ 0 & 3 & 1 \\ -5 & 5 & -1 \end{pmatrix}$

ii $\mathbf{T} = \begin{pmatrix} a & -2 & 4 \\ 3 & 4-a & 2a \\ -2a & 0 & 0 \end{pmatrix}$ for $a > 0$

b Explain which of the transformations involve(s) a reflection.

5 Find the inverse, if it exists, of each of these matrices. If it does not exist, explain why not.

a $\begin{pmatrix} 3 & 2 & -1 \\ 4 & 0 & 5 \\ 2 & 1 & -3 \end{pmatrix}$ **b** $\begin{pmatrix} 6 & -2 & -3 \\ 1 & 0 & -1 \\ 0 & 2 & -3 \end{pmatrix}$

6 Use matrices to find a solution to the simultaneous equations

$3x - 2y + z = 6$

$x - y + 3z = -23$

$5x - 3y + 2z = 5$

7 Three planes have equations

$6x - 2y + 4z = 3$

$(k+4)x - 3y + z = -5$

$-3x + ky - 2z = 7$

a Find the values of k for which the system of equations does not have a unique solution.

b Describe the arrangement of the three planes for each value of k found in part **a**.

8 Multiply out the determinant $\begin{vmatrix} 3 & -2 & 5 \\ -6 & 4 & 0 \\ 5 & -1 & -3 \end{vmatrix}$ using the 3rd row and show that the solution is the same as the result of multiplying out using the 3rd column.

9 Given that $\begin{vmatrix} a & 3 & 2b \\ -6 & 0 & 4b \\ a & 2 & -5b \end{vmatrix} = 90$, state the values of

a $\begin{vmatrix} a & 2b & 3 \\ -6 & 4b & 0 \\ a & -5b & 2 \end{vmatrix}$ **b** $\begin{vmatrix} a & 3 & 2b \\ 3 & 0 & -2b \\ a & 2 & -5b \end{vmatrix}$

c $\begin{vmatrix} a & -6 & a \\ 3 & 0 & 2 \\ 2b & 4b & -5b \end{vmatrix}$ **d** $\begin{vmatrix} 0 & 1 & 7 \\ 3 & 0 & -2 \\ a & 2 & -5 \end{vmatrix}$

10 a Show that

$\begin{vmatrix} 1 & -1 & 1 \\ a & a^2 & -a^2 \\ b & b^2 & -b^2 \end{vmatrix} = ab(a+1)(b+1) \begin{vmatrix} 1 & -1 \\ 1 & -1 \end{vmatrix}$

b Hence state the value of $\begin{vmatrix} 1 & -1 & 1 \\ a & a^2 & -a^2 \\ b & b^2 & -b^2 \end{vmatrix}$

11 Fully factorise the determinant $\begin{vmatrix} bc & a & a^2 \\ ac & b & b^2 \\ ab & c & c^2 \end{vmatrix}$

12 Find the eigenvalues and corresponding eigenvectors of

a $\begin{pmatrix} -3 & 5 \\ 4 & -2 \end{pmatrix}$ **b** $\begin{pmatrix} -15 & 12 \\ 6 & -21 \end{pmatrix}$

13 Find the eigenvalues and corresponding eigenvectors of

a $\begin{pmatrix} 2 & 0 & 0 \\ 2 & 3 & 2 \\ 7 & -4 & -3 \end{pmatrix}$ **b** $\begin{pmatrix} 4 & 2 & 9 \\ 0 & 9 & 0 \\ 2 & 1 & 7 \end{pmatrix}$

14 a Show that the characteristic equation of the matrix $A = \begin{pmatrix} a & 3 \\ -2 & 2a \end{pmatrix}$ is

$$\lambda^2 - 3a\lambda + 2a^2 + 6 = 0$$

b Demonstrate that the matrix **A** satisfies its own characteristic equation.

15 Given that $M = \begin{pmatrix} -1 & 0 & 2 \\ -1 & 2 & 0 \\ -4 & 0 & 3 \end{pmatrix}$

a Show that $\lambda = 2$ is the only real eigenvalue of **M**

b Find the eigenvector corresponding to $\lambda = 2$

16 $A = \begin{pmatrix} -3 & -2 & 0 \\ 0 & 2 & 0 \\ 3 & 1 & 0 \end{pmatrix}$

State a matrix **P** and a diagonal matrix **D** such that $A = PDP^{-1}$

17 $M = \begin{pmatrix} 2 & 2 & 0 \\ 1 & 1 & 0 \\ 0 & 2 & 1 \end{pmatrix}$

Show that $M^n = \begin{pmatrix} 2(3^{n-1}) & 2(3^{n-1}) & 0 \\ 3^{n-1} & 3^{n-1} & 0 \\ 3^{n-1}-1 & 3^{n-1}+1 & 1 \end{pmatrix}$

18 The matrix **A** can be written

$$A = \begin{pmatrix} 3 & 2 \\ -1 & 3 \end{pmatrix} \begin{pmatrix} 1 & 0 \\ 0 & -2 \end{pmatrix} \begin{pmatrix} 3 & 2 \\ -1 & 3 \end{pmatrix}^{-1}$$

Write down the vector equation of

a A line of invariant points,

b An invariant line which is not a line of invariant points.

19 a Find the eigenvalues of the matrix

$$B = \begin{pmatrix} -3 & 5 & -10 \\ 10 & -8 & 20 \\ 5 & -5 & 12 \end{pmatrix}$$

b For each eigenvalue, find the full set of corresponding eigenvectors.

c Describe the geometrical significance of the eigenvectors of **B** in relation to the transformation it represents.

Did you know?

Internet search engines use algorithms that rank the importance of pages according to an eigenvector of a weighted link matrix.

Google

Google Search   I'm Feeling Lucky

Research

There are several alternative methods for solving systems of linear equations. Find out about:

- Cramer's rule
- Cholesky decomposition
- Quantum algorithm for linear systems of equations.

Research

Find the eigenvalues and the eigenvectors of the matrix $A = \begin{bmatrix} 2 & 1 \\ 1 & 2 \end{bmatrix}$.

Now plot the four points $(0,0)$, $(1,0)$, $(0, 1)$ and $(1,1)$ and draw in the three vectors as shown below.
Transform each of the four points using matrix A. Draw in the new vectors created by the transformed points.

For each of the vectors consider the following questions.
- Which vectors have had their direction maintained and why?
- Which vectors have been enlarged, by how much, and why?
- How do these observations relate to the eigenvalues?
- How do these observations relate to the eigenvectors?

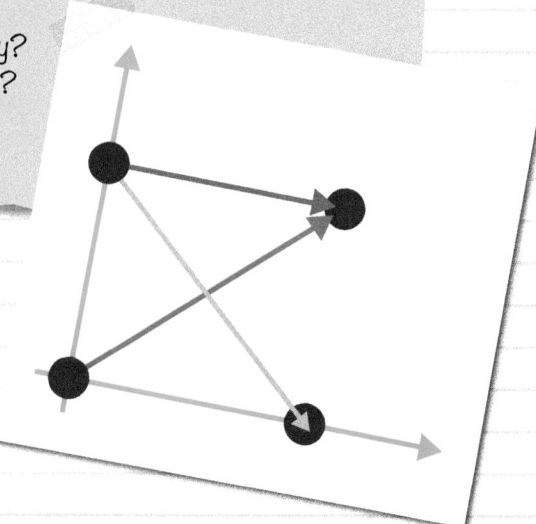

22   Assessment

1   Find the coordinates of the point of intersection between the three planes with equations

$2x+y-z=21$

$x+2y-3z=-14$

$5x-y+z=-7$ **[5 marks]**

2   A linear transformation is defined by the matrix $\mathbf{T}=\begin{pmatrix} 1 & 0 & 0 \\ 0 & 0.5 & -a \\ 0 & a & 0.5 \end{pmatrix}$

When the transformation is applied to a cube, the volume of the image is the same as the volume of the image.

a   Calculate the possible values of a **[3]**

b   Describe the possible transformation represented by $\mathbf{T}$ **[2]**

3   Find the range of values of k for which the transformation $\begin{pmatrix} 3-k & 2 \\ k & k+1 \end{pmatrix}$ involves a reflection.   **[3]**

4   a   Find the eigenvalues and corresponding eigenvectors of the matrix $\mathbf{A}=\begin{pmatrix} 2 & -3 \\ -8 & 4 \end{pmatrix}$ **[5]**

b   Verify that A satisfies its own characteristic equation.   **[3]**

5   $\mathbf{M}=\begin{pmatrix} 1 & k-1 & -k \\ k+2 & 3 & 0 \\ -2k & 2 & 8 \end{pmatrix}$

a   Calculate the values of k for which $\mathbf{M}$ is singular.   **[4]**

b   Three planes have equations

$x-z=5$

$3x+3y=2$

$-2x+2y+8z=-1$

Explain how your answer to part **a** guarantees that the three planes have a unique point of intersection.   **[3]**

6   Given that $\mathbf{A}=\begin{pmatrix} 2 & 0 & 1 \\ 0 & 1 & 0 \\ 1 & 0 & 2 \end{pmatrix}$, prove that $\mathbf{A}^n=\dfrac{1}{2}\begin{pmatrix} 3^n+1 & 0 & 3^n-1 \\ 0 & 2 & 0 \\ 3^n-1 & 0 & 3^n+1 \end{pmatrix}$ for all $n\in\mathbb{N}$ **[5]**

7   Use row and column operations to show that $\begin{vmatrix} 2 & 1 & 2 \\ 4 & 1 & 2 \\ 1 & -1 & 1 \end{vmatrix}=A\begin{vmatrix} 0 & 1 & 0 \\ 1 & 1 & 0 \\ 1 & 0 & 1 \end{vmatrix}$, where A is a constant to be found. Explain each step clearly.   **[4]**

8   The 3×3 matrices **A** and **B** are such that $\mathbf{AB}=\begin{pmatrix} 6 & 1 & -2 \\ 14 & 2 & 4 \\ 2 & 0 & -1 \end{pmatrix}$ and $\mathbf{B}=\begin{pmatrix} a & 0 & 0 \\ 0 & 2 & 2 \\ 1 & 0 & 1 \end{pmatrix}$

Find the matrix **A** in terms of a **[5]**

9 Given that the determinant of $\begin{pmatrix} a & b & c \\ d & e & f \\ g & h & i \end{pmatrix}$ is 30, write down the determinants of

a $\begin{pmatrix} a & b & 3c \\ 2d & 2e & 6f \\ g & h & 3i \end{pmatrix}$ **b** $\begin{pmatrix} g & h & i \\ d+a & e+b & f+c \\ a & b & c \end{pmatrix}$ **[2]**

10 a Use at least 2 row or column operations to fully factorise $\begin{vmatrix} a & b & c \\ a^2 & b^2 & c^2 \\ 1 & 1 & 1 \end{vmatrix}$ **[4]**

b Hence or otherwise, calculate the value of $\begin{vmatrix} k & k+1 & k-1 \\ k^2 & (k+1)^2 & (k-1)^2 \\ 1 & 1 & 1 \end{vmatrix}$ **[2]**

11 $\mathbf{T} = \begin{pmatrix} 1 & 2 & 0 \\ 0 & 0 & 3 \\ 2 & 0 & 1 \end{pmatrix}$

a Show that $\lambda = 3$ is the only real eigenvalue of $\mathbf{T}$ **[6]**

b Find the unit eigenvector corresponding to $\lambda = 3$ **[5]**

c Hence write down the vector equation of an invariant line under the transformation represented by matrix $\mathbf{T}$ **[2]**

12 Use row and column operations to solve the equation **[6]**

$\begin{vmatrix} 1 & x+2 & x-1 \\ x^2 & 1 & 4 \\ x & -1 & 2 \end{vmatrix} = 0$

13 Given that $\mathbf{M} = \begin{pmatrix} 0 & 0 & 2 \\ -2 & 1 & 0 \\ 2 & 0 & 0 \end{pmatrix}$

a Find matrix $\mathbf{P}$ and diagonal matrix $\mathbf{D}$ such that $\mathbf{M} = \mathbf{PDP}^{-1}$ **[8]**

b Hence find $\mathbf{M}^4$. You must show how you used your answer to part **a**. **[3]**

c Explain how you know from part **a** that the y-axis is a line of invariant points. **[2]**

14 Three planes have equations

$x + 2y - 3z = -2$

$4x - z = 3$

$2x - 4y + 5z = 1$

Identify the geometric configuration of the planes. Fully justify your answer. **[4]**

15 The matrix $\mathbf{A}$ can be written $\mathbf{A} = \begin{pmatrix} 1 & 0 & 1 \\ 1 & -1 & 2 \\ 0 & 1 & -2 \end{pmatrix} \begin{pmatrix} 3 & 0 & 0 \\ 0 & -7 & 0 \\ 0 & 0 & -7 \end{pmatrix} \begin{pmatrix} 1 & 0 & 1 \\ 1 & -1 & 2 \\ 0 & 1 & -2 \end{pmatrix}^{-1}$

Write down three eigenvectors of $\mathbf{A}$ and explain their geometrical significance in relation to the transformation represented by $\mathbf{A}$ **[4]**

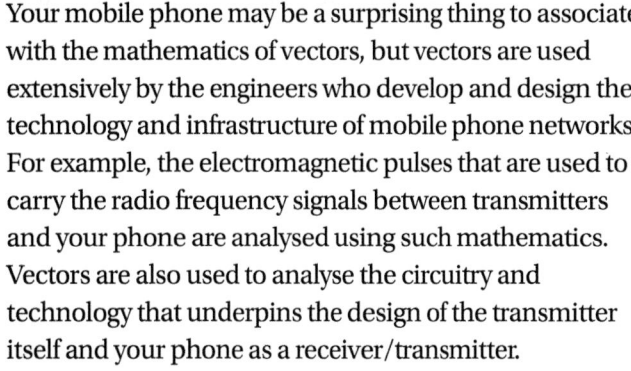

23 Vectors 2

Your mobile phone may be a surprising thing to associate with the mathematics of vectors, but vectors are used extensively by the engineers who develop and design the technology and infrastructure of mobile phone networks. For example, the electromagnetic pulses that are used to carry the radio frequency signals between transmitters and your phone are analysed using such mathematics. Vectors are also used to analyse the circuitry and technology that underpins the design of the transmitter itself and your phone as a receiver/transmitter.

The use of vectors by engineers is widespread and of great importance in many aspects of our lives, particularly in the technology that we associate with modern living such as televisions, radios, microwave ovens and more. The development of our modern-day transportation systems including electric trains, electric cars and aircraft has all been achieved by our increasing knowledge and application of science, together with our use of vector mathematics.

Orientation

What you need to know	What you will learn	What this leads to
Ch6 Vectors 1	• To calculate the vector product of two vectors. • To use the vector product to find angles and areas. • To find the vector equation of a plane. • To calculate the distance between two lines. • To calculate the distance between a line and a plane.	**Careers** • Electrical engineering. • Mechanical engineering.

23.1 The vector product

Fluency and skills

Vectors can be multiplied by calculating the scalar product. Another sort of vector multiplication is the vector product.

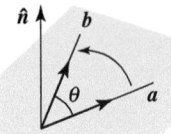

> **Key point**
>
> The **vector product** $\mathbf{a} \times \mathbf{b}$ of vectors $\mathbf{a}$ and $\mathbf{b}$ is defined as $\mathbf{a} \times \mathbf{b} = |\mathbf{a}||\mathbf{b}|\sin\theta\hat{\mathbf{n}}$ where θ is the angle between the vectors $\mathbf{a}$ and $\mathbf{b}$ and $\hat{\mathbf{n}}$ is a unit vector perpendicular to both $\mathbf{a}$ and $\mathbf{b}$

The vector product is sometimes called the **cross product**.

Suppose you have a plane containing the vectors $\mathbf{a}$ and $\mathbf{b}$. An anticlockwise angle from vector $\mathbf{a}$ to vector $\mathbf{b}$ implies that $\hat{\mathbf{n}}$ is directed 'upwards', whereas a clockwise angle implies that $\hat{\mathbf{n}}$ is directed in the opposite direction.

Therefore, you can see that the vector product is not commutative.

> The 'right-hand rule' can be used to determine the direction of the cross product. Use your index finger for vector $\mathbf{a}$ and middle finger for vector $\mathbf{b}$; then your thumb will indicate the direction of $\mathbf{a} \times \mathbf{b}$

> The convention is to use 'right-handed' axes as shown here

Example 1

Show that $\mathbf{i} \times \mathbf{j} = \mathbf{k}$ and $\mathbf{j} \times \mathbf{i} = -\mathbf{k}$

$\mathbf{i} \times \mathbf{j} = |1||1|\sin 90 \mathbf{k}$

$\quad = \mathbf{k}$

$\mathbf{j} \times \mathbf{i} = |1||1|\sin(-90)\mathbf{k}$

$\quad = -\mathbf{k}$

> Since $\mathbf{i}, \mathbf{j}$ and $\mathbf{k}$ are unit vectors which are perpendicular to each other.

> The angle between $\mathbf{i}$ and $\mathbf{j}$ is now 90° clockwise which you can write as −90° or +270°

Similarly, you can show that

> **Key point**

$\mathbf{i} \times \mathbf{j} = \mathbf{k}, \mathbf{j} \times \mathbf{k} = \mathbf{i}$ and $\mathbf{k} \times \mathbf{i} = \mathbf{j}$

And that, in general

> **Key point**
>
> For any vectors $\mathbf{a}$, $\mathbf{b}$ and $\mathbf{c}$
>
> - $\mathbf{a} \times \mathbf{b} = -\mathbf{b} \times \mathbf{a}$ (anticommutative property)
> - $\mathbf{a} \times (\mathbf{b}+\mathbf{c}) = \mathbf{a} \times \mathbf{b} + \mathbf{a} \times \mathbf{c}$ (distributive property)

From the definition $\mathbf{a} \times \mathbf{b} = |\mathbf{a}||\mathbf{b}|\sin\theta\,\hat{\mathbf{n}}$, it is clear that if vectors $\mathbf{a}$ and $\mathbf{b}$ are parallel, then $\mathbf{a} \times \mathbf{b} = \mathbf{0}$ since the angle between parallel vectors is either $0°$ or $180°$ and $\sin 0 = \sin 180 = 0$

> **Key point**
>
> For any non-zero vectors $\mathbf{a}$ and $\mathbf{b}$,
>
> $\mathbf{a} \times \mathbf{b} = \mathbf{0}$ if and only if $\mathbf{a}$ and $\mathbf{b}$ are parallel

An important consequence of this fact is that $\mathbf{a} \times \mathbf{a} = \mathbf{0}$ for any vector $\mathbf{a}$

Example 2

Given that $\mathbf{a}$ and $\mathbf{b}$ and $\mathbf{c}$ are non-parallel vectors and $\mathbf{b} \neq \mathbf{0}$

a Write $(\mathbf{a} + 3\mathbf{b}) \times (\mathbf{a} - \mathbf{b})$ in its simplest form,

b If $\mathbf{a} \times \mathbf{b} = \mathbf{b} \times \mathbf{c}$, show that $\mathbf{a} + \mathbf{c} = \lambda\mathbf{b}$ for some scalar λ

a $(a + 3b) \times (a - b) = a \times a + a \times (-b) + 3b \times a + (3b) \times (-b)$

$= a \times a - a \times b + 3b \times a - 3b \times b$

$= -a \times b + 3b \times a$ Since $a \times a = 0$ and $b \times b = 0$

$= b \times a + 3b \times a$ Since $a \times b = -b \times a$

$= 4b \times a$

b If $a \times b = b \times c$

Then $b \times c - a \times b = 0$

$b \times c + b \times a = 0$ Use the anticommutative property.

Therefore $b \times (a + c) = 0$ Use the distributive property.

So $a + c$ must be parallel to b

Therefore $a + c = \lambda b$ for some scalar λ Since $b \neq 0$ and $a \neq -c$ (since not parallel).

If you have two vectors $\mathbf{a} = a_1\mathbf{i} + a_2\mathbf{j} + a_3\mathbf{k}$ and $\mathbf{b} = b_1\mathbf{i} + b_2\mathbf{j} + b_3\mathbf{k}$ then you can find the vector product

$\mathbf{a} \times \mathbf{b} = (a_1\mathbf{i} + a_2\mathbf{j} + a_3\mathbf{k}) \times (b_1\mathbf{i} + b_2\mathbf{j} + b_3\mathbf{k})$

$= a_1 b_1 (\mathbf{i} \times \mathbf{i}) + a_1 b_2 (\mathbf{i} \times \mathbf{j}) + a_1 b_3 (\mathbf{i} \times \mathbf{k}) + a_2 b_1 (\mathbf{j} \times \mathbf{i}) + a_2 b_2 (\mathbf{j} \times \mathbf{j}) + a_2 b_3 (\mathbf{j} \times \mathbf{k})$

$\quad + a_3 b_1 (\mathbf{k} \times \mathbf{i}) + a_3 b_2 (\mathbf{k} \times \mathbf{j}) + a_3 b_3 (\mathbf{k} \times \mathbf{k})$

$= a_1 b_2 \mathbf{k} - a_1 b_3 \mathbf{j} - a_2 b_1 \mathbf{k} + a_2 b_3 \mathbf{i} + a_3 b_1 \mathbf{j} - a_3 b_2 \mathbf{i}$

$= (a_2 b_3 - a_3 b_2)\mathbf{i} - (a_1 b_3 - a_3 b_1)\mathbf{j} + (a_1 b_2 - a_1 b_2)\mathbf{k}$

> Since $\mathbf{i} \times \mathbf{j} = \mathbf{k}$, $\mathbf{j} \times \mathbf{k} = \mathbf{i}$ and $\mathbf{k} \times \mathbf{i} = \mathbf{j}$, and also $\mathbf{i} \times \mathbf{i} = \mathbf{j} \times \mathbf{j} = \mathbf{k} \times \mathbf{k} = 0$

> **Key point**
>
> If you have two vectors $\mathbf{a} = a_1\mathbf{i} + a_2\mathbf{j} + a_3\mathbf{k}$ and $\mathbf{b} = b_1\mathbf{i} + b_2\mathbf{j} + b_3\mathbf{k}$, then the vector product is defined as
>
> $\mathbf{a} \times \mathbf{b} = (a_2 b_3 - a_3 b_2)\mathbf{i} - (a_1 b_3 - a_3 b_1)\mathbf{j} + (a_1 b_2 - a_1 b_2)\mathbf{k}$

Example 3

Given $\mathbf{a} = \begin{pmatrix} 1 \\ -3 \\ -2 \end{pmatrix}$ and $\mathbf{b} = \begin{pmatrix} -2 \\ 0 \\ 4 \end{pmatrix}$,

a Find a vector perpendicular to both $\mathbf{a}$ and $\mathbf{b}$

b Calculate the acute angle between $\mathbf{a}$ and $\mathbf{b}$

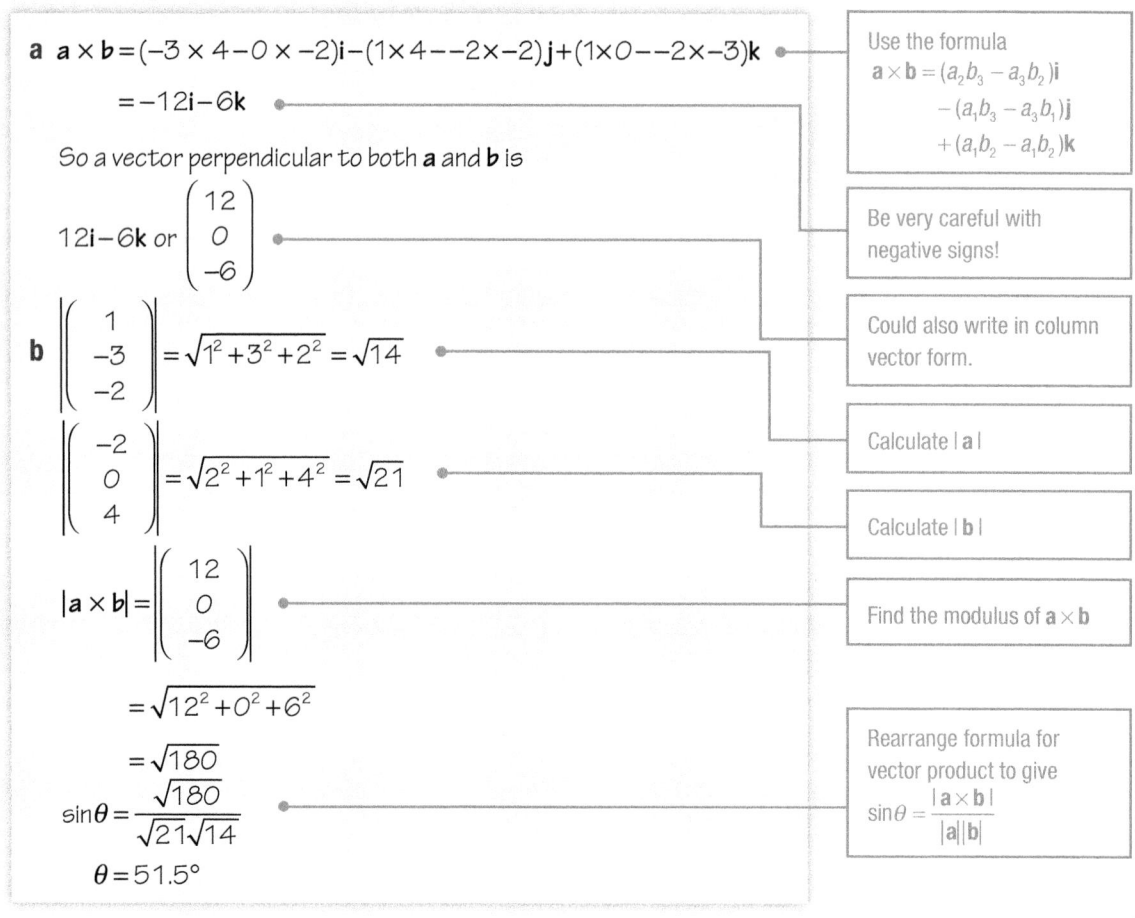

a $\mathbf{a} \times \mathbf{b} = (-3 \times 4 - 0 \times -2)\mathbf{i} - (1 \times 4 - -2 \times -2)\mathbf{j} + (1 \times 0 - -2 \times -3)\mathbf{k}$

$= -12\mathbf{i} - 6\mathbf{k}$

So a vector perpendicular to both $\mathbf{a}$ and $\mathbf{b}$ is

$12\mathbf{i} - 6\mathbf{k}$ or $\begin{pmatrix} 12 \\ 0 \\ -6 \end{pmatrix}$

b $\left| \begin{pmatrix} 1 \\ -3 \\ -2 \end{pmatrix} \right| = \sqrt{1^2 + 3^2 + 2^2} = \sqrt{14}$

$\left| \begin{pmatrix} -2 \\ 0 \\ 4 \end{pmatrix} \right| = \sqrt{2^2 + 1^2 + 4^2} = \sqrt{21}$

$|\mathbf{a} \times \mathbf{b}| = \left| \begin{pmatrix} 12 \\ 0 \\ -6 \end{pmatrix} \right|$

$= \sqrt{12^2 + 0^2 + 6^2}$

$= \sqrt{180}$

$\sin\theta = \dfrac{\sqrt{180}}{\sqrt{21}\sqrt{14}}$

$\theta = 51.5°$

Use the formula
$\mathbf{a} \times \mathbf{b} = (a_2 b_3 - a_3 b_2)\mathbf{i}$
$\quad - (a_1 b_3 - a_3 b_1)\mathbf{j}$
$\quad + (a_1 b_2 - a_1 b_2)\mathbf{k}$

Be very careful with negative signs!

Could also write in column vector form.

Calculate $|\mathbf{a}|$

Calculate $|\mathbf{b}|$

Find the modulus of $\mathbf{a} \times \mathbf{b}$

Rearrange formula for vector product to give
$\sin\theta = \dfrac{|\mathbf{a} \times \mathbf{b}|}{|\mathbf{a}||\mathbf{b}|}$

See Ch22.1
For a remind of how to calculate determinants.

From your study of determinants of 3×3 matrices, you will be able to use the fact that

$\mathbf{a} \times \mathbf{b} = \left| \begin{pmatrix} \mathbf{i} & \mathbf{j} & \mathbf{k} \\ a_1 & a_2 & a_3 \\ b_1 & b_2 & b_3 \end{pmatrix} \right|$

Example 4

Find the vector product of $2\mathbf{i} + \mathbf{j} - \mathbf{k}$ and $\mathbf{i} - 3\mathbf{j} - 2\mathbf{k}$

$\left| \begin{pmatrix} \mathbf{i} & \mathbf{j} & \mathbf{k} \\ 2 & 1 & -1 \\ 1 & -3 & -2 \end{pmatrix} \right| = \mathbf{i}(-2 - 3) - \mathbf{j}(-4 - -1) + \mathbf{k}(-6 - 1)$

$= -5\mathbf{i} + 3\mathbf{j} - 7\mathbf{k}$

Use $\mathbf{a} \times \mathbf{b} = \left| \begin{pmatrix} \mathbf{i} & \mathbf{j} & \mathbf{k} \\ a_1 & a_2 & a_3 \\ b_1 & b_2 & b_3 \end{pmatrix} \right|$

with the coefficients of $\mathbf{i}$, $\mathbf{j}$ and $\mathbf{k}$ you are given.

Try it on your calculator

You can use a calculator to find the vector product.

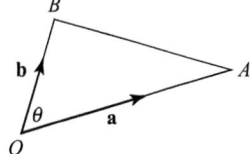

VctA × VctB
[13 -3 -2]

Activity

Find out how to work out

$\begin{pmatrix} 1 \\ 5 \\ -1 \end{pmatrix} \times \begin{pmatrix} 0 \\ -2 \\ 3 \end{pmatrix}$ on your

calculator.

The vector product can also be used to calculate areas. For example, the area of the triangle shown can be found using Area $= \frac{1}{2}|\mathbf{a}||\mathbf{b}|\sin\theta$ which, by definition, is $\frac{1}{2}|\mathbf{a}\times\mathbf{b}|$

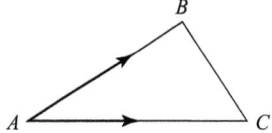

Key point

Area of triangle $OAB = \frac{1}{2}|\mathbf{a}\times\mathbf{b}|$

You may have to first work out vectors **a** and **b** from the coordinates of the vertices of the triangle. So, in general,

Key point

Area of triangle $ABC = \frac{1}{2}|\overrightarrow{AB}\times\overrightarrow{AC}|$

Example 5

Calculate the area of the triangle ABC with vertices at $A(2, 5, -1)$, $B(-4, 9, -6)$ and $C(3, -4, 8)$

$\overrightarrow{AB} = \begin{pmatrix} -4 \\ 9 \\ -6 \end{pmatrix} - \begin{pmatrix} 2 \\ 5 \\ -1 \end{pmatrix} = \begin{pmatrix} -6 \\ 4 \\ -5 \end{pmatrix}$

$\overrightarrow{AC} = \begin{pmatrix} 3 \\ -4 \\ 8 \end{pmatrix} - \begin{pmatrix} 2 \\ 5 \\ -1 \end{pmatrix} = \begin{pmatrix} 1 \\ -9 \\ 9 \end{pmatrix}$

Calculate $\overrightarrow{AB}$ and $\overrightarrow{AC}$ by subtracting position vectors, e.g. $\overrightarrow{AB} = \overrightarrow{OB} - \overrightarrow{OA}$

Area of triangle $= \frac{1}{2}\left|\begin{pmatrix} -6 \\ 4 \\ -5 \end{pmatrix} \times \begin{pmatrix} 1 \\ -9 \\ 9 \end{pmatrix}\right|$

Use Area $= \frac{1}{2}|\overrightarrow{AB}\times\overrightarrow{AC}|$

$= \frac{1}{2}\left|\begin{pmatrix} -9 \\ 49 \\ 50 \end{pmatrix}\right|$

$= \frac{1}{2}\sqrt{(-9)^2 + 49^2 + 50^2}$

Find magnitude of vector.

$= 35.3$ square units

1 Simplify each of these vector products.

  a $\mathbf{j} \times \mathbf{k}$ b $\mathbf{i} \times \mathbf{k}$

  c $\mathbf{k} \times \mathbf{k}$ d $(\mathbf{i}+\mathbf{j}) \times \mathbf{j}$

2 Fully simplify each of these expressions involving vectors $\mathbf{a}$, $\mathbf{b}$ and $\mathbf{c}$

  a $\mathbf{a} \times (\mathbf{a}+\mathbf{b})$ b $\mathbf{a} \times (\mathbf{b}+2\mathbf{c})+\mathbf{c} \times \mathbf{a}$

  c $(\mathbf{a}+\mathbf{b}) \times (\mathbf{a}-\mathbf{b})$ d $(\mathbf{a}+\mathbf{b}+\mathbf{c}) \times (\mathbf{a}-\mathbf{b}+\mathbf{c})$

3 Use the distributive and anticommutative properties of the vector product to show that

  a $\mathbf{a} \times \mathbf{b}+\mathbf{b} \times \mathbf{c}+\mathbf{a} \times \mathbf{c}=(\mathbf{b}+\mathbf{a}) \times (\mathbf{c}-\mathbf{a})$

  b $\mathbf{b} \times (2\mathbf{a}+\mathbf{c})-\mathbf{c} \times (\mathbf{b}-\mathbf{c})=2\mathbf{b} \times (\mathbf{a}+\mathbf{c})$

4 The non-zero vectors $\mathbf{a}$ and $\mathbf{b}$ are such that $(\mathbf{a}+2\mathbf{b}) \times (\mathbf{a}-3\mathbf{b})=0$

  Deduce the possible sizes of the angle between $\mathbf{a}$ and $\mathbf{b}$

5 For the vectors $\mathbf{a}=2\mathbf{i}-3\mathbf{j}+\mathbf{k}$, $\mathbf{b}=\mathbf{i}+5\mathbf{j}-2\mathbf{k}$ and $\mathbf{c}=-\mathbf{i}+3\mathbf{j}+4\mathbf{k}$, calculate

  a $\mathbf{a} \times \mathbf{b}$ b $\mathbf{a} \times \mathbf{c}$ c $\mathbf{b} \times \mathbf{c}$

  d $\mathbf{b} \times \mathbf{a}$ e $\mathbf{c} \times \mathbf{a}$ f $\mathbf{c} \times \mathbf{b}$

  g $\mathbf{a} \times \mathbf{a}$ h $\mathbf{b} \times \mathbf{b}$

6 Calculate each of these vector products.

  a $\begin{pmatrix} 0 \\ 2 \\ -5 \end{pmatrix} \times \begin{pmatrix} 3 \\ -1 \\ 4 \end{pmatrix}$

  b $(2\mathbf{i}+3\mathbf{j}+\mathbf{k}) \times (6\mathbf{i}-4\mathbf{j}+7\mathbf{k})$

  c $(7\mathbf{j}-2\mathbf{k}) \times (-4\mathbf{i}-9\mathbf{j}+3\mathbf{k})$

  d $\begin{pmatrix} 1 \\ -8 \\ -2 \end{pmatrix} \times \begin{pmatrix} -4 \\ -11 \\ 0 \end{pmatrix}$

  e $\begin{pmatrix} \sqrt{3} \\ 3 \\ 1 \end{pmatrix} \times \begin{pmatrix} 0 \\ \sqrt{3} \\ -\sqrt{3} \end{pmatrix}$

  f $(a\mathbf{i}-2a\mathbf{j}-\mathbf{k}) \times (a\mathbf{i}+\mathbf{j}-5a\mathbf{k})$

7 Find a vector which is perpendicular to both $\begin{pmatrix} 1 \\ 3 \\ -2 \end{pmatrix}$ and $\begin{pmatrix} -1 \\ 4 \\ 7 \end{pmatrix}$

8 Find a vector which is perpendicular to both $6\mathbf{i}+3\mathbf{j}-\mathbf{k}$ and $7\mathbf{i}+3\mathbf{k}$

9 Find a unit vector which is perpendicular to both $\begin{pmatrix} -5 \\ 0 \\ 0 \end{pmatrix}$ and $\begin{pmatrix} -2 \\ 1 \\ -3 \end{pmatrix}$

10 Find a unit vector which is perpendicular to both $2\mathbf{i}-\mathbf{j}$ and $\mathbf{j}-2\mathbf{k}$

11 Use the vector product to show that the vectors $-2\mathbf{i}+6\mathbf{j}-4\mathbf{k}$ and $3\mathbf{i}-9\mathbf{j}+6\mathbf{k}$ are parallel.

12 Use the vector product to find the acute angle between these pairs of vectors.

  a $\begin{pmatrix} 9 \\ -2 \\ 0 \end{pmatrix}$ and $\begin{pmatrix} 4 \\ 1 \\ 1 \end{pmatrix}$

  b $2\mathbf{i}-\mathbf{j}$ and $\mathbf{j}-6\mathbf{k}$

  c $3\mathbf{i}+2\mathbf{j}+\mathbf{k}$ and $\mathbf{i}+\mathbf{k}$

  d $\begin{pmatrix} 5 \\ 3 \\ \sqrt{2} \end{pmatrix}$ and $\begin{pmatrix} 7 \\ -\sqrt{2} \\ \sqrt{2} \end{pmatrix}$

13 Find, in surd form, the sine of the angle between $3\mathbf{i}+\mathbf{j}+2\mathbf{k}$ and $\mathbf{i}+2\mathbf{j}$

14 Find the values of a, b and c given that $\begin{pmatrix} 1 \\ a \\ -2 \end{pmatrix} \times \begin{pmatrix} b \\ 0 \\ 3 \end{pmatrix} = \begin{pmatrix} 12 \\ -13 \\ c \end{pmatrix}$

15 Find the values of a, b and c given that $\begin{pmatrix} 3 \\ -4 \\ a \end{pmatrix} \times \begin{pmatrix} 0 \\ 2 \\ b \end{pmatrix} = \begin{pmatrix} 10 \\ 9 \\ c \end{pmatrix}$

16 A triangle ABC is such that $AB=2\mathbf{i}-\mathbf{j}+\mathbf{k}$ and $AC=7\mathbf{i}+\mathbf{j}-4\mathbf{k}$. Calculate the exact area of the triangle.

17 A triangle DEF is such that $ED=6\mathbf{i}+5\mathbf{j}-\mathbf{k}$ and $EF=-4\mathbf{i}-2\mathbf{j}+3\mathbf{k}$. Calculate the area of the triangle. Give your answer to 3 significant figures.

18 Calculate the exact area of the triangles with vertices at

  a $(0, 0, 0)$, $(3, -1, 4)$ and $(2, -3, -5)$

  b $(1, 1, -2)$, $(0, 1, -1)$ and $(-2, 0, 1)$

Reasoning and problem-solving

Two forms for the equation of a line in 3D are the Cartesian form
$\dfrac{x-a_1}{b_1} = \dfrac{y-a_2}{b_2} = \dfrac{z-a_3}{b_3}$ and the vector form $\mathbf{r} = \mathbf{a} + \lambda\mathbf{b}$

A third possibility is the **vector product form**.

> **Key point**
>
> A straight line passing through the point with position vector $\mathbf{a}$ and parallel to the vector $\mathbf{b}$ has equation $(\mathbf{r}-\mathbf{a}) \times \mathbf{b} = \mathbf{0}$

This follows from the fact that any vector on the line will be parallel to the vector $\mathbf{b}$ and the vector product of two parallel vectors is zero.

You can expand the brackets of $(\mathbf{r}-\mathbf{a}) \times \mathbf{b} = \mathbf{0}$ to give $\mathbf{r} \times \mathbf{b} - \mathbf{a} \times \mathbf{b} = \mathbf{0}$
which rearranges to $\mathbf{r} \times \mathbf{b} = \mathbf{a} \times \mathbf{b}$

You should calculate the result of $\mathbf{a} \times \mathbf{b}$ using the vectors given.

Strategy

To find the equation of a line in vector product form

(1) Find a vector parallel to the line by subtracting the position vectors of two points on the line.

(2) Use the equation $(\mathbf{r}-\mathbf{a}) \times \mathbf{b} = \mathbf{0}$ where $\mathbf{a}$ lies on the line and $\mathbf{b}$ is parallel to the line.

(3) Expand the brackets and rearrange if required.

Example 6

Find the equation of the line through the points $(2, 0, -1)$ and $(-3, 1, 4)$ in the form $\mathbf{r} \times \mathbf{b} = \mathbf{c}$

A vector parallel to the plane is given by $\begin{pmatrix} 2 \\ 0 \\ -1 \end{pmatrix} - \begin{pmatrix} -3 \\ 1 \\ 4 \end{pmatrix} = \begin{pmatrix} 5 \\ -1 \\ -5 \end{pmatrix}$

> **(1)** Since these are the position vectors of two points on the line.

So the equation can be written $\left(\mathbf{r} - \begin{pmatrix} 2 \\ 0 \\ -1 \end{pmatrix}\right) \times \begin{pmatrix} 5 \\ -1 \\ -5 \end{pmatrix} = \mathbf{0}$

> **(2)** Use $(\mathbf{r} - \mathbf{a}) \times \mathbf{b} = \mathbf{0}$

$\mathbf{r} \times \begin{pmatrix} 5 \\ -1 \\ -5 \end{pmatrix} - \begin{pmatrix} 2 \\ 0 \\ -1 \end{pmatrix} \times \begin{pmatrix} 5 \\ -1 \\ -5 \end{pmatrix} = \mathbf{0}$

> **(3)** Expand the brackets.

Which can be rearranged to give

$\mathbf{r} \times \begin{pmatrix} 5 \\ -1 \\ -5 \end{pmatrix} = \begin{pmatrix} 2 \\ 0 \\ -1 \end{pmatrix} \times \begin{pmatrix} 5 \\ -1 \\ -5 \end{pmatrix}$

$\mathbf{r} \times \begin{pmatrix} 5 \\ -1 \\ -5 \end{pmatrix} = \begin{pmatrix} -1 \\ 5 \\ -2 \end{pmatrix}$

> By calculating the vector product on the right of the equation.

1 Show that $\mathbf{a} \times (\mathbf{b}+\mathbf{c}) = \mathbf{a} \times \mathbf{b} + \mathbf{a} \times \mathbf{c}$ for any 3D vectors $\mathbf{a}$, $\mathbf{b}$ and $\mathbf{c}$

2 Show that $\mathbf{a} \times \mathbf{b} = -\mathbf{b} \times \mathbf{a}$

3 Find the equation of the line passing through the point $(2, 5, 0)$ and parallel to the vector $\mathbf{i}+\mathbf{j}-\mathbf{k}$ in the form $(\mathbf{r}-\mathbf{a}) \times \mathbf{b} = \mathbf{0}$

4 Find the equation of the line that intercepts the y-axis at $y = 3$ and is parallel to the vector $\mathbf{k}$ in the form $(\mathbf{r}-\mathbf{a}) \times \mathbf{b} = \mathbf{0}$

5 Find the equation of the line that intercepts the z-axis at $z = -2$ and is parallel to the vector $\mathbf{i}+\mathbf{k}$ in the form $\mathbf{r} \times \mathbf{b} = \mathbf{c}$

6 Find the equation of the line that passes through the points $(1, 3, 5)$ and $(4, 2, 7)$ in the form $(\mathbf{r}-\mathbf{a}) \times \mathbf{b} = \mathbf{0}$

7 Find two different versions of the equation of the line that passes through the points $(0, 3, -2)$ and $(1, 2, -1)$ in the form $\mathbf{r} \times \mathbf{b} = \mathbf{c}$

8 Convert each of these equations of lines to the form $(\mathbf{r}-\mathbf{a}) \times \mathbf{b} = \mathbf{0}$

  a $\mathbf{r} = 2\mathbf{i}-\mathbf{k}+s(\mathbf{j}+5\mathbf{k})$

  b $\mathbf{r} = \begin{pmatrix} 3 \\ 1 \\ -4 \end{pmatrix} + t \begin{pmatrix} 5 \\ 3 \\ -6 \end{pmatrix}$

  c $\dfrac{x-1}{-2} = \dfrac{y+3}{4} = \dfrac{z}{-2}$

  d $x = \dfrac{y}{-3} = \dfrac{z-1}{2}$

9 Show that these two equations represent the same line.

  A: $\left(\mathbf{r} - \begin{pmatrix} 1 \\ 3 \\ 5 \end{pmatrix} \right) \times \begin{pmatrix} 4 \\ 3 \\ -1 \end{pmatrix} = \mathbf{0}$

  B: $\mathbf{r} = \begin{pmatrix} 5 \\ 6 \\ 4 \end{pmatrix} + \lambda \begin{pmatrix} -8 \\ -6 \\ 2 \end{pmatrix}$

10 Do these equations represent the same line? Explain how you know.

  A: $\dfrac{x-3}{-2} = \dfrac{y+1}{4} = \dfrac{z-5}{-6}$

  B: $\left(\mathbf{r} - \begin{pmatrix} 4 \\ -3 \\ 8 \end{pmatrix} \right) \times \begin{pmatrix} -3 \\ 6 \\ -9 \end{pmatrix} = \mathbf{0}$

11 Find, in Cartesian form, a possible equation of the line given by $\mathbf{r} \times \begin{pmatrix} 1 \\ 0 \\ -1 \end{pmatrix} = \begin{pmatrix} -3 \\ 0 \\ -3 \end{pmatrix}$

12 Find, in the form $\mathbf{r} = \mathbf{a}+t\mathbf{b}$, an equation of the line given by $\mathbf{r} \times \begin{pmatrix} -2 \\ -1 \\ 0 \end{pmatrix} = \begin{pmatrix} 1 \\ -2 \\ 0 \end{pmatrix}$

13 The lines l_1 and l_2 have equations

  $l_1: \mathbf{r} \times (2\mathbf{i}+\mathbf{j}-2\mathbf{k}) = (\mathbf{i}-2\mathbf{j}+\mathbf{k})$ and

  $l_2: \mathbf{r} = (-3\mathbf{i}-11\mathbf{j}+5\mathbf{k}) + t(-\mathbf{i}+3\mathbf{j}+\mathbf{k})$

  a Show that l_1 and l_2 intersect and find their point of intersection.

  b Calculate the sine of the angle between l_1 and l_2

14 A line l_1 has equation $\left(\mathbf{r} - \begin{pmatrix} 0 \\ 4 \\ -2 \end{pmatrix} \right) \times \begin{pmatrix} 1 \\ -6 \\ 5 \end{pmatrix} = \mathbf{0}$

  a Verify that the point $A(-2, 16, -12)$ lies on l_1

  A second line, l_2 has equation

  $\mathbf{r} = \begin{pmatrix} 2 \\ 0 \\ -3 \end{pmatrix} + \lambda \begin{pmatrix} 1 \\ 2 \\ -6 \end{pmatrix}$

  The lines l_1 and l_2 intersect at the point B

  b Find the coordinates of B

  c Calculate the area of triangle OAB

15 The lines l_1 and l_2 have equations

  $l_1: \mathbf{r} \times (5\mathbf{i}+\mathbf{k}) = (3\mathbf{j}-7\mathbf{k})$ and

  $l_2: \mathbf{r} = (\mathbf{i}+\mathbf{j}-\mathbf{k}) + t(9\mathbf{i}+8\mathbf{j}+\mathbf{k})$

  Find a unit vector which is perpendicular to both l_1 and l_2

Fluency and skills

A plane is a 2D surface, extending infinitely far in both directions. You can write the equation of a plane in Cartesian, vector and scalar product form.

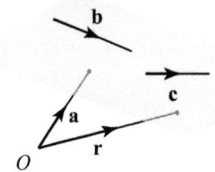

Suppose you wish to find the equation of the plane containing a point with position vector **a** and the non-parallel vectors **b** and **c**. Any other vector on the plane can be described as a multiple of **b** plus a multiple of **c**. Therefore the position vector of any point on the plane is given by $\mathbf{r} = \mathbf{a} + s\mathbf{b} + t\mathbf{c}$ for some values of s and t

> **Key point**
>
> The vector equation of the plane containing the point with position vector **a** and the non-parallel vectors **b** and **c** is $\mathbf{r} = \mathbf{a} + s\mathbf{b} + t\mathbf{c}$

Example 1

Find the equation of the plane containing the points $(1, 4, -2)$, $(0, 3, 2)$ and $(-5, 0, 3)$

$\mathbf{a} = \mathbf{i} + 4\mathbf{j} - 2\mathbf{k}$

A vector on the plane is $(\mathbf{i} + 4\mathbf{j} - 2\mathbf{k}) - (3\mathbf{j} + 2\mathbf{k}) = \mathbf{i} + \mathbf{j} - 4\mathbf{k}$

Another vector on the plane is $(\mathbf{i} + 4\mathbf{j} - 2\mathbf{k}) - (-5\mathbf{i} + 3\mathbf{k}) = 6\mathbf{i} + 4\mathbf{j} - 5\mathbf{k}$

So the vector equation of the plane is
$\mathbf{r} = \mathbf{i} + 4\mathbf{j} - 2\mathbf{k} + s(\mathbf{i} + \mathbf{j} - 4\mathbf{k}) + t(6\mathbf{i} + 4\mathbf{j} - 5\mathbf{k})$

Alternatively, you can write this in column vector form:

$$\mathbf{r} = \begin{pmatrix} 1 \\ 4 \\ -2 \end{pmatrix} + s\begin{pmatrix} 1 \\ 1 \\ -4 \end{pmatrix} + t\begin{pmatrix} 6 \\ 4 \\ -5 \end{pmatrix}$$

> You could use the position vector of any of the three points.

> Any combination of the position vectors of the three points can be used as long as they do not give parallel vectors.

Consider a plane containing the point with position vector **a** and with a perpendicular vector **n**. Any vector on this plane will be perpendicular to the vector **n**. Using the definition of the scalar product, any point on the plane with position vector **r**, satisfies $(\mathbf{r} - \mathbf{a}) \cdot \mathbf{n} = 0$. Expanding the bracket gives $\mathbf{r} \cdot \mathbf{n} - \mathbf{a} \cdot \mathbf{n} = 0$, which you can rearrange to give $\mathbf{r} \cdot \mathbf{n} = \mathbf{a} \cdot \mathbf{n}$

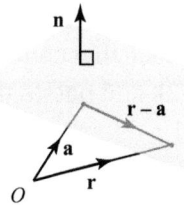

> **Key point**
>
> The scalar product equation of the plane perpendicular to the vector **n** and passing through the point with position vector **a** is $\mathbf{r} \cdot \mathbf{n} = p$, where $p = \mathbf{a} \cdot \mathbf{n}$

You can convert this into Cartesian form by writing the vector **r** in component form: $x\mathbf{i}+y\mathbf{j}+z\mathbf{k}$ and calculating the scalar product.

> **Key point**
>
> The vector perpendicular to a plane is called a **normal** to the plane.

Since any vectors on the plane will be perpendicular to the normal, you can use the vector product to find the normal vector.

Example 2

The plane Π contains the vectors $\mathbf{a}=\mathbf{i}-\mathbf{j}+3\mathbf{k}$ and $\mathbf{b}=3\mathbf{i}+\mathbf{j}-2\mathbf{k}$ and the point $(1, 0, -2)$. Find the equation of Π in

a Scalar product form,   **b** Cartesian form.

> The capital Greek letter Π is often used to denote a plane.

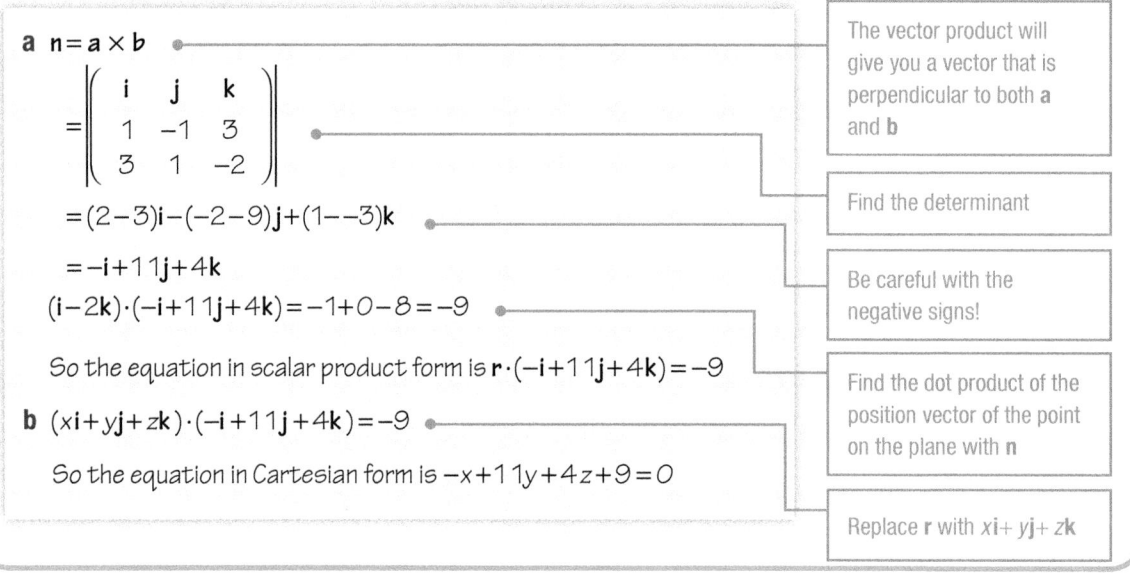

a $\mathbf{n}=\mathbf{a}\times\mathbf{b}$

$$=\begin{vmatrix} \mathbf{i} & \mathbf{j} & \mathbf{k} \\ 1 & -1 & 3 \\ 3 & 1 & -2 \end{vmatrix}$$

$$=(2-3)\mathbf{i}-(-2-9)\mathbf{j}+(1--3)\mathbf{k}$$

$$=-\mathbf{i}+11\mathbf{j}+4\mathbf{k}$$

$(\mathbf{i}-2\mathbf{k})\cdot(-\mathbf{i}+11\mathbf{j}+4\mathbf{k})=-1+0-8=-9$

So the equation in scalar product form is $\mathbf{r}\cdot(-\mathbf{i}+11\mathbf{j}+4\mathbf{k})=-9$

b $(x\mathbf{i}+y\mathbf{j}+z\mathbf{k})\cdot(-\mathbf{i}+11\mathbf{j}+4\mathbf{k})=-9$

So the equation in Cartesian form is $-x+11y+4z+9=0$

- The vector product will give you a vector that is perpendicular to both **a** and **b**
- Find the determinant
- Be careful with the negative signs!
- Find the dot product of the position vector of the point on the plane with **n**
- Replace **r** with $x\mathbf{i}+y\mathbf{j}+z\mathbf{k}$

> **Key point**
>
> The Cartesian equation of a plane is $ax+by+cz+d=0$, where the vector $a\mathbf{i}+b\mathbf{j}+c\mathbf{k}$ is perpendicular to the plane.

Exercise 23.2A Fluency and skills

1   Find, in the form $\mathbf{r}=\mathbf{a}+s\mathbf{b}+t\mathbf{c}$, the equation of the plane that contains the vectors $\mathbf{i}+\mathbf{j}$ and $\mathbf{j}+2\mathbf{k}$ and passes through the point $(1, 5, 2)$

2   Find, in vector form, the equation of the plane that contains the vectors $\begin{pmatrix} 5 \\ -2 \\ 4 \end{pmatrix}$ and $\begin{pmatrix} 2 \\ 0 \\ -3 \end{pmatrix}$ and the point $(0, 6, 2)$

3   A plane contains the points $(3, 1, 0)$, $(2, 4, -2)$ and $(-5, 0, 4)$

Find the equation of the plane in

a   Vector form,

b   Scalar product form,

c   Cartesian form.

4 A plane contains the points $(5, 1, 1)$, $(-2, 0, 0)$ and $(6, 2, -5)$

Find the equation of the plane in

a Vector form,   **b** Scalar product form,

c Cartesian form.

5 A plane contains the points A, B and C with position vectors $\mathbf{a} = \mathbf{i} - 2\mathbf{j}$, $\mathbf{b} = 3\mathbf{j} + 2\mathbf{k}$ and $\mathbf{c} = -\mathbf{i} + 2\mathbf{j} - \mathbf{k}$ respectively. Find the equation of the plane in

a Vector form,

b Scalar product form,

c Cartesian form.

6 Find, in the form $\mathbf{r} \cdot \mathbf{n} = p$, the equation of the plane that passes through the point $(3, -1, 4)$ and is perpendicular to the vector $\mathbf{i} + 2\mathbf{k}$

7 Find, in the form $\mathbf{r} \cdot \mathbf{n} = p$, the equation of the plane that passes through the point $(6, 0, -2)$ and has normal $3\mathbf{i} - \mathbf{j} + 4\mathbf{k}$

8 Find the scalar product equation of the plane that passes through the point with position vector $\mathbf{i} + \mathbf{j} + \mathbf{k}$ and is perpendicular to the vector $2\mathbf{j} - 3\mathbf{k}$

9 Find the Cartesian equation of the plane that passes through the point $(0, 3, -3)$ and is perpendicular to the vector $5\mathbf{i} + 4\mathbf{j} - 2\mathbf{k}$

10 Find the Cartesian equation of the plane that passes through the point $(1, -1, 2)$ and is perpendicular to the vector $2\mathbf{i} - 5\mathbf{k}$

11 Find the Cartesian equation of the plane that passes through the point with position vector $3\mathbf{i} + 7\mathbf{k}$ and is perpendicular to the vector $\mathbf{i} - \mathbf{j} + 4\mathbf{k}$

12 The plane Π is perpendicular to the line with equation $\mathbf{r} = \mathbf{i} - \mathbf{j} + t(2\mathbf{i} + \mathbf{j} - 3\mathbf{k})$ and passes through the point $(4, -1, 3)$. Find the equation of Π in

a Scalar product form,

b Cartesian form.

13 The plane Π is perpendicular to the line with equation $\dfrac{x+1}{3} = \dfrac{y-4}{-2} = \dfrac{z}{5}$ and passes through the point $(4, 3, -2)$. Find the equation of Π in

a Scalar product form,

b Cartesian form.

14 Convert each of these equations of planes into Cartesian form.

a $\mathbf{r} \cdot \begin{pmatrix} 3 \\ 5 \\ -1 \end{pmatrix} = -2$

b $\mathbf{r} \cdot (7\mathbf{i} - \mathbf{j} + 8\mathbf{k}) = 3$

c $\mathbf{r} = (2\mathbf{i} + \mathbf{j} - 5\mathbf{k}) + s(-\mathbf{i} + 3\mathbf{j} + \mathbf{k}) + t(6\mathbf{i} - 2\mathbf{j} + 7\mathbf{k})$

d $\mathbf{r} = \begin{pmatrix} 0 \\ 4 \\ -2 \end{pmatrix} + s\begin{pmatrix} 3 \\ -2 \\ 5 \end{pmatrix} + t\begin{pmatrix} 1 \\ 0 \\ 4 \end{pmatrix}$

15 Convert each of these equations of planes into scalar product form.

a $9x + 3y - z = 5$

b $2x - 7y - 15z + 4 = 0$

c $\mathbf{r} = (5\mathbf{i} - 2\mathbf{k}) + s(-4\mathbf{i} + \mathbf{j}) + t(8\mathbf{i} - 3\mathbf{j} + 4\mathbf{k})$

d $\mathbf{r} = \begin{pmatrix} 6 \\ 2 \\ -5 \end{pmatrix} + s\begin{pmatrix} 0 \\ 1 \\ -3 \end{pmatrix} + t\begin{pmatrix} 5 \\ -8 \\ 0 \end{pmatrix}$

Reasoning and problem-solving

The vector equation of a plane is not unique, as you can chose any two vectors and any point on the plane to build the equation. In the case of the scalar product equation and the Cartesian equation, all vectors that are perpendicular to the plane are parallel to each other, so these forms of the equation will all simplify to a unique equation. You need to be able to tell if two equations represent the same plane.

To check whether two equations represent the same plane

(1) Check whether the normals are parallel.

(2) Choose a point that you know is on one of the planes and check whether it also lies on the other plane.

(3) Conclude whether the planes are the same, parallel or intersecting.

Example 3

Parallel planes will have parallel normals but will not have a point in common.

The planes Π_1 and Π_2 have equations $\mathbf{r}\cdot(\mathbf{i}-6\mathbf{j}+2\mathbf{k})=1$ and $\mathbf{r}=\mathbf{i}+\mathbf{j}+3\mathbf{k}+s(\mathbf{j}+3\mathbf{k})+t(2\mathbf{i}-\mathbf{k})$ respectively.

Show that Π_1 and Π_2 are, in fact, the same plane.

The normal to Π_2 is given by

$(\mathbf{j}+3\mathbf{k})\times(2\mathbf{i}-\mathbf{k})=-\mathbf{i}+6\mathbf{j}-2\mathbf{k}$

$-\mathbf{i}+6\mathbf{j}-2\mathbf{k}=-(\mathbf{i}-6\mathbf{j}+2\mathbf{k})$

The point (1, 1, 3) lies on Π_2

$(\mathbf{i}+\mathbf{j}+3\mathbf{k})\cdot(\mathbf{i}-6\mathbf{j}+2\mathbf{k})=1-6+6$

$=1$

Therefore Π_1 and Π_2 represent the same plane.

Calculate the normal vector to Π_2

(1) So their normal vectors are parallel.

Substitute into the equation of Π_1

(2) So the point (1, 1, 3) lies on Π_1

(3) Since their normals are parallel and they have a point in common.

Consider the intersection of a plane $\mathbf{r}\cdot\mathbf{n}=p$ and a line $\mathbf{r}=\mathbf{a}+\lambda\mathbf{b}$.

You know that the acute angle between the vectors $\mathbf{n}$ and $\mathbf{b}$ is given by formula $\cos\alpha=\left|\dfrac{\mathbf{b}\cdot\mathbf{n}}{\|\mathbf{b}\|\|\mathbf{n}\|}\right|$. Therefore, the angle θ is given by

$\theta=90-\alpha$ which implies that $\sin\theta=\left|\dfrac{\mathbf{b}\cdot\mathbf{n}}{\|\mathbf{b}\|\|\mathbf{n}\|}\right|$ since $\sin(90-\alpha)=\cos\alpha$

Key point

The acute angle, θ, between the plane $\mathbf{r}\cdot\mathbf{n}=p$ and the line $\mathbf{r}=\mathbf{a}+\lambda\mathbf{b}$ is given by $\sin\theta=\left|\dfrac{\mathbf{b}\cdot\mathbf{n}}{\|\mathbf{b}\|\|\mathbf{n}\|}\right|$

Now consider the intersection of two planes, $\mathbf{r}\cdot\mathbf{n}_1=p$ and $\mathbf{r}\cdot\mathbf{n}_2=p_2$. To find the angle, α, between the normals you can use the formula $\cos\alpha=\dfrac{\mathbf{n}_1\cdot\mathbf{n}_2}{\|\mathbf{n}_1\|\|\mathbf{n}_2\|}$

Using the quadrilateral formed from the planes and their normals you can see that $\theta=180-\alpha$ which implies that $\cos\theta=\left|\dfrac{\mathbf{n}_1\cdot\mathbf{n}_2}{\|\mathbf{n}_1\|\|\mathbf{n}_2\|}\right|$ since $|\cos(180-\alpha)|=|\cos\alpha|$

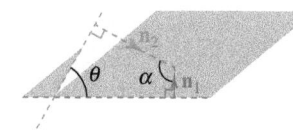

See C
For a reminc of how calcula the ang betwee vectors

> **Key point**
>
> The acute angle, θ, between the planes $\mathbf{r} \cdot \mathbf{n}_1 = p_1$
> and $\mathbf{r} \cdot \mathbf{n}_2 = p_2$ is given by $\cos\theta = \left| \dfrac{\mathbf{n}_1 \cdot \mathbf{n}_2}{\|\mathbf{n}_1\| \|\mathbf{n}_2\|} \right|$

Strategy 2

To calculate the angle between two planes or between a line or a plane

(1) Identify the normal vector/s.

(2) Identify the direction vector of the line.

(3) Use $\cos\theta = \left| \dfrac{\mathbf{n}_1 \cdot \mathbf{n}_2}{\|\mathbf{n}_1\| \|\mathbf{n}_2\|} \right|$ to find the acute angle between two planes.

(4) Use $\sin\theta = \left| \dfrac{\mathbf{b} \cdot \mathbf{n}}{\|\mathbf{b}\| \|\mathbf{n}\|} \right|$ to find the acute angle between a line $r = \mathbf{a} + t\mathbf{b}$ and a plane.

(5) Subtract from $180°$ to give an obtuse angle if necessary.

Example 4

Calculate the obtuse angle between the planes $\mathbf{r} \cdot (13\mathbf{i} - 9\mathbf{j} + 5\mathbf{k}) = 8$ and $\mathbf{r} \cdot (2\mathbf{j} + 3\mathbf{k}) = 4$

The normal to the first plane is $\mathbf{n}_1 = 13\mathbf{i} - 9\mathbf{j} + 5\mathbf{k}$

Similarly the normal to the second plane is $\mathbf{n}_2 = 2\mathbf{j} + 3\mathbf{k}$

(1) Identify the normals.

$\cos\theta = \left| \dfrac{(13\mathbf{i} - 9\mathbf{j} + 5\mathbf{k}) \cdot (2\mathbf{j} + 3\mathbf{k})}{|13\mathbf{i} - 9\mathbf{j} + 5\mathbf{k}||2\mathbf{j} + 3\mathbf{k}|} \right|$

(3) Using $\cos\theta = \left| \dfrac{\mathbf{n}_1 \cdot \mathbf{n}_2}{\|\mathbf{n}_1\| \|\mathbf{n}_2\|} \right|$

$= \left| \dfrac{-3}{\sqrt{275}\sqrt{13}} \right|$

$\theta = 87.1°$

So the obtuse angle is

$180 - 87.1 = 92.9°$

(5) Subtract from $180°$ since the question asked for an obtuse angle.

Example 5

Calculate the acute angle between the plane $2x + 3y - z = 1$ and the line $\dfrac{x}{4} = \dfrac{y+1}{-2} = \dfrac{z-2}{7}$

The normal to the plane is $\mathbf{n} = 2\mathbf{i} + 3\mathbf{j} - \mathbf{k}$

The direction vector of the line is $\mathbf{b} = 4\mathbf{i} - 2\mathbf{j} + 7\mathbf{k}$

(1) Since the Cartesian equation of a plane is $ax + by + cz = d$ where $\mathbf{n} = a\mathbf{i} + b\mathbf{j} + c\mathbf{k}$

$\sin\theta = \left| \dfrac{(4\mathbf{i} - 2\mathbf{j} + 7\mathbf{k}) \cdot (2\mathbf{i} + 3\mathbf{j} - \mathbf{k})}{|4\mathbf{i} - 2\mathbf{j} + 7\mathbf{k}||2\mathbf{i} + 3\mathbf{j} - \mathbf{k}|} \right|$

$= \left| \dfrac{-5}{\sqrt{69}\sqrt{14}} \right|$

$\theta = 9.26°$

(4) Using $\sin\theta = \left| \dfrac{\mathbf{b} \cdot \mathbf{n}}{\|\mathbf{b}\| \|\mathbf{n}\|} \right|$

(2) Since the Cartesian equations of a line are $\dfrac{x - x_1}{b_1} = \dfrac{y - y_1}{b_2} = \dfrac{z - z_1}{b_3}$ where $\mathbf{b} = b_1\mathbf{i} + b_2\mathbf{j} + b_3\mathbf{k}$ is the direction vector of the line.

1 Which of these planes does the point $(-7, 10, 0)$ lie on? Show your working.

a $\mathbf{r} = 2\mathbf{i} - 3\mathbf{j} - 7\mathbf{k} + s(-\mathbf{i} + 5\mathbf{j} + 3\mathbf{k}) + t(7\mathbf{i} - 3\mathbf{j} - \mathbf{k})$

b $\mathbf{r} = \begin{pmatrix} 2 \\ 0 \\ 8 \end{pmatrix} + s\begin{pmatrix} 1 \\ 3 \\ -2 \end{pmatrix} + t\begin{pmatrix} -2 \\ 4 \\ 0 \end{pmatrix}$

c $\mathbf{r} \cdot (\mathbf{i} - \mathbf{j} + 3\mathbf{k}) = 17$

d $2x + 3y - 8z = 16$

2 Verify whether each pair of equations represent the same plane.

a $2x + 3y - z - 2 = 0$ and $\mathbf{r} \cdot (2\mathbf{i} + 3\mathbf{j} - \mathbf{k}) = 2$

b $\mathbf{r} = 3\mathbf{i} - \mathbf{k} + s(4\mathbf{i} + \mathbf{j} + 2\mathbf{k}) + t(5\mathbf{i} + 3\mathbf{j})$ and
$\mathbf{r} = (\mathbf{i} - 4\mathbf{j} + 4\mathbf{k}) + \lambda(2\mathbf{i} + 4\mathbf{j} + 5\mathbf{k}) + \mu(5\mathbf{i} + 3\mathbf{j})$

c $\mathbf{r} = \begin{pmatrix} 1 \\ 0 \\ -7 \end{pmatrix} + s\begin{pmatrix} 5 \\ -2 \\ 4 \end{pmatrix} + t\begin{pmatrix} 0 \\ 0 \\ 1 \end{pmatrix}$ and

$\mathbf{r} = \begin{pmatrix} -4 \\ 2 \\ -7 \end{pmatrix} + \lambda\begin{pmatrix} 0 \\ 1 \\ 3 \end{pmatrix} + \mu\begin{pmatrix} 5 \\ -2 \\ 0 \end{pmatrix}$

d $\mathbf{r} = \begin{pmatrix} -2 \\ 1 \\ 0 \end{pmatrix} + s\begin{pmatrix} 4 \\ 3 \\ 2 \end{pmatrix} + t\begin{pmatrix} -5 \\ 0 \\ 1 \end{pmatrix}$ and

$\mathbf{r} \cdot \begin{pmatrix} 6 \\ -2 \\ -1 \end{pmatrix} = -14$

e $\mathbf{r} \cdot (2\mathbf{i} - 7\mathbf{j} - 3\mathbf{k}) = -2$ and
$\mathbf{r} = (5\mathbf{i} + 4\mathbf{k}) + \lambda(3\mathbf{j} - 7\mathbf{k}) + \mu(2\mathbf{i} + \mathbf{j} - \mathbf{k})$

f $x + 3y + z = -1$ and
$\mathbf{r} = (\mathbf{i} - 2\mathbf{k}) + \lambda(\mathbf{i} - 9\mathbf{j} + 6\mathbf{k}) + \mu(4\mathbf{i} + 3\mathbf{j} - 3\mathbf{k})$

3 Calculate the acute angle between the line $\mathbf{r} = 2\mathbf{i} - 8\mathbf{j} + \mathbf{k} + t(\mathbf{j} + \mathbf{k})$ and the plane $\mathbf{r} \cdot (6\mathbf{i} - \mathbf{j} + 5\mathbf{k}) = 9$

4 Calculate the obtuse angle between the line $\mathbf{r} = 9\mathbf{i} - 5\mathbf{j} + 3\mathbf{k} + \lambda(\mathbf{i} - \mathbf{j} - 4\mathbf{k})$ and the plane $6x - y + z = 24$

5 Calculate the acute angle between the line $\dfrac{x-1}{5} = \dfrac{y-2}{-2} = \dfrac{z+3}{-2}$ and the plane $2x + 7y - z = 5$

6 Calculate the obtuse angle between the line $\mathbf{r} = \begin{pmatrix} -5 \\ 0 \\ 1 \end{pmatrix} + \lambda\begin{pmatrix} 7 \\ 2 \\ -3 \end{pmatrix}$ and the plane $\mathbf{r} \cdot \begin{pmatrix} -2 \\ 4 \\ -9 \end{pmatrix} = 1$

7 Calculate the acute angle between the plane $\mathbf{r} \cdot (2\mathbf{i} + \mathbf{j} - 3\mathbf{k}) = 2$ and the plane $\mathbf{r} \cdot (6\mathbf{i} - 4\mathbf{j} - 7\mathbf{k}) = 5$

8 Calculate the acute angle between the plane $\mathbf{r} \cdot (7\mathbf{i} - \mathbf{j} - 5\mathbf{k}) = 9$ and the plane $2x - y + z = 4$

9 Calculate the obtuse angle between the plane $x + y - 7z - 3 = 0$ and the plane $8x - 3y + 4z = 15$

10 Find the scalar product equation of the plane that contains the vectors $\begin{pmatrix} 1 \\ 0 \\ -3 \end{pmatrix}$ and $\begin{pmatrix} 2 \\ 6 \\ -1 \end{pmatrix}$ and passes through the point $(3, -1, 0)$

11 Convert this equation of a plane into scalar product form.
$\mathbf{r} = \begin{pmatrix} 0 \\ 3 \\ -2 \end{pmatrix} + \mu\begin{pmatrix} 6 \\ -2 \\ 2 \end{pmatrix} + \lambda\begin{pmatrix} -1 \\ 4 \\ 7 \end{pmatrix}$

12 Find the Cartesian equation of the plane that passes through the points $(3, -5, 1)$, $(2, 4, 5)$ and $(1, 0, 3)$

13 The plane Π has vector equation
$\mathbf{r} = \begin{pmatrix} 3 \\ 0 \\ 1 \end{pmatrix} + \lambda\begin{pmatrix} 1 \\ -2 \\ 4 \end{pmatrix} + \mu\begin{pmatrix} 6 \\ 1 \\ -3 \end{pmatrix}$

a Write the equation of Π in scalar product form.

b The line l passes through the point $P(1, 4, -2)$ and is perpendicular to Π
Find the vector product equation of l

A line will intersect any given plane unless the line and the plane are parallel.

You can find the point of intersection between a line and a plane by using the equation of the line to substitute for **r** in the equation of the plane.

Example 1

Find the point of intersection between the line with equation $\mathbf{r} = -4\mathbf{j}+2\mathbf{k}+\lambda(3\mathbf{i}+\mathbf{j}+\mathbf{k})$ and the plane with equation $\mathbf{r} = \mathbf{i}-2\mathbf{j}+s(\mathbf{j}+3\mathbf{k})+t(2\mathbf{i}+2\mathbf{j}+\mathbf{k})$

At the point of intersection
$-4\mathbf{j}+2\mathbf{k}+\lambda(3\mathbf{i}+\mathbf{j}+\mathbf{k}) = \mathbf{i}-2\mathbf{j}+s(\mathbf{j}+3\mathbf{k})+t(2\mathbf{i}+2\mathbf{j}+\mathbf{k})$

*Use the equation of the line to substitute for **r** in the equation of the plane.*

$\mathbf{i}: 3\lambda = 1+2t$

$\mathbf{j}: -4+\lambda = -2+s+2t \Rightarrow \lambda = 2+s+2t$

*Consider the **i** components.*

Subtracting these two equations gives: $2\lambda = -1-s$

Which can be rearranged to give $s = -1-2\lambda$

$\mathbf{k}: 2+\lambda = 3s+t$

Substitute $s = -1-2\lambda$

$\Rightarrow 2+\lambda = 3(-1-2\lambda)+t$

$\Rightarrow t = 5+7\lambda$

Solve this together with $3\lambda = 1+2t$

to give $\lambda = -1$ and $t = -2$

By substituting back into $s = -1-2\lambda$

$\Rightarrow s = 1$

Substitute $\lambda = -1$ into equation of the line.

$\mathbf{r} = -4\mathbf{j}+2\mathbf{k}-1(3\mathbf{i}+\mathbf{j}+\mathbf{k})$

$= -3\mathbf{i}-5\mathbf{j}+\mathbf{k}$

So the point of intersection is $(-3, -5, 1)$

Could also find by substituting $t = -2$ and $s = 1$ into equation of the plane.

Example 2

Find the point of intersection between the line with equation $\mathbf{r} = \mathbf{i}-\mathbf{k}+t(\mathbf{i}-2\mathbf{j}+3\mathbf{k})$ and the plane with equation $2x-y-3z = 20$

At the point of intersection $2(1+t)-(-2t)-3(-1+3t) = 20$

Substituting $x = 1+t$, $y = -2t$, $z = -1+3t$ from the equation of the line.

$2+2t+2t+3-9t = 20$

$5t = -15$

$t = -3$

Substitute $t = -3$ into equation of the line.

So $\mathbf{r} = \mathbf{i}-\mathbf{k}-3(\mathbf{i}-2\mathbf{j}+3\mathbf{k})$

$= -2\mathbf{i}+6\mathbf{j}-10\mathbf{k}$

So the point of intersection is $(-2, 6, -10)$

You can find the shortest distance from a point to a plane.

Suppose you have the plane Π with equation $\mathbf{r} \cdot \mathbf{n} = p$ and the point P. The shortest distance from P to the plane will be perpendicular to the plane. Therefore you need the length of $\overrightarrow{PQ}$

Q is the point of intersection of the plane and the line through P and Q. You can easily write down the equation of this line using the fact that it passes through P and has direction vector parallel to the vector $\mathbf{n}$ which is normal to the plane.

Once you've found the point of intersection Q you can find the length of the line segment PQ and hence the shortest distance from the point P to the plane Π

Example 3

Find the shortest distance from the plane Π with equation

$$\mathbf{r} \cdot \begin{pmatrix} 1 \\ 3 \\ -2 \end{pmatrix} = -4 \text{ to the point } P(4, 0, -7)$$

The equation of the line though P and the plane is

$$\mathbf{r} = \begin{pmatrix} 4 \\ 0 \\ -7 \end{pmatrix} + \lambda \begin{pmatrix} 1 \\ 3 \\ -2 \end{pmatrix}$$

> Since the line is perpendicular to the plane.

At Q, $\begin{pmatrix} 4+\lambda \\ 3\lambda \\ -7-2\lambda \end{pmatrix} \cdot \begin{pmatrix} 1 \\ 2 \\ -2 \end{pmatrix} = -4$

> Find the point of intersection between the plane and the line.

$$\Rightarrow 4 + \lambda + 2(3\lambda) - 2(-7-2\lambda) = -4$$
$$\Rightarrow 11\lambda = -22$$
$$\Rightarrow \lambda = -2$$

Therefore, the position vector of Q is

$$\overrightarrow{OQ} = \begin{pmatrix} 4 \\ 0 \\ -7 \end{pmatrix} - 2 \begin{pmatrix} 1 \\ 3 \\ -2 \end{pmatrix}$$

> Substitute the value of λ found into the equation of the line.

$$\overrightarrow{PQ} = \overrightarrow{OQ} - \overrightarrow{OP}$$

$$= \begin{pmatrix} 4 \\ 0 \\ -7 \end{pmatrix} - 2 \begin{pmatrix} 1 \\ 3 \\ -2 \end{pmatrix} - \begin{pmatrix} 4 \\ 0 \\ -7 \end{pmatrix}$$

$$= \begin{pmatrix} -2 \\ -6 \\ 4 \end{pmatrix}$$

> This is just the vector $\lambda\mathbf{n}$ where $\lambda = -2$

$$\left| \overrightarrow{PQ} \right| = \sqrt{(-2)^2 + (-6)^2 + 4^2}$$
$$= 2\sqrt{14}$$

So the shortest distance from the point P to the plane Π is $2\sqrt{14}$

> Notice that the distance of the point from the plane was given by $|\lambda\mathbf{n}|$

1 Find the shortest distance between each point and plane.

   a $(2, 5, 1)$ and $2x-4y+4z+5=0$

   b $(3, 0, -4)$ and $3x+y-2z=6$

   c $(-1, 5, 4)$ and $\mathbf{r}\cdot(\mathbf{i}-7\mathbf{j}+5\mathbf{k})=-3$

   d $(6, 1, 4)$ and $\mathbf{r}\cdot\begin{pmatrix}0\\-8\\6\end{pmatrix}=12$

   e $(-2, -5, -1)$ and $\mathbf{r}\cdot(\mathbf{i}-\mathbf{j}-3\mathbf{k})=1$

2 Show that the line $\mathbf{r}=\begin{pmatrix}-11\\-4\\10\end{pmatrix}+t\begin{pmatrix}4\\-2\\0\end{pmatrix}$ and

   the plane $\mathbf{r}=\begin{pmatrix}3\\-5\\-10\end{pmatrix}+\mu\begin{pmatrix}7\\-10\\0\end{pmatrix}+\lambda\begin{pmatrix}1\\-3\\4\end{pmatrix}$

   intersect and find their point of intersection.

3 Find the point of intersection of the line with equation $\mathbf{r}=-6\mathbf{i}+2\mathbf{j}+16\mathbf{k}+\lambda(3\mathbf{i}-\mathbf{j}+\mathbf{k})$ and the plane with equation $\mathbf{r}=-7\mathbf{i}-9\mathbf{j}+\mathbf{k}+s(5\mathbf{i}+\mathbf{j}+6\mathbf{k})+t(-2\mathbf{i}+4\mathbf{j}+\mathbf{k})$

4 Find the point of intersection of the line with

   equation $\mathbf{r}=\begin{pmatrix}1\\-2\\5\end{pmatrix}+t\begin{pmatrix}-5\\1\\2\end{pmatrix}$ and the plane

   with equation $\mathbf{r}\cdot\begin{pmatrix}1\\6\\-5\end{pmatrix}=0$

5 Find the point of intersection of the line with

   equation $\dfrac{x}{5}=\dfrac{y+2}{-1}=\dfrac{3-z}{4}$ and the plane

   with equation $\mathbf{r}\cdot(5\mathbf{i}-\mathbf{j}+4\mathbf{k})=-6$

6 Find the point of intersection of the

   line with equation $\dfrac{x-3}{4}=\dfrac{y-2}{5}=\dfrac{z-1}{-3}$

   and the plane with equation

   $\mathbf{r}=13\mathbf{i}+7\mathbf{j}-6\mathbf{k}+s(\mathbf{i}+2\mathbf{j}-5\mathbf{k})+t(-\mathbf{i}+\mathbf{j}+2\mathbf{k})$

7 The plane Π is perpendicular to the y-axis and passes through the point $(1, 2, 0)$

   Calculate the shortest distance of this plane from the point $(3, 4, -1)$

8 Show that the shortest distance from the origin to the plane $\mathbf{r}\cdot\mathbf{n}=p$ is $\dfrac{p}{|\mathbf{n}|}$

9 For each pair of lines, decide whether they are parallel, skew or intersecting. If they are intersecting, find their point of intersection.

   a $\mathbf{r}=\begin{pmatrix}7\\-6\\-2\end{pmatrix}+\lambda\begin{pmatrix}3\\0\\-1\end{pmatrix}$ and $\mathbf{r}=\begin{pmatrix}3\\4\\0\end{pmatrix}+\mu\begin{pmatrix}-2\\5\\1\end{pmatrix}$

   b $\mathbf{r}=\mathbf{i}+2\mathbf{k}+\lambda(3\mathbf{i}+4\mathbf{j}-2\mathbf{k})$ and $\mathbf{r}=3\mathbf{i}+\mathbf{j}-\mathbf{k}+\mu(-6\mathbf{i}-8\mathbf{j}+4\mathbf{k})$

   c $\dfrac{x+5}{-1}=\dfrac{y}{2}=\dfrac{z-1}{2}$ and $\dfrac{x-2}{1}=\dfrac{y+9}{3}=\dfrac{z-5}{-2}$

   d $\dfrac{-4-x}{3}=\dfrac{y-2}{-1}=\dfrac{z}{5}$ and $\dfrac{x-6}{2}=\dfrac{y+8}{-2}=\dfrac{5-z}{-1}$

10 Two lines, L_1 and L_2, have equations

   $\dfrac{x+10}{5}=\dfrac{y-5}{-1}=\dfrac{z-5}{2}$ and $\mathbf{r}=\begin{pmatrix}3\\3\\5\end{pmatrix}+\lambda\begin{pmatrix}-3\\0\\4\end{pmatrix}$

   respectively.

   Given that L_1 and L_2 intersect at the point A, calculate the length of $\overline{OA}$

11 The points A, B, C are defined as $(2, -1, 4)$, $(0, -2, 4)$ and $(1, 0, 5)$ respectively.

   Calculate the shortest distance from the origin to the plane containing points A, B and C

Reasoning and problem-solving

See Ch6.3

For a reminder of finding the distance between skew lines.

You can find the distance between lines by finding the length of the perpendicular to both lines. It is also possible to use the vector product to find this distance.

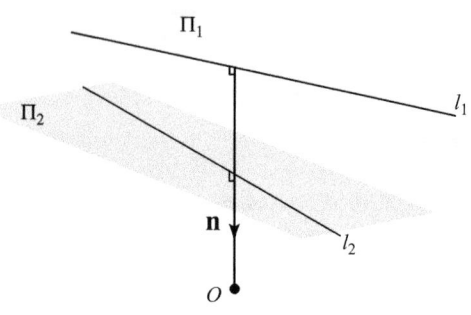

Consider the lines l_1 and l_2 with equations $\mathbf{r} = \mathbf{a} + \lambda\mathbf{b}$ and $\mathbf{r} = \mathbf{c} + \mu\mathbf{d}$ respectively. Since the lines do not intersect they must lie on parallel planes Π_1 and Π_2

The normal to the planes Π_1 and Π_2 will be the same since they are parallel and is given by the vector product of the two direction vectors: $\mathbf{n} = \mathbf{b} \times \mathbf{d}$

So the equations of the planes are $\mathbf{r} \cdot \mathbf{n} = \mathbf{a} \cdot \mathbf{n}$ and $\mathbf{r} \cdot \mathbf{n} = \mathbf{c} \cdot \mathbf{n}$

In Section 23.3A you saw how to find the distance of a point from a plane.

The equation of the line through the origin which is perpendicular to a plane $\mathbf{r} \cdot \mathbf{n} = p$ is $\mathbf{r} = \lambda\mathbf{n}$

Therefore, at the point of intersection, $\lambda\mathbf{n} \cdot \mathbf{n} = p$ which implies that $\lambda|\mathbf{n}|^2 = p$ so $\lambda = \dfrac{p}{|\mathbf{n}|^2}$

The distance of a point from a plane is given by $|\lambda\mathbf{n}|$. So, using $\lambda = \dfrac{p}{|\mathbf{n}|^2}$ gives

$$\text{distance from plane to origin} = \left|\frac{p}{|\mathbf{n}|^2}|\mathbf{n}|\right| = \left|\frac{p}{|\mathbf{n}|}\right|$$

The sign of p indicates on which side of the origin the plane lies.

You can then find the distance between the two planes, and hence the shortest distance between the lines, by adding or subtracting the two distances.

Strategy 1

To find the distance between skew lines equations with $\mathbf{r} = \mathbf{a} + \lambda\mathbf{b}$ and $\mathbf{r} = \mathbf{c} + \mu\mathbf{d}$

(1) Find the vector perpendicular to both lines using $\mathbf{n} = \mathbf{b} \times \mathbf{d}$

(2) Find parallel planes on which the lines lie: $\mathbf{r} \cdot \mathbf{n} = \mathbf{a} \cdot \mathbf{n}$ and $\mathbf{r} \cdot \mathbf{n} = \mathbf{c} \cdot \mathbf{n}$

(3) Find the distance of each of these planes from the origin.

(4) Find the distance between the planes and hence the shortest distance between the lines.

Example 4

Find the shortest distance between the lines l_1 and l_2 with equations

$l_1 : \mathbf{r} = 2\mathbf{i} - \mathbf{j} + s(\mathbf{i} + \mathbf{j} - 3\mathbf{k})$ and $l_2 : \mathbf{r} = 4\mathbf{i} + \mathbf{j} - \mathbf{k} + t(2\mathbf{i} - \mathbf{k})$

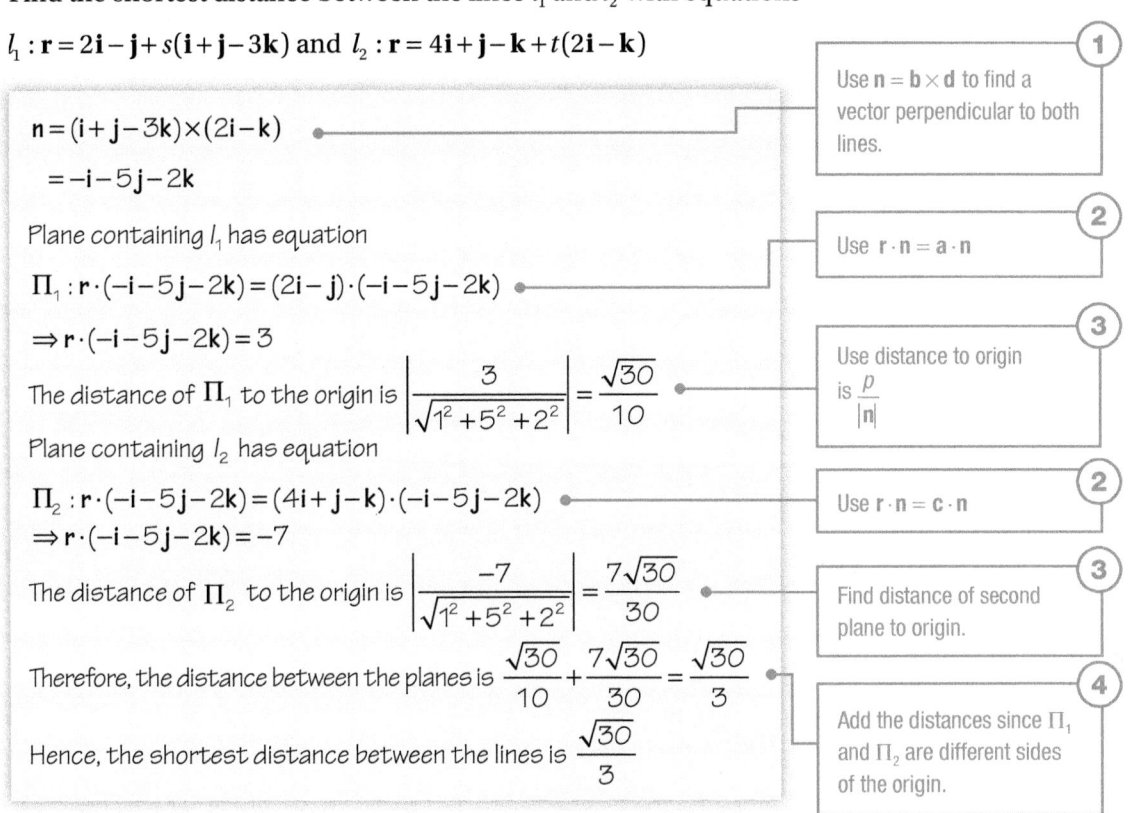

$\mathbf{n} = (\mathbf{i} + \mathbf{j} - 3\mathbf{k}) \times (2\mathbf{i} - \mathbf{k})$

$\quad = -\mathbf{i} - 5\mathbf{j} - 2\mathbf{k}$

(1) Use $\mathbf{n} = \mathbf{b} \times \mathbf{d}$ to find a vector perpendicular to both lines.

Plane containing l_1 has equation

$\Pi_1 : \mathbf{r} \cdot (-\mathbf{i} - 5\mathbf{j} - 2\mathbf{k}) = (2\mathbf{i} - \mathbf{j}) \cdot (-\mathbf{i} - 5\mathbf{j} - 2\mathbf{k})$

$\Rightarrow \mathbf{r} \cdot (-\mathbf{i} - 5\mathbf{j} - 2\mathbf{k}) = 3$

(2) Use $\mathbf{r} \cdot \mathbf{n} = \mathbf{a} \cdot \mathbf{n}$

The distance of Π_1 to the origin is $\left| \dfrac{3}{\sqrt{1^2 + 5^2 + 2^2}} \right| = \dfrac{\sqrt{30}}{10}$

(3) Use distance to origin is $\dfrac{p}{|\mathbf{n}|}$

Plane containing l_2 has equation

$\Pi_2 : \mathbf{r} \cdot (-\mathbf{i} - 5\mathbf{j} - 2\mathbf{k}) = (4\mathbf{i} + \mathbf{j} - \mathbf{k}) \cdot (-\mathbf{i} - 5\mathbf{j} - 2\mathbf{k})$

$\Rightarrow \mathbf{r} \cdot (-\mathbf{i} - 5\mathbf{j} - 2\mathbf{k}) = -7$

(2) Use $\mathbf{r} \cdot \mathbf{n} = \mathbf{c} \cdot \mathbf{n}$

The distance of Π_2 to the origin is $\left| \dfrac{-7}{\sqrt{1^2 + 5^2 + 2^2}} \right| = \dfrac{7\sqrt{30}}{30}$

(3) Find distance of second plane to origin.

Therefore, the distance between the planes is $\dfrac{\sqrt{30}}{10} + \dfrac{7\sqrt{30}}{30} = \dfrac{\sqrt{30}}{3}$

Hence, the shortest distance between the lines is $\dfrac{\sqrt{30}}{3}$

(4) Add the distances since Π_1 and Π_2 are different sides of the origin.

Another application of the vector product is to find the image of a point reflected in a plane.

You know that the reflection will be along the line perpendicular to the plane and can use this fact to find the image.

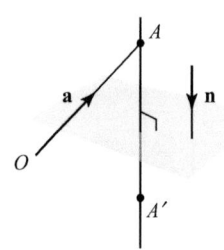

You need to find a vector, $\mathbf{n}$, which is perpendicular to the plane, then use this to write the equation of a line passing through the point you wish to reflect. The line will have equation $\mathbf{r} = \mathbf{a} + t\mathbf{n}$

By solving simultaneously, you can find the value of t where the line intersects the plane.

Then, since the image and the object are equidistant from the plane, the coordinates of the image can be found by doubling this value of t and substituting it back into the equation of the line.

Strategy 1

To find the reflection of a point in a plane

(1) Find a vector, $\mathbf{n}$, that is perpendicular to the plane.

(2) Write the equation of a line perpendicular to the plane in the form $\mathbf{r} = \mathbf{a} + t\mathbf{n}$, where $\mathbf{a}$ is the point to be reflected.

(3) Calculate the value of t where the line intersects the plane.

(4) Multiply the value of t by two and substitute back into the equation to find the coordinates of the image.

Example 5

Find the image of the point $A(1, 0, 2)$ in the plane $\mathbf{r} = \begin{pmatrix} 1 \\ 0 \\ -2 \end{pmatrix} + \mu \begin{pmatrix} -1 \\ 3 \\ 2 \end{pmatrix} + \lambda \begin{pmatrix} 0 \\ 2 \\ 1 \end{pmatrix}$

$\mathbf{n} = \begin{pmatrix} -1 \\ 3 \\ 2 \end{pmatrix} \times \begin{pmatrix} 0 \\ 2 \\ 1 \end{pmatrix} = \begin{pmatrix} -1 \\ 1 \\ -2 \end{pmatrix}$

(1) Use the vector product to find a vector perpendicular to the plane.

$\mathbf{r} = \begin{pmatrix} 1 \\ 0 \\ 2 \end{pmatrix} + t \begin{pmatrix} -1 \\ 1 \\ -2 \end{pmatrix}$

(2) This is the equation of line through A which is perpendicular to the plane.

$\begin{pmatrix} 1 \\ 0 \\ 2 \end{pmatrix} + t \begin{pmatrix} -1 \\ 1 \\ -2 \end{pmatrix} = \begin{pmatrix} 1 \\ 0 \\ -2 \end{pmatrix} + \mu \begin{pmatrix} -1 \\ 3 \\ 2 \end{pmatrix} + \lambda \begin{pmatrix} 0 \\ 2 \\ 1 \end{pmatrix}$

$1 - t = 1 - \mu$ (1)

$t = 3\mu + 2\lambda$ (2)

$2 - 2t = -2 + 2\mu + \lambda$ (3)

(3) You need to solve these simultaneously to find the value of t

Equation 1 gives $t = \mu$

Substitute into equations (2) and (3)

$-t = \lambda$ and

$\lambda = 4 - 4t$

$So -t = 4 - 4t \Rightarrow t = \dfrac{4}{3}$

So

$OA' = \begin{pmatrix} 1 \\ 0 \\ 2 \end{pmatrix} + \dfrac{8}{3} \begin{pmatrix} -1 \\ 1 \\ -2 \end{pmatrix}$

(4) $2 \times \dfrac{4}{3} = \dfrac{8}{3}$, substitute this for t in the equation of the line to find the position vector of A'

$= \begin{pmatrix} -\dfrac{5}{3} \\ \dfrac{8}{3} \\ -\dfrac{10}{3} \end{pmatrix}$

So coordinates of A' are $\left(-\dfrac{5}{3}, \dfrac{8}{3}, -\dfrac{10}{3} \right)$

State the actual coordinates.

1 The line l passes through the points $A\,(1, 4, -2)$ and $B\,(0, 2, -7)$

The plane Π has equation $5x - 5y + z = 3$

  a Find the shortest distance from the plane to the point

    i A **ii** B

  b Does the line intersect the plane? Explain your answer.

2 **a** Find the perpendicular distance between each of these pairs of planes. Give your answers to 3 significant figures.

    i $3x - y + 9z = 15$ and $6x - 2y + 18z - 3 = 0$

    ii $\mathbf{r} \cdot \begin{pmatrix} 4 \\ 1 \\ -2 \end{pmatrix} = 2$ and $\mathbf{r} \cdot \begin{pmatrix} -6 \\ -\dfrac{3}{2} \\ 3 \end{pmatrix} = 5$

    iii $\mathbf{r} \cdot (\mathbf{i} + 3\mathbf{j} - 5\mathbf{k}) = 12$ and $2x - 6y + 10z + 13 = 0$

  b Explain why the planes in part **a** do not meet.

3 The plane Π is perpendicular to the x-axis and passes through the point $(5, 4, -2)$. The line L is parallel to the vector $\mathbf{i} - \mathbf{k}$ and passes through the point $(1, 5, 2)$. Π and L intersect at the point A. Calculate the length $\overrightarrow{OA}$

4 The lines $\mathbf{r} = \begin{pmatrix} 1 \\ 0 \\ -2 \end{pmatrix} + s \begin{pmatrix} 5 \\ -2 \\ -4 \end{pmatrix}$ and

$\mathbf{r} = \begin{pmatrix} 16 \\ 10 \\ 1 \end{pmatrix} + t \begin{pmatrix} -1 \\ 6 \\ 6 \end{pmatrix}$ intersect the plane

$\mathbf{r} \cdot \begin{pmatrix} 1 \\ -3 \\ 2 \end{pmatrix} = 9$ at the points A and B respectively.

Calculate the area of triangle OAB

5 Three lines have equations as follows:

$L_1 \colon \mathbf{r} = 6\mathbf{i} - 3\mathbf{j} + \lambda(\mathbf{i} + \mathbf{k})$,

$L_2 \colon \mathbf{r} = s(3\mathbf{i} - \mathbf{j} + \mathbf{k})$ and

$L_3 \colon \mathbf{r} = t(\mathbf{j} + 2\mathbf{k})$

The lines L_1 and L_2 intersect at the point A, L_1 and L_3 intersect at the point B and L_2 and L_3 intersect at the point C, as shown.

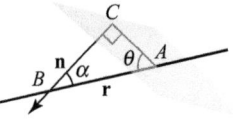

Calculate the exact area of triangle ABC

6 Find the shortest distance between these pairs of skew lines.

  a $\mathbf{r} = 2\mathbf{i} + \mathbf{j} + s(\mathbf{i} - 2\mathbf{k})$ and $\mathbf{r} = 8\mathbf{i} + 2\mathbf{j} - 3\mathbf{k} + t(\mathbf{i} - \mathbf{j})$

  b $\mathbf{r} = \begin{pmatrix} 2 \\ 1 \\ 1 \end{pmatrix} + \lambda \begin{pmatrix} 4 \\ 0 \\ -3 \end{pmatrix}$ and $\mathbf{r} = \begin{pmatrix} 2 \\ 7 \\ 0 \end{pmatrix} + \mu \begin{pmatrix} 5 \\ 3 \\ -1 \end{pmatrix}$

  c $\dfrac{x}{-1} = \dfrac{y+4}{0} = \dfrac{z-1}{2}$ and $\dfrac{x-2}{2} = \dfrac{y}{-1} = z - 6$

  d $\mathbf{r} = \begin{pmatrix} 0 \\ 0 \\ 2 \end{pmatrix} + \lambda \begin{pmatrix} -1 \\ 1 \\ 1 \end{pmatrix}$ and $\dfrac{x-4}{2} = 4 - y = \dfrac{z+4}{-2}$

  e $\mathbf{r} \times \begin{pmatrix} 2 \\ 0 \\ 1 \end{pmatrix} = \begin{pmatrix} 2 \\ 5 \\ -4 \end{pmatrix}$ and $\mathbf{r} \times \begin{pmatrix} 0 \\ -1 \\ 0 \end{pmatrix} = \begin{pmatrix} 2 \\ 0 \\ -4 \end{pmatrix}$

7 Calculate the shortest distance between the x-axis and the line with equation

$\mathbf{r} = \begin{pmatrix} 3 \\ -1 \\ -1 \end{pmatrix} + \lambda \begin{pmatrix} 2 \\ -1 \\ 1 \end{pmatrix}$

8 Find the image of the point $(1, 4, -7)$ when reflected in the plane $\mathbf{r} \cdot \begin{pmatrix} 1 \\ 3 \\ 2 \end{pmatrix} = 5$

9 The point $A(0, 2, -1)$ is reflected in the plane $\mathbf{r} = (2\mathbf{i} - \mathbf{k}) + s(\mathbf{i} + \mathbf{j}) + t(-3\mathbf{j} + \mathbf{k})$

Find the coordinates of the image of A

Chapter summary

- The vector product is defined as $\mathbf{a} \times \mathbf{b} = |\mathbf{a}||\mathbf{b}|\sin\theta\,\hat{\mathbf{n}}$, where θ is the angle between the vectors $\mathbf{a}$ and $\mathbf{b}$ and $\hat{\mathbf{n}}$ is a unit vector perpendicular to both $\mathbf{a}$ and $\mathbf{b}$

- $\mathbf{i} \times \mathbf{j} = \mathbf{k}$, $\mathbf{j} \times \mathbf{k} = \mathbf{i}$ and $\mathbf{k} \times \mathbf{i} = \mathbf{j}$

- For any vectors $\mathbf{a}$ and $\mathbf{b}$, $\mathbf{a} \times \mathbf{b} = -\mathbf{b} \times \mathbf{a}$

- For two vectors $\mathbf{a} = a_1\mathbf{i} + a_2\mathbf{j} + a_3\mathbf{k}$ and $\mathbf{b} = b_1\mathbf{i} + b_2\mathbf{j} + b_3\mathbf{k}$, the vector product is defined as
  $\mathbf{a} \times \mathbf{b} = (a_2 b_3 - a_3 b_2)\mathbf{i} - (a_1 b_3 - a_3 b_1)\mathbf{j} + (a_1 b_2 - a_1 b_2)\mathbf{k}$. You can use the determinant to calculate

$$\mathbf{a} \times \mathbf{b} = \left| \begin{pmatrix} \mathbf{i} & \mathbf{j} & \mathbf{k} \\ a_1 & a_2 & a_3 \\ b_1 & b_2 & b_3 \end{pmatrix} \right|$$

- Area of triangle $ABC = \dfrac{1}{2}\left| \overrightarrow{AB} \times \overrightarrow{AC} \right|$

- A straight line passing through the point with position vector $\mathbf{a}$ and parallel to the vector $\mathbf{b}$ has equation $(\mathbf{r} - \mathbf{a}) \times \mathbf{b} = \mathbf{0}$

- The vector equation of a plane is $\mathbf{r} = \mathbf{a} + s\mathbf{b} + t\mathbf{c}$ where $\mathbf{a}$ is the position vector of a point on the plane, and $\mathbf{b}$ and $\mathbf{c}$ are non-parallel vectors on the plane.

- The scalar product equation of a plane is $\mathbf{r} \cdot \mathbf{n} = p$ where $\mathbf{n}$ is perpendicular to the plane and $p = \mathbf{a} \cdot \mathbf{n}$ for $\mathbf{a}$ the position vector of a point on the plane.

- The Cartesian equation of a plane is $ax + by + cz = d$ where the vector $x\mathbf{i} + y\mathbf{j} + z\mathbf{k}$ is perpendicular to the plane.

- The acute angle, θ, between the plane $\mathbf{r} \cdot \mathbf{n} = p$ and the line $\mathbf{r} = \mathbf{a} + \lambda\mathbf{b}$ is given by $\sin\theta = \left| \dfrac{\mathbf{b} \cdot \mathbf{n}}{|\mathbf{b}||\mathbf{n}|} \right|$

- The acute angle, θ, between the planes $\mathbf{r} \cdot \mathbf{n}_1 = p_1$ and $\mathbf{r} \cdot \mathbf{n}_2 = p_2$ is given by $\cos\theta = \left| \dfrac{\mathbf{n}_1 \cdot \mathbf{n}_2}{|\mathbf{n}_1||\mathbf{n}_2|} \right|$

- The shortest distance between the skew lines with equations $\mathbf{r} = \mathbf{a} + \lambda\mathbf{b}$ and $\mathbf{r} = \mathbf{c} + \mu\mathbf{d}$ is given by $\left| \dfrac{(\mathbf{a} - \mathbf{c}) \cdot (\mathbf{b} \times \mathbf{d})}{|\mathbf{b} \times \mathbf{d}|} \right|$

Check and review

You should now be able to...	Try Questions
✓ Calculate the vector product of two vectors.	1
✓ Use the vector product to calculate the size of the angle between two vectors.	2
✓ Find a unit vector that is perpendicular to two other vectors.	3
✓ Use the vector product to calculate areas.	4

Write the equation of a line using vector product form.	5, 6
Write the equation of a plane in Cartesian, vector and scalar product form.	7, 8
Find the point of intersection between a line and a plane.	9, 10
Calculate the angle between a line and a plane.	9, 10
Calculate the angle between two planes.	11, 12
Calculate the shortest distance from a point to a plane.	13, 14
Calculate the shortest distance between skew lines.	15

PURE

1 Calculate these vector products.

   a $(\mathbf{i}+5\mathbf{j}-2\mathbf{k})\times(3\mathbf{i}-3\mathbf{j}-4\mathbf{k})$

   b $(2\mathbf{i}-3\mathbf{j}+8\mathbf{k})\times(4\mathbf{i}-2\mathbf{k})$

2 Find the acute angle between the pairs of vectors in question **1**.

3 Find a unit vector which is perpendicular to both $\mathbf{a}=\mathbf{i}-3\mathbf{k}$ and $\mathbf{b}=-\mathbf{j}+5\mathbf{k}$

4 A triangle has vertices at $(1, 3, -2)$, $(2, 3, 0)$ and $(4, 2, -3)$. Calculate its area.

5 A line is parallel to the vector $-5\mathbf{i}+3\mathbf{j}-\mathbf{k}$ and passes through the point with position vector $2\mathbf{j}+3\mathbf{k}$. Find the equation of the line in the form $\mathbf{r}\times\mathbf{a}=\mathbf{b}$

6 Write the equation of the line $\mathbf{r}=(\mathbf{i}-2\mathbf{j})+t(3\mathbf{j}-4\mathbf{k})$ in the form $\mathbf{r}\times\mathbf{a}=\mathbf{b}$

7 Write down the vector equation of the plane that contains the vectors $-2\mathbf{i}+3\mathbf{k}$ and $\mathbf{i}-4\mathbf{j}-\mathbf{k}$ and passes through the point $(2, 0, -3)$

8 A plane is perpendicular to the vector $7\mathbf{i}-5\mathbf{j}+\mathbf{k}$ and contains the point with position vector $2\mathbf{j}-5\mathbf{k}$. Write down the equation of the plane in

   a The form $\mathbf{a}\cdot\mathbf{n}=p$ **b** Cartesian form.

9 A plane has equation $\mathbf{r}\cdot(\mathbf{i}-2\mathbf{j}+2\mathbf{k})=5$

   The line with equation $\mathbf{r}=\mathbf{i}+t(2\mathbf{i}+3\mathbf{j}-\mathbf{k})$ intersects the plane at the point A

   a Calculate the acute angle between the plane and the line at A

   b Find the coordinates of A

   c Calculate the length OA

10 The line with equation $\dfrac{x-1}{2}=y=\dfrac{z+2}{4}$ intersect the plane $8x+6y-3z-4=0$ at the point A

   a Find the coordinates of A

   b Calculate the obtuse angle between the line and the plane at point A

11 The plane Π_1 has equation $\mathbf{r}\cdot(\mathbf{i}-4\mathbf{k})=7$ and the plane Π_2 has equation $\mathbf{r}\cdot(-3\mathbf{i}+\mathbf{j}+2\mathbf{k})$

   Given that the acute angle between the two planes is θ, find the exact value of $\cos\theta$

12 Calculate the size of the obtuse angle between the planes $x+y-5z=4$ and $2x-y+7z+1=0$

13 Calculate the shortest distance from the point $(5, 2, 1)$ to the plane $\mathbf{r}\cdot(12\mathbf{i}+7\mathbf{j}-14\mathbf{k})=9$

14 Calculate the shortest distance from the point $(-12, -7, 4)$ to the plane $2x-9y-12z=-2$

15 Calculate the shortest distance between the lines $\mathbf{r}=(2\mathbf{i}-\mathbf{k})+t(\mathbf{i}+\mathbf{j}-2\mathbf{k})$ and $\mathbf{r}\times(\mathbf{i}-\mathbf{k})=-2\mathbf{i}+\mathbf{j}-2\mathbf{k}$

Research

Galileo (1564–1642) was a major figure in the Scientific Renaissance, making significant contributions to a number of important areas in science. In order to investigate the motion of objects due to gravity, he rolled a ball along a sloping plane and recorded a notable result. The distance traversed *horizontally* by the ball was uniform, that is, in equal time intervals the ball travelled equal distances, but the distance travelled *vertically* was proportional to the square of the time taken.

Research Galileo's work in this field and use vectors to find an equation for the position of such an object measured from the point of projection.

History

Although mathematicians and scientists worked with a conceptual understanding of vectors prior to the 19th century, it was not until that time that vectors first began to be formalised. A number of mathematicians including Wessel, Argand and Gauss, explored complex numbers as vectors having two dimensions: a real and imaginary part. These new complex numbers provided many avenues to explore and helped mathematicians and scientists to think 'vectorially'.

In the mid-nineteenth century the mathematician Hamilton extended his thinking into four-dimensions developing a system that he called quaternions. Maxwell, a Scottish scientist, found that he could use this new way of thinking to develop understanding and insight into physical quantities that he classified as being either scalar or vector in nature.

Investigation

Consider geometrically how three planes in three-dimensional space might intersect each other.

Find a set of equations that illustrate each geometrical situation.

Try to solve your equations and consider how the algebra and geometry of the situation complement each other.

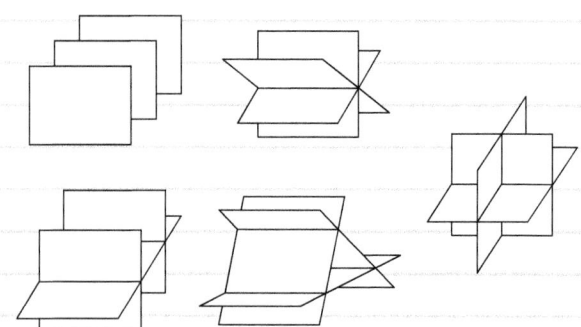

1 a Calculate the vector product $(\mathbf{i}-\mathbf{j}-\mathbf{k})\times(\mathbf{i}+\mathbf{k})$ **[2]**

  b Find a unit vector which is perpendicular to both $\mathbf{i}-\mathbf{j}-\mathbf{k}$ and $\mathbf{i}+\mathbf{k}$ **[2]**

2 The point $(2, 7, -3)$ lies on the line with equation $\dfrac{x-a}{2}=\dfrac{y-b}{-3}=\dfrac{z-c}{1}$

  a Write down the values of a, b and c **[1]**

  b Find the equation of the line in

    i Vector form,   ii Vector product form. **[4]**

3 Give the equation of the straight line $\mathbf{r}=\mathbf{i}+2\mathbf{j}-3\mathbf{k}+t(2\mathbf{i}-\mathbf{k})$ in the form $(\mathbf{r}-\mathbf{a})\times\mathbf{b}=0$ **[2]**

4 Show that the point $(3, 1, -2)$ lies on each of these planes.

  a $\mathbf{r}\cdot\begin{pmatrix}0\\4\\-2\end{pmatrix}=8$ b $2x+3y-z=11$ **[4]**

5 A triangle has vertices at $A(3, 5, -1)$, $B(4, 0, 2)$ and $C(-3, 2, -4)$

Use the vector product to calculate the area of the triangle. **[5]**

6 A plane passes through the point with position vector $\mathbf{i}-5\mathbf{j}+2\mathbf{k}$ and contains the vectors $-3\mathbf{j}-\mathbf{k}$ and $-2\mathbf{i}+4\mathbf{j}$

  a Find the equation of the plane in vector form. **[3]**

  b Show that the vector $2\mathbf{i}+\mathbf{j}-3\mathbf{k}$ is perpendicular to the plane. **[4]**

  c Find the equation of the plane in

    i Scalar product form,   ii Cartesian form. **[4]**

7 a Find the equation of the plane that passes through the points $(7, 12, -14)$, $(5, 4, -10)$ and $(1, 9, -5)$ in the form $\mathbf{r}=\mathbf{a}+s\mathbf{b}+t\mathbf{c}$ **[3]**

  b Does the point $(-1, 0, 4)$ lie on the plane? Explain how you know. **[4]**

8 The lines l_1 and l_2 have equations $l_1:\mathbf{r}=(\mathbf{i}-\mathbf{j}+3\mathbf{k})+s(2\mathbf{i}-5\mathbf{k})$ and $l_2:\mathbf{r}=(3\mathbf{j}+\mathbf{k})+t(\mathbf{i}+3\mathbf{j})$

  a Find the shortest distance between l_1 and l_2. Give your answer correct to 3 significant figures.

  The plane Π has Cartesian equation $3x+y-z=2$

  The line l_1 intersects the plane Π at the point A **[6]**

  b Find the exact coordinates of A **[2]**

  c Calculate the acute angle between l_1 and Π, giving your answer to 3 significant figures. **[2]**

9 The point A has coordinates $(0, 3, -1)$. Find the image of A after reflection in the plane $\mathbf{r}\cdot(\mathbf{i}+\mathbf{j}-2\mathbf{k})=3$ **[6]**

10 Given $L_1: \mathbf{r} \times \begin{pmatrix} 1 \\ 0 \\ -2 \end{pmatrix} = \begin{pmatrix} -4 \\ 5 \\ -2 \end{pmatrix}$ and $L_2: \mathbf{r} = \begin{pmatrix} 5 \\ 5 \\ 2 \end{pmatrix} + \lambda \begin{pmatrix} -8 \\ -6 \\ 2 \end{pmatrix}$

   a Show that L_1 and L_2 intersect.   **[3]**

   b Find the point of intersection between L_1 and L_2

11 The plane Π_1 has equation $\mathbf{r} \cdot (2\mathbf{i} - 3\mathbf{j}) = 4$ and the plane Π_2 has equation $\mathbf{r} \cdot (\mathbf{i} + 2\mathbf{j} - \mathbf{k}) = 7$

   Calculate the acute angle between planes Π_1 and Π_2 **[3]**

12 A plane Π_1 has equation $\mathbf{r} \cdot \begin{pmatrix} -1 \\ 3 \\ -3 \end{pmatrix} = -2$ and a line l_1 has equation $\mathbf{r} = \begin{pmatrix} 5 \\ -7 \\ 1 \end{pmatrix} + t \begin{pmatrix} 0 \\ -4 \\ 2 \end{pmatrix}$

   a Calculate the acute angle between the plane Π_1 and the line l_1 **[3]**

   b Find the point of intersection between the plane Π_1 and the line l_1 **[4]**

13 Find the perpendicular distance from the point $(5, 0, -2)$ to the plane with equation $2x + y - 3y = 9$ **[3]**

14 Find the equation of the plane that contains the point $(5, -2, 7)$ and the line with equation $\dfrac{x-2}{4} = \dfrac{y+1}{5} = \dfrac{z-1}{-2}$. Give your answer in the form $\mathbf{r} = \mathbf{a} + \lambda\mathbf{b} + \mu\mathbf{c}$ **[6]**

15 The planes Π_1 and Π_2 have equations $\mathbf{r} \cdot \begin{pmatrix} 2 \\ 0 \\ -3 \end{pmatrix} = -11$ and $x - 2y - z = 1$ respectively.

   The line with equation $\dfrac{x+2}{4} = \dfrac{y+1}{-3} = \dfrac{z-4}{1}$ intersects Π_1 at the point A and Π_2 at the point B

   Calculate the length of AB **[9]**

16 The lines L_1 and L_2 have equations $\mathbf{r} = 8\mathbf{i} - 14\mathbf{j} + 13\mathbf{k} + s(-4\mathbf{i} + 7\mathbf{j} - 6\mathbf{k})$ and $\dfrac{x}{2} = \dfrac{y-17}{5} = \dfrac{z+7}{-1}$ respectively. The plane Π contains both L_1 and L_2

   a Find the vector equation of the plane.   **[5]**

   b Calculate the distance of the plane from the point $(16, 11, -13)$ **[5]**

17 Find a vector equation of the line with vector product equation $\mathbf{r} \times \begin{pmatrix} 3 \\ -1 \\ 2 \end{pmatrix} = \begin{pmatrix} 2 \\ 6 \\ 1 \end{pmatrix}$ **[5]**

18 Show that the line $\mathbf{r} = \begin{pmatrix} -11 \\ -4 \\ 10 \end{pmatrix} + t \begin{pmatrix} 4 \\ -2 \\ 0 \end{pmatrix}$ and the plane

   $\mathbf{r} = \begin{pmatrix} 3 \\ -5 \\ -10 \end{pmatrix} + \mu \begin{pmatrix} 7 \\ -10 \\ 0 \end{pmatrix} + \lambda \begin{pmatrix} 1 \\ -3 \\ 4 \end{pmatrix}$ intersect and find their point of intersection.   **[5]**

19 The planes Π_1 and Π_2 have equations $\mathbf{r} = \begin{pmatrix} 5 \\ 3 \\ -2 \end{pmatrix} + s \begin{pmatrix} 4 \\ 2 \\ -1 \end{pmatrix} + t \begin{pmatrix} 0 \\ -5 \\ 1 \end{pmatrix}$ and $\mathbf{r} \cdot \begin{pmatrix} 3 \\ 4 \\ 20 \end{pmatrix} = 12$

   a Show that Π_1 and Π_2 are parallel planes.   **[6]**

   b Find the perpendicular distance between Π_1 and Π_2 **[2]**

24 Circular motion 2

In sports such as cycling, it is important to consider the mechanics of circular motion to ensure both speed and safety. In a velodrome the circular parts of the track, at either end of two straight sections, are steeply banked. This banking helps cyclists to develop force towards the centre of their motion (modelled as being part of a circle) at speed. The maximum angle of banking in velodromes that are used to host world championship and Olympic events is forty-five degrees. This appears very steep when you ride a bike around such a track for the first time.

The same principles apply in the design of surfaces used by other vehicles, such as roads. Busy main roads must be designed to ensure that vehicles can take corners, and use roundabouts, safely. Engineers involved in road design, railway track engineering, and so on, use the principles of the mechanics of circular motion to make sure that their designs are safe under all conditions.

Orientation

What you need to know

Ch9 Circular motion 1
- Kinematics of circular motion.
- Horizontal circular motion.

What you will learn

- How to analyse circular motion.
- How to analyse the conical pendulum.
- How to analyse vertical circular motion.

What this leads to

Careers
Civil engineering.
Architecture.

Fluency and skills

There are equations that describe the motion of a point mass moving in a horizontal circle at constant speed.

> **Key point**
>
> A body moving on a circular path of radius r with constant angular velocity ω about the centre has
>
> - a constant linear speed $v = r\omega$ along the tangent
> - an acceleration $a = r\omega^2 = \dfrac{v^2}{r}$ towards the centre
> - a centripetal force $F = mr\omega^2$ or $\dfrac{mv^2}{r}$ towards the centre
> - a time period $T = \dfrac{2\pi}{\omega}$ to make one revolution.

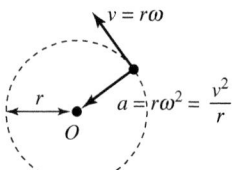

You can also use vectors to find some of these equations using a more mathematically elegant method.

Let the circle have a radius r and a centre O

If the mass starts at point A and, after a time t seconds, it has reached point P, then $\angle AOP = \theta = \omega t$, where ω is the angular velocity in rad s^{-1}.

Let point P have coordinates (x, y) which, using trigonometry, can also be written as $(r\cos \omega t, r\sin \omega t)$.

So, the position vector of P is

$$\mathbf{r} = r\cos \omega t\,\mathbf{i} + r\sin \omega t\,\mathbf{j}$$

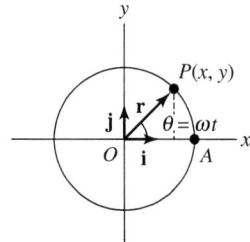

> $\mathbf{i}$ and $\mathbf{j}$ are unit vectors parallel to the x-axis and y-axis respectively.

When you differentiate with respect to t, you get

$$\mathbf{v} = -r\omega\sin \omega t\,\mathbf{i} + r\omega\cos \omega t\,\mathbf{j}$$

When you differentiate again,

$$\mathbf{a} = -r\omega^2 \cos \omega t\,\mathbf{i} - r\omega^2 \sin \omega t\,\mathbf{j}$$
$$= -\omega^2(r\cos \omega t\,\mathbf{i} + r\sin \omega t\,\mathbf{j})$$

> ω is a constant, so
> $$\dfrac{d(\cos \omega t)}{dt} = -\omega \sin \omega t$$
> and $\dfrac{d(\sin \omega t)}{dt} = \omega \cos \omega t$

which you can write as

$$\mathbf{a} = -\omega^2 \mathbf{r}$$

The negative sign in $\mathbf{a} = -\omega^2 \mathbf{r}$ indicates that $\mathbf{a}$ is always *towards* the centre, in the opposite direction to $\mathbf{r}$.

So, although there is no linear acceleration along the tangent, circular motion requires a linear acceleration towards the centre of the circle of magnitude $r\omega^2 = \dfrac{v^2}{r}$

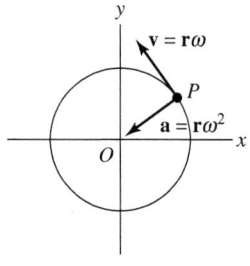

You can think of the acceleration as changing the direction of the particle, rather than its speed.

Since there is an acceleration, Newton's 2nd law requires that there is a force $\mathbf{F} = m\mathbf{a} = -m\omega^2\mathbf{r}$ towards the centre. This force is called the **centripetal force**.

Example 1

A point mass of 5 kg has a position vector $\mathbf{r} = 8\sin\left(\frac{1}{2}t\right)\mathbf{i} + 8\cos\left(\frac{1}{2}t\right)\mathbf{j}$ metres at a time t seconds.

a Find the magnitude in newtons of the force $\mathbf{F}$ acting on the mass when $t = \frac{\pi}{3}$ seconds.

b Show that the force $\mathbf{F}$ and position vector $\mathbf{r}$ are parallel at all times.

a $\mathbf{r} = 8\sin\left(\frac{1}{2}t\right)\mathbf{i} + 8\cos\left(\frac{1}{2}t\right)\mathbf{j}$ [1]

$\mathbf{v} = \dfrac{d\mathbf{r}}{dt} = 8 \times \dfrac{1}{2}\cos\left(\frac{1}{2}t\right)\mathbf{i} - 8 \times \dfrac{1}{2}\sin\left(\frac{1}{2}t\right)\mathbf{j}$ ← Differentiate with respect to t

$= 4\cos\left(\frac{1}{2}t\right)\mathbf{i} - 4\sin\left(\frac{1}{2}t\right)\mathbf{j}$

$\mathbf{a} = \dfrac{d\mathbf{v}}{dt} = -4 \times \dfrac{1}{2}\sin\left(\frac{1}{2}t\right)\mathbf{i} - 4 \times \dfrac{1}{2}\cos\left(\frac{1}{2}t\right)\mathbf{j}$ ← Differentiate again with respect to t

$= -2\sin\left(\frac{1}{2}t\right)\mathbf{i} - 2\cos\left(\frac{1}{2}t\right)\mathbf{j}$

Newton's 2nd law gives

$\mathbf{F} = m\mathbf{a} = -10\sin\left(\frac{1}{2}t\right)\mathbf{i} - 10\cos\left(\frac{1}{2}t\right)\mathbf{j}$ [2]

When $t = \dfrac{\pi}{3}$

$\mathbf{F} = -10\left(\sin\dfrac{\pi}{6}\mathbf{i} + \cos\dfrac{\pi}{6}\mathbf{j}\right) = -5\mathbf{i} - 5\sqrt{3}\mathbf{j}$

Magnitude of $\mathbf{F} = |\mathbf{F}| = \sqrt{(-5)^2 + (-5\sqrt{3})^2}$

$= \sqrt{25 + 75} = 10\,\text{N}$

b From [1] $\mathbf{r} = 8\left[\sin\left(\frac{1}{2}t\right)\mathbf{i} + \cos\left(\frac{1}{2}t\right)\mathbf{j}\right]$

From [2] $\mathbf{F} = -10\left[\sin\left(\frac{1}{2}t\right)\mathbf{i} + \cos\left(\frac{1}{2}t\right)\mathbf{j}\right]$

$= -10 \times \dfrac{1}{8}\mathbf{r} = -1.25\mathbf{r}$

Hence, $\mathbf{F}$ and $\mathbf{r}$ are parallel but have opposite directions.

1   A point mass of 10 kg has a position vector $\mathbf{r} = 2\sin 4t\mathbf{i} + 2\cos 4t\mathbf{j}$ metres at time t seconds.

Find vector expressions for

a   The acceleration of the mass,

b   The force acting on the mass.

2   A 2 kg particle P moves on a smooth horizontal plane containing x- and y-axes. Its velocity $\mathbf{v}$ is given by $\mathbf{v} = 6\cos 2t\mathbf{i} - 6\sin 2t\mathbf{j}\,\mathrm{m\,s^{-1}}$.

When $t = 0$, P has the position vector $2\mathbf{i} + 4\mathbf{j}$.

a   Find the position vector, $\mathbf{r}$, of P at time t

b   Use the components of $\mathbf{r}$ to show that P moves on a circular path and find the equation of the path.

3   A mass, M, of 3 kg at the point (x, y) on coordinate axes moves on a circular path with the equation $x^2 + y^2 = 16$. The line OM rotates about the origin O with a constant angular speed of $0.5\,\mathrm{rad\,s^{-1}}$, starting in line with the x-axis when $t = 0$

a   Write expressions for

   i   The position vector $\mathbf{r}$ of M at time t

   ii   The acceleration of M as a vector at time t

b   Find the time for M to make one revolution about O

c   Calculate the magnitude of the force acting on M

4   The position vector of a particle at time t seconds is given by $\mathbf{r} = 2\sin 3t\mathbf{i} + 2\cos 3t\mathbf{j}$ metres.

a   Express, as vectors, the velocity and acceleration of the particle.

b   Show that the direction of the acceleration is

   i   Parallel to the direction of the position vector,

   ii   Perpendicular to the direction of the velocity.

5   A particle of mass 4 kg is acted on by a force $\mathbf{F}$ newtons and it moves in a horizontal plane with a velocity $\mathbf{v} = 4\cos 2t\mathbf{i} + 4\sin 2t\mathbf{j}\,\mathrm{m\,s^{-1}}$ at a time t seconds.

a   Find an expression for force $\mathbf{F}$ in terms of t and find its magnitude when $t = \pi$

b   When $t = 0$, the particle is at a point with a position vector $3\mathbf{i} - 10\mathbf{j}$ metres. Find the position vector, $\mathbf{r}$ of the particle at time t seconds. Describe the path in which the particle moves.

6   A particle of mass m kg is acted upon by a force of $\mathbf{F}$ newtons acting through the origin, O. At time t seconds, the particle has position $\mathbf{r}$ in the x–y plane and velocity $\mathbf{v} = a\sin \omega t\mathbf{i} + b\cos \omega t\mathbf{j}\,\mathrm{m\,s^{-1}}$.

Prove that $\mathbf{F} = k\mathbf{r}$ and find k in terms of the constants m and ω

Reasoning and problem-solving

Strategy

To solve problems involving motion in a horizontal circle

(1) Draw and label a diagram showing all forces and other key variables.

(2) Use Newton's 2nd law to write an equation of motion to the centre.

(3) Write other equations involving forces, as necessary, and solve them.

Example 2

A car travels round a bend in a smooth road of radius 60 m which is banked at 10° to the horizontal.

a Find the only safe speed (in $km\,h^{-1}$) at which the car can travel.

b What assumption in this model is unrealistic?

a
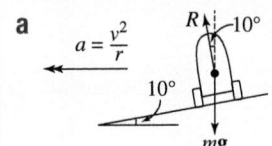

① Draw a labelled diagram.

Let the mass, velocity and acceleration of the car be m, v and a

Let the normal reaction of the ground on the car be R

The car does not move vertically.

Resolve vertically

$$R\cos 10° = mg \qquad [1]$$

③ Vertical forces balance.

Horizontal equation of motion is

$$R\sin 10° = m \times \frac{v^2}{60} \qquad [2]$$

② Use Newton's 2nd law.

Divide [2] by [1] to eliminate R

$$\tan 10° = \frac{mv^2}{60} \div mg = \frac{v^2}{60g}$$

The only safe speed, $v = \sqrt{60g\tan 10°} = 10.1\,m\,s^{-1} = 36.6\,km\,h^{-1}$

If $v > 10.1\,m\,s^{-1}$, the car skids up the slope.

If $v < 10.1\,m\,s^{-1}$, the car skids down the slope.

b The model is unrealistic as it takes no account of friction acting along the slope and a speed of $10.1\,m\,s^{-1}$ could not be maintained precisely.

Example 3

The bend in Example 2 is resurfaced so the coefficient of friction is μ

Find the least value of μ for the car to round the bend at a speed of $26\,m\,s^{-1}$ without any side-slip up the slope.

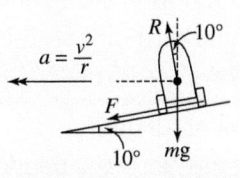

① At its greatest safe speed, the car is on the point of slipping up the slope, so friction acts down the slope.

Equation of motion down the slope is

② Resolving parallel to the slope means that R does not appear in the equation, so it is easy to calculate F

$$F + mg\sin 10° = \frac{m \times 26^2}{60} \times \cos 10°$$

(*Continued on the next page*)

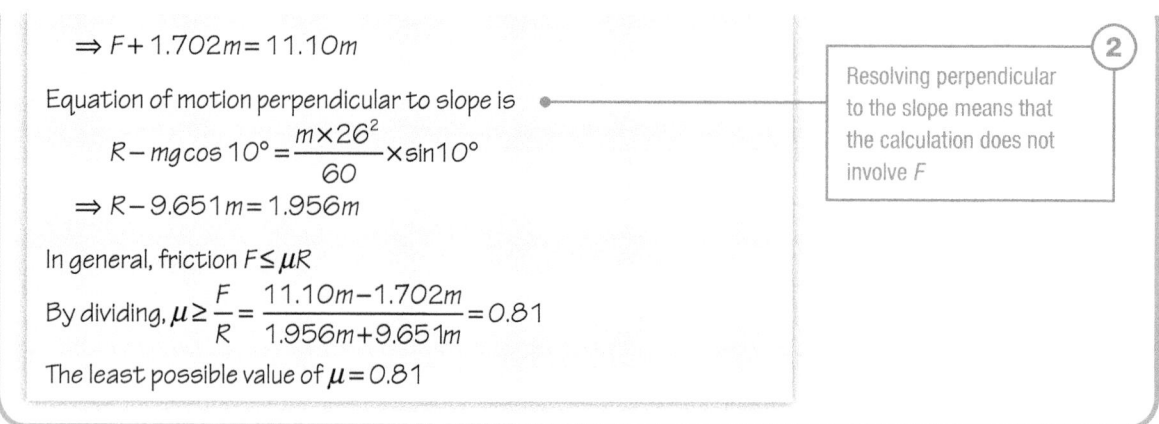

$$\Rightarrow F + 1.702m = 11.10m$$

Equation of motion perpendicular to slope is

$$R - mg\cos 10° = \frac{m \times 26^2}{60} \times \sin 10°$$

$$\Rightarrow R - 9.651m = 1.956m$$

In general, friction $F \leq \mu R$

By dividing, $\mu \geq \dfrac{F}{R} = \dfrac{11.10m - 1.702m}{1.956m + 9.651m} = 0.81$

The least possible value of $\mu = 0.81$

Resolving perpendicular to the slope means that the calculation does not involve F

Example 3 could also be solved by writing a horizontal equation of motion and by resolving vertically for vertical equilibirum (as in Example 2). F and R are then found by solving the equations simultaneously.

Exercise 24.1B Reasoning and problem-solving

1 The smooth road surface on a corner of radius 80 m is designed so that a car can travel without sideways movement when driven at a certain speed. If the road is banked at an angle of 15°, find the speed.

2 A lorry is driven round a bend in a smooth road of radius 100 m which is banked at an angle of 12°. At what speed should it be driven so that there is no sideways force on its tyres?

3 A smooth road is banked so that a car can travel round a circular bend without any skidding if its speed is 20 m s⁻¹. If the radius of the bend is 120 m, calculate the angle of the banking.

4 A circular bend in a road has a radius of 180 m and it is banked at 45° to the horizontal. The coefficient of friction between the road and a car is 0.5

   a At what speed would a car rounding the bend have no sideways friction on its tyres?

   b What are the car's maximum and minimum speeds that are possible without any sideways slipping?

5 A car is just on the point of skidding when it goes round a bend of radius 30 m at a speed of 54 km h⁻¹ on a level racing track.

   a At what angle to the horizontal should the track be banked so the car can round the same bend at 108 km h⁻¹ without skidding up the slope?

   b State an assumption you have made in your solution.

6 Find the maximum speed at which a car can be driven round a bend of radius 120 m if the coefficient of friction between the car and road is 0.4 and the road surface is

   a Horizontal,

   b Banked at an angle of 10°.

7 A bend in a road has a radius of 80 m and is banked at 14°. A car is on the point of sliding up the slope when it travels at 25 m s⁻¹

   a Find the coefficient of friction, μ between the road and car at the bend.

   b On an icy morning when $\mu = 0.1$, what is the maximum speed of the car for it not to slide upwards?

8 Part of a railway track is a circular arc of radius 250 m. The track is banked so that there is no side-thrust on the flanges of the train's wheels when it travels at 50 km h^{-1}. Calculate the angle of the banking of the track.

9 A railway track curves with a radius of 60 m and it is banked so that an engine travelling at 15 m s^{-1} exerts no sideways force on the track. What sideways force would there be when a 50 tonne engine

  a Stands at rest on the bend,

  b Moves at 30 m s^{-1} round the bend?

10

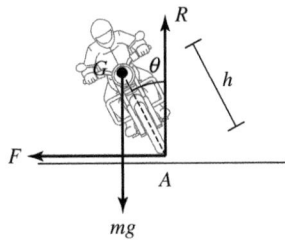

  a A motorcyclist rounds a curve of radius 12 m on a level road at 5 m s^{-1}. By considering the forces on the bike and the moments about the centre of mass G, find the angle at which he and his bike are inclined to the vertical and the least value of the coefficient of friction for the bike not to side-slip.

  b Another corner on a level road has a radius of 30 m. Find the greatest speed in km h^{-1} at which he can round the corner if the coefficient of friction with the road is 0.4

11 A particle moves in a horizontal circle on the smooth inner surface of an inverted hollow cone. The circle is at a height h above the vertex of the cone. Prove that the particle's linear velocity is $\sqrt{hg}$

12 Four masses of 3 kg each are connected by four strings 10 cm long so they form a square with the strings as its sides. The square is placed with its centre at the centre of a turntable that is rotating with an angular speed of 1.5 rad s^{-1}. Find the tension in each of the strings.

13 A smooth bowl is formed from a segment of a sphere of radius $2x$ such that the bowl has a depth x in the middle. The bowl is rotated about its vertical axis with angular velocity ω such that a small particle is at rest relative to the bowl when placed just within the rim. Find the value of ω in terms of x

14 A car rounds a bend of radius r on a road banked at an angle θ to the horizontal, where the coefficient of friction $\mu = \tan \lambda$ Prove that the car can drive round the bend at a speed v without skidding provided that

$$\theta - \lambda < \arctan\left(\frac{v^2}{rg}\right) < \theta + \lambda$$

Fluency and skills

An ordinary pendulum swings to and fro with all the motion in the same vertical plane. A **conical pendulum**, however, has its bob moving in a horizontal circle so that the string of the pendulum traces out a hollow vertical cone.

Example 1

A conical pendulum has a bob P of mass 2 kg hanging at the lower end of a light, inextensible string of length 2 m. The upper end of the string is fixed to a point O. The bob moves in a horizontal circle, centre C and radius 1.2 m, at constant speed.

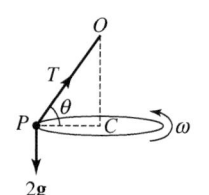

Find the tension in the string, the angular speed of the bob and the time taken for the bob to make one revolution.

In $\triangle OPC$, $OC = \sqrt{2^2 - 1.2^2} = 1.6\,\text{m}$

Resolve vertically $T \sin \theta = 2g$ | P is in vertical equilibrium.

$$T \times \frac{1.6}{2} = 2 \times 9.8$$

Tension, $T = \dfrac{4 \times 9.8}{1.6} = 24.5\,\text{N}$

Equation of motion of P towards C is

$$T \cos \theta = mr\omega^2$$ | Resolve horizontally and use Newton's 2nd law.

$$24.5 \times \frac{1.2}{2} = 2 \times 1.2 \times \omega^2$$

$$\omega^2 = 6.125$$

Angular speed, $\omega = 2.47\,\text{rad s}^{-1}$

Time for one revolution $= \dfrac{2\pi}{\omega} = \dfrac{2\pi}{2.47} = 2.5$ seconds | Time $= \dfrac{\text{angular distance}}{\text{angular speed}}$

Exercise 24.2A Fluency and skills

In this exercise, all strings are light and inextensible.

1 A particle of mass 2 kg is attached to one end of a string 0.5 m long, the other end of which is fixed to a point O. The particle rotates in a horizontal circle about a point vertically below O so that the string makes an angle of $60°$ with the vertical. Find the tension in the string and the time taken for the particle to make one revolution.

2 The bob of a conical pendulum moves in a horizontal circle at a steady speed of 80 revolutions per minute (rpm). If the string of the pendulum is 0.25 m long, show that it makes an angle of approximately $56°$ with the vertical.

3 The string of a conical pendulum is 80 cm long and makes an angle of $60°$ with the vertical. Show that the bob of the pendulum makes about 8 revolutions every 10 seconds.

4 A mass of 0.5 kg is attached to a string of length 1.5 m. The mass acts as conical pendulum and moves in a horizontal circle at a steady speed of 60 rpm. Calculate the radius of the circle and the tension in the string.

5 A conical pendulum has a bob with a mass of 1.2 kg which makes a full rotation every second. If the length of the pendulum is 50 cm, calculate the tension in the string and the angle which the string makes with the vertical.

6 A small object is attached to a fixed point by a string of length l. It describes a horizontal circle with an angular speed of ω rad s^{-1}. Prove that its vertical distance below the fixed point is independent of the length of the string.

Reasoning and problem-solving

To solve problems involving motion in horizontal circles

1. Draw a clear diagram showing all the forces acting on the moving parts.

2. Write equations of motion to the centre of the circle using Newton's 2nd law.

3. Write other equations involving forces, as necessary, and solve them simultaneously.

Example 2

A 0.5 kg mass P is tied to the midpoint of a 2 m string XY with $X \sqrt{3}$ m vertically above Y

P rotates in a horizontal circle so that the string stays taut.

Find the least possible angular velocity of P. Take g as 9.81 m s^{-2}.

(1) Draw a labelled diagram. Let T and U be tensions, as shown.

$\triangle XYP$ is isosceles, so $\cos \theta = \dfrac{\sqrt{3}}{2}$ and $\theta = 30°$

Vertical equilibrium:

(3) Write an equilibrium equation for vertical forces.

$T\cos \theta = U\cos \theta + 0.5g$

giving $T - U = 5.664$ [1]

Horizontal equation of motion for P is

(2) Use Newton's 2nd law for horizontal forces.

$T\sin \theta + U\sin \theta = mr\omega^2 = 0.5 \times 1 \times \sin \theta \times \omega^2$

$T + U = 0.5\omega^2$ [2]

The string can only go slack between P and Y

The string stays taut if $U \geq 0$

From [1][2], $2U = 0.5\omega^2 - 5.664$

(3) Solve simultaneously to find an expression for U

$U \geq 0$ if $\omega^2 \geq 11.33$,

so least angular velocity, ω is $\sqrt{11.33} = 3.4$ rad s^{-1}

If mass P were a smooth ring threaded onto string XY, then tensions T and U would be equal on either side of P. P will adjust its position and may not stay at the midpoint M

Example 3

Particle Y is connected to two light rods XY and YZ, both of length l, such that end X pivots about a fixed point.

End Z is attached to a ring which slides on a smooth vertical rod XZ. Y and Z have the same mass m

The system rotates so that Y performs horizontal circles with constant speed ω

Prove that, if $\angle ZXY = \theta$, then $\cos\theta = \dfrac{3g}{l\omega^2}$

$\triangle XYZ$ is isosceles. T and U are forces in the rods. R is the reaction of rod XZ on Z.

Z is in vertical equilibrium: $U\cos\theta = mg$ [1]

Y is in vertical equilibrium $T\cos\theta = U\cos\theta + mg$

$= 2mg$ [2]

Horizontal equation of motion for Y is

$T\sin\theta + U\sin\theta = m \times l\sin\theta \times \omega^2$ [3]

Substituting [1] and [2] into [3]

$\dfrac{2mg}{\cos\theta} \times \sin\theta + \dfrac{mg}{\cos\theta} \times \sin\theta = m \times l\sin\theta \times \omega^2$

Simplify to give $\cos\theta = \dfrac{3g}{l\omega^2}$

> ② ③ Write equations for Y and Z

> Substitute from [1]

> Divide by $m\sin\theta$ and rearrange.

Exercise 24.2B Reasoning and problem-solving

1   a   A 2 kg mass is fixed to the midpoint of a string XY of length 10 m, with X 6 m vertically above Y. The mass describes a horizontal circle with an angular speed ω. Find the minimum value of ω so that both parts of XY remain taut.

   b   The mass in part **a** is changed to a smooth 2 kg ring R which can slide freely along the same string XY. Find the angular speed ω of the ring for it to rotate in a horizontal circle about Y with the horizontal distance $RY = 3.2$ m.

   c   Give two assumptions that you have made about this mathematical model.

2   A light, inextensible string PQ passes through a smooth fixed ring R in the ceiling. A 2 kg mass at Q hangs at rest vertically below R as a 1 kg mass at P acts as a conical pendulum moving in a horizontal circle with an unknown height and radius. What is the radius of the circle if the linear speed of the 1 kg mass is 7 m s^{-1}?

3   A light rod BC of length l is pivoted at a fixed point C and held horizontal by a string AB where A is a fixed point at a distance h vertically above C. A mass m attached to B rotates about C with angular velocity ω so that the rod BC remains horizontal. Find the force exerted on the mass by the rod in terms of m, l, h, ω and g.

4   A particle moves horizontally on a circular path of radius r inside a smooth hemispherical bowl of radius R. Find the period of its rotation in terms of r, R and g

5   The ends of an inextensible string of length 0.4 m are attached to a mass m on a smooth

horizontal table and a fixed point above the table. The mass moves in a horizontal circle of radius 0.2 m in contact with the table, with the string taut, at an angular speed ω. Find the reaction between the mass and table in terms of m and ω and show that the maximum value of ω to keep in contact with the table is 5.3 rad s^{-1}

6   The ends A and B of an inextensible string are fastened to a vertical rod with A above B A smooth ring R of mass 3 kg slides on the string and rotates with the string taut in a horizontal circle of radius 0.7 m about the rod below the level of B. If $\angle RAB = 28°$ and $\angle RBA = 144°$, find the angular speed of R.

7   An elastic string of natural length l and modulus of elasticity $2mg$ has its upper end fixed and its lower end attached to a particle of mass m which makes horizontal circles with angular velocity ω

    a   When the extension in the string is $\frac{3}{5}l$
        i   Show that $\omega^2 = \dfrac{3g}{4l}$
        ii   Find the exact value of the angle that the string makes with the vertical.

    b   State two assumptions you have made in this mathematical model.

8   A light, inextensible string has identical particles of mass m attached at each end. The string is threaded through a hole in the vertex V of a smooth, hollow cone of semi-vertical angle 30°. The particle inside the cone hangs at rest. The other particle describes horizontal circles on the outer surface of the cone at a distance of 0.5 metres from V. Show that the angular speed of the moving particle is 3.24 rad s^{-1}

9   A smooth vertical hoop of radius r rotates with angular speed ω about its vertical diameter. Two beads, A of mass m and B of mass M (with $M > m$), are threaded on the hoop with B vertically below A. They are joined by a light rigid rod of length r

    a   Show that the reaction R of the hoop on A is given by $R = mr\omega^2$

    b   If $m = 2$ kg, $M = 4$ kg and $r = 1.5$ m, find the value of ω

10   A particle P of mass m moves in a horizontal circle, centre C and radius 12 cm, on the inside surface of a smooth hemispherical bowl of radius 13 cm.

C is 8 cm vertically above the lowest point H of the bowl. A light, inextensible string attached to P passes through a small smooth hole at H with its lower end attached to an identical particle Q suspended at rest. Find the angular speed of P

11   Steam engines have a mechanism, called a governor, which rotates so that it opens or closes a valve to adjust the speed and so keep it constant.

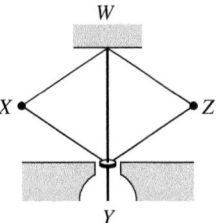

Four equal light rods are hinged to form a rhombus $WXYZ$ with W fixed. Y is a collar of mass m which slides smoothly on a vertical fixed rod WY. Two identical metal balls of mass M are attached at X and Z The mechanism rotates about rod WY at an angular speed ω

    a   Find the distance WY in terms of M, m and ω

    b   Describe three ways in which this model of a governor can be refined.

12   An elastic string of natural length l is fixed at one end. A mass m hangs from the other end and produces an extension e when in equilibrium. When the same mass acts as a conical pendulum and describes horizontal circles with angular velocity ω, the same string makes an angle θ with the vertical. Find an expression for $\cos\theta$ in terms of l, e, ω and g

24.3 Vertical circular motion

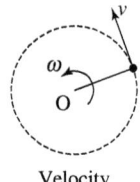

Fluency and skills

In the circle problems you have studied so far, angular velocity, ω, and tangential velocity, v, have both been constant. However, when circular motion is in a vertical plane, ω and v both vary, so $\dfrac{d\omega}{dt}$ and $\dfrac{dv}{dt}$ are not zero.

Velocity

The acceleration of a particle now has two components:

- A **radial component** towards the centre of the circle, O, of
  magnitude $r\omega^2 = r\left(\dfrac{d\theta}{dt}\right)^2 = r\dot{\theta}^2 = \dfrac{v^2}{r}$
  where $\dot{\theta}$ is used as a shorthand for $\dfrac{d\theta}{dt}$

- A **tangential component** along the tangent of magnitude
  $\dfrac{dv}{dt} = r\dfrac{d\omega}{dt} = r\dfrac{d^2\theta}{dt^2} = r\ddot{\theta}$ where $\ddot{\theta}$ is used as a shorthand for $\dfrac{d^2\theta}{dt^2}$

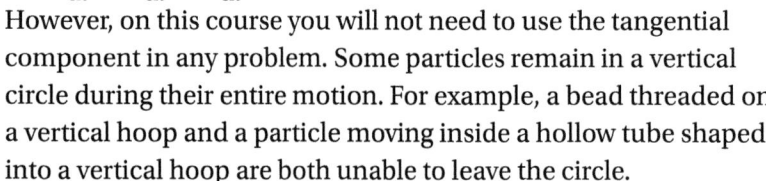

Acceleration

However, on this course you will not need to use the tangential component in any problem. Some particles remain in a vertical circle during their entire motion. For example, a bead threaded on a vertical hoop and a particle moving inside a hollow tube shaped into a vertical hoop are both unable to leave the circle.

High-energy particles in these, or similar, situations will be able to make complete circles, but low-energy particles will merely oscillate in the lower part of the circle.

Consider a particle moving smoothly in a vertical circle under its own weight. It slows down as it travels towards the top of the circle and speeds up as it moves down towards the bottom. There are changes in its kinetic and gravitational potential energy.

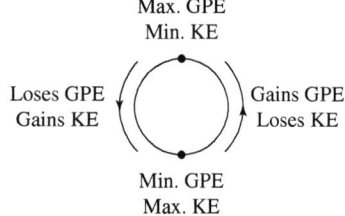

An **energy equation** can state *either* that the total energy of the particle is constant *or* that a gain in one form of energy is balanced by a loss in another form.

The energy equation involves velocities and heights, but not forces.

You can also use Newton's 2nd law to write an equation of motion towards the centre of the circle. This equation will involve forces and accelerations.

> **Key point**
>
> For a particle moving in a vertical circle of radius r with a variable velocity, v, the radial component of acceleration is
> $$r\omega^2 = r\dot{\theta}^2 = \dfrac{v^2}{r}$$
> Newton's 2nd law can be used for motion towards the centre of the circle.
> An energy equation can also be used.

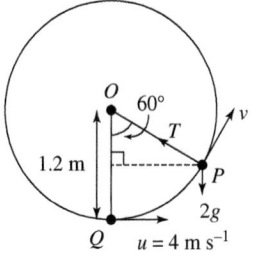

Example 1

A 2 kg mass, *P* is attached to a string *OP* of length 1.2 m where *O* is a fixed point. *P* is given a horizontal velocity, *u* of 4 m s⁻¹ when hanging vertically below *O* at point *Q*

When $\angle QOP = 60°$, find

a The velocity, *v*, of *P*

b The tension, *T* in the string.

---

a Let *P* have velocity *v* when $\angle QOP = 60°$

Vertical height of *P* above *Q* is *h*

$h = OQ - OP\cos 60° = 1.2 - 1.2\cos 60°$

$\qquad = 1.2(1 - \cos 60°)$

$\qquad = 0.6\,m$

> Finding a vertical height as $r(1 - \cos\theta)$ is very common in these questions.

Energy equation from *Q* to *P* is

$\quad$ KE lost = GPE gained

$\dfrac{1}{2}mu^2 - \dfrac{1}{2}mv^2 = mgh$

> There is no friction or other external force in the system, so energy is conserved.

$\dfrac{1}{2}\times 2\times 4^2 - \dfrac{1}{2}\times 2\times v^2 = 2\times 9.8\times 0.6$

$v^2 = 16 - 11.76 = 4.24$

Velocity, $v = 2.06\,m\,s^{-1}$

b Equation of motion of *P* along radius is

$T - mg\cos 60° = m\times\dfrac{v^2}{r}$

> Resultant force = mass × acceleration

$T = 2\times\dfrac{4.24}{1.2} + 2\times 9.8\times\cos 60°$

Tension, $T = 16.9\,N$

In Example 1, the energy equation gave you the velocity, *v*, which you then used in the equation of motion to the centre. This link between the two equations is a common strategy.

Exercise 24.3A Fluency and skills

In this exercise, use g = 9.8 m s⁻².

1 A 5 kg mass, *P*, is attached to a string *OP* of length 1.5 m where *O* is a fixed point. *P* is given a horizontal velocity, *u*, of 10 m s⁻¹ when hanging vertically below *O* at point *Q*

Find the velocity of *P* and the tension in the string when

a $\angle QOP = 60°$,

b *OP* is horizontal.

2 A 10 kg mass *P* is attached to a light rigid rod *OP* of length 2 m such that the rod can rotate about the fixed point *O*

P is given a horizontal velocity, *u*, of 8 m s⁻¹ when *OP* is vertical with *P* below *O* at point *Q*

Find the velocity of *P* and the force in the rod when

a $\angle QOP = 60°$, **b** $\angle QOP = 120°$

In each case, state whether the force in the rod is a tension or a thrust.

3 A 250 gram bead B is threaded on a fixed, smooth, vertical hoop of radius 2 m and centre O. It is projected horizontally from the lowest point A of the hoop with a velocity, u, of 10 m s^{-1}.

   a Will it travel in complete circles of the hoop? Explain your answer.

   b If its initial speed is reduced to 7 m s^{-1}, find the greatest value of $\angle AOB$ in the subsequent motion.

4 A particle P of mass 0.5 kg rotates on the inside of a smooth circular surface, in the vertical plane, of radius 1.4 m and centre O. It leaves the lowest point Q of the surface with a horizontal velocity of 6 m s^{-1}.

   a When $\angle QOP = 45°$, find

   i The velocity of P

   ii The reaction between P and the surface.

   b Will P rise to the same horizontal level as O? Explain your answer.

5 **a** A 5 kg particle P is attached to a string OP of length 1.5 m where O is a fixed point. P is given a horizontal velocity, u, of 8 m s^{-1} when hanging vertically below O at point Q

   i Find the velocity, v, of P and the tension, T in OP when $\angle QOP = 30°$.

   ii Give a reason why the method used in part **i** might not apply when $\angle QOP$ is an obtuse angle.

   b P is now connected to O by a light rod rather than a string.

   i Will your answer to part **a i** still be valid?

   ii Find the velocity, v, of P and the force in the rod when $\angle QOP = 150°$, stating whether the force is a tension or a thrust.

Reasoning and problem-solving

When a particle P rotates into the upper half of a vertical circle, centre O, it may not have enough energy to complete a full circle. The subsequent motion will vary depending on the particular situation. For example:

- If P is attached to O by a light rod, the rod will come to instantaneous rest when the particle's kinetic energy (and therefore its velocity) is reduced to zero. P will then oscillate to and fro on a circular arc, but never make a full rotation.
- If P is attached to O by a string, P will leave the circle if the tension in the string reduces to zero and the string becomes slack. P will fall inside the circle under gravity.
- If the particle P is sliding up the inside or down the outside of a circular surface, circular motion is broken when the reaction with the surface becomes zero. P will leave the circle and fall away under gravity.

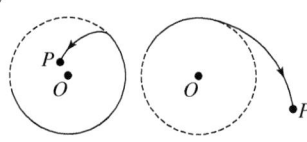

Example 2

A particle P of mass m at point A hangs vertically from a fixed point O where OA is a light rod of length $r = 1.5$ m. P is given a horizontal speed, u, at A

Find the least value of u for P to perform a complete circle about O

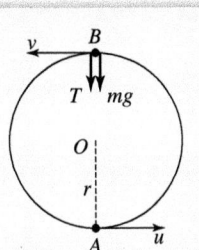

P completes a circle if it reaches the topmost point B with a speed $v \geq 0$

Gain in GPE from A to B is $m \times g \times 2r = 2mgr$

Loss of KE from A to B is $\dfrac{1}{2}mu^2 - \dfrac{1}{2}mv^2$

So the energy equation is $2mgr = \dfrac{1}{2}mu^2 - \dfrac{1}{2}mv^2$

giving $v^2 = u^2 - 4gr$

So $v \geq 0$ if $u^2 \geq 4gr$

Minimum speed, $u = \sqrt{4 \times 9.8 \times 1.5} = 7.67$ m s^{-1} for a complete circle for P

> ① ②
> Draw a clear diagram.
>
> State the condition for reaching the top.

> ③
> Write the energy equation.

Example 3

The light rod OA in Example 2 is replaced by a string of the same length. Find the new least value of u for P to perform a complete circle about O

P now completes a circle if the string is still taut at B; that is, if its tension $T \geq 0$ at B

The energy equation is unchanged, so $v^2 = u^2 - 4gr$ [1]

From Newton's 2nd law, the equation of motion of P at B towards O is

$$T + mg = m \times \dfrac{v^2}{r} \implies T = \dfrac{mv^2}{r} - mg$$

So, $T \geq 0$ if $\dfrac{mv^2}{r} \geq mg$ or $v^2 \geq rg$ [2]

From [1] and [2], $u^2 - 4gr \geq rg \implies u^2 \geq 5gr$

Minimum speed, $u = \sqrt{5 \times 9.8 \times 1.5} = 8.57$ m s^{-1} for a complete circle for P

> ②
> State the condition for reaching the top.

> ③
> Write an energy equation and an equation of motion using Newton's 2nd law.

Example 4

A particle P of mass m is at the lowest point A on the inside of a thin, smooth sphere of radius 1.2 m with centre O. P is projected along the surface from A with a velocity, u, of 6 m s^{-1}. Find $\angle AOP$ when P is about to lose contact with the sphere.

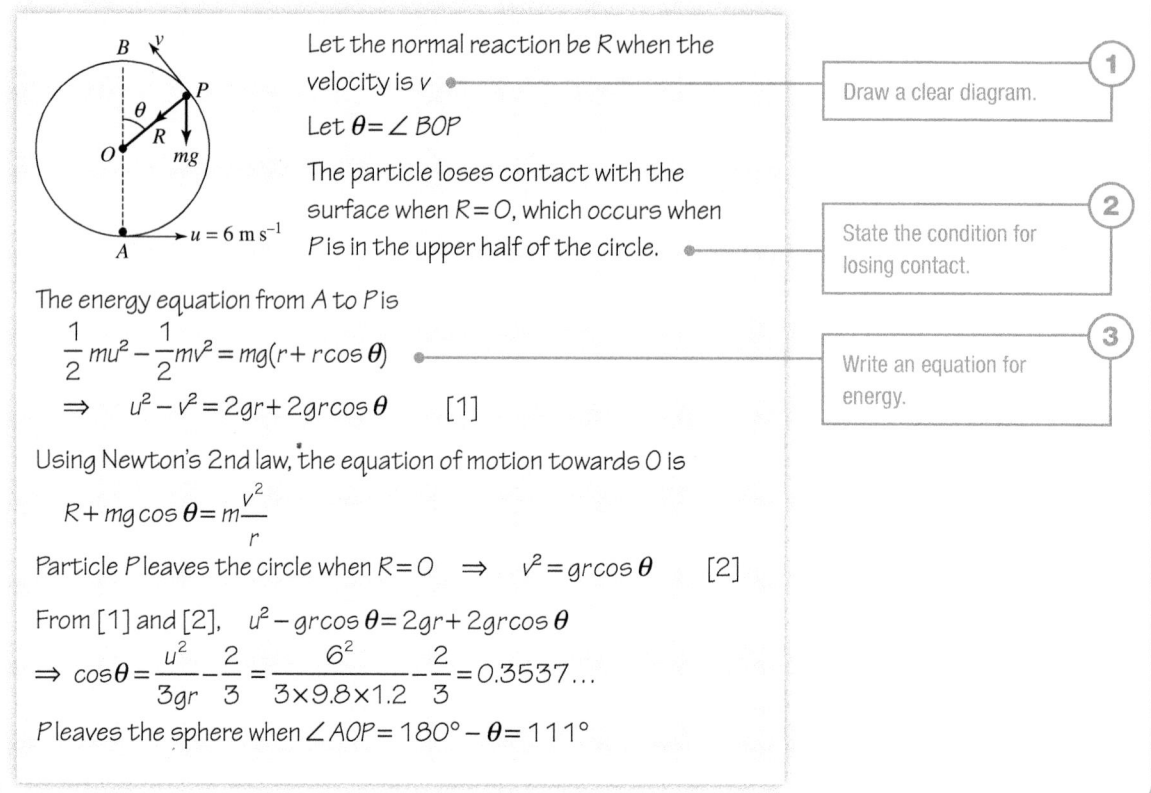

Let the normal reaction be R when the velocity is v

Let $\theta = \angle BOP$

The particle loses contact with the surface when $R = 0$, which occurs when P is in the upper half of the circle.

The energy equation from A to P is

$$\frac{1}{2}mu^2 - \frac{1}{2}mv^2 = mg(r + r\cos\theta)$$

$$\Rightarrow \quad u^2 - v^2 = 2gr + 2gr\cos\theta \qquad [1]$$

Using Newton's 2nd law, the equation of motion towards O is

$$R + mg\cos\theta = m\frac{v^2}{r}$$

Particle P leaves the circle when $R = 0 \quad \Rightarrow \quad v^2 = gr\cos\theta \qquad [2]$

From [1] and [2], $\quad u^2 - gr\cos\theta = 2gr + 2gr\cos\theta$

$$\Rightarrow \quad \cos\theta = \frac{u^2}{3gr} - \frac{2}{3} = \frac{6^2}{3 \times 9.8 \times 1.2} - \frac{2}{3} = 0.3537\ldots$$

P leaves the sphere when $\angle AOP = 180° - \theta = 111°$

- **1** Draw a clear diagram.
- **2** State the condition for losing contact.
- **3** Write an equation for energy.

Example 5

A particle P of mass 4 kg rests at the topmost point A of the surface of a smooth sphere of radius 1.5 m, fixed to a horizontal plane at point B. P is slightly disturbed from rest and moves down the surface of the sphere before leaving it at C. If P strikes the plane at D, find the distance BD

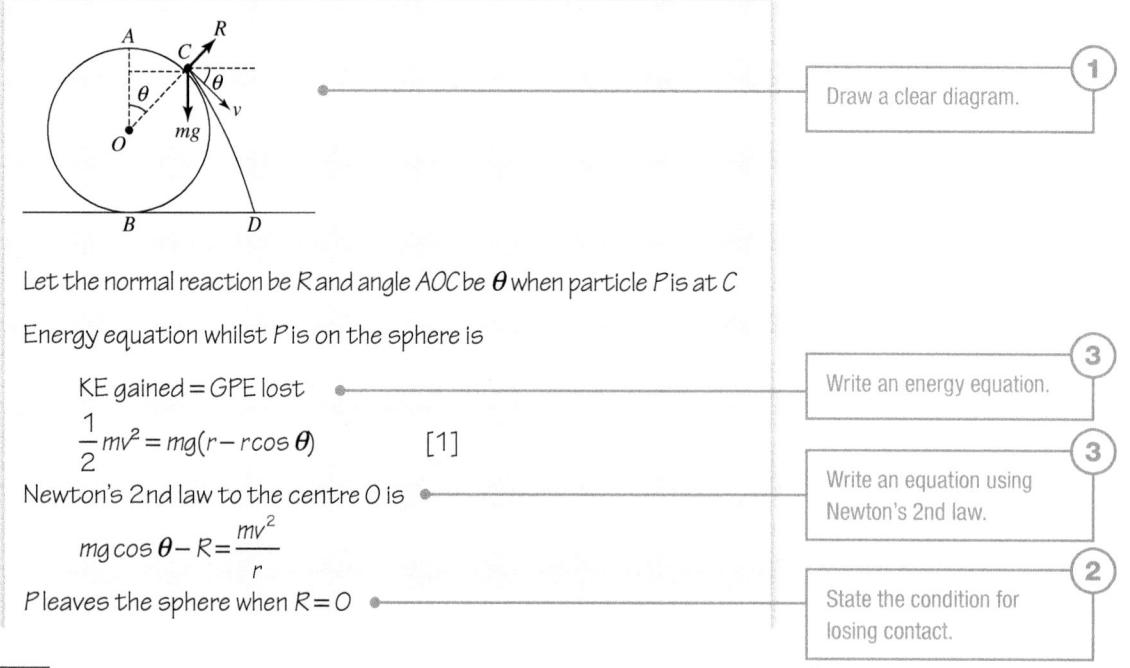

Let the normal reaction be R and angle AOC be θ when particle P is at C

Energy equation whilst P is on the sphere is

KE gained = GPE lost

$$\frac{1}{2}mv^2 = mg(r - r\cos\theta) \qquad [1]$$

Newton's 2nd law to the centre O is

$$mg\cos\theta - R = \frac{mv^2}{r}$$

P leaves the sphere when $R = 0$

- **1** Draw a clear diagram.
- **3** Write an energy equation.
- **3** Write an equation using Newton's 2nd law.
- **2** State the condition for losing contact.

(Continued on the next page)

So $v^2 = gr\cos\theta$ [2]

[1] and [2] give $\dfrac{1}{2} mgr\cos\theta = mgr - mgr\cos\theta$

$\Rightarrow \cos\theta = \dfrac{2}{3}$ so $\theta = 48.2°$

$\Rightarrow v^2 = \dfrac{2gr}{3} = \dfrac{2 \times 9.8 \times 1.5}{3}$ and $v = 3.13\,\text{m s}^{-1}$

P leaves the sphere at a distance of $r + r\cos\theta = \dfrac{5r}{3} = 2.5\,\text{m}$ above the plane.

The vertical component of velocity is
$v\sin\theta = 3.13 \times \sin 48.2° = 2.33\,\text{m s}^{-1}$

The horizontal component of velocity is
$v\cos\theta = 3.13 \times \cos 48.2° = 2.09\,\text{m s}^{-1}$

Consider P as a projectile, taking a time t to reach the plane at D

$$s = ut + \dfrac{1}{2}at^2$$
$$2.5 = 2.33t + \dfrac{1}{2} \times 9.8 \times t^2$$

Take downwards as positive and write a kinematic equation to find t

$$\Rightarrow 9.8t^2 + 4.66t - 5 = 0$$

Solve the quadratic equation; t must be positive.

$$\Rightarrow t = 0.515\,\text{s} \ (t > 0)$$

Horizontal distance travelled in this time is

$$v\cos\theta \times t = 2.09 \times 0.515 = 1.08\,\text{m}$$

Horizontal velocity × time

Impact with plane occurs at a distance
$BD = BC + CD = 1.5\sin 48.2° + 1.08 = 2.20\,\text{m}$

These problems emphasise that you should take care to state the correct condition for completing a full circle or breaking away from the circle.

Exercise 24.3B Reasoning and problem-solving

In this exercise, assume that all strings are light and inextensible and use $\mathbf{g} = 9.8\,\text{m s}^{-2}$.

1
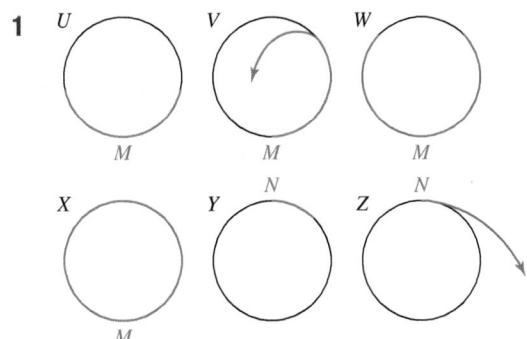

The loci in red show the possible paths traced out by particles or beads which move in a vertical circle centred at O after being given a starting velocity at point M or N. Match the possible paths U to Z with descriptions **a** to **f**. Each description may have more than one matching diagram.

a   A bead threaded onto a smooth vertical hoop.

b   A particle attached to a string OP

c   A particle attached to a light rod OP

d A particle moving from the topmost point of a solid, smooth sphere.

e A particle moving from the topmost point of a smooth sphere.

f A particle moving on the inside of a smooth circular surface.

2 A 3 kg mass P hangs freely from a fixed point O on a string OP of length 2 m.

    a Find the least velocity that P must be given so that it moves in a vertical circle about O

    b Also find the radial component of its acceleration as the string OP passes through the horizontal, given the velocity calculated in part **a**.

3 A 2 kg particle P is placed at the lowest point inside the surface of a smooth hollow cylinder of radius 1.5 m and axis O. Find the velocity that P is given to move up the line of greatest slope if it

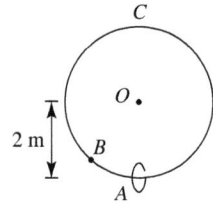

    a Just reaches the horizontal through O

    b Just makes a complete vertical circle about O

4 A particle P of mass m kg is at rest on the highest point of a smooth sphere of radius r and centre O. It is disturbed slightly so that it begins to move down the sphere's surface. Find the angle that OP makes with the vertical at the point where P leaves the surface. If a refined model takes air resistance into account, how would this affect your answer?

5 A 0.5 kg mass P hangs freely from a fixed point O on a string OP of length 2 m. It is given a horizontal velocity of 8 m s^{-1}. Find the angle through which OP has rotated when the string goes slack.

6 A hollow cylinder of radius r is fixed with its axis O horizontal. A particle of mass m is placed on the smooth inner surface of the cylinder and made to oscillate through 180°. Show that, when v is the speed of the particle, the reaction between the particle and the surface is $\dfrac{3mv^2}{2r}$

7 A road over a bridge has the shape of a circular arc of radius 20 m. Find the greatest speed in km h^{-1} at which a car can travel over the bridge without leaving the ground at the highest point of the road. State an assumption that you have made in your solution.

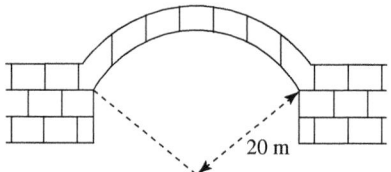

8 A 0.25 kg bead B is threaded onto a fixed vertical hoop with centre O and radius 2 m.

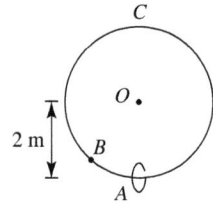

    a B is given a horizontal velocity of 8 m s^{-1} at the lowest point, A, of the hoop. When $\angle AOB = 45°$, find

      i The reaction between B and the hoop,

      ii The radial component of its acceleration.

    b If B is given a horizontal velocity of 8 m s^{-1} when at the highest point C of the hoop, find the radial component of the resultant acceleration of B when $\angle COB = 60°$

9 A 0.5 kg particle, P is at rest at the highest point inside a narrow hollow vertical circular tube of radius 2 m and centre O. It moves from rest and slides down inside the tube.

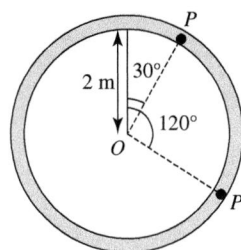

a Find the reaction R between P and the tube when OP has rotated through an angle of

   i 30°,   **ii** 120°.

b What angle must OP rotate through for the reaction R to be zero?

10 A smooth sphere of radius 2 m is fixed to a horizontal plane at point B. A 3 kg mass P at rest on the highest point A of the sphere is slightly disturbed and moves down the sphere's surface and leaves it. How far from B does P hit the plane?

11 The ends of a string OM of length 2 m are attached to a 3 kg mass M and a fixed point O. OM is held taut and horizontal and M is released from rest. When OM is vertical, it catches on a small fixed peg P which is a distance x below O

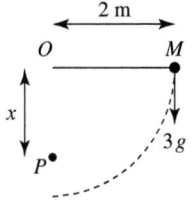

a Find x if M just completes a circle with P as centre.

b Find the ratio of the tensions in the string just before and just after making contact with P

12 A mass m, hanging at point A from a fixed point O on a string of length a, is projected from A with a horizontal velocity of $\sqrt{\dfrac{7ag}{2}}$ Show that the string becomes slack and that the mass, on leaving the circle, moves freely under gravity back to A

13 A 10 kg mass P hangs from a 2 m string attached to a fixed point O which is 3 m above point A on a horizontal plane.

P receives a horizontal impulse of 50 N s. When the string has rotated through 60° about O, it breaks. P now travels as a projectile and strikes the plane at B

Calculate the distance AB

Chapter summary

- For a particle moving in a circle, with centre O and radius r, at **constant** speed,
  - Its linear tangential speed, v, and angular speed, ω, are related by $v = r\omega = r\dot{\theta}$
  - Its acceleration, a, to the centre O is $r\omega^2 = \dfrac{v^2}{r}$
  - The centripetal force, F is $mr\omega^2 = \dfrac{mv^2}{r}$
  - The time, T, for one full revolution is $\dfrac{2\pi}{\omega}$
- A conical pendulum has a bob which moves in a horizontal circle at constant speed as its length traces out a hollow cone.
- For a particle moving in a vertical circle, centre O and radius r, at **variable** speed,
  - Its linear tangential speed, v, and angular speed, ω, are related by $v = r\omega = r\dot{\theta}$
  - Its acceleration has a radial component of $r\omega^2 = r\dot{\theta}^2 = \dfrac{v^2}{r}$
  - The centripetal force $F = mr\omega^2 = \dfrac{mv^2}{r}$
- Newton's 2nd law can give an equation of motion along the radius.
- An energy equation can be written in terms of *either* the total energy of the particle *or* its gains and losses of energy.

Check and review

You should now be able to...	Try Questions
✔ Use vectors to solve certain problems with circular motion.	1–4
✔ Solve problems when the motion is in a horizontal circle.	5–10, 14–15
✔ Solve problems when the motion is in a vertical circle.	11–13

1 A mass of 2 kg has a position vector $\mathbf{r} = 3\sin 2t\,\mathbf{i} + 3\cos 2t\,\mathbf{j}$ metres at a time t seconds.

    a Calculate the magnitude of the force $\mathbf{F}$ acting on the mass when $t = \dfrac{\pi}{3}$ seconds.

    b Find the equation of the circle on which the point moves in the x–y plane.

2 The position vector of a particle at time t seconds is $\mathbf{r} = (2\cos 3t + 5)\mathbf{i} + (2\sin 3t - 1)\mathbf{j}$ metres.

    a What is the speed of the particle?

    b Find the radius and centre of the circle on which the particle is moving.

3 A 2 kg mass M moves in a circle with the equation $x^2 + y^2 = 9$ at an angular speed of $2\,\text{rad s}^{-1}$, starting on the x-axis when $t = 0$. Find its position vector $\mathbf{r}$ at time t and the force $\mathbf{F}$ acting on M when $t = \dfrac{\pi}{4}$ seconds.

4 A particle P of mass 4 kg has a velocity $\mathbf{v} = 5\sin\left(\dfrac{1}{2}t\right)\mathbf{i} + 5\cos\left(\dfrac{1}{2}t\right)\mathbf{j}\,\text{m s}^{-1}$ at a time t s. If its initial position is the point $(-2, 1)$, find its position vector at time t and the force acting on it when $t = \dfrac{\pi}{2}$ seconds.

5   A car of mass 900 kg rounds a bend of radius 40 m at 54 km h^{-1}. Find the smallest value of μ if the road is banked at an angle of 10° to the horizontal and there is no sideways slipping up the slope.

6   A motorcyclist goes round a curve of radius 20 m on a level road at 27 km h^{-1}. Find the angle at which she and her bike are inclined to the vertical and the least value of μ between the bike and the road for the bike not to side-slip.

7   A light, taut string connects particle P of mass m_1 and particle Q of mass m_2 as they lie on a rough horizontal surface, where the coefficient of friction for P is μ_1 and for Q is μ_2. P is given a velocity, u, perpendicular to PQ and rotates about Q which does not move. Find the greatest possible angle through which PQ can rotate in terms of m_1, m_2, μ_1 and μ_2.

8   A conical pendulum has a bob with a 2 kg mass making a full rotation every 2 seconds. If the pendulum's length is 1.5 m, calculate the tension in the string and the angle of the string to the vertical.

9   An elastic string has a natural length of 0.8 m and modulus of elasticity $\lambda = 100$ N. Its ends are attached to a fixed point O and to a 2 kg mass. Find the extension in the string when the mass rotates at 4 rad s^{-1} as a conical pendulum suspended from O

10   A ball of mass 2 kg rests on a smooth horizontal table and is attached to one end of a string of length 3 m. The other end is tied to a fixed point 1 m above the table. The ball moves in a horizontal circle while maintaining contact with the table. Show that the greatest linear speed for the ball to stay in contact with the table is $\sqrt{8g}$ m s^{-1}

11   A 1 kg mass slides down the surface of a smooth, fixed hemisphere of radius 2 m from being at rest at the topmost point. Find the distance it travels along the surface before contact is lost.

12   A 3 kg mass, P, is tied to the midpoint M of a string of length 2.6 m with its ends A and B fixed 2.4 m apart horizontally. Initially, P hangs at rest.

   a   Find the minimum speed it must be given to make complete vertical circles with the string taut.

   b   Find the maximum tension in the string during the motion.

13   A pendulum, P has a light rod OP 1 m in length which is fixed at O with a bob of 2 kg attached at P. When P is at rest vertically below O, the bob is struck so it has an initial velocity u. Find the angle through which the rod has turned when the bob reaches its highest point if

   a $u = 3\,\text{ms}^{-1}$ b $u = 5\,\text{ms}^{-1}$

14   Each end of a light, inextensible string is attached to a mass of 2 kg. The string passes through a hole in the vertex of a smooth, hollow cone of semi-vertical angle 45°, so that one mass hangs at rest inside the cone and the other mass moves in a horizontal circle on the outer surface of the cone. If the string is 2 m long and the two masses are at the same horizontal level, show that the angular velocity of the moving mass is $\sqrt{\dfrac{g}{2}}\,\text{rad s}^{-1}$

15   a   A bead of mass m is threaded on a smooth string QR of length 3 m where Q is fixed a distance of 2 m vertically above R. The bead rotates in a horizontal circle about QR. Find the angular velocity ω of the bead for it to position itself at point P on the string where $PQ = 2$ m.

   b   If the bead in part **a** is tied to the string at point P, find the minimum angular velocity, ω, which keeps both PQ and PR taut during the motion.

History

In the mid-nineteenth century, the French physicist Foucault realised that the plane of oscillation of a pendulum stays constant, even when the pivot point moves. This allowed him to demonstrate the rotation of the Earth because the plane of the swinging pendulum was seen to rotate over time. Many museums of science now include a 'Foucault pendulum' to illustrate the rotation of the Earth.

Research

The 'Wall of Death' motorcycle ride involves a motorcyclist riding a bike around a (near) vertical wooden circular wall inside a wooden cylinder. To do this safely, the motorcyclist needs to go at a speed that is sufficient to stop him or her from slipping down the wall and, because (s)he together with the motorcycle cannot be modelled as a particle, (s)he must also lean at an angle to the normal to the wall to ensure that the motorcycle remains in contact with the wall.

Carry out some research to find typical values of dimensions, mass and speed of a 'Wall of Death' ride and riders. Model and explore the situation.

Investigation

Explore the forces of attraction acting on planets on their paths around the Sun. You will need to research planetary data to do this.

1   A particle of mass 2 kg is moving in a horizontal plane under the action of a force **F** N so that at time t s its position vector is given by

$$\mathbf{r} = (0.3\cos 10t\,\mathbf{i} + 0.3\sin 10t\,\mathbf{j})\,\text{m}$$

a   Show that the particle is moving in a circle. [3]

b   Find an expression for **v**, its velocity vector, and hence show that it is moving with constant speed. [4]

c   Find the magnitude of the force **F** and state its direction. [4]

2   A particle moves as a conical pendulum at the end of a light, inextensible string of length 40 cm. If the string makes an angle of 30° with the horizontal, find the angular speed of the particle. [6]

3   A car travels round a roundabout of radius 10 m at a uniform speed of 6 m s⁻¹. Taking the centre of the roundabout as the origin, the car is initially at the point with position vector 10**i** and is moving in an anticlockwise direction.

a   Find an expression for the position vector **r** at time t [3]

b   Find an expression for the acceleration of the car at time t and explain how you know that the acceleration is directed towards the origin. [3]

4   A car travels round a curve of radius 150 m on a track which is banked at 25° to the horizontal. The coefficient of friction between the wheels and the road is 0.4

a   At what speed is the car travelling if there is no frictional force acting? [5]

b   What is the maximum speed at which the car can travel without slipping up the slope? [6]

c   What is the minimum speed at which the car can travel without slipping down the slope? [5]

5   A pendulum consists of a rod of length 1 m with a bob of mass 2 kg attached to one end. The rod is freely pivoted at the other end, O, so that it can rotate in a vertical circle. Initially, the bob is vertically below O when it is given an impulse so that it starts to move with speed 6.5 m s⁻¹. Assuming that the rod is light and that the bob can be modelled as a particle, calculate

a   The speed of the bob, [5]

b   The force in the rod, [5]

when the pendulum makes an angle of   i   30°,   ii   150° with the downward vertical.

6   A pendulum of length a has a bob of mass m. The speed of the bob at the lowest point of its path is U. Find the condition which U must satisfy for the bob to make complete revolutions if the length of the pendulum consists of

a   A rod, [4]

b   A string. [4]

7 A particle of mass m hangs at rest, suspended from a point, O, by a light, inextensible string of length a. The particle receives an impulse so that it starts moving with speed $\sqrt{3ga}$ m s^{-1}. Find the angle between the string and the upward vertical when the string goes slack. **[6]**

8 An engine of mass 60 tonnes is travelling on rails 1.5 m apart around a curve of radius 1500 m. The outer rail is raised a distance h m above the inner rail, so that at a speed of 50 km h^{-1} there is no sideways force on the rails.

   a Find the value of h **[4]**

   b Given that the maximum safe sideways force on the rails is 100 kN, what is the maximum safe speed of the engine round the curve? **[5]**

9 A ball of mass 3 kg is fastened to one end of a string of length 0.5 m. The other end of the string is fixed to a point A and the ball revolves as a conical pendulum at a rate of 5 rad s^{-1}.

   a How far below A is the centre of the circle traced out by the ball? **[4]**

   b What assumptions have you made in your answer? **[2]**

10 A mass of 0.5 kg, suspended by a light, inextensible string of length 1.5 m, revolves as a conical pendulum at 30 rev min^{-1}. Find the radius of the circle in which it travels and the tension in the string. **[6]**

11 A particle moves as a conical pendulum at the end of a light, inextensible string, which has its fixed end at point A. The angular speed of the particle is ω. The centre of the circle in which the particle moves is O

   a Show that $AO = \dfrac{g}{\omega^2}$ **[5]**

   b Explain why the string cannot be horizontal. **[2]**

12 A particle of mass 0.1 kg is attached by a string of length 1.5 m to a fixed point, and is made to travel in a vertical circle about that point.

   a Find the minimum velocity the particle must have at the lowest point of the circle if it is to make complete revolutions. **[6]**

   b For this velocity, find the tension in the string when the particle is at point A, a distance of 75 cm above the lowest point. **[5]**

13 A particle of mass 0.01 kg is placed on the topmost point, A, of a smooth sphere of centre O and radius 0.5 m. It is slightly displaced. When it reaches point B, it is about to leave the surface of the sphere. Calculate the angle AOB **[5]**

14 A ball of mass 1 kg is fastened to one end, B, of a light, inextensible string AB of length 1 m. The ball is placed on a smooth, horizontal table. The end A is attached to a fixed point above the table so that the particle moves as a conical pendulum whilst staying in contact with the table. The radius of the circle in which it travels is 0.5 m.

   a If the speed of the particle is 1.5 m s^{-1}, what is the normal reaction between the ball and the table? **[4]**

   b What is the maximum speed at which the ball could travel without lifting off the table? **[4]**

15 A stone of mass 0.5 kg performs complete revolutions in a vertical circle on the end of a light, inextensible string of length 1 m. Show that the string must be strong enough to support a tension of at least 29.4 N. **[6]**

16 A smooth hollow cone with semi-vertical angle θ is fixed with its axis vertical and its vertex V at the bottom, as shown. A particle P travels with angular speed ω in a horizontal circle inside the cone about a point A on its axis, where $AV = h$. Show that

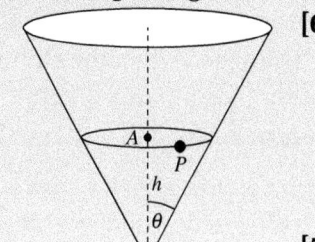

$$h = \frac{g}{\omega^2 \tan^2 \theta}$$ **[5]**

17 A bead of mass m is threaded onto a smooth, circular hoop of radius a, which is fixed in a vertical plane. The bead is displaced from rest at the top of the hoop. Find, in terms of g, the resultant acceleration of the bead when it has reached a point which is a vertical distance $\frac{3}{4}a$ below its starting point. **[9]**

18 A particle of mass m is projected horizontally with speed v from the topmost point, A, of a sphere of radius a and centre O. It remains in contact with the sphere until leaving the surface at point B. If angle AOB is 30°, find v **[6]**

19 A hemispherical bowl of radius 13 cm is fixed with its rim horizontal. A ball-bearing of negligible diameter is made to travel in a horizontal circle inside the bowl at a speed of $1.68\,\mathrm{m\,s^{-1}}$. How far is the centre of the circle above the bottom of the bowl? **[8]**

20 A ball, B, of mass 2 kg is attached to one end of a light, inextensible string. The string passes through a smooth, fixed ring, O, and a second ball, A, of mass 4 kg, is attached to the other end. B is made to move as a conical pendulum while A hangs vertically below the ring, as shown. If the speed of B is $7\,\mathrm{m\,s^{-1}}$, calculate the distance BO **[4]**

21 Two points, A and B, are on a vertical pole, 9 m apart with A above B, as shown. A rope of length 27 m is fastened at its ends to A and B. A smooth, heavy, metal ring, S, of mass m is threaded onto the rope. The ring is made to move in a horizontal circle about the pole. The upper section, AS, of the rope makes an angle θ with the vertical, as shown. Find the speed of the ring and the tension in the rope when

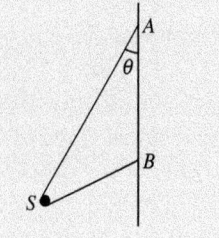

a $\tan\theta = \frac{8}{15}$, **[6]**

b $\tan\theta = \frac{4}{3}$ **[5]**

22 A smooth wire has a bead of mass 0.005 kg threaded onto it. The wire is bent to form a circular hoop of radius 0.2 m. The hoop is fastened in a horizontal position, whilst the bead travels round it at a constant speed of $1\,\mathrm{m\,s^{-1}}$. Find the magnitude and direction of the reaction force between the hoop and the bead. **[6]**

23 A particle of mass m travels in complete vertical circles on the end of a light, inextensible string of length a. If the maximum tension in the string is three times the minimum tension, find the speed of the particle as it passes through the lowest point on the circle. **[6]**

24 A particle of mass 2 kg is attached to the end of a light, inextensible string of length 1 m, the other end of which is attached to a fixed point, O The particle is held with the string taut and horizontal, and is released from rest. When the string reaches the vertical position, it meets a fixed pin, A, a distance x below O. Given that the particle just completes a circle about A, find the value of x 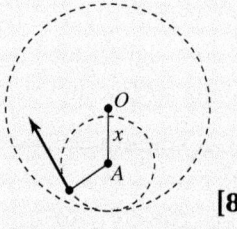 [8]

25 A pendulum bob of mass m is fastened to one end of a light, inextensible string of length r whose other end is fixed at a point O. The bob is at rest in its lowest position when it is set in motion with an initial speed of $\sqrt{\dfrac{7gr}{2}}$. As it swings upwards, the string meets a small, fixed peg, P, on the same level as O

The string then wraps round P. What is the closest that P can be to O so that the bob makes a complete revolution about P? [8]

26 A ring of mass 5 kg is threaded onto a rope of length 10 m, whose ends are attached to two fixed points 6 m apart and on the same level. The ring hangs at rest. It is then set in motion so that it travels on a circular path whose plane is perpendicular to the line joining the two ends of the rope. Given that the ring can just make complete revolutions, find the maximum tension in the rope. State any modelling assumptions that you have made in reaching your answer. [10]

27 Particles A and B, of masses m and $2m$, respectively, are connected by a light, inextensible string of length πa. The particles are placed symmetrically, and with the string taut, on the smooth outer surface of a cylinder of radius $3a$, as shown, and the system is released from rest. Find the reactions between the cylinder and the particles at the moment when A reaches the topmost point. [9]

28 A pendulum bob, P, of mass 1.2 kg, hangs at one end of a light, inextensible string which passes through a smooth hole in a table at a point O. The length of OP is 0.7 m. The other end of the string is attached to a particle, Q, of mass 5.2 kg, which is resting on the rough horizontal surface of the table. The coefficient of friction between Q and the table is 0.25 The bob, P, is made to move as a conical pendulum below O. Find the maximum angular speed at which it can move without making Q slip. [5]

29 The diagram shows a loop-the-loop on a roller-coaster ride. The car approaches the loop on a horizontal track. The maximum speed at which the car can enter the loop is 80 km h^{-1}. What is the greatest radius with which the loop can be constructed if the car is not to leave the track?  [6]

25 Centres of mass and stability

The centre of mass of a car affects how it behaves when it is driven.

This is important for all vehicles. It is particularly significant in racing cars, such as those used in Formula 1, where small differences in performance have a big effect on outcomes. Designers of racing cars make sure a car is stable by ensuring that the centre of mass of the car is as close to the ground as possible. When the brakes on a standard car are applied suddenly, the car might feel as though it is diving forwards towards the road. This happens when forces are unbalanced about the centre of mass. Such behaviour needs to be minimised in racing cars.

The position of the centre of mass of many other vehicles must also be considered at the design stages. For example, because of the uneven weight distribution of passengers in a double decker bus, the vehicle could quite easily become unstable when cornering on a banked section of road, unless its centre of mass is positioned correctly.

Orientation

What you need to know

Maths Ch19
- Moments.

What you will learn

- To calculate the moment about a point.
- How to find the centre of mass for point masses and laminas.
- How to find the centre of mass for solids.

What this leads to

Careers
Mechanical engineering.
Architecture.

Fluency and skills

The magnitude of the turning effect or **moment** of a force F about a fixed point A depends on the perpendicular distance between F and A and its direction can be clockwise (negative) or anticlockwise (positive).

> **Key point**
>
> The moment of a force F about a point A is $F \times x$, where x is the perpendicular distance of the force from A

> A moment is sometimes called a **torque**.

When several forces act on the same object at the same time, each force has its own turning effect on the object.

> **Key point**
>
> The **resultant moment** of several forces acting on an object equals the sum of the moments of the individual forces, taking into account their directions.

> Moment and work have the same dimensions and thus the same units (Nm). Whereas Nm are called Joules for work, moments are always measured in Nm.

For an object that can rotate about a fixed point A, it is in rotational equilibrium if the total moment acting on it is zero. (That is, when the clockwise moments balance the anticlockwise moments exactly.)

> **Key point**
>
> If a system of forces is in equilibrium, the resultant moment about any point is zero.

> The fixed point A is called a **pivot**.

A light rectangular lamina L, can rotate about point A

Find the resultant moment of the three forces about A

Total (anticlockwise) moment $= (6 \times 5) + (7 \times 0) - (8 \times 3)$

$$= 30 - 24 = 6 \, \text{Nm}$$

> The line of action of the 7 N force passes through A, so it produces a zero moment. The 8 N force produces a negative moment.

The net horizontal force on the object shown in the diagram is zero (as $F - F = 0$) and there is no vertical force. However, the object is not in equilibrium as the two parallel forces cause the object to rotate with a clockwise moment of $F \times d$. Two forces like these produce a **couple**.

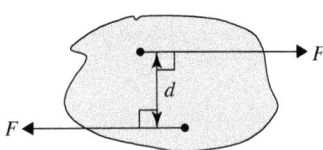

> **Key point**
>
> A couple comprises two equal and opposite forces which do not act in the same straight line. The moment of a couple, about any point, equals $F \times d$, where d is the perpendicular distance between the two forces. This is independent of the point about which the moment is taken.

The forces used to unscrew a bottle top or turn a doorknob cause rotation without translation. Such forces also form a couple which is measured by its moment.

Example 2

A cantilever is formed by clamping a uniform horizontal beam AB at A. The beam is 4 m long, weighs 20 N and has a load of 15 N hanging vertically at B

Find the force and couple which must act at A to keep the beam in equilibrium.

Let the force at A have components X and Y
and the couple at A have a moment M •——

Resolve horizontally $X = 0$

Resolve vertically $Y = 20 + 15 = 35$

Take moments about A $20 \times 2 + 15 \times 4 = M$ •——

$\Rightarrow M = 100 \, Nm$

The force at A is 35 N upwards and the couple at A is 100 Nm anticlockwise.

The components of the force at A prevent translation and the couple prevents rotation.

Clockwise moment = anti-clockwise moment

Exercise 25.1A Fluency and skills

1 Find the total moment about point P of these systems of forces.

a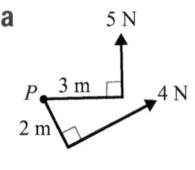

b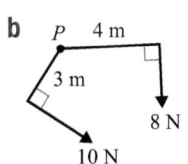

c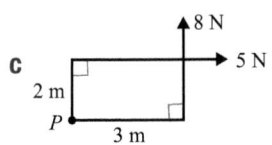

2 a Force $\mathbf{F} = 2\mathbf{i} + 3\mathbf{j}$ N acts at point $(4, 3)$. Find its moment about point $(2, 1)$.

b Force $\mathbf{F} = 5\mathbf{i} - 2\mathbf{j}$ N acts at point $(2, 2)$. Find its moment about point $(1, 4)$.

3 Rectangle $PQRS$ with sides 6 m by 4 m and centre O is a light lamina. Find the resultant moment of the forces about

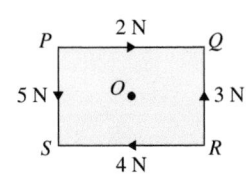

a Point Q **b** Point O

4 A uniform horizontal cantilever AB of mass 20 kg is 6 m long. It is clamped at A and carries a mass 10 kg at its midpoint. Find the force and couple acting at A to keep it in equilibrium.

5 A square $ABCD$ of side 2 m has forces of 10 N along AB and CD and forces of 6 N along CB and AD Find the total moment acting on the square about A.

6 Find the resultant moment of the two forces about point A. Hence show that the moment of a couple is independent of the position of A

Reasoning and problem-solving

In situations where you apply a force F to an object at an angle θ which is not a right angle, you can use one of two methods to calculate its moment about a point, as illustrated in Example 3

The first method in Example 3 resolves force F into two components, $F\cos\theta$ and $F\sin\theta$, and finds the moment of each component about point A

The second method uses a right-angled triangle to find the perpendicular distance from the line of action of force F to point A

In each problem you meet, you can choose which method is more appropriate.

Strategy

To solve problems involving moments and couples

(1) Draw a clear diagram, marking on all the forces and distances.

(2) Take account of the directions of moments when finding their resultant.

Example 3

A force F of 10 N acts at end B of a light rod AB of length l, making an angle θ with the rod.

If $l = 3\,\text{m}$ and $\theta = 30°$, find the moment of F about A

First method

Moment of F about $A = F\cos\theta \times 0 + F\sin\theta \times l$

$\qquad = 0 + 10\sin 30° \times 3$

$\qquad = 15\,\text{Nm}$

(1) Draw a diagram and resolve the force F into components $F\cos\theta$ along AB and $F\sin\theta$ perpendicular to AB

Second method

Moment of F about $A = F \times x$

$\qquad = F \times l\sin\theta$

$\qquad = 10 \times 3\sin 30° = 15\,\text{Nm}$

(1) Draw a diagram and extend the line of the force to show the perpendicular distance, x from A to the line of action of the force.

Notice that both methods give the moment of F as $F \times l \times \sin\theta$

Example 4

A light square lamina $ABCD$ of side 10 cm can rotate about point A

The diagram shows three forces acting.

Find the magnitude of force F for the lamina to be in equilibrium.

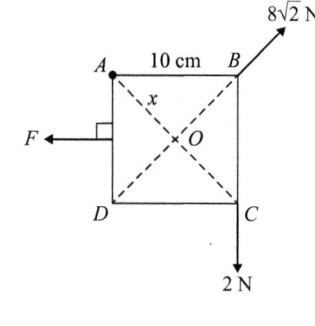

In $\triangle AOB$, $x^2 + x^2 = 10^2$, so $x = \sqrt{50}\,\text{cm} = 5\sqrt{2}\,\text{cm}$

Take moments about point A

For equilibrium:

$\qquad$ Anticlockwise moments = Clockwise moments

$\qquad 8\sqrt{2} \times 5\sqrt{2} = (2 \times 10) + (F \times 5)$

$\qquad 80 = 20 + 5F$

$\qquad$ Force $F = 12\,\text{N}$

(Continued on the next page)

Or

Total moment about $A = 0$

$(8\sqrt{2} \times 5\sqrt{2}) - (2 \times 10) - (F \times 5) = 0$

$80 - 20 - 5F = 0$

Force $F = 12\,\text{N}$

② Take anticlockwise as positive.

The reaction at point A is not on the diagram and it does not appear in the equation because moments are taken about A. This reaction exists because a balance of forces is needed for equilibrium as well as a balance of moments.

Exercise 25.1B Reasoning and problem-solving

1 Find the resultant moment about point A in each case.

a

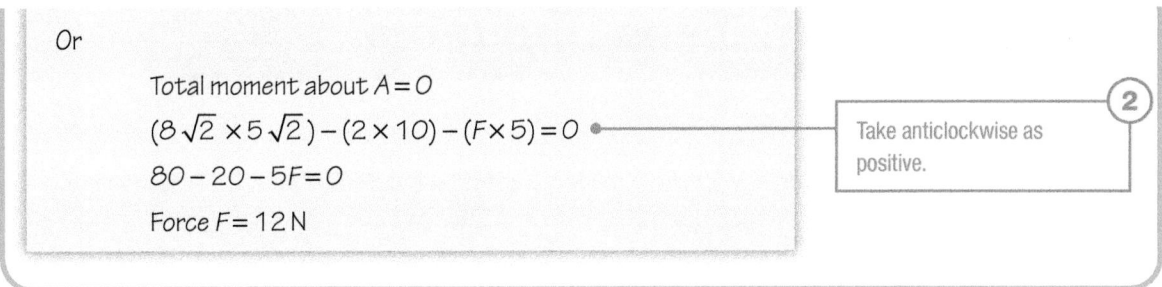

b

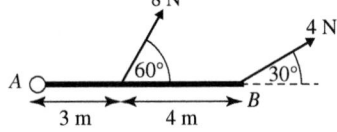

c

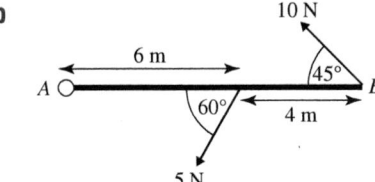

2 The equilateral triangle ABC of side $4\,\text{m}$ and right-angled triangle PQR with $PQ = 3\,\text{m}$ and $QR = 4\,\text{m}$ are two light laminas. M_1 and M_2 are midpoints of sides AB and PR, respectively. Find the resultant moment of the given forces about points

a B **b** M_1 **c** Q **d** M_2

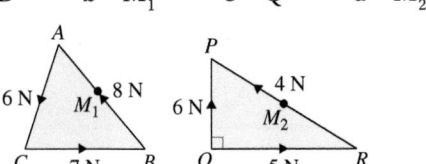

3 A uniform cantilever AB of length $2\,\text{m}$ and weight $10\,\text{N}$ is held horizontal by a tension of $45\,\text{N}$ in a string BC when $\angle ABC$ is $30°$. Find the resultant moment about point A

4 A uniform beam AB of length $8\,\text{m}$ and weight $10\,\text{N}$ is held by two strings AC and BD such that $\angle BAC = \angle ABD = 46°$. If the tension in the two strings is $7\,\text{N}$, find the total moment about A and also about a point on the beam $2\,\text{m}$ from A

5 A rusty trapdoor of weight $40\,\text{N}$ is made of a uniform material and has centre of mass at the point indicated on the diagram. It is held partially open in equilibrium by a vertical force of $30\,\text{N}$ as shown.

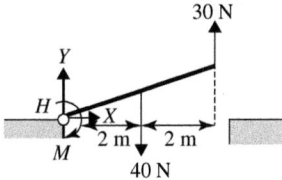

The reaction at the hinge H has components X and Y. The rust at the hinge provides a resisting couple of moment M. Find the values of X, Y and M

6 A horizontal uniform cantilever YZ of length $8\,\text{m}$, weight $30\,\text{N}$ and midpoint M is fixed to a vertical wall PYQ with P above Q such that $PY = QY = 5\,\text{m}$. Two light rods PM and QZ support YZ with a tension of $10\,\text{N}$ in PM and a thrust of $12\,\text{N}$ in QZ. Find the resultant moment about Y

Fluency and skills

If you place n masses $m_1, m_2, m_3, ..., m_n$ along the x-axis at distances $x_1, x_2, x_3, ..., x_n$ from the origin O, the total mass M is given by

$$M = m_1 + m_2 + m_3 + ... = \sum_{i=1}^{n} m_i$$

This system of masses behaves as if its total weight Mg acts at a point, called the **centre of mass**, which is a distance $\bar{x}$ from O

To find $\bar{x}$, take moments about O and equate the moment of the total weight to the sum of the moments of the individual weights.

$$Mg \times \bar{x} = m_1 g \times x_1 + m_2 g \times x_2 + m_3 g \times x_3 + ...$$

giving

$$M \times \bar{x} = \sum_{i=1}^{n} m_i x_i$$

So, the centre of mass is at the point $(\bar{x}, 0)$ where $\bar{x} = \dfrac{\sum_{i=1}^{n} m_i x_i}{\sum_{i=1}^{n} m_i}$

> You can think of $\bar{x}$ as the weighted mean of the various x-values where the weight of each x-value is the mass at that point.

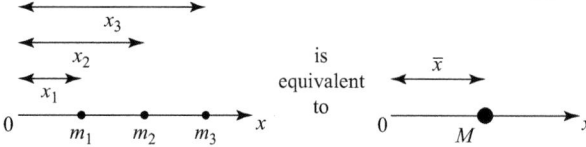

You can extend this method to two dimensions where m_1 is at the point (x_1, y_1), m_2 at (x_2, y_2), m_3 at (x_3, y_3), ... and the total mass M is at the point $(\bar{x}, \bar{y})$.

By taking moments about the x-axis, you get $M \times \bar{y} = \sum_{i=1}^{n} m_i \times y_i$

which can be written as $\bar{y} = \dfrac{\sum_{i=1}^{n} m_i \times y_i}{\sum_{i=1}^{n} m_i}$

Key point

The centre of mass of a system of particles is at the point $(\bar{x}, \bar{y})$ where $\bar{x} = \dfrac{\sum_{i=1}^{n} m_i \times x_i}{\sum_{i=1}^{n} m_i}$ and $\bar{y} = \dfrac{\sum_{i=1}^{n} m_i \times y_i}{\sum_{i=1}^{n} m_i}$

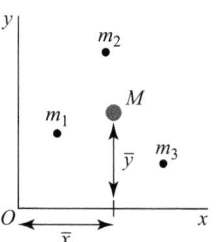

> Remember that $M = \sum_{i=1}^{n} m_i$

Example 1

Four masses of 3 kg, 2 kg, 4 kg and 1 kg lie at the points $(2, 3)$, $(5, 1)$, $(8, 4)$ and $(9, 6)$, respectively. Calculate the position of their centre of mass.

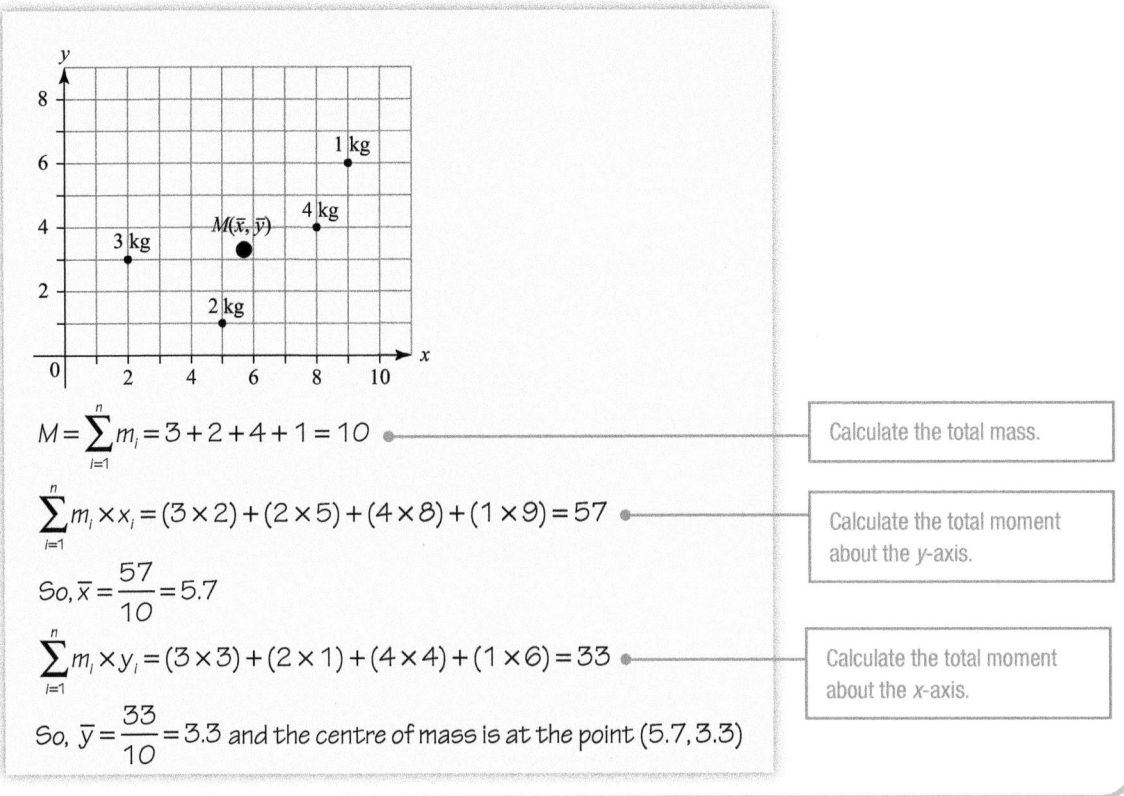

$M = \sum_{i=1}^{n} m_i = 3 + 2 + 4 + 1 = 10$ ———— Calculate the total mass.

$\sum_{i=1}^{n} m_i \times x_i = (3 \times 2) + (2 \times 5) + (4 \times 8) + (1 \times 9) = 57$ ———— Calculate the total moment about the y-axis.

So, $\bar{x} = \dfrac{57}{10} = 5.7$

$\sum_{i=1}^{n} m_i \times y_i = (3 \times 3) + (2 \times 1) + (4 \times 4) + (1 \times 6) = 33$ ———— Calculate the total moment about the x-axis.

So, $\bar{y} = \dfrac{33}{10} = 3.3$ and the centre of mass is at the point $(5.7, 3.3)$

Exercise 25.2A Fluency and skills

1   Find the coordinates of the centres of mass of these systems of masses.

  a   2 kg at the point $(4, 0)$, 3 kg at $(6, 0)$ and 5 kg at $(8, 0)$,

  b   3 kg at the point $(1, 0)$, 5 kg at $(3, 0)$, 4 kg at $(4, 0)$ and 8 kg at $(6, 0)$,

  c   4 kg at the point $(0, 2)$, 6 kg at $(0, 5)$ and 10 kg at $(0, 4)$,

  d   6 kg, 12 kg and 15 kg at the points $(4, -4)$, $(10, 1)$ and $(0, -5)$, respectively.

2   Masses of 7 kg, 8 kg and 5 kg are placed at points with position vectors $3\mathbf{i} + 6\mathbf{j}$, $4\mathbf{i} - 2\mathbf{j}$ and $6\mathbf{i} + \mathbf{j}$, respectively. Find the position vector of their centre of mass.

3   Masses of 2 kg and 3 kg are placed at the points $(6, 0)$ and $(2, 0)$. Where should a mass of 4 kg be placed on the x-axis so that the centre of mass of all three masses is at the point $(4, 0)$?

4   Masses of 6 kg, 4 kg and 10 kg are placed at the points $(5, 0)$, $(0, 6)$ and (a, b), respectively. If their centre of mass is at the point $(3.5, 2.2)$, find the values of a and b

5   Four masses of 0.5 kg, 1.5 kg, 1 kg and 2 kg are placed on coordinates axes at points with position vectors $4\mathbf{i} + 4\mathbf{j}$, $4\mathbf{i} + 8\mathbf{j}$, $10\mathbf{i} + 2\mathbf{j}$ and $a\mathbf{i} + b\mathbf{j}$, respectively. The position vector of their centre of mass is $6\mathbf{i} + 6\mathbf{j}$. Find the values of a and b

6   Four masses have a centre of mass at the point $(0, -1)$. A 4 kg mass is at the point $(5, 3)$, 9 kg is at $(6, -2)$ and 6 kg is at $(-1, 4)$. Find the position of the final mass of 5 kg.

To solve problems involving a centre of mass

(1) Draw a diagram showing the information and construct a table to summarise the data.

(2) Take moments to calculate unknown values. Make sure you take distances from the correct axes when finding moments.

When working with any simple or composite system of point masses or laminas, you should consider the symmetry of both the shape and the distribution of masses. Symmetry may make your solution to a problem easier.

A light, rectangular, rigid framework $OABC$ has side lengths $OA = 3\,\text{m}$ and $OC = 2\,\text{m}$.

D and E are the midpoints of AB and BC. Masses of $4\,\text{kg}$, $3\,\text{kg}$, $1\,\text{kg}$ and $5\,\text{kg}$ are fixed to A, C, D and E, respectively, and it is suspended by a string from O

a Find the position of the centre of mass, G

b Find the angle which OA makes with the vertical.

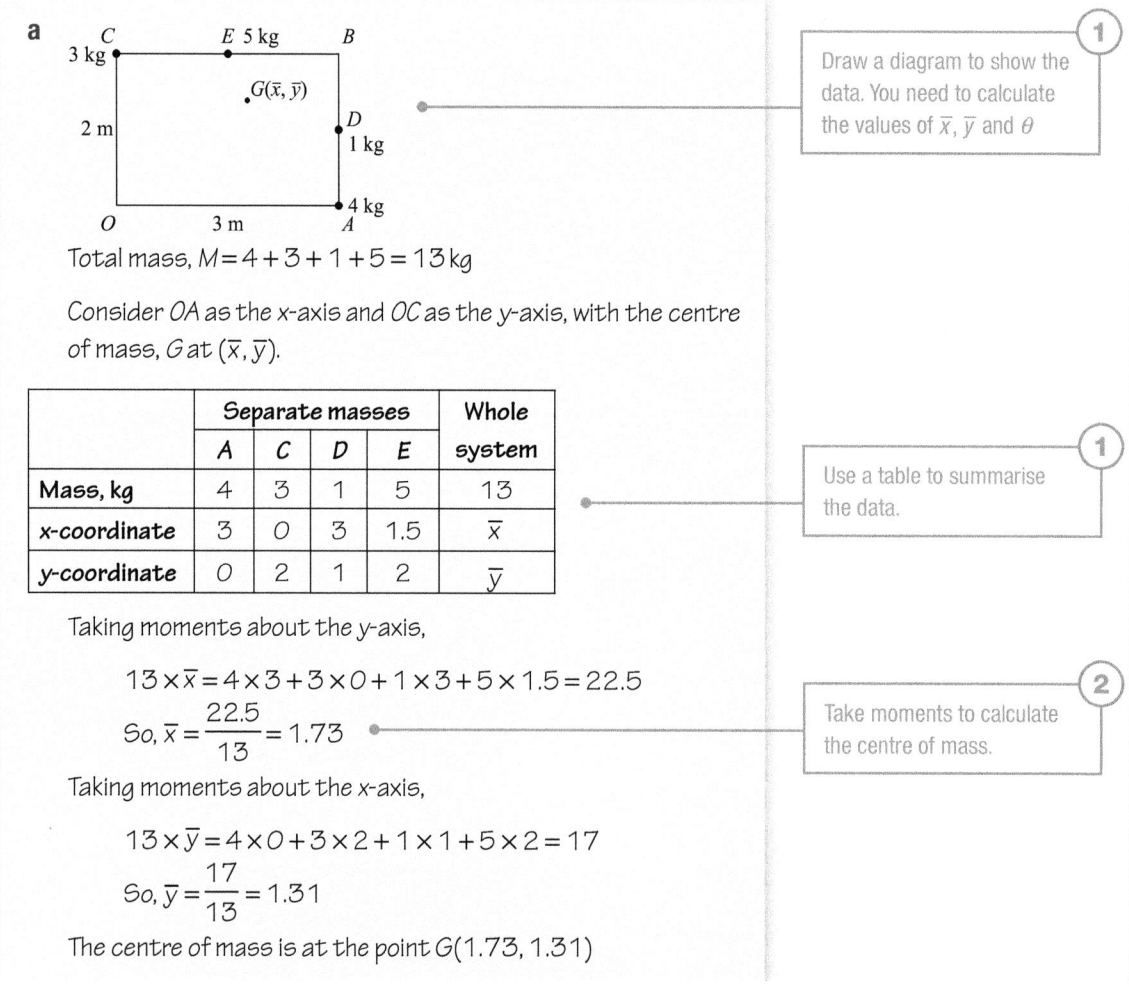

a

Total mass, $M = 4 + 3 + 1 + 5 = 13\,\text{kg}$

Draw a diagram to show the data. You need to calculate the values of $\bar{x}$, $\bar{y}$ and θ ①

Consider OA as the x-axis and OC as the y-axis, with the centre of mass, G at $(\bar{x}, \bar{y})$.

	Separate masses				Whole
	A	C	D	E	system
Mass, kg	4	3	1	5	13
x-coordinate	3	0	3	1.5	$\bar{x}$
y-coordinate	0	2	1	2	$\bar{y}$

Use a table to summarise the data. ①

Taking moments about the y-axis,

$$13 \times \bar{x} = 4 \times 3 + 3 \times 0 + 1 \times 3 + 5 \times 1.5 = 22.5$$

$$\text{So, } \bar{x} = \frac{22.5}{13} = 1.73$$

Take moments to calculate the centre of mass. ②

Taking moments about the x-axis,

$$13 \times \bar{y} = 4 \times 0 + 3 \times 2 + 1 \times 1 + 5 \times 2 = 17$$

$$\text{So, } \bar{y} = \frac{17}{13} = 1.31$$

The centre of mass is at the point $G(1.73, 1.31)$

(Continued on the next page)

b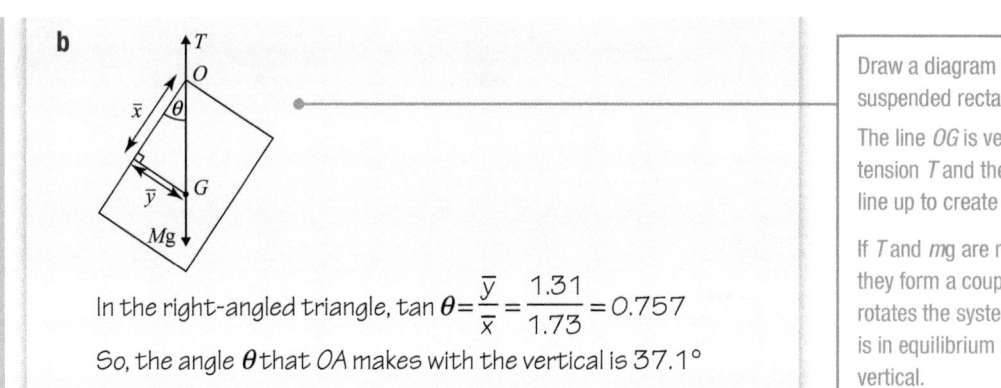

In the right-angled triangle, $\tan \theta = \dfrac{\bar{y}}{\bar{x}} = \dfrac{1.31}{1.73} = 0.757$

So, the angle θ that OA makes with the vertical is $37.1°$

Draw a diagram of the suspended rectangle.

The line OG is vertical as the tension T and the weight Mg line up to create equilibrium.

If T and mg are not in line, they form a couple which rotates the system until it is in equilibrium with OG vertical.

Exercise 25.2B Reasoning and problem-solving

1 Masses of 2 kg, 4 kg, 6 kg and 9 kg are placed at the vertices of a light, rigid, rectangular framework $ABCD$, where $AB = 5$ m and $BC = 3$ m. Find the position of the centre of mass of the system from AB and AD

2 A light, rectangular lamina $OPQR$ has a mass of 5 kg fixed to P and 4 kg fixed to Q. A third mass of 8 kg is fixed to the centre C of the rectangle. $OP = 10$ m and $OR = 5$ m. Taking OP and OR as the x- and y-axes, find the position of the centre of mass of the system.

3 A rigid, square framework of light rods has particles fixed to its four corners and the centre of the square, as shown in the diagram.
The square has sides 2.4 m long.

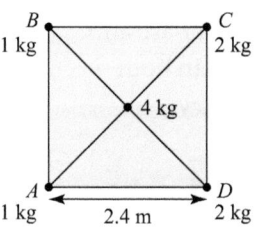

a Explain, without any calculation, why the distance of the centre of mass, G from side AD is 1.2 m.

b Find the distance of G from side AB and the angle of AB to the vertical when the framework is suspended from A

4 A light lamina $ABCDEF$ has the shape of a letter L made from two identical rectangles 10 cm by 4 cm and a square of side 4 cm.

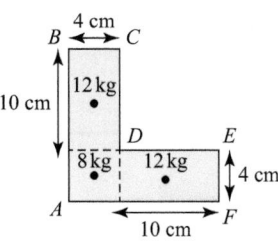

Three masses are attached to the lamina, two of 12 kg to the centre of each rectangle and one of 8 kg to the centre of the square.

a Explain why the centre of mass of the three masses must lie on the line AD

b Find the distances of the centre of mass from the sides AB and AF

5 A light, rectangular framework $ABCD$ has $AB = 4$ m and $BC = 3$ m. Masses of 5 kg, 4 kg, 2 kg and 3 kg are placed at A, B, C and D, respectively. A fifth mass m kg is placed at a point E on CD so that the centre of mass of the whole system is at the centre of the rectangle. Find the value of m and the position of E

6 A square, metal framework $OXZY$ of side 2 m is made from uniform rods of different densities. The density of OX is 1.2 kg per metre; the density of XZ and YZ is 2 kg per metre; and the density of OY is 1.8 kg per metre. Find the distance of the centre of mass from OX and OY

7 A uniform piece of wire has a mass of 10 grams per cm of length. It is bent into the shape of a triangle OXY, which is right-angled at O. If $OX = 24$ cm and $OY = 10$ cm, find the distance of the centre of mass from OX and OY

8 A rectangular framework $OXPY$ with $OX = YP = 1.2$ m is made from two bent uniform rods. The rods are both 2 m but rod OXP is twice as heavy as rod OYP. If OYP has a density of 4 kg per metre length, find the distance of the centre of mass from O

Fluency and skills

You can use symmetry to find the centres of mass of some common shapes, provided their mass is uniform over their area. Two simple cases are a rectangular lamina and a circular lamina. Symmetry gives their centres of mass G at their centre points.

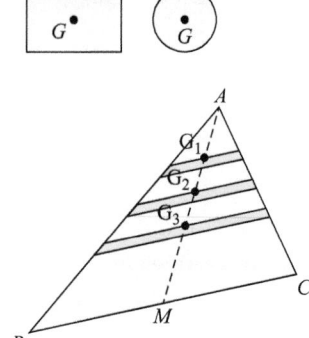

Most triangular laminas are not symmetrical. But you can think of a triangle as being made up of an infinite number of uniform rods of negligible thickness which are parallel to one side. The symmetry of the rods gives their centres of mass at their midpoints G_1, G_2, G_3, ... so the centre of mass G of the whole triangle lies on the median AM When you repeat this method for the other two medians, you find point G where all the medians intersect.

This point of intersection is called the **centroid** of the triangle.

Key point

The centre of mass, G, of a triangular lamina is at the point of intersection of its medians which is $\frac{2}{3}$ of the way along each median from the vertex.

The formulae for the centre of mass for a uniform circular arc and for a sector of a circle are listed in your formula booklet. In both cases, their centres of mass are on their lines of symmetry.

Key point

A circular arc of radius r and angle 2α at the centre has
$$OG = \frac{r\sin\alpha}{\alpha}$$

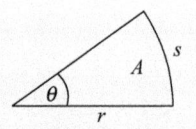

Key point

A sector of a circle of radius r and angle 2α at the centre has
$$OG = \frac{2r\sin\alpha}{3\alpha}$$

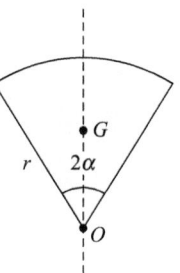

In both formulae, α must be in radians, not degrees.

Remember, for a sector with angle theta radians,
arc length, $s = r\theta$
area, $A = \frac{1}{2}r^2\theta$

Example 1

A uniform wire of length 2.1 m is bent to make a circular arc of radius 1.4 m.

Find the position of its centre of mass.

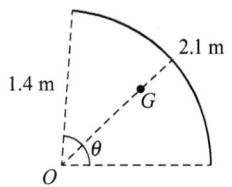

The length of a circular arc, $s = r\theta$

So angle θ at the centre $O = \dfrac{s}{r} = \dfrac{2.1}{1.4} = 1.5$ radians

As $\theta = 2\alpha$, angle $\alpha = 0.75$ radians

The centre of mass, G, is on the wire's line of symmetry such that

$OG = \dfrac{r\sin\alpha}{\alpha} = \dfrac{1.4 \times \sin 0.75}{0.75} = 1.27\,\text{m}$

When working with any simple or composite system of laminas and point masses, you should consider the symmetry of both the shape and the distribution of any masses.

Exercise 25.3A Fluency and skills

1   Find the position of the centre of mass of a uniform wire bent into a circular arc with

   a A radius of 2.4 m and a length of 3.6 m,

   b A radius of 5 m and an angle at the centre of 135°,

   c A semicircular shape and a diameter of 3.6 m.

2   Find the position of the centre of mass of a uniform sector of a circle with

   a A radius of 2 m and an angle at the centre of 1.2 radians,

   b A radius of 4 m and an area of 16 m²,

   c An area of 7.2 m² and an angle at the centre of 1.6 radians.

3   Find the position of the centre of mass of a uniform lamina which is

   a A semicircle of diameter 1.6 m,

   b A quadrant of a circle of radius 1.4 m.

4   Find the coordinates of the centre of mass of a uniform triangular lamina with its vertices at the points:

   a $(1, 1)$, $(10, 1)$ and $(10, 7)$,

   b $(0, 0)$, $(9, 0)$ and $(0, 6)$,

   c $(2, 6)$, $(8, 6)$ and $(8, 0)$.

5   a A uniform lamina has the shape of an isosceles triangle ABC with $AB = AC$. If $AB = 13$ cm and $BC = 10$ cm, calculate the position of its centre of mass.

   b Calculate the position of the centre of mass of an equilateral triangular lamina of side 10 cm.

6   A uniform triangular lamina OXY is right-angled at O with its two other vertices at $(a, 0)$ and $(0, b)$. If the triangle's centre of mass is at the point $(5, 4)$, find the values of a and b

7   Prove that the centre of mass of a uniform lamina in the shape of a parallelogram is at the intersection of the two lines which join the midpoints of pairs of opposite parallel sides.

To solve problems involving the centre of mass of shapes

1. Draw a diagram showing the data and summarise the data in a table.

2. Use symmetry and standard results for the centre of mass of common shapes.

3. Take moments to find the centre of mass of a shape.

When working with laminas which have a mass, you can use the relationship $mass = area \times density$, where the units of density ρ are $kg\,m^{-2}$. You need to be very careful as, in some problems, the area density across the shape may vary.

a A uniform lamina $OABCD$ consists of a rectangle, a semicircle and a right-angled triangle with the dimensions shown. Taking OB and OD as axes, find the point $G(\bar{x}, \bar{y})$ of the lamina's centre of mass.

b If the rectangle is made from a material with twice the density of that for the semicircle and triangle, find the new position $G(\bar{x}, \bar{y})$ of the lamina's centre of mass.

a Let the area density of the lamina be $\rho\,kg\,m^{-2}$.

	Area, m^2	Mass, kg	x-value	y-value
Rectangle	$3 \times 6 = 18$	18ρ	3	1.5
Semicircle	$\frac{1}{2}\pi \times 3^2 = 14.1$	14.1ρ	3	$3 + \dfrac{2 \times 3\sin\frac{\pi}{2}}{3 \times \frac{\pi}{2}} = 4.27$
Triangle	$\frac{1}{2} \times 3 \times 3 = 4.5$	4.5ρ	$6 + \frac{1}{3} \times 3 = 7$	$\frac{1}{3} \times 3 = 1$
Whole lamina	36.6	36.6ρ	$\bar{x}$	$\bar{y}$

① Summarise the data in a table.

Taking moments about OD

$36.6\rho \times \bar{x} = (18\rho \times 3) + (14.1\rho \times 3) + (4.5\rho \times 7)$

$\qquad = 127.8\rho$

giving $\bar{x} = 3.49$

Taking moments about OB

$36.6\rho \times \bar{y} = (18\rho \times 1.5) + (14.1\rho \times 4.27) + (4.5\rho \times 1)$

$\qquad = 91.7\rho$

giving $\bar{y} = 2.51$

Centre of mass is $G(3.49, 2.51)$

③ Take moments to find $\bar{x}$ and $\bar{y}$

(Continued on the next page)

b The density of the rectangle is now $2\rho\,\mathrm{kg\,m^{-2}}$.

So the new total mass $= 18 \times 2\rho + 14.1\rho + 4.5\rho = 54.6\rho$

Taking moments about OD and OB gives

$54.6\rho \times \bar{x} = (18 \times 2\rho \times 3) + (14.1\rho \times 3) + (4.5\rho \times 7)$

$\qquad\qquad = 181.8\rho$

giving $\quad \bar{x} = 3.33$

$54.6\rho \times \bar{y} = (18 \times 2\rho \times 1.5) + (14.1\rho \times 4.27) + (4.5\rho \times 1)$

$\qquad\qquad = 118.7\rho$

giving $\quad \bar{y} = 2.17$

So the new centre of mass is $G(3.33, 2.17)$

Example 3

A uniform rectangular lamina $ABCD$, 4 m by 6 m, has a rectangular hole $PQRS$, 2 m by 1 m, removed, such that S coincides with the lamina's centre, G_1. Taking AB and AD as the x- and y-axes, find the distances $(\bar{x}, \bar{y})$ from A of the centre of mass of the remaining part.

Let the density of the lamina be $\rho\,\mathrm{kg\,m^{-2}}$.

For $ABCD$

mass $= 6 \times 4 \times \rho = 24\rho$ with centre of mass G_1 at $(3, 2)$ •⎯⎯⎯⎯⎯ ② Work out the mass and centre of mass for each of the separate parts.

For $PQRS$

mass $= 2 \times 1 \times \rho = 2\rho$ with centre of mass G_2

at $(3 + 1, 1 + 0.5) = (4, 1.5)$

For the remainder

mass $= 24\rho - 2\rho = 22\rho$ with centre of mass at $(\bar{x}, \bar{y})$

Taking moments about AD

Moment of $ABCD =$ moment of $PQRS +$ moment of remainder •⎯⎯⎯⎯⎯ ③ Take moments to find $\bar{x}$ and $\bar{y}$

$\qquad 24\rho \times 3 = (2\rho \times 4) + (22\rho \times \bar{x})$

giving $\qquad 72 = 8 + 22\bar{x}$

$\qquad\qquad \bar{x} = 2.91$

Taking moments about AB

$\qquad 24\rho \times 2 = (2\rho \times 1.5) + (22\rho \times \bar{y})$

giving $\qquad 48 = 3 + 22\bar{y}$

$\qquad\qquad \bar{y} = 2.05$

So, the centre of mass of the remaining part is $(2.91, 2.05)$ metres from A

MECH

The formulae booklet also provides some standard results for the centre of mass of 3D shapes which you can use without proof, as in Example 4

Example 4

A child's toy is made from a hollow, hemispherical shell attached to a hollow, conical shell of the same material and radius. The cone is 12 cm tall with a radius of 5 cm.

Find the position of the toy's centre of mass G from point O in the centre of the base of the hemisphere and cone.

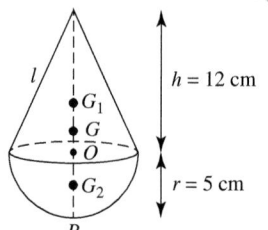

Let the area density be ρ and points G_1 and G_2 be the centres of mass of the two parts of the toy.

By symmetry, G lies on the line of symmetry through O

Using standard results, $OG_1 = \dfrac{1}{3} \times 12 = 4\,\text{cm}$

and $OG_2 = \dfrac{1}{2} \times 5 = 2.5\,\text{cm}$

> The results for the centres of mass of a conical shell and a hemispherical shell are given in the formulae booklet.

Total mass = (area of cone + area of hemisphere) $\times \rho$

$= (\pi r l + \dfrac{1}{2} \times 4\pi r^2)\rho = (\pi \times 5 \times \sqrt{5^2 + 12^2} + 2 \times \pi \times 25)\rho$

$= (65\pi + 50\pi)\rho = 115\pi\rho$

1st method

Taking moments about point P

$115\pi\rho \times PG = 65\pi\rho \times PG_1 + 50\pi\rho \times PG_2$

$= 65\pi\rho \times (5+4) + 50\pi\rho \times (5-2.5) = 710\pi\rho$

$\Rightarrow PG = 6.2\,\text{cm and } OG = 6.2 - 5 = 1.2\,\text{cm}$

2nd method

Taking moments about point O

$115\pi\rho \times OG = 65\pi\rho \times OG_1 + 50\pi\rho \times OG_2$

$= 65\pi\rho \times 4 + 50\pi\rho \times (-2.5)$

$= 135\pi\rho$

> Distances can be negative. Note the negative sign as G_2 is on the opposite side of O

$\Rightarrow OG = 1.2\,\text{cm}$

The toy's centre of mass G is 1.2 cm from point O within the conical shell.

Exercise 25.3B Reasoning and problem-solving

1 Find the coordinates of the centres of mass of these uniform laminas.

a An L-shape $(1, 7)$, $(1, 9)$, $(8, 9)$, $(8, 3)$, $(4, 3)$, $(4, 7)$.

b A trapezium $(2, 1)$, $(8, 1)$, $(8, 8)$, $(2, 5)$.

c A rectangle $(1, 1)$, $(1, 6)$, $(9, 6)$, $(9, 1)$ with

i A semicircle attached to the side joining $(1, 6)$ and $(9, 6)$,

ii A circular hole, centre $(5, 4)$ and radius 2 units, cut out.

2 For each uniform lamina, find the centre of mass taking O as the origin of x- and y-axes.

a

5 m   3 m

2 m

O

b

4 m

4 m   4 m

2 m

8 m

O 10 m

3 a A uniform wire of length 3 m is bent into a triangle ABC where $AB = 0.5$ m and $AC = 1.2$ m. Find the distances of the centre of mass from AB and AC

b The same wire is now bent to form the perimeter of a sector of a circle of radius 1 m. Find the distance of the centre of mass from the centre of the circle.

4 If the child's toy in Example 4 above were made from a solid cone and a solid hemisphere of the same material, use standard results to find the new position of the toy's centre of mass.

5 A circular hole of radius 3 cm is cut from a circular, metal disc of radius 9 cm. If the centre of mass of the remaining piece is 0.5 cm from the centre of the disc, find the position of the centre of the hole.

6 a A uniform square lamina of side 5 m has a square of side 2 m cut from one corner. Find the distance of the centre of mass of the remainder from that corner.

b A uniform semicircular lamina, diameter 24 cm, has a circular hole, diameter 6 cm, cut out of it. The hole's centre is 4 cm from its straight edge on the semicircle's line of symmetry. Find the exact position of the centre of mass of the remainder.

7 A child's building block is a prism with a symmetrical cross-section formed from a rectangle 10 cm by 5 cm with a semicircle of diameter 6 cm cut away. If the prism is uniformly dense, how far is its centre of mass from the baseline AB?

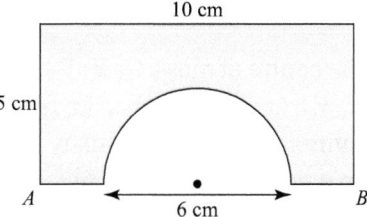

10 cm

5 cm

A 6 cm B

8 A uniform metal sheet $PQRST$ has the shape of a square $PQRT$ of side 6 cm joined to an isosceles triangle RST of height 4.5 cm with $SR = ST$. The triangle is twice the thickness of the square. How far is the centre of mass of the sheet from side PQ?

9 A toy rocket is made by joining the plane faces of a solid cone and a solid cylinder, both of radius 5 cm and height 9 cm. The density of the cylinder is twice that of the cone. How far is the rocket's centre of mass from the cone's vertex?

10 A frustum is made from a solid cone of height $6a$ and base radius $2a$ by removing its conical top so that the exposed plane surface has a radius of a. Find how far, in terms of a, the frustum's centre of mass is from the midpoint of the base.

11 An isosceles trapezium $ABCD$ has height h and parallel sides $AB = a$ and $CD = b$
Prove that its centre of mass is a distance of $\dfrac{h(a+2b)}{3(a+b)}$ from AB. Explain why this result applies to trapezia which are not isosceles.

Fluency and skills

You can find the centre of mass, G($\bar{x}$, $\bar{y}$) of n masses m_1, m_2, m_3, ... m_n at points (x_1, y_1), (x_2, y_2), (x_3, y_3), ... (x_n, y_n). You can apply the same method to a lamina with an area density ρ and mass M by dividing the lamina into n thin strips of mass $\delta m = \rho \times y \delta x$, where the base of each strip is on the x-axis so that the centre of mass of each strip is at the point $(x, \frac{1}{2}y)$.

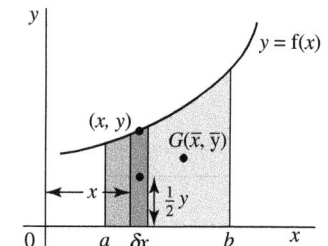

Suppose a lamina is bounded by the curve $y = f(x)$, the x-axis and the lines $x = a$ and $x = b$

The total mass $M = \sum_{x=a}^{b} \delta m = \sum_{x=a}^{b} \rho y \delta x$

By taking moments about the y-axis and the x-axis,

$$M \times \bar{x} = \sum_{x=a}^{b} \delta m \times x = \sum_{x=a}^{b} \rho x y \delta x$$

and $\quad M \times \bar{y} = \sum_{x=a}^{b} \delta m \times \frac{1}{2}y = \sum_{x=a}^{b} \frac{1}{2}\rho y^2 \delta x$

As the number of strips increases, $\delta x \to 0$

In the limit, as $\delta x \to 0$, these summations combine strips across the whole area from $x = a$ to $x = b$ to give exact values for $\bar{x}$ and $\bar{y}$

So, in the limit,

$$M = \rho \int_a^b y \, dx$$

$$M \times \bar{x} = \rho \int_a^b xy \, dx$$

and $\quad M \times \bar{y} = \rho \int_a^b \frac{1}{2}y^2 \, dx$

By substituting for M and cancelling ρ, you can find $\bar{x}$ and $\bar{y}$

Key point

A uniform lamina with an area bounded by the curve $y = f(x)$, the x-axis and the lines $x = a$ and $x = b$ has a centre of mass $G(\bar{x}, \bar{y})$, where

$$\bar{x} = \frac{\displaystyle\int_a^b xy \, dx}{\displaystyle\int_a^b y \, dx} \quad \text{and} \quad \bar{y} = \frac{\displaystyle\int_a^b \frac{1}{2}y^2 \, dx}{\displaystyle\int_a^b y \, dx}$$

You can *either* use the formulae for $\bar{x}$ and $\bar{y}$ as given in the key point, *or* set out your solution by taking moments as in Example 1. You have a choice.

Find the centre of mass, G of the lamina formed from the area under the curve $y = \dfrac{12}{x}$ for $2 \leq x \leq 6$

Let the area density be ρ and G be at the point $(\bar{x}, \bar{y})$

Total mass $M = \rho \displaystyle\int_2^6 y\,dx = \rho \int_2^6 \dfrac{12}{x}\,dx = 12\rho \int_2^6 \dfrac{1}{x}\,dx = 12\rho[\ln x]_2^6 = 12\rho \ln 3$

Taking moments about the y-axis

$12\rho \ln 3 \times \bar{x} = \rho \displaystyle\int_2^6 xy\,dx = \rho \int_2^6 x\dfrac{12}{x}\,dx = 12\rho \int_2^6 1\,dx = 12\rho[x]_2^6 = 48\rho$

$\Rightarrow \bar{x} = \dfrac{48}{12 \ln 3} = 3.64$ to 3 sf

Taking moments about the x-axis

$12\rho \ln 3 \times \bar{y} = \rho \displaystyle\int_2^6 \dfrac{1}{2}y^2\,dx = \rho \int_2^6 \dfrac{1}{2}\left(\dfrac{12}{x}\right)^2 dx = 72\rho \int_2^6 \dfrac{1}{x^2}\,dx = 72\rho\left[-\dfrac{1}{x}\right]_2^6 = 24\rho$

$\Rightarrow \bar{y} = \dfrac{24}{12 \ln 3} = 1.82$ to 3 sf

The centre of mass, G, is at the point $(3.64, 1.82)$

Note that the lamina in Example 1 has the *x*-axis as a boundary. Some laminas have a different shape. In the one shown here, the lamina is symmetrical about the *x*-axis, the *x*-axis is not a boundary and the vertical strips have a length $2y$, so you have to take moments about the *y*-axis to find the centre of mass noting that, by symmetry, $\bar{y} = 0$

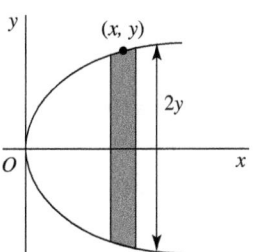

You can use a similar method for finding the centre of mass of a 3D solid with a volume density ρ. When the area under the curve $y = f(x)$ for $a \leq x \leq b$ is rotated about the *x*-axis, a thin strip of width δx generates a thin disc of mass $\delta m = \rho \times \pi y^2 \times \delta x$

In the limit as $\delta x \to 0$, a summation of these discs gives the total

mass $M = \displaystyle\lim_{\delta x \to 0} \sum_{x=a}^{b} \delta m = \lim_{\delta x \to 0} \sum_{x=a}^{b} \rho \pi y^2 \delta x = \rho \int_a^b \pi y^2\,dx$

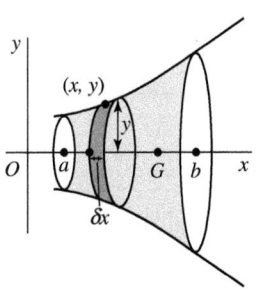

Taking moments about the *y*-axis, $M \times \bar{x} = \rho \displaystyle\int_a^b x \times \pi y^2\,dx$, which gives

you the value of $\bar{x}$. Considering symmetry about the *x*-axis, you have $\bar{y} = 0$

A uniform solid of revolution formed by rotating the **Key point**
area between the curve $y = f(x)$, the x-axis and the lines
$x = a$ and $x = b$ through $360°$ about the x-axis, has a centre
of mass $G(\bar{x}, \bar{y})$, where

$$\bar{x} = \dfrac{\displaystyle\int_a^b xy^2\,dx}{\displaystyle\int_a^b y^2\,dx} \quad \text{and } \bar{y} = 0$$

As before, you can choose whether to use the formula given in this key
point or set out your solution by taking moments about the y-axis, as
in Example 2. You have a choice.

Example 2

Find the centre of mass, G, of the solid of revolution formed
by rotating the area under the curve $y = \dfrac{12}{x}$ for $2 \le x \le 6$
through 2π radians about the x-axis. Give an exact answer.

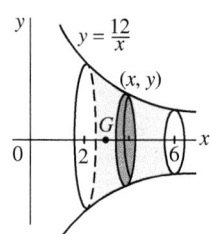

Let the volume density be ρ and G be at the point $(\bar{x}, \bar{y})$

Total mass $M = \rho \displaystyle\int_2^6 \pi y^2\,dx = \rho\pi \int_2^6 \dfrac{144}{x^2}\,dx = 144\rho\pi\left[-\dfrac{1}{x}\right]_2^6 = 48\rho\pi$

Taking moments about the y-axis

$48\rho\pi \times \bar{x} = \rho\displaystyle\int_2^6 x \times \pi y^2\,dx = \rho\pi\int_2^6 \dfrac{144}{x}\,dx = 144\rho\pi[\ln x]_2^6 = 144\rho\pi\ln 3$

$\Rightarrow \bar{x} = \dfrac{144\rho\pi\ln 3}{48\rho\pi} = 3\ln 3$

By symmetry, $\bar{y} = 0$

The solid's centre of mass, G, is at the point $(3\ln 3, 0)$

Exercise 25.4A Fluency and skills

1 Find the centre of mass of a uniform lamina
defined by the area between the curve $y = f(x)$
and the x-axis for the given range, when

   a $y = \dfrac{9}{x}$ for $1 \le x \le 3$

   b $y = x^2$ for $0 \le x \le 2$

   c $y = 1 + x^2$ for $0 \le x \le 1$

   d $y = x(1 - x)$ for $0 \le x \le 1$

2 Find the centre of mass of a uniform lamina
defined by the area between

   a the curve $y^2 = x$ and the line $x = 4$

   b the curve $y = 9 - x^2$ where $y \ge 0$

   c the curve $y^2 = x^3$ and the line $x = 4$

3 Find the centre of mass of a uniform solid
of revolution generated by rotating the area
between the curve and the x-axis through 2π
about the x-axis.

   a $y = 2x$ for $0 \le x \le 4$

   b $y = \dfrac{9}{x}$ for $1 \le x \le 3$

   c $y = x^2$ for $0 \le x \le 2$

   d $y^2 = x$ and the line $x = 4$, where $y \ge 0$

4 Find the centre of mass of a uniform
semicircular lamina bounded by these curves.

   a $x^2 + y^2 = 4$, where $x \ge 0$

   b $x^2 + y^2 = r^2$, where $x \ge 0$

Centres of mass and stability Further topics with centres of mass

5 a A uniform solid hemisphere is generated by rotating a semicircle given by $x^2 + y^2 = 1$, where $x \geq 0$, through $180°$ about the x-axis. Find the coordinates of its centre of mass.

b Repeat for the hemisphere generated by $x^2 + y^2 = r^2$, where $x \geq 0$

6 Find the position of the centre of mass of a solid cone generated by a rotation of $360°$ about the x-axis of the straight line $y = mx$ for $0 \leq x \leq h$

Reasoning and problem-solving

To solve problems involving the centre of mass of shapes

(1) Draw a diagram showing the information you are given.

(2) Work from first principles or, if you understand the method and can recall accurately, you may decide to quote a standard formula.

(3) Take moments to find the centre of mass of a shape, using symmetry where possible.

A lamina can be suspended in equilibrium *either* by a string attached to its perimeter *or* from a nail or other smooth, horizontal axis through any point on its surface.

Only two forces act on the lamina: its weight and *either* the tension in the string *or* the reaction at the axis.

The lamina adjusts its position until the two forces act in the same vertical line and it is then in equilibrium. If the forces were not in the same vertical line, there would be a couple that would tend to rotate the suspended object. The same applies when a solid is suspended in equilibrium.

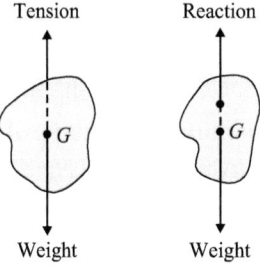

Key point

When a lamina or solid is freely suspended in equilibrium, its centre of mass is vertically below its point of suspension.

A uniform semicircular lamina of centre O with diameter $PQ = 2a$ is suspended from P and hangs freely. Calculate the angle θ that PQ makes with the vertical.

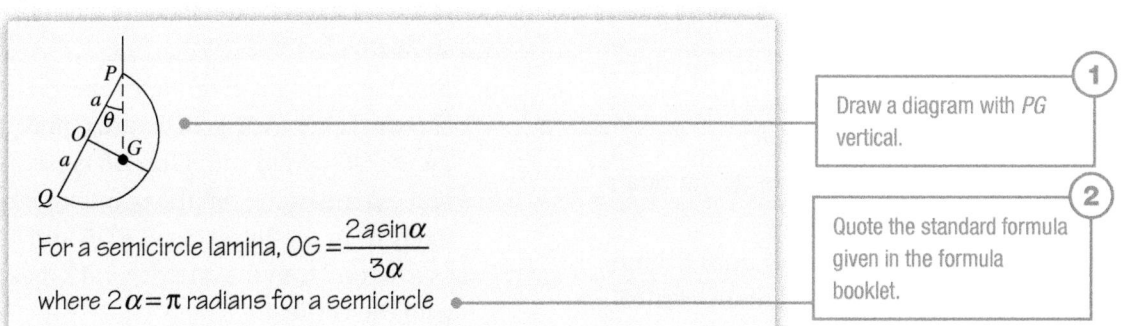

For a semicircle lamina, $OG = \dfrac{2a\sin\alpha}{3\alpha}$

where $2\alpha = \pi$ radians for a semicircle

(1) Draw a diagram with PG vertical.

(2) Quote the standard formula given in the formula booklet.

(Continued on the next page)

$$\Rightarrow OG = \frac{2a\sin\frac{\pi}{2}}{3\times\frac{\pi}{2}} = \frac{4a}{3\pi}$$

In $\triangle OPG$, $\tan\theta = \dfrac{OG}{OP} = \dfrac{4}{3\pi}$

$\Rightarrow$ Angle of PQ to vertical is $\theta = 23.0°$

Exercise 25.4B Reasoning and problem-solving

Where possible, you may quote standard results from the formula booklet for the centres of mass of uniform bodies.

1 Each of these uniform objects is suspended by a string from any point, P, on its circular edge, centre O. Find the angle that OP makes with the vertical.

   a   A solid hemisphere, radius a

   b   A solid cone, radius 3 cm, height 8 cm,

   c   A hemispherical shell, radius a

   d   A conical shell, radius 3 cm, height 8 cm.

2 A lamina in the shape of an isosceles triangle LMN with $MN = LN = 13$ cm, is suspended by a string to hang freely from L. If $LM = 10$ cm, find the angle that LM makes with the vertical.

3 A lamina in the shape of a sector of a circle, centre O and radius a with an arc PQ, is suspended to hang freely from point P. Find the angle that OP makes with the vertical when $\angle POQ = \dfrac{\pi}{2}$

4 One end of a solid cylinder is joined to the base of a solid cone of the same material. The object formed is hung from a point on the edge of the other end of the cylinder. The cylinder and cone have the same radius, r and height, h. Find the angle between the string and the object's axis in terms of r, and h

5 A uniform plane lamina is made by joining a semicircle, centre O and diameter $PQ = 2a$, to an equilateral triangle PQR

   a   Find the exact distance of the lamina's centre of mass from O

   b   The lamina is hung freely from P. Find the angle between OP and the vertical.

6 A vertical uniform rectangular laminar board $OABC$ of mass 2 kg is designed to hold boxes on its surface. Taking OA and OC as axes with $OA = 1$ m and $OC = 0.5$ m, model three boxes as particles A, B and C with masses 2 kg at point $(0.2, 0.4)$, 1 kg at $(0.4, 0.3)$ and 3 kg at $(0.6, 0.4)$. Find the position G of the centre of mass. If the board is hung freely from a nail at the midpoint M of BC, find the angle that BC makes with the horizontal.

7 Two uniform semicircular laminas of masses m and $2m$ have the same radius r They combine to form a circle, centre O and centre of mass G. Calculate the distance OG The circle is suspended from one end P of their common diameter. Show that the angle that this diameter makes with the vertical is equal to $\tan^{-1}\left(\dfrac{4}{9\pi}\right)$

8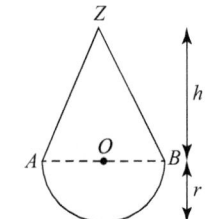

   a   A pendant is made from a uniform lamina consisting of a semicircle, centre O and radius r, attached to an isosceles triangle of base $2r$ and height h. If the centre of mass of the pendant is at O, show that $h = r\sqrt{2}$

   b   Another pendant of the same design has $2r = h = 15$ cm. A hole of centre C and radius $OC = 5$ cm with C on AB is cut out of the pendant. When the pendant hangs freely from the vertex Z, take OB and OZ as axes and find the angle that AZ makes with the vertical.

Fluency and skills

When an object is subjected to a system of forces and couples which all act in the same plane, it will be in equilibrium only if

(a) The resultant of all forces must be zero.

(b) The total moment of all forces and couples must be zero.

The sum of the components of all forces must be zero in any two different directions to ensure that there is no linear movement. You can write three equations to describe the equilibrium and then solve them simultaneously.

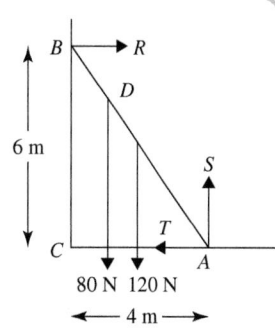

> **Key point**
>
> For an object to be in equilibrium under a system of coplanar forces, these conditions must be met:
>
> - The sum of the components of all forces must be zero in any *two different* directions
> - The total moment of all forces and couples must be zero.

Coplanar means 'in the same plane'.

Sometimes you can use equivalent conditions: for example, resolve in one direction and take moments about two different points.

Take care when choosing the directions in which you resolve and also when choosing the point to take moments about. Try to keep the algebra as simple as possible.

Example 1

A uniform ladder AB weighing 120 N rests on a smooth floor with its upper end B on a smooth wall.

End A is held by a string AC fixed to the base of the wall at C

A weight of 80 N is attached three-quarters of the way up the ladder at D

If $AC = 4$ m and $BC = 6$ m, find the tension T in the string and the reaction S at the ground.

As both surfaces are smooth, the reactions R and S are normal to the wall and ground.

For vertical equilibrium,

resultant vertical force is zero: $S - 120 - 80 = 0$

or   upward force balances downward force: $S = 120 + 80$ [1]

$\Rightarrow$ Reaction $S = 200$ N

(*Continued on the next page*)

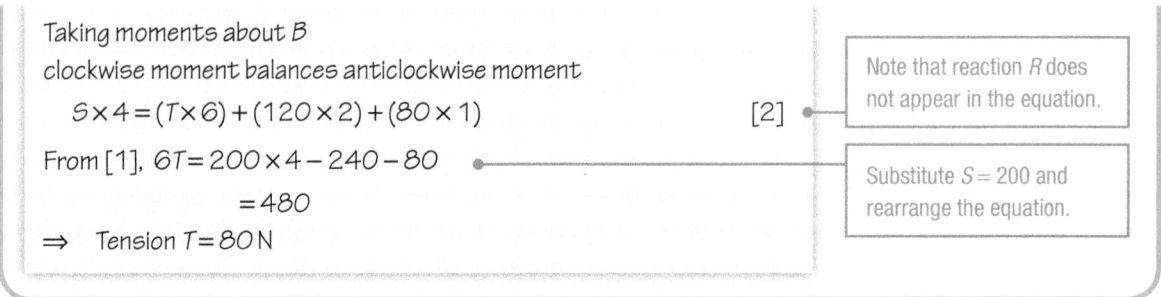

Taking moments about B
clockwise moment balances anticlockwise moment

$$S \times 4 = (T \times 6) + (120 \times 2) + (80 \times 1) \qquad [2]$$

From [1], $6T = 200 \times 4 - 240 - 80$

$$= 480$$

$$\Rightarrow \quad \text{Tension } T = 80\,\text{N}$$

Note that reaction R does not appear in the equation.

Substitute $S = 200$ and rearrange the equation.

The method in Example 1 used only two equations because it involved only two unknowns (T and S). Taking moments about B to give equation [2] was chosen because it did not involve R and so it made the algebra easier.

Instead of taking moments about B, you could have used the condition for horizontal equilibrium (giving $R = T$) and then taken moments about A. This method, however, involves three unknowns (R, T and S) and three equations. The algebra is not as easy.

You should think about your method before you start, so that you minimise the number of equations and unknowns.

Exercise 25.5A Fluency and skills

1   A uniform ladder PQ weighs 200 N. It stands on a smooth floor with its upper end Q against a smooth vertical wall. The lower end P is joined to the foot of the wall R by a string PR. If $PR = 6$ m and $QR = 8$ m, find the tension in the string and the reactions at the ground and at the wall.

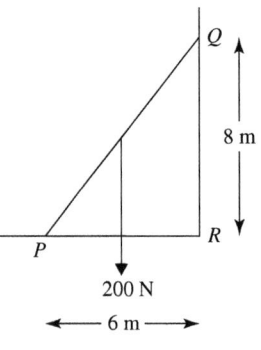

2   A uniform ladder AB weighing 100 N rests with end A on rough horizontal ground where the coefficient of friction, μ is $\dfrac{3}{4}$. End B rests against a smooth vertical wall. If A is 4 m from the foot of the wall, find the reaction at B and the frictional force at A if the ladder is on the point of slipping.

3   A uniform rod PQ weighs 8 N. It slopes up at 45° from a hinge at P and has a weight of 10 N hanging from Q. It is held in equilibrium by a horizontal string attached to Q so that Q is 4 m higher than P. Find the horizontal and vertical components of the reaction at the hinge and hence the total reaction at the hinge.

4   A uniform square lamina $EFGH$ of mass 4 kg and side 3 m is hinged at H with its upper edge GH horizontal. It is held in equilibrium by a horizontal string attached to F. Find the tension in the string and the exact value of the total reaction at the hinge.

5   A man of mass 75 kg climbs two-thirds of the way up a ladder AB of mass 20 kg which rests on a smooth floor at A and against a smooth wall at B. A string joins A to the foot of the wall at C such that $AC = 3$ m and $BC = 6$ m. Find the tension in the string in terms of g

6   A rectangular shop-sign $ABCD$, such that $AB = CD = 1$ m and $AD = BC = 0.5$ m, with a mass of 4 kg is hinged to a smooth wall at A. Its upper side AB is kept horizontal by corner D making contact with the wall. A 2 kg mass hangs from C. Find

   a   The reaction with the wall at D

   b   The total reaction at the hinge A

Reasoning and problem-solving

When an object is placed on an inclined plane, it may rest in equilibrium, slide down the plane or topple over. If there is enough friction to prevent sliding, there are three possible outcomes depending on the angle of inclination, θ

Consider a rectangular lamina $ABCD$ of weight W and centre of mass G.

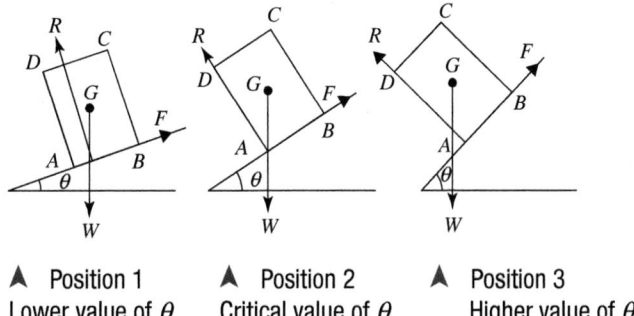

▲ Position 1
Lower value of θ

▲ Position 2
Critical value of θ

▲ Position 3
Higher value of θ

In position 1, the rectangle and plane make contact all along AB producing a normal reaction R. The moments of R and W about A are balanced and the rectangle is in equilibrium.

In position 2, θ has increased and the rectangle is about to topple. R, F and W all act through the only point of contact, A. The rectangle is in limiting equilibrium.

In position 3, θ has increased further. R and F still act through A, but W has an anticlockwise moment about A. The rectangle is not in equilibrium and it topples.

> **Key point**
>
> An object on an inclined plane is in equilibrium without toppling if the line of action of its weight falls within the side of the object that makes contact with the plane.

To solve problems involving the equilibrium of objects

(1) Draw a diagram showing the information you are given and the information you need to find.

(2) Devise a strategy to minimise the number of unknowns and equations.

(3) Write equations to satisfy the conditions for equilibrium, along with any other associated equations, and solve them simultaneously.

Rectangle $ABCD$ with $AD = 6\,\text{cm}$ and $CD = 4\,\text{cm}$ rests on an inclined plane as shown. If it does not slide down the plane, find the greatest possible angle of inclination θ

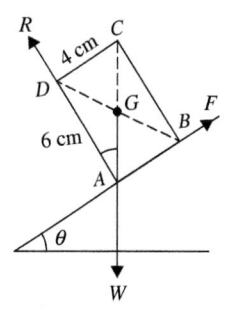

Let G be the centre of mass and the weight be W.

The rectangle is on the point of toppling when G is vertically above A

Geometry gives $\angle GAD = \theta$

In $\triangle ACD$, $\tan\theta = \dfrac{4}{6}$

So the greatest possible angle of inclination is $\theta = 33.7°$

Example 3

A cylindrical roller, centre O and radius 63 cm, is fixed to smooth, horizontal ground with a uniform plank resting against it at point C

One end, A, of the plank lies on the ground and the other end, B, projects beyond the roller. The plank is perpendicular to the horizontal axis of the roller.

A light string AO of length 1.05 m keeps the plank from slipping.

If the plank has weight W and length 90 cm, find the tension T in the string in terms of W.

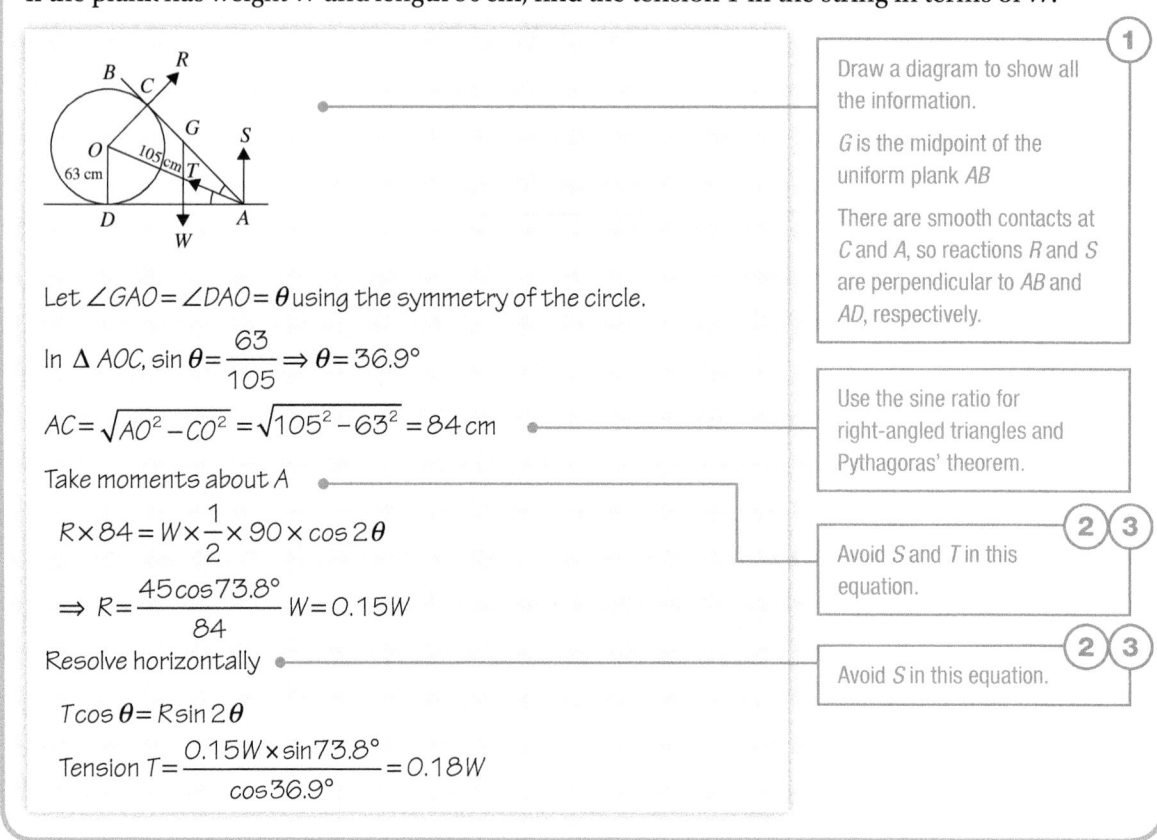

Let $\angle GAO = \angle DAO = \theta$ using the symmetry of the circle.

In $\triangle AOC$, $\sin \theta = \dfrac{63}{105} \Rightarrow \theta = 36.9°$

$AC = \sqrt{AO^2 - CO^2} = \sqrt{105^2 - 63^2} = 84$ cm

Take moments about A

$R \times 84 = W \times \dfrac{1}{2} \times 90 \times \cos 2\theta$

$\Rightarrow R = \dfrac{45 \cos 73.8°}{84} W = 0.15W$

Resolve horizontally

$T \cos \theta = R \sin 2\theta$

Tension $T = \dfrac{0.15W \times \sin 73.8°}{\cos 36.9°} = 0.18W$

Draw a diagram to show all the information.

G is the midpoint of the uniform plank AB

There are smooth contacts at C and A, so reactions R and S are perpendicular to AB and AD, respectively.

Use the sine ratio for right-angled triangles and Pythagoras' theorem.

Avoid S and T in this equation.

Avoid S in this equation.

A third equation (from resolving vertically) was not needed as you had to find only the tension T.

Example 4

A uniform rod AB of weight 10 N and length $2a$ is hinged at A Its end B is held by a string so that the rod and string are both inclined at 30° to the horizontal. Find the tension in the string and the total reaction at the hinge.

Resolve horizontally

$X = T \cos 30° \qquad \Rightarrow \qquad 2X = \sqrt{3}\, T \qquad$ [1]

Resolve vertically

$Y + T \sin 30° = 10 \quad \Rightarrow \quad 2Y + T = 20 \qquad$ [2]

(*Continued on the next page*)

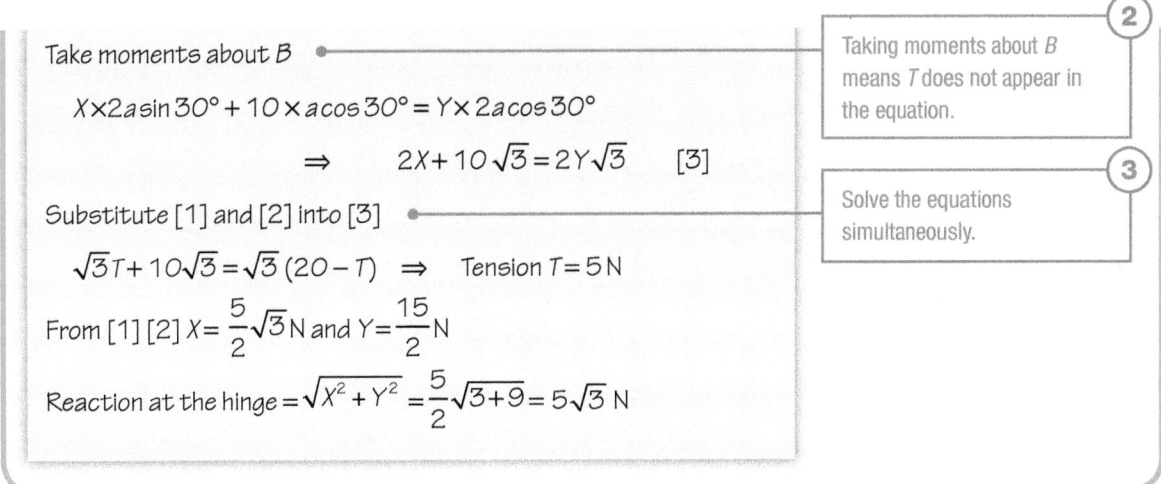

Take moments about B

$X \times 2a \sin 30° + 10 \times a \cos 30° = Y \times 2a \cos 30°$

$$\Rightarrow \quad 2X + 10\sqrt{3} = 2Y\sqrt{3} \quad [3]$$

Substitute [1] and [2] into [3]

$\sqrt{3}T + 10\sqrt{3} = \sqrt{3}(20 - T) \quad \Rightarrow \quad$ Tension $T = 5\,\text{N}$

From [1] [2] $X = \dfrac{5}{2}\sqrt{3}\,\text{N}$ and $Y = \dfrac{15}{2}\,\text{N}$

Reaction at the hinge $= \sqrt{X^2 + Y^2} = \dfrac{5}{2}\sqrt{3+9} = 5\sqrt{3}\,\text{N}$

> **2**
> Taking moments about B means T does not appear in the equation.

> **3**
> Solve the equations simultaneously.

The algebra could be reduced by taking moments about A with T resolved into two components. This means that X and Y do not appear in the equation.

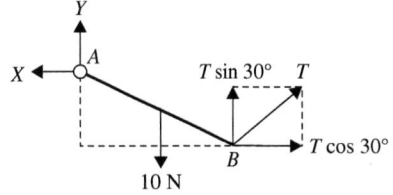

$$10 \times a \cos 30° = T \sin 30° \times 2a \cos 30° + T \cos 30° \times 2a \sin 30°$$

$$\Rightarrow T = 5\,\text{N}$$

Using the vector product in mechanics

The vector product of two vectors **a** and **b** is given by $\mathbf{a} \times \mathbf{b} = |\mathbf{a}||\mathbf{b}|\sin\theta\,\hat{\mathbf{n}}$ and can be used to find the moment about the origin O of a force **F** acting at point R with position vector **r**

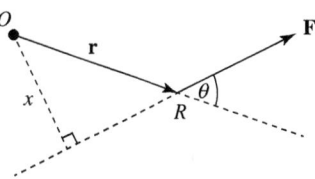

Let x be the shortest distance of O from the line of action of **F** and θ be the acute angle between the directions of **r** and **F**

The moment of **F** about O has a magnitude of $F \times x = F \times r \sin\theta = |\mathbf{r}||\mathbf{F}|\sin\theta$

Using the definition of a vector product and the right-hand rule, you can express this moment as a vector **M** in the direction of the axis of rotation where $\mathbf{M} = \mathbf{r} \times \mathbf{F}$

> The moment about O is actually the moment about an axis through O perpendicular to the plane containing O, R and **F**

> **Key point**
> In general, a force **F** acting at point P has a moment **M** about point Q given by $\mathbf{M} = \mathbf{r} \times \mathbf{F}$, where $\mathbf{r} = \mathbf{QP}$

Example 5

Two forces $\mathbf{F}_1 = 2\mathbf{i} + 4\mathbf{j} - \mathbf{k}\,\text{N}$ and $\mathbf{F}_2 = 3\mathbf{i} - 3\mathbf{j} - 2\mathbf{k}\,\text{N}$ act at points $P(2, -2, 3)$ and $Q(0, 4, -3)$ respectively. Their resultant forms a couple with a third force $\mathbf{F}_3$ which acts at the origin O

Find

a The force $\mathbf{F}_3$ and its magnitude,

b The moment **M** of the couple and its magnitude.

(*Continued on the next page*)

a $F_1 + F_2 = (2i + 4j - k) + (3i - 3j - 2k) = 5i + j - 3k$ N

As the overall resultant is a couple and not a single force,

Force $F_3 = -(F_1 + F_2) = -5i - j + 3k$

Magnitude of $F_3 = \sqrt{(25 + 1 + 9)} = \sqrt{35}$ N

b Moment **M** of couple = Total moment of all three forces about any point

Taking moments about O

Moment $M = (p \times F_1) + (q \times F_2) + (O \times F_3)$

$$\text{Moment } M = \begin{vmatrix} i & j & k \\ 2 & -2 & 3 \\ 2 & 4 & -1 \end{vmatrix} + \begin{vmatrix} i & j & k \\ 0 & 4 & -3 \\ 3 & -3 & -2 \end{vmatrix} + 0 = -27i + 17j + 24k$$

Magnitude of the moment of the
couple $= |M| = \sqrt{(-27)^2 + 17^2 + 24^2} = 39.9$ Nm

Exercise 25.5B Reasoning and problem-solving

1 A uniform rectangular lamina $ABCD$ stands in a vertical plane with AB in contact with a rough plane inclined at an angle θ to the horizontal. Sides AD and BC are perpendicular to the slope with AD on the downside of BC. The rectangle does not slide. If $AD = 10$ cm and $CD = 5$ cm, find the value of θ when the lamina is on the point of toppling.

2 A lamina has the shape of an isosceles triangle ABC with $AC = BC = 12$ cm and $AB = 6$ cm. It stands in a vertical plane with AB in contact with a rough plane inclined at angle θ to the horizontal. Given that the lamina does not slide, find the greatest value of θ for it not to topple.

3 A triangular lamina PQR has $PR = 12$ cm, $PQ = 6$ cm and $\angle RPQ = 90°$. It rests in a vertical plane on a rough slope inclined at angle θ to the horizontal with PQ in contact with the slope and P lower than Q. Given that the lamina does not slide, find the minimum value of θ for it to topple.

Also find the least value of the coefficient of friction, μ, for the lamina not to slide when it is about to topple.

4 A cone of height h and radius r rests in equilibrium with its plane face on a rough slope which makes an angle θ with the horizontal.

   a Calculate the maximum possible value of θ before the cone topples, without sliding, if

      i The cone is solid,

      ii The cone is hollow.

   b What must be the least value of the coefficient of friction, μ, between the hollow cone and the slope for the cone to topple before it slides?

5 **a** A uniform L-shaped lamina is made by removing the square $OCDE$ of side 2 cm from the square $ABDF$. If $AB = 4$ m, find the position of the centre of mass, G, of the lamina from AB and AF

b The lamina stands at rest with AB on a rough inclined plane. What is the greatest possible angle of inclination of the plane if the lamina is not to topple over for each of the two positions shown?

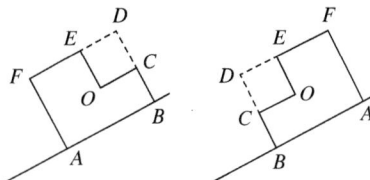

6 a A uniform horizontal beam PQ with mass 8 kg and length 2 m acts as a cantilever. It is clamped at P and has a 10 kg mass hanging from Q. Find the force and couple which acts at P to maintain the beam in equilibrium.

b The same beam and 10 kg mass are supported additionally by a sloping string QR fixed to a point R which is 2 m vertically above P. The string breaks if the tension in it exceeds 100 N. When the tension in the string is at its maximum, find the force and couple acting at P to keep the beam in equilibrium.

7 A light, triangular lamina ABC has $AB = 3$ cm, $BC = 4$ cm and $CA = 5$ cm. Forces of 15 N, 20 N and 25 N act along the sides AB, BC and CA, respectively. What force and couple, if any, must act at A to keep the lamina in equilibrium?

8 A uniform ladder of length $2a$ and weight W stands on a smooth floor against a smooth wall so that it is inclined at an angle α to the floor. A weight w is placed on the ladder at a distance x from the lower end. The ladder is held in equilibrium by a couple of moment M. Prove that $M = (Wa + wx)\cos\alpha$

9 A uniform rod PQ of mass 10 kg and length 4 m is hinged at P. A horizontal string attached to Q holds the rod at rest such that PQ makes 30° with the horizontal. Find the tension in the string if

a Q is higher than P

b Q is lower than P

10 A uniform rod AB of mass 8 kg and length 4 m rests at an angle θ to rough, level ground with A on the ground and B on a smooth peg 2 m above the ground. Find the total reaction at A and the least possible value of the coefficient of friction, μ, at A

11 a The area between the curve $y = f(x)$ and the x-axis for $0 \le x \le 2$ is rotated 360° about the x-axis to form a uniform solid. Find the exact coordinates of the centre of mass of the solid, when

 i $f(x) = \dfrac{1}{2}x^2$

 ii $f(x) = \sqrt{x}$

b Each solid is placed with its plane face on a rough plane inclined at an angle θ to the horizontal and it does not slide. Find, in each case, the exact value of θ when the solid is just about to topple.

12 a The area between the curve $y = \dfrac{12}{x}$ and the x-axis for $4 \le x \le 6$ is rotated through an angle of 2π about the x-axis to form a uniform solid. Find the exact coordinates of the centre of mass of the solid.

b The solid is placed with its smaller plane surface in contact with a rough plane that is inclined at an angle θ to the horizontal and has coefficient of friction, μ. As θ increases, find the value of θ at which it topples, but does not slide.

13 A uniform beam FH, of length 4 m and weight 50 N, is hinged to a vertical wall at H It is held at rest by a horizontal rope attached to the beam at F so that the beam slopes down at 20° from H to F. Find the tension in the rope and the magnitude and direction of the reaction at the hinge.

14 A horizontal light rod PQ of length 6 m is held in equilibrium by two strings PR and QS which slope upwards away from the rod such that $\angle QPR = 120°$ and $\angle PQS = 135°$. A weight of 10 N hangs from a point on the rod x m from P. Calculate the tension in each string and the value of x

15 A smooth cylinder, radius a, is fixed to rough, level ground at point C with its axis horizontal. A uniform rod AB of weight W rests at right angles to the axis with A on the ground and with B on the cylinder's surface such that $\angle BAC = 60°$. If the rod is about to slide, find the exact value of μ

16 A uniform ladder PQ of weight W and length $2a$ leans at an angle of $60°$ against a rough, vertical wall with its lower end Q on a rough level floor. The coefficient of friction at both ends is 0.8

What is the maximum weight of a person, in terms of W, who can climb to the top of the ladder without the ladder slipping?

17 A heavy, horizontal, uniform rod CD of mass 75 kg and length 90 cm is in contact with a rough vertical wall at C. A string DE holds the rod in limiting equilibrium, where E is on the wall 72 cm above C

A mass 100 kg hangs freely from D. Calculate the tension in the string, the thrust in the beam and the coefficient of friction at C

18 A uniform, square lamina $PQRS$ of weight W has its plane at right angles to a rough, vertical wall with P in contact with the wall and side PQ sloping upwards from P to Q at an angle θ to the vertical. It is held in equilibrium by a horizontal force W acting at R. Show that the coefficient of friction must be at least 1. Find the value of θ

19 A toy is made from two solid pieces with the same density ρ: a hemisphere and a cone with the same radius r. Their circular faces are joined together and have a common centre O. The toy stands with the hemisphere on a rough level surface.

a Show that the toy's centre of mass, G is a distance $\bar{x} = \dfrac{h^2 - 3r^2}{8r + 4h}$ from O, where h is the height of the cone.

b The toy stands with its axis at an angle to the vertical and with its hemisphere on a rough, level surface. By taking moments about its point of contact with the

surface, explain what happens after it is released if G

i Lies within the hemisphere,

ii Lies within the cone,

iii Coincides with O

20 A uniform footbridge AB of weight 400 N and length 4 m is rigidly fixed at end A and hinged at end B. When a person of weight 800 N stands on it 1 m from B, the vertical force on the bridge at B is three times that at A. Find the magnitude and sense of the couple which acts at A to keep the bridge in equilibrium.

21 A uniform rod AB of weight W and length 4 m is acted on by two forces T and U as shown. If the three forces are equivalent to a couple when $U = \sqrt{2}$ N, find the rod's weight and the magnitude and sense of the couple.

22 A light rectangular lamina has vertices $A(6, 0)$, $B(6, 5)$, $C(0, 5)$ and the origin O, with distances in m. Three forces $\mathbf{F}$, $\mathbf{G}$ and $\mathbf{H}$ act on the rectangle where $\mathbf{F} = 10\mathbf{i}$ N at the point $(2, 1)$, $\mathbf{G} = -3\mathbf{j}$ N at $(5, 4)$, $\mathbf{H} = 2\mathbf{i} + 8\mathbf{j}$ N at $(0, 3)$. Show that this system is equivalent to

a A single force $\mathbf{R}$ acting along the line $12y - 5x = 31$. Find the magnitude of $\mathbf{R}$

b A force $\mathbf{S}$ acting at the origin and a couple $\mathbf{C}$. Find the magnitudes of $\mathbf{S}$ and $\mathbf{C}$

23 Two equally rough planes are both inclined at $20°$ to the horizontal. Their lower edges meet in a horizontal line to form a V-shape in which an oil drum, of weight 800 N and radius 25 cm, rests symmetrically in equilibrium with its axis horizontal. If the coefficient of friction between the drum and planes is 0.5, what is the maximum external couple which can act on the drum in a vertical plane at right-angles to its axis if it is to stay in equilibrium?

24 A force $\mathbf{F} = 3\mathbf{i} - \mathbf{j} + 2\mathbf{k}$ newtons acts on an object at the point R(4, 3, −1). Find the magnitude of the moment of $\mathbf{F}$ about

a The origin,

b The point P(2, −2, 0).

25 Three forces $(4n\mathbf{i} + 2\mathbf{j} + 4\mathbf{k})$N, $(2n\mathbf{i} - 4\mathbf{j} + 6\mathbf{k})$N and $(4\mathbf{i} - 2n\mathbf{j} + 4n\mathbf{k})$N act on a light object at the points A(n, 0.5, 0), B(0.5, 1, 0) and C(−1, −0.5, 1) respectively, where n is a constant.

a Show that the total moment of the three forces about the origin O is independent of n

b Find the magnitude of the moment of the force $(4n\mathbf{i} + 2\mathbf{j} + 4\mathbf{k})$N about the point B when $n = 2$

26 Two forces $\mathbf{F}_1 = 2\mathbf{i} + \mathbf{j} - \mathbf{k}$ N and $\mathbf{F}_2 = \mathbf{i} - 2\mathbf{j} - 2\mathbf{k}$ N act at points P(1, −2, 3) and Q(0, 2, −3) respectively. The resultant of $\mathbf{F}_1$ and $\mathbf{F}_2$ forms a couple with a third force $\mathbf{F}_3$ which acts at the origin O

Find

a The resultant of $\mathbf{F}_1$ and $\mathbf{F}_2$

b The force $\mathbf{F}_3$ and its exact magnitude,

c The magnitude of the moment of the couple.

27 Three forces, in N, act in the x-y plane at the given points and they form a couple.

$\mathbf{F}_1 = \mathbf{i} + 2m\mathbf{j}$ at (3, 1)

$\mathbf{F}_2 = 4n\mathbf{i} + 11\mathbf{j}$ at (−2, −5)

$\mathbf{F}_3 = -2m\mathbf{i} + 8n\mathbf{j}$ at (4, −1)

a Find the constants m and n.

b Calculate the couple and its magnitude.

28 The force $\mathbf{F} = 2\mathbf{i} + 5\mathbf{j} - \mathbf{k}$ newtons acts at the point A(3, −2, 1) and the force $\mathbf{G} = \mathbf{i} - \mathbf{j} - 2\mathbf{k}$ newtons acts at the point B(4, 2, −4). A third force $\mathbf{H}$ acts at the origin and forms a couple with the resultant of $\mathbf{F}$ and $\mathbf{G}$. Find the exact values of

a The force $\mathbf{H}$ and its magnitude,

b The moment of the couple and its magnitude.

29 Three forces $\mathbf{F}$, $\mathbf{G}$ and $\mathbf{H}$, form a system of forces which is equivalent to a single force $\mathbf{Z}$ acting at the origin together with a couple of moment 96 N m. If $\mathbf{F} = 4\mathbf{i} + 2\mathbf{j}$, $\mathbf{G} = 10\mathbf{i} - 8\mathbf{j}$ and $\mathbf{H} = n\mathbf{j}$ in newtons act at the points (6, 8), (−4, −6) and (4, 0) respectively, find

a Two possible values of n

b The force $\mathbf{Z}$ in each case.

30 A rigid light rod AB has a force $\mathbf{F} = -\mathbf{i} + 3\mathbf{j} + 4\mathbf{k}$ N acting at its midpoint M

Given that A and B are at the points (−10, 8, 2) and (−2, −4, 6) respectively, find

a The vector $\mathbf{AM}$

b The moment of $\mathbf{F}$ about A and its exact magnitude,

c The acute angle between the direction of $\mathbf{F}$ and the rod by using

  i A vector product,

  ii A scalar product.

Chapter summary

- The moment of a force F about a point A is $F \times d$, where d is the perpendicular distance of F from A. Moments are measured in N m. They can be positive (anticlockwise) or negative (clockwise).
- The resultant moment of several forces acting on an object equals the sum of the moments of the individual forces, taking into account their directions.
- A couple comprises two equal and opposite forces, F, which are not acting in the same straight line. The moment of a couple, about any point, equals $F \times d$
- When several masses m_1, m_2, m_3,... at the points (x_1, y_1), (x_2, y_2), (x_3, y_3),... have their centre of mass G at the point $(\bar{x}, \bar{y})$, then
  - Their total mass $M = \sum\limits_{i=1}^{n} m_i$
  - Taking moments about the axes, $M \times \bar{x} = \sum\limits_{i=1}^{n} m_i \times x_i$ and $M \times \bar{y} = \sum\limits_{i=1}^{n} m_i \times y_i$
- You can use standard formulae to find the centre of mass of some laminas and solid bodies.
- When a uniform lamina with area density ρ has an area bounded by the curve $y = f(x)$, the x-axis and the lines $x = a$ and $x = b$, it has a centre of mass $G(\bar{x}, \bar{y})$, where
$$M = \rho \int_a^b y\,dx \,, \; M \times \bar{x} = \rho \int_a^b xy\,dx \text{ and } M \times \bar{y} = \rho \int_a^b \frac{1}{2}y^2\,dx$$
- When a uniform solid is formed by rotating the region between the curve $y = f(x)$, the x-axis and the lines $x = a$ and $x = b$ through $360°$ about the x-axis, it has a centre of mass $G(\bar{x}, \bar{y})$, where
$$\bar{y} = 0, \; M = \rho \int_a^b \pi y^2\,dx \text{ and } M \times \bar{x} = \rho \int_a^b x \times \pi y^2\,dx$$
- For an object to be in equilibrium under a system of coplanar forces
  - The sum of the components of all forces must be zero in any *two different* directions
  - The total moment of all forces and couples must be zero.
- When a plane shape is suspended from a point or can rotate about a horizontal axis, its centre of mass must be vertically below the point or the axis for it to be in equilibrium.
- An object on an inclined plane is in equilibrium without toppling if the line of action of its weight falls within the side of the object that makes contact with the plane.
- In general, a force F acting at point P has a moment M about point Q given by $M = \mathbf{r} \times \mathbf{F}$, where $\mathbf{r} = \mathbf{QP}$

Check and review

You should now be able to...	Review Questions
✓ Find the centres of mass for systems of point masses, frameworks, laminas and solids of revolution.	1–4, 6
✓ Explore the stability of suspended objects and objects on inclined planes.	2, 4–7
✓ Use the conditions required for an object to be in equilibrium.	8–10
✓ Find forces and couples acting on a system	11–13

1 Masses of 2 kg, 4 kg, 6 kg and 8 kg are placed respectively at the vertices of a light, rigid rectangular framework $ABCD$, where $AB = 5$ m and $BC = 3$ m. Find the position of the centre of mass of the system from AB and AD

2 A uniform rectangular lamina $ABCD$ with centre O, $AB = 12$ cm and $AD = 8$ cm has the rectangle $OMCN$ cut away, where M and N are midpoints of BC and CD. Find the centre of mass, G, of the remaining piece and the angle that AB makes with the vertical when the piece is hung by a string from A

3 A uniform circular lamina, centre O and radius 12 cm, has a triangular hole OMN cut out of it, where M and N are midpoints of two radii with $\angle MON = 120°$. Find the position of the centre of mass of the remainder.

4 Find the centre of mass of a uniform lamina defined by the region between the curve $y^2 = x$ and the line $x = 4$. The lamina hangs freely by a thread from the point $(4, 2)$. Find the angle between the thread and the x-axis.

5 A solid hemisphere of radius 24 cm is placed with its plane face on a rough slope inclined at an angle θ to the horizontal. It does not slide. Find the greatest value of θ for the hemisphere not to topple.

6 A graph is defined by

$y = x^2$ for $0 \leq x \leq 1$,
$y = 1$ for $1 \leq x \leq 3$

and $y = 0$ for all other values of x

The region between the graph and the x-axis is rotated through $360°$

a Find the centre of mass of the solid generated.

b The solid is placed without sliding with its plane surface on a slope inclined at angle θ to the horizontal. Find the maximum value of θ for it not to topple.

7 A solid cone of height h, radius r and semi-vertical angle θ is cut to make a frustum by a plane parallel to its base at a distance of $\frac{1}{2}h$ from its vertex. The frustum is placed on its curved face on a horizontal plane. Prove that it will not topple if $\cos \theta \geq \sqrt{\dfrac{28}{45}}$

8 A uniform ladder of length 8 m and mass 24 kg rests on rough, level ground, where $\mu = 0.3$, against a smooth, vertical wall. The angle, θ, between the ladder and the ground is given by arctan 2. How far up the ladder can an object of mass 24 kg be placed without causing the ladder to slip?

9 A uniform rod AB of length 4 m and mass 10 kg is in equilibrium at $60°$ to the horizontal with its lower end A on a rough, level floor where the coefficient of friction, μ, is 0.5. The rod rests against a smooth peg P which moves along the length of AB so that $AP = 2 + x$ metres. Calculate the value of x for the rod to be on the point of moving.

10 A semicircular prism of radius r has its plane face fixed to a level floor. A uniform rod PQ of weight W and length $2r$ rests against the curved surface of the prism so that end P is on the floor and end Q extends beyond the point of contact with the prism. Both contacts are rough with the same coefficient of friction, μ. Prove that, if the rod is on the point of moving when inclined at an angle θ,
$\sin^2 \theta = \dfrac{\mu}{\mu^2 + 1}$

11 A uniform horizontal cantilever AB of mass 20 kg and length 10 m is in equilibrium rigidly fixed to a wall at A. End B is on a support with a vertical reaction of 30 N. Find the magnitude and direction of the force and the couple acting at A

12 A light, equilateral triangular lamina OAB lies in the first quadrant with $A(5, 0)$ and O the origin. Forces of 6 N act along OA, 10 N along AB and 8N along BO. This system is equivalent to

a A single force R. Find the equation of its line of action.

b A force S acting at O and a couple C. Find the moment of the couple.

13 A light rod OA of length 6 m lies along the x-axis. Force $\mathbf{T} = a\mathbf{i} + b\mathbf{j}$ acts at the origin O at an angle of $30°$ with OA. Force $\mathbf{U} = -c\mathbf{j}$ acts at the point $(4, 0)$. Force $\mathbf{V}$ of magnitude 4 N acts at A at an angle of $60°$ with AO This system is equivalent to a single couple C. Find the values of a, b and c and the magnitude and sense of $\mathbf{C}$

Investigation

A castle drawbridge can be simplified and modelled as shown in the diagram.

Investigate how the tension in the rope varies as the drawbridge is raised from the horizontal.
Draw a graph that shows this.

Investigate how the tension in the rope varies as a person or vehicle crosses the drawbridge.
Again, draw a graph that shows this.

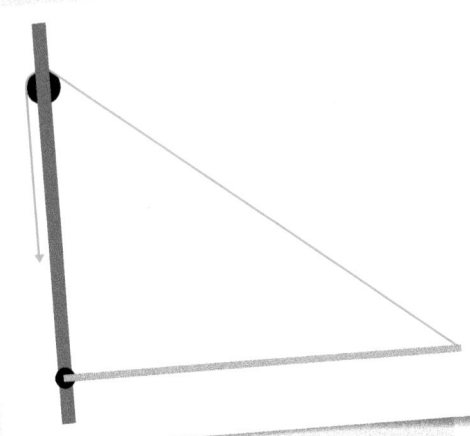

Applications

When designing buildings, an understanding of how forces act on, and within, frameworks is essential. Steel frames are often used when large buildings are built. For example, the Turning Torso skyscraper in Malmo, Sweden, has an external steel frame. It was designed by the famous Spanish architect Santiago Calatrava. The building is in nine segments made of irregular pentagons that twist, so that the top floor is rotated by 90° with respect to the ground floor. Calatrava has designed many iconic buildings around the world. Investigate some of these, and see his use of steel frames, which often add beauty to buildings as well as being an integral part of the structure.

Applications

Understanding the concept of centre of mass is important in field events in athletics, as well as in gymnastics. For example, the American high jumper Dick Fosbury developed the 'Fosbury flop'. In this technique, the athlete jumps over the bar with an arched back, passing as close as possible to the bar. Using this technique, the jumper's centre of mass passes under the bar by up to 20 centimetres.

Find out more about the technique and consider the advantages of a jump where the athlete's centre of mass is under the bar.

In what other sports is the centre of mass of a person important?

1   Particles of mass 1 kg, 2 kg, 3 kg and 4 kg are attached in that order to a rod AB of length 1.5 m at distances of 0.3 m, 0.6 m, 0.9 m and 1.2 m from A

    a   Assuming that the rod is of negligible mass, find the distance from A of the centre of mass of the system.   **[3]**

    b   In fact, the rod is uniform and of mass m kg. The centre of mass of the system is 0.85 m from A Find the value of m **[3]**

2   ABC is a triangle formed of three uniform rods. AB has length 4 m and mass 4 kg, AC has length 3 m and mass 2 kg and BC has length 5 m and mass 4 kg. Find the distance of the centre of mass of the triangle from

    a AB **[3]**

    b AC **[3]**

3   Masses of 2 kg, 4 kg, 6 kg and 9 kg are placed, respectively, at the vertices A, B, C and D of a light rectangular framework $ABCD$, where $AB = 5$ cm and $BC = 3$ m. Find the angle that AB makes with the vertical when the framework is suspended from A **[6]**

4   A uniform lamina consists of the region enclosed by $y = \sqrt{x}$, the x-axis and the lines $x = 1$ and $x = 4$

    a   Find the coordinates of the centre of mass of the lamina.   **[8]**

    b   Find the coordinates of the uniform solid formed by rotating the region about the x-axis.   **[5]**

5   A bridge on a model railway is a prism with a symmetrical cross-section in the form of a rectangle with a semicircular arch cut out of it. The whole structure is 28 cm high and 36 cm wide, and the radius of the arch is 10 cm.
Assuming that the bridge is made of a uniformly dense material, find the height of its centre of mass above the base.   **[5]**

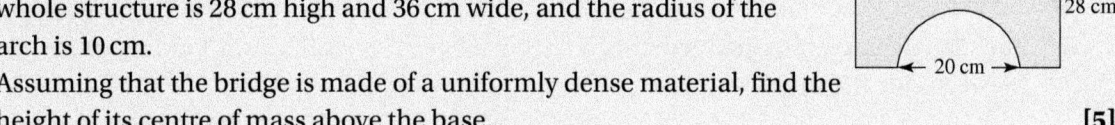

6   A uniform lamina is in the form of the minor segment of a circle of radius 60 cm cut off by a chord AB of length 60 cm. Find the distance of the centre of mass of the segment from the centre of the circle.   **[8]**

7   Find the centre of mass of this uniform lamina relative to the origin O and the axes shown.   **[6]**

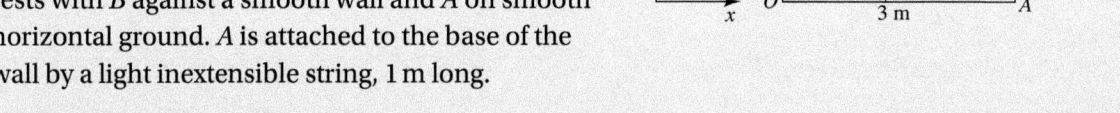

8   A uniform ladder AB of mass 20 kg and length 3 m rests with B against a smooth wall and A on smooth horizontal ground. A is attached to the base of the wall by a light inextensible string, 1 m long.

    a   Find the tension in the string.   **[4]**

    b   If the breaking strain of the string is 250 N, find how far up the ladder a man of mass 80 kg can safely ascend.   **[5]**

9 A uniform wire AB is bent to form the arc of a semicircle. The wire is then suspended freely from A. Find the angle between the diameter AB and the vertical. **[4]**

10 Masses of 5 kg, 4 kg, 2 kg and 3 kg are placed at A, B, C and D, respectively, on a light rectangular framework $ABCD$, where $AB = 4$ m and $BC = 3$ m. A mass m kg is placed at a point E on CD so that the centre of mass of the system is the centre of the rectangle. Find m and the position of E **[7]**

11 The diagram shows a carpenter's square, with a rectangular handle $PQUV$, of length 0.2 m and width 0.1 m, and a rectangular blade $RSTU$ of length 0.3 m and width 0.1 m. The blade is made from metal of density 1 kg m^{-2}, and the handle from material of density m kg m^{-2}.

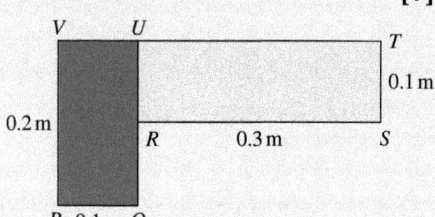

a Find the value of m if the centre of mass of the object lies on the line QU **[3]**

b For this value of m, find the distance of the centre of mass from PQ **[2]**

12 The triangular lamina with vertices at $(0, 0)$, $(h, 0)$ and (h, r) is rotated about the x-axis. Show, by integration, that the centre of mass of the resulting uniform cone is at $(0.75h, 0)$ **[6]**

13 Particles of mass 3 kg, 5 kg, m_1 kg and m_2 kg are placed, respectively, at points $(1, 4)$, $(4, 1)$, $(2, 1)$ and $(4, 2)$. Find the values of m_1 and m_2 if the centre of mass of the system is $(3, 2)$ **[8]**

14 $ABCD$ is a uniform rectangular lamina with $AB = 60$ cm and $BC = 30$ cm. E is the midpoint of CD The triangle BCE is removed from the lamina, and the remainder is suspended from E. Find the angle that the direction of AD makes with the vertical when the lamina hangs in equilibrium. **[7]**

15 A uniform rod of length 3 m is bent to form triangle ABC, with angle $BAC = 90°$, $AB = 1.2$ m and $AC = 0.5$ m.

a Taking AB and AC as the x- and y-axes, find the coordinates of the centre of mass of the triangle. **[4]**

b The triangle is freely suspended from B so that it hangs in equilibrium. Find the angle between side AB and the vertical. **[3]**

16 A uniform lamina consists of the region enclosed by the curve $y = e^x$, the axes and the line $x = 2$

a Find the exact coordinates of the centre of mass of the lamina. **[8]**

b Find the exact centre of mass of the uniform solid produced by rotating the region about the x-axis **[5]**

17 The diagram shows a rod AB of length 2 m and mass 4 kg. The rod is freely hinged at A to a point on a vertical wall and makes an angle of 50° with the wall. The rod is held in equilibrium by a horizontal force of TN applied at B

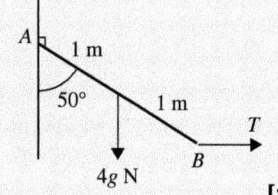

a Calculate the value of T. **[3]**

b Find the magnitude and direction of the reaction at the hinge. **[7]**

18 A uniform ladder AB has length 8 m and mass 20 kg. It leans against a rough wall with end A on rough, horizontal ground. The coefficient of friction at both contacts is 0.3 and the ladder makes an angle of 65° with the horizontal. A woman of mass 50 kg starts to climb the ladder.

a How far can she ascend up the ladder before the ladder is on the point of slipping? [7]

b Her daughter, of mass m kg, then stands on the ladder at A. This enables the woman to climb right to the top of the ladder. Find the least value of m [6]

19 A uniform rod, AB, of length 1.6 m and mass 6 kg, is freely hinged at A to a vertical wall. The rod rests in a horizontal position, supported by a light, inextensible string BC attached to the wall at the point C, which is 1.2 m vertically above A. The string has a breaking strain of 40 N, and equilibrium is maintained by a couple of moment P N m applied at A. Take $g = 9.8$ m s^{-2}.

a Find the greatest and least possible values of P. [6]

b Find the magnitude and direction of the reaction force at the hinge in each case. [8]

20 A uniform right circular cone has height 1 m, base radius 0.6 m and density 1 kg m^{-3}. A uniform hemisphere has radius 0.6 m and density ρ kg m^{-3}. The two are joined together at their circular plane faces to form a symmetrical solid.

a Find the position of the centre of mass of the solid relative to O, the centre of the common plane face. Give your answer in terms of ρ [7]

b The solid is placed on a horizontal plane surface. The curved surface of the hemisphere is in contact with the horizontal plane and the solid remains in equilibrium whatever the angle between its axis and the vertical. Find the value of ρ [4]

21 The diagram shows a lamina in the form of a sector of a circle, centre O, of radius 6 cm. The angle of the sector is 60°. A uniform rod is bent so that it attaches exactly to the arc of the sector, as shown. The mass of the lamina is twice the mass of the rod. Show that the centre of mass of the combined object is $\dfrac{14}{\pi}$ cm from O [7]

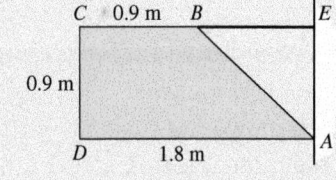

22 The diagram shows a sign outside a shop. It is made of a uniform lamina $ABCD$ in the shape of a square attached to a triangle. The mass of the lamina is 6 kg. The sign is mounted in a vertical plane with AD horizontal. It is smoothly hinged to the wall at A and is held in equilibrium by a horizontal string BE attached to the wall at E

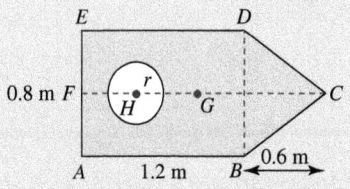

a Calculate the tension in the string. [5]

b Find the magnitude and direction of the reaction at the hinge. [5]

23 The diagram shows a lamina comprising a rectangle $ABDE$ of length 1.2 m and width 0.8 m together with an isosceles triangle BCD of height 0.6 m. F is the midpoint of AE. H is a point on CF, with $FH = 0.4$ m. A hole, centre H and radius r cm, is cut in the lamina. The centre of mass of the object is at G

a Explain why G lies on CF [1]

b If $r = 0.2$, find the distance FG [4]

c If $FG = 1$ m, find the value of r [3]

d Explain why FG cannot be 1.1 m. [2]

24 A uniform ladder AB of length $2a$ and weight W rests with B against a smooth, vertical wall and A on rough, horizontal ground. The ladder makes an angle of $60°$ with the horizontal. A child of weight W can just climb to the top of the ladder without causing it to slip. Find how far up the ladder a man of weight $4W$ could safely climb. **[7]**

25 The diagram shows the cross-section of a prism consisting of a rectangle $ABCD$ of length 50 cm and width 30 cm from which a triangle BCE has been removed so that DE has length x cm. Find the minimum value of x for which the prism will stand on a horizontal surface with DE in contact with the surface. **[8]**

26 A uniform lamina comprises the region in the first quadrant enclosed by the ellipse $\dfrac{x^2}{a^2} + \dfrac{y^2}{b^2} = 1$
The area of the ellipse is πab

a Find the coordinates of the centre of mass of the lamina. **[8]**

b Find the centre of mass of the uniform solid formed by rotating the region about the x-axis. **[5]**

27 The cross-section of a uniform prism is an isosceles trapezium of height 6 cm with parallel sides of length 4 cm and 10 cm. The prism is placed on a rough inclined plane, as shown. If the prism does not slip, but is on the point of toppling, find the angle of the sloping plane. **[7]**

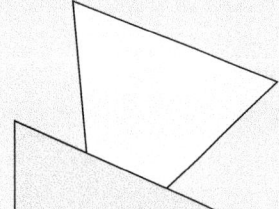

28 A uniform rod of length 1 m and mass 1 kg is bent to form the framework shown, where angle $B =$ angle $C = 90°$. A particle of mass m kg is attached to the framework at D and the framework is suspended from A. If the point C hangs vertically below A, find the value of m **[9]**

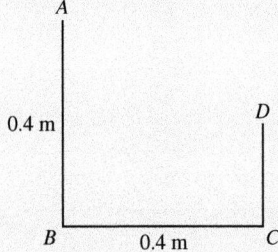

29 Two forces $\mathbf{F}_1 = 3\mathbf{i} + 2\mathbf{j} - \mathbf{k}$ N and $\mathbf{F}_2 = 2\mathbf{i} - \mathbf{j} - 2\mathbf{k}$ N act at points A(1, 0, 3) and B(0, 2, −3) respectively. Their resultant forms a couple $\mathbf{M}$ with a third force $\mathbf{F}_3$ which acts at the origin O. Find the force $\mathbf{F}_3$, the couple $\mathbf{M}$ and their magnitudes. **[4]**

26 Random processes

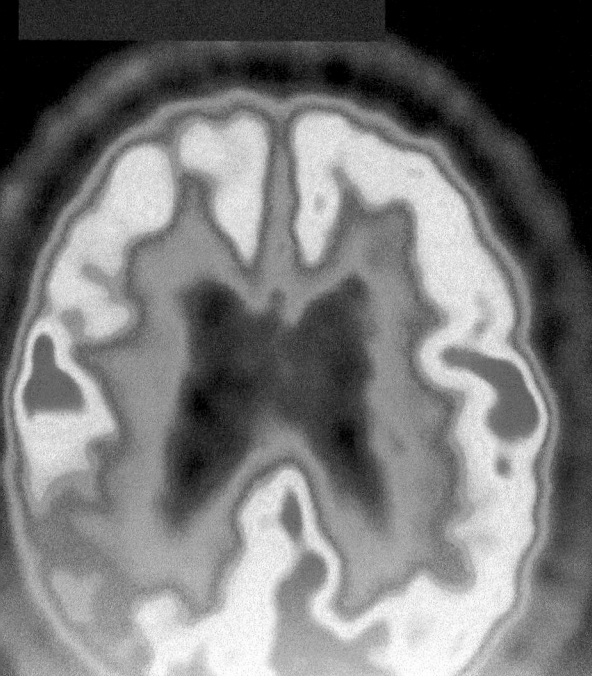

Radiation can help medical professionals investigate patients' bodies without resorting to invasive surgery. A radioactive tracer is ingested by a patient, and a radiographer detects the beta particles and gamma rays emitted by the patient's body. Due to the potentially harmful nature of the radiation, the amount used must be measured carefully to ensure that it will decay to a certain level after a short period of time. An exponential distribution provides a good model for radioactive disintegrations, and allows medical professionals to estimate the amount of radiation at any point in the process.

Continuous probability distributions are used to model a wide variety of continuous random variables, from waiting times on a public transport network to the frequency of flaws in a length of fibre-optic cable.

Orientation

What you need to know

Maths Ch15
- Differentiating exponential functions.

Maths Ch16
- Integrating exponential functions.

Ch10 Discrete and continuous random variables

What you will learn

- To calculate the cumulative distribution function and probability density function for a continuous random variable.
- To calculate the mean and variance of mixed random variables.
- To use rectangular and exponential probability models.

What this leads to

Careers
- Medicine.
- Quality assurance.
- Insurance.

Fluency and skills

See Ch10.3

For a reminder of continuous distributions.

If X is a continuous random variable with probability density function $f(x)$, then

$$P(a < X < b) = \int_a^b f(x)\,dx$$

The mean and variance of X are $\mu = E(X) = \int_{-\infty}^{+\infty} x f(x)\,dx$ and $\sigma^2 = \text{Var}[X] = E[X^2] - \mu^2 = \int_{-\infty}^{\infty} x^2 f(x)\,dx - \mu^2$

The cumulative distribution function, $F(x)$, of a random variable is useful when calculating probabilities as well as values of the median and other quartiles and percentiles. It is defined by the equation $F(x) = P(X < x)$

But, for a continuous random variable, $P(a < X < b) = \int_{x=a}^{x=b} f(x)\,dx$

Therefore

> **Key point**
>
> The cumulative distribution function of a continuous random variable with probability density function $f(t)$ is given by $F(x) = P(X < x) = \int_{-\infty}^{x} f(t)\,dt$

Example 1

A continuous random variable X has probability density function given by

$$f(x) = \begin{cases} 4x^a; & 0 < x < 1 \\ 0; & \text{otherwise} \end{cases}$$

Find:

a The value of a

b The cumulative distribution function $F(x)$

c The median value, M, of X

a $\int_0^1 4x^a \, dx = 1$ •   Total probability equals 1

$\left[\dfrac{4x^{a+1}}{a+1}\right]_0^1 = 1 \rightarrow \dfrac{4}{a+1} = 1$ Therefore $a = 3$

b $F(x) = \int_0^x 4x^3 \, dx = \left[x^4\right]_0^x = x^4$

c $\int_0^M 4x^3 \, dx = \dfrac{1}{2}$ •   Definition of the median.

$M^4 = \dfrac{1}{2} \rightarrow M = 0.841$ (3dp)

Using the definition of integration as the inverse of differentiation, the probability density function of X can be found from the cumulative distribution function.

If $F(x)$ and $f(x)$ are the cumulative distribution function and the probability density function of X, respectively, then

$$f(x) = \frac{d}{dx}F(x)$$

Example 2

The cumulative distribution function of a random variable X is given by $F(x) = 1 - e^{-\lambda x}$

a Find the probability density function of X

b Write the modal value of X

c Prove that, when two X values are chosen at random, the probability that one is less than x_1 and the other is more than x_1 is given by $2e^{-\lambda x_1}(1 - e^{-\lambda x_1})$

a $f(x) = \dfrac{d}{dx}F(x) = \dfrac{d}{dx}(1 - e^{-\lambda x}) = \lambda e^{-\lambda x}$ • ——— Use the chain rule for differentiation.

b Mode $= 0$ • ——— The maximum value of $f(x)$ for $x \geq 0$ occurs when $x = 0$. Visualise the graph.

c $P(X < x_1) = F(x_1) = 1 - e^{-\lambda x_1}$. Therefore $P(X > x_1) = e^{-\lambda x_1}$

P(one x-value is less than x_1 and the other is more than x_1)
$= 2e^{-\lambda x_1}(1 - e^{-\lambda x_1})$ • ——— The outcome can occur in two ways.

Example 3

A continuous random variable X has a probability density function given by

$$f(x) = \begin{cases} ax^2; & 0 < x \leq 1 \\ \dfrac{-9x + 15}{4}; & 1 < x < \dfrac{5}{3} \\ 0; & \text{otherwise} \end{cases}$$

a Show that $a = \dfrac{3}{2}$

b Find

i The cumulative distribution function $F(x)$

ii The median and the interquartile range of X

a $\displaystyle\int_1^{\frac{5}{3}} \frac{-9x + 15}{4}dx = \left[\frac{1}{4}\left(\frac{-9x^2}{2} + 15x\right)\right]_1^{\frac{5}{3}} = \frac{1}{2}$

$\displaystyle\int_0^1 ax^2 dx = \left[\frac{ax^3}{3}\right]_0^1 = \frac{a}{3} = 1 - \frac{1}{2} = \frac{1}{2} \rightarrow a = \frac{3}{2}$

(*Continued on the next page*)

b i $F(x) = \begin{cases} \displaystyle\int_{0}^{x}\frac{3x^2}{2}dx; & 0 < x \le 1 \\[2ex] \dfrac{1}{2}+\displaystyle\int_{1}^{x}\frac{-9x+15}{4}dx; & 1 < x \le \dfrac{5}{3} \\[2ex] 1; & x > \dfrac{5}{3} \end{cases}$

Therefore

$F(x) = \begin{cases} \dfrac{x^3}{2}; & 0 < x \le 1 \\[2ex] \dfrac{-9x^2}{8}+\dfrac{15x}{4}-\dfrac{17}{8}; & 1 < x \le \dfrac{5}{3} \\[2ex] 1; & x > \dfrac{5}{3} \end{cases}$

ii Median $= 1$

If Q_1 and Q_3 are the first and third quartile, respectively, then

$F(Q_1) = \dfrac{1}{4}$ and $F(Q_3) = \dfrac{3}{4} \rightarrow$

$\dfrac{Q_1^3}{2} = \dfrac{1}{4} \rightarrow Q_1 = \dfrac{1}{\sqrt[3]{2}} = 0.79 \,(2dp)$

and

$\dfrac{-9Q_3^2}{8}+\dfrac{15Q_3}{4}-\dfrac{17}{8} = \dfrac{3}{4} \rightarrow Q_3 = 1.20 \,(2dp)$

$IQR = 0.41$

The rectangular distribution models many real-life situations where, within the domain of the function, all small intervals of width Δx are equally likely to occur when an experiment is performed. It is therefore characterised by having a probability density function equal to a constant over its entire domain. The value of the constant is chosen so that the total area under the function is equal to 1

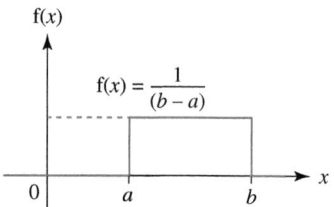

Key point

If R is a random variable with a rectangular probability distribution defined on the interval (a, b) (that is, $R \sim \text{Rectangular}(a,b)$), then its probability density function is given by $f(r) = \dfrac{1}{b-a}, a < r < b$

This probability density function can be used to find the mean and variance of the distribution.

$$E(R) = \int_{-\infty}^{+\infty} rf(r)dr = \int_{a}^{b} r\frac{1}{b-a}dr$$

$$= \frac{1}{b-a} \times \frac{b^2-a^2}{2} = \frac{a+b}{2}$$

$$\mathrm{Var}(R) = \int_{-\infty}^{+\infty} r^2 \, \mathrm{f}(r)\,\mathrm{d}r - \{\mathrm{E}(R)\}^2$$

$$= \int_{r=a}^{b} r^2 \frac{1}{b-a}\,\mathrm{d}r - \left(\frac{a+b}{2}\right)^2$$

$$= \frac{1}{b-a} \times \frac{b^3 - a^3}{3} - \left(\frac{a+b}{2}\right)^2$$

$$= \frac{1}{b-a} \times \frac{(b-a)(b^2 + ab + a^2)}{3} - \left(\frac{a+b}{2}\right)^2$$

$$= \frac{(b^2 + ab + a^2)}{3} - \left(\frac{a+b}{2}\right)^2$$

$$= \frac{4b^2 + 4ab + 4a^2 - 3a^2 - 6ab - 3b^2}{12}$$

$$= \frac{(b-a)^2}{12}$$

$$= \frac{(b-a)^2}{12}$$

Key point

If R has a rectangular probability distribution then the mean and variance of R are given by $\mathrm{E}(X) = \dfrac{a+b}{2}$ and $\mathrm{Var}(X) = \dfrac{(b-a)^2}{12}$

The rectangular distribution can be used to model 'round off' values as in Example 4

Exercise 26.1A Fluency and skills

1   A random variable, X, has a probability density function given by

$$\mathrm{f}(x) = \begin{cases} \dfrac{1}{4}x; & 1 < x < k \\[2mm] 0; & \text{otherwise} \end{cases}$$

where k is a positive constant.

   a   Prove that $k = 3$

   b   Show that the cumulative distribution function is given by $\dfrac{1}{8}(x^2 - 1)$

   c   Find the median value of X

2   A random variable, X, has probability density function $\mathrm{f}(x) = \begin{cases} ax^2(4x-1); & 0 < x < \dfrac{1}{4} \\[2mm] 0; & \text{otherwise} \end{cases}$

where a is a negative constant.

   a   Find the value of a and the cumulative distribution function $\mathrm{F}(x)$

   b   Use your answer in part **a** to show that the median of X is approximately 0.15

3   A random variable, X, has probability density function $\mathrm{f}(x) = \begin{cases} -kx(x-3); & 0 < x < 3 \\[2mm] 0; & \text{otherwise} \end{cases}$

where k is some positive constant.

   a   Find the value of k

   b   Find the cumulative distribution function of X and hence find $P(X > 2)$

4   **a**   Sketch the probability density function of X, given by

$$\mathrm{f}(x) = \begin{cases} 1 - |x|; & -1 < x < 1 \\[2mm] 0; & \text{otherwise} \end{cases}$$

b Show that $P(S) = 1$, where S is a sample space for X

c By finding the cumulative probability distribution $F(x)$, prove that the exact value of the upper quartile is given by

$$Q_3 = \frac{\sqrt{2}-1}{\sqrt{2}}$$

5 The random variable X has a cumulative probability distribution given by

$$F(x) = \frac{1}{k}(x^2 + x) \text{ for } 0 < x < 5$$

a Find the value of k

b Show that the lower quartile is 2.28 (2 dp) and find the upper quartile.

c Show that the probability density function of X is $f(x) = \frac{2x+1}{30}$ and hence write down the modal value of X

6 A random variable, X, has probability density function $f(x) = \begin{cases} \dfrac{1}{k}; & a < x < b \\ 0; & \text{otherwise} \end{cases}$

a Sketch the graph of the probability density function of X and show that $k = b - a$

b Find $E(X)$ and $Var(X)$ when $a = 3$ and $b = 5$

c Show that the cumulative probability distribution function of X is given by

$$F(x) = \frac{x-3}{2}$$

and hence find the probability that $P(X < 3.5)$

7 A random variable, X, has probability density function given by $f(x) = \begin{cases} \dfrac{1}{4}; & -1 < x < a \\ 0; & \text{otherwise} \end{cases}$

a Find the value of a

b Sketch the probability density function of X and shade the area representing $P(X > 1.5)$

c Use your diagram to write down the value of $P(X > 1.5)$

d Find the cumulative distribution function of X and use it to show that the median value of X is 1

8 A continuous random variable X has probability density function given by

$$f(x) = \begin{cases} x; & 0 < x \le 1 \\ 2-x; & 1 < x < 2 \\ 0; & \text{otherwise} \end{cases}$$

a Show that $f(x)$ is a valid probability density function.

b Find

i The cumulative probability distribution function $F(x)$

ii The median and the interquartile range of X

Reasoning and problem-solving

Strategy

To solve a problem using continuous distributions

1. Integrate the probability density function to find probabilities and the cumulative probability distribution function.

2. Recognise continuous variables that follow a rectangular distribution and identify values for the parameters a and b

3. Calculate the mean and variance of rectangularly distributed random variables.

4. Recognise random variables that are partly discrete and partly continuous and use appropriate formulae for mean and variance.

These ideas can be applied to the continuous rectangular distribution and to mixed distributions that are partly discrete and partly continuous.

Example 4

The length of a line segment is measured to the nearest millimetre. Calculate the mean and variance of the error in any recorded length.

X – a random variable for the error (mm) in any recorded length

$X \sim$ Rectangular$(-0.5, 0.5)$

Therefore mean is $\dfrac{0.5 + (-0.5)}{2} = 0$ and variance is $\dfrac{(0.5 - (-0.5))^2}{12} = \dfrac{1}{12}$

Defining the error as 'true' – recorded value.

Example 5

A circle of radius R is drawn, where R is a rectangularly distributed random variable defined on the interval 1 to 5

The random variable A is the area of the circle. Find the cumulative distribution function of A and its mean.

If f(r) is the probability density function of R, then $f(r) = \dfrac{1}{4}, 1 < x < 5$

and $F(r) = \dfrac{r-1}{4}$

$A = \pi R^2 \rightarrow R = \sqrt{\dfrac{A}{\pi}}$

Therefore $\Phi(a) = P(A < a) = P(\pi R^2 < a) = P\left(R < \sqrt{\dfrac{a}{\pi}}\right) = \dfrac{\sqrt{\dfrac{a}{\pi}} - 1}{4}$

$$= \dfrac{1}{4}\left(\sqrt{\dfrac{a}{\pi}} - 1\right)$$

$\varphi(a) = \dfrac{d}{da}\Phi(a) = \dfrac{d}{da}\dfrac{1}{4}\left(\sqrt{\dfrac{a}{\pi}} - 1\right) = \dfrac{1}{8\sqrt{a\pi}}$

$\Phi(a)$ and $\varphi(a)$ are the cumulative distribution function and probability density function of A, respectively.

$E(A) = \displaystyle\int_{\pi}^{25\pi} a\dfrac{1}{8\sqrt{a\pi}}\,da = \dfrac{1}{8}\int_{\pi}^{25\pi}\sqrt{\dfrac{a}{\pi}}\,da = \dfrac{1}{8\sqrt{\pi}}\int_{\pi}^{25\pi} a^{\frac{1}{2}}\,da = \left[\dfrac{1}{12\sqrt{\pi}}a^{\frac{3}{2}}\right]_{\pi}^{25\pi}$

$= \dfrac{\pi}{12}(25^{\frac{3}{2}} - 1) = \dfrac{31}{3}\pi$

Some random variables are partly discrete and partly continuous. For example, queuing times may take the value zero exactly (discrete) if there is no one else queuing, or they may be a set of positive values (continuous).

Example 6

Supermarket checkout waiting times, T minutes, are found to have the following distribution:
$P(T=0)=c$; $f(t)=kt(t-3)$, $0<t\leq 3$

a Prove that $k=\dfrac{2(c-1)}{9}$

b If the probability of zero waiting time is 0.4, show that the expected waiting time is 54 seconds and find the probability that the waiting time exceeds one minute.

a $k\displaystyle\int_0^3 t(t-3)\,dt = 1-c$ ●———— Total probability in a sample space $=1$

$\left[k\left(\dfrac{t^3}{3}-\dfrac{3t^2}{2}\right)\right]_0^3 = 1-c \rightarrow k\left(9-\dfrac{27}{2}\right)=1-c \rightarrow k=\dfrac{2(c-1)}{9}$

b $c=0.4 \rightarrow k=\dfrac{2(0.4-1)}{9}=-\dfrac{2}{15}$

$E(T)=0\times 0.4 - \dfrac{2}{15}\displaystyle\int_0^3 t^2(t-3)\,dt=\left[\dfrac{2}{15}\left(t^3-\dfrac{t^4}{4}\right)\right]_0^3 = \dfrac{9}{10}=54s$

$P(T>1) = -\dfrac{2}{15}\displaystyle\int_1^3 t(t-3)\,dt=\left[\dfrac{2}{15}\left(\dfrac{3t^2}{2}-\dfrac{t^3}{3}\right)\right]_1^3 = \dfrac{4}{9}$

Exercise 26.1B Reasoning and problem-solving

1 Buses arrive at a bus stop on the hour and the half hour throughout the day. Isaac has no idea when the buses are due and arrives at the stop to catch a bus.

    a If T is a random variable for the number of minutes since the last bus, use an appropriate distribution to calculate the mean and variance of T

    b Using these results, find the mean and variance of the variable, S, the number of minutes Isaac has to wait for the next bus.

2 The random variable Y is part discrete, part continuous. The probability density function of Y is given by

$$f(y)=\begin{cases} c; & y=0 \\ k(y-1)(y-3); & 1<y<3 \\ 0; & \text{otherwise} \end{cases}$$

    a Find k in terms of c

    b Given that $c=0.1$, find the expected value of $\sqrt{Y}$

3 A random variable X takes the exact value 1 or any value in the interval $2\leq x\leq 4$
Its distribution is given by $P(X=1)=0.2$, $P(X<x)=k(-x^2+8x-11)$ for $2\leq x\leq 4$

    a Prove that $k=0.2$

    b Find the probability density function, $f(x)$, of X in the interval $2\leq x\leq 4$

c Find

    i The expected value of X **ii** $P(X<3)$

4 A continuous random variable X, defined on the domain $x>0$, has cumulative probability distribution function given by

$$F(x)=\begin{cases} ax; & 0<x\leq 2 \\ -2a+3ax-\dfrac{1}{2}ax^2; & 2<x\leq 3 \\ 1; & x>3 \end{cases}$$

a By calculating in terms of a an expression for F(3), find the value of a

b Find the value of F(2) and hence explain why the median, M, of X satisfies $M<2$

c Find the median of X

d By differentiation, find the density function, f(x)

5 A continuous random variable X, defined on the domain $x>0$, has cumulative probability distribution function given by

$$F(x)=\begin{cases} \dfrac{x^2}{k}; & 0<x\leq 3 \\ -3+2x-\dfrac{1}{4}x^2; & 3<x\leq 4 \\ 1; & x>4 \end{cases}$$

a Find the value of k

b By differentiation, find the density function, f(x) and hence sketch the density function.

c Find the median and show that $P(X>3.5)=\dfrac{1}{16}$

Fluency and skills

The exponential distribution is used to model continuous random variables such as the waiting times between random events. These can include radioactive disintegrations, the failure of electronic components or even the lengths of certain telephone calls.

> **Key point**
>
> If X is a random variable with an exponential probability distribution with parameter λ, then its probability density function is given by $f(x) = \lambda e^{-\lambda x}, x > 0$

The probability density function has a positive skew and is defined for positive values of x

Its graph is shown for several values of λ

The exponential cumulative distribution function, F(x), can be found from the probability density function by integration:

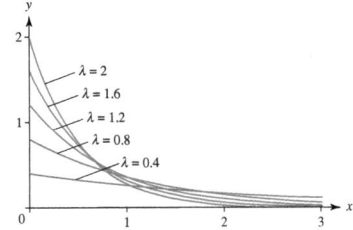

$$F(x) = P(X < x) = \int_0^x f(t)\,dt$$

$$= \int_0^a \lambda e^{-\lambda t}\,dt = \left[\lambda \frac{e^{-\lambda t}}{-\lambda} \right]_0^x$$

$$= 1 - e^{-\lambda x}$$

> **Key point**
>
> If X is a random variable with an exponential probability distribution, then its cumulative probability distribution function is given by $F(x) = 1 - e^{-\lambda x}, x > 0$

Example 1

A random variable X has an exponential distribution with parameter $\lambda = \dfrac{1}{4}$

a Write down the cumulative distribution function and then use it to find

   i $P(X < 1)$ **ii** $P(X > 3)$ **iii** $P(1 < X < 3)$

   Use a calculator to check your answers.

b If three x-values are chosen at random, find the probability that one will be less than 1 and the other two will be greater than 1

---

a $F(x) = 1 - e^{(-\lambda x)}$

  i $P(X < 1) = 1 - e^{-0.25} = 0.221$

  ii $P(X > 3) = 1 - P(X < 3) = 1 - (1 - e^{-0.75}) = 0.472$

  iii $P(1 < X < 3) = 1 - (0.221 + 0.472) = 0.31$ (2 dp)

b $Y -$ number less than 1 out of 3 values. $Y \sim B(3, 0.221)$

      $P(Y = 1) = 0.40$

The probability density function is used to calculate the mean value and variance of the distribution.

$$E(X) = \int_0^\infty x\,f(x)\,dx = \lambda \int_0^\infty x e^{-\lambda x}\,dx$$

$$= \lambda \left\{ \left[x \frac{e^{-\lambda x}}{-\lambda} \right]_0^\infty - \int_0^\infty \frac{e^{-\lambda x}}{-\lambda}\,dx \right\}$$

$$= \lambda \left\{ 0 - \left[\frac{e^{-\lambda x}}{\lambda^2} \right]_0^\infty \right\} = \lambda \left(\frac{1}{\lambda^2} \right)$$

$$= \frac{1}{\lambda}$$

Using integration by parts.

$$E(X^2) = \int_0^\infty x^2\,f(x)\,dx = \lambda \int_0^\infty x^2 e^{-\lambda x}\,dx$$

$$= \lambda \left\{ \left[x^2 \frac{e^{-\lambda x}}{-\lambda} \right]_0^\infty + \int_0^\infty \frac{e^{-\lambda x}}{\lambda} 2x\,dx \right\}$$

$$= \lambda \left\{ 0 - 2 \left(\left[x \frac{e^{-\lambda x}}{\lambda^2} \right]_0^\infty - \int_0^\infty \frac{e^{-\lambda x}}{\lambda^2}\,dx \right) \right\}$$

$$= \lambda \left\{ 0 - 2 \left(0 + \left[\frac{e^{-\lambda x}}{\lambda^3} \right]_0^\infty \right) \right\} = \lambda \frac{2}{\lambda^3} = \frac{2}{\lambda^2}$$

$$\mathrm{Var}(X) = E(X^2) - \{E(X)\}^2 = \frac{2}{\lambda^2} - \left(\frac{1}{\lambda} \right)^2 = \frac{1}{\lambda^2}$$

Using integration by parts.

Using integration by parts.

> **Key point**
>
> If X has a probability density function given by $f(x) = \lambda e^{-\lambda x}$, $x > 0$, then the mean and variance of X are given by $E(X) = \dfrac{1}{\lambda}$ and $\mathrm{Var}(X) = \dfrac{1}{\lambda^2}$

The exponential distribution can be used in calculations on the waiting times between random events.

$X \sim \text{Poisson}(\mu)$

Suppose that random events occur at a mean rate of μ per unit time. Let X be a random variable for the number of occurrences in a random unit interval and let T be the time between successive events.

$P(T < t) = 1 - P(T > t) = 1 - P(\text{no occurrences in interval } t) = 1 - e^{-\mu t}$, which is the cumulative distribution function of an exponential random variable, parameter μ and mean $\dfrac{1}{\mu}$

These events form a **Poisson process** with mean μ

If the mean number of occurrences in unit time is Poisson (μ), the mean number in time t is Poisson (μt)

> **Key point**
>
> If T is a random variable for the time between successive random events that occur at a mean rate μ per unit time, then T follows an exponential distribution, parameter μ, with mean $\dfrac{1}{\mu}$ and variance $\dfrac{1}{\mu^2}$

Remember that μ is the mean of the underlying Poisson process, not the waiting time mean.

Example 2

a Write down the mean and variance of a random variable, X, that has an exponential distribution with $\lambda = 2$

b Find the probability that a randomly chosen value of X will be within 1 standard deviation of the mean.

a $X \sim \text{exponential}(\lambda = 2) \rightarrow$ mean $\dfrac{1}{2}$, variance $\dfrac{1}{2^2} = \dfrac{1}{4}$

b Standard deviation $\sqrt{\dfrac{1}{4}} = \dfrac{1}{2}$

$P\left(\dfrac{1}{2} - \dfrac{1}{2} < X < \dfrac{1}{2} + \dfrac{1}{2}\right) = P(0 < X < 1) = 0.865 \ (3dp)$

Exercise 26.2A Fluency and skills

1 A random variable X has an exponential distribution with a mean of 3

Find the probability that X takes values

  a Greater than 3   **b** Between 1 and 3

2 A random variable X has an exponential distribution with parameter $\lambda = 2$

  a Using the probability density function of X, copy and complete this table

x	0	0.2	0.4	0.6	0.8	1.0	1.2	1.4
$f(x)$	2				0.40	0.27	0.18	0.12

  b Sketch the probability density function of X for $0 < x < 1.4$

  c Write the modal value of X

3 **a** Write the cumulative distribution function of a random variable X that follows an exponential distribution with parameter λ

  b Hence prove that M, the median of X, satisfies the equation $M\lambda = \ln 2$

4 An exponentially distributed random variable, X, has parameter λ equal to 0.25

  a Find the mean and standard deviation of X

  b Using the definition of skewness,

$\dfrac{\text{mean} - \text{mode}}{\text{standard deviation}}$, calculate the

skewness of X

5 A random variable X has an exponential distribution, parameter $\lambda = 2.5$

  a Write the probability density function of X

  b Show that its cumulative distribution function is $F(x) = 1 - e^{-2.5x}$, $x > 0$

  c Find the values of

    i $P(X < 0.5)$

    ii $P(X > 0.2)$

    iii $P(0.30 < X < 0.65)$

  d Using the general formulae for the mean and variance of a random variable, prove that the mean and variance of X are 0.4 and 0.16, respectively.

6 **a** State the condition for waiting times between events to follow an exponential distribution with parameter $\dfrac{1}{2}$

  b Requests for cash withdrawals arrive at a bank as a Poisson process with mean 12 per second.

    i What is the mean waiting time between requests?

    ii What is the probability that, from a randomly chosen time, there is a gap of more than 0.2 s before the first request?

7 A random variable X has an exponential distribution with parameter $\lambda = 4$

    a Write the cumulative distribution function of X and hence use differentiation to derive its probability density function.

    b Sketch the probability density function of X

    c Write the modal value of X

    d Use your answer in part **a** to find $P\left(X > \dfrac{1}{2}\right)$

8 A random variable X has an exponential distribution, parameter $\lambda = 5$

    a Write the probability density function of X and sketch this function, showing where it cuts the vertical axis.

    b Using integration, show that the cumulative distribution function of X is $F(x) = 1 - e^{-5x}$

    c Show that the mean and variance of X are 0.2 and 0.04, respectively. You may quote standard results for these quantities.

Reasoning and problem-solving

Strategy

To solve a problem using the exponential distribution

  ① Ensure that the conditions for the exponential distribution apply to the problem.

  ② Use a calculator, statistical tables or the exponential distribution function to find probabilities.

  ③ Where required, use standard results to find the mean and variance of the distribution.

Example 3

Random events occur at a rate of 4 per minute.

  a Write the probability density function, $f(t)$, and the cumulative distribution function, $F(t)$, of the random variable T, the waiting time in minutes between events.

  b Find the probability that, from the occurrence of one event, the waiting time until the next event will be greater than 15 seconds.

  c Calculate the mean and variance of the waiting time.

> **a** $f(t) = 4e^{-4t}$, $t \geq 0$. $F(t) = 1 - e^{-4t}$, $t \geq 0$
>
> **b** $P(T > 0.25) = 1 - P(T < 0.25) = e^{-4 \times 0.25} = e^{-1} = 0.37$ (2dp)
>
> **c** Mean $= \dfrac{1}{\mu} = \dfrac{1}{4}$; variance $= \dfrac{1}{\mu^2} = \dfrac{1}{16}$, where μ is the distribution parameter.

Example 4

Radioactive decays occur randomly in time with a mean of 3 per minute and T is a random variable for the waiting time in seconds between events.

  a Specify fully the distribution of T

  b Find the probability that, in the t_1 seconds after a clock is started, no decays occurred.

  c Given that there were no decays in the first t_0 seconds after the clock started, find the probability that there were no decays in a further t_1 seconds.

  d Comment on the results in parts **b** and **c**

(Continued on the next page)

a Exponential, parameter 0.05

b $P(\text{no decays in } t_1 \text{ s}) = P(T > t_1) = 1 - P(T < t_1) = 1 - (1 - e^{-0.05t_1}) = e^{-0.05t_1}$

c $P(\text{no decays in further } t_1 \text{ s} \mid \text{no decays in } t_0 \text{ s})$

$$= \frac{P(\text{no decays in } t_0 + t_1)}{P(\text{no decays in } t_0)} = \frac{e^{-0.05(t_0 + t_1)}}{e^{-0.05t_0}} = \frac{e^{-0.05t_0} \times e^{-0.05t_1}}{e^{-0.05t_0}} = e^{-0.05t_1}$$

d The answers are equal. The probability that there were no decays in t_1 seconds is unaffected by the absence of decays during any previous time interval.

3 per minute is $\dfrac{3}{60} = 0.05$ per second

Or, substitute $x = 0$ into a Poisson distribution function, mean $0.05t_1$

Exercise 26.2B Reasoning and problem-solving

1 Random events occur at a mean rate of 6 per minute.

   a Prove that the cumulative distribution function, $F(t)$, of the random variable T, the waiting time in minutes between events, is given by $1 - e^{-6t}$

   b By differentiation, find the probability density function, $f(t)$

   c Show that the median waiting time between events, M_T, is 0.12 (2 dp)

   A clock is started at some randomly chosen time.

   d Find the probability that the waiting time until the next event will be greater than 20 seconds.

2 The mean number of customers who arrive at a supermarket checkout during a 6 minute period is 12

   a Assuming that their arrivals constitute a Poisson process, what is the probability that a period of at least two minutes will occur without any customer appearing?

   b Find the expected number of customers arriving during a 10-hour day and hence the mean number of gaps of at least 2 minutes between customers.

3 Telephone calls to a call centre arrive randomly with a mean of 4 calls per minute. The centre has 10 operators and calls last at least 4 minutes. Lines open at 9 am.

   a Give the distribution of the waiting time between calls and hence find the probability that the first call arrives after 09:01

   b Given that there are no calls by 09:01, with no further calculation, write down the probability that the operators wait until after 09:02 before the first call. Give a reason for your answer.

   c Find the probability that at 09:03 at least half of the operators are busy.

4 A firm produces microchips and has found that the mean lifetime for these components is 2.1 years, with the exponential distribution providing a good model for the lifetime.

   a Specify completely the distribution of the lifetime for a randomly chosen component and find the probability that its lifetime is less than one year.

   The firm guarantees the components for one year. If a failure occurs within the first year, the component is replaced and a new guarantee for one year then applies to the new component. Subsequent replacements are not guaranteed.

b What is the probability that for a randomly chosen component, a buyer will apply for exactly one replacement under guarantee?

c Explain why it is unlikely that failure is generally due to wear and tear factors.

5 a T, the waiting times between random events, follows an exponential distribution with parameter $\dfrac{1}{3}$

Write the probability density function and the mean and variance of T

b Cars arriving at a traffic checkpoint can be considered as a Poisson process with mean 14 per hour.

   i What is the mean waiting time between car arrivals?

   ii What is the probability that, from the start of counting, there is a five-minute period during which no cars arrive?

6 In the production of fibre-optic cable, flaws occur at random with a mean of 0.21 per 10 metres.

a What is the probability that, from the start of production one day, a length of at least 80 metres will be produced without any flaws?

b In 10 000 metres of cable produced, find the mean number of flawless lengths of at least 80 metres.

7 Random events occur at a rate of 8 per hour.

a Write the probability density function, $f(t)$, and the cumulative distribution function, $F(t)$, of the random variable T, the waiting time in minutes between events.

b Find the probability that, from the occurrence of one event, the waiting time until the next event will be greater than 12 minutes.

c Calculate the mean and variance of the waiting time.

Chapter summary

- The cumulative distribution function of a continuous random variable with probability density function f(t) is given by $F(x) = P(X < x) = \int_{-\infty}^{x} f(t)dt$

- If F(x) and f(x) are the cumulative distribution function and the probability density function of X, respectively, then $f(x) = \dfrac{d}{dx}F(x)$

- A mixed distribution occurs when the corresponding random variable is partly discrete and partly continuous.

- If R is a random variable with a rectangular probability distribution defined on the interval (a, b) (that is, $R \sim \text{Rectangular}(a,b)$), then its probability density function is given by $f(r) = \dfrac{1}{b-a}; a < r < b$

- If X has a rectangular probability distribution, then the mean and variance of X are, respectively, given by $E(X) = \dfrac{a+b}{2}$ and $\text{Var}(X) = \dfrac{(b-a)^2}{12}$

- If X is a random variable with an exponential probability distribution, parameter λ, then its probability density function is given by $f(x) = \lambda e^{-\lambda x}; x > 0$

- If X is a random variable with an exponential probability distribution, parameter λ, then its cumulative probability distribution function is given by $F(x) = 1 - e^{-\lambda x}; x > 0$

- If X has a probability density function given by $f(x) = \lambda e^{-\lambda x}; x > 0$, then the mean and variance of X are, respectively, given by $E(X) = \dfrac{1}{\lambda}$ and $\text{Var}(X) = \dfrac{1}{\lambda^2}$

- If T is a random variable for the time between successive random events which occur at a mean rate μ, then T follows an exponential distribution, parameter μ, with mean $\dfrac{1}{\mu}$ and variance $\dfrac{1}{\mu^2}$

Check and review

You should now be able to...	Review Questions
✔ Calculate the cumulative distribution function and probability density function for continuous random variables.	1, 2, 3
✔ Recognise and calculate the mean and variance of mixed random variables.	4
✔ Apply rectangular and exponential probability models in different circumstances.	5, 6, 7, 8
✔ Prove and use formulae for the mean and variance of rectangular and exponential random variables.	3, 5, 6, 7, 8

1 A random variable, X, has a probability density function given by $f(x) = \begin{cases} ax(x-3)(x+1); & -1 < x < 0 \\ 0; & \text{otherwise} \end{cases}$.

   a Find the value of a and the cumulative distribution function $F(x)$.

   b Hence show that the median of the distribution is approximately -0.52

2 A continuous random variable X has pdf given by $f(x) = \begin{cases} \dfrac{(1-3x)}{2}; & a < x < \dfrac{1}{3} \\ 0; & \text{otherwise} \end{cases}$

   a Show that $a = \dfrac{1-2\sqrt{3}}{3}$

   b Find the cumulative distribution function of X

3 A random variable, X, has probability density function $f(x) = \begin{cases} \dfrac{1}{k}; & a < x < b \\ 0; & \text{otherwise} \end{cases}$

   a Show that $k = b - a$

   b Find $E(X)$ and $Var(X)$ when $a = -2$ and $b = 2$

   c Show that the cumulative distribution function of X is given by $F(x) = \dfrac{(x+2)}{4}$

4 A random variable X takes the exact value 0 or any value in the interval $2 \le x \le 4$

   Its distribution is given by $P(X=0) = 0.1$, $P(2 < X < x) = \left(\dfrac{k}{10}\right)(-x^2 + 10x - 16)$ for $2 < x < 4$

   a Prove that $k = 1.125$

   b Find the probability density function, $f(x)$, of X in the interval $2 \le x \le 4$

   c Find

      i The expected value of X **ii** $P(X < 3)$

5 Radioactive decay can be modelled as a Poisson process with mean 12 nuclei decaying per second.

   a What is the mean waiting time between disintegrations?

   b What is the probability that, from a randomly chosen time, there is a gap of more than 0.2 s before the first disintegration?

6 Cars pass a traffic checkpoint at a rate of 4 per minute.

   a Find the probability that the interval between successive cars will be at least 20 seconds. You should state any distributional assumptions you make.

   b Using the same assumptions as in part **a**, calculate how many cars you would expect to pass the checkpoint in 5 hours.

   c How many intervals greater than 20 seconds would you expect during 5 hours?

7 Random events occur at a mean rate of 10 per minute.

   a Prove that the cumulative distribution function, $F(t)$, of the random variable T, the waiting time in minutes between events, is given by $1 - e^{-10t}$

   b Show that the median waiting time between events, M_T, is 0.069 (3 dp)

   c Find the probability that the waiting time between successive events will be greater than 20 seconds.

8 On a quiet country road, cars pass a given point randomly in time with a mean of 6 every 10 minutes. Let T be a random variable for the waiting time in minutes between successive cars.

   a Specify fully the distribution of T

An observer records car arrivals.

   b Find the probability that she has to wait longer than 5 minutes before the first car arrives.

   c Given that there were no cars in the first 2 minutes, use your answer to part **b** to find the probability that she has to wait more than a further 5 minutes until the first car. You should explain your answer.

   d Given that observations start at 9 am, find the probability that the first car recorded arrives before 9:02 am and the second after 9:04 am.

Did you know?

The exponential distribution and the geometric distribution are the only probability distributions with the memoryless property:

$$P(x > a + b | x > b) = P(x > a)$$

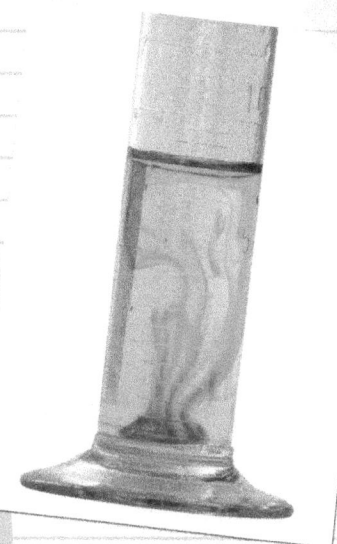

Research

The arcsine distribution has a cumulative distribution function of $F(x) = \dfrac{2}{\pi} \arcsin(\sqrt{x})$

Can you show that this is equivalent to

$$F(x) = \frac{1}{\pi} \arcsin(2x - 1) + \frac{1}{2}?$$

Can you find the corresponding probability distribution function through differentiating $F(x)$?

Find out about the three arcsine laws that cover one-dimensional random walks and Brownian motion.

Research

There are many applications of the exponential function that you could find out about. These include:

- Queuing theory
- The Barometric formula used in physics
- Daily rainfall calculations
- Dead time in particle detection analysis

1 The continuous random variable X has probability density function given by

$$f(x) = \begin{cases} \dfrac{1}{4}(x-1); & 2 \le x \le 4 \\ 0; & \text{otherwise} \end{cases}$$

  a Find $P(2.5 < X < 3)$ [3]

  b Calculate the expectation and variance of X [6]

  c Calculate $E(2X+1)$ and $\text{Var}(2X+1)$ [4]

2 The continuous random variable X has probability density function given by

$$f(x) = \begin{cases} k(1+3x^2); & 0 \le x \le 2 \\ 0; & \text{otherwise} \end{cases}$$

  a Find the value of k [2]

  b Sketch the probability density function of X [2]

  c Write the mode of X [1]

  d Show that the median, m, of X satisfies the equation $m^3 + m - 5 = 0$ [3]

3 The cumulative distribution function of the continuous random variable X is given by

$$F(x) = \begin{cases} 0; & x < 1 \\ k(x^3 - 1); & 1 \le x \le 2 \\ 1; & x > 2 \end{cases}$$

  a Work out the value of k [2]

  b Calculate $P(X < 1.5)$ [2]

  c Work out the probability density function of X [3]

4 The continuous random variable U is uniformly distributed over the interval $[1, 5]$

  a Sketch the probability density function of U [2]

  b Write $P(X > 2)$ [1]

  a Calculate the mean and variance of U [4]

5 The lifetime, X, of a component, measured in thousands of days, can be modelled by an

  exponential distribution with probability density function $f(x) = \begin{cases} \dfrac{1}{4}e^{-\frac{x}{4}}; & x > 0 \\ 0; & \text{otherwise} \end{cases}$

  a State the values of the mean and the standard deviation of the lifetime. [2]

  b Find the probability that the lifetime of the component is less than 2400 days. [3]

6 A random variable, R, is exponentially distributed with $\lambda = 5$

   a Write the probability density function of R **[2]**

   b Write the cumulative distribution function of R **[2]**

   c Calculate the probability that R lies within one standard deviation of its mean.   **[3]**

7 The continuous random variable X has probability density function given by

$$f(x) = \begin{cases} \dfrac{4}{3}(x^3 + x); & 0 \le x \le 1 \\ 0; & \text{otherwise} \end{cases}$$

   a Find the cumulative distribution function of X **[3]**

   b Calculate

      i $E(5X - 3)$ **ii** $\text{Var}(5X - 3)$ **[8]**

8 The continuous random variable X has probability density function given by

$$f(x) = \begin{cases} \dfrac{x}{24} + \dfrac{1}{12}; & 2 \le x \le 6 \\ 0; & \text{otherwise} \end{cases}$$

   a Find the cumulative distribution function of X **[4]**

   b Find the exact value of the median of X **[3]**

   c Calculate $P(\text{mean} < X < \text{median})$.   **[5]**

9 The continuous random variable U is uniformly distributed over the interval $[4, 10]$

   Work out the cumulative distribution function of U **[4]**

10 The duration, in minutes, of a phone call to a bank is modelled by a uniform distribution on the interval $[1, 10]$

   a Use this model to calculate the probability that the length of a call is between 4 and 8 minutes.   **[2]**

   b Give a reason why this model may not be appropriate.   **[1]**

   It is decided instead to model the duration as an exponential distribution, using the mean of the uniform distribution as the parameter.

   c Use this model to calculate the probability that a call is

      i Between 4 and 8 minutes,

      ii Over 15 minutes given it has already been over 10 minutes.   **[5]**

11 Prove that the variance of an exponential random variable with parameter 2 is $\dfrac{1}{4}$ **[5]**

12 A continuous random variable, U, is uniformly distributed over the interval $[k, 2k]$

   Prove that the variance of U is given by $\dfrac{1}{12}k^2$ **[4]**

13 The continuous random variable X has probability density function given by

$$f(x) = \begin{cases} \dfrac{1}{x}; & 1 \le x \le e \\ 0; & \text{otherwise} \end{cases}$$

   a Work out the cumulative distribution function of X **[3]**

   b Find the standard deviation of X **[6]**

14 The continuous random variable X has cumulative distribution function given by

$$F(x) = \begin{cases} 0; & x < 2\sqrt{2} \\ \dfrac{x^2}{8} - 1; & 2\sqrt{2} \le x \le 4 \\ 1; & x > 4 \end{cases}$$

   a Calculate the median of X **[2]**

   b Calculate the exact values of $E(2X^{-1})$ and $\text{Var}(2X^{-1})$.   **[6]**

15 The continuous random variable X has probability density function given by

$$f(x) = \begin{cases} k; & -1 \le x < 1 \\ k(x-2)^2; & 1 \le x \le 2 \\ 0; & \text{otherwise} \end{cases}$$

   a Sketch the probability density function of X **[4]**

   b Calculate the value of k **[4]**

   c Fully define the cumulative distribution function of X **[5]**

   d Find the median of X **[3]**

   e Calculate $E(4X)$ and $\text{Var}(4X)$ **[8]**

16 The rate of customers arriving at a bank is modelled by a Poisson distribution with a rate of 20 people per half-hour.

What is the probability that there is a gap of less than 30 seconds until the next customer? State clearly the distribution you are using.   **[4]**

27 Hypothesis testing and the *t*-test

In 1972, researchers at Stanford University famously conducted the 'Marshmallow Experiment' to study whether deferred gratification is an indicator of success later in life. In the experiment, a sample of children aged between 4 and 6 were each given a marshmallow and asked to wait alone in a room for 15 minutes for the experimenter to return. The children were told that if the marshmallow was still uneaten when the experimenter returned, then they would be given a second marshmallow. One third of the children resisted eating the marshmallow for the full 15 minutes. Follow up experiments showed that the children who resisted eating the first marshmallow had higher SAT scores, thereby supporting the hypothesis that deferred gratification is an indicator of later success.

Hypothesis testing is used widely in psychology, and the ability to identify the risk of a Type I or Type II error is crucial to avoid incorrect conclusions from being widely accepted in the discipline.

Orientation

What you need to know

Ch11 Hypothesis testing and contingency tables

- Type I/II errors.

What you will learn

- To find the probability of a Type I error.
- To find the probability of a Type II error.
- Calculate the power of a hypothesis test.
- Use a *t*-test to evaluate a possible population mean for a Normal distribution.

What this leads to

Careers

- Scientific research.
- Quality control.
- Psychology.

27.1 Type II errors and power

Fluency and skills

See Ch11.1 For a reminder of Hypothesis testing. In hypothesis testing, the null hypothesis is always assumed to be true when determining the critical region. The probability of a result in the critical region should be less than or equal to the test's significance level.

> **Key point**
>
> A **type I error** is when a null hypothesis which is true is rejected.
>
> A **type II error** is when a null hypothesis which is false is accepted.

The probability of making a type I error is equal to the probability of a result in the critical region. It is commonly denoted by α

The probability of a type II error is more difficult to calculate as it depends on the true value of the population parameter being tested. If this was known already, then a test would be pointless. Taking some example true values of the parameter can provide an idea of the probability of making a type II error. The probability of making a type II error is commonly denoted by β

The probabilities of making type I and a type II errors are generally not independent; decreasing one is likely to increase the other. The probability of making a type I error is much easier to control via the significance level.

Example 1

A Poisson distribution is being tested with null hypothesis $\lambda = 1.9$ and alternative hypothesis $\lambda > 1.9$

The critical value at the 5% significance level is 5

a Find the probability of a type I error.

The true parameter value is $\lambda = 3.1$

b Find the probability of a type II error.

> The probabilities of the Poisson distribution are given by the formula $P(X = k) = \dfrac{\lambda^k e^{-\lambda}}{k!}$ where k is the number of events in the interval.

See Ch10.2 For a reminder of the Poisson distribution.

a The probability of a type I error is the probability of getting a result in the critical region. $P(X \geq 5) = 4.41\%$

b The probability of a type II error is the probability of getting a result outside of the critical region when the parameter value is 3.1. $P(X \leq 4) = 79.8\%$

> This is very large. This means that the situations $\lambda = 1.9$ and $\lambda = 3.1$ are difficult to distinguish. To overcome this, it is often a good idea to use a larger sample.

The **power** of a hypothesis test is the probability that a false null hypothesis is rejected. It is equal to $1 - \beta$

Key point

This depends on the true parameter value and is rarely a fixed value. The larger the power, the better the test is at avoiding accepting a false null hypothesis by mistake.

Example 2

The mean of a Normal distribution with standard deviation 2.2 is being tested at the 10% level. The null hypothesis is $\mu = -0.8$ and the alternative hypothesis is $\mu \neq -0.8$

The true mean is $\mu = -3.2$

a Find the probability of a type II error.

b Calculate the power of the test.

a For this two-tailed test, the critical region is $X < -4.42$ and $X > 2.82$

When the mean is -3.2, the probability of obtaining a value outside that region is 70.7%

b The power is $1 - 0.707 = 0.293$

This was found using the Normal distribution with mean -0.8 and standard deviation 2.2. Each tail is 5% of the distribution.

For $\mu = -3.2$,
$P(-4.42 < X < 2.82)$
$= 1 - P(X < -4.42) - P(X > 2.82)$
$= 1 - 0.2898 - 0.0031$

Exercise 27.1A Fluency and skills

1   With reference to the correct hypotheses, explain the difference between a type I and a type II error.

2   A Normal distribution with standard deviation 6.25 is being tested at the 5% significance level. The null hypothesis is H_0: $\mu = 15.7$ and the alternative is H_1: $\mu > 15.7$

   a   State the probability of a type I error.

   b   Find the probability of a type II error if actually $\mu = 5.7$

   c   Find the probability of a type II error if actually $\mu = 25.7$

3   A Poisson distribution is tested at the 5% level with hypotheses H_0: $\lambda = 6.8$ and H_1: $\lambda > 6.8$

   a   Find the critical region,

   b   Calculate the probability of a type I error.

   The true value is $\lambda = 9.7$

   c   Calculate the probability of a type II error,

   d   Calculate the power.

4   For a binomial distribution, the hypotheses H_0: $p = 0.2$ and H_1: $p \neq 0.2$ are tested at the 10% level. 20 trials are performed and the critical region is $X = 0$ or $X > 7$

   Calculate the probability of a type II error and the power when the true value of p is

   a   0.3   **b**   0.4   **c**   0.6

5   A Normal distribution is being tested with hypotheses $H_0: \mu = 0$ and $H_1: \mu > 0$
    The significance level is 5%. The standard deviation is 4.5

    a   Calculate the power if the true mean is 12.1

    b   Explain why the power tends to 1 as the true mean tends to infinity.

6   A binomial distribution is being tested at the 5% level with the hypotheses $H_0: p = 0.4$ and
    $H_1: p > 0.4$

    A sample of size 16 is taken.

    a   Find the critical region,

    b   Calculate the power if the true value of p is 0.65

7   A crisp manufacturer wants to check that the mean weight of a packet of crisps is not too
    different from the advertised weight. They perform a two-tailed test.

    a   Explain what a type I error would mean in context.

    Actually, the mean weight is above the advertised weight but the crisp manufacturer believes that
    it is not.

    b   Decide whether or not they have made a type II error.

8   A binomial distribution $X \sim B(50, p)$ is tested at the 10% significance level to decide between the
    hypotheses $H_0: p = 0.4$ and $H_1: p < 0.4$

    a   Explain why the critical region is $X \le 15$, showing supporting calculations,

    b   State the probability of making a type I error,

    c   Calculate the probability of making a type II error if actually $p = 0.35$

    A common procedure to ease calculations is to approximate a binomial distribution with a
    Normal distribution.

    d   Explain why the approximating Normal distribution is $N(20, 12)$ and explain why the new
        hypotheses are $H_0: \mu = 20$ and $H_1: \mu < 20$

    e   Determine the new critical value at the 10% level,

    f   Determine the approximating distribution if actually $p = 0.35$, and hence calculate the
        probability of making a type II error.

9   Two hypothesis tests are considered for determining the mean of a distribution.
    A statistician wishes to know which is likely to be a better test.

    For both tests the probability of making a type I error is 5%

    a   Explain what a type I error is and hence why these probabilities might be made the same by the
        statistician to choose a better test.

    For one test the probability of making a type II error is larger than for the other test.

    b   Explain what a type II error is and hence decide which test should be used by the
        statistician.

Reasoning and problem-solving

Example 3

Strategy

(1) When deciding on the significance level for a test you should consider whether it is more important to reduce the probability of making a type I error or of making a type II error.

(2) The sample size should be large enough so that if the true value is very different from the null hypothesis, then it is likely to be inside the critical region.

A Normal distribution with standard deviation 4 is being tested with hypotheses $H_0: \mu = 7.7$ and $H_1: \mu < 7.7$. Calculate the critical value and hence the probability of making a type II error if the true mean is 2.5 when the significance level is

a 10%   **b** 5%   **c** 1%

Cumulative probabilities for the Normal distribution can be found using your calculator.

a The critical value is 2.57. $P(X < 2.57 \mid \mu = 2.5) = 40.3\%$

b The critical value is 1.12. $P(X < 1.12 \mid \mu = 2.5) = 63.5\%$

c The critical value is -1.61. $P(X < -1.61 \mid \mu = 2.5) = 84.8\%$

Note that relatively small changes to the significance level have brought large changes to the probability of making a type II error.

Consider a Poisson distribution being tested with hypotheses
$H_0: \lambda = 10$ and $H_1: \lambda > 10$

At the 5% significance level, the critical region is $X \geq 16$

The graph shows the power for a range of true parameter values.

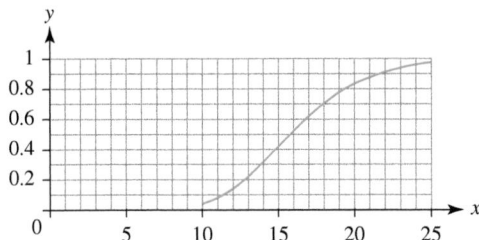

The power exceeds 0.5 near to 15.7, meaning that if the true value of the parameter is much smaller than the critical value of the test, then there is a better-than-evens chance that the false null hypothesis is accepted.

When the true value is much higher than the critical value, the power becomes close to 1. Note that the power isn't technically defined when the null hypothesis really is the true value.

This is because power is the probability that a false null hypothesis is rejected. A true null hypothesis is not false.

Example 4

A new medical test for detecting a rare illness is being trialled. The null hypothesis is that a person being tested does not have the illness.

a Explain why the probability of a type II error should be made as small as possible.

b This makes the probability of a type I error larger. Why might this not be a problem?

> **a** A type II error is when a false null hypothesis is accepted. A medical test should aim to not overlook somebody who genuinely has the illness.
>
> **b** A type I error is when a true null hypothesis is rejected. If the test claims that a person has the illness when they don't, then a simple repeat of the test will help to decide.

Exercise 27.1B Reasoning and problem-solving

1 A Normal distribution with standard deviation 0.6 is being tested with the hypotheses $H_0: \mu = 0.3$ and $H_1: \mu > 0.3$

Calculate the critical value, the probability of making a type II error if the true mean is 1.3 and the power when the significance level is

    a 10%

    b 5%

    c 1%

2 A Poisson distribution is being tested with hypotheses $H_0: \lambda = 1.7$ and $H_1: \lambda \neq 1.7$ The true value of the mean is 2.3. For the significance level

    a 10%

    b 5%

    calculate

      i The critical region,

      ii The probability of a type II error,

      iii The power.

3 For a binomial distribution, the hypotheses

$H_0: p = \dfrac{1}{3}$ and $H_1: p \neq \dfrac{1}{3}$ are tested at the 2% level. 20 trials are performed and the critical region is $X < 2$ or $X > 12$

The true value of p is 0.4

a State how many successes you would expect to see (using the true value of p),

b Calculate the probability of a type II error and the power.

Instead of 20 trials, 2000 trials are performed. The new critical region is $X < 618$ or $X > 716$

c State how many successes you would expect to see,

d Giving a reason, explain how you would expect the power of this test to compare with the 20-trial test.

4 A treasure hunter is configuring their metal detector to find buried valuables.

a Explain why the probability of a type II error should be made as small as possible,

b This makes the probability of a type I error larger. Why might this not be a problem?

5 The hypotheses H_0: $\lambda = 4.25$ and H_1: $\lambda < 4.25$ are tested at the 5% level, where λ is the parameter of a population Poisson distribution.

    a Find the critical region,

    b Find the probability of a type I error.

The true mean is 9.25

    c Calculate the probability of a type II error and the power of the test.

6 A binomially distributed population is tested with the hypotheses H_0: $p = 0.2$ and H_1: $p > 0.2$ at the 5% level. Different-sized samples are taken to test this. The true value of p is 0.35. Calculate the critical region and hence the power when the sample size is

    a 10

    b 20

    c 30

7 The number of goals scored by a team playing away from home in a football match can be modelled by a Poisson distribution with parameter $\lambda = 1.2$

A team has undergone a training regime that it is hoped will increase the number of goals they score away. The next 5 away games are played and the hypotheses H_0: $\lambda = 6$ and H_1: $\lambda > 6$ are tested at the 5% significance level.

    a Explain why the hypotheses take this form,

    b Find the critical region and calculate the probability of making a type I error.

After training, the number of goals scored away can now be modelled by a Poisson distribution with parameter $\lambda = 1.4$

    c Calculate the probability of making a type II error,

    d Explain why this error wouldn't be very concerning to a fan of the team,

    e Explain why this error would be very concerning to a sports-betting company, who gamble on the results of matches.

27.2 *t*-tests

Fluency and skills

You can use a hypothesis test to investigate the mean of a Normal distribution with known variance. When the variance is unknown, more work is required.

Suppose you believe that a process follows a Normal distribution and expect the mean to be 12.0

You take a sample of size 10 and get the following results:

11.9	8.37	14.8	14.6	11.2
11.1	14.9	12.2	10.7	15.7

The mean of these values is 12.547, which could be close to 12.0 or it could be far from it, depending on the size of the variance. You need to know the distribution of the sample mean.

Key point

The **sample mean**, $\overline{X}$, itself has a probability distribution. The mean of this distribution (i.e. the mean sample mean) is the same as the population mean.

Just as there is a standard Normal distribution that other Normal distributions can be related to, there are standard distributions for $\overline{X}$. These ***t*-distributions** describe the distribution of a sample mean for varying numbers of data points.

If you know the variance of the population distribution, σ, then the distribution of the test statistic, $\dfrac{\overline{x}-\mu_0}{\dfrac{\sigma}{\sqrt{n}}}$, is Normal with mean 0 and variance 1

If you do not know the variance of the population distribution, then you use the variance of the sample as an estimate but it will be inaccurate if the sample size is small.

For the sample of size 10 above, the sample variance is

$$\frac{1}{10-1}((11.9^2+8.37^2+\ldots+15.7^2)-10\times12.547^2)=5.56$$

For a sample of size n the sample variance is

$$S^2=\frac{1}{n-1}\left(\sum X_i^2-n\overline{X}^2\right)=\frac{1}{n-1}\sum(X_i-\overline{X})^2$$

Key point

The ***t*-test statistic** is $t=\dfrac{\overline{x}-\mu_0}{\dfrac{S}{\sqrt{n}}}$

It follows the *t*-distribution with $n-1$ degrees of freedom.

It is well worth stressing here that a particular sample's mean is very unlikely to be the same as that of the population, but the average of the means of many samples will be the same as the population mean.

The example t-test statistic is $\dfrac{12.547-12.0}{\dfrac{\sqrt{5.56}}{\sqrt{10}}} = 0.734$

Using the t-distribution with $10 - 1 = 9$ degrees of freedom, the probability of getting a sample mean this big or larger is 24.1%

No significance level was stated, but it is reasonable to say that the sample has come from a Normal distribution with mean 12.0

Example 1

A population Normal distribution with unknown variance is being tested at the 10% significance level with hypotheses $H_0: \mu_0 = 3.2$ and $H_1: \mu_0 \neq 3.2$. A sample of size 8 is taken, which has a sample mean of 2.39 and a sample variance of 1.12

a Calculate the t-test statistic,

b State the number of degrees of freedom,

c Calculate the p-value of the statistic,

d Determine the conclusion of the test.

a $t = \dfrac{2.39-3.2}{\dfrac{\sqrt{1.12}}{\sqrt{8}}} = -2.165$

b There are $8 - 1 = 7$ degrees of freedom.

c $P(T < -2.165) = 3.36\%$

Your calculator can find these cumulative probabilities from the t-test statistic.

d This is a two-tailed test and $3.36\% < 5\%$. There is significant evidence to reject the null hypothesis. You conclude that the sample is not likely to come from a population Normal distribution with mean 3.2

Key point

For a hypothesis test using the t-distribution to be valid, the population should be Normally distributed. As with all hypothesis tests, the sample must be taken at random

Example 2

A student wants to test whether the average number of devices that can connect to the internet per UK household is five or greater. They ask seventeen friends to record their values and decide, due to the sample size, to use a t-test. Give two reasons why this test will not give a valid result.

1 The sample isn't random; it is likely to be biased due to the student asking friends.

2 The values aren't even approximately Normally distributed so a t-test is unsuitable.

1 Calculate the test statistics for the following tests.

	Sample size	Sample mean	Sample variance	Hypothesis mean
a	16	2.51	1.08	1.92
b	4	−0.121	3.56	1.35
c	7	10.49	18.65	21.3
d	11	−21.7	5.16	−18.3

2 For the following data

   a Calculate the sample mean,

   b Calculate the sample variance,

   c Calculate the test statistic if the hypothesis mean is 0.2

0.342	−0.348	0.692	3.845
−0.361	0.998	0.681	2.029

3 A population Normal distribution with unknown variance is being tested at the 5% level with hypotheses $H_0: \mu_0 = 1.61$ and $H_1: \mu_0 < 1.61$

   A sample of size 12 is taken, which has sample mean −2.56 and sample variance 29.3

   a Calculate the t-test statistic,

   b State the number of degrees of freedom,

   c Calculate the p-value of the statistic,

   d Determine the conclusion of the test.

4 A population Normal distribution with unknown variance is being tested at the 5% significance level with hypotheses $H_0: \mu_0 = -2.1$ and $H_1: \mu_0 < -2.1$

   A sample of size 17 is taken, which has sample mean −2.49 and sample variance 18.7

   a Find the critical region,

   b Calculate the t-test statistic,

   c Determine the conclusion of the test.

5 The hypotheses $H_0: \mu_0 = 3.17$ and $H_1: \mu_0 \neq 3.17$ are tested at the 2% level for a population Normal distribution with unknown variance. A sample of size 14 is taken, which has a sample mean of 0.31 and a sample variance of 14.0

   Showing your reasoning, determine the conclusion of the test.

6 The hypotheses $H_0: \mu_0 = 18.2$ and $H_1: \mu_0 < 18.2$ are tested at the 5% level for a population Normal distribution with unknown variance. The following sample is taken.

0.387	4.23	20.2	−15.3
22.9	26.5	−6.39	13.5

   a Calculate the t-test statistic,

   b State the number of degrees of freedom,

   c Calculate the p-value of the statistic,

   d Determine the conclusion of the test,

   e State two assumptions made about the sample that are required for the test to be valid.

7 A population Normal distribution with unknown variance is being tested at the 5% significance level with hypotheses $H_0: \mu_0 = 129.1$ and $H_1: \mu_0 < 129.1$

   This sample is taken.

153.9	26.53	120.8
183.5	75.95	

   a Find the critical region,

   b Calculate the t-test statistic,

   c Determine the conclusion of the test.

8 A student wants to see whether the average height of a student in their school has changed over the last ten years. They obtain a list of all the students currently in the school and the previous average value.

   a Could a t-test be suitable for this investigation?

   b What would the student need to do with the sample to ensure that a valid test is performed?

Reasoning and problem-solving

The *t*-distribution can also be used to generate confidence intervals for the mean of a Normal distribution when the sample size is small and the population variance is unknown. A rule of thumb is that a sample of size at least 30 is 'large enough' for the standard Normal distribution to be used. For a sample of size n taken from a Normal distribution $N(\mu, \sigma^2)$, the p%-confidence interval is

See Ch11.3
For a reminder of confidence intervals.

$$\bar{x} - t \times \frac{s}{\sqrt{n}} < \mu < \bar{x} + t \times \frac{s}{\sqrt{n}}$$

where t is drawn from the *t*-distribution with $n-1$ degrees of freedom, $\bar{x}$ is the sample mean and s^2 is the sample variance. For a p%-confidence interval $t = t_{n-1}^{-1}\left(\dfrac{1+p}{2}\right)$, you can work these out yourself using your calculator.

Example 3

A sample of size 16 is taken, whose mean is 13.6 and whose standard deviation is 20.4, in order to generate a 95% confidence interval for the mean of the population. Find the confidence interval.

For $p = 95\%$ and with $16-1 = 15$ degrees of freedom, $t_{15}^{-1}(0.975) = 2.13$

The interval $13.6 - 2.13 \times \dfrac{20.4}{\sqrt{16}} < \mu < 13.6 + 2.13 \times \dfrac{20.4}{\sqrt{16}}$

simplifies to $2.74 < \mu < 24.46$

Strategy

When using a *t*-test

(1) Use a *t*-test if a sample comes from a Normal population with unknown variance.

(2) For better estimates of population parameters, use a larger sample.

Key point

If the population variance is known, then a *t*-test is unsuitable. Instead, use a regular Normal distribution test.

The *t*-distribution is used when only a small sample is available. As the sample size becomes larger, the distribution approaches the standard Normal distribution. In the limit as the number of degrees of freedom tends to infinity, the *t*-distribution does become the standard Normal distribution, as the sample approaches becoming the whole population.

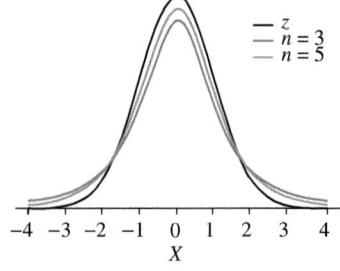

A good rule of thumb is that if the sample size is larger than 30, then the *t*-distribution is numerically close enough to the standardised Normal distribution, Z, so that Z can be used instead.

STATS

Example 4

Two samples are taken from a Normal distribution with unknown mean and variance. The first sample is of size 12; it has sample mean 3.75 and sample variance 2.9. The second sample is of size 7; it has sample mean 5.1 and sample variance 4.3. The two samples are combined. The samples are tested to see whether or not the population mean is 4.5

a Calculate the t-test statistic for each sample,

b State the combined sample size,

c Calculate the combined sample mean,

d Calculate the combined sample variance,

e Calculate the t-test statistic for the combined sample.

a $t = \dfrac{3.75 - 4.5}{\dfrac{\sqrt{2.9}}{\sqrt{12}}} = -1.526$ for sample 1; $t = \dfrac{5.1 - 4.5}{\dfrac{\sqrt{4.3}}{\sqrt{7}}} = 0.766$ for sample 2

b 19

c $\dfrac{(12 \times 3.75 + 7 \times 5.1)}{19} = 4.25 = \overline{X}$

d The combined sample variance is $\dfrac{1}{19-1}\left(\sum X_i^2 - 19 \times \overline{X}^2 \right)$. To find $\sum X_i^2$ you need the contributions from each sample, which can be found by rearranging the variance formula. For the first sample it is $(12-1) \times 2.9 + 12 \times 3.75^2 = 200.65$
For the second sample it is $(7-1) \times 4.3 + 7 \times 5.1^2 = 207.87$. Their sum is 408.52 which, along with $\overline{X} = 4.25$, gives the combined variance as 3.63

e The combined sample t-test statistic is $t = \dfrac{4.25 - 4.5}{\dfrac{\sqrt{3.63}}{\sqrt{19}}} = -0.572$

Exercise 27.2B Reasoning and problem-solving

1 A sample of size 1000 is taken from a Normally distributed population with unknown mean and variance. The sample has mean 14.8 and variance 91.6

The hypotheses $H_0: \mu_0 = 12.1$ and $H_1: \mu_0 > 12.1$ are tested at the 5% level.

a Explain why a Normal test is required instead of a t-test,

b Calculate the test statistic,

c Determine the conclusion of the test.

2 A sample is taken from a Normally distributed population with unknown mean and variance. The hypotheses $H_0: \mu_0 = 13.4$ and $H_1: \mu_0 < 13.4$ are tested at the 10% level.

a Perform the test using the first row of five values,

b Perform the test using the first two rows (ten values total),

c Perform the test using all fifteen values.

−16.8	86.1	15.1	−149	96.8
−62.7	−10.3	−114	−52.5	−42.0
32.6	51.4	3.01	27.9	42.4

3 Two samples are taken from a Normal distribution with unknown mean and variance. The first sample is of size 8; it has sample mean 14.3 and sample variance 18.5

The second sample is of size 11; it has sample mean 12.7 and sample variance 22.3 The two samples are combined. The samples are tested with the hypotheses $H_0: \mu_0 = 9.1$ and $H_1: \mu_0 > 9.1$ at the 10% level.

a Calculate the combined sample mean,

b Calculate the combined sample variance,

c Calculate the t-test statistic for the combined sample.

4 The amount of sugar in grams found in a can of drink is measured. A sample of size 8 has mean 5.19 g and variance 2.34 g².

a Describe how to take such a sample,

b Give one assumption that must be made about the sample in order for a t-test to be appropriate,

c The intended sugar level is 4.26 g. Perform a t-test at the 5% level and state the conclusion of the test.

5 The distance at which a professional darts player throws a dart away from the bullseye is believed to have mean 0.31 cm. The hypotheses $H_0: \mu_0 = 0.31$ and $H_1: \mu_0 > 0.31$ are tested at the 5% level. They throw 12 darts and the sample has mean 0.68 cm and standard deviation 0.539 cm.

a Calculate the p-value of the statistic,

b Determine the conclusion of the test,

c Discuss the claim that the dart thrower's mean value being higher than the null hypothesis mean suggests that they throw inconsistently.

6 Two samples are taken from a Normal distribution with unknown mean and variance. The first sample is of size 10; it has sample mean 12.5 and sample variance 14.3

The second sample is of size 15; it has sample mean 12.3 and sample variance 27.3 The two samples are combined. The samples are tested with the hypotheses $H_0: \mu_0 = 14.7$ and $H_1: \mu_0 < 14.7$ at the 5% level.

a Calculate the t-test statistic for each sample and determine the conclusion of the test for each individual sample,

b Calculate the t-test statistic for the combined sample and determine the conclusion of the test for the combined sample.

7 The lengths of pipes produced by machines are Normally distributed with unknown variance. An engineer has two sets of pipes but has forgotten whether they came from the same machine. The machines are supposed to produce pipes with a mean length of 30.0 cm.

a The first sample of 14 pipes has sample mean 29.91 cm and sample variance 0.0576 cm². Test at the 10% level the hypotheses $H_0: \mu_0 = 30.0$ and $H_1: \mu_0 \neq 30.0$

b The second sample of 12 pipes has sample mean 30.07 cm and sample variance 0.0289 cm². Test at the 10% level the hypotheses $H_0: \mu_0 = 30.0$ and $H_1: \mu_0 \neq 30.0$

c Should the engineer conclude that the two sets of pipes have come from the same machine?

8 For each of these samples find a 90%-confidence interval for the population mean using an appropriate t-distribution.

a 11.6, 12.1, 12.5, 12.7, 13.0, 13.2, 14.8

b 3.2, 7.1, 8.8, 9.0, 9.5, 10.1, 10.9, 11.8, 12.1, 12.1, 12.7, 13.1, 15.3, 16.5

9 Find 99%-confidence intervals for samples with these summary statistics.

a $n = 9, \sum x = 83.5, \sum x^2 = 1296.9,$

b $n = 24, \sum x = 260, \sum x^2 = 3000$

10 Find the 95%-confidence interval for a sample with mean 25 and variance 25 when n takes the following values.

a $n = 5$ **b** $n = 10$ **c** $n = 25$

Chapter summary

- A **type I error** is when a null hypothesis that is true is rejected. The probability of such an error is denoted by α
- A **type II error** is when a null hypothesis that is false is accepted. The probability of such an error is denoted by β
- The **power** of a hypothesis test is the probability that a false null hypothesis is rejected. It is equal to $1 - \beta$
  This depends on the true parameter value and is rarely a fixed value.
- For a given sample size, decreasing the significance level will decrease α and increase β
- For a sample of size n taken from a Normally distributed population, the distribution of the sample mean follows a t-distribution with $n - 1$ degrees of freedom.
- The mean of the distribution of the sample mean is likely to be similar to the mean of the population, but the variance will be larger.
- A t-test may be used when the population is Normally distributed but only a small sample can be obtained; smaller than 30 is a good rule of thumb.
- The t-test determines whether a suggested value appears to be a suitable description of the population mean.

Check and review

You should now be able to...	Review Questions
✔ Calculate the probabilities of making type I and type II errors.	1, 2, 3
✔ Calculate the power of a hypothesis test.	1, 2
✔ Calculate the t-test statistic from a sample.	4, 5, 7
✔ Identify when a t-test is appropriate.	5
✔ Determine the conclusion of a t-test.	6, 7

1 A test is performed at the 5% significance level.

   a Write down an inequality involving α, the probability of making a type I error.

   The null hypothesis is H_0: $p = 15$ and the test has power P_1 when the true value is $p = k$, where $k > 15$

   b Write down an inequality for the power P_2 of the test when the true value is $p = q$, where $q > k$

2 A Poisson distribution is tested at the 10% level with hypotheses H_0: $\lambda = 12.4$ and H_1: $\lambda > 12.4$

   a Find the critical region,

   b Calculate the probability of a type I error.

   The true value is actually $\lambda = 18.3$

   c Calculate

   i The probability of making a type II error,

   ii The power.

3 A Normal distribution with standard deviation 18.2 is being tested at the 2% significance level. The null hypothesis is $H_0: \mu = 30.1$ and the alternative is $H_1: \mu < 30.1$

  a State the probability of a type I error,

  b Find the probability of a type II error if actually $\mu = 15.2$

  c Find the probability of a type II error if actually $\mu = -15.2$

4 Calculate the t-test statistics for the following samples. The mean is 26.7 in each case.

  a 16.6 24.6 27.9 29.3 49.9

  b 12.4 15.3 18.1 21.0 28.1 31.0 33.9

  c −18.2 −8.83 6.13 23.0 24.8 41.7 47.3 49.1 60.4

5 A sample of size 23 is taken from a Normal distribution with sample mean 106.7 and sample variance 533

  The hypotheses $H_0: \mu_0 = 104.5$ and $H_1: \mu_0 \neq 104.5$ are tested at the 5% level.

  a Explain why a t-test is appropriate for this situation,

  b Calculate the t-test statistic.

6 A population Normal distribution with unknown variance is being tested at the 5% level with hypotheses $H_0: \mu_0 = -5.14$ and $H_1: \mu_0 < -5.14$

  For a sample of size 7, the t-test statistic is −2.87

  a State the number of degrees of freedom,

  b Calculate the p-value of the statistic,

  c Determine the conclusion of the test.

7 Two samples are taken from a Normal distribution with unknown mean and variance. The first sample is of size 4; it has sample mean 71.6 and sample variance 23.8 The second sample is of size 6; it has sample mean 72.5 and sample variance 31.9

  The two samples are combined. The samples are tested with the hypotheses $H_0: \mu_0 = 105$ and $H_1: \mu_0 < 105$ at the 5% level.

  a Calculate the combined sample mean,

  b Calculate the combined sample variance,

  c Calculate the t-test statistic for the combined sample.

Did you know?

Student's t-distribution was developed by William Sealy Gosset. Gosset worked for Guinness who had banned publications by employees to prevent competing brewers from using their techniques. To get around this problem Gosset published under the pseudonym 'Student' – hence the name of the test.

   Gosset was known for his modesty and once remarked that 'Fisher would have discovered it all anyway' in reference to his friend Ronald Aylmer Fisher.

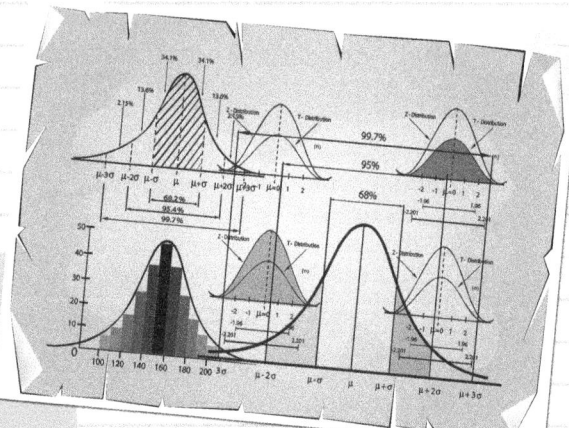

Quotation

With regard to the ubiquity of the Normal curve, Poicare has been attributed as saying 'Everyone believes in it: experimentalists believing that it is a mathematical theorem, mathematicians believing that it is an empirical fact.'

Did you know?

You have studied type I and type II errors but you can also have type III errors. There are a number of interpretations for these errors of the third kind that you might like to look up. One popular interpretation is when you correctly reject a null hypothesis but for the wrong reasons!

27 Assessment

1 A Poisson distribution is being tested with null hypothesis $\lambda = 4.2$ and alternative hypothesis $\lambda > 4.2$

    a   Find the critical value for the test at a 5% significance level, [1]

    b   Find the probability of a type I error. [1]

    The true parameter value is $\lambda = 8$

    c   Find the probability of a type II error. [2]

2 A biased coin is thrown 20 times and the number of heads noted. It is suggested that the probability of a head is 0.3

    a   State the null and alternative hypotheses in a test of this claim, [1]

    b   Find the critical region for the test using a significance level of 10% [3]

    c   Assuming that the probability of obtaining a head is actually 0.4, find the probability of making a type II error, [2]

    d   Write down the power of the test. [2]

3 The mean of a Normal distribution with standard deviation 12.1 is being tested at the 5% level. A sample of size 15 is used. The null hypothesis is $\mu = 24$ and the alternative hypothesis is $\mu > 24$. The true mean is 30

    a   Find the critical x-value for this test, [7]

    b   Find the probability of a type II error, [4]

    c   Calculate the power of the test. [1]

4   a   Explain the meaning of the terms 'type II error' and 'the power of a hypothesis test'. [2]

    b   In the manufacture of soft drinks, a machine fills cans with cola. The nominal volume of cola in a can is 330 ml and it is known that the standard deviation of the volume per can is 4 ml. The manufacturer claims that the machine dispenses a mean volume greater than 330 ml. To test this claim, 10 cans are chosen at random and the mean volume calculated. You should assume that the quantity of drink in a randomly chosen can follows a Normal distribution.

       i   State the null and alternative hypotheses for this test, [1]

       ii   Find the set of sample mean values which would lead to the null hypothesis being accepted. [6]

    It turns out that the population mean volume per can was 331.4 ml.

       iii   Find the probability of making a type II error, [3]

       iv   Write down the power of the test. [1]

5 A sample of size 12 is taken from a population modelled by a Normal distribution with unknown mean and variance. The sample values are:

54.1 50.9 46.3 49.2 56.3 51.1 50.6 48.3 47.2 51.4 52.0 50.3

The population mean is believed to be 52.0 and a 5% significance test is to be performed.

a Find the sample mean and variance, [3]

b Write the test statistic and calculate its value, [3]

c Calculate an appropriate p-value and state clearly your conclusion. [2]

6 a A sample of size 10 is taken from a population with unknown mean μ and variance σ^2

Explain why you should **not** assume that the test statistic $\dfrac{\overline{X}-\mu}{\frac{s}{\sqrt{n}}}$, where s is the sample

standard deviation, follows a standard Normal distribution, [1]

b A sample of size 10 is taken from a Normally distributed population and has mean 12.1 and variance 4.4

The population mean was known to be 10.2 but is now thought to have increased. You wish to test this belief.

i Write down the null and alternative hypotheses for this test, [1]

ii Write down the test statistic and calculate its value, [2]

iii By calculating the appropriate p-value, perform the test at a significance level of 5% [2]

7 In a trial on the effectiveness of drugs designed to reduce cholesterol levels, 9 participants received a course of statins. The reduction in cholesterol level (mmol L^{-1}) for each person is as follows (– denotes an increase in level):

0.3 1.1 1.2 1.4 −0.3 0.5 −0.2 0.4 −0.5

a State the null and alternative hypotheses for a test of the effectiveness of the treatment, [1]

b Test at a 5% significance level whether the statin treatment affects cholesterol levels. [7]
   You should state any assumptions you make.

c Comment on the confidence with which your final conclusion can be drawn. [1]

8 A sample of size 8 was taken from a Normal population and the values were as follows:

$$16.3 \quad 18.2 \quad 13.4 \quad 11.1 \quad 21.5 \quad 15.3 \quad 18.1 \quad 12.4$$

Find a 95% confidence interval for the mean. [7]

9 A sample of size 14 was taken from a Normal population and gave the following statistics:

$$\sum x = 94.2, \ \sum x^2 = 822.3$$

Find a 98% confidence interval for the mean. [7]

10 In an investigation into crop yield, 13 identical test beds are sown with a crop and the yield from each is noted. A 95% confidence interval for the population mean yield is (43.41, 44.99)

a State a necessary assumption related to the crop yield distribution, [1]

b Explain why it is not acceptable to use a Normal distribution to investigate the confidence interval for the population mean, [1]

c Find the sample mean and standard deviation. [5]

28 Graphs and networks 2

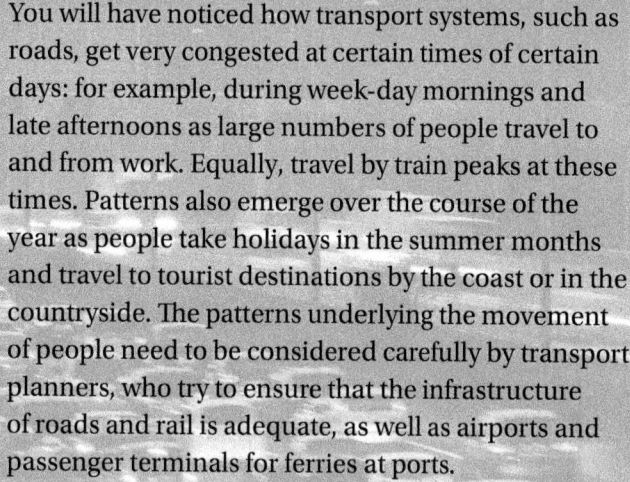

You will have noticed how transport systems, such as roads, get very congested at certain times of certain days: for example, during week-day mornings and late afternoons as large numbers of people travel to and from work. Equally, travel by train peaks at these times. Patterns also emerge over the course of the year as people take holidays in the summer months and travel to tourist destinations by the coast or in the countryside. The patterns underlying the movement of people need to be considered carefully by transport planners, who try to ensure that the infrastructure of roads and rail is adequate, as well as airports and passenger terminals for ferries at ports.

Transport planners use the mathematics of flow in such strategic planning to make sure that they can cater for the transportation expectations of workers, holidaymakers and other travellers. For example, it informs how many trains can travel on any section of the rail network and this, in turn, informs how many trains need to be bought. On the roads it informs the development of the 'smart motorway' network that seeks to optimise the flow of traffic along the busiest roads in the country.

Orientation

What you need to know	What you will learn	What this leads to
Ch12 Graphs and networks 1	• To construct the complement of a graph. • To identify isomorphic graphs. • To use Euler's formula. • To use Kuratowski's theorem. • To find the maximum flow through a network. • To solve problems involving multiple sources/sinks.	**Careers** • Transport industry. • Operational research.

Fluency and skills

See Ch12
For a reminder on graphs.

This section expands on some of the properties of graphs that were covered in the first year of this course. Recall that:

- A graph is a set of points (**vertices**) connected by a set of lines (**edges**).
- A graph is **connected** if, for every two vertices, there is a **trail** (a sequence of edges) joining one to the other.
- A graph may have loops (edges starting and ending at the same vertex) or multiple edges (two or more edges connecting a pair of vertices). A **simple graph** has no loops or multiple edges.
- A graph with 'one-way streets' is a **directed graph** or **digraph**.
- A simple graph with n vertices in which there is an edge connecting every pair of vertices is the **complete graph K_n** that has $\dfrac{n(n-1)}{2}$ edges.
- A graph with two distinct sets of vertices, in which every edge connects a vertex from one set with a vertex from the other, is a **bipartite graph**. If every vertex in a set of m vertices connects with every vertex in a set of n vertices, you have the **complete bipartite graph $K_{m,n}$** that has mn edges.
- The number of edges meeting at a vertex is its **degree**.
- The sum of the degrees $= 2 \times$ the number of edges (this is the **handshaking lemma** – each edge 'shakes hands' with two vertices).
- A trail which doesn't visit any vertex more than once (except perhaps the start/finish) is a **path**.
- A path which starts and ends at the same vertex is a **cycle**.
- A graph in which there is a closed trail that uses every edge of the graph only once is **traversable** or **Eulerian**. In an Eulerian graph, all vertices are of even degree.
- A cycle which visits every vertex of the graph once and only once and which returns to the starting vertex is a **Hamiltonian cycle**.
- A connected graph with no cycles is a **tree**. A **spanning tree** for a graph is a subgraph, which is a tree connecting all vertices. The spanning tree for a graph with n vertices has $(n-1)$ edges.
- A **subgraph** of a graph G is a graph formed by some of the vertices and edges of G
- A **subdivision** of a graph G is a graph formed by adding vertices of degree 2 along edges of G

Consider Graph G

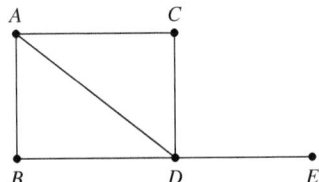

To describe G you can list the **vertex set** $\{A, B, C, D, E\}$ together with the **edge set** $\{AB, AC, AD, BD, CD, DE\}$ of G

Alternatively, you can construct the **adjacency matrix** of G. This shows the number of connections between each pair of vertices (an undirected loop would be recorded as 2).

	A	B	C	D	E
A	0	1	1	1	0
B	1	0	0	1	0
C	1	0	0	1	0
D	1	1	1	0	1
E	0	0	0	1	0

The **complement** of a simple graph G, denoted by G' (in some books by $\overline{G}$), and has the same vertices as G but contains only the edges which are *not* in G. It looks like this.

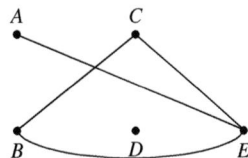

You find the adjacency matrix for G' by changing 1s to 0s and vice versa (ignoring the leading diagonal).

	A	B	C	D	E
A	0	0	0	0	1
B	0	0	1	0	1
C	0	1	0	0	1
D	0	0	0	0	0
E	1	1	1	0	0

If you combine G and G' you get the complete graph K_5

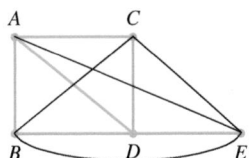

In general, if a simple graph has n vertices and you combine it with its complement, you get the complete graph K_n

Graphs G_1 and G_2 are **isomorphic** if they have the same structure. This means that for each vertex in G_1 there is a corresponding vertex in G_2, so that if an edge joins two vertices in G_1, then there is an edge between the corresponding vertices in G_2, and vice versa. Provided you list corresponding vertices in the same order, the adjacency matrices will look identical.

For example, look at these graphs.

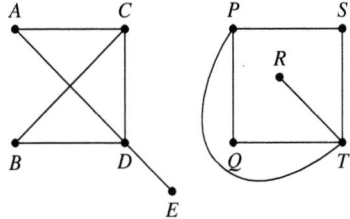

If you pair vertices A, B, C, D and E with vertices Q, S, P, T and R, respectively, then the edge set $\{AC, AD, BC, BD, CD, DE\}$ of the first graph corresponds to the edge set $\{QP, QT, SP, ST, PT, TR\}$ of the second graph. The adjacency matrices are

	A	B	C	D	E
A	0	0	1	1	0
B	0	0	1	1	0
C	1	1	0	1	0
D	1	1	1	0	1
E	0	0	0	1	0

	Q	S	P	T	R
Q	0	0	1	1	0
S	0	0	1	1	0
P	1	1	0	1	0
T	1	1	1	0	1
R	0	0	0	1	0

When looking for corresponding vertices, it can help to use the degrees of the vertices – for example, in this case, you would expect to pair D with T because each has degree 4. However, just because two graphs have the same set of degrees you cannot assume they are isomorphic. For example, look at these graphs.

They each have one vertex of degree 3, three vertices of degree 2 and three of degree 1. However, they are clearly not isomorphic.

Example 1

The diagram shows the graph G

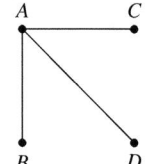

a Construct the adjacency matrix for G

b Find the adjacency matrix for the complement, G' and sketch G'

c Show that G and G' are not isomorphic.

d Find a graph H with 4 vertices such that H and H' are isomorphic.

a

	A	B	C	D
A	0	1	1	1
B	1	0	0	0
C	1	0	0	0
D	1	0	0	0

b

	A	B	C	D
A	0	0	0	0
B	0	0	1	1
C	0	1	0	1
D	0	1	1	0

Change 1s to 0s and 0s to 1s except on the leading diagonal.

(Continued on the next page)

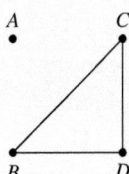

c G has vertices with degrees 3, 1, 1, 1. G' has degrees 2, 2, 2, 0, so they are not isomorphic.

d K_4 has $\dfrac{4 \times 3}{2} = 6$ edges and therefore the sum of the degrees of the vertices is 12

If H has 4 vertices and H and H' are isomorphic, then each has 3 edges, the degrees of their vertices are identical and each total 6. The degrees must be 2, 2, 1, 1 because the degrees of the complement would then be 1, 1, 2, 2

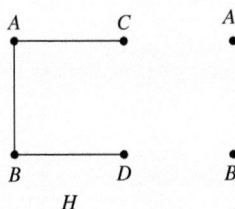 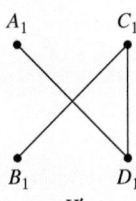

These are isomorphic. If you pair $\{C, A, B, D\}$ with $\{A_1, D_1, C_1, B_1\}$ in that order, then the edges $\{CA, AB, BD\}$ correspond to $\{A_1D_1, D_1C_1, C_1B_1\}$.

A graph is **planar** if it can be drawn without edges intersecting except at a vertex.

> **Key point**
>
> If a simple, connected graph drawn in the plane has V vertices, E edges and F faces (including the 'infinite face' surrounding the graph), then $V + F - E = 2$ (Euler's formula). (If a graph is drawn in the plane then it is planar.)

A face in a plane drawing is a region bounded by edges.

- K_n is non-planar if $n \geq 5$
- $K_{m,n}$ is non-planar if $m \geq 3$ and $n \geq 3$

There are some useful corollaries (consequences) of Euler's formula.

- If a graph drawn in the plane has no vertices of degree 1, loops or multiple edges then every edge borders two faces and every face has at least three edges.

Edges leading to vertices of degree 1, multiple edges and loops can all be temporarily removed from a graph and then added back once the graph has been drawn in the plane.

It follows that $2E \geq 3F$. Substituting this into Euler's formula gives $E \leq 3V - 6$. If $V = 5$, then $E \leq 9$. As K_5 has 10 edges, it follows that K_5 is non-planar.

- If you have a bipartite graph, every edge is associated with two faces and every face has at least four edges. It follows that $2E \geq 4F$. Substituting this into Euler's formula gives $E \leq 2V - 4$. If $V = 6$, then $E \leq 8$. As $K_{3,3}$ has 9 edges, it follows that $K_{3,3}$ is non-planar.

> Notice that if $E \leq 3V - 6$, it does not mean that the graph is planar. However, if $E > 3V - 6$, then the graph is non-planar.

Key point

A graph is non-planar if and only if it has a subgraph which is a subdivision of K_5 or $K_{3,3}$ (**Kuratowski's theorem**).

For example:

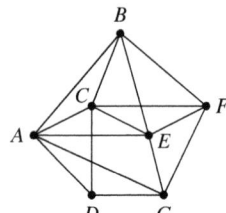 can be drawn as 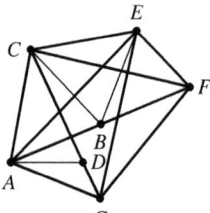 with subgraph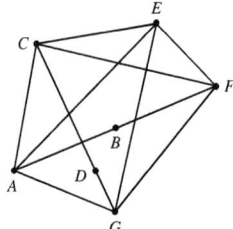

This is a subdivision (by vertices B and D) of the K_5 graph formed by A, C, E, F and G. The original graph is therefore non-planar.

Exercise 28.1A Fluency and skills

1

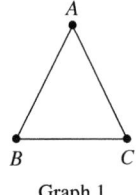

Graph 1

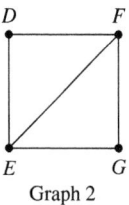

Graph 2

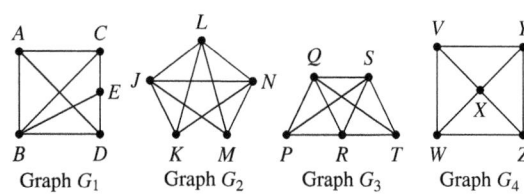

Graph G_1 Graph G_2 Graph G_3 Graph G_4

Identify

a A sub-graph of graph 2 which is isomorphic to Graph 1

b A sub-graph of graph 2 which is isomorphic to a subdivision of Graph 1

2 Identify a sub-graph of this graph which is a subdivision of the complete graph K_4

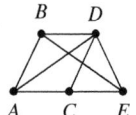

3 Identify a sub-graph of this graph which is isomorphic to the complete bipartite graph $K_{3,2}$

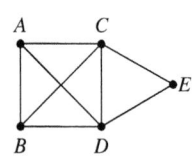

4 For graphs G_1, G_2, G_3 and G_4 shown, identify any pairs which are isomorphic. For each isomorphic pair you find, state a correspondence between their vertices and edges.

5 State whether each of these graphs is planar or non-planar. If the graph is planar, show this by redrawing it. If it is non-planar, explain using Kuratowski's theorem why this is the case.

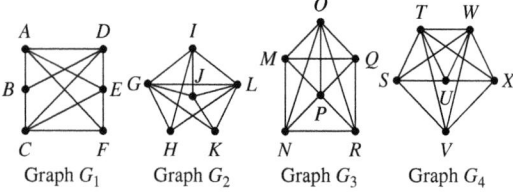

Graph G_1 Graph G_2 Graph G_3 Graph G_4

6 a Draw a graph to show that $K_{4,2}$ is planar.

b Draw the complement of $K_{4,2}$

7 This table is the adjacency matrix for the graph G

	A	B	C	D	E
A	0	0	1	0	1
B	0	0	0	1	1
C	1	0	0	0	0
D	0	1	0	0	1
E	1	1	0	1	0

a Construct the adjacency matrix of the graph G', the complement of G

b Draw the graphs G and G'

c State, with reasons, whether G and G' are isomorphic.

8 Show, by finding a sub-graph which is a subdivision of $K_{3,3}$, that this graph is non-planar.

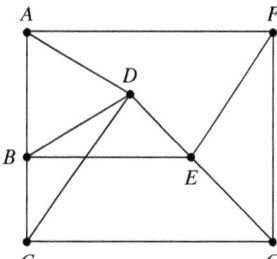

Reasoning and problem-solving

To solve problems in graph theory

1. Draw clear, labelled diagrams.

2. Use the correct terminology.

3. Use theorems and rules to prove results.

4. Answer the question.

Example 2

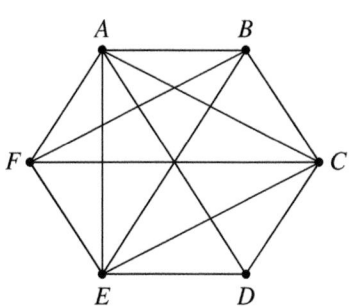

a By considering the number of edges and vertices, show that the graph is non-planar,

b Use Kuratowski's theorem to show that if CE is removed the graph is still non-planar.

(*Continued on the next page*)

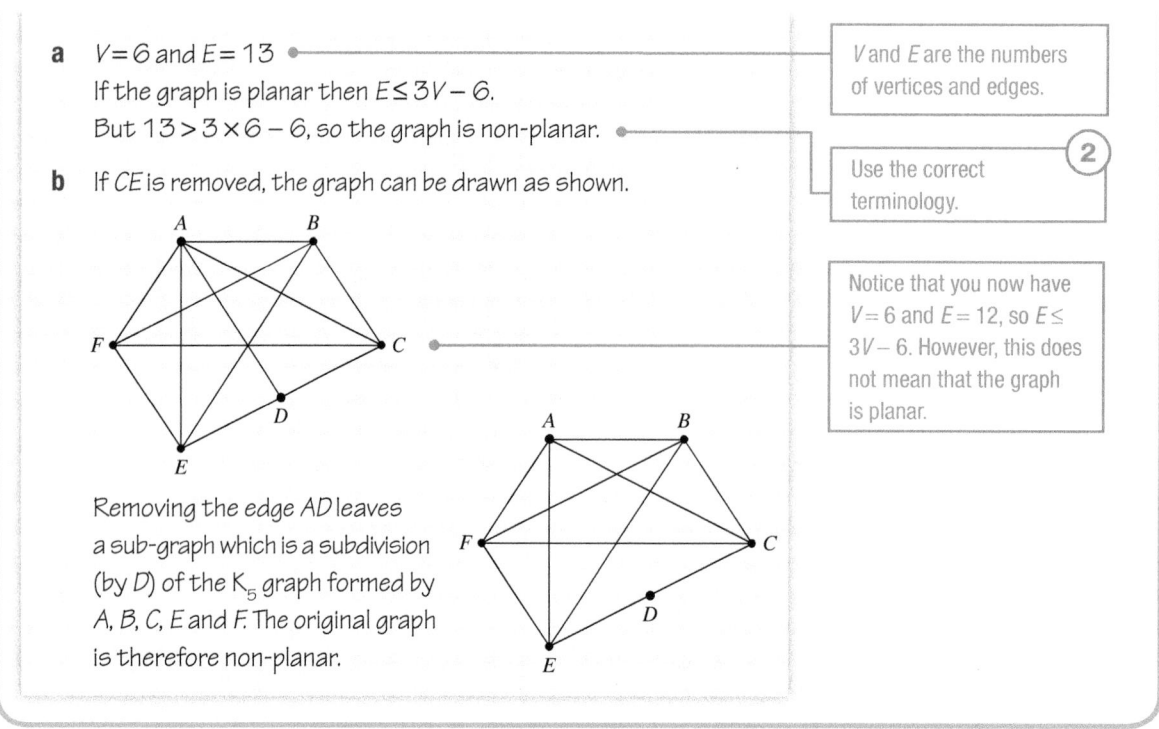

a $V = 6$ and $E = 13$

If the graph is planar then $E \leq 3V - 6$.

But $13 > 3 \times 6 - 6$, so the graph is non-planar.

> V and E are the numbers of vertices and edges.

> ② Use the correct terminology.

b   If CE is removed, the graph can be drawn as shown.

> Notice that you now have $V = 6$ and $E = 12$, so $E \leq 3V - 6$. However, this does not mean that the graph is planar.

Removing the edge AD leaves a sub-graph which is a subdivision (by D) of the K_5 graph formed by A, B, C, E and F. The original graph is therefore non-planar.

Exercise 28.1B Reasoning and problem-solving

1 T is a spanning tree of K_n

   a   State the number of edges in T

   b T' is the complement of T. For what value(s) of n could T and T' be isomorphic? Draw an example where this is true.

2   **a**   Find the number of distinct, simple graphs with 3 vertices (ignoring isomorphisms).

   b   Find the number of distinct, simple, connected graphs with 4 vertices.

   c   A graph has 3 vertices, each of degree 4 (the graph cannot be simple).

      i   If the graph is connected, how many distinct graphs are possible?

     ii   How many more are there if they do not have to be connected?

3   A simple, connected graph has n vertices, each of degree d

   a   If the graph is non-Eulerian, what can you say about n?

   b   If $n = 6$, for what values of d is the graph planar?

4   A coding system uses three-digit binary codes $A(000)$, $B(001)$, $C(010)$, $D(011)$, $E(100)$, $F(101)$, $G(110)$ and $H(111)$. A single-bit transmission error could change, for example, A into E

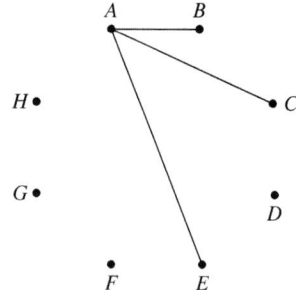

a Copy and complete this graph in which an edge connects codes which are just one error apart.

b How long is the shortest path from

    i A to D

    ii A to H?

c Show by drawing that the graph is planar.

5 The diagram shows the Petersen graph, which occurs in many mathematical topics related to graph theory.

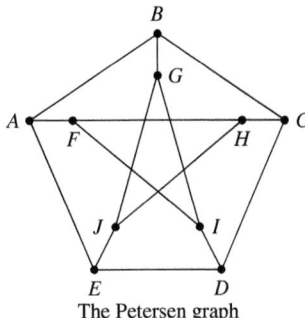

The Petersen graph

By considering the vertex sets $\{A, C, G\}$ and $\{B, E, F\}$, make a drawing to show that the Petersen graph has a subgraph which is a subdivision of $K_{3,3}$ and hence is non-planar.

6 a Draw a graph with four vertices which is isomorphic to its complement.

b Draw two distinct graphs with five vertices which are each isomorphic to their complement.

c A graph G has n vertices.

    i If $n = 6$, explain why it is not possible for G and G' to be isomorphic.

    ii For what values of n could G and G' be isomorphic?

7 The diagram shows a graph G which is not connected.

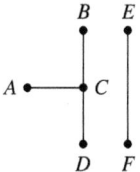

a Draw the complement, G'

b The vertices of a disconnected graph F are in two separate disconnected sets, $\{A_1, A_2, ... A_m\}$ and $\{B_1, B_2, ..., B_n\}$, each of which forms a connected subgraph of F

For what values of m and n is F' bound to be non-planar?

c In **a**, the complement of a disconnected graph was connected. Prove that this is always the case, that is, for any disconnected graph H, the complement H' is connected.

8 Use Kuratowski's theorem to show that this graph is non-planar.

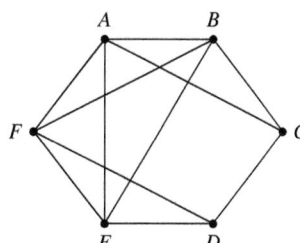

Fluency and skills

See Ch12.8
For a
reminder on
networks.

In Chapter 12 you met the idea of a commodity flowing through a directed network from a **source** (usually called S) to a **sink** (usually called T).

Each arc has a **capacity** and the **flow** along it cannot be more than this. (This is the **feasibility condition**.) Together these flows form the **flow in the network**.

If the flow in an arc equals its capacity you say it is **saturated**.

At every node of the network the total inflow equals the total outflow. (This is the conservation condition.)

The total outflow from S equals the total inflow at T. This quantity is the **value of the flow**. The **capacity of the network** is the value of the maximum possible flow.

You describe a **cut** either by listing the arcs in the cut (the cut set), or by listing the nodes in the source set X and in the sink set Y

For this network, cut 2 has cut set $= \{AC, BC, BD\}$, source set $X = \{S, A, B\}$ and sink set $Y = \{C, D, T\}$.

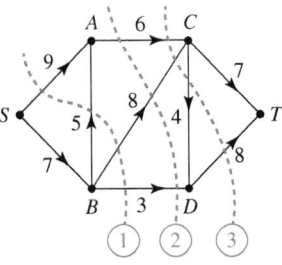

The capacity of a cut is the sum of the capacities of those arcs of the cut which are directed from X to Y. In this network, cut 1 has a capacity of 25, cut 2 a capacity of 17 and cut 3 a capacity of 22

The flow in the network cannot be more than the capacity of any cut. This gives the **maximum flow, minimum cut theorem**.

> **Key point**
>
> The value of the maximal flow = the capacity of a minimum cut
> It follows that if you find a flow and a cut such that
> $$(\text{value of flow}) = (\text{capacity of cut})$$
> then the flow is a maximum and the cut is a minimum.

You need a systematic way of finding the maximal flow. You start with an initial flow of some sort (which could just be zero, although you will usually do better than this) and look for ways to **augment** (increase) it. This **flow augmentation** process may happen several times until you have the maximal flow.

For example, look at this diagram.

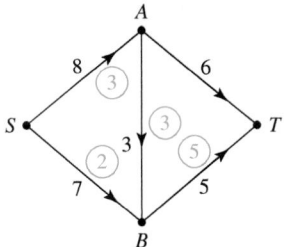

It shows a network with an initial flow of 3 along *SABT* and 2 along *SBT*. (The circled values show the flow in each arc.)

You subtract these flows from the capacities of the arcs to see how much spare capacity remains, as shown in this diagram. You can see that a flow of 5 is possible along *SAT*. *SAT* is called a **flow-augmenting path**.

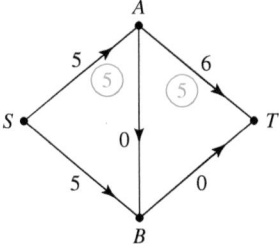

You again subtract this flow to see what capacity remains.

There is no obvious flow-augmenting path, but you have not yet reached the maximal flow.

If you increase the flow in *SB* and *AT* by 1, and **reduce** the flow in *AB* by 1, the overall flow increases by 1

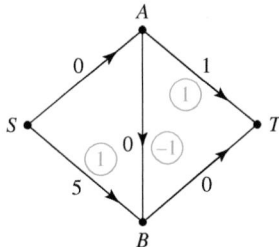

You now have an overall flow of 11, as shown.

The arcs *AT* and *BT* are saturated, so the flow is maximal.

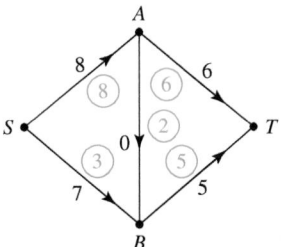

It is not easy to spot flow-augmenting paths, especially if they involve reducing a flow. It is also tedious to keep redrawing the diagram. The **labelling procedure** overcomes these problems.

Suppose an arc AB of capacity 12 has a flow of 8 along it. So far you have shown this as

$$A \bullet \xrightarrow{\quad 12 \quad} \bullet B$$
$$(8)$$

You could increase the flow by 4, this is the **potential flow**.

You could decrease the flow by 8, this is the **potential backflow**.

You show these as

$$A \bullet \xrightarrow{\quad 4 \rightarrow \quad} \bullet B$$
$$\leftarrow 8$$

The forward arrow represents the spare capacity while the backward arrow represents the actual flow. The total of the two gives the capacity of the arc.

To find a flow-augmenting path you look for a route from S to T where all the potential flows (arrows pointing forwards along the route) are greater than zero.

> **Key point**
>
> A flow-augmenting path is a route from S to T where the **arrows pointing forward** all have **non-zero values**.
>
> The lowest of these values gives the possible extra flow along that path.

> Notice that the labelling procedure automatically reduces the flow along AB by 1 in the second stage of this solution.

Example 1

Starting with an initial flow of 3 along the route $SABT$ and 2 along SBT, find a maximal flow through this network.

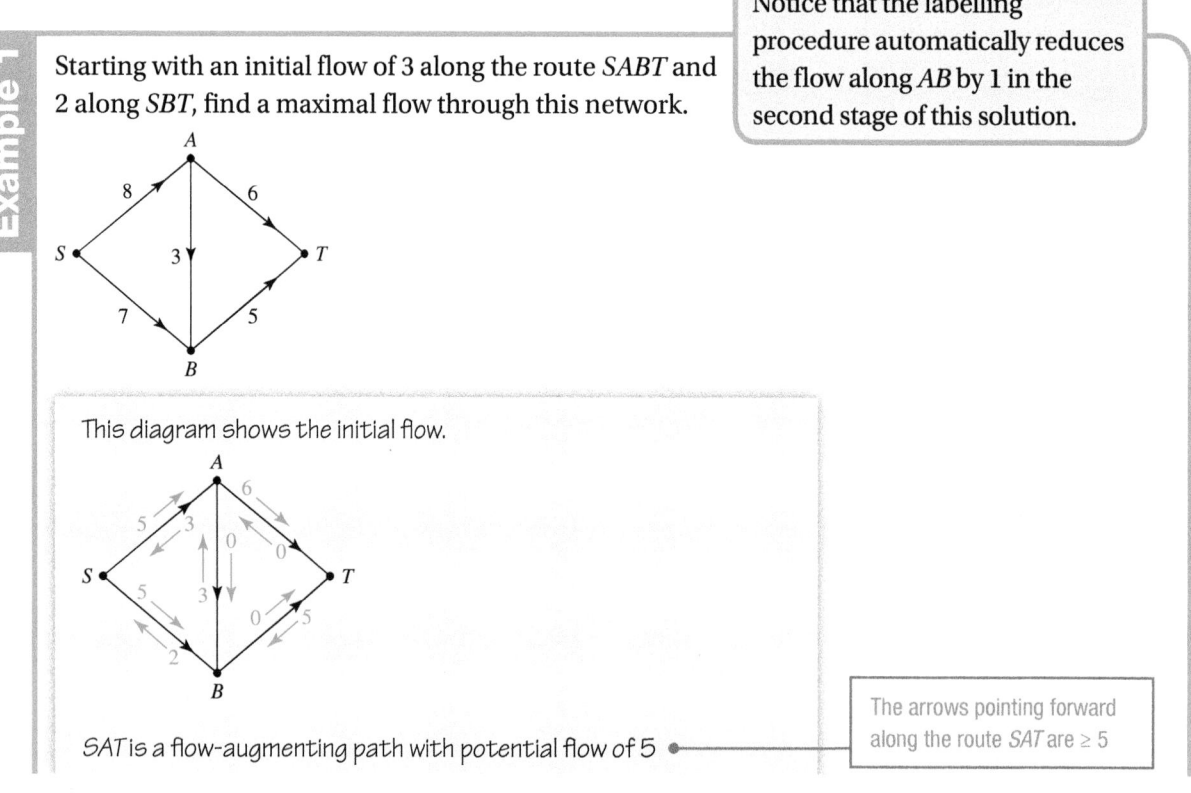

This diagram shows the initial flow.

SAT is a flow-augmenting path with potential flow of 5

> The arrows pointing forward along the route SAT are ≥ 5

(Continued on the next page)

Updating the labels gives this flow.

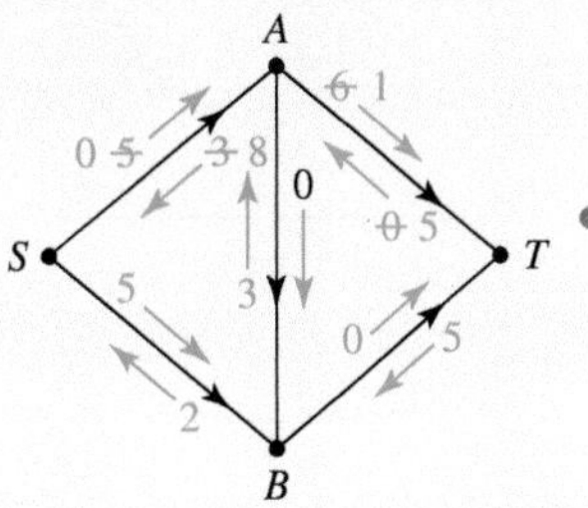

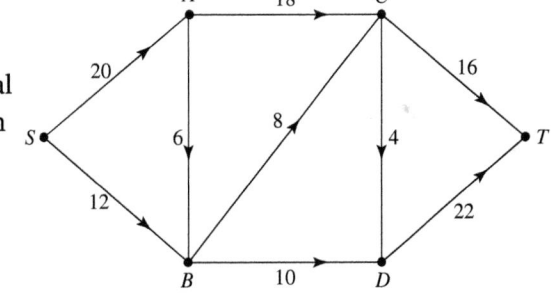

Reduce forward arrows and increase backward arrows by 5

SBAT is a flow-augmenting path with potential flow of 1

Updating the labels gives this flow.

The arrows pointing forward along the route *SBAT* are all ≥ 1

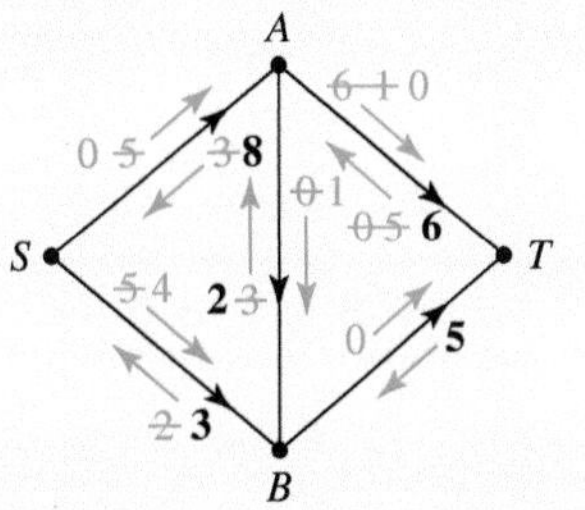

Reduce forward arrows and increase backward arrows by 1

AT and *BT* are now saturated (potential flow is 0), so the flow is maximal. The potential backflows (shown in black) give the flow in each arc.

Example 2

Taking a flow of 16 along *SACT* and 10 along *SBDT* as the initial flow:

a Use flow-augmenting paths to find a maximal flow for this network. Record your working in this table.

Augmenting path	Flow

b Use the maximum flow, minimum cut theorem theorem to show that your flow is maximal.

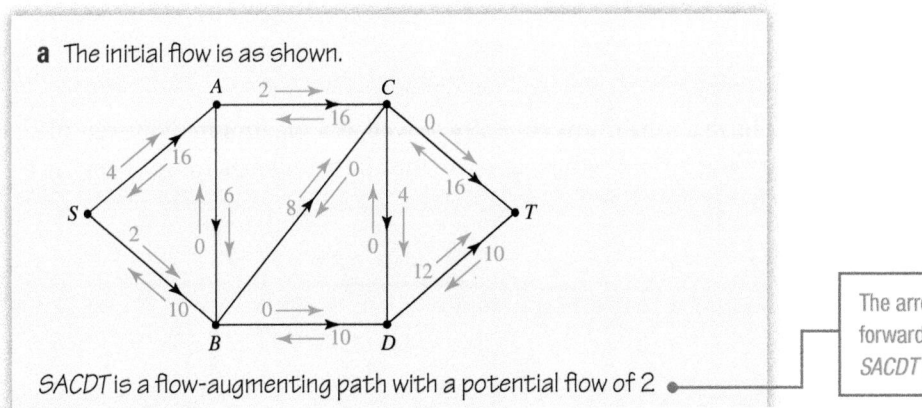

a The initial flow is as shown.

SACDT is a flow-augmenting path with a potential flow of 2

The arrows pointing forward along the route *SACDT* are ≥ 2

(*Continued on the next page*)

Augmenting path	Flow
SACDT	2

Update the labels to include this flow, as shown.

Reduce forward arrows and increase backward arrows by 2

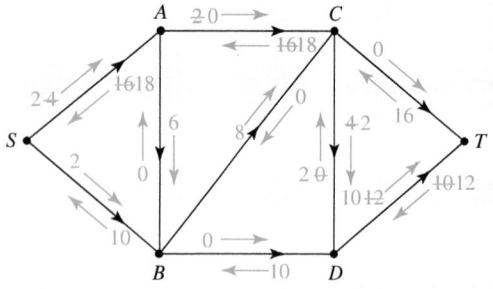

SBCDT is a flow-augmenting path with a potential flow of 2

Augmenting path	Flow
SACDT	2
SBCDT	2

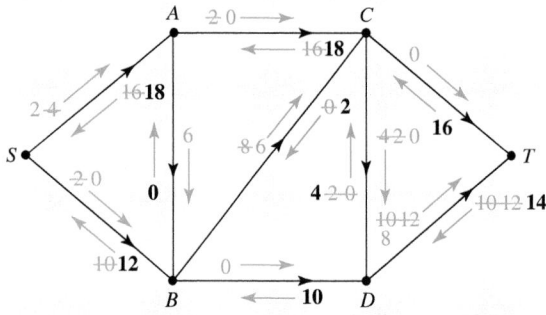

Update the labels to include this flow, as shown.

There is no flow-augmenting path available, so the flow (shown in black) is maximal.

The flow has been increased by a total of 4, so the value of the maximal flow is 30

In some circumstances arcs may have a **minimum capacity** as well as a maximum. For example, if the arcs represent pipelines carrying inflammable gas there may be safety reasons for keeping a certain amount of gas flowing.

The first task is to find an initial feasible flow. Previously you could start with an initial flow of zero, but this is not possible now. The flow through each node must be compatible with both the minimum and maximum total inflows and outflows.

Look at this partial network.

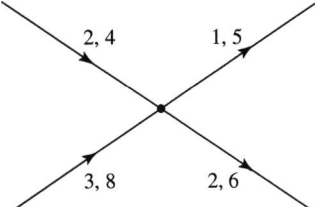

The values shown on each arc give the minimum and maximum permissible flow in that arc. At least 5 units must flow through this node because of the minimum incoming capacities. No more than 11 units can flow because of the maximum outgoing capacities.

In rare cases it is not possible to find a flow. This happens if the maximum outflow is less than the minimum inflow, or if the maximum inflow is less than the minimum outflow.

Once you have found an initial flow, you use the labelling procedure as before to find flow-augmenting paths. The only difference is that the potential flow and backflow in an arc must be compatible with the minimum and maximum flows in that arc.

For example, this diagram shows a flow of 6 units. The flow could be increased by 2 or decreased by 3

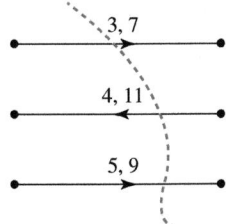

You can find the flow (6) from the diagram as (min capacity (3) + potential backflow (3)) or as (maximum capacity (8) − potential flow(2)).

To find the capacity of a cut you need to allow for the minimum flow for any arc directed from the sink set to the source set.

For example, $7 + 9 = 16$ could flow left to right through this cut, but at least 4 must flow right to left. The capacity of the cut is therefore $16 - 4 = 12$

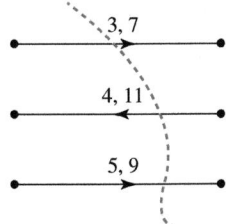

Capacity of cut = (maximum capacities from S to T) − (minimum capacities from T to S) **Key point**

Example 3

For this network, an initial feasible flow is *SAT* 3, *SABT* 4, *SBT* 5

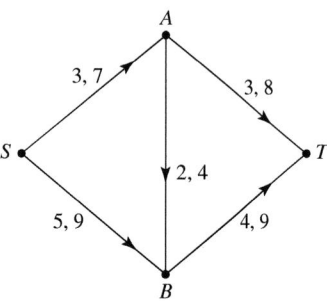

a Use flow augmentation to find a maximal flow,

b Confirm the result using the maximum flow, minimum cut theorem.

a This labelling shows the initial flow.

> Check that you can see how all these values arise.

There is a flow-augmenting path *SBAT* 2
Update the labels to add this flow.

> The forward arrows have minimum value 2

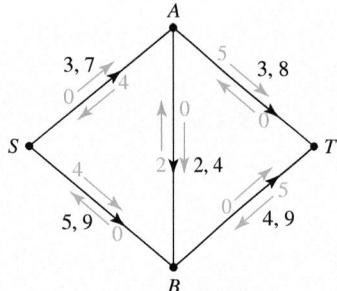

There are now no flow-augmenting paths, so the flow is maximal.
The total flow is 14. This diagram shows the flow in each arc.

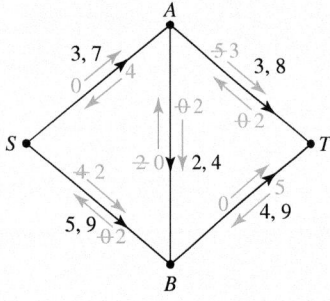

(*Continued on the next page*)

b The cut {SA, AB, BT} shown in this diagram has capacity
$$7 + 9 - 2 = 14$$

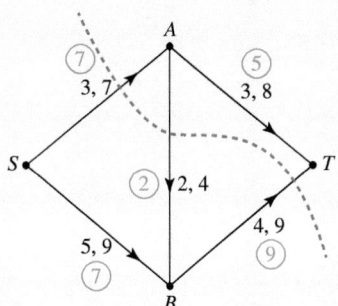

There is a flow of 14 and a cut of 14, so by the maximum flow, minimum cut theorem this flow is maximal.

Answer sheet available

1 For this network, take an initial flow comprising a flow of 15 along *SAT*, 14 along *SBT* and 10 along *SCT*

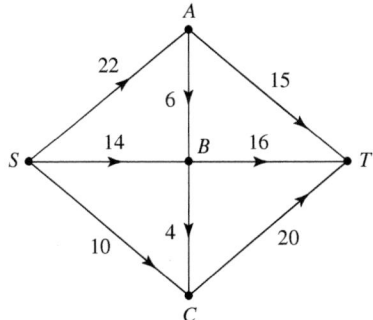

a Use the labelling procedure to augment the flow until a maximal flow is obtained.

b Confirm that your flow is maximal by using the maximum flow, minimum cut theorem.

2 For this network, take an initial flow comprising a flow of 20 along *SADT*, 20 along *SBCET* and 40 along *SBFT*

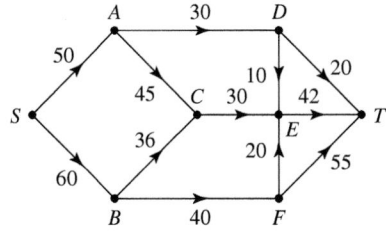

a Use the labelling procedure to augment the flow until a maximal flow is obtained.

b Confirm that your flow is maximal by using the maximum flow, minimum cut theorem.

3 Repeat question **2a** starting with a flow of 20 along *SADT*, 30 along *SACET* and 40 along *SBFT*

4 In this network, the maximum possible outflow from *S* and the maximum possible inflow to *T* are both 16

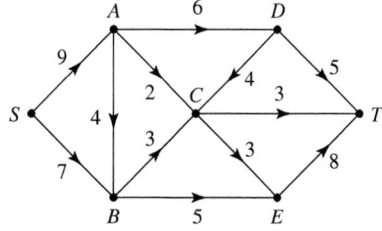

By starting with a flow of 5 along *SADT* and 5 along *SBET*, use flow augmentation to obtain a flow pattern with this maximum value.

5 For this network

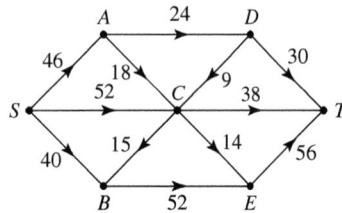

a Use an initial flow and flow augmentation to find the maximal flow.

b Use the maximum flow, minimum cut theorem to confirm that your flow is maximal.

6 Find a feasible flow, if one exists, for each of these networks.

a

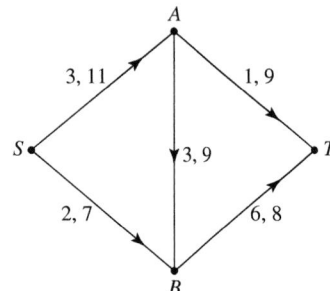

b

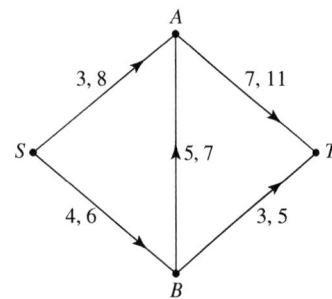

c

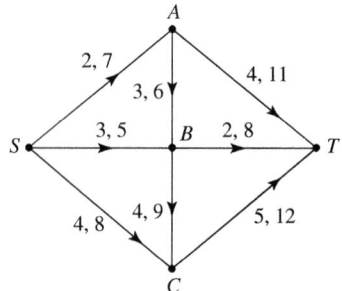

d

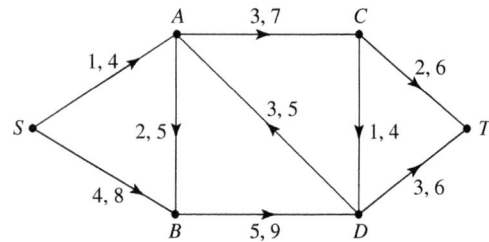

7 Find a minimum cut for each of the networks in question **6** for which a feasible flow exists.

8 This network shows the minimum and maximum flows along each arc and the circled values show a feasible flow.

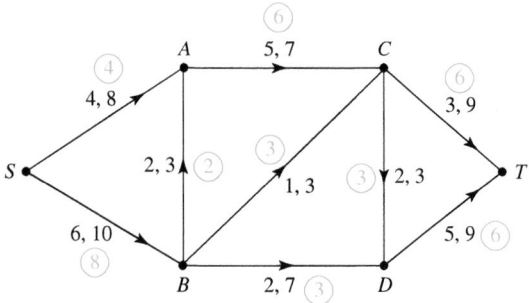

a Use the labelling procedure to find a maximum flow for the network.

b Confirm that it is a maximum by using the maximum flow, minimum cut theorem.

Reasoning and problem-solving

If there is more than one source or sink, you introduce a dummy supersource or supersink.

Example 4

Use an initial flow and flow augmentation to find a maximal flow through this network.

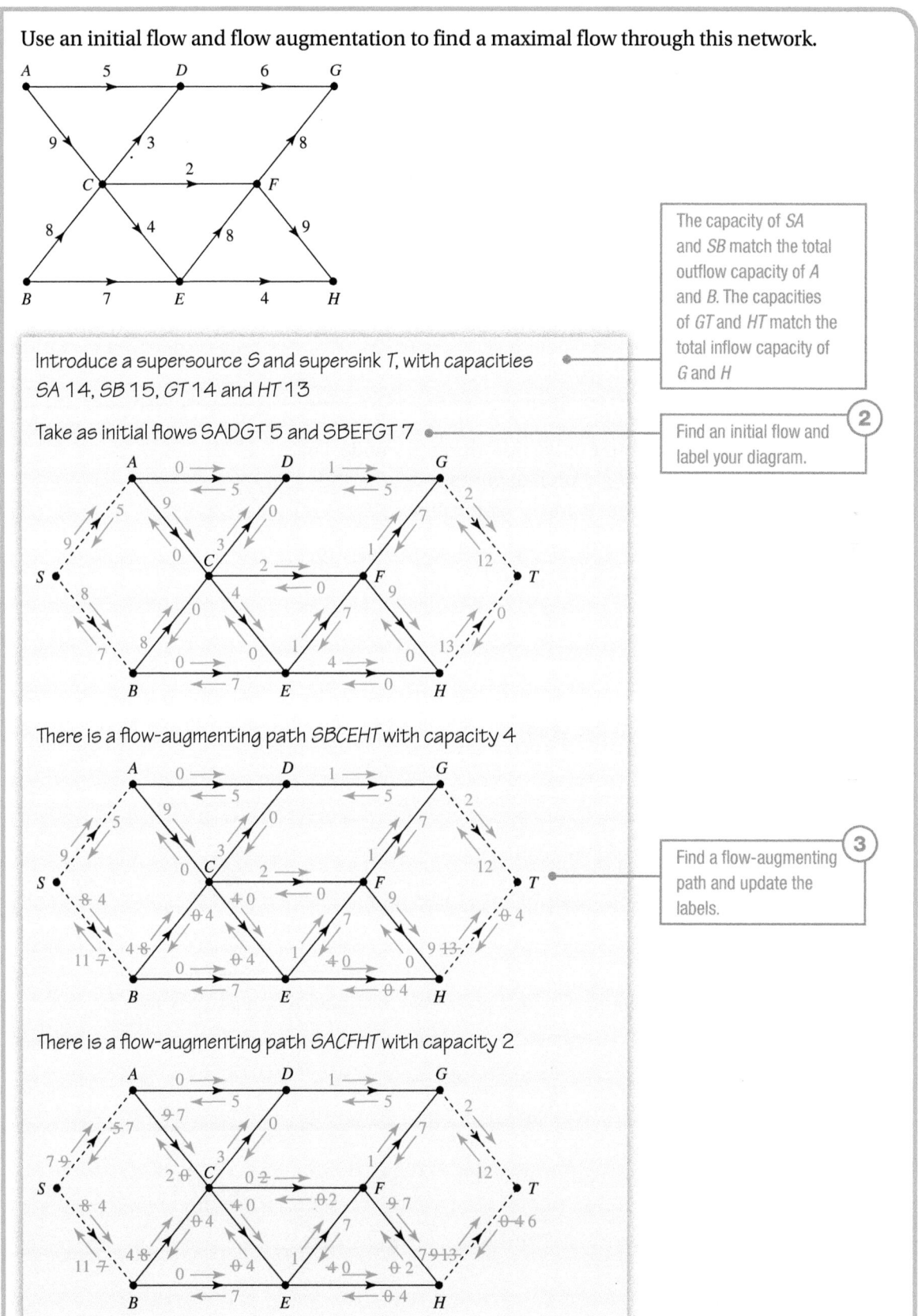

The capacity of *SA* and *SB* match the total outflow capacity of *A* and *B*. The capacities of *GT* and *HT* match the total inflow capacity of *G* and *H*

Introduce a supersource *S* and supersink *T*, with capacities *SA* 14, *SB* 15, *GT* 14 and *HT* 13

Take as initial flows *SADGT* 5 and *SBEFGT* 7

② Find an initial flow and label your diagram.

There is a flow-augmenting path *SBCEHT* with capacity 4

③ Find a flow-augmenting path and update the labels.

There is a flow-augmenting path *SACFHT* with capacity 2

(Continued on the next page)

There is a flow-augmenting path *SBCDGT* with capacity 1

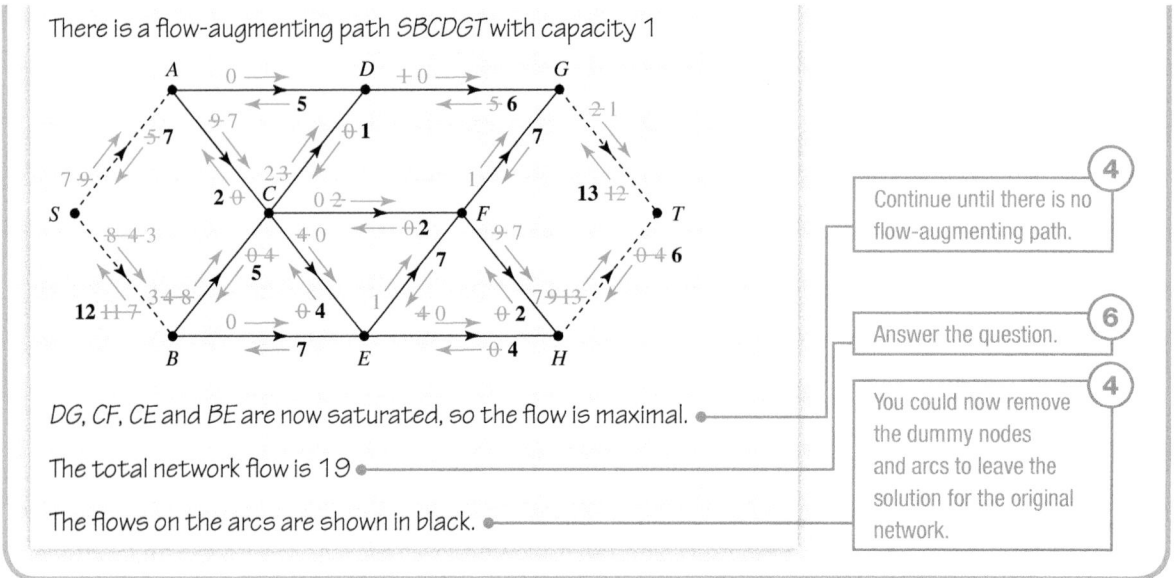

4 Continue until there is no flow-augmenting path.

DG, *CF*, *CE* and *BE* are now saturated, so the flow is maximal.

The total network flow is 19

The flows on the arcs are shown in black.

6 Answer the question.

4 You could now remove the dummy nodes and arcs to leave the solution for the original network.

There can be a restriction on the amount of flow through a node. For example, the capacity of a pumping station to pump water may be less than the capacity of the incoming and outgoing pipes.

To deal with this you draw a modified network, with an extra arc to represent the flow through the restricted node.

For example, if this node has a capacity of 4

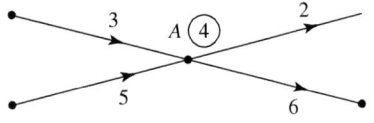

you redraw it as

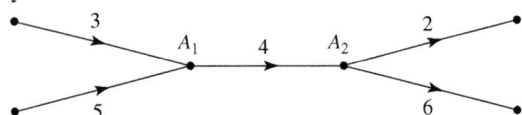

Notice that all inflows to A go to the first node A_1 and all outflows emerge from the second node A_2

Example 5

Find the maximal flow through this network, given that the nodes A and B have capacities of 5 and 4, respectively.

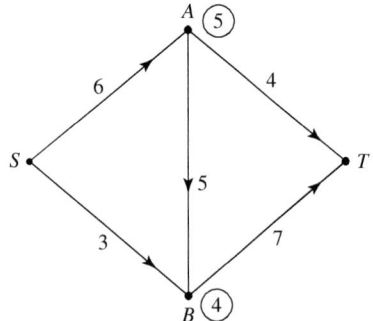

(*Continued on the next page*)

The modified network is as shown.

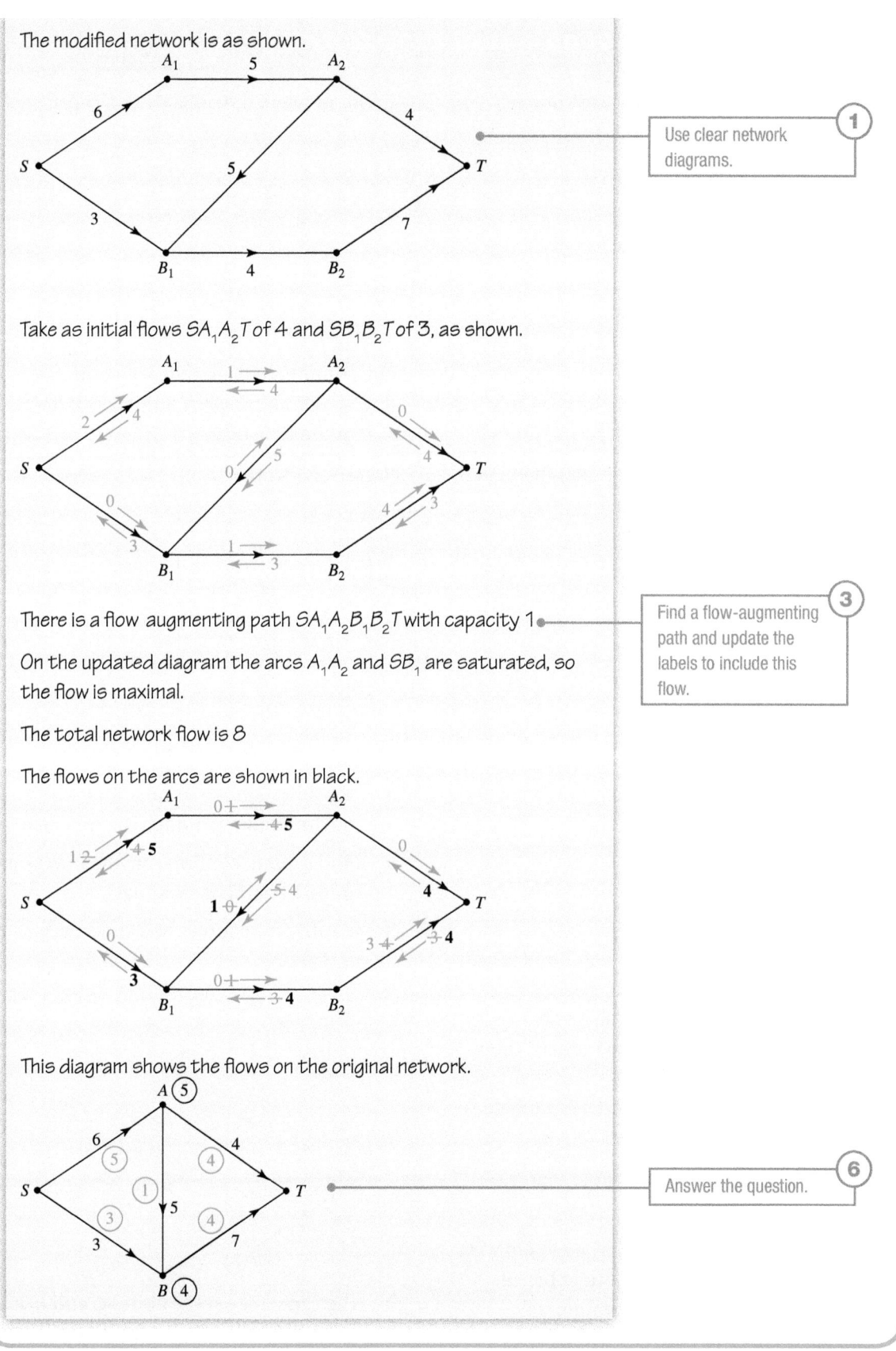

Take as initial flows SA_1A_2T of 4 and SB_1B_2T of 3, as shown.

There is a flow augmenting path $SA_1A_2B_1B_2T$ with capacity 1

On the updated diagram the arcs A_1A_2 and SB_1 are saturated, so the flow is maximal.

The total network flow is 8

The flows on the arcs are shown in black.

This diagram shows the flows on the original network.

1 Use clear network diagrams.

3 Find a flow-augmenting path and update the labels to include this flow.

6 Answer the question.

DISCRETE

1 This network has a sink and two sources.

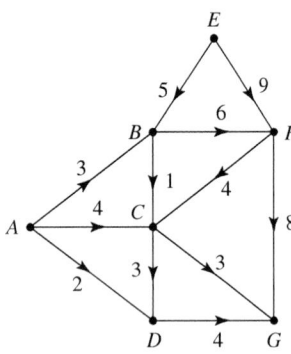

 a Introduce a supersource S

 b Use an initial flow and flow augmentation to find the maximal flow though the network.

 c Draw the original network to show the flow you have found, together with a minimum cut.

2 This network represents a system of one-way streets.

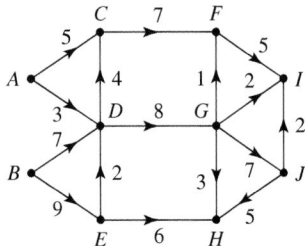

Traffic enters the system at A and B and leaves at I and H. The weights are the maximum traffic flows, in hundreds of cars per hour, which can safely pass along the streets. There is an environmental charge of £4 for all diesel vehicles using these roads, and it is estimated that 40% of vehicles are diesel.

 a Draw a diagram with a supersource, S, and a supersink, T

 b Use an initial flow and flow augmentation to find the maximum flow and hence estimate the maximum amount of environmental charge that could be collected in a peak hour.

3 In this network the node B has a capacity of 7

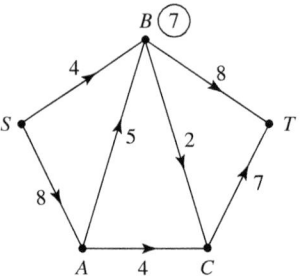

On a modified diagram use an initial flow and flow augmentation to find the maximum flow from S to T

4 For this network

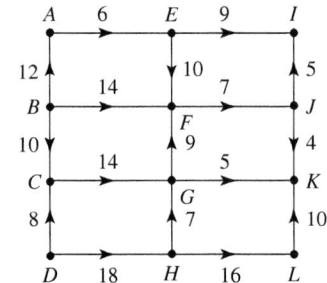

 a Identify the source(s) and sink(s).

 b Use flow augmentation to find the maximum flow through the network.

 c Find a cut to confirm that the flow you found in part **b** is maximal.

5 a Identify the sources and sinks in this network.

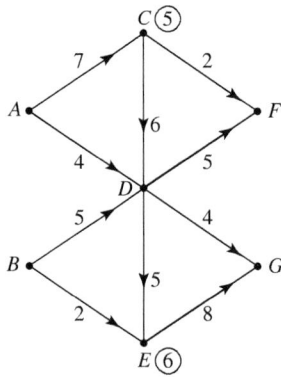

 b The capacities at the nodes C and E are restricted to 5 and 6, respectively. Draw a modified network to allow for this and find an initial flow with a total capacity of 10

c Starting with your flow from part **b**, use flow augmentation to find a maximal flow through the network.

6 Find the maximum flow through this network and confirm your result using a minimum cut.

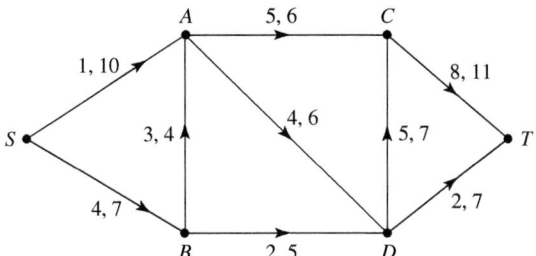

7 In a network of pipes used to deliver gas from S to T, each pipeline has a maximum flow, but for safety reasons it is also necessary to maintain a minimum flow in each pipe. This table shows both the minimum and the maximum number of units that can flow in each pipe.

		To				
		A	B	C	D	T
	S	6,16	2,12	–	–	–
	A	–	3,4	4,12	1,2	–
From	B	–	–	–	6,10	–
	C	–	–	–	2,8	4,12
	D	–	–	–	–	8,18

Find the maximum network flow consistent with these safety conditions and investigate whether it differs from the maximum flow possible if the safety restrictions were removed.

8 The diagram shows the minimum and maximum flows through the arcs of a network.

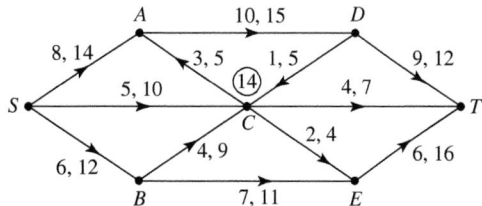

In addition, the node C has a maximum capacity of 14. Find the maximum flow through the network and confirm your result by finding a suitable cut.

Chapter summary

- The complement of a simple graph G, denoted by G', has the same vertices as G but contains only the edges which are *not* in G
- You find the adjacency matrix for G' by changing 1s to 0s and vice versa (ignoring the leading diagonal).
- If a graph has n vertices and you combine it with its complement, you get the complete graph K_n
- Graphs G_1 and G_2 are isomorphic if, for each vertex in G_1, there is a corresponding vertex in G_2 so that if an edge joins two vertices in G_1, then there is an edge between the corresponding vertices in G_2, and vice versa. If corresponding vertices are listed in the same order, the adjacency matrices are identical.
- A subdivision of a graph G is a graph formed by adding vertices of degree 2 along edges of G
- A graph is planar if it can be drawn without edges intersecting except at a vertex.
- If a simple, connected, planar graph has V vertices, E edges and F faces (including the 'infinite face' surrounding the graph) then $V + F - E = 2$ (Euler's formula).
- It follows from Euler's formula that, for all planar graphs, $E \leq 3V - 6$. (Be aware that this is also true for some non-planar graphs.) If $V = 5$, then $E \leq 9$. As K_5 has 10 edges, it follows that K_5 is non-planar.
- For a bipartite graph, it follows from Euler's formula that $E \leq 2V - 4$. If $V = 6$, then $E \leq 8$. As $K_{3,3}$ has 9 edges, it follows that $K_{3,3}$ is non-planar.
- A graph is non-planar if and only if it has a subgraph which is a subdivision of K_5 or $K_{3,3}$ (Kuratowski's theorem).
- For a network flow from source S to sink T
  - The flow in each arc cannot be more than its capacity (feasibility condition).
  - At each node, the total inflow equals the total outflow (conservation condition).
  - A cut is a set of arcs whose removal disconnects the network into two parts X and Y, with X containing S and Y containing T
  - The value of the maximal flow = the capacity of a minimum cut.
  - If you have a flow and a cut such that (value of flow) = (capacity of cut), then the flow is a maximum and the cut is a minimum.
- Flow augmentation seeks to improve on an existing feasible flow.
  - The labelling procedure labels each arc with a forward arrow showing spare capacity and a backward arrow showing actual flow.
  - A flow-augmenting path is a route from S to T where the arrows pointing forwards along the route all have non-zero values. The lowest of these values gives the possible extra flow along that path.
- A network with more than one source or sink is modified by introducing a dummy supersource and/or supersink.
- If a node has a restricted capacity, you modify the network by replacing that node with an inflow node and an outflow node connected by an arc with the restricted capacity.

- If arcs have a minimum required flow, then the initial feasible flow must be consistent with these. You can still use the labelling procedure and flow augmentation.
- For a network with minimum and maximum capacities,
  capacity of cut = (maximum flows from S to T) – (minimum flows from T to S)

Check and review

You should now be able to ...	Review Questions
✔ Construct the complement of a graph given in the form of a drawing or an adjacency matrix.	1
✔ Decide whether two given graphs are isomorphic.	2
✔ Use Euler's formula in relation to planar graphs.	3
✔ Use Kuratowski's theorem to determine whether a graph is non-planar.	4
✔ Use the labelling procedure and flow-augmentation to find the maximum flow through a network.	5, 6, 7, 8
✔ Use the maximum flow, minimum cut theorem to show that a flow is maximal.	5, 8
✔ Work with multiple sources/sinks using a supersource/supersink.	6
✔ Solve problems in which there is a restricted flow through one or more nodes.	7
✔ Solve problems where arcs have a minimum flow as well as a maximum flow.	8

1 For the adjacency matrix shown:

	A	B	C	D	E	F
A	0	0	1	0	0	0
B	0	0	1	1	0	0
C	1	1	0	0	1	1
D	0	1	0	0	1	1
E	0	0	1	1	0	0
F	0	0	1	1	0	0

a Draw the corresponding graph, G

b Construct the adjacency table for the complement, G'

c Draw the graph G'

2 Two of these graphs are isomorphic. Identify these and list corresponding vertices.

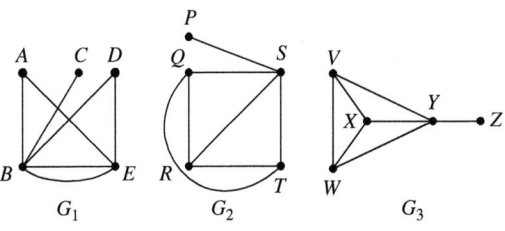

G_1 G_2 G_3

3 a A planar graph has 6 vertices and 9 edges. Calculate the number of faces in the graph and draw a graph which fits these facts.

b A graph has 7 vertices and 16 edges. Use a corollary of Euler's formula to show that the graph is non-planar.

4 Use Kuratowski's theorem to show that this graph is non-planar.

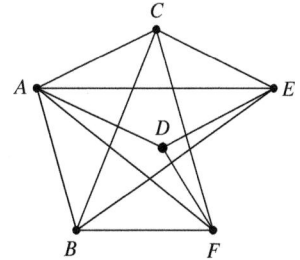

5 For this capacitated network:

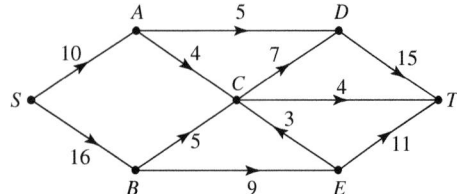

a Explain why a flow of 16 in *SB* is not possible.

b State the maximum flow along

    i *SADT*

    ii *SBET*

c Taking your results from part **b** as the initial flow, use the labelling procedure to find the maximum flow through the network.

d Prove that your flow is maximal.

6 For this capacitated network.

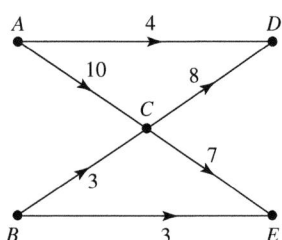

a Identify the source(s) and sink(s).

b Draw a diagram to include a dummy source and/or sink as necessary.

c Use the labelling procedure to find a maximal flow through the network.

7 In the network from question **6** the node *C* is then restricted to a flow of 8

a Draw a modified diagram to allow for this restriction.

b Use the labelling procedure to find a maximal flow through this network.

8

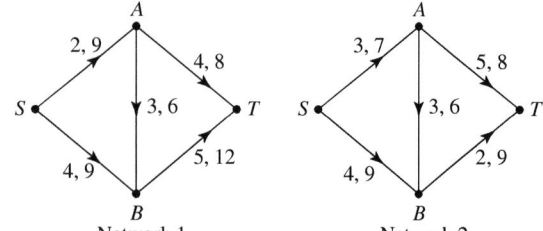

a For one of these networks, no feasible flow exists. Identify the network and explain why no flow is possible.

b Consider the other network.

    i Find the capacity of the cut {*SA*, *AB*, *BT*}.

    ii An initial feasible flow has 7 on *SA* and 4 on *SB*. Starting with this, use the labelling procedure to find the maximum flow.

    iii Explain how you know it is a maximum.

Investigation

The complete bipartite graph on six vertices $K_{3,3}$ is also known as the utility graph. This comes from the puzzle of trying to connect three utility points to 3 buildings with no connection lines crossing.

$K_{3,3}$ is also a toroidal graph as it can be embedded on a torus without any edges crossing. This offers a potential solution for the utility problem – can you find it?

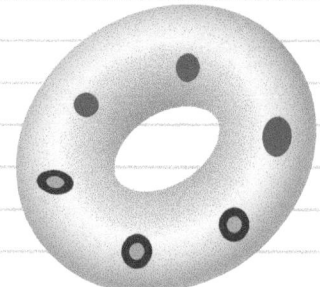

Research

Image segmentation is the process of segmenting a digital image to make it easier to analyse. Applications include video surveillance and magnetic resonance imaging. Research how the maximum flow problem can be used to assign pixels to the background or foreground of a picture.

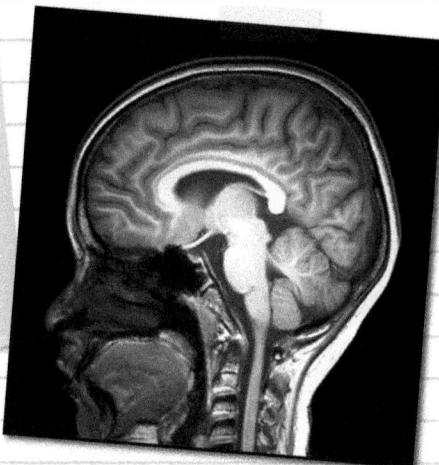

Did you know?

The first maximum flow problem was formulated in 1954 as a simplified model of Soviet railway traffic flow.

Research

Algorithms designed to find the maximum flow of a network include Ford-Fulkerson and push-relabel. The Ford-Fulkerson algorithm augments flows along paths from the source all the way to the sink. In contrast, the push-relabel algorithm gradually finds a maximum flow by moving flow locally between neighbouring nodes.

1 Using Kuratowski's theorem, determine whether or not this graph is planar. **[2]**

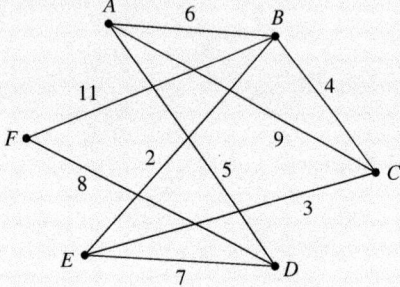

2 G_1 G_3

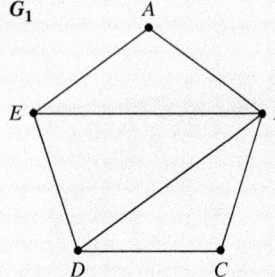 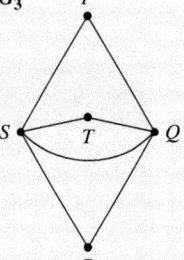

G_2

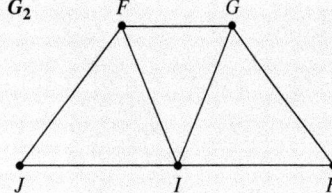

   a Construct the adjacency matrix for the complement of G_1 **[3]**

   b Explain whether or not each of G_2 and G_3 are isomorphic to G_1 **[3]**

3 **a** Using an initial flow of 7 along $SADT$ and 9 along $SCFT$, augment the flow until a maximal flow is obtained. **[5]**

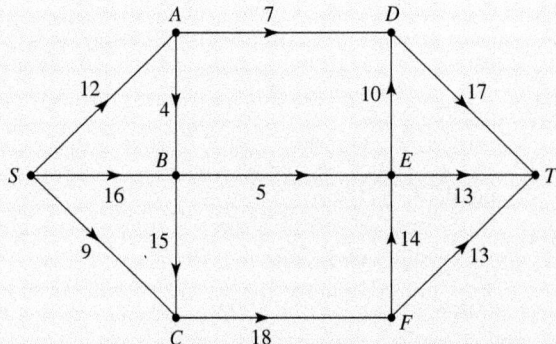

b Verify that your flow is maximal, stating the name of the theorem used. **[3]**

c Explain which edge in the network could be removed without affecting the maximum flow. **[2]**

d Which of the edges are saturated? **[2]**

4 The edges on this network represent road with the capacities shown.

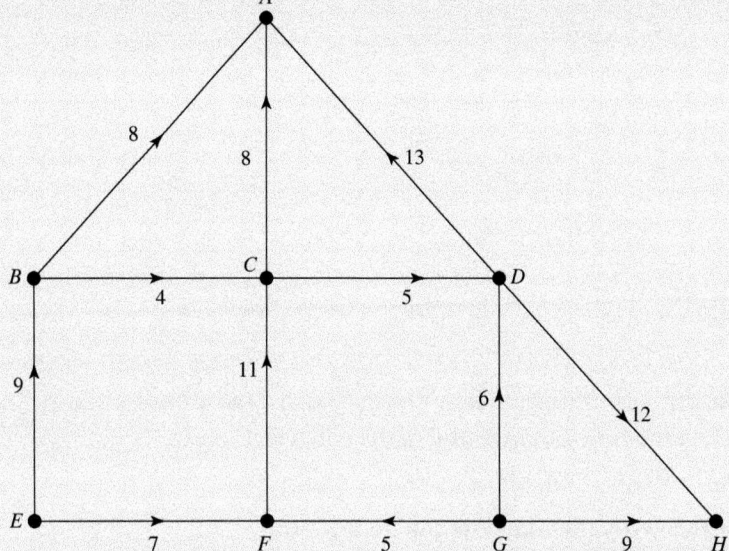

a Copy the network, introducing a supersource and/or supersink if needed. **[3]**

b Clearly stating your initial flows, use flow augmentation to obtain the maximal flow. **[5]**

c Draw your maximal flow onto the copy of the network. **[3]**

d Verify that the flow you have found is maximal. **[2]**
Additional capacity is to be added either at *EB* or *GF*

e Where would you suggest it is placed and how much additional capacity should be added? Explain your answer fully. **[3]**

5 This network represents a system of pipes.

There is a blockage at *C* that restricts the flow to a maximum of 16

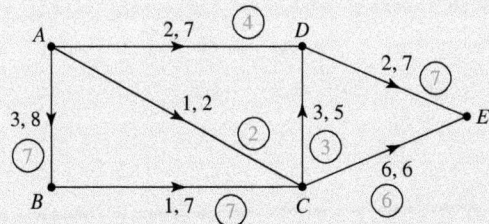

a Draw a modified network to account for this restricted node. **[3]**

b Find the maximal flow through this network and verify that it is a maximum. **[5]**

6 The numbers on the arcs of this network give the minimum and maximum permissible flow in that arc. A initial feasible flow is $ADE = 2$, $ACDE = 2$, $ABCE = 6$

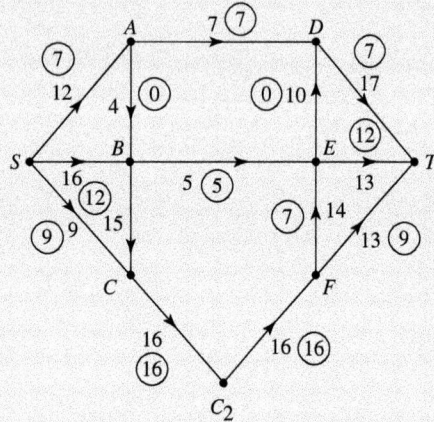

a Using arrows for potential flow and backflow, illustrate the initial flow on a copy of this diagram. **[3]**

b Use flow augmentation, starting with the initial flow given, to find a maximum flow for this network, and state the flow-augmenting paths you use. Clearly illustrate your flow on the copy of the network. **[4]**

c Use the cut $\{DE, DC, AC, BC\}$ to verify that the flow you have found is maximal. **[2]**

29 Critical path analysis 2

Natural disasters such as floods, volcanic eruptions, earthquakes and spread of a major disease, require a rapid response from disaster relief agencies. Once an incident occurs, such organisations are well prepared to move people to safety, arrange medical aid, transport food supplies and set up sanitation. The mathematics of critical path analysis is a crucial tool in ensuring that support reaches disaster victims as quickly as possible. Through this, agencies are able to coordinate large-scale operations using advance planning and variable factors, which can be adjusted using information from on-the-ground experts.

Critical path analysis is at the heart of project planning and implementation, from major building projects such as building a new school, airport terminal or retail distribution centre, through to holding a major event such as a summer festival or sports tournament.

Orientation

What you need to know	What you will learn	What this leads to
Ch13 Critical path analysis 1	• To calculate earliest start times and latest finish times for an activity network. • To use Gantt charts. • To use resource histograms to schedule activities.	**Careers** • Project management. • Construction. • Event planning.

Fluency and skills

See Ch13.1

For a reminder on activity networks and critical paths.

A project can be described by a precedence table showing each activity involved and the order in which they must be completed.

You draw an activity network and use a forward and backward pass to find the earliest possible start time and latest possible finish time for each activity. The time available for an activity is the gap between these. If the time available is greater than the duration of the activity, then the activity has float.

> **Key point**
>
> Float = (latest finish time − earliest start time) − duration

An activity with zero float is a critical activity. The critical activities form a critical path. Some projects may have more than one critical path.

With non-critical activities, you have some flexibility about when to start them. They will have a latest possible start time and an earliest possible finish time.

> **Key point**
>
> Earliest finish time = earliest start time + duration
> Latest start time = latest finish time − duration

For example, look at this activity network.

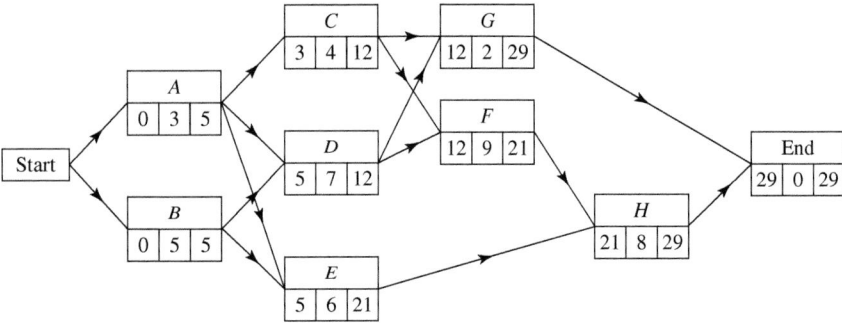

The table on the next page shows the complete information for this network.

For example, for activity C

Earliest start time = 3

Earliest finish time = 3 + duration of C = 7

Latest finish time = 12

Latest start time = 12 − duration of C = 8

Activity	Duration (days)	Start time Earliest	Start time Latest	Finish time Earliest	Finish time Latest	Float
A	3	0	2	3	5	2
B	5	0	0	5	5	0
C	4	3	8	7	12	5
D	7	5	5	12	12	0
E	6	5	15	11	21	10
F	9	12	12	21	21	0
G	2	12	27	14	29	15
H	8	21	21	29	29	0

You can now illustrate this information on a **Gantt chart** or **cascade chart**. Activities are shown as bars against a time scale. The critical activities are fixed, so you insert those first.

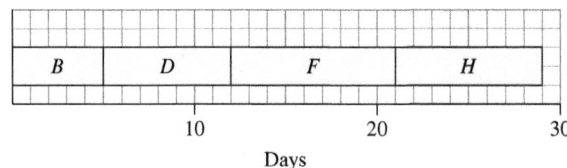

You now place each non-critical activity at its earliest start time, and show the boundaries within which it can be moved.

Activity A could start at 0 and has a latest finish of 5

Activity C has earliest start 3 and latest finish 12

Activity E has earliest start 5 and latest finish 21

Activity G has earliest start 12 and latest finish 29

Although C has 5 days float, 2 of these depend on when A starts. You can put a 'fence' on the chart to show that C cannot start until A is complete.

The finished Gantt chart looks like this.

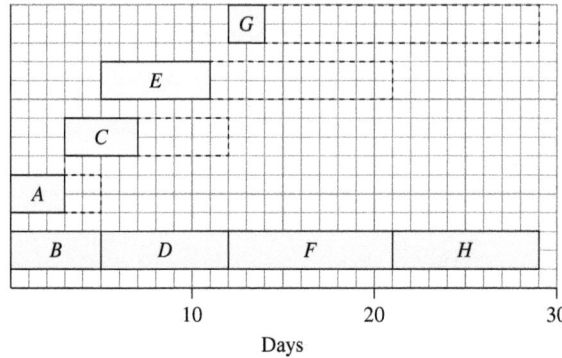

You may see variants of this layout. The most common is to put each activity on a separate line.

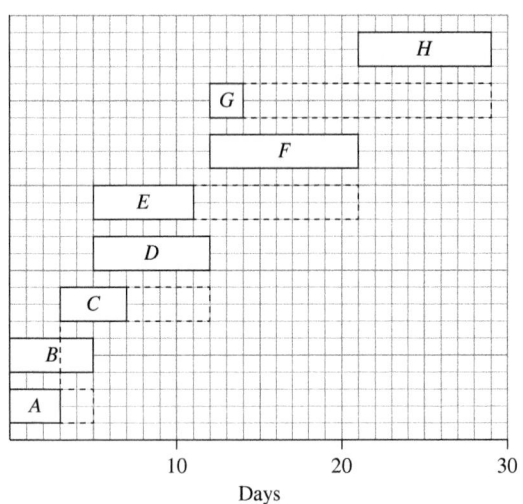

You will also see them drawn 'upside-down' with the early activities at the top of the diagram.

Example 1

Draw a Gantt chart corresponding to this activity network.

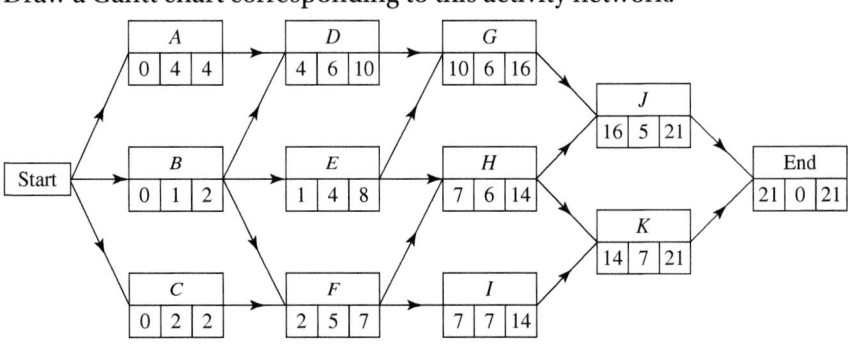

Activity	Duration (hours)	Start time		Finish time		Float
		Earliest	Latest	Earliest	Latest	
A	4	0	0	4	4	0
B	1	0	1	1	2	1
C	2	0	0	2	2	0
D	6	4	4	10	10	0
E	4	1	4	5	8	3
F	5	2	2	7	7	0
G	6	10	10	16	16	0
H	6	7	8	13	14	1
I	7	7	7	14	14	0
J	5	16	16	21	21	0
K	7	14	14	21	21	0

First, create a table to show the earliest and latest times.

The critical activities are A, C, D, F, G, I, J, K

They form two critical paths through the network – ADGJ and CFIK

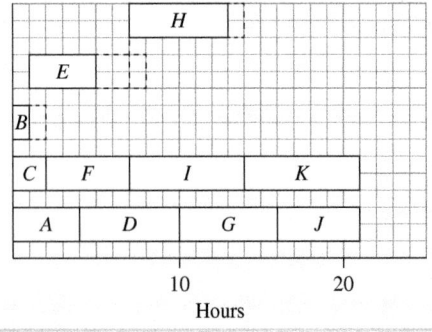

Draw the Gantt chart.

1 Complete the following table. Hence draw a Gantt chart to show the project.

Activity	Duration (hours)	Start		Finish		Float
		Earliest	Latest	Earliest	Latest	
A	6	0			6	
B	8	0			15	
C	14	6			22	
D	12	6			18	
E	3	8			18	
F	4	18			22	
G	3	22			25	

2 Draw a Gantt chart for each of these activity networks.

a

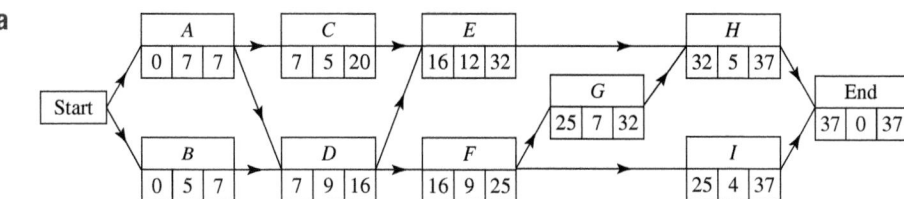

b

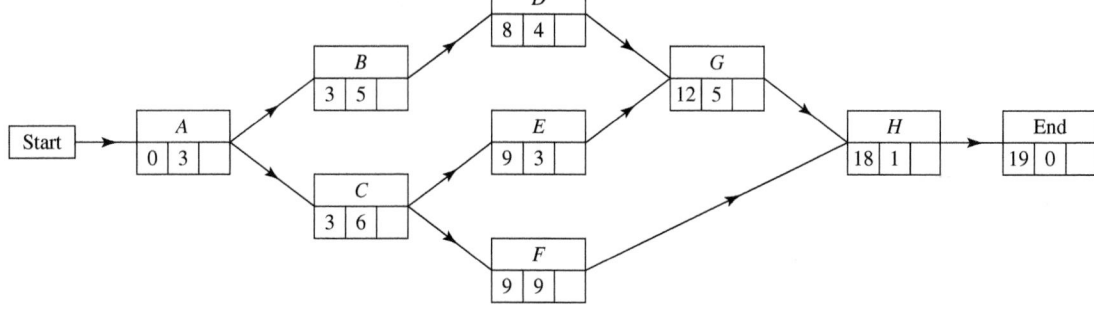

c

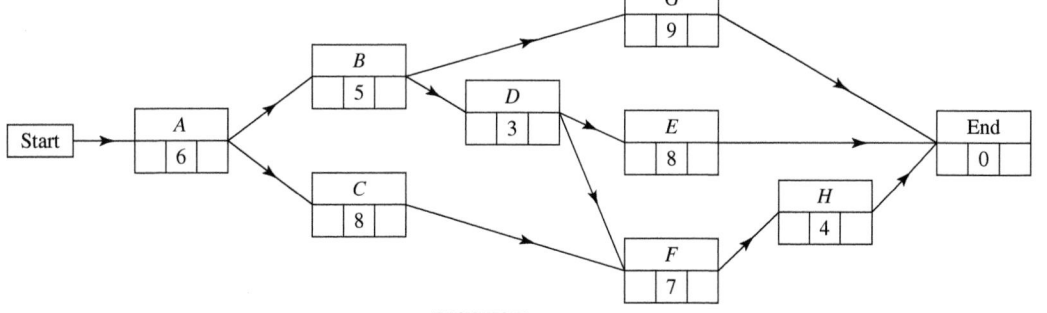

d

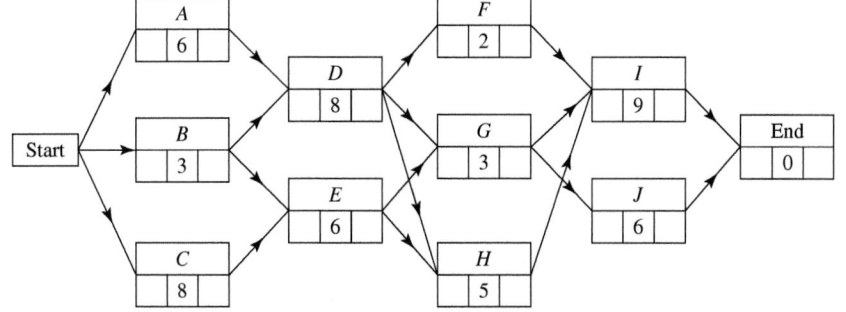

Strategy

To solve problems involving Gantt charts

① Find the earliest and latest start and finish times for each activity.

② Identify the critical activities and critical path(s).

③ Draw the Gantt chart, entering critical activities first.

④ Answer the question.

Example 2

a Draw a Gantt chart to show this project.

Activity	Depends on:	Duration (days)
A	–	8
B	–	4
C	A	10
D	B	7
E	C	2
F	C, D	6
G	E, F	3

Explain what would happen if

b Activity B overran by 3 days,

c Activity F overran by 3 days.

Assume that, in each case, they started on time.

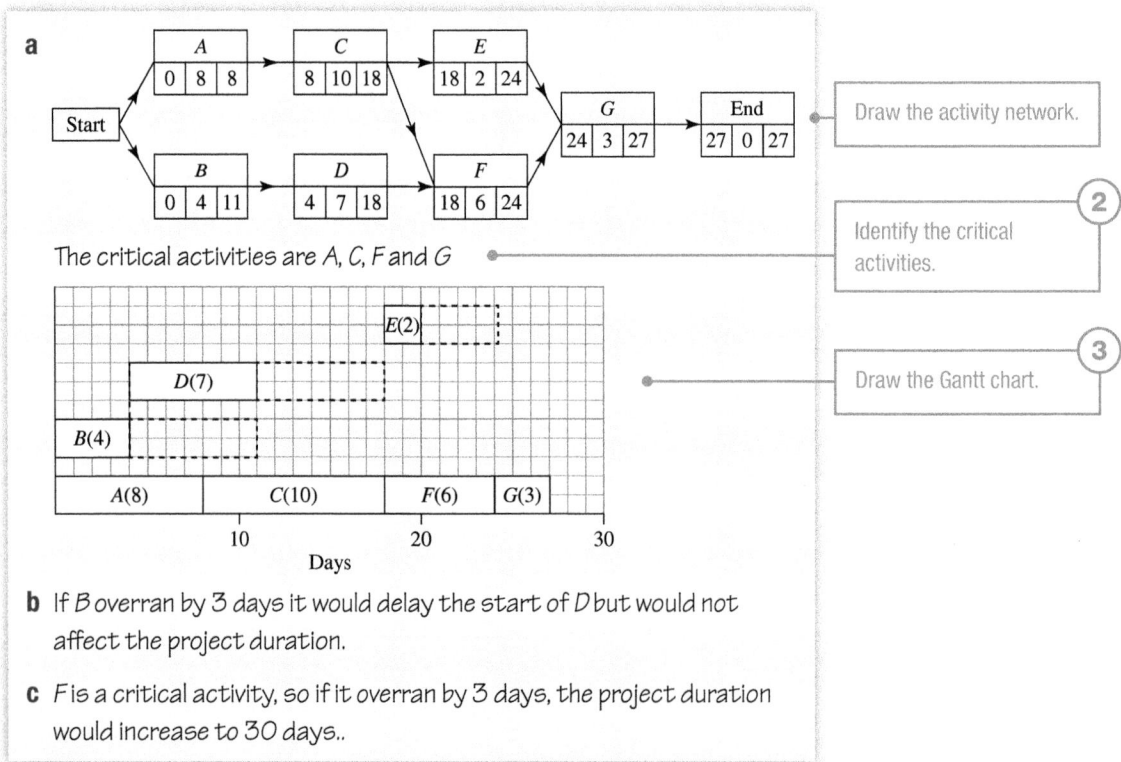

a Draw the activity network.

The critical activities are A, C, F and G

② Identify the critical activities.

③ Draw the Gantt chart.

b If B overran by 3 days it would delay the start of D but would not affect the project duration.

c F is a critical activity, so if it overran by 3 days, the project duration would increase to 30 days..

1 Draw a Gantt chart for the project described by this precedence table.

Activity	Depends on:	Duration (hours)
A	–	9
B	–	7
C	A	4
D	B	5
E	C, D	6
F	C, D	4
G	E	7
H	E, F	3

2 a Draw a Gantt chart for the project described by this precedence table.

Activity	Depends on:	Duration (days)
A	–	4
B	–	6
C	–	5
D	A, B	3
E	B, C	2
F	D	6
G	E	7
H	D, G	3
I	F, H	3
J	F	3
K	G	4

b Activities D and J overrun by 2 days each. What effect will this have on the project duration?

3 A gardener plans to make a pond. The table shows the tasks involved. Each task requires one worker.

	Activity	Depends on:	Duration (hours)
A	Clear site	–	2
B	Dig hole	A	10
C	Clear soil	B	3
D	Line hole	B	2
E	Fill pond	D	3
F	Install pump	D	2
G	Test pump	E, F	1
H	Put in fish	E	1
I	Put in plants	E	2
J	Landscape site	C, D	5

a How long will the project take if the gardener works alone?

b What is the least time needed for the complete project?

c Draw a Gantt chart for the project.

d Find the least workforce needed to finish the project in the minimum time.

4 The table shows a project for renovating a flat. Draw a Gantt chart for this project.

	Activity	Depends on:	Duration (hours)
A	Remove furniture	–	1
B	Remove old carpet	A	1
C	Remove curtains	–	1
D	Strip wallpaper	B, C	3
E	Sand down paintwork	B, C	2
F	Rewire	D	4
G	Install central heating	F	4
H	Repair plaster	G	2
I	Wash ceiling and walls	E, H	1
J	Paint ceiling	I	1
K	Paint walls	J	2
L	Paint woodwork	H	2
M	Install new carpet	K, L	1
N	Replace furniture	M	1
O	Replace curtains	K, L	1
P	Clean	N, O	2

DISCRETE

Fluency and skills

You use the term **resource** to refer to the workforce needed for a project. The different activities may require different numbers of workers. You will need to

- decide how many workers are required at each stage if every activity starts as early as possible,
- reschedule activities to minimise the number of workers needed; this is known as **resource levelling**,
- extend the duration of the project to allow for the available number of workers.

To illustrate the number of workers required at each stage you use a **resource histogram**.

Example 1

This Gantt chart shows the activities in a project. The figures in brackets show the number of workers needed for the activity.

a Draw a resource histogram assuming that each activity starts as early as possible. How many workers would be needed?

b By rescheduling activities, find the least number of workers needed to complete the project in the minimum time.

c If only 8 workers are available, find the length of time needed for the project.

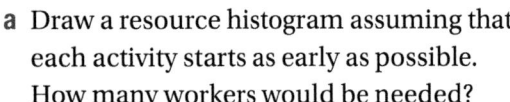
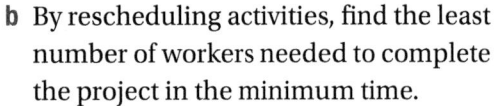
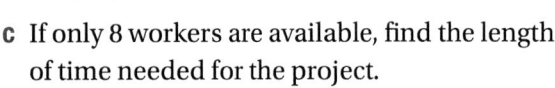

a

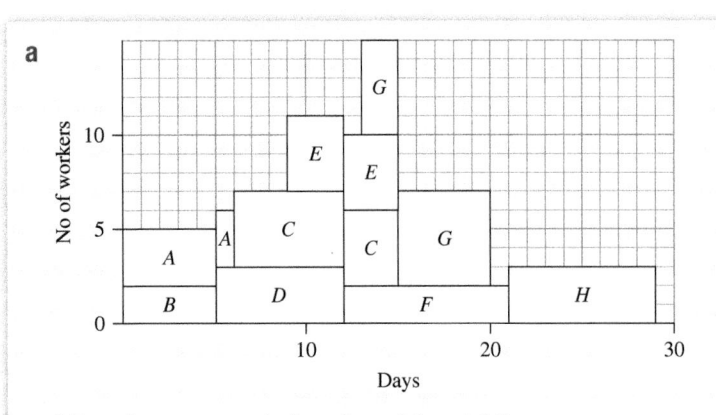

15 workers are needed on days 14 and 15

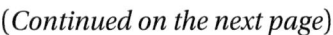

(*Continued on the next page*)

b There is enough float to delay the start of E until D has finished, and G until E has finished.

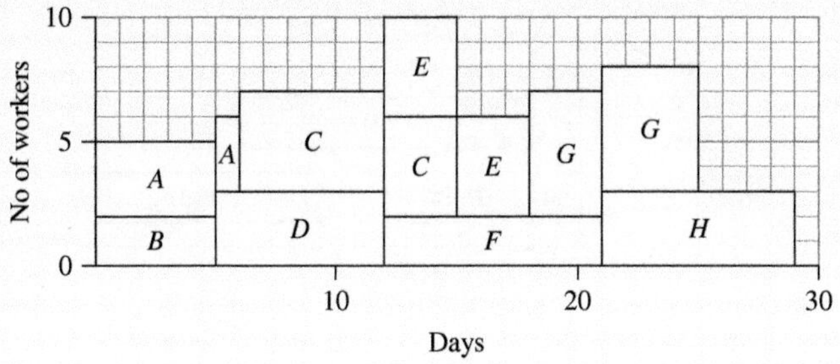

No of workers
Days
A
B
A
C
D
E
C
E
F
G
G
H

This is the revised histogram.

The project now requires 10 workers.

c The problem occurs on days 13–15. It will be necessary to delay F by 3 days. F is a critical activity, so the project now takes 32 days.

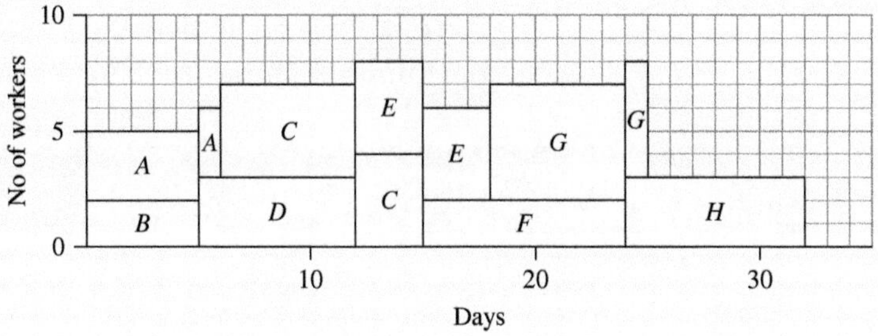

No of workers
Days
A
B
A
C
D
E
C
E
G
F
G
H

This is the resource histogram.

Exercise 29.2A Fluency and skills

1 On each of these Gantt charts the values in brackets give the number of workers needed for the activity.

a Draw a resource histogram assuming that every activity starts as early as possible. How many workers are needed in this case?

b Use resource levelling to construct a revised resource histogram to minimise the workforce needed without extending the project duration. How many workers are now required?

i

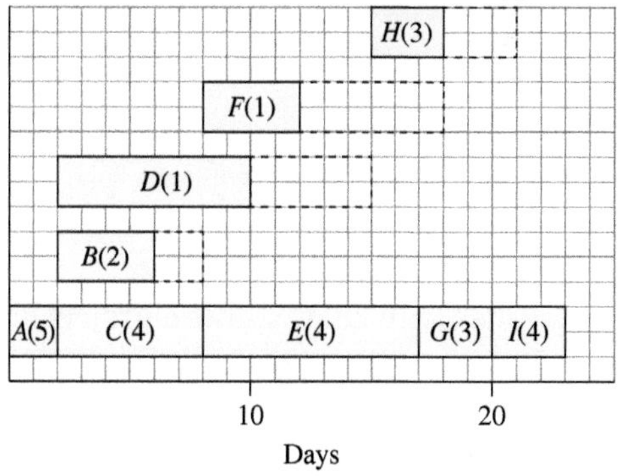

H(3)
F(1)
D(1)
B(2)
A(5)
C(4)
E(4)
G(3)
I(4)
10
20
Days

ii

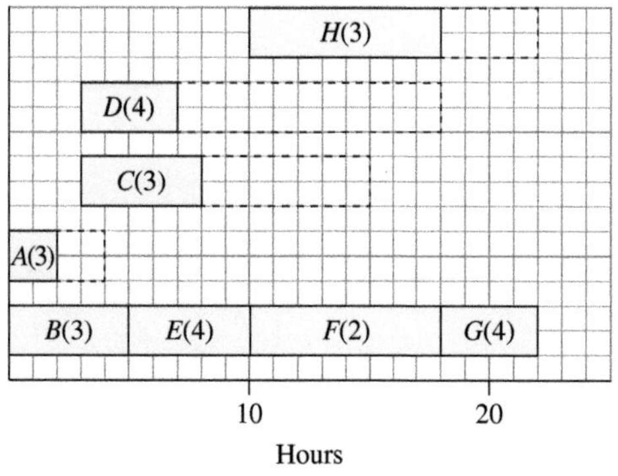

H(3)
D(4)
C(3)
A(3)
B(3)
E(4)
F(2)
G(4)
10
20
Hours

iii

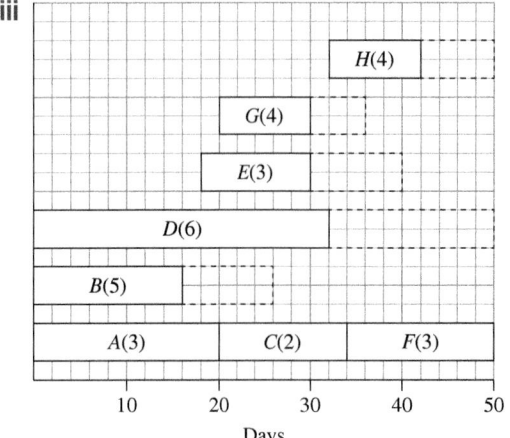

2 Draw a resource histogram for the project shown in this Gantt chart, assuming that every activity starts as early as possible.

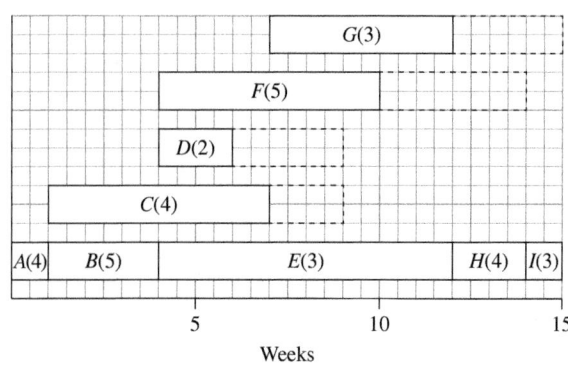

Weeks

a Use resource levelling to show that the project can be completed in the minimum time using 12 workers.

b Show that, if each activity does not have to be completed in a single block of time, it is possible to complete the project in the minimum time using just 11 workers.

Reasoning and problem-solving

To solve a problem involving resource levelling

1. If necessary, find the earliest start time, latest finish time and float for each activity.

2. Sketch a Gantt chart if one is not given.

3. Draw a resource histogram assuming that all activities start as early as possible.

4. Reschedule activities within their float times to reduce the size of the workforce needed.

5. If there is a given maximum number of workers available, it may be necessary to extend the duration of the project.

6. Answer the question.

Example 2

This Gantt chart shows a project lasting 27 days. The values in brackets give the number of workers needed for each activity.

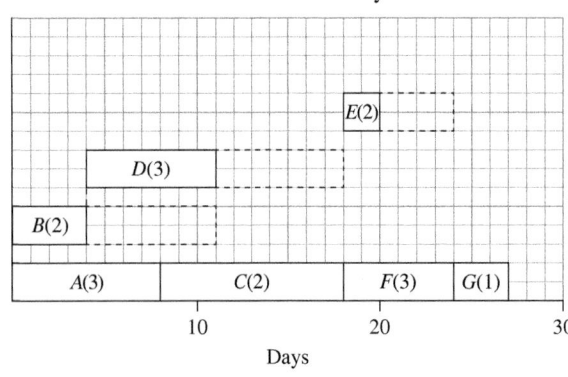

Days

(*Continued on the next page*)

a Draw a resource histogram assuming that each activity starts as soon as possible. How many workers are needed?

b Show that the project can be completed on time if only 5 workers are available.

c Find the effect on the project duration if it has to be completed using just 4 workers.

a

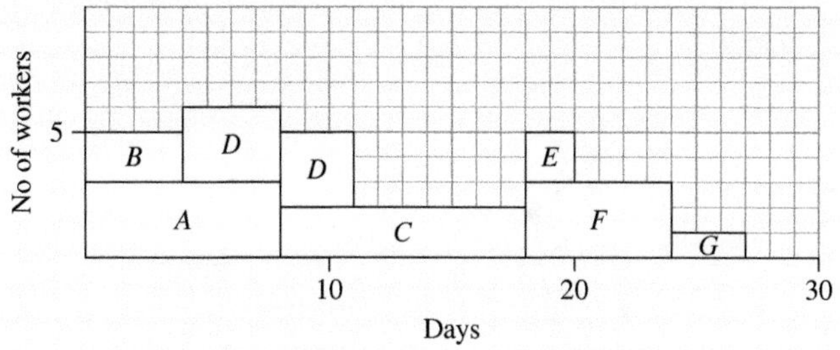

The project needs 6 workers.

> ③ Draw the resource histogram.

b By delaying the start of D, the project can be done by 5 workers.

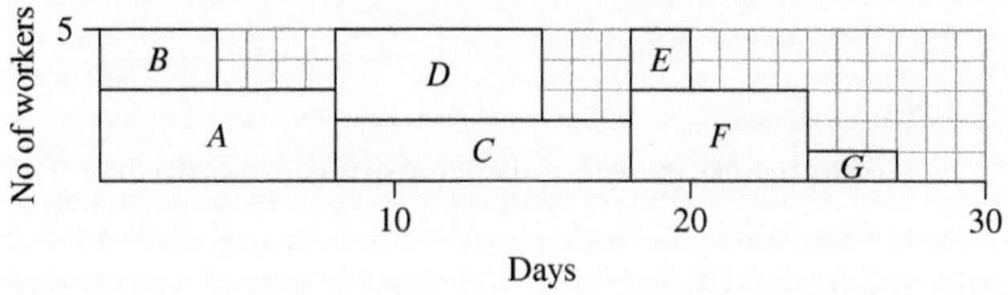

> ④ Reschedule activities within their float times to reduce the size of the workforce needed.

c If only 4 workers are available, the only activities which can be carried out in parallel are B and C

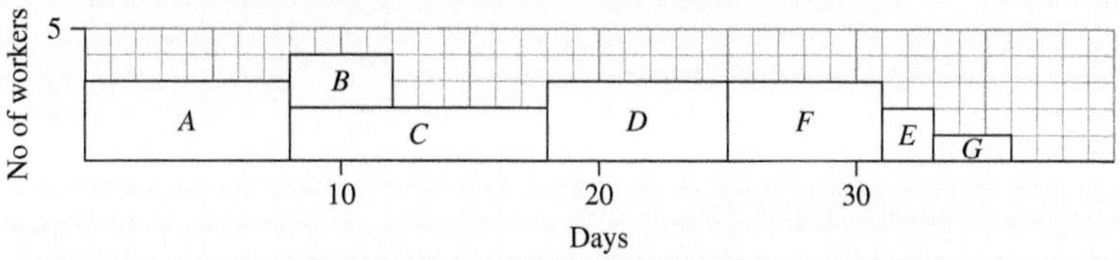

The resource histogram shows that the project now takes 36 days.

> ④ Extend the duration of the project.

Answer
sheet
available

1   The table shows the requirements for a given project.

Activity	Preceded by	Duration (days)	Number of workers
A	–	5	3
B	A	3	5
C	A	4	2
D	B	6	4
E	B, C	4	4
F	C	2	3
G	C	5	5
H	D, E	2	4
I	E, F	2	2

DISCRETE

a Draw a resource histogram for this project, assuming that all activities start as early as possible. State the number of workers needed.

b Show that the number of workers can be reduced by 3 without extending the duration of the project.

c How long will the project take if there are just 11 workers available?

2

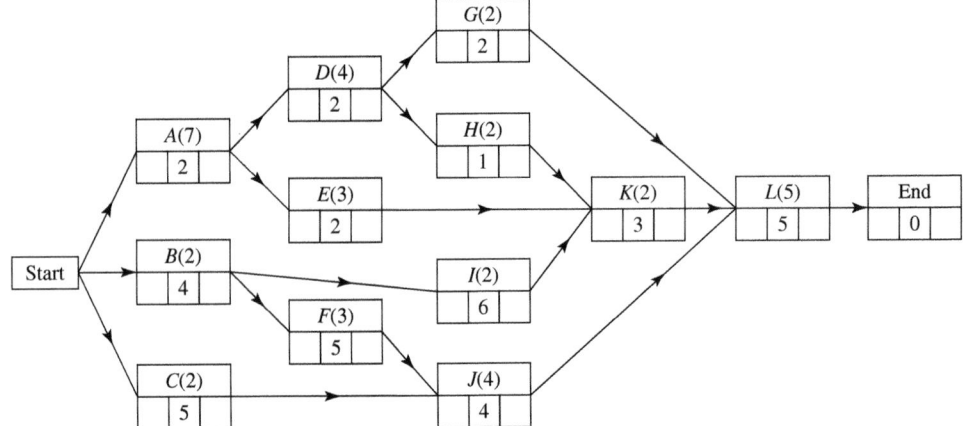

a Draw a resource histogram for the project shown in this activity network assuming that each activity starts as early as possible. (The values in brackets are the number of workers needed for the activity and the non-bracketed numbers are durations, in days, of each activity.) State the number of workers the project manager needs to be able to call on.

b By rescheduling activities, find the minimum number of workers required if the project is to be completed in the minimum time.

c By how many days would the project be extended if only 8 workers were available?

3 The table shows the requirements for a given project.

Activity	Preceded by	Duration (days)	Number of workers
A	–	3	2
B	A	5	1
C	A	3	2
D	B	3	3
E	B	4	3
F	C	3	2
G	D	2	4
H	E	4	2
I	E, F	5	2
J	G, H	3	1
K	I, J	3	4

a Draw a resource histogram for this project assuming that all activities start as early as possible. State the number of workers needed.

b Show, by resource levelling, that this number of workers can be reduced.

c By how many days would the project have to be extended if there were only

   i 5 workers,

   ii 4 workers available?

4

Activity	Preceded by	Duration (days)	Number of workers
A	–	2	5
B	A	4	2
C	A	5	4
D	A	7	2
E	B, C	8	3
F	C, D	4	2
G	E, F	2	3
H	G	2	3
I	G	4	4
J	H, I	2	2
K	I	2	2
L	H, J	1	3

a Draw a resource histogram for this project assuming that all activities start as early as possible. State the number of workers needed.

b In fact, there are only 6 workers available. Show how the project can be completed and state the number of extra days required.

Chapter summary

- From an activity network you find the earliest start time and latest finish time for each activity.
- Float = (latest finish time − earliest start time) − duration.
- Critical activities have zero float. They form one or more critical paths.
- Earliest finish time = earliest start time + duration.
  Latest start time = latest finish time − duration.
- You show this information on a Gantt chart or cascade chart. Activities are shown as bars against a time scale, starting as early as possible and with the float shown dotted.
- A resource histogram shows the number of workers needed at each stage of the project.
- Resource levelling involves rescheduling activities to reduce the number of workers needed.
- If the number of workers is limited, this may involve extending the duration of the project.

Check and review

You should now be able to...	Review Questions
✔ Calculate the earliest and latest start and finish times for an activity network.	1
✔ Draw a Gantt (cascade) chart for a project.	2
✔ Draw a resource histogram with every activity starting as early as possible.	3a
✔ Use resource levelling to reschedule activities to minimise the number of workers needed.	3b
✔ Use resource levelling to reschedule a project with extended duration when the number of workers is limited.	3c

1   Construct a table showing the earliest and latest start and finish times for the activities in the project shown in this network.

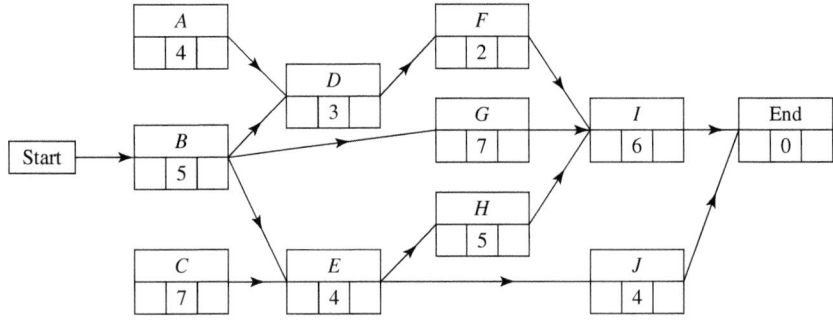

2   Draw a Gantt chart for the table you made in question **1**.

3

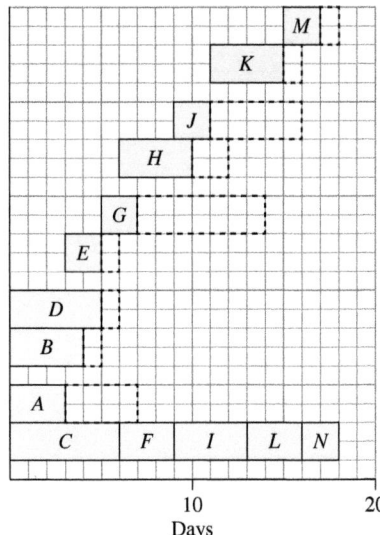

10 20
Days

a All the activities in this Gantt chart require one worker, with the exception of $D(2)$, $G(3)$, $H(4)$, $I(2)$ and $K(2)$. Draw a resource histogram with every activity starting as early as possible. State the number of workers needed.

b Use resource levelling to reduce the number of workers needed. State the minimum workforce needed if the project is to be completed in 18 days.

c If the available workforce is 5 workers, show how the project can be completed with just one extra day.

Research

Critical Path Drag is defined to be the amount of time that an activity on the critical path adds to the project duration. An important alternative way to understand this is to see it as the maximum amount of time that an activity can be shortened before it is no longer on the critical path.

For the activity network shown below: identify the floats, critical activities and hence the critical path drag.

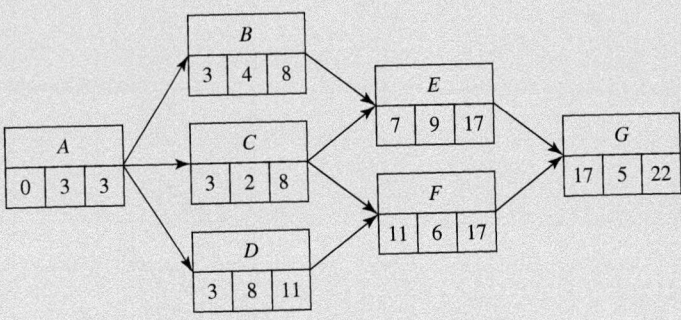

Can you explain why the critical path drag is equal to the minimum of either the remaining duration or (if there are parallel activities) the total float of the parallel activity that has the least total float?

Research

There are many criticisms (and offered solutions) of the critical path method that are worth finding out about.

These include:

1. The difficulty of estimating completion times – particularly for new projects.
2. Defining links for large projects becomes complicated very quickly.
3. The reality of resources (employees) shifting and changing the plan.
4. Identifying single critical paths.

Look into some suggested solutions to these problems by finding out a little more on probabilistic critical paths, critical chain project management and crash durations.

Did you know?

One modern critic of Gannt carts is Professor Edward Tufte. He argues that graphics should provide viewers with the greatest number of ideas in the shortest time, with the least ink, in the smallest space. He feels that the over simplistic Gannt charts have 'regressed to Microsoft mediocrity'.

29 Assessment

1 a Draw a Gantt chart for this activity network. **[5 marks]**

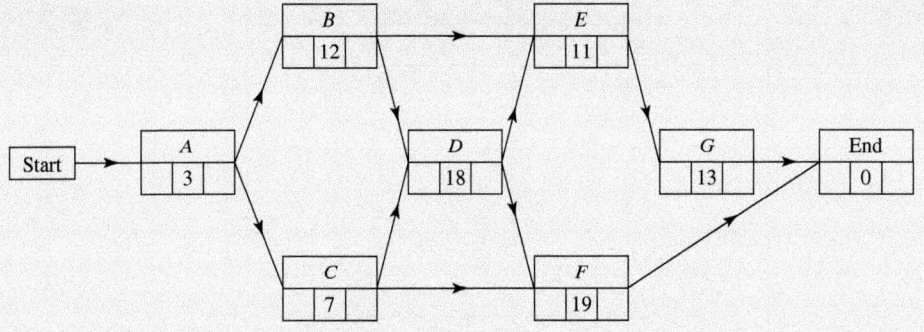

b Given that each activity requires 3 workers, how many workers are required to complete the project in the minimum time? **[1]**

2 The activity network for a project is shown below. The values given are the duration of each activity in hours.

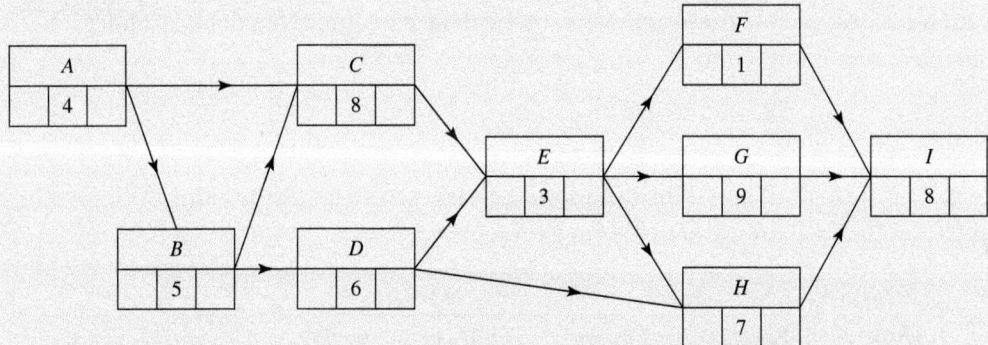

a Find the earliest start time and latest finish time for each activity and complete the activity network. **[4]**

b State the critical path. **[1]**

c Calculate the float of activity *H* **[2]**

d Explain how a delay of 3 hours for activity *D* will affect the minimum completion time for the project. **[2]**

3 The precedence table shows the dependencies and duration of the activities in a project.

Activity	Preceded by	Duration (days)
A	–	4
B	–	2
C	A	5
D	A, B	7
E	B	6
F	E	1
G	D, E	4
H	F, G	5
I	F	2
J	C, H, I	3
K	I, J	4

a Use the precedence table to draw an activity network. Include the duration, the earliest start time and the latest finish time for each activity. **[6]**

b Find the critical path for the network. **[1]**

c Explain how long each of these activities can be delayed without having an impact on the minimum completion time for the project.

    i Activity G   **ii** Activity C **[3]**

4 The table gives information about the duration, the earliest and latest start and finish times and the float for the activities in a project.

Activity	Duration (hours)	Start Earliest	Start Latest	Finish Earliest	Finish Latest	Float
A	1.5	0			1.5	
B	2.4				3.9	0
C		3.9	3.9		10	0
D	1.8	3.9				4.6
E	0.3	10				0
F	3	3.9			10.3	
G	2.6				12.9	0
H	5.2	12.9			18.1	

a Copy and complete the table. **[8]**

b Draw a Gantt chart to show the project. **[3]**

c Use your Gantt chart to list which activities

    i Are definitely taking place at time 2.5, **[1]**   **ii** May be happening at time 9 **[3]**

5 A project taking place in a workshop is shown in the Gantt chart.

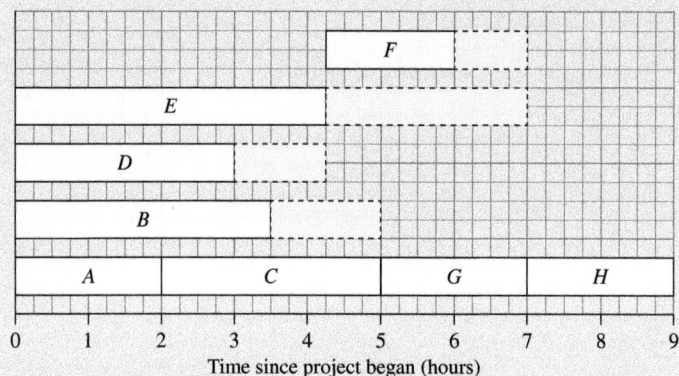

Work on the project begins at 8:00 am. An inspector visits at 11:45 am.

a Which activities will the inspector definitely see happening? [1]

Each activity requires two people to complete it and every person is capable of carrying out every activity.

b What is the lower bound for the number of people required to complete the project in the minimum time? [2]

c Draw a resource histogram to show how the project can be completed in the minimum time by 8 people. [3]

The machine needed for activity *F* is broken and will not be fixed until 2:30 pm.

d What is the earliest time at which the whole project can now end? [2]

6 On this Gantt chart, the values in brackets show the number of workers needed for the activity.

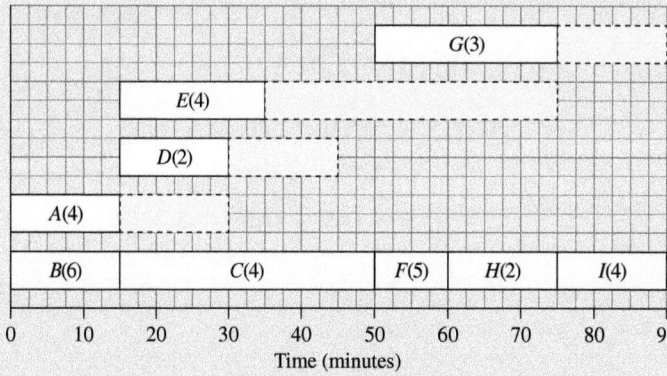

a Assuming that every activity starts as early as possible, draw a resource histogram and state the number of workers required. [4]

b Use resource levelling to show how the project can be completed in the minimum time by just 9 workers. [3]

Only 8 workers are now available.

c Draw a resource histogram and give the new minimum length of the project. [4]

7 The table gives the duration, dependencies and number of workers required for the activities in a project.

Activity	Depends on:	Duration (hours)	Number of workers
A	–	5	4
B	–	2	2
C	–	4	5
D	A, B	6	1
E	A, B, C	2	6
F	B, C	2	3
G	F	5	4
H	E, F	8	2
I	G	3	7
J	I	5	3
K	J	6	4
L	I	8	3

a Calculate a lower bound for the number of workers required to complete the project in 26 hours. **[2]**

b Draw a resource histogram for the project, assuming that all activities start as early as possible. **[4]**

It is possible for 4 workers to complete activity C in 5 hours.

c Would you recommend this approach? Fully explain your answer. **[3]**

8 This activity network has the number of workers required for each activity shown in brackets. The duration of each activity is given in days.

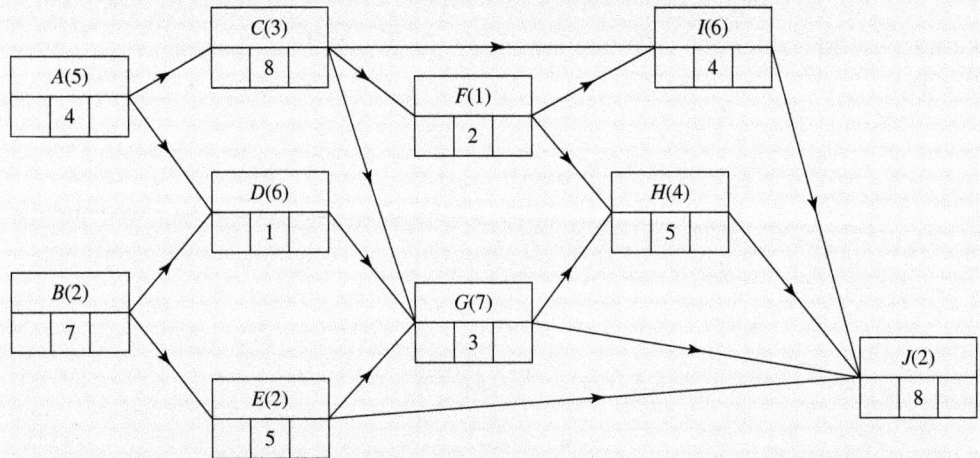

a Copy and complete the activity network with the earliest start and latest finish times. **[4]**

b Find the critical paths. **[2]**

c Draw a resource histogram, assuming that each activity starts as early as possible. State the number of workers required. **[5]**

d Explain how to reschedule the activities to find the minimum number of workers required to complete the project in the minimum time. **[2]**

e By how many days would the project have to be extended if only 10 workers were available? Explain your answer. **[2]**

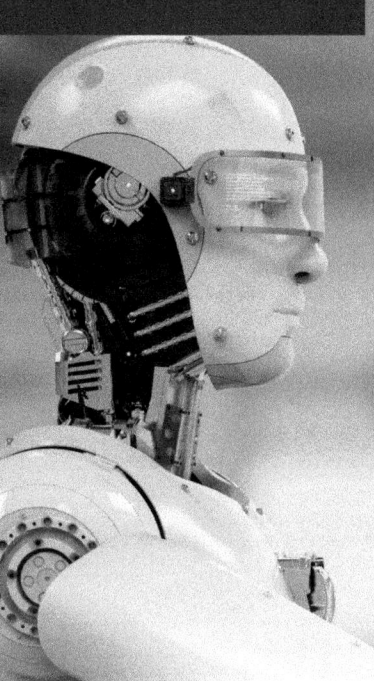

30 Linear programming and game theory 2

Making decisions about situations that involve cooperation, or non-cooperation, with outcomes that are either advantageous or non-advantageous, is fundamental to **game theory**. Game theory has applications in a variety of areas including economics, business, political science and biology. The behaviours and decisions of humans, or animals, can lead to desirable or less desirable outcomes. Game theory can be used to model likely patterns of behaviour.

Until recently, computer programs have been able to take decisions only if every eventuality was pre-programmed. If certain conditions were not met, then the programming did not allow a decision to be made. Game theory has been used to advance programming for artificial intelligence (AI) systems. In future, AI systems may generate payoff matrices that are based on observed stimuli and experience. This will mean that, when decisions need to be made, the system responds more like a human.

Orientation

What you need to know	What you will learn	What this leads to
Ch14 Linear programming and game theory 1	• How to use the simplex algorithm. • How to analyse zero-sum games. • How to analyse mixed-strategy games.	**Careers** • Business. • Economics. • Political science. • Biology. • Computer programming.

In linear programming, you aim to find the best combination of a number of quantities (the decision variables) to maximise or minimise a given quantity (the objective function), subject to a number of constraints (usually inequalities).

To solve a problem with just two decision variables using a graph, you need to identify the feasible region. This is the set of points on the graph that satisfy all the constraints. You then draw an objective line. This is a line joining all points for which the objective function takes the same value.

By moving the objective line you can find the optimal value of the objective function for which the line intersects the feasible region. The optimal solution corresponds to a vertex of the feasible region.

If the problem has three or more decision variables, a graphical approach may be difficult or infeasible, so it is necessary to find a solution by algebraic methods. The simplex algorithm is an algebraic method that tests whether a vertex gives the optimal solution and, if not, moves systematically to a better vertex until the problem is solved.

You first write the problem using equations rather than inequalities. You do this by introducing **slack variables**.

Look at this linear programming formulation. This has only two variables, so you can compare the simplex method with the graphical method.

Maximise $\quad P = 6x + 8y$

subject to $\quad 4x + 3y \leq 1500$

$\quad\quad\quad\quad x + 2y \leq 500$

$\quad\quad\quad\quad x \geq 0, y \geq 0$

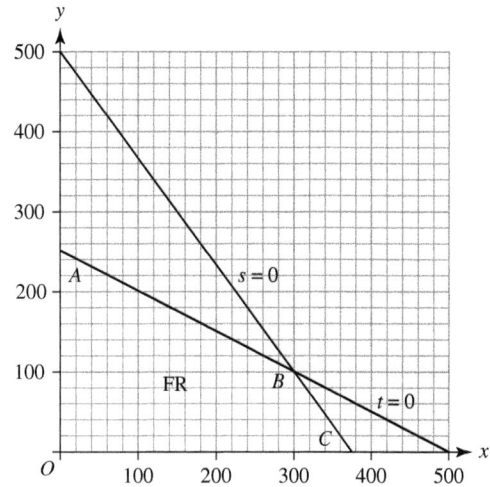

You use slack variables s and t to show the difference (the slack) between the two sides of the inequalities. You write the formulation like this:

Maximise $\quad P = 6x + 8y$

subject to $\quad 4x + 3y + s = 1500$

$\quad\quad\quad\quad x + 2y + t = 500$

$\quad\quad\quad\quad x \geq 0, y \geq 0, s \geq 0, t \geq 0$

The line $4x + 3y = 1500$ is equivalent to $s = 0$

and the line $x + 2y = 500$ to $t = 0$

At each of the vertices O, A, B and C of the feasible region, two of the variables x, y, s and t are zero.

The algebraic method effectively sets two of the variables to zero, finds the values of the other two and evaluates P

For example, setting $x = 0$, $t = 0$ (that is vertex A) gives

$3y + s = 1500$

$2y = 500$

Solving gives $y = 250$, $s = 750$ and $P = 6 \times 0 + 8 \times 250 = 2000$

The complete table is

Vertex	x	y	s	t	P
O	0	0	1500	500	0
A	0	250	750	0	2000
B	300	100	0	0	2600
C	375	0	0	125	2250
–	0	500	0	–500	–
–	500	0	–500	0	–

The optimal solution is $x = 300$, $y = 100$ $P = 2600$

In each row of the table, the non-zero variables form the basis of a possible solution. They are called the **basic variables** and the solution is a **basic solution**. Each of the first four rows is a **basic feasible solution** because it corresponds to a point of the feasible region.

You are going to use the **simplex algorithm** to search the basic solutions for the optimal solution.

You write each constraint as an equation, using a slack variable, in **standard form**. All the variables should be on the left-hand side and a non-negative value should be on the right. You also write the objective function in standard form, though the right-hand side can be negative.

You then enter the coefficients into a table called a **simplex tableau** (plural tableaux)

So for the above example

$P - 6x - 8y - 0s - 0t = 0$

$4x + 3y + s + 0t = 1500$

$x + 2y + 0s + t = 500$

and the simplex tableau is

> The last two rows do not correspond to vertices of the feasible region. This is shown by the negative values taken by the variables.

DISCRETE

907

Basic variable	P	x	y	s	t	Value	Row
	1	−6	−8	0	0	0	R1
s	0	4	3	1	0	1500	R2
t	0	1	2	0	1	500	R3

This starting tableau corresponds to the origin O in the feasible region, with non-basic variables $x = 0$, $y = 0$, giving $P = 0$

The basic variables are $s = 1500$ and $t = 500$

Because $P - 6x - 8y = 0$, increasing either x or y will make P increase. In general, a negative number in the objective row tells you that the corresponding variable must be increased to reach the optimal solution.

> The basic variables, s and t, always have a coefficient of 1 in their row, and zeros elsewhere in their column. The 'Row' column is for recording your working. The left-hand column is not really needed, but you might find it helpful to keep track of the process.

Test for optimality

> **Key point**
> The table shows the optimal solution when the objective (top) row contains no negative coefficients.

You move to a different vertex of the feasible region. This is called **changing the basis** of the solution.

This means either x or y becomes a basic variable. It is said to **enter the basis** as you increase its value. In turn, either s or t must become a non-basic variable. It is said to **leave the basis** as you make it zero.

You first have to choose between x and y for the new basic variable. You could choose either, but with coefficients −6 and −8 it is clear that increasing y will have a greater effect on P than increasing x

> **Key point**
> Choose the variable with the 'most negative' entry in the objective row to enter the basis. The corresponding column is called the **pivot column**.

Next you choose which of s and t should become zero. You divide each *positive* entry in the pivot column into the corresponding entry in the 'Value' column. These ratios are called θ-**values**.

> **Key point**
> The row giving the smallest θ-value is the **pivot row**.
> The entry in the pivot row and pivot column is called the **pivot**.
> The basic variable in the pivot row is the one that will leave the basis.

In the example, the y-column is the pivot column.

The θ-value for $s = \dfrac{1500}{3} = 500$, and for $t = \dfrac{500}{2} = 250$

It is smaller for t, so this is the pivot row, and t will leave the basis.

The pivot is the 2 in the pivot column and pivot row.

Basic variable	P	x	y	s	t	Value	Row
	1	−6	−8	0	0	0	R1
s	0	4	3	1	0	1500	R2
t	0	1	2	0	1	500	R3

Pivot column (above y)

Pivot row → t

Pivot (the 2 in R3, y-column)

You divide the pivot row by the pivot to give a 1 in the pivot position. In this case, you divide row R3 by 2

You now want to get zeros in the rest of the pivot column by combining the other rows with a suitable multiple of the pivot row.

In the y-column, change the 3 to a 0 by subtracting $3 \times R6$

In the y-column, change the −8 to 0 by adding $8 \times R6$

> Remember that basic variables have a 1 in their row and zeros in the rest of their column. You need to make this true for y

Basic variable	P	x	y	s	t	Value	Row
	1	−2	0	0	4	2000	$R4 = R1 + 8 \times R6$
s	0	$2\frac{1}{2}$	0	1	$-1\frac{1}{2}$	750	$R5 = R2 - 3 \times R6$
y	0	$\frac{1}{2}$	1	0	$\frac{1}{2}$	250	$R6 = \dfrac{R3}{2}$

> Use the 'Row' column to record what has happened, as shown. Always use fractions, not decimals, to avoid introducing rounding errors.

At this stage, $s = 750$ and $y = 250$, $x = 0$, $t = 0$, giving $P = 2000$

You apply the optimality test. There is −2 in the x-column, so you need to do another change of basis.

The x-column is the pivot column, so x will enter the basis.

The θ-value for $s = \dfrac{750}{2\frac{1}{2}} = 300$, and for $t = \dfrac{250}{\frac{1}{2}} = 500$

> Remember that the solution is optimal if there are no negative coefficients in the objective row.

It is smaller for s, so this is the pivot row, and s will leave the basis.

Basic variable	P	x	y	s	t	Value	Row
	1	−2	0	0	4	2000	R4
s	0	$2\frac{1}{2}$	0	1	$-1\frac{1}{2}$	750	R5
y	0	$\frac{1}{2}$	1	0	$\frac{1}{2}$	250	R6

Divide the pivot row by $2\frac{1}{2}$ to give row R8. The basic variable will change from s to x

In the x-column on the y-row, change the $\frac{1}{2}$ to a 0 by subtracting $\frac{1}{2} \times R8$

In the x-column on the objective row, change the −2 to 0 by adding $2 \times R8$

DISCRETE

Basic variable	P	x	y	s	t	Value	Row
	1	0	0	$\dfrac{4}{5}$	$2\dfrac{4}{5}$	2600	$\text{R7} = \text{R4} + 2 \times \text{R8}$
x	0	1	0	$\dfrac{2}{5}$	$-\dfrac{3}{5}$	300	$\text{R8} = \dfrac{\text{R5}}{2\dfrac{1}{2}}$
y	0	0	1	$-\dfrac{1}{5}$	$\dfrac{4}{5}$	100	$\text{R9} = \text{R6} - \dfrac{1}{2} \times \text{R8}$

You now have $x = 300$, $y = 100$, $s = 0$, $t = 0$, giving $P = 2600$

There are no negative coefficients in the objective row, so the solution is optimal.

The simplex algorithm

Key point

Step 1 Write the constraints and the objective function as equations in standard form, using slack variables.

Step 2 Transfer the data to a simplex tableau. At this stage the slack variables form the basis.

Step 3 Choose the column with the most negative coefficient in the objective row. This is the **pivot column**.

Step 4 If the **positive** numbers in the pivot column are p_1, p_2, ... and the corresponding numbers in the 'Value' column are v_1, v_2, ..., calculate $\theta_1 = \dfrac{v_1}{p_1}$, $\theta_2 = \dfrac{v_2}{p_2}$, .

The row giving the smallest θ-value is the **pivot row**. (If there is a 'tie', choose at random). The number in the pivot column and pivot row is the **pivot**.

Step 5 Divide the pivot row by the pivot. Replace the basic variable for that row by the variable for the pivot column.

Step 6 Combine suitable multiples of the new pivot row with the other rows to give zeros in the pivot column.

Step 7 If there are no negative coefficients in the objective row, the solution is optimal. Otherwise, go to Step 3

The main advantage of the simplex algorithm is that you can use it when there are three or more decision variables.

Example 1

Maximise $\qquad P = x + 2y + z$

subject to $\qquad x + 3y + 2z \le 60$

$\qquad\qquad\quad 2x + y \le 40$

$\qquad\qquad\quad x + 3z \le 30$

$\qquad\qquad\quad x, y, z \ge 0$

Step 1 In standard form with slack variables s, t and u

$\qquad$ Maximise $\qquad P - x - 2y - z = 0$

$\qquad$ subject to $\qquad x + 3y + 2z + s = 60$

$\qquad\qquad\qquad\quad 2x + y + t = 40$

$\qquad\qquad\qquad\quad x + 3z + u = 30$

$\qquad\qquad\qquad\quad x, y, z, s, t, u \ge 0$

Step 2

Basic variables.

B.V.	P	x	y	z	s	t	u	Value	Row
	1	−1	−2	−1	0	0	0	0	R1
s	0	1	3	2	1	0	0	60	R2
t	0	2	1	0	0	1	0	40	R3
u	0	1	0	3	0	0	1	30	R4

This is the simplex tableau.

Step 3 The y-column is the pivot column.

Step 4 The θ-values are: R2: $60 \div 3 = 20$, R3: $40 \div 1 = 40$

These give R2 as the pivot row and the 3 as the pivot, as shown.

Step 5 Divide R2 by 3

y becomes the basic variable.

Step 6

B.V.	P	x	y	z	s	t	u	Value	Row
	1	$-\dfrac{1}{3}$	0	$\dfrac{1}{3}$	$\dfrac{2}{3}$	0	0	40	$R5 = R1 + 2 \times R6$
y	0	$\dfrac{1}{3}$	1	$\dfrac{2}{3}$	$\dfrac{1}{3}$	0	0	20	$R6 = \dfrac{R2}{3}$
t	0	$1\dfrac{2}{3}$	0	$-\dfrac{2}{3}$	$-\dfrac{1}{3}$	1	0	20	$R7 = R3 - R6$
u	0	1	0	3	0	0	1	30	$R8 = R4$

Combine the rows to get zeros in the y-column.

Step 7 The solution is not optimal because there is a negative value in the top row.

(Continued on the next page)

DISCRETE

Step 3 The x-column is the pivot column.

Step 4 The θ-values are: R6: $20 \div \dfrac{1}{3} = 60$

$$\text{R7: } 20 \div 1\dfrac{2}{3} = 12$$

$$\text{R8: } 30 \div 1 = 30$$

> These give R7 as the pivot row and $1\dfrac{2}{3}$ as the pivot, as shown.

Step 5 Divide R7 by $1\dfrac{2}{3}$

> x becomes the basic variable.

Step 6

B.V.	P	x	y	z	s	t	u	Value	Row
	1	0	0	$\dfrac{1}{5}$	$\dfrac{3}{5}$	$\dfrac{1}{5}$	0	44	$R9 = R5 + \dfrac{R11}{3}$
y	0	0	1	$\dfrac{4}{5}$	$\dfrac{2}{5}$	$-\dfrac{1}{5}$	0	16	$R10 = R6 - \dfrac{R11}{3}$
x	0	1	0	$-\dfrac{2}{5}$	$-\dfrac{1}{5}$	$\dfrac{3}{5}$	0	12	$R11 = \dfrac{R7}{1\frac{2}{3}}$
u	0	0	0	$3\dfrac{2}{5}$	$\dfrac{1}{5}$	$-\dfrac{3}{5}$	1	18	$R12 = R8 - R11$

> Combine the rows to get zeros in the x-column.

Step 7 There are no negative numbers in the top row so the solution is optimal.

The optimal solution is $y = 16$, $x = 12$, $u = 18$, $z = 0$, $s = 0$, $t = 0$, giving $P = 44$

Exercise 30.1A Fluency and skills

1 Use the simplex method to solve the following linear programming problems.

a Maximise $P = x + y$

  subject to $x + 2y \le 40$

  $3x + 2y \le 60$

  $x \ge 0, y \ge 0$

b Maximise $P = 2x + y$

  subject to $4x + 3y \le 170$

  $5x + 2y \le 160$

  $x \ge 0, y \ge 0$

c Maximise $P = 4x + 5y$

  subject to $2x + 3y \le 30$

  $x + 3y \le 24$

  $4x + 3y \le 48$

  $x \ge 0, y \ge 0$

d Maximise $P = x + 2y$

  subject to $x + y \le 6$

  $2x + y \le 9$

  $3x + 2y \le 15$

  $x \ge 0, y \ge 0$

e Maximise $P = 3x + 4y + 2z$

  subject to $8x + 5y + 2z \le 7$

  $x + 2y + 3z \le 4$

  $x \ge 0, y \ge 0, z \ge 0$

f Maximise $P = 7x + 6y + 4z$

  subject to $2x + 4y - z \le 7$

  $5x + 6y + 2z \le 16$

  $7x + 7y + 4z \le 25$

  $x \ge 0, y \ge 0, z \ge 0$

2

P	x	y	z	s	t	Value
1	-3	-1	-2	0	0	0
0	2	3	1	1	0	20
0	1	2	1	0	1	12

a Write the objective function and the constraints (in inequality form) corresponding to this simplex tableau.

b Perform one iteration of the simplex algorithm. Explain how you know that the solution is not yet optimal. State the values of the variables at this stage.

c Complete the solution of the problem, stating the final values of the variables.

3 Consider the following linear programming problem.

Maximise $P = x + 3y$
subject to $x + y \leq 6$
$x + 4y \leq 12$
$x + 2y \leq 7$
$x \geq 0, y \geq 0$

a Write the problem in terms of equations with slack variables.

b Enter the data into a simplex tableau and perform one iteration of the simplex algorithm. Explain how you know that the optimal solution has not yet been reached.

c Perform a second iteration of the algorithm to obtain the optimal solution.

4 A linear programming problem is written as a simplex tableau as follows.

P	x	y	s	t	Value
1	-5	-6	0	0	0
0	3	3	1	0	40
0	1	2	0	1	25

a Write down the original linear programming formulation, stating the objective function and showing the constraints as inequalities. Explain the meaning of the variables s and t in the tableau.

b Perform one iteration of the simplex algorithm. State the values of x, y, s, t and P at this stage, and explain how you know that the solution is not optimal.

c Perform a second iteration of the algorithm to obtain the optimal solution.

Reasoning and problem-solving

Strategy

To solve a problem using the simplex algorithm

1. Write the problem as a linear programming formulation.

2. Express the constraints as equations by using slack variables.

3. Apply the algorithm until there are no negative coefficients in the objective row.

4. Read off the values of the basic variables and the objective function.

5. State the solution in context, including units where appropriate.

6. If the problem is non-standard, modify it until you can apply the simplex algorithm.

7. Answer the question in context.

You will usually need to express a problem as a linear programming formulation before solving it.

Example 2

A farmer has 65 hectares of land on which to grow a mixture of wheat and potatoes. The costs and profits involved are shown in the table.

	Labour (man-hours per ha)	Fertiliser (kg per ha)	Profit (£ per ha)
Wheat	30	700	80
Potatoes	50	400	100

There are 2800 man-hours of labour and 40 tonnes of fertiliser available. Construct a simplex tableau and solve it to find the optimal planting scheme. Explain the meaning of slack variables in the final solution.

Plant x ha of wheat and y ha of potatoes.

Maximise $P = 80x + 100y$

subject to $30x + 50y \leq 2800$ and so $3x + 5y \leq 280$

$700x + 400y \leq 40\,000$ and so $7x + 4y \leq 400$

$x + y \leq 65$

$x \geq 0, y \geq 0$

$P - 80x - 100y = 0$

$3x + 5y + s = 280$

$7x + 4y + t = 400$

$x + y + u = 65$ ● — — — — — — — — — Write as equations.

P	x	y	s	t	u	Value	Row
1	−80	−100	0	0	0	0	R1
0	3	5	1	0	0	280	R2
0	7	4	0	1	0	400	R3
0	1	1	0	0	1	65	R4

Apply the simplex algorithm using the pivot shown.

P	x	y	s	t	u	Value	Row
1	−20	0	20	0	0	5600	$R5 = R1 + 100 \times R6$
0	$\frac{3}{5}$	1	$\frac{1}{5}$	0	0	56	$R6 = \dfrac{R2}{5}$
0	$4\frac{3}{5}$	0	$-\frac{4}{5}$	1	0	176	$R7 = R3 - 4 \times R6$
0	$\frac{2}{5}$	0	$-\frac{1}{5}$	0	1	9	$R8 = R4 - R6$

Not yet optimal, so apply the simplex algorithm using the pivot shown.

(*Continued on the next page*)

P	x	y	s	t	u	Value	Row
1	0	0	10	0	50	6050	$R9 = R5 + 20 \times R12$
0	0	1	$\frac{1}{2}$	0	$-1\frac{1}{2}$	$42\frac{1}{2}$	$R10 = R6 - R12 \times \frac{3}{5}$
0	0	0	$1\frac{1}{2}$	1	$-11\frac{1}{2}$	$72\frac{1}{2}$	$R11 = R7 - R12 \times \frac{23}{5}$
0	1	0	$-\frac{1}{2}$	0	$2\frac{1}{2}$	$22\frac{1}{2}$	$R12 = R8 \times \frac{5}{2}$

The solution is now optimal.

Maximum profit = £6050, by planting 22.5 ha of wheat and 42.5 ha of potatoes.

The other non-zero value is $t = 72.5$. This corresponds to 7250 kg of fertiliser left over.

Be careful to allow for any simplification made in the constraints before the slack variables were introduced.

You can use the standard simplex algorithm provided:

- you need to maximise the objective function
- every non-trivial constraint is an inequality using ≤
- all variables are ≥ 0
- the origin is a vertex of the feasible region.

Provided these conditions are met, you can write the problem as equations in standard form using slack variables, with the right-hand sides all being non-negative (except perhaps for the objective function).

You may need to deal with non-standard situations in which the aim is to minimise the objective function, or when one or more of the constraints involves an inequality using ≥

You convert a minimisation problem into a related maximisation problem.

Key point

Minimising the objective function C is equivalent to maximising the objective function $P = -C$

You can rewrite an inequality involving ≥ so that it uses ≤ by multiplying through by −1

If the resulting problem can be written in standard form, you can go ahead with the simplex algorithm.

Example 3

Minimise	$C = 3x - 4y - 3z$
subject to	$x + y - z \geq -2$
	$x + 2y + z \leq 3$
	$x \geq 0, y \geq 0, z \geq 0$

(*Continued on the next page*)

Maximise $P = -3x + 4y + 3z$

subject to $\quad -x - y + z \le 2$

$\qquad\qquad x + 2y + z \le 3$

$\qquad\qquad x, y, z \ge 0$

First restate the problem.

$P + 3x - 4y - 3z = 0$

$-x - y + z + s = 2$

$x + 2y + z + t = 3$

$x, y, z, s, t \ge 0$

Write in standard form with slack variables.

P	x	y	z	s	t	Value	Row
1	3	-4	-3	0	0	0	R1
0	-1	-1	1	1	0	2	R2
0	1	2	1	0	1	3	R3

Complete the simplex tableau.

P	x	y	z	s	t	Value	Row
1	5	0	-1	0	2	6	$R4 = R1 + 4 \times R6$
0	$-\dfrac{1}{2}$	0	$1\dfrac{1}{2}$	1	$\dfrac{1}{2}$	$3\dfrac{1}{2}$	$R5 = R2 + R6$
0	$\dfrac{1}{2}$	1	$\dfrac{1}{2}$	0	$\dfrac{1}{2}$	$1\dfrac{1}{2}$	$R6 = \dfrac{R3}{2}$

Apply the simplex algorithm with the pivot shown.

There is still a negative coefficient in the objective row, so apply the simplex algorithm again with the pivot shown.

P	x	y	z	s	t	Value	Row
1	$4\dfrac{2}{3}$	0	0	$\dfrac{2}{3}$	$2\dfrac{1}{3}$	$8\dfrac{1}{3}$	$R7 = R4 + R8$
0	$-\dfrac{1}{3}$	0	1	$\dfrac{2}{3}$	$\dfrac{1}{3}$	$2\dfrac{1}{3}$	$R8 = R5 \times \dfrac{2}{3}$
0	$\dfrac{2}{3}$	1	0	$-\dfrac{1}{3}$	$\dfrac{1}{3}$	$\dfrac{1}{3}$	$R9 = R6 - \dfrac{R8}{2}$

The solution is now optimal.

Maximum $P = 8\dfrac{1}{3}$ when $x = 0, y = \dfrac{1}{3}, z = 2\dfrac{1}{3}$

As $C = -P$, minimum $C = -8\dfrac{1}{3}$ when $x = 0, y = \dfrac{1}{3}, z = 2\dfrac{1}{3}$

An extra problem arises if you have a constraint such as $2x - 3y \geq 4$

The simplex method is based on the origin being a vertex of the feasible region, but clearly $(0, 0)$ does not satisfy this inequality. Attempting to write the constraint as an equation with a slack variable would give $-2x + 3y + s = -4$, which if $x = 0$ and $y = 0$ violates the need to have $s \geq 0$

There are some methods beyond the scope of this syllabus for dealing with the situation, but it is sometimes possible to modify the simplex tableau to give a basic feasible solution (a vertex of the feasible region) from which to start the simplex algorithm. To do this you first pivot on a negative coefficient in the row with the negative value.

Example 4

Maximise $\quad P = x + 2y$

subject to $\quad 2x + y \leq 6$

$\qquad\qquad x + y \geq 2$

$\qquad\qquad x, y \geq 0$

First write the second constraint as $-x - y \leq -2$

$P - x - 2y = 0$

$2x + y + s = 6$

$-x - y + t = -2$

> Write the problem as equations with slack variables.
>
> The last constraint is not in standard form.

P	x	y	s	t	Value	Row
1	−1	−2	0	0	0	R1
0	2	1	1	0	6	R2
0	−1	−1	0	1	−2	R3

> Before starting the simplex algorithm, pivot on the cell shown to produce +2 in the value column.

P	x	y	s	t	Value	Row
1	0	−1	0	−1	2	R4 = R1 + R6
0	0	−1	1	2	2	R5 = R2 − 2 × R6
0	1	1	0	−1	2	R6 = − R3

This gives $x = 2$, $s = 2$ ($y = 0$, $t = 0$) as an initial feasible solution.

> There are negative coefficients in the objective row, so apply the simplex algorithm with the pivot shown.

P	x	y	s	t	Value	Row
1	1	0	0	−2	4	R7 = R4 + R9
0	1	0	1	1	4	R8 = R5 + R9
0	1	1	0	−1	2	R9 = R6

> The solution is not optimal, so apply the simplex algorithm with the pivot shown.

P	x	y	s	t	Value	Row
1	3	0	2	0	12	R10 = R7 + 2 × R11
0	1	0	1	1	4	R11 = R8
0	2	1	1	0	6	R12 = R9 + R11

> This is now optimal.

Maximum $P = 12$ when $x = 0$, $y = 6$ ($t = 4$)

1 a Use the simplex method to minimise $C = x - y$

subject to $y \le 2x$

$2y \ge x$

$2x + y \le 12$

$x \ge 0, y \ge 0$

b Use the simplex method to minimise $C = 4x - 3y - 5z$

subject to $x + y - z \ge -2$

$x + 2y + z \le 4$

$x \ge 0, y \ge 0, z \ge 0$

c Use the simplex method to minimise $C = x - 2y + z$

subject to $2x - y + z \ge 0$

$2x - 3y \ge -6$

$4y + 3z \le 10$

$x \ge 0, y \ge 0, z \ge 0$

d Use the simplex method to maximise $P = 2x - 3y + z$

subject to $3y \ge x + 2z - 5$

$y - x - z \ge -8$

$x + y + z \ge -2$

$x \ge 0, y \ge 0, z \ge 0$

2 A situation involving variables x, y and z is subject to these constraints:

$x - 4y + z \ge -4$

$2x - 3y + z \le 33$

$x \ge 10, y \ge 3, z \ge 8$

You need to minimise the cost function $C = x + 3y - 2z$

At present, the origin is not in the feasible region. Substitute $X = x - 10$, $Y = y - 3$ and $Z = z - 8$ and solve the resulting problem using the simplex method. Hence state the solution to the original problem.

3 A garden centre produces and sells two grades of grass seed, each a mixture of perennial ryegrass (PR) and creeping red fescue (CRF). Regular Lawn Mix is 70% PR and 30% CRF; Luxury Lawn Mix is 50% of each.

They buy PR at £4 per kg and CRF at £5 per kg. They sell regular mix at £6 per kg and luxury mix at £7 per kg. They have 8000 kg of PR and 6000 kg of CRF in stock.

Use the simplex algorithm to decide what quantities of regular and luxury mix they should make and state the profit they achieve.

4 An electric bicycle has two settings. On setting A the rider pedals, with the motor providing assistance. On setting B the motor does all the work. Setting A gives a speed of $4\,\mathrm{m\,s^{-1}}$, using 6 J of battery energy per metre. Setting B gives a speed of $6\,\mathrm{m\,s^{-1}}$ and uses 9 J per metre. The battery can store 45 000 J. The rider wishes to travel as far as possible in 20 minutes.

Given that the rider travels x metres on setting A and y metres on setting B, express the problem in linear programming terms and use the simplex algorithm to find the rider's furthest distance and the settings that should be used to achieve it.

5 A company makes three models of bicycle. In each week they make x bicycles of type A, y of type B and z of type C. The weekly profit, in £hundreds, is P. In a certain week there are constraints on resources and manpower. They need a production plan to maximise the profit. The simplex tableau corresponding to this problem is as shown.

P	x	y	z	s	t	Value
1	-4	-5	-2	0	0	0
0	5	6	3	1	0	90
0	2	4	1	0	1	42

a How much profit is made on a bicycle of type A?

b Write down the two constraints as inequalities.

c Use the simplex algorithm to solve this linear programming problem. Explain how you know when you have reached the optimal solution.

d Explain why the solution to part **c** is not practical.

e Find a practical solution which gives a profit of £7200

Verify that it is feasible.

6 Moltobuono Meals Ltd make three different pasta sauces. Each is a blend of tomato paste and onion (of which they have unlimited quantities), with the addition of varying amounts of garlic paste, oregano and basil. The table below gives details of the recipes (for 1 kg of sauce), together with the availability of ingredients and the profit on each type of sauce.

Sauce	Garlic (kg)	Oregano (kg)	Basil (kg)	Profit (£ per kg)
Assolato	0.03	0.02	0.01	0.60
Buona Salute	0.02	0.03	0.01	0.50
Contadino	0.03	0.04	0.02	0.90
Availability	50	60	40	

They wish to maximise their profit.

a Write the problem as a linear programming formulation, simplifying the constraints so they have integer coefficients.

b Use the simplex algorithm to find the most profitable production plan.

7 A manufacturer makes two fruit-based drinks. Froo-T is 50% grape juice, 30% peach juice and 20% cranberry juice. Joo-C is 40% grape juice, 20% peach juice and 40% cranberry juice. The profit made is 20p per litre for Froo-T and 25p per litre for Joo-C. The manufacturer has 2100 litres of grape juice, 1200 litres of peach juice and 1500 litres of cranberry juice in stock. Find the amount of each drink they should make to maximise their total profit, and the amount of raw materials left over.

8 A vegetable-box delivery scheme offers four different boxes, containing different proportions of baking potatoes, cabbages, swedes and butternut squash. The table below shows the proportions of these, the availability of the different vegetables and the profit made on the boxes.

Box	Potatoes	Cabbages	Swedes	Squash	Profit (p per box)
A	6	2	1	1	50
B	4	1	1	2	30
C	8	2	2	2	80
D	5	1	2	1	60
Availability	800	240	300	320	

a Express the problem as a simplex tableau and solve it to find the optimal solution. (Assume that all the boxes will be sold.)

b Explain why the result you obtained in **a** does not provide a solution to the original problem. Use your result to arrive at a solution to the problem.

9 a Use the simplex method to maximise $P = 3x + 2y$

subject to $2x + y \le 6$

$3x - 2y \ge 4$

$x, y \ge 0$

b Use the simplex method to minimise $C = x + 2y$

subject to $2x \ge 7$

$x + y \ge 4$

$x, y \ge 0$

10 Part of an area of land is being made into a garden, some of which will be covered with turf and the rest with paving slabs. The aim is to make as large a garden as possible for a maximum outlay of £2000

It costs £5 per m² to lay turf and £10 per m² to lay slabs. There should be at least 50 m² of slabs and at least half of the garden will be turf.

Let there be x m² of turf and y m² of slabs. Find the size of the garden and the areas of turf and slabs to be laid.

11 A linear programming problem is stated as

Maximise $P = 5x + 2y - z$

subject to $x + 2y + z \leq 14$

$\qquad 2x - y + 2z \leq 16$

$\qquad 2x + y - z = 10$

$\qquad x, y, z \geq 0$

a Use the equality constraint to restate the problem as a two-variable problem. Solve it using the simplex method.

b Restate the original problem by replacing the equality constraint with the two constraints $2x + y - z \leq 10$ and $2x + y - z \geq 10$ and then solve it.

12 Three students, A, B and C, attempt to solve this linear programming problem using a simplex tableau:

Maximise $P = x + 2y$

subject to $x + y \leq 28$

$\qquad 3x + y \geq 30$

$\qquad 2x + 3y \geq 60$

$\qquad x, y \geq 0$

a Student A completes the exercise successfully. Show her solution and state the final result.

b Student B miscopies the 28 in the first constraint as 18, and then complains that he can't find a solution. Show how his solution breaks down and explain, by means of a sketch graph or otherwise, why there has been a problem.

c Student C miscopies the first constraint as $x - y \leq 28$

She also complains that there are problems with the solution. Show how her solution breaks down and explain, by means of a sketch graph or otherwise, why there has been a problem.

Games as linear programming problems

Fluency and skills

Here is a brief summary of two-player zero-sum games.

See Ch14.2
For a reminder on zero-sum games.

The table always shows the pay-offs for the row player. In this table, if player A plays A_1 they will lose 3 or gain 5 depending on what player B does.

		B		Row minimum	
		B_1	B_2		
A	A_1	−3	5	−3	
	A_2	3	−1	−1	← max. = −1
Column maximum		3	5		

↑
min. = 3

You find the play-safe strategy for A by finding the maximum of the row minima, and for B by finding the minimum of the column maxima. In this table, the play-safe strategies are A_2 and B_1

If they play a **pure strategy game** (both always play safe), A will win 3.

This is the value of the game.

If the solution is stable, there is no advantage in moving from the play-safe strategy.

> **Key point**
>
> A game has a stable solution if
> maximum of row minima = minimum of column maxima

In the table shown the solution is not stable. It is best for each player to play each strategy for some proportion of the time, i.e play a **mixed strategy game**. They choose the proportions so as to maximise their expected pay-off.

> **Key point**
>
> If pay-offs $x_1, x_2, ..., x_n$ occur with probabilities $p_1, p_2, ..., p_n$, the **expectation** or **expected (mean) pay-off** $E(x)$ is given by
>
> $$E(x) = x_1 p_1 + x_2 p_2 + ... + x_n p_n = \sum_{i=1}^{n} x_i p_i$$

DISCRETE

If A plays A_1 with probability p and B plays B_1, the pay-offs and probabilities are as shown.

	A_1	A_2
Pay-off	−3	3
Probability	p	$(1-p)$

The expected pay-off is $(-3) \times p + 3 \times (1-p) = 3 - 6p$

If B plays B_2, the possible pay-offs for A are as shown.

	A_1	A_2
Pay-off	5	−1
Probability	p	$(1-p)$

The expected pay-off is $5 \times p + (-1) \times (1-p) = 6p - 1$

The value, v, of the game cannot be more than either of these expected pay-offs, so

$v \le 3 - 6p$ and $v \le 6p - 1$

Player A needs to maximise v subject to $v \le 3 - 6p$ and $v \le 6p - 1$

This is a linear programming problem that can be solved graphically if there are just two strategies available to A

If more than two strategies are involved, you need to use the simplex algorithm.

You will need to modify the pay-off matrix if any of the entries are negative, because the simplex method requires all variables to be non-negative including, in this case, the value v of the game. You do this by adding a constant to each entry in the matrix – once the problem has been solved, you then subtract this constant from the value you found.

Example 1

Use the simplex method to solve this zero-sum game to find the optimal strategy for players A and B and the value of the game.

		B	
		B_1	B_2
	A_1	1	−2
A	A_2	0	−1
	A_3	−1	0

Add 2 to each table entry.

		B	
		B_1	B_2
	A_1	3	0
A	A_2	2	1
	A_3	1	2

All entries must be non-negative.

(*Continued on the next page*)

Let A play A_1, A_2 and A_3 with probabilities p_1, p_2 and p_3

Maximise $\quad\quad P = v$

subject to $\quad\quad v \le 3p_1 + 2p_2 + p_3$

$\quad\quad\quad\quad\quad v \le p_2 + 2p_3$

$\quad\quad\quad\quad\quad p_1 + p_2 + p_3 \le 1$

$\quad\quad\quad\quad\quad v, p_1, p_2, p_3 \ge 0$

$P - v = 0$

$v - 3p_1 - 2p_2 - p_3 + s = 0$

$v - p_2 - 2p_3 + t = 0$

$p_1 + p_2 + p_3 + u = 1$

> A's expected profit, P, is the value, v, of the game.

> You replace $p_1 + p_2 + p_3 = 1$ with the inequality shown so that you can introduce a slack variable for the simplex tableau. The slack variable will be found to be zero.

> Introduce slack variables.

P	v	p_1	p_2	p_3	s	t	u	Value	Row
1	−1	0	0	0	0	0	0	0	R1
0	1	−3	−2	−1	1	0	0	0	R2
0	1	0	−1	−2	0	1	0	0	R3
0	0	1	1	1	0	0	1	1	R4
1	0	−3	−2	−1	1	0	0	0	R5 = R1 + R6
0	1	−3	−2	−1	1	0	0	0	R6 = R2
0	0	3	1	−1	−1	1	0	0	R7 = R3 − R6
0	0	1	1	1	0	0	1	1	R8 = R4
1	0	0	−1	−2	0	1	0	0	R9 = R5 + R11 × 3
0	1	0	−1	−2	0	1	0	0	R10 = R6 + R11 × 3
0	0	1	$\frac{1}{3}$	$-\frac{1}{3}$	$-\frac{1}{3}$	$\frac{1}{3}$	0	0	$R11 = \dfrac{R7}{3}$
0	0	0	$\frac{2}{3}$	$1\frac{1}{3}$	$\frac{1}{3}$	$-\frac{1}{3}$	1	1	R12 = R8 − R11
1	0	0	0	0	$\frac{1}{2}$	$\frac{1}{2}$	$1\frac{1}{2}$	$1\frac{1}{2}$	R13 = R9 + R16 × 2
0	1	0	0	0	$\frac{1}{2}$	$\frac{1}{2}$	$1\frac{1}{2}$	$1\frac{1}{2}$	R14 = R10 + R16 × 2
0	0	1	$\frac{1}{2}$	0	$-\frac{1}{4}$	$\frac{1}{4}$	$\frac{1}{4}$	$\frac{1}{4}$	$R15 = R11 + \dfrac{R16}{3}$
0	0	0	$\frac{1}{2}$	1	$\frac{1}{4}$	$-\frac{1}{4}$	$\frac{3}{4}$	$\frac{3}{4}$	$R16 = R12 \times \dfrac{3}{4}$

> This is the simplex solution.

DISCRETE

(Continued on the next page)

This gives $v = 1.5$ when $p_1 = \dfrac{1}{4}$, $p_2 = 0$ and $p_3 = \dfrac{3}{4}$

So the value of the game is -0.5

Subtract the 2 you added initially.

A's optimal strategy is to play A_1 and A_3 with probabilities 0.25, 0.75

and never play A_2

The value of the game to B is 0.5. Suppose B plays B_1 and B_2 with probabilities

q_1 and q_2

If A plays A_1, the value to B is $-q_1 + 2q_2$, so $\quad -q_1 + 2q_2 = 0.5 \quad$ [1]

Similarly, for A_2 $\hspace{5cm} q_2 = 0.5 \quad$ [2]

and A_3 $\hspace{5.5cm} q_1 = 0.5 \quad$ [3]

[1], [2] and [3] are all satisfied by $q_1 = 0.5$ and $q_2 = 0.5$, so B

plays B_1 and B_2 with equal probability.

Of course, instead of using all three probabilities, you could rewrite the problem
by replacing p_3 by $(1 - p_1 - p_2)$. The solution then looks like this.

Example 2

Maximise $\quad P = v$

subject to $\quad v \le 3p_1 + 2p_2 + (1 - p_1 - p_2) \quad \rightarrow \quad v \le 2p_1 + p_2 + 1$

$\hspace{2.3cm} v \le p_2 + 2(1 - p_1 - p_2) \quad\quad \rightarrow \quad v \le 2 - 2p_1 - p_2$

$\hspace{2.3cm} v, p_1, p_2 \ge 0$

Introduce slack variables:

$\hspace{2cm} P - v = 0$

$\hspace{2cm} v - 2p_1 - p_2 + s = 1$

$\hspace{2cm} v + 2p_1 + p_2 + t = 2$

Here is the simplex solution.

P	v	p_1	p_2	s	t	Value	Row
1	-1	0	0	0	0	0	R1
0	1	-2	-1	1	0	1	R2
0	1	2	1	0	1	2	R3

(Continued on the next page)

924 **Linear programming and game theory 2** Games as linear programming problems

P	v	p_1	p_2	s	t	Value	Row
1	0	−2	−1	1	0	1	R4 = R1+R5
0	1	−2	−1	1	0	1	R5 = R2
0	0	4	2	−1	1	1	R6 = R3−R5
1	0	0	0	$\frac{1}{2}$	$\frac{1}{2}$	$1\frac{1}{2}$	R7 = R4+2 × R9
0	1	0	0	$\frac{1}{2}$	$\frac{1}{2}$	$1\frac{1}{2}$	R8 = R5+2 × R9
0	0	1	$\frac{1}{2}$	$-\frac{1}{4}$	$\frac{1}{4}$	$\frac{1}{4}$	R9 = R$\frac{6}{4}$

This gives $v = 1.5 - 2 = -0.5$ when $p_1 = \frac{1}{4}, p_2 = 0$ and $p_3 = 1 - p_1 - p_2 = \frac{3}{4}$, as before.

Exercise 30.2A Fluency and skills

1 For each of these games, use the simplex method to find the optimal mixed strategy for each player and the value of the game.

a

		B	
		B_1	B_2
A	A_1	2	−2
	A_2	−2	3

b

		B	
		B_1	B_2
A	A_1	7	6
	A_2	5	8

c

		B	
		B_1	B_2
A	A_1	5	2
	A_2	−1	3

d

		B	
		B_1	B_2
A	A_1	−2	1
	A_2	3	0

2 For each of these zero-sum games, use the simplex method to find the optimal strategy for A and the value of the game.

a

		B		
		B_1	B_2	B_3
A	A_1	1	−1	0
	A_2	−2	0	−1

b

		B	
		B_1	B_2
	A_1	1	−2
A	A_2	−1	1
	A_3	0	−1

3 Consider each of the following games. If there is a stable solution, find the play-safe strategies and the value of the game. Otherwise use the simplex method to find the optimal strategy for A and the value of the game. (Simplify the table using dominance where possible.)

a

		B	
		B_1	B_2
	A_1	−3	−3
A	A_2	1	−3
	A_3	−2	1

b

		B	
		B_1	B_2
	A_1	−1	2
A	A_2	−3	4
	A_3	2	1

c

		B		
		B_1	B_2	B_3
	A_1	−2	5	3
A	A_2	2	3	2
	A_3	−1	0	1

d

		B		
		B_1	B_2	B_3
	A_1	−4	−1	1
A	A_2	0	−3	−1
	A_3	−1	2	0

e

		B			
		B_1	B_2	B_3	B_4
	A_1	3	2	−1	3
	A_2	−1	0	3	1
A	A_3	4	3	2	5
	A_4	2	−1	−2	1

f

		B			
		B_1	B_2	B_3	B_4
	A_1	−3	−6	2	−11
	A_2	13	−10	−4	8
A	A_3	5	8	4	7
	A_4	4	9	−5	0

Reasoning and problem-solving

To solve a zero-sum game using the simplex algorithm

1. If necessary construct a pay-off matrix.

2. If possible, use dominance to reduce the table.

3. Check that the game does not have a stable solution.

4. Write the problem as a linear programming formulation.

5. Express the constraints as equations by using slack variables.

6. Apply the algorithm until there are no negative coefficients in the objective row.

7. Read off the values of the probabilities and the value of the game.

8. Answer the question in context.

Example 3

Show that this zero-sum game requires a mixed strategy, and find A's best strategy and the value of the game.

		B		
		B_1	B_2	B_3
	A_1	0	−1	−1
A	A_2	−1	0	1
	A_3	−2	1	2

2 If possible use dominance to reduce the table.
$-1 = -1, 0 < 1, 1 < 2$
(Remember: these are A's gains.)

Column 2 dominates column 3, so delete column 3

The revised table is

		B	
		B_1	B_2
	A_1	0	−1
A	A_2	−1	0
	A_3	−2	1

The row minima are $-1, -1, -2$, with a maximum value of -1

The column maxima are $0, 1$, with a minimum of 0

These are not equal, so the solution is not stable.

3 Check that the game does not have a stable solution.

Add 2 to all the pay-offs. Let the value of the revised game be v

So there are no negative entries.

		B	
		B_1	B_2
	A_1	2	1
A	A_2	1	2
	A_3	0	3

Suppose A plays A_1, A_2 and A_3 with probabilities p_1, p_2 and p_3

If B plays B_1 then A's expected pay-off is $\quad 2p_1 + p_2$

If B plays B_2 then A's expected pay-off is $\quad p_1 + 2p_2 + 3p_3$

The linear programming formulation is

Maximise $\qquad P = v$

subject to $\qquad v \le 2p_1 + p_2$

$\qquad\qquad v \le p_1 + 2p_2 + 3p_3$

$\qquad\qquad p_1 + p_2 + p_3 \le 1$

$\qquad\qquad v, p_1, p_2, p_3 \ge 0$

4 Write the problem as a linear programming formulation.

(Continued on the next page)

Write as equations:

$$P - v = 0$$

$$v - 2p_1 - p_2 + s = 0$$

$$v - p_1 - 2p_2 - 3p_3 + t = 0$$

$$p_1 + p_2 + p_3 + u = 1$$

⑤ Express the constraints as equations by using slack variables.

Enter these in a tableau and apply the simplex algorithm.

P	v	p_1	p_2	p_3	s	t	u	Value	Row
1	−1	0	0	0	0	0	0	0	R1
0	1	−2	−1	0	1	0	0	0	R2
0	1	−1	−2	−3	0	1	0	0	R3
0	0	1	1	1	0	0	1	1	R4
1	0	−2	−1	0	1	0	0	0	R5 = R1 + R6
0	1	−2	−1	0	1	0	0	0	R6 = R2
0	0	1	−1	−3	−1	1	0	0	R7 = R3 − R6
0	0	1	1	1	0	0	1	1	R8 = R4
1	0	0	−3	−6	−1	2	0	0	R9 = R5 + R11 × 2
0	1	0	−3	−6	−1	2	0	0	R10 = R6 + R11 × 2
0	0	1	−1	−3	−1	1	0	0	R11 = R7
0	0	0	2	4	1	−1	1	1	R12 = R8 − R11
1	0	0	0	0	$\frac{1}{2}$	$\frac{1}{2}$	$1\frac{1}{2}$	$1\frac{1}{2}$	R13 = R9 + R16 × 6
0	1	0	0	0	$\frac{1}{2}$	$\frac{1}{2}$	$1\frac{1}{2}$	$1\frac{1}{2}$	R14 = R10 + R16 × 6
0	0	1	$\frac{1}{2}$	0	$-\frac{1}{4}$	$\frac{1}{4}$	$\frac{3}{4}$	$\frac{3}{4}$	R15 = R11 + R16 × 3
0	0	0	$\frac{1}{2}$	1	$\frac{1}{4}$	$-\frac{1}{4}$	$\frac{1}{4}$	$\frac{1}{4}$	$R16 = \dfrac{R12}{4}$

This gives $v = 1.5$ and so the value of the original game is −0.5

Remember to subtract the 2 from v

The optimal strategy for A is to play A_2 and A_3 with probabilities 0.75 and 0.25

Example 4

DISCRETE

Solve this zero-sum game to find the optimal strategy for player A and the value of the game.

	B		
	B_1	B_2	B_3
A_1	1	−1	0
A_2	0	2	1
A_3	1	1	−1

(A on left spanning A_1, A_2, A_3)

> Entries must be non-negative.

The game cannot be reduced using dominance.

Add 1 to all the entries.

	B		
	B_1	B_2	B_3
A_1	2	0	1
A_2	1	3	2
A_3	2	2	0

Suppose A plays A_1, A_2, A_3 with probabilities p_1, p_2, p_3

Let the value of the revised game be v

The linear programming formulation is

Maximise $P = v$

subject to $v \leq 2p_1 + p_2 + 2p_3$

$v \leq 3p_2 + 2p_3$

$v \leq p_1 + 2p_2$

$p_1 + p_2 + p_3 \leq 1$

$v, p_1, p_2, p_3 \geq 0$

Introduce slack variables:

$P - v = 0$

$v - 2p_1 - p_2 - 2p_3 + s_1 = 0$

$v - 3p_2 - 2p_3 + s_2 = 0$

$v - p_1 - 2p_2 + s_3 = 0$

$p_1 + p_2 + p_3 + s_4 = 1$

Here is the simplex solution.

P	v	p_1	p_2	p_3	s_1	s_2	s_3	s_4	Value	Row
1	−1	0	0	0	0	0	0	0		R1
0	1	−2	−1	−2	1	0	0	0		R2
0	1	0	−3	−2	0	1	0	0		R3
0	1	−1	−2	0	0	0	1	0		R4
0	0	1	1	1	0	0	0	1	1	R5

(Continued on the next page)

1	0	-2	-1	-2	1	0	0	0	0	R6 = R1 + R7
0	1	-2	-1	-2	1	0	0	0	0	R7 = R2
0	0	2	-2	0	-1	1	0	0	0	R8 = R3 − R7
0	0	1	-1	2	-1	0	1	0	0	R9 = R4 − R7
0	0	1	1	1	0	0	0	1	1	R10 = R5
1	0	0	-3	-2	0	1	0	0	0	R11 = R6 + 2 × R13
0	1	0	-3	-2	0	1	0	0	0	R12 = R7 + 2 × R13
0	0	1	-1	0	$-\frac{1}{2}$	$\frac{1}{2}$	0	0	0	$R13 = R\frac{8}{2}$
0	0	0	0	2	$-\frac{1}{2}$	$-\frac{1}{2}$	1	0	0	R14 = R9 − R13
0	0	0	2	1	$\frac{1}{2}$	$-\frac{1}{2}$	0	1	1	R15 = R10 − R13
1	0	0	0	$-\frac{1}{2}$	$\frac{3}{4}$	$-\frac{1}{4}$	0	$1\frac{1}{2}$	$1\frac{1}{2}$	R16 = R11 + 3 × R20
0	1	0	0	$-\frac{1}{2}$	$\frac{3}{4}$	$-\frac{1}{4}$	0	$1\frac{1}{2}$	$1\frac{1}{2}$	R17 = R12 + 3 × R20
0	0	1	0	$\frac{1}{2}$	$-\frac{1}{4}$	$\frac{1}{4}$	0	$\frac{1}{2}$	$\frac{1}{2}$	R18 = R13 + R20
0	0	0	0	2	$-\frac{1}{2}$	$-\frac{1}{2}$	1	0	0	R19 = R14
0	0	0	1	$\frac{1}{2}$	$\frac{1}{4}$	$-\frac{1}{4}$	0	$\frac{1}{2}$	$\frac{1}{2}$	$R20 = R\frac{15}{2}$
1	0	0	0	0	$\frac{5}{8}$	$\frac{1}{8}$	$\frac{1}{4}$	$1\frac{1}{2}$	$1\frac{1}{2}$	$R21 = R16 + R\frac{24}{2}$
0	1	0	0	0	$\frac{5}{8}$	$\frac{1}{8}$	$\frac{1}{4}$	$1\frac{1}{2}$	$1\frac{1}{2}$	$R22 = R17 + R\frac{24}{2}$
0	0	1	0	0	$-\frac{1}{8}$	$\frac{3}{8}$	$-\frac{1}{4}$	$\frac{1}{2}$	$\frac{1}{2}$	$R23 = R18 - R\frac{24}{2}$
0	0	0	0	1	$-\frac{1}{4}$	$-\frac{1}{4}$	$\frac{1}{2}$	0	0	$R24 = R\frac{19}{2}$
0	0	0	1	0	$\frac{3}{8}$	$-\frac{1}{8}$	$-\frac{1}{4}$	$\frac{1}{2}$	$\frac{1}{2}$	$R25 = R20 - R\frac{24}{2}$

This gives $v = 1.5$, so the value of the original game is 0.5

The optimal strategy is for A to play A_1 and A_2 each with a probability 0.5

> Remember to subtract the 1 from v

1 The table shows the pay-offs in a game between A and B.

		B		
		B_1	B_2	B_3
	A_1	−1	2	1
A	A_2	1	−1	−2
	A_3	0	−1	−1

a Show that A must play a mixed strategy in this game.

b Find A's optimal strategy and the value of the game.

c Find B's optimal strategy.

2 Analyse the games shown in these tables to find the optimal strategy for each player and the values of the games.

a

		B		
		B_1	B_2	B_3
A	A_1	0	2	−1
	A_2	2	1	3

b

		B	
		B_1	B_2
	A_1	−3	1
A	A_2	−1	−4
	A_3	0	−5

c

		B	
		I	II
	I	−2	4
A	II	−1	2
	III	4	1

d

		B		
		B_1	B_2	B_3
	A_1	−8	−2	1
A	A_2	−3	1	4
	A_3	0	−4	−2

e

		B		
		B_1	B_2	B_3
	A_1	2	1	5
A	A_2	1	−4	4
	A_3	5	2	1

f

		N			
		I	II	III	IV
	I	−2	−1	3	−1
M	II	2	−1	1	2
	III	−3	−2	1	−2
	IV	−3	−3	2	−1

3 You may be familiar with the game 'rock, paper, scissors' where the rock blunts the scissors, the scissors cut the paper and the paper wraps round the rock. Awarding 1 point for a win and 0 for a draw, construct a pay-off table for the game and analyse it to find the best strategy.

4 An entertainment company plans to run nine outdoor events in different localities on the same day. If the weather at a given locality is fine they expect to make £1200 for that event, but in a place where it rains they will lose £400

They could choose to take out one of two possible insurance policies. The basic policy costs £500 and pays £1300 if it rains. The comprehensive policy costs £800 and pays £2400 if it rains.

a Their strategies are A_1 no insurance, A_2 basic, A_3 comprehensive. Analyse this as a 'game against the weather' (the weather's strategies are rain or no rain) and advise the company on their best strategy.

b On what assumptions is your advice based?

Chapter summary

- A linear programming problem can be solved using the simplex algorithm if:
  - the aim is to maximise the objective function,
  - every constraint can be written in standard form, that is, as an equation with a slack variable ≥ 0 and a non-negative value on the right-hand side.
- To minimise an objective function, C, you maximise the objective function $P = -C$.

  You rewrite inequalities involving $\geq$ by multiplying through by -1, so that they involve $\leq$.

  If the problem can then be written in standard form, you use the simplex algorithm.
- If introducing a slack variable gives a negative value on the right-hand side, then the origin is not in the feasible region. You pivot on a negative coefficient in the row with the negative value, to generate a basic feasible solution (a vertex of the feasible region) from which to start the simplex algorithm.
- In a pure strategy game, each player always plays the option that maximises their pay-off. This corresponds to the row with the greatest minimum or the column with the least maximum entry.
- A game has a stable solution (a saddle point) if

  maximum of row minima = minimum of column maxima
- This gives the value of the game.
- For an unstable situation, each player should play a mixed-strategy game, that is, play each strategy some of the time.
- In a mixed-strategy game, you aim to maximise the expected pay-off, where for pay-offs $x_1, x_2, ..., x_n$ occuring with probabilities $p_1, p_2, ..., p_n$, the expected (mean) pay-off $E(x)$ is given by

$$E(x) = x_1 p_1 + x_2 p_2 + ... + x_n p_n = \sum_{i=1}^{n} x_i p_i$$

- To solve a mixed-strategy game you assume that the row player plays their strategies with probabilities $p_1, p_2, ...$ and write inequalities connecting v, their expected value, with the ps for each of the column player's strategies. For a 2×2 game you can solve the resulting linear programming problem graphically, but for larger games you need to use the simplex algorithm.

Check and review

You should now be able to...	Review Questions
✔ Use slack variables to write a linear programming problem as a simplex tableau.	1
✔ Solve a standard maximisation problem using the simplex algorithm.	1
✔ Convert a minimisation problem to a maximisation problem and solve it using the simplex algorithm.	2
✔ Solve a non-standard linear programming problem by pivoting on negative coefficients to generate a starting feasible solution.	3
✔ Decide whether a game has a stable solution.	4
✔ Formulate a game as a linear programming problem, and solve it by using the simplex method to find the optimal mixed strategy.	4

1 A linear programming problem is stated as follows:

Maximise $P = x + 2y - z$

subject to $x + y + 2z \le 14$

$2x + y - z \le 8$

$x, y, z \ge 0$

a Rewrite the problem as equations in standard form using slack variables.

b Construct a simplex tableau and apply the simplex algorithm to solve the problem.

2 Use the simplex algorithm to solve this linear programming problem:

Minimise $C = 2x - y - 2z$

subject to $2x + y - 2z \ge -10$

$y \ge 2x$

$z \ge 3x + y$

$x, y, z \ge 0$

3 A linear programming problem is stated as follows:

Maximise $P = 5x + 4y$

subject to $3x + y \ge 10$

$4x + 3y \le 24$

$x, y, z \ge 0$

a Explain why the problem cannot be written directly in standard form using equations and slack variables.

b Construct a tableau and pivot on a suitable coefficient to produce a basic feasible solution. State the values of the variables at this stage.

c Use the simplex algorithm to complete the solution of the problem.

4 The table shows the pay-offs in a game between A and B

		B	
		B_1	B_2
A	A_1	−3	2
	A_2	1	−2
	A_3	0	1

a Show that the game does not have a stable solution.

b Use the simplex method to find the value of this game and the optimal strategy for player A

DISCRETE

933

Research

One of the most famous games analysed in game theory is that of the Prisoner's dilemma. It was formalised by Albert Tucker as follows:

Two members of a criminal gang are arrested and imprisoned. Each prisoner is in solitary confinement with no means of communicating with the other. The prosecutors lack sufficient evidence to convict the pair on the principal charge. They hope to get both sentenced to a year in prison on a lesser charge. Simultaneously, the prosecutors offer each prisoner a bargain. Each prisoner is given the opportunity either to betray the other by testifying that the other committed the crime, or to cooperate with the other by remaining silent. The offer is:

- If A and B each betray the other, each of them serves 2 years in prison,
- If A betrays B but B remains silent, A will be set free and B will serve 3 years in prison (and vice versa),
- If A and B both remain silent, both of them will only serve 1 year in prison (on the lesser charge).

Using this scenario, try to determine the dominant strategy, and then explain why this is considered a dilemma.

You could also find out about iterated versions of the game where the prisoners repeatedly play the game and are able to base their decisions on previous actions.

Did you know?

Le Her is a simple French card game based on a standard 52 card deck dating from the 16th century. A minimax mixed strategy game based around Le Her is described by James Waldergrave in a 1713 letter.

Research

The Klee-Minty cube is a system of linear inequalities for which the simplex algorithm has poor worst-case performance when initialised at the origin.

Try solving the following using the simplex algorithm.

Maximise $\quad 4x + 2y + z$

Subject to $\quad x \le 5, 4x + y \le 25, 8x + 4y + z \le 125,$
$\quad\quad\quad\quad x, y, z \ge 0$

Find out why this is called the Klee-Minty cube problem.

1 A linear programming problem is defined as

Maximise $P = x + y$

subject to $3x + 4y \le 120$

$4x + y \le 80$

$x \ge 0, y \ge 0$

The constraints are rewritten as

$3x + 4y + s = 120$

$4x + y + t = 80$

$x \ge 0, \ y \ge 0, \ s \ge 0, \ t \ge 0$

a State the name given to s and t and rewrite the objective function as an equation involving s and t [2]

b Display the linear programming problem in a simplex tableau. [2]

The first pivot is chosen from the x-column.

c State a pivot and explain why this value is chosen. [2]

d Perform one iteration of the simplex algorithm. [4]

e Explain how you know that this solution is not optimal. [1]

f Find the optimal solution to this problem. [5]

2 Use the simplex method to solve this linear programming problem.

Maximise $P = 2x + y + 3z$

subject to $x + 2y + 3z \le 10$

$3x + 4y + 2z \le 22$

$5x + y + z \le 15$

$x, y, z \ge 0$

Show your method clearly. [10]

3 Sarina and Jeremy play a zero-sum game. The pay-off matrix for Sarina is given.

Strategy	J_1	J_2	J_3
S_1	−4	9	−1
S_2	3	4	4
S_3	3	5	5

a Work out the play-safe strategies for Sarina and for Jeremy and state the value of the game. You must show your method. [5]

b Explain whether or not this is a stable solution. [2]

c Write down the pay-off matrix for Jeremy. [2]

4 Three types of drink, X, Y and Z, are to be produced. They all require orange, pineapple, grapefruit juice and water in different quantities. The amount of each type of juice (in litres) required per litre of each drink along with the availability of each juice and the profit per litre of each drink produced is shown in the table. You can assume that there is an unlimited supply of water.

	Orange	Pineapple	Grapefruit	Profit
X	0.4	0.1	0.1	£1
Y	0.6	0.2	0.1	£1.50
Z	0.5	0.3	0.05	£1.20
Availability	550	160	100	

a Given that you wish to maximise the profit, formulate this scenario as a linear programming problem. Ensure that you define your variables clearly. [4]

b Show the simplex tableau after one iteration of the simplex algorithm. [6]

c Write down the solution illustrated by this tableau and explain how you know whether or not this solution is optimal. [3]

A second iteration gives the solution

$P = £1300$, $x = 400$, $y = 600$, $z = 0$, $r = 30$, $s = 0$, $t = 0$

d Explain the significance of the values of r, s and t in this solution. [2]

5 The table shows a zero-sum game between two players, R and C

Strategy	C_1	C_2	C_3	C_4
R_1	7	2	−4	−5
R_2	−2	−3	2	3
R_3	5	4	−7	6
R_4	4	3	−3	5

a Use dominance arguments to explain why the matrix can be written as [3]

Strategy	C_2	C_3
R_2	−3	2
R_3	4	−7
R_4	3	−3

b Use a graphical method to find the optimal strategy for both players and work out the value of the game. [10]

6 A zero-sum game between two players, A and B, is represented by this pay-off matrix.

Strategy	B_1	B_2	B_3	B_4
A_1	−3	2	3	−1
A_2	6	−1	6	7
A_3	5	−3	4	−1

a Explain why player A should never play A_3 [2]

b Simplify the pay-off matrix as far as possible. [3]

c Verify that there is not a stable solution to the game. [2]

d Find the optimal mixed-strategy for both players. [7]

7 This optimal mixed-strategy problem for Alex in the zero-sum game shown in this pay-off matrix is to be solved using the simplex algorithm.

		Becky		
		A	**B**	**C**
Alex	**I**	3	−1	2
	II	−2	0	3
	III	0	1	−1

a Add 2 to every value in the pay-off matrix and then express the game as a linear programming problem for Alex. Write the constraints as equations involving slack variables. [5]

b Why was it necessary to add 2 to every value in the matrix? [1]

c Reduce the number of decision variables to three. [3]

d Construct a simplex tableau and carry out one iteration of the simplex algorithm. [5]

After a further two iterations, the tableau becomes

P	V	p_1	p_2	r	s	t	Value
1	0	0	0	$\dfrac{1}{6}$	$\dfrac{3}{5}$	$\dfrac{7}{30}$	$\dfrac{71}{30}$
0	0	0	1	$\dfrac{1}{6}$	0	$-\dfrac{1}{6}$	$\dfrac{1}{6}$
0	0	1	0	$-\dfrac{1}{6}$	$\dfrac{1}{5}$	$-\dfrac{1}{30}$	$\dfrac{7}{30}$
0	1	0	0	$\dfrac{1}{6}$	$\dfrac{3}{5}$	$\dfrac{7}{30}$	$\dfrac{71}{30}$

e State the optimal solution for Alex and the value of the game. [4]

8 Hayley and Sanjit play a zero-sum game. The pay-off matrix for the game is given.

	Sanjit		
Strategy	S_1	S_2	S_3
H_1	4	2	1
H_2	1	1	3
H_3	2	3	2

(The left column is labelled **Hayley** spanning rows H_1, H_2, H_3.)

Hayley choses to play strategy H_i with probability p_i, $i = 1, 2, 3$

a Formulate the problem of finding the value, V, of the game as a linear programming problem.

Give the constraints as equations. [4]

b Perform two iterations of the simplex algorithm and indicate your pivots. [8]

9 Greg and Aidan play a zero-sum game. The pay-off matrix for the game is given.

	Aidan			
Strategy	**W**	**X**	**Y**	**Z**
A	1	3	−2	−3
B	2	−1	3	1
C	3	1	0	2
D	4	4	−1	3

(The left column is labelled **Greg** spanning rows A, B, C, D.)

a Verify that this game does not have a stable solution. [3]

b Reduce this 4×4 game to a 3×3 game. Justify your solution. [3]

c Formulate the 3×3 game as a linear programming problem, giving your constraints as inequalities. [5]

31

Group theory

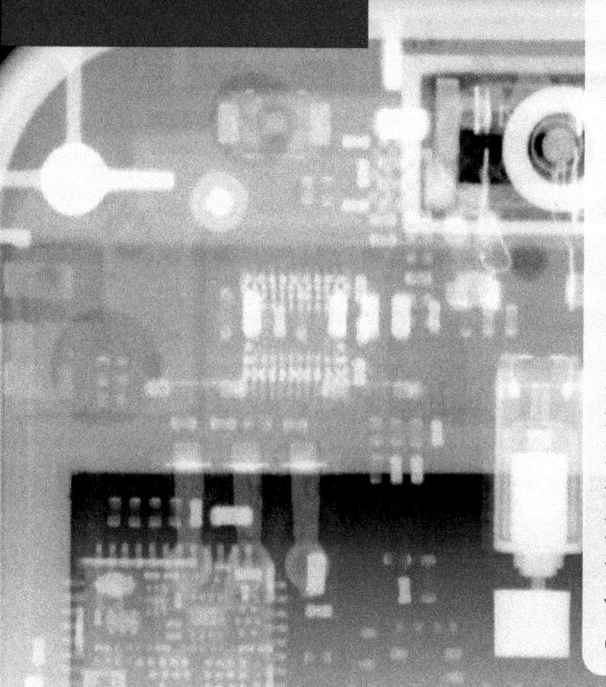

Materials scientists use experimentation and the mathematical analysis of molecular symmetries to ensure that a material has particular properties. Symmetries of molecules are classified using **groups**, and the group provides an indication of the physical properties of its elements. The materials that are developed are used in products such as satellites. They may also be used to manufacture artificial limbs, and implants used in the human body.

Materials scientists are also involved in developing materials used in the manufacture of mobile phones. Mobile phones need screens that are made of a material that is resistant to breakages, sensitive to the touch, and has high levels of clarity. Inside the phone, the materials that are used in the complex circuitry must work efficiently, whilst being on a very small scale to ensure that the phone can be as small and powerful as possible

Orientation

What you need to know	What you will learn	What this leads to
KS4 • Sets. **Ch15 Abstract algebra** • Binary operations	• How to analyse groups mathematically. • How to analyse subgroups. • How to recognise and find isomorphisms.	**Careers** • Materials science.

31.1 Groups

Fluency and skills

A group is a collection of mathematical objects that can be combined subject to a set of basic rules, or axioms.

An example of a group is the set of permutations of the numbers 1, 2 and 3

You can list these permutations: (1 2 3), (1 3 2), (2 1 3), (2 3 1), (3 1 2) and (3 2 1). You can also write the permutations using notation which indicates how the original order of the numbers changes under each permutation.

$$e = \begin{pmatrix} 1 & 2 & 3 \\ 1 & 2 & 3 \end{pmatrix}, a = \begin{pmatrix} 1 & 2 & 3 \\ 1 & 3 & 2 \end{pmatrix}, b = \begin{pmatrix} 1 & 2 & 3 \\ 2 & 1 & 3 \end{pmatrix}, c = \begin{pmatrix} 1 & 2 & 3 \\ 2 & 3 & 1 \end{pmatrix}, d = \begin{pmatrix} 1 & 2 & 3 \\ 3 & 1 & 2 \end{pmatrix}, f = \begin{pmatrix} 1 & 2 & 3 \\ 3 & 2 & 1 \end{pmatrix}$$

The six permutations are called **elements**. The first of these permutations leaves the order of the numbers unchanged. This element, *e*, is known as the **identity**.

In order to be a group, the set of elements must contain an identity element.

A second condition for the set of objects to be a group is that any combination of the elements is contained within the original set. This condition is known as **closure**.

If you apply permutation *b* and then permutation *c*, then under the permutation *cb*: 1 maps to 2 maps to 3, 2 maps to 1 maps to 2 and 3 maps to 3 maps to 1

Overall, the result is permutation *f*

All of the possible combinations of two permutations can be shown in a **Cayley table**.

	followed by	\multicolumn{6}{c}{First permutation}

Second permutation	followed by	*e*	*a*	*b*	*c*	*d*	*f*
	e	*e*	*a*	*b*	*c*	*d*	*f*
	a	*a*	*e*	*d*	*f*	*b*	*c*
	b	*b*	*c*	*e*	*a*	*f*	*d*
	c	*c*	*b*	*f*	*d*	*e*	*a*
	d	*d*	*f*	*a*	*e*	*c*	*b*
	f	*f*	*d*	*c*	*b*	*a*	*e*

You can see from the Cayley table that for any combination of two permutations, the result is one of the original set of permutations.

In addition to this, you can see that every element has another element which, when combined, leads to the identity element *e*

For example, $aa = e$ and $cd = e$

These elements are known as **inverse** elements. Since *a*, *b*, *e* and *f* combine with themselves to produce the identity element they are known as **self-inverses**, while *d* is the inverse of *c* and vice versa.

The final condition for a set of elements to be a group is known as **associativity**. Three elements are associative if $a(bc) = (ab)c$, that is, bc followed by a gives the same result as c followed by ab

From the table on the previous page, you can see that $a(bc) = aa$ and $(ab)c = dc$ and both aa and dc give e

Hence a, b and c are associative. This can be shown for any other combination of the elements in the same way.

Since this final condition is met, you can say that the set of permutations of the numbers 1, 2 and 3 form a **group**.

> **Key point**
>
> The conditions under which a set of mathematical objects form a group under a given binary operation are known as the **axioms**. These are **closure**, the existence of an **identity element**, **associativity** and the existence of an **inverse** element for each member of the set.

Using formal notation, a group can be defined as follows.

> **Key point**
>
> A group $(S, \circ)$ is a non-empty set S with a binary operation $\circ$ such that:
>
> $\circ$ is closed in S
>
> $\circ$ is associative
>
> there is an identity element such that $x \circ e = e \circ x = x$ for all x
>
> each element has an inverse x^{-1} such that $x \circ x^{-1} = x^{-1} \circ x = e$

> A binary operation is an operation that combines two elements of a set according to a given rule.

> **See Ch15**
> For a reminder on binary operations.

The most difficult axiom to test for is associativity. However, there are certain binary operations that are known to be associative. For example, modular multiplication and addition are associative, as is matrix multiplication. Also, any binary operation that can be interpreted as the composition of mappings is associative. You can use these facts without proof.

DISCRETE

Example 1

a Draw a Cayley table for the binary operation addition modulo 4 ($+_4$) on the set $S = \{0, 1, 2, 3\}$

b Is the set closed under $+_4$?

c State the element that is the identity element.

d For each element, state its inverse.

a

$+_4$	0	1	2	3
0	0	1	2	3
1	1	2	3	0
2	2	3	0	1
3	3	0	1	2

b Yes, because every combination of elements is in the original set of elements.

c 0

d 0 and 2 are self-inverse, 3 is the inverse of 1 and 1 is the inverse of 3

> Combining 0 with any other element leaves that element unchanged.

There are several terms that you need to know.

> **Key point**
>
> The **order** of a group is equal to the number of elements in the group.

The **period** (or order) of a particular element, x, of a group is the smallest non-negative integer n such that $x^n = e$, where e is the identity.

In the permutations example described at the start of this section, the order of the group is 6 since there are 6 different permutations.

Element a has period 2 since $a^2 = e$ and element c has period 3 since
$$c^3 = c(cc) = cd = e$$

> **Key point**
>
> An **abelian** group is a group with the additional property of **commutativity** between the elements of the group.

Commutativity is the property that, for all elements of a group, $xy = yx$ under the given binary operation.

In the permutations example, $ba \neq ab$; therefore this group is not an abelian group.

In Example 1, $1 + 3 = 3 + 1$ and so on, so this group *is* an abelian group.

Abelian groups are characterised by a line of symmetry down the leading diagonal.

Exercise 31.1A Fluency and skills

1 a Draw a Cayley table for the binary operation multiplication modulo 5 ($\times_5$) on the set $S = \{0, 1, 2, 3, 4\}$

b Is the set closed under $\times_5$?

c State the element that is the identity element.

d For each element, write down its inverse.

e State, with reasons, whether or not S forms a group.

2 a Draw a Cayley table for the binary operation addition modulo 6 ($+_6$) on the set $S = \{0, 1, 2, 3, 4, 5\}$

b Is the set closed under $+_6$?

c State the element that is the identity element.

d Show that S forms a group under $+_6$

e Is the group formed an abelian group? Give a reason for your answer.

3 The binary operation $x \bullet y$ is defined as $|x - y|$

a Draw a Cayley table for the binary operation when applied to the set $S = \{0, 1, 2, 3\}$

b Is the set closed for $x \bullet y$?

c Identify the identity element.

d Fahima says that the set S forms a group under $\bullet$. Is Fahima correct? Give a reason for your answer.

4 Show that the set of matrices, S, of the form $\begin{pmatrix} 1 & p \\ 0 & 1 \end{pmatrix}$, $p \in \mathbb{Z}$ forms an abelian group under the operation of matrix multiplication.

5 Prove that the set of natural numbers, $\mathbb{N}$, does not form a group under the operation of subtraction.

6 The set S, defined as $S = \{1, 2, 3, 4, 5, 6\}$, forms a group G under the binary operation $\times_7$

    a State the order of the group G

    b Determine the period of the element 6 in the group G

7 **a** Show that the set $S = \{\mathbb{Z}\}$ forms a group under the binary operation $\bullet$ where $x \bullet y = x + y - 2$

    b Explain why the set $T = \{\mathbb{Z}^+\}$ does not form a group under the same binary operation.

Reasoning and problem-solving

The set of all symmetries of a regular polygon also form a group.

These groups are called **dihedral groups** (denoted by $\boldsymbol{D}_n$ where n is the number of sides in the regular n-gon).

A **cyclic group** is formed if, for example, only rotational symmetries are considered. Cyclic groups are groups that can be generated by a single element.

Strategy

To show that a given set forms a group under a given binary operation

(**1**) Write down the elements of the set.

(**2**) Produce a Cayley table to show the combinations of each element under the given binary operation.

(**3**) Check that the binary operation meets the axioms necessary to be a group.

DISCRETE

Example 2

a Show that the set of all symmetries of an equilateral triangle forms a group.

b Write down the cyclic group, C_3, represented by the set of rotational symmetries of an equilateral triangle and state an element that can be used as a generator.

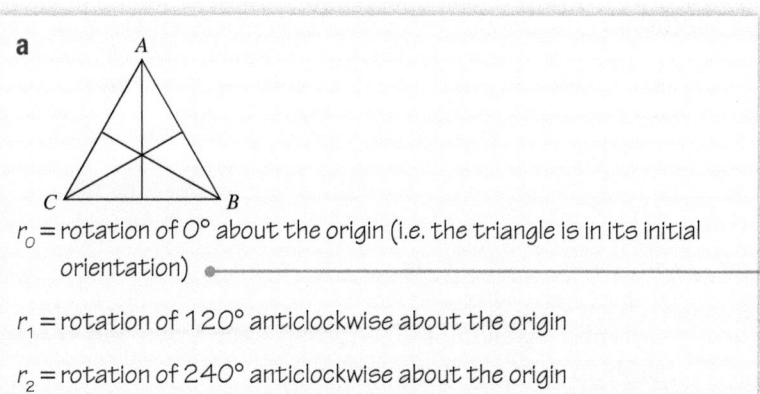

a

r_0 = rotation of 0° about the origin (i.e. the triangle is in its initial orientation)

r_1 = rotation of 120° anticlockwise about the origin

r_2 = rotation of 240° anticlockwise about the origin

Define the point of intersection of the lines of symmetry as the origin and then list the symmetries in turn.

(1)

(continued on the next page)

m_1 = reflection in the mirror line through A

m_2 = reflection in the mirror line through B

m_3 = reflection in the mirror line through C

	r_0	r_1	r_2	m_1	m_2	m_3
r_0	r_0	r_1	r_2	m_1	m_2	m_3
r_1	r_1	r_2	r_0	m_2	m_3	m_1
r_2	r_2	r_0	r_1	m_3	m_1	m_2
m_1	m_1	m_3	m_2	r_0	r_2	r_1
m_2	m_2	m_1	m_3	r_1	r_0	r_2
m_3	m_3	m_2	m_1	r_2	r_1	r_0

> **2** Draw up a Cayley table to show the combinations of the different symmetries.

Axioms

Identity: r_0 is the identity element.

Inverses: r_0, m_1, m_2 and m_3 are self-inverses; r_1 is the inverse of r_2 and vice versa.

Closure: Every combination of symmetries is in the original set of symmetries.

> **3** Check that the binary operation representing the combination of two symmetries satisfies the axioms.

Associativity: Since the binary operation is a composition of mappings, the operation is associative.

Hence the set of symmetries of an equilateral triangle form a group (this is the dihedral group D_3)

b

	r_0	r_1	r_2
r_0	r_0	r_1	r_2
r_1	r_1	r_2	r_0
r_2	r_2	r_0	r_1

> This is the Cayley table for the rotational symmetries.

The group C_3 is therefore $\{r_0, r_1, r_2\}$

The group can be generated by successively applying symmetry r_1

$r_1^1 = r_1$

$r_1^2 = r_2$

$r_1^3 = r_0$

Hence r_1 is a generator of the group which can be denoted by $\langle r_1 \rangle$

> r_2 is also a generator of the group since $r_2^2 = r_1$ and $r_2^3 = r_0$
> Either r_1 or r_2 are suitable answers.

1 a Show that the set of all symmetries of a square form a group and state the order of the group.

b Write the cyclic group, C_4, represented by the set of rotational symmetries of a square and state two different elements that are generators.

2 Jeremiah defines the cyclic group represented by the set of rotational symmetries of a regular hexagon as
$$C_6 = \{r_0, r_1, r_2, r_3, r_4, r_5\}$$

a Using a Cayley table, or otherwise, find all of the generators of Jeremiah's group.

He claims that the group C_6 is abelian.

b Is Jeremiah correct?

3 a Give a geometric interpretation of the dihedral group D_5

b Explain why the order of D_5 is 10

Philomena claims that the order of any dihedral group D_n is equal to $2n$

c Explain why Philomena is correct.

4 The group G is defined as $G = (\langle 5 \rangle, \times_7)$

a Find the set of all elements contained within G

b Is G an abelian group?

5 The group R is defined as $R = (\langle 3 \rangle, +_7)$

a Draw a Cayley table for R

b Write the order of the group.

Joanna claims that 3 is not the unique generator of the group.

c Assess Joanna's claim.

6 The group G is defined as
$$G = \left(\left\langle \begin{pmatrix} 0 & 1 \\ -1 & 0 \end{pmatrix} \right\rangle, \text{matrix multiplication} \right)$$

a Find all of the elements of G and state its order.

b Give a geometrical interpretation of G

7 A cyclic group G under the binary operation • has the Cayley table

•	a	b	c	d
a	a	b	c	d
b	b	a	d	c
c	c	d	b	a
d	d	c	a	b

a Write the identity element.

b Find a generator of G

8 The set of symmetries of a circle form an infinite group, O_2

a Write the identity element.

b State why the associativity property holds for this group.

c Write the inverse of each:

   i A rotation through $32.1°$

   ii A reflection in a line of symmetry inclined at $\theta°$

The table shows the effects of combining two symmetries, 'rot' (a rotation) and 'ref' (a reflection):

	rot	ref
rot	?	ref
ref	ref	rot

d What is represented by '?' Give a reason for your answer.

e Use this table to explain why the set of symmetries of a circle has the closure property.

[Note: The group O_2 is known as the **orthogonal** group and the subscript '2' tells you that you are dealing with a two-dimensional object.]

Fluency and skills

A **subgroup** of a group G is any subset H of G such that H itself is also a group under the same binary operation as G

> **Key point**

The **trivial subgroup** consisting of just the identity element is a subgroup of the (parent) group.

The group itself is also a subgroup of the group.

A **non-trivial subgroup** is any subgroup that is not the trivial subgroup.

> **Key point**

A **proper subgroup** is any subgroup that is not the group itself.

You need to be able to show that a stated group H is a subgroup of another stated group G. You do this by checking four conditions:

- That H is non-empty
- That the identity element in G exists in H
- That H is closed under the binary operation for G
- That the inverse of each element of H belongs to H

> $\mathbb{R} \setminus \{0\}$ denotes the set of real numbers *excluding zero.*

Example 1

a Show that $H = (\mathbb{Q}, +)$ is a subgroup of $G = (\mathbb{R}, +)$

b Show that the set $J = \{x \in \mathbb{R} : x \geq 1\}$ is not a subgroup of $K = (\mathbb{R} \setminus \{0\}, \times)$

a H is non-empty since there exist (an infinite number of) elements in H

The identity element of G under the binary operation addition is 0 and $0 \in \mathbb{Q}$

H is closed since if $x, y \in \mathbb{Q}$, then $x + y \in \mathbb{Q}$

The inverse of an element x is x^{-1}, which is $-x$ under the binary operation of addition, and if $x \in \mathbb{Q}$, then $-x \in \mathbb{Q}$

Hence H is a subgroup of G

b J is non-empty since there exist (an infinite number of) elements in J

The identity element of K under the binary operation multiplication is 1, and 1 is in J

(continued on the next page)

J is closed since if *x* and *y* are in *J*, their product *xy* is in *J*

The inverse of an element *x* is x^{-1}, which is $\dfrac{1}{x}$ under the binary operation of multiplication, but if $x > 1$, $x^{-1} < 1$, so then *J* does not contain the inverse of *x*

Hence *J* is not a subgroup of *K*

If you can show that any given condition is false, you do not have to check all of the conditions, since a single counter-example is enough.

You also need to be able to find subgroups for a given group.

Example 2

A group *G* is formed by the set $S = \{a, b, c, d\}$ under the binary operation $\bullet$

A Cayley table is drawn to show the outcome for each pair of elements under $\bullet$

$\bullet$	*a*	*b*	*c*	*d*
a	*a*	*b*	*c*	*d*
b	*b*	*a*	*d*	*c*
c	*c*	*d*	*b*	*a*
d	*d*	*c*	*a*	*b*

Find all the non-trivial subgroups of *G*

Since *a* is the identity element, the group {*a*} is not a non-trivial subgroup, so this should not be included.

Since subgroups are groups under the same binary operation as the original group, all subgroups must contain the identity element *a*.

Hence the options are {*a, b*}, {*a, c*}, {*a, d*}, {*a, b, c*}, {*a, b, d*}, {*a, c, d*} and {*a, b, c, d*}

Checking each one:

{*a, b*} is non-empty, closed and since *a* and *b* are self-inverse, {*a, b*} is a subgroup of *G*

{*a, c*} is non-empty but since $c^2 = b$, it is not closed.

{*a, d*} is non-empty but since $d^2 = b$, it is not closed.

{*a, b, c*} is non-empty but since $bc = d$, it is not closed.

{*a, b, d*} is non-empty but since $bd = c$, it is not closed.

{*a, c, d*} is non-empty but since $cd = b$, it is not closed.

{*a, b, c, d*} is the group itself so is, by definition, a non-trivial subgroup of *G*

Hence the non-trivial subgroups of *G* are {*a, b*} and {*a, b, c, d*}

1 Show that the group $H = (\mathbb{Z}, +)$ is a subgroup of $G = (\mathbb{R}, +)$

2 Show that the group of rotations of an equilateral triangle is a subgroup of the group of all symmetries of an equilateral triangle.

3 A group $G = (\{1, 2, 3, 4\}, \times_5)$. Find all the proper subgroups of G

4 A group $H = (\{0, 1, 2, 3\}, +_4)$. Find all the non-trivial subgroups of H

5 **a** Show that the group C_4 contains only one proper subgroup other than the trivial subgroup.

b Explain how group H in question **4** and C_4 are related.

6 The Klein group $V = (\{a, b, c, d\}, \bullet)$ has the corresponding Cayley table

$\bullet$	a	b	c	d
a	a	b	c	d
b	b	a	d	c
c	c	d	a	b
d	d	c	b	a

Find all the proper subgroups of V

7 M is the set of all non-singular real 2 by 2 matrices such that

$$M = \left\{ \begin{pmatrix} a & b \\ c & d \end{pmatrix} : a, b, c, d \in \mathbb{R}, ad - bc \neq 0 \right\}$$

Given that M forms a group under the binary operation of matrix multiplication, show that the set $N = \left\{ \begin{pmatrix} 1 & 0 \\ 0 & 1 \end{pmatrix}, \begin{pmatrix} -1 & 0 \\ 0 & -1 \end{pmatrix} \right\}$ is a subgroup of M

Reasoning and problem-solving

So far you have been identifying subgroups by checking all possible combinations of elements. If the number of elements is large, however, this becomes an arduous task. Imagine the group of all symmetries of a regular octagon. This would have order 16 and you would have to check a substantial number of cases.

Lagrange's theorem allows you to reduce the number of cases significantly.

> **Key point**
>
> **Lagrange's theorem** states that for any finite group G, the order of every subgroup of G divides the order of G

> Also by Lagrange's theorem, the period of any element in the group G is a factor of the order of G

The consequence of this is that, for the example of the symmetries of the regular octagon, you would only need to look for subgroups of order 1 (the trivial subgroup containing just the identity which is always a subgroup), order 16 (the group itself), order 2, order 4 and order 8. For a group of order 16, this is still quite a few, but significantly fewer than if you had to check for groups of order 3, 5, 7, etc. as well.

Strategy

To find all the possible subgroups of a particular group

1. Use Lagrange's theorem to find the order of the possible subgroups.

2. Use a systematic process to identify the possibilities.

3. Check each possibility against the conditions for being a subgroup.

Example 3

The group $G = (\{0, 1, 2, 3, 4, 5\}, +_6)$

a State the order of G

b Write down the order of possible subgroups of G. Justify your answer.

c Use your answer to **b** to find all the proper subgroups of G

a There are six elements in G so the order is 6

b 1, 2, 3 and 6 are factors of 6 so these are the orders of possible subgroups. •————————

> **1** Use Lagrange's theorem.

c List all the possibilities:

$\{0\}$ – this is the trivial subgroup.

$\{0, 1, 2, 3, 4, 5, 6\}$ – this is the group itself so is *not* a proper subgroup.

Other possibilities: $\{0, 1\}, \{0, 2\}, \{0, 3\}, \{0, 4\}, \{0, 5\}, \{0, 1, 2\},$
$\{0, 1, 3\}, \{0, 1, 4\}, \{0, 1, 5\}, \{0, 2, 3\}, \{0, 2, 4\}, \{0, 2, 5\}, \{0, 3, 4\},$
$\{0, 3, 5\}$ and $\{0, 4, 5\}$ •————————

> **2** List the possibilities.

For the two-element options, the element other than the identity must be self-inverse.

$2 + 2 \neq 0 \pmod 6$, $3 + 3 = 0 \pmod 6$, $4 + 4 \neq 0 \pmod 6$,
$5 + 5 \neq 0 \pmod 6$

Hence $\{0, 3\}$ is a subgroup.

For the three-element options, both elements other than the identity must be self-inverse, or they must be inverse pairs.

Since 3 is the only self-inverse element other than the identity, $\{0, 1, 3\}, \{0, 2, 3\}, \{0, 3, 4\}$ and $\{0, 3, 5\}$ can be discounted.

> **3** Check each possibility against the conditions. Checking for a closed group using inverses is a good way of doing this.

$2 + 4 = 0 \pmod 6$ and $1 + 5 = 0 \pmod 6$, so $\{0, 1, 5\}$ and $\{0, 2, 4\}$ are also subgroups. •————————

DISCRETE

Exercise 31.2B Reasoning and problem-solving

1 Determine the only possible orders of the subgroups of a group which has an order of 90. Fully justify your answer.

2 Which of the following groups *cannot* be a subgroup of G, which has order 30

    a A, order 5

    b B, order 3

    c C, order 12

    d D, order 10

3 Prove that no subgroups of order 6 exist for a group of order 92. Fully justify your answer.

4 A group

$H = (\{0, 1, 2, 3, 4, 5, 6, 7, 8, 9, 10\}, +_{11})$

Explain why H has only two subgroups, the trivial subgroup and the group itself.

5 A group $G = (\{e, p, q, r, s, t, u, v\}, \bullet)$ has the following Cayley table.

$\bullet$	e	p	q	r	s	t	u	v
e	e	p	q	r	s	t	u	v
p	p	q	u	t	v	s	e	r
q	q	u	e	s	r	v	p	t
r	r	t	s	u	p	e	v	q
s	s	v	r	p	u	q	t	e
t	t	s	v	e	q	p	r	u
u	u	e	p	v	t	r	q	s
v	v	r	t	q	e	u	s	p

a State the order of G

b Write down the orders of possible subgroups of G. Justify your answer.

c Use your answer to **b** to find all the non-trivial proper subgroups of G

6 A group $J = (\{1, -1, i, -i\}, \times)$. This is a subgroup of the complex numbers under multiplication.

Carmelita says that $(\{1, i, -i\}, \times)$ is a proper subgroup of J

a Explain why Carmelita is wrong.

b Identify all the proper subgroups of J

7 A group $G = (\{0, 1, 2, 3, 4, 5, 6, 7, 8\}, +_9)$. Explain why there are exactly three subgroups of G

8 The group, G, of all operations on a standard Rubik's Cube has 43 252 003 274 489 856 000 elements. Use Lagrange's theorem to work out the number of orders of possible subgroups of G

Hint: Try writing the number as the product of prime factors. You do not need to write out all the possible orders.

Fluency and skills

Isomorphism is an important concept in group theory. Suppose that you prove certain results for a group of order 4. If you encounter another group of order 4, your first thought is to start to prove these same results again for this different group. However, if you can show that the two groups are **isomorphic**, then the results proved for the first group automatically hold for the second group.

> **Key point**
>
> Two groups are **isomorphic** if there is a one-to-one mapping (an **isomorphism**) which associates each of the elements of one group with one of the elements of the other such that:
>
> If p maps to a and q maps to b, then the result of combining p and q under the binary operation of the first set maps to the result of combining a and b under the binary operation of the second set.

Example 1

A group G under the binary operation $\bullet$ has the Cayley table

$\bullet$	p	q	r	s
p	p	q	r	s
q	q	p	s	r
r	r	s	q	p
s	s	r	p	q

A second group $H = (\{i, -i, 1, -1\}, \times)$

Show that H is isomorphic to G and state the corresponding elements in each group.

The Cayley table for H is

$\times$	i	$-$i	1	-1
i	-1	1	i	$-$i
$-$i	1	-1	$-$i	i
1	i	$-$i	1	-1
-1	$-$i	i	-1	1

The identity element for H is 1 since $1 \times a = a \times 1 = a$

Elements 1 and -1 are self-inverse, so 1 corresponds to p and -1 to q

(continued on the next page)

DISCRETE

Reorder the columns and rows in the Cayley table:

×	1	−1	i	−i
1	1	−1	i	−i
−1	−1	1	−i	i
i	i	−i	−1	1
−i	−i	i	1	−1

By reordering the columns and rows in the Cayley table, you can see that the pattern of entries here is the same as in the Cayley table for G

Hence i corresponds to r and $−i$ corresponds to s

The groups are isomorphic.

The corresponding elements for each group are

$1 \leftrightarrow p, −1 \leftrightarrow q, i \leftrightarrow r$ and $−i \leftrightarrow s$

Note that this correspondence is not unique. Interchanging i and −i would also give the same pattern of entries as in G

The notation $H \cong G$ is used to denote isomorphism between groups.

Exercise 31.3A Fluency and skills

1 Show that the group formed by the set of rotational symmetries of a square is also isomorphic to the groups in Example 1

2 **a** Show that the groups $A = (\{1, 2, 3, 4\}, \times_5)$ and $B = (\{0, 1, 2, 3\}, +_4)$ are isomorphic.

   b Are groups A and B also isomorphic to those in Example 1?

3 $M = \left\{ \begin{pmatrix} 1 & 0 \\ 0 & 1 \end{pmatrix}, \begin{pmatrix} 1 & 0 \\ 0 & -1 \end{pmatrix}, \begin{pmatrix} -1 & 0 \\ 0 & 1 \end{pmatrix}, \begin{pmatrix} -1 & 0 \\ 0 & -1 \end{pmatrix} \right\}$,

matrix multiplication)

$N = (\{1, 3, 5, 7\}, \times_8)$

   a Show that M and N are isomorphic.

   b Are M and N also isomorphic to those in Example 1?

4 Show that the group formed by the set of rotational symmetries of an equilateral triangle is isomorphic to the group $R = (\langle 4 \rangle, \times_7)$

Reasoning and problem-solving

Showing that two groups of order 3 or 4 are isomorphic (or not) is quite straightforward since it is simple to rearrange the columns and rows of one of the Cayley tables to see if the pattern of entries matches the other table. For groups of a larger order, a systematic approach is advised.

Strategy 1

To show that two groups are isomorphic

  (1) Identify the identity element in each group.

  (2) Find the self-inverse elements in each group.

  (3) Identify the elements in each group that are not self-inverse and choose an arbitrary mapping.

  (4) Choose an arbitrary pairing of the self-inverse elements and write down a possible mapping.

  (5) Rewrite the Cayley table for one of the groups using the possible mapping and check that the pattern of entries is the same as for the Cayley table of the other group.

Example 2

Groups G and H have Cayley tables as shown.

G

	a	b	c	d	e	f
a	a	b	c	d	e	f
b	b	c	a	f	d	e
c	c	a	b	e	f	d
d	d	e	f	a	b	c
e	e	f	d	c	a	b
f	f	d	e	b	c	a

H

	P	Q	R	S	T	U
P	S	U	T	P	R	Q
Q	T	R	S	Q	U	P
R	U	S	Q	R	P	T
S	P	Q	R	S	T	U
T	Q	P	U	T	S	R
U	R	T	P	U	Q	S

Show that H is isomorphic to G

In G, the identity element is a

In H, the identity element is S ●—————————————————

> ① Identify the identity element in each group.

The self-inverse elements in G are d, e and f

The self-inverse elements in H are P, T and U ●—————

> ② Identify the self-inverse elements in each group.

In G, b and c are not self-inverse.

In H, Q and R are not self-inverse.

Choose an arbitrary mapping, say $b \leftrightarrow Q$

This means that $c \leftrightarrow R$ ●—————————————————

> ③ Identify the elements that are not self-inverse and choose an arbitrary mapping.

Choose an arbitrary mapping for the self-inverse elements:

Let $d \leftrightarrow P$; hence $e \leftrightarrow U$ and $f \leftrightarrow T$

A possible mapping is therefore $[b, c, d, e, f] \leftrightarrow [Q, R, P, U, T]$ ●——

> ④ Choose an arbitrary pairing of the self-inverse elements and write down a possible mapping.

The Cayley table for H is now

	S	Q	R	P	U	T
S	S	Q	R	P	U	T
Q	Q	R	S	T	P	U
R	R	S	Q	U	T	P
P	P	U	T	S	Q	R
U	U	T	P	R	S	Q
T	T	P	U	Q	R	S

> ⑤ Redraw the Cayley table for one of the groups using the possible mapping and check the pattern of entries.

The pattern of entries in the redrawn Cayley table for H now matches that for the Cayley table for G ●—————

Hence $H \cong G$

To show that two groups are *not* isomorphic, you need to establish at least one of the following

(1) That there are a different number of self-inverse elements in the two groups.

(2) That some of the elements in one group do not have the same period as in the other.

(3) That one group is cyclic and the other is not.

Example 3

Group *S* has a Cayley table as shown.

Show that *S* is not isomorphic to the two groups in Example 1

	1	2	3	4	5	6
1	1	2	3	4	5	6
2	2	4	6	1	3	5
3	3	6	2	5	1	4
4	4	1	5	2	6	3
5	5	3	1	6	4	2
6	6	5	4	3	2	1

The identity element is 1

The only other self-inverse element in *S* is 6

G and *H* both have three self-inverse elements other than the identity.

Hence *S* is not isomorphic to *G* nor to *H*

(1) There are a different number of self-inverse elements.

You could also have shown that *S* has elements of period 6 (the elements 3 and 5) whereas *G* and *H* do not.

Exercise 31.3B Reasoning and problem-solving

1   Explain why all groups of order 2 are isomorphic to each other.

2   Explain why all groups of order 3 are isomorphic to each other.

3   Brice claims that every group of order 4 is isomorphic to exactly one of four particular groups, *A, B, C, D* of order 4.
    Group *A* has three elements of period 2
    Group *B* has two elements of period 2 and one of period 4
    Group *C* has one element of period 2 and two of period 4
    Group *D* has four elements of period 4.
    Explain why Brice's claim is incorrect and show that every group of order 4 is isomorphic to exactly one of two of the groups Brice has listed.

4   Leona says that if *p* is a prime number, all groups of order *p* are isomorphic, i.e. all groups of order 7, for example, are isomorphic to each other. Explain why Leona is correct in her assertion.

5   Explain why it is impossible for an abelian group to be isomorphic to a non-abelian group.

6   The group $G = (\{0, 1, 2, 3, 4, 5\}, +_6)$
    The group *H* consists of the set of rotational symmetries of a regular hexagon.
    Show that $G \cong H$

7   The group $S = (\{1, 2, 4, 7, 8, 11, 13, 14\}, \times_{15})$
    Another group *M* has elements consisting of the set of rotations of angle $\dfrac{k\pi}{4}$, $k = 1, 2, 3, 4,$ 5, 6, 7, 8, of a unit square about the origin.
    Show that *S* is not isomorphic to *M*

8   Show that the group of matrices of the form $\begin{pmatrix} 1-a & a \\ -a & 1+a \end{pmatrix}$ under the operation matrix multiplication is isomorphic to the group $G = (\mathbb{Z}, +)$

Hint: Since the groups are infinite, you need to specify a mapping connecting each element in one group with an element in the other group.

Chapter summary

- The conditions under which a set of mathematical objects form a group under a given binary operation are known as the axioms.
- These are closure, the existence of an identity element, associativity and the existence of an inverse element for each member of the set.
- A group $(S, \odot)$ is a non-empty set S with a binary operation $\odot$ such that:
  - $\odot$ is closed in S
  - $\odot$ is associative
  - there is an identity element such that $x \odot e = e \odot x = x$ for all x
  - each element has an inverse, x^{-1}, such that $x \odot x^{-1} = x^{-1} \odot x = e$
- The order of a group is equal to the number of elements in the group.
- The period (or order) of a particular element, x, of a group is the smallest non-negative integer n such that $x^n = e$, where e is the identity.
- An abelian group is a group with the additional property of commutativity between the elements of the group.
- To show that a given set forms a group under a given binary operation:
  - Write down the elements of the set.
  - Produce a Cayley table to show the combinations of each element under the given binary operation.
  - Check that the binary operation meets the axioms necessary to be a group.
- A subgroup of a group G is any subset H of G such that H itself is also a group under the same binary operation as G
- A non-trivial subgroup is any subgroup that is not the trivial subgroup.
- A proper subgroup is any subgroup that is not the group itself.
- Lagrange's theorem states that, for any finite group G, the order of every subgroup of G divides the order of G
- To find all the possible subgroups of a particular group:
  - Use Lagrange's theorem to find the order of the possible subgroups.
  - Use a systematic process to identify the possibilities.
  - Check each possibility against the conditions for being a subgroup.
- Two groups are isomorphic if there is a one-to-one mapping (an isomorphism) which associates each of the elements of one group with one of the elements of the other such that if p maps to a and q maps to b, then the result of combining p and q under the binary operation of the first set maps to the result of combining a and b under the binary operation of the second set.

- To show that two groups are isomorphic:
  - Identify the identity element in each group.
  - Find the self-inverse elements in each group.
  - Identify the elements in each group that are not self-inverse and choose an arbitrary mapping.
  - Choose an arbitrary pairing of the self-inverse elements and write down a possible mapping.
  - Rewrite the Cayley table for one of the groups using the possible mapping and check that the pattern of entries is the same as for the Cayley table of the other group.
- To show that two groups are not isomorphic, you need to establish at least one of the following:
  - That there are a different number of self-inverse elements in the two groups.
  - That some of the elements in one group do not have the same period as in the other.
  - That one group is cyclic and the other is not.

Check and review

You should now be able to...	Review Questions
✓ Understand and use the language of groups including: order, period, subgroup, proper, trivial, non-trivial.	1, 3
✓ Understand and use the group axioms: closure, identity, inverses and associativity, including use of Cayley tables.	1, 2
✓ Recognise and use finite and infinite groups and their subgroups, including: groups of symmetries of regular polygons, cyclic groups and abelian groups.	1, 3, 4
✓ Understand and use Lagrange's theorem.	3
✓ Identify and use the generators of a group.	1
✓ Recognise and find isomorphism between groups of finite order.	4

1 $G = (\{1, 2, 4, 5, 7, 8\}, \times_9)$

   a State the order of G

   b Draw a Cayley table for G

   c State the period of the element '5'

   d Explain why your answer to c shows that G is cyclic.

   e Is G an abelian group? Explain your answer.

2   It is suggested that the set $S = \{1, 5, 7, 11\}$ forms a group under $\times_{12}$

By drawing a Cayley table, or otherwise,

a   Write down the identity element,

b   Show that S is closed under $\times_{12}$

c   Show that every element in S has an inverse,

d   Give at least two examples of the axiom of associativity holding for S

3   The group, S, formed by the set of all symmetries of an equilateral triangle has Cayley table

	r_0	r_1	r_2	m_1	m_2	m_3
r_0	r_0	r_1	r_2	m_1	m_2	m_3
r_1	r_1	r_2	r_0	m_2	m_3	m_1
r_2	r_2	r_0	r_1	m_3	m_1	m_2
m_1	m_1	m_3	m_2	r_0	r_2	r_1
m_2	m_2	m_1	m_3	r_1	r_0	r_2
m_3	m_3	m_2	m_1	r_2	r_1	r_0

a   State the order of possible subgroups of S

b   Write down the trivial subgroup of S

c   Find all proper subgroups of S

4   a   Show that the groups $G = (\{1, 3, 7, 9\}, \times_{10})$ and $H = (\{0, 1, 2, 3\}, +_4)$ are isomorphic and write down a possible mapping.

b   Are the groups cyclic? Explain your reasoning.

c   Are the groups abelian? Explain your reasoning.

DISCRETE

Did you know?

An automorphism is an isomorphism from a mathematical object to itself. One of the earliest group automorphisms was found by William Rowan Hamilton as he developed icosian calculus. This resulted in the development of the popular icosian game. The aim the game is to find a Hamiltonian cycle along the edges of a dodecahedron, visiting every vertex only once.

Research

The three general isomorphism theorems were formulated by Emmy Noether in 1927. Find out about them, particularly in the context of groups. You may also like to extend this research by looking at the lattice theorem and the Butterfly lemma.

Research

You can use the theories around zero knowledge proofs to discover whether two graphs are isomorphic.

Start by finding out about zero knowledge proofs. For example, look at the research paper *How to Explain Zero-Knowledge Protocols to Your Children* by Quisquater. In particular, well-known examples are the Ali Baba cave, and two-coloured balls with a colour-blind friend.

1 The group G has the following Cayley table.

*	a	b	c	d
a	c	a	d	b
b	a	b	c	d
c	d	c	b	a
d	b	d	a	c

    a State, with a reason, which element is the identity element. **[2]**

    b Write down all of the elements that are self-inverse. **[1]**

    c Explain why G is an abelian group. **[1]**

2 A finite group H has proper subgroups $\{a, b\}$, $\{b, d, e\}$ and $\{b, f, g, h, k\}$

    a State, with a reason, which element of H is the identity. **[2]**

    b State, with a reason, the smallest possible order of H **[2]**

3 The group $G = (\{1, 3, 5, 7\}, \times_8)$

  The group $H = (\{1, 3, 7, 9\}, \times_{10})$

  By drawing Cayley tables for G and H, or otherwise, show that G and H
  are not isomorphic. **[4]**

4 The cyclic group $C_n = (\langle 9 \rangle, \times_{64})$

    a Explain what is meant by $\langle 9 \rangle$ **[1]**

    b Find the value of n and justify your answer. **[3]**

    c State the order of all of the possible subgroups, giving a reason for your answer. **[2]**

5 The binary operation $\odot$ is defined as $a \odot b = a + b + 3 \, [\text{mod } 6]$, where $a, b \in \mathbb{Z}$

    a Show that the set $\{0, 1, 2, 3, 4, 5\}$ forms a group G under $\odot$ **[5]**

    b Find all of the proper subgroups of G **[2]**

    c Determine whether G is isomorphic to the group $K = (\langle 2 \rangle, \times_9)$ **[3]**

6 The group G has Cayley table

	a	b	c	d	e	f	g	h
a	c	e	b	f	a	h	d	g
b	e	c	a	g	b	d	h	f
c	b	a	e	h	c	g	f	d
d	f	g	h	a	d	c	e	b
e	a	b	c	d	e	f	g	h
f	h	d	g	c	f	b	a	e
g	d	h	f	e	g	a	b	c
h	g	f	d	b	h	e	c	a

a Explain why element e is the identity element. [1]

Emelia says that $\{e, d, g\}$ is a proper subgroup of G

b Explain, with a reason, whether Emelia is correct. [2]

c Show that G is cyclic and hence explain why G is not isomorphic to the group of symmetries of a square. [5]

7 The set P consists of the set of all integers under the binary operation • such that

$x • y = x + y - 1$

a Show that • is associative. [2]

b Hence, show that P forms a group under • [4]

c State, with a reason, whether the set of positive integers forms a group under • [2]

Pure Mathematics

Quadratic equations

$ax^2 + bx + c = 0$ has roots $\dfrac{-b \pm \sqrt{b^2 - 4ac}}{2a}$

Laws of indices

$a^x a^y \equiv a^{x+y}$

$a^x \div a^y \equiv a^{x-y}$

$(a^x)^y \equiv a^{xy}$

Laws of logarithms

$x = a^n \Leftrightarrow n = \log_a x$ for $a > 0$ and $x > 0$

$\log_a x + \log_a y \equiv \log_a xy$

$\log_a x - \log_a y \equiv \log_a\left(\dfrac{x}{y}\right)$

$k\log_a x \equiv \log_a (x)^k$

Coordinate geometry

A straight-line graph, gradient m passing through (x_1, y_1) has equation $y - y_1 = m(x - x_1)$

Straight lines with gradients m_1 and m_2 are perpendicular when $m_1 m_2 = -1$

Sequences

General term of an arithmetic progression: $u_n = a + (n-1)d$

General term of a geometric progression: $u_n = ar^{n-1}$

Trigonometry

In the triangle ABC

Sine rule: $\dfrac{a}{\sin A} = \dfrac{b}{\sin B} = \dfrac{c}{\sin C}$

Cosine rule: $a^2 = b^2 + c^2 - 2bc\cos A$

Area $= \dfrac{1}{2}ab\sin C$

$\cos^2 A + \sin^2 A \equiv 1$

$\sec^2 A \equiv 1 + \tan^2 A$

$\operatorname{cosec}^2 A \equiv 1 + \cot^2 A$

$\sin 2A \equiv 2\sin A \cos A$

$\cos 2A \equiv \cos^2 A - \sin^2 A$

$\tan 2A \equiv \dfrac{2\tan A}{1 - \tan^2 A}$

Mensuration

Circumference and Area of circle, radius r and diameter d:

$$C = 2\pi r = \pi d \qquad A = \pi r^2$$

Pythagoras' Theorem:

In any right-angled triangle where a, b and c are the lengths of the sides and c is the hypotenus:
$$c^2 = a^2 + b^2$$

Area of a trapezium $= \dfrac{1}{2}(a+b)h$, where a and b are the lengths of the parallel sides and h is their perpendicular separation.

Volume of a prism = area of cross section $\times$ length

For a circle of radius r, where an angle at the centre of θ radians subtends an arc of length s and encloses an associated sector of area A:

$$s = r\theta \qquad A = \dfrac{1}{2}r^2\theta$$

Complex numbers

For two complex numbers $z_1 = r_1 e^{i\theta_1}$ and $z_2 = r_2 e^{i\theta_2}$:

$$z_1 z_2 = r_1 r_2\, e^{i(\theta_1 + \theta_2)}$$

$$\dfrac{z_1}{z_2} = \dfrac{r_1}{r_2}\, e^{i(\theta_1 - \theta_2)}$$

Loci in the Argand diagram:

$|z - a| = r$ is a circle radius r centred at a

$\arg(z - a) = \theta$ is a half line drawn from a at angle θ to a line parallel to the positive real axis.

Exponential form: $e^{i\theta} = \cos\theta + i\sin\theta$

Matrices

For a 2 by 2 matrix $\begin{pmatrix} a & b \\ c & d \end{pmatrix}$ the determinant $\Delta = \begin{vmatrix} a & b \\ c & d \end{vmatrix} = ad - bc$

the inverse is $\dfrac{1}{\Delta}\begin{pmatrix} d & -b \\ -c & a \end{pmatrix}$

The transformation represented by matrix $\mathbf{AB}$ is the transformation represented by matrix $\mathbf{B}$ followed by the transformation represented by matrix $\mathbf{A}$.

For matrices $\mathbf{A}$, $\mathbf{B}$:

$$(\mathbf{AB})^{-1} = \mathbf{B}^{-1}\mathbf{A}^{-1}$$

Algebra

$$\sum_{r=1}^{n} r = \frac{1}{2}n(n+1)$$

For $ax^2 + bx + c = 0$ with roots α and β:

$$\alpha + \beta = \frac{-b}{a} \qquad \alpha\beta = \frac{c}{a}$$

For $ax^3 + bx^2 + cx + d = 0$ with roots α, β and γ:

$$\sum \alpha = \frac{-b}{a} \qquad \sum \alpha\beta = \frac{c}{a} \qquad \alpha\beta\gamma = \frac{-d}{a}$$

Hyperbolic functions

$$\cosh x \equiv \frac{1}{2}(e^x + e^{-x}) \qquad \sinh x \equiv \frac{1}{2}(e^x - e^{-x}) \qquad \tanh x \equiv \frac{\sinh x}{\cosh x}$$

Calculus and differential equations

Differentiation

Function	Derivative	Function	Derivative
x^n	nx^{n-1}	e^{kx}	ke^{kx}
$\sin kx$	$k\cos kx$	$\ln x$	$\dfrac{1}{x}$
$\cos kx$	$-k\sin kx$	$f(x) + g(x)$	$f'(x) + g'(x)$
		$f(x)g(x)$	$f'(x)g(x) + f(x)g'(x)$
		$f(g(x))$	$f'(g(x))g'(x)$

Integration

Function	Integral		Function	Integral
x^n	$\dfrac{1}{n+1}x^{n+1}+c,\ n\neq -1$		e^{kx}	$\dfrac{1}{k}e^{kx}+c$
$\cos kx$	$\dfrac{1}{k}\sin kx+c$		$\dfrac{1}{x}$	$\ln|x|+c,\ x\neq 0$
			$f'(x)+g'(x)$	$f(x)+g(x)+c$
$\sin kx$	$-\dfrac{1}{k}\cos kx+c$		$f'(g(x))g'(x)$	$f(g(x))+c$

Area under a curve $= \displaystyle\int_a^b y\,\mathrm{d}x\ (y\geq 0)$

Volumes of revolution about the x and y axes:

$$V_x = \pi\int_a^b y^2\,\mathrm{d}x \qquad V_y = \pi\int_c^d x^2\,\mathrm{d}y$$

Simple Harmonic Motion: $\ddot{x} = -\omega^2 x$

Vectors

$$\left|\,x\mathbf{i}+y\mathbf{j}+z\mathbf{k}\,\right| = \sqrt{(x^2+y^2+z^2)}$$

Scalar product of two vectors $\mathbf{a} = \begin{pmatrix} a_1 \\ a_2 \\ a_3 \end{pmatrix}$ and $\mathbf{b} = \begin{pmatrix} b_1 \\ b_2 \\ b_3 \end{pmatrix}$ is

$$\begin{pmatrix} a_1 \\ a_2 \\ a_3 \end{pmatrix} \cdot \begin{pmatrix} b_1 \\ b_2 \\ b_3 \end{pmatrix} = a_1 b_1 + a_2 b_2 + a_3 b_3 = |\mathbf{a}|\,|\mathbf{b}|\cos\theta$$

where θ is the acute angle between the vectors $\mathbf{a}$ and $\mathbf{b}$.

The equation of the line through the point with position vector $\mathbf{a}$ parallel to vector $\mathbf{b}$ is:

$$\mathbf{r} = \mathbf{a} + t\mathbf{b}$$

The equation of the plane containing the point with position vector $\mathbf{a}$ and perpendicular to vector $\mathbf{n}$ is:

$$(\mathbf{r} - \mathbf{a}) \cdot \mathbf{n} = 0$$

Mechanics

Forces and equilibrium

Weight $= \text{mass} \times g$

Friction: $F \le \mu R$

Newton's second law in the form: $F = ma$

Kinematics

For motion in a straight line with variable acceleration:

$$v = \frac{dr}{dt} \qquad a = \frac{dv}{dt} = \frac{d^2r}{dt^2}$$

$$r = \int v\, dt \qquad v = \int a\, dt$$

Statistics

The mean of a set of data: $\bar{x} = \dfrac{\sum x}{n} = \dfrac{\sum fx}{\sum f}$

The standard Normal variable: $Z = \dfrac{X - \mu}{\sigma}$ where $X \sim N(\mu, \sigma^2)$

Mathematical notation

For A Level Further Maths

Set Notation

Notation	Meaning
$\in$	is an element of
$\notin$	is not an element of
$\subseteq$	is a subset of
$\subset$	is a proper subset of
$\{x_1, x_2, \dots\}$	the set with elements $x_1, x_2, \dots$
$\{x : \dots\}$	the set of all x such that ...
$n(A)$	the number of elements in set A
$\varnothing$	the empty set
ε	the universal set
A'	the complement of the set A
$\mathbb{N}$	the set of natural numbers, $\{1, 2, 3, \dots\}$
$\mathbb{Z}$	the set of integers, $\{0, \pm1, \pm2, \pm3, \dots\}$
$\mathbb{Z}^+$	the set of positive integers, $\{1, 2, 3, \dots\}$
$\mathbb{Z}_0^+$	the set of non-negative integers, $\{0, 1, 2, 3, \dots\}$
$\mathbb{R}$	the set of real numbers
$\mathbb{Q}$	the set of rational numbers, $\left\{ \dfrac{p}{q} : p \in \mathbb{Z}, \ q \in \mathbb{Z}^+ \right\}$
$\cup$	union
$\cap$	intersection
(x, y)	the ordered pair x, y
$[a, b]$	the closed interval $\{x \in \mathbb{R} : a \le x \le b\}$
$[a, b)$	the interval $\{x \in \mathbb{R} : a \le x < b\}$
$(a, b]$	the interval $\{x \in \mathbb{R} : a < x \le b\}$
(a, b)	the open interval $\{x \in \mathbb{R} : a < x < b\}$
$\mathbb{C}$	the set of complex numbers

Miscellaneous Symbols

Notation	Meaning
$=$	is equal to
$\neq$	is not equal to
$\equiv$	is identical to or is congruent to
$\approx$	is approximately equal to
∞	infinity
$\propto$	is proportional to
$\therefore$	therefore
$\because$	because
$<$	is less than
$\leqslant, \le$	is less than or equal to; is not greater than
$>$	is greater than
$\geqslant, \ge$	is greater than or equal to; is not less than
$p \Rightarrow q$	p implies q (if p then q)
$p \Leftarrow q$	p is implied by q (if q then p)
$p \Leftrightarrow q$	p implies and is implied by q (p is equivalent to q)

Mathematical notation – for A Level Further Maths

Notation	Meaning
a	first term for an arithmetic or geometric sequence
l	last term for an arithmetic sequence
d	common difference for an arithmetic sequence
r	common ratio for a geometric sequence
S_n	sum to n terms of a sequence
S_∞	sum to infinity of a sequence
$\cong$	is isomorphic to

Operations

Notation	Meaning
$a + b$	a plus b
$a - b$	a minus b
$a \times b, ab, a \cdot b$	a multiplied by b
$a \div b, \dfrac{a}{b}$	a divided by b
$\displaystyle\sum_{i=1}^{n} a_i$	$a_1 + a_2 + \ldots + a_n$
$\displaystyle\prod_{i=1}^{n} a_i$	$a_1 \times a_2 \times \ldots \times a_n$
$\sqrt{a}$	the non-negative square root of a
$|a|$	the modulus of a
$n!$	n factorial: $n! = n \times (n-1) \times \ldots \times 2 \times 1$, $n \in \mathbb{N}$; $0! = 1$
$\dbinom{n}{r}, {}^nC_r, {}_nC_r$	the binomial coefficient $\dfrac{n!}{r!(n-r)!}$ for $n, r \in \mathbb{Z}_0^+, r \le n$ or $\dfrac{n(n-1)\ldots(n-r+1)}{r!}$ for $n \in \mathbb{Q}, r \in \mathbb{Z}_0^+$
$a \times_n b$	multiplication modulo n of a by b
$a +_n b$	addition modulo n of a and b
$G = (<n>,*)$	n is the generator of a given group G under the operation*

Functions

Notation	Meaning
$f(x)$	the value of the function f at x
$f: x \mapsto y$	the function f maps the element x to the element y
f^{-1}	the inverse function of the function f
gf	the composite function of f and g which is defined by $gf(x) = g(f(x))$
$\lim\limits_{x \to a} f(x)$	the limit of $f(x)$ as x tends to a
$\Delta x, \delta x$	an increment of x
$\dfrac{\mathrm{d}y}{\mathrm{d}x}$	the derivative of y with respect to x
$\dfrac{\mathrm{d}^n y}{\mathrm{d}x^n}$	the nth derivative of y with respect to x
$f'(x), f''(x), \ldots, f^{(n)}(x)$	the first, second, ..., nth derivatives of $f(x)$ with respect to x
$\dot{x}, \ddot{x}, \ldots$	the first, second, ... derivatives of x with respect to t
$\displaystyle\int y \, \mathrm{d}x$	the indefinite integral of y with respect to x

Mathematical notation – for A Level Further Maths

Notation	Meaning
$\displaystyle\int_a^b y\,dx$	the definite integral of y with respect to x between the limits $x = a$ and $x = b$

Exponential and Logarithmic Functions

Notation	Meaning
e	base of natural logarithms
e^x, exp x	exponential function of x
$\log_a x$	logarithm to the base a of x
$\ln x$, $\log_e x$	natural logarithm of x

Trigonometric Functions

Notation	Meaning
sin, cos, tan, cosec, sec, cot	the trigonometric functions
$\sin^{-1}$, $\cos^{-1}$, $\tan^{-1}$ arcsin, arccos, arctan	the inverse trigonometric functions
$^\circ$	degrees
rad	radians
cosec^{-1}, $\sec^{-1}$, $\cot^{-1}$, arccosec, arcsec, arccot	the inverse trigonometric functions
sinh, cosh, tanh, cosech, sech, coth	the hyperbolic functions
$\sinh^{-1}$, $\cosh^{-1}$, $\tanh^{-1}$ cosech^{-1}, sech^{-1}, $\coth^{-1}$ arcsinh, arccosh, arctanh, arccosech, arcsech, arccoth, arsinh, arcosh, artanh, arcosech, arsech, arcoth	the inverse hyperbolic functions

Complex numbers

Notation	Meaning
i, j	square root of -1
$x + iy$	complex number with real part x and imaginary part y
$r(\cos\theta + i\sin\theta)$	modulus argument form of a complex number with modulus r and argument θ
z	a complex number, $z = x + iy = r(\cos\theta + i\sin\theta)$
$\text{Re}(z)$	the real part of z, $\text{Re}(z) = x$
$\text{Im}(z)$	the imaginary part of z, $\text{Im}(z) = y$
$|z|$	the modulus of z, $|z| = r = \sqrt{x^2 + y^2}$
$\arg(z)$	the argument of z, $\arg(z) = \theta$, $-\pi < \theta \leq \pi$
z^*	the complex conjugate of z, $x - iy$

Matrices

Notation	Meaning
$\mathbf{M}$	a matrix $\mathbf{M}$
$\mathbf{0}$	zero matrix
$\mathbf{I}$	identity matrix
$\mathbf{M}^{-1}$	the inverse of the matrix $\mathbf{M}$
$\mathbf{M}^{T}$	the transpose of the matrix $\mathbf{M}$
Δ, det $\mathbf{M}$ or $|\mathbf{M}|$	the determinant of the square matrix $\mathbf{M}$
$\mathbf{Mr}$	Image of the column vector $\mathbf{r}$ under the transformation associated with the matrix $\mathbf{M}$

Vectors

Notation	Meaning
$\mathbf{a}$, $\underline{a}$, $\underset{\sim}{a}$	the vector $\mathbf{a}$, $\underline{a}$, $\underset{\sim}{a}$ these alternatives apply throughout section 9
$\overrightarrow{AB}$	the vector represented in magnitude and direction by the directed line segment AB
$\hat{\mathbf{a}}$	a unit vector in the direction of $\mathbf{a}$
$\mathbf{i}, \mathbf{j}, \mathbf{k}$	unit vectors in the directions of the Cartesian coordinate axes
$|\mathbf{a}|$, a	the magnitude of $\mathbf{a}$
$|\overrightarrow{AB}|$, AB	the magnitude of $\overrightarrow{AB}$
$\begin{pmatrix} a \\ b \end{pmatrix}$, $a\mathbf{i} + b\mathbf{j}$	column vector and corresponding unit vector notation
$\mathbf{r}$	position vector
$\mathbf{s}$	displacement vector
$\mathbf{v}$	velocity vector
$\mathbf{a}$	acceleration vector
$\mathbf{a} \cdot \mathbf{b}$	the scalar product of $\mathbf{a}$ and $\mathbf{b}$

Differential equations

Notation	Meaning
ω	angular speed

Probability and statistics

Notation	Meaning
A, B, C, etc.	events
$A \cup B$	union of the events A and B
$A \cap B$	intersection of the events A and B
$\mathrm{P}(A)$	probability of the event A
A'	complement of the event A
$\mathrm{P}(A \mid B)$	probability of the event A conditional on the event B
X, Y, R etc.	random variables
x, y, r etc.	values of the random variables X, Y, R etc.
$x_1, x_2, \ldots$	values of observations

$f_1, f_2, \dots$	frequencies with which the observations x_1, x_2, ... occur
$\mathrm{p}(x)$, $\mathrm{P}(X = x)$	probability function of the discrete random variable X
$p_1, p_2, \dots$	probabilities of the values x_1, x_2, ... of the discrete random variable
X	
$\mathrm{E}(X)$	expectation of the random variable X
$\mathrm{Var}(X)$	variance of the random variable X
$\sim$	has the distribution
$\mathrm{B}(n, p)$	binomial distribution with parameters n and p, where n is the number of trials and p is the probability of success in a trial
q	$q = 1 - p$ for binomial distribution
$\mathrm{N}(\mu, \sigma^2)$	Normal distribution with mean μ and variance σ^2
$Z \sim \mathrm{N}(0, 1)$	standard Normal distribution
ϕ	probability density function of the standardised Normal variable with distribution $\mathrm{N}(0, 1)$
Φ	corresponding cumulative distribution function
μ	population mean
σ^2	population variance
σ	population standard deviation
$\bar{x}$	sample mean
s^2	sample variance
s	sample standard deviation
H_0	null hypothesis
H_1	alternative hypothesis
r	product moment correlation coefficient for a sample
ρ	product moment correlation coefficient for a population

Mechanics

Notation	Meaning
kg	kilograms
m	metres
km	kilometres
m/s, $\mathrm{m\,s^{-1}}$	metre(s) per second (velocity)
m/s², $\mathrm{m\,s^{-2}}$	metre(s) per second square (acceleration)
F	Force or resultant force
N	newton
Nm	newton metre (moment of a force)
t	time
s	displacement
u	initial velocity
v	velocity or final velocity
a	acceleration
g	acceleration due to gravity
μ	coefficient of friction

Chapter 1

Exercise 1.1A

1 a $x = \pm 5i$ **b** $x = \pm 11i$
  c $x = \pm 2\sqrt{5}i$ **d** $x = \pm 2\sqrt{2}i$
  e $z = \pm 3i$ **f** $z = \pm 2\sqrt{3}i$

2 a $7 - 6i$ **b** $-7 - 10i$
  c $18 - 27i$ **d** $26 + 25i$
  e $-41 - 63i$ **f** $27 - 10i$

3 a $7 + 17i$ **b** $39 - 27i$
  c $3 + 8i$ **d** $65 - 72i$

4 a $-i$ **b** 1
  c i **d** $-8i$
  e 81 **f** $-112 + 180i$

5 a $\dfrac{6}{5} - \dfrac{3}{5}i$ **b** $-\dfrac{5}{13} + \dfrac{1}{13}i$

  c $-\dfrac{2}{5} + \dfrac{11}{5}i$ **d** $-\dfrac{1}{5} - \dfrac{7}{5}i$

  e $(1 - 2\sqrt{2}) + (-2 - \sqrt{2})i$ **f** $-\sqrt{2}$

6 a $2 + 4i$ **b** $-11 + 10i$

  c $-\dfrac{5}{17} + \dfrac{14}{17}i$ **d** $-\dfrac{5}{13} - \dfrac{14}{13}i$

Exercise 1.1B

1 $a = \pm 3$, $b = \pm 12$

2 $b = 7$, $a = 34$

3 $z = 4 + i$

4 $x = 2 - i$

5 $z_2 = 8 + 2i$, $z_1 = 3 - 5i$

6 $w = -2 - \dfrac{1}{2}i$, $z = 4 + 7i$

7 $z_2 = -3 - 2i$, $z_1 = \dfrac{1}{2} + i$

8 a $w = 3 + 5i$ or $w = -3 - 5i$
  b $w = 1 - 2i$ or $w = -1 + 2i$
  c $w = 5 - 2i$ or $w = -5 + 2i$

9 $-2 + \sqrt{2}i$ and $2 - \sqrt{2}i$

10 $w = -3 \pm i$, $z = 1 \pm 3i$

11 a $z = 1 + 7i$ or $z = 1 - 7i$
  b $z = 2 - 3i$ or $z = -2 + 3i$

12 a $-7 - 24i$
  b $4282 + 1475i$
  c $44 + 8i$

13 $a = \pm 3$, $b = \pm 96$

Exercise 1.2A

1 a $5 + 2i$ **b** $8 - i$
  c $-5i - 6$ **d** $\sqrt{2} + i\sqrt{3}$

  e $\dfrac{1}{3} - 4i$ **f** $-\dfrac{2}{3}i - 5$

2 a 85 **b** 18

  c $-4i$ **d** $\dfrac{77}{85} - \dfrac{36}{85}i$

  e $9 - 2i$ **f** $\dfrac{77}{85} + \dfrac{36}{85}i$

3 a 8 **b** $-2\sqrt{6}$

  c $2\sqrt{2}i$ **d** $\dfrac{1}{2} - \dfrac{\sqrt{3}}{2}i$

  e 8 **f** 24

4 a $x = -\dfrac{5}{2} \pm \dfrac{\sqrt{3}}{2}i$ **b** $x = \dfrac{3}{2} \pm \dfrac{\sqrt{11}}{2}i$

  c $x = -\dfrac{7}{4} \pm \dfrac{\sqrt{7}}{4}i$ **d** $x = \dfrac{5}{3} \pm \dfrac{\sqrt{2}}{3}i$

5 a $x^2 - 3x - 28 = 0$ **b** $x^2 - 6x + 34 = 0$
  c $x^2 + 2x + 82 = 0$ **d** $x^2 + 10x + 41 = 0$

6 a $\sqrt{3} - i$ **b** $z^2 - 2\sqrt{3}z + 4 = 0$

7 a $x^2 - 4x + 5 = 0$ **b** $x^2 - 8x + 25 = 0$
  c $x^2 + 2x + 50 = 0$ **d** $x^2 + 10x + 29 = 0$
  e $x^2 - 2ax + (9 + a^2) = 0$ **f** $x^2 - 10x + (25 + b^2) = 0$

8 a $x^3 + x^2 - 7x + 65 = 0$
  b $x^3 - 2x - 4 = 0$
  c $x^3 - 2\sqrt{3}x^2 + 7x = 0$
  d $x^3 + (2\sqrt{2} - 3)x^2 + (3 - 6\sqrt{2})x - 9 = 0$

9 a $6 + 2i$
  b $2z^3 - 23z^2 + 68z + 40 = 0$

10 a $x^3 + 9x^2 + 25x + 25 = (x + 5)(x^2 + 4x + 5)$
    $f(x) = 0 \Rightarrow x^2 + 4x + 5 = 0$
  b $x = -5$, $-2 \pm i$

11 a $1 - 8i$ is also a root giving quadratic factor $x^2 - 2x + 65$
    $x^4 + 4x^3 + 66x^2 + 364x + 845$
    $= (x^2 - 2x + 65)(x^2 + 6x + 13) = 0$
    $\Rightarrow x^2 + 6x + 13 = 0$
  b $x = 1 \pm 8i$, $-3 \pm 2i$

12 a $a = 1$, $b = 4$, $c = 30$
  b $a = 1$, $b = -3$, $c = -185$

13 a $a = -2$, $b = 1$, $c = -20$
  b $a = 27$, $b = -38$, $c = 26$

14 $(x - 7i)(x + 7i) = x^2 + 49$
  $x^4 - x^3 + 43x^2 - 49x - 294 = (x^2 + 49)(x^2 - x - 6) = 0$
  So can be written $(x^2 + 49)(x - 3)(x + 2) = 0$
  $(A = 49, B = -3, C = 2)$

Exercise 1.2B

1 a Let $z = a + bi$, $w = c + di$, $a, b, c, d \in \mathbb{R}$
    $\begin{aligned}(zw)^* &= ((a + bi)(c + di))^* \\ &= ((ac - bd) + (bc + ad)i)^* \\ &= (ac - bd) - (bc + ad)i \\ &= (a - bi)(c - di) \\ &= z^* w^* \end{aligned}$

  b $\begin{aligned}(z^*)^* &= ((a + bi)^*)^* \\ &= (a - bi)^* \\ &= a + bi \\ &= z \end{aligned}$

c $\left(\dfrac{z}{w}\right)^* = \left(\dfrac{a+bi}{c+di}\right)^*$

$= \left(\dfrac{(a+bi)(c-di)}{(c+di)(c-di)}\right)^*$

$= \left(\dfrac{(ac+bd)+(bc-ad)i}{c^2-d^2i^2}\right)^*$

$= \dfrac{ac+bd-(bc-ad)i}{c^2+d^2}$

$\dfrac{z^*}{w^*} = \dfrac{a-bi}{c-di}$

$= \dfrac{(a-bi)(c+di)}{(c-di)(c+di)}$

$= \dfrac{ac+bd-(bc-ad)i}{c^2+d^2}$

$= \left(\dfrac{z}{w}\right)^*$

2 a Let $z = a+bi$, $a, b \in \mathbb{R}$
$z+z^* = (a+bi)+(a-bi)$
$= 2a$ so a real number

b $z-z^* = (a+bi)-(a-bi)$
$= 2bi$ so an imaginary number

c $zz^* = (a+bi)(a-bi)$
$= a^2 - abi + abi - b^2i^2$
$= a^2 + b^2$ so a real number

3 $z = 3 \pm 7i$

4 $w = 2+9i$ or $w = -2+9i$

5 $z = \sqrt{3} - i$

6 $w = 4+2i$

7 a $k = -119$ **b** $x = (7), 1 \pm 4i$

8 $k = -8$

$x = -\dfrac{3}{2} \pm \dfrac{\sqrt{23}}{2}i, 1$

9 $x = 1, 3, \pm i$

10 $x = -3, 7, 1+\sqrt{2}i$ and $1-\sqrt{2}i$

11 a $k = 10$

b $x = -6 \pm i, 3, -1$

12 $A = -10$
$x = 5 \pm i, \pm \sqrt{2}i$

13 a e.g. $x^4 - 4x^3 + 24x^2 - 40x + 100 = 0$

b e.g. $x^4 + 12x^3 + 62x^2 + 156x + 169 = 0$

14 a $(x+1)(x^2 - 20x + 109)$

b

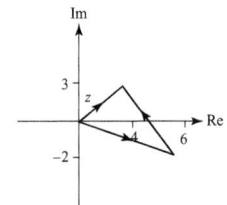

15 a $4x^4 + 12x^3 - 35x^2 - 300x + 625 = (4x^2 - 20x + 25)(x^2 + 8x + 25)$

b

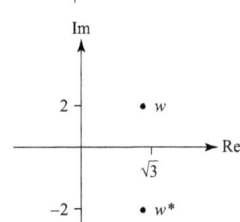

16 a Two real and two complex roots

b $x = -5$ (repeated), $2+i, 2-i$

17 $x^3 - (a+b+c)x^2 + (ab+ac+bc)x - abc = 0$

18 a -5 **b** -12

19 a $-\alpha - \beta - \gamma - \delta$ **b** $\alpha\beta\gamma\delta$

20 $a = 2, b = -41, c = 336, d = -1318, e = 2262$

21 a Order 5 (quintic)

b $x^5 - 6x^4 + 10x^3 - 20x^2 + 9x + 306 = 0$

Exercise 1.3A

1 $\overrightarrow{OA} = 5+3i$

$\overrightarrow{OB} = -3-6i$

$\overrightarrow{OC} = 6$

$\overrightarrow{OD} = -1+2i$

$\overrightarrow{OE} = -4i$

$\overrightarrow{OF} = -6+5i$

2 $u = -3+5i, v = 2-7i, w = 4i, z = -4+i$

3

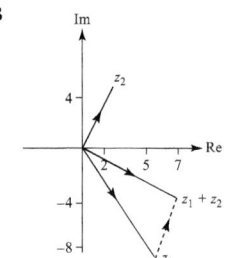

4

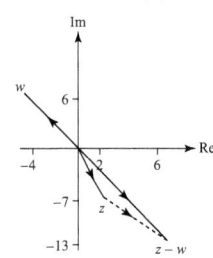

5

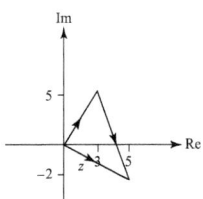

6

7

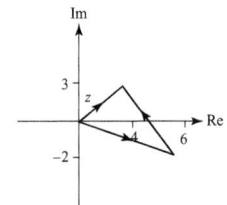

w^* is a reflection of w in the real axis.

8

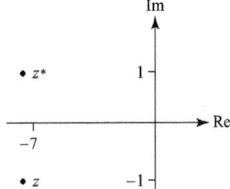

z is a reflection of z^* in the real axis.

9 **a** $x = \pm 4i$

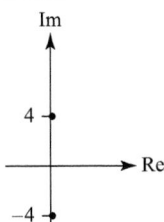

b $x = \pm 4\sqrt{5}i$

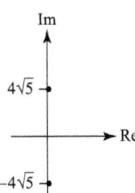

In both cases the points are reflections of each other in the real axis

10 a $z = -1 \pm \sqrt{3}i$

b

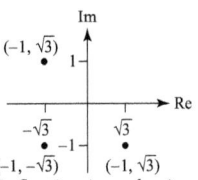

c Reflection in real axis.

11 a $z = 1 \mp 5i$

b

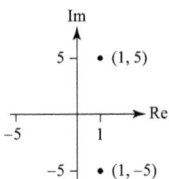

c Reflection in real axis.

12 a

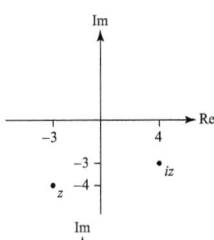

b

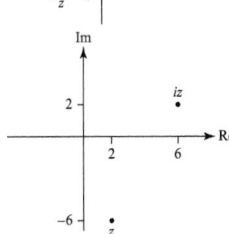

In both cases iz is the image of z rotated $\dfrac{\pi}{2}$ radians (anti–clockwise) about the origin.

13 a

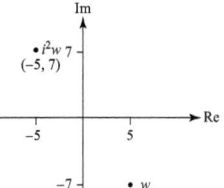

b

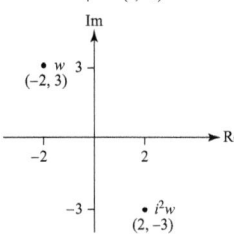

In both cases, i^2w is w rotated π radians around the origin.

Exercise 1.3B

1 **a** $z_2 = 2 + 5i$

b

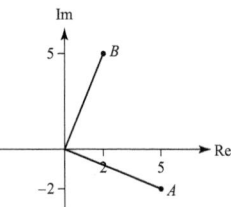

c Gradient of OA is $-\dfrac{2}{5}$, gradient of OB is $\dfrac{5}{2}$

$-\dfrac{2}{5} \times \dfrac{5}{2} = -1$ so OA and OB are perpendicular

$\therefore AOB$ is a right angle

2 **a** $x = -1 \pm \sqrt{7}i, 1$

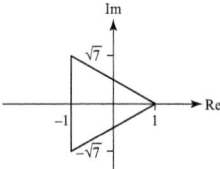

b Isosceles triangle

c $2\sqrt{7}$

3 6 square units

4 **a** $A = -2, B = 3$

b $x = -\dfrac{3}{2} \pm \dfrac{\sqrt{3}}{2}i, 1 \pm \sqrt{2}i$

c

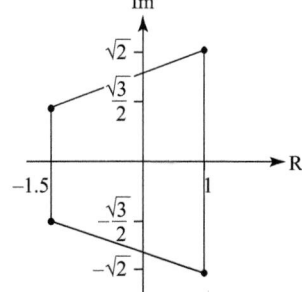

d A trapezium

e 5.70 square units

5 **a** $(-2)^4 + 8(-2)^3 + 40(-2)^2 + 96(-2) + 80 = 0$
$x^4 + 8x^3 + 40x^2 + 96x + 80 = (x+2)^2(x^2+4x+20) = 0$,
so $x = -2$ is repeated

b $x = -2 \pm 4i$

c

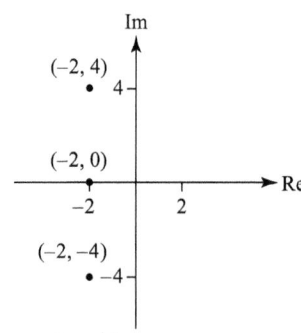

6 **a** $x = -1$ and 2
    b 9 square units

7 **a** $x^3 - x^2 + 9x - 9 = (x-1)(x^2+9) = 0$
$\Rightarrow x = 1, \pm 3i$
$A(1, 0)$, $B(0, 3)$, $C(0, -3)$
$|AB| = \sqrt{1^2 + 3^2} = \sqrt{10}$
$|AC| = \sqrt{1^2 + 3^2} = \sqrt{10}$
$|BC| = \sqrt{0^2 + 6^2} = 6$
So ABC is isosceles

    b 3 square units

8 **a** $A = (5, -2)$
$B = (7, 3)$
$C = (2, 5)$
Length $OA = \sqrt{5^2 + 2^2} = \sqrt{29}$
Length $AB = \sqrt{(7-5)^2 + (3+2)^2} = \sqrt{29}$
Length $BC = \sqrt{(7-2)^2 + (5-3)^2} = \sqrt{29}$
Length $OC = \sqrt{29}$
Gradient $OA = -\dfrac{2}{5}$
Gradient $AB = \dfrac{5}{2}$ So OA and AB are perpendicular
Gradient $BC = -\dfrac{2}{5}$ so AB and BC are perpendicular
Gradient $OC = \dfrac{5}{2}$ so OC and OA are perpendicular.
Therefore it is a square.

    b 29 square units

9 **a** Q represents $2 + 4i$, R represents $6 - 4i$,

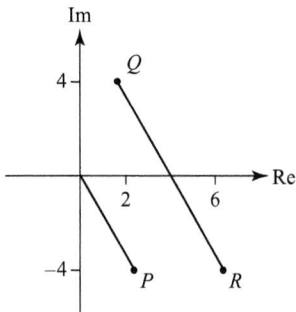

    b OP has been enlarged by scale factor 2 centre the origin then translated by the vector $\begin{pmatrix} 2 \\ 4 \end{pmatrix}$.

Or alternatively: OP has been enlarged by scale factor 2 centre $(-2, -4)$.

10 **a** $-2 + 9i$

    b CB is an enlargement of OA centre the origin, scale factor 2, then translated by the vector $\begin{pmatrix} -2 \\ 9 \end{pmatrix}$

11 $|a - c|(b + d)$

12

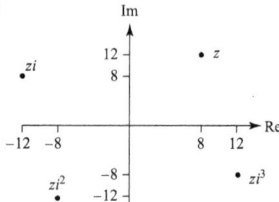

z is rotated around the origin by $\dfrac{\pi}{2}$ radians (anticlockwise) for zi, by π radians for zi^2 and by $\dfrac{3\pi}{2}$ radians (anticlockwise) for zi^3

13 Let $z = a + bi$
Then $iz = ai + bi^2 = -b + ai$
Gradient of $OA = \dfrac{b}{a}$
Gradient of $OB = \dfrac{a}{-b}$
$\dfrac{b}{a} \times \dfrac{a}{-b} = -1$ so OA is perpendicular to OB

14 17 square units

15 3 square units

16 10

17 **a** $\dfrac{z}{w} = \dfrac{12 - 5i}{3 + 2i} = \dfrac{(12 - 5i)(3 - 2i)}{(3 + 2i)(3 - 2i)}$
$= \dfrac{36 - 15i - 24i - 10}{9 + 4}$
$= \dfrac{26 - 39i}{13} = 2 - 3i$

    b

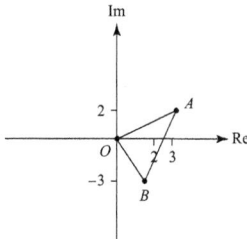

    c Gradient of $OB = -\dfrac{3}{2}$
Gradient of $OA = \dfrac{2}{3}$
$-\dfrac{3}{2} \times \dfrac{2}{3} = -1$ therefore OB and OA are perpendicular so OAB is a right-angled triangle.

    d $\dfrac{13}{2}$ square units

18

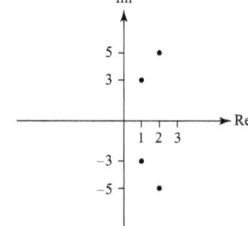

19 a i A kite

  ii An isosceles triangle

  iii A trapezium

 b i $a=1, b=0, c=0, d=0, e=-16$

  ii $a=1, b=-4, c=6, d=-4, e=-80$

Exercise 1.4A

1 a Modulus $=13$
  Argument $=0.395^c$

 b Modulus $=5$
  Argument $=-0.644^c$

 c Modulus $=3$
  Argument $=-\dfrac{\pi}{2}$

 d Modulus $=10$
  Argument $=-2.21^c$.

 e Modulus $=5\sqrt{2}$
  Argument $=1.71$

 f Modulus $=\sqrt{2^2+1^2}=\sqrt{5}$
  Argument $=-2.68^c$

 g Modulus $=2$
  Argument $=-\dfrac{3\pi}{4}$

 h Modulus $=3$
  Argument $=0.955^c$

2 a $zw=(1+3i)(-5+2i)=-11-13i$

$$|zw|=\sqrt{11^2+13^2}=\sqrt{290}$$

$$|z|=\sqrt{1^2+3^2}=\sqrt{10}$$

$$|w|=\sqrt{5^2+2^2}=\sqrt{29}$$

$$|z||w|=\sqrt{10}\sqrt{29}=\sqrt{290}=|zw|\text{ as required.}$$

$$\frac{z}{w}=\frac{1+3i}{-5+2i}$$

$$=\frac{(1+3i)(5+2i)}{(-5+2i)(5+2i)}=\frac{1}{29}-\frac{17}{29}i$$

$$\left|\frac{z}{w}\right|=\sqrt{\left(\frac{1}{29}\right)^2+\left(\frac{17}{29}\right)^2}=\sqrt{\frac{10}{29}}$$

$$\frac{|z|}{|w|}=\frac{\sqrt{10}}{\sqrt{29}}=\sqrt{\frac{10}{29}}=\left|\frac{z}{w}\right|\text{ as required}$$

 b $zw=(-2-i)(\sqrt{5}i)=\sqrt{5}-2\sqrt{5}i$

$$|zw|=\sqrt{(\sqrt{5})^2+(2\sqrt{5})^2}=5$$

$$|z|=\sqrt{2^2+1^2}=\sqrt{5}$$

$$|w|=\sqrt{5}$$

$$|z||w|=\sqrt{5}\sqrt{5}=5=|zw|\text{ as required.}$$

$$\frac{z}{w}=\frac{-2-i}{\sqrt{5}i}$$

$$=\frac{(-2-i)i}{\sqrt{5}i\cdot i}=-\frac{1}{\sqrt{5}}+\frac{2}{\sqrt{5}}i$$

$$\left|\frac{z}{w}\right|=\sqrt{\left(\frac{1}{\sqrt{5}}\right)^2+\left(\frac{2}{\sqrt{5}}\right)^2}=1$$

$$\frac{|z|}{|w|}=\frac{\sqrt{5}}{\sqrt{5}}=1=\left|\frac{z}{w}\right|\text{ as required}$$

 c $zw=(-\sqrt{3}+6i)(1-\sqrt{3}i)=5\sqrt{3}+9i$

$$|zw|=\sqrt{(5\sqrt{3})^2+(9)^2}=2\sqrt{39}$$

$$|z|=\sqrt{(\sqrt{3})^2+6^2}=\sqrt{39}$$

$$|w|=\sqrt{1^2+(\sqrt{3})^2}=2$$

$$|z||w|=2\sqrt{39}=|zw|\text{ as required.}$$

$$\frac{z}{w}=\frac{-\sqrt{3}+6i}{1-\sqrt{3}i}$$

$$=\frac{(-\sqrt{3}+6i)(1+\sqrt{3}i)}{(1-\sqrt{3}i)(1+\sqrt{3}i)}=-\frac{7\sqrt{3}}{4}+\frac{3}{4}i$$

$$\left|\frac{z}{w}\right|=\sqrt{\left(\frac{7\sqrt{3}}{4}\right)^2+\left(\frac{3}{4}\right)^2}=\frac{\sqrt{39}}{2}$$

$$\frac{|z|}{|w|}=\frac{\sqrt{39}}{2}=\left|\frac{z}{w}\right|\text{ as required}$$

3 a $zw=(1+i)(3+\sqrt{3}i)=(3-\sqrt{3})+(3+\sqrt{3})i$

$$\arg(zw)=\tan^{-1}\left(\frac{3+\sqrt{3}}{3-\sqrt{3}}\right)=\frac{5\pi}{12}$$

$$\arg z=\tan^{-1}1=\frac{\pi}{4}$$

$$\arg w=\tan^{-1}\frac{\sqrt{3}}{3}=\frac{\pi}{6}$$

$$\arg z+\arg w=\frac{\pi}{4}+\frac{\pi}{6}=\frac{5\pi}{12}\text{ as required}$$

$$\frac{z}{w}=\frac{1+i}{3+\sqrt{3}i}$$

$$=\frac{(1+i)(3-\sqrt{3}i)}{(3+\sqrt{3}i)(3-\sqrt{3}i)}=\frac{3+\sqrt{3}}{12}+\frac{3-\sqrt{3}}{12}i$$

$$\arg\left(\frac{z}{w}\right)=\tan^{-1}\frac{3-\sqrt{3}}{3+\sqrt{3}}=\frac{\pi}{12}$$

$$\arg z-\arg w=\frac{\pi}{4}-\frac{\pi}{6}=\frac{\pi}{12}\text{ as required}$$

 b $zw=i(2-2i)=2+2i$

$$\arg(zw)=\tan^{-1}\left(\frac{2}{2}\right)=\frac{\pi}{4}$$

$$\arg z=\frac{\pi}{2}$$

$$\arg w=-\tan^{-1}\frac{2}{2}=-\frac{\pi}{4}$$

$$\arg z+\arg w=\frac{\pi}{2}+\frac{-\pi}{4}=\frac{\pi}{4}\text{ as required}$$

$$\frac{z}{w}=\frac{i}{2-2i}$$

$$=\frac{i(2+2i)}{(2-2i)(2+2i)}=-\frac{1}{4}+\frac{1}{4}i$$

$$\arg\left(\frac{z}{w}\right)=\pi-\tan^{-1}\frac{0.25}{0.25}=\frac{3\pi}{4}$$

$$\arg z-\arg w=\frac{\pi}{2}-\frac{-\pi}{4}=\frac{3\pi}{4}\text{ as required}$$

 c $zw=-2i(\sqrt{3}-3i)=-6-2\sqrt{3}i$

$$\arg(zw)=-\pi+\tan^{-1}\left(\frac{2\sqrt{3}}{6}\right)=-\frac{5\pi}{6}$$

$$\arg z=-\tan^{-1}\frac{3}{\sqrt{3}}=-\frac{\pi}{3}$$

$$\arg w=-\frac{\pi}{2}$$

$$\arg z+\arg w=-\frac{\pi}{3}+\frac{-\pi}{2}=-\frac{5\pi}{6}\text{ as required}$$

$$\frac{z}{w}=\frac{\sqrt{3}-3i}{-2i}$$

$$=\frac{i(\sqrt{3}-3i)}{-2i\cdot i}=\frac{3}{2}+\frac{\sqrt{3}}{2}i$$

$$\arg\left(\frac{z}{w}\right)=\tan^{-1}\frac{\sqrt{3}}{3}=\frac{\pi}{6}$$

$$\arg z-\arg w=-\frac{\pi}{3}-\frac{-\pi}{2}=\frac{\pi}{6}\text{ as required}$$

4 $|w| = \sqrt{3}$

$\arg(w) = \dfrac{2\pi}{3}$

5 $|z_2| = \dfrac{\sqrt{3}}{6}$

$\arg(z_2) = -\dfrac{11}{12}\pi$

6 $|w| = 8\sqrt{3}$

$\arg(w) = \dfrac{\pi}{2}$

7 **a** $3i$ **b** -5 **c** $-5\sqrt{3} + 5i$ **d** $-\dfrac{\sqrt{3}}{2} - \dfrac{3}{2}i$

8 **a** $z = 3\sqrt{2}\left(\cos\dfrac{\pi}{4} + i\sin\dfrac{\pi}{4}\right)$

b $z = 2\left(\cos\left(-\dfrac{\pi}{3}\right) + i\sin\left(-\dfrac{\pi}{3}\right)\right)$

c $z = 4\left(\cos\left(-\dfrac{5\pi}{6}\right) + i\sin\left(-\dfrac{5\pi}{6}\right)\right)$

d $z = \sqrt{97}\left(\cos(1.99) + i\sin(1.99)\right)$

9 a

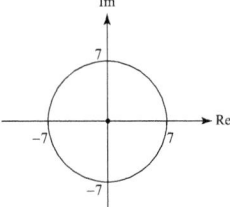

b

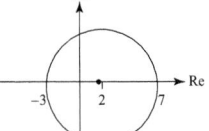

c

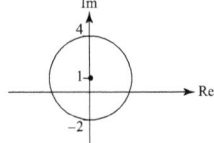

d

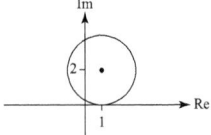

e

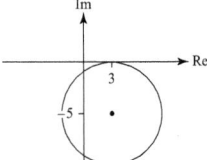

f

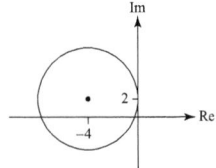

10 a

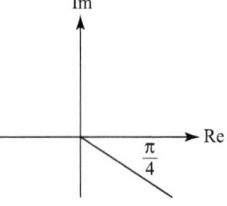

b

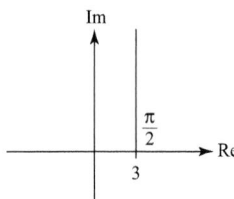

c

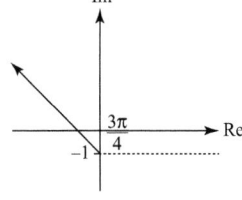

d

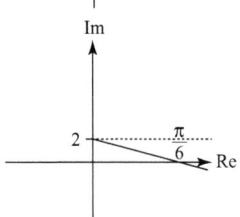

e

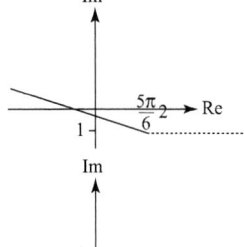

f

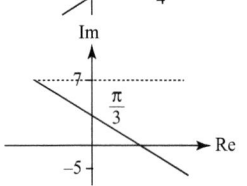

g

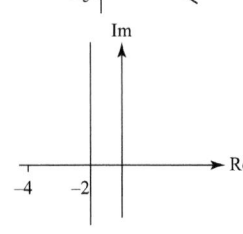

11 a

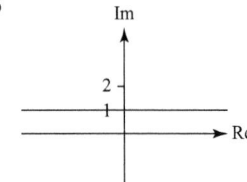

b

c

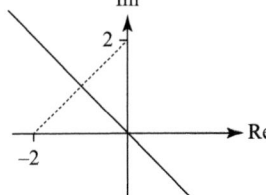

d

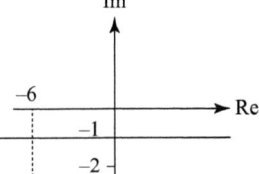

e

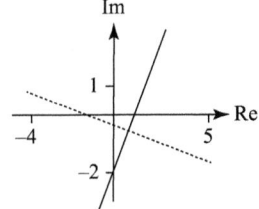

Exercise 1.4B

1 Let $z = |z|(\cos A + i \sin A)$ and $w = |w|(\cos B + i \sin B)$

Then $zw = |z|(\cos A + i \sin A)|w|(\cos B + i \sin B)$

$\qquad = |z||w|(\cos A \cos B + i \sin A \cos B + i \sin B \cos A$
$\qquad\quad + i^2 \sin A \sin B)$

$\qquad = |z||w|(\cos A \cos B - \sin A \sin B$
$\qquad\quad + i(\sin A \cos B + \sin B \cos A))$

$\qquad = |z||w|(\cos(A + B) + i \sin(A + B))$

So $|zw| = |z||w|$

And $\arg(zw) = A + B = \arg z + \arg w$ as required

2 Let $z = x + iy$

Then $|x + iy + 3 - 2i| = 4$

$(x + 3)^2 + (y - 2)^2 = 16$

Circle centre $(-3, 2)$ and radius 4

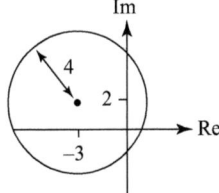

3 a Let $z = x + iy$

$|x + iy - 2 - i| = 2$

$(x - 2)^2 + (y - 1)^2 = 4$

Circle centre $(2, 1)$ and radius 2

  b $\sqrt{3}$ square units

4 a $x = 2.5$ **b** $y = -x$

  c $x + 2y = 3$ **d** $8x - 2y = 7$

  e $2x = 7$ **f** $13x - 7y = 10$

5 a $y = x + 3$ **b** $x = -5$

  c $y = \sqrt{3}x + 1 + 2\sqrt{3}$ **d** $y = -\sqrt{3}x - 1 + 4\sqrt{3}$

6 $y = 5x + 3$

7 $y = \sqrt{2} - x$

8 $x - 2y + 3 = 0$

9 a

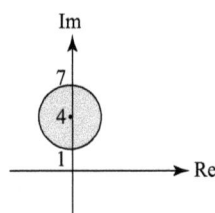

  b

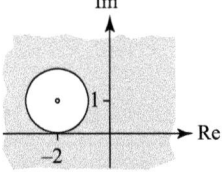

  c

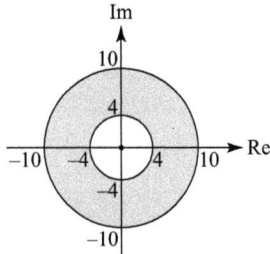

  d

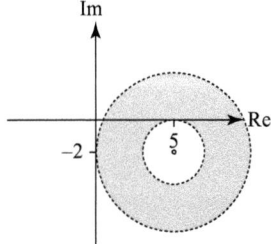

10 42π square units

11 a

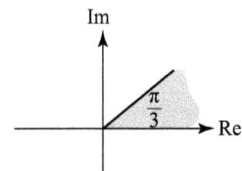

  b

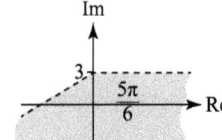

  c

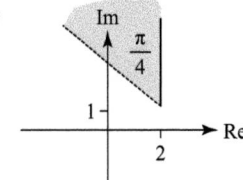

12 16π square units

13 a

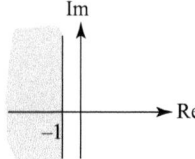

b

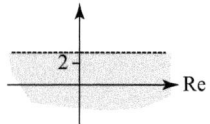

c

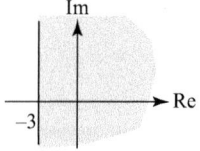

d

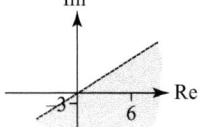

e

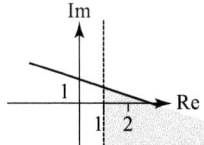

14

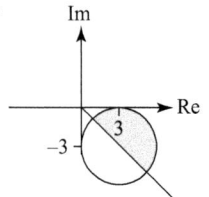

15

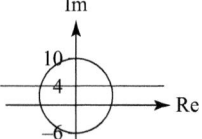

16

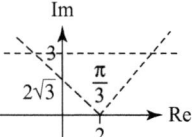

17

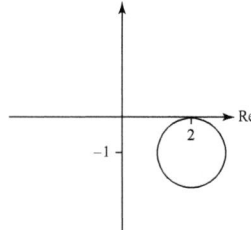

18

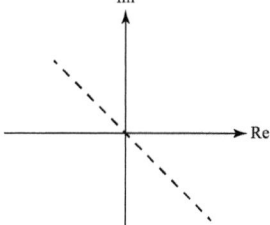

19

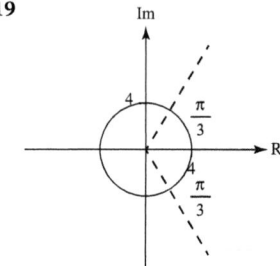

20

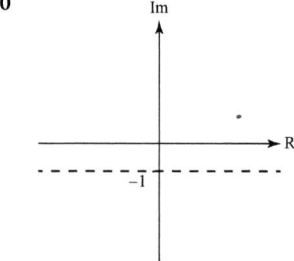

21 a Let $z = x + iy$

$$\left|x + iy + 3\right| = 2\left|x + iy - 6i\right|$$
$$\left|x + iy + 3\right|^2 = 4\left|x + iy - 6i\right|^2$$
$$(x + 3)^2 + y^2 = 4x^2 + 4(y - 6)^2$$
$$x^2 + 6x + 9 + y^2 = 4x^2 + 4y^2 - 48y + 144$$
$$3x^2 - 6x + 3y^2 - 48y + 135 = 0$$
$$x^2 - 2x + y^2 - 16y + 45 = 0$$
$$(x - 1)^2 - 1 + (y - 8)^2 - 64 + 45 = 0$$
$$(x - 1)^2 + (y - 8)^2 = 20$$

Circle centre (1, 8) radius $2\sqrt{5}$

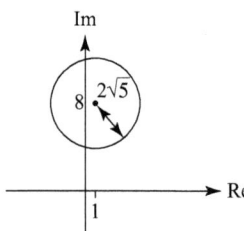

b

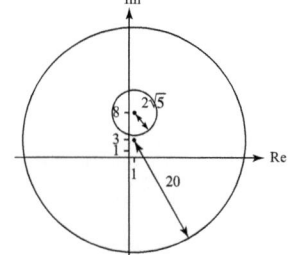

22 a $(x-3)^2+(y+1)^2=9$

b $x=y$

c $z=\dfrac{2+\sqrt{2}}{2}+\dfrac{2+\sqrt{2}}{2}i$ or $z=\dfrac{2-\sqrt{2}}{2}+\dfrac{2-\sqrt{2}}{2}i$

23 $z=2\sqrt{2}+(3-2\sqrt{2})i$

Review exercise 1

1 a $3-7i$ **b** $-6-9i$ **c** $12-5i$

d $-22-7i$ **e** $4-28i$ **f** $-40-42i$

g $5+4i$ **h** $-2+3i$ **i** $\dfrac{2}{13}+\dfrac{23}{13}i$

j $\dfrac{3}{41}-\dfrac{69}{82}i$ **k** $\dfrac{10}{41}+\dfrac{8}{41}i$ **l** $1+\dfrac{2}{3}i$

2 $a=1,\ b=14,\ c=53$

3 a -16 **b** $z=7\pm4i,\ 2$

4 $x=3\pm2i$ and $x=-2\pm3i$

5 a Modulus $=\sqrt{85}$ **b** Modulus $=3\sqrt{2}$
   Argument $=1.35^c$ Argument $=-\dfrac{\pi}{4}$

c Modulus $=7$ **d** Modulus $=2$
   Argument $=\dfrac{\pi}{2}$ Argument $=-\dfrac{\pi}{2}$

e Modulus $=\sqrt{17}$ **f** Modulus $=5$
   Argument $=1.82^c$ Argument $=-2.21^c$

6 a $10(\cos(0.644)+i\sin(0.644))$

b $13(\cos(2.75)+i\sin(2.75))$

c $2\sqrt{2}\left(\cos\left(-\dfrac{3\pi}{4}\right)+i\sin\left(-\dfrac{3\pi}{4}\right)\right)$

d $2\left(\cos\left(-\dfrac{\pi}{6}\right)+i\sin\left(-\dfrac{\pi}{6}\right)\right)$

e $\sqrt{5}\left(\cos\left(-\dfrac{\pi}{3}\right)+i\sin\left(-\dfrac{\pi}{3}\right)\right)$

7 a $\sqrt{3}+i$ **b** $\dfrac{\sqrt{6}}{2}-\dfrac{\sqrt{6}}{2}i$

8 a $3\sqrt{2}\left(\cos\left(-\dfrac{\pi}{6}\right)+i\sin\left(-\dfrac{\pi}{6}\right)\right)$

b $\sqrt{2}\left(\cos\left(-\dfrac{\pi}{2}\right)+i\sin\left(-\dfrac{\pi}{2}\right)\right)$

c $\dfrac{\sqrt{2}}{2}\left(\cos\left(\dfrac{\pi}{2}\right)+i\sin\left(\dfrac{\pi}{2}\right)\right)$

9

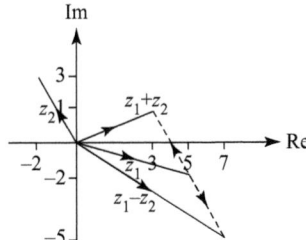

10 a

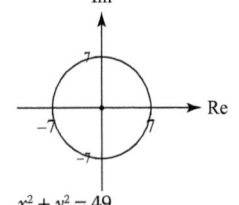

$x^2+y^2=49$

b

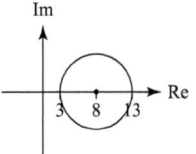

$(x-8)^2+y^2=25$

c

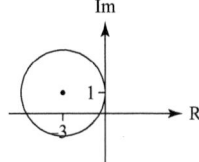

$(x+3)^2+(y-1)^2=9$

d

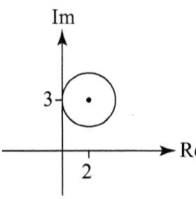

$(x-2)^2+(y-3)^2=4$

11 a

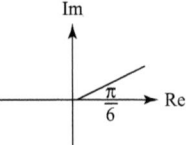

b

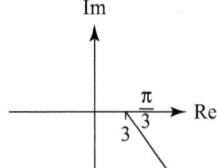

c

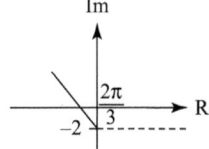

d

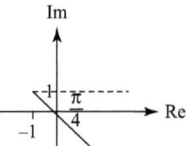

12 a

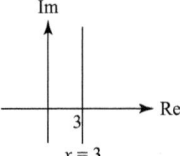

$x=3$

b

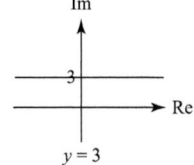

$y=3$

c

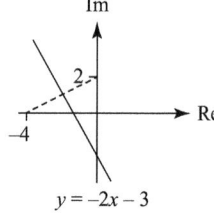

$y = -2x - 3$

d

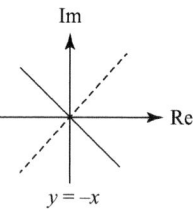

$y = -x$

13 a

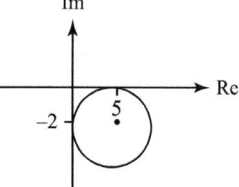

b

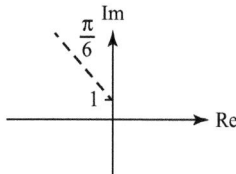

c

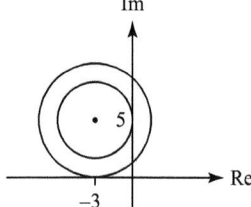

d

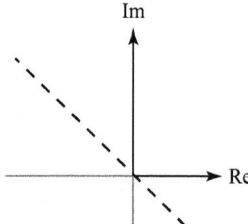

14

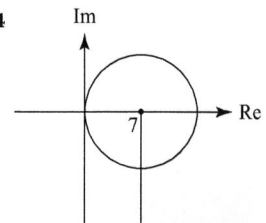

Assessment 1

1 a i $-2 + 14i$ **ii** $96 - 50i$ **iii** $\dfrac{15}{101} + \dfrac{52}{101}i$
  b $|z| = 14.2$, $\arg z = -0.885^c$

2 a $-1 + 2i, -1 - 2i$

b

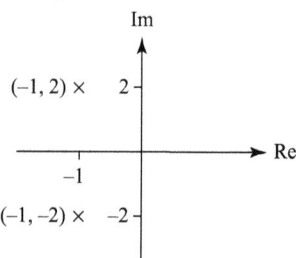

c Reflection in real axis

3 a $z = \dfrac{(5 + 2i)(3 - i)}{(3 + i)(3 - i)}$

$= \dfrac{15 - 5i + 6i - 2i^2}{9 - i^2}$

$= \dfrac{17 + i}{10}$

$= \dfrac{17}{10} + \dfrac{1}{10}i$

$\left(a = \dfrac{17}{10}, b = \dfrac{1}{10} \right)$

b i $\dfrac{17}{5}$ **ii** $\dfrac{1}{5}i$ **iii** $\dfrac{29}{10}$

c i

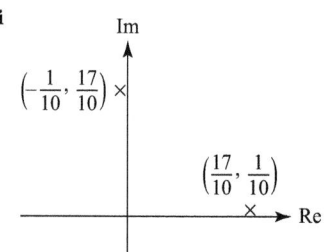

  ii iz is a rotation of z 90° anticlockwise around origin.

4 $b = \pm 4$, $a = \pm 1$

5 $b = \pm 1$, $a = \pm\sqrt{2}$

6 a $2 \pm 3i, -2$

  b 12 square units

7 a $5 + i$, 7

  b $a = -17$, $b = 96$

8 $z = 6 + 9i$

9 $w = -3 - 7i$, $z = 2 - 9i$

10 a $|z| = 6$, $\arg z = \dfrac{\pi}{3}$

  b $|w| = 2$, $\arg z = \dfrac{3\pi}{4}$

  c i $|zw| = 12$

   ii $\left| \dfrac{z}{w} \right| = 3$

   iii $\arg(zw) = -\dfrac{11\pi}{12}$

   iv $\arg\left(\dfrac{z}{w} \right) = -\dfrac{5\pi}{12}$

11 a $w = \sqrt{3}\left(\cos(-0.615) + i\sin(-0.615) \right)$

  b $\sqrt{33}$

12 a $z = 2\left(\cos\left(-\dfrac{\pi}{6} \right) + i\sin\left(-\dfrac{\pi}{6} \right) \right)$

  b i $\arg(zw) = -\dfrac{\pi}{4}$ **ii** $\arg\left(\dfrac{z}{w} \right) = -\dfrac{\pi}{12}$

  c $|w| = 5$

13 a   i

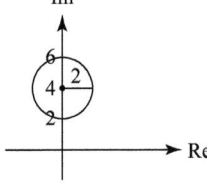

ii

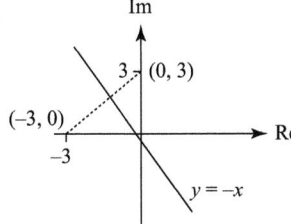

b   i $x^2 + (y-4)^2 = 4$
   ii $y = -x$

14 a $|x + iy + 3 - 4i| = 4$
   $(x+3)^2 + (y-4)^2 = 16$

   Therefore a circle (with centre $(-3, 4)$ and radius 4)

b, c

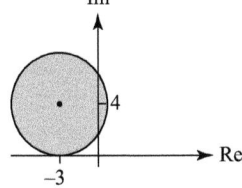

15 a

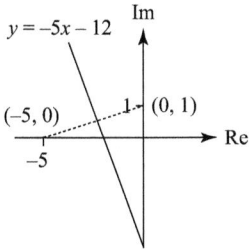

b $y = -5x - 12$

16 $-8 + 5i$ and $8 - 5i$

17 $z = \sqrt{5} - i$ or $-\sqrt{5} + i$

18 28 square units

19 a $-i - 4$
   b $x^3 + 11x^2 + 41x + 51 = 0$

20 a $z = -2 - i$ and $z = 3 + 5i$
   b $z^4 - 2z^3 + 15x^2 + 106z + 170 = 0$

21 So $a = 1$, $b = -4$, $c = 56$, $d = -104$, e $= 676$

22 $w = 2 \pm i$, $z = -1 \mp 2i$

23 $w = 6 \pm i$, $z = 2 \mp 3i$

24 a $2(3i)^5 + (3i)^4 + 36(3i)^3 + 18(3i)^2 + 162(3i) + 81$
   $= 486i + 81 - 972i - 162 + 486i + 81$
   $= 0$ so $3i$ is a solution
   b $(x^2 + 9)^2(2x + 1)$

25 $x = -1 \pm i$, 3 (repeated)

26 a $\beta = 3\left(\cos\left(-\dfrac{5\pi}{6}\right) + i\sin\left(-\dfrac{5\pi}{6}\right)\right)$

   b   i $|\alpha\beta| = 9$ **ii** $\left|\dfrac{\alpha}{\beta}\right| = 1$ **iii** $\arg(\alpha\beta) = 0$

   c $x^2 + 3\sqrt{3}x + 9 = 0$

27 a $\dfrac{z_1}{z_2} = \dfrac{(6-a)}{5} + \dfrac{(3+2a)}{5}i$

   b $a = \pm 9$

28 a $w = \dfrac{1}{2}\left(\cos\left(-\dfrac{\pi}{3}\right) + i\sin\left(-\dfrac{\pi}{3}\right)\right)$

   b $z = \dfrac{\sqrt{3}}{24} + \dfrac{1}{24}i$

29 Circle, centre $(3, -1)$, radius 1

30 a

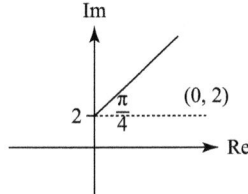

   b $y = x + 2$

31

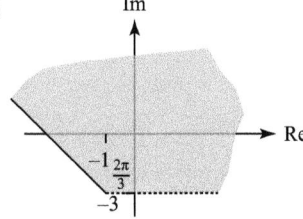

32

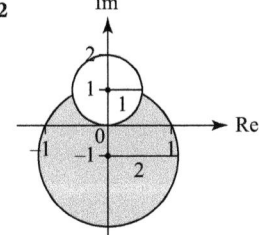

33 5 square units

34 $\arg(z - (\sqrt{3} + i)) = -\dfrac{5\pi}{6}$

35 $|x + yi + 2| = 3|x + yi - 2i|$
   $(x+2)^2 + y^2 = 9(x^2 + (y-2)^2)$
   $x^2 + 4x + 4 + y^2 = 9x^2 + 9y^2 - 36y + 36$
   $8x^2 - 4x + 8y^2 - 36y + 32 = 0$
   $x^2 - \dfrac{x}{2} + y^2 - \dfrac{9}{2}y + 4 = 0$
   $\left(x - \dfrac{1}{4}\right)^2 + \left(y - \dfrac{9}{4}\right)^2 = \dfrac{9}{8}$

   So a circle, centre $\left(\dfrac{1}{4}, \dfrac{9}{4}\right)$ and radius $\dfrac{3}{4}\sqrt{2}$

36 a

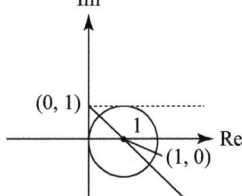

   b $z = \left(\dfrac{2-\sqrt{2}}{2}\right) + \dfrac{\sqrt{2}}{2}i$ or $z = \left(\dfrac{2+\sqrt{2}}{2}\right) - \dfrac{\sqrt{2}}{2}i$

37 a $z = 6 - i$ or $z = -2 - i$

b

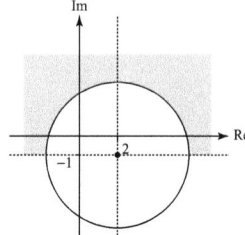

38 a

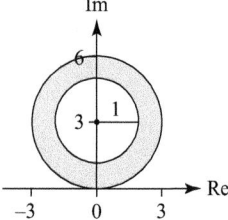

b 8π

39 a

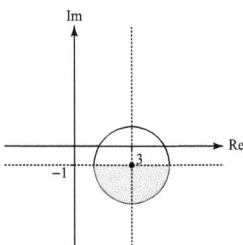

b 2π

Chapter 2
Exercise 2.1A

1 The sums and products are:

  a 5 and 9   **b** −6 and 7   **c** 8 and −12

  d −10 and −5   **e** 4 and $\dfrac{8}{3}$ **f** $-\dfrac{1}{4}$ and $\dfrac{3}{2}$

2 The sum, sum of the products in pairs and the product of the roots are:

  a −4, −9 and 14   **b** 7, −11 and −12

  c 13, 22 and 26   **d** $\dfrac{-5}{2}, \dfrac{17}{2}$ and $\dfrac{21}{2}$

  e $\dfrac{1}{4}, \dfrac{3}{4}$ and −2   **f** $\dfrac{-3}{4}, \dfrac{-3}{2}$ and $\dfrac{-5}{8}$

3 a $-\dfrac{k}{3}$ **b** $-\dfrac{k}{3} - \dfrac{1}{3}$ **c** $\dfrac{k^2}{9} + \dfrac{8}{3}$

  d $\dfrac{44}{3} + \dfrac{4k}{3}$ **e** $-\dfrac{k^3}{27} - \dfrac{4k}{3}$ **f** $\dfrac{k}{4}$

4 $x^3 - 5x^2 + 4x + 2 = 0$

5 a i 26   **ii** 0

  b i 30   **ii** $-\dfrac{11}{12}$

6 a $-\dfrac{a}{2}$ **b** $\dfrac{b}{2}$ **c** $-\dfrac{c}{2}$ **d** $\dfrac{a^2}{4} - b$

  e $-\dfrac{a^3}{8} + \dfrac{3ab}{4} - \dfrac{3c}{2}$ **f** $1 - \dfrac{a}{2} + \dfrac{b}{2} - \dfrac{c}{2}$ **g** $-\dfrac{b}{c}$ **h** $\dfrac{b^2}{4} - \dfrac{ac}{2}$

7 $ax^4 + bx^3 + cx^2 + dx + e = a(x - \alpha)(x - \beta)(x - \gamma)(x - \delta)$

$= a(x^4 - \alpha x^3 - \beta x^3 - \gamma x^3 - \delta x^3 + \alpha\beta x^2 + \alpha\gamma x^2 + \alpha\delta x^2$
$+ \beta\gamma x^2 + \beta\delta x^2 + \gamma\delta x^2 - \alpha\beta\gamma x - \alpha\beta\delta x - \alpha\gamma\delta x - \beta\gamma\delta x + \alpha\beta\gamma\delta)$

$= a(x^4 - (\alpha + \beta + \gamma + \delta)x^3 + (\alpha\beta + \alpha\gamma + \alpha\delta + \beta\gamma + \beta\delta + \gamma\delta)x^2$
$- (\alpha\beta\gamma + \alpha\beta\delta + \alpha\gamma\delta + \beta\gamma\delta)x + \alpha\beta\gamma\delta)$

Therefore,

  a $-a(\alpha + \beta + \gamma + \delta) = b \Rightarrow \alpha + \beta + \gamma + \delta = -\dfrac{b}{a}$

  b $a(\alpha\beta + \alpha\gamma + \alpha\delta + \beta\gamma + \beta\delta + \gamma\delta) = c$

    $\Rightarrow \alpha\beta + \alpha\gamma + \alpha\delta + \beta\gamma + \beta\delta + \gamma\delta = \dfrac{c}{a}$

  c $-a(\alpha\beta\gamma + \alpha\beta\delta + \alpha\gamma\delta + \beta\gamma\delta) = d$

    $\Rightarrow \alpha\beta\gamma + \alpha\beta\delta + \alpha\gamma\delta + \beta\gamma\delta = -\dfrac{d}{a}$

  d $a(\alpha\beta\gamma\delta) = e \Rightarrow \alpha\beta\gamma\delta = \dfrac{e}{a}$

8 a $\dfrac{k^2}{9} - \dfrac{14}{3}$

  b $\dfrac{2}{k}$

Exercise 2.1B

1 a $y^3 + 5y^2 + 3y + 1 = 0$

  b $y^3 + 9y^2 + 45y + 27 = 0$

2 a $y^3 + 4y^2 + 3y = 0$

  b $y = 0, -1$ or −3

  c $x = 2, 1$ or −1

3 $8y^3 - 4y^2 - 14y + 9 = 0$

4 $y^4 + 16y^3 + 42y^2 + 96y + 197 = 0$

5 a $y^3 + 3y^2 - 18y - 56 + k = 0$

  b $x = 4, -2, 7$ (since $x = y + 4$)

6 a $x = 3 \pm i, 5$

  b $m = -11$

    $n = 40$

7 a 24

  b $4 + 24y - 11y^2 + 119y^3 = 0$

8 a $x = -\dfrac{7}{2}, 9 \pm 2i$

  b $p = -115$

    $q = 1498$

9 $b = 5$

10 $\alpha^3 = 4\alpha + 3; \beta^3 = 4\beta + 3$ and $\gamma^3 = 4\gamma + 3$

  Hence $\alpha^3 + \beta^3 + \gamma^3 = 4(\alpha + \beta + \gamma) + 9$

  But from $x^3 = 4x + 3$ or $x^3 - 4x - 3 = 0$, $\alpha + \beta + \gamma = 0$

  Hence $\alpha^3 + \beta^3 + \gamma^3 = 4(0) + 9 = 9$

11 $2y^3 - 4y^2 - 4y + 9 = 0$

12 a 4

  b $-y^3 + 8y^2 - 15y + 3 = 0$ or $y^3 - 8y^2 + 15y - 3 = 0$

  c 3

Exercise 2.2A

1 a If $a = -3$ and $b = 4$ then

    $-\dfrac{1}{3} < \dfrac{1}{4}$

    but if $a = 3$ and $b = 4$ then $\dfrac{1}{3} > \dfrac{1}{4}$

  b If $a = 2, b = 4, x = 5$ and $y = 6$ then

    $2 - 4 < 5 - 6$ (−2 < −1)

    but if $a = 2, b = 3, x = 5$ and $y = 7$

    then $2 - 3 > 5 - 7$ (−1 > −2)

  c If $a = 2, b = 4, x = 5$ and $y = 6$ then $2 \times 4 < 5 \times 6$ (8 < 30),

    but if $a = -2, b = -4, x = -1$ and $y = 6$ then

    $-2 \times -4 > -1 \times 6$ (8 > −6)

2 a $x > 0$ **b** $x > 0$ or $x \leq -3$

  c $x < -2$ or $x \geq -\dfrac{1}{2}$ **d** $2 < x < 11$

3 a

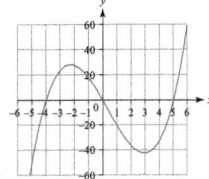

$-4 < x < 0 \text{ or } x > 5$

b

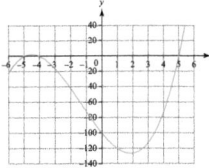

$x < -5 \text{ or } -4 < x < 5$

c

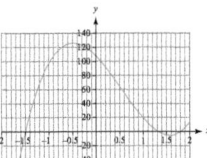

$x \le -\dfrac{3}{2} \text{ or } \dfrac{4}{3} \le x \le \dfrac{9}{5}$

d

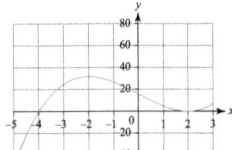

$x \ge -4$

e

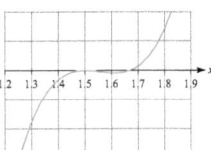

$x > \dfrac{5}{3}$

f

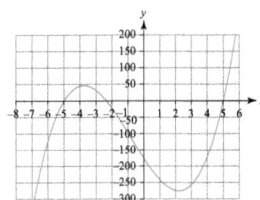

$x < -5 \text{ or } -\dfrac{7}{3} < x < 5$

g

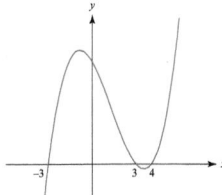

$x \le -3 \text{ or } 3 \le x \le 4$

4 a $x < -1 \text{ or } 0 < x < 2 \text{ or } x > 9$

b $-2 < x < -1 \text{ or } 3 < x < 7$

c $-\dfrac{1}{2} \le x \le \dfrac{5}{4} \text{ or } \dfrac{4}{3} \le x \le 11$

d $(x-1)^2(x-3)^2 \ge 0 \rightarrow \{ x : x \in \mathrm{R} \}$

e $(2x-7)^2(3x+2)^2 > 0 \rightarrow \{ x : x \in \mathrm{R} \}$

f $-7 < x < -\sqrt{5} \text{ or } \sqrt{5} < x < 7$

g $-3 < x < -2 \text{ or } 4 < x < 5$

5 a $x > 2 \text{ or } x < 0$

b $x > 2 \text{ or } x < -3$

c $x > 5 \text{ or } x < 2$

d $-2 < x < -1 \text{ or } x > 0$

e $-5 < x < -3 \text{ or } x > 4$

f $-\dfrac{4}{3} < x < \dfrac{3}{2} \text{ or } x > 7$

g $-\dfrac{3}{2} < x < 4 \text{ or } x < -\dfrac{5}{2}$

h $-2 < x < \dfrac{2}{3} \text{ or } x > \dfrac{3}{2}$

6 a $-1 < x < 0$

b $-4 < x < -3$

c $-2 < x < 3$

d $x < -3 \text{ or } 0 < x < 5$

e $x < -7 \text{ or } -6 < x < -4$

f $x < -6 \text{ or } -\dfrac{5}{2} < x < \dfrac{3}{4}$

g $-\dfrac{6}{5} < x < \dfrac{1}{2} \text{ or } x > \dfrac{9}{2}$

h $x < -\dfrac{5}{2} \text{ or } -1 < x < 2$

Exercise 2.2B

1 a $(-3,0), (2.5,0), (7,0) \text{ and } (0,105)$

b $(x+3)(2x-5)(x-7) = 0$
$\rightarrow 2x^3 - 13x^2 - 22x + 105 = 0$

c $A = 2, B = -13, C = -22 \text{ and } D = 105$

d $(-0.7, 115) \text{ and } (5.1, -80)$

e $x < -3 \text{ or } 2.5 < x < 7$

2 a $(-3,0), (2,0), (7,0) \text{ and } (0,-84)$
The turning point at $(2,0)$ means there is a repeated root.

b The equation of the function is $(x+3)(x-7)(x-2)^2 = 0$
$x^4 - 8x^3 - x^2 + 68x - 84 = 0$

c $A = 1, B = -8, C = -1, D = 68 \text{ and } E = -84$

d $(-1.5, -156), (2,0) \text{ and } (5.5, -156)$

e $x < -3 \text{ or } x > 7$

3 $0 \le x \le 2 \text{ and } 6 \le x$

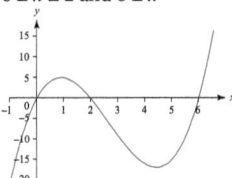

4 $-4 \le x \le -2; 2 \le x \le 4$

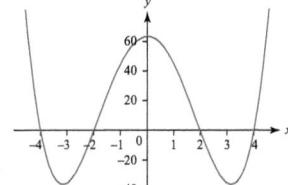

5 a $x > 3 \text{ or } -3 < x < 2$

b $-\dfrac{11}{4} < x < 0; x < -4$

c $x \ge 5 \text{ or } 3 \le x < 4$

d $-6 \le x \le -2; x > -1$

6 $-3 < x < 2 \text{ or } x > 7$

7 $-5 \le x \le -3; 3 \le x \le 5$

8 $0 < x < 1; x > 4$

9 a $y \geq -3 + \sqrt{8}; y \leq -3 - \sqrt{8}$

b
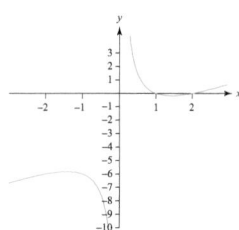

10 a $-2 < x < -1.5$ or $x > 4$

b
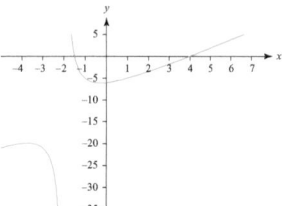

11 a $x \geq 0$

b
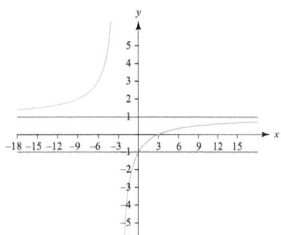

12 Let the minimum value of $x + \dfrac{1}{x}$ be k. Thus $x + \dfrac{1}{x} = k$, so

$x^2 - kx + 1 = 0$

If x is real then $k^2 - 4 \times 1 \times 1 \geq 0$, i.e. $k^2 \geq 4$ or $k \leq -2$ or $k \geq 2$

Since x is positive, $k \geq 2$

Alternatively:

Find x such that $x + \dfrac{1}{x} < 2$

$x + \dfrac{1}{x} - 2 < 0$

$\dfrac{x^2 + 1 - 2x}{x} < 0$

$\dfrac{(x-1)^2}{x} < 0$

The numerator is squared so is positive for all x.

So, the denominator must be negative, but this is not possible since $x > 0$

So there is no value of x for which $x + \dfrac{1}{x}$ is less than 2

Exercise 2.3A

1 a $\displaystyle\sum_{1}^{5} r^3 \equiv 1^3 + 2^3 + 3^3 + 4^3 + 5^3$

b $\displaystyle\sum_{4}^{n} r^2 \equiv 4^2 + 5^2 + \dots + n^2$

c $\displaystyle\sum_{1}^{n} (r^2 - 2r) \equiv (1^2 - 2 \times 1) + (2^2 - 2 \times 2) + (3^2 - 2 \times 3)$
$+ (4^2 - 2 \times 4) + \dots + (n^2 - 2 \times n)$
$= -1 + 0 + 3 + 8 + \dots + (n^2 - 2 \times n)$

d $\displaystyle\sum_{1}^{n} \dfrac{1}{m+2} \equiv \dfrac{1}{3} + \dfrac{1}{4} + \dfrac{1}{5} + \dfrac{1}{6} + \dots + \dfrac{1}{n+2}$

e $\displaystyle\sum_{1}^{6} (-1)^r r^3 \equiv -1 + 8 - 27 + 64 - 125 + 216$

f $\displaystyle\sum_{n-3}^{n} r(r+1) \equiv (n-3)(n-2) + (n-2)(n-1) + (n-1)(n)$
$+ (n)(n+1)$

2 a $\displaystyle\sum_{1}^{16} (2r - 1)$ **b** $\displaystyle\sum_{1}^{n} r^5$

c $\displaystyle\sum_{1}^{n+1} \dfrac{1}{r}$ **d** $\displaystyle\sum_{1}^{14} (-1)^{r+1} 3r$

e $\displaystyle\sum_{1}^{n} r(r+2)$ **f** $\displaystyle\sum_{1}^{n} (-1)^{r+1} \dfrac{(2r-1)(r+3)}{(r+2)}$

3 a $\displaystyle\sum_{r=1}^{n} (r^2 + 6r - 3) = \dfrac{1}{6} n(n+1)(2n+1) + 6 \times \dfrac{1}{2} n(n+1) - 3n$

$= n\left[\dfrac{1}{6}(n+1)(2n+1) + 3(n+1) - 3 \right]$

$= \dfrac{1}{6} n\left[(n+1)(2n+1) + 18(n+1) - 18 \right]$

$= \dfrac{1}{6} n(2n^2 + 3n + 1 + 18n + 18 - 18)$

$= \dfrac{1}{6} n(2n^2 + 21n + 1)$ as required

b $\displaystyle\sum_{r=1}^{n} (2r^2 - 7r) = 2 \times \dfrac{1}{6} n(n+1)(2n+1) - 7 \times \dfrac{1}{2} n(n+1)$

$= \dfrac{1}{3} n(n+1)(2n+1) - \dfrac{7}{2} n(n+1)$

$= n(n+1)\left[\dfrac{1}{3}(2n+1) - \dfrac{7}{2} \right]$

$= \dfrac{1}{6} n(n+1)\left[2(2n+1) - 21 \right]$

$= \dfrac{1}{6} n(n+1)(4n - 19)$ as required

c $\displaystyle\sum_{r=1}^{n} (8 - 5r^2) = 8n - 5 \times \dfrac{1}{6} n(n+1)(2n+1)$

$= n\left[8 - \dfrac{5}{6}(n+1)(2n+1) \right]$

$= \dfrac{1}{6} n\left[48 - 5(n+1)(2n+1) \right]$

$= \dfrac{1}{6} n(48 - 10n^2 - 15n - 5)$

$= -\dfrac{1}{6} n(10n^2 + 15n - 43)$ as required

d $\displaystyle\sum_{r=1}^{n} (r+1)^2 = \sum_{r=1}^{n} (r^2 + 2r + 1)$

$= \dfrac{1}{6} n(n+1)(2n+1) + 2 \times \dfrac{1}{2} n(n+1) + n$

$$= n\left[\frac{1}{6}(n+1)(2n+1)+(n+1)+1\right]$$

$$=\frac{1}{6}n\left[(n+1)(2n+1)+6(n+1)+6\right]$$

$$=\frac{1}{6}n(2n^2+3n+1+6n+6+6)$$

$$=\frac{1}{6}n(2n^2+9n+13)\text{ as required}$$

e $\displaystyle\sum_{r=1}^{n}(r^3+2r+3)=\frac{1}{4}n^2(n+1)^2+2\times\frac{1}{2}n(n+1)+3n$

$$=n\left[\frac{1}{4}n(n+1)^2+(n+1)+3\right]$$

$$=\frac{1}{4}n\left[n(n+1)^2+4(n+1)+12\right]$$

$$=\frac{1}{4}n(n^3+2n^2+n+4n+4+12)$$

$$=\frac{1}{4}n(n^3+2n^2+5n+16)\text{ as required}$$

f $\displaystyle\sum_{r=1}^{n}(2r^3+3r^2)=2\times\frac{1}{4}n^2(n+1)^2+3\times\frac{1}{6}n(n+1)(2n+1)$

$$=\frac{1}{2}n^2(n+1)^2+\frac{1}{2}n(n+1)(2n+1)$$

$$=\frac{1}{2}n(n+1)\left[n(n+1)+(2n+1)\right]$$

$$=\frac{1}{2}n(n+1)(n^2+n+2n+1)$$

$$=\frac{1}{2}n(n+1)(n^2+3n+1)\text{ as required}$$

g $\displaystyle\sum_{r=1}^{n}(r^3-2r^2+3r)$

$$=\frac{1}{4}n^2(n+1)^2-2\times\frac{1}{6}n(n+1)(2n+1)+3\times\frac{1}{2}n(n+1)$$

$$=n(n+1)\left[\frac{1}{4}n(n+1)-\frac{1}{3}(2n+1)+\frac{3}{2}\right]$$

$$=\frac{1}{12}n(n+1)\left[3n(n+1)-4(2n+1)+18\right]$$

$$=\frac{1}{12}n(n+1)(3n^2+3n-8n-4+18)$$

$$=\frac{1}{12}n(n+1)(3n^2-5n+14)\text{ as required}$$

h $\displaystyle\sum_{r=1}^{n}(r+2)^3=\sum_{r=1}^{n}r^3+6r^2+12r+8$

$$=\frac{1}{4}n^2(n+1)^2+6\times\frac{1}{6}n(n+1)(2n+1)+12\times\frac{1}{2}n(n+1)+8n$$

$$=n\left[\frac{1}{4}n(n+1)^2+(n+1)(2n+1)+6(n+1)+8\right]$$

$$=\frac{1}{4}n\left[n(n+1)^2+4(n+1)(2n+1)+24(n+1)+32\right]$$

$$=\frac{1}{4}n(n^3+2n^2+n+8n^2+12n+4+24n+24+32)$$

$$=\frac{1}{4}n(n^3+10n^2+37n+60)\text{ as required}$$

4 a There are $3n$ terms. **b** The $(2n+1)$th term is
$$(2n+1)^2-(2n+1)+3$$
$$=4n^2+2n+3$$

5 a $\dfrac{3n(3n+1)}{2}$ **b** $\dfrac{n(2n-1)(4n-1)}{3}$

c $n^2(2n-1)^2$ **d** n^2

e $\dfrac{(n+1)(n+2)}{3}$ **f** $\dfrac{n(n+1)(n-1)}{3}$

6 a $\dfrac{n}{3}(n^2-1)$ **b** 330

7 a rth term $=r(r+1)(r+2)$

Sum to n terms $=\dfrac{n(n+1)(n+2)(n+3)}{4}$

b rth term $=(2r-1)(r+3)$

Sum to n terms $=\dfrac{n}{6}(4n^2+21n-1)$

c rth term $=r(r+2)(r+4)$

Sum to n terms $=\dfrac{n(n+1)(n+4)(n+5)}{4}$

d rth term $=4r-2$

Sum to n terms $=2n^2$

e rth term $=r^2+1$

Sum to n terms $=\dfrac{n(2n^2+3n+7)}{6}$

8 a $n^2+4n-32$

b $\dfrac{(n-2)(n-1)(n-3)}{3}-2$

c $(n+1)(2n+1)(2n^2+3n+4)-\dfrac{n(n-1)(n^2-n+6)}{4}$

9 a $\dfrac{n(1-3n)}{2}$

b $1-\dfrac{1}{n+1}\equiv\dfrac{n}{n+1}$

c $\dfrac{n(3n+5)}{2(n+1)(n+2)}$

d $\dfrac{11}{6}-\dfrac{3n^2+12n+11}{(n+1)(n+2)(n+3)}$

e $\dfrac{2n}{2n+1}$

f $\dfrac{n(n+7)}{12(n+3)(n+4)}$

Exercise 2.3B

1 a First term is $2(1)^2+4(1)=6$
Sum of first two terms is $2(2)^2+4(2)=16$, so second term is $16-6=10$
Sum of first three terms is $2(3)^2+4(3)=30$, so third term is $30-16=14$

b The nth term is $4n+2$

2 a $\dfrac{n(n+1)}{2}$

b Altogether there are $\dfrac{n(n+1)(n+2)}{6}$

c 599 kg (3sf)

3 Number of cannon balls $=\dfrac{n(2n+1)(7n+1)}{6}$

4 a

$$\frac{1}{1^2} - \frac{1}{2^2} = \frac{3}{4}$$
$$\frac{1}{2^2} - \frac{1}{3^2} = \frac{5}{36}$$
$$\frac{1}{3^2} - \frac{1}{4^2} = \frac{7}{144}$$
$$\frac{1}{4^2} - \frac{1}{5^2} = \frac{9}{400}$$
$$\frac{1}{n^2} - \frac{1}{(n+1)^2}$$

b $\displaystyle\sum_1^n \left(\frac{1}{r^2} - \frac{1}{(r+1)^2}\right) \equiv \frac{n(n+2)}{(n+1)^2}$

The sum to infinity $\displaystyle\lim_{n \to \infty}\left(1 - \frac{1}{(n+1)^2}\right) = 1$

5 $\dfrac{m(m+1)(m+2)}{6}$

6 a $\dfrac{1}{(2r-1)^2} - \dfrac{1}{(2r+1)^2} \equiv \dfrac{(2r+1)^2 - (2r-1)^2}{(2r-1)^2(2r+1)^2}$

$\equiv \dfrac{8r}{[(2r-1)(2r+1)]^2}$

$\equiv \dfrac{8r}{((2r)^2-1)^2}$

b $\displaystyle\sum_1^n \frac{8r}{((2r)^2-1)^2} \equiv \sum_1^n \frac{1}{(2r-1)^2} - \frac{1}{(2r+1)^2}$

$\equiv \left(\dfrac{1}{1^2} - \dfrac{1}{3^2}\right) + \left(\dfrac{1}{3^2} - \dfrac{1}{5^2}\right) + \left(\dfrac{1}{5^2} - \dfrac{1}{7^2}\right) + ... +$

$\left(\dfrac{1}{(2n-3)^2} - \dfrac{1}{(2n-1)^2}\right) + \left(\dfrac{1}{(2n-1)^2} - \dfrac{1}{(2n+1)^2}\right)$

$\equiv 1 - \dfrac{1}{(2n+1)^2}$

$$\sum_1^n \frac{r}{((2r)^2-1)^2} = \frac{n(n+1)}{2(2n+1)^2}$$

7 a $r(r+1) - r(r-1) \equiv r^2 + r - r^2 + r \equiv 2r$

b $\Sigma r(r+1) - r(r-1) \equiv$

$$\begin{array}{lll}
\cancel{1\times 2} & - & 1\times 0 \\
\cancel{2\times 3} & - & 2\times 1 \\
\cancel{3\times 4} & - & 3\times 2 \\
& \Downarrow & \\
\cancel{(n-2)\times(n-1)} & - & \cancel{(n-3)\times(n-2)} \\
\cancel{(n-1)\times n} & - & \cancel{(n-2)\times(n-1)} \\
n\times(n+1) & - & \cancel{(n-1)\times n}
\end{array}$$

Hence $2\Sigma r \equiv \Sigma r(r+1) - r(r-1) \equiv 1\times 0 + n\times(n+1)$

$2\Sigma r \equiv n(n+1)$

$$\sum_1^n r \equiv \frac{n(n+1)}{2}$$

8 a $(2r+1)^3 - (2r-1)^3 \equiv 8r^3 + 12r^2 + 6r + 1 - (8r^3 - 12r^2 + 6r - 1)$

$\equiv 24r^2 + 2$

b $\Sigma[(2r+1)^3 - (2r-1)^3] = 24\Sigma r^2 + 2n$

$\Sigma[(2r+1)^3 - (2r-1)^3]$

$$\begin{array}{lll}
3^3 & - & 1^3 \\
5^3 & - & 3^3 \\
7^3 & - & 5^3 \\
\downarrow & & \downarrow \\
\cancel{[(2n-3)^3} & - & (2n-5)^3] \\
\cancel{[(2n-1)^3} & - & \cancel{(2n-3)^3]} \\
[(2n+1)^3 & - & \cancel{(2n-1)^3]} \\
\equiv (2n+1)^3 & - & 1
\end{array}$$

Hence $24\displaystyle\sum_1^n r^2 + 2n \equiv (2n+1)^3 - 1$

$24\displaystyle\sum_1^n r^2 \equiv (2n+1)^3 - 2n - 1$

$\equiv 8n^3 + 12n^2 + 4n$

$\equiv 4n(n+1)(2n+1)$

Hence $\displaystyle\sum_1^n r \equiv \frac{n(n+1)(2n+1)}{6}$

9 a $\Sigma(2r-1)^2 = 4\Sigma r^2 - 4\Sigma r + \Sigma 1$

$= \dfrac{4n(n+1)(2n+1)}{6} - \dfrac{4n(n+1)}{2} + n \to$

$\dfrac{n[4(n+1)(2n+1) - 12(n+1) + 6]}{6}$

$= \dfrac{n(8n^2 - 2)}{6} \to \dfrac{n(2n-1)(2n+1)}{3}$

b $1^2 + 3^2 + 5^2 + ... + (2r-1)^2 \equiv \displaystyle\sum_1^{2n} r^2 - 4\sum_1^n r^2$

$= \dfrac{2n(2n+1)(4n+1)}{6} - \dfrac{4n(n+1)(2n+1)}{6}$

$= \dfrac{n(2n+1)[(4n+1) - 2(n+1)]}{3}$

$= \dfrac{n(2n+1)(2n-1)}{3}$

10 $-\dfrac{1}{x} + \dfrac{3}{x+2} - \dfrac{2}{x+4} \equiv$

$\dfrac{-(x+2)(x+4) + 3x(x+4) - 2x(x+2)}{x(x+2)(x+4)}$

$\equiv \dfrac{-x^2 - 6x - 8 + 3x^2 + 12x - 2x^2 - 4x}{x(x+2)(x+4)}$

$\equiv \dfrac{2x-8}{x(x+2)(x+4)} \equiv \dfrac{2(x-4)}{x(x+2)(x+4)}$

11 $\left(2r^2 + 3r + \dfrac{1}{r} - \dfrac{1}{r+1}\right) \equiv \dfrac{2r^3(r+1) + 3r^2(r+1) + (r+1) - r}{r(r+1)}$

$\equiv \dfrac{2r^4 + 2r^3 + 3r^3 + 3r^2 + r + 1 - r}{r(r+1)}$

$\equiv \dfrac{2r^4 + 5r^3 + 3r^2 + 1}{r(r+1)}$

$\displaystyle\sum \frac{2r^4 + 5r^3 + 3r^2 + 1}{r(r-1)} \equiv \sum_1^n \left(2r^2 + 3r + \frac{1}{r} - \frac{1}{r+1}\right)$

$\equiv 2\displaystyle\sum_1^n r^2 + 3\sum_1^n r + \sum_1^n \left(\frac{1}{r} - \frac{1}{r+1}\right)$

$\equiv 2\dfrac{n(n+1)(2n+1)}{6} + 3\dfrac{n(n+1)}{2} + \left(\dfrac{1}{1} - \dfrac{1}{2}\right) + \left(\dfrac{1}{2} - \dfrac{1}{3}\right)$

$+ \left(\dfrac{1}{3} - \dfrac{1}{4}\right) + \left(\dfrac{1}{4} - \dfrac{1}{5}\right) + ... + \left(\dfrac{1}{n-1} - \dfrac{1}{n}\right) + \left(\dfrac{1}{n} - \dfrac{1}{n+1}\right)$

$\equiv \dfrac{n(n+1)(2n+1)}{3} + \dfrac{3n(n+1)}{2} + 1 - \dfrac{1}{n+1}$

$\equiv \dfrac{2n(n+1)(2n+1)(n+1) + 9n(n+1)(n+1) + 6(n+1) - 6}{6(n+1)}$

$\equiv \dfrac{n(4n^3 + 19n^2 + 26n + 17)}{6(n+1)}$

Exercise 2.4A

1 a When $n = 1$, $\displaystyle\sum_{r=1}^{n} 1 = 1$

and $n = 1$ so true for $n = 1$

Assume true for $n = k$ and consider $n = k + 1$:

$$\sum_{r=1}^{k+1} 1 = \sum_{r=1}^{k} 1 + 1$$

$$= k + 1$$

So true for $n = k + 1$

The statement is true for $n = 1$ and by assuming it is true for $n = k$ it is shown to be true for $n = k + 1$, therefore, by mathematical induction, it is true for all $n \in \mathbb{N}$

b When $n = 1$, $\displaystyle\sum_{r=1}^{n} r = 1$

and $\dfrac{1}{2} n(n+1) = \dfrac{1}{2} \times 1(1+1)$

$$= \dfrac{1}{2}(2)$$

$$= 1$$

so true for $n = 1$

Assume true for $n = k$ and consider $n = k + 1$:

$$\sum_{r=1}^{k+1} r = \sum_{r=1}^{k} r + (k+1)$$

$$= \dfrac{1}{2} k(k+1) + k + 1$$

$$= \dfrac{1}{2}(k+1)(k+2)$$

$$= \dfrac{1}{2}(k+1)(k+1+1)$$

So true for $n = k + 1$

The statement is true for $n = 1$ and by assuming it is true for $n = k$ it is shown to be true for $n = k + 1$, therefore, by mathematical induction, it is true for all $n \in \mathbb{N}$

c When $n = 1$, $\displaystyle\sum_{r=1}^{n} (2r+3) = 2 \times 1 + 3$

$$= 5$$

and $n(n+4) = 1(1+4)$

$$= 5$$

so true for $n = 1$

Assume true for $n = k$ and consider $n = k + 1$:

$$\sum_{r=1}^{k+1} (2r+3) = \sum_{r=1}^{k} (2r+3) + 2(k+1) + 3$$

$$= k(k+4) + 2k + 5$$

$$= k^2 + 6k + 5$$

$$= (k+1)(k+5)$$

$$= (k+1)(k+1+4)$$

So true for $n = k + 1$

The statement is true for $n = 1$ and by assuming it is true for $n = k$ it is shown to be true for $n = k + 1$, therefore, by mathematical induction, it is true for all $n \in \mathbb{N}$

d When $n = 1$, $\displaystyle\sum_{r=1}^{n} r(r+1) = 1(1+1)$

$$= 2$$

and $\dfrac{1}{3} n(n+1)(n+2) = \dfrac{1}{3} \times 1(1+1)(1+2)$

$$= \dfrac{1}{3}(2)(3)$$

$$= 2$$

so true for $n = 1$

Assume true for $n = k$ and consider $n = k + 1$:

$$\sum_{r=1}^{k+1} r(r+1) = \sum_{r=1}^{k} r(r+1) + (k+1)(k+2)$$

$$= \dfrac{1}{3} k(k+1)(k+2) + (k+1)(k+2)$$

$$= \dfrac{1}{3}(k+1)(k+2)(k+3)$$

$$= \dfrac{1}{3}(k+1)(k+1+1)(k+1+2)$$

So true for $n = k + 1$

The statement is true for $n = 1$ and by assuming it is true for $n = k$ it is shown to be true for $n = k + 1$, therefore, by mathematical induction, it is true for all $n \in \mathbb{N}$

e When $n = 1$, $\displaystyle\sum_{r=1}^{n} (r-1)^2 = (1-1)^2$

$$= 0$$

and $\dfrac{1}{6} n(n-1)(2n-1) = \dfrac{1}{6} \times 1(1-1)(2 \times 1 - 1)$

$$= \dfrac{1}{6}(0)(1)$$

$$= 0$$

so true for $n = 1$

Assume true for $n = k$ and consider $n = k + 1$:

$$\sum_{r=1}^{k+1} (r-1)^2 = \sum_{r=1}^{k} (r-1)^2 + (k+1-1)^2$$

$$= \dfrac{1}{6} k(k-1)(2k-1) + k^2$$

$$= \dfrac{1}{6} k[(k-1)(2k-1) + 6k]$$

$$= \dfrac{1}{6} k[2k^2 - 2k - k + 1 + 6k]$$

$$= \dfrac{1}{6} k(2k^2 + 3k + 1)$$

$$= \dfrac{1}{6} k(2k+1)(k+1)$$

$$= \dfrac{1}{6}(k+1)(k+1-1)(2(k+1)-1)$$

So true for $n = k + 1$

The statement is true for $n = 1$ and by assuming it is true for $n = k$ it is shown to be true for $n = k + 1$, therefore, by mathematical induction, it is true for all $n \in \mathbb{N}$

f When $n = 1$, $\displaystyle\sum_{r=1}^{n} (r+1)(r-1) = (1+1)(1-1)$

$$= 0$$

and $\dfrac{1}{6} n(2n+5)(n-1) = \dfrac{1}{6} \times 1(2 \times 1 + 5)(1-1)$

$$= \dfrac{1}{6}(7)(0)$$

$$= 0$$

so true for $n = 1$

Assume true for $n = k$ and consider $n = k + 1$:

$$\sum_{r=1}^{k+1}(r+1)(r-1) = \sum_{r=1}^{k}(r+1)(r-1)+(k+1+1)(k+1-1)$$

$$= \frac{1}{6}k(2k+5)(k-1)+k(k+2)$$

$$= \frac{1}{6}k[(2k+5)(k-1)+6(k+2)]$$

$$= \frac{1}{6}k(2k^2-2k+5k-5+6k+12)$$

$$= \frac{1}{6}k(2k^2+9k+7)$$

$$= \frac{1}{6}k(2k+7)(k+1)$$

$$= \frac{1}{6}(k+1)(2(k+1)+5)(k+1-1)$$

So true for $n = k + 1$

The statement is true for $n = 1$ and by assuming it is true for $n = k$ it is shown to be true for $n = k + 1$, therefore, by mathematical induction, it is true for all $n \in \mathbb{N}$

2 a When $n = 1$, $\displaystyle\sum_{r=1}^{n}(r+1)^2 = 2^2 = 4$

and $\dfrac{1}{6}n(2n^2+9n+13) = \dfrac{1}{6}(1)(24) = 4$

So the statement is true when $n = 1$

Assume statement is true for $n = k$ and substitute $n = k+1$ into the formula:

$$\sum_{r=1}^{k+1}(r+1)^2 = \sum_{r=1}^{k}(r+1)^2 + (k+1+1)^2$$

$$= \frac{1}{6}k(2k^2+9k+13)+(k+2)^2$$

$$= \frac{1}{6}(2k^3+9k^2+13k+6k^2+24k+24)$$

$$= \frac{1}{6}(2k^3+15k^2+37k+24)$$

$$= \frac{1}{6}(k+1)(2k^2+13k+24)$$

$$= \frac{1}{6}(k+1)(2(k+1)^2+9(k+1)+13)$$

So the statement is true when $n = k + 1$

The statement is true for $n = 1$ and by assuming it is true for $n = k$ it is shown to be true for $n = k + 1$, therefore, by mathematical induction, it is true for all $n \in \mathbb{N}$

b When $n = 1$, $\displaystyle\sum_{r=1}^{n}5^{n-1} = 5^0 = 1$

and $\dfrac{1}{4}(5^n-1) = \dfrac{1}{4}(4) = 1$

So the statement is true when $n = 1$

Assume statement is true for $n = k$ and substitute $n = k+1$ into the formula:

$$\sum_{r=1}^{k+1}5^{r-1} = \sum_{r=1}^{k}5^{r-1}+5^{k+1-1}$$

$$= \frac{1}{4}(5^k-1)+5^k$$

$$= \frac{1}{4}(5^k-1+4(5^k))$$

$$= \frac{1}{4}(5(5^k)-1)$$

$$= \frac{1}{4}(5^{k+1}-1)$$

So the statement is true when $n = k + 1$

The statement is true for $n = 1$ and by assuming it is true for $n = k$ it is shown to be true for $n = k + 1$, therefore, by mathematical induction, it is true for all $n \in \mathbb{N}$

3 When $n = 1$, $\displaystyle\sum_{r=1}^{n}r^3 = 1^3$

$$= 1$$

and $\dfrac{1}{4}n^2(n+1)^2 = \dfrac{1}{4}\times1^2(1+1)^2$

$$= \frac{1}{4}(1)(2)^2$$

$$= 1$$

so true for $n = 1$

Assume true for $n = k$ and consider $n = k + 1$:

$$\sum_{r=1}^{k+1}r^3 = \sum_{r=1}^{k}r^3+(k+1)^3$$

$$= \frac{1}{4}k^2(k+1)^2+(k+1)^3$$

$$= \frac{1}{4}(k+1)^2[k^2+4(k+1)]$$

$$= \frac{1}{4}(k+1)^2[k^2+4k+4]$$

$$= \frac{1}{4}(k+1)^2(k+2)^2$$

$$= \frac{1}{4}(k+1)^2(k+1+1)^2$$

So true for $n = k + 1$

The statement is true for $n = 1$ and by assuming it is true for $n = k$ it is shown to be true for $n = k + 1$, therefore, by mathematical induction, it is true for all $n \in \mathbb{N}$

4 When $n = 1$, $\displaystyle\sum_{r=1}^{2n}r = 1+2$

$$= 3$$

and $n(2n+1) = 1(2\times1+1)$

$$= 3$$

so true for $n = 1$

Assume true for $n = k$ and consider $n = k + 1$:

$$\sum_{r=1}^{2(k+1)}r = \sum_{r=1}^{2k}r+(2k+1)+(2k+2)$$

$$= k(2k+1)+4k+3$$

$$= 2k^2+k+4k+3$$

$$= 2k^2+5k+3$$

$$= (2k+3)(k+1)$$

$$= (k+1)(2(k+1)+1)$$

So true for $n = k + 1$

The statement is true for $n = 1$ and by assuming it is true for $n = k$ it is shown to be true for $n = k + 1$, therefore, by mathematical induction, it is true for all $n \in \mathbb{N}$

5 When $n=1$, $\displaystyle\sum_{r=1}^{2n} r^2 = 1^2 + 2^2 = 5$

and $\dfrac{1}{3}n(2n+1)(4n+1) = \dfrac{1}{3}(1)(3)(5) = 5$

So the statement is true when $n=1$

Assume statement is true for $n=k$ and substitute $n=k+1$ into the formula.

$\displaystyle\sum_{r=1}^{2(k+1)} r^2 = \sum_{r=1}^{2k} r^2 + (2k+1)^2 + (2k+2)^2$

$\quad = \dfrac{1}{3}k(2k+1)(4k+1) + (4k^2+4k+1) + (4k^2+8k+4)$

$\quad = \dfrac{1}{3}(8k^3 + 6k^2 + k + 12k^2 + 12k + 3 + 12k^2 + 24k + 12)$

$\quad = \dfrac{1}{3}(8k^3 + 30k^2 + 37k + 15)$

$\quad = \dfrac{1}{3}(k+1)(8k^2 + 22k + 15)$

$\quad = \dfrac{1}{3}(k+1)(2k+3)(4k+5)$

$\quad = \dfrac{1}{3}(k+1)(2(k+1)+1)(4(k+1)+1)$

So statement is true when $n=k+1$

The statement is true for $n=1$ and by assuming it is true for $n=k$ it is shown to be true for $n=k+1$, therefore, by mathematical induction, it is true for all $n \in \mathbb{N}$

6 a When $n=1$, $\displaystyle\sum_{r=1}^{n} 2^r = 2^1$

$\quad = 2$

and $2(2^n - 1) = 2(2^1 - 1)$

$\quad = 2$

so true for $n=1$

Assume true for $n=k$ and consider $n=k+1$:

$\displaystyle\sum_{r=1}^{k+1} 2^r = \sum_{r=1}^{k} 2^r + 2^{k+1}$

$\quad = 2(2^k - 1) + 2^{k+1}$

$\quad = 2(2^k) - 2 + 2(2^k)$

$\quad = 4(2^k) - 2$

$\quad = 2(2^{k+1} - 1)$

So true for $n=k+1$

The statement is true for $n=1$ and by assuming it is true for $n=k$ it is shown to be true for $n=k+1$, therefore, by mathematical induction, it is true for all $n \in \mathbb{N}$

b When $n=1$, $\displaystyle\sum_{r=1}^{n} 3^r = 3^1$

$\quad = 3$

and $\dfrac{3}{2}(3^n - 1) = \dfrac{3}{2}(3^1 - 1)$

$\quad = \dfrac{3}{2}(2)$

$\quad = 3$

so true for $n=1$

Assume true for $n=k$ and consider $n=k+1$:

$\displaystyle\sum_{r=1}^{k+1} 3^r = \sum_{r=1}^{k} 3^r + 3^{k+1}$

$\quad = \dfrac{3}{2}(3^k - 1) + 3^{k+1}$

$\quad = \dfrac{3}{2}\left(3^k - 1 + \dfrac{2}{3}(3^{k+1})\right)$

$\quad = \dfrac{3}{2}\left(3^k - 1 + \dfrac{2}{3}(3)3^k\right)$

$\quad = \dfrac{3}{2}(3^k - 1 + 2(3^k))$

$\quad = \dfrac{3}{2}(3(3^k) - 1)$

$\quad = \dfrac{3}{2}(3^{k+1} - 1)$

So true for $n=k+1$

The statement is true for $n=1$ and by assuming it is true for $n=k$ it is shown to be true for $n=k+1$, therefore, by mathematical induction, it is true for all $n \in \mathbb{N}$

c When $n=1$, $\displaystyle\sum_{r=1}^{n} 4^r = 4^1$

$\quad = 4$

and $\dfrac{4}{3}(4^n - 1) = \dfrac{4}{3}(4^1 - 1)$

$\quad = \dfrac{4}{3}(3)$

$\quad = 4$

so true for $n=1$

Assume true for $n=k$ and consider $n=k+1$:

$\displaystyle\sum_{r=1}^{k+1} 4^r = \sum_{r=1}^{k} 4^r + 4^{k+1}$

$\quad = \dfrac{4}{3}(4^k - 1) + 4^{k+1}$

$\quad = \dfrac{4}{3}\left(4^k - 1 + \dfrac{3}{4}(4^{k+1})\right)$

$\quad = \dfrac{4}{3}\left(4^k - 1 + \dfrac{3}{4}(4)4^k\right)$

$\quad = \dfrac{4}{3}(4^k - 1 + 3(4^k))$

$\quad = \dfrac{4}{3}(4(4^k) - 1)$

$\quad = \dfrac{4}{3}(4^{k+1} - 1)$

So true for $n=k+1$

The statement is true for $n=1$ and by assuming it is true for $n=k$ it is shown to be true for $n=k+1$, therefore, by mathematical induction, it is true for all $n \in \mathbb{N}$

d When $n=1$, $\displaystyle\sum_{r=1}^{n} 2^{r-1} = 2^{1-1}$

$\quad = 1$

and $2^n - 1 = 2^1 - 1$

$\quad = 1$

so true for $n=1$

Assume true for $n = k$ and consider $n = k + 1$:

$$\sum_{r=1}^{k+1} 2^{r-1} = \sum_{r=1}^{k} 2^{r-1} + 2^{k+1-1}$$
$$= 2^k - 1 + 2^k$$
$$= 2(2^k) - 1$$
$$= 2^{k+1} - 1$$

So true for $n = k + 1$

The statement is true for $n = 1$ and by assuming it is true for $n = k$ it is shown to be true for $n = k + 1$, therefore, by mathematical induction, it is true for all $n \in \mathbb{N}$

e When $n = 1$, $\displaystyle\sum_{r=1}^{n} 3^{r-1} = 3^{1-1}$

$$= 1$$

and $\dfrac{1}{2}(3^n - 1) = \dfrac{1}{2}(3^1 - 1)$

$$= 1$$

so true for $n = 1$

Assume true for $n = k$ and consider $n = k + 1$:

$$\sum_{r=1}^{k+1} 3^{r-1} = \sum_{r=1}^{k} 3^{r-1} + 3^{k+1-1}$$
$$= \frac{1}{2}(3^k - 1) + 3^k$$
$$= \frac{1}{2}(3^k - 1 + 2(3^k))$$
$$= \frac{1}{2}(3(3^k) - 1)$$
$$= \frac{1}{2}(3^{k+1} - 1)$$

So true for $n = k + 1$

The statement is true for $n = 1$ and by assuming it is true for $n = k$ it is shown to be true for $n = k + 1$, therefore, by mathematical induction, it is true for all $n \in \mathbb{N}$

f When $n = 1$, $\displaystyle\sum_{r=1}^{n} \left(\frac{1}{2}\right)^r = \left(\frac{1}{2}\right)^1$

$$= \frac{1}{2}$$

and $1 - \left(\dfrac{1}{2}\right)^n = 1 - \left(\dfrac{1}{2}\right)^1$

$$= \frac{1}{2}$$

so true for $n = 1$

Assume true for $n = k$ and consider $n = k + 1$:

$$\sum_{r=1}^{k+1} \left(\frac{1}{2}\right)^r = \sum_{r=1}^{k} \left(\frac{1}{2}\right)^r + \left(\frac{1}{2}\right)^{k+1}$$
$$= 1 - \left(\frac{1}{2}\right)^k + \left(\frac{1}{2}\right)^{k+1}$$
$$= 1 - \left(\frac{1}{2}\right)^k \left(1 - \frac{1}{2}\right)$$
$$= 1 - \left(\frac{1}{2}\right)^k \left(\frac{1}{2}\right)$$
$$= 1 - \left(\frac{1}{2}\right)^{k+1}$$

So true for $n = k + 1$

The statement is true for $n = 1$ and by assuming it is true for $n = k$ it is shown to be true for $n = k + 1$, therefore, by mathematical induction, it is true for all $n \in \mathbb{N}$

7 When $n = 1$, $\displaystyle\sum_{r=1}^{n} \frac{1}{r(r+1)} = \frac{1}{1(1+1)}$

$$= \frac{1}{2}$$

and $\dfrac{n}{n+1} = \dfrac{1}{1+1}$

$$= \frac{1}{2}$$

so true for $n = 1$

Assume true for $n = k$ and consider $n = k + 1$:

$$\sum_{r=1}^{k+1} \frac{1}{r(r+1)} = \sum_{r=1}^{k} \frac{1}{r(r+1)} + \frac{1}{(k+1)(k+1+1)}$$
$$= \frac{k}{k+1} + \frac{1}{(k+1)(k+2)}$$
$$= \frac{k(k+2)+1}{(k+1)(k+2)}$$
$$= \frac{k^2 + 2k + 1}{(k+1)(k+2)}$$
$$= \frac{(k+1)^2}{(k+1)(k+2)}$$
$$= \frac{k+1}{k+2}$$
$$= \frac{k+1}{k+1+1}$$

So true for $n = k + 1$

The statement is true for $n = 1$ and by assuming it is true for $n = k$ it is shown to be true for $n = k + 1$, therefore, by mathematical induction, it is true for all $n \in \mathbb{N}$

8 When $n = 2$, $\displaystyle\sum_{r=2}^{n} \frac{1}{r(r-1)} = \frac{1}{2(2-1)}$

$$= \frac{1}{2}$$

and $\dfrac{n-1}{n} = \dfrac{2-1}{2}$

$$= \frac{1}{2}$$

so true for $n = 2$

Assume true for $n = k$ and consider $n = k + 1$:

$$\sum_{r=2}^{k+1} \frac{1}{r(r-1)} = \sum_{r=2}^{k} \frac{1}{r(r-1)} + \frac{1}{(k+1)(k+1-1)}$$
$$= \frac{k-1}{k} + \frac{1}{k(k+1)}$$
$$= \frac{(k-1)(k+1)+1}{k(k+1)}$$
$$= \frac{k^2 - 1 + 1}{k(k+1)}$$
$$= \frac{k^2}{k(k+1)}$$
$$= \frac{k}{k+1}$$
$$= \frac{(k+1)-1}{k+1}$$

So true for $n = k + 1$

The statement is true for $n = 2$ and by assuming it is true for $n = k$ it is shown to be true for $n = k + 1$, therefore, by mathematical induction, it is true for all $n \in \mathbb{N}$

9 When $n = 1$, $\displaystyle\sum_{r=1}^{n} \frac{1}{r^2 + 2r} = \frac{1}{1^2 + 2}$

$$= \frac{1}{3}$$

and $\dfrac{n(3n+5)}{4(n+1)(n+2)} = \dfrac{1(3+5)}{4(1+1)(1+2)}$

$$= \frac{8}{24}$$

$$= \frac{1}{3}$$

so true for $n = 1$

Assume true for $n = k$ and consider $n = k + 1$:

$$\sum_{r=1}^{k+1} \frac{1}{r^2 + 2r} = \sum_{r=1}^{k} \frac{1}{r^2 + 2r} + \frac{1}{(k+1)^2 + 2(k+1)}$$

$$= \frac{k(3k+5)}{4(k+1)(k+2)} + \frac{1}{k^2 + 2k + 1 + 2k + 2}$$

$$= \frac{k(3k+5)}{4(k+1)(k+2)} + \frac{1}{k^2 + 4k + 3}$$

$$= \frac{k(3k+5)}{4(k+1)(k+2)} + \frac{1}{(k+3)(k+1)}$$

$$= \frac{k(3k+5)(k+3) + 4(k+2)}{4(k+1)(k+2)(k+3)}$$

$$= \frac{k(3k^2 + 14k + 15) + 4(k+2)}{4(k+1)(k+2)(k+3)}$$

$$= \frac{3k^3 + 14k^2 + 15k + 4k + 8}{4(k+1)(k+2)(k+3)}$$

$$= \frac{(k+1)(3k^2 + 11k + 8)}{4(k+1)(k+2)(k+3)}$$

$$= \frac{(3k+8)(k+1)^2}{4(k+1)(k+2)(k+3)}$$

$$= \frac{(k+1)(3k+8)}{4(k+2)(k+3)}$$

$$= \frac{(k+1)(3(k+1)+5)}{4(k+1+1)(k+1+2)}$$

So true for $n = k + 1$

The statement is true for $n = 1$ and by assuming it is true for $n = k$ it is shown to be true for $n = k + 1$, therefore, by mathematical induction, it is true for all $n \in \mathbb{N}$

Exercise 2.4B

1 a When $n = 1$, $n^2 + 3n = 1 + 3$

$$= 4$$

$$= 2 \times 2$$

so true for $n = 1$

Assume true for $n = k$ and consider $n = k + 1$:

$(k+1)^2 + 3(k+1) = k^2 + 2k + 1 + 3k + 3$

$$= k^2 + 5k + 4$$

$$= (k^2 + 3k) + 2k + 4$$

$$= 2A + 2k + 4 \text{ since } k^2 + 3k \text{ divisible by 2}$$

$$= 2(A + k + 2)$$

So true for $n = k + 1$

The statement is true for $n = 1$ and by assuming it is true for $n = k$ it is shown to be true for $n = k + 1$, therefore, by mathematical induction, it is true for all $n \in \mathbb{N}$

b When $n = 1$, $5n^2 - n = 5 - 1$

$$= 4$$

$$= 2 \times 2$$

so true for $n = 1$

Assume true for $n = k$ and consider $n = k + 1$:

$5(k+1)^2 - (k+1) = 5k^2 + 10k + 5 - k - 1$

$$= 5k^2 + 9k + 4$$

$$= (5k^2 - k) + 10k + 4$$

$$= 2A + 10k + 4 \text{ since } 5k^2 - k \text{ divisible by 2}$$

$$= 2(A + 5k + 2)$$

So true for $n = k + 1$

The statement is true for $n = 1$ and by assuming it is true for $n = k$ it is shown to be true for $n = k + 1$, therefore, by mathematical induction, it is true for all $n \in \mathbb{N}$

c When $n = 1$, $8n^3 + 4n = 8 + 4$

$$= 12$$

$$= 12 \times 1$$

so true for $n = 1$

Assume true for $n = k$ and consider $n = k + 1$:

$8(k+1)^3 + 4(k+1) = 8(k^3 + 3k^2 + 3k + 1) + 4k + 4$

$$= 8k^3 + 24k^2 + 24k + 8 + 4k + 4$$

$$= 8k^3 + 24k^2 + 28k + 12$$

$$= (8k^3 + 4k) + 24k^2 + 24k + 12$$

$$= 12A + 12(2k^2 + 2k + 1) \text{ since } 8k^3 + 4k$$
$$\text{divisible by 12}$$

$$= 12(A + 2k^2 + 2k + 1)$$

So true for $n = k + 1$

The statement is true for $n = 1$ and by assuming it is true for $n = k$ it is shown to be true for $n = k + 1$, therefore, by mathematical induction, it is true for all $n \in \mathbb{N}$

d When $n = 1$, $11n^3 + 4n = 11 + 4$

$$= 15$$

$$= 3 \times 5$$

so true for $n = 1$

Assume true for $n = k$ and consider $n = k + 1$:

$11(k+1)^3 + 4(k+1) = 11(k^3 + 3k^2 + 3k + 1) + 4k + 4$

$$= 11k^3 + 33k^2 + 33k + 11 + 4k + 4$$

$$= 11k^3 + 33k^2 + 37k + 15$$

$$= (11k^3 + 4k) + 33k^2 + 33k + 15$$

$$= 3A + 3(11k^2 + 11k + 5) \text{ since } 11k^3 + 4k$$
$$\text{divisible by 3}$$

$$= 3(A + 11k^2 + 11k + 5)$$

So true for $n = k + 1$

The statement is true for $n = 1$ and by assuming it is true for $n = k$ it is shown to be true for $n = k + 1$, therefore, by mathematical induction, it is true for all $n \in \mathbb{N}$

2 When $n = 1$, $7n^2 + 25n - 4$

$$= 7 + 25 - 4$$

$$= 28$$

$$= 2 \times 14$$

so true for $n = 1$

Assume true for $n = k$ and consider $n = k + 1$:

$7(k+1)^2 + 25(k+1) - 4 = 7(k^2 + 2k + 1) + 25k + 25 - 4$

$$= 7k^2 + 14k + 7 + 25k + 25 - 4$$

$$= 7k^2 + 39k + 28$$

$$= (7k^2 + 25k - 4) + 14k + 32$$

$$= 2A + 2(7k + 16) \text{ since } 7k^2 + 25k - 4$$
$$\text{divisible by 2}$$

$$= 2(A + 7k + 16)$$

So true for $n = k + 1$

The statement is true for $n = 1$ and by assuming it is true for $n = k$ it is shown to be true for $n = k + 1$, therefore, by mathematical induction, it is true for all $n \in \mathbb{N}$

3 When $n = 2$, $n^3 - n = 2^3 - 2$

$$= 6$$

$$= 3 \times 2$$

so true for $n = 2$

Assume true for $n = k$ and consider $n = k + 1$:

$(k+1)^3 - (k+1) = (k^3 + 3k^2 + 3k + 1) - k - 1$

$\qquad = k^3 + 3k^2 + 2k$

$\qquad = (k^3 - k) + 3k^2 + 3k$

$\qquad = 3A + 3(k^2 + k)$ since $k^3 - k$ divisible by 3

$\qquad = 3(A + k^2 + k)$

So true for $n = k + 1$

The statement is true for $n = 2$ and by assuming it is true for $n = k$ it is shown to be true for $n = k + 1$, therefore, by mathematical induction, it is true for all $n \in \mathbb{N}, n \geq 2$

4 When $n = 1, 10n^3 + 3n^2 + 5n - 6$

$\qquad = 10 + 3 + 5 - 6$

$\qquad = 12$

$\qquad = 6 \times 2$

so true for $n = 1$

Assume true for $n = k$ and consider $n = k + 1$:

$10(k+1)^3 + 3(k+1)^2 + 5(k+1) - 6$

$\qquad = 10(k^3 + 3k^2 + 3k + 1) + 3(k^2 + 2k + 1) + 5(k+1) - 6$

$\qquad = 10k^3 + 30k^2 + 30k + 10 + 3k^2 + 6k + 3 + 5k + 5 - 6$

$\qquad = 10k^3 + 33k^2 + 41k + 12$

$\qquad = (10k^3 + 3k^2 + 5k - 6) + 30k^2 + 36k + 18$

$\qquad = 6A + 30k^2 + 36k + 18$ since $10k^3 + 3k^2 + 5k - 6$

$\qquad$ divisible by 6

$\qquad = 6(A + 5k^2 + 6k + 3)$

So true for $n = k + 1$

The statement is true for $n = 1$ and by assuming it is true for $n = k$ it is shown to be true for $n = k + 1$, therefore, by mathematical induction, it is true for all $n \in \mathbb{N}$

5 a When $n = 1, 6^n + 9 = 6^1 + 9$

$\qquad = 15$

$\qquad = 5 \times 3$

so true for $n = 1$

Assume true for $n = k$ and consider $n = k + 1$:

$6^{k+1} + 9 = 6(6^k) + 9$

$\qquad = (6^k + 9) + 5(6^k)$

$\qquad = 5A + 5(6^k)$ since $6^k + 9$ divisible by 5

$\qquad = 5(A + 6^k)$

So true for $n = k + 1$

The statement is true for $n = 1$ and by assuming it is true for $n = k$ it is shown to be true for $n = k + 1$, therefore, by mathematical induction, it is true for all $n \in \mathbb{N}$

b When $n = 1, 3^{2n} - 1 = 3^2 - 1$

$\qquad = 8$

$\qquad = 8 \times 1$

so true for $n = 1$

Assume true for $n = k$ and consider $n = k + 1$:

$3^{2(k+1)} - 1 = 3^{2k}3^2 - 1$

$\qquad = 9(3^{2k}) - 1$

$\qquad = (3^{2k} - 1) + 8(3^{2k})$

$\qquad = 8A + 8(3^{2k})$ since $3^{2k} - 1$ divisible by 8

$\qquad = 8(A + 3^{2k})$

So true for $n = k + 1$

The statement is true for $n = 1$ and by assuming it is true for $n = k$ it is shown to be true for $n = k + 1$, therefore, by mathematical induction, it is true for all $n \in \mathbb{N}$

c When $n = 1, 2^{3n+1} - 2 = 2^4 - 2$

$\qquad = 14$

$\qquad = 7 \times 2$

so true for $n = 1$

Assume true for $n = k$ and consider $n = k + 1$:

$2^{3(k+1)+1} - 2 = 2^{3k+4} - 2$

$\qquad = 2^3(2^{3k+1}) - 2$

$\qquad = 8(2^{3k+1}) - 2$

$\qquad = (2^{3k+1} - 2) + 7(2^{3k+1})$

$\qquad = 7A + 7(2^{3k+1})$ since $3^{2k} - 1$ divisible by 7

$\qquad = 7(A + 2^{3k+1})$

So true for $n = k + 1$

The statement is true for $n = 1$ and by assuming it is true for $n = k$ it is shown to be true for $n = k + 1$, therefore, by mathematical induction, it is true for all $n \in \mathbb{N}$

6 When $n = 1, 5^n - 4n + 3 = 5^1 - 4 + 3$

$\qquad = 4$

$\qquad = 4 \times 1$

so true for $n = 1$

Assume true for $n = k$ and consider $n = k + 1$:

$5^{k+1} - 4(k+1) + 3 = 5(5^k) - 4k - 4 + 3$

$\qquad = (5^k - 4k + 3) + 4(5^k) - 4$

$\qquad = 4A + 4(5^k - 1)$ since $5^k - 4k + 3$ divisible by 4

$\qquad = 4(A + 5^k - 1)$

So true for $n = k + 1$

The statement is true for $n = 1$ and by assuming it is true for $n = k$ it is shown to be true for $n = k + 1$, therefore, by mathematical induction, it is true for all $n \in \mathbb{N}$

7 When $n = 1, 3^n + 2n + 7 = 3^1 + 2 + 7$

$\qquad = 12$

$\qquad = 4 \times 3$

so true for $n = 1$

Assume true for $n = k$ and consider $n = k + 1$:

$3^{k+1} + 2(k+1) + 7 = 3(3^k) + 2k + 9$

$\qquad = 3(3^k + 2k + 7) - 4k - 12$

$\qquad = 3(4A) - 4(k+3)$ since $3^{k+1} + 2(k+1) + 7$

$\qquad$ divisible by 4

$\qquad = 4(3A - k - 3)$

So true for $n = k + 1$

The statement is true for $n = 1$ and by assuming it is true for $n = k$ it is shown to be true for $n = k + 1$, therefore, by mathematical induction, it is true for all $n \in \mathbb{N}$

8 When $n = 1, 7^n - 3n + 5 = 7 - 3 + 5$

$\qquad = 9$

$\qquad = 3 \times 3$

so true for $n = 1$

Assume true for $n = k$ and consider $n = k + 1$:

$7^{k+1} - 3(k+1) + 5 = 7(7^k) - 3k + 2$

$\qquad = (7^k - 3k + 5) + 6(7^k) - 3$

$\qquad = 3A + 3(2(7^k) - 1)$ since $7^k - 3k + 5$

$\qquad$ divisible by 3

$\qquad = 3(A + 2(7^k) - 1)$

So true for $n = k + 1$

The statement is true for $n = 1$ and by assuming it is true for $n = k$ it is shown to be true for $n = k + 1$, therefore, by mathematical induction, it is true for all $n \in \mathbb{N}$

9 a When $n = 1, \displaystyle\sum_{r=n+1}^{2n} r^2 = \sum_{r=2}^{2} r^2$

$\qquad\qquad = 2^2$

$\qquad\qquad = 4$

and $\dfrac{1}{6}n(2n+1)(7n+1) = \dfrac{1}{6}(3)(8)$

$\qquad\qquad\qquad = 4$

so true for $n = 1$

Assume true for $n = k$ and consider $n = k + 1$:

$$\sum_{r=(k+1)+1}^{2(k+1)} r^2 = \sum_{r=k+1}^{2k} r^2 + (2k+1)^2 + (2k+2)^2 - (k+1)^2$$

$$= \frac{1}{6}k(2k+1)(7k+1) + 4k^2 + 4k + 1 + 4k^2$$
$$\quad + 8k + 4 - k^2 - 2k - 1$$

$$= \frac{1}{6}(14k^3 + 9k^2 + k + 42k^2 + 60k + 24)$$

$$= \frac{1}{6}(14k^3 + 51k^2 + 61k + 24)$$

$$= \frac{1}{6}(k+1)(2k+3)(7k+8)$$

$$= \frac{1}{6}(k+1)(2(k+1)+1)(7(k+1)+1)$$

So true for $n = k + 1$

The statement is true for $n = 1$ and by assuming it is true for $n = k$ it is shown to be true for $n = k + 1$, therefore, by mathematical induction, it is true for all $n \in \mathbb{N}$

b When $n = 1$, $\displaystyle\sum_{r=n}^{2n} r^3 = \sum_{r=1}^{2} r^3$

$$= 1 + 8$$
$$= 9$$

and $\dfrac{3}{4}n^2(5n+1)(n+1) = \dfrac{3}{4}(1)(6)(2)$

$$= 9$$

so true for $n = 1$

Assume true for $n = k$ and consider $n = k + 1$:

$$\sum_{r=(k+1)}^{2(k+1)} r^3 = \sum_{r=k}^{2k} r^3 + (2k+1)^3 + (2k+2)^3 - k^3$$

$$= \frac{3}{4}k^2(5k+1)(k+1) + 8k^3 + 12k^2 + 6k + 1$$
$$\quad + 8k^3 + 24k^2 + 24k + 8 - k^3$$

$$= \frac{3}{4}(5k^4 + 6k^3 + k^2 + 20k^3 + 48k^2 + 40k + 12)$$

$$= \frac{3}{4}(5k^4 + 26k^3 + 49k^2 + 40k + 12)$$

$$= \frac{3}{4}(k+1)(5k^3 + 21k^2 + 28k + 12)$$

$$= \frac{3}{4}(k+1)(k+1)(k+2)(5k+6)$$

$$= \frac{3}{4}(k+1)^2(k+1+1)(5(k+1)+1)$$

So true for $n = k + 1$

The statement is true for $n = 1$ and by assuming it is true for $n = k$ it is shown to be true for $n = k + 1$, therefore, by mathematical induction, it is true for all $n \in \mathbb{N}$

c When $n = 1$, $8^n - 5^n = 8^1 - 5^1$
$$= 3$$
$$= 3 \times 1$$

so true for $n = 1$

Assume true for $n = k$ and consider $n = k + 1$:
$$8^{k+1} - 5^{k+1} = 8(8^k) - 5(5^k)$$
$$= 5(8^k - 5^k) + 3(8^k)$$
$$= 5(3A) + 3(8^k) \text{ since } 8^k - 5^k \text{ divisible by 3}$$
$$= 3(5A + 8^k)$$

So true for $n = k + 1$

The statement is true for $n = 1$ and by assuming it is true for $n = k$ it is shown to be true for $n = k + 1$, therefore, by mathematical induction, it is true for all $n \in \mathbb{N}$

Exercise 2.5A

1 a $1 + 2x + 2x^2 + \dfrac{4x^3}{3} + \dfrac{2x^4}{3} + \dots$

b $1 - x + \dfrac{1}{2}x^2 - \dfrac{x^3}{6} + \dfrac{x^4}{24} + \dots$

c $1 - 3x + \dfrac{9}{2}x^2 - \dfrac{9x^3}{2} + \dfrac{27x^4}{8} + \dots$

d $1 + \dfrac{x}{2} + \dfrac{1}{8}x^2 + \dfrac{x^3}{48} + \dfrac{x^4}{384} + \dots$

e $1 - \dfrac{x}{3} + \dfrac{1}{18}x^2 + \dfrac{x^3}{162} + \dfrac{x^4}{1944} + \dots$

2 a i $-x - \dfrac{x^2}{2} - \dfrac{x^3}{3} - \dfrac{x^4}{4} - \dfrac{x^5}{5} \dots - \dfrac{x^r}{r} + \dots$

   ii $-1 \leq x < 1$

b i $2x - 2x^2 + \dfrac{8x^3}{3} - 4x^4 + \dfrac{32x^5}{5} \dots + \dfrac{(2x)^r}{r} + \dots$

   ii $-\dfrac{1}{2} < x \leq \dfrac{1}{2}$

c i $-3x - \dfrac{9x^2}{2} - 9x^3 - \dfrac{81x^4}{4} - \dfrac{243x^5}{5} \dots + \dfrac{(-3x)^r}{r} + \dots$

   ii $-\dfrac{1}{3} \leq x < \dfrac{1}{3}$

d i $\dfrac{x}{2} - \dfrac{x^2}{8} + \dfrac{x^3}{24} - \dfrac{x^4}{64} + \dfrac{x^5}{180} \dots + \dfrac{x^r}{(2^r)r} + \dots$

   ii $-2 < x \leq 2$

e i $\dfrac{-x}{3} - \dfrac{x^2}{18} - \dfrac{x^3}{81} - \dfrac{x^4}{324} - \dfrac{x^5}{1215} \dots - \dfrac{x^r}{(3)r} + \dots$

   ii $-3 \leq x < 3$

3 a $2x - \dfrac{4x^3}{3} + \dfrac{4x^5}{15} - \dfrac{8x^7}{315} + \dots + \dfrac{(-1)^{r+1}(2x)^{2r-1}}{(2r-1)!} + \dots$

b $\dfrac{x}{2} - \dfrac{x^3}{48} + \dfrac{x^5}{3840} - \dfrac{x^7}{645\,120} + \dots + \dfrac{(-1)^{r+1}(x)^{2r-1}}{2^{2r-1}(2r-1)!} + \dots$

c $-x + \dfrac{x^3}{3!} - \dfrac{x^5}{5!} + \dfrac{x^7}{7!} + \dots + \dfrac{(-1)^r(x)^{2r-1}}{(2r-1)!} + \dots$

d $-3x + \dfrac{9x^3}{2} - \dfrac{81x^5}{40} + \dfrac{243x^7}{560} + \dots + \dfrac{(-1)^r(x)^{2r-1}}{3^{2r-1}(2r-1)!} + \dots$

e $\dfrac{3x}{2} - \dfrac{9x^3}{16} + \dfrac{81x^5}{1280} - \dfrac{243x^7}{71\,680} + \dots + \dfrac{(-1)^{r+1}(3x)^{2r-1}}{2^{2r-1}(2r-1)!} + \dots$

4 a $1 - 8x^2 + \dfrac{32x^4}{3} - \dfrac{256x^6}{45} + \dots + \dfrac{(-1)^{r+1}(x)^{2r}}{4^{2r}(2r)!} + \dots$

b $1 - \dfrac{x^2}{18} + \dfrac{x^4}{1944} - \dfrac{x^6}{524\,880} + \dots + \dfrac{(-1)^{r+1}(x)^{2r}}{3^{2r}(2r)!} + \dots$

c $1 - \dfrac{x^2}{2!} + \dfrac{x^4}{4!} - \dfrac{x^6}{6!} + \dots + \dfrac{(-1)^{r+1}(x)^{2r}}{(2r)!} + \dots$

d $1 - 2x^2 + \dfrac{2x^4}{3} - \dfrac{4x^6}{45} + \dots + \dfrac{(-1)^{r+1}(x)^{2r}}{2^{2r}(2r)!} + \dots$

e $1 - \dfrac{x^2}{8} + \dfrac{x^4}{384} - \dfrac{x^6}{46\,080} + \dots + \dfrac{(-1)^{r+1}(x)^{2r}}{2^{2r}(2r)!} + \dots$

5 a i $1 + \dfrac{x}{2} - \dfrac{x^2}{8} + \dfrac{x^3}{16} + \dots$

   ii $-1 < x < 1$

b i $1 - \dfrac{x}{2} - \dfrac{x^2}{8} - \dfrac{x^3}{16} + \dots$

   ii $-1 < x < 1$

c i $1 + \dfrac{x}{3} - \dfrac{x^2}{9} + \dfrac{5x^3}{81} + \dots$

   ii $-1 < x < 1$

d i $1 + 3x + \dfrac{3x^2}{2} - \dfrac{x^3}{2} + \dots$

ii $-\dfrac{1}{2} < x < \dfrac{1}{2}$

e i $1 - \dfrac{15x}{2} + \dfrac{135x^2}{8} - \dfrac{135x^3}{16} + \dots$

ii $-\dfrac{1}{3} < x < \dfrac{1}{3}$

f i $1 + x + x^2 + x^3 + \dots$

ii $-1 < x < 1$

g i $3 - 3x + 3x^2 - 3x^3 + \dots$

ii $-1 < x < 1$

6 a $x^2 - \dfrac{x^6}{3!} + \dfrac{x^{10}}{5!} - \dfrac{x^{14}}{7!} + \dots$

b $x^{\frac{1}{2}} - \dfrac{x^{\frac{3}{2}}}{3!} + \dfrac{x^{\frac{5}{2}}}{5!} - \dfrac{x^{\frac{7}{2}}}{7!} + \dots$

c $1 - \dfrac{x^4}{2!} + \dfrac{x^8}{4!} - \dfrac{x^{12}}{6!} + \dots$

d $1 - \dfrac{x}{2!} + \dfrac{x^2}{4!} - \dfrac{x^3}{6!} + \dots$

Exercise 2.5B

1 a $x + x^2 + \dfrac{x^3}{3} + \dots$

b $1 + x + \dfrac{x^2}{2} - \dots$

c $1 + x - \dfrac{x^2}{2} + \dots$

d $1 - x^2 + \dfrac{x^4}{3} + \dots$

2 $x - \dfrac{x^3}{2!} + \dfrac{x^5}{4!} - \dots$ Valid for all x.

3 a $1 + \dfrac{3x}{2} - \dfrac{9x^2}{8} + \dots$ Valid when $-\dfrac{1}{3} < x < \dfrac{1}{3}$

b $-2x - \dfrac{2x^3}{3} - \dfrac{2x^5}{5} + \dots$ Valid when $-1 < x < 1$

c $x + x^2 + x^3 + \dots$ Valid when $-1 < x < 1$

d $1 - x - x^2 + \dots$ Valid when $-1 < x < 1$

4 a i $1 + \dfrac{1}{2x} - \dfrac{1}{8x^2} + \dfrac{1}{16x^3} + \dots$

ii $\dfrac{1}{x} - \dfrac{1}{2x^2} + \dfrac{1}{3x^3} - \dfrac{1}{4x^4} \dots$

b $x \geq 1$ or $x < -1$

5 a i $x + \dfrac{x^3}{3!} + \dfrac{x^5}{5!} + \dots$

ii $1 - \dfrac{x^2}{2} + \dfrac{x^4}{24}$

iii $\dfrac{5}{6} - \dfrac{x^2}{2} + \dfrac{x^4}{6} - \dfrac{x^6}{16} + \dfrac{13x^8}{1152} - \dfrac{x^{10}}{1152} + \dfrac{x^{12}}{41472}$

b The range of values of x for which $\ln(1 + \cos x)$ is valid is
$-1 < \cos x \leq 1$
or $\cos^{-1}(-1) < x \leq \cos^{-1}(1)$
or $-\pi < x \leq 0$

6 a 0.5488 (4sf)

b i 1.2528 (5 sf)

ii The series is only valid for $-1 < x \leq 1$
$x = 2.5$ is outside this range, so you have to make x in this range.

7 a $f(x) = e^{2x}$ so $f(0) = e^0 = 1$
$f'(x) = 2e^{2x}$ so $f'(0) = 2e^0 = 2$
$f''(x) = 4e^{2x}$ so $f''(0) = 4e^0 = 4$
$f'''(x) = 8e^{2x}$ so $f'''(0) = 8e^0 = 8$
$f^{iv}(x) = 16e^{2x}$ so $f^{iv}(0) = 16e^0 = 16$

Hence $e^{2x} \equiv 1 + x(2) + \dfrac{x^2}{2!}(4) + \dfrac{x^3}{3!}(8) + \dfrac{x^4}{4!}(16) + \dots$

$\equiv 1 + 2x + 2x^2 + \dfrac{4x^3}{3} + \dfrac{2x^4}{3} + \dots$

b $e^x = 1 + x + \dfrac{x^2}{2!} + \dfrac{x^3}{3!} + \dots$

so $e^{2x} \equiv 1 + 2x + \dfrac{4x^2}{2!} + \dfrac{8x^3}{3!} + \dfrac{16x^3}{4!} + \dots$

$\equiv 1 + 2x + 2x^2 + \dfrac{4x^3}{3} + \dfrac{2x^4}{3} + \dots$

c $\sqrt{e} \equiv 1 + 2\left(\dfrac{1}{4}\right) + 2\left(\dfrac{1}{4}\right) + \dfrac{4\left(\frac{1}{4}\right)^3}{3} + \dfrac{2\left(\frac{1}{4}\right)^4}{3} + \dots$

$\approx 1 + 0.5 + 0.125 + 0.02083333333\dots + 0.002604166666\dots$
$\approx 1.6484375\dots$

8 a $f(x) = \sin x$ so $f(0) = \sin 0 = 0$
$f'(x) = \cos x$ so $f'(0) = \cos 0 = 1$
$f''(x) = -\sin x$ so $f''(0) = -\sin 0 = 0$
$f'''(x) = -\cos x$ so $f'''(0) = -\cos 0 = 1$
$f^{iv}(x) = \sin x$ so $f^{iv}(0) = \sin 0 = 0$

Hence $\sin x \equiv 0 + x(1) + \dfrac{x^2}{2!}(0) + \dfrac{x^3}{3!}(-1) + \dfrac{x^4}{4!}(0)$

$+ \dfrac{x^5}{5!}(-1) + \dfrac{x^6}{6!}(0) + \dfrac{x^7}{7!}(-1) + \dots$

$\equiv x - \dfrac{x^3}{3!} + \dfrac{x^5}{5!} - \dfrac{x^7}{7!} + \dots$

b Hence $\sin \dfrac{\pi}{2} \approx \left(\dfrac{\pi}{2}\right) - \dfrac{\left(\frac{\pi}{2}\right)^3}{3!} + \dfrac{\left(\frac{\pi}{2}\right)^5}{5!} - \dfrac{\left(\frac{\pi}{2}\right)^7}{7!} + \dots$

$\approx 0.99984\dots$

Hence percentage error between 0.99984 and the exact value of $1 = \dfrac{(0.99984 - 1) \times 100}{1}$

$\approx -0.016\%$

Review exercise 2

1 a Sum of roots $= -5$
Sum of the products in pairs $= 2$
Product of the roots $= -1$

b Sum of roots $= -9$
Sum of the products in pairs $= -11$
Product of the roots $= 8$

2 $\left(\dfrac{y}{2}\right)^3 - 4\left(\dfrac{y}{2}\right)^2 + 6\left(\dfrac{y}{2}\right) - 2 = \dfrac{y^3}{8} - y^2 + 3y - 2$

3 a 102

b $510y^3 - 1428y^2 + 1291y - 371 = 0$

4 a $\pm i, -2$

b $a = 2$
$b = 1$

5 a -1

b i $y^3 + 2y^2 - 2y - 2 = 0$ **ii** 2

6 a -3 **b** $-\dfrac{3}{2}$ **c** 6

7 a $x \leq 0$ or $x \geq 5$

b $0 < x < 4$

c $-3 < x < -2$

d $x \leq \dfrac{5}{3}$ or $x > \dfrac{5}{2}$

e $2 < x < 3$ or $x < 1$

f $-15 \leq x < -4$

8 **a** $x < -\dfrac{3}{2}$ or $0 < x < \dfrac{2}{3}$

   b $-6 < x < -1$ or $x > 8$

   c $-\dfrac{8}{3} \le x \le 1$ or $\dfrac{5}{2} \le x \le 8$

   d $-5 \le x \le -1$ or $3 \le x \le 7$

9 **a** $x \le 1$ or $x \ge 3$

   b $-3 < x \le 7$

10 **a** $27 + 64 + 125 + 216 + 343 + 512 + 729 + 1000 = 3016$

   b $n^2 + (n+1)^2 + (n+2)^2 + (n+3)^2 + (n+4)^2$

   c $6 + 14 + 24 + \ldots + ((n-1)^2 + 5 \times (n-1)) + (n^2 + 5n)$

   d $-\dfrac{1}{2} - 1 + \infty + \ldots + \dfrac{1}{(n-4)} + \dfrac{1}{n-3}$

   e 56

   f $2(n-1)(3(n-1)-2) + 2(n)(3n-2) + 2(n+1)(3(n+1)-2)$
     $+ 2(n+2)(3(n+2)-2) + 2(n+3)(3(n+3)-2)$
     $= 2(n-1)(3n-5) + 2n(3n-2) + 2(n+1)(3n+1)$
     $+ 2(n+2)(3n+4) + 2(n+3)(3n+7)$

11 **a** $n(2-n)$

   b $\dfrac{n(n+1)(2n-5)}{3}$

   c $\dfrac{n(n+1)(n^2+n+10)}{4}$

12 **a** $(2n+1)^2(n+1)^2 - \dfrac{n^2(n-1)^2}{4} + 2n + 4$

   b When $n = 6$, $\displaystyle\sum_{n}^{2n+1}(r^3 + 2) = 8072$

13 **a** $\dfrac{1}{2} - \dfrac{1}{n+2}$

   b $\dfrac{1}{3} - \dfrac{1}{n+3}$

14 1

15 $t^2(t+1)^2 - t^2(t-1)^2 \equiv t^2(t^2 + 2t + 1) - t^2(t^2 - 2t + 1)$
     $\equiv t^4 + 2t^3 + t^2 - t^4 + 2t^3 - t^2$
     $\equiv 4t^3$

   $\Sigma\, t^2(t+1)^2 - t^2(t-1)^2$
   $\equiv [1^2(2)^2 - 1^2(0)^2] + [2^2(3)^2 - 2^2(1)^2]$
     $+ [3^2(4)^2 - 3^2(2)^2] + \ldots + (n-1)^2(n)^2 - (n-1)^2(n-2)^2$
     $+ n^2(n+1)^2 - n^2(n-1)^2$
   $\equiv (4+0) + (36-4) + (144-36) + \ldots + n^2(n+1)^2$

   Hence $\Sigma 4t^3 \equiv n^2(n+1)^2$

   So $\displaystyle\sum_{1}^{n} t^3 = \dfrac{n^2(n+1)^2}{4}$

16 $\displaystyle\sum_{1}^{k}\left(\dfrac{1}{n^2} - \dfrac{1}{(n+1)^2}\right) = \dfrac{1}{1} - \dfrac{1}{4}$
     $+ \dfrac{1}{4} - \dfrac{1}{9}$
     $+ \dfrac{1}{9} - \dfrac{1}{16}$
     $\downarrow$

     $\downarrow$

     $+ \dfrac{1}{(k-1)^2} - \dfrac{1}{(k)^2}$
     $+ \dfrac{1}{(k)^2} - \dfrac{1}{(k+1)^2}$

   Thus $\displaystyle\sum_{1}^{k}\left(\dfrac{1}{n^2} - \dfrac{1}{(n+1)^2}\right) = \dfrac{1}{1} - \dfrac{1}{(k+1)^2} =$

   $\dfrac{(k+1)^2 - 1}{(k+1)^2} = \dfrac{k^2 + 2k}{(k+1)^2} = \dfrac{k(k+2)}{(k+1)^2}$

17 When $n = 1$, $3^{2n+1} + 1 = 3^{2+1} + 1$
                     $= 28$
                     $= 4 \times 7$

   so true for $n = 1$

   Assume true for $n = k$ and consider $n = k + 1$:

   $3^{2(k+1)+1} + 1 = 3^{2k+3} + 1$
            $= 3^2 3^{2k+1} + 1$
            $= 9(3^{2k+1}) + 1$
            $= (3^{2k+1} + 1) + 8(3^{2k+1})$
            $= 4A + 8(3^{2k+1})$ since true for $n = k$
            $= 4(A + 2(3^{2k+1}))$

   So true for $n = k + 1$

   The statement is true for $n = 1$ and by assuming it is true for $n = k$ it is shown to be true for $n = k + 1$, therefore, by mathematical induction, it is true for all $n \in \mathbb{N}$

18 When $n = 1$, $2^{2n} - 3n + 2 = 2^2 - 3 + 2$
                       $= 3$

   so true for $n = 1$

   Assume true for $n = k$ and consider $n = k + 1$:

   $2^{2(k+1)} - 3(k+1) + 2 = 2^{2k+2} - 3k - 3 + 2$
                  $= 2^2 2^{2k} - 3k + 2 - 3$
                  $= (2^{2k} - 3k + 2) + 3(2^{2k}) - 3$
                  $= 3A + 3(2^{2k} - 1)$ since true for $n = k$
                  $= 3(A + 2^{2k} - 1)$

   So true for $n = k + 1$

   The statement is true for $n = 1$ and by assuming it is true for $n = k$ it is shown to be true for $n = k + 1$, therefore, by mathematical induction, it is true for all $n \in \mathbb{N}$

19 Let $n = 1$, then $6^n - 1 = 5$ which is divisible by 5

   So true for $n = 1$

   Assume true for $n = k$ and let $n = k + 1$

   $6^{k+1} - 1 = 6(6^k) - 1$
          $= 5(6^k) + 6^k - 1$
          $= 5(6^k) + 5A$ for some integer A since $6^k - 1$
          is divisible by 5
          $= 5\left(6^k + A\right)$

   So divisible by 5

   True for $n = 1$ and true for $n = k$ implies true for $n = k + 2$ therefore true for all $n \in \mathbb{N}$

20 **a** The validity of $(1 + x^2)^{-\frac{1}{2}}$ is $-1 < x^2 < 1$;
     Since x^2 cannot be negative the validity becomes $-1 < x < 1$

   b The validity of $\ln\left(1 - \dfrac{x}{2}\right)$ is $-1 < -\dfrac{x}{2} \le 1$ and hence $-2 < -x < 2$
     Or, multiplying through by -1, $2 > x > -2$ or, as is more commonly written $-2 < x < 2$

21 **a** $x^2 - \dfrac{x^4}{2} + \dfrac{x^6}{3} - \ldots$

   b $2x - \dfrac{2x^2}{3} - \dfrac{8x^3}{9} + \ldots$

22 $1 + \dfrac{x}{3} + \dfrac{x^2}{18} + \dfrac{x^3}{162} + \ldots$

   Hence $\sqrt[3]{e} \equiv e^{\frac{1}{3}}$ so, substituting $x = 1$,

   $\sqrt[3]{e} \approx 1 + \dfrac{1}{3} + \dfrac{1}{18} + \dfrac{1}{162} + \ldots \approx \dfrac{113}{81} \approx 1.3951\ (4\,\text{dp})$

Assessment 2

1 $\alpha + \beta = -\dfrac{4}{3}$

2 $\displaystyle\sum \alpha\beta\delta = -1$

3 $x < -3$ or $x > 1$

4 a $1 - 15x + 150x^2 - 1250x^3 + \dots$

   b $|x| < \dfrac{1}{5}$

5 a $= 1 - \dfrac{x}{2} + \dfrac{x^2}{8} - \dfrac{x^3}{48} + \dots$

   b 1.05123

6 $2n(n+1)^2$

7 a $u^3 = 1$ b $u = 1$ c $x = 1$

8 $x < -\dfrac{1}{13}$

9 a $\dfrac{n}{2}(2n+3)(n-1)$ b 4200

10 a $\dfrac{n}{6}(4n^2 + 21n - 1)$ b 3035   c 2012.5

11 a $f(x) = \dfrac{1}{2}(x-2)(x-3)(x+4)$

   b $g(x) = \dfrac{1}{2}\left(\dfrac{1}{2}x+1-2\right)\left(\dfrac{1}{2}x+1-3\right)\left(\dfrac{1}{2}x+1+4\right)$

      $= \dfrac{1}{16}(x-2)(x-4)(x+10)$

12 a $\alpha\beta\gamma = -1$

      $\sum \alpha\beta = -\dfrac{3}{2}$

      $\sum \alpha = \dfrac{1}{2}$

   b $u^3 + 2u^2 - 5u + 2 = 0$

13 a $-2x^2 - 2x^4 - \dfrac{8}{3}x^6 - 4x^8 - \dots$

   b $-\dfrac{1}{\sqrt{2}} \le x < \dfrac{1}{\sqrt{2}}$

14 $\dfrac{x}{3} - \dfrac{x^2}{3} + \dfrac{13}{81}x^3 - \dfrac{4}{81}x^4$

15 a $f(1) = 0$

      Therefore, by factor theorem, $x - 1$, is a factor.

   b $2x^3 + 3x^2 - 8x + 3 = (x-1)(2x^2 + 5x - 3)$

      $= (x-1)(2x-1)(x+3)$

   c $x < -3$ or $\dfrac{1}{2} < x < 1$

16 When $n = 1$, $\displaystyle\sum_{r=1}^{n} 4^{r-1} = 1$

   $\dfrac{1}{3}(4^n - 1) = \dfrac{1}{3}(4-1) = 1$

   So true for $n = 1$

   Assume true for $n = k$

   $\displaystyle\sum_{r=1}^{k+1} 4^{r-1} = 4^{(k+1)-1} + \sum_{r=1}^{k} 4^{r-1}$

      $= 4^k + \dfrac{1}{3}(4^k - 1)$

      $= \dfrac{4}{3}(4^k) - \dfrac{1}{3}$

      $= \dfrac{1}{3}(4^{k+1} - 1)$

   True for $n = 1$ and assuming true for $n = k$ implies true for $n = k+1$, hence true for all $n \in \mathbb{N}$

17 When $n = 1$, $4n^3 + 8n = 12$ which is divisible by 12 so true for $n = 1$

   Assume true for $n = k$

   $4(k+1)^3 + 8(k+1)$

      $= 4(k^3 + 3k^2 + 3k + 1) + 8(k+1)$

      $= 4k^3 + 8k + 12k^2 + 12k + 4 + 8$

      $= 4k^3 + 8k + 12k^2 + 12k + 12$

      $= 4k^3 + 8k + 12(k^2 + k + 1)$

   So divisible by 12 since $4k^3 + 8k$ and $12(k^2 + k + 1)$ are divisible by 12

   True for $n = 1$ and assuming true for $n = k$ implies true for $n = k + 1$, hence true for all $n \in \mathbb{N}$

18 When $n = 1$, $4^{2n+1} - 1 = 4^3 - 1 = 63$

   $63 = 21 \times 3$ hence a multiple of 3

   Assume true for $n = k$

   $4^{2(k+1)+1} - 1 = 4^{2k+3} - 1$

      $= 4^2 4^{2k+1} - 1$

      $= 16(4^{2k+1} - 1) + 15$

   So a multiple of 3 since $4^{2k+1} - 1$ assumed to be a multiple of 3 and $15 = 5 \times 3$ so a multiple of 3

   True for $n = 1$ and assuming true for $n = k$ implies true for $n = k + 1$, hence true for all $n \in \mathbb{N}$

19 a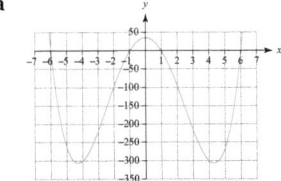

      $-4 < x < 3$ or $x > 5$

   b $x \le -13$ or $4 < x < 5$

20 a $\dfrac{1}{x} - \dfrac{1}{x+1} = \dfrac{x+1-x}{x(x+1)}$

         $= \dfrac{1}{x(x+1)}$

   b $1 - \dfrac{1}{n+1}$

   c $\dfrac{99}{100}$

   d $\dfrac{1}{100}$

21 a $\dfrac{1}{2}\left(\dfrac{1}{x-1} - \dfrac{1}{x+1}\right) = \dfrac{1}{2}\left(\dfrac{(x+1)-(x-1)}{(x-1)(x+1)}\right)$

      $\dfrac{1}{2}\left(\dfrac{2}{x^2-1}\right) = \dfrac{1}{x^2-1}$

   b $\dfrac{1}{2} + \dfrac{1}{4} - \dfrac{1}{2n} - \dfrac{1}{2n+2} = \dfrac{3}{4} - \dfrac{1}{2n} - \dfrac{1}{2n+2}$

   c $\dfrac{(3\times 20 + 2)(20-1)}{4\times 20 \times(20+1)} = \dfrac{589}{840}$

   d $\displaystyle\sum_{r=2}^{\infty} \dfrac{1}{r^2-1} = \dfrac{3}{4}$

22 $\displaystyle\sum_{r=1}^{n} r^2 = \dfrac{1}{6}(n+1)(2n^2 + n) = \dfrac{n}{6}(n+1)(2n+1)$

23 a $u^4 + 27u + 81 = 0$ b $u^4 + 2u^2 - u + 1 = 0$

24 $1 < x < 1.25$ and $x \le -1$

25 a $k \le -3, k \ge 5$

b $f(x) \le -3$ or $f(x) \ge 5$

$-1 \le \sin\theta \le 1$

Which is not in $f(x) \ge 5$ or $f(x) \le -3$

26 a $5x - 7x^2 + \dfrac{68}{3}x^3$ **b** $-\dfrac{1}{4} < x \le \dfrac{1}{4}$

27 a $\dfrac{1}{2}\left(\dfrac{(x+1)(x+2) - 2x(x+2) + x(x+1)}{x(x+1)(x+2)} \right)$

$= \dfrac{x^2 + 3x + 2 - 2x^2 - 4x + x^2 + x}{2x(x+1)(x+2)}$

$= \dfrac{1}{x(x+1)(x+2)}$

b $= \dfrac{n^2 + 3n + 2 - 2n - 4 + 2n + 2}{4(n+1)(n+2)} = \dfrac{n(n+3)}{4(n+1)(n+2)}$

c $\dfrac{5}{1848}$

28 a $\dfrac{2x(x(x+2)) + 3(x+2) - 3x}{x(x+2)}$

$= \dfrac{2x^3 + 4x^2 + 3x + 6 - 3x}{x(x+2)}$

$= \dfrac{2x^3 + 4x^2 + 6}{x(x+2)}$

b $\dfrac{n(2n^3 + 8n^2 + 19n + 19)}{2(n+1)(n+2)}$

29 When $n = 2$, $\displaystyle\sum_{r=1}^{n} \dfrac{1}{r^2 - 1} = \dfrac{1}{3}$

$\dfrac{3n^2 - n - 2}{4n(n+1)} = \dfrac{12 - 2 - 2}{4(2)(3)} = \dfrac{8}{24} = \dfrac{1}{3}$

So true for $n = 2$

Assume true for $n = k$

$\displaystyle\sum_{r=1}^{k+1} \dfrac{1}{r^2 - 1} = \dfrac{1}{(k+1)^2 - 1} + \sum_{r=1}^{k} \dfrac{1}{r^2 - 1}$

$= \dfrac{1}{(k+1)^2 - 1} + \dfrac{3k^2 - k - 2}{4k(k+1)}$

$= \dfrac{1}{k^2 + 2k} + \dfrac{3k^2 - k - 2}{4k(k+1)}$

$= \dfrac{4(k+1)}{4k(k+2)(k+1)} + \dfrac{(3k^2 - k - 2)(k+2)}{4k(k+1)(k+2)}$

$= \dfrac{4(k+1) + (3k^2 - k - 2)(k+2)}{4k(k+1)(k+2)}$

$= \dfrac{4k + 4 + 3k^3 + 6k^2 - k^2 - 2k - 2k - 4}{4k(k+1)(k+2)}$

$= \dfrac{3k^3 + 5k^2}{4k(k+1)(k+2)}$

$= \dfrac{k^2(3k + 5)}{4k(k+1)(k+2)}$

$= \dfrac{k(3k + 5)}{4(k+1)(k+2)}$

$\dfrac{3(k+1)^2 - (k+1) - 2}{4(k+1)(k+1+1)} = \dfrac{3k^2 + 6k + 3 - k - 1 - 2}{4(k+1)(k+2)}$

$= \dfrac{3k^2 + 5k}{4(k+1)(k+2)}$ as required]

True for $n = 2$ and assuming true for $n = k$ implies true for $n = k + 1$, hence true for all $n \in \mathbb{N}$. Since $\dfrac{1}{r^2 - 1}$ not defined at $r = 1$

Chapter 3

Exercise 3.1A

1 a i $\left(0, -\dfrac{1}{2}\right)$ $(-1, 0)$

ii $x = 2$ $y = 1$

iii

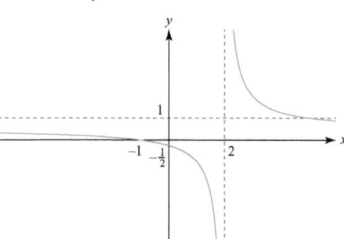

b i $\left(0, -\dfrac{1}{2}\right)$ $(1, 0)$

ii $x = -2$ $y = 1$

iii

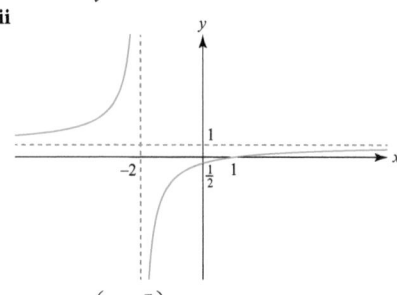

c i $(-5, 0)$ $\left(0, -\dfrac{5}{3}\right)$

ii $x = 3$ $y = 1$

iii

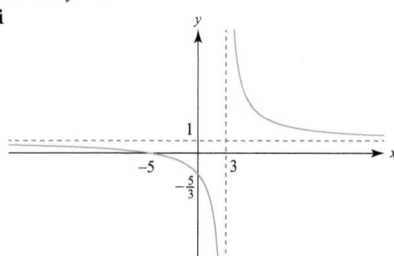

d i $\left(0, -\dfrac{5}{3}\right)$ $\left(\dfrac{5}{6}, 0\right)$

ii $x = -\dfrac{3}{8}$ $y = \dfrac{3}{4}$

iii

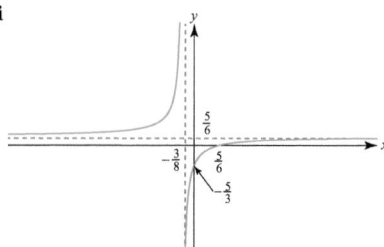

e i $(0, -1)$ $(-12, 0)$

ii $x = \dfrac{12}{5}$ $y = \dfrac{1}{5}$

iii

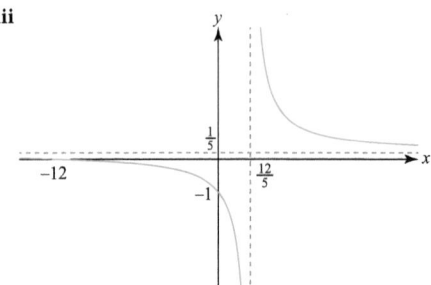

f i $\left(0, \dfrac{1}{2}\right)$ $\left(\dfrac{10}{7}, 0\right)$

ii $x = \dfrac{20}{3}$ $y = \dfrac{7}{3}$

iii

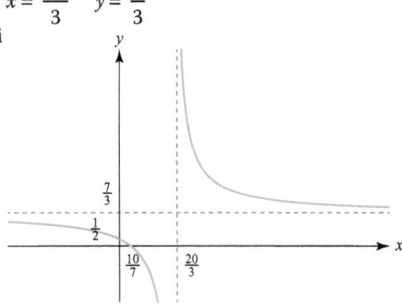

2 a intercepts: $(0, -a); (-a, 0)$
asymptotes: $x = \dfrac{1}{2}$; $y = \dfrac{1}{2}$

b intercepts: $\left(0, \dfrac{4}{b}\right); \left(-\dfrac{4}{3}, 0\right)$
asymptotes: $x = -b$; $y = 3$

c intercepts: $\left(0, \dfrac{a}{5}\right); \left(\dfrac{a}{2}, 0\right)$
asymptotes: $x = \dfrac{5}{3}$; $y = \dfrac{2}{3}$

d intercepts: $\left(0, \dfrac{-a}{b}\right); (a, 0)$
asymptotes: $x = -\dfrac{b}{2}$; $y = \dfrac{1}{2}$

3 a intercepts: $(-3, 0); \left(0, -\dfrac{3}{4}\right)$
asymptotes: $x = 4$; $y = 1$
points of intersection: $(11, 2)$

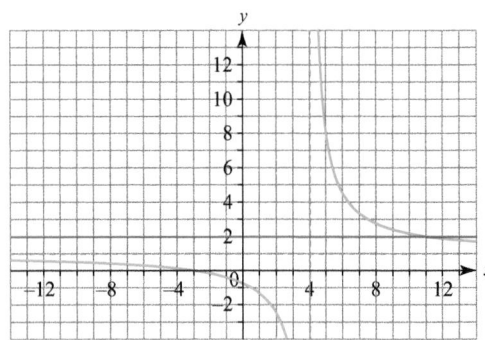

b intercepts: $\left(0, -\dfrac{3}{2}\right); (-6, 0)$
asymptotes: $x = 4$; $y = 1$
points of intersection: $(6, 6)$

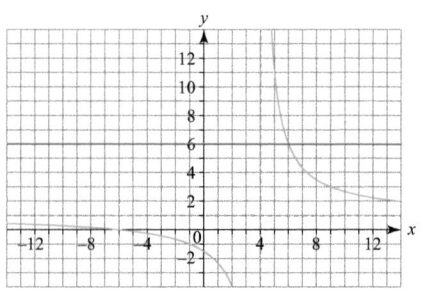

c intercepts: $(0, -1); (-12, 0)$
asymptotes: $x = \dfrac{12}{5}; y = \dfrac{1}{5}$
points of intersection: $(-12, 0)$

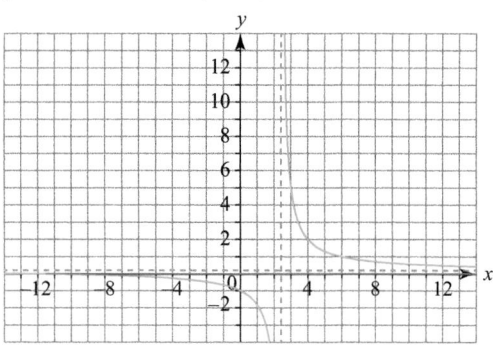

d intercepts: $\left(0, -\dfrac{4}{3}\right); \left(\dfrac{4}{3}, 0\right)$
asymptotes: $x = \dfrac{3}{2}; y = -\dfrac{3}{2}$
points of intersection: $\left(\dfrac{20}{13}, -8\right)$

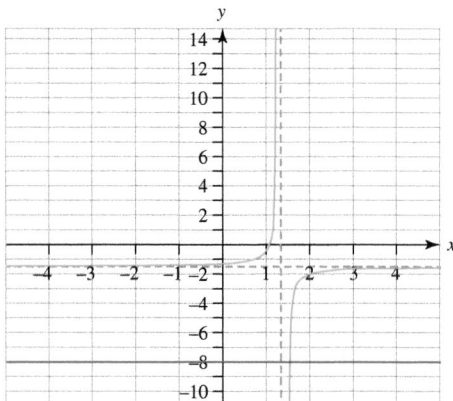

e intercepts: $(-5, 0); (0, 5); \left(-\dfrac{1}{2}, 0\right); (0, 1)$
asymptotes: $x = -1; y = 1$
points of intersection: $(-2, -3); (1, 3)$

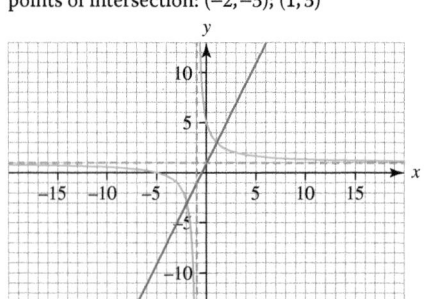

f intercepts: $(3, 0)$; $\left(0, -\dfrac{6}{5}\right)$; $(0, 3)$

asymptotes: $x = -5$; $y = 2$
points of intersection: $(-7, 10)$; $(3, 0)$

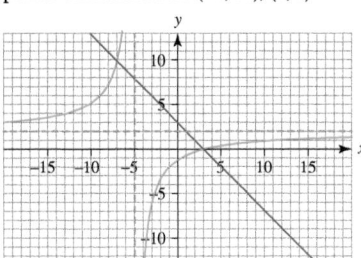

g intercepts: $\left(-\dfrac{15}{7}, 0\right)$; $\left(\dfrac{5}{3}, 0\right)$; $\left(0, \dfrac{10}{3}\right)$; $(0, 15)$

asymptotes: $x = -3$; $y = -6$
points of intersection: $(-5, -20)$; $(-1, 8)$

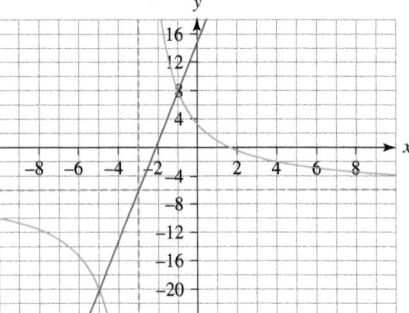

h intercepts: $(-6, 0)$; $(24, 0)$; $(0, 24)$; $(0, -6)$

asymptotes: $x = \dfrac{8}{3}$; $y = \dfrac{8}{3}$

points of intersection: $(4, 20)$; $(20, 4)$

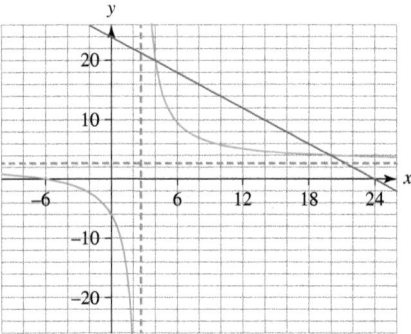

4 a intercepts: $(2, 0)$; $\left(0, -\dfrac{2}{5}\right)$

asymptotes: $x = -5$; $y = 1$
points of intersection: $(-6, 8)$

$\dfrac{x - 2}{x + 5} < 8$ when $x < -6$ or $x > -5$

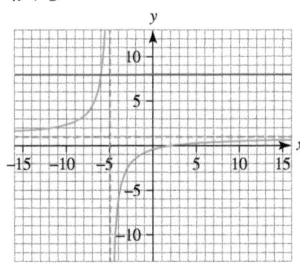

b intercepts: $\left(0, -\dfrac{5}{2}\right)$; $(15, 0)$

asymptotes: $x = -6$; $y = 1$
points of intersection: $(-13, 4)$

$\dfrac{x - 15}{x + 6} \le 4$ when $x \le -13$ or $x > -6$

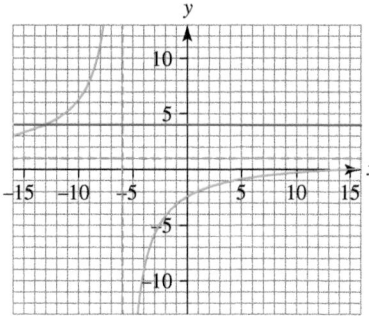

c intercepts: $\left(0, -\dfrac{1}{4}\right)$; $\left(\dfrac{1}{4}, 0\right)$

asymptotes: $x = -\dfrac{4}{3}$; $y = \dfrac{4}{3}$

points of intersection: $\left(-\dfrac{9}{2}, 2\right)$

$\dfrac{4x - 1}{3x + 4} < 2$ when $x < -\dfrac{9}{2}$ or $x > \dfrac{-4}{3}$

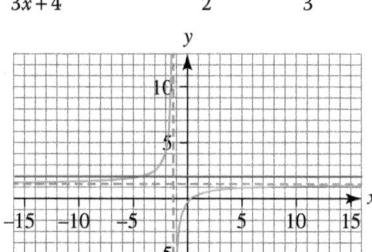

d intercepts: $\left(0, -\dfrac{5}{3}\right)$; $\left(-\dfrac{5}{3}, 0\right)$

asymptotes: $x = \dfrac{3}{2}$; $y = \dfrac{3}{2}$

points of intersection: $(2, 11)$

$\dfrac{3x + 5}{2x - 3} \ge 11$ when $\dfrac{3}{2} < x \le 2$

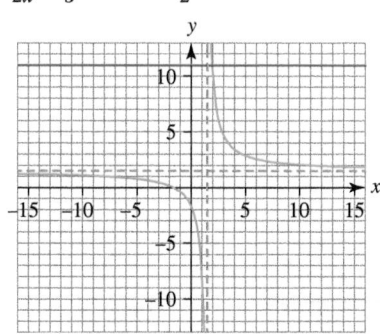

Exercise 3.1B

1 a intercepts: $(-5, 0)$; $(0, 5)$; $(0, 1)$; $\left(\dfrac{-1}{2}, 0\right)$

asymptotes: $x = -1$; $y = 1$

points of intersection: $(-2, -3)$; $(1, 3)$

$\dfrac{x + 5}{x + 1} > 2x + 1$ when $x < -2$ or $-1 < x < 1$

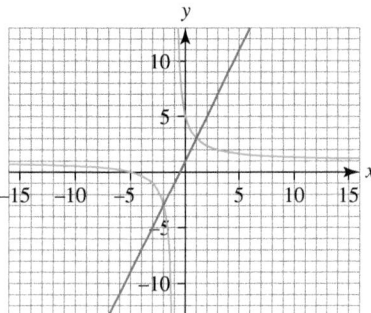

b intercepts: $(3, 0)$; $\left(0, -\dfrac{6}{5}\right)$; $(0, 3)$

asymptotes: $x = -5$; $y = 2$

points of intersection: $(-7, 10)$; $(3, 0)$

$\dfrac{2x - 6}{x + 5} < 3 - x$ when $x < -7$ or $-5 < x < 3$

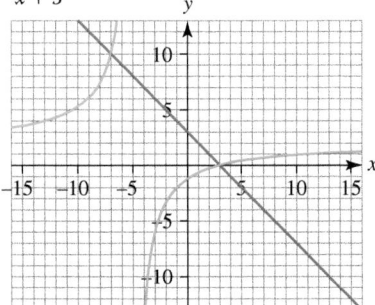

c intercepts: $\left(\dfrac{9}{2}, 0\right)$; $(-8, 0)$; $(0, -2)$; $(0, -9)$

asymptotes: $x = 4$; $y = 1$

points of intersection: $(2, -5)$; $(7, 5)$

$\dfrac{x + 8}{x - 4} \geq 2x - 9$ when $x \leq 2$, $4 < x \leq 7$

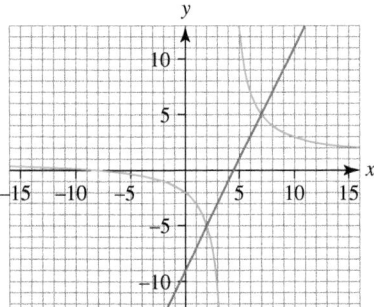

d intercepts: $(-3, 0)$; $(7, 0)$; $(0, -2)$; $(0, 14)$

asymptotes: $x = \dfrac{3}{2}$; $y = 1$

points of intersection: $(2, 10)$; $(6, 2)$

$\dfrac{2x + 6}{2x - 3} \geq 14 - 2x$ when $\dfrac{3}{2} < x \leq 2$, $x \geq 6$

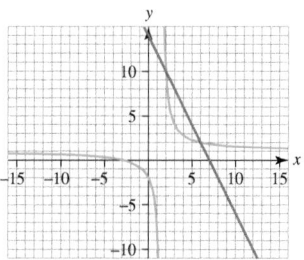

e intercepts: $\left(-\dfrac{15}{7}, 0\right)$; $\left(\dfrac{5}{3}, 0\right)$; $\left(0, \dfrac{10}{3}\right)$; $(0, 15)$

asymptotes: $x = -3$; $y = -6$

points of intersection: $(-5, -20)$; $(-1, 8)$

$\dfrac{10 - 6x}{x + 3} \leq 7x + 15$ when $-5 \leq x < -3$, $x \geq -1$

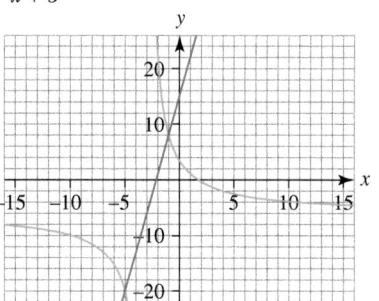

f intercepts: $(-6, 0)$; $(24, 0)$; $(0, 24)$; $(0, -6)$

asymptotes: $x = \dfrac{8}{3}$; $y = \dfrac{8}{3}$

points of intersection: $(4, 20)$; $(20, 4)$

$\dfrac{8x + 48}{3x - 8} > 24 - x$ when $\dfrac{8}{3} < x < 4$ or $x > 20$

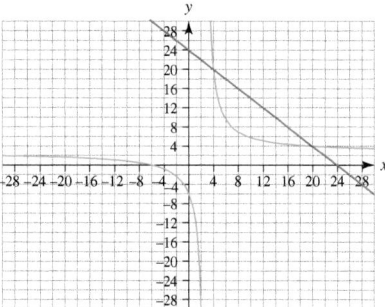

2 a intercepts: $\left(0, -\dfrac{1}{2}\right)$; $(-2, 0)$

asymptotes: $x = 4$; $y = 1$

points of intersection: $(2, -2)$; $(7, 3)$

$-2 < \dfrac{x + 2}{x - 4} < 3$ when $x < 2$ or $x > 7$

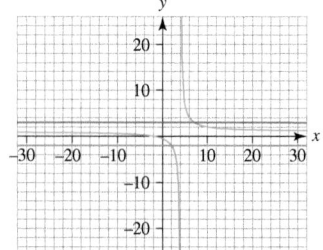

b intercepts: $(0,-1)$; $\left(-\dfrac{5}{2},0\right)$

asymptotes: $x=\dfrac{5}{2}$; $y=1$

points of intersection: $\left(\dfrac{-5}{2},0\right)$; $(2,-9)$

$0 \geq \dfrac{2x+5}{2x-5} \geq -9$ when $\dfrac{-5}{2} \leq x \leq 2$

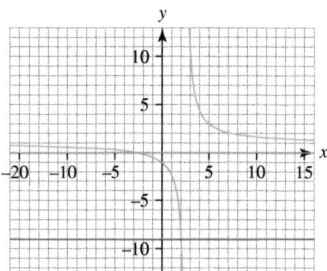

c intercepts: $(0,1)$; $\left(\dfrac{5}{3},0\right)$

asymptotes: $x=\dfrac{5}{2}$; $y=\dfrac{3}{2}$

points of intersection: $(0,1)$; $(2,-1)$

$-1 \leq \dfrac{3x-5}{2x-5} \leq 1$ when $0 \leq x \leq 2$

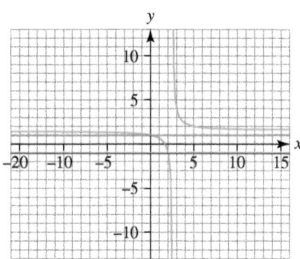

d intercepts: $(12,0)$; $(2,0)$; $(0,2)$; $(0,6)$
asymptotes: $x=6$; $y=1$
points of intersection between curve and lines:
$(0,2)$; $(4,4)$; $(7,-5)$; $(12,0)$

$2-x < \dfrac{x-12}{x-6} \leq 6-\dfrac{x}{2}$ when $0 < x \leq 4$ or $7 < x \leq 12$

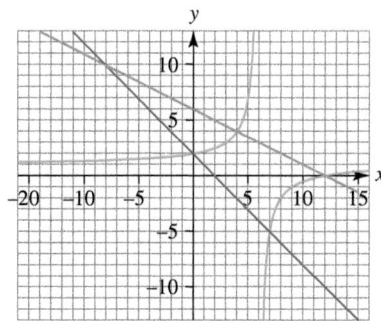

e intercepts: $(9,0)$; $\left(\dfrac{11}{3},0\right)$; $(0,11)$; $(0,6)$; $(0,3)$

asymptotes: $x=3$; $y=2$
points of intersection between curve and lines:
$(-3,4)$; $(1,8)$; $(5,-4)$; $(9,0)$

$11-3x \leq \dfrac{2x-18}{x-3} \leq \dfrac{9-x}{3}$ when $5 \leq x \leq 9$

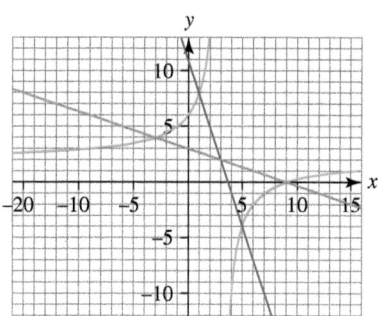

f intercepts: $\left(0,\dfrac{2}{3}\right)$; $(0,-3)$;

asymptotes: $x=1$; $x=-3$; $y=0$

points of intersection: $\left(-11,-\dfrac{1}{4}\right)$

$\dfrac{2}{x+3} > \dfrac{3}{x-1}$ when $x<-11$ or $-3<x<1$

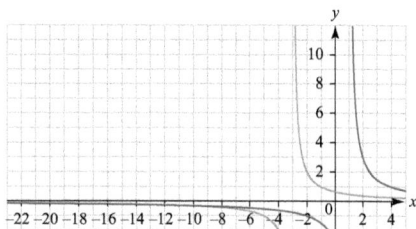

g intercepts: $\left(0,\dfrac{5}{4}\right)$; $\left(0,-\dfrac{3}{2}\right)$

asymptotes: $x=-4$; $x=2$; $y=0$
points of intersection: $\left(11,\dfrac{1}{3}\right)$

$\dfrac{5}{x+4} \leq \dfrac{3}{x-2}$ when $x<-4$, $2<x\leq 11$

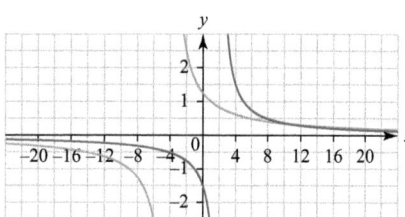

3 The graph shows that $\dfrac{2x}{x+2}$ is only less than 1 in the range $-2<x<2$

At all other times the function is greater than or equal to 1

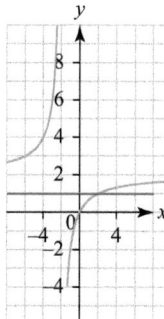

4 The two appropriate graphs are $y = \dfrac{3}{x-1}$ and $y = x+1$

The inequality is thus $\dfrac{3}{x-1} \geq x+1$

The curve is 'above' the line in the ranges
$x \leq -2$ and $1 < x \leq 2$

Therefore the solution is $x \leq -2,\, 1 < x \leq 2$

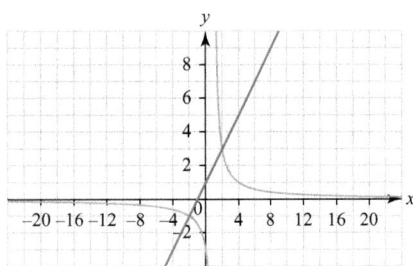

5 The inequality is $-1 \leq \dfrac{x-5}{x+5} \leq 1$

The curve lies between the straight lines only when $x \geq 0$

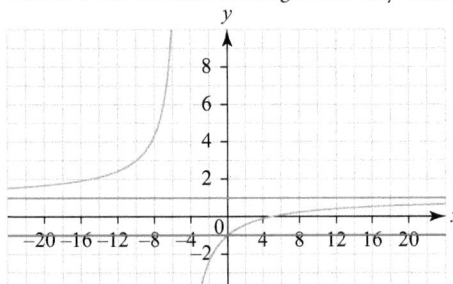

6 The inequality is $-1 \leq \dfrac{x+1}{1-x} \leq 1$

The curve lies between the straight lines only when $x \leq 0$

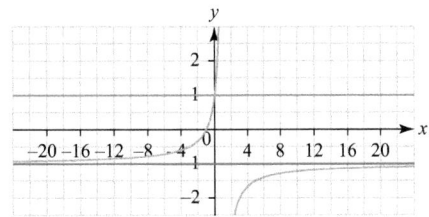

7 a $y = \dfrac{2x}{x+1} - \dfrac{x-3}{x-5}$

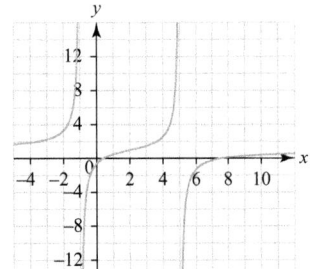

b The graph crosses the y-axis when $x = 0$

At this point, $y = 0 - \dfrac{-3}{-5} = \dfrac{3}{5}$

c The equations of its vertical asymptotes are where
denominators of the fractions are 0 i.e. where $x = -1$ and $x = 5$

Its horizontal asymptote occurs when x gets very large,

and y approaches $\dfrac{2}{1} - \dfrac{1}{1} = 1$

8 $\dfrac{x}{x+1}$ is the blue graph and $\dfrac{x+1}{x}$ is the red graph.

$\dfrac{x}{x+1} \geq \dfrac{x+1}{x}$ when $x < -1$,

$-\dfrac{1}{2} \leq x < 0$

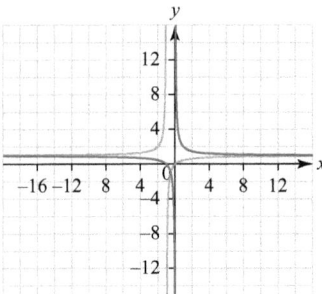

9 a $x^2 - 2x - 15 = 0 \rightarrow x^2 - 2x = 15 \rightarrow x - 2 = \dfrac{15}{x}$

Thus $x^2 - 2x - 15 = 0$ is equivalent to $x - 2 = \dfrac{15}{x}$

$x - 2 = \dfrac{15}{x}$ when $x = -3$ or 5

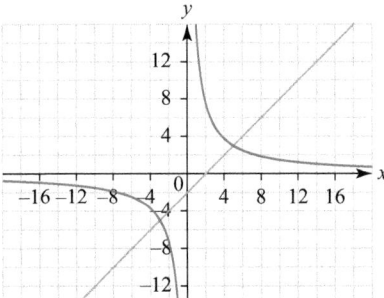

Alternatively, $x^2 - 2x - 15 = 0 \rightarrow x - 2 + x^2 - 3x - 13 = 0$
$\rightarrow x - 2 = -x^2 + 3x + 13$ when $x = -3$ or 5

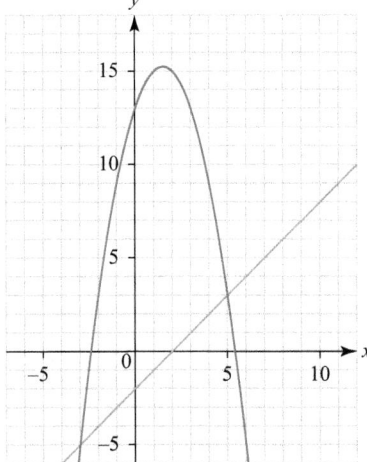

A similar alternative is $x^2 - x - 17 = x - 2$

b $0 = x^2 - 2x - 15 = (x-5)(x+3) \Rightarrow x = 5, -3$

10 $x^3 - x^2 - 6x = 0 \rightarrow x^2(x-1) = 6x \rightarrow x^2 = \dfrac{6x}{x-1}$
Thus $A = 6$ and $B = -1$

x^2 is the blue graph and $\dfrac{6x}{x-1}$ is the red graph.

$x^2 = \dfrac{6x}{x-1}$ when $x = -2$ or $x = 0$ or $x = 3$

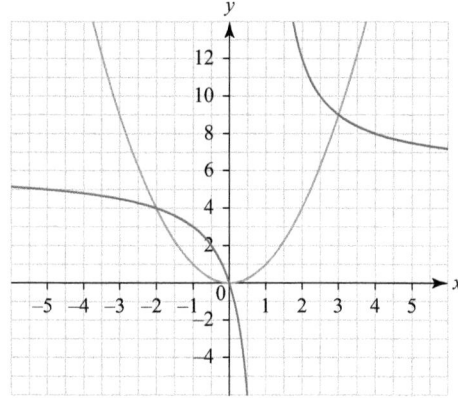

11 $\dfrac{1}{x}$ is the blue graph, $\dfrac{x}{2}$ is the red graph and $\dfrac{3}{x+2}$ is the green graph.

Blue meets green when $\dfrac{1}{x} = \dfrac{3}{x+2} \rightarrow x = 1$

Green has an asymptote at $x = -2$, and blue has an asymptote at $x = 0$

Hence green is the largest in the range $-2 < x < 0$

Blue meets red when $x^2 = 2$, i.e. $x = \pm\sqrt{2}$

Thus $\dfrac{1}{x}$ is the largest when $x < -2$ or $0 < x < 1$;

Red meets green when $\dfrac{x}{2} = \dfrac{3}{x+2} \rightarrow x^2 + 2x - 6 = 0$

$\rightarrow (x+1)^2 - 1 - 6 = 0 \rightarrow x = -1 \pm\sqrt{7}$

Thus $\dfrac{x}{2}$ is the largest when $x > -1 + \sqrt{7}$

$\dfrac{3}{x+2}$ is the largest when $-2 < x < 0$ or $1 < x < -1 + \sqrt{7}$

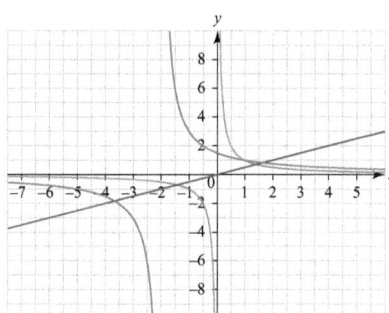

Exercise 3.2A

1 a Intercepts are $(0, -1)$, $(3, 0)$ and $(-3, 0)$
There are no vertical asymptotes as denominator never zero.
As $x \rightarrow \infty$, $y \rightarrow 1$ so $y = 1$ is a horizontal asymptote.
b Intercept is $(0, -1)$
Denominator $= 0$ when $x = \pm 3$ so vertical asymptotes at $x = 3$ and $x = -3$
As $x \rightarrow \infty$, $y \rightarrow 1$ so $y = 1$ is a horizontal asymptote.
c Intercept is $(0, 0)$
There are no vertical asymptotes as denominator never zero.
As $x \rightarrow \infty$, $y \rightarrow 1$ so $y = 1$ is a horizontal asymptote.

d Intercepts are $(0, 0)$ and $(-5, 0)$
There are no vertical asymptotes as denominator never zero.
As $x \rightarrow \infty$, $y \rightarrow 1$ so $y = 1$ is a horizontal asymptote.
e When $x = 0$ denominator is zero so $x = 0$ is vertical asymptote.
Denominator also zero when $x = -4$ so this is another vertical asymptote.
Numerator never equal to zero line doesn't cross x axis.
As $x \rightarrow \infty$, $y \rightarrow 1$ so $y = 1$ is a horizontal asymptote.
f When $x = 0$ denominator is zero so $x = 0$ is vertical asymptote.
Denominator also zero when $x = -4$ so this is another vertical asymptote.
When $y = 0$, $x^2 - 3x - 4 = 0$
$(x - 4)(x + 1) = 0$ so $x = 4$ or $x = -1$
So intercepts are $(4, 0)$ and $(-1, 0)$
As $x \rightarrow \infty$, $y \rightarrow 1$ so $y = 1$ is a horizontal asymptote.

2 a The denominator $= 0$ when $x = -2 \pm\sqrt{7}$ so the graph has asymptotes and does not exist for these values of x. Otherwise the graph is defined. Hence it has range

$x \neq -2 \pm\sqrt{7}$, $y \geq \dfrac{3}{7}$ or $y \leq 0$

b The denominator $= 0$ when $x = 1$ or $x = 2$ so the graph has asymptotes and does not exist for these values of x. Otherwise the graph is defined, and hence has range
$x \neq 1, 2$
$y < 1$ or $y \geq 9$

c The denominator $= 0$ when $x = \dfrac{3 \pm\sqrt{5}}{2}$ so the graph has asymptotes and does not exist for these values of x

The graph exists when $y \neq 1$ and $y < 1$ or $y \geq 5$

When $y = 1$, the equation $1 = \dfrac{x^2 - 3x - 4}{x^2 - 3x + 1}$ has no real solutions, so $y = 1$ is excluded from the range.

d The denominator is never zero so the graph exists for all values of x.
$-0.112 \leq y \leq 1.11$ (3sf)
The function takes the value $y = 0$ at $x = 2$, so this is included in the range.

3 a $(-2, -4)$ is a local maximum point.
$(0, 0)$ is a local minimum point.
b $(2, 3)$ is a local maximum point.
$(-2, -3)$ is a local minimum point.

c $\left(0, \dfrac{3}{2}\right)$ is a local maximum point.

d $\left(-2, -\dfrac{5}{4}\right)$ is a local maximum point.

4 a

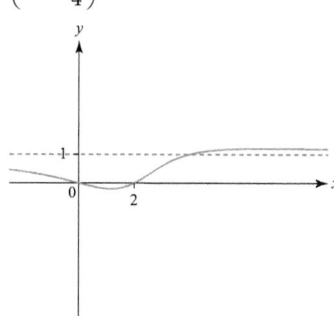

b

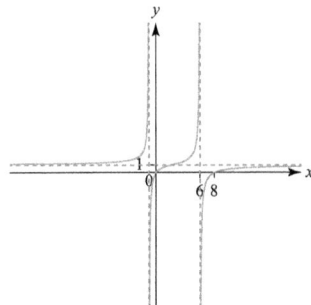

c

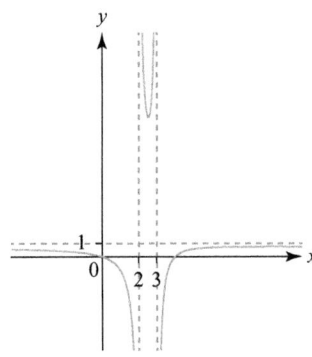

d

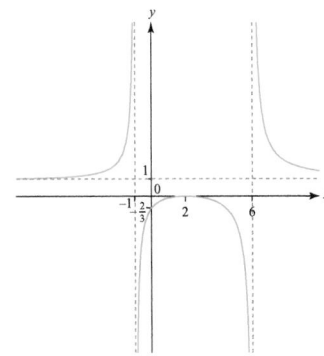

Exercise 3.2B

1 $\dfrac{x}{x^2 + 1} = y$

$y(x^2 + 1) = x$

$yx^2 - x + y = 0$

$y = 0$ at $x = 0$; otherwise y is real if $1 - 4y(y) \geq 0$

$1 - 4y^2 \geq 0$

$4y^2 - 1 \leq 0$

$y^2 \leq \dfrac{1}{4}$

$-\dfrac{1}{2} \leq y \leq \dfrac{1}{2}$

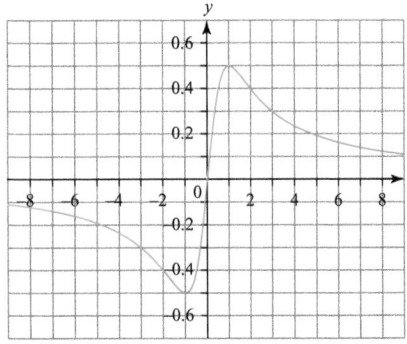

2 $\dfrac{x^2}{x^2 + 1} = y$

$y(x^2 + 1) = x^2$

$(y - 1)x^2 + y = 0$

$y = 0$ at $x = 0$; otherwise y has real roots if $0 - 4y(y - 1) \geq 0$

$4y(y - 1) \leq 0$

$0 \leq y \leq 1$

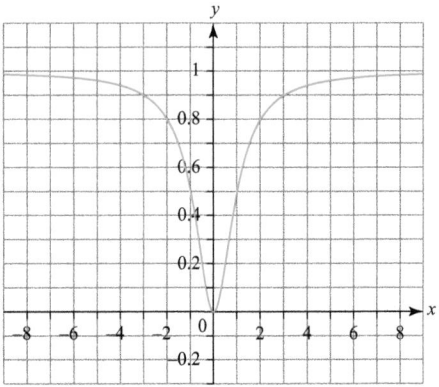

3 $\dfrac{x^2}{x^2 - 1} = y$

$y(x^2 - 1) = x^2$

$(y - 1)x^2 - y = 0$

There are no values of x for which $y = 1$; otherwise y has no real roots if $0 + 4y(y - 1) < 0$

$4y(y - 1) < 0$

$0 < y < 1$

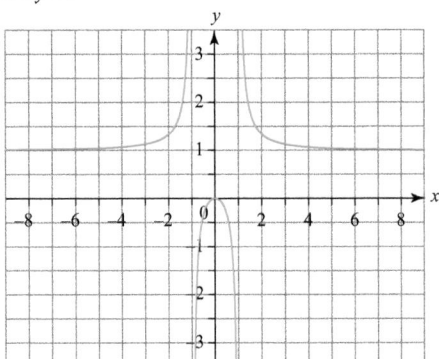

4 $y = \dfrac{x^2 + 5}{x - 2}$

$y(x - 2) = x^2 + 5$

$x^2 - yx + (2y + 5) = 0$

y has real roots if $y^2 - 4(2y + 5) \geq 0$

$y^2 - 8y - 20 \geq 0$

$(y - 10)(y + 2) \geq 0$

$y < -2$ or $y > 10$ At $y = -2$, we have the equation

$0 = x^2 + 2x + 1 = (x + 1)^2$

hence we have a local maximum at $(-1, -2)$

At $y = 10$, we have the equation

$0 = x^2 - 10x + 25 = (x - 5)^2$

hence we have a local minimum at $(5, 10)$

This apparent contradiction is explained by the fact that these are local maxima and minima.

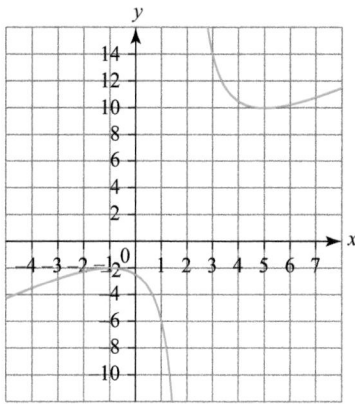

5

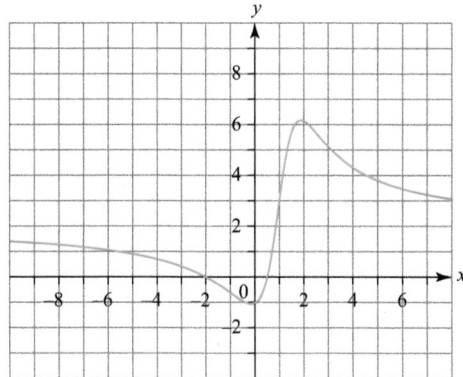

6 a $c \le 1$

Hence greatest possible value of c is 1

b When $c = 1$

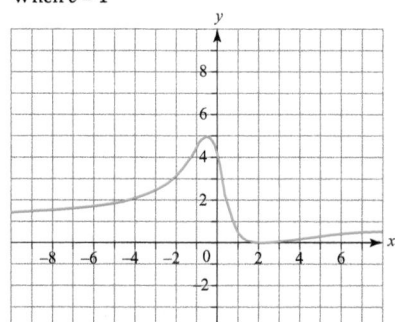

7 $C \le -9, C \ge -1$

local maximum at $\left(-\dfrac{1}{3}, -9 \right)$

local minimum at $(1, -1)$

8 $y = \dfrac{x^2 + A}{x^2 - B^2}$

$y(x^2 - B^2) = x^2 + A$

$(y - 1)x^2 - (B^2 y + A) = 0$

The value $y = 1$ cannot be attained for any value of x

Otherwise, y has no real roots if $0 + 4(y - 1)(B^2 y + A) < 0$

$4(y - 1)(B^2 y + A) < 0$

$(y - 1)(B^2 y + A) < 0$

$-\dfrac{A}{B^2} < y < 1$

For $y = \dfrac{x^2 + 4}{x^2 - 3^2}$, $A = 4$ and $B = 3$, and so y has no real roots in the range

$-\dfrac{4}{9} < y < 1$

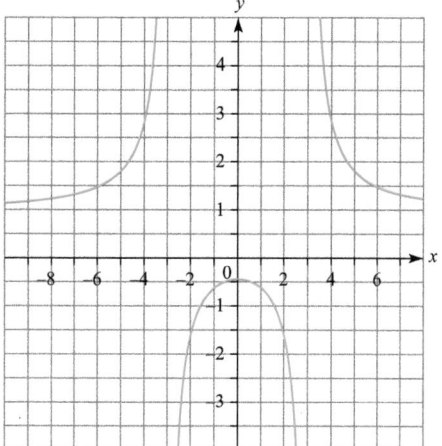

Max turning point has coordinates $\left(0, \dfrac{-4}{9} \right)$

9 a $y = \dfrac{4}{x^2 - x}$ is the blue graph and $y^2 = \dfrac{4}{x^2 - x}$ is the red graph.

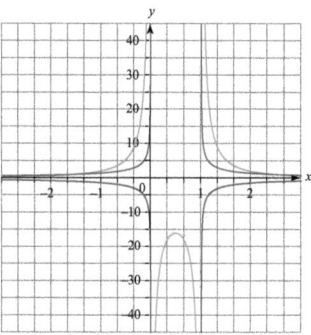

b The graphs intersect at $x = \dfrac{1}{2}(1 \pm \sqrt{17})$ Hence the inequality holds for

$\dfrac{1}{2}(1 - \sqrt{17}) < x < 0$ and $1 < x < \dfrac{1}{2}(1 + \sqrt{17})$

c $y^2 = \dfrac{4}{x^2 - x}$ is negative between $x = 0$ and $x = 1$ and the square root of a negative number does not have any real values.

10 a $y = \dfrac{x^2 - A}{x^2 - 4x - A}$

$y(x^2 - 4x - A) = x^2 - A$

$(y - 1)x^2 - 4yx - A(y - 1) = 0$

$y = 1$ is attained at $x = 0$ otherwise y has real roots if

$16y^2 + 4A(y - 1)^2 \ge 0$

$16y^2$ and $4(y - 1)^2$ are both ≥ 0, this holds when $A \ge 0$

b i The graph shows there is no turning point.

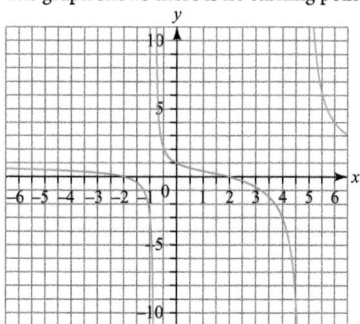

ii $x = \pm 2$

11 If $y = \dfrac{kx}{x^2 - x + 1}$

$yx^2 - (k+y)x + y = 0$

$y = 0$ is attained at $x = 0$ otherwise y has no real roots if
$(k+y)^2 - 4y^2 < 0$

$k^2 + 2ky - 3y^2 < 0$

$y^2 - \dfrac{2k}{3}y - \dfrac{k^2}{3} > 0$

$\left(y - \dfrac{k}{3}\right)^2 - \dfrac{k^2}{3} - \dfrac{k^2}{9} > 0$

$\left(y - \dfrac{k}{3}\right)^2 - \dfrac{4k^2}{9} > 0$

$\Rightarrow y$ has no real roots if $y < \dfrac{-k}{3}$, $y > k$ for no real roots

Hence the minimum and maximum values of y are $-\dfrac{k}{3}$ and k

When $k = 6$, the minimum and maximum values of y are -2 and 6

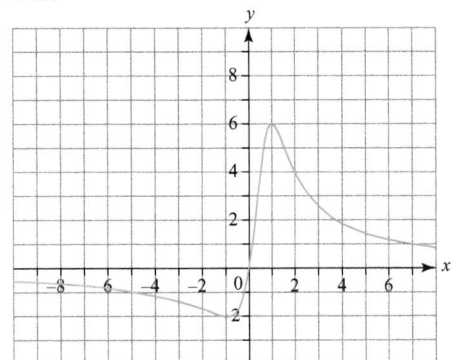

Exercise 3.3A

1

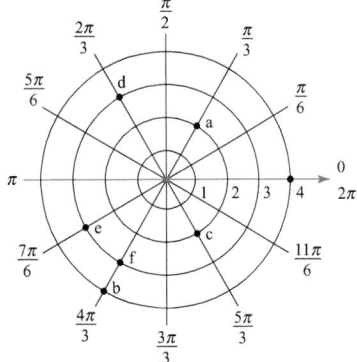

a $(1, \sqrt{3})$ **b** $(-2, -2\sqrt{3})$

c $(1, -\sqrt{3})$ **d** $\left(\dfrac{-3}{2}, \dfrac{3\sqrt{3}}{2}\right)$

e $\left(-\dfrac{3\sqrt{3}}{2}, -\dfrac{3}{2}\right)$ **f** $\left(-\dfrac{3}{2}, -\dfrac{3\sqrt{3}}{2}\right)$

2 **a** $\left(4, \dfrac{\pi}{6}\right)$ **b** $\left(6, \dfrac{2\pi}{3}\right)$

   c $\left(8, \dfrac{5\pi}{3}\right)$ **d** $(13, 4.32)$

3 **a** $\left(4, -\dfrac{5\pi}{6}\right)$ **b** $\left(10, -\dfrac{\pi}{3}\right)$

   c $(5, 2.21)$ **d** $(\sqrt{37}, 1.41)$

4 $r = \sqrt{5}$, circle centre origin, radius $\sqrt{5}$

5 Circle, centre origin, radius 4; $x^2 + y^2 = 16$

6 **a** Sketch should show graph of $y = x$

$\theta = \dfrac{\pi}{4} \Rightarrow \tan\theta = \tan\dfrac{\pi}{4} = 1 \Rightarrow \dfrac{y}{x} = 1 \Rightarrow y = x$

   b Sketch should show graph $x = 0$

$\theta = \dfrac{\pi}{2} \Rightarrow \tan\theta = \tan\dfrac{\pi}{2}$

so gradient is undefined, hence lies on $x = 0$

   c Sketch should show $y = -x$

$\theta = \dfrac{3\pi}{4} \Rightarrow \tan\theta = \tan\dfrac{3\pi}{4} = -1 \Rightarrow \dfrac{y}{x} = -1 \Rightarrow y = -x$

   d Sketch should show $y = x\dfrac{\sqrt{3}}{3}$

$\theta = -\dfrac{\pi}{6} \Rightarrow \dfrac{y}{x} = \tan\left(-\dfrac{\pi}{6}\right)$ so $y = -\dfrac{\sqrt{3}}{3}x$

7 **a**

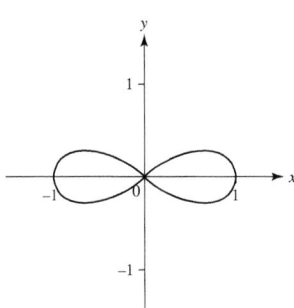

   b

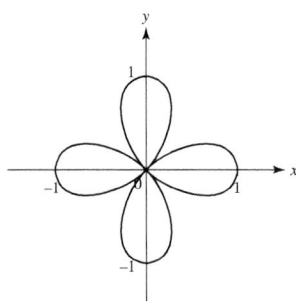

8 **a** **i** maximum = 1, minimum = 0

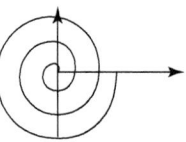

   ii maximum = 1, minimum = -1

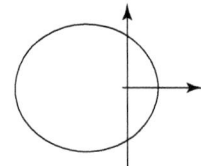

   b maximum = 1, minimum = 0

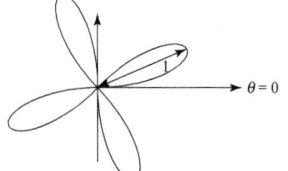

c **i** maximum = 2, minimum = 0

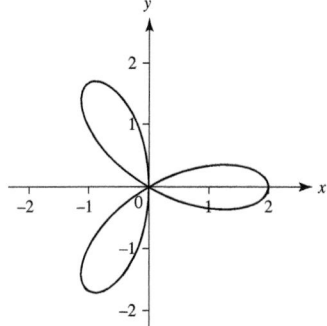

ii maximum = 2, minimum = −2

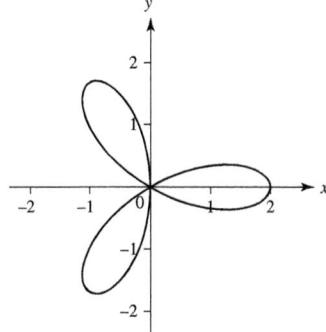

d **i** maximum = 4, minimum = 0

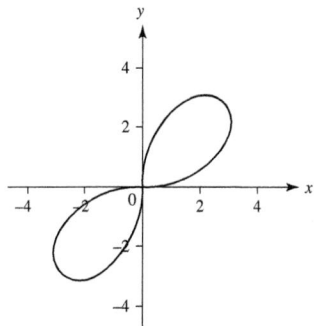

ii maximum = 4, minimum = −4

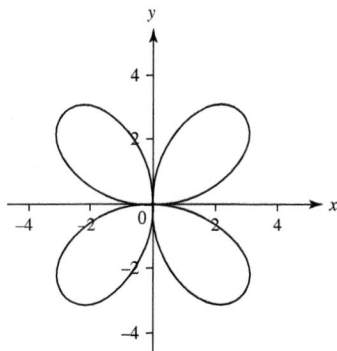

e Max = 2, min = 0

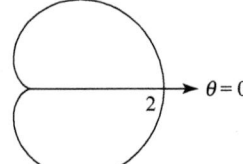

f Max = 5, min = 3

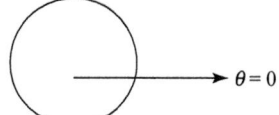

g Max = 5, min = 1

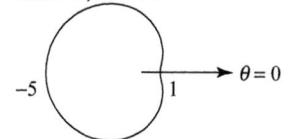

h **i** maximum = 8, minimum = 0

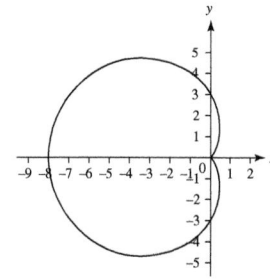

ii maximum = 8, minimum = −2

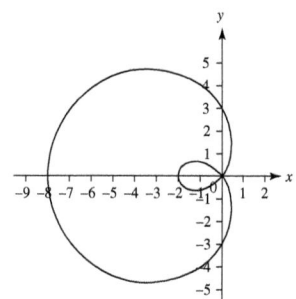

i Max = 4π, min = 0

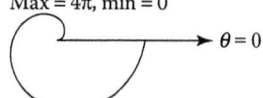

j **i** maximum = 2, minimum = 0

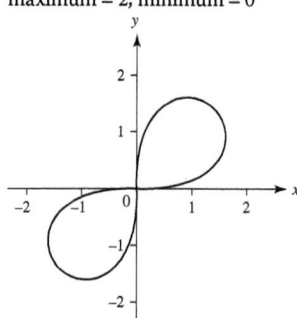

ii maximum = 2, minimum = −2

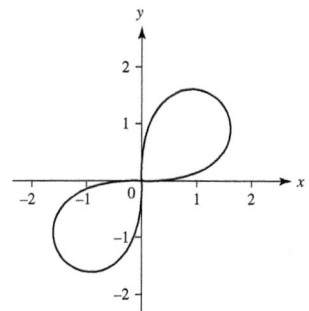

Exercise 3.3B

1 a $\theta = \dfrac{\pi}{4}$ **b** $\theta = \arctan 2$

c $\theta = \dfrac{\pi}{2}, r \geq 0$ **d** $\theta = 0 \quad r \geq 0$

e $r = \dfrac{5}{\sin \theta}$ **f** $r = \dfrac{3}{\cos \theta}$

g $r(\sin \theta - \cos \theta) = 1 \Rightarrow r$

$= \dfrac{1}{\sin \theta - \cos \theta} \quad \theta \neq n\pi + \dfrac{\pi}{4}$

h $r = \dfrac{c}{\sin \theta - m \cos \theta} \quad \theta \neq n\pi + \arctan(m)$

i $r = 5$

j $r = 10 \cos \theta + 24 \sin \theta$ **k** $r = 6 \cos \theta$

l $r = 8 \sin \theta$ **m** $r = 10(\cos \theta + \sin \theta)$

n $r = \dfrac{\sin \theta}{\cos^2 \theta}$

o $r^2 = \dfrac{5}{1 + \sin 2\theta} \quad \theta \neq n\pi - \dfrac{\pi}{4}$

2 a $y = 1 - x$
   a straight line, gradient -1, y-intercept 1.

b $y = 2x + 5$
   a straight line, gradient 2, y-intercept 5.

c $x = 2$
   a line parallel to y-axis cutting x-axis at 2.

d $y = -\dfrac{2}{3}$
   a line parallel to x-axis cutting y-axis at $-\dfrac{2}{3}$

e $x^2 + (y - 2)^2 = 2^2$
   a circle centre $(0, 2)$ radius 2.

f $(x - 4)^2 + (y - 3)^2 = 25 = 5^2$
   a circle centre $(4, 3)$ radius 5

3 a $\left(2, \dfrac{\pi}{6}\right), \left(2, \dfrac{5\pi}{6}\right)$

b $\left(\dfrac{1}{2}, \pm\dfrac{\pi}{12}\right), \left(\dfrac{1}{2}, \pm\dfrac{5\pi}{12}\right), \left(\dfrac{1}{2}, \pm\dfrac{7\pi}{12}\right), \left(\dfrac{1}{2}, \pm\dfrac{11\pi}{12}\right)$

c $\left(\dfrac{\sqrt{2}}{2}, \dfrac{\pi}{8}\right), \left(\dfrac{\sqrt{2}}{2}, -\dfrac{7\pi}{8}\right),$

4 Spiral has equation $r = a\theta$ and circle has equation $r = b$

Intersect when $a\theta = b \Rightarrow \theta = \dfrac{b}{a}$

So single point of intersection at $\left(b, \dfrac{b}{a}\right)$

5 a i

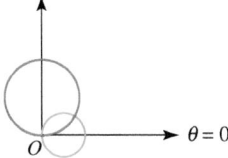

ii $\left(\dfrac{\sqrt{3}}{2}, \dfrac{\pi}{6}\right)$

b i

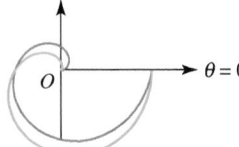

ii $(0, 0), (\pi, \pi), (2\pi, 2\pi)$

c i

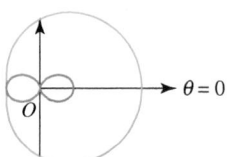

ii $(1, \pi)$

d i

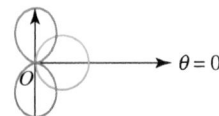

ii $\left(\dfrac{-1+\sqrt{5}}{2}, 0.905\right), \left(\dfrac{-1+\sqrt{5}}{2}, 5.38\right)$

e

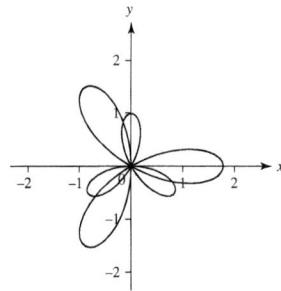

$\left(\dfrac{\sqrt{3}}{2}, \dfrac{5\pi}{9}\right), \left(\dfrac{\sqrt{3}}{2}, \dfrac{11\pi}{9}\right), \left(\dfrac{\sqrt{3}}{2}, \dfrac{17\pi}{9}\right)$

f $\left(\dfrac{\sqrt{3}}{2}, \dfrac{2\pi}{9}\right), \left(\dfrac{\sqrt{3}}{2}, \dfrac{5\pi}{9}\right), \left(\dfrac{\sqrt{3}}{2}, \dfrac{8\pi}{9}\right), \left(\dfrac{\sqrt{3}}{2}, \dfrac{11\pi}{9}\right),$

$\left(\dfrac{\sqrt{3}}{2}, \dfrac{14\pi}{9}\right), \left(\dfrac{\sqrt{3}}{2}, \dfrac{17\pi}{9}\right)$

6 a $-1 \leq \cos \theta \leq 1$
   $-4 \leq 4 \cos \theta \leq 4$
   $a - 4 \leq a + 4 \cos \theta \leq a + 4$
   So if $a > 4$ then $a + 4 \cos \theta > 0$ for all θ. So $r > 0$. To pass through the pole $r = 0$.
   So curve never passes through pole.

b $r(\theta) = a + 4 \cos \theta$
   $r(-\theta) = a + 4 \cos(-\theta) = a + 4 \cos \theta = r(\theta)$
   Since $r(-\theta) = r(\theta)$ curve is symmetrical about the initial line.

c i

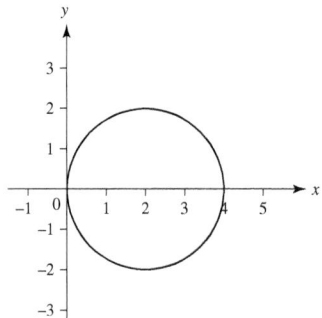

ii

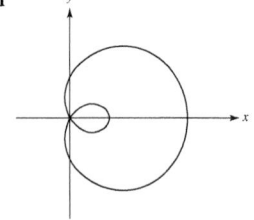

iii

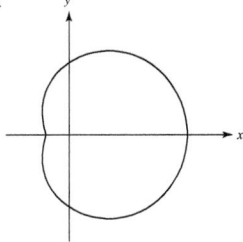

iv

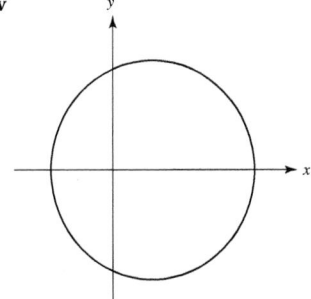

d $-1 \le \cos\theta \le 1$

$\Rightarrow a - 4 \le r \le a + 4$

7 $\sec^2\theta = 2\tan\theta$

$\Rightarrow 1 = 2\sin\theta\cos\theta = \sin 2\theta$

$\Rightarrow 2\theta = \dfrac{\pi}{2}, \dfrac{5\pi}{2}$

$\Rightarrow \theta = \dfrac{\pi}{4}, \dfrac{5\pi}{4}$

$\theta = \dfrac{\pi}{4}, r = \sqrt{2}$

$\theta = \dfrac{5\pi}{4}, r = -\sqrt{2}$

So they intersect at $\left(\sqrt{2}, \dfrac{\pi}{4}\right)$ t and $\left(-\sqrt{2}, \dfrac{5\pi}{4}\right)$.

But these are the same point, so they only intersect at one point.

8   **a**

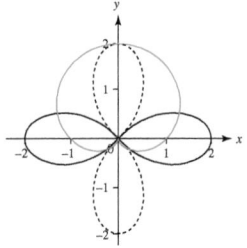

$(1.390, 0.401), (1.390, 2.741), (0.360, -0.695), (0.360, -2.45)$

Since $1 + \sin\theta$ is always non-negative, extending r to permit negative values does not add any additional solutions.

b

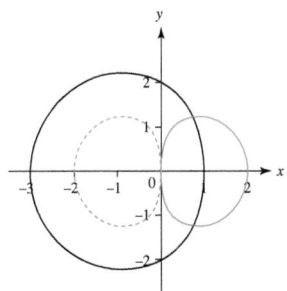

Points of intersection are $(1.464, 1.005), (1.464, -1.005)$
Since $2 - \cos\theta$ is always non-negative, extending r to permit negative values does not add any additional solutions.

9 $2\sin^2\theta = \cos 2\theta = 1 - 2\sin^2\theta$

$4\sin^2\theta = 1$

$\sin^2\theta = \dfrac{1}{4}$

$\sin\theta = \pm\dfrac{1}{2}$

$\theta = \pm\dfrac{\pi}{6}, \pm\dfrac{5\pi}{6}$

$r = 2\sin^2\theta = \dfrac{1}{2}$ so points of intersection are

$\left(\dfrac{1}{2}, \pm\dfrac{\pi}{6}\right), \left(\dfrac{1}{2}, \pm\dfrac{5\pi}{6}\right)$

The second pair of points are reflections of the first pair through the y-axis, so these form a rectangle, with Cartesian

coordinates $\left(\pm\dfrac{\sqrt{3}}{4}, \pm\dfrac{1}{4}\right)$,

$\text{Area} = 2 \times \dfrac{1}{4} \times \dfrac{\sqrt{3}}{4} = \dfrac{\sqrt{3}}{4}$

Exercise 3.4A

1   **a**

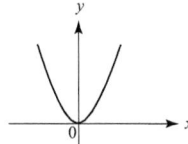

b

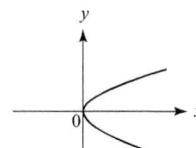

c

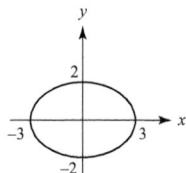

d

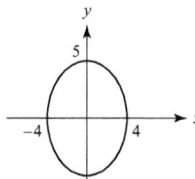

e

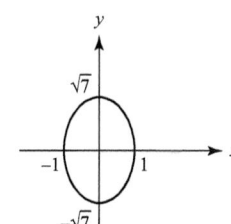

f

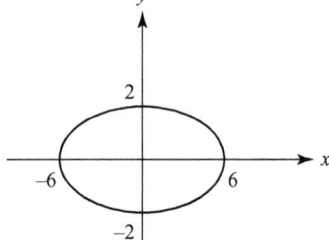

g

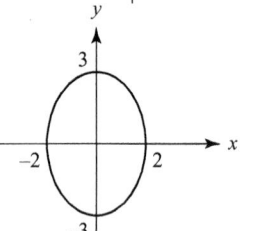

h

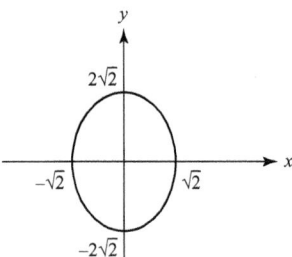

2 a

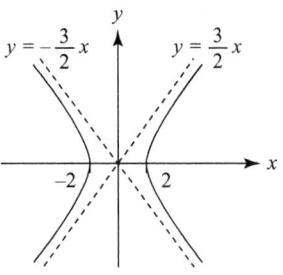

Asymptotes: $y = \pm\dfrac{3}{2}x$

b

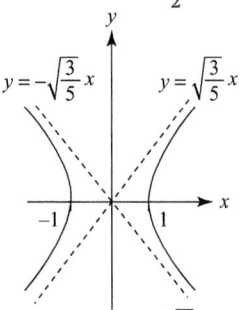

Asymptotes: $y = \pm\sqrt{\dfrac{3}{5}}x$

c

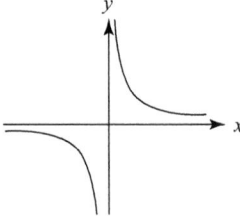

Asymptotes $x = 0$, $y = 0$

d

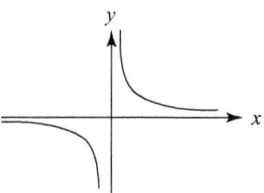

Asymptotes $x = 0$, $y = 0$

e

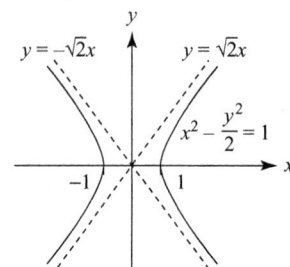

Asymptotes: $y = \pm\sqrt{2}x$

f

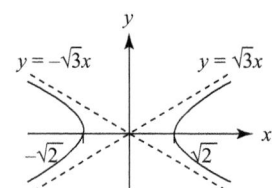

Asymptotes: $y = \pm\sqrt{3}x$

g

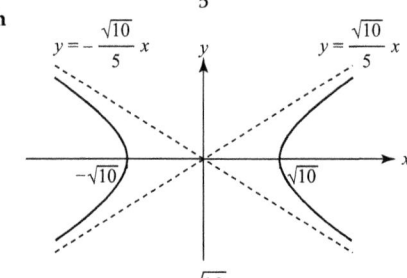

Asymptotes: $y = \pm\dfrac{\sqrt{5}}{5}x$

h

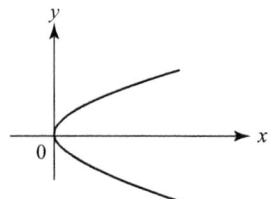

Asymptotes: $y = \pm\dfrac{\sqrt{10}}{5}x$

3 a

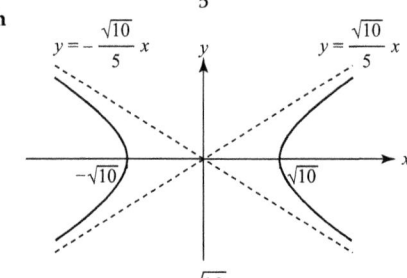

b (0, 0) and (5, 10)

4 a

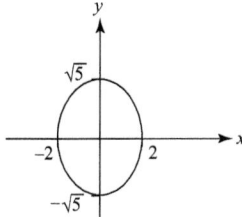

b $(-2, 0)$ and $\left(\dfrac{2}{9}, \dfrac{20}{9} \right)$

5 a

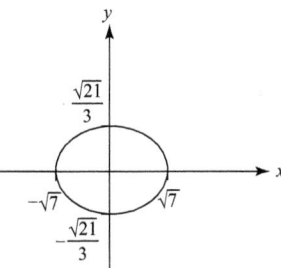

b $x = \dfrac{7 - 3y}{2}$

Substitute to give $\left(\dfrac{7 - 3y}{2} \right)^2 + 3y^2 = 7$

$\Rightarrow 49 - 42y + 9y^2 + 12y^2 = 28$

$\Rightarrow 21y^2 - 42y + 21 = 0$

$\Rightarrow y^2 - 2y + 1 = 0$

$\Rightarrow (y - 1)^2 = 0$

So only one solution: $y = 1$, $x = 2$

Therefore must be a tangent

6 a Ellipse $\dfrac{x^2}{27} + \dfrac{y^2}{8} = 1$

b Hyperbola $\dfrac{x^2}{256} - \dfrac{y^2}{128} = 1$

7 $\dfrac{9x^2}{8} - \dfrac{y^2}{8} = 1$

8 a Rectangular hyperbola

b

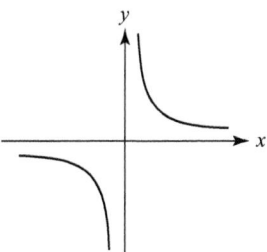

Asymptotes at $x = 0$, $y = 0$

c $\dfrac{14}{3}\sqrt{10}$ (or 14.8)

9 a

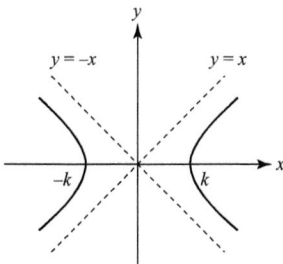

b The tangent lines $y = \pm x$ both make an angle $\dfrac{\pi}{4}$ with the x and y axes, and thus make an angle $\dfrac{\pi}{2}$ with each other, which is a right angle.

Exercise 3.4B

1 a

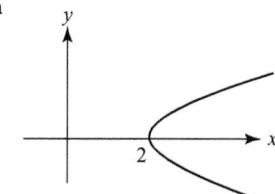

b

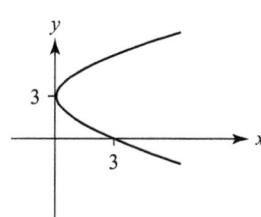

c

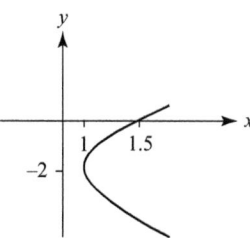

d

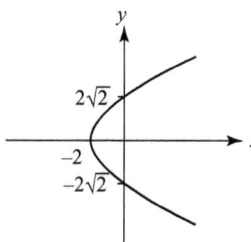

e

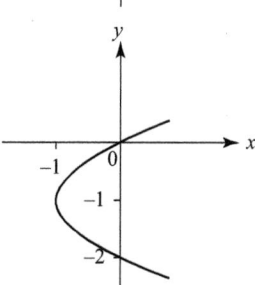

f

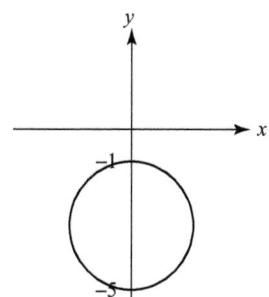

g

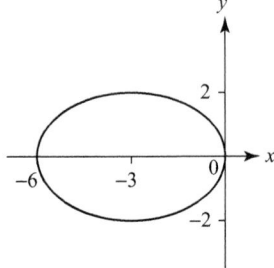

h

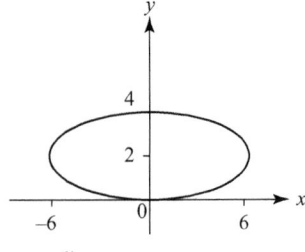

i

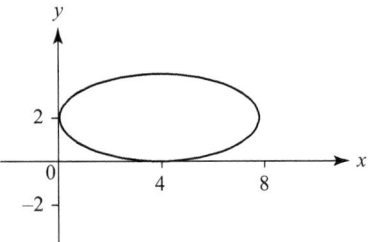

j

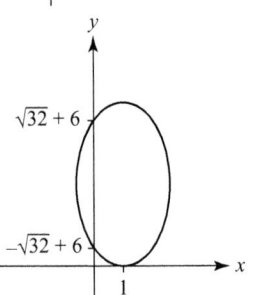

2 a

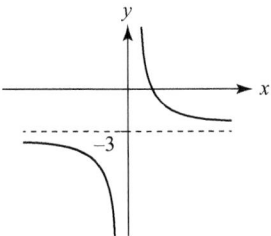

Asymptotes $x = 0$, $y = -3$

b

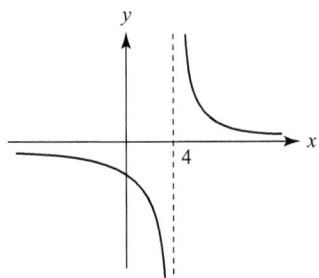

Asymptotes $x = 4$, $y = 0$

c

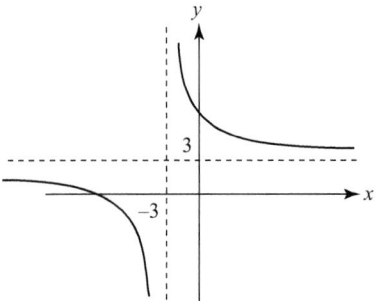

Asymptotes $x = -3$, $y = 3$

d $x(y-2) = 4$

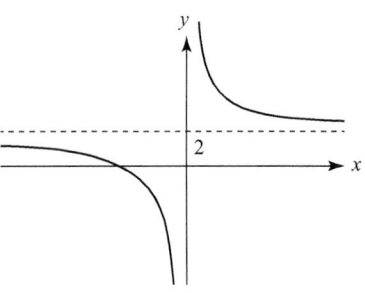

Asymptotes $x = 0$, $y = 2$

e $y(x-5) = 2$

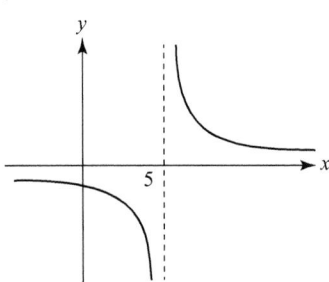

Asymptotes $x = 5$, $y = 0$

f

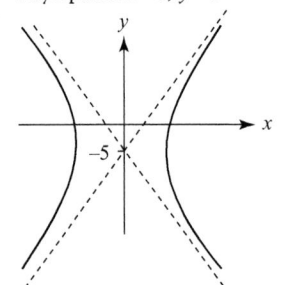

Asymptotes $y + 5 = \pm\dfrac{3}{\sqrt{2}}x$

g

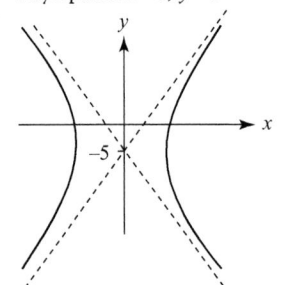

Asymptotes $y = \pm\dfrac{5}{4}(x+2)$

h

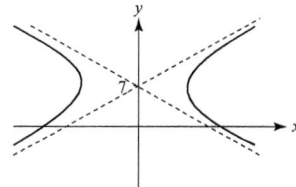

Asymptotes $y - 7 = \pm\dfrac{1}{2}x$

i

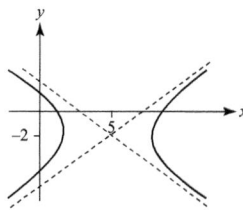

Asymptotes $y + 2 = \pm\dfrac{\sqrt{2}}{2}(x - 5)$

j

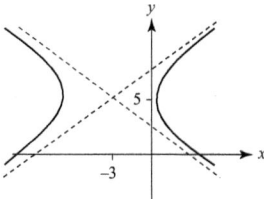

Asymptotes $y - 5 = \pm\dfrac{\sqrt{3}}{3}(x + 3)$

3 a

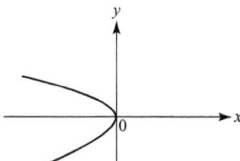

b

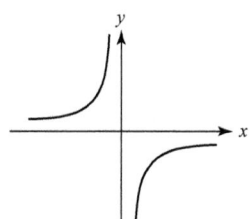

4 a $y^2 = 8\left(\dfrac{x}{3}\right) \Rightarrow y^2 = \dfrac{8}{3}x$

b $(4y)^2 = 8x \Rightarrow y^2 = \dfrac{x}{2}$

c $(-y)^2 = 8x \Rightarrow y^2 = 8x$

d $y^2 = 8(-x) \Rightarrow y^2 = -8x$

5 a $(y + 3)^2 = -4(x - 2)$ **b** $x^2 = -4y$

6 a $\dfrac{x^2}{12} + \dfrac{y^2}{16} = 1$ **b** $\dfrac{x^2}{300} + \dfrac{y^2}{4} = 1$

c $\dfrac{x^2}{12} + \dfrac{y^2}{4} = 1$ **d** $\dfrac{x^2}{4} + \dfrac{y^2}{12} = 1$

7 $\dfrac{(x - 3)^2}{8} + \dfrac{(y + 1)^2}{4} = 1$

8 a i $-x(y - 1) = 4$ **ii** $y = 1, x = 0$
    b i $(x - 5)(y - 3) = 4$ **ii** $y = 3, x = 5$
    c i $y(x - 1) = 4$ **ii** $x = 1, y = 0$
    d i $y(x + 1) = 4$ **ii** $y = 0, x = -1$

9 a i $\dfrac{x^2}{9} - \dfrac{y^2}{729} = 1$ **ii** $y = \pm 9x$

    b i $\dfrac{x^2}{9} - \dfrac{y^2}{81} = 1$ **ii** $y = \pm 3x$

    c i $\dfrac{(x + 1)^2}{9} - \dfrac{(y - 4)^2}{81} = 1$ **ii** $y - 4 = \pm 3(x + 1)$

    d i $\dfrac{y^2}{9} - \dfrac{x^2}{81} = 1$ **ii** $x = \pm 3y \Rightarrow y = \pm\dfrac{1}{3}x$

10 a $y + k = \pm\dfrac{1}{a}(x - k)$

    b Asymptotes $x - k = \pm\dfrac{1}{a}(y + k)$

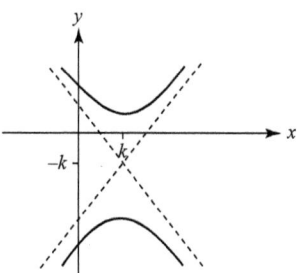

Exercise 3.5A

1 a 0 **b** 1.54 **c** $\dfrac{3}{4}$

    d $\dfrac{5}{3}$ **e** 1 **f** $\dfrac{3}{5}$

2 a

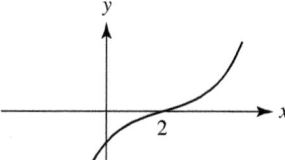

b

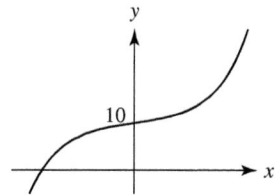

c

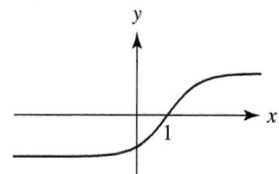

d

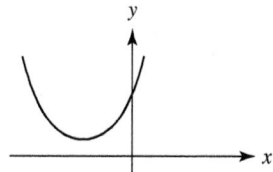

e

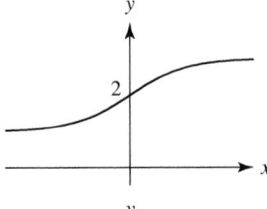

f

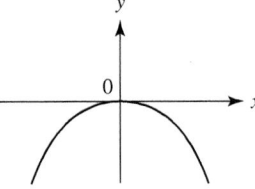

3 a $y = f(x)$

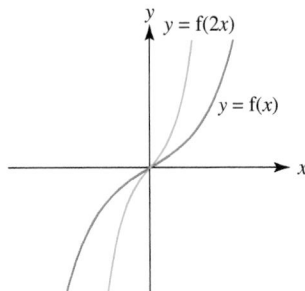

b $\sinh(2x) = 2 \Rightarrow x = 0.722$

4

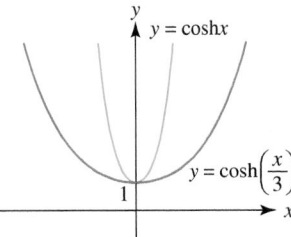

5 a

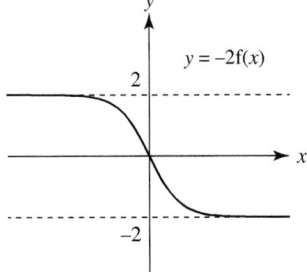

b $-2\tanh(x) = 1 \Rightarrow \tanh(x) = -\dfrac{1}{2}$
$\Rightarrow x = -0.549$

6 a

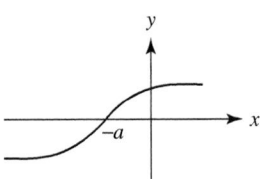

b $y = \pm 1$

7 a

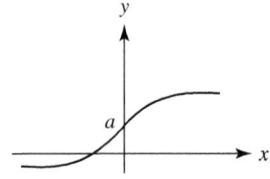

b $y = a+1; y = a-1$

8 a

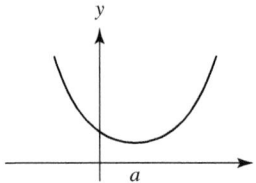

b

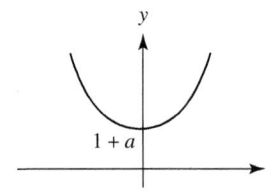

c

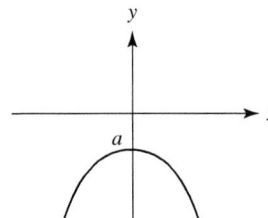

9 a 2.31   **b** 1.32, −1.32   **c** −0.549

   d 0.763, −2.76   **e** 0.698   **f** 0.549

10 a 0   **b** 0   **c** ln(2)

   d $\ln(4), \ln\left(\dfrac{1}{4}\right)$ **e** ln(2)   **f** ln(9)

11 a $\ln\dfrac{1}{2}$ **b** $0, \ln\left(\dfrac{1}{5}\right)$

   c $\ln\left(\dfrac{3 \pm \sqrt{6}}{3}\right)$ **d** $\dfrac{1}{4}\ln 2$

Exercise 3.5B

1 a $\ln(2+\sqrt{5}), -\ln(3+\sqrt{10})$ **b** $0, \pm\ln\left(\dfrac{3+\sqrt{5}}{2}\right)$

   c $\pm\dfrac{1}{2}\ln(5+2\sqrt{6})$ **d** $\pm\dfrac{1}{2}\ln(3+2\sqrt{2})$

2 a 0   **b** $\ln(-4+\sqrt{17}), \ln(1+\sqrt{2})$

   c $0, \ln\left(\dfrac{-2+\sqrt{13}}{3}\right)$ **d** $\dfrac{1}{2}\ln 3$

   e $\dfrac{1}{2}\ln(3+\sqrt{10})$ **f** $\pm\dfrac{1}{3}(\ln(4+\sqrt{15}))$

   g $\dfrac{1}{6}\ln(3)$

3 $LHS = \left(\dfrac{e^x + e^{-x}}{2}\right)^2 - \left(\dfrac{e^x - e^{-x}}{2}\right)^2$

$\qquad = \dfrac{1}{4}(e^{2x} + 2 + e^{-2x}) - \dfrac{1}{4}(e^{2x} - 2 + e^{-2x})$

$\qquad = \dfrac{1}{4}(2+2)$

$\qquad = 1 = RHS$

4 a $\sinh(A)\cosh(B) + \sinh(B)\cosh(A)$

$$\equiv \left(\frac{e^A - e^{-A}}{2}\right)\left(\frac{e^B + e^{-B}}{2}\right) + \left(\frac{e^B - e^{-B}}{2}\right)\left(\frac{e^A + e^{-A}}{2}\right)$$

$$\equiv \frac{e^{A+B} + e^{A-B} - e^{-(A-B)} - e^{-(A+B)}}{4}$$

$$+ \frac{e^{A+B} + e^{-(A-B)} - e^{A-B} - e^{-(A+B)}}{4}$$

$$\equiv \frac{2e^{A+B} - 2e^{-(A+B)}}{4}$$

$$\equiv \frac{e^{A+B} - e^{-(A+B)}}{2}$$

$$\equiv \sinh(A+B) \text{ as required}$$

b $\sinh(A)\cosh(B) - \sinh(B)\cosh(A)$

$$= \left(\frac{e^A - e^{-A}}{2}\right)\left(\frac{e^B + e^{-B}}{2}\right) - \left(\frac{e^B - e^{-B}}{2}\right)\left(\frac{e^A + e^{-A}}{2}\right)$$

$$= \frac{e^{A+B} + e^{A-B} - e^{-(A-B)} - e^{-(A+B)}}{4}$$

$$- \frac{e^{A+B} - e^{A-B} + e^{-(A-B)} - e^{-(A+B)}}{4}$$

$$= \frac{2e^{A-B} - 2e^{-(A-B)}}{4}$$

$$= \frac{e^{A-B} - e^{-(A-B)}}{2}$$

$$= \sinh(A-B) \text{ as required}$$

c $2\sinh(x)\cosh(x) = 2\left(\frac{e^x + e^{-x}}{2}\right)\left(\frac{e^x - e^{-x}}{2}\right)$

$$= \frac{(e^{2x} + 1 - 1 - e^{-2x})}{2}$$

$$= \frac{(e^{2x} - e^{-2x})}{2}$$

$$= \sinh(2x) \text{ as required}$$

5 a $\tanh x = \dfrac{\sinh x}{\cosh x}$

$$= \frac{\dfrac{e^x - e^{-x}}{2}}{\dfrac{e^x + e^{-x}}{2}}$$

$$= \frac{e^x - e^{-x}}{e^x + e^{-x}}$$

$$= \frac{e^{2x} - 1}{e^{2x} + 1}$$

b $\dfrac{2\tanh x}{1 + \tanh^2 x} \equiv \dfrac{2\left(\dfrac{e^{2x} - 1}{e^{2x} + 1}\right)}{1 + \left(\dfrac{e^{2x} - 1}{e^{2x} + 1}\right)^2}$

$$\equiv \frac{2(e^{2x} - 1)(e^{2x} + 1)}{(e^{2x} + 1)^2 + (e^{2x} - 1)^2}$$

$$\equiv \frac{2(e^{4x} - 1)}{(e^{4x} + 2e^{2x} + 1) + (e^{4x} - 2e^{2x} + 1)}$$

$$\equiv \frac{2(e^{4x} - 1)}{2e^{4x} + 2}$$

$$\equiv \frac{e^{4x} - 1}{e^{4x} + 1}$$

$$\equiv \tanh(2x) \text{ as required}$$

6 a $2\cosh^2 x - 1 \equiv 2\left(\dfrac{e^x + e^{-x}}{2}\right)^2 - 1$

$$\equiv \frac{2(e^{2x} + 2 + e^{-2x})}{4} - 1$$

$$\equiv \frac{e^{2x} + e^{-2x} + 2}{2} - \frac{2}{2}$$

$$\equiv \frac{e^{2x} + e^{-2x}}{2}$$

$$\equiv \cosh(2x) \text{ as required}$$

b $2\cosh^2 x - 1 + \cosh x = 5$

$$2\cosh^2 x + \cosh x - 6 = 0$$

$$\Rightarrow \cosh x = \frac{3}{2}, (-2)$$

$$\Rightarrow \cosh x = \pm\ln\left(\frac{3}{2} + \sqrt{\left(\frac{3}{2}\right)^2 - 1}\right)$$

$$= \pm\ln\left(\frac{3 + \sqrt{5}}{2}\right)$$

7 Let $y = \sinh^{-1} x$ then $x = \sinh y = \dfrac{e^y - e^{-y}}{2}$

$$2x = e^y - e^{-y}$$

$$e^{2y} - 2xe^y - 1 = 0$$

$$(e^y - x)^2 = x^2 + 1$$

$$e^y = x \pm \sqrt{x^2 + 1}$$

$e^y > 0$ therefore, $x - \sqrt{x^2 + 1}$ not valid as $x^2 + 1 > x^2$

so $\sqrt{x^2 + 1} > x \Rightarrow x - \sqrt{x^2 + 1} < 0$

So $y = \ln(x + \sqrt{x^2 + 1})$ as required

8 a Let $y = \tanh^{-1} x$ then $x = \tanh y = \dfrac{e^y - e^{-y}}{e^y + e^{-y}}$

$$xe^y + xe^{-y} = e^y - e^{-y}$$

$$(1 + x)e^{-y} = (1 - x)e^y$$

$$\frac{1 + x}{1 - x} = e^{2y}$$

$$2y = \ln\left(\frac{1 + x}{1 - x}\right)$$

So $\operatorname{artanh} x = \dfrac{1}{2}\ln\left(\dfrac{1 + x}{1 - x}\right)$ as required

b The range of $y = \tanh x$ is $-1 < y < 1$

9 $\pm\ln\left(\dfrac{5}{2} + \sqrt{\dfrac{21}{4}}\right) = \pm\ln\left(\dfrac{5 + \sqrt{21}}{2}\right)$

10 $\ln\left(\dfrac{1 + \sqrt{5}}{2}\right), \ln(-1 + \sqrt{2})$

11 $\pm\ln(2 + \sqrt{3}), \pm\ln\left(\dfrac{3 + \sqrt{5}}{2}\right)$

12 a $3\sinh\left(\dfrac{x}{2}\right) - \sinh(x) \equiv 3\sinh\left(\dfrac{x}{2}\right) - 2\sinh\left(\dfrac{x}{2}\right)\cosh\left(\dfrac{x}{2}\right)$

$\equiv \sinh\left(\dfrac{x}{2}\right)\left(3 - 2\cosh\left(\dfrac{x}{2}\right)\right)$ as required

b $0, \pm 2\ln\left(\dfrac{3+\sqrt{5}}{2}\right)$

13 a $\dfrac{1}{1+\cosh x} + \dfrac{1}{1-\cosh x} \equiv \dfrac{(1-\cosh x)+(1+\cosh x)}{(1+\cosh x)(1-\cosh x)}$

$\equiv \dfrac{2}{1-\cosh^2(x)}$

$\equiv -\dfrac{2}{\sinh^2(x)}$ as required

b $\ln\left(\dfrac{1+\sqrt{5}}{2}\right), \ln\left(\dfrac{-1+\sqrt{5}}{2}\right)$

14 a $2\sinh x\cosh x = \cosh^2 x$

$\cosh^2 x - 2\sinh x\cosh x = 0$

$\cosh x(\cosh x - 2\sinh x) = 0$

$\cosh x \neq 0$ so $\cosh x - 2\sinh x = 0 \Rightarrow \cosh x = 2\sinh x$

$\Rightarrow \dfrac{1}{2} = \dfrac{\sinh x}{\cosh x} = \tanh x$ as required

b $x = \dfrac{1}{2}\ln(3)$

15 1, 1.84

16 a $\sqrt{5}$

b $\dfrac{2\sqrt{5}}{5}$

17 a $2\sqrt{2}$

b $\dfrac{2\sqrt{2}}{3}$

18 a $\dfrac{\sqrt{3}}{3}$

b $\dfrac{2\sqrt{3}}{3}$

19 a $\dfrac{d}{dx}(\sinh x) = \dfrac{d}{dx}\left(\dfrac{e^x - e^{-x}}{2}\right) = \dfrac{e^x + e^{-x}}{2} = \cosh x$

$\dfrac{d}{dx}(\cosh x) = \dfrac{d}{dx}\left(\dfrac{e^x + e^{-x}}{2}\right) = \dfrac{e^x - e^{-x}}{2} = \sinh x$

b $\dfrac{d}{dx}(\sinh x)_{x=0} = \dfrac{e^0 + e^0}{2} = \cosh 0 = 1$

$\dfrac{d}{dx}(\cosh x)_{x=0} = \dfrac{e^0 - e^0}{2} = \sinh 0 = 0$

c When $x > 0$, $e^{-x} > 0 \Rightarrow e^x + e^{-x} > e^x > e^x - e^{-x}$

$\Rightarrow e^x + e^{-x} > e^x - e^{-x} \Rightarrow \dfrac{e^x + e^{-x}}{2} > \dfrac{e^x - e^{-x}}{2} \Rightarrow \cosh x > \sinh x$

So the gradient of $y = \sinh x$ is greater than the gradient of $y = \cosh x$ when $x > 0$.

d 1

Review exercise 3

1

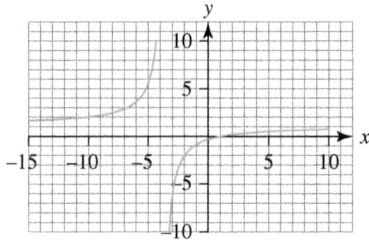

2 a

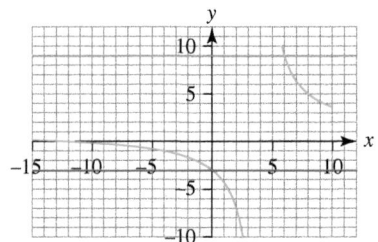

Intercepts are $(-12, 0)$ and $(0, -3)$
Asymptotes are $x = 4$ and $y = 1$
Solution to the inequality is $x < 0$ or $x > 6$

b

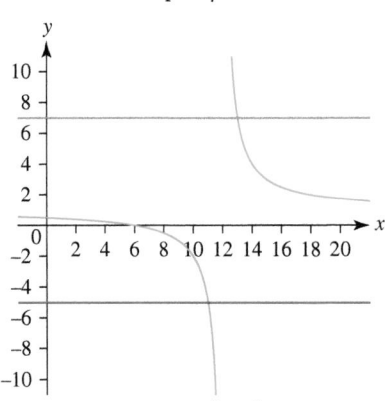

Intercepts are $(6, 0)$ and $\left(0, \dfrac{1}{2}\right)$

Asymptotes are $x = 12$ and $y = 1$
Solution to the inequality is $x \leq 11$ or $x \geq 13$

3 a

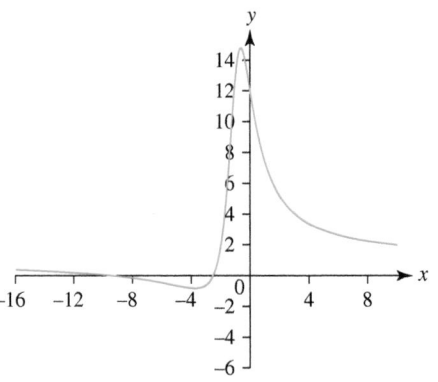

Intercepts are $(-6 \pm 2\sqrt{3}, 0)$ (i.e. $\approx (-9.46, 0)$ and $(-2.54, 0)$)
and $(0, 12)$
Asymptotes are $y = 1$ and there are no vertical asymptotes;

b $y = \dfrac{x^2 + 12x + 24}{x^2 + 2x - 3}$

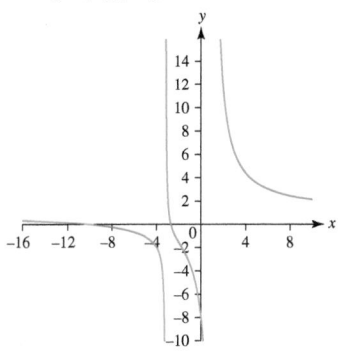

Intercepts are $(-6 \pm 2\sqrt{3}, 0)$ (i.e. $\approx (-9.46, 0)$ and $(-2.54, 0))$ and $(0, -8)$

Asymptotes are $x = -3$ and $x = 1$ and $y = 1$

4 $y < 1$ or $y \geq 5$.

5 $\left(\dfrac{5}{2}, 5\right)$.

no maximum.

6 a $\left(\dfrac{5}{2}, \dfrac{5\sqrt{3}}{2}\right)$ **b** $\left(\dfrac{5}{2}, -\dfrac{5\sqrt{3}}{2}\right)$

7 a $\left(4, \dfrac{\pi}{6}\right)$ **b** $\left(4, \dfrac{5\pi}{6}\right)$

  c $\left(4, \dfrac{7\pi}{6}\right)$ **d** $\left(4, \dfrac{11\pi}{6}\right)$

8 a $\tan^{-1} 3$ **b** $\dfrac{1}{\sin\theta - 2\cos\theta}$

  c $r = 4$ **d** $2(\cos\theta + \sin\theta)$

9 a centre is $(0, 2)$ radius 2

  b $(x^2 + y^2 + 4y) = 4(x^2 + y^2)$

10 a i

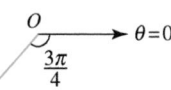

  ii $y = x$

  b i

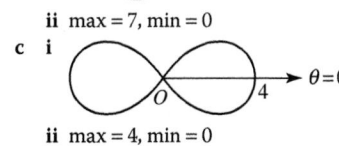

  ii $y = \sqrt{3}x$

11 a i

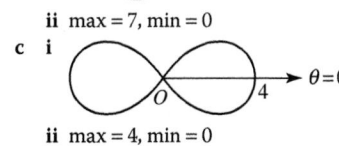

  ii max = 3, min = 3

  b i

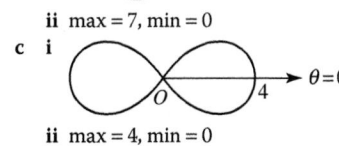

  ii max = 7, min = 0

  c i

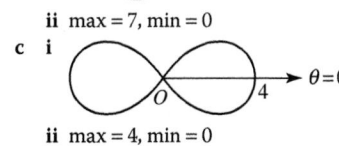

  ii max = 4, min = 0

d i

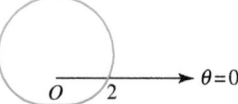

  ii max = 3, min = 1

e i

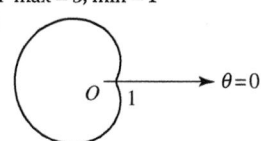

  ii max = 7, min = 1

f i

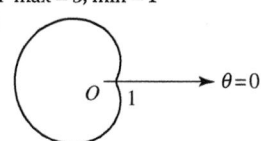

  ii max = 6π, min = 0

12 $\left(2, \dfrac{\pi}{3}\right), \left(2, \dfrac{5\pi}{3}\right)$

13 If $r \geq 0$ on both curves, then solutions are

$$\left(\dfrac{\sqrt{3}}{2}, \dfrac{\pi}{6}\right), \left(\dfrac{\sqrt{3}}{2}, \dfrac{7\pi}{6}\right)$$

If $r < 0$ on both curves, then there are 4 additional solutions:

$$\left(\dfrac{\sqrt{3}}{2}, \dfrac{\pi}{3}\right), \left(\dfrac{\sqrt{3}}{2}, \dfrac{5\pi}{6}\right), \left(\dfrac{\sqrt{3}}{2}, \dfrac{4\pi}{3}\right), \left(\dfrac{\sqrt{3}}{2}, \dfrac{11\pi}{6}\right)$$

14 a

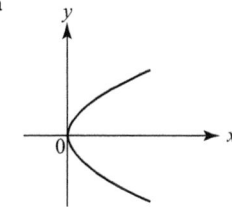

  b $(9, 6)$ and $(1, -2)$

15 a

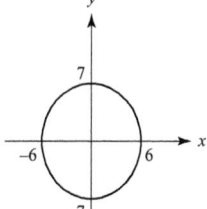

  b

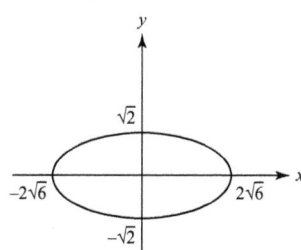

16

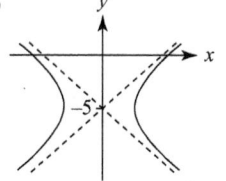

Asymptotes $y = \pm \dfrac{\sqrt{2}}{2} x$

17 a

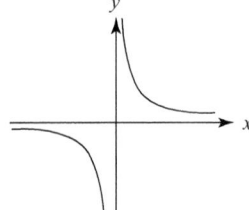

Asymptotes, $x = 0$, $y = 0$

b $y = 1 - x$

Substitute to give $x(1-x) = 5 \Rightarrow -x^2 + x - 5 = 0$

$b^2 - 4ac = 1^2 - 4(-1)(-5) = -19$

$b^2 - 4ac < 0$ so no solutions

Hence they do not intersect.

18 a

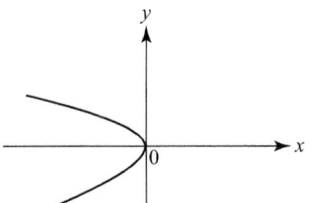

b

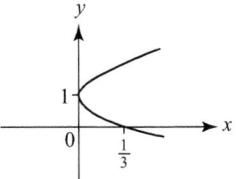

19 a

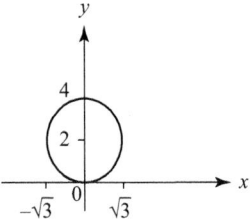

b

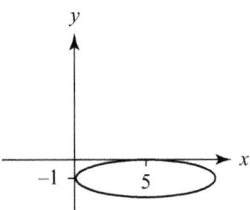

20 a Asymptotes at $y = -5 \pm \dfrac{2}{\sqrt{3}} x$

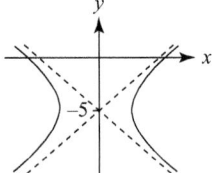

b Asymptotes at $y - 2 = \pm \dfrac{2}{3}(x+3)$

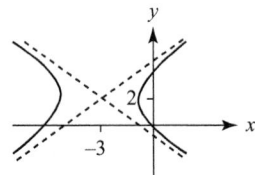

c Asymptotes at $x = 0$, $y = 0$

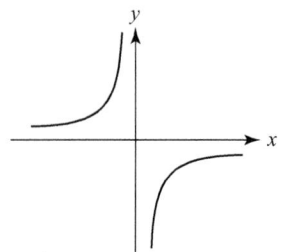

d Asymptotes at $x = -3$, $y = 4$

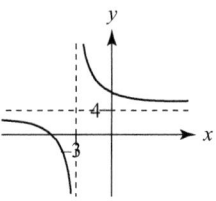

21 a $\dfrac{y^2}{8} + x^2 = 2 \Rightarrow \dfrac{x^2}{2} + \dfrac{y^2}{16} = 1$

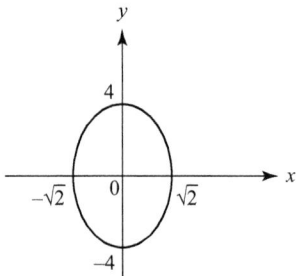

b $6(-y)^2 - 2(-x)^2 = 6 \Rightarrow y^2 - \dfrac{x^2}{3} = 1$

Asymptotes are at $y = \pm \dfrac{x}{\sqrt{3}}$

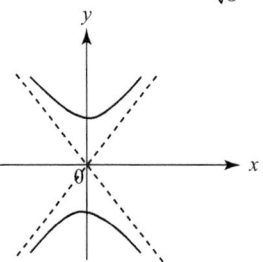

22 a $\dfrac{13}{5}$

b $-\dfrac{3}{5}$

23 a

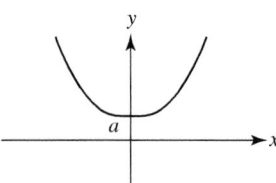

b

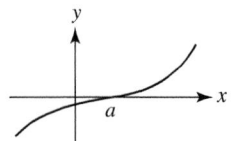

c

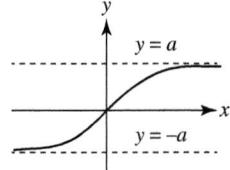

d

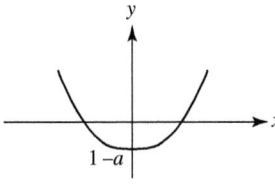

24 $x = \ln(-2+\sqrt{5}), \ln(1+\sqrt{2})$

25 a $x = \ln 3$

   b $x = \ln\left(1+\sqrt{2}\right)$

26 Let $y = \cosh^{-1} 2x$

   Then $2x = \cosh y = \dfrac{e^y + e^{-y}}{2}$

   $4x = e^y + e^{-y} \Rightarrow e^{2y} - 4xe^y + 1 = 0$

   $e^y = 2x + \sqrt{4x^2-1} \Rightarrow y = \ln(2x + \sqrt{4x^2-1})$

27 a $\sinh^{-1}(2) = \ln\left(2+\sqrt{5}\right)$

   b $\ln\left(\dfrac{\sqrt{3}}{3}\right)$

Assessment 3

1 a

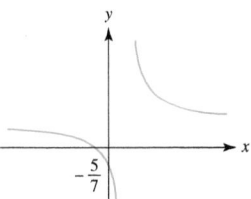

   b $x = 7$

       $y = 1$

   c $(1, -1), (5, -5)$

2 a

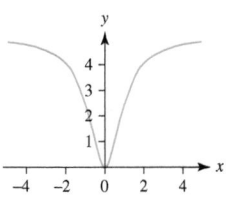

   b $\left(\pm\dfrac{1}{2}, 1\right)$

3 a $\left(2\sqrt{3}, -\dfrac{\pi}{3}\right)$

   b i

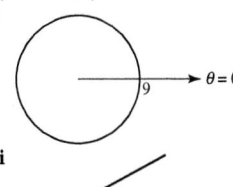

     ii

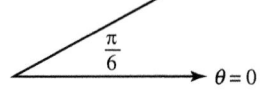

c i $x^2 + y^2 = 81$ **ii** $y = \dfrac{1}{\sqrt{3}}x$

4 a $\left(-2\sqrt{3}, -2\right)$ **b** $x^2 + y^2 = 2y$

   c

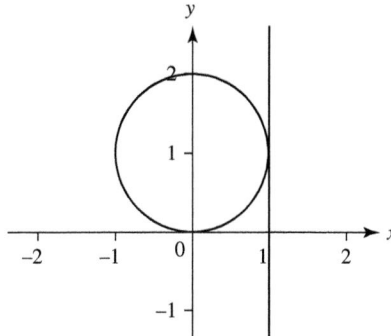

5 a

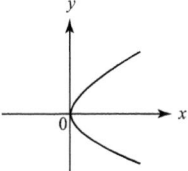

   Parabola

   b $\dfrac{3}{4}$

6 a

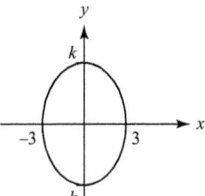

   b

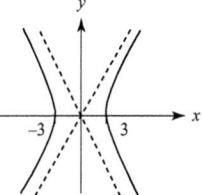

   Asymptotes $y = \pm\dfrac{k}{3}x$

7 a

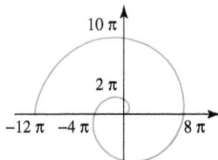

   b

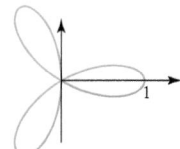

   c

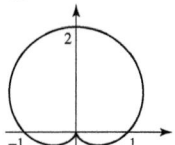

d

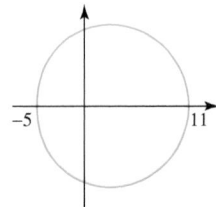

8 a $\sinh(\ln 2) = \dfrac{e^{\ln 2} - e^{-\ln 2}}{2}$

$= \dfrac{e^{\ln 2} - e^{\ln\left(\frac{1}{2}\right)}}{2}$

$= \dfrac{2 - \dfrac{1}{2}}{2}$

$= \dfrac{3}{4}$

b $x = 1.8$

9 a

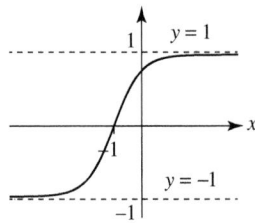

b $y = 1$

$y = -1$

c $\tanh x = \dfrac{\sinh x}{\cosh x}$

$= \dfrac{e^x - e^{-x}}{2} \div \dfrac{e^x + e^{-x}}{2}$

$= \dfrac{2(e^x - e^{-x})}{2(e^x + e^{-x})}$

$= \dfrac{e^x - e^{-x}}{e^x + e^{-x}}$

$= \dfrac{e^{2x} - 1}{e^{2x} + 1}$ as required

d $x = \dfrac{1}{2}\ln 3$

10 a

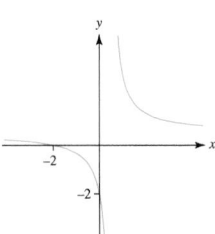

b $-4 < x \le -1$ or $\dfrac{1}{2} < x \le 8$

11 a

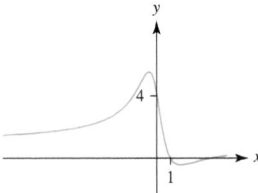

b $x > \dfrac{\sqrt{13}-1}{2}$ or $\dfrac{-1-\sqrt{13}}{2} < x < 1$

12 a $x = -1 \pm \sqrt{7}$

b $x = 1$ or $-\dfrac{8}{3}$, $y = \dfrac{4}{3}$

c $x > -1 + \sqrt{7}, -2 < x < \dfrac{1}{2}, x > -1 + \sqrt{7}$

13 a $(x-5)(y+1) = xy - 5y + x - 5$
So $xy - 5y + x - 9 = 0$
$\Rightarrow (x-5)(y+1) - 4 = 0$
$\Rightarrow (x-5)(y+1) = 4$
$(k = 4)$

b

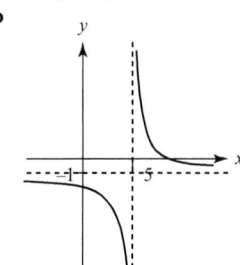

Asymptotes $x = 5$, $y = -1$

14

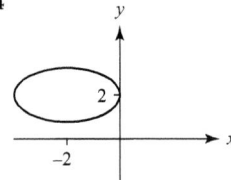

15 $r\sin\theta = \dfrac{1}{r\cos\theta}$

$r^2\sin\theta\cos\theta = 1$

$r^2\left(\dfrac{1}{2}\sin 2\theta\right) = 1$

$r^2 = \dfrac{2}{\sin 2\theta}$

$r^2 = 2\operatorname{cosec} 2\theta$

16 a $(x^2 + y^2)^3 = 36x^2y^2$

b

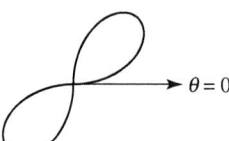

c Max value of r is 3
This occurs at $\theta = \dfrac{\pi}{4}, \dfrac{5\pi}{4}$

17 a $r = 7$ is maximum
$r = 1$ is minimum

b

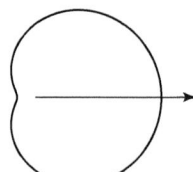

18 a

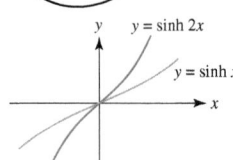

b $y = \dfrac{e^x - e^{-x}}{2}$

$2y = e^x - e^{-x}$

$e^{2x} - 2ye^x - 1 = 0$

$(e^x - y)^2 - y^2 - 1 = 0$

$(e^x - y)^2 = y^2 + 1$

$e^x - y = \pm\sqrt{y^2 + 1}$

$e^x = y \pm \sqrt{y^2 + 1}$

$e^x > 0 \therefore e^x = y + \sqrt{y^2 + 1}$ as $(y - \sqrt{y^2 + 1}) < 0$

$x = \ln(y + \sqrt{y^2 + 1})$ as required.

19 a $\dfrac{y^2}{10} - \dfrac{x^2}{5} = 1$

b

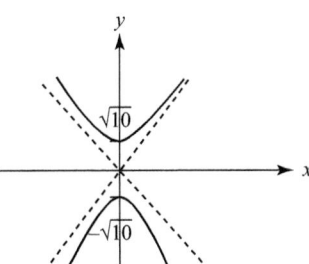

Asymptotes at $x = \pm\dfrac{1}{\sqrt{2}} y$

(or $y = \pm\sqrt{2}x$)

20 a $y^2 = 48x$

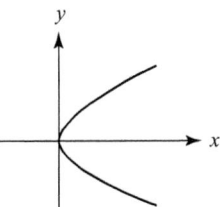

b $(y+8)^2 = 12(x-3)$

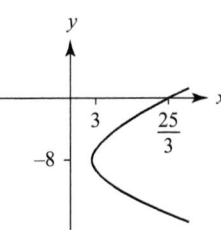

c $y = -\dfrac{x^2}{12}$

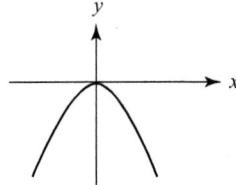

21 a i and ii

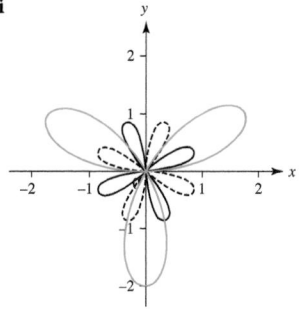

b Dotted sections only included when r can take negative values.

22 a $2\sinh x + \cosh x \equiv 2\left(\dfrac{e^x - e^{-x}}{2}\right)\left(\dfrac{e^x + e^{-x}}{2}\right)$

$\equiv 2\left(\dfrac{e^{2x} - e^{-2x}}{4}\right)$

$\equiv \dfrac{e^{2x} - e^{-2x}}{2}$

$\equiv \sinh(2x)$ as required

b i $x = \dfrac{1}{2}\ln 5, 0$ **ii** $x = \dfrac{1}{2}\ln(2+\sqrt{5}), \dfrac{1}{2}\ln\left(\dfrac{-3+\sqrt{13}}{2}\right) 1$

Chapter 4

Exercise 4.1A

1 a $\dfrac{45}{4}$ (11.25) **b** $\dfrac{15}{4}$ (3.75)

2 a

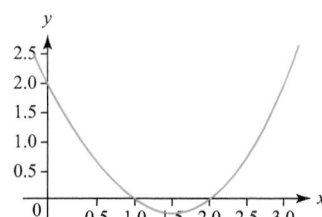

b $\dfrac{1}{6}$

c i $-\dfrac{1}{6}$ **ii** $\dfrac{5}{6}$

3 a

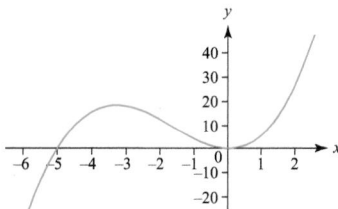

b $\dfrac{625}{12}$

c i $\dfrac{125}{12}$ **ii** $\dfrac{26}{3}$ (8.67)

4 a 256 **b** 121 **c** 11 **d** $\dfrac{1}{16}$

5 a 9 **b** $\dfrac{113}{3}$ (37.7)

c 5 **d** $\dfrac{23}{3}$ (7.67)

6 a $-\dfrac{325}{2}$ (−162.5) **b** $-\dfrac{99}{2}$ (−49.5)

  c 0 **d** $\dfrac{245}{2}$ (122.5)

7 $\dfrac{1}{5-2}\displaystyle\int_2^5 x^{-2}\,dx = \dfrac{1}{3}\Big[-x^{-1}\Big]_2^5$

$$= \dfrac{1}{3}\left(-\dfrac{1}{5}\right) - \dfrac{1}{3}\left(-\dfrac{1}{2}\right)$$

$$= \dfrac{1}{10}\ (0.1)$$

8 $\dfrac{1}{4-1}\displaystyle\int_1^4 x^{\frac{1}{2}}\,dx = \dfrac{1}{3}\left[\dfrac{2}{3}x^{\frac{3}{2}}\right]_1^4$

$$= \dfrac{1}{3}\left(\dfrac{16}{3}\right) - \dfrac{1}{3}\left(\dfrac{2}{3}\right)$$

$$= \dfrac{14}{9}\ (1.56)$$

9 $\dfrac{1}{9-4}\displaystyle\int_4^9 x^{-\frac{1}{2}}\,dx = \dfrac{1}{5}\left[2x^{\frac{1}{2}}\right]_4^9$

$$= \dfrac{1}{5}(6) - \dfrac{1}{5}(4)$$

$$= \dfrac{2}{5}\ (0.4)$$

10 a $A = 1, B = \dfrac{3}{2}$ and $c = -\dfrac{1}{2}$

  b $\dfrac{1}{9-1}\displaystyle\int_1^9 x^{-\frac{1}{2}} + \dfrac{3}{2}\,dx = \dfrac{1}{8}\left[2x^{\frac{1}{2}} + \dfrac{3}{2}x\right]_1^9$

$$= \dfrac{1}{8}\left(6 + \dfrac{27}{2}\right) - \dfrac{1}{8}\left(2 + \dfrac{3}{2}\right)$$

$$= 2$$

11 a $A = \dfrac{3}{5}, B = -\dfrac{1}{5}, c = \dfrac{1}{2}, d = \dfrac{3}{2}$ **b** $\dfrac{8}{25} - \dfrac{6}{25}\sqrt{2}$

12 a $-\dfrac{13}{3}$ (−4.33) **b** $-\dfrac{229}{200}$ (−1.145)

  c $\dfrac{1}{12}$ (0.0833) **d** $\dfrac{28}{5}$ (5.6)

13 a $1 - 2T + \dfrac{4}{3}T^2$ **b** $4T^2 + 4T + \dfrac{7}{3}$

14 $\dfrac{26}{9}\sqrt{2}$

15 $\dfrac{45}{4}a^3 - \dfrac{3}{8a^3}$

16 $= -\dfrac{1}{6}X^4$

Exercise 4.1B

1 a 3.6 ms⁻¹

  b $\dfrac{1}{2}\left[\dfrac{2}{5}t^3 - \dfrac{1}{5}t\right]_1^3 = \dfrac{1}{2}\left(\dfrac{54}{5} - 3\right) - \dfrac{1}{2}\left(\dfrac{2}{5} - \dfrac{1}{5}\right)$

$$= 5\ \text{ms}^{-2}\ \text{as required.}$$

2 a $\dfrac{1}{5-0}\displaystyle\int_0^5 \dfrac{3t^2 - 5}{5}\,dt = \dfrac{1}{25}[t^3 - 5t]_0^5$

$$= \dfrac{1}{25}(125 - 25) - 0$$

$$= 4\ \text{ms}^{-1}$$

  b 3 ms⁻¹

3 $s = 2t^{\frac{3}{2}}$

$v = 3t^{\frac{1}{2}}$

$\left(a = \dfrac{3}{2}t^{-\frac{1}{2}}\right)$

Mean acceleration $= \dfrac{1}{1 - \dfrac{1}{2}}\displaystyle\int_1^4 \dfrac{3}{2}t^{-\frac{1}{2}}\,dt$

$$= 2\left[3t^{\frac{1}{2}}\right]_{0.5}^1$$

$$= 2\left(3 - \dfrac{3\sqrt{2}}{2}\right)$$

$$= 6 - 3\sqrt{2}\ \text{ms}^{-2}$$

4 a $\dfrac{1}{3-1}\displaystyle\int_1^3 \dfrac{t}{10} - 10t^{-3}\,dt = \dfrac{1}{2}\left[\dfrac{t^2}{20} + 5t^{-2}\right]_1^3$

$$= \dfrac{1}{2}\left(\dfrac{9}{20} + \dfrac{5}{9}\right) - \dfrac{1}{2}\left(\dfrac{1}{20} + 5\right)$$

$$= -\dfrac{91}{45}\ \text{ms}^{-2}$$

  b $\dfrac{11}{6}$ (1.83) ms⁻¹

5 Mean value $= \dfrac{1}{b-a}\displaystyle\int_a^b mx + c\,dx$

$$= \dfrac{1}{b-a}\left[\dfrac{mx^2}{2} + cx\right]_a^b$$

$$= \dfrac{1}{b-a}\left(\dfrac{mb^2}{2} + cb\right) - \dfrac{1}{b-a}\left(\dfrac{ma^2}{2} + ca\right)$$

$$= \dfrac{1}{b-a}\left(\dfrac{m}{2}(b^2 - a^2) + c(b-a)\right)$$

$$= \dfrac{m(b+a)(b-a)}{2(b-a)} + \dfrac{c(b-a)}{b-a}$$

$$= \dfrac{m(b+a)}{2} + c\ \text{as required}$$

6 Mean value $= \dfrac{1}{a-0}\displaystyle\int_0^a x^2\,dx$

$$= \dfrac{1}{a}\left[\dfrac{x^3}{3}\right]_0^a$$

$$= \dfrac{1}{a}\dfrac{a^3}{3}$$

$$= \dfrac{a^2}{3}\ \text{as required}$$

7 a

  b $\dfrac{443}{6}$ (73.8)

  c i $\dfrac{175}{12}$ **ii** $= -\dfrac{11}{12}$

8 $\dfrac{49}{36}$

9 -8

10 7

11 $6, -9$

12 $b = 4$
$a = -1$

13 $\dfrac{1}{3-1}\displaystyle\int_1^3 x^4 - 2x^3 + 3x - 5 \; dx = \dfrac{1}{2}\left[\dfrac{x^5}{5} - \dfrac{x^4}{2} + \dfrac{3}{2}x^2 - 5x\right]_1^3$

$= \dfrac{1}{2}\left(\dfrac{243}{5} - \dfrac{81}{2} + \dfrac{27}{2} - 15\right) - \dfrac{1}{2}\left(\dfrac{1}{5} - \dfrac{1}{2} + \dfrac{3}{2} - 5\right)$

$= \dfrac{26}{5}$

$\dfrac{1}{-1--2}\left[\dfrac{x^5}{5} - \dfrac{x^4}{2} + \dfrac{3}{2}x^2 - 5x\right]_{-2}^{-1}$

$= \left(-\dfrac{1}{5} - \dfrac{1}{2} + \dfrac{3}{2} + 5\right) - \left(-\dfrac{32}{5} - 8 + 6 + 10\right)$

$= \dfrac{21}{5}$

So [1, 3] is bigger by 1.

14 Mean speed of A: $\dfrac{1}{\frac{1}{2}}\displaystyle\int_0^{1.5} t^2 + t \; dt = 2\left[\dfrac{t^3}{3} + \dfrac{t^2}{2}\right]_0^{0.5}$

$= \dfrac{1}{3}$

Mean speed of B: $\dfrac{1}{\frac{1}{2}}\displaystyle\int_0^{0.5} t^{\frac{1}{2}} dt = 2\left[\dfrac{2}{3}t^{\frac{3}{2}}\right]_0^{0.5}$

$= \dfrac{\sqrt{2}}{3}$

B is $\dfrac{1}{3}(\sqrt{2} - 1)$ faster than A

15 $a = \dfrac{2}{5}$, $b = 27\sqrt{2}$, $c = -1$

16 $\dfrac{1}{8}\ln\left(\dfrac{4k^2 + 8k + 1}{4k^2 - 3}\right)$

17 a $= 2 + \dfrac{1}{2}\ln 3$

b No as function not continuous in this range.

18 a $\dfrac{3}{2\pi}$ **b** 0

c $\dfrac{1}{3}(e^3 - 1)$ **d** $\dfrac{4}{3}e^3 + \dfrac{2}{3}$

Exercise 4.2A

1 $\dfrac{9207}{5}\pi$ (5785)

2 a $\displaystyle\int_0^2 8 - x^3 \; dx = \left[8x - \dfrac{x^4}{4}\right]_0^2$
$= (16 - 2) - 0$
$= 12$ as required

b $\dfrac{576}{7}\pi$ (259)

3 a $\dfrac{1}{20}$ **b** $\dfrac{\pi}{252}$ (0.0125)

4 $\dfrac{59}{30}\pi$ (6.18)

5 $\dfrac{5261}{105}\pi$ (157)

6 $\pi\displaystyle\int_{0.5}^1 \left(\dfrac{1}{x^2}\right)^2 dx = \pi\displaystyle\int_{0.5}^1 x^{-4} dx$

$= \pi\left[-\dfrac{1}{3}x^{-3}\right]_{0.5}^1$

$= \pi\left(-\dfrac{1}{3} - -\dfrac{8}{3}\right)$

$= \dfrac{7}{3}\pi$

7 a $\displaystyle\int_1^3 \dfrac{2}{5}x^{-2} + \dfrac{1}{5}x^{-\left(\frac{3}{2}\right)} \; dx = \left[-\dfrac{2}{5}x^{-1} - \dfrac{2}{5}x^{-\frac{1}{2}}\right]_1^3$

$= -\dfrac{2}{15} - \dfrac{2}{5\sqrt{3}} - \left(-\dfrac{2}{5} - \dfrac{2}{5}\right)$

$= -\dfrac{2}{15} - \dfrac{2\sqrt{3}}{15} + \dfrac{4}{5}$

$= \dfrac{1}{15}(-2 - 2\sqrt{3} + 12)$

$= \dfrac{10 - 2\sqrt{3}}{15}$

b i 0.405 **ii** 0.101

8 a $\dfrac{256}{15}\pi$

b 8π

9 $x = \left(\dfrac{y}{4}\right)^{\frac{1}{4}}$

$\pi\displaystyle\int_2^8 \left(\left(\dfrac{y}{4}\right)^{\frac{1}{4}}\right)^2 dy = \pi\displaystyle\int_2^8 \dfrac{1}{2}y^{\frac{1}{2}} \; dy$

$= \pi\left[\dfrac{1}{3}y^{\frac{3}{2}}\right]_2^8$

$= \pi\left(\dfrac{16}{3}\sqrt{2} - \dfrac{2}{3}\sqrt{2}\right)$

$= \dfrac{14}{3}\sqrt{2} \; \pi$

10 $\dfrac{36}{35}\sqrt{3}\pi$

Exercise 4.2B

1 a $\dfrac{19}{12}$ **b** $\dfrac{109}{30}\pi$

2 a 12 **b** $\dfrac{912}{7}\pi$

3 a $\displaystyle\int_0^{\sqrt{3}} 16 - x^4 dx = \left[16x - \dfrac{x^5}{5}\right]_0^{\sqrt{3}}$

$= 16\sqrt{3} - \dfrac{9}{5}\sqrt{3} - 0$

$= \dfrac{71}{5}\sqrt{3}$

Area of R $= \dfrac{71}{5}\sqrt{3} - 7 \times \sqrt{3}$

$= \dfrac{36}{5}\sqrt{3}$ as required

b $\dfrac{792}{5}\sqrt{3}\pi$ (862)

4

a $\displaystyle\int_0^4 8+2x-x^2\,dx = \left[8x+x^2-\frac{x^3}{3}\right]_0^4$

$$= 32+16-\frac{64}{3}$$

$$= \frac{80}{3}$$

Area required $= \dfrac{80}{3}+\dfrac{1}{2}\times 4\times 4$

$$= \frac{104}{3} \text{ as required.}$$

b $\dfrac{1696}{15}\pi$ (355)

5 a $\dfrac{10}{3}$ square units   **b** $\dfrac{8}{3}\pi$

6 $V = \pi\displaystyle\int_0^b (ax)^2\,dx$

$$= \pi\int_0^b a^2x^2\,dx$$

$$= \pi\left[\frac{a^2}{3}x^3\right]_0^b$$

$$= \frac{\pi a^2}{3}(b^3-0)$$

$$= \frac{\pi(ab)^2}{3}b$$

$$= \frac{1}{3}\pi r^2 h \text{ where } r = ab \text{ and } h = b$$

7 $V = \pi\displaystyle\int_0^h r^2\,dx$

$$= \pi\left[r^2 x\right]_0^h$$
$$= \pi(r^2 h - 0)$$
$$= \pi r^2 h \text{ as required}$$

8 $V = \pi\displaystyle\int_0^r \sqrt{r^2-x^2}^{\,2}\,dx$

$$= \pi\int_0^r r^2 - x^2\,dx$$

$$= \pi\left[r^2 x - \frac{x^3}{3}\right]_0^r$$

$$= \pi\left(r^3 - \frac{r^3}{3} - 0\right)$$

$$= \pi\left(\frac{2}{3}r^3\right)$$

$$= \frac{2}{3}\pi r^3$$

9 a $A = \dfrac{\pm\sqrt{6}}{3}$

b $\dfrac{\pi}{15}(27\sqrt{2}-\sqrt{6})$

10 a A has coordinates $(7, 14)$, B has coordinates $(7, -14)$

b $686\,\pi$

11 a $\dfrac{\pi}{4}\ln\left(\dfrac{5}{2}\right)$ **b** $\dfrac{2}{3}\sqrt{3}\pi$

c $\dfrac{127}{14}\pi$ **d** $\dfrac{9}{4}\pi\left(1-\dfrac{1}{e}\right)$ or $\dfrac{9\pi(e-1)}{4e}$

12 $\dfrac{48\pi}{5}$

Review exercise 4

1 $\dfrac{241}{3}$

2 $\dfrac{1}{8-2}\displaystyle\int_2^8 \frac{1+x}{\sqrt{x}}\,dx = \frac{1}{6}\int_2^8 x^{-\frac{1}{2}}+x^{\frac{1}{2}}\,dx$

$$= \frac{1}{6}\left[2x^{\frac{1}{2}}+\frac{2}{3}x^{\frac{3}{2}}\right]_2^8$$

$$= \frac{1}{6}\left(4\sqrt{2}+\frac{32}{3}\sqrt{2}\right)-\frac{1}{6}\left(2\sqrt{2}+\frac{4}{3}\sqrt{2}\right)$$

$$= \frac{17}{9}\sqrt{2}$$

3 $-\dfrac{75}{8}$

4 $\dfrac{1}{1-a}\displaystyle\int_a^1 x^{-3}-\frac{1}{2}x^{-2}\,dx = \frac{1}{1-a}\left[-\frac{1}{2}x^{-2}+\frac{1}{2}x^{-1}\right]_a^1$

$$= \frac{1}{1-a}\left(-\frac{1}{2}+\frac{1}{2}\right)-\frac{1}{1-a}\left(-\frac{1}{2a^2}+\frac{1}{2a}\right)$$

$$= \frac{1}{2(a-1)}\frac{(a-1)}{a^2}$$

$$= \frac{1}{2a^2} \text{ as required}$$

5 a 12 ms^{-1}

b 15.4 ms^{-2}

6 291π (914)

7 a $\displaystyle\int_{\frac{1}{\sqrt{3}}}^2 3-x^{-2}\,dx = \left[3x+\frac{1}{x}\right]_{\frac{1}{\sqrt{3}}}^2$

$$= \left(6+\frac{1}{2}\right)-\left(\frac{3}{\sqrt{3}}+\sqrt{3}\right)$$

$$= \frac{13}{2}-2\sqrt{3}$$

$$= \frac{13-4\sqrt{3}}{2}$$

b $\dfrac{503-192\sqrt{3}}{24}\pi$ (22.3)

8 50π

9 $\pi\displaystyle\int_0^{\frac{1}{2}} 16y^4\,dy = \pi\left[\frac{16}{5}y^5\right]_0^{\frac{1}{2}}$

$$= \pi\left(\frac{1}{10}-0\right)$$

$$= \frac{\pi}{10}$$

10 a
$$\int_0^2 y^4 + 1\, dy = \left[\frac{y^5}{5} + y\right]_0^2$$
$$= \frac{32}{5} + 2 - 0$$
$$= \frac{42}{5} \text{ as required}$$

b $\dfrac{3226}{45}\pi$ (225)

11 a

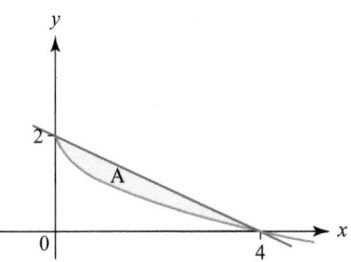

b $\dfrac{8}{3}\pi$

c $\dfrac{64}{15}\pi$ (13.4)

12 a $\dfrac{49}{48}$

b $\dfrac{901}{960}\pi$ (2.95)

13 a $\dfrac{32}{3}\pi$

b $\dfrac{128}{15}\pi$

Assessment 4

1 a $\dfrac{16}{3}$ **b** $-\dfrac{1}{7}$

2 a $\dfrac{1}{2}x - \dfrac{3}{2}x^{-\frac{3}{2}}$ **b** $\dfrac{3}{4}$

3 $\dfrac{9}{5}\sqrt{3} + \dfrac{4}{5}$

4 a 36 ms^{-1} **b** 18

5 6

6 a 2 **b** $\dfrac{20}{3}$ **c** $\dfrac{496}{15}\pi$ or 103.9

7 18π

8 $\dfrac{619}{21}\pi$ or 92.6

9 1170

10 9.44 ms^{-1}

11 $a = 6, b = 2$

12 $k = 2$ or $-\dfrac{2}{5}$

13 128

14 a $a = 3, b = 2$

b 4π

15 a

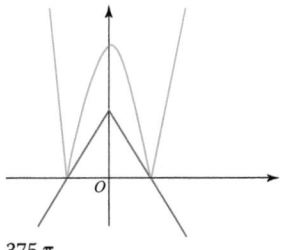

b 375π

16 $\dfrac{1}{2-0}\int_0^1 e^{2x} + x^3 + 2\,dx$
$$= \frac{1}{2}\left[\frac{1}{2}e^{2x} + \frac{x^4}{4} + 2x\right]_0^2$$
$$= \frac{1}{2}\left(\frac{1}{2}e^4 + 4 + 4\right) - \frac{1}{2}\left(\frac{1}{2} + 0 + 0\right)$$
$$= \frac{1}{4}e^4 + \frac{15}{4}$$

17 $\dfrac{1}{2(3-a)}\ln\left(\dfrac{3}{2a-3}\right)$

18 a $\dfrac{3}{2\pi}$ or 0.477

b $\dfrac{1}{2} + \dfrac{3\sqrt{3}}{4\pi}$ or 0.913

19 $-\dfrac{3}{4} + \dfrac{7}{2}\ln 2$

20 a 4

b $2\pi(e^2 - 1)$

21 177π

22 a

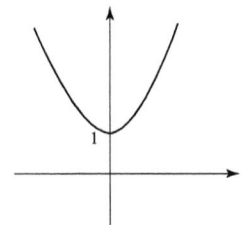

b $\dfrac{\pi}{32}\left(e^4 - 8 - e^{-4}\right)$

23 $\dfrac{2}{3}$

24 $2\pi - \dfrac{\pi^2}{2}$

Chapter 5

Exercise 5.1A

1 a i 3×2 **ii** 3×1 **iii** 2×2 **iv** 2×4

b the matrix in part **iii** is a square matrix

2 a $\begin{pmatrix} 4 & 3 & 2 \\ 5 & -1 & 6 \end{pmatrix}$ **b** $\begin{pmatrix} -4 & -6 \\ 1 & -3 \end{pmatrix}$

c $\begin{pmatrix} -4 & 0 & 16 \\ 8 & 20 & -12 \\ 32 & -24 & 0 \end{pmatrix}$ **d** $\begin{pmatrix} -2 & 12 \\ 6 & 10.5 \\ 1 & -4 \end{pmatrix}$

3 a i $\begin{pmatrix} 9 & -3 \\ -5 & 2 \end{pmatrix}$ **ii** Not possible as different order matrices

iii $\begin{pmatrix} 13 & 6 \\ -5 & -6 \\ -5 & -1 \end{pmatrix}$ **iv** $\begin{pmatrix} 24 & 12 \\ 6 & -18 \\ -6 & 0 \end{pmatrix}$

b i $\begin{pmatrix} 45 & -20 \\ 0 & -10 \end{pmatrix} - \begin{pmatrix} 0 & 1 \\ -5 & 4 \end{pmatrix} = \begin{pmatrix} 45 & -21 \\ 5 & -14 \end{pmatrix}$

ii $\begin{pmatrix} -10 & -4 \\ 14 & 0 \\ 6 & 2 \end{pmatrix} + \begin{pmatrix} 56 & 28 \\ 14 & -42 \\ -14 & 0 \end{pmatrix} = \begin{pmatrix} 46 & 24 \\ 28 & -42 \\ -8 & 2 \end{pmatrix}$
$$= 2\begin{pmatrix} 23 & 12 \\ 14 & -21 \\ -4 & 1 \end{pmatrix}$$

4 **a** $\begin{pmatrix} -6 & -15 & 3 & -18 \\ -10 & 9 & -1 & -10 \end{pmatrix}$ **b** $\begin{pmatrix} 16 & 2 \\ 36 & 72 \\ 30 & -15 \end{pmatrix}$

  c $\begin{pmatrix} 0 \\ 13 \end{pmatrix}$ **d** $(-8 \quad -5 \quad -10)$

  e $\begin{pmatrix} -4 & 8 & 20 \\ -3 & 6 & 15 \end{pmatrix}$ **f** $\begin{pmatrix} -5 & 6 & 4 \\ -2 & -12 & 4 \\ 6 & -3 & -5 \end{pmatrix}$

5 $\begin{pmatrix} 3 & 2 & 0 \\ -1 & 0 & -2 \end{pmatrix}\begin{pmatrix} 4 & 1 \\ 0 & 3 \\ -3 & 0 \end{pmatrix}$

  $= \begin{pmatrix} (3\times4)+(2\times0)+(0\times-3) & (3\times1)+(2\times3)+(0\times0) \\ (-1\times4)+(0\times0)+(-2\times-3) & (-1\times1)+(0\times3)+(-2\times0) \end{pmatrix}$

  $= \begin{pmatrix} 12+0+0 & 3+6+0 \\ -4+0+6 & -1+0+0 \end{pmatrix}$

  $= \begin{pmatrix} 12 & 9 \\ 2 & -1 \end{pmatrix}$

6 **a** Not possible since 1 column in **A** but 2 rows in **B**

  b $\begin{pmatrix} 51 \\ -7 \end{pmatrix}$

  c $\begin{pmatrix} -14 \\ -72 \end{pmatrix}$

  d Not possible since 3 columns in **C** but 2 rows in **B**

  e $\begin{pmatrix} 36 \\ 40 \end{pmatrix}$

  f Not possible as **C** not a square matrix.

7 **a** $\begin{pmatrix} 7a & 8a+6 & 3a \\ 2-a & 2 & 3+a \end{pmatrix}$ **b** $\begin{pmatrix} 3a & -a^2 \\ 6 & -2a \\ -3a & a^2 \end{pmatrix}$

  c $(a \quad a)$

  d $\begin{pmatrix} a^2-6 & 6a \\ -4a & a^2-6 \end{pmatrix}$

8 $\mathbf{A}^3 = \begin{pmatrix} 2 & 5 \\ -1 & 2 \end{pmatrix}\begin{pmatrix} 2 & 5 \\ -1 & 2 \end{pmatrix}\begin{pmatrix} 2 & 5 \\ -1 & 2 \end{pmatrix}$

  $= \begin{pmatrix} 2 & 5 \\ -1 & 2 \end{pmatrix}\begin{pmatrix} 4-5 & 10+10 \\ -2-2 & -5+4 \end{pmatrix}$

  $= \begin{pmatrix} 2 & 5 \\ -1 & 2 \end{pmatrix}\begin{pmatrix} -1 & 20 \\ -4 & -1 \end{pmatrix}$

  $= \begin{pmatrix} -2-20 & 40-5 \\ 1-8 & -20-2 \end{pmatrix}$

  $= \begin{pmatrix} -22 & 35 \\ -7 & -22 \end{pmatrix}$ $k=35$

9 $\begin{pmatrix} 3k & 2k \\ 12-2k & 8 \end{pmatrix}$

10 $\begin{pmatrix} 20+2k & 4 & 8 \\ 4+k^2 & -1 & k \\ 12+2k & -6 & 0 \end{pmatrix}$

Exercise 5.1B

1 $a=-3$
  $b=-5$
  $c=-1$
  $d=2$
2 $a=-3$
  $b=4$
  $c=-3$
3 **a** $x=-1, y=4$
  b $x=-2, y=4$
4 **a** $x=1, -\dfrac{3}{5}$
  b $x=-2, -5$
  c $x=\pm 3$
  d $x=-2$
5 $a=7, b=-3, c=2$
6 $x=3, y=4, z=-1$
7 $a=-1, b=-2, c=6$
8 **a** $\begin{pmatrix} 2 & 5 \\ 6 & -1 \end{pmatrix}\begin{pmatrix} x \\ y \end{pmatrix}=\begin{pmatrix} 6 \\ 3 \end{pmatrix}$

  b $\begin{pmatrix} 1 & 1 & -1 \\ 2 & 1 & 1 \\ 3 & 2 & 2 \end{pmatrix}\begin{pmatrix} x \\ y \\ z \end{pmatrix}=\begin{pmatrix} -4 \\ 4 \\ 10 \end{pmatrix}$

9 $\begin{pmatrix} 1 & 5 \\ 0 & 1 \end{pmatrix}^1=\begin{pmatrix} 1 & 5 \\ 0 & 1 \end{pmatrix}$ and $\begin{pmatrix} 1 & 5\times1 \\ 0 & 1 \end{pmatrix}=\begin{pmatrix} 1 & 5 \\ 0 & 1 \end{pmatrix}$ so true for $n=1$

  Assume true for $n=k$

  $\begin{pmatrix} 1 & 5 \\ 0 & 1 \end{pmatrix}^{k+1}=\begin{pmatrix} 1 & 5 \\ 0 & 1 \end{pmatrix}^k\begin{pmatrix} 1 & 5 \\ 0 & 1 \end{pmatrix}$

  $= \begin{pmatrix} 1 & 5k \\ 0 & 1 \end{pmatrix}\begin{pmatrix} 1 & 5 \\ 0 & 1 \end{pmatrix}$

  $= \begin{pmatrix} 1 & 5+5k \\ 0 & 1 \end{pmatrix}$

  $= \begin{pmatrix} 1 & 5(k+1) \\ 0 & 1 \end{pmatrix}$ so true for $n=k+1$

  Since true for $n=1$ and assuming true for $n=k$ implies true for $n=k+1$, therefore true for all positive integers n

10 $\begin{pmatrix} 5 & 4 \\ 0 & 1 \end{pmatrix}^1=\begin{pmatrix} 5 & 4 \\ 0 & 1 \end{pmatrix}$ and $\begin{pmatrix} 5^1 & 5^1-1 \\ 0 & 1 \end{pmatrix}=\begin{pmatrix} 5 & 4 \\ 0 & 1 \end{pmatrix}$ so true for $n=1$

  Assume true for $n=k$

  $\begin{pmatrix} 5 & 4 \\ 0 & 1 \end{pmatrix}^{k+1}=\begin{pmatrix} 5 & 4 \\ 0 & 1 \end{pmatrix}^k\begin{pmatrix} 5 & 4 \\ 0 & 1 \end{pmatrix}$

  $= \begin{pmatrix} 5^k & 5^k-1 \\ 0 & 1 \end{pmatrix}\begin{pmatrix} 5 & 4 \\ 0 & 1 \end{pmatrix}$

  $= \begin{pmatrix} 5(5^k) & 4(5^k)+5^k-1 \\ 0 & 1 \end{pmatrix}$

  $= \begin{pmatrix} 5^{k+1} & 5(5^k)-1 \\ 0 & 1 \end{pmatrix}$

  $= \begin{pmatrix} 5^{k+1} & 5^{k+1}-1 \\ 0 & 1 \end{pmatrix}$ so true for $n=k+1$

  Since true for $n=1$ and assuming true for $n=k$ implies true for $n=k+1$, therefore true for all positive integers n

11 $\begin{pmatrix} -2 & -1 \\ 9 & 4 \end{pmatrix}^1 = \begin{pmatrix} -2 & -1 \\ 9 & 4 \end{pmatrix}$ and $\begin{pmatrix} 1-3\times1 & -1 \\ 9\times1 & 3\times1+1 \end{pmatrix} = \begin{pmatrix} -2 & -1 \\ 9 & 4 \end{pmatrix}$

so true for $n = 1$

Assume true for $n = k$

$\begin{pmatrix} -2 & -1 \\ 9 & 4 \end{pmatrix}^{k+1} = \begin{pmatrix} -2 & -1 \\ 9 & 4 \end{pmatrix}^k \begin{pmatrix} -2 & -1 \\ 9 & 4 \end{pmatrix}$

$= \begin{pmatrix} 1-3k & -k \\ 9k & 3k+1 \end{pmatrix} \begin{pmatrix} -2 & -1 \\ 9 & 4 \end{pmatrix}$

$= \begin{pmatrix} -2+6k-9k & -1+3k-4k \\ -18k+27k+9 & -9k+12k+4 \end{pmatrix}$

$= \begin{pmatrix} -2-3k & -1-k \\ 9k+9 & 3k+4 \end{pmatrix}$

$= \begin{pmatrix} 1-3(k+1) & -(k+1) \\ 9(k+1) & 3(k+1)+1 \end{pmatrix}$ so true for $n = k+1$

Since true for $n = 1$ and assuming true for $n = k$ implies true for $n = k+1$, therefore true for all positive integers n

12 $\begin{pmatrix} 1 & 4 \\ 0 & 2 \end{pmatrix}^1 = \begin{pmatrix} 1 & 4 \\ 0 & 2 \end{pmatrix}$ and $\begin{pmatrix} 1 & 4(2^1-1) \\ 0 & 2^1 \end{pmatrix} = \begin{pmatrix} 1 & 4 \\ 0 & 2 \end{pmatrix}$ so true

for $n = 1$

Assume true for $n = k$

$\begin{pmatrix} 1 & 4 \\ 0 & 2 \end{pmatrix}^{k+1} = \begin{pmatrix} 1 & 4 \\ 0 & 2 \end{pmatrix}^k \begin{pmatrix} 1 & 4 \\ 0 & 2 \end{pmatrix}$

$= \begin{pmatrix} 1 & 4(2^k-1) \\ 0 & 2^k \end{pmatrix} \begin{pmatrix} 1 & 4 \\ 0 & 2 \end{pmatrix}$

$= \begin{pmatrix} 1 & 4+8.2^k-8 \\ 0 & 2.2^k \end{pmatrix}$

$= \begin{pmatrix} 1 & 4(2.2^k-1) \\ 0 & 2.2^k \end{pmatrix}$

$= \begin{pmatrix} 1 & 4(2^{k+1}-1) \\ 0 & 2^{k+1} \end{pmatrix}$ so true for $n = k+1$

Since true for $n = 1$ and assuming true for $n = k$ implies true for $n = k+1$, therefore true for all positive integers n

13 Let $A = \begin{pmatrix} a_1 & a_2 \\ a_3 & a_4 \\ a_5 & a_6 \end{pmatrix}$, $B = \begin{pmatrix} b_1 & b_2 \\ b_3 & b_4 \\ b_5 & b_6 \end{pmatrix}$, $C = \begin{pmatrix} c_1 & c_2 \\ c_3 & c_4 \\ c_5 & c_6 \end{pmatrix}$

$A + (B + C) = \begin{pmatrix} a_1 & a_2 \\ a_3 & a_4 \\ a_5 & a_6 \end{pmatrix} + \begin{pmatrix} b_1+c_1 & b_2+c_2 \\ b_3+c_3 & b_4+c_4 \\ b_5+c_5 & b_6+c_6 \end{pmatrix}$

$= \begin{pmatrix} a_1+b_1+c_1 & a_2+b_2+c_2 \\ a_3+b_3+c_3 & a_4+b_4+c_4 \\ a_5+b_5+c_5 & a_6+b_6+c_6 \end{pmatrix}$

$= \begin{pmatrix} (a_1+b_1)+c_1 & (a_2+b_2)+c_2 \\ (a_3+b_3)+c_3 & (a_4+b_4)+c_4 \\ (a_5+b_5)+c_5 & (a_6+b_6)+c_6 \end{pmatrix}$

$= \begin{pmatrix} a_1+b_1 & a_2+b_2 \\ a_3+b_3 & a_4+b_4 \\ a_5+b_5 & a_6+b_6 \end{pmatrix} + \begin{pmatrix} c_1 & c_2 \\ c_3 & c_4 \\ c_5 & c_6 \end{pmatrix}$

$= (A + B) + C$ as required

14 Let $A = \begin{pmatrix} a_1 & a_2 \\ a_3 & a_4 \end{pmatrix}$, $B = \begin{pmatrix} b_1 & b_2 \\ b_3 & b_4 \end{pmatrix}$

$A + B = \begin{pmatrix} a_1 & a_2 \\ a_3 & a_4 \end{pmatrix} + \begin{pmatrix} b_1 & b_2 \\ b_3 & b_4 \end{pmatrix}$

$= \begin{pmatrix} a_1+b_1 & a_2+b_2 \\ a_3+b_3 & a_4+b_4 \end{pmatrix}$

$= \begin{pmatrix} b_1+a_1 & b_2+a_2 \\ b_3+a_3 & b_4+a_4 \end{pmatrix}$

$= \begin{pmatrix} b_1 & b_2 \\ b_3 & b_4 \end{pmatrix} + \begin{pmatrix} a_1 & a_2 \\ a_3 & a_4 \end{pmatrix}$

$= B + A$ as required

15 e.g. let $A = \begin{pmatrix} 2 & 0 \\ 1 & 1 \end{pmatrix}$, $B = \begin{pmatrix} 0 & 1 \\ 2 & 2 \end{pmatrix}$

$AB = \begin{pmatrix} 2 & 0 \\ 1 & 1 \end{pmatrix} \begin{pmatrix} 0 & 1 \\ 2 & 2 \end{pmatrix}$

$= \begin{pmatrix} 0 & 2 \\ 2 & 3 \end{pmatrix}$

$BA = \begin{pmatrix} 0 & 1 \\ 2 & 2 \end{pmatrix} \begin{pmatrix} 2 & 0 \\ 1 & 1 \end{pmatrix}$

$= \begin{pmatrix} 1 & 1 \\ 6 & 2 \end{pmatrix}$

So $AB \neq BA$

16 Let $A = \begin{pmatrix} a_1 & a_2 \\ a_3 & a_4 \end{pmatrix}$, $B = \begin{pmatrix} b_1 & b_2 \\ b_3 & b_4 \end{pmatrix}$, $C = \begin{pmatrix} c_1 & c_2 \\ c_3 & c_4 \end{pmatrix}$

$A(B + C)$

$= \begin{pmatrix} a_1 & a_2 \\ a_3 & a_4 \end{pmatrix} \left[\begin{pmatrix} b_1 & b_2 \\ b_3 & b_4 \end{pmatrix} + \begin{pmatrix} c_1 & c_2 \\ c_3 & c_4 \end{pmatrix} \right]$

$= \begin{pmatrix} a_1 & a_2 \\ a_3 & a_4 \end{pmatrix} \begin{pmatrix} b_1+c_1 & b_2+c_2 \\ b_3+c_3 & b_4+c_4 \end{pmatrix}$

$= \begin{pmatrix} a_1(b_1+c_1)+a_2(b_3+c_3) & a_1(b_2+c_2)+a_2(b_4+c_4) \\ a_3(b_1+c_1)+a_4(b_3+c_3) & a_3(b_2+c_2)+a_4(b_4+c_4) \end{pmatrix}$

$= \begin{pmatrix} (a_1b_1+a_2b_3)+(a_1c_1+a_2c_3) & (a_1b_2+a_2b_4)+(a_1c_2+a_2c_4) \\ (a_3b_1+a_4b_3)+(a_3c_1+a_4c_3) & (a_3b_2+a_4b_4)+(a_3c_2+a_4c_4) \end{pmatrix}$

$= \begin{pmatrix} a_1b_1+a_2b_3 & a_1b_2+a_2b_4 \\ a_3b_1+a_4b_3 & a_3b_2+a_4b_4 \end{pmatrix} + \begin{pmatrix} a_1c_1+a_2c_3 & a_1c_2+a_2c_4 \\ a_3c_1+a_4c_3 & a_3c_2+a_4c_4 \end{pmatrix}$

$= \begin{pmatrix} a_1 & a_2 \\ a_3 & a_4 \end{pmatrix} \begin{pmatrix} b_1 & b_2 \\ b_3 & b_4 \end{pmatrix} + \begin{pmatrix} a_1 & a_2 \\ a_3 & a_4 \end{pmatrix} \begin{pmatrix} c_1 & c_2 \\ c_3 & c_4 \end{pmatrix}$

$= AB + AC$

17 Let $A = \begin{pmatrix} a_1 & a_2 \\ a_3 & a_4 \end{pmatrix}$

$A \begin{pmatrix} 1 & 0 \\ 0 & 1 \end{pmatrix} = \begin{pmatrix} a_1 & a_2 \\ a_3 & a_4 \end{pmatrix} \begin{pmatrix} 1 & 0 \\ 0 & 1 \end{pmatrix}$

$= \begin{pmatrix} a_1 & a_2 \\ a_3 & a_4 \end{pmatrix}$

$$\begin{pmatrix} 1 & 0 \\ 0 & 1 \end{pmatrix}\mathbf{A} = \begin{pmatrix} 1 & 0 \\ 0 & 1 \end{pmatrix}\begin{pmatrix} a_1 & a_2 \\ a_3 & a_4 \end{pmatrix}$$

$$= \begin{pmatrix} a_1 & a_2 \\ a_3 & a_4 \end{pmatrix}$$

So $\mathbf{A}\begin{pmatrix} 1 & 0 \\ 0 & 1 \end{pmatrix} = \begin{pmatrix} 1 & 0 \\ 0 & 1 \end{pmatrix}\mathbf{A}$ as required

18 Let $\mathbf{B} = \begin{pmatrix} b_1 & b_2 & b_3 \\ b_4 & b_5 & b_6 \\ b_7 & b_8 & b_9 \end{pmatrix}$

$$\mathbf{B}\begin{pmatrix} 1 & 0 & 0 \\ 0 & 1 & 0 \\ 0 & 0 & 1 \end{pmatrix} = \begin{pmatrix} b_1 & b_2 & b_3 \\ b_4 & b_5 & b_6 \\ b_7 & b_8 & b_9 \end{pmatrix}\begin{pmatrix} 1 & 0 & 0 \\ 0 & 1 & 0 \\ 0 & 0 & 1 \end{pmatrix}$$

$$= \begin{pmatrix} b_1 & b_2 & b_3 \\ b_4 & b_5 & b_6 \\ b_7 & b_8 & b_9 \end{pmatrix}$$

$$\begin{pmatrix} 1 & 0 & 0 \\ 0 & 1 & 0 \\ 0 & 0 & 1 \end{pmatrix}\mathbf{B} = \begin{pmatrix} 1 & 0 & 0 \\ 0 & 1 & 0 \\ 0 & 0 & 1 \end{pmatrix}\begin{pmatrix} b_1 & b_2 & b_3 \\ b_4 & b_5 & b_6 \\ b_7 & b_8 & b_9 \end{pmatrix}$$

$$= \begin{pmatrix} b_1 & b_2 & b_3 \\ b_4 & b_5 & b_6 \\ b_7 & b_8 & b_9 \end{pmatrix}$$

So $\mathbf{B}\begin{pmatrix} 1 & 0 & 0 \\ 0 & 1 & 0 \\ 0 & 0 & 1 \end{pmatrix} = \begin{pmatrix} 1 & 0 & 0 \\ 0 & 1 & 0 \\ 0 & 0 & 1 \end{pmatrix}\mathbf{B}$ as required

19 a $\begin{pmatrix} 8 & 14 & 15 & 17 \\ 6 & 11 & 7 & 9 \\ 9 & 18 & 19 & 12 \end{pmatrix}$

$\begin{pmatrix} 5 \\ 2 \\ 3 \\ 1 \end{pmatrix}$ or $\begin{pmatrix} 0.05 \\ 0.02 \\ 0.03 \\ 0.01 \end{pmatrix}$

b $\begin{pmatrix} 8 & 14 & 15 & 17 \\ 6 & 11 & 7 & 9 \\ 9 & 18 & 19 & 12 \end{pmatrix}\begin{pmatrix} 0.05 \\ 0.02 \\ 0.03 \\ 0.01 \end{pmatrix}$

$= \begin{pmatrix} 1.3 \\ 0.82 \\ 1.5 \end{pmatrix}$

c £3620

20 a $\begin{pmatrix} 5 & 20 & 7 & 6 & 50 \\ 4 & 15 & 15 & 8 & 100 \\ 12 & 4 & 2 & 30 & 50 \end{pmatrix}\begin{pmatrix} 12 \\ 8 \\ 18 \\ 30 \\ 1 \end{pmatrix}$

$\begin{pmatrix} 576 \\ 778 \\ 1162 \end{pmatrix}$

b £25.16

21 a $ae + bg = a\left(\dfrac{d}{ad - bc}\right) + b\left(-\dfrac{c}{ad - bc}\right)$

$= \dfrac{ad}{ad - bc} - \dfrac{bc}{ad - bc}$

$= \dfrac{ad - bc}{ad - bc}$

$= 1$ as required.

b $h = \dfrac{a}{ad - bc}, \ f = -\dfrac{b}{ad - bc}$

22 When $n = 1$, $\begin{pmatrix} 0 & 1 \\ 2 & 0 \end{pmatrix}^{2n} = \begin{pmatrix} 0 & 1 \\ 2 & 0 \end{pmatrix}^{2}$

$= \begin{pmatrix} 2 & 0 \\ 0 & 2 \end{pmatrix}$

$2^n\mathbf{I} = 2\begin{pmatrix} 1 & 0 \\ 0 & 1 \end{pmatrix}$

$= \begin{pmatrix} 2 & 0 \\ 0 & 2 \end{pmatrix}$

So true for $n = 1$

Assume true for $n = k$ and consider $n = k + 1$:

$\begin{pmatrix} 0 & 1 \\ 2 & 0 \end{pmatrix}^{2(k+1)} = \begin{pmatrix} 0 & 1 \\ 2 & 0 \end{pmatrix}^{2k}\begin{pmatrix} 0 & 1 \\ 2 & 0 \end{pmatrix}^{2}$

$= 2^k\begin{pmatrix} 1 & 0 \\ 0 & 1 \end{pmatrix}\begin{pmatrix} 2 & 0 \\ 0 & 2 \end{pmatrix}$

$= 2^k\begin{pmatrix} 2 & 0 \\ 0 & 2 \end{pmatrix}$

$= 2^k \cdot 2\begin{pmatrix} 1 & 0 \\ 0 & 1 \end{pmatrix}$

$= 2^{k+1}\mathbf{I}$

So true for $n = k + 1$

The statement is true for $n = 1$ and by assuming it is true for $n = k$ it is shown to be true for $n = k + 1$, therefore, by mathematical induction, it is true for all $n \in \mathbb{N}$

Exercise 5.2A

1 a

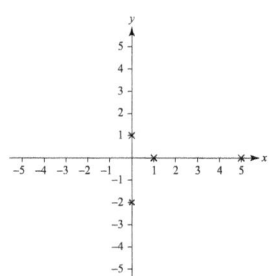

b $\begin{pmatrix} 5 & 0 \\ 0 & -2 \end{pmatrix}$

2 a

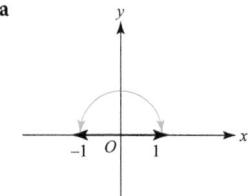

b $\begin{pmatrix} 0 \\ -1 \end{pmatrix}$

c $\begin{pmatrix} -1 & 0 \\ 0 & -1 \end{pmatrix}$

3 a $\begin{pmatrix} 0 & 1 \\ 1 & 0 \end{pmatrix}$ **b** $\begin{pmatrix} \dfrac{\sqrt{3}}{2} & -\dfrac{1}{2} \\ \dfrac{1}{2} & \dfrac{\sqrt{3}}{2} \end{pmatrix}$

c $\begin{pmatrix} 5 & 0 \\ 0 & 5 \end{pmatrix}$ **d** $\begin{pmatrix} 1 & 0 \\ 0 & 4 \end{pmatrix}$

4 a $(0, -2)$, $(4, 3)$ and $(-8, 3)$
b Stretch parallel to x-axis of scale factor 2

5 $\theta = 315°$

6 a $\begin{pmatrix} 6 \\ -1 \end{pmatrix}, \begin{pmatrix} -1 \\ 2 \end{pmatrix}$
b Stretch scale factor 0.5 parallel to y-axis

7 a Enlargement scale factor 3 centre the origin
b Stretch scale factor 2 parallel to y-axis
c Rotation by 135° anticlockwise about origin
d Rotation by 270° anticlockwise about origin
e Reflection in y-axis
f Rotation by 20° clockwise about origin or 340° anticlockwise.

8 a i $\mathbf{BA} = \begin{pmatrix} 0 & -1 \\ -1 & 0 \end{pmatrix}$ **ii** $\mathbf{AB} = \begin{pmatrix} -1 & 0 \\ -1 & 1 \end{pmatrix}$
b $\mathbf{BA} \neq \mathbf{AB}$ since matrix multiplication is not commutative.

9 a $\mathbf{A}$ is reflection in x-axis, $\mathbf{B}$ is rotation about origin of 90° clockwise or 270° anticlockwise
b i $\mathbf{BA} = \begin{pmatrix} 0 & -1 \\ -1 & 0 \end{pmatrix}$ **ii** $\mathbf{AB} = \begin{pmatrix} 0 & 1 \\ 1 & 0 \end{pmatrix}$
c i Reflection in $y = -x$ **ii** Reflection in $y = x$

10 a $\begin{pmatrix} -\dfrac{\sqrt{2}}{2} & \dfrac{\sqrt{2}}{2} \\ \dfrac{\sqrt{2}}{2} & \dfrac{\sqrt{2}}{2} \end{pmatrix}$ **b** $\begin{pmatrix} 0 & -2 \\ -2 & 0 \end{pmatrix}$ **c** $\begin{pmatrix} 0 & 1 \\ 2 & 0 \end{pmatrix}$

11 $\left(\dfrac{3}{2}\sqrt{2}, 7\sqrt{2}\right), \left(-\dfrac{\sqrt{2}}{2}, 3\sqrt{2}\right), (2\sqrt{2}, -2\sqrt{2}), (4\sqrt{2}, 2\sqrt{2})$

12 a $\begin{pmatrix} -\dfrac{\sqrt{2}}{2} & -\dfrac{\sqrt{2}}{2} \\ \dfrac{\sqrt{2}}{2} & -\dfrac{\sqrt{2}}{2} \end{pmatrix}$

b $\mathbf{A}^2 = \begin{pmatrix} -\dfrac{\sqrt{2}}{2} & -\dfrac{\sqrt{2}}{2} \\ \dfrac{\sqrt{2}}{2} & -\dfrac{\sqrt{2}}{2} \end{pmatrix}\begin{pmatrix} -\dfrac{\sqrt{2}}{2} & -\dfrac{\sqrt{2}}{2} \\ \dfrac{\sqrt{2}}{2} & -\dfrac{\sqrt{2}}{2} \end{pmatrix}$

$= \begin{pmatrix} \dfrac{1}{2} - \dfrac{1}{2} & \dfrac{1}{2} + \dfrac{1}{2} \\ -\dfrac{1}{2} - \dfrac{1}{2} & -\dfrac{1}{2} + \dfrac{1}{2} \end{pmatrix}$

$= \begin{pmatrix} 0 & 1 \\ -1 & 0 \end{pmatrix}$

c $\sin\theta = -1 \Rightarrow \theta = -90$
and $\cos(-90) = 0$
So a clockwise rotation of 90°,
or show geometrically.
d An anticlockwise rotation by 45° about the origin.

13 a $\begin{pmatrix} -1 & 0 \\ 0 & -1 \end{pmatrix}$; rotation of 180° around origin

b $\begin{pmatrix} -\dfrac{1}{2} & -\dfrac{\sqrt{3}}{2} \\ \dfrac{\sqrt{3}}{2} & -\dfrac{1}{2} \end{pmatrix}$; rotation of 120° anticlockwise

around origin

14 a $\begin{pmatrix} 1 & 0 \\ 0 & -3 \end{pmatrix}$
b A stretch of scale factor -3 parallel to the y-axis.

15 a $\mathbf{M} = \begin{pmatrix} k\cos\theta & -k\sin\theta \\ k\sin\theta & k\cos\theta \end{pmatrix}$

b $\theta = (-45°), 135°, k = 2\sqrt{2}$

16 a $(4, 1, 0)$ and $(2, -5, 3)$
b Reflection in $y = 0$

17 a $\begin{pmatrix} 1 & 0 & 0 \\ 0 & \dfrac{\sqrt{3}}{2} & -\dfrac{1}{2} \\ 0 & \dfrac{1}{2} & \dfrac{\sqrt{3}}{2} \end{pmatrix}$ **b** $\begin{pmatrix} 1 & 0 & 0 \\ 0 & 1 & 0 \\ 0 & 0 & -1 \end{pmatrix}$

c $\begin{pmatrix} \dfrac{\sqrt{2}}{2} & 0 & \dfrac{\sqrt{2}}{2} \\ 0 & 1 & 0 \\ -\dfrac{\sqrt{2}}{2} & 0 & \dfrac{\sqrt{2}}{2} \end{pmatrix}$

18 a Rotation around x-axis, $\theta = 300°$

b $\begin{pmatrix} 3 \\ 2 - \dfrac{\sqrt{3}}{2} \\ -\dfrac{1}{2} - 2\sqrt{3} \end{pmatrix}$

19 Anticlockwise rotation of 300° around z-axis.
20 a Reflection in plane $z = 0$

b $\mathbf{M} = \begin{pmatrix} 1 & 0 & 0 \\ 0 & 1 & 0 \\ 0 & 0 & -1 \end{pmatrix}$

21 a Rotation of 90° (anticlockwise) around the x-axis.
b Reflection in $x = 0$
c Rotation of 180° around the z-axis.
d Rotation of 120° anticlockwise around y-axis.

22 a $\begin{pmatrix} -\dfrac{\sqrt{2}}{2} & 0 & -\dfrac{\sqrt{2}}{2} \\ 0 & 1 & 0 \\ \dfrac{\sqrt{2}}{2} & 0 & -\dfrac{\sqrt{2}}{2} \end{pmatrix}$ **b** $\left(-1 - \dfrac{\sqrt{2}}{2}, 0, 1 - \dfrac{\sqrt{2}}{2}\right)$

23 a $\begin{pmatrix} \dfrac{\sqrt{3}}{2} & \dfrac{1}{2} & 0 \\[2mm] -\dfrac{1}{2} & \dfrac{\sqrt{3}}{2} & 0 \\[2mm] 0 & 0 & 1 \end{pmatrix}$ **b** $\left(\dfrac{3}{2}, -\dfrac{\sqrt{3}}{2}, 1 \right)$

Exercise 5.2B

1 a $y = 0$
  b $y = -2x$

2 a $\begin{pmatrix} 3 & -2 \\ 2 & 3 \end{pmatrix}\begin{pmatrix} x \\ y \end{pmatrix} = \begin{pmatrix} x \\ y \end{pmatrix}$

  $3x - 2y = x \Rightarrow x = y$
  $2x + 3y = y \Rightarrow x = -y$
  (0, 0) is only point that satisfies both of these equations, therefore (0, 0) is the only invariant point.

  b $\begin{pmatrix} 2 & 0 & 1 \\ 0 & 3 & -2 \\ 1 & 0 & -4 \end{pmatrix}\begin{pmatrix} x \\ y \\ z \end{pmatrix} = \begin{pmatrix} x \\ y \\ z \end{pmatrix}$

  $2x + z = x \Rightarrow z = -x$
  $3y - 2z = y \Rightarrow y = z$
  $x - 4z = z \Rightarrow x = 5z$
  (0, 0, 0) is only point that satisfies all of these equations, therefore (0, 0, 0) is the only invariant point.

3 a No difference, e.g. reflection in y-axis affects only the x-coordinates and stretch parallel to y-axis affects only the y-coordinates. Or use a sketch. Or show that combined transformation is $\begin{pmatrix} -1 & 0 \\ 0 & k \end{pmatrix}$ for both orders.

  b No difference, e.g. use a sketch or show that combined transformation is $k\begin{pmatrix} \cos\theta & -\sin\theta \\ \sin\theta & \cos\theta \end{pmatrix}$ in both cases.

  c Different. Reflection in $y = x$ followed by stretch along x-axis will be $\begin{pmatrix} 0 & k \\ 1 & 0 \end{pmatrix}$, other way around is $\begin{pmatrix} 0 & 1 \\ k & 0 \end{pmatrix}$.

4 a Stretch parallel to x-axis of scale factor 3 and stretch parallel to y-axis of scale factor -2.
  b Enlargement of scale factor 4 centre the origin and reflection in $y = x$.

5 a $y = \dfrac{1}{5}x$ **b** $y = 3x$ and $y = -x$

6 a e.g. A stretch parallel to the y-axis and a reflection in the y-axis (or rotation by 180° or 360° about origin)
  b Reflection in y-axis
7 e.g. Reflection in line $y = x$ (or $y = x + k$)
8 Enlargement, centre the origin

9 a $y = \left(\dfrac{-1 \pm \sqrt{5}}{2}\right)x$ **b** $y = mx$ for any m

  c $y = 2x$ and $y = -\dfrac{1}{2}x + c$ for any c.
10 ai; bi $y = 0$; line of invariant points
  aii; bii (0, 0)
  aiii; biii $x = 0$; line of invariant points
  aiv; biv $y = x + c$ for any c; not lines of invariant points, only (0, 0) is invariant. $y = -x$; line of invariant points.
11 a $y = x$

b $\begin{pmatrix} -2 & 3 \\ -3 & 4 \end{pmatrix}\begin{pmatrix} x \\ x+c \end{pmatrix} = \begin{pmatrix} -2x + 3(x+c) \\ -3x + 4(x+c) \end{pmatrix}$

$= \begin{pmatrix} x + 3c \\ x + 4c \end{pmatrix}$

$= \begin{pmatrix} x + 3c \\ (x + 3c) + c \end{pmatrix}$

So invariant lines are $y = x + c$

12 $\begin{pmatrix} 1 & 0 \\ 0 & -1 \end{pmatrix}$

13 a Plane $x = 0$ **b** Plane $y = z$
  c z-axis   **d** Plane $x = y$

Exercise 5.3A

1 $\det\begin{pmatrix} -7 & 3 \\ 5 & -4 \end{pmatrix} = -7 \times -4 - 3 \times 5$

                          $= 28 - 15$
                          $= 13$ as required

2 a -18 **b** -3
  c $-5a$ **d** 2

3 $\det\begin{pmatrix} a+b & 2a \\ 2b & a+b \end{pmatrix} = (a+b)(a+b) - (2a)(2b)$

                        $= a^2 + 2ab + b^2 - 4ab$
                        $= a^2 - 2ab + b^2$
                        $= (a-b)^2$ as required

4 $k = 3, 4$
5 a Non-singular
  b Singular
  c Non-singular unless $a = 0$
  d Singular
6 a $x = \dfrac{5}{2}$

  b $x = \pm\sqrt{3}$

  c $x = 0, \dfrac{-3}{2}$

  d $x = 0, 4$

7 $y = -1, -2$
8 a i -30 **ii** 1

  b i $AB = \begin{pmatrix} -1 & 2 \\ 0 & 5 \end{pmatrix}\begin{pmatrix} 0 & -3 \\ 2 & 4 \end{pmatrix} = \begin{pmatrix} 4 & 11 \\ 10 & 20 \end{pmatrix}$

    $\det(AB) = 4 \times 20 - 11 \times 10 = -30$ as required

  ii $A + B = \begin{pmatrix} -1 & 2 \\ 0 & 5 \end{pmatrix} + \begin{pmatrix} 0 & -3 \\ 2 & 4 \end{pmatrix} = \begin{pmatrix} -1 & -1 \\ 2 & 9 \end{pmatrix}$

    $\det(A + B) = -1 \times 9 - -1 \times 2 = -7$

9 a -2

  b $\dfrac{1}{9}$

10 a -8

  b 16

11 $\dfrac{1}{7}$

12 a 5 **b** $\dfrac{5}{3}$

13 a $\dfrac{1}{27}\begin{pmatrix} 5 & -2 \\ -4 & 7 \end{pmatrix}$ **b** $\dfrac{1}{20}\begin{pmatrix} 1 & 3 \\ 8 & 4 \end{pmatrix}$

  c $-\dfrac{1}{x}\begin{pmatrix} -3 & -5 \\ x & 2x \end{pmatrix}$ **d** $\dfrac{1}{50}\begin{pmatrix} 2\sqrt{5} & \sqrt{5} \\ -2\sqrt{5} & 4\sqrt{5} \end{pmatrix}$

14 $\begin{pmatrix} 6 & 1 \\ -9 & -1 \end{pmatrix}^{-1} = \dfrac{1}{6\times -1 - 1\times -9}\begin{pmatrix} -1 & -1 \\ 9 & 6 \end{pmatrix}$

$= \dfrac{1}{3}\begin{pmatrix} -1 & -1 \\ 9 & 6 \end{pmatrix}$

$= \begin{pmatrix} -\dfrac{1}{3} & -\dfrac{1}{3} \\ 3 & 2 \end{pmatrix}$ as required.

Alternatively show that

$\begin{pmatrix} 6 & 1 \\ -9 & -1 \end{pmatrix}\begin{pmatrix} -\dfrac{1}{3} & -\dfrac{1}{3} \\ 3 & 2 \end{pmatrix} = \mathbf{I}$

15 a $\begin{pmatrix} \dfrac{3}{b} & -\dfrac{2}{b} \\ -\dfrac{1}{a} & \dfrac{1}{a} \end{pmatrix}$

b $\det(\mathbf{A}^{-1}) = \dfrac{3}{b}\cdot\dfrac{1}{a} - \left(-\dfrac{2}{b}\right)\left(-\dfrac{1}{a}\right)$

$= \dfrac{3}{ab} - \dfrac{2}{ab}$

$= \dfrac{1}{ab}$

$= \dfrac{1}{\det(A)}$ as required

16 $\begin{pmatrix} 3 & -2 \\ 4 & -3 \end{pmatrix}\begin{pmatrix} 3 & -2 \\ 4 & -3 \end{pmatrix} = \begin{pmatrix} 3\times 3 + -2\times 4 & 3\times -2 + -2\times -3 \\ 4\times 3 + -3\times 4 & 4\times -2 + -3\times -3 \end{pmatrix}$

$= \begin{pmatrix} 1 & 0 \\ 0 & 1 \end{pmatrix}$ therefore it is self-inverse.

17 $a = \pm\sqrt{7}$

18 $a = 3$

19 a $x = \dfrac{5}{3}, -2$

b $\dfrac{1}{10}\begin{pmatrix} 2x & x \\ 1-x & x+1 \end{pmatrix}$

20 a $\mathbf{T}^{-1} = \dfrac{1}{4}\begin{pmatrix} 2 & 0 \\ 0 & 2 \end{pmatrix}$

$= \dfrac{1}{2}\begin{pmatrix} 1 & 0 \\ 0 & 1 \end{pmatrix}$

$= \dfrac{1}{2}\mathbf{I}$ as required

b Enlargements scale factor $2\left(\dfrac{1}{2}\right)$ centre the origin.

21 a $\dfrac{\sqrt{2}}{2}\begin{pmatrix} -1 & -1 \\ 1 & -1 \end{pmatrix}$

b $\mathbf{A}^{-1}\mathbf{A} = \dfrac{1}{\sqrt{2}}\begin{pmatrix} -1 & 1 \\ -1 & -1 \end{pmatrix}\dfrac{\sqrt{2}}{2}\begin{pmatrix} -1 & -1 \\ 1 & -1 \end{pmatrix}$

$= \dfrac{1}{2}\begin{pmatrix} 1+1 & 1-1 \\ 1-1 & 1+1 \end{pmatrix}$

$= \dfrac{1}{2}\begin{pmatrix} 2 & 0 \\ 0 & 2 \end{pmatrix}$

$= \begin{pmatrix} 1 & 0 \\ 0 & 1 \end{pmatrix}$ as required, therefore the inverse

is as given.

22 a Always self-inverse.
b Only self-inverse if enlargement of scale factor ± 1
c Only self-inverse when rotated by a multiple of $180°$
d Always self-inverse.

Exercise 5.3B

1 a $x = 2, y = -3$ **b** $x = -5, y = -1$
c $x = 3, y = \dfrac{1}{2}$ **d** $x = -\dfrac{1}{4}, y = \dfrac{1}{3}$

2 a $x = 2, y = -5, z = -3$
b $x = \dfrac{1}{4}, y = -\dfrac{1}{2}, z = 1$

3 $x = -\dfrac{1}{a} - 6, y = 2 + 2a$

4 $x = \dfrac{3k-1}{3(k+1)}, y = \dfrac{4k}{3(k+1)}$

5 $\begin{pmatrix} -2 & 4 \\ -1 & -2 \end{pmatrix}$

6 $\begin{pmatrix} 4 & -2 \\ 0 & -1 \end{pmatrix}$

7 $\dfrac{b+2a}{b}$

8 $\begin{pmatrix} 17 & 16 \\ -9 & -8 \end{pmatrix}$

9 $\begin{pmatrix} 5 & 0 \\ 4 & -1 \end{pmatrix}$

10 e.g. $\sqrt{2}\begin{pmatrix} 1 & 0 \\ 0 & 1 \end{pmatrix}$, or $\begin{pmatrix} 0 & \sqrt{2} \\ \sqrt{2} & 0 \end{pmatrix}$

11 a $\mathbf{B}^{-1}$ **b** $\mathbf{A}$
12 $\mathbf{A} = \mathbf{C}^{-1}\mathbf{BC} \Rightarrow \mathbf{CA} = \mathbf{CC}^{-1}\mathbf{BC}$
$\Rightarrow \mathbf{CA} = \mathbf{BC}$
$\Rightarrow \mathbf{CAC}^{-1} = \mathbf{BCC}^{-1}$
$\Rightarrow \mathbf{CAC}^{-1} = \mathbf{B}$ as required
13 $\mathbf{ABA}^{-1} = \mathbf{I} \Rightarrow \mathbf{A}^{-1}\mathbf{ABA}^{-1} = \mathbf{A}^{-1}\mathbf{I}$
$\Rightarrow \mathbf{BA}^{-1} = \mathbf{A}^{-1}$
$\Rightarrow \mathbf{BA}^{-1}\mathbf{A} = \mathbf{A}^{-1}\mathbf{A}$
$\Rightarrow \mathbf{B} = \mathbf{I}$
14 $(\mathbf{ABC})^{-1}(\mathbf{ABC}) = \mathbf{I}$
$\Rightarrow (\mathbf{ABC})^{-1}\mathbf{ABCC}^{-1} = \mathbf{IC}^{-1}$
$\Rightarrow (\mathbf{ABC})^{-1}\mathbf{AB} = \mathbf{C}^{-1}$
$\Rightarrow (\mathbf{ABC})^{-1}\mathbf{ABB}^{-1} = \mathbf{C}^{-1}\mathbf{B}^{-1}$
$\Rightarrow (\mathbf{ABC})^{-1}\mathbf{A} = \mathbf{C}^{-1}\mathbf{B}^{-1}$
$\Rightarrow (\mathbf{ABC})^{-1}\mathbf{AA}^{-1} = \mathbf{C}^{-1}\mathbf{B}^{-1}\mathbf{A}^{-1}$
$\Rightarrow (\mathbf{ABC})^{-1} = \mathbf{C}^{-1}\mathbf{B}^{-1}\mathbf{A}^{-1}$ as required
15 $\mathbf{P}$ self-inverse $\Rightarrow \mathbf{P} = \mathbf{P}^{-1}$
$\Rightarrow \mathbf{PP} = \mathbf{P}^{-1}\mathbf{P}$
$\Rightarrow \mathbf{P}^2 = \mathbf{I}$
$\mathbf{I} = \mathbf{I}^{-1} \Rightarrow \mathbf{P}^2 = \mathbf{I}^{-1}$ as required
16 $\mathbf{PQP} = \mathbf{I} \Rightarrow \mathbf{P}^{-1}\mathbf{PQP} = \mathbf{P}^{-1}\mathbf{I}$
$\Rightarrow \mathbf{QP} = \mathbf{P}^{-1}$
$\Rightarrow \mathbf{QPP}^{-1} = \mathbf{P}^{-1}\mathbf{P}^{-1}$
$\Rightarrow \mathbf{Q} = (\mathbf{P}^{-1})^2$ as required
17 $(5, -1)$
18 Let $\mathbf{A} = \begin{pmatrix} a_1 & a_2 \\ a_3 & a_4 \end{pmatrix}$ and $\mathbf{B} = \begin{pmatrix} b_1 & b_2 \\ b_3 & b_4 \end{pmatrix}$

$\mathbf{AB} = \begin{pmatrix} a_1 & a_2 \\ a_3 & a_4 \end{pmatrix}\begin{pmatrix} b_1 & b_2 \\ b_3 & b_4 \end{pmatrix} = \begin{pmatrix} a_1b_1 + a_2b_3 & a_1b_2 + a_2b_4 \\ a_3b_1 + a_4b_3 & a_3b_2 + a_4b_4 \end{pmatrix}$

Then
$\det(\mathbf{AB}) = (a_1b_1 + a_2b_3)(a_3b_2 + a_4b_4) - (a_1b_2 + a_2b_4)(a_3b_1 + a_4b_3)$
$= (a_1b_1a_3b_2 + a_1b_1a_4b_4 + a_2b_3a_3b_2 + a_2b_3a_4b_4) -$
$(a_1b_2a_3b_1 + a_1b_2a_4b_3 + a_2b_4a_3b_1 + a_2b_4a_4b_3)$
$= a_1b_1a_4b_4 - a_1b_2b_3a_4 - a_2b_1a_3b_4 + a_2b_2a_3b_3$
$= (a_1a_4 - a_2a_3)(b_1b_4 - b_2b_3)$
$= \det(\mathbf{A})\det(\mathbf{B})$ as required

19 Let $\mathbf{A} = \begin{pmatrix} a_1 & a_2 \\ a_3 & a_4 \end{pmatrix}$ then $\mathbf{A}^{-1} = \dfrac{1}{a_1a_4 - a_2a_3}\begin{pmatrix} a_4 & -a_2 \\ -a_3 & a_1 \end{pmatrix}$

$$\det(\mathbf{A}^{-1}) = \dfrac{a_4}{a_1a_4 - a_2a_3}\cdot\dfrac{a_1}{a_1a_4 - a_2a_3} - \dfrac{-a_2}{a_1a_4 - a_2a_3}\cdot\dfrac{-a_3}{a_1a_4 - a_2a_3}$$

$$= \dfrac{a_4a_1 - a_2a_3}{(a_1a_4 - a_2a_3)^2}$$

$$= \dfrac{1}{a_1a_4 - a_2a_3}$$

$$= \dfrac{1}{\det(\mathbf{A})}$$

$$= [\det(\mathbf{A})]^{-1} \text{ as required}$$

20 a   i $\begin{pmatrix} \cos\theta & -\sin\theta \\ \sin\theta & \cos\theta \end{pmatrix}$ **ii** $\begin{pmatrix} \cos\theta & \sin\theta \\ -\sin\theta & \cos\theta \end{pmatrix}$

b $\begin{pmatrix} \cos\theta & -\sin\theta \\ \sin\theta & \cos\theta \end{pmatrix}\begin{pmatrix} \cos\theta & \sin\theta \\ -\sin\theta & \cos\theta \end{pmatrix} = \begin{pmatrix} 1 & 0 \\ 0 & 1 \end{pmatrix}$

$\Rightarrow \cos\theta\cos\theta - \sin\theta(-\sin\theta) \equiv 1$

$\Rightarrow \cos^2\theta + \sin^2\theta \equiv 1$ as required.

21 a   Square with vertices at $(0, 0)$, $(x, 0)$, $(0, y)$ and (x, y)

$\mathbf{T} = \begin{pmatrix} a & 0 \\ 0 & b \end{pmatrix}$ is a stretch of a in x-direction and b in y-direction.

$\begin{pmatrix} a & 0 \\ 0 & b \end{pmatrix}\begin{pmatrix} 0 & x & 0 & x \\ 0 & 0 & y & y \end{pmatrix} = \begin{pmatrix} 0 & ax & 0 & ax \\ 0 & 0 & by & by \end{pmatrix}$

So a rectangle with vertices at $(0, 0)$, $(ax, 0)$, $(0, by)$ and (ax, by).

b   Area of original square is xy

Area of image is $(ax)(by) = abxy$

$= \det(\mathbf{T}) \times xy$ as required since

$\det(\mathbf{T}) = ab - 0 = ab$

Review exercise 5

1 a   A has order 2×2, B has order 2×3, C has order 3×2, D has order 2×3

b   i $\begin{pmatrix} 8 & 0 & 2 \\ 5 & 3 & 0 \end{pmatrix}$ **ii** $\begin{pmatrix} -3 & -15 \\ 12 & 0 \\ 0 & 6 \end{pmatrix}$

iii $\begin{pmatrix} 8 & -6 & 5 \\ -10 & -9 & 0 \end{pmatrix}$ **iv** $\begin{pmatrix} 3 & 1 \\ -2 & 6 \end{pmatrix}$

c   i $\begin{pmatrix} 10 & 20 & -6 \\ 60 & 40 & 4 \end{pmatrix}$

ii **BA** not possible as 3 columns in **B** but 2 rows in **A**

iii $\begin{pmatrix} 8 & -2 \\ 11 & -25 \end{pmatrix}$ **iv** $\begin{pmatrix} 14 & -62 \\ 24 & 8 \\ -8 & 24 \end{pmatrix}$

v **AC** not possible as 2 columns in **A** but 3 rows in **C**

vi $\begin{pmatrix} -8 & 7 & -3 \\ 32 & -8 & 12 \\ 0 & -2 & 0 \end{pmatrix}$ **vii** $\begin{pmatrix} 28 & 36 \\ -72 & 136 \end{pmatrix}$

viii $\mathbf{C}^2$ not possible as **C** not a square matrix

2 a $\begin{pmatrix} a^2 - a & 3a - 2 \\ 3a & 6 \\ 2a^2 - 2a & 4a - 6 \end{pmatrix}$

b $\begin{pmatrix} -a - 6 & 1 + 9a \\ -2a & 3 + 2a^2 \\ 2a^2 & -2a - 9 \end{pmatrix}$

c $\begin{pmatrix} a^2 + 3a & 3a + 6 \\ a^2 + 2a & 3a + 4 \end{pmatrix}$

d $\begin{pmatrix} 1 + 6a & -9 & a^2 - 3 \\ 2a^2 & 1 - 3a & a \\ -2a & 2a^2 - 3 & 3a \end{pmatrix}$

3 a   i $(2, -1)$ reflection in line $y = x$

ii $(-7, 2)$ stretch scale factor 7 parallel to x-axis

iii $(-2, -1)$ rotation of $90°$ anticlockwise about origin

iv $\dfrac{1}{2}(2 + \sqrt{3}), \dfrac{1}{2}(1 - 2\sqrt{3})$; rotation of $210°$ anticlockwise about origin

b   i 5   **ii** 35   **iii** 5   **iv** 5

c   i   determinant negative so does not preserve orientation

ii, iii, iv   have positive determinants so do preserve orientation

4 a $(3, -3\sqrt{2}, \sqrt{2})$; rotation of $45°$ anticlockwise around x-axis.

b $(3, -2, -4)$; reflection in plane $z = 0$

5 a $\begin{pmatrix} -1 & 0 \\ 0 & 1 \end{pmatrix}$

b $\begin{pmatrix} \dfrac{\sqrt{2}}{2} & \dfrac{\sqrt{2}}{2} \\ -\dfrac{\sqrt{2}}{2} & \dfrac{\sqrt{2}}{2} \end{pmatrix}$

c $\begin{pmatrix} -3 & 0 \\ 0 & -3 \end{pmatrix}$

6 a $\begin{pmatrix} \dfrac{\sqrt{2}}{2} & 0 & \dfrac{\sqrt{2}}{2} \\ 0 & 1 & 0 \\ -\dfrac{\sqrt{2}}{2} & 0 & \dfrac{\sqrt{2}}{2} \end{pmatrix}$

b $\begin{pmatrix} 1 & 0 & 0 \\ 0 & -1 & 0 \\ 0 & 0 & 1 \end{pmatrix}$

7 a $\begin{pmatrix} 0 & 1 \\ -2 & 0 \end{pmatrix}$

b $\begin{pmatrix} 0 & -1 \\ 1 & 0 \end{pmatrix}$

c $\begin{pmatrix} -\dfrac{\sqrt{3}}{2} & \dfrac{1}{2} \\ \dfrac{1}{2} & \dfrac{\sqrt{3}}{2} \end{pmatrix}$

8 a   Invariant point is $(0, 0)$

b   Line of invariant points is $x = 3y$

9 a $y = -x$ and $y = \dfrac{3}{2}x$

b $y = 3x$ and $y = -\dfrac{1}{3}x + c$

10 a $\det\begin{pmatrix} 2 & -1 \\ 4 & -3 \end{pmatrix} = 2 \times -3 - -1 \times 4 = -2$

b $\det\begin{pmatrix} 3a & b \\ -2a & b \end{pmatrix} = 3ab - -2ab = 5ab$

11 a $x = \dfrac{4}{3}$

b $x = \pm\sqrt{2}$

12 a $-\dfrac{1}{9}\begin{pmatrix} -2 & -3 \\ 1 & 6 \end{pmatrix}$

b $-\dfrac{1}{3}\begin{pmatrix} -2 & -1 \\ -3 & 0 \end{pmatrix}$

c Does not exist since $\det\begin{pmatrix} 3 & -6 \\ -4 & 8 \end{pmatrix} = 3 \times 8 - -6 \times -4 = 0$

d $\dfrac{1}{ab}\begin{pmatrix} 3b & -2b \\ -7a & 5a \end{pmatrix}$

13 $\begin{pmatrix} 2 & 3 \\ 0 & 1 \end{pmatrix}$

14 $\begin{pmatrix} 2 & -7 \\ -5 & -3 \end{pmatrix}$

15 a $x = -5, y = 2$ **b** $x = \dfrac{1}{5}, y = -1$

Assessment 5

1 a $\begin{pmatrix} 15 & 6 \\ -9 & 3 \end{pmatrix}$

b $\begin{pmatrix} 8 & -6 \\ 0 & -4 \end{pmatrix}$

c $\begin{pmatrix} 9 & -1 \\ -3 & -1 \end{pmatrix}$

d $\begin{pmatrix} -20 & 19 \\ 12 & -7 \end{pmatrix}$

2 a $\begin{pmatrix} 12 & 9 \\ -7 & 25 \end{pmatrix}$

b $\begin{pmatrix} 30 & -17 \\ 9 & 7 \end{pmatrix}$

c **M** and **N** do not commute

3 a $\begin{pmatrix} 2 & 7 \\ 9 & 11 \end{pmatrix}$ **b** $\begin{pmatrix} 4 & 14 \\ 18 & 22 \end{pmatrix}$ **c** $\begin{pmatrix} 67 & 91 \\ 117 & 184 \end{pmatrix}$

4 $a = 4, b = 1, c = 16$

5 a Enlargement, 6
b the origin, 90 degrees
c reflection, the x-axis

6 a $\mathbf{M} = \begin{pmatrix} -\dfrac{\sqrt{2}}{2} & -\dfrac{\sqrt{2}}{2} \\ \dfrac{\sqrt{2}}{2} & -\dfrac{\sqrt{2}}{2} \end{pmatrix}$

b $\mathbf{N} = \begin{pmatrix} \dfrac{\sqrt{2}}{2} & \dfrac{\sqrt{2}}{2} \\ \dfrac{-\sqrt{2}}{2} & \dfrac{\sqrt{2}}{2} \end{pmatrix}$

c $\begin{pmatrix} 0 & -1 \\ 1 & 0 \end{pmatrix}$

A rotation by 90 degrees anticlockwise about the origin. This the combined effect of the two rotations represented by **M** and **N**.

7 a The point $(0, 0)$
b Any point $(\lambda, -2\lambda)$
c Any point $(-3\lambda, \lambda)$

8 a The y-axis
b The z-axis

9 $\begin{pmatrix} x \\ y \end{pmatrix} = \begin{pmatrix} \dfrac{57}{25} \\ \dfrac{22}{25} \end{pmatrix}$

10 $\begin{pmatrix} x \\ y \end{pmatrix} = \begin{pmatrix} 4 \\ 3 \end{pmatrix}$

11 a 2 **b** $(0, 0)$
12 a 12 **b** 0

13 a $\begin{pmatrix} 9 & -16 \\ -5 & 9 \end{pmatrix}$ **b** $\dfrac{1}{22}\begin{pmatrix} 3 & 2 \\ 5 & 4 \end{pmatrix}$

14 a $\mathbf{A}^{-1} = \begin{pmatrix} 2 & -1 \\ -5 & 3 \end{pmatrix}$

b $\mathbf{B}^{-1} = \dfrac{1}{2}\begin{pmatrix} 1 & 4 \\ 1 & 2 \end{pmatrix}$

c $\begin{pmatrix} -9 & 5.5 \\ -4 & 2.5 \end{pmatrix}$

d $\begin{pmatrix} -5 & 11 \\ -8 & 18 \end{pmatrix}$

$(\mathbf{B}^{-1}\mathbf{A}^{-1})^{-1} = \mathbf{AB}$

Chapter 6
Exercise 6.1A

1 $\mathbf{r} = \begin{pmatrix} 0 \\ 4 \\ 3 \end{pmatrix} + \lambda\begin{pmatrix} 1 \\ -2 \\ 0 \end{pmatrix}$

2 $\mathbf{r} = 7\mathbf{i} - \mathbf{j} + 2\mathbf{k} + \lambda(5\mathbf{i} - 3\mathbf{j} + \mathbf{k})$

3 $\mathbf{r} = \begin{pmatrix} -7 \\ 2 \\ -5 \end{pmatrix} + \lambda\begin{pmatrix} 11 \\ -11 \\ 3 \end{pmatrix}$ or alternatively

$\mathbf{r} = \begin{pmatrix} 4 \\ -9 \\ -2 \end{pmatrix} + \lambda\begin{pmatrix} 11 \\ -11 \\ 3 \end{pmatrix}$

4 $\mathbf{r} = 2\mathbf{i} + 3\mathbf{k} + \lambda(\mathbf{j} - 4\mathbf{k})$ or alternatively
$\mathbf{r} = 2\mathbf{i} + \mathbf{j} - \mathbf{k} + \lambda(\mathbf{j} - 4\mathbf{k})$

5 $x = 5; \dfrac{y-2}{4} = \dfrac{z+7}{-1}$

6 $\dfrac{x-4}{8} = \dfrac{y+2}{-2} = \dfrac{z}{3}$

7 $\dfrac{x-9}{2} = \dfrac{y-4}{1} = \dfrac{z+3}{-2}$ or alternatively $\dfrac{x-1}{2} = \dfrac{y-0}{1} = \dfrac{z-5}{-2}$

8 $\dfrac{x-0}{-2} = \dfrac{y-5}{3} = \dfrac{z-1}{1}$ or alternatively $\dfrac{x-4}{-2} = \dfrac{y+1}{3} = \dfrac{z+1}{1}$

9 a $\dfrac{x-0}{-4} = \dfrac{z+2}{8}$ and $y = 1$ **b** $\dfrac{x+3}{5} = \dfrac{y-0}{-1} = \dfrac{z-7}{3}$

10 a $\mathbf{r} = \begin{pmatrix} 5 \\ 1 \\ -3 \end{pmatrix} + \lambda \begin{pmatrix} 2 \\ -3 \\ 1 \end{pmatrix}$

b $\mathbf{r} = \begin{pmatrix} -2 \\ 0 \\ 5 \end{pmatrix} + \lambda \begin{pmatrix} 4 \\ -2 \\ -3 \end{pmatrix}$

11 a (3, 2, 1) lies on the line.

    b (3, 2, 1) does not lie on the line

12 a (−2, 0, 4) does not line on the line

    b (−2, 0, 4) lies on the line.

Exercise 6.1B

1 a Not the same line.

    b The equations represent the same line.

    c The same line.

2 a Not the same line.

    b Not the same line.

    c Represent the same line.

    d Represent the same line.

    e Not the same line.

3 a Intersect at (−11, 16)

    b Intersect at (−2, 10)

    c Intersect at (1, −1)

4 a Intersect at (7, −6, −2)

    b lines are parallel

    c skew lines.

    d intersect at (−4, 2, 0)

5 $3\sqrt{10}$

6 $\mathbf{r} = \begin{pmatrix} 5 \\ -6 \end{pmatrix} + \lambda \begin{pmatrix} -3 \\ -2 \end{pmatrix}$

7 $\mathbf{r} = \begin{pmatrix} 1 \\ 2 \end{pmatrix} + \lambda \begin{pmatrix} -1 \\ 2 \end{pmatrix}$

8 a $\mathbf{r} = \begin{pmatrix} 3 \\ 1 \\ 0 \end{pmatrix} + \lambda \begin{pmatrix} 1 \\ 2 \\ -1 \end{pmatrix}$

    b $\mathbf{r} = \begin{pmatrix} -3 \\ 0 \\ 1 \end{pmatrix} + \lambda \begin{pmatrix} -1 \\ -1 \\ 2 \end{pmatrix}$

    c $\mathbf{r} = \begin{pmatrix} 3 \\ 0 \\ 1 \end{pmatrix} + \lambda \begin{pmatrix} 1 \\ -1 \\ 2 \end{pmatrix}$

9 $\mathbf{r} = \begin{pmatrix} 1 \\ 3 \\ 5 \end{pmatrix} + \lambda \begin{pmatrix} 2 \\ 1 \\ 4 \end{pmatrix}$

Exercise 6.2A

1 a 1   **b** 0   **c** 0   **d** 1

    e 2   **f** 0   **g** 1   **h** 3

2 a 4   **b** −5

    c $-\sqrt{2}$ **d** −3

3 $6a^2 - 4a$

4 k −6

5 a 46.4°   **b** 26.5°

    c 63.9°   **d** 101.6°

6 a $\dfrac{a}{a^2 + 1}$ **b** $\dfrac{5 - 4a^2}{5a^2 + 5}$

7 $\begin{pmatrix} 6 \\ -4 \end{pmatrix} \cdot \begin{pmatrix} -8 \\ -12 \end{pmatrix} = -48 + 48 = 0$ so perpendicular

8 $\begin{pmatrix} 4 \\ -1 \\ 5 \end{pmatrix} \cdot \begin{pmatrix} -2 \\ -3 \\ 1 \end{pmatrix} = (4 \times -2) + (-1 \times -3) + (5 \times 1)$

                 = 0 therefore they are perpendicular

9 $3 \times 2 - 5 \times 4 - 2 \times -7 = 0$ so perpendicular

10 $2 \times 4 + 0 + 1 \times -3 = 5$ so not perpendicular

11 $a = 1$

12 $b = 2$

13 $c = 1, 2$

Exercise 6.2B

1 $\begin{pmatrix} -1 \\ 3 \end{pmatrix} \cdot \begin{pmatrix} 6 \\ 2 \end{pmatrix} = -6 + 6 = 0$ so perpendicular

2 $\begin{pmatrix} 6 \\ 7 \\ -5 \end{pmatrix} \cdot \begin{pmatrix} 2 \\ -1 \\ 1 \end{pmatrix} = 12 - 7 - 5 = 0$ so perpendicular

3 $\begin{pmatrix} 3 \\ -2 \\ 6 \end{pmatrix} \cdot \begin{pmatrix} 2 \\ 6 \\ 1 \end{pmatrix} = 6 - 12 + 6 = 0$ therefore they are perpendicular.

4 $\theta = 45°$

5 $2 = 5 + 3t$ ①

    $1 + 2s = 11 - 4t$ ②

    $-3 + s = 3 - t$ ③

    Solving ① and ② gives $s = 7$ and $t = -1$

    Check in ③

    $-3 + 7 = 4$

    $3 - t = 4$

    So the lines intersect

$$\cos\theta = \dfrac{\begin{pmatrix} 0 \\ 2 \\ 1 \end{pmatrix} \cdot \begin{pmatrix} 3 \\ -4 \\ -1 \end{pmatrix}}{\left|\begin{pmatrix} 0 \\ 2 \\ 1 \end{pmatrix}\right|\left|\begin{pmatrix} 3 \\ -4 \\ -1 \end{pmatrix}\right|} \frac{-8-1}{\sqrt{5}\sqrt{26}} = \frac{-9}{\sqrt{5}\sqrt{26}}$$

    $\theta = 180 - 142.1 = 38°$ to the nearest degree

6 93°

7 $-1 + \lambda = -2 \Rightarrow \lambda = -1$

    $2 + \lambda = 2 - \mu \Rightarrow \mu = 1$

    Check: $3 - \lambda = 4$

    $1 + 3\mu = 4$ hence they intersect

    $\theta = 43.1°$

8 $a = \pm 2\sqrt{2}$

9 $a = \pm 2$

10 a $\left(2, 3, \dfrac{7}{2}\right)$ **b** $\cos\theta = \dfrac{67}{9\sqrt{161}}$

11 $k = 2 \pm \sqrt{3}$

12 $k = -\dfrac{1}{4}$

13 $\dfrac{\sqrt{89}}{2}$

14 $\dfrac{\sqrt{11}}{2}$

15 $\mathbf{a} \cdot \mathbf{b} = (a_1\mathbf{i} + a_2\mathbf{j} + a_3\mathbf{k}) \cdot (b_1\mathbf{i} + b_2\mathbf{j} + b_3\mathbf{k})$

$\quad = a_1b_1\mathbf{i} \cdot \mathbf{i} + a_1b_2\mathbf{i} \cdot \mathbf{j} + a_1b_3\mathbf{i} \cdot \mathbf{k} + a_2b_1\mathbf{j} \cdot \mathbf{i} + a_2b_2\mathbf{j} \cdot \mathbf{j}$
$\quad\quad + a_2b_3\mathbf{j} \cdot \mathbf{k} + a_3b_1\mathbf{k} \cdot \mathbf{i} + a_3b_2\mathbf{k} \cdot \mathbf{j} + a_3b_3\mathbf{k} \cdot \mathbf{k}$

$\quad = a_1b_1\mathbf{i} \cdot \mathbf{i} + a_2b_2\mathbf{j} \cdot \mathbf{j} + a_3b_3\mathbf{k} \cdot \mathbf{k}$ Using fact that $\mathbf{i}$, $\mathbf{j}$ and $\mathbf{k}$ are perpendicular

$\quad = a_1b_1 + a_2b_2 + a_3b_3$ as required since $\mathbf{i} \cdot \mathbf{i} = \mathbf{j} \cdot \mathbf{j} = \mathbf{k} \cdot \mathbf{k} = 1$

16 Let $\mathbf{a} = a_1\mathbf{i} + a_2\mathbf{j} + a_3\mathbf{k}$, $\mathbf{b} = b_1\mathbf{i} + b_2\mathbf{j} + b_3\mathbf{k}$ and $\mathbf{c} = c_1\mathbf{i} + c_2\mathbf{j} + c_3\mathbf{k}$

$\quad \mathbf{b} + \mathbf{c} = (b_1\mathbf{i} + b_2\mathbf{j} + b_3\mathbf{k}) + (c_1\mathbf{i} + c_2\mathbf{j} + c_3\mathbf{k})$

$\quad = (b_1 + c_1)\mathbf{i} + (b_2 + c_2)\mathbf{j} + (b_3 + c_3)\mathbf{k}$

$\quad \mathbf{a} \cdot (\mathbf{b} + \mathbf{c}) = (a_1\mathbf{i} + a_2\mathbf{j} + a_3\mathbf{k})((b_1 + c_1)\mathbf{i} + (b_2 + c_2)\mathbf{j} + (b_3 + c_3)\mathbf{k})$

$\quad = a_1(b_1 + c_1)\mathbf{i} \cdot \mathbf{i} + a_1(b_2 + c_2)\mathbf{i} \cdot \mathbf{j} + a_1(b_3 + c_3)\mathbf{i} \cdot \mathbf{k}$
$\quad\quad + a_2(b_1 + c_1)\mathbf{j} \cdot \mathbf{i} + a_2(b_2 + c_2)\mathbf{j} \cdot \mathbf{j} + a_2(b_3 + c_3)\mathbf{j} \cdot \mathbf{k}$
$\quad\quad + a_3(b_1 + c_1)\mathbf{k} \cdot \mathbf{i} + a_3(b_2 + c_2)\mathbf{k} \cdot \mathbf{j} + a_3(b_3 + c_3)\mathbf{k} \cdot \mathbf{k}$

$\quad = a_1(b_1 + c_1)\mathbf{i} \cdot \mathbf{i} + a_2(b_2 + c_2)\mathbf{j} \cdot \mathbf{j} + a_3(b_3 + c_3)\mathbf{k} \cdot \mathbf{k}$

$\quad = a_1(b_1 + c_1) + a_2(b_2 + c_2) + a_3(b_3 + c_3)$

$\quad = a_1b_1 + a_1c_1 + a_2b_2 + a_2c_2 + a_3b_3 + a_3c_3$

$\quad = (a_1b_1 + a_2b_2 + a_3b_3) + (a_1c_1 + a_2c_2 + a_3c_3)$

$\quad = \mathbf{a} \cdot \mathbf{b} + \mathbf{a} \cdot \mathbf{c}$ (using qu 8 or prove as above) as required

17 a Let $\mathbf{a} = \begin{pmatrix} a_1 \\ a_2 \\ a_3 \end{pmatrix}$

$\quad$ Then $\mathbf{a} \cdot \mathbf{a} = \begin{pmatrix} a_1 \\ a_2 \\ a_3 \end{pmatrix} \cdot \begin{pmatrix} a_1 \\ a_2 \\ a_3 \end{pmatrix}$

$\quad = a_1^2 + a_2^2 + a_3^2$

$\quad = |\mathbf{a}|^2$

$\quad$ since $|\mathbf{a}| = \sqrt{a_1^2 + a_2^2 + a_3^2}$

b $\mathbf{a} = \mathbf{b} - \mathbf{c}$ using the 'triangle rule'

$\quad |\mathbf{a}|^2 = \mathbf{a} \cdot \mathbf{a}$

$\quad = (\mathbf{b} - \mathbf{c}) \cdot (\mathbf{b} - \mathbf{c})$

$\quad = \mathbf{b} \cdot \mathbf{b} + \mathbf{c} \cdot \mathbf{c} - 2\mathbf{b} \cdot \mathbf{c} \cos\theta$

$\quad |\mathbf{b}|^2 + |\mathbf{c}|^2 - 2|\mathbf{b}||\mathbf{c}|\cos\theta$ as required

18 a When $y = 0$, $\dfrac{y - 5}{5} = -1$

$\quad$ So $\dfrac{z - 3}{3} = -1 \Rightarrow z = 0$

$\quad$ Therefore it cuts through x-axis

$\quad \dfrac{x + 2}{-4} = -1 \Rightarrow x = 2$

$\quad$ So intercept is $(2, 0, 0)$

b $55.6°$

19 $\dfrac{27}{2}\sqrt{6}$

Exercise 6.3A

1 6

2 3

3 $2\sqrt{2}$ (2.83)

4 $\dfrac{\sqrt{114}}{3}$ (3.56)

5 1

6 a $\sqrt{66}$ (8.12)　　**b** $3\sqrt{10}$ (9.49)

$\quad$ **c** $\dfrac{4}{3}\sqrt{30}$ (7.30)　　**d** $\sqrt{14}$ (3.74)

$\quad$ **e** $\sqrt{134}$ (11.58)

7 a $\dfrac{5}{2}\sqrt{2}$　　　　(3.54 to 3 significant figures)

$\quad$ **b** $\dfrac{5}{6}\sqrt{6}$　　　　(2.04 to 3 significant figures)

$\quad$ **c** $\dfrac{4\sqrt{5}}{5}$　　　　(1.79 to 3 significant figures

Exercise 6.3B

1 $\dfrac{\sqrt{110}}{5}$ (2.10)

2 $\dfrac{\sqrt{30}}{6}$ (0.913)

3 $\dfrac{\sqrt{329}}{10}$ (5.74)

4 a $\dfrac{7\sqrt{3}}{3}$　　　　(4.04 to 3 sig. fig.)

$\quad$ **b** $(7, 2, -8)$

$\quad$ **c** $\sqrt{14}$　　　　·(3.74 to 3 sig. fig.)

5 $(-2, 5, 2)$

6 $\left(\dfrac{5}{3}, \dfrac{2}{3}, -\dfrac{16}{3}\right)$

7 $\sqrt{\dfrac{19}{7}}$, (1.65)

8 For L_1 :

$$x = \left| \begin{pmatrix} 5 \\ -2 \\ -4 \end{pmatrix} + \lambda \begin{pmatrix} 1 \\ 2 \\ 1 \end{pmatrix} \right|$$

$\quad x^2 = (5 + \lambda)^2 + (-2 + 2\lambda)^2 + (-4 + \lambda)^2$

$\quad \dfrac{\mathrm{d}(x^2)}{\mathrm{d}\lambda} = 2(5 + \lambda) + 4(-2 + 2\lambda) + 2(-4 + \lambda)$

$\quad \dfrac{\mathrm{d}(x^2)}{\mathrm{d}\lambda} = 0 \Rightarrow \lambda = \dfrac{1}{2}$

$\quad$ Therefore, $x = \sqrt{\left(5 + \dfrac{1}{2}\right)^2 + \left(-2 + 2\left(\dfrac{1}{2}\right)\right)^2 + \left(-4 + \dfrac{1}{2}\right)^2}$

$\quad = \dfrac{\sqrt{174}}{2}$ (6.60)

$\quad$ For L_2

$$x = \left| \begin{pmatrix} -1 \\ 3 \\ 2 \end{pmatrix} + \lambda \begin{pmatrix} 1 \\ 1 \\ 0 \end{pmatrix} \right|$$

$\quad x^2 = (-1 + \lambda)^2 + (3 + \lambda)^2 + 2^2$

$\quad \dfrac{\mathrm{d}(x^2)}{\mathrm{d}\lambda} = 2(-1 + \lambda) + 2(3 + \lambda)$

$\quad \dfrac{\mathrm{d}(x^2)}{\mathrm{d}\lambda} = 0 \Rightarrow \lambda = -1$

$\quad$ Therefore, $x = \sqrt{(-1 - 1)^2 + (3 - 1)^2 + 2^2}$

$\quad = 2\sqrt{3}$ (3.46)

$\quad$ Therefore L_2 comes closer to the origin (by 3.13 units) than L_1.

9 For point A:

$$x = \left| \begin{pmatrix} 4 \\ -3 \\ 7 \end{pmatrix} + \lambda \begin{pmatrix} 1 \\ 3 \\ -2 \end{pmatrix} - \begin{pmatrix} 4 \\ 1 \\ 6 \end{pmatrix} \right|$$

$$x^2 = (\lambda)^2 + (-4 + 3\lambda)^2 + (1 - 2\lambda)^2$$

$$\frac{d(x^2)}{dt} = 2\lambda + 6(-4 + 3\lambda) - 4(1 - 2\lambda)$$

$$\frac{d(x^2)}{dt} = 0 \Rightarrow \lambda = 1$$

Therefore, $x = \sqrt{(1)^2 + (-4 + 3)^2 + (1 - 2)^2}$

$$= \sqrt{3} \ (1.73)$$

For point B:

$$x = \left| \begin{pmatrix} 4 \\ -3 \\ 7 \end{pmatrix} + \lambda \begin{pmatrix} 1 \\ 3 \\ -2 \end{pmatrix} - \begin{pmatrix} 6 \\ -2 \\ 6 \end{pmatrix} \right|$$

$$x^2 = (-2 + \lambda)^2 + (-1 + 3\lambda)^2 + (1 - 2\lambda)^2$$

$$\frac{d(x^2)}{dt} = 2(-2 + \lambda) + 6(-1 + 3\lambda) - 4(1 - 2\lambda)$$

$$\frac{d(x^2)}{dt} = 0 \Rightarrow \lambda = \frac{1}{2}$$

Therefore, $x = \sqrt{\left(-2 + \frac{1}{2}\right)^2 + \left(-1 + \frac{3}{2}\right)^2 + (1 - 1)^2}$

$$= \frac{\sqrt{10}}{2} \ (1.58)$$

Therefore point B comes closer to the line (by 0.151) than line A.

10 $\dfrac{1}{\sqrt{2}}$ (= 0.71)

Review exercise 6

1 a $\dfrac{x-1}{2} = \dfrac{y+1}{-3} = \dfrac{z-4}{1}$

b $r = i - j + 4k + s(2i - 3j + k)$

2 a $\dfrac{x-3}{5} = \dfrac{y}{-4} = \dfrac{z-5}{-2}$

b $r = 3i + 5k + s(5i - 4j - 2k)$

3 a Skew
b Parallel
c Intersect at (17, 9, −2)
d Intersect at (10, −7, 7)

4 37

5 a −13   **b** 60.6°

6 109.1°

7 $\begin{pmatrix} -2 \\ 8 \\ -4 \end{pmatrix} \cdot \begin{pmatrix} -4 \\ 2 \\ 6 \end{pmatrix} = 8 + 16 - 24$

$= 0$ therefore they are perpendicular

8 $\sqrt{3}$

9 a $-2 (3i - 6k) = -6i + 12k = 3(-2i + 4k)$ so the lines are parallel

b 1.67 (3 sf)

10 $\dfrac{\sqrt{46}}{\sqrt{5}}$ (= 3.03)

11 a they do not intersect, and they are not parallel, therefore they must be skew

b $\dfrac{3}{11}\sqrt{11}$ (0.905 to 3 sf)

Assessment 6

1 a $r = 2i + 4j - k + \lambda(i + j - 2k)$

b $-i + j + 5k = 2i + 4j - k + \lambda(i + j - 2k)$
Coefficients of **i**: $-1 = 2 + \lambda \Rightarrow \lambda = -3$
Check coefficients of **j**: $4 - 3 = 1$
Check coefficients of **k**: $-1 - 3(-2) = 5$
So (−1, 1, 5) lies on the line.

2 a $r = \begin{pmatrix} 3 \\ 0 \\ 1 \end{pmatrix} + \lambda \begin{pmatrix} 1 \\ -1 \\ 1 \end{pmatrix}$

b $3 + \lambda = 8 \Rightarrow \lambda = 5$
Check: $0 - 5 = -5$ as required
$1 + 5 = 6$ not 7 so (8, −5, 7) does not lie on the line.

3 $r = \begin{pmatrix} 2 \\ 1 \\ 0 \end{pmatrix} + \lambda \begin{pmatrix} 8 \\ -5 \\ 3 \end{pmatrix}$

4 a −3   **b** $\theta = 87.1°$

5 $\dfrac{1}{6}\sqrt{3}$

6 $(-4i + k).(2i - j + 8k) = -8 + 0 + 8 = 0$
Therefore they are perpendicular.

7 4

8 $\begin{pmatrix} -12 \\ -14 \\ 18 \end{pmatrix}$

9 $\dfrac{x-2}{-1} = \dfrac{y-0}{4} = \dfrac{z+3}{3}$

10 a $a = 2, b = 7, c = -3$

b $r = \begin{pmatrix} 2 \\ 7 \\ -3 \end{pmatrix} + t \begin{pmatrix} 2 \\ -3 \\ 1 \end{pmatrix}$

11 $\dfrac{\sqrt{42}}{14}$

12 So first line closer

13 a Intersect at (1, −2, 6)
b parallel
c skew

14 a $\dfrac{x-2}{-1} = \dfrac{y-0}{4} = \dfrac{z+3}{3}$

b They are not parallel as the direction vectors are not equal, so it is not the same line.

15 B and C

16 $\dfrac{1}{3}\sqrt{137}$

17 $\dfrac{\sqrt{42}}{14}$

18 a 48.2°

b $\left(\dfrac{-1}{3}, \dfrac{19}{3}, \dfrac{-4}{3}\right)$

19 3.32 square units

20 a they do not intersect, and they are not parallel, therefore they must be skew

b $\dfrac{4}{3}\sqrt{3}$
(2.31 to 3 sig. fig)

c (−4, 2, −3)

d $\dfrac{\sqrt{165}}{2}$

Chapter 7

Exercise 7.1A

1. **a** 80 J   **b** 600 J   **c** 150 kJ
2. C does the most work
3. A exerted the most power
4. At start, 16 000 W, at end 24 000 W
5. Initial power 700 W final power 150 W
6. Work: 50 000 J

   Power P = 5000 W

   Power Q = $3333\frac{1}{3}$ W
7. **a** 1000 N   **b** 600 N
8. First: 30 N   Second: 20 N
9. **a** 720J, 180 W   **b** 60J, 40 W
10. Box A
    a 12g N   **b** 36g J
    c 9g W

    Box B
    a 4g N   **b** 2g J
    c 0.5g W

    Box C
    a 0.5g N   **b** 0.1g J
    c 0.05g W
11. **a** 16 J   **b** $42\frac{2}{3}$ J   **c** $21\frac{1}{3}$ J
12. **a** 30 J   **b** 46 J
13. **a** 40 J   **b** 14.1 J   **c** 0.104 J
14. **a** 10 W   **b** 17.7 W   **c** 0.693 W
15. 39.0 W

Exercise 7.1B

1. **a** 15.3 m s^{-1} **b** 6.57 m s^{-1}
2. **a** 29.4 m   **b** 27.2 m
3. 62.5 N
4. **a** 72 J   **b** 24 J   **c** 24 W
5. **a** **i** 10^5 J   **ii** 10^5 J   **iii** 250 N
   b 120000 J, 300 N
6. **a** 7.5×10 J   **b** 8.5×10^6 J, 17000 N
   c 170000 W, 340000 W
7. **a** **i** 352.8 J, 352.8 J   **ii** 24.2 m s^{-1}
   b 470.4 J, 28 m s^{-1}
   c 28 m s^{-1}
   d 20.9 m s^{-1}
8. **a** 375 kW   **b** $\frac{1}{4}$ m s^{-2}
9. 435 000 W
10. 3.55 m s^{-1}
11. 3640 J, 13.2 N
12. 26.7 m s^{-1}
13. 3560 W
14. 1134 W
15. **a** 30 m s^{-1} **b** $\frac{2}{3}$ m s^{-2}
16. 21.5 m s^{-1}
17. $\frac{81}{320}$
18. **a** 3.95   **b** 0.147 m s^{-2}
19. 3.32 m s^{-1}, 66.5 W
20. **a** **i** 1764000 J   **ii** 5880 N   **iii** 147 kW
    b 8880 N, 222 kW
21. **a** 8.24 m s^{-1} **b** 7.38 m s^{-1}
22. 20 m s^{-1}
23. **a** 0.391 m s^{-2} **b** 207 N
24. **a** Energy equation from bottom to top of loop is:
    GPE gained = KE lost
    $$m \times g \times 2r = \frac{1}{2}mu^2$$
    $$u^2 = 4gr$$

b energy equation from bottom to point P is
GPE gained = KE lost
$$mgr + mgr\cos\theta = \frac{1}{2}mu^2 - \frac{1}{2}m\left(\frac{u}{2}\right)^2$$

$$gr(1+\cos\theta) = \frac{3u^2}{8}$$

$$\cos\theta = \frac{3u^2}{8gr} - 1$$

θ is acute, so cos θ
$$0 < \cos\theta < 1$$
$$0 < \frac{3u^2}{8gr} - 1 < 1$$

$$8gr \times 1 < 3u^2 < 2 \times 8gr$$
$$\frac{8gr}{3} < u^2 < \frac{16gr}{3}$$

25. 1.05°
26. 65.3 m s^{-1}
27. 2.08 m s^{-1}, 41.5 W
28. **a** 1500×10^4 J   **b** 8.40 m s^{-1}, 756 312 W

Exercise 7.2A

1. **a** A: 2 N, 0.5 J   **B:** 2 N, 1 J
   C: 0.4 m, 0.8 J   **D:** 9 N, 1.35 J
   E: 0.8 m, 3.2 N
   b A 4 N m^{-1} B 2 N m^{-1}
   C 10 N m^{-1} D 7.5 N m^{-1}
   E 4 N m^{-1}
2. 0.4 m
3. 2.5 m
4. **a** 0.8 m   **b** 8 J   **c** 8 J
5. **a** 0.72 m   **b** 5.4 J
6. **a** 78.4 N   **b** 0.98 m
7. $x_1 = 1$ m, $x_2 = 0.5$ m, $26\frac{2}{3}$ N
8. $x_1 = \frac{4}{9}$ m, $x_2 = \frac{5}{9}$ m, 4.17 J
9. 0.260 m, 8.33 J
10. For vertical equilibrium
    $$T_1 + T_2 = mg$$
    $$\frac{\lambda_1}{l}x + \frac{\lambda_2}{l}x = mg$$
    $$(\lambda_1 + \lambda_2)x = mgl$$
    Extension $x = \dfrac{mgl}{\lambda_1 + \lambda_2}$
    $$\frac{mg\lambda_1}{\lambda_1 + \lambda_2} \text{ N}, \quad \frac{mg\lambda_2}{\lambda_1 + \lambda_2} \text{ N}$$

Exercise 7.2B

1. $16\frac{2}{3}$ J
2. **a** 39.2 N in both **b** 0.441 m   **c** 8.64 J
3. **a** 0.65 m   **b** 15 N, 34 N **c** 15.6 J
4. **a** 980, 1.36 m   **b** 3a
5. 6.62 m s^{-2}, 1.94 m
6. 5.77 m s^{-1}, 3.46 m s^{-1}
7. Either using calculus
   $$\text{Work done} = \int_{x_2}^{x_1} T \, dx = \frac{\lambda}{l}\left[\frac{x^2}{2}\right]_{x_1}^{x_2}$$
   $$= \frac{1}{2} \times \frac{\lambda}{l} \times (x_2^2 - x_1^2)$$
   $$= \frac{1}{2} \times \frac{\lambda}{l}(x_2 + x_1)(x_2 - x_1)$$
   $$= \frac{T_1 + T_2}{2}(x_2 - x_1)$$
   Or using the graph of $T = \dfrac{\lambda x}{l}$

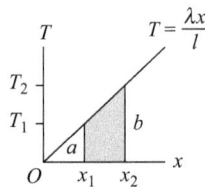

Work done = shaded area on T–x graph

= area of a trapezium

$$= \tfrac{1}{2}h(a+b)$$

$$= \tfrac{1}{2}(T_1 + T_2)(x_2 - x_1)$$

8 0.8 m

9 a 0.3 m

b 1.71 m s^{-1}, 0.3 m above P

The mass now falls back to Q and continues to oscillate about P, 0.3 m above and below.

c 1.23 m below A

d The mass will oscillate about P, 0.2 m above and below. 1.3 m below A

10 a Energy equation gives:

KE gained + EPE gained = GPE lost

$$\tfrac{1}{2} \times 60 \times v^2 + \tfrac{1}{2}\,\frac{1200}{25}(x-25)^2 = 60 \times g \times x$$

$$30v^2 + 24(x^2 - 50x + 625) = 588x;$$

$$30v^2 = 1788x - 24x^2 - 15000$$

25, 24.7 m s^{-1}

b 64.9 m

11 Max extension occurs when m_1 and m_2 have the same velocity v and string is no longer extending.

energy equation at this point is

Loss of KE of m_1 = Gain in KE of m_2 + Gain in EPE

$$\tfrac{1}{2}m_1u^2 - \tfrac{1}{2}m_1v^2 = \tfrac{1}{2}m_1v^2 + \tfrac{1}{2}\times\frac{\lambda}{a}\times x^2$$

$$\frac{\lambda}{a}x^2 = m_1u^2 - (m_1 + m_2)v^2 \qquad (1)$$

There are no external impacts or impulses so momentum is conserved.

Momentum equation gives

$$m_1u = (m_1 + m_2)v \qquad (2)$$

Substitute (2) in (1) to eliminate v.

$$\frac{\lambda}{a}x^2 = m_1u^2 - (m_1 + m_2)\left(\frac{m_1u}{m_1 + m_2}\right)^2$$

$$= \frac{m_1(m_1 + m_2)u^2 - (m_1u)^2}{m_1 + m_2}$$

$$= \frac{m_1m_2u^2}{m_1 + m_2}$$

Max extension, $x = \sqrt{\dfrac{axm_1m_2u^2}{\lambda(m_1 + m_2)}}$

12 a 2.235 m   **b** 0.509 m

13 a 3.13 m s^{-1}

b Resolving vertically for equilibrium

$2T\cos\theta = 15g$

where $\theta = \tan^{-1}\left(\dfrac{2}{1.1}\right) = 61.1892...°$

Hooke's law gives $T = \dfrac{30g}{1.5}(\sqrt{2^2 + 1.1^2} - 1.5)$

$= 153.3783...$ N

$2T\cos\theta = 2 \times 153.3783... \times \cos 61.1892...° = 147.8317$ N

$15g = 15 \times 9.8 = 147$ N

$147 \approx 147.8$, mass hangs in equilibrium

14 a 10.2 kg   **b** 117°

Exercise 7.3A

1 b, c, e, f, g

2 a [velocity] = $[L][T]^{-1}$ [acceleration] = $[L][T]^{-2}$

b [work] = $[M][L]^2[T]^{-2}$

c [momentum] = $[M][L][T]^{-1}$

3 a $[L]^3$ **b** $[M][L]^{-3}$ **c** $[M][L]^2[T]^{-3}$

d $[M][L]^2[T]^{-2}$ **e** $[M][L][T]^{-1}$ **f** $[M][L]^{-1}[T]^{-2}$

4 $2as = [a]\,[s]$

$= [L][T]^{-2}[L]$

$= [L]^2\,[T]^{-2}$

$v^2 = u^2 = ([L][T]^{-1})^2$

$= [L]^2\,[T]^{-2}$

5 $ut = [L][T]^{-1}[T] = [L]$

$\tfrac{1}{2}at^2 = [L][T]^{-2}[T]^2 = [L]$

6 ML^2T^{-2}

7 $E = k \times m$

$[E] = ML^2T^{-2}$

$[m] = M$

$[k] = \dfrac{ML^2T^{-2}}{M} = L^2T^{-2}$

velocity = distance/time

$[\text{velocity}]^2 = \left(\dfrac{L}{T}\right)^2 = L^2T^{-2}$

8 a i $k = \dfrac{A}{t}$ **ii** $[L]^2[T]^{-1}$

b i $k = \dfrac{F}{v^2}$ **ii** $[M][L]^{-1}$

c i $k = \dfrac{F}{xt}$ **ii** $[M][T]^{-3}$

d i $k = \dfrac{a}{mu}$ **ii** $[M]^{-1}[T]^{-1}$

e i $k = \dfrac{vma}{xF}$ **ii** $[T]^{-1}$

f i $k = \dfrac{u}{at^2}$ **ii** $[T]^{-1}$

9 $\dfrac{2mMg}{m+M} = \dfrac{[M]^2[L][T]^{-2}}{[M]}$

$= [M][L][T]^{-2}$

$= [\text{force}]$

10 $[M][T]^{-1}$

11 $M^{-1}L^3T^{-2}$

12 $[M][L]^2$

13 a Inconsistent

b Inconsistent

c Inconsistent

d Consistent

e Consistent

f Consistent

Exercise 7.3B

1 P might be a velocity, Q might be an angular velocity

2 Consistent

3 $T = \dfrac{kmv^2}{r}$

4 $v = k\rho^{-\frac{1}{2}}P^{\frac{1}{2}} = k\sqrt{\dfrac{P}{\rho}}$

5 $v = kv^2Ar$

6 Henry's as v does not depend on ρ

7 I is dimensionless

8 $v = k\sqrt{\dfrac{TL}{m}}$

9 $V = k\sqrt{\lambda g}$

10 a $F = k \times R^4 L^{-1} \eta^{-1} P$

   b Substitute

     $0.15 = \dfrac{k \times 0.05^4 \times 3600}{6 \times 0.01}$

     $k = 0.4$

Review exercise 7

1 1080 N

2 a 150 N

   b 15 000 J

   c 15 000 J

   d 1500 Watts

3 a 0.24 m s^{-2}

   b 2.4 m s^{-1}

   c i 576 J   **ii** 23 520 J

   d 24 096 J

   e 2008 N

   f 4819 Watts

4 802.5 Watts

5 63 N

6 Speed $v = 7.43$ m s^{-1}

   156 Watts

7 Newton's 2nd Law gives:

   down slope

   $\dfrac{p}{v} + mg\sin\alpha - kv = 0$

   up slope

   $\dfrac{2p}{v} - mg\sin\alpha - k\dfrac{v}{2} = 0$

   add equations

   $\dfrac{3p}{v} - 3\dfrac{kv}{2} = 0$

   $v^2 = 2\dfrac{p}{k}$

   $v = \sqrt{\dfrac{2p}{k}}$

8 a 118 N   **b** 1.47 m   **c** 86.4 J

9 $x = 0.652$ m

   $y = 0.348$ m

   17.4 N

10 a Consistent   **b** Inconsistent

   c Consistent   **d** Inconsistent

11 a Dimensions are consistent

   b Dimensions are consistent

12 $T = 0.0282\, Av^2\rho$

13 Total increase in energy $= 705\,000$ J to 3 sf

   tractive force $T = 1410$ N

   Power at end $= 35250$ W

14 2.48 m s^{-1}

Assessment 7

1 a 1176 J   **b** 6.53 s

2 a 8.85 m s^{-1}

   b i 530 J   **ii** 88.3 N

3 a 33 N   **b** 528 J

4 $AB = 1.65$ m, $BC = 0.745$ m

5 4.70 m s^{-1}

6 3.83 m

7 a 60 J   **b** 35 J

8 0.221 m s^{-2}

9 $[\text{LHS}] \equiv T^2$

   $[\text{RHS}] \equiv \dfrac{L^3}{LT^{-2} \times L^2} \equiv T^2$

   So formula is dimensionally consistent.

10 Driving force $= \dfrac{\text{power}}{\text{velocity}}$

   Against the wind: $\dfrac{P}{V} - kV - W = 0$ ....[1]

   With the wind: $\dfrac{P}{2V} - 2kV + W = 0$ ....[2]

   Add [1] and [2]: $\dfrac{3P}{2V} - 3kV = 0 \;\Rightarrow\; V = \sqrt{\dfrac{P}{2k}}$

11 a 7.67 m s^{-1} **b** 4.17 m

12 a 1.1 m   **b** 0.6 m below A **c** 0.2 m

13 a 4 m   **b** 8.94 m s^{-1}

14 $f = \dfrac{k}{l^2}\sqrt{\dfrac{T}{\rho}}$

15 $M^{\frac{1}{2}}$

16 28.5 m s^{-1}

17 a 31.25 J   **b** 9.08 m s^{-1}

18 formula is dimensionally consistent

19 $P_1 = 8P$ W

20 400 N

21 Let speed at top of tube be v, and zero PE be at ground level.

   Energy at top of tube $= \dfrac{1}{2}mv^2 + mgh$

   Energy at highest point $= 4mgh$

   These are equal, so $v^2 = 6gh$

   Let time in tube be t and acceleration be a

   $v = at$ and $v^2 = 2ah$

   These give $t = \dfrac{v}{3g} = \dfrac{\sqrt{6gh}}{3g} = \sqrt{\dfrac{2h}{3g}}$

   Total work done $= 4mgh$, so rate of working

   $= \dfrac{4mgh}{t} = 4mgh\sqrt{\dfrac{3g}{2h}} = 2m\sqrt{6g^3h}$

22 $h = \dfrac{kv^2}{g}$

23 a Three dimensions give three equations, so can't find four unknowns.

   b $V = \dfrac{kr^4 p}{\eta l}$

24 0.25 m s^{-2}

25 a 3.3 MJ

   b 43.5 kW to 3 s.f.

26 17.9 N

27 25.9 MW to 3 s.f.

28 a 26.4 m

   b 23.9 m

   c 29.7 m s^{-1}

29 a $k \geq 0.5$

   b $d = \dfrac{1-k}{k}$

30 a 12.5 m

   b 10.6 m s^{-1}

   c 20 m

Chapter 8
Exercise 8.1A

1 a $2\dfrac{1}{2}$ m s^{-1} **b** 1 m s^{-1}

   c $4\dfrac{1}{2}$ m s^{-1} **d** -2 m s^{-1}

   e $-1\dfrac{1}{2}$ m s^{-1}

2 a 3.6 m s^{-1} **b** 1.5 m s^{-1} **c** $= -1.4$ m s^{-1}

3 10 kg

4 $\dfrac{3.2}{2.04} = 1.57 \text{ m s}^{-1}$

5 $\dfrac{5}{2}\mathbf{i} - 2\mathbf{j} \text{ m s}^{-1}$

6 $\begin{pmatrix} 6 \\ 1 \end{pmatrix} \text{m s}^{-1}$

7 $\begin{pmatrix} 6 \\ 5.5 \end{pmatrix} \text{m s}^{-1}$

8 $\begin{pmatrix} 1 \\ -3 \end{pmatrix} \text{m s}^{-1}$

9 $\begin{pmatrix} 2 \\ 2\frac{1}{4} \end{pmatrix} \text{m s}^{-1}$ and $\begin{pmatrix} 4 \\ 4\frac{1}{2} \end{pmatrix} \text{m s}^{-1}$

$2\mathbf{i} + 2\frac{1}{4}\mathbf{j}$ and $4\mathbf{i} + 4\frac{1}{2}\mathbf{j}$

10 1.25 kg
11 Masses are $2\frac{1}{7}$ kg and $\frac{6}{7}$ kg

Exercise 8.1B

1 20 m s^{-1}
2 0.676 m s^{-1} to 3 sf in the direction of motion of the arrow before the collision
3 6.43 km h^{-1} to 3 sf

4 $\dfrac{24}{2.04} = 11.8 \text{ m s}^{-1}$

5 a 3.2 m s^{-1} **b** 0.8 m s^{-1}
6 7.2 m s^{-1}
7 6.25 m s^{-1}
Taking gun and bullet together, the explosion is not an external impulse, so does not affect the total momentum.
8 227 m s^{-1} to 3sf
9 $a = 18$, $b = 23$
10 3.2 m s^{-1} in the direction Q was originally moving

11 $\dfrac{90}{12} = 7\frac{1}{2} \text{ m s}^{-1}$

12 $u = 5 \text{ m s}^{-1}$

$v = -\dfrac{10}{6} = -1\frac{2}{3} \text{ m s}^{-1}$ There are no more collisions as X and Y together are moving in the opposite direction to Z.

13 a $3\frac{1}{3} \text{ m s}^{-1}$

  b 2 m s^{-1}
  R is now moving towards S at rest.
  So there is another collision between R and S.

14 $\dfrac{2}{5}u$ and $\dfrac{4}{5}u$

15 a Momentum equations gives
  $10 \times 5 + m \times (-2) = 10 \times 3 + mv$
  $mv = 20 - 2m$
  b No further collisions if
  $v \geq 3$
  $\dfrac{20 - 2m}{m} \geq 3$
  $20 \geq 5m$
  $m \leq 4$
16 $m = 2$
  $a = -5$
17 a $m = 6$
  b $V = 3\sqrt{2} \text{ m s}^{-1}$

18 $v = \dfrac{12\sqrt{2}}{5} \text{ m s}^{-1}$

19 $v_A = \begin{pmatrix} 0 \\ 3\sqrt{3} \end{pmatrix}$, $v_B = \begin{pmatrix} 6 \\ 0 \end{pmatrix}$

Exercise 8.2A

1 a $v_2 = 4 \text{ m s}^{-1}$
  $v_1 = 3 \text{ m s}^{-1}$
  b $v_2 = \dfrac{8}{3} \text{ m s}^{-1}$

  $v_1 = \dfrac{10}{3} \text{ m s}^{-1}$
  c $v_2 = -9 \text{ m s}^{-1}$
  $v_1 = -3 \text{ m s}^{-1}$
  d $v_2 = 0 \text{ m s}^{-1}$
  $v_1 = -3 \text{ m s}^{-1}$

  e $v_2 = 2\frac{1}{6} \text{ m s}^{-1}$

  $v_1 = 2\frac{1}{6} - 2\frac{1}{2} = -\frac{1}{3} \text{ m s}^{-1}$

2 a 11 m s^{-1}, $\dfrac{3}{4}$ **b** 3.25 m s^{-1}, $\dfrac{1}{4}$

  c 0 m s^{-1}, $\dfrac{1}{5}$ **d** 2 m s^{-1}, $\dfrac{4}{5}$

3 0.18 m s^{-1} and 0.15 m s^{-1}
4 $v_2 = 12 \text{ m s}^{-1}$
  $v_1 = 3 \text{ m s}^{-1}$
5 0.733 to 3 sf
6 a $\mathbf{v} = 4\mathbf{i} + 3\mathbf{j} \text{ m s}^{-1}$

  at $\tan^{-1}\left(\dfrac{3}{4}\right) = 36.9°$

   to x-axis
  b $\mathbf{v} = 3\mathbf{i} + \mathbf{j} \text{ m s}^{-1}$

  at $\tan^{-1}\left(\dfrac{1}{3}\right) = 18.4°$

  to x-axis
  c $\mathbf{v} = 2\mathbf{i} + 2\mathbf{j} \text{ m s}^{-1}$

  at $\tan^{-1}\left(\dfrac{2}{2}\right) = 45°$

  to x-axis
  d $\mathbf{v} = 1.2\mathbf{i} + 7.2\mathbf{j} \text{ m s}^{-1}$

  at $\tan^{-1}\left(\dfrac{7.2}{1.2}\right) = 80.5°$

7 $\mathbf{v} = 3\mathbf{i} + 2\mathbf{j} \text{ m s}^{-1}$
  at an angle of
  86.8°

Exercise 8.2B

1 a 8.85 m s^{-1} (3 sf), 7.67 m s^{-1} (3 sf) **b** 0.866 (3sf)

2 $3\frac{1}{3} \text{ m s}^{-1}$

3 2 kg

4 $\dfrac{1}{2}$

5 $= 5.55 \text{ m s}^{-1}$
  25.6° to the plane.
6 1.5 kg

7 a $\dfrac{9}{16}u$ and $\dfrac{5}{16}u$

  b $\dfrac{3}{8}u$

8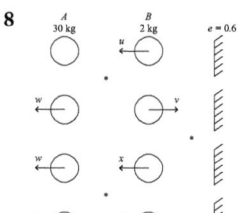

For 1st impact A and B
Momentum equation is
$2u + 0 = 30w - 2v$
$u = 15w - v$ (1)
Newton's equation is
$\dfrac{w+v}{u} = 0.6$
$0.6u = w + v$ (2)
Solve (1) + (2)
$1.6\,u = 16w$

$w = 0.1u$

$v = 0.5u$

For impact of B on wall, Newton's equation
$x = 0.6v = 0.3u$
$x > w$ so B impacts on A
Momentum equation is
$30w + 2x = 30y + 2z$
$3u + 0.6u = 30y + 2z$
$1.8\,u = 15y + z$ (3)
Newton's equation is
$\dfrac{y-z}{x-w} = 0.6$
$y - z = 0.12u$ (4)
Solve (3) + (4)
$1.8u + 0.12u = 16y$
$y = \dfrac{1.92u}{16} = 0.12u$
From (4)
$z = y - 0.12u = 0$
Hence, sphere B is at rest.

9

For P, Q impact
Momentum equation
$6 \times 15 = 6u + 4v$
$45 = 3u + 2v$ (1)
Newton's equation
$\dfrac{v-u}{15} = \dfrac{3}{5}$

$v - u = 9$ (2)

Solve (1) + 3 × (2)
$5v = 45 + 27$
$v = 14.4$
$u = v - 9 = 5.4$

For Q, R impact
Momentum equation
$4 \times 14.4 = 4w + 8x$
$14.4 = w + 2x$ (3)

Newton's equation
$\dfrac{x-w}{14.4} = \dfrac{3}{5}$
$x - w = 8.64$ (4)
Solve (3) + (4)
$3x = 23.04$

$x = 7.68$
$w = v - 2x = -0.96$
For 2nd P, Q impact
Momentum equation is
$6 \times 5.4 + 4 \times (-0.96) = 6y + 4z$
$28.56 = 6y + 4z$ (5)
Newton's equation is
$\dfrac{z-y}{5.4 - (-0.96)} = \dfrac{3}{5}$
$z - y = 3.816$ (6)
Solve (5) + 6 × (6)
$10z = 51.456$
$z = 5.1456$
$y = z - 3.816 = 1.3296$
Final velocities are 1.33 ms^{-1} to 3 sf, 5.15 ms^{-1} to 3 sf and 7.68 ms^{-1} so $v_P < v_Q < v_R$ so there are no more collisions.

10 3 s

11 a 4 m s^{-1}
   b 2 m s^{-1}
     As z is moving faster than X and Y together and they are moving in the same direction, there are no more collisions.

12 $\dfrac{1}{3}$

13 10.2 m s^{-1}
   33.7° below the horizontal.

Exercise 8.3A

1 a −40 N s   **b** −210 N s
2 a 18 N s   **b** 45 N s
3 a 960 m s^{-1} **b** 232 m s^{-1}
4 a 181 N   **b** 283 N
5 0.12 N
6 a 32 N s
     8.4 m s^{-1}
   b 4 N s
     2.8 m s^{-1}

Exercise 8.3B

1 7.2 N s
2 $\mathbf{v} = 2\mathbf{i} + 0.6\mathbf{j}$ m s^{-1}

   $\mathbf{I} = 18\mathbf{j}$ N s

3 $\mathbf{v} = 2.4\mathbf{i} + 3\mathbf{j}$ m s^{-1}

   $\mathbf{I} = 38.4\mathbf{i}$ N s

4 a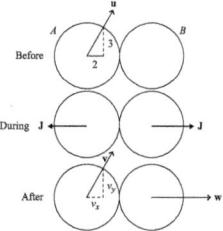

No impulse perpendicular to x–axis. Momentum conserved so B does not change its velocity in the y–direction.
   b 4$\mathbf{i}$ N s

5 Final velocities are
$4\mathbf{i} + 0.4\mathbf{j}$ m s^{-1} for P
and $2.8\mathbf{j}$ m s^{-1} for Q
$11.2\mathbf{j}$ N s

6 a 135 Ns
  b i 45 ms^{-1} **ii** 41 ms^{-1}

7 a 120 Ns   **b** 26 ms^{-1} **c** 6.1s

8 a 2.6 Ns   **b** 3 ms^{-1}

9 a 72 Ns   **b** 96 ms^{-1} **c** 3s

10 Impulse $= 42\dfrac{2}{3}$ N s

Final velocity $= 20\dfrac{2}{3}$ m s^{-1}

11 240 N s
Final velocity $v = 5.6$ m s^{-1}

12 a 48 N s
  b 2.4 m s^{-1}
  c $\sqrt{8}$ secs

13 $J = \dfrac{Mm(u-v)}{M+m}$

14 12 m s^{-1}
36 N s

15 a 475 N s
9.5 m s^{-1}
  b 2260 N (3 sf)

16 Velocities are 5 m s^{-1} and
5.20 m s^{-1}
Impulse $= 60$ N s

17 Final velocities are
$\dfrac{4\sqrt{3}}{3}\mathbf{i} + 4\mathbf{j}$ and $\dfrac{7\sqrt{3}}{3}\mathbf{i} + \mathbf{j}$
Impulse $= \dfrac{16\sqrt{3}}{3}$ N s

18 After impact the velocities are
$\sqrt{v_1^2 + v_2^2} = \sqrt{\dfrac{u^2}{4} + \dfrac{u^2}{4}} = \dfrac{u}{\sqrt{2}}$
and
$\sqrt{w_1^2 + w_2^2} = \sqrt{\dfrac{3u^2}{4} + \dfrac{3u^2}{4}} = \sqrt{\dfrac{3}{2}}\, u$
Impulse
$J = m\dfrac{u\sqrt{3}}{2} - m\dfrac{u}{2} = \dfrac{\sqrt{3}-1}{2}mu$

19 a

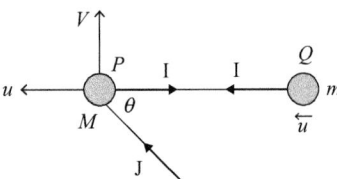

Let velocity of Q be $\mathbf{u}$ and components of velocity of P be u and v.
Impulse equation perpendicular to the rod is
$J\sin\theta = Mv$
Impulse equation along the rod for P and Q together is
$J\cos\theta = Mu + mu$

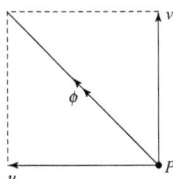

If angle of final velocity of P to rod $= \phi$
$\tan\phi = \dfrac{v}{u} = \dfrac{J\sin\theta}{M} \times \dfrac{M+m}{J\cos\theta}$
$\tan\phi = \dfrac{M+m}{M}\tan\theta$

  b i $M \gg m$ $\tan\phi$ grows very large $\Rightarrow \phi$ tends to 90°
Q effectively becomes fixed and P swings around Q in a circle.

  b ii $M = 0$ $\tan\phi = \tan\theta \Rightarrow \phi = \theta$
A massless Q has no effect on the motion of P which moves in the direction of the impulse.

20 a No impulse perpendicular to AB, so component $\dfrac{u}{2}\mathbf{j}$ is unchanged

  b Magnitude of velocity of $B = \dfrac{u}{2}$
Magnitude of final velocity of
$B = \sqrt{\dfrac{3}{16}u^2 + \dfrac{1}{4}u^2}$
$= \dfrac{\sqrt{7}}{4}u$
Impulse $= \dfrac{mu\sqrt{3}}{4}$

Review exercise 8

1 -6 m s^{-1}

2 speed $= \sqrt{v_1^2 + v_2^2} = \sqrt{10}$ m s^{-1}
at $\tan^{-1}\left(\dfrac{-1}{3}\right) = -18.4°$ (3 sf) to the x-direction.

3 $m = 10$,
so final velocity $= \begin{pmatrix} \dfrac{4}{3} \\ \dfrac{4}{3} \end{pmatrix}$ ms^{-1}

4 $9\dfrac{1}{4}$ m s^{-1} and $3\dfrac{3}{4}$ m s^{-1}

5 $e = \dfrac{2\sqrt{5}}{6\sqrt{5}} = \dfrac{1}{3}$
$11\dfrac{2}{9}$ m

6 $v = 2$ m s^{-1}
10 000 N s

7 $m = 10\dfrac{1}{2}$ kg
$\dfrac{6+2}{12-2} = e \Rightarrow e = \dfrac{4}{5}$
42 N s

8 Final velocity $= 5\mathbf{i} + \mathbf{j}$ m s^{-1}
$\mathbf{I} = 12\mathbf{j}$ N s

9 18 750 N

10 6 N s
12 m s^{-1}

11 Velocities are
8 m s^{-1} for A to the left
4 m s^{-1} for B to the left
1 m s^{-1} for C to the right
As $8 > 4 > -1$ there are no more collisions.

12 Velocity of P is $\dfrac{10}{3}\mathbf{i}$
Velocity of Q is $\dfrac{10}{3}\mathbf{i} + 6\mathbf{j}$
Impulse in string $= \dfrac{20}{3}$ N s

13 13.0°, $\mathbf{J} = 14\mathbf{j}$ N s

14 Assume no friction and that the balls are modelled as particles.

$$0, \frac{I}{2\sqrt{2}m}\mathbf{i}, \frac{I}{2\sqrt{2}m}(\mathbf{i}+\mathbf{j}), \frac{I}{2\sqrt{2}m}\mathbf{j}$$

15 **a** 32 Ns **b** 64 m s⁻¹ **c** 2.47 s

Assessment 8

1 **a** 42 Ns **b** 120 N
2 4 s
3 $\mathbf{v} = (4.4\mathbf{i} - 0.4\mathbf{j})\,\text{m s}^{-1}$
4 4.5 s
5 **a** 3 m s⁻¹ **b** 4.76 m s⁻¹

6 **a** $V = \dfrac{10v}{7}$ **b** $V = \dfrac{2v}{7}$

7 **a** $V = \dfrac{14}{3}$ m s⁻¹

   b $50 - 2m = 30 + mv$ gives $v = \dfrac{20 - 2m}{m}$

     B must travel at least as fast as A

     $\dfrac{20 - 2m}{m} \geq 3 \implies m \leq 4$

8 **a** A has velocity $(4\mathbf{i} - 1.8\mathbf{j})\,\text{m s}^{-1}$, B has velocity $(-\mathbf{i} + 2.2\mathbf{j})\,\text{m s}^{-1}$

   b $14.4\mathbf{j}$ Ns **c** $\tan^{-1}\left(\dfrac{-1.8}{4}\right) - \tan^{-1}\left(\dfrac{3}{4}\right) = -61.1°$

9 **a** 15 m s⁻¹ **b** 6000 N **c** 5882
10 $\mathbf{v} = (2.5\mathbf{i} + 0.5\mathbf{j})\,\text{m s}^{-1}$
11 **a** **i** $v_A = 4\,\text{m s}^{-1}$, $v_B = 9\,\text{m s}^{-1}$ **ii** $v = 6\,\text{m s}^{-1}$

   b **i** No momentum is lost, so final momentum $5v$ equals initial momentum 30 provided that e > 0

     ii $e = 0$ means the particles would not separate after collision, so the string would never become taut. The particles would move together at $6\,\text{m s}^{-1}$

12 **a** $v_A = 7.5\,\text{m s}^{-1}$, $v_B = 10\,\text{m s}^{-1}$

   b $v_A = 0$, $v_B = 7.5\,\text{m s}^{-1}$

   c B is brought to rest, so there will be no more collisions.

13 **a** A $u_1 = 2.2\,\text{m s}^{-1}$, for B $v_1 = 3.2\,\text{m s}^{-1}$

   b $v_2 = \dfrac{7.4 - 2.2e}{3}$

   c If there are no more collisions, B must be moving faster than A

     $\dfrac{7.4 - 2.2e}{3} \geq 2.2 \implies e \leq \dfrac{4}{11}$

14 **a** **i** The cushion is smooth, so no impulse and hence no change of momentum in the $\mathbf{i}$-direction

     ii 4.9 m s⁻¹ **iii** 98.9°

   b 2380

15 **a** 4 m s⁻¹ **b** 6 Ns

   c The total momentum of A and B is now $1 \times 4 + 2 \times 1 = 6\,\text{Ns}$. This will be unchanged by the collision between A and B and the string again becomes taut. Their common velocity is then v, where $3v = 6$, $v = 2\,\text{m s}^{-1}$. They are moving slower than C, so there are no more collisions.

16 $\dfrac{u}{4}$ and $\dfrac{u}{2}$

17 4.25 or 0.3125
18 **a** 6 **b** 1.8 m s⁻¹
19 7.2 N s
20 2 m s⁻¹
21 **a** When A and B collide the momentum equation is $4 - 8 = 4v_A + 2v_B$ .....[1] The restitution equation is $0.2 \times 5 = 1 = v_B - v_A$ ....[2] Solving [1] and [2] $v_B = 0$

b 1 m s⁻¹
22 $-8\mathbf{j}\,\text{m s}^{-1}$
23 **a** 5.6 m s⁻¹ **b** 0.2
24 $u = 7.38\,\text{m s}^{-1}$, $v = 4.62\,\text{m s}^{-1}$
25 $I = (10\mathbf{i} + 20\mathbf{j})\,\text{Ns}$

   Magnitude:

   $|I| = 10\sqrt{5}$ Ns

26 **a** 68 Ns **b** 10.5 m s⁻¹

Chapter 9
Exercise 9.1A

1

	rpm	rad min⁻¹	rad s⁻¹
a	33	207	3.5
b	9	57	0.94
c	95	600	10
d	2480	15600	260

2 0.0039 rad s⁻¹
  0.23 m s⁻¹
3 Hour hand: 0.0004 m s⁻¹; 0.00015 rad s⁻¹
  Minute hand: 0.0073 m s⁻¹; 0.0017 rad s⁻¹
4 7.3×10^{-5} rad s⁻¹, 1670 km h⁻¹
5 4 rad s⁻¹, 24 m s⁻²

6 Linear speed $= \dfrac{60\pi}{60} = \pi$ m s⁻¹

  Acceleration to centre $= \dfrac{\pi^2}{3} = 3.3$ m s⁻²

7 Linear speed $= \dfrac{120\pi}{60} = 2\pi$ m s⁻¹

  Acceleration to centre $= \dfrac{(2\pi)^2}{0.03}$

                  $= 1320$ m s⁻²

8 Linear speed $= \dfrac{576\pi}{60} = 30.2$ m s⁻¹

  Acceleration to centre $= \dfrac{30.2^2}{0.24}$

  $= 3800$ m s⁻²

Exercise 9.1B

1 12 N
2 66.6 N
3 1.05 N
4 27.4 N
5 0.118 rad s⁻¹
  8.5 km h⁻¹
6 47.7 rpm
7 4170 N
8

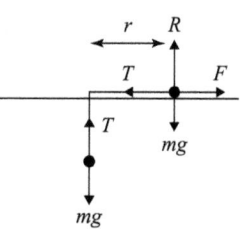

Let linear speed of mass on table $= v$
Equation of motion of this mass is

Tension $T = m \times \dfrac{v^2}{r}$

Resolve vertically for the hanging mass
Tension $= mg$

So $\dfrac{mv^2}{r} = mg$

Required velocity, $v = \sqrt{rg}$

Exercise 9.2A

1 94.7 N

2 $v = \sqrt{\dfrac{0.5 \times 0.8}{0.2}} = \sqrt{2}\ \text{m s}^{-1}$

3 It completes 86 revs per minute

4 0.45 N

5 Tension $T = 3ma\omega^2$
   Tension $S = 2ma\omega^2$

6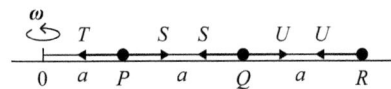

Let tensions be T, S and U as shown
Equation of motion about 0

For R is:	$U = m \times 3a \times \omega^2$	(1)
For S is:	$S - U = m \times 2a \times \omega^2$	(2)
For T is:	$T - S = m \times a \times \omega^2$	(3)
(1) gives	$U = 3ma\omega^2$	
Add (1)(2)	$S = 5ma\omega^2$	
Add (1)(2)(3)	$T = 6ma\omega^2$	

Hence, ratio of tension $U{:}S{:}T$
is 3:5:6

7 $\dfrac{\omega}{\sqrt{2}}$

Exercise 9.2B

1 0.13 m

2 3.125 N

3
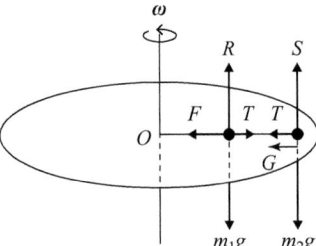

Let vertical reactions be R and S and frictional forces be
F and G
Tension in string $= T$
Resolve vertically
$R = m_1 g$
$S = m_2 g$
On point of slipping,
$F = \mu R = \mu m_1 g$
and
$G = \mu S = \mu m_2 g$
Horizontal equations of motion to centre are
$F - T = m_1 \times x \times \omega^2$
and
$G + T = m_2 \times 2x \times \omega^2$
Add to eliminate T
$G + F = \omega^2(m_1 x + 2m_2 x)$
Substitute for F and G
$(m_1 + m_2)\mu g = x\omega^2(m_1 + 2m_2)$
Hence,
$\omega = \sqrt{\dfrac{\mu g(m_1 + m_2)}{x(m_1 + 2m_2)}}$

4 The tension in the rim is at right angle to that in the spokes
   and thus has no effect.

5 2.53 m s^{-1}
   11.2 J

6 17.1 cm

7 Maximum speed is 15.3 m s^{-1}
   The speed v does not depend on the mass of the car.

8 $\mu \geq \dfrac{320}{800 \times 9.8} = 0.04$

9 0.15

10
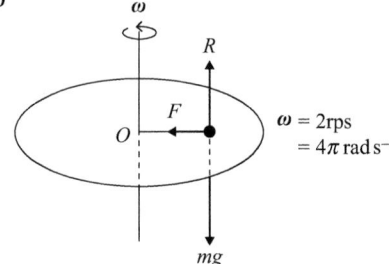

Resolve vertically
$R = mg$
Equation of motion to centre is
$F = mr(4\pi)^2$
For limiting friction,
$F = \mu R$
$\ = \mu mg$
So
$\mu mg = mr(4\pi)^2$
Greatest distance,
$r = \mu \times \dfrac{9.8}{(4\pi)^2}\ \text{m}$
$\ = \mu \times 6.2\ \text{cm}$

11 a $F \alpha \dfrac{1}{(R^2)}$

   $F = \dfrac{k}{(R^2)}$

   Constant
   $k = mgR^2$

b Angular velocity
   $= 0.00116\ \text{rad s}^{-1}$ to 3sf
   Height
   $= 227\ \text{km}$

c Height of geostationary satellite
   $= 35800\ \text{km}$ to 3sf.

Review exercise 9

1 $800\ \text{rpm} = \dfrac{800 \times 2\pi}{60} = 83.8\ \text{rad s}^{-1}$

   $1200\ \text{rpm} = \dfrac{1200 \times 2\pi}{60} = 125.7\ \text{rad s}^{-1}$

2 0.0015 rad s^{-1}

3 $v = r\omega$ so $\omega = \dfrac{v}{r} = \dfrac{0.6}{0.8} = \dfrac{3}{4}\ \text{rad s}^{-1}$

   Acceleration $= \dfrac{v^2}{r} = \dfrac{0.6^2}{0.8} = 0.45\ \text{rad s}^{-2}$

4 $\dfrac{1}{30}\ \text{rad s}^{-1}$
   31.4 sec

5 118 kJ to 3sf

6 822 N

7 Angular speed $= 1.99 \times 10^{-7}\ \text{rad s}^{-1}$
   Force $= 3.56 \times 10^{19}\ \text{N}$

8 22.5 N

9 28.3 m s^{-1}

10 16.7 s

11 $3\,\text{m s}^{-1}$

12 $23.2\,\text{J}$

13 Reaction $= m\sqrt{g^2 + r^2\omega^4}$

$$\theta = \tan^{-1}\left(\frac{r\omega^2}{g}\right)$$

14 89.3 minutes

15 0.34

16 0.574

Assessment 9

1 $10.4\,\text{N}$

2 $149\,\text{rev min}^{-1}$

3 $60\,\text{cm}$

4 $70\,\text{cm}$

5 $15.7\,\text{m s}^{-1}$

6 Magnitude $= 0.055\,\text{N}$
Direction $= 63°$

7 $6.02 \times 10^{24}\,\text{kg}$

8 **a** $1.2\,\text{m}$ **b** $6\,\text{m s}^{-1}$

9 $3\,mga$

10 **a** $\dfrac{12.5^2 m}{50} = \dfrac{25m}{8}\,\text{N}$

     b $75.6\,\text{km h}^{-1}$

11 **a** $5.07\,\text{rad s}^{-1}$ **b** $0.932\,\text{m}$

12 $3.13\,\text{rad s}^{-1}$

13 $\omega = \sqrt{\dfrac{\mu g}{r}}$

14 $r = \dfrac{\mu g}{16\pi^2}$

15 Maximum friction is μmg.

For maximum speed $mg + \mu mg = ma\omega_1^2$

$$\Rightarrow \omega_1^2 = \frac{g(1+\mu)}{a}$$

For minimum speed $mg - \mu mg = ma\omega_2^2$

$$\Rightarrow \omega_2^2 = \frac{g(1-\mu)}{a}$$

Hence $\dfrac{\omega_1}{\omega_2} = \sqrt{\dfrac{1+\mu}{1-\mu}}$

Chapter 10

Exercise 10.1A

1 $E(X) = 1 \times \dfrac{1}{6} + 4 \times \dfrac{1}{6} + \ldots + 36 \times \dfrac{1}{6} = \dfrac{91}{6}$

$\text{Var}(X) = 1^2 \times \dfrac{1}{6} + 4^2 \times \dfrac{1}{6} + \ldots + 36^2 \times \dfrac{1}{6} - \left(\dfrac{91}{6}\right)^2 = 149.14$ (2 d.p.)

$E[Y] = -1 \times \dfrac{1}{3} + 1 \times \dfrac{1}{6} + 2 \times \dfrac{1}{6} = \dfrac{1}{6}$

$\text{Var}[Y] = 1 \times \dfrac{1}{3} + 1 \times \dfrac{1}{6} + 4 \times \dfrac{1}{6} - \left(\dfrac{1}{6}\right)^2 = \dfrac{41}{36}$

2 **a** $E[X] = 1 \times \dfrac{1}{4} + 2 \times \dfrac{1}{4} + \ldots + 4 \times \dfrac{1}{4} = \dfrac{5}{2}$

$\text{Var}(X) = 1^2 \times \dfrac{1}{4} + 2^2 \times \dfrac{1}{4} + \ldots + 4^2 \times \dfrac{1}{4} - \left(\dfrac{5}{2}\right)^2 = \dfrac{5}{4}$

     b $E[X] = 1 \times \dfrac{1}{3} + 2 \times \dfrac{2}{3} = \dfrac{5}{3}$

$\text{Var}(X) = 1^2 \times \dfrac{1}{3} + 2^2 \times \dfrac{2}{3} - \left(\dfrac{5}{3}\right)^2 = \dfrac{2}{9}$

c $E[X] = 2 \times \dfrac{1}{2} + 3 \times \dfrac{1}{3} + 4 \times \dfrac{1}{6} = \dfrac{8}{3}$

$\text{Var}(X) = 2^2 \times \dfrac{1}{2} + 3^2 \times \dfrac{1}{3} + 4^2 \times \dfrac{1}{6} - \left(\dfrac{8}{3}\right)^2 = \dfrac{5}{9}$

3 $E(X) = 1 \times \dfrac{1}{10} + 2 \times \dfrac{1}{10} + \ldots + 10 \times \dfrac{1}{10} = \dfrac{11}{2}$

$\text{Var}(X) = 1^2 \times \dfrac{1}{10} + 2^2 \times \dfrac{1}{10} + \ldots + 10^2 \times \dfrac{1}{10} - \left(\dfrac{11}{2}\right)^2 = \dfrac{33}{4}$

4 $E[T] = \dfrac{150}{25} = 6$

5 $E(\text{score}) = 1 \times \dfrac{1}{3} + 2 \times \dfrac{1}{6} + 3 \times \dfrac{1}{3} + 4 \times \dfrac{1}{6} = \dfrac{7}{3}$

$\text{Var}(\text{score}) = 1^2 \times \dfrac{1}{3} + 2^2 \times \dfrac{1}{6} + 3^2 \times \dfrac{1}{3} + 4^2 \times \dfrac{1}{6} - \left(\dfrac{7}{3}\right)^2 = \dfrac{11}{9}$

6 $E(X) = 1 \times \dfrac{1}{9} + 2 \times \dfrac{1}{9} + \ldots + 9 \times \dfrac{1}{9} = \dfrac{45}{9} = 5$

$\text{Var}(X) = 1^2 \times \dfrac{1}{9} + 2^2 \times \dfrac{1}{9} + \ldots + 9^2 \times \dfrac{1}{9} - 5^2 = \dfrac{285}{9} - 25 = \dfrac{20}{3}$

7 $E(X^2) = 1^2 \times \dfrac{1}{6} + 2^2 \times \dfrac{1}{6} + \ldots + 6^2 \times \dfrac{1}{6} = \dfrac{91}{6}$

8

s	1	2	3	4	5	6
P(S = s)	k	4k	9k	16k	25k	36k

$k = \dfrac{1}{91} \Rightarrow E[S] = 1 \times \dfrac{1}{91} + 2 \times \dfrac{4}{91} + \ldots + 6 \times \dfrac{36}{91} = 4.85$ (2 d.p.)

9 $E[R] = \dfrac{1}{1} \times \dfrac{1}{6} + \dfrac{1}{2} \times \dfrac{1}{6} + \ldots + \dfrac{1}{6} \times \dfrac{1}{6} = 0.41$ (2 d.p.)

10 **a**

x	P(X = x)	xP(X = x)	x²P(X = x)
1	0.4	0.4	0.4
2	0.2	0.4	0.8
3	0.2	0.6	1.8
4	0.2	0.8	3.2
		2.2	6.2

If the mean and variance of X are μ and σ^2, then

$\mu = \displaystyle\sum_{\text{all } x} x\text{P}(X = x) = 2.2$ and $\sigma^2 = \displaystyle\sum_{\forall x} x^2\text{P}(X = x) - \mu^2$

$= 6.2 - 2.2^2 = 1.36$

b $E(Y) = 2E(X^2) = 2 \times 6.2 = 12.4$; $\text{Var}(Y) = \text{Var}(2X^2) = 4\text{Var}(X^2)$
$= 4E(X^4) - 4\{E(X^2)\}^2$
$= 4(1 \times 0.4 + 16 \times 0.2 + 81 \times 0.2 + 256 \times 0.2) - 4 \times 6.2^2$
$= 130.24$

11 **a** Expected value $0 \times \dfrac{2}{3} + 1 \times \dfrac{1}{3} = \dfrac{1}{3}$

b If the experiment is performed a large number of times, approximately $\dfrac{1}{3}$ of the results will be prime and, with high probability, the proportion will get closer to $\dfrac{1}{3}$ as number of trials increases.

12 **a**

x	1	2	3	4
P(X = x)	$\dfrac{1}{2}$	$\dfrac{1}{4}$	$\dfrac{1}{8}$	$\dfrac{1}{8}$

b $E[X] = 1 \times \dfrac{1}{2} + 2 \times \dfrac{1}{4} + 3 \times \dfrac{1}{8} + 4 \times \dfrac{1}{8} = \dfrac{15}{8}$

13 Mode $= 1$ Median $= 1$

14 **a** $\dfrac{1}{6}$ **b** 3

15 $\mu = E(U) = \sum_{u=1}^{n} \dfrac{u}{n} = \dfrac{1}{n}\sum_{u=1}^{n} u = \dfrac{1}{n} \times \dfrac{n(n+1)}{2} = \dfrac{n+1}{2}$

$\sigma^2 = \text{Var}(U) = \sum_{i=1}^{n} \dfrac{u^2}{n} - \mu^2 = \dfrac{1}{n} \times \dfrac{1}{6} n(n+1)(2n+1) - \left(\dfrac{n+1}{2}\right)^2$

$= \dfrac{2(n+1)(2n+1) - 3(n+1)^2}{12} = \dfrac{n^2 - 1}{12}$

Exercise 10.1B

1 $a = \dfrac{1}{5}, b = \dfrac{2}{15}$

2 a By symmetry, the mean of this distribution is 0

r	$P(R = r)$	$rP(R = r)$	$r^2 P(R = r)$
-3	$\dfrac{1}{7}$	$-\dfrac{3}{7}$	$\dfrac{9}{7}$
-2	$\dfrac{1}{7}$	$-\dfrac{2}{7}$	$\dfrac{4}{7}$
-1	$\dfrac{1}{7}$	$-\dfrac{1}{7}$	$\dfrac{1}{7}$
0	$\dfrac{1}{7}$	0	0
1	$\dfrac{1}{7}$	$\dfrac{1}{7}$	$\dfrac{1}{7}$
2	$\dfrac{1}{7}$	$\dfrac{2}{7}$	$\dfrac{4}{7}$
3	$\dfrac{1}{7}$	$\dfrac{3}{7}$	$\dfrac{9}{7}$
		0	4

$\sigma^2 = \sum_{\forall r} r^2 P(R = r) - \mu^2 = 4$. Therefore $\sigma = 2$.

b The probability that it will be one standard deviation or closer to the mean is therefore $\dfrac{5}{7}$.

3 a BBBB GBBB BGBB BBGB BBBG GGBB GBGB GBBG
BGGB BGBG BBGG GGGB GGBG GBGG BGGG GGGG

x	$P(X = x)$
0	$0.51^4 = 0.07$
1	$4 \times 0.51^3 \times 0.49 = 0.26$
2	$6 \times 0.51^2 \times 0.49^2 = 0.37$
3	$4 \times 0.51 \times 0.49^3 = 0.24$
4	$0.49^4 = 0.06$

b $E[X] = 0 \times 0.07 + 1 \times 0.26 + ... + 4 \times 0.06 = 1.96 = (2 \text{ d.p.})$
$\text{Var}[X] = 0 \times 0.07 + 1^2 \times 0.26 + ... + 4^2 \times 0.06 - 1.96^2 = 1.02$ (2 d.p.)

4 a $P(\text{John wins}) = P(X \geq 1) = 1 - P\{X = 0\} = 1 - \left(\dfrac{5}{6}\right)^4 = 0.518$ (3 d.p.)

b Jenny should pay John 48p to enter the game.

5

b	1	2
$P(B = b)$	$\dfrac{2}{3} \times \dfrac{1}{2} + \dfrac{1}{3} \times 1 = \dfrac{2}{3}$	$\dfrac{2}{3} \times \dfrac{1}{2} = \dfrac{1}{3}$

$E[B] = 1 \times \dfrac{2}{3} + 2 \times \dfrac{1}{3} = \dfrac{4}{3}$

6 a $P(X \geq 1) = 1 - P(X = 0) = 1 - \left(\dfrac{35}{36}\right)^{24} = 0.491$ (3 d.p.)

b 1 counter

7 a $E(X) = \dfrac{1}{5n} \sum_{x=1}^{5n} x = \dfrac{5n(5n+1)}{2 \times 5n} = \dfrac{1+5n}{2}$

b $\text{Var}(X) = \left(\dfrac{1}{5n} \sum_{x=1}^{5n} x^2\right) - \left(\dfrac{1+5n}{2}\right)^2 = \dfrac{1}{5n} \times \dfrac{5n}{6} \times$
$(5n+1)(10n+1) - \dfrac{(1+5n)^2}{4} = \dfrac{25n^2 - 1}{12}$

8 a $P(\text{one dart landing in circle}) = \dfrac{\pi r^2}{4r^2} = \dfrac{\pi}{4}$

$E[D] = 0 \times \left(1 - \dfrac{\pi}{4}\right)^3 + 1 \times \left(1 - \dfrac{\pi}{4}\right)^2 \times 3 + 2 \times \left(1 - \dfrac{\pi}{4}\right)\left(\dfrac{\pi}{4}\right)^2$

$\times 3 + 3 \times \left(\dfrac{\pi}{4}\right)^3 = 2.36$ (2 d.p.)

b Estimate of $\dfrac{\pi}{4} = 0.71$. π estimated as 2.84

9 $E(X + Y) = E(X) + E(Y) = 3.2 + 22 = 25.2$
$\text{Var}(X + Y) = \text{Var}(X) + \text{Var}(Y) = 12.1 + 9 = 21.1$
Standard deviation of $X + Y = 4.59$ (2 d.p.)

10 $\text{Var}(X + Y) = E\left((X+Y)^2\right) - [E(X+Y)]^2$

$= E(X^2) + E(Y^2) + 2E(XY) - \left\{\{E(X)\}^2 + \{E(Y)\}^2 + 2E(X)E(Y)\right\}$

$= \text{Var}(X) + \text{Var}(Y) + 2\left\{E(XY) - E(X)E[Y]\right\}$

$= \text{Var}(X) + \text{Var}(Y)$

because $E[XY] = E(X)E(Y)$ for independent X and Y.

11 a i $a\mu, a^2\sigma^2$ **ii** $b\mu - c, b^2\sigma^2$

b X – score on a randomly chosen game. W – Net winnings. W = 5X – 20.

$E(X) = 3.5 \; \text{Var}(X) = \dfrac{105}{36}$
$E(W) = 5E(X) - 20 = 5 \times 3.5 - 20 = -2\dfrac{1}{2}$
$\text{Var}(W) = 25\text{Var}(X) = \dfrac{2625}{36} = 72.9$ (1 dp)

12 a $\dfrac{5}{24}$

b $E(S) = \dfrac{11}{6}$; $\text{Var}(S) = \dfrac{5}{9}$
$E(R) = \dfrac{13}{3}$; $\text{Var}(R) = \dfrac{80}{9}$

13 9 and 20; −1, 20.

14 $E(X) = 2 \times \dfrac{1}{6} + 4 \times \dfrac{1}{6} + ... + 12 \times \dfrac{1}{6} = 7$

$\text{Var}(X) = 2^2 \times \dfrac{1}{6} + 4^2 \times \dfrac{1}{6} + ... + 12^2 \times \dfrac{1}{6} - 7^2 = \dfrac{35}{3}$

$E(Y) = 1 \times \dfrac{1}{8} + 2 \times \dfrac{1}{8} + ... + 8 \times \dfrac{1}{8} = \dfrac{9}{2}$

$\text{Var}(Y) = 1^2 \times \dfrac{1}{8} + 2^2 \times \dfrac{1}{8} + ... + 8^2 \times \dfrac{1}{8} - \left(\dfrac{9}{2}\right)^2 = \dfrac{21}{4}$

$E(X + Y) = \dfrac{23}{2}$

$\text{Var}(X + Y) = \dfrac{203}{12}$

15 Let X be the score obtained.

$E(X) = 1 \times \dfrac{1}{10} + 2 \times \dfrac{1}{10} + ... + 10 \times \dfrac{1}{10} = 5.5$

$\text{Var}(X) = 1^2 \times \dfrac{1}{10} + 2^2 \times \dfrac{1}{10} + ... + 10^2 \times \dfrac{1}{10} - 5.5^2 = 8.25$

$E(X_1 + X_2) = 11$

$\text{Var}(X_1 + X_2) = 16.5$

Exercise 10.2A

1 $P(X = x) = \dfrac{e^{-\mu}\mu^x}{x!}$. 0.01832, 0.07326, 0.1465, 0.1954, 0.5665

2 0.8009

3 0.39 (2 d.p.)

4

x	0	1	2	3	4	>4
P(X = x)	0.333	0.366	0.201	0.074	0.020	0.006

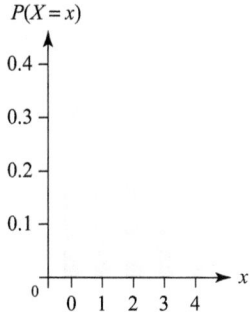

5 0.0028 (2 s.f.).

6 0.35 (2 d.p.).

7 0.98168. Assume independence from one day to the next. $0.98168^7 = 0.88$ (2 d.p.)

8 0.76

9 $P(X \geq 1) = 0.982$. Number in any bar independent of any other bar. $0.982^5 = 0.91$ (2 d.p.)

10 0.0011 (2s.f.)

11 0.20 (2 d.p.)

12 $\mu = 2.29$ $\sigma^2 = 2.80$

Data likely to be approximately Poisson distributed → cars passing randomly. In addition, the mean and variance of the data are fairly similar which is consistent with a Poisson distribution (in which the mean and variance are equal).

Exercise 10.2B

1 0.347 (3 d.p.)

2 0.195 (3 d.p.)

3 0.841 (3 d.p.)

4 0.142 (3 d.p.)

5 0.97214

6 0.374 (3 d.p.)

7 a 0.1215 (4 d.p.)   b 0.9314

8 Drops arrive independently and at a constant average rate. 0.0067(2 s.f.)

9 0.145 (3 d.p.)

10 0.20 (2dp)

11 0.14 (2dp)

Exercise 10.3A

1 a

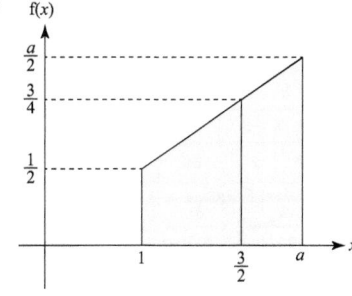

b $a = \sqrt{5}$, $P(X > 1.5) = \dfrac{11}{16}$

2 $k = \dfrac{3}{8}$. $P(X < 1) = \dfrac{3}{8}\displaystyle\int_0^1 x^2 dx = \dfrac{1}{8}$

3 $\dfrac{1}{36}$

4 a $\displaystyle\int_0^{\frac{1}{2}} ax(2x-1)dx = 1$, $\displaystyle\int_0^{\frac{1}{2}} (2ax^2 - ax)dx = 1$,

$\dfrac{2a}{24} - \dfrac{a}{8} = 1$, $a = -24$

b i $-24\displaystyle\int_0^M (2x^2 - x)dx = \dfrac{1}{2}$ → $-32M^3 + 24M^2 - 1 = 0$ which

is satisfied by $M = \dfrac{1}{4}$ or by symmetry.

ii $-24\displaystyle\int_0^{Q_1} (2x^2 - x)dx = \dfrac{1}{4}$ → $-64Q_1^{3} + 48Q_1^{2} - 1 = 0$ which

is satisfied by $Q_1 = 0.16318$ (5 d.p.).

$-24\displaystyle\int_0^{Q_3} (2x^2 - x)dx = \dfrac{3}{4}$ → $-64Q_3^{3} + 48Q_3^{2} - 3 = 0$ which

is satisfied by $Q_3 = 0.33682$ (5 d.p.).

iii IQR = 0.3362 − 0.16318 = 0.1736 (4 d.p.)

5 a

b $\dfrac{1}{2} \times 2a \times a = 1$, $a = \pm 1$. $a \neq -1$ (f(x) > 0 for all x). $a = 1$

c $Q_1 = \dfrac{\sqrt{2}-2}{2}$, $Q_3 = \dfrac{-\sqrt{2}+2}{2}$ by symmetry. IQR = $2 - \sqrt{2}$

6 a $k = 20$

b Area under pdf between $y = 0$ and $y = Q_1$ is

$\dfrac{1}{20}(1 + 2Q_1 + 1) \times \dfrac{Q_1}{2} = \dfrac{1}{4}$

$Q_1^2 + Q_1 - 5 = 0$, $Q_1 = 1.79$ (3 sf). Area under pdf between

$y = 0$ and $y = M$ is

$\dfrac{1}{20}(2 + 2M) \times \dfrac{M}{2} = \dfrac{1}{2}$

$M^2 + M - 10 = 0$, $M = 2.70$ (3 sf)

c $E(Y) = \dfrac{1}{20}\displaystyle\int_0^4 y(2y + 1)dy = \dfrac{1}{20}\left(\dfrac{128}{3} + \dfrac{16}{2}\right) = \dfrac{38}{15}$

$E(Y^2) = \dfrac{1}{20}\displaystyle\int_0^4 y^2(2y + 1)dy = \dfrac{1}{20}\left(128 + \dfrac{64}{3}\right) = \dfrac{112}{15}$

$Var(Y) = \dfrac{112}{15} - \left(\dfrac{38}{15}\right)^2 = \dfrac{236}{225}$

7 a $\dfrac{a}{4}\displaystyle\int_0^5 (2x + 1)dx = \dfrac{a}{4}(25 + 5) = \dfrac{15}{2}a = 1$ → $a = \dfrac{2}{15}$

$E(X) = \dfrac{1}{30}\displaystyle\int_0^5 x(2x + 1)dx = \dfrac{1}{30}\left(\dfrac{250}{3} + \dfrac{25}{2}\right) = \dfrac{115}{36}$

$Var(X) = \dfrac{1}{30}\displaystyle\int_0^5 x^2(2x + 1)dx - \left(\dfrac{115}{36}\right)^2$

$= \dfrac{1}{30}\left(\dfrac{625}{2} + \dfrac{125}{3}\right) - \left(\dfrac{115}{36}\right)^2 = 1.60$ (2 d.p.).

b $P(X > 4) = \dfrac{1}{30}\displaystyle\int_4^5 (2x+1)\mathrm{d}x = \dfrac{1}{30}(25+5-16-4) = \dfrac{1}{3}$

P(one value above 4) $= 3 \times \dfrac{1}{3} \times \left(\dfrac{2}{3}\right)^2 = \dfrac{4}{9}$

8 $E(X^2) = \dfrac{33}{4}$

$E\left(X + \dfrac{1}{X^3}\right) = 2.86$ (2 dp)

9 a $-\dfrac{6}{125}$, 0.54 (2dp)

  b $P(X = 2) = 0$. Exact values of continuous random variables have probability zero.

  c $\dfrac{5}{2}$

10 a 4.395 (3 dp)   **b** 4.481   **c** 0.241 (3 dp)

11 a

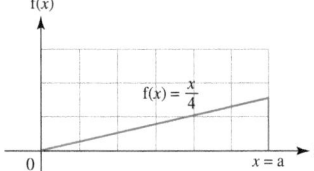

  b Since the area under the graph equals 1,

$\left(\dfrac{a}{2}\right)\dfrac{a}{4} = 1$

$a^2 = 8$

$a = \sqrt{8} = 2\sqrt{2}$ (since $a > 0$)

For the median M,

$\dfrac{M}{2}\left(\dfrac{M}{4}\right) = \dfrac{1}{2}$

$M^2 = 4$

$M = 2$ (since $M > 0$)

12 a

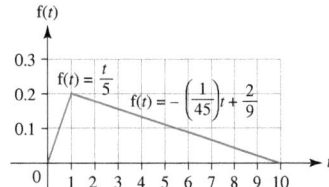

  b $f(t) \geq 0$ for all t and area under the graph is

$\left(\dfrac{1}{2}\right) \times 10 \times 0.2 = 1$

  c 1

  d 3.29 (2dp)

Exercise 10.3B

1 a $k = \dfrac{3}{28}$, $E(X) = \dfrac{15}{7}$, $Var(X) = \dfrac{156}{245}$

  b $E(Y) = -\dfrac{23}{7}$, $Var(Y) = \dfrac{624}{245}$

2 a $k = \dfrac{1}{36000}$, $P(X > 50) = \dfrac{2}{27}$

  b 50: $0.9 \times 50 + 3 = 48$. 10: $0.9 \times 10 + 3 = 12$

  c Original mean 30, by symmetry. New mean: $0.9 \times 30 + 3$ $= 30$. Zero change.

3 a $a = -\dfrac{1}{9}$, $b = 4$

  b $-\dfrac{1}{9}\displaystyle\int_0^3 x^3(x-4)\mathrm{d}x - \left(\dfrac{7}{4}\right)^2 = -\dfrac{1}{9}\left(\dfrac{3^5}{5} - 3^4\right) - \left(\dfrac{7}{4}\right)^2 = \dfrac{43}{80}$

  c 0.61

4 $f(x) = \dfrac{3x(100-x)}{500\,000}$ for $0 < x < 100$

$\dfrac{3}{500\,000}\displaystyle\int_0^M (100x - x^2)\mathrm{d}x = \dfrac{1}{2}$

$\dfrac{3}{500\,000}\left(50M^2 - \dfrac{M^3}{3}\right) = \dfrac{1}{2}$. $M = 50$.

5 a $a = \dfrac{3}{4}$

  b $\dfrac{3}{4}\displaystyle\int_0^M x(2-x)\mathrm{d}x = \dfrac{3}{4}\left[x^2 - \dfrac{x^3}{3}\right]_0^M = \dfrac{3}{4}\left(M^2 - \dfrac{M^3}{3}\right) = \dfrac{1}{2} \rightarrow$

$M^3 - 3M^2 + 2 = 0$

  c $1^3 - 3 \times 1^2 + 2 = 0$. Hence $M = 1$

6 a $\dfrac{3}{26}$

  b $E(3X + 2Y) = 6.62$ (2 dp)
     $Var(3X + 2Y) = 1.54$ (2 dp)

7 a 0.952   **b i** 5.461   **ii** 0.305 (3 dp)

8 a $a = 2^{\frac{1}{4}}$

  b $E(X) = 0.951\cdots$
     $Var(X) = 0.037\ldots$

  c $E(2X - 3) = -1.10$
     $Var(2X - 3) = 0.15$ (2dp)

9 $E(X + Y) = \mu_X + \mu_Y$
$Var(X + Y) = \sigma_X^2 + \sigma_Y^2 \rightarrow sd(X + Y) = \sqrt{(\sigma_x^2 + \sigma_y^2)}$
Assume X and Y are independent.

10 a $a = -\dfrac{1}{288}$

  b $E(X) = 6$
     $Var(X) = 7.2$

  c $T = X + Y + 2$ where Y is cycle time.
     $E(Y) = 7$, $Var(Y) = 4$
     $E(T) = E(X + Y + 2) = 6 + 7 + 2 = 15$;
     $Var(X + Y + 2) = 7.2 + 4 = 11.2$

Review exercise 10

1 $E(X) = 1 \times \dfrac{1}{6} + 3 \times \dfrac{1}{6} + \ldots + 11 \times \dfrac{1}{6} = 6$

$Var(X) = 1^2 \times \dfrac{1}{6} + 3^2 \times \dfrac{1}{6} + \ldots + 11^2 \times \dfrac{1}{6} - 6^2 = \dfrac{35}{3}$

2

c	1	2	3	4	5	6
P(C = c)	k	8k	27k	64k	125k	216k

$k = \dfrac{1}{441} \Rightarrow E[C] = 1 \times \dfrac{1}{441} + 2 \times \dfrac{8}{441} + \ldots + 6 \times \dfrac{216}{441} = 5.16$
(2 d.p.)

3 $P(X = x) = \dfrac{e^{-\mu}\mu^x}{x!}$; $\mu = 5$. $P(X < 2) = 0.04$ (2 d.p.).

4 a Let N be a random variable for the number of letters in one day. Assume random arrivals in time and therefore a Poisson distribution. $N \sim \text{Po}(18)$. $P(N \geq 12) = 0.945$. Assume independence from one day to the next. $0.945^5 = 0.75$ (2 d.p.)

b Let M be a random variable for the number of letters in five days. $M \sim \text{Po}(90)$. $P(M \geq 90) = 0.51$ (2 d.p.).

5 a $\dfrac{a}{5}\displaystyle\int_0^2 (x+3)\,dx = \dfrac{a}{5}(2+6) = \dfrac{8}{5}a = 1 \rightarrow a = \dfrac{5}{8}$

$E(X) = \dfrac{1}{8}\displaystyle\int_0^2 x(x+3)\,dx = \dfrac{1}{8}\left(\dfrac{8}{3}+6\right) = \dfrac{13}{12}$;

$\text{Var}(X) = \dfrac{1}{8}\displaystyle\int_0^2 x^2(x+3)\,dx - \left(\dfrac{13}{12}\right)^2 = \dfrac{1}{8}(4+8) - \left(\dfrac{13}{12}\right)^2$

$= 0.33$ (2 d.p.).

b $P(X>1) = \dfrac{1}{8}\displaystyle\int_1^2 (x+3)\,dx = \dfrac{1}{8}\left(2+6-\dfrac{1}{2}-3\right) = \dfrac{9}{16}$

$P(\text{one value above } 1) = 4 \times \dfrac{9}{16} \times \left(\dfrac{7}{16}\right)^3 = 0.19$ (2 d.p.)

6 a $\dfrac{2}{3}$

b $E(S+2T) = \dfrac{23}{9}$

$\text{VarE}(S + 2T) = 0.28$ (2 dp)

7 a 11, 33   **b** $\dfrac{9}{2}, \dfrac{33}{4}$

Assessment 10

1 a $k = 10$
$b = 0.4$
$a = 0.15$

b $E(X) = 1 \times \dfrac{1}{10} + 2 \times \dfrac{2}{10} + 3 \times \dfrac{3}{10} + 4 \times \dfrac{4}{10}$
$= 3$
$E(Y) = 1 \times 0.3 + 2 \times 0.15 + 3 \times 0.15 + 4 \times 0.4$
$= 2.65$
So X is higher on average

2 a $k = \dfrac{1}{20}$

b Median is 2

Mode is 3

c $E(N) = 1 \times \dfrac{5}{20} + 2 \times \dfrac{6}{20} + 3 \times \dfrac{7}{20} + 4 \times \dfrac{1}{20} + 5 \times \dfrac{1}{20}$
$= \dfrac{47}{20}$

d $\text{Var}(N) = \dfrac{133}{20} - \left(\dfrac{47}{20}\right)^2$
$= \dfrac{451}{400}$ or 1.1275

3 a i 0.9161   **ii** 0.224   **iii** 0.1728   **iv** 0.4232
b 3
4 a Assume apples are sold individually and at a constant rate.
b i 0.0821   **ii** 0.5438   **iii** 0.1088
c 0.1378

5 a $E(X^2) = 25$ **b** $\sqrt{2.5}$

6 a $k = \dfrac{1}{18}$

b i $P(X<3) = \left[\dfrac{1}{36}x^2\right]_0^3$
$= \dfrac{1}{36}(9-0)$
$= \dfrac{1}{4}$

ii $P(X>1) = \left[\dfrac{1}{36}x^2\right]_1^6$
$= \dfrac{1}{36}(36-1)$
$= \dfrac{35}{36}$

iii $P(1.5<X<5) = \left[\dfrac{1}{36}x^2\right]_{1.5}^5$
$= \dfrac{91}{144}$

c 4
d 2

7 a

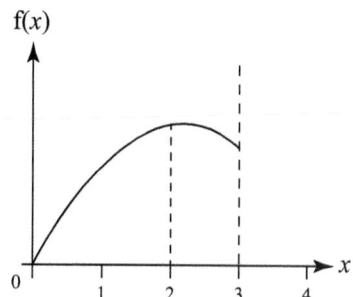

b 2
c 1.79

8 a

t	11	21	12	22	13	23
$P(T=t)$	0.2	0.1	0.15	0.05	0.3	0.2

b $E(T) = 15.7$
c $E(T) = E(X) + E(Y)$ for any discrete random variables X and Y where $T = X + Y$
but $\text{Var}(T) = \text{Var}(X) + \text{Var}(Y)$ only if X and Y are independent.
In this case, X and Y are not independent, we know this since, e.g.
$P(X=2) = 0.2$, $P(Y=10) = 0.6$,
$P(X=2 \text{ and } Y=10) = 0.15$
$P(X=2) \times P(Y=10) = 0.2 \times 0.6 = 0.12 \neq 0.15$

9 a i 10   **ii** 10
b 0.6672   **c** 0.0286

10 a $E(U) = 4.5$
$\text{Var}(U) = \dfrac{1}{12}(6^2 - 1)$
$= \dfrac{35}{12}$ or 2.92

b $P(X>3.5) = P(X \geq 4)$
$= \dfrac{2}{3}$

11 a 3

b $E(5V+3) = 5E(V)+3$

$$= 5\left(\frac{10}{3}\right)+3$$

$$= \frac{59}{3} \text{ or } 19.7$$

$$\text{Var}(5V+3) = 5^2\text{Var}(V)$$

$$= 25\left(\frac{31}{45}\right)$$

$$= \frac{155}{9} \text{ or } 17.2$$

12 a $a > 7$ or $a \geq 8$

b $b > 8$ or $b \geq 9$

c 0.6078

d i 0.6620   **ii** 0.8641   **iii** 0.4126

13 a $\dfrac{8}{75}$ **b** 0.45

c $E(W) = 4E(Y)-3$

$$= 4\left(\frac{8}{5}\right)-3$$

$$= \frac{17}{5}$$

$$\text{Var}(W) = 16\text{Var}(Y)$$

$$= 16\left(\frac{8}{75}\right)$$

$$= \frac{128}{75} \text{ or } 1.71$$

14 a $P(T > 3.92) = P(T=4)+P(T=5)$

$$= \frac{4}{12}+\frac{5}{12}$$

$$= \frac{3}{4}$$

b $\text{Var}(7-3T) = 9\text{Var}(T)$

$$= \frac{227}{16} \text{ or } 14.2$$

15 a $E(X) = \dfrac{n+1}{2}$

b $E(X^2) = \dfrac{1}{n}+\dfrac{4}{n}+\dfrac{9}{n}+\ldots+\dfrac{n^2}{n}$

$$= \frac{1}{n}\left(1+4+9+\ldots+n^2\right)$$

$$= \frac{1}{n}\left(\frac{n}{6}(n+1)(2n+1)\right)$$

$$= \frac{1}{6}(n+1)(2n+1)$$

$$\text{Var}(X) = \frac{1}{6}(n+1)(2n+1)-\left(\frac{n+1}{2}\right)^2$$

$$= \frac{1}{6}(n+1)(2n+1)-\frac{1}{4}(n+1)^2$$

$$= \frac{1}{24}(n+1)\left[4(2n+1)-6(n+1)\right]$$

$$= \frac{1}{24}(n+1)(2n-2)$$

$$= \frac{1}{12}(n+1)(n-1)$$

$$= \frac{1}{12}(n^2-1) \text{ as required}$$

16 a 0.0119   **b** 0.401   **c** 0.0368

17 a $a = \dfrac{1}{8}$, $b = \dfrac{3}{8}$

b $m = -3+\sqrt{17}$ (1.12)

c $E(Y) = \dfrac{13}{12}A+2$

d 12

18 a 1.5   **b** 3.89

19 a $\displaystyle\int_0^3 k(9-x^2)dx = 1$

$$k\left[9x-\frac{x^3}{3}\right]_0^3 = 1$$

$$k(27-9) = 1$$

$$18k = 1$$

$$k = \frac{1}{18}$$

b $= \dfrac{117}{40}$ (2.925)

20 $= \dfrac{7}{4}+\dfrac{2}{3}\ln 2$ (2.21)

21 a $k = \dfrac{2}{\sqrt{3}}$

b $E(\sqrt{3}R+2) = \sqrt{3}\dfrac{2}{\sqrt{3}}\left(\dfrac{\sqrt{3}}{3}\pi-\dfrac{\pi^2}{36}-1\right)+2$

$$= 2\left(\frac{\sqrt{3}}{3}\pi-\frac{\pi^2}{36}-1\right)+2$$

$$= \frac{2\sqrt{3}}{3}\pi-\frac{\pi^2}{18}$$

22 a $\displaystyle\int_1^3 \frac{1}{8}\left(s^2-\frac{1}{s^2}\right)ds = \frac{1}{8}\left[\frac{s^3}{3}+\frac{1}{s}\right]_1^3$

$$= \frac{1}{8}\left(\left(9+\frac{1}{3}\right)-\left(\frac{1}{3}+1\right)\right)$$

$$= \frac{1}{8}\left(\frac{28}{3}-\frac{4}{3}\right) = 1 \text{ as required}$$

$$A(1^2+2^2+3^2+4^2) = 1$$

$$A = \frac{1}{30}$$

b 28

Chapter 11

Exercise 11.1A

1 a $H_0: \lambda = 5.31$, $H_1: \lambda < 5.31$

b 53 as $P(X \leq 53) = 0.0976$ and $P(X \leq 54) = 0.1224$

2 a The critical values are 21 and 41

b $3 \times 9 = 27$ is between the critical values so the null hypothesis is accepted.

3 a The test is to see if the parameter has changed, but we don't know if we expect it to be higher or lower.

b The statistic is larger than the upper critical value, so the result is significant. There is sufficient evidence to suggest that the parameter is not 72.

c Actually the parameter is still 72, but an extreme result has incorrectly led us to reject that conclusion.

4 a The parameter is suspected to have decreased, so the alternative hypothesis takes the form given.

b The p-value is larger than the significance level, so the result is not significant. The null hypothesis is not rejected and you conclude that there is insufficient evidence, at the 5% level, to suggest that the parameter is less than 12.5.

c Actually the parameter has decreased, but the result was not extreme enough to lead us to that conclusion.

5 a $H_0 : \lambda = 8.3$, $H_1 : \lambda \neq 8.3$

b 0.0885

c Since the p-value is larger than half the significance level (2-tailed test) there is insufficient evidence to reject the null hypothesis.

6 a 163

b 0.04505

c Since the p-value is less than the significance level there is sufficient evidence to reject the null hypothesis.

7 a 72 as $P(X \geq 72) = 0.0482$ and $P(X \geq 71) = 0.0617$

b 0.0482

8 a 0.0568

b Since the p-value is larger than the significance level there is insufficient evidence to reject the null hypothesis.

c 0.0469 (as 107 is the critical value)

9 a A false positive can be treated by re-testing, which hopefully will avoid error and show the person as disease-free.

b A false negative means a person with the disease would go untreated for longer and potentially suffer as a result.

10 a Since re-designs are costly, they wouldn't want to repeatedly go through the process unnecessarily.

b If they are convinced that the dice are fair from their testing then probably they are close enough to being fair that people can't tell the difference even when they aren't fair.

11 a and b Students' own answers and reasons.

Exercise 11.1B

1 a 34 and 63 as $P(X \leq 34) = 0.0243$, $P(X \leq 35) = 0.0350$, $P(X \geq 63) = 0.0188$, and $P(X \geq 62) = 0.0256$.

b The result is outside of the critical region so there is insufficient evidence to reject the null hypothesis. You conclude that the average rate of phone calls to the call centre is reasonably estimated at 6.8 per 10 minutes.

c A type I error is when a true null hypothesis is falsely rejected. It occurs when a result in the critical region is obtained.

d The probability is $P(X \leq 34) + P(X \geq 62) = 0.0429$ (to 3 s.f.)

2 a The critical value is 2122 points.

b As $2098 < 2122$ there is insufficient evidence to reject the null hypothesis. You conclude that the average number of points scored per game has not increased.

3 a $H_0 : \lambda = 41$, $H_1 : \lambda < 41$

b $P(X \leq 38) = 35.6\% > 10\%$. There is insufficient evidence to suggest that the number of accidents has been reduced.

c It would mean that you believe the rate of accidents has not reduced when actually it has.

4 a 28

b $H_0 : \lambda = 28$, $H_1 : \lambda > 28$.

c $P(X \geq 37) = 5.89\% < 10\%$. There is sufficient evidence to suggest that the number of goals on the final day was exceptionally large.

d It would mean that the rate of goal scoring on the final day was not exceptionally large despite our belief that it

was.

5 a $P(X \leq 5) = 8.38\%$ (3s.f.)

b Since the p-value is larger than the significance level, the result is not significant. There is insufficient evidence to suggest that there are fewer accidents per month at the sports centre.

6 a 0.0420

b The p-value is smaller than the significance level. There is sufficient evidence to reject the null hypothesis. You conclude that the average rate of customers coming into the book shop has increased.

7 a 0.117

b The p-value is larger than half the significance level (2-tailed test). There is insufficient evidence to reject the null hypothesis.

8 a 3 and 13 as $P(X \leq 3) = P(X \geq 13) = 0.0106$ and $P(X \leq 4) = P(X \geq 12) = 0.0384$

b $0.0213 = P(X \leq 3) + P(X \geq 13)$ (3s.f.)

9 a i The critical region is 0, 1, 2, 9, 10, 11, 12. $\alpha = 0.0777$

ii The critical region is 0, 1, 10, 11, 12 $\alpha = 0.0162$

iii The critical region is 0, 11, 12 $\alpha = 0.00185$

b i $\beta = 0.772$ **ii** $\beta = 0.916$ **iii** $\beta = 0.980$

Exercise 11.2A

1 a

Expected	A	B	C
X	53.0	85.4	15.6
Y	17.9	28.8	5.3
Z	68.1	109.8	20.1

b

X^2	A	B	C
X	0.30	0.51	7.23
Y	6.63	0.35	11.3
Z	0.69	0.87	0.42

X^2 Total	28.3

c There are 4 degrees of freedom, so the critical value is 9.49.

d The p-value is 0.00001

e The test statistic is larger than the critical value (or the p-value is less than the significance level), so there is sufficient evidence to reject the null hypothesis.

2 a

Expected	A	B	C	D
W	12.4	12.1	14.9	15.6
X	10.3	10.1	12.4	13.1
Y	13.0	12.8	15.7	16.5
Z	13.3	13.0	16.0	16.8

b

X^2	A	B	C	D
W	1.07	0.07	0.08	2.04
X	0.53	0.81	1.59	1.17
Y	1.20	2.61	0.11	0.02
Z	2.09	0.31	0.26	0.09

X^2 Total	14.04275

c There are 9 degrees of freedom, so the critical value is 16.9.

d The p-value is 0.120839

e The test statistic is smaller than the critical value (or the p-value is larger than the significance level), so there is insufficient evidence to reject the null hypothesis.

3 a

Expected	A	B	C
X	74.4	68.5	78.1
Y	65.6	60.5	68.9

b

X^2	A	B	C
X	0.29	1.94	0.61
Y	0.33	2.20	0.69

X^2 Total	6.055767

c There are 2 degrees of freedom so the critical value is 5.99.

d The p-value is 0.04842

e The test statistic is larger than the critical value (or the p-value is less than the significance level), so there is sufficient evidence to reject the null hypothesis.

4 a

Expected	A	B
X	80.6	113.4
Y	39.4	55.6

b

X^2	A	B
X	1.25	0.89
Y	2.56	1.82

Yates:	6.52749

c There is 1 degree of freedom so the critical value is 3.84.

d The p-value is 0.010623

e The test statistic is larger than the critical value (or the p-value is less than the significance level), so there is sufficient evidence to reject the null hypothesis.

5 a

E_i	0–1	2–3	4+
Brown	8.8	12.7	13.5
White	14.2	20.3	21.5

b

X^2	0–1	2–3	4+
Brown	0.916	0.007	0.479
White	0.57	0.005	0.30

The total X^2 statistic is 2.27802

c There are $(3 - 1) \times (2 - 1) = 2$ degrees of freedom so the p-value of the test statistic is 0.320139

d The p-value is larger than the significance level so there is insufficient evidence to reject the null hypothesis.

6 a

E_i	Chicken	BLT	Ham & Cheese	Egg Mayo
Orange	26.9	28.4	20.4	8.4
Apple	23.4	24.6	17.7	7.3
Water	23.7	25.0	17.9	7.4

b

X^2	Chicken	BLT	Ham & Cheese	Egg Mayo
Orange	0.622	0.005	0.091	0.668
Apple	0.822	0.768	0.027	0.074
Water	0.004	0.636	0.237	0.362

The total statistic X^2 is 4.31525

c The statistic is less than the critical value so there is insufficient evidence to reject the null hypothesis.

7 a

E_i	Hamburger	Cheeseburger	Pizza	Hot Dog
Class A	4.6	4.6	7.6	6.2
Class B	4.8	4.8	7.9	6.5
Class C	4.6	4.6	7.6	6.2

b Group hamburger and cheeseburger data together. 4 degrees of freedom.

c The critical value is 9.49, so the result is not significant. There is insufficient evidence to say there is any association between which class a student is in and their food preference.

8 a

Expected	Chocolate Bars	Apples	Packets of Crisps
Frieda	10.7	4.0	13.3
Helga	11.1	4.2	13.8
Hans	15.3	5.8	19.0

b They are twins and so it isn't unreasonable to suggest that they might have similar preferences.

c There are $(3 - 1) \times (2 - 1) = 2$ degrees of freedom so the critical value is 5.99. The result is significant; there is sufficient evidence to suggest there is association between the person and the snacks they eat.

Exercise 11.2B

1 a

Expected	Blue Eyes	Green Eyes	Brown Eyes
Blue	79.1	47.9	47.0
Red	43.6	26.4	25.9
Yellow	42.3	25.6	25.1

X^2	Blue Eyes	Green Eyes	Brown Eyes
Blue	5.01	2.49	1.71
Red	2.13	0.25	1.94
Yellow	2.50	2.74	0.143

The total X^2 statistic is 18.9. For 4 degrees of freedom at the 5% significance level the critical value is 9.49. The test statistic is larger than the critical value so the result is significant at the 5% level. There is sufficient evidence to reject the null hypothesis.

b The largest contribution is from people with blue eyes who say blue is their favourite colour, significantly more than expected. The number of people with green eyes whose favourite colour is red and those with brown eyes whose favourite colour is yellow is very close to what is expected.

2 a

Expected	Axe	Sword	Bow	Spear
Celts	8.6	26.1	23.7	34.6
Vikings	7.1	21.6	19.7	28.7
Huns	6.4	19.3	17.6	25.7

X^2	Axe	Sword	Bow	Spear
Celts	0.242	0.329	0.588	0.0125
Vikings	3.40	0.270	4.741	0.189
Huns	6.351	1.476	10.179	0.112

The total X^2 statistic is 27.9. For 6 degrees of freedom at the 5% significance level the critical value is 12.6. The test statistic is larger than the critical value so the result is significant at the 5% level. There is sufficient evidence to reject the null hypothesis.

b The association suggests that people who play the different tribes choose their weapons accordingly and not independently. The largest contributions are from people who play as Huns, using far fewer Axes than expected and far more bows than expected. People who play as Vikings use more Axes than expected and fewer Bows than expected. Spears are the most popular weapon but they occur around the expected frequency for each tribe.

3 a

Expected	Grey	Brown
Beach	83.7	56.3
River	99.3	66.7

X^2	Grey	Brown
Beach	0.62	0.93
River	0.53	0.78

Using Yates' correction the total X^2 statistic is 2.86. At the 5% significance level with one degree of freedom the critical value is 3.84. The statistic is lower than the critical value so there is insufficient evidence to reject the null hypothesis.

b Although in the observed data there are more brown stones on the beach than by the river, the numbers are not dissimilar enough from the expected values to suggest there is any association.

4 a

Expected	Pizza	Pasta
Soup	97.4	110.6
Salad	93.6	106.4

X^2	Pizza	Pasta
Soup	1.001	0.881
Salad	1.041	0.916

With Yates' correction the test statistic is 3.839. The critical value with 1 degree of freedom at the 10% level is 2.706. The test statistic is larger than the critical value so there is sufficient evidence to reject the null hypothesis. You conclude that there is association between choice of starter and main meal.

b No individual contribution is very large or small but together they suggest association. Pasta + Soup and Pizza + Salad go together slightly more than you would expect, and the other pairs go together slightly less than you would expect.

5 a

Expected	Red	Blue	Yellow
Boys	67.5	76.1	59.4
Girls	65.5	73.9	57.6

X^2	Red	Blue	Yellow
Boys	0.834	0.345	0.095
Girls	0.859	0.356	0.098

The total X^2 test statistic is 2.59. The critical value with 2 degrees of freedom at the 1% level is 9.21. The test statistic is less than the critical value so there is insufficient evidence to reject the null hypothesis. You conclude that there is no association between child gender and monster colour preference.

b The numbers of boys and girls who prefer the yellow monster match the expected numbers well. There are no large contributions to the test statistic.

6 a

Observed	Dark	Colourful	Light	Totals
Black	40	32	24	96
Brown	66	42	12	120
Blonde	34	34	34	102
Totals	140	108	70	318

b

Expected	Dark	Colourful	Light
Black	42	32.6	21.1
Brown	52.8	40.8	26.4
Blonde	44.9	34.6	22.5

X^2	Dark	Colourful	Light
Black	0.121	0.011	0.389
Brown	3.28	0.038	7.87
Blonde	2.65	0.01	5.9

There are 4 degrees of freedom. The critical value at the 5% level is 9.49. The test statistic is 20.3. This is larger than the critical value so the result is significant. There is sufficient evidence to suggest that there is association between hair colour and choice of clothing colour.

c People with black hair have broadly typical choices of clothing. Light-coloured clothes are particularly favoured by those with blonde hair and not favoured by those with brown hair. The reverse is true for dark-coloured clothes.

7 a

Observed	0–59	60–69	70–79	80+	Totals
London	30	14	10	1	55
Midlands	17	16	13	2	48
North-East	5	12	7	2	26
North-West	30	21	10	0	61
South	8	2	0	0	10
Totals	90	65	40	5	200

b The expected frequencies for 80+ are below 5 and the most-similar grouping is with 70–79.

c The South should be excluded as the expected frequencies are below 5. It is not clear which other region is similar enough to group them with.

d 6

e The North East teams earned below 60 yellow cards much less often than expected. The Midlands teams earned between 60 and 69 yellow cards almost exactly as often as expected.

8 Check correct from Ex 11.2A Q4 and Ex 11.2B Qs 3 and 4

Exercise 11.3A

1 **a** $90.43 \le \mu \le 92.37$

  b $0.09903 \le \mu \le 0.144497$

  c $10.53 \le \mu \le 14.87$

2 **a** $31.95 \le \mu \le 51.05$

  b $-10.19 < \mu < -1.33$

3 **a** Since the population variance is known, the sample variance does not need to be used to estimate it.

  b 95%: $55.87 \le \mu \le 69.13$ 99%: $53.79 \le \mu \le 71.21$

4 **a** 187.25cm **b** $183.0 \le \mu \le 191.5$

5 $208.5 \le \mu \le 215.8$

6 **a** 0.939

  b $2655 \le \mu \le 4745$

7 **a** 2.5%

  b 2.5% $213.12 \le \mu \le 420.88$

8 Need $25 \ge z \times \dfrac{\sigma}{\sqrt{n}}$, so $n \ge \left(\dfrac{98z}{25}\right)^2$.

  a 102 **b** 60 **c** 42

Exercise 11.3B

1 **a** $58.67 \le \mu \le 63.93$, $57.28 \le \mu \le 63.72$

  b Size = 90, mean = 60.95. $58.91 \le \mu \le 62.99$

  c The sample is larger and there is no reason to believe the two samples are significantly different. The combined sample should give better estimates for the true mean.

2 **a** $1063.80 \le \mu \le 1078.20$ from 432, and $1055.88 \le \mu \le 1086.12$ from 1905.

  b The sample variance is very different from the population variance but the sample size is large enough that the sample variance should be used. The confidence intervals themselves aren't too dissimilar and if it was really important to tell the difference then likely a larger sample should be taken.

3 **a** mean estimate is −0.05

  b standard deviation estimate is 0.119

  c $-0.0799 \le \mu \le -0.020$

4 **a** mean estimate is 82.8, standard deviation estimate is 11.7

  b Expected frequencies are 4.5, 13.7, 23.6, 20.2, 8.7 and 1.4, which are very close to the recorded values. The assumption of a normal distribution seems justified.

  c $80.1 \le \mu \le 85.5$

5 **a** estimate of population variance is 35.41

  b Although 95% of all intervals generated in this way are expected to contain the true mean value, it is impossible to say for certain what the probability is that a generated interval contains the true mean.

Review exercise 11

1 **a** The null hypothesis is $H_0: \lambda = 1.9$ and the alternative hypothesis is $H_1: \lambda = 1.9$

  b $P(X \ge 48) = 0.0332$ (3s.f.).

  c The p-value is less than the significance level so there is $H_0: \lambda = 1.9$ sufficient evidence to reject the null hypothesis.

2 **a** The critical value is 13 as $P(X \le 13) = 3.11\%$ and $P(X \le 14) = 5.29\%$ (both 3s.f.).

  b The probability of making a type I error is 0.0311 (3s.f.) as that is the probability of obtaining a result in the critical region if the parameter actually is 0.75.

3 **a** The critical value is 5 as $P(X \le 5) = 5.34\%$ and $P(X \le 14) = 5.29\%$ (both 3s.f.).

  b 7 is outside the critical region so there is insufficient evidence to reject the null hypothesis. You conclude that there are not fewer accidents per month than previously, and that 1.3 is a reasonable estimate of the rate of accidents per month.

4 **a**

E_i	A	B	C
X	100.0	68.7	40.3
Y	105.7	72.6	42.6
Z	104.3	71.7	42.1

  b

X^2	A	B	C
X	0.091	0.077	0.701
Y	0.895	1.775	0.062
Z	0.432	2.603	1.148

The X^2 test statistic is 7.78 (3s.f.).

5 **a** For a 5×3 table there are $4 \times 2 = 8$ degrees of freedom.

  b The critical value at the 5% significance level is 15.5 (3s.f.).

6 The p-value is 0.0697 (3s.f.). This is larger than the significance level so there is insufficient evidence to reject the null hypothesis. You conclude that there is no association between height and birthplace of UK adults.

7 **a**

E_i	A	B	C	D
Red	86.4	85.9	67.2	84.5
Blue	67.7	67.3	52.7	66.2
Silver	30.9	30.8	24.1	30.3

  b

X^2	A	B	C	D
Red	2.151	0.042	0.777	0.240
Blue	2.385	0.166	0.350	2.086
Silver	0.028	0.894	0.357	1.739

The X^2 statistic is 11.22.

  c The test statistic is larger than the critical value so there is sufficient evidence to reject the null hypothesis. You conclude that there is association between the public's preference for the four car models and their choice of colour. For model A there is positive association with Red and negative association with Blue. For model D there is positive association with Blue and negative association with Silver. Models B and C appear to have colour preferences in roughly the expected proportion.

8 **a** $14.7 - \phi^{-1}(0.975) \times \dfrac{\sqrt{29.5}}{\sqrt{36}} \le \mu \le 14.7 + \phi^{-1}(0.975) \times \dfrac{\sqrt{29.5}}{\sqrt{36}}$, $12.93 \le \mu \le 16.47$

  b The test statistic is $\dfrac{14.7 - 16.5}{\sqrt{\dfrac{29.5}{36}}} = -1.988$. $\phi(-1.988) = 0.0234$

This is smaller than the significance level so the result is significant. There is sufficient evidence to suggest that the population mean is not 16.5

  c The size of the test is the probability of making a type I error, which for a continuous distribution is simply the significance level, 5%.

9 **a** $1.2282 < \mu < 6.7718$

  b $4.9066 < \mu < 4.9623$

10 The X^2 statistic using Yates' correction is 12.96. The contributions are shown in this table.

X^2	A	B
X	3.15	4.41
Y	2.25	3.15

Assessment 11

1 a $H_0: \lambda = 2.7$
$H_1: \lambda > 2.7$
b 0.057. No reason to reject the null hypothesis that the mean rate is 2.7

2 a $H_0: \lambda = 6.4$ $H_1: \lambda < 6.4$
b $P(X \le 2) = 0.046$ (3dp)
For a 5% one-tailed test, reject the null hypothesis; evidence suggests that the number of flaws has been reduced.
c Rejecting the null hypothesis when it is true.
$P(\text{Type I error}) = 0.046$ Rejecting the null hypothesis when it is true. $1 - 0.046 = 0.954$

3 a For a 5% one-tailed test, reject the null hypothesis; evidence suggests that the number of calls has increased.
b 0.047. This is the probability that the null hypothesis is rejected given that it is true.

4 a The Poisson distribution models the occurrence of random events in time; road traffic accidents can be modelled as random events.
b $H_0: \lambda = 3.2$ $H_1: \lambda < 3.2$
Under H_0, $P(X \le 2) = 0.380$ (3dp)
For a 5% one-tailed test, no reason to reject the null hypothesis; no evidence to suggest that the number of accidents has been reduced.
c Only zero accidents would result in a conclusion that the rate of accidents had been reduced.

5 a

		Pastime					
		Reading	Sport	Online	Television	Other	Total
Gender	Male	7.67	17.11	17.7	7.67	8.85	59
	Female	5.33	11.89	12.3	5.33	6.15	41
	Total	13	29	30	13	15	100

b Reject null hypothesis; evidence suggests that there is an association between pastime and gender.

6 a

		Result			
		Fail	Pass	Distinction	Total
Exam	A	7.7	17.6	7.7	33
	B	6.3	14.4	6.3	27
	Total	14	32	14	60

b No reason to reject teacher's belief.

7 There is no reason to reject the hypothesis that weekdays and weekends result in the same proportion of the different errors.

8 There is no reason to reject the hypothesis that there is no association between grade of oil and price.

9 a $H_0: p = 0.3$; $H_1: p > 0.3$
b Under H_0

x	6	7	8	9
$P(X \ge x)$	0.1179	0.0386	0.0095	0.0017

c $x = 7, 8, 9, 10, 11, 12$
d A type I error occurs when the null hypothesis is rejected given that it is true.
0.0386

10 a $4.4 \pm 1.96 \dfrac{0.6}{\sqrt{15}} = (4.10, 4.70)$ (2 d.p.)
b No reason to reject the hypothesis as 4.6 is within the interval (4.10, 4.70).

11 a Large sample, which suggests sample variance close to population variance.
b $14 \pm 1.645 \sqrt{\dfrac{36}{60}} = (12.73, 15.27)$ (2 d.p.)

Chapter 12

Exercise 12.1A

1 (these are examples – others may be possible)
a
b
c
d

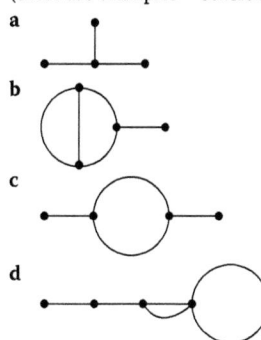

2

	V	E	F
a	5	7	4
b	8	12	6
c	4	9	7

a

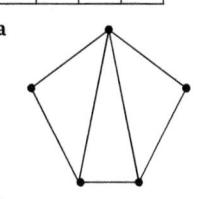

b

c

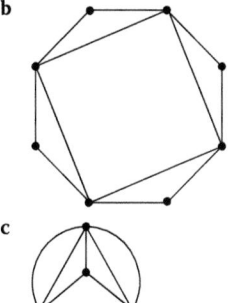

3 a Graph B is the odd one out.
b For graph A: {A, B, C, D, E, F}, {AD, DF, FC, FE, EB} other sets are possible
For graph C: {A, B, C, D, E, F}, {AD, DF, FE, EB, BC, CA} other sets are possible

4 Graph 1, the extra vertices are A, E and F.

5

	A	B	C	D	E
A	0	1	1	1	1
B	1	0	1	0	1
C	1	1	0	2	0
D	1	0	2	0	1
E	1	1	0	1	0

6 a

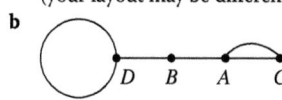

(your layout may be different)
b

(your layout may be different)

7 a

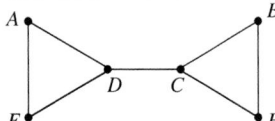

	A	B	C	D	E	F
A	0	1	1	0	1	0
B	1	0	0	1	0	1
C	1	0	0	0	0	1
D	0	1	0	0	1	0
E	1	0	0	1	0	1
F	0	1	1	0	1	0

b

	A	B	C	D	E	F
A	0	0	0	1	0	1
B	0	0	1	0	1	0
C	0	1	0	1	1	0
D	1	0	1	0	0	1
E	0	1	1	0	0	0
F	1	0	0	1	0	0

8 a

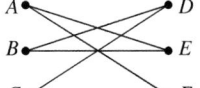

b

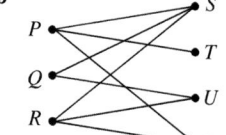

9

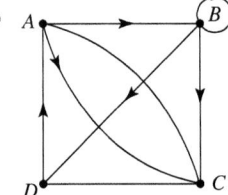

10

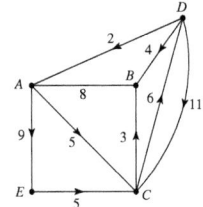

Exercise 12.1B

1 a

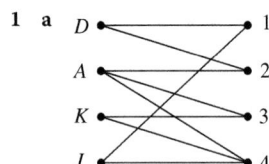

b $D1\ A2\ K3\ J4$ or $D2\ A3\ K4\ J1$ or $D2\ A4\ K3\ J1$

2 a

	P	E	F	G	A
P	–	7	22	–	–
E	7	–	–	8	15
F	22	–	–	20	–
G	–	8	20	–	4
A	–	15	–	4	–

b $EG + GA < EA$

c

	P	E	F	G	A
P	–	7	22	15	19
E	7	–	28	8	12
F	22	28	–	20	24
G	15	8	20	–	4
A	19	12	24	4	–

d

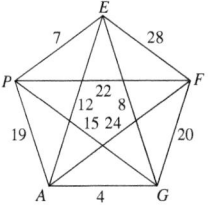

3 a $n-1$

b Sum of degrees = twice number of edges

Number of edges $= \dfrac{1}{2}n(n-1)$

4 a

	A	B	C	D	E
A	0	1	1	1	1
B	1	0	1	0	1
C	1	1	0	1	1
D	1	0	1	0	0
E	1	1	1	0	0

b

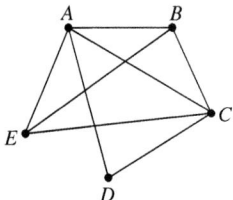

c

	A	B	C	D	E
A	0	0	0	0	0
B	0	0	0	1	0
C	0	0	0	0	0
D	0	1	0	0	1
E	0	0	0	1	0

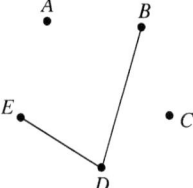

This graph shows which members of the group do not know each other.'

d Yes, $\{A, B, C, E\}$, these four all know each other.

5 a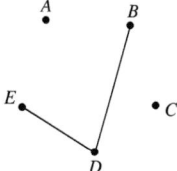

b 3

c Must have *fcd* (the only triple containing *d*), *afe* (the only one with *e*) and *abc* (the only one with *b*).

Exercise 12.2A

1 a i *ADFC*
  ii *ABEDFC* (these are just examples)
  b i *BACB*
  ii *BADFEB* (these are just examples)
  c For example *BADEFCB*
  d It visits *C* twice.

2 Graphs **a**, **d** and **e**

3 a neither **b** Eulerian **c** semi–Eulerian
  d Eulerian **e** Eulerian **f** neither

4 (These are just examples)

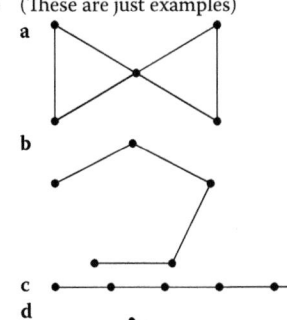

5 a 10

b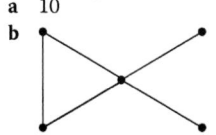
  Semi–Eulerian

c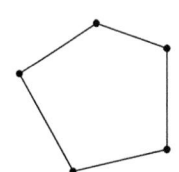

Exercise 12.2B

1 a

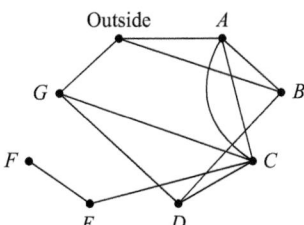

b non–Eulerian because 6 vertices of odd order

2 For example

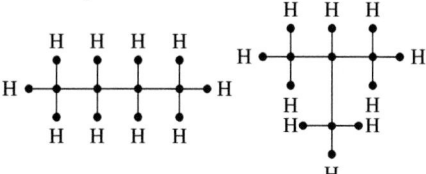

10 hydrogen in each case.

3 a Two odd order vertices – *B* and *D*
  b *BD*
  c For example *ABFCAEGFEDBCDA*

4

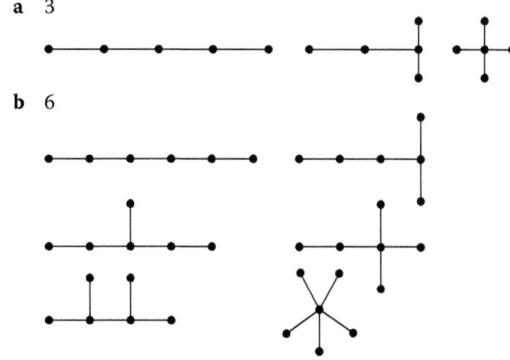

All vertices odd, so non–Eulerian.

5 a 3

b 6

6 K_n is traversable if and only if *n* is odd.

7 $K_{m,n}$ is Hamiltonian if and only if *m* = *n*

Exercise 12.3A

1 a *AC*(2), *AD*(2), *DF*(3), *BE*(4), *BD*(5) or *AD*(2), *AC*(2), *DF*(3), *BE*(4), *BD*(5)
  Total 16.

  b *CD* (4), *AB* (5), *BD* (6), *BE* (7) or *CD* (4), *AB* (5), *BD* (6), *CE* (7)
  Total 22

  c *BC* (26), *DE* (29), *AB* (37), *EF* (38), *GI* (40), *HI* (40), *AH* or *BH* (42), *FI* (55) or *BC* (26), *DE* (29), *AB* (37), *EF* (38), *HI* (40), *GI* (40), *AH* or *BH* (42), *FI* (55)
  Total 307

  d *BD*/*CD* (2), *CD*/*BD* (2), *FG* (3), *EG*/*DF* (4) *DF*/*EG* (4), *AF* (5)
  Total 20

  e *DE* (15), *HG*/*HI* (17), *HI*/*HG* (17), *AH* (18), *CI*(24), *AB* (25), *CD* (26), *EF* (36)
  Total 178

2 *AB*/*BD* (2), *BD*/*AB* (2), *AC* (3) or *AB*/*BD* (2), *BD*/*AB* (2), *BC* (3)
  Total 7

3 There are 4 trees. The order of choosing them is
  BD/*CD*(3), *CD*/*BD*(3), *AB*(4), *CE*(5)
  BD/*CD*(3), *CD*/*BD*(3), *AB*(4), *ED*(5)
  BD/*CD*(3), *CD*/*BD*(3), *AC*(4), *CE*(5)
  BD/*CD*(3), *CD*/*BD*(3), *AC*(4), *ED*(5)

4 a

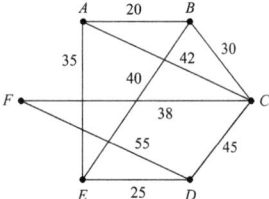

b AB (20), DE (25), BC (30), AE (35), CF (38)
(Total 148)

c

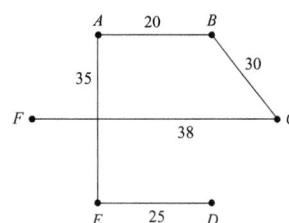

Exercise 12.3B

1

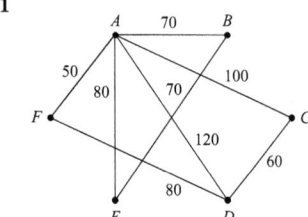

AF (50), CD (60), AB/BE (70), BE/AB (70), DF (80)
Total 330

2 a BG/EG (3), EG/BG (3), AC/BD (4), BD/AC (4), CD/DF (5), DF/CD (5)
Total 24 miles

b CD

3 a From a sketch of the network: IS (20), ES/IF (22), IF/ES (22), PS (23), GE (30)

b Starting with German: GE (30), ES (22), SP (23), SI (20), IF (22)
Total cost £1170

c GE (30), GF (35), GI (35), GP (40), GS (32)
Total cost £1720
May be better because successive translations can gradually change the meaning of the text. May also be quicker because translators can work at the same time.

4 a and b

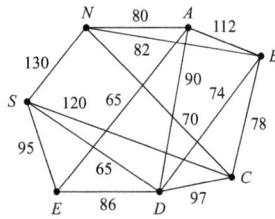

Exercise 12.4A

1 a AB (5), BD (6), CD (4), BE (7)
Total 22

b AB (37), BC (26), AH/BH (42), HI (40), GI (40), FI (55), EF (38), DE (29)
Total 307

c AF (5), FG (3), DF (4), CD/BD (2), BD/CD (2), EG (4)
Total 20

d AH (18), HG/HI (17), HI/HG (17), CI (24), AB (25), CD (26), DE (15), EF (36)
Total 178

2

	1	3	5	4	2
	A	B	C	D	E
A	–	5	9	6	3
B	(5)	–	11	4	7
C	9	11	–	(5)	8
D	6	(4)	5	–	12
E	(3)	7	8	12	–

AE (3), AB (5), BD (4), CD (5)
Total 17

3 a Assume start from P (but you could start from anywhere, which would change the labeling of the columns).
Circle, in order, PQ (2), RQ (1), RS (4), ST (3), TU (4) total 14, with columns labelled 1, 2, 3, 4, 5, 6 **or** circle PQ (2), RQ (1), QT (4), ST (3), UT (4) with columns labelled 1, 2, 3, 4, 5, 6

b

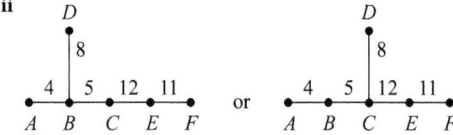

4 Circle, in order, AD (8), DG (6), AB (9), BC (6), CF (8), ED (10)
Total 47 with columns labelled 1, 4, 5, 2, 7, 6, 3

Exercise 12.4B

1 Circle, in order, AC (20), CH (30), AB (56), BD (15), DG (25), EG (40), EF (30)
Total 216, with columns labelled 1, 4, 2, 5, 7, 8, 6, 3

2 a Because of the difficulty of checking for cycles

b i Circle, in order, AB (4), BC (5), BD (8), CE (12), EF (11) total 40 with columns labelled 1, 2, 3, 4, 5, 6
or
circle AB (4), BC (5), CD (8), CE (12), EF (11) with same labelling and total

ii

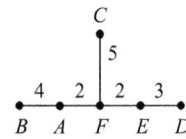

c i $4 + 5 + 12 + 11 = 32$ miles
ii C, gives lowest maximum distance 23 miles

3 a Circle, in order, AF (2), EF (2), DE (3), AB (4), CF (5), with columns labelled 1, 5, 6, 4, 3, 2

b

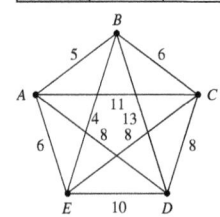

c 11 mins, between B and C, or between B and D.

4 a

	A	B	C	D	E
A	–	5	11	8	6
B	5	–	6	13	4
C	11	6	–	8	8
D	8	13	8	–	10
E	6	4	8	10	–

b

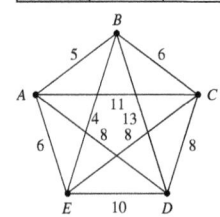

c AB (5), BE (4), BC (6), AD/CD (8)
Total 23 with columns labelled 1, 2, 4, 5, 3

d Because, for example, BE and BC might share a stretch of road

e Close AB, CD with all others open.

5 At each stage choose the road with the greatest capacity
BC (12), CH (11), AB (10), CD (10), AG (12), DJ (10), FJ (10), EF (13)
A lorry of 10 tonnes can go everywhere.

Exercise 12.5A

1 a There are two nodes, B and E, of odd degree, so the route must travel twice between these.

b The shortest route between these is $BCDE = 9$, so repeat arcs BC, CD and DE.

c Total length $= 41 + 9 = 50$

2 a Odd nodes are A and D. Least distance is $ACD = 19$, so repeat AC and CD.

b Total weight $= 130$, so route $= 130 + 19 = 149$

c For example, $ABCACDCEBDEA$

3 a Odd nodes are A, B, C, D
$AB(AEB) + CD = 7 + 7 = 14$. $AC(AEC) + BD(BCD)$
$= 6 + 13 = 19$
$AD + BC(BEC) = 13 + 5 = 18$.
Repeat AE, EB, CD. Total weight $= 56$,
so route $= 56 + 14 = 70$

b Odd nodes are A, B, C, F
$AB + CF(CDGF) = 14 + 14 = 28$. $AC(AGDC) + BF(BGF)$
$= 16 + 13 = 29$
$AF(AGF) + BC(BGDC) = 14 + 15 = 29$
Repeat AB, CD, DG, GF. Total weight $= 135$,
so route $= 135 + 28 = 163$

c Odd nodes are A, B, C, E
$AB + CE = 7 + 10 = 17$. $AC + BE(BCE) = 5 + 14 = 19$.
$AE(ADE) + BC = 14 + 4 = 18$
Repeat AB, CE. Total weight $= 59$, so route $= 59 + 17 = 76$

d Odd nodes are B, C, D, F
$BC + DF = 5 + 18 = 23$. $BD(BCD) + CF = 13 + 11 = 24$
$BF(BCF) + CD = 16 + 8 = 24$
Repeat BC, DF
Total weight $= 86$, so route $= 86 + 23 = 109$

4 a The odd nodes are A, B, D and F.
The pairings are $AB + DF = 14 + 8 = 22$, $AD + BF = 7 + 15$
$= 22$, and $AF + BD = 15 + 11 = 26$.
Either repeat AB, CD and CF or AD, BC and CF.

b Total length $= 82 + 22 = 104$

Exercise 12.5B

1 a Odd nodes A, F, H, I.
$AF + HI = 740 + 140 = 880$. $AH + FI = 740 + 340 = 1080$
$AI + FH = 600 + 200 = 800$.
Repeat (AJ, JI or AK, KI) and (FG, GH or FL, LH)
Total of weights $= 3240$, so route $= 3240 + 800 = 4040$ m

b Need to repeat route from A to F (AK, KL, LF)
Length of route $= 3240 + 740 = 3980$ m

c Least distance between odd vertices is $HI = 140$, so repeat this. Start at A and finish at F (or vice versa).
Total route $= 3240 + 140 = 3380$ m

2 a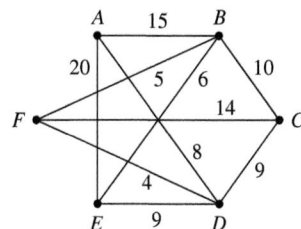

b Odd nodes A, C, E, F
$AC(ADC) + EF(EBF) = 17 + 11 = 28$. $AE(ADE) + CF(CDF)$
$= 17 + 13 = 30$
$AF(ADF) + CE(CBE) = 12 + 16 = 28$
Repeat AD, DF, CB, BE or repeat AD, DE, EB, BF
Total weight $= 100$, so route $= 100 + 28 = 128$ km

3 a If all arcs are to be repeated then every node becomes even.
The network is Eulerian and distance $= 86 \times 2 = 172$ km

b Odd nodes A, B, D, E
$AB + DE = 20 + 6 = 26$. $AD + BE = 14 + 10 = 24$
$AE(ACE) + BD(BED) = 19 + 16 = 35$
Repeat AD, BE. Distance $= 86 + 24 = 110$ km.

4 a Odd nodes A, C, D, E
$AC(AFGC) + DE = 600 + 100 = 700$. $AD + CE(CGE)$
$= 200 + 550 = 750$
$AE(ADE) + CD = 300 + 450 = 750$
Repeat AF, FG, GC, DE
Total distance $= 3350 + 700 = 4050$ m

b Removing BG would create two more odd nodes, so there would be more repeats and he would go further.

c Pick up at C, so the only repeat needed is $DE = 100$, Total distance $= 3450$ m.

5 a There are 6 odd nodes (A, B, D, E, F, G), but it is clear the least pairing is
$AB + EF + DG = 200$ or the symmetrical combination.
Repeat AB, EF, DC, CG

b Distance $= 680 + 200 = 880$ m.

c With repeats C has order 8, so route passes through 4 times.

6 a Odd nodes B, D, E, G, I, J.
Discounting symmetries and the obvious bad pairings such as DI, consider
$BG + DE + IJ = 440$ and $BJ + GI + DE = 400$
Repeat BA, AJ, DE, GH, HI (or symmetry) gives rope length 1800 cm.

b Repeat DE and IJ. Start at B, end at G (or vice versa). Rope length 1640 cm.

Exercise 12.6A

1 a

	A	B	C	D
A	–	10	16	2
B	10	–	6	8
C	16	6	–	14
D	2	8	14	–

b From A: $ADBCA = 32$ is the same tour as from C:
$CBDAC = 32$
From B: $BCDAB = 32$ is the same tour as from D:
$DABCD = 32$

c Both tours would visit $ADBCBDA$.

2 a $AEBDCA = 38$

b $BDAECB = 35$

3 a MST is AB, AC, AD with total length 42. Upper bound $= 84$ min.

b For example, replace CAB by CB and BAD by BD. Route is $ACBDA = 67$ min.

4 a MST is AC, BC, BE, CF, DE with total length 34. Upper bound $= 68$ km.

b For example, replace ACB by AB and $DEBCF$ by DF. Route is $ABEDFCA = 50$ km.

5 a MST is AE, BC, BE, DE or AE, BC, CE, DE with total length 74.

b 148

c For example, from the second MST replace AED by AD and $DECB$ by DB.
Route is $ADBCEA = 103$.

6 a MST is AC, BD, CD, CF, DE
Total 28
b 56
c Start, for example, with ACDBDEDCFCA. Then replace DCF by DF, $ACDB$ by AB, BDE by BE. Upper bound = 41
7 a Arcs $(AC + AD) + (BC + CD) = 14$.
Not possible because a tour of length 14 would have to use AC, CD, DA which form a cycle ACD.
b Arcs $(BC + BD) + (AC + AD) = 16$
These form a cycle $ACBDA$, so this must be the optimal tour.
8 The bounds by deleting A, B, C, D, E, F in turn give lower bounds of 63, 59, 72, 60, 55, 57. The best is 72.
9 The bounds by deleting A, B, C, D, E in turn are 74, 72, 72, 77, 72. The best is 77.

Exercise 12.6B

1 a
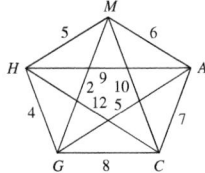
b $AGMH(G)CA = 31$ min. On original network the route is $HGMACGH$.
c $HGCA(G)MH = 30$ min, so $HGMACH$ is not optimal.
2 a

	A	B	C	D
A	–	5	9	6
B	5	–	10	7
C	9	10	–	3
D	6	7	3	

b $ABDCA$, 24 miles. Actual route $ABDCDA$.
3 a $(AB + AD) + (BC + CD + CE) = £950$
b MST is AD, AB, BC, CE, total 690. Upper bound is £1380 using $ADABCECBA$ but replacing $ECBA$ by EA gives a tour of £1240.
c $950 \leq C \leq 1240$
4 a
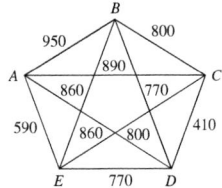
b Lower bound $= (AD + AE) + (BD + CD + DE) = 3400\,\text{m}$
$ABCDEA = 3520$
so $3400 \leq$ optimal tour ≤ 3520, so maximum saving is 120 m
5 a MST is AE, BC, BD, BE, BG, EF, length 29 km. Upper bound = 58 km
b For example, replace $DBEF$ by DF GBC by GC and AEF by AF.
c Route $AFDBGCBEA = 48$ km
d Lower bound is $(AB + AE) + (BC + BD + BE + BG + EF) = 34$ km
$34 \leq T \leq 48$
6 a
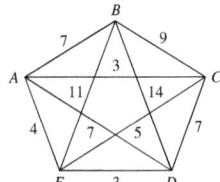
b $ACEDBA = 32$ miles. Actual order is $ACEDEABA$.

7 a

	A	B	C	D	E	F
A	–	4	2	4	3	5
B	4	–	2	4	3	5
C	2	2	–	4	3	3
D	4	4	4	–	3	3
E	3	3	3	3	–	2
F	5	5	3	3	2	–

b Upper bound 16 (for example, using nearest neighbour starting at A, $ACBDFEA$), lower bound 15 (for example, remove C, and choose $AE, BE, EF, DF + AC, BC$)

Exercise 12.7A

1 a Cut 1: **i** $\{SA, SB, SC\}$
  ii $X = \{S\}, Y = \{A, B, C, T\}$
  iii 41
Cut 2: **i** $\{AT, AB, SB, SC\}$
  ii $X = \{S, A\}, Y = \{B, C, T\}$
  iii 52
Cut 3: **i** $\{SA, AB, BT, CT\}$
  ii $X = \{S, B, C\}, Y = \{A, T\}$
  iii 38 (not 46, because AB is directed from the sink to the source)
b Cut 1: **i** $\{SB, AB, BC, DC, CF, EF, ET\}$
  ii $X = \{S, A, C, E\}, Y = \{B, D, F, T\}$
  iii 90
Cut 2: **i** $\{SA, AB, BC, BD\}$
  ii $X = \{S, B\}, Y = \{A, C, D, E, F, T\}$
  iii 70
Cut 3: **i** $\{AC, BC, DC, CF, EF, FT\}$
  ii $X = \{S, A, B, D, F\}, Y = \{C, E, T\}$
  iii 69
c Cut 1: **i** $\{SA, AC, BC, SB, CE\}$
  ii $X = \{S, C\}, Y = \{A, B, D, E, F, T\}$
  iii 144
Cut 2: **i** $\{BF, CE, DE, DT\}$
  ii $X = \{S, A, B, C, D\}, Y = \{E, F, T\}$
  iii 106
Cut 3: **i** $\{AD, DE, ET, FT\}$
  ii $X = \{S, A, B, C, E, F\}, Y = \{D, T\}$
  iii 117
d Cut 1: **i** $\{SB, SC, AC, AD\}$
  ii $X = \{S, A\}, Y = \{B, C, D, E, T\}$
  iii 104
Cut 2: **i** $\{SC, AC, BE, CB, DC, DT\}$
  ii $X = \{S, A, B, D\}, Y = \{C, E, T\}$
  iii 131
Cut 3: **i** $\{AD, DC, CT, ET\}$
  ii $X = \{S, A, B, C, E\}, Y = \{D, T\}$
  iii 118
2 a $\{SA, SB, BC, CT\}$
b SAT (12), SBT (9), SCT (14)
c Optimal flow = 35 (max flow = min cut)
3 Cut $\{CE, EF, FT\} = 24$, flow for example $SACET$ (10) and $SBDFT$ (14)
4 Cut $\{SA, AB, BC, BE\} = 17$, flow for example $SADT$ (5), $SACT$ (4), $SBCT$ (2), $SBCET$ (1), $SBET$ (5)
5 Cut $\{SA, AC, CD, ET\} = 12$
Flow 4 along $SADT$, 5 along $SCET$ and 3 along $SBET$

Exercise 12.7B

1 a Sources A, E
b SE (14) and SA (9)
c 15
d $\{DG, CG, FG\} = 15$

2
  a Add the arcs *SA* (8), *SB* (16), *HT* (14), *IT* (9)
  b *ACFI* (5), *BDGJH* (5), *BEH* (6)
  c Those in **b** plus *ADGH* (3)
  d {*EH, DG, GF, FI*} = 19 so maximal (max flow = min cut}

3
  a Sources *B, D* and sinks *I, K*
  b Add arcs *SB* (36), *SD* (26), *IT* (14), *KT* (19)
  c Cut {*AE, EF, FJ, GK, KL*} = 28
  d For example *BAEI* (6), *BFJI* (5), *BFJK* (2), *BCGK* (5), *DHLK* (10)

4
  a Sources *A, E* and sinks *D, H*
  b
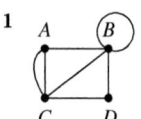
  c Cut {*SA, SE*} = 44
  d For example *ABD* (10), *ABFD* (2), *ACBFD* (8), *EGH* (15), *ECGH* (3), *ECGFH* (6)

Review exercise 12

1
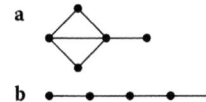

2
  a
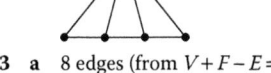
  b
  c
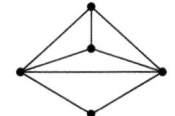

3
  a 8 edges (from $V + F - E = 2$)
  b

4 *FG* (3), *AC* (4), *CF* (5), *CD* (6), *BC* (9), *BE* (10) Total 37.

5 *BC, AC, CF, FG, CD, BE*

6

	1	5	3	2	6	7	4
	A	*B*	*C*	*D*	*E*	*F*	*G*
A	–	9	12	4	6	8	14
B	9	–	10	13	8	11	③
C	12	10	–	②	9	15	7
D	④	13	2	–	13	14	4
E	⑥	8	9	13	–	15	8
F	⑧	11	15	14	15	–	13
G	14	3	7	④	8	13	–

AD, CD, DG, BG, AE, AF

7
  a Odd nodes *A, B, C, G*
     AB (*ACB*) + *CG* (*CFG*) = 7 + 14 = 21. *AC* + *BG* (*BEG*) = 5 + 14 = 19
     AG (*ADG*) + *BC* = 16 + 2 = 18
     Repeat arcs *AD, DG, BC*. Length of route = 104 + 18 = 122
  b *D* now has degree 6, so route passes through *D* 3 times.

8
  a

	A	*B*	*C*	*D*	*E*
A	–	14	17	11	29
B	14	–	9	16	25
C	17	9	–	25	34
D	11	16	25	–	18
E	29	25	34	18	–

  b *ADBCEA* = 99. Actual route is *ADBCBEDA*. Upper bound because the optimal tour must be ≤ any known tour.
  c and d MST is *CB, BA, AD, DE*, total 52. So 104 is an upper bound. Shortcut replacing *EDABC* by *EC* gives upper bound 86.

9
  a *SADT* (5), *SACT* (4), *SBCET* (2) *SBET* (7)
  b $C_1 = 19$, $C_2 = 18$
  c From max flow– min cut theorem, the max flow = 18

Assessment 12

1
  a A connected graph with no cycles
  b
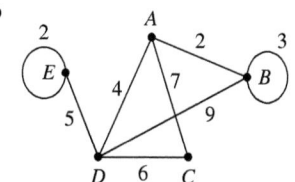
  c For example, *EDAC*
  d Any cycle would need to visit *D* twice as it is the only vertex that is directly connected to *E*. So there is no Hamiltonian cycle, and the graph is not Hamiltonian.
  e There are two vertices with odd order (*A* and *E*) (it is semi-Eulerian).
  f Add the edge *AE*.

2
  a Yes. For example, demonstrate by labelling and listing edges, using a table, or redrawing graph 1 with the right-most arc redrawn further to the left so that it looks like graph 2.
  b Semi-traversable since it has two odd order vertices.
  c For example,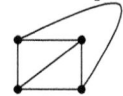
  d   Yes,

    For example, it can be drawn like this:

3
  a

From	To				
	A	*B*	*C*	*D*	*E*
A	0	1	1	1	0
B	1	0	1	1	0
C	1	2	0	1	0
D	0	1	1	0	1
E	0	0	0	1	2

  b There is no Eulerian trail. There is, however, a semi-Eulerian trail of *EDCBCACADBA* with a total weight of 53.
  c This cannot be done since one arc (either *BD* or *AC*) has no weight.
  d Yes,
    BD can be moved to around the outside so that none of the edges cross.

6 a For example, *ABCDA*
  For example, 19

b
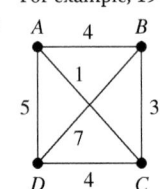

c *ACBDA*

d *ACBCDCA*

7 a Vertex *E*: $9 = x + 2 + 5 \Rightarrow x = 2$
  Vertex *C*: $2 + 2 = z \Rightarrow z = 4$
  Vertex *A*: $10 = y + 2 + y \Rightarrow y = 4$
  Or:
  Vertex *B*: $y + 5 = 9$ so $y = 4$
  Vertex *D*: $4 + z + 2 = 10$ so $z = 4$
  Vertex *C*: $2 + x = z$ so $x = 2$

b *AC*, *BE*, *ET*
  EC, *AD*

c {*AD*, *AC*, *EC*, *ED*, *ET*}
  Capacity of cut $= 4 + 2 + 2 + 6 + 5$
  $= 19$

8 a
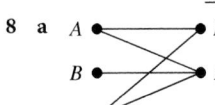

b bipartite

c *AF*, *BD*, *BF* to form $K_{3,3}$ then additional vertex and 3 edges to expand to $K_{4,3}$.
  So 1 vertex
  6 edges

d Non-Eulerian, as all four of the vertices in the set of 4 will have order 3.

9 a

	A	B	C	D	E	F
A	–	12	9	–	7	–
B	12	–	4	5	–	1
C	9	4	–	3	8	–
D	–	5	3	–	6	–
E	7	–	8	6	–	2
F	–	1	–	–	2	–

  AE, *EF*, *BF*, *BC*, *CD*
  Weight $= 17$

b
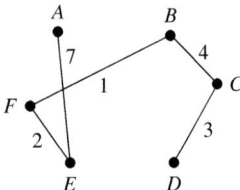

  Add *BF*, *EF*, *CD*, *BC*
  Reject *BD*, *DE*
  Add *AE*

c Vertices with odd degree: *A*, *D*
  Shortest route between *A* and *D* is *ACD* so repeat *AC*, *CD*.
  Possible Route: *ABCDCEDBFEACA*
  Length $= 57 + 12 = 69$

10 a a Minimum spanning tree is
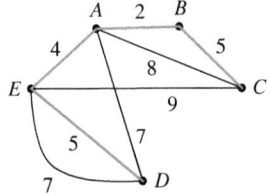

  So upper bound $= 2 \times 16$
  $= 32$ minutes

b For example, a short cut is *AD* instead of *AED*
  So improved upper bound is 30

c Removing *A* gives MST of weight 19
  So lower bound is $19 + 2 + 4 = 25$

  Removing *B* gives MST of weight 16
  So lower bound is $16 + 2 + 5 = 23$

  Removing *C* gives MST of weight 11
  So lower bound is $11 + 5 + 8 = 24$

  Removing *D* gives MST of weight 11
  So lower bound is $11 + 5 + 7 = 23$

  Removing *E* gives MST of weight 14
  So lower bound is $14 + 4 + 5 = 23$

  So best lower bound is 25 minutes

d

	A	B	C	D	E
A	–	2	7	7	4
B	2	–	5	9	6
C	7	5	–	14	9
D	7	9	14	–	5
E	4	6	9	5	–

e *CBAEDC*
  Length of tour is $5 + 2 + 4 + 5 + 14 = 30$ minutes
  Tour on original network is:
  CBAEDEC

f For example, a shorter tour is *ABCEDA* with length 28 minutes

11 a 4

b **i** {*AD*, *CD*, *CE*, *BE*}
  ii {*S*, *A*, *B*, *C*}
  iii {*D*, *E*, *T*}

c Minimum cut: {*DT*, *ET*}
  has capacity 18
  The maximum flow is therefore 18.

12 a For example, Kruskals: Add *CE*, *DE*,
  Reject *CD*
  Add *BC*
  Reject *BD*
  Add *EF*, *AB*
  Length $= 2 + 3 + 5 + 9 + 11 = 30\,\text{m}$
  Kruskal's or Prim's

b Use algorithm to find minimum spanning tree.
  Redraw the network with weight 0 on *AC* and *BD*. Then use algorithm as before
  Using Kruskal's,
  Add *AC*, *BD*, *CE*, *DE*
  Reject *CD*, *BC*
  Add *EF*
  Length reqired is $2 + 3 + 9 = 14\,\text{m}$

c Upper bound $= 2 \times 30 = 60$
ABCEFEDECBA (in either direction)
Short cuts from *A* to *F* and *D* to *C* give the route
AFEDCBA with weight 47

d removing *A* gives MST of weight 19. So lower bound is
$19 + 11 + 12 = 42$
removing *B* gives MST of weight 26
So lower bound is $26 + 5 + 8 = 39$

Removing *C* gives MST of weight 31
So lower bound is $31 + 2 + 4 = 37$

Removing *D* gives MST of weight 27
So lower bound is $27 + 3 + 4 = 34$

Removing *E* gives MST of weight 33
So lower bound is $33 + 2 + 3 = 38$

Greatest lower bound is 42
e $42 \le$ length of optimal tour ≤ 47
f Matrix/Complete graph of shortest lengths:

	A	*B*	*C*	*D*	*E*	*F*
A	–	11	12	16	14	15
B	11	–	5	8	7	16
C	12	5	–	4	2	11
D	16	8	4	–	3	12
E	14	7	2	3	–	9
F	15	16	11	12	9	–

ABCEDFA = 48
BCEDFAB = 48
CEDBAFC = 50
DECBAFD = 48
ECDBAFE = 49
FECDBAF = 49
None are in the range so not optimal.

13 a

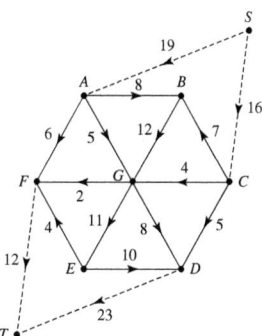

b 4
c Either cut {*AF, GF, GE, GD, CD*} or cut
{*AF,AG,BG,CG,CD*} have capacity 32.
Therefore since there exists a possible flow of size 32, the
max flow-min cut theorem tells us this is the maximal
flow.

Chapter 13

Exercise 13.1A

1

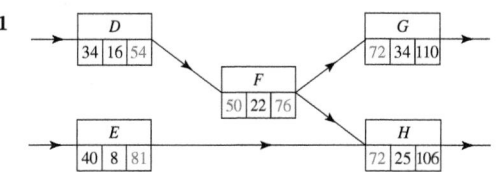

2 a

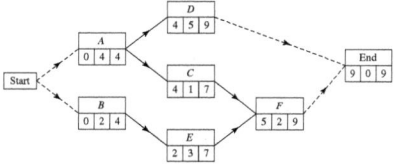

Project duration = 9

b

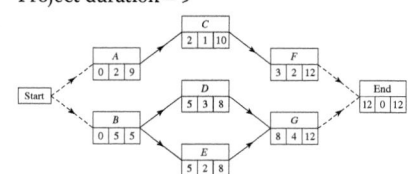

Project duration = 12

c

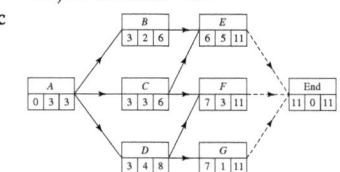

Project duration = 11

d

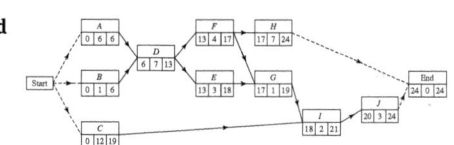

Project duration = 24

e

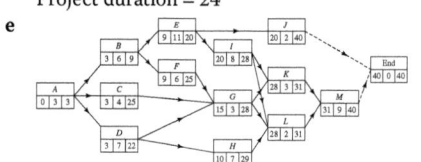

Project duration = 40

Exercise 13.1B

1 a

Activity	Description	Depends on
A	Purchase wood	–
B	Purchase wheels	–
C	Purchase cord	–
D	Cut out sides	*A*
E	Cut out base	*A*
F	Attach sides to base	*D, E*
G	Attach wheels to base	*B, E*
H	Attach cord	*C, D*

b

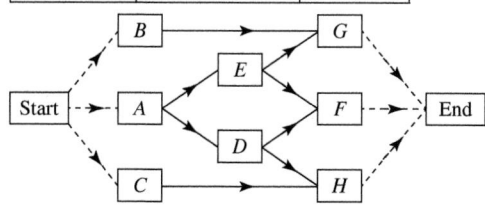

(you may have decided on slightly different precedence)

2

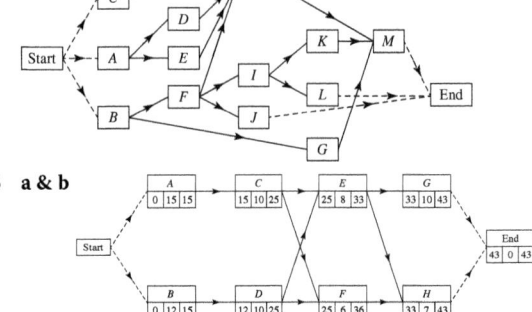

3 a & b

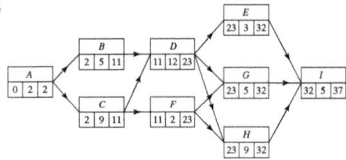

Project duration = 43 days

c i Project duration would be 45 days

ii Project duration would be unchanged at 43 days

4 a & b

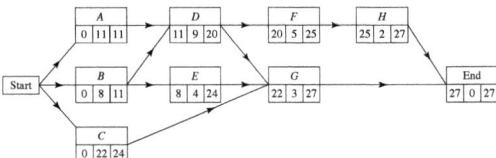

Project duration = 37 hours

c 12 hours

Exercise 13.2A

1 Floats are A 0, B 2, C 4, D 0, E 7, F 10, G 0, H 4, I 12, J 0, K 7, L 9, M 0

Critical activities are A, D, G, J, M

2 a

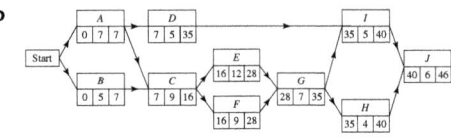

Project duration = 27

Floats are A 0, B 3, C 2, D 0, E 12, F 0, G 2, H 0

Critical activities are A, D, F, H

b

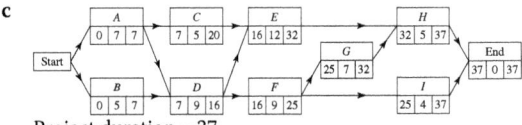

Project duration = 46

Floats are A 0, B 2, C 0, D 23, E 0, F 3, G 0, H 1, I 0, J 0

Critical activities are A, C, E, G, I, J

c

Project duration = 37

Floats are A 0, B 2, C 8, D 0, E 4, F 0, G 0, H 0, I 8

Critical activities are A, D, F, G, H

3

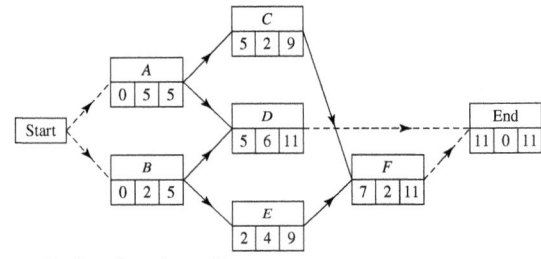

Project duration = 11

Critical activities are A and D

4 a

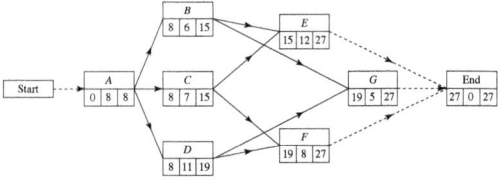

Duration = 27 hours

b Critical activities are A, C, D, E and F, forming critical paths A-C-E and A-D-F.

Exercise 13.2B

1 a

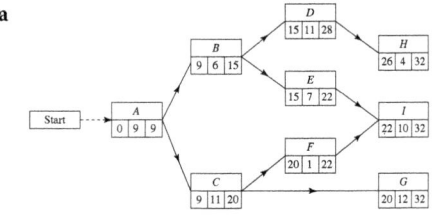

Duration = 32 hours.

b Critical paths A-B-E-I and A-C-G.

c Reduce C and E by 3 hours and D by 1 hour. Total cost £350.

2 a

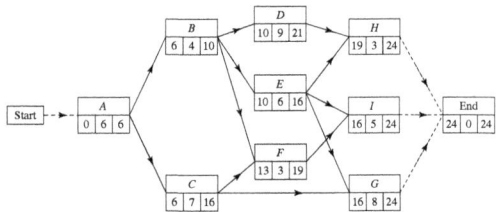

Duration = 24 days

b A, B, E, G

c i D has 2 days float, so 4 days delay extends the project by 2 days.

ii If F is delayed by 2 days then I cannot start until 18. If I is then delayed by 2 days it means the project is extended by 1 day.

3 a

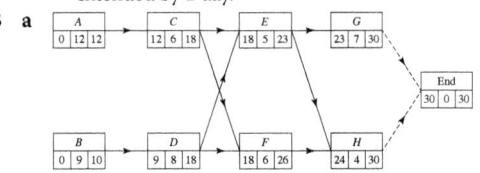

Duration = 30 days

b Critical activities A, C, E, G

Floats B 1, D 1, F 2, H 2

c 1 day

4 a

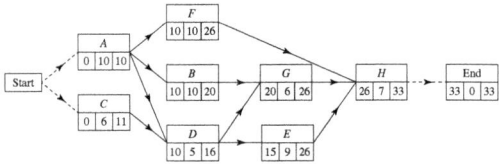

Duration = 33 hours

Critical activities A, B, G, H

b Split B into B₁ in its current place and B₂ following F. The duration is then 32 hours. The new critical activities are A, F, B₂, H

Review exercise 13

1

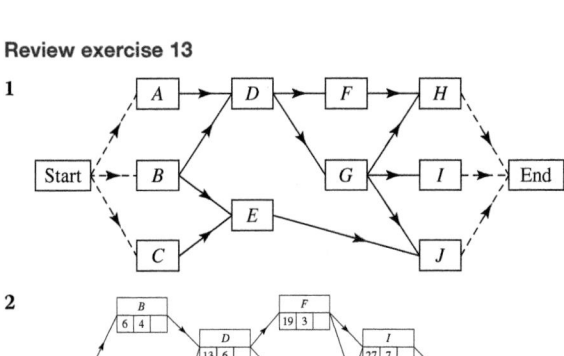

2

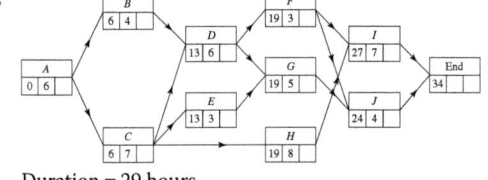

Duration = 29 hours

3

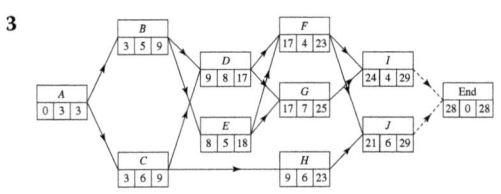

4 a A 2, B 0, C 5, D 0, E 10, F 0, G 15, H 0
  b B, D, F, H

Assessment 13

1

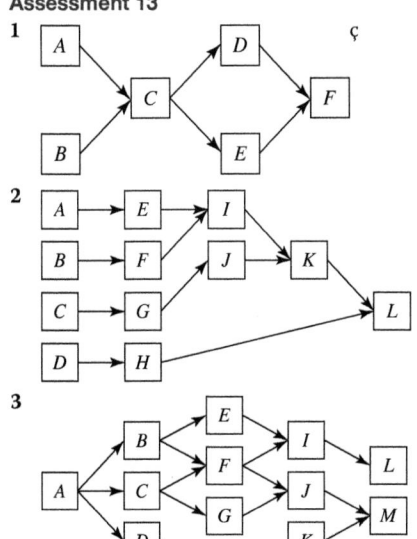

ç

2

3

4

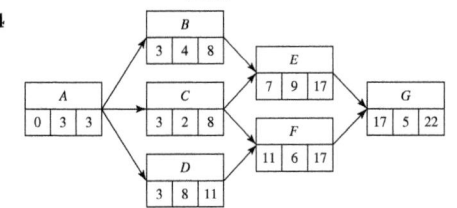

5

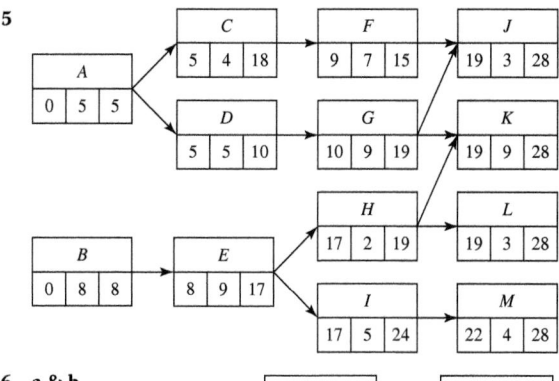

6 a & b

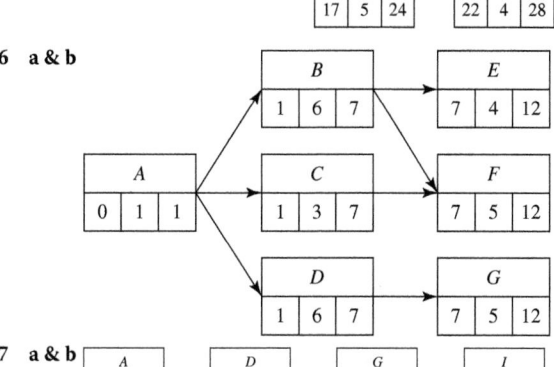

7 a & b

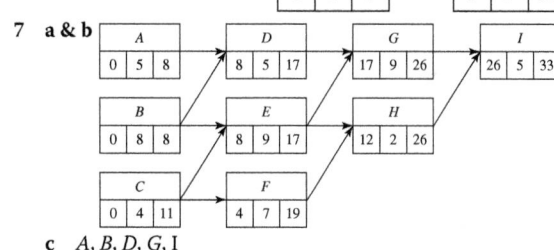

  c A, B, D, G, I

8 a & b

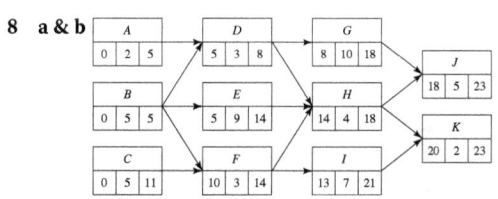

  c $BDGJ, BEHJ$

9 a & b

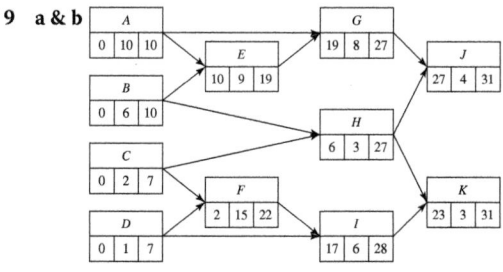

  c $AEGJ$
  d $x \leq 11$

10 a & b

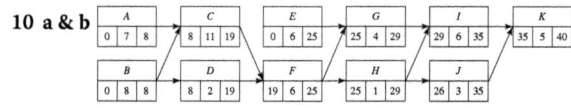

  c $BCFGIK$
  d $ACFGIK$

11 a & b

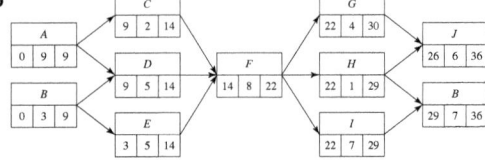

c *ADFIK*

d The earliest start time for J becomes 31 days and the earliest finish time becomes 40 days.

Chapter 14

Exercise 14.1A

1 If profit is £P, then

Maximise $P = 0.1x + 0.15y$

subject to $0.2x + 0.4y \leq 8000$ (or $x + 2y \leq 40\,000$)

$0.3x + 0.1y \leq 6000$ (or $3x + y \leq 60\,000$)

$x \geq 0, y \geq 0$

2 a The graph shows the objective line at $P = 50$

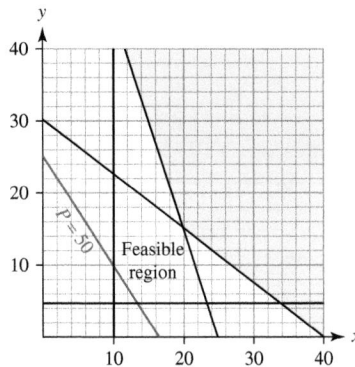

Optimal value $P = 90$, $x = 20$, $y = 15$

b The graph shows the objective line at $P = 8$.

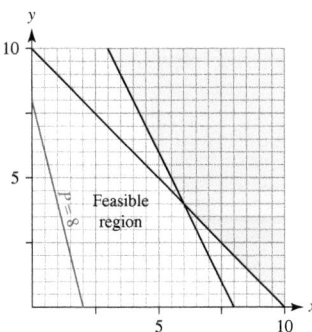

Optimal value $P = 32$, $x = 8$, $y = 0$

c The graph shows the objective line at $C = 50$.

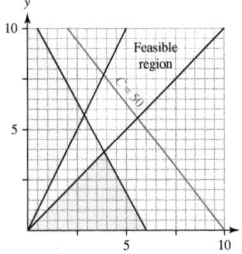

Optimal value $C = 39$, $x = 3$, $y = 6$

d The graph shows the objective line at $R = 15$.

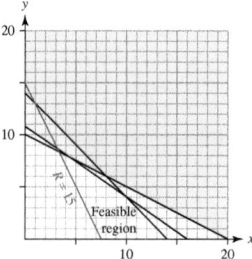

Optimal value $R = 28$, $x = 14$, $y = 0$.

3 x litres of Econofruit, y litres of Healthifruit, profit £P

Maximise $P = 0.3x + 0.4y$

subject to $0.2x + 0.4y \leq 20\,000$ (or $x + 2y \leq 100\,000$)

$0.5x + 0.3y \leq 30\,000$ (or $5x + 3y \leq 300\,000$)

$x \geq 0, y \geq 0$

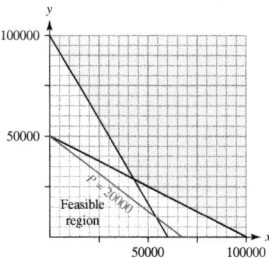

Objective line drawn at $P = 20\,000$

Optimal value $P = 24\,285\frac{5}{7}$ when $x = 42\,857\frac{1}{7}$, $y = 28\,571\frac{3}{7}$

Make 42 857 litres of Econofruit and 28 571 litres of Healthifruit, making £24 285 profit.

4 x ha of wheat, y ha of potatoes. Profit £P

Maximise $P = 80x + 100y$

subject to $3x + 5y \leq 280$

$7x + 4y \leq 400$

$x + y \leq 65$

$x \geq 0, y \geq 0$

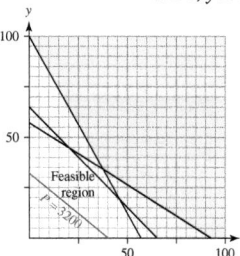

Objective line drawn at $P = 3200$

Optimal value $P = 6050$

when $x = 22.5$, $y = 42.5$

Plant 22.5 ha wheat and 42.5 ha of potatoes. making £6050 profit.

Exercise 14.1B

1 So hire 2 cars and 6 minibuses at a cost of £400.

2 So mix $28\frac{4}{7}$ ml of whisky with $71\frac{3}{7}$ ml ginger wine at a cost of £0.70.

3 Buy 13 cases of shampoo and 47 cases of cleaner, making a profit of £918.

4 Invest £8000 in bonds and £12 000 in shares.

5 Build 13 houses and 12 bungalows, making a profit of £560 000

6 x senior staff, y trainees, z children. Cost is £C.

a Minimise $\quad C = 20x + 15y + 12z$

subject to $\quad x + y + z = 50$

$2x + 2y \geq z$

$x \geq y$

$x \geq 10, y \geq 5, z \geq 0$

x, y, z are integers

b Substitute $z = 50 - (x + y)$ in the objective function and constraints.

$C = 20x + 15y + 12(50 - x - y) = 8x + 3y + 600$

$2x + 2y \geq z$ becomes $3x + 3y \geq 50$

Complete LP statement is

Minimise $\quad C = 8x + 3y + 600$

subject to $\quad 3x + 3y \geq 50$

$x \geq y$

$x \geq 10, y \geq 5$

x, y are integers

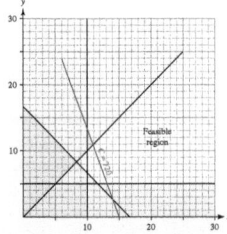

c Take 10 senior staff, 7 trainees and 33 children at a cost of £701.

7 The possible solutions are:

10 chew A, 20 chew B, 30 chew C

9 chew A, 23 chew B, 28 chew C

8 chew A, 26 chew B, 26 chew C

7 chew A, 29 chew B, 24 chew C

all of which give a cost of 95p.

Exercise 14.2A

1 a

		B			Row minima	Max of row minima
		B_1	B_2	B_3		
A	A_1	5	3	9	3	
	A_2	4	7	6	4	4
	A_3	2	4	5	2	
	Column maxima	5	7	9		
	Min of column maxima	5				

Play-safe strategies A_2 and B_1. Not a stable solution. Value = 4.

b

		B			Row minima	Max of row minima
		B_1	B_2	B_3		
A	A_1	9	-2	3	-2	
	A_2	-1	-4	2	-4	
	A_3	4	6	3	3	3
	Column maxima	9	6	3		
	Min of column maxima			3		

Play-safe strategies A_3 and B_3. Stable solution. Value = 3.

c

		B			Row minima	Max of row minima
		B_1	B_2	B_3		
A	A_1	6	-1	4	-1	-1
	A_2	-1	-3	5	-3	
	Column maxima	6	-1	5		
	Min of column maxima		-1			

Play–safe strategies A_1 and B_2. Stable solution. Value = -1.

d

		B		Row minima	Max of row minima
		B_1	B_2		
A	A_1	-1	3	-1	
	A_2	2	7	2	2
	A_3	6	-2	-2	
	Column maxima	6	7		
	Min of column maxima	6			

Play–safe strategies A_2 and B_1. Not a stable solution. Value = 2.

e

		B				Row minima	Max of row minima
		B_1	B_2	B_3	B_4		
A	A_1	4	-2	5	9	-2	
	A_2	2	1	3	5	1	1
	A_3	3	-1	-8	4	-8	
	Column maxima	4	1	5	9		
	Min of column maxima		1				

Play-safe strategies A_2 and B_2. Stable solution. Value = 1.

f

		B				Row minima	Max of row minima
		B_1	B_2	B_3	B_4		
A	A_1	3	2	7	10	2	
	A_2	5	5	7	8	5	5
	A_3	-3	2	-5	6	-5	
	Column maxima	5	5	7	10		
	Min of column maxima	5	5				

Play-safe strategies A_2 and B_1 or B_2. Stable solution. Value = 5.

2

		A		
		A_1	A_2	A_3
B	B_1	1	-2	-6
	B_2	-3	-7	2

3 a Row 2 dominates row 1.

		B	
		B_1	B_2
A	A_2	2	3
	A_3	4	−2

b Row 1 dominates row 2. After row 2 deleted, column 2 dominates columns 1 and 3, leaving this matrix.

		B
		B_2
A	A_1	−1

This is stable. Play A_1 and B_2. Value = −1

c Row 2 dominates row 3, then column 1 dominates column 3.

		B	
		B_1	B_2
A	A_1	3	1
	A_2	2	5

d Row 3 dominates row 1, then column 3 dominates column 2.

		B	
		B_1	B_3
A	A_2	3	−2
	A_3	−4	1

e Row 2 dominates rows 1 and 3, then column 1 dominates columns 2 and 3.

		B	
		B_1	B_4
A	A_2	−1	−1

This is stable. Play A_2 and B_1 or B_4. Value = −1.

f Row 1 dominates rows 2 and 4, then column 1 dominates column 2.

		B	
		B_1	B_3
A	A_1	3	2
	A_3	1	4

Exercise 14.2B

1 a Dominance eliminates row 3 then column 3.

		B			
		B_1	B_2	Row minima	Max of row minima
A	A_1	−4	6	−4	−4
	A_2	−2	−8	−8	
	Column maxima	−2	6		
	Min of column maxima	−2			

Play-safe strategies are A_1 and B_1. Value = −4. Unstable solution.

b Stable solution where A plays A_1 or A_3 and B plays B_2. Value −2.

c Dominance eliminates row 4 then column 2.

		B				
		B_1	B_2	B_3	Row minima	Max of row minima
A	A_1	4	1	−2	−2	
	A_2	3	−3	−1	−3	
	A_3	−1	3	0	−1	−1
	Column maxima	4	3	0		
	Min of column maxima			0		

Play-safe strategies are A_3 and B_4. Value = −1. Unstable solution.

d Stable solution where A plays A_2 and B plays B_3. Value 1.

e Stable solution where A plays A_3 and B plays B_3. Value −2.

f Stable solution where A plays A_4 and B plays B_3 or B_5. Value 0.

2 a

		Y		
		Choose 2	Choose 6	Choose 7
X	Choose 3	−2	3	4
	Choose 5	3	−2	−4
	Choose 9	5	3	−4

b Play-safe strategies are A_1 and B_2. The solution is unstable as the max of the row minima (−2) does not equal the min of the column maxima (3).

c For example, change Y's set to {3, 6, 7}

3 a

		B		
		B_1	B_2	B_3
A	A_1	1	0	−1
	A_2	−1	−3	0
	A_3	3	1	2

b

		B				
		B_1	B_2	B_3	Row minima	Max of row minima
A	A_1	1	0	−1	−1	
	A_2	−1	−3	0	−3	
	A_3	3	1	2	1	1
	Column maxima	3	1	2		
	Min of column maxima		1			

Play-safe strategies A_3 and B_2. Max of row minima = min of column maxima = 1, so stable solution (saddle point).

c Player A would win 10 − 0.

Exercise 14.3A

1 a Max of row mins = 2
Min of column maxs = 3
Solution is not stable
Let A play A1 with probability p
$v \le -2p + 3(1 - p) = 3 - 5p$
$v \le 4p + 2(1 - p) = 2p + 2$
Graph as shown.

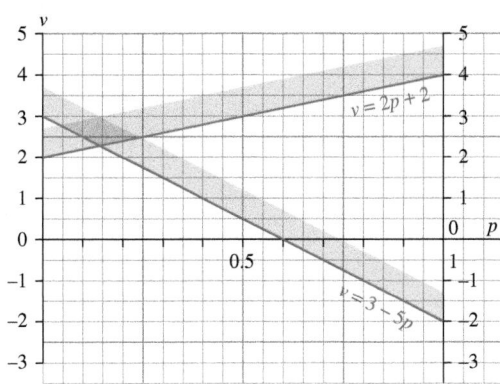

Max v when $3 - 5p = 2p + 2$, so $p = \dfrac{1}{7}$

This gives $v = 2\dfrac{2}{7}$

A plays A_1, A_2 with probabilities $\dfrac{1}{7}$, $\dfrac{6}{7}$

If B plays B_1 with probability q then

$v = -2q + 4(1 - q) = 2\dfrac{2}{7}$

This gives $q = \dfrac{2}{7}$, so B plays B_1, B_2

with probabilities $\dfrac{2}{7}$, $\dfrac{5}{7}$

b Max of row mins = 2
Min of column maxs = 3
Solution is not stable
Let A play A_1 with probability p
$v \le 4p - (1 - p) = 5p - 1$
$v \le 2p + 3(1 - p) = 3 - p$
Graph as shown.

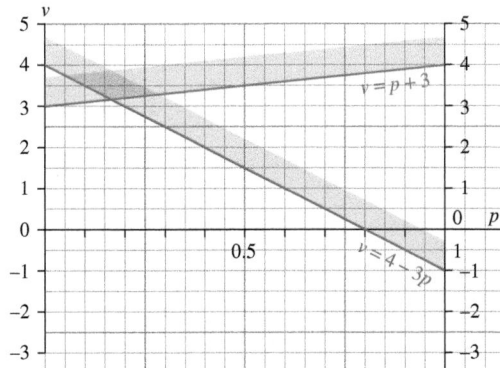

Max v when $5p - 1 = 3 - p$, so $p = \dfrac{2}{3}$

This gives $v = 2\dfrac{1}{3}$

A plays A_1, A_2 with probabilities $\dfrac{2}{3}$, $\dfrac{1}{3}$

If B plays B_1 with probability q then

$v = 4q + 2(1 - q) = 2\dfrac{1}{3}$

This gives $q = \dfrac{1}{6}$, so B plays B_1, B_2 with probabilities $\dfrac{1}{6}$, $\dfrac{5}{6}$

c Max of row mins = 2, min of column maxs = 2
Solution is stable. A plays A_2 and B plays B_2. Value = 2.

d Max of row mins = 3
Min of column maxs = 4
Solution is not stable
Let A play A_1 with probability p

$v \le p + 4(1 - p) = 4 - 3p$
$v \le 4p + 3(1 - p) = p + 3$
Graph as shown.

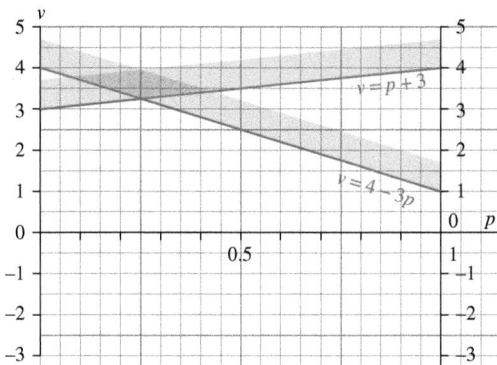

Max v when $4 - 3p = p + 3$, so $p = \dfrac{1}{4}$

This gives $v = 3\dfrac{1}{4}$

A plays A_1, A_2 with probabilities $\dfrac{1}{4}$, $\dfrac{3}{4}$

If B plays B_1 with probability q then $v = q + 4(1 - q) = 3\dfrac{1}{4}$

This gives $q = \dfrac{1}{4}$, so B plays B_1, B_2 with probabilities $\dfrac{1}{4}$, $\dfrac{3}{4}$

2 a Max of row mins = −2
Min of column maxs = 2
Solution is not stable
Let A play A_1 with probability p
$v \le 2p - 2(1 - p) = 4p - 2$
$v \le -2p + 3(1 - p) = 3 - 5p$
Graph as shown.

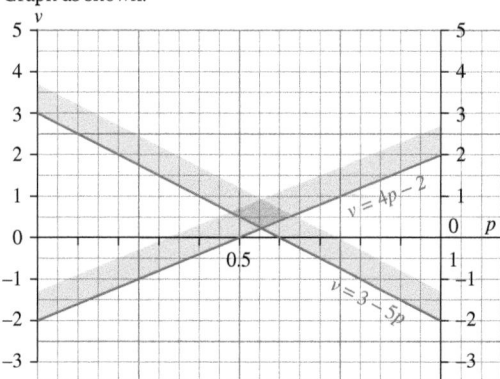

Max v when $4p - 2 = 3 - 5p$, so $p = \dfrac{5}{9}$

This gives $v = \dfrac{2}{9}$

A plays A_1, A_2 with probabilities $\dfrac{5}{9}$, $\dfrac{4}{9}$

If B plays B_1 with probability q then $v = 2q - 2(1 - q) = \dfrac{2}{9}$

This gives $q = \dfrac{5}{9}$, so B plays B_1, B_2 with probabilities $\dfrac{5}{9}$, $\dfrac{4}{9}$

b Max of row mins = 6
Min of column maxs = 7
Solution is not stable
Let A play A_1 with probability p
$v \le 7p + 5(1 - p) = 2p + 5$
$v \le 6p + 8(1 - p) = 8 - 2p$
Graph as shown.

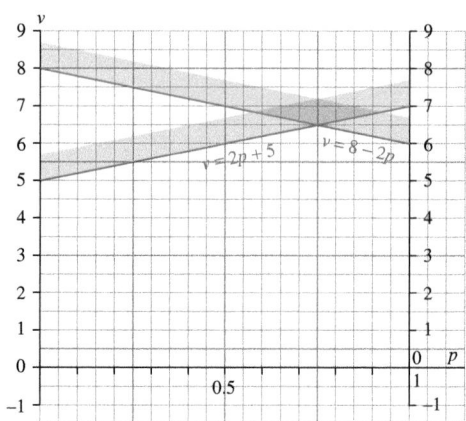

Max v when $2p+5=8-2p$, so $p=\dfrac{3}{4}$

This gives $v=6\dfrac{1}{2}$

A plays A_1, A_2 with probabilities $\dfrac{3}{4}$, $\dfrac{1}{4}$

If B plays B_1 with probability q then $v=7q+6(1-q)=6\dfrac{1}{2}$

This gives $q=\dfrac{1}{2}$, so B plays B_1, B_2 with probabilities $\dfrac{1}{2}$, $\dfrac{1}{2}$

c Max of row mins $=1$
Min of column maxs $=3$
Solution is not stable
Let A play A_1 with probability p
$v \le 5p-(1-p)=6p-1$
$v \le 2p+3(1-p)=3-p$
   Graph as shown.

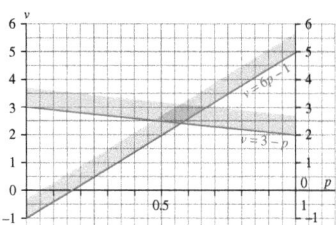

Max v when $6p-1=3-p$, so $p=\dfrac{4}{7}$

This gives $v=2\dfrac{3}{7}$

A plays A_1, A_2 with probabilities $\dfrac{4}{7}$, $\dfrac{3}{7}$

If B plays B_1 with probability q then $v=5q+2(1-q)=2\dfrac{3}{7}$

This gives $q=\dfrac{1}{7}$, so B plays B_1, B_2 with probabilities $\dfrac{1}{7}$, $\dfrac{6}{7}$

d Max of row mins $=-1$
Min of column maxs $=1$
Solution is not stable
Let A play A_1 with probability p
$v \le -2p+2(1-p)=2-4p$
$v \le p$
Graph as shown.

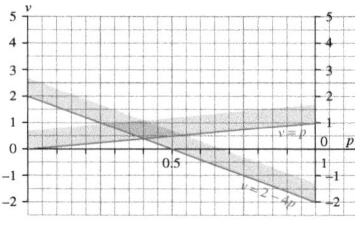

Max v when $2-4p=p$, so $p=\dfrac{2}{5}$

This gives $v=\dfrac{2}{5}$

A plays A_1, A_2 with probabilities $\dfrac{2}{5}$, $\dfrac{3}{5}$

If B plays B_1 with probability q then $v=-2q+(1-q)=\dfrac{2}{5}$

This gives $q=\dfrac{1}{5}$, so B plays B_1, B_2 with probabilities $\dfrac{1}{5}$, $\dfrac{4}{5}$

3 a Let B play B_1 with probability q
$(-v) \le 6q-2(1-q)$ and so $v \ge 2-8q$
$(-v) \le -q+2(1-q)$ and so $v \ge 3q-2$
Graph as shown.

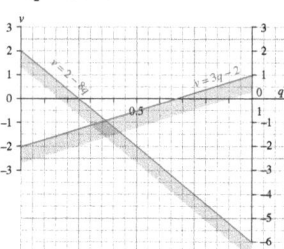

Max $(-v)$ when $2-8q=3q-2$, so $q=\dfrac{4}{11}$

This gives $v=-\dfrac{10}{11}$

B plays B_1, B_2 with probabilities $\dfrac{4}{11}$, $\dfrac{7}{11}$

b Let B play B_1 with probability q
$(-v) \le -9q-7(1-q)$ and so $v \ge 2q+7$
$(-v) \le -3q-10(1-q)$ and so $v \ge 10-7q$
Graph as shown.

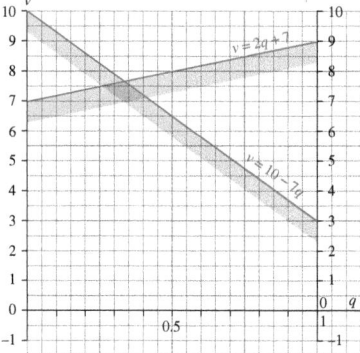

Max $(-v)$ when $2q+7=10-7q$, so $q=\dfrac{1}{3}$

This gives $v=7\dfrac{2}{3}$

B plays B_1, B_2 with probabilities $\dfrac{1}{3}$, $\dfrac{2}{3}$

c Let B play B_1 with probability q
$(-v) \le 5q-2(1-q)$ and so $v \ge 2-7q$
$(-v) \le -q+4(1-q)$ and so $v \ge 5q-4$
Graph as shown.

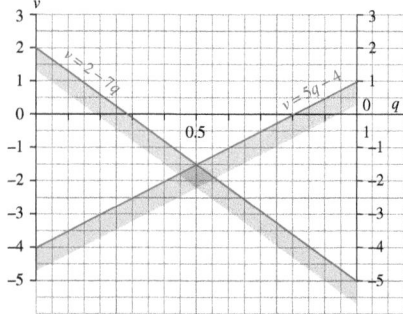

Max $(-v)$ when $2 - 7q = 5q - 4$, so $q = \dfrac{1}{2}$

This gives $v = -1\dfrac{1}{2}$

B plays B_1, B_2 with probabilities $\dfrac{1}{2}$, $\dfrac{1}{2}$

4 a

		B				Row minima	Max of row minima
		B_1	B_2	B_3	B_4		
A	A_1	−1	3	7	−1	−1	
	A_2	1	5	4	2	1	
	A_3	3	2	3	3	2	2
	Column maxima	3	5	7	3		
	Min of column maxima	3			3		

Max of row minima ≠ min of column maxima, so no stable solution.

b Column 4 dominates column 3, so remove column 3
Then row 2 dominates row 1, so remove row 1.
Then column 3 dominates column 1, so remove column 1.
The table is now:

		B	
		B_2	B_4
A	A_2	5	2
	A_3	2	3

c Let A play A_2 with probability p
$v \le 5p + 2(1 - p) = 3p + 2$
$v \le 2p + 3(1 - p) = 3 - p$
Graph as shown.

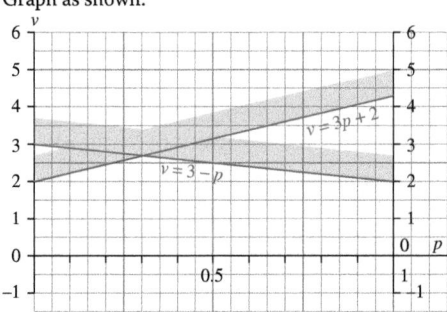

Max v when $3p + 2 = 3 - p$, so $p = \dfrac{1}{4}$

This gives $v = 2\dfrac{3}{4}$

A plays A_2, A_3 with probabilities $\dfrac{1}{4}$, $\dfrac{3}{4}$

d If B plays B_2 with probability q then $v = 5q + 2(1 - q) = 2\dfrac{3}{4}$

This gives $q = \dfrac{1}{4}$, so B plays B_2, B_4 with probabilities $\dfrac{1}{4}$, $\dfrac{3}{4}$

Exercise 14.3B

1 a Row 2 dominates row 1, so remove row 1
Column 2 dominates column 3, so removes column 3
This leaves:

		B	
		B_1	B_2
A	A_2	−3	3
	A_3	0	−4

Max of row mins = −3
Min of column maxs = 0

Solution is not stable
Let A play A_2 with probability p
$v \le -3p$
$v \le 3p - 4(1 - p) = 7p - 4$
Graph as shown.

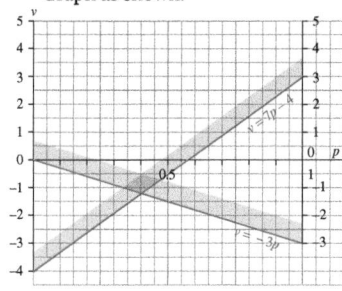

Max v when $-3p = 7p - 4$, so $p = \dfrac{2}{5}$

This gives $v = -1\dfrac{1}{5}$

A plays A_2, A_3 with probabilities $\dfrac{2}{5}$, $\dfrac{3}{5}$

If B plays B_1 with probability q then $v = -3q + 3(1 - q) = -1\dfrac{1}{5}$

This gives $q = \dfrac{7}{10}$, so B plays B_1, B_2 with probabilities $\dfrac{7}{10}$, $\dfrac{3}{10}$

b Column 2 dominates column 1, so remove column 1
Row 1 dominates row 2, so remove row 2
This leaves:

		B	
		B_2	B_3
A	A_1	1	5
	A_3	2	1

Max of row mins = 1
Min of column maxs = 2
Solution is not stable
Let A play A_1 with probability p
$v \le p + 5(1 - p)$ and so $v \le 5 - 4p$
$v \le 2p + (1 - p)$ and so $v \le p + 1$
Graph as shown.

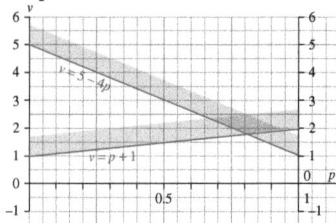

Max v when $5 - 4p = p + 1$, so $p = \dfrac{4}{5}$

This gives $v = 1\dfrac{4}{5}$

A plays A_1, A_3 with probabilities $\dfrac{4}{5}$, $\dfrac{1}{5}$

If B plays B_2 with probability q then $v = q + 2(1 - q) = 1\dfrac{4}{5}$

This gives $q = \dfrac{1}{5}$, so B plays B_2, B_3 with probabilities $\dfrac{1}{5}$, $\dfrac{4}{5}$

c Max of row mins = 2
Min of column maxs = 3
Solution is not stable
Let A play A_1 with probability p
$v \le 5p + (1 - p)$ and so $v \le 4p + 1$
$v \le 2p + 4(1 - p)$ and so $v \le 4 - 2p$
$v \le 3p + 2(1 - p)$ and so $v \le p + 2$

Graph as shown.

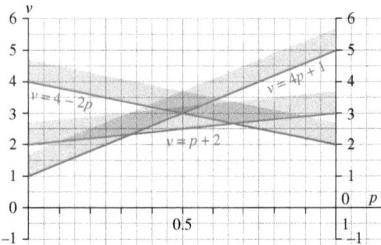

B would never play B_1

Max v when $p + 2 = 4 - 2p$, so $p = \dfrac{2}{3}$

This gives $v = 2\dfrac{2}{3}$

A plays A_1, A_2 with probabilities $\dfrac{2}{3}$, $\dfrac{1}{3}$

If B plays B_2 with probability q then $v = 2q + 3(1 - q) = 2\dfrac{2}{3}$

This gives $q = \dfrac{1}{3}$, so B plays B_2, B_3 with probabilities $\dfrac{1}{3}$, $\dfrac{2}{3}$

d Max of row mins $= -1$

Min of column maxs $= 2$

Solution is not stable

Let A play A_1 with probability p

$v \le 3p - 2(1 - p)$ and so $v \le 5p - 2$

$v \le -p + 4(1 - p)$ and so $v \le 4 - 5p$

$v \le p + 2(1 - p)$ and so $v \le 2 - p$

Graph as shown.

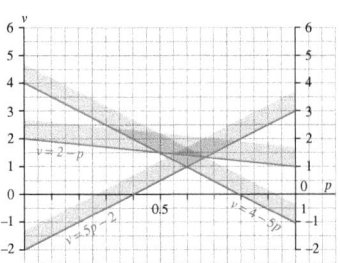

B would never play B_3

Max v when $5p - 2 = 4 - 5p$, so $p = \dfrac{3}{5}$

This gives $v = 1$

A plays A_1, A_2 with probabilities $\dfrac{3}{5}$, $\dfrac{2}{5}$

If B plays B_1 with probability q then $v = 3q - (1 - q) = 1$

This gives $q = \dfrac{1}{2}$, so B plays B_1, B_2 with probabilities $\dfrac{1}{2}$, $\dfrac{1}{2}$

2 a

		Y				Row minima	Max of row minima
		I	II	III	IV		
	I	1	4	3	2	1	1
X	II	−1	3	0	3	−1	
	III	4	2	5	1	1	1
	IV	−2	6	−3	1	−3	
	Column maxima	4	6	5	3		
	Min of column maxima				3		

Max of row minima $\ne$ min of column maxima, so no stable solution.

b Column 4 dominates column 2, so remove column 2

Then row 1 dominates row 4, so remove row 4

Then column 1 dominates column 3. so remove column 3.

This leaves:

		Y	
		I	IV
	I	1	2
X	II	−1	3
	III	4	1

Let Y play I with probability q

$(-v) \le -q - 2(1 - q)$ and so $v \ge 2 - q$

$(-v) \le q - 3(1 - q)$ and so $v \ge 3 - 4q$

$(-v) \le -4q - (1 - q)$ and so $v \ge 3q + 1$

Graph as shown.

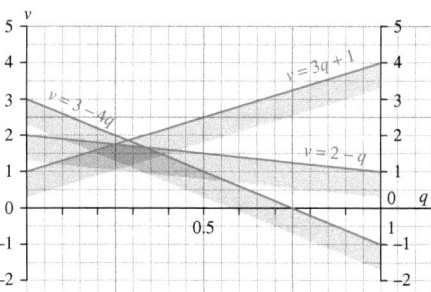

X would never play I

Max $(-v)$ when $3 - 4q = 3q + 1$, so $q = \dfrac{2}{7}$

This gives $v = 1\dfrac{6}{7}$

Y plays I, IV with probabilities $\dfrac{2}{7}$, $\dfrac{5}{7}$

If X plays II with probability p then $v = -p + 4(1 - p) = 1\dfrac{6}{7}$

This gives $p = \dfrac{3}{7}$, so X plays II, III with probabilities $\dfrac{3}{7}$, $\dfrac{4}{7}$

3 $V \le -p_1 + p_2$

$V \le -2p_2 - (1 - p_1 - p_2)$, giving $V \le p_1 - p_2 - 1$

$V \le -2p_1 - p_2 + (1 - p_1 - p_2)$, giving $V \le 1 - 3p_1 - 2p_2$

4 a Let A play A_1 with probability p

$v \le 2p + 5(1 - p) = 5 - 3p$

$v \le 5p + 2(1 - p) = 3p + 2$

$v \le 6p + (1 - p) = 5p + 1$

Graph as shown.

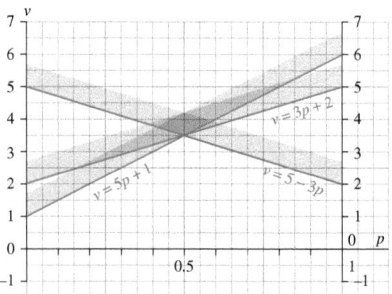

Max v when $5 - 3p = 3p + 2$, so $p = \dfrac{1}{2}$

A plays A_1, A_2 with probabilities $\dfrac{1}{2}$, $\dfrac{1}{2}$

b This gives $v = 3\dfrac{1}{2}$

c All three line pass through the same point, so none of the strategies is worse than the others.

d $q_1 + q_2 + q_3 = 1$ [1]

$2q_1 + 5q_2 + 6q_3 = 3\frac{1}{2}$ [2]

$5q_1 + 2q_2 + q_3 = 3\frac{1}{2}$ [3]

e From part **d**, [3] − [1], $q_2 = 2\frac{1}{2} - 4q$. Sub in [1] gives

$q_3 = 3q - 1\frac{1}{2}$

f $q_2 \geq 0$, which gives $q \leq \frac{5}{8}$

$q_3 \geq 0$, which gives $\frac{1}{2} \leq q$

Hence $\frac{1}{2} \leq q \leq \frac{5}{8}$

Max $q_2 = 2\frac{1}{2} - 4 \times \frac{1}{2} = \frac{1}{2}$. Max $q_3 = 3 \times \frac{5}{8} - 1\frac{1}{2} = \frac{3}{8}$

So $0 \leq q_2 \leq \frac{1}{2}$, $0 \leq q_3 \leq \frac{3}{8}$

Review exercise 14

1 a, b

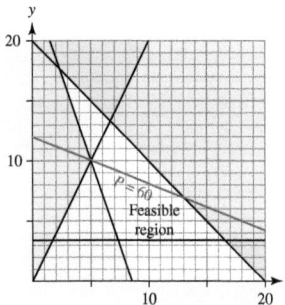

c $x = 6\frac{2}{3}$, $y = 13\frac{1}{3}$

2 a x = Value, y = Luxury, profit P = pence.

Maximise $P = 10x + 12y$

subject to $2x + 3y \leq 2400$

             $4x + 3y \leq 4050$

             $x \geq 0, y \geq 0$

b

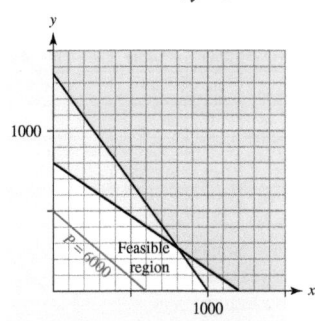

Maximum $P = 11250$, $x = 825$, $y = 250$

Make 825 kg of Value and 250 kg of Luxury

Profit = £112.50

3 a Play-safe strategy for A is A_2, with value 2.

b Play-safe strategy for B is B_2

c

		A	
		A_1	A_2
B	B_1	1	−4
	B_2	−3	−2

4 a

		B			Row minima	Max of row minima
		I	II	III		
A	I	1	0	−2	−2	
	II	−1	4	0	−1	
	III	2	3	5	2	2
	Column maxima	2	4	5		
	Min of column maxima	2				

Play-safe strategies are III for A and I for B.

Max of row minima = min of column maxima, so the solution is stable, that is a saddle point.

Neither player can gain by changing from their play-safe strategy.

b Value = 2

5 a Row 2 dominates row 3, so eliminate row 3

Column 3 dominates column 1 so eliminate column 1.

This leaves:

		B	
		II	III
A	I	3	9
	II	7	6

No further reduction is possible.

b The game does not have a stable solution because if it did the table could be reduced to a single value using dominance.

6 a Max of row mins = −1

Min of column maxs = 2

Solution is not stable

b Column 2 dominates column 3, so table becomes:

		B	
		B_1	B_2
A	A_1	3	−1
	A_2	−2	4

c $v \leq 3p - 2(1 - p) = 5p - 2$

$v \leq -p + 4(1 - p) = 4 - 5p$

d Graph as shown.

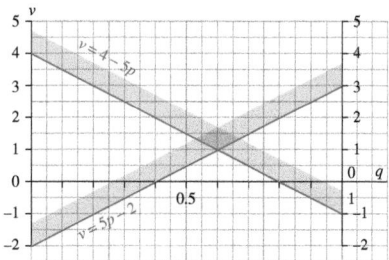

Max v when $5p - 2 = 4 - 5p$ giving $p = 0.6$.

A plays A_1, A_2 with probabilities 0.6, 0.4

When $p = 0.6$, $v = 1$

e If B plays B_1 with probability q then

$v = 3q - 2(1 - q) = 1$

This gives $q = 0.6$

B plays B_1, B_2 with probabilities 0.6, 0.4.

Assessment 14

1 a

		B				
		B_1	B_2	B_3	Row minima	Max of row minima
A	A_1	6	4	9	4	
	A_2	5	8	7	5	5
	A_3	3	5	6	3	
	Column maxima	6	8	9		
	Min of column maxima	6				

Play-safe strategy for A is A_2 and for B is B_1

b Game not stable as max and min values are different.

c Value of the game is 5

2 a Row M_1 dominates M_2 so M_2 can be removed.

		N		
		N_1	N_2	N_3
M	M_1	6	6	8
	M_3	4	2	−1

Column N_2 dominates N_1 so N_1 can be removed.

		N	
		N_2	N_3
M	M_1	6	8
	M_3	2	−1

Row M_1 dominates M_2 so M_2 can be removed.

		N	
		N_2	N_3
M	M_1	6	8

Column N_2 dominates N_3 so N_3 can be removed.

		N
		N_2
M	M_1	6

b You can deduce that there is a stable solution to the game. M should always play M_1 and N should always play N_2.
The value of the game is 6.

3 a Attempt to draw graph of $ax + by = c$
Correct plot of $x + y = 30$
Correct plot of $y = 2x$
Correct plot of $2y = 9$
Correct plot of $y + 2x = 28$
Correct shading of feasible region and identification in some suitable way.

b Attempt to draw graph of $ax + by = c$
Correct plot of $2x + 3y = m$

4 a

		B				Row minima
		B_1	B_2	B_3	B_4	
A	A_1	14	2	2	7	2
	A_2	6	2	2	9	2
	A_3	15	−1	1	0	−1
	Column maxima	15	2	2	9	

Maximum of row minima = 2
Minimum of column maxima = 2
The game is therefore stable as these values are equal.

b The value of the game is 2.

5 a 5

b Column I_3 is dominant over column I_2 (all values are lower) so Isla would never use this strategy.

c Table can be reduced to

	I_1	I_2	Row minima
C_1	2	−5	−5
C_2	−1	−3	−3
C_3	3	−5	−5
C_4	3	−2	−2
Column maxima	3	−2	

Play-safe for Connor is $C_4 (= −2)$
Play-safe for Isla is $I_2 (= −2)$
The game is stable as these values are equal.

6 a $x + y \leq 22$

b Correct plot of $x = 12$
Correct plot of $y = 9$
Correct plot of $x + y = 22$
Correct plot of $x + 3y = 45$
Clear identification of feasible region.

7 a

		Danny			
		Red	Blue	Purple	Row minima
Sanjay	Yellow	2	−4	−1	−4
	Orange	−4	1	3	−4
	Green	1	−1	1	−1
	Column maxima	2	1	3	

Play-safe strategy for Sanjay is Green
Play-safe strategy for Danny is Blue

b Sanjay should play Orange.

8 a Assume x = number of hectares of cauliflowers and y = number of hectares of sprouts
Maximise $P = 100x + 125y$
Subject to
$x + y \leq 20$ (land area)
$300x + 200y \leq 4800$ (cost per hectare)
$10x + 20y \leq 360$ (crop care)
$x \geq 0$
$y \geq 0$

b Farmer should plan 4 hectares of cauliflowers and 16 hectares of sprouts
Maximum Profit is £2400

9 a Let x = number of Supervit tablets and y = number of Extravit tablets

minimise $C = 15x + 20y$

Subject to

$15x + 20y \leq 600$ (cost)

$200x + 135y \leq 5400$ (Vit C)

$30x + 20y \geq 700$ (Niacin)

$20x + 40y \leq 800$

$x \geq 0, y \geq 0$

b Jenny should buy 16 tablets of Supervit and 12 tablets of Extravit.

Total cost is £4.80

Chapter 15

Exercise 15.1A

1 a $\circ$ is not a binary operation on $\mathbb{N}$ since when $a \leq b$, then $a - b \leq 0$ so $a \circ b$ is not a member of $\mathbb{N}$.

b $*$ is a binary operation on $\mathbb{N}$ since $a \times b$ is a natural number for all natural numbers a and b.

2 a $*$ is a binary operation on $\mathbb{Z}$ since $a - b$ is an integer for all integers a and b.

b $\lozenge$ is not a binary operation on $\mathbb{Z}$ since $\dfrac{a}{b^2 + 1}$ is not

necessarily an integer, for example $\dfrac{3}{2^2 + 1} = \dfrac{3}{5}$ which is not an integer.

3 a $\lozenge$ is a binary operation on $\mathbb{Z}^+$ since the max of a and b will be either a or b, therefore a positive integer.

b $*$ is a binary operation on $\mathbb{Z}^+$ since $2a + b$ is a positive integer for all integers a and b.

4 a $\circ$ is not a binary operation on $\mathbb{Q}$ since a^b not

always rational, for example if $a = 2$ and $b = \dfrac{1}{2}$ then

$a \circ b = 2^{\frac{1}{2}} = \sqrt{2}$ which is irrational.

b $*$ is not a binary operation on $\mathbb{Q}$ since $\dfrac{a}{b}$ is not defined for $b = 0$.

5 a $*$ is not a binary operation on the set of even integers

since $\dfrac{a}{2}$ could be odd so $\dfrac{a}{2} + b$ would be odd, e.g. $a = 6$,

$b = 8 \Rightarrow a * b = \dfrac{6}{2} + 8 = 11$ which is not in the set.

b $*$ is a binary operation on the set of even integers since if a and b are even then $3a$ and $7b$ must also be even so $3a - 7b$ is even.

6 a Is not commutative, e.g. $5 - 3 \neq 3 - 5$

Is not associative since $(a - b) - c = a - b - c$ but $a - (b - c) = a - b + c$

b Is commutative and associative since addition of numbers is commutative: $a + b = b + a$ and associative: $a + (b + c) = (a + b) + c$

c Is commutative and associative and since $a * b$ is the maximum of the two values so the order they are written makes no difference.

d Is not commutative: $a^2 + b \neq b^2 + a$ or associative: $a * (b * c) = a * (b^2 + c) = a^2 + b^2 + c$ but $(a * b) * c = (a^2 + b) * c = (a^2 + b)^2 + c = a^4 + 2a^2b + b^2 + c$

e Is commutative and associative since multiplication of numbers is commutative: $a \times b = b \times a$ and associative: $a \times (b \times c) = (a \times b) \times c$

f Is commutative and associative and since $a * b$ is the minimum of the two values so the order they are written makes no difference.

g Is commutative and associative since multiplication of numbers is commutative: $a \times b = b \times a$ and associative: $a \times (b \times c) = (a \times b) \times c$

7 a i identity element does not exist: $a - 0 = a$ but $0 - a \neq a$

b i identity element is 0

ii All elements have inverses; the inverse of a is $-a$

c i Identity element is 1

ii The only element with an inverse is 1, which is self-inverse.

d i Identity element does not exist

e i Identity element is 1

ii The only element with an inverse is 1, which is self-inverse.

f i The identity element is -1

ii The only element with an inverse is -1, which is self-inverse.

g i Identity element is 1.

ii All elements except 0 have an inverse; the inverse of a is $\dfrac{1}{a}$

8 a Is not commutative: $\begin{pmatrix} a_1 & a_2 \\ a_3 & a_4 \end{pmatrix} \begin{pmatrix} b_1 & b_2 \\ b_3 & b_4 \end{pmatrix}$

$= \begin{pmatrix} a_1 b_1 + a_2 b_3 & a_1 b_2 + a_2 b_4 \\ a_3 b_1 + a_4 b_3 & a_3 b_2 + a_4 b_4 \end{pmatrix}$

but $\begin{pmatrix} b_1 & b_2 \\ b_3 & b_4 \end{pmatrix} \begin{pmatrix} a_1 & a_2 \\ a_3 & a_4 \end{pmatrix}$

$= \begin{pmatrix} a_1 b_1 + a_3 b_2 & a_2 b_1 + a_4 b_2 \\ a_1 b_3 + a_3 b_4 & a_2 b_3 + a_4 b_4 \end{pmatrix}$ and in general $a_1 b_1 + a_3 b_2$

$\neq a_1 b_1 + a_2 b_3$ etc.

b $\begin{pmatrix} a_1 & a_2 \\ a_3 & a_4 \end{pmatrix} + \left(\begin{pmatrix} b_1 & b_2 \\ b_3 & b_4 \end{pmatrix} + \begin{pmatrix} c_1 & c_2 \\ c_3 & c_4 \end{pmatrix} \right)$

$= \begin{pmatrix} a_1 & a_2 \\ a_3 & a_4 \end{pmatrix} + \begin{pmatrix} b_1 + c_1 & b_2 + c_2 \\ b_3 + c_3 & b_4 + c_4 \end{pmatrix}$

$= \begin{pmatrix} a_1 + b_1 + c_1 & a_2 + b_2 + c_2 \\ a_3 + b_3 + c_3 & a_4 + b_4 + c_4 \end{pmatrix}$ so associative.

$\begin{pmatrix} a_1 & a_2 \\ a_3 & a_4 \end{pmatrix} + \begin{pmatrix} b_1 & b_2 \\ b_3 & b_4 \end{pmatrix} = \begin{pmatrix} a_1 + b_1 & a_2 + b_2 \\ a_3 + b_3 & a_4 + b_4 \end{pmatrix}$

$= \begin{pmatrix} b_1 & b_2 \\ b_3 & b_4 \end{pmatrix} + \begin{pmatrix} a_1 & a_2 \\ a_3 & a_4 \end{pmatrix}$

since addition of real numbers is commutative.

c i Multiplicative identity is $\begin{pmatrix} 1 & 0 \\ 0 & 1 \end{pmatrix}$ and addition identity

is $\begin{pmatrix} 0 & 0 \\ 0 & 0 \end{pmatrix}$

ii Multiplicative inverse of $\begin{pmatrix} a & b \\ c & d \end{pmatrix}$ is

$\dfrac{1}{ad - bc} \begin{pmatrix} d & -b \\ -c & a \end{pmatrix}$, it exists only when $ad - bc \neq 0$

Addition inverse of $\begin{pmatrix} a & b \\ c & d \end{pmatrix}$ is of $\begin{pmatrix} -a & -b \\ -c & -d \end{pmatrix}$ and

always exists.

9 a The identity element is 1; the numbers 1 and -1 are both self-inverse but 0 does not have an inverse.

b The identity element is 2, which is also self-inverse but the other elements do not have an inverse

c The identity element is 0, all elements have an inverse; the inverse of a is $-a$.

d The identity element is 1, which is also self-inverse but the other elements do not have an inverse.

Exercise 15.1B

1 i a Identity element is c

 b Inverse of a is b and vice-versa; c is self-inverse.

 ii a Identity element is a

 b Inverse of b is d and vice-versa; a and c are self-inverse.

 iii a Identity element is B

 b All elements are self-inverse.

 iv a Identity element is A

 b Inverse of B is D and vice-versa; inverse of C is E and vice-versa; A is self-inverse.

2 a $a*(b*c) = a*b = c$ and $(a*b)*c = c*c = c$
$b*(a*c) = b*a = c$ and $(b*a)*c = c*c = c$
$c*(a*b) = c*c = c$ and $(c*a)*b = a*b = c$
No need to check, e.g. $a*(c*b)$ as function is commutative so $b*c = c*b$
Therefore it is associative.

 b $a*(b*c) = a*d = d$ and $(a*b)*c = b*c = d$
$a*(b*d) = a*a = a$ and $(a*b)*d = b*d = a$
$a*(d*c) = a*b = b$ and $(a*d)*c = d*c = b$
$b*(a*c) = b*c = d$ and $(b*a)*c = b*c = d$
$b*(a*d) = b*d = a$ and $(b*a)*d = b*d = a$
$b*(c*d) = b*b = c$ and $(b*c)*d = d*d = c$
$c*(a*b) = c*b = d$ and $(c*a)*b = c*b = d$
$c*(a*d) = c*d = b$ and $(c*a)*d = c*d = b$
$c*(b*d) = c*a = c$ and $(c*b)*d = d*d = c$
No need to check, e.g. $a*(c*b)$ as function is commutative so $b*c = c*b$
Therefore it is associative.

3 a

$\times$	1	-1	i	$-i$
1	1	-1	i	$-i$
-1	-1	1	$-i$	i
i	i	$-i$	-1	1
$-i$	$-i$	i	1	-1

 b The identity element is 1

 c The inverse of i is $-i$ and vice-versa; 1 and -1 are self-inverse.

 d The operation is associative since multiplication of complex numbers is associative.

4 a $S = \left\{ \begin{pmatrix} 1 & 0 \\ 0 & 1 \end{pmatrix}, \begin{pmatrix} 0 & -1 \\ 1 & 0 \end{pmatrix}, \begin{pmatrix} -1 & 0 \\ 0 & -1 \end{pmatrix}, \begin{pmatrix} 0 & 1 \\ -1 & 0 \end{pmatrix} \right\}$

 b Not a binary operation since A + B not necessarily a member of S, for example $\begin{pmatrix} 1 & 0 \\ 0 & 1 \end{pmatrix} + \begin{pmatrix} 1 & 0 \\ 0 & 1 \end{pmatrix} = \begin{pmatrix} 2 & 0 \\ 0 & 2 \end{pmatrix}$ which is not in S.

 c

	0	1	2	3
0	0	1	2	3
1	1	2	3	0
2	2	3	0	1
3	3	0	1	2

 d i Identity is $\begin{pmatrix} 1 & 0 \\ 0 & 1 \end{pmatrix}$ or 0

 ii 0 and 2 are self-inverses, 1 is inverse of 3 and vice-versa.

 e Represent rotations of $0, \dfrac{\pi}{2}, \pi$ and $\dfrac{3\pi}{2}$ radians anti-clockwise around the origin.

5 a Minimum value of k is 5

 b

	1	2	3	4	5
1	2	3	4	5	1
2	3	4	5	1	2
3	4	5	1	2	3
4	5	1	2	3	4
5	1	2	3	4	5

 c 1 is inverse of 4 and vice-versa; 2 is inverse of 3 and vice-versa; 5 is a self-inverse.

6 a A, B, C, D, F, H are self-inverses; E is the inverse of G and vice-versa

 b $A = \begin{pmatrix} 1 & 0 \\ 0 & -1 \end{pmatrix}$, $B = \begin{pmatrix} -1 & 0 \\ 0 & 1 \end{pmatrix}$, $C = \begin{pmatrix} 0 & 1 \\ 1 & 0 \end{pmatrix}$,

 $D = \begin{pmatrix} 0 & -1 \\ -1 & 0 \end{pmatrix}$, $E = \begin{pmatrix} 0 & -1 \\ 1 & 0 \end{pmatrix}$, $F = \begin{pmatrix} -1 & 0 \\ 0 & -1 \end{pmatrix}$,

 $G = \begin{pmatrix} 0 & 1 \\ -1 & 0 \end{pmatrix}$, $H = \begin{pmatrix} 1 & 0 \\ 0 & 1 \end{pmatrix}$

 c

$*$	A	B	C	D	E	F	G	H
A	H	F	E	G	C	B	D	A
B	F	H	G	E	D	A	C	B
C	G	E	H	F	B	D	A	C
D	E	G	F	H	A	C	B	D
E	D	C	A	B	F	G	H	E
F	B	A	D	C	G	H	E	F
G	C	D	B	A	H	E	F	G
H	A	B	C	D	E	F	G	H

 d e.g. A * C = E but C * A = G

7 a All the possible combinations of matrices of this form are in S so it is a binary operation

 b $\begin{pmatrix} 1 & 0 & 0 \\ 0 & 1 & 0 \\ 0 & 0 & 1 \end{pmatrix}$ since this is the identity matrix, I, and

 AI = IA = A for all matrices A

 c $\begin{pmatrix} a & 0 & 0 \\ 0 & b & 0 \\ 0 & 0 & c \end{pmatrix}^2 = \begin{pmatrix} a^2 & 0 & 0 \\ 0 & b^2 & 0 \\ 0 & 0 & c^2 \end{pmatrix}$ if $a, b, c = \pm 1$ then

 $a^2, b^2, c^2 = 1$ therefore all the matrices are self-inverse

 d $\begin{pmatrix} a & 0 & 0 \\ 0 & b & 0 \\ 0 & 0 & c \end{pmatrix} \begin{pmatrix} d & 0 & 0 \\ 0 & e & 0 \\ 0 & 0 & f \end{pmatrix} = \begin{pmatrix} ad & 0 & 0 \\ 0 & be & 0 \\ 0 & 0 & cf \end{pmatrix}$

 and $\begin{pmatrix} d & 0 & 0 \\ 0 & e & 0 \\ 0 & 0 & f \end{pmatrix} \begin{pmatrix} a & 0 & 0 \\ 0 & b & 0 \\ 0 & 0 & c \end{pmatrix} = \begin{pmatrix} ad & 0 & 0 \\ 0 & be & 0 \\ 0 & 0 & cf \end{pmatrix}$

 So matrix multiplication is commutative on S

Exercise 15.2A

1 e.g.
  a 7 (mod 3) = 4
  b 12 (mod 7) = 5
  c 19 (mod 4) = 3
  d 27 (mod 9) = 0
  e 25 (mod 6) = 1
  f −5 (mod 8) = 3
  g −55 (mod 11) = 0
  h 48 (mod 13) = 9

2
  a 21 + 19 (mod 3) = 0 + 1 (mod 3) = 1 (mod 3)
  b 4 + 25 (mod 7) = 4 + 4 (mod 7) = 8 (mod 7) = 1 (mod 7)
  c 68 − 34 (mod 4) = 0 − 2 (mod 4) = −2 (mod 4) = 2 (mod 4)
  d 79 − 94 (mod 9) = 7 − 4 (mod 9) = 3(mod 9)
  e 5 × 14 (mod 6) = 5 × 2 (mod 6) = 12 (mod 6) = 4 (mod 6)
  f 25 × 37 (mod 8) = 1 × 5 (mod 8) = 5 (mod 8)
  g 82 × −16 (mod 11) = 5 × 6 (mod 11) = 30 (mod 11) = 8 (mod 11)
  h 29 × 75 (mod 13) = 3 × 10 (mod 13) = 30 (mod 13) = 4 (mod 13)

3
  a $37951 \equiv 1^{37} \,(\text{mod } 10)$
    $\equiv 1(\text{mod } 10)$
  b $1736 \equiv 1^{26}(\text{mod } 5)$
    $\equiv 1(\text{mod } 5)$
  c $982 \equiv 2^{9}(\text{mod } 5)$
    $\equiv 2\,(\text{mod } 5)$
  d $3219 \equiv 0^{48}(\text{mod } 3)$
    $\equiv 0\,(\text{mod } 3)$
  e $3268 \equiv 1^{84}(\text{mod } 3)$
    $\equiv 1(\text{mod } 3)$
  f $12653 \equiv 5(\text{mod } 6)$
    $\equiv (-1)(\text{mod } 6)$
    so $12653^{11} \equiv (-1)^{11}(\text{mod } 6)$
    $\equiv (-1)(\text{mod } 6)$
    $\equiv 5(\text{mod } 6)$

Exercise 15.2B

1 a

*	0	1
0	0	1
1	1	0

2 a

*	0	1	2
0	0	1	2
1	1	2	0
2	2	0	1

b

*	0	1	2	3
0	0	1	2	3
1	1	2	3	0
2	2	3	0	1
3	3	0	1	2

c

*	1	2	3	4
1	1	2	3	4
2	2	4	1	3
3	3	1	4	2
4	4	3	2	1

d

*	1	2	3	4	5	6
1	1	2	3	4	5	6
2	2	4	6	1	3	5
3	3	6	2	5	1	4
4	4	1	5	2	6	3
5	5	3	1	6	4	2
6	6	5	4	3	2	1

3 In question 1
  i identity element is 0
  ii both elements are self-inverse
  In question 2
  a i identity element is 0
    ii 1 is the inverse of 2 and vice-versa; 0 is self-inverse
  b i identity element is 0
    ii 1 is the inverse of 3 and vice-versa; 0 and 2 are self-inverse
  c i identity element is 1
    ii 2 is the inverse of 3 and vice-versa; 1 and 4 are self-inverse
  d i identity element is 1
    ii 2 is the inverse of 4 and vice-versa; 3 is the inverse of 5 and vice-versa; 1 and 6 are self-inverse

4 a

*	0	1	2	3
0	0	1	2	3
1	1	3	0	2
2	2	0	3	1
3	3	2	1	0

b i Identity is 0
  ii 0 and 3 are self-inverses, 1 is inverse of 2 and vice-versa.

5 a

◆	0	1	2	3	4	5
0	4	5	0	1	2	3
1	5	0	1	2	3	4
2	0	1	2	3	4	5
3	1	2	3	4	5	0
4	2	3	4	5	0	1
5	3	4	5	0	1	2

b i Identity is 2
  ii 2 and 5 are self-inverses, 0 is inverse of 4 and vice-versa; 1 is inverse of 3 and vice-versa.

Review exercise 15

1 a Is a binary operation since $a + 3b$ always a natural number if a and b are natural numbers.
  b Is not a binary operation, if $b = 1$ then $(a+1)(b-1) = (a-1) \times 0 = 0$ which is not a positive integers.
  c Is a binary operation: if $a, b \in \mathbb{Q}$ then they can be written $a = \dfrac{p}{q}, b = \dfrac{m}{n}$ for some integers p, q, m, n.
    So $\dfrac{b}{a^2+1} = \dfrac{\dfrac{m}{n}}{\left(\dfrac{p}{q}\right)^2 + 1} = \dfrac{mq^2}{n(p^2+q^2)} \in \mathbb{Q}$ since p, q, m, n are integers.
  d Is not a binary operation, if one of a and b is negative then ab will be negative so $\sqrt{ab}$ will not be a real number, e.g. $a = 1, b = -1, \sqrt{ab} = \sqrt{-1} = i$.

2 **a** Not associative since $\dfrac{a}{\frac{b}{c}}=\dfrac{ac}{b}$ but $\dfrac{\frac{a}{b}}{c}=\dfrac{a}{bc}$

   b Is not associative: $(a^2+b^2)^2+c^2=a^4+2a^2b^2+b^4+c^2$

     $a^2+(b^2+c^2)^2=a^2+b^4+2b^2c^2+c^4\neq a^4+2a^2b^2+b^4+c^2$

   c Associative since $\dfrac{a\left(\frac{bc}{b+c}\right)}{a+\left(\frac{bc}{b+c}\right)}=\dfrac{abc}{a(b+c)+bc}=\dfrac{abc}{ab+ac+bc}$

     and $\dfrac{\left(\frac{ab}{a+b}\right)c}{\left(\frac{ab}{a+b}\right)+c}=\dfrac{abc}{ab+c(a+b)}=\dfrac{abc}{ab+ac+bc}$

3 **a** Not commutative, in general $\dfrac{a}{b}\neq\dfrac{b}{a}$

   b Commutative since $a^2+b^2=b^2+a^2$

   c Commutative since $\dfrac{ab}{a+b}=\dfrac{ba}{b+a}$

4 **a** **i** Identity element does not exist (since not commutative).

     ii Therefore no inverse elements.

   b **i** Identity element is 0 since $0^2+a^2=a^2+0^2=a^2$

     ii 0 is self-inverse but no other elements have an inverse in $\mathbb{Z}$ since $a^2+(a^{-1})^2=0\Rightarrow(a^{-1})^2\leq0$ which is only possible if a^{-1} is imaginary.

   c **i** No identity element.

     ii So no inverse elements.

5 **a** Identity element is E

   b Inverse of A is D and vice-versa; inverse of B is C and vice-versa; E is self-inverse.

   c Cayley table is symmetric along leading diagonal therefore * is commutative.

6 **a** $379+612 \pmod 5\equiv 4+2\pmod 5\equiv 6\pmod 5\equiv 1\pmod 5$

   b $1079-351\pmod 3\equiv 2-0\pmod 3\equiv 2\pmod 3$

   c $326\times249\pmod 8\equiv 6\times 1\pmod 8\equiv 6\pmod 8$

   d $-532\times249\pmod{10}\equiv-2\times 9\pmod{10}\equiv 8\times 9\pmod{10}\equiv 72\pmod{10}\equiv 2\pmod{10}$

7 **a**

*	0	2	4	6	8
0	0	2	4	6	8
2	2	4	6	8	0
4	4	6	8	0	2
6	6	8	0	2	4
8	8	0	2	4	6

   b Identity element is 0

   c The inverse of 2 is 8 and vice-versa; the inverse of 4 is 6 and vice-versa; 0 is self-inverse.

8 **a**

	A	B	C
A	B	C	A
B	C	A	B
C	A	B	C

   b **i** identity is $\begin{pmatrix}1 & 0\\0 & 1\end{pmatrix}$ or C

     ii C is self-inverse, A is inverse of B and vice-versa.

   c Rotations of 0°, 120°, 240° clockwise around origin.

Assessment 15

1 **a** 52

   b e.g. $2\blacksquare(3\blacksquare1)=2+(3+1^2)^2=18$

     $(2\blacksquare3)\blacksquare1=(2+3^2)+1^2=12$

   So not associative

   e.g. $2\blacksquare3=2+3^2=11$

   $3\blacksquare2=3+2^2=7$ so not commutative

2 **a** $3+7\equiv0\pmod{10}$

   b $125-621\equiv5-1\pmod{10}$

     $\equiv4\pmod{10}$

   c $358\times6914\equiv8\times4\pmod{10}$

     $32\equiv2\pmod{10}$

   d $2563^4\equiv3^4\pmod{10}$

     $81\equiv1\pmod{10}$

3 **a** **i** If b is a fraction, then a^b may not be rational, e.g. if $b=\dfrac{1}{2}$, $a=2$ then $2^{\frac{1}{2}}=\sqrt2$ which is irrational.

     So not a binary operation on $\mathbb{Q}$.

     ii If a is negative then a^b may not be a real number, e.g. $a=-1$, $b=\dfrac{1}{2}$. Then $(-1)^{\frac{1}{2}}=i$ which is an imaginary number

     So not a binary operation on $\mathbb{R}$

     iii If a and b are natural numbers then a^b will also be a natural number so it is a binary operation on $\mathbb{N}$.

   b In general $a^b\neq b^a$ so operation is not commutative, therefore no identity element

   c $(a*b)*c=(a^b)^c=a^{bc}$

     $a*(b*c)=a^{(b^c)}$

     so not associative

4 **a**

*	−2	2
−2	2	−2
2	−2	2

   b 2

   c both elements are self-inverses

5 **a** the identity element is U

   b $V^{-1}=X$

   c U and W are self-inverses

6 **a** **i**

*	0	1	2	3	4
0	0	1	2	3	4
1	1	2	3	4	0
2	2	3	4	0	1
3	3	4	0	1	2
4	4	0	1	2	3

     ii identity element is 0

     iii inverse of 1 is 4 and vice-versa; inverse of 2 is 3 and vice-versa; 0 is self-inverse

   b e.g. $18^{12}\equiv3^{12}\pmod 5$

     $\equiv(3^4)^3\pmod 5$

     $\equiv81^3\pmod 5$

     $\equiv1^3\pmod 5$

     $\equiv1\pmod 5$

7 **a** C

   b The operation is not commutative

   There is no element, e, such that $x\blacklozenge e=e\blacklozenge x=x$

   Therefore no identity element

   c e.g. $(A\blacklozenge B)\blacklozenge C=B\blacklozenge C=C$

   but $A\blacklozenge(B\blacklozenge C)=A\blacklozenge C=E$

   so not associative

8 a

	A	B	C	D
A	B	C	D	A
B	C	D	A	B
C	D	A	B	C
D	A	B	C	D

b D and B are self-inverses

A is inverse of C and vice-versa

9 Let $a = 2m + 1$, $b = 2n + 1$ where m and n are integers.

Then $a * b = 2(2m + 1) - (2n + 1)$

$$= 4m + 2 - 2n - 1$$

$$= 2(2m - n) + 1$$

Which is an odd number since $2(2m - n)$ is even.

Hence it is a binary operation

10 a

×	M_1	M_2	M_3	M_4	M_5	M_6
M_1	M_2	M_3	M_4	M_5	M_6	M_1
M_2	M_3	M_4	M_5	M_6	M_1	M_2
M_3	M_4	M_5	M_6	M_1	M_2	M_3
M_4	M_5	M_6	M_1	M_2	M_3	M_4
M_5	M_6	M_1	M_2	M_3	M_4	M_5
M_6	M_1	M_2	M_3	M_4	M_5	M_6

b M_6

c M_4

d M_3

Chapter 16

Exercise 16.1A

1 **a** $5e^{i0.927}$ **b** $\sqrt{5}e^{-i0.464}$ **c** $10e^{i0}$

     d $5e^{i\pi}$ **e** $2e^{i\frac{\pi}{2}}$ **f** $6e^{-i\frac{\pi}{2}}$

     g $13e^{i1.97}$ **h** $4\sqrt{5}e^{-i2.03}$ **i** $2e^{i\frac{\pi}{6}}$ **j** $5\sqrt{2}e^{-i\frac{\pi}{4}}$

2 **a** $2e^{\frac{\pi}{12}i}$ **b** $4e^{-\frac{2\pi}{3}i}$ **c** $3e^{-\frac{5\pi}{6}i}$

     d $6e^{-\frac{\pi}{7}i}$ **e** $e^{-\frac{3\pi}{5}i}$ **f** $\sqrt{2}e^{\frac{\pi}{8}i}$

     g $\sqrt{3}e^{\frac{5\pi}{6}i}$ **h** $8e^{-\frac{7\pi}{12}i}$

3 **a** $2i$ **b** $\dfrac{7}{2}-\dfrac{7\sqrt{3}}{2}i$

     c $1+i$ **d** $\dfrac{\sqrt{3}}{2}-\dfrac{1}{2}i$

     e $-2\sqrt{2}$ **f** $-\dfrac{3}{2}+\dfrac{\sqrt{3}}{2}i$

4 **a** 6   **b** $\dfrac{2}{3}$ **c** 0   **d** $\dfrac{2\pi}{3}$

5 **a** 1   **b** 25   **c** $\dfrac{\pi}{7}$ **d** $\dfrac{3\pi}{7}$

6 **a** $3\sqrt{2}$ **b** $\sqrt{2}$ **c** $\dfrac{11\pi}{12}$ **d** $\dfrac{7\pi}{12}$

Exercise 16.1B

1 $e^{i\theta} = \cos\theta + i\sin\theta$

$e^{-i\theta} = \cos(-\theta) + i\sin(-\theta) = \cos\theta - i\sin\theta$

$e^{i\theta} - e^{-i\theta} = \cos\theta + i\sin\theta - (\cos\theta - i\sin\theta)$

$\qquad = 2i\sin\theta$

$\sin\theta = \dfrac{e^{i\theta} - e^{-i\theta}}{2i}$ as required

2 $z = e^{\theta i}$

   a $z^n + \dfrac{1}{z^n} = (e^{\theta i})^n + \dfrac{1}{(e^{\theta i})^n}$

$\qquad\qquad = e^{n\theta i} + e^{-n\theta i}$

$\qquad\qquad = 2\cos(n\theta)$ as required

   b $z^n - \dfrac{1}{z^n} = (e^{\theta i})^n - \dfrac{1}{(e^{\theta i})^n}$

$\qquad\qquad = e^{n\theta i} - e^{-n\theta i}$

$\qquad\qquad = 2i\sin(n\theta)$ as required

3 **a** $z_1 z_2 = r_1 e^{\theta_1 i} \times r_2 e^{\theta_2 i} = r_1 r_2 e^{\theta_1 i} e^{\theta_2 i} = r_1 r_2 e^{(\theta_1+\theta_2)i}$

So $|z_1 z_2| = r_1 r_2 = |z_1||z_2|$

and $\arg(z_1 z_2) = \theta_1 + \theta_2 = \arg z_1 + \arg z_2$

   b $\dfrac{z_1}{z_2} = \dfrac{r_1 e^{\theta_1 i}}{r_2 e^{\theta_2 i}} = \dfrac{r_1}{r_2} e^{(\theta_1-\theta_2)i}$

So $\left|\dfrac{z_1}{z_2}\right| = \dfrac{r_1}{r_2} = \dfrac{|z_1|}{|z_2|}$

and $\arg\left(\dfrac{z_1}{z_2}\right) = \theta_1 - \theta_2 = \arg z_1 - \arg z_2$

4 **a** $\text{RHS} = \left(\dfrac{e^{iA}-e^{-iA}}{2i}\right)\left(\dfrac{e^{iB}+e^{-iB}}{2}\right) + \left(\dfrac{e^{iB}-e^{-iB}}{2i}\right)\left(\dfrac{e^{iA}+e^{-iA}}{2}\right)$

$= \dfrac{\left(e^{i(A+B)}+e^{i(A-B)}-e^{-i(A-B)}-e^{-i(A+B)}\right)}{4i}$

$\quad + \dfrac{\left(e^{i(A+B)}+e^{-i(A-B)}-e^{i(A-B)}-e^{-i(A+B)}\right)}{4i}$

$= \dfrac{2e^{i(A+B)}-2e^{-i(A+B)}}{4i}$

$= \dfrac{e^{i(A+B)}-e^{-i(A+B)}}{2i} = \sin(A+B)$ as required

   b $\text{RHS} = \left(\dfrac{e^{iA}+e^{-iA}}{2}\right)\left(\dfrac{e^{iB}+e^{-iB}}{2}\right) - \left(\dfrac{e^{iA}-e^{-iA}}{2i}\right)\left(\dfrac{e^{iB}-e^{-iB}}{2i}\right)$

$= \dfrac{e^{i(A+B)}+e^{i(A-B)}+e^{-i(A-B)}+e^{-i(A+B)}}{4}$

$\quad - \dfrac{e^{i(A+B)}-e^{i(A-B)}-e^{-i(A-B)}+e^{-i(A+B)}}{4i^2}$

$= \dfrac{e^{i(A+B)}+e^{i(A-B)}+e^{-i(A-B)}+e^{-i(A+B)}}{4}$

$\quad + \dfrac{e^{i(A+B)}-e^{i(A-B)}-e^{-i(A-B)}+e^{-i(A+B)}}{4}$

$= \dfrac{2e^{i(A+B)}+2e^{-i(A+B)}}{4}$

$= \dfrac{e^{i(A+B)}+e^{-i(A+B)}}{2} = \cos(A+B)$ as required

5 **a** $\text{RHS} = \left(\dfrac{e^{ix}+e^{-ix}}{2}\right)^2 - \left(\dfrac{e^{ix}-e^{-ix}}{2i}\right)^2$

$= \dfrac{e^{2ix}+2+e^{-2ix}}{4} - \dfrac{e^{2ix}-2+e^{-2ix}}{4i^2}$

$= \dfrac{e^{2ix}+2+e^{-2ix}}{4} + \dfrac{e^{2ix}-2+e^{-2ix}}{4}$

$= \dfrac{2e^{2ix}+2e^{-2ix}}{4}$

$= \dfrac{e^{2ix}+e^{-2ix}}{2} = \cos 2x$ as required

   b $\text{LHS} = \left(\dfrac{e^{ix}+e^{-ix}}{2}\right)^2 + \left(\dfrac{e^{ix}-e^{-ix}}{2i}\right)^2$

$= \dfrac{e^{2ix}+2+e^{-2ix}}{4} + \dfrac{e^{2ix}-2+e^{-2ix}}{4i^2}$

$= \dfrac{e^{2ix}+2+e^{-2ix}}{4} - \dfrac{e^{2ix}-2+e^{-2ix}}{4}$

$= \dfrac{4}{4} = 1$ as required

6 a LHS $= \left(\dfrac{e^{i\theta} + e^{-i\theta}}{2} + i \left(\dfrac{e^{i\theta} - e^{-i\theta}}{2i} \right) \right)^2$

$= \left(\dfrac{e^{i\theta} + e^{-i\theta}}{2} + \dfrac{e^{i\theta} - e^{-i\theta}}{2} \right)^2$

$= \left(\dfrac{2e^{i\theta}}{2} \right)^2 = (e^{i\theta})^2$

$= e^{2i\theta}$

RHS $= \dfrac{e^{2i\theta} + e^{-2i\theta}}{2} + i \left(\dfrac{e^{2i\theta} - e^{-2i\theta}}{2i} \right)$

$= \dfrac{e^{2i\theta} + e^{-2i\theta}}{2} + \dfrac{e^{2i\theta} - e^{-2i\theta}}{2}$

$= \dfrac{2e^{2i\theta}}{2} = e^{2i\theta} =$ LHS

so $(\cos\theta + i\sin\theta)^2 \equiv \cos 2\theta + i\sin 2\theta$

b LHS $= \left(\dfrac{e^{i\theta} + e^{-i\theta}}{2} + i \left(\dfrac{e^{i\theta} - e^{-i\theta}}{2i} \right) \right)^n$

$= \left(\dfrac{e^{i\theta} + e^{-i\theta}}{2} + \dfrac{e^{i\theta} - e^{-i\theta}}{2} \right)^n$

$= \left(\dfrac{2e^{i\theta}}{2} \right)^n = (e^{i\theta})^n$

$= e^{in\theta}$

RHS $= \dfrac{e^{in\theta} + e^{-in\theta}}{2} + i \left(\dfrac{e^{in\theta} - e^{-in\theta}}{2i} \right)$

$= \dfrac{e^{in\theta} + e^{-in\theta}}{2} + \dfrac{e^{in\theta} - e^{-in\theta}}{2}$

$= \dfrac{2e^{in\theta}}{2} = e^{in\theta} =$ LHS

so $(\cos\theta + i\sin\theta)^n \equiv \cos(n\theta) + i\sin(n\theta)$

7 $zw = (4 \times 3) \left(\cos \left(\dfrac{\pi}{9} + \dfrac{2\pi}{9} \right) + i\sin \left(\dfrac{\pi}{9} + \dfrac{2\pi}{9} \right) \right)$

$= 12 \left(\cos \left(\dfrac{\pi}{3} \right) + i\sin \left(\dfrac{\pi}{3} \right) \right)$

$= 12 \left(\dfrac{1}{2} + i \left(\dfrac{\sqrt{3}}{2} \right) \right)$

$= 6 + 6\sqrt{3}i$ as required

8 $z^2 = \left(8e^{i\frac{5\pi}{12}} \right)^2 = 64e^{i\frac{5\pi}{6}}$

$= 64 \left(\cos \left(\dfrac{5\pi}{6} \right) + i\sin \left(\dfrac{5\pi}{6} \right) \right)$

$= 64 \left(\left(-\dfrac{\sqrt{3}}{2} \right) + i \left(\dfrac{1}{2} \right) \right) = -32\sqrt{3} + 32i$

9 $|w| = \sqrt{1^2 + 1^2} = \sqrt{2}$ $\arg(w) = -\dfrac{\pi}{4}$

a $|zw| = |z||w| = \sqrt{2}k$

$\arg(zw) = \arg(z) + \arg(w) = \theta - \dfrac{\pi}{4}$

b $\left| \dfrac{z}{w} \right| = \dfrac{|z|}{|w|} = \dfrac{k}{\sqrt{2}} = \dfrac{\sqrt{2}}{2}k$

$\arg \left(\dfrac{z}{w} \right) = \arg(z) - \arg(w) = \theta + \dfrac{\pi}{4}$

Exercise 16.2A

1 a 1   **b** $-\dfrac{\sqrt{3}}{2} + \dfrac{1}{2}i$ **c** $\dfrac{1}{2} - \dfrac{\sqrt{3}}{2}i$ **d** $-i$

e $-\dfrac{\sqrt{2}}{2} - \dfrac{\sqrt{2}}{2}i$

2 a $\dfrac{81}{2}\sqrt{3} + \dfrac{81}{2}i$ **b** $\dfrac{\sqrt{2}}{1458} - \dfrac{\sqrt{2}}{1458}i$

3 a $4\sqrt{2} + 4\sqrt{2}i$ **b** $\dfrac{\sqrt{3}}{8} - \dfrac{1}{8}i$

c $64i$ **d** $\dfrac{1}{32} - \dfrac{\sqrt{3}}{32}i$

4 a -16 **b** $64i$

c $\dfrac{1}{4}i$ **d** $\dfrac{1}{16}$

5 a $-2 - 2i$ **b** $8 + 8i$

c $\dfrac{1}{8} - \dfrac{1}{8}i$ **d** $-\dfrac{1}{8}i$

6 a -9 **b** $-\dfrac{1}{3}i$

c $\dfrac{1}{27}i$ **d** $\dfrac{1}{9}i$

7 a $8 - 8\sqrt{3}i$ **b** $-\dfrac{1}{8}i$

c $\dfrac{1}{8} + \dfrac{\sqrt{3}}{8}i$ **d** $-\dfrac{1}{8}$

Exercise 16.2B

1 a $2\cos^2\theta \equiv \dfrac{1}{2}(2\cos\theta)^2$

$\equiv \dfrac{1}{2}(e^{i\theta} + e^{-i\theta})^2$

$\equiv \dfrac{1}{2}(e^{2i\theta} + 2 + e^{-2i\theta})$

$\equiv \dfrac{1}{2}(2\cos 2\theta + 2)$

$\equiv \cos 2\theta + 1$

b $8\sin^3\theta \equiv \dfrac{1}{i^3}(2i\sin\theta)^3$

$\equiv \dfrac{1}{-i}(e^{i\theta} - e^{-i\theta})^3$

$\equiv \dfrac{-1}{i}(e^{3i\theta} - e^{-3i\theta} - 3e^{i\theta} + 3e^{-i\theta})$

$\equiv -\dfrac{1}{i}(2i\sin 3\theta - 6i\sin\theta)$

$\equiv 6\sin\theta - 2\sin 3\theta$

c $4\sin^4\theta \equiv \dfrac{1}{4}(2i\sin\theta)^4$

$\equiv \dfrac{1}{4}(e^{i\theta} - e^{-i\theta})^4$

$\equiv \dfrac{1}{4}(e^{4i\theta} + e^{-4i\theta} - 4e^{2i\theta} - 4e^{-2i\theta} + 6)$

$\equiv \dfrac{1}{4}(2\cos 4\theta - 8\cos 2\theta + 6)$

$\equiv \dfrac{1}{2}\cos 4\theta - 2\cos 2\theta + \dfrac{3}{2}$

2 a $\cos^5\theta \equiv \dfrac{1}{32}(2\cos\theta)^5$

$\qquad \equiv \dfrac{1}{32}(e^{i\theta}+e^{-i\theta})^5$

$\qquad \equiv \dfrac{1}{32}(e^{5i\theta}+e^{-5i\theta}+5e^{3i\theta}+5e^{-3i\theta}+10e^{i\theta}+10e^{-i\theta})$

$\qquad \equiv \dfrac{1}{32}(2\cos5\theta+10\cos3\theta+20\cos\theta)$

$\qquad \equiv \dfrac{1}{16}(10\cos\theta+5\cos3\theta+\cos5\theta)$

$\quad$ So $A=\dfrac{1}{16}$

b $\dfrac{1}{16}\left(10\sin\theta+\dfrac{5}{3}\sin3\theta+\dfrac{1}{5}\sin5\theta\right)+c$

3 a $\sin^6\theta \equiv \dfrac{1}{-64}(2i\sin\theta)^6 \equiv -\dfrac{1}{64}(e^{i\theta}-e^{-i\theta})^6$

$\qquad \equiv -\dfrac{1}{64}(e^{6i\theta}+e^{-6i\theta}-6e^{4i\theta}-6e^{-4i\theta}+15e^{2i\theta}+15e^{-2i\theta}-20)$

$\qquad \equiv -\dfrac{1}{64}(2\cos6\theta-12\cos4\theta+30\cos2\theta-20)$

$\qquad \equiv -\dfrac{1}{32}(15\cos2\theta-6\cos4\theta+\cos6\theta-10)$

$\quad$ So $B=-\dfrac{1}{32}$

b $-\dfrac{1}{32}\left(\dfrac{15}{2}\sin2\theta-\dfrac{3}{2}\sin4\theta+\dfrac{1}{6}\sin6\theta-10\theta\right)+c$

4 a $2\sin^3\theta = \dfrac{1}{-4i}(2i\sin\theta)^3 = -\dfrac{1}{4i}(e^{i\theta}-e^{-i\theta})^3$

$\qquad = -\dfrac{1}{4i}\left(e^{3i\theta}-e^{-3i\theta}-3e^{i\theta}+3e^{-i\theta}\right)$

$\qquad = -\dfrac{1}{4i}(2i\sin3\theta-6i\sin\theta) = \dfrac{3}{2}\sin\theta-\dfrac{1}{2}\sin3\theta$

b $\theta=\dfrac{\pi}{6},\dfrac{5\pi}{6}$

5 a $5\cos^4\theta = \dfrac{5}{16}(2\cos\theta)^4 = \dfrac{5}{16}(e^{i\theta}+e^{-i\theta})^4$

$\qquad = \dfrac{5}{16}(e^{4i\theta}+e^{-4i\theta}+4e^{2i\theta}+4e^{-2i\theta}+6)$

$\qquad = \dfrac{5}{16}(2\cos4\theta+8\cos2\theta+6)$

$\qquad = \dfrac{5}{8}\cos4\theta+\dfrac{5}{2}\cos2\theta+\dfrac{15}{8}$

$\quad$ So $A=\dfrac{5}{8}$, $B=\dfrac{5}{2}$ and $C=\dfrac{15}{8}$

b $\theta=\pm\dfrac{\pi}{4},\pm\dfrac{3\pi}{4}$

6 a $\cos2\theta+i\sin2\theta=(\cos\theta+i\sin\theta)^2$

$\qquad = \cos^2\theta+2i\cos\theta\sin\theta+i^2\sin^2\theta$

$\qquad = \cos^2\theta+2i\cos\theta\sin\theta-\sin^2\theta$

$\quad$ Im $:\sin2\theta=2\cos\theta\sin\theta$

b $\cos3\theta+i\sin3\theta=(\cos\theta+i\sin\theta)^3$

$\qquad = \cos^3\theta+3\cos^2\theta(i\sin\theta)+3\cos\theta(i\sin\theta)^2$

$\qquad +(i\sin\theta)^3$

$\qquad = \cos^3\theta+3i\cos^2\theta\sin\theta-3\cos\theta\sin^2\theta$

$\qquad -i\sin^3\theta$

$\quad$ Im $:\sin3\theta=3\cos^2\theta\sin\theta-\sin^3\theta$

$\qquad = 3(1-\sin^2\theta)\sin\theta-\sin^3\theta$

$\qquad = 3\sin\theta-3\sin^3\theta-\sin^3\theta$

$\qquad = 3\sin\theta-4\sin^3\theta$

c Re $:\cos3\theta=\cos^3\theta-3\cos\theta\sin^2\theta$

$\qquad = \cos^3\theta-3\cos\theta(1-\cos^2\theta)$

$\qquad = \cos^3\theta-3\cos\theta+3\cos^3\theta$

$\qquad = 4\cos^3\theta-3\cos\theta$

d $\cos4\theta+i\sin4\theta=(\cos\theta+i\sin\theta)^4$

$\qquad = \cos^4\theta+4\cos^3\theta(i\sin\theta)$

$\qquad +6\cos^2\theta(i\sin\theta)^2$

$\qquad +4\cos\theta(i\sin\theta)^3+(i\sin\theta)^4$

$\qquad = \cos^4\theta+4i\cos^3\theta\sin\theta-6\cos^2\theta\sin^2\theta$

$\qquad -4i\cos\theta\sin^3\theta+\sin^4\theta$

$\quad$ Im $:\sin4\theta=4\cos^3\theta\sin\theta-4\cos\theta\sin^3\theta$

$\qquad = 4\cos\theta\sin\theta(\cos^2\theta-\sin^2\theta)$

$\qquad = 4\cos\theta\sin\theta(1-2\sin^2\theta)$

$\qquad = 4\cos\theta\sin\theta-8\cos\theta\sin^3\theta$

7 $\cos6\theta+i\sin6\theta=(\cos\theta+i\sin\theta)^6$

$\qquad = \cos^6\theta+6i\cos^5\theta\sin\theta+15i^2\cos^4\theta\sin^2\theta$

$\qquad +20i^3\cos^3\theta\sin^3\theta+15i^4\cos^2\theta\sin^4\theta$

$\qquad +6i^5\cos\theta\sin^5\theta+i^6\sin^6\theta$

$\qquad = \cos^6\theta+6i\cos^5\theta\sin\theta-15\cos^4\theta\sin^2\theta$

$\qquad -20i\cos^3\theta\sin^3\theta+15\cos^2\theta\sin^4\theta$

$\qquad +6i\cos\theta\sin^5\theta-\sin^6\theta$

a Re $:\cos6\theta=\cos^6\theta-15\cos^4\theta\sin^2\theta$

$\qquad +15\cos^2\theta\sin^4\theta-\sin^6\theta$

$\qquad = \cos^6\theta-15\cos^4\theta(1-\cos^2\theta)$

$\qquad +15\cos^2\theta(1-\cos^2\theta)^2-(1-\cos^2\theta)^3$

$\qquad = \cos^6\theta-15\cos^4\theta+15\cos^6\theta+15\cos^2\theta$

$\qquad -30\cos^4\theta+15\cos^6\theta-1+3\cos^2\theta$

$\qquad -3\cos^4\theta+\cos^6\theta$

$\qquad = 32\cos^6\theta-48\cos^4\theta+18\cos^2\theta-1$

b Im $:\sin6\theta=6\cos^5\theta\sin\theta-20\cos^3\theta\sin^3\theta+6\cos\theta\sin^5\theta$

$\qquad = 2\sin\theta\cos\theta(3\cos^4\theta-10\cos^2\theta\sin^2\theta+3\sin^4\theta)$

$\qquad = 2\sin\theta\cos\theta(3(1-\sin^2\theta)^2$

$\qquad -10(1-\sin^2\theta)\sin^2\theta+3\sin^4\theta)$

$\qquad = 2\sin\theta\cos\theta(3-6\sin^2\theta+3\sin^4\theta-10\sin^2\theta$

$\qquad +10\sin^4\theta+3\sin^4\theta)$

$\qquad = 2\sin\theta\cos\theta(16\sin^4\theta-16\sin^2\theta+3)$

8 a $\cos5\theta+i\sin5\theta=(\cos\theta+i\sin\theta)^5$

$\qquad = \cos^5\theta+5\cos^4\theta(i\sin\theta)+10\cos^3\theta(i\sin\theta)^2$

$\qquad +10\cos^2\theta(i\sin\theta)^3$

$\qquad +5\cos\theta(i\sin\theta)^4+(i\sin\theta)^5$

$\qquad = \cos^5\theta+5i\cos^4\theta\sin\theta-10\cos^3\theta\sin^2\theta$

$\qquad -10i\cos^2\theta\sin^3\theta+5\cos\theta\sin^4\theta+i\sin^5\theta$

Re: $\cos 5\theta = \cos^5\theta - 10\cos^3\theta\sin^2\theta + 5\cos\theta\sin^4\theta$

$\qquad = \cos^5\theta - 10\cos^3\theta(1-\cos^2\theta) + 5\cos\theta(1-\cos^2\theta)^2$

$\qquad = \cos^5\theta - 10\cos^3\theta + 10\cos^5\theta + 5\cos\theta$

$\qquad\qquad - 10\cos^3\theta + 5\cos^5\theta$

$\qquad = 16\cos^5\theta - 20\cos^3\theta + 5\cos\theta$

b $x = 1, 0.309, -0.809$

9 a $\cos 4\theta + i\sin 4\theta = (\cos\theta + i\sin\theta)^4$

$\qquad = \cos^4\theta + 4\cos^3\theta(i\sin\theta) + 6\cos^2\theta(i\sin\theta)^2$

$\qquad\qquad + 4\cos\theta(i\sin\theta)^3 + (i\sin\theta)^4$

$\qquad = \cos^4\theta + 4i\cos^3\theta\sin\theta - 6\cos^2\theta\sin^2\theta$

$\qquad\qquad - 4i\cos\theta\sin^3\theta + \sin^4\theta$

Re: $\cos 4\theta = \cos^4\theta - 6\cos^2\theta(1-\cos^2\theta) + (1-\cos^2\theta)^2$

$\qquad = \cos^4\theta - 6\cos^2\theta + 6\cos^4\theta + 1 - 2\cos^2\theta + \cos^4\theta$

$\qquad = 8\cos^4\theta - 8\cos^2\theta + 1$

b $x = \pm 0.966, \pm 0.259$

10 $\cos 2\theta + i\sin 2\theta = (\cos\theta + i\sin\theta)^2 = \cos^2\theta + 2i\cos\theta\sin\theta - \sin^2\theta$

Re: $\cos 2\theta = \cos^2\theta - \sin^2\theta$

Im: $\sin 2\theta = 2\cos\theta\sin\theta$

$\tan 2\theta \equiv \dfrac{\sin 2\theta}{\cos 2\theta}$

$\qquad = \dfrac{2\cos\theta\sin\theta}{\cos^2\theta - \sin^2\theta}$

$\qquad = \dfrac{\dfrac{2\cos\theta\sin\theta}{\cos^2\theta}}{\dfrac{\cos^2\theta}{\cos^2\theta} - \dfrac{\sin^2\theta}{\cos^2\theta}}$

$\qquad = \dfrac{2\tan\theta}{1-\tan^2\theta}$

11 $\left[r(\cos\theta + i\sin\theta)\right]^1 = r(\cos\theta + i\sin\theta)$

and $r^1(\cos(1\theta) + i\sin(1\theta)) = r(\cos\theta + i\sin\theta)$

So true for $n = 1$

Assume true for $n = k$

$\left[r(\cos\theta + i\sin\theta)\right]^{k+1} = \left[r(\cos\theta + i\sin\theta)\right]^k\left[r(\cos\theta + i\sin\theta)\right]$

$\qquad = r^k(\cos k\theta + i\sin k\theta)\left[r(\cos\theta + i\sin\theta)\right]$

$\qquad = r^{k+1}(\cos k\theta\cos\theta + i\cos k\theta\sin\theta$

$\qquad\qquad + i\cos\theta\sin k\theta + i^2\sin k\theta\sin\theta)$

$\qquad = r^{k+1}(\cos k\theta\cos\theta - \sin k\theta\sin\theta$

$\qquad\qquad + i(\cos k\theta\sin\theta + \cos\theta\sin k\theta))$

$\qquad = r^{k+1}(\cos(k\theta + \theta) + i\sin(k\theta + \theta))$

$\qquad = r^{k+1}[\cos(k+1)\theta + i\sin(k+1)\theta]$ as required

Since true for $n = 1$ and assuming true for $n = k$ implies true for $n = k + 1$ hence true for all positive integers n.

12 a $z^n = (\cos\theta + i\sin\theta)^n = \cos(n\theta) + i\sin(n\theta)$

$\dfrac{1}{z^n} = z^{-n} = (\cos\theta + i\sin\theta)^{-n}$

$\qquad = \cos(-n\theta) + i\sin(-n\theta)$

$\qquad = \cos(n\theta) - i\sin(n\theta)$

Therefore $z^n + \dfrac{1}{z^n} = \cos(n\theta) + i\sin(n\theta)$

$\qquad\qquad + (\cos(n\theta) - i\sin(n\theta))$

$\qquad = 2\cos(n\theta)$ as required

b $4\cos\theta\sin^2\theta \equiv 4\cos\theta(1-\cos^2\theta)$

$\qquad \equiv 4\cos\theta - 4\cos^3\theta$

$\qquad \equiv 4\cos\theta - 4\left(\dfrac{1}{2}\right)^3\left(z + \dfrac{1}{z}\right)^3$

$\qquad \equiv 4\cos\theta - \dfrac{1}{2}\left(z^3 + \dfrac{1}{z^3} + 3\left(z + \dfrac{1}{z}\right)\right)$

$\qquad \equiv 4\cos\theta - \dfrac{1}{2}(2\cos(3\theta) + 6\cos\theta)$

$\qquad \equiv \cos\theta - \cos(3\theta)$

13 a $z^n = (\cos\theta + i\sin\theta)^n$

$\qquad = \cos(n\theta) + i\sin(n\theta)$

$\dfrac{1}{z^n} = z^{-n} = (\cos\theta + i\sin\theta)^{-n}$

$\qquad = \cos(-n\theta) + i\sin(-n\theta)$

$\qquad = \cos(n\theta) - i\sin(n\theta)$

Therefore $z^n - \dfrac{1}{z^n} = \cos(n\theta) + i\sin(n\theta)$

$\qquad\qquad - (\cos(n\theta) - i\sin(n\theta))$

$\qquad = 2i\sin(n\theta)$ as required

b $16\sin^3\theta\cos^2\theta \equiv 16\sin^3\theta(1-\sin^2\theta)$

$\qquad \equiv 16(\sin^3\theta - \sin^5\theta)$

$\qquad \equiv 16\left(\dfrac{1}{2i}\right)^3\left(z - \dfrac{1}{z}\right)^3 - 16\left(\dfrac{1}{2i}\right)^5\left(z - \dfrac{1}{z}\right)^5$

$\qquad \equiv 2i\left(z^3 - \dfrac{1}{z^3} - 3\left(z - \dfrac{1}{z}\right)\right)$

$\qquad\qquad + \dfrac{i}{2}\left(z^5 - \dfrac{1}{z^5} - 5\left(z^3 - \dfrac{1}{z^3}\right) + 10\left(z - \dfrac{1}{z}\right)\right)$

$\qquad \equiv 2i(2i\sin(3\theta) - 6i\sin\theta) + \dfrac{i}{2}(2i\sin(5\theta)$

$\qquad\qquad - 10i\sin(3\theta) + 20i\sin\theta)$

$\qquad \equiv -4\sin(3\theta) + 12\sin\theta - \sin(5\theta)$

$\qquad\qquad + 5\sin(3\theta) - 10\sin\theta$

$\qquad \equiv 2\sin\theta + \sin(3\theta) - \sin(5\theta)$

14 $\displaystyle\sum_{r=1}^{n}\cos(r\theta) + i\sin(r\theta) = \sum_{r=1}^{n}(\cos(r\theta) + i\sin(r\theta))^r$

$\qquad = (\cos\theta + i\sin\theta) + (\cos\theta + i\sin\theta)^2 + \ldots$

$\qquad\qquad + (\cos\theta + i\sin\theta)^n$

$\qquad = \dfrac{(\cos\theta + i\sin\theta)(1 - (\cos(\theta) + i\sin(\theta))^n)}{1 - (\cos(\theta) + i\sin(\theta))}$

$\qquad = \dfrac{(\cos\theta + i\sin\theta)(1 - (\cos(n\theta) + i\sin(n\theta)))}{1 - (\cos(\theta) + i\sin(\theta))}$

$\qquad = \dfrac{(\cos\theta + i\sin\theta)\left(2\sin^2\left(\dfrac{n\theta}{2}\right) - 2i\sin\left(\dfrac{n\theta}{2}\right)\cos\left(\dfrac{n\theta}{2}\right)\right)}{2\sin^2\left(\dfrac{\theta}{2}\right) - 2i\sin\left(\dfrac{\theta}{2}\right)\cos\left(\dfrac{\theta}{2}\right)}$

$$= \frac{(\cos\theta + i\sin\theta)2\sin\left(\frac{n\theta}{2}\right)\left(\sin\left(\frac{n\theta}{2}\right) - i\cos\left(\frac{n\theta}{2}\right)\right)}{2\sin\left(\frac{\theta}{2}\right)\left(\sin\left(\frac{\theta}{2}\right) - i\cos\left(\frac{\theta}{2}\right)\right)}$$

$$= \frac{(\cos\theta + i\sin\theta)\sin\left(\frac{n\theta}{2}\right)\left(\sin\left(\frac{n\theta}{2}\right) - i\cos\left(\frac{n\theta}{2}\right)\right)\left(\sin\left(\frac{\theta}{2}\right) + i\cos\left(\frac{\theta}{2}\right)\right)}{\sin\left(\frac{\theta}{2}\right)\left(\sin\left(\frac{\theta}{2}\right) - i\cos\left(\frac{\theta}{2}\right)\right)\left(\sin\left(\frac{\theta}{2}\right) + i\cos\left(\frac{\theta}{2}\right)\right)}$$

$$= \frac{(\cos\theta + i\sin\theta)\left(\begin{array}{c}\sin\left(\frac{n\theta}{2}\right)\sin\left(\frac{\theta}{2}\right) + \cos\left(\frac{n\theta}{2}\right)\cos\left(\frac{\theta}{2}\right) \\ + i\left(\cos\left(\frac{\theta}{2}\right)\sin\left(\frac{n\theta}{2}\right) - \cos\left(\frac{n\theta}{2}\right)\sin\left(\frac{\theta}{2}\right)\right)\end{array}\right)\sin\left(\frac{n\theta}{2}\right)}{\sin\left(\frac{\theta}{2}\right)\left(\sin^2\left(\frac{\theta}{2}\right) + \cos^2\left(\frac{\theta}{2}\right)\right)}$$

$$= \frac{(\cos\theta + i\sin\theta)\left(\begin{array}{c}\cos\left(\frac{(n-1)\theta}{2}\right) \\ + i\sin\left(\frac{(n-1)\theta}{2}\right)\end{array}\right)\sin\left(\frac{n\theta}{2}\right)}{\sin\left(\frac{\theta}{2}\right)}$$

Considering only the real parts of this sum gives

$$\sum_{r=1}^{n}\cos(r\theta) = \frac{\left(\cos\theta\cos\left(\frac{(n-1)\theta}{2}\right) - \sin\left(\frac{(n-1)\theta}{2}\right)\sin\theta\right)\sin\left(\frac{n\theta}{2}\right)}{\sin\left(\frac{\theta}{2}\right)}$$

$$= \frac{\cos\left(\frac{(n+1)\theta}{2}\right)\sin\left(\frac{n\theta}{2}\right)}{\sin\left(\frac{\theta}{2}\right)}$$

Exercise 16.3A

1 a $1, -\frac{1}{2} + \frac{\sqrt{3}}{2}i, -\frac{1}{2} - \frac{\sqrt{3}}{2}i$

b

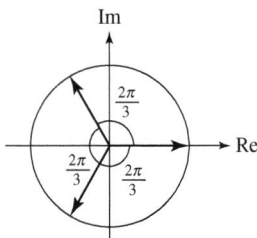

2 a $1, -1, i, -i$

b

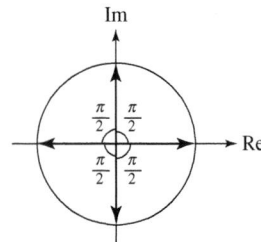

3 a $1, e^{\frac{2\pi i}{5}}, e^{\frac{4\pi i}{5}}, e^{\frac{6\pi i}{5}}, e^{\frac{8\pi i}{5}}$

b

Im(x) graph on unit circle with axes Re(x) from −1.0 to 1.0 and Im(x) from −1.0 to 1.0.

4 a $2, -1 + \sqrt{3}i, -1 - \sqrt{3}i$

b $\frac{\sqrt{3}}{2} + \frac{1}{2}i, -\frac{\sqrt{3}}{2} + \frac{1}{2}i, -i$

c $\frac{\sqrt{14}}{2} - \frac{\sqrt{14}}{2}i, \frac{\sqrt{14}}{2} + \frac{\sqrt{14}}{2}i, -\frac{\sqrt{14}}{2} + \frac{\sqrt{14}}{2}i,$
$-\frac{\sqrt{14}}{2} - \frac{\sqrt{14}}{2}i$

d $\frac{3\sqrt{3}}{2} - \frac{3}{2}i, 3i, -\frac{3\sqrt{3}}{2} - \frac{3}{2}i$

e $\frac{\sqrt{6}}{2} - \frac{\sqrt{2}}{2}i, \frac{\sqrt{6}}{2} + \frac{\sqrt{2}}{2}i, \sqrt{2}i, -\frac{\sqrt{6}}{2} + \frac{\sqrt{2}}{2}i, -\frac{\sqrt{6}}{2} - \frac{\sqrt{2}}{2}i,$
$-\sqrt{2}i$

5 a $2e^{-\frac{\pi}{8}i}, 2e^{\frac{3\pi}{8}i}, 2e^{\frac{7\pi}{8}i}, 2e^{-\frac{5\pi}{8}i}$

b $2e^{\frac{\pi}{10}i}, 2e^{\frac{\pi i}{2}}, 2e^{\frac{9\pi}{10}i}, 2e^{-\frac{7\pi}{10}i}, 2e^{-\frac{3\pi}{10}i}$

6 a $2e^{\frac{\pi}{12}i}, 2e^{\frac{3\pi}{4}i}, 2e^{-\frac{7\pi}{12}i}$

b $2e^{\frac{\pi}{4}i}, 2e^{\frac{11\pi}{12}i}, 2e^{-\frac{5\pi}{12}i}$

c $2e^{-\frac{\pi}{4}i}, 2e^{\frac{5\pi}{12}i}, 2e^{-\frac{11\pi}{12}i}$

d $2e^{-\frac{\pi}{12}i}, 2e^{\frac{7\pi}{12}i}, 2e^{-\frac{3\pi}{4}i}$

7 a $\sqrt{3}(\cos(0.18) + i\sin(0.18)), \sqrt{3}(\cos(1.75) + i\sin(1.75)),$
$\sqrt{3}(\cos(-1.39) + i\sin(-1.39)), \sqrt{3}(\cos(-2.96) + i\sin(-2.96))$

b $\sqrt{3}(\cos(0.60) + i\sin(0.60)), \sqrt{3}(\cos(2.17) + i\sin(2.17))$
$\sqrt{3}(\cos(-0.97) + i\sin(-0.97)),$
$\sqrt{3}(\cos(-2.54) + i\sin(-2.54))$

c $\sqrt{3}(\cos(-0.58) + i\sin(-0.58)), \sqrt{3}(\cos(1.00) + i\sin(1.00))$
$\sqrt{3}(\cos(2.57) + i\sin(2.57)), \sqrt{3}(\cos(-2.14) + i\sin(-2.14))$

d $\sqrt{3}(\cos(-0.21) + i\sin(-0.21)), \sqrt{3}(\cos(1.36) + i\sin(1.36))$
$\sqrt{3}(\cos(2.93) + i\sin(2.93)), \sqrt{3}(\cos(-1.78) + i\sin(-1.78))$

8 a $3e^{-\frac{\pi}{5}i}, 3e^{\frac{\pi}{5}i}, 3e^{\frac{3\pi}{5}i}, 3e^{-\frac{3\pi}{5}i}, 3e^{\pi i}$

b $\frac{1}{2}e^{-\frac{\pi}{5}i}, \frac{1}{2}e^{\frac{\pi}{5}i}, \frac{1}{2}e^{\frac{3\pi}{5}i}, \frac{1}{2}e^{-\frac{3\pi}{5}i}, \frac{1}{2}e^{\pi i}$

c $2e^{-\frac{\pi}{12}i}, 2e^{\frac{\pi}{4}i}, 2e^{\frac{7\pi}{12}i}, 2e^{\frac{11\pi}{12}i}, 2e^{-\frac{5\pi}{12}i}, 2e^{-\frac{3\pi}{4}i}$

d $1.06e^{-0.161i}, 1.06e^{1.41i}, 1.06e^{2.98\,i}, 1.06e^{-1.73\,i}$

9 a $\dfrac{\sqrt{2}}{2}\left(1+\sqrt{3}+\left(\sqrt{3}-1\right)i\right)$, $\dfrac{\sqrt{2}}{2}\left(1-\sqrt{3}+\left(\sqrt{3}+1\right)i\right)$

$\dfrac{\sqrt{2}}{2}\left(\sqrt{3}-1-\left(\sqrt{3}+1\right)i\right)$, $\dfrac{\sqrt{2}}{2}\left(-1-\sqrt{3}+\left(-\sqrt{3}+1\right)i\right)$

b $\sqrt{3}+i,\ -1+\sqrt{3}i,\ -\sqrt{3}-i,\ 1-\sqrt{3}i$

10 $2-2\sqrt{3}\,i,\ -4,\ 2+2\sqrt{3}\,i$

11 a $\sqrt{6}+\sqrt{2}i,\ -\sqrt{2}+\sqrt{6}i,\ -\sqrt{6}-\sqrt{2}i,\ \sqrt{2}-\sqrt{6}i$

b

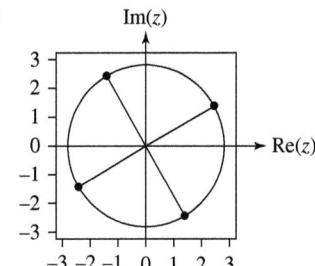

12 a $2e^{\frac{\pi i}{18}},\ 2e^{\frac{13\pi i}{18}},\ 2e^{-\frac{11\pi i}{18}}$

b $\sqrt[4]{6}e^{-\frac{\pi i}{16}},\ \sqrt[4]{6}e^{\frac{7\pi i}{16}},\ \sqrt[4]{6}e^{\frac{15\pi i}{16}},\ \sqrt[4]{6}e^{-\frac{9\pi i}{16}}$

13 a $\sqrt{2}\left(\cos\left(-\dfrac{\pi}{9}\right)+i\sin\left(-\dfrac{\pi}{9}\right)\right),\ \sqrt{2}\left(\cos\left(\dfrac{5\pi}{9}\right)+i\sin\left(\dfrac{5\pi}{9}\right)\right),$

$\sqrt{2}\left(\cos\left(-\dfrac{7\pi}{9}\right)+i\sin\left(-\dfrac{7\pi}{9}\right)\right)$

b $\sqrt{2}\left(\cos\left(\dfrac{5\pi}{36}\right)+i\sin\left(\dfrac{5\pi}{36}\right)\right),\ \sqrt{2}\left(\cos\left(\dfrac{17\pi}{36}\right)+i\sin\left(\dfrac{17\pi}{36}\right)\right)$

$\sqrt{2}\left(\cos\left(\dfrac{29\pi}{36}\right)+i\sin\left(\dfrac{29\pi}{36}\right)\right),$

$\sqrt{2}\left(\cos\left(-\dfrac{31\pi}{36}\right)+i\sin\left(-\dfrac{31\pi}{36}\right)\right)$

$\sqrt{2}\left(\cos\left(-\dfrac{19\pi}{36}\right)+i\sin\left(-\dfrac{19\pi}{36}\right)\right),$

$\sqrt{2}\left(\cos\left(-\dfrac{7\pi}{36}\right)+i\sin\left(-\dfrac{7\pi}{36}\right)\right)$

Exercise 16.3B

1 a $1+\omega+\omega^{2}=\dfrac{(1-\omega^{3})}{1-\omega}$

$=\dfrac{(1-1)}{1-\omega}=0$

b i 0   **ii** 1   **iii** −1   **iv** 0

2 a $1+\omega+\omega^{2}+\omega^{3}+\omega^{4}=\dfrac{(1-\omega^{5})}{1-\omega}$

$=\dfrac{(1-1)}{1-\omega}=0$

b −1   **ii** 0   **iii** −1

3 a $\dfrac{3\sqrt{3}}{2}-\dfrac{3}{2}i,\ 3i,\ -\dfrac{3\sqrt{3}}{2}-\dfrac{3}{2}i$

b i $\dfrac{27\sqrt{3}}{4}$ square units   **ii** $9\sqrt{3}$ units

4 a $z^{3}=125e^{(-\pi+2k\pi)i}$

$z=(125e^{(2k-1)\pi i})^{\frac{1}{3}}$

$=5e^{\frac{(2k-1)\pi i}{3}}$

$k=0:z=5e^{-\frac{\pi i}{3}}=\dfrac{5}{2}-\dfrac{5\sqrt{3}}{2}i$

$k=1:z=5e^{\frac{\pi i}{3}}=\dfrac{5}{2}+\dfrac{5\sqrt{3}}{2}i$

$k=2:z=5e^{\pi i}=-5$

$\text{Area}=\dfrac{1}{2}\times\left(5+\dfrac{5}{2}\right)\times\left(\dfrac{5\sqrt{3}}{2}+\dfrac{5\sqrt{3}}{2}\right)$

$=\dfrac{75\sqrt{3}}{4}$ square units $\left(k=\dfrac{75}{4}\right)$

b $15\sqrt{3}$ units

5 a Square   **b** 4 square units

6 a $\dfrac{\sqrt{3}}{3},\ \dfrac{\sqrt{3}}{3}i,\ -\dfrac{\sqrt{3}}{3},\ -\dfrac{\sqrt{3}}{3}i$

b $A\left(\dfrac{\sqrt{3}}{3},0\right),\ B\left(0,\dfrac{\sqrt{3}}{3}\right),\ C\left(-\dfrac{\sqrt{3}}{3},0\right),\ D\left(0,-\dfrac{\sqrt{3}}{3}\right)$

$\text{gradient } AB=\dfrac{\dfrac{\sqrt{3}}{3}}{-\dfrac{\sqrt{3}}{3}}=-1$

$\text{gradient } BC=-\dfrac{\dfrac{\sqrt{3}}{3}}{-\dfrac{\sqrt{3}}{3}}=1$

$-1\times1=-1$ so a right-angle at B

$\text{gradient } CD=-\dfrac{\dfrac{\sqrt{3}}{3}}{\dfrac{\sqrt{3}}{3}}=-1$

$-1\times1=-1$ so a right-angle at C etc

Hence all angles are right angles.

The length of all four sides is given by

$$\sqrt{\left(\dfrac{\sqrt{3}}{3}\right)^{2}+\left(\dfrac{\sqrt{3}}{3}\right)^{2}}=\dfrac{\sqrt{6}}{3}$$

Hence all sides have the same length, and so the shape formed is a square.

c $\dfrac{2}{3}$ square units

7 a $n=6$

b $\left(-\dfrac{1}{2},\dfrac{\sqrt{3}}{2}\right)$ **c** $\dfrac{3\sqrt{3}}{2}$

Review exercise 16

1 a $3e^{-\frac{\pi}{2}i}$ **b** $\sqrt{2}e^{\frac{\pi}{4}i}$ **c** $5e^{0}$

d $2e^{-\frac{5\pi}{6}i}$ **e** $\sqrt{3}e^{-0.615i}$ **f** $2e^{\frac{2\pi}{3}i}$

2 a $3e^{\frac{\pi}{7}i}$ **b** $\sqrt{2}e^{\frac{\pi}{9}i}$ **c** $\sqrt{3}e^{-\frac{\pi}{8}i}$ **d** $5e^{\frac{\pi}{5}i}$

3 a $7i$ **b** $3+3\sqrt{3}i$ **c** $\dfrac{3}{2}-\dfrac{\sqrt{3}}{2}i$

d $-1+i$ **e** $-\dfrac{\sqrt{6}}{2}-\dfrac{3\sqrt{2}}{2}i$ **f** $-3+\sqrt{3}i$

4 a $|zw| = 15 \arg(zw) = \dfrac{12}{35}\pi$

  b $|zw| = 2\sqrt{3} \arg(zw) = \dfrac{5}{24}\pi$

  c $|zw| = 6\sqrt{2} \arg(zw) = \dfrac{\pi}{12}$

  d $|zw| = 2 \arg(zw) = \dfrac{5}{9}\pi$

5 a $\left|\dfrac{z}{w}\right| = \dfrac{1}{2} \arg\left(\dfrac{z}{w}\right) = \dfrac{\pi}{4}$

  b $\left|\dfrac{z}{w}\right| = \sqrt{3} \arg\left(\dfrac{z}{w}\right) = -\dfrac{5\pi}{8}$

  c $\left|\dfrac{z}{w}\right| = \dfrac{5}{6}\sqrt{3} \arg\left(\dfrac{z}{w}\right) = -\dfrac{11\pi}{12}$

  d $\left|\dfrac{z}{w}\right| = 8\sqrt{2} \arg\left(\dfrac{z}{w}\right) = -\dfrac{7\pi}{11}$

6 a -64 **b** $-512i$

7 $8 + 8\sqrt{3}i$

8 a
$$\cos 5\theta + i\sin 5\theta = (\cos\theta + i\sin\theta)^5$$
$$= \cos^5\theta + 5\cos^4\theta(i\sin\theta)$$
$$+ 10\cos^3\theta(i\sin\theta)^2$$
$$+ 10\cos^2\theta(i\sin\theta)^3$$
$$+ 5\cos\theta(i\sin\theta)^4 + (i\sin\theta)^5$$
$$= \cos^5\theta + 5i\cos^4\theta\sin\theta - 10\cos^3\theta\sin^2\theta$$
$$- 10i\cos^2\theta\sin^3\theta + 5\cos\theta\sin^4\theta + i\sin^5\theta$$
$$\text{Im}: \sin 5\theta \equiv 5\cos^4\theta\sin\theta - 10\cos^2\theta\sin^3\theta + \sin^5\theta$$
$$= 5(1 - \sin^2\theta)^2\sin\theta - 10(1 - \sin^2\theta)\sin^3\theta + \sin^5\theta$$
$$= 5\sin\theta - 10\sin^3\theta + 5\sin^5\theta - 10\sin^3\theta$$
$$+ 10\sin^5\theta + \sin^5\theta$$
$$= 5\sin\theta - 20\sin^3\theta + 16\sin^5\theta$$

  b $x = 0, 0.588, 0.951$

9
$$\cos^3\theta = \frac{1}{8}(2\cos\theta)^3$$
$$= \frac{1}{8}(e^{i\theta} + e^{-i\theta})^3$$
$$= \frac{1}{8}(e^{3i\theta} + 3e^{i\theta} + 3e^{-i\theta} + e^{-3i\theta})$$
$$= \frac{1}{8}(2\cos 3\theta + 6\cos\theta)$$
$$= \frac{1}{4}(\cos 3\theta + 3\cos\theta) \text{ as required } \left(A = \frac{1}{4}\right)$$

10 a $1, \dfrac{\sqrt{2}}{2} + \dfrac{\sqrt{2}}{2}i, i, -\dfrac{\sqrt{2}}{2} + \dfrac{\sqrt{2}}{2}i, -1, -\dfrac{\sqrt{2}}{2} - \dfrac{\sqrt{2}}{2}i,$ $-i, \dfrac{\sqrt{2}}{2} - \dfrac{\sqrt{2}}{2}i$

  b

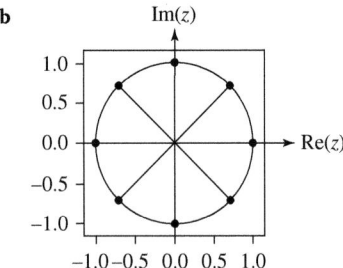

11 a $\sqrt{2}, 1+i, \sqrt{2}i, -1+i, -\sqrt{2}, -1-i, -\sqrt{2}i, 1-i$

  b $\dfrac{\sqrt{3}}{2} + \dfrac{1}{2}i, -\dfrac{\sqrt{3}}{2} + \dfrac{1}{2}i, -i$

  c $\dfrac{3\sqrt{2}}{2} - \dfrac{3\sqrt{2}}{2}i, -\dfrac{3\sqrt{2}}{2} + \dfrac{3\sqrt{2}}{2}i$

  d $\dfrac{\sqrt{15}}{2} + \dfrac{\sqrt{5}}{2}i, \sqrt{5}i, -\dfrac{\sqrt{15}}{2} + \dfrac{\sqrt{5}}{2}i, -\dfrac{\sqrt{15}}{2} - \dfrac{\sqrt{5}}{2}i, -\sqrt{5}i,$ $\dfrac{\sqrt{15}}{2} - \dfrac{\sqrt{5}}{2}i$

12 $\sqrt{2}e^{\frac{-3\pi i}{16}}, \sqrt{2}e^{\frac{5\pi i}{16}}, \sqrt{2}e^{\frac{13\pi i}{16}}, \sqrt{2}e^{\frac{-11\pi i}{16}}$

Assessment 16

1 a $2\sqrt{3}e^{-\frac{\pi}{6}i}$

  b $144\left(\cos\left(-\dfrac{2\pi}{3}\right) + i\sin\left(-\dfrac{2\pi}{3}\right)\right)$

2 a $\text{mod}(z) = 2k \arg(z) = \dfrac{2\pi}{3}$

  b $-128 + 128\sqrt{3}i$

3 a $e^{-\frac{\pi}{6}i}, e^{\frac{\pi}{2}i}, e^{-\frac{5\pi}{6}i}$

  b

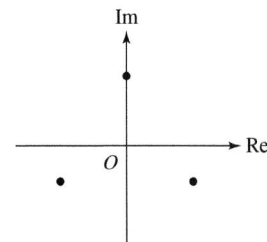

4 a $2\sqrt{2}\left(\cos\left(\dfrac{3\pi}{4}\right) + i\sin\left(\dfrac{3\pi}{4}\right)\right)$ **b** $\dfrac{1}{32} - \dfrac{1}{32}i$

5 $-8 - 8\sqrt{3}i$

6 $(\cos\theta - i\sin\theta)^n \equiv (\cos(-\theta) + i\sin(-\theta))^n$

  since $\sin(-\theta) \equiv -\sin\theta$ and $\cos(-\theta) \equiv \cos\theta$

  Therefore

  $(\cos\theta - i\sin\theta)^n \equiv \cos(-n\theta) + i\sin(-n\theta)$ by de Moivre's theorem

  $\equiv \cos(n\theta) - i\sin(n\theta)$

7 a $\cos(5\theta) + i\sin(5\theta) \equiv (\cos\theta + i\sin\theta)^5$
$$\equiv \cos^5\theta + 5i\cos^4\theta\sin\theta - 10\cos^3\theta\sin^2\theta$$
$$- 10i\cos^2\theta\sin^3\theta + 5\cos\theta\sin^4\theta + i\sin^5\theta$$
$$\sin(5\theta) \equiv 5\cos^4\theta\sin\theta - 10\cos^2\theta\sin^3\theta + \sin^5\theta$$
$$\equiv 5(1 - \sin^2\theta)^2\sin\theta - 10(1 - \sin^2\theta)\sin^3\theta + \sin^5\theta$$
$$\equiv 5\sin\theta - 10\sin^3\theta + 5\sin^5\theta - 10\sin^3\theta$$
$$+ 10\sin^5\theta + \sin^5\theta$$
$$\equiv 5\sin\theta - 20\sin^3\theta + 16\sin^5\theta$$

  b $\theta = 6°, 30°, 78°, 102°, 150°, 174°$

8 a $32\cos^5 x \equiv (2\cos x)^5$

$$\equiv (e^{ix} + e^{-ix})^5$$
$$\equiv (e^{ix})^5 + 5(e^{ix})^4(e^{-ix}) + 10(e^{ix})^3(e^{-ix})^2$$
$$+ 10(e^{ix})^2(e^{-ix})^3 + 5(e^{ix})(e^{-ix})^4 + (e^{-ix})^5$$
$$\equiv e^{5ix} + 5e^{3ix} + 10e^{ix} + \frac{10}{e^{ix}} + \frac{5}{e^{3ix}} + \frac{1}{e^{5ix}}$$
$$\equiv 2\cos(5x) + 10\cos(3x) + 20\cos(x)$$

b $\dfrac{1}{80}\sin(5x) + \dfrac{5}{48}\sin(3x) + \dfrac{5}{8}\sin(x) + c$

9 a $\cos 3\theta + i\sin 3\theta \equiv (\cos\theta + i\sin\theta)^3$

$$\equiv \cos^3\theta + 3i\cos^2\theta\sin\theta$$
$$- 3\cos\theta\sin^2\theta - i\sin^3\theta$$

Im : $\sin 3\theta \equiv 3\cos^2\theta\sin\theta - \sin^3\theta$ as required

b Re : $\cos 3\theta \equiv \cos^3\theta - 3\cos\theta\sin^2\theta$

$$\tan 3\theta \equiv \frac{\sin 3\theta}{\cos 3\theta}$$
$$\equiv \frac{3\cos^2\theta\sin\theta - \sin^3\theta}{\cos^3\theta - 3\cos\theta\sin^2\theta}$$
$$\equiv \frac{3\tan\theta - \tan^3\theta}{1 - 3\tan^2\theta} \text{ as required}$$

c $-\dfrac{18}{35}\sqrt{3}$

10 a 6

b $\sqrt{2}\left(\cos\left(-\dfrac{\pi}{6}\right) + i\sin\left(-\dfrac{\pi}{6}\right)\right)$,

$\sqrt{2}\left(\cos\left(\dfrac{\pi}{6}\right) + i\sin\left(\dfrac{\pi}{6}\right)\right)$,

$\sqrt{2}\left(\cos\left(\dfrac{\pi}{2}\right) + i\sin\left(\dfrac{\pi}{2}\right)\right)$,

$\sqrt{2}\left(\cos\left(\dfrac{5\pi}{6}\right) + i\sin\left(\dfrac{5\pi}{6}\right)\right)$,

$\sqrt{2}\left(\cos\left(-\dfrac{\pi}{2}\right) + i\sin\left(-\dfrac{\pi}{2}\right)\right)$,

$\sqrt{2}\left(\cos\left(-\dfrac{5\pi}{6}\right) + i\sin\left(-\dfrac{5\pi}{6}\right)\right)$

11 RHS $\equiv 1 - 2\left(\dfrac{(e^{ix} - e^{-ix})}{2i}\right)^2$

$$\equiv 1 - \frac{2(e^{2ix} + e^{-2ix} - 2)}{4i^2}$$
$$\equiv 1 - \frac{e^{2ix} + e^{-2ix} - 2}{-2}$$
$$\equiv 1 + \frac{e^{2ix} + e^{-2ix} - 2}{2}$$
$$\equiv \frac{2 + e^{2ix} + e^{-2ix} - 2}{2}$$
$$\equiv \frac{e^{2ix} + e^{-2ix}}{2} \equiv \cos 2x$$

12 a $\dfrac{3\pi}{4}$ **b** $\dfrac{\pi}{4}$ **c** $\dfrac{n}{4}\pi$ **d** $-\dfrac{3\pi}{4}$

13 a $\cos 7\theta + i\sin 7\theta \equiv (\cos\theta + i\sin\theta)^7$

Re : $\cos 7\theta \equiv \cos^7\theta - 21\cos^5\theta\sin^2\theta$
$$+ 35\cos^3\theta\sin^4\theta - 7\cos\theta\sin^6\theta$$
$$\equiv \cos^7\theta - 21\cos^5\theta(1 - \cos^2\theta)$$
$$+ 35\cos^3\theta(1 - \cos^2\theta)^2$$
$$- 7\cos\theta(1 - \cos^2\theta)^3$$
$$\equiv \cos^7\theta - 21\cos^5\theta + 21\cos^7\theta + 35\cos^3\theta$$
$$- 70\cos^5\theta + 35\cos^7\theta - 7\cos\theta$$
$$+ 21\cos^3\theta - 21\cos^5\theta + 7\cos^7\theta$$
$$\equiv 64\cos^7\theta - 112\cos^5\theta + 56\cos^3\theta - 7\cos\theta$$
$$\equiv \cos\theta(64\cos^6\theta - 112\cos^4\theta + 56\cos^2\theta - 7)$$

as required

b 0.975, 0.782, 0.434, −0.434, −0.782, −0.975

14 $\sqrt{2}e^{-\frac{5}{48}\pi i},\ \sqrt{2}e^{\frac{7}{48}\pi i},\ \sqrt{2}e^{\frac{19}{48}\pi i},\ \sqrt{2}e^{\frac{31}{48}\pi i},\ \sqrt{2}e^{\frac{43}{48}\pi i},$

$\sqrt{2}e^{-\frac{17}{48}\pi i},\ \sqrt{2}e^{-\frac{29}{48}\pi i},\ \sqrt{2}e^{-\frac{41}{48}\pi i}$

15 a $(\sin\theta)^4 \equiv \left(\dfrac{e^{i\theta} - e^{-i\theta}}{2}\right)^4$

$$\equiv \frac{1}{16}\left[\begin{array}{c}(e^{i\theta})^4 - 4(e^{i\theta})^3(e^{-i\theta}) + 6(e^{i\theta})^2(e^{-i\theta})^2 \\ - 4(e^{i\theta})(e^{-i\theta})^3 + (e^{-i\theta})^4\end{array}\right]$$
$$\equiv \frac{1}{16}\left[e^{4i\theta} + e^{-4i\theta} - 4(e^{2i\theta} + e^{-2i\theta}) + 6\right]$$
$$\equiv \frac{1}{16}\left[2\cos(4\theta) - 8\cos(2\theta) + 6\right]$$
$$\equiv \frac{1}{8}\left(\cos(4\theta) - 4\cos(2\theta) + 3\right) \text{ as required}$$

b $\dfrac{1}{4}\sin(4\theta) - 2\sin(2\theta) + 3\theta + c$

16 a $1.86\left(\cos\left(-\dfrac{\pi}{6}\right) + i\sin\left(-\dfrac{\pi}{6}\right)\right)$,

$1.86\left(\cos\left(\dfrac{\pi}{3}\right) + i\sin\left(\dfrac{\pi}{3}\right)\right)$,

$1.86\left(\cos\left(\dfrac{5\pi}{6}\right) + i\sin\left(\dfrac{5\pi}{6}\right)\right)$,

$1.86\left(\cos\left(-\dfrac{2\pi}{3}\right) + i\sin\left(-\dfrac{2\pi}{3}\right)\right)$

b $2.6 - 0.93i$ $1.9 + 1.6i$
$-0.61 + 0.93i$ $0.069 - 1.6i$

17 a $\dfrac{1}{z^n} \equiv z^{-n} \equiv (\cos\theta + i\sin\theta)^{-n}$

$$\equiv \cos(-n\theta) + i\sin(-n\theta)$$
$$\equiv \cos(n\theta) - i\sin(n\theta)$$
$$z^n \equiv (\cos\theta + i\sin\theta)^n \equiv \cos(n\theta) + i\sin(n\theta)$$
$$z^n - \frac{1}{z^n} \equiv \cos(n\theta) + i\sin(n\theta) - (\cos(n\theta) - i\sin(n\theta))$$
$$\equiv 2i\sin(n\theta) \text{ as required}$$

b $4\sin^3\theta \equiv \dfrac{1}{-2i}(2i\sin\theta)^3$

$$\equiv \frac{1}{2}i\left(z-\frac{1}{z}\right)^3$$

$$\equiv \frac{1}{2}i\left[(z)^3-3(z)^2\left(\frac{1}{z}\right)+3(z)\left(\frac{1}{z}\right)^2-(z)^3\right]$$

$$\equiv \frac{1}{2}i\left[z^3-\frac{1}{z^3}-3\left(z-\frac{1}{z}\right)\right]$$

$$\equiv \frac{1}{2}i\left(2i\sin(3\theta)-6i\sin\theta\right)$$

$$\equiv 3\sin\theta-\sin(3\theta) \text{ as required}$$

c $\theta=\dfrac{7\pi}{6},\dfrac{11\pi}{6}$

18 a

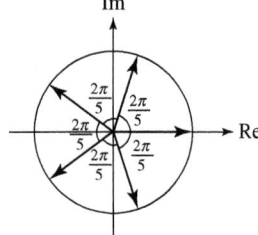

b i -1 **ii** 4

19 a $\cos(5\theta)+i\sin(5\theta)\equiv(\cos\theta+i\sin\theta)^5$

$$\equiv \cos^5\theta+5i\cos^4\theta\sin\theta$$
$$-10\cos^3\theta\sin^2\theta$$
$$-10i\cos^2\theta\sin^3\theta$$
$$+5\cos\theta\sin^4\theta+i\sin^5\theta$$

$\text{Re}:\cos(5\theta)\equiv\cos^5\theta-10\cos^3\theta\sin^2\theta+5\cos\theta\sin^4\theta$

$$\equiv \cos^5\theta-10\cos^3\theta(1-\cos^2\theta)$$
$$+5\cos\theta(1-\cos^2\theta)^2$$

$$\equiv \cos^5\theta-10\cos^3\theta+10\cos^5\theta$$
$$+5\cos\theta-10\cos^3\theta+5\cos^5\theta$$

$$\equiv 16\cos^5\theta-20\cos^3\theta+5\cos\theta$$
$$\text{as required}$$

b $\theta=\dfrac{\pi}{2}(2n-1)$ for $n\in\mathbb{Z}$

20 a $\omega^7=\cos\left(\dfrac{2\pi}{7}\right)+i\sin\left(\dfrac{2\pi}{7}\right)$

$$=\left(e^{\frac{2\pi i}{7}}\right)^7=e^{2\pi i}$$

$$=\cos(2\pi)+i\sin(2\pi)$$
$$=1 \text{ as required}$$

b $\omega^2,\omega^3,\omega^4,\omega^5,\omega^6$

c -1

21 a $\left(\dfrac{1}{2}+\dfrac{\sqrt{3}}{2}i-1\right)^3=\left(-\dfrac{1}{2}+\dfrac{\sqrt{3}}{2}i\right)^3$

$$=\left(-\frac{1}{2}\right)^3+3\left(-\frac{1}{2}\right)^2\left(\frac{\sqrt{3}}{2}i\right)$$

$$+3\left(-\frac{1}{2}\right)\left(\frac{\sqrt{3}}{2}i\right)^2+\left(\frac{\sqrt{3}}{2}i\right)^3$$

$$=-\frac{1}{8}+\frac{3\sqrt{3}}{8}i+\frac{9}{8}-\frac{3\sqrt{3}}{8}i$$

$$=1 \text{ as required}$$

b $\dfrac{1}{2}-\dfrac{\sqrt{3}}{2}i \quad 2$

c

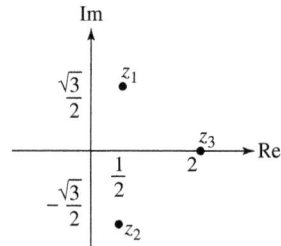

d $(1,0),1$

22 a $\sqrt{2}+\sqrt{2}i \qquad \sqrt{2}-\sqrt{2}i$

$-\sqrt{2}-\sqrt{2}i \qquad -\sqrt{2}+\sqrt{2}i$

b 8 square units

23 a $\cos5\theta+i\sin5\theta=(\cos\theta+i\sin\theta)^5$

$$=\cos^5\theta+5i\cos^4\theta\sin\theta-10\cos^3\theta\sin^2\theta$$
$$-10i\cos^2\theta\sin^3\theta+5\cos\theta\sin^4\theta+i\sin^5\theta$$

Imaginary parts give:

$$\sin5\theta=5\cos^4\theta\sin\theta-10\cos^2\theta\sin^3\theta+\sin^5\theta$$

$$=5(1-\sin^2\theta)^2\sin\theta-10\left(1-\sin^2\theta\right)\sin^3\theta+\sin^5\theta$$

$$=5\sin\theta-10\sin^3\theta+5\sin^5\theta-10\sin^3\theta$$
$$+10\sin^5\theta+\sin^5\theta$$

$$=16\sin^5\theta-20\sin^3\theta+5\sin\theta \text{ as required}$$

b $\sin\left(\dfrac{\pi}{10}\right),\sin\left(\dfrac{\pi}{2}\right),\sin\left(\dfrac{9\pi}{10}\right),\sin\left(\dfrac{13\pi}{10}\right),\sin\left(\dfrac{17\pi}{10}\right)$

24 a $3 \qquad \sqrt{3}i \qquad -\sqrt{3}i$

b

Im

$\sqrt{3}\ \bullet\,z_1$

z_1

$\bullet$ — Re
$\quad 3$

$-\sqrt{3}\ \bullet\,z_2$

c $3\sqrt{3}$ square units

d $\dfrac{3}{2}+\dfrac{3\sqrt{3}}{2}i \qquad -\dfrac{3}{2}+\dfrac{\sqrt{3}}{2}i$

$\dfrac{3}{2}-\dfrac{\sqrt{3}}{2}i$

25 a

$$\left(\dfrac{\frac{1}{2}(1+i)-1}{\frac{1}{2}(1+i)}\right)^6 = \left(\dfrac{1+i-2}{1+i}\right)^6$$

$$= \left(\dfrac{i-1}{1+i}\right)^6$$

$$= \left(\dfrac{(i-1)(1-i)}{(1+i)(1-i)}\right)^6$$

$$= \left[-\dfrac{1}{2}(i-1)^2\right]^6$$

$$= \dfrac{1}{64}(i-1)^{12}$$

$$= \dfrac{1}{64}\left(\sqrt{2}\left(\cos\left(\dfrac{3\pi}{4}\right)+i\sin\left(\dfrac{3\pi}{4}\right)\right)\right)^{12}$$

$$= \dfrac{64}{64}\left(\cos\left(\dfrac{36\pi}{4}\right)+i\sin\left(\dfrac{36\pi}{4}\right)\right)$$

$$= (-1+0)$$

$$= -1 \text{ as required}$$

b $\dfrac{1}{2}+\dfrac{1}{2}i$

$\dfrac{1}{2}+\dfrac{2-\sqrt{3}}{2}i$

$\dfrac{1}{2}+\dfrac{-2+\sqrt{3}}{2}i$

$\dfrac{1}{2}-\dfrac{1}{2}i$

$\dfrac{1}{2}+\dfrac{-2-\sqrt{3}}{2}i$

c $\left(\dfrac{1}{4}(1+\sqrt{3}),\ \dfrac{1}{4}(-1+\sqrt{3})\right)$

26 a $2e^{-\frac{\pi i}{9}}$ $2e^{\frac{5\pi i}{9}}$ $2e^{-\frac{7\pi i}{9}}$

b $3\sqrt{3}$ square units

c Sum of roots $= 0$

So sum of imaginary parts $= 0$

Therefore $\sin\left(-\dfrac{\pi}{9}\right)+\sin\left(\dfrac{5\pi}{9}\right)+\sin\left(-\dfrac{7\pi}{9}\right)=0$

$\Rightarrow -\sin\left(\dfrac{\pi}{9}\right)+\sin\left(\dfrac{5\pi}{9}\right)-\sin\left(\dfrac{7\pi}{9}\right)=0$

$\Rightarrow \sin\left(\dfrac{\pi}{9}\right)-\sin\left(\dfrac{5\pi}{9}\right)+\sin\left(\dfrac{7\pi}{9}\right)=0$

27 First consider the case when n is a positive integer.

Let $n = 1$ then $(\cos\theta+i\sin\theta)^1 = \cos(1\theta)+i\sin(1\theta)$ so true for $n = 1$

Assume true for $n = k$ and consider when $n = k+1$

$(\cos\theta+i\sin\theta)^{k+1} = (\cos\theta+i\sin\theta)^k (\cos\theta+i\sin\theta)$

$= (\cos(k\theta)+i\sin(k\theta))(\cos\theta+i\sin\theta)$

$= \cos(k\theta)\cos\theta - \sin(k\theta)\sin\theta$

$\quad + i(\cos(k\theta)\sin\theta + \sin(k\theta)\cos\theta)$

$= \cos((k+1)\theta)+i\sin((k+1)\theta)$

So true for $n = k+1$

Hence, since true for $n = 1$ and by assuming it is true for $n = k$ we have proved it is true for $n = k+1$, therefore it is true for all positive integers n.

When $n = 0$, $(\cos\theta+i\sin\theta)^0 = 1$ and $\cos(0)+i\sin(0)=1$ so true for $n = 0$

Consider when n is a negative integer, it can be written as $n = -m$ where m is a positive integer.

So $(\cos\theta+i\sin\theta)^n = (\cos\theta+i\sin\theta)^{-m}$

$$= \dfrac{1}{(\cos\theta+i\sin\theta)^m}$$

$$= \dfrac{1}{(\cos(m\theta)+i\sin(m\theta))}$$

since m is a positive integer and we have just proved de Moivre's holds for positive integers.

$$= \dfrac{\cos(m\theta)-i\sin(m\theta)}{(\cos(m\theta)+i\sin(m\theta))(\cos(m\theta)-i\sin(m\theta))}$$

$$= \dfrac{\cos(m\theta)-i\sin(m\theta)}{\cos^2(m\theta)+\sin^2(m\theta)}$$

$$= \cos(m\theta)-i\sin(m\theta)$$

$$= \cos(-m\theta)+i\sin(-m\theta)$$

$$= \cos(n\theta)+i\sin(n\theta) \text{ as required}$$

So de Moivre's theorem holds for all integers n

Chapter 17
Exercise 17.1A

1 a $\dfrac{2}{x+5}+\dfrac{4}{x-5}$ **b** $\dfrac{3}{x-3}-\dfrac{7}{x-7}$

   c $\dfrac{9}{1-x}+\dfrac{2}{x+6}$ **d** $\dfrac{2}{x+2}+\dfrac{5}{x-2}-\dfrac{7}{x+3}$

2 a $f(x)-f(x+1) \equiv \dfrac{1}{x(x-1)}-\dfrac{1}{x(x+1)} \equiv \dfrac{(x+1)-(x-1)}{x(x-1)(x+1)}$

$$\equiv \dfrac{2}{x(x-1)(x+1)}$$

b $\dfrac{1}{2}\left[\dfrac{n^2+n-2}{2n(n+1)}\right] \equiv \left[\dfrac{(n+2)(n-1)}{4n(n+1)}\right]$

3 a $\left[\dfrac{5n^2+13n}{6(n+2)(n+3)}\right]$ **b** $\dfrac{2n}{(2n+1)}$ **c** $\left[\dfrac{8(n+1)(n+3)}{3(2n+3)(2n+5))}\right]$

4 a $\dfrac{n-1}{n}$ **b** $\dfrac{(n-1)}{2(n+1)}$

   c $\dfrac{5n^2+13n}{12(n+2)(n+3)}$ **d** $\dfrac{(3n+2)(n-1)}{4n(n+1)}$

Exercise 17.1B

1 a $r(r+1)(r+2)-(r-1)r(r+1)$
$\equiv (r^3+3r^2+2r)-(r^3-r) \equiv 3r^2+3r \equiv 3r(r+1)$

b $\dfrac{n(n+1)(n+2)}{3}$

2 a $\dfrac{1}{2}\left[\dfrac{1}{2r-1}-\dfrac{1}{2r+1}\right]$ **b** $\dfrac{n}{(2n+1)}$ **c** $\dfrac{1}{2}$

3 $f(r) = \dfrac{1}{(r+1)(r+2)}$

a $f(r) - f(r+1) \equiv \dfrac{1}{(r+1)(r+2)} - \dfrac{1}{(r+2)(r+3)}$

$$\equiv \dfrac{(r+3)-(r+1)}{(r+1)(r+2)(r+3)} \equiv \dfrac{2}{(r+1)(r+2)(r+3)}$$

b $\dfrac{n(n+5)}{12(n+2)(n+3)}$ **c** $\dfrac{1}{12}$

4 a $4r^3$ **b** $\dfrac{1}{4}\left[n^2(n+1)^2\right]$

5 a $\dfrac{(3n+2)(n-1)}{n(n+1)}$ **b** $\dfrac{3}{4}$

6 $\dfrac{2r-1}{r(r+1)(r+2)} \equiv \dfrac{A}{r} + \dfrac{B}{(r+1)} + \dfrac{C}{(r+2)}$

$\Rightarrow 2r-1 \equiv A(r+1)(r+2) + B(r)(r+2) + C(r)(r+1)$

When $r=0$, $-1 = 2A$, so $A = -\dfrac{1}{2}$

When $r=-1$, $-3 = -B$, so $B = 3$

When $r=-2$, $-5 = 2C$, so $C = -\dfrac{5}{2}$

Hence $\dfrac{2r-1}{r(r+1)(r+2)} \equiv \dfrac{1}{2}\left[\dfrac{-1}{r} + \dfrac{6}{(r+1)} - \dfrac{5}{(r+2)}\right]$

$\displaystyle\sum_1^n \dfrac{1}{2}\left[\dfrac{-1}{r} + \dfrac{6}{(r+1)} - \dfrac{5}{(r+2)}\right]$

$$\equiv \dfrac{1}{2}\begin{bmatrix}\dfrac{-1}{1} + \dfrac{6}{2} - \dfrac{5}{3} \\ \dfrac{-1}{2} + \dfrac{6}{3} - \dfrac{5}{4} \\ \dfrac{-1}{3} + \dfrac{6}{4} - \dfrac{5}{5} \\ \dfrac{-1}{4} + \dfrac{6}{5} - \dfrac{5}{6} \\ - \ldots \\ \dfrac{-1}{r-2} + \dfrac{6}{(r-1)} - \dfrac{5}{(r)} \\ \dfrac{-1}{r-1} + \dfrac{6}{(r)} - \dfrac{5}{(r+1)} \\ \dfrac{-1}{r} + \dfrac{6}{(r+1)} - \dfrac{5}{(r+2)}\end{bmatrix}$$

Hence $\displaystyle\sum_1^n \dfrac{1}{2}\left[\dfrac{-1}{r} + \dfrac{6}{(r+1)} - \dfrac{5}{(r+2)}\right]$

$$\equiv \dfrac{1}{2}\left[-1 + 3 - \dfrac{1}{2} - \dfrac{5}{(r+1)} + \dfrac{6}{(r+1)} - \dfrac{5}{(r+2)}\right]$$

$$\equiv \dfrac{1}{2}\left[\dfrac{3}{2} + \dfrac{1}{(r+1)} - \dfrac{5}{(r+2)}\right]$$

Hence $\displaystyle\sum_1^\infty \dfrac{1}{2}\left[\dfrac{-1}{r} + \dfrac{6}{(r+1)} - \dfrac{5}{(r+2)}\right] = \dfrac{3}{4} + \dfrac{1}{\infty} - \dfrac{5}{\infty} = \dfrac{3}{4}$

7 a $\dfrac{x-1}{x} - \dfrac{x-2}{x-1} \equiv \dfrac{(x-1)^2 - x(x-2)}{x(x-1)}$

$$\equiv \dfrac{(x^2 - 2x + 1 - x^2 + 2x)}{x(x-1)} \equiv \dfrac{1}{x(x-1)}$$

b $\dfrac{n-2}{2n}$ **c** $\dfrac{1}{2}$

8 $\ln(n)$

9 a $\dfrac{-1}{(x-1)} + \dfrac{3}{(x)} - \dfrac{2}{(x+1)}$ **b** $\dfrac{1}{n} - \dfrac{2}{(n+1)}$

c $\displaystyle\sum_2^\infty\left[\left(\dfrac{-1}{(x-1)} + \dfrac{3}{(x)} - \dfrac{2}{(x+1)}\right)\right] = \lim_{n\to\infty} \dfrac{1}{n} - \dfrac{2}{(n+1)} = 0$

10 $\dfrac{1}{r(r+1)(r+2)(r+3)} \equiv \dfrac{A}{(r)} + \dfrac{B}{(r+1)} + \dfrac{C}{(r+2)} + \dfrac{D}{(r+3)}$

$\Rightarrow 1 \equiv A(r+1)(r+2)(r+3) + B(r)(r+2)(r+3) + C(r)(r+1)(r+3) + D(r)(r+1)(r+2)$

When $r=0$, $1 = 6A$ so $A = \dfrac{1}{6}$; When $r=-1$, $1 = -2B$ so $B = -\dfrac{1}{2}$

When $r=-2$, $1 = 2C$ so $C = \dfrac{1}{2}$; When $r=-3$, $1 = -6D$ so $D = -\dfrac{1}{6}$

Hence $\dfrac{1}{r(r+1)(r+2)(r+3)} \equiv \dfrac{1}{(6r)} - \dfrac{1}{2(r+1)} + \dfrac{1}{2(r+2)} - \dfrac{1}{6(r+3)}$

Hence $\displaystyle\sum_1^n \dfrac{1}{r(r+1)(r+2)(r+3)} \equiv$

$$\dfrac{1}{6}\sum_1^n \dfrac{1}{(r)} - \dfrac{3}{(r+1)} + \dfrac{3}{(r+2)} - \dfrac{1}{(r+3)}$$

$$\equiv \dfrac{1}{6}\begin{bmatrix}\dfrac{1}{1} - \dfrac{3}{2} + \dfrac{3}{3} - \dfrac{1}{4} \\ + \dfrac{1}{2} - \dfrac{3}{3} + \dfrac{3}{4} - \dfrac{1}{5} \\ + \dfrac{1}{3} - \dfrac{3}{4} + \dfrac{3}{5} - \dfrac{1}{6} \\ + \dfrac{1}{4} - \dfrac{3}{5} + \dfrac{3}{6} - \dfrac{1}{7} \\ + \dfrac{1}{5} - \dfrac{3}{6} + \dfrac{3}{7} - \dfrac{1}{8} \\ + \ldots \\ + \dfrac{1}{(n-3)} - \dfrac{3}{(n-2)} + \dfrac{3}{(n-1)} - \dfrac{1}{(n)} \\ + \dfrac{1}{(n-2)} - \dfrac{3}{(n-1)} + \dfrac{3}{(n)} - \dfrac{1}{(n+1)} \\ + \dfrac{1}{(n-1)} - \dfrac{3}{(n)} + \dfrac{3}{(n+1)} - \dfrac{1}{(n+2)} \\ + \dfrac{1}{(n)} - \dfrac{3}{(n+1)} + \dfrac{3}{(n+2)} - \dfrac{1}{(n+3)}\end{bmatrix}$$

Hence $\displaystyle\sum_1^n \dfrac{1}{r(r+1)(r+2)(r+3)}$

$$\equiv \dfrac{1}{6}\left[\dfrac{1}{1} - \dfrac{3}{2} + \dfrac{3}{3} + \dfrac{1}{2} - \dfrac{3}{2} + \dfrac{1}{3} - \dfrac{1}{(n+1)} + \dfrac{3}{(n+1)} - \dfrac{1}{(n+2)} \right.$$

$$\left. - \dfrac{3}{(n+1)} + \dfrac{3}{(n+2)} - \dfrac{1}{(n+3)}\right]$$

$$\equiv \dfrac{1}{6}\left[\dfrac{1}{3} - \dfrac{1}{(n+1)} + \dfrac{2}{(n+2)} - \dfrac{1}{(n+3)}\right]$$

Hence $\displaystyle\sum_1^\infty \dfrac{1}{r(r+1)(r+2)(r+3)} = \dfrac{1}{6}\left[\dfrac{1}{3} - \dfrac{1}{(\infty+1)} + \dfrac{2}{(\infty+2)}\right.$

$$\left. - \dfrac{1}{(\infty+3)}\right] = \dfrac{1}{18}$$

11 a $\dfrac{2}{3(1+2x)} + \dfrac{1}{3(1-x)}$

b $1 - \dfrac{1}{2}x + \dfrac{19}{8}x^2 - \dfrac{53}{16}x^3 + \ldots$

Exercise 17.2A

1 a $\dfrac{(-x)^r}{r!}$ or $\dfrac{(-1)^r x^r}{r!}$

b $(-1)^{r+1}\dfrac{(-2x)^r}{r}$ or $-\dfrac{2^r x^r}{r}$

c $(-1)^r\dfrac{\left(\dfrac{x}{5}\right)^{2r+1}}{(2r+1)!}$ or $(-1)^r\dfrac{x^{2r+1}}{5^{2r+1}(2r+1)!}$

d $\dfrac{n(n-1)\ldots(n-1+r)(3x)^r}{r!}$ or $\dfrac{n(n-1)\ldots(n-1+r)(3^r)(x)^r}{r!}$

2 $\sin x \equiv x - \dfrac{x^3}{3!} + \dfrac{x^5}{5!} - \dfrac{x^7}{7!} + \ldots$

The general term is $(-1)^r\dfrac{x^{2r+1}}{(2r+1)!}$

3 $\cos x \equiv 1 - \dfrac{x^2}{2!} + \dfrac{x^4}{4!} - \dfrac{x^6}{6!} + \ldots$

The general term is $(-1)^r\dfrac{x^{2r}}{(2r)!}$

4 $-4x - \dfrac{16x^3}{3} - \dfrac{64x^5}{5} - \ldots$

5 $1 - x^2 - \dfrac{x^4}{3} - \ldots$

6 $1 - \dfrac{3x^2}{2} - \dfrac{9x^4}{8} - \ldots$

7 $x + 2x^2 + \dfrac{11x^3}{6} - \ldots$

8 $x + \dfrac{x^2}{2} - \dfrac{x^3}{3} - \ldots$

9 a 1   **b** -1 **c** 3   **d** $\dfrac{1}{2}$

e $\dfrac{1}{4}$ **f** $\dfrac{1}{3}$ **g** -1 **h** -11

10 a 1.5   **b** $\dfrac{1}{3}$ **c** $-\dfrac{5}{8}$

d 2   **e** $\dfrac{\sqrt{2}}{2}$

11 a 1   **b** 1   **c** 0   **d** 1

e 2   **f** $\dfrac{29}{12}$ **g** $\dfrac{5}{2}$

Exercise 17.2B

1 a $4 - 16x + 64x^2 - 256x^3 + \ldots$

b $4x - 8x^2 + \dfrac{64}{3}x^3 - 64x^4 + \ldots$

c $x - \dfrac{x^2}{2} + \dfrac{x^3}{3} - \dfrac{x^4}{4} + \dfrac{x^5}{5} \ldots$

So $\ln(1+4x) = 4x - \dfrac{16x^2}{2} + \dfrac{64x^3}{3} - \dfrac{256x^4}{4} + \ldots$

$\equiv 4x - 8x^2 + \dfrac{64}{3}x^3 - 64x^4 + \ldots$

2 a -6 **b** $\dfrac{1}{2}$ **c** 8   **d** -1

3 a 2   **b** 0   **c** 0

d $-\dfrac{1}{5}$ **e** $\dfrac{1}{4}$

4 a $x + \dfrac{x^2}{2} + \dfrac{5x^3}{6} + \dfrac{7x^4}{12} \ldots$

b i $\sqrt{2}\left(1 - \dfrac{9x}{4} + \dfrac{143x^2}{32} - \dfrac{1145x^3}{128} + \ldots\right)$

ii $\sqrt{2}$

When $x = 0$, $\sqrt{2}\left(1 - \dfrac{9x}{4} + \dfrac{9x^2}{2} - \dfrac{1145x^3}{128} + \ldots\right)$

$= \sqrt{2}\left(1 - \dfrac{0}{4} + \dfrac{0}{2} - \dfrac{0}{128} + \ldots\right) = \sqrt{2}$

5 a $\dfrac{9}{(x+1)} - \dfrac{15}{(x+2)}$

b $\dfrac{3}{2} - \dfrac{21x}{4} + \dfrac{57x^2}{8} - \dfrac{129x^3}{16} + \ldots$

c $\dfrac{3}{2}$

When $x = 0$, $\dfrac{3}{2} - \dfrac{21x}{4} + \dfrac{57x^2}{8} - \dfrac{129x^3}{16} + \ldots = \dfrac{3}{2}$

6 a -1

b $\lim\limits_{x\to 0}\dfrac{e^x - e^{2x}}{x} = \dfrac{0}{0} \Rightarrow \lim\limits_{x\to 0}\dfrac{e^x - 2e^{2x}}{1} = -1$

7 a $e^2(2 - 3x + 2x^2 - \ldots)$; $2e^2$ **b** $2e^2$

8 a $1 - \dfrac{x}{2} + \dfrac{x^2}{12} + \ldots$ 1

b $\lim\limits_{x\to 0}\dfrac{x}{e^x - 1} = \dfrac{0}{0} \Rightarrow \lim\limits_{x\to 0}\dfrac{1}{e^x} = 1$

9 a $2 - x + \dfrac{x^2}{2} + \ldots$

b $x + \dfrac{9x^2}{8}$ **c** $\dfrac{1}{4}$

10 a $2x - \dfrac{4x^3}{3} + \dfrac{4x^5}{15}$

b $-\dfrac{2x^2}{3} - \dfrac{4x^4}{45}$ **c** $-\dfrac{2}{3}$

11 3

12 a $\ln 2 + \dfrac{3}{2}x + \dfrac{3}{8}x^2 + \dfrac{3}{8}x^3 + \ldots$

b $\ln 2$

When $x = 0$, $\ln 2 + \dfrac{3}{2}x + \dfrac{3}{8}x^2 + \dfrac{3}{8}x^3 + \ldots = \ln 2$

13 a $\ln(\cos x) \equiv \ln(1 + (\cos x - 1)) \equiv \ln(1 + [1 - \dfrac{x^2}{2!} + \dfrac{x^4}{4!} - \dfrac{x^6}{6!} \ldots - 1])$

$\equiv \ln\left(1 + \left[-\dfrac{x^2}{2} + \dfrac{x^4}{24} - \dfrac{x^6}{6!}\right]\right)$

$\equiv \left[-\dfrac{x^2}{2} + \dfrac{x^4}{24} - \dfrac{x^6}{6!}\right] - \dfrac{1}{2}\left[-\dfrac{x^2}{2} + \dfrac{x^4}{24} - \dfrac{x^6}{6!}\right]^2$

$+ \dfrac{1}{3}\left[-\dfrac{x^2}{2} + \dfrac{x^4}{24} - \dfrac{x^6}{6!}\right]^3 + \ldots$

$\equiv -\dfrac{x^2}{2} + \dfrac{x^4}{24} - \dfrac{x^6}{6!} - \dfrac{1}{2}\left[\dfrac{x^4}{4} - \dfrac{x^6}{24} + \ldots\right]$

$+ \dfrac{1}{3}\left[-\dfrac{x^6}{8} + \ldots\right] + \ldots$

$\equiv -\dfrac{x^2}{2} + \dfrac{x^4}{24} - \dfrac{x^4}{8} - \dfrac{x^6}{720} + \dfrac{x^6}{48} - \dfrac{x^6}{24} + \ldots$

$\equiv -\dfrac{x^2}{2} - \dfrac{x^4}{12} - \dfrac{x^6}{45} + \ldots$

b $\dfrac{1}{2}$

14 0

Review exercise 17

1 $\left(\dfrac{5}{(2x+9)} - \dfrac{4}{(3x-1)}\right)$

2 a $\dfrac{4}{x(x+4)} \equiv \dfrac{1}{x} - \dfrac{1}{x+4}$

b $\dfrac{25}{12} - \left[\dfrac{1}{n+1} + \dfrac{1}{n+2} + \dfrac{1}{n+3} + \dfrac{1}{n+4}\right]; \dfrac{25}{12}$

3 a $\dfrac{1}{r^2} - \dfrac{2}{(r+1)^2} + \dfrac{1}{(r+2)^2}$

$= \dfrac{(r+1)^2(r+2)^2 - 2r^2(r+2)^2 + r^2(r+1)^2}{r^2(r+1)^2(r+2)^2}$

$= \dfrac{(r^2+2r+1)(r^2+4r+4) - 2r^2(r^2+4r+4) + r^2(r^2+2r+1)}{r^2(r+1)^2(r+2)^2}$

$= \dfrac{6r^2+12r+4}{r^2(r+1)^2(r+2)^2}$

$= \dfrac{2(3r^2+6r+2)}{r^2(r+1)^2(r+2)^2}$

b $\dfrac{3}{4} + \dfrac{1}{(n+2)^2} - \dfrac{1}{(n+1)^2}$ **c** $\dfrac{3}{4}$

4 $2x + \dfrac{5x^3}{3} - 2x^4 \ldots$

5 a $\dfrac{x^r}{3^r r!}$ **b** $((-1)^r)\dfrac{x^{2r}}{r!}$

c $((-1)^r)\dfrac{(x)^{2r}}{(3)^{2r}(2r)!}$

d $((-1)^r)\dfrac{(4x+5)^{(2r+1)}}{(2r+1)!}$

e $\dfrac{n(n-1)(n-2)(n-3)\ldots(n-r+1)}{r!}\left(\dfrac{-x}{6}\right)^r$

f $\dfrac{x^{(r+1)}}{r!}$

6 a $-\sqrt{2} \le x < \sqrt{2}$ **b** $-2 \le x \le 2$

7 a $-6 < x < 6$ **b** $-\dfrac{1}{3} \le x < \dfrac{1}{3}$ **c** $-\dfrac{1}{3} \le x < \dfrac{1}{3}$

8 a $x + 2x^2 + \dfrac{11x^3}{6} + \ldots$

b $4 - 8x + 16x^2 + \ldots$

c $\dfrac{4}{\ln 2} + x\left(\dfrac{-2}{(\ln 2)^2}\right) + \dfrac{x^2}{2!}\left(\dfrac{1}{(\ln 2)^2} + \dfrac{2}{(\ln 2)^3}\right) + \ldots$

d $1 - 2x - 2x^2 + \ldots$

e $1 + x\ln 3 + \dfrac{(x\ln 3)^2}{2!} + \dfrac{(x\ln 3)^3}{3!} + \ldots$

f $\dfrac{2}{5} - \dfrac{2x}{25} - \dfrac{23x^2}{125} + \ldots$

9 a $\dfrac{4}{3}$ **b** ∞ **c** 4 **d** $\dfrac{1}{4}$

e 0 **f** $\dfrac{64}{3}$ **g** 2

Assessment 17

1 a $\dfrac{n}{2(n+2)}$ **b** 50

---

2 e.g. $(r+1)^3 - (r-1)^3$

$= r^3 + 3r^2 + 3r + 1 - (r^3 - 3r^2 + 3r - 1)$

$= 6r^2 + 2$

$\displaystyle\sum_{r=1}^{n}(6r^2+2) = 2^3 - 0^3$

$+3^3 - 1^3$
$+4^3 - 2^3$
$+5^3 - 3^3$
$+\ldots$
$+n^3 - (n-2)^3$
$+(n+1)^3 - (n-1)^3$

$= -1 + n^3 + (n+1)^3$

$= -1 + n^3 + n^3 + 3n^2 + 3n + 1$

$= 2n^3 + 3n^2 + 3n$

$= n(2n^2 + 3n + 3)$

$\displaystyle\sum_{r=1}^{n}6r^2 + 2 = 6\sum_{r=1}^{n}r^2 + \sum_{r=1}^{n}2$

Therefore, $\displaystyle\sum_{r=1}^{n}r^2 = \dfrac{n(2n^2+3n+3) - 2n}{6}$

$= \dfrac{n}{6}(2n^2 + 3n + 3 - 2)$

$= \dfrac{n}{6}(2n^2 + 3n + 1)$

$= \dfrac{n}{6}(n+1)(2n+1)$ as required

3 a $2x^2 - \dfrac{4}{3}x^6 + \dfrac{4}{15}x^{10} - \dfrac{8}{315}x^{14} + \ldots$

b Valid for all real values of x

c $\dfrac{1}{2}$

4 a $-5x - \dfrac{25}{2}x^2 - \dfrac{125}{3}x^3 - \dfrac{625}{4}x^4 - \ldots$

b $-\dfrac{1}{5} \le x < \dfrac{1}{5}$ **c** $-\dfrac{1}{5}$

5 a $x + x^2 + \dfrac{1}{3}x^3 - \dfrac{1}{30}x^5 + \ldots$

b $0.398\,919$

6 a $\dfrac{3n}{n+1}$

b $\displaystyle\sum_{r=n}^{2n}\dfrac{3}{r(r+1)} = \dfrac{3(2n)}{2n+1} - \dfrac{3(n-1)}{n-1+1}$

$= \dfrac{6n}{2n+1} - \dfrac{3(n-1)}{n}$

$= \dfrac{6n^2 - 3(n-1)(2n+1)}{n(2n+1)}$

$= \dfrac{6n^2 - 6n^2 + 3n + 3}{n(2n+1)}$

$= \dfrac{3n+3}{n(2n+1)}$

$= \dfrac{3(n+1)}{n(2n+1)}$

c $\dfrac{30}{31}$

7 a $\displaystyle\sum_{r=2}^{n}\frac{2}{r^2-1}=1-\frac{1}{3}$

$+\dfrac{1}{2}-\dfrac{1}{4}$

$+\dfrac{1}{3}-\dfrac{1}{5}$

$+\dfrac{1}{4}-\dfrac{1}{6}$

$+...$

$+\dfrac{1}{n-1-1}-\dfrac{1}{n-1+1}$

$+\dfrac{1}{n-1}-\dfrac{1}{n+1}$

$=1+\dfrac{1}{2}-\dfrac{1}{n}-\dfrac{1}{n+1}=\dfrac{3}{2}-\dfrac{n+1+n}{n(n+1)}$

$=\dfrac{3(n^2+n)-2(2n+1)}{2n(n+1)}$

$=\dfrac{3n^2-n-2}{2n(n+1)}=\dfrac{(3n+2)(n-1)}{2n(n+1)}$

So $\displaystyle\sum_{r=2}^{n}\frac{1}{r^2-1}=\dfrac{(3n+2)(n-1)}{2n(n+1)}\div 2$

$=\dfrac{(3n+2)(n-1)}{4n(n+1)}$ as required

b $\displaystyle\sum_{r=2}^{\infty}\frac{1}{r^2-1}=\lim_{n\to\infty}\dfrac{(3n+2)(n-1)}{4n(n+1)}$

$=\lim_{n\to\infty}\dfrac{3n^2-n-2}{4(n^2+n)}$

$=\lim_{n\to\infty}\dfrac{3-\dfrac{1}{n}-\dfrac{2}{n^2}}{4\left(1+\dfrac{1}{n}\right)}=\dfrac{3}{4}$

8 a

$f(x)=\cos 3x \qquad f(0)=1$

$f'(x)=-3\sin 3x \qquad f'(0)=0$

$f''(x)=-9\cos x \qquad f''(0)=-9$

$f'''(x)=27\sin x \qquad f'''(0)=0$

$f^{iv}(x)=81\cos x \qquad f^{iv}(0)=81$

So $\cos 3x=1+0x-\dfrac{9}{2!}x^2+0x^3+\dfrac{81}{4!}x^4+...$

$=1-\dfrac{9}{2}x^2+\dfrac{27}{8}x^4-...$

b $\dfrac{(-1)^{r-1}(3x)^{2r-2}}{(2r-2)!}$

9 a $1-\dfrac{3}{2}x-\dfrac{9}{8}x^2-\dfrac{27}{16}x^3+...$ **b** $\dfrac{4}{3}$

10 $1+\dfrac{1}{2}x^2$

11 $1+2x+2x^2+...$

12 a $\dfrac{2}{3}$

b $\displaystyle\lim_{x\to 0}2x=0$ and $\displaystyle\lim_{x\to 0}e^{3x}-1=1-1=0$

13 a $\dfrac{1}{3}$ **b** 2

14 a $\dfrac{1}{r!}-\dfrac{1}{(r+1)!}=\dfrac{(r+1)!-r!}{r!(r+1)!}$

$=\dfrac{r!(r+1)-r!}{r!(r+1)!}$

$=\dfrac{r!(r+1-1)}{r!(r+1)!}$

$=\dfrac{r}{(r+1)!}$ as required

b $\displaystyle\sum_{r=1}^{n}\frac{r}{(r+1)!}=1-\dfrac{1}{2!}$

$+\dfrac{1}{2!}-\dfrac{1}{3!}$

$+\dfrac{1}{3!}-\dfrac{1}{4!}$

$+...$

$+\dfrac{1}{n!}-\dfrac{1}{(n+1)!}$

$=1-\dfrac{1}{(n+1)!}$

c $\dfrac{(2n+1)!-n!}{n!(2n+1)!}$

15 a $\dfrac{1}{r+2}-\dfrac{2}{r+1}+\dfrac{1}{r}$

b $\displaystyle\sum_{r=1}^{n}\frac{2}{r(r+1)(r+2)}=\dfrac{1}{3}-1+1$

$+\dfrac{1}{4}-\dfrac{2}{3}+\dfrac{1}{2}$

$+\dfrac{1}{5}-\dfrac{2}{4}+\dfrac{1}{3}$

$+\dfrac{1}{6}-\dfrac{2}{5}+\dfrac{1}{4}$

$+...$

$+\dfrac{1}{n+1}-\dfrac{2}{n}+\dfrac{1}{n-1}$

$+\dfrac{1}{n+2}-\dfrac{2}{n+1}+\dfrac{1}{n}$

$=-1+1+\dfrac{1}{2}+\dfrac{1}{n+1}+\dfrac{1}{n+2}-\dfrac{2}{n+1}$

$=\dfrac{1}{2}+\dfrac{1}{n+2}-\dfrac{1}{n+1}$

$=\dfrac{(n+2)(n+1)+2(n+1)-2(n+2)}{2(n+2)(n+1)}$

$=\dfrac{n^2+3n+2+2n+2-2n-4}{2(n+2)(n+1)}$

$=\dfrac{n^2+3n}{2(n+2)(n+1)}$

$=\dfrac{n(n+3)}{2(n+2)(n+1)}$

So $\displaystyle\sum_{r=1}^{n}\frac{1}{r(r+1)(r+2)}=\dfrac{n(n+3)}{4(n+2)(n+1)}$ as required

c $=\dfrac{1}{4}$

16 $\lim\limits_{x\to 0}\dfrac{3-3\cos x}{2x^2}=\lim\limits_{x\to 0}\dfrac{3\sin x}{4x}$

Applicable since $\lim\limits_{x\to 0}(3-3\cos x)=\lim\limits_{x\to 0}2x^2=0$

$\qquad\qquad\qquad=\lim\limits_{x\to 0}\dfrac{3\cos x}{4}$

Applicable since $\lim\limits_{x\to 0}(3\sin x)=\lim\limits_{x\to 0}4x=0=\dfrac{3}{4}$

17 $\lim\limits_{x\to 0}x^2\ln(x^2)=\lim\limits_{x\to 0}\dfrac{\ln(x^2)}{\dfrac{1}{x^2}}$

$\lim\limits_{x\to 0}\ln(x^2)=-\infty$ and $\lim\limits_{x\to 0}\dfrac{1}{x^2}=\infty$ so possible to use l'Hopital's rule

$\lim\limits_{x\to 0}\dfrac{\ln(x^2)}{\dfrac{1}{x^2}}=\lim\limits_{x\to 0}\dfrac{\left(\dfrac{2x}{x^2}\right)}{\left(-\dfrac{2}{x^3}\right)}$

$\qquad\qquad=\lim\limits_{x\to 0}-x^2$

$\qquad\qquad=0$

18 $-\dfrac{5}{6}$

19 -1

20 $\dfrac{1}{r(r+1)(r+2)}=\dfrac{A}{r}+\dfrac{B}{r+1}+\dfrac{C}{r+2}$

$1=A(r+1)(r+2)+Br(r+2)+Cr(r+1)$

$r=-1\Rightarrow 1=-B\Rightarrow B=-1$

$r=0\Rightarrow 1=2A\Rightarrow A=\dfrac{1}{2}$

$r=-2\Rightarrow 1=2C\Rightarrow C=\dfrac{1}{2}$

Multiply through by 2

$\dfrac{2}{r(r+1)(r+2)}=\dfrac{1}{r}-\dfrac{2}{r+1}+\dfrac{1}{r+2}$

$\sum\limits_{r=1}^{n}\left(\dfrac{1}{r}-\dfrac{2}{r+1}+\dfrac{1}{r+2}\right)$

$=\cancel{\dfrac{1}{1}}-\cancel{\dfrac{2}{1}}+\dfrac{1}{3}$

$+\dfrac{1}{2}-\cancel{\dfrac{2}{3}}+\cancel{\dfrac{1}{4}}$

$+\cancel{\dfrac{1}{3}}-\cancel{\dfrac{2}{4}}+\cancel{\dfrac{1}{5}}$

$+\cancel{\dfrac{1}{4}}-\cancel{\dfrac{2}{5}}+\cancel{\dfrac{1}{6}}$

$+\ldots$

$+\cancel{\dfrac{1}{n-1}}-\cancel{\dfrac{2}{n}}+\dfrac{1}{n+1}$

$+\cancel{\dfrac{1}{n}}-\dfrac{2}{n+1}+\dfrac{1}{n+2}$

$=\dfrac{1}{2}-\dfrac{1}{n+1}+\dfrac{1}{n+2}$

$=\dfrac{n^2+3n}{2n^2+6n+4}$

$=\dfrac{1+\dfrac{3}{n}}{2+\dfrac{6}{n}+\dfrac{4}{n^2}}\to\dfrac{1}{2}$ as $n\to\infty$

$\sum\limits_{r=1}^{\infty}\dfrac{1}{r(r+1)(r+2)}=\dfrac{1}{2}\div 2=\dfrac{1}{4}$

21 $\dfrac{1}{5}$

Chapter 18
Exercise 18.1A

1 In each case the graph of f(x) is blue and either $\dfrac{1}{f(x)}$ or $|f(x)|$ is red.

a i

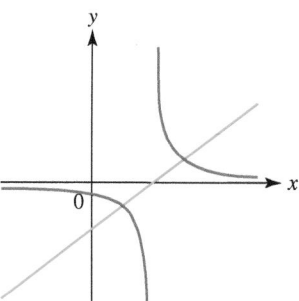

ii

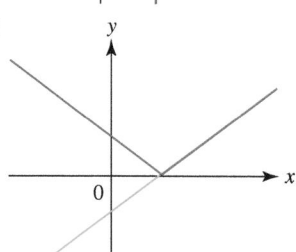

b i

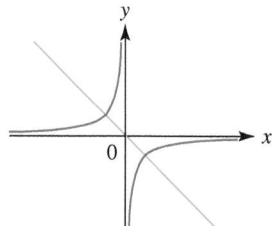

ii

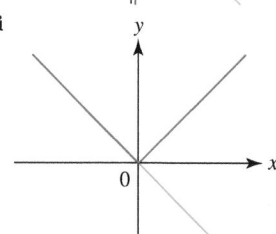

c i

ii

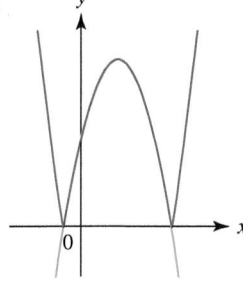

d i

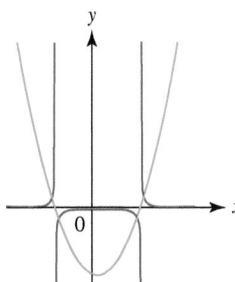

ii

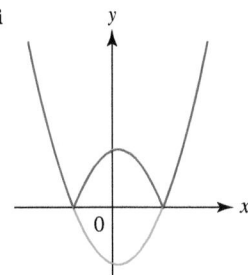

2 In each case the graph of f(x) is blue and $\dfrac{1}{f(x)}$ is red.

a

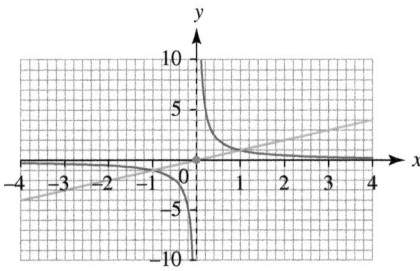

Asymptotes at $x = 0$ and at $y = 0$
Intersections at $(-1, -1)$ and $(1, 1)$

b

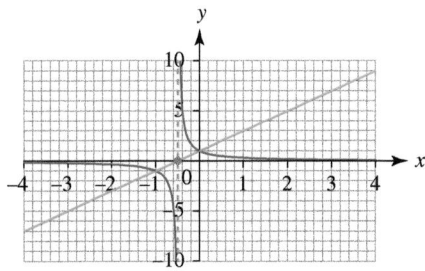

Asymptotes at $x = \dfrac{1}{2}$ and at $y = 0$
Intersections at $(-1, -1)$ and $(0, 1)$

c

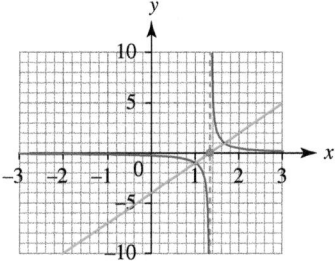

Asymptotes at $x = \dfrac{4}{3}$ and at $y = 0$
Intersections at $(1, -1)$ and $(\dfrac{5}{3}, 1)$

d

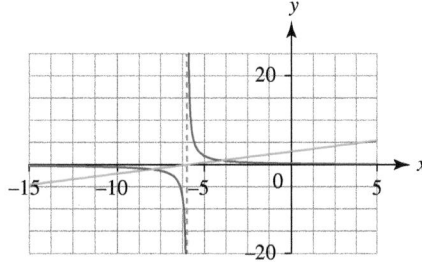

Asymptotes at $x = -6$ and at $y = 0$
Intersections at $(-4, 1)$ and $(-8, -1)$

e

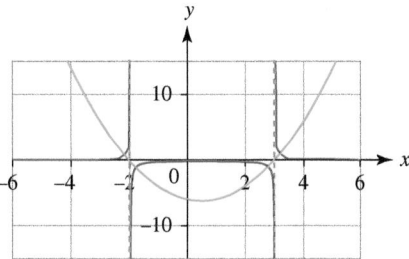

Asymptotes at $x = -2$, $x = 3$ and at $y = 0$
Intersections at $(-2.193, 1)$ and $(-1.791, -1)$ and
$(2.791, -1)$ and $(3.193, 1)$

f

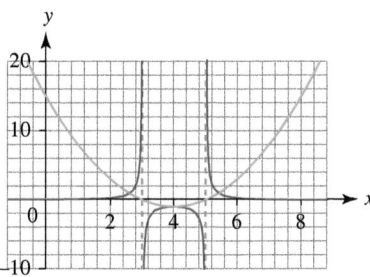

Asymptotes at $x = 3$, $x = 5$ and at $y = 0$
Intersections at $(2.586, 1)$ and $(4, -1)$ and $(5.414, 1)$

g

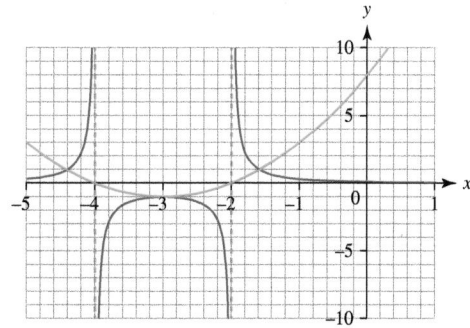

Asymptotes at $x = -4$, $x = -2$ and at $y = 0$
Intersections at $(-4.414, 1)$ and $(-3, -1)$ and $(-1.586, 1)$

h

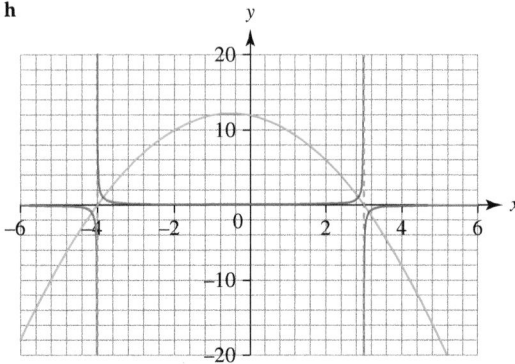

Asymptotes at $x = -4$, $x = 3$ and at $y = 0$
Intersections at $(-4.14, -1)$ and $(-3.854, 1)$ and $(2.854, 1)$
and $(3.14, -1)$

3 In each case the graph of $f(x)$ is blue and $|f(x)|$ is red.

a

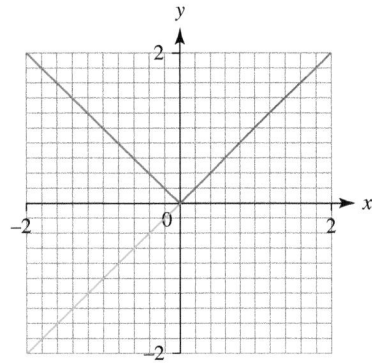

b

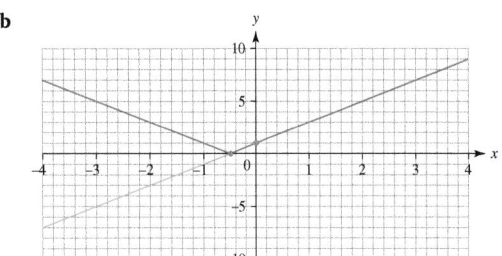

c

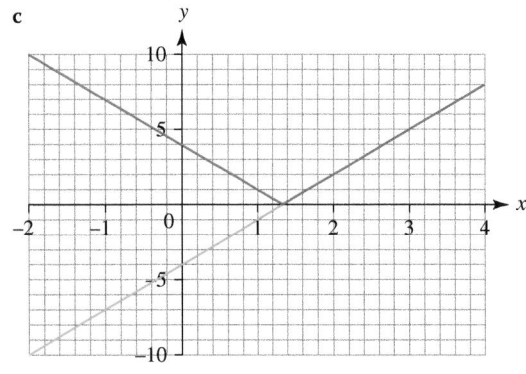

d

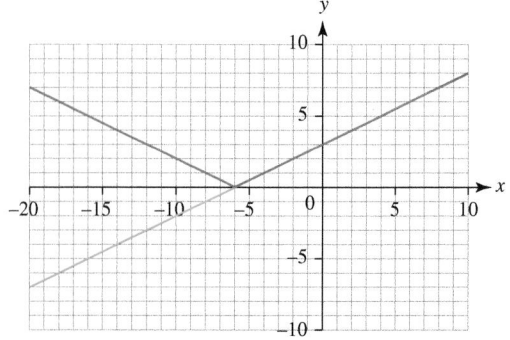

e

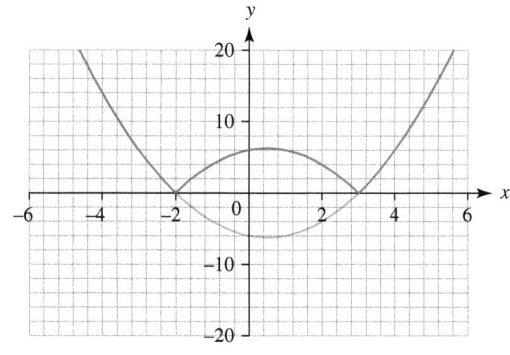

f

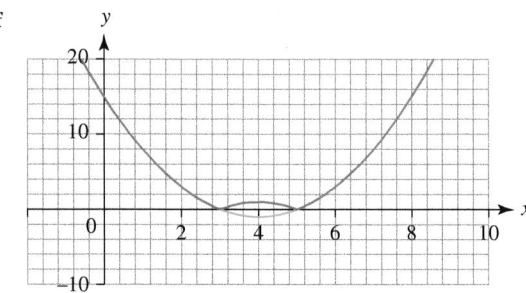

g

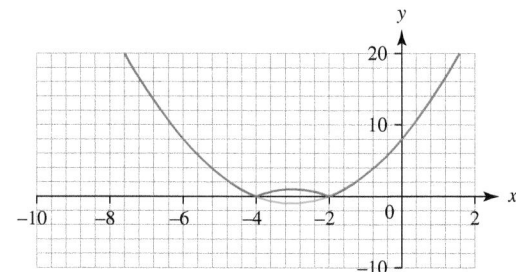

h

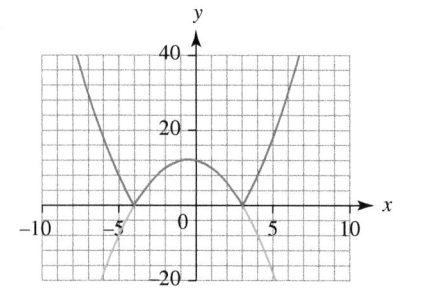

4 f(x) is blue and |f(x)| and $\dfrac{1}{f(x)}$ are red.

a i

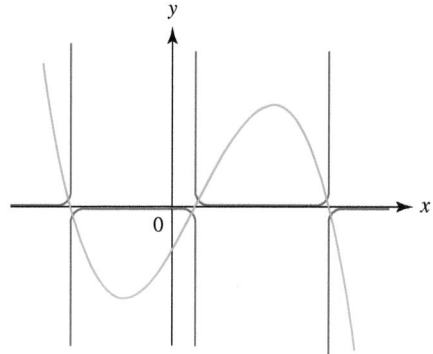

ii

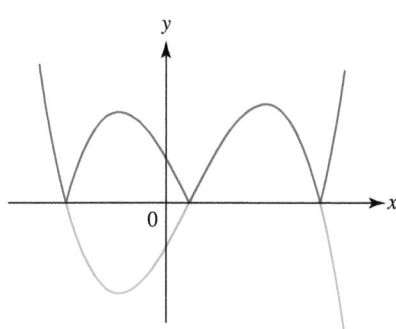

b i

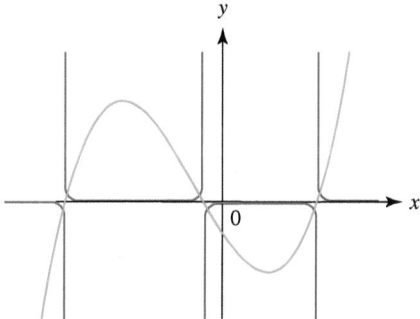

ii

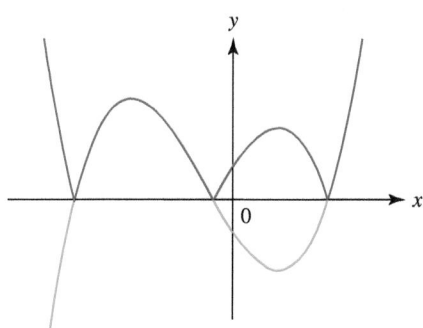

5 a i
ii

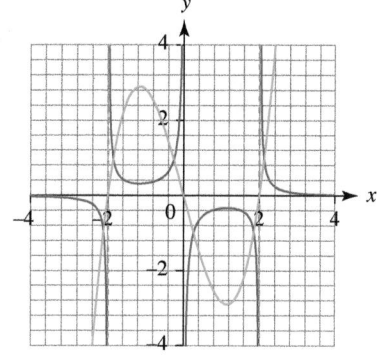

Asymptotes at $x = -2$, $x = 0$, $x = 2$ and at $y = 0$
Intersections at $(-2.115, -1)$ and $(-1.861, 1)$ and
$(-0.254, 1)$ and $(0.254, -1)$ and $(1.861, -1)$ and $(2.115, 1)$

iii

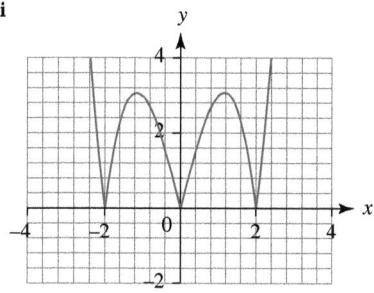

b i
ii

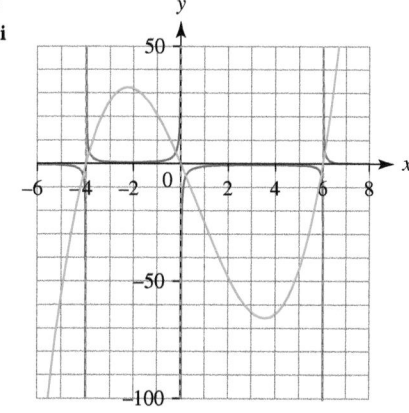

Asymptotes at $x = -4$, $x = 0$, $x = 6$ and at $y = 0$
Intersections at $(-4.025, -1)$ and $(-3.975, 1)$ and
$(-0.042, 1)$ and $(0.042, -1)$ and $(5.983, -1)$ and $(6.017, 1)$

iii

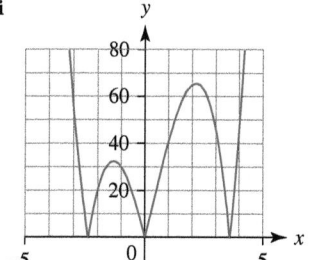

c i

ii

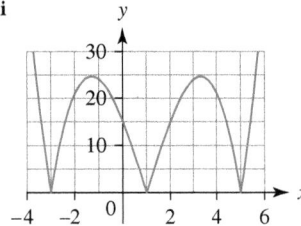

Asymptotes at $x = -3$, $x = 1$, $x = 5$
Intersections at $(-3.031, -1)$ and $(-2.968, 1)$ and
$(0.937, 1)$ and $(1.063, -1)$ and $(4.968, -1)$ and $(5.031, 1)$

iii

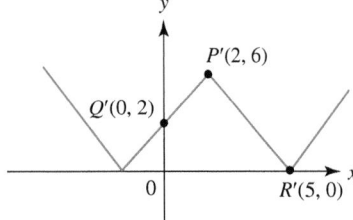

6

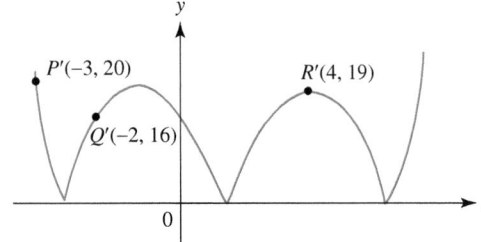

7

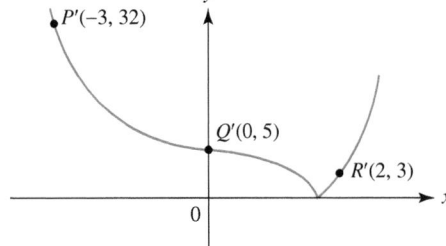

8

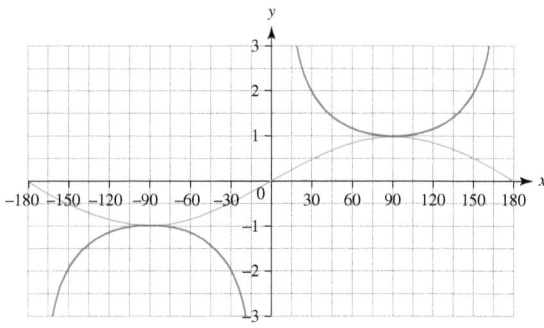

Exercise 18.1B

1 a i $y = \sin x$ and $y = \operatorname{cosec} x$

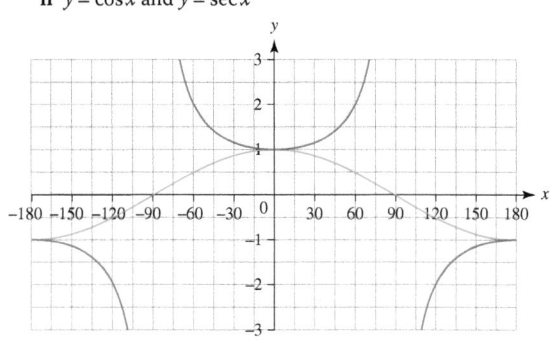

$\sin x$ is the blue graph and $y = \operatorname{cosec} x$ the red graph

ii $y = \cos x$ and $y = \sec x$

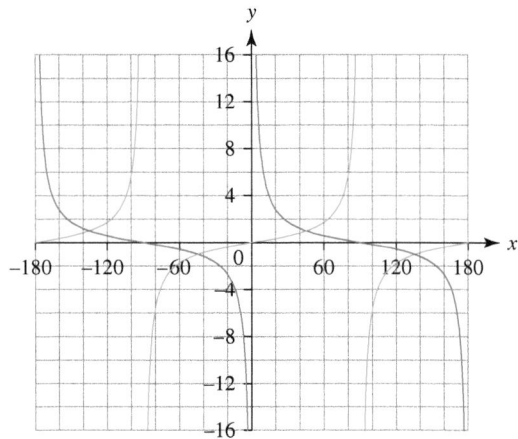

$\cos x$ is the blue graph and $y = \sec x$ the red graph

iii $y = \tan x$ and $y = \cot an x$

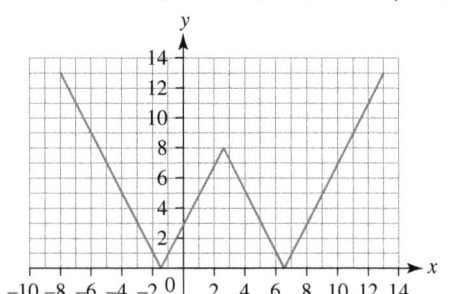

$\tan x$ is the blue graph and $y = \cot an x$ the red graph

b i $0 < x < 180$ **ii** $-90 < x < 90$
iii $-180 < x < -90$ and $0 < x < 90$

2 a P becomes $(0, 3)$, Q becomes $(2.5, 8)$ and R stays as $(6.5, 0)$

b $a = 2$, $b = -2.5$ and $c = -8$

3 a A' has coordinates $\left(-\dfrac{1}{2}, \dfrac{4}{25}\right)$ B' has coordinates $\left(3, -\dfrac{1}{6}\right)$

b f(x) is the blue graph and $\dfrac{1}{f(x)}$ is the red graph

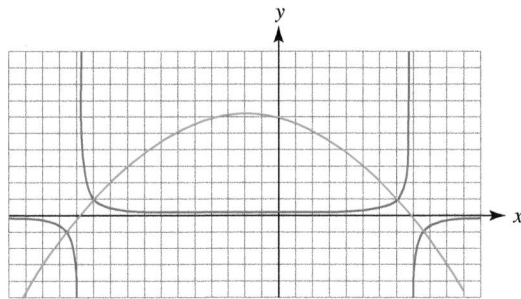

$\dfrac{1}{f(x)} > 0$ when x is between the values of the two roots of f(x)

4

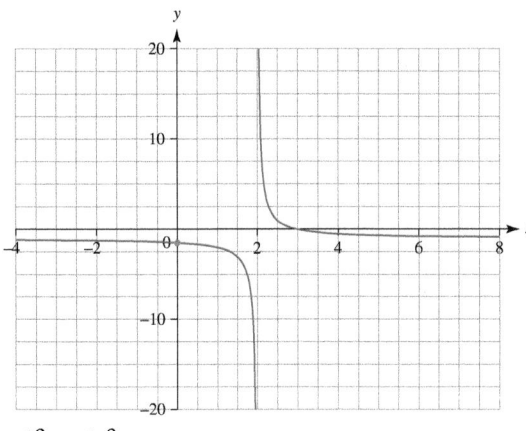

$x < 2$ or $x > 3$

5 a i $x = 1$; $y = 2$; $(0, 2)$; $(2, 0)$

ii

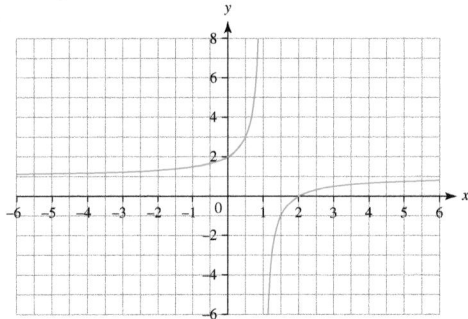

b

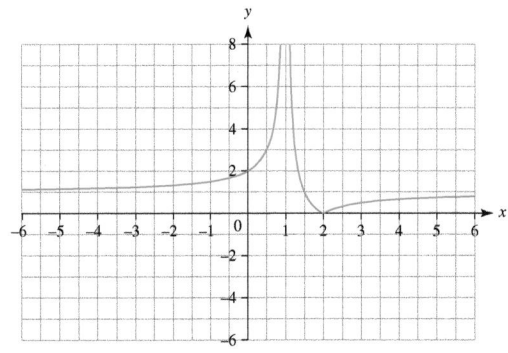

c

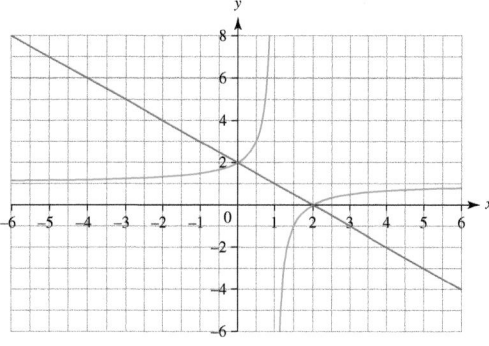

$x < 0$ or $1 < x < 2$

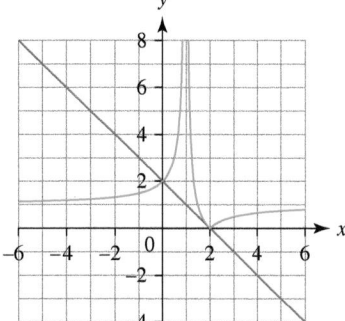

$x < 0$

6 a

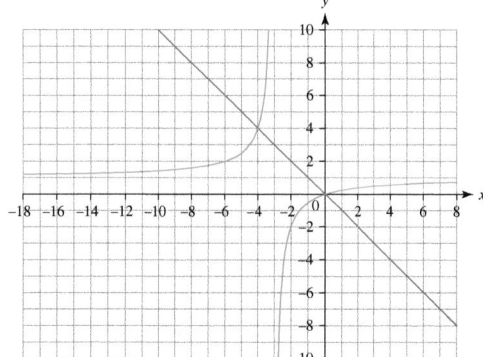

$x < -4$ or $-3 < x < 0$

b

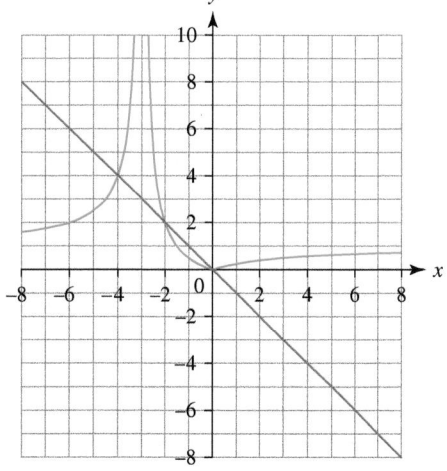

$x < -4$ or $-2 < x < 0$

Exercise 18.2A

1 a $\left(\dfrac{y}{2}\right)^2 = 12\left(\dfrac{x}{2}\right) \Rightarrow y^2 = 24x$

b $(3y)^2 = 12(3x) \Rightarrow y^2 = 4x$

c $(-x)^2 = 12y \Rightarrow x^2 = 12y$

d $(-y)^2 = 12(-x) \Rightarrow y^2 = -12x$

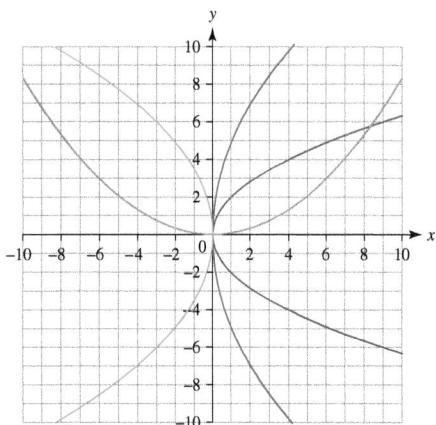

2 a $\dfrac{(2x)^2}{16} + \dfrac{(2y)^2}{25} = 1 \Rightarrow \dfrac{x^2}{4} + \dfrac{4y^2}{25} = 1$

Intercepts at $(\pm 2, 0)$ and $\left(0, \pm\dfrac{5}{2}\right)$

b $\dfrac{\left(\frac{x}{3}\right)^2}{16} + \dfrac{\left(\frac{y}{3}\right)^2}{25} = 1 \Rightarrow \dfrac{x^2}{144} + \dfrac{y^2}{225} = 1$

Intercepts at $(\pm 12, 0)$ and $(0, \pm 15)$

c $\dfrac{(-y)^2}{16} + \dfrac{x^2}{25} = 1 \Rightarrow \dfrac{x^2}{25} + \dfrac{y^2}{16} = 1$.

Intercepts at $(\pm 5, 0)$ and $(0, \pm 4)$

d $\dfrac{(-x)^2}{16} + \dfrac{(-y)^2}{25} = 1 \Rightarrow \dfrac{x^2}{16} + \dfrac{y^2}{25} = 1$.

Intercepts at $(\pm 4, 0)$ and $(0, \pm 5)$

Graphs: In order - red, purple, blue, green

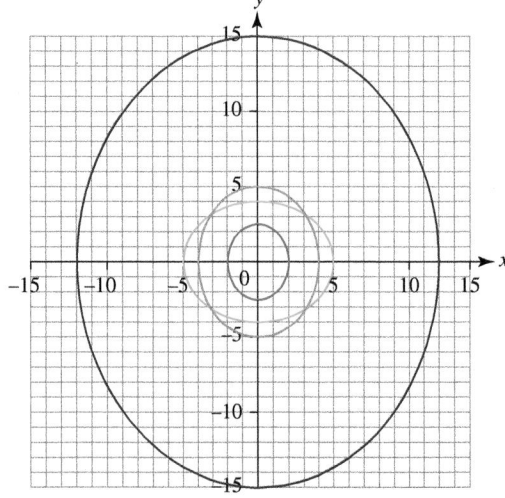

3 a $\dfrac{\left(\frac{x}{4}\right)^2}{9} - \dfrac{\left(\frac{y}{4}\right)^2}{4} = 1 \Rightarrow \dfrac{x^2}{144} - \dfrac{y^2}{64} = 1$

Asymptotes at $y = \pm\dfrac{2}{3}x$

b $\dfrac{\left(\frac{2x}{3}\right)^2}{9} - \dfrac{\left(\frac{2y}{3}\right)^2}{4} = 1 \Rightarrow \dfrac{4x^2}{81} - \dfrac{y^2}{9} = 1$

Asymptotes at $y = \pm\dfrac{2}{3}x$

c $\dfrac{(y)^2}{9} - \dfrac{(-x)^2}{4} = 1 \Rightarrow \dfrac{y^2}{9} - \dfrac{x^2}{4} = 1$ Asymptotes at $y = \pm\dfrac{3}{2}x$

d $\dfrac{(-y)^2}{9} - \dfrac{(x)^2}{4} = 1 \Rightarrow \dfrac{y^2}{9} - \dfrac{x^2}{4} = 1$ Asymptotes at $y = \pm\dfrac{3}{2}x$

Graphs: In order - red, blue, green

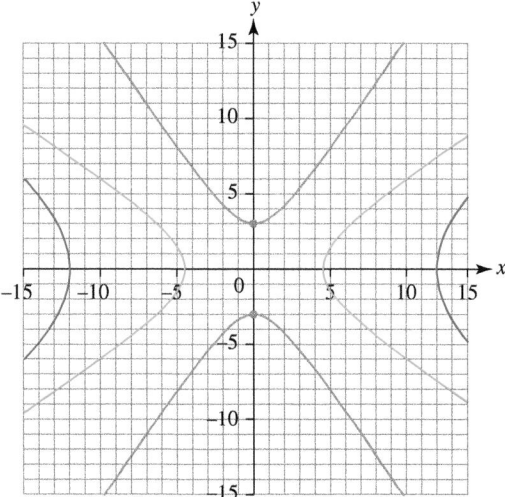

4 a $\left(\dfrac{3x}{2}\right)\left(\dfrac{3y}{2}\right) = 16 \Rightarrow 9xy = 64$

b $\left(\dfrac{3x}{4}\right)\left(\dfrac{3y}{4}\right) = 16 \Rightarrow 9xy = 256$

c $(-x)(-y) = 16 \Rightarrow xy = 16$

d $y(-x) = 16 \Rightarrow xy = -16$

Asymptotes in all cases are the lines $x = 0$ and $y = 0$
Graphs: In order - red, blue, purple, green

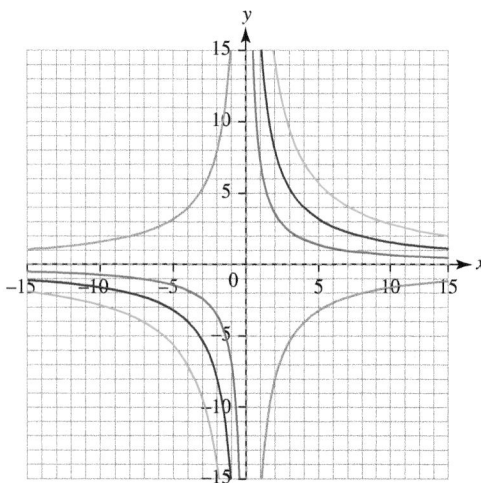

5 $\dfrac{2}{3}$ **6** $\dfrac{6}{5}$ **7** $\theta = \dfrac{3\pi}{2}$

Exercise 18.2B

1 a $9y^2 - 4x^2 = 324$ **b** Asymptotes: $y = \pm\dfrac{2}{3}x$

2 a $y^2 + 4y + 12x = 8$ **b** $(1, -2)$

c

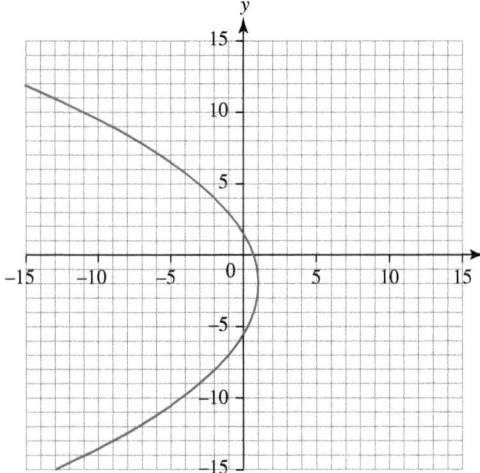

3 a Apply the translation: $\dfrac{(x-3)^2}{2} + \dfrac{(y+1)^2}{3} = 1$

Apply the rotation: $\dfrac{(y-3)^2}{2} + \dfrac{(-x+1)^2}{3} = 1$

Apply the stretch: $\dfrac{(y-3)^2}{2} + \dfrac{\left(-\dfrac{x}{2}+1\right)^2}{3} = 1$

Expand and simplify:

$3(y^2 - 6y + 9) + 2\left(\dfrac{x^2}{4} - x + 1\right) = 6$

$3y^2 - 18y + 27 + \dfrac{1}{2}x^2 - 2x + 2 - 6 = 0$

$x^2 + 6y^2 - 4x - 36y + 46 = 0$

b $3 - \sqrt{2} < k < 3 + \sqrt{2}$

4 Translation with vector $\begin{pmatrix} 2 \\ -2 \end{pmatrix}$ followed by a stretch in the y-direction, scale factor $\dfrac{1}{2}$

5 Translated by vector $\begin{pmatrix} 1 \\ 3 \end{pmatrix}$ and then enlarged by scale factor 3

6 a

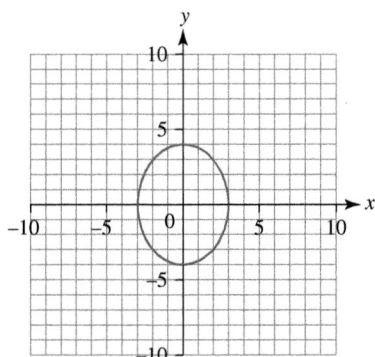

Intercepts at $(-3, 0)$ and $(3, 0)$ and $(0, 4)$ and $(0, -4)$

b $\dfrac{x^2}{9} + \dfrac{(c-x)^2}{16} = 1$

$16x^2 + 9(c-x)^2 = 144$

$25x^2 - 18cx + 9c^2 - 144 = 0$

Solve discriminant > 0

$(-18c)^2 - 4 \times 25 \times (9c^2 - 144) > 0 \Rightarrow 576c^2 < 14\,400 \Rightarrow c^2 < 25$

Hence $-5 < c < 5$

c $a = 4$ $b = -1$ $d = 121$

d $y = 8 - x$ and $y = -2 - x$

7 a $x = 2 - \dfrac{3}{2}y$

Substitute into curve: $\dfrac{\left(2 - \dfrac{3}{2}y\right)^2}{4} - \dfrac{y^2}{5} = 1$

Expand and simplify:

$5\left(2 - \dfrac{3}{2}y\right)^2 - 4y^2 = 20$

$\dfrac{45}{4}y^2 - 30y + 20 - 4y^2 = 20$

$29y^2 - 120y = 0$

$y(29y - 120) = 0$

Hence $y = 0$ or $\dfrac{120}{29}$ giving coordinates $(2, 0)$ and $\left(\dfrac{-122}{29}, \dfrac{120}{29}\right)$

b $\dfrac{y^2}{4} - \dfrac{x^2}{5} = 1$

$\dfrac{y^2}{4} - \dfrac{\left(2 - \dfrac{3}{2}y\right)^2}{5} = 1$

$5y^2 - 4\left(2 - \dfrac{3}{2}y\right)^2 = 20$

$5y^2 - 9y^2 + 24y - 16 = 20$

$y^2 - 6y + 9 = 0$

$(y-3)^2 = 0$

Hence $y = 3$. Repeated solution implies tangent.

Exercise 18.3A

1 a $\dfrac{4}{5}$ **b** $\dfrac{5}{12}$ **c** 3

d $\dfrac{32}{257}$ **e** $\dfrac{\sqrt{5}}{2}$ **f** $\dfrac{5}{4}$

2 a

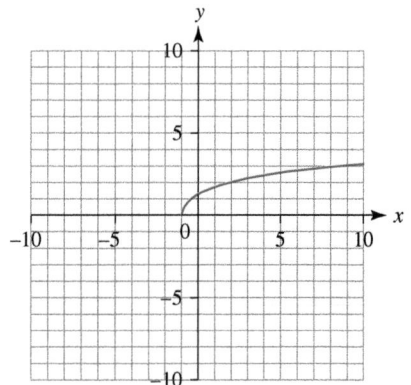

domain is $x \in \mathbb{R}$, $x \geq -1$; range is $y \in \mathbb{R}$

b

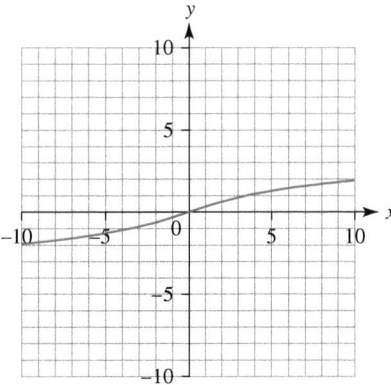

domain is $x \in \mathbb{R}$; range is $y \in \mathbb{R}$

c

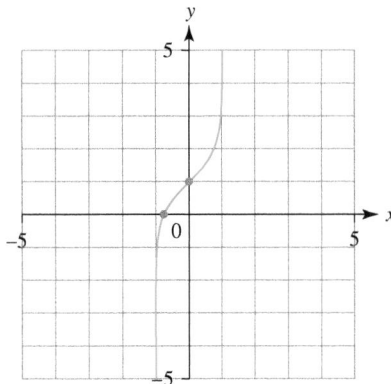

domain is $x \in \mathbb{R}$, $-1 < x < 1$; range is $y \in \mathbb{R}$

d

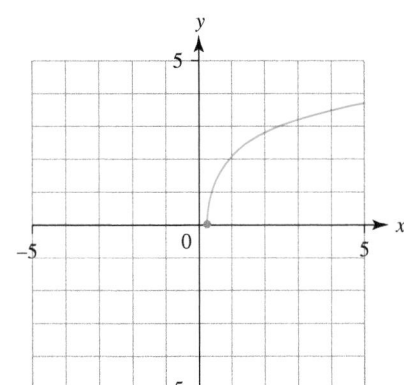

domain is $x \in \mathbb{R}$, $x \geq \dfrac{1}{4}$; range is $y \in \mathbb{R}$

e

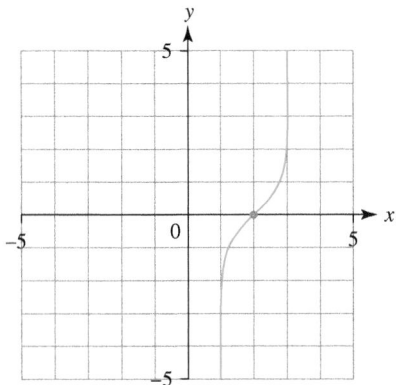

domain is $x \in \mathbb{R}$, $1 < x < 3$; range is $y \in \mathbb{R}$

f

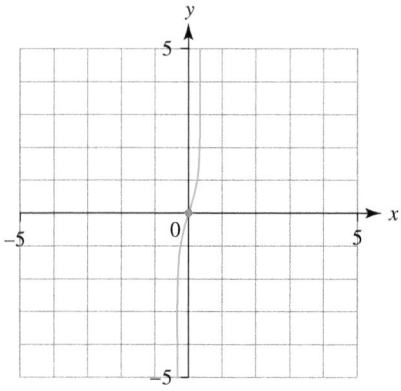

domain is $x \in \mathbb{R}$, $-\dfrac{1}{3} < x < \dfrac{1}{3}$; range is $y \in \mathbb{R}$

3 a

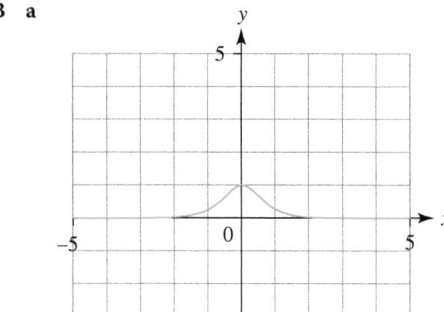

domain is $x \in \mathbb{R}$, range is $y \in \mathbb{R}$, $0 < y \leq 1$

b

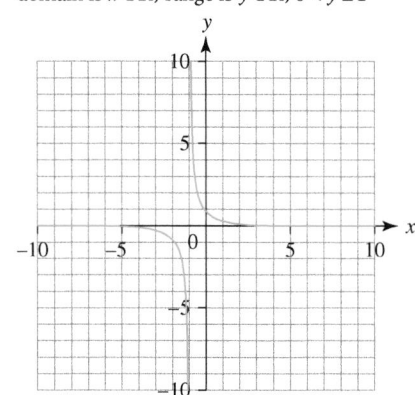

domain is $x \in \mathbb{R}$, $x \neq -1$; range is $y \in \mathbb{R}$, $y \neq 0$

c

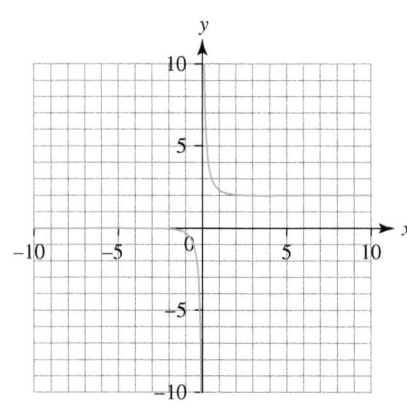

domain is $x \in \mathbb{R}$, $x \neq 0$; range is $y \in \mathbb{R}$, $y < 0$, $y > 2$

d

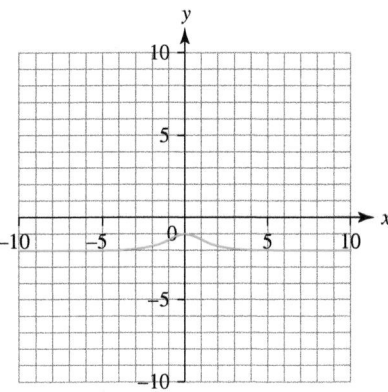

domain is $x \in \mathbb{R}$, range is $y \in \mathbb{R}$, $-2 < y \le -1$

e

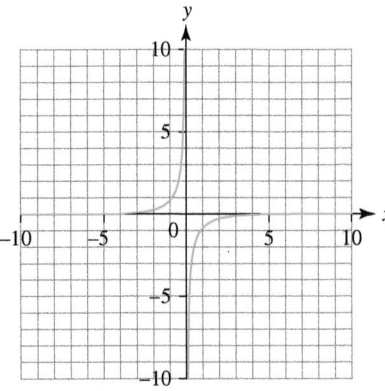

domain is $x \in \mathbb{R}$, $x \ne 0$; range is $y \in \mathbb{R}$, $y \ne 0$

f

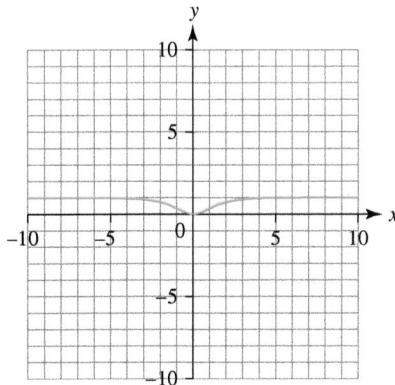

domain is $x \in \mathbb{R}$; range is $y \in \mathbb{R}$, $0 \le y < 1$

4 a $\ln\left(\dfrac{1+\sqrt{5}}{2}\right)$ **b** $\ln(\sqrt{2})$ **c** $\pm\ln(1+\sqrt{2})$

d $\ln\left(\sqrt{\dfrac{7}{3}}\right)$ **e** $\ln\left(\dfrac{\sqrt{17}-1}{4}\right)$ **f** $\pm\ln(2+\sqrt{3})$

5 a $\dfrac{7}{25}$ **b** $\dfrac{16}{63}$ **c** $\dfrac{2}{3}$ **d** $\dfrac{8}{85}$

6 a $\pm\ln(\sqrt{3})$ **b** $\ln(\sqrt{3}+2)$ **c** $\dfrac{1}{2}\ln(2\pm\sqrt{3})$

Exercise 18.3B

1 a $1 + \operatorname{cosech}^2 x \equiv 1 + \left(\dfrac{2}{e^x - e^{-x}}\right)^2$

$\equiv 1 + \dfrac{4}{(e^x - e^{-x})^2}$

$\equiv \dfrac{(e^x - e^{-x})^2 + 4}{(e^x - e^{-x})^2}$

$\equiv \dfrac{e^{2x} - 2 + e^{-2x} + 4}{(e^x - e^{-x})^2}$

$\equiv \dfrac{e^{2x} + 2 + e^{-2x}}{(e^x - e^{-x})^2} \equiv \dfrac{(e^x + e^{-x})^2}{(e^x - e^{-x})^2}$

$\equiv \left(\dfrac{e^x + e^{-x}}{e^x - e^{-x}}\right)^2 \equiv \left(\dfrac{e^{2x} + 1}{e^{2x} - 1}\right)^2$

$\equiv \coth^2 x$

b $\dfrac{1}{2}\operatorname{cosech} x \operatorname{sech} x \equiv \dfrac{1}{2}\left(\dfrac{2}{e^x - e^{-x}}\right)\left(\dfrac{2}{e^x + e^{-x}}\right)$

$\equiv \dfrac{1}{2}\left(\dfrac{4}{(e^x - e^{-x})(e^x + e^{-x})}\right)$

$\equiv \dfrac{2}{e^{2x} - e^{-2x}} \equiv \operatorname{cosech} 2x$

c $\dfrac{\coth^2\left(\dfrac{x}{2}\right) + 1}{2\coth\left(\dfrac{x}{2}\right)} \equiv \dfrac{\left(\dfrac{e^x + 1}{e^x - 1}\right)^2 + 1}{2\left(\dfrac{e^x + 1}{e^x - 1}\right)}$

$\equiv \dfrac{(e^x + 1)^2 + (e^x - 1)^2}{2(e^x + 1)(e^x - 1)}$

$\equiv \dfrac{(e^{2x} + 2e^x + 1) + (e^{2x} - 2e^x + 1)}{2(e^{2x} - 1)}$

$\equiv \dfrac{2(e^{2x} + 1)}{2(e^{2x} - 1)} \equiv \dfrac{e^{2x} + 1}{e^{2x} - 1}$

$\equiv \coth x$

d $\dfrac{1 - \operatorname{sech} x}{1 + \operatorname{sech} x} \equiv \dfrac{1 - \dfrac{2}{e^x + e^{-x}}}{1 + \dfrac{2}{e^x + e^{-x}}}$

$\equiv \dfrac{e^x + e^{-x} - 2}{e^x + e^{-x} + 2}$

$\equiv \dfrac{\left(e^{\frac{x}{2}} - e^{-\frac{x}{2}}\right)^2}{\left(e^{\frac{x}{2}} + e^{-\frac{x}{2}}\right)^2}$

$\equiv \left(\dfrac{e^{\frac{x}{2}} - e^{-\frac{x}{2}}}{e^{\frac{x}{2}} + e^{-\frac{x}{2}}}\right)^2$

$\equiv \tanh^2\left(\dfrac{x}{2}\right)$

2 a $\cosh x + \dfrac{2}{\cosh x} = 3$

$\cosh^2 x + 2 = 3\cosh x$

$\cosh^2 x - 3\cosh x + 2 = 0$

b $x = 0$, $\pm\ln(2+\sqrt{3})$

3 a $2\sinh x\cosh x = 2\left(\dfrac{e^x - e^{-x}}{2}\right)\left(\dfrac{e^x + e^{-x}}{2}\right)$

$$= \dfrac{2(e^{2x} - e^{-2x})}{4}$$

$$= \dfrac{e^{2x} - e^{-2x}}{2} = \sinh(2x)$$

b $\coth x - \tanh x = \dfrac{\cosh x}{\sinh x} - \dfrac{\sinh x}{\cosh x}$

$$= \dfrac{\cosh^2 x - \sinh^2 x}{\sinh x \cosh x}$$

$$= \dfrac{1}{\frac{1}{2}\sinh(2x)} = \dfrac{2}{\sinh(2x)}$$

$$= 2\operatorname{cosech}(2x)$$

c $x = \dfrac{1}{2}\ln(\sqrt{5} - 2)$

4 $x = \ln\sqrt{3}$

5 a $\sinh^2 x = \left(\dfrac{e^x - e^{-x}}{2}\right)^2$

$$= \dfrac{e^{2x} - 2 + e^{-2x}}{4}$$

$\dfrac{1}{2}(\cosh(2x) - 1) = \dfrac{1}{2}\left(\dfrac{e^{2x} + e^{-2x}}{2} - 1\right)$

$$= \dfrac{1}{2}\left(\dfrac{e^{2x} + e^{-2x} - 2}{2}\right)$$

$$= \dfrac{e^{2x} - 2 + e^{-2x}}{4} = \sinh^2 x$$

b $x = 0$ or $x = \ln\left(\dfrac{-1 + \sqrt{5}}{2}\right)$

6 a Let $y = \operatorname{arcosech} x$ then $x = \operatorname{cosech} y$

$$x = \dfrac{2}{e^y - e^{-y}}$$

$$xe^y - xe^{-y} = 2$$

$$xe^{2y} - 2e^y - x = 0$$

$$e^y = \dfrac{2 \pm \sqrt{4 + 4x^2}}{2x}$$

$$e^y = \dfrac{1}{x} \pm \sqrt{\dfrac{1}{x^2} + 1}$$

$$y = \ln\left(\dfrac{1}{x} + \sqrt{\dfrac{1}{x^2} + 1}\right)$$

$$\dfrac{1}{x} - \sqrt{\dfrac{1}{x^2} + 1} < 0 \text{ since } \sqrt{\dfrac{1}{x^2} + 1} > \dfrac{1}{x}$$

so $\ln\left(\dfrac{1}{x} - \sqrt{\dfrac{1}{x^2} + 1}\right)$ is not a solution.

b $\operatorname{arcosech} x + \operatorname{arcosech}(-x)$

$= \ln\left(\dfrac{1}{x} + \sqrt{\dfrac{1}{x^2} + 1}\right) + \ln\left(\dfrac{1}{(-x)} + \sqrt{\dfrac{1}{(-x)^2} + 1}\right)$

$= \ln\left(\dfrac{1}{x} + \sqrt{\dfrac{1}{x^2} + 1}\right) + \ln\left(-\dfrac{1}{x} + \sqrt{\dfrac{1}{x^2} + 1}\right)$

$= \ln\left[\left(\dfrac{1}{x} + \sqrt{\dfrac{1}{x^2} + 1}\right)\left(-\dfrac{1}{x} + \sqrt{\dfrac{1}{x^2} + 1}\right)\right]$

$$= \ln\left[-\dfrac{1}{x^2} + \left(\dfrac{1}{x^2} + 1\right)\right]$$

$$= \ln 1 = 0$$

7 Let $y = \operatorname{arsech} x$ then $x = \operatorname{sech} y$

$$x = \dfrac{2}{e^y + e^{-y}}$$

$$x(e^y + e^{-y}) = 2$$

$$xe^{2y} - 2e^y + x = 0$$

$$e^y = \dfrac{2 \pm \sqrt{4 - 4x^2}}{2x}$$

$$= \dfrac{1 \pm \sqrt{1 - x^2}}{x} = \dfrac{1}{x} \pm \sqrt{\dfrac{1}{x^2} - 1}$$

$$y = \ln\left(\dfrac{1}{x} \pm \sqrt{\dfrac{1}{x^2} - 1}\right)$$

$$= \pm\ln\left(\dfrac{1}{x} + \sqrt{\dfrac{1}{x^2} - 1}\right)$$

since $\left(\dfrac{1}{x} - \sqrt{\dfrac{1}{x^2} - 1}\right)\left(\dfrac{1}{x} + \sqrt{\dfrac{1}{x^2} - 1}\right) = \dfrac{1}{x^2} - \left(\dfrac{1}{x^2} - 1\right) = 1$

so $\left(\dfrac{1}{x} - \sqrt{\dfrac{1}{x^2} - 1}\right) = \left(\dfrac{1}{x} + \sqrt{\dfrac{1}{x^2} - 1}\right)^{-1}$

$\ln\left(\dfrac{1}{x} - \sqrt{\dfrac{1}{x^2} - 1}\right) = \ln\left(\dfrac{1}{x} + \sqrt{\dfrac{1}{x^2} - 1}\right)^{-1}$

$$= -\ln\left(\dfrac{1}{x} + \sqrt{\dfrac{1}{x^2} - 1}\right)$$

8 a $\dfrac{\sqrt{6}}{2}$ **b** $\dfrac{\sqrt{3}}{3}$

9 a $\sqrt{2}$ **b** $\dfrac{\sqrt{6}}{2}$

10 a $\tanh(2x) = \dfrac{\sinh(2x)}{\cosh(2x)}$

$$= \dfrac{2\sinh x \cosh x}{\cosh^2 x + \sinh^2 x}$$

$$= \dfrac{\dfrac{2\sinh x \cosh x}{\cosh^2 x}}{\dfrac{\cosh^2 x}{\cosh^2 x} + \dfrac{\sinh^2 x}{\cosh^2 x}}$$

$$= \dfrac{2\tanh x}{1 + \tanh^2 x}$$

b $\coth(2x) = \dfrac{1 + \tanh^2 x}{2\tanh x}$

$$= \dfrac{1}{2\tanh x} + \dfrac{\tanh^2 x}{2\tanh x}$$

$$= \dfrac{1}{2}(\coth x + \tanh x)$$

c $\coth(2x) = 2$

$$\tanh(2x) = \dfrac{1}{2}$$

$$2x = \dfrac{1}{2}\ln 3$$

$$x = \dfrac{1}{4}\ln 3$$

11 a

$$\frac{1}{1+\coth x}+\frac{1}{1-\coth x}\equiv\frac{(1-\coth x)+(1+\coth x)}{(1+\coth x)(1-\coth x)}$$

$$\equiv\frac{2}{1-\coth^2 x}$$

$$\equiv\frac{2}{-\operatorname{cosech}^2 x}$$

$$\equiv-2\sinh^2 x$$

b $x=\ln\left(\dfrac{3+\sqrt{13}}{2}\right)$ or $x=\ln\left(\dfrac{-3+\sqrt{13}}{2}\right)$

12 a

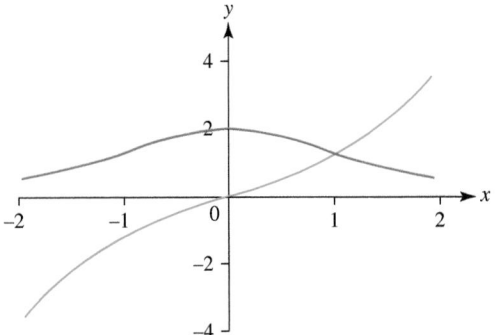

b one solution since they intersect once only

c $x=\dfrac{1}{2}\ln\left(4+\sqrt{17}\right)$

13 a

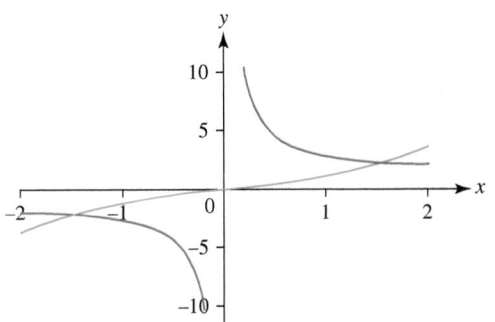

Asymptotes $y=\pm 2$ and $x=0$

b $x=\pm 1.53$

14 $x=\ln\left(1+\sqrt{2}\right),\ \ln\left(\dfrac{-1+\sqrt{10}}{3}\right)$

15 $x=\pm\ln\left(2+\sqrt{3}\right)$

16 $x=\dfrac{1}{2}\ln 3$ Or $x=\dfrac{1}{2}\ln\dfrac{1}{5}$

17 $\ln\sqrt{5}$

18 $x=\dfrac{1}{2}\ln\left(\dfrac{3+\sqrt{5}}{2}\right)$ or $x=\dfrac{1}{2}\ln\left(\dfrac{3-\sqrt{5}}{2}\right)$

Exercise 18.4A

1 a $y=x+7$ and $x=5$ **b** $y=-2x-2$ and $x=1$
  c $y=-3x-3$ and $x=1$ **d** $y=x+3$ and $x=2$

2 a Asymptotes at $x=-3$ and $y=x-3$

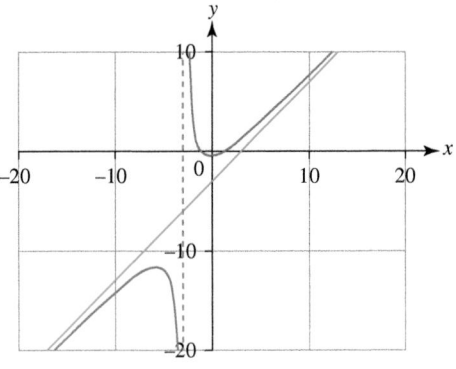

b Asymptotes at $x=-\dfrac{3}{2}$ and $y=x-\dfrac{3}{2}$

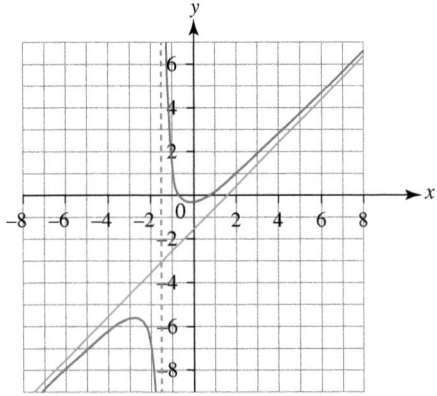

c Asymptotes at $x=4$ and $y=-x-2$

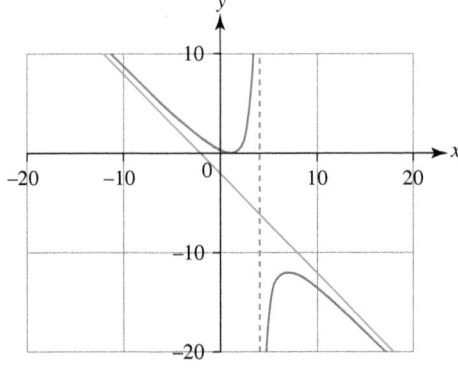

d Asymptotes at $x=\dfrac{3}{2}$ and $y=-\dfrac{1}{2}x+\dfrac{5}{4}$

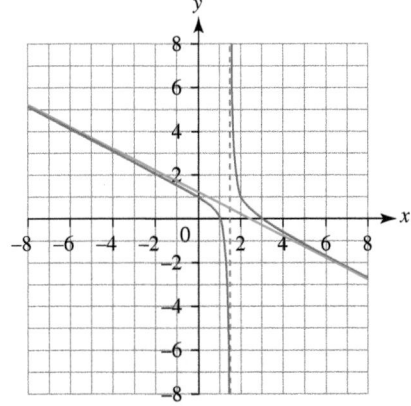

3 **a** Asymptotes at $x = \pm\dfrac{1}{\sqrt{2}}$ and $y = \dfrac{1}{2}x$

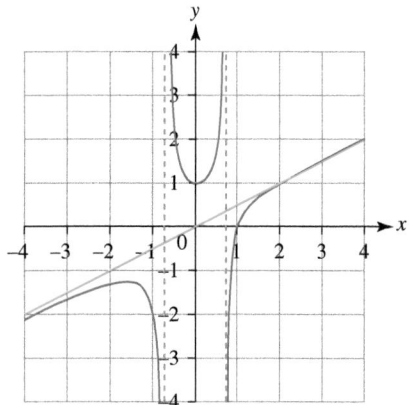

b Asymptotes at $x = \pm 2$ and $y = x$

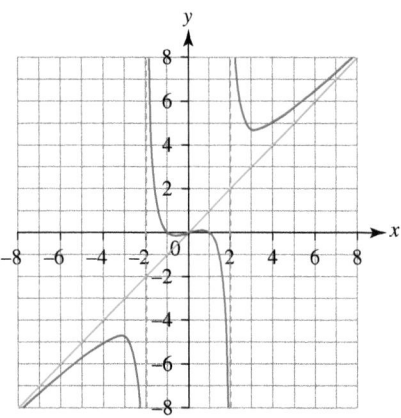

Exercise 18.4B

1 **a** $a = 4, b = 2$ **b** $x = -1$

    c $\left(-\dfrac{1}{2}, -2\right), \left(-\dfrac{3}{2}, -10\right)$

    d

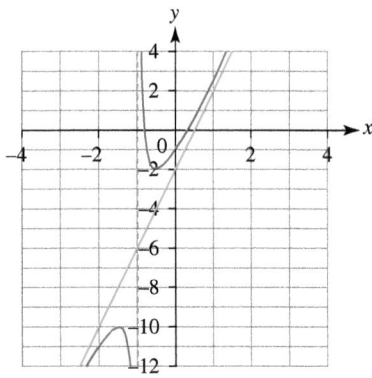

2 **a** $a = -1, b = -3$ **b** $x = 2$

    c $(-1, -1), (5, -13)$

d

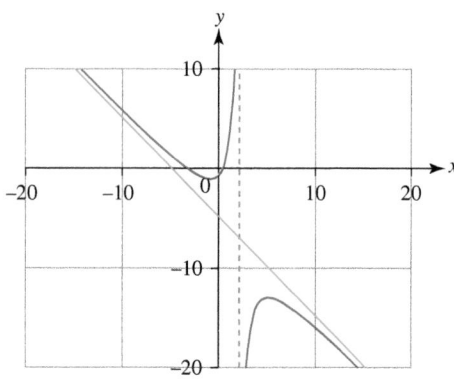

3 **a** $\dfrac{x^3 - 1}{2(x^2 - 1)} = \dfrac{(x-1)(x^2 + x + 1)}{2(x-1)(x+1)} = \dfrac{x^2 + x + 1}{2x + 2} = \dfrac{1}{2}x + \dfrac{1}{2(x+1)}$

    b $x = -1$ and $y = \dfrac{1}{2}x$ **c** $\left(0, \dfrac{1}{2}\right), \left(-2, -\dfrac{3}{2}\right)$

    d

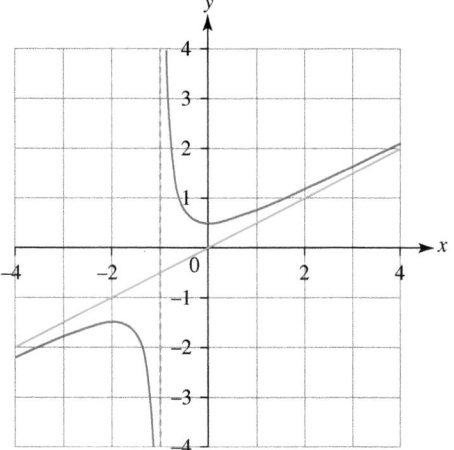

4 **a** $x^2 + x + 1 + \dfrac{1}{x-1}$

    b $y = x^2 + x + 1$ and $x = 1$

    c

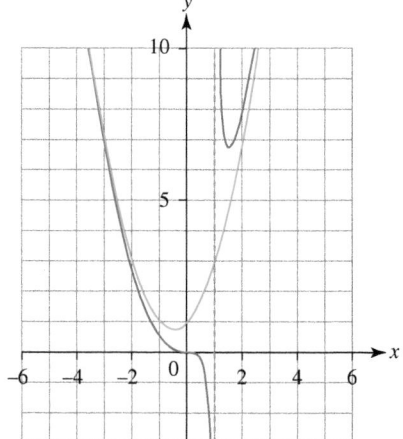

Review exercise 18

1

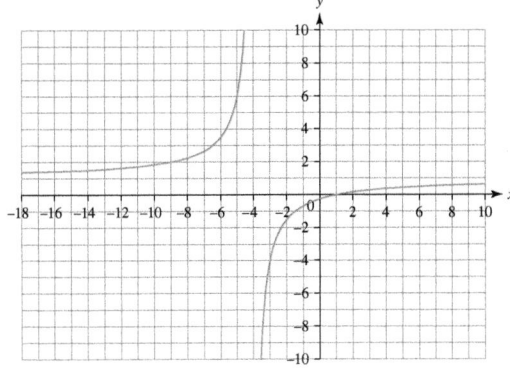

2 a

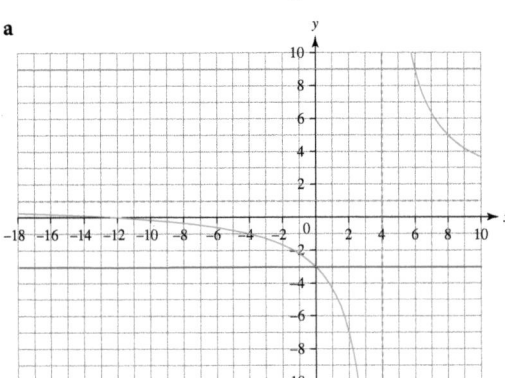

Intercepts are $(-12, 0)$ and $(0, -3)$
Asymptotes are $x = 4$ and $y = 1$
Solution to the inequality is $x < 0$ or $x > 6$

b

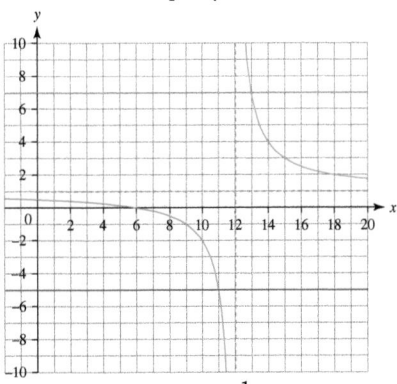

Intercepts are $(6, 0)$ and $(0, \frac{1}{2})$
Asymptotes are $x = 12$ and $y = 1$
Solution to the inequality is $x < 11$ or $x > 13$

3 a

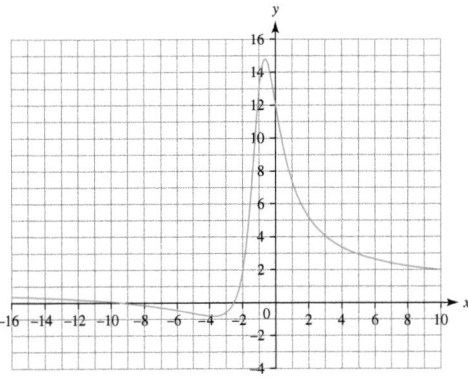

b

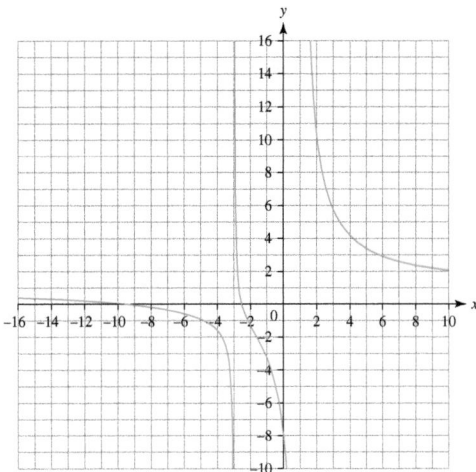

4 $\left(\frac{5}{2}, 5\right)$ Maximum value of f(x) is infinite.

5

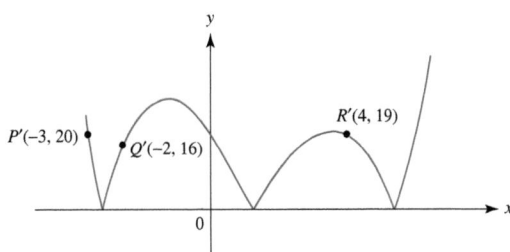

6 a

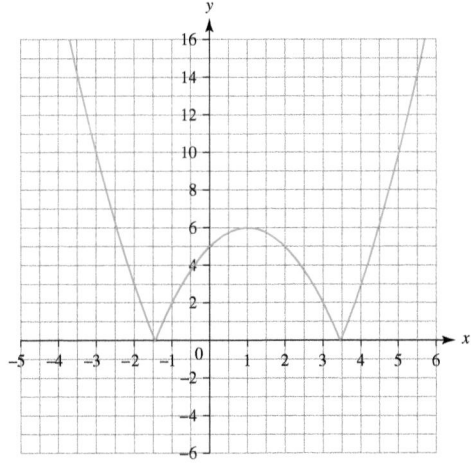

b

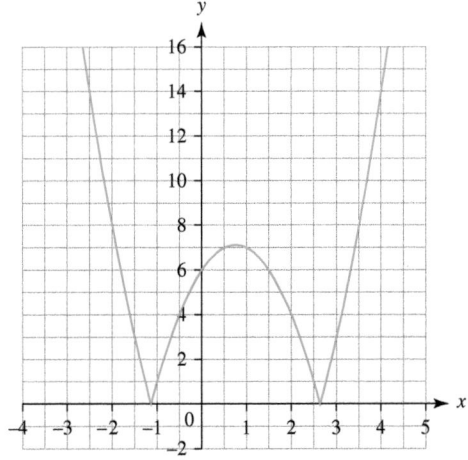

7 The blue line is f(x) and the red line is $\dfrac{1}{f(x)}$

a

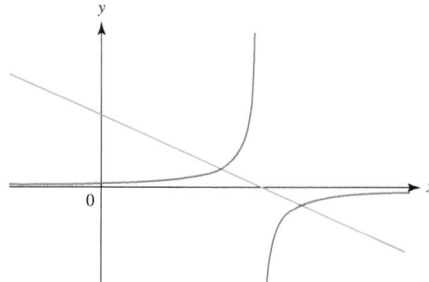

b

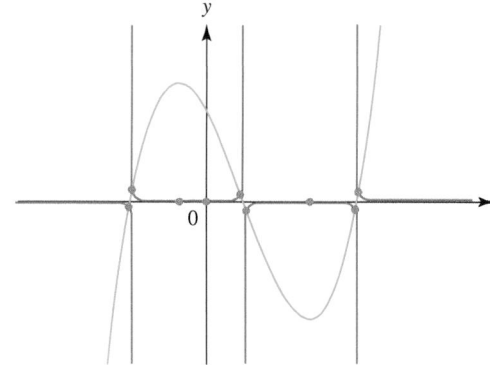

8 The blue line is f(x) and the red line is $\dfrac{1}{f(x)}$

a

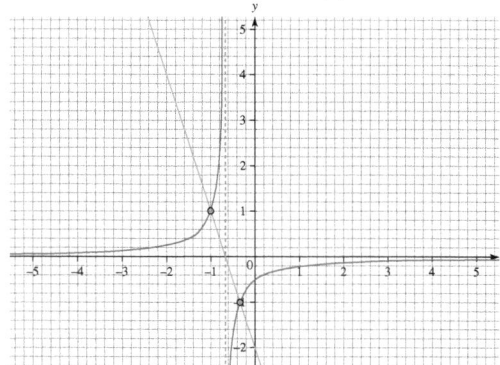

asymptotes: $x = -\dfrac{2}{3}$; $y = 0$

points of intersection: $(-1, 1)$, $(-\dfrac{1}{3}, -1)$

b asymptotes: $x = -5$; $x = -1$; $y = 0$

points of intersection: $(-3 - \sqrt{5}, 1)$, $(-3 + \sqrt{5}, 1)$,

$(-3 - \sqrt{3}, -1)$, $(-3 + \sqrt{3}, -1)$

or $(-5.24, 1)$, $(-0.764, 1)$, $(-4.73, -1)$, $(-1.27, -1)$ to 3 sf

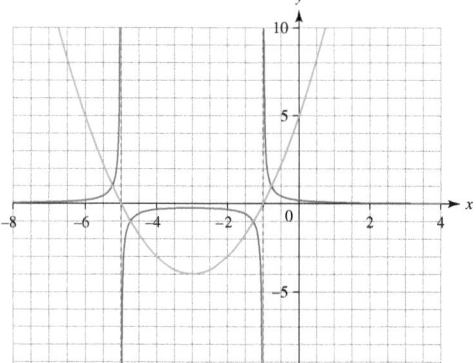

c asymptotes: $x = \sqrt[3]{-4}$ and $y = 0$

points of intersection: $(\sqrt[3]{-5}, -1)$, $(\sqrt[3]{-3}, 1)$ or $(-1.71, -1)$,

$(-1.44, 1)$ to 3 sf

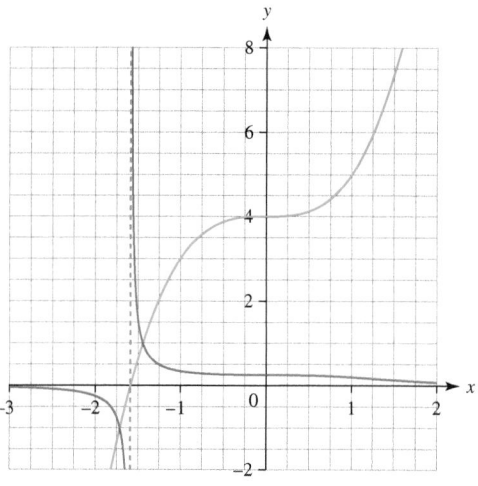

9 a i $x = 2$; $y = -1$; $(-4, 0)$, $(0, 2)$

ii

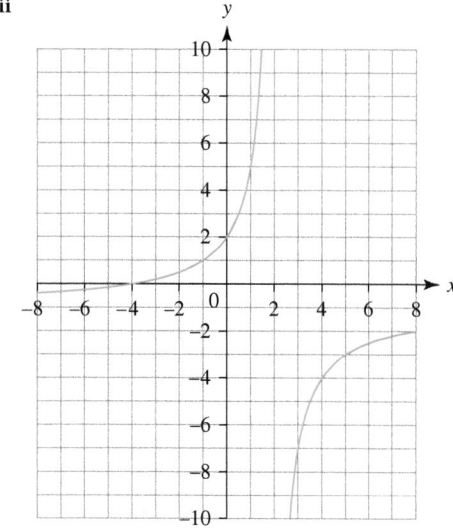

Intersections at $(-4, 0)$ and $(0, 2)$

b

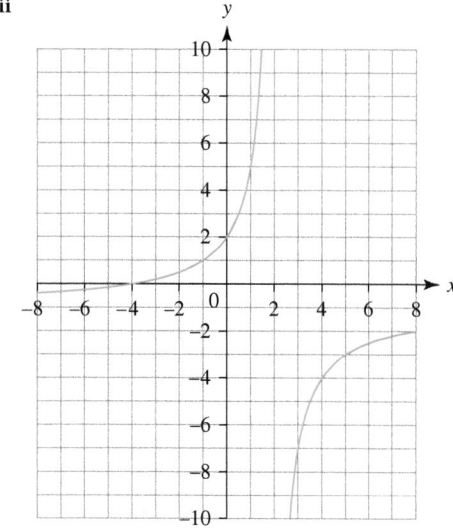

c

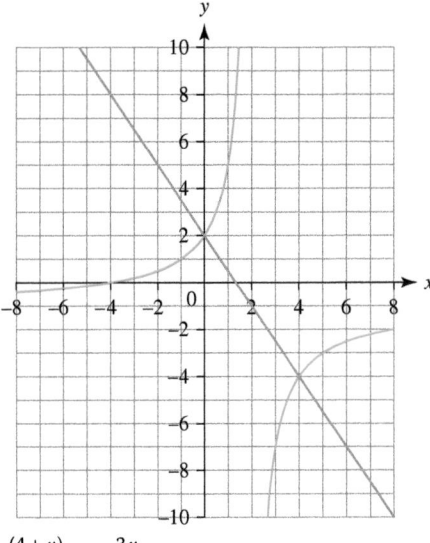

$\dfrac{(4+x)}{(2-x)} > 2 - \dfrac{3x}{2}$ when $0 < x < 2$ and when $4 < x$

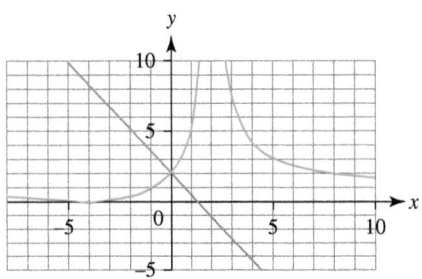

$0 < x < 2$ and $x > 2$

10 a $\dfrac{x^2}{45} + \dfrac{y^2}{81} = 1$

b

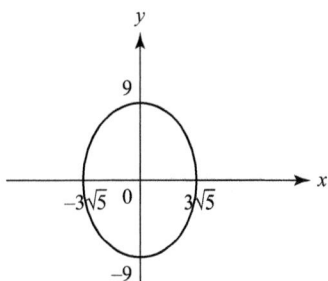

11 a $\dfrac{x^2}{16} - \dfrac{y^2}{9} = 1$

b asymptotes at $y = \pm\dfrac{3}{4}x$

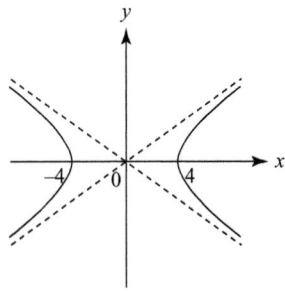

12 a $y = \dfrac{x^2}{18}$ **b** $(y-2) = \dfrac{1}{18}(x+1)^2$

c

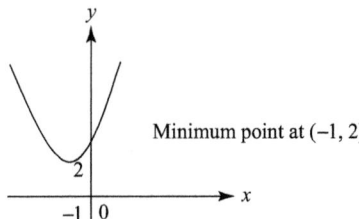

Minimum point at $(-1, 2)$

13 a $\dfrac{x^2}{4} + \dfrac{y^2}{9} = 1$

b $\dfrac{(x-2)^2}{4} + \dfrac{(y-3)^2}{9} = 1$

c

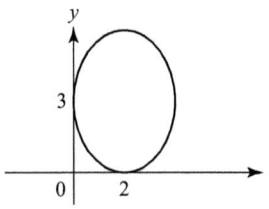

14 a $\dfrac{(x-1)^2}{100} + \dfrac{(y+4)^2}{75} = 1$

b

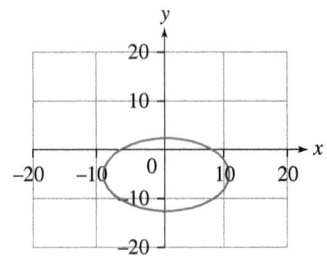

15 a $y^2 = 7x$

b

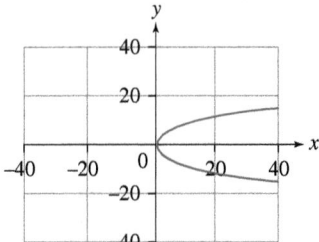

16 a

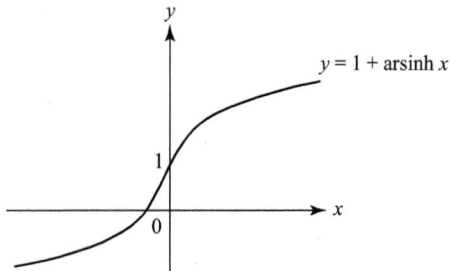

$y = 1 + \operatorname{arsinh} x$

domain $x \in \mathbb{R}$; range $y \in \mathbb{R}$

b

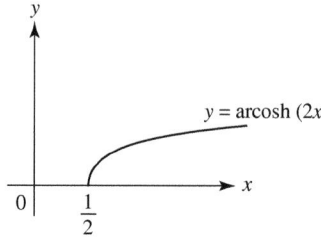

domain $x \in \mathbb{R}$, $x \geq \dfrac{1}{2}$; range $y \in \mathbb{R}$, $y \geq 0$

c

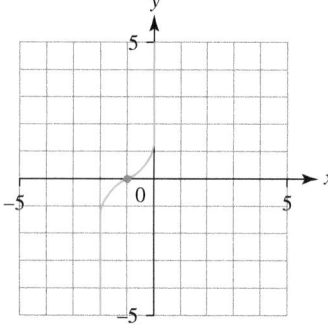

domain $x \in \mathbb{R}$, $-2 < x < 0$; range $y \in \mathbb{R}$

17

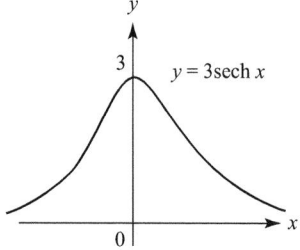

domain, $x \in \mathbb{R}$; range $y \in \mathbb{R}$, $0 < y \leq 3$

18 a $\dfrac{1}{2}\ln(2)$ **b** $\dfrac{1}{2}\ln\left(\dfrac{1}{3}\right)$

c $\dfrac{1}{2}\ln(1+\sqrt{2})$

19 a $\operatorname{cosech}(2x) = \dfrac{2}{e^{2x} - e^{-2x}}$

$\dfrac{1}{2}\operatorname{cosech}(x)\operatorname{sech}(x) = \dfrac{1}{2}\left(\dfrac{2}{e^{x} - e^{-x}}\right)\left(\dfrac{2}{e^{x} + e^{-x}}\right)$

$= \dfrac{1}{2}\left(\dfrac{4}{e^{2x} - e^{-2x}}\right)$

$= \dfrac{2}{e^{2x} - e^{-2x}} = \operatorname{cosech}(2x)$

b $x = \dfrac{1}{2}\ln(-1+\sqrt{2})$

20 a $X = \ln(1+\sqrt{2})$ or $X = \ln\left(\dfrac{1}{4}\left(\sqrt{17}-1\right)\right)$

b $X = \dfrac{1}{2}\ln\left(\dfrac{3}{2}\right)$

21 $\tanh^{2} x = 1 - \operatorname{sech}^{2} x$

$1 - \operatorname{sech}^{2}x + \operatorname{sech} x + 5 = 0$

$\operatorname{sech}^{2}x - \operatorname{sech} x - 6 = 0$

$\operatorname{sech} x = 3,\ -2$

But the range of $y = \operatorname{sech} x$ is $0 < y \leq 1$ so there are no solutions to the equation.

22 a $y = x + 3$ and $x = 5$

b

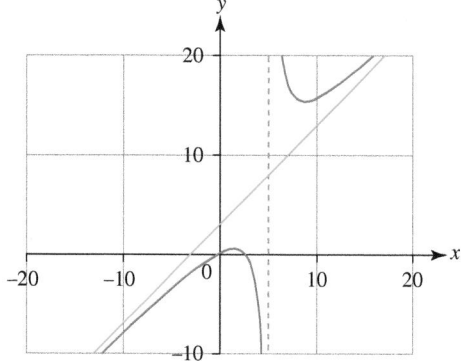

23 a $a = 3$, $b = -2$ **b** $x = \dfrac{1}{3}$

c

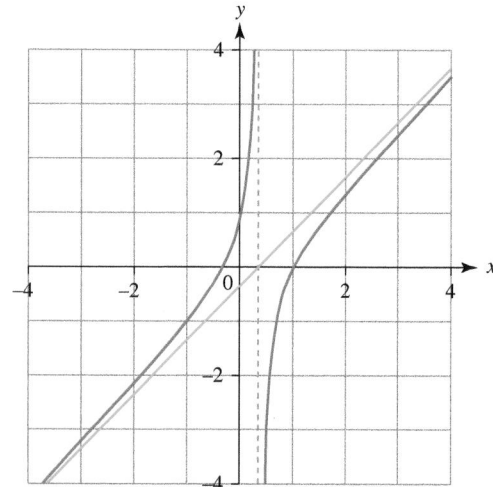

Assessment 18

1 a

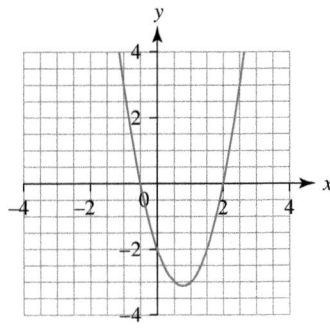

Intersections at $(-0.5, 0)$, $(2, 0)$ and $(0, -2)$

b

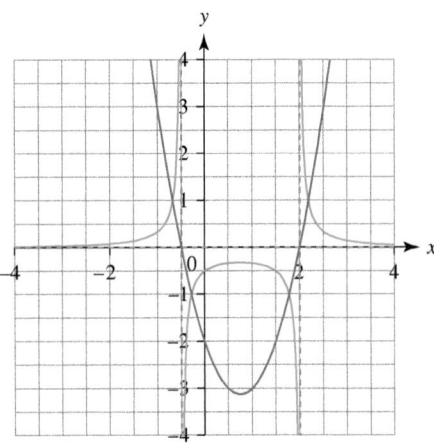

Asymptotes at $x = -0.5$, $x = 2$, $y = 0$

c $\left\{ \left(\dfrac{3 - \sqrt{33}}{4} < x < -\dfrac{1}{2} \right) \cup \left(\dfrac{3 - \sqrt{17}}{4} < x < \dfrac{3 + \sqrt{17}}{4} \right) \right.$

$\left. \cup \left(2 < x < \dfrac{3 + \sqrt{33}}{4} \right) \right\}$

2 a

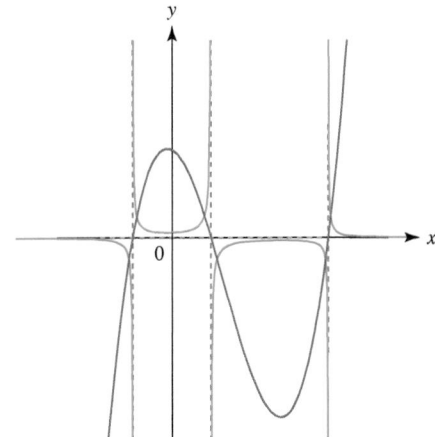

b

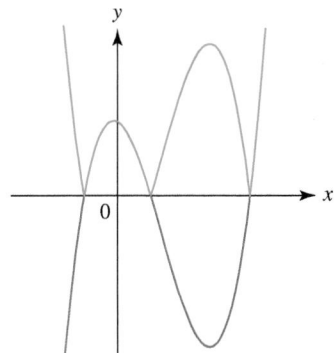

3 a $x = -1$ $y = -1$ $(0, 3)$, $(3, 0)$

b

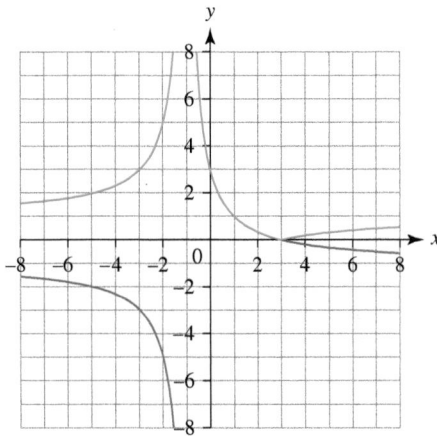

c $x < -3$, $x > 0$

4 a $\dfrac{\left(\dfrac{x}{2}\right)^2}{4} + \dfrac{\left(\dfrac{y}{2}\right)^2}{2} = 1$

$\dfrac{x^2}{16} + \dfrac{y^2}{8} = 1$

$\dfrac{(x-2)^2}{16} + \dfrac{(y+3)^2}{8} = 1$

$x^2 - 4x + 4 + 2(y^2 + 6y + 9) = 16$

$x^2 + 2y^2 - 4x + 12y + 6 = 0$ $(k = 6)$

b $y = -3 \pm 2\sqrt{2}$

5 $(x-1)^2 - 1 = -16y - 33$

$(x-1)^2 = -16(y + 2)$

Recognise second transformation is translation vector $\begin{pmatrix} 1 \\ -2 \end{pmatrix}$

$x^2 = -16y \Rightarrow$ rotation $\dfrac{3\pi}{2}$ radians about $(0, 0)$

Must state: rotation *followed by* translation.

6 a $x = 3y - 3$ and substitute:

$\dfrac{(3y-3)^2}{9} - \dfrac{y^2}{4} = 1$

$4(9y^2 - 18y + 9) - 9y^2 = 36$

$3y^2 - 8y = 0$

$y = 0$ or $y = \dfrac{8}{3}$

Coordinates (both): $(-3, 0)$ and $\left(5, \dfrac{8}{3} \right)$

b

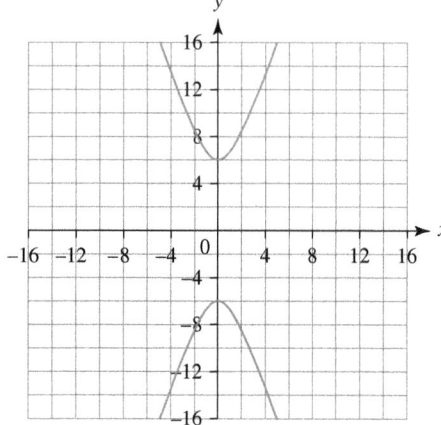

c $y = \pm 3x$

7 a $\left(\dfrac{e^x + e^{-x}}{2}\right)^2 - \left(\dfrac{e^x - e^{-x}}{2}\right)^2$

$= \dfrac{e^{2x} + 2 + e^{-2x}}{4} - \dfrac{e^{2x} - 2 + e^{-2x}}{4}$

$= \dfrac{2 - (-2)}{4} = 1$

b $x = \ln\dfrac{1}{2}$ or $x = -\ln 2$

8 a $2\cosh^2 x = 2\left(\dfrac{e^x + e^{-x}}{2}\right)^2$

$= \dfrac{e^{2x} + 2 + e^{-2x}}{2}$

$= \dfrac{e^{2x} + e^{-2x}}{2} + 1 = \cosh 2x + 1$

b $x = \ln(3 + \sqrt{8})$

9 a $\sinh x + \dfrac{3}{\sinh x} = 4 \Rightarrow \sinh^2 x + 3 = 4\sinh x$

$\sinh^2 x - 4\sinh x + 3 = 0$

b $x = \ln(3 + \sqrt{10})$ or $x = \ln(1 + \sqrt{2})$

10 a

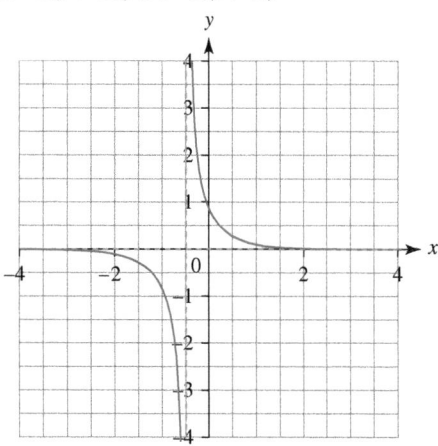

b Domain: $x \in \mathbb{R}, x \neq -\dfrac{1}{2}$

Range: $f(x) \in \mathbb{R}, f(x) \neq 0$

c $y = \text{cosech}\,1$

$\dfrac{2}{e^1 - e^{-1}} \times \dfrac{e^1}{e^1} = \dfrac{2e}{e^2 - 1}$

11 a $y = 2x - 2$ **b** $a = 4$ and $b = -6$

c

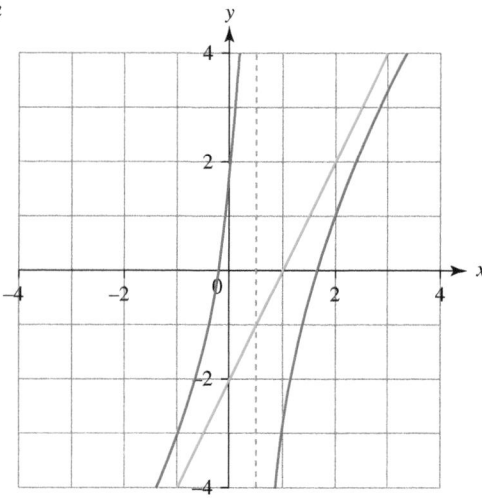

Chapter 19

Exercise 19.1A

1 a Not an improper integral, fully defined in interval and limits finite

b An improper integral since $\ln x$ undefined at $x = 0$

c An improper integral since one of the limits is ∞

d An improper integral since the limits are $\pm\infty$

e Not an improper integral, fully defined in interval and limits finite

f An improper integral since $\tan x$ undefined at $x = \dfrac{\pi}{2}$

2 a 1   **b** $\dfrac{1}{8}$ **c** 1

   d $\dfrac{1}{6}$ **e** $\dfrac{1}{8}$ **f** $\dfrac{1}{3}$

   g $\dfrac{1}{4}$ **h** 1

3 $\dfrac{1}{e^3}$

4 2

5 a 6   **b** $\dfrac{27}{2}$ **c** $2\sqrt{2}$

   d 3   **e** -4 **f** 2

   g 2   **h** $\sqrt{\dfrac{1}{2}}$

6 a $\dfrac{1}{4}e^2$ **b** $\dfrac{2}{9}e^3$

7 $\displaystyle\int_{1-e}^{a} \ln(1-x)\,dx = [-x + x\ln(1-x) - \ln(1-x)]_{1-e}^{a}$

$= (-a + a\ln(1-a) - \ln(1-a)) - (-(1-e) + (1-e)\ln e - \ln e)$

$= -a + (a-1)\ln(1-a) + (1-e) + e\ln e$

$= -a + (a-1)\ln(1-a) + (1-e) + e$

$= -a + (a-1)\ln(1-a) + 1$

$\to -1 + 1 = 0$ as $a \to 1$ since $-a \to -1$ and $(a-1)\ln(1-a) \to 0$ as $a \to 1$

8 $\dfrac{4}{9}e^3$

Exercise 19.1B

1 a $\displaystyle\int_a^1 x^{-4}\,dx=\left[-\frac{1}{3}x^{-3}\right]_a^1$

$\displaystyle=-\frac{1}{3}\left(1-\frac{1}{a^3}\right)$

As $a\to 0,\dfrac{1}{a^3}\to\infty$ so integral does not converge.

b Integral converges and has value $\dfrac{1}{3}$

c Integral has value $\ln\sqrt{2}$

d $\displaystyle\int_0^a \tan x\,dx=[-\ln\cos x]_0^a$

$=-(\ln\cos a-\ln 1)$

$=-\ln\cos a$

As $a\to\dfrac{\pi}{2},\cos a\to 0$ so $\ln\cos a\to\infty$

So integral does not converge.

e $\displaystyle\int_0^a \cos x\ dx=[\sin x]_0^a$

$=\sin a-\sin 0$

$=\sin a$

As $a\to\infty$, $\sin a$ does not converge so integral does not converge.

f $\displaystyle\int_1^a \frac{1}{x}\,dx=[\ln x]_1^a$

$=\ln a-\ln 1$

$=\ln a$

As $a\to\infty$, $\ln a\to\infty$ so integral does not converge.

g $\displaystyle\int_a^0 \frac{1}{3-x}\,dx=[-\ln(3-x)]_a^0$

$=-\ln 3+\ln(3-a)$

As $a\to-\infty,\ln(3-a)\to\infty$ so integral does not converge.

h Integral converges and has value $2\sqrt{7}$

i $\displaystyle\int_0^7 (7-x)^{-2}\,dx=[(7-x)^{-1}]_0^a$

$=\dfrac{1}{7-a}-\dfrac{1}{7}$

As $a\to 7,\dfrac{1}{7-a}\to\infty$ so integral does not converge.

j Integral converges and has value $\ln\left(\dfrac{3}{2}\right)$

k $\displaystyle\int_1^a \frac{1}{x}-\frac{2x}{x^2+1}\,dx=[\ln x-\ln(x^2+1)]_1^a$

$=\left[\ln\frac{x}{x^2+1}\right]_1^a$

$=\ln\dfrac{a}{a^2+1}-\ln\dfrac{1}{2}$

$\dfrac{a}{a^2+1}=\dfrac{\dfrac{1}{a}}{1+\dfrac{1}{a^2}}$

As $a\to\infty,\dfrac{\dfrac{1}{a}}{1+\dfrac{1}{a^2}}\to 0$ so $\ln\dfrac{a}{a^2+1}\to-\infty$

So integral does not converge.

i Integral converges and has value $\ln\left(\dfrac{\sqrt{3}}{2}\right)$

2 a $\displaystyle\int \frac{x}{x^2+3}-\frac{2}{2x+3}=\frac{1}{2}\ln(x^2+3)-\ln(2x+3)$

$=\dfrac{1}{2}\ln(x^2+3)-\dfrac{1}{2}\ln(2x+3)^2$

$=\dfrac{1}{2}\ln\left(\dfrac{x^2+3}{(2x+3)^2}\right)$

$=\dfrac{1}{2}\ln\dfrac{x^2+3}{4x^2+12x+9}$

b $\displaystyle\int_0^a \frac{x}{x^2+3}-\frac{2}{2x+3}=\frac{1}{2}\left[\ln\frac{x^2+3}{4x^2+12x+9}\right]_0^a$

$=\dfrac{1}{2}\ln\dfrac{a^2+3}{4a^2+12a+9}-\dfrac{1}{2}\ln\dfrac{3}{9}$

$=\dfrac{1}{2}\ln\dfrac{1+\dfrac{3}{a^2}}{4+\dfrac{12}{a}+\dfrac{9}{a^2}}-\dfrac{1}{2}\ln\dfrac{1}{3}$

$\to\dfrac{1}{2}\ln\dfrac{1}{4}-\dfrac{1}{2}\ln\dfrac{1}{3}$ as $a\to\infty$ since

$\dfrac{1+\dfrac{3}{a^2}}{4+\dfrac{12}{a}+\dfrac{9}{a^2}}\to\dfrac{1}{4}$ as $a\to\infty$

$=\dfrac{1}{2}\ln\left(\dfrac{\dfrac{1}{4}}{\dfrac{1}{3}}\right)$

$=\dfrac{1}{2}\ln\left(\dfrac{3}{4}\right)$

$=\ln\left(\dfrac{\sqrt{3}}{2}\right)\qquad\left(k=\dfrac{\sqrt{3}}{2}\right)$

3 a $\displaystyle\int_{a}^{0}\frac{6}{3x-2}-\frac{2x}{x^2+4}\,dx=[2\ln(3x-2)-\ln(x^2+4)]_a^0$

$$=\left[\ln\frac{(3x-2)^2}{x^2+4}\right]_a^0$$

$$=\ln 1-\ln\frac{(3a-2)^2}{a^2+4}$$

$$=-\ln\frac{9a^2-12a+4}{a^2+4}$$

$$\frac{9a^2-12a+4}{a^2+4}=\frac{9-\dfrac{12}{a}+\dfrac{4}{a^2}}{1+\dfrac{4}{a^2}}$$

As $a\to-\infty$, $\dfrac{9-\dfrac{12}{a}+\dfrac{4}{a^2}}{1+\dfrac{4}{a^2}}\to 9$, so $-\ln\dfrac{9a^2-12a+4}{a^2+4}\to-\ln 9$

So integral converges and value is $-\ln 9$ or $\ln\dfrac{1}{9}$

b $\displaystyle\int_{a}^{0}\frac{3-x}{x^2-6x+1}-\frac{8}{5-8x}\,dx$

$$=\left[-\frac{1}{2}\ln(x^2-6x+1)+\ln(5-8x)\right]_a^0$$

$$=\left[\frac{1}{2}\ln(5-8x)^2-\frac{1}{2}\ln(x^2-6x+1)\right]_a^0$$

$$=\left[\frac{1}{2}\ln\frac{(5-8x)^2}{x^2-6x+1}\right]_a^0$$

$$=\frac{1}{2}\ln 25-\frac{1}{2}\ln\frac{(5-8a)^2}{a^2-6a+1}$$

$$\frac{(5-8a)^2}{a^2-6a+1}=\frac{25-80a+64a^2}{a^2-6a+1}$$

$$=\frac{\dfrac{25}{a^2}-\dfrac{80}{a}+64}{1-\dfrac{6}{a}+\dfrac{1}{a^2}}\to 64\text{ as }a\to-\infty$$

So integral converges and is given by:

$$\frac{1}{2}\ln 25-\frac{1}{2}\ln 64=\frac{1}{2}\ln\left(\frac{25}{64}\right)=\ln\left(\frac{5}{8}\right)$$

4 Integral converges when $p<1$ and has value $\dfrac{1}{1-p}$

5 Integral converges when $p>1$ and has value $\dfrac{1}{p-1}$

Exercise 19.2A

1 a $-\dfrac{1}{\sqrt{1-x^2}}$

b $\dfrac{1}{1+x^2}$

c $\dfrac{2}{\sqrt{1-4x^2}}$

d $-\dfrac{5}{\sqrt{1-25x^2}}$

e $\dfrac{1}{x^2-2x+2}$

f $\dfrac{2}{\sqrt{1-x^2}}$

g $-\dfrac{3}{\sqrt{9-x^2}}$

h $-\dfrac{3}{\sqrt{4x-x^2-3}}$

i $-\dfrac{2x}{\sqrt{1-x^4}}$

j $\dfrac{x}{\sqrt{1-x^2}}+\arcsin x$

2 Let $y=\operatorname{arcsec}x\Rightarrow x=\sec y=(\cos y)^{-1}$

$$\frac{dx}{dy}=-(\cos y)^{-2}(-\sin y)$$

$$=\frac{\sin y}{\cos^2 y}$$

$$=\tan y\sec y$$

$$=\sec y\sqrt{\sec^2 y-1}$$

$$=x\sqrt{x^2-1}$$

$$\frac{dy}{dx}=\frac{1}{x\sqrt{x^2-1}}\text{ as required}$$

3 Let $y=\operatorname{arccosec}x\Rightarrow x=\operatorname{cosec}y=(\sin y)^{-1}$

$$\frac{dx}{dy}=-(\sin y)^{-2}(\cos y)$$

$$=-\frac{\cos y}{\sin^2 y}$$

$$=-\cot y\operatorname{cosec}y$$

$$=-\operatorname{cosec}y\sqrt{\operatorname{cosec}^2 y-1}$$

$$=-x\sqrt{x^2-1}$$

$$\frac{dy}{dx}=-\frac{1}{x\sqrt{x^2-1}}\text{ as required}$$

4 Let $y=\operatorname{arccot}x\Rightarrow x=\cot y=(\tan y)^{-1}$

$$\frac{dx}{dy}=-(\tan y)^{-2}(\sec^2 y)$$

$$=-\frac{\sec^2 y}{\tan^2 y}$$

$$=-\operatorname{cosec}^2 y$$

$$=-(1+\cot^2 y)$$

$$=-(1+x^2)$$

$$\frac{dy}{dx}=-\frac{1}{1+x^2}\text{ as required}$$

5 a $\dfrac{dy}{dx}=e^x\arctan x+\dfrac{e^x}{1+x^2}$

b $\dfrac{dy}{dx}=-\dfrac{1}{\sqrt{1-(3x^2-1)^2}}(6x)$

$$=-\frac{6x}{\sqrt{6x^2-9x^4}}$$

c $\dfrac{dy}{dx} = -\dfrac{2\sin x}{\sqrt{1-4x^2}} + \cos x \arccos 2x$

d $\dfrac{dy}{dx} = \dfrac{2(\arcsin x)}{\sqrt{1-x^2}}$

e $\dfrac{dy}{dx} = \dfrac{e^x}{\sqrt{1-e^{2x}}}$

f $\dfrac{dy}{dx} = \dfrac{2}{1+4x^2}e^{\arctan 2x}$

6 $x = \cos u \Rightarrow \dfrac{dx}{du} = -\sin u$

$$\int \frac{1}{\sqrt{1-x^2}}\,dx = \int \frac{1}{\sqrt{1-\cos^2 u}}(-\sin u)\,du$$

$$= -\int \frac{\sin u}{\sin u}\,du$$

$$= -\int 1\,du$$

$$= -u + c$$

$$= -\arccos x + c \text{ as required}$$

7 $x = \tan u \Rightarrow \dfrac{dx}{du} = \sec^2 u$

$$\int \frac{1}{1+x^2}\,dx = \int \frac{1}{1+\tan^2 u}(\sec^2 u)\,du$$

$$= \int \frac{\sec^2 u}{\sec^2 u}\,du$$

$$= \int 1\,du$$

$$= u + c$$

$$= \arctan x + c \text{ as required}$$

8 $\dfrac{1}{3}\arcsin 3x + c$

9 $-5\arccos\left(\dfrac{x}{5}\right) + c$

10 $\dfrac{1}{3}\arctan\left(\dfrac{x}{3}\right) + c$

Exercise 19.2B

1 a $\arcsin\left(\dfrac{x}{3}\right) + c$ **b** $\arcsin\left(\dfrac{x}{10}\right) + c$

c $-2\arcsin\left(\dfrac{x}{6}\right) + c$ **d** $-4\arcsin\left(\dfrac{x}{2\sqrt{2}}\right) + c$

e $\dfrac{1}{5}\arctan\left(\dfrac{x}{5}\right) + c$ **f** $\dfrac{1}{7}\arctan\left(\dfrac{x}{7}\right) + c$

g $-\dfrac{3\sqrt{2}}{2}\arctan\left(\dfrac{\sqrt{2}}{2}x\right) + c$ **h** $\sqrt{6}\arctan\left(\dfrac{\sqrt{6}}{6}x\right) + c$

i $\dfrac{1}{2}x\sqrt{1-x^2} + \dfrac{1}{2}\arcsin x + c$ **j** $\dfrac{1}{2}x\sqrt{16-x^2} + 8\arcsin\left(\dfrac{x}{4}\right) + c$

2 a $8\arctan\left(\dfrac{x}{8}\right) + c$ **b** $5\arctan(5x) + c$

c $\sqrt{2}\arctan\left(\dfrac{\sqrt{2}}{4}x\right) + c$ **d** $\dfrac{2\sqrt{3}}{3}\arctan\left(\dfrac{\sqrt{3}}{3}x\right) + c$

e $3\arcsin\left(\dfrac{x}{3}\right) + c$ **f** $\dfrac{1}{2}\arcsin\left(\dfrac{x}{2}\right) + c$

g $-6\arcsin\left(\dfrac{x}{6}\right) + c$ **h** $\arcsin\left(\dfrac{\sqrt{2}}{2}x\right) + c$

i $\dfrac{1}{18}x\sqrt{81-x^2} + \dfrac{9}{2}\arcsin\left(\dfrac{x}{9}\right) + c$

j $\dfrac{3}{2}x\sqrt{4-x^2} + 6\arcsin\left(\dfrac{x}{2}\right) + c$

3 a $\dfrac{1}{2}\arctan\left(\dfrac{x+1}{2}\right) + c$ **b** $\arctan(x-3) + c$

c $\dfrac{1}{2}\arctan\left(\dfrac{x-7}{2}\right) + c$ **d** $\sqrt{2}\arctan\left(\dfrac{x+6}{\sqrt{2}}\right) + c$

e $\arcsin\left(\dfrac{x+4}{6}\right) + c$ **f** $\arcsin\left(\dfrac{x-1}{\sqrt{2}}\right) + c$

g $-2\arcsin\left(\dfrac{x-4}{4}\right) + c$ **h** $\arcsin\left(\dfrac{x+2}{2\sqrt{2}}\right) + c$

i $\dfrac{x-2}{2}\sqrt{16-(x-2)^2} + 8\arcsin\left(\dfrac{x-2}{4}\right) + c$

j $\dfrac{1}{2}(x+3)\sqrt{25-(x+3)^2} + \dfrac{25}{2}\arcsin\left(\dfrac{x+3}{5}\right) + c$

4 $\dfrac{\pi}{12}$

Exercise 19.3A

1 $\cosh x = \dfrac{1}{2}(e^x + e^{-x})$

So $\dfrac{d(\cosh x)}{dx} = \dfrac{1}{2}(e^x - e^{-x})$

$= \sinh x$

2 a $\tanh x = \dfrac{(e^x - e^{-x})}{e^x + e^{-x}}$

So $\dfrac{d(\tanh x)}{dx} = \dfrac{(e^x + e^{-x})(e^x + e^{-x}) - (e^x - e^{-x})(e^x - e^{-x})}{(e^x + e^{-x})^2}$

$= \dfrac{(e^{2x} + 2 + e^{-2x}) - (e^{2x} - 2 + e^{-2x})}{(e^x + e^{-x})^2}$

$= \dfrac{4}{(e^x + e^{-x})^2}$

$= \left(\dfrac{2}{e^x + e^{-x}}\right)^2$

$= \operatorname{sech}^2 x$

b $\coth x = (\tanh x)^{-1}$

So $\dfrac{d(\coth x)}{dx} = -(\tanh x)^{-2}\operatorname{sech}^2 x$

$= -\dfrac{\operatorname{sech}^2 x}{\tanh^2 x}$

$= -\dfrac{1}{\cosh^2 x} \div \dfrac{\sinh^2 x}{\cosh^2 x}$

$= -\dfrac{1}{\sinh^2 x}$

$= -\operatorname{cosech}^2 x$

3 $\operatorname{cosech} x = \dfrac{2}{e^x - e^{-x}} = 2(e^x - e^{-x})^{-1}$

$\dfrac{d(\operatorname{cosech} x)}{dx} = -2(e^x - e^{-x})^{-2}(e^x + e^{-x})$

$= -\dfrac{2(e^x + e^{-x})}{(e^x - e^{-x})^2}$

$= -\dfrac{2}{e^x - e^{-x}} \cdot \dfrac{e^x + e^{-x}}{e^x - e^{-x}}$

$= -\dfrac{2}{e^x - e^{-x}} \cdot \dfrac{e^{2x} + 1}{e^{2x} - 1}$

$= -\operatorname{cosech} x \coth x$

4 a $-3\coth x \operatorname{cosech} x$ **b** $-\operatorname{cosech}^2(x+1)$

c $2\cosh 2x$ **d** $\dfrac{1}{3}\sinh\left(\dfrac{x}{3}\right)$

e $\operatorname{sech}^2(x-2)$ **f** $2x\cosh x^2$

g $2\sinh(2x-3)$ **h** $2\sinh x \cosh x$

i $\tanh x + x\operatorname{sech}^2 x$ **j** $\sqrt{\operatorname{sech} x} + \dfrac{1}{2}x\tanh x\sqrt{\operatorname{sech} x}$

5 a $\tanh x + c$

b $-\operatorname{cosech} x + c$

c $\dfrac{1}{2}\sinh 2x + c$

d $3\cosh\left(\dfrac{x}{3}\right) + c$

e $\dfrac{1}{3}\sinh^3 x + c$

f $\dfrac{1}{4}\sinh 2x + \dfrac{1}{2}x + c$

g $\dfrac{1}{4}\ln\sinh 4x + c$

h $\dfrac{1}{5}\coth 5x + c$

6 a $\dfrac{9}{16}$

b $\dfrac{4}{3}$

c $\dfrac{27}{65}$

d $\dfrac{13}{240}$

7 Let $y = \operatorname{arsinh} x$ then $x = \sinh y$

$$\frac{dx}{dy} = \cosh y$$

$$\frac{dy}{dx} = \frac{1}{\cosh y}$$

$$= \frac{1}{\sqrt{1+\sinh^2 y}}$$

$$= \frac{1}{\sqrt{1+x^2}} \text{ as required}$$

8 Let $y = \operatorname{artanh} 2x$ then $x = \dfrac{1}{2}\tanh y$

$$\frac{dx}{dy} = \frac{1}{2}\operatorname{sech}^2 y$$

$$\frac{dy}{dx} = \frac{2}{\operatorname{sech}^2 y}$$

$$= \frac{2}{1-\tanh^2 y}$$

$$= \frac{2}{1-(2x)^2}$$

$$= \frac{2}{1-4x^2} \text{ as required}$$

9 Let $y = \operatorname{arcoth} x$ then $x = \coth y$

$$\frac{dx}{dy} = -\operatorname{cosech}^2 y$$

$$\frac{dy}{dx} = \frac{1}{-\operatorname{cosech}^2 y}$$

$$= \frac{1}{-(\coth^2 y - 1)}$$

$$= \frac{1}{-(x^2 - 1)}$$

$$= \frac{1}{1-x^2} \text{ as required}$$

10 Let $y = \operatorname{arcosech} x$ then $x = \operatorname{cosech} y$

$$\frac{dx}{dy} = -\coth y\operatorname{cosech} y$$

$$\frac{dy}{dx} = \frac{1}{-\coth y\operatorname{cosech} y}$$

$$= \frac{1}{-\operatorname{cosech} y\sqrt{\operatorname{cosech}^2 y + 1}}$$

$$= \frac{1}{-x\sqrt{x^2+1}} \text{ as required}$$

11 a $\dfrac{4}{\sqrt{16x^2-1}}$

b $\dfrac{1}{\sqrt{x^2+25}}$

c $\dfrac{2x}{\sqrt{x^4-1}}$

d $\dfrac{e^x}{\sqrt{e^{2x}+1}}$

e $\dfrac{\cosh x}{\sqrt{\sinh^2 x - 1}}$

f $e^x\operatorname{arsinh}(x^2-1) + \dfrac{2xe^x}{\sqrt{x^4-2x^2+2}}$

Exercise 19.3B

1 a $\operatorname{arsinh}\left(\dfrac{x}{7}\right) + c$

b $\operatorname{arcosh}\left(\dfrac{x}{9}\right) + c$

c $\operatorname{arsinh}\left(\dfrac{x}{4}\right) + c$

d $-3\operatorname{arcosh}\left(\dfrac{x}{3}\right) + c$

e $2\operatorname{arsinh}\left(\dfrac{x+3}{4}\right) + c$

f $\operatorname{arsinh}(x-5) + c$

g $\operatorname{arcosh}\left(\dfrac{x+7}{5}\right) + c$

h $3\operatorname{arcosh}\left(\dfrac{x-12}{10}\right) + c$

2 a 0.398

b 23.7

3 a $\ln(1+\sqrt{2})$

b $\dfrac{1}{\sqrt{3}}\ln 2$

4 a $\dfrac{1}{9}\sqrt{16+9x^2} + \dfrac{1}{3}\operatorname{arsinh}\left(\dfrac{3x}{4}\right) + c$

b $6\operatorname{arcosh}\left(\dfrac{x}{\sqrt{12}}\right) - 4\sqrt{\dfrac{x^2}{4}-3} + c$

c $\sqrt{8}\arcsin\left(\dfrac{x}{\sqrt{8}}\right) - \dfrac{1}{2}\sqrt{16-2x^2} + c$

d $-\dfrac{35}{2}\ln(x^2+49) + 2\arctan\left(\dfrac{x}{7}\right) + c$

5 $$\int \operatorname{arsinh} x\,dx = \int 1\times\operatorname{arsinh} x\,dx$$

$$= x\operatorname{arsinh} x - \int \frac{x}{\sqrt{x^2+1}}\,dx$$

$$= x\operatorname{arsinh} x - \sqrt{x^2+1} + c$$

6 a $x\operatorname{arcosh} x - \sqrt{x^2-1} + c$

b $x\operatorname{arcoth} x + \dfrac{1}{2}\ln(1-x^2) + c$

7 Let $x = a\cosh u$, $\dfrac{dx}{du} = a\sinh u$

$$\int \frac{1}{\sqrt{x^2-a^2}}\,dx = \int \frac{1}{\sqrt{a^2\cosh^2 u - a^2}}\,a\sinh u\,du$$

$$= \int \frac{a\sinh u}{\sqrt{a^2\sinh^2 u}}\,du$$

$$= \int 1\,du$$

$$= u\ (+c)$$

$$= \operatorname{arcosh}\left(\frac{x}{a}\right)\ (+c)$$

A: $$\left[\operatorname{arcosh}\left(\frac{x}{a}\right)\right]_t^{2a} = \operatorname{arcosh}(2) - \operatorname{arcosh}\left(\frac{t}{a}\right)$$

$$= \ln(2+\sqrt{4-1}) - \ln\left(\frac{t}{a} + \sqrt{\left(\frac{t}{a}\right)^2 - 1}\right)$$

As $t \to a$, $\dfrac{t}{a} \to 1$ so $\dfrac{t}{a} + \sqrt{\left(\dfrac{t}{a}\right)^2 - 1} \to 1$ so

$$\ln\left(\frac{t}{a} + \sqrt{\left(\frac{t}{a}\right)^2 - 1}\right) \to \ln 1 = 0$$

So the integral converges and is equal to $\ln(2+\sqrt{3})$

B: $$\left[\operatorname{arcosh}\left(\frac{x}{a}\right)\right]_{2a}^t = \operatorname{arcosh}\left(\frac{t}{a}\right) - \operatorname{arcosh}(2)$$

$$= \ln\left(\frac{t}{a} + \sqrt{\left(\frac{t}{a}\right)^2 - 1}\right) - \ln(2 + \sqrt{3})$$

As $t \to \infty$, $\dfrac{t}{a} \to \infty$ so $\dfrac{t}{a} + \sqrt{\left(\dfrac{t}{a}\right)^2 - 1} \to \infty$ so

$$\ln\left(\frac{t}{a} + \sqrt{\left(\frac{t}{a}\right)^2 - 1}\right) \to \infty$$

So the integral does not converge.

Exercise 19.4A

1 a $5x + \ln(x-2) + 3\ln(x+4) + c$

b $2x + 3\ln(x+7) + \dfrac{4}{x+7} + c$

c $\dfrac{5}{2}x^2 + x + 2\ln(x+3) - 3\ln(x-3) + c$

d $8x + 3\ln(x+3) - 2\ln(x+8) + 3\ln x + c$

e $\dfrac{x^3}{3} + \dfrac{x^2}{2} - 3x + \ln(x-9) - 2\ln(x+1) + c$

f $\dfrac{9}{2}x^2 + 27x + \ln(x+1) + 80\ln(x-2) - \dfrac{48}{x-2} + c$

g $-\dfrac{x^3}{3} - x + \dfrac{5}{3}\ln(3x+4) + \dfrac{3}{4}\ln(3-4x) + c$

2 a $3\ln\left(\dfrac{x+3}{\sqrt{x^2+1}}\right) + 2\arctan x + c$

b $\ln\left(\dfrac{x-3}{\sqrt{x^2+25}}\right) + \arctan\left(\dfrac{x}{5}\right) + c$

c $\ln\left(\dfrac{\sqrt{x^2+36}}{6-x}\right) + \dfrac{5}{6}\arctan\left(\dfrac{x}{6}\right) + c$

d $\ln\left(\dfrac{x^2+2}{(1-x)^2}\right) - \dfrac{5}{\sqrt{2}}\arctan\left(\dfrac{x}{\sqrt{2}}\right) + c$

e $\ln(x+1) + 2\ln(x-2) - \dfrac{3}{2}\ln(x^2+9) - \arctan\left(\dfrac{x}{3}\right) + c$

f $-2\ln(x-1) - \dfrac{3}{x-1} + \ln(x^2+16) - \dfrac{1}{4}\arctan\left(\dfrac{x}{4}\right) + c$

3 a $\ln x + \dfrac{3}{7}\arctan\left(\dfrac{x}{7}\right) - \dfrac{1}{2}\ln(x^2+49) + c$

b $-\dfrac{3}{2}\ln x + \dfrac{3}{4}\ln(x^2+64) + \dfrac{1}{16}\arctan\left(\dfrac{x}{8}\right) + c$

c $\ln(3x-1) + \dfrac{3}{2}\arctan\left(\dfrac{x}{2}\right) - \dfrac{1}{2}\ln(x^2+4) + c$

4 a $\dfrac{1}{x-3} + \dfrac{1}{(x-3)^2} - \dfrac{x+4}{x^2+5}$

b $\ln(x-3) - \dfrac{1}{x-3} - \dfrac{1}{2}\ln(x^2+5) - \dfrac{4}{\sqrt{5}}\arctan\left(\dfrac{x}{\sqrt{5}}\right) + c$

Exercise 19.4B

1 a $\ln\left(\dfrac{4}{3}\right)$

b $\ln\left(\dfrac{1}{6}\right)$

c $2 + \ln\left(\dfrac{5}{2}\right)$

d $\dfrac{1}{2}\ln 10$

---

2 $\dfrac{-2x-12}{2x^3 - x^2 + 6x - 3} = \dfrac{Ax+B}{x^2+3} + \dfrac{C}{2x-1}$

$-2x - 12 = (Ax+B)(2x-1) + C(x^2+3)$

Let $x = \dfrac{1}{2}$ then $-13 = \dfrac{13}{4}C \Rightarrow C = -4$

$x^2 : 0 = 2A + C \Rightarrow A = 2$

$1 : -12 = -B + 3C \Rightarrow B = 0$

A:

$$\int_1^a \frac{2x}{x^2+3} - \frac{4}{2x-1}\,dx = [\ln(x^2+3) - 2\ln(2x-1)]_1^a$$

$$= \left[\ln\left(\frac{x^2+3}{(2x-1)^2}\right)\right]_1^a$$

$$= \ln\left(\frac{a^2+3}{(2a-1)^2}\right) - \ln(4)$$

$$\frac{a^2+3}{(2a-1)^2} = \frac{a^2+3}{4a^2 - 4a + 1}$$

$$= \frac{1 + \dfrac{3}{a^2}}{4 - \dfrac{4}{a} + \dfrac{1}{a^2}}$$

$$\frac{1 + \dfrac{3}{a^2}}{4 - \dfrac{4}{a} + \dfrac{1}{a^2}} \to \frac{1}{4} \text{ as } a \to \infty \text{ since } \frac{1}{a}, \frac{1}{a^2} \to 0$$

So $\ln\left(\dfrac{a^2+3}{(2a-1)^2}\right) \to \ln\left(\dfrac{1}{4}\right) = -\ln 4$

So integral converges and is $-2\ln 4$

B:

Point of discontinuity at $x = \dfrac{1}{2}$

$\dfrac{a^2+3}{(2a-1)^2} \to \infty$ as $x \to \dfrac{1}{2}$ since $2a - 1 \to 0$

So $\ln\left(\dfrac{a^2+3}{(2a-1)^2}\right) \to \infty$

Hence integral does not converge.

3 a $\dfrac{2-x}{x^2+a} + \dfrac{1}{x}$

b $\displaystyle\int_0^a \frac{1}{x}\,dx$ does not converge.

$$\int_0^a \frac{2-x}{x^2+a}\,dx = \left[\frac{2}{\sqrt{a}}\arctan\left(\frac{x}{\sqrt{a}}\right) - \frac{1}{2}\ln(x^2+a)\right]_0^a$$

$$= \frac{2}{\sqrt{a}}\arctan(\sqrt{a}) - \frac{1}{2}\ln(a^2+a) - 0 + \frac{1}{2}\ln(a)$$

converges to a finite value.

Therefore $\displaystyle\int_0^a \frac{2x+a}{x^3+ax}\,dx$ does not converge.

4 $\dfrac{3-a}{(x+a)(x+3)} = \dfrac{1}{x+a} - \dfrac{1}{(x+3)}$

$$\int \frac{3-a}{(x+a)(x+3)}\,dx = \int \frac{1}{x+a} - \frac{1}{(x+3)}\,dx$$

$$= \ln(x+a) - \ln(x+3)$$

$$= \ln\left(\frac{x+a}{x+3}\right)$$

a $\dfrac{1}{(x+3)}$ is discontinuous at $x=-3$

When $x \to -3$, $\dfrac{x+a}{x+3}$ does not converge (as the limit is different when approaching from the left and from the right). Hence $\ln\left(\dfrac{x+a}{x+3}\right)$ does not have a limit as $x \to -3$ so the integral is undefined for all values of a

b $\dfrac{1}{x+a}$ is discontinuous at $x=-a$, so if $a \leq 0$ then this is between 0 and ∞ and as $x \to -a$ the limit of $\dfrac{1}{x+a}$ does not exist, hence the integral does not converge.

If $a > 0$ then there are no points of discontinuity in the specified range.

$$\ln\left(\dfrac{x+a}{x+3}\right) \to \ln 1 = 0 \text{ as } x \to \infty$$

Hence $\displaystyle\int_0^t \dfrac{3-a}{(x+a)(x+3)}\,dx = \ln\left(\dfrac{t+a}{t+3}\right) - \ln\left(\dfrac{a}{3}\right) \to \ln\left(\dfrac{3}{a}\right)$

as $t \to \infty$

Thus the integral converges precisely when $a > 0$

Exercise 19.5A

1 a $I_n = \displaystyle\int x^n e^x \, dx = x^n e^x - \int n x^{n-1} e^x \, dx$

$= x^n e^x - n I_{n-1}$

b $x^4 e^x - 4x^3 e^x + 12x^2 e^x - 24 x e^x + 24 e^x + c$

2 a $I_n = \displaystyle\int x^n e^{-\frac{x}{2}}\,dx = -2x^n e^{-\frac{x}{2}} + 2\int n x^{n-1} e^{-\frac{x}{2}}\,dx$

$= -2x^n e^{-\frac{x}{2}} + 2n I_{n-1}$

b $-2x^3 e^{-\frac{x}{2}} - 12x^2 e^{-\frac{x}{2}} - 48 x e^{-\frac{x}{2}} - 96 e^{-\frac{x}{2}} + c$

3 a $I_n = \displaystyle\int_0^1 x^n e^{3x}\,dx = \left[\dfrac{1}{3} x^n e^{3x}\right]_0^1 - \dfrac{1}{3} n \int_0^1 x^{n-1} e^{3x}\,dx$

$= \dfrac{1}{3}(e^3 - 0) - \dfrac{1}{3} n I_{n-1}$

$= \dfrac{1}{3} e^3 - \dfrac{n}{3} I_{n-1}$

b $\dfrac{11}{81} e^3 - \dfrac{8}{81}$

4 a $\displaystyle\int x(\ln x)^n \, dx = \dfrac{x^2}{2}(\ln x)^n - n \int \dfrac{x^2}{2} \dfrac{1}{x}(\ln x)^{n-1}\,dx$

$= \dfrac{x^2}{2}(\ln x)^n - \dfrac{n}{2} \int x(\ln x)^{n-1}\,dx$

$= \dfrac{x^2}{2}(\ln x)^n - \dfrac{n}{2} I_{n-1}$

b $\dfrac{x^2}{2}(\ln x)^2 - \dfrac{x^2}{2}\ln x + \dfrac{x^2}{4} + c$

5 a $I_n = \displaystyle\int_0^1 \dfrac{x^n}{\sqrt{1-x}}\,dx = \left[-2x^n(1-x)^{\frac{1}{2}}\right]_0^1$

$- \displaystyle\int_0^1 -2n x^{n-1}(1-x)^{\frac{1}{2}}\,dx$

$= 0 + 2n \displaystyle\int_0^1 x^{n-1}(1-x)^{\frac{1}{2}}\,dx$

$= 2n \displaystyle\int_0^1 x^{n-1}(1-x)(1-x)^{-\frac{1}{2}}\,dx$

$= 2n \displaystyle\int_0^1 x^{n-1}(1-x)^{-\frac{1}{2}} - x^n(1-x)^{-\frac{1}{2}}\,dx$

$= 2n I_{n-1} - 2n I_n$

$$I_n = \dfrac{2n}{1+2n} I_{n-1}$$

b $\dfrac{2048}{3003}$

6 a $I_n = \displaystyle\int x^n \sin x \, dx = -x^n \cos x - \int -nx^{n-1}\cos x \, dx$

$= -x^n \cos x + n \displaystyle\int x^{n-1}\cos x \, dx$

$= -x^n \cos x + n\left(x^{n-1}\sin x - (n-1)\displaystyle\int x^{n-2}\sin x\,dx\right)$

$= -x^n \cos x + nx^{n-1}\sin x - n(n-1)I_{n-2}$

b i $-x^4 \cos x + 4x^3 \sin x + 12x^2 \cos x - 24 x \sin x$
$-24\cos x + c$

ii $-x^3 \cos x + 3x^2 \sin x + 6x \cos x - 6 \sin x + c$

7 a $I_n = \left[\dfrac{1}{2} x^n \sin 2x\right]_0^{\frac{\pi}{4}} - \dfrac{n}{2}\displaystyle\int_0^{\frac{\pi}{4}} x^{n-1}\sin 2x \, dx$

$= \dfrac{1}{2}\left(\dfrac{\pi}{4}\right)^n - \dfrac{n}{2}\left(\left[-\dfrac{1}{2}x^{n-1}\cos 2x\right]_0^{\frac{\pi}{4}} + \dfrac{n-1}{2}\displaystyle\int_0^{\frac{\pi}{4}} x^{n-2}\cos 2x\,dx\right)$

$= \dfrac{1}{2}\left(\dfrac{\pi}{4}\right)^n - \dfrac{n}{2}\left(\dfrac{n-1}{2}\right)I_{n-2}$

$= \dfrac{1}{2}\left(\dfrac{\pi}{4}\right)^n - \dfrac{n(n-1)}{4}I_{n-2}$ as required

b i $\dfrac{1}{8192}\pi^6 - \dfrac{15}{1024}\pi^4 + \dfrac{45}{64}\pi^2 - \dfrac{45}{8}$

ii $\dfrac{1}{2048}\pi^5 - \dfrac{5}{128}\pi^3 + \dfrac{15}{16}\pi - \dfrac{15}{8}$

Exercise 19.5B

1 a $I_n = \displaystyle\int_0^{\frac{\pi}{2}} \cos x \cos^{n-1} x \, dx$

$= \left[\sin x \cos^{n-1} x\right]_0^{\frac{\pi}{2}} - (n-1)\displaystyle\int_0^{\frac{\pi}{2}} -\sin^2 x \cos^{n-2} x \, dx$

$= 0 + (n-1)\displaystyle\int_0^{\frac{\pi}{2}}(1-\cos^2 x)\cos^{n-2} x \, dx$

$= (n-1)\displaystyle\int_0^{\frac{\pi}{2}} \cos^{n-2} x - \cos^n x \, dx$

$= (n-1)(I_{n-2} - I_n)$

$I_n = \dfrac{n-1}{n} I_{n-2}$

b i $\dfrac{8}{15}$

ii $\dfrac{5}{32}\pi$

2 a $I_n = \displaystyle\int_{\frac{\pi}{2}}^{\pi} \sin x \sin^{n-1} x\, dx$

$$= \left[-\cos x \sin^{n-1} x\right]_{\frac{\pi}{2}}^{\pi} - (n-1)\int_{\frac{\pi}{2}}^{\pi} -\cos^2 x \sin^{n-2} x\, dx$$

$$= 0 + (n-1)\int_{\frac{\pi}{2}}^{\pi} (1 - \sin^2 x)\sin^{n-2} x\, dx$$

$$= (n-1)\int_{\frac{\pi}{2}}^{\pi} \sin^{n-2} x - \sin^n x\, dx$$

$$= (n-1)(I_{n-2} - I_n)$$

$$I_n = \frac{n-1}{n} I_{n-2}$$

b i $\dfrac{16}{35}$

ii $\dfrac{35}{256}\pi$

3 a $I_n = \dfrac{\tan^{n-1} x}{n-1} - I_{n-2}$

b i $\dfrac{1}{5}\tan^5 x - \dfrac{1}{3}\tan^3 x + \tan x - x + c$

ii $\dfrac{5}{12} - \dfrac{1}{2}\ln 2$

4 a $I_n = \displaystyle\int_0^1 x^n\sqrt{1+x^2}\, dx = \int_0^1 x^{n-1} x\sqrt{1+x^2}\, dx$

$$= \left[\frac{1}{3}x^{n-1}(1+x^2)^{\frac{3}{2}}\right]_0^1 - \frac{n-1}{3}\int_0^1 x^{n-2}(1+x^2)^{\frac{3}{2}}\, dx$$

$$= \left(\frac{1}{3}2\sqrt{2} - 0\right) - \frac{n-1}{3}\int_0^1 x^{n-2}(1+x^2)(1+x^2)^{\frac{1}{2}}\, dx$$

$$= \frac{2}{3}\sqrt{2} - \frac{n-1}{3}\int_0^1 x^{n-2}(1+x^2)^{\frac{1}{2}} + x^n(1+x^2)^{\frac{1}{2}}\, dx$$

$$= \frac{2}{3}\sqrt{2} - \frac{n-1}{3}(I_{n-2} + I_n)$$

$$3I_n = 2\sqrt{2} - (n-1)I_{n-2} - (n-1)I_n$$

$$I_n = \frac{2\sqrt{2} - (n-1)I_{n-2}}{2+n}$$

b i $\dfrac{22\sqrt{2} - 8}{105}$

ii $\dfrac{7\sqrt{2} + 3\ln(1+\sqrt{2})}{48}$

5 a $\displaystyle\int \frac{\cos(nx)}{\cos x}\, dx$

$$= \int \frac{\cos((n-1)x + x)}{\cos x}\, dx$$

$$= \int \frac{\cos(n-1)x\,\cos x - \sin(n-1)x\,\sin x}{\cos x}\, dx$$

$$= \int \cos(n-1)x - \frac{\sin(n-1)x\,\sin x}{\cos x}\, dx$$

$$= \frac{1}{n-1}\sin(n-1)x - \frac{1}{2}\int \frac{\cos(n-2)x - \cos nx}{\cos x}\, dx$$

$$= \frac{1}{n-1}\sin(n-1)x - \frac{1}{2}\int \frac{\cos(n-2)x}{\cos x} - \frac{\cos nx}{\cos x}\, dx$$

$$I_n = \int_0^{\pi} \frac{\cos(nx)}{\cos x}\, dx = \left[\frac{1}{(n-1)}\sin(n-1)x\right]_0^{\pi} - \frac{1}{2}I_{n-2} + \frac{1}{2}I_n$$

$$= -\frac{1}{2}I_{n-2} + \frac{1}{2}I_n$$

So $\dfrac{1}{2}I_n = -\dfrac{1}{2}I_{n-2}$, so $I_n = -I_{n-2}$ as required

b $-\pi$

c i $-\pi$
ii π
d Even values of n will reduce to I_0

$$I_0 = \int_0^{\pi} \frac{\cos 0}{\cos x}\, dx$$

$$= \int_0^{\pi} \frac{1}{\cos x}\, dx$$

Discontinuity at $x = \dfrac{\pi}{2}$

$$\int_0^a \frac{1}{\cos x}\, dx = \int_0^a \sec x\, dx$$

$$= \ln(\tan x + \sec x)$$

When $a \to \dfrac{\pi}{2}$ from the left, $\tan a \to \infty$ and $\sec a \to \infty$

When $a \to \dfrac{\pi}{2}$ from the right, $\tan a \to -\infty$ and $\sec a \to -\infty$

Hence the limit as $a \to \dfrac{\pi}{2}$ of $\tan x + \sec x$ does not exist, and hence nor does the limit of $\ln(\tan x + \sec x)$. Therefore the integral does not converge. So I_n will not exist for even values of n

Exercise 19.6A

1 a

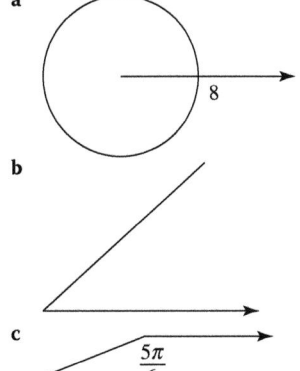

b

c $\dfrac{5\pi}{6}$

d

2 **a** max = 2, min = 0

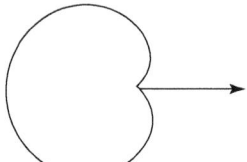

b max = 3, min = 1

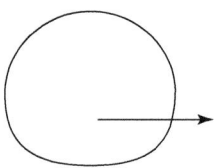

c max = 4, min = 0

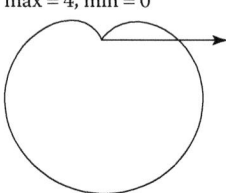

d max = 9, min = 3

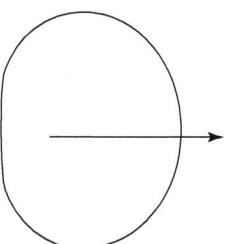

e max = 6, min = 4

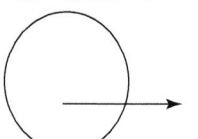

f max = 1, min = 0

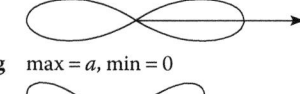

g max = a, min = 0

h max = b, min = 0

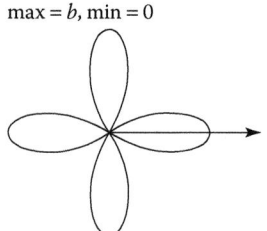

i max = c, min = 0

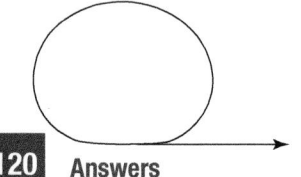

3 $\dfrac{4}{3}\pi + \sqrt{3}$

4 $\dfrac{\pi^3}{48}$

5 $\dfrac{\pi}{3}$

6 $2\sqrt{2}$

7 **a** $\dfrac{19}{2}\pi$ **b** 27π

    c $\dfrac{3}{2}\pi$ **d** 11π

8 **a** $\dfrac{\pi}{4}$ **b** $\dfrac{\pi}{4}$

    c 4π **d** $\dfrac{25}{4}\pi$

    e 2 **f** 5π

Exercise 19.6B

1 **a** If there exists θ such that $\cos\theta + \sin\theta = 2 + \sin\theta$
Then $\cos\theta = 2$ which has no solutions so they do not
intersect

    b 4π

2 **a** $(0, 0)$ and $\left(\dfrac{\pi}{3}, \dfrac{\sqrt{3}}{2}\right)$.

    b $\dfrac{3\sqrt{3}}{16} - \dfrac{\pi}{12}$

3 $\dfrac{3\pi}{4} + \dfrac{3\sqrt{3}}{16}$

4 **a** $\dfrac{5}{24}\pi - \dfrac{\sqrt{3}}{4}$ **b** $\dfrac{9}{4}\pi - 3\sqrt{3}$

    c $\dfrac{7}{4}\pi$ **d** $\dfrac{\pi}{2} - \dfrac{3}{4}\sqrt{3}$ **e** $\dfrac{59}{3}\pi - \dfrac{19}{2}\sqrt{3}$

5 **a** 2.514 units   **b** 8.473 square units

Exercise 19.7A

1 $\dfrac{dy}{dx} = x^{\frac{1}{2}}$

$$s = \int_0^1 \sqrt{1 + \left(x^{\frac{1}{2}}\right)^2}\, dx$$

$$= \int_0^1 \sqrt{1 + x}\, dx$$

$$= \left[\dfrac{2}{3}(1 + x)^{\frac{3}{2}}\right]_0^1$$

$$= \dfrac{2}{3}(2\sqrt{2} - 1)$$

$$= \dfrac{4}{3}\sqrt{2} - \dfrac{2}{3}$$

2 19.7

3 $\dfrac{9}{8}$

4 $\dfrac{15}{16}$

5 $\dfrac{56}{27}$

6 **a** $\dfrac{d(\ln(\sec x + \tan x))}{dx} = \dfrac{\sec x \tan x + \sec^2 x}{\sec x + \tan x}$

$$= \dfrac{\sec x(\tan x + \sec x)}{\sec x + \tan x}$$
$$= \sec x \text{ as required}$$

b $\ln(\sqrt{3})$

7 $\dfrac{8\pi}{3}(5\sqrt{5}-1)$

8 $\dfrac{61}{6}$

9 a $\dfrac{\pi}{56}$ **b** $\dfrac{\pi}{108}(10\sqrt{10}-1)$

10 a $\dfrac{\pi}{2}\ln 3+\dfrac{10\,\pi}{9}$ **b** $\pi\ln 3+\dfrac{20\,\pi}{9}$

Exercise 19.7B

1 a $\displaystyle\int \sqrt{1+16\,x^2}\,dx$

Let $x=\dfrac{1}{4}\sinh u$, then $\dfrac{dx}{du}=\dfrac{1}{4}\cosh u$

$s=\displaystyle\int \sqrt{1+16\left(\dfrac{1}{16}\sinh^2 u\right)}\,\dfrac{1}{4}\cosh u\ du$

$=\dfrac{1}{4}\displaystyle\int \sqrt{1+\sinh^2 u}\ \cosh u\ du$

$=\dfrac{1}{4}\displaystyle\int \sqrt{\cosh^2 u}\ \cosh hu\ du$

$=\dfrac{1}{4}\displaystyle\int \cosh u\cosh u\ du$

$=\dfrac{1}{4}\displaystyle\int \cosh^2 u\ du$

$=\dfrac{1}{8}\displaystyle\int 1+\cosh 2u\ du$

$=\dfrac{1}{8}\left(u+\dfrac{1}{2}\sinh 2u\right)+c$

$=\dfrac{1}{8}(\operatorname{arsinh}4x+\sinh(\operatorname{arsinh}4x)\cosh(\operatorname{arsinh}4x))+c$

$=\dfrac{1}{8}\operatorname{arsinh}(4x)+\dfrac{1}{2}x\cosh(\operatorname{arsinh}4x)+c$

b $\dfrac{1}{8}\ln(1+\sqrt{2})+\dfrac{\sqrt{2}}{8}$

2 $\ln(1+\sqrt{2})+\sqrt{2}$

3 a $\dfrac{dy}{dx}=x^{-\frac{1}{2}}$

$s=\displaystyle\int_0^4 \sqrt{1+\left(x^{-\frac{1}{2}}\right)^2}\,dx$

$=\displaystyle\int_0^4 \sqrt{1+\dfrac{1}{x}}\,dx$

$=\displaystyle\int_0^4 \sqrt{\dfrac{x+1}{x}}\,dx$ as required

b $\ln(2+\sqrt{5})+2\sqrt{5}$

4 $\dfrac{3}{2}$

5 a $\dfrac{3\pi}{2}$

b $\dfrac{dy}{dx}=-2e^{-x}$

$S=2\pi\displaystyle\int_0^{\ln 2} 2e^{-x}\sqrt{1+(-2e^{-x})^2}\,dx$

$=2\pi\displaystyle\int_0^{\ln 2} 2e^{-x}\sqrt{1+4e^{-2x}}\,dx$

$=4\pi\displaystyle\int_0^{\ln 2} e^{-x}\sqrt{1+4e^{-2x}}\,dx$

c $\pi(\ln(2+\sqrt{5})+2\sqrt{5}-\ln(1+\sqrt{2})-\sqrt{2})$

6 a $\dfrac{dx}{d\theta}=-r\sin\theta,\ \dfrac{dy}{d\theta}=r\cos\theta$

$s=\displaystyle\int_0^{2\pi} \sqrt{(-r\sin\theta)^2+(r\cos\theta)^2}\,d\theta$

$=\displaystyle\int_0^{2\pi} \sqrt{r^2\sin^2\theta+r^2\cos^2\theta}\,d\theta$

$=\displaystyle\int_0^{2\pi} \sqrt{r^2}\,d\theta$

$=\displaystyle\int_0^{2\pi} r\,d\theta$

$=[r\theta]_0^{2\pi}$

$=2\pi r-0=2\pi r$ as required

b $V=\pi\displaystyle\int_0^{\pi}(r\sin\theta)^2(-r\sin\theta)d\theta=\pi\displaystyle\int_0^{\pi}-r^3\sin^3\theta d\theta$

$=\pi\displaystyle\int_0^{\pi}-r^3\sin\theta\sin^2\theta d\theta$

$=\pi\displaystyle\int_0^{\pi}-r^3\sin\theta(1-\cos^2\theta)\,d\theta$

$=-\pi r^3\displaystyle\int_0^{\pi}\sin\theta-\sin\theta\cos^2\theta\,d\theta$

$=-\pi r^3\left[-\cos\theta+\dfrac{1}{3}\cos^3\theta\right]_0^{\pi}$

$=-\pi r^3\left(1-\dfrac{1}{3}\right)-\pi r^3\left(-1+\dfrac{1}{3}\right)$

$=-\dfrac{4}{3}\pi r^3$, so volume $=\dfrac{4}{3}\pi r^3$ as required

c $S=2\pi\displaystyle\int_0^{\pi}r\sin\theta\sqrt{(-r\sin\theta)^2+(r\cos\theta)^2}\,d\theta=2\pi\displaystyle\int_0^{\pi}r^2\sin\theta d\theta$

$=2\pi r^2\left[-\cos\theta\right]_0^{\pi}$

$=2\pi r^2(1--1)$

$=4\pi r^2$ as required

7 $\dfrac{2\pi}{3}\left(\dfrac{13\sqrt{13}}{8}-1\right)$

8 **a** $\sqrt{1-\cos\theta}=\sqrt{2\sin^2\dfrac{\theta}{2}}$

$$=\sqrt{2}\sin\dfrac{\theta}{2}$$

b 4

c $\dfrac{32}{3}\pi$

Review exercise 19

1 A is an improper integral as integrand is undefined at $x=1$ which is within the limits
B is not an improper integral as it is fully defined within the limits
C is an improper integral as one of its limits is ∞

2 **a** 1
   b 12

3 $\displaystyle\int_a^{2e}x^3\ln x\,dx=\left[\dfrac{1}{4}x^4\ln x\right]_a^{2e}-\int_a^{2e}\dfrac{1}{4}x^3$

$$=\left[\dfrac{x^4}{4}\left(\ln x-\dfrac{1}{4}\right)\right]_a^{2e}$$

$$=\dfrac{(2e)^4}{4}\left(\ln 2+1-\dfrac{1}{4}\right)-\dfrac{a^4}{4}\left(\ln a-\dfrac{1}{4}\right)$$

$\to e^4(\ln 16+3)$ as $a\to 0$ since $\ln a-\dfrac{1}{4}\to\dfrac{1}{4}$ and so
$\dfrac{a^4}{4}\left(\ln a-\dfrac{1}{4}\right)\to 0$ as $a\to 0$

4 $\displaystyle\int_1^a\dfrac{2}{x}\,dx=[2\ln x]_1^a$

$$=2\ln a-2\ln 1$$

$\to\infty$ as $a\to\infty$ so integral does not converge.

5 Let $y=\arccos x$ then $x=\cos y\Rightarrow\dfrac{dx}{dy}=-\sin y$

$$\dfrac{dy}{dx}=-\dfrac{1}{\sin y}$$

$$=-\dfrac{1}{\sqrt{1-\cos^2 y}}$$

$$=-\dfrac{1}{\sqrt{1-x^2}}$$

6 **a** $2x\arctan x+\dfrac{x^2}{1+x^2}$ **b** $\dfrac{x}{\sqrt{1-\left(\dfrac{x^2}{2}\right)^2}}$ or $\dfrac{2x}{\sqrt{4-x^4}}$

7 **a** $\arcsin\left(\dfrac{x}{2}\right)+c$

   b $\dfrac{1}{4}\arctan\left(\dfrac{x}{4}\right)+c$

8 **a** $\sinh A\cosh B-\sinh B\cosh A=\left(\dfrac{e^A-e^{-A}}{2}\right)\left(\dfrac{e^B+e^{-B}}{2}\right)$

$$-\left(\dfrac{e^B-e^{-B}}{2}\right)\left(\dfrac{e^A+e^{-A}}{2}\right)$$

$$=\dfrac{1}{4}(e^{A+B}+e^{A-B}-e^{-(A-B)}-e^{-(A+B)})$$

$$-\dfrac{1}{4}(e^{A+B}+e^{-(A-B)}-e^{(A-B)}-e^{-(A+B)})$$

$$=\dfrac{1}{4}(2e^{A-B}-2e^{-(A-B)})$$

$$=\dfrac{1}{2}(e^{A-B}-e^{-(A-B)})$$

$$=\sinh(A-B)\text{ as required}$$

b $\cosh^2 x=\left(\dfrac{e^x+e^{-x}}{2}\right)^2$

$$=\dfrac{1}{4}(e^{2x}+2+e^{-2x})$$

$$=\dfrac{1}{2}+\dfrac{1}{4}(e^{2x}+e^{-2x})$$

$$=\dfrac{1}{2}\left(1+\dfrac{1}{2}(e^{2x}+e^{-2x})\right)$$

$$=\dfrac{1}{2}(1+\cosh 2x)\text{ as required}$$

9 **a** $2\sinh 2x$ **b** $2x\sinh x+x^2\cosh x$

10 Let $y=\operatorname{artanh}x$ then $x=\tanh y\Rightarrow\dfrac{dx}{dy}=\operatorname{sech}^2 y$

$$\dfrac{dy}{dx}=\dfrac{1}{\operatorname{sech}^2 y}$$

$$=\dfrac{1}{1-\tanh^2 y}$$

$$=\dfrac{1}{1-x^2}\text{ as required}$$

11 **a** $\dfrac{4x}{\sqrt{(2x^2)^2+1}}=\dfrac{4x}{\sqrt{4x^4+1}}$

   b $\operatorname{arcosh}(x-1)+\dfrac{x}{\sqrt{(x-1)^2-1}}$

12 **a** $\ln(1+\sqrt{2})$

   b $\dfrac{\sqrt{2}}{2}\ln(3+2\sqrt{2})$

13 **a** $2x-3+\dfrac{4}{x-2}-\dfrac{1}{x+3}$

   b $4+\ln\left(\dfrac{96}{7}\right)$

14 $\ln 640+3\arctan(3)$

15 **a** $I_n=\displaystyle\int x^n e^{-2x}\,dx=-\dfrac{1}{2}x^n e^{-2x}+\dfrac{n}{2}\int x^{n-1}e^{-2x}\,dx$

$$=-\dfrac{1}{2}x^n e^{-2x}+\dfrac{n}{2}I_{n-1}$$

   b $-\dfrac{1}{8}e^{-2x}(4x^3+6x^2+6x+3)+c$

16 **a** $I_n=\displaystyle\int_0^{\pi}x^n\cos x\,dx=[x^n\sin x]_0^{\pi}-\int_0^{\pi}nx^{n-1}\sin x\,dx$

$$=0-n\left([-x^{n-1}\cos x]_0^{\pi}-(n-1)\int_0^{\pi}-x^{n-2}\cos x\,dx\right)$$

$$=-n(-\pi^{n-1}(-1)-0+(n-1)I_{n-2})$$

$$=-n(\pi^{n-1}+(n-1)I_{n-2})=-n\pi^{n-1}-n(n-1)I_{n-2}$$
as required

b i $-6\pi^5 + 120\pi^3 - 720\pi$

ii $-5\pi^4 + 60\pi^2 - 240$

17 a i
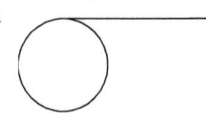

ii Max = 8, min = 0

b i
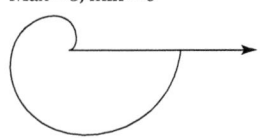

ii Max = 3, min = 0

c i
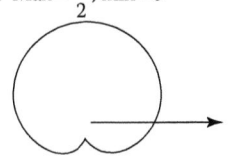

ii Max $= \dfrac{\pi}{2}$, min $= 0$

d i
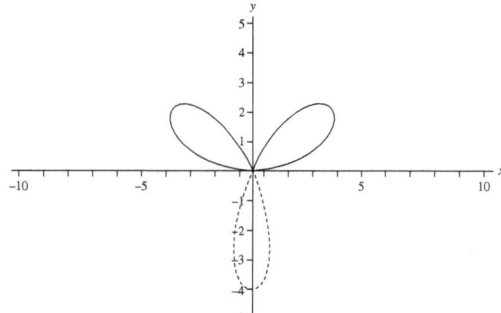

ii Max = 12, min = 2

18 a $\dfrac{19}{2}\pi$

b π

19 $\dfrac{7\pi}{4}$

20 $\dfrac{488}{27}$

21 $\ln(2+\sqrt{5})+2\sqrt{5}$

22 $\dfrac{\pi}{9}(2\sqrt{2}-1)$

23 $\dfrac{dx}{d\theta}=-\sin\theta, \dfrac{dy}{d\theta}=\cos\theta$

$$S = 2\pi \int_{\frac{\pi}{2}}^{\frac{3\pi}{4}} \sin\theta\sqrt{(-\sin\theta)^2+(\cos\theta)^2}\,d\theta$$

$$= 2\pi \int_{\frac{\pi}{2}}^{\frac{3\pi}{4}} \sin\theta\sqrt{\sin^2\theta+\cos^2\theta}\,d\theta$$

$$= 2\pi \int_{\frac{\pi}{2}}^{\frac{3\pi}{4}} \sin\theta\,d\theta$$

$$= 2\pi\left[-\cos\theta\right]_{\frac{\pi}{2}}^{\frac{3\pi}{4}}$$

$$= 2\pi\left(\frac{\sqrt{2}}{2}-0\right)$$

$$= \sqrt{2}\pi$$

Assessment 19

1 $\ln\left(\dfrac{e^x-3}{e^x+3}\right)+c$

2 a $I_n = \displaystyle\int 1\times(\ln x)^n\,dx$

$$v' = \frac{n(\ln x)^{n-1}}{x}$$

$$I_n = x(\ln x)^n - \int x\,\frac{n(\ln x)^{n-1}}{x}\,dx$$

$$I_n = x(\ln x)^n - n I_{n-1}$$

b $4(\ln 4)^3 - 12(\ln 4)^2 + 24(\ln 4) - 18$

3 a $P\left(\dfrac{3}{2},\dfrac{\pi}{3}\right)$ **b** $\dfrac{\pi}{2}-\dfrac{9\sqrt{3}}{16}$

4 a $\dfrac{\pi}{2}$ **b** $\cosh^{-1} 2$

5 a $\dfrac{a^2}{3}$ **b** $\dfrac{2\pi a^3}{15}$

6 a $I_n = \displaystyle\int \tanh^n x\,dx = \int (1-\operatorname{sech}^2 x)\tanh^{n-2} x\,dx$

$$= \int \tanh^{n-2} x\,dx - \int \operatorname{sech}^2 x\tanh^{n-2}x\,dx$$

$$= I_{n-2} - \frac{1}{n-1}\tanh^{n-1} x$$

b $I_8 = x - \tanh x - \dfrac{1}{3}\tanh^3 x - \dfrac{1}{5}\tanh^5 x - \dfrac{1}{7}\tanh^7 x\,(+c)$

7 a

(graph)

b $\dfrac{4\pi}{3}$

8 a $\sinh y = x$

$$\cosh y\,\frac{dy}{dx}=1$$

$$\sqrt{1+\sinh^2 y}\,\frac{dy}{dx}=1$$

$$\frac{dy}{dx}=\frac{1}{\sqrt{1+x^2}}$$

b $\displaystyle\int_0^2 1\times\sinh^{-1} x\,dx$

$$= \left[x\sinh^{-1} x\right]_0^2 - \int_0^2 \frac{x}{\sqrt{1+x^2}}\,dx$$

$$= \left[x\sinh^{-1} x - \sqrt{1+x^2}\right]_0^2$$

$$2\sinh^{-1} 2 - (\sqrt{5}) - (-1)$$

$$2\ln(2+\sqrt{5})+1-\sqrt{5}$$

9 a $\dfrac{3a}{2}$ **b** $\dfrac{6\pi a^2}{5}$ square units

10 $x^2 - 2x + 10 \equiv (x-1)^2 + 9$

$$\int_1^4 \frac{2x+1}{\sqrt{x^2-2x+10}}\,dx = \int_1^4 \frac{2x+1}{\sqrt{(x-1)^2+9}}\,dx$$

$$x - 1 = 3\sinh\theta$$

$$\frac{dx}{d\theta} = 3\cosh\theta$$

Limits of 0 and arsinh 1

$$\int_0^{\sinh^{-1}1} \frac{6\sinh\theta+3}{\sqrt{9\sinh^2\theta+9}}\,3\cosh\theta\,d\theta$$

$$= \int_0^{\text{arsinh}\,1} 6\sinh\theta + 3\theta\,d\theta$$

$$= \left[6\cosh\theta + 3\right]_0^{\text{arsinh}\,1}$$

$$= \left[6\sqrt{1+\sinh^2\theta} + 3\theta\right]_0^{\text{arsinh}\,1}$$

$$= 6\sqrt{2} + 3\,\text{arsinh}\,1 - 6$$

$$= 6\sqrt{2} + 3\ln(1+\sqrt{2}) - 6$$

11 $\dfrac{\pi(\pi-2)}{2}$

12 a $\dfrac{1}{3}\tan^{-1}\left(\dfrac{x-2}{3}\right) + c$

b 2

13 a $\left(r = \dfrac{\sqrt{3}}{2}, \theta = \dfrac{\pi}{6}\right)$

b $\dfrac{1}{2}\displaystyle\int_0^{\frac{\pi}{6}} (\sin 2\theta)^2\,d\theta$

$$= \frac{1}{2}\int_0^{\frac{\pi}{6}} \frac{1-\cos 4\theta}{2}\,d\theta$$

$$= \left[\frac{\theta}{4} - \frac{\sin 4\theta}{16}\right]_0^{\frac{\pi}{6}}$$

$$= \frac{\pi}{24} - \frac{\sqrt{3}}{32}$$

$$\frac{1}{2}\int_{\frac{\pi}{6}}^{\frac{\pi}{2}} (\cos\theta)^2\,d\theta$$

$$= \frac{1}{2}\int_{\frac{\pi}{6}}^{\frac{\pi}{2}} \frac{1+\cos 2\theta}{2}\,d\theta$$

$$= \left[\frac{\theta}{4} + \frac{\sin 2\theta}{8}\right]_{\frac{\pi}{6}}^{\frac{\pi}{2}}$$

$$= \frac{\pi}{12} - \frac{\sqrt{3}}{16}$$

Total area $= \dfrac{\pi}{8} - \dfrac{3\sqrt{3}}{32}$

14 a $= 1 - \dfrac{1}{\sqrt{3}}$

b $\dfrac{3x-x^2}{(x+1)(x^2+3)} \equiv \dfrac{A}{x+1} + \dfrac{Bx+C}{x^2+3}$

$$3x - x^2 \equiv A(x^2+3) + (Bx+C)(x+1)$$

$$-3 - 1 = 4A \implies A = -1$$

$$-1 = A + B \implies B = 0$$

$$0(x^2)$$

$$0 = 3A + C \implies C = 3$$

$$\int_1^3 -\frac{1}{x+1} + \frac{3}{x^2+3}\,dx$$

$$\left[-\ln(x+1) + \frac{3}{\sqrt{3}}\tan^{-1}\left(\frac{x}{\sqrt{3}}\right)\right]_1^3$$

$$= -\ln 4 + \frac{3}{\sqrt{3}}\tan^{-1}\left(\sqrt{3}\right) - \left(-\ln 2 + \frac{3}{\sqrt{3}}\tan^{-1}\left(\frac{1}{\sqrt{3}}\right)\right)$$

$$-\ln\left(\frac{4}{2}\right) + \frac{3}{\sqrt{3}}\left(\frac{\pi}{3} - \frac{\pi}{6}\right)$$

$$= \frac{\pi}{2\sqrt{3}} - \ln 2$$

15 a $2\pi\displaystyle\int_0^2 y\sqrt{\left(\dfrac{dy}{dt}\right)^2 + \left(\dfrac{dx}{dt}\right)^2}\,dt$

$$= 2\pi\int_0^2 2at\sqrt{(2a)^2 + (2at)^2}\,dt$$

$$= 8\pi a^2\int_0^2 t\sqrt{1+t^2}\,dt$$

b $8\pi a^2\left(\dfrac{5\sqrt{5}-1}{3}\right)$

c $\displaystyle\int_0^2 \sqrt{\left(\dfrac{dy}{dt}\right)^2 + \left(\dfrac{dx}{dt}\right)^2}\,dt$

$$= \int_0^2 \sqrt{(2a)^2 + (2at)^2}\,dt$$

$$= 2a\int_0^2 \sqrt{1+t^2}\,dt$$

d $a(\sinh^{-1}2 + 2\sqrt{5})$

16 a $I_n = \displaystyle\int_0^{\frac{\pi}{2}} \sin^n x\,dx = \int_0^{\frac{\pi}{2}} \sin x\,\sin^{n-1}x\,dx,$

$$v' = (n-1)\sin^{n-2}x\cos x$$

$$\left[-\cos x\,\sin^{n-1}x\right]_0^{\frac{\pi}{2}} - \int_0^{\frac{\pi}{2}} -\cos x \times (n-1)\sin^{n-2}x\cos x\,dx$$

$$\left[-\cos x\,\sin^{n-1}x\right]_0^{\frac{\pi}{2}} - \int_0^{\frac{\pi}{2}} -(1-\sin^2 x) \times (n-1)\sin^{n-2}x\,dx$$

$$= 0 + (n-1)\int_0^{\frac{\pi}{2}} \sin^{n-2}x \, dx - (n-1)\int_0^{\frac{\pi}{2}} \sin^n x \, dx$$

$$I_n = (n-1)I_{n-2} - (n-1)I_n$$

$$nI_n = (n-1)I_{n-2}$$

b $\dfrac{3\pi}{16}$

c $\dfrac{35}{256}\pi^2$

17 a $\left(\dfrac{3}{2}, \dfrac{\pi}{3}\right)$ or $\left(\dfrac{3}{2}, -\dfrac{\pi}{3}\right)$

b

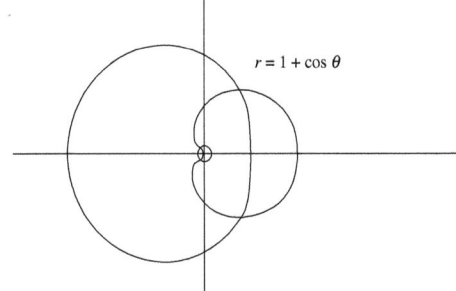

$r = 1 + \cos\theta$

c $3\sqrt{3} - \pi$

18 a $\dfrac{1}{2}x^2 + \dfrac{1}{2}\ln(3x-2) - \dfrac{1}{2}\ln(3x+2) + c$

b Limits 0 and $\dfrac{\pi}{6}$

$$dx = \cos\theta \, d\theta$$

$$\int_0^{\frac{\pi}{6}} \frac{24\sin^2\theta}{\sqrt{1-\sin^2\theta}} \cos\theta \, d\theta$$

$$= \int_0^{\frac{\pi}{6}} 24\sin^2\theta \, d\theta$$

$$= \int_0^{\frac{\pi}{6}} 12 - 12\cos 2\theta \, d\theta$$

$$= \left[12\theta - 6\sin 2\theta\right]_0^{\frac{\pi}{6}}$$

$$= \frac{12\pi}{6} - \frac{6\sqrt{3}}{2} - (0)$$

$$= 2\pi - 3\sqrt{3}$$

19 a Consider

$$I_{n-1} - I_n = \int_0^{\frac{\pi}{4}} \frac{\sin^{2n-1}x}{\cos x} - \int_0^{\frac{\pi}{4}} \frac{\sin^{2n+1}x}{\cos x} \, dx$$

$$= \int_0^{\frac{\pi}{4}} \frac{(1-\sin^2 x)\sin^{2n-1}x}{\cos x} \, dx$$

$$= \int_0^{\frac{\pi}{4}} \frac{\cos^2 x \sin^{2n-1}x}{\cos x} \, dx = \int_0^{\frac{\pi}{4}} \cos x \sin^{2n-1}x \, dx$$

$$= \left[\frac{\sin^{2n}x}{2n}\right]_0^{\frac{\pi}{4}}$$

$$= \frac{1}{2n}\left(\frac{1}{\sqrt{2}}\right)^{2n} = \frac{1}{2^{n+1}n}$$

Hence $I_n = I_{n-1} - \dfrac{1}{2^{n+1}n}$

b $\dfrac{1}{2}\ln 2 - \dfrac{1}{3}$

20 a $P\left(4a, \dfrac{\pi}{3}\right)$ and $Q\left(4a, -\dfrac{\pi}{3}\right)$

b $\dfrac{1}{2}\displaystyle\int_0^{\frac{\pi}{3}} a^2(5-2\cos\theta)^2 \, d\theta$

$$= \frac{a^2}{2}\int_0^{\frac{\pi}{3}} 25 - 20\cos\theta + 4\cos^2\theta \, d\theta$$

$$= \frac{a^2}{2}\int_0^{\frac{\pi}{3}} 27 - 20\cos\theta + 2\cos 2\theta \, d\theta$$

$$\frac{a^2}{2}\left[27\theta - 20\sin\theta + \sin 2\theta\right]_0^{\frac{\pi}{3}}$$

$$= \frac{a^2}{2}\left(9\pi - 10\sqrt{3} + \frac{\sqrt{3}}{2}\right) = \frac{a^2}{2}\left(9\pi - \frac{19\sqrt{3}}{2}\right)$$

$$\frac{1}{2}\int_{\frac{\pi}{3}}^{\pi} a^2(3+2\cos\theta)^2 \, d\theta$$

$$\frac{a^2}{2}\int_{\frac{\pi}{3}}^{\pi} 9 + 12\cos\theta + 4\cos^2\theta \, d\theta$$

$$= \frac{a^2}{2}\int_{\frac{\pi}{3}}^{\pi} 11 + 12\cos\theta + 2\cos 2\theta \, d\theta$$

$$= \frac{a^2}{2}\left[11\theta + 12\sin\theta + \sin 2\theta\right]_{\frac{\pi}{3}}^{\pi}$$

$$= \frac{a^2}{2}\left(11\pi - \left(\frac{11\pi}{3} + 6\sqrt{3} + \frac{\sqrt{3}}{2}\right)\right) = \frac{a^2}{2}\left(\frac{22\pi}{3} - \frac{13\sqrt{3}}{2}\right)$$

$$\left(\frac{a^2}{2}\left(9\pi - \frac{19\sqrt{3}}{2}\right) + \frac{a^2}{2}\left(\frac{22\pi}{3} - \frac{13\sqrt{3}}{2}\right)\right)$$

Total area $= 2 \times \left(\dfrac{a^2}{2}\left(9\pi - \dfrac{19\sqrt{3}}{2}\right) + \dfrac{a^2}{2}\left(\dfrac{22\pi}{3} - \dfrac{13\sqrt{3}}{2}\right)\right)$

$$= a^2\left(\frac{49\pi}{3} - 16\sqrt{3}\right) = a^2\left(\frac{49\pi - 48\sqrt{3}}{3}\right)$$

21 a $(1,0)$

b $6y\left(\dfrac{dy}{dx}\right) = (1-x)^2 - 2x(1-x)$

$\left(\dfrac{dy}{dx}\right) = \dfrac{(1-x)(1-3x)}{6y}$

$1 + \left(\dfrac{dy}{dx}\right)^2 = \dfrac{36y^2 + (1-x)^2(1-3x)^2}{36y^2}$

$= \dfrac{12x(1-x)^2 + (1-x)^2(1-3x)^2}{12x(1-x)^2}$

$= \dfrac{(1-x)^2(1+6x+9x^2)}{12x(1-x)^2} = \dfrac{(1+3x)^2}{12x}$

So $2\times\displaystyle\int_0^1 \sqrt{1+\left(\dfrac{dy}{dx}\right)^2}\,dx = 2\times\int_0^1 \sqrt{\dfrac{(1+3x)^2}{12x}}\,dx$

$= 2\times\displaystyle\int_0^1 \dfrac{1+3x}{2\sqrt{3x}}\,dx = \int_0^1 \dfrac{1+3x}{\sqrt{3x}}\,dx$

c $= \dfrac{4\sqrt{3}}{3}$

d $2\pi\displaystyle\int_0^1 y\sqrt{1+\left(\dfrac{dy}{dx}\right)^2}\,dx$

$= 2\pi\displaystyle\int_0^1 y\sqrt{\dfrac{(1-x)^2(1+6x+9x^2)}{36y^2}}\,dx$

$= \dfrac{\pi}{3}\displaystyle\int_0^1 (1-x)(1+3x)\,dx$

$= \dfrac{\pi}{3}\displaystyle\int_0^1 1+2x-3x^2\,dx$

$= \dfrac{\pi}{3}\left[x+x^2-x^3\right]_0^1 = \dfrac{\pi}{3}$

22 a i $= \sin^{-1}\left(\dfrac{x+2}{4}\right) + c$

ii $x\,\mathrm{artanh}\,(2x) + \dfrac{1}{4}\ln(1-4x^2) + c$

b Let $x = 4\,\mathrm{cosech}\,u$, so $\dfrac{dx}{du} = -4\coth u\,\mathrm{cosech}\,u$

At $x=1$, $u = \mathrm{arsinh}\,4$; at $x=a$, $u = \mathrm{arsinh}\,\dfrac{4}{a}$

$\displaystyle\int_1^a \dfrac{1}{x\sqrt{x^2+16}}\,dx$

$= \displaystyle\int_{\mathrm{arsinh}\,4}^{\mathrm{arsinh}\,\frac{4}{a}} \dfrac{1}{4\,\mathrm{cosech}\,u\,\sqrt{16\,\mathrm{cosech}^2 u+16}}\,(-4\,\mathrm{cosech}\,u\coth u)\,du$

$= -\displaystyle\int_{\mathrm{arsinh}\,4}^{\mathrm{arsinh}\,\frac{4}{a}} \dfrac{1}{4}\,du$

$= \dfrac{1}{4}\mathrm{arsinh}\,4 - \dfrac{1}{4}\mathrm{arsinh}\,\dfrac{4}{a}$

$= \dfrac{1}{4}\ln\left(4+\sqrt{17}\right) - \dfrac{1}{4}\ln\left(\sqrt{\dfrac{16}{a^2}+1}+\dfrac{4}{a}\right)$

As $a \to \infty$, $\sqrt{\dfrac{16}{a^2}+1}+\dfrac{4}{a} \to 1$ so $\dfrac{1}{4}\ln\left(\sqrt{\dfrac{16}{a^2}+1}+\dfrac{4}{a}\right) \to 0$

and so $\displaystyle\int_1^\infty \dfrac{1}{x\sqrt{x^2+16}}\,dx = \dfrac{1}{4}\ln 4 + \sqrt{17}$

23 a $I_{2n+1} = \displaystyle\int_0^1 x^{2n}\,x\sqrt{1-x^2}\,dx$

$= \left[-\dfrac{1}{3}x^{2n}\left(1-x^2\right)^{\frac{3}{2}}\right]_0^1 - \displaystyle\int_0^1 -\dfrac{2n}{3}x^{2n-1}\left(1-x^2\right)^{\frac{3}{2}}\,dx$

$= \displaystyle\int_0^1 \dfrac{2n}{3}x^{2n-1}(1-x^2)^{\frac{3}{2}}\,dx$

$= \dfrac{2n}{3}\displaystyle\int_0^1 x^{2n-1}(1-x^2)(1-x^2)^{\frac{1}{2}}\,dx$

$= \dfrac{2n}{3}I_{2n-1} - \dfrac{2n}{3}I_{2n+1}$

So $(2n+3)I_{2n+1} = 2n\,I_{2n-1}$

ie $I_{2n+1} = \dfrac{2n}{2n+3}I_{2n-1}$

b $I_1 = \displaystyle\int_0^1 x\sqrt{1-x^2}\,dx$

$= \left[-\dfrac{1}{3}(1-x^2)^{\frac{3}{2}}\right]_0^1 = \dfrac{1}{3}$

$I_{2n+1} = \dfrac{2n}{2n+3}I_{2n-1} = \dfrac{2n}{2n+3}\times\dfrac{2(n-1)}{2n+1}I_{2n-3}$

$= \dfrac{2n}{2n+3}\times\dfrac{2(n-1)}{2n+1}\times\dfrac{2(n-2)}{2n-1}\times.....\times\dfrac{2(1)}{5}\,I_1$

$= 2^n\,n!\times\dfrac{1}{(2n+3)\times(2n+1)\times(2n-1)\times.....\times 5\times 3\times 1}$

$= 2^n\,n!\times\dfrac{1}{(2n+3)\times(2n+1)\times(2n-1)\times.....\times 5\times 3\times 1}$

$\times\dfrac{(2n+2)\times(2n)\times(2n-2)\times...\times 4\times 2}{(2n+2)\times(2n)\times(2n-2)\times...\times 4\times 2}$

$= 2^n\,n!\times\dfrac{2^{n+1}(n+1)!}{(2n+3)!}$

$= \dfrac{2^{2n+1}\,n!\,(n+1)!}{(2n+3)!}$

24 a $Q\left(\dfrac{3}{2}, \dfrac{\pi}{6}\right)$ and $P\left(\dfrac{3}{2}, \dfrac{5\pi}{6}\right)$

b $\dfrac{1}{2}\displaystyle\int_{\frac{\pi}{6}}^{\frac{5\pi}{6}} (1+\sin\theta)^2\,d\theta$

$= \dfrac{1}{2}\displaystyle\int_{\frac{\pi}{6}}^{\frac{5\pi}{6}} 1 + 2\sin\theta + \sin^2\theta\,d\theta$

$= \dfrac{1}{2}\displaystyle\int_{\frac{\pi}{6}}^{\frac{5\pi}{6}} 1 + 2\sin\theta + \dfrac{1-\cos 2\theta}{2}\,d\theta$

$$= \frac{1}{2}\left[\frac{3\theta}{2} - 2\cos\theta - \frac{\sin 2\theta}{4}\right]_{\frac{\pi}{6}}^{\frac{5\pi}{6}}$$

$$= \frac{1}{2}\left(\frac{5\pi}{4} + \sqrt{3} + \frac{\sqrt{3}}{8}\right) - \frac{1}{2}\left(\frac{\pi}{4} - \sqrt{3} - \frac{\sqrt{3}}{8}\right) = \frac{4\pi + 9\sqrt{3}}{8}$$

$$\frac{1}{2}\int_{\frac{\pi}{6}}^{\frac{5\pi}{6}} 3^2(1-\sin\theta)^2\,d\theta$$

$$= \frac{9}{2}\int_{\frac{\pi}{6}}^{\frac{5\pi}{6}} 1 - 2\sin\theta + \sin^2\theta\,d\theta$$

$$= \frac{9}{2}\int_{\frac{\pi}{6}}^{\frac{5\pi}{6}} 1 - 2\sin\theta + \frac{1-\cos 2\theta}{2}\,d\theta$$

$$= \frac{9}{2}\left[\frac{3\theta}{2} + 2\cos\theta - \frac{\sin 2\theta}{4}\right]_{\frac{\pi}{6}}^{\frac{5\pi}{6}}$$

$$= \frac{9}{2}\left(\frac{5\pi}{4} - \sqrt{3} + \frac{\sqrt{3}}{8}\right) - \frac{9}{2}\left(\frac{\pi}{4} + \sqrt{3} - \frac{\sqrt{3}}{8}\right) = \frac{36\pi - 63\sqrt{3}}{8}$$

$$\frac{4\pi + 9\sqrt{3}}{8} - \frac{36\pi - 63\sqrt{3}}{8}$$

$$= 9\sqrt{3} - 4\pi$$

An alternative method is to work out

$$\frac{1}{2}\int_{\frac{\pi}{6}}^{\frac{5\pi}{6}} (1+\sin\theta)^2 - 9(1-\sin\theta)^2\,d\theta$$

25 a $\sinh^{-1}\left(\dfrac{x-1}{3}\right) + c$

b Let $x = \tan u$, so $\dfrac{dx}{du} = \sec^2 u$

At $x = a$, $u = \arctan a$

$$\int_a^b \frac{1}{1+x^2}\,dx = \int_{\arctan a}^{\arctan b} \frac{\sec^2 u}{1+\tan^2 u}\,du$$

$$= \int_{\arctan a}^{\arctan b} 1\,du$$

$$= \arctan b - \arctan a$$

As $a \to -\infty \arctan a \to -\dfrac{\pi}{2}$, so $\displaystyle\int_{-\infty}^b \frac{1}{1+x^2}\,dx = \arctan b + \frac{\pi}{2}$

As $b \to \infty \arctan b \to \dfrac{\pi}{2}$, so $\displaystyle\int_{-\infty}^{\infty} \frac{1}{1+x^2}\,dx = \frac{\pi}{2} + \frac{\pi}{2} = \pi$

c The function $\dfrac{1}{(1+x)^2}$ is undefined at $x = -1$

In order to find $\displaystyle\int_{-\infty}^{\infty} \frac{1}{(1+x)^2}\,dx$ you need to work out

$\displaystyle\int_{-\infty}^{-1-\alpha} \frac{1}{(1+x)^2}\,dx + \int_{-1+\alpha}^{\infty} \frac{1}{(1+x)^2}\,dx$, and then let $\alpha \to 0$

26 a $A(2\pi,0), B(4\pi,0), C(6\pi,0)$

b $\left(\dfrac{dx}{d\theta}\right)^2 + \left(\dfrac{dy}{d\theta}\right)^2 = (1-\cos\theta)^2 + (\sin\theta)^2$

$$= 2(1-\cos\theta)$$

$$\int_0^{6\pi} \sqrt{\left(\frac{dx}{d\theta}\right)^2 + \left(\frac{dy}{d\theta}\right)^2}\,d\theta = 3\int_0^{2\pi} \sqrt{2(1-\cos\theta)}\,d\theta$$

$$= 3\int_0^{2\pi} \sqrt{2\left(2\sin^2\left(\frac{\theta}{2}\right)\right)}\,d\theta$$

c 24

d $2\pi\displaystyle\int_0^{6\pi} y\sqrt{\left(\dfrac{dx}{d\theta}\right)^2 + \left(\dfrac{dy}{d\theta}\right)^2}\,d\theta$

$$= 2\pi\int_0^{6\pi} (1-\cos\theta)\sqrt{2(1-\cos\theta)}\,d\theta$$

$$= 6\pi\int_0^{2\pi} (1-\cos\theta)\sqrt{2(1-\cos\theta)}\,d\theta$$

$$= 6\pi\int_0^{2\pi} 2\sin^2\frac{\theta}{2}\,2\sin\frac{\theta}{2}\,d\theta$$

$$= 24\pi\int_0^{2\pi} \sin^3\frac{\theta}{2}\,d\theta$$

$$= 24\pi\int_0^{2\pi} \sin\frac{\theta}{2} - \sin\frac{\theta}{2}\cos^2\frac{\theta}{2}\,d\theta$$

$$= 24\pi\left[-2\cos\frac{\theta}{2} + \frac{2}{3}\cos^3\frac{\theta}{2}\right]_0^{2\pi}$$

$$= 24\pi\left(2 - \frac{2}{3}\right) - 8\pi\left(-2 + \frac{2}{3}\right) = 64\pi$$

27 a $\dfrac{\pi}{4}$ **b** $\dfrac{\pi}{4}\ln\left(\dfrac{2+\sqrt{2}}{2-\sqrt{2}}\right)$

c $2\pi(2-\sqrt{2})$ cubic units

Chapter 20

Exercise 20.1A

1 a $y = \dfrac{e^{-x}}{2} + ce^{-3x}$ **b** $y = 4 + ce^{-x^2}$

c $y = \dfrac{\sin^3 x\cos x}{3} + c\cos x$

d $y = x\ln[A(x-5)]$ **e** $y = \dfrac{2}{5}x^{\frac{1}{2}} + cx^{-\frac{1}{3}}$

f $y = x^2\ln x - x^2 + cx$

g $y = 4\sec x\tan x + c\sec x$

h $y = \dfrac{1}{2}e^{(x^2-2x+1)} + ce^{-(x+1)^2}$

2 a $y = 2xe^{3x} - 2e^{3x} + e^{2x}$

b $y = \dfrac{2}{x} + \dfrac{3}{x^3}$ **c** $y = \dfrac{11-16\cos^4 x}{16\sin x}$

d $y = e^{\sin 2x}(2x+1)$ **e** $y = \dfrac{14x+31}{(x+1)^2}$

f $y = \dfrac{\sqrt{2} + 4\cos^3 x}{2\sin x}$

g $y = \dfrac{x}{4}(2x^2 \ln x - x^2 + 9)$

h $y = \dfrac{3 + e^{5x}}{\cosh x}$

3 a $y \sec^2 x = \tan x - x + c$

b $y \sec^2 x = \tan x - x + \dfrac{\pi}{4}$

4 $y = \dfrac{x^3 + 2}{\cos x}$

5 $y = x^3 \ln x - x^3 + 6x^2$

6 a $y = 2\sin x - \cos x + ce^{-2x}$

b $y = 2\sin x - \cos x + 2e^{-2x}$

7 a By the chain rule $\dfrac{du}{dx} = \dfrac{du}{dy} \times \dfrac{dy}{dx} = -\dfrac{2}{y^3} \times \dfrac{dy}{dx}$

so $\dfrac{dy}{dx} = -\dfrac{y^3}{2}\dfrac{du}{dx}$

Substituting for $\dfrac{dy}{dx}$ gives

$-\dfrac{y^3}{2}\dfrac{du}{dx} = y + 2xy^3 \Rightarrow -\dfrac{1}{2}\dfrac{du}{dx} = \dfrac{1}{y^2} + 2x$

or $-\dfrac{1}{2}\dfrac{du}{dx} = u + 2x$ or $\dfrac{du}{dx} + 2u = -4x$

b $y^2(1 - 2x + 3e^{-2x}) = 1$

8 a $y = \cos^2 x(e^x + c)$ **b** $y = \cos^2 x(e^x + 4)$

9 a $y \sin x = 8\sin^3 x + c$

b $2\sin\left(\dfrac{\pi}{4}\right) = 8\sin^3\left(\dfrac{\pi}{4}\right) + c$

$\sqrt{2} = 2\sqrt{2} + c \Rightarrow c = -\sqrt{2}$

$y \sin x = 8\sin^3 x - \sqrt{2}$

10 a $x^3 e^x - 3x^2 e^x + 6xe^x - 6e^x + c$

b $x\dfrac{dy}{dx} + (x + 2)y = 2x^2$

i.e. $\dfrac{dy}{dx} + \left(1 + \dfrac{2}{x}\right)y = 2x$

Integrating factor $e^{\int\left(1 + \frac{2}{x}\right)dx} = e^{x + 2\ln x} = e^x e^{2\ln x} = x^2 e^x$

So $x^2 e^x \dfrac{dy}{dx} + e^x(x^2 + 2x)y = 2x^3 e^x$

$\dfrac{d}{dx}\left[x^2 e^x y\right] = 2x^3 e^x$

$x^2 e^x y = 2\displaystyle\int x^3 e^x\, dx$

$x^2 e^x y = 2e^x(x^3 - 3x^2 + 6x - 6) + c$

or $x^2 y = 2(x^3 - 3x^2 + 6x - 6) + ce^{-x}$

11 a $\displaystyle\int \sec^3 x\, dx = \int \sec x \sec^2 x\, dx = \sec x \tan x -$

$\displaystyle\int (\sec x \tan x)\tan x\, dx$

$= \sec x \tan x - \displaystyle\int \sec x(\sec^2 x - 1)\, dx$

$= \sec x \tan x - \displaystyle\int \sec^3 x\, dx + \int \sec x\, dx$

$= \sec x \tan x - \displaystyle\int \sec^3 x\, dx + \ln(\sec x + \tan x) + c$

So $2\displaystyle\int \sec^3 x\; dx = \sec x \tan x + \ln(\sec x + \tan x) + c$

$\displaystyle\int \sec^3 x\; dx = \dfrac{1}{2}\left[\sec x \tan x + \ln(\sec x + \tan x)\right] + c$

b $x\dfrac{dy}{dx} + (1 - x\tan x)y = 2\sec^4 x$

i.e. $\dfrac{dy}{dx} + \left(\dfrac{1}{x} - \tan x\right)y = \dfrac{2\sec^4 x}{x}$

Integrating factor $e^{\int\left(\frac{1}{x} - \tan x\right)dx} = e^{(\ln x + \ln \cos x)} = x\cos x$

So $x\cos x\dfrac{dy}{dx} + (\cos x - x\sin x)y = 2\sec^3 x$

$\dfrac{d}{dx}\left[y(x\cos x)\right] = 2\sec^3 x$

$y(x\cos x) = \sec x \tan x + \ln(\sec x + \tan x) + c$

$y = \dfrac{\sec x \tan x + \ln(\sec x + \tan x) + c}{x\cos x}$

12 $x = \dfrac{e^{2t} + c}{\ln t}$

13 a $x = \dfrac{3\ln(1 + t) - 1}{9} + \dfrac{c}{(1 + t)^3}$

b $\dfrac{8}{9} = -\dfrac{1}{9} + c \Rightarrow c = 1$

$x = \dfrac{3\ln(1 + t) - 1}{9} + \dfrac{1}{(1 + t)^3}$

14 a $y = z^2 \Rightarrow \dfrac{dy}{dx} = 2z\dfrac{dz}{dx}$

$\dfrac{dz}{dx} - 3z\tan x = \dfrac{3}{2}$

b $y = \left(\dfrac{\dfrac{3}{2}\sin x - \dfrac{1}{2}\sin^3 x + c}{\cos^3 x}\right)^2$

15 a $\dfrac{dv}{dt} - \dfrac{1}{t}v = 5t$

b $x = \dfrac{5}{8}t^4 + ct^2 + d$

Exercise 20.1B

1 a $\dfrac{dp}{dt} = 20t - P$ so $\dfrac{dP}{dt} + P = 20t$

$e^{\int F(t)dt} = e^{\int 1\,dt} = e^t$

$e^t \dfrac{dP}{dt} + e^t P = 20te^t$

$\dfrac{d}{dt}\left[e^t P\right] = 20te^t$

$e^t P = \displaystyle\int 20te^t\, dt$

$e^t P = 20te^t - \displaystyle\int 20e^t\, dt = 20te^t - 20e^t + c$

$40 = -20 + c \Rightarrow c = 60$

$e^t P = 20te^t - 20e^t + 60$

$P = 20t - 20 + 60e^{-t} = 20(t + 3e^{-t} - 1)$

b 260

c $\dfrac{dP}{dt} = 20(1 - 3e^{-t}) = 0$

$1 - 3e^{-t} = 0$ gives $t = \ln 3$

$P = 20(\ln 3 + 1 - 1)$

By inspection this is a minimum, so the value of P never falls below $20\ln 3 \approx 21.97$

2 a The integrating factor is $e^{\int 2\,dt} = e^{2t}$

$$e^{2t}\frac{dI}{dt} + 2e^{2t}I = 6e^{2t}$$

so $\dfrac{d}{dt}\left[e^{2t}I\right] = 6e^{2t}$

$e^{2t}I = 3e^{2t} + c$ or $I = 3 + ce^{-2t}$

$t = 0, I = 8$ gives $8 = 3 + c$, so $c = 5$

so $I = 3 + 5e^{-2t}$

b $e^{-2t} > 0$, so $I > 3$

c
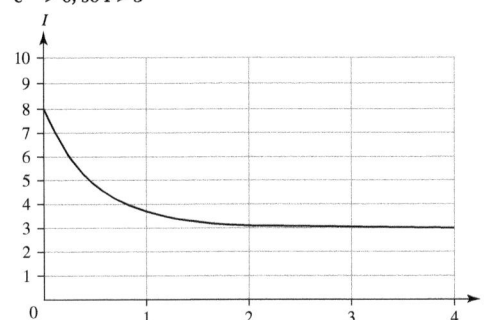

3 a $\dfrac{dx}{dt} = 0 - 8 + 4 = -4 < 0$ so population starts to decline initially

b $x = t + 3 + 5e^{-t}$ **c** 5.61 million (3 sf)

d Population continues to increase but at a constant rate in the long term.

4 a $v = \dfrac{10t(1+t)}{1+2t}$ **b** 4.4 s

c In this model, as $t \to \infty, v \to \infty$. In practice the hailstone will approach a terminal velocity.

5 a Integrating factor $e^{\int \frac{1}{20+t}\,dt} = e^{\ln(20+t)} = (20+t)$

$$(20+t)\frac{dC}{dt} + C = 4(20+t)$$

So $\dfrac{d}{dt}\left[(20+t)C\right] = 4(20+t)$

$(20+t)C = 2(20+t)^2 + K$

$t = 0, C = 10$ gives $200 = 800 + K$, so $K = -600$

so $(20+t)C = 2(20+t)^2 - 600$

or $C = 2(20+t) - \dfrac{600}{20+t}$

b 10 minutes

6 a $x = \dfrac{1}{2}e^{-\frac{1}{2}t}(4 - \sin t)$ **b** 0.204 m (3 sf)

7 a Multiplying throughout by $(1+2x)$ gives

$$2v\frac{dv}{dx}(1+2x) + 2v^2 = -4(1+2x)$$

Notice that $\dfrac{d}{dx}\left[v^2(1+2x)\right] = 2v\dfrac{dv}{dx}(1+2x) + 2v^2$

so $\dfrac{d}{dx}\left[v^2(1+2x)\right] = -4(1+2x)$

Integrating gives $v^2(1+2x) = -(1+2x)^2 + c$

$x = 0, v = 13$, gives $13^2 = -1 + c$, so $c = 170$

$v^2(1+2x) = 170 - (1+2x)^2$

b 6.02 m (3 sf)

8 a $\dfrac{dP}{dt} - \dfrac{P}{1+t} = 2$

Integrating factor is $e^{\int -\frac{1}{1+t}\,dt} = e^{-\ln(1+t)} = \dfrac{1}{1+t}$

$$\frac{1}{1+t}\frac{dP}{dt} - \frac{1}{(1+t)^2}P = \frac{2}{1+t}$$

$$\frac{d}{dt}\left(\frac{P}{1+t}\right) = \frac{2}{1+t}$$

$$\frac{P}{1+t} = 2\ln(1+t) + c$$

$10000 = 2\ln 1 + c \Rightarrow c = 10000$

$$\frac{P}{1+t} = 2\ln(1+t) + 10000$$

$P = 2(1+t)(5000 + \ln(1+t))$

b 70027

9 a Divide through by x to give:

$$\frac{dy}{dx} + \frac{4}{x}y = x^3y^2$$

which is in the form of a Bernoulli equation $\dfrac{dy}{dx} + py = qy^n$

with $p = \dfrac{4}{x}$ and $q = x^3$

b $y = \dfrac{1}{-x^4\ln x + \left(\ln 2 - \dfrac{1}{32}\right)x^4}$ (or equivalent)

Exercise 20.2A

1 a $y = Ae^{2x} + Be^{4x}$ **b** $y = (Ax+B)e^{-4x}$

c $y = e^{2x}(A\cos x + B\sin x)$

d $y = A + Be^{-3x}$ **e** $y = Ae^{3x} + Be^{-4x}$

f $y = A\cos x + B\sin x$

g $y = Ae^{\frac{1}{2}x} + Be^{-3x}$

h $y = e^{-\frac{1}{2}x}(A\cos x + B\sin x)$

2 a $y = Ae^{3x} + Be^{-5x} + 2x - 1$

b $y = (Ax+B)e^{3x} + 3e^{2x}$

c $y = e^{-x}(A\cos 4x + B\sin 4x) + \cos 3x$

d $y = Ae^{4x} + Be^{-4x} - 2x + 3$

e $y = Ae^x + Be^{-4x} - e^{-3x}$

f $y = (Ax+B)e^{-x} + x^2 + 5$

g $y = Ae^{-\frac{1}{3}x} + Be^{3x} - \dfrac{5}{7}e^{2x}$

h $y = (Ax+B)e^{-\frac{1}{2}x} + 5x - 2$

3 a $\dfrac{d^2y}{dx^2} - 2a\dfrac{dy}{dx} + a^2y = \dfrac{d^2y}{dx^2} - a\dfrac{dy}{dx} - a\dfrac{dy}{dx} + a^2y$

$$= \frac{d}{dx}\left[\frac{dy}{dx} - ay\right] - a\left[\frac{dy}{dx} - ay\right]$$

Let $u = \dfrac{dy}{dx} - ay$, then $\dfrac{du}{dx} - au = 0$ (1)

b Multiplying (1) throughout by the integrating factor of e^{-ax}:

$$e^{-ax}\frac{du}{dx} - ae^{-ax}u = 0$$

$$\frac{d}{dx}\left[ue^{-ax}\right] = 0$$

so $ue^{-ax} = A$

$u = Ae^{ax}$

c $u = \dfrac{dy}{dx} - ay$, so $\dfrac{dy}{dx} - ay = Ae^{ax}$ (2)

Multiplying (2) throughout by the integrating factor of e^{-ax}

$$e^{-ax}\frac{dy}{dx} - ae^{-ax}y = A$$

$$\frac{d}{dx}\left[ye^{-ax}\right] = A$$

$ye^{-ax} = Ax + B$

$y = (Ax+B)e^{ax}$

4 $Ae^{(m+in)x} + Be^{(m-in)x} = Ae^{mx}e^{inx} + Be^{mx}e^{inx}$
$= Ae^{mx}(\cos nx + i\sin nx) + Be^{mx}(\cos nx - i\sin nx)$
$= e^{mx}\left[(A+B)\cos nx + (Ai - Bi)\sin nx\right]$
Let $A + B = \alpha$, and $Ai - Bi = \beta$
Then $Ae^{(m+in)x} + Be^{(m-in)x} = e^{mx}(\alpha\cos nx + \beta\sin nx)$

Exercise 20.2B

1 a $y = 2e^x + e^{6x}$ **b** $y = (4x+1)e^{-2x}$
  c $y = 3\cos 5x + 6\sin 5x$
  d $y = 3e^{2x} - 2e^{-3x}$ **e** $y = (5-12x)e^{3x}$
  f $y = 6(e^{4x} + e^{-3x})$ **g** $y = 12e^{\frac{2}{3}x}$
  h $y = 2e^{\frac{1}{2}x}(3\cos x - 2\sin x)$
2 a $y = 15e^{2x} - 5e^{5x} + 3e^x$
  b $y = (1-8x)e^{6x} + 2$
  c $y = e^{-2x}(3\cos x - 2\sin x) + 2(\cos x + \sin x)$
  d $y = \frac{1}{6}(13e^{2x} + 11e^{-4x}) + 3x - 3$
  e $y = 6e^{\frac{2}{3}x} - 9e^{-\frac{2}{3}x} - x^2 - 5$
  f $y = e^{2x}(8\cos x + 4\sin x) + \cos 2x$
  g $y = 3e^{\frac{3}{2}x} + 7e^{-\frac{3}{2}x} + x^2 - 3x - 2$
  h $y = (x+7)e^{\frac{1}{4}x} + 3e^{-\frac{1}{4}x}$
3 $a = -2$, $y = Ae^{4x} + Be^{-2x} - 2xe^{-2x}$
4 $a = 3$, $y = (Ax+B)e^x + 3x^2e^x$
5 $y = 2e^x - 2e^{-6x} + 3xe^x - 2$
6 $y = (2x+3)e^{3x} + 17x^2e^{3x}$
7 $y = 3e^{2x} - e^{3x} + 2e^{4x}$
8 a Given $x = e^u, \dfrac{dx}{du} = e^u = x$

By the chain rule $\dfrac{dy}{dx} = \dfrac{dy}{du} \times \dfrac{du}{dx}$

$\dfrac{dy}{dx} = \dfrac{dy}{du} \times \dfrac{1}{e^u} = \dfrac{1}{x}\dfrac{dy}{du}$

$\dfrac{d^2y}{dx^2} = -\dfrac{1}{x^2}\dfrac{dy}{du} + \dfrac{1}{x}\dfrac{d^2y}{du^2} \times \dfrac{du}{dx}$

$\dfrac{d^2y}{dx^2} = -\dfrac{1}{x^2}\dfrac{dy}{du} + \dfrac{1}{x^2}\dfrac{d^2y}{du^2}$

Substitute $\dfrac{dy}{dx} = \dfrac{1}{x}\dfrac{dy}{du}$ and $\dfrac{d^2y}{dx^2} = -\dfrac{1}{x^2}\dfrac{dy}{du} + \dfrac{1}{x^2}\dfrac{d^2y}{du^2}$ into

$x^2\dfrac{d^2y}{dx^2} - 4x\dfrac{dy}{dx} + 6y = 12$

$x^2\left(-\dfrac{1}{x^2}\dfrac{dy}{du} + \dfrac{1}{x^2}\dfrac{d^2y}{du^2}\right) - 4x\left(\dfrac{1}{x}\dfrac{dy}{du}\right) + 6y = 12$

$-\dfrac{dy}{du} + \dfrac{d^2y}{du^2} - 4\dfrac{dy}{du} + 6y = 12$

$\dfrac{d^2y}{du^2} - 5\dfrac{dy}{du} + 6y = 12$

  b $y = 7x^2 - 2x^3 + 2$

Exercise 20.3A

1 a $x = 7\cos 5t + 5\sin 5t$
  b $x = -4\cos 2t + 8\sin 2t$
  c $x = -5\cos 10t + 3\sin 10t$
  d $x = -\cos t + 9\sin t$
  e $x = 5\cos\dfrac{t}{2} - 3\sin\dfrac{t}{2}$
  f $x = 6\sqrt{3}\cos\dfrac{t}{3} - 4\sin\dfrac{t}{3}$

2 a $x = -3\cos 4t + 2\sin 4t + 3$
  b $x = 5\cos 5t + \sin 5t + 4$
  c $x = -5\cos 3t + 4\sin 3t + 4$
  d $x = -6\cos 2t + 4\sin 2t + 3$
  e $x = 2\sqrt{2}\cos\dfrac{t}{4} + 5\sqrt{2}\sin\dfrac{t}{4} + 1$
  f $x = -2\cos\dfrac{5t}{2} + 6\sin\dfrac{5t}{2} - 2$
3 $x = 4\cos 5t + 3\sin 5t$

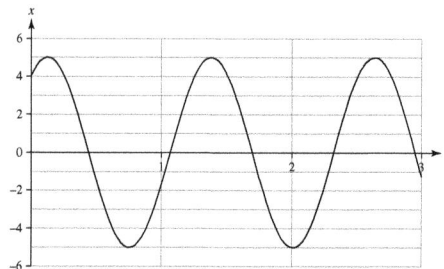

4 $x = 5\cos 3t - 12\sin 3t + 2$

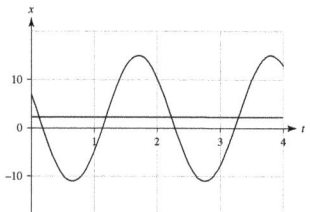

5 a $T = \dfrac{2\pi}{\omega} = \dfrac{2\pi}{3}$ s **b** $6\ \text{ms}^{-1}$
6 a $T = \dfrac{2\pi}{\omega} = \dfrac{2\pi}{2} = \pi$ s **b** $a = 6$ metres
7 a $T = \dfrac{2\pi}{\omega} = \dfrac{2\pi}{0.2} = 10\pi$ s
  b $0.3 = \sqrt{a^2 - 0.4^2} \Rightarrow a = 0.5$ metres
  c $v = \omega\sqrt{a^2 - x^2} \Rightarrow v = 0.2\sqrt{0.5^2 - 0.3^2} = 0.08\ \text{ms}^{-1}$
8 a $v = \omega\sqrt{a^2 - x^2} \Rightarrow v = 2\sqrt{0.9^2 - 0^2} \Rightarrow v = 1.8\ \text{m s}^{-1}$
  b 0.446 s (3 sf)
9 a $v = \omega\sqrt{a^2 - x^2} \Rightarrow v = 4\sqrt{0.6^2 - 0^2} \Rightarrow v = 2.4\ \text{m s}^{-1}$
  b $0.6\sin 4t = 0.3 \Rightarrow \sin 4t = \dfrac{1}{2} \Rightarrow 4t = \dfrac{\pi}{6} \Rightarrow t = \dfrac{\pi}{24}$ s
10 $v = \omega\sqrt{a^2 - x^2} \Rightarrow 4 = 8\sqrt{1.3^2 - x^2} \Rightarrow x = 1.2$ m
11 2.04 s
12 a $a^2 = 2.4^2 + 1 \Rightarrow a = 2.6$ m **b** $T = \dfrac{2\pi}{\omega} = \dfrac{2\pi}{3}$ s
13 a 50 cm **b** 100π s

Exercise 20.3B

1 a $-20x = 5\dfrac{d^2x}{dt^2}$

  $\dfrac{d^2x}{dt^2} = -4x$, which is SHM with $\omega = 2$

  b $T = \dfrac{2\pi}{\omega} = \dfrac{2\pi}{2} = \pi$ secs

  c $v = \omega\sqrt{a^2 - x^2} = 2\sqrt{5^2 - 3^2} = 8\ \text{m s}^{-1}$

2 a $-5x = 0.2\dfrac{d^2x}{dt^2}$

  $\dfrac{d^2x}{dt^2} = -25x$, which is SHM with $\omega = 5$

b The auxiliary equation is $m^2 + 25 = 0$
$m^2 = -25$, so $m = \pm 5i$
So the general solution is $x = A\cos 5t + B\sin 5t$
Use $x = 0.3$, $\dfrac{dx}{dt} = 0$ at $t = 0$ to get $A = 0.3$, $B = 0$
$\Rightarrow x = 0.3\cos 5t$

c 15 cm   **d** 1.3 m s^{-1}

3 a $-12x = \dfrac{d^2x}{dt^2}$

$\dfrac{d^2x}{dt^2} = -4x$, which is SHM with $\omega = 2$

b $T = \dfrac{2\pi}{\omega} = \dfrac{2\pi}{2} = \pi$ s

c $v = \omega\sqrt{a^2 - x^2} = 2\sqrt{0.2^2 - 0.16^2} = 0.24$ m s^{-1}

d No friction or air resistance. The body is a particle. The spring has no mass.

4 a $-10x = 0.1\dfrac{d^2x}{dt^2}$

$\dfrac{d^2x}{dt^2} = -100x$ which is SHM with $\omega = 10$

b 12 cm

c $v = \omega\sqrt{a^2 - x^2} = 10\sqrt{0.12^2 - 0.06^2} = \dfrac{3\sqrt{3}}{5} \approx 1.04$ m s^{-1}

5 a $-48x - 48x = 1.5\dfrac{d^2x}{dt^2}$

$\dfrac{d^2x}{dt^2} = -64x$, which is SHM with $\omega = 8$

b $T = \dfrac{2\pi}{8} = \dfrac{\pi}{4}$ s

c The auxiliary equation is $m^2 + 64 = 0$
$m^2 = -64$, so $m = \pm 8i$
So the general solution is $x = A\cos 8t + B\sin 8t$
And $\dfrac{dx}{dt} = -8A\sin 8t + 8B\cos 8t$
Use $x = 0$, $\dfrac{dx}{dt} = 4$ at $t = 0$ to get $A = 0$, $B = 0.5$
$\Rightarrow x = 0.5\sin 8t$
$0.5 = 0.5\sin 8t \Rightarrow \sin 8t = 1 \Rightarrow t = \dfrac{\pi}{16}$
(Or for a more elegant solution, notice that this is a quarter period.)

d $v = \omega\sqrt{a^2 - x^2} = 8\sqrt{0.5^2 - 0.3^2} = 3.2$ m s^{-1}

6 a $-0.32x = 0.5\dfrac{d^2x}{dt^2}$

$\dfrac{d^2x}{dt^2} = -0.64x$, which is SHM with $\omega = 0.8$

b The auxiliary equation is $m^2 + 0.64 = 0$
$m^2 = -0.64$, so $m = \pm 0.8i$
So the general solution is $x = A\cos 0.8t + B\sin 0.8t$
Use $x = 0.9$, $\dfrac{dx}{dt} = 0$ at $t = 0$ to get $A = 0.9$, $B = 0$
$\Rightarrow x = 0.9\cos 0.8t$

c 0.624 m s^{-1}

d If the body is pulled 120 cm in the direction OB, then one side of the string will be slack for part of the motion, and the model will need to be applied to two different situations – one when both parts of the string are taut, and one where one part of the string is slack.

7 a $-mx - 3mx = m\dfrac{d^2x}{dt^2}$

$\dfrac{d^2x}{dt^2} = -4x$, which is SHM with $\omega = 2$

b $T = \dfrac{2\pi}{2} = \pi$ s   **c** 0.346 m s^{-1}

d No friction or air resistance, the box is a particle, the spring has no mass.

8 a $-\dfrac{2(x-2)}{5} = 0.1\dfrac{d^2x}{dt^2}$

$\dfrac{d^2x}{dt^2} + 4x = 8$

b The auxiliary equation is $m^2 + 4 = 0$
$m^2 = -4$, so $m = \pm 2i$
So the complementary function is $x = A\cos 2t + B\sin 2t$
By inspection the particular integral is $x = 2$
So the general solution is $x = 2 + A\cos 2t + B\sin 2t$
Use $x = 2.4$, $\dfrac{dx}{dt} = 0$ at $t = 0$ to get $A = 0.4$, $B = 0$
$x = 2 + 0.4\cos 2t$

c 0.8 m s^{-1}

d $\cos 2t = -1 \Rightarrow 2t = \pi \Rightarrow t = \dfrac{\pi}{2}$

9 a $-15(2x - 3) = 0.3\dfrac{d^2x}{dt^2}$

$\dfrac{d^2x}{dt^2} + 100x = 150$

b $x = 1.5 - 1.5\cos 10t + 2\sin 10t$

c $\dfrac{dx}{dt} = 15\sin 10t + 20\cos 10t$

d 25 m s^{-1}

10 a Apply $\mathbf{F} = m\mathbf{a}$ perpendicular to the string to get

$-mg\sin\theta = ml\dfrac{d^2\theta}{dt^2}$

b Use $\sin\theta \approx \theta$ to get

$-mg\theta \approx ml\dfrac{d^2\theta}{dt^2}$

i.e. $\dfrac{d^2\theta}{dt^2} \approx -\dfrac{g}{\ell}\theta$

This is the equation of SHM with $\omega = \sqrt{\dfrac{g}{\ell}}$

$T = \dfrac{2\pi}{\omega} = 2\pi\sqrt{\dfrac{\ell}{g}}$

Exercise 20.4A

1 a $x = -e^{-3t}(5\cos 2t + 6\sin 2t)$, light damping

b $x = 3e^{-2t} - e^{-4t}$, heavy damping

c $x = (3t - 1)e^{-2t}$, critical damping

d $x = e^{-5t}(2\cos 4t + \sin 4t)$, light damping

e $x = e^{-2.5t}(6\cos 0.5t + 14\sin 0.5t)$, light damping

f $x = 6t\, e^{-\frac{1}{3}t}$, critical damping

2 a $x = 3e^{-t} - e^{-9t} - 2$

b $x = (5t + 2)e^{-4t} + 3$

c $x = e^{-5t}(5\cos 2t + 2\sin 2t) + 1$

d $x = 4 - e^{-2t}\cos t$

e $x = (11t + 1)e^{-2t} + 3$

f $x = e^{-t}(\cos t + \sin t) + 5$

3 $x = 4e^{-t} - e^{-3t} + 2$

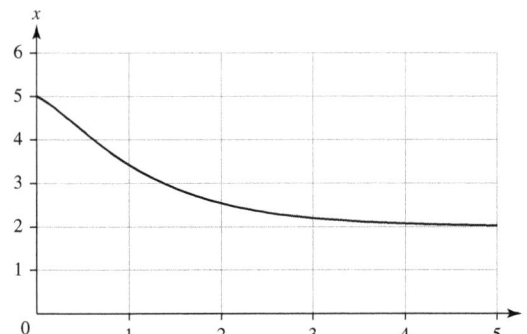

4 $x = 4 + e^{-t}\sin 2t$

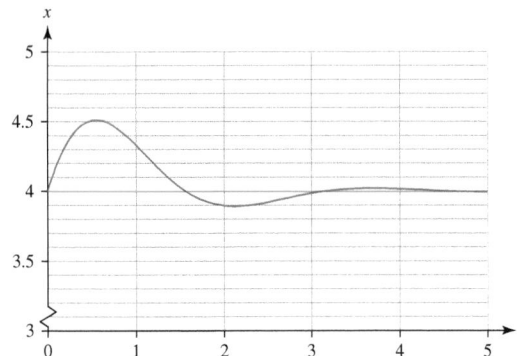

5 a $x = 0.4e^{-2t}\sin 2t$
   b

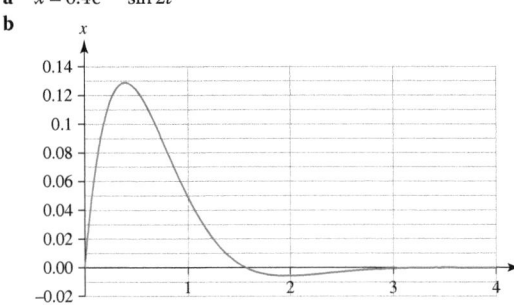

   c This is light damping.

6 a $x = 4e^{-3t} - 3.5e^{-4t} + 0.5$ **b** 0.154 s (3 sf)
   c

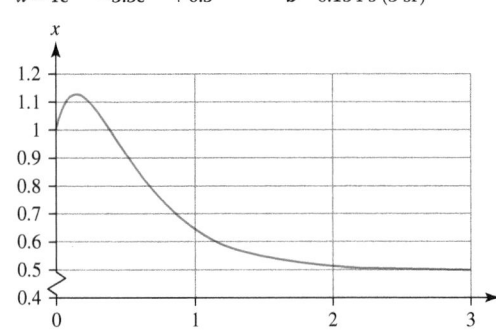

   d This is heavy damping.

7 a $x = ute^{-4\omega t}$ **b** $t = \dfrac{1}{4\omega}$
   c The damping is critical.

8 a $I = 2\sin 4t + 3\sin 2t$ **b** 1.21 s (3 sf)

9 a $\theta = \dfrac{\pi e^{-3t}}{8}(2\cos 2t + 3\sin 2t)$

b

c As, $t \to \infty$, $\theta \to 0$, i.e. the door settles down towards a closed position.

10 a $x = e^{-\frac{1}{2}t}\left(3 + 2\cos\dfrac{1}{2}t + 2\sin\dfrac{1}{2}t\right)$

   b $x = e^{-\frac{1}{2}t}\left(3 + 2\cos\dfrac{1}{2}t + 2\sin\dfrac{1}{2}t\right)$

      $= e^{-\frac{1}{2}t}\left(3 + 2\sqrt{2}\,\sin\left(\dfrac{1}{2}t + \dfrac{\pi}{4}\right)\right) > 0$ for all values

      of t since $3 > 2\sqrt{2}$ and $\sin\left(\dfrac{1}{2}t + \dfrac{\pi}{4}\right) > -1$. So the particle

      never reaches the origin.

   c 7.98 s

11 a $x = -\dfrac{4}{3}e^{-\frac{1}{2}t} + \dfrac{4}{3}e^{-2t} + 2\sin t$ **b** 1.35 cm above A

12 a $x = \dfrac{1}{4}(2t + 1 - \cos 2t - \sin 2t)$ **b** $t = \dfrac{3\pi}{4}$

   c

(graph)

13 a $x = e^{-t}(\cos t + \sin t)$ **b** $t = \pi, 2\pi, 3\pi, \ldots$

   c $x(\pi) = -e^{-\pi}, x(2\pi) = e^{-2\pi}, x(3\pi) = -e^{-3\pi}, \ldots$

      So total distance $= 1 + 2(e^{-\pi} + e^{-2\pi} + e^{-3\pi} + \ldots)$

      $= 1 + \dfrac{2e^{-\pi}}{1 - e^{-\pi}}$, summing an infinite GP

      $= \dfrac{1 + e^{-\pi}}{1 - e^{-\pi}} = \dfrac{e^{\frac{\pi}{2}} + e^{-\frac{\pi}{2}}}{e^{\frac{\pi}{2}} - e^{-\frac{\pi}{2}}}$

      $= \coth\left(\dfrac{\pi}{2}\right)$

Exercise 20.4B

1 a Using $\mathbf{F} = m\mathbf{a}$ gives $-36x - 45\dfrac{dx}{dt} = 9\dfrac{d^2x}{dt^2}$

      So $\dfrac{d^2x}{dt^2} + 5\dfrac{dx}{dt} + 4x = 0$

   b $x = e^{-t} - 3e^{-4t}$ **c** heavy damping
   d 0.37 s

2 a Using $\mathbf{F} = m\mathbf{a}$ gives $-25mn^2x - km\dfrac{dx}{dt} = m\dfrac{d^2x}{dt^2}$

      So $\dfrac{d^2x}{dt^2} + k\dfrac{dx}{dt} + 25n^2x = 0$

   b $k = 10n$ **c** $x = (At + B)e^{-5nt}$

3 a $\frac{4m}{45}$ might be friction between the body and the table,

and $\frac{4m\frac{dx}{dt}}{3}$ might be air resistance.

b $9\frac{d^2x}{dt^2}+12\frac{dx}{dt}+4x=\frac{4}{5}$

c The auxiliary equation is $9m^2+12m+4=0$

$(3m+2)^2=0$, so $m=-\frac{2}{3}$ (repeated)

So the complementary function is $x=(At+B)e^{-\frac{2}{3}t}$

By inspection the particular integral is $x=\frac{1}{5}$

The general solution is $x=(At+B)e^{-\frac{2}{3}t}+\frac{1}{5}$

$x=\frac{2}{5}$ when $t=0$ gives $B+\frac{1}{5}=\frac{2}{5}$, so $B=\frac{1}{5}$

$\frac{dx}{dt}=Ae^{-\frac{2}{3}t}-\frac{2}{3}\left(At+\frac{1}{5}\right)e^{-\frac{2}{3}t}$

$\frac{dx}{dt}=0$ when t = 0 gives $A-\frac{2}{15}=0$ so $A=\frac{2}{15}$

Particular solution is $x=\left(\frac{2}{15}t+\frac{1}{5}\right)e^{-\frac{2}{3}t}+\frac{1}{5}$

$=\frac{1}{15}\left[(2t+3)e^{-\frac{2}{3}t}+3\right]$

d The damping is critical, so the body moves directly towards an equilibrium position in which the spring is stretched by 20 cm.

4 a Using $\mathbf{F}=m\mathbf{a}$ gives $24e^{-t}-15x-6\frac{dx}{dt}=3\frac{d^2x}{dt^2}$

So $\frac{d^2x}{dt^2}+2\frac{dx}{dt}+5x=8e^{-t}$

b $x=e^{-t}(2+3\cos 2t+4\sin 2t)$

c 1.45 s (3 sf)

5 a Using $\mathbf{F}=m\mathbf{a}$ gives $m\omega^2\ell e^{-\omega t}-5m\omega\frac{dx}{dt}-6m\omega^2\frac{dx}{dt}=m\frac{d^2x}{dt^2}$

So $\frac{d^2x}{dt^2}+5\omega\frac{dx}{dt}+6\omega^2 x=\omega^2\ell e^{-\omega t}$

b The auxiliary equation is $m^2+5\omega m+6\omega^2=0$

$(m+2\omega)(m+3\omega)=0$, so $m=-2\omega$ or $m=-3\omega$

So the complementary function is $x=Ae^{-2\omega t}+Be^{-3\omega t}$

For the particular integral try $x=ae^{-\omega t}$

Then $a\omega^2 e^{-\omega t}-5a\omega^2 e^{-\omega t}+6a\omega^2 e^{-\omega t}\equiv\omega^2\ell e^{-\omega t}$, so $a=\frac{\ell}{2}$

The particular integral is $x=\frac{\ell}{2}e^{-\omega t}$ and the general

solution is $x=Ae^{-2\omega t}+Be^{-3\omega t}+\frac{\ell}{2}e^{-\omega t}$

$x=0$ when $t=0$ gives $A+B+\frac{\ell}{2}=0$ (1)

$\frac{dx}{dt}=-2A\omega e^{-2\omega t}-3B\omega e^{-3\omega t}-\frac{\ell}{2}\omega e^{-\omega t}$

$\frac{dx}{dt}=0$ when t = 0 gives $-2A\omega-3B\omega-\frac{\ell}{2}\omega=0$ ie

$-2A-3B-\frac{\ell}{2}=0$ (2)

Solving (1) and (2) gives $A=-\ell$ $B=\frac{\ell}{2}$

Particular solution is $x=-\ell e^{-2\omega t}+\frac{\ell}{2}e^{-3\omega t}+\frac{\ell}{2}e^{-\omega t}$

$=\frac{\ell}{2}e^{-\omega t}(1-2e^{-\omega t}+e^{-2\omega t})=\frac{\ell}{2}e^{-\omega t}(1-e^{-\omega t})^2$

c $\frac{dx}{dt}=-\frac{\ell}{2}\omega e^{-\omega t}(1-e^{-\omega t})^2+\ell\omega e^{-\omega t}(1-e^{-\omega t})$

$=\frac{\ell}{2}\omega e^{-\omega t}(1-e^{-\omega t})(-1+e^{-\omega t}+2e^{-\omega t})$

$=\frac{\ell}{2}\omega e^{-\omega t}(1-e^{-\omega t})(-1+3e^{-\omega t})$

So $\frac{dx}{dt}=0$ when $-1+3e^{-\omega t}=0$ ie when $e^{-\omega t}=\frac{1}{3}$

And $x=\frac{\ell}{2}\times\frac{1}{3}\left(1-\frac{1}{3}\right)^2=\frac{2\ell}{27}$

6 a Using $\mathbf{F}=m\mathbf{a}$ gives $-m\omega^2 x-2mk\frac{dx}{dt}=m\frac{d^2x}{dt^2}$

So $\frac{d^2x}{dt^2}+2k\frac{dx}{dt}+\omega^2 x=0$

b $x=Ute^{-\omega t}$ **c** critical

d $\frac{dx}{dt}=Ue^{-\omega t}-U\omega te^{-\omega t}=Ue^{-\omega t}(1-\omega t)$

Tub comes to rest when $1-\omega t=0\Rightarrow t=\frac{1}{\omega}$

$t<2\Rightarrow\frac{1}{\omega}<2\Rightarrow\omega>\frac{1}{2}$

e Tub is moving horizontally as it is brought to rest, and the spring has no mass.

7 a Using $\mathbf{F}=m\mathbf{a}$ gives $5g-25x-10\frac{dx}{dt}=5\frac{d^2x}{dt^2}$

So $5\frac{d^2x}{dt^2}+10\frac{dx}{dt}+25x=50$ or $\frac{d^2x}{dt^2}+2\frac{dx}{dt}+5x=10$

b $x=\frac{e^{-t}}{2}(\sin 2t-4\cos 2t)+2$

c 1.22 s (3 sf)

8 a Using $\mathbf{F}=m\mathbf{a}$ gives $80g-10x-40\frac{dx}{dt}=80\frac{d^2x}{dt^2}$

So

$80\frac{d^2x}{dt^2}+40\frac{dx}{dt}+10x=800$ or $8\frac{d^2x}{dt^2}+4\frac{dx}{dt}+x=80$

b $x=80-20e^{-\frac{1}{4}t}\left(4\cos\left(\frac{1}{4}t\right)+\sin\left(\frac{1}{4}t\right)\right)$

c 10.4 secs (3 sf)

d The man is a particle and the rope is light.

9 a i After t seconds the end A has moved a displacement of $3t$ metres, and box B has moved a displacement of y metres.

Since the spring is initially in equilibrium, the extension of the spring, x metres, is given by $x=3t-y$, ie $x+y=3t$

ii Differentiating $x+y=3t$ gives $\frac{dx}{dt}+\frac{dy}{dt}=3$ (1) and

$\frac{d^2x}{dt^2}+\frac{d^2y}{dt^2}=0$ (2)

Applying $\mathbf{F}=m\mathbf{a}$ to B gives $\frac{90x}{2}-30v=5\frac{d^2y}{dt^2}$

So $45x-30\frac{dy}{dt}=5\frac{d^2y}{dt^2}$ ie $9x-6\frac{dy}{dt}=\frac{d^2y}{dt^2}$

Substituting for $\frac{dy}{dt}$ and $\frac{d^2y}{dt^2}$ from (1) and (2):

$9x-6\left(3-\frac{dx}{dt}\right)=-\frac{d^2x}{dt^2}$

So $\frac{d^2x}{dt^2}+6\frac{dx}{dt}+9x=18$

iii Auxiliary equation is $m^2+6m+9=0$ so

$(m+3)^2=0 \Rightarrow m=-3$ (repeated)

Complementary function is $x=(At+B)e^{-3t}$

Particular integral is $x=2$, so general solution is

$x=(At+B)e^{-3t}+2$

$\dfrac{dx}{dt}=A\,e^{-3t}-3(At+B)e^{-3t}$

At $t=0$, $x=y=0$, and $\dfrac{dy}{dt}=0$. Using (1) gives $\dfrac{dx}{dt}=3$

$t=0$, $x=0$ and $\dfrac{dx}{dt}=3$ give $B+2=0$, and $A-3B=3$

$\Rightarrow B=-2$ and $A=-3$

So particular solution is $x=2-(2+3t)e^{-3t}$

b For large values of t the oscillations damp down, the box moves with a near constant velocity of 3 m s^{-1}, and the spring stretched by a near constant 2 metres.

Exercise 20.5A

1 a $y=Ae^{2t}+Be^{3t}$

b $y=A\cos 3t+B\sin 3t$

c $y=Ae^{\frac{16}{3}t}+Be^t$

d $y=Ae^{3t}+Be^{-2t}+\dfrac{4}{3}$

e $y=e^{-5t}(A\cos t+B\sin t)$

f $y=e^{-t}(A\cos 2t+B\sin 2t)+\frac{1}{2}\sin t-\frac{1}{2}\cos t$

2 a $y=2e^{3t}+e^{-2t}$, $x=8e^{3t}-e^{-2t}$

b $y=e^t+3e^{-3t}-5$, $x=e^t+\dfrac{3}{5}e^{-3t}-4$

c $y=e^{-t}(3\cos t-2\sin t)$, $x=e^{-t}(4\cos t-7\sin t)$

d $y=3t\,e^{-t}-5t+9$, $x=9t-11-\dfrac{3}{2}e^{-t}(2t+1)$

e $y=e^{-2t}(5\cos t+12\sin t)$, $x=\dfrac{e^{-2t}}{2}(7\cos t-17\sin t)-1$

f $y=\left(2t+\dfrac{5}{3}\right)e^t+\dfrac{1}{3}e^{-2t}$, $x=3e^t(t+1)$

3 $x=e^t+2e^{3t}$, $y=-e^t+2e^{3t}$

4 $x=17-7t-(4t+9)e^{-t}$, $y=11(t-2)+(4t+11)e^{-t}$

5 a From (2): $\dfrac{d^2y}{dt^2}=\dfrac{dy}{dt}-\dfrac{dz}{dt}$,

so $\dfrac{d^2y}{dt^2}=\dfrac{dy}{dt}+x$ (4)

from (1): $\dfrac{d^2x}{dt^2}=\dfrac{dx}{dt}+2\dfrac{dy}{dt}$ (5)

differentiating (5): $\dfrac{d^3x}{dt^3}=\dfrac{d^2x}{dt^2}+2\dfrac{d^2y}{dt^2}$

from (4): $\dfrac{d^3x}{dt^3}=\dfrac{d^2x}{dt^2}+2\left(\dfrac{dy}{dt}+x\right)$

so $\dfrac{d^3x}{dt^3}=\dfrac{d^2x}{dt^2}+2\dfrac{dy}{dt}+2x$

substituting $2\dfrac{dy}{dt}=\dfrac{d^2x}{dt^2}-\dfrac{dx}{dt}$ from (5):

$\dfrac{d^3x}{dt^3}=\dfrac{d^2x}{dt^2}+\dfrac{d^2x}{dt^2}-\dfrac{dx}{dt}+2x$

$\Rightarrow \dfrac{d^3x}{dt^3}-2\dfrac{d^2x}{dt^2}+\dfrac{dx}{dt}-2x=0$

b $x=e^{2t}-\cos t+2\sin t$

Exercise 20.5B

1 a $\dfrac{dx}{dt}=-3y$

$\dfrac{dy}{dt}=x-4y$

b $y=20e^{-t}-20e^{-3t}$ $\qquad x=60e^{-t}-20e^{-3t}$

c $t=\dfrac{1}{2}\ln 3$

2 a $\dfrac{dx}{dt}=3y-2x$ $\qquad \dfrac{dy}{dt}=x-4y$

b $y=3e^{-t}+2e^{-5t}$ $\qquad x=9e^{-t}-2e^{-5t}$

c $x-y=9e^{-t}-2e^{-5t}-(3e^{-t}+2e^{-5t})$

$=6e^{-t}-4e^{-5t}$

$=2e^{-5t}(3e^{4t}-2)$

For $t>0$, $3e^{4t}-2>0$ and $e^{-5t}>0$

$\Rightarrow x>y$, as required

3 a $\dfrac{dx}{dt}=3(x-y)$ $\qquad \dfrac{dy}{dt}=5y-x$

$y=20e^{2t}+10e^{6t}$ $\qquad x=60e^{2t}-10e^{6t}$

b X becomes extinct when

$t=\dfrac{1}{4}\ln 6$ months

4 a $\dfrac{dx}{dt}=y-4x+7$ $\qquad \dfrac{dy}{dt}=6x-5y$

b $x=\dfrac{5e^{-2t}-4e^{-7t}+5}{2}$

$y=5e^{-2t}+6e^{-7t}+3$

c As $t\to\infty$, $e^{-kt}\to 0$ for $k>0$

$\Rightarrow x\to\dfrac{5}{2}$ and $y\to 3$

$\therefore x:y\to 5:6$

5 a $\dfrac{dx}{dt}=x-y$ $\qquad \dfrac{dy}{dt}=3x+5y$

b $x=220e^{2t}-100e^{4t}$

$y=-220e^{2t}+300e^{4t}$ $\qquad$ **c** 0.4 years

6 a $\dfrac{dx}{dt}=6x-y$ $\qquad \dfrac{dy}{dt}=3x+2y$

$x=30e^{3t}+70e^{5t}$ $\qquad y=90e^{3t}+70e^{5t}$

b $\dfrac{x}{y}=\dfrac{90e^{3t}+70e^{5t}}{30e^{3t}+70e^{5t}}$

$=\dfrac{90e^{-2t}+70}{30e^{-2t}+70}\to\dfrac{70}{70}=1$ as $t\to\infty$

$\therefore$ Over time, the number of foxes will be approximately the same as the number of rabbits.

7 $t=\dfrac{60}{7}=8\dfrac{4}{7}$ years

8 a $\dfrac{dx}{dt}=0.4x+0.1y$

$\dfrac{dy}{dt}=-0.2x+0.2y$

$\dfrac{d^2y}{dt^2}=-0.2\dfrac{dx}{dt}+0.2\dfrac{dy}{dt}$

$=-0.2(0.4x+0.1y)+0.2\dfrac{dy}{dt}$

$=0.4\left(\dfrac{dy}{dt}-0.2y\right)-0.02y+0.2\dfrac{dy}{dt}$

$10\dfrac{d^2y}{dt^2}-6\dfrac{dy}{dt}+y=0$

b $x=[4\cos(0.1t)+1004\sin(0.1t)]e^{0.3t}$

$y=[1000\cos(0.1t)-1008\sin(0.1t)]e^{0.3t}$

c 7.81 years (3 sf)

d Numbers of bears would grow without limit even after there are no fish left. This is unrealistic.

9 a $L = 0$: $\dfrac{dH}{dt} = 1.2H$ In the absence of lynx, hares increase

at $+120\% = +1.2$ per year

$L' \neq 0$: $\dfrac{dH}{dt} = 1.2H - 1.15L$ On average, each lynx eats 1.15 hares; this additional term reduces the rate of growth of the hare population.

b $\dfrac{dL}{dt} = 0.05H$

$\dfrac{d^2H}{dt^2} = 1.2\dfrac{dH}{dt} - 1.15\dfrac{dL}{dt}$

$= 1.2\dfrac{dH}{dt} - 1.15 \times 0.05H$

$\dfrac{d^2H}{dt^2} - 1.2\dfrac{dH}{dt} + 0.0575H = 0$

c $L = 20e^{1.15t} + 40e^{0.05t}$ $H = 460e^{1.15t} + 40e^{0.05t}$

d According to the model, the populations of both hares and lynx grow without limit. This is unrealistic since food and territory are both finite which would limit the growth.

10 a $\dfrac{dL}{dt} = 0.2Z$ $\dfrac{dZ}{dt} = 1.3Z - 1.1L$

$\dfrac{d^2Z}{dt^2} = 1.3\dfrac{dZ}{dt} - 1.1\dfrac{dL}{dt} = 1.3\dfrac{dZ}{dt} - 1.1 \times 0.2Z$

$\dfrac{d^2Z}{dt^2} + 1.3\dfrac{dZ}{dt} + 0.22Z = 0$

b $Z = 1100e^{1.1t} - 100e^{0.2t}$ $L = 200e^{1.1t} - 100e^{0.2t}$

c $\dfrac{Z}{L} = \dfrac{1100e^{1.1t} - 100e^{0.2t}}{200e^{1.1t} - 100e^{0.2t}}$

$= \dfrac{1100 - 100e^{-0.9t}}{200 - 100e^{-0.9t}} \to \dfrac{1100}{200} = \dfrac{11}{2}$ as $t \to \infty$

$\therefore$ For large values of t, $L : Z = 2 : 11$

11 a The rate of change of X is $\dfrac{dx}{dt}$

It is decaying so $\dfrac{dx}{dt} < 0$

The rate is 0.2 times the amount of X present: $\dfrac{dx}{dt} = -0.2x$

X changes into Y, so the $0.2x$ decrease in X is a $0.2x$ increase in Y.

In addition, Y decreases at a rate of 0.1 times the amount of Y present: $\dfrac{dy}{dt} = 0.2x - 0.1y$

Y changes into Z, so the $0.1y$ decrease in Y is a $0.1y$ increase in Z: $\dfrac{dz}{dt} = 0.1y$

b $x = 100e^{-0.2t}$

$y = -200e^{-0.2t} + 200e^{-0.1t}$

$z = 100e^{-0.2t} - 200e^{-0.1t} + 100$

c $x + y + z = 100e^{-0.2t} - 200e^{-0.2t} + 200e^{-0.1t} +$

$100e^{-0.2t} - 200e^{-0.1t} + 100$

$= 100$ as required

OR $\dfrac{d(x+y+z)}{dt} = -0.2x + 0.2x - 0.1y + 0.1y$

$= 0$

$\Rightarrow x + y + z = \text{constant} = 100$

Review exercise 20

1 $y = \dfrac{-\frac{1}{4}\cos^4 x + 1}{\sin x}$

2 a $S = \dfrac{t(300 - t)}{600}$ **b** $37.5\,\text{kg}$

3 a $\dfrac{dT}{dt} = -kT$

$\displaystyle\int \dfrac{1}{T}\,dT = \int -k\,dt$

$\ln(cT) = -kt$

$t = 0$, $T = 100$ giving $\ln(100c) = 0$, so $c = \dfrac{1}{100}$

$\ln\left(\dfrac{T}{100}\right) = -kt$

$T = 100e^{-kt}$

b $T = 25$ when $t = 6$ gives $25 = 100e^{-6k}$

$\ln\left(\dfrac{1}{4}\right) = -6k$

$k = \dfrac{\ln 4}{6} = \dfrac{\ln 2}{3}$

c 13.0 minutes (3 sf)

4 a Using $\mathbf{F} = m\mathbf{a}$ gives $600 - 6v^2 = 120\dfrac{dv}{dt} \Rightarrow 100 - v^2 = 20\dfrac{dv}{dt}$

b 2.94 seconds (3 sf)

c 16.6 metres (3 sf)

5 a $y = e^{-2x}(A\cos 3x + B\sin 3x) + 2e^{4x}$

b $y = e^{-2x}(4\cos 3x - \sin 3x) + 2e^{4x}$

6 a $y = -\dfrac{1}{8}\sin 4t + \dfrac{1}{2}t$ **b** $\dfrac{\pi}{2}$ seconds

c

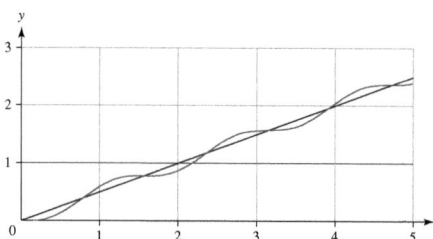

7 a Using $\mathbf{F} = m\mathbf{a}$ gives $-9x - 12v = 4a$

$\Rightarrow -9x - 12\dfrac{dx}{dt} = 4\dfrac{d^2x}{dt^2}$ or $4\dfrac{d^2x}{dt^2} + 12\dfrac{dx}{dt} + 9x = 0$

b $4m^2 + 12m + 9 = 0 \Rightarrow (2m+3)^2 = 0$, so $m = -\dfrac{3}{2}$ (repeated)

$x = (At + B)e^{-\frac{3}{2}t}$

$\dfrac{dx}{dt} = -\dfrac{3}{2}e^{-\frac{3}{2}t}(At + B) + Ae^{-\frac{3}{2}t}$

$x = 6$ and $\dfrac{dx}{dt} = -11$ when $t = 0$ give $B = 6$ and

$-\dfrac{3}{2}B + A = -11$,, so $A = -2$

The particular solution is $x = e^{-\frac{3}{2}t}(6 - 2t)$

c 3 s   **d** $0.022\,\text{m\,s}^{-1}$ (3 dp)

8 a $\dfrac{dx}{dt} = x - y - 3t$ $\dfrac{dy}{dt} = y - 4x - 3$

$\dfrac{d^2x}{dt^2} = \dfrac{dx}{dt} - \dfrac{dy}{dt} - 3$

$= \dfrac{dx}{dt} - y + 4x + 3 - 3$

$= \dfrac{dx}{dt} + \dfrac{dx}{dt} - x + 3t + 4x$

$\dfrac{d^2x}{dt^2} - 2\dfrac{dx}{dt} - 3x = 3t$

b $x = \dfrac{1}{12}e^{3t} + \dfrac{5}{4}e^{-t} + \dfrac{2}{3} - t$

$y = -\dfrac{1}{6}e^{3t} + \dfrac{5}{2}e^{-t} + \dfrac{5}{3} - 4t$

1 $y = \ln \dfrac{2}{k - e^x(\cos x + \sin x)}$

2 $y = \ln \dfrac{1}{x \ln x - x + 2}$

3 a $y^2 = x^2 \ln x - \dfrac{x^2}{2} + k$

  b $y^2 = x^2 \ln x - \dfrac{x^2}{2} + \dfrac{33}{2}$

4 a $P = 50 + ce^{-4t}$ **b** $P = 50 + 500e^{-4t}$
  c The model predicts that the population will fall to 50, and will then remain constant.
  d Because the exponential term is negative.
    The model does not take into account external factors, so it is possible that the population will fall to zero, which this model does not predict.

5 $y = \dfrac{x^2}{4} + \dfrac{c}{x^2}$

6 $y = 4x^2 \tan x + 8x - 8\tan x + \sec x$

7 a $y = \ln x - 2 + \dfrac{c}{\sqrt{x}}$

  b $y = \ln x - 2 + \dfrac{1}{\sqrt{x}}$

8 a $y = Ae^{-x} + Be^{-3x}$
  b $y = Ae^{-2x} + Bxe^{-2x}$
  c $y = e^{2x}(A\cos x + B\sin x)$

9 $y = 2e^{\frac{1}{2}x} + 3e^{-2x}$

10 $y = 6e^{\frac{x}{3}} + 2xe^{\frac{x}{3}}$

11 $y = e^{-x}(10\cos 2x + 6\sin 2x)$

12 $y = Ae^{3x} + Be^{-4x} - x - \dfrac{1}{6}$

13 $y = 4e^{2x}(\cos 3x - \sin 3x) + 3\sin x + \cos x$

14 a $0.5\dfrac{d^2x}{dt^2} = -\dfrac{7}{2}\dfrac{dx}{dt} - (5x - 6e^{-t})$

  $\dfrac{d^2x}{dt^2} = -\dfrac{7}{2 \times 0.5}\dfrac{dx}{dt} - \dfrac{1}{0.5}(5x - 6e^{-t})$

  $\dfrac{d^2x}{dt^2} + 7\dfrac{dx}{dt} = -10x + 12e^{-t}$

  $\dfrac{d^2x}{dt^2} + 7\dfrac{dx}{dt} + 10x = 12e^{-t}$

  b $x = \dfrac{1}{3}e^{-5t} - \dfrac{10}{3}e^{-2t} + 3e^{-t}$
  c For the auxiliary equation

  $b^2 - 4ac = 7^2 - 4 \times 1 \times 10$

    > 0 so the damping is heavy
  d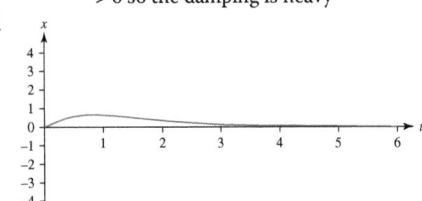

15 a $x = \sin 3t + \sin t$
  b $\cos t = 0 \Rightarrow t = \dfrac{\pi}{2}, \dfrac{3\pi}{2}, \dfrac{5\pi}{2}, \dots$

16 a $\dfrac{dA}{dt} = kA$; $A(0) = 1000$; $A(1) = 1005$
  b $A = 1000\left(\dfrac{1005}{1000}\right)^t$
  c The account has £1025.25 after 5 years.

17 a $V = RI + L\dfrac{dI}{dt}$

  $\dfrac{V}{L} = \dfrac{RI}{L} + \dfrac{L}{L}\dfrac{dI}{dt}$

  $\therefore \dfrac{dI}{dt} + \dfrac{RI}{L} = \dfrac{V}{L}$
  b $I = \dfrac{V}{R} + \dfrac{V}{R}e^{-\frac{R}{L}t}$ $I = \dfrac{V}{R}\left(1 - e^{-\frac{R}{L}t}\right)$
  c 0.787 amps (3 sf)
  d The current reaches a steady value of 2 amps.
  e

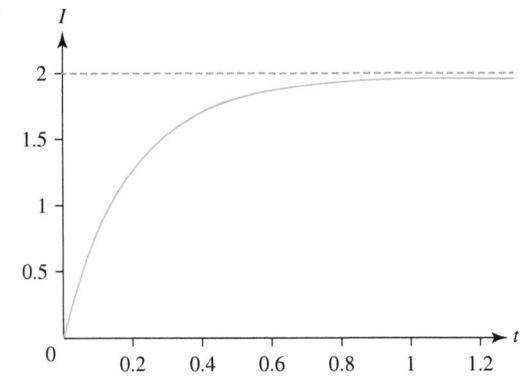

18 a $x = 4\sin \dfrac{1}{2}t$ **b** 2

19 a Auxiliary equation has equal roots so the damping is critical. The oscillations will die away quickly.
  b $x = 0.5e^{-0.15t}(1 + t)$
  c

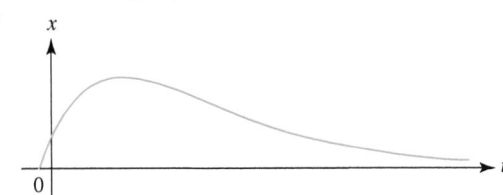

20 a $m\dfrac{d^2x}{dt^2} = -5m\dfrac{dx}{dt} - 6mx + 3m\sin 2t$

  $\dfrac{d^2x}{dt^2} + 5\dfrac{dx}{dt} + 6x = 3\sin 2t$
  b $x = Ae^{-2t} + Be^{-3t}$
  c $x = -\dfrac{15}{52}\cos 2t + \dfrac{3}{52}\sin 2t$
  d $x = \dfrac{7}{4}e^{-2t} - \dfrac{19}{13}e^{-3t} + \dfrac{3}{52}(\sin 2t - 5\cos 2t)$

21 a $x = 9(2e^{-t} - e^{-3t} - 1)$ $y = \dfrac{1}{2}(9e^{-3t} - 9e^{-t} + 2)$
  b As $t \to \infty$, e^{-t} and $e^{-3t} \to 0$ so $x \to -9$ and $y \to 1$ so the system settles around the point $(-9, 1)$.

Chapter 21

Exercise 21.1A

1.
   a. 12.9 square units
   b. π square units
   c. 4.07 square units
   d. 29.0 square units
   e. 0.654 square units
   f. 2.42 square units
   g. 3.71 square units
   h. 0.211 square units

2.
   a. 12.99 square units (4 sf)
   b. 1.303 square units (4 sf)
   c. 1.764 square units (4 sf)
   d. 0.7751 square units (4 sf)
   e. 0.7819 square units (4 sf)
   f. 0.3227 square units (4 sf)
   g. 202.1 square units (4 sf)

3.
   a. 165.333 square units
      Area $\approx$ 165.0 square units   Relative error = 0.002016
   b. 2 square units
      Area $\approx$ 2.052 square units   Relative error = 0.0262
   c. 114.583 square units
      Area $\approx$ 116.9 square units   Relative error = 0.0200
   d. 32.47 square units (4 sf)
      Area $\approx$ 32.51 square units   Relative error = 0.0011

4.
   a. 3.343 square units (4 sf).
      Area $\approx$ 3.353 (4 sf) square units   Relative error = 0.00299
   b. 8.959 square units (4 sf).
      Area $\approx$ 8.966 square units (4 sf)   Relative error = 0.000781
   c. 6.220 square units (4 sf)
      Area $\approx$ 6.340 square units (4 sf)   Relative error = 0.0193
   d. 36.92 square units (4 sf).
      Area $\approx$ 36.84 square units (4 sf)   Relative error = 0.00217

Exercise 21.1B

1. 1190 m (3 sf)

2.
   a. $1 + \dfrac{x^2}{2} + \dfrac{x^4}{24} + \ldots$
   b. 0.63665 (5 dp)

3.
   a. $x - \dfrac{x^3}{3!} + \dfrac{x^5}{5!} - \dfrac{x^7}{7!} + \ldots$
   b. 0.4597 (4 d.p)

4.
   a. $\displaystyle\int_0^{\frac{\pi}{4}} \tan x \, dx = \left[\ln|\sec x|\right]_0^{\frac{\pi}{4}} = \ln\left(\sec x\left(\frac{\pi}{4}\right)\right)$
      $- \ln(\sec x (0)) = \ln\sqrt{2}$ (= 0.3466 (4 dp)).
   b. Area $\approx$ 0.3466 square units (4 dp)   Relative error = 0

5. $(1+x^2)^{\frac{1}{3}} \equiv 1 + \dfrac{1}{3}(x)^2 + \dfrac{\left(\frac{1}{3}\right)\left(\frac{-2}{3}\right)}{2!}(x)^4 + \dfrac{\left(\frac{1}{3}\right)\left(\frac{-2}{3}\right)\left(\frac{-5}{3}\right)}{3!}(x)^6 +$
   .... $\equiv 1 + \dfrac{x^2}{3} - \dfrac{x^4}{9} + \dfrac{5x^6}{81} - \ldots\ldots$
   a. 0.6225 (4 dp)
   b. 0.6225 (4 dp)

6. Area $\approx$ 26.61237 square units (5 dp)

7.
   a. 2.68 square units (3 sf)
   b. 2.68 (3 sf)
   c. 0.000486 (3 sf) and 0.00000350 (3 sf)

8. 1.09894 (6 sf)

9. $\bar{x} = \dfrac{\displaystyle\int xy \, dx}{\displaystyle\int y \, dx} = \dfrac{16}{9.3636} \approx 1.709$ (3 sf) and $\bar{y} = \dfrac{\displaystyle\int \dfrac{y^2}{2} \, dx}{\displaystyle\int y \, dx}$
   $= \dfrac{12}{9.3636} \approx 1.282$ (3 sf)

Exercise 21.2A

1.

	$\dfrac{dy}{dx}$	x_0	y_0	h	End point
1	$2x + 3y$	0	1	0.1	$y(0.1)$

A	B	C	D		
r	x	y(formula)	y	$y_{r+1} = y_r + hf(x_r, y_r)$	
0	0		1.0000000	$f(x, y) =$	$2x + 3y$
1	0.1	(D3) + 0.1*(2(B3) + 3(D3))	1.3	$h =$	0.1

2.

$\dfrac{dy}{dx}$	x_0	y_0	h	End point
$3 + xy$	1	1	0.1	$y(1.2)$

A	B	C	D		
r	x	y(formula)	y	$y_{r+1} = y_r + hf(x_r, y_r)$	
0	1		1.0000000	$f(x, y) =$	$3 + xy$
1	1.1	(D3) + 0.1*(3 + (B3)*(D3))	1.4	$h =$	0.1
2	1.2	(D4) + 0.1*(3 + (B4)*(D4))	1.854		

3.

$\dfrac{dy}{dx}$	x_0	y_0	h	End point
$x^2 y$	0	1	0.1	$y(0.3)$

A	B	C	D		
r	x	y(formula)	y	$y_{r+1} = y_r + hf(x_r, y_r)$	
0	0		1.0000000	$f(x, y) =$	$x^2 y$
1	0.1	(D3) + 0.1*((B3)^2 *(D3))	1	$h =$	0.1
2	0.2	(D4) + 0.1*((B4)^2 *(D4))	1.001		
3	0.3	(D5) + 0.1*((B5)^2 *(D5))	1.005004		

4

$\dfrac{dy}{dx}$	x_0	y_0	h	End point
$\cos x$	0	0	0.1	$y(0.5)$

A	B	C	D	
r	x	y(formula)	y	$y_{r+1} = y_r + hf(x_r, y_r)$
0	0		0.0000000	$f(x, y) =$ $\cos x$
1	0.1	(D3) + 0.1*(cos(B3))	0.1	$h =$ 0.1
2	0.2	(D4) + 0.1*(cos(B4))	0.199500417	
3	0.3	(D5) + 0.1*(cos(B5))	0.297507074	
4	0.4	(D6) + 0.1*(cos(B6))	0.393040723	
5	0.5	(D7) + 0.1*(cos(B7))	0.485146823	

5

$\dfrac{dy}{dx}$	x_0	y_0	h	End point
$\sqrt{x}$	0	5	1	$y(10)$

A	B	C	D	
r	x	y(formula)	y	$y_{r+1} = y_r + hf(x_r, y_r)$
0	0		5.0000000	$f(x, y) =$ $\sqrt{x}$
1	1	(D3) + 1*($\sqrt{}$(B3))	5	$h =$ 1
2	2	(D4) + 1*($\sqrt{}$(B4))	6	
3	3	(D5) + 1*($\sqrt{}$(B5))	7.414213562	
4	4	(D6) + 1*($\sqrt{}$(B6))	9.14626437	
5	5	(D7) + 1*($\sqrt{}$(B7))	11.14626437	
6	6	(D8) + 1*($\sqrt{}$(B8))	13.38233235	
7	7	(D9) + 1*($\sqrt{}$(B9))	15.83182209	
8	8	(D10) + 1*($\sqrt{}$(B10))	18.4775734	
9	9	(D11) + 1*($\sqrt{}$(B11))	21.30600053	
10	10	(D12) + 1*($\sqrt{}$(B12))	24.30600053	

6

$\dfrac{dy}{dx}$	x_0	y_0	h	End point
$\ln(\sqrt{x} + \sqrt{y})$	1	2	1	$y(5)$

r	x	y (formula)	y	k_1 (formula)	k_1	k_2 (formula)	K_2
0	1		2				
1	2	(D2) + 0.5*((F3) + (H3))	3.008267	1*LN(SQRT (B2) +SQRT(D2))	0.881373587	1*(LN(SQRT(B3) + SQRT((D2) + (F3))))	1.135161001
2	3	(D3) + 0.5*((F4) + (H4))	4.245357	1*LN(SQRT (B3) +SQRT(D3))	1.146973564	1*(LN(SQRT(B4) + SQRT((D3) + (F4))))	1.32720539
3	4	(D4) + 0.5*((F5) + (H5))	5.648316	1*LN(SQRT (B4) +SQRT(D4))	1.333019411	1*(LN(SQRT(B5) + SQRT((D4) + (F5))))	1.472898258
4	5	(D5) + 0.5*((F6) + (H6))	7.181608	1*LN(SQRT(B5) +SQRT(D5))	1.4762764	1*(LN(SQRT(B6) + SQRT((D5) + (F6))))	1.590308326

7

$\dfrac{dy}{dx}$	x_0	y_0	h	End point
$\sin(x^2)$	1	1	0.1	$y(1.4)$

r	x	y (formula)	y	k_1 (formula)	k_1	k_2 (formula)	K_2
0	1		1				
1	1.1	(D2) + 0.5*((F3) + (H3))	1.088854	0.1*(SIN((B2)^2))	0.084147098	0.1*(SIN((B3)^2))	0.0935616
2	1.2	(D3) + 0.5*((F4) + (H4))	1.185208	0.1*(SIN((B3)^2)	0.0935616	0.1*(SIN(B4)^2))	0.099145835
3	1.3	(D4) + 0.5*((F5) + (H5))	1.284426	0.1*(SIN(B4)^2)	0.099145835	0.1*(SIN(B5)^2)	0.099290365
4	1.4	(D5) + 0.5*((F6) + (H6))	1.380332	0.1*(SIN(B5))^2	0.099290365	0.1*(SIN(B6))^2	0.092521152

8

$\dfrac{dy}{dx}$	x_0	y_0	h	End point
$\sin^2 x$	1	1	0.1	$y(1.4)$

r	x	y (formula)	y	k_1 (formula)	k_1	k_2 (formula)	K_2
0	1		1				
1	1.1	(D2) + 0.5*((F3) + (H3))	1.075116	0.1*(SIN(B2))^2	0.070807342	0.1*(SIN(B3))^2	0.079425056
2	1.2	(D3) + 0.5*((F4) + (H4))	1.158264	0.1*(SIN((B3)^2))	0.079425056	0.1*(SIN(B4)^2	0.086869686
3	1.3	(D4) + 0.5*((F5) + (H5))	1.248121	0.1*(SIN((B4)^2))	0.086869686	0.1*(SIN(B5)^2))	0.092844438
4	1.4	(D5) + 0.5*((F6) + (H6))	1.343098	0.1*(SIN((B5)^2))	0.092844438	0.1*(SIN(B6)^2))	0.097111117

9

$\dfrac{dy}{dx}$	x_0	y_0	h	End point
$(x - y)\mathrm{e}^{-x}$	0	5	1	$y(5)$

r	x	y (formula)	y	k_1 (formula)	k_1	k_2 (formula)	k_2
0	0		5				
1	1	(D2) + 0.5*((F3) + (H3))	2.68394	1*(((B2) − (D2))/ EXP(B2))	−5	1*(((B3) − ((D2) + (F3)))/EXP(B3))	0.367879441
2	2	(D3) + 0.5*((F4) + (H4))	2.369835	1*(((B3) − (D3))/ EXP(B3))	−0.619486803	1*(((B4) − ((D3) + (F4)))/EXP(B4))	−0.008722754
3	3	(D4) + 0.5*((F5) + (H5))	2.361742	1*((B4) − (D4))/ EXP(B4))	−0.050051717	1*(((B5) − ((D4) + (F5)))/EXP(B5))	0.033865999
4	4	(D5) + 0.5*((F6) + (H6))	2.392342	1*(((B5) − (D5))/ EXP(B5))	0.031776991	1*(((B6) − ((D5) + (F6)))/EXP(B6))	0.029423725
5	5	(D6) + 0.5*((F7) + (H7))	2.415751	1*(((B6) − (D6))/ EXP(B6))	0.029445275	1*(((B7) − ((D6) + (F7)))/EXP(B7))	0.017371858

10

$\dfrac{dy}{dx}$	x_0	y_0	h	End point
$ye^{\tan x}$	0	5	0.25	$y(1)$

r	x	y (formula)	y	k_1 (formula)	k_1	k_2 (formula)	r
	0		5				
0	0.25	(D2) + 0.5*((F3) + (H3))	6.633518	0.25*((D2)*EXP(TAN(B2)))	1.25	0.25*(((D2) + (F3))EXP(TAN(B3)))	2.017035832
1	0.5	(D3) + 0.5*((F4) + (H4))	9.597921	0.25*((D3)*EXP(TAN(B3)))	2.140806933	0.25*(((D3) + (F4))EXP(TAN(B4)))	3.787999165
2	0.75	(D4) + 0.5*((F5) + (H5))	16.03014	0.25*((D4)*EXP(TAN(B4)))	4.143557165	0.25*(((D4) + (F5))EXP(TAN(B5)))	8.720887048
3	1	(D5) + 0.5*((F6) + (H6))	36.6637	0.25*((D5)*EXP(TAN(B5)))	10.17336459	0.25*(((D5) + (F6))EXP(TAN(B6)))	31.09374413

11

$\dfrac{dy}{dx}$	x_0	y_0	h	End point
$2x - 3y$	0	1	0.1	$y(0.3)$

r	x	y (formula)	y	$y_1 = y_0 + hf(x_0, y_0)$	
0	0		1	$y_{r+1} = y_{r-1} + 2hf(x_r, y_r)$	
1	0.1	(D2) + 0.1*(2*(B2) − 3*(D2))	0.7	$f(x, y) =$	$2x - 3y$
2	0.2	(D2) + 0.2*(2*(B3) − 3*(D3))	0.62	$h =$	0.1
3	0.3	(D3) + 0.2*(2*(B4) − 3*(D4))	0.408		

12

$\dfrac{dy}{dx}$	x_0	y_0	h	End point
$3x + xy - 4$	1	1	0.1	$y(1.3)$

r	x	y (formula)	y	$y_1 = y_0 + hf(x_0, y_0)$	
0	1		1.000000	$y_{r+1} = y_{r-1} + 2hf(x_r, y_r)$	
1	1.1	(D2) + 0.1*(3*(B2) + (B2)*(D2) − 4)	1.000000	$f(x, y) =$	$3x + xy - 4$
2	1.2	(D2) + 0.2*(3*(B3) + (B3)*(D3) − 4)	1.080000	$h =$	0.1
3	1.3	(D3) + 0.2*(3*(B4) + (B4)*(D4) − 4)	1.179200		

13

$\dfrac{dy}{dx}$	x_0	y_0	h	End point
x^2y^2	0	1	0.1	$y(0.3)$

r	x	y (formula)	y	$y_1 = y_0 + hf(x_0, y_0)$	
0	0		1.000000	$y_{r+1} = y_{r-1} + 2hf(x_r, y_r)$	
1	0.1	(D2) + 0.1*((B2)^2 + (D2)^2)	1.0	$f(x, y) =$	x^2y^2
2	0.2	(D2) + 0.2*((B3)^2 + (D3)^2)	1.002	$h =$	0.1
3	0.3	(D3) + 0.2*((B4)^2 + (D4)^2)	1.008032		

14

$\frac{dy}{dx}$	x_0	y_0	h	End point
$\tan x$	0	0	0.1	$y(0.5)$

r	x	y (formula)	y	$y_1 = y_0 + hf(x_0, y_0)$	
0	0		0.000000	$y_{r+1} = y_{r-1} + 2hf(x_r, y_r)$	
1	0.1	(D2) + 0.1*(TAN(B2))	0.000000	f(x, y) =	$\tan x$
2	0.2	(D2) + 0.2*(TAN(B3))	0.020067	h =	0.1
3	0.3	(D3) + 0.2*(TAN(B4))	0.040542		
4	0.4	(D4) + 0.2*(TAN(B5))	0.081934		
5	0.5	(D5) + 0.2*(TAN(B6))	0.125101		

15

$\frac{dy}{dx}$	x_0	y_0	h	End point
$x^{\frac{-1}{2}}$	1	5	1	$y(7)$

r	x	y (formula)	y	$y_1 = y_0 + hf(x_0, y_0)$	
0	1		5.000000	$y_{r+1} = y_{r-1} + 2hf(x_r, y_r)$	
1	2	(D2) + 1*(1/SQRT(B2))	6.000000	f(x, y) =	$\dfrac{1}{\sqrt{x}}$
2	3	(D2) +2*(1/SQRT(B3))	6.414214	h =	0.1
3	4	(D3) +2*(1/SQRT(B4))	7.154701		
4	5	(D4) +2*(1/SQRT(B5))	7.414214		
5	6	(D5) +2*(1/SQRT(B6))	8.049128		
6	7	(D6) +2*(1/SQRT(B7))	8.230710		

Exercise 21.2B

1 a

A	B	C	D		
r	x	y(formula)	y	$y_{r+1} = y_r + hf(x_r, y_r)$	
0	0		150.0000000	f(x, y) =	$2x - 1$
1	0.5	(D3) + 0.5*(2*(B3) −1)	149.5	h =	0.5
2	1	(D4) + 0.5*(2*(B4) −1)	149.5		
3	1.5	(D5) + 0.5*(2*(B5) −1)	150		
4	2	(D6) + 0.5*(2*(B6) −1)	151		
5	2.5	(D7) + 0.5*(2*(B7) −1)	152.5		
6	3	(D8) + 0.5*(2*(B8) −1)	154.5		
7	3.5	(D9) + 0.5*(2*(B9) −1)	157		
8	4	(D10) + 0.5*(2*(B10) −1)	160		

b $\dfrac{ds}{dt} = v = 2t - 1 \rightarrow s = \displaystyle\int (2t - 1)dt = t^2 - t + c;$ $s = 150$ when $t = 0$, so $s = t^2 - t + 150$

Hence when $t = 4$, $s = 4^2 - 4 + 150 = 162$

2

A	B	C	D		
r	x	y (formula)	y	$y_{r+1} = y_r + hf(x_r, y_r)$	
0	0		0.0000000	f(x, y) =	10 − 0.5y
1	0.1	(D3) + 0.1*(10 − 0.5*(D3))	1.000000	h =	0.1
2	0.2	(D4) + 0.1*(10 − 0.5*(D4))	1.950000		
3	0.3	(D5) + 0.1*(10 − 0.5*(D5))	2.852500		
4	0.4	(D6) + 0.1*(10 − 0.5*(D6))	3.709875		
5	0.5	(D7) + 0.1*(10 − 0.5*(D7))	4.524381		
6	0.6	(D8) + 0.1*(10 − 0.5*(D8))	5.298162		
7	0.7	(D9) + 0.1*(10 − 0.5*(D9))	6.033254		
8	0.8	(D10) + 0.1*(10 − 0.5*(D10))	6.731591		
9	0.9	(D11) + 0.1*(10 − 0.5*(D11))	7.395012		
10	1	(D12) + 0.1*(10 − 0.5*(D12))	8.025261		

3

A	B	C	D		
r	x	y (formula)	y	$y_{r+1} = y_r + hf(x_r, y_r)$	
0	0		0.0000000	f(x, y) =	π(20x − x^2)
1	0.5	(D3) + 0.5*π*(20(B3) − (B3)^2)	0.000000	h =	0.5
2	1	(D4) + 0.5*π*(20(B4) − (B4)^2)	15.315251		
3	1.5	(D5) + 0.5*π*(20(B5) − (B5)^2)	45.160356		
4	2	(D6) + 0.5*π*(20(B6) − (B6)^2)	88.749918		

4 a

A	B	C	D		
r	x	y (formula)	y	$y_{r+1} = y_r + hf(x_r, y_r)$	
0	0		0.0000000	f(x,y) =	$3x^2 - 10x + 18$
1	1	(D3) + 1*(3*(B3)^2 − 10*(B3) + 18)	18.000000	h =	0.5
2	2	(D4) + 1*(3*(B4)^2 − 10*(B4) + 18)	29.000000		
3	3	(D5) + 1*(3*(B5)^2 − 10*(B5) + 18)	39.000000		
4	4	(D6) + 1*(3*(B6)^2 − 10*(B6) + 18)	54.000000		
5	5	(D7) + 1*(3*(B7)^2 − 10*(B7) + 18)	80.000000		
6	6	(D8) + 1*(3*(B8)^2 − 10*(B8) + 18)	123.000000		

b $\dfrac{\mathrm{d}s}{\mathrm{d}t} = 3t^2 - 10t + 18.$ $s = 0$ when $t = 0 \rightarrow s = \displaystyle\int (3t^2 - 10t + 18)\,\mathrm{d}t = t^3 - 5t^2 + 18t$

Hence when $t = 6$, $s = 6^3 - 5(6)^2 + 18(6) = 144$

5 a

A	B	C	D		
r	x	y (formula)	y	$y_{r+1} = y_r + hf(x_r, y_r)$	
0	1		4.0000000	f(x,y) =	$\dfrac{4}{\sqrt{t}}$
1	3	(D3) + 2*(4/sqrt(B3))	12.000000	h =	2
2	5	(D4) + 2*(4/sqrt(B4))	16.618802		
3	7	(D5) + 2*(4/sqrt(B5))	20.196511		
4	9	(D6) + 2*(4/sqrt(B6))	23.220227		

5	11	(D7) + 2*(4/sqrt(B7))	25.886893		
6	13	(D8) + 2*(4/sqrt(B8))	28.298941		
7	15	(D9) + 2*(4/sqrt(B9))	30.517785		
8	17	(D10) + 2*(4/sqrt(B10))	32.583376		
9	19	(D11) + 2*(4/sqrt(B11))	34.523661		
10	21	(D12) + 2*(4/sqrt(B12))	36.358987		

b $\dfrac{dv}{dt} = \dfrac{4}{\sqrt{t}}$, so $v = \displaystyle\int \dfrac{4+c}{\sqrt{t}} = 8\sqrt{t}$ When $t = 1$, $v = 4$ so $c = -4$. Therefore $v = 8\sqrt{t} - 4$

so when $t = 21$, $v = 8\sqrt{21} - 4 = 32.6606056$

relative error $= \dfrac{36.358987 - 32.6606056}{32.6606056} = 0.1132$ (4sf) percentage error $= 11.32\%$ (4sf)

6 a

A	B	C	D		
r	x	y (formula)	y	$y_{r+1} = y_r + hf(x_r, y_r)$	
0	0		0.0000000	f(x, y) =	$\dfrac{x(60 - 3x)}{1000}$
1	1	(D3) + 1*((B3)*((60 − 3*(B3)))/1000	0.000000	h =	1
2	2	(D4) + 1*((B4)*((60 − 3*(B4)))/1000	0.057000		
3	3	(D5) + 1*((B5)*((60 − 3*(B5)))/1000	0.165000		
4	4	(D6) + 1*((B6)*((60 − 3*(B6)))/1000	0.318000		
5	5	(D7) + 1*((B7)*((60 − 3*(B7)))/1000	0.510000		
6	6	(D8) + 1*((B8)*((60 − 3*(B8)))/1000	0.735000		
7	7	(D9) + 1*((B9)*((60 − 3*(B9)))/1000	0.987000		
8	8	(D10) + 1*((B10)*((60 − 3*(B10)))/1000	1.260000		
9	9	(D11) + 1*((B11)*((60 − 3*(B11)))/1000	1.548000		
10	10	(D12) + 1*((B12)*((60 − 3*(B12)))/1000	1.845000		

b $\dfrac{dv}{dt} = \dfrac{t(60 - 3t)}{1000} = \dfrac{3t}{50} - \dfrac{3t^2}{1000}$. Therefore $v = \displaystyle\int \dfrac{3t}{50} - \dfrac{3t^2}{1000}\,dt$. When $t = 10$, $v = \left[\dfrac{3t^2}{100} - \dfrac{t^3}{1000}\right] = 3 - 1 = 2\,\mathrm{m\,s^{-1}}$

7 a

A	B	C	D		
r	x	y (formula)	y	$y_{r+1} = y_r + hf(x_r, y_r)$	
0	0		10.0000000	f(x, y) =	$-\dfrac{x}{y}$
1	1	(D3) + 1*((− (B3)/(D3)))	10.000000	h =	1
2	2	(D4) + 1*((− (B4)/(D4)))	9.900000		
3	3	(D5) + 1*((− (B5)/(D5)))	9.697980		
4	4	(D6) + 1*((− (B6)/(D6)))	9.388637		
5	5	(D7) + 1*((− (B7)/(D7)))	8.962590		

b $v\dfrac{dv}{dx} = -x \to \displaystyle\int v\,dv = \int -x\,dx ; \to \dfrac{v^2}{2} = -\dfrac{x^2}{2} + C; \to$ Curve passes through $(0, 10)$ so $\dfrac{10^2}{2} = -\dfrac{0^2}{2} + C \to C = 50$

Hence $\dfrac{v^2}{2} = -\dfrac{x^2}{2} + 50; \to v^2 = 100 - x^2; \to$ When $x = 5$, $v^2 = 100 - 5^2 = 75; \to v = \sqrt{75} = 8.66$

8 a

r	x	y (formula)	y	k_1 (formula)	k_1	k_2 (formula)	k_2
0	0		3				
1	0.5	(D2) + 0.5*((F3) + (H3))	2.976563	0.5*(–9*(B2)/(16*(D2)))	0	0.5*(– 9*(B3)/(16*((D2) + F3))))	–0.046875
2	1	(D3) + 0.5*((F4) + (H4))	2.904934	0.5*(–9*(B3)/(16*(D3)))	–0.047244094	0.5*(– 9*(B4)/(16*((D3) + F4))))	–0.096012096
3	1.5	(D4) + 0.5*((F5) + (H5))	2.781408	0.5*(–9*(B4)/(16*(D4)))	–0.096818021	0.5*(– 9*(B5)/(16*((D4) + F5))))	–0.150234158
4	2	(D5) + 0.5*((F6) + (H6))	2.59862	0.5*(–9*(B5)/(16*(D5)))	–0.151676759	0.5*(– 9*(B6)/(16*((D5) + F6))))	–0.21390016
5	2.5	(D6) + 0.5*((F7) + (H7))	2.342808	0.5*(–9*(B6)/(16*(D6)))	–0.216461057	0.5*(– 9*(B7)/(16*((D6) + F7))))	–0.295162942
6	3	(D7) + 0.5*((F8) + (H8))	1.986218	0.5*(–9*(B7)/(16*(D7)))	–0.300120643	0.5*(– 9*(B8)/(16*((D7) + F8))))	–0.413058835

b Accurate value when $x = 3$: $3\sqrt{\left(1 - \dfrac{3^2}{16}\right)} = 1.984313$ (6 dp) so relative error $= \dfrac{1.986218 - 1.984313}{1.984313} = 0.000960$ (3 sf) and

percentage error $= 0.0960\%$ (3 sf)

9

r	t	v (formula)	v	k_1 (formula)	k_1	k_2 (formula)	k_2
0	1		4.000000				
1	3	(D2) + 0.5*((F3) + (H3))	7.386723	2*(2/(B2)^(1/3))	4.000000	2*(2/(B3)^(1/3))	2.773445
2	5	(D3) + 0.5*((F4) + (H4))	9.943052	2*(2/(B3)^(1/3))	2.773455	2*(2/(B4)^(1/3))	2.339214
3	7	(D4) + 0.5*((F5) + (H5))	12.158175	2*(2/(B4)^(1/3))	2.339214	2*(2/(B5)^(1/3))	2.091032
4	9	(D5) + 0.5*((F6) + (H6))	14.165191	2*(2/(B5)^(1/3))	2.091032	2*(2/(B6)^(1/3))	1.923014
5	11	(D6) + 0.5*((F7) + (H7))	16.025979	2*(2/(B6)^(1/3))	1.922999	2*(2/(B7)^(1/3))	1.798592
6	13	(D7) + 0.5*((F8) + (H8))	17.775849	2*(2/(B7)^(1/3))	1.798577	2*(2/(B8)^(1/3))	1.701176
7	15	(D8) + 0.5*((F9) + (H9))	19.437390	2*(2/(B8)^(1/3))	1.701161	2*(2/(B9)^(1/3))	1.621935
8	17	(D9) + 0.5*((F10) + (H10))	21.026172	2*(2/(B9)^(1/3))	1.621921	2*(2/(B10)^(1/3))	1.555659
9	19	(D10) + 0.5*((F11) + (H11))	22.553507	2*(2/(B10)^(1/3))	1.555644	2*(2/(B11)^(1/3))	1.499039
10	21	(D11) + 0.5*((F12) + (H12))	24.027939	2*(2/(B11)^(1/3))	1.499025	2*(2/(B12)^(1/3))	1.449855

10 a

A	B	C	D		
r	x	y (formula)	y	$y_{r+1} = y_r + hf(x_r, y_r)$	
0	1		2.0000000	f(x,y) =	$\dfrac{y}{(x^2 + 1)}$
1	1.2	(D3) + 0.2*((D3)/((B3)^2 + 1))	2.200000	h =	0.2
2	1.4	(D4) + 0.2*((D4)/((B4)^2 + 1))	2.380328		
3	1.6	(D5) + 0.2*((D5)/((B5)^2 + 1))	2.541161		
4	1.8	(D6) + 0.2*((D6)/((B6)^2 + 1))	2.683923		
5	2	(D7) + 0.2*((D7)/((B7)^2 + 1))	2.810523		

b $\dfrac{dy}{dx} = \dfrac{y}{x^2 + 1} \rightarrow \int \dfrac{dy}{y} = \dfrac{dx}{x^2 + 1} \rightarrow \ln(y) = \tan^{-1}(x) + C$; Curve passes through $(1, 2)$ so $\ln(2) = \tan^{-1}(1) + C \rightarrow C = \ln 2 - \dfrac{\pi}{4}$

$\rightarrow \ln(y) = \tan^{-1}(x) + \ln 2 - \dfrac{\pi}{4}$; $\rightarrow \ln\left(\dfrac{y}{2}\right) = \tan^{-1}(x) - \dfrac{\pi}{4}$; $\rightarrow y = 2e^{(\tan^{-1}(x) - \frac{\pi}{4})}$

Hence when $x = 2$, $y = 2e^{(\text{Tan}^{-1}(2) - \frac{\pi}{4})} = 2.76$ (3 sf)

11 a

r	x	y(formula)	y
0	0		0.000000
1	1	(D2) + 1*((B2)^2 − 5*(B2) + 8)	8.000000
2	2	(D2) + 2*((B3)^2 − 5*(B3) + 8)	8.000000
3	3	(D3) + 2*((B4)^2 − 5*(B4) + 8)	12.000000
4	4	(D4) + 2*((B5)^2 − 5*(B5) + 8)	12.000000
5	5	(D5) + 2*((B6)^2 − 5*(B6) + 8)	20.000000

b $\dfrac{ds}{dt} = t^2 - 5t + 8$. $s = 0$ when $t = 0 \Rightarrow s = \displaystyle\int (t^2 - 5t + 8)\, dt$

$$= \frac{1}{3}t^3 - \frac{5}{2}t^2 + 8t$$

hence when $t = 5$, $s = \dfrac{1}{3}5^3 - \dfrac{5}{2}5^2 + 8 \times 5 = 19.17$ (2 dp)

Review exercise 21

1 a 5.72 square units (3 sf)
b Area ≈ 12.5 square units (3 sf)
c Area ≈ 6.80 square units (3 sf)

2 a Area ≈ 12.7 square units (3 sf)
b Area ≈ 0.687 square units (3 sf)
c Area ≈ 1.38 (3 sf)

3 a Relative error   −0.0336364
  Percentage error   −3.3636364
b Relative error   0.01186
  Percentage error   1.18644
c Relative error   −0.77
  Percentage error   −77
d Relative error   0.01394
  Percentage error   1.394

4 a

$\dfrac{dy}{dx}$	x_0	y_0	H	End point
$10 - 3x$	0	1	0.1	$y(0.4)$

r	x	y(formula)	y
0	0		1.0000000
1	0.1	(D2) + 0.1*(10 − 3*(B2))	2.000000
2	0.2	(D3) + 0.1*(10 − 3*(B3))	2.970000
3	0.3	(D4) + 0.1*(10 − 3*(B4))	3.910000
4	0.4	(D5) + 0.1*(10 − 3*(B5))	4.820000

b

$\dfrac{dy}{dx}$	x_0	y_0	H	End point
$10 - 3y$	1	1	0.1	$y(1.5)$

r	x	y(formula)	y
0	1		1.0000000
1	1.1	(D2) + 0.1*(10 − 3*(D2))	1.700000
2	1.2	(D3) + 0.1*(10 − 3*(D3))	2.190000
3	1.3	(D4) + 0.1*(10 − 3*(D4))	2.533000
4	1.4	(D5) + 0.1*(10 − 3*(D5))	2.773100
5	1.5	(D2) + 0.1*(10 − 3*(D6))	2.941170

c

$\dfrac{dy}{dx}$	x_0	y_0	H	End point
$x^2\sqrt{y}$	3	2	0.01	$y(3.05)$

r	x	y(formula)	y
0	3		2.0000000
1	3.01	(D2) + 0.01*((B2)^2)*SQRT(D2)	2.12727922
2	3.02	(D3) + 0.01* ((B3)^2)*SQRT(D3)	2.25942255
3	3.03	(D4) + 0.01* ((B4)^2)*SQRT(D4)	2.39651471
4	3.04	(D5) + 0.01* ((B5)^2)*SQRT(D5)	2.53864129
5	3.05	(D6) +0.01* ((B6)^2)*SQRT(D6)	2.68588876

5 a

$\dfrac{dy}{dx}$	x_0	y_0	h	End point
$2x - y + xy$	3	4	0.1	$y(3.6)$

r	x	y (formula)	y	k_1 (formula)	k_1	k_2 (formula)	k_2
0	3		4.000000				
1	3.1	(D2) + 0.5*((F3) + (H3))	5.577000	0.1*(2*(B2) − (D2) + (B2)*(D2))	1.400000	0.1*(2*(B3) − ((D2)+(F3)) + (B3)*((D2) + (F3)))	1.754000
2	3.2	(D3) + 0.5*((F4) + (H4))	7.603084	0.1*(2*(B3) − (D3) + (B3)*(D3))	1.789700	0.1*(2*(B4) − ((D3)+(F4)) + (B4)*((D3) + (F4)))	2.260997
3	3.3	(D4) + 0.5*((F5) + (H5))	10.229736	0.1*(2*(B4) − (D4) + (B4)*(D4))	2.31268	0.1*(2*(B5) − ((D4)+(F5)) + (B5)*((D4) + (F5)))	2.940625
4	3.4	(D5) + 0.5*((F6) + (H6))	13.665264	0.1*(2*(B5) − (D5) + (B5)*(D5))	3.01284	0.1*(2*(B6) − ((D5)+(F6)) + (B6)*((D5) + (F6)))	3.858218
5	3.5	(D6) + 0.5*((F7) + (H7))	18.198212	0.1*(2*(B6) − (D6) + (B6)*(D6))	3.95967	0.1*(2*(B7) − ((D6)+(F7)) + (B7)*((D6) + (F7)))	5.106232
6	3.6	(D7) + 0.5*((F8) + (H8))	24.231198	0.1*(2*(B7) − (D7) + (B7)*(D7))	5.24955	0.1*(2*(B8) − ((D7)+(F8)) + (B8)*((D7) + (F8)))	6.816419

b

dy/dx	x_0	y_0	h	End point
$\sqrt{xy}$	0	36	0.01	$y(0.04)$

r	x	y (formula)	y	k_1 (formula)	k_1	k_2 (formula)	k_2
0	0		36.000000				
1	0.01	(D2) + 0.5*((F3) + (H3))	36.003000	0.01*(SQRT((B2)*(D2)))	0.000000	0.01*(SQRT((B3)*((D2) + (F3))))	0.006000
2	0.02	(D3) + 0.5*((F4) + (H4))	36.010243	0.01*(SQRT((B3)*(D3)))	0.006000	0.01*(SQRT((B4)*((D3) + (F4))))	0.008486
3	0.03	(D4) + 0.5*((F5) + (H5))	36.019684	0.01*(SQRT((B4)*(D4)))	0.008486	0.01*(SQRT((B5)*((D4) + (F5))))	0.010395
4	0.04	(D5) + 0.5*((F6) + (H6))	36.030884	0.01*(SQRT((B5)*(D5)))	0.010395	0.01*(SQRT((B6)*((D5) + (F6))))	0.012005
5	0.05	(D6) + 0.5*((F7) + (H7))	36.043599	0.01*(SQRT((B6)*(D6)))	0.012005	0.01*(SQRT((B7)*((D6) + (F7))))	0.013424
6	0.06	(D7) + 0.5*((F8) + (H8))	36.057665	0.01*(SQRT((B7)*(D7)))	0.013425	0.01*(SQRT((B8)*((D7) + (F8))))	0.014709

c

dy/dx	x_0	y_0	h	End point
$\left(\dfrac{e^{x^2}}{y}\right)$	1	4	1	$y(5)$

r	x	y (formula)	y	k_1 (formula)	k_1	k_2 (formula)	k_2
0	1		4.000000				
1	2	(D2) + 0.5*((F3) + (H3))	10.173456	1*(EXP((B2)^2)/(D2))	0.679570	1*(EXP((B3)^2)/((D2) + (F3)))	11.66734223
2	3	(D3) + 0.5*((F4) + (H4))	273.570750	1*(EXP((B3)^2)/(D3))	5.366726	1*(EXP((B4)^2)/((D3) + (F4)))	521.4278625
3	4	(D4) + 0.5*((F5) + (H5))	14942.718275	1*(EXP((B4)^2)/(D4))	29.619701	1*(EXP((B5)^2)/((D4) + (F5)))	29308.67535
4	5	(D5) + 0.5*((F6) + (H6))	2332388.201382	1*(EXP((B5)^2)/(D5))	594.678315	1*(EXP((B6)^2)/((D5) + (F6)))	4634296.288

6 a 21.8 minutes.

b If $m = 10e^{-0.0318t}$, then $\dfrac{dm}{dt} = -0.318e^{-0.0318t}$ hence using a starting point of $(0, 10)$ with a step length of 1, you get:

r	x	y (formula)	y	k_1 (formula)	k_1	k_2 (formula)	k_2
0	0		10.000000				
1	1	(D2) + 0.5*((F3) + (H3))	9.686977	1*(-0.318*EXP(-0.0318*(B2)))	-0.318000	1*(-0.318*EXP(-0.0318*(B3)))	-0.308046696
2	2	(D3) + 0.5*((F4) + (H4))	9.383751	1*(-0.318*EXP(-0.0318*(B3)))	-0.308047	1*(-0.318*EXP(-0.0318*(B4)))	-0.298404928
3	3	(D4) + 0.5*((F5) + (H5))	9.090016	1*(-0.318*EXP(-0.0318*(B4)))	-0.298405	1*(-0.318*EXP(-0.0318*(B5)))	-0.289064944
4	4	(D5) + 0.5*((F6) + (H6))	8.805475	1*(-0.318*EXP(-0.0318*(B5)))	-0.289065	1*(-0.318*EXP(-0.0318*(B6)))	-0.280017299
5	5	(D6) + 0.5*((F7) + (H7))	8.529840	1*(-0.318*EXP(-0.0318*(B6)))	-0.280017	1*(-0.318*EXP(-0.0318*(B7)))	-0.271252842
6	6	(D7) + 0.5*((F8) + (H8))	8.262832	1*(-0.318*EXP(-0.0318*(B7)))	-0.271253	1*(-0.318*EXP(-0.0318*(B8)))	-0.26276271
7	7	(D8) + 0.5*((F9) + (H9))	8.004181	1*(-0.318*EXP(-0.0318*(B8)))	-0.262763	1*(-0.318*EXP(-0.0318*(B9)))	-0.254538317
8	8	(D9) + 0.5*((F10) + (H10))	7.753627	1*(-0.318*EXP(-0.0318*(B9)))	-0.254538	1*(-0.318*EXP(-0.0318*(B10)))	-0.246571345
9	9	(D10) + 0.5*((F11) + (H11))	7.510914	1*(-0.318*EXP(-0.0318*(B10)))	-0.246571	1*(-0.318*EXP(-0.0318*(B11)))	-0.238853736
10	10	(D11) + 0.5*((F12) + (H12))	7.275798	1*(-0.318*EXP(-0.0318*(B11)))	-0.238854	1*(-0.318*EXP(-0.0318*(B12)))	-0.231377687
11	11	(D12) + 0.5*((F13) + (H13))	7.048042	1*(-0.318*EXP(-0.0318*(B12)))	-0.231378	1*(-0.318*EXP(-0.0318*(B13)))	-0.224135635
12	12	(D13) + 0.5*((F14) + (H14))	6.827414	1*(-0.318*EXP(-0.0318*(B13)))	-0.224136	1*(-0.318*EXP(-0.0318*(B14)))	-0.217120258
13	13	(D14) + 0.5*((F15) + (H15))	6.613691	1*(-0.318*EXP(-0.0318*(B14)))	-0.217120	1*(-0.318*EXP(-0.0318*(B15)))	-0.210324459

(Continued)

r	x	y (formula)	y	k_1 (formula)	k_1	k_2 (formula)	k_2
14	14	(D15) + 0.5*((F16) + (H16))	6.406658	1*(−0.318*EXP(−0.0318*(B15)))	−0.210324	1*(−0.318*EXP(−0.0318*(B16)))	−0.203741367
15	15	(D16) + 0.5*((F17) + (H17))	6.206106	1*(−0.318*EXP(−0.0318*(B16)))	−0.203741	1*(−0.318*EXP(−0.0318*(B17)))	−0.197364324
16	16	(D17) + 0.5*((F18) + (H18))	6.011830	1*(−0.318*EXP(−0.0318*(B17)))	−0.197364	1*(−0.318*EXP(−0.0318*(B18)))	−0.191186881
17	17	(D18) + 0.5*((F19) + (H19))	5.823635	1*(−0.318*EXP(−0.0318*(B18)))	−0.191187	1*(−0.318*EXP(−0.0318*(B19)))	−0.185202789
18	18	(D19) + 0.5*((F20) + (H20))	5.641331	1*(−0.318*EXP(−0.0318*(B19)))	−0.185203	1*(−0.318*EXP(−0.0318*(B20)))	−0.179405998
19	19	(D20) + 0.5*((F21) + (H21))	5.464732	1*(−0.318*EXP(−0.0318*(B20)))	−0.179406	1*(−0.318*EXP(−0.0318*(B21)))	−0.173790645
20	20	(D21) + 0.5*((F22) + (H22))	5.293662	1*(−0.318*EXP(−0.0318*(B21)))	−0.173791	1*(−0.318*EXP(−0.0318*(B22)))	−0.16835105
21	21	(D22) + 0.5*((F23) + (H23))	5.127945	1*(−0.318*EXP(−0.0318*(B22)))	−0.168351	1*(−0.318*EXP(−0.0318*(B23)))	−0.163081713
22	22	(D23) + 0.5*((F24) + (H24))	4.967416	1*(−0.318*EXP(−0.0318*(B23)))	−0.163082	1*(−0.318*EXP(−0.0318*(B24)))	−0.157977305
23	23	(D24) + 0.5*((F25) + (H25))	4.811911	1*(−0.318*EXP(−0.0318*(B24)))	−0.157977	1*(−0.318*EXP(−0.0318*(B25)))	−0.153032663

(using formulae in a spreadsheet)

The half life of Francium 223 is 22 minutes (nearest minute)

  c   6.83 g (3 sf)

7 a

r	x	y (formula)	y	k_1 (formula)	k_1	k_2 (formula)	k_2
0	0		1.000000				
1	0.1	(D2) + 0.5*((F3) + (H3))	1.105556	0.1*((D2)/(1 − (B2)^2))	0.100000	0.1*(((D2) + (F3))/((1−(B3)^2)))	0.111111
2	0.2	(D3) + 0.5*((F4) + (H4))	1.224789	0.1*((D3)/(1 − (B3)^2))	0.111672	0.1*(((D3) + (F4))/((1−(B4)^2)))	0.126795
3	0.3	(D4) + 0.5*((F5) + (H5))	1.362886	0.1*((D4)/(1 − (B4)^2))	0.127582	0.1*(((D4) + (F5))/((1−(B5)^2)))	0.148612
4	0.4	(D5) + 0.5*((F6) + (H6))	1.527809	0.1*((D5)/(1 − (B5)^2))	0.149768	0.1*(((D5) + (F6))/((1−(B6)^2)))	0.180078
5	0.5	(D6) + 0.5*((F7) + (H7))	1.732729	0.1*((D6)/(1 − (B6)^2))	0.181882	0.1*(((D6) + (F7))/((1−(B7)^2)))	0.227959
6	0.6	(D7) + 0.5*((F8) + (H8))	2.001663	0.1*((D7)/(1 − (B7)^2))	0.231031	0.1*(((D7) + (F8))/((1−(B8)^2)))	0.306837

  b   2

8 a

$\frac{dy}{dx}$	x_0	y_0	h	End point
$2xy$	1	1	0.1	$y(1.4)$

r	x	y(formula)	Y
0	1		1.000000
1	1.1	(D2) + 0.1*2*(B2)*(D2)	1.2
2	1.2	(D2) + 0.2*2*(B3)*(D3)	1.528
3	1.3	(D3) + 0.2*2*(B4)*(D4)	1.93344
4	1.4	(D4) + 0.2*2*(B5)*(D5)	2.5333888

 c

$\frac{dy}{dx}$	x_0	y_0	h	End point
$2x + \ln(1 + x)$	1	2	0.01	$y(1.05)$

r	x	y(formula)	Y
0	1		2.000000
1	1.01	(D2)+0.01*(2*(B2) + LN(1+(B2)))	2.026931
2	1.02	(D2)+0.02*(2*(B3) + LN(1+(B3)))	2.054363
3	1.03	(D2)+0.02*(2*(B4) + LN(1+(B4)))	2.081793
4	1.04	(D4)+0.02*(2*(B5) + LN(1+(B5)))	2.109723
5	1.05	(D5)+0.02*(2*(B6) + LN(1+(B6)))	2.137652

9 a   12 383, and a baby on the way.

  b

r	t	n(formula)	n
0	0		1000.000000
1	1	(D2) +700*1.5^(B2)	1700
2	2	(D2) +1400*1.5^(B3)	3100
3	3	(D3) +1400*1.5^(B4)	4850
4	4	(D4) +1400*1.5^(B5)	7825
5	5	(D5) +1400*1.5^(B6)	11937.5

 b

$\frac{dy}{dx}$	x_0	y_0	h	End point
$\tan^2 x$	0	0	0.2	$y(1)$

r	x	y(formula)	Y
0	0		0.000000
1	0.2	(D2) + 0.2*(TAN(B2) * TAN(B2))	0.000000
2	0.4	(D2) + 0.4*(TAN(B3) * TAN(B3))	0.016437
3	0.6	(D3) + 0.4*(TAN(B4) * TAN(B4))	0.071502
4	0.8	(D4) + 0.4*(TAN(B5) * TAN(B5))	0.203654
5	1	(D5) + 0.4*(TAN(B6) * TAN(B6))	0.495564

Assessment 21

1 **a** 0.5775 square units (4 dp)
   b 0.5760 square units
2 **a** 3.307 square units (4 sf)
   b Use more strips
3 **a** 50.8 cm² (3 sf)
   b $\dfrac{25}{48}\pi^3 + 100\ln(\sqrt{2})$
   c Answers agree to 3 sf so approximation in part **a** is suitable, accurate, good, etc.
4 **a** 4.75
   b 5.5
   c 13.6% (3 sf)
5 £580.61
6 **a** $k_1 = 0.25 \times (2 - (1^2) \times 1) = \frac{1}{4}$
       $k_2 = 0.25 \times (2 - (1.25^2) \times (1 + 1/4)) = 3/256$
       $y_1 = 1 + (1/4 + 3/256)/2 = 1 + (67/256)/2 = 579/512$
   b 1.0755 (4dp)
7 **a** 6
   b 8.25
   c 0.273 (3 sf)
8 $v = 5.087\ \text{m s}^{-1}$

Chapter 22
Exercise 22.1A

1 **a** 2   **b** $5a + b$ **c** 50   **d** $12b - 15ab$

2 **a** $x = -6$ **b** $x = 0, -5$ **c** $x = \dfrac{4}{3}$ **d** $x = \pm 2\sqrt{\dfrac{6}{19}}$

3 **a** **i** 21 square units   **ii** 7 square units
   b **i** does not involve a reflection
       ii involves a reflection
4 **a** 1.5 square units
   b triangle has not been reflected
5 **a** 80 cube units
   b involves a reflection

6 **a** $\dfrac{1}{3}\begin{pmatrix} 0 & -3 \\ 1 & 2 \end{pmatrix}$ **b** $\dfrac{1}{a^2}\begin{pmatrix} a-2 & 2 \\ a & -a \end{pmatrix}$

   c $\dfrac{1}{6}\begin{pmatrix} -2 & 3 & -2 \\ -2 & -6 & 4 \\ 4 & -3 & 4 \end{pmatrix}$ **d** $\begin{pmatrix} -5 & 8 & 11 \\ 18 & -28 & -39 \\ 2 & -3 & -4 \end{pmatrix}$

   e $\begin{pmatrix} \dfrac{2}{a} & -\dfrac{3}{a} & \dfrac{4}{a} \\ 5 & -7 & 9 \\ -1 & 2 & -2 \end{pmatrix}$

   f $\dfrac{1}{9a^2 + 11a}\begin{pmatrix} 3a-7 & 6a+2 & -6-2a \\ 3a^2-9a & a+6a^2 & a^2-3a \\ 6a & 3a & 2a \end{pmatrix}$

7 $\begin{pmatrix} -\dfrac{1}{k^2} & \dfrac{k^2+1}{k^3} & -\dfrac{1}{k^3} \\ -\dfrac{1}{k^2} & \dfrac{1}{k^3} & \dfrac{k^2-1}{k^3} \\ \dfrac{1}{k} & -\dfrac{1}{k^2} & \dfrac{1}{k^2} \end{pmatrix}$

8 $(1, -2, 0)$

Exercise 22.1B

1 **a** $(2, -5, -3)$
   b $\left(\dfrac{1}{4}, -\dfrac{1}{2}, 1\right)$

2 $\begin{pmatrix} -3 & -1 & -4 \\ 0 & 0 & -5 \\ 2 & 3 & -1 \end{pmatrix}$

3 $\begin{pmatrix} -6 & 3 & 3 \\ 2 & -1 & 0 \\ 4 & -5 & -3 \end{pmatrix}$

4 $\begin{pmatrix} 17 & 9 & -1 \\ 22 & 0 & 4 \\ 15 & 6 & 2 \end{pmatrix}$

5 **a** If we multiply the first equation by 2 we get $6x - 4y + 2z = 14$ so this plane is parallel to the second as they are the same except the constant term. Therefore, the three planes do not meet.
   b Multiplying the first equation by -3 gives $6x - 9y + 3z = -12$ so this plane is parallel to the second as they are the same except the constant term. Therefore the three planes do not meet.

6 **a** $k > \dfrac{1+\sqrt{7}}{2}$ or $k < \dfrac{1-\sqrt{7}}{2}$
   b $k = -1$

7 **a** $\det\begin{pmatrix} -7 & -3 & 5 \\ 1 & 1 & 1 \\ 1 & 2 & 4 \end{pmatrix} = -7(4-2)+3(4-1)+5(2-1) = 0$

   Therefore they do not intersect at a unique point.
   b Let $x = 2\lambda + 1$ then second equation gives:
       $2\lambda + 1 + y + z = 0$
       Multiply by 2 to give $4\lambda + 2 + 2y + 2z = 0$
       Third equation becomes: $2\lambda + 2y + 4z = 0$
       Subtract to eliminate y: $2\lambda + 2 - 2z = 0 \Rightarrow z = \lambda + 1$
       Substitute back into second equation to give:
       $2\lambda + 1 + y + \lambda + 1 = 0 \Rightarrow y = -3\lambda - 2$
       So general point on line is $(2\lambda+1, -3\lambda-2, \lambda+1)$
       Check first equation:
       $-7(2\lambda + 1) - 3(-3\lambda - 2) + 5(\lambda + 1) = 4$

8 **a** $\det\begin{pmatrix} 1 & 0 & 2 \\ -2 & 1 & 4 \\ 9 & -2 & 2 \end{pmatrix} = 1(12 - -8) + 2(4 - 9)$
       $= 0$ so not a unique solution
       First equation gives $x = 1 - 2z$
       Substitute into second equation to give
       $-2(1 - 2z) + y + 4z = 0 \Rightarrow y = 2 - 8z$
       Check in third equation:
       $9(1 - 2z) + 2(2 - 8z) + 2z = 5$
       So they form a sheaf (the line has equations such as $x = 1 - 2z, y = 2 - 8z$)
   b $\det\begin{pmatrix} 5 & -1 & 1 \\ 2 & 1 & -2 \\ 36 & 11 & -24 \end{pmatrix}$
       $= 5(-24 - -22) - -1(-48 - -72) + 1(22 - 36)$
       $= 0$ so not a unique solution
       Adding first two equations gives $7x - z = 7 \Rightarrow z = 7x - 7$
       Substitute into second equation to give
       $2x + y - 2(7x - 7) = 7 \Rightarrow y = 12x - 7$
       Check in third equation:
       $36x + 11(12x - 7) - 24(7x - 7) = 91$
       So they form a sheaf (the line had equations such as $z = 7x - 7, y = 12x - 7$)

9 $a = -3, 2$

10 a $\begin{pmatrix} 1 & 1 & 1 \\ 1 & -1 & 0 \\ 0 & 1 & -1 \end{pmatrix} \begin{pmatrix} r \\ h \\ f \end{pmatrix} = \begin{pmatrix} 17 \\ 1 \\ -10 \end{pmatrix}$

b 3 rabbits, 2 hamsters and 12 fish

11 a $\begin{pmatrix} 1 & 1 & 1 \\ 2 & 3 & 4 \\ 1 & 0 & -1 \end{pmatrix} \begin{pmatrix} x \\ y \\ z \end{pmatrix} = \begin{pmatrix} 12 \\ 36 \\ 0 \end{pmatrix}$

b $\det \begin{pmatrix} 1 & 1 & 1 \\ 2 & 3 & 4 \\ 1 & 0 & -1 \end{pmatrix}$

$= 1(-3-0) - 1(-2-4) + 1(0-3)$

$= 0$ so not a unique solution

c Equations are: $x + y + z = 12$
$$2x + 3y + 4z = 36$$
$$x - z = 0$$
Final equation gives $x = z$
Substitute into first equation to give
$z + y + z = 12 \Rightarrow y = 12 - 2z$
Check in second equation:
$2z + 3(12 - 2z) + 4z = 36$
So solutions lie on line $x = z$, $y = 12 - 2z$
But x, y, z must all be positive integers so possible solutions are:
$(1, 10, 1), (2, 8, 2), (3, 6, 3), (4, 4, 4), (5, 2, 5)$
So 5 possibilities (not 6 as $(6, 0, 6)$ not a solution since has no three-bed houses)

12 a $\dfrac{16}{k}$ square units **b** $k < 0$

c $A\left(-\dfrac{6}{k}, 4\right), B(0,0), C\left(\dfrac{2}{k}, 4\right)$ **d** $k = -2$

13 a $\det \begin{pmatrix} k & -1 & -1 \\ 2 & k & -1 \\ 3 & 2 & -2 \end{pmatrix} = k(-2k+2) + (-4+3) - (4-3k)$

$= -2k^2 + 2k - 1 - 4 + 3k$

$= -2k^2 + 5k - 5$

$= -[2k^2 - 5k + 5]$

$= -\left[2\left(k - \dfrac{5}{4}\right)^2 + \dfrac{15}{8}\right] < -\dfrac{15}{8}$

therefore determinant $\neq 0$

So the planes intersect at a single, unique point for all possible values of k

b $\left(\dfrac{13}{3}, \dfrac{1}{3}, \dfrac{16}{3}\right)$

14 a $k = 4$

b $x = \dfrac{8-k}{k-4}$ $y = \dfrac{10+k}{4-k}$ $z = \dfrac{6}{k-4}$

15 a $a = 2$ **b** $b = 4$

Exercise 22.2A

1 a $\begin{vmatrix} 2 & 5 & 1 \\ 4 & 3 & -1 \\ -3 & 2 & 0 \end{vmatrix} = 1 \begin{vmatrix} 4 & 3 \\ -3 & 2 \end{vmatrix} - 1 \begin{vmatrix} 2 & 5 \\ -3 & 2 \end{vmatrix} + 0$

$= 1(8 - -9) + 1(4 - -15) = 36$

b $\begin{vmatrix} 0 & 14 & 3 \\ 0 & -1 & 2 \\ 8 & 3 & 1 \end{vmatrix} = 0 - 0 + 8 \begin{vmatrix} 14 & 3 \\ -1 & 2 \end{vmatrix}$

$= 8(28 - -3) = 248$

c $\begin{vmatrix} 8 & 13 & -2 \\ 0 & 1 & 5 \\ 2 & 5 & 7 \end{vmatrix} = -0 + 1 \begin{vmatrix} 8 & -2 \\ 2 & 7 \end{vmatrix} - 5 \begin{vmatrix} 8 & 13 \\ 2 & 5 \end{vmatrix}$

$= 1(56 - -4) - 5(40 - 26)$

$= 60 - 70 = 10$

d $\begin{vmatrix} 12 & 4 & 6 \\ 8 & -2 & 5 \\ 0 & 9 & 0 \end{vmatrix} = 0 - 9 \begin{vmatrix} 12 & 6 \\ 8 & 5 \end{vmatrix} + 0$

$= -9(60 - 48)$

$= -108$

e $\begin{vmatrix} 0 & b & 2a \\ a & a & 1 \\ 0 & b & 0 \end{vmatrix} = 0 - a \begin{vmatrix} b & 2a \\ b & 0 \end{vmatrix} + 0$

$= -a(0 - 2ab) = 2a^2 b$

f $\begin{vmatrix} b & a & -b \\ 0 & -3 & 0 \\ a & 2 & a \end{vmatrix} = -0 - 3 \begin{vmatrix} b & -b \\ a & a \end{vmatrix} - 0$

$= -3(ab - -ab) = -6ab$

2 a For example, $\begin{vmatrix} 2 & 1 & -3 \\ 4 & 2 & 1 \\ 1 & 1 & 2 \end{vmatrix} = \begin{vmatrix} 0 & 1 & -3 \\ 0 & 2 & 1 \\ -1 & 1 & 2 \end{vmatrix}$ C1 − 2C2

b -7

3 a For example, $\begin{vmatrix} 6 & 3 & -3 \\ 9 & 2 & -1 \\ -5 & -7 & -2 \end{vmatrix} = \begin{vmatrix} 0 & 3 & -3 \\ 5 & 2 & -1 \\ 9 & -7 & -2 \end{vmatrix}$ C1 − 2C2

$= \begin{vmatrix} 0 & 0 & -3 \\ 5 & 1 & -1 \\ 9 & -9 & -2 \end{vmatrix}$ C2 + C3

$= \begin{vmatrix} 0 & 0 & -3 \\ 6 & 1 & -1 \\ 0 & -9 & -2 \end{vmatrix}$ C1 + C2

$= \begin{vmatrix} 0 & 0 & -3 \\ 6 & 1 & 0 \\ 0 & -9 & -11 \end{vmatrix}$ C2 + C3

$= 162$

b 162

4 For example, $\begin{vmatrix} 5 & -2 & 12 \\ -2 & -8 & 24 \\ 4 & 1 & 16 \end{vmatrix} = \begin{vmatrix} 5 & -2 & 12 \\ -12 & -4 & 0 \\ 4 & 1 & 16 \end{vmatrix}$ R2 − 2R1

$= \begin{vmatrix} 11 & -2 & 12 \\ 0 & -4 & 0 \\ 1 & 1 & 16 \end{vmatrix}$ C1 − 3C2

5 a −128 since rows 1 and 2 swapped
b 128 since transpose
c 128 since row 2 added to row 1
d 128 since 4 × column 1 subtracted from column 3
e 128 × 3 = 384 since column 2 multiplied by 3
f (128 ×−1) ÷ 2 = −64 since column 2 divided by 2 and column 3 multiplied by −1
g 256 since columns 1 and 3 then columns 2 and 3 swapped and column 2 doubled
h (128 ÷ 2) × 4 = 256 since row 3 divided by 2 and column 2 multiplied by 4

1 a $\begin{vmatrix} a & b & c \\ b & c & a \\ c & a & b \end{vmatrix} = \begin{vmatrix} a+b & b+c & c+a \\ b & c & a \\ c & a & b \end{vmatrix}$ R1+R2

$= \begin{vmatrix} a+b+c & b+c+a & c+a+b \\ b & c & a \\ c & a & b \end{vmatrix}$ R1+R3

$= (a+b+c) \begin{vmatrix} 1 & 1 & 1 \\ b & c & a \\ c & a & b \end{vmatrix}$

$= (a+b+c)[(bc-a^2)-(b^2-ac)+(ab-c^2)]$

$= (a+b+c)(ab+ac+bc-a^2-b^2-c^2)$
as required

b $\begin{vmatrix} a^2 & b^2 & c^2 \\ bc & ca & ab \\ 1 & 1 & 1 \end{vmatrix} = \begin{vmatrix} a^2 & b^2-a^2 & c^2 \\ bc & ca-bc & ab \\ 1 & 0 & 1 \end{vmatrix}$ C2−C1

$= \begin{vmatrix} a^2 & b^2-a^2 & c^2-a^2 \\ bc & ca-bc & ab-bc \\ 1 & 0 & 0 \end{vmatrix}$ C3−C1

$= \begin{vmatrix} a^2 & (b-a)(b+a) & c^2-a^2 \\ bc & -c(b-a) & ab-bc \\ 1 & 0 & 0 \end{vmatrix}$

$= (b-a) \begin{vmatrix} a^2 & b+a & c^2-a^2 \\ bc & -c & ab-bc \\ 1 & 0 & 0 \end{vmatrix}$

$= (b-a) \begin{vmatrix} a^2 & b+a & (c-a)(c+a) \\ bc & -c & -b(c-a) \\ 1 & 0 & 0 \end{vmatrix}$

$= (b-a)(c-a) \begin{vmatrix} a^2 & b+a & c+a \\ bc & -c & -b \\ 1 & 0 & 0 \end{vmatrix}$

$= (b-a)(c-a)[-b(b+a)--c(c+a)]$

$= (b-a)(c-a)(-b^2-ab+c^2+ac)$

$= (b-a)(c-a)(c-b)(a+b+c)$ as required

c $\begin{vmatrix} a+b & a+c & b+c \\ c & b & a \\ c^2 & b^2 & a^2 \end{vmatrix} = \begin{vmatrix} a+b+c & a+c+b & b+c+a \\ c & b & a \\ c^2 & b^2 & a^2 \end{vmatrix}$ R1+R2

$= (a+b+c) \begin{vmatrix} 1 & 1 & 1 \\ c & b & a \\ c^2 & b^2 & a^2 \end{vmatrix}$

$= (a+b+c) \begin{vmatrix} 0 & 1 & 1 \\ c-b & b & a \\ c^2-b^2 & b^2 & a^2 \end{vmatrix}$ C1−C2

$= (a+b+c)(c-b) \begin{vmatrix} 0 & 1 & 1 \\ 1 & b & a \\ c+b & b^2 & a^2 \end{vmatrix}$

$= (a+b+c)(c-b) \begin{vmatrix} 0 & 0 & 1 \\ 1 & b-a & a \\ c+b & b^2-a^2 & a^2 \end{vmatrix}$ C2−C3

$= (a+b+c)(c-b)(b-a)[(a+b)-(c+b)]$

$= (a+b+c)(c-b)(b-a)(a-c)$
as required

d $\begin{vmatrix} a & -b & c \\ c-b & a+c & a-b \\ -bc & ac & -ab \end{vmatrix} = \begin{vmatrix} a+b & -b & c \\ -b-a & a+c & a-b \\ -bc-ac & ac & -ab \end{vmatrix}$ C1−C2

$= \begin{vmatrix} a+b & -b & c \\ -(a+b) & a+c & a-b \\ -c(a+b) & ac & -ab \end{vmatrix}$

$= (a+b) \begin{vmatrix} 1 & -b & b+c \\ -1 & a+c & -b-c \\ -c & ac & -ab-ac \end{vmatrix}$ C3−C2

$= (a+b) \begin{vmatrix} 1 & -b & b+c \\ -1 & a+c & -(b+c) \\ -c & ac & -a(b+c) \end{vmatrix}$

$= (a+b)(b+c) \begin{vmatrix} 1 & -b & 1 \\ -1 & a+c & -1 \\ -c & ac & -a \end{vmatrix}$

$= (a+b)(b+c) \begin{vmatrix} 0 & -b & 1 \\ 0 & a+c & -1 \\ -c+a & ac & -a \end{vmatrix}$ C1−C3

$= (a+b)(b+c)(c-a) \begin{vmatrix} 0 & -b & 1 \\ 0 & a+c & -1 \\ -1 & ac & -a \end{vmatrix}$

$= (a+b)(b+c)(c-a)[-1(b-(a+c))]$

$= (a+b)(b+c)(c-a)(a-b+c)$
as required

2 $\begin{vmatrix} 1 & 1 & 1 \\ a & a^2 & a^3 \\ b & b^2 & b^3 \end{vmatrix} = a \begin{vmatrix} 1 & 1 & 1 \\ 1 & a & a^2 \\ b & b^2 & b^3 \end{vmatrix}$

$= ab \begin{vmatrix} 1 & 1 & 1 \\ 1 & a & a^2 \\ 1 & b & b^2 \end{vmatrix}$

$= ab \begin{vmatrix} 1 & 1 & 1 \\ 0 & a-1 & a^2-1 \\ 1 & b & b^2 \end{vmatrix}$ R2−R1

$= ab \begin{vmatrix} 1 & 1 & 1 \\ 0 & a-1 & a^2-1 \\ 0 & b-1 & b^2-1 \end{vmatrix}$ R3−R1

$= ab(a-1) \begin{vmatrix} 1 & 1 & 1 \\ 0 & 1 & a+1 \\ 0 & b-1 & b^2-1 \end{vmatrix}$

$= ab(a-1)(b-1) \begin{vmatrix} 1 & 1 & 1 \\ 0 & 1 & a+1 \\ 0 & 1 & b+1 \end{vmatrix}$

$$= ab(a-1)(b-1)\begin{vmatrix} 1 & 1 & 1 \\ 0 & 1 & a+1 \\ 0 & 1 & b+1 \end{vmatrix}$$

$$= ab(a-1)(b-1)[(b+1)-(a+1)]$$

$$= ab(a-1)(b-1)(b-a)$$

3 a $\begin{vmatrix} a & b & c \\ a^2 & b^2 & c^2 \\ a^3 & b^3 & c^3 \end{vmatrix} = a\begin{vmatrix} 1 & b & c \\ a & b^2 & c^2 \\ a^2 & b^3 & c^3 \end{vmatrix}$

$$= ab\begin{vmatrix} 1 & 1 & c \\ a & b & c^2 \\ a^2 & b^2 & c^3 \end{vmatrix}$$

$$= abc\begin{vmatrix} 1 & 1 & 1 \\ a & b & c \\ a^2 & b^2 & c^2 \end{vmatrix}$$

$$= abc\begin{vmatrix} 1 & 1 & 1 \\ a & b & c \\ 0 & b^2-ab & c^2-ac \end{vmatrix} \quad R3-aR2$$

$$= abc\begin{vmatrix} 1 & 1 & 1 \\ 0 & b-a & c-a \\ 0 & b^2-ab & c^2-ac \end{vmatrix} \quad R2-aR1$$

$$= abc\begin{vmatrix} 1 & 1 & 1 \\ 0 & b-a & c-a \\ 0 & 0 & c^2-ac-b(c-a) \end{vmatrix} \quad R3-bR2$$

since $c^2-ac-b(c-a)=c(c-a)-b(c-a)$
$$=(c-b)(c-a)$$

$$= abc\begin{vmatrix} 1 & 1 & 1 \\ 0 & b-a & c-a \\ 0 & 0 & (c-b)(c-a) \end{vmatrix}$$

as required

b i 48   **ii** −1680

4 a $(a-b)(a^2+ab+b^2)=a^3+a^2b+ab^2-a^2b-ab^2-b^3$
$$=a^3-b^3 \text{ as required}$$

b $-(a-b)(b-c)(c-a)(a+b+c)$

c Unique solution if: $a \neq b$, $a \neq c$, $b \neq c$, $a+b+c \neq 0$

5 a $(a-b)(b-c)(a-c)(ab+ac+bc)$

b $a \neq b$, $a \neq c$ and $b \neq c$

6 a $(a-b)(a+b)(a+b-2)$ **b** $a \neq b$, $a \neq -b$, $a+b \neq 2$

7 e.g. $\begin{vmatrix} a^2 & b^2 & a \\ 1 & c & b \\ a & ac & ab \end{vmatrix} = a\begin{vmatrix} a^2 & b^2 & a \\ 1 & c & b \\ 1 & c & b \end{vmatrix}$

$$= a\begin{vmatrix} a^2 & b^2 & a \\ 0 & 0 & 0 \\ 1 & c & b \end{vmatrix} \quad R2-R3$$

$$= a \times 0$$
$$= 0 \quad \text{so singular}$$

8 $x = 0, a, b$

9 $x = a, b, -a-b$

10 $\det A = \begin{vmatrix} a & a^2-a & a^2+b^2+c^2 \\ b & b^2-b & a^2+b^2+c^2 \\ c & c^2-c & a^2+b^2+c^2 \end{vmatrix}$

$$= (a^2+b^2+c^2)\begin{vmatrix} a & a^2-a & 1 \\ b & b^2-b & 1 \\ c & c^2-c & 1 \end{vmatrix}$$

$$= (a^2+b^2+c^2)\begin{vmatrix} a & a^2-a & 1 \\ b-a & b^2-b-a^2+a & 0 \\ c-a & c^2-c-a^2+a & 0 \end{vmatrix}$$

subtract R1 from R2 & R3

$$= (a^2+b^2+c^2)\begin{vmatrix} a & a^2-a & 1 \\ b-a & (b-a)(b+a-1) & 0 \\ c-a & (c-a)(c+a-1) & 0 \end{vmatrix}$$

$$= (a^2+b^2+c^2)(b-a)(c-a)\begin{vmatrix} a & a^2-a & 1 \\ 1 & b+a-1 & 0 \\ 1 & c+a-1 & 0 \end{vmatrix}$$

$$= (a^2+b^2+c^2)(b-a)(c-a)(c+a-1-b-a+1)$$
$$= (a^2+b^2+c^2)(b-a)(c-a)(c-b)$$

$a \neq b \neq c$. so $(b-a)(c-a)(c-b) \neq 0$

a, b, c are distinct and real so $a^2+b^2+c^2 > 0$

Therefore $\det A \neq 0$ so **A** is non-singular for all real, distinct values of a, b and c.

Exercise 22.3A

1 a i $\lambda = 2, -4$, corresponding eigenvectors are $\begin{pmatrix} 1 \\ 0 \end{pmatrix}, \begin{pmatrix} 1 \\ -2 \end{pmatrix}$

ii $\lambda = 2, -1$, corresponding eigenvectors are $\begin{pmatrix} 1 \\ 1 \end{pmatrix}, \begin{pmatrix} 1 \\ 4 \end{pmatrix}$

iii $\lambda = 3, 7$, corresponding eigenvectors are $\begin{pmatrix} -4 \\ 3 \end{pmatrix}, \begin{pmatrix} 0 \\ 1 \end{pmatrix}$

iv $\lambda = -1, 5$, corresponding eigenvectors are $\begin{pmatrix} 1 \\ 1 \end{pmatrix}, \begin{pmatrix} -5 \\ 1 \end{pmatrix}$

b i Characteristic equation is $\lambda^2+2\lambda-8=0$

Substitute matrix **A**:

$$A^2+2A-8I = \begin{pmatrix} 2 & 3 \\ 0 & -4 \end{pmatrix}^2 + 2\begin{pmatrix} 2 & 3 \\ 0 & -4 \end{pmatrix} - 8\begin{pmatrix} 1 & 0 \\ 0 & 1 \end{pmatrix}$$

$$= \begin{pmatrix} 4 & -6 \\ 0 & 16 \end{pmatrix} + \begin{pmatrix} 4 & 6 \\ 0 & -8 \end{pmatrix} - \begin{pmatrix} 8 & 0 \\ 0 & 8 \end{pmatrix}$$

$$= \begin{pmatrix} 0 & 0 \\ 0 & 0 \end{pmatrix} \text{ as required}$$

ii Characteristic equation is $\lambda^2 - \lambda - 2 = 0$
Substitute matrix **A**:

$$\mathbf{A}^2 - \mathbf{A} - 2\mathbf{I} = \begin{pmatrix} 3 & -1 \\ 4 & -2 \end{pmatrix}^2 - \begin{pmatrix} 3 & -1 \\ 4 & -2 \end{pmatrix} - 2\begin{pmatrix} 1 & 0 \\ 0 & 1 \end{pmatrix}$$

$$= \begin{pmatrix} 5 & -1 \\ 4 & 0 \end{pmatrix} - \begin{pmatrix} 3 & -1 \\ 4 & -2 \end{pmatrix} - \begin{pmatrix} 2 & 0 \\ 0 & 2 \end{pmatrix}$$

$$= \begin{pmatrix} 0 & 0 \\ 0 & 0 \end{pmatrix} \text{ as required}$$

iii Characteristic equation is $\lambda^2 - 10\lambda + 21 = 0$
Substitute matrix **A**:

$$\mathbf{A}^2 - 10\mathbf{A} + 21\mathbf{I} = \begin{pmatrix} 3 & 0 \\ 3 & 7 \end{pmatrix}^2 - 10\begin{pmatrix} 3 & 0 \\ 3 & 7 \end{pmatrix} + 21\begin{pmatrix} 1 & 0 \\ 0 & 1 \end{pmatrix}$$

$$= \begin{pmatrix} 9 & 0 \\ 30 & 49 \end{pmatrix} - \begin{pmatrix} 30 & 0 \\ 30 & 70 \end{pmatrix} + \begin{pmatrix} 21 & 0 \\ 0 & 21 \end{pmatrix}$$

$$= \begin{pmatrix} 0 & 0 \\ 0 & 0 \end{pmatrix} \text{ as required}$$

iv Characteristic equation is $\lambda^2 - 4\lambda - 5 = 0$
Substitute matrix **A**:

$$\mathbf{A}^2 - 4\mathbf{A} - 5\mathbf{I} = \begin{pmatrix} 4 & -5 \\ -1 & 0 \end{pmatrix}^2 - 4\begin{pmatrix} 4 & -5 \\ -1 & 0 \end{pmatrix} - 5\begin{pmatrix} 1 & 0 \\ 0 & 1 \end{pmatrix}$$

$$= \begin{pmatrix} 21 & -20 \\ -4 & 5 \end{pmatrix} - \begin{pmatrix} 16 & -20 \\ -4 & 0 \end{pmatrix} - \begin{pmatrix} 5 & 0 \\ 0 & 5 \end{pmatrix}$$

$$= \begin{pmatrix} 0 & 0 \\ 0 & 0 \end{pmatrix} \text{ as required}$$

2 a $\lambda = -3,\ 1,\ 5$, corresponding eigenvectors are

$$\begin{pmatrix} 5 \\ -3 \\ 4 \end{pmatrix}, \begin{pmatrix} 1 \\ -1 \\ 0 \end{pmatrix}, \begin{pmatrix} 0 \\ 1 \\ 0 \end{pmatrix}$$

b $\lambda = -2,\ 2,\ 3$, corresponding eigenvectors are

$$\begin{pmatrix} 1 \\ -1 \\ 0 \end{pmatrix}, \begin{pmatrix} 3 \\ 1 \\ -4 \end{pmatrix}, \begin{pmatrix} 3 \\ 2 \\ 0 \end{pmatrix}$$

c $\lambda = 5,\ 2,\ -4$, corresponding eigenvectors are

$$\begin{pmatrix} 1 \\ 0 \\ 0 \end{pmatrix}, \begin{pmatrix} 6 \\ 6 \\ -5 \end{pmatrix}, \begin{pmatrix} 0 \\ 0 \\ 1 \end{pmatrix}$$

d $\lambda = -7,\ 4,\ 9$, corresponding eigenvectors are

$$\begin{pmatrix} -44 \\ 19 \\ 11 \end{pmatrix}, \begin{pmatrix} 0 \\ 1 \\ 0 \end{pmatrix}, \begin{pmatrix} 0 \\ 1 \\ 5 \end{pmatrix}$$

3 a $\det\begin{pmatrix} 1-\lambda & 3 & -1 \\ -2 & 3-\lambda & 2 \\ 1 & 0 & 2-\lambda \end{pmatrix}$

$$= 1\big[6 - -1(3-\lambda)\big] + (2-\lambda)\big[(1-\lambda)(3-\lambda) - -6\big]$$

$$= 9 - \lambda + (2-\lambda)(\lambda^2 - 4\lambda + 9)$$

$$= 9 - \lambda + 2\lambda^2 - 8\lambda + 18 - \lambda^3 + 4\lambda^2 - 9\lambda$$

$$= -\lambda^3 + 6\lambda^2 - 18\lambda + 27$$

When $\lambda = 3$, $-\lambda^3 + 6\lambda^2 - 18\lambda + 27 = 0$ so 3 is an eigenvalue.

$$-\lambda^3 + 6\lambda^2 - 18\lambda + 27 = (\lambda - 3)(-\lambda^2 + 3\lambda - 9)$$

$-\lambda^2 + 3\lambda - 9 = 0$ has no real solutions since
$b^2 - 4ac = 9 - 36 < 0$

An eigenvector is $\begin{pmatrix} 1 \\ 1 \\ 1 \end{pmatrix}$

b Characteristic equation is $-\lambda^3 + 6\lambda^2 - 18\lambda + 27 = 0$
Substitute matrix **A**:

$$-\mathbf{A}^3 + 6\mathbf{A}^2 - 16\mathbf{A} + 9\mathbf{I} = -\begin{pmatrix} 1 & 3 & -1 \\ -2 & 3 & 2 \\ 1 & 0 & 2 \end{pmatrix}^3$$

$$+ 6\begin{pmatrix} 1 & 3 & -1 \\ -2 & 3 & 2 \\ 1 & 0 & 2 \end{pmatrix}^2 - 18\begin{pmatrix} 1 & 3 & -1 \\ -2 & 3 & 2 \\ 1 & 0 & 2 \end{pmatrix} + 27\begin{pmatrix} 1 & 0 & 0 \\ 0 & 1 & 0 \\ 0 & 0 & 1 \end{pmatrix}$$

$$= -\begin{pmatrix} -27 & 18 & 36 \\ 0 & -9 & 36 \\ 0 & 18 & 9 \end{pmatrix} + 6\begin{pmatrix} -6 & 12 & 3 \\ -6 & 3 & 12 \\ 3 & 3 & 3 \end{pmatrix}$$

$$-18\begin{pmatrix} 1 & 3 & -1 \\ -2 & 3 & 2 \\ 1 & 0 & 2 \end{pmatrix} + 27\begin{pmatrix} 1 & 0 & 0 \\ 0 & 1 & 0 \\ 0 & 0 & 1 \end{pmatrix}$$

$$= -\begin{pmatrix} -27 & 18 & 36 \\ 0 & -9 & 36 \\ 0 & 18 & 9 \end{pmatrix} + \begin{pmatrix} -36 & 72 & 18 \\ -36 & 18 & 72 \\ 18 & 18 & 18 \end{pmatrix}$$

$$-\begin{pmatrix} 18 & 54 & -18 \\ -36 & 54 & 36 \\ 18 & 0 & 36 \end{pmatrix} + \begin{pmatrix} 27 & 0 & 0 \\ 0 & 27 & 0 \\ 0 & 0 & 27 \end{pmatrix}$$

$$= \begin{pmatrix} 0 & 0 & 0 \\ 0 & 0 & 0 \\ 0 & 0 & 0 \end{pmatrix} \text{ as required}$$

4 a $\det\begin{pmatrix} -2-\lambda & 6 & -3 \\ 3 & 2-\lambda & 6 \\ 2 & 6 & -3-\lambda \end{pmatrix}$

$$= (-2-\lambda)\,[(2-\lambda)(3-\lambda) - 36] - 6[3(-3-\lambda) - 12]$$
$$-3[18 - 2(2-\lambda)]$$

$$= -\lambda^3 - 3\lambda^2 + 52\lambda + 168$$

When $\lambda = -7$, $-\lambda^3 - 3\lambda^2 + 52\lambda + 168 = 0$ so -7 is an eigenvalue and $\lambda = 2 \pm 2\sqrt{7}$

b $\begin{pmatrix} 21 \\ -13 \\ 9 \end{pmatrix}$

c Characteristic equation is
$-\lambda^3 - 3\lambda^2 + 52\lambda + 168 = 0$
Substitute matrix **A**:
$-\mathbf{A}^3 - 3\mathbf{A}^2 + 52\mathbf{A} + 168\mathbf{I}$

$$= -\begin{pmatrix} -2 & 6 & -3 \\ 3 & 2 & 6 \\ 2 & 6 & -3 \end{pmatrix}^3 - 3\begin{pmatrix} -2 & 6 & -3 \\ 3 & 2 & 6 \\ 2 & 6 & -3 \end{pmatrix}^2$$

$$+ 52\begin{pmatrix} -2 & 6 & -3 \\ 3 & 2 & 6 \\ 2 & 6 & -3 \end{pmatrix} + 168\begin{pmatrix} 1 & 0 & 0 \\ 0 & 1 & 0 \\ 0 & 0 & 1 \end{pmatrix}$$

$$= -\begin{pmatrix} 16 & 366 & -309 \\ 120 & 98 & 357 \\ 80 & 294 & -105 \end{pmatrix} - 3\begin{pmatrix} 16 & -18 & 51 \\ 12 & 58 & -15 \\ 8 & 6 & 39 \end{pmatrix}$$

$$+ 52\begin{pmatrix} -2 & 6 & -3 \\ 3 & 2 & 6 \\ 2 & 6 & -3 \end{pmatrix} + 168\begin{pmatrix} 1 & 0 & 0 \\ 0 & 1 & 0 \\ 0 & 0 & 1 \end{pmatrix}$$

$$= -\begin{pmatrix} 16 & 366 & -309 \\ 120 & 98 & 357 \\ 80 & 294 & -105 \end{pmatrix} - \begin{pmatrix} 48 & -54 & 153 \\ 36 & 174 & -45 \\ 24 & 18 & 117 \end{pmatrix}$$

$$+ \begin{pmatrix} -104 & 312 & -156 \\ 156 & 104 & 312 \\ 104 & 312 & -156 \end{pmatrix} + \begin{pmatrix} 168 & 0 & 0 \\ 0 & 168 & 0 \\ 0 & 0 & 168 \end{pmatrix}$$

$$= \begin{pmatrix} 0 & 0 & 0 \\ 0 & 0 & 0 \\ 0 & 0 & 0 \end{pmatrix} \text{ as required}$$

5 $\mathbf{P} = \begin{pmatrix} 0 & 1 \\ 1 & 0 \end{pmatrix}$, $\mathbf{D} = \begin{pmatrix} 7 & 0 \\ 0 & 2 \end{pmatrix}$

6 $\mathbf{B} = \begin{pmatrix} -1 & 6 \\ 1 & 1 \end{pmatrix}\begin{pmatrix} -4 & 0 \\ 0 & 3 \end{pmatrix}\begin{pmatrix} -\dfrac{1}{7} & \dfrac{6}{7} \\ \dfrac{1}{7} & \dfrac{1}{7} \end{pmatrix}$

7 $\mathbf{P} = \begin{pmatrix} 0 & 0 & -30 \\ -2 & 3 & 1 \\ 3 & 2 & 24 \end{pmatrix}$, $\mathbf{D} = \begin{pmatrix} -7 & 0 & 0 \\ 0 & 6 & 0 \\ 0 & 0 & -4 \end{pmatrix}$

8 $\mathbf{T} = \begin{pmatrix} -1 & 0 & 1 \\ 1 & 0 & 1 \\ 0 & 1 & 0 \end{pmatrix}\begin{pmatrix} 8 & 0 & 0 \\ 0 & -7 & 0 \\ 0 & 0 & 6 \end{pmatrix}\begin{pmatrix} -\dfrac{1}{2} & \dfrac{1}{2} & 0 \\ 0 & 0 & 1 \\ \dfrac{1}{2} & \dfrac{1}{2} & 0 \end{pmatrix}$

Exercise 22.3B

1 a $\begin{pmatrix} 1 & 0 \\ 189 & 64 \end{pmatrix}$

b $\begin{pmatrix} 1 & 0 \\ 3(2^n) - 3 & 2^n \end{pmatrix}$

2 a $\begin{pmatrix} 5^4 & -2(5^4) \\ -2(5^4) & 4(5^4) \end{pmatrix}$

b $5^{n-1}\begin{pmatrix} 1 & -2 \\ -2 & 4 \end{pmatrix}$

3 a $\begin{pmatrix} -1 & 1 \\ 1 & 1 \end{pmatrix}$, $\begin{pmatrix} 3 & 0 \\ 0 & 7 \end{pmatrix}$

b $3^3 = 27$ and $7^3 = 343$

c $\begin{pmatrix} -1 \\ 1 \end{pmatrix}$ and $\begin{pmatrix} 1 \\ 1 \end{pmatrix}$

4 a $\mathbf{r} = s\begin{pmatrix} 1 \\ -3 \end{pmatrix}$ and $\mathbf{r} = t\begin{pmatrix} 2 \\ 1 \end{pmatrix}$

b Only eigenvalue is $7^2 = 49$, eigenvectors are $\begin{pmatrix} 1 \\ -3 \end{pmatrix}$ and $\begin{pmatrix} 2 \\ 1 \end{pmatrix}$

5 a Eigenvalues are $3^4 = 81$, $2^4 = 16$, $(-4)^4 = 256$

Eigenvectors are $\begin{pmatrix} 1 \\ 0 \\ 1 \end{pmatrix}$, $\begin{pmatrix} 2 \\ -3 \\ 0 \end{pmatrix}$, $\begin{pmatrix} 1 \\ 1 \\ 2 \end{pmatrix}$

b $\mathbf{r} = s\begin{pmatrix} 1 \\ 0 \\ 1 \end{pmatrix}$, $\mathbf{r} = t\begin{pmatrix} 2 \\ -3 \\ 0 \end{pmatrix}$ and $\mathbf{r} = u\begin{pmatrix} 1 \\ 1 \\ 2 \end{pmatrix}$

6 a $\begin{pmatrix} 101 & -37 & 0 \\ 74 & -10 & 0 \\ 0 & 0 & 1 \end{pmatrix}$

b $\begin{pmatrix} 2(4^n) - 3^n & 3^n - 4^n & 0 \\ 2(4^n) - 2(3^n) & 2(3^n) - 4^n & 0 \\ 0 & 0 & 1 \end{pmatrix}$

7 a $x = -2y$ or $y = -\dfrac{1}{2}x$, $y = x$

b $y = x$ is a line of invariant points since corresponding eigenvalue is one.

8 $\lambda = a \pm b$, corresponding eigenvectors are $\begin{pmatrix} 1 \\ 1 \end{pmatrix}$, $\begin{pmatrix} 1 \\ -1 \end{pmatrix}$

9 a $\lambda = -3, 1$, corresponding set of eigenvectors are $\alpha\begin{pmatrix} 2 \\ 1 \end{pmatrix}$, $\beta\begin{pmatrix} 3 \\ 2 \end{pmatrix}$

b $\begin{pmatrix} 2 \\ 1 \end{pmatrix}$ gives the direction of an invariant line through the origin and $\begin{pmatrix} 3 \\ 2 \end{pmatrix}$ gives the direction of a line of invariant points.

10 a $\lambda = -1, 3, 1$

b Eigenvectors are of the form $\alpha\begin{pmatrix} 2 \\ 1 \\ -1 \end{pmatrix}$, $\beta\begin{pmatrix} 1 \\ 0 \\ 0 \end{pmatrix}$, $\gamma\begin{pmatrix} 1 \\ 0 \\ 1 \end{pmatrix}$

c The eigenvector $\begin{pmatrix} 1 \\ 0 \\ 1 \end{pmatrix}$ gives the direction of a line of invariant points through the origin.

The eigenvectors $\begin{pmatrix} 1 \\ 0 \\ 0 \end{pmatrix}$ and $\begin{pmatrix} 2 \\ 1 \\ -1 \end{pmatrix}$ gives the directions of two invariant lines through the origin.

11 a $\lambda = 2, 2, -1$

b Set of eigenvectors of the form $\alpha\begin{pmatrix} 0 \\ 1 \\ 2 \end{pmatrix} + \beta\begin{pmatrix} 1 \\ 0 \\ -1 \end{pmatrix}$, $\lambda\begin{pmatrix} 1 \\ 1 \\ -2 \end{pmatrix}$

c The set of eigenvectors $\alpha\begin{pmatrix} 0 \\ 1 \\ 2 \end{pmatrix} + \beta\begin{pmatrix} 1 \\ 0 \\ -1 \end{pmatrix}$ represent an invariant plane.

The eigenvector $\begin{pmatrix} 1 \\ 1 \\ -2 \end{pmatrix}$ gives the direction of an invariant line through the origin.

12 a $\det\begin{pmatrix} 1-\lambda & 1 & 2 \\ 0 & 1-\lambda & -1 \\ 0 & 3 & -\lambda \end{pmatrix}=0$

$\Rightarrow (1-\lambda)(-\lambda(1-\lambda)+3)=0$

$\Rightarrow (1-\lambda)(\lambda^2-\lambda+3)=0$

$\Rightarrow 1-\lambda=0$ or $\lambda^2-\lambda+3=0$

$\lambda=1$ is a real eigenvalue

$\lambda^2-\lambda+3=0$ has no real solutions since discriminant
$=(-1)^2-4(1)(3)=-11<0$

So only one real eigenvalue

b $\begin{pmatrix} 1 & 1 & 2 \\ 0 & 1 & -1 \\ 0 & 3 & 0 \end{pmatrix}\begin{pmatrix} x \\ y \\ z \end{pmatrix}=\begin{pmatrix} x \\ y \\ z \end{pmatrix}$

$x+y+2z=x \Rightarrow y=-2z$

$y-z=y \Rightarrow z=0$

$\Rightarrow y=0$

So eigenvector is $\begin{pmatrix} 1 \\ 0 \\ 0 \end{pmatrix}$

So, since eigenvalue is 1, $\mathbf{r}=s\begin{pmatrix} 1 \\ 0 \\ 0 \end{pmatrix}$ is a line of invariant

points under the transformation represented by **M**, and this is the equation of the x-axis.

13 a $\mathbf{M}=\mathbf{U}\mathbf{D}\mathbf{U}^{-1}\Rightarrow \mathbf{M}^n=(\mathbf{U}\mathbf{D}\mathbf{U}^{-1})^n$

$=(\mathbf{U}\mathbf{D}\mathbf{U}^{-1})(\mathbf{U}\mathbf{D}\mathbf{U}^{-1})(\mathbf{U}\mathbf{D}\mathbf{U}^{-1})\ldots(\mathbf{U}\mathbf{D}\mathbf{U}^{-1})$

$=(\mathbf{U}\mathbf{D})(\mathbf{U}^{-1}\mathbf{U})\mathbf{D}(\mathbf{U}^{-1}\mathbf{U})\mathbf{D}\ldots(\mathbf{U}^{-1}\mathbf{U})(\mathbf{D}\mathbf{U}^{-1})$

$=(\mathbf{U}\mathbf{D})\mathbf{I}\mathbf{D}\mathbf{I}\mathbf{D}\ldots\mathbf{I}(\mathbf{D}\mathbf{U}^{-1})$

$=(\mathbf{U}\mathbf{D})\mathbf{D}\mathbf{D}\ldots(\mathbf{D}\mathbf{U}^{-1})$

$=\mathbf{U}\mathbf{D}^n\mathbf{U}^{-1}$ as required

b $\begin{pmatrix} 0 & -\dfrac{1}{2} \\ 0 & 1 \end{pmatrix}$

14 a $\lambda=\pm k$, corresponding eigenvectors are $\begin{pmatrix} k+3 \\ k-3 \end{pmatrix}, \begin{pmatrix} 1 \\ -1 \end{pmatrix}$

b $\begin{pmatrix} k+3 & 1 \\ k-3 & -1 \end{pmatrix}^{-1}=\dfrac{1}{2k}\begin{pmatrix} 1 & 1 \\ k-3 & -k-3 \end{pmatrix}$

$\mathbf{M}=\dfrac{1}{2k}\begin{pmatrix} 3+k & 1 \\ k-3 & -1 \end{pmatrix}\begin{pmatrix} k & 0 \\ 0 & -k \end{pmatrix}\begin{pmatrix} 1 & 1 \\ k-3 & -k-3 \end{pmatrix}$

Therefore,

$\mathbf{M}^{2n+1}=\dfrac{1}{2k}\begin{pmatrix} k+3 & 1 \\ k-3 & -1 \end{pmatrix}\begin{pmatrix} k & 0 \\ 0 & -k \end{pmatrix}^{2n+1}\begin{pmatrix} 1 & 1 \\ k-3 & -k-3 \end{pmatrix}$

$=\dfrac{1}{2k}\begin{pmatrix} k+3 & 1 \\ k-3 & -1 \end{pmatrix}\begin{pmatrix} k^{2n+1} & 0 \\ 0 & (-k)^{2n+1} \end{pmatrix}\begin{pmatrix} 1 & 1 \\ k-3 & -k-3 \end{pmatrix}$

$=\dfrac{k^{2n+1}}{2k}\begin{pmatrix} k+3 & 1 \\ k-3 & -1 \end{pmatrix}\begin{pmatrix} 1 & 0 \\ 0 & -1 \end{pmatrix}\begin{pmatrix} 1 & 1 \\ k-3 & -k-3 \end{pmatrix}$

$=\dfrac{k^{2n}}{2}\begin{pmatrix} k+3 & -1 \\ k-3 & 1 \end{pmatrix}\begin{pmatrix} 1 & 1 \\ k-3 & -k-3 \end{pmatrix}$

$=\dfrac{k^{2n}}{2}\begin{pmatrix} (k+3)-(k-3) & (k+3)-(-k-3) \\ (k-3)+(k-3) & (k-3)+(-k-3) \end{pmatrix}$

$=\dfrac{k^{2n}}{2}\begin{pmatrix} 6 & 2(k+3) \\ 2(k-3) & -6 \end{pmatrix}$

$=k^{2n}\begin{pmatrix} 3 & k+3 \\ k-3 & -3 \end{pmatrix}$

$=k^{2n}\mathbf{M}$ as required

Review exercise 22

1 a i 200 square units **ii** 80 square units

  iii 8 square units **iv** $128a^2$ square units

b ii, iv determinants negative

2 a -63 **b** $-44a-21b$

3 a $x=-\dfrac{4}{3}$ **b** $x=\dfrac{-5\pm\sqrt{37}}{2}$

4 a i 4 cubic units **ii** $\dfrac{7}{4a}$ cubic units

b Determinant negative

5 a $\dfrac{1}{25}\begin{pmatrix} -5 & 5 & 10 \\ 22 & -7 & -19 \\ 4 & 1 & -8 \end{pmatrix}$

b $\det\begin{pmatrix} 6 & -2 & -3 \\ 1 & 0 & -1 \\ 0 & 2 & -3 \end{pmatrix}=6(0--2)+2(-3-0)-3(2-0)=0$

therefore matrix is singular so inverse does not exist.

6 $x=2, y=-5, z=-10$

7 a $k=-\dfrac{5}{2}, 1$

b When $k=1$, the first and third planes are parallel hence do not intersect.

When $k=-\dfrac{5}{2}$, there is an inconsistency, hence the three planes form a triangular prism.

8 $\begin{vmatrix} 3 & -2 & 5 \\ -6 & 4 & 0 \\ 5 & -1 & -3 \end{vmatrix}=5(0-20)--1(0--30)+(12--12)$

$=-100+30+70=-70$

$\begin{vmatrix} 3 & -2 & 5 \\ -6 & 4 & 0 \\ 5 & -1 & -3 \end{vmatrix}=5(6-20)-0+3(12-12)$

$=-70+0=-70$

9 a -90 **b** -45 **c** 90 **d** $\dfrac{-45}{b}$

10 a $\begin{vmatrix} 1 & -1 & 1 \\ a & a^2 & -a^2 \\ b & b^2 & -b^2 \end{vmatrix}=\begin{vmatrix} 1 & -1 & 1 \\ a+1 & a^2-1 & 1-a^2 \\ b+1 & b^2-1 & 1-b^2 \end{vmatrix}$

$=(a+1)(b+1)\begin{vmatrix} 1 & -1 & 1 \\ 1 & a-1 & 1-a \\ 1 & b-1 & 1-b \end{vmatrix}$

$=(a+1)(b+1)\begin{vmatrix} 1 & -1 & 1 \\ 0 & a & -a \\ 0 & b & -b \end{vmatrix}$

$=ab(a+1)(b+1)\begin{vmatrix} 1 & -1 & 1 \\ 0 & 1 & -1 \\ 0 & 1 & -1 \end{vmatrix}$

$=ab(a+1)(b+1)\left(1\begin{vmatrix} 1 & -1 \\ 1 & -1 \end{vmatrix}-0+0\right)$

$=ab(a+1)(b+1)\begin{vmatrix} 1 & -1 \\ 1 & -1 \end{vmatrix}$

b 0

11 $(a-b)(c-a)(b-c)(ac+ab+bc)$

12 a $\lambda = 2,-7$, corresponding eigenvectors are $\begin{pmatrix} 1 \\ 1 \end{pmatrix}, \begin{pmatrix} 5 \\ -4 \end{pmatrix}$

b $\lambda = -9,-27$, corresponding eigenvectors are $\begin{pmatrix} 2 \\ 1 \end{pmatrix}, \begin{pmatrix} -1 \\ 1 \end{pmatrix}$

13 a $\lambda = 2, 1,-1$, corresponding eigenvectors are

$$\begin{pmatrix} 1 \\ 8 \\ -5 \end{pmatrix}, \begin{pmatrix} 0 \\ 1 \\ -1 \end{pmatrix}, \begin{pmatrix} 0 \\ 1 \\ -2 \end{pmatrix}$$

b $\lambda = 1, 9, 10$, corresponding eigenvectors are

$$\begin{pmatrix} -3 \\ 0 \\ 1 \end{pmatrix}, \begin{pmatrix} 13 \\ -8 \\ 9 \end{pmatrix}, \begin{pmatrix} 3 \\ 0 \\ 2 \end{pmatrix}$$

14 a Characteristic equation is $\det(\mathbf{A} - \lambda\mathbf{I}) = 0$

So, in this case: $(a-\lambda)(2a-\lambda)+6 = 0$

$\Rightarrow \lambda^2 - 3a\lambda + 2a^2 + 6 = 0$ as required

b $\mathbf{A}^2 - 3a\mathbf{A} + (2a^2+6)\mathbf{I}$

$$= \begin{pmatrix} a & 3 \\ -2 & 2a \end{pmatrix}^2 - 3a\begin{pmatrix} a & 3 \\ -2 & 2a \end{pmatrix} + (2a^2+6)\begin{pmatrix} 1 & 0 \\ 0 & 1 \end{pmatrix}$$

$$= \begin{pmatrix} a^2-6 & 9a \\ -6a & 4a^2-6 \end{pmatrix} - \begin{pmatrix} 3a^2 & 9a \\ -6a & 6a^2 \end{pmatrix}$$

$$+ \begin{pmatrix} 2a^2+6 & 0 \\ 0 & 2a^2+6 \end{pmatrix}$$

$$= \begin{pmatrix} a^2-6-3a^2+2a^2+6 & 9a-9a+0 \\ -6a+6a+0 & 4a^2-6-6a^2+2a^2+6 \end{pmatrix}$$

$$= \begin{pmatrix} 0 & 0 \\ 0 & 0 \end{pmatrix}$$ as required

15 a $\det\begin{pmatrix} -1-\lambda & 0 & 2 \\ -1 & 2-\lambda & 0 \\ -4 & 0 & 3-\lambda \end{pmatrix} = (2-\lambda)[(-1-\lambda)(3-\lambda)--8]$

$(2-\lambda)(\lambda^2-2\lambda+5) = 0 \Rightarrow \lambda = 2 \text{ or } \lambda^2-2\lambda+5 = 0$

Discriminant is $(-2)^2 - 4(5) = -16 < 0$ so no real solutions.

b $\begin{pmatrix} 0 \\ 1 \\ 0 \end{pmatrix}$

16 $\mathbf{P} = \begin{pmatrix} 0 & 4 & -1 \\ 0 & -10 & 0 \\ 1 & 1 & 1 \end{pmatrix}$, $\mathbf{D} = \begin{pmatrix} 0 & 0 & 0 \\ 0 & 2 & 0 \\ 0 & 0 & -3 \end{pmatrix}$

17 Eigenvalues are $\lambda = 1, 0, 3$

Eigenvectors are $\begin{pmatrix} 0 \\ 0 \\ 1 \end{pmatrix}, \begin{pmatrix} 1 \\ -1 \\ 2 \end{pmatrix}, \begin{pmatrix} 2 \\ 1 \\ 1 \end{pmatrix}$

$$\mathbf{M}^n = \frac{1}{3}\begin{pmatrix} 0 & 1 & 2 \\ 0 & -1 & 1 \\ 1 & 2 & 1 \end{pmatrix}\begin{pmatrix} 1 & 0 & 0 \\ 0 & 0 & 0 \\ 0 & 0 & 3 \end{pmatrix}^n \begin{pmatrix} -3 & -3 & 3 \\ 1 & -2 & 0 \\ 1 & 1 & 0 \end{pmatrix}$$

$$= \frac{1}{3}\begin{pmatrix} 0 & 1 & 2 \\ 0 & -1 & 1 \\ 1 & 2 & 1 \end{pmatrix}\begin{pmatrix} 1 & 0 & 0 \\ 0 & 0 & 0 \\ 0 & 0 & 3^n \end{pmatrix}\begin{pmatrix} -3 & -3 & 3 \\ 1 & -2 & 0 \\ 1 & 1 & 0 \end{pmatrix}$$

$$= \frac{1}{3}\begin{pmatrix} 0 & 0 & 2(3^n) \\ 0 & 0 & 3^n \\ 1 & 0 & 3^n \end{pmatrix}\begin{pmatrix} -3 & -3 & 3 \\ 1 & -2 & 0 \\ 1 & 1 & 0 \end{pmatrix}$$

$$= \frac{1}{3}\begin{pmatrix} 2(3^n) & 2(3^n) & 0 \\ 3^n & 3^n & 0 \\ 3^n-3 & 3^n+3 & 3 \end{pmatrix}$$

$$= \begin{pmatrix} 2(3^{n-1}) & 2(3^{n-1}) & 0 \\ 3^{n-1} & 3^{n-1} & 0 \\ 3^{n-1}-1 & 3^{n-1}+1 & 1 \end{pmatrix}$$

18 a $\mathbf{r} = s\begin{pmatrix} 3 \\ -1 \end{pmatrix}$

b $\mathbf{r} = t\begin{pmatrix} 2 \\ 3 \end{pmatrix}$

19 a $\lambda = -3, 2$ (repeated)

b Eigenvectors are of the form $\alpha\begin{pmatrix} -1 \\ 2 \\ 1 \end{pmatrix}$ or the form

$\beta\begin{pmatrix} 2 \\ 0 \\ -1 \end{pmatrix} + \gamma\begin{pmatrix} 1 \\ 1 \\ 0 \end{pmatrix}$

c The eigenvector $\begin{pmatrix} -1 \\ 2 \\ 1 \end{pmatrix}$ gives the direction of an invariant line through the origin.

Eigenvectors of the form $\beta\begin{pmatrix} 2 \\ 0 \\ -1 \end{pmatrix} + \gamma\begin{pmatrix} 1 \\ 1 \\ 0 \end{pmatrix}$ define an invariant plane.

Assessment 22

1. $(2, 67, 50)$

2 a $\pm\dfrac{\sqrt{3}}{2}$ **b** Rotation of $\pm 60°$ around the x-axis

3 $k < -\sqrt{3}, k > \sqrt{3}$

4 a $-2, 8$; corresponding eigenvectors are $\begin{pmatrix} 3 \\ 4 \end{pmatrix}, \begin{pmatrix} 1 \\ -2 \end{pmatrix}$

b $\begin{pmatrix} 2 & -3 \\ -8 & 4 \end{pmatrix} - 6\begin{pmatrix} 2 & -3 \\ -8 & 4 \end{pmatrix} - 16\begin{pmatrix} 1 & 0 \\ 0 & 1 \end{pmatrix}$

$= \begin{pmatrix} 28 & -18 \\ -48 & 40 \end{pmatrix} - \begin{pmatrix} 12 & -18 \\ -48 & 24 \end{pmatrix} - \begin{pmatrix} 16 & 0 \\ 0 & 16 \end{pmatrix}$

$= \begin{pmatrix} 0 & 0 \\ 0 & 0 \end{pmatrix}$ as required

5 a 1.25 and -2

b The equations are represented by the matrix $\mathbf{M}$ with $k = 1$. When $k \neq 1.25$ or -2, $\mathbf{M}$ is not singular. Therefore the equations have a unique solution so the planes have a unique point of intersection.

6 When $n=1$, $A^n = \begin{pmatrix} 2 & 0 & 1 \\ 0 & 1 & 0 \\ 1 & 0 & 2 \end{pmatrix}^1 = \begin{pmatrix} 2 & 0 & 1 \\ 0 & 1 & 0 \\ 1 & 0 & 2 \end{pmatrix}$

and $\dfrac{1}{2}\begin{pmatrix} 3^1+1 & 0 & 3^1-1 \\ 0 & 2 & 0 \\ 3^1-1 & 0 & 3^1+1 \end{pmatrix} = \begin{pmatrix} 2 & 0 & 1 \\ 0 & 1 & 0 \\ 1 & 0 & 2 \end{pmatrix}$

So true for $n=1$

Assume true for $n=k$ and let $n=k+1$

$A^{k+1} = \begin{pmatrix} 2 & 0 & 1 \\ 0 & 1 & 0 \\ 1 & 0 & 2 \end{pmatrix}^{k+1}$

$= \begin{pmatrix} 2 & 0 & 1 \\ 0 & 1 & 0 \\ 1 & 0 & 2 \end{pmatrix}^{k}\begin{pmatrix} 2 & 0 & 1 \\ 0 & 1 & 0 \\ 1 & 0 & 2 \end{pmatrix}$

$= \dfrac{1}{2}\begin{pmatrix} 3^k+1 & 0 & 3^k-1 \\ 0 & 2 & 0 \\ 3^k-1 & 0 & 3^k+1 \end{pmatrix}\begin{pmatrix} 2 & 0 & 1 \\ 0 & 1 & 0 \\ 1 & 0 & 2 \end{pmatrix}$

$= \dfrac{1}{2}\begin{pmatrix} 2(3^k+1)+3^k-1 & 0 & 3^k+1+2(3^k-1) \\ 0 & 2 & 0 \\ 2(3^k-1)+3^k+1 & 0 & 3^k-1+2(3^k+1) \end{pmatrix}$

$= \dfrac{1}{2}\begin{pmatrix} 3(3^k)+1 & 0 & 3(3^k)-1 \\ 0 & 2 & 0 \\ 3(3^k)-1 & 0 & 3(3^k)+1 \end{pmatrix}$

$= \dfrac{1}{2}\begin{pmatrix} 3^{k+1}+1 & 0 & 3^{k+1}-1 \\ 0 & 2 & 0 \\ 3^{k+1}-1 & 0 & 3^{k+1}+1 \end{pmatrix}$

So true for $n=k$

Since true for $n=1$ and assuming true for $n=k$ implies true for $n=k+1$, therefore true for all $n \in \mathbb{N}$

7 For example, add C3 to C2

$\begin{vmatrix} 2 & 1 & 2 \\ 4 & 1 & 2 \\ 1 & -1 & 1 \end{vmatrix} = \begin{vmatrix} 2 & 3 & 2 \\ 4 & 3 & 2 \\ 1 & 0 & 1 \end{vmatrix}$

multiply R3 by 2

$= \dfrac{1}{2}\begin{vmatrix} 2 & 3 & 2 \\ 4 & 3 & 2 \\ 2 & 0 & 2 \end{vmatrix}$

subtract R3 from R1

$= \dfrac{1}{2}\begin{vmatrix} 0 & 3 & 0 \\ 4 & 3 & 2 \\ 2 & 0 & 2 \end{vmatrix}$

divide C2 by 3

$= \dfrac{3}{2}\begin{vmatrix} 0 & 1 & 0 \\ 4 & 1 & 2 \\ 2 & 0 & 2 \end{vmatrix}$

divide C1 by 2

$= 3\begin{vmatrix} 0 & 1 & 0 \\ 2 & 1 & 2 \\ 1 & 0 & 2 \end{vmatrix}$

divide C3 by 2

$= 6\begin{vmatrix} 0 & 1 & 0 \\ 2 & 1 & 1 \\ 1 & 0 & 1 \end{vmatrix}$

subtract R3 from R2

$= 6\begin{vmatrix} 0 & 1 & 0 \\ 1 & 1 & 0 \\ 1 & 0 & 1 \end{vmatrix}$

8 $\begin{pmatrix} \dfrac{9}{a} & \dfrac{1}{2} & -3 \\ \dfrac{12}{a} & 1 & 2 \\ \dfrac{3}{a} & 0 & -1 \end{pmatrix}$

9 a 180 **b** -30

10 a $(b-a)(c-a)(c-b)$ **b** 2

11 a $\det\begin{pmatrix} 1-\lambda & 2 & 0 \\ 0 & -\lambda & 3 \\ 2 & 0 & 1-\lambda \end{pmatrix} = 0$

$(1-\lambda)(-\lambda)(1-\lambda)-2(0-6)+0=0$

$-\lambda^3+2\lambda^2-\lambda+12=0$

$(\lambda-3)(-\lambda^2-\lambda-4)=0$

$\lambda=3$ or $-\lambda^2-\lambda-4=0$

$b^2-4ac = (-1)^2-4(-1)(-4)=-15$

So no further real solutions.

Hence 3 is the only real eigenvalue.

b $\begin{pmatrix} \dfrac{1}{\sqrt{3}} \\ \dfrac{1}{\sqrt{3}} \\ \dfrac{1}{\sqrt{3}} \end{pmatrix}$ **c** $\mathbf{r}=s\begin{pmatrix} 1 \\ 1 \\ 1 \end{pmatrix}$

12 $2, -1$ (repeated)

13 a $\mathbf{P}=\begin{pmatrix} 1 & 3 & 0 \\ -2 & 2 & 1 \\ 1 & -3 & 0 \end{pmatrix}$ $\mathbf{D}=\begin{pmatrix} 2 & 0 & 0 \\ 0 & -2 & 0 \\ 0 & 0 & 1 \end{pmatrix}$

b $\mathbf{P}^{-1}=\dfrac{1}{6}\begin{pmatrix} 3 & 0 & 3 \\ 1 & 0 & -1 \\ 4 & 6 & 8 \end{pmatrix}$

$\mathbf{M}^4 = \mathbf{P}\mathbf{D}^4\mathbf{P}^{-1}$

$= \begin{pmatrix} 1 & 3 & 0 \\ -2 & 2 & 1 \\ 1 & -3 & 0 \end{pmatrix}\begin{pmatrix} 2^4 & 0 & 0 \\ 0 & (-2)^4 & 0 \\ 0 & 0 & 1^4 \end{pmatrix}\dfrac{1}{6}\begin{pmatrix} 3 & 0 & 3 \\ 1 & 0 & -1 \\ 4 & 6 & 8 \end{pmatrix}$

$= \begin{pmatrix} 16 & 0 & 0 \\ -10 & 1 & -20 \\ 0 & 0 & 16 \end{pmatrix}$

c The eigenvector $\begin{pmatrix} 0 \\ 1 \\ 0 \end{pmatrix}$ which represents the direction of the y-axis corresponds to an eigenvalue of 1

14 First equation $\times 2$ gives: $2x + 4y - 6z = -4$

Add to third equation to give: $4x - z = -3$

But $4x - z = 3$ so inconsistent.

Hence they do not intersect.

The three planes form a triangular prism.

15 $\begin{pmatrix} 1 \\ 1 \\ 0 \end{pmatrix}$ gives the direction of an invariant line, $\begin{pmatrix} 0 \\ -1 \\ 1 \end{pmatrix}, \begin{pmatrix} 1 \\ 2 \\ -2 \end{pmatrix}$

define an invariant plane.

Chapter 23

Exercise 23.1A

1 **a** $\mathbf{j} \times \mathbf{k} = \mathbf{i}$

   b $\mathbf{i} \times \mathbf{k} = -\mathbf{k} \times \mathbf{i} = -\mathbf{j}$

   c $\mathbf{k} \times \mathbf{k} = 0$

   d $(\mathbf{i} + \mathbf{j}) \times \mathbf{j} = \mathbf{i} \times \mathbf{j} + \mathbf{j} \times \mathbf{j} = \mathbf{k}$

2 **a** $\mathbf{a} \times \mathbf{b}$

   b $\mathbf{a} \times (\mathbf{b} + \mathbf{c})$

   c $-2\mathbf{a} \times \mathbf{b}$

   d $2\mathbf{b} \times (\mathbf{a} + \mathbf{c})$

3 **a** $\mathbf{a} \times \mathbf{b} + \mathbf{b} \times \mathbf{c} + \mathbf{a} \times \mathbf{c} = -\mathbf{b} \times \mathbf{a} + \mathbf{b} \times \mathbf{c} + \mathbf{a} \times \mathbf{c}$

   $= \mathbf{b} \times (\mathbf{c} - \mathbf{a}) + \mathbf{a} \times \mathbf{c}$

   $= (\mathbf{b} + \mathbf{a}) \times (\mathbf{c} - \mathbf{a})$ (since $\mathbf{a} \times \mathbf{a} = 0$)

   b $\mathbf{b} \times (2\mathbf{a} + \mathbf{c}) - \mathbf{c} \times (\mathbf{b} - \mathbf{c}) = \mathbf{b} \times (2\mathbf{a}) + \mathbf{b} \times \mathbf{c} - \mathbf{c} \times \mathbf{b} - \mathbf{c} \times (-\mathbf{c})$

   $= 2\mathbf{b} \times \mathbf{a} + \mathbf{b} \times \mathbf{c} + \mathbf{b} \times \mathbf{c} + \mathbf{c} \times \mathbf{c}$

   $= 2\mathbf{b} \times \mathbf{a} + 2\mathbf{b} \times \mathbf{c}$

   $= 2\mathbf{b} \times (\mathbf{a} + \mathbf{c})$

4 $0°$ or $180°$

5 **a** $\mathbf{i} + 5\mathbf{j} + 13\mathbf{k}$

   b $-15\mathbf{i} - 9\mathbf{j} + 3\mathbf{k}$

   c $26\mathbf{i} - 2\mathbf{j} + 8\mathbf{k}$

   d $-\mathbf{i} - 5\mathbf{j} - 13\mathbf{k}$

   e $15\mathbf{i} + 9\mathbf{j} - 3\mathbf{k}$

   f $-26\mathbf{i} + 2\mathbf{j} - 8\mathbf{k}$

   g 0

   h 0

6 **a** $\begin{pmatrix} 3 \\ -15 \\ -6 \end{pmatrix}$

   b $25\mathbf{i} - 8\mathbf{j} - 26\mathbf{k}$

   c $3\mathbf{i} + 8\mathbf{j} + 28\mathbf{k}$

   d $\begin{pmatrix} -22 \\ 8 \\ -43 \end{pmatrix}$

   e $\begin{pmatrix} -4\sqrt{3} \\ 3 \\ 3 \end{pmatrix}$

   f $\begin{pmatrix} 10a^2 + 1 \\ 5a^2 - a \\ 2a^2 + a \end{pmatrix}$

7 $\begin{pmatrix} 29 \\ -5 \\ 7 \end{pmatrix}$

8 $9\mathbf{i} - 25\mathbf{j} - 21\mathbf{k}$

9 $\begin{pmatrix} 0 \\ -\dfrac{3}{\sqrt{10}} \\ -\dfrac{1}{\sqrt{10}} \end{pmatrix}$

10 $\begin{pmatrix} \dfrac{1}{\sqrt{6}} \\ \dfrac{2}{\sqrt{6}} \\ \dfrac{1}{\sqrt{6}} \end{pmatrix}$

11 $(-2\mathbf{i} + 6\mathbf{j} - 4\mathbf{k}) \times (3\mathbf{i} - 9\mathbf{j} + 6\mathbf{k})$

$= (6 \times 6 - -4 \times -9)\mathbf{i} - (-2 \times 6 - -4 \times 3)\mathbf{j} + (-2 \times -9 - 6 \times 3)\mathbf{k}$

$= 0\mathbf{i} - 0\mathbf{j} + 0\mathbf{k}$

$= 0$ so parallel as required

12 **a** $29.6°$

   b $85.8°$

   c $40.9°$

   d $\theta = 41.4$

13 $\dfrac{3}{\sqrt{14}}$

14 $a = 4$

   $b = 5$

   $c = -20$

15 $b = -3$

   $a = 1$

   $c = 6$

16 $\dfrac{3}{2}\sqrt{35}$ square units

17 10.4 square units

18 **a** $\dfrac{17\sqrt{3}}{2}$

   b $\dfrac{\sqrt{2}}{2}$

Exercise 23.1B

1 Let $\mathbf{a} = \begin{pmatrix} a_1 \\ a_2 \\ a_3 \end{pmatrix}, \mathbf{b} = \begin{pmatrix} b_1 \\ b_2 \\ b_3 \end{pmatrix}, \mathbf{c} = \begin{pmatrix} c_1 \\ c_2 \\ c_3 \end{pmatrix}$

$\mathbf{a} \times (\mathbf{b} + \mathbf{c}) = \begin{pmatrix} a_1 \\ a_2 \\ a_3 \end{pmatrix} \times \begin{pmatrix} b_1 + c_1 \\ b_2 + c_2 \\ b_3 + c_3 \end{pmatrix}$

$= \begin{pmatrix} a_2(b_3 + c_3) - a_3(b_2 + c_2) \\ a_3(b_1 + c_1) - a_1(b_3 + c_3) \\ a_1(b_2 + c_2) - a_2(b_1 + c_1) \end{pmatrix}$

$= \begin{pmatrix} (a_2 b_3 - a_3 b_2) + (a_2 c_3 - a_3 c_2) \\ (a_3 b_1 - a_1 b_3) + (a_3 c_1 - a_1 c_3) \\ (a_1 b_2 - a_2 b_1) + (a_1 c_2 - a_2 c_1) \end{pmatrix}$

$= \begin{pmatrix} a_2 b_3 - a_3 b_2 \\ a_3 b_1 - a_1 b_3 \\ a_1 b_2 - a_2 b_1 \end{pmatrix} + \begin{pmatrix} a_2 c_3 - a_3 c_2 \\ a_3 c_1 - a_1 c_3 \\ a_1 c_2 - a_2 c_1 \end{pmatrix}$

$= \mathbf{a} \times \mathbf{b} + \mathbf{a} \times \mathbf{c}$ as required

2 $\mathbf{a} \times \mathbf{b} = \begin{pmatrix} a_1 \\ a_2 \\ a_3 \end{pmatrix} \times \begin{pmatrix} b_1 \\ b_2 \\ b_3 \end{pmatrix}$

$= \begin{pmatrix} a_2 b_3 - a_3 b_2 \\ a_3 b_1 - a_1 b_3 \\ a_1 b_2 - a_2 b_1 \end{pmatrix}$

$= -\begin{pmatrix} a_3 b_2 - a_2 b_3 \\ a_1 b_3 - a_3 b_1 \\ a_2 b_1 - a_1 b_2 \end{pmatrix}$

$= -\begin{pmatrix} b_1 \\ b_2 \\ b_3 \end{pmatrix} \times \begin{pmatrix} a_1 \\ a_2 \\ a_3 \end{pmatrix}$

$= -\mathbf{b} \times \mathbf{a}$ as required

3 $\left(\mathbf{r} - \begin{pmatrix} 2 \\ 5 \\ 0 \end{pmatrix} \right) \times \begin{pmatrix} 1 \\ 1 \\ -1 \end{pmatrix} = 0$

4 $\left(\mathbf{r} - \begin{pmatrix} 0 \\ 3 \\ 0 \end{pmatrix} \right) \times \begin{pmatrix} 0 \\ 0 \\ 1 \end{pmatrix} = 0$

5 $\mathbf{r} \times \begin{pmatrix} 1 \\ 0 \\ 1 \end{pmatrix} = \begin{pmatrix} 0 \\ -2 \\ 0 \end{pmatrix}$

6 $\left(\mathbf{r} - \begin{pmatrix} 4 \\ 2 \\ 7 \end{pmatrix} \right) \times \begin{pmatrix} 3 \\ -1 \\ 2 \end{pmatrix} = 0$ or $\left(\mathbf{r} - \begin{pmatrix} 1 \\ 3 \\ 5 \end{pmatrix} \right) \times \begin{pmatrix} 3 \\ -1 \\ 2 \end{pmatrix} = 0$

7 e.g. $\begin{pmatrix} 0 \\ 3 \\ -2 \end{pmatrix} - \begin{pmatrix} 1 \\ 2 \\ -1 \end{pmatrix} = \begin{pmatrix} -1 \\ 1 \\ -1 \end{pmatrix}$ (or could subtract other way round)

$\mathbf{r} \times \begin{pmatrix} -1 \\ 1 \\ -1 \end{pmatrix} = \begin{pmatrix} 0 \\ 3 \\ -2 \end{pmatrix} \times \begin{pmatrix} -1 \\ 1 \\ -1 \end{pmatrix}$

$\mathbf{r} \times \begin{pmatrix} -1 \\ 1 \\ -1 \end{pmatrix} = \begin{pmatrix} -1 \\ 2 \\ 3 \end{pmatrix}$

Alternatively: $\mathbf{r} \times \begin{pmatrix} 1 \\ -1 \\ 1 \end{pmatrix} = \begin{pmatrix} 1 \\ -2 \\ -3 \end{pmatrix}$

8 a $\left(\mathbf{r} - \begin{pmatrix} 2 \\ 0 \\ -1 \end{pmatrix} \right) \times \begin{pmatrix} 0 \\ 1 \\ 5 \end{pmatrix} = 0$

b $\left(\mathbf{r} - \begin{pmatrix} 3 \\ 1 \\ -4 \end{pmatrix} \right) \times \begin{pmatrix} 5 \\ 3 \\ -6 \end{pmatrix} = 0$

c $\left(\mathbf{r} - \begin{pmatrix} 1 \\ -3 \\ 0 \end{pmatrix} \right) \times \begin{pmatrix} -2 \\ 4 \\ -2 \end{pmatrix} = 0$

d $\left(\mathbf{r} - \begin{pmatrix} 0 \\ 0 \\ 1 \end{pmatrix} \right) \times \begin{pmatrix} 1 \\ -3 \\ 2 \end{pmatrix} = 0$

9 $\begin{pmatrix} -8 \\ -6 \\ 2 \end{pmatrix} = -2 \begin{pmatrix} 4 \\ 3 \\ -1 \end{pmatrix}$ so they are parallel

$\begin{pmatrix} 1 \\ 3 \\ 5 \end{pmatrix}$ is on line A so check B: $\begin{pmatrix} 1 \\ 3 \\ 5 \end{pmatrix} = \begin{pmatrix} 5 \\ 6 \\ 4 \end{pmatrix} + \lambda \begin{pmatrix} -8 \\ -6 \\ 2 \end{pmatrix}$

$1 = 5 - 8\lambda \Rightarrow \lambda = \dfrac{1}{2}$

$3 = 6 - 6\lambda \Rightarrow \lambda = \dfrac{1}{2}$

$5 = 4 + 2\lambda \Rightarrow \lambda = \dfrac{1}{2}$

Therefore (1, 3, 5) satisfies both equations so they do represent the same line.

10 $\begin{pmatrix} -3 \\ 6 \\ -9 \end{pmatrix} = \dfrac{3}{2} \begin{pmatrix} -2 \\ 4 \\ -6 \end{pmatrix}$ so they are parallel

(4, −3, 8) satisfies B so check A:

$\dfrac{4-3}{-2} = -\dfrac{1}{2}$

$\dfrac{-3+1}{4} = -\dfrac{1}{2}$

$\dfrac{8-5}{-6} = -\dfrac{1}{2}$ so (4, −3, 8) satisfies both equations therefore they represent the same line.

11 $\dfrac{x-1}{1} = \dfrac{y-3}{0} = \dfrac{z+1}{-1}$

12 $\mathbf{r} = \begin{pmatrix} 2 \\ 1 \\ 1 \end{pmatrix} + t \begin{pmatrix} -2 \\ -1 \\ 0 \end{pmatrix}$

13 a $\begin{pmatrix} 1 \\ -2 \\ 1 \end{pmatrix} + s \begin{pmatrix} 2 \\ 1 \\ -2 \end{pmatrix} = \begin{pmatrix} -3 \\ -11 \\ 5 \end{pmatrix} + t \begin{pmatrix} -1 \\ 3 \\ 1 \end{pmatrix}$

$i : 1 + 2s = -3 - t \Rightarrow t = -4 - 2s$

$j : -2 + s = -11 + 3t \Rightarrow 3t = 9 + s$

$\Rightarrow 3(-4 - 2s) = 9 + s$

$\Rightarrow -12 - 6s = 9 + s$

$\Rightarrow 7s = -21$

$\Rightarrow s = -3$

$\Rightarrow t = -4 - 2(-3) = 2$

$k : 1 - 2s = 1 - 2(-3) = 7$

$5 + t = 5 + 2 = 7$ so they intersect when $t = 2$

$\begin{pmatrix} -3 \\ -11 \\ 5 \end{pmatrix} + 2 \begin{pmatrix} -1 \\ 3 \\ 1 \end{pmatrix} = \begin{pmatrix} -5 \\ -5 \\ 7 \end{pmatrix}$

They intersect at (−5, −5, 7)

b $\dfrac{7}{33}\sqrt{22}$

14 a $\left(\begin{pmatrix} -2 \\ 16 \\ -12 \end{pmatrix} - \begin{pmatrix} 0 \\ 4 \\ -2 \end{pmatrix}\right) \times \begin{pmatrix} 1 \\ -6 \\ 5 \end{pmatrix} = \begin{pmatrix} -2 \\ 12 \\ -10 \end{pmatrix} \times \begin{pmatrix} 1 \\ -6 \\ 5 \end{pmatrix}$

$= \begin{pmatrix} 60-60 \\ 10-10 \\ 12-12 \end{pmatrix} = \begin{pmatrix} 0 \\ 0 \\ 0 \end{pmatrix}$

as required so point A lies on the line

b $(1, -2, 3)$

c $3\sqrt{21}$

15 $\dfrac{-2}{\sqrt{105}}\mathbf{i} + \dfrac{1}{\sqrt{105}}\mathbf{j} + \dfrac{10}{\sqrt{105}}\mathbf{k}$

Exercise 23.2A

1 $\mathbf{r} = \begin{pmatrix} 1 \\ 5 \\ 2 \end{pmatrix} + s\begin{pmatrix} 1 \\ 1 \\ 0 \end{pmatrix} + t\begin{pmatrix} 0 \\ 1 \\ 2 \end{pmatrix}$

2 $\mathbf{r} = \begin{pmatrix} 0 \\ 6 \\ 2 \end{pmatrix} + s\begin{pmatrix} 2 \\ 0 \\ -3 \end{pmatrix} + t\begin{pmatrix} 5 \\ -2 \\ 4 \end{pmatrix}$

3 a $\mathbf{r} = \begin{pmatrix} 3 \\ 1 \\ 0 \end{pmatrix} + s\begin{pmatrix} 1 \\ -3 \\ 2 \end{pmatrix} + t\begin{pmatrix} 8 \\ 1 \\ -4 \end{pmatrix}$

b $\mathbf{n} = \begin{pmatrix} 1 \\ -3 \\ 2 \end{pmatrix} \times \begin{pmatrix} 8 \\ 1 \\ 4 \end{pmatrix} = \begin{pmatrix} 10 \\ 20 \\ 25 \end{pmatrix} = 5\begin{pmatrix} 2 \\ 4 \\ 5 \end{pmatrix}$

$\begin{pmatrix} 3 \\ 1 \\ 0 \end{pmatrix} \cdot \begin{pmatrix} 2 \\ 4 \\ 5 \end{pmatrix} = 6+4 = 10$

$\mathbf{r} \cdot (2\mathbf{i} + 4\mathbf{j} + 5\mathbf{k}) = 10$

c $2x + 4y + 5z - 10 = 0$

4 a $\mathbf{r} = \begin{pmatrix} -2 \\ 0 \\ 0 \end{pmatrix} + s\begin{pmatrix} 7 \\ 1 \\ 1 \end{pmatrix} + t\begin{pmatrix} 8 \\ 2 \\ -5 \end{pmatrix}$

b $\mathbf{n} = \begin{pmatrix} 7 \\ 1 \\ 1 \end{pmatrix} \times \begin{pmatrix} 8 \\ 2 \\ -5 \end{pmatrix} = \begin{pmatrix} -7 \\ 43 \\ 6 \end{pmatrix}$

$\begin{pmatrix} -2 \\ 0 \\ 0 \end{pmatrix} \cdot \begin{pmatrix} -7 \\ 43 \\ 6 \end{pmatrix} = 14$

$\mathbf{r} \cdot (-7\mathbf{i} + 43\mathbf{j} + 6\mathbf{k}) = 14$

c $-7x + 43y + 6z - 14 = 0$

5 a $\mathbf{r} = \begin{pmatrix} 0 \\ 3 \\ 2 \end{pmatrix} + s\begin{pmatrix} 1 \\ -5 \\ -2 \end{pmatrix} + t\begin{pmatrix} -1 \\ -1 \\ -3 \end{pmatrix}$

b $\mathbf{n} = \begin{pmatrix} 1 \\ -5 \\ -2 \end{pmatrix} \times \begin{pmatrix} -1 \\ -1 \\ -3 \end{pmatrix} = \begin{pmatrix} 13 \\ 5 \\ -6 \end{pmatrix}$

$\begin{pmatrix} 0 \\ 3 \\ 2 \end{pmatrix} \cdot \begin{pmatrix} 13 \\ 5 \\ -6 \end{pmatrix} = 3$

$\mathbf{r} \cdot (13\mathbf{i} + 5\mathbf{j} - 6\mathbf{k}) = 3$

c $13x + 5y - 6z - 3 = 0$

6 $\mathbf{r} \cdot \begin{pmatrix} 1 \\ 0 \\ 2 \end{pmatrix} = 11$

7 $\mathbf{r} \cdot \begin{pmatrix} 3 \\ -1 \\ 4 \end{pmatrix} = 10$

8 $\begin{pmatrix} 1 \\ 1 \\ 1 \end{pmatrix} \cdot \begin{pmatrix} 0 \\ 2 \\ -3 \end{pmatrix} = 0 + 2 - 3 = -1$

$\mathbf{r} \cdot \begin{pmatrix} 0 \\ 2 \\ -3 \end{pmatrix} = -1$

9 $5x + 4y - 2z = 18$

10 $2x - 5z = -8$

11 $x - y + 4z = 31$

12 a $\mathbf{r} \cdot \begin{pmatrix} 2 \\ 1 \\ -3 \end{pmatrix} = -2$

b $2x + y - 3z = -2$

13 a $\mathbf{r} \cdot \begin{pmatrix} 3 \\ -2 \\ 5 \end{pmatrix} = -4$

b $3x - 2y + 5z = -4$

14 a $3x + 5y - z = -2$

b $7x - y + 8z = 3$

c $23x + 13y - 16z - 139 = 0$

d $-8x - 7y + 2z + 32 = 0$

15 a $\mathbf{r} \cdot \begin{pmatrix} 9 \\ 3 \\ -1 \end{pmatrix} = 5$

b $\mathbf{r} \cdot \begin{pmatrix} 2 \\ -7 \\ -15 \end{pmatrix} = -4$

c $\mathbf{r} \cdot \begin{pmatrix} 4 \\ 16 \\ 4 \end{pmatrix} = 12$

d $\mathbf{r} \cdot \begin{pmatrix} -24 \\ -15 \\ -5 \end{pmatrix} = -149$

Exercise 23.2B

1 a If there exist s and t such that

$\begin{pmatrix} 2 \\ -3 \\ -7 \end{pmatrix} + s\begin{pmatrix} -1 \\ 5 \\ 3 \end{pmatrix} + t\begin{pmatrix} 7 \\ -3 \\ -1 \end{pmatrix} = \begin{pmatrix} -7 \\ 10 \\ 0 \end{pmatrix}$

Then, taking $\mathbf{i}$: $2 - s + 7t = -7 \Rightarrow s = 7t + 9$
$\mathbf{j}$: $-3 + 5s - 3t = 10 \Rightarrow 5s = 13 + 3t$
Substitute for s to give $5(7t + 9) = 13 + 3t$
$\Rightarrow 32t = -32$
$\Rightarrow t = -1, s = 2$
$\mathbf{k}$: $-7 + 2 \times 3 - 1 \times -1 = 0$ as required
So $(-7, 10, 0)$ does lie on this plane.

b If there exist s and t such that

$$\begin{pmatrix} 2 \\ 0 \\ 8 \end{pmatrix} + s \begin{pmatrix} 1 \\ 3 \\ -2 \end{pmatrix} + t \begin{pmatrix} -2 \\ 4 \\ 0 \end{pmatrix} = \begin{pmatrix} -7 \\ 10 \\ 0 \end{pmatrix}$$

Then, taking $\mathbf{k}$: $8 - 2s = 0 \Rightarrow s = 4$

$\mathbf{j}$: $3 \times 4 + 4t = 10 \Rightarrow t = -\dfrac{1}{2}$

$\mathbf{i}$: $2 + 4 - \dfrac{1}{2} \times -2 = 7$ not -7 so $(-7, 10, 0)$ not on the plane.

c $\begin{pmatrix} -7 \\ 10 \\ 0 \end{pmatrix} \cdot \begin{pmatrix} 1 \\ -1 \\ 3 \end{pmatrix} = -7 \times 1 + 10 \times -1 + 0 \times 3 = -17$

So $(-7, 10, 0)$ not on the plane.

d $\begin{pmatrix} -7 \\ 10 \\ 0 \end{pmatrix} \cdot \begin{pmatrix} 2 \\ 3 \\ -8 \end{pmatrix} = -7 \times 2 + 10 \times 3 + 0 \times 8 = 16$ as required

So $(-7, 10, 0)$ does lie on this plane.

2 a $\begin{pmatrix} x \\ y \\ z \end{pmatrix} \cdot \begin{pmatrix} 2 \\ 3 \\ -1 \end{pmatrix} = 2x + 3y - z = 2$ so they represent the same plane.

b Check if $(3, 0, -1)$ satisfies second equation:

$$\begin{pmatrix} 1 \\ -4 \\ 4 \end{pmatrix} + \lambda \begin{pmatrix} 2 \\ 4 \\ 5 \end{pmatrix} + \mu \begin{pmatrix} 5 \\ 3 \\ 0 \end{pmatrix} = \begin{pmatrix} 3 \\ 0 \\ -1 \end{pmatrix}$$

$\mathbf{k}$: $4 + 5\lambda = -1 \Rightarrow \lambda = -1$

$\mathbf{i}$: $1 - 2 + 5\mu = 3 \Rightarrow \mu = \dfrac{4}{5}$

$\mathbf{j}$: $-4 - 4 + 3\left(\dfrac{4}{5}\right) = -\dfrac{28}{5}$ (not 0) so not the same plane.

c Check if $(1, 0, -7)$ satisfies second equation:

$$\begin{pmatrix} -4 \\ 2 \\ -7 \end{pmatrix} + \lambda \begin{pmatrix} 0 \\ 1 \\ 3 \end{pmatrix} + \mu \begin{pmatrix} 5 \\ -2 \\ 0 \end{pmatrix} = \begin{pmatrix} 1 \\ 0 \\ -7 \end{pmatrix}$$

$\mathbf{i}$: $-4 + 5\mu = 1 \Rightarrow \mu = 1$
$\mathbf{j}$: $2 + \lambda - 2 = 0 \Rightarrow \lambda = 0$
$\mathbf{k}$: $-7 + 0 + 0 = -7$ so on the same plane.
Check if normals are parallel.
Normal to first plane:

$$\begin{pmatrix} 5 \\ -2 \\ 4 \end{pmatrix} \times \begin{pmatrix} 0 \\ 0 \\ 1 \end{pmatrix} = \begin{pmatrix} -2 \\ -5 \\ 0 \end{pmatrix}$$

Normal to second plane:

$$\begin{pmatrix} 0 \\ 1 \\ 3 \end{pmatrix} \times \begin{pmatrix} 5 \\ -2 \\ 0 \end{pmatrix} = \begin{pmatrix} 6 \\ 15 \\ -5 \end{pmatrix}$$

This is not a multiple of normal to first plane so not the same plane.

d Check if $(-2, 1, 0)$ satisfies second equation:

$$\begin{pmatrix} -2 \\ 1 \\ 0 \end{pmatrix} \cdot \begin{pmatrix} 6 \\ -2 \\ -1 \end{pmatrix} = -12 - 2 = -14$$

For first plane $\mathbf{n} =$

$$\begin{pmatrix} 4 \\ 3 \\ 2 \end{pmatrix} \times \begin{pmatrix} -5 \\ 0 \\ 1 \end{pmatrix} = \begin{pmatrix} 3 \\ -14 \\ 15 \end{pmatrix}$$

which is not equal to $\lambda \begin{pmatrix} 6 \\ -2 \\ -1 \end{pmatrix}$ for any λ

So not the same plane.

e Check if $(5, 0, 4)$ satisfies first equation:

$$\begin{pmatrix} 5 \\ 0 \\ 4 \end{pmatrix} \cdot \begin{pmatrix} 2 \\ -7 \\ -3 \end{pmatrix} = 10 + 0 - 12 = -2$$

For second plane $\mathbf{n} = \begin{pmatrix} 0 \\ 3 \\ -7 \end{pmatrix} \times \begin{pmatrix} 2 \\ 1 \\ -1 \end{pmatrix} = \begin{pmatrix} 4 \\ -14 \\ -6 \end{pmatrix} = 2 \begin{pmatrix} 2 \\ -7 \\ -3 \end{pmatrix}$

so the planes are parallel and share a point
so the equations represent the same plane.

f Check if $(1, 0, -2)$ satisfies first equation:

$$\begin{pmatrix} 1 \\ 0 \\ -2 \end{pmatrix} \cdot \begin{pmatrix} 1 \\ 3 \\ 1 \end{pmatrix} = 1 + 0 - 2 = -1$$

Setting $\lambda = 0$ and $\mu = 1$ gives point $(5, 3, -5)$ on second equation
Check if $(5, 3, -5)$ satisfies first equation:

$$\begin{pmatrix} 5 \\ 3 \\ -5 \end{pmatrix} \cdot \begin{pmatrix} 1 \\ 3 \\ 1 \end{pmatrix} = 5 + 9 - 5 = 9 \text{ (not } -1)$$

Not the same plane.

3 $21.1°$

4 $173°$

5 $2.72°$

6 $164.6°$

7 $39.5°$

8 $61.9°$

9 $110°$

10 $\mathbf{r} \cdot \begin{pmatrix} 18 \\ -5 \\ 6 \end{pmatrix} = 59$

11 $\mathbf{r} \cdot \begin{pmatrix} -1 \\ -2 \\ 1 \end{pmatrix} = -8$

12 $2x + 6y - 13z = -37$

13 a $\mathbf{r} \cdot \begin{pmatrix} 2 \\ 27 \\ 13 \end{pmatrix} = 19$

b $\left(\mathbf{r} - \begin{pmatrix} 1 \\ 4 \\ -2 \end{pmatrix} \right) \times \begin{pmatrix} 2 \\ 27 \\ 13 \end{pmatrix} = 0$

Exercise 23.3A

1 a $\dfrac{7}{6}$ units

b $\dfrac{11}{14}\sqrt{14}$ units

c $\dfrac{13}{15}\sqrt{3}$ units

d 0.4 units

e $\dfrac{5}{11}\sqrt{11}$ units

2 $\begin{pmatrix} -11 \\ -4 \\ 10 \end{pmatrix} + t\begin{pmatrix} 4 \\ -2 \\ 0 \end{pmatrix} = \begin{pmatrix} 3 \\ -5 \\ -10 \end{pmatrix} + \mu\begin{pmatrix} 7 \\ -10 \\ 0 \end{pmatrix} + \lambda\begin{pmatrix} 1 \\ -3 \\ 4 \end{pmatrix}$

$\mathbf{k}: 10 = -10 + 4\lambda \Rightarrow \lambda = 5$

$\mathbf{i}: -11 + 4t = 3 + 7\mu + 5$

$\Rightarrow 4t = 19 + 7\mu$

$\mathbf{j}: -4 - 2t = -5 - 10\mu - 15$

$\Rightarrow 2t = 16 + 10\mu$

$\times 2$ to give $4t = 32 + 20\mu$

Substitute into $4t = 19 + 7\mu$ to give $32 + 20\mu = 19 + 7\mu$

$\Rightarrow \mu = -1$

$\Rightarrow t = 3$

$\begin{pmatrix} -11 \\ -4 \\ 10 \end{pmatrix} + 3\begin{pmatrix} 4 \\ -2 \\ 0 \end{pmatrix} = \begin{pmatrix} 1 \\ -10 \\ 10 \end{pmatrix}$

and $\begin{pmatrix} 3 \\ -5 \\ -10 \end{pmatrix} - 1\begin{pmatrix} 7 \\ -10 \\ 0 \end{pmatrix} + 5\begin{pmatrix} 1 \\ -3 \\ 4 \end{pmatrix} = \begin{pmatrix} 1 \\ -10 \\ 10 \end{pmatrix}$

So they intersect at the point $(1, -10, 10)$

3 $(6, -2, 20)$

4 $(21, -6, -3)$

5 $(-10, 0, 11)$

6 $(11, 12, -5)$

7 2 units

8 Distance $= \left| \dfrac{n_1\alpha + n_2\beta + n_3\gamma + d}{\sqrt{n_1^2 + n_2^2 + n_3^2}} \right|$

$= \left| \dfrac{n_1 \times 0 + n_2 \times 0 + n_3 \times 0 - p}{\sqrt{n_1^2 + n_2^2 + n_3^2}} \right|$

$= \left| -\dfrac{p}{|\mathbf{n}|} \right|$

$= \dfrac{p}{|\mathbf{n}|}$ as required

9 a So they intersect when $\lambda = 0$

   Which is point $(7, -6, -2)$

 b Lines are parallel

 c Skew lines

 d Intersect at $(-4, 2, 0)$

10 $3\sqrt{10}$

11 $\dfrac{16}{\sqrt{14}}$

Exercise 23.3B

1 a i $\left| \dfrac{-20\sqrt{51}}{51} \right|$

   ii $\left| \dfrac{-20\sqrt{51}}{51} \right|$

 b Since A and B are the same distance from Π and are the same side of Π the line l must be parallel to the plane therefore it doesn't intersect it.

2 a i 1.42 units

   ii 0.291 units

   iii 3.13 units

 b Each pair of planes is parallel, hence they do not meet and we can find the perpendicular distance between them since it is constant.

3 $\sqrt{54}$ (7.35)

4 11.67 square units

5 $= \dfrac{27}{2}\sqrt{6}$ square units

6 a $\dfrac{11}{3}$ (3.67)

 b $\dfrac{78}{\sqrt{346}}$ (4.19)

 c $\dfrac{29}{\sqrt{30}}$ (5.29)

 d $\sqrt{2}$ (1.41)

 e $\sqrt{5}$ (2.24)

7 $\sqrt{2}$ (1.41)

8 $\left(\dfrac{13}{7}, \dfrac{46}{7}, -\dfrac{37}{7} \right)$

9 $\left(\dfrac{8}{11}, \dfrac{14}{11}, -\dfrac{35}{11} \right)$

Review exercise 23

1 a $\begin{pmatrix} -26 \\ -2 \\ -18 \end{pmatrix}$

 b $\begin{pmatrix} 6 \\ 36 \\ 12 \end{pmatrix}$

2 a $82.8°$

 b $78.2°$

3 $\pm\begin{pmatrix} \dfrac{3}{\sqrt{35}} \\ \dfrac{5}{\sqrt{35}} \\ \dfrac{1}{\sqrt{35}} \end{pmatrix}$

4 $\dfrac{3}{2}\sqrt{6}$ units^2

5 $\mathbf{r} \times \begin{pmatrix} -5 \\ 3 \\ -1 \end{pmatrix} = \begin{pmatrix} -11 \\ -15 \\ 10 \end{pmatrix}$

6 $\mathbf{r} \times \begin{pmatrix} 0 \\ 3 \\ -4 \end{pmatrix} = \begin{pmatrix} 8 \\ 4 \\ 3 \end{pmatrix}$

7 $\mathbf{r} = \begin{pmatrix} 2 \\ 0 \\ -3 \end{pmatrix} + s\begin{pmatrix} -2 \\ 0 \\ 3 \end{pmatrix} + t\begin{pmatrix} 1 \\ -4 \\ -1 \end{pmatrix}$

8 a $\mathbf{r} \cdot \begin{pmatrix} 7 \\ -5 \\ 1 \end{pmatrix} = -15$

 b $7x - 5y + z = -15$

9 a $32.3°$

 b $\left(-\dfrac{1}{3}, -2, \dfrac{2}{3} \right)$

 c $\dfrac{\sqrt{41}}{3}$ (2.13)

10 a $(-1, -1, -6)$

  b $168°$

Answers

11 $-\dfrac{11}{\sqrt{238}}$

12 $153°$

13 2.59 units

14 0.463 units

15 $\dfrac{2\sqrt{3}}{3}$

Assessment 23

1 a $-\mathbf{i}-2\mathbf{j}+\mathbf{k}$

  b $\dfrac{1}{\sqrt{6}}(-\mathbf{i}-2\mathbf{j}+\mathbf{k})$

2 a $a=2, b=7, c=-3$

  b i $\mathbf{r}=\begin{pmatrix}2\\7\\-3\end{pmatrix}+t\begin{pmatrix}2\\-3\\1\end{pmatrix}$

  ii $\left(\mathbf{r}-\begin{pmatrix}2\\7\\-3\end{pmatrix}\right)\times\begin{pmatrix}2\\-3\\1\end{pmatrix}=0$

3 $\left(\mathbf{r}-\begin{pmatrix}1\\2\\-3\end{pmatrix}\right)\times\begin{pmatrix}2\\0\\-1\end{pmatrix}=0$

4 a $\begin{pmatrix}3\\1\\-2\end{pmatrix}\cdot\begin{pmatrix}0\\4\\-2\end{pmatrix}=0+4+4$

   $=8$ so lies on this plane

  b $2(3)+3(1)-(-2)=11$
   so lies on this plane

5 21.7 square units

6 a $\mathbf{r}=\begin{pmatrix}1\\-5\\2\end{pmatrix}+s\begin{pmatrix}0\\-3\\-1\end{pmatrix}+t\begin{pmatrix}-2\\4\\0\end{pmatrix}$

  b $\begin{pmatrix}2\\1\\-3\end{pmatrix}\cdot\begin{pmatrix}0\\-3\\-1\end{pmatrix}=0-3+3=0$

   So perpendicular to $\begin{pmatrix}0\\-3\\-1\end{pmatrix}$

   $\begin{pmatrix}2\\1\\-3\end{pmatrix}\cdot\begin{pmatrix}-2\\4\\0\end{pmatrix}=-4+4+0=0$

   So perpendicular to $\begin{pmatrix}-2\\4\\0\end{pmatrix}$

   Hence perpendicular to plane

  c i $\mathbf{r}\cdot\begin{pmatrix}2\\1\\-3\end{pmatrix}=-9$ **ii** $2x+y-3z=-9$

7 a $\mathbf{r}=\begin{pmatrix}7\\12\\-14\end{pmatrix}+s\begin{pmatrix}2\\8\\-4\end{pmatrix}+t\begin{pmatrix}2\\1\\-3\end{pmatrix}$

b If there exist s and t such that

$\begin{pmatrix}-1\\0\\4\end{pmatrix}=\begin{pmatrix}7\\12\\-14\end{pmatrix}+s\begin{pmatrix}2\\8\\-4\end{pmatrix}+t\begin{pmatrix}2\\1\\-3\end{pmatrix}$ then

$7+2s+2t=-1\Rightarrow s+t=-4$
$12+8s+t=0\Rightarrow 8s+t=-12$

$\Rightarrow s=-\dfrac{8}{7}$

$\Rightarrow t=-\dfrac{20}{7}$

Check for $\mathbf{k}$ component:

$-14-\dfrac{8}{7}(-4)-\dfrac{20}{7}(-3)=-\dfrac{6}{7}$ not 4

So does not lie on the plane

8 a 2.78

  b $\left(\dfrac{17}{11},-1,\dfrac{18}{11}\right)$

  c 0.664 rad (or $38.0°$)

9 $\left(-\dfrac{2}{3},\dfrac{7}{3},\dfrac{1}{3}\right)$

10 a $\begin{pmatrix}5-8\lambda\\5-6\lambda\\2+2\lambda\end{pmatrix}\times\begin{pmatrix}1\\0\\-2\end{pmatrix}$

$=\begin{pmatrix}-2(5-6\lambda)-0\\(2+2\lambda)--2(5-8\lambda)\\0-(5-6\lambda)\end{pmatrix}$

$\begin{pmatrix}12\lambda-10\\12-14\lambda\\6\lambda-5\end{pmatrix}=\begin{pmatrix}-4\\5\\-2\end{pmatrix}$

$12\lambda-10=-4\Rightarrow\lambda=\dfrac{1}{2}$

$12-14\lambda=5\Rightarrow\lambda=\dfrac{1}{2}$

$6\lambda-5=-2\Rightarrow\lambda=\dfrac{1}{2}$

Therefore they intersect

  b $\begin{pmatrix}1\\2\\3\end{pmatrix}$

11 $63.1°$

12 a $67.4°$

  b $\begin{pmatrix}5\\-1\\-2\end{pmatrix}$

13 $\dfrac{16}{\sqrt{14}}$

14 $\mathbf{r}=\begin{pmatrix}5\\-2\\7\end{pmatrix}+\lambda\begin{pmatrix}3\\-1\\6\end{pmatrix}+\mu\begin{pmatrix}-1\\-6\\8\end{pmatrix}$

15 $\dfrac{4}{9}\sqrt{26}$ or 2.27

16 a $\mathbf{r}=\begin{pmatrix}-4\\7\\-5\end{pmatrix}+\lambda\begin{pmatrix}-4\\7\\-6\end{pmatrix}+\mu\begin{pmatrix}2\\5\\-1\end{pmatrix}$

  b 15.2 units

17 $\mathbf{r} = \begin{pmatrix} -3 \\ 1 \\ 0 \end{pmatrix} + \lambda \begin{pmatrix} 3 \\ -1 \\ 2 \end{pmatrix}$

18 $\begin{pmatrix} 3 \\ -5 \\ -10 \end{pmatrix} + \mu \begin{pmatrix} 7 \\ -10 \\ 0 \end{pmatrix} + \lambda \begin{pmatrix} 1 \\ -3 \\ 4 \end{pmatrix} = \begin{pmatrix} -11 \\ -4 \\ 10 \end{pmatrix} + t \begin{pmatrix} 4 \\ -2 \\ 0 \end{pmatrix}$

k: $-10 + 4\lambda = 10 \Rightarrow \lambda = 5$
i: $3 + 7\mu + 5 = -11 + 4t \Rightarrow 7\mu + 19 = 4t$
j: $-5 - 10\mu - 15 = -4 - 2t \Rightarrow 10\mu + 16 = 2t$
$\Rightarrow 20\mu + 32 = 7\mu + 19$
$\Rightarrow \mu = -1$
$\Rightarrow t = 3$

$\begin{pmatrix} -11 \\ -4 \\ 10 \end{pmatrix} + 3 \begin{pmatrix} 4 \\ -2 \\ 0 \end{pmatrix} = \begin{pmatrix} 1 \\ -10 \\ 10 \end{pmatrix}$

So they intersect at (1, −10, 10)

19 a $\begin{pmatrix} 4 \\ 2 \\ -1 \end{pmatrix} \times \begin{pmatrix} 0 \\ -5 \\ 1 \end{pmatrix} = \begin{pmatrix} -3 \\ -4 \\ -20 \end{pmatrix} = -1 \begin{pmatrix} 3 \\ 4 \\ 20 \end{pmatrix}$

Therefore planes are parallel
We know they aren't the same plane as (5, 3, −2) does not

lie on Π_2, since $\begin{pmatrix} 5 \\ 3 \\ -2 \end{pmatrix} \cdot \begin{pmatrix} 3 \\ 4 \\ 20 \end{pmatrix} = 15 + 12 - 40 = -16$ (not 12).

b $\dfrac{5\sqrt{17}}{17}$ (1.23)

Chapter 24
Exercise 24.1A

1 a $\mathbf{a} = -32\sin 4t\,\mathbf{i} - 32\cos 4t\,\mathbf{j}\ \mathrm{ms}^{-2}$

b $-320\sin 4t\,\mathbf{i} - 320\cos 4t\,\mathbf{j}\ \mathrm{N}$

2 a $(3\sin 2t + 2)\mathbf{i} + (3\cos 2t + 1)\mathbf{j}\ \mathrm{m}$

b $x = 3\sin 2t + 2$
$y = 3\cos 2t + 1$
$\cos^2 2t + \sin^2 2t = 1$
$\left(\dfrac{y-1}{3}\right)^2 + \left(\dfrac{x-2}{3}\right)^2 = 1$
$(y-1)^2 + (x-2)^2 = 9$
which is a circle, radius 3 and centre (2, 1).

3 a i $\mathbf{OM} = 4\cos\frac{1}{2}t\,\mathbf{i} + 4\sin\frac{1}{2}t\,\mathbf{j}\ \mathrm{m}$

ii $-\cos\frac{1}{2}t\,\mathbf{i} - \sin\frac{1}{2}t\,\mathbf{j}\ \mathrm{ms}^{-2}$

b 4π seconds
c 3 N

4 a $\mathbf{v} = 6\cos 3t\,\mathbf{i} - 6\sin 3t\,\mathbf{j}\ \mathrm{ms}^{-1}$

$\mathbf{a} = -18\sin 3t\,\mathbf{i} - 18\cos 3t\,\mathbf{j}\ \mathrm{ms}^{-2}$

b i Gradient of $\mathbf{a} = \dfrac{-18\cos 3t}{-18\sin 3t} = \cot 3t$

Gradient of $\mathbf{r} = \dfrac{2\cos 3t}{2\sin 3t} = \cot 3t$

Gradients are equal, so **a** and **r** are parallel
Using vectors:
$\mathbf{a} = -9(2\sin 3t\,\mathbf{i} + 2\cos 3t\,\mathbf{j}) = -9\mathbf{r}$

So, **a** and **r** are parallel.
They are in opposite directions.

ii Gradient of $\mathbf{r} = \dfrac{-6\sin 3t}{6\cos 3t} = -\tan 3t$

As $\cot 3t \times (-\tan 3t) = -1$,
gradients of **a** and **v** multiply to −1.
So **a** and **v** are perpendicular.
Using vectors:

$\mathbf{a} \times \mathbf{v} = -18\begin{pmatrix} \sin 3t \\ \cos 3t \end{pmatrix} \times 6\begin{pmatrix} \cos 3t \\ -\sin 3t \end{pmatrix}$
$= -108\,(\sin 3t\cos 3t - \cos 3t\sin 3t)$
$= 0$
So **a** and **v** are perpendicular.

5 a $-32\sin 2t\,\mathbf{i} + 32\cos 2t\,\mathbf{j},\ 32\mathrm{N}$

b $(2\sin 2t + 3)\mathbf{i} - (2\cos 2t + 8)\mathbf{j}\ \mathrm{m}$,
circle radius 2 and centre (3, −8)

6 $\mathbf{r} = \displaystyle\int \mathbf{v}\,dt = -\dfrac{a}{\omega}\cos\omega t\,\mathbf{i} + \dfrac{b}{\omega}\sin\omega t\,\mathbf{j}$

$\mathbf{F} = m\mathbf{a} = m\dfrac{d\mathbf{v}}{dt} = m(a\omega\cos\omega t\,\mathbf{i} - b\omega\sin\omega t\,\mathbf{j})$

$= -m\omega^2\left(-\dfrac{a}{\omega}\cos\omega t\,\mathbf{i} + \dfrac{b}{\omega}\sin\omega t\,\mathbf{j}\right)$

$= -m\omega^2\mathbf{r}$

So $\mathbf{F} = k\mathbf{r}$ where $k = -m\omega^2$.

Exercise 24.1B

1 $14.5\ \mathrm{ms}^{-1}$
2 $14.4\ \mathrm{ms}^{-1}$
3 $18.8°$
4 a $42.0\ \mathrm{ms}^{-1}$ **b** $24.2\ \mathrm{ms}^{-1}, 72.7\ \mathrm{ms}^{-1}$
5 a $34.5°$
  b Assuming that $108\ \mathrm{km\,h}^{-1}$ is the maximum speed around the bend.
  Assume that the limiting friction force and hence the coefficient of friction remain the same.
6 a $21.7\ \mathrm{ms}^{-1}$ **b** $27.0\ \mathrm{ms}^{-1}$
7 a 0.46 (2 sf) **b** $16.8\ \mathrm{ms}^{-1}$
8 $4.50°$
9 a 175 kN **b** 525 kN
10 a $12.0°, 0.21$ **b** $39\ \mathrm{km\,h}^{-1}$
11

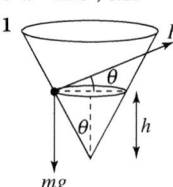

Let angle of cone be 2θ
Vertical equilibrium:
$R\sin\theta = mg$
Horizontal equation of motion:
$R\cos\theta = m \times \dfrac{v^2}{h\tan\theta}$
$v^2 = \dfrac{mg}{\sin\theta} \times \cos\theta \times \dfrac{h\tan\theta}{m}$
$= hg$
Velocity $v = \sqrt{hg}$

12 $0.34\,\text{N}$

13 $\sqrt{\dfrac{g}{x}}$

14 Maximum speed occurs when on point of moving up slope.

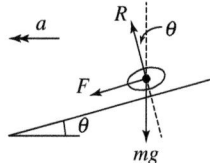

Equation of motion down slope:

$$F + mg\sin\theta = \dfrac{mv^2}{r}\cos\theta$$

Equation of motion $\perp$ slope:

$$R - mg\cos\theta = \dfrac{mv^2}{r}\sin\theta$$

Friction: $F \leq \mu R$

$$\mu \geq \dfrac{F}{R} = \dfrac{mv^2\cos\theta - mgr\sin\theta}{mv^2\sin\theta + mgr\cos\theta}$$

$$= \dfrac{v^2 - gr\tan\theta}{r^2\tan\theta + gr}$$

$\mu v^2\tan\theta + \mu gr \geq v^2 - gr\tan\theta$
and $\mu = \tan\lambda$
So $v^2(1 - \tan\lambda + \tan\theta) \leq gr(\tan\lambda + \tan\theta)$

$$\dfrac{v^2}{gr} \leq \dfrac{\tan\lambda + \tan\theta}{1 - \tan\lambda\tan\theta} = \tan(\theta + \lambda)$$

Minimum speed occurs when on point of moving down slope.

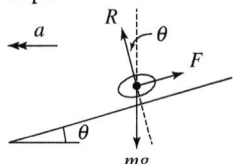

Equations are now:

$$mg\sin\theta - F = \dfrac{mv^2}{r}\cos\theta \text{ and}$$

$$R - mg\cos\theta = \dfrac{mv^2}{r}\sin\theta$$

Friction $F \leq \mu R$

$$mg\sin\theta - \dfrac{mv^2}{r}\cos\theta \leq \tan\lambda\left(mg\cos\theta + \dfrac{mv^2}{r}\sin\theta\right)$$

$$\dfrac{mv^2}{r}(\tan\lambda\sin\theta + \cos\theta) \geq mg(\sin\theta - \tan\lambda\cos\theta)$$

$$\dfrac{v^2}{rg}(\tan\lambda\tan\theta + 1) \geq \tan\theta - \tan\lambda$$

$$\dfrac{v^2}{rg} \geq \tan(\theta - \lambda)$$

So $\tan(\theta - \lambda) \leq \dfrac{v^2}{rg} \leq \tan(\theta + \lambda)$

or $\theta - \lambda \leq \tan^{-1}\left(\dfrac{v^2}{rg}\right) \leq \theta + \lambda$

Exercise 24.2A

1 $4g\,\text{N}$, $1.0\,\text{s}$

2

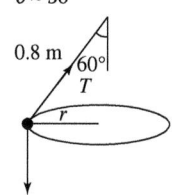

$$80\,\text{rpm} = \dfrac{80 \times 2\pi}{60} = 8.38\,\text{rad s}^{-1}$$

Vertical equilibrium: $T\cos\theta = mg$
Horizontal equation of motion:

$$T\sin\theta = m \times 0.25\sin\theta \times \omega^2$$

$$\Rightarrow \dfrac{mg}{\cos\theta} = m \times 0.25 \times 8.38^2$$

$\cos\theta = 0.559$
$\theta \approx 56°$

3

$T\cos 60° = mg$
$T\sin 60° = mr\omega^2$

$$\Rightarrow \tan 60° = \dfrac{r\omega^2}{9}$$

$$\omega^2 = \dfrac{g\sqrt{3}}{0.8 \times \sin 60°}$$

$$\Rightarrow \omega = 4.95\,\text{rad s}^{-1}$$
$$= 0.788\,\text{rad s}^{-1}$$

In 10 seconds, the number of revolutions $= 7.88 \approx 8$

4 $1.48\,\text{m}$, $29.6\,\text{N}$

5 $23.7\,\text{N}$, $60.2°$

6

mg
Vertically: $T\cos\theta = mg$
Horizontally: $T\sin\theta = m \times l\sin\theta \times \omega^2$
Eliminating T: $ml\omega^2 \times \cos\theta = mg$

Height $h = l\cos\theta = \dfrac{g}{\omega^2}$ which is independent of l

Exercise 24.2B

1 **a** $1.8\,\text{rad s}^{-1}$ **b** $2.26\,\text{rad s}^{-1}$
  c Any two from:
  String is light and inextensible.
  There is no air resistance to slow the mass as it rotates.
  Masses are modelled as point masses or particles.

2 $2.89\,\text{m}$

3 $\dfrac{ml}{h}(\omega^2 h - g)$

4 $2\pi\sqrt{\dfrac{\sqrt{R^2 - r^2}}{g}}\,\text{s}$

5 $mg - \dfrac{m\omega^2\sqrt{3}}{5}$

$R \geq 0$ implies $mg \geq \dfrac{m\omega^2\sqrt{3}}{5}$

$5g \geq \omega^2 \times \sqrt{3}$ So maximum $\omega^2 = \dfrac{5g}{\sqrt{3}}$

$\omega = \sqrt{28.29} = 5.3 \,\text{rad s}^{-1}$

6 $3.0 \,\text{rad s}^{-1}$

7 a

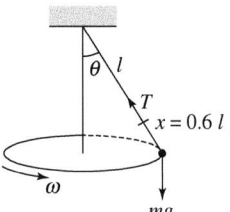

  i Vertical equilibrium:
  $T \cos\theta = mg$ (1)
  Horizontal equation of motion:
  $T \sin\theta = m \times 1.6l \sin\theta \times \omega^2$ (2)
  Hooke's law:
  $T = \dfrac{2mg}{l} \times 0.6l$

  $T = 1.2\,mg$ (3)
  From (2) and (3): $1.2\,mg = m \times 1.6l \times \omega^2$

  $\omega^2 = \dfrac{1.2\,mg}{1.6\,ml} = \dfrac{3g}{4l}$

  ii $\cos^{-1}\left(\dfrac{5}{6}\right)$

  b String is light.
  There is no air resistance to slow the mass as it rotates.

8

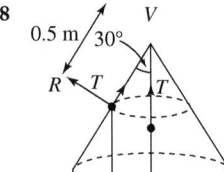

Cone is smooth, so tension T is constant along all the length.
For hanging mass in equilibrium: $T = mg$ (1)
Vertical equilibrium for moving mass:
$T \cos 30° + R \cos 60° = mg$ (2)
Horizontal equation of motion:
$T \sin 30° - R \sin 60° = mr\omega^2$ (3)
where $r = 0.5 \sin 30° = \dfrac{1}{4}\text{metre}$

From (1) and (2): $T\dfrac{\sqrt{3}}{2} + R \times \dfrac{1}{2} = mg$

$R = 2mg - 2 \times mg\dfrac{\sqrt{3}}{2}$

$R = 0.268\,mg$

Substitute into (3): $mg \times \dfrac{1}{2} - 0.268\,mg \times \sin 60° = m \times \dfrac{1}{4}\omega^2$

$\dfrac{1}{4}\omega^2 = 2.625$

Angular speed $\omega = 3.24 \,\text{rad s}^{-1}$

9

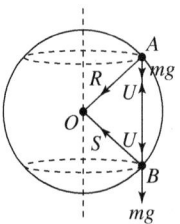

$\triangle OAB$ is equilateral with sides of length r
Assume directions of forces R, S, U as shown.
  a For A, vertical equilibrium:
  $R \cos 60° + mg = U$ (1)
  Horizontal equation of motion:
  $R \cos 30° = m \times r \sin 60° \times \omega^2$

  $R \times \dfrac{\sqrt{3}}{2} = m \times r \times \dfrac{\sqrt{3}}{2} \times \omega$

  $R = mr\,\omega^2$ (2)
  b $6.3 \,\text{rad s}^{-1}$
10 $19.3 \,\text{rad s}^{-1}$
11 a $\dfrac{2(M+m)g}{M\omega^2}$

  b Let the rods have a mass that is not negligible.
  Include friction at the contact between the collar and the rod.
  Take air resistance into account.

12 $\dfrac{g - e\omega^2}{l\omega^2}$

Exercise 24.3A

1 a $9.24 \,\text{m s}^{-1}$, 309 N   **b** $8.40 \,\text{m s}^{-1}$, 235 N
2 a $6.66 \,\text{m s}^{-1}$, 271 N, tension   **b** $2.28 \,\text{m s}^{-1}$, -23 N, thrust
3 a Find out whether $\dfrac{1}{2}mu^2 > mgh$

  $\dfrac{1}{2}u^2 > gh$ where $h = 2r = 4$ metres

  Is $u^2 > 8g$?
  $100 > 78.4$ so yes, it will complete the circle.
  b $104°$
4 a i $5.29 \,\text{m s}^{-1}$ **ii** 13.45 N

  b P rises to level of O if $\dfrac{1}{2}mu^2 > mgr$

  $\dfrac{1}{2}mu^2 = 9.0 \,\text{J}$

  $mgr = \dfrac{1}{2} \times 9.8 \times 1.4 = 6.9 \,\text{J}$
  P rises to level of O and goes higher.
5 a i $7.75 \,\text{m s}^{-1}$, 243 N
  ii The string may go slack, so that P leaves the circle.
  b i Yes; the same equations apply.
  ii $30.2 \,\text{m s}^{-1}$, 12 N, a thrust

Exercise 24.3B

1 a UWX **b** UVX **c** UWX
  d YZ **e** Z **f** UVX
2 a $10 \,\text{m s}^{-1}$ **b** $-9.8 \,\text{m s}^{-2}$
3 a $5.42 \,\text{m s}^{-1}$ **b** $8.57 \,\text{m s}^{-1}$
4 $48.2°$
If air resistance is taken into account, the particle, P, would be slowed down and, travelling more slowly, it would leave the surface lower down, or at an angle to the vertical greater than $48.2°$.

5 115°

6

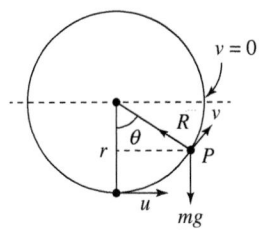

Energy equation to horizontal is

$\frac{1}{2}mu^2 - 0 = mgr$

$u^2 = 2gr$

Energy equation to point P is

$\frac{1}{2}\cancel{m}u^2 - \frac{1}{2}\cancel{m}v^2 = \cancel{m}g(r - r\cos\theta)$

$\cancel{gr} - \frac{v^2}{2} = \cancel{gr} - gr\cos\theta$

$\cos\theta = \frac{v^2}{2gr}$

Equation of motion at P towards centre is

$R - mg\cos\theta = m \times \frac{v^2}{r}$

Reaction $R = \frac{mv^2}{r} + m\cancel{g} \times \frac{v^2}{2g\cancel{r}}$

$= \frac{3mv^2}{2r}$

7 50.4 km h^{-1}

Car is a point mass, so no air resistance.

8 a i 8.3 N (3 sf)   **ii** 26.3 m s^{-2}

   b 41.8 m s^{-2}

9 a i 2.93 N   **ii** 17.2 N

   b 48.2°

10 2.92 m

11 a 1.2 m   **b** 1 : 2

12

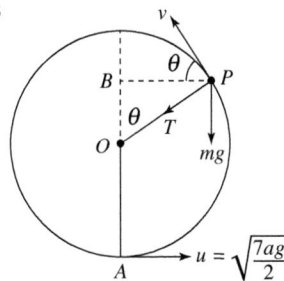

Energy equation for A to P: $\frac{1}{2}\cancel{m}u^2 - \frac{1}{2}\cancel{m}v^2 = \cancel{m}ga(1 + \cos\theta)$

$\frac{7ag}{4} - \frac{v^2}{2} = ag(1 + \cos\theta)$ (1)

Equation of motion along PO: $T + mg\cos\theta = m\frac{v^2}{a}$

Leaves circle when $T = 0$:

$v^2 = ag\cos\theta$ (2)

(1) and (2) give: $\frac{7ag}{4} - \frac{ag\cos\theta}{2} = ag + ag\cos\theta$

$\frac{3}{4}ag = \frac{3}{2}ag\cos\theta$

$\cos\theta = \frac{1}{2} \Rightarrow \theta = 60°$, so $v^2 = \frac{ag}{2}$

Time taken to travel PB horizontally

$= \frac{BP}{v\cos\theta} = \frac{a \times \frac{\sqrt{3}}{2}}{\sqrt{ag \times \frac{1}{2} \times \frac{1}{2}}} = \sqrt{\frac{6a}{g}}$

Using $s = ut + \frac{1}{2}at^2$ vertically from P:

$s = v\sin\theta \times t - \frac{1}{2} \times g \times t^2$

$= \sqrt{\frac{ag}{2}} \times \frac{\sqrt{3}}{2} \times \sqrt{\frac{6a}{g}} - \frac{1}{2} \times g \times \frac{6a}{g}$

$= \frac{3}{2}a - 3a$

$= -\frac{3a}{2}$, which is a distance of $\frac{3a}{2}$ below B

$AB = a + a\cos\theta = 3\frac{a}{2}$

So, mass passes through A

13 2.75 m

Review exercise 24

1 a 24 N   **b** $x^2 + y^2 = 9$

2 a 6 m s^{-1} **b** 2 m, (5, −1)

3 $3\cos 2t\mathbf{i} + 3\sin 2t\mathbf{j}, -24\mathbf{j}$

4 $\left(8 - 10\cos\frac{t}{2}\right)\mathbf{i} + \left(1 + 10\sin\frac{t}{2}\right)\mathbf{j}, \frac{10}{\sqrt{2}}(\mathbf{i} - \mathbf{j})$

5 0.36

6 16°, 0.287

7 $\frac{\mu_2 m_2}{2\mu_1 m_1}$

8 29.6 N, 48.6°

9 0.28 m (28 cm)

10

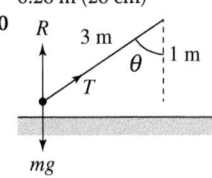

Vertical equation: $R + T\cos\theta = mg$

$R + \frac{T}{3} = mg$

Horizontal equation of motion:

$T\sin\theta = m \times 3\sin\theta \times \omega^2$

$T = 3m\omega^2$

Eliminating T

$R = mg - \frac{1}{3} \times 3m\omega^2 = mg - m\omega^2$

Contact occurs if $R > 0$

$mg > m\omega^2$

$\omega < \sqrt{g}$

Greatest linear velocity, $v = r\omega$

$= \sqrt{9 - 1} \times \omega = \sqrt{8g}$ m s^{-1}

11 1.68 m

12 a 4.95 m s^{-1} **b** 229 N

13 a 57°   **b** 106°

14

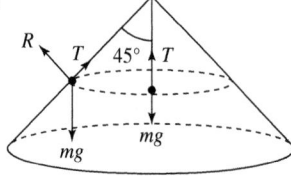

For hanging mass, $T = mg$
Tension constant as hole is smooth.
For outside mass,
vertical equilibrium means:
$T\cos 45° + R\cos 45° = mg$
$T + R = \sqrt{2} \times mg$ (1)
Horizontal equation of motion:
$T\sin 45° - R\sin 45° = m \times r\omega^2$
$T - R = \sqrt{2} \times mr\omega^2$ (2)

To find r

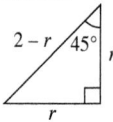

$\cos 45° = \dfrac{r}{2-r}$
$2 - r = \sqrt{2}r$
$r = \dfrac{2}{\sqrt{2}+1}$
Eliminate R from (1) and (2)
$2T = \sqrt{2}mg + \sqrt{2}m\omega^2 \times \dfrac{2}{\sqrt{2}+1}$
$\dfrac{2\sqrt{2}}{\sqrt{2}+1}\cancel{m}\omega^2 = 2\cancel{m}g - \sqrt{2}\cancel{m}g$
      $= \sqrt{2}g(\sqrt{2}-1)$
$2\omega^2 = (\sqrt{2}-1)(\sqrt{2}+1)g = 1g$
Angular velocity $\omega = \sqrt{\dfrac{g}{2}}$

15 a 4.85 rad s^{-1} **b** 2.37 rad s^{-1}

Assessment 24

1 a $|\mathbf{r}| = \sqrt{(0.3\cos 10t)^2 + (0.3\sin 10t)^2}$
         $= \sqrt{0.09(\cos^2 10t + \sin^2 10t)} = 0.3$
     The particle moves at a constant distance from O
     The path is a circle of radius 0.3 m about O
b $|\mathbf{v}| = \sqrt{(-3\sin 10t)^2 + (3\cos 10t)^2} = 3$
     so speed is a constant 3 m s^{-1}
c   60 N directed towards O
2 7 rad s^{-1}
3 a $\mathbf{r} = 10\cos 0.6t\mathbf{i} + 10\sin 0.6t\mathbf{j}$
b $\mathbf{a} = -3.6\cos 0.6t\mathbf{i} - 3.6\sin 0.6t\mathbf{j}$
     Hence $\mathbf{a} = -0.36\mathbf{r}$
     $\mathbf{r}$ is directed away from the origin, so $\mathbf{a}$ is directed towards it.
4 a $v = 26.2$ m s^{-1} **b** $v = 39.6$ m s^{-1} **c** $v = 9.06$ m s^{-1}
5 a i $v = 6.29$ m s^{-1} **ii** $v = 2.38$ m s^{-1}
   b i $T = 96.2$ N (tension)   **ii** $T = -5.62$ N (thrust)
6 a $U \geq 2\sqrt{ga}$ **b** $U \geq \sqrt{5ga}$
7 $70.5°$
8 a 0.0193 m   **b** 51.9 m s^{-1} or 187 km h^{-1}
9 a 0.392 m
   b   Assume ball is a particle, string is light and inextensible, no air resistance.

10 7.4 N, 1.12 m
11 a   Let length of string be l, let AO be given by h
     Resolve horizontally: $T\sin\theta = mr\omega^2$, where $r = l\sin\theta$
     $T = ml\omega^2$
     Resolve vertically: $T\cos\theta = mg$, where $\cos\theta = \dfrac{h}{l}$
     $ml\omega^2 = \dfrac{h}{l} = mg \Rightarrow h = \dfrac{g}{\omega^2}$
   b $\dfrac{g}{\omega^2} > 0$ for all values of ω, so $h > 0$ and string is not horizontal
     or for a horizontal string there is no vertical component of T to balance the weight of the particle.
12 a 8.57 m s^{-1} **b** $T = 4.41$ N
13 $48.2°$
14 a 2.01 N   **b** 1.68 m s^{-1}
15   Let speeds at bottom and top be u and v
     Resolve vertically at top: $T + 0.5g = \dfrac{0.5v^2}{1}$
     For a complete circle $T \geq 0$, so minimum $v^2 = g$
     From conservation of energy: $\dfrac{1}{2}mu^2 = \dfrac{1}{2}mv^2 + 2mg$
     $u^2 = v^2 + 4g$, so minimum $u^2 = 5g$
     Resolve vertically at bottom: $T_1 - 0.5g = 0.5\,u^2 = 2.5g$
     $T_1 = 3g = 29.4$ N
16   Radius of circle $= h\tan\theta$
     Resolve vertically: $R\sin\theta = mg$ (1)
     Resolve horizontally: $R\cos\theta = m\omega^2 h\tan\theta$ (2)
     Divide (1) by (2): $\tan\theta = \dfrac{mg}{mh\omega^2\tan\theta} \Rightarrow h = \dfrac{g}{\omega^2\tan^2\theta}$
17 $\dfrac{g\sqrt{51}}{4}$ m s^{-2}
18 $v = \sqrt{\dfrac{ga(3\sqrt{3}-4)}{2}}$ m s^{-1}
19 8 cm
20 $3\dfrac{1}{3}$ m
21 a $\dfrac{85\,mg}{126}$ N, 8.20 m s^{-1} **b** $\dfrac{5\,mg}{3}$ N, 18.8 m s^{-1}
22 $27.0°$, 0.055 N
23 $\sqrt{8ga}$
24 0.6 m
25 $0.5r$
26 183.75 N
     Assumes light, inextensible rope, ring is a particle, no friction/air resistance.
27 $R_A = \dfrac{mg(10-3\sqrt{3})}{6}$ and $R_B = \dfrac{mg(14-6\sqrt{3})}{6}$
28 3.89 rad s^{-1}
29 10.08 m

Chapter 25
Exercise 25.1A

1 a 23 m   **b** -2 N m   **c** 14 N m
2 a 2 N m   **b** 8 N m
3 a 14 N m   **b** 12 N m
4 $30g$ N, $90g$ N m anticlockwise
5 8 N m anticlockwise
6   Take moments about A clockwise.
     Resultant clockwise moment $= F \times (d + x) - F \times x = F \times d$
     which is independent of x and the position of point A

Exercise 25.1B

1   **a**   34.8 Nm   **b**   44.7 Nm   **c**   5 Nm anticlockwise

2   **a**   20.8 Nm   **b**   22.5 Nm   **c**   9.6 Nm   **d**   4.5 Nm

3   35 Nm

4   0.010 47 Nm, 0.005 24 Nm

5   0 N, 10 N, –40 Nm

6   –37.9 Nm

Exercise 25.2A

1   **a**   (6.6, 0)   **b**   (4.1, 0)   **c**   (0, 3.9)   **d**   (4.36, –2.64)

2   $r = 4.15i + 1.55j$

3   (4.5, 0)

4   $a = 4, b = 2$

5   $a = 6, b = 7$

6   (–13.6, –8.4)

Exercise 25.2B

1

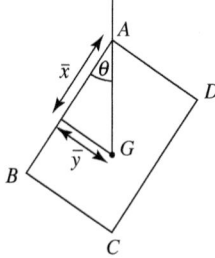

$$\bar{x} = \frac{50}{21}, \bar{y} = \frac{45}{21}, \theta = 42.0°$$

2   $\bar{x} = \dfrac{130}{17}, \bar{y} = \dfrac{40}{17}$

3   **a**

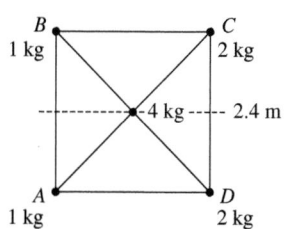

The masses are positioned, in position and in size, symmetrically about the dotted line.

So centre of mass, G, lies on this line.

  b   1.44m, 50.2°

4   **a**

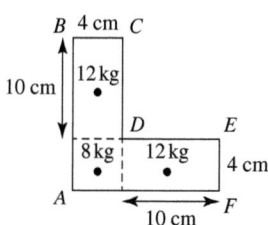

The line AD is a line of geometric symmetry and mass symmetry.

So centre of mass lies on AD (or AD extended).

  b $\bar{x} = 4\dfrac{5}{8}$ cm, $\bar{y} = 4\dfrac{5}{8}$ cm

5   $m = 4, DE = 3$ metres

---

6   Centre of mass is 1.03 m from OY and 1.11 m from OX.

7   Centre of mass is 3 cm from OX and 10 cm from OY.

8   0.75 m

Exercise 25.3A

1   **a** OG is 2.18 m on line of symmetry of arc.

  b OG is 3.9 m on line of symmetry of arc.

  c OG is 1.15 m on line of symmetry of arc.

2   **a** OG is 1.24 m on line of symmetry.

  b OG is 2.24 m on line of symmetry.

  c OG is 1.79 m on line of symmetry.

3   **a** OG is 0.34 m on line of symmetry.

  b OG is 0.84 m on line of symmetry.

4   **a**   (7, 5)   **b**   (3, 2)   **c**   (6, 4)

5   **a** AG is 8 cm on line of symmetry.

  b AG is $\dfrac{10\sqrt{3}}{3}$ cm on line of symmetry.

6   $a = 15, b = 12$

7

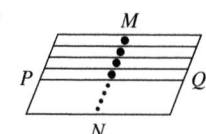

If P, Q are midpoints then mass of parallelogram above PQ equals mass below PQ, hence the centre of mass lies on PQ Repeat the process with MN and the question is answered.

Exercise 25.3B

1   **a**   (5.3, 6.4)   **b**   (5.27, 3.82)

  c   **i**   (5, 5.12)   **ii**   (5, 3.27)

2   **a**   (3.31, 1)   **b**   (4.625, 3.375)

3   **a**   Centre of mass is 0.5 m from AB and 0.15 m from AC

  b   0.612 m from centre on line of symmetry

  4   0.78 cm above O on axis of symmetry

5

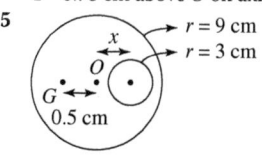

$x = 4$ cm

6   **a**

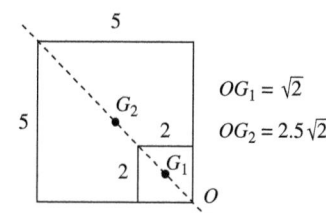

3.94 m from O

  b   5.25 cm from the straight edge of the semicircle and along the line of symmetry of the lamina

7   2.98 cm

8   4.93 cm

9   12.5 cm

10   $\dfrac{33a}{28}$

11

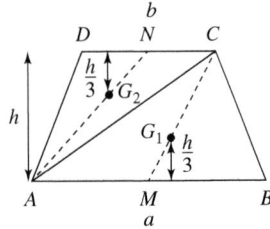

M, N are midpoints, so G_1 is $\dfrac{h}{3}$ from AB and G_2 is $h - \dfrac{h}{3} = \dfrac{2}{3}h$ from AB.

	Mass	C of M from AB
$\triangle ABC$	$\dfrac{1}{2}\,ah\rho$	$\dfrac{h}{3}$
$\triangle ACD$	$\dfrac{1}{2}\,bh\rho$	$\dfrac{2h}{3}$
Trapezium	$\dfrac{1}{2}h(a+b)\rho$	$\bar{y}$

Taking moments about AB

$$\frac{1}{2}h(a+b)\bar{y} = \frac{1}{2}ah \times \frac{h}{3} + \frac{1}{2}bh \times \frac{2}{3}h$$

$$(a+b)\bar{y} = (a+2b)\frac{h}{3}$$

Distance of centre of mass from $AB = \bar{y} = \dfrac{(a+2b)h}{3(a+b)}$

The symmetry of the isosceles trapezium has not been used in this proof, so the proof applies to all trapezia.

[NB: If the trapezium has been dissected like this, symmetry would been used.

The chosen method is more general.]

Exercise 25.4A

1 a $\left(\dfrac{2}{\ln 3}, \dfrac{3}{\ln 3}\right)$ **b** $\left(\dfrac{3}{2}, \dfrac{6}{5}\right)$

   c $\left(\dfrac{9}{16}, \dfrac{7}{10}\right)$ **d** $\left(\dfrac{11}{8}, -18\dfrac{2}{5}\right)$

2 a $(2.4, 0)$ **b** $(1.8, 1.8)$ **c** $\left(2\dfrac{6}{7}, 0\right)$

3 a $(3, 0)$ **b** $\left(\dfrac{3}{2}\ln 3, 0\right)$

   c $\left(1\dfrac{2}{3}, 0\right)$ **d** $\left(2\dfrac{2}{3}, 0\right)$

4 a $\left(\dfrac{8}{3\pi}, 0\right)$ **b** $\left(\dfrac{4r}{3\pi}, 0\right)$

5 a $\left(\dfrac{3}{8}, 0\right)$ **b** $\left(\dfrac{3}{8}r, 0\right)$

6 $\left(\dfrac{3\pi}{4}, 0\right)$

Exercise 25.4B

1 a $20.6°$ **b** $33.7°$

   c $26.6°$ **d** $41.6°$

2 $38.7°$

3 $36.4°$

4 $\tan^{-1}\left(\dfrac{16r}{11h}\right)$

5 a $\dfrac{2a}{3(\pi + 2\sqrt{3})}$ **b** $5.8°$

6 $(0.45, 0.35)$, $18.4°$

7 Let G_1 and G_2 be centres of mass of the two semicircles.

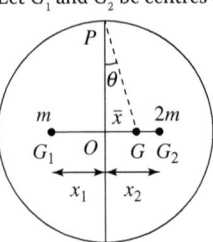

By symmetry, G lies on the line $G_1 G_2$ which passes through O
Take moments about O
$$3m_r OG = 2m_x x_2 - m_r x_1$$
Move negative sign as G_1 is on opposite side of O

$$OG = \frac{2x_2 - x_1}{3} \text{ where } x_1 = \frac{2r\sin\dfrac{\pi}{2}}{3\dfrac{\pi}{2}} = \frac{4r}{3\pi} \text{ and } x_2 = x_1$$

So $OG = \dfrac{4r}{9\pi}$

When suspended from P, PG is vertical and the required angle θ is given by $\tan^{-1}\left(\dfrac{OG}{OP}\right) = \tan^{-1}\left(\dfrac{4}{9\pi}\right)$

8 a Take OB and OZ as x- and y- axes.

   Centre of mass of pendant is at O, therefore taking moments about OB

$$0 = \frac{1}{2}(2r)h \times \frac{h}{3} - \frac{4r}{3\pi} \times \frac{1}{2}\pi r^2$$

$$\frac{h^2}{3} = \frac{2r^2}{3} \text{ therefore } h = r\sqrt{2}$$

   b $47.6°$

Exercise 25.5A

1 Reactions are $75\,\text{N}$, $200\,\text{N}$
   Tension $T = 75\,\text{N}$

2 Reaction at $B = S = 75\,\text{N}$
   Frictional force $F = 75\,\text{N}$

3 Components $k = 14\,\text{N}$
   $y = 18\,\text{N}$, $22.8\,\text{N}$ at angle of $7.1°$ above rod

4 $2g\,\text{N}$, $2\sqrt{5}\,g\,\text{N}$

5 $30g\,\text{N}$

6 a $8g\,\text{N}$ **b** $98.1\,\text{N}$

Exercise 25.5B

1 $26.6°$

2 $37.8°$

3 $26.6°$

4 a i $\tan^{-1}\left(\dfrac{4r}{h}\right)$ **ii** $\tan^{-1}\left(\dfrac{3r}{h}\right)$

   b $\tan^{-1}\left(\dfrac{3r}{h}\right)$

5 **a** $\dfrac{5}{3}m$ **b** 45°, 54.5°

6 **a** $18g$ N, $28g$ N m anticlockwise

    b 127.3 N, 35.2 N m

7 0 N, 60 N m clockwise

8

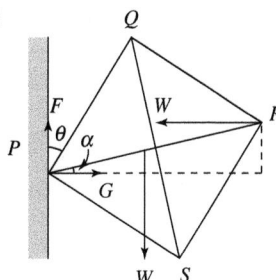

Let R and S be normal reactions at the two ends.
For equilibrium,
resolving horizontally: $S = 0$
resolving vertically: $R = W + w$
Take moments about lower end:
$Wa \cos \alpha + wx \cos \alpha = 0 + M$
Movement of couple, $M = (Wa + wx) \cos \alpha$

9 **a** 84.9 N   **b** 84.9 N

10 $2\sqrt{7g}\ N$ at an angle 70.9°, $\dfrac{\sqrt{3}}{5}$

11 **a** **i** $\left(1\dfrac{2}{3}, 0\right)$ **ii** $\left(1\dfrac{1}{3}, 0\right)$

    b **i** 80.5°   **ii** 64.8°

12 **a** $(12 \ln 1.5, 0)$ **b** 60.4°

13 68.7 N, 85.0 N at 36° to horizontal

14 $T_1 = 7.32$ N and $T_2 = 5.18$ N, 2.2

15 $\dfrac{\sqrt{3}}{7}$

16 0.52 W (to 2 d.p.)

17 2160 N (3 s.f.), 1680 N, 0.218

18

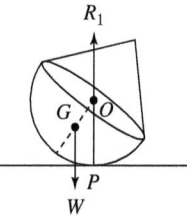

Let diagonals have length $2a$ and PR make angle α with the horizontal.
Let F and G be the friction and normal reaction at P
Resolving horizontally: $G = W$
Resolving vertically: $F = W$
At P, $F \le \mu G$

$$\mu \ge \dfrac{F}{G} = \dfrac{W}{W}$$

So $\mu \ge 1$ and the coefficient must be at least 1
$\theta = 18.4°$

19 a

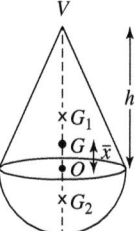

Centre of mass of cone is at G_1
Centre of mass of hemisphere is at G_2
Centre of mass of whole toy is at G.
Standard results give:

$$OG_1 = \dfrac{1}{4}h,\ OG_2 = \dfrac{3}{8}r$$

	Mass	Distance of C of M from O
Cone	$\dfrac{1}{3}\pi r^2 h p$	$\dfrac{1}{4}h$
Hemisphere	$\dfrac{1}{2}\left(\dfrac{4}{3}\pi r^2 p\right)$	$\dfrac{3}{8}r$
Whole	$\dfrac{1}{3}\pi r^2 p(h+2r)$	$\bar{x}$

Take moments about O

$$\dfrac{1}{3}\pi r^2 h\rho \times \dfrac{1}{4}h - \dfrac{1}{2} \times \dfrac{4}{3}\pi r^3 \rho \times \dfrac{3}{8}r$$

$$= \dfrac{1}{3}\pi r^2 \rho\,(h+2r) \times \bar{x}$$

Cancel $\dfrac{1}{3}\pi r^2 \rho$

$$\dfrac{1}{4}h^2 - \dfrac{3}{4}r^2 = (h+2r)\bar{x}$$

$$\bar{x} = \dfrac{h^2 - 3r^2}{4h+8r}$$

b **i** G lies within the hemisphere.
    $\bar{x}$ is negative.
    $h^2 < 3r^2$
    $h < r\sqrt{3}$

Normal reaction always passes through O
The anticlockwise moment about P returns the toy to the vertical position.
It is in stable equilibrium when it is vertical.

ii

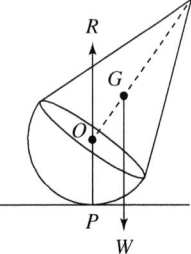

As before, R always passes through O. The clockwise moment about P turns the toy even further away from the vertical and the toy will always topple until the vertex of the cone lies on the surface.

The equilibrium when the toy is vertical is unstable for $h > r\sqrt{3}$

iii

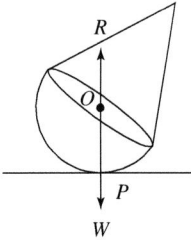

When G and O coincide, reaction R and weight W are equal and opposite.

There is no moment about P in any position. So the toy is in equilibrium in all positions (the equilibrium is said to be 'neutral') when $h = r\sqrt{3}$

20 400 Nm clockwise

21 $1 + \sqrt{3}\,N$, anticlockwise of magnitude $2\sqrt{3} - 2$ Nm

22 a 13 N   **b** 13 N, 31 Nm

23 142 Nm

24 a 17.7 Nm   **b** 20.5 Nm

25 a A total moment about O

$$= \begin{vmatrix} \mathbf{i} & \mathbf{j} & \mathbf{k} \\ n & \frac{1}{2} & 0 \\ 4n & 2 & 4 \end{vmatrix} + \begin{vmatrix} \mathbf{i} & \mathbf{j} & \mathbf{k} \\ \frac{1}{2} & 1 & 0 \\ 2n & -4 & 6 \end{vmatrix} + \begin{vmatrix} \mathbf{i} & \mathbf{j} & \mathbf{k} \\ -1 & -\frac{1}{2} & 1 \\ 4 & -2n & 4n \end{vmatrix}$$

$= (2\mathbf{i} - 4n\,\mathbf{j} + 0\mathbf{k})$

$\quad + (6\mathbf{i} - 3\mathbf{j} + (-2 - 2n)\mathbf{k})$

$\quad + (0\mathbf{i} - (4n - 4)\mathbf{j} + (2n + 2)\mathbf{k})$

$= 8\mathbf{i} + \mathbf{j} + 0\mathbf{k}$, which is independent of n

  b 9.4 Nm

26 a $\begin{pmatrix} 3 \\ -1 \\ -3 \end{pmatrix}$ N   **b** $\begin{pmatrix} -3 \\ 1 \\ 3 \end{pmatrix}$ N, $\sqrt{19}$ N   **c** 12.1 Nm

27 a $m = -1\frac{1}{2}, n = -1$ **b** 81 Nm anticlockwise

28 a $-3\mathbf{i} - 4\mathbf{j} + 3\mathbf{k}, \sqrt{34}$ N   **b** $-11\mathbf{i} - 9\mathbf{j} + 13\mathbf{k}, \sqrt{371}$ Nm

29 a $n = 6$ or -42 **b** The force $\mathbf{Z}$ is either $\begin{pmatrix} 14 \\ 0 \end{pmatrix}$ N or $\begin{pmatrix} 14 \\ 48 \end{pmatrix}$ N

30 a $\begin{pmatrix} 4 \\ -6 \\ 2 \end{pmatrix}$ **b** $-30\mathbf{i} - 18\mathbf{j} + 6\mathbf{k}, 6\sqrt{35}$ Nm

  c i 68.5°   **ii** 68.5°

Review exercise 25

1

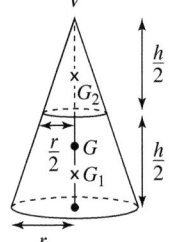

(2.5, 2.1)

2 $\left(5, 3\frac{1}{3}\right)$ cm, 33.7°

3 Centre of mass of remainder is 0.07 cm from O on opposite side of O to the triangle, on the line of symmetry.

4 51.3°

5 69.4°

6 a (1.894, 0)   **b** 42.1°

7

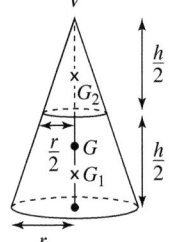

	Mass	Distance of C of M from vertex V
Cone removed	$\frac{1}{3}\pi\left(\frac{r}{2}\right)^2\left(\frac{h}{2}\right) = \frac{\pi r^2 h}{24}$	$VG_2 = \frac{3}{4}\left(\frac{h}{2}\right) = \frac{3h}{8}$
Whole cone	$\frac{1}{3}\pi r^2 h$	$VG_1 = \frac{3h}{4}$
Frustum	$\pi r^2 h\left(\frac{1}{3} - \frac{1}{24}\right) = \frac{7\pi r^2 h}{24}$	$\bar{x}$

Take moments about V

$$\frac{1}{3}\pi r^2 h \times \frac{3h}{4} - \frac{\pi r^2 h}{24} \times \frac{3h}{8} = \frac{7\pi r^2 h}{24} \times \bar{x}$$

Divide by $\pi r^2 h$

$$\frac{1}{4}h - \frac{1}{64}h = \frac{7}{24}\bar{x} \implies \bar{x} = \frac{15h}{64} \times \frac{24}{7} = \frac{45h}{56}$$

It is on the point of toppling when G is vertically above P

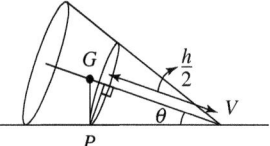

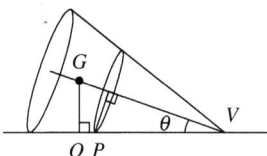

Will not topple if GQ is vertical such that $VQ > VP$

or $VG\cos\theta > \dfrac{\dfrac{h}{2}}{\cos\theta}$

$$\cos^2\theta > \frac{h}{2}\times\frac{1}{\bar{x}} = \frac{h}{2}\times\frac{56}{45h}$$

$$\cos^2\theta > \frac{28}{45}$$

$$\cos\theta > \sqrt{\frac{28}{45}}$$

8 5.6 m

9 0.23 m

10

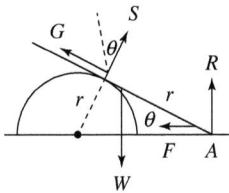

Resolving vertically:

$W = S\cos\theta + G\sin\theta + R$ (1)

Resolving horizontally:
$G\cos\theta + F = S\sin\theta$ (2)

Taking moments about A

$$S\times\frac{r}{\tan\theta} = W\times r\cos\theta$$

$S = W\sin\theta$ (3)

Friction is limiting.

$F = \mu R$ (4)
$G = \mu S$ (5)

[5 equations: 5 unknowns, R, F, S, G, θ]

Substitute from (4) and (5) into (2):

$\mu S\cos\theta + \mu R = S\sin\theta$

Substitute from (3):
$\mu R = W\sin^2\theta - \mu W\sin\theta\cos\theta$ (6)
Substitute from (6) and (5) into (1):

$$W = W\sin\theta\cos\theta + \mu W\sin^2\theta + \frac{W\sin^2\theta}{\mu} - W\sin\theta\cos\theta$$

$$\mu = \mu^2\sin^2\theta + \sin^2\theta$$

$$\sin^2\theta = \frac{\mu}{\mu^2+1}$$

11 166 N vertically downwards, 680 N m

12 a $\sqrt{3}y + x = 25$ **b** $25\sqrt{3}$ N m

13 $a = 2, b = \frac{2\sqrt{3}}{3}, c = \frac{8}{\sqrt{3}}, \frac{4}{\sqrt{3}}$ N m in an anticlockwise direction

Assessment 25

1 a 0.9 m   **b** 5 kg
2 a 0.9 m   **b** 1.6 m
3 42.0°
4 a (2.66, 0.804)   **b** (2.8, 0)
5 15.8 cm
6 55.2 cm
7 (1.54, 0.641)
8 a 34.6 N   **b** 2.33 m
9 32.5°
10 $m = 4, DE = 3$ m
11 a 4.5 kg m^{-2} **b** 0.1125 m from PQ

12 The region is bounded by the x-axis, the line $x = h$ and the line $y = \dfrac{rx}{h}$

By symmetry the centre of mass is on the x-axis.

$$\bar{x} = \frac{\displaystyle\int_0^h xy^2\,dx}{\displaystyle\int_0^h y^2\,dx} = \frac{\dfrac{r^2}{h^2}\displaystyle\int_0^h x^3\,dx}{\dfrac{r^2}{h^2}\displaystyle\int_0^h x^2\,dx} = \frac{\left[\dfrac{1}{4}x^4\right]_0^h}{\left[\dfrac{1}{3}x^3\right]_0^h} = 0.75h$$

13 $m_1 = 1, m_2 = 2$

14 21.8°

15 a (0.5, 0.15)   **b** 12.1°

16 a $\left(\dfrac{e^2+1}{e^2-1}, \dfrac{1}{4e^2}+\dfrac{1}{4}\right)$ **b** $\dfrac{3e^4+1}{\dfrac{1}{2}(e^4-1)}$

17 a 23.4 N
   b The resultant reaction has magnitude 45.6 N and makes an angle of 59.2° with the horizontal (into the wall).

18 a 5.94 m   **b** 19.2 kg

19 a 47.04 N m, 8.64 N m
   b P is a maximum when the reaction is $6g$ N vertically.
     47.3 N at an angle of 47.4° to the horizontal

20 a $\bar{x} = \dfrac{25-27\rho}{100+120\rho}$ m

   $\left(\text{or equivalent, for example, } \dfrac{0.03-0.0324\rho}{0.12+0.648\rho}\right)$

   below point O on the line of symmetry
   b $\dfrac{25}{27}$

21 In relation to the centre, the centres of mass are:

   rod: $\dfrac{6\sin\dfrac{\pi}{6}}{\dfrac{\pi}{6}} = \dfrac{18}{\pi}$

   lamina: $\dfrac{2\times6\sin\dfrac{\pi}{6}}{3\times\dfrac{\pi}{6}} = \dfrac{12}{\pi}$

   Hence $2m\times\dfrac{12}{\pi} + m\times\dfrac{18}{\pi} = 3m\bar{x}$

   This gives $\bar{x} = \dfrac{14}{\pi}$

22 a 71.9 N   **b** 92.9 N at 39.3° to the horizontal

23 a G lies on CF because it is the line of symmetry.
   b 0.802 m   **c** 0.391 m
   d Solving (1) when $\bar{x} = 1.1$ m gives $r = 0.43$ m, but as $FH = 0.4$ m the hole would overlap the edge of the lamina.

24 $\dfrac{13a}{8}$ m

25 18.3 cm

26 a $\left(\dfrac{4a}{3\pi}, \dfrac{4b}{3\pi}\right)$ **b** $\dfrac{3a}{8}$

27 30.3°

28 0.7 kg

29 $\begin{pmatrix}-5\\-1\\3\end{pmatrix}\sqrt{35}$ N, $-13\mathbf{i}+4\mathbf{j}-2\mathbf{k}$, $\sqrt{189}$ N m

Chapter 26
Exercise 26.1A

1 a $\displaystyle\int_1^k \frac{1}{4}x\,dx = 1 \rightarrow \left[\frac{x^2}{8}\right]_1^k = 1$

$\frac{1}{8}(k^2 - 1) = 1 \rightarrow k = 3$

b $\displaystyle F(x) = \int_1^x \frac{1}{4}x\,dx = \left[\frac{x^2}{8}\right]_1^x = \frac{1}{8}(x^2 - 1)$

c $M = \sqrt{5}$

2 a $a = -768$

$F(x) = -768\left(x^4 - \frac{x^3}{3}\right)$

b $F(0.15) = -768\left(0.15^4 - \frac{0.15^3}{3}\right) = 0.48\ (2\text{dp}) \approx 0.5$

Therefore $M \approx 0.15$

3 a $k = \dfrac{2}{9}$

b $F(x) = -\dfrac{2}{9}\left(\dfrac{x^3}{3} - 3\dfrac{x^2}{2}\right)$

$P(X > 2) = \dfrac{7}{27}$

4 a
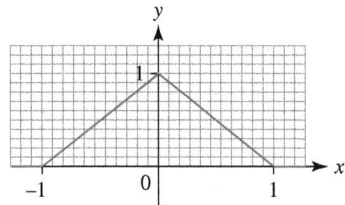

b Area under f(x) is $\dfrac{1}{2} \times 2 \times 1 = 1$

c $F(x) = \begin{cases} \dfrac{1}{2}(x+1)^2; & -1 < x < 0 \\[2mm] 1 - \dfrac{1}{2}(1-x)^2; & 0 < x < 1 \end{cases}$

$1 - \dfrac{1}{2}(1 - Q_3)^2 = \dfrac{3}{4} \rightarrow Q_3 = \dfrac{\sqrt{2} - 1}{\sqrt{2}}$

5 a $k = 30$

b $\dfrac{1}{30}(x^2 + x) = \dfrac{1}{4}$

$2x^2 + 2x - 15 = 0 \rightarrow x = 2.28\ (2\ \text{d.p.})$

$\dfrac{1}{30}(x^2 + x) = \dfrac{3}{4}$

$2x^2 + 2x - 45 = 0 \rightarrow x = 4.27\ (2\ \text{d.p.})$

c $f(x) = \dfrac{1}{30}\dfrac{d}{dx}(x^2 + x) = \dfrac{2x + 1}{30}$

From graph of f(x), modal value is 5

6 a
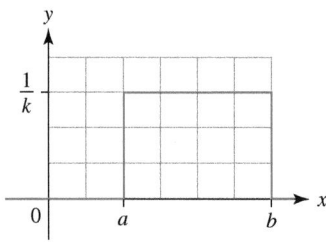

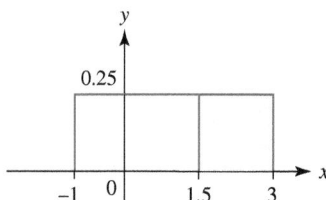

$(b - a) \times \dfrac{1}{k} = 1$

$k = b - a$

b $E(X) = 4$ and $Var(X) = \dfrac{1}{3}$

c $F(x) = P(X < x) = (x - 3) \times \dfrac{1}{2} = \dfrac{x - 3}{2}$

$P(X < 3.5) = (3.5 - 3) \times \dfrac{1}{2} = \dfrac{0.5}{2} = 0.25$

7 a $a = 3$

b

(graph: y-axis value 0.25; x-axis marks at −1, 0, 1.5, 3)

c $P(X > 1.5) = \dfrac{3}{8}$

d $P(X < x) = \dfrac{X + 1}{4}$

$M = 1$

8 a $f(x) \geq 0$ for $0 < x \leq 2$ and area under $f(x) = \dfrac{1}{2} \times 1 + \dfrac{1}{2} \times 1 = 1$

b i $F(x) = \begin{cases} \displaystyle\int_0^x x\,dx; & 0 < x \leq 1 \\[3mm] \dfrac{1}{2} + \displaystyle\int_1^x 2 - x\,dx; & 1 < x \leq 2 \\[3mm] 1; & x > 2 \end{cases}$

Therefore:

$F(x) = \begin{cases} \dfrac{x^2}{2}; & 0 < x \leq 1 \\[3mm] -1 + 2x - \dfrac{x^2}{2}; & 1 < x \leq 2 \\[3mm] 1; & x > 2 \end{cases}$

ii Median $= 1$

IQR $= 0.58$

Exercise 26.1B

1 a $E[T] = 15$ and $Var[T] = 75$ **b** $E(S) = 30 - E(T) = 15$; $Var(S) = Var(T) = 75$

2 a $k = \dfrac{3(c - 1)}{4}$ **b** $E(\sqrt{Y}) = 1.26\ (2\ \text{d.p.})$

3 a $P(X \leq x) = k(-x^2 + 8x - 11)$ for $2 \leq x \leq 4$

$P(2 < X < 4) = P(X < 4) - P(X < 2)$

$= k(-4^2 + 8 \times 4 - 11) - k(-2^2 + 8 \times 2 - 11) = 4k = 1 - 0.2$

$k = 0.2$

b $f(x) = -0.4x + 1.6$

c i $E(X) = \dfrac{7}{3}$ **ii** $P(X < 3) = \dfrac{4}{5}$

4 a $a = \dfrac{2}{5}$

b $F(2) = 0.8$

$P(X < 2) = \dfrac{4}{5} > \dfrac{1}{2} \Rightarrow M < 2$

c $M = \dfrac{5}{4}$

d $f(x) = \begin{cases} \dfrac{2}{5}; & 0 < x \le 2 \\[2mm] \dfrac{6}{5} - \dfrac{2}{5}x; & 2 < x < 3 \\[2mm] 0; & \text{otherwise} \end{cases}$

5 a $k = 12$

b $f(x) = \begin{cases} \dfrac{x}{6}; & 0 < x \le 3 \\[2mm] 2 - \dfrac{x}{2}; & 3 < x \le 4 \\[2mm] 0; & \text{otherwise} \end{cases}$

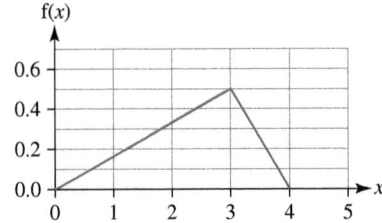

c $M = \sqrt{6}$

$P(X > 3.5) = 1 - F(3.5) = 1 - \left(-3 + 7 - \dfrac{49}{16}\right) = \dfrac{1}{16}$

Exercise 26.2A

1 a $P(X > 3) = 0.37$ **b** $P(1 < X < 3) = 0.35$ (2 dp)

2 a

x	0	0.2	0.4	0.6	0.8	1.0	1.2	1.4
f(x)	2	1.34	0.90	0.60	0.40	0.27	0.18	0.12

b

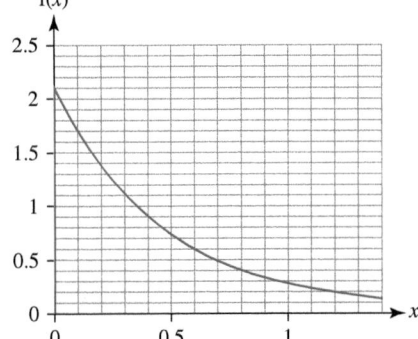

c $x = 0$

3 a $F(x) = 1 - e^{-\lambda x}$

b $F(M) = \dfrac{1}{2} \Rightarrow 1 - e^{-\lambda M} = \dfrac{1}{2}$

$e^{-\lambda M} = \dfrac{1}{2} \Rightarrow -\lambda M = \ln\dfrac{1}{2}$

$\lambda M = \ln 2$

4 a Mean = standard deviation = 4   **b** skewness = 1

5 a $f(x) = 2.5e^{-2.5x}$ **b** $F(x) = 1 - e^{-2.5x}$

c i 0.71   ii 0.61   iii 0.28

d $E(X) = 2.5\displaystyle\int_0^\infty xe^{-2.5x}\,dx = 2.5\left\{\left[\dfrac{-xe^{-2.5x}}{2.5}\right]_0^\infty - \displaystyle\int_0^\infty \dfrac{-e^{-2.5x}}{2.5}\,dx\right\}$

$= 2.5\left\{0 - \left[\dfrac{e^{-2.5x}}{2.5^2}\right]_0^\infty\right\} = \dfrac{2.5}{2.5^2} = 0.4$

$E(X^2) = \displaystyle\int_0^\infty x^2 f(x)\,dx = 2.5\int_0^\infty x^2 e^{-2.5x}\,dx$

$= 2.5\left\{\left[x^2\dfrac{e^{-2.5x}}{-2.5}\right]_0^\infty + \displaystyle\int_0^\infty \dfrac{e^{-2.5x}}{2.5}2x\,dx\right\}$

$= 2.5\left\{0 - 2\left(\left[x\dfrac{e^{-2.5x}}{2.5^2}\right]_0^\infty - \displaystyle\int_0^\infty \dfrac{e^{-2.5x}}{2.5^2}\,dx\right)\right\}$

(Using integration by parts.)

$= 2.5\left\{0 - 2\left(0 + \left[\dfrac{e^{-2.5x}}{2.5^3}\right]_0^\infty\right)\right\} = 2.5\dfrac{2}{2.5^3} = 0.32$

$\mathrm{Var}(X) = E(X^2) - \{E(X)\}^2 = 0.32 - 0.4^2 = 0.16$

6 a Events occur at random with mean 2 per unit time.

b i $\dfrac{1}{12}$ s   ii 0.09

7 a $F(x) = 1 - e^{-4x}$

$f(x) = \dfrac{dF(x)}{dx} = -(-4e^{-4x}) = 4e^{-4x}$

b

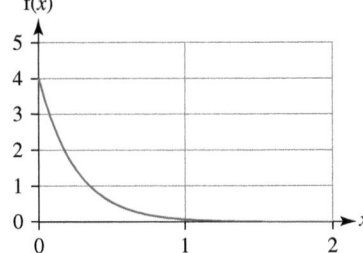

c 0   **d** 0.14

8 a $f(x) = 5e^{-5x}$

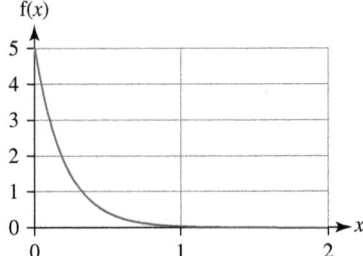

b $F(x) = \displaystyle\int_0^x 5e^{-5x}\,dx = [e^{-5x}]_x^0 = 1 - e^{-5x}$

c Mean $= \dfrac{1}{\lambda} = \dfrac{1}{5} = 0.2$; variance $= \dfrac{1}{\lambda^2} = \dfrac{1}{25} = 0.04$

Exercise 26.2B

1 a $F(t) = P(T < t) = 1 - P(T > t) = 1 - P(\text{no events in } t)$

$= 1 - e^{-6t}$

b $f(t) = 6e^{-6t}$

c $1 - e^{-6M_T} = \dfrac{1}{2} \Rightarrow e^{-6M_T} = \dfrac{1}{2}$

$6M_T = \ln 2 \Rightarrow M_T = \dfrac{1}{6}\ln 2 = 0.12$ (2 d.p.)

d 0.14

2 a 0.018   **b** 22.0

3 a $T \sim \mathrm{Exp}(4)$

0.02

b 0.02. Exponential distribution is memoryless.

c 1.00

4 a $T \sim \text{Exp}\left(\dfrac{1}{2.1} = 0.476\ldots\right)$

0.38 (2 d.p.)

b 0.24

c Failure occurs at random, which is not a characteristic of wear and tear.

5 a $f(t) = \dfrac{1}{3}e^{-\frac{1}{3}t}$; 3, 9

b i $\dfrac{1}{14}$ hour **ii** 0.311

6 a 0.186 **b** 39.06

7 a Mean $\dfrac{2}{15}$ per minute

$f(t) = \dfrac{2}{15}e^{-\frac{2}{15}t}$; $t \geq 0$. $F(t) = 1 - e^{-\frac{2}{15}t}$; $t \geq 0$.

b 0.20

c Mean = 7.5; variance = 56.25

Review exercise 26

1 a $= \dfrac{12}{7}$

$F(x) = \dfrac{12}{7}\left\{\dfrac{x^4}{4} - \dfrac{2x^3}{3} - \dfrac{3x^2}{2} + \dfrac{7}{12}\right\}$

b $F(-0.52) = \dfrac{12}{7}\left\{\dfrac{0.52^4}{4} + \dfrac{2\times0.52^3}{3} - \dfrac{3\times0.52^2}{2} + \dfrac{7}{12}\right\} = 0.50$ (2 d.p.)

$M \approx -0.52$

2 $f(x) = \begin{cases} \dfrac{(1-3x)}{2}; & a < x < \dfrac{1}{3} \\ 0; & \text{otherwise} \end{cases}$

$\dfrac{1}{2}\displaystyle\int_{a}^{\frac{1}{3}}(1-3x)\,dx = 1$

$\dfrac{1}{2}\left[x - \dfrac{3x^2}{2}\right]_{a}^{\frac{1}{3}} = 1$

$\dfrac{1}{2}\left\{\left(\dfrac{1}{6}\right) - \left(a - \dfrac{3a^2}{2}\right)\right\} = 1$

$a = \dfrac{1-2\sqrt{3}}{3}$

b $F(x) = \left(\dfrac{x}{2} - \dfrac{3x^2}{4} + \dfrac{11}{12}\right)$

3 a $(b-a) \times \dfrac{1}{k} = 1$

$k = b - a$

b $E(X) = 0$ and $\text{Var}(X) = \dfrac{4}{3}$

c $F(x) = P(X < x) = (x+2) \times \dfrac{1}{4} = \dfrac{(x+2)}{4}$

4 a $P(2 < X < x) = \left(\dfrac{k}{10}\right)(-x^2 + 10x - 16)$ for $2 < x < 4$.

$P(2 < X < 4) = \left(\dfrac{k}{10}\right)(-16 + 40 - 16) = 1 - 0.1$

$\Rightarrow 0.8k = 0.9$

$k = 1.125$

b $f(x) = -0.225x + 1.125$

c i $E(X) = 2.55$

ii $P(X < 3) = 0.6625$

5 a $\dfrac{1}{12}$ s **b** 0.09

6 a Assume that cars arrive randomly and therefore constitute a Poisson process.

0.264

b 1200 **c** 316.8

7 a $F(t) = P(T < t) = 1 - P(T > t) = 1 - P(\text{no events in } t)$
$= 1 - e^{-10t}$

b $1 - e^{-10M_T} = \dfrac{1}{2} \rightarrow e^{-10M_T} = \dfrac{1}{2}$

$10M_T = \ln 2 \rightarrow M_T = \dfrac{1}{10}\ln 2 = 0.069$ (3 d.p.)

c 0.036

8 a $T \sim \text{Exp}(0.6)$

b 0.050

c 0.050 (3dp). Exponential distribution is memoryless; outcome of first two minutes is not 'remembered'.

d 0.11

Assessment 26

1 a $\dfrac{7}{32}$

b $E(X) = \dfrac{19}{6}$, $\text{Var}(X) = \dfrac{11}{36}$

c $E(2X+1) = \dfrac{22}{3}$, $\text{Var}(2X+1) = \dfrac{11}{9}$

2 a $k = \dfrac{1}{10}$

b Section of a positive quadratic curve between $x = 0$ and $x = 2$. y-intercept is 0.1

c 2

d $\dfrac{1}{10}\displaystyle\int_{0}^{m}(1+3x^2)\,dx = \dfrac{1}{2} \Rightarrow \left[x + x^3\right]_0^m = 5$

$m + m^3 - 0 = 5$

$m^3 + m - 5 = 0$ as required

3 a $k = \dfrac{1}{7}$ **b** $\dfrac{19}{56}$

c $f(x) = \begin{cases} \dfrac{3}{7}x^2; & 1 \leq x \leq 2 \\ 0, & \text{otherwise} \end{cases}$

4 a Rectangular distribution drawn on axes, with boundaries $x = 1$, $x = 5$, $y = 0.25$; graph continues along x-axis for $x \leq 1$ and $x \geq 5$

b $P(X > 2) = \dfrac{3}{4}$

c $E(X) = 3$, $\text{Var}(X) = \dfrac{4}{3}$

5 a Mean = 4000 days, standard deviation = 4000 days

b 0.451

6 a $f(x) = \begin{cases} 5e^{-5x}; & x > 0 \\ 0; & \text{otherwise} \end{cases}$

b $F(x) = \begin{cases} 0; & x < 0 \\ 1 - e^{-5x}; & x > 0 \end{cases}$

c 0.865

7 a $F(x) = \begin{cases} 0; & x < 0 \\ \dfrac{4}{3}\left(\dfrac{x^4}{4} + \dfrac{x^2}{2}\right); & 0 \leq x \leq 1 \\ 1; & x > 1 \end{cases}$

b $E(5X - 3) = \dfrac{5}{9}$, $\text{Var}(5X - 3) = \dfrac{101}{81}$

8 a $F(x) = \begin{cases} 0; & x < 2 \\ \dfrac{x^2}{48} + \dfrac{1}{12}x - \dfrac{1}{4}; & 2 \le x \le 6 \\ 1; & x > 6 \end{cases}$

b $-2 + 2\sqrt{10}$ **c** $\dfrac{13}{486}$

9 $F(x) = \begin{cases} 0; & x < 4 \\ \dfrac{x}{6} - \dfrac{2}{3}; & 4 \le x \le 10 \\ 1; & x > 10 \end{cases}$

10 a $\dfrac{4}{9}$

b Phone calls may last longer than 10 minutes.

c i 0.2497   **ii** 0.403

11 $\displaystyle\int_0^a 2x^2 e^{-2x}\,dx = \left[\left(-x^2 - x - \dfrac{1}{2}\right)e^{-2x}\right]_0^a$

$= \left(-a^2 - a - \dfrac{1}{2}\right)e^{-2a} - \left(0 - 0 - \dfrac{1}{2}\right)e^0$

$= -\left(a^2 + a - \dfrac{1}{2}\right)e^{-2a} + \dfrac{1}{2}$

$\displaystyle\int_0^\infty 2x^2 e^{-2x}\,dx = \lim_{a\to\infty}\left(\dfrac{1}{2} - \left(a^2 + a - \dfrac{1}{2}\right)e^{-2a}\right)$

$= \dfrac{1}{2}$

$\displaystyle\int_0^a 2x e^{-2x}\,dx = \left[\left(-x - \dfrac{1}{2}\right)e^{-2x}\right]_0^a$

$= \left(-a - \dfrac{1}{2}\right)e^{-2a} - \left(-0 - \dfrac{1}{2}\right)e^0$

$= -\left(a + \dfrac{1}{2}\right)e^{-2a} + \dfrac{1}{2}$

$\displaystyle\int_0^\infty 2x e^{-2x}\,dx = \lim_{a\to\infty}\left(\dfrac{1}{2} - \left(a + \dfrac{1}{2}\right)e^{-2a}\right)$

$= \dfrac{1}{2}$

$\mathrm{Var}(X) = \dfrac{1}{2} - \left(\dfrac{1}{2}\right)^2$

$= \dfrac{1}{4}$

12 pdf is $f(u) = \dfrac{1}{k}$ for $a \le u \le b$

$\mathrm{Var}(U) = \displaystyle\int_a^b \dfrac{u^2}{k}\,du - \left(\int_a^b \dfrac{u}{k}\,du\right)^2$

$= \left[\dfrac{u^3}{3k}\right]\dfrac{2k}{k} - \left(\left[\dfrac{u^2}{2k}\right]\dfrac{2k}{k}\right)^2$

$= \dfrac{8k^3 - k^3}{3k} - \left(\dfrac{4k^2 - k^2}{2k}\right)^2$

$= \dfrac{7k^2}{3} - \dfrac{9k^2}{4}$

$= \dfrac{28k^2 - 27k^2}{12}$

$= \dfrac{1}{12}k^2$ as required

13 a $F(x) = \begin{cases} 0; & x < 1 \\ \ln x; & 1 \le x \le e \\ 1; & x > e \end{cases}$

b Standard deviation is $\dfrac{1}{2}\sqrt{8e - 2e^2 - 6}$

14 a $2\sqrt{3}$

b $E(2X^{-1}) = 2 - \sqrt{2}$, $\mathrm{Var}(2X^{-1}) = \ln 2 - 6 + 4\sqrt{2}$

15 a

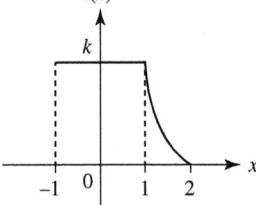

b $k = \dfrac{3}{7}$

c $F(x) = \begin{cases} 0; & x < -1 \\ \dfrac{3}{7}x + \dfrac{3}{7}; & -1 \le x < 1 \\ \dfrac{1}{7}(x-2)^3 + 1; & 1 \le x \le 2 \\ 1; & x > 2 \end{cases}$

d $\dfrac{1}{6}$

e $E(4X) = \dfrac{5}{7}$, $\mathrm{Var}(4X) = \dfrac{3039}{245}$

16 0.283

Exponential distribution with $\lambda = \dfrac{2}{3}$

Chapter 27
Exercise 27.1A

1 While both concern mistakes, type I errors occur when the null hypothesis is true and type II errors occur when the alternative hypothesis is true.

2 a 5%   **b** 48.2%   **c** 0.0588%

3 a $X \ge 12$ **b** 4.48%   **c** 73.0%   **d** 0.27

4 a 77.1%, 0.229   **b** 41.6%, 0.584   **c** 2.10%, 0.979

5 a 0.852

b As the true mean gets further from the hypothesis mean, the smaller the probability of obtaining an outcome not in the critical region, so the probability of making a type II error decreases.

6 a 11 or more   **b** 0.490

7 a They would get a result inside the critical region and determine that the mean weight is not the advertised weight, when actually it is.

b They have accepted a null hypothesis which is false, so they have made a type II error.

8 a Because $P(X \le 15) = 9.55\% < 10\%$ and $P(X \le 16) = 15.61\% > 10\%$

b 9.55%   **c** 71.99%

d $N(50 \times 0.4, 50 \times 0.4 \times 0.6)$. For a Normal distribution you can test the mean value, which corresponds to the expected number of outcomes $(n \times p)$.

e 15.56

f $N(17.5, 11.375)$. $P(X \geq 15.56 \mid X \sim N(17.5, 11.375)) = 71.74\%$

9 a A type I error occurs when a null hypothesis that is true is rejected. If both tests have the same probability of making such an error then a fair comparison can be made. It is easy to control the probability of making a type I error.

b A type II error occurs when a null hypothesis that is false is rejected. The statistician should use the test that has a lower probability of making such an error, since overall it produces a lower probability of error.

Exercise 27.1B

1 a Critical value = 1.07, P(type II error) = 35%, power = 0.650

b Critical value = 1.29, P(type II) = 49.1%, power = 0.509

c Critical value = 1.70, P(type II) = 74.5%, power = 0.255

2 a i $X \geq 5$ **ii** 91.6% **iii** 0.0838

b i $X \geq 6$ **ii** 97.0% **iii** 0.0300

3 a 8

b P(type II error) = 97.8%. Power = 0.022

c 800

d This expected number is inside the critical region. You should expect the power to be much closer to 1 than for the 20-trial test, due to the larger sample size.

4 a The treasure hunter does not want to miss any valuables.

b If the hunter digs a little and finds nothing valuable then they can keep trying other locations.

5 a $X = 0$ **b** 1.43%

c P(type II error) = 99.99038%, power = 0.00961%

6 a $X \geq 5$; 0.249 **b** $X \geq 8$; 0.399 **c** $X \geq 11$; 0.492

7 a Since 5 games are played, 1.2 goals per game leads to 6 goals per 5 games. The alternative is > 6 because the team thinks they might be scoring more goals. A two-tailed test would be inappropriate here as the team is not concerned with the probability of the training leading to fewer goals.

b $X \geq 11$; 4.26% **c** 90.1%

d Most fans probably wouldn't notice an extra goal every 5 away matches, and it would have a small impact on overall performance, game by game or in the league.

e Bookmakers offer odds on specific scores in matches and this is a popular type of bet. An extra goal every 5 away games will change the odds enough that the company could lose a lot of money making bad predictions.

Exercise 27.2A

1 a 2.271 **b** −1.559 **c** −6.623 **d** −4.964

2 a 0.985 **b** 1.92 **c** 1.60

3 a −2.669 **b** 11 **c** 0.0109

d This is smaller than the significance level so there is sufficient evidence to reject the null hypothesis. You conclude that 1.61 is not a reasonable value for the mean of the parent population.

4 a $T < -1.75$ **b** $t = -0.372$

c This is not in the critical region; there is insufficient evidence to reject the null hypothesis. You conclude that −2.1 is a reasonable value for the mean of the parent population.

5 t-test statistic is −2.860. The p-value is 0.670%, or the critical region is $T < -2.65$ or $T > 2.65$. The result is significant, so 3.17 is not a reasonable value for the mean of the parent population.

6 a $t = -1.88$ **b** 7 **c** 5.10%

d The p-value is larger than the significance level so there is insufficient evidence to reject the null hypothesis. You conclude that 18.2 is a reasonable value for the mean of the parent population.

e The population must be normally-distributed and the sample must be random.

7 a $T < -2.13$ **b** −0.608

c The t-test statistic is smaller than the critical value. There is insufficient evidence to reject the null hypothesis. You conclude that 129.1 is a reasonable value for the mean of the parent population.

8 a Yes as heights are reasonably approximated by a Normal distribution and the sample can be taken at random.

b The student would need to ensure that the sample is taken at random from the population, therefore avoiding possible bias.

Exercise 27.2B

1 a When the sample size is large, the distribution approaches a Normal distribution. When the sample size is 30 the distribution is approximately a Normal distribution, so when it is 1000 the difference will be negligible and it is acceptable to use a Normal distribution.

b 8.92

c The critical value is 1.6463803. 8.92 > 1.646… so reject H_0 in favour of H_1 and conclude that there is sufficient evidence at the 5% level of significance to suggest that the true mean is greater than 12.1

2 a The t-test statistic is −0.157, which has a p-value of 44.1%. This is larger than the significance level so there is insufficient evidence to reject the null hypothesis.

b The t-test statistic is −1.553, which has a p-value of 7.74%. This is smaller than the significance level so there is sufficient evidence to reject the null hypothesis.

c The t-test statistic is −1.095, which has a p-value of 14.6%. This is larger than the significance level so there is insufficient evidence to reject the null hypothesis.

3 a 13.4 **b** 20.2 **c** 4.14

4 a The cans should not have been produced together in the factory. Cans should be purchased from different locations if possible, and over a period of time.

b The sample must come from a Normal distribution for a t-test to be appropriate.

c The p-value of the sample is 6.46%. This is larger than the significance level so there is insufficient evidence to suggest that the sample has not come from a parent Normal distribution with mean 4.26 g.

5 a 1.83%

b The p-value is smaller than the significance level so there is sufficient evidence to reject the null hypothesis. You conclude that the dart player is throwing more than 0.31 cm away from the bullseye, on average.

c While the mean is higher, it could be that the dart thrower is consistently getting the darts to land 0.68 cm away at a specific point and not all around the target radially.

6 a For sample 1 the t-test statistic is −1.840. This has a p-value of 4.95%, which is smaller than the significance level. There is sufficient evidence with sample 1 to suggest that the true population mean is less than 14.7.
For sample 2 the t-test statistic is −1.779. This has a p-value of 4.85%, which is smaller than the significance level. There is sufficient evidence with sample 2 to suggest that 14.7 is not a suitable mean value of the population Normal distribution.

b For the combined sample the t-test statistic is -2.51. This has a p-value of 0.96%, which is smaller than the significance level. There is sufficient evidence with the combined sample to suggest that 14.7 is not a suitable mean value of the population Normal distribution.

7 a The t-test statistic is -1.403, which has a p-value of 9.20%. The significance level for this tail is 5%, so the result is not significant. There is insufficient evidence to reject the null hypothesis.

b The t-test statistic is 1.426, which has a p-value of 9.09%. The significance level for this tail is 5%, so the result is not significant. There is insufficient evidence to reject the null hypothesis.

c Each set of pipes could reasonably have come from a machine with mean 30.0 cm, so the engineer should believe that the two sets of pipes have come from the same machine. They might want larger samples to get a better idea though.

8 a (12.1, 13.6)   **b** (9.28, 12.46)
9 a (0.241, 18.3)   **b** (9.22, 12.5)
10 a (18.8, 31.7)   **b** (21.4, 28.6)   **c** (22.9, 27.1)

Review exercise 27

1 a $\alpha \le 5$ **b** $P_2 > P_1$
2 a $X \ge 18$ **b** 7.96%
   c i 44.1%   **ii** 0.559
3 a 2%   **b** 89.2%   **c** 33.2%
4 a 0.536   **b** -1.24 **c** -0.182
5 a The sample size isn't too large.   **b** 0.457
6 a 6   **b** 0.0142
   c Since the p-value is lower than the significance level there is sufficient evidence to reject the null hypothesis. You conclude that the mean is lower than -5.14.
7 a 72.14   **b** 25.87   **c** -20.43

Assessment 27

1 a 9   **b** 0.028   **c** 0.593
2 a $H_0: p = 0.3$, $H_1: p \ne 0.3$ **b** $X \le 2, X \ge 10$
   c 0.751   **d** 0.249
3 a 29.14   **b** 0.39   **c** 0.61
4 a A type II error occurs when the null hypothesis is accepted when it is false. The power of a test is the probability of the complementary event when the null hypothesis is rejected when it is false.
   b i $H_0: \mu = 330$, $H_1: \mu > 330$ **ii** $\bar{X} < 332.08$
    iii 0.70   **iv** 0.30
5 a 7.7317   **b** -1.692
   c p value $= 0.1187$, therefore do not reject null hypothesis
6 a Sample is not large enough to assume that sample and population standard deviations are approximately equal.
   b i $\mu = 10.2$, $H_1: \mu > 10.2$ **ii** 2.8644
    iii p value $= 0.0093$, therefore reject the null hypothesis.
7 a $\mu = 0$, $H_1: \mu > 0$
   b X is normally distributed; $t = 1.8861$, p value $= 0.048$, therefore reject null hypothesis.
   c Result is only just significant; small sample suggests further research would be a good idea.
8 (13.39, 18.18)
9 (4.03, 9.43)
10 a Crop yield is normally distributed.
   b For a Normal distribution with unknown population variance, sample size must be large.
   c 44.2 and 1.31

Chapter 28
Exercise 28.1A

1 a Either DEF or EFG. For example $\{AB, AC, BC\}$ correspond to $\{DE, DF, EF\}$.
   b The sub-graph formed by removing EF can be draw like this: This is a subdivision (by G) of DEF

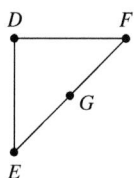

2 Removing the edge CD gives a subdivision (by C) of the K_4 graph $ABDE$

3 Removing edges AB and CD gives the complete bipartite graph joining the vertex sets $\{A, B, E\}$ and $\{C, D\}$.

4 G_1 and G_4 are isomorphic – $\{A, B, C, D, E\}$ corresponds to $\{V, X, Y, W, Z\}$
G_2 and G_3 are isomorphic – $\{J, K, L, M, N\}$ corresponds to $\{Q, P, R, T, S\}$.

5 G_1 is planar:

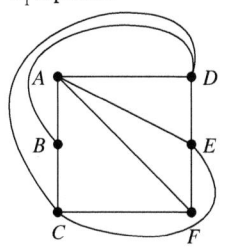 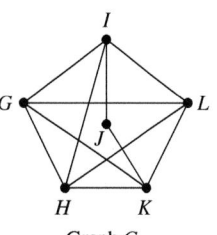

Graph G_2

G_2 contains a subdivision (by J) of the K_5 graph formed by G, H, I, K and L, so it is non-planar.
G_3 contains the complete bipartite graph $K_{3,3}$ joining $\{M, P, R\}$ and $\{N, O, Q\}$, so is non-planar.

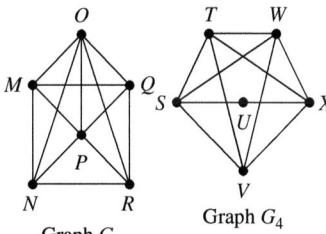

Graph G_3 Graph G_4

G_4 contains a subdivision (by U) of the K_5 graph formed by S, T, V, W and X, so it is non-planar.

6 a   **b**

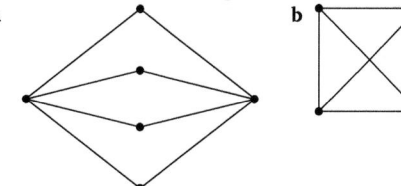

7 a

	A	B	C	D	E
A	0	1	0	1	0
B	1	0	1	0	0
C	0	1	0	1	1
D	1	0	1	0	0
E	0	0	1	0	0

b

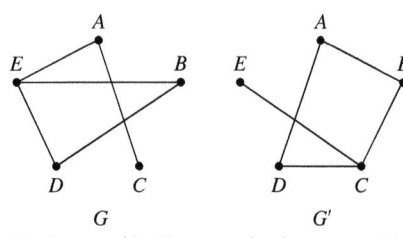

G G'

c Not isomorphic. For example, the vertex with degree 1 in G connects to a vertex of degree 2, but in G' it connects to a vertex with degree 3

8 The graph can be drawn as shown. Removing EG and BD gives a subdivision (by G) of $K_{3,3}$. Hence by Kuratowski's theorem the graph is non-planar.

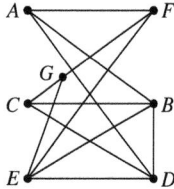

Exercise 28.1B

1 a $n-1$

b If T and T' are isomorphic they have the same number of edges.

K_n has $\dfrac{n(n-1)}{2}$ edges, so T and T' must each have $\dfrac{n(n-1)}{4}$

Hence $\dfrac{n(n-1)}{4} = n-1 \Rightarrow n^2 - 5n + 4 = 0$, giving

$n = 4$ (or $n = 1$ which is trivial).

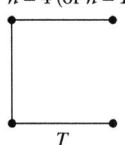   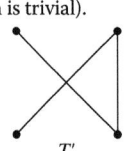

T T'

2 a There are 4

b There are 6

c i There are 4

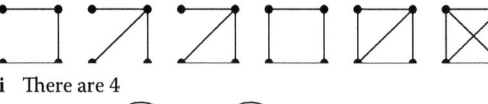

ii There are 3 more:

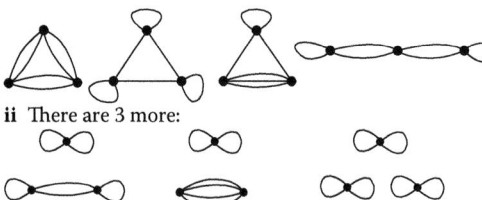

3 a d is odd but these odd vertices occur in pairs, so n is even.

b $d = 2$ or 4 ($d = 3$ gives $K_{3,3}$ and $d = 5$ gives K_5 as a subgraph)

4 a

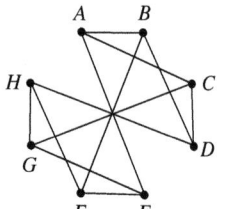

b i 2   **ii** 3

c AC and EG 'inside', BF and DH 'outside', so planar plane drawing:

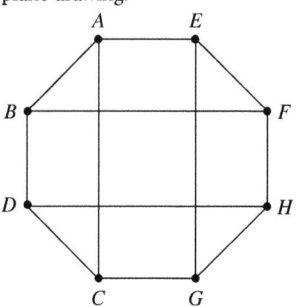

(The plane drawing is the cube graph, which you may have realised because the 'codes' are the 3D coordinates of the vertices of a unit cube.)

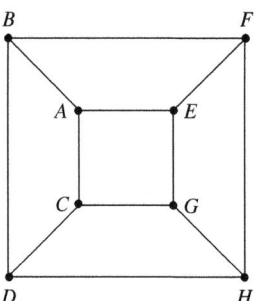

5

6 a

b   and

c i K_6 has 15 edges. As G and G' combine to form K_6, they must have different numbers of edges, so they are not isomorphic.

ii $\dfrac{n(n-1)}{2}$ must be even, so either n or $(n-1)$ is a multiple of 4

So $n = 4m$ or $n = 4m + 1$

7 a

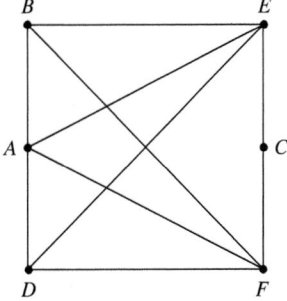

b $m \geq 3$ and $n \geq 3$

c Suppose the vertices of H are in two sets, $\{A_1, A_2, ...A_m\}$ and $\{B_1, B_2, ..., B_n\}$.

For vertices A_i and B_j there will be an edge A_iB_j in H' Whether or not A_i and A_k were connected in H, there will be a trail $A_iB_jA_k$ between them in H'. Similarly, there will be a trail joining any two B vertices.

Hence there is a trail connecting any two vertices in H', so H' is a connected graph.

If H consists of more than two disjoint sets of vertices, the same argument applies between any two sets, so H' will be connected.

8 The graph does not contain a sub-division of K_5 (although you could generate one by "collapsing" C down onto D. The resulting graph is called a minor. Try Googling Wagner's theorem. A similar process is possible for the Petersen graph in question 5).

However, the graph does contain $K_{3,3}$ as is shown in this diagram, so the graph is non-planar.

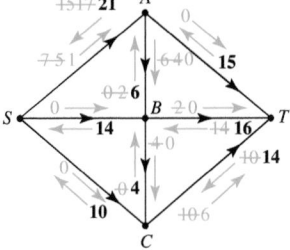

Exercise 28.2A

(Final flows are shown in black. There may be other possible flow patterns.)

1 a Flow-augmenting paths
SABT 2 and SABCT 4
Maximum flow = 45

b There is a cut $\{SC, BC, BT, AT\}$ with capacity 45. By the maximum flow-minimum cut theorem flow = cut = 45 means 45 is maximal.

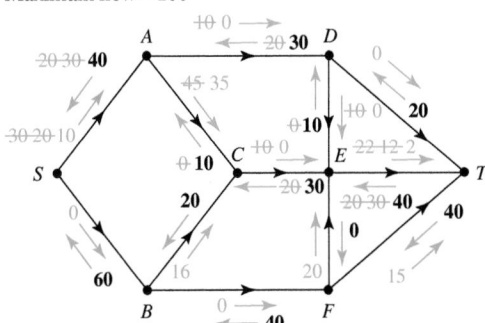

2 a Flow-augmenting paths SADET 10 and SACET 10
Maximum flow = 100

b Cut $\{AD, CE, BF\} = 100$
Flow = cut = 100 means 100 is maximal.

3 Flow-augmenting path SBCADET 10
Maximum flow = 100

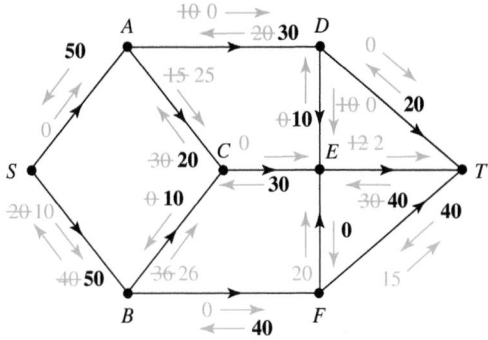

4 Flow-augmenting paths
SBCET 2, SACT 2, SADCT 1 and SABCET 1

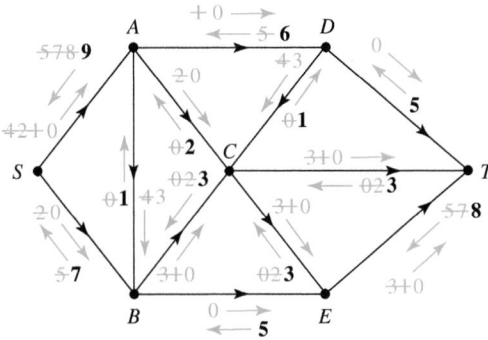

5 a Take, for example, initial flow SADT 24, SCT 38 and SBET 40
Flow augmenting paths SCET 14 and SACBET 2
Maximum flow = 118

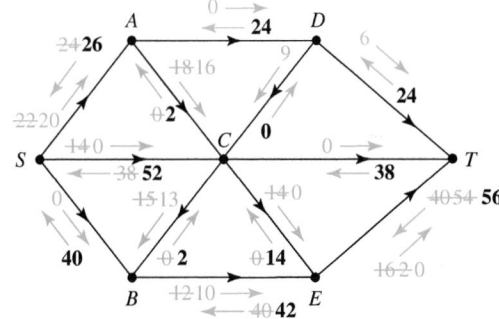

b Cut $\{AD, CD, CT, ET\} = 118$
Flow = cut = 118 means 118 is maximal.

6 These are just examples.

a

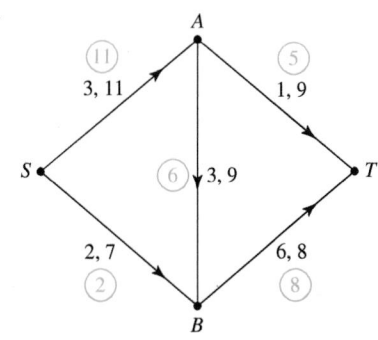

b No flow exists, because at B maximum inflow = 6 but minimum outflow = 5 + 3 = 8

c

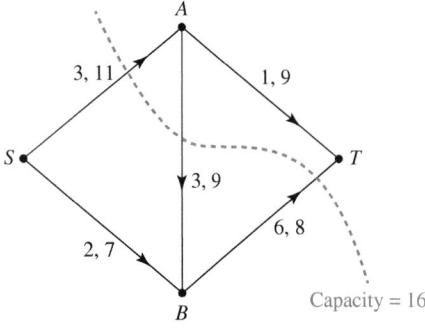

d

7 a

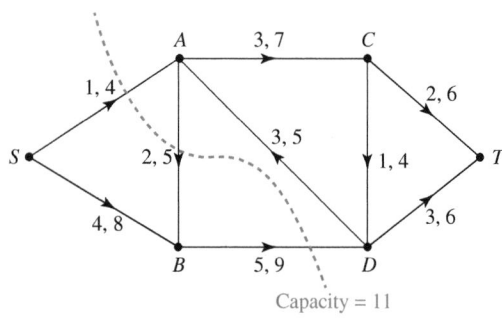

Capacity = 16

b No feasible flow

c

Capacity = 20

d

Capacity = 11

8 a Flow augmenting paths $SACT$ 1 and $SBDT$ 2
Maximum flow = 15
Final flows as in second diagram.

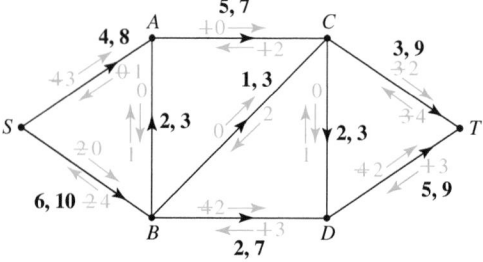

b Cut $\{SB, AB, AC\} = 15$
Flow = cut, so maximal.

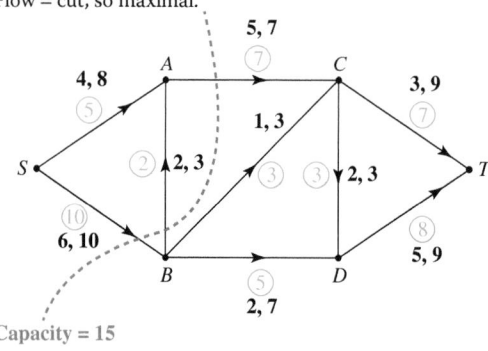

Capacity = 15

Exercise 28.2B

1 a, b Take, for example, initial flow
$SEFG$ 8, $SACG$ 3 and $SADG$ 2
Flow-augmenting paths $SACDG$ 1 and
$SABCDG$ 1
Maximum flow = 15

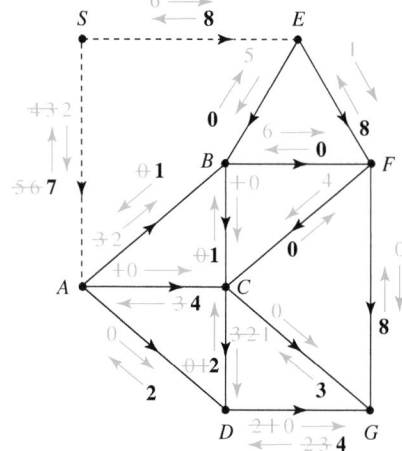

c

2 a, b Take, for example, initial flow *SACFIT* 5 and *SBEHT* 6
Flow-augmenting paths *SBDGJHT* 5, *SBDGIT* 2 and
SBEDGHT 1

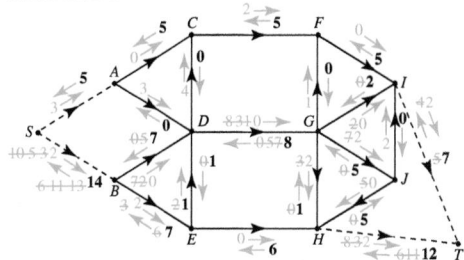

Maximum flow = 19
This is confirmed by cut {*EH, DG, FG, FI*} = 19
So expected revenue = 1900 × 0.4 × 4 = £3040

3 Convert *B* to B_1B_2 with capacity 7
Take, for example, initial flow SB_1B_2T 4 and *SACT* 4
Flow-augmenting path SAB_1B_2T 3
Maximum flow = 11
Confirmed by cut {B_1B_2, AC} = 11

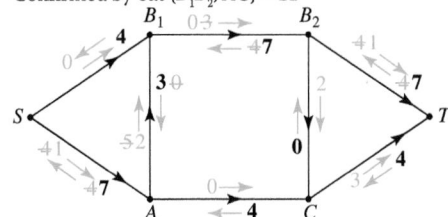

4 a Sources *B* and *D*, sinks *I* and *K*
  b Take, for example, initial flow *SBAEIT* 6, *SBFJIT* 5,
SBCGKT 5 and *SDHLKT* 10
Flow-augmenting path *SBFJKT* 2
Maximum flow = 28
  c Cut {*AE, EF, FJ, GK, KL*} = 28

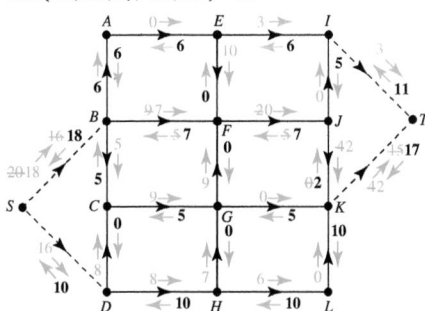

5 a Sources *A* and *B*, sinks *F* and *G*
  b, c Take, for example, initial flow SAC_1C_2DFT 5 and
$SBDE_1E_2GT$ 5
Flow-augmenting paths *SADGT* 4, SBE_1E_2GT 1 and
SBE_1DC_2FT 1
Maximum flow = 16
Confirmed by cut (C_1C_2, AD, BD, BE_1) = 16

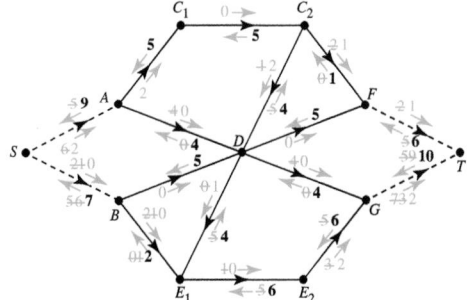

6 The first diagram shows working starting from a total initial
flow of 13

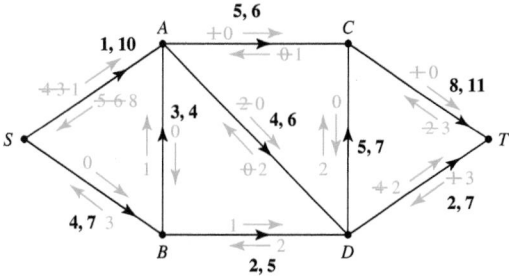

The second diagram shows final flows for a maximum flow of
16, with a cut of 16 to confirm this.

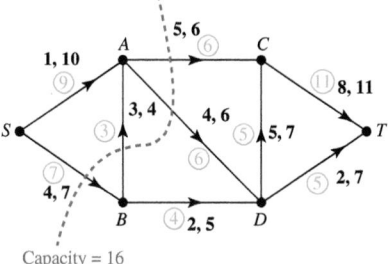

Capacity = 16

7 The first diagram shows the network with an initial feasible
flow of 22

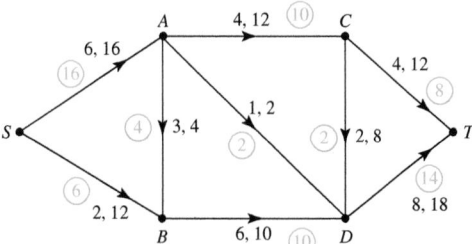

The second diagram shows the working using a flow-
augmenting path *SBACT* 1. The flow (23) is then maximal.
There is a cut {*SA, AB, BD*} with a capacity 23

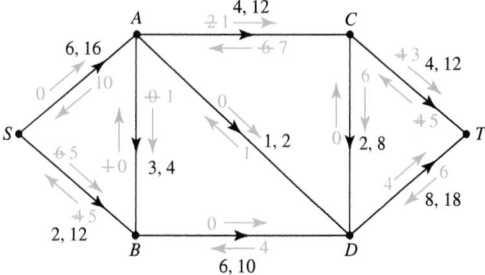

If there were no minimum requirements there is a flow of
24 (*SACT* 12, *SBDT* 10 *SADT* 2) and a cut of 24 {*AC, AD, BD*},
so the maximum flow is then 24

8 An initial feasible flow is as shown (others are possible).

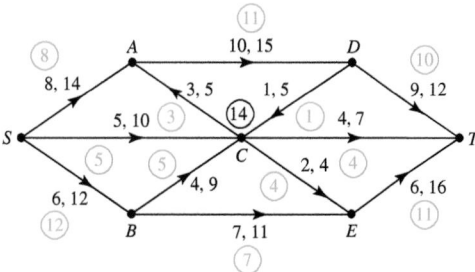

Split C into incoming node C_1 and outgoing node C_2. C_1C_2 has maximum capacity 14, and can be assumed to have minimum capacity 10 to be consistent with incoming values.

The second diagram shows the working using flow-augmenting paths SC_1C_2T 3, $SADT$ 2 and SC_1BET 1. This gives a total flow of 31, consistent with the cut shown.

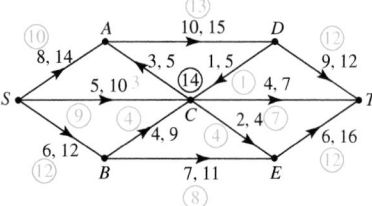

The third diagram shows the final flows.

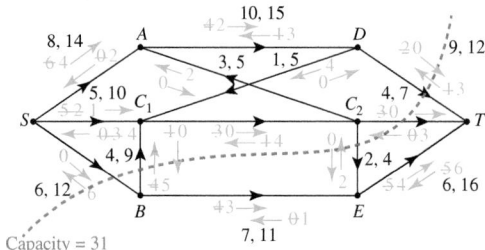

Capacity = 31

Review exercise 28

1 **a**

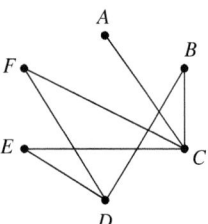

b

	A	B	C	D	E	F
A	0	1	0	1	1	1
B	1	0	0	0	1	1
C	0	0	0	1	0	0
D	1	0	1	0	0	0
E	1	1	0	0	0	1
F	1	1	0	0	1	0

c

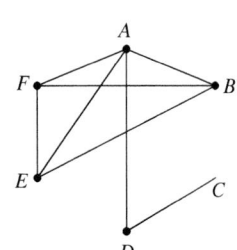

2 G_2 and G_3 are isomorphic. P corresponds to Z, S to Y and the remaining three can be in any order.

3 **a** $V = 6$ and $E = 9$. $V + F - E = 2$, so $F = 5$
These are two possible graphs.

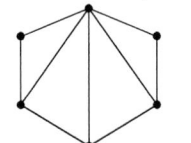 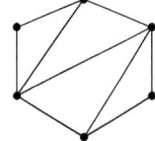

b $V = 7$, $E = 16$. From Euler a planar graph satisfies $E \le 3V - 6$. Here $3V - 6 = 15$, so the graph is non-planar.

4 The graph contains a sub-graph which is a sub-division of K_5, as shown, so by Kuratowski's theorem it is non-planar.

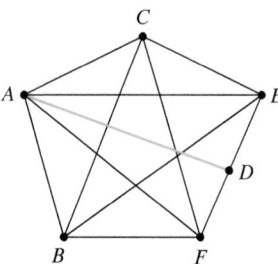

5 **a** The maximum outflow from B is 14, so the inflow ≤ 14
b **i** $SADT$ 5 **ii** $SBET$ 9
c Flow-augmenting paths $SBCT$ 4, $SACDT$ 4 and $SBCDT$ 1
Maximum flow = 23

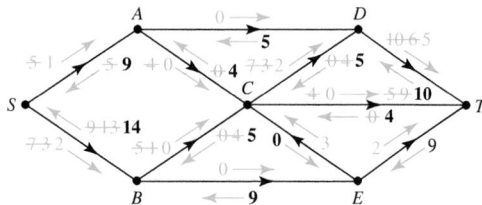

d Cut $\{AD, AC, BC, BE\} = 23$
Flow = cut = 23 means maximum flow is 23 (by the maximum flow-minimum cut theorem).

6 **a** Sources A and B, sinks D and E
b, c Take, for example, initial flow $SADT$ 4 and $SBET$ 3
Flow-augmenting paths $SACDT$ 8, $SBCET$ 3 and $SACET$ 2

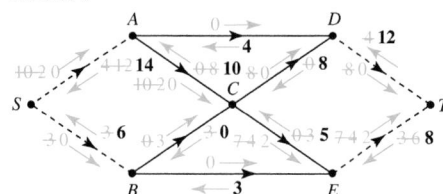

Maximum flow = 20
(You know because SA and SB are saturated.)

7 **a, b** Take, for example, initial flow $SADT$ 4 and $SBET$ 3
Flow-augmenting path SAC_1C_2DT 8
Maximum flow = 15
Confirmed by cut $(AD, C_1C_2, DE\} = 15$

8 **a** There is no feasible flow for Network 2
Node A has a maximum inflow of 7, but needs a minimum outflow of 8
b **i** $9 + 12 - 3 = 18$
ii This is the initial flow pattern.

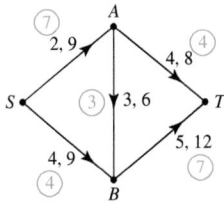

Using the labelling procedure with flow-augmenting paths *SAT* 2 and *SBT* 5 gives:

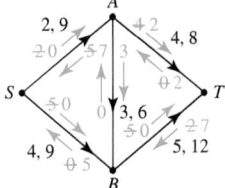

These are the final flows. Total flow = 18

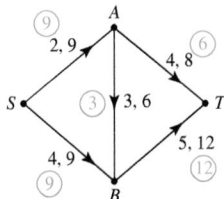

iii Flow = 18 and there is a cut of 18, so the flow is maximal by the maximum flow-minimum cut theorem.

Assessment 28

1 a The graph does not contain K_5 or $K_{3,3}$ as a subgraph therefore Kuratowski's theorem tells us it is planar.

2 a

	A	B	C	D	E
A	0	0	1	1	0
B	0	0	0	0	0
C	1	0	0	0	1
D	1	0	0	0	0
E	0	0	1	0	0

b i G_2 is isomorphic.
$J = A, F = E, I = B, G = D, H = C$

ii G_3 is not isomorphic: order of vertices is 2, 4, 2, 4, 2 not 2, 4, 2, 3, 3

3 a Flow-augmenting paths *SBCFES* (9)
SBEDS (5)
Final network:

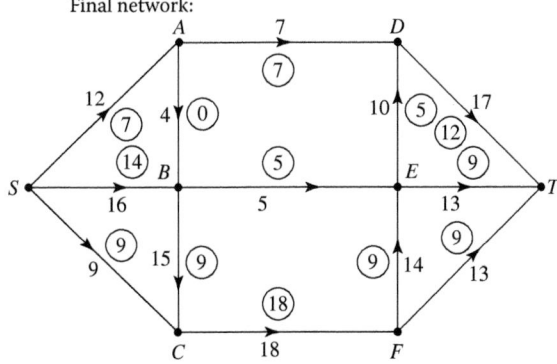

Maximum flow is 30

b A possible cut is {*AD, BE, CF*} of size 30 so flow of 30 is maximal using the maximum flow-minimum cut theorem.
c *AB* could be removed as no flow along this edge
d *SC, CF, AD* are saturated.

4 a, c

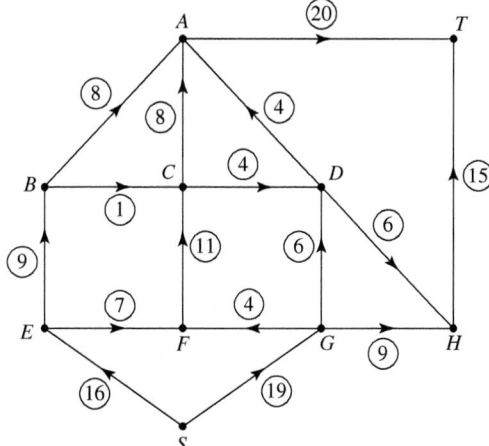

b For example, initial flow of 8 through *SEBA*
initial flow of 9 through *SGHT*
Flow-augmenting paths
SEFCAT (7)
SBCAT (1)
SGFCDAT (4)
SGDHT (6)

Maximum flow is 35

d Cut {*GH, GD, FC, EB*} has capacity 35
So 35 is maximum-flow using maximum flow-minimum cut theorem.

e *EB* as all edges from *F* are saturated
Only worth adding additional capacity of 1 as *BA* saturated so will go along *BC* then *CA* saturated so will go along *CD* which only has a spare capacity of 1. *DH* and *DA* both have plenty of spare capacity so can reach sink node.

5 a

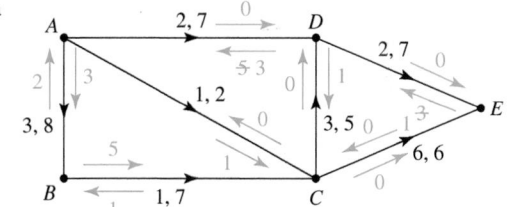

b Flow augmenting paths
ADE (+2)
and *ABCDE* (+1)

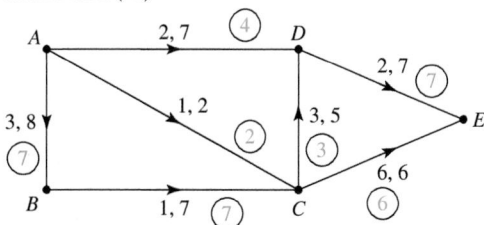

c Capacity of cut = 7 + 2 + 7 − 3 = 13
Therefore 13 is maximum-flow using maximum flow-minimum cut theorem.

6 a

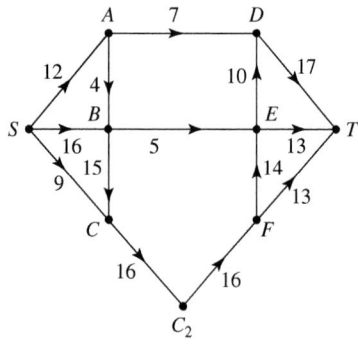

b e.g. initial feasible flow of 9 along SCC_2FT, 7 along $SADT$ and 5 along $SBET$

Flow augmenting path For example, $SBCC_2FET$ (+7)

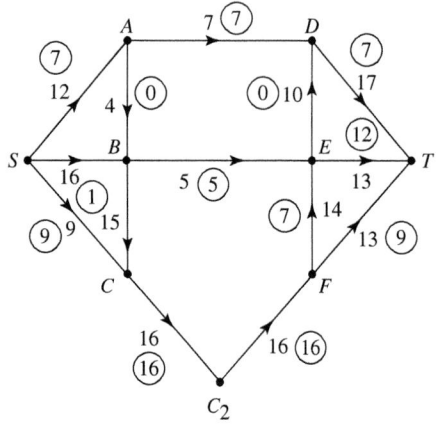

Min-cut e.g. $\{AD, BE, FE, FT\} = 7 + 5 + 7 + 9 = 28$

So flow of 28 is maximum by maximum flow-minimum cut theorem.

Chapter 29
Exercise 29.1A

1

Activity	Duration (hours)	Start		Finish		Float
		Earliest	Latest	Earliest	Latest	
A	6	0	0	6	6	0
B	8	0	7	8	15	7
C	14	6	8	20	22	2
D	12	6	6	18	18	0
E	3	8	15	11	18	7
F	4	18	18	22	22	0
G	3	22	22	25	25	0

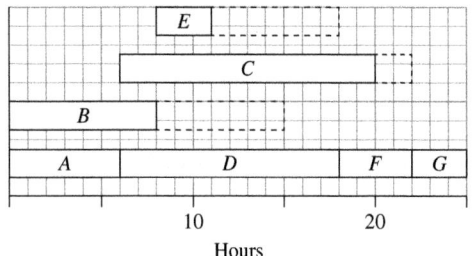

2 a

Activity	Duration	Start		Finish		Float
		Earliest	Latest	Earliest	Latest	
A	7	0	0	7	7	0
B	5	0	2	5	7	2
C	5	7	15	12	20	8
D	9	7	7	16	16	0
E	12	16	20	28	32	4
F	9	16	16	25	25	0
G	7	25	25	32	32	0
H	5	32	32	37	37	0
I	4	25	33	29	37	8

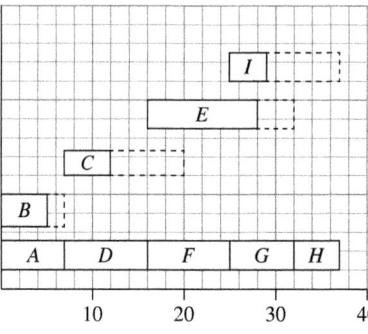

b

Activity	Duration	Start		Finish		Float
		Earliest	Latest	Earliest	Latest	
A	3	0	0	3	3	0
B	5	3	4	8	9	1
C	6	3	3	9	9	0
D	4	8	9	12	13	1
E	3	9	10	12	13	1
F	9	9	9	18	18	0
G	5	12	13	17	18	1
H	1	18	18	19	19	0

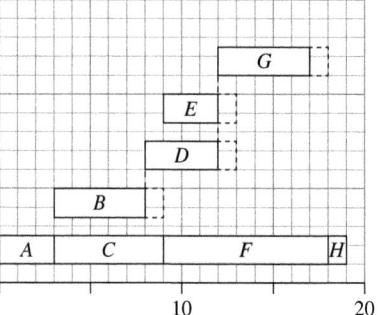

c

Activity	Duration	Start Earliest	Start Latest	Finish Earliest	Finish Latest	Float
A	6	0	0	6	6	0
B	5	6	6	11	11	0
C	8	6	6	14	14	0
D	3	11	11	14	14	0
E	8	14	17	22	25	3
F	7	14	14	21	21	0
G	9	11	16	20	25	5
H	4	21	21	25	25	0

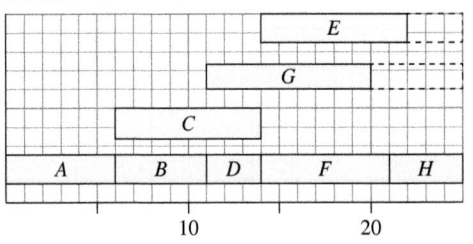

Activity	Duration	Start Earliest	Start Latest	Finish Earliest	Finish Latest	Float
A	6	0	2	6	8	2
B	3	0	3	3	6	3
C	8	0	0	8	8	0
D	8	6	6	14	14	0
E	6	8	8	14	14	0
F	2	14	17	16	19	3
G	3	14	16	17	19	2
H	5	14	14	19	19	0
I	9	19	19	28	28	0
J	6	17	22	23	28	5

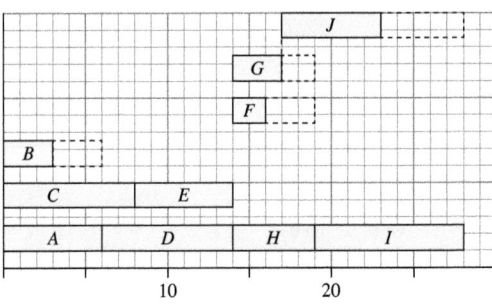

Exercise 29.1B

1

Activity	Duration (hours)	Start Earliest	Start Latest	Finish Earliest	Finish Latest	Float
A	9	0	0	9	9	0
B	7	0	1	7	8	1
C	4	9	9	13	13	0
D	5	7	8	12	13	1
E	6	13	13	19	19	0
F	4	13	19	17	23	6
G	7	19	19	26	26	0
H	3	19	23	22	26	4

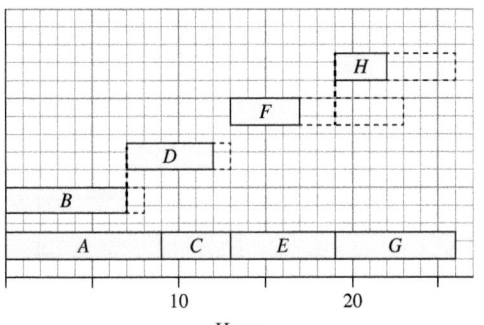

Hours

2 a

Activity	Duration	Start Earliest	Start Latest	Finish Earliest	Finish Latest	Float
A	4	0	5	4	9	5
B	6	0	0	6	6	0
C	5	0	1	5	6	1
D	3	6	9	9	12	3
E	2	6	6	8	8	0
F	6	9	12	15	18	3
G	7	8	8	15	15	0
H	3	15	15	18	18	0
I	3	18	18	21	21	0
J	3	15	18	18	21	3
K	4	15	17	19	21	2

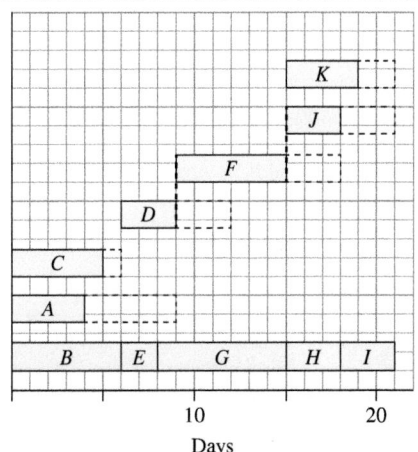

Days

b *D* has 3 days of float. If *D* continues until 11 days, *F* cannot start until then, so *J* cannot start until 17 days. It can therefore be started on day 18 while *J* has 3 days float and will **still** finish at 20 days, so the project duration is not affected.

3 a 31 hours (the total of all the tasks)

b

Activity	Duration	Start		Finish		Float
		Earliest	Latest	Earliest	Latest	
A	2	0	0	2	2	0
B	10	2	2	12	12	0
C	3	12	12	15	15	0
D	2	12	13	14	15	1
E	3	14	15	17	18	1
F	2	14	17	16	19	3
G	1	17	19	18	20	2
H	1	17	19	18	20	2
I	2	17	18	19	20	1
J	5	15	15	20	20	0

Minimum project duration = 20 hours

c

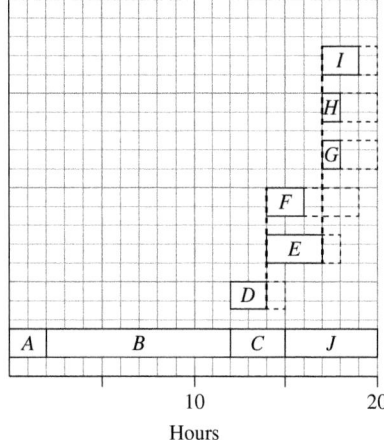

d At least 3 workers needed (once *D* is complete at least 2 of *E*, *F*, *G*, *H* and *I* must be taking place in addition to the critical tasks *C* and *J*).

4

Activity	Duration	Start		Finish		Float
		Earliest	Latest	Earliest	Latest	
A	1	0	0	1	1	0
B	1	1	1	2	2	0
C	1	0	1	1	2	1
D	3	2	2	5	5	0
E	2	2	13	4	15	11
F	4	5	5	9	9	0
G	4	9	9	13	13	0
H	2	13	13	15	15	0
I	1	15	15	16	16	0
J	1	16	16	17	17	0
K	2	17	17	19	19	0
L	2	15	17	17	19	2
M	1	19	19	20	20	0
N	1	20	20	21	21	0
O	1	19	20	20	21	1
P	2	21	21	23	23	0

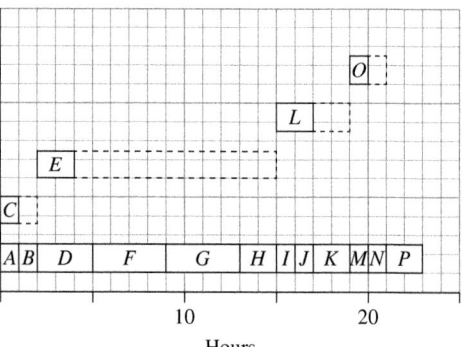

Exercise 29.2A

1 a i

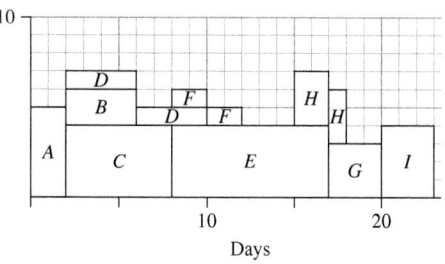

7 workers needed.

ii

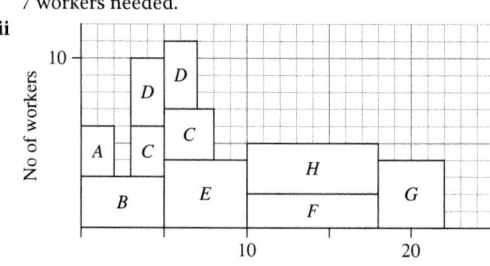

11 workers needed.

iii

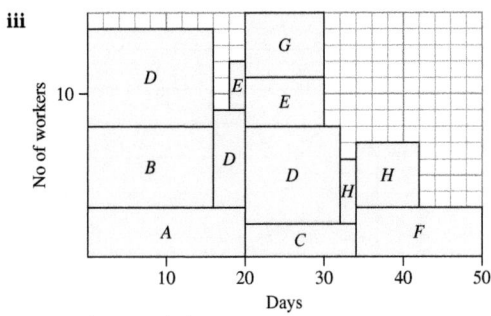

15 workers needed.

b i

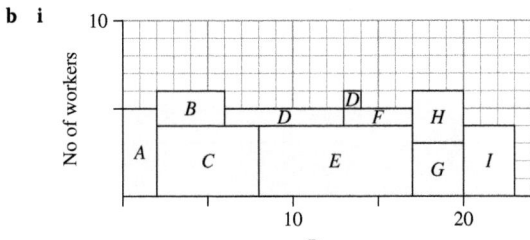

6 workers needed.

ii

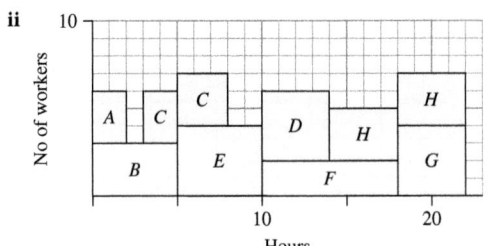

7 workers needed.

iii

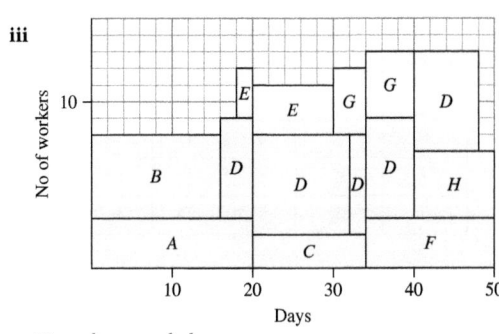

13 workers needed.

2

a

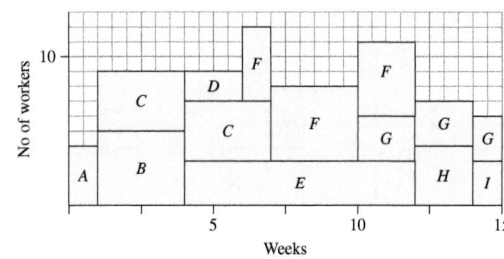

b

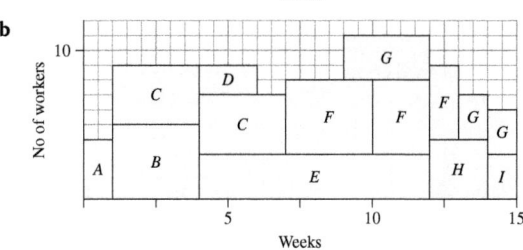

Exercise 29.2B

1 a

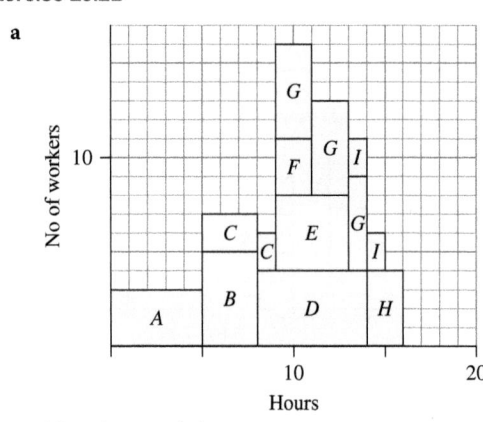

16 workers needed.

b

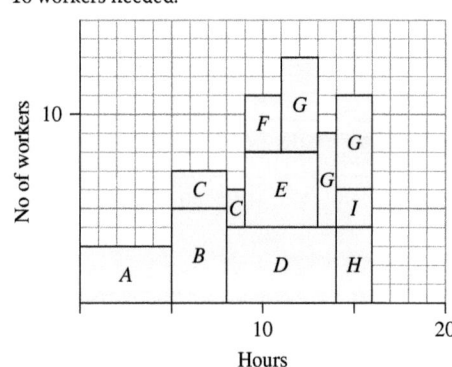

13 workers needed.

c

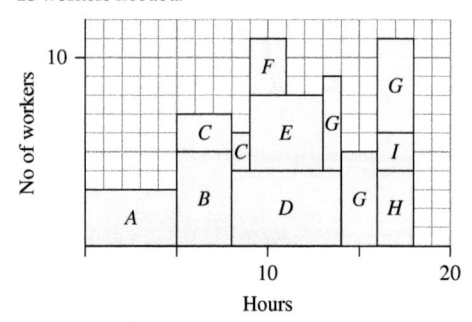

18 hours needed.

2 a

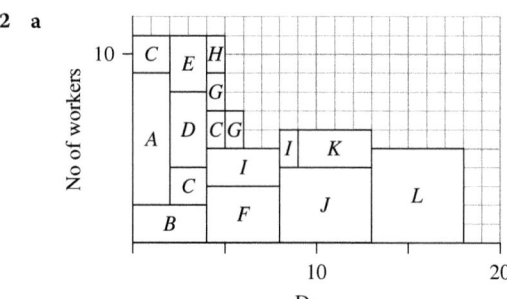

11 workers needed.

b

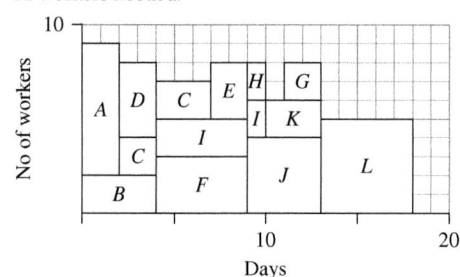

9 workers needed.

c

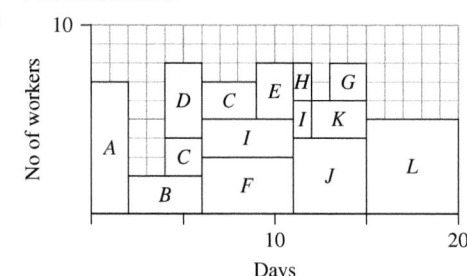

2 extra days needed.

3 a

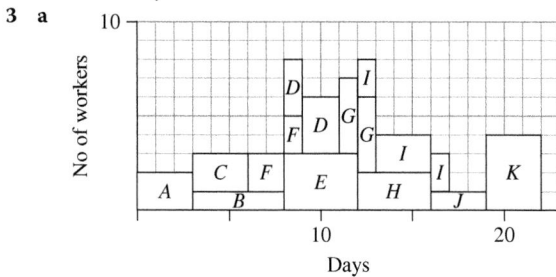

8 workers needed.

b

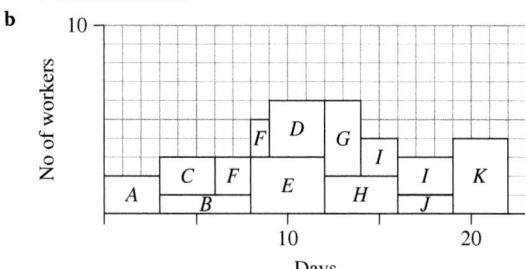

6 workers needed.

c i

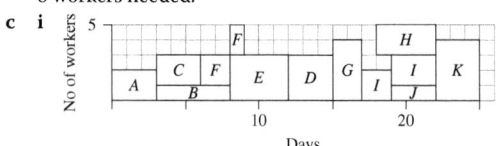

4 extra days

ii

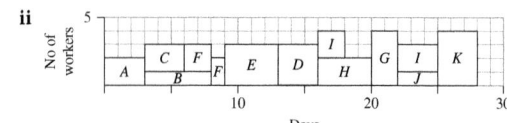

4 a

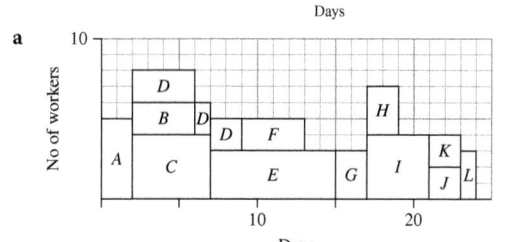

8 workers needed.

b

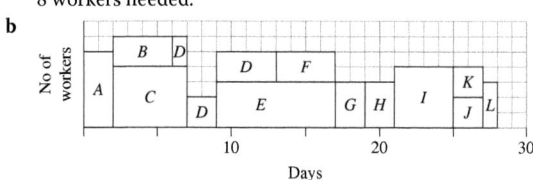

4 extra days needed.

Review exercise 29

1

Activity	Duration	Start		Finish		Float
		Earliest	Latest	Earliest	Latest	
A	4	0	7	4	11	7
B	5	0	2	5	7	2
C	7	0	0	7	7	0
D	3	5	11	8	14	6
E	4	7	7	11	11	0
F	2	8	14	10	16	6
G	7	5	9	12	16	4
H	5	11	11	16	16	0
I	6	16	16	22	22	0
J	4	11	18	15	22	7

2

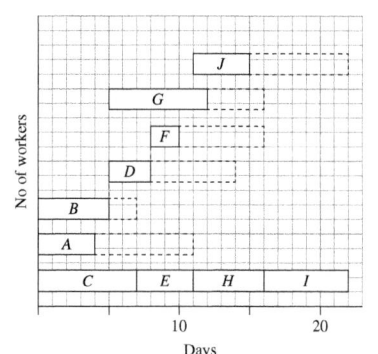

3 a

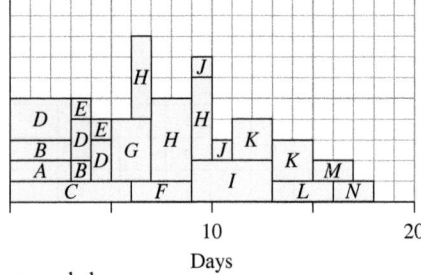

8 workers needed.

b

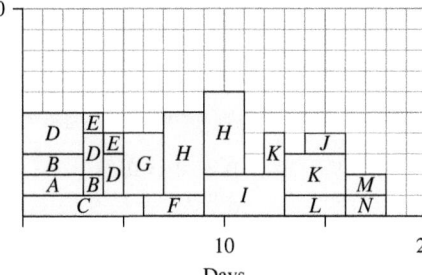

6 workers needed.

c

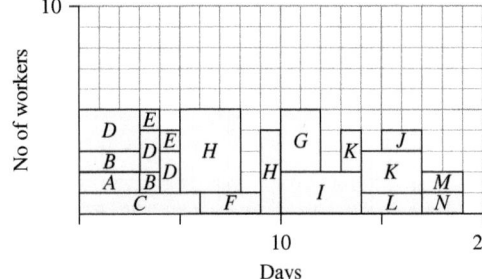

Assessment 29

1 a

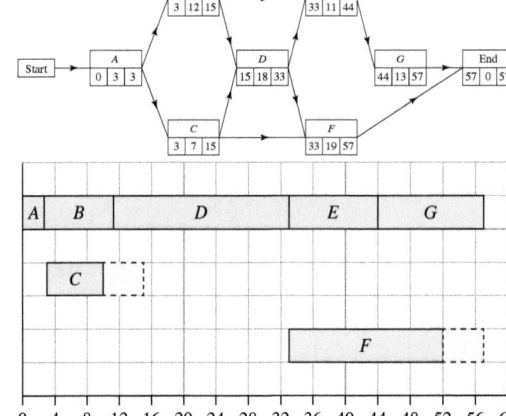

b 6

2 a

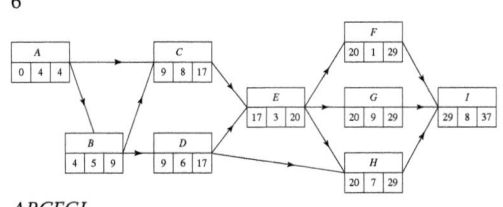

b *ABCEGI*

c $29 - 20 - 7$
   $= 2$

d Float of *D* is $17 - 9 - 6 = 2$
   Increase of 1 hour

3 a

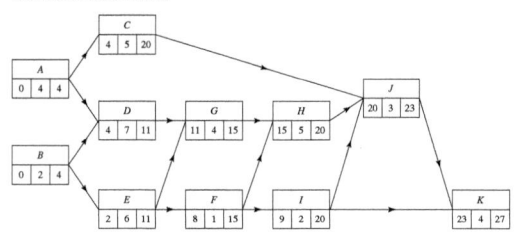

b *ADGHJK*

c i *G* is on the critical path so any delay will delay the project.
   ii Float of *C* is $20 - 4 - 5 = 11$
   So can delay by 11 days without delaying project.

4 a

Activity	Duration (hours)	Start		Finish		Float
		Earliest	Latest	Earliest	Latest	
A	1.5	0	0	1.5	1.5	0
B	2.4	1.5	1.5	3.9	3.9	0
C	6.1	3.9	3.9	10	10	0
D	1.8	3.9	8.5	5.7	10.3	4.6
E	0.3	10	10	10.3	10.3	0
F	3	3.9	7.3	6.9	10.3	3.4
G	2.6	10.3	10.3	12.9	12.9	0
H	5.2	12.9	12.9	18.1	18.1	0

b

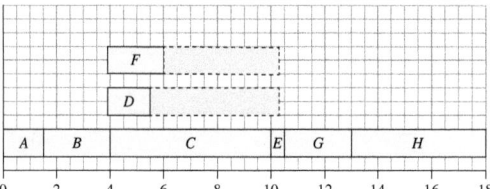

c i Only *B* is definitely happening.
   ii *C* definitely happening, *D* and *F* may be happening.

5 a *C* and *E*

b 5

c

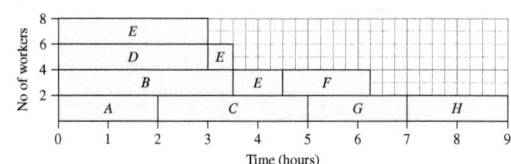

d Delay of 2.25 hours, float of 1 hour so delay of 1.25 hours.
   6:15 pm

6 a

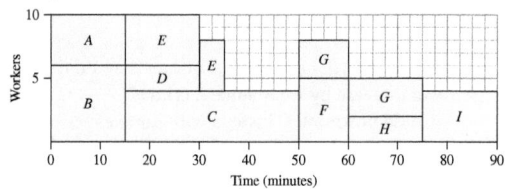

b For example:

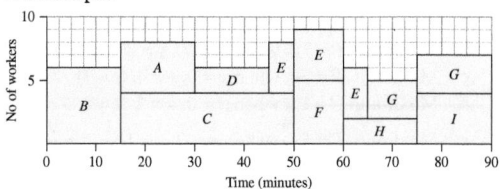

c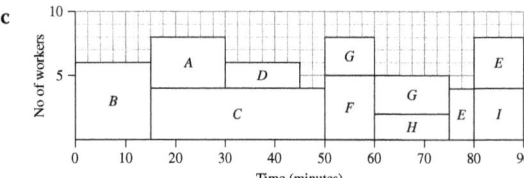

95 minutes

7 a
$$\frac{5\times4+2\times2+4\times5+6\times1+2\times6+3\times2+5\times4+8\times2+3\times7+5\times3+6\times4+8\times3}{26}$$
$$= 7.23$$

So 8 workers needed.

b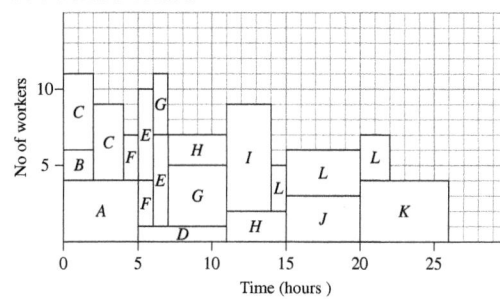

c Activity D can move (e.g. to time 15) since nothing depends on it. This will mean only 10 workers on the 7th day.
Completing 4 workers completing C in 5 hours would mean only 10 workers required on 1st day. Activities G, I, L will be postponed by a day, but this will not increase the overall length of the project
So do recommend this approach as 10 workers will be required overall instead of 11

8 a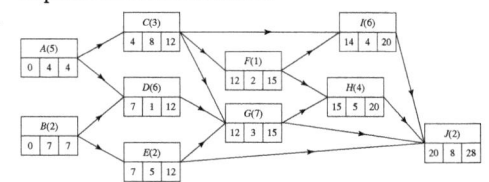

b Critical paths are: $ACGHJ$ and $BEGHJ$

c e.g.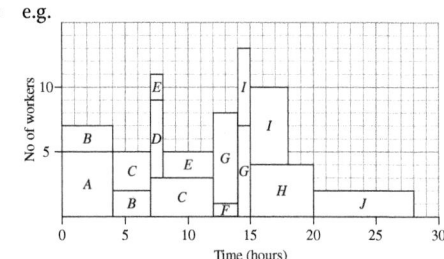

13 workers required.

d Activity I can start at time 15 instead without impacting the total time for the project.
So 11 workers required.

e Activity E will need to be postponed by 1 day so project will increase by 1 day since E critical.
(Could postpone D instead with same effect.)

Chapter 30
Exercise 30.1A

1 a Maximum $P = 25$, when $x = 10$, $y = 15$
b Maximum $P = 70$, when $x = 20$, $y = 30$
c Maximum $P = 56$, when $x = 9$, $y = 4$, $(t = 3)$
d Maximum $P = 12$, when $x = 0$, $y = 6$, $(t = 3, u = 3)$
e Maximum $P = 5\frac{9}{11}$ when $x = 0$, $y = 1\frac{2}{11}$, $z = \frac{6}{11}$
f Maximum $P = 25$ when $x = 2\frac{1}{3}$, $y = 0$, $z = 2\frac{1}{6}$, $\left(s = 4\frac{1}{2}\right)$

2 a $P = 3x + y + 2z$, $2x + 3y + z \le 20$, $x + 2y + z \le 12$, $x \ge 0$, $y \ge 0$, $z \ge 0$

b

P	x	y	z	s	t	Value	Row
1	−3	−1	−2	0	0	0	R1
0	2	3	1	1	0	20	R2
0	1	2	1	0	1	12	R3
1	0	$3\frac{1}{2}$	$-\frac{1}{2}$	$1\frac{1}{2}$	0	30	R4 = R1 + 3 × R5
0	1	$1\frac{1}{2}$	$\frac{1}{2}$	$\frac{1}{2}$	0	10	R5 = $\frac{R2}{2}$
0	0	$\frac{1}{2}$	$\frac{1}{2}$	$-\frac{1}{2}$	1	2	R6 = R3 − R5

Not optimal because there is still a negatve entry in the objective row.
$P = 30$, $x = 10$, $y = 0$, $z = 0$, $s = 0$, $t = 2$

c Maximum $P = 32$, when $x = 8$, $y = 0$, $z = 4$

3 a $P - x - 3y = 0$, $x + y + s = 6$, $x + 4y + t = 12$, $x + 2y + u = 7$

b

P	x	y	s	t	u	Value	Row
1	−1	−3	0	0	0	0	R1
0	1	1	1	0	0	6	R2
0	1	4	0	1	0	12	R3
0	1	2	0	0	1	7	R4
1	$-\frac{1}{4}$	0	0	$\frac{3}{4}$	0	9	R5 = R1 + 3 × R7
0	$\frac{3}{4}$	0	1	$-\frac{1}{4}$	0	3	R6 = R2 − R7
0	$\frac{1}{4}$	1	0	$\frac{1}{4}$	0	3	R7 = $\frac{R3}{4}$
0	$\frac{1}{2}$	0	0	$-\frac{1}{2}$	1	1	R8 = R4 − 2 × R7

Not optimal because there is still a negative entry in the objective row.

c Maximum $P = 9\frac{1}{2}$, when $x = 2$, $y = 2\frac{1}{2}$ $\left(s = 1\frac{1}{2}\right)$

4 a Maximise $P = 5x + 6y$
subject to $3x + 3y \le 40$
$x + 2y \le 25$
$x \ge 0$, $y \ge 0$ s, t are slack variables.

b

P	x	y	s	t	Value	Row
1	−5	−6	0	0	0	R1
0	3	3	1	0	40	R2
0	1	2	0	1	25	R3
1	−2	0	0	3	75	R4 = R1 + 6 × R6
0	$1\frac{1}{2}$	0	1	$-1\frac{1}{2}$	$2\frac{1}{2}$	R5 = R2 − 3 × R6
0	$\frac{1}{2}$	1	0	$\frac{1}{2}$	$12\frac{1}{2}$	R6 = $\frac{R3}{2}$

$P = 75$, $x = 0$, $y = 12.5$, $s = 2.5$, $t = 0$
Not optimal because there is still a negative entry in the objective row.

c Maximum $P = 78\frac{1}{3}$, $x = 1\frac{2}{3}$, $y = 11\frac{2}{3}$

Exercise 30.1B

1 a Maximum $P = 3$, so minimum $C = -3$, when $x = 3$, $y = 6$, $(t = 9)$

b Maximum $P = 15\frac{1}{3}$, so minimum $C = -15\frac{1}{3}$, when $x = 0$, $y = \frac{2}{3}$, $z = 2\frac{2}{3}$

c Maximum $P = 3\frac{3}{4}$, so minimum $C = -3\frac{3}{4}$, when $x = 1\frac{1}{4}$, $y = 2\frac{1}{2}$, $z = 0$, $(t = 1)$

d Maximum $P = 14\frac{1}{2}$, when $x = 9\frac{1}{2}$, $y = 1\frac{1}{2}$ $z = 0$, $(u = 13)$

2 Maximum $P = 97$, so minimum $C = -97$, when $X = 0$, $Y = 24$, $Z = 86$, that is when $x = 10$, $y = 27$, $z = 94$

3 Maximum profit = £31 000. Make 5000 kg Regular and 9000 kg Luxury.

4 Travel x m on setting A and y m on setting B.
Maximise $\qquad D = x + y$
subject to $\qquad 3x + 2y \leq 14400$
$\qquad\qquad 2x + 3y \leq 15000$
$\qquad\qquad x, y \geq 0$
Maximum $D = 5880$ m. Travel 2640 m on setting A and 3240 m on setting B.

5 a £400

b $5x + 6y + 3z \leq 90$
$2x + 4y + z \leq 42$

c Maximum $P = 72.75$, when $x = 13.5$, $y = 3.75$, $z = 0$ No negatives in the objective row.

d Cannot make fractions of a bicycle.

e 13 type A, 4 type B, no type C. Satisfies both constraints.

6 a Make x kg of A, y kg of B and z kg of C.
Maximise profit $\quad P = 0.6x + 0.5y + 0.9z$
subject to $\;0.03x + 0.02y + 0.03z \leq 50$, so $3x + 2y + 3z \leq 5000$
$\qquad\quad 0.02x + 0.03y + 0.04z \leq 60$, so $2x + 3y + 4z \leq$ 6000
$\qquad\quad 0.01x + 0.01y + 0.02z \leq 40$, so $x + y + 2z \leq 4000$
$\qquad\quad x, y, z \geq 0$

b $P = 1400$, $x = 333\frac{1}{3}$, $y = 0$, $z = 1333\frac{1}{3}$, so make $333\frac{1}{3}$ kg Assolato and $1333\frac{1}{3}$ kg Contadino (and no Buona Salute), giving £1400 profit.

7 Maximum profit = £1087.50. Make 2000 litres of Froo-T and 2750 litres of Joo-C.
500 litres of peach juice left over.

8 a Maximum $P = 9333\frac{1}{3}$ when $w = 0$, $x = 0$, $y = 16\frac{2}{3}$, $z = 133\frac{1}{3}$

b This is only an approximate answer, as the variables should be integers. Trial and error suggests that $y = 16$, $z = 134$, giving $P = 9320$, is the best result.
So produce 16 box C and 134 box D, profit £93.20

9 a Maximum $P = 9\frac{5}{7}$, when $x = 2\frac{2}{7}$, $y = 1\frac{3}{7}$

b Maximum $P = -4$, so minimum $C = 4$, when $x = 4$, $y = 0$, $(s = 1)$

10 Garden is 350 m², with 300 m² turf and 50 m² slabs.

11 a Substitute $z = 2x + y - 10$
$P = 3x + y + 10$, $x + y \leq 8$, $6x + y \leq 36$
$P - 3x - y = 10$, $x + y + s = 8$, $6x + y + t = 36$
Maximum $P = 29.2$, when $x = 5.6$, $y = 2.4$, which gives $z = 3.6$

b $P - 5x - 2y + z = 0$, $x + 2y + z + s = 14$, $2x - y + 2z + t = 16$
$2x + y - z + u = 10$. $-2x - y + z + v = -10$
Maximum $P = 29.2$, when $x = 5.6$, $y = 2.4$, $z = 3.6$

12 a

P	x	y	s	t	u	Value	Row
1	−1	−2	0	0	0	0	R1
0	1	1	1	0	0	28	R2
0	−3	−1	0	1	0	−30	R3
0	−2	−3	0	0	1	−60	R4
1	$\frac{1}{3}$	0	0	0	$-\frac{2}{3}$	40	R5 = R1 + 2 × R8
0	$\frac{1}{3}$	0	1	0	$\frac{1}{3}$	8	R6 = R2 − R8
0	$-2\frac{1}{3}$	0	0	1	$-\frac{1}{3}$	−10	R7 = R3 + R8
0	$\frac{2}{3}$	1	0	0	$-\frac{1}{3}$	20	$R8 = -\frac{R4}{3}$
1	0	0	0	$\frac{1}{7}$	$-\frac{5}{7}$	$38\frac{4}{7}$	$R9 = R5 - \frac{R11}{3}$
0	0	0	1	$\frac{1}{7}$	$\frac{2}{7}$	$6\frac{4}{7}$	$R10 = R6 - \frac{R11}{3}$
0	1	0	0	$-\frac{3}{7}$	$\frac{1}{7}$	$4\frac{2}{7}$	$R11 = -R7 \times \frac{3}{7}$
0	0	1	0	$\frac{2}{7}$	$-\frac{3}{7}$	$17\frac{1}{7}$	$R12 = R8 - R11 \times \frac{2}{3}$
1	0	0	$2\frac{1}{2}$	$\frac{1}{2}$	0	55	$R13 = R9 - R14 \times \frac{5}{7}$
0	0	0	$3\frac{1}{2}$	$\frac{1}{2}$	1	23	$R14 = R10 \times \frac{7}{2}$
0	1	0	$-\frac{1}{2}$	$-\frac{1}{2}$	0	1	$R15 = R11 - \frac{R14}{7}$
0	0	1	$1\frac{1}{2}$	$\frac{1}{2}$	0	27	$R16 = R12 - R14 \times \frac{3}{7}$

Maximum $P = 55$, when $x = 1$, $y = 27$, $(u = 23)$

b

P	x	y	s	t	u	Value	Row
1	−1	−2	0	0	0	0	R1
0	1	1	1	0	0	18	R2
0	−3	−1	0	1	0	−30	R3
0	−2	−3	0	0	1	−60	R4
1	$\frac{1}{3}$	0	0	0	$-\frac{2}{3}$	40	R5 = R1 + 2 × R8
0	$\frac{1}{3}$	0	1	0	$\frac{1}{3}$	−2	R6 = R2 − R8
0	$-2\frac{1}{3}$	0	0	1	$-\frac{1}{3}$	−10	R7 = R3 + R8
0	$\frac{2}{3}$	1	0	0	$-\frac{1}{3}$	20	R8 = $-\frac{R4}{3}$
1	0	0	0	$\frac{1}{7}$	$-\frac{5}{7}$	$38\frac{4}{7}$	R9 = R5 − $\frac{R11}{3}$
0	0	0	1	$\frac{1}{7}$	$\frac{2}{7}$	$-3\frac{3}{7}$	R10 = R6 − $\frac{R11}{3}$
0	1	0	0	$-\frac{3}{7}$	$\frac{1}{7}$	$4\frac{2}{7}$	R11 = R7 × $\frac{(-3)}{7}$
0	0	1	0	$\frac{2}{7}$	$-\frac{3}{7}$	$17\frac{1}{7}$	R12 = R8 − R11 × $\frac{2}{3}$

This can go no further because R10 has a negative right–hand side but no negative coefficients on the left–hand side on which to pivot.

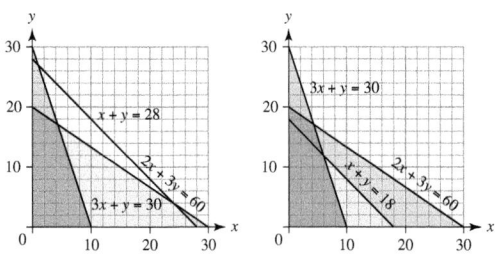

On the correct graph there is a feasible region, but putting 18 instead of 28 means that no point satisfies all three inequalities.

c

P	x	y	s	t	u	Value	Row
1	−1	−2	0	0	0	0	R1
0	1	−1	1	0	0	28	R2
0	−3	−1	0	1	0	−30	R3
0	−2	−3	0	0	1	−60	R4
1	$\frac{1}{3}$	0	0	0	$-\frac{2}{3}$	40	R5 = R1 + 2 × R8
0	$1\frac{2}{3}$	0	1	0	$-\frac{1}{3}$	48	R6 = R2 + R8
0	$-2\frac{1}{3}$	0	0	1	$-\frac{1}{3}$	−10	R7 = R3 + R8
0	$\frac{2}{3}$	1	0	0	$-\frac{1}{3}$	20	R8 = $-\frac{R4}{3}$
1	0	0	0	$\frac{1}{7}$	$-\frac{5}{7}$	$38\frac{4}{7}$	R9 = R5 − $\frac{R11}{3}$
0	0	0	1	$\frac{5}{7}$	$\frac{4}{7}$	$40\frac{6}{7}$	R10 = R6 − R11 × $\frac{5}{3}$
0	1	0	0	$-\frac{3}{7}$	$\frac{1}{7}$	$4\frac{2}{7}$	R11 = R7 × $\frac{3}{7}$
0	0	1	0	$\frac{2}{7}$	$-\frac{3}{7}$	$17\frac{1}{7}$	R12 = R8 − R11 × $\frac{2}{3}$
1	5	0	0	−2	0	60	R13
0	4	0	1	−1	0	58	R14
0	7	0	0	−3	1	30	R15 = R11 × 7
0	3	1	0	−1	0	30	R16

This breaks down because there are no positive values in the t column on which to pivot.

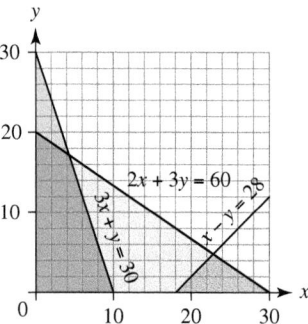

The feasible region is "open–ended" – there is no limit to how large P can be.

Exercise 30.2A

1 a Add 2 to all entries. A plays A_1, A_2 with probabilities p_1, p_2

$P = v$, $v \le 4p_1$, $v \le 5p_2$, $p_1 + p_2 \le 1$

$P - v = 0$, $v - 4p_1 + s = 0$, $v - 5p_2 + t = 0$, $p_1 + p_2 + u = 1$

Value is $2\frac{2}{9} - 2 = \frac{2}{9}$ when $p_1 = \frac{5}{9}$, $p_2 = \frac{4}{9}$

If B plays B_1, B_2 with probabilities q_1, q_2 then $4q_1 = 5q_2$

$= 2\frac{2}{9}$ so $q_1 = \frac{5}{9}$, $q_2 = \frac{4}{9}$

b $P - v = 0$, $v - 7p_1 - 5p_2 + s = 0$, $v - 6p_1 - 8p_2 + t = 0$,
$p_1 + p_2 + u = 1$
Value is 6.5 when $p_1 = 0.75$, $p_2 = 0.25$
For B, $7q_1 + 6q_2 = 5q_1 + 8q_2 = 6.5$ and solving gives $q_1 = q_2 = 0.5$

c Add 1 to all entries.
$P - v = 0$, $v - 6p_1 + s = 0$, $v - 3p_1 - 4p_2 + t = 0$, $p_1 + p_2 + u = 1$
Value $= 3\dfrac{3}{7} - 1 = 2\dfrac{3}{7}$ when $p_1 = \dfrac{4}{7}$, $p_2 = \dfrac{3}{7}$
For B, $6q_1 + 3q_2 = 4q_2 = 3\dfrac{3}{7}$ so $q_1 = \dfrac{1}{7}$, $q_2 = \dfrac{6}{7}$

d Add 2 to all entries.
$P - v = 0$, $v - 5p_2 + s = 0$, $v - 3p_1 - 2p_2 + t = 0$,
$p_1 + p_2 + p_3 + u = 1$
Value is $2.5 - 2 = 0.5$ when $p_1 = p_2 = 0.5$
For B, $3q_2 = 5q_1 + 2q_2 = 2.5$, giving $q_1 = \dfrac{1}{6}$, $q_2 = \dfrac{5}{6}$

2 a Add 2 to all entries.
$P - v = 0$, $v - 3p_1 + s_1 = 0$, $v - p_1 - 2p_2 + s_2 = 0$
$v - 2p_1 - p_2 + s_3 = 0$, $p_1 + p_2 + s_4 = 1$
Value $= 1.5 - 2 = -0.5$ when $p_1 = p_2 = 0.5$

b Add 2 to all entries.
$P - v = 0$, $v - 3p_1 - p_2 - 2p_3 + s = 0$, $v - 3p_2 - p_3 + t = 0$,
$p_1 + p_2 + p_3 + u = 1$
Value $= 1.8 - 2 = -0.2$ when $p_1 = 0.4$, $p_2 = 0.6$, $p_3 = 0$

3 a Maximum of row minima $= -2$, minimum of column maxima $= 1$, no stable solution.
Row 2 dominates row 1, so table becomes

		B	
		B_1	B_2
A	A_2	1	−3
	A_3	−2	1

Add 3 to all entries.
$P - v = 0$, $v - 4p_1 - p_2 + s = 0$, $v - 4p_2 + t = 0$, $p_1 + p_2 + u = 0$
Value is $q_1 = \dfrac{1}{6}$, $q_2 = \dfrac{5}{6}$. $2\dfrac{2}{7} - 3 = -\dfrac{5}{7}$ when $p_1 = \dfrac{3}{7}$, $p_2 = \dfrac{4}{7}$.
$2\dfrac{2}{7} - 3 = -\dfrac{5}{7}$ when $p_1 = \dfrac{3}{7}$, $p_2 = \dfrac{4}{7}$

b Maximum of row minima $= 1$, minimum of column maxima $= 2$ so no stable solution.
Add 3 to all entries.
$P - v = 0$, $v - 2p_1 - 5p_3 + s = 0$, $v - 5p_1 - 7p_2 - 4p_3 + t = 0$.
$p_1 + p_2 + p_3 + u = 1$
Value is $4\dfrac{3}{8} - 3 = 1\dfrac{3}{8}$ when $p_1 = 0$, $p_2 = \dfrac{1}{8}$, $p_3 = \dfrac{7}{8}$

c Maximum of row minima $=$ minimum of column maxima $= 2$, so stable solution.
Play-safe strategies are A_2 and B_1. The value $= 2$

d Maximum of row minima $= -2$, minimum of column maxima $= 0$, no stable solution.
Add 4 to all entries.
$P - v = 0$, $v - 4p_2 - 3p_3 + s_1 = 0$, $v - 3p_1 - p_2 - 6p_3 + s_2 = 0$
$v - 5p_1 - 3p_2 - 4p_3 + s_3 = 0$, $p_1 + p_2 + p_3 + s_4 = 1$
Value $= 3.5 - 4 = -0.5$ when $p_1 = 0$, $p_2 = p_3 = 0.5$

e Maximum of row minima $= 2$, minimum of column maxima $= 3$, no stable solution.
Row 3 dominates rows 1 and 4, then column 1 dominates column 4.
Table becomes

		B		
		B_1	B_2	B_3
A	A_2	−1	0	3
	A_3	4	3	2

Add 1 to all entries.
$P - v = 0$, $v - 5p_2 + s_1 = 0$, $v - p_1 - 4p_2 + s_2 = 0$
$v - 4p_1 - 3p_2 + s_3 = 0$, $p_1 + p_2 + s_4 = 1$
Value $= 3.25 - 1 = 2.25$ when $p_1 = 0.25$, $p_2 = 0.75$

f Maximum of row minima $=$ minimum of column maxima $= 4$, so stable solution.
Play–safe strategies are A_3 and B_3. Value $= 4$.

Exercise 30.2B

1 a Maximum of row minima $= -1$, minimum of col maxima $= 1$, so no stable solution.

b Col 3 dominates col 2, so table becomes

		B	
		B_1	B_3
A	A_1	−1	1
	A_2	1	−2
	A_3	0	−1

Add 2 to all entries.
$P - v = 0$, $v - p_1 - 3p_2 - 2p_3 + s = 0$, $v - 3p_1 - p_3 + t = 0$,
$p_1 + p_2 + p_3 + u = 1$
Value $= 1.8 - 2 = -0.2$ when $p_1 = 0.6$, $p_2 = 0.4$, $p_3 = 0$

c A never plays A_3. If B plays B_1, B_3 with probs q_1, q_3 then $-q_1 + q_3 = q_1 - 2q_3 = -0.2$, which gives $q_1 = 0.6$, $q_3 = 0.4$

2 a Maximum of row minimums $= 1$, minimum of col maximums $= 2$, no stable solution.
Add 1 to all entries.
$P - v = 0$, $v - p_1 - 3p_2 + s_1 = 0$, $v - 3p_1 - 2p_2 + s_2 = 0$
$v - 4p_2 + s_3 = 0$, $p_1 + p_2 + s_4 = 1$
Value is $2\dfrac{1}{3} - 1 = 1\dfrac{1}{3}$ when $p_1 = \dfrac{1}{3}$, $p_2 = \dfrac{2}{3}$
A plays A_1, A_2 with probabilities $\dfrac{1}{3}$, $\dfrac{2}{3}$
If B plays B_1, B_2, B_3 with probabilities q_1, q_2, q_3
then $2q_2 - q_3 = 2q_1 + q_2 + 3q_3 = 1\dfrac{1}{3}$ and $q_1 + q_2 + q_3 = 1$.
Solving gives $q_1 = \dfrac{1}{3}$, $q_2 = \dfrac{2}{3}$, $q_3 = 0$

b Maximum of row minima $= -3$, minimum of column maxima $= 0$, no stable solution.
Add 5 to all entries.
$P - v = 0$, $v - 2p_1 - 4p_2 - 5p_3 + s = 0$, $v - 6p_1 - p_2 + t = 0$,
$p_1 + p_2 + p_3 + u = 1$
Value is $3\dfrac{1}{3} - 5 = -1\dfrac{2}{3}$ when $p_1 = \dfrac{5}{9}$, $p_2 = 0$, $p_3 = \dfrac{4}{9}$
A never plays A_2.
For B, $-3q_1 + q_2 = -5q_2 = -1\dfrac{2}{3}$
Solving gives $q_1 = \dfrac{2}{3}$, $q_2 = \dfrac{1}{3}$

c Maximum of row minima $= 1$, minima of column maxima $= 4$, no stable solution.
Add 2 to all entries.
$P - v = 0$, $v - p_2 - 6p_3 + s = 0$, $v - 6p_1 - 4p_2 - 3p_3 + t = 0$,
$p_1 + p_2 + p_3 + u = 1$
Value $= 4 - 2 = 2$ when $p_1 = \dfrac{1}{3}$, $p_2 = 0$, $p_3 = \dfrac{2}{3}$
A never plays II.
For B, $-2q_1 + 4q_2 = 4q_1 + q_2 = 2$
Solving gives $q_1 = \dfrac{1}{3}$, $q_2 = \dfrac{2}{3}$

d Maximum of row minima $= -3$, minimum of column maxima $= 0$, no stable solution.
Row 2 dominates row 1, column 2 dominates column 3.

Table becomes

		B	
		B_1	B_2
A	A_2	-3	1
	A_3	0	-4

Add 4 to all entries.
$P - v = 0$, $v - p_1 - 4p_2 + s = 0$, $v - 5p_1 + t = 0$, $p_1 + p_2 + u = 1$
Value is $2.5 - 4 = -1.5$ when $p_1 = p_2 = 0.5$
For B, $-3q_1 + q_2 = -4q_2 = -1.5$ which gives $q_1 = \frac{5}{8}$, $q_2 = \frac{3}{8}$

e Maximum of row minima = 1, minimum of column maxima = 2, no stable solution.
Row 1 dominates row 2, column 2 dominates column 1
Table becomes

		B	
		B_2	B_3
A	A_1	1	5
	A_3	2	1

$P - v = 0$, $v - p_1 - 2p_2 + s = 0$, $v - 5p_1 - p_2 + t = 0$, $p_1 + p_2 + u = 1$
Value is 1.8 when $p_1 = 0.2$, $p_2 = 0.8$
For B, $q_1 + 5q_2 = 2q_1 + q_2 = 1.8$, giving $q_1 = 0.8$, $q_2 = 0.2$

f Maximum of row minima = minimum of column maxima = -1, so stable solution.
M plays II, N plays II, value = -1

3

		B		
		R	P	S
A	R	0	-1	1
	P	1	0	-1
	S	-1	1	0

Add 1 to all entries.
$P - v = 0$, $v - p_1 - 2p_2 + s_1 = 0$, $v - p_2 - 2p_3 + s_2 = 0$
$v - 2p_1 - p_3 + s_3 = 0$, $p_1 + p_2 + p_3 + s_4 = 0$
Value is $1 - 1 = 0$ when all three strategies are equally probable. The same is true for player B. This accords with the symmetry of the situation.

4 a Allowing for the cost of insurance the pay-off matrix (in £00) is

		Weather	
		Rain	No rain
Acme	A_1	-4	12
	A_2	4	7
	A_3	12	4

Add 4 to all entries.
$P - v = 0$, $v - 8p_2 - 16p_3 + s = 0$,
$v - 16p_1 - 11p_2 - 8p_3 + t = 0$, $p_1 + p_2 + p_3 + u = 1$
Value is $10\frac{2}{3} - 4 = 6\frac{2}{3}$ when $p_1 = \frac{1}{3}$, $p_2 = 0$, $p_3 = \frac{2}{3}$
Advise them to avoid the basic insurance and to buy comprehensive insurance for a randomly chosen six of the nine events. They would expect to make a total of $9 \times 6\frac{2}{3} \times 100 = £6000$

b This assumes that the weather is independent at every location and it is equally likely to be wet at all the events.

Review exercise 30

1 a $P - x - 2y + z = 0$, $x + y + 2z + s = 14$, $2x + y - z + t = 8$
  b

P	x	y	z	s	t	Value	Row
1	-1	-2	1	0	0	0	R1
0	1	1	2	1	0	14	R2
0	2	1	-1	0	1	8	R3
1	3	0	-1	0	2	16	R4 = R1 + 2 × R6
0	-1	0	3	1	-1	6	R5 = R2 − R6
0	2	1	-1	0	1	8	R6 = R3
1	$2\frac{2}{3}$	0	0	$\frac{1}{3}$	$1\frac{2}{3}$	18	R7 = R4 + R8
0	$-\frac{1}{3}$	0	1	$\frac{1}{3}$	$-\frac{1}{3}$	2	R8 = $\frac{R5}{3}$
0	$1\frac{2}{3}$	1	0	$\frac{1}{3}$	$\frac{2}{3}$	10	R9 = R6 + R8

Maximum $P = 18$, when $x = 0$, $y = 10$, $z = 2$

2 Maximise $P = (-C) = -2x + y + 2z$
$P + 2x - y - 2z = 0$, $-2x - y + 2z + s = 10$, $2x - y + t = 0$,
$3x + y - z + u = 0$
Maximum $P = 30$, so minimum $C = -30$, when $x = 0$, $y = 10$, $z = 10$, $(t = 10)$

3 a As equations the problem is like this:
$P - 5x - 4y = 0$, $-3x - y + s = -10$, $4x + 3y + t = 24$
This is not standard form because of the -10. The initial solution $x = 0$, $y = 0$ would give a negative value for s, which violates $s \geq 0$
  b

P	x	y	s	t	Value	Row
1	-5	-4	0	0	0	R1
0	-3	-1	1	0	-10	R2
0	4	3	0	1	24	R3
1	0	$-\frac{1}{4}$	0	$1\frac{1}{4}$	30	R4 = R1 + 5 × R6
0	0	$1\frac{1}{4}$	1	$\frac{3}{4}$	8	R5 = R2 +3 × R6
0	1	$\frac{3}{4}$	0	$\frac{1}{4}$	6	R6 = $\frac{R3}{4}$

This is now a feasible solution with $x = 6$, $y = 0$, $s = 8$, $t = 0$
  c Maximum $P = 31.6$, when $x = 1.2$, $y = 6.4$

4 a Maximum of row minima = 0, minimum of column maxima = 1, so solution is not stable.
  b Add 3 to all entries.
$P - v = 0$, $v - 4p_2 - 3p_3 + s = 0$, $v - 5p_1 - p_2 - 4p_3 + t = 0$,
$p_1 + p_2 + p_3 + u = 1$
Value is $3.25 - 3 = 0.25$ when $p_1 = 0$, $p_2 = 0.25$, $p_3 = 0.75$

Assessment 30

1 a s and t are slack variables;
$13P + 8t - 2s = 1000$
  b

P	x	y	s	t	Value
1	-1	-1	0	0	0
0	3	4	1	0	120
0	4	1	0	1	80

c The pivot is the 4 in the x–column;
since the θ-value of row 2 is $120 \div 3 = 40$
but the θ-value of row 3 is $80 \div 4 = 20$
$20 < 40$ so chose row 3/pivot is 4

d

P	x	y	s	t	Value
1	0	−0.75	0	0.25	20
0	0	3.25	1	−0.75	60
0	1	0.25	0	0.25	20

e Not optimal since one of the values in the objective row is negative

f $P_{max} = \dfrac{440}{13}$ when $x = \dfrac{200}{13}, y = \dfrac{140}{13}$

2 $P_{max} = \dfrac{25}{2}, x = \dfrac{5}{2}, y = 0, z = \dfrac{5}{2}$

3 a Play-safe strategy for Sarina is S_2 or S_3
Play-safe strategy for Jeremy is J_1
Value of game is 3

b Maximum of row minimums = minimum of column maxima, so a stable solution.

c

Strategy	S_1	S_2	S_3
J_1	4	−3	−3
J_2	−9	−4	−5
J_3	1	−4	−5

4 a x = number of litres of type X
y = number of litres of type Y
z = number of litres of type Z
Maximise $P = x + 1.5y + 1.2z$
subject to $0.4x + 0.6y + 0.5z \leq 550$
$0.1x + 0.2y + 0.3z \leq 160$
$0.1x + 0.1y + 0.05z \leq 100$

b

P	x	y	z	r	s	t	Value
1	−0.25	0	1.05	0	7.5	0	1200
0	0.1	0	−0.4	1	−3	0	70
0	0.5	1	1.5	0	5	0	800
0	0.05	0	−0.1	0	−0.5	1	20

c $P = £1200, x = 0, y = 800, z = 0, \quad (r = 70, s = 0, t = 20)$
Not optimal since negative value in objective function row

d There is $30l$ of orange juice remaining but no pineapple or grapefruit juice.

5 a C_2 dominates C_1 so delete C_1
R_4 dominates R_1 so delete R_1
C_3 dominates C_4 so delete C_4

b C should chose strategies C_2 and C_3 with probabilities $\dfrac{5}{11}$
and $\dfrac{6}{11}$ respectively
R should chose strategies R_2 and R_4 with probabilities
$\dfrac{6}{11}$ and $\dfrac{5}{11}$ respectively
Value of game is $-\dfrac{3}{11}$ (to player C or $\dfrac{3}{11}$ to player R)

6 a Row 2 dominates row 3 since $6 > 5, -1 > -3, 6 > 4$ and $7 > -1$. So player A should never play strategy A_3

b

Strategy	B_1	B_2
A_1	−3	2
A_2	6	−1

c Max of row minima $= -1$
Min of column maxima $= 2$
So no stable solution since $-1 \neq 2$

d A should play A_1 with probability $\dfrac{7}{12}$ and A_2 with probability $\dfrac{5}{12}$.
Value of the game is $-9 \times \dfrac{7}{12} + 6 = \dfrac{3}{4}$ to A
B should play B_1 with probability $\dfrac{1}{4}$ and B_2 with probability $\dfrac{3}{4}$
Value of the game is $-5 \times \dfrac{1}{4} + 2 = -\dfrac{3}{4}$ to B

7 a Let p_1 be the probability of playing strategy I, p_2 be the probability of playing strategy II, p_3 be the probability of playing strategy III,
Let V = value of the game
Maximise $P - V = 0$
Subject to $V - 5p_1 - 2p_3 + r = 0$
$V - p_1 - 2p_2 - 3p_3 + s = 0$
$V - 4p_1 - 5p_2 - p_3 + t = 0$
$p_1 + p_2 + p_3 = 1$
$v, p_1, p_2, p_3 \geq 1$

b Since the simplex algorithm requires decision variables to be non-negative

c For example:
$1 - p_1 - p_2 = p_3$
$V - 3p_1 + 2p_2 + r = 2$
$V + 2p_1 + p_2 + s = 3$
$V - 3p_1 - 4p_2 + t = 1$

d

P	V	p_1	p_2	r	s	t	Value
1	0	−3	−4	0	0	1	1
0	0	0	6	1	0	−1	1
0	0	5	5	0	1	−1	2
0	1	−3	−4	0	0	1	1

e Alex should chose strategy I $\dfrac{7}{30}$ of the time, II $\dfrac{1}{6}$ of the time and III $\dfrac{3}{5}$ of the time.
Value of game for Alex is $\dfrac{11}{30}$

8 a Maximise $P = V$
subject to: $V - 4p_1 - p_2 - 2p_3 + r = 0$
$V - 2p_1 - p_2 - 3p_3 + s = 0$
$V - p_1 - 3p_2 - 2p_3 + t = 0$
$p_1 + p_2 + p_3 + u = 1$
$p_1, p_2, p_3, r, s, t, u \geq 0$

b Set up table:

	V	p_1	p_2	p_3	r	s	t	u	Value
r	①	−4	−1	−2	1	0	0	0	0
s	1	−2	−1	−3	0	1	0	0	0
t	1	−1	−3	−2	0	0	1	0	0
u	0	1	1	1	0	0	0	1	1
P	−1	0	0	0	0	0	0	0	0

	V	p_1	p_2	p_3	s_1	s_2	s_3	s_4	Value
V	1	−4	−1	−2	1	0	0	0	0
s	0	②	0	−1	−1	1	0	0	0
t	0	3	−2	0	−1	0	1	0	0
u	0	1	1	1	0	0	0	1	1
P	0	−4	−1	−2	1	0	0	0	0

	V	P_1	P_2	P_3	s_1	s_2	s_3	s_4	Value
V	1	0	−1	−4	−1	2	0	0	0
P_1	0	1	0	−0.5	−0.5	0.5	0	0	0
t	0	0	−2	1.5	0.5	−1.5	1	0	0
u	0	0	1	1.5	0.5	−0.5	0	1	1
P	0	0	−1	−4	−1	2	0	0	0

9 a Maximum of row minima = $\max\{-3, -1, 0, -1\} = 0$
Minimum of column maxima = $\min\{4, 4, 3, 3\} = 3$
$0 \neq 3$ so no stable solution

b Strategy W is dominated by strategy Z and
strategy A is dominated by strategy D.
So 3×3 pay-off matrix is

	Strategy	X	Y	Z
Greg	**B**	−1	3	1
	C	1	0	2
	D	4	−1	3

c

	Strategy	X	Y	Z
Greg	**B**	0	4	2
	C	2	1	3
	D	5	0	4

Greg plays B with probability p, C with probability q and
D with probability r
Maximise $P = V - 1$
subject to: $V - 2p - 3q - 4r \leq 0$
$V - 2q - 5r \leq 0$
$V - 4p - q \leq 0$
$p + q + r \leq 1$
$V, p, q, r \geq 0$

Chapter 31
Exercise 31.1A

1 a

$\times_5$	0	1	2	3	4
0	0	0	0	0	0
1	0	1	2	3	4
2	0	2	4	1	3
3	0	3	1	4	2
4	0	4	3	2	1

b Yes   **c** 1
d 1 and 4 are self-inverse, 2 is the inverse of 3 and vice
versa. 0 has no inverse.
e S does not form a group since not every element has an
inverse.

2 a

$+_6$	0	1	2	3	4	5
0	0	1	2	3	4	5
1	1	2	3	4	5	0
2	2	3	4	5	0	1
3	3	4	5	0	1	2
4	4	5	0	1	2	3
5	5	0	1	2	3	4

b Yes   **c** 0
d S forms a group since modular addition is associative,
every element has an inverse (e.g. the inverse of 5 is 1),
and parts **b** and **c** give us that the operation is closed and
has an identity element

e Yes: $x +_6 y = y +_6 x$ for all x, y. (The Cayley table has a line
of symmetry down the leading diagonal.)

3 a

•	0	1	2	3
0	0	1	2	3
1	1	0	1	2
2	2	1	0	1
3	3	2	1	0

b Yes   **c** 0
d No: inverse axiom is ok but $(2 \cdot 1) \cdot 3 \neq 2 \cdot (1 \cdot 3)$ so not
associative.

4 Closed: $\begin{pmatrix} 1 & p \\ 0 & 1 \end{pmatrix}\begin{pmatrix} 1 & q \\ 0 & 1 \end{pmatrix} = \begin{pmatrix} 1 & p+q \\ 0 & 1 \end{pmatrix}$ which is also a
member of the set S

Identity: $\mathbf{I} = \begin{pmatrix} 1 & 0 \\ 0 & 1 \end{pmatrix}$ is a member of the set S, and for any
element s of S, $s\mathbf{I} = \mathbf{I}s = s$
Associativity: Matrix multiplication is associative.

Inverse: The inverse of $\begin{pmatrix} 1 & p \\ 0 & 1 \end{pmatrix}$ is $\begin{pmatrix} 1 & -p \\ 0 & 1 \end{pmatrix}$ which is also a
member of set S

Commutativity: $\begin{pmatrix} 1 & p \\ 0 & 1 \end{pmatrix}\begin{pmatrix} 1 & q \\ 0 & 1 \end{pmatrix} = \begin{pmatrix} 1 & p+q \\ 0 & 1 \end{pmatrix}$ and

$\begin{pmatrix} 1 & q \\ 0 & 1 \end{pmatrix}\begin{pmatrix} 1 & p \\ 0 & 1 \end{pmatrix} = \begin{pmatrix} 1 & p+q \\ 0 & 1 \end{pmatrix}$ so the set forms an Abelian
group under the operation of matrix multiplication.

5 For example, $1, 2 \in \mathbb{N}$ but $1 - 2 = -1 \notin \mathbb{N}$

6 a 6   **b** 2

7 a Closed: if x and y are integers then $x + y - 2$ is an integer.
Identity: $x \cdot e = x \Rightarrow x + e - 2 = x \Rightarrow e = 2$. Conversely,
$2 \cdot x = 2 + x - 2 = x$
Associativity: $(x \cdot y) \cdot z = (x + y - 2) + z - 2 = x + y + z - 4$;
$x \cdot (y \cdot z) = x + (y + z - 2) - 2 = x + y + z - 4$
Inverse: $x \cdot x^{-1} = e \Rightarrow x + x^{-1} - 2 = 2 \Rightarrow x^{-1} = 4 - x$ which is
also an integer.
b $1 \in \mathbb{Z}^+$, but $1 \cdot 1 = 1 + 1 - 2 = 0 \notin \mathbb{Z}^+$. Hence T is not closed
under •.

Exercise 31.1B

1 a Define the symmetries:
Given that the centre of the square is the origin,
r_0 = rotation of 0° about the origin (i.e. the square is in its
initial orientation)
r_1 = rotation of 90° anti-clockwise about the origin
r_2 = rotation of 180° anti-clockwise about the origin
r_3 = rotation of 270° anti-clockwise about the origin
m_1 = reflection in the x-axis
m_2 = reflection in the line $y = x$
m_3 = reflection in the y-axis
m_4 = reflection in the line $y = -x$

Draw up a Cayley table:

	r_0	r_1	r_2	r_3	m_1	m_2	m_3	m_4
r_0	r_0	r_1	r_2	r_3	m_1	m_2	m_3	m_4
r_1	r_1	r_2	r_3	r_0	m_2	m_3	m_4	m_1
r_2	r_2	r_3	r_0	r_1	m_3	m_4	m_1	m_2
r_3	r_3	r_0	r_1	r_2	m_4	m_1	m_2	m_3
m_1	m_1	m_4	m_3	m_2	r_0	r_3	r_2	r_1
m_2	m_2	m_1	m_4	m_3	r_1	r_0	r_3	r_2
m_3	m_3	m_2	m_1	m_4	r_2	r_1	r_0	r_3
m_4	m_4	m_3	m_2	m_1	r_3	r_2	r_1	r_0

Identity element: r_0

Inverses: The mirror lines are self-inverse, as are r_0 and r_2. r_1 is the inverse of r_3 and vice versa.

Associativity: Since the binary operation is a composition of mappings, the operation is associative.

Closure: Every combination of symmetries is in the original set of symmetries.

The order of the group is 8

b $C_4 = \{r_0, r_1, r_2, r_3\}$
Generators are r_1 and r_3

2 a

	r_0	r_1	r_2	r_3	r_4	r_5
r_0	r_0	r_1	r_2	r_3	r_4	r_5
r_1	r_1	r_2	r_3	r_4	r_5	r_0
r_2	r_2	r_3	r_4	r_5	r_0	r_1
r_3	r_3	r_4	r_5	r_0	r_1	r_2
r_4	r_4	r_5	r_0	r_1	r_2	r_3
r_5	r_5	r_0	r_1	r_2	r_3	r_4

Generators are r_1 and r_5

b Yes

3 a The set of all symmetries of a regular pentagon

b There is the identity element, r_0 plus four other rotations through multiples of 72° and five lines of symmetry, one through each vertex and the midpoint of its opposite side. Hence the group has at least 10 elements.

Conversely, select one vertex v of the pentagon, and consider an adjacent vertex w. Any symmetry of the pentagon maps v to one of 5 vertices of the pentagon, 5 possibilities.

Following this mapping, w must still be adjacent to v, either immediately clockwise, or immediately anticlockwise, so 2 possibilities.

These 5×2 possibilties uniquely determine the action of the symmetry on all vertices; hence there can be at most 10 elements of the group. Hence the group has 10 elements.

c For each group of symmetries there will be n rotations through multiples of $\dfrac{360°}{n}$ plus n lines of symmetry. For odd n, these will be through each vertex and the midpoint of its opposite side and for even n there will be $\dfrac{n}{2}$ lines through pairs of opposite vertices and $\dfrac{n}{2}$ lines through the midpoints of opposite sides. Hence the group has at least $2n$ elements.

Conversely, select one vertex v of the n-gon, and consider an adjacent vertex w. Any symmetry of the n-gon maps v to one of n vertices of the n-gon, so n possibilities.

Following this mapping, w must still be adjacent to v, either immediately clockwise or immediately anticlockwise, so two possibilities. These two possiilities uniquely determine the action of the symmetry on all vertices; hence there can be at most $2n$ elements of the group. Hence the group has exactly $2n$ elements.

4 a $\{1, 2, 3, 4, 5, 6\}$ **b** Yes

5 a

$+_7$	0	1	2	3	4	5	6
0	0	1	2	3	4	5	6
1	1	2	3	4	5	6	0
2	2	3	4	5	6	0	1
3	3	4	5	6	0	1	2
4	4	5	6	0	1	2	3
5	5	6	0	1	2	3	4
6	6	0	1	2	3	4	5

b 7

c She is correct. Any element coprime to 7 is a generator under modular addition so in this case all of the elements except 0 are possible generators of the group.

6 a $\left\{\begin{pmatrix} 0 & 1 \\ -1 & 0 \end{pmatrix}, \begin{pmatrix} -1 & 0 \\ 0 & -1 \end{pmatrix}, \begin{pmatrix} 0 & -1 \\ 1 & 0 \end{pmatrix}, \begin{pmatrix} 1 & 0 \\ 0 & 1 \end{pmatrix}\right\}$, 4

b The group of rotational symmetries of the unit square.

7 a a **b** c or d

8 a A rotation through 0° about the centre of the circle

b Since the binary operation is a composition of mappings, the operation is associative.

c i A rotation through 327.9°

  ii A reflection in a line of symmetry inclined at $\theta°$

d rot since a combination of any two rotations is another rotation.

e Any combination of rotations and reflections leads to either another rotation or reflection and since the set of elements contains every rotation and every reflection, the resulting combination must be contained within the set.

Exercise 31.2A

1 H is non-empty since there exist (infinitely many) elements in H

The identity element of G under the binary operation addition is 0 and $0 \in \mathbb{Z}$

H is closed since if x and y are integers, then so is their sum $x + y$. Hence if $x, y \in \mathbb{Z}$ then $x + y \in \mathbb{Z}$

The inverse of an element x is x^{-1} which is $-x$ under the binary operation of addition. Since the negation of an integer is also an integer, it follows that if $x \in \mathbb{Z}$ then $x^{-1} \in \mathbb{Z}$

Hence H is a subgroup of G

2 The Cayley table of all symmetries of an equilateral triangle is:

	r_0	r_1	r_2	m_1	m_2	m_3
r_0	r_0	r_1	r_2	m_1	m_2	m_3
r_1	r_1	r_2	r_0	m_2	m_3	m_1
r_2	r_2	r_0	r_1	m_3	m_1	m_2
m_1	m_1	m_3	m_2	r_0	r_2	r_1
m_2	m_2	m_1	m_3	r_1	r_0	r_2
m_3	m_3	m_2	m_1	r_2	r_1	r_0

The group of rotations is given by:

	r_0	r_1	r_2
r_0	r_0	r_1	r_2
r_1	r_1	r_2	r_0
r_2	r_2	r_0	r_1

This contains the identity element r_0 and is non-empty. r_0 is self-inverse and r_1 and r_2 form an inverse pair. The set of rotations is closed since every combined element is one of the original elements, therefore the group of rotations is a subgroup of the group of all symmetries.

3 {1} and {1, 4}

4 {0, 2} and {0, 1, 2, 3}

5 a The Cayley table for C_4

	c_0	c_1	c_2	c_3
c_0	c_0	c_1	c_2	c_3
c_1	c_1	c_2	c_3	c_0
c_2	c_2	c_3	c_0	c_1
c_3	c_3	c_0	c_1	c_2

{c_0, c_1}, {c_0, c_3}, {c_0, c_1, c_2}, {c_0, c_1, c_3} and {c_0, c_2, c_3} are not closed, e.g. $c_3 c_2 = c_1$

{c_0, c_2} is closed, contains the identity element c_0 and both elements are self-inverse so this is the only non-trivial proper subset of C_4

b The Cayley tables for **Q4** and **Q5** have exactly the same structure indicating that H is equivalent to C_4

6 {a}, {a, b}, {a, c} and {a, d}

7 $N = \left\{ \begin{pmatrix} 1 & 0 \\ 0 & 1 \end{pmatrix} \begin{pmatrix} -1 & 0 \\ 0 & -1 \end{pmatrix} \right\}$ contains the identity matrix and is

non-empty.

$$\begin{pmatrix} 1 & 0 \\ 0 & 1 \end{pmatrix} \begin{pmatrix} -1 & 0 \\ 0 & -1 \end{pmatrix} = \begin{pmatrix} -1 & 0 \\ 0 & -1 \end{pmatrix}$$

$$\begin{pmatrix} -1 & 0 \\ 0 & -1 \end{pmatrix} \begin{pmatrix} 1 & 0 \\ 0 & 1 \end{pmatrix} = \begin{pmatrix} -1 & 0 \\ 0 & -1 \end{pmatrix}$$

$$\begin{pmatrix} -1 & 0 \\ 0 & -1 \end{pmatrix} \begin{pmatrix} -1 & 0 \\ 0 & -1 \end{pmatrix} = \begin{pmatrix} 1 & 0 \\ 0 & 1 \end{pmatrix}$$

Hence the set is closed under matrix multiplication. Both matrices are self-inverse, hence N is a subgroup of M

Exercise 31.2B

1 By Lagrange's theorem, only groups with order that is a factor of 90 can be subgroups so the only possibilities are: 1, 2, 3, 5, 6, 9, 10, 15, 18, 30, 45, 90

2 c C, order 12

Since 12 is not a factor of 30

3 6 does not divide 92, therefore, by Lagrange's theorem, 6 cannot be the order of a possible subgroup.

4 There are 11 elements in H and since 11 is prime, the only orders of possible subgroups are 1 and 11 The only subgroup of order 1 is the trivial subgroup, since any subgroup must contain the identity element, and the only subgroup of order 11 is the group itself, since it must contain every element. Hence the only subgroups are the trivial subgroup and the group itself.

5 a 8

b Orders of possible subgroups, by Lagrange's theorem, are 1, 2, 4, 8 (the factors of 8)

c {e, q} and {e, q, p, u}

6 a There are three elements in Carmelita's proposed subgroup and since 3 is not a factor or 4, by Lagrange's theorem, Carmelita must be wrong. (Also accept that $-i \times -i = 1$ and 1 is not an element of Carmelita's proposed subgroup, so it's not closed.)

b {1} and {1, −1}

7 There are nine elements so the order of G is 9. Hence possible subgroups should be of orders 1, 3 or 9

Order 1: The trivial subgroup {0}

Order 9: The group itself, G

Order 3: There are no self-inverse elements under $+_9$ so a subgroup of order 3 will consist of the identity, an element and its inverse. There are four such sets: {0, 1, 8}, {0, 2, 7}, {0, 3, 6} and {0, 4, 5}. However, all except {0, 3, 6} are not closed under the group operation. Hence there is a single subgroup of order 3. Hence there are three subgroups in total.

8 $43\,252\,003\,274\,489\,856\,000 = 2^{27} \times 3^{14} \times 5^3 \times 7^2 \times 11$

$(27 + 1)(14 + 1)(3 + 1)(2 + 1)(1 + 1) = 10\,080$ so $43\,252\,003\,274\,489\,856\,000$ has 10 080 factors and, by Lagrange's theorem, this gives the number of orders of possible subgroups.

Exercise 31.3A

1 Draw a Cayley table for the group:

	r_0	r_1	r_2	r_3
r_0	r_0	r_1	r_2	r_3
r_1	r_1	r_2	r_3	r_0
r_2	r_2	r_3	r_0	r_1
r_3	r_3	r_0	r_1	r_2

r_0 is the identity element. r_2 is self-inverse. Rearranging the columns and rows gives

	r_0	r_2	r_1	r_3
r_0	r_0	r_2	r_1	r_3
r_2	r_2	r_0	r_3	r_1
r_1	r_1	r_3	r_2	r_0
r_3	r_3	r_1	r_0	r_2

Now it is clear to see that the pattern of entries is the same as for the groups in example 1, for instance, via the mapping $r_0 \mapsto 1$, $r_2 \mapsto -1$, $r_1 \mapsto i$, $r_3 \mapsto -i$, hence the groups are isomorphic.

2 a Write out the Cayley tables for each group:

$\times_5$	1	2	3	4
1	1	2	3	4
2	2	4	1	3
3	3	1	4	2
4	4	3	2	1

$+_4$	0	1	2	3
0	0	1	2	3
1	1	2	3	0
2	2	3	0	1
3	3	0	1	2

The self-inverse element in $+_4$ is 2 so rearrange the columns and rows:

$+_4$	0	1	3	2
0	0	1	3	2
1	1	2	0	3
3	3	0	2	1
2	2	3	1	0

Now it is clear to see the pattern of entries is the same for both groups, for instance via the mapping between A and B given by $1 \mapsto 0, 2 \mapsto 1, 3 \mapsto 3, 4 \mapsto 2$ hence the groups are isomorphic.

b Yes

3 a For group M, let the four matrices be a, b, c and d respectively. Hence the Cayley table for M is

$\times$	a	b	c	d
a	a	b	c	d
b	b	a	d	c
c	c	d	a	b
d	d	c	b	a

For group N, the Cayley table is

	1	3	5	7
1	1	3	5	7
3	3	1	7	5
5	5	7	1	3
7	7	5	3	1

The identity elements are a and 1 respectively and since all elements in both groups are self-inverse, the groups are isomorphic to each other, for instance via the mapping $a \mapsto 1, b \mapsto 3, c \mapsto 5, d \mapsto 7$.

b No

4 The Cayley table for the rotations of the equilateral triangle is

	r_0	r_1	r_2
r_0	r_0	r_1	r_2
r_1	r_1	r_2	r_0
r_2	r_2	r_0	r_1

For group R the Cayley table is

$\times_7$	4	2	1
4	2	1	4
2	1	4	2
1	4	2	1

Since the identity elements are r_0 and 1, rearrange the rows/columns in the table for $\times_7$

$\times_7$	1	4	2
1	1	4	2
4	4	2	1
2	2	1	4

Now it is clear to see the matching patterns of entries in the two tables, for instance via the mapping $r_0 \mapsto 1, r_1 \mapsto 4, r_2 \mapsto 2$, hence the two groups are isomorphic.

Exercise 31.3B

1 All groups of order 2 consist of the identity element and one other self-inverse element of period 2 so they must be isomorphic.

2 All groups of order 3 consist of the identity element and two other elements of period 3 (period 2 is not allowed since 2 does not divide into 3). The groups are therefore cyclic with elements that map to e, a and a^2.

3 Period 4 elements must occur in pairs, paired with their inverse. Hence one or three elements of period 4 cannot happen. So every group of order 4 is isomorphic to exactly one of two particular groups of order 4, Group A with all elements other than the identity of period 2 (the Klein 4-group) and Group C with two elements of period 4 (the C_4 group).

4 Groups of order p, where p is prime, have an identity element of period 1 and all remaining elements must have period p, since p has just two factors, 1 and p. Let G and H be two such groups of order p, and select an element $g \in G$ and $h \in H$, both of order p. Then the mapping $a^i \mapsto b^i$ gives a one-to-one mapping that respects the operations of G and H, hence giving an isomorphism. Hence all groups of order p are isomorphic to the cyclic group C_p

5 Let G be a non-abelian group, and H an abelian group. Since G is not abelian there are elements a, $b \in G$ such that $ab \neq ba$. Suppose there were an isomorphism between G and H with $a \mapsto x, b \mapsto y$. Then $ab \mapsto xy, ba \mapsto yx$, and since H is abelian $xy = yx$. This means that ab and ba map to the same element of H despite being different elements of G, so the mapping cannot have been one-to-one. Hence no isomorphism exists.

6 The Cayley table for G

$+_6$	0	1	2	3	4	5
0	0	1	2	3	4	5
1	1	2	3	4	5	0
2	2	3	4	5	0	1
3	3	4	5	0	1	2
4	4	5	0	1	2	3
5	5	0	1	2	3	4

The Cayley table for H

	r_0	r_1	r_2	r_3	r_4	r_5
r_0	r_0	r_1	r_2	r_3	r_4	r_5
r_1	r_1	r_2	r_3	r_4	r_5	r_0
r_2	r_2	r_3	r_4	r_5	r_0	r_1
r_3	r_3	r_4	r_5	r_0	r_1	r_2
r_4	r_4	r_5	r_0	r_1	r_2	r_3
r_5	r_5	r_0	r_1	r_2	r_3	r_4

It is evident that the two Cayley tables has the same pattern of entries, for instance via the mapping $i \mapsto r_i$, therefore the two groups are isomorphic.

7 M is a cyclic group since when $k = 1$, the period of the element is 8 (8 successive rotations of $\dfrac{\pi}{4}$ will map the unit square back to itself). In particular, the only self-inverse element of M is the rotation by π, so M only has one self-inverse element. However, in S the elements 4, 11 and 14 are all self-inverse. Hence the groups cannot be isomorphic.

8 0, which is the identity element of G

Consider a mapping between the sets, say a maps to

$\begin{pmatrix} 1-a & a \\ -a & 1+a \end{pmatrix}$ for all $a \in \mathbb{Z}$. This means that p maps to

$\begin{pmatrix} 1-p & p \\ -p & 1+p \end{pmatrix}$ and q maps to $\begin{pmatrix} 1-q & q \\ -q & 1+q \end{pmatrix}$. This mapping is clearly one-to-one.

You must now show that the product of these two matrices

is $\begin{pmatrix} 1-(p+q) & (p+q) \\ -(p+q) & 1+(p+q) \end{pmatrix}$, i.e. $p+q$ maps to this matrix product.

$$\begin{pmatrix} 1-p & p \\ -p & 1+p \end{pmatrix}\begin{pmatrix} 1-q & q \\ -q & 1+q \end{pmatrix}$$

$$=\begin{pmatrix} (1-p)(1-q)-pq & (1-p)q+p(1+q) \\ -p(1-q)-q(1+p) & -pq+(1+p)(1+q) \end{pmatrix}$$

$$=\begin{pmatrix} 1-p-q & p+q \\ -p-q & 1+p+q \end{pmatrix}=\begin{pmatrix} 1-(p+q) & (p+q) \\ -(p+q) & 1+(p+q) \end{pmatrix}$$

as required.

Hence the two groups are isomorphic.

Review exercise 31

1 a 6

b

$\times_9$	1	2	4	5	7	8
1	1	2	4	5	7	8
2	2	4	8	1	5	7
4	4	8	7	2	1	5
5	5	1	2	7	8	4
7	7	5	1	8	4	7
8	8	7	5	4	7	1

c 6

d 5 has period equal to the order of the group; hence G is the group generated by 5 and hence the group is cyclic.

e Yes: The Cayley table has a line of symmetry down the leading diagonal. (Or G is a cyclic group and cyclic groups are abelian).

2 a 1

b Since all of the entries in the Cayley table are elements of S, the set is closed under the operation $\times_{12}$.

c All of the elements are self-inverse since for $a \in S$, $a^2 = 1$

d For instance:
$(5 \times 7) \times 11 = 11 \times 11 = 1 \pmod{12}$ and $5 \times (7 \times 11) = 5 \times 5 = 1 \pmod{12}$
$(11 \times 7) \times 5 = 5 \times 5 = 1 \pmod{12}$ and $11 \times (7 \times 5) = 11 \times 11 = 1 \pmod{12}$

3 a 1, 2, 3 and 6

b $\{r_0\}$

c $\{r_0\}$, $\{r_0, m_1\}$, $\{r_0, m_2\}$, $\{r_0, m_3\}$ and $\{r_0, r_1, r_2\}$.

4 a The Cayley tables are

$\times_{10}$	1	3	7	9
1	1	3	7	9
3	3	9	1	7
7	7	1	9	3
9	9	7	3	1

$\times_4$	0	1	2	3
0	0	1	2	3
1	1	2	3	0
2	2	3	0	1
3	3	0	1	2

The self-inverse element in $+_4$ is 2 so swapping the order of columns and rows gives

$+_4$	0	1	3	2
0	0	1	3	2
1	1	2	0	3
3	3	0	2	1
2	2	3	1	0

Now the pattern of elements in the two groups are the same, hence $G \cong H$.

A possible mapping is $[1, 3, 7, 9] \leftrightarrow [0, 1, 3, 2]$.

b Yes: 1 and 3 are generators of H and 3 and 7 are generators of G.

c Yes: The Cayley tables have a line of symmetry along the leading diagonal. Also the groups are cyclic, and cyclic groups are abelian.

Assessment 31

1 a b
Since $b * x = x * b = x$ for all elements x

b b, c

c Symmetric in the leading diagonal.

2 a b
Since it is the only element in all the subgroups.

b 30
Since it is the lowest common multiple of 2, 3 and 5

3

$\times_8$	1	3	5	7
1	1	3	5	7
3	3	1	7	5
5	5	7	1	3
7	7	5	3	1

$\times_{10}$	1	3	7	9
1	1	3	7	9
3	3	9	1	7
7	7	1	9	3
9	9	7	3	1

All elements in $\times_8$ are self-inverse, $\times_{10}$ has only two self-inverse elements, therefore not isomorphic.

4 a It means that 9 is a *generator* of the group.

b $9^2 = 17 \pmod{64}$
$9^3 = 25 \pmod{64}$
$9^4 = 33 \pmod{64}$
$9^5 = 41 \pmod{64}$
$9^6 = 49 \pmod{64}$
$9^7 = 57 \pmod{64}$
$9^8 = 1 \pmod{64}$
Since 1 is the multiplicative identity,
$n = 8$

c 1, 2, 4, 8
Since by Lagrange's theorem, order of subgroup must be a factor of the group order.

5 a

$\odot$	0	1	2	3	4	5
0	3	4	5	0	1	2
1	4	5	0	1	2	3
2	5	0	1	2	3	4
3	0	1	2	3	4	5
4	1	2	3	4	5	0
5	2	3	4	5	0	1

Since every element generated under $\odot$ is an element of the set, closure under $\odot$.

Identity element is 3

5, 1 and 4, 2 are inverse pairs and 0, 3 are self-inverse so every element has an inverse.

$(a \odot b) \odot c = (a + b + 3) + c + 3 = a + b + c + 6$

$a \odot (b \odot c) = a + (b + c + 3) + 3 = a + b + c + 6$

Therefore associative.

Since the four axioms of being a group are satisfied, G forms a group under $\odot$

b $\{3\}, \{1, 3, 5\}, \{0, 3\}$

c $G = (\langle 2 \rangle, \odot)$

or $K = (\{1, 2, 4, 5, 7, 8\}, \times_9)$

Produce a mapping of G to K, e.g.

$3 \mapsto 1$

$0 \mapsto 8$

$1 \mapsto 4$

$2 \mapsto 2$

$4 \mapsto 5$

$5 \mapsto 7$

As there is a one-to-one mapping between the elements of G and the elements of K which preserves the group operation, $G \cong K$

6 a Since $ex = xe = x$ for all x

b 3 does not divide into 8 so no, she is not correct.

c $d^2 = a, d^4 = c,$

$d^8 = e$

Since period of d is 8, d is a generator and group is cyclic.

Rotations have period 2 or 4 and reflections have period 2, hence no element of order 8, hence not isomorphic.

7 a $(x \cdot y) \cdot z = (x + y - 1) + z - 1 = x + y + z - 2$

$x \cdot (y \cdot z) = x + (y + z - 1) - 1 = x + y + z - 2$

Therefore associative.

b $x \cdot 1 = x + 1 - 1 = x, 1 \cdot x = x$ hence 1 is the identity.

$x \cdot x^{-1} = 1 \Rightarrow x + x^{-1} - 1 = 1 \Rightarrow x^{-1} = 2 - x$, and conversely $x \cdot (2 - x) = x + 2 - x - 1 = 1$.

$2 - x$ is an integer, therefore elements have inverse

$x + y - 1$ is an integer therefore closed.

Hence P forms a group under $\cdot$

c No.

Since inverse $2 - x$ is not a positive integer for all $x > 1$

For instance the inverse of 3 is -1, which is not a positive integer.

Index

principle of conservation of mechanical
energy 210, 220
probability, confidence intervals 333
probability density function (pdf) 816–823
continuous distributions 295–306
exponential distributions 824–829
rectangular probability distributions
818–819
probability distribution functions 276–292
project management 409–430
activity networks 886–891
critical path analysis 885–904
resourcing 892–897
proof
characteristics of matrices 170
methods 59
scalar products of vectors 190
proof by induction 59–63
properties of algebraic expressions
61–63
sums of series 59–61
using matrices 150–151
proper subgroups, group theory 946–950
pure strategy games 445, 450, 921

Q

quadratic equations
roots and transformations 36–37, 40–41
using complex numbers 8, 10–12
using hyperbolic functions 115–116
quadratic formula, complex number
solutions 8
quadratic function curves 82–87, 668–669
quartic equations
roots and transformations 38
using complex numbers 9, 11

R

radians, Argand diagrams 19, 21
radioactive disintegrations 288
random processes 275–324, 815–836
rational expression, improper integrals 567
rational functions
curve sketching 76–87
oblique asymptotes 548–551
real axis, Argand diagrams 14
real coefficients, polynomial equations 8
real roots
auxiliary equations 621, 639
inequalities 48–49
reciprocal functions, curve sketching
528–530, 533–535
rectangular hyperbolas 103
rectangular probability distribution 818–819
reducing flows, flow-augmenting network
paths 865–866
reduction formulae, integration 583–587
reflections
conic sections 107–108
points in lines 197–198
using matrices 154, 157–158, 162
repeated roots, auxiliary equations 621, 622,
623, 625
representative samples 335, 336
resisting forces 207, 208, 210–211, 220, 264,
266–267
resource histograms 892–895
resource levelling 892
resourcing, project workforce 892–897

restricted nodes, network flow 874
resultant moments 780
revolution of curves 132–137, 140
Riemann hypothesis 71
right-hand rule, vector products 728
roots of a function, graphs 528
roots of polynomials
algebra 36–42
versus factors 8
roots of unity, complex numbers 495–500
rotational symmetry, hyperbolic function
graphs 111
rotations
curve sketching 537
using matrices 154–159, 163
route inspection problem/algorithm
374–379, 404
row and column operations, matrices
702–707

S

sample mean, t-distributions 844
sample mean and variance, relationship
to population mean and variance
332–333
sample size 316, 332, 333, 336, 841
sampling, confidence intervals 332–337
saturated arcs, network flow 393
scalar product equation form, planes 735–736
scalar products, vectors 187–191
second order differential equations 620–627
coupled equations 651, 653
harmonic motion 632–649
self-inverse elements
group theory 940, 941, 944, 947, 949,
951–954
sets 469
self-inverse matrices 167
semi-Eulerian (semi-traversable) graphs/
networks 356–357, 359, 377–378
separating the variables, first order
differential equations 610, 629
series 507–526
convergent series 513
general terms 514–515
Maclaurin series 513–517
summing 51–58, 508–512
sets, binary operations 468–476
sheaf of planes 694
SHM *see* simple harmonic motion
shortest distances, points to planes 742
sign matrix, finding determinants 692, 701
significance levels, hypothesis testing 839
simple graphs, networks 346
simple harmonic motion (SHM) 628–637,
638, 639
simplex algorithm
linear programming 464, 906–920
zero-sum games 922–931
simplex tableaux 907–920
Simpson's rule, numerical integration
668–669
simultaneous equations, using matrices 143,
165, 169–171
singular matrices 165, 166, 692
sink set 394
sinks, flow in networks 392, 394, 397
sketching curves *see* curve sketching

skew lines, vectors 184, 193–194
slack variables, linear programming 906–920
solids, centres of mass 784–793, 795–796
source set 394
sources, flow in networks 392, 394, 397
spanning trees 360
see also minimum spanning trees
speed
approach and separation in collisions
242–243
dimensional analysis 224
see also velocity
splitting integrals 563–566, 586–587
spreadsheets 310, 464, 677, 682
springs, elasticity 217–223
square matrices 146–147, 165–167
stability, centres of mass 779–814
stable solutions, zero-sum games 446–448,
450, 921, 926, 927
standard deviation of discrete random
variables 279, 284
standard deviation of sample mean 333, 336
standard error, confidence intervals 333, 336
standard form, variables in linear
programming 907
stiffness, Hooke's law 217–223
stochastic processes, matrices 176
strategies in zero-sum games 445–448,
921–931
stretches 107–108, 111, 155–156
strings, elasticity 217–223
subdivisions of graphs, networks 347
subgraphs, networks 346, 360–373
subgroups, group theory 946–950
substitutions, differential equations
612–613
sum of cubes 52
sum of independent random variables 284,
304–305
sum of integers 51
sum of squares 52
summing series 51–58, 59–61, 508–512
supersinks/supersources, network flow 397
surface area of revolution 140
symmetry
Abelian groups 942
centres of mass 786, 788–789
hyperbolic function graphs 111
systems of linear simultaneous equations,
matrices 143, 165, 169–171

T

t-tests 844–849
table problems, matrices 151–152
tangential velocity, circular motion 262–263,
266
tension, Hooke's law 217–223
test statistics, hypothesis testing 316
timings, activity networks 410–423
torque 780
total probability rule 295–296, 302
tractive forces 207, 211–212
trails, graphs 356–357, 404
transcript notation, elements of a matrix 149
transformations
conic sections 105–108
curve sketching 536–541
hyperbolic functions 111